Nanotechnology 2008: Materials, Fabr~~ication~~ Particles, and Characterization

Technical Proceedings of the 2008 NSTI Nanotechnology Conference and Trade Show

Boston, June 1 – 5, 2008

NSTI Nanotech

The Nanotechnology Conference and Trade Show

An Interdisciplinary Integrative Forum on Nanotechnology, Biotechnology and Microtechnology

June 1-5, 2008
Hynes Convention Center
Boston, Massachusetts, U.S.A.
www.nsti.org

NSTI
Nano Science and
Technology Institute

CRC Press
Taylor & Francis Group

Caustic I (cover):
Here we focus on the origin of caustics as projections of three- dimensional transparent sheets onto two dimensions. The sheets consist of colored panels; the color combine with each other (by color subtraction for example) in unexpected ways.

Light from a point source is passed through two wavy surfaces, each like the surface of a swimming pool with a few people in it. This focuses the light randomly in a characteristic caustic pattern, here projected by intersection with a plane surface, as on a pool bottom.

Caustics are places where things accumulate; in this case it is light which is accumulating. We often think of focal points as where light gathers after passing through a lens, but more generally for "random" lenses there are much more interesting patterns to examine. In Caustic I, rays of light from a point source have passed through two imaginary, successive layers of water, each with a wavy surface, i.e. a random lens refracting the light. A "sea bottom" ultimately interrupts the refracted rays, which is the plane of the image. The bright patches, or "caustics", are similar to those produced when sunlight shines on a swimming pool, giving the familiar ribbed pattern on the bottom made famous by the painter David Hockney. The swimming pool provides only one refracting surface, but what if there are more? We were led to this question by recent work on the motion of electrons in random landscapes where they are deflected by hills and valleys. The electrons are moving in a thin, two-dimensional layer (see my Transport series), whereas the rays here are moving in three dimensions, and shown interrupted by a two dimensional surface. The electrons are refracted many times, so we wanted to look at light refracted by many wavy surfaces.

Nanotechnology 2008: Materials, Fabrication, Particles, and Characterization

Technical Proceedings of the 2008 NSTI Nanotechnology Conference and Trade Show

NSTI-Nanotech 2008, Vol. 1

*An Interdisciplinary Integrative Forum on
Nanotechnology, Biotechnology and Microtechnology*

June 1-5, 2008
Boston, Massachusetts, U.S.A.
www.nsti.org

NSTI Nanotech 2008 Joint Meeting

The 2008 NSTI Nanotechnology Conference and Trade Show includes:
2008 NSTI Bio Nano Conference and Trade Show, Bio Nano 2008
11th International Conference on Modeling and Simulation of Microsystems, MSM 2008
8th International Conference on Computational Nanoscience and Technology, ICCN 2008
7th Workshop on Compact Modeling, WCM 2008
NSTI Nanotech Ventures 2008
2008 TechConnect Summit
Clean Technology 2008

NSTI Nanotech 2008 Proceedings Editors:

Matthew Laudon
mlaudon@nsti.org

Bart Romanowicz
bfr@nsti.org

Nano Science and Technology Institute
Boston • Geneva • San Francisco

Nano Science and Technology Institute
One Kendall Square, PMB 308
Cambridge, MA 02139
U.S.A.

The papers in this book comprise the proceedings of the 2008 NSTI Nanotechnology Conference and Trade Show, Nanotech 2008, Boston, Massachusetts, June 1-5 2008. They reflect the authors' opinions and, in the interests of timely dissemination, are published as presented and without change. Their inclusion in this publication does not necessarily constitute endorsement by the editors, Nano Science and Technology Institute, or the sponsors. Dedicated to our knowledgeable network of willing and forgiving friends.

ISBN 978-1-4200-8503-7
ISBN 978-1-4200-8507-5 (Volumes 1-3 SET)
ISBN 978-1-4200-8511-2 (Volumes 1-3 CD-ROM)

Additional copies may be ordered from:

CRC Press
Taylor & Francis Group
an informa business
www.taylorandfrancisgroup.com

6000 Broken Sound Parkway, NW
Suite 300, Boca Raton, FL 33487
270 Madison Avenue
New York, NY 10016
2 Park Square, Milton Park
Abingdon, Oxon OX14 4RN, UK

Printed in the United States of America

Nanotech 2008 Proceeding Editors

VOLUME EDITORS

Matthew Laudon
Nano Science and Technology Institute, USA

Bart Romanowicz
Nano Science and Technology Institute, USA

TOPICAL EDITORS

Nanotechnology
Clayton Teague
National Nanotechnology Coordination Office, USA
Wolfgang Windl
Ohio State University, USA
Nick Quirke
Imperial College, London, UK
Philippe Renaud
Swiss Federal Institute of Technology, Switzerland
Mihail Roco
National Science Foundation, USA

Carbon Nanotech
Wolfgang S. Bacsa
Université Paul Sabatier, France
Anna K. Swan
Boston University, USA

Biotechnology
Srinivas Iyer
Los Alamos National Laboratory, USA
Gabriel A. Silva
University of California, San Diego, USA

Micro-Bio Fluidics
Daniel Attinger
Columbia University, USA
Hang Lu
Georgia Institute of Technology, USA
Steffen Hardt
Leibniz Universität Hannover, Germany

Pharmaceutical
Kurt Krause
University of Houston, USA
Mansoor M. Amiji
Northeastern University, USA

Soft Nanotechnology
Fiona Case
Case Scientific, USA

Composites & Interfaces
Thomas E. Twardowski
Twardowski Scientific, USA

Microtechnology
Narayan R. Aluru
University of Illinois Urbana-Champaign, USA
Bernard Courtois
TIMA-CMP, France
Anantha Krishnan
Lawrence Livermore National Laboratory, USA

Sensors
Elena Gaura
Coventry University, UK

Semiconductors
David K. Ferry
Arizona State University, USA
Andreas Wild
Freescale Semiconductor, France

Nano Particles
Sotiris E. Pratsinis
Swiss Federal Institute of Technology, Switzerland

Nano Fabrication
Warren Y.C. Lai
Alcatel-Lucent, USA
Leonidas E. Ocola
Argonne National Laboratory, USA
Stanley Pau
University of Arizona, USA

Characterization
Pierre Panine
European Synchrotron Radiation Facility, France
Greg Haugstad
University of Minnesota, USA

Workshop on Compact Modeling
Xing Zhou
Nanyang Technological University, Singapore

Sponsors

Accelrys Software, Inc.
Ace Glass, Inc.
Advance Reproductions Corporation
Advanced Materials Technologies Pte Ltd
Agilent Technologies
AIXTRON AG
AJA International, Inc.
ALM
Amuneal Manufacturing Corporation
ANSYS, Inc.
Anton Paar USA
Applied MicroStructures, Inc.
Applied Surface Technologies
Arkalon Chemical Technologies, LLC
Arkema Group
Artech House Publishers
Asemblon, Inc.
ASML
Asylum Research
AZoNano
Banner & Witcoff, Ltd.
Beckman Coulter, Inc.
Bio Nano Consulting
BioForce Nanosciences, Inc.
Bioneer Corporation
Birck Nanotechnology Center - Purdue University Discovery Park
Brookhaven Instruments Corporation
Buchanan Ingersoll & Rooney PC
Buhler AG
BusinessWeek
California NanoSystems Institute
Capovani Brothers, Inc.
Center for Functional Nanomaterials, Brookhaven National Laboratory
Center for Integrated Nanotechnologies, Los Alamos and Sandia National Laboratories
Center for Nanophase Materials Sciences at Oak Ridge National Laboratory
Center for Nanoscale Materials, Argonne National Laboratory
Cheap Tubes, Inc.
Clean Technology and Sustainable Industries Organization (CTSI)
COMSOL, Inc.
Cooley Godward Kronish LLP
Core Technology Group, Inc.
CRESTEC Corporation
CSEM
CVI Melles Griot
Department of Innovation, Industry, Science and Research
Digital Matrix
Draiswerke, Inc.
ETH Zurich
Eulitha AG

Marks & Clerk
Marubeni Techno-Systems Corporation
Massachusetts Technology Transfer Center
Materials Research Society
Merck & Co., Inc.
Microfluidics
Micromeritics Instrument Corporation
Microtrac, Inc.
MINAT 2008, Messe Stuttgart
Minus K Technology, Inc.
Misonix, Inc.
Molecular Foundry at Lawrence Berkeley National Laboratory
Motorola, Inc.
Nano Korea 2008
Nano Science and Technology Institute
nano tech 2009 Japan
Nano Technology Research Association of Korea (NTRA)
NanoAndMore USA, Inc.
NanoDynamics, Inc.
NanoEurope 2008
NanoInk, Inc.
Nanomotion, Inc.
Nanonics Imaging Ltd.
NANOSENSORS™
Nanosystems Initiative Munich (NIM)
Nanotech Northern Europe 2008
nanoTox, Inc.
NanoWorld AG
National Cancer Institute
National Institute of Standards and Technology (NIST)
National Institutes of Health
National Nanomanufacturing Network (NNN)
National Nanotechnology Infrastructure - GATech - Microelectronics Research Center
Natural Nano
Nature Publishing Group
NETZSCH Fine Particle Technology LLC
NIL Technology ApS
Novomer
NTT Advanced Technology Corporation
Olympus Industrial America
OSEC - Business Network Switzerland
Oxford Instruments
Particle Technology Labs, Ltd.
Pennsylvania State University
Photonics Spectra
PI (Physik Instrumente) L.P.
picoDrill SA
Piezo Institute
PVD Products, Inc.
Q-Sense, Inc.
Quantum Analytics
Raith USA, Inc.

European Patent Office (EPO)
Evans Analytical Group
EXAKT Technologies, Inc.
First Nano
Flow Science, Inc.
Food and Drug Administration
Goodwin Procter LLP
Greater Houston Partnership
Greenberg Traurig, LLP
Halcyonics
Headwaters Technology Innovation, LLC
Heidelberg Instruments
Hielscher USA, Inc.
HighNanoAnayltics
Hiscock & Barclay, LLP
Hitachi High Technologies America, Inc.
Hochschule Offenburg University of Applied Sciences
HOCKMEYER Equipment Corporation
HORIBA Jobin Yvon, Inc.
IBU-tec advanced materials GmbH
IDA Ireland
ImageXpert, Inc.
Inspec, Inc.
Institute for Nanoscale and Quantum Scientific and Technological Advanced Research (nanoSTAR)
Institute of Electrical and Electronics Engineers, Inc. (IEEE)
IntelliSense Software Corp.
Invest in Germany
IOP Publishing
Italian Trade Commission
JENOPTIK
Justus-Liebig-University (JLU) in Giessen
KAUST - King Abdullah University of Science and Technology
Keithley Instruments
Kelvin Nanotechnology Ltd.
Kodak
Kotobuki Industries Co., Ltd.
Lake Shore Cryotronics, Inc.
MACRO-M / NanoClay

Research in Germany Land of Ideas
RKS Legal Solutions, LLC
Sandia National Laboratories
Scottish Enterprise
SEMTech Solutions, Inc.
Serendip
Silvix Corporation
SNS Nano Fiber Technology, LLC
SoftMEMS LLC
Sonics & Materials, Inc.
SouthWest NanoTechnologies, Inc.
Specialty Coating Systems
Spectrum Laboratories, Inc.
SPEX SamplePrep, LLC
Springer, Inc.
Sterne, Kessler, Goldstein & Fox P.L.L.C.
Strem Chemicals, Inc.
Sukgyung A-T Co., Ltd.
Surrey NanoSystems
SUSS MicroTec
Swissnanotech Pavilion
Taylor & Francis Group LLC - CRC Press
TechConnect
Technovel Corporation
Tekna Plasma Systems, Inc.
Thinky Corporation
Thomas Swan & Co. Ltd.
UGL Unicco
UK Trade & Investment
UniJet
University Muenster (WWU)
University of Applied Sciences (TFH) Berlin
University of Duisburg-Essen / Center for Nanointegration Duisburg-Essen (CeNIDE)
Veeco Instruments
Wasatch Molecular Incorporated
Weidmann Plastics Technology
Whiteman Osterman & Hanna LLP
Willy A. Bachofen AG
WITec GmbH
World Gold Council

Supporting Organizations

AllConferences.Com
Asia Pacific Nanotechnology Forum (APNF)
AZoNano
Battery Power Products & Technology
Berkeley Nanotechnology Club
BioExecutive International
Biolexis.com
Biophotonics International
BioProcess International
BioTechniques
Biotechnology Industry Organization (BIO)
Business Wire
BusinessWeek
CMP: Circuits Multi-Projets
Drug & Market Development Publications
EurekAlert! / AAAS
Food and Drug Administration (FDA)
Foresight Institute
Fuel Cell Magazine
GlobalSpec
Greenberg Traurig, LLP
IEEE CiSE
IEEE San Francisco Bay Area Nanotechnology Council
ivcon.net
Journal of Experimental Medicine
JSAP - the Japan Society of Applied Physics
KCI Investing
Laser Focus World
Materials Today
MEMS Investor Journal
Microsystem Technologies - Micro- and Nanosystems
mst|news

Nano Today
NanoBiotech News
Nanomedicine: Nanotechnology, Biology and Medicine
NanoNow!
nanoparticles.org
NanoSPRINT
Nanotechnology Law & Business
Nanotechnology Now
nanotechweb
nanotimes
Nanovip.com
Nanowerk
National Cancer Institute (NCI)
National Institutes of Health (NIH)
Nature Publishing Group
Photonics Spectra
PhysOrg.com
R&D Magazine
Red Herring, Inc.
Science Magazine
SciTechDaily Review
Small Times
Solid State Technology
Springer
Springer Micro and Nano Fluidics
The Journal of BioLaw & Business (JB&B)
The Journal of Cell Biology
The Real Nanotech Investor
The Scientist
tinytechjobs
Understanding Nanotechnology
Virtual Press Office

Table of Contents

Nano Materials & Composites

Nano Surfaces & Interfaces

Nanofabrication & Direct-Write Nanolithography

Nanoparticles & Applications

Characterization

Initiatives, Education & Policy

NSTI Nanotech 2008 Program Committee

TECHNICAL PROGRAM CO-CHAIRS

Matthew Laudon — *Nano Science and Technology Institute, USA*
Bart Romanowicz — *Nano Science and Technology Institute, USA*

TOPICAL AND REGIONAL SCIENTIFIC ADVISORS AND CHAIRS

Nanotechnology

Clayton Teague	*National Nanotechnology Coordination Office, USA*
Wolfgang Windl	*Ohio State University, USA*
Nick Quirke	*Imperial College, London, UK*
Philippe Renaud	*Swiss Federal Institute of Technology, Switzerland*
Mihail Roco	*National Science Foundation, USA*

Carbon Nanotech

Wolfgang S. Bacsa	*Université Paul Sabatier, France*
Bennett Goldberg	*Boston University, USA*
Anna K. Swan	*Boston University, USA*

Biotechnology

Srinivas Iyer	*Los Alamos National Laboratory, USA*
Gabriel A. Silva	*University of California, San Diego, USA*

Micro-Bio Fluidics

Daniel Attinger	*Columbia University, USA*
Hang Lu	*Georgia Institute of Technology, USA*
Steffen Hardt	*Leibniz Universität Hannover, Germany*

Pharmaceutical

Kurt Krause	*University of Houston, USA*
Mansoor M. Amiji	*Northeastern University, USA*

Soft Nanotechnology

Fiona Case	*Case Scientific, USA*

Composites & Interfaces

Thomas E. Twardowski	*Twardowski Scientific, USA*

Microtechnology

Narayan R. Aluru	*University of Illinois Urbana-Champaign, USA*
Bernard Courtois	*TIMA-CMP, France*
Anantha Krishnan	*Lawrence Livermore National Laboratory, USA*

Sensors

Elena Gaura	*Coventry University, UK*

Semiconductors

David K. Ferry	*Arizona State University, USA*
Andreas Wild	*Freescale Semiconductor, France*

Nano Particles

Sotiris E. Pratsinis	*Swiss Federal Institute of Technology, Switzerland*

Nano Fabrication

Warren Y.C. Lai	*Alcatel-Lucent, USA*
Leonidas E. Ocola	*Argonne National Laboratory, USA*
Stanley Pau	*University of Arizona, USA*

Characterization

Pierre Panine	*European Synchrotron Radiation Facility, France*
Greg Haugstad	*University of Minnesota, USA*

Workshop on Compact Modeling

Xing Zhou	*Nanyang Technological University, Singapore*

NANOTECH PHYSICAL SCIENCES COMMITTEE

M.P. Anantram	*NASA Ames Research Center, USA*
Phaedon Avouris,	*IBM, USA*
Wolfgang S. Bacsa	*Université Paul Sabatier, France*
Gregory S. Blackman	*DuPont, USA*
Alexander M. Bratkovsky	*Hewlett-Packard Laboratories, USA*
Roberto Car	*Princeton University, USA*
Fiona Case	*Case Scientific, USA*
Franco Cerrina	*University of Wisconsin - Madison, USA*
Alex Demkov	*University of Texas at Austin, USA*
David K. Ferry	*Arizona State University, USA*
Lynn Foster	*Greenberg Traurig L.L.P., USA*
Toshio Fukuda	*Nagoya University, Japan*
Sharon Glotzer	*University of Michigan, USA*
William Goddard	*California Institute of Technology, USA*
Gerhard Goldbeck-Wood	*Accelrys, Inc., UK*
Bennett Goldberg	*Boston University, USA*
Niels Gronbech-Jensen	*UC Davis and Berkeley Laboratory, USA*
Jay T. Groves	*University of California at Berkeley, USA*
James R. Heath	*California Institute of Technology, USA*
Karl Hess	*University of Illinois at Urbana-Champaign, USA*
Christian Joachim	*CEMES-CNRS, France*
Hannes Jonsson	*University of Washington, USA*
Anantha Krishnan	*Lawrence Livermore National Laboratory, USA*
Kristen Kulinowski	*Rice University, USA*
Alex Liddle	*Lawrence Berkeley National Laboratory, USA*
Shenggao Liu	*ChevronTexaco, USA*
Lutz Mädler	*University of California, Los Angeles, USA*
Chris Menzel	*Nano Science and Technology Institute, USA*
Meyya Meyyappan	*National Aeronautics and Space Agency, USA*
Martin Michel	*Nestlé, Switzerland*
Sokrates Pantelides	*Vanderbilt University, USA*
Philip Pincus	*University of California at Santa Barbara, USA*
Joachim Piprek	*University of California, Santa Barbara, USA*
Sotiris E. Pratsinis	*Swiss Federal Institute of Technology (ETH Zürich), Switzerland*
Serge Prudhomme	*University of Texas at Austin, USA*
Nick Quirke	*Imperial College, London, UK*
PVM Rao	*IIT Delhi, India*
Mark Reed	*Yale University, USA*
Philippe Renaud	*Swiss Federal Institute of Technology of Lausanne, Switzerland*
Doug Resnick	*Molecular Imprints, USA*
Mihail Roco	*National Science Foundation, USA*
Rafal Romanowicz	*CTSI, Switzerland*
Robert Rudd	*Lawrence Livermore National Laboratory, USA*
Brent Segal	*Nantero, USA*
Douglas Smith	*University of San Diego, USA*
Donald C. Sundberg	*University of New Hampshire, USA*
Anna K. Swan	*Boston University, USA*
Clayton Teague	*National Nanotechnology Coordination Office, USA*
Loucas Tsakalakos	*GE Global Research, USA*
Arthur Voter	*Los Alamos National Laboratory, USA*
Wolfgang Windl	*Ohio State University, USA*
Xiaoguang Zhang	*Oakridge National Laboratory, USA*

NANOTECH LIFE SCIENCES COMMITTEE

Mansoor M. Amiji	*Northeastern University, USA*
Mostafa Analoui	*Pfizer, USA*
Amos Bairoch	*Swiss Institute of Bioinformatics, Switzerland*
Jeffrey Borenstein	*Draper Laboratory, USA*
Stephen H. Bryant	*National Institute of Health, USA*
Dirk Bussiere	*Chiron Corporation, USA*
Fred Cohen	*University of California, San Francisco, USA*
Tejal Desai	*University of California, San Francisco, USA*
Daniel Davison	*Bristol Myers Squibb, USA*
Tejal Desai	*University of California, San Francisco, USA*
Robert S. Eisenberg	*Rush Medical Center, Chicago, USA*
Michael N. Helmus	*Advance Nanotech, USA*
Andreas Hieke	*GEMIO Technologies, Inc., USA*
Leroy Hood	*Institute for Systems Biology, USA*
Sorin Istrail	*Brown University, USA*
Srinivas Iyer	*Los Alamos National Laboratory, USA*
Brian Korgel	*University of Texas-Austin, USA*
Kurt Krause	*University of Houston, USA*
Daniel Lacks	*Case Western ReserveUniversity, USA*
Jeff Lockwood	*Novartis, USA*
Hang Lu	*Georgia Institute of Technology, USA*
Atul Parikh	*University of California, Davis, USA*
Andrzej Przekwas	*CFD Research Corporation, USA*
Don Reed	*Ecos Corporation, Australia*
George Robillard	*BioMade Corporation, Netherlands*
Jonathan Rosen	*Center for Integration of Medicine & Innovative Technology, USA*
Gabriel A. Silva	*University of California, San Diego, USA*
Srinivas Sridhar	*Northeastern University, USA*
Sarah Tao	*The Charles Stark Draper Laboratory, Inc., USA*
Tom Terwilliger	*Los Alamos National Laboratory, USA*
Vladimir Torchilin	*Northeastern University, USA*
Michael S. Waterman	*University of Southern California, USA*
Thomas J. Webster	*Brown University, USA*
Steven T. Wereley	*Purdue University, USA*

NANOTECH MICROSYSTEMS COMMITTEE

Narayan R. Aluru	*University of Illinois Urbana-Champaign, USA*
Daniel Attinger	*Columbia University, USA*
Xavier J. R. Avula	*Washington University, USA*
Stephen F. Bart	*Bose Corporation, USA*
Bum-Kyoo Choi	*Sogang University, Korea*
Bernard Courtois	*TIMA-CMP, France*
Peter Cousseau	*Honeywell, USA*
Robert W. Dutton	*Stanford University, USA*
Gary K. Fedder	*Carnegie Mellon University, USA*
Edward P. Furlani	*Eastman Kodak Company, USA*
Elena Gaura	*Coventry University, UK*
Steffen Hardt	*Leibniz Universität Hannover, Germany*
Eberhard P. Hofer	*University of Ulm, Germany*
Michael Judy	*Analog Devices, USA*
Yozo Kanda	*Toyo University, Japan*
Jan G. Korvink	*University of Freiburg, Germany*
Mark E. Law	*University of Florida, USA*
Mary-Ann Maher	*SoftMEMS, USA*
Kazunori Matsuda	*Tokushima Bunri University, Japan*
Tamal Mukherjee	*Carnegie Mellon University, USA*
Andrzej Napieralski	*Technical University of Lodz, Poland*

Ruth Pachter	*Air Force Research Laboratory, USA*
Michael G. Pecht	*University of Maryland, USA*
Marcel D. Profirescu	*Technical University of Bucharest, Romania*
Marta Rencz	*Technical University of Budapest, Hungary*
Siegfried Selberherr	*Technical University of Vienna, Austria*
Sudhama Shastri	*ON Semiconductor, USA*
Armin Sulzmann	*Daimler-Chrysler, Germany*
Mathew Varghese	*MEMSIC, Inc., USA*
Dragica Vasilesca	*Arizona State University, USA*
Gerhard Wachutka	*Technical University of Münich, Germany*
Jacob White	*Massachusetts Institute of Technology, USA*
Thomas Wiegele	*Goodrich, USA*
Andreas Wild	*Freescale Semiconductor, France*
Cy Wilson	*North American Space Agency, USA*
Wenjing Ye	*Georgia Institute of Technology, USA*
Xing Zhou	*Nanyang Technological University, Singapore*

NANO FABRICATION COMMITTEE

Adekunle Adeyeye	*National University of Singapore, Singapore*
Ronald S. Besser	*Stevens Institute of Technology, USA*
Gregory R. Bogart	*Symphony Acoustics, USA*
Chorng-Ping Chang	*Applied Materials, Inc., USA*
Charles Kin P. Cheung	*National Institute of Standards and Technology, USA*
Seth B. Darling	*Argonne National Laboratory, USA*
Guy A. DeRose	*California Institute of Technology, USA*
Zhixiong Guo	*Rutgers University, USA*
Takamaro Kikkawa	*Hiroshima University and National Institute of Advanced Industrial Science and Technology, Japan*
Jungsang Kimg	*Duke University, USA*
Uma Krishnamoorthy	*Sandia National Laboratory, USA*
Andres H. La Rosa	*Portland State University, USA*
Warren Y.C. Lai	*Lucent Technologies, USA*
Sergey D. Lopatin	*Applied Materials, Inc., USA*
Pawitter Mangat	*Motorola, USA*
Omkaram Nalamasu	*Applied Materials, Inc., USA*
Vivian Ng	*National University of Singapore, Singapore*
Leonidas E. Ocola	*Argonne National Laboratory, USA*
Sang Hyun Oh	*University of Minnesota, USA*
Stanley Pau	*University of Arizona, USA*
John A. Rogers	*University of Illinois at Urbana-Champaign, USA*
Nicolaas F. de Rooij	*University of Neuchâtel, Switzerland*
Aaron Stein	*Brookhaven National Laboratory, USA*
Vijay R. Tirumala	*National Institute of Standards and Technology, USA*
Gary Wiederrecht	*Argonne National Laboratory, USA*

SOFT NANOTECHNOLOGY CONFERENCE COMMITTEE

Fiona Case	*Case Scientific, USA*
Greg Haugstad	*University of Minnesota, USA*
Pierre Panine	*European Synchrotron Radiation Facility, France*
Peter Schurtenberger	*University of Fribourg, Switzerland*
Patrick Spicer	*The Procter & Gamble Company, USA*
Donald C. Sundberg	*University of New Hampshire*
Krassimir Velikov	*Unilever Research Vlaardingen, Netherland*

CONFERENCE OPERATIONS MANAGER

| Sarah Wenning | *Nano Science and Technology Institute, USA* |

NSTI Nanotech 2008 Proceedings Topics

Nanotechnology 2008: Materials, Fabrication, Particles, and Characterization,

NSTI-Nanotech 2008, Vol. 1, ISBN: 978-1-4200-8503-7:

1. Carbon Nano Structures & Applications
2. Nano Materials & Composites
3. Nano Surfaces & Interfaces
4. Nanofabrication & Direct-Write Nanolithography
5. Nanoparticles & Applications
6. Characterization
7. Initiatives, Education & Policy

Nanotechnology 2008: Life Sciences, Medicine, and Bio Materials,

NSTI-Nanotech 2008, Vol. 2, ISBN: 978-1-4200-8504-4:

1. Cancer Diagnostics, Imaging & Treatment
2. Environment, Health & Toxicology
3. Biomarkers, Nano Particles & Materials
4. Drug & Gene Delivery Systems
5. Phage Nanobiotechnology
6. Nano Medicine & Neurology
7. Bio & Chem Sensors
8. Soft Nanotechnology & Polymers

Nanotechnology 2008: Microsystems, Photonics, Sensors, Fluidics, Modeling, and Simulation,

NSTI-Nanotech 2008, Vol. 3, ISBN: 978-1-4200-8505-1:

1. Photonics & Nanowires
2. Sensors & Systems
3. Lab-on-a-Chip, Micro & Nano Fluidics
4. MEMS & NEMS
5. Modeling & Simulation of Microsystems
6. Computational Nanoscience
7. Compact Modeling

Nanotechnology 2008, Vol. 1-3, ISBN: 978-1-4200-8507-5 (hardcopy)

Nnaotechnology 2008, Vol. 1-3 CDROM, ISBN: 978-1-4200-8511-2

Diameter Selective Growth of Vertically Aligned Single Walled Carbon Nanotubes by Ethanol Flow Control

M. G. Hahm[*], D. Pina[**], Y. K. Kwon[**] and Y. J. Jung[*]

[*]Northeastern University, Boston, MA, USA, jungy@coe.neu.edu
[**]University of Massachusetts Lowell, Lowell, MA, USA, YoungKyun_Kwon@uml.edu

ABSTRACT

Nanotechnologies based on single walled carbon nanotubes (SWNTs) are developing very rapidly because of their outstanding mechanical, electrical, and optical properties. However, large scale synthesis of SWNT with desired structures still have many difficulties. In particular, controlling the diameter and chirality of SWNT is one of the biggest challenges that need to be solved. In this presentation, we introduce the study on role of ethanol flow rate in a chemical vapor deposition (CVD) for the selective diameter distribution on vertically aligned single walled carbon nanotubes (VA-SWNTs) and selective synthesis of desired structure (single walled, double walled and multi walled carbon nanotube) by systematically deposited catalyst particles.

Keywords: carbon nanotube, diameter control, raman, ethanol flow rate

1 ITRODUCTION

Carbon nanotubes (CNTs) are unique nanostructures with remarkable electronic and mechanical properties, some stemming from the close relation between CNTs and graphite, and from their one-dimensional aspects. As other intriguing properties have been discovered, such as their remarkable electronic transport properties[1], their unique Raman spectra[2], and their unusual mechanical properties[3], interest has grown in their potential use in nanometer sized electronics and in a variety of other applications. The full technological potential of CNTs has been hindered somewhat by the difficulty associated with the control of their properties such as structure, diameter, and chirality. In recent years, significant research efforts have concentrated on overcoming these barriers[4-7]. Chemical vapor deposition (CVD), which offers versatile control and the possibility of scaling-up, is the most attractive method of producing CNTs because of the critical role played by catalyst nanoparticles. Although the growth mechanism of CNTs in CVD is still unclear, the size of the metal catalyst nanoparticles approximately determines the eventual structure and diameter of the CNTs. We have shown that the structures of CNTs can be controlled by controlling the metal catalyst nanoparticles. Additionally, we introduce the study on role of ethanol flow rate in a CVD for the selective diameter distribution of VA-SWNTs.

2 EXPERIMENTAL

For synthesis of structure selective VA-CNTs growth, we first prepared new catalyst system using cobalt (Co) as a catalyst. For catalyst system, Al/SiO_2 multilayered substrate was used. And Co was deposited on Al/SiO_2 multilayer using sputter. In order to grow CNTs selectively desired structure, Co catalyst film was deposited by different sputter current at the same sputter time (3 sec/5mA and 3 sec/25mA). The growth of CNTs was performed in a 1.25-in. quartz tube and a furnace. $Co/Al/SiO_2$ substrate was loaded in a quartz tube. A loaded wafer was baked at 400 °C for 10 min and inside of the quartz tube was evacuated by a rotary pump, heated at 850 °C for growth CNTs. During the heat-up the furnace, argon-hydrogen mixture gas was supplied so that the pressure was 700 torr. And ethanol vapor was supplied as a carbon source for 30 min with determined flow rate. In order to investigate the role of ethanol flow rate in CVD for the selective diameter distribution on VA-SWNTs, the ethanol flow rate is controlled by systematically varying while keeping all other CVD parameters, such as growth pressure, growth temperature, and growth time.

3 RESULT AND DISCUSSION

Figure 1a, 1b, 1c, and 1d show low magnification and high magnification SEM images of vertically aligned carbon nanotubes. Ethanol CVD results in the growth of dense and vertically aligned CNTs with several hundred height in a 30-min growth time. The best result to date is 1 mm in 30-min. First sample (Figure 1a, 1c, and 1e) was synthesized using Co catalyst system was deposited by 3sec/5mA. VA-SWNTs were up to 98% on substrate as

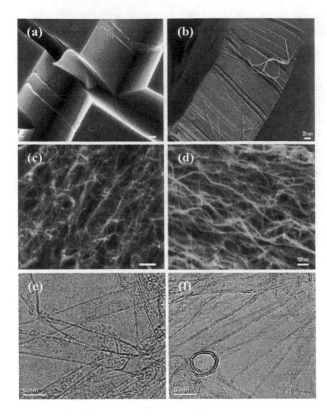

Figure 1: SEM and TEM images of vertically aligned CNTs synthesized with different sputter conditions. (a) low magnification SEM image of VA-SWNTs, (b) low magnification SEM image of VA-DWNTs, (c) high magnification SEM image of VA-SWNTs, (d) high magnification SEM image of VA-SWNTs, (e) HR-TEM image of VA-SWNTs, and (f) HR-TEM image of VA-DWNTs. Scale bars of a, b are 20 μm, scale bars of c, d are 100 nm.

shown Figure 1e. In case of second sample (Figure 1b, 1d, and 1f), CNTs were synthesized using Co catalyst system was deposited by 3sec/25mA. On second sample, VA-DWNTs were up to 97% as shown Figure 1f. Using different catalyst deposition condition, we obtained the structure selective growth of VA-SWNTs and VA-DWNTs.

The diameter selective growth of highly dense and VA-SWNTs were synthesized by three different flow rate of ethanol vapor (50, 100 and 200 sccm), the carbon source, in a CVD as shown Figure 2. It is clearly seen that SWNTs are vertically aligned for all ethanol flow rate such as 50, 100 and 200 sccm indicating that our ethanol CVD process is effective for SWNTs growth. In this experiment, one interesting feature was found in Raman studies, especially in Radial Breathing Mode (RBM) mapping process. Raman map and image score gives us spatial distribution of SWNTs diameter. Therefore, statistical treatment of Raman images can quantify the diameter distribution on substrates.

RBM of SWNTs gives information on the diameter and chirality of SWNTs. The diameter of SWNTs can be calculated from the equation; $d(nm) = A/[\omega_R(cm^{-1})]$, where d is the diameter of SWNT, A is a proportionality constant (248), and ω is the RBM frequency[8]. In order to investigate the large scale diameter distribution of three different VA-SWNTs, Raman maps were recorded using Raman microscope and mapping stage. The excitation laser was 785 nm, the Raman mapping area was 10 μm by 10 μm at 0.3 μm laser steps, the exposure time was 5 second/spectrum, and the number of accumulations is 14. The 600 gr/mm grating was used, and the confocal hole diameter was set to 200 μm. Rman maps were processed using Modeling function in LabSPEC 5 (HORIBA Jobin Yvon).

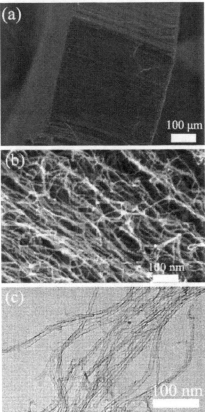

Figure 2: SEM and TEM images of vertically aligned SWNTs. (a) low magnification SEM image of VA-SWNTs synthesized with 50 sccm flow rate of ethanol, (b) high magnification SEM image, (C) TEM image shows that VA-CNTs are single walled carbon nanotubes

Raman spectra models and score images of SWNTs grown with three different flow rate of ethanol vapor are shown Figure 3. A multivariate analysis algorithm (direct classical least square, DCLS) was applied in an unsupervised mode to extract significantly different Raman spectra (models) and to calculate scores of individual

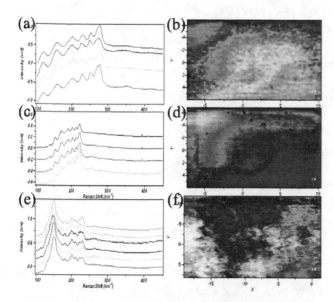

Figure 3: (a), (c) and (e) show models produced from Raman maps over $10 \times 10 \ \mu m^2$ area (b), (d) and (f) show corresponding score images. (a) and (b) are results from VA-SWNTs grown with 50 sccm ethanol flow rate, (c) and (d) 100 sccm, (e) and (f) 200 sccm.

spectra with respect to these models. The RBM models recorded from VA-SWNTs synthesized with 50 sccm ethanol flow rate (Figure 3 a) produced four model spectra with the RBM peak positions ranging between 159 cm^{-1} and 291 cm^{-1}. These RBM frequencies correspond to VA-SWNTs diameter in the range of 0.89 nm – 1.55 nm. The corresponding Raman image shows the diameter distribution of the VA-SWNTs with the four factors represented with corresponding colors. For example, green areas in the Raman image represent the spatial diameter distribution of VA-SWNTs corresponding to the green factor (spectrum). Vertically grown SWNTs with 100 sccm ethanol flow rate also yielded 4 different RBM models whose frequencies range from 143 cm^{-1} to 231 cm^{-1} (Figure 3 c). The diameter distribution of VA-SWNTs grown with 100 sccm ethanol flow rate is 1.07 nm to 1.73 nm. In the case of the 200 sccm ethanol flow rate, VA-SWNTs produced 6 different RBM models in the range of 119 cm^{-1} to 243 cm^{-1} (Figure 3 e). The observed RBM frequencies correspond to VA-SWNTs diameters from 1.02 nm to 2.08 nm. From images (Figure 3 b, d and f), we can determine RBM spectrum among the several RBM spectra on 10 by 10 μm^2 region of each sample. From above results, as the ethanol vapor flow rate increased from 50 sccm to 200 sccm, the diameter distributions of VA-SWNTs were enlarged. The diameter distribution of VA-SWNTs changed from 0.85 nm – 1.55 nm with 50 sccm ethanol flow rate to 1.02 nm – 2.08 nm with 200 sccm ethanol flow rate.

In summary, using different catalyst deposition condition, CNTs were synthesized selectively single walled, double walled and multi walled carbon nanotubes. Additionally, our results show that the diameter distribution selective synthesis of SWNTs was closely correlated with flow rate of ethanol vapor.

REFERENCES

[1] Z. Yao, C. L. Kane and C. Dekker, Phys. Rev. Lett. 84, 2941-2944, 2000.
[2] M. S. Dresselhaus, G. Dresselhaus, R. Saito and A. Jorio, Phys. Rep. 10, 2004.
[3] R. S. Ruoff and D. C. Lorents, Carbon, 33, 925, 2000.
[4] S. Maruyama, E. Einarsson, Y. Murakami, T. Edamura, Chem. Phys. Lett. 2005, 403, 320.
[5] K. Hata, D. N. Futaba, K. Mizuno, T. Namai, S. Iijima, Science 2004, 306, 1362.
[6] Y. Murakami, S. Chiashi, Y. Miyauchi, M. Hu, M. Ogura, T. Okubo, S. Maruyama, Chem. Phys. Lett. 2004, 385, 298.
[7] G. Zhang, D. Mann, L. Zhang, A. Javey, Y. Li, E. Yenilmez, Q. Wang, J. P. McVitte, Y. Nishi, J. Gibbons, H. Dai, PNAS 2005, 102, 16141.
[8] A. Jorio, R. Saito, J. H. Hanfner, C. M. Lieber, M. Hunter, T. McClure, G. Dresselhaus and M. S. Dresselhaus, Phys. Rev. Lett. 86, 1118, 2001

Growth of Nanostructured Diamond, Diamond-Like Carbon, and Carbon Nanotubes in a Low Pressure Inductively Coupled Plasma

Katsuyuki Okada

National Institute for Materials Science, 1-1 Namiki, Tsukuba, Ibaraki 305-0044, Japan,
okada.katsuyuki@nims.go.jp

ABSTRACT

A 13.56 MHz low pressure inductively coupled $CH_4/CO/H_2$ plasma has been applied to prepare nanocrystalline diamond particles of 200-700 nm diameter. The minimum diameter of the particles was found to be 5 nm. Two-dimensional platelet-like graphite and carbon nanotubes also were deposited with different conditions. The characterizations were performed with transmission electron microscopy (TEM) and electron energy loss spectroscopy (EELS). The TEM observations have revealed that the two-dimensional platelet-like deposits consist of disordered microcrystalline graphite, whereas the particles are composed of only diamond nanocrystallites. The high-resolution TEM images clearly show that each particle is composed of small particles of about several ten nm in diameter. The mapping of sp^2 bonding by the $\pi*$ image reveals that sp^2-bonded carbons are localized in the grain boundaries of 20-50 nm sub-grains of nanocrystalline diamond particles at approximately 1 nm width.

Keywords: nanocrystalline diamond, nanostructured carbon, inductively coupled plasma, electron energy loss spectroscopy, sp^2/sp^3 bonding,

1 INTRODUCTION

Nanocrystalline diamond and nanostructured carbon films have attracted considerable attention because they have a low coefficient of friction and a low electron emission threshold voltage [1]. The small grain size (approximately 5-100 nm) gives films valuable tribological and field-emission properties comparable to those of conventional polycrystalline diamond films. Furthermore, applications for micro-electro-mechanical systems (MEMS) devices, metal-semiconductor field effect transistors (MESFETs), electrochemical electrodes, and biochemical devices have been proposed that take advantage of these excellent properties [2-4].

A 13.56 MHz low pressure inductively coupled $CH_4/CO/H_2$ plasma has been applied to prepare nanocrystalline diamond particles of 200-700 nm diameter. The minimum diameter of the particles was found to be 5 nm. Two-dimensional platelet-like graphite and carbon nanotubes also were deposited with different conditions. The characterizations were performed with transmission electron microscopy (TEM) and electron energy loss spectroscopy (EELS).

2 EXPERIMENT

The schematic view of the low pressure ICP-CVD system is illustrated in Figure 1. The detailed description and deposition procedures were reported previously [5]. To be brief, a low pressure ICP was generated in a growth chamber by applying 13.56 MHz rf powers of 1 kW to a three-turn helical antenna. The flow rates of CH_4 and H_2 were kept at 4.5 and 75 sccm, respectively, whereas the flow rate of CO ([CO]) was varied between 0, 1.0, and 10 sccm, respectively. The total gas pressure was accordingly varied from 45 to 50 mTorr. Silicon (100) wafers (10 mm in diameter) were used as a substrate. The substrate temperature was kept at 900 ℃. The deposition duration was 2 h.

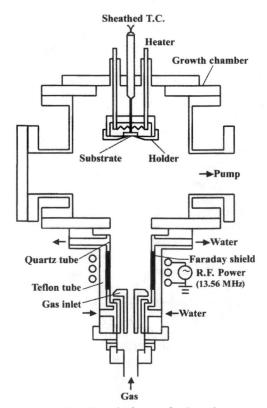

Fig. 1. Schematic description of the low pressure inductively coupled rf plasma CVD system.

EELS measurements [6] were carried out by using a post-column energy filter (GATAN, GIF2002) equipped with a transmission electron microscope (TEM; Hitachi HF-3000) at 297 keV.

3 RESULTS & DISCUSSION

3.1 Morphology

Figure 2 shows SEM photographs of the resultant deposits on a Si(100) substrate. Figs. 2(a), 2(b), and 2(c) correspond to [CO]=0, 1.0, and 10 sccm, which are referred to as samples A, B, and C, respectively. The morphology of sample A was platelet-like, as shown in Fig. 2(a), and no crystal facets were clearly seen. When CO was added to the CH_4/H_2 plasma, particles of 200-300 nm diameter as well as platelet-like deposits appeared, as shown in Fig. 2(b). With the increase in [CO], only particles were deposited on the Si substrate, as shown in Fig. 2(c). The diameters of the particles were 200-700 nm. Detailed observation reveals that the particles consist of small particles of about 20-50 nm diameter, and that the particle size remains are almost the same regardless of increasing [CO]. It is therefore speculated that increasing [CO] results in a large supersaturation degree of carbon; thus, the number of encounters between particles is increased.

The previous TEM observations have revealed [5] that the two-dimensional platelet-like deposits consist of disordered microcrystalline graphite, whereas the particles are composed of only diamond nanocrystallites. The high-resolution TEM images clearly show that each particle is composed of small particles of about several ten nm in diameter. The X-ray diffraction pattern for the sample C exhibits the diffraction peaks of diamond (111) and (220) planes [7]. The crystallite size was estimated to be approximately 20 nm from the full width at half maximum (FWHM) of the diamond peaks by using the Scherrer's equation. It is consistent with the TEM observations.

3.2 EELS

The EEL spectrum of the outer part of a nanocrystalline diamond particle is shown in Figure 3. It exhibits a peak at 290 eV corresponding to $\sigma*$ states and a small peak appears at ~285 eV corresponding to $\pi*$ states. The ELNES above 290 eV is similar to that of diamond [8] and is clearly different from that of graphite or sp^3-rich tetrahedral amorphous carbon [9]. The intensity of the $\pi*$ peak is much lower than that of the $\sigma*$ peak. Although the $\sigma*$ peak in general includes contributions from both sp^2 and sp^3 bonding, the $\sigma*$ peak of the EEL spectrum in Fig. 3 is considered to be mainly due to sp^3 bonding. The use of a narrow energy window positioned on the ELNES signal allows the mapping of the variation in intensity as a function of position within the microstructure. The conventional so-called three-window method [10] was employed to remove the background contribution. Two pre-edge images (272-277 and 277-282 eV), indicated by a_1 and a_2 in Fig. 3, were used to obtain an extrapolated background image. The subtraction of the extrapolated background image from the postedge images (282-287 and 287-292 eV) indicated by b and c produces the $\pi*$ and $\sigma*$ images, respectively, with the background contribution removed.

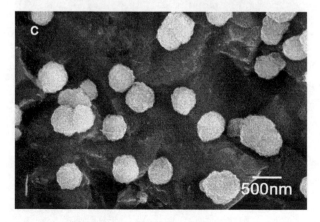

Fig. 2. SEM micrographs of obtained deposits: (a) [CH_4/CO]=4.5/0 sccm, (b) [CH_4/CO]=4.5/1.0 sccm, and (c) [CH_4/CO]=4.5/10 sccm.

The $\pi*$ and $\sigma*$ images of the outer part of a nanocrystalline diamond particle are shown in Figures 4(a) and 4(b), respectively. Since the intensity of the $\pi*$ image was weak compared with that of the $\sigma*$ image, the former was increased by a factor of 5. The $\pi*$ image reveals that the intensity is strong around the subgrains, whereas the $\sigma*$ image shows that the intensity is strong within the subgrains. Although sp^2-bonded graphitic layers do not clearly appear because of the limitation at the resolution of the HR-TEM image, these energy-filtered $\pi*$ and $\sigma*$ images imply that sp^2 bonding is localized around 20-50 nm subgrains. The sp^2 bonding around the subgrains is considered to contribute to the small peak at ~285 eV in the ELNES. The width of the sp^2 bonding is estimated to be approximately 1 nm from the $\pi*$ image in Fig. 4(a). Fallon and Brown [11] reported the presence of amorphous carbon at the grain boundaries of CVD diamond films by TEM observation and EELS analysis. The amorphous carbon is shown to contain almost exclusively sp^2 bonding and to be nonhydrogenated. It was also demonstrated from the theoretical point of view [12] that sp^2 bonding is energetically stable in the grain boundaries of nanocrystalline diamond. It is consequently considered that the sp^2 bonding is localized in the grain boundaries of 20-50 nm subgrains.

Figures 5(a) and 5(b) show the low loss region of EEL spectra of a nanocrystalline diamond particle in sample C and a graphite-like platelet in sample A, respectively. The low loss region, extending from 0 to ~50 eV, corresponds to the excitation of electrons in the outermost atomic orbitals and reflects the solid state character of the sample. The low loss region is dominated by collective, resonant oscillations of the valence electrons known as plasmons [10]. In Fig. 5(a), the peak at 33 eV is assigned to the bulk plasmon, E_B of diamond, and the shoulder at 23 eV is attributed to the surface plasmon, E_S of diamond. The E_S is almost equal to the $E_B/\sqrt{2}$, which is consistent with the previous report on diamond [13]. It was also reported [13] that the intensity of surface plasmon peak to that of bulk plasmon peak increases with decreasing crystallite size of nanocrystalline diamond.

On the other hand, the low loss spectrum of a graphite-like platelet as shown in Fig. 5(b) consists of two peaks at 27 eV and 6 eV. The former corresponds to the main plasmon peak of graphite, and the latter to a π to $\pi*$ interband transition, which reflects the excitation of valence electrons to low-energy unoccupied electronic states above the Fermi level [10]. The whole spectrum shape is similar to that of graphite reported previously [10].

The author has demonstrated the sp^2 bonding distributions in the nanocrystalline diamond particles by the $\pi*$ peak of the high loss region of the EEL spectrum. The mapping of sp^2 states reveals that sp^2 bondings are localized in the grain boundaries of 20-50 nm subgrains. Thus one can say that the combination of low loss and high loss regions of EEL spectra enables one to perform extensive

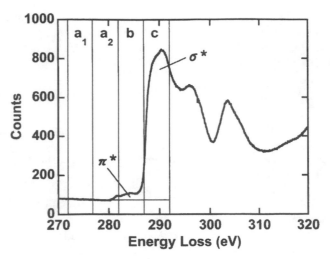

Fig. 3. EEL spectrum of the outer part of a nanocrystalline diamond particle.

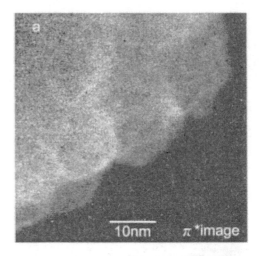

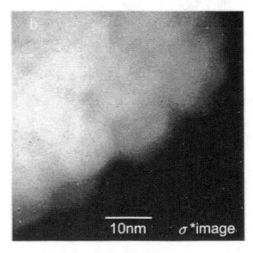

Fig. 4. (a) $\pi*$ image and (b) $\sigma*$ image of the outer part of a nanocrystalline diamond particle.

characterizations of nanocrystalline diamond and nanostructured carbon.

The nanocrystalline diamond particles of 5 nm diameter as shown in Figure 6 are expected to exhibit an emission in UV region and could be used for an UV-light emitting nanodevice.

4 SUMMARY

Nanocrystalline diamond particles of 200-700 nm diameter, two-dimensional platelet-like graphite, and carbon nanotubes were prepared in a 13.56 MHz low pressure inductively coupled $CH_4/CO/H_2$ plasma. The minimum diameter of the particles was found to be 5 nm. The mapping of sp^2 bonding by the $\pi *$ image derived from the EEL spectrum reveals that sp^2-bonded carbons are localized in the grain boundaries of 20-50 nm sub-grains of nanocrystalline diamond particles at approximately 1 nm width. The low loss region of EEL spectra exhibits a bulk plasmon peak at 33 eV and a surface plasmon peak at 23 eV. It is concluded in this study that the combination of low loss and high loss regions of EEL spectra enables one to perform extensive characterizations of nanocrystalline diamond and nanostructured carbon.

REFERENCES

[1] D. M. Gruen, Annu. Rev. Mater. Sci. **29**, 211 (1999).

[2] J. Philip, P. Hess, T. Feygelson, J. E. Butler, S. Chattopadhyay, K. H. Chen, and L. C. Chen, J. Appl. Phys. **93**, 2164 (2003).

[3] J. A. Carlisle, J. Birrell, J. E. Gerbi, O. Auciello, J. M. Gibson, and D. M. Gruen, 8th Inter. Conf. New Diamond Sci. Tech., Melbourne, p. 129, (2002).

[4] G. M. Swain, A. B. Anderson, J. C. Angus, MRS Bull. p.56, (1998).

[5] K. Okada, S. Komatsu, and S. Matsumoto, J. Mater. Res. **14,** 578 (1999).

[6] K. Okada, K. Kimoto, S. Komatsu, and S. Matsumoto, J. Appl. Phys. **93,** 3120 (2003).

[7] K. Okada, H. Kanda, S. Komatsu, and S. Matsumoto, J. Appl. Phys. **88,** 1674 (2000).

[8] R.F. Egerton, M.J. Whelan, J. Elect. Spect. Relat. Phenom. **3,** 232 (1974).

[9] J. Bruley, D.B. Williams, J.J. Cuomo, D.P. Pappas, J. Microscopy **180**, 22 (1995).

[10] R. Brydson, *Electron Energy Loss Spectroscopy*, Springer-Verlag, New York, (2001).

[11] P.J. Fallon, L.M. Brown, Diamond Relat. Mater. **2**, 1004 (1993).

[12] P. Keblinski, D. Wolf, S.R. Phillpot, H. Gleiter, J. Mater. Res. **13**, 2077 (1998).

[13] S. Prawer, K. W. Nugent, D. N. Jamieson, J. O. Orwa, L. A. Bursill, and J. L. Peng, Chem. Phys. Lett. **332**, 93 (2000).

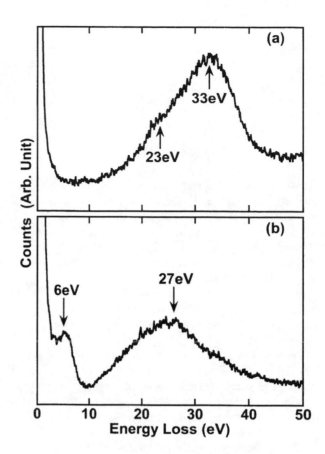

Fig. 5. Low loss spectra of (a) nanocrystalline diamond particle and (b) graphite-like platelet.

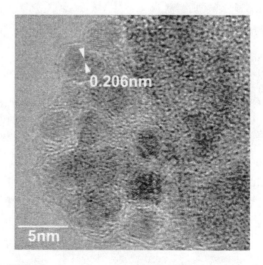

Fig. 6. TEM image of nanocrystalline diamond particles of 5 nm diameter.

New developments in the purification, filling and functionalization of carbon nanotubes

Russell Clarke*

*Thomas Swan & Co Ltd., Consett, County Durham, DH8 7ND, United Kingdom
rclarke@thomas-swan.co.uk

ABSTRACT

A vast amount of research is currently being done on the purification, filling and functionalization of carbon nanotubes (CNTs) due to the potential applications of these materials. Here we report on the use of steam for the purification and opening of CNTs, which allows the removal of the amorphous carbon and the graphitic layers surrounding the metal particles.

The steam purified CNTs can be filled by solutions of the desired compounds. Fullerenes can be used as corks for the containment of the encapsulated materials inside the open-ended CNTs.

The removal of amorphous carbon is proven to be a key step for the sidewall functionalization of CNTs. Finally, we show that the direct visualization of organic molecules covalently bonded to the carbon nanotubes by electron microscopy techniques can be achieved by first labelling them with a high scattering element.

SWAN CNTs were used in all the reported experiments.

Keywords: carbon nanotubes, purification, filling, functionalization, corking

1 CARBON NANOTUBE PURIFICATION BY STEAM TREATMENT

The main impurities in as-made CNTs are typically amorphous carbon, graphitic particles and metal particles. Although several procedures have been used to remove them off, the nitric acid has become the standard reagent for purification of CNTs and constitutes one of the steps in many different purification schemes [1]. Recent studies have shown that acid purification also leads to partial oxidation of CNT themselves and sometimes to an extensive disruption of the tubular structure, especially in the case of single-walled carbon nanotubes (SWCNTs) [2, 3]. On the contrary, the use of the mild oxidising agent steam can remove the amorphous carbon present in samples of as-made CNTs without introducing defects and functional groups, as confirmed by IR and Raman spectroscopies [4]. Raman spectra (Figure 1) show a decrease in the ratio of D/G band intensity (I_D/I_G) from 8.2 ±1.4% for the as-made SWCNTs to 4.6 ± 0.4% after steam treatment at 900 °C for 4 h. On the contrary, when refluxing the as-made SWCNTs in 3 M nitric acid for 45 h, which is the most commonly used method reported in the literature, the D/G band ratio increases from 8.2 ± 1.4% to 44.7 ±

0.5%. Both, sp^3-hybridized carbon present as impurities and SWCNT defects contribute to the intensity of the D-band [5]. If compared to the as-made SWCNTs, the increase in D/G band ratio for the nitric acid treated SWCNTs indicates a large amount of defects, which is consistent with the studies showing that nitric acid purification alters the SWCNT structure. In the case of steam treated samples, the observed decrease in I_D/I_G can be attributed to the removal of carbonaceous impurities during purification and to a self-healing process of the defect sites resulting in an enhanced graphitization. The preferential removal of the more defective SWCNTs could also play a role in the decrease of the D-band intensity. There is no significant change in I_D/I_G when a sample of as-made SWCNTs is heated at 900 °C for 4 h under argon ($I_D/I_G = 7.7 \pm 1.3\%$; spectrum not shown). The radial breathing modes (RBM) show the same features before and after the steam purification, which indicates that the tubular structures of the SWCNTs are preserved during the steam treatment.

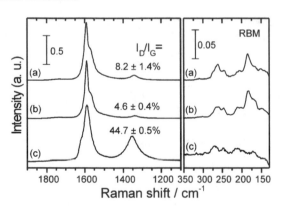

Figure 1. Raman spectra of (a) as-made SWCNTs, (b) steam-purified SWCNTs at 900 °C for 4 h, (c) as-made SWCNTs after treatment with 3 M HNO₃ for 45 h.

We observed that amorphous carbon was much more reactive with the steam than carbon nanotubes and graphitic particles. No amorphous carbon was present after 2 h treatment. Although graphitic particles cannot be removed from the sample without consuming SWCNTs, we found that 4 h treatment was a good compromise for purifying the as-made SWCNTs. Figure 2 shows HRTEM pictures of as-made SWCNTs and after steam treatment at 900 °C for 4 h. The purification of the sample with steam clearly removes amorphous carbon and some graphitic particles entangling the as-made SWCNTs (Figure 2a) leaving behind cleaner SWCNTs (Figure 2b).

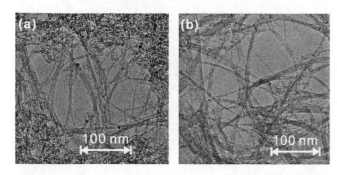

Figure 2. HRTEM images of (a) as-made SWCNTs and (b) SWCNTs after steam treatment at 900°C for 4h

Recently we have proved that steam can also remove the graphitic shells coating the metal particles present in samples of as-made SWCNTs [6]. These now exposed catalytic particles can be easily dissolved by hydrochloric acid treatment.

In order to get the minimum amount of metal catalyst particles, the steam treatment time needs to be optimized. Therefore we treated SWCNTs for different times with steam at 900°C and we subsequently washed the steam treated samples with HCl. Figure 3 summarises the metal content (% of Fe) in each steam and acid washed sample. The values were obtained by carrying out TGA experiments under flowing air.

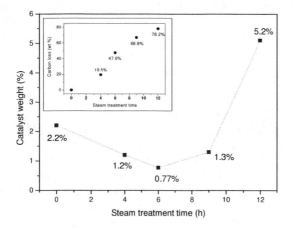

Figure 3. Summary of the amount of iron present in the as-made SWCNTs and the different steam treated samples after the HCl wash. *Inset:* Weight percentage of carbon loss with respect to the amount of carbon in the starting material after different steam treatment times.

The as-prepared sample has a metal content of 2.2 %wt. Lower metal contents with respect to the starting material can be obtained after 4 h, 6 h and 9 h steam treatment. However, the lowest percentage of catalytic particles is observed after 6 h treatment (0.77%). Longer treatments result in a relative increase of the metal content (1.3% after 9 h and 5.2% after 12 h) and therefore are not recommended. The carbon loss in weight % with respect to the initial amount of carbon present in the sample is

depicted in the inset in Figure 3. As expected, the longer the steam treatment the more carbon in its different forms (amorphous carbon, graphitic particles or SWCNTs) is removed from the sample. Therefore there is a clear compromise between the quality of the sample and the weight loss.

Steam purification has the advantages of being an economic, easily scalable and non-toxic process. The purification time needs to be adjusted depending on the carbon nanotube source. For instance, arc-discharged SWCNTs are more reactive towards steam compared to SWAN CVD SWCNTs [7].

2   FILLING AND CORKING OF CARBON NANOTUBES

As-made SWCNTs, normally closed at both ends, may be readily filled by direct heating to about 700-900 °C in the presence of any material which is liquid at that temperature and of suitable surface tension [8]. On cooling the reaction mixture, the ends of the filled SWCNTs are still closed [9]. Therefore, the excess of material external to the SWCNTs may be dissolved away by choice of a suitable solvent. This high temperature filling method is limited mainly by the requirement that the chosen filling material is thermally stable as a melt. For example, organic molecules can not be filled using this approach.

The alternative methods of filling SWCNTs require opened ends. This can readily be accomplished by use of an oxidising agent such as steam [4]. The resulting open-ended SWCNTs may then be filled by solution, vapour or melt of the chosen materials [10]. A high resolution transmission electron microscopy (HRTEM) image of uranyl acetate solution-filled carbon nanotubes is shown in Figure 4. The encapsulated uranium compound is presented in the form of short crystals along the SWCNTs, characteristic of the solution filling method.

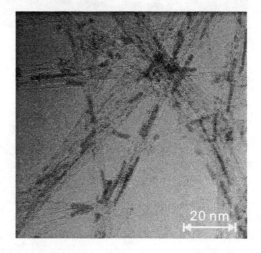

Figure 4. HRTEM image of an uranium compound encapsulated inside steam opened SWCNTs

The solution filling experiment was carried out by stirring 10 mg of steam-purified carbon nanotubes at 70 °C for 2 days in a saturated solution of uranyl acetate $\{UO_2(CH_3CO_2)_2(H_2O)_2\}$. Uranyl acetate was chosen since the heavy element component uranium can be easily observed by HRTEM.

Bulk filling of carbon nanotubes always results in a large amount of unwanted external material. However, when using solution filling it is not possible to remove the material external to the filled SWCNT by washing, as this also removes the material encapsulated in the open-ended SWCNTs. To solve this problem, fullerenes have been used as corks (plugs) to block the opened ends of filled SWCNTs [11]. The resulting "nanocapsules" can then be readily purified from the external material by washing in a suitable solvent.

Following the procedure described by Iijima *et al.* for filling C_{60} into the SWCNTs (C_{60}@SWCNTs) [12], the fullerene C_{60} in ethanol was added to the uranyl acetate filled SWCNTs (prepared as described above). The mixture was then stirred in aqueous hydrochloric acid, in which uranyl acetate is very soluble. After filtering and drying, the sample was examined by HRTEM and the continuing presence of uranyl acetate inside the SWCNTs was confirmed whilst the uranium material outside the SWCNTs had been removed. Also, the presence of both C_{60} molecules and uranyl acetate crystals could be observed along individual SWCNTs (Figure 5). A blank experiment with uranyl acetate was made by repeating all the experimental procedures but without addition of C_{60}. After the final washing step with hydrochloric acid, no uranyl acetate could be detected neither inside nor outside the SWCNTs and only empty SWCNTs were observed by HRTEM. The fact that filled tubes with uranyl acetate after the HCl wash can only be seen when C_{60} are also present, confirms the ability of C_{60} to seal materials inside the SWCNTs.

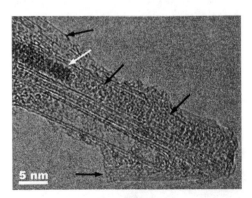

Figure 5. HRTEM images of C_{60} corked uranyl acetate filled SWCNTs after being washed in HCl. As guides to the eye, white arrows point the uranium compound, and black arrows point some of the encapsulated C_{60}.

The methodology described here is simple and effective way to seal soluble materials inside open-ended SWCNTs, using C_{60} molecules, and allow the washing of the external undesired material. Samples free of material outside the SWCNTs are of great interest for the bulk characterization and application of filled SWCNTs.

3 FUNCTIONALIZATION OF CARBON NANOTUBES

The use of CNTs for materials science and biomedical applications is limited by their poor solubility and processability. An approach to overcome these problems consists in the chemical functionalization of the sidewalls of the carbon nanotubes. The most common functional group that has been discussed in the literature is the covalently bonded COOH group, by treatment of CNT samples with nitric acid or a mixture of sulfuric and nitric acids [13].

The nitric acid treatment of as-produced SWCNTs generates contaminating debris, which can be easily removed by base wash leaving behind CNT samples with low degree of functionalization. The sidewall functionalization of carbon nanotubes using nitric acid treatment can be greatly enhanced by first removing the amorphous carbon present in the sample [7].

As-made SWCNTs (sample A) were refluxed in 3 M nitric acid for 24 h, followed by filtration and rinsing with water. The remaining solid (sample B) was treated with 4 M NaOH, filtered and rinsed with water leading to a dark filtrate due to the presence of oxidation debris. The solid sample on the filter membrane was further washed with water (sample C). The IR spectrum of sample A (Figure 6a) shows an absence (low degree) of functional groups. After nitric acid treatment (sample B) strong peaks appear at 1735 cm^{-1}, 1585 cm^{-1} and 1200 cm^{-1}, due to the C=O, C=C and C–O stretching transitions respectively, most likely due to formation of carboxylic acid groups. The peak at 1200 cm^{-1} could also be assigned to O–H bending. The peaks at 1735 cm^{-1} and 1200 cm^{-1} vanish after NaOH wash (sample C) due to the removal of the oxidation debris, which contains the majority of COOH functionality created by the acid treatment.

In another set of experiments, as-made SWCNTs were treated with steam in order to remove the amorphous carbon [4]. The same reactions performed on sample A were then carried out on this purified material (sample A'), namely treatment with nitric acid (sample B') and washing with NaOH (sample C'). Only a small peak at 1575 cm^{-1} due to C=C bonds can be observed in the IR spectrum after steam purification (sample A', Figure 6b), confirming that this treatment does not introduce functional groups. The IR spectrum of sample B' shows the appearance of the C=O band corresponding to the formation of carboxylic acid groups in the sample. This peak remains unchanged on the IR spectrum of the powder (sample C') obtained after washing the sample with NaOH, confirming that for the purified sample the majority of the carboxylic acid groups are present on the SWCNT walls rather than on oxidation debris. The band around 1200 cm^{-1} due to C–O stretching becomes more visible after the NaOH wash.

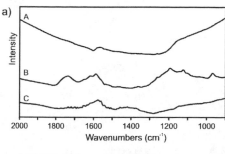

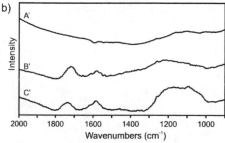

Figure 6. FTIR spectra of (a) as-made SWCNTs (sample A), after acid treatment (sample B) and base wash (sample C) and (b) steam purified SWCNTs (sample A'), after acid treatment (sample B') and base wash (sample C').

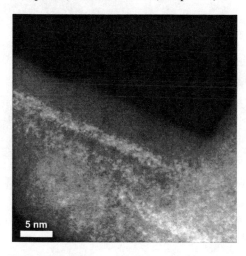

Figure 7. Z-contrast dark field STEM images of SWCNTs functionalized with iodide labelled carbohydrates. Individual iodides (white dots) can be clearly seen.

However, these do not provide direct evidence of the attachment of individual molecules. The use of electron microscopy techniques such as HRTEM and scanning transmission electron microscopy (STEM) for characterization of organic compounds is typically limited

by the lack of structural periodicity and the presence of low scattering elements (H, C, N, O). We have been able to directly visualise organic molecules covalently bonded to the carbon nanotubes by first labelling them with a high scattering element, such as iodine [14] (Figure 7).

REFERENCES

[1] E. Dujardin, T. W. Ebbesen, A. Krishnan, M. M. J. Treacy, Adv. Mater. 10, 611, 1998.

[2] H. Hu, B. Zhao, M. E. Itkis, R. C. Haddon, J. Phys. Chem. B 107, 13838, 2003.

[3] M. Monthioux, B. W. Smith, B. Burteaux, A. Claye, J. E. Fischer, D. E. Luzzi, Carbon 39, 1251, 2001.

[4] G. Tobias, L. Shao, C. G. Salzmann, Y. Huh, M. L. H. Green, J. Phys. Chem. B 110, 22318, 2006.

[5] R. Saito, G. Dresselhaus, M. S. Dresselhaus in *Physical Properties of Carbon Nanotubes*, Imperial College Press, Singapore 1998.

[6] B. Ballesteros, G. Tobias, L. Shao, E. Pellicer, J. Nogués, E. Mendoza, M. L. H. Green, Small doi: 10.1002/smll.200701283, 2008.

[7] L. Shao, G. Tobias, C. G. Salzmann, B. Ballesteros, S. Y. Hong, A. Crossley, B. G. Davis, M. L. H. Green, Chem. Comm. 5090, 2007.

[8] E. Dujardin, T. W. Ebbesen, H. Hiura, K. Tanigaki, Science 265, 1850, 1994.

[9] L.D. Shao, G. Tobias, Y. Huh, M. L. H. Green, Carbon 44, 2855, 2006.

[10] M. Monthioux, Carbon 40, 1809, 2002.

[11] L. Shao, T.-W. Lin, G. Tobias, M. L. H. Green, Chem. Comm., doi: 10.1039/b800881g, 2008.

[12] M. Yudasaka, K. Ajima, K. Suenaga, T. Ichihashi, A. Hashimoto, S. Iijima, Chem. Phys. Lett. 380, 42, 2003.

[13] J. Liu, A. G. Rinzler, H. Dai, J. H. Hafner, R. K. Bradley, P. J. Boul, A. Lu, T. Iverson, K. Shelimov, C. B. Huffman, F. J. Rodriguez-Macias, Y.-S. Shon, T. R. Lee, D. T. Colbert and R. E. Smalley, Science 280, 1253, 1998.

[14] S. Y. Hong, G. Tobias, B. Ballesteros, F. El Oualid, J. C. Errey, K. J. Doores, A. I. Kirkland, P. D. Nellist, M. L. H. Green, B. G. Davis, J. Am. Chem. Soc. 129, 10966, 2007.

Linear scaling techniques for first-principle calculations of large nanowire devices

D. Zhang and E. Polizzi

Department of Electrical and Computer Engineering
University of Massachusetts, 01003 Amherst, USA
dzhang@ecs.umass.edu, polizzi@ecs.umass.edu

ABSTRACT

The purpose of this paper is to improve performances for the calculation of the electron density at a given step of the self-consistent first-principle procedure (for one given atomistic potential configuration). We propose to use a combination of numerical techniques and demonstrate their robustness and scalability for simulating a 3D Carbon nanotube (CNT) device.

Keywords: First-principle calculations, DFT/Kohn-Sham, Carbon nanotube, FEM, Electronic structure, Mode approach, Contour integration

1 Introduction

In nanoelectronics, large scale 'ab-initio' simulations could significantly enhance our understanding of nanoscale physics and engineering related issues of materials and transistor devices. At large scale (scale of a transistor), first-principle atomistic simulations of devices are in need of appropriate efficient modeling strategies and innovative numerical algorithms.

In this study, we propose to make use of a full 3D atomistic real-space mesh technique to discretize the DFT/Kohn-Sham equations for an entire CNT device. Real-space mesh techniques for first-principle electronic and transport calculations (finite element method- FEM- in our case) can exhibit significant numerical advantages as compared to other traditional discretization techniques such as plane waves expansion schemes or linear combination of atomic orbitals [1]. However, they have been so far limited to the simulation of small devices containing only few number of atoms. One limiting factor is concerned with the high numerical costs that are required for calculating the electron density at each step of the DFT/Kohn-Sham self-consistent iterations. In [2], we have presented an effective combination of the following three $O(n)$ numerical techniques for addressing these problems for nanowire-type devices: (i) the electron density is calculated by performing a contour integration of the Green's function along the complex energy plane [3], [4], (ii) a mode approach is used to reduce the size of the discretized problem (typically from $O(10^8)$ to $O(10^5)$) while producing very narrow banded matrices [5], (iii) the diagonal elements of the Green's

function are obtained using our in-house $O(n)$ banded system solver. The summary chart of our numerical approach is showed in Figure 1.

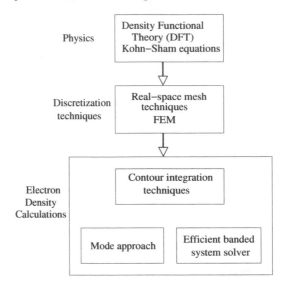

Figure 1: Summary chart of our combined modeling techniques for computing the electron density using first principle calculations.

Our results have shown that the time needed to compute the electron density in electronic structure calculations using our contour integration technique, is now order of magnitude faster than the one obtained using other traditional approaches such as solving eigenpairs. Here, we propose to discuss how we can improve the robustness and scalability of those three techniques and their combination while performing numerical simulations on 3D Carbon nanotube (CNT).

2 Numerical scaling techniques for electronic density calculations

2.1 The contour integration technique

For isolated systems, the contour integration technique allows solving a large number of eigenpairs of the Kohn-Sham equation to be replaced by computing the diagonal elements of few Green's function matrices ($O(10)$) in the complex plane[3]. The expression of the electron density is then given by:

$$n(\mathbf{r}) = -\frac{1}{\pi}\left(\int_C dz\mathrm{Im}(G(\mathbf{r},\mathbf{r},Z_E))f_{FD}(Z_E - E_F)\right)$$
$$-\frac{1}{\pi}\sum_{poles,n}\mathrm{Im}(-2i\pi k_B T G(\mathbf{r},\mathbf{r},Z_{E_n})) \quad (1)$$

where $G(\mathbf{r},\mathbf{r},Z_E)$ represents the diagonal elements of the Green's function for the complex energy Z_E, the complex contour C contains all the resonances on the real axis (eigenvalue solutions of the Kohn-Sham equation), and the second term of the right hand side corresponds to the contribution of the residues of all poles on the imaginary axis that are introduced by the Fermi-Dirac distribution function f_{FD}. One should note that $Z_{E_n} = E_F + i(2n+1)\pi k_B T$. For open systems (transport problem), this approach can be used as well to perform the integration on the equilibrium part of the electron density while avoiding solving a prohibitive number of linear systems[4].

Since, in practice, only the diagonal elements of few Green's functions need to be calculated along the complex plane (this number is independent of the size of the system), the contour integration technique represents an attractive alternative approach to the traditional eigenvalue problem. In order to take fully advantage of the contour integration technique, however, we need to overcome the major numerical difficulties in inverting directly the Hamiltonian matrix (to obtain the Green's function) that can reach size of millions (for large atomistic systems). For matrices that are banded, which are naturally obtained after discretization with nanotubes or any type of nanowire devices, it is indeed possible to make use of an efficient algorithm to compute only the diagonal elements of the inverse of a banded matrix. This algorithm, which will be presented in section 2.3, exhibits a linear scaling performance $O(b^2 n)$ with the size of the system n and a quadratic scaling with the bandwidth b. Therefore this scheme, and then the contour integration technique, is particularly well adapted for narrow banded systems. In order to reduce the bandwidth of the Hamiltonian matrix obtained by the finite element method (FEM), we make use in the following of a subband decomposition technique (mode approach) [5].

2.2 The mode approach

One of the underlying advantage of using an atomistic real-space mesh discretization, is to be able to rigorously perform a mode approach where the quantum confinement and transport problems are treated separately. The $3D$ wave-function are expanded by

$$\Psi_{3D}(x,y,z) = \sum_{n=1}^{\infty}\psi_n(x)\chi_n(y,z), \quad (2)$$

where y,z are the coordinates in the cross section of the wire. and the calculation procedure is as follows:
(a) Solving only one $2D$ Schrödinger equation with closed (Dirichlet) boundary condition to obtain the eigenvalues E_n and corresponding eigenfunctions $\chi_n(y,z)$ in the cross section:

$$-\frac{\hbar^2}{2m}\left(\frac{\partial^2}{\partial y^2} + \frac{\partial^2}{\partial z^2}\right)\chi_n(y,z) + U_{2D}\chi_n(y,z) = E_n\chi_n(y,z).$$
$$(3)$$

The same mathematical complete basis set $\{\chi_n(y,z)\}$ is then used for all cross sections along the nanotube ($\forall x$). Therefore, a suitable choice for the 2D potential U_{2D} can be defined by [5], [6]

$$U_{2D}(y,z) = \frac{\int_x U(x,y,z)n(x,y,z)}{\int_x n(x,y,z)} \quad (4)$$

where the regions "seen by the electrons" are captured in average along the longitudinal x direction.
(b) Solving the following 1D coupled Schrödinger equation (with appropriate boundary conditions):

$$-\frac{\hbar^2}{2m}\frac{\partial^2}{\partial x^2}\psi_n(x) + \sum_{m=1}^{\infty}U_{mn}\psi_n(x) = (E - E_n)\psi_n(x) \quad (5)$$

$$U_{mn}(x) = \int_{y,z}(U - U_{2D})\chi_n\chi_m dydz \quad (6)$$

When this decomposition is accounting for all coupling between modes and with a sufficient number of modes, the mode approach is equivalent to obtaining the full 3D real-space solutions. If only M modes are taken into account in the calculations, the size of this system matrix and bandwidth can be both reduced by a factor $N_{Y,Z}/M$ (where $N_{Y,Z}$ is the number of nodes in the cross section and $M << N_{Y,Z}$ in practice). Using an appropriate reordering, the obtained matrix is block tridiagonal, each $M \times M$ blocks being completely dense. Using this approach, however, all the diagonal blocks of size $M \times M$ of the obtained mode approach Green's function matrix g are needed in the calculation of the density since:

$$G(\mathbf{r},\mathbf{r},Z_E) = \sum_{m,n}\chi_m(y,z)\chi_n(y,z)g_{m,n}(x,x,Z_E).$$

where $g_{m,n}(x,x',Z_E)$ denotes the matrix elements of g.

2.3 Banded system solver

Since the obtained mode approach matrices are narrow banded and block tridiagonal (as described above), in Figure 2 we show that it is possible to make use of a LU factorization and modified forward and backward sweeps which focus only in computing the diagonal blocks of the Green's function. This banded system solver for diagonal elements is the numerical algorithm underneath the recursive Green's function technique[7]. The computational cost of this technique is

$O(b^2N)$ which is linear in N, and takes advantage of the small bandwidth b obtained using the mode approach. In addition, our diagonal system solver has been implemented using LAPACK and BLAS level 3 routines to minimize memory references and increase performances.

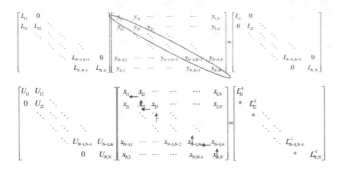

Figure 2: Overview of the different solve steps for obtaining the diagonal blocks of the inverse of a symmetric and non singular banded matrix after a LU factorization without pivoting (i.e. obtaining the diagonal blocks of X where AX=I and A=LU). For the forward sweep $LY = I$, we compute only the diagonal blocks of Y which corresponds to the inverse of each local diagonal blocks of L. Then for the backward sweep we can proceed as follows: (i) solve $U_{n,n}X_{n,n} = L_{n,n}^{-1}$ to obtain $X_{n,n}$ (ii) solve $U_{n-1,n-1}X_{n-1,n} = -U_{n-1,n}X_{n,n}$ to obtain $X_{n-1,n}$, (iii) since the matrix is symmetric it comes $X_{n,n-1} = X_{n-1,n}$, (iv) Solve $U_{n-1,n-1}X_{n-1,n-1} = L_{n-1,n-1}^{-1} - U_{n-1,n}X_{n,n-1}$ to obtain $X_{n-1,n-1}$, (v) repeat recursively from step (ii) with $n <= n-1$ until $n = 1$.

3 Results

In order to illustrate the efficiency of the above techniques, one needs to consider two parameters: M, the number of modes that will be considered, and N_E, the number of points of discretization for the complex energy contour. For a given atomistic potential configuration (the Hartree and exchange-correlation potentials are given), we propose to compute the electron density of an isolated (13,0) CNT with arbitrary length by solving the Schrödinger-type equation with different values for M and N_E.

3.1 Validity of the contour integration technique

In table 1, we propose first to fix the number of modes M and compare the results obtained on the electron density with the contour integration technique and different N_E, with the ones obtained with a traditional eigenvalue solver (used as reference). The results indicate that using an uniform repartition of the energy points along the contour, the poles Z_{E_n}, which appear in equation (1) at room temperature, may not be captured appropriately. This gives rise to inconsistency in

N_E	1-unit ($T = 0K$)	1-unit, uniform ($T = 300K$)	1-unit, non-uniform ($T = 300K$)
40	0.85%	1.3%	0.50%
60	0.34%	7.1%	0.34%
80	0.19%	0.6%	0.22%
100	0.14%	1.5%	0.22%
200	0.12%	0.3%	0.18%

Table 1: Relative residual (error) $||n_e - n_c||_\infty / ||n_e||_\infty$ for different calculations of the electron density in function of N_E by two techniques: eigenvalue problem (n_e) and contour integration (n_c), for 1-unit CNT with 50 modes included. A non-uniform repartition of energy points becomes necessary at room temperature to capture the abrupt variation of the Green's function caused by the poles Z_{E_n}.

the error while varying the number of energy points N_E. In order to obtain an identical accuracy at both zero and room temperatures, we propose in Figure 3 to make use of a non-uniform repartition of energy points allowing a much larger concentration of contour points around $Re(Z_{E_n})$.

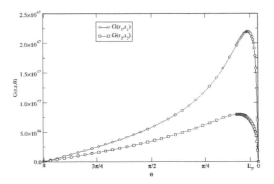

Figure 3: Variation of the Green's function at a given position in space along the complex contour of integration going from $\theta = 0$ to $\theta = \pi$ (half-circle), and using a non-uniform repartition of the energy points. For the purpose of illustrating a simple case, we select two particular real space points $\mathbf{r_1}$ and $\mathbf{r_2}$ inside the CNT, one at the position of a center of an atom and the other one between two atoms. The contour integration of these two curves give the electron density at position $\mathbf{r_1}$ and $\mathbf{r_2}$. Using a non-uniform mesh (here 60 points), one can see that the abrupt variation of the Green's function is well captured, and in table 1 we obtain a 0.34% error accuracy as compared to solving the eigenvalue problem.

3.2 Validity of the mode approach

In order to illustrate the robustness of the mode approach, we propose to fix the number of energy points along the contour integration to $N_E = 100$ and vary the number of modes from $M = 1$ to $M = 100$. In Figure 4, the case $M = 100$ is also used as the reference solution to calculate the error, since for this large number of modes and coupling between modes the mode approach and the full 3D solutions can be considered equivalent. The curves show that the error decays exponentially as we increase the number of modes and for $M = 50$, for example, the error is already below 2%.

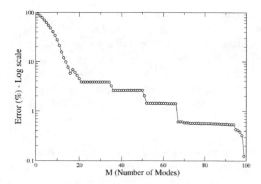

Figure 4: Error on the electron density obtained using different number of modes M and compared to a reference solution where $M = 100$.

3.2.1 Scalability of the proposed techniques

Table 2 summarizes the time obtained by the combination of our proposed technique for computing the electron density of CNT with different lengths. One can

#unit cells	1-unit	3-unit	6-unit	24-unit	48-unit
$L(nm)$	0.57	1.42	2.70	10.37	20.59
#atoms	52	156	312	1248	2498
$T_{eigen}(s)$	283	5806	N/A	N/A	N/A
$T_{contour}(s)$ $N_E = 100$	33	86	165	629	1245
$T_{contour}(s)$ $N_E = 60$	20	52	100	381	754

Table 2: Times obtained using our contour integration technique ($N_E = 60$ and 100, $M = 50$) and solving the eigenvalue problem for obtaining the electron density of a (13,0) CNT up to 48-unit cell (2498 atoms). The calculations are performed on only one core of a Clovertown 2.66Ghz and 16Gb. The eigenvalue problem is solved using the direct banded eigenvalue solver in LAPACK and exhibits memory limitations beyond 3-unit cells.

see that the time for the contour integration technique scales linearly with the length of the device, and it is directly proportional to the number of energy points in the contour N_E. If we compare these results with those obtained with a more traditional eigenvalue solver

technique (which takes also advantage of the mode approach), one observe a $\sim \times 100$ speed up improvement for 3-unit cells.

4   Discussions

The proposed numerical scheme is applicable to any nanowire type device independently of their size (as long as the cross section of the wire is much smaller than its length). While increasing the length of the device, we have demonstrated in Table 2 that the computational cost of the our numerical scheme is linear. The efficiency of this technique will eventually exhibit limitations if the cross section of the nanowire does not contain less than ~ 100 atoms (in order to minimize the number of modes that are needed). It would be possible, however, to address these limitations by coupling the proposed approach with an outer layer domain decomposition technique.

It is also possible to perform transport calculations within the framework of the proposed scheme. To account for the flow of electrons, small dense blocks (self-energy matrices) need to be added at the edges of the the obtained block tridiagonal Hamiltonian matrix with the mode approach. Since all the blocks are already treated as dense blocks, it is possible to solve the resulting linear systems (resulting from the NEGF transport equation [8]) with at least the same efficiency than for the isolated systems (the efficiency will be in practice much higher since the diagonal elements are not needed here and the complexity of the forward and backward sweeps will be reduced).

Acknowledgment

This material is based upon work supported by the National Science Foundation under Grant No. CCF 0635196 and EEC 0725613.

REFERENCES

[1] T. Beck, Rev. of Modern Phys., 72, 1041, (2000).

[2] D. Zhang, E. Polizzi, Journal of Comput. Electronics, to appear(2008).

[3] R. Zeller, J. Deutz, and P. H. Dederichs, Solid State Commun. 44, 993-997 (1982).

[4] M. Brandbyge, J.-L. Mozos, P. Ordejn, J. Taylor, and K. Stokbro, Phys. Rev. B 65, 165401 (2002).

[5] E. Polizzi, N. Ben Abdallah, J. of Comput. Phys., 202, 150-180 (2005).

[6] E. Polizzi, N. Ben Abdallah Phys. Rev. B 66, 245301 (2002).

[7] F. Sols, M. Macucci, U. Ravaioli, J. Appl. Phys. 66 (8), 3892 (1989).

[8] E. Polizzi, A. Sameh, H. Sun, 2004 NSTI Nanotechnology Conference and Trade Show. Technical Proceedings, Vol. 2, Chapter 8, pp403-406 (2004)

Controlled Assembly of High Density SWNT Networks on a Flexible Parylene-C Substrate

C.-L. Chen[1,*], X. Xiong[2,*], A. Busnaina[2], and M. R. Dokmeci[1]

[1] ECE Department, NSF Center for High Rate Nanomanufacturing, Northeastern University, Boston, MA
[2] MIE Department, NSF Center for High Rate Nanomanufacturing, Northeastern University, Boston, MA
chen.ch@neu.edu

ABSTRACT

In this paper, we present a directed assembly technique for controlled micro-patterning of Single-Walled Carbon Nanotubes (SWNTs) on a flexible parylene-C substrate for electronic applications. The presented large scale fabrication of ordered carbon nanotube arrays and networks is achieved by performing site-selective fluidic assembly of SWNTs on a plasma treated parylene-C substrate. Parylene-C, which is lightweight, mechanically strong and stress-free material deposited at room temperature, is an emerging substrate material for flexible devices. The uniformly deposited nanotube lateral structures are formed directly on the parylene-C substrate without utilizing printing or transfer techniques. Both electrical and structural characterizations are performed on the SWNT-based devices on the flexible substrate. The developed nanotube patterning on polymeric substrates has immediate applications in wearable electronics and sensors, flexible field effect transistors (FETs) and lateral interconnects.

Keywords: Flexible Parylene-C substrates, Single-Walled Carbon Nanotubes, Nanoscale patterning, Dip coating.

1 INTRODUCTION

Single-Walled Carbon Nanotubes (SWNTs) with their attractive properties such as large surface-to-volume ratio, high packing density and long-range order may serve as the potential building blocks for the next generation of nanoscale devices. [1, 2] The integration of ordered arrays of carbon nanotubes on to rigid as well as flexible substrates offers many opportunities for realizing novel multifunctional devices. [3-5] The transfer of vertically or horizontally aligned carbon nanotube structures are often realized utilizing complicated steps of site-selective CVD nanotube growth or conformal contact printing. [6, 7] The major problem of CVD based approach is the requirement of high processing temperatures (~800°C) which is not CMOS compatible and also can't be applied to most polymeric devices. PDMS stamps are also utilized as an intermediate carrier to transfer-print SWNTs on to different substrates including plastic sheets. [8] Though challenging, it is highly desirable to fabricate integrated nanotube-polymer flexible electronic devices at room temperature in a simple and cost-effective way. Here, we present a novel technique for localized patterning of SWNT networks on a flexible Parylene-C substrate by utilizing surface controlled microfluidic assembly technique (Fig.1).

Unlike most rigid substrates which can be chemically functionalized for large-scale assembly of carbon nanotubes, [9, 10] the soft polymer surfaces are hydrophobic and their properties cannot be easily altered chemically. The low surface energy makes the direct assembly of nanotubes on a hydrophobic surface a challenging task. To overcome these challenges, we developed a plasma treatment method to modify the surface properties of the polymer for effective direct assembly of carbon nanotubes onto its surface. The previous SWNT patterning approach used a PDMS substrate which was also flexible. [11] A major limitation of the previous approach, however, was the requirement of a shadow mask which limited the flexibility in pattern dimensions. In this study, we utilized photolithography for the large-scale assembly of SWNT arrays on parylene-C substrates.

This versatile technology has direct applications in the realization of nanoscale devices on flexible substrates potentially useful in numerous fields including flexible electronics, wearable nanosensors, CNT-field effect transistors and lateral CNT interconnects.

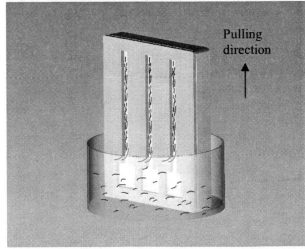

Fig.1 Schematic drawing of the surface controlled microfluidic assembly technique.

2 FABRICATION PROCESS

The fabrication process is shown in Fig.2 and starts with the deposition of a 10μm parylene-C layer on top of a

---

*These authors contributed equally.

15mm×15mm silicon substrate. Next, a brief O_2 plasma treatment is utilized to render the as-deposited hydrophobic parylene-C surface hydrophilic. Parylene-C, is known to have very low permeability to moisture and corrosive gases and has been historically utilized in encapsulation applications. Along with its ability to create a true pin-hole free insulation, the high resistivity, $6×10^{16}\Omega$-cm, and high breakdown voltage (300Volts/μm), all indicate that Parylene-C has excellent dielectric properties. The physical properties such as its high tensile strength (10,000psi) and mechanical strength (Young's modulus of 400Kpsi) [12] suggest that it is also a very promising candidate amongst materials for flexible substrates. The surface property of the plasma treated parylene-C substrates was studied by measuring contact angles in a dynamic mode (Phoenix 300 Plus, SEO). The contact angle before oxygen plasma treatment of parylene-C was 97.2°±4.2° [13]. After the oxygen plasma treatment, the parylene-C surface became hydrophilic (contact angle was around 4.82°±0.57°). Fig. 3 shows the contact angle measurements from the parylene-C surface. The hydrophilic property of parylene-C surface is relatively stable after plasma modification (for at least 2 hours following the plasma treatment). This characteristic will be utilized for the SWNT fluidic assembly.

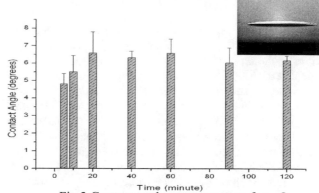

Fig.3 Contact angle measurements from O_2 plasma treated Parylene-C substrates.

3 RESULTS

3.1 Dip Coating Process

Commercially available SWNTs (Nantero Inc.) dispensed in an aqueous solution are used in our experiments. The SWNT solutions consist of about 0.23wt% SWNTs with lengths ranging from 1 to 5μm. The SWNTs are terminated with carboxylic acid groups, which adsorb ions such as H^+ and OH^- from the aqueous solution resulting in the presence of a net charge on the surface of the SWNTs. The patterned substrate was first vertically submerged into the solution containing SWNTs. The substrate was next gradually pulled out of the solution with a constant pulling speed of 0.1mm/min using a dip coater. Fig.4 shows the SEM micrographs of the assembled SWNT patterns, and the inset shows the details of the assembled SWNT networks inside the trenches. The assembled nanotube patterns retain their original shape after the removal of the photoresist layer. The assembled SWNT structures are dense and show complete coverage over the parylene-C surface. In addition, the boundaries of the nanotube micropatterns are well defined.

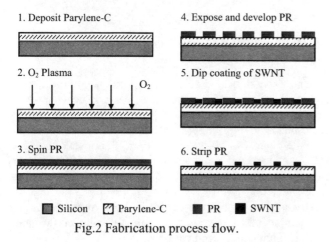

1. Deposit Parylene-C
2. O_2 Plasma
3. Spin PR
4. Expose and develop PR
5. Dip coating of SWNT
6. Strip PR

☐ Silicon ▨ Parylene-C ▨ PR ■ SWNT

Fig.2 Fabrication process flow.

We then performed photolithography to create trenches (30μm×1.1mm) on the silicon die. The patterned substrate was first vertically submerged into an aqueous solution containing SWNTs using a dip coater (KSV Instruments). The substrate was then gradually pulled upward from the solution with a constant pulling speed of 0.1 mm/min. Due to the hydrophobic nature of the top photoresist surface, surface controlled microfluidic assembly technique can be utilized to selectively assemble SWNTs inside the hydrophilic trenches. After assembly of the SWNTs, the photoresist was next removed in acetone in a gentle manner, leaving the assembled SWNT arrays intact on desired parylene-C sites. The thin (10μm) layer of parylene-C film with SWNT patterns was next peeled off from the silicon substrate and is ready for characterization.

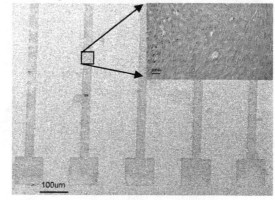

Fig.4 shows the SEM micrograph of the assembled SWNT patterns, and the inset shows the details of the assembled SWNT networks inside the trenches.

3.2 I-V Characterization

A potential application of the assembled SWNT microstructures is in macroelectronics, specifically in flexible devices. We next conduct electrical characterization of the assembled nanotube microstructures on Parylene-C substrates. The dimensions of the tested nanotube microlines are 30 μm in width and 1mm in length. The measured I-V results are shown in Fig. 5. The two-terminal resistance measurements from the nanotube strips ranged from 15KΩ to 21KΩ. To test the reliability of the assembled SWNTs on the defined trenches, a repeatability test was performed on the same sample which demonstrated a variation of less than 1% and is shown in Fig. 6.

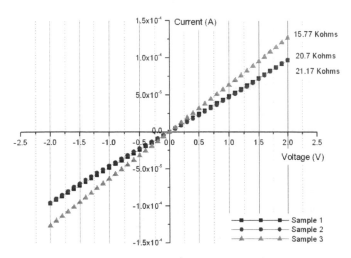

Fig.5 I-V measurements from assembled SWNT networks.

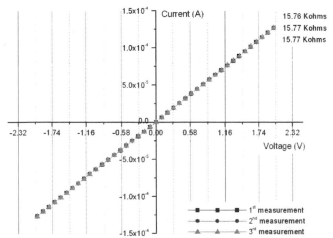

Fig.6 Repeatability test on the same sample where the variation is found to be less than 1%.

3.3 Raman Spectroscopy Measurements

To further characterize the structural properties of the assembled SWNTs, the Raman spectra analysis was performed. We obtained an estimation of the SWNT diameter distribution using Raman spectroscopy with a 783 nm excitation. The diameter, chirality and phonon structure are reflected in the first and second order Raman frequencies. For example, the radial breathing mode (RBM) is usually located between 75 and 300 cm^{-1}. [14] The diameter of the assembled SWNT is calculated from equation d (nm) = 248/[ω(cm^{-1})], [15] where d is the diameter of the assembled SWNTs, and ω is the RBM frequency. Fig. 7 shows the spectral range between 100 to 450 cm^{-1}, with radial breathing mode (RBM) of SWNT peaks at frequencies around 160, 210, and 270 cm^{-1} (taken from 4 different spot areas). RBM analysis revealed that the observed frequencies of the RBM peaks correspond to SWNT diameters in the range of 0.9 – 1.55 nm. The peak at 210 cm^{-1} indicated a dominant SWNT of approximately 1.18 nm.

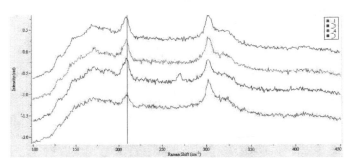

Fig.7 Raman spectra measured from 4 different points of the same SWNT networks.

The D band located around 1300 cm^{-1} is due to defects or disordered features in the nanotubes. The G mode is usually located between 1500 and 1600 cm^{-1}, which corresponds to the tangential C-C bond stretching in the graphite plane. As observed in Fig. 8, the observed peak is located at 1585 cm^{-1}, and a minor shoulder shifted to lower frequencies by around 25 cm^{-1}, indicating the presence of semiconducting SWNTs. [16] Our results show that the electrical behavior of high-density SWNT micro-patterns is linear or metallic, which is suitable for high-performance flexible interconnect applications.

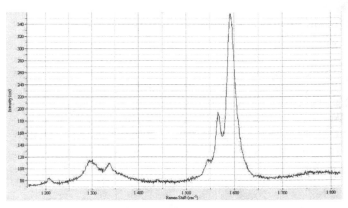

Fig.8 Raman spectra measurement showing G and D band of SWNT networks.

4 CONCLUSIONS

In summary, we report the patterned assembly of SWNT microscale networks with controllable dimensions on a flexible parylene-C substrate using surface controlled microfluidic assembly. The surface properties of the oxygen plasma treated parylene-C were characterized by contact angle measurements which indicated that following the O_2 plasma treatment, the parylene-C surface became hydrophilic with a contact angle of about $4.82°\pm0.57°$ and stayed hydrophilic for at least 2 hours. The two-terminal resistance of the SWNT microlines ranged from $15K\Omega$ to $21K\Omega$. Repeatability tests were performed on the sample which had a variation in resistance of less than 1%. Raman spectroscopy analysis revealed that the observed frequencies of the RBM peaks correspond to SWNTs with diameters ranging $0.9 - 1.55$ nm.

This successful assembly of SWNT networks opens a possibility to study their intrinsic properties and anisotropic behavior at the microscale. The assembled SWNT networks on the parylene-C substrate show desirable electrical properties that are promising candidates for the next generation of flexible devices. This work provides a simple and economical approach to realize patterned SWNT microstructures on flexible polymer substrates. The presented method is also scalable to wafer scale fabrication of SWNT-based flexible electronics including high performance flexible interconnects, sensors and FETs.

ACKNOWLEDGMENTS

This work was supported by the National Science Foundation Nanoscale Science and Engineering Center (NSEC) for High-rate Nanomanufacturing (NSF grant-0425826).

REFERENCES

[1] J. Kong, N. R. Franklin, C. Zhou, M. G. Chapline, S. Peng, K. Cho, and H. Dai, "Nanotube molecular wires as chemical sensors," *Science,* vol. 287, pp. 622-5, 2000.

[2] R. Krupke, F. Hennrich, M. M. Kappes, and H. v. Lohneysen, "Surface conductance induced dielectrophoresis of semiconducting single-walled carbon nanotubes," *Nano Lett,* vol. 4, pp. 1395-99, 2004.

[3] H. Ko and V. V. Tsukruk, "Liquid-crystalline processing of highly oriented carbon nanotube arrays for thin-film transistors," *Nano Lett,* vol. 6, pp. 1443-8, Jul 2006.

[4] S.-H. Hur, O. O. Park, and J. A. Rogers, "Extreme bendability of single-walled carbon nanotube networks transferred from high-temperature growth substrates to plastic and their use in thin-film transistors," *Appl. Phys. Lett.,* vol. 86, p. 243502, 2005.

[5] S. J. Kang, C. Kocabas, T. Ozel, M. Shim, N. Pimparkar, M. A. Alam, S. V. Rotkin, and J. A. Rogers, "High-performance electronics using dense, perfectly aligned arrays of single-walled carbon nanotubes," *Nature Nanotechnology,* vol. 2, p. 230, 2007.

[6] Y. J. Jung, S. Kar, S. Talapatra, C. Soldano, G. Viswanathan, X. Li, Z. Yao, F. S. Ou, A. Avadhanula, R. Vajtai, S. Curran, O. Nalamasu, and P. M. Ajayan, "Aligned carbon nanotube-Polymer hybrid architectures for diverse flexible electronic applicatons," *Nano Lett.,* vol. 6, pp. 413-418, 2006.

[7] S. J. Kang, C. Kocabas, H. S. Kim, Q. Cao, M. A. Meitl, D. Y. Khang, and J. A. Rogers, "Printed multilayer superstructures of aligned single-walled carbon nanotubes for electronic applications," *Nano Lett,* vol. 7, pp. 3343-8, 2007.

[8] M. A. Meitl, Y. Zhou, A. Gaur, S. Jeon, M. L. Usrey, M. S. Strano, and J. A. Rogers, "Solution casting and transfer printing single-walled carbon nanotube films," *Nano Lett,* vol. 4, pp. 1643-47, 2004.

[9] S. G. Rao, L. Huang, W. Setyawan, and S. Hong, "Nanotube electronics: large-scale assembly of carbon nanotubes," *Nature,* vol. 425, pp. 36-7, Sep 4 2003.

[10] V. V. Tsukruk, H. Ko, and S. Peleshanko, "Nanotube surface arrays: weaving, bending, and assembling on patterned silicon," *Phys Rev Lett,* vol. 92, p. 065502, 2004.

[11] X. Xiong, L. Jaberansari, M. G. Hahm, A. Busnaina, and Y. J. Jung, "Building highly organized single-walled-carbon-nanotube networks using template-guilded fluidic assembly," *Small,* vol. 12, pp. 2006-2010, 2007.

[12] J. Frados, "Modern Plastics Encyclopedia." vol. 45 NY, USA: McGraw Hill, 1968.

[13] T. Chang, V. Yadav, S. Deleo, A. Mohedas, B. Rajalingam, C.-L. Chen, S. Selvarasah, M. R. Dokmeci, and A. Khademhosseini, "Cell and protein compatibility of parylene-C surfaces," *Langmuir,* vol. 23, pp. 11718-11725, 2007.

[14] S. Rols, A. Righi, L. Alvarez, E. Anglaret, R. Almairac, C. Journet, P. Bernier, J. L. Sauvajol, A. M. Benito, W. K. Maser, E. Muñoz, M. T. Martinez, G. F. d. l. Fuente, A. Girard, and J. C. Ameline, "Diameter distribution of single wall carbon nanotubes in nanobundles," *Eur Phys J B,* vol. 18, pp. 201-05, 2000.

[15] A. Jorio, R. Saito, J. H. Hafner, C. M. Lieber, M. Hunter, T. McClure, G. Dresselhaus, and M. S. Dresselhaus, "Structural (n, m) determination of isolated single-wall carbon nanotubes by resonant Raman scattering," *Phys Rev Lett,* vol. 86, pp. 1118-21, 2001.

[16] M. S. Dresselhaus, G. Dresselhaus, R. Saito, and A. Jorio, "Raman spectroscopy of carbon nanotubes," *Phys. Rep.,* vol. 409, p. 47, 2005.

First Disclosure of a Viable Semi-Commercial Single CNT-Based FE Device

S. Daren[*], I. Kalifa and S. Peretz

El-Mul Technologies Ltd., Yavne, Israel

ABSTRACT

Several disclosures covering Carbon Nanotube (CNT) based devices for Field Emitter (FE) applications have been published, none of which have shown commercial viability. As a result of our research and development, El-Mul Technologies has recently gained knowledge of some critical factors which enable semi-commercial manufacturing of single CNT-based FE devices that are useful for a wide variety of applications. We present here this novel device and discuss its key characteristics. Our proprietary technology combines conventional semiconductor manufacturing processes with nanoscale innovations to yield scalable, high performance electron micro-gun emitters and emitter arrays.

Keywords: carbon nanotube, CNT, field emitter, FE, electron micro-gun array, commercial

1 MANUFACTURE OF CNT-BASED FIELD EMITTERS

Although there have been numerous disclosures covering Carbon Nanotube (CNT) based devices for Field Emitter (FE) applications have been published[1][2], none of these have shown commercial viability. As a result of our research and development, El-Mul Technologies has recently gained knowledge of some critical factors which enable semi-commercial manufacturing of single CNT-based FE devices that are useful for a wide variety of applications.

The fabrication methods we have developed overcome the difficulties of two critical capabilities for CNT integration inside a gating structure: (1) mastering growth of single, vertically aligned CNTs and (2) patterning techniques at the nanoscale. We are now able to produce micro-gun based devices such as single beam emitters with well-characterized fine beams or multi-beam emitter arrays with high current broad beams.

2 METHODOLOGY

As the basis for our technology, we employ a stack of silicon, silicon oxide and conductive poly silicon layers. A cathode well (a cavity measuring 2 μm in diameter and depth) is etched through the poly and oxide layers to the silicon substrate, using conventional semiconductor processing techniques, resulting in formation of a capacitor-like gating structure.

This structure is further processed to place either single or multiple CNTs in each well, depending on the catalyst seed pattern which is deposited during the pre-growth stage. Additional etching into the silicon is possible in order to form a structure which enables the creation of a narrow-spreading beam.

The resulting proprietary, patented structure[3][4] creates a focusing field-line regime which, when applied to the CNT apex, produces a narrow electron beam, as demonstrated in the 2D simulation presented in Figure 1.

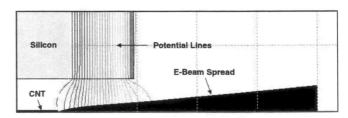

Figure 1: Simulation of electron trajectories focused by field line regime, based on our proprietary structure. Note that the diagram is rotated clockwise here.

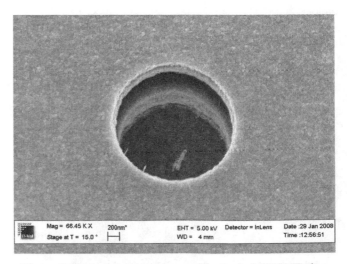

Figure 2: Completed structure with single MWCNT that results in a narrow beam spread. In this version, gate electrode and CNT apex are 2 μm apart.

Figure 2 presents a manufactured instance of the proprietary structure that is described in Figure 1. Our work confirms prior efforts[5] that report the need for a catalyst site in the diameter range of 200-300 nm in order to yield single MWCNT growth.

In our work Ni catalyst is deposited by means of electron beam evaporation and is confined to form such sites on the well bottom. Then, a Plasma Enhanced Chemical Vapor Deposition (PECVD) process is applied to yield a vertically aligned CNT with the desired dimensions.

We have already achieved good control over our patterning and growth processes, enabling us fully reproducible manufacturing of cathodes with a ±200 nm precision in CNT centricity and height. A top-down image of our CNT is presented in Figure 3.

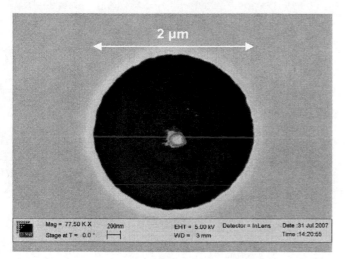

Figure 3: Cathode well with single MWCNT, as viewed directly from above.

3 CHARACTERIZATION

CNT cathode devices are characterized by two principle parameters: current characteristics and beam optics behavior.

3.1 Device Current

Direct current measurements are performed, including I-V profiling, beam stability and gate leakage testing. Our devices are characterized by a low turn-on voltage of (a) 20-50 V for cathodes where the gate electrode and CNT apex are on the same level, as presented in Figure 4; and (b) 100-150 V for cathodes where the gate electrode and CNT apex are 2 μm apart, as already presented in Figure 2.

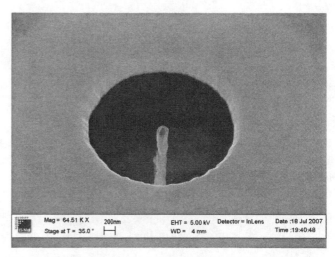

Figure 4: Cathode well with single MWCNT, as viewed at slight angle from above.

All cathodes show typical field emission behavior expressed by an exponential rise of current versus voltage and a good fit to the Fowler-Nordheim model, as depicted in Figure 5.

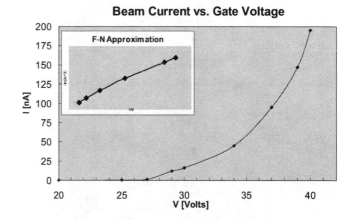

Figure 5: Sampled CNT emission data showing the exponential behavior which characterizes field emitters. Inset shows data coordinates fit to Fowler-Nordheim model, resulting in a straight line.

Currents of 200 nA per source are easily surpassed for extended time periods, without causing any apparent damage to the CNTs. Stability tests, performed at ultra high vacuum levels of 1E-9 Torr show an unsteady behavior expressed by current fluctuations sometimes exceeding 50%. (Measures to resolve this undesired behavior are being taken currently.)

Gate-to-source leakages during operation measure as high as 10-20% of the beam current. Such leakages do not affect operating lifetimes, as no visual damage to the source or performance degradation has been identified over time. However, these leakages may still affect the beam stability.

3.2 Beam Optics

Characterization is performed using Field Emission Microscopy (FEM) where a Multichannel Plate (MCP) and phosphor screen are placed in front of the emitter chip to facilitate imaging of the beam profile. Our electrical design allows imaging at conditions where voltages on the face of the detector (MCP input) and the gate electrode are equal, thus imposing a "field free zone" for electrons during flight. These conditions enable measurement of the intrinsic emitting angle of the device without the distortion that would be imposed by external electrical fields. Typical divergence angles vary from 10-15 degrees for emitters where the CNT tip is located below the oxide level, to 30-40 degrees for emitters where the CNT extends to the gate level.

4 DEVICE YIELDS IN SEMI-COMMERCIAL CONFIGURATIONS

As stated above, our current process successfully addresses critical demands of semi-commercial manufacturing. As presented in Figure 6, we currently produce CNT FE devices in various array sizes with high yields, achieved on the die level. For purposes of our target applications, *e.g.*, next generation e-beam tools, microscopy and other electron field emission systems, our preliminary results confirm the potential for high volume device production with very good yields and repeatability.

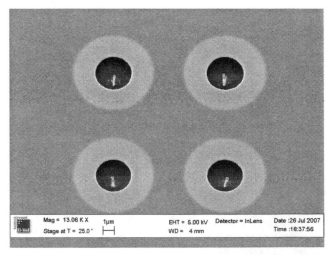

| Mag = 13.06 K X | 1µm | EHT = 5.00 kV | Detector = InLens | Date :26 Jul 2007 |
| Stage at T = 25.0 ° | | WD = 4 mm | | Time :18:37:56 |

Figure 6: Sample micro-gun emitter array manufactured using our proprietary structure and conventional semi-conductor manufacturing processes.

REFERENCES

[1] W.I. Milne, K.B.K. Teo, M. Chhowalla, *etal*, "Electrical and Field Emission Investigation of Individual Carbon Nanotubes," Diamond and Related Materials, 12, 422–428, 2003.

[2] L. Gangloff, E. Minoux, K.B.K. Teo, *etal*, "Self-Aligned, Gated Arrays of Individual Nanotube and Nanowire Emitters," Nano Letters, Volume 4 Number 9, 1575-1579, 2004.

[3] J.G. Leopold, O. Zik, E. Cheifetz, *etal*, "Carbon Nanotube-Based Electron Gun for Electron Microscopy," Journal of Vacuum Science & Technology A: Vacuum, Surfaces, and Films, Volume 19 Issue 4, 1790-1795, 2001.

[4] "Nanotube-based electron emission device and systems using the same," U.S. Patent No. 6,512,235, Issued January 2003.

[5] V. Golovko, M. Cantoro, S. Hofmann, *etal*, "Selective growth of vertically aligned carbon nanofibres in sub-micron patterning and Raman mapping of produced arrays," Diamonds and Related Materials, 15, 1023, 2005.

* Correspondence may be addressed to Mr. Sagi Daren of El-Mul Technologies Ltd., sagi.daren@el-mul.com.

Chemical doping by sulfuric acid in double wall carbon nanotubes

P Puech[*], A Ghandour[***], A Sapelkin[***], C Tinguely[**], E Flahaut[**], D Dunstan[***] and W Bacsa[*]

[*]CEMES CNRS UPR, 29 rue Jeanne Marvig, Univ. Toulouse, 31055 Toulouse France,
wolfgang.bacsa@cemes.fr
[**] CIRIMAT-LCMIE, CNRS UMR, 118 route de Narbonne, Univ. Toulouse, 31062 Toulouse, France
[***] Physics Department, Queen Mary, University of London, London E1 4NS, United Kingdom

ABSTRACT

Charge transfer due to chemical doping in carbon nanotubes can be detected through changes in the band shape and spectral shifts of the Raman G-band. In double wall carbon nanotubes, the inner tube is well protected from the environment and contributions of the inner tube to the Raman G-band can be detected when applying hydrostatic pressure. We find that by combining doping with sulfuric acid and high hydrostatic pressure, we can determine the ratio of single to double wall carbon nanotubes and we propose empirical parameters to fit the G-band line shape. We observe a spectral band at 1560 cm^{-1} which shifts with pressure at the same rate as the outer tube and which attribute to electronic coupling of the two tube walls.

Keywords: Raman spectroscopy, carbon nanotubes, double wall, chemical doping, hydrostatic pressure

1 INTRODUCTION

Double wall carbon nanotubes (DWs) are the simplest form of multi-wall carbon nanotubes. While single wall carbon nanotubes (SW) can either be semi-conducting or metallic depending on the way the graphene sheet is rolled up, multi-wall carbons nanotubes (MWs) are electrical conductors due to their larger diameter. DWs are the ideal system to study the inter-wall coupling. The electrical conductivity perpendicular to the graphene layer is less than 1% of the in-plane electrical conductivity [1]. Electronic conductivity has been extensively studied, and inter-shell conductance in MWs is consistent with tunneling through orbitals of neighboring walls [2].

Two main synthesis methods for DWs are known to day: conversion of peapods into DWs leading to DWs with uniform diameter distribution [3] and the use of the catalytical chemical vapor deposition (CCVD) method resulting in 80−100% of DWs with a larger diameter distribution [4, 5]. Peapods or single wall carbon nanotubes filled with C_{60} molecules. Raman spectroscopy is routinely used to screen the diameter distribution using the low frequency radial breathing mode, by changing the excitation wavelength and by measuring defect induced scattering (D band) [6]. The G band in DWs contains contributions from the internal and external tubes which depend on external parameters such as pressure,

temperature and applied electric field [7]. We combine the influence of the G band shape as a function chemical doping and hydrostatic pressure to separate contributions from inner and outer tubes. The internal tube does not experience any pressure from the inside and is only slightly affected by doping (10%) [8,9].

FIG. 1: Transmission electron microscopy images of a bundle of DWs of uniform diameter distribution.

The G band frequency of the outer tube in DWs is about the same as that for SWs making it impossible to separate both G band contributions applying pressure or changing the temperature. Kim et al [10] have recently proposed a scheme to determine the purity of the sample using chemical doping. Chemical doping with sulphuric acid has a large effect on the G band of SWs [11] depending on the excitation wavelength, while the shape change is more subtle in the case of DWs as will be shown in our study. Raman scattering of DW doped with H_2SO_4 reported in literature [12] have not been able to discriminate contribution form the inner and outer tubes to the Raman G band. Hydrostatic pressure experiments give us the opportunity to separate contributions from the inner and outer tubes [13, 14].

2 EXPERIMENTAL

The DWs were prepared by CCVD [15]. High-resolution transmission electron microscopy images show the presence of individual and small bundles of DWs with diameters ranging from 0.6 to 3 nm (see figure1). The tubes are single(15%), double (80%) or triple walled (< 5%). For the high pressure experiment Raman spectra were recorded at room temperature using a Renishaw Raman microprobe instrument. The high-pressure Raman measurements were performed in a diamond anvil cell. Raman spectra were also recored using a XY-Dilor spectrometer. All spectra have been recorded in air using 1 mW before entering the optical microscope and spectrometer for the high pressure experiment to prevent any heating of the tubes. We can estimate that the tubes are heated less than 10 K using Stokes and Anti-Stokes scattering and G, D band shifts.

3 RAMAN BANDS OF DOPED DWS

Kim et al [10] propose to use chemical doping with sulphuric acid to determine the composition of DW samples. Here we conduct hydrostatic pressure experiment with sulphuric acid as a medium to identify a set of empirical parameters that can be used to determine the fraction of DW to SW. Zhou et al [11] has shown that the G band frequency is diameter dependent. In figure 2, we show Raman spectra excited at 633 nm in the spectral region of the D and G band and the G' 2D band of SWs of 1.4 nm and 0.8 nm diameter and DWs.

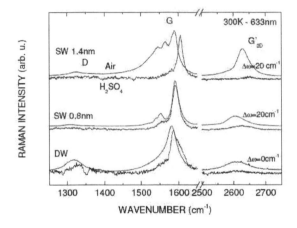

FIG. 2: First and second order Raman spectra of SWs and DWs recorded at 633nm in air and doped with sulphuric acid.

We note that with diameters larger than 1.4nm, chemical doping leads to a considerable upshift of the G band. Contribution from the G− band in SWs, as well as a band associated with electronic coupling are strongly reduced in intensity when doped. As the mean tube diameters of SWs and DWs are larger than 1.4nm in our DW sample, a large upshift is expected as a result of the chemical doping. This results in the change of the G band shape of our DW sample containing 15% of SWs

4 HYDROSTATIC PRESSURE

Figure 3 shows on the left side the G band of DWs at 5 GPa using four different pressure media. A clear difference in the G band shape is observed when using argon or sulphuric acid as pressure medium.

The G band splits when increasing pressure due to the different pressure experienced by the inner and outer tubes. The differences as a function of pressure between the media demonstrate that the pressure experienced by the tube depends on them. At normal pressure, ie without anvil cell, the signal of the outer tubes on the higher energy side of the G band is not present. The signal from the outer tube increases with pressure and its intensity is comparable to the G band of the inner tube. The signal from the remaining SWs in the DW is also present but is less intense. At around 5GPa we observe a decrease in the intensity of the entire G band.

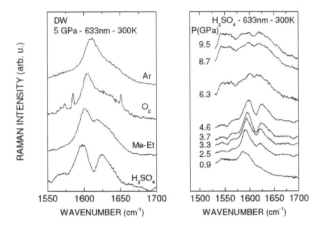

FIG. 3: Left side: Raman G band of DWNTs at 5 GPa for 4 different pressure transmitting media. Right side: G band of DWNTs and pressures upto 9.5 GPa for H2SO4 as pressure transmitting medium.

We used the spectrum recorded without anvil cell for the value at zero pressure. In graphite intercalation compounds the G band shifts by 16 cm^{-1} for the first H_2SO_4 intercalation stage and shifts two or three times more for the second and third intercalation stage [16]. Doping of graphite induces strain [17]. It has been found from high pressure experiments [18] that the lattice parameter a/a is 8×10^{-4} for each stage leading to a total shift of 4 cm^{-1}. Consequently, strain alone cannot explain the observed shift in DWs upon doping. The remaining shift of 12 cm^{-1} has clearly a different origin. This effect is related to electron-phonon interaction [19].

In table I, we report the spectral G band position. The spectral G band position at zero pressure depends also on the medium. The larger effect of oxygen on the G band position compared to alcohol or argon can be explained by p-doping of the tubes by oxygen which is expected to up shift the G band. It is important to notice that a shoulder is observed on the lower side of the G band at 1560 cm^{-1}. This band persists with increasing pressure for DWs in contrast with what is observed for SWs [20-22].

It is observed that D band is more intense after pressure loading. Doping has clearly the effect of reducing the

NSTI-Nanotech 2008, www.nsti.org, ISBN 978-1-4200-8503-7 Vol. 1

intensity of the G band. Reduction of the G band intensity through doping of up to 50% has been reported in the literature [10].

TABLE I: G band position of inner and outer tube of DWs, pressure coefficients for four different pressure transmitting media (i: inner, o: outer).

Medium	$\omega_i(P=0)$ (cm^{-1})	$\omega_o(P=0)$ (cm^{-1})
Me-Et	1582	1594
O_2	1584	1598
Argon	1581	1592
H_2SO_4	1587	1618

The splitting of the G band with pressure allows to determine contributions of inner and outer tubes and to extrapolate the G band frequencies at zero pressure. In the low pressure regime (< 3GPa) the two bands overlap and the numerical fitting is not stable. A change in shape can be either due to intensity variation or change of spectral position. There are clear differences seen between the DWs obtained from peapods and DWs grown with CCVD. The spectral position of the inner tube at zero pressure deduced from linear fitting is at 1579 cm^{-1} [14] for the DW from peapods but at 1581 cm^{-1} for DW grown by CCVD and we find that the shifting of the G band of the inner tube with pressure is delayed. This implies that the coupling of the two walls is not the same which is consistent with differences observed for the band at 1560 cm^{-1}.

5 SPECTRAL ANALYSIS

Figure 4 shows DW G band spectra recorded at three different locations (A, B, C) and recorded at different laser power levels (right hand side). The left hand side shows Stokes and anti-Stokes spectra at location B and a spectrum of DWs in H_2SO_4. To obtain the same background level for Stokes and anti Stokes spectra, we have corrected the Anti-Sokes part by the factor ω^4 and the Bose-Einstein factor corresponding to T = 775K. This high Temperature can be attributed to the single particle excitations [23]. On the right side of figure 4 we have subtracted a linear background for each spectra for the spectra of the three locations A, B and C. To fit the data, only the 4 intensities of the four main contributions are taken as free parameters. Even if a small spectral shift for the four bands is added the intensity ratio remains unchanged. For location B we show spectra from two different laser powers. The fit is stable and higher power reduces the spectral noise. We associate the two intense bands to the G band of SWs and the inner tubes of the DWs and we correlate the intensity ratio with experimentally observed SW/DW ratio.

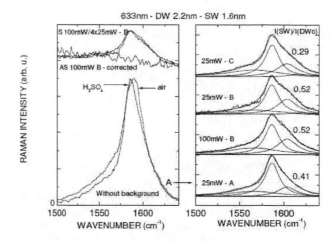

FIG. 4: Fraction of SW in our sample: 3 places (A,B and C), S for Stokes, AS for Anti-Stokes.

Using the determined purity (80%) using transmission electron microscopy, and the average value of I_S/I_D reported in figure 4, we find an empirical relation: $N_{SW}/N_{DW}=0.3I_S/I_D$. We note that with increasing pressure and doping the intensity decreases and increases the contribution of the outer tube. We use the parameters deduced on the CCVD grown DWs to test the consistency of our approach. We consider 3 bands at fixed spectral position and fixed HWHM for a given temperature. The data used for the fitting are reported in the table II.

We have fitted our data using the set of parameters determined in the first part. We use 2 parameters for the linear background and 3 parameters for the intensities and include a small shift for all associated to temperature increase. Two sets of spectra have been used to test the scheme and the results are reported in figure 5.

Figure 5 shows the G band as a function of laser power using two different microscope objectives.

TABLE II: Parameters for fitting G bands of SWs and DWs in H_2SO_4 using 647 nm excitation wavlength.

Wavenumber (cm^{-1})	HWHM (cm^{-1})
$\omega_{DW-inner} = 1587$	$\Gamma = 10$
$\omega_{DW-outer} = 1618$	$\Gamma = 10$
$\omega_{SW} = 1606$	$\Gamma = 10$
$\omega_{DW_{ei}} = 1568$	$\Gamma = 35$

6 CONCLUSION

We find that chemical doping with H_2SO_4 allows us to accurately determine the composition of CCVD DW samples by using a set of parameters obtained from hydrostatic pressure experiments to fit the Raman G band by keeping the spectral position of contributions from the inner and outer tubes fixed.

ACKNOWLEDGMENT

We thank Jenny Patterson, Intel Ireland, for stimulating discussions.

REFERENCES

[1] K. Matsubara, K. Sugihara, and T. Tsuzuku, Phys. Rev. B 41, 969 (1990)

[2] B. Bourlon, C. Miko, L. Forro, D. C. Glattli, and A. Bachtold, Phys. Rev. Lett. 93, 176806 (2004)

[3] S. Bandow, M. Takizawa, K. Hirahara, M. Yudasaka, S. Iijima, Chem. Phys. Lett. 337, 48 (2001)

[4] E. Flahaut, A. Peigney, Ch. Laurent, A. Rousset, J. Mater. Chem. 10, 249 (2000)

[5] M. Endo, H. Muramatsu, T. Hayashi, Y. A. Kim, M. Terrones, and M. S. Dresselhaus, Nature London 433, 476 (2005)

[6] C. Thomsen and S. Reich, Light Scattering in Solids IX, Ed. M. Cardona and R. Merlin (2005)

[7] Ladislav Kavan , Lothar Dunsch, Chemphyschem.8, 974 (2007)

[8] G. Chen, S. Bandow, E.R. Margine, C. Nisoli, A.N. Kolmogorov, V.H. Crespi, R. Gupta, G.U. Sumanasekera, S. Iijima,and P.C. Eklund, Phys. Rev. Lett. 90, 257403 (2003)

[9] P. Puech, E. Flahaut, A. Sapelkin, H. Hubel, D.J. Dunstan, G. Landa, and W.S. Bacsa, Phys. Rev. B73, 233408 (2006)

[10] Y.A. Kim et al Chem. Phys. Lett. 420, 377 (2006)

[11] W. Zhou, J. Vavro, N.M. Nemes, J.E. Fischer, F.Borondics, K. Kamaras and D.B. Tanner, Phys. Rev. B 71, 205423 (2005)

[12] E.B. Barros et al Phys. Rev. B 76, 045425 (2007)

[13] P. Puech, H. Hubel, D.J. Dunstan, R.R. Bacsa, C. Laurent and W.S. Bacsa, Phys. Rev. Lett. 93, 095506 (2004)

[14] J. Arvanitidis, D. Christofilos, K. Papagelis, K.S. Andrikopoulos, T. Takenobu, Y. Iwasa, H. Kataura, S. Ves, and G.A. Kourouklis, Phys. Rev. B 71, 125404 (2005)

[15] E. Flahaut, R. Bacsa, A. Peigney and Ch. Laurent, Chemical Communication 12, 1442 (2003)

[16] R. Nishitani, Y. Sasaki and Y. Nishina, Phys. Rev. B37, 3141 (1988)

[17] C.T. Chan, W.A. Kamitakahara, K.M. Ho and P.C. Eklund, Phys. Rev. Lett.58, 1528 (1987)

[18] M. Hanfland, H. Beister, and K. Syassen, Phys. Rev. B 39, 12598 (1989)

[19] S. Piscanec, M. Lazzeri, J. Robertson, A.C. Ferrari, and F. Mauri, Phys. Rev. B 75, 035427 (2007)

[20] A. Merlen, N. Bendiab, P. Toulemonde, A. Aouizerat, A. San Miguel, J. L. Sauvajol, G. Montagnac, H. Cardon, and P. Petit, Phys. Rev. B 72, 035409 (2005)

[21] R. Pfeiffer, F. Simon, H. Kuzmany and V.N. Popov, Phys. Rev. B72, 161404 (2005)

[22] P. Puech, A. Bassil, J. Gonzalez, Ch. Power, E. Flahaut, S. Barrau, Ph. Demont, C. Lacabanne, E. Perez, and W. S. Bacsa, Phys. Rev. B 72, 155436 (2005)

[23] D.S. Kim and P.Y. Yu, Phys. Rev. B43, 4158 (1991)

The Influence of DNA Wrapping on SWNT Optical Absorption in Perpendicular Polarization

S. Snyder* and S. V. Rotkin*,**

* Lehigh University Department of Physics
16 Memorial Drive East, Bethlehem, PA 18015, USA, ses7@lehigh.edu
** Center for Advanced Materials and Nanotechnology
Lehigh University, Bethlehem, PA, USA, rotkin@lehigh.edu

ABSTRACT

Numerous electronics applications have been proposed for discrete single-walled carbon nanotubes (SWNTs) due to their exceptional electronic properties, which vary with diameter and chirality of the tube. Any solution processing to sort as-grown SWNTs by type requires a robust dispersion method that must overcome nanotube insolubility and strong van der Waals attraction between tubes. Functionalization of SWNTs with helically wrapped single-stranded DNA yields stable hybrid structures that disperse in aqueous solution.[1], [2] Since these hybrids are intended for use in optical and electronic devices, we investigate whether the optical properties of DNA-wrapped SWNTs are different than those of bare nanotubes. We find changes in the optical absorption spectra in perpendicular polarization, including circular dichroism of achiral nanotubes. These effects may serve as characterization tools to identify the presence of a DNA wrap.

Keywords: carbon nanotube, optical absorption, DNA, circular dichroism

1 INTRODUCTION

Single-walled carbon nanotubes are being studied as potential components in of the next generation of transistors, field emitters, photodetectors, and biological sensors [3], to name a few potential applications, but to realize these devices, methods must be developed to gain control over the material. Currently all nanotube fabrication processes produce tubes of various structures and, hence, various electronic properties. If we wish to take advantage of the unique properties of a specific SWNT structure, starting with currently available material, SWNTs must first be dispersed in solution. Single-stranded DNA can form a stable hybrid with a SWNT, allowing dispersion. Ion exchange chromatography has been used with success to help sort DNA-SWNT hybrids by structure and electronic properties [1], [2], and methods are being developed for controlled placement of the hybrids. Since a DNA-SWNT hybrid is not readily dismantled, and indeed the DNA wrap may prove useful for some applications, it is important to determine whether the DNA changes optical or electronic properties of SWNTs.

For the case of optical absorption of light parallel to the SWNT axis, the direct optical bandgap of a semiconducting SWNT decreases only slightly due to DNA wrapping. This result of our earlier work [4] is in agreement with experiments using parallel polarization.[5] For the present work, we consider optical absorption with perpendicular (or circular) polarization. Although perpendicular absorption is weaker than parallel absorption, it has been successfully measured for surfactant dispersed SWNTs.[6]–[8] We find changes in absorption spectra upon hybridization with DNA, including new transitions prohibited for pristine nanotubes in the same polarization. In addition, we find circular dichroism for non-chiral SWNTs. These optical effects may be used to verify the presence of DNA-wrapping. Understanding these effects could also aid in the characterization and identification of nanotubes by type.

2 RESULTS

The ionized DNA backbone has a net negative charge, and this charged helical wrap breaks the symmetry of a bare SWNT. The potential of the DNA is too strong for a perturbative approach, so we numerically solve the joint Schrödinger-Poisson equations beyond the perturbation approximation to determine the changes in optical absorption resulting from hybridization.

As observed previously, absorption of light with *parallel* polarization should not change significantly upon DNA hybridization.[4] In contrast, Figure 1 shows that for a semiconducting zigzag (5,0) DNA-SWNT hybrid, the optical absorption coefficients for light polarized *perpendicularly to the SWNT axis* (dashed orange curve) drastically differ from the bare tube absorption in the same polarization (solid black curve). The first absorption peak in cross-polarization for the bare SWNT corresponds to E_{12} and E_{21} transitions. This peak is also present for the DNA-SWNT hybrid, although it is shifted to higher frequency. In addition, a peak at lower frequency near that of the bare E_{11} transitions appears as a consequence of the lifting of selection rules. The upper insets show absorption-luminescence maps for the bare (left) and DNA-wrapped (right) (5,0) nanotube. Additional transitions can clearly be seen for the hy-

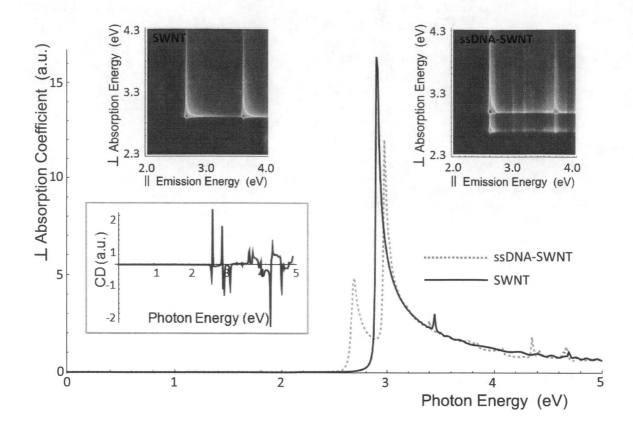

Figure 1: Calculated absorption spectrum in perpendicular polarization for a (5,0) SWNT with (dashed orange) and without (solid black) a DNA wrap. The upper left inset shows an absorption-luminescence map for the bare (5,0) SWNT, while the upper right inset shows the same for the DNA-wrapped SWNT. The lower left inset shows circular dichroism for the DNA-wrapped (5,0) SWNT.

brid.

For cross-polarized transitions, the selection rule for angular momentum is $\Delta m = +1$ for the bare tube. However, angular momentum is not a good quantum number for the subbands of the wrapped SWNT, due to the helical polarization of the electron (and hole) wave functions by the Coulomb potential of the DNA, as shown in Figure 2. This symmetry breaking results in allowed perpendicularly polarized optical transitions at lower frequency near that of the $\sim E_{11}$, which is prohibited for the bare tube. The corresponding physical explanation is that the polarization of the electron (hole) due to the transverse electric field of the DNA creates a permanent dipole across the nanotube, which can then be excited by the perpendicularly polarized incident light.

3 CALCULATION METHODS

A semi-empirical orthogonal tight-binding approach was used to calculate optical absorption of several of DNA-SWNT complexes. This simple numerical approach is chosen to capture the essential physics of the problem with low computational cost. The DNA backbone is modeled as a regular, infinite helix of point charges representing the phosphate groups wrapped around the tube (as seen in Figure 2). The angle of wrap, its position with respect the underlying graphene lattice of the nanotube, the distance between the tube and the wrap, and linear charge density are parameters that can be obtained from molecuar dynamics simulations or adjusted to the experimental data. For a broad range of these parameters, we observe similar symmetry breaking effects.

In order to break the perpendicular selection rules for angular momentum, one must mix subbands with $\Delta m = \pm 1$. The most effective wrap is along the zigzag direction of the SWNT lattice (at an angle of 60° to the circumference). Such a wrap effectively mixes the lowest conduction and valence bands of different angular momentum as a result of the phase matching [15] of the perturbation.

The partial absorption coefficient is calculated for a

NSTI-Nanotech 2008, www.nsti.org, ISBN 978-1-4200-8503-7 Vol. 1

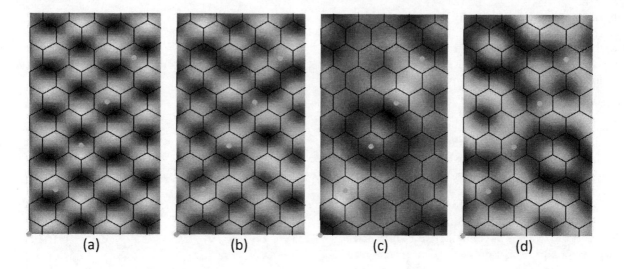

Figure 2: Wavefunctions for the unscrolled (5,0) SWNT: (a) Unperturbed (bare) SWNT, (b) LUMO, (c) HOMO-2, and (d) HOMO-6. The projected charges of the idealized helical wrap can be seen as diagonal lines of dots.

$$\alpha_{\pm}(\hbar\omega, k) \propto \sum_i \sum_f \frac{q^2 \left|\langle\psi(k, m_f, \lambda_f)|\vec{p}\cdot\vec{e}_{\pm}|\psi(k, m_i, \lambda_i)\rangle\right|^2}{m_0\omega} \times \frac{f(E_i(k, m_i, \lambda_i))[1 - f(E_f(k, m_f, \lambda_f))]\,\Gamma}{[E_f(k, m_f, \lambda_f) - E_i(k, m_i, \lambda_i) - \hbar\omega]^2 + \Gamma^2} \qquad (1)$$

bare SWNT as shown in Equation (1), where $\hbar\omega$ is the energy of the incident light, k is the electron wavevector, q is the magnitude of the electron charge, m_i and m_f refer to the initial and final angular momentum quantum numbers, λ_i and λ_f label the band (conduction or valence) of the initial and final states, m_0 is the free electron mass, $f(E)$ is the Fermi-Dirac function, and Γ is a phenomenological broadening parameter, which was set to 0.008 eV for these simulations. $|p_{\pm}| = qR|\langle\psi(k, m_i, \lambda_i)|e^{\pm i\theta}|\psi(k, m_i \pm 1, \lambda_f)\rangle|$ is the dipole matrix element for a transition from an initial state (with energy E_i) in the valence band, $|k, m_i, \lambda_i\rangle$, to a final state (of energy E_f) in the conduction band, $|k, m_f, \lambda_f\rangle$, for incident light with the circular polarization $\vec{e}_{\pm}$, with the selection rule for angular momentum $\Delta m = m_f - m_i = \pm 1$. One can exclude the DNA from the spectra analysis because the DNA absorption edge [9], [10] is at approximately 4.1 eV, which is well above the main IR spectral features of the SWNT itself.

The total absorption spectrum is determined by integrating the partial absorption coefficient Equation (1) over the wavevector inside the first Brillouin zone, $\alpha_{\pm}(\hbar\omega) = \int_{BZ} \alpha_{\pm}(\hbar\omega, k)\,dk$, and is plotted in Figure 1. Absorption of cross-polarized light drastically differs for the hybrid and the bare tube.

The circular dichroism (CD) spectrum of the DNA-SWNT hybrid is given by the difference in absorption for right and left circularly polarized ligh: $CD = \alpha_+ - \alpha_-$ (lower left inset of Figure 1). This CD is not the result of any chirality of the DNA, because DNA absorption is not included in this work. Then the optical activity must be fully attributed to the nanotube itself and the "natural" helicity of the electron states. This result is further supported by the experimental data in Ref.[11]. Similar results were seen for a variety of achiral as well as chiral tubes.

4 DISCUSSION AND CONCLUSIONS

The exact geometry of the DNA wrap for an arbitrary tube is not yet known, and may not be uniform across the length of a SWNT. The effect of variation of the parameters of the wrap and/or tube diameters should be considered. Our simulations showed that CD and the cross-polarized absorption spectrum do not change significantly when a given helical wrap is displaced along the length of the tube or rotated about the axis of the tube.[14] Neither of these shifts changes the helical angle of the DNA backbone. In contrast, absorption spectra did change when the helical angle changed, due to the importance off symmetry matching between the SWNT lattice and the DNA backbone helical angle. Despite these changes, a general qualitative feature of the helical symmetry breaking remains: new van Hove singularities appear in the optical data. A random coating of DNA with no particular symmetry is expected to yield at most in a broad, featureless background.

In summary, we predict that SWNT optical absorption in perpendicular polarization can be used for optical detection of regular helical DNA wrapping. We calculated optical absorption spectra, as well as circular dichroism for DNA-wrapped single-wall nanotubes. The Coulomb potential of the ionized DNA wrap polarizes the electronic structure of the tube and results in helical symmetry breaking in an intrinsically achiral SWNT. We predict qualitative changes in the absorption of circularly or linearly polarized light with its electric field perpendicular to the nanotube axis. Namely, new transitions appear in the cross-polarized absorption of the SWNT at wavelengths substantially larger than those of all allowed perpendicular E_{12} transitions in the bare tube. We propose to use the splitting and shift of the lowest perpendicular absorption peak to experimentally identify the existence and symmetry of the wrapping.

Acknowledgements: This work was partially supported by DoD-ARL (W911NF-07-2-0064), NSF (CMS-0609050), and PA Infrastructure Technology Fund (PIT-735-07).

REFERENCES

[1] M. Zheng, A. Jagota, M. S. Strano, A. P. Santos, P. Barone, S. G. Chou, B. A. Diner, M. S. Dresselhaus, Mildred S., R. S. Mclean, G. B. Onoa, G. G. Samsonidze, E. D. Semke, M. Usrey, D. J. Walls, *Science* **2003**, *302*, 1545.

[2] M. Zheng, A. Jagota, E. D. Semke, B. A. Diner, R. S. Mclean, S. R. Lustig, R. E. Richardson, N. G. Tassi, *Nat. Mater.* **2003**, *2*, 338.

[3] Ph. Avouris, *Physics World* **March 2007**, 40.

[4] S. E. Snyder, S. V. Rotkin, *JETP Letters* **2006**, *84*, 348.

[5] S. G. Chou, H. B. Ribeiro, E. B. Barros, A. P. Santos, D. Nezich, Ge. G. Samsonidze, C. Fantini, M. A. Pimenta, A. Jorio, F. Plentz Filho, M. S. Dresselhaus, G. Dresselhaus, R. Saito, M. Zheng, G. B. Onoa, E. D. Semke, A. K. Swan, M. S. Uenlue, B. B. Goldberg, *Chem. Phys. Lett.* **2004**, *397*, 296.

[6] Y. Miyauchi, M. Oba, S. Maruyama, *Phys. Rev. B* **2006**, *74*, 205440.

[7] M. F. Islam, D. E. Milkie, C. L. Kane, A. G. Yodh, J. M. Kikkawa, *Phys. Rev. Lett.* **2004**, *93*, 037404.

[8] J. Lefebvre, P. Finnie, *Phys. Rev. Lett.* **2007**, *98*, 167406.

[9] T. Okada, T. Kaneko, R. Hatakeyama, K. Tohji, *Chem. Phys. Lett.* **2006**, *417*, 288.

[10] M. E. Hughes, E. Brandin, J. A. Golovchenko, *Nano Lett.* **2007**, *7*, 1191.

[11] G. Dukovic, M. Balaz, P. Doak, N. D. Berova, M. Zheng, R. S. Mclean, L. E. Brus, *J. Am. Chem. Soc.* **2006**, *128*, 9004.

[12] M. Zheng, E. D. Semke, *J. Am. Chem. Soc.* **2007**, *129*, 6084.

[13] L. Zhang, S. Zaric, X. Tu, X. Wang, W. Zhao, H. Dai *J. Am. Chem. Soc.* **2008**, *130*, 2686.

[14] S. E. Snyder, S. V. Rotkin, *Small* **2008**, accepted.

[15] T. Ando, T. Nakanishi, R. Saito, *J. Phys. Soc. Jpn.* **1998**, *67*, 2857.

Modeling and Simulations of Adhesion between Carbon Nanotubes and Surfaces

N. R. Paudel[1], A. Buldum[1,2], T. Ohashi[3] and L. Dai[4]
[1]Department of Physics, The University of Akron
[2]Department of Chemistry, The University of Akron,
[3]Honda Research Institute USA Inc.
[4]Department of Chemical and Materials Engineering, University of Dayton

ABSTRACT

Recent experiments showed that there are very strong adhesion forces between nanotubes and surfaces. These forces were much stronger then the adhesion forces of Gecko's foot hairs on substrates. Thus, nanotubes become candidates to be used in dry adhesives. Here, we present theoretical investigations on the adhesion between nanotubes and graphite surfaces. Molecular dynamics simulations and energy minimization calculations were performed. Layer by layer deformations of the nanotube tips were observed. Parallel and perpendicular components of the forces were calculated for different contact angles of nanotubes. The adhesion forces are found to be maximized at 15° angles with respect to surface normal.

Keywords: Nanotubes, Adhesion, Gecko's Foot, Molecular Dynamics

1 INTRODUCTION

Carbon nanotubes (CNTs) have attracted great interest in the field of nanotechnology because of their remarkable mechanical, electrical and transport properties [1-5]. The mechanical properties of nanotubes are very important as many potential applications depend on these properties. Over the past couple of years, there have been many studies on mechanical properties and deformation on CNTs such as bending, buckling, twisting and curvature effects using molecular dynamics (MD) simulations [4]. With the development of new experimental techniques, the buckling behavior of the CNTs has been observed under large deformation [6].

Recent experiments showed that there are very strong adhesion forces between nanotubes and surfaces. Yurdumakan et. al. [7] measured the adhesion force on multiwalled carbon nanotubes. In the experiment, they formed MWNT brushes on PMMA polymer surfaces. Then, scanning probe microscopy was used to measure the strong adhesion forces of the MWNT brushes. They found that the forces were 200 times higher than that of gecko's foot hairs[8]. Thus, nanotubes are good candidates to be used in dry adhesives.

In this research study, we perform theoretical investigations of CNTs interacting with surfaces. To study the deformation behavior and adhesion of CNTs, atomistic simulations of capped armchair (10,10) nanotubes with two different lengths are performed on rigid and relaxed graphite surfaces. There had been many theoretical studies of CNT tips interacting with surfaces[9-13]. These studies were mainly focused on employing CNTs as SPM tips. We started with a similar system, however, beside studying the deformation of CNTs, we also focused on parallel and perpendicular components of the forces for different contact angles to investigate the adhesive behavior of nanotubes.

2 MODEL

Atomic models of (10, 10) armchair single-wall nanotubes were created and combined with atomic models of graphite surfaces which are multilayer periodic graphite lattice structures. Two capped nanotubes of different lengths were selected such that the shorter tube had 790 atoms and the longer tube had 1990 atoms. The lengths between two extreme ends of the tubes were 46.5 Å and 122 Å, respectively. These tubes were then combined with graphite surfaces consisting 10 layers. In the longer tube case, the lattice constants of the overall structure were $a = 49.2$ Å, $b = 49.2$ Å and $c = 163$ Å where, $a = 36.9$ Å, $b = 36.9$ Å and $c = 86$ Å were chosen for the shorter tube case. 120 atoms of the nanotube from the uncapped end were chosen to be fixed. The nanotubes were initially positioned at 6 Å above the top layer of the graphite surface. The simulations were performed by using Cerius^2 with the Universal Force Field [10].

During the simulations, the fixed nanotube atoms were moved towards the surface in the increments of 0.1 Å for energy minimization calculations and 0.2 Å for the molecular dynamics simulations. In MD simulations, the entire system was allowed to equilibrate for 5000 steps initially and then moved towards the surface. There were 500 MD steps in between each increment of the displacement of the tip. All MD simulations were performed at constant NVT. The temperature was selected as 10 K and the time step was 0.001 PS.

3 DEFORMATION OF THE NANOTUBES AND ADHESION FORCES

The first series of simulations were energy minimization calculations and nanotubes were facing hollow, top1 and top2 atomic sites of graphite surface. These atomic sites are shown in figure 1.

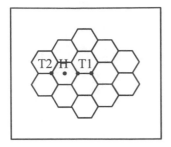

Figure 1: The hollow (H), top1 (T1) and top2 (T2) sites on the top-most layer of the graphite surface.

Energy minimization calculations were performed for both short and long nanotubes on rigid and relaxed surfaces. The total potential energy (E) was calculated during the simulations and the variation of E as a function of z displacement for a short nanotube on three different atomic sites is presented in figure 2. In the figure, the first major peak represents the critical position of the tip in which the repulsive interaction between the graphite surface and tip reaches a maximum. When the tip was moved closer to the surface, a layer of carbon atoms from the tip moved into the tube which resulted with the first deformation of the tip and caused the energy to drop. Snapshots of the tip deformation for the hollow atomic site case are shown in figure 3.

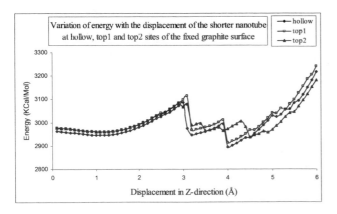

Figure 2: Variation of the total potential energy as a function of tip displacement for the shorter nanotube on a fixed graphite surface. The nanotube tip was initially 6A above the surface. Positive z-direction is towards the surface.

When the tip moves toward the graphite surface, the nearest atoms of the capped end are first attracted by the graphite layers up to a critical position then repulsive forces push them back along the tube axis. When the tip moves further, the repulsive forces are increased that causes layer by layer the compression of the cap into the interior of the nanotube. The sharp end of the cap changes gradually from an original concave shape to a convex shape which also was observed in previous simulations [11]. The symmetrical deformation of the inversed cap of the tip was observed for the hollow site but not necessarily for the top1 and the top2 sites. Two results from these calculations appeared to be interesting at this point. One is the layer by layer compression of the tip surface which is related with the discrete nature of the tip deformation. Another is the atomic surface site dependence of the energy variation and deformation. How atoms are positioned with respect to each other is important for the energy variation and deformation of the nanotube cap.

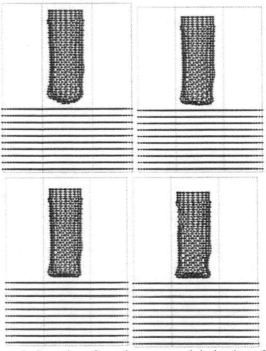

Figure 3: Snapshots from the energy minimization of the interaction of a capped [10, 10] nanotube with fixed graphite layers at hollow position for (a) $\Delta Z = 2$ Å (b) $\Delta Z = 3.5$ Å (c) $\Delta Z = 4.3$ Å and (d) $\Delta Z = 6$ Å.

After further compression, the structural deformation of the side wall increases gradually and then changes into buckling of the nanotube. The large deformation of the longer nanotube with buckling, bending and slipping from MD simulations is presented in figure 4. As it can be seen in the figure, the front part of the nanotube after the onset of buckling makes an angle with the graphite surface. After buckling and bending, the slipping of the tip on the graphite surface was observed.

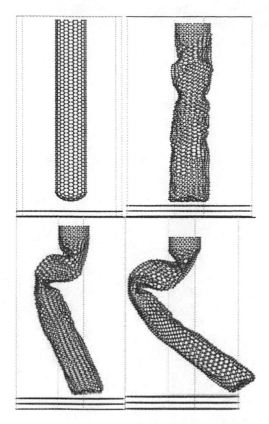

Figure 4: Snapshots from the molecular dynamics simulations of a long capped (10, 10) nanotube on a fixed graphite surface at hollow position for (a) $\Delta Z = 0$ Å (b) $\Delta Z = 10$ Å (c) $\Delta Z = 20$ Å and (d) $\Delta Z = 40$ Å.

Another series of simulations were performed for different contact angles of nanotubes on graphite surfaces. The contact angle (or impact angle) here is the angle between the nanotube axis and the normal of the graphite surface as shown in the figure 5.

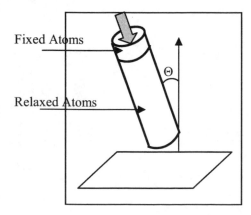

Figure 5: Schematic diagram of the tilted nanotube. The contact or impact angle Θ is defined with respect to the surface normal.

Simulations were performed for seven impact angles in intervals of 5^0 from 0 to 30 degrees. The nanotubes were initially positioned 6 Å above the graphite surface and moved along the axes of the nanotubes (the direction of the motion is shown by an arrow in figure 5) up to 20 Å axial displacement. One important quantity for the adhesion is the length of front part of the nanotube after buckling (i.e. buckling length). The front part would become parallel to the surface after further compression and the main contribution to the adhesion force would be coming from this part. The buckling length of a buckled nanotube is from the point of buckling to the capped end of the nanotube. The buckling length as a function of impact angles is presented in figure 6.

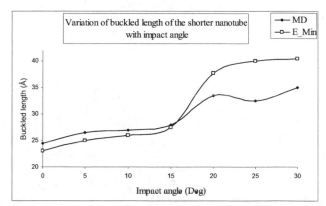

Figure 6: Impact angle dependence of the buckling length of the shorter tube determined from MD simulations and energy minimization calculations.

In the figure, we can see the variation of buckling length of the shorter tube, which increases slowly at first with the impact angle, and then increases rapidly between 15 and 20 degrees and finally becomes nearly constant after 25 degree. Beyond 30 degree slipping occurs quickly so that buckling of the nanotube could be observed rarely beyond this point. One can clearly seen in the figure that the buckling length change appears to be significantly larger in the energy minimization case. This is due to the temperature dependent relaxation of the atoms in the MD simulations. Similar results were also obtained for the longer nanotube.

By performing MD simulations, the maximum force on the nanotubes for different impact angles were determined. Maximum force values for different impact angles are presented in figure 7. The results show that the maximum force depends on the impact angle and has a maximum at 15 degrees. This angle coincides with the angles where sudden change in the buckling length was occurred. The components of the adhesion forces were investigated also. The parallel components of forces were much weaker than the perpendicular components. On the other hand, for tilted orientations of the nanotubes, the parallel component increases with the increase of the impact angle.

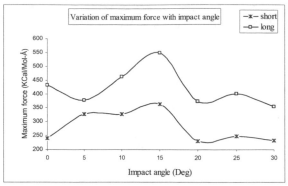

Figure 7: Impact angle dependence of the maximum force experienced by the atoms of tip surface.

4   CONCLUSIONS

Theoretical investigations on the adhesion between nanotubes and graphite surfaces were performed. Atomistic simulations of capped armchair (10,10) nanotubes with two different lengths on rigid and relaxed graphite surfaces were carried out. The layer by layer compression of the tip surface was observed which is related with the discrete nature of the tip deformation. The total potential energy and deformation of the nanotube cap were dependent on the local atomic surface sites. The buckling length increased slowly at first with the impact angle, and then increased rapidly between 15 and 20 degrees. The maximum force on the nanotube depends on the impact angle and has a maximum at 15 degree.

REFERENCES

[1] S. Iijima, Nature, 354 56, 1991.
[2] J. P. Lu, Phys. Rev. Lett., 79, 1297, 1997
[3] B. I. Yakobson, C. J. Brabec, and J. Bernholc, Phys. Rev. Lett., 76, 2511, 1996.
[4] J. Mintmire, B. Dunlap and C. White, Phys. Rev. Let., 68, 631, 1992.
 [5] A. Buldum and J. P. Lu, Phys. Rev. Lett., 91, 236801, 2003.
[6] G. Cao and X. Chen, Phys. Rev. B, 74, 165422, 2006
[7] B. Yurdumakan, N. R. Raravikar, P. M. Ajayan and A. Dhinojwala, Chem. Communication, 3799, 2005
[8] K. Autumn, Y. A. Liang, S. T. Hsieh, W. Zesch, W. P. Chan, T. W. Kenny, R. Fearing and R. J. Full, Nature, 405, 681, 2000.
[9] N. Yao and V. Lordi, Phys. Rev. B, 58, 12649, 1998.
[10] A. K Rappe, C. J. Caeswit, K. S. Colwell, W. A. Goddard III and W. M. Skiff,, J. Am. Chem. Soc., 114, 10024, 1992.

[11] S. P. Ju, C. L. Weng and C. H. Lin, J. Appl. Phys., 95, 5703, 2004
[12] J. A. Harrison, S. J. Stuart, D. H. Robertson and C. T. White, J. Phys. Chem. B, 101, 9682, 1997.

[13] A. Grag and S. B. Sinnott, Phys. Rev. B, 60, 13786, 1999

Improve Contacts in Carbon Nanotube Networks by In situ Polymerization of Thin Skin of Self-Doped Conducting Polymer

Yufeng Ma, William Cheung, Dongguang Wei, Albert Bogozi, Pui Lam Chiu, Lin Wang, Francesco Pontoriero, Richard Mendelsohn, and Huixin He*

*Chemistry Department, Rutgers University, Newark, NJ 07102
Phone: 973-353-1254; Fax: 973-353-1264
Email: huixinhe@newark.rutgers.edu

ABSTRACT

The overall conductivity of SWNT networks is dominated by the existence of high resistance and tunneling/Schottky barriers at the intertube junctions in the network. Here we report that *in-situ* polymerization of a highly conductive self-doped conducting polymer "skin" around and along single stranded DNA dispersed- and functionalized- single wall carbon nanotubes can greatly decrease the contact resistance. The polymer skin also acts as "conductive glue" effectively assembling the SWNTs into a conductive network, which decreases the amount of SWNTs needed to reach the high conductive regime of the network. The conductance of the composite network after the percolation threshold can be two orders of magnitude higher than the network formed from SWNTs alone.

Keywords: carbon nanotubes, DNA, self-doped polyaniline, composite

1 INTRODUCTION

There is increasing enthusiasm for the use of single walled carbon nanotube (SWNT) networks as conductive flexible electrodes and sensing materials. However, the experimentally measured conductivities of the SWNT networks are significantly lower than the conductivity of a SWNT rope. Previous studies demonstrate the existence of high resistance and tunneling/Schottky barriers at the intertube junctions, which dominates the overall film conductivity in the network.

Herein, we report that the conductivity of SWNT networks can be dramatically improved by *in-situ* polymerization of a thin layer of self-doped conducting polymer (polyaniline boronic acid, PABA) around and along the carbon nanotubes. The formed conducting polymer improves the contacts between the SWNTs and it also acts as a "conductive glue or zipper", which effectively assembles the SWNTs into a conductive network and decreases the amount of SWNTs needed to reach the high conductive regime of the network. The conductance of the composite network beyond the percolation threshold can be two orders of magnitude higher than the network formed from SWNTs alone. In addition, the thin layer of conducting polymer provides a powerful functionality for a variety of potential applications, including flexible sensors.[1] We also found that the enhancement highly depends on the methods to coat the thin layer of the polymer onto the carbon nanotubes.

2 EXPERIMENTAL SECTION

2.1 *In-situ* fabrication of a self-doped polyaniline/ss-DNA-SWNTs nanocomposite

The materials used and the protocol for the dispersion of SWNT into water solution are described in our previous work.[2] A typical procedure for the preparation of a solution of ss-DNA/SWNTs/PABA nanocomposite by *in-situ* polymerization approach is as follows: 50 μL of ABA solution (50 mM) and KF (40 mM) in 0.05 M H_2SO_4 was added to 2.5 mL of the ss-DNA/SWNTs solution (70 mg/L) in 0.05 M H_2SO_4. The solution was bubbled with nitrogen for 30 min at 0 °C to remove the dissolved oxygen. The chemical polymerization of ABA was then initiated by adding 11.34 μL of 37.5 mM $(NH_4)_2S_2O_8$ (APS) (in 0.05 M H_2SO_4) drop-wise to the mixture. It is important to note that the amounts of ABA and APS were determined from titration experiments to make sure only a thin layer of PABA is produced around and along the carbon nanotubes. The polymerization was carried out at 0 °C with nitrogen bubbling for 7 h and another 43 h in a refrigerator (4 °C). The obtained composite solution is referred to as "in-situ polymerized composite". For the "seed" approach, the polymerization conditions were kept the same, except 11.34 μL of 37.5 mM APS was first added to pre-oxidize the same amount of ss-DNA/SWNTs, followed by addition of 50 μL of ABA solution (50 mM). The obtained composite solution is termed "seed composite". A neat poly (aniline boronic acid) (PABA) was fabricated by a recipe described in our previous work, which was demonstrated to produce PABA with longer conjugation length.[2] A composite of ss-DNA/SWNT/PABA was prepared by mixing 50 μL of the preformed neat polymer solution with 2.5 mL of the ss-DNA/SWNTs (70 mg/L) in 0.05 M H_2SO_4. The resulting solution is so called "postmixture composite". All the composites, neat PABA, and ss-DNA/SWNTs solutions were dialyzed to remove excess salts before conductance measurements.

2.2 Characterization

Percolation-like conductive behaviors of the composites and ss-DNA/SWNT alone were prepared and studied by measuring the conductance of the films in a layer-by-layer approach on a pre-patterned Si chip. Each layer was prepared by adding 2 µL of the dialyzed solution (10 mg/L of SWNTs) onto a Si chip and dried under vacuum. The Si chip was fabricated at Air Force Research Labs and the distance between two facing gold electrodes is 2 µm. The conductance of the composites was measured with an Electrochemical Workstation CHI 760C.

The conductance was also studied by a four point-probe approach. Films with different thickness were prepared from the corresponding composite and ss-DNA/SWNT solutions by vacuum filtration using Anodisc 47 inorganic membranes with 200nm pores (Whatman Ltd). To evaluate the impact of the conducting polymer skin on the conductivity of the carbon nanotube network, the thickness of each film were prepared with different composite solutions while maintaining the concentration of the carbon nanotubes. After filtration, the thin films were dried in vacuum for 15 – 20 minutes. The sheet conductance was determined by a 302 manual Four Point Resistivity Probe (Lucas Labs).

The morphology of the resulting composites, neat polymer, and the ss-DNA/SWNT films was characterized by a Nanoscope III A (Digital Instruments) operating with tapping mode in ambient air. The thickness of PABA on the carbon nanotubes was measured from high resolution transmission electron microscope (TEM) (Libra 120 Energy Filtering TEM, Zeiss) operated at 200kV. Samples were prepared for imaging by placing a drop of aqueous composite solution on TEM grids and wicking away the liquid after 2 minutes. Electronic and molecular structures of the PABA in the pure polymer and the composites were measured by a Spectrum Spotlight FTIR Imaging System (Perkin Elmer instruments) and a Cary 500 UV-Vis-NIR Spectrophotometer operating in double beam mode.

3 RESULTS AND DISCUSSION

3.1 Electrical properties of the composite and ss-DNA/SWNT films

Figure 1a shows the room-temperature percolation behavior of the ss-DNA/SWNTs network and the composite films. At the third layer, the *in-situ* fabricated composite reaches its percolation threshold while the ss-DNA/SWNT film just begins to have measurable current. At this percolation point, the conductance of the *in-situ* polymerized composite is 5 orders of magnitude higher than that of the ss-DNA/SWNT film. There is no detectable current for the composites prepared by post-mixing and "seed" method. According to the percolation mechanism, multiple conducting channels have formed in the *in-situ* polymerized composite film and the tube-tube junctions

begin to dominate its overall resistance, whereas space between conducting sticks is still the governing factor for the other three samples.

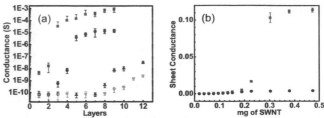

Figure 1. (a) Conductance of in-situ polymerized composite (■), ss-DNA/SWNTs (●), post mixture (▲), and "seed" composite (▼) as a function of layers of the composites and ss-DNA-SWNT. Each layer corresponds to 2 µL of solution with a concentration of SWNT at 10mg/mL. The conductance was measured by a two probe approach. Each data points presented here was an average of 18 pair of electrodes on five silicon chips. (b) Conductance of *in-situ* polymerized composite (■), ss-DNA/SWNTs (●) measured by a four probe approach. Each data point presented here was an average of 10 measurements.

The films reached their percolation threshold at 3, 6, 10 and 11 layers for in-situ polymerized composite, ss-DNA/SWNTs, post-mixture and "seed" composite, respectively. After percolation, the conductance of the *in-situ* polymerized composite is $\sim10^2$-fold, $\sim10^5$-fold, and $\sim10^6$-fold higher than that of the ss-DNA/SWNTs, post-mixture and "seed" composite, respectively. Based on the percolation theory for randomly arranged conducting sticks, above the percolation threshold, the overall resistance of the film is dominated by intertube junctions in the conducting channels. The significant difference in conductance illustrates that only the *in-situ* polymerized PABA remarkably improved the intertube contacts, while the pre-formed PABA and the *in-situ* polymerized PABA by "seed" approach did not.

To eliminate the influence of electronic contacts between the electrodes and the composite on the electrical measurements, the percolation behavior was also studied by a four probe approach with a probe distance of 1mm. The conductance of the composite prepared by the "seed" method and postmixture approach is beyond the sensitivity of the measurement setup. Figure 1b shows that the sheet conductance of the *in-situ* polymerized composite network increased 40 times compared to that of ss-DNA/SWNT alone. In contrast to two-probe measurements, it seems that the in-situ polymerized conducting polymer skin did not decrease the amount of SWNTs needed to reach the high conductive regime of the network. This discrepancy may be related to the geometry-dependent percolation behavior of the carbon nanotube networks, which has been studied by Kumar et al and Ural groups recently. With larger distances between the source and drain electrodes, the percolation probability of the carbon nanotubes, defined as the probability of finding at least one conducting path between the source and drain electrodes, decreased more dramatically compared to the small ones. In this work, the distance between the two electrodes is around 2 μm in the two-probe measurement, 500 times shorter than the distance between the probes in the four-probe measurements (1mm). It is possible that the length of the "glued" tubes by self-assembling during the polymerization is still too short compared to 1 mm, while it is comparable or longer than 2μm.

Molecular structures of PABA in the composites. We used Fourier Transform Infrared (FTIR) spectroscopy to systematically study the molecular structures of PABA in the composites fabricated from different approaches. Figure 2 shows the FTIR spectra of the four samples: in-situ polymerized composite, neat PABA, "seed"-composite, and postmixture. IR characteristic bands of polyaniline are observed at 1574, 1478, and 1120 cm^{-1}, corresponding to quinoid, benzenoid, and C=N stretching (-N=quinoid=N-, "electron-like band") modes, respectively. The peak at 1210 cm^{-1} is related to the antisymmetric stretching vibration of the PO^{2-} in DNA. The spectra also exhibit characteristic vibrations at 1510 and 1340 cm^{-1}, assigned to the B-N and B-O stretching mode. The calculated ratio for the absorption intensity of quinoid to benzenoid ring modes (I_{1574}/I_{1478}) in the pure PABA is 3.5, which suggests that the percentage of quinoid units is much higher than that of benzenoid units for the neat PABA film. However, the ratio of I_{1574} to I_{1478} decreases to 1.5 for PABA in the in-situ fabricated composite, indicating that the relative amount of quinoid units decreased in the PABA when polymerized in the presence of ss-DNA/SWNTs. This result is consistent with our previous work,[2] indicating that the PABA exists more in the fully oxidized pernigraniline state in pure PABA, and more in the conductive emeraldine state in the composite.

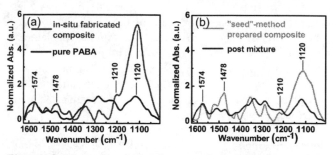

Figure 2. Normalized Fourier-transform IR spectra of (a) in-situ polymerized composite (red) and pure PABA (blue); (b) "seed" composite (green) and postmixture (purple). After the baseline correction at two points (1626 and 1010 cm^{-1}), all IR spectra are normalized as the standard peak (quinoind) at 1574 cm^{-1}.

The ratio of relative intensities of quinoid to benzenoid ring modes (I_{1574}/I_{1478}) in the post mixture is 3.4, which is similar to the neat PABA (3.5), indicating the neat PABA

weakly interact with the ss-DNA/SWNTs in the composites. The results demonstrate that the PABA in the postmixture composite exists in the non-conductive pernigraniline state.

A striking difference among the four spectra in Figure 2 is found in -N=quinoid=N- stretching at ~1120 cm^{-1}. The peak has been described by MacDiarmid et al. as the "electronic-like band" and is considered to be a measure of the degree of delocalization of electrons along polyaniline backbone and thus it is a characteristic peak of polyaniline conductivity. There is a dramatic intensity increase of the "electronic-like band" in the in-situ polymerized composite, as shown in Figure 2a. This remarkable increase suggests that PABA in the in-situ polymerized composite has higher conductivity compared to the neat PABA. The composite fabricated by the postmixture approach shows similar low intensity of the "electron-like band" as the neat PABA, indicating that conductivity of the PABA in the postmixture composite may be similar to the neat PABA.

Unexpectedly, the relative intensity of the "electron-like band" in the composite prepared by the "seed" approach is 2.9, much higher than that of the neat PABA and the postmixture composite (1.3). The ratio of relative intensities of quinoid to benzenoid ring modes (I_{1574}/I_{1478}) in the composite prepared the "seed"-method is 0.9, much lower than that of the neat PABA and the postmixture composite (3.5, and 3.4 respectively). The results indicated that the PABA in the composite fabricated by the "seed" approach also existed in the conductive emeraldine states. Therefore, the conductivity of the PABA should be higher than the PABA in the postmixture composite, even though lower than that of the composite prepared by in-situ polymerization with the intact ss-DNA/SWNTs. However, the conductance measurements described above did not show this trend, indicating that there are other uncovered factors that determine the overall macroscopic conductivity of the films in addition to the modified tube-tube contacts.

Atomic force microscope (AFM) and high resolution transmission electron microscope (HRTEM) images in Figure 3 shows the third layer of the different samples (ss-DNA/SWNTs, in-situ polymerized composite, postmixture and "seed" composite) that have been prepared layer-by-layer from the corresponding solutions with a SWNT concentration of 10 mg/L. At this percolation point, as shown in Figure 1a, only the composite fabricated by in-situ polymerization with the intact ss-DNA/SWNTs reached the percolation threshold studied by two-probe measurements. The composites fabricated by postmixture and "seed" method showed no detectable current by our detection instruments. Figure 3a shows the morphology of the films prepared from ss-DNA/SWNT alone. Most of the SWNTs are individual (~2 nm diameter), with some bundled structures (~17 nm). The nanotubes appear randomly oriented. Most of the tubes remain isolated from each other, with some jointed tubes. Figure 3b is a typical AFM image of the in-situ polymerized composite. Remarkably, individual ss-DNA/SWNTs are replaced by long fibers (most of the fibers are longer than 4 μm). These fibers are randomly arranged and self-assembled into a network. The diameter of the fibers ranges from 3 nm to 20 nm measured by high resolution TEM (Figure 3c). From the TEM images it is noted that some of the fibers are composed of individual nanotubes with a polymer coating of 1 to 3 nm in thickness. Some of the fibers are carbon nanotubes bundles, which are composed of carbon nanotubes with polymer coating.

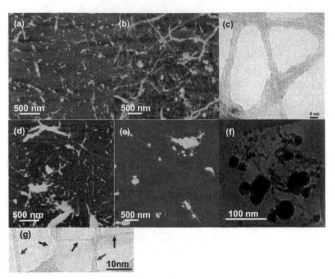

Figure 3. AFM images of the third layers of the films prepared from (a) ss-DNA/SWNTs, (b) *in-situ* polymerized composite, (d) postmixture and (e) "seed"composite. The concentration of SWNT in all these samples is 10 mg/L. TEM images of *in-situ* polymerized composite (c), "seed" composite (f) and postmixture (g).

The pre-polymerized PABA didn't show the tendency to self-assemble the carbon nanotubes into networks (Figure 3d). The arrangement and the spatial distribution of the carbon nanotubes in the postmixture composite are very similar to that of ss-DNA/SWNT alone. In the AFM image, there are large bright regions (~60 nm in height), which are suspected to be the neat PABA without being uniformly mixed with the SWNTs. From the study by high resolution TEM (Figure 3g), we found that while some of the tubes were not coated, some were coated with a 1-3 nm layer of polymer. In the conductance experiment, we found the conductance of the postmixture network dramatically decreased and the percolation threshold of the SWNT networks largely increased (3 fold) (Figure 1). After the percolation threshold, the conductance of the postmixture composite is three orders of magnitude lower than the network prepared from SWNT alone, and is five orders of magnitude lower than the network formed from the in-situ polymerized PABA composite. Combined with the FTIR results, showing that the PABA exists in the non-conductive pernigraniline state in the postmixture composite and conductive emeraldine state in the *in-situ*

polymerized composite, this morphological study strongly suggest that the electronic and molecular structure, and therefore the conductivity of the interfacial PABA on the SWNTs can modulate the overall electronic performance and percolation behavior of the SWNT films. Compared to the simple postmixture process, the *in-situ* polymerization process also facilitates SWNTs self-assembling to highly conducting networks, which also largely contributes to the highly improved conductance and the low percolation threshold of the *in-situ* polymerized composite. This result soundly supports the hypothesis discussed earlier in this work.

The morphological study of the "seed" composite (Figure 3e, f) revealed our curiosity pertaining to why the conductance is even lower than the postmixture. In contrast to the PABA *in-situ* polymerized in the presence of the intact ss-DNA/SWNTs, the PABA in the "seed" composite did not interlink the nanotubes into a conductive network. Instead, the PABA induced severe aggregation of the nanotubes into large particles (as large as 1μm). This aggregation dramatically changed the effective length/diameter aspect ratio of the carbon nanotubes, which is known to impact the conductivity and percolation behavior of the carbon nanotube films. Therefore, even though the PABA in the "seed" composite may has higher conductivity compared to that of PABA in the postmixture, the aggregated carbon nanotubes makes the conductivity of the "seed" composite even lower than the postmixture composite. We conclude that not only do the molecular structure of the polymer, but also the arrangement or distribution of carbon nanotubes in the composites determines the overall macroscopic electronic property and percolation behavior of the composites.

Summary. The electrical performance of SWNT network can be significantly improved by *in-situ* polymerization of a thin layer of PABA on the intact ss-DNA/SWNTs. However, the fabrication process rigorously impacts the electronic and molecular structure of the produced PABA in the composites, and also the arrangement or lateral distribution of the carbon nanotubes in the composites. Understanding these reaction characteristics is important to effectively optimize the fabrication parameters and ensure the formation of SWNT networks in a controllable fashion for a variety of potential applications.

Acknowledgement. This material is based upon work supported by the National Science Foundation under CHE-0750201 and Petroleum Research Fund.

Reference:

(1) Ali, S. R.; Ma, Y. F.; Parajuli, R. R.; Balogan, Y.; Lai, W. Y.-C.; He, H. X., *Anal. Chem.* **2007,** 79, 2583-2587.
(2) Ma, Y. F.; Ali, S. R.; Wang, L.; Chiu, P. L.; Mendelsohn, R.; He, H. X., *J. Am. Chem. Soc.* **2006,** 128, 12064-12065.

Deagglomeration and Dispersion of Carbon Nanotubes Using Microfluidizer® High Shear Fluid Processors

Thomai Panagiotou*, John Michael Bernard** and Steven Vincent Mesite***

Microfluidics, 33 Ossipee Rd., Newton, MA, USA

* mimip@mfics.com, ** mbernard@mfics.com and *** smesite@mfics.com

ABSTRACT

For many carbon nanotube (CNT) applications, paramount performance can be achieved if the CNTs get deagglomerated, frequently shortened, and uniformly dispersed in liquid media. Microfluidizer high shear fluid processors were used successfully to disperse single- and multi- wall CNTs in various liquid media including polymer resins, organic solvents and water. Processing deagglomerated the CNTs and caused them to form networks inside the liquid media. Length reduction of CNTs was possible and controlled based on the processing conditions. During processing, CNTs detached from the catalyst substrate, facilitating purification. Finally, the electrical resistivity of polymer/CNT composites was measured as a function of CNT concentration. For samples processed with the Microfluidizer® processor, the volume resistivity of the composites decreased by seven orders of magnitude as the CNT concentration increased from 0.008wt% to 0.06wt%. There was no change in resistivity for the baseline samples.

Keywords: dispersion, carbon nanotubes, nanocomposites microfluidizer

1 INTRODUCTION

Deagglomeration and dispersion of carbon nanotubes (CNTs) in various media has been recognized as one of the major challenges in utilizing CNTs in commercial applications. For many CNT applications, paramount performance can be achieved if the carbon nanotubes get deagglomerated, frequently shortened and uniformly dispersed in media such as organic solvents, polymer resins, water, etc.[1-3]. Deagglomerated CNTs form networks, and such networks are responsible for enhanced strength and electrical conductivity of polymer composites containing CNTs.

Any method used to de-agglomerate CNTs should be efficient, economical, and scalable. Sonication is a method that is often used in lab scale to de-agglomerate and shorten the CNTs. This method does not scale up and may contaminate the CNTs with metal particles from the sonication probe. In addition, it may change the chiral properties of the nanotubes[3]. Functionalization of CNT surfaces to improve dispersion may be effective but does not solve the problem fully and is still in the research stage[2].

Microfluidizer high shear fluid processors have been used extensively for particle deagglomeration, dispersion and size reduction. Processing of multi-phase liquids takes place as the fluids flow at velocities of up to 500 m/s in microchannels of various geometries, thus exposing uniformly the liquids to high shear stresses. The process pressure and the channel geometry control the velocities inside the channels, and therefore the energy dissipation.

Microfluidizer processors were used to process various formulations of single- and multi- wall CNTs. These were dispersed in water, high viscosity mineral oil, polymer resins and organic solvents. Process pressures in the range of 41 to 158 MPa were used in which the CNT dispersion is forced through specially designed microchannels called interaction chambers. Carbon nanotube concentrations varied in the range of 0.008 to 1.9 wt%. The effect of processing was assessed using FESEM (Field Emmision Scanning Electron Microscope) imaging, particle size analysis and bulk resistivity measurements.

2 EXPERIMENTAL APPARATUS

The Microfluidizer processor model M-110EH, equipped with two 'Z' type interaction chambers, was used to process the CNT dispersions. Figure 1 shows schematics of a 'Z' type microchannel design.

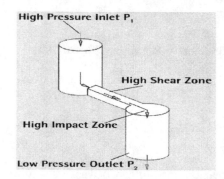

Figure 1: Schematic of a "Z" chamber. Copyright by Microfluidics, 2007.

Typically, the CNT bundles, dispersed in a liquid media, are forced through this type of configuration that has been specially designed to optimize shear. This consists of sharp turns and a very narrow rectangular cross-section; as small as 50μm across the smallest dimension. This combination is referred to as the interaction chamber.

Process scale up simply means adding multiple chambers or microchannels in parallel. Two interaction chambers were used in series for these experiments, the H30Z(200μm) and the G10Z(87μm).

The intensification of the process fluid involves the use of electro-hydraulics. An electric motor may drive a hydraulic pump, which in turn forces hydraulic fluid over the area of a piston. In all cases this piston is coupled to a plunger which forces the process fluid through the interaction chamber. The ratio of the areas or the piston to the plunger defines the amplification of pressure. This is schematically shown in Figure 2.

The premixes were obtained using either a Talboy 102 propeller stirrer, or an IKA T-25 shear mixer. The processed materials were analyzed using an Olympus BH-2 optical microscope and a Horiba LA-910 particle size analyzer. Samples were imaged by Hitachi S-4800, a Zeisis Supra™ 25 and Jeol JSM 6340F FESEMs. Volume resistivity testing was also conducted referencing ASTM-D257-99 using a Quadtech 1825 Tester.

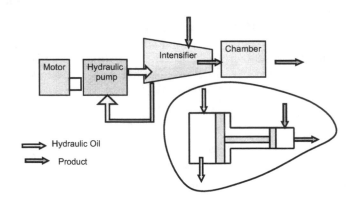

Figure 2: Principal of operation of a Microfluidizer processor. Copyright by Microfluidics, 2007.

3 PROCEDURE

The respective concentrations of the CNT dispersions were prepared by adding specific quantities (by weight) to the desired solvent or matrix material. Upon formulating, each dispersion mix of CNT was then stirred or shear mixed before addition to the reservoir of the Microfluidizer processor.

The test to de-agglomerate in a liquid media was such that all materials in the formulation were mixed well before processing. Table 1 shows a list of carbon nanotube formulations that were processed. Subsequently, they were analyzed by a combination of techniques, including optical microscopy, field emission scanning electron microscopy and particle size analysis.

The resistivity procedure involved first weighing the SWNTs (Single Wall carbon Nanotube) and then adding them to the Epon™ 862 [4-6] epoxy resin, which was preheated to approximately 40ºC and was continuously being stirred in a water bath using a Talboy model 102

Sample	Solid Content	Nanotube Type	Dispersant
1	1.90 %	SWNT	mineral oil
2	0.38 %	SWNT	mineral oil
3	1.00 %	SWNT	Polymer resin
4	1.00 %	SWNT	Water
5	0.70 %	SWNT	Toluene
6	6.00 %	MWNT	Methanol
7	6.00 %	MWNT	Methanol
8	0.60 %	MWNT	Methanol
9	0.06%	SWNT	Epoxy resin

Table 1: Typical CNT formulations processed with the Microfluidizer

propeller mixer. Batches of up to 255mL were used at each SWNT concentration which were 0.008, 0.02, 0.04 and 0.06 wt%. The SWNT-resin mixture was then processed at 158 MPa for 1 pass using an H30Z (200μm)-G10Z (87μm) chamber combination. The temperature of the material exiting the processor was maintained between 45 and 57ºC. This mixture was then combined with the curing agent (Epicure™ 3282) [4-6] within 5 minutes and allowed to cure in 4" x 4"x 1/8" moulds for 2 hours. The SWNT-epoxy tiles thus formed, were then allowed to finish curing in an oven at 120 ºC for 4 hours.

4 RESULTS

4.1 Deagglomeration in Liquid Media

The effects of processing 0.38% SWNT carbon nanotubes dispersed in high density mineral oil can be seen in the FESEM images of Figure 3. The unprocessed nanotubes, Figure 3a, are densely packed and the individual nanotube strands are not discernable. After a single pass, individual long and thin tubes can be clearly seen; as in Figure 3b. The CNTs appear much more de-agglomerated and forming a network. They are also uniformly dispersed throughout the media. It is not clear that these are individual nanotubes or thin strands but their diameter appears to be fairly uniform.

Additional passes shorten such strands, without necessarily decreasing the diameter of the strands; as seen Figures 3c and 3d.

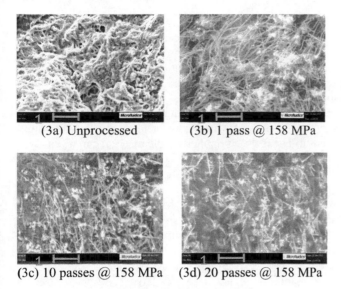

(3a) Unprocessed   (3b) 1 pass @ 158 MPa

(3c) 10 passes @ 158 MPa   (3d) 20 passes @ 158 MPa

Figure 3: Images captured with an FESEM at 20,000x Magnification. Each picture corresponds to different number of passes through the Microfluidizer processor. Copyright by Microfluidics, 2007.

4.2 Length Reduction

Results were obtained using a laser scattering particle size analyzer for the samples that had water or ethanol as the diluent. One of the assumptions that the particle size analyzer makes is that the particles are spherical. Since the carbon nanotubes have an extremely high aspect ratio, the numbers that are reported may not necessarily correspond to any dimensions of the nanotubes themselves. These particle size distributions merely act as a comparison between unprocessed and processed samples. The particle size analysis of the MWNTs (Multi Wall carbon Nanotubes) are shown in Figure 4. The particle size data shown was measured using 60 seconds of sonication in the flow cell of the particle size analyzer.

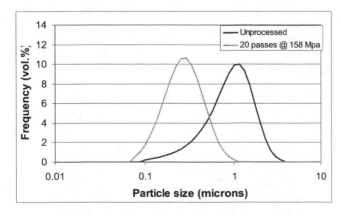

Figure 4: Particle size distributions of MWNTs dispersed in water before and after processing with the Microfluidizer processor. The median particle size went from 0.892μm down to 0.254μm after 20 passes at 158 MPa.

4.3 Impurity Stripping

Catalyst used during the reaction to from the CNTs becomes a residual impurity. After processing in the Microfluidizer processor, the residual catalyst that were entangled with the CNTs are broken and detached from the CNTs. This is shown in FESEM images of Figure 5.

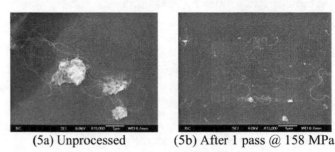

(5a) Unprocessed   (5b) After 1 pass @ 158 MPa

Figure 5: FESEM (20,000x) images of catalyst clumps being stripped from MWNTs (5a) before and (5b) catalyst impurity segregated after processing. Images provided by Nanolab Inc., Newton, MA.

4.4 Control Conductivity

A distinct change in appearance between the processed and unprocessed epoxy tiles is shown in Figure 6.

Figure 6: SWNT-Epoxy tiles during the initial curing state in the mold. Starting from the 1st column from the lower left, 0.008% at 0 and after 1 pass above, 2nd column 0.02% at 0 and 1 pass, 3rd column 0.04% at 0 and 1 pass and finally the 4th column 0.06% at 0 and 1 pass.

Comparisons of the unprocessed and the processed SWNT-epoxy systems are shown in Figure 7 show a dramatic difference between the resistivity value measured both before and after processing, and also between the resistivity values measured at increased concentrations of SWNT. At 0.008 wt%, there was a 23% reduction in resistivity after processing. At 0.02 wt%, there was a 6 order of magnitude drop in resistivity when the volume resistivity dropped from 7.28×10^{15} ohm-cm down to 1.21×10^{9} ohm-cm. Similarly at 0.04 wt%, there was a 7 order of magnitude drop in resistivity when the volume resistivity dropped from 7.22×10^{15} ohm-cm down to 1.34×10^8 ohm-cm. At the maximum SWNT loading of 0.06 wt% there was also a large drop in resistivity from 7.16×10^{15} ohm-cm down to 2.35×10^8 ohm-cm. The electrical behavior exhibited suggests that the critical loading required for a significant change in resistivity lies between 0.008 wt% and 0.02 wt% of SWNT. It also suggests that beyond the critical loading, enhanced dispersion of the SWNTs, ultimately resulted in

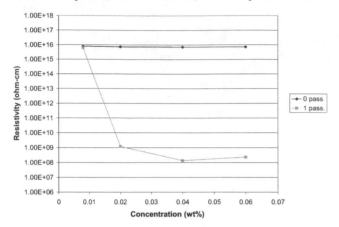

Figure 7: Resistivity vs concentration of SWNT after processing for 1 pass at 158 MPa using Microfluidics technology.

significant drops is resistivity. There was no significant change in resistivity for the SWNT-epoxy samples that was not processed with the Microfluidizer processor. The images of Figure 8 show the topical characteristics of the 0.06 wt% SWNT-epoxy specimen at a magnification of about 4,000x before and after processing with the Microfluidizer processor.

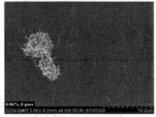 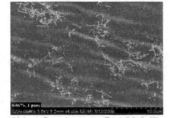
(8a) Unprocessed   (8b) After 1 pass @ 158 MPa

Figure 8: FESEM images of the 0.06 wt% SWNT-epoxy composites, (8a) before and (8b) after processing with a Microfluidizer processor.

These FESEM images suggest that for each concentration range, the SWNT clumps were dispersed by processing. This may have contributed to the enhanced conductivity of the SWNT-epoxy composite at loadings above 0.008 wt%.

5 SUMMARY

Microfluidizer technology was successful in de-agglomerating SWNT and MWNT, and creating stable suspensions of nanotubes in various liquid media. These

included polymer resins, organic solvents, mineral oil and water, with concentrations of carbon nanotubes varied in the range of 0.38-6%. Microscopy, both electron and optical, confirmed that the nanotubes were de-agglomerated to a large extent, and dispersed uniformly within in the dispersion media in a single pass. Additional passes resulted in controlled length reduction of the nanotubes. A distinct advantage of Microfluidizer technology is that the amount of shear that is applied on the material can be easily varied, allowing the nanotubes to detangle and be shortened to the extent required by any application. Microfluidizer processor technology was successful in reducing the particle size distribution of the MWNTs' dispersion in all four of the diluents that were analyzed.

The Microfluidizer processor provided the dispersion necessary for enhanced electrical conductivity as well. There was no significant change in resistivity at the lowest SWNT loading of 0.008 wt%. Similarly, there was also no significant change in resistivity for the specimens with loadings above 0.008 wt% that were not processed with the Microfluidizer processor. If good dispersion is initiated, the critical loading level above which there is a significant drop in resistivity lies between 0.008 and 0.02 wt% for this SWNT-epoxy system. This was also observed by Andrews et al [4-6].

REFERENCES

[1] Colbert, Daniel T, *Single-wall Nanotubes: A New Option for Conductive Plastics and Engineering Polymers*. Plastics Additives & Compounding. January/February 2003.

[2] "Carbon Nanotubes: A Small Scale Wonder". Chemical Engineering, February 2003.

[3] "Sonication-induced changes in chiral distribution: A complication in the use of single-walled carbon nanotube fluorescence for determining species distribution". Heller, D.A., Barone, P.W. and Strano, M.S., *Carbon*, **43**, 651-653 (2005).

[4] "Preparation of SWNT-reinforced composites by a continuous mixing process". Subramanian, Gopinath, Andrews, Malcolm J., Nanotechnology, **16**, 836- 840 (2005).

[5] "On SWNT-reinforced composites from a continuous mixing process". Subramanian, Gopinath, Andrews, Malcolm J., Nanotechnology, **18**, 345703 (9pp) (2007).

[6] "Carbon Nanotubes-Polymer Composites: A Study on Electrical Conductivity". Barrera, E.V., Bisch, C., Chakravarthi, Khabashesku, V., Lozano, K., Pena-Paras, L., Vaidyanathan, R., Zeng, Q. 38th SAMPE Fall Technical Conference November 2006.

A High-Flux, Flexible Membrane with Parylene-encapsulated Carbon Nanotubes

H. G. Park[*], J. In[**], S. Kim[***], F. Fornasiero[*], J. K. Holt[*],
C. P. Grigoropoulos[**], A. Noy[*] and O. Bakajin[*,***]

[*]Lawrence Livermore National Laboratory, LLNS LLC, Livermore, CA, USA, bakajin1@llnl.gov
[**]Mechanical Engineering, University of California, Berkeley, CA, USA,
[***]NSF Center for Biophotonics Science & Technology, Univ. of California Davis, Sacramento, CA, USA

ABSTRACT

We present fabrication and characterization of a membrane based on carbon nanotubes (CNTs) and parylene. Carbon nanotubes have shown orders of magnitude enhancement in gas and water permeability compared to estimates generated by conventional theories [1, 2]. Large area membranes that exhibit flux enhancement characteristics of carbon nanotubes may provide an economical solution to a variety of technologies including water desalination [3] and gas sequestration [4]. We report a novel method of making carbon nanotube-based, robust membranes with large areas. A vertically aligned dense carbon nanotube array is infiltrated with parylene. Parylene polymer creates a pinhole free transparent film by exhibiting high surface conformity and excellent crevice penetration. Using this moisture-, chemical- and solvent-resistant polymer creates carbon nanotube membranes that promise to exhibit high stability and biocompatibility. CNT membranes are formed by releasing a free-standing film that consists of parylene-infiltrated CNTs, followed by CNT uncapping on both sides of the composite material. Thus fabricated membranes show flexibility and ductility due to the parylene matrix material. These membranes have a potential for applications that may require high flux, flexibility and durability.

Keywords: membrane, carbon nanotube, parylene, high-flux

1 INTRODUCTION

Carbon nanotubes are a unique platform for studies of nanoscale mass transport and flow phenomena. Membranes that have carbon nanotubes as pores can be employed in a variety of important applications. Having nanometer-scale diameters of innermost walls whose graphitic surface is atomically smooth, they give rise to newly discovered phenomena of ultra-efficient transport of water through these ultra-narrow molecular pipes [5]. According to the experiments, water and gas flows through carbon nanotubes are enhanced by 3-5 and 1-2 orders of magnitudes, respectively, compared to conventional theories [1, 2]. This newly discovered unique nanoscale phenomena has applications in energy efficient filtration such as water desalination [3] and gas sequestration [4].

Use of CNT-based membranes in practical applications, requires scale up of the membrane area. It is also important to make sure that the large-area membranes are flexible and durable. When choosing materials for fabrication of carbon nanotube membranes it is important to consider chemical, biological and mechanical compatibility, ease of fabrication, and durability.

In an effort toward scale up the size of the carbon nanotube-based membrane, we explore a parylene-CNT composite. We describe fabrication of the parylene/CNT composite membrane. In section 2, we first address the currently available CNT-based membrane techniques and give an overview of simaulations and experiments that describe flow through CNTs. Properties of parylene and parylene-CNT composite follow in section 3. Then, we present fabrication of a parylene-CNT composite membrane.

2 FLOW THROUGH CNTS

2.1 Water Transport - Simulation

The task of observing and understanding fluid and gas flows in CNT pores raises a set of unique fundamental questions [6]. First, it is surprising that hydrophilic liquids, especially water, enter and fill very narrow and hydrophobic CNTs. If water does enter CNTs, what influence does extreme confinement have on the water structure and properties? It is important to evaluate how these changes in structure influence the rates, efficiency, and selectivity of the transport of liquids and gases through CNTs. As is often the case, MD (molecular dynamics) simulations have provided some of the first answers to these questions. Hummer et al. [7] have used MD simulations to observe the filling of a (6,6) CNT (0.81 nm in diameter and 1.34 nm in length) with water molecules. Surprisingly, they find that water fills the empty cavity of a CNT within a few tens of picoseconds and the filled state continues over the entire simulation time (66 ns). More importantly, the water molecules confined in such a small space form a single-file configuration that is unseen in the bulk water. Several experimental studies also provide some evidence of water filling of CNTs [8,9]. Further analysis of the simulation results of Hummer et al. shows that water molecules inside and outside a nanotube are in thermodynamic equilibrium. This observation illustrates one of the more important and counterintuitive phenomena associated with nanofluidic systems: nanoscale confinement leads to a narrowing of the interaction energy distribution, which lowers the chemical potential [7]. In other words, confining a liquid inside a nanotube channel actually lowers its free energy. Further simulations by the same group have shown that the filling equilibrium is very sensitive to water-nanotube interaction parameters: a 40% reduction in the carbon-water interaction potential results in the emptying of the CNT cavity, while a 25% reduction results in a fluctuation between filled and empty states (bi-stable states) [7,10]. This sharp transition between the two states has been observed for other hydrophobic nanopores as well [11,12]. MD simulations have also studied the dependence of CNT hydration on other properties of CNTs such as nanotube wall flexibility, charge, chirality, length, and diameter [5].

2.2 Gas Transport - Simulation

MD simulations also provide an indication that the intrinsic smoothness of the graphitic wall of CNTs is a defining feature for the gas phase transport in these channels. Sholl and colleagues [13] have shown that gas flux in CNTs reaches a value almost three orders of magnitude higher than in zeolites with equivalently sized pores. In these simulations, transport diffusivities of light gases such as hydrogen and methane reach almost the same level as that of the bulk gas diffusivity ($O(10^{-1})$ cm2/s). Depending on the nanotube size, the diffusivity is as large as 10 cm^2/s for hydrogen. In a follow-up set of simulations, the same group presents a more detailed picture of gas molecule behavior inside a SWNT pore by displaying a density profile and molecular trajectories [14]. The picture of gas transport through CNTs that has emerged from these simulations centers on the predominantly specular nature of the molecule-wall collisions inside CNTs. As a result, a CNT is remarkably effective in allowing gas transport since most of the gas molecules travel along the tube in an almost billiard-balllike manner [13-15]. Sokhan, Quirke, and others [16-18] have performed a series of MD simulations and reached a similar conclusion to Sholl and coworkers. They simulated both an atomic network of carbons and a smooth imaginary surface with a given slip boundary condition. Maxwell's coefficient or tangential momentum accommodation coefficient (TMAC) determined by the simulation are of the order of 10^{-3} (these values indicate that statistically only 0.1% of gas molecules in the nanotube are thermalized by the wall and randomize their reflection velocities). At the same time, Sholl's group [14] indicates that some of the gas molecules adsorb on the nanotube walls. Therefore, such behavior could give rise to the alternative mechanism of transport based on two-dimensional surface diffusion.

2.3 Membrane Fabrication

Testing these seemingly exotic predictions of fast transport through CNTs, which have emerged from MD simulations, has required the fabrication of a robust test platform. Practical considerations dictate that experimental efforts focus on a geometry that would allow observation of transport through multiple nanotube pores, such as a CNT membrane geometry. These membranes typically consist of an aligned array of CNTs encapsulated by a filler (matrix) material, with the nanotube ends open at the top and bottom. While there are many ways to produce such a structure, (a notable early result by Martin and coworkers was based on fabrication of amorphous CNTs within porous alumina membrane template [19]), the most fruitful approach to date involves growing an aligned array of CNTs, followed by infiltration of a matrix material in between the CNTs. The extremely high aspect ratio of the gaps between the nanotubes in the array (length/diameter on the order of 1000 or larger) presents a fabrication challenge. Fortunately, researchers have developed successful strategies to overcome this issue [1,5,20-22].

2.4 Water and Gas Flow Measurement

We have measured the bulk flow rate of air through a SiN$_x$/DWNT membrane by mounting it in an O-ring sealed flow cell and applying a pressure gradient of ~1 atm [1]. To compare experimental data with simulations, we need to calculate the flow rate per nanotube and, therefore, we have to estimate the density of open pores. This estimate introduces the single largest uncertainty in the measurement of flow per pore. Estimation obtained by plan-view TEM images of the membrane gives an upper limit to the membrane pore density (this number provides an upper limit because not all of the imaged pores span the entire 2–4 μm thickness of the membrane). This pore density estimate from TEM of 2.5×10^{11} cm^{-2} agrees well with the density of nanoparticles that catalyze nanotube synthesis. Using this number as an upper boundary, we can estimate a lower limit to the measured flow per nanotube. Even using this lower limit, it is readily apparent that there is a significant transport enhancement; the measured gas flows are up to 100 times greater than the predictions of the conventional Knudsen model for gas transport in rarefied environments (Fig. 1c). These experiments cannot definitively determine the cause of the observed gas flow enhancement. However, a change in the nature of molecule-wall collisions from diffuse to specular, as indicated by the MD simulations, provides a very likely explanation [13]. This conclusion is supported by the observation that nonhydrocarbon gas species tested in these experiments exhibit Knudsen-like $M^{-1/2}$ scaling of their membrane permeability (Fig. 1a), where M is the molecular weight. Another interesting observation is that the sub-2-nm CNT membranes exhibit some gas selectivity for hydrocarbon transport [1]. Hydrocarbon gases show a deviation from Knudsen scaling and instead exhibit slightly higher permeabilities (Fig. 1a). This deviation in scaling may reflect additional surface adsorption and diffusion of these molecules along the surface, although more detailed measurements are needed to establish this as a viable enhancement mechanism.

We have also observed high rates of water transport through sub-2-nm DWNT membranes using pressure-driven flow [1]. Similarly high rates have also been observed by Majumder et al. [2] using MWNT membranes with larger pore diameters. As previously discussed, the single largest uncertainty in quantifying the flux through individual pores lies in determination of the active pore density. Majumder et al. estimate the active pore densities by quantifying the diffusion of small molecules through CNTs. They report enhancements of 4– 5 orders of magnitude compared with the Hagen– Poiseuille formalism. As described above, we estimate the upper bounds of the pore densities so our measurements represent lower boundary estimates. The transport rates we measure reveal a flow enhancement that is at least 2– 3 orders of magnitude faster than no-slip, hydrodynamic flow calculated using the Hagen– Poiseuille equation (Fig. 1b). The calculated slip length for sub-2-nm CNTs is as large as hundreds of nanometers, which is almost three orders of magnitude larger than the pore size. In contrast, a polycarbonate membrane with a pore size of 15 nm has a slip length of just 5 nm. This suggests that slip flow formalism may not be applicable to water flow through CNTs, possibly because of length scale confinement [2,23] or partial wetting between the water and the CNT surface [24]. Interestingly, the measured water flux does not only compare well with that predicted by MD simulations, but it is also similar to the water flow through aquaporin (3.9 molecules/pore) [25]. The comparison to the aquaporins is not straightforward since the diameters of our CNTs are twice that of aquaporins and are considerably longer, to name just a few differences. Therefore, we cannot yet imply that the same mechanism is responsible for transport in our CNTs and aquaporins. Nevertheless, our experiments demonstrate that water transport through CNTs starts to approach the efficiency of biological channels.

3 PARYLENE-CNT COMPOSITE

3.1 Properties of parylene

Parylene is a generic name for a series of polymers. The basic member of the series, called Parylene N, is poly-para-xylylene, a completely linear and highly crystalline material. Parylene C is basically the same form as Parylene N with one chlorine atom replacing an aromatic hydrogen atom. Parylene polymer is deposited as a vapor phase, thus forming a continuous film following the shape of a target substrate is possible. Due to the uniqueness of the vapor phase deposition together with good dielectric material properties and chemical inertness, parylene has been a material of choice for pinhole-free coating of critical electronic assemblies.

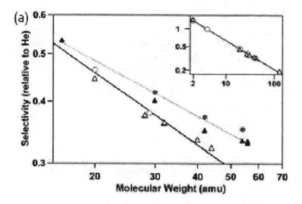

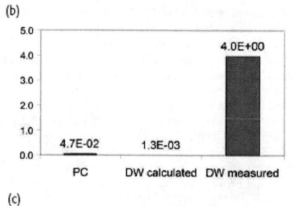

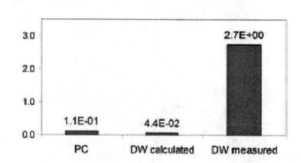

Figure 1: Water and gas transport in sub-2-nm CNT membranes. (a) Gas selectivity (defined as the permeability relative to that of He) data for sub-2-nm CNT (triangle) and MWNT (circles) membranes. Open symbols denote nonhydrocarbon gases, filled symbols hydrocarbon gases. The solid line is a power law fit of the nonhydrocarbon gas selectivity data showing a scaling predicted by the Knudsen diffusion model (exponent of – 0.49 ± 0.01). The dashed line is a power law fit of the hydrocarbon gas data showing a deviation from the Knudsen model (exponent of – 0.37 ± 0.02). The inset shows the full mass range of nonhydrocarbon gas data, illustrating agreement with the Knudsen model scaling. (b) Comparison of the water flux predicted for a polycarbonate (PC) membrane (left) and a DWNT (center) membrane with the flux measured for the DWNT membrane. (c) Comparison of the air flux predicted for a PC membrane (left) and a DWNT membrane (center) with the flux measured for the DWNT membrane (right).

Recently, researchers have used parylene for coating nanowires and large diameter CNTs (MWNT). Yang and Majumdar and their colleagues [26] deposited parylene onto their vertically aligned Si nanowire mat as a chemical etch mask. Upon completion of reactions, they could easily remove parylene with embedded silica nanotubes intact. Miserendino et al. [27] used parylene to coat MWNT arrays for a nanoelectrode application. In both nanowire and nanotube cases, the size and gaps between individual pillars were on the order of 100 nm.

3.2 Parylene-CNT composite and membrane fabrication

For producing flexible CNT membrane, we deposited parylene onto our CNT mat. Parylene dimmer (di-para-xylylene) was thermally evaporated at $160^{\circ}C$, dissociated at $670^{\circ}C$, and then deposited onto a CNT mat on top of a Si substrate at room temperature, when it forms parylene polymer (poly-para-xylylene). Although major deposition reaction control factor is temperature, depending on the chamber pressure, the deposition rate may vary. Unlike the previous efforts, the gaps between nanotubes are very small, and the conformal filling may become challenging. If the deposition is too fast, parylene dimers can deposit only on top and hamper an access of fresh dimers further down to the CNT array, thereby coating only a few micrometers from the top of a mat. Parameters influencing conformal coatings of CNTs include (1) vertically aligned but short enough CNTs that do not suffer from fast deposition and (2) deposition speed control. After releasing the composite film from the substrate, the free standing film was dipped in HCl for 24 hrs at room temperature to remove catalyst layers (Fe and Mo). Then, we mounted the freestanding composite on a thin Al plate with a millimeter size hole in the middle. Overly deposited parylene was etched out with gentle O2 plasma. Once CNTs were exposed, we used water plasma to uncap the carbon nanotubes.

4 CONCLUSION

We fabricated a flexible membrane based on the parylene-CNT composite. Many benign properties of parylene such as environmental friendliness, biocompatibility, chemical inertness, no swelling upon water exposure, and good mechanical and dielectric properties may allow unique properties of this composite material.

5 ACKNOWLEDGEMENT

This work was partially supported by DARPA DSO & Lawrence Livermore National Laboratory. Lawrence Livermore National Laboratory is operated by Lawrence Livermore National Security, LLC, for the U.S. Department of Energy, National Nuclear Security Administration under Contract DE-AC52-07NA27344. CG & OB were partially supported by NSF NER 0608964. AN, CG & OB acknowledge support by NSF NIRT CBET-0709090. OB also acknowledges support by the Center for Biophotonics, an NSF Science and Technology Center, managed by the University of California, Davis, under Cooperative Agreement PHY 0120999.

REFERENCES

[1] J.K. Holt et al., Science, **312**, 1034, 2006.
[2] M. Majumder et al., Nature, **438** ,44, 2005.
[3] F. Fornasiero et al., Proc. Natl. Acad. Sci., 2008 (in press).
[4] D.S. Sholl, and J.K. Johnson, Science, **312**, 1033, 2006
[5] A. Noy et al., Nano Today, **2 (6)**, 22, 2007.
[6] G. Hummer, Mol. Phys., **105**, 201, 2007.
[7] G. Hummer et al., Nature, **414**, 188, 2001.
[8] A.I. Kolesnikov et al., Phys, Rev. Lett., **93**, 35503, 2004.
[9] N. Naguib et al., Nano Lett., **4**, 2237, 2004.
[10] A. Waghe et al., J. Chem. Phys., 117, 10789, 2002.
[11] O. Beckstein and M.S.P. Sansome, Proc. Natl. Acad. Sci., **100**, 7063, 2003.
[12] R. Allen et al., J. Chem. Phys., **119**, 3905, 2003.
[13] A.I. Skoulidas et al., Phys. Rev. Lett., **89**, 185901, 2002.
[14] A.I. Skoulidas et al., J. Chem. Phys., **124**, 054708, 2006
[15] H. Chen et al., J. Phys. Chem. B, **110**, 1971, 2006.
[16] V.P. Sokhan et al., J.Chem. Phys., **120**, 3855, 2004.
[17] V.P. Sokhan et al., J.Chem. Phys., **115**, 3878, 2001.
[18] V.P. Sokhan et al., J.Chem. Phys., **117**, 8531, 2002.
[19] G. Che et al., Chem. Mater., 10, 260, 1998.
[20] B.J. Hinds et al., Science, **303**, 62, 2004.
[21] J. Holt et al., Nano Lett., **4**, 2245, 2004.
[22] S. Kim et al., Nano Lett., **7**, 2806, 2007.
[23] C. Cottin-Bizonne et al., Euro. Phys, J. E **9**, 47, 2002.
[24] V.S.J. Craig et al., Phys. Rev. Lett., **87**, 54504, 2001.
[25] A. Kalra et al., Proc. Natl. Acad. Sci., **100**, 10175, 2003
[26] R. Fan et al., J. Am. Chem. Soc., **125**, 5254, 2003.
[27] S. Miserendino et al., Nanotechnology, **17**, S23, 2006.

Arkema Graphistrength® Multi-Walled Carbon Nanotubes

T. Page McAndrew, Pierre Laurent, Mickael Havel, and Chris Roger

Arkema Inc., Research Center, 900 First Ave., King of Prussia, PA 19406
page.mcandrew@arkemagroup.com

ABSTRACT

Arkema Inc. produces multi-walled carbon nanotubes (CNT's) under the trade name *Graphistrength®*. Typical dimensions are 10-15 nm in diameter (5-15 walls) and 1-10 microns in length. Commercial and developmental products comprise plain CNT's (*Graphistrength®* C100), high-purity CNT's (*Graphistrength®* U100), thermoplastic masterbatches, water dispersions and epoxy pre-mix. Masterbatches, which comprise CNT's blended in a resin (up to 50%) have advantages – CNT's in a resin are easier to handle and dispersed already. Shown is the value of the masterbatch approach. Electrical conductivity data are presented for a variety of polymers containing CNT's – polyamides, polycarbonate, polyesters, and fluoropolymers. Mechanical properties data are presented as well. For epoxy, use of a masterbatch enables properties above that possible with plain CNT's – and simultaneous increase in both strength and toughness – an excellent combination.

Keywords: multi-walled carbon nanotube, masterbatch, Arkema, Graphistrength, composite

1 INTRODUCTION

Carbon nanotubes (CNT's) are subject of enormous interest world-wide – based on extraordinary properties:

- electrical – conductivity to 10^5 ohm^{-1}cm^{-1}
- mechanical – tensile modulus approximately 1,000 GPa
- thermal – conductivity approximately 3,000 watts/meter-K (exceeds diamond)

The general structure of a CNT is depicted in Figure 1. It can be regarded as a rolled graphene sheet. There are three types – single-walled, double-walled, and multi-walled. Double- and multi-walled systems comprise tubes concentrically nested – depicted in Figure 2. Electrical properties depend on the angle at which graphene sheet is rolled. See Figure 3. *Zig-zag* and *chiral* structures are semiconductive. *Arm-chair* structure is metallic. [1] Because of larger diameters, and thus more graphite-like structure, this construct does not apply to multi-walled carbon nanotubes, which are metallic. [2]

CNT's have been discovered in Middle Ages artifacts – the legendary *Damascus swords*. [3] However, not until the early 1990's was there structure elucidation and recognition as a distinct allotrope of carbon. [4,5,6]

Figure 1. Structure of Single-Walled Carbon Nanotube

Figure 2. Depiction of Nesting of Tubes in a Multi-Walled Carbon Nanotube

One area of interest in CNT's is incorporating them into a polymer to produce a composite having enhanced properties, specifically:

- <u>Improved electrical conductivity</u>. This is at incorporation levels far below that needed with other conductive fillers, such as carbon black or metal powders. Said lower levels enable preservation of other polymer properties, such as strength, permeability, cost, etc. End-use applications include coatings for electromagnetic shielding, composites for automotive fuel systems, and composites amenable to electrostatic painting.
- <u>Improved mechanical properties</u>. This is primarily for thermosets like epoxies and polyesters. End-use

applications include aeronautical products and sporting goods.

In addition to improved electrical and mechanical properties, recent studies have shown exceptional performance in flame retardancy – at CNT concentrations ≤ 0.5%. [7]

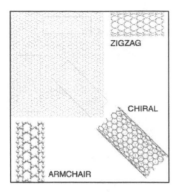

Figure 3. Depiction of Rolling of a Graphene Sheet

2 GRAPHISTRENGTH® PRODUCTS

Arkema Inc. manufactures multi-walled CNT's at Lacq, France. Process is chemical vapor deposition at elevated temperature of ethylene on metal/ceramic catalyst. Arkema CNT's and CNT-containing products are trademarked *Graphistrength®*.

SEM and TEM images of plain CNT's, i.e., Graphistrength C100, are presented in Figure 4. CNT's exist in bundles – median diameter approximately 10-20 microns. Diameter of individual CNT's is approximately 10-15 nanometers – corresponding to approximately 5-15 concentric tubes, with lengths approximately 1-10 microns.

As shown, multi-walled CNT's exist in bundles. A key point concerns how to prepare said CNT's for best use. Improving electrical conductivity is a crucial area. Electrical conductivity in a composite containing CNT's relies on percolation, as it does with other conductive fillers. Percolation is the state at which conductive filler particles interconnect to form a continuous network in which there can be current flow.

If bundles are not dispersed in composite at all, there would be low electrical conductivity – as all are surrounded by insulating polymer. Likewise if all CNT's are removed from bundles and perfectly dispersed, there would be low electrical conductivity – as all CNT's are surrounded by insulating polymer. [8] Needed is dispersion that enables percolation, i.e., connections of CNT's. Perfect dispersion is not needed, but rather that in which CNT layers and seams can connect throughout.

To this end, Arkema offers masterbatch products, in addition to plain CNT's. Masterbatches are composites comprising CNT's already introduced into a resin. Benefits are:

- easier handling – density of approximately 1 g/cc, versus approximately 0.1 g/cc for plain CNT's
- good dispersion established already

There are two types of masterbatches:

- specific – CNT's in a resin (e.g., polyamide-6) matching end-use polymer
- general – CNT's in a general-purpose resin

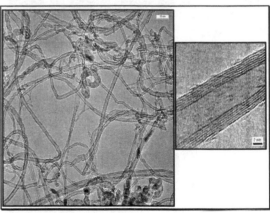

Figure 4. Graphistrength C100 Images - SEM (upper) and TEM (lower)

Current Graphistrength products are:

- *CNT's*
  - C100   plain carbon nanotubes (>90%)
  - U100   high-purity plain carbon nanotubes (> 97%)
- *Specific Masterbatches*
  - CM1-20   20% CNT, 80% polyamide-12
  - CM3-20   20% CNT, 80% polyamide-11
  - CM6-20   20% CNT, 80% polyamide-6
  - CS1-015   1.5% CNT, 98.5% epoxy (diglycidyl ether bisphenol A)
- *General Masterbatches*
  - 200 P50   50% CNT, 50% block copolymer
  - CM12-25   25% CNT, 75% general-purpose resin (developmental)

3 EXPERIMENTAL

Production of CNT-containing thermoplastic composites was with a DSM Research B.V. MIDI 2000 twin-screw extruder. Production of CNT-containing epoxies employed only stirring/shaking. Electrical conductivity was by 2-probe on freshly-fractured ends coated by silver paint. Mechanical properties were measured with a Zwick/Roell Z005. SEM imaging was with a Gemini LEO 1530, and optical imaging with a Nikon Eclipse ME 600.

4 PERFORMANCE

Consider the utility of the general masterbatch – Graphistrength CM12-25. In Figure 5 is shown the electrical conductivity of several composites comprising 2% CNT's. CNT's were introduced plain, and in CM12-25, by melt mixing. Performance with CM12-25 is as good or better than achievable with plain CNT's. This highlights general utility, especially considering that masterbatch is easier to use.

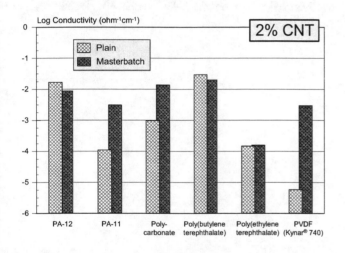

Figure 5. Electrical Conductivity of CNT-Containing Composites. Concentration of CNT's is 2% by weight. Masterbatch is CM12-25.

Performance in polycarbonate was examined in greater detail. In Figure 6 is shown electrical conductivity as a function of CNT concentration. CNT's were introduced plain, and in CM12-25. Note that CM12-25 provides for better conductivity, and a lower percolation level.

Mechanical performance in polyamide-11 was examined. See Figure 7. Tensile modulus increases and elongation-at-break decreases with increasing CNT concentration. CNT's were introduced in CM12-25.

Mechanical performance in epoxy was examined. See Figure 8. Flexural modulus data were compared for two systems comprising 0.5% CNT's. CNT's were introduced plain and in CM12-25. Note the system incorporating

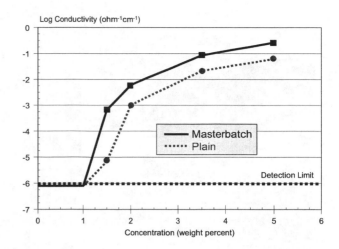

Figure 6. Electrical Conductivity of CNT-Containing Polycarbonate Composites. Masterbatch is CM12-25.

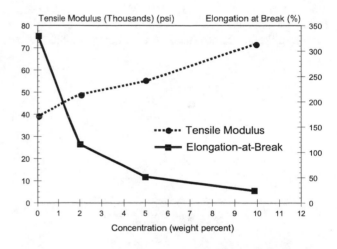

Figure 7. Mechanical Properties of CNT-Containing Polyamide-11 Composites. CNT's introduced in CM12-25

CM12-25 not only has improved strength (i.e., higher flexural modulus), but substantially increased toughness. This is a very desirable combination. Possibly interaction of CNT's with epoxy matrix transfers load to CNT's, resulting in improved modulus, and well-dispersed CNT's serve as crack stoppers, resulting in improved toughness. [9]

5 SUMMARY

Arkema produces multi-walled carbon nanotubes – plain and in masterbatches. Masterbatch products have advantages – easy handling and good performance in a wide range of polymers. Moreover in epoxy, materbatch product enables increase in strength and toughness – a very desirable combination.

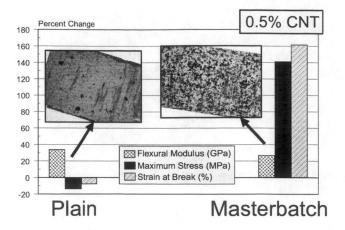

Figure 8. Mechanical Properties and Transmission Optical Images of CNT-Containing Epoxy Composites. Concentration of CNT's is 0.5% by weight. Data are plotted as change versus plain epoxy (Flexural Modulus = 2,950 MPa, Maximum Stress = 41.9 MPa, Strain-at-Break = 1.3%). Magnification of photographs is 15X. Epoxy is diglycidyl ether of bisphenol A and dicyandiamide.

6 ACKNOWLEDGEMENTS

The following Arkema Inc. personnel are thanked for support: P. Piccione and P. Miaudet (Groupement de Recherches, Lacq); N. Mekhilef, Z. Cherian, and G. Moeller (Analytical & Systems Research); G. Jones and R. Barsotti (Functional Additives).

REFERENCES

1. R. Baughman, et al. Science, 297, 787 (8/2/02)
2. R. Martel, et al. Applied Physics Letters, 73 (17), 2447 (1998)
3. M. Inman. National Geographic News (11/16/06). http://news.nationalgeographic.com/news/2006/11/061 116-nanotech-swords.html
4. S. Ijimia. Nature, 356, 56 (1991)
5. S. Iijima, T. Ichihashi. Nature, 363, 603 (1993)
6. D.S. Bethune, et al. Nature, 363, 605 (1993)
7. J. Gilman. Polymer Nanocomposites 2008, Bethlehem, PA (3/4/08)
8. F.H. Gojny, et al. Polymer, 47, 2036 (2006)
9. B. Fiedler, et al. Composites Science and Technology, 66, 3115 (2006)

Resistance measurements and weak localization in long SWNTs

P. G. Gabrielli[*], S. Gabrielli[**] and N. Lisi[*]

[*]ENEA, Centro Ricerche Casaccia, via Anguillarese 301 Rome Italy
[**]Dip. di fisica, Univ. degli Studi di Roma "Tor Vergata" Rome Italy

ABSTRACT

In many IV measurements an evidence has emerged pointing towards the existence of Luttinger liquid behavior in metallic SWNT, as expected for strongly interacting electrons in 1-d, such behavior was observed via the power law temperature and bias-voltage dependence of the current through tunneling contacts attached to the nanotubes. In particular recent advances in the growth of extremely long nanotube (>1 mm) have allowed for experimental measurements on the scaling behavior of resistance in individual, millimeter long SWNT for the temperature range of 1.6-300 K. From the linear scaling of resistance, the temperature dependent electron mean free path has been calculated for each temperature and, beyond the linear scaling regime, it has been observed that the resistance increases exponentially with length, indicating localization behavior. In this work we analyse the results of the resistance measurements of different lengths SWNT indicating the weak localization behavior.

Keywords: quantum transport, carbon nanotube, phase coherence, weak localization

1   INTRODUCTION

The rapidly advancing technology of nanometric devices has led to the production of smaller and smaller systems: among the main items in the design of these electronic devices there are the measurement and understanding of the current-voltage response of electronic circuits in which carbon nanotubes   act as conducting elements. These devices, often called mesoscopic systems, are large on the atomic scale but sufficiently small that the electron wavefunction is coherent over the entire sample. The condition for coherence is that the electron traverse the wire without undergoing any inelastic collision: indeed in perfect single-walled carbon nanotube (SWNT) electrons propagate ballistically if the inelastic scattering can be neglected. On the other hand it is also well known that at low-temperature charge transport in any disordered conductor is governed by the interplay between inelastic scattering and elastic scattering off static disorder (impurities and defects) of electrons and, in low dimensions systems like SWNTs, an arbitrarily weak disorder localizes [1,2] all single-electron states and there would be no transport without inelastic processes (Anderson localization[1] of electronic states leads to metal-insulator transition at zero temperature [3]: note

that, for electrons in a given conduction band, strong enough disorder can localize the whole band. The metal – insulator transition induced by disorder is called Anderson transition.).

2   QUANTUM TRANSPORT IN SWNT

Coherent quantum transport in low dimensional systems can be investigated with either the Kubo or the Landauer-Buttiker formalism [4]. The first approach, which derives from the fluctuation-dissipation theorem, allows one to evaluate the intrinsic conduction regime within the linear response and gives direct access to the fundamental transport length scales[2] such as the elastic mean free path l_0 and the localization length ξ. While l_0 results from elastic backscattering by static disorder, ξ denotes the scale beyond which quantum conductance decays exponentially with the system length L driving the system from weak to strong localization. The localization length gives the scale beyond which localization effects are fully suppressed owing to mechanisms such as electron-phonon (e-ph)   or electron-electron (e-e) coupling: weak localization regime. When l_0 becomes longer than the length of the nanotube between the leads, the carriers propagate ballistically and contact effects prevail. In such a situation, the Landauer-Buttiger formalism becomes more appropriate, since it rigorously treats transmission properties for open systems and arbitrary interface geometries.

2.1 Conductivity and Transport

The more general formula for actual local current measured by experimentalists (generalized Ohm's law) for conductivity in infinite length system is

$$J_\alpha(\mathbf{r},t) = \int d\mathbf{r}' \sum_\beta \sigma_{\alpha\beta}(\mathbf{r},\mathbf{r}',\omega)E_\beta(\mathbf{r}',t) \tag{1}$$

---

[1]The localization is a property of the states in random QM systems and can be interpreted by total back-reflection of particles from potential barriers so that they become localized in a single potential well.

[2] Important length scales: the coherence length l_ϕ, the energy relaxation length l, the elastic mean free length l_0, the Fermi wave length λ_F of the electron, the sample size L: in mesoscopic systems it will be $\lambda_F \leq l_0 < L < l_\phi \leq l$. Note that $k_F l \geq 1$ is called the Ioffe-Regel limit and the atomic Bohr radius $a_0 << \lambda_F$.

where the (non-local) conductivity $\sigma_{\alpha\beta}$ is response to actual (external + induced) electric field and is given by the Kubo formula [5] (where periodic boundary conditions and coupling of only the charge degrees of freedom, none spin degree of freedom is considered, to external $\mathbf{E}$-field are assumed)

$$\sigma_{\alpha\beta}(\mathbf{r},\mathbf{r}',\omega) = \frac{e^2}{\hbar\omega} \int_0^\infty dt e^{i\omega t} \left\langle \psi \left| \left[j_\alpha^+(\mathbf{r},t), j_\beta(\mathbf{r}',0) \right] \right| \psi \right\rangle + \frac{ne^2}{m\omega} i\delta_{\alpha\beta}\delta(\mathbf{r}-\mathbf{r}') \quad (2)$$

The wave function ψ is the ground state of the many-body Hamiltonian which contains all possible interactions in the solid (except the interaction between the total electric field and the particles of the system) and the first term in (2) is called retarded current-current correlation function. Kubo first derived the equations for electrical conductivity in the solid and Kubo formulas are the name applied to the correlation function which describes the linear response.

But this procedure is physically incorrect as a way of defining the conductivity in a finite system in which electrons enter from an external electrode at one end and are removed at the other end collected by another external electrode. In a finite (ideal) sample if the chemical potential is higher at one lead (electrons of the large reservoir with constant μ_1) than at the other lead (with constant μ_2) the current is the response to the gradient of chemical potential for electrons not to the electric field; in other words, if $\mu_1 > \mu_2$ and e is the absolute value of the electronic charge, the voltage difference ΔV between the two baths due to flow across the sample of the current I is

$$\Delta V = \frac{I}{G_c} = \frac{\mu_1 - \mu_2}{e} \quad (3)$$

where

$$G_c = \frac{e^2}{\pi\hbar} T(E_f) \quad (4)$$

is the irreducible conductance measured between the two outside reservoirs being T the transmission probability for channel (to go from electrode 1 to electrode 2). The inelastic processes (which break the time-reversal invariance and the phase coherence of the states at the two extremities, dissipate energy and restore equilibrium) in this case are assumed to exist only in the two electrons baths, so that the ramdomized phase of the injected and absorbed electrons through these processes results in no phase relation between particles.

At low temperature, in presence of only elastic scattering for the electrons at the Fermi surface with a linear series of random scatterers connecting the two reservoirs, the true conductance G due to barrier (including spin degeneracy) is correctly given by the transmission and reflection coefficients of the sample by the Landauer-Buttiker, not the Kubo, formula [4,5]

$$G = \frac{e^2}{\pi\hbar} \frac{T(E_F)}{1 - T(E_F)} \quad (5)$$

Electronic transport measurements [6] on individual SWNT demonstrate that, in the absence of scattering (then the transmission probability is $T = 1$), the momentum relaxation length and the localization length ξ are much larger than the wire length and the transport in these systems is ballistic: the wavefunction of the electron is extended over the total length of the nanotube and there are only two channels which contribute to the electronic transport giving $G = 2G_c$. However, as already outlined, in presence of some mechanism of scattering the conductance is described by the Landauer formula (4) and the conductance is no longer exactly quantized.

Because the electron can lose energy and equilibrate with heat bath only via inelastic collisions it is necessary to reexamine the conventional concept of energy dissipation in a quasi 1-d resistor systems. In the theory of the 1-d electron systems, called Luttinger-liquid (LL)[7,8] (see AppendixA), the correlated electron state is characterized by a parameter g that measures the strength of the interaction between the electrons ($g=1$ for non-interacting electrons gas). The most important feature of the LL, in contrast to Landau Fermi-Liquid theory[3] (FL), is the absence of the fermion quasiparticle branch at low energy: excited states of the system must be described by the bosonic fluctuations of the charge and spin densities dispersing with different velocities, which correspond to many-body electron state with a huge number of the electron-hole pairs. This have a pronounced effect on the tunneling into a LL conductor: the IV curve of a tunnel junction between a normal FL and a LL conductor is expected to be non-ohmic and described by a power law with an exponent depending on interaction strength.

Not only, in [9] Bockrath et al. observe a Luttinger-liquid behavior in (rope of) SWNT's in measurements of the electrical transport as a function of temperature resulting in a power laws for linear response conductance

$$G(T) \propto T^\alpha \quad (6)$$

where $\alpha = f(g)$ and $g = \left[1 + \frac{2U}{\Delta}\right]^{-1/2}$ (U is the charging energy of the tube and Δ [10] is the single-particle level spacing).

---

[3] Landau Fermi-Liquid theory is concerned with the properties of many-fermion system at low temperature (much lower than Fermi energy) in the normal state, i.e. in the absence or at least at temperatures above any symmetry breaking phase transition (superconduction, etc).

NSTI-Nanotech 2008, www.nsti.org, ISBN 978-1-4200-8503-7 Vol. 1

Weak localization

When the temperature is so high that the conductivity can be treated as local quantity, like in Anderson localization which deals with wave function of the single electron in presence of impurities, the conductance is obtained by combination of smaller parts of materials. But when the temperature is low (or the sample is small) the dephasing length l_φ is greater than the sample linear dimensions L so that the quantum corrections to the conductivity are non-local and the conductance can no longer be treated as a self-averaging quantity. Since the lack of self-averaging of the conductance is a feature of mesoscopic conductors, in SWNT it is necessary to analyze the effects of weak disorder and the inelastic scatterings on charges transport. Due to the existence of the impurities the transport is more diffusive than ballistic (the existence of scattering is possible in both regimes but in diffusive one we have that $l = v_F \tau << L$ so that the material is characterized by a relatively low mobility) though the elastic scattering of the electrons, if these impurities are equivalent to static defects, can modify the interference terms but does not cause decoherence [4]. At low temperature the conduction take place mainly with electrons at Fermi energy and, due to some gate potential, the Fermi point upon which the electrons travel can be shifted slightly, therefore it is possible that an electron that moved on one side (path) of an impurity begins to move on the other side (path) after the shift. This process (analogous to Aharonov-Bohm effect in the presence of some magnetic field) induced a quantum fluctuation of the conductance of the order of $2e^2/h$ and depends on the exact configuration of scattering centers within the sample: these two paths are time-reversed with respect to one another and since the electron return to its original position it can interfere with itself creating an additional resistance called weak localization [11,12]. Then the weak localization is caused by the quantum interference effect on the diffusive motion of a single electron.

From semiclassical point of view it is possible to calculate this additional resistance considering that the conductivity is related to the current-current correlation function, as in eq. (2), and being $D = v_F^2 \tau$ the diffusion coefficient and $l_\varphi = \sqrt{D\tau_\varphi}$ for 1D we get

$$\Delta\sigma_{WL} \approx -\frac{e^2}{\pi\hbar} l_\varphi \left[1 - \left(1 + \frac{\tau_\varphi}{\tau}\right)^{-1/2}\right] \qquad (7)$$

Even though in [13] it has been suggested that zero-point fluctuations cause the dephasing in one dimensional quantum wire at low temperature (ascribed to finite broadening of Fermi surface) and presenting a zero-point-limited dephasing time τ_0 in good agreement with the measured saturation values τ_φ found in many experiments,

in [14] this hypothesis is rejected using purely physical arguments.

A more sophisticated theory [15] tells to us that in presence of a vector potential **A**, for $\omega\tau << 1$, the probability of return path in a disordered SWNT can be conveniently obtained by calculating the Cooperon $C_\omega(\mathbf{r},\mathbf{r}')$, which is a retarded classical electron-electron propagator satisfying a modified diffusion equation in the frequency domain [16]

$$\left[-D\left(\nabla_r - \frac{2ie}{\hbar}\mathbf{A}\right)^2 + i\omega + \frac{1}{\tau_\varphi}\right]C_\omega(\mathbf{r},\mathbf{r}') = \delta(\mathbf{r}-\mathbf{r}') \qquad (8)$$

The WL correction to the conductivity is due to enhanced probability to return, so that

$$\Delta\sigma_{WL}(\omega) = -\frac{2e^2 D}{\pi\hbar} C_\omega(\mathbf{r},\mathbf{r}') \qquad (9)$$

3 RESISTANCE IN LONG SWNT

The conductance of metallic SWNT has been shown to depend strongly on the nature of the contacts between the nanotube and the leads. In a typical experimental setup [17] a bias voltage is applied across a nanotube connected to metallic leads, while a gate voltage applied to a third electrode acts as a chemical potential and modulates the charge on the nanotube (see [17] fgg.1,3 and 4). At room temperature the main origin of the resistivity at low bias in high-quality metallic SWNT is believed to be inelastic scattering by acoustic phonon: the scattering is weak resulting in long mean-free path in a range from few hundred nanometers to several micrometers both in the measurements and in the calculations. Then at low bias regime we get ballistic transport. When sufficiently large bias are applied to drive the electric current, higher energy vibrational modes are activated and e-ph coupling limits ballistic transport: electrons gain enough energy to emit optical or zone-boundary phonons leading to a saturation of the current, in [18] indicated at ~20μA. The effect of electron-(optical) phonon coupling was found to strongly affect electronic conductance and to induce some energy dependence of the coherence length scale, completely similar to the experimental data obtained in the weak localization regime [19,20].

4 CONCLUSION

We study the suppression of the quasi-ballistic conduction in long (many l_φ) SWNT noting that some environmental conditions also at low temperature introduce some dynamic disorder which involves, by means inelastic scattering [19-21], a weak localization correction to the conductance.

Appendix A LUTTINGER LIQUID: VERY BRIEF REVIEW

A Luttinger liquid (LL) is a one-dimensional (Fermi liquid) correlated electron state characterized by a parameter g that measures the strength of the interaction between electrons: strong repulsive interactions have $g \ll 1$, whereas $g = 1$ for the non-interacting electron gas (remembering that weakly interacting electrons in normal metal are described by quasiparticles of the Fermi liquid). The LL's are very special in that they retain a Fermi surface enclosing the same k-space volume as that of free fermions, but there are no fermionic quasi-particles (like in normal Fermi liquids), their elementary excitations are bosonic collective charge and spin fluctuations dispersing with different velocities. An incoming electron decays into such charge and spin excitations which then spatially separate with time (charge-spin separation): the correlations between these excitations are anomalous and show up as interaction-dependent non-universal power laws in many physical properties where those of ordinary metals are characterized by universal (interaction-independent) powers. A list of such properties includes: 1) a continuous momentum distribution function $n(k)$, varying with as $\left| k - k_F \right|^\alpha$ with an interaction-dependent exponent α, and a pseudogap in the single-particle density of states $\propto \left| \omega \right|^\alpha$, consequences of the non-existence of fermionic quasi-particles; 2) similar power-law behavior in all correlation functions (in those for charge or spin density wave fluctuations) with universal scaling relations between the different non-universal exponents, which depend only on one effective coupling constant per degree of freedom; 3) finite spin and charge response at small wave vectors and finite Drude weight in the conductivity;   4) spin-charge separation; persistent currents quantized in units of $2k_F$.

REFERENCES

[1] P.W.Anderson, Phys.Rev. **109**, 1492, 1958.

[2]   E.Abrahams,   P.W.Anderson,   D.C.Licciardello   and T.V.Ramakrishnan, Phys.Rev.Lett. **42**,673 ,1979.

[3]D.M.Basko, I.L. Aleiner and B.L. Altshuler., Ann.Phys. **321**, 1126, 2006

[4] Y.Imry, "Introduction to Mesoscopic Physics", Oxford, 1997;

[5]D.K.Ferry and S.M.Goodnick, "Transport in Nanostructures", Cambridge, 1997.

[6] R.Saito, G.Dresselhaus and M.S. Dresselhaus, "Physical Properties of Carbon Nanotube" Imperial College Press, 1998.

[7] T.Ando, J.Phys.Soc.Jpn,**74**,777,2005; J.-C.Charlier, X.Blase and S.Roche, Rev.Mod.Phys. **79**,677,2007.

[8] T.Giamarchi, "Quntum Physics in One Dimension", Oxford, 2004

[9] M.Bockrath, D.H.Cobden, J.Lu, A.G.Rinler, R.E.Smalley, L.Balents and P.L.McEuen, Natur*e* **397**,598,1999; J.Nygard, D.H.Cobden, M.Bockrath, P.L.McEuen, P.E.Lindelof, Appl.Phys.A 69,297,1999.

[10]J.Cao, Q.Wang and H.Dai, Nature Materials, **4**,745,2005

[11]G.Bergmann, Phy.Rep. **107**,1,1984

[12]I.V.Gornyi, A.D.Mirlin and D.G.Poliakov, Phys.Rev.B,**75**,085421, 2007.

[13] P.Mohanty and R.A.Webb, Phys.Rev.B **55**,R13452,1997

[14] I.L. Aleiner and B.L. Altshuler.and M.E.Gershenson, cond-mat/9808053

[15] H.Bruus and K.Flensberg, "Many-Boby Quantum Theory in Condensed Matter Physics, An Introduction", Oxford, 2004.

[16] B.L.Altshuler, A.G.Aronov and D.E.Khmelnitskii, J.Phys. C **15**,7367, 1982; C.-K. Lee, J.Cho, J.Ihm and K.-H.Ahn, Phys.Rev. **69**,205404,2004.

[17]J.Cao, Q.Wang and H.Dai, Nature Materials, **4**,745,2005

[18] Z.Yao, C.L.Kane and C.Dekker, Phys.Rev.Lett. **84**,2941,2000; Ji-Y.Park et al,Nano Letters, **4**,517,2004

[19] S.Roche, J.Jiang, F.Triozon and R.Saito, Phys.Rev.Lett. **95**, 076803, 2005

[20]B.Sojetz, c.Miko, L.Forrò and C.Strunk, Phys.Rev.Lett. **94**, 186802, 2005

[21]Y.Imry,   O.Entin-Wohlman   and   A.Aharony,   Europhys.Lett. **72**,263,2005.

Design Optimization of Field Emission From A Stacked Carbon Nanotube Array

D. Roy Mahapatra*, N. Sinha**, S.V. Anand***, R. Krishnan***, Vikram N.V.***,
R.V.N. Melnik**** and J.T.W. Yeow**

* Department of Aerospace Engineering, Indian Institute of Science, Bangalore, India
** Department of Systems Design Engineering, University of Waterloo, Waterloo, ON, Canada
*** Department of Mechanical Engineering, R.V. College of Engineering, Bangalore, India
**** M^{2}NeT Lab, Wilfrid Laurier University, Waterloo, ON, Canada

ABSTRACT

Fluctuation of field emission in carbon nanotubes (CNTs) is not desirable in many applications and the design of biomedical x-ray devices is one of them. In these applications, it is of great importance to have precise control of electron beams over multiple spatio-temporal scales. In this paper, a new design is proposed in order to optimize the field emission performance of CNT arrays. A diode configuration is used for analysis, where arrays of CNTs act as cathode. The results indicate that the linear height distribution of CNTs, as proposed in this study, shows more stable performance than the conventionally used unifrom distribution.

Keywords: carbon nanotube, field emission, biomedical x-ray devices, electron-phonon, array, electro-mechanical fatigue, optimization.

1 INTRODUCTION

The quest for building functional devices has led to the development of new fabrication techniques and materials, which have been scaled to nano level. The extensive research on nanomaterials can be traced back to the discovery of carbon nanotubes (CNTs) by Iijima in 1991 [1]. Since then newly proposed applications of CNTs such as field emitters, chemical and biological sensors have been successfully demonstrated. With significant improvement in synthesis techniques, CNTs are currently ranked among the best field emitters. CNTs grown on substrates are used as electron sources in field emission applications. Field emission from CNTs is difficult to characterize using simple formulae or data fitting, which is due to (1) electron-phonon interaction; (2) electromechanical force field leading to stretching of CNTs; and (3) ballistic transport induced thermal spikes, coupled with high dynamic stress, leading to degradation of emission performance at the device scale. Fairly detailed physics-based models of CNTs considering the aspects (1) and (2) above have already been developed by the authors [2]-[5]. However, design optimization issues aimed at better field emission devices to reduce the extent of electro-mechanical fatigues and to improve spatio-temporal localization of emitted electrons remain open and important areas of research. With

due success in designing such devices, various applications such as in-situ biomedical x-rays probes and thin film pixel based imaging technology, to name just a few, are of great significance. The authors' interest towards this study stems from the problem of precision biomedical x-ray generation. In this paper, we focus on the device-level performance of CNTs grown on a metallic surface in the form of an array (for field emission) under diode configuration. We analyze a new design concept, wherein (a) the electrodynamic force field leading to strong electron-phonon interaction during ballistic transport and also (b) the usually observed reorientation of the CNT tips and instability due to coulomb repulsions, can be harnessed optimally.

2 MODEL FORMULATION

Let N_T be the total number of carbon atoms (in CNTs and in cluster form) in a representative volume element ($V_{\text{cell}} = \Delta A d$), where ΔA is the cell surface interfacing the anode and d is distance between the inner surfaces of cathode substrate and the anode. Let N be the number of CNTs in the cell, and N_{CNT} be the total number of carbon atoms present in the CNTs. We assume that during field emission some CNTs are decomposed and form clusters. Such degradation and fragmentation of CNTs can be treated as the reverse process of CVD or a similar growth process used for producing the CNTs on a substrate. Hence,

$$N_T = N N_{CNT} + N_{\text{cluster}} , \qquad (1)$$

where N_{cluster} is the total number of carbon atoms in the clusters in a cell at time t and is given by

$$N_{\text{cluster}} = V_{\text{cell}} \int_0^t dn_1(t) , \qquad (2)$$

where n_1 is the concentration of carbon clusters in the cell. By combining Eqs. (1) and (2), one has

$$N = \frac{1}{N_{CNT}} \left[N_T - V_{\text{cell}} \int_0^t dn_1(t) \right] . \qquad (3)$$

The number of carbon atoms in a CNT is proportional to its length. Let the length of a CNT be a function of time, denoted as $L(t)$. Therefore, one can write

$$N_{CNT} = N_{\text{ring}} L(t) . \qquad (4)$$

where N_{ring} is the number of carbon atoms per unit length of a CNT and can be determined from the geometry of the hexagonal arrangement of carbon atoms in the CNT. By combining Eqs. (3) and (4), one can write

$$N = \frac{1}{N_{\text{ring}}L(t)}\left[N_T - V_{\text{cell}}\int_0^t dn_1(t)\right]. \quad (5)$$

In order to determine $n_1(t)$ phenomenologically, we employ a nucleation coupled model developed by us previously [2]. Based on the model, the rate of degradation of CNTs (v_{burn}) is defined as

$$v_{\text{burn}} = V_{\text{cell}}\frac{dn_1(t)}{dt}\left[\frac{s(s-a_1)(s-a_2)(s-a_3)}{n^2a_1^2 + m^2a_2^2 + nm(a_1^2 + a_2^2 - a_3^2)}\right]^{1/2}, \quad (6)$$

where a_1, a_2, a_3 are lattice constants, $s = \frac{1}{2}(a_1+a_2+a_3)$, n and m are integers ($n \geq |m| \geq 0$). The pair (n, m) defines the chirality of the CNT. Therefore, at a given time, the length of a CNT can be expressed as $h(t) = h_0 - v_{\text{burn}}t$, where h_0 is the initial average height of the CNTs and d is the distance between the cathode substrate and the anode.

In the absence of electronic transport within a CNT and field emission from its tip, the background electric field is simply $E_0 = -V_0/d$, where $V_0 = V_d - V_s$ is the applied bias voltage, V_s is the constant source potential on the substrate side, V_d is the drain potential on the anode side and d, as before, is the clearance between the electrodes. The total electrostatic energy consists of a linear drop due to the uniform background electric field and the potential energy due to the charges on the CNTs. Therefore, the total electrostatic energy can be expressed as

$$\mathcal{V}(x,z) = -eV_s - e(V_d - V_s)\frac{z}{d} + \sum_j G(i,j)(\hat{n}_j - n), \quad (7)$$

where e is the positive electronic charge, $G(i,j)$ is the Green's function [6] with i indicating the ring position and $\hat{n}_j$ describing the electron density at node position j on the ring. In the present case, while computing the Green's function, we also consider the nodal charges of the neighboring CNTs. This essentially introduces non-local contributions due to the CNT distribution in the film. We compute the total electric field $\mathbf{E}(z) = -\nabla\mathcal{V}(z)/e$, which is expressed as

$$E_z = -\frac{1}{e}\frac{d\mathcal{V}(z)}{dz}. \quad (8)$$

The current density (J) due to field emission is obtained by using the Fowler-Nordheim (FN) equation [7]

$$J = \frac{BE_z^2}{\Phi}\exp\left[-\frac{C\Phi^{3/2}}{E_z}\right], \quad (9)$$

where Φ is the work function of the CNT, and B and C are constants. Computation is performed at every time step, followed by update of the geometry of the CNTs. As a result, the charge distribution among the CNTs also changes and such a change affects Eq. (7). The field emission current (I_{cell}) from the anode surface corresponding to an elemental volume V_{cell} of the film is then obtained as

$$I_{\text{cell}} = A_{\text{cell}}\sum_{j=1}^N J_j, \quad (10)$$

where A_{cell} is the anode surface area and N is the number of CNTs in the volume element. The total current is obtained by summing the cell-wise current (I_{cell}). This formulation takes into account the effect of CNT tip orientations, and one can perform statistical analysis of the device current for randomly distributed and randomly oriented CNTs.

3 RESULTS AND DISCUSSIONS

In the proposed design, we introduce two additional gates on the edges of the cathode substrate. An array of stacked CNTs is considered on the cathode substrate. The height of the CNTs is such that a symmetric force field is maintained in each pixel with respect to the central axis parallel to z-axis (see Fig. 1). As a result, it is expected that a maximum current density and well-shaped beam can be produced under DC voltage across the cathode-anode structure. In the present design, the anode is assumed to be simply a uniform conducting slab. However, such an anode can be replaced with a porous thin film along with MEMS-based beam control mechanism. Figure 1 shows the transverse electric field distribution in the pixel, which directly influences the field emission current.

In the simulation and analysis, the distance between the cathode substrate and anode surface was taken as 34.7 μm. The height of CNTs in arrays was varied between 6 μm to 12 μm. The constants B and C in Eq. (9) were taken as $(1.4 \times 10^{-6}) \times \exp((9.8929) \times \Phi^{-1/2})$ and 6.5×10^7, respectively [8]. It has been reported in the literature (e.g., [8]) that the work function Φ for CNTs is smaller than the work functions for metal, silicon, and graphite. However, there are significant variations in the experimental values of Φ depending on the types of CNTs (i.e., SWNT/MWNT) and geometric parameters. The type of substrate materials have also significant influence on the electronic band-edge potential. The results reported in this paper are based on a representative value of $\Phi = 2.2 eV$.

The first step in our computation is to obtain the value of n_1 (the carbon cluster concentration) at a given time step from the nucleation coupled model. In this paper, it has been assumed that at $t = 0$, the diode contains minimal amount of carbon cluster in plasma. The CNTs degrade over time (due to both fragmentation

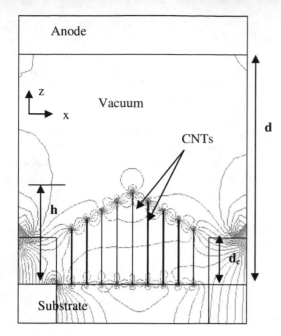

Figure 1: Contour plots of electric field E_z showing the concentration near the CNT tips under symmetric lateral force field. $V_0 = 650V$ and the side-wise gates are shorted with the substrate.

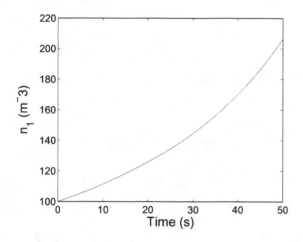

Figure 2: Variation of carbon cluster concentration with time.

and self-assembly) and the carbon cluster concentration in each cell also changes accordingly. Based on this assumption, the value of n_1 was computed. Fig. 2 shows the $n_1(t)$ history over a small time duration ($50s$). Such evolution indicates that the rate of decay is very slow, which in turn implies longer lifetime of cathodes.

Next, we simulate the field emission current histories for two different parametric variations: diameter and spacing between CNTs at the cathode substrate. The current histories are simulated for a constant bias

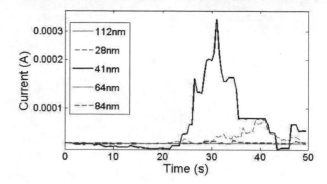

Figure 3: Simulated field emission current histories for varying diameters of CNTs under a DC voltage of 650 V.

voltage of 650 V. In the first case, the spacing between neighboring CNTs is kept constant, while the diameter is varied. The current histories for different values of diameters are shown in Fig. 3. As evident from the figure, the output current is low at large diameter values. This is due to the fact that current amplification is less with large diameter of CNTs.

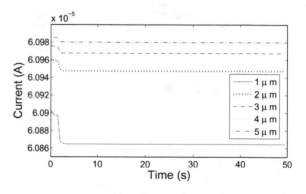

Figure 4: Simulated field emission current histories for varying spacing between neighboring CNTs under a DC voltage of 650 V.

In the second case, the diameter is kept constant, while the spacing between neighboring CNTs is varied. Following five values of spacing between neighboring CNTs have been considered: 1 μm, 2 μm, 3 μm, 4 μm and 5 μm. The current histories for all these cases are shown in Fig. 4. The trends in five curves in Fig. 4 tell us the following: (1) the current in all cases decreases initially and then becomes constant afterwards. This may be due to realignment of CNTs in the array when voltage is applied; (2) as the spacing between neighboring CNTs increases, the output current increases, which is physically consistent because the screening effect becomes less pronounced. These results are in agreement with a previously reported study [9].

Next, we simulate the current-voltage (I-V) charac-

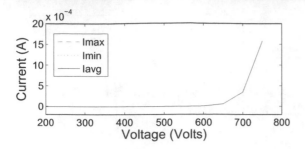

Figure 5: Simulated current-voltage characteristics.

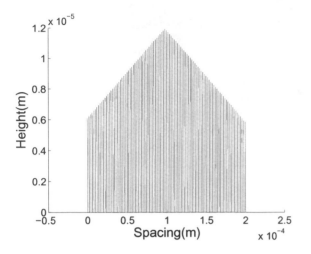

Figure 6: Initial and deflected shape of an array of 100 CNTs at t=50 s of field emission.

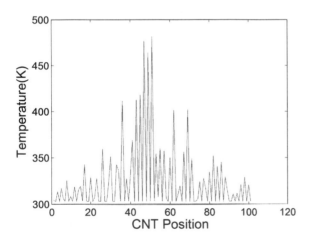

Figure 7: Maximum temperature of CNT tips during 50 s of field emission.

I-V response drastically as compared to uniform height reported in the previous studies. The variation in I-V characteristics for the three cases in this new design is negligible. This is further confirmed by analyzing the simulated results of CNT tip deflections in Fig. 6, in which there is no significant deflection of CNT tips after 50 s.

Finally, in Fig. 7, we plot the maximum tip temperature distribution over an array of 100 CNTs during field emission at a bias voltage of 650 V over 50 s duration. This result indicates a temperature rise of up to $\approx$ 480 K.

4 CONCLUSION

In this paper, a new design concept for the stable performance of CNT arrays in a diode configuration has been proposed. In this design, the CNT height distribution is linearly distributed as opposed to the uniform distribution, considered in most of the previous studies. The results reveal highly stable performance of CNT cathodes in this new design.

REFERENCES

[1] S. Iijima, Nature 354, 56, 1991.

[2] N. Sinha, D. Roy Mahapatra, J.T.W. Yeow, R.V.N. Melnik and D. A. Jaffray, Proc. IEEE Int. Conf. Nanotech., 673, 2006.

[3] N. Sinha, D. Roy Mahapatra, J.T.W. Yeow, R.V.N. Melnik and D.A. Jaffray, J. comp. Theor. Nanosci. 4, 535, 2007.

[4] N. Sinha, D. Roy Mahapatra, Y. Sun, J.T.W. Yeow, R.V.N. Melnik and D.A. Jaffray, Nanotechnology 19, 25710, 2008.

[5] N. Sinha, D. Roy Mahapatra, J.T.W. Yeow and R.V.N. Melnik, Proc. IEEE Int. Conf. Nanotech., 961, 2007.

[6] A. Svizhenko, M.P. Anantram and T.R. Govindan, IEEE Trans. Nanotech. 4, 557, 2005.

[7] R.H. Fowler and L. Nordheim, Proc. Royal Soc. London A 119, 173, 1928.

[8] Z.P. Huang, Y. Tu, D.L. Carnahan and Z.F. Ren, "Field emission of carbon nanotubes," Encyclopedia of Nanoscience and Nanotechnology (Ed. H.S. Nalwa) 3, 401-416, 2004.

[9] L. Nilsson, O. Groening, C. Emmeneger, O. Kuettel, E. Schaller, L. Schlapbach, H. Kind, J.M. Bonard and K. Kern, Appl. Phys. Lett. 76, 2071, 2000.

teristics for this arrangement. The minimum, average and maximum values of current were computed. The simulated results are shown in Fig. 5. These results give some interesting insigths. For example, as evident from Fig. 5, the linear height distribution has stabilized the

High Current Cold Electron Source Based on Carbon Nanotube Field Emitters and Electron Multiplier Microchannel Plate

Raghunandan Seelaboyina and Wonbong Choi[1]

Nanomaterials and Device Lab, Department of Mechanical and Materials Engineering,
Florida International University, Miami, FL 33174, USA

ABSTRACT

In this work, we report the synthesis and field emission properties of carbon nanotube multistage emitter arrays, which were grown on porous silicon by catalytic thermal chemical vapor deposition. The emitter structure consisted of arrays of multiwall nanotubes (MWNTs) on which single/thin-multiwall nanotubes were grown. The structure was confirmed by TEM and Raman analysis. Higher field emission current ~ 32 times and low threshold field ~1.5 times were obtained for these structures in comparison to only MWNT arrays. The enhanced field emission results for these multistage emitters are a consequence of higher field concentration, which was ~3 times more than only MWNTs. In this work we also report a novel method to amplify electron emission from field emitters by electron multiplication. A microchannel plate positioned between the emitter cathode and the anode was utilized to enhance the field emission current.

Keywords: Carbon nanotubes, Multistage, Field emission, Microchannelplate, Electron source.

1 INTRODUCTION

Field emission (FE) is one of the most promising applications of carbon nanotubes (CNTs). The high aspect ratio (~1000) and atomically sharp apex of CNTs enhances the local field and lowers the threshold field for electron emission (~0.5-1V/μm) [1-2]. In addition to the geometrical features they posses high electrical, thermal conductivity and high chemical, temperature stability. All these unique properties make CNT a robust and stable field emitter [3]. Field emission displays (FED) [4], electron guns for next generation scanning electron microscopes (SEM) and transmission electron microscopes (TEM) [5], miniature x-ray tubes [6] and source for high powered microwave (HPM) devices [7] are some of the various CNT FE applications that have been demonstrated. For these applications various CNT synthesis methods and structures have been investigated. One of the factors for efficient field emission is the distance between CNTs, which should be greater than their height to minimize electric-field screening effect [8]. This has lead to the growth of vertically and spaced CNTs by application of electric field [9], plasma enhanced chemical vapor deposition (PECVD) [10] and nano template assisted thermal CVD [11]. To further improve the emitter performance i.e. low threshold electric field and high emission current, recent works [12-17] have demonstrated a novel emitter design consisting of smaller emitters on a larger one. The enhanced field emission properties of such structures termed as multistage are attributed to the enhanced field enhancement factor (β). A ratio of local field (E_L) around the emitter tip to applied electric field (E_a) and the overall β for multistage emitters can be expressed as a product of their respective field enhancement factors [12-13, 17].

The emitters in the previous studies consisted of CNTs grown on carbon cloth [12], silicon posts [14], porous silicon pillars [15], ZnO nanorods [16] and tungsten oxide nanowires on tungsten tip [13]. To the best of our knowledge this is the first report on field emission on CNT-CNT multistage structures i.e. single wall nanotubes (SWNTs) or thin-multiwall nanotubes (thin-MWNTs) on MWNTs. Here we report the synthesis of vertically oriented CNT multistage emitter periodic arrays by catalytic thermal CVD method and its field emission properties. The emitter structure consisted of arrays of MWNTs on which SWNTs/thin-MWNTs were grown, and the distance between each multistage pillar was controlled to have higher field emission performance.

Here we also report on a method for achieving high emission current from CNTs and electron multiplier microchannel plate (MCP). Micro channel plates are used in several applications including X-ray, astronomy, e-beam fusion studies, nuclear science [18] and high efficiency field emission displays [19]. Our novel approach of combining CNT emitters and MCP will provide several benefits. First MCP amplifies the emission current and second it also protects the CNTs from irreversible damage during vacuum arcing [20]. Operation of MCP is based on avalanche multiplication of secondary electrons, which are generated when incident electrons strike the channel walls of a MCP. A voltage applied across the ends of the MCP creates a field which accelerates the secondary electrons along the channel leading to avalanche multiplication.

2 EXPERIMENTAL

The porous-Si substrates for CNT growth were prepared by anodization of 2 inch diameter and 300 μm thick p-type Si<100> with a resistivity of 0.008-0.02 ohm-cm. Anodization was performed in an electrolyte under galvanostatic i.e. current-controlled conditions in a simple

---

[1]Corresponding author Tel: (305)348-1973, Fax (305)348-1932
E-mail address choiw@fiu.edu

o-ring Teflon cell in the dark [21]. The electrolyte consisted of a 1:1 mixture (by volume) of 48 wt% HF and 100% ethanol. An aluminum foil pressed on the backside of the sample served as an ohmic contact and a platinum wire served as counter electrode. Anodization was carried out at a rate of 10 mA/cm² for ~5 minutes resulting in a nano porous-Si layer with a pore diameter of ~15-20 nm. On the resulting porous-Si substrates iron (Fe) catalyst thin film (~10 nm) was sputtered through a shadow mask followed by annealing for ~12 hours at 300⁰C. Annealing under these conditions was observed to improve the contact of Fe catalyst with the nanopores. In addition annealing was also observed to relieve the stress in the porous film thus avoiding cracking during the nanotube growth [11]. For the nanotube synthesis the annealed substrates were placed in the cylindrical quartz tube of the CVD system, followed by heating to 700⁰C in Ar (~1000 sccm). After the temperature was stabilized at 700⁰C, Ar was replaced with C_2H_2 (~1000 sccm, 40 minutes) precursor gas, followed by cooling to room temperature for ~2 hours. For SWNT/thin-MWNT nanotube growth on top of MWNTs, 'Fe' catalyst was deposited for ~5 minutes through a shadow mask aligned with the previously grown MWNT arrays. Resulting substrates were then placed in a thermal CVD and nanotubes were grown by flowing a mixture of CH_4 (~1000 sccm) and C_2H_4 (~5 sccm) for ~5 minutes at 900⁰C. Morphology of the synthesized nanotubes was determined by scanning electron microscopy (SEM) and transmission electron microscopy (TEM). Raman spectroscopy was used to determine the degree of graphitization. Field emission measurements for the multistage emitters were performed with and without MCP. In all the experiments indium tin oxide (ITO) coated quartz screen printed with conventional green phosphor paste was used as anode and the vacuum level was maintained at ~5E-7 Torr. The inter-electrode distance (D) i.e. cathode-anode in experiments without MCP and cathode-MCP in experiments with MCP was maintained at ~900 μm.

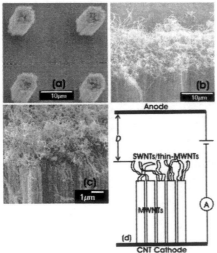

Figure 1. SEM images of (a) vertically oriented MWNT arrays grown on porous silicon; (b) MWNTs; (c) CNT

multistage structure, i.e. thin MWNTs and SWNTs on top of MWNTs; (d) schematic of the multistage structure and field emission measurement schematic.

3   RESULTS AND DISCUSSION

Figure 1(a) shows the SEM image of vertically grown MWNTs on porous-Si nanotemplate. The growth mode of nanotubes on the porous-silicon is base growth mode similar to the one demonstrated by Fan et al [11]. This was verified by completely removing the nanotubes by reactive ion etching (RIE). Nanotube growth was observed on the resulting etched substrates without catalyst deposition, thus confirming the base growth (images not shown here). Vertical alignment for nanotubes grown on porous-Si nanotemplates is achieved by van der Walls force interaction between the adjacent nanotubes forming bundles with high rigidity from densely packed catalyst particles. The rigidity enables nanotubes to keep growing along the original direction i.e. normal to the substrate and even outermost nanotubes are held by inner nanotubes without branching away [11]. Length and diameter of the nanotubes were controlled by varying the precursor gas concentration and flow during the CVD growth. The average length and diameter of MWNT arrays grown for 45 minutes were ~55 μm and ~15-20 nm, respectively. The multistage structure of the CNTs seen in SEM image of figure 1c and schematic figure 1d can be distinguished by comparing with the top surface of MWNTs shown in figure 1b. The presence of smaller diameter nanotubes i.e. SWNT and thin-MWNT on top of the MWNT arrays i.e. multistage structure was determined from TEM and Raman spectroscopy. For TEM analysis few drops of suspension obtained from nanotube blocks sonicated in dichloroethane was deposited on a TEM grid. Figure 2a, 2b show the TEM images consisting of MWNT and thin-MWNT. The defective structure of thin-MWNT seen in figure 2b may be a result of lower growth time of ~5 minutes [16]. From the TEM analysis presence of SWNT was not found this may be due the modest amount of SWNT existence in the total sample, however their presence was confirmed from Raman spectroscopy. Usually radial breathing mode (RBM) peaks are not observed in MWNTs [22-23], however in our samples they were observed due to the presence of SWNT and thin-MWNT on top of MWNTs (figure 1c, d). The diameters (d_t) of nanotubes were calculated from the relation $d_t = 248/\omega_{RBM}$ [19], where ω_{RBM} is RBM frequency. The d_t values ranging from ~2.76-3.52 nm calculated for ω_{RBM} 50-150 cm⁻¹ correspond to thin-MWNTs [23], d_t values ranging from ~1.28-1.96 nm for ω_{RBM} 125-200 cm⁻¹ correspond to SWNTs [22] (figure 2a-c). The calculated d_t of nanotubes ω_{RBM} 50-200 cm⁻¹ agreed well with internal diameters obtained from TEM analysis. The G band at ~1570 cm⁻¹ of Raman spectra (figure 2c) indicates formation of graphene sheets and D-band peak at ~1344 cm⁻¹ indicates defects or impurities [22] in the sample. The average length and diameter of the SWNT and thin-MWNTs were ~10-15 μm and ~2-10 nm, respectively.

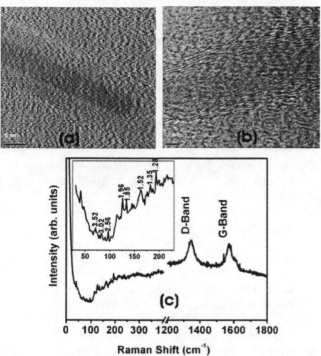

Figure 2. HRTEM images of (a) MWNTs and (b) thin MWNTs; (c) Raman spectrum (λ: 514 nm) of multistage CNTs with argon ion (Ar+) laser; inset, radial breathing mode (RBM) peaks, where inset numbers correspond to inner tube diameters of thin MWNTs and SWNT. Usually radial breathing mode (RBM) peaks are not observed in MWNTs [15]: however, in our samples they were observed due to the presence of SWNTs and thin MWNTs on top of MWNTs.

FE measurements of as-grown MWNT and multistage arrays at ~1 × 10⁻⁶ Torr are shown in figure 3. Obtained data was analyzed by Fowler-Nordheim (FN) equation [24], a relation between current (I) and applied electric field

$$I = \left(aA\beta^2 E^2 / \phi \right) exp\left(-b\phi^{3/2} / \beta E \right) \qquad (1)$$

Where $a=1.54x10^{-6} A\ eV\ V^{-2}$ and $b=6.83x10^7\ eV^{3/2}\ V\ cm^{-1}$, respectively. A is emission area, β is field enhancement factor, E is applied electric field in $V\ cm^{-1}$ and ϕ is work function in eV. The turn-on field (E_t) and current at a field of 1 V/μm as seen in figure 3 for the multistage and MWNT arrays was ~0.4 V/μm, 0.6 V/μm and ~450 μA, 14 μA respectively. The lower turn-on field (~1.5 times) and higher emission current (~32 times) for multistage arrays can be attributed to its geometry, i.e. to higher field enhancement at the smallest nanotube tip on the vertically aligned MWNTs [12-13, 17]. The calculated β value from the slope of FN plot $(-b\phi^{3/2}D\beta^{-1})$ assuming ϕ 5 eV

similar to graphite was ~ 26200 and ~8400 for multistage and as-grown MWNT arrays, respectively. The higher β (~3 times) value may also have resulted in higher field emission in multistage MWNT arrays [12-13]. Inset of figure 3 shows a fairly uniform emission image of the multistage arrays and the detailed investigation about their emission stability will be performed in future studies.

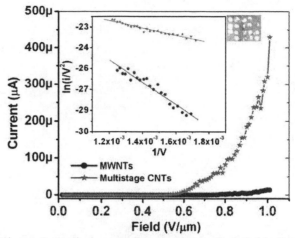

Figure 3. Emission current versus applied field plot, for multistage and MWNT arrays. Turn-on field (Et) was ~ 1.5 times lower and emission current was ~32 times higher for multistage in comparison to only MWNT arrays. Inset shows the corresponding FN plot and a fairly uniform emission image.

Electron multiplication was characterized by placing a commercial MCP (Hamamatsu) on top of the CNT cathode as represented by the drawing in figure 4b. The characterization was achieved by measuring the I-V data and by electron emission imaging. Schematic of the measurement circuit with and without MCP is shown in figure 4c. The 7.5 times higher (figure 4a) current with MCP is attributed to electron multiplication. So by placing an MCP on top of CNT emitters higher emission current could be achieved at moderate conditions. In both cases the turn-on voltages were approximately same. Enhanced emission current which was ~7.5 times higher with MCP can be attributed to electron multiplication. The commercial MCP plate placed on top CNTs amplified current by few micro amperes. However they are not suitable for achieving higher currents (few amperes) because of the lower secondary yield of the constituent materials. So a new MCP with stable and high secondary emission materials has been designed and fabricated which could be an efficient electron multiplier to achieve higher currents (data not shown here). This unique approach of placing the MCP over the field emitters could provide a more consistent and reliable cold electron sources operating at moderate power.

In summary, we have demonstrated synthesis and the field emission of vertically oriented multistage CNT arrays i.e. SWNTs/thin MWNTs on MWNTs on porous-Si and a reliable method for current amplification using a microchannel plate. The enhanced field emission properties of CNTs in this study were a result of the multistage structure which produced ~3 times higher field concentration compared to only MWNTs. Current was amplified by 7.5 times with the microchannel plate positioned between the cathode and the anode.

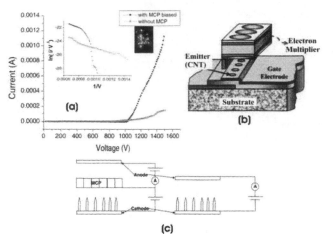

Figure 4(a) Emission current vs. applied voltage plot with and without MCP, (inset) FN plot and emission image with well defined spots (b) Schematic of a proposed novel cold cathode for high current density including carbon nanotube emitters and electrons multiplier microchannel plate, field emitted electron will be enhanced due to electro multiplication from MCP channels [patent pending] (c) schematic of measurement circuit with and without MCP.

REFERENCES

[1] Rinzler A G, Hafner J H, Nikolaev P, Lou L, Kim S G, Tomanek D, Nordlander P, Colbert D T, and Smalley R E , Science, 269,1550, 1995.

[2] Utsumi, T, IEEE Trans. Electron. Devices, 38, 2276, 1991.

[3] Seelaboyina R and Choi W B, Recent Patents on Nanotechnology, 1, 238, 2007.

[4] Choi W B et al, Appl. Phys. Lett., 75, 3129, 1999.

[5] De Heer W A, Chatelain A. and Ugarte D, Science, 270, 1179, 1995.

[6] Zhang J, Yang G, Lee Y Z, Chang S, Lu J P, and Zhou O, Appl. Phys. Lett., 89, 064106-1, 2006.

[7] Teo K B K, Minoux E, Hudanski L, Peauger F, Schnell J-P,Gangloff L, Legagneux P, Dieumegard D,Amaratunga G A J and Milne W I, Nature, 437 968, 2005.

[8] Nilsson L, Groening O, Emmenegger C, Kuettel O,Schaller E,Schlapbach L, Kind H, Bonard J-M and Kern K, Appl. Phys. Lett., 76, 2071, 2000.

[9] Chen L H, AuBuchon J F, Gapin A, Daraio C, Bandaru P, Jin S, Kim D W, Yoo I K and Wang C M, Appl. Phys. Lett., 85, 5373, 2004.

[10]Meyyappan M, Delzeit L, Cassell A, and David H, Plasma Sources Sci. Technol., 12, 205, 2003.

[11]Fan S S, Chapline M G, Franklin N R., Tombler T W, Cassell A M and Dai H, Science 283, 512, 1999.

[12]Huang J Y, Kempa K, Jo S H, Chen S and Ren Z F , Appl.Phys. Lett., 87, 053110, 2005.

[13]Seelaboyina R., Huang J, Park J, Kang D H, and Choi W B, Nanotechnology, 17, 4840, 2006.

[14]Hsu D S Y, and Shaw J, Appl.Phys.Lett., 80, 118, 2002.

[15]Li J X, and Jiang W F, Nanotechnology 18, 065203, 2007.

[16]Li C, Fang G, Yuan L, Liu N, Ai L, Xiang Q, Zhao D, Pan C, and Zhao X, Nanotechnology 18, 155702, 2007.

[17]Niemann D L, Ribaya D P, Gunther N, Rahman M, Leung J and Nguyen C V, Nanotechnology 18, 485702, 2007.

[18] Joseph Ladislas Wiza, Nuclear Instruments and Methods, 162, 587, 1979.

[19]Whikun Yi, Sunghwan Jin, Taewon Jeong, Jeonghee Lee, SeGi Yu, Yongsoo Choi, and J. M. Kim, Appl. Phys. Lett., 77, 1716, 2000.

[20]W. P. Dyke, J. K. Trolan, E. E. Martin, and J. P. Barbour, Phys. Rev., 91, 1043, 1953.

[21]Lehmanna V and Rönnebeck, S, Journal of The Electrochemical Society, 146, 2968, 1999.

[22]Dresselhausas M S, Dresselhaus G, Saito R, and Jorio A 2005 Physics Reports 409 47

[23]Seelaboyina R, Huang J and Choi W B 2006 Appl. Phys. Lett.88 194104

[24]Fowler R H and Nordheim L W 1928 Proc. R. Soc. A 119 173

Advantages and Limitations of Diamond-Like Carbon as a MEMS Thin Film Material

P. Ohlckers*,**, T. Skotheim**, V. K. Dmitriev*** and G. G. Kirpilenko***

*Vestfold University College, Raveien 197, 3184 Borre, Norway, Per.Ohlckers@hive.no
**INTEX, Tucson, AZ, USA, terje.skotheim@gmail.com
**Patinor Coatings, Moscow, Russia, pt_ire@mail.compnet.ru

ABSTRACT

Combining Bulk Silicon Micromachining (BSM) with Diamond-Like Carbon (DLC) thin film technology can be favourable used to make high performance MEMS devices. We highlight the versatility of BSM combined with the unique features of our proprietary DLC thin film technology to make high performance MEMS devices at favourable cost. A high performance infrared emitter has been designed and commercialised, with the most distinctive features being high speed with a modulation depth of more than 100 HZ, broadband IR emission from 1 to 20 micrometers, more than 10% power efficiency, and a lifetime beyond 100,000 hours. These emitters are already in use in system applications like non-dispersive infrared gas sensors.

1 INTRODUCTION: MEMS STRUCTURES WITH DIAMOND-LIKE THIN FILMS

We assume here that the versatility of bulk silicon micromachining is well know, and in addition we can observe from the material properties of Diamond-Like Carbon (DLC) that one or more of the following characteristics can be favourable exploited to make MEMS structures and/or devices with DLC thin films:

- Extraordinary Yield Strength of up towards 30 times better than stainless steel and up to 5 times better than silicon.
- Extraordinary stiffness with Young Modulus of Elasticity of around 8 times stiffer than silicon and around 7 times stiffer than steel.
- Indentation hardness and wear resistance approaching diamond, the best among any other materials.
- High thermal conductivity.
- Superior chemical and corrosion resistance
- Processing of DLC films compatible with most silicon processes up to 500 °C ((short pulsing to ~800°C).
- DLC thin films can be made by different Physical Vapor Deposition and Chemical Deposition Methods, or combined methods.

- Combination of silicon MEMS with DLC thin films can be used to make devices combining the versatility of silicon processing with the unique features of DLC thin film

If one of more of these features can be exploited together with silicon MEMS technology such as Bulk Silicon Micromachining, the resulting BSM/DLC mixed technology can be used to make very competitive MEMS devices, such as the infrared emitter described here and shown in Figure 1.

Figure 1: Picture of the BSM/DLC based infrared emitter packaged in a metal can transistor header. The picture is taken during operation, showing the visible part of radiation from the emitting membrane.

2 DIAMOND-LIKE THIN FILM PROCESSES

We have developed the following DLC thin film processes that can be combined with silicon MEMS technologies:

A Pulsed Cathodic Arc (PCA) process, a physical vapor deposition process for producing ultra-hard amorphous diamond (AD) carbon coatings. These materials, also called tetrahedral amorphous carbon (ta-C), consist essentially of pure carbon, with only a trace of hydrogen (~90 % sp^3 bonding). The AD coatings have extreme hardness (70-80 GPa), close to that of crystalline diamond. The process can be scaled to coat substrates of an arbitrary size at high rates and low cost. The first application is as protective coatings in places of severe wear or corrosion. The process is flexible—multiple independent sources can be mounted in any direction—and has high deposition rate capability. The films produced with PCA are generally denser and of higher quality than films produced with other vacuum deposition technologies. Exceptional adhesion is achieved with a proprietary process. Intex has developed a deposition process that results in stress-free AD films for use as structural elements such as membranes and cantilevers. Figure 2 shows a micrograph of an AD film conformally coated on a WC substrate.

A NanoAmorphous Carbon (NAC) process, giving a new class of multi-functional electronic materials, coatings, with conductivity that can be varied from dielectric to metallic. The films are obtained by Plasma Enhanced Chemical Vapor Deposition (PECVD). The PECVD-produced film is an amorphous dielectric with a composition consisting of a substantially sp^3-bonded carbon network that also contains silicon and oxygen. NAC films can be made electrically conducting by incorporating metals into the carbon matrix by a simultaneous sputtering process, or Plasma Enhanced Physical Vapor Deposition (PVD) The resistivity of the film is controlled by controlling the metal concentration. The conductivity reaches a maximum of $\sim 10^4$ S/cm. Dielectric films (without conducting additives) have conductivities in the 10^{-10} S/cm range. This is a larger range of conductivity than has been observed with any other known material.

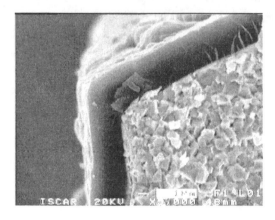

Figure 2: Micrograph of a ~3 micrometer amorphous diamond carbon on sintered tungsten carbide.

Processing in air is normally limited to 350°C - 400°C. NAC, which is a mixture of SiO_x and DLC-type of material is more stable. Short pulsing up to ~800°C is possible, and sustained processing can be done at 500°C. The principle of operation for NAC deposition system with combined Plasma Enhanced Vapor Deposition (PECVD) and Physical Vapor Deposition (PVD) is shown in Figure 3.

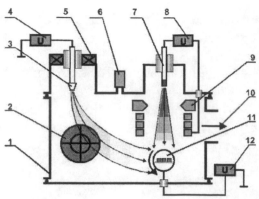

Figure 3: Schematic representation of the Intex Pulsed Cathodic Arc deposition system for producing strongly adhering AD films. (1) Vacuum chamber; (2) deflecting magnetic system; (3) metal plasma source; (4) power supply unit; (5) focusing magnetic system; (6) manometer; (7) pulsed plasma accelerator cathode; (8) power supply unit; (9) pulsed plasma accelerator anode; (10) vacuum pumping; (11) carousel with substrates; (12) substrate bias voltage.

The conductivity of the NAC film can be controlled by the atomic fraction of a metal as the conductive additive, as shown in Figure 4 with tungsten (W) as the metal additive.

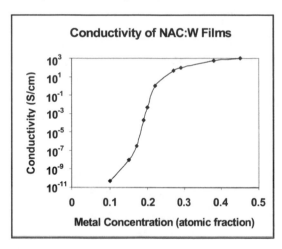

Figure 4: Conductivity of the NAC film as a function of the atomic fraction of tungsten as the metal additive cosputtered during the NAC deposition process.

3 THE EMITTER MADE BY SILICON BULK MICROMACHINING AND DIAMOND-LIKE CARBON THIN FILM TECHNOLOGY

The versatility of the NAC process in combination with silicon MEMS process technology is demonstrated by a broadband infrared emitter. Intex and its subsidiary Patinor Coatings have developed a micromachined pulsed infrared light source with high intensity and ability to pulse at high frequencies. The IR emitter incorporates a NAC film as a thermoresistive element in the form of a free-hanging thin multilayer membrane supported by silicon (Figure 5). The emission spectrum is that of a greybody and provides a wide spectral output, from 1 μm up to 20 μm. The high emissivity, high thermal conductivity, low thermal mass and high strength of the NAC material allow rapid heating and cooling by passing an alternating current through the resistive membrane. The resistivity of the NAC material is controlled by cosputtering of a metal additive.

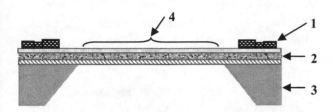

Figure 5: Cross section view of the emitter chip: 1) Bonding pads; 2) NAC multilayer membrane, 3) Silicon support, 4) Active emitter area.

The source can pulse at frequencies up to 100Hz at ~50% modulation depth. High frequency pulsed sources are important for achieving good signal-to-noise ratios (high sensitivity) in IR gas sensors. The IR emitters are fabricated using MEMS technology and can be tailored to the customer's requirements over a wide range.

The emitter is now in production with the following most distinctive typical specifications:
Chip Size: 3.7 mm x 3.7 mm
Spectral Output Range 1.0 – 20 micrometer
Emitter Surface Area 1.7mm x 1.7 mm^2
Resistance 50 Ω
Drive Voltage (pulsed, bi-polar or DC) 6.5 V
Drive Current 135 mA
Working Temperature 750 °C
Modulation frequency 0 – 100 Hz
Maximum Frequency at 50% Modulation 100 Hz
Power Consumption 900 mW
Integrated Power Emission 90 mW
Warm-up Time <30 msec
Decay time <5 msec
Lifetime >5,000 hours at 750°C
>25,000 hours at 600°C
>100,000 hours at 500°C

In Figure 1, the emitter is shown packaged in I TO-5 type of metal can header, with an open header cap. The cap can alternatively be sealed in nitrogen with a filter window like calcium fluoride, sapphire, silicon etc., depending upon the spectral properties wanted for the cap window.

In Figure 6, the pulsing speed of the emitter is shown as the modulation depth at 50% duty cycle, showing that the emitter can be modulated beyond 100 Hz.

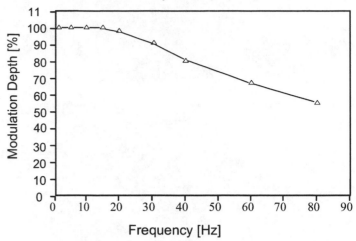

Figure 6: Modulation depth measurement for the infrared emitter at 50% duty cycle.

The temperature distribution across the area of the emitter membrane is shown in Figure 7 at a working temperature of approximately 750 °C at the centre of the membrane, showing the effect of sideways conduction cooling to the silicon frame of the emitter.

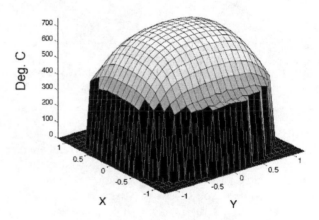

Figure 7: The temperature distribution across the emitter membrane. X and y coordinates in mm.

The emitter output can be focused with a parabolic reflector mounted on the header cap, as shown in Figure 8, Figure 9 and Figure 10. In this way, improved signal-to-noise ratio can be achieved for the same power input to the emitter.

Figure 8: Picture of the emitter with the parabolic reflector mounted on the header cap.

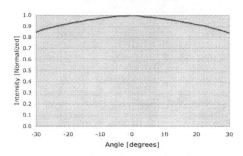

Figure 9: Angular distribution of infrared radiation without parabolic reflector.

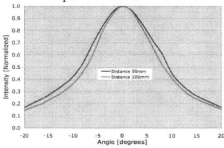

Figure 10: Angular distribution of infrared radiation with parabolic reflector.

The IR emitters can be used as the light source in infrared gas sensors for industrial, consumer, research and medical applications, mostly based upon the non-dispersive infrared sensing principle (ND-IR) Application examples are: Explosive gas detection systems (methane); combustion efficiency and stack emissions monitoring systems (CO, , CO_2); toxic emission systems (SO_x, NO_x, NH_3); air quality monitoring systems; HVAC efficiency (controlling airflow by measuring CO_2 concentration); spectrophotometers; patient bedside monitoring systems; anesthesia gas monitoring systems (CO_2 and anesthetic gases); noninvasive glucose measurements; automotive engine control and exhaust monitoring; and chemical warfare agent detection.

4 CONCLUSIONS AND FURTHER WORK

We have explained and demonstrated that Bulk Silicon Micromachining and Diamond-Like Carbon Thin Film Technology can be combined to make versatile MEMS devices. The demonstrator device is an infrared emitter with high speed, broadband infrared emission spectrum and high power efficiency as distinctive features. This emitter is already in production, to be ramped up to high volumes the coming years. In the future, other MEMS devices will be developed, which can benefit from these mixed technologies.

REFERENCES

[1] V. Dorfman and B. Pypkin, "Method for Forming Diamond-like Nano-structured or Doped Nanostructured Films," US Patent No. 5,352,493 (1994)

[2] P.E. Nordal and T. Skotheim, "Infrared Emitter and Methods for Fabrication the Same," US Patent 6,031,970 (2000)

[3] V.P. Goncharenko, A.Y. Kolpakov and A.I. Maslov, "Method of Forming Diamond-like Carbon Coating in Vacuum," US Patent 6,261,424 (2001)

[4] A. Kolpakov, V.N. Inkin and G.G. Kirpilenko, "Vacuum Coating Apparatus," US Patent 6,692,624 (2004)

[5] L. P. Sidorova, V. K. Dmitriev and V. N. Inkin, "Method for Producing a Conducting Doped Diamond-like Nano-structured Film and a Conducting Doped Nano-structured Diamond-like Film," US Patent (2004) Publication date 02/05/2004.

[6] A. Kolpakov, V. N. Inkin and G. G. Kirpilenko, "Pulsed Carbon Plasma Apparatus," US Patent (2004) Publication date 01/29/04.

First-principles studies of electrophilic molecules on the carbon nanotubes

Sohee Park, Ki-jeong Kong, Hye-Mi So, Gyoung-Ho Buh, Jeong-O Lee, and Hyunju Chang*

Fusion Biotechnology Research Center, Advanced Materials Division
Korea Research Institute of Chemical Technology
P. O. Box 107, Daejeon, 305-600, Korea, hjchang@krict.re.kr

ABSTRACT

Various approaches have been made to separate semiconducting carbon nanotubes (CNTs) from metallic ones including chemical treatments. Recently we have demonstrated that some electrophilic molecules, such as 2,4,6-triphenylpyrylium tetrafluoroborate (2,4,6-TPPT) and 1,3-benzodithiolylium tetrafluoroborate (1,3-BDYT), preferentially interact with metallic CNTs. In this study, we present systematic theoretical studies on the interactions between the CNTs and these electrophilic molecules, using density functional theory. It was found that 2,4,6-TPPT and 1,3-BDYT interact with SWNT via noncovalent and covalent bindings, respectively. We have also found that 1,3-BDYT tends to be more strongly bound to the metallic SWNT of the smaller diameter. However, 2,4,6-TPPT tends to interact with CNTs via π-π interaction with very small binding energies, regardless of chiralities and diameters.

Keywords: carbon nanotube (CNT), electrophilic molecule, electronic structure

1 INTRODUCTION

Since the first finding on carbon nanotubes (CNTs) [1], there have been extensive investigations on the applications of CNTs. CNTs are fascinating because of their diversity of electronic structure, from metal to semiconductor, depending on their chirality. However, this diversity becomes a major disadvantage in applications of CNTs for real electronic devices. There have been many attempts to separate metallic nanotubes from semiconducting ones, including electrophoresis[2], DNA wrapping[3], electrical breakdown of metallic nanotubes [4], and chemical treatment with various molecules[5-8]. Among them, chemical treatment seems very attractive for a large-scale electronic device fabrication. It was reported that the diazonium compound and nitronium ions selectively attack metallic nanotubes resulting in sorting out semiconducting nanotubes only[5-7]. Recently, our group also reported that some electrophilic molecules can play a similar role to metallic nanotubes to change their electronic structures from the investigations on the electrical transport properties of single-walled CNT field effect transistor reacted with several electrophilic molecules [8]. However, no rigorous explanation has not been suggested for these electrophilic molecules how to affect on the electronic structure of the carbon nanotubes. A few theoretical approaches have been reported to explain how nitronium ion and diazonium compounds to affect the electronic structures of the carbon nanotubes [9,10]. They suggested that those molecules selectively attack the sidewall of the metallic nanotubes due to the abundant presence of the electron density at the Fermi level. Another theoretical study reported that some large nearly neutral aromatic molecules and some small charge transfer aromatic molecules interact more strongly metallic nanotubes than the semiconducting ones [11].

In the present study, we investigated the interaction of the electrophilic molecules, specifically 1,3-benzodithiolylium tetrafluoroborate (1,3-BDYT) and 2,4,6-triphenylpyrylium tetrafluoroborate (2,4,6-TPPT), with both metallic and semiconducting CNTs in various diameters.

2 COMPUTATIONAL DETAILS

First-principles calculations were performed for the models of 2 electrophilic molecules, 1,3-BDYT and 2,4,6-TPPT on the CNTs of various chiralities. 1,3-BDYT and 2,4,6-TPPT molecules are shown in Fig. 1. The density-functional theory (DFT) was employed using Vienna Ab initio Simulation Package (VASP)[12], and some calculations were done with Dmol^3[13]. Generalized gradient approximation (GGA) was adopted for the exchange and correlation with Vanderbilt ultrasoft pseudopotentials. The plan-wave basis sets were used with the energy cutoff as 396 eV. Only Γ-point in k-space is considered for the calculations here.

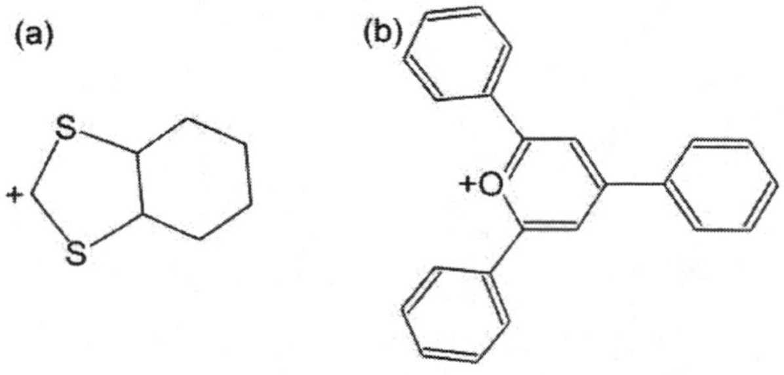

Fig. 1 Electrophilic molecules we considered. (a) 1,3-benzodithiolylium tetrafluoroborate (1,3-BDYT), (b) 2,4,6-triphenylpyrylium tetrafluoroborate (2,4,6-TPPT)

For the small-radius nanotube models, armchair (5, 5) metallic CNT and zigzag (10, 0) semiconducting CNT were chosen, while (10, 10) metallic tube and (17, 0) semiconducting tube were chosen for the large-radius cases. Periodic supercells were taken as containing 3 unitcells for semiconducting tubes and 6 unit cells for the metallic tubes. The sizes of the supercells for the calculations are $24 \times 24 \times 12.68$ Å^3 and $24 \times 24 \times 14.64$ Å^3, for semiconducting tubes and metallic tubes, respectively. Within these supercells, the electrophilic molecule was attached to the sidewall of CNTs, one by one, and the geometric structures were fully relaxed. The atomic relaxation is terminated if the each component of the Hellmann-Feynman force on each atom is reduced to within 0.02 eV/Å. After structural relaxation of each molecule on the CNT, the binding energy was calculated as the following equation, $E_{bind} = E_{tot}$(CNT + mol)-E_{tot}(CNT)-E_{tot}(mol), where E_{tot} is the total energy of a given system

3 RESULTS AND DISCUSSTIONS

3.1 1,3-BDYT on various CNTs

We first studied the single 1,3-BDYT molecule on the side wall of CNT. We have selected the most stable geometry of 1,3-BDYT on metallic or semiconducting CNT, in aqueous solution model. For experimental chemical functionalization, CNTs are immerged into the solution of given electrophilic molecules [8]. In the solution, 1,3-BDYT is dissolved into $(C_7H_5S_2)^+$ and $(BF_4)^-$ ions. We can suppose that firstly $(C_7H_5S_2)^+$ ion attacks C' atoms of the sidewall of CNT, then the C atom between S atoms of $(C_7H_5S_2)^+$ is covalently bonded to C' atom of the CNT. After C-C' bonding between the electrophilic ion and the CNT, another C1 atom of CNT, which is neighboring to C-

C' bond, becomes active. It is because that the original C1-C' bond becomes weak after C-C' bonding. One can imagine that H$^+$ ions, which usually exist in aqueous solution, easily attack the active C1 atom, then the H$^+$-C1 bond can be formed. From several model calculations of the possible pathways of C-C' bonding, that will be published elsewhere, we have obtained the most stable structure of the covalent bonding between $(C_7H_5S_2)^+$ and CNT with additional H$^+$ ion adsorption, as shown in Fig. 2-(a)

After adsorption structures were fully relaxed for 4 different CNTs, such as (5,5), (10,0), (17,0) and (10,10), we have calculated binding energies of 1,3-BDYT for each case. The calculated binding energies are plotted with respect to the diameter in Fig. 3. The solid line connects binding energies of 1,3-BDYT on metallic CNTs, whereas the dashed line connects the binding energies on semiconducting CNTs. As shown in Fig. 3, 1,3-BDYT is more strongly bound to the metallic nanotube than the semiconducting nanotube for the smaller diameter.

On the contrary to that, for the larger diameter, such as (10,10) or (17,0) CNT, the binding energies are almost the same for both metallic and semiconducting ones. It infers that 1,3-BDYT is not preferentially bound to the metallic nanotubes anymore for larger diameters, more than 13 Å. Our previous experiments reported that the yield of selective suppression of metallic nanotubes seems to depend on the diameters of the nanotubes. The best efficiency of selective suppression lies around 10 ~20 Å [8]. Our calculation results can explain the observed diameter-dependency of selective suppression of metallic nanotubes. The strong adsorption of 1,3-BDYT on metallic CNT can result in changing electronic structure of the metallic nanotubes.

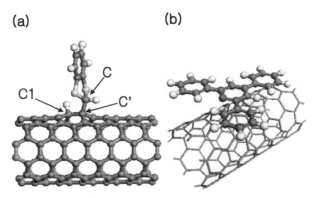

Fig. 2 The optimized structures after full relaxation of (a) 1,3-BDTY with additional H$^+$ ion, and (b) 2,4,6-TPPT, on (5,5) CNT. Grey balls and sticks are carbons. White ball, yellow ball and red ball represent hydrogen, sulfur and oxygen respectively

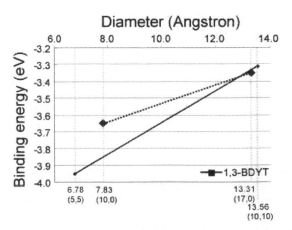

Fig. 3 Binding energies of 1,3-BDYT on CNT are plotted with respect to the diameters and chiralites of CNTs.

3.2 2,4,6-TPPT on various CNTs

Another molecule we have calculated is 2,4,6-TPPT. It consists of 4 aromatic rings, as shown in Fig. 1-(b). 2,4,6-TPPT is dissolved into $(C_{23}H_{17}O)^+$ and $(BF_4)^-$ ions in the solution. We set up the adsorption models of 2,4,6-TPPT on (5,5) and (10,0) nanotubes.

Unlike 1,3-BDYT, 2,4,6-TPPT is expected to be attached to the sidewall of CNT via noncovalent π interaction, because of its planar structure. Moreover, it doesn't need to consider additional H^+ ion adsorption for 2,4,6-TPPT solution, because noncovalent π interaction does not seem to induce any active C site of CNT.

The optimized stable structure of adsorption of 2,4,6-TPPT on CNT is shown in Fig. 2-(b). After adsorption structures were fully relaxed, the binding energies were calculated as -70 meV and -60 meV, for metallic (5,5) nanotube and semiconducting (10,0) nanotube, respectively. There is no difference between metallic and semiconducting nanotubes in the binding energies of 2,4,6-TPPT. However, the experimental effect in selective suppression of metallic nanotubes using 2,4,6-TPPT is very similar to that of 1,3-BDYT [8]. For the case of 2,4,6-TPPT, we need another explanation rather than binding energy difference to explain the experimental results. Recently, Lu et al.[11] suggested that effective contact area and atomistic correlation between aromatic molecules and CNT can lead chiral selectivity. Atomistic correlation between aromatic rings of 2,4,6-TPPT and the hexagonal rings of CNT should be investigated more rigorously to explain chiral selectivity of 2,4,6-TPPT.

4 RAMAN SPECTRA ANALYSIS

In order to investigate the effect of 1,3-BDYT and 2,4,6-TPPT on the CNTs , we had previously performed a Raman spectrum analysis [8]. Fig. 5 shows the Raman spectra of the SWNT-FET device, after reaction with 1,3-BDYT and 2,4,6-TPPT solutions, respectively. The Raman spectra show two peaks, G band and D band, for 1,3-BDYT treatment, while only G band for 2,4,6-TPPT treatment. It is known that the G band near 1600 cm^{-1} is a characteristic peak of pristine CNTs, while the D band near 1350 cm^{-1} is a defect peak. In order to explain this difference between 1,3-BDYT and 2,4,6-TPPT, we have examined the structure of adsorption geometry and have calculated charge density of the given system. Figure 6 shows the charge density plot of the final geometry containing (a)2,4,6-TPPT and (b) 1,3-BDYT on (5,5) CNT, respectively. 1,3-BDYT forms chemical bonding with C atom of CNT, whereas 2,4,6-TPPT lies above the sidewall of CNT at a distance of 3.43 Å , where is van der Waals interaction region. 2,4,6-TPPT interacts with CNT non-covalently via π- π interaction. Thus there is no significant structural distortion of CNT. It explains the only G band appears in the Raman spectra of 2,4,6-TPPT treatment. On the contrary, 1,3-BDYT is

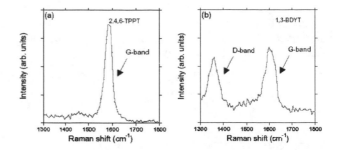

Fig.4 D, G band of Raman spectra of the SWNT-FET device after reaction with (a) 2,4,6-TPPT and (b) 1,3-BDYT treatment.

covalently bound to CNT and forms *sp3*-like bond, which is the origin of D band in Fig. 6-(b).

5 CONCLUTIONS

In this study, we present systematic theoretical studies on the interactions between the CNTs and the electrophilic molecules, such as 2,4,6-TPPT and 1,3-BDYT. We have used the first principle method based on density functional theory. Firstly, we have calculated the binding energies of these two electrophilic molecules on the several CNTs including both metallic and semiconducting ones with different diameters, such as (5,5), (10,0), (17,0) and (10,10). We have found that 1,3-BDYT tends to be more strongly bound to the metallic SWNT in the smaller diameter less than 13 Å. However, 2,4,6-TPPT tends to interact with CNTs via π-π interaction with very small binding energies about 60-70 meV, regardless of chiralities and diameters. We have also shown that 2,4,6-TPPT and 1,3-BDYT interact with CNT via noncovalent and covalent bindings, respectively. It can explain the observed Raman spectra of CNTs with 2,4,6-TPPT and 1,3-BDYT treatment. Additional D band observed in 1,3-BDYT treatment is well explained with *sp3* bond formation between 1,3-BDYT and CNT. From these calculations, we can conclude that electrophilic molecule tends to more strongly bind to the

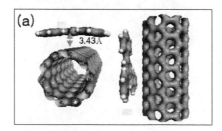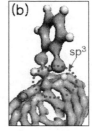

Fig.5 Optimized structures of SWNT and (a) 2,4,6-TPPT , and (b) 1,3-BDYT. The calculated total electron densities are also shown in blue and gray for (a) 2,4,6-TPPT and (b) 1,3-BDYT ,respectively

metallic CNT than to the semiconducting CNT, at the small diameter region, when it forms covalent bonding. Its strong binding leads to change in electronic structures of metallic nanotube via forming *sp3* configuration. It seems to result in selective suppression of metallic nanotubes.

ACKNOWLEDGEMENTS

This work was supported by the Korea Research Council for Industrial Science & Technology.

REFERENCES

[1] Ijima, S. Nature, **1991**, 354, 56.

[2] Krupke, R.; Hennrich, F.; von Lohneysen, H.; Kappes, M. M. Science, **2003**, 301, 344.

[3] Zeng, M.; Jagota, A.; Semke, E. D.; Diner, B. A.; Mclean, R. S.; Rustig, S. R.; Richardson, R. E.; Tassi, N. G. Nat. Mater. **2003**, 2, 338.

[4] Collins, P. G.; Arnold, M. S.; Avouris, P. Science **2001**, 292, 706.

[5] Strano, M. S.; Dyke, C. A.; Usrey, M. L.; Baron, P. W.; Allen, M. J.; Shan, H.; Kittrell, C.; Hauge, R. H.; Tour, J. M.; Smalley, R. E. Science **2003**, 301, 1519.

[6] Maeda, Y. et al. J. Am. Chem. Soc. **2005**, 127, 10287.

[7] An, K. H.; Park, J. S.; Yang, C.-M.; Jeong, S. Y.; Lim, S. J.; Kang, C.; Son, J.-H.; Jeong, M. S.; Lee, Y. H. J. Am. Chem. Soc. **2005**, 127, 5196.

[8] So, H.-M.; Kim, B.-K.; Park, D.-W.; Kim, B. S.; Kim, J.-J.; Kong K.-j.; Chang, H.; Lee, J.-O. J. Am. Chem. Soc., **2007**,129, 4866.

[9] An, K. H.; Yang, C.-M.;Seo, K.; Park, K. A.; Lee, Y. H. Current Applied Physics, **2006**, 6S1, e99.; Yang, C.-M.; Park, J. S.; An, K. H.; Seong, C. L.; Seo, K.; Kim, B.; Park, K. A.; Han, S.; Park, C. Y.; Lee, Y. H. J. Phys.Chem. B. **2005**, 109, 19242.

[10] Du, A.J.; Smith, S. C. Molecular Simulation, **2006**, 32, 1212.

[11] Lu, J.; Nagase, S.; Zhang, X.; Wang, D.; Ni, M.; Maeda, Y.; Wakahara, T.; Nakahodo, T.; Tsuchiya, T.; Akasaka, T.; Gao, Z.; Yu, D.; Ye, H.;Mei, W. N.; Zhou, Y. J. Am. Chem. Soc. **2006**, 128, 5114.

[12] G. Kresse and J. Hafner, Phys. Rev. B, **1993**, 47, 558,; ibid., **1994**, 49, 14251.; G. Kresse and J. Furthmueller, ibid., **1996,** 54, 11169.

[13] DMol^3 is a registered software product of Accelrys, Inc. Delley, B. J. Chem. Phys. **1990**, 92, 508.

Carbon nanotube interfaces for single molecular level bio sensing

H. Vedala*, S. Roy*, T-H. Kim*, W.B. Choi*

*Department of Mechanical and Materials Engineering, Florida International University, Miami, FL 33174, USA, choiw@fiu.edu

ABSTRACT

Single molecule detection, aimed both at fundamental investigation and applications, has recently been attracting a lot of attentions. The present set of studies is focused on the design, fabrication and optimization of novel carbon nanotube (CNT) probes for recognition of biomolecules (target species) with the accuracy down to molecular level. The sensor platforms are fabricated with aligned grown CNTs on desired substrates with high level of control. We have demonstrated excellent electrochemical sensing of functionalized array of CNT for quantitative and selective detection of a range of metabolites including cholesterol, ascorbic acid and uric acid, in buffer solution as well as in human plasma and blood. In addition, a label-free detection scheme for DNA hybridization as well as environmental gas has recently been demonstrated (Nano Letters vol 8, 26 2008). In this presentation the charge transport of bio-molecular binding at the CNT-transducer will be presented and discussed.

Keywords: Carbon Nanotube, DNA, Hybridization, electric conductance, bio sensors

1 INTRODUCTION

Electronic detection of biomolecules at single molecular level has several advantages as compared to detection of ensemble of molecules. Single molecules studies unravel the intrinsic properties of these molecules which are essential for both fundamental studies and various technological applications [1]. Most of the current available single molecule detection techniques use spectroscopic properties of the molecule which require optical labeling. This adds more complexity to the spectroscopic detection systems. On the other hand use of electronic detection techniques based on nanomaterials provide a direct and label free alternative method.

In this communication we report the development of a novel nanoelectronic platform for measuring direct electrical transport in single-molecule DNA of genomic significance. We have used single-walled carbon nanotube (SWNT) electrodes for anchoring a DNA molecule of compatible diameter (1-2 nm). Characterization of DNA using carbon nanotubes (CNT) has been pursued in the past, motivated by the prospects of CNT as a unique electrode material[2,3]. A couple of recent reports detailed the techniques of creating a nanogap in a SWNT and bridging the gap by organic molecules[4,5]. The preset study extends this concept to overcome the challenge of anchoring and electrically characterizing single-molecule DNA. We also report the influence of local environmental factors such as counterion variation, pH, temperature, ionic strength on charge transport (CT) of double-stranded (ds) DNA molecule at single molecule-level.

2 EXPERIMENTAL

2.1 Nanoelectrode fabrication

SWNTs were synthesized by chemical vapor deposition technique and were suspended in isopropyl alcohol by ultrasonication. Initially a 2 µl droplet of SWNT suspension was spun on an thermally oxidized (500 nm) silicon substrate having photolithographically patterned microelectrodes and bonding pad [6,7]. Electrical contacts to individual SWNT were made by first locating them with respect to prepatterned index marks using field emission scanning electron microscope (FESEM) imaging. That was followed by making contact leads using e-beam lithography and sputtering of 50 nm of Au on 10 nm Ti adhesion layer. To fabricate a pair of nanoelectrodes, Focused Ion Beam (FIB) was used for etching near the center of an individual SWNT segment between the metal electrodes. FIB etching parameters (beam current, exposure time) were optimized to obtain a uniform gap in accordance with the length of the DNA strands.

2.2 Electrode functionalization and DNA attachment

Electrical conductivity of an 80 base–pair (bp) in denatured (ssDNA) and hybridized (dsDNA) form (contour length ~27nm), encoding a portion of the *H5N1* gene of avian *Influenza A* virus (AIV) was measured. The template strand obtained with amine modifications at the 5′ and 3′ ends was hybridized with the unmodified complementary strand at 90°C for 5 min in 10 mM NaAc buffer (pH 5.8) at equimolar concentrations. To measure the electrical conductivity of the dsDNA molecules, the SWNT nanoelectrodes were first functionalized with COOH groups to form a strong covalent bond with amine terminated DNA molecule. This was performed by chemical oxidation of SWNT as reported in [8]. In short the SWNTs were treated with HNO_3 for 1 hour followed by rinsing with DI water and vacuum drying. The sample was then incubated for 30 min in 2 mM 1-ethyl-3-(3-dimethylaminopropyl)carbodiimide hydrochloride and 5 mM *N*-hydroxysuccinimide (NHS) to convert carboxyl groups to amine-reactive Sulfo-NHS esters. Amine terminated dsDNA molecules from a diluted solution (10 nM) were deposited on the electrodes and a.c. dielectrophoresis technique was used to align and immobilize DNA molecule between the electrodes. The devices were then washed with corresponding buffer solutions to remove non specifically attached DNA molecules. The samples were then blow dried with nitrogen stream.

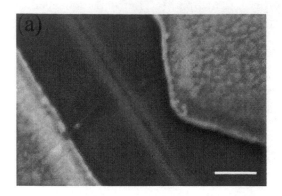

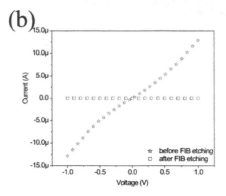

Figure 1: (a) SEM image of a pair of FIB-etched SWNT electrodes connected by Ti/Au micro-contact leads. Scale bar = 100 nm. (b) Room temperature *I-V* characteristics of SWNT before and after FIB etching (30 nm gap)

3 ELECTRICAL CHARACTERIZATION OF SINGLE DNA MOLECULE

3.1 Nanoelectrode electrical characterization

Figure 1 (a) illustrates the SEM image of a single SWNT nanoelectrodes with a gap of 27 nm. Figure 1 (b) depicts *I-V* characteristics of a typical SWNT before and after the etching process. The SWNT exhibited resistance in the range of kΩ. After FIB etching the current decreased from several micro-amperes to a few femto-amperes (noise range of instrument) indicating the nanogap formation. FIB etching also resulted in the formation of a trench in the oxide layer beneath the gap. The presence of the nanotrench

(width: 27±3 nm) helps in minimizing the interaction between the anchored DNA molecule and the substrate surface, which, otherwise causes a strong compression deformation of the immobilized DNA molecule and hence a perturbation of CT through it.

As the DNA molecules are coiled in an aqueous medium due to thermal agitation, a strong electric field gradient is essential to straighten and attach them between the electrodes. We used a.c. dielectrophoresis to trap and align the DNA molecule between the SWNT nanoelectrodes with a peak-to-peak voltage of 0.1-1 V and for a frequency range between 0.01-10 MHz. The applied field (40 MV/m) is sufficiently high to overcome Brownian motion, which is dominant in nanoscale objects [9]. Figure 2 shows the AFM image of the DNA molecule attached between the nanoelectrodes after dielectrophoresis.

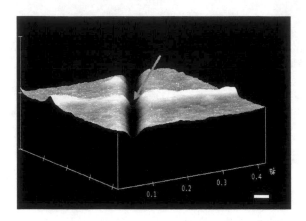

Figure 2: High resolution AFM image of the DNA molecule between the SWNT electrodes. Scale bar = 30 nm.

3.2 Electrical conductivity of single DNA molecule

The I-V characteristics of the double helix, hybridized in NaAc buffer, dried and measured at ambient and in high vacuum (10^{-5} Torr) conditions is shown in Figure 3 (a). At ambient conditions a current signal of 30 pA for a 1 V bias was observed while at high vacuum condition the signal decreased by 33 %. This decrease may be ascribed to the partial removal of water molecules from the proximity of the DNA. In fact, in ambient condition the proton-transfer process in the hydration layer surrounding the DNA promotes the electrical conductivity but diminishes in high vacuum [10].

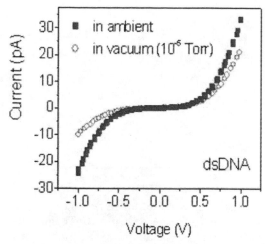

Figure 3: High resolution AFM image of the DNA molecule between the SWNT electrodes. Scale bar = 30 nm.

3.3 Influence of environmental factors on DNA conductivity

The structural and electronic properties of DNA molecule are significantly effect by its local environment. Several earlier reports have shown the strong influence of humidity on the conductivity of DNA molecules. However other environmental factors such as counterion variation, pH and temperature which have substantial effect on DNA conductivity were not studied in detail. The effect of ionic variation on the electrical conductivity of dry DNA molecule was studied by measuring I-V characteristics of dsDNA previously suspended either in sodium acetate (NaAc) or Tris(hydroxymethyl) aminomethane (TE). Figure 4 shows a comparison of I-V characteristics of hybridized DNA (in dry state) previously suspended either in NaAc or TE buffer.

A current signal in the range of 25-50 pA (at 1 V bias) was observed for the DNA molecule in the case of sodium acetate buffer. A nonlinear I-V characteristic was observed indicating a semiconducting behavior of the trapped DNA fragment encoding a specific gene but devoid of any periodic arrangement of the base pairs. In comparison, we observed that dsDNA in TE buffer exhibit almost two orders of magnitude higher current. To further investigate the cause of differences in magnitude, control experiments for both the buffer solutions were performed. A comparison of I-V characteristics after spotting a droplet of DNA-free NaAc and TE buffer, followed by drying in nitrogen stream showed that after drying, TE buffer exhibits about two orders of magnitude higher current as compared to that for sodium acetate. This observation is in accordance reports wherein a large current was observed for a similar control experiment with TE buffer[10]. Hence it is apparent that, unlike sodium acetate, the intrinsic conductivity of TE buffer strongly influences the measured conductivity of

DNA and thus can be misled as the intrinsic conductivity of the dsDNA even in the dry state.

Figure 5 (a) shows the influence of pH variation on the conductivity of the DNA molecule. After the I-V measurement of DNA molecule at pH 5.8, the substrates were washed with NaAc buffer (10 mM) with pH values from 3.5 to 9.3 using the same experimental process mentioned earlier. It can be observed that as the pH is increased there is gradual increase in current signal. After measuring conductivity at pH 9.3, I-V measurements were repeated by washing the devices with descending values of pH. Current signals were similar to the values obtained for the corresponding pH values during the increase of pH. A similar trend in pH dependence of DNA conductivity was reported by Lee et al [11]. Further studies have to be performed to understand the true mechanism of pH influence on DNA conductivity.

Figure 5 (b) depicts the influence of variation of temperature on the conductivity of the dsDNA molecule attached between the nanoelectrodes. As the temperature was increased from 25°C (in high vacuum) the current signal decreased gradually. We believe that eventual evaporation of water molecules from the hydration shell surrounding the DNA molecule, and subsequent change in DNA conformation became predominant factors in this case. Further increase in temperature above the melting temperature (T_M = 75.6°C) resulted in complete loss of signal possibly due to the thermal denaturation of DNA. I-V measurements after cooling the device to room temperature yielded current in the range of few pA ($\leq$10 pA) confirming the above assumption.

The influence of ionic strength of buffer on DNA conductivity was also studied. It was found that for buffer concentrations between 10 mM and 100 mM, the current signal varied from 40 - 100 pA. However at high salt concentration (>100 mM) there was a drastic increase in current signal. This could be attributed to the condensation of salt residue on DNA, which might have not been removed by the washing procedure followed in this work. On the other hand at very low ion concentration (<1 mM) the current signal diminished to few pA ranges indicating probable denaturation of dsDNA.

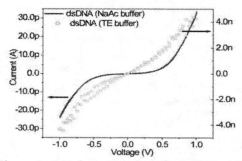

Figure 4 : *I-V* characteristics of a dsDNA molecule previously suspended either in NaAc and TE buffer. The two orders of higher magnitude in current signal observed for DNA in TE is attributed to the high intrinsic conductivity of the buffer

(a)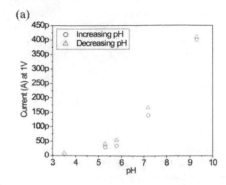

(b)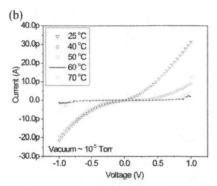

Figure 5 : (a) Effect of pH variation on conductivity of DNA measured at 1V bias. *I-V* of the same device washed with buffer solution of different pH values is shown both for gradual increase and decrease in pH. (b) Temperature effect on DNA conductivity. A gradual increase in the device temperature resulted in a diminishing current signal. Above the melting temperature of the DNA molecule (75.6 °C), there was no detectable signal. (c) Effect of ionic strength (NaAc) variation on DNA conductivity measured at 1V bias. At very low ionic strength (<1 mM), the signal diminished.

To confirm that the current signal was from the immobilized DNA molecule the following control experiment was performed. The devices were treated with DNase I enzyme (37°C, 30 min), followed by cleaning with NaAc and DI water. DNAse I enzyme cuts the DNA regardless of sequence. I –V measurements taken after the cleaning step showed no current signal (noise in fA range) indicating that the current signal obtained earlier was indeed from the DNA molecule bridging the SWNT electrodes.

4 CONCLUSION

In conclusion, we have developed a novel platform based on SWNT nanoelectrodes for directly probing the dc conductivity in DNA at the single molecule level. A n application of DEP in our system causes stretching of the DNA molecules and positioning them between the electrodes. Statistically, for majority of the devices, we observed a current value in the range of 25-40 pA when a dsDNA molecule bridges the SWNT electrodes. Influence of various local environmental factors on DNA conductivity was investigated. A reversible shift in the current signal was observed for pH variation. Influence of ionic strength on DNA conductivity was significant especially at very high (<100 mM) concentrations due to the condensation of salt residues. The present study demonstrates that SWNT can be employed as efficient nanoelectrodes for direct measurements of charge transport in DNA at the single molecule level.

REFERENCES

[1] Singh-Zocchi M, Dixit S, Ivanov V and Zocchi G 2003 *Proceedings of the National Academy of Sciences* 100 7605

[2] Roy S, Vedala H and Choi W 2006 *Nanotechnology* 17 S14-S18

[3] Vedala H, Huang J, Zhou X Y, Kim G, Roy S and Choi W B 2006 *Applied Surface Science* 252 7987

[4] Qi P, Javey A, Rolandi M, Wang Q, Yenilmez E and Dai H 2004 *J Am Chem Soc* 126 11774

[5] Guo X, et al. 2006 *Science* 311 356

[6] Kim D H, Huang J, Shin H K, Roy S and Choi W 2006 *Nano Lett.* 6 2821

[7] Roy S, et al. 2008 *Nano Lett.* 8 26

[8] Chen J, Hamon M A, Hu H, Chen Y, Rao A M, Eklund P C and Haddon R C 1998 *Science* 282 95

[9] Tuukkanen S, Toppari J J, Kuzyk A, Hirviniemi L, Hytonen V P, Ihalainen T and Torma P 2006 *Nano Lett.* 6 1339

[10] Kleine-Ostmann T, Jordens C, Baaske K, Weimann T, de Angelis M H and Koch M 2006 *Applied Physics Letters* 88 102102

[11] Lee J M, Ahn S K, Kim K S, Lee Y and Roh Y 2006 *Thin Solid Films* 515 818

Osteogenic Induction on Single-Walled Carbon Nanotube Scaffolds

Laura P. Zanello, Parul Sharma[#], Renzo Corzano, and Peter Hauschka[#]

Department of Biochemistry, University of California-Riverside, CA 92521, laura.zanello@ucr.edu;
[#]Department of Orthopaedic Surgery, Children's Hospital Boston, Harvard Medical School, Boston, MA 02115, peter.hauschka@childrens.harvard.edu

ABSTRACT

We demonstrated recently that CNT scaffolds support proliferation of mature osteoblasts as well as production of mineralized bone in vitro. More specifically, we showed that osteoblasts grown on single-walled (SW) CNTs retain plasma membrane electrical functions involved in secretory activities. Here we studied osteoinductive properties of SWCNT scaffolds as they support the growth of osteoblastic precursors, their differentiation into mature osteoblasts, and expression of membrane proteins involved in secretory processes. We used electrically neutral ("as prepared", AP-) SWCNTs, and SWCNTs chemically modified with carboxyl (-COOH, net negative electric charge), polyethylene glycol (PEG, electrically neutral), and poly-(m-amino-benzene sulfonic acid, PABS, net positive and negative electric charges) functional groups. We found that PEG-SWCNT and PABS-SWCNT showed the highest upregulation of ALP activity as a measure of osteoinduction. In addition, we found that AP-SWCNTs induced the expression of voltage-gated chloride channels ClC-3 and ClC-5 involved in secretory activities in hFOB osteoblasts. SWCNT preparations might be seen as osteoinductive materials with great potential for use in bone regeneration and repair.

1 INTRODUCTION

One main goal in bone tissue engineering is to create bone substitutes that will be incorporated successfully to native bones. Surgical procedures for bone repair utilize a variety of materials as prosthesis, grafts, and fillers. Common problems leading to multiple surgeries arise from quick degradation, poor contact, or rejection of the implant. Development of new biomaterials for bone tissue regeneration with improved biocompatibility, physical properties, and resistance to resorption is of crucial importance for the treatment of bone defects.

We propose the use of carbon nanotube (CNT)(1) scaffolds to stimulate the growth of osteoblasts and production of bone matrix in vitro(2). Production of bone materials is the result of secretory activities by osteoblasts, the bone-forming cells found on bone surfaces(3). In the fractured and wounded bone, successful repair and/or regeneration depends in large part on the capacity of osteoblast precursors to proliferate and fully differentiate into mature, secretory osteoblasts(4). In cases of large

affected areas, the use of an adequate scaffold material that facilitates osteoblast proliferation, differentiation and secretion is of crucial importance(5). Compatibility of the regenerated bone with the existing bone will be determined by the chemical composition, physical properties, and nanotopography of the scaffold used to induce bone formation.

The unique physical-chemical characteristics of carbon nanotubes (CNT) make them a potentially ideal material for applications in bioengineering(6). With the purpose to identify CNT preparations with osteoinductive properties, we measured the proliferation and differentiation of mouse MC3T3 E1 and human hFOB preosteoblasts cultured on glass coverslips coated with SWCNTs.

2 MATERIALS AND METHODS

2.1 Materials

Bone morphogenetic protein-2 (BMP-2), ascorbic aid, and β-glycerophosphate were obtained from Sigma.

2.2 Cell culture

MC3T3 E1 preosteoblastic cells we grown for 20 days in differentiating and mineralizing medium containing 10 mM β-glycerophosphate, 50 µg/ml ascorbic acid, and 100 ng/ml BMP-2 in DMEM (Sigma), with the addition of 5% fetal bovine serum (Sigma) and antibiotics, at 37°C in a 5 % CO_2 humidified incubator. Human FOB preosteoblast differentiation was induced in DMEM medium at 39°C. Cells were grown on glass coverslips and CNT-sprayed coverslips in 35-mm plastic culture dishes. The culture medium was changed every 3-4 days.

We studied osteoblast proliferation on CNTs in 5 day-old cultures. By day 5, control cells grown on glass reached confluence. Cell growth was calculated on the basis of number of cells per field of observation at a magnification of x 200 with an Olympus IX50 inverted microscope. Cell counts were performed with phase contrast and fluorescence microscopy.

2.3 CNT preparations

AP-SWNTs and nitric acid-treated SWCNTs (SWCNT-COOH) were obtained from Carbon Solution

Inc. (Riverside, CA), and SWCNTs chemically functionalized with poly-(*m*-aminobenzene sulfonic acid) (SWCNT-PABS) and poly ethylene glycol (SWCNT-PEG), were obtained by previously published synthetic methods.(7) SWCNT-COOH, SWCNT-PABS, and SWCNT-PEG were chosen for the present study on the basis of their net negative, zwitterionic, and neutral electric charge, respectively, at the pH of the experiment. Carbon nanotube coated glass coverslips were prepared as described previously.(8) Briefly, CNT samples (100 µg/mL) were sonicated in solvent (water for functionalized CNTs, and 95 % ethanol for AP-SWCNTs and AP-MWCNTs, multi-walled CNTs) for about 2 hours, and the resulting dispersion was sprayed onto pre-heated (ca. 80°C) glass coverslips. Sprayed coverslips were allowed to dry in air and used for cell culture after a sterilization procedure with UV irradiation overnight.

2.4 Alkaline phosphatise (ALP) activity

Cell cultures were assessed for total ALP expression using the pNPP substrate (Sigma) according to the manufacturer's protocol.

2.5 Chloride channel (ClC) expression

ClC expression was detected with standard Western blotting, using anti-ClC-3 and anti-ClC-5 antibodies from Santa Cruz Biotechnology.

3 RESULTS AND CONCLUSIONS

Osteoblast preparations reached confluency by day 4 (Fig. 1). We assessed osteogenic differentiation of MC3 E1 cells by measuring alkaline phosphatase (ALP) activ after 20 days in culture. We found that PEG-SWCNT a PABS-SWCNT showed the highest upregulation of A activity as a measure of osteoinduction. This suggests t electrically neutral and zwitterionic CNTs are bet inducers of bone formation than negatively charg scaffolds.

AP-SWCNT scaffolds as well as the control on plas dishes equally supported the growth of differentiated hF(osteoblasts for up to 7 days (Fig. 2). We found that A SWCNTs induced the expression of voltage-gated chlor channels ClC-3 and ClC-5 involved in secretory activities hFOB osteoblasts(9). We conclude that AP-SWCNT as w as some chemically modified SWCNT scaffolds posses osteoinductive properties that induce differentiation preosteoblasts. Electrically neutral AP-SWCr preparations support the proliferation of preosteoblas cells and their differentiation into mature osteoblasts, well as the expression of the molecular machinery involv in secretory activities.

SWCNT preparations might be seen as osteoinduct materials with great potential for use in bone regenerati and repair.

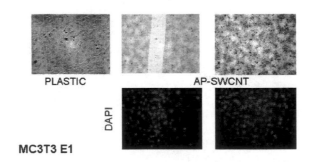

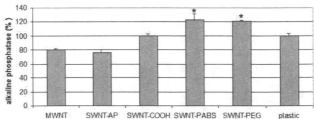

Figure 1. MC3T3 E1 osteoblasts cultured on AP-SWCNT and plastic as control. DAPI was used to stain cell nuclei. Alkaline phosphatase activity was measured with pNPP substrate on differentiated osteoblasts grown on different SWCNT preparations.

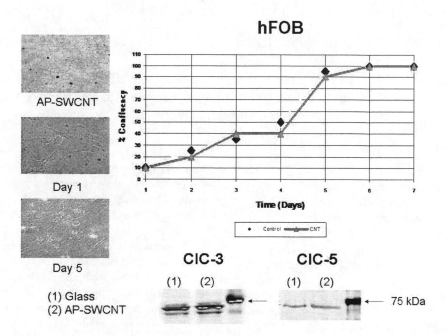

Figure 2. Growth curve of differentiated hFOB osteoblasts grown on plastic (control) and AP-SWCNTs. Western blots for the expression of chloride channels ClC-3 and ClC-5.

References

1. Ajayan PM 1999 Nanotubes from carbon. Chem Rev 99: 1787-1799

2. Zanello LP, Zhao B, Hu H, Haddon RC 2006 Bone cell proliferation on carbon nanotubes. Nano Lett 6: 562-567

3. Zanello LP, Norman AW 2004 Electrical responses to $1\alpha,25(OH)_2$-vitamin D_3 and their physiological significance in osteoblasts. Steroids 69: 561-565

4. Beck GR, Jr., Zerler B, Moran E 2001 Gene array analysis of osteoblast differentiation. Cell Growth Differ 12:61-83

5. Shea LD, Wang D, Franceschi RT, Mooney DJ 2000 Engineered bone development from a pre-osteoblast cell line on three-dimensional scaffolds. Biomaterials 6: 605-617

6. Baughman RH, Zakhidov AA, de Heer WA. 2002 Carbon nanotubes--the route toward applications. 297: 787-792

7. Zhao B, Yu A, Perea D, Haddon RC 2005 Synthesis and characterization of water soluble single-walled carbon nanotube graft copolymers. JACS 127: 8197-8203

8. Hu H, Ni Y, Montana V, Haddon RC, Parpura V 2004 Chemically functionalized carbon nanotubes as substrates for neuronal growth. 4: 507-511

9. Zanello LP 2006 Electrical properties of osteoblasts cultured on carbon nanotubes. IEE Micro and Nano Lett 1: 19-22

SWNTs Inhibit Normal Physiological Function of Calcium Ion Channels through Yttrium Release

L.M. Jakubek, J. Raingo, S. Marangoudakis, D. Lipscombe and R.H. Hurt

Brown University, Providence, RI, USA
Lorin_Jakubek@brown.edu, Jesica_Raingo@Brown.edu, Spiro@Brown.edu
Diane_Lipscombe@brown.edu, Robert_Hurt@brown.edu

ABSTRACT

The calcium ion channel is a voltage gated channel that enables calcium to enter electrically active cells. Proper function of these channels is essential for gene expression, neuronal excitability, muscle contraction and the release of neurotransmitters and hormones. As carbon nanotubes are electrically conductive and hydrophobic, they may interact with cell membranes and ion channels and adversely affect cells that rely on membrane potential for proper function. In this study, tsA201 cells were transfected with the CaV2.2 neuronal calcium ion channel and exposed to sulfonated SWNTs prior to electrophysiological characterization. As-produced SWNTs were found to strongly inhibit calcium channel function. Surprisingly, the same inhibitory dose response curve was seen in supernatant solutions after the nanotubes were removed by centrifugation. Careful study of this system reveals that the inhibitory effect is due to soluble yttrium ion that is mobilized from residual catalyst nanoparticles imbedded in these arc-synthesized SWNTs.

Keywords: SWNT, yttrium, calcium, ion channel

1 INTRODUCTION

Single walled nanotubes are being developed for various biomedical applications including cancer therapeutics, biomedical sensors and scaffolds for tissue engineering. The nanomaterials are well suited for these applications given their small aspect ratio, thermal and electrical conductance, and strength.

Films of carbon nanotubes have demonstrated capacitance in aqueous solutions(6) and the ability to conductively stimulate neural cells grown upon them(3) making carbon nanotube films a promising scaffold or implant for healing brain related injuries. Additionally, the surface texture provided by carbon nanotube composites results in increased cell adhesion and subsequently the potential for more durable prosthetics. However, the qualities that make these materials valuable for these applications may also result in unique and potentially detrimental interactions with biological systems if the nanomaterials are released from the films or composites.

In cell studies on electrically active cardiomyocytes, short-term assays suggest that SWNTs are biocompatible, however long term studies indicate that the SWNTs bind to the membrane of the cells resulting in a physically induced change in cell morphology, altered proliferation and increase in cell death.(2) Given the nanotubes' affinity for the cellular membrane and the potential hydrophobic interactions that may result combined with the conductive capabilities of the nanomaterial, it is possible that the carbon nanotubes may have a unique effect on cells which rely on membrane potential for proper function.

In this paper we investigate the interaction of SWNT with electrically active cells with the specific aim of determining the direct effect of SWNTs on the calcium ion channel. The calcium ion channel is a voltage gated channel that enables calcium to enter when the cell membrane undergoes depolarization. Proper function of these channels is essential for gene expression, neuronal excitability, muscle contraction and the release of neurotransmitters and hormones. In order to make this assessment, tsA201 cells, a modified human embryonic kidney (HEK) cell line, were transfected with the CaV2.2 neuronal calcium ion channel and exposed to aryl-sulfonated SWNTs prior to electrophysiological characterization. Metal leaching data was obtained in order to assess the concentrations of soluble metals in each sample and a comparison was drawn between the electrophysiological effect of the SWNTs and that of the supernatant. Based upon these experiments, it was determined that release of yttrium from catalyst residues in the SWNT is the main mechanism behind calcium channel inhibition.

2 MATERIALS

Preparation of SWNT: As Produced (AP) and vendor purified SWNTs were obtained from Carbolex. In order to suspend the hydrophobic SWNT in the cell media, the SWNTs were made hydrophilic through aryl-sulfonation(11). SWNTs were immersed in 8.4mM sulfanilic acid solution at 70ºC. While maintaining constant temperature and agitation, 1.5mL of .2M sodium nitrite solution was added and allowed to incubate for 2 hours. The SWNTs were subsequently washed with distilled water six times and dried at 100 ºC for 8 hours. Aryl-sulfonated SWNTs were then suspended in external solution (see below) through bath sonication.

NSTI-Nanotech 2008, www.nsti.org, ISBN 978-1-4200-8503-7 Vol. 1

Transient expression of $Ca_V2.2$ calcium channels in tsA201 cell line. Calcium channel subunits $Ca_V2.2$ $Ca_V2.2e[37b]$ (AF055477(5)) together with $Ca_V\beta_3$ (sequence homologous to M88751), $Ca_V\alpha_2\delta_1$ (AF286488((4)), and enhanced green fluorescent protein cDNAs (eGFP; BD Bioscience) were transiently expressed in tsA201 cells as described previously using Lipofectamine 2000 (Invitrogen)(10).

Electrophysiology: We performed standard whole cell patch clamp recording as described previously(10). External solution contained: 1 mM $CaCl_2$, 4 mM $MgCl_2$, 10 mM HEPES, 135 mM choline chloride, pH adjusted to 7.2 with CsOH. Control internal solution contained: 126 mM CsCl, 10 mM EGTA, 1 mM EDTA, 10 mM HEPES, 4 mM MgATP, pH 7.2 with CsOH. Recording electrodes had resistances of 2-4 MΩ when filled with internal solution and were coated with Sylgard (Dow Corning) to reduce capacitance. Series resistances (< 6 MΩ for whole cell recording) were compensated 70-80% with a 10 μs lag time. Calcium currents were evoked by voltage-steps and currents leak subtracted on-line using a P/-4 protocol. Data were sampled at 20 kHz and filtered at 10 kHz (-3 dB) using pClamp V8.1 software and the Axopatch 200A amplifier (Molecular Devices). All recordings were obtained at room temperature. Cells were typically held at -100 mV to remove closed-state inactivation(10). Test potentials 20-25 ms in duration were applied every 6 seconds.

Metal Mobilization: Metal mobilization assays were performed by sonicating aryl-sulfonated SWNTs in a bath sonicator for 2 hours in external solution. The suspensions were then transferred to a 5000 NMWL Amicon Ultra Centrifuge tube (Millipore, MA) and centrifuged at 4500 RPM and 4°C for 30 min. The resultant supernatant contained soluble metals. Metal content was determined through Inductively Coupled Plasma Atomic Emission Spectroscopy (ICP).

3 RESULTS AND DISCUSSION

Voltage-gated calcium ion channels are essential for coupling membrane depolarization to an increase in intracellular calcium. They control a large number of essential calcium-dependent cellular events including muscle contraction, neurotransmitter release, hormone release, gene expression, neurite extension, and overall cell excitability. CaV2-type calcium channels are prototypic presynaptic calcium channels which control calcium entry that triggers neurotransmitter release. Synaptic transmission in all parts of the nervous system therefore relies on normal functioning of CaV2 channels.

Upon first exposure to the aryl-sulfonated AP SWNTs, the tsA201 cells expressing the Ca V2 neuronal calcium channel exhibited signs of extreme and rapid channel inhibition through electrophysiological assays. Surprisingly, after the nanotubes were removed through centrifugation, the cells demonstrated the same channel

inhibitory response. This suggested that the primary mechanism was related to changes in the fluid medium caused by the release of some substance from the nanotubes, or depletion of some substance through adsorption on or destruction by the nanotubes. Auxiliary experiments were needed to test this hypothesis. The SWNTs utilized in this study are made through an arc-synthesis procedure that relies on the incorporation of nickel and yttrium catalytic components to the original graphite target. Based on our previous work with nickel (8,7), we hypothesized that metal may be mobilized into solution from these imbedded catalyst residues in concentrations sufficient for channel inhibition. In order to ascertain the concentration of nickel and yttrium present in the extracellular solution an ICP based bioavailability assay was conducted. In this assay, solutions of purified and AP SWNTs were suspended in external solution in concentrations typical of cell studies (i.e. 5-100 μg SWNT/ml external solution.) The samples underwent centrifugation to remove the nanotubes leaving the supernatant containing the bioavailable metals. Inductively coupled plasma atomic emission spectroscopy (ICP) was then used to simultaneously analyze the supernatants for nickel and yttrium content. ICP is accurate at low concentrations down to 10 PPB.

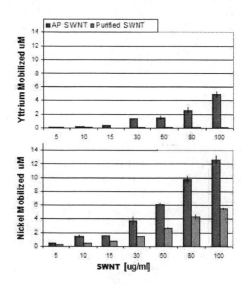

Figure 1. Dose-dependent release of metals from SWNTs in external solution. Concentration of nickel and yttrium available in the extracellular media (μM) as a function of the concentration of AP and purified SWNTs (μg/ml) present in suspension before centrifugation.

The metal mobilization results presented in Figure 1 indicate that the AP SWNTs contain soluble nickel and yttrium, whereas the purified SWNTs only contain a reduced amount of nickel. The combination of sulfonation and vendor purification removed roughly 60% of the bioavailable nickel and all of the bioavailable yttrium. In addition to defining the metal composition of the supernatant, this information also enables the generation of metal salt solutions commensurate with the concentrations leached from the SWNTs.

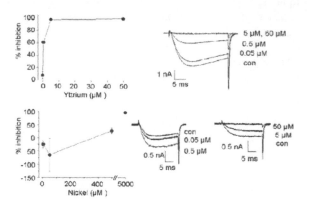

Figure 2: Calcium channel inhibitory response generated from the addition of varying concentrations of nickel and yttrium salts (left) and representative current-voltage curves (right).

In order to simulate and isolate the individual effects of CNT nickel and yttrium release on calcium channel dynamics, current-voltage analyses were conducted on cells exposed to varying amounts of fully soluble nickel and yttrium salts respectively. The results, presented in Figure 2, demonstrate that yttrium is a significantly more effective channel blocker than nickel. Yttrium produces a 50% inhibition of CaV2 currents at 0.5 µM and completely blocks at 5 µM. Nickel was about 100-fold less sensitive, 500 µM inhibited 30% of current and 5000 µM induced 100% inhibition. These results are comparable to previously published data. (9,1)

It is possible for the supernatant to contain another component in addition to the metals that may interact with the channel. In order to address this concern, physiological time course traces were performed on a solution of purified SWNTs. In this assay, the supernatant, the solution containing the purified SWNTs and a nickel salt solution (commensurate with the concentration of metal leaching from the purified SWNTs as determined by ICP) are serially added while current readings are recorded from the cell undergoing a uniform voltage step. If any of these components is an inhibitor than the inward current (pA) would decrease resulting in a sloping line. If, however, these components are not inhibitory to the channel, then a horizontal line representing no change in inward calcium current would result. The assay was performed in triplicate and presented in Figure 3.

The three traces presented in Figure 3, are recorded from three independent cells. While all three traces are relatively horizontal, with minor wavering caused by the physical action of changing the samples, the third appears to have a slight slope. This slope is steady across the samples and is considered to be due to cellular run down.

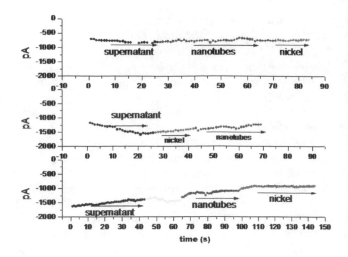

Figure 3. Electrophysiological effect of serial addition of supernatant, SWNTs and nickel salt solutions. The three traces are recorded from three individual cells. The y-axis is pA of inward calcium current under constant voltage step. The arrows indicate when each sample was applied to the cell. Samples included 100 µg/ml purified SWNTs, associated supernatant and 5.5µM nickel salt solution all in extracellular solution.

The cell requires nutrients in order to open channels. When a cell is "run down" these nutrients are depleted, resulting in fewer open channels and a decrease in inward calcium current.

It can therefore be concluded that the high concentration of purified SWNTs, the associated supernatant and the commensurate nickel salt solution do not have any appreciable inhibitory effect on the calcium ion channel. Additionally, the similarity in inhibitory response among the supernatant and the nickel salt solution at such a high concentration of SWNTs suggests that there is no previously unforeseen element in the supernatant that is acting independently of the known metals to cause inhibition.

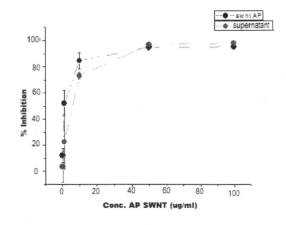

Figure 4 Dose response curve for AP SWNTs and the corresponding supernatant on calcium channel inhibition.

A dose response curve was generated incorporating samples of AP SWNTs and the associated supernatant (Figure 4). The channels are completely blocked by 50ug/ml AP SWNTs. The amount of metal leached from this sample is 6.1µM Ni and 1.5 µM Y. Based on the results presented in Figure 2, 5,000 µM Ni would be required for complete channel block whereas 0.5 µM Y would be required. Therefore, it can be concluded that the inhibitory response observed from the AP SWNT samples is caused essentially entirely by yttrium release. The result emphasizes the need to carefully investigate the mechanistic origin and material basis for biological impacts of complex nanostructures.

4 ACKNOWLEDGEMENT

We would like to thank Xinyuan Liu, David Murray and Joseph Orchardo for their experimental assistance. This work supported by NSF NIRT 0506661 and the Brown Superfund Basic Research Project.

REFERENCES

1. Beedle, A. M. and Zamponi, G. Inhibition of Transiently Expressed Low- and High- Voltage-Activated Calcium Channels by Trivalent Metal Cations. Journal of Membrane Biology. 187, 225-238, 2002.

2. Garibaldi, S., Brunelli, C., Bavastrello, V., Ghigliotti, G., and Nicolini, C. Carbon nanotube biocompatibility with cardiac muscle cells. Nanotechnology. 17, 391-397, 2006.

3. Gheith, M.K., Pappas, T.C., Liopo, A.V., Sinani, V.A., Shim, B. S., Motamedi, M., Wicksted, J.P., and Kotov, N.A. Stimulation of Neural Cells by Lateral Currents in Conductive Layer-by-Layer Films of Single-Walled Carbon Nanotubes. Advanced Materials. 18, 2975-2979, 2006.

4. Lin, Y., McDonough, S.I., and Lipscombe, D. Alternative Splicing in the Volateg-Seneing Region of N-Type Cav2.2 Channels Modulates Channel Kinetics. Journal of Neurophysiology. 92, 2820-2830, 2004

5. Lin, Z., Haus, S., Edgerton, J., and Lipscombe, Diane. Identification of Functionally Distinct Isoforms of the N-Type Ca 2+ Channel in Rat Sympathetic Ganglia and Brain. Neuron. 18, 1, 153-166, 1997.

6. Liu, H. and Zhu, G. The electrochemical capacitance of nanoporous carbons in aqueous and ionic liquids. Journal of Power Sources. 171, 1054-1061, 2007.

7. Liu, X., Guo, L., Morris, D., Kane, A.B., and Hurt, R.H. Targeted Removal of Bioavailable Metal as a Detoxification Strategy for Carbon Nanotubes. Carbon. in press 2008.

8. Liu, X., Gurel, V., Morris, D., Murray, D.W., Zhitkovich, A., Kane, A.B., and Hurt, R.H. Bioavailability of Nickel in Single-Wall Carbon Nanotubes. Advanced Materials. 19, 2790-2796, 2007.

9. Nachshen, D.A. Selectivity of the Ca Binding Site in Synaptosome Ca Channels. Journal of General Physiology. 83, 941-967, 1984.

10. Thaler, C., Gray, A.C., and Lipscombe, D. Cumulative inactivation of N-type Cav2.2 calcium channels modified by alternative splicing. Proceedings of the National Academy of Sciences. 101, 15, 5675-5679, 2004.

11. Yan, A., Xiao, X., Külaots, I., Sheldon, B.W., and Hurt, R.H. Controlling water contact angle on carbon surfaces from 5° to 167°. Carbon. 44:3113-3148, 2006.

Carbon nanotube superlattices- An oscillatory metallic behaviour

B. K. Agrawal and A. Pathak

Physics Department, Allahabad University, Allahabad - 211002, India.
E-mail : balkagl@rediffmail.com, balkagr@yahoo.co.in

ABSTRACT

We study the structural, electronic and optical properties of the (n,n)/(2n,0); n = 3 and 6 superlattices of the carbon nanotubes (CN's) by employing the first–principle pseudopotential method within density functional theory (DFT) in generalized gradient approximation (GGA). The role of the enhanced size of the superlattice unit cell on the electronic and optical properties has been investigated.The curvature effects on the various properties are also investigated. The heterojuction of the small diameter n(3,3)/n(6,0) superlattice which possesses a three-fold rotational symmetry exhibits an oscillatory behavior in terms of the fundamental energy band gap which vanishes whenever the integer 'n' is a multiple of 3. A similar behaviour having a periodicity of six may be observed in the case of the large diameter n(6,6)/n(12,0) superlattice whose heterojunction reveals a six-fold symmetry. The electronic structure and optical absorption of a superlattice are quite different from those of its constituent carbon nanotubes. The present results obtained after employing all the s-, p- and d-orbitals of the atoms are quite different from the findings of the earlier workers who have employed a phenomenological tight binding formulation considering only one π orbital or four orbitals. We find that most of the states are extended resonance states and are quite delocalized in contrast to the earlier finding of the occurrence of the completely localized states in the sections of the constituent nanotube. The metallic superlattices exhibit a high density of states (DOS) at the Fermi level (E_F). For the large diameter n(6,6)/n(12,0) superlattices, the electron energy gap vanishes for n=1 and 2 but for n=3 increases up to a maximum value and decreases thereafter for larger 'n', a result which is in disagreement with the earlier workers. These new facts have not been reported in the literature.
Keywords: Carbon nanotubes, electronic structure, optical absorption.

INTRODUCTION

The present day need of devices miniaturization in nanoelectronics can be met by employing the molecules as functional devices. The promising candidates are the single-walled carbon nanotubes (SWCNT's) which behave like one-dimensional metals or semiconductors.

The metallic and the semiconducting carbon nanotubes (CNT's) have been used to create single electron transistors [1,2] and the field effect transistors, respectively. Different kinds of deformations or defects in single nanotubes [3,4], junctions between the materials and the CNT's [5], junctiuon between different nanotubes are the basis in many of these applications.

We study the role of the length of the superlattice unit cell on the electronic and optical properties of small and large diameters (n,n)/(2n,0); n = 3 and 6 superlattices of the (CNT's). The curvature effects on the various properties are also discussed.

CALCULATION AND RESULTS

For calculations, the pseudopotential and the plane waves within the DFT are employed [6]. We consider the exchange correlation potential of Perdew et al [7] in GGA which is generated by the FHI code [8].

n(3,3)/n(6,0) superlattices

The junctions built of the armchair (3,3) and the zigzag (6,0) tubes are connected by a ring of three sets of the pentagon and heptagon defects. One observes a three-fold rotational symmetry at the junction.

In our future discussion, we name E = 0.0 as the Fermi energy which is the correct one for the metallic systems but not true for the semiconducting systems where it appears in the band gap.

The electronic structure and the DOS for the n(3,3)/n(6,0) superlattices for n = 1- 9 have been shown in Figs. 1 and 2. The change in the scale of DOS with 'n' should be noted.

The electronic structure and DOS for the smallest period 1(3,3)/1(6,0) superlattice [Fig. 1(a)] are totally different from those of its constituent nanotubes. Two hybridized (s-p-d-orbitals) conduction states descend into the valence band and make the superlattice metallic. Two valence states are quite flat in the vicinity of the boundary of the Brillouin zone (BZ) that gives rise high DOS near E_F.

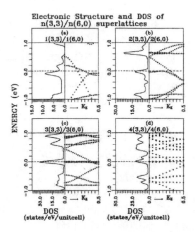

Electronic Structure and DOS of
n(3,3)/n(6,0) superlattices

Fig. 1 *Electronic structure and density of electron states (DOS) for the n(3,3)/n(6,0) superlattices for n=1 to 4. . The Fermi level for each nanotube has been fixed at the valence band maximum and is taken as the origin of the energy.*

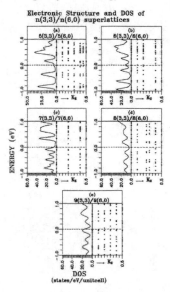

Electronic Structure and DOS of
n(3,3)/n(6,0) superlattices

Fig.2 *Same as for Fig.1 but for the n(3,3)/n(6,0) superlattices for n=5 to 9*

.

For the 2(3,3)/2(6,0) superlattice [Fig. 1(b)], one conduction state originates at $k_z = 0.0$ and descends into the valence band making the superlattice conducting. The electronic structures of 3(3,3)/3(6,0) and 4(3,3)/4(6,0) superlattices are quite similar [Fig. 1(c), Fig. 1(d)] and are metallic.

In 5(3,3)/5(6,0) superlattice [Fig. 2(a)], the valence and conduction state at E_F show a direct small band gap of 51 meV. On the other hand, for the 6(3,3)/6(6,0) superlattice [Fig. 2(b)], a valence state crosses the E_F and the superlattice is metallic. For n = 7 and 8 superlattices, again small gaps appear. Finally for the 9(3,3)/9(6,0) superlattice is seen to be metallic.

We do not find any pure s- or p-state as reported by Jaskolski and Chico [9].

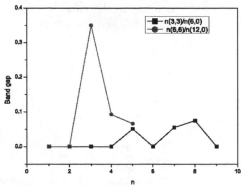

Fig. 3 *Variation of the electron energy band gap(eV) with 'n' in the various superlattices. The curves marked by squares and dots depict the band gaps for the small diameter n(3,3)/n(6,0) and large diameter n(6,6)/n(12,0) superlattices, respectively.*

We find that most of the states observed in the electronic structures of the various small diameter n(3,3)/n(6,0) superlattices are s-p hybridized states containing the major contributions from the p-orbitals in the vicinity of the E_F. In some states, the s-type orbitals dominate over the p-type orbitals and show large dispersion. We do not find any pure s or pure p-state. Further, we do not find the localized states as seen by Jaskolski and Chico [9]. The variation of the band gap with 'n' has been depicted in Fig. 3.

The small diameter n(3,3)/n(6,0) superlattices are metallic for n = 1 to 4. The band gap is opened up for n ≥ 5 except when the index 'n' is a perfect multiple of 3. Thus, a notable feature seen for the n(3,3)/n(6,0) superlattices is that for a value of 'n' which is an exact multiples of 3 i.e., the 3(3,3)/3(6,0), 6(3,3)/6(6,0) and 9(3,3)/9(6,0) superlattices are metallic. The small diameter n(3,3)/n(6,0) superlattice, thus, exhibit an oscillatory behavior in terms of the fundamental energy band gap. The gap vanishes whenever the integer 'n' is a multiple of 3. A high DOS at the E_F is observed in these superlattices.

n(6,6)/n(12,0) superlattices

The (6,6) and (12,0) nanotubes are connected by a ring of six pairs of pentagon/ heptagon defects which have also been studied earlier by Jaskolski and Chico [9].

A minimum energy is observed for the 3(6,6)/3(12,0) superlattice which as we will see later shows a semiconducting behaviour.

The electronic structure and the DOS for the n(6,6)/n(12,0) superlattices (n= 1 to 4) are depicted in Fig.4.

In the electronic structure of the shortest period 1(6,6)/1(12,0) superlattice [Fig. 4(a)], the crossing of the valence and conduction states at E_F is retained and the superlattice is metallic.

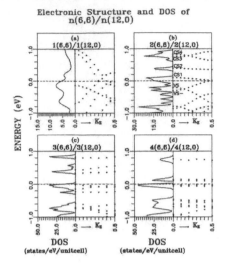

Electronic Structure and DOS of n(6,6)/n(12,0)

Fig. 4 *Same as for Fig.1 but for the n(6,6)/n(12,0) superlattices for n= 1 to 4. In Fig. 4(b), the highest occupied valence state is denoted by VS and the one below the VS is named as VS- . CS1, CS2, CS3 and CS4 depict the different conduction states.*

The electronic structure and DOS of the larger period 2(6,6)/2(12,0) superlattice are shown in Fig. 4(b). The top most filled doubly-degenerate state at $k_z = 0.0$ named as VS in Fig. 4(b) descends into the lower part of the valence band whereas the next valence state, VS- rises and crosses the E_F at the boundary of the BZ making the superlattice metallic. The first doubly-degenerate flat conduction state nearest to the E_F named as CS1 in Fig. 4(b) descends with the wave vector and crosses the E_F at the boundary of the BZ. This flat state shows a dispersion of only 0.1 eV and incurs a high DOS in the vicinity of the E_F. The two doubly-degenerate valence VS and the conduction CS1 states are widely separated at the center of the BZ but approach each other for the larger values of the wave vector. The next singlet conduction state called as CS2 in Fig. 4(b) is very flat and lies at about 0.4 eV above the E_F and shows dispersion of 39 meV only. Going up in the conduction band region, the next states occur at about 0.9 eV which are marked as CS3 and CS4 at $k_z = 0.0$, respectively. They are interchanged before the BZ boundary at $k_z = 0.5$. These conduction states have also appreciable contributions of the s-like orbitals and show therefore comparatively larger dispersion.

Out of them, the state CS3 is the doubly-degenerate one whereas the state CS4 is a singlet.

The wavefunctions are comparatively large at the heterojuction in the (12,0) section of the superlattice but fluctuate inside both the nanotubes. This variation in the wavefunction near the heterojunction reveals the resonance character of the wavefunctions. These resonance wavefunctions are broad in the case of the valence VS and the lowest conduction CS1 states. However, for the higher conduction state CS2, the resonance may be considered as sharp resonance. These resonant wavefunctions will give rise to non-zero electric conduction.

The variation of the band gap with the superlattice period for the n(6,6)/n(12,0) superlattices has already been shown in Fig. 3. For n = 3, there appears a large band gap. It may be noted that the energy of this superlattice is minimum and this superlattice is the most favoured one.

The optical absorption for the n(3,3)/n(6,0), n = 1 to 3 superlattices is depicted in Figs. 5(a), 5(b) and 5(c), respectively. For comparison, we have included in the figures the optical absorption of the constituent nanotubes.

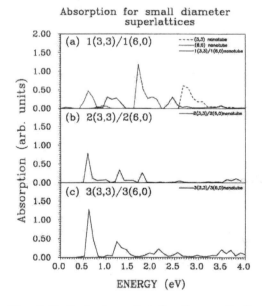

Absorption for small diameter superlattices

Fig. 5 *Optical absorption for the small diameter 1(3,3)/1(6,0), 2(3,3)/2(6,0) and 3(3,3)/3(6,0) superlattices denoted by the continuous curves. The optical absorption for the infinitely long constituent (6,0) and (3,3) nanotubes has been denoted by the dotand dashed curves, respectively.*

The optical absorption obtained for the infinitely long (3, 3) nanotube formed by a single unit cell [Fig. 5(a)] shows a broad peak at about 2.82 eV. The optical absorption of the infinitely long (6,0) nanotube of the single unit cell contains three peaks at 0.61, 1.70 and 2.0 eV, the peak at

NSTI-Nanotech 2008, www.nsti.org, ISBN 978-1-4200-8503-7 Vol. 1

1.70 eV being the strongest one. No experimental data is available for comparison

The optical absorption of the infinitely long 1(3,3)/1(6,0) superlattice is totally different from those of its constituent nanotubes. There appear only three main peaks at 1.03, 1.23 and 2.45 eV at the energy locations, which are totally different from those of the constituent nanotubes.

For the 2(3, 3)/2(6, 0) long period superlattice, there appears appreciable absorption in the energy range, 0.5- 2.0 eV and the weak absorption above it. There occurs broad absorption showing three peaks at 0.60, 1.30 and 1.80 eV which are located at different energies from those of the small period 1(3,3)/1(6,0) superlattice. An absorption peak at 0.60 eV was also observed in the case of the isolated 1(6,0) nanotube.

In the optical absorption of the larger period 3(3, 3)/3(6, 0) superlattice, one observes continuous absorption up to 4 eV with several peaks in different energy regions. The two lowest energy peaks seen earlier at 0.60 and 1.30 eV in the optical absorption of the 2(3,3)/2(6,0) superlattice appears here also approximately at the same energy locations (0.63, 1.24 eV). The other high-energy peaks occur at 2.16, 3.49 and 3.70 eV. Again, the absorption of the superlattice is quite different from its parent nanotubes.

One observes that in the optical absorption of the larger period 2(3, 3)/2(6, 0) and 3(3, 3)/3(6, 0) superlattices, two lowest peaks appear at quite similar energy locations. This behaviour may also be observed in the larger period n(3,3)/n(6,0) n> 3 superlattices.

The optical absorption of each superlattice is different and it may, thus be used to characterize the lattice parameter of a superlattice.

Conclusions

The electrons having the weakly localized character or the extended states will mean the occurrence of mobile delocalized electrons which will lead to appreciable electrical conduction in the metallic heterojunctions of the carbon nanotubes.

In the case of the large diameter n(6,6)/n(12,0) superlattices, for n = 1 and 2, the band gap is zero and for n > 2, the band gap increases up to a certain limit but decreases thereafter, a finding not reported so far. These superlattices might also reveal an oscillatory behavior in terms of the fundamental energy band gap as seen in the small diameter superlattices. The optical absorption for the superlattice is seen to be completely different from those of the constituent nanotubes. Each superlattice of a particular periodicity shows its own characteristic absorption.

In a single uniform nanotube kinks have been seen to form the metal-semiconductor junction showing rectifying behaviour [10]. With the use of heterojunctions formed between the carbon nanotubes of different orientations and sizes, one may obtain several types of junctions for their applications in the development of SWNT devices.

REFERENCES

[1]. S. J. Trans et al., Nature **386**, 474 (1997).

[2]. M. Bockrath et al, Science **275**, 1922 (1997).

[3]. J. -Q Lu, J.Wu, W.Duan, and B.-L Gu, Appl. Phys. Lett. **84**, 4203 (2004).

[4]. A. G. Petrov and S. V Rotkin, Phys. Rev. B **70**, 035408 (2004).

[5]. M. Terrones, F. Banhart, N. Grobert, J.-C. Charlier, H. Terrones, and P. M. Ajayan, Phys. Rev. Lett. **89**, 075505 (2002).

[6]. ABINIT code is a common project of the University Catholique de Louvain, Corning incorporated and other Contributors.

[7]. J. P. Perdew, K. Burke and M. Ernzerhof, Phys. Rev. Lett. **77**, 3865 (1996).

[8]. M. Fuchs and M. Scheffler, Comput. Phys. Commun. **119**, 67 (1999).

[9]. W. Jaskolski and L. Chico, Phys. Rev.B **71**, 155405 (2005).

[10]. Z. Yao, H. W. C. Postma, L. Balents and C. Dekker, Nature **402**, 273 (1999).

4

AQUEOUS DISPERSION OF SINGLE-WALLED CARBON NANOTUBES BY USING SELF-DOPED POLYANILINE AS DISPERSANT

J. H. Hwang[*], H.-J. Lee[**], M. S. Hong[***], H. Lee[***], and M.-H. Lee[*,****]

[*]Chonbuk National University, Jeonju, Korea, jhh3945@naver.com
[**]Korea Basic Science Institute, Jeonju Center, Jeonju, Korea, hajinlee@kbsi.re.kr
[***]Jeonju University, Jeonju, Korea, haeseong@jj.ac.kr
[****]Chonbuk National University, Jeonju, Korea, mhlee2@chonbuk.ac.kr

ABSTRACT

Single-walled carbon nanotubes (SWCNTs) functionalized with poly(aniline N-butylsulfonate) (PAnBuS), a water-soluble self-doped polyaniline, were prepared by mixing solutions in ex-situ. The interaction between poly(aniline N-butylsulfonate) (PAnBuS) and SWCNTs was investigated by XPS and UV-vis spectroscopy. The morphology of SWCNTs-PAnBus nanocomposites was investigated by FE-SEM. The SWCNTs-PAnBuS dispersion was exhibited long-term stability in aqueous environments. The sheet resistance of SWCNTs-PAnBuS thin film on a PET substrate was 1~2 x 10^4 Ω /sq using four-point probe measurement.

Keywords: self-doped polyaniline, single-walled carbon nanotubes, dispersion

1 INTRODUCTION

Since the discovery of single-walled carbon nanotubes (SWCNTs) in 1991, they have attracted great interest due to their high aspect ratio, small diameter, light weight, and high Young's modulus.[1,2] However, poor solubility of SWCNTs in both aqueous and common organic solvents is a major hindrance for their potential applications. The dispersion and dissolution of SWNTs are recognized to be an essential step in utilizing them. To obtain the good dispersion, various techniques have been developed to debundle SWCNTs. Recently, carbon nanotubes-polymer composites by using *in-situ* polymerization of inherently conducting polymer (ICP) such as polyaniline have been reported to enhance a dispersion property of carbon nanotubes and allow their new electrical applications.[3,4] Due to their unique properties ICPs have been widely stuied as a working component in many fields including rechargeable batteries, light emitting diodes, transistors, molecular sensors, nonlinear optical devices, corrosion protection, and electrochromic display. However, from a practical view point the *in-situ* method has limitations in controlling the ratio of both components easily and in completing functionalizing of SWCNTs in the molecular level. Among ICPs, self-doped polyanilines can be a good candidate to enhance the dispersibility of SWCNT in aqueous media because of their water solubility, electroactivity, conductivity, and redox activity in a wider pH range.[5] Previously, Kim et al reported the synthesis of poly(aniline N-butylsulfonate) (PAnBuS) *via* either by electrochemical or chemical oxidations of 4-anilino-1-butane sulfonic acid monomer.[6] The PAnBuS shows reversible electrochemical redox behavior, moderate conductivity in the range of ~10^{-3} S/cm, and most of all, excellent solubility in water.

In this paper, a water-soluble self-doped polyaniline, poly(aniline N-butylsulfonate) (PAnBuS), is going to be introduced to disperse SWCNTs. (Figure 1) The SWCNTs functionalized with PAnBuS are expected to exhibit high dispersibility in water, and to form a stable and homogeneous dispersion. Preparation of transparent conductive films (TCF) from the aqueous dispersion of SWCNTs functionalized with PAnBuS will be attempted.[7] Electrical conductivity and other physical properties of the resulting conductive film will be also discussed.

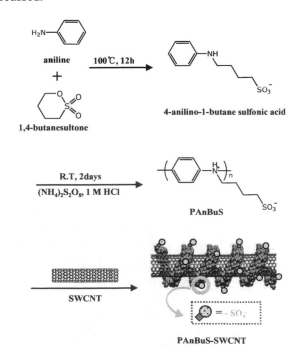

Figure 1 : Preparation route for PAnBuS polymer and a dispersion process of SWCNTs using PAnBuS.

2 EXPERIMENTAL

2.1 Materials

The purified SWCNTs were purchased from Carbon Nanotechnologies Inc. (HiPco SWCNTs, Lot #PO304) and used without further purification. Aniline (Aldrich), 1,4-butane sultone (99+%, Aldrich) and ammonium persulfate (98+%, Aldrich) were used as received.

2.2 Synthesis

A water-soluble self-doped polyaniline, poly(aniline N-butylsulfonate) (PAnBuS) was synthesized by the electrochemical oxidations of aniline N-butylsulfonate, which was readily obtained from aniline and 1,4-butane sultone following a reported procedure.[6] The aniline monomer (0.6 mg) was added to a solution of 1 M HCl (35 %, 0.45 mL) and stirred. Ammonium persulfate (0.66 g) dissolved in deionized water (2.5 mL) solution was drop-wisely added to the aniline/HCl solution with stirring After a few minutes, the dark suspension became green, indicating polymerization of aniline has been started. The solution was further stirred for 48 h at room temperature. The dark-green PAnBuS product was purified by dialysis in deionized water for 3 days, and dried under vacuum for 48 h at 40 ℃.

2.3 Dispersion of SWCNTs

As-received SWCNTs were placed in an aqueous solution of PAnBuS, followed by sonication in a bath-type sonicator (300 W, Power Sonic 405, Hwashin Tec.) for 2 h at room temperature. We also isolated solid products of SWCNTs-PAnBuS nanocomposite in order to characterize using XPS and SEM. An aqueous dispersion of SWCNTs-PAnBuS was filtered under 0.2 µm pore membrane filter and sufficiently rinsed with deionized water until the filtrate was colorless. The collected solids were dried under vacuum for 48 h at 60 ℃, and stored in a closed container minimizing exposure to moisture in the air.

2.4 Characterization

The UV-vis spectrum of SWCNT dispersed in aqueous PAnBus solution was recorded using Scinco Co., S-1500 and the result was compared with that of aqueous PAnBus solution itself. Measurements of the sheet resistance were carried out by four-point probe method (Keithley, 3220/2182A) at room temperature.The morphology studies of the SWCNTs-PAnBuS were performed with field emission scanning electron microscopy (FE-SEM, HITACHI S-4800, Japan) operating at 30 kV and the samples for the measurement were prepared as follows: bath sonicate SWCNTs-PAnBus in methanol for 20 min,

place one drop of the suspension on the silicon wafer and dry on the hot plate for 80 ℃ for 10 min. The X-ray photoelectron spectroscopy (XPS) was conducted on an AXIS-NOVA (Kratos Inc.) at using monochromatized Al $K\alpha$ radiation ($h\nu = 1486.6$ eV).

3 RESULTS AND DISCUSSION

A poly(aniline N-butylsulfonate) (PAnBuS) has been successfully synthesized by the electrochemical oxidations of aniline N-butylsulfonate. The dispersion of SWCNTs in water containing PAnBuS was performed by using a typical procedure as descried in the experimental section. We observed the dispersed SWCNTs in aqueous PAnBuS solution had not been aggregated even for a month. The pictures of as-received SWCNT and SWCNT-PAnBuS dispersed in water shown in Figure 2 were taken after 30 days since they had been treated under ultrasonic agitation for 1 hour.

Figure 2 : Aqueous solutions of (a) as-received SWCNTs and (b) SWCNT-PAnBuS.

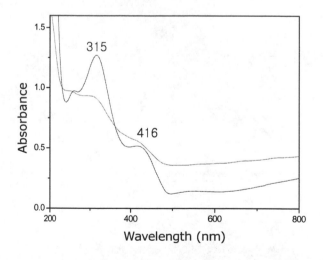

Figure 3 : UV-vis absorption spectra of SWCNT-PAnBuS (red line) and PAnBuS (blue line).

Figure 3 shows the UV-vis spectra of aqueous PAnBuS solution itself (blue line) and dispersed SWCNTs in PAnBuS solution (red line). Two absorption bands at 315 nm and 416 nm of aqueous PAnBuS correspond to π -π * transitions [8]. The two absorption bands, especially the band at 315 nm, were significantly decreased upon forming dispersion with SWCNTs. It suggests a change in conformation of the PAnBus polymer backbone due to the interaction with added SWCNTs. [9]

The formation of SWCNTs-PAnBuS nanocomposites was investigated by X-ray photoelectron spectroscopy (XPS). Since nitrogen atom in PAnBuS would be most highly affected when the PAnBuS molecules were interacted with SWCNTs, we chose the spectra of N(1s) for both PAnBus and SWCNTs-PAnBuS for XPS study. The XPS spectra of N(1s) of them were depicted in Figure 4. Two peaks of N(1s) (Figure 4(a)) were observed at 399 eV and 401 eV in PAnBuS which corresponded to benzenoid amine (-NH-) and positively charged nitrogen atoms (N^+), respectively.[10] These peaks were not shown in as-received SWCNTs. (Figure 4(c) After the formation of the SWCNTs-PAnBuS nanocomposites, the peak from positively charged nitrogen atoms was dramatically reduced and the peak from the amine was dominantly appeared. (Figure 4(b) It implies that the SWCNTs-PAnBuS nano-composites was fabricated *via* a redox reaction.

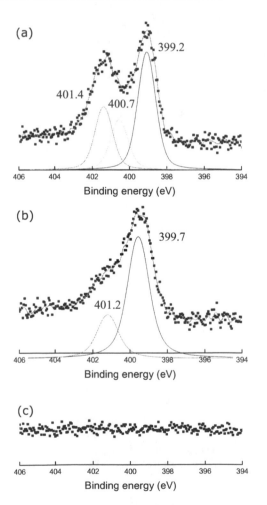

Figure 4 : XPS N(1s) spectra of (a) PAnBuS, (b) SWCNTs-PAnBuS, and (c) as-received SWCNTs

The morphology of SWCNTs-PAnBuS was compared with that of SWCNTs using a field emission scanning electron microscope (FE-SEM). In order to investigate the difference in the surface of SWCNTs and the of SWCNTs-PAnBuS nanocomposites, FE-SEM images were obtained at large area. The surface of the latter (Fig. 5(b)) was seemed to be covered with polymer thin films which were not observed in that of the former (Fig. 5(a). In addition, bundle diameters of SWCNTs were in the 20 - 50 nm range. By comparison, small bundle sized SWCNTs (less than 20 nm) were frequently observed for SWCNTs-PAnBuS nanocomposites. This explains that the bundled SWCNTs become dispersed into small dimeters through strong chemical interaction between PAnBuS with side-wall of SWCNTs.

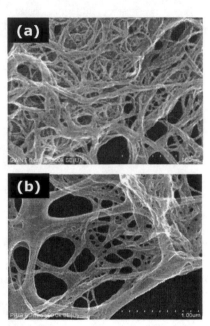

Figure 5 : SEM images of (a) as-received SWCNTs and (b) SWCNTs-PAnBuS.

We attempted to fabricate electrically conducting thin film using the SWCNTs-PAnBuS nanocomposites in order to trace the application possibility for new electrode materials. 3 mg of SWCNTs was added to 10 mL of aqueous solution containing 15 mg PAnBuS under ultrasonic agitation for 10 h. The dark solution was centrifuged at 10,000 g for 10 minutes.[7] The SWCNTs – PanBuS supernatant was sprayed onto polyethylene terephthalate (PET) substrate using an air brush pistol to form a uniformed thin film. The sheet resistance of TCFs prepared from PAnBuS dispersed SWCNTs are in the range of $1\sim2 \times 10^4$ Ω /sq. This values are slightly higher than that of films prepared SDS/SWNT dispersion ($2\sim8 \times 10^3$ Ω /sq) at the same transparency. Even though a conducting polymer was employed to fabricate thin film of SWCNTs, the electrical property was not improved or even detracted.. However, there is more room for enhancing conductivity of thin films, e.g. adding other dopants into conducting

NSTI-Nanotech 2008, www.nsti.org, ISBN 978-1-4200-8503-7 Vol. 1

polymers, and improving SWCNT networks in polymer matrix, etc, so that the conductivity of the composite can be enhanced. This study is still on-going for achieving suitable conductivity.

4 CONCLUSIONS

Dispersed and debundled SWCNTs were prepared in an aqueous solution of PAnBuS using a simple mixing procedure under ultrasonic treatment. Unlike the as-received SWCNTs, the synthesized PAnBuS-SWCNTs were completely soluble in water at least for 1 month. This fact implies that the interaction between polymer and carbon nanotubes is strong enough to overcome the van der Waals interaction among SWCNTs. This functionalized complex can be utilized in many applications, especially in fabricating transparent conducting films.

ACKNOWLEDGMENT

This work is outcome of the fostering project of the Specialized Graduate School supported financially by the Ministry of Commerce, Industry and Energy (MOCIE), Korea. We thank Mr. C. H. Choi and Mr. B.C. Son for FE-SEM and XPS measurement, respectively.

REFERENCES

[1] S. Iijima, "Helical Microtubules of Graphitic Carbon," Nature, 354, 56-58, 1991.

[2] M. S. Dresselhaus, G. Dresselhaus, Ph. "Introduction to Carbon Materials Research," Avouris (Eds.) in Carbon Nanotubes, Topics Appl. Phys., Springer-Verlag, Berlin, 80, 1-9, 2001.

[3] H. Zengin, W. Zhou, J. Jin, R. Czerw, D. W. Smith, Jr., L. Echegoyen, D. L Carroll, S. H. Foulger, J. Ballato, "Carbon Nanotubes Doped Polyaniline," Adv. Mater., 14, 1480, 2002.

[4] H. Zhang, H. X. Li, H. M. Cheng, "Water-Soluble Multiwalled Carbon Nanotubes Functionalized with Sulfonated Polyaniline," J. Phys. Chem. B, 110, 9095-9099, 2006.

[5] J. Yue, Z. H. Wang, K. R. Cromack, A. J. Epstein, A. G. McDiarmid, "Effect of Sulfonic Acid Group on Polyaniline Backbone," J. Am. Chem. Soc., 113, 2665, 1991.

[6] E. K. Kim, M.-H. Lee, B. S. Moon, S. B. Rhee, "Redox Cyclability of a Self-Doped Polyaniline," J. Electrochem. Soc., 141, 26-28, 1994.

[7] H.-Z. Geng, K. K. Kim, K. P. So, Y. S. Lee, Y. Chang, Y. H. Lee, "Effect of Acid Treatment on Carbon Nanotube-Based Flexible Transparent Conducting Films," J. Am. Chem. Soc., 129, 7758-7759, 2007.

[8] M. in het Panhuis, L. A. Kane-Maguire, S. E,. Moulton, P. C. Innis, G. G. Wallace, "Stabilization of Single-Wall Carbon Nanotubes in Fully Solfonated Polyaniline," J. Nanosci. Nanotech., 4, 976-981, 2004.

[9] B. McCarthy, J. N. Coleman, R. Czerw, A. B. Dalton, M. in het Panhuis, A. Mait, A. Drury, H. J. Byrne, D. L. Caroll, W. J. Blau, "A Microscopic and Spectrscopic Study of Interactions between Carbon Nanotubes and a Conjugated Polymer," J. Phys. Chem. B., 106, 2210-2216, 2002.

[10] H. Zhang, H. X. Li, H. Cheng, J. Phys. Chem. B., "Water-Soluble Multiwalled Carbon Nantubes Functionalized with Sulfonated Polyaniline," 110, 9095-9099, 2006.

Department of Polymer Nano Science and Technology, Chonbuk Nat'l University, Jeonju, 561-756, Korea, Ph: +82-63-270-2337, Fax: +82-63-270-2341
mhlee2@chonbuk.ac.kr

Nanocatalysts Assisted Growth of Diamond Films on Si by Hot Filament CVD

C. C. Teng,[*] F. C. Ku,[**] J. P. Deng,[***] S. F. Chien,[***] C. M. Sung[**] and C. T. Lin[*]

[*]Department of Chemistry and Biochemistry,
Northern Illinois University, DeKalb, IL, USA, ctlin@niu,edu
[**]Kinik Company, Ying-Ko, Taipei, Taiwan, sung@kinik.com.tw
[***]Tamkang University, Tamsui, Taiwan, jpdeng@mail.tku.edu.tw

ABSTRACT

Four different catalysts, nano-Ni, diamond powder, mixture of nano-Ni/diamond powder, and ultrasonicated nanodiamond seed, were used to activate Si wafers for diamond film growth by hot filament CVD (HFCVD). Diamond crystals were shown to grow directly on both large diamond powder and small nanodiamond seed, but a better crystallinity of diamond film was observed on the ultrasonicated nanodiamond seeded Si substrate. On the other hand, nano-Ni nanocatalysts seem to promote the formation of amorphous and/or graphitic-like carbon at the initial growth of diamond films. The subsequent nucleation and growth of diamond crystals on top of the amorphous carbon layer were followed to generate the spherical diamond particles and clusters prior to coalescence into continuous diamond films. Moreover, the mixture of nano-Ni/diamond powder catalyst tend to promote the formation of a mixed of amorphous and crystallite diamond structures as characterized by TEM, Raman, and XRD techniques.

Keywords: Ni nanoparticle, nanodiamond seed, nanocatalyst, growth mechanism, diamond film, HFCVD

1 INTRODUCTION

Recently, nanocrystalline diamond films have been considered as the attractive materials for solid-state electronic applications; due to their hardness, thermal conductivity, electrical resistivity, UV transparency, etc. Different surface modification techniques, such as mechanical abrasion, ultrasonication and bias, have been used to enhance the initial nucleation density for diamond film growth [1]. In ultrasonication seeding, diamond powders and/or mixed with transition metal powders (Ni of 3~5 µm in size, Ti, Cu, Fe etc.) were processed as catalysts. However, the catalytic effect of Ni has been illustrated to inhibit CVD diamond nucleation and growth, though diamond might nucleate and grow on the amorphous carbon or graphitic interlayer initially formed on Ni in a low pressure methane-hydrogen environment [2-5]. For this reason, Ni nanoparticles were widely reported for synthesis of carbon nanotube (CNT) because the graphitic layer covering Ni particles of proper size in nanoscale would transform into CNT by CVD under a controlled temperature [6,7]. Under these conditions, CNT and diamond can be selectively grown on Ni coated and diamond powder abraded Si substrates [8]. Therefore, the

better understanding of the detailed chemistry of diamond film growth assisted by Ni is essential to the better control of CVD diamond film or CNT growth in relation to the growth of initial species, chemical mechanisms and transformations to the final crystalline phase. In this paper, we employed four different catalysts, nano-Ni, diamond powder (250 ~ 350 nm), mixture of nano-Ni/diamond powder, and ultrasonicated nanodiamond (50 nm) seed, on Si wafers to investigate diamond film growth by HFCVD method. A growth mechanism is suggested to explain the effect of Ni nanoparticles on the initial growth stage of CVD diamond film formation.

2 EXPERIMENTAL

2.1 Preparation of Nanocatalysts

Nano-Ni nanocatalysts Nano-Ni nanocatalysts were prepared by a chemical reduction method and stabilized in ethylene glycol (EG) solutions containing poly(vinylpyrrolidone) (PVP) polymer stabilizers. First, a 10mL of 10mM $NaH_2PO_2 \cdot H_2O$ (sodium hypophosphite, J. T. Baker 3740) in an EG solution and a 10mL of 4mM $NiSO_4 \cdot 6H_2O$ (nickel (II) sulfate hexahydrate, Fisher N73-100) in an EG solution were freshly prepared, and then well mixed by a magnetic stirrer for 30 minutes. Second, about 0.74 wt% of PVP (polyvinylpyrrolidone, SIGMA-ALDRICH 437190-500G, $M_w \approx 1,300,000$ (LS)) powder was added into the $NaH_2PO_2 \cdot H_2O/NiSO_4 \cdot 6H_2O/EG$ solution and mixed for at least 1.5 hour. Third, the final solution was heated under microwave (2.45 GHz, 900W) for 3 minutes until the color of the solution changed from light green to black brown, indicating the formation of nano-Ni nanoparticle in solutions.

Diamond powders solution First, about 0.37wt% of PVP was added into a 46g EG solution and mixed with a magnetic stirrer for at least 1.5 hour. Second, 1wt% of fluorosurfactant (FS, Dupont FS-510) was added into the PVP/EG solution and mixed for 30 minutes. Third, 0.005wt% of diamond powder (SP-DP MapleCanada Group) was added into the FS/PVP/EG solution and well dispersed by a sonicationb dismembrator (Fisher Scientific Model-100) with 15W power for 30 seconds.

2.2 Deposition of Diamond Films by HFCVD

The polished (100) Si substrates were pretreated with either (a) nanodiamond seeds, ND (4~50 nm, Carbo-Tec Dynaget M3D) by ultrasonication, (b) diamond powders, NDP (250~350 nm, SP-DP MapleCanada Group), (c) nano-Ni nanoparticles or (d) a mixture of nano-Ni and NDP. The surface modification processes, (b), (c) and (d) were done by spin-coating (Photo-resist spinner EC101D-R485, Headway Res. Inc.). The pretreated Si substrates were used for diamond films deposition by HFCVD system (Sp3 Inc., Model 500). The deposition pressure was controlled at 18~20 Torr, containing a gas mixture of 2.3 vol% methane (99.995% purity) and hydrogen (99.95% purity) under a flow rate of 3000 sccm. The distance between the tungsten (W) filament and Si substrates was adjusted at 10 mm. The W filament temperature was controlled at 2000-2200 °C. The temperature of Si substrates was about 650~700°C in the seeding stage and about 800-850°C in the growing stage. In this work, an initial nucleation time of 1.5 hrs and a total deposition time of 48 hrs were employed.

2.3 Characterization Methods

A dynamic light scattering (DLS) goniometer equipped with a HeNe laser at 630 nm (Brookhaven Instruments, Model BI-200SM) was used to characterize the size and size distribution of Ni nanoparticle solutions. The diamond films were analyzed by X-ray powder diffraction measurements (XRD, Rigaku Corp., Model MiniFlex). The lattice structures were examined for the front surface diamond film (front-side) and also the interface between diamond film and Si substrate (backside). Renishaw Raman Scattering Noodles System 2000, equipped with a microscope and a HeNe red (632.8 nm) excitation laser, was used to examine the molecular properties and quality of diamond films from both the front and backsides. Scanning electron microscopy (SEM, JEOL JSM-5600) micrographs were taken to illustrate the surface morphology of diamond films on Si substrates. Atomic force microscopy (AFM, Quesant Q-scope 350) measurements were scanned for the surface pretreated Si substrates and also the front-side and backside of diamond films. Plan-view Transmission electron microscopy (TEM, JEOL, JSM-1200EX II) pictures were taken to identify the possible growing structures and/or species for the formation of diamond films.

3 RESULTS AND DISCUSSIONS

3.1 Size Distribution of Nano-Ni Catalysts with PVP in EG by DLS and AFM

DLS measurements gave a mean diameter of around 50 nm for both 1 mM and 2 mM of nano-Ni catalysts with 0.37 wt% PVP in EG solutions. The AFM pictures for the spin-coated nano-Ni on Si substrates were also shown to have the particle sizes of around 50 ~ 70 nm. (Both DLS and AFM figures are not shown here)

3.2 XRD Patterns of Diamond Films

Figure 1 shows XRD patterns of the front-side diamond films (catalyzed by ND – blue, nano-Ni – red, NDP - pink, and a mixture of nano-Ni/NDP – brown), which are labeled as CVDD-ND, CVDD-Ni, CVDD-NDP, and CVDD-NDP-Ni, respectively. The XRD spectra display three characteristic 2θ peaks around 44.3, 75.6 and 91.8 degrees, which may be assigned to <111>, <220> and <311> lattice planes of diamond crystal, respectively. In Figure 2(b), the XRD patterns for the backsides (B-CVDD) of diamond films are also presented, where only a small bump at 2θ around 75.5 degree (or <220> plane) of diamond crystal has started to come out at the initial growth stage. On the backside of diamond film, both <111> and <311> were not observed. This observation may suggest the possible phase transition from the nondiamond to diamond <111> faces during the growth of diamond films. From the crystallographic diagram, the dominant growth in <110> (or its parallel <220> face) should give diamond crystals of pyramidal <111> facets.[12] Thus, the backside shoulder band of 2θ around 75.8 degree is probably resulted from the stacking faults along <220> direction. All XRD peaks observed in Figure 1 are quite broad (FWHM 0.3 ~ 0.5 degrees in 2θ), suggesting the formation of small diamond crystals. The observed XRD peaks displayed a peak shift of 2θ about 0.3~0.4 degrees relative to the theoretical positions that indicates the inhomogeneous stresses existed in diamond films. A peak splitting for <111> peak at 2θ = 44.3 degrees was observed for nano-Ni assisted diamond films as shown in Figure 1(a). The reason is not yet known, but it may be resulted from the lattice mismatching for the growth of diamond crystals on nano-Ni particle-like catalysts.

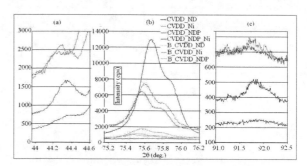

Figure 1: XRD of diamond films: (a) D<111>, (b) D<220>, front & backside and (c) D<311> lattice planes

3.3 Raman Spectra of Diamond Films

Figure 2 shows Raman spectra of diamond films catalyzed by ND (blue), nano-Ni (pink), and NDP (green). Both front and backsides of diamond films were examined, where the backside film served as the film growth at the initial stage. In all cases, the crystalline diamond peak at 1332 cm^{-1} is only weakly observed (the inserted in Figure 2), suggesting the % and size of diamond in film composition is low and small. However, an amorphous phase starts to form at the initial stage as observed by a small broad band at 1520 cm^{-1} on the backside. This is then transformed to give the diamond structure as displayed by a small peak around 1334 cm^{-1} on the front side.

All spectra showed an intense band at 2270 cm^{-1} and a shoulder band at 2800 cm^{-1}, which have been attributed to several origins: (1) a photoluminescence peak or a color center due to silicon defects in the growth of diamond films [10,11], and (2) the overtone Raman band of a short conjugation chain of trans-polyacetylene (t-PA), which may grow around the grain boundaries of diamond films [12-16]. The origin (2) is preferred since these Raman bands have also been observed for the diamond films grown on non-silicon substrates [17]. Moreover, in Figure 2, the intensity of 2270 cm^{-1} peak is even stronger on the front than on the backsides of diamond films grown on Si by HFCVD. This indicates that the 2270 cm^{-1} peak is not due the substrate defects, but the t-PA formed around the grain boundaries of small (or nano) diamonds. It is worthwhile to mention, the backside diamond film catalyzed by nano-Ni gives the lowest peak intensity of Raman band at 2270 cm^{-1}, indicating the production of t-PA is low in the initial growth of diamond phase on nano-Ni catalysts. This may be resulted from the catalytic dehydrogenation effects that inhibit the formation of t-PA.

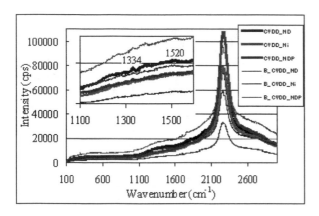

Figure 2: Raman spectra of diamond films

3.4 TEM Study of Diamond Films

TEM pictures are very informative for identifying the initial growth species of diamond films as shown in Figure 3 on (a) ND, (b) nano-Ni, (c) NDP, and (d) nano-Ni/NDP mixture. For diamond film grown on diamond seed/powder, such as pictures (a), (c) & (d-1), the nanocrystalline diamond clusters were shown to be the dominant structures of several overlapping hexagonal forms. Picture d-1, in particular, a well-defined crystalline structure is clearly seen. On the surface of nano-Ni catalyst, however, some short tube-like structures (picture b-2) and a 2-5 nm thin

graphitic-like layer (picture d-2) were shown to nucleate on nano-Ni catalysts of various sizes, which may be identified in picture b-2 (10-15 nm), b-3 and d-2 (30 nm). An interesting observation in picture b-1 should be mentioned, where diamond crystals seem to be formed on amorphous/graphitic-like interlayers initially grown on nano-Ni catalysts without diamond seeding. Moreover, in picture d-3, nano-Ni catalysts clusters were shown to embed in amorphous phase of diamond films. The observed tube-like and graphitic-like layer forms on nano-Ni surface may lead us to speculate that those are the initial growth species before a transformation to amorphous phase and then to the crystalline diamond structures.

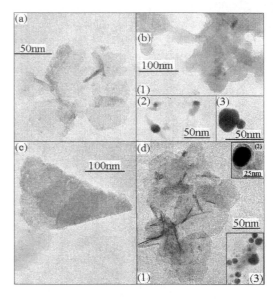

Figure 3: TEM of diamond films (a) CVDD-ND, (b) CVDD-Ni, (c) CVDD-NDP, (d) CVDD-NDP-Ni

3.5 SEM and AFM Study of Diamond Films

SEM pictures of diamond films grown on the pretreated Si substrates (a - nano-Ni and b - NDP) are shown in Figure 4. The pyramidal <111> facet of diamond crystals are clearly observed in the inserted pictures under high resolution. It is interesting to point out that the continuous polycrystalline diamond films were formed by the coalescence of the spherical cauliflower-like microcrystalline diamond particles about 20 μm in size. These diamond structures are particularly striking as seen in Figure 4(b), and their coalescing boundaries of the resulting diamond films are pointed by arrows. The formation of spherical cauliflower-like microcrystalline diamond particles may result from the isotropic nucleation and growth of nanodiamond crystals on large nanodiamond powders. On the surface of nano-Ni catalyst, however, amorphous carbon (or graphitic-like layer as shown in TEM) may be formed initially at lower temperature of 700°C and followed by the nucleation of nanodiamond crystals and then transformed to the spherical cauliflower-like microcrystalline diamond particles.

The continuous faceted diamond films grown on different pretreated Si substrates are quite similar at first glance. The detailed examination shows that the diamond films grown on ND surface gave the most crystalline diamond particles as demonstrated in XRD analysis. The diamond films grown on nano-Ni catalysts seem to have the smaller size and less crystalline diamond particles as shown in Figure 4(a). This can be viewed from the right-bottom inserts of Figure 4 (a) & (b), the diamond crystals on the surface of the single spherical cauliflower-like microcrystalline diamond particle assisted by nano-Ni have less crystallinity than those seeded with NDP and ND.

The smaller diamond particles grown on nano-Ni (a), as compared to those on ND, can also be seen in AFM scans in Figure 5. The surface roughness (Rt) of diamond films were measured as 1.552 µm and 0.905 µm for Figure 5 (a) and (b), respectively. A 40% reduction in surface roughness for diamond film grown on nano-Ni is a surprise and quite significant.

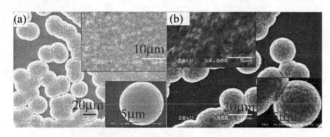

Figure 4: SEM of diamond films (a) CVDD-Ni (b) CVDD-NDP

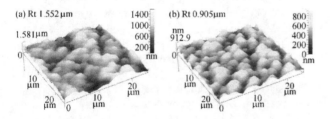

Figure 5: AFM of diamond films (a) CVDD-ND and (b) CVDD-Ni

4 CONCLUSIONS

We have successfully demonstrated the diamond films grown on ND, nano-Ni, NDP, and nano-Ni/NDP mixture by HFCVD. The initial growth seems to have the <220> (or <110>) facets and then leads to diamond crystals of pyramidal <111> facets as shown in XRD. The Raman spectral band at 2270 cm^{-1} has been verified to originate from the short chain t-PA formed around the grain boundaries of small (or nano) diamonds. The SEM and TEM results indicated that diamond crystals were grown directly on both large NDP sites and small ND seeds, but latter displayed a better crystallinity of diamond film. It was

clearly illustrated that nano-Ni nanocatalysts can promote the formation of amorphous and/or graphitic-like carbon phases at the initial growth of diamond films, and the subsequent nucleation and growth of diamond crystals on top of the amorphous carbon layer to generate the spherical diamond particles and clusters prior to coalescence into continuous diamond films. Moreover, a 40% reduction in surface roughness for diamond film grown on nano-Ni is a surprise and quite significant.

ACKNOWLEDGMENT

This work was supported by Kinik Company, Taiwan and Institute for Nano Science, Engineering and Technology, NIU. The technical assistances from Dr. Laurence Lurio, Dr. Haji-Shiekh and Dr. Chong Zheng are greatly appreciated.

REFERENCES

[1] Y. Chakk, R. Brener and A. Hoffman, Appl. Phys. Lett. 66, 2819, 1995.; Diamond Relat. Mater. 5, 286, 1996.; Diamond Relat. Mater. 6, 681, 1997.

[2] D. N. Belton and S. J. Schieg, J. Appl. Phys. 66, 4223, 1989.

[3] P. C. Yang, W. Zhu and J. T. Glass, J. Mater. Res. 9, 1063, 1994.; J. Mater. Res. 8, 1773, 1993.

[4] E. Johansson, P. Skytt, J.-O. Carlsson, N. Wassdahl and J. Nordgren, J. Appl. Phys. 79, 7248, 1996.

[5] W. Zhu, P. C. Yang, J. T. Glass, Appl. Phys. Lett. 63, 1640, 1993.

[6] M. Yudasaka, R. Kikuchi, T. Matsui, Y. Ohki and S. Yoshimura, Appl. Phys. Lett. 67, 2477, 1995.

[7] M. Mauger, V. T. Binh, A. Levesque and D. Guillot Phys. Lett. 85, 305, 2004.

[8] Q. Yang, C. Xiao, W. Chen and A. Hirose, Diamond Relat. Mater. 13, 433, 2004.

[9] M. N. R. Ashfold, P. W. May and C. A. Rego, Chem. Soc. Rev. 23, 21, 1994.

[10] J. Birrell, J. E. Gerbi, O. Auciello, J. M. Gibson, J. Johnson and J. A. Carlisle, Diamond Relat. Mater. 14, 86, 2005.

[11] D. V. Musale, S. R. Sainkar, S. T. Kshirsagar, Diamond Relat. Mater. 11, 75, 2002

[12] I. I. Vlasov, V. G. Ralchenko, E. Goovaerts, A. V. Savelie and M. V. Kanzyuba, Phys. Stat. Sol. (a) 203, 3028, 2006.

[13] Sh. Michaelson, O. Ternyak and A. Hoffman, Appl. Phys. Lett. 89, 131918, 2006.

[14] D. Roy, Z. H. Barber and T. W. Clyne, J. Appl. Phys. 91, 6085, 2002.

[15] A. C. Ferrari and J. Robertson, Phys. Rev. B. 63, 121405-1, 2001

[16] T. López-Ríos, É. Sandré, S. Leclercq and É. Sauvain, Phys. Rev. Lett. 76, 4935, 1996.

[17] L.-T. S. Lin, G. Popovici, M. A. Prelas, S. Khasawinah and T. Sung, J. Chem. Vap. Dep. 3, 102, 1994.

Single-Walled Carbon Nanotube Network based Biosensors using Aptamers and its Characteristics

Dong Wan Kim*, Sung Min Seo*, Young June Park*

*School of Electrical Engineering and Nano-Systems Institute (NSI-NCRC), Seoul National University, Seoul 151-742, Korea, dwkim@isis.snu.ac.kr

ABSTRACT

We have successfully demonstrated the single-walled carbon nanotube (SWNT) network based biosensor using aptamers as a protein recognition site. Aluminum was first patterned on the substrate with CVD-grown oxide. Then, gold was electrolessly plated on the Al electrodes and SWNTs were dip-coated on the substrate. Electrical pulses were applied through the electrodes to reduce a contact resistance. Finally, aptamers were attached on the surface of SWNTs and were used as a recognition site of human serum albumin (HSA). Carbon nanotube shows that it can be a powerful candidate for the next-generation sensor element for the biological/chemical detection.

Keywords: Single-walled carbon nanotube network, Biosensor, Aptamer, Human Serum Albumin

1 INTRODUCTION

Since its discovery in 1991 by Iijima, carbon nanotube (CNT) has attracted a lot of attention from researchers [1]. Much effort has been made to fabricate biosensors with carbon nanotubes as sensor elements due to their several advantages including large surface-to-volume ratio, one-dimensional electronic structure, and a molecular composition consisting of only surface atoms.

In this paper, we present a SWNT network based biosensor using aptamer and its experimental results. CNT network was used in the expectation that it will show a better performance and consistency in comparison with the single nanotube in that they contain a number of carbon nanotubes with various chirality and diameter making them have better uniformity in electrical properties [2]. Aptamers are artificial oligonucleotides (DNA or RNA) that can bind to a variety of materials with high selectivity, specificity. Since aptamers (1-2 nm) are much smaller than proteins, it is likely that aptamer-protein binding occurs inside the electrical double layer, which is characterized by Debye length (~3 nm in 10 mM ionic concentration), of solution resulting in higher sensitivity. The possibility to fabricate biosensors using SWNT network and aptamers is investigated.

2 DEVICE FABRICATION AND EXPERIMENTAL RESULTS

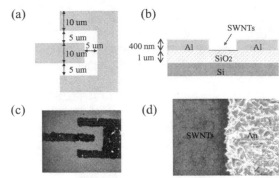

Fig. 1. (a) Semi-concentric electrode structure. (b) Cross-sectional view of the device. (c) Microscopic image of the device after Au electroless plating (ELP). (d) Scanning electron microscopy (SEM) image of the single-walled carbon nanotubes (SWNTs) deposited on the device.

Carbon nanotube biosensors were constructed as shown in Fig. 1. A 1000 nm thick tetra-ethyl-ortho-silicate (TEOS) oxide layer was deposited on the silicon wafer using chemical vapor deposition (CVD), followed by a 400 nm thick aluminum (Al) metal that was sputtered and patterned as the source/drain electrode. It is compatible with the conventional CMOS process so its structure can easily be incorporated to existing CMOS circuits. The device geometry was varied with the spacing between the source and drain electrode ranging from 5 to 20 um, and the channel width ranging from 77 to 106 um. To improve the contact resistance between the metal electrodes and the SWNTs, yet without the need of the additional mask, an electroless plating (ELP) can be adopted. In our case, gold (Au) was electrolessly plated on the Al electrodes using palladium (Pd) as a catalyst. The 0.05 g SWNTs were immersed in 1L 1,2-dichlorobenzene undergoing wet-oxidation to introduce defects such as carboxylic acid on the surface of the SWNTs followed by ultra-sonication for dispersion and were spread on the substrate using the dip-coating method. SEM image revealed the deposited SWNTs are 1.6 um long with diameter of 1.4 nm on average. Fig. 2 shows the change in 2-terminal resistance before and after Au electroless plating [3].

Thereafter, electrical pulses of 10 V were applied through the electrodes to reduce the contact resistance [4]. Fig. 3 shows a series of I_{DS}-V_{DS} curves of SWNTs at room temperature as an electrical pulse is applied with an incremental height and fixed duration (16 msec) through electrodes. The statistics show that the 2-terminal resistance becomes more uniform after pulse annealing compared with those before pulse was applied.

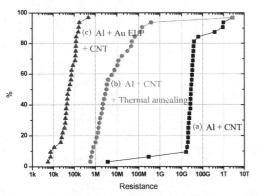

Fig. 2. Change in 2-terminal resistance (a) Resistance after SWNTs are deposited on Al electrodes. (b) After thermal annealing process. (c) After Au ELP.

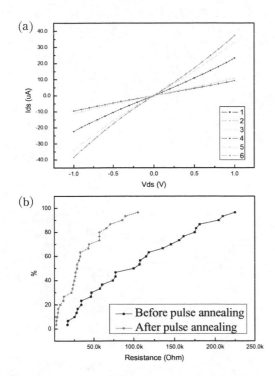

Fig. 3. (a) The evolution of the I_{DS}-V_{DS} characteristics of SWNTs with an incremental height of the applied pulse. Curves 1–6 correspond to the following: 1 – without applying pulse, 2–after applying a pulse of 2 V for 16 msec, 3 – 4 V, 4 – 6 V, 5 – 8 V, 6 – 10 V. (b) The statistics of the changes in 2-terminal resistance after pulse annealing.

Then, SWNTs were treated with carbodiimidazole-activated Tween 20 (CDI-tween 20) which serves as a linker between SWNTs and aptamers. While the Tween-20 component which has a hydrocarbon chain in it was bound to the carbon nanotube side wall through hydrophobic interactions, the carbodiimidazole was used to covalently

attach the 3'-amine group of Aptamer. Aptamers that bind specifically to human serum albumin (HSA) were attached onto CDI-Tween 20 [5]. The electrical transfer characteristics were measured at each process stage.

The real-time measurement of conductance from HSA aptamer immobilized SWNTs were performed. First, 5 uL droplet of diethylpyrocabonate (DEPC)-treated water was placed on the aptamer-modified SWNTs. Then, 5 ul of

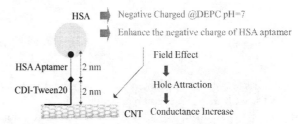

Fig. 4. Schematic diagram of binding of human serum albumin (HSA) on a aptamer immobilized SWNT-based biosensor.

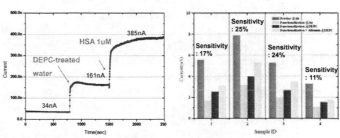

Fig. 5. (a) A real-time conductance measurement from the SWNT-based biosensor. (b) Statistics of the changes after aptamer immobilization and introducing HSA on biosensor.

HSA with concentration of 1.5 uM was added to the DEPC-treated water droplet. In Fig. 4, we show a schematic diagram of HSA binding on HSA aptamer immobilized SWNTs.

As shown in Fig. 5a, the conductance increased as the DEPC-treated water droplet was placed. After the initial increase, addition of HSA caused the sharp increase in conductance until it reached the saturation point. Fig. 5b displays the resistance at each process and shows that the sensitivity of the aptamer-immobilized SWNTs biosensor reaches up to 25%. The HSA molecules are negatively charged in DEPC-treated water (pH 7.0), since the isoelectric point of HSA is rather low (pI 4.7). The increase in conductance after introduction of HSA could be attributed to these negatively charged HSA molecules enhancing the negative charge of HSA aptamers when the binding occurs. These negative charges are expected to give a field effect on SWNTs, thereby attracting holes in SWNTs which results in increase in conductance, while CDI-Tween 20 acts as an insulator.

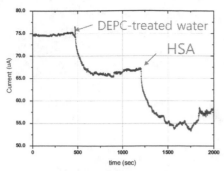

Fig. 6. Control experiment using SWNTs with no HSA aptamers attached.

As a control, we conducted the same experiment using SWNTs with no HSA aptamers attached. In this case, decrease in conductance was observed upon introducing HSA as shown in Fig. 6. This sudden drop in conductance could be explained by amine group in HAS aptamers which donates its unshared electron pair to SWNTs leading to decrease in hole concentration in SWNTs.

3 SUMMARY

We have fabricated SWNT-based biosensors, in which SWNT channels were modified with aptamers and detected human serum albumin (HSA). Aptamer modified SWNT network displayed a good performance for the detection of HSA. Our aptamer immobilized SWNT-based biosensor is a promising candidate for the development of an integrated, high-throughput, real-time biosensor. As the development and optimization studies continue, we expect that sensitive detection of numerous important bio-molecules using SWNT-based biosensors would become possible.

ACKNOWLEDGEMENT

This work was supported by the Nano System Institute-National Core Research Center (NSI-NCRC) program of KOSEF, Korea.

REFERENCES

[1] Iijima, Nature 354, 56–58, 1991.
[2] E. S. Snow, et al., Appl. Phys. Lett., Vol. 82, No. 13, 2003.
[3] M. Liebau, et al., Appl. Phys. A 77, p.731–734, 2003.
[4] Yunsung Woo, G. S. Duesberg, S. Roth, Nanotechnology 18, 095203, 2007.
[5] Hye-Mi So, et al., J. AM. CHEM. SOC., Vol. 127, No. 34, 11906–11907, 2005.

Alcohol Vapor Sensors Using Multiple Spray-Coated SWCNTs

S. J. Kim[*], Y. M. Choi[**], G. W. Lee[***]

[*]Kyungnam University, Department of Electronic Engineering
449 Wolyoung-dong, Masan, Korea, 631-701, sjk1216@kyungnam.ac.kr
[**] Kyungnam University, Masan, Korea, fululu@hanmail.net
[***]KERI, Changwon, Korea, gwleephd@keri.re.kr

ABSTRACT

We suggest a gas sensor using single-walled carbon nanotubes (SWCNTs) thin film for alcohol vapor detection. The SWCNTs thin film were deposited by cost-effective and scalable spray method on flexible PES (polyethersulfone) polymer substrates. From the fabricated sensors, conductivity response properties were measured and discussed. Although alcohol sensors are currently studied for the use of various purposes such as bio-sensing and detection of VOCs (volatile organic compounds), they have been most widely commercially used as a meter for breath alcohol measurement which is applicable to checking whether car drivers are drinking-driving or not. Our sensors showed good sensitivity and linearity in conductance response.

Keywords: alcohol sensors, spray, SWCNTs, conductance

1 INTRODUCTION

The main requirements of a good sensor are high sensitivity, fast response, low cost, high mass production and high reliability. Sensors continue to make significant impact in everyday life with application spreading from biomedical to automotive industry. This has led to intensive research activities across the world in developing new sensing materials and technologies. Currently, the discovery of carbon nanotubes has generated outstanding interest among researchers to develop carbon nanotube(CNT)-based sensors for many applications[1]. Their peculiar properties, especially high surface area and atomic structure, make carbon nanotubes promising candidates for nanoscale sensing materials. Nanotube sensors offer potential and significant advantages over traditional sensor materials (mainly semiconducting metal oxides) in terms of sensitivity, operation at room temperature, small sizes for device miniaturization, massive sensor arrays, and it has been experimentally demonstrated that the electrical conductance of carbon nanotubes can be modulated upon exposure of gaseous molecules of CO_2, NO_2 and NH_3, etc.[2-3]

Currently alcohol sensors using CNTs have been developed for the use of various purposes such as bio-

sensing [4] and detection of VOCs (volatile organic compounds)[5]. Among them, alcohol vapor sensors have been most widely commercially used as a meter for breath alcohol measurement which is applicable to checking whether car drivers are drinking-driving or not, and also they are expected as alcohol detecting devices attached to a dashboard for traffic safety. In general, alcohol gas sensors for breath alcohol measurement should be able to measure samples at low concentrations of the order of a few hundred parts per million. To do this, several types of alcohol vapor sensors have been developed, that is, fuel-cell, semiconductor and infrared absorption. But there are a few drawbacks in the present alcohol gas sensors, so we suggest a new type of SWCNTs-based alcohol gas sensors showing potential and significant advantages in terms of sensitivity, operation at room temperature and small sizes for device miniaturization.

2 EXPERIMENTAL

2.1 Device Fabrication

The gas sensors studied in our experiment were comprised of bundles of single-walled carbon nanotubes (SWCNTs). Although dispersion and distribution of CNTs in aqueous media are proved to be challenging, some organic solvents cause less coagulation of the carbon nanotubes and thus permit greater extent of dispersion. In this work, we used SWCNTs solution dispersed in ethanol solvent with epoxy resin, and SWCNTs thin films were formed by multiple spray-coating with the SWCNTs solution where the SWCNTs were modified and functionalized by carboxyl groups during acid treatment for high solubility and gas selectivity. Many current sensors using CNTs follow this approach because of easiness in sensor fabrication.

To fabricate the gas sensors at first, 100 mg of unpurified HiPco SWCNTs (CNI, 35% metal catalyst) was used. In general bundle of CNTs is purified and dispersed by acid treatment with sonication. The unpurified SWCNTs were dispersed in 50 ml of 30 vol% nitric acid and subsequently sonicated for 1 hour in a water bath sonicator, and then the sonicated suspensions were refluxed at boiling condition for 1 hour. The suspensions were filtered through filter paper (Fisher Scientific, No. 1), 40mg of the SWCNT suspensions were added with 40mg of epoxy resin into 100ml of ethanol to complete final SWCNTs solution, and then were bath sonicated for 6 hours to be dispersed clearly.

Next, Al metallization was carried out on the surface of the SWCNTs thin films with a shadow mask to form two electrodes, and then wire bonding with silver paste was followed. Figure 1 indicates a diagram of our alcohol vapor sensors.

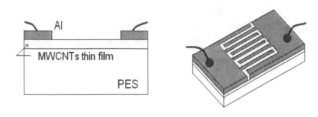

Figure 1: A diagram of our alcohol vapor sensors

2.2 Measurements

Electrical measurements for the sensors were carried out with a semiconductor device analyzer and a HP-4280A in a chamber. We used industrial ethanol diluted in water instead of drinkable alcohol. To adjust the experiment to normal breath alcohol measurement, we injected ethanol molecules vaporized from different vol% ethanol solutions at 36°C, close to the temperature of human body, into the surface of the sensors with N_2 carrier gas. The sensors were first flushed with clean nitrogen gas before exposure to vapors, and were measured 30 seconds after exposure to gases. Figure 2 shows a set of equipment for testing alcohol gas sensors. Current-voltage (I-V) measurement was carried out with a dc power meter to observe the current under a bias to Al electrodes from -5 to 5 V, and electrical conductance and capacitance characteristics were examined.

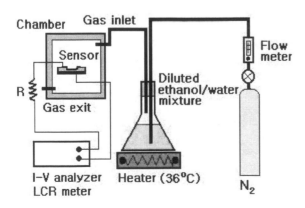

Figure 2: A set of equipment for testing alcohol gas sensors

3 RESULTS AND DISCUSSION

We used PES (polyethersulfone) polymer instead of typical silicon or glass as substrate in this experiment. Since PES polymer has excellent characteristics in flexibility, thermal stability (Tg: 225°C) and surface roughness, it is expected as substrate applicable for flexible sensors. Figure 3 shows an image of multiple spray-coated SWCNTs thin film on a flexible PES substrate

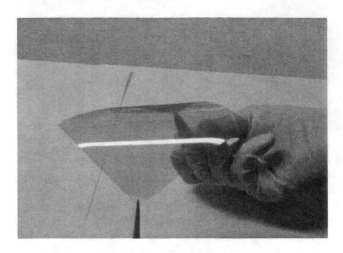

Figure 3: multiple spray-coated SWCNTs thin film on a flexible PES substrate

As the electrical properties of SWCNTs are a critical function of their atomic structure, any gas adsorption can induce great changes in conductance. The promotion or mediation of charge transfer in CNTs thin film by adsorption of electroactive molecules is a main mechanism in most CNT-based chemical sensors application. When gas molecules are adsorbed and then transfer charges on SWCNTs, the molecules may be operated as an electron-acceptor such as NO_2 and O_2, or an electron donor such as NH_3 and H_2O[6]. If an electron-acceptor type of gas molecules is adsorbed at the surface of SWCNTs, the SWCNTs thin films will show p-type semiconducting property of increasing conductance while showing decreasing conductance against an electron-donor type of gas molecules. Figure 4 (a), (b) and (c) show variations of conductance (G) for the three different sensors. Before exposure to alcohol vapor, the conductance was about 3, 15 and 50[μS], respectively, from the 20, 40 and 60 times-coated sensors at room temperature. When the sensors were exposed to alcohol molecules vaporized from 0 to 0.3 vol% ethanol solutions, the conductance of all sensors was increased considerably while the capacitance was nearly invariable. The average value of $\Delta G/Go$ per 0.1 vol% alcohol concentration (where Go indicates the value of conductance at 0.0vol% and ΔG, the variation of conductance) was about 4.0%, while the capacitance is little changed. The invariance of capacitance is due to low face-to-face area between the electrodes in structure.

In result, the increase of conductance seems to be related to the charge transfer due to the electron-acceptor characteristic of alcohol molecules and physical adsorption of these molecules in the tube wall when the SWCNTs functionalized by carboxyl groups are exposed to alcohol molecules. Nearly linear relationship between the change in conductance and alcohol concentration was observed and considerable change in electrical conductivity was found in response to alcohol gas adsorption at room temperature.

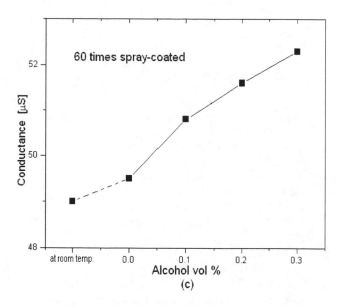

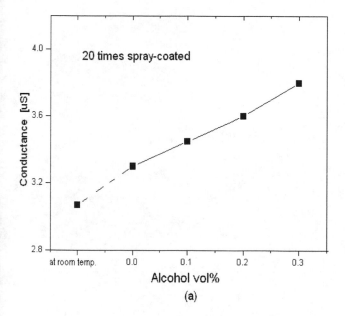

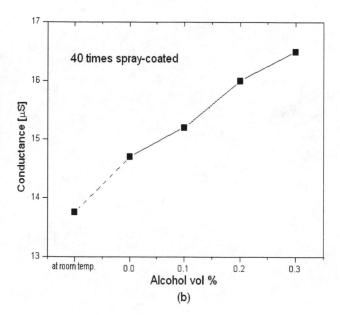

Figure 4: The dependence of conductance on different vol% alcohol solutions for the three different sensors

4 CONCLUSIONS

With the advent of nanotechnology, research is underway to create miniaturized sensors. Miniaturized sensors can lead to reduced weight, lower power consumption and low cost. Carbon nanotubes have been known as one of the most proper materials for miniaturized sensors due to their unique electronic, mechanic, thermal and chemical properties.

The aim of this paper is to present alcohol gas-sensing results for single-walled carbon nanotubes (SWCNTs) thin film prepared by multiple spray-coating on PES (polyethersulfone) substrate. CNTs thin films can be formed simply with the dispersed CNTs solution by screen printing, spray coating, spin coating, imprinting or ink-jet printing onto various substrates and subject to solvent evaporation. In this work, SWCNTs thin films were formed by multiple spray-coating. The multiple spray-coating procedure is very reliable on relatively large area (maximum 20x20cm2), leading to cost reduction and it provides easiness in controlling thickness and conductivity of SWCNTs thin film.

In conclusion, it is expected that our sensors are superior to present alcohol sensors in respects of simple process, low power consumption and good flexibility as well as working at room temperature without heating. Besides, they showed good sensitivity and linearity in conductance response.

ACKNOWLEDGEMENTS

This work was supported by Kyungnam University Foundation Grant, 2007-00039.

REFERENCES

[1] N. Sinha, J. Ma and J. Yeow, J. Nanosci. Nanotechnol. 6, 573, 2006.

[2] K. Ong, K. Zeng and C. Grimes, IEEE Sensor J., 2, 82, 2002.

[3] P. Qi, O. Vermesh, M. Grecu, A. Javey, Q. Wang, H. Dai, S. Peng and K. Cho, Nano Lett. 3, 347, 2003.

[4] J. Wang and M. Musameh, Anal. Chem. 75, 2075, 2003.

[5] M. Penza, F. Antolini and M. Antisari, Sens. Actuators B, 100, 47, 2004.

[6] Y. G. Lee, W. S. Cho, S. I. Moon, Y. H. Lee, J. K. Kim, S. Nahm and B. K. Ju, Chem. Phys. Lett. 433, 105, 2006.

NSTI-Nanotech 2008, www.nsti.org, ISBN 978-1-4200-8503-7 Vol. 1

Single Walled Carbon Nanotubes based Ionic Building Blocks for Nanoelectronic Devices

Yamini Yadav and Shalini Prasad*

Department of Electrical and Computer Engineering, Portland State University
Post Office Box 751, Portland, OR 97207-0751, *prasads@ece.pdx.edu

ABSTRACT

We present the electrical characterization of crossbar ionic nanoscale devices built using surfactant coated single walled carbon nanotubes (SWCNTs) via micro contact printing. These electro-ionic building blocks were synthesized by doping intrinsic semi conducting SWCNT doped with surfactant molecules, thereby altering the Fermi energy levels of SWCNTs and their electrical properties. The surfaces of these SWCNTs, were modified by two types of surfactants; sodium docedyl sulfate and cetyl trimethylammonium bromide, to produced anionic SWCNT (P-type) and cationic SWCNT (N-type) respectively. Using dual micro-patterning process, anionic and cationic SWCNT ionic blocks were alternatively symmetrically patterned into a parallel array to form crossbar p-n junctions. Functionality of the nanodevices was demonstrated by studying the current – voltage (I-V) characteristics that shows promise towards the formation of nanoelectronic diode arrays.

Keywords: Carbon nanotubes, microcontact printing, nanoelectronics, ionic junction, surfactants

1   INTRODUCTION

Fundamental need for high throughput and improved device performance has lead to the investigation of incorporating both organic and inorganic materials in multi-scale architectures[1]. Secondly, organic chemical doping of nanomaterials enables the modulation of nanomaterial electrical properties by nanoionic transport[2]. These capabilities have potential for developing nanoelectronic device building blocks that are suitable for a wide range of device application. Numerous nanoscale semiconductor fabrication techniques and materials engineering processes have guided the development of rapid synthesis of extrinsic nanomaterial, assembly and patterning of nanomaterials to form electronic nanodevice components[3]. Development of these nanoelectronic systems has wide range of applications in semiconductor business as well as sensors based biomedical and defense industries [4].

Various techniques such as Langmuir Blodgett method, superlattice nanowire pattern transfer (SNAP) and fluidic alignment for positioning these surface modified nanomaterials have been studies [5, 6]. These methods are complex, difficult to implement and require complex laboratory equipment for nanomaterial patterning[6]. In contrast, in our research we have adopted and illustrated an integrated approach of simple, hard and soft lithography techniques to fabricate and assemble crossbar ionic junctions comprising of bundles of extrinsically functionalized SWCNTs.

For building junction devices, chemical properties of nanomaterials are typically modulated by adding dopants using most common techniques such as diffusion and ion implantation methods [7, 8]. These techniques show crystallographic damage. Amorphization and sputtering have also been used to dope nanomaterial, but these techniques use hazardous materials such as antimony, arsenic, phosphorus, and boron and require high voltages [9]. In contrast, in the current paper, we have chemically doped SWCNTs using non-hazardous organic surfactants. Also, control over the dopant concentration is achieved by varying concentration of surfactant.

2   MATERIALS AND METHODS

Fabrication and assembly of cross bar ionic junction based nanodevices consist of following five steps: (i) Synthesis of extrinsic SWCNT by chemically doping using surfactant (ii) fabrication of microelectrode arrays using photolithography techniques, (iii) fabrication of polymer PDMS stamps using rapid soft lithography techniques, (iv) patterning of cationic and anionic SWCNTs using PDMS stamp in crossbar fashion and (v) Electrical characterization of crossbar ionic junctions.

2.1   Synthesis of extrinsic SWCNTs

The charge transport in the ionic p-n junctions is by ion molecules in contrast to semiconductor p-n junction where the flow of current is by holes and electrons. The extrinsic SWCNTs were synthesized using surfactants having positive and negative charge ions at its hydrophilic ends. Two types of surfactants were used; sodium dodecyl sulfate (SDS) and cetyl trimethylammonium bromide (CTAB) having having Na+ positive ionic charge (anions) and Br- negative ionic charge (cations) on its hydrophilic ends were used respectively to achieve this doping. The intrinsic SWCNTs were dispersed into surfactant, by forming an encapsulate over the SWCNTs by generating a micelle layer around SWCNTs. Figure 1 shows orientation of the micelles in perpendicular and parallel onto the surface of the SWCNTs. Doping SWCNT with SDS and CTAB produces anionic SWCNT and cationic SWCNT respectively.

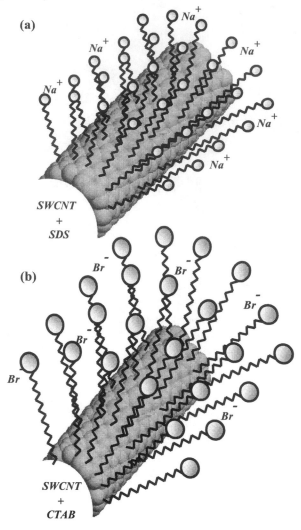

Figure 1: Surfactant molecules on the SWCNT. (a) and (b) SDS and CTAB form a uniform layer or a micelle around nanotubes with hydrophobic group adsorbed onto SWCNT surface and hydrophilic heads having Na^+ and Br^- ions

2.2 Microelectrode array and PDMS fabrication

The microelectrode platform consists of 2x2 array of metallic gold microfingers. The microfingers were placed opposite to each other in cross bar junction fashion to obtain electrical I-V measurements. Traditional, simple and standard photolithography techniques were used for fabrication of the microelectrode array. The technique follows the following process sequence; (a) substrate preparation, (b) positive photoresist application, (c) exposure and pattern development and (d) finally metal sputtering and lift off.

For crossbar pattering of the extrinsically doped SWCNTs in a parallel symmetric array, primarily a master template having a micro- channel surface topology was fabricated using the previously described photolithography process. Later, softlithography techniques were incorporated for fabricating soft, flexible PDMS stamps having micro-relief structures using the master templates. This technique involved mixing of elastomer and curing agent, followed by curing of the mixture and finally peeling off the flexible PDMS stamp, having the negative replica of the master template for stamping of anionic and cationic SWCNTs.

2.3 Patterning nanomaterial

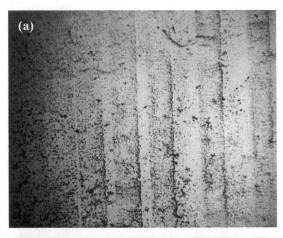

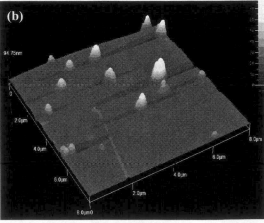

Figure 2: (a) Optical micrograph and (b) AFM image representing patterned SWCNT in parallel array by imprinting inked PDMS stamps with surfactant dispersed SWCNT

Double microimprinting techniques were involved for pattering of extrinsic SWCNTs. Due to the parallel relief structure on the PDMS stamps, repetitive parallel pattering of SWCNT was achieved which is shown in figure 2. The PDMS stamp was inked alternatively with anionic and cationic SWCNTs and these were transferred in a grid like manner onto the base microelectrode platform to form crossbar p-n ionic junction comprising of clusters of SWCNTs. Finally, the assembled crossbar ionic junction was electrically characterized to demonstrate the formation of the ionic junction.

3   RESULTS AND DISCUSSION

In our earlier research we have shown that randomly dispersed CNT's have a resistive lumped element characteristics. These randomly dispersed SWCNT when physically assembled in well-ordered patterns demonstrate device configuration such as transmission lines and transistors. The P-type and N-type SWCNT in this application were stamped in a crossbar manner to form p-n ionic junction nanodevices. Figure 3 demonstrates the forward and reverse bias junction characteristics.

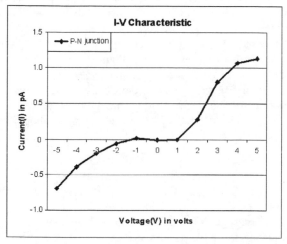

Figure 3: Represents I-V characteristics of P_N ionic junction. The graph shows rectifying forward junction with a turn ON voltage of 1V.

In this paper, micro contact patterning techniques associated with soft lithography polymer based processes were implemented for the simultaneous and rapid assembly of nanomaterial. This technique has helped to overcome the issues associated with the use of standard electron beam lithography techniques that are currently being adopted for individual patterning of nanomaterial, namely time and cost per structure. We have demonstrated the formation of crossbar p-n junction arrays using ion doped SWCNT by adopting micro contact printing techniques. We have currently established that integrated SWCNTs with microstructures in a patterned manner is a nanoelectronic tool for understanding nanoscale electro-ionic transport, as these tools can be easily probed for electrical characterization. This device is a good example of multi-dimensional integration of SWCNTs with metallic microelectrodes. These microelectrodes can be probed using standard lead based address lines associated with IC technology.

Analogous to the electrons and holes transport in a semiconductor p-n junction, ionic charge transfer occurs through the ionic junction in chemically doped nanomaterial. When anionic and cationic SWCNTs are patterned to form crossbar ionic junctions across base electrodes, ions transfer between two electrodes. We observe this behavior in the electro-ionic tool that we have developed by using the dually doped SWCNT grid. We anticipate that ionic tunneling and ionic charge transfer occurs between two adjacent SWCNTs. This behavior when cumulated over clusters of SWCNTs in the ionic unction result in a potential difference across the nanomaterial junction resulting in tuning of Fermi energy level. Hence, rectification is observed in an ionic junction similar to the rectification behavior due to electron-hole transport in a semi conducting junction.

4   CONCLUSIONS AND FUTURE WORK

In summary, we have demonstrated a method for chemically modulating the electrical properties of nanomaterial and implemented simple, low cost and rapid prototyping technique for developing crossbar SWCNT p-n junction arrays. The electrical rectification characteristics show promise towards assembling of junction array for building highly integrated circuits. Future work involves building of complex functional devices by assembly ionic nanodiode as principal building blocks.

REFERENCES

[1]   Y. Oaki and H. Imai, Chem. Commun., 6011-6013, 2005.
[2]   T. Schenkel, Nature Materials 4, 799–800 2005.
[3]   D. Whang, S. Jin, Y. Wu, and C. M. Lieber, Nano Letters 3, 1255-1259, 2003.
[4]   B. L. Allen, P. D Kichambare and A. Star, Adv. Mat. 19, 1439-1451, 2007.
[5]   N. A. Melosh, A. Boukai, F. Diana, B. Gerardot, A. Badolato, P. M. Petroff, and J. R. Heath, Science 300, 112-115, 2003.
[6]   P. Yang, Nature 425, 243 - 244, 2003.
[7]   P. L. Degen, Physica Status Solidi (a) 16, 1973, 9-42.
[8]   K.-S. Kim, Y.-H. Song, K.-T. Park, H. Kurino, T. Matsuura, K. Hane, and M. Koyanagi, Thin Solid Films 369, 207-212, 2000.
[9]   S. G. Tavakoli, S. Baek, and H. Hwang, Materials Science and Engineering B 114-115, 376-380, 2004.

Novel Branched Nanostructures of Carbon Nanotubes on Si Substrates Suitable for the Realization of Gas Sensors

Y. Abdi, S. Mohajerzadeh, M.H. Sohrabi, M. Fathipour and M. Araghchini

Nano-Electronic Center of Excellence
Thin film Lab. and Nanoelectronic lab, ECE Department,
University of Tehran, Tehran, Iran, Email: mohajer@ut.ac.ir

ABSTRACT

A novel tree-like nanostructure of carbon nanotubes is reported through the multi-stage growth of vertically aligned nanotubes. The growth of branched structures is feasible by means of a hydrogenation treatment of already grown carbon nanotubes on silicon substrates. Subsequent growth of CNTs would lead to the evolution of tree-like nano-structures. Such branched nano-structures are useful for the formation of gas sensors as well as field emission devices and displays.

Keywords: hydrogenation, carbon nanotubes, nickel seed, tree-like, vertical growth.

1   INTRODUCTION

Carbon nanotubes are of great importance due to their exceptional electrical and mechanical properties. Such one dimensional structures have been considered as promising candidates for ballistic field effect transistor fabrication, gas sensors, field emission devices and displays and also as additive for improving the mechanical properties of polymers and cements. They can be used as sharp tips for AFM applications as well as for nano-writing and nano-lithography [1-2].

The growth of carbon nanotube has been feasible using a variety of techniques among which one can identify chemical vapor deposition and specially, plasma enhanced chemical vapor deposition as one of the most promising methods to realize electronic-grade structures. The presence of plasma during the growth, not only lowers the processing temperature, it also leads to a vertical alignment of CNTs which can be further used for the formation of composite structures using an oxide or polymer matrix.

In this paper we report a new structure in which the growth of carbon nanotubes is achieved on top of already grown CNTs. The initial growth happens on (100) silicon substrate whereas the subsequent growth is achieved on top of the vertically grown nanotubes, using the trapped nickel at the top side of CNTs as the seed for the second and third growth. Scanning electron microscopy has been widely employed to study the growth. Also a preliminary sensing element for oxygen has been realized. Apart from gas sensing elements, the newly presented nano-structures can find applications for the formation of nano-actuators and motors where the branched structure assimilates the rotating element of the nano-metric system.

2   EXPERIMENTAL

The fabrication of branched structures started with the deposition of 5-8 nm thickness of Ni layer by e-beam evaporation system onto a (100) silicon substrate with a P-type doping of 1×10^{15} cm^{-3}. Ni was used as a catalyst of growth and can be patterned using standard photo-lithography. If features smaller than 0.5μm are needed standard optical lithography cannot be exploited. To achieve such features we have developed a hydrogenation assisted nano-island formation where small clusters of nanotubes can be formed with no need to a nanolithography technique [3].

As the second step, vertical nanotubes were grown using plasma enhanced chemical vapor deposition (PECVD) on Si wafers. This process was carried out at the temperature of 650ºC and at a pressure of 1.6 torr of acetylene/hydrogen gases. The ratio of acetylene/hydrogen flows during the process was kept at 1 to 4. Well-aligned nanotubes with a diameter of 50 to 150 nm are regularly obtained. Most recently we have been able to achieve nanotubes with diameters below 10nm using a reduced plasma power during the hydrogenation step and by increasing the hydrogen percentage during the growth of CNTs in the C_2H_2/H_2 mixture.

In Figure 1 we have collected some of the SEM images of the samples prepared for this study. Parts (a) and (b) of Fig. 1 depict two of the scanning electron microscope images of the patterned and grown nanotubes on Si with desired features. The evolution of a circular ring as seen in part (a) of this image is a result of direct photo-lithography of a small round and solid spot of nickel as the seed for the growth of CNTs and a proper hydrogenation to hollow out the circle and to form a ring. To obtain the image of Fig.(1.b), we have used a back-scattering mode of SEM imaging. As shown in this image, Ni particles are placed on top of nanotubes, as expected from a tip-growth mechanism. The idea of the growth of branched-nanostructures originates from using such Ni particles for the subsequent growth of CNT branches on top of the original nanotubes. The process flow is schematically shown in Figure2. Hydrogen plasma was used to create small Ni nano-particles on tip of nanotubes. To protect the nanotubes, they were coated by TiO_2 layer. Deposition of TiO_2 was carried out in CVD reactor in presence of oxygen gas and $TiCl_4$ vapor at a temperature of 220°C and at atmospheric pressure.

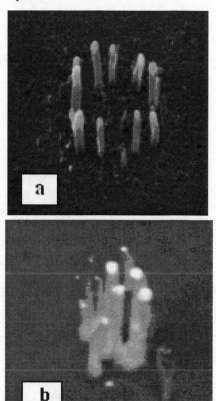

Figure 1: A collection of various CNTs grown on Si substrate patterned prior to the growth. (a) Circular growth, (b) a small cluster of CNTs where its image indicate the presence of Ni at the tip side.

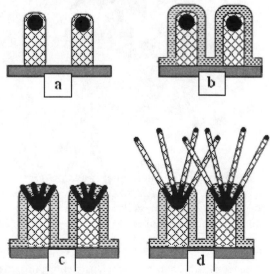

Figure 2: The evolution of branched structure from vertically aligned carbon nanotubes (a). Coating with titanium oxide (b), hydrogenation bombardment (c) and the subsequent growth (d).

TiO_2-coated nanotubes were placed in the PECVD reactor and at a temperature of 650°C and a plasma power density of 5.5 W/cm^2, the hydrogenation processing of the sample has been carried out. The pressure of the chamber was kept at 1.5 torr during the hydrogenation step. After 15 minutes of hydrogenation, the tip side of nanotubes is expanded and small particles of Ni appear. These Ni nano-particles act as catalyst seed for the growth of secondary nanotubes on original structures. The hydrogenated samples were immediately exposed to the acetylene gas in the same reactor and the growth of branches on nanotubes was achieved.

3 RESULTS AND DISCUSSION

Figure 3 demonstrates the SEM results of the process which was schematically presented in the previous section as a step by step manner. Parts (a) and (b) of this figure correspond to the growth of individually placed nanotubes before and after coating with the TiO_2 layer. Part (c) of this figure shows the TiO_2-coated nanotube after hydrogen bombardment. As shown in this part, Ni catalyst comes out through the titanium-oxide layer. Part (d) shows an image of the nanotubes after subsequent growth has been accomplished.

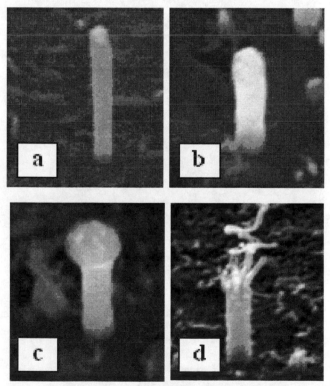

Figure 3: (a) Individually placed CNTs on Si, (b) coated with TiO2, (c) hydrogenation of the specimen leading to an expansion of the tip side and (d) the growth of a second growth.

Because the size of newly evolved nickel spots is much smaller than the original seeds, the secondary nanotubes with smaller diameters can be grown on top of initial nanotube, as shown in part (d).

A collection of tree-like nano-structures is also presented in Fig. 4. Two top images in this figure show the creation of individually placed branched nanotubes on top of individual CNTs on the silicon substrate. The image at the bottom depicts a corner of a large cluster of CNTs formed in a branched like manner. By controlling the plasma power density in hydrogenation step, samples with different diameter of branches were obtained.

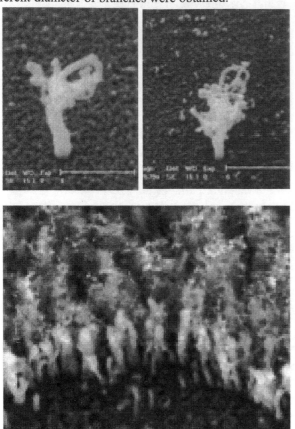

Figure 4: SEM images of tree-like structures with different diameter of branches. The bottom image shows a cluster of CNTs in a branched-like structure.

3.1 GAS SENSORS

For the fabrication of the gas sensor based on such nano-structures, Si substrates were first doped by phosphorous in designated areas to form P-N junctions at desired places. Then Ni layer was deposited and patterned to form interdigital structures suitable for gas sensor fabrication. The exposed areas of the silicon substrate were vertically etched away to create recession in those regions and to avoid short-circuiting the parallel fingers of the interdigital structure. This step is carried out using a reactive ion etching technique assisted by hydrogen/oxygen passivation. By using the process which was mentioned before, tree-like structures were grown on these interdigital patterns.

Figure 5 represents the SEM images of the sensor structure. Parts (a) and (b) show nanotubes and tree-like structures on comb-like fingers, respectively. As shown in the figure in some places, branches were joined together. So the fingers of the interdigital structure are electrically connected together through these branches. It must be born in mind that the electrical isolation of parallel fingers in the original interdigital structure has been ensured by a P-N junction isolation.

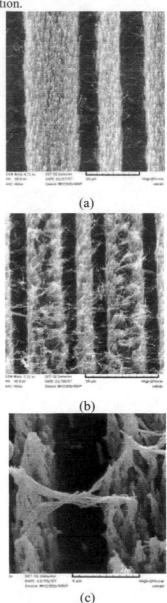

(a)

(b)

(c)

Figure 5: The formation of interdigital structures for the gas sensor fabrication. (a) as-grown CNTs, (b) tree-like formation and (c) a higher magnification view showing the connection of neighboring fingers.

In order to perform the gas sensing experiment the sample is placed in a chamber which is being purged by nitrogen as an inert ambient. By introducing oxygen gas to the chamber, the electrical conductivity of the structure shows a considerable reduction. Oxygen incorporation can passivate the carbon bonds and lower the electrical

NSTI-Nanotech 2008, www.nsti.org, ISBN 978-1-4200-8503-7 Vol. 1

conductivity of the sensor. Since the sensing mechanism is mainly due to the oxygen presence at the surface of the device, this structure can also be used for sensing the reducing gases such as hydrogen and carbon monoxide.

Figure 6 represents the sensing characteristic of the structure. This measurement has been carried out at room temperature and oxygen gas was introduced after 20s. As seen from this figure, not a significant increase is observed in the electrical resistance of the sample.

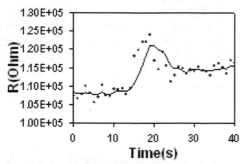

Figure 6: The resistance-time behavior of the sample at room temperature. Not a significant result is observed under such conditions.

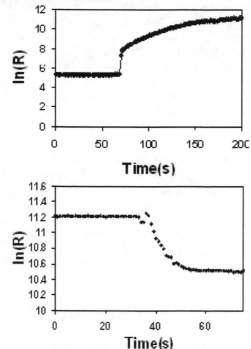

Figure 7: The response of the sensor to gas incorporation at a temperature of 200°C (top) evidencing a sharp and quick response. (bottom) The recovery of the sensor at the same temperature.

For better functionalizing the nanostructures the measurement was carried out again at 200 ºC. Results of this measurement are shown in figure 7. Part (a) of the figure demonstrates that by introducing the gas, the resistance of the sensor was increased. Part (b) shows the electrical characteristic of the sample during the recovery period.

4 CONCLUSIONS

We have successfully fabricated new branch-like nano-structures of carbon nanotubes on silicon substrates using a hydrogenation-assisted sequential growth. The evolution of such nano-structures has been possible by a second growth of carbon nanotubes on top of the original vertically aligned structures. The coating of the nickel seed which is placed at the tip side of the CNTs by a titanium oxide layer and its expansion by a hydrogenation process leads to a leach of nickel through the protective layer and allows the formation of newly evolved nano-size islands of nickel on the very tip side of the original CNTs. These new sites can be used as the seed for the growth of thinner nanotube and in the form of a branched tree-like structure. The realization of such nano-structures on parallel lines of an interdigital structure has been used as a device for gas sensing application.

This work has been supported with a grant from Ministry of Industry and Mines of Iran and partial support from Research Council of the University of Tehran. Technical assistances of Mr. J. Koohsorkhi and Dr. M.D. Robertson are greatly acknowledged.

REFERENCES

[1] J.-M. Bonard, M. Croci, C. Klinke, F. Conus, I. Arfaoui, T. Stockli, A. Chatelaine, Carbon 40 (10) (2002) 1715.
[2] M.I. Milne, K.B. Teo, G.A. Amaratunga, P. Legaganeux, J. Mater. Chem. 14 (6) (2004) 933.
[3] Y. Abdi, S. Mohajerzadeh, H. Hoseinzadegan, J. Koohsorkhi, Appl. Phys. Lett. 88 (2006) 053124.

Electronic structure of single wall carbon nanotubes under transverse external electric field, radial deformations and defects

Y. Shtogun and L. M. Woods

University of South Florida, yshtogun@cas.usf.edu

ABSTRACT

We present density functional theory and tight binding calculations of a radially deformed (8,0) single walled carbon nanotube under transverse external electric fields. Density functional theory calculations are also presented for radially deformed and defective (8,0) nanotube. Three types of single defects are considered – Stone-Wales, single Nitrogen impurity, and a mono-vacancy. The electronic structure and energy gap changes are explained in terms of orbital admixture in the total density of states and energy band structure as a function of deformation and fields strengths, types of defects, and/or curvature. Our results can be used to understand and interpret experimental data of realistic nanotubes under extreme conditions, as well as to engineer new or build on existing nanoelectronic devices.

Keywords: carbon nanotubes, electronic structure, electric field, defects, deformations

1 INTRODUCTION

Carbon nanotubes (CNT) are cylindrical quasi-one dimensional graphitic structures characterized with a chiral index *(n,m)* with diverse properties and many potential applications [1]. There has been significant interest in modulating different CNT properties in order to improve our fundamental understanding of these systems as well as to explore possibilities for new devices.

CNT properties can be modified in various ways. One way is to apply a transverse external electric field (TEEF) to the radial direction of the CNT. Strong enough TEEF can couple different energy bands in the nanotube electronic structure resulting in possible semiconductor-metal transitions [2,3]. Understanding the CNT electronic structure as a function of TEEF is an important element in nanotube based field-effect transistors, rectifiers, or p-n junctions [1].

CNT properties can also be modified by incorporating mechanical defects or deformations in their structure. Defects are always present in nanotubes during the stages of synthesis and purification [1], or later on during device production [4]. Defects can also be engineered using electron, ion irradiation [5] or chemical methods. In addition, radial deformation of CNTs has been shown to lead to significant changes in their properties and semiconductor- metals transitions can be achieved. Such deformations can be achieved by applying external

hydrostatic pressures [6] or by squashing the nanotube using an AFM tip [7].

Nevertheless, modifying the CNT properties can be challenging experimentally. For example, transforming semiconducting CNTs into metallic ones by radial deformation requires the application of pressures on the order of several ~GPa [8]. Also, relatively strong TEEF ~0.5-0.8 eV/Å are needed to close the gaps for most semiconducting CNTs [3,9]. The large deformation and electric field strengths have led to the search for different ways to modulate the CNT properties. Recently, researchers have shown that weaker TEEF fields can introduce dramatic changes in a defective nanotube resistance and gap as compared to the resistance and gap of a perfect one [10].

Here we investigate additional ways for modulating the CNT characteristics by considering a carbon nanotube under the influence of two external factors. The first case we investigate is related to understanding the combined effect of radial deformation and TEEF on the nanotube electronic structure, and the second one is the combined effect of various mechanical defects and radial deformations. This work presents a fundamental theoretical study of the electronic structure changes as a function of such external factors. Our results reveal that CNT metal-semiconductor transitions can be achieved for various combinations of electric fields and mechanical modifications. Therefore, greater experimental capabilities for new devices are possible when two external factors are applied.

The rest of the paper is organized as follows. In Section 2 the calculation methods are presented. In Section 3 results and discussions are given for deformed CNT under TEEF and for deformed CNT with a defect.

2 CALCULATION METHODS

To obtain and analyze the electronic structure as a function of external electric fields, deformations, and defects, we apply *ab initio* density functional theory (DFT) and tight binding (TB) methods. To illustrate our results we take a single walled nanotube (SWNT) with a chiral index (8,0). The circular perfect structure is simulated first using the DFT VASP package [11] within the local density approximation for the exchange correlation function. A 1x1x7 Monhorst-Pack k grid sampling of the Brillouin zone was taken with an energy cutoff of 420 eV. We construct a super-cell consisting of 4 unit cells along the axial direction of the nanotube with length 17.03 Å after relaxation. The length in the transverse direction is 22.12 Å after the ionic relaxation. The convergence criteria are taken as 10^{-5} eV for the energy and 0.005 eV/Å for the force.

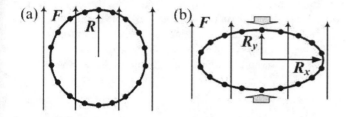

Fig.1 (a) - Perfect and (b) - Radially deformed SWNT under a transverse external electric field F. $2R_x$ is the major axis connecting the highest curvature and $2R_y$ is the minor axis connecting the lowest curvature.

The radially deformed (8,0) nanotube was also calculated using VASP with the same super-cell under the same convergence criteria. The tube is squeezed in the y-direction and elongated in the x-direction. This is characterized by a dimensionless parameter defined as $\eta = (R - R_y)/R$, where R is the radius of the perfect nanotube, and R_y is the semi-minor axis connecting the SWNT regions with lowest curvature – Fig. 1. For all values of η the system is relaxed by freezing only the y-coordinate of the atoms on the top and bottom rows of the flattened regions while all others are let free.

After the relaxed structure of the (8,0) SWNT for various η was obtained, various defects were also introduced. We consider three types of defects – Stone-Wales defect, a single vacancy, and a N substitution impurity. Each defect is calculated separately. For a given η, the Stone-Wales defect is simulated on the highest curvature region in the (8,0) supercell. The same is done for the vacancy and the N impurity. Such procedure corresponds to first deforming the nanotube and then introducing the defect in the structure. The vacancy and impurity defects are simulated with the inclusion of spin polarization effects, while the Stone-Wales defect is done without spin polarization.

In addition, the structure of the perfect and radially deformed (8,0) nanotube is calculated using a four orbital $\sigma-\pi$ non-orthogonal tight binding model [12]. This is motivated by the fact that the application of an electric field breaks the periodicity of the system. Since DFT methods are applicable only to periodic system, the TB method is adapted to determine the TEEF effect on the nanotube electronic structure.

The TB model involves solving the non-orthogonal eigenvalue problem for the matrix $(H_{TB} - ES)$, where H_{TB} is the tight-binding Hamiltonian for the (8,0) tube including the $\sigma-\pi$ orbitals for each atom, E is the energy, and S is the overlap matrix written in the Slater-Koster scheme. The positions of the atoms for the perfect and radially deformed (8,0) tubes are taken to correspond to the results from the DFT calculations. Also, the TB parameters were adjusted in order to have a good agreement with the DFT results for the electronic structure in each case.

The TEEF provides an additional contribution H^F to the tight-binding Hamiltonian matrix H_{TB}, [13] with diagonal elements $H^F{}_{ii} = -e F \mathbf{R}_i \cos(\varphi_i + \theta)$ and off-diagonal elements $H^F{}_{ij} = -e S_{ij} F \mathbf{R}_{ij} \cos(\varphi_{ij} + \theta)$. The notations are as follows: e is electron charge, F is the strength of the external electric field, θ is initial phase of TEEF with respect to the x-axis, S_{ij} are the overlap matrix elements between two nearest neighbor atoms, $\mathbf{R}_i$ is the distance from the ith atom to the center of the nanotube, $\mathbf{R}_{ij} = (\mathbf{R}_i + \mathbf{R}_j)/2$ is the distance to the center of nanotubes for the center of mass of the two atoms, φ_i and φ_{ij} are the angles between the direction of F and $\mathbf{R}_i$ and $\mathbf{R}_{ij}$, respectively. The inclusion of the $\sigma-\pi$ orbitals and the non-orthogonality condition is necessary to describe correctly the electronic structure changes due to the various energy bands admixture from the electric field and from to the high curvature regions for the radial deformation cases.

3 RESULTS AND DISCUSSIONS

3.1 Radially deformed (8,0) SWNT under transverse external electric field

The electronic structure of the (8,0) SWNT with various degrees of radial deformation η is calculated using VASP first. The results are summarized in Fig. 2. The perfect (8,0) tube is semiconducting with a relatively large band gap $E_g=0.55$ eV. Fig. 2a) shows that increasing η results in decreasing of the band gap. Eventually the gap closes at $\eta=0.25$. Further increase in the deformation keeps the band gap closed. This behavior is due to the hybridization of the $\sigma-\pi$ orbitals mainly from the regions with higher curvature. We find that if all atomic coordinates of the deformed tube are allowed to relax, the tube returns to its perfect circular form. In fact, this elastic type of deformation is found for all $\eta \leq 0.75$. If $\eta > 0.75$, the tube breaks down.

The DFT structure and band gap evolution are also calculated and reproduced using the non-orthogonal TB model. Next, TB calculations are performed when TEEF is applied to the (8,0) tube with different degrees of deformation. Fig. 2b) shows how the band gap changes as a function of the electric field strength for several deformations η. The band gap for the perfect tube does not change much for smaller valeus of F, and it eventaully shows oscillatory-like behavior as F is increased indicating several metal-semiconducor transitions. As the radial deformation is increased, E_g first increases and then decreases following similar oscillations as in the case for $\eta=0.0$. Even when the gap is closed, the electric field can induce several metal-semiconductor transitions. This is characteristic for a metallic nanotube response to an external electric filed. In fact, we examined several $(n,0)$ SWNTs, and we found that for all tubes (perfect or deformed) with small or no band gaps, relatively small electric fields can open larger band gaps. For nanotubes with relatively large energy gaps, stronger fields are needed to decrease E_g [3].

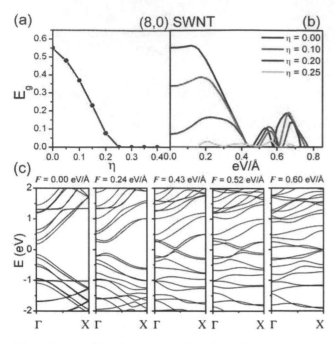

Fig.2 (a) – (8,0) SWNT E_g as a function of deformation η; (b) – (8,0) SWNT E_g as the function of F for different value of η; (c) – deformed (8,0) SWNT ($\eta = 0.10$) energy bands for several TEEF strengths.

The origin of the semiconductor-metal transitions are analyzed by calculating the energy band structure. In Fig. 2c) we show the energy bands for $\eta=0.10$ and several values of the TEEF strength. The electric field couples the states according to the selection rule $\Delta J = \pm 1$. Note that for a perfect tube, each energy level is characterized by the angular momentum quantum number J. The radial deformation couples states according to the selection rule $\Delta J = 0, \pm 2$. When both F and η are present, the band gap changes are due to $\Delta J = 0, \pm 1, \pm 2$ as well as to the increased $\sigma-\pi$ hybridization from the higher curvature regions. Thus the band gap opening or closure happens as a result of the competing effects between different energy levels admixture originating from the increased curvature and/or the strength of the applied electric field.

3.2 Radially deformed (8,0) SWNT with defects

Here we calculate and analyze the electronic structure of radially deformed (8,0) SWNT with various defects imbedded in its structure. Three single defects are considered – a Stone-Wales (SW) defect, a N atom

substitution, and a single vacancy. For discussion purposes the cases with $\eta=0.0$ and $\eta=0.20$ are taken as examples. The bare perfect and deformed (8,0) nanotubes are calculated first using the DFT-VASP code as specified in Section 2. After the relaxation was completed, each defect is introduced on one of the highest curvature sides for the deformed tube. The deformed and defective tube atoms are also allowed to relax. This procedure corresponds to first deforming the nanotube and then making the defect.

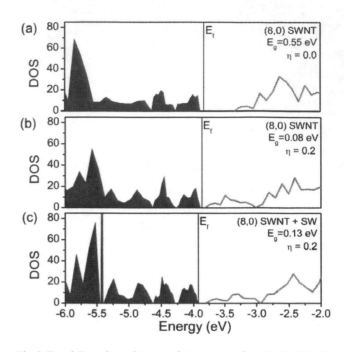

Fig.3 Total Density of States for (a) – perfect (8,0) SWNT, (b) – radially deformed (8,0) SWNT for $\eta = 0.20$, (c) – radially deformed (8,0) SWNT with a Stone-Wales defect on the high curvature side for $\eta = 0.20$.

We consider first the SW defect. It is characterized by rotating one C-C bond in the tube structure by 90^0. Since the chemical structure of the nanotube is not changed in this case, the SW defective tube is a closed-shell system as in the case of a perfect nanotube. Thus our calculations do not involve the inclusion of a spin polarization effects. We find that the most energetically stable SW defect is when the bond parallel to the nanotube axis is rotated. In Fig. 3, the total DOS is shown for the perfect, radially deformed $\eta=0.20$, and radially deformed $\eta=0.20$ with a SW defect (8,0) carbon nanotube. The radial deformation decreases the original $E_g=0.55$ eV (Fig. 3a) to $E_g=0.08$ eV (Fig. 3b), while the SW defect increases the gap to $E_g=0.64$ eV (DOS not shown in here). The SW defect also causes increasing in the gap to $E_g=0.13$ eV (Fig. 3c) for the deformed tube with $\eta=0.20$. The electronic structure reveals that the SW effect affects the deeper valence region of the deformed (8,0) tube $\sim$-5.4 eV, where a sharp peak is found, and states around the Fermi level are not affected much by it.

Next we consider the defect created by substituting one of the nanotube C atoms on the highest curvature side for the deformed case $\eta=0.20$ with a N atom. This results in obtaining a partially filled electron shell system since the N atom has an extra e as compared to C. Therefore, we include spin polarization in the calculation procedure. The results are shown in Fig. 4a). It is evident that $E_g=0.0\ eV$ for the radially deformed (8,0) tube with $\eta=0.20$ as compared to $E_g=0.08\ eV$ for the defect free deformed tube. In addition, small difference in the spin "up" and spin "down" states is found 0.4 eV below the Fermi level indicating that some magnetism can exist in a N-doped carbon nanotube. The peak at -4.15 eV is mainly due to the N impurity in the structure.

Finally, a mono-vacancy is created by removing a C atom on the highest curvature side of the deformed nanotube with $\eta=0.20$. The results are shown in Fig. 4b). One sees that there is no difference in the spin "up" and "down" species due to the spin-polarization effects. The band gap of 0.08 eV for the defect-free (8,0) tube with $\eta=0.20$ is reduced to 0.03 eV for the defective deformed tube. The main contribution from the vacancy is seen mainly in the conduction region ~ -3.5 eV as compared to the DOS of the defect free nanotube (Fig.3a).

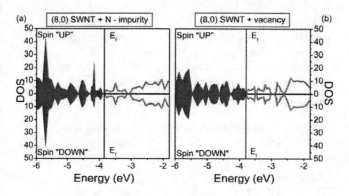

Fig. 4 a) Total density of states for spin "up" and "down" for (8,0) nanotube with $\eta=0.20$ and a N impurity; b) Total density of states for spin "up" and "down" for (8,0) nanotube with mono-vacancy with $\eta=0.20$.

In summary, our comprehensive study shows that by combining different strengths of deformations and electric field, one can achieve several insulator-metal transitions in a single deformed nanotube. In addition, depending on the type of single defect introduced in the deformed tube, various energy gap and electronic structure modulations can be achieved.

REFERENCES

[1] R. Saito, G. Dresselhaus, and M. S. Dresselhaus, "Physical Properties of Carbon Nanotubes," Imperial College Press, London, 1998.

[2] Y. H. Kim and K.J. Chang, Phys Rev. B 64, 153404, 2001.

[3] Y.Shtogun, and L.Woods, Appl. Phys. Lett. submitted for publication, 2007.

[4] D.B.Mawhinney, V. Naumenko, A. Kuznetsova, J. T. Yates Jr., J. Liu, and R. E. Smalley, Chem Phys. Lett. 324, 213, 2000.

[5] P.M. Ajayan, V. Ravikumar, and J.-C. Charlier, Phys.Rev. Lett. 81, 1437, 1998.

[6] S. A. Chesnokov, V. A. Nalimova, A. G. rinzler, R. E. Smalley, and J. E. Fisher, Phys. Rev. Lett. 82, 343, 1999.

[7] T.W. Tombler, C. Zhou, L. Alexseyev, J. Kong, H. Dai, L. Liu, C. S. Jayanthi, M. Tang, and S. Wu. Nature 405, 769, 2000.

[8] M. Mazzoni and H. Chacham, Appl. Phys. Lett. 76, 1561, 2000.

[9] C.W. Chen, M. H. Lee, and S. J. Clark, Nanotechnology 15, 1837, 2004.

[10] J. Y. Park, Appl. Phys. Lett. 90, 023112, 2007.

[11] G. Kressen and J. Futhmuller, Phys. Rev. B 54, 11169, 1996.

[12] V. N. Popov and L. Henrard, Phys. Rev. B 70, 115407, 2004.

[13] X. Zhou, H. Chen, and O. Y. Zhong-can, J. Phys.: Cond. Mat. 13, L635, 2001.

Multiwalled carbon nanotube films as temperature nano-sensors

A. Di Bartolomeo[*,***], F. Giubileo[*,****], M. Sarno[**,***], C. Altavilla[**,***], D. Sannino[**,***], L. Iemmo[*],
F. Bobba[*,***], S. Piano[*], A. M. Cucolo[*,***] and P. Ciambelli[**,****]

[*] Department of Physics, University of Salerno, via S. Allende, 84081 Baronissi (SA), Italy
[**] Department of Chemical and Food Engineering, University of Salerno, 84084 Fisciano (SA), Italy
[***] NANO_MATES, Research Centre for NANOMAterials and nanoTEchnology at Salerno University,
c/o Department of Physics, University of Salerno, via S. Allende, 84081 Baronissi (SA), Italy
[****] CNR-INFM Regional Laboratory SUPERMAT, via S. Allende, 84081 Baronissi (SA), Italy

ABSTRACT

Most of the basic research on the electrical behavior of CNTs has been carried out on individual or bundle nanotubes. More recently random or oriented CNT networks (CNTN) are emerging as new material for electronic application. We present the fabrication of thick and dense CNTNs, in the form of freestanding films, and the study of their electric resistance as a function of the temperature, from -200 to +150 °C. A non-metallic behavior has been observed with a monotonic R(T). A good long-term stability and a behavioral accordance with the temperature measured Si or Pt thermistor are demonstrated. We underline that a transition from non-metallic to metallic can take place at few degrees below 0°C. A model involving regions of highly anisotropic metallic conduction separated by tunneling barrier regions can explain the non-metallic to metallic crossover, based on the competing mechanisms of the metallic resistance rise and the barrier resistance lowering.

Keywords: carbon nanotubes, bucky paper, temperature nano-sensor, non-metallic behavior

1 INTRODUCTION

The temperature dependence of the electric resistance of carbon nanotubes is an important topic for technological applications. A through understanding of it is relevant for the utilization of carbon nanotube networks as sensing element in temperature nano-sensors. A sensor of nanometric size can provide local accurate measurements of a rapidly changing temperature, while reducing the possibility of disturbing the neighboring environment. In addition, the small size sensor implies a very low power consumption.

For single and multi-wall CNTs, both a non-metallic (with negative dR/dT) and a metallic (with positive dR/dT) temperature dependence of the electric resistance has been reported [1-6] as well as a mixed behavior with a transition from non-metal to metal occurring a few tens degree below 0 °C [7-10]. A remarkable similarity between the temperature behavior of CNTNs and of highly conducting polymers (as for example blends of polyaniline dispersed in non conducting PMMA) has been pointed out and attributed to a common feature in the conduction mechanism, namely the presence of metallic conducting regions separated by insulating barriers. A heterogeneous model involving regions of highly anisotropic metallic conduction separated by tunneling barrier regions can explain the non-metallic to metallic transition by competing mechanism of the metallic resistance augmentation and the barrier resistance lowering [11].

Since most of the basic research on the electrical behavior of carbon nanotubes has been carried out on individual or bundle nanotubes, it is interesting to study other forms of interacting nanotubes, such as random or oriented networks in freestanding thick films.

In this paper we present the fabrication of thick and dense multiwalled carbon nanotube (MWCNT) freestanding films (often referred as bucky paper) and the study of their electric resistance as a function of the temperature, from -200 to +150 °C. been sometimes observed.

2 EXPERIMENTAL

2.1 CNT fabrication process

MWCNTs have been synthesized by ethylene catalytic chemical vapour deposition (CCVD) on Co/Fe-Al_2O_3 catalyst, prepared by wet impregnation of gibbsite (γ-Al(OH)$_3$) powder with cobalt acetate (2.5 wt%) and iron acetate (2.5 wt %) ethanol solution [12]. The catalyst was dried at 393 K for 720 min and preheated before synthesis at 70 K/min up to 973 K under N_2 flow. For the CNT synthesis a mixture of ethylene 10% v/v in helium was fed to a continuous flow microreactor at 973 K, with a runtime of 30 min. Gas flow rate and catalyst mass were 120 (stp)cm^3/min and 400 mg. The nanotubes were obtained by a very effective synthesis, yielding more than 95% conversion of the injected carbon.

The MWCNTs selectivity was about 100%, while to remove catalyst impurities the sample was treated with HF (46% aqueous solution), and the solid residue was washed with distilled water, centrifuged and finally dried at 353 K for 12 h. High purity multiwalled carbon nanotubes (>97%) were obtained.

To prepare a BuckyPaper free-standing sheet, 0.5 g of MWCNTs were suspended in 100 g of water in presence of 0.1 mg of sodium dodecyl sulfate, sonicated and then vacuum filtered onto a membrane support. After drying, a bucky paper was removed from the support as MWCNT films of different thickness and density.

The paper can be folded and cut with scissor and was sufficiently robust to let stable silver paint contacts to be formed and to withstand long thermal stresses.

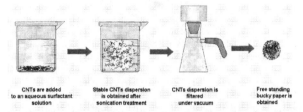

Figure 1: Scheme of the different steps for the production of the CNTNs.

2.2 Setup for electrical measurements

A 4-probe method was adopted to measure the low resistance of the CNT films, typically of ~1 Ω, and to overcome the problem of the comparatively high contact resistance. The electrical contacts were improved by making pads with smeared silver paint. Films of different but approximately rectangular shape, dimensions up to 3mm × 6mm and thicknesses between 300 and 500 μm, were measured by forcing a current of 1-10 mA through the outer probes and measuring the voltage thus developed between the inner ones. For this purpose a Keithley 4200 SCS was used as source and measurement unit (SMU). The temperature was monitored through a fast silicon temperature sensor (Infineon KT–11-6) and/or a platinum PT100 thermistor, mounted very close to the bucky paper and read by additional SMUs of the same Keithley 4200 SCS. The temperature cycles, with warming or cooling sweeps, were performed while operating the device in constant current mode and with a power consumption as low as 1 μW.

The temperature of the air-filled and high thermal capacity chamber, which was housing the sample, was varied by means of an external resistive heater; the measurements below ambient temperature were performed by flowing nitrogen vapours in the chamber or by inserting the sample in a dewar containing liquid nitrogen or helium.

Since considerably room temperature resistance drift was observed in non-treated samples, before systematic measurements, a few stabilising thermal annealing cycles, from room temperature up to about 100 °C, were performed. Thermal annealing is believed to make the

connections between the CNTs and of the CNTs with the silver paint more robust and to evaporate adsorbates.

3 RESULTS AND DISCUSSION

3.1 Bucky Paper Characterization

Figure 2a shows a typical SEM picture of the as produced bundle of nanotubes grown from the catalyst, with a length in the range 100-200 μm. Bundle organisation, constituted of entangled nanotubes, is more clearly visible in the SEM picture of Figure 2b, and in TEM Figure 2c. Nanotubes are multiwalled with a diameter ranging from 10 to 30 nm, while the internal diameter varies between 5 and 10 nm (see Figure 2d).

In Figure 3 a SEM image of a final film is shown.

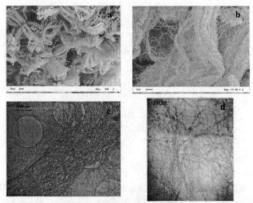

Figure 2: SEM image of as produced CNT bundles (a), of a particular of a bundle (b); TEM image of an as produced bundle (c), of CNTs at high resolution (d).

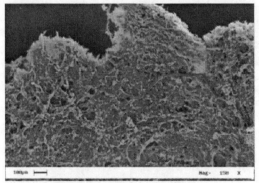

Figure 3: SEM image of the bucky paper

3.2 Electrical characterization

As expected, all the MWCNT freestanding films were highly conductive with a resistance around 1 Ω.

Figure 4 shows a typical result with a linearly decreasing resistance for raising temperature on the whole range investigated (from -40°C to +150°C). Several temperature sweeps, corresponding to warming up and cooling down cycles, were measured. A good reproducibility and a low hysteresis were obtained. These measurements also demonstrate that the CNT sensor and its contacts are not damaged by the temperature variations, as a consequence of a possible mismatch of the coefficients of thermal expansion at the interfaces of the device.

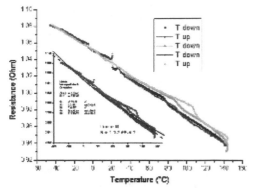

Figure 4: R-T characteristics of a bucky paper. The sample was heated and cooled down several times in the range -50 to 145 °C. The fits in the insert considers all the data together.

A linear fit to the data of figure 4, taken all together, can be used to estimate the temperature coefficient of resistance (TCR), defined as TCR = $1/R_0 \cdot dR/dT$ (where R is the resistance at temperature T and R_0 is the resistance at the standard temperature of 0°C). A negative TCR = -0.0007 is obtained and is consistent with values reported by other authors [6,13-15]. Such behavior is expected when the CNT network becomes thinner and less denser, i.e. with reduced metallic percolating paths, making a barrier tunneling mechanism dominant (see following).

The behavior of the bucky paper at lower temperatures was further tested with a different experimental setup, consisting of a Keithley 2400 used as source and meter unit connected to an insert that was slowly driven in or out of a dewar containing liquid helium. The curve shown in figure 5 confirms the non-metallic behavior, with lesser linearity.

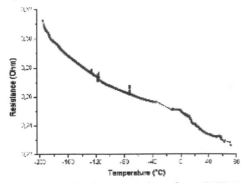

Figure 5: R-T characteristic of a CNTN at low temperature.

A different behavior is shown in figure 6. A sample, from the same batch of that used in the measurements of figures 4/5 was slowly heated from -45 till +140 °C (in a time of more than three hours). The curve obtained shows a non-metallic behavior at low temperatures which turns into a metallic one at T=-15°C.

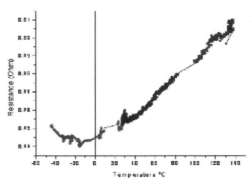

Figure 6: R-T characteristic of a bucky paper sample from the same batch of that used in the measurements of figures 4/5 and slowly heated from -45 to +140 °C.

A model of interrupted metallic conduction, with temperature dependent tunneling through thin electrical barriers separating metallic regions, has been suggested to account for the mixed non metallic-metallic behavior in quasi one-dimensional conductors, in which carriers cannot circumvent defects or other barriers to conduction [9,11]. If the barriers between metallic regions, intertube or inter-rope contacts or due to tubule defects are thin, a substantial conductance can still be seen in the absolute zero-temperature limit. As temperature increases, thermal fluctuations assist the tunnel and reduce the resistance (yielding a negative sign for the temperature coefficient, dR/dT<0). At higher temperatures, the usual increase in resistivity due to scattering of carriers by phonons may dominate the T dependence, thus leading to a change of behavior with a crossover and a valley in the R(T) curve. Backscattering by phonons is the main cause of the changeover to a metallic sign of dR/dT.

Such model can easily fit R-T experimental data with a non-metallic to metallic crossover.

3.3   CNTNs thermal response

Samples with a monotonic R(T) can be used to realize miniaturized temperature sensors, with fast response, repeatability, durability, etc. For an ideal sensor repeatability means to recover the same minimum/maximum resistance when temperature reaches the same minimum/maximum values.

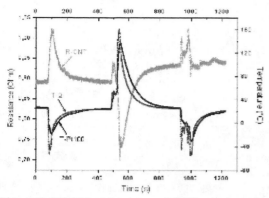

Figure 7: CNTN resistance and Si/Pt temperature vs time for a sample under heating/cooling fast cycles.

From figure 7 it is clearly seen that the resistance of our CNT sensor returns back to the same minimum/maximum value, with the same or an higher speed than the Si/Pt sensors.

To check the reliability of our MWCNT freestanding film we monitored its long-term stability by measuring the room temperature for a time of 24 hours. The result is shown in figure 8.

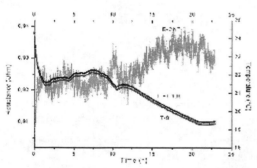

Figure 8: Room temperature monitored by the CNT film and Si/Pt thermistors.

A good accordance with the measurements performed by the Si/Pt thermistor are observed. The CNTN resistance fluctuation have successively discovered to be reduced by increasing the operating current.

4 CONCLUSIONS

In this paper, a study of the electric resistance vs temperature of freestanding MWCNT films has been reported in the aim of a possible application of carbon nanotubes as sensing element in temperature nano-sensors. A monotonic R(T) has been demonstrated, making the CNT films suitable as temperature sensors. Once protected from moisture and contaminants in an operational environment, CNTN films would have wide operating range, fast time response, low size and power consumption.

The temperature coefficient of resistance may not be so good as compared with the ones for platinum temperature sensors. However the nano-size of CNTNs can result in a very high sensitivity to the environmental temperature change and in an excellent time response, which is highly desirable for local measurements in systems with very rapid temperature variations and where the perturbation introduced by the thermometer has to be reduced as much as possible.

REFERENCES

[1] Langer L., Bayot V., Grivei E., Issi, J.P., Heremans J. P., Olk C. H., Stockaman L., VanHaesendonck C., Bruyinseraede Y., Phys. Rev. Lett., 76, 479, 1996.

[2] Song S. N., Wang X. K., Chang R. P. H., Ketterson J. B., Phys. Rev. Lett., 72, 697, 1994.

[3] Jang J. W., Lee D. K., Lee C. E., Lee T. J., Lee C. J., Noh S. J., Solid State Comm., 124, 147, 2002.

[4] Zhang H.L., Li J. F., Zhang B. P, Yao K. F, Liu W. S., Wang H., Phys. Rev. B 75, 205407, 2007.

[5] Deheer H., Bacsa W. S., Chatelain A., Gerfin T., Humphreybacker R, Forro L., Ugarte D., Science 268, 845, 1995.

[6] Kuo C. Y., Chan C. L., Ghau C., Liu C. W., Shiau S. H., Ting J. H., IEEE Transactions on Nanotechnology, 6, 63-69, 2007.

[7] Fisher J.E., Dai H., Thess A., Lee R., Hanjani N. M., Dehaas D. L., Smalley R. E, Phys. Rev. B, 55, R4921, 1997.

[8] Home J., LLaguno M. C., Nemes N. M., Johnson A. T., Fisher J. E., Walters D. A., Casawant M. J., Schmidt J., Smalley R. E., Appl. Phys. Lett., 77, 666, 2000.

[9] Skakalova V., Kaiser A. B., Woo Y.-S., Roth S., Phys. Rev. B, 74, 085403, 2006.

[10] Sun Y., Miyasato T., Kirimoto K., Kusunoki M., Appl. Phys. Lett., 86, 223108, 2005.

[11] Kaiser A. B., Dusberg G., Roth S., Phys. Rev. B, 57, 1814-1821, 1998.

[12] Ciambelli P., Sannino D., Sarno M., Leone C., Lafont U., Diamond & Related Materials 16, 1144–1149, 2007.

[13] Yosida Y., Journal of Physics and Chemistry of Solids, 60, 1-4, 1999.

[14] Fung C. K. M., Wong T. S., Chan R. H. M., Li W. J., IEEE Transactions on Nanotechnology, 3, 395-403, 2004.

[15] Bachtold A., Henny M., Terrier C., Strunk C., Schonenberger C., Salvetat J.P, Bonard J. M., L Forrò, Appl. Phys. Lett. 73, 274-276, 1998.

NANO STRUCTURE OF CARBON NANO TUBE PRODUCTS SYNTHESIZED IN SOLID PHASE

KC Nguyen*, S.T.Do*

*Saigon Hi Tech Park Research Laboratories, Hochiminh City, Vietnam **KTube Technology LLC, San Jose CA, USA

ABSTRACT

Solid phase synthetic process of carbon nano tube using solid precursor comprised of solid state carbon source, metallic catalyst and tube control agent has been reported in the last Nanotech 2007 (Santa Clara).

In the present study, we discovered that a tube control agent molecule having carbonitrile –CN functionality, is an effective hook which links carbon source molecule to metallic atoms to form tube shape products in the pyrrolysis process. Certain carbonitrile compounds show adequate interaction with specific carbon sources to give rise to unique tube structure and eliminate non-tube components in the product. A harmonica precursor with above mentioned 3 components can give rise to various tube shapes including fat tubes (outside diameter ≈ 600nm, inside diameter ≈ 400nm), thin tubes (diameter ≈ 20-30nm), ultra thin tube (diameter ≈1nm), twisting tubes etc. These tubes are suitable for applications in nanofiltration, nanocomposites for EM shielding, nanocomposites for light structural materials and even for energy devices.

Keywords: solid phase synthesis, solid precursor, tube control agent

5. INTRODUCTION

We have reported in Nanotech 2007 a novel solid phase synthetic process of carbon nano tube [1]. In this process, the tube growing process is controlled by a tube control agent mixed together with a solid carbon source instead of gas phase raw material conventionally utilized in catalytic growth process [2] . It is likely that somehow the tube control molecule must play a role of a connector slighly linking carbon source molecule with metallic atoms. The need of a chemical link between carbon source and metallic element has been report in a polymer template technique [3] utilized for the synthesis of SWCNT. In this technique, a copolymer of vinyl pyridine capable of forming complex with metal element becomes polymer template for synthesis of CNT using ethanol as carbon source. In the present study, we found that a molecule carrying carbonitrile -CN group even in a polymeric format or in a non-polymeric molecule can enhance the formation of tube shape products in our solid phase synthetic process. Furthermore, specific additives can dramatically change the tube geometry such as tube length, tube diameter as well as tube shape. Straight CNT, coiled CNT, Y-form CNT, zigzag CNT products also obtained from this process.

2. EXPERIMENTAL PROCEDURE

In the present study, the CNT was prepared by the pyrolysis of solid CS in a reactor described in Fig .1. The reactor is an oven 1 equipped with high heat resistant ceramic materials including oven cover 1.1, heat resistant layer 1.2, heat resistant ceramic tube 1.3 , coil heater 1.4 , heat controller 1.5 and a Pyrex glass reactor tube 2 . The Pyrex glass tube is connected with 3 neck connector in one end where suitable inert gases can be fed in or the air can be succeed out to form unoxidizing environment in the reaction chamber . The diameter of the Pyrex glass tube is about 25mm, active heating length is about 40 cm and the entire length of the tube is about 70cm. For the larger scale production, larger diameter (Þ=100 mm) and longer length tube will be used. The heating system (heater and

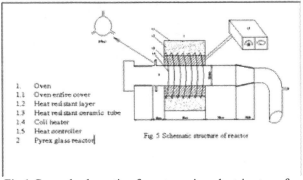

Fig 1 General schematic of reactor using electric stove for the synthesis of carbon nanotubes in solid phase

controller) can provide a well controlled temperature to the reactor chamber up to 1000C.

5. SOLID PRECURSOR AND TUBE FORMING PROCESS

The solid precursor is mainly composed of a) flammable solid carbon source b) hooks or tube control

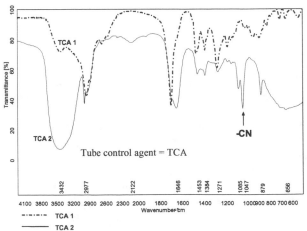

Fig. 2 FtIR chart of two different TCA in which TCA2 shows absorption peak due to the–CN group

agent having specific chemical functional groups of carbonitrile derivatives c) metal salts as metal source.

Fig.2 exhibits FtIR data of two different kinds of tube controls agents (TCA) (1) and (2) and one can see that only the additive (2) does contain carbonitrile –CN group but the first one doesn't .

Fig.3 exhibits FE-SEM image of the product of the solid precursor containing additive (1) and no tube shape product was achieved besides round shape particle

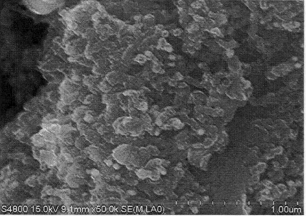

Fig.3 FE-SEM image of product using TCA 1 containing no-CN group. Absolutely, there is no tube observed.

agglomerate .

On the other hand, Fig. 4 shows the FE-SEM image of the product of the solid precursor containing additive (2). It is obvious that the tube shape product was successfully formed, confirming the role of the precursor carbonitrile – CN in the solid synthetic process. In this case, the tube is relatively straight with uniform diameter. This is the proof of carbonitrile –CN additives in the tube forming mechanism.

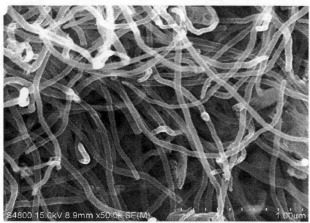

Fig. 4 FE-SEM image product using TCA2 containing CN group. Absolutely, tube shape product is obtained.

Next, Fig. 5, 6, 7 are FE-SEM image of solid phase synthesized products at three different process (A), (B), (C) showing three different shapes : fiber like (process A) , fat tube (process B) and coiled tube (process C) . It should be noted that fiber like product (process A) was made at high level of metal (carbon source CS/metal source MS > 0.5) showing tube diameter in the range of a few nm . On the other hand , the fat tube (outer diameter $\cong$ 600nm, inner

diameter $\cong$400nm) was formed when the unpurified carbon nano tube product in the solid phase , was used as catalyst. The fat tube product can be used as membrane for nanofiltration of waste water, blood .... The coiled product is achieved by a combination of process B with gaseous additive during burning period. The coiled product seems to be adequate for nano composite.

5. DISCUSSION

In the art of making carbon nano tube, the solid phase process reveals major advantages over the gas phase process such as a) well handled feed stock raw material (well controlled CS / MS ratio) b) suitable for large scale production c) better uniformity and higher purity d) more easily handled products ; for example, 100g of gas phase product has to be packed in a 100,000ml volume bag while solid phase products are easily contained in a 500 ml plastic jar e) well controlled morphologies with various additives .

In general, the solid phase synthesized products exhibit shorter length than gas phase products. Thus, it is ready to use and there is no need to cut long tubes into short tube using expensive and complicated tools such as E-beam or X-ray which may cause the damages for the tube. The short tube tends to show more porous than the very long tube produced in gas phase process. This unique feature of solid phase synthesized products makes it more compatible with polymeric binders in nano composite process.

In the solid phase synthetic process, the raw materials which are solid precursors need to have some hooks which can effectively generate the chemical links between free radicals generated in early stage of heating. We found that the effective hook is selected from group of chemicals having carbonitrile -CN substituents such as phthalonitrile, polymeric molecules containing –CN such as poly acrylonitrile and its copolymers including ABS resin (copolymer of acrylonitrile, butadiene, and styrene). Phthalonitrile has been known as precursor for metal chelates such as metal phthalocyanine pigments synthesis. Using a mixture of phthalonitrile and transitional metallic salts as a solid precursor in a oxygen free chamber of a furnace heated up at 1000C shows the formation of multiwalled carbon nano tube (MWCNT) and singlewalled carbon nano tube (SWCNT) upon the variation of (CS/MS) ratio. In the solid phase synthetic process, TGA data shows that the pure carbon nano tube content occupies up to 75% wt in the product in-situ. The impurities mainly are metals and non tubular carbon including amorphous carbon and the like. Re-heating the product in-situ in ambient at 1000C will eliminate the non tubular products. Otherwise the non-tubular products can be also eliminated by chemical process.

5. CONCLUSION

MWCNT product has been achieved with solid phase synthetic process using additive containing carbonitrile – CN functionality group .Changing the chemistry and the amount of additive (or tube control agent) will change the tube shape . Further works will be done with SWCNT in future .

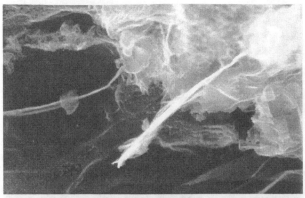

Fig.6 FE-SEM image of process (B)'s product
(Unpurified CNT product was used as catalyst)

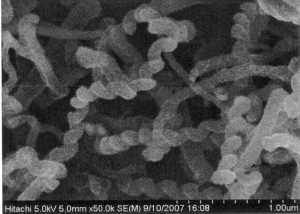

Fig.7 FE-SEM image of process (C)'s product
(Process B + gaseous additive)

REFERENCES

[1] K.C. Nguyen, S.T. Do T.V.Le, S.T. Nguyen, Nanotech 2007 Proceeding, and p. 61

[2] Jos-Yacamn, M. (1993). "Catalytic growth of carbon microtubules with fullerene structure". *Appl. Phys. Lett.* 62: 657

[3] J.Lu, Q.Fu, C.Lu, and J. Liu, Nanotech 2007 Proceeding, p.25

Formation of Single Walled Carbon Nanotube Via the Interaction of Graphene Nanoribbons: Molecular Dynamics Simulation

A. J. Du*,**, Sean C. Smith *,** and G. Q. Lu**

*Centre for Computational Molecular Science, Australian Institute for Bioengineering and Nanotechnology, The University of Queensland, QLD 4072, Brisbane, Australia, a.du@uq.edu.au and s.smith@uq.edu.au

**ARC Centre for Functional Nanomaterials, Australian Institute for Bioengineering and Nanotechnology, The University of Queensland, QLD 4072, Brisbane, Australia, maxlu@cheque.uq.edu.au

ABSTRACT

Classical Molecular Dynamics simulations were carried out to study the interaction of two zigzag graphene nanoribbons (GNRs). Remarkably, single walled armchair carbon nanotube could be formed via two zigzag GNRs at room temperature. The reaction process strongly depends on the distances between two ZGNRs and the widths of ZGNR. Our results suggest an effective route for the controllable growth of specific armchair nanotube.

Keywords: Graphene Nanoribbon, Single walled carbon nanotube, Molecular Dynamics

1 INTRODUCTION

Since the discovery of fullerenes and carbon nanotubes, low-dimensional nanoscale carbon materials has been the subject of intensive research during the past two decades due to the peculiar electronic structures that are expected to be important for practical applications in nanoelectronics [1-2]. Recently, single graphite layers, referred to as Graphene nanoribbons (GNR) have been prepared experimentally by using conventional device set up [3-4]. Such findings have opened up exciting opportunities for the design of novel electronic devices and interconnects, e.g., quantum information processing [5] and tiny transistors [4, 6]. Theoretically, the energy gaps and optical properties have been predicted with various widths by Son et al. and Barone et al [7-8], respectively. These provided a qualitative way of determining the electronic properties of ribbons with widths of practical significance. More recently, results obtained in the Berkeley lab show that zigzag graphene nanoribbons are magnetic and can carry a spin current response to the external electric field. This opens a new path to the application of spintronics [9].

Generally, SWCNTs are typically grown as mixtures of metallic and semiconducting tubes, depending on the arrangement of the hexagonal rings along the tubular surface [10]. However, this actually constitutes one of the notable obstacles to the widespread application of this unique material, since metallic and semiconducting materials have very different functions in nano-devices.

Hence, separating them has become a central issue in terms of effective fabrication of high performance electronic devices. Currently, many physical and chemical methods have been developed for the separation according to the respective electronic properties by using dielectrophoresis [11-12], selective flocculation [13], selective adsorption of the functional group [14-15], and density gradient induced centrifugation [16] so on. However, none of these is satisfactory from the point of view of high throughput, better selectivity and yield, and more favorable scalability [17].

Bare GNR also has unsaturated dangling bonds at zigzag. Clearly a single graphene sheet is very difficult to roll into a SWCNT without any catalyst. An intriguing question, however, is whether it is possible to form SWCNT via the interaction of bare ZGNRs? To explore this question, we report below a molecular dynamics (MD) simulation to study the interaction of bare GNRs with zigzag shaped edges. We found that two bare nanoribbons (N_z=8) with zigzag shaped edge could form a (8, 8) single-wall armchair carbon nanotube at room temperature, suggesting a possible route for selective synthesis and growth of armchiair nanotubes via the interaction of ZGNRs.

2 COMPUTATIONAL DETAILS

MD simulation was performed by putting two 8-ZGNRs at natural separation distance of graphite layers (3.3 Å). The second-generation reactive empirical bond order potential (REBO) developed by Brenner was used to describe the C-C and C-H interaction [18]. Long distance van der Waals forces expressed in 6-12 Lennard-Jones form were also taken into account. The simulated system contains 1600 Carbon and 32 H atoms. Temperature was controlled by Langevin scheme at room temperature.

3 RESULTS AND DISCUSSION

ZGNRs are classified by the number of zigzag chains (N_z) across the ribbon width as shown in Fig 1. 8-ZGNR represent GNR with 8 zigzag chains. First, geometry optimization for 8-ZGNR with and without H-termination were performed utilising the conjugate gradient method.

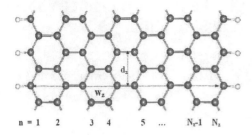

Figure 1 Geometry structure for hydrogen terminated graphene nanoribbons (GNRs) with zigzag shaped edges. The grey and white balls represent C and H atoms, respectively. The 1D unit cell distance and the ribbon width are denoted by $d_z(W_z)$ for 8-ZGNR.

Apparently, the existence of dangling carbon bonds in the bared GNR should offer a much higher chemical reactivity to manipulate the interaction of ZGNRs. Here we performed MD simulation at 300K for two 8-ZGNRs positioned at a distance of 3.3 Å, which is very close to the natural separation of two graphite layers. Our MD simulation was based on the canonical (NVT) ensemble. Figure 2 present the energy profile along the MD trajectory. It can be seen that a (8,8) armchair SWCNT was formed around 160 ps.

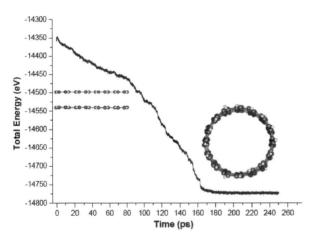

Figure 2 the variation of total energy along MD trajectory for the interaction of two 8-ZGNRs.

Then we turn to study the effect of the distance between two 8-ZGNRs. Two distances, 3.2 and 5.0 Å were set for two separated MD run. Figure 3a present their energy profiles along the MD trajectory. Width could be also very important during the formation of SWCNT via the interaction of ZGNRs. Here we performed another two MD runs at different ribbon widths, i.e. 6-ZGNR and 18 ZGNR. The energy profile along MD trajectory was presented in Figure 3b. Clearly, the formation of SWCNT is faster

when distance between two 8-ZGNRs is short (3.2 Å) and the width of ZGNR is narrow (6-ZGNR).

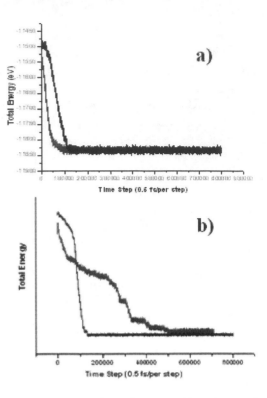

Figure 3 a) Energy profiles along the MD trajectory for the distance of two ZGNRs are 3.2 (red) and 5.0 Å (blue), respectively. b) for the 6-ZGNR (blue) and 18-ZGNR (red), respectively

In order to build reliable microelectronic nano-devices based on SWCNTs, selective processes must be developed to create SWCNTs with specific electronic and mechanical properties. Despite many experimental efforts directed towards the separation of metallic and semiconducting nanotubes, most of them are actually not selective for the diameter and chirality [19]. Furthermore, techniques such as chemical functionalization may also introduce defects and require further processing to restore the useful electronic properties of SWCNT. Hence, improved techniques are still needed. More recently, Smalley's group has developed a cloning method to cut up individual nanotubes into short segments that act as the seeds for regrowing entire tubes [20]. In light of exciting recent progress on the controllable growth of two dimensional GNRs with various widths, these results suggest an alternative approach to the selective synthesis of SWCNTs with specific diameter and chiarility that may provide an interesting avenue for future investigations.

4 CONCLUSIONS

In summary, the interactions of two bare 8-ZGNRs with were investigated by *a classical Molecular Dyanmics based on the*

Brenner potential. Remarkably, two bare 8-ZGNRs could form a (8, 8) armchair SWCNT even at room temperature. Additionally, the interaction also strongly depends on the distance and width of ZGNRs. Considering the challenges in separating metallic and semiconducting SWCNT experimentally, our results suggest a possible alternative route to selectively synthesize specific types of nanotubes via GNRs.

5 ACKNOWLEDGEMENT

We acknowledge generous grants of high-performance computer time from both the Computational Molecular Science cluster computing facility at The University of Queensland and the Australian Partnership for Advanced Computing (APAC) National Facility. The authors also greatly appreciate the financial support of the Australian Research Council and The University of Queensland through the ARC Centre for Functional Nanomaterials.

REFERENCES

[1] Dresselhaus, G.; Dresselhaus, M. S.; Eklund, P. C*Science of Fullerenes and Carbon Nanotubes: Their Porperties and Applications* (Academic, New York, 1996).

[2] Chico, L. *Phys. Rev. Letts.* **1996**, 76, 971.

[3] Novoselov, K. S.; Geim, A. K.; Morozov, S. V.; [4] Jiang, D.; Zhang, Y.; Dubonos, S. V.; Grigorieva, I. V.; Firsob, A. A. *Science* **2004**, *306*, 666.

[4] Geim, A. K.; Novoselov, K. S. *Nature Materials* **2007**, *6*, 183.

[5] Falko, V. *Nature Physics*, **2007**, 3, 151.

[6] Yan, Q. M.; Huang, B.; Yu, J.; Zheng, F. W.; Zang, J.; Wu, J.; Gu, B. L.; Liu, F.; Duan, W. H. *Nano Letts.* **2007**, 7, 1469.

[7] Son, Y. W.; Cohen, M. L.; Louie, S. G. *Phys. Rev. Letts* **2006**, 97, 216803.

[8] Barone, B.; Hod, O.; Scuseria, G. E. *Nano letts* **2006**, 6, 2748.

[9] Son, Y. W.; Cohen, M. L., Louie, S. G.; *Nature* **2006**, 444, 347.

[10] R. Saito, G. Dresselhaus, M.S. Dresselhaus. Physical Properties of Carbon Nanotubes, Imperial College Press, London (1998).

[11] Peng, H.; Alvarez, N. T.; Kittrell, C.; Hauge, R. H.; Schmidt, H. K. *J. Am. Chem. Soc.* **2006**, 128, 8396.

[12] Krupke, R.; Hennrich, F.; von Lohneysen, H.; Kappes, M. M. *Science* **2003**, *301*, 344.

[13] D.Chattopadhyay L. Galeska, F. Papadimitrakopoulos, *J. Am. Chem Soc.* **2003**, *125*, 3370.

[14] Strano, M. S.; Dyke, C. A.; Usrey, M. L.; Barone, P. W., Allen, M. J.; Shan, H.; Kittrell, C.; Hauge, R. H.; Tour, J. M.; Smalley, R. E. *Science* **2003**, *301*, 1519.

[15] Hudson, J. L.; Jian, H.; Leonard, A. D.; Stephenson, J. J.; Tour, J. M. *Chem. Mater.* **2006**, *18*, 2766.

[16] Arnold, M. S.; Green, A. A.; Hulvat, J. F.; Stupp, S. I.; Hersam, M. C. *Nature Nanotechnology* **2006**, *1*, 60-65.

[17] Rinzler, A. G. *Nature Nanotechnology* **2006**, *1*, 17. 702.

[18] Brenner, D. W.; Shenderova, O. A.; Harrison, J. A.; Stuart, S. J.; Ni, B.; Sinnott, S. B. *J. Phys. Condensed matter* **2002**, 14, 783.

[19] Ren, Z. F. Nature Nanotechnology 2006, 2, 17.

[20] Smalley, R. E.; Li, Y. B, Moore, V. C.; Katherine Price, B.; Colorado, R.; Schmidt, H. K.; Hauge, R. H.; Barron, A. R.; Tour, J. M. J. Am. Chem. Soc. **2006**, 128, 158234.

Defect Reduction of Multi-walled Carbon Nanotubes by Rapid Vacuum Arc Annealing

Jeff T. H. Tsai*, Jason S. Li, and Anders A. P. Tseng

*Graduate Institute of Electro-Optical Engineering,
Tatung University, Taipei 10452, Taiwan, thtsai@ttu.edu.tw

ABSTRACT

A rapid thermal annealing process is demonstrated for defect reduction of multiwalled carbon nanotubes (MWCNTs) using a DC vacuum arc discharge system. A vacuum arc discharge system was used to generate high temperatures (~ 1800 °C) followed by rapid cooling. The MWCNTs were rapidly annealed in this system by several cycles of fast heating (300 °C per sec.) and cooling (100 °C per sec.). The annealed samples were characterized by Raman spectroscopy and transmission electron microscopy. It was found that the defect density was reduced effectively by rapid thermal anneal. When inlet water vapor to the annealing chamber, the more defects were removed which indicate the oxygen may play an important role in combustion of the imperfect structures and removing the weakly bonded defects during the rapid heating cycles. This method produces effective defect reduction of the MWCNTs.

Keywords: carbon nanotubes, raman spectroscopy

1 INTRODUCTION

CNTs is a unique material because of its novel physical and mechanical properties. Although the MWNTs fabricated by CVD can be manufactured a large-scale amounts, the highly disordered MWNTs are obtained. Dai *et al.* reported that defects in MWNT increase its resistivity dramatically [1]. In this sense, structural improvement of MWNTs for the progress of conductivitiy is necessary. Andrews *et al.* reported the thermal treatment of nanotubes synthesized by CVD [2]. The effect of high temperature annealing CNTs has been extensively studied [3]. In additional, it also shows that the etching rate between CNTs and amorphous carbon by oxidation is different. It revealed the oxygen play an important role to modify the structures of CNTs.

2 EXPERIMENTAL

We adapt the vacuum arc system into a vacuum arc rapid thermal annealing apparatus (VARTA) to carry out the defect annealing of MWNTs. The graphite anode of VARTA is a process chamber with 5 mm hole filled with the CVD grown MWNTs were linked together with a graphitic segment. The annealing process was controlled by arc current and the purging Ar or air flow. In the heating cycle, the heating rate was furnished with dc power supply in different current magnitude. Temperature of the process chamber was measured by an infrared thermometer and the maximum value reached to 1800 ℃ in 6 sec by the arc current of 60 A in the chamber based pressure of 2 x 10-2 Torr. Fig. 1(a) shows the heating rate and the saturated temperature in different arc current. The cooling rate was achieved at ~170℃ /sec. by injected 4000 sccm high flow Ar direct to the process chamber surface, as seen in Fig. 1(b).

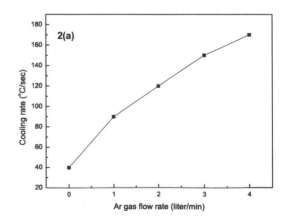

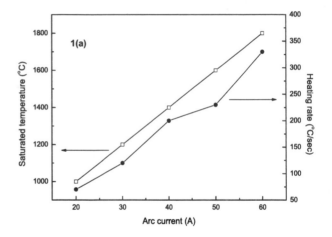

Fig. 1(a) The diagram of the terminal temperature and heating rate in different arc current. 1(b) The diagram of the cooling rate by Argon purging in different flow rate.

The raw MWNTs with diameters in 20-150 nm fabricated by CVD in relative low process temperature (~700 ℃) were used in our VARTA experiment. In this work, we modulated the different arc current for the observation of annealed MWNTs. The raw MWNTs underwent RTA for 20 cycles in 10 min from 1000 ℃ to 1800 ℃ and high flow Ar purging of 4000 sccm in the vacuum chamber at a pressure of 2×10^{-2} Torr. Furthermore, we also constructed the humid environment to perform VARTA process at pressure of 1 Torr. The change in structure of samples was observable by Raman spectroscopy analysis, TEM, and SEM.

3 RESULTS AND DISCUSSION

In the Raman measurements, the raw and annealed MWNTs for 20, 30, 40, 50, and 60 A in VARTA process, as shown in Fig. 2(a). The result of Raman analysis indicates that the ratios of the intensity G-band peak to that of D-band peak (I_G/I_D) ascend with the increasing arc current. It represents that the higher the I_G/I_D reveals, the better the graphitic structure obtained. The TEM images, shown in Fig. 2(b) and 2(c), for raw and optimum annealing temperature of 1800 ℃. The main characteristics of raw samples are relatively crooked and highly disordered morphology. Through VARTA at 1800 ℃, the crooked tubes convert into straight structures.

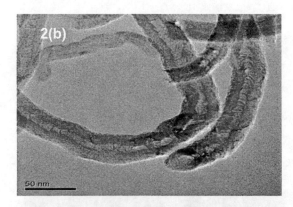

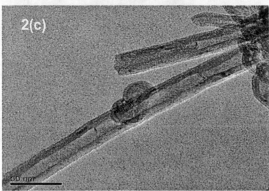

Fig. 2(a) Raman spectra of raw and annealed nanotubes. 2(b) TEM image of raw MWNTs synthesized by CVD shows imperfect structures. 2(c) After the optimum VARTA temperature of 1800 ℃, the less defects are obtained.

Fig. 3(a) shows the relationship between graphitization degrees and arc currents. We found that the differences between the samples which were annealed by standard VARTA process and in humid environment. The humid environment provide oxygen molecules in VARTA process. Hence, the oxygen attack week C-C bonds in the defects of the CNT to help forward to reconstruct MWNTs. It is clear to see that the effect of oxidation or even combustion on the CNTs by wet-oxygen VARTA from the SEM images, the CNTs present less disorder by removing the amorphous carbon coating on the outer shell as seen in Fig. 3(b) and 3(c).

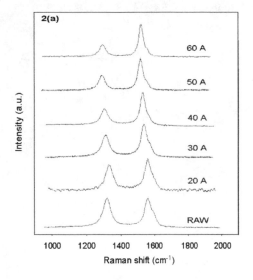

4 CONCLUSION

In summary, we have demonstrated the VARTA process is a reliable method for defect annealing. From rapid thermal spiking enforcing the carbon atoms rearranged into more graphitic structure to improve the imperfection of the MWNTs. Most importantly, the VARTA process on defected MWNTs promotes the efficiency of annealing in time greatly. By wet-oxidation assistants, oxygen play an important role to attack the week bonds of defect structures. This may originated from combustion of amorphous carbon coating or thermal assisted oxidation inside the defects.

ACKNOWLEDGEMENT

The authors gratefully acknowledge the National Science Council Taiwan under the grants 95-2112-M-036-001-MY2.

REFERENCES

[1] Hongjie Dai, Eric W. Wong, Charles M. Lieber, Science **272**, 523 (1996)
[2] R. Andrews, D. Jacques, D. Qian, E.C. Dickey, Carbon **39**, 1681 (2001)
[3] Y.A. Kim, T. Hayashi, K. Osawa, M.S. Dresselhaus, M. Endo, Chem. Phys. Lett. **380,** 319 (2003)

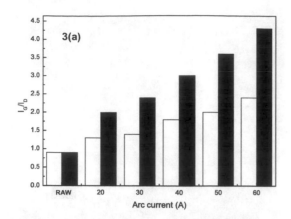

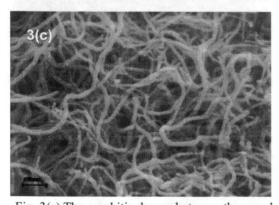

Fig. 3(a) The graphitic degree between the samples annealed by VARTA process and wet-oxidation (White bar: VARTA, Black bar: Oxidation).
3 (b) SEM image of entangled raw MWNTs. 3(c) After the wet-oxidation, MWNTsappear less disorder.

Biocompatibility of Carbon Nanotubes for Cartilage Tissue Engineering

Nadeen O. Chahine [*,#], Nicole M. Collette [**], Heather Thompson [*], Gabriela G. Loots [*,**,^]

[*] Lawrence Livermore National Laboratory, Livermore, CA.
[**] University of California Berkeley, Berkeley, CA.
[#] chahine2@llnl.gov   [^] loots1@llnl.gov

ABSTRACT

Carbon nanotubes (CNT) have the potential to become an important component of scaffolding in tissue engineering, due to their unique physical properties. However, one major limitation of CNTs that must be overcome is their inherent cytotoxicity. In this study, we assessed the long-term biocompatibility of CNTs for chondrocyte growth. The effect of CNTs on chondrocyte viability and biochemical deposition has been examined in two dimensional (2D) cultures, and in three dimensional (3D) composite materials consisting of hydrogels and CNTs. The exposure of chondrocytes to CNTs was carried out up to 14 days in 2D culture and 21 days in 3D agarose composites. Our results suggest that functionalized CNTs alter the viability and metabolic response of cells over the 2 to 3 week duration. Interestingly, the results of this study also suggest the dose dependent effect of CNTs on cellular responses vary between 2D and 3D cultures, suggesting that chondrocytes tolerate the presence of CNTs at greater concentrations in 3D more than in 2D cultures. In addition, CNTs appear to have a stimulatory metabolic effect on chondrocytes in 3D cultures, reflected by enhanced production of glycosaminoglycans (GAGs) and collagen deposition. These findings support the notion that optimization of the use of nanotubes in cell-based therapies should be performed in 3D systems directly, and that CNTs appear to promote cellular growth and metabolic activity of chondrocytes.

Keywords: Carbon Nanotubes, Cytotoxicity, Tissue Engineering, Biocompatibility, Cartilage.

1 INTRODUCTION

Carbon nanotubes (CNTs) are cylindrical allotropes of carbon that are nanometers in diameter and posses unique physical properties, positioning them as ideal materials for studying physiology at a single cell level. CNTs have the potential to become a very important component of medical therapeutics, such as in (a) drug delivery system [1], (b) existing as an interfacial layer in surgical implants [2,3], or (c) acting as scaffolding in tissue engineering [4,8]. While some studies have explored the use of CNTs as a novel material in regenerative medicine, they have not yet been fully evaluated in cellular systems. One major limitation of CNTs that must be overcome is their inherent cytotoxicity. The goal of this study is to assess the long-term biocompatibility of CNTs for chondrocyte growth. We hypothesize that CNT-based material in tissue engineering can provide an improved molecular sized substrate for stimulation of cellular growth, and structural reinforcement of the scaffold mechanical properties. Here we present data on the effects of CNTs on chondrocyte viability and biochemical deposition examined in composite materials of hydrogels + CNTs mixtures that supports our hypothesis. Also, the effects of CNTs surface functionalization with polyethlyne glycol (PEG) or carboxyl groups (COOH) were examined.

2 METHODS

Purified single wall carbon nanotubes (SWNTs) functionalized with terminal carboxylic acid (SWNT-COOH) or covalent polyethylene glycol (SWNT-PEG) (Carbon Solutions Inc., Riverside, CA) were UV-sterilized, suspended in DI water, and sonicated for ~3 hours to arrive at a stable, homogenous suspension at room temperature (RT). Chondrocytes were isolated from immature bovine articular cartilage (1-3 days old) using enzymatic digestion, as previously described [5].

2.1 Effect of CNTs on cell viability (2D Culture)

Chondrocytes were plated at 1×10^6 cells/cm^2 and cultured in high glucose DMEM + 10% FBS + 1X penicillin for up to 14 days. Cells were exposed to media containing one of the following treatment groups: a) Control, (b) Sonicated Media (SM), (c) 0.1 mg/ml SWNT+COOH, (d) 0.01 mg/ml SWNT+COOH, (e) 0.1 mg/ml SWNT+PEG, (f) 0.01 mg/ml SWNT+PEG. Chondrocyte viability was measured with a fluorescence microplate assay, using a Live-Dead assay kit (Molecular Probes, Portland, OR). Cell viability was reported as % live cells, on day 3 (D3), 7 (D7) and 14 (D14). Absolute cell percentages were acquired by killing all cells in 1 well per group using 0.25% digitonin. Fluorescence microscopy images of the cells were also taken using an inverted scope.

2.2 CNT-Hydrogel Constructs (3D Culture)

Chondrocytes were seeded at a final density of 10×10^6 cells/ml in composite mixtures of 2% low melt agarose ± SWNTs. Agarose/SWNT mixtures were sonicated prior to addition of cells, and final mixtures were cast between 2 glass plates. Cylindrical constructs were prepared (∅5mm/~1.7mm thick) from 4 treatment groups: (a) Control, (b) 0.1 mg/ml SWNT+PEG, (c) 1.0 mg/ml SWNT+PEG, (d) 0.1 mg/ml SWNT+COOH. Disks were cultured up to 21 days in high glucose DMEM supplemented with 10 ng/ml TGF-β3, 1% ITS, 100nM dexamethasone, 50μg/ml L-Proline, 100 μg/ml sodium pyruvate, and antibiotics (penicillin, streptomycin).

Water content of the disks was measured by lypholization. In addition, the biochemical content of molecules typically synthesized by chondrocytes was measured, as an indication of biological synthesis. Glycosaminoglycans (GAG) are extracellular matrix molecules typically found in cartilage, and are indicative of the presence of differentiated chondrocytes in culture. The GAG content in these constructs was assessed using a calorimetric assay (Blyscan Assay; Biocolor, U.K.). Double stranded DNA content in the samples was also measured using Quant-IT assay (Molecular Probes), against calf thymus DNA standards. Cellular composition was also assessed histologically on paraffin embedded sections to determine glyocoaminoglycan (GAG), (Alcian blue) and collagen deposition (picrosirius red) differences among the various culture groups. Cellular viability in the constructs was also assessed using the Live/Dead viability assay, and imaged on an inverted fluorescence microscope.

At each time point, the biomechanical properties of constructs were measured under unconfined compression using an Instron testing frame (equipped with 50N load cell), similar to previously described protocols [5]. Three stress relaxation compressions were applied (5% strain per ramp), and the Young's modulus (E_Y) was determined as the slope of the stress-strain curve.

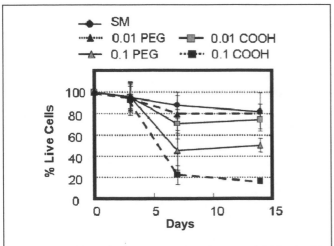

Figure 1: Effect of SWNTs on chondrocyte viability in 2D culture over 2 week period.

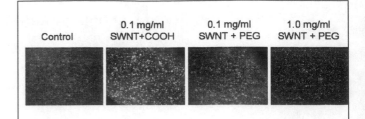

Figure 2: Viability of chondrocytes at Day 7 in varying CNT hybrid constructs. Green indicates live cells, red indicates dead cells.

3 RESULTS

3.1 Effect of SWNTs on cell viability (2D Culture)

Chondrocytes maintained comparable viability in all groups, up to Day 3 in culture (Figure 1). Cells cultured in sonicated media (SM) exhibited limited loss in viability by Day 14 (82 ± 18% live). The addition of CNTs at 0.01 mg/ml resulted in comparable viability for SWNT+PEG (83±6%) and SWNT+COOH (76±9%). However by Day 7, the percentage of live cells decreased in the presence of higher concentrations of SWNTs, with 45±13% and 22±8% of cells remaining viable in SWNT+PEG & SWNT+COOH, respectively (Figure 1). No additional loss of cells was seen between Day 7 and Day 14.

3.2 CNT-Hydrogel Constructs (3D Culture)

The presence of carbon nanotubes altered the viability and biochemical synthesis of chondrocytes in the hydrogel composites in a dose dependent manner. No significant loss of cell viability was seen for cells cultured in the presence of 0.1 mg/ml SWNT, same concentration shown to be detrimental to cells in 2D cultures. Even at concentrations 10 fold higher, the presence of SWNT+PEG at 1.0 mg/ml resulted in limited cell death by Day 7 (Figure 2).

The GAG content of the constructs increased over the culture duration for all groups, demonstrating 2- to 4-fold increases in the control and 0.1 mg/ml CNT groups. In the presence of 1.0 mg/ml CNTs, GAG content increased significantly over all groups, demonstrating a 10- to 13-fold increases over Day 0 (Figure 3).

The compressive modulus (E_Y) of agarose ± SWNT constructs also increased with time in culture and with the addition of CNTs (Figure 4). At D0, the addition of 0.1 mg/ml SWNT+COOH increased E_Y by ~53% and that of 1.0 mg/ml SWNT+PEG by 32% (Figure 4). By D21, E_Y increased by 45% in the control group, where as the presence of 0.1 mg/ml SWNT+COOH increased E_Y by 73%. The presence of SWNT+PEG demonstrated transient effects on E_Y, with no significant increase over control (Figure 4).

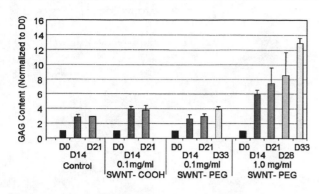

Figure 3: GAG content is composite constructs at various times in culture in the presence or absence of SWNTs (content is reported normalized to Day 0 values).

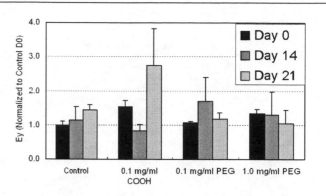

Figure 4: Young's Modulus (E_Y) of agarose constructs ± SWNTs (normalized to Day 0 control values)

Histological analysis revealed comparable GAG deposition for all groups by D21 (Figure 5). However, an increased deposition of collagen matrix was seen for chondrocytes in 1.0 mg/ml SWNT+PEG at D21. The water content of the constructs at D0 was comparable for all groups (0.89±0.04), and decreased slightly over the culture duration, reaching 0.84±0.03 by Day 21.

4 DISCUSSION

The goal of this study was to assess the long-term biocompatibility of SWNTs for chondrocyte growth. The exposure of chondrocytes to SWNTs was carried out for up to 14 days in 2D and 21 days in 3D cultures. This represents a significantly longer culture-time than previous reports examining the effects of CNTs on cell responses (such as adhesion, etc) [7]. Our results suggest that functionalized CNTs alter the viability, metabolic and mechanical environment of cells over the 2 to 3 week course of these experiments.

In 2D cultures, chondrocytes experienced a dose dependent loss in viability in the presence of SWNT+COOH or SWNT+PEG (Figure 1), where low concentrations of SWNT+PEG were as benign as sonicated medial alone (Figure 1). However, the behavior of chondrocytes in 2D can only be limitedly applicable to longer cultures, due the propensity of cells to dedifferentiate in this environment [9]. Morphological examination of cells revealed that a significant number of cells acquired fibroblast like morphology by day 14, in contrast to control samples where cells continued to maintain chondrocyte-like phenotype.

In 3D cultures, chondrocytes demonstrated viability and biochemical deposition through 21 days of exposure to SWNTs. Despite some loss of viability in the presence of SWNTs at higher concentrations, the biochemical deposition through 21 days of exposure to CNTs was found to be significantly greater than control groups (Figure 3 and 5). The presence of CNTs did not diminish the biochemical deposition of GAGs nor of collagen, compared to the

control (Figure 5), suggesting that the presence of CNTs in 3D allows chondrocytes to maintain their phenotype. Moreover, 1.0 mg/ml SWNT+PEG resulted in increased deposition of collagen by Day 21. This increased collagen deposition suggests that SWNT composite structures can act as molecular sized stimulants of cellular growth and promoters of biochemical synthesis.

The presence of SWNTs also altered the biomechanical properties of these composite structures, though no obvious dose dependent behavior was observed in the equilibrium compressive modulus. The presence of SWNT+COOH increased E_Y initially (Day 0 constructs) by 53%, suggesting a reinforcement behavior prior to matrix deposition, *de novo*. Interestingly, the presence of SWNT+PEG at the same concentration did not result in a similar reinforcement behavior in compression. This finding suggests that the presence of the negatively charged carboxylic acid groups may have mediated this increase in stiffness, possibly by increasing osmotic effects. These findings are consistent with previous studies, where CNTs introduced into synthetic biopolymers were found to improve the mechanical properties of the nanocomposites [8].

Interestingly, the results of this study also suggest the dose dependent effects of CNTs on cellular responses vary between 2D and 3D cultures. Chondrocytes appear to tolerate the presence of SWNTs at 10-fold higher concentrations in 3D than concentrations shown to be lethal in 2D cultures. This may be due to the interaction of disbursed SWNTs with cellular structures, either through adhesion to the cell membrane or via endocytosis into the intracellular space. Though no specific cross links between the SWNTs and the hydrogel were targeted, the entrapment of the CNTs in the hydrogel during the curing process may have prevented cellular uptake of CNTs. These findings support the notion that optimization of the use of nanotubes or particles in cell-based therapies should be performed in the intended 3D systems directly, thus alleviating complications unique to other models in use.

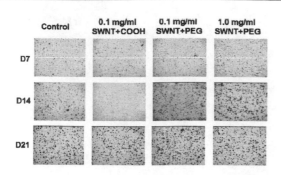 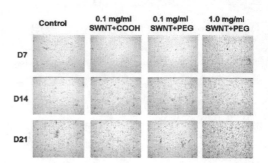

Figure 5: Histological evidence of GAG (left) and collagen (right) deposition in chondrocyte seeded agarose + SWNT composites at days 7, 14 and 21 in culture (Magnification: 10X).

In conclusion, the results of this study indicate that SWNTs offer a unique potential for cartilage tissue engineering, where functionalization with bioactive molecules may provide an ideal substrate for stimulation of cellular growth and repair. Future studies will examine the effect of growth factor conjugation to CNTs on cell growth, as well as the effect of CNTs on the tensile properties in composite structures.

ACKNOWLEDGMENTS

This work performed under the auspices of the U.S. Department of Energy by Lawrence Livermore National Laboratory under Contract DE-AC52-07NA27344. LLNL-ABS-400929.

REFERENCES

[1] Kam et al. , 2005, Prooceedings of the National Academy of Science. Vol. 102 (33).
[2] Popat et al., 2007, Biomaterials, Vol. 28 (32).
[3] Yao et al., 2008, Journal of Biomedical Material Reseearch, Part A. Vol. 85(1).
[4] Shi et al., 2007, Journal of Biomedical Material Reseearch, Part A.
[5] Mauck et al., 2000, Journal of Biomechanical Engineering. Vol. 122 (3).
[6] Mauck et al., 2006, Osteoarthritis & Cartilage Vol. 14 (2).
[7] Price et al., 2003, Biomaterials, Vol. 24 (11).
[8] Shi et al., 2006, Biomacromolecules Vol. 7 (7).
[9] Benya and Shafer, 1982, Cell, Vol. 30 (1).

Exact field enhancement factor, electrostatic force and field emission current from a long carbon nanotube

A.I. Zhbanov*, E.G. Pogorelov** and Y.-C. Chang***

*Research Center for Applied Science (RCAS), Academia Sinica, Taiwan,
128, Section 2, Academia Road, Nankang, Taipei, 11529, azhbanov@gate.sinica.edu.tw
** RCAS, evgenypogorelov@gmail.com
*** RCAS, yiachang@gate.sinica.edu.tw

ABSTRACT

Field enhancement factor, electrostatic force and field emission current from carbon nanotube film are theoretically investigated. In our model the cathode resembles a dense carpet consisting from randomly twisted nanotubes. Under the action of electric field the free ends of nanotube are built in parallel lines of intensity of a field. Emitting tubes come off at the certain electrostatic forces. The further voltage increase causes other tubes to emit till the force is still insufficient to tear them off of a substrate.

Keywords: carbon nanotubes, field emission, electrostatic force

1 INTRODUCTION

After experimental discovering of fullerenes [1] and carbon tubes [2] many researches are devoted to studying of their properties and possible applications. One is easier to imagine nanotube by mental folding of a graphite layer. As a result of such action the seamless open cylindrical tube should turn out. Closed nanotubes are ended with hemispherical caps.

Remarkable properties of carbon nanotube films is their ability to act as field emission electron sources at low applied voltages. Now field emission arrays on the basis of carbon nanotube films for field emission displays are developing.

The key problems restricting large-scale industrial use of cathodes on carbon nanotube films is the problem of lifetime and reliability.

We assume one of aspects of this problem consists that during emission in strong electric fields the part of tubes is pulled out from a film and adheres to the anode [3].

2 EXPERIMENTAL DATA

We investigated films obtained by various technologies of synthesis. Fig. 1 shows the I-V characteristic of the carbon film synthesized by a method of plasma-chemical decomposition of spirit C_2H_5OH in presence of iron catalyst at temperature 800 ºC. It is visible that at a direct current *250 µA* a reverse current appears and at a direct current *500*

µA it reaches *20 µA*. Occurrence of a reverse current also confirms that nanotubes come off from the cathode and adhere to the anode. The assumption about pulling out of nanotubes was reported elsewhere [2-3].

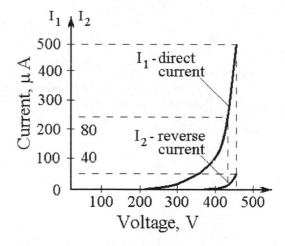

Figure 1: I-V characteristic of the film.

3 THEORETICAL CALCULATIONS

Let's consider how to explain tubes detaching from a substrate and what forces hold these tubes.

3.1 Electrostatic forces acting on a nanotube

During field emission electrostatic force acts on a tube and aspires to tear off it from a substrate. For numerical calculation of a field emission from aligned carbon nanotube we use the following model:

1. The nanotube is a cylinder with length H and diameter d that is closed with a hemispherical cap with the same diameter d.

2. Nanotube obeys the laws of continuous medium, it is an ideally conducting material and the cathode potential is maintained at its whole surface.

For numerical calculation we used special finite-element technique [3].

In case $L \gg H$ the field amplification factor is well described by the formula

$$\beta = 3 + Bx^D \qquad (1)$$

where $x = H / d$; $B = 2.372$; $D = 0.8562$. Parameters B and D were evaluated using least squares method.

Distribution of the electrical field intensity on the hemispherical nanotube cap was fitted as following:

$$E = E_{max}(1 - A_\alpha \alpha^2) \qquad (2)$$

where $E_{max} = \beta E_0$ is the maximum electrostatic field at the apex of the nanotube and E_0 is the applied macroscopic field, i.e., the far-field., α is a minimal angle between radius to a point on hemisphere and nanotube axis, parameter $A_\alpha = 0.1288$. Angle α is measured in radians $0 \le \alpha \le \pi / 2$. Noticeable field emission may be observed from approximately half of nanotube cap, the rest of emission current is negligible.

Consider force stretching vertically oriented nanotube.

Using the formula (2) it is possible to find the electrostatic force stretching a single tube, perpendicularly oriented to a substrate [3]

$$F = 8.26 \varepsilon_0 E_{max}^2 \pi r^2 / 2 \qquad (3)$$

where E_{max} is maximal electric field (V/µm); r is tube radius (Å); ε_0 is electric constant (F/m); F is force (nN), and 8.26 is dimensionless factor.

As schematically shown in Fig. 2 our carbon films represent dense "carpet" from randomly oriented nanotubes. We suppose that under the action of electric field the free ends of nanotube are built in parallel lines of intensity of a field. Thus a film from randomly oriented nanotubes converts into the array of aligned nanotubes (see Fig. 2). Force F (3) acts to nanotube apex and aspires to extract a tube from "carpet". The tube is kept by van der Waals interaction with other tube.

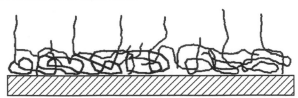

Figure 2: "Carpet" from randomly oriented nanotubes.

3.2 Van der Waals forces for nanotubes

For van der Waals forces calculation we use Lennard-Jones potential well describing interatomic interactions.

$$\varphi_a(r) = B / r^{12} - A / r^6 \qquad (4)$$

Parameters of potential $A = 15.2$ eV·Å⁶ and $B = 24.1 \cdot 10^3$ eV·Å¹² for C-C interactions in graphite planes are taken from work [7]. As we believe, nanotube is kept due to interaction with other tubes, thus using of carbon-carbon potential in calculations is quite justified.

We used method [8] of potential integration over tube surfaces to get single wall tubes van der Waals potential. Then it can be generalized by summation over different interacting walls to potential of multi-wall nanotubes. We was able to get exact analytical formula for van der Waals potential and force $F_w(t_1,t_2,x)$ between carbon multi-wall nanotubes at radius t_1, t_2 and distance x. Full analytical formula and derivations are quite huge, so we would like present here only potential between two single wall tube at the same radius.

$$\varphi_t(r,t) = -\pi^2\left(-\frac{Af_a}{r^2} + \frac{Bf_b}{r^8}\right), \qquad (5)$$

$$\lambda \equiv \frac{2\sqrt{t_1 t_2}}{\sqrt{r + t_1 - t_2}\sqrt{r + t_2 - t_1}}, \qquad (6)$$

$$f_i = f_{ik}K(\lambda) + f_{ie}E(\lambda), \qquad (7)$$

where K and E are complete elliptical integrals of first and second kind.

$$f_{ak} = -\frac{2(5\chi^2 - 4)}{3(\chi - 2)^2(\chi + 2)^2}, \quad \chi \equiv \frac{r}{t}, \qquad (8)$$

$$f_{ae} = \frac{2(32 - 20\chi^2 + 11\chi^4)}{3(\chi + 2)^3(\chi - 2)^3}, \qquad (9)$$

$$f_{bk} = -\left(\begin{array}{l} 4609\chi^{14} + 56038\chi^{12} + 321132\chi^{10} - \\ 473632\chi^8 + 1885952\chi^6 - 3867648\chi^4 + \\ 4510720\chi^2 - 2293760 \end{array}\right) /$$

$$/ 1575(\chi - 2)^8(\chi + 2)^8, \qquad (10)$$

$$f_{be} = -\left(\begin{array}{l} 7129\chi^{16} + 97220\chi^{14} + 763489\chi^{12} - \\ 1533424\chi^{10} + 7790944\chi^8 - \\ 21756160\chi^6 + 38781184\chi^4 - \\ 40099840\chi^2 + 18350080 \end{array}\right) /$$

$$/ 1575(\chi - 2)^9(\chi + 2)^9. \qquad (11)$$

Potential derivation gives required van der Waals force. Investigating this derivative on extreme, it is possible to find distance on which van der Waals interaction will be maximal.

$$F(r) = \max_x F_w(r, r, x) \qquad (12)$$

In figure 3 we show dependence of maximal force (nN) on tubes radius (Å).

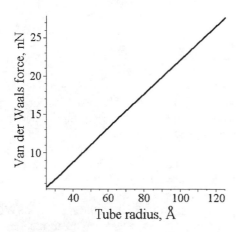

Figure 3: Dependence of maximal van der Waals force on tube radius.

Using (1), (3) and $F(r)$ (Fig. 3) we get

$$H(E, r) = 2r \left(\frac{1}{B} \left[\frac{1}{Er} \sqrt{\frac{2F(r)}{8.26\pi\varepsilon_0}} - 3 \right] \right)^{1/D} \qquad (13)$$

In figure 4 for different values of applied electrical field we draw $H(r)$ divided space (H,r) into two regions. In the upper region tubes having one contact with other tube at the same radius will be detached. Otherwise it will be hold by van der Waals force for the down region.

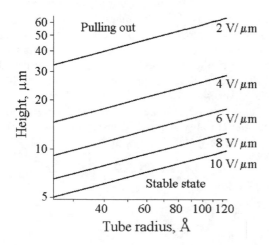

Figure 4: Critical curve $H(r)$ for given electrical field.

In figure 5 we have similar "break lines" for different given tube diameters that predict dependence of critical height on applied electrical field.

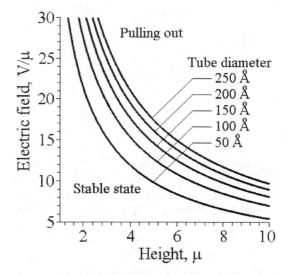

Figure 5: "Break line" for emitting nanotube

4 CONCLUSION

In this work the electrostatic forces acting on a nanotube are calculated. Using the Lennard–Jones potential the van der Waals forces keeping tubes on a substrate are estimated. Tubes detachment conditions are being analyzed.

REFERENCES

[1] H.W. Kroto, J.R. Heath, S.C. O'Brien, R.F. Curl, and R.E. Smalley, Nature, 318, 162-164, 1985.
[2] S. Iijima, Nature, 354, 1991, 56-58, 1991.
[3] O.E. Glukhova, A.I. Zhbanov, I.G. Torgashov, N.I. Sinitsyn, and G.V. Torgashov, Appl. Surf. Sci., 215, 149-159, 2003.
[4] A.I. Zhbanov, Y.-C. Chang, N.I. Sinitsyn, and G.V. Torgashov, Eurodisplay-2007, Moscow, Russia, September 18-20, 77-80, 2007.
[5] Y. Tzeng, Y. Chen, and C. Liu, Diamond and Related Materials, 12, 442, 2003.
[6] R. Rosen, W. Simendinger, C. Debbault, H. Shimoda, L. Fleming, B. Stoner, and O. Zhou, Applied Physics Letters, 76, 1668, 2000.
[7] L.A. Girifalco, M. Hodak, and L.S. Roland, Phys. Rev. B, 62, 13104-13110, 2000.
[8] L.A. Girifalco, J. Phys. Chem., 96, 858-861, 1992.

Solubilization of Single Wall Carbon Nanotubes with Salmon Sperm DNA

S. N. Kim[*], K. M. Singh[*], F. Ouchen[†‡], J. G. Grote[†] and R. R. Naik[*]

[*] Air Force Research Laboratory, RXBP, WPAFB, OH, USA, Rajesh.Naik@wpafb.af.mil

[†] Air Force Research Laboratory, RXPS, WPAFB, OH, USA, James.Grote@wpafb.af.mil

[‡] University of Dayton, 300 College Park, Dayton, OH 45469

ABSTRACT

Single wall carbon nanotubes (SWCNTs) have been highlighted among scientific communities due to their potential to advance numerous application areas, such as in nanoscale circuits, ultrathin, flexible, and transparent conductors, supercapacitors, field emitters, actuators, nanosized electrochemical probes, transistors, photovoltaic devices, and nanoscale sensors. Aside from their immense technological importance, enhancing the structural purity and homogeneity by obtaining well dispersed and fractionated samples could also enable us to better characterize and model the SWCNTs. Recently, it has been possible to separate and/or enrich fractions of SWCNTs according to metallicity and diameter (d_t). Deoxyribonucleic acids (DNAs), specifically in synthetic oligomeric form, have been playing an important role in the exfoliation and subsequent d_t and metallicity dependent separation of SWCNTs. However, their extreme high price for the synthesis needs to be overcome for the cost-effective and bulk scale solubilization and separation of SWCNTs. In this paper, we present a solubilization study of commercially available SWCNTs by using the DNA extracted from waste materials of the salmon fishing industry through an enzyme isolation process. The optical properties, including NIR and circular dichroism spectra, are also presented.

Keywords: single wall carbon nanotubes, DNA, dispersion, solubilization, spectroscopy

1 INTRODUCTION

Single wall carbon nanotubes can be pictorially presented as the cylindrical roll-up of a simple flat graphene sheet containing carbon atoms in a hexagonal lattice, with a circumferential rolling vector of $C_h = na_1 + ma_2$, where n and m are integer and a_1 and a_2 are unit vectors.[1] Typically, their diameter varies from 0.4 to 3 nm and length ranges from tens of nanometers to micrometers by controlling the growth conditions and/or applying appropriate post-growth chemical treatments, such as oxidative shortening.[2, 3] The large scale growth of structurally perfect single wall carbon nanotubes has been typically performed via arc discharge, laser ablation, chemical vapor deposition (CVD), and plasma enhanced chemical vapor deposition (PECVD) methods. Among the growth processes, CVD and PECVD methods have appeared to be particularly important processes for the production of SWCNTs that contain minimal concentrations of graphitic nanoparticle contaminations. However, the extreme SWCNTs inter-tube aggregation forces, along with inhomogeneity in the chirality, pose significant obstacle in the achievement for technological breakthroughs where nanotubes of precisely defined length, diameter, and chirality are used.

By wrapping[4, 5] and/or groove-binding[6] with SWCNTs via hydrophobic interactions, DNA has been recognized as one of the most efficient dispersion media that enables us to acquire both individually exfoliated samples and chirality-fractionated carbon nanotubes according to their diameter and metallicity.[5, 7] From a series of experiments, single stranded $d(GT)_n$ DNA has been recognized to exhibit not only individual-level nanotube dispersion, but also effective SWCNT chirality separation when eluted from an anion exchange column at various salt concentrations.[5, 7] However, the typical synthetic oligomer price for a $d(GT)_{20}$, which is used mostly in the SWCNT dispersion and separation experiments, is ~25,000\$/g. Usual oligo-DNA assisted SWCNT dispersion experiments are carried out with a DNA:SWCNT weight ratio of 1:1 and the majority of non-interacting DNAs are discarded. This poses a high price of \$25 for oligo-DNA in treating every mg of carbon nanotubes, and thus, a cost-effective nucleic acid system is highly in demand. Here we present a solubilization and separation study of SWCNTs by using salmon sperm DNA (SaDNA), which is a byproduct of the fishing industry and costs 20\$/g from the preparation method used in this work, as a dispersion media in water or D_2O. We show that the SaDNA exfoliates and disperses SWCNTs on an individual level. The analysis on NIR and resonance Raman spectra from the solubilized nanotube samples will be presented along with the chirality dependent separation results from SaDNA mediated SWCNT dispersion.

2 EXPERIMENTS

2.1 DNA preparation

Synthetic $d(GT)_{20}$ DNA was purchased from IDT and used to compare the solubility of SWCNTs to SaDNA solution. The SaDNA used for this research was purified DNA provided by the Chitose Institute of Science and Technology (CIST)[8], through an Asian Office of Aerospace Research and Development (AOARD) supported effort. It is marine-based, extracted from frozen salmon milt and roe sacs through a homogenization process. It goes through an enzymatic treatment to degrade the proteins by protease. Proteins are then removed by controlling the pH level. The saDNA undergoes a carbon treatment for decolourization, is filtered, and precipitated by adding acetone. The purified saDNA is finally filtered from the acetone and freeze dried. The molecular weight of the purified saDNA typically measures >8,000,000 Daltons (Da), via gel phase electrophoresis. The saDNA material purity measures ~96% and the protein content measures ~2%.

2.2 SWCNT dispersion and spectroscopy

Commercial SWCNTs (HipCo, CNI) were ultrasonicated with a horn sonicator at the concentration of 1mg/ml in D_2O with 1mg of DNA for each sample. The dispersed SWCNTs were further centrifuged at 14,000g for 90 min, and the supernatants were carefully taken for further spectroscopic measurements.

A Renishaw inVia Raman microscope was used for resonance Raman spectroscopy (RRS) measurements, equipped with a 1.96 eV excitation laser line focused on a 1 μm spot by using a 50x objective. All samples were subjected to the same laser intensity, focus, exposure time, and collection scan number for the given laser line. To enhance the sampling signal and uniformity, the RRS data were collected and averaged out from 30 different individual spots.

A JASCO J-815 circular dichroism (CD) spectrometer equipped with 0.2mm path length sample holder was used for circular dichroism measurements.

3   RESULTS AND DISCUSSION

The first electronic transition energy ($E_{ii}^{M/S}$) of van Hove singularity (VHS) in *sem*-SWCNTs (E_{11}^S) have been well characterized by both theory and experiments. Since this distinctive excitation energy is not overlapped by other higher electronic transitions, such as E_{22}^S and E_{11}^M, and the SWCNT peaks are widely spread for HipCo nanotube samples, the detailed feature of the NIR peak can not only provide the degree of solubilization, but also reveal the relative enrichment information about the dispersed tubes. Figure 1 shows normalized NIR absorption spectra obtained from oligo d(GT)$_{20}$ DNA and SaDNA dispersed HipCo SWCNTs. Characteristic E_{11}^S peaks from {(9,7)}, {(11,1), (10,3)}, {(11,3)}, {(8,4), (9,2), (9,4)}, {(10,2)}, {(7,5), (11,0)}, and {(8,3), (6,5)} SWCNTs were observed (marked as diamond shape from left to right in each spectrum).[9] SaDNA dispersed SWCNTs show relatively comparable peak features, and even better isolated peaks from {(11,3)} and {(8,4), (9,2), (9,4)} groups of tubes than

the spectrum from d(GT)$_{20}$ oligomer dispersed SWCNTs. This indicates that the natural SaDNA product enables us to obtain individually dispersed SWCNTs in solution at a comparable or even enhanced level than d(GT)$_{20}$ DNA oligomer.

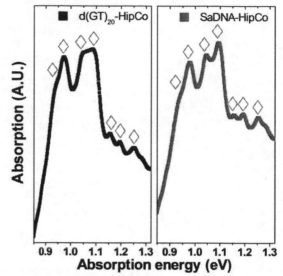

Figure 1: Normalized NIR absorption spectra from oligo d(GT)$_{20}$ DNA and SaDNA dispersed HipCo SWCNTs.

Figure 2 shows NIR spectra profile changes from SaDNA dispersed HipCo SWCNTs at various ultrasonication times. A Sigmoid-like 0.97 eV NIR absorption change was observed (Figure 2a). By extrapolating this curve with a sigmoid function (not shown) fitted from the above profile, we could estimate a saturation ultrasonication time of 90 min with an absorption

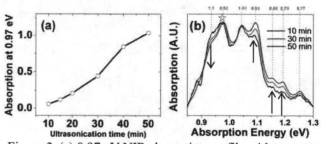

Figure 2: (a) 0.97 eV NIR absorption profile with respect to sonication time, and (b) Normalized NIR absorption spectra from SaDNA dispersed HipCo SWCNTs at various ultrasonication times. The star indicates the absorption from the {(11,1), (10,3)} SWCNT group. The up arrows indicate the increase of the absorption from the {(8,4), (9,2), (9,4)}, {(11,3)} and {(7,5), (11,0)} SWCNT groups (from left to right). The down arrow indicates the decrease of the absorption from {(9,7)} SWCNTs. The numbers on the top *x*-axis indicate the average diameter (nm) from each SWCNT group.

of 0.173. When the NIR spectra from 10, 30, and 50 min sonicated samples were normalized with respect to the absorption from {(11,1), (10,3)} tubes (indicated as a star in Figure 2b) and baseline-corrected by subtracting the plasmonic and scattering lines, the absorption from {(8,4), (9,2), (9,4)}, {(11,3)} and {(7,5), (11,0)} (indicated as arrows) SWCNTs showed an increase in relative intensity, implying their enrichment in the dispersion as the ultrasonication time increased. Meanwhile, the relative absorption from {(9,7)} tubes diminished indicating their decrease in abundance.

RRS is particularly important to analyze SWCNTs with respect to their chirality.[1] When the electronic transitions between the VHSs of a given SWCNT are within the resonance window (~ 0.2 eV) of excitation laser energy,[1] the phonons from this one dimensional system are strongly coupled with electrons, generating clear and distinctive SWCNT peaks in RRS. When SWCNTs d_t is correlated with the lower frequency radial breathing mode (RBM)

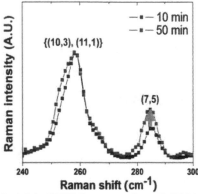

Figure 3: 1.96 eV RRS spectra from SaDNA dispersed HipCo SWCNTs at 10 and 50 min ultrasonication time. The spectra are normalized with the peak intensity at 258 cm^{-1}.

signal (ω_{RBM}) via the relation, $\omega_{RBM} = \alpha / d_t + \beta$, where the values of α and β depend on SWCNT synthesis condition and environments,[10] we can easily estimate the chirality of nanotubes from the Kataura plot.[11] Based on this theoretical model and other experimental observation,[12] (7,5) tubes and {(11,1), (10,3)} SWCNT groups were determined to be exhibiting the RBM peaks at 258 cm^{-1} and 285 cm^{-1}, respectively, in 1.96 eV excitation energy RRS spectra. The enrichment profile from NIR absorption, showing an increase in small diameter (7,5)[12] tubes, as oppose to the {(11,1), (10,3)}[12] SWCNT group, was consistently observed from the RRS RBM spectra (Figure 3) obtained from 10 min and 50 min sonicated SaDNA-HipCo SWCNT samples.

The ideal d_t of a SWCNT can be calculated from a simple relation, $d_t = (a/\pi)(n^2 + m^2 + nm)^{1/2}$,[1] where a is an average carbon-carbon distance (1.44Å in this paper). It is
clear from the NIR absorption profile that smaller diameter ($d_t<0.9$ nm) SWCNTs are becoming more abundant as the ultrasonication time increases, meanwhile the larger

diameter ($d_t>1.0$ nm) tubes are becoming less stable in the dispersion. Further CD study from these samples has shown a matching trend in the helicity of the SaDNA, where small d_t tubes are stabilized as the DNA configures from high near-complete A- (NCA-) form to an intermedia or mixture of A- and B- (A&B-) form. Circular dichroism has been typically utilized to estimate the representative secondary structures of DNA, A- and B-form, by monitoring the intensity changes from the negative and positive peaks at 290-260 nm and 260-230 nm, respectively.[13] The SWCNT-induced changes of DNA CD characteristics have been investigated by several groups.[14, 15] By using heavily acid-treated commercial SWCNTs, Li et al. reported carbon nanotubes induce a B-A transition of DNA. In Figure 4, this stabilization of A-form DNA

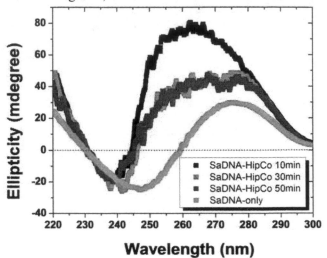

Figure 4: CD spectra from SaDNA dispersed HipCo SWCNTs at 10, 30 and 50 min ultrasonication time. SaDNA solution without SWCNT is shown as reference B-DNA spectrum.

structure is clearly shown in the CD measurement from SaDNA dispersed HipCo SWCNTs. At the initial ultrasonication of 10 min, SaDNA is in NCA-form, and moves to A&B-form as the sonication time increases. Typically, DNA helical structure exhibits the diameters of 2.6 nm and 2.0 nm at A- and B- forms, respectively. Assuming the SaDNA are wrapping around SWCNTs, the enrichment trend of small d_t nanotubes at prolonged sonication (50 min) matches surprisingly well with the decrease in diameters of surfactant DNA.

4 CONCLUSION

The NIR study revealed that single wall carbon nanotubes were dispersed into the D_2O media at a level comparable to or better than that of d(GT)$_{20}$ DNA oligomer. Moreover, selective stabilization of smaller d_t SWCNTs was observed as the sonication time increased. 1.96 eV RRS measurement on (7,5) and {(11,1), (10,3)} tubes showed agreeing enrichment trend with the observation from NIR study. CD measurements on the SaDNA dispersed SWCNTs and SaDNA revealed that SaDNA preferred A-form when sonicated with SWCNTs. Also the NCA to A&B transition trend matches well with the small d_t SWCNT stabilization at longer sonication time. More precise determination of solubilization and separation degree in SaDNA dispersion is under progress by using a UV-VIS-NIR fluorometer. Ion exchange, gel permeation, density gradient and agarose-gel electrophoresis can be good candidates for further SWCNT separation studies using this SaDNA. In addition, with optimally chirality-separated nanotube samples, the immense number of possible DNA oligomeric-structures available in SaDNA will enable us to investigate the kind of DNA strands that are interacting with each specific (n,m) SWCNT. For this, a conventional separation and amplification technique from DNA research, such as polymerase chain reaction, can be used to identify the sequences of preferentially interacting DNAs.

ACKNOWLEDGEMENT

This research was performed while Dr. Kim held a National Research Council Research Associateship Award at AFRL. Financial support from AFSOR is greatly appreciated.

REFERENCES

1. Dresselhaus, M. S.; Eklund, P. C. *Adv. Phys.* **2000,** 49, (6), 705.
2. Liu, J.; Rinzler, A. G.; Dai, H.; Hafner, J. H.; Bradley, R. K.; Boul, P. J.; Lu, A.; Iverson, T.; Shelimov, K.; Huffman, C. B.; Rodriguez-Macias, F.; Shon, Y.-S.; Lee, T. R.; Colbert, D. T.; Smalley, R. E. *Science* **1998,** 280, (5367), 1253.
3. Gooding, J. J.; Wibowo, R.; Liu, J.; Yang, W.; Losic, D.; Orbons, S.; Mearns, F. J.; Shapter, J. G.; Hibbert, D. B. *J. Am. Chem. Soc.* **2003,** 125, (30), 9006.
4. Zheng, M.; Jagota, A.; Semke, E. D.; Diner, B. A.; McLean, R. S.; Lustig, S. R.; Richardson, R. E.; Tassi, N. G. *Nat. Mater.* **2003,** 2, 338.
5. Zheng, M.; Jagota, A.; Strano, M. S.; Santos, A. P.; Barone, P.; Chou, S. G.; Diner, B. A.; Dresselhaus, M. S.; Mclean, R. S.; Onoa, G. B.; Samsonidze, G. G.; Semke, E. D.; Usrey, M.; Walls, D. J. *Science* **2003,** 302, (5650), 1545.Saito, R.; Dresselhaus, M. F.; Dresselhaus, M. S. *Appl. Phys. Lett.* **1992,** 60, (18), 2204.
6. Lu, G.; Maragakis, P.; Kaxiras, E. *Nano Lett.* **2005,** 5, (5), 897.
7. Zheng, M.; Semke, E. D. *J. Am. Chem. Soc.* **2007,** 129, (19), 6084.
8. Wang, L.; Yoshida, J.; Ogata, N.; Sasaki, S.; Kajiyama, T. *Chem. Mater.* **2001,** 13, (4), 1273.
9. Luo, Z.; Pfefferle, L. D.; Haller, G. L.; Papadimitrakopoulos, F. *J. Am. Chem. Soc.* **2006,** 128, (48), 15511.
10. H. Kuzmany, W. P., M. Hulman, Ch. Kramberger, A. Gruneis, Th. Pichler, H. Peterlik, H. Kataura, Y. Achiba. *Eur. Phys. J. B* **2001,** 22, (3), 307.
11. Kataura, H.; Kumazawa, Y.; Maniwa, Y.; Umezu, I.; Suzuki, S.; Ohtsuka, Y.; Achiba, Y. *Syn. Met.* **1999,** 103, (1-3), 2555.
12. Menna, E.; Della Negra, F.; Dalla Fontana, M.; Meneghetti, M. *Phys. Rev. B* **2003,** 68, (19), 193412/1.
13. Baase, W. A.; Johnson, W. C., Jr. *Nucl. Acids Res.* **1979,** 6, (2), 797.
14. Heller, D. A.; Jeng, E. S.; Yeung, T.-K.; Martinez, B. M.; Moll, A. E.; Gastala, J. B.; Strano, M. S. *Science* **2006,** 311, (5760), 508.
15. Li, X.; Peng, Y.; Qu, X. *Nucl. Acids Res.* **2006,** 34, (13), 3670.

Studies on Carbon Nanotubes – Polyaniline based Composites

Chhotey Lal, Tejendra Kumar Gupta, Anil Kumar and Manoj Kumar

National Physical Laboratory (NPL),

Dr. K.S.Krishnan Road

New Delhi - 110012

Email-: clal@mail.nplindia.ernet.in

ABSTRACT

Carbon Nanotubes (CNTs) have become leading material for tremendous strategic applications. We have prepared CNT composites with the Polyaniline by solvent and powder methodology. The adhesion, alignment and the effect of composite nature have been studied by systematic characterization employing Scanning Electron Microscope (SEM) and thermal behaviour (TGA). The Polyaniline has effectively surface attachment and influenced binding properties, which have been observed on different types of Nanotubes precursor materials, prepared by different roots of synthesis. The CNTs have been synthesized at National Physical Laboratory (NPL), New Delhi by Chemical Vapour Deposition (CVD) from the organic substrates and arc discharge process. The CNTs have been accompanied with the iron material as impurities along with some amorphous carbons.

The Functionalization of CNTs has been done with varying concentration of oxidizing agents. The possibility of the attachment of Polyaniline with CNTs may be either covalent or non-covalent which require detailed studies. The Polyaniline matrix is suitable for CNTs addition & the composites may be useful for various applications.

1. INTRODUCTION

Conducting polymers have found different applications in energy storage, photovoltaics, transistors, sensors etc. PANI is one of the first intrinsic conducting polymer which is used in these applications. PANI was discovered in 1934. The conductivity range of PANI mixture is between 10^{-9} Scm^{-1} and 10 Scm^{-1}.it is green in colour and redox active material.

It is made by the polymerization of aniline monomers by the oxidation of aniline monomers. Due to these excellent properties, PANI is used for making CNT based composites.

The MWNTs synthesized by catalytic chemical vapour deposition (CVD), by taking 8 times concentration of toluene & ferrocene. The solution of these mixtures is injected by uniform flow rate of 0.1 ml/min in to the reactor furnance at 750^{0C} and 200 torr pressure of Ar gas.

To increase the conducting properties of polymers, composites of these polymers with other phase are more effective than the pure polymers because by doping of high conducting materials in the polymer matrix increases the conductivity of the composites.

A composite material may be defines as a material system consisting of a mixture of two or more microphases, which are mutually insoluble and differing in composition.

[1]Some advantages of composite materials over the bulk material which we are using like matrix, polymers and ceramics are:

- High specific strength

- Low specific gravity

- High stiffness

- They maintain the strength at high temperature

- Better toughness

- Easily fabricated.

2. EXPERIMENTAL

2.1 Synthesis of MWNTs :

MWNTs are synthesized by catalytic chemical vapour deposition method at Carbon Technology Unit, NPL, New Delhi, India by taking 8 times concentration of toluene and ferrocene at 750^{0C} temperature.

Conditions & percentage conversion of toluene with ferrocene at 750^{0C} temperature is manipulated by Carbon Technology Unit, NPL, New Delhi and values are given as:

Conditions:

- Flow rate: 0.1ml/min.

- Zone temperature: 750^{0C}

- Reactor diameter: 42mm

- Toluene is taken as carbon source.

- Ferrocene is taken as catalyst.

Synthesized as such carbon Nanotubes are impure and called carbon soot.

These CNTs soot have been purified by acidic treatment taking 1N-HNO_3 followed washing with the distilled water and dried at $100^{0}C$ in electric oven, the purified material was further heat treated by air oxidation process at 350^{0C}.

3. CHARACTERIZATION

Some techniques are used for the characterization of these composites.

1. Thermo gravimetric analysis (TGA)

2. FT-RAMAN

3. Scanning Electronic Microscope(SEM)

4. Electrical Conductivity Measurement

1. TGA

➢ The decomposition of PANI starts above $300^{0}C$.

➢ TGA shows the higher thermal stability of PANI-CNT Composites as compared to pure PANI.

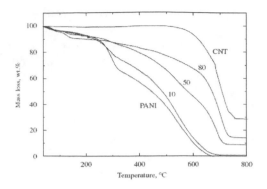

Fig.1: TGA Graph

2. FT-RAMAN

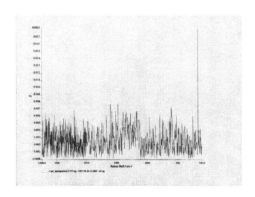

Fig.2: FT-RAMAN Spectra for 5% CNT in PANI

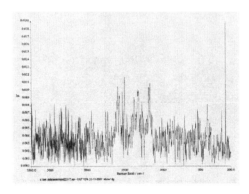

Fig.3: FT-RAMAN Spectra for 10% CNT in PANI

3. SEM IMAGES

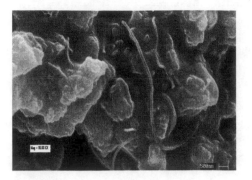

Fig.5: Composite of 10 % CNT in PANI

Fig.6: Composite of 20 % CNT in PANI

Fig.7: Composite of 30 % CNT in PANI

NSTI-Nanotech 2008, www.nsti.org, ISBN 978-1-4200-8503-7 Vol. 1

Fig.8: Composite of 40 % CNT in PANI

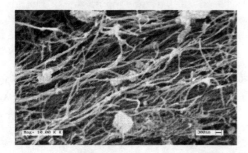

Fig.9: Composite of 50 % CNT in PANI

RESULTS AND DISCUSSION

SEM shows uniform coating of CNT with PANI (Shown in fig. 4, 5, 6, 7, 8, 9 and 10). The coated CNT become thicker as the amount of deposited PANI increases. Uniformity increases with increasing the fraction of CNTs in PANI.

Increasing the fraction of CNT, electrical conductivity of the composite increases.

This shows the successful fabrication of CNT-PANI composites with varying the fraction of CNTs.

TGA graph shows that thermal stability of PANI in the composites with CNTs is better than the stability of neat PANI.

REFERENCES

[1] *S.Iijima, Nature 1991, 354, 56.*

[2] *Carbon Nanotubes: Synthesis, Structure, Properties, and Applications (Eds: M.S. Dresselhaus, G. Dresselhaus, Ph. Avouris). Topics in Applied Physics, Vol. 80, Springer, Berlin, Germany 2001.*

[3] *Soluble Self- Aligned Carbon Nanotube/PANI Composites, By Raquel Sainz, Ana M. Benito, M. Teresa Martinez, Juan F. Galindo, Javier Sotres, Arturo M. Baro, Benoit Corraze, Olivier Chauvet, and Wolfgang K. Maser.*

[4] *Multi-wall Carbon Nanotubes Coated With Polyaniline By Elena N. Konyushenko, Jaroslav Stejskal, Miroslava Trchova, Jiri Hradil, Jana Kovarova, Jan Prokes, Miroslav Cieslar, Jeon- Yuan Hwang, Kuei-Hsein Chen, Irina Sapurina, Polymer 47 (2006) 5715-5723.*

In situ Synthesis of Cu Nanoparticles on MWCNTS using Microwave Irradiation

Sanchita Dey, Vijaya K Rangari[*] and Shaik Jeelani

Materials Science and Engineering, Center for Advanced Materials Tuskegee University

ABSTRACT

Carbon nanotubes are one of the most extensively studied nanostructured materials. In particular, carbon nanotubes are unique and ideal templates onto which to immobilze nanoparticles allowing the construction of designed nanoarchitectures that are extremely attractive as supports for heterogeneous catalysts, for use in fuel cells and multifunctional composite structural applications. This paper describes a simple and efficient way of deposition of Cu nanoparticle on multiwalled carbon nanotube (MWCNTS) using microwave irradiation technique. Cu/MWCNTS composites were prepared by microwave heating of ethylene glycol (EG) solutions of Cu acetate precursor salt. 250 gm copper(II)acetate dissolved in 100 ml of ethylene glycol in a round bottom flask by using magnetic stirrer. Cetylmethyl Ammonium Bromide (CTAB) used as a surfacent. Then 50mg MWCNTS were dispersed. The flask was placed in the center of a microwave oven (SHARP 1000V/R21HT) for 10-20 min under microwave power of 60 W. The products obtained were centrifuged and washed thoroughly with ethanol and vacuum dried at room temp for overnight. For comparison similar experiments were carried out with and without CTAB and MWCNTS. In this method, metal acetate acts as source of metal and ethylene glycol acts as a solvent as well as reducing agent. The as-prepared nanocomposites were structurally and morphologically characterized by X-ray diffraction (XRD), Transmission electron microscopy (TEM). These results clearly show that the MWCNTS are covered by crystalline Cu nanoparticles. The copper nanoparticles are uniform in size and shape.

Keywords: MWCNTS, microwave synthesis, copper nanoparticles, and coating

1. INTRODUCTION

Multi-walled carbon nanotube (MWCNT) is an ideal raw material for various applications due to its outstanding mechanical characteristics such as high tensile strength and high elastic modulus, high thermal conductivity and electric conductivity [1]. Recently, there has been great interest in the metal coating of MWCNTs [2] for creating new metal-matrix-based carbon tube composites. The metallization process is a kind of surface modification of MWCNTS. This kind of modification not only can increase the surface active sites to improve bonding between nanotube and resin or ceramic [3], but also can preserve the superior performance and excellent intrinsic properties of MWCNTS in the composites. Furthermore, this metal coating of MWCNTs has been shown to have significant potential for the fabrication of new powder MWCNTS-metal composites [4] thus extending the application fields of MWCNTS. In the past, some successful attempts have been made to synthesize nanoparticles of copper and copper oxides, using various method including sonochemical, microwave irradiation photochemical, hydrothermal, solvothermal, electrochemical, sol-gel methods, solid-state reactions, chemical reduction and decomposition route and so on [5].

Microwaves are a portion of the electromagnetic spectrum with frequencies in the range of 300 MHz to 300 GHz. The corresponding wavelengths of these frequencies are 1 m to 1 mm. The most commonly used frequency is 2.45 GHz. The degree of interaction of microwaves with a dielectric medium is related to the material's dielectric constant and dielectric loss. [6] When microwaves penetrate and propagate through a dielectric solution or suspension, the internal electric fields generated within the affected volume induce translation motions of free or bound charges such as electrons or ions and rotate charged complexes such as dipoles. [6]

The resistance of these induced motions due to inertial, elastic, and frictional force, which are frequency dependent, causes losses and attenuates the electric field. The main advantages of microwave-assisted reactions over conventional methods in synthesis are: (a) the kinetics of the reaction are increased by one to two orders of magnitude, (b) novel phases are formed, (c) the initial heating is rapid, which can lead to energy savings, and (d) selective formation of one phase over another often occurs. [7] One possible hypothesis for these microwave-induced effects is the generation of localized high temperatures at the reaction sites to enhance reaction rates in an analogous manner to that of ultrasonic waves, where both high temperatures and pressures have been reported during

NSTI-Nanotech 2008, www.nsti.org, ISBN 978-1-4200-8503-7 Vol. 1

reactions. The enhanced kinetics of crystallization can lead to energy savings of up to 90%. [8]

The polyol method is a low-temperature process that is environmentally friendly because the reactions are carried out under closed-system conditions. It was first introduced to produce metal submicron-sized powders. In this method, a suitable solid metal salt is suspended in a liquid polyol. The suspension is stirred and heated to a certain temperature; the reduction of the starting compound yields fine metal powders. The polyol itself acts not only as a solvent in the process but also as a stabilizer, limiting particle growth and restricting agglomeration. Recently, this method has also been extended to the preparation of metal oxides and metal chalcogenides. [9]

In this paper, we present a microwave-induced polyol route to synthesize and characterization of cu coated carbon nanotube.

2. EXPERIMENTAL

Cu/MWCNTS composites were prepared by microwave heating of ethylene glycol (EG, Aldrich) solutions of Cu acetate (Aldrich) precursor salts. 250 gm copper(II)acetate dissolved in 100 ml of ethylene glycol in a round bottom flask by using magnetic stirrer. Cetylmethyl ammonium bromide (CTAB, 500 mg) or polyvinyl alcohol (PVA, 5 mg) used as a stabilizing agent in a different reactions. All solutions were prepared using reagent grade chemicals. Then 50 mg MWCNTS were dispersed in the solution. The flask was placed in the center of a microwave oven (SHARP 1000V/R21HT) and irradiated for 10 min under microwave power of 60 W. After the reaction completed, the solution was cooled to room temperature, and the products obtained were separated from the liquid by centrifugation and followed by repeated washing with absolute ethanol several times and vacuum dried at room temperature overnight.

Similarly the Cu nanoparticles were also synthesized using the same method as described earlier. Dissolving Cu acetate (250 mg) and stabilizing agent (CTAB or PVA) in Ethylene glycol and then irradiated 20 min with microwave. Then particles the obtained were separated from the liquid by centrifugation and then repeated washing with absolute ethanol and vacuum dried at room temperature overnight

The XRD measurements were carried out using a Rigaku, D/Max 2200 instrument. Transmission electron microscopy (TEM) examinations of the samples were carried out with a JEOL-2010 microscope. The powdered samples were dispersed in ethanol and subjected to ultrasonic treatment and dropped on to a conventional carbon coated molybdenum grid.

3. RESULT AND DISCUSSION

The x-ray diffraction patterns reveal that the synthesized copper particles have high crystalinity and high purity. Figure 1 depicts the powder XRD patterns of the powder XRD patterns of: (a) Cu/MWCNT without CTAB, (b) Cu/MWCNT with CTAB (c) Copper nanoparticles with CTAB. Figures 1(a) and 1(b) indicate that the Cu/MWCNT composite particles are crystalline and all the peaks match with the standard MWCNT and copper JCPDS file number 4-0836. In figure 2 represents the TEM pictures of (a) the as prepared Cu nanoparticle with CTAB, and (b) the as prepared Cu coated MWCNT. The as prepared Cu particles show a polydispersion and the particles sizes measured are about size of 200-400 nm. Figure 2(b) also shows that the Cu particles coated on carbon nanotube shows much smaller size (~ 50nm) nanoparticles compared to the copper nanoparticles prepared without MWCNT. The shapes of these nanoparticles are similar to the nanoparticles prepared without MWCNT. It has been found that if the irradiated for longer time the Cu particles are prone to merge in to large particles to reduce surface energy.

Figure 3 represents the TEM pictures of (a) the as prepared Cu nanoparticle with PVA and (b) the as prepared Cu coated MWCNT with PVA. The as prepared Cu particles show a monodispersion and the particles sizes measured are about size of 20 nm. Figure 2(b) also shows that the Cu particles coated on carbon nanotube also of similar dimension and almost round shape. It has been found that if the irradiated for longer time the Cu particles are prone to merge in to large particles to reduce surface energy as in the case of Cu/MWCNT with CTAB. The interaction and percentage of coating on MWCNTS are under investigation.

The initial dispersion of carbon nanotube in the polyol solution is also an important factor to produce uniform coating of copper nanoparticles on MWCNT. The better the dispersion of MWCNT in initial solution is the better the coating. The reaction scheme for producing fine and dispersed Cu particles using polyol process involves the following reactions: reduction of soluble copper (II) acetate by ethylene glycol, nucleation of metallic Cu and growth of individual nuclei in the presence of a protective agent. The nucleation of Cu particles involves intermediate solid phase formed between the starting material and the final metal powder is formed; during the second stage of the reaction the re-dissolution of the intermediate solid phase takes place. The reduction reaction of ethylene glycol is due to diacetyl, which is formed by a duplicative oxidation of acetaldehyde previously produced by dehydration of ethylene glycol. The reaction can be summarized as follows:

$$2CH_2OH\text{-}CH_2OH \xrightarrow{-H_2O} 2\ CH_3CHO$$

$$\xrightarrow{\text{Intermediate stage}} CH_3\text{-}CO\text{-}CO\text{-}CH_3 + H_2O + 2Cu \quad (1)$$

Since Carbon nanotubes absorb maximum microwave [10] (graphite powder of 1mm can reach 1072 °C in 1.76 min) it require less time to coat MWCNT with similar size Cu particles.

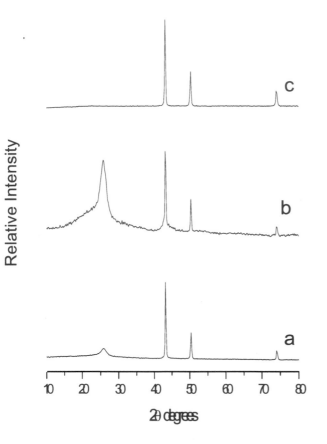

Figure 1. The powder XRD patterns of: (a) Cu/MWCNT without CTAB, (b) Cu/MWCNT with CTAB (c) Copper nanoparticles with CTAB

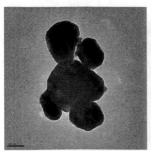 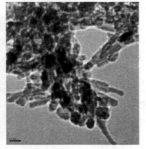

Figure 2. The Transmission electron micrograph of (a) As prepared Cu nanoparticle with CTAB (b) As-prepared Cu coated MWCNT with CTAB

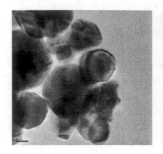 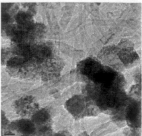

Figure 3. The Transmission electron micrograph of (a) As prepared Cu nanoparticle with PVA (b) As-prepared Cu coated MWCNT with PVA.

4. CONCLUSIONS

Microwave irradiation technique is successfully used to synthesize the copper coated MWCNTs in a one-pot reaction with and without surfactant. Since the graphite carbon absorbs the (in 1.76min 1072°C) maximum microwave radiation had led to the quick and uniform copper nanoparticles coating on MWCNTs. The advantage of this method is particles sizes and extent of coating can be controlled by varying the concentration of precursor and surfactant. Besides this the method is also can be used to synthesize or coating of other transition metals from their metals acetates. The method can be scalable for bulk production, which is the key element in the usage of multifunctional nanostructural materials for structural composite applications.

5. ACKNOWLEDGEMENTS

The authors gratefully acknowledge the support of this research by NSF-PREM, RISE, and CREST grants.

REFERENCES

[1] Oberlin, M. Endo and T. Koyama, J. Cryst. Growth 32 (1976), p. 335.

[2] Chand, J. Mater. Sci. 35 (2000), p. 1303.

[3] S.J. Park, Y.S. Jang and J. Kawasaki, Carbon Lett. 3 (2002), p. 77.

[4] S. Arai and M. Endo, Electrochem. Commun. 5 (2003), p. 797.

[5] Georgakilas, D. Gournis, V. Tzitzios, L. Pasquato, D. M. Goldi and M. Prato" Decorating carbon nanotubes with metal oxides W. H. Sutton, Am. Ceram. Soc. Bull. 1989, 68, 376.

[6] W. H. Sutton, Am. Ceram. Soc. Bull. 1989, 68, 376.

[7] H. Katsuki, S. Komarneni, J. Am. Ceram. Soc. 2001, 84, 313_2317.

[8] Brochure on Aultra CLAVE by Milestone, Microwave Laboratory Systems (Monroe, CT).

[9] Y. Zhao, J. Zhu, J. Hong, N. Bian, and H. Chen, "Microwave-induced polyol-process synthesis of Copper and Copper Oxide Nanocrystals with Controllable Morphology", Eur. J. Inorg. Chem. 2004, 4072-4080.

[10] K. J. Rao, B. Vaidhyanathan, M. Ganguli, and P. A. Ramakrishnan "Synthesis of Inorganic Solids Using Microwaves" Chem. Mater. 1999, 11, 882-895

*Dr.Vijaya Kumar Rangari
Email: rangariv@tuskegee.edu
Fax: 334 727 2286

DNA Functionalized Carbon Nanotubes as Active Stabilizers: Enhanced Stability of Conducting Polymer Composites

William Cheung, Yufeng Ma, Guangru Mao, and Huixin He*

*Chemistry Department, Rutgers University, Newark, NJ 07102
Phone: 973-353-1254; Fax: 973-353-1264
Email: huixinhe@newark.rutgers.edu

ABSTRACT

Short lifetime has been a thorny problem for chemical and biosensors, and light emitting devices consisting of organic (polymer) materials. In this work, a *water-soluble* self-doped polyaniline nanocomposite was fabricated by *in-situ* polymerization of 3-aminophenylboronic acid monomers in the presence of single-stranded DNA dispersed- and functionalized- single-walled carbon nanotubes. For the first time, we found that carbon nanotubes act as novel *active stabilizers*. This is possibly due to DNA functionalization: they reduced the polyaniline backbone from the unstable, degradable, fully oxidized pernigraniline state to the stable, conducting emeraldine state, which significantly improves the chemical stability of the self-doped polyaniline against the harsh UV irradiation.

Keywords: carbon nanotubes, DNA, self-doped polyaniline, composite, UV irradiation, stability

1   INTRODUCTION

Conjugated polymers received a renewed intense interest in the fabrication of numerous light and/or foldable electronic devices, such as light emitting devices, electrochromic displays, rechargeable batteries, microelectronic devices, protection coatings and chemical/biosensors. However, the relative low conductivity, mechanical strength, and stability severely limit conducting polymers for practical applications. Environmental stability is an essential determinant of a device's lifetime. Currently, very little effort has been expended on this subject.

Inspired by the remarkable electronic and thermal conductivity and the superior mechanical properties of carbon nanotubes (CNTs), tremendous efforts have been made over the past decade to prepare polymer and CNT composites with an aim of synergistically combining the merits of each individual component. Herein, for the first time, we exploit the special redox chemistry of carbon nanotubes for improving the stability of conducting polymers under harsh UV irradiation. We fabricated a *water-soluble* self-doped polyaniline/carbon nanotube nanocomposite by *in-situ* polymerization of the 3-aminophenylboronic acid monomers in the presence of single-stranded DNA dispersed- and functionalized- single-walled carbon nanotubes (SWNTs). We found that not only was the electrical performance of the conducting polymer

dramatically improved as predicted, the composite also shows remarkable enhanced stabilization under UV irradiation. Such stabilization effect is of academic interest and practical importance. Short lifetime has been a significant problem in devices consisting of organic (polymer) materials. Incorporation of carbon nanotubes into such devices may help develop organic photonic systems with longer life spans and thus commercial value.

2   EXPERIMENTAL SECTION

2.1   Materials

Purified HiPco single-walled carbon nanotubes (SWNT) were purchased from Carbon Nanotechnologies. Single stranded DNA (ssDNA) with sequence $d(T)_{30}$ was purchased from Integrated DNA Technologies. Ammonium persulfate (APS; 98%), potassium fluoride, 3 – aminophenylboronic acid hemisulfate salt (ABA; $\geq$ 95%) were purchased from Aldrich. Sulfuric acid was purchased from Pharmco. All chemicals were used as received without further purification.

2.2   Dispersion of SWNTs into water solution

Single-walled carbon nanotubes were dispersed in water using a method previously described by Zheng et al. Briefly, 11 mg of purified HiPco SWNT was suspended in aqueous ssDNA solution. This mixture was kept at 0°C with an ice-water bath and sonicated with a Sonics Vibracell (at 30% amplitude) for 30 minutes. After sonication, the sample was centrifuged with a Beckman centrifuge at 5000g to remove undispersed SWNT. After centrifugation, the sample was also dialyzed several times with a Centricon centrifugal filter unit with a molecular weight cutoff of 50kDa to remove free ssDNA. The resulting solution contains DNA – dispersed carbon nanotubes (ssDNA/SWNT) at a mass concentration of 200 – 500 mg/L.

2.3   *In-situ* fabrication of a self-doped polyaniline/ss-DNA/SWNTs nanocomposites

A typical synthetic procedure for the preparation of a water soluble poly (anilineboronic acid)/ssDNA/SWNT nanocomposite (ssDNA/SWNT/PABA) in the presence of fluoride is as follows: An aqueous solution of 40mM ABA and 40mM potassium fluoride was prepared with 0.05M sulfuric acid. A known quantity of ssDNA/SWNT (178.15μl for 0.2%, 890.75μl for 1%) with a concentration of 250.35 mg/L was added to the ABA solution bringing the total volume to 3mL. The quantity of ssDNA/SWNT added is the weight percent based on the amount of ABA monomers in the mixture. The mixture was then purged with nitrogen for 30 minutes to remove dissolved oxygen. 0.38mL of 40mM APS, the oxidizing agent, was slowly added to this mixed solution over a period of 70 minutes to initiate the polymerization process. This reaction was carried out at 0°C under nitrogen bubbling for an additional 5 hrs. The mixture was left to react overnight in the refrigerator at 4°C. The same protocol was applied for the synthesis of pure PABA except ssDNA/SWNT was not added into the polymerization solution. The formed polymer was centrifuged at 4000g for 30 minutes to remove the water soluble oligomers and monomers from the mixture. This process was repeated several times. Dialysis was then performed to remove free DNA and other salts from the solution. This process was repeated until the pH of the formed polymer solution was adjusted to approximately 3.5.

2.4 Ultraviolet irradiation of self-doped polyaniline/ss-DNA/SWNTs nanocomposites

The following steps were taken to expose the polymer nanocomposites with ultraviolet (UV) light: A known quantity of the polymer solutions were dissolved in de-ionized water. Using UV-Visible spectroscopy, the absorbance intensity of the polymer peak for the three polymer samples (pure PABA, 0.2% ssDNA/SWNT/PABA composite, and 1% ssDNA/SWNT/PABA composite) were adjusted to approximately the same intensity to keep the polymer concentration the same. The three polymer solutions, in quartz cuvettes, were then placed 4cm from each other and 7cm from a 3UV multi-wavelength lamp light source (Ultraviolet Products). The three samples were then irradiated for different lengths of time and characterized using UV-Visible spectroscopy (UV-Vis), Fourier Transform Infrared spectroscopy (FTIR), and conductance measurements. All UV-Vis spectra were obtained using a Cary 500 UV-Vis-NIR Spectrophotometer in double beam mode. All FTIR spectra were obtained using a Spectrum Spotlight FTIR Imaging System with a spectral resolution of 4 cm^{-1}. All conductance measurements were obtained using an Electrochemical Workstation CHI 760C.

3 RESULTS AND DISCUSSION

3.1 Improved stability of ss-DNA/SWNT/PABA nanocomposite

Oxidation of the PABA backbone causes a change in its structure. This change is represented by a decrease in conductance due to the conversion of benzenoid units to the more reactive quinoid units in the PABA backbone. Therefore, the conductance can be monitored as a function of time using an Electrochemical Workstation CHI 760C.

Thin films were prepared by adding 5ul of the polymer composites to pre-patterned silicon chips in a layer-by-layer fashion and the conductance was measured until the percolation point was reached. After percolation, the three polymers on the silicon chips were subjected to UV irradiation light at 365nm, the wavelength where the π- π* electron orbital transition occurs along the backbone of the PABA polymer chain. The conductance was monitored for pure PABA, 0.2% ssDNA/SWNT/PABA, and 1% ssDNA/SWNT/PABA as a function of UV irradiation time. Figure 1a shows a typical I-V curve for pure PABA after 150 minutes of UV light irradiation.

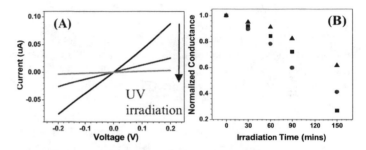

Figure 1: (a) I-V curve of pure PABA irradiated for 0 (black), 60 (red), and 150 mins (green). (b) Normalized relative conductance decrease for pure PABA (black), 0.2% ssDNA/SWNT/PABA (red), 1% ssDNA/SWNT/PABA (blue). Larger percentage of SWNT in the in-situ polymerized solution yields more stable composite.

The normalized relative conductance decrease is shown in Figure 1b. Pure PABA showed the largest relative decrease in conductance (73.31%) compared to 1% ssDNA/SWNT/PABA (38.40%) and 0.2% ssDNA/SWNT/PABA (58.79%) after UV light irradiation. These results reveal that the amount of ssDNA/SWNT added to the polymerization process affects the stability of the formed polymer nanocomposite. In order to understand this phenomenon further, we used UV-Visible spectroscopy and FTIR spectroscopy to study the electronic and molecular structures of the formed polymers before and after UV irradiation.

A typical UV – Vis spectrum of PABA in the emeraldine oxidation state contains three absorption bands of interest. The first absorption peak at ~266nm is characteristic of ABA monomers, free ssDNA, and certain polyaniline degradation products such as hydroquinone, p-benzoquinone, and p-aminophenol.[1] The second peak at ~390nm originates from the π- π* electron orbital transition in the benzenoid rings along the backbone of the PABA chain and also from oligomers. The last absorption peak of interest is at ~800nm. This peak is caused by the intrachain electron excitation of the polymer.

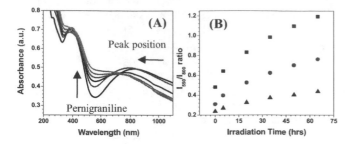

Figure 2: (a) UV – Visible spectra of pure PABA after exposure to UV light irradiation for 0 (black), 5 (red), 20 (blue), 35 (green), 50 (pink), and 65 hours (olive). (b) Pernigraniline to emeraldine ratio at different irradiation time periods for pure PABA (black), 0.2% ssDNA/SWNT/PABA (red), and 1% ssDNA/SWNT/PABA (blue). The values for pernigraniline and emeraldine were obtained by fitting Gaussian curves and taking the total area under these curves. Higher percentage of SWNT used during the in-situ polymerization process results in less pernigraniline state produced under UV irradiation.

Figure 2a shows the UV-Vis spectra of pure PABA at different UV irradiation times. As shown in the figure, UV irradiation induces significant changes in the UV-Vis spectra. The increase in absorbance at ~266nm is due to the oxidation of the PABA backbone, resulting in an increase in concentration of ABA monomers and PABA degraded products such as hydroquinone, p-benzoquinone, and p-aminophenol. Similarly, the increase in absorbance at ~550nm, attributed to the non-conductive pernigraniline oxidation state of PABA and to quinoneimines, indicates oxidation of the emeraldine state of PABA to the pernigraniline state and thus the degradation of the polymer chain. Furthermore, irradiation of the PABA nanocomposite shows a decrease in absorbance and a blue shift of the absorption band at ~800nm, attributed to the emeraldine state of PABA. The decrease in absorption at ~800nm is consistent with the increase in absorption at ~550nm, verifying that the emeraldine state is oxidized to the pernigraniline state. The conjugation length of the conducting PABA polymer can also be qualitatively determined from the UV spectrum by examining the emeraldine peak position. It is well documented that longer wavelengths of the emeraldine peak position indicates longer PABA conjugation length and thus higher conductivity.

The emeraldine peak position of pure PABA blue shifts (149nm) much more compared with the PABA nanocomposites containing 0.2% wt SWNT (45nm) and 1% wt SWNT (24nm) after 65 hours of UV light irradiation. The conjugation length of the polymer produced and thus

the conductivity is preserved more efficiently in the presence of harsh UV light with higher concentration of SWNT.

In addition, the stability of the produced polymer can be determined by analyzing the amount of pernigraniline (~550nm) to emeraldine (~800nm) in the solution, as shown in figure 2b. The increasing pernigraniline to emeraldine ratio for the three polymers verify the degradation of the PABA polymer chain with UV light irradiation. In addition, it shows that the amount of quinoid units in the polymer chain increases as irradiation time increases. The percent increase of the pernigraniline to emeraldine ratio for pure PABA, 0.2% ssDNA/SWNT/PABA, and 1% ssDNA/SWNT/PABA are 148.0%, 127.6%, and 84.9% respectively. The UV-Vis data shown are consistent with conductance measurements; increasing the concentration of ssDNA/SWNT in the in-situ polymerization process yields a polymer nanocomposite with higher stability.

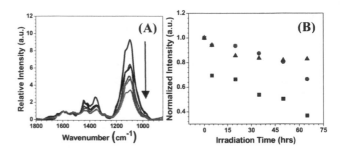

Figure 3: (a) FTIR spectra of pure PABA after exposure to UV light irradiation for 0 (black), 5 (red), 20 (blue), 35 (green), 50 (pink), and 65 hours (olive). (b) Relative decrease in intensity of the "electronic-like" peak (~1100nm) at different irradiation times for pure PABA (black), 0.2% ssDNA/SWNT/PABA (red), 1% ssDNA/SWNT/PABA (blue).

Furthermore, FTIR spectroscopy was used to study and characterize molecular structural changes of the polymers after exposure to UV irradiation. Figure 3a shows the FTIR spectra of pure PABA at different UV irradiation times. The strong absorption band at 1100 cm^{-1} can be attributed to the "electronic-like band" and is a measure of the degree of delocalization of electrons, and thus conductivity. The absorption band at 1596 cm^{-1} can be attributed to the C=C stretching in the quinoid type ring and the absorption band at 1440 cm^{-1} is from mixed C-C stretching, C-H, and N-H bending vibrations of the benzenoid ring.

A general relative decrease in the "electronic-like" peak is seen, indicating that the conductivity of the polymers decreases with increasing UV irradiation time. In addition, the amount of benzenoid units in the polymer chain, represented by the intensity band at 1440 cm^{-1}, shows a decrease with increasing irradiation time. This indicates the oxidization of PABA from the emeraldine state, which has more benzenoid units than quinoid units, to the pernigraniline state. The relative intensity decrease of the "electronic-like band" for the three polymers was calculated. After 65 hours of UV light irradiation, pure PABA showed the greatest relative decrease in intensity (63.2%) compared to 0.2% ssDNA/SWNT/PABA (33.6%) and 1% ssDNA/SWNT/PABA (17.0%). This reveals that pure PABA experiences the greatest conductivity loss compared to the composites containing SWNT. Furthermore, as the concentration of SWNT increases, the polymer produced has greater stability.

The absorption bands corresponding to the quinoid ring at 1596 cm^{-1} and the benzenoid ring at 1440 cm^{-1} can be expressed as an intensity ratio (I_{1596}/I_{1440}). This quinoid to benzenoid ratio can provide information on the degree of oxidation of the polymer. Figure 3b shows the quinoid to benzenoid intensity ratio of pure PABA, 0.2% ssDNA/SWNT/PABA, and 1% ssDNA/SWNT/PABA at different irradiation times. 1% ssDNA/SWNT/PABA showed almost no change (~0.9% increase) in the quinoid to benzenoid ratio, indicating that its molecular structure is relatively unaffected by UV irradiation even after 65 hours of exposure. The quinoid to benzenoid ratio, however, increased 1.9 times for 0.2% ssDNA/SWNT/PABA and over 2.5 times for pure PABA. This shows that as the concentration of ssDNA/SWNT used in the in-situ polymerization decreases the amount of quinoid units formed in the polymer chain increases under UV irradiation and thus makes the formed polymer less stable.

3.2 Proposed Mechanistic Aspect and Role of the ssDNA/SWNTs

The mechanism of the chemical degradation of polyaniline induced by UV irradiation has been discussed in literature.[2] In the presence of dissolved oxygen, the amine units in the polyaniline backbone were oxidized to imine units. The polyaniline is converted from the conductive, half oxidized emeraldine state to the non-conductive and fully oxidized pernigraniline state. Under UV light irradiation, this oxidation process is accelerated. In the pernigraniline form, polyaniline contains quinoid units which are easily hydrolyzed by water, producing various degraded products such as hydroquinone and p-aminophenol. To our knowledge, there have not been any studies on the degradation mechanism of PABA; we assume that PABA experience similar degradation process as polyaniline. The results from our UV-Vis and FTIR study of the molecular structure of PABA after different UV irradiation time periods support this assumption.

However, the data shown from the conductance measurements, UV-Vis, and FTIR spectra all lead to the conclusion that PABA/SWNT nanocomposites are much more stable as the concentration of ssDNA/SWNT is increased during the in-situ polymerization process. The exact mechanism of how the ssDNA/SWNTs enhance the chemical stability of the PABA under UV irradiation is still under investigation in our group. We hypothesis that this is due to the reductive ability of ssDNA/SWNTs, which can reduce the nonconductive pernigraniline to the stable and highly conductive emeraldine state (Figure 4) and making the polymer more stable and less susceptible to hydrolysis. Therefore, the molecular structure and thus the conductivity of the polymer are largely preserved. As the concentration of SWNT is increased, this stabilizing effect is more prominent, as supported by the conductance, UV-Vis, and FTIR data. In this process, the SWNTs are also oxidized. Reduction of the oxidized carbon nanotubes by water or hydroxide ions would complete the cycle.[3, 4]

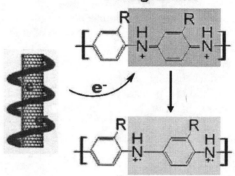

Figure 4: Proposed stabilization mechanism of ssDNA/SWNT during the in-situ polymerization process.

In summary, not only was the electrical performance of the conducting polymer dramatically improved by in-situ polymerization of 3-aminophenylboronic acid monomers in the presence of single-stranded DNA dispersed- and functionalized- single-walled carbon nanotubes (SWNTs), the composite also showed remarkable enhanced stabilization under UV irradiation. This may possibly be due to the reductive capability of the carbon nanotubes functionalized with DNA. The enhanced stability would greatly help develop organic photonic systems with longer life spans and thus commercial value.

Acknowledgement. This material is based upon work supported by the National Science Foundation under CHE-0750201 and Petroleum Research Fund.

References.
(1) Rannou, P.; Nechtschein, M.; Travers, J.P.; Berner, D.; Wolter, A.; Djurado, D., *Synthetic Metals*. **1999**, 101, 734-737.
(2) Neoh, K.G.; Kang, E. T.; Tan, K.L., *Polymer*, **1992**, 33, 2292.
(3) Zheng, M.; Rostovtsev, V. V., *J. Am. Chem. Soc.* **2006**, 128, 7702-7703.
(4) Zheng, M.; Diner, B. A., *J. Am. Chem. Soc.* **2004**, 126, 15490-15494.

Simultaneous tip and base growth mechanism in carbon nanotubes produced by PECVD

J. García-Céspedes, E. Pascual and E. Bertran

FEMAN Group, IN$_2$UB, Departament de Física Aplicada i Òptica, Universitat de Barcelona.
C/ Martí i Franquès, 1, E-08028, Barcelona, Catalonia, Spain

ABSTRACT

Carbon nanotubes (CNTs) with curly and straight segments were synthesized by means of radio-frequency plasma enhanced chemical vapour deposition (RF-PECVD). The origin of this peculiar combination of aligned and misaligned structures within a single CNT is attributed to a simultaneous growth process from the tip and the base of the CNT. Evidences for supporting such a growth mechanism were found by scanning and transmission electron microscopy (SEM and TEM) studies. In addition, energy dispersive X-ray analysis (EDX) provided information about the chemical purity of the active catalyst. The growth kinetics is affected by the particular local conditions surrounding the base and the tip of the CNTs, such as the degree of ion bombardment originated by the presence of plasma.

Keywords: PECVD, VACNTs, VACNFs, CNT growth mechanism, catalyst–substrate interaction.

1 INTRODUCTION

Plasma enhanced chemical vapour deposition (PECVD) has been demonstrated over the years as the preferred method to obtain free-standing vertically aligned carbon nanotubes and nanofibers (VACNTs and VACNFs) [1,2]. This kind of structures show an excellent behaviour in applications such as field emission [3,4] or electrochemistry [5,6]. A full understanding of the growth and alignment mechanisms of these structures is crucial in order to achieve their optimal performance.

Merkulov and co-authors [7] were who firstly proposed a consistent alignment mechanism for CNFs and CNTs grown by PECVD. This model is based on the force that the electric field generated at the plasma sheath exerts on the tip of the growing CNFs. In Merkulov's article, only CNTs following a tip-growth mechanism grew vertically aligned, while CNFs grown from the base showed random orientations. In contrast, Bower's article [8] showed VACNTs with their catalyst particle at their base. Although the authors of this work claimed that the electric field was clearly responsible for the alignment, experimental evidences seem ambiguous and unclear presented.

Here, we present CNTs grown from tip and base simultaneously by PECVD. This new kind of structure has been found to provide valuable information on the growth mechanism of these structures, as well as on the factors that determine their alignment with respect to the substrate.

2 EXPERIMENTAL

CNTs/CNFs have been grown on c-Si (100) substrates with a native oxide layer, using a gas mixture of NH$_3$:C$_2$H$_2$=2:1. Ni, Fe and Co have been used as the catalyst materials in our experiments. For this, a 4-5nm thin layer is first deposited by RF magnetron sputtering on top of the substrate. The PECVD system used in this work has been powered by a 300W RF source (Hüttinger) working at 13.56MHz. The deposition temperature ranged from 650ºC to 750ºC. Details of the deposition procedure can be found elsewhere [9].

SEM micrographs of the nanotubes were obtained on a Hitachi H-4100FE operated at 30kV. For TEM analysis, samples were prepared scrapping the films off the substrate, dispersing them in a hexane ultrasonic bath, and placing them onto holey carbon Cu grids. Bright field pictures were taken on a PHILIPS CM30 electron microscope operating at 300kV, while EDX microanalysis was performed on a Hitachi 800MT operated at 200kV equipped with an EDX microprobe.

3 RESULTS AND DISCUSSION

Figure 1a shows SEM and TEM images obtained from various CNT and CNF samples. In figure 1a (sample A) there are mostly VACNFs, although there is a small portion of spaghetti-like CNTs grown among them (marked by arrows). These nanotubes show particular characteristics, as greater length and smaller diameter with respect to the aligned ones. In figure 1b (sample A), TEM pictures of VACNFs reveal a typical bamboo-like structure of their inner cavity, which provides, in addition, a fingerprint of growth direction. Note that in this picture, two CNFs show a particle at both ends. This means that the original Ni catalyst nanoislands broke into two separate fragments during growth, one remaining at the base and the other at the tip. However, only one of the particles appears to be active. EDX analysis performed on both ends of several CNFs revealed a significant Si content, which might be incorporated by temperature-activated diffusion from the silicon substrate. The proportion of Si with respect to Ni was around 10% at the active particles located at the tips, while the percentage at the bottom fragments was slightly over 30%. Hence, the Si contamination threshold for the inactivation of a Ni catalyst must be within this range (10-30%). Regarding the spaghetti-like CNTs, they were not found in the TEM prepared sample, probably due to its extremely low concentration, although it would be interesting to check the Si content at the active catalyst fragments of these CNTs. In figure 1c, a new kind of structure is shown (sample B). Here, we present nanotubes composed of a partially aligned segment (>5µm), with a VACNF at their tip (~1µm). These two segments are associated to a simultaneous growth from the tip and the base of the nanotube. In figure 1d (Sample C), a TEM

micrograph shows clearly a junction of a short VACNF (~200nm) and its spaghetti-like "tail". The tip segment resembles a CNF grown by a typical PECVD process, while the tail shows a very common structure obtained by thermal CVD. Furthermore, the bamboo structure concavities present inverted orientations, thus highlighting opposite growth start points along the tubules (see inset for the long segment). The great morphological difference found between the two segments of the carbon nanostructure suggests an important variation of the surrounding conditions at their tip and base. On one hand, we must take into account the electric field created at the

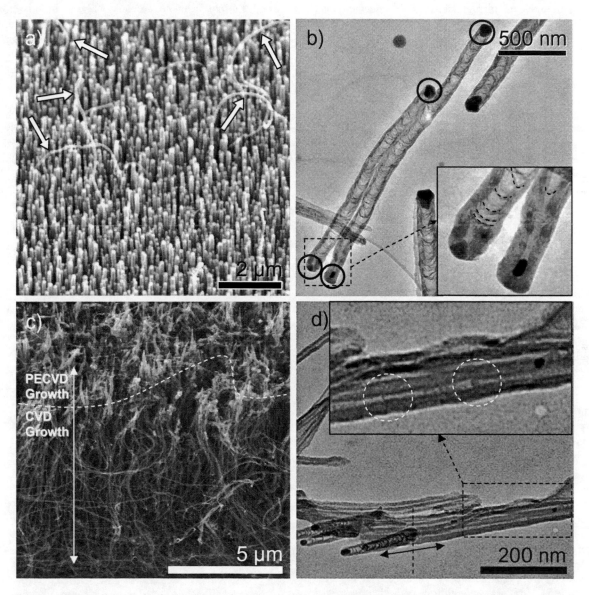

Figure 1. a) SEM micrograph of VACNFs. A few CNT (pointed by arrows) grew in a curly fashion, and show a greater length. b) TEM images of the catalyst material found at both edges of the VACNFs. The inset shows their base edges, where the typical bamboo-like growth fingerprints presumably correspond to the tip particles. c) SEM micrograph of CNTs with straight and curly segments. d) TEM image showing a junction which divide a CNTs into two segments: one grown from the tip (left) and one grown from the base (right). The inset highlights the orientation of the bamboo-like structure. Pictures a) and b) corresponds to sample A; picture c) corresponds to sample B and picture d) corresponds to Sample C.

sheath between the plasma bulk and the substrate (cathode). Given that the distance between adjacent tubes is very low compared to their length (<1:10) and the plasma sheath is in the order of several mm, almost all the electric field lines are collected by the tips of the growing nanotubes. Positively ionized species will be electrically attracted by these tips and accelerated against them. It is well established that the role of NH_3 is to etch preferentially the amorphous carbon during CNT deposition [10]. Nevertheless, the ion bombardment can also remove building atoms of the nanotube, which results in a slowed down growth process [1]. On the other hand, we can consider that neutral species reach equally the base and the tip of the nanotubes, in this geometry. Consequently, the base of the growing tubules is shielded from high energetic ions but benefits from the extra carbonaceous radicals supplied by the plasma discharge, as CN, CH or C_2 [9]. This would explain why in sample B and C the nanotube tails show greater lengths than the tips. Furthermore, it is likely that the small portion of spaghetti-like CNTs in sample A is the result of a base-growth mechanism as well. From a different approach, all these results highly supports Merkulov's model for the alignment mechanism in CNFs grown by PECVD [7]. SEM images in figure 1c allow us to extract a direct estimation of the growth rates at the tip with respect to the base. If we approximate a constant growth rate with time and consider no deactivation of the catalyst particle during the deposition process, then we obtain that the base-growth is at least 7 times faster in our system under the specified deposition conditions for sample B. This estimation takes into account, for the spaghetti-like segment, the distance from the base of the VACNF to the substrate, which is a lower limit, given that this fragments are not totally aligned. If we apply the same criteria to sample C, then the ratio between base and tip growth rate is as high as 35 (the length of the total nanotube was estimated from another SEM picture). In parallel, this large difference between growth rates gives an idea about how intense is the etching process that takes place on the nanotube tip.

In figure 2, it can be seen possible scenarios for CNT/CNF growth. On step I, the precursor species present in the plasma reach the catalyst nanoisland formed after the annealing process. On step II, carbon atoms from catalytic dissociation of precursor species diffuse along the particle surface and precipitate to form a MWCNT or MWCNF. At this point, the catalyst particle will be lifted-up by the growing tubule ("tip-growth" mechanism) or will remain anchored to the substrate surface because of a strong interaction with it ("base-growth" mechanism). In addition, the particle can fragment into two (or more) parts as a consequence of the tensile stress established along the particle. On one side, the interaction with the substrate retains the particle anchored; and on the other side, the continuous incorporation of new carbon atoms at the catalyst-wall frontiers of the CNT/CNF lifts the particle upwards. This last supposition is highly supported by CNF

in-situ growth TEM observations carried out by Helveg S. and co-authors [11]. In the supplementary movie of this reference, it can be clearly seen how a Ni particle attached to the $MgAl_2O_4$ substrate is firstly elongated because of the incorporation of new C atoms, and finally it releases the accumulated stress detaching itself from the surface and recovering its initial shape. Finally, on step III, we consider the most feasible processes that take place in our samples. If the catalyst particle keeps its original volume during growth, it will yield a pure base or tip growth (cases 1 and 2, respectively), depending on the interaction between the catalyst and the substrate. If the particle breaks into two or more fragments, the resulting nanotubes will present several morphologies. Case 3 is observed in our sample A, where the CNFs grow mainly from the tip. Finally, the most complex case (4) is when the catalyst material breaks into two active fragments, one located at the tip and the other one at the base of the nanostructure, giving rise to the CNTs shown in samples B and C (figures 1c and 1d). Moreover, these two fragments are continuously subjected to the tensile stress mentioned in step II, so it is possible that new fragments originate and rest encapsulated in the body of the nanotube. In fact, this is the mechanism proposed in [12] for the formation of metallic nanowires encapsulated by CNTs.

A possible cause of the simultaneous growth from the tip and the base described in the case 4 is the effect of temperature on the catalyst particles. At ~700°C, their plastic deformation and further fragmentation are favoured, which does not necessarily imply a melting of the metal, even if we take into account the size correction for its

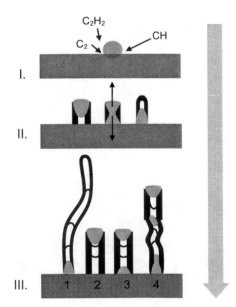

Figure 2. Different possibilities of growth. Step I: metal particle obtained from the annealed precursor film. Step II: Nucleation. Step III: Fragmentation of the catalyst particle and growth possibilities.

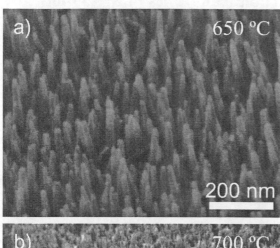

Figure 3. SEM images obtained from CNTs grown at 650°C (a) and 700°C (b), using Fe as the catalyst. The inset is shown for comparison, due to the great difference in length. Picture (a) correspond to sample D and picture (b) correspond to sample E.

melting point [13]. In figure 3, it can be seen the transition of the growth mechanism from the case 2/3 to case to the case 4, due to a change in growth temperature, keeping the rest of the deposition parameters unaltered. The length of the nanostructures is found to increase one order of magnitude (~500nm compared more than 5000nm!), and again we found long and partially aligned CNTs with short straight CNFs over them (see inset of figure 3b).

4 CONCLUSIONS

A new kind of CNT-based structure has been presented, which results from a simultaneous tip and base growth mechanism during a PECVD process. SEM and TEM micrographs provided morphological evidences of this kind of growth. EDX elemental analysis was performed directly on the catalyst particles located at both ends of straight Ni-catalysed CNFs. This allowed us to estimate the Si atomic content necessary to inactivate completely Ni as a catalyst, which is between 10 and 30%. From the SEM images obtained of our samples, it has been possible to compare the growth rates of the segments obtained by the tip-growth mechanism and the base-growth mechanism. The highest growth rate ratio found until now in our samples is $R_{tip}:R_{base}\sim35$. Possible growth scenarios were proposed and discussed according to experimental evidences. The main causes of this new kind of simultaneous growth are the high temperature conditions in combination with the catalyst-substrate interaction. In order to obtain VACNFs (regardless of any crowding effect), growth must take place exclusively from their tip, given that the alignment driving force caused by the electric field is only applied at this point.

ACKNOLEDGEMENTS

This study was supported by MEDU of Spain (project No. DPI2006-03070) and The Generalitat de Catalunya (project No. SGR2005-00666). The authors thank Serveis Cientifico-tècnics of the Universitat de Barcelona (SCT-UB) for measurement facilities.

REFERENCES

1 M. Meyyappan, L. Delzeit, A. Cassell, and D. Hash, Plasma Sources Science & Technology, 12, 205-216, 2003.

2 M. Meyyappan, Carbon Nanotubes: Science and Applications, 99-116, 2005.

3 W. I. Milne, K. B. K. Teo, G. A. J. Amaratunga, P. Legagneux, L. Gangloff, J. P. Schnell, V. Semet, V. T. Binh, and O. Groening, Journal of Materials Chemistry, 14, 933-943, 2004.

4 E. Minoux, O. Groening, K. B. K. Teo, S. H. Dalal, L. Gangloff, J. P. Schnell, L. Hudanski, I. Y. Y. Bu, P. Vincent, P. Legagneux, G. A. J. Amaratunga, and W. I. Milne, Nano Letters, 5, 2135-2138, 2005.

5 J. Li, H. T. Ng, A. Cassell, W. Fan, H. Chen, Q. Ye, J. Koehne, J. Han, and M. Meyyappan, Nano Letters, 3, 597-602, 2003.

6 W. Ke, H. A. Fishman, D. Hongjie, and J. S. Harris, Nano Letters|Nano Letters, 6, 2043-8, 2006.

7 V. I. Merkulov, A. V. Melechko, M. A. Guillorn, D. H. Lowndes, and M. L. Simpson, Applied Physics Letters, 79, 2970-2972, 2001.

8 C. Bower, W. Zhu, S. H. Jin, and O. Zhou, Applied Physics Letters, 77, 830-832, 2000.

9 J. Garcia-Cespedes, M. Rubio-Roy, M. C. Polo, E. Pascual, U. Andujar, and E. Bertran, Diamond and Related Materials, 16, 1131-1135, 2007.

10 M. Chhowalla, K. B. K. Teo, C. Ducati, N. L. Rupesinghe, G. A. J. Amaratunga, A. C. Ferrari, D. Roy, J. Robertson, and W. I. Milne, Journal of Applied Physics, 90, 5308-5317, 2001.

11 S. Helveg, C. Lopez-Cartes, J. Sehested, P. L. Hansen, B. S. Clausen, J. R. Rostrup-Nielsen, F. Abild-Pedersen, and J. K. Norskov, Nature, 427, 426-429, 2004.

12 W. Z. Qian, F. Wei, T. Liu, and Z. W. Wang, Solid State Communications, 126, 365-367, 2003.

13 O. A. Louchev, T. Laude, Y. Sato, and H. Kanda, Journal of Chemical Physics, 118, 7622-7634, 2003.

Ways to Increase the Length of Single Wall Carbon Nanotubes in a Magnetically Enhanced Arc Discharge

M. Keidar[*], I. Levchenko[**], A. Shashurin[*], A.M. Waas[***] and K. Ostrikov[**]

[*]The George Washington University, Washington DC 20052, keidar@gwu.edu
[**] School of Physics, The University of Sydney, Sydney NSW 2006, Australia
[***] University of Michigan, Ann Arbor MI

ABSTRACT

Ability to control the properties of single-wall nanotubes produced in the arc discharge is important for many practical applications. Our experiments suggest that the length and purity of single-wall nanotubes significantly increase when the magnetic field is applied to the arc discharge. A model of a single wall carbon nanotube interaction and growth in the thermal plasma was developed which considers several important effects such as anode ablation that supplies the carbon plasma in an anodic arc discharge technique, and the momentum, charge and energy transfer processes between nanotube and plasma. The numerical simulations based on Monte-Carlo technique were performed, which explain an increase of the nanotubes produced in the magnetic field – enhanced arc discharge.

Keywords: single wall carbon nanotubes, arc discharge

1 INTRODUCTION

Despite significant progress in synthesis techniques the nucleation and growth of carbon nanotubes (CNT) are not completely understood. Several techniques were developed for CNT synthesis such as arc discharge, chemical vapor deposition (CVD), and laser ablation [1, 2, 3, 4]. Plasma-enhanced methods of CNT synthesis are one of the most efficient and precise tools of fabrication of the carbon-based nanostructures [5, 6, 7]. Among other techniques, arc discharge is the most practical method of CNT synthesis.

Main feature of the arc discharge is that the carbon nanotubes produced by the arc discharge technique have fewer structural defects than those produced by low temperature techniques probably due to fast growth that prevents defect formation. In addition it was shown that among several method of CNT production, nanotubes produced by the arc discharge have lowest time degradation of emission capability [8] that is very important for field emission applications.

One important issue related to SWNT synthesis is ability to control SWNT properties, such as radius, chirality and length. It was demonstrated that SWNT radius can be controlled by type of the gas in the chamber while gas pressure leads to fairly constant radius (though some tendency of radius increase with the pressure was recently reported [9]). Recently some ideas regarding control of the chirality were suggested [10]. There is tremendous interest in synthesis long SWNT, which will enable new types of MEMS/NEMS systems, such as micro-electric motors and can act as a nanoconducting cable [11]. Recently growth of 4 cm long SWNT was reported [11]. While, in general, arc discharge technique is considered to offer poor flexibility, it is primarily result of limited understanding of the SWNT synthesis mechanism [12]. In this paper we show that the magnetic field provides a considerable increase in the plasma density and electron temperature, and also ensures a significant enlargement of the high-density plasma area. As a result, the longer SWNT of better quality (lower density of the structural defects) can be synthesized.

2 EXPERIMENTAL SET UP

The arc discharge system consists of anode-cathode assembly installed in a stainless steel flanged chamber capped at both ends (Figure 1). A linear drive connected to the bottom of the chamber acts as the anode feed system. Two portholes on the vertical sides of the chamber are connected to a digital pressure transducer and a constant pressure control system. The arc discharge is sustained with a constant power supply, using a LabView feedback program connected to the linear drive of the anode and the power supply generating the arc.

Figure 1: Photo of GWU set up.

The anode is the pure carbon rod while the cathode is stainless steel rod, with the anode being hollow and the cathode being solid. The cathode has a length and diameter of 1.5 in and .5 in, respectively, while the anode has a length of 3 in., and an outer and inner diameter of .25 in and .125 in, respectively. The anode hole is packed with various metal catalysts. Previous quanta sizing and

NSTI-Nanotech 2008, www.nsti.org, ISBN 978-1-4200-8503-7 Vol. 1

microscope examinations of arc-discharge products for equal arc runtime has revealed that the catalyst combination yielding the largest amount of nanotubes was Y-Ni in a 1-4 ratio [13].

The nanotube samples were produced at constant helium pressures ranging from 150 to 750 Torr. The magnetic field was applied to confine the discharge plasma. Samples containing SWNTs were collected after 180 s run of the arc discharge under various conditions – with and without magnetic field. The samples produced were examined under SEM and High Resolution Transmission Electron Microscope (HRTEM). The average ablation rate of the anode was determined by measuring the initial and final anode geometry. From these measurements, the dependency of the anode material consumption on the arc current with and without magnetic field was determined.

3. RESULTS

The current-voltage characteristic of the discharge (without magnetic field) were measured for helium pressures in the range from 150 to 750 Torr, interelectrode gap ~0.5-1 mm and for 2 anode compositions (C:Ni=15:1 and C:Ni:Y=10.4:4:1 wt. % ratio). Typical dependence of arc voltage (U_{arc}) on arc current (I_{arc}) is shown in Fig. 2 (He pressure –750 Torr, anode composition - C:Ni:Y=10.4:4:1). It was found that current-voltage characteristics were slightly depended on the He pressure and anode composition.

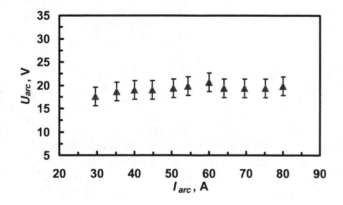

Figure 2: Current-voltage characteristic of the discharge (interelectrode gap ~0.5-1 mm, anode composition - C:Ni:Y=10.4:4:1, He pressure – 750 Torr).

Below we present experimental results related to study of the effect of a magnetic field on SWNT synthesis. Several interesting effects were observed with application of the magnetic field.

The magnetic field strongly confines the plasma causing brighter discharge in smaller zone. We recall here that the application of the magnetic field to the similar discharges usually causes a strong increase in the plasma density [14]. In our previous works we also have demonstrated that the application of magnetic field to the arc discharge leads to

the significant change in the cathode and anode erosion rate [15]. It is natural that the carbon deposit produced in the magnetic field – applied discharge is different of those produced without the magnetic field. A detailed analysis and high-resolution TEM images of the carbon deposits have demonstrated that the samples produced in the magnetic field consist mainly of isolated single-wall nanotubes and bundles. In Fig. 3 we show TEM images of various magnification that demonstrate a 6-nm bundle of several SCWNTs (a), bundle and isolated SWCNT (b), as well as a large bundle where the individual nanotubes are perfectly visible (c). The detailed studies, along with the radius measurements, have proved that the nanotubes produced in the magnetic field-assisted discharge are mostly single-walled.

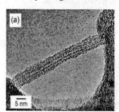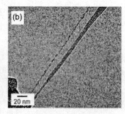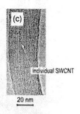

Figure 3: (a) TEM image of the bundle of single-wall carbon nanotubes at high magnification; (b) TEM image of bundle of carbon nanotubes and individual nanotube in parallel the bundle at lower magnification; and (c) TEM image of large bundle of single-wall carbon nanotubes.

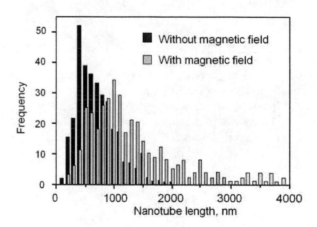

Figure 4: Distribution of SWNT lengths in deposits produced with and without magnetic field. Magnetic field 0.4 T.

We recall again that just the length of the SWCNT are the main our focus; thus we have made a measurement of the nanotubes collected from the deposits produces with and without the magnetic field, and the results obtained are presented in Fig. 4 where we show the distribution of SWCNT length. It can be seen from this graph that the maximum of the distribution of SWCNTs produced without the magnetic field corresponds to 400 nm; for the SWCNTs produced in the magnetic field – enhanced discharge, the maximum of the length distribution corresponds to 1000 nm, and the deposit contains nanotubes of 4 μm length, with the maximum length of 2 μm found in the deposits produced without magnetic field.

4. SIMULATION

Now we attempt to explain the features of SWNTs growth in the magnetically-enhanced arc discharges. We recall that the application of magnetic field, first of all, strongly increases the plasma density and electron temperature of the discharge. Together with the electric field related effects, this should strongly affect the SWNT growth [16]. We have already shown in our previous works that the ion focusing in nano-scaled systems can play a very important role [17, 18]. Now we present a model that helps to describe a SWNT growth in the dense plasma.

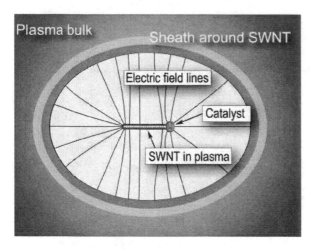

Figure 5: SWNT on metal catalyst particle in plasma. In the practically interesting condition of large sheath ($\Delta \ll \delta$), the shape of the sheath weakly depend on the SWNT length. Electric field is applied between SWNT/C and plasma bulk border. Carbon ion flux is deposited mainly on SWNT tip and on catalyst particle.

Let us first consider in details the SWNT – plasma interaction. As any object immersed in the plasma, SWNT esquires an electric charge (dependent on plasma density as shown in Fig. 5) and eventually encloses by the plasma – surface sheath which thickness can be estimated as few Debye lengths. The plasma is quasi-neutral outside of the sheath, and the electrical field is close to zero. In the sheath there is a non-compensated electric charge which created an electric field applied to the SWNT surface (Fig. 5).

It was shown that in the process with pure helium the SWNT length reaches several μm [19]; so in our calculations we have assumed the maximum SWNT length of 5 μm. Thus, for the typical plasma density (10^{17} m^{-3} – 10^{18} m^{-3}) and electron temperature (1-5 eV) in the arc discharge we can estimate the sheath thickness in the range of 15 to 50 μm. This estimate shows that the sheath

thickness δ well exceeds the SWNT length Δ, and hence the shape of the sheath envelope does not depend significantly on it.

In the vicinity of SWNT, the ion motion is determined by the electrical field between SWNT surface and plasma bulk boundary. The electric field is described by the Poisson equation for the electric potential $\Delta \varphi = \rho_e / \varepsilon_0$, where ρ_e is the density of electrical charge on the sheath. As a boundary condition for the Poisson equation, we assumed the equi-potentiality of the entire SWNT surface (thus we assumed that the SWNT is well conductive): $\varphi(x, r, \alpha)\big|_{(x, R, \alpha)} = \Psi_{SWNT}$, where Ψ_{SWNT} is the electric potential of the SWNT/C system and R is the SWNT radius. With the electric field in the sheath calculated, the ion trajectories can be obtained by integrating the motion equation. More details on the electric field and ion motion calculations on nanostructures can be found elsewhere [20, 21].

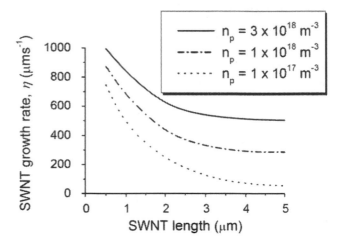

Figure 6. Dependence of SWNT growth rate η on SWNT length with plasma density as a parameter. SWNT diameter 2 nm, catalyst particle diameter is 10 nm. The graph shows strong decrease of SWNT growth rate with the SWNT length.

In this work, we have implemented the following scenario of the SWNT growth in plasma. We assume that the SWNT grows on the partially molten metal catalyst particle supplied to the plasma from ablated electrode (below, we will term the SWNT growing on catalyst particle as SWNT/C system). In the plasma, the metal catalyst particle is a subject to the additional heating and ablation, which reduce the catalyst size, and then condition and molten the external layer creating an external liquid shell. The carbon atom flux gets to the catalyst surface, diffuse through it and eventually incorporate into the SWNT structure. The ion flux at the sheath border – SWNT/C surface supplies carbon atoms to the SWNT/C system. Upon recombination, carbon adatoms migrate about the SWNT/C surface and eventually reach the molten catalyst shell or re-evaporate to the plasma bulk. Today, the two main growth scenarios are mostly accepted: the vapor–liquid-solid (VLS) [22, 23] and solid-liquid-solid (SLS) [24]. In despite of the different initial stages, both scenarios involve the carbon atom diffusion in the molten metal of the catalyst particle, and thus the process of the carbon supply to the external catalyst surface is a decisive factor that determines the SWNT growth kinetics. To calculate the carbon supply to the catalyst surface, we implemented a diffusion model which was used for simulation of the diffusion-driven growth of carbon nanostructures on surface [20, 21].

The above described model was used to simulate the SWNT growth on metal catalyst article in plasma. For the ion motion calculations, we used a Monte-Carlo (MC) technique to obtain the ion flux distribution over the SWNT-catalyst surface [25]. The adatom migration in the collisionless approximation about SWNT surface was simulated also by the MC method, and the carbon atoms diffusion in the molten catalyst was calculated by the diffusion equation [25]. The detailed description of the numerical method and boundary conditions can be found elsewhere [25].

The results of the growth rate calculations are shown in Fig. 6, with the plasma density as a parameter. We should point out that the SWNT growth rate strongly decreases with the SWNT length, and increases with the pressure.

REFERENCES

[1] S. Iijima, Nature **354**, 56 (1991).

[2] Z. Huang, J. Xu, Z. Ren, J. Wang, M. Siegal, and P. Provencio, Appl. Phys. Lett. **73**, 3845 (1998).

[3] Y.H. Wang, M.J. Kim, H.W. Shan, C. Kittrell, H. Fan, L.M. Ericson, W.F. Hwang, S. Arepalli, R.H. Hauge, R.E. Smalley, Nano Lett. **5**, 997 (2005).

[4] S. Arepalli, J. Nanosci. Nanotech. **4**, 317 (2004).

[5] C. D. Scott. J. Nanosci. Nanotech. **4**, 377 (2004).

[6] Z. Markovic, B. Todorovic-Markovic, I. Mohai, Z. Farkas, E. Kovats, J. Szepvolgyi, D. Otaševic, P. Scheier, S. Feil, and N. Romcevic. J. Nanosci. Nanotechnol. **7**, 1357 (2007).

[7] K. Ostrikov, Rev. Mod. Phys. **77**, 489 (2005).

[8] Y. Okawa, S. Kitamura, Y.Iseki, 9th Spacecraft charging technology conference, Japan, April 2005.

[9] E.I. Waldorff, A. M. Waas, P.P. Friedmann, and M. Keidar, J. Appl. Phys. **95**, 2749 (2004).

[10] J. Kunstmann, A. Quanddt, and I. Boustani, Nanotechnology **18**, 155703, (2007).

[11] L.X. Zheng, M.J. O'Connell, S.K. Doorn, X.Z. Liao, Y.H. Zhao, E.A. Akhadov, M.A. Hoffbauer, B.J. Roop, Q.X. Jia, R.C. Dye, D.E. Peterson, S.M. Huang, J. Liu, and Y.T. Zhu, Nature Materials **3**, 673 (2004).

[12] M. Keidar, J. Phys. D: Appl. Phys. **40**, 2388-2393 (2007).

[13] M. Keidar, Y. Raitses, A. Knapp, A.M. Waas, Carbon **44**, 1022 (2006).

[14] I. Levchenko, M. Romanov, Surf. Coat. Technol. **184**, 356-360 (2004).

[15] M. Keidar, I. Levchenko, T. Arbel, M. Alexander, A.M. Waas, K. Ostrikov, Appl. Phys. Lett. (in press)

[16] I. Levchenko, K. Ostrikov, M. Keidar and S. Xu. J. Appl. Phys. **98**, 064304 (2005).

[17] I. Levchenko, K. Ostrikov, M. Keidar and S. Xu. Appl. Phys.Lett. **89**, 033109 (2006).

[18] I. Levchenko, K. Ostrikov, E. Tam, Appl. Phys. Lett. **89**, 223108 (2006)

[19] F. Du, Y. Ma, Xin Lv, Yi Huang, F. Li, Y. Chen. Carbon **44**, 1298 (2006).

[20] I. Levchenko, K. Ostrikov. J. Phys. D: Appl. Phys. **40**, 2308-2319 (2007).

[21] I. Levchenko, K. Ostrikov. Phys. Plasmas **14**, 063502 (2007).

[22] J. Gavillet , J. Thibault, O. Stephan, H. Amara , A. Loiseau , C. Bichara et al. J. Nanosci. Nanotechnol. **4**, 346 (2004).

[23] B.B. Wang, Soonil Lee, X.Z. Xu, Seungho Choi, H. Yan, B. Zhang, W. Hao. Appl. Surf. Sci. **236**, 6 (2004).

[24] R. Sen, S. Suzuki, H. Kataura, Y. Achiba. Chem. Phys. Lett. 349, 383 (2001).

[25] I. Levchenko, A. E. Rider, K. Ostrikov, Appl. Phys. Lett. **90**, 193110 (2007).

One-Step Flame-Synthesis of Carbon-Embedded and -Supported Platinum Clusters

F. Ernst, R. Büchel, R. Strobel, S.E. Pratsinis

Department of Mechanical and Process Engineering, ETH Zurich, Switzerland

ABSTRACT

Carbon-embedded or -supported Pt-clusters were made by a scalable, single-step flame spray pyrolysis (FSP) process. Pt-containing precursors was dissolved in xylene and sprayed and combusted in a controlled oxidation atmosphere resulting in nanostructured, carbon-embedded Pt-clusters. Combustion of xylene alone and subsequent addition of Pt-precursor downstream of the flame (onto the freshly-made carbon particles) led to carbon-supported Pt-clusters.

The majority of the Pt clusters are in the range of 2 – 5 nm. The presence of Pt decreased the carbon yield by catalytic burn-off. Changing the Pt loading, however, had only little effect. The presence of Pt on the surface was measured by CO-chemisorption. For carbon embedded Pt particles no CO was adsorbed and therefore also no catalytic activity for hydrogenation of cyclohexene was observed.

Carbon-supported Pt-clusters on the other hand were made by adding Pt downstream when carbon formation was finished. These carbon-supported Pt-clusters exhibited catalytic activity

Both carbon-embedded and -supported Pt-clusters possessed the self-preserving size distribution of aerosols grown by coagulation in the free-molecular regime. This indicates that homogeneous gas-phase formation rather than a heterogeneous pathway is present.

Keywords: Carbon black,, Flame spray synthesis, co-sythesis, catalytic soot combustion

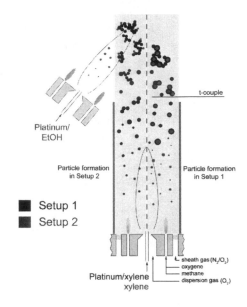

Figure 1: The Experimental setup consists of a flame spray pyrolysis (FSP) enclosed by a quartz glass tube, FSP1. C-embedded Pt clusters were made by dissolving Pt-precursor in xylene and co-oxidizing them. An additional FSP unit (FSP2) was used for synthesis of C-supported Pt clusters. The Pt precursor was dissolved in ethanol/water solution, sprayed, and combusted through the FSP2 unit at the top of the FSP1 chimney into the carbon-loaded effluents of FSP1.

1 INTRODUCTION

Most platinum-based catalysts consist of nanometer-sized metal particles embedded in or dispersed on high-surface-area supports [1].

Carbon embedded Pt nanoparticles find application in sensing applications such as for hydrogen peroxide or biomolecules such as glucose or choline [2]. They are produced by thermolysis incorporating them in glassy carbon [3] or by co-sputtering [4].

Carbon supported Pt nanoparticles are used in many catalytic processes, such as hydrogenation, oxidation, and reforming, but also in electrodes for fuel cells [5,6]. These particles are mainly produced in wet-phase processes [7].

Recently it has been shown that FSP is a suitable process for one-step flame-synthesis of carbon-embedded and -supported platinum clusters [8]

2 EXPERIMENTAL

Figure 1 shows the experimental setup for synthesis of carbon-embedded and -supported Pt clusters by FSP. Xylene (Riedel-de Haën, >96%) was fed into the nozzle (of FSP1) by a syringe pump (Inotech R232) at 5 mL/min and dispersed by 5 L/min nitrogen (Pan Gas, >99.95%) into fine droplets. The spray was ignited and maintained by a premixed flame ring surrounding the spray capillary. This premixed methane/oxygen supporting flame ring was fed by 1.63 L/min CH_4 and 3.88 L/min O_2 throughout all experiments. The reactor is surrounded by a 50 mm outer diameter and 400 mm long quartz glass tube (wall thickness 2 mm). Oxygen and nitrogen were fed as sheath gas with a constant total flow of 7 L/min through a sinter metal ring. The stoichiometry during combustion was controlled by varying the sheath nitrogen to oxygen ratio. With aid of a vacuum pump the product particles were collected on a glass fiber filter (GF/D Whatman, 257 mm in diameter). For synthesis of carbon-embedded Pt clusters, platinum acetyl-acetonate (Strem Chemicals, 98%) was dissolved in xylene and simultaneously fed into the flame, referred to as setup A in Figure 1 (only FSP1 operated). The Pt-precursor concentrations were chosen as to result in 1 to 5 wt % Pt loading of the final product, whereas the total precursor flow rate was in the range from 2.2 to 5 mL/min. For synthesis of carbon-supported Pt clusters, FSP1 (Figure 1) was used as carbon source while a second FSP unit (FSP2) delivering the Pt precursor was at the top end of the quartz glass tube at an angle of 45° (setup B). In FSP2, oxygen was used as dispersion gas at 3 L/min while the liquid precursor consisted of a mixture of ethanol (Fluka), water (deionized), and platinum acetyl-acetonate. The ethanol (EtOH) fraction in the EtOH/H_2O solvent was varied from 0.5 to 1.0 and platinum acetyl-acetonate concentrations were adjusted to obtain 2.7 to 12 wt % Pt in the product powder.

3 RESULTS AND DISCUSSION

3.1 Pt Clusters Embedded in Carbon.

Two conditions for carbon black synthesis were chosen to analyze formation of carbon-embedded Pt-clusters using only FSP1 (Figure 1). At a Equivalence ratio ϕ =1.31 and 1.10 high yields of 2 - 5% of the total C in the precursor and the SSAs of 25 and 105 m²/g, respectivel. Different Pt-contents in the final product (1 to 5 wt-%) were achieved by controlling the Pt concentration in xylene. The Pt clusters seem well dispersed; however, a few single Pt clusters in the size range of tens of nanometers can be observed in STEM pictures. This is attributed to incomplete Pt precursor droplet evaporation at short residence time at high temperatures. Increasing the Pt content also increased Pt cluster size as expected by coagulation and condensation. .

Setup [a]	Pt content [b] / wt%	Surface area / m²g⁻¹	Pt disp.[c] / %
1	2.6	110	0
2	10	247	5.3
2	12	223	14.0

Table 1: Characteristics of selected Pt/C materials prepared by one-step flame synthesis. The absence of platinum on the surface results in a Pt dispersion of 0%.
[a]See Fig. 1 [b]Thermograviometrically determined [c]Derived from CO chemisorption measurements.

The Pt loading of the product depends not only on the Pt content in the precursor but also on the carbon yield. If the latter is reduced, the product Pt-loading increases without changing the Pt concentration in the precursor. Obviously some Pt clusters serve as nucleation sites for the carbon to grow layer-wise on their surfaces (surface growth) as seen in figure 2. No chemisorption of CO on Pt could be observed for all samples synthesized using only FSP1. Together with the inactivity of these samples for catalytic hydrogenation of cyclohexene, this indicates complete or hermetic coating of all Pt clusters with carbon.

3.2 Carbon-Supported Pt Clusters.

Using the FSP1 combined with the FSP2 carbon-supported Pt clusters were FSP2 provides the Pt while FSP1 served as a carbon black source. Lowering the temperature of the FSP2 unit (by using lower ethanol/water fractions) may have led to incomplete combustion and dissociation of the Pt precursor. This is supported by the lack of CO-adsorption (Table 1) and catalytic activity for this powder (as discussed further down). Only little, if any, Pt surface seemed to be available.

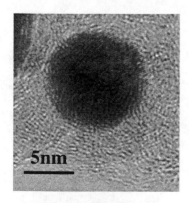

Figure 2: HR-TEM image of carbon embedded Pt nanoparticles. Crystalline Pt particles are surrounded by amorphous, layerwise built carbon.

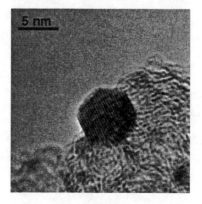

Figure 3:HRTEM image of carbon supported Pt nanoparticles with Pt particles sitting on the carbon surface.

Figure 3 shows a typical HR-TEM image of these Pt/C particles made with pure ethanol as solvent for the Pt-precursor in FSP2. In contrast to carbon embedded Pt clusters (Figure 2), well-developed Pt clusters are located on the carbon surface being well-attached with a good fraction of Pt surface exposed for reaction.

4 CONCLUSIONS

The production of carbon black with FSP was investigated where the maximum carbon yield was observed at the minimum of the process temperature and the highest SSA was observed close to stoichiometric combustion accompanied with a small yield.

0.2-5.0 wt-% Pt nanoparticles embedded in carbon black were synthesized by introducing the respective amounts of Pt precursor to the xylene fuel (Fig. 1 setup 1). Co-formation of Pt nanoparticles and carbon results in carbon condensation on the freshly formed Pt particles. Pt particles were completely enclosed by carbon as shown by CO chemisorption. Few large Pt particles were detected by XRD and STEM, the majority though was in the range of 2 – 5 nm.

5-12 wt-% carbon black supported Pt nanoparticles were produced by using the above mentioned flame spray combusted xylene as the carbon source and introducing the Pt by aid of an additional gas assist nozzle. These particles were catalytically active for the hydrogenation of cyclohexene showing the accessibility of the Pt surface.

REFERENCES

[1] A. T. Bell, *Science* **299** (2003).

[2] T. You, O. Niwa, M. Tomita, S.Hirono, *Anal-Chem.* **75** (2003).

[3] O.J.A. Schueller, N.L. Pocard, M.E. Huston, R.J. Spontak, T.X. Neenan, M.R. Callstrom, *Chem. Mater.* **5** (1993).

[4] T. You, O. Niwa, T. Horiuchi, M. Tomita, Y. Iwasaki, Y. Ueno, S. Hirono, *Chem. Mater.* **14** (2002).

[5] E. Auer, A. Freund, J. Pietsch and T. Tacke, *Appl. Cat. A* **173** (1998).

[6] M. Hogarth and T. Ralph, *Plat. Met. Rev.* **46** (2002).

[7] D. Thompsett, in Fuel cell technology handbook, CRC Press, Boca Raton, FL, USA (2003)

[8] F. Ernst, R. Büchel, R. Strobel, S.E. Pratsinis, *Chem. Mater.* DOI: 10.1021/cm702023n (2008).

Reactor Design for Low-Temperature Growth of Vertically Aligned Carbon Nanotubes

A. V. Vasenkov[*], D. Carnahan[**], D. Sengupta[*] and M. Frenklach[***]

[*]CFD Research Corporation, 215 Wynn Drive, Huntsville, AL 35805
[**]NanoLab, Inc., 55 Chapel Street, Newton, MA 02458
[***]Department of Mechanical Engineering,
University of California, Berkeley, CA 94720-1740

ABSTRACT

This paper reports on the design of novel reactor for low-temperature growth of Vertically Aligned Carbon Nanotubes (VACNTs). Two mechanisms for selective heating catalytic nanoparticles were investigated: (i) a heating from exothermic reaction of catalytic oxidation and (ii) an induction heating. The requirements for efficient induction heating catalytic particles of size varying from micron to 100s nanometers are discussed.

Keywords: CNT, low-temperature growth, selective heating nanoparticles, multiscale modeling

1 INTRODUCTION

For many applications, varying from field emitters, tweezers, electrical interconnects, and antennas to thermal management materials, there is a need for well aligned arrays of Carbon Nanotubes (CNTs) that can be patterned both in fields and as single, stand-alone structures. CFDRC and NanoLab has recognized the need for a new, lower temperature synthesis method to produce these arrays. The standard temperature for the growth of Vertically Aligned CNTs (VACNTs) is about 600°C, which can easily damage many materials commonly found in integrated circuits, such as aluminum, photoresists, etc. The high growth temperature is a limitation to designers who seek to create CNT- based devices with lithographically defined features, in low temperature and lower cost materials. If the growth temperature could be reduced at least down to 400°C, the spectrum of applications for in-situ grown VACNTs would be substantially widened.

Recently, we have reported results of feasibility studies for low-temperature growth of CNTs by selective heating catalyst [1, 2]. This paper continues the study of previously reported selective heating catalytic particles and outlines reactor design for low-temperature growth of CNTs.

2 EFFECT OF OXYGEN ON VACNT GROWTH

The CNT growth involves catalytic decomposition of hydrocarbons and carbon incorporation into CNT. The results of computational studies presented in Sec. 4 show that catalytic decomposition of hydrocarbons can efficiently occur even at room temperature. Consequently, high temperature of CNT synthesis is required only for carbon incorporation into CNT. We attempted to reduce this temperature via addition of oxygen to the gaseous feedstock in thermal Chemical Vapor Deposition (CVD) reactor. Results of our previous experiments indicated that heating from exothermic reactions of oxidation can be significant in plasma-enhanced CVD reactor [2]. A series of experiments were conducted and yield data obtained in thermal CVD reactor are summarized in Fig. 1. Results show that at all temperatures, the addition of oxygen reduced the observed yield. At 500°C, there was essentially no growth for either condition, indicating that the catalyst is not active at this temperature. At 600°C, the carbon yield from the 20 sccm oxygen test was approximately half of the yield without oxygen. At 700°C, the yield with oxygen was ~80% of the yield without oxygen. We found that the positive effect of oxygen was an improved morphology of CNTs at temperatures below 700°C as shown in Fig. 2.

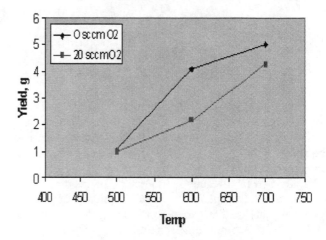

Figure 1: The thermal CVD tests of oxygen addition.

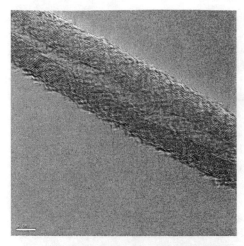

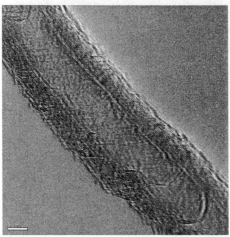

Figure 2: Morphology of CNTs grown at 600°C without (top) and with (bottom) gaseous oxygen. The graphitization was improved with the oxygen addition.

3 INDUCTION HEATING NANOPARTICLES

Results of our multi-scale (spanning 100s nm to 10 cm ranges) simulations demonstrated that particles of specific sizes could be efficiently heated within certain ranges of frequency [1]. Particularly, we found for nickel nanoparticles that micron scale particles are heated efficiently by MHz frequency irradiation, but as the particle size is reduced, the optimal frequencies quickly reached the GHz range [1]. Here, we report results of experimental validation of computational trends. In our tests, nanoparticles were prepared using our sphere masking technique. Al particles of various sizes were deposited on a Corning 1737F glass substrate. These substrates were sent to Duke University, where the frequency response of the structures was investigated between 8 and 12 GHz. In these investigations, a microwave waveguide setup was used to analyze absorption of the signal by measuring both the reflected and transmitted power as a function of frequency. The total of these two is constant, and equal to the original transmitted power, unless the substrate is an absorptive. It was found that the substrate was capable of absorbing power through inductive heating of the metal particles as was computationally predicted. In the series of experiments, three sizes of aluminum metal particles were investigated and the smallest of these (590 nm) has a resonant frequency at about 8 GHz range.

Aluminum is not catalytic to CNT growth, but it was useful to show experimentally that the nanoparticles made by sphere masking showed a resonance. Nickel, with its higher resistance and therefore smaller skin depth as was predicted by our calculations, will have a higher frequency than aluminum.

Based on the obtained results, we formulated requirements for induction heating catalyst nanoparticles during VACNT that are very unique. Typically, industrial applications use RF induction heating in the 450 KHz range. For these applications, water-cooled copper coils are used as inductors. For plasma CVD in semiconductor industry, typical supplies are at 13.56 MHz frequency. Our inductive heating concept points in a very rare frequency range. For example, our results indicated that for catalyst particle of 100 nm in size, the most efficient RF heating for CNT growth is near 20 GHz. Smaller catalyst particles are typically required for single-wall VACNT growth. The use of these particles would require even higher frequencies. Based on our discussion with industrial engineers we identified a specific RF source for novel reactor capable of manufacturing VACNTs at low temperature. The assembly and testing of the reactor are in progress.

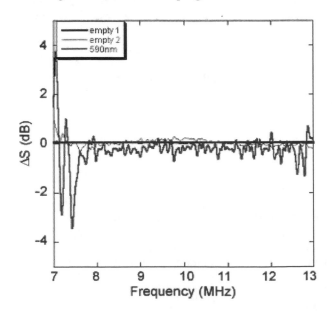

Figure 3: Resonances between 7 and 8 GHZ are clear in the 590 nm aluminum nanoparticles deposited on the glass substrate.

NSTI-Nanotech 2008, www.nsti.org, ISBN 978-1-4200-8503-7 Vol. 1

4 ATOMISTIC SIMULATION OF HYDROCARBON DECOMPOSITION

To investigate the effect of temperature on catalytic decomposition of hydrocarbons, we have integrated Molecular Dynamics (MD) code developed by Prof. Frenklach group at UCB [3] with Mopac, a general-purpose semiempirical quantum mechanics package for the study of chemical properties and reactions in gas, solution or solid-state. The integrated multi-scale simulator was used to investigate the dissociation of C_2H_2 on the surface of iron nanoparticle used as a catalyst during the CNT synthesis. This process is known to be a major path for carbon supply during CNT growth [4].

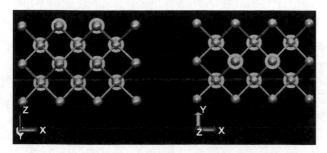

Figure 4: Simulation domain represents a small fraction of the BCC Fe crystal nanoparticle. Large spheres depict quantum atoms.

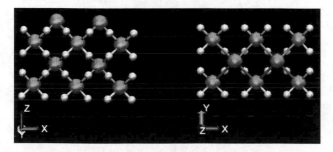

Figure 5: Quantum atoms were passivated with hydrogen at the boundaries for computing semiempirical quantum potential.

Prior to running simulations, we have prepared a simulation domain by truncating the surface of Fe nanoparticle used in CNT growth experiments to a size that is a computationally tractable by the developed multi-scale simulator. The constructed simulation domain, shown in Fig. 4, represents a small fraction of the Body Centered Cubic (BCC) Fe crystal. As a result of the truncation, artificial boundaries with periodic boundary conditions were introduced. Since Mopac 7.2 does not support the calculation of gradients for a 2-dimensional periodic system we considered two groups of atoms as shown in Fig. 4: fixed empirical atoms, located at the bottom and side boundaries, and quantum atoms forming a cluster.

Interactions between empirical atoms and between empirical atoms and quantum cluster were simulated using the empirical Morse potential [5] and interactions of atoms in the quantum cluster were modeled using semiempirical quantum mechanics potential computed by Mopac after each time step in the MD solver [3]. To perform Mopac calculations, the quantum cluster was saturated with "ghost" hydrogen atoms as shown in Fig. 5. The distance between the ghost hydrogen atoms and the quantum cluster was determined by scaling distances between boundary atoms from the quantum and empirical regions [3]. The force matrix computed by Mopac was transferred to MD code. Here, the contributions from the ghost hydrogen atoms were zeroed and empirical atoms with Morse potential were added to quantum atoms.

To validate the developed solver we considered the dissociation of C_2H_2 on the surface of iron catalyst which is known as a major mechanism for carbon supply to growing CNT [4]. In the beginning of simulation, C_2H_2 was introduced 5 Å above the Fe surface as shown in the top panel of Fig. 6 with a velocity directed towards the surface. We found that the dissociation of C_2H_2 involves the formation of C_2H-Fe_n complex (see bottom panel in Fig. 6) due to hydrogen removal reaction:

$$C_2H_2 + Fe_n \rightarrow C_2H\text{-}Fe_n + H\text{-}Fe_n \qquad (R1)$$

C_2H-Fe_n was not stable and quickly dissociated:

$$C_2H\text{-}Fe_n \rightarrow CH\text{-}Fe_n + C\ Fe_n \qquad (R2)$$

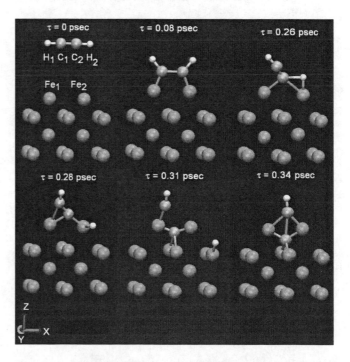

Figure 6: Computational results illustrating dissociation of C2H2 on the surface of Fe at 700°C.

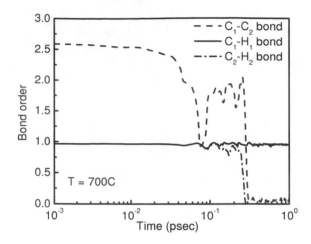

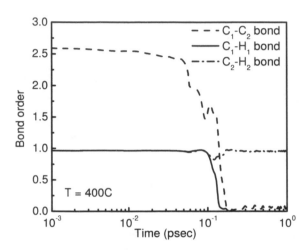

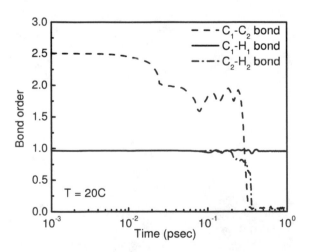

Figure 7: Time-dependent bond order profiles at different temperatures: 700°C (a typical temperature for CNT growth), 400°C (desirable temperature for our CNT experiments), and 20°C (room temperature).

Reactions (R1) and (R2) were investigated at three different temperatures: 700°C (a typical temperature for CNT growth), 400°C (desirable temperature for our CNT experiments), and 20°C (room temperature). Time-dependent bond order profiles for the cases with different temperatures are given in Fig. 7. At the beginning of simulations, when C_2H_2 was 5 Å above the Fe surface, C-C bond order was slightly less than 3 due to the strain resulting from Fe cluster. For $t < 3 \times 10^{-2}$ psec, C-C bond order slowly decreased with time at all temperatures. We observed a sharp minimum for C-C bond order at about 0.08 psec (see Fig. 6 for corresponding space configuration). This minimum resulted in braking C-C bond at T= 400°C. C-C bond at 700°C and 20°C recovered from the minimum at 0.08 psec and was broken at the later time. C-H bond always preceded braking C-C bond. The obtained results demonstrated that breaking C-C bond during interaction of C_2H_2 molecule with Fe cluster can happen even at room temperature. Consequently, carbon supply for growing CNT can be successfully generated even at room temperature.

The further work is in progress. For example, we plan to investigate how the presence of oxygen affects the catalytic decomposition of hydrocarbons (C_2H_2, CH_4 etc.).

5 CONCLUSIONS

We obtained mixed experimental results for reducing growth temperature of VACNTs using oxygen in CVD reactor. Oxygen improved the morphology of CNTs, but diminished the yield of CNTs. The combined experimental and theoretical results demonstrated that 100s nm scale metal particles are heated efficiently by GHz frequency irradiation. The results of atomistic simulations showed that decomposition of hydrocarbons can efficiently occur on the surface of catalyst even at room temperature. The assembly and testing of the reactor for low-temperature VACNTs are in progress.

REFERENCES

[1] A. V. Vasenkov, J. Comput. Theor. Nanosci. 5, 48, 2008.
[2]. A. V. Vasenkov, V. I. Kolobov, and A. V. Melechko, Technical Proceedings of the 2007 Nanotechnology Conference and Trade Show, Santa Clara, CA. Vol. 1, p. 29, 2007.
[3] M. Frenklach and C. S. Carmer,. Advances in Classical Trajectory Methods 4, 27, 1999.
[4] A. V. Melechko, et al., J. Appl. Phys., 97, 041301-1, 2005.
[5] H. O. Pamur, and T. Halicioglu, Phys. Stat. Sol. A 37, 695, 1976.

MWCNTs Production by means of Pyrolysis of Polyethylene-Terephtalate in a Bubbling Fluidized Bed

M.L. Mastellone, U. Arena

Department of Environmental Sciences – Second University of Naples, Via Vivaldi 43, Caserta (Italy)
mlaura.mastellone@unina2.it; umberto.arena@unina2.it

ABSTRACT

A bubbling fluidized bed reactor 102mmID has been used as pyrolyser by feeding polyethylene-terephtalate at two different reactor temperatures. Experiments were carried out with the aim to quantify the yield and composition of gas, liquids and to characterize the solid phase. The solid phase has been characterized by means of different methods: TG-DTG allowed to obtain a preliminary indication about the nature of the different compounds present in the solid sample by means of determination of the thermal stability of their structures; SEM and TEM microscopy, coupled with EDAX analysis, allowed to investigate the morphology of solid structures and to recognize the presence of some specific elements. Different nanostructures, having different degradation temperatures, have been obtained at 600°C and at 800°C. Moreover, the effect of metals extracted by reactor walls on MWCNTs production and the activation/deactivation of this "in-situ" catalyst during pyrolysis of PET is described and supported by experimental evidences.

Keywords: bubbling fluidized bed, carbon nanotubes, polymers, pyrolysis

1   RESULTS AND DISCUSSION

1.1   Effect of reactor temperature: experiments at 600°C.

Pyrolysis of PET in BFB reactors at temperatures lower than 800°C presents some troubles due to risk of defluidization as a consequence of the accumulation on the bed particles of the solid-liquid products of polymer cracking [1]. In fact, even if the thermo gravimetrical studies demonstrated that the temperature for which there is the onset of PET degradation was about 370°C and that at 550°C degradation was totally completed, in the actual reaction conditions a high viscosity, very stable product formed and covered the bed particles so causing a "layering" process [2] that induced defluidization. This occurrence did not allow having stable operation for time longer than 1100s, as showed by Figure 1. During the steady-state regime the gas products have been measured. The composition in term of gas concentration (molar fraction) and in term of yields (mass of gas compound/mass of PET fed during the test) is reported in table 1. Gas composition obtained by PET pyrolysis is strongly related to the monomer structure of PET that contains two carbonyl groups, one aromatic ring and an oxidized ethylene radical. As expected the experimental results showed the presence of hydrogen, methane, ethylene, carbon monoxide, benzene, ethylene glycol and other heavier hydrocarbons. In particular, the GC-MS of a gas sample collected during a test at 600°C allowed to identify also acetaldehyde, methyl-benzene, naphthenol, phenil-naphtenol, and phenanthrene.

Once the feeder was turned off the accumulated residue in the reactor (bed, wall and connected lines) was retrieved in order to carry out TG-DTG analysis and SEM and TEM observations. The reactor wall appeared as completely

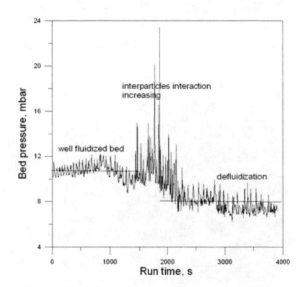

Figure 1 Variation of the bed pressure with the run time

Gas compound	Mol fraction,	Yield, g/g
Hydrogen	20	0.0024
Methane	14	0.014
Carbon monoxide	63	0.11
Ethylene	2.7	0.0045

Table 1 Gas composition and yields for tests at 600°C

covered by a non compact layer, 2-5mm thick, made of bundles of fibers. The yield of this solid fraction was about 0.05g/g. On the basis of previous study on pyrolysis of polymers this residue was expected to be amorphous carbon or microfibers [3] or MWCNTs [4]. TG-DTG gave first indication about the nature of the carbonaceous solid adhering to the bed retrieved after experiments at 600°C that resulted to be stable until 580°C under an oxidizing environment (Figure 2). This degradation temperature is

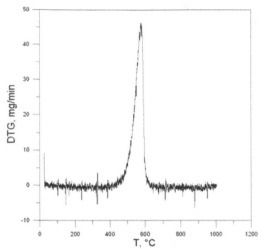

Figure 2 DTG curve of the sample obtained for the run with PET at 600°C.

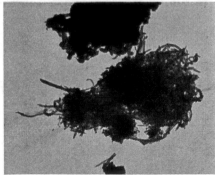

Figure 3. SEM and TEM photographs of samples obtained by the test at 600°C.

coherent with a CNTs structure, as reported in previous studies carried out with other polymers [4] and in other studies carried out by pyrolysing pure methane [5].

In order to verify the actual nature of these structures SEM and TEM investigations have also been made (Figure 3). The photos show bundles of CNTs present in the solids as collected by the reactor wall and by exit lines.

Effect of reactor temperature: experiments at 800°C.

The second series of tests have been carried out at 800°C. At this temperature the progressing of pyrolysis was completely different from that obtained at the lower temperature. Firstly, the pressure continuously increased with time due to a large amount of solids produced in the bed and along the reactor wall; this solid fraction, partially transported out from the reactor, occluded the downstream pipelines so causing pressure drop increasing. Secondly, gas yield is higher (from 0.13g/g at 600°C to 0.31g/g at 800°C) and gas composition was that reported in Table 2. It is evident, by comparing table 1 and 2, that the higher temperature led to have almost the same hydrogen yield but that of methane and ethylene was strongly reduced. In particular, carbon monoxide reaches the maximum yield that can be produced by chain scission. In fact, the maximum yield of carbon monoxide obtainable from PET monomer scission is 28*2/194=0.29.

The part of carbon that is not contained in the gas/liquid fraction constitutes the solid phase that was totally recovered at the end of each test from the reactor and connected pipelines. After collection the solids were weighted, analyzed and completely characterized. The yield of solid phase produced at 800°C was about 0.06g/gPET.

At this temperature the structure of produced solids was different than that obtained at 600°C. CNTs were present in the collected samples only in a low quantity while the main part was composed by nanostructures having a quasi-spherical shape. The possible reason can be a carbon deposition rate over catalysts surface too high to allow the CNT growing. Analyses of TG (Figure 4 and 5) and SEM/TEM photos (Figure 6) revealed the presence of these structures that are absolutely predominant respect to MWCNTs. The degradation temperature found for these structures was equal to 680°C that is greater respect to that found for samples obtained at 600°C. Samples of solids collected from the wall (Figure 4A) are more heterogeneous compared with those obtained by the bed

Gas compound	Mol fraction, %	Yield, g/g
Hydrogen	49.7	0.0024
Methane	0.047	0.0017
Carbon monoxide	46	0.29
Ethylene	2.7	0

Table 2 Gas composition and yields for tests at 800°C

NSTI-Nanotech 2008, www.nsti.org, ISBN 978-1-4200-8503-7 Vol. 1

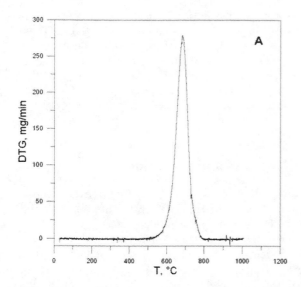 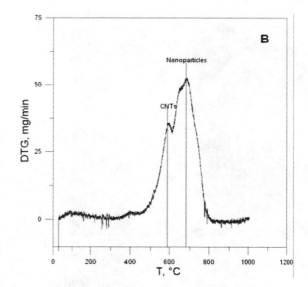

Figure 4 DTG curves – (A): a single well defined peak for bed samples obtained at 800°C. (B): two peaks for the samples obtained by the reactor wall.

zone of the reactor as demonstrated by TG-DTG of Figure 4B. The reason of simultaneous presence of CNTs and other structures, recognized by the two peaks of DTG curve, is related to the temperature gradient of last part of freeboard that is at lower temperature in proximity of the reactor top. This feature is confirmed by observing the results of TG-DTG of a sample collected at the top of reactor (Figure 5). The degradation temperature obtained for this sample is that typical of CNTs; this is in agreement with the very low temperature of the external wall of this zone that is cooled by a water jacket.

1.2   Catalytic effect of metals extracted by the reactor wall.

The internal reactor wall is constituted by stainless steel AISI 316L that contains 13%Ni, 2%Mo, 17% Cr, 0.02%C and 68% Fe. Nichel and ferrous are also the main constituents of catalysts used to produce CNTs by means of CVD method. Several studies demonstrated, in fact, that these metals can catalyze the synthesis of CNTs by acting as a support around that the nanotube can grow if the metal catalyst have nanometric dimension [6; 7; 8].

During the previously described experiments no catalyst has been added into the reactor and the bed material was fresh quartz sand that does not contain Fe or Ni-based compounds. EDAx analysis carried out on fresh sand demonstrated the presence of Si, O and C on the sand surface. On the contrary, the samples retrieved after the pyrolysis tests (sand of bed retrieved after the test and carbonaceous sample collected by reactor wall, condenser, filter, etc.) contained Fe, Cr and Ni. Therefore, it can be deduced that an interaction between the pyrolysis gas and these metals, extracted by the reactor wall, occurred during the process. The presence of hydrogen (acting as reducing

agent) can allow the extraction of metals over that the radicals produced by pyrolysis underwent graphitization and CNT formation. Therefore, the effect of Fe is that of enhancing (respect to thermal cracking) dehydrogenation of radicals, so allowing the carbon deposition and the formation of CNT [5; 7; 8].

1.3   Effect of carbon monoxide on MWCNTs formation.

A peculiar aspect of CNTs production by carrying out the pyrolysis of PET is the progressive deactivation of the catalyst. In particular, during a test 3-4 hours long carried out at 800°C it has been noted a dramatic change in the gas

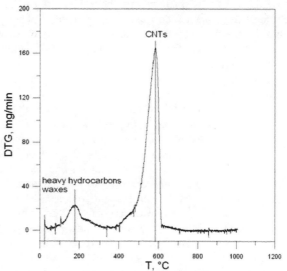

Figure 5 TG-DTG – A single peak is obtained for sample taken in the cold region of freeboard.

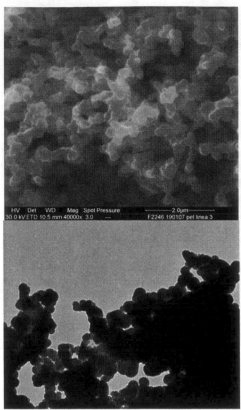

Figure 6 SEM e TEM Spherical nanostructures obtained at 800°C

composition and the complete disappearance of solid phase. After this test, all the other experiments at 600°C and 800°C showed to have solids yield and gas composition totally different from the first ones. The main difference was the totally absence of solids: no solids in the bed, over the wall or in the filters were found. Moreover, the hydrogen yield decreases from 2.3% to 0.28% and CO from 29% to 13%. In order to verify the reliability of these experiments the reactor tube was substituted with a new one and the experiments were repeated with the same schedule. All the results were confirmed: solid phase containing CNTs were produced only for a certain time (from 2 to 4 hours depending on temperature). After this period the solid phase disappeared and a liquid phase was produced.

A reliable hypothesis to explain this behavior is a chemical reaction between CO and Fe that, at high temperature and in presence of a reducing agent as hydrogen, can form metals carbonyls [9]. These iron compounds are complex structures that do not allow any catalytic action since the link between the iron and the CO molecule is preferential. The formation of these kinds of compounds can explain: the progressively reduction of CO amount (reacting with Fe) during the tests; the parallel reduction of hydrogen yield (dehydrogenation is less favored); the formation of liquids instead of solids and, as a consequence, the absence of nanostructures at all. The formation of carbonyls is also supported by experimental evidence that was the presence, in the reactor, of a large amount of white, transparent and crystalline flakes. These compounds covered the reactor wall and filled the pipes and seem like small flakes of "ice". After few hours under ambient temperature they disappeared leaving a thin yellow-orange layer in the container. All these characteristics are in accordance with compounds derived by the further reduction of carbonyls i.e. with their hydrides [9].

1.4   Remarks and conclusion

Pyrolysis of PET in a bubbling fluidized bed allows obtaining good yield of MWCNTs at 600°C. This production is allowed by the catalytic effect of metals (Fe and Ni) extracted by the reactor wall. These metals promote the dehydrogenation and carbonization of the hydrocarbons produced by the PET chain cracking occurring in the fluidized bed. The CNT formation is affected by the reactor temperature and by the gas composition. In particular, the presence of carbon monoxide in the gas phase leads to a chemical deactivation of metals with the consequent total inhibition of CNTs production.

REFERENCES

[1] Mastellone, M.L. and U. Arena, "Bed Defluidization During the Fluidised Bed Pyrolysis of Plastic Waste Mixtures", Polymer Degradation and Stability, 85/3:1051-1058 (2004)

[2] Ennis B.J., Litster, J., "The science and engineering of granulation processes", Kluver Academic Publishers, ISBN: 1-6020-1877-0, (2004)

[3] Dente M, Bozzano G, Faravelli T, Marongiu A, Pierucci S and Ranzi E. Kinetic modeling of pyrolysis processes in gas and condensed phase. In: Guy Marin. Advances in Chemical Engineering. Elsevier Inc, 2007, 32:52-166.

[4] Arena U., M.L. Mastellone, G. Camino, E. Boccaleri, "An innovative process for mass production of multi-wall carbon nanotubes by means of low-cost pyrolysis of polyolefins", Polymer Degradation & Stability, 2006

[5] Chen C.-M., Dai Y.-M., Huang J.G., Jehng J.-M.. Intermetallic catalyst for carbon nanotubes (CNTs) growth by thermal chemical vapor deposition method. Carbon, 44: 1808-1820 (2006).

[6] Moisala, A., Nasibulin, A. G., Brown, D. P., Jiang, H., Khriachtchev, L. and Kauppinen, E. I. "Single-walled carbon nanotube synthesis using ferrocene and iron pentacarbonyl in a laminar flow reactor", Chemical Engineering Science 61, 4393 – 4402, (2006)

[7] Vander Wal R.L. and Lee J. Hall "Ferrocene as a precursor reagent for metal-catalyzed carbon nanotubes: competing effects", Combustion and flame, 130:27–36 (2002a)

[8] Vander Wal R.L., "Fe-Catalyzed Single-Walled Carbon Nanotube Synthesis within a Flame Environment", Combustion And Flame 130:37–47 (2002b)

[9] Cotton, F. A. & Wilkinson G. Advanced Inorganic Chemistry, 6th edition J. Wiley & Sons Inc. (1999)

Growth and characterization of CNT Forests using Bimetallic Nanoparticles as Catalyst

K.H. Lee[*], A. Sra[**], H.S. Jang[*], B.J. Lee[***], L. Overzet[*], D.J. Yang[*], and G.S. Lee[*]

[*]University of Texas, Dallas, TX, USA, khlee@student.utdallas.edu
[**]University of Texas Southwestern Medical Center, Dallas, TX, USA
[***]Yeungnam University, Kyungsan, Kyungbuk, Rep. of KOREA

ABSTRACT

The synthesis of Fe, Fe-Pt and Fe-Co nanoparticles were carried out using the bottom-up polyol process using standard ari free techniques. Using these nanoparticles, multiwalled carbon nanotubes were synthesized in an atmospheric pressure plasma enhanced chemical vapor deposition(APPECVD) process with a mixture of helium and acetylene gases. The APPECVD process produced dense carbon nanotube forests that were 30µm for Fe, 5µm for Fe-Pt and 80~100µm for Fe-Co nanoparticles.

Keywords: bometallic nanoparticles, iron-platinium, iron-cobalt, carbon nanotubes

1 INTRODUCTION

Multiwalled carbon nanotubes (MWCNTs) promise to be useful many applications of great technological importance as they have excellent behaviors as nano-electronics [1], field electron emitters [2], contact electrode [3], sensors [4], etc. Current efforts have been concentrated on using zerovalent metal ions such as Fe, Ni, Co as catalysts, we have expanded our horizons to using bimetallic nanoparticles as catalyst materials. Bimetallic nanoparticles have great improved catalytic properties of the original single metal catalyst and create new properties which may not be achieved by monometallic catalysts [5]. One of the main advantages of using bimetallic nanoparticles is that both the external (size and shape) and internal composition (atomic ordering) can be well controlled.

Fe-Pt and Fe-Co nanoparticles have been demonstrated to be suitable catalysts for growth of carbon nanotubes [6], with some growth recipes reported to yield single-walled nanotubes (SWNTs) [7]. Nanoparticles make for attractive catalysts because, in addition to the possibility of making high-quality CNTs, they are suitable for a variety of growth conditions [8] and applications [6, 9-11]. It is widely believe that the diameter of the catalyst form affects the diameter of carbon nanotubes, the nanoparticle, which can be controlled for size is a good candidate for selective growth of CNTs [10].

Here, we report a simple and efficient way of producing iron-platinum and iron-cobalt (Fe-Pt, Fe-Co) bimetallic nanoparticles which are subsequently used to catalyze continuous growth of multi-walled carbon nanotube (MWCNTs) forests. We compared growth behaviors of MWCNTs grown on monometallic nanoparticles and bimetallic nanoparticles.

2 EXPERIMENT

2.1 Synthesis & Preparation of Nanoparticles

The synthesis of Fe, Fe-Pt, and Fe-Co nanoparticles were carried out using the bottom up polyol process using standard air free techniques [11]. Fe, Fe-Pt, and Fe-Co nanocrystals were prepared by thermal decomposition at 280 °C for 30 minutes of iron acetylacetonate, platinum acetylacetone, and cobalt acetylacetonate with octyl ether in the presence of 1,2-hexadecanediol, oleic acid, and lleyamine. The ratio of platinum acetylacetonate to iron acetylacetonate and cobalt acetylacetone to iron acetylacetonate was 3:1 and 2:1 respectively. After 30minutes, the solution was cooled to room temperature, and the nanoparticles were precipitated using ethanol. The nanoparticles was isolated by centrifuge, were re-dispersed in hexane and then spin-coated onto a silicon substate. Previous studies have demonstrated that the synthesized nanoparticles are typically surrounded by the organic stabilizer shell [12], which in our case is oleic acid. Additionally, some nanoparticles spin coated on the Si wafer were first heated to 680 °C and then treated with oxygen plasma, in order to expose the nanoparticles.

2.2 Growth of Carbon Nanotubes

Subsequent growth of MWCNTs forests was accomplished with an atmospheric pressure plasma jet(APPJ) procedure previously published by our group [13].

The plasma jet was maintained bt 13.56 MHz and the RF power was limited to 30W. Nanoparticles were spin-coated on the SiO_2 covered Si wafer. Acetylene as the carbon-precursor gas was delivered through a capacitively coupled plasma source onto the growth surface. The substrate with spin-coated nanoparticles was heated to the growth temperature on the Cu heating block, after which acetylene carried in the helium plasma was supplied to the surface. Growths were preformed at 680 °C for 3 min with RF power at 30W. After 3 min, the substrate was cooled under helium environment to the room temperature.

2.3 Characterization

Samples for transmission electron microscopy (TEM) were prepared by sonicating the nanoparticles in hexane, dropping a dilute suspension on a copper grid, and then drying. TEM images and energy-dispersive X-ray analysis (EDS) were acquired using a JEOL JEM-2100F TEM. Scanning electron microscopy (SEM) images were obtained at 10kV using the LEO 1350V. For carbon nanotube growths, SEM and TEM images were obtained using the same equipment.

In observing growth products, the CNT forests were checked for long range order at low magnification to check for nanotubes, The forest contains disordered content such as estimate of the quality of these nanotube forests. The Raman spectrometer used in this study had a laser at 633 nm spectral detection.

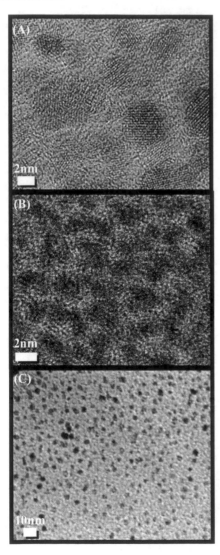

Fig. 1. TEM Images of nanoparticles (A) Fe NPs, (B) Fe-Pt NPs, (C) Fe-Co NPs

3 RESULTS AND DISCUSSION

3.1 Imaging - Nanoparticles

The ability to control composition and size in the synthesis of bimetallic Fe-Pt and Fe-Co nanoparticles is important for the exploitation of their catalytic properties. The Fe, Fe-Pt, and Fe-Co nanoparticles used in the process were examined under TEM and found to have an average diameter of 2~3nm (Fig. 1).

EDS analysis reveals the composition of bimetallic nanoparticles. The molar ratio of iron to platinum and iron to cobalt is 4:1 and 1.2:1, repectively. This composition of iron to platinum in a 4:1 ratio is important for the growth of carbon nanotubes, as both 1:1 and 2:1 molar ratios resulted in no growths. The effects of the composition of the nanoparticles may be related to their ability to dissolve carbon atoms and form nanotubes [7, 14].

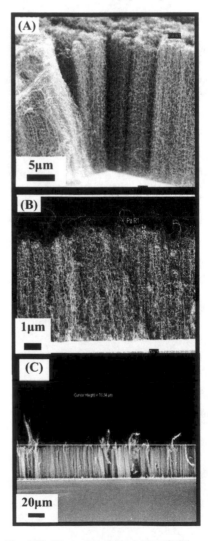

Fig. 2. The SEM Images of the MWCNTs grown on (A) the Fe nanoparticles, (B) the Fe-Pt nanoparticles and (C) the Fe-Co nanoparticles.

NSTI-Nanotech 2008, www.nsti.org, ISBN 978-1-4200-8503-7 Vol. 1

3.2 Imaging – Growths in PECVD

Growths were performed, using the atmospheric plasma jet as previously described, at 680°C for 3min.

Fig. 2. showes MWCNT forests which were dense, uniform, and well-aligned on Fe, Fe-Pt and Fe-Co nanoparticles, respectively. The forest's height varied from 30 μm for Fe to 5 μm for Fe-Pt and 80~100 μm for Fe-Co nanoparticles. The MWCNT forests produced on Fe-Co particles was taller than those grown on Fe and Fe-Pt nanoparticles. We think that the advantages of using Fe-Co nanoparticles over Fe alone is the increase in the MWCNT forest's height from 30 μm to 80~100 μm. The addition of Co prevents the catalyst from poisoning and enhance the growth of the MWCNT forests. But Pt does not improve the catalytic properties.

Fig. 3. is the TEM Images of MWCNTs grown on different particles (Fe, Fe-Pt, and Fe-Co). According to TEM results, the diameter of MWCNTs grown on Fe, Fe-Pt, and Fe-Co is in the range 10~25nm, 7~22nm, and 7~23nm respectively. This result indicates that bimetallic nanoparticles decrease diameter of MWCNTs slightly, as compare to the MWCNTs grown on Fe NPs.

3.3 Spectroscopy

Raman spectroscopy was performed on separate samples to provide another measure of the quality of the carbon nanotubes. Three scans reveal similar spectroscopic profiles. Fig. 4. showed Raman spectra in the range from 1200 cm^{-1} to 1700 cm^{-1} for MWCNTs grown on different nanoparticles. As shown Fig. 4, D and G peaks appeared at 1321 cm^{-1} and 1573 cm^{-1} for Fe nanoparticles, 1327 cm^{-1} and 1580 cm^{-1} for Fe-Pt nanoparticles, and 1324 cm^{-1} and 1577 cm^{-1} for Fe-Co nanoparticles, respectively. The D peak is generally caused by defects in the curved graphite sheet and by the finite sizes of graphite crystallites. The G peak corresponds to the tangential stretching mode of graphite and indicates the presence of a crystalline graphitic structure for MWCNT forests.

The quality of carbon nanotubes can be estimated as the G/D peak ratio in the Raman spectra.

The G/D peak ratio of the MWCNTs grown on Fe-Co nanoparticles is 1.16. In case of MWCNT forests grown on Fe and Fe-Pt catalysts are 0.99 and 0.92, respectively. From the Raman results, the quality of the MWCNT forests grown on the Fe-Co nanoparticles as a catalyst is ~17% better.

4 CONCLUSION

We demonstrate bimetallic nanoparticles using simple and efficient ways of producing bimetallic nanoparticles that can be used as catalysts for the continuous growth of MWCNTs forests by APPECVD. MWCNTs have been successfully grown on catalyst nanoparticles at atmospheric pressure using an atmospheric pressure plasma jet (APPJ)

system. Bimetallic nanoparticles decrease the diameter of MWCNTs compared with monometallic nanoparticles. Although the height of the forests on Fe-Pt is less than that of the forests on Fe, the Fe-Co nanoparticles increase the forests height significantly. In addition, Raman G/D peak ratio of MWCNTs grown on Fe-Co indicate that the quality of forests is better than that of Fe and Fe-Pt. The results from these CNT growths on bimetallic particles are promising for large scale production with controllable nanotube features. Careful study of bimetallic nanoparticles which offer ease in both synthesis and dissemination, and how they catalyze carbon nanotube growth, promises development toward superior and reliable results [15, 16]. Further detailed studies are under way to fully understand the catalytic activity of the iron platinum nanoparticles and to elucidate the mechanism of the carbon nanotube growths from these particles.

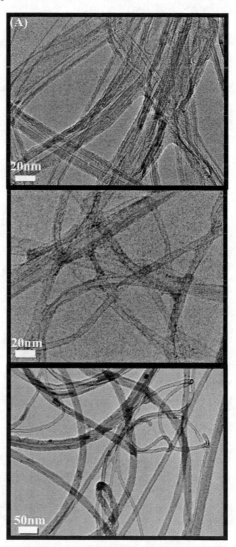

Fig. 3. The TEM Images of the CNTs grown on (A) the Fe nanoparticles and (B) the Fe-Pt nanoparticles, and (C) Fe-Co nanoparticles

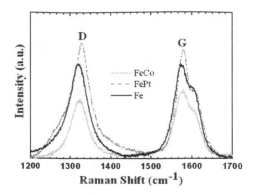

Fig. 4. Raman spectra of MWCNTs grown on Fe, Fe-Pt, and Fe-Co nanoparticles

REFERENCES

[1] W. B. Choi, E. Bae, D. Kang, S. Chae, B. H. Cheong, J. H. Ko, E. Lee and W. Park, "Aligned carbon nanotubes for nanoelectronics", Nanotechnology, 15, 512-516, 2004

[2] W. A. de Heer, A. Chatelain and D. Ugarte, "A Carbon Nanotube Field-Emission Electron Source," Science, 270, 1179, 1995

[3] P. Qi. O. Vermech, M. Grecu, A. Javey, Q. Wang, H. Dai, S. Peng and K. J. Cho, "Toward large arrays of multiplex functionalized carbon nanotube sensors for highly sensitive and selective molecular detection," Nano Lett., 3, 347-351, 2003

[4] Y. Tzeng, Y. Chen and C. Liu, Diam. Relat. Mater., 12, 774-779, 2003

[5] N. Toshima and Tetsu Yonezawa, "Bimetallic nanoparticles – novel materials for chemical and physical applications," New J. Chem., 22, 1179-1201, 1998

[6] S. Huang and A. W. H. Mau, "3D carbon nanotube architectures on galss substrate by stamp printing bimetallic FePt/polymer catalyst," J. Phys. Chem. B, 107, 8285-8288, 2003

[7] X. Wang, W. Yue, M. He, M. Liu, J. Zhang and Z. Liu,. "Bimetallic catalysts for the efficient growth of SWNTs on surfaces," Chem. Mater., 16, 799-805, 2004

[8] D. C. Lee, F. V. Mikulec and B. A. Korgel, "Carbon nanotube synthesis in supercritical toluene," J. Am. Chem. Soc., 126, 4951-4957, 2004

[9] A. Javey and H. Dai, "Regular arrays of 2nm metal nanoparticles for deterministic synthesis of nanomaterials," J. Am. Chem. Soc., 127, 11942-11943, 2005

[10] C. L. Cheung, A. Kurtz, H. Park and C. M. Lieber, "Diameter controlled synthesis of carbon nanotubes," J. Phys. Chem. B, 106, 2429-2433, 2002

[11] A. K. Sra, T. D. Ewers, Q. Xu., H. zandbergenb and R. E. Schaak, "one pot synthesis of bi-disperse FePt nanoparticles and size selective self assembly into AB2, AB5, and AB13 superlattices," Chem. Comm., 750-752, 2006

[12] E. V. Shevchenko, D. V. Talapin, C. B. Murray and S. O'Brien, "Structure characterization of self-assembled multifunctional binary nanoparticle superlattices," J. Am. Soc., 128, 3620-3637, 2006

[13] A. Chandrashekar, J. Lee, G. S. Lee, M. J. Goeckner and L. J. Overzet, " Gas phase and sample characterizations of multiwall carbon nanotube growth using an atmospheric pressure plasma," J. Vac. Sci. Technol. A, 24, 1812-1817, 2006

[14] E. W. Wong, M. J. Bronikowski, M. E. Hoenk, R. S. Kowalczyk and B. D. Hunt, "Submicron patterning of iron nanoparticl monolayers for carbon nanotube growth," Chem. Mater., 17, 237-241, 2005

[15] W. Deng, X. Xu and W. A. Goddard III , " A two stage mechanism of bimetallic catalyzed growth of single walled carbon nanotubes," Nano Letters, 4, 2331-2335, 2004

[16] V. Brotons, B. Coq and J. M. Planeix, "Catalytic influence of bimetallic phases for the synthesis of single walled carbon nanotubes," J. Molecular Catalysis A, 116, 397-403, 1997

Characterization of Activated Carbon particles for nanocomposite synthesis.

L.P. Terrazas-Bandala*, E.A. Zaragoza-Contreras*, G. González-Sánchez*, M.L. Ballinas -Casarrubias **.

*Centro de Investigación en Materiales Avanzados, S.C. Miguel de Cervantes 120 Compl. Ind. Chih. 31109, Chihuahua, México, piroshka.terrazas@cimav.edu.mx
**Fac. Ciencias Químicas. Universidad Autónoma de Chihuahua, Circuito Universitario S/N, Chihuahua, México.

ABSTRACT

Activated carbon (AC) micro and nano particles were produced from Granular AC. Particle size distribution was measured before and after THF solvation by light dispersion. Particles were incorporated into a matrix of polystyrene via in situ miniemulsion polymerization. The nanostructured latexes were characterized by TEM and by SEM. Triacetate cellulose (TAC) and AC particles composite membranes were produced and analyzed by AFM tapping® for height and face contrast. Particle size after solvation shows a significant decrease in size. We obtained stable latex with particles of less than 100 nm diameter, TEM micrographs show the encapsulation of the AC nanoparticles into the polystyrene matrices. Integration of AC particles into TAC membranes was successfully done, and AFM images show the nanodispersion of these particles.

Keywords: Activated carbon, miniemulsion polymerization, nanoparticles, AFM.

1 INTRODUCTION

Activated carbon from inexpensive sources has been of great interest for separation systems, especially in the field of water treatment [1]. Efforts had been made to produce AC nanoparticles by different means, in the aim to integrate them to composite materials to enhance their characteristic efficiency [2,3,4]. The success of these applications strongly depends on the availability of colloidal particles with tightly controlled size and surface properties and therefore there is a continuously increase research effort in the synthesis of colloidal carbon micro and nano particles and the methods for characterize them [5]. Numerous reports on the problems to precisely determine the size of these materials have been made [5,6,7,8]. Depending on composition, solvent used or matrix where they are dispersed, these nanosize materials tend to agglomerate in different proportions.

Recently p-aminophenol was synthesized from p-nitrophenol over nanosize nickel catalysts. Size characterization was made by SEM and Mastersizer 2000. Results obtained by light dispersion showed sizes up to six times bigger than the size observed by SEM. One of the reasons for larger particle size measured by Mastersizer 2000 is the presence of extensive twinning and agglomeration in the samples. In this case, Mastersizer measured the agglomerations instead of crystallites [5].

Miniemulsion polymerization has been successfully used to stabilize different types of nanoparticles, as carbon nanotubes [9], ceramics [10,11], magnetic particles [6, 12] and carbon black [13]. This technique allows the distribution of small concentration of nanoparticles into nanoscale-independent matrices.

Composite materials that incorporate inorganic fillers to polymeric matrices have been prepared for many different applications, specifically the use of activated carbon as filler has drown attention due to its high adsorbent capacity. ABS copolymer-activated carbon mixed matrix membranes for CO_2/CH_4 separation were produce by casting–evaporation process at 45% relative humidity and showed a simultaneous increase of CO_2 gas permeability (40–600%) and CO_2/CH_4 selectivity (40–100%) by increasing the percentage of carbon loaded in the mixed matrix composite membrane [14].

The adhesion between the polymer phase and the external surface of the particles appeared to be a major problem when glassy polymers are used in the preparation of composite membranes. It seems that the weak polymer–filler interaction makes the filler tend to form voids in the interface between the polymer and the filler [15]. Various techniques have been employed to improve the polymer–filler contact, among them sonicated baths [7,16] and particle surface modification [7] are widely used.

Recently a very inexpensive method to produce AC nanoparticles by solvation was proposed [3]. The objective of this work is to study this production of AC micro and nanoparticles and to study different methods to characterize their average particle size and the possibility to fully disperse them into a polymeric matrix.

2 METHODOLOGY

2.1 Materials

Granular AC (Carbochem LQ 1,000; 870 m2/g surface area, 5.75 Å average pore size) was triturated by steel balls mill (8000 Mixer/Mill,Spex Cetriprep) for 10 min. Powder obtained was sieved to separate particles with less than 53 microns, AC microparticles were agitated in tetrahydrofurane (THF, Merck 99%) 0.1g in 100ml for 96 hours for solvation, and finally dried in an oven for 24 hours[3].

For miniemulsion polymerization, the initiator azobisisobutyronitrile (AIBN; Aldrich) was freshly recrystallized from methanol and kept at 5°C until use, sodium lauril sulfate (SLS; Aldrich) and hexadecane (HD; Aldrich) were used as received. The water was of tridistilled quality.

Triacetate cellulose (TAC; Aldrich) and AC Particles composite membranes were produce by previously reported method, using methylene chloride (Aldrich) [3]. Vapor induced phase separation was carried out at 70% relative humidity at 35 and 45°C.

2.2 Characterization

Particle size distribution was measured before and after the solvation with THF by light dispersion (Malvern Instruments, Mastersizer 2000), with three different solvents (ethanol, water, and water + sodium hexametaphosphate as surfactant).

AC particles were incorporated, into a matrix of polystyrene via in situ miniemulsion polymerization following the procedure describe by Lopez et al. [4]. A mixture formed of 0.05 g of AC particles, 20 g of styrene, 0.18 g of SLS, 0.2 g of AIBN, and 0.83 g of HD was added to a glass flask and sonicated for 15 min to disperse the particles. Some drops of this dispersion were studied by scanning electron microscope (SEM). Afterward 80 g of tridistilled water were introduced into a 250-mL, three-necked, round-bottom glass reactor; the sonicated mixture was fed to the reactor to complete the polymerization system, which was sonicated for 45 min to obtain a stable miniemulsion. It is important to mention that ice was added to the sonication bath during the time of dispersion to avoid the occurrence of free-radical polymerization induced by the high shear strength. Finally, the reactor was immersed into a heating bath (at 60°C). The polymerization was allowed to react for 180 min.

The nanostructured latexes were characterized by TEM with a Philips CM200 operated at a 200-kV accelerating voltage, and by SEM with a JEOL, JSM 5800-LV. Triacetate cellulose (TAC) and AC particles composite membranes were analyzed by AFM tapping® (Nanoscope IV, Digital Instruments) for height and face contrast characteristics. Silica coated phosphorus tips model RTESP (Veeco®), and scan rates of 2 –3 Hz were used. Set point and the gains were adjusted to obtain the best image resolution

3 RESULTS
3.1 Light dispersion

The average particle size measured by light dispersion after THF solvation with the use of a surfactant shows a significant decrease in size (Table 1), but still these results do not fall into the nano scale (below 100 nm). Previous results of TEM micrograph analysis (images not shown) suggest that some of these particles detected by laser diffraction are in fact particles agglomerates. This is the reason for the larger size detected in the same instrument when different solutions are used (water, and ethanol).

	Suspension	Particle Size distribution (80%) μm	mean size μm
Before THF	Surfactant	1.89 - 33.58	8.52
After THF	Surfactant	0.6 - 25.48	1.6
	Ethanol	33.35- 89.76	42.6
	Water	45.8 - 220	98.4

Table 1: AC Particle size distribution, measure by light dispersion.

3.2 Miniemulsion polymerization

AC particles were successfully dispersed in the mixture of styrene and surfactants by 15 minutes sonication. On SEM micrographs we observed different sizes of particles, ranging from micrometers (Fig.1) to less than 400 nm (Fig.2).

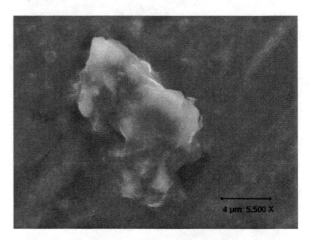

Figure 1: SEM micrograph of AC particles after THF solvation.

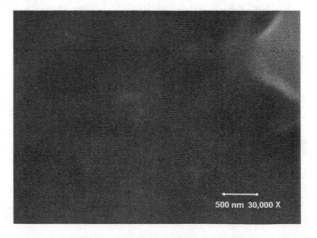

Figure 2: SEM micrograph of AC particles after THF solvation.

Tetrahydrofuran (IUPAC name: Oxacyclopentane) with the formula $(CH2)_4O$ is a heterocyclic organic compound, a strong solvent that dissolves a wide range of non-polar and polar compounds. Activated carbon is a highly porous non-polar solid that can be easily penetrated by THF and dissociated into smaller particles by means of solvation. Previous studies reported the effect of strong solvents like dimethylformamide (DMF) on the casting of composite membranes with activated carbon particles; SEM images of the AC particles indicated an increase in particle channelling and size decrement due to the close contact of solvent and particles during polymer dissolution [15].

During our experiment the contact time between THF and AC particles was enough to let the solvent penetrate into the porous structure and led to partial fragmentation of the particles. Nevertheless this fragmentation was by non means uniform since SEM micrographs clearly shows a very wide particle size dispersion.

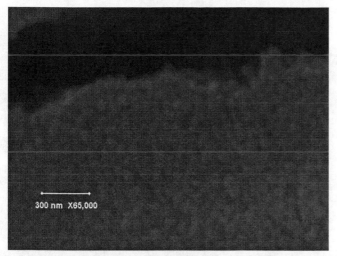

Figure 3: SEM micrograph of AC/polystyrene nanocomposite.

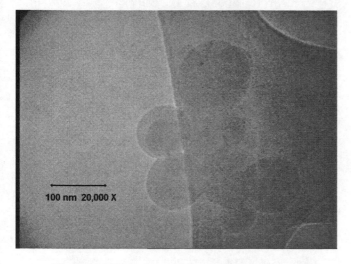

Figure 4: TEM micrograph of AC/polystyrene nanocomposite.

By miniemulsion polymerization stable latex with particles of less than 100 nm diameter was obtained (Fig.3), and TEM micrograph shows the encapsulation of the AC nanoparticles into the polystyrene particles (Fig.4). It is important to realize that some AC particles were not integrated into the miniemulsion, and were precipitated during the polymerization, which indicate that there were still a good amount of AC particles that were bigger than 100 nm.

3.3 Membrane characterization by AFM

Integration of AC particles into TAC membranes was macroscopically homogenous. With the aid of an optical microscope integrated to the AFM the scanning of the surface was made only were there were not micrometric AC particles. Fig. 5(a) shows an AFM topographic image obtained from the height signal, scan area of 2 μm^2. Valleys and peaks are clearly distingue by color pallets, darkest areas represent porous formed in the membrane during the casting process; light color areas indicate either nodules formed by the polymer or presence of AC particles.

Face contrast images obtained from AFM are a powerful extension of the tapping mode® (Digital Instruments, Santa Barbara, CA), since they provide information at nanometric scale about the surface structure and some characteristics that are not regularly revealed by other microscopic techniques. By register the delay on the cantilever oscillation on tapping mode, the obtained images detect variations on composition, adhesion, friction, viscoelasticity among others.

Face contrast image of the exact same area (Fig. 5b) reveals the presence of two different materials, corresponding to the polymer TAC (dark homogenous areas) and AC particles (light, granular shape areas).

AC particles can be distinguished and size range is from 4 nm to 220 nm. Grain analysis from each image was made with software Nanoscope 5.30r3sr3®, Fig. 6 shows the particle size distribution obtained. Data from this analysis indicate a particle mean size of 19.53 nm and a standard deviation of 1.87 nm Nanodispersion can be seen on every image obtained of all the membranes prepared.

4 CONCLUSIONS

Activated Carbon solvation in THF produced micro and nano-particles due to fragmentation, resulting in wide particle size dispersion. Results obtained by light dispersion shows a decrement of particle size, but the use of different solvents in the experiment can lead to confusion since particles tend to agglomerate.

AC particles disperse well in styrene and surfactants, SEM micrographs show particles of different sizes in both micro and nano scale. Stable latex with particles of less than 100 nm diameter was obtained, TEM micrographs show the encapsulation of the AC nanoparticles into the polystyrene particles. Integration of AC particles into TAC membranes was successfully done, and AFM images show the nanodispersion of these particles.

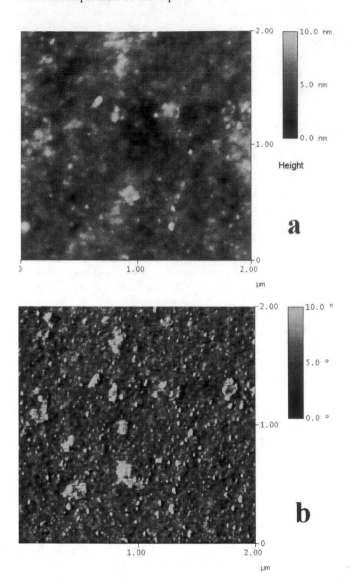

a

b

Figure 5: AFM images of CTA/AC composite membranes obtained at 70%RH. (a) topographic image, (b) phase contrast image. Scanning area of 2 μ².

REFERENCES

[1] Pattanayak, J., Mondal, K., Mathew, S. and Lalvani, S.B. Carbon, 38: 589–596, 2000.

[2] Shing-Dar W., Ming-Hao Ch., Jen-Jui, Ch., Hsun-Kai, Ch. and Ming-Der K,. Carbon 43: 1322–1325, 2005.

[3] Ballinas-Casarrubias, M., Terrazas-Bandala, L.P., Ibarra-Gómez, R., Mendoza-Duarte, M.E., Manjarrez -Nevárez, L. and González-Sánchez, G. Polymers for Adv. Tech. 17: 991-999. 2006.

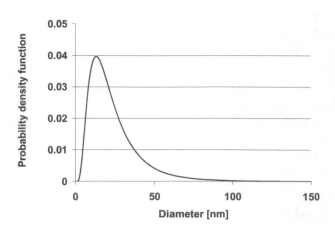

Figure 6: Particle size distribution from AFM images analysis.

[4] López-Martínez, E.I., Márquez-Lucero, A., Hernández -Escobar, C.A., Flores-Gallardo, S. G., Ibarra-Gómez, R., Yacaman, M.J. and Zaragoza-Contreras, E.A. Journal of Polymer Science: Part B: Polymer Physics, 45: 511- 518, 2007.

[5] Du, Yan, Hongling Ch., Rizhi Ch., and Nanping X. Applied Catalysis A: General 277: 259- 264, 2004.

[6] Liu. J., Wang, H., and Zhang, L. Chem. Mater. 16: 4205-4207, 2004.

[7] Garrigue P., Delville, M.H., Labrugere, Ch., Cloutet, E., Kulesza, P.J., Morand, J. P., and Kuhn, A. Chem. Mater. 16: 2984-2986, 2004.

[8] Bandyopadhyaya, R., Rong, W., and Friedlander, S. K. Chem. Mater. 16: 3147-3154, 2004.

[9] Barraza, H. J., Pompeo, F., O'Rear, E.A. and Resasco, D.E. Nano Letters. 2: 797–802, 2002.

[10] Erdem, B., Sudol, E. D., Dimonie, V.L. and El-Aasser, M.S. Macromol Symp. 155:181–198, 2000.

[11] Erdem, B., Sudol, E. D., Dimonie, V.L. and El-Aasser J. Polymer Science Part A: Polym Chem 38: 4431–4440, 2000.

[12] Xu, Z., Wang, C.C., Yang, W.L., Deng, Y.H. and Fu, S.K. J Magn Magn Mater. 277: 136–143, 2004.

[13] Bechthold, N., Tiarks, F., Willert, M., Landfester, K., and Antonietti, M. M. Macromol Symp. 51:549–555, 2000.

[14] Anson, M., Marchese, J., Garis, E., Ochoa, N., Pagliero, C. Journal of Membrane Science 243 (2004) 19–28.

[15] Ballinas, L., Torras, C., Fierro, V. and Garcia-Valls. R. Journal of Physics and Chemistry of Solids 65: 633–637, 2004.

[16] Garcia, M., Barsema, J., Galindo, R.E., Cangialosi, D., Garcia-Turiel, J., Van Zyl, W.E., Verweij, H. and Blank D.H.A. Polymer Engineering and Science, 44: 1240-126,2004.

Multi-Functional Carbon Nanotube Based Filtration Material

A. Cummings

Seldon Technologies, Inc.
Windsor, VT, USA, acummings@seldontech.com

ABSTRACT

As the fibers making up a filter become smaller in diameter, the more efficient the filter is at depth filtration. Seldon Technologies has developed carbon nanotube (CNT) based filtration media (Nanomesh™) that takes advantage of the inherent properties of CNTs (high strength, high surface area, electrical conductivity, etc.). Seldon's Nanomesh™ filters are technically disruptive because the carbon nanotubes they contain represent the ultimate limit in fiber diameter.

Over the past six years, Seldon has demonstrated the feasibility of using its material to clean water, air and fuels. In 2007 Seldon was issued a patent [1] for the filtration of fluids using CNTs. Additionally Seldon has demonstrated that it is possible to produce large quantities of Nanomesh™ on papermaking machines largely using off-the-shelf processing equipment.

Keywords: carbon nanotubes, filtration, water, air, fuel

1 INTRODUCTION

It is understood that due to the hydrodynamics of flow around fibers, the efficiency of depth filters improves as the diameter of the fibers making up the filter element decreases. This is due primarily to two effects: 1) smaller fibers possess larger surface area and can therefore capture and hold more contaminants; and 2) smaller fibers disrupt the fluid stream to a lesser degree – the particles get less "advance warning" that they are about to strike the capturing fiber. However, if the fibers are too small in diameter they may lack sufficient strength to avoid breakage under the stress induced by fluid flow. CNTs possess exceptional mechanical strength (tensile strength ~100 GPa, modulus ~1000 GPa) [2] making breakage nearly impossible. Additionally, due to their very small diameters (on the order of the mean free path of air molecules), carbon nanotubes present a relatively low resistance to the flow of fluids. This, combined with their high surface area, means that highly-efficient, bio-contaminant removals are possible with a thin filtration media possessing a relatively low pressure drop.

For the past five years, Seldon Technologies has focused primarily on developing CNT-based water filtration media. Seldon's water filter material removes bacteria, viruses, spores, cysts, total organic carbons and inorganic contaminants from water using adsorption and sieving. Additionally, the pressure drop across the filter is much less than what is typically seen with nanofiltration and reverse osmosis systems (1 – 2 bar compared to 3 – 20 bar for nanofiltration and 5 – 120 bar for RO [3]) and is capable of working for well over 1,000 gallons of influent. This level of filtration is accomplished passively without the use of chemicals.

More recently Seldon has worked on developing Nanomesh™ that can be used in air and fuel filtration applications. In a 2005 DARPA sponsored project, Seldon demonstrated its air filter Nanomesh™ removed over 99% of an anthrax surrogate. This has lead to a larger research effort that in the past year developed material with near-HEPA filtration performance at a lower pressure drop. In the area of fuel filtration, internal testing showed that Seldon's material was effective at removing bacteria from contaminated jet fuel. Additional testing by a third party has suggested that the Nanomesh™ fuel filter media will be very effective at cleaning contaminated bio-fuels, a market that will be very important in the coming years.

2 WATER FILTRATION

2.1 Background

Over 1 billion people lack access to adequately clean water. Additionally, an estimated 2.5 million people die each year from diarrheal diseases and millions more are chronically ill due to a lack of portable water. In the United States and throughout the world, the number of identified contaminants is increasing, the water distribution infrastructure is aging and the costs of upgrading and repairing it are beyond the capability of many municipalities. Hence, there is a growing concern that the water being delivered to people's houses contains harmful impurities.

Common illness-causing organisms other than viruses are 1 - 5 microns long and can typically be removed by size-exclusion filtration. However, the removal of viruses (on the order of 20 nanometers) by size exclusion is impractical as the material would present a very high-resistance to flow and could not be used in small filtration systems. Seldon's approach uses chemically activated carbon nanotubes as an adsorptive surface for the attraction of viruses and other microorganisms. The very small size of CNTs creates an enormous removal capacity in the Nanomesh™ filtration media which equates to the ability to purify large volumes of water.

2.2 Work to Date

The bulk of Seldon's development work has focused on testing the technology's effectiveness in removing bioburden (i.e. bacteria, viruses, etc.) from water. Seldon's 2" x 9" water filters now reliably remove bacteria and virus from more than 1,000 gallons of water and some tests show that filters can purify up to 3,000 gallons of water.

Seldon has successfully completed third party testing of its Nanomesh™ filter at the University of New Hampshire's Water Treatment Technology Assistance Center. Using specific protocols from NSF P231, UNH demonstrated that while there was a very high biological challenge upstream of the filter, none of the contaminants made it through the filter. The third party test of Seldon's Nanomesh™ filter was terminated at 600 gallons. At Seldon, duplicating the third party's test conditions as closely as possible, the filter testing continued to operate up to 1,000 gallons.

2.3 Large Scale Production

Since May 2007 Seldon has conducted ten large scale production runs of its water filtration material (this includes producing both Nanomesh™ and pre-filtration material). It has been an on-going process at Seldon's facility in Vermont to evaluate this material's biological removal performance. While not all runs of the material have shown a high level of performance (filtration of 3000$^+$ gallons), the material does perform relatively well and shows a higher level of consistency than batch processes used previously. As a result of this work, Seldon has demonstrated it is possible to produce large quantities of a product incorporating a nanoscale material on a papermaking machine (see Figure 1).

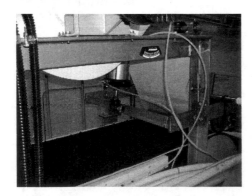

Figure 1: Wet lay deposition of Seldon's Nanomesh™ water filtration media.

2.4 Water Filtration in the Field

Throughout 2007, a Dian Fossey Gorilla Fund International field project located in Bisate, Rwanda used prototypes of Seldon's water purification products. Users from the NGO have provided valuable feedback to Seldon on how the units operate in the field. Additionally, all users have regularly commented on how good the water tastes when filtered through Seldon's system as compared to other water purification methods.

3 AIR FILTRATION

3.1 Background

There is a widespread pressing need for the effective filtration of airborne biological and chemical agents. Applications range from building ventilation systems, public transportation to personal health protection. While some very efficient air filtration technologies exist (e.g., HEPA and ULPA) they typically are restricted in their use by their bulk and/or the energy required to achieve adequate flow rates.

Air filtration presents challenges that are somewhat different than those for liquids. Because of the inherent limitations in air handling systems and human lung performance, air filtration must be carried out at a considerably lower pressure drop than liquid filtration. The most significant technical concern, then, is being able to achieve sufficient biological and chemical removal levels at the required flow rates without undue pressure drops. Fortunately, the low density/viscosity of air leads to higher levels of diffusion and easier deflection of small particles allowing for a thinner material and lower pressure drops as compared to liquid filtration media. Further, because of the absence of surface tension issues it is easier for contaminants to interact with the fibers within the filter.

Early air filtration tests showed that several non-optimized designs operated in the performance range of HEPA filters for bio-contaminants (see Figure 2). Samples of Seldon's candidate air filters achieved a log 2 reduction (~98%) for a weaponized form of *Bacillus subtilus*, the common surrogate for the bio-warfare agent, *Bacillus anthracis*. This level of removal is comparable to the early results achieved in the development of a water filtration product that currently has log 7 bacteria and log 6 virus reductions in water (99.99999% and 99.9999%, respectively).

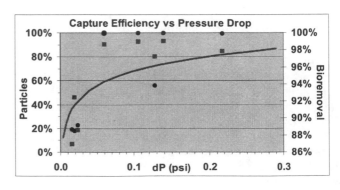

Figure 2: Neutral particle (circles) and surrogate bio-warfare agent (square) removal efficiency for several Seldon air filters. Solid curve is typical filtration relationship.

3.2 Recent Work

Following a typical design of experiment (DOE) methodology, researchers tested over 300 individual filters in the past 18 months. Careful analysis of early results suggested that the properties of the carbon nanotubes were not being fully leveraged to optimize a filter's performance. The key element was that the CNTs significantly contributed to the pressure drop across the filter, but did not lead to a concomitant improvement in the particle removal efficiency. This is typically indication that the specific fibers are not being adequately dispersed and exist in the media as tight agglomerations that block flow but do not effectively capture contaminants.

By closely examining the processing and mixing of each component fiber, Seldon's team managed to ease the constraints driving the efficiency-permeability relationship. SEM images show that the CNTs span the large pores created by the scaffold fibers effectively capturing particles but causing very little impediment to the flow of air. As a result, the overall media performance has significantly improved. Comparing the performance of Seldon's media to that of a flat sheet of HEPA media under the test conditions used in Seldon's lab, researchers have shown that several of Seldon's media designs achieve higher capture efficiency than HEPA media at nearly *three times* the permeability (Figure 3). Researchers are in the process of getting the comparison of Seldon's materials to HEPA media validated by a third party lab to confirm the improved performance.

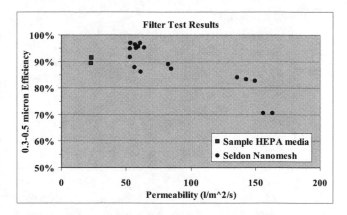

Figure 3: Filter performance comparison between several of Seldon's improved Nanomesh™ air media (circles) and a sample of HEPA media (squares) received from a third party supplier.

4 FUEL FILTRATION

4.1 Background

Although freshly refined oil is pure, bacterial growth can occur at the interface layer of fuel and water in storage tanks. (The water needed for the growth of microorganisms can enter the fuel tank by condensation due to temperature changes.) When disturbed, this growth enters vehicles and can quickly clog fuel filters and interfere with engine operation. The importance of microbial contamination in fuels was realized in 1958 when a B-52 crash was directly related to the plugging of an in-line fuel filter. Together with improved housekeeping, toxic chemicals like ethylene glycol monomethyl ether (EGME) and di-ethylene glycol monomethyl ether (di-EGME) were used as icing inhibitor additives to reduce the microbial infestation in the fuel. Although these measures reduced biological contamination, a complete reduction of microbial growth in aviation fuel was never achieved. A recent microbial study [4] shows that there is a variety of microbial organisms found in aviation fuel tanks throughout the United States. Moreover, the high toxicity [5] to mammals and aquatic animals of the chemicals used to treat this problem demands an alternate solution.

In 2004, at the request of the AFRL, Seldon tested the feasibility of its Nanomesh™ filtration media in removing bacteria from contaminated JP-8 jet fuel. Figure 4 shows a result from these feasibility tests. Figure 4a shows a culture of a sample taken from the fuel-water interface of contaminated JP-8. Figure 4b is a culture of the fuel after it has been passed through a Nanomesh™ filter. The Nanomesh™ was clearly capable of removing the bacteria from the fuel.

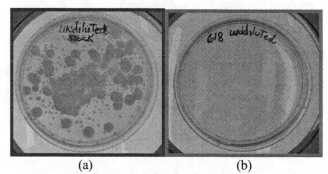

(a)   (b)

Figure 4: (a) Culture of sample from fuel-water interface showing considerable bacterial growth. (b) Culture of sample filtered through Seldon's Nanomesh™ showing total absence of bacteria

4.2 Recent Work

An outside laboratory has conducted numerous tests showing the feasibility of using Seldon's fuel filtration material in a variety of applications. In a qualitative test, Seldon's Nanomesh™ fuel filtration media was shown to significantly reduce the amount of fuel degradation products (FDPs) from biodiesel as well as commercial diesel products. Figure 5 compares a samples of biodiesel and regular diesel before and after filtration. The clarity of the post-filtration samples indicates the removal of fuel degradation products. For diesel fuel (based on the clarity of the filtrate), Seldon's sample appears to outperform the commercial filter.

Figure 5: (a) Image of thermally degraded biodiesel fuel before and after removal of FDP using Seldon's Nanomesh™ filter; (b) Image comparing the effect of a traditional filter against the effect of Seldon's Nanomesh™ filter for the removal of FDP from commercial diesel fuel.

In another test, the laboratory demonstrated the removal of small particulate matter using Seldon's media. In this measurement, test filters were placed upstream of a sacrificial "filter patch" (see inset in Figure 6) and the change in pressure across the patch was then monitored. Figure 6 shows the results of flowing ultra-low sulfur diesel fuel containing 0.14% of dimethyl-phenanthrene (DMP) particulate (less than 3 μm) through a traditional fuel filter, a sample of Seldon's Nanomesh™ fuel filter and when no test filter is in place. As seen in the figure, with no test filter in place, the sacrificial patch quickly becomes clogged. No increase in pressure drop was observed in both the Nanomesh™ and the traditional filter; indicating that they work with similar removal efficiencies. However, because Seldon's Nanomesh™ fuel filter media is slightly electrically conductive, it has superior electrostatic discharge capabilities as compared to the traditional filter. At high flow rates, the build up of large electrostatic charges causes electrical sparks within the filter. These discharges cause damage to the filter and hence reduce filter efficiency and life.

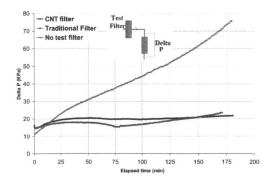

Figure 6: Plot depicting change in pressure drop across a downstream filter when Seldon's Nanomesh™ pre-filter, a traditional pre-filter and no pre-filter is used for removal

of 0.14% (DMP) fine particulates (< 3 micron) from ultralow sulfur diesel(ULSD).

Internal testing has shown that swatches of Seldon's fuel filtration media is capable of removing over 7 logs of bacteria from emulsions of made from 85% fuel surrogate and 15% bacteria laden water.

5 CONCLUSIONS

Seldon Technologies has successfully applied its concept of filtration using nanoscale particles (specifically carbon nanotubes) to the areas of water, air and fuel filtration. Seldon's water filtration technology has matured to the point where a single filter cartridge can potentially clean 3000 gallons of water heavily contaminated with bio-burden. Seldon's air filtration media appears to perform on par with some types of HEPA media in terms of efficiency but at a lower pressure drop. Finally, fuel filtration media produced by Seldon could prove very useful in the remediation of contaminated fuel stocks greatly reducing waste.

6 ABOUT SELDON TECHNOLOGIES

Founded in 2002, Seldon Technologies has received $15 M in government sponsored funding from AFRL, DARPA and NASA. Additionally, Seldon has entered into collaborative research agreements worth $2.4 M with corporate sponsors. In addition to its patent for the purification of fluids using carbon nanotubes, Seldon has 11 patents pending. The company currently employees 35 people in eastern Vermont.

REFERENCES

[1] US Patent #7,211,320

[2] Yakobson, B.I. and L.S. Couchman, 2004: Carbon Nanotube: Supermolecular Mechanics, in: *Encyclopedia of Nanoscience and Nanotechnology*, Ed. J.A. Schwartz *et al.* (Marcel Dekker, New York), 508-601pp

[3] Van der Brugen, Bart, *et al*, "A Review of Pressure-Driven Membrane Processes in Wastewater Treatment and Drinking Water Production", *Environmental Progress*, 22(1), 2003, 46 – 56.

[4] Rauch, M. E., Harold, H. W., Rozenzhak, S. M., "Characterization of microbial contamination in United States Air Force aviation fuel tanks", *J Ind. Microbiol Biotechnol*ogy, 33, 2006, 29-36.

[5] D. A. Pillard "Comparative Toxicity of Formulated Glycol Deicers and Pure Ethylene and Propylene Glycol to Ceriodaphnia Dubia and Pimephales Promelas" Environmental Toxicology and Chemistry, (1995), 14 (2); 311-315.

Separation Mechanisms for Nanoscale Spheres and Rods in Field-Flow Fractionation

Frederick R. Phelan Jr.[*] and Barry J. Bauer[**]

[*]Polymers Division, NIST, Gaithersburg, MD, USA, frederick.phelan@nist.gov
[**]Polymers Division, NIST, Gaithersburg, MD, USA, barry.bauer@nist.gov

ABSTRACT

Separation mechanisms for spheres and rodlike particles in classical field-flow fractionation (FFF) are studied using a Brownian dynamics simulation. For spheres, simulation results for mean elution time are found to be in good agreement with experimental data and the steric inversion theory of Giddings [1]. Modeling of particle separation for rods is compared with spheres of equal diffusivity. The simulation shows that nanotube scale particles elute by a normal mode mechanism up to aspect ratios of about 500, based on a particle diameter of 1 nm. At larger sizes, the rods also show a steric deviation from normal mode elution, but in the opposite sense as for spheres. The different behavior is attributable to the effect of particle shape. Extension of the steric mode theory for spheres to rods illustrates a potential steric mode separation that can be used to separate rods based on chirality.

Keywords: Brownian Dynamics, Field-Flow Fractionation, Nanotubes, Separations, SWNT

1 INTRODUCTION

For nanotubes to achieve their full potential in applications, it is desirable to be able to separate them according to their various physical properties. One possible technique for achieving this is field-flow fractionation (FFF), depicted in Figure 1.

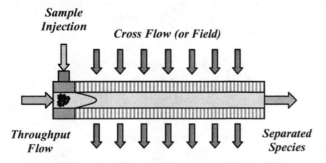

Sample Injection

Cross Flow (or Field)

Throughput Flow

Separated Species

Figure 1: Schematic of frit-inlet symmetrical, field-flow fractionation. In the asymmetric variation, only the lower wall is porous.

Classical flow-FFF is a separation technique in which a perpendicular cross flow is imposed upon a channel flow of dilute particulates [1-3]. The cross flow exits through a porous accumulation wall which is impermeable to the particulates. Competition between various flow mechanisms drives particles of unlike size to different average positions in the cross flow direction. Separation is achieved due to the different residence times of the particles based upon their position in the parabolic velocity profile in the throughput direction. FFF can also be combined with other techniques such as dielectrophoresis to produce separations based on electronic, as well as size based properties.

A number of different mechanisms can be exploited to achieve separation in flow-FFF. Normal mode separation applies to particles which are small enough to undergo significant Brownian motion, and whose size is small compared to the cross flow gap size. In this case, smaller particles, which are more diffusive, have an average position closer to the centerline and elute faster than larger particles. Steric mode separation occurs when the particle layer in FFF is strongly compressed along the accumulation wall. In this case for spheres, larger particles are more highly entrained by the throughput flow and elute more quickly than smaller ones. This turnaround is called steric inversion.

Particle separation in flow-FFF can be described by a theoretical variable called the retention, R, which represents the ratio of the average residence time of non-retained tracers, t_0, to the average retention time of the particles, t_r, *i.e.*, $t_r = t_0 / R$, where $0 \leq R \leq 1$. For the case of normal mode separation, the retention is given by

$$R = 6\lambda \left[\coth\left(\frac{1}{2\lambda}\right) - 2\lambda \right] \quad (1.1)$$

where λ is an inverse Peclet Number (Pe) given by

$$\lambda = \frac{D}{|v_c| H} = \frac{1}{Pe} \quad (1.2)$$

v_c is the cross flow velocity, D is the diffusion coefficient of the particle in the cross flow direction, and H is the cross-flow thickness. A model which takes into account both normal mode diffusion and steric effects for spheres in FFF derived by Giddings [1]

$$R = 6\left(\alpha - \alpha^2\right) + 6\lambda\left(1 - 2\alpha\right)\left[\coth\left(\frac{1-2\alpha}{2\lambda}\right) - \frac{2\lambda}{1-2\alpha} \right] (1.3)$$

where α is the ratio of the particle radius to gap thickness, *i.e.*, $\alpha = r / H$. For the case of negligible particle size, Eq.

(1.3) reduces to Eq. (1.1). It can be seen that, theoretically, normal mode elution is independent of the particle shape and depends only on the particle diffusion coefficient. Steric mode separations become independent of diffusion, but little is known about how steric mode separations are effected by different particle shapes.

In what follows, a Brownian dynamics simulation is used to investigate the separation of spheres and rodlike particles in flow-FFF. First, we study the case of spheres to validate the simulation method by comparison with experimental data and existing theory. We then compare simulation results for rods with spheres of equal diffusivity and show that the difference in particle shape leads to divergent trends at large particle size in the steric mode regime. Finally, we look at the elution of oriented rods and elucidate a potential steric mode separation that can be used to separate rods based on chirality.

2   SIMULATION

In this work, we have developed a Brownian dynamics simulation to investigate the separation of spheres and rodlike particles in flow-FFF. A summary of pertinent details of the numerical method are discussed below. Full details are found in references [4-6].

2.1   Model Equations

For spheres, the particle motions are governed by a Langevin equation which takes into account the drag force due to fluid flow and the Brownian force [7,8]. The linear momentum balance for an ensemble of spheres in a viscous flow, individually denoted by the superscript (i), is given by the Langevin equation

$$\frac{d}{dt}\left(\underline{R}^{(i)}\right) = \underline{v}\left(\underline{R}^{(i)}\right) + \frac{\underline{F}_B^{(i)}(t)}{\zeta^{(i)}} \quad (1.4)$$

where $\underline{R}$ is the position vector of the particle, $\underline{v}(\underline{R})$ is the unperturbed velocity of the fluid evaluated at the particle position, $\underline{F}_B$ is the random force due to Brownian motion, and ζ is the Stokes' law drag coefficient given by

$$\zeta = 6\pi r \eta \quad (1.5)$$

where r is the particle radius, and η is the fluid viscosity.

Rods are modeled as prolate ellipsoids. The particle motions are governed by a similar Langevin equation with orientation dependent drag and diffusion coefficients, and the Jeffrey equation with rotational diffusion [4,7,8]

$$\frac{d}{dt}\left(\underline{R}^{(i)}\right) = \underline{v}\left(\underline{R}^{(i)}\right) + \left[\underline{\underline{\zeta}}^{(i)}\right]^{-1} \cdot \underline{F}_B^{(i)} \quad (1.6)$$

$$\frac{d}{dt}\left(\underline{p}^{(i)}\right) = -\underline{\underline{W}} \cdot \underline{p}^{(i)} + \lambda_p^{(i)}\left(\underline{\underline{D}} \cdot \underline{p}^{(i)} - \underline{\underline{D}} : \underline{p}^{(i)}\underline{p}^{(i)}\underline{p}^{(i)}\right)$$
$$+ \underline{p}^{(i)} \times \left[\underline{\underline{\xi}}^{(i)}\right]^{-1} \cdot \underline{T}_B^{(i)} \quad (1.7)$$

where $\underline{R}$ and $\underline{p}$ are the position and orientation vectors of the particles, $\underline{F}_B$ and $\underline{T}_B$ are the Brownian force and torque, $\underline{\underline{\zeta}}$ and $\underline{\underline{\xi}}$ are hydrodynamic resistance tensors, $\underline{\underline{D}}$ and $\underline{\underline{W}}$ are the stretching and vorticity tensors, respectively, $\lambda_p = \dfrac{\mathfrak{R}^2 - 1}{\mathfrak{R}^2 + 1}$ and $\mathfrak{R} = \dfrac{a}{b}$.

2.2   Boundary Conditions

An important part of the scheme is the interaction of the particles with the boundaries. The cross flow velocity continually drives the particles towards the accumulation wall and the diffusion step being random may also cause the particles to collide with or overstep this boundary. To handle this, a no-penetration boundary condition is used. For spheres, particles may only approach within a distance equal to one particle radius of the upper and lower boundaries. For rods, a similar criterion based on the rod size and 3-D orientation is used [6].

3   RESULTS

3.1   Spheres

Modeling and experimentation of particle separation for spheres under conditions spanning the normal to steric transition was examined. The goal of this was to test the simulation model by comparison with experimental data and existing theory. The experimental procedure is described in reference [5]. Results are shown in Figure 2 and Table 1. Both the simulation results and experimental data for mean elution time are in good agreement with the steric inversion theory of Giddings, i.e., Eq. (1.3). For the given experimental conditions, the steric transition occurs when the particle diameter is in the range 300 to 600 nm, with full steric inversion at 600 nm.

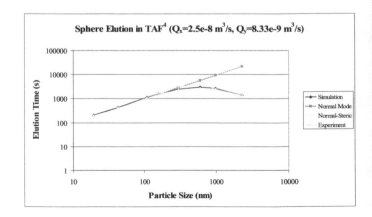

Figure 2: Comparison of simulation elution times with experimental data and theory [5].

3.2 Brownian Rodlike Particles

In this section we consider the modeling of rodlike particles which are subject to both translational and rotational diffusion, shown schematically in Figure 3.

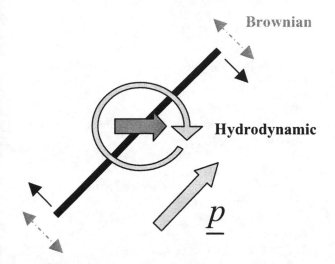

Figure 3: Forces on a rodlike particle subject to both shear flow and translational and rotational Brownian motion.

Particle separation for rods is compared with spheres of equal diffusivity as under normal mode conditions such particles should elute at the same rate. The rods are assumed to be 1 nm in diameter. The equivalent diffusion sphere size is given by the expression [6]

$$r_{eq} = a \frac{16}{3} \frac{s^3}{2s + (3s^2 - 1)Z} \quad (1.8)$$

where a and b are the major and minor axes of the ellipse, respectively, $s = \sqrt{a^2 - b^2}/a$, and $Z = \ln\left[(1+s)/(1-s)\right]$.

Results are shown in Figure 4, which plots the deviation from normal mode theory for the two different particle types as a function of equivalent sphere size. The data show that rods and spheres of equal diffusivity elute at the same rate up to an equivalent sphere size of 90 nm, which corresponds to a rod size of about 500 nm. At larger sizes, the rods begin to deviate from normal mode theory, but in the opposite sense as for spheres. While the steric effect for spheres causes larger spheres to elute faster than predicted by normal mode theory, an inverse steric effect occurs for rods in which larger rods move increasing slower than that predicted by the theory. This occurs due to alignment of the larger, less Brownian rods in low velocity region along the accumulation wall. It can be seen that past sphere:rod sizes of approximately 90 nm:500 nm, the negative steric deviations for the spheres and the positive steric deviations for rods lead to increasingly greater differences in mean elution times for the two different particle types. This can

be viewed as a positive result as it shows that nanotubes can be increasingly fractionated by size in flow-FFF.

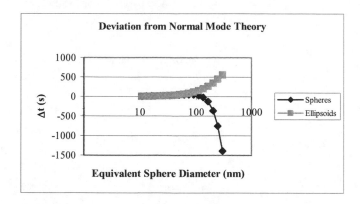

Figure 4: Comparison of the deviation of elution time from normal mode theory for spheres and rods of equal diffusivity.

3.3 Oriented Rodlike Particles

The steric inversion of spheres suggests that an even stronger steric effect might be obtained for rodlike particles due to their great length, if they could be oriented normal to the flow direction. This could be useful in the context of nanotube separation if either metallic or semi-conducting types could be preferentially oriented relative to the other, and could be accomplished by use of an AC electric field to induce alignment, as shown in Figure 5. The torque due to the electric field would have to be strong enough to resist the torque due to shear and the Brownian force.

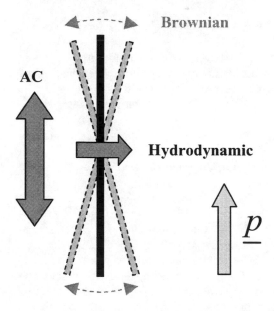

Figure 5: Forces on a rodlike particle subject to both shear flow, translational and rotational Brownian motion, and an AC electric field to induce "wobbly" alignment.

We investigate this in a straightforward manner by modifying Eq. (1.3) with $\alpha' = a/H$, where a is once again the length of the ellipsoid major axis. Results are shown in Figure 6.

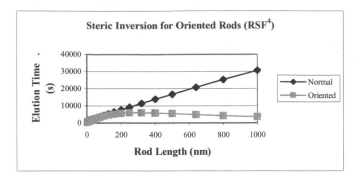

Figure 6: Comparison of elution time for rodlike particles under normal mode conditions, and for rods oriented perpendicular to flow direction.

The results show that oriented rods will exhibit steric inversion. Thus, in the case of mixtures of oriented and Brownian rods, the species can be separated into two fractions. The first fraction will contain oriented rods of all sizes, and Brownian rods whose size is such that their elution time is less than the elution time corresponding to the steric inversion point for the oriented rods. The second fraction will contain the longer Brownian rods. For the case shown in Figure 6, the cutoff size in the first fraction is 150 nm; smaller cutoff sizes can be obtained by optimizing parameter conditions. The first fraction may be further refined using size separation only.

4 SUMMARY AND CONCLUSIONS

We have investigated particle separation mechanisms in flow-FFF using Brownian dynamics method, and comparison with experimental results and theory. The following conclusions can be drawn from the present work.

- Under normal mode conditions, spheres and rods of equal diffusivity elute at the same rate and small particles elute faster than large particles.
- For large sphere sizes, steric inversion occurs due to size exclusion of the particles at the boundary. Past the steric inversion point, large particles elute faster than small particles. The Brownian dynamics results show very good agreement with both theory and modeling on this point.
- Large rods show an inverse steric effect relative to spheres, in which larger rods move increasing slower than that predicted by normal mode theory due to alignment of the rods in low velocity region along the accumulation wall. This can be viewed as a positive result as it shows that nanotubes can be increasingly fractionated by size in flow-FFF.
- Rods oriented normal to the flow direction can be expected to exhibit steric inversion. This suggests a potential mechanism to separate nanotubes based on chirality, if nanotubes of different type can be preferentially oriented in an AC electric field.

REFERENCES

1. J. C. Giddings, Separation Science and Technology **13,** 241 (1978).
2. J. Janca, in *Field-Flow Fractionation*, (Marcel Dekker, New York, 1987).
3. in *Field-Flow Fractionation Handbook*, (Wiley-Interscience, 2000).
4. F. R. Phelan Jr. and B. J. Bauer, Chemical Engineering Science **62,** 4620 (2007).
5. F. R. Phelan Jr. and B. J. Bauer, Analytical Chemistry **submitted,** (2008).
6. F. R. Phelan Jr. and B. J. Bauer, AICHE Journal **submitted,** (2008).
7. A. Satoh, in *Introduction to Molecular-Microsimulation of Colloidal Dispersions*, (Elsevier Science B.V., Amsterdam, The Netherlands, 2003).
8. S. Kim and S. J. Karrila, in *Microhydrodynamics: Principles and selected applications*, (Butterworth-Heinemann, Stoneham, 1991).

Official contribution of the National Institute of Standards and Technology; not subject to copyright in the United States.

5 TABLES

Elution Data for Spheres			
d (nm)	Experiment Mean (s)	Simulation Mean (s)	% diff
19	195	208.6	6.5
42	389	439.6	11.6
110	1080	1121.6	4.0
160	1490	1581.1	5.6
300	2990	2532.7	-17.9
600	4200	3067.3	-36.9
984	2510	2639.3	5.0
2300	1380	1404.2	1.6

Table 1: Comparison of simulation and experiment for spheres [5]. The relative uncertainties reported are one standard deviation, based on the goodness of the fit or from multiple runs.

Development of nanocomposites with carbon nanotubes in a thermoplastic elastomer matrix

Georg Broza[1], Zbigniew Roslaniec[2] and Karl Schulte[1]

[1]·Technical University Hamburg-Harburg,
Denickestrasse 15, D-21073 Hamburg, Germany,
broza@tuhh.de, fax.: +4940428782002
[2] Technical University of Szczecin, Al. Piastów 17
PL-37301 Szczecin, Poland

ABSTRACT

We have developed a polycondensation process, which allows the manufacturing of thermoplastic elastomers with carbon nanotubes (CNTs). The nanocomposites are obtained by introducing the CNTs into the reaction mixture whilst the synthesis of polybutylene terephthalate/polyoxytetramethylene (PBT/PTMO) blockcopolymers. In the first step the CNTs were dispersed in dimethyl terephthalate (DMT) and 1,4-butanediol (BD) by ultrasonication and ultrahigh speed stirring. The nanocomposites were characterized by transmission, scanning electron microscopy, atomic force niscroscopy, tensile testing and dynamic-mechanical analysis. Only a small amount of CNTs was needed to improve the mechanical and dynamical properties, reversibility and energy absorption of the nanocomposites.

Keywords: carbon nanotubes, thermoplastic elastomers, nanocomposites

1 INTRODUCTION

Carbon nanotubes (CNTs) are a promising new class of material with exceptional mechanical, electrical and thermal properties [1-3]. Their presence in polymer at very low concentrations may improve mechanical properties and electrical conductivity [4-7]. However strong interfacial bonding between filler and polymer matrix as well as good wetting of the filler surface by the polymer and its homogenous dispersion in the polymer system have to be ensured in order to obtain a maximum effect of reinforcement in composites [8].

We have obtained polymer matrix + CNT nanocomposites by introducing CNTs into a reaction mixture during the synthesis of poly(ether-ester) block copolymers (PBT/PTMO). The polymers without and with carbon nanotubes were synthesised using an in situ polycondensation reaction process [9-10]. After polycondensation the nanocomposites were extruded and then injection moulded. The Young`s modulus, tensile strength, and also strain to failure go all up with increasing amounts of CNTs in the PBT/PTMO matrix. However, an addition of only a small amounts of CNTs to the matrix is sufficient to improve the mechanical and dynamical mechanical (DMA) properties.

2 EXPERIMENTAL

2.1 Materials

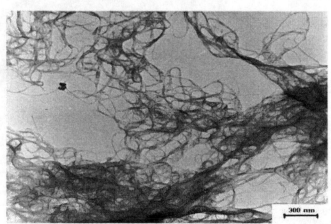

Polybutylene terephthalate/polyoxytetramethylene (PBT/PTMO) was used and synthesised in a two-stage process: transesterification followed by polycondensation in the melt as described in [11,12], with a content PTMO of 55 wt%.

2.1.1 Characterization of carbon nanotubes (CNTs)

The multi walled nanotubes (MWCNT) used for the experiments were purchased from Nanocyl S.A. (Belgium) and used without further purification process.For the experiments very thin multi-wall carbon nanotubes (MWCNT) were used. Typically, their outer diameter range is 3-15 nm, length up to 50 μm. The CNTs were produced by catalytic chemical vapour deposition method (CCVD) [13]. The structure of the bundles of the multi wall nanotubes was investigated using TEM (Fig. 1).
The single walled carbon nanotubes (SWCNT) were ready supplied by CNI Technology Co., TX, USA, synthesised by the HIPCO method [14]. The diameter of the SWCNT was 0.7-1.2 nm with a length of a few μm. A TEM image of the SWCNT is shown in Fig. 2.

Figure 1: TEM image of the multi walled nanotubes

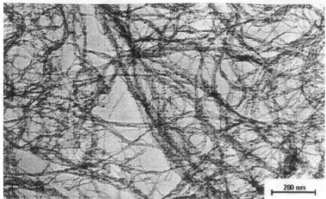

Figure 2: TEM image of the single walled nanotubes, SWCNTs

2.2 Processing of PBT/PTMO with CNTs

2.2.1 Dispersion of CNTs in DMT

Dimethyl terephthalate (DMT) was melt in butanediol (BD), at 140^0C. The CNTs were dispersed in this mixture. The mixture was ultrahigh speed stirred for 5 min. at 20.000 rpm and sonicated for 5 min. at room temperature, applying a SONOPLUS-Homogenisator HD 2200 with a titanium sonotrode made by BANDELIN ELECTRONIK GmbH, Berlin, Germany. The first step and the second step were repeated six times. The bundles of nanotubes are suspended in DMT+BD.

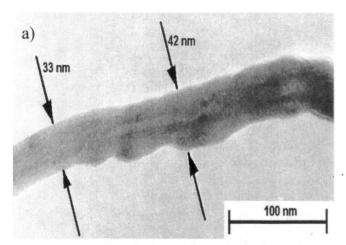

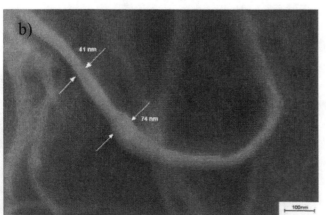

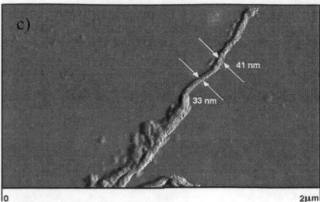

Figure 3: (a) TEM image, (b) SEM image, (c) AFM image showing the dispersion of MWCNTs in DMT+BD. The different diameters of MWCNT are due to the surrounding of the surface of nanotubes with DMT+BD layer.

The molten dispersion was casted onto a plate, and after cooling down to room temperature the material crystallized. It was then crushed and added to the reactor. The TEM micrograph confirms the covering of nanotubes by a layer of DMT+BD.

Fig. 3 shows a) the TEM micrographs b) SEM micrograph and c) AFM micrograph of the crystallized powder of DMT+BD containing MWCNTs.

The TEM characterization in Fig. 3(a) provides a clearer view of the nanotube surface covering due to the DMT+BD in crystallized powder. In this figure changes in diameter of the nanotubes are in the range of 33 to 42 nm, which is comparable to the diameter range taken from AFM showed in Fig. 3(c). Only SEM image in Fig. 3(b) shows the different diameters of the nanotubes, which change in the range from 33 to 74 nm.

2.2.2 Schematic illustration of the PBT/PTMO+CNT nanocomposites processing

In a steel reactor (Autoclave Eng. Inc., USA) dimethyl terephthalate (DMT) and catalyst were mixed; CNT+BD system was slowly added to this mixture and was mechanically stirred to obtain a homogeneous dispersed system. The nanocomposite formed were extruded from the reactor by compressed nitrogen and cooled down to room temperature. All nanocomposites were granulated and dried before processing. In the present investigation the nanocomposites containing 0.1 wt% of SWCNT, and 0.5 wt% of MWCNT were included. The samples of neat PBT/PTMO were synthesized as a reference.

The process is schematically shown in Fig. 4

NSTI-Nanotech 2008, www.nsti.org, ISBN 978-1-4200-8503-7 Vol. 1

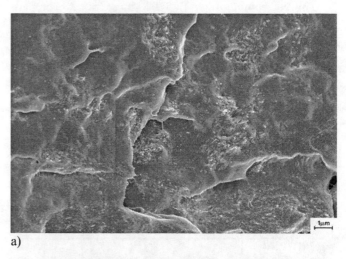

Figure 4: Schematic illustration of the PBT/PTMO+CNT composites processing

3 RESULTS AND DISCUSSION

3.1 Morphology

A LEO 1530 SEM FEG (Field Scanning Electron Microscope) was used to characterize the morphology of the failure surfaces of the PBT/PTMO+CNT nanocomposites. As shown in Fig. 5, the nanotubes which protrude from the fracture surface appear to be coated by a thin layer of the PBT/PTMO matrix. The coating layer was found to have a thickness of 80 to 200 nm.

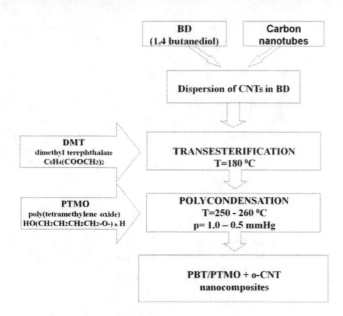

Figure 5: SEM micrograph of the fracture surface of the granulated PBT/PTMO nanocomposite with MWCNTs dispersed in DMT+BD before injection moulding

Fig.6 a,b shows the distribution of carbon nanotubes after injection molding. It is quite clear that the CNTs are uniformly distributed in the PBT/PTMO phase, even though the nanocomposite was prepared via in situ polykondensation reaction.

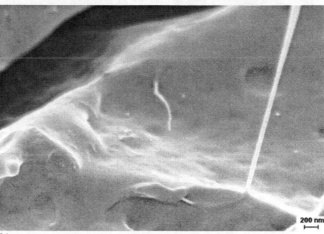

a)

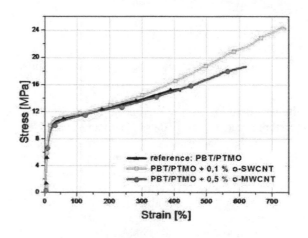

b)

Figure 6 a,b: SEM micrographs of the fracture surface of the granulated PBT/PTMO nanocomposite with MWCNTs dispersed in DMT+BD after injection moulding

3.2 Mechanical and dynamical properties

Figure 7: Stress-strain curves of pure PBT/PTMO and PBT/PTMO nanocomposites with different weight fraction and type of carbon nanotubes.

Table 1 lists the average values for Young's modulus, tensile strength, and fracture strain for SWCNT and MWCNT content in the PBT/PTMO matrix. In comparison to the neat PBT/PTMO tensile tests results show that nanotubes dispersed via in situ polycondensation in PBT/PTMO have an improved tensile strength and the tensile strain (Fig.7).

Sample	E [MPa]	σ [MPa]	ε [%]	E' at temp. 50^{0}C [MPa]
Reference: PBT/PTMO (45% of hard segments)	88,1	15,4	433,61	77,44
PBT/PTMO +0,1wt% SWCNT	109,9	24,4	735,69	88,58
PBT/PTMO +0,5wt% MWCNT	93,5	18,7	620,13	84,11

Table 1: The mechanical and dynamical-mechanical properties of PBT/PTMO+CNT nanocomposites

The dynamic mechanical properties of the neat matrix and the nanocomposites were studied by DMTA. Fig.8 shows the temperature dependent storage modulus E` of PBT/PTMO and the nanocomposites.

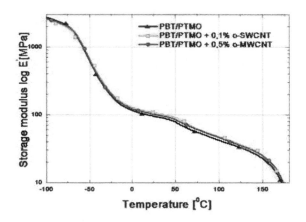

Figure 8: Storage modulus – temperature curves of pure PBT/PTMO and PBT/PTMO nanocomposites with different weight fraction and type of carbon nanotubes

In the nanocomposites the incorporation of CNTs causes a measurable increase in stiffness. The storage modulus E` of PBT/PTMO is increased by the stiffenning effect of the nanotubes, which is partycularly significant at temperatures between T= -50 and + 150 ^{0}C. This is a hint that the rein-forcement effect of CNTs is mainly active not only in the soft segments (PTMO) but also in the amorphous phase of PBT, which leads to the suggestion that the nanotubes are only present in the amorphous phases.

The two-stage polycondensation method in molten state was successfully applied to obtain block ether-ester elastomers with carbon nanotubes dispersed in dimethyl terephthalate (DMT) and 1,4-butanodiol (BD) and added to the reaction mixture. The essential element of obtaining the nanocomposites in such a way the total dispersion of nanotubes in BD just before starting the synthesis. The CNTs were relatively uniformly dispersed with a minor trend for agglomeration, and incorporated in thermoplastic elastomers in order to assess the effectiveness of the polycondensations process. Tensile test results showed that nanotubes dispersed via in situ polycondensation in PBT/PTMO could improve the tensile strength and the tensile strain concomitantly. The E-modulus of the nano-composite with SWCNTs reached slightly higher values as MWCNT.

ACKNOWLEDGEMENTS

The authors gratefully acknowledge the support by the Deutsche Forschungsgemeinschaft (DFG), Contract SCHU 926/14-1 and by Polish Ministry of Science and High Education, contract 2/DFG/2007/02.

REFERENCES

[1] M.S. Dresselhaus, G. Dresselhaus and Ph. Avouris ,Top. Appl. Phys. 80, 2001.
[2] A.G. Mamalis, L.O.G. Vogtlander and A. Markopolous, Precis. Eng. 28, 16, 2004.
[3] E.T. Thostenson, Z. Ren and T.W. Chou, Compos. Sci. Technol. 61, 1899, 2001.
[4] R. Andrews and M.C. Weisenberger, Curr. Opin. Solid St. M. 8, 31, 2004.
[5] W.D. Zhang, L.Shen, I.Y. Phang and T.Liu, Macromolecules 37 ,256, 2004.
[6] J.K.W. Sandler, J.E. Kirk, I.A. Kinloch, M.S.P. Shaffer and A.H. Windle, Polymer 44, 5893, 2003.
[7] S.J. park, S.T. Lim, M.S. Cho, H.M. Kim, J. Joo and H.J. Hoi, Curr. Apl. Phys. 5, 302, 2005.
[8] E.T. Thostenson, Ch. Li and T.W. Chou, Compos. Sci. Technol. 65, 491, 2005.
[9] G. Broza, M. Kwiatkowska, Z. Rosłaniec, K. Schulte, Polymer 46, 5860-67, 2005.
[10] Z. Rosłaniec, G. Broza, K. Schulte, Composite Interfaces 10, 95-102, 2003.
[11] A. Szepke-Wrobel, J. Slonecki, H. Wojcikiewicz, Polimery 27, 400, 1982.
[12] R. Ukielski, H. Wojcikiewicz, Polimery 23, 48, 1978.
[13] D. Lupu, A.R. Biris, A. Jianu, C. Bunescu, E. Burkel, E. Indrea, G. Mihailescu, S. Pruneanu, L. Olenic and I. Misan, "Carbon nanostructures produced by CCVD with induction heating", Carbon 42, 503-507, 2004.
[14] P. Nikolaev, M.J. Bronikowski, B.R. Kelly, F. Rohmund, D.T. Colbert, K.A. Smith, et all, Chem. Phys. Lett. 313, 91-97, 1999.

4   CONCLUSIONS

Halloysite Nanotubes in Polymers

Robert C. Daly, Cathy A. Fleischer, Aaron L. Wagner and Michael Duffy

NaturalNano, Inc., 15 Schoen Place, Pittsford, NY, 14534
daly@naturalnano.com

Abstract

Halloysite Nanotubes (HNT[TM]) provide a new avenue for the preparation of nanocomposites. Halloysite is a naturally occurring member of the kaolin family of aluminosilicate clays. Its uniqueness is that it exists predominantly in a tubular form with lengths of up to 10 microns and diameters of up to 400 nm rather than the deck of cards platy form of essentially all other clays.

Well behaved dispersions of HNT in nylon-6, polypropylene, TPO and several varieties of PE have been obtained by standard melt processes. In all cases, improved physical performance has been realized for molded parts.

HNT have also been dispersed in polymer latexes and dispersions at quite high loadings. These polymer and clay dispersions have been coated and produce coatings with useful performance properties.

Key words: halloysite, nanocomposite, nanotube, clay

Introduction

Polymer nanocomposites are an area of intense research and development due to their potential to provide stronger and lighter materials and parts. Unfortunately, it has proven hard to produce finished parts that deliver this potential with a cost and the robust productivity required for critical uses. Clays are a class of nanomaterials that have received a great deal of attention. [1] While the particles of a typical clay are far from nano-sized, the particle can be separated into its individual component sheets by a combination of chemical, thermal and mechanical steps. These two dimensional sheets provide a very high surface area to weight ratio which improved the performance of polymers that they were compounded with, particularly nylon. Despite the improved performance obtained in typical platy clay nanocomposites; their utility has continued to be limited. Separating the platy clay particles into their individual sheets is a difficult process that can alter the materials balance of the melt formulations or in the extreme requires in situ polymerization of the monomer in the presence of the clay.

Halloysite is a naturally occurring member of the Kaolin family of aluminosilicate clays. Halloysite can occur in several structures but predominantly exists as a tubular structure believed to be the result of hydrothermal alteration, or surface weathering of other aluminosilicate minerals. [2], [3] Halloysite nanotubes (HNT[TM]) do not require large amounts of chemical modification or complex chemical

processes such as intercalation and exfoliation in order to produce stable nanoparticle clay dispersions in the melt. This makes it possible to obtain performance improvements without the complexity and processing cost associated with platy clays.

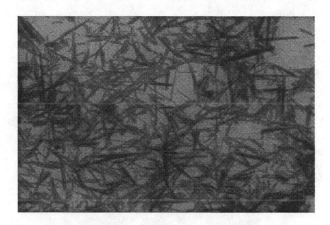

Figure 1 TEM of halloysite at 2600 X magnification.

Figure 2 TEM of halloysite at 25000X magnification

Our results for nylon/HNT and polypropylene/HNT polymer nanocomposites indicate that an increase in modulus and strength is obtained with little reduction in elongation. NaturalNano has prepared well-dispersed HNT polymer concentrates which are commercially available under the name Pleximer[TM] at HNT concentrations of up to 50%.

Additionally, stable acrylic polymer latex/HNT dispersions were prepared to concentrations as high as 30% HNT. Coatings made from these latex formulations exhibit a ten fold increase in storage modulus, while maintaining key properties such as transparency and tack.

Experimental Work

The scope and property improvements seen by Toyota in their work with platy clays in nylon for injection molded parts [4] caused us to begin our nanocomposite work with nylon. The composites were typically prepared by melting the nylon pellets in the early part of the extruder and feeding the desired amount of halloysite onto the melt at a later hopper. The halloysite was milled to an extremely fine, near white powder that flows and handles like flour. Both the polymer and the halloysite were dried to discourage degradation of the nylon. The composite strand was passed through a water bath and chopped into pellet form for direct use or to be let down for injection molding.

Figure 3 is a 20,000X magnified SEM image of the end of a frozen strand of a 30% halloysite composite in nylon-6. Individual tubes can be seen as well as small clusters of tubes.

Figure 3 SEM of 30% halloysite in nylon-6.

Figure 4 shows that the tubes are oriented within the flow while remaining dispersed. Micrograph 4a was taken across the strand while 4b is taken along the length of a fractured strand.

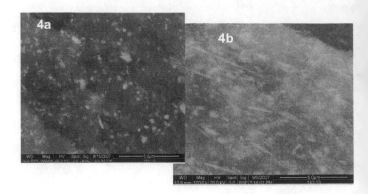

Figure 4 SEM of 20% halloysite in nylon-6 when fractured in the cross direction (4a) or along the strand (4b).

Properties \ Composite	Nylon 0% HNT	Nylon 2% HNT	Nylon 4.1% HNT	Nylon 5.1% HNT	Nylon 10.2% HNT	Nylon 3.8% HNT2	Nylon 4.7% HNT2	Nylon 9.4% HNT2
Modulus of Elasticity, psi	322000	438000	455000	456000	466000	448000	474000	516000
Elongation at Break, % (2in/min)	91.1%	35.8%	31.3%	41.4%	16.9%	68.5%	42.5%	39.0%
Tensile Modulus, psi	407000	457000	464000	481000	495000	477000	502000	572000
Gardner Impact, in.lb	>160	140	140	120	120	>160	>160	80

Table 1 HNT-Nylon Nanocomposite Physicals.

Testing bars were prepared from the extruded pellets by injection molding. HNT generally improved physical performance but it was especially effective at producing parts which have good ductility, impact resistance and resistance to flex fatigue. The improvement in physical performance can be obtained at low levels of incorporation and while maintaining good elongation properties.

Two different halloysite samples are shown in Table 1. HNT1 is a shorter tube than HNT 2. HNT1 has an average tube length of about 1.2 microns while HNT2 averages slightly less than 5 microns. The tensile performance of the long tubes is preferred but at some expense in impact and fatigue.

A potential disadvantage for HNT is that the addition of HNT increases the mold shrinkage for nylon injection molded parts compared to the nylon itself. However, this brings a significant opportunity for mixing HNT with the more normal talc or glass fiber fillers in the composites.

Composites with talc or glass fibers, will have large improvements in tensile properties but may suffer from very low elongation at break, poor Gardner impact and too little shrinkage. A composite containing both fillers in low amounts can produce an excellent set of final properties as shown in Table 2

% HNT in B3K nylon	% GF in nylon	tensile mod (psi)	flex modulus (psi)	mold shrink (%)
0	0	407000	322000	0.92
5.1	0	481000	456000	1.60
10.2	0	495000	466000	1.76
10	5	600000	449000	1.00
5	7.5	581000	415000	0.65
0	15	731000	524000	0.30

Table 2 Mixed Composites of HNT and glass fibers in nylon.

The mold shrinkage can be dialed in by controlling the ratio and amount of the two fillers. Reducing glass fiber loadings can dramatically improve the processing and esthetics of the finished parts while the HNT keeps the mechanical properties quite high.

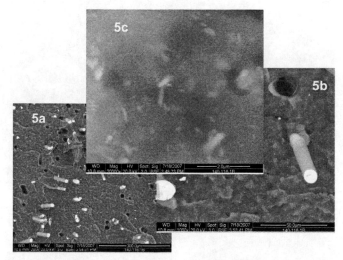

Figure 5   SEM images of mixed composites of halloysite and glass fibers in nylon-6. The large difference in size can be recognized by the different magnifications needed to see the two types of fillers. 5a at 200X, 5b at 1000X and 5c at 20000X.

Polyolefins similarly form composites using melt technology with improved physical performance but the impact is not as large as for nylon. As with platy clays it is difficult to get complete dispersion and more clay is needed to get large effects. Figure 6 includes data for HNT

composites with a polypropylene homopolymer and a propylene copolymer. In both cases the flex modulus improves while some elongation to break remains. Similar results have been observed for a variety of polyethylenes and TPO's.

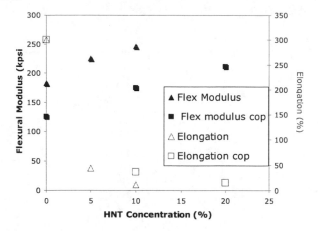

Figure 6 Flexural Modulus and Elongation at break for HNT composites in polypropylene and a polypropylene copolymer.

As the polypropylene extrusions were being made it became obvious that the addition of halloysite to the polymer melt made it easier to process. Temperatures could be dropped or outputs increased. Dry blended samples of halloysite in INEOS H12F-00 were prepared and extruded under the same input and output conditions using the same screw and temperature profile. As seen in Table 3, less torque was required as more HNT was added.

Percent HNT	0	10	20	30
Extruder Torque (%)	85.8	69.5	64.1	63.2

Table 3 Extruder torque as a function of adding HNT to polypropylene.

Composites of HNT have also been prepared by dispersing the milled and treated halloysite with polymer latexes and emulsions. Significant mixing is required but not enough to harm the polymer system. The subsequent filled fluids may be coated and dried in the normal way to give coatings which have improved physical properties. Figure 7 contains information about the change in modulus produced by putting various amounts of HNT and platy clay into an acrylic polymer dispersion and coating it onto a substrate. When HNT were added to the acrylic dispersion, the modulus of the coating rose significantly while it dropped when even a small amount of a platy clay was added. In fact 5% platy clay was the maximal amount that could be placed in a stable dispersion while the HNT went easily to 20% percent of the total solids.

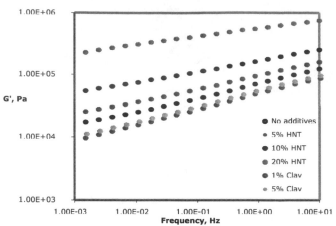

Figure 7 G' modulus for MG 0580 latex coatings with varying amounts of HNT or platy clay.

Since the coatings of this polymer latex could function as a pressure sensitive adhesive, the HNT filled coatings were also evaluated as an adhesive. Formulations were prepared containing 40% polymer and 10% clay within the latex. Coatings were drawn down onto PET using a 20 mil coating knife and air dried. Remarkably, the coatings remained tacky even at higher levels of HNT up to 20% HNT. A measured strip of the coated PET was cut and pressed with a weighted roller onto an uncoated strip of PET. The strips were peeled apart on a Tinius Olsen H5KT. The peel forces reported in Table 4 for composite coatings on PET show that the compositions were not only tacky but they formed bonds that were considerably stronger than the polymer itself. On the other hand, one of the two platy clays tested crashed out and the other made adhesion much worse.

Clay added to Rhoderm 5600 Latex	Strength (N)	Failure Mode
HNT1	31.8	cohesive
HNT2	28	cohesive
Latex alone	26.9	cohesive
Veegum T	N/A	N/A
Bentonite	4.1	adhesive to uncoated PET

Table 4 Peel strengths for composite coatings from acrylic polymer latexes with HNT and platy clays

Summary

Halloysite nanotubes were formulated in the melt and via polymer dispersions to form polymer nanocomposites which produced objects with useful and unusual properties.

Acknowledgements

We would like to thank Bo Wang and Tim Patterson of Noble Polymers for their partnership in much of the melt composite work, Dr Robert Corkery of YKI for his work with adhesives and Donald Black for his SEM and TEM microscopy.

References

[1] T. Pinnavaia, G. Beall, "Polymer-Clay Nanocomposites" Wiley Series in Polymer Science, John Wiley and Sons Ltd., Chitchester, Eng, 2000.
[2] E. Joussein, S. Petit, G. Churchman, B. Theng, D. Righi, B. Delvaux, Clay Minerals **40**, 383-426, 2005.
[3] B. Theng, M. Russell, G. Churchman, R.Parfitt, Clays and Clay Minerals **30,** 143-149, 1982.
[4] A. Okana, A. Usuki, Macromolecular Materials and Engineering, **291**, 1449-1476, 2006.

Adapting carbon nanotubes for fine composite structures

M. Shaffer*, R. Verdejo*, M. Tran**, A. Menner**, A. Bismarck**

*Department of. Chemistry, Imperial College, London SW7 2AZ, UK, m.shaffer@imperial.ac.uk
**Department of. Chemical Engineering, Imperial College, London SW7 2AZ, UK

ABSTRACT

Although defective, commercially-available, CVD-grown, multi-walled carbon nanotubes are widely used for polymer nanocomposites due to their availability and low cost. However, to use these materials effectively, they must be disentangled, and chemically modified in order to produce high quality composites. This paper shows that such nanotubes can be disentangled by cutting; abrupt, repeated exposures to oxidising conditions in air is a clean, convenient and efficient means of producing material with open ends, moderate functionalisation, and enhanced solvent dispersibility; the surface character can easily be tuned to acidic or basic. These approaches could be deliberately integrated into conventional CVD processes, but also have implications for existing products. Matrix-filler interactions can be quantified by examining the contact angle of individual polymer droplets on single nanotubes, and correlated to the mechanical performance of corresponding macroscopic nanocomposites.

Keywords: nanotubes, nanocomposites, surface modification, cutting, wetting, CVD synthesis

1 INTRODUCTION

There is enormous interest in the use of carbon nanotubes as fillers in composite materials. The intrinsic properties of individual carbon nanotubes are remarkable; they have unmatched strength and thermal conductivity, as well as high stiffness and electrical conductivity, all at low density. In addition, their high surface area enables an intimate interaction with polymer matrices, potentially influencing matrix characteristics such as glass transition temperature and crystallinity, as well as functional properties inducing operating temperature, solvent resistance, and tribology. Although encouraging results have been obtained for nanotube-filled polymers, significant improvements over conventional fillers have proved elusive (a recent review summarises progress in this area [1]). A critical factor is the reliance on CVD-grown material, which is available in large quantity at reasonable purity, but defective and often entangled. However, a range of steps can be taken to improve the applicability of such materials, and to employ them in situations where they can provide a unique benefit. Although the intrinsic properties of nanotubes grown by other techniques, such as arc or laser evaporation are better, these methods only produce gram-scale yields, often with intractable carbonaceous by-products. In contrast, the annual production capacity for CVD nanotubes is now measured in thousands of tonnes; scalable methods for improving the usability of these materials are therefore of great interest.

One promising avenue is to exploit CVD nanotubes in fine structures where other reinforcements cannot be accommodated; examples include fibres [2][3], foams [4][5][6] and the matrices of conventional fibre composites [7]. Although nanotubes tend to raise viscosity, thereby introducing processing difficulties, in certain circumstances, the change in rheology can be beneficial even enabling [6]. In these cases, the nanofiller can simultaneously aid processing and improve properties.

2 CUTTING

Multiwall carbon nanotubes grown by CVD are usually inherently entangled, which hinders efforts to produce composites containing the well-dispersed nano-reinforcement that is required for effective stress transfer. Most nanotube composite manufacturing, therefore, includes either an explicit cutting step, or implicitly involves breakage during shear or ultrasonic processing. Explicit cutting is more reliable, but most methods both cause damage to the nanotubes, and involve a length work-up procedure; examples include mechanical ball-milling or grinding, and sonication or irradiation in aggressive acidic environments. In our work [8], the use of abrupt, repeated exposure to thermally oxidising conditions proved to be an efficient means of producing material with open ends, moderate levels of functionalisation, and, most importantly, enhanced dispersibility in organic solvents.

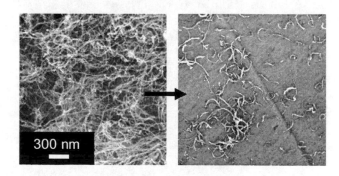

Figure 1:Scanning electron micrograph showing the effect of cutting the original entangled nanotubes (left) into shorter fragments (right)

The key is to cycle the oxidation conditions rapidly and repeatedly in order to localize the reaction at defect sites, to encourage cutting rather than the thinning process usually associated with gas phase oxidation. In our experiments, cycling was achieved by moving the sample between room temperature and a furnace at 600°C; but an alternative strategy would be to control the supply of oxygen in short bursts. After treatment the material is much more readily dispersed in solvents, such as DMF.

Figure 2: Photographs of dispersions of carbon nanotubes in DMF after 5 min of mild sonication and settling overnight. Left to right: as-grown nanotubes, 3 cycles of thermal cutting, 6 cycles of thermal cutting.

Interestingly, heating the nanotubes to 800°C in an inert atmosphere and cooling to room temperature before exposing them to air, changes the surface chemistry. The oxygen-containing surface groups switch from acid character (carboxylic acids, phenols etc) to basic groups (pyrones), as confirmed by XPS. The nanotubes with acidic character disperse best in solvents with Lewis base character (eg DMF), whilst the nanotubes with basic character disperse best in solvents with Lewis acid character (eg chloroform). It is worth noting that this type of change in character may happen inadvertently if the procedure for removing nanotubes from the original growth furnace is not standardized. It may also go unnoticed, as typical quality control techniques (EM, Raman, TGA) do not reveal the nature of the surface chemistry. Such effects may help to explain apparent variability in performance between nanotube batches. On the other hand, by using these effects deliberately, nanotubes can be adjusted for different applications. The basic equipment for these cutting and modification processes is similar or identical to typical CVD growth equipment. There is, therefore, considerable scope for integrating such processes into production facilities.

3 WETTING

The surface interactions of (modified) carbon nanotubes with a given polymer matrix has an important role in determining composite performance. Characterization of wetting and adhesion interactions at the nanoscale is challenging. However, we have developed a new methodology to create polymer droplets on individual nanotubes which allows quantitative determination of contact angle. The contact angle can be directly correlated to the mechanical effectiveness of the filler; the lower the angle the greater the stiffness. This methodology allows the surface modification of the nanotube to be linked to the interaction with the polymer, and hence to the performance of the macroscopic composite. Optimizing these factors is a crucial step to improving nanocomposite performance.

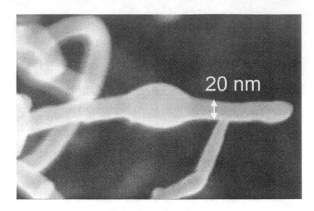

Figure 3: Scanning electron micrograph of a polymer droplet on a carbon nanotube. The shape of the droplet can be used to calculate the contact angle.

REFERENCES

[1] M Shaffer & J Sandler, Carbon Nanotube/Nanofibre Polymer Composites, Chapter 1, pages 1-59, in Processing and Properties of Nanocomposites, ed S Advani, World Scientifc, Dec 2006, ISBN 978-981-270-390-3. Currently available online at http://www.worldscibooks.com/nanosci/6317.html

[2] J Sandler, S Pegel, M Cadek, F. Gojny, M. van Es, J Lohmar, WJ Blau, K Schulte, AH Windle, MSP Shaffer, A comparative study of melt spun polyamide-12 fibres reinforced with carbon nanotubes and nanofibres, Polymer, 45/6, 2001-2015 2004

[3] J. Sandler, P. Werner, V. Denchuk, V. Altstädt, A.H. Windle, M. Shaffer, Carbon-nanofibre-reinforced poly(ether ether ketone) fibres, J. Mat. Sci., 38, 2135 – 2141, 2003

[4] Carbon nanotube enhanced polyurethane scaffolds fabricated by thermally induced phase separation, G Jell, R Verdejo, L Safinia, M. Shaffer, M Stevens, A Bismarck, J. Mat. Chem., In Press, 2008, DOI: 10.1039/b716109c

[5] Reactive polyurethane carbon nanotube foams and their interactions with osteoblasts, R Verdejo, G Jell, L Safinia, M Stevens, A Bismarck, M Shaffer,

J Biomed. Res., In Press, DOI: 10.1002/jbm.a.31698, 2008

[6] P. Werner, F Wőllecke, V. Altstädt, J K.W. Sandler, M Shaffer, Carbon Nanofibres allow foaming of semicrystalline Poly (ether ether ketone), Advanced Materials 17, 2864 – 2869, 2005

[7] H Qian, A Bismarck, E Greenhalgh, G Kalinka, M Shaffer, Hierarchical composites reinforced with carbon nanotube grafted fibres: the potential assessed at the single fibre level, Chem. Mat., In Press, DOI: 10.1021/cm702782j, 2008

[8] M Tran; C Tridech; A Alfrey; A Bismarck and M S P Shaffer, Thermal Oxidative Cutting of Multi-walled Carbon Nanotubes, Carbon, 45, 2341, 2007

Nano-engineered Composites Reinforced with Aligned Carbon Nanotubes (CNTs)

S. Wicks[*], N. Yamamoto, R. Guzman de Villoria, K. Ishiguro, E.J. Garcia, H. Cebeci, A.J. Hart, and B. L. Wardle

Massachussetts Institute of Technology
Room 41-317, Cambridge, MA 02139
[*] swicks@mit.edu

ABSTRACT

We present the implementation of aligned carbon nanotubes (CNTs) as a method to enhance properties of traditional advanced composites. The hybrid composites are 3-dimensional architectures of aligned CNTs, and existing advanced fibers and polymeric resins creating nano-engineered composites. Our work to date has focused on interlaminar strength and toughness, and electrical and thermal conductivities of two laminate-level architectures. The first approach utilizes aligned CNT forests placed between ply layers perpendicular to the fiber direction to create "nano-stitches". The second approach involves growing the CNTs directly on woven alumina fibers, so that the CNTs extend radially outward from every fiber forming "fuzzy fibers". Three standard processing routes for are reviewed: nano-stitching of graphite/epoxy prepreg, nano-stitching of graphite/epoxy cloth in a resin-infusion process, and hand layup of fuzzy fiber ceramic cloth.

Keywords: carbon nanotube, nanocomposite, composite, laminate

1 INTRODUCTION

Ever since the initial investigation of carbon nanotubes [1, 2], the multi-functional properties of CNTs have inspired research not only into their mechanical attributes, such as tensile stiffness and strength, but also into their electrical and thermal properties as well. Numerous researchers have confirmed the high tensile stiffness and strength of CNTs [3]. Other theoretical and experimental observations on CNTs include tunable electrical conductivities (from semiconducting to metallic), excellent thermal conductivity, minimal thermal expansion, and low density [4].

These exceptional properties are only beneficial if they can be implemented at engineering-relevant lengthscales. Successful examples in engineering applications include micro-probes and data storage elements [5], and are generally limited to micro and nano-scale applications. This stems from the difficulties inherent in fabricating with nanoscale constituents, scaling production to useable levels, and other issues that arise during integration. Recently, our group has implemented CNTs into macroscopic structures such as aerospace composites, where minimizing weight is

a great priority [6, 7]. Laminate mechanical, electrical, and thermal properties could be significantly enhanced by CNT implementation without significant weight addition. Two main approaches to incorporating CNTs in a macro scale have been explored in this work. Embedding VA-CNTs in the interply region focuses on enhancement in interlaminar properties. Another, more aggressive method to implement CNTs is by growing them directly on all fibers, including the ones inside the weaves, providing both interlaminar and intralaminar reinforcement. This paper summarizes work to date fabricating and testing both nano-engineered composite architectures.

2 EXPERIMENTAL APPROACH

2.1 Design of Nano-engineered Composites

The two main architectures developed for introducing aligned carbon nanotube into traditional composites are shown in Fig. 1 [6, 7]. In the first architecture, aligned carbon nanotube forests are introduced between ply layers perpendicular to the fiber direction to create "nano-stitches". Vertically aligned CNT forests are grown on a flat substrate using chemical vapor deposition (CVD) as described below. The CNTs are placed between layers of either carbon/epoxy prepreg or dry carbon weave and cured with conventional methods. The resin-rich region between traditional composite plies is a source of weak interlaminar properties. This particular approach allows the carbon nanotubes to effectively reinforce this region facilitating load transfer between plies. Such stitching has been shown to have substantial (>100%) improvement in critical Mode I and II strain-energy release rates. The second, fuzzy fiber architecture, addresses similar interlaminar weaknesses, but incorporates CNT forests that are radially grown on the fiber surfaces. By spanning both the region between fibers and the interply region, CNTs grown on adjacent fibers can form mechanical connections and electrical/thermal conductive pathways throughout the structure.

2.2 CNT Growth

Multi-walled carbon nanotubes to be implemented into nano-engineered composites mentioned above are grown with a modified chemical vapor deposition (CVD) method developed at MIT [8, 9]. The substrate is first coated with a

194

catalyst precursor, then is treated with heat and a reduction gas. This pre-conditions the catalyst into nano-particles, which nucleate CNTs when a gaseous carbon source is introduced. Our existing setup consists of an atmospheric pressure quartz tube furnace (Lindberg, 22 mm inner diameter, 300 mm heated length).

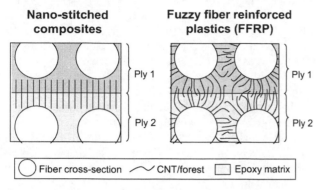

Figure 1: Illustration of idealized micro-structure of the nano-engineered composites reinforced with CNTs (after [6] and [7]). Not to scale.

2.3 FFRP Fabrication

The FFRP fabrication process (Fig. 2) involves growth of aligned CNTs on woven cloth, and then hand layup and curing with thermoset epoxy resin. Alumina (Al_2O_3) fiber cloth was selected as an advanced fiber substrate, as Al_2O_3 can regulate surface diffusion of an iron (Fe) catalyst, and also since Al_2O_3 composites are used in some demanding applications like armor [10]. As-obtained cloth (McMaster-Carr) is cut into swatches, soaked in a solution of iron nitrate dissolved in 2-propanol, and subsequently dried by hanging vertically in ambient air. The samples are placed in the quartz tube furnace. After purging the furnace with helium (Airgas), the furnace is heated up to 750 °C, at which temperature the Fe layer is pre-conditioned with hydrogen (H_2, Airgas, UHP grade), and then CNTs are grown with ethylene (C2H4) gas. The grown CNTs show radial alignment, uniform distribution, and high density as revealed in scanning electron microscopy (SEM, JOEL 5910 and FEI/Philips XL30) images, which fulfill the desired micro-structure in Fig. 1. The grown CNTs are measured to have 7-8 walls with an inner diameter of ~9.7 nm (standard deviation of 1.8 nm) and an outer diameter of ~17.1 nm (standard deviation of 3.3 nm) averaged over 30 CNTs when inspected under transmission electron microscope (TEM, JOEL 2011). With these diameters, the CNT density grown on the fibers is calculated as ~0.8 mg/mm³, based on 1.4 mg/mm³ density of 1-nm-diameter SWCNT [11], and an assumed wall thickness of 0.34 nm (interlayer spacing of graphite). Each cloth sample is weighed with a microbalance (Sartorius) after catalyst coating and after CNT growth. The catalyst mass change during the CNT growth is not considered. Since decomposition of organic elements in the catalyst layer is

likely, the estimated CNT mass yield is a conservative value.

After CNT growth, Al_2O_3 cloth layers with CNTs are embedded in epoxy to fabricate the nano-engineered composites based on the traditional lay-up method of dry fibers. First, epoxy (West Systems) is poured into cork dams (I G. Marston) on a guaranteed non-porous teflon (GNPT) covered vacuum table. Cloth plies with CNTs are placed in the pool of epoxy. Enough time is allowed for the epoxy to wet the fibers and CNTs, facilitated by the strong capillary forces generated by the aligned CNTs small spacing [12, 13]. When wetting on the top cloth surface is observed, extra epoxy is poured, and a new cloth was applied on the top. This process is repeated until all the plies are assembled. After introducing cure materials, samples are sealed in with a vacuum bag and cured under vacuum (~88 kPa below the atmospheric pressure) at ~60 °C to draw out voids and enhance the curing process. Cured composites are trimmed using a carbide-grit blade on a water-assisted cutting wheel.

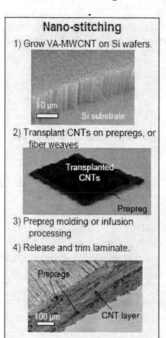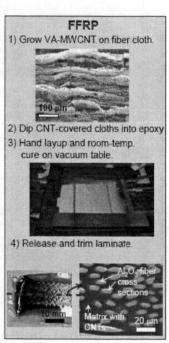

Figure 2: Fabrication and characterization of two nano-engineering architectures

2.4 Nano-stitched Composite Fabrication

The following section describes two different methods of fabricating the nano-stitched composite architecture based on standard prepreg and infusion manufacturing techniques for advanced composites.

2.4.1. Prepreg Method

Chemical Vapor Deposition (CVD) on the nano-stitched composites employs the same catalyst (Fe), reduction gas (H_2), and carbon source gas (C_2H_4) as FFRP fabrication [8, 9]. The main differences, however, are the substrate, catalyst deposition method, and the growth method. Silicon substrates are e-beam deposited first with 10 nm Al_2O_3 layer as a diffusion barrier, and then with 1 nm Fe catalyst layer on top. These wafers are cut into pieces and are placed inside the quartz tube furnace mentioned above. In this growth method, the H_2 gas is introduced from the start of temperature ramp. The average CNTs diameter produced is ~8 nm, with an aligned-CNT volume fraction of 1%.

The CNT forests are transplanted onto the surface of a prepreg ply with the following method. First, unidirectional prepreg (Cytec IM7/977-3 or Hexcel AS4/8552) is cut and placed on a metal cylinder with the tacky prepreg surface exposed. Then, with pressure, the cylinder is rolled over a CNT-forest-covered Si wafer. As the prepreg contacts the forest, CNTs adhere to the ply, and are removed from the Si wafers. SEM has confirmed that the CNT alignment is maintained during the transplantation process. All of the layers above and below the CNT forest are laid up and cured following the recommended standard method of the material supplier. In our case, the curing cycle is the following: 1 atm of vacuum, 7 atm of total pressure, heating ramp of 2.8 °C/min until 180 °C, hold for 2.5 h, cooled to 60 °C at 2.8 °C/min, and then left until room temperature.

2.4.2. Infusion Method

The CNTs for this application are grown following the same method as the prepreg method described above. The length of the CNT forests is ~100 μm, similar to the ply spacing between the woven carbon cloths. After growth, CNT forests are extracted from the substrate using a razor blade. The laminates are prepared with the delaminated CNT forests sandwiched by carbon fiber (CF) woven cloth (Tenax-J G40-800 24K EP03) layers. A stainless steel plate is used as a mold. Teflon mesh flow media sheets are placed on the top and the bottom of the assembled laminates. The entire assembly is enclosed in a vacuum bag and connected to a vacuum pump on one side, with a opening on the opposite side for resin (Hexcel RTM6) introduction. After moving the setup in a heating chamber, RTM6 is heated up to 90 °C to achieve a viscosity appropriate for infusion processing. The curing cycle is the following: hold at 160 °C for 1h and hold at 180 °C for 2h.

3 RESULTS

In the following section, we review the composite quality and laminate-level property enhancement as observed by standard testing.

3.1 FFRP Composites

The fabricated FFRP composites are inspected under an optical microscope (Axiotech) for micron-scale void fraction and scanning electron microscope (SEM, JOEL 5910) for epoxy wetting of CNTs. Wetting of fibers and CNTs is overall successful, with void fractions estimated at <1% for both baseline (no CNTs) and FFRP laminates. TEM imaging of the composite cross sections is in progress to inspect micro-structure morphology of CNTs (alignment, bonding to the fibers) inside the hybrid composite.

Ignoring void fractions, Al_2O_3 weight fraction is evaluated from the conservative measured CNT weight, the measured composite weight, and Al_2O_3 cloth area density. With a composite thickness set by the cork dam height, the Al_2O_3 weight fraction is estimated as ~60%, while the CNT weight fraction varied from ~0.05 to 5 % depending on the growth time. Epoxy weight fraction also varied accordingly.

These hybrid woven composites reinforced by aligned CNTs have been measured for property enhancement. The shear strength was measured by three point bending with short span following the ASTM standard D 2344-00, and ~31% improvement in interlaminar shear strength was observed with ~1 wt% of CNT fraction [7]. The samples were also tested for electrical conductivities with four probes, and the side planes of the samples were coated with silver paint (Structures Probe Inc.). DC electrical conductivity showed $x10^6$-10^8 increase with ~0.5-3 wt% CNT fraction [7]. AC impedance was also acquired by applying an AC test signal with 10 mV over the frequency range of 10 Hz–40 MHz [14]. While the baseline composites without CNTs, consisting of dielectric epoxy/alumina sandwiched by conductive silver plates, exhibited resistive and capacitive behavior, the nano-engineered composites with ~2-7 wt% CNT fraction showed behavior dominated by a small resistance. Thermal property measurements are in progress with the laser-flash method.

3.2 Nano-stitched Composites

3.2.1. Prepreg

Fabricated prepreg samples reinforced with CNTs were tested for Mode I and II interlaminar fracture toughness with an Instron 1332 testing machine, following ASTM standard 471 and the 4ENF method, respectively [7]. Two kinds of specimens were produced of standard size following the manufacturing procedure described above: 24 layer carbon fiber (CF) epoxy prepreg, and 24 layer CF/epoxy/CNT prepreg, with a layer of CNTs at the mid-plane.

The fracture surface was inspected by SEM (JOEL). CNTs have been observed on both sides of the crack, indicating a bridging-type toughening of the interface due to the aligned-CNT nanostitches in the z-direction. Surface or functionalization treatments could improve toughness of the CNTs due to bridging/pullout.

3.2.2. Infusion

Samples were fabricated by sandwiching a 1cm x 3cm forest of CNTs between 5 layers of carbon fiber cloth on top and bottom, producing a 10 CF layers plus 1 CNT/epoxy layer composite. This infusion process was a demonstration of the technique only to this point in time. Although the CNTs are oriented transversely to the resin flow, they remain aligned as evidenced in SEM imaging. No significant differences are observed in the position of the CNT forests. The low viscosity of the resin and the capillary effects of the CNTs resulted in wetting of the forests. No air voids were been observed in the nanostitched center CNT/epoxy layer.

4 CONCLUSIONS

Two nano-engineered composite architectures reinforced with carbon nanotubes have been designed to enhance the multi-functional properties of aerospace structural composites. These two composites were successfully fabricated to achieve the designed nano/micro-structure in laminates suitable for standard mechanical and other testing. Electrical and fracture properties have been tested to demonstrate improved performance with the inclusion of vertically aligned carbon nanotubes. Future work on these architectures will include further optimization of the fabrication processes and a more complete (and expanded, e.g., in-plane laminate strength) characterization of the enhanced multi-functional properties.

REFERENCES

1. Dresselhaus, M.S., G. Dresselhaus, and P.C. Eklund, *Science of fullerenes and carbon nanotubes.* 1996, San Diego: Academic Press. xviii, 965 p.
2. Iijima, S., *Helical Microtubules of Graphitic Carbon.* Nature, 1991. **354**(6348): p. 56-58.
3. Wong, S.S., et al., *Covalently functionalized nanotubes as nanometre-sized probes in chemistry and biology.* Nature, 1998. **394**(6688): p. 52-55.
4. Ebbesen, T.W., et al., *Electrical conductivity of individual carbon nanotubes.* Nature, 1996. **382**(6586): p. 54-56.
5. Rice, P., et al., *Broadband electrical characterization of multiwalled carbon nanotubes and contacts.* Nano Letters, 2007. **7**(4): p. 1086-1090.
6. Garcia, E.J., Wardle, B. L., Hart, A. J., and Yamamoto, N., *Fabrication and Multifunctional Properties of a Hybrid Laminate with Aligned Carbon Nanotubes Grown In Situ.* Composites Science & Technology, 2008.
7. Garcia, E.J., Wardle, B.L., Hart, A.J., *Stitching Prepreg Composite Interfaces with Aligned Carbon Nanotubes.* Composites Part A, 2008.
8. Hart, A.J. and Massachusetts Institute of Technology. Dept. of Mechanical Engineering., *Chemical, mechanical, and thermal control of substrate-bound carbon nanotube growth.* 2006. p. 357 p.
9. Hart, A.J. and A.H. Slocum, *Rapid growth and flow-mediated nucleation of millimeter-scale aligned carbon nanotube structures from a thin-film catalyst.* Journal of Physical Chemistry B, 2006. **110**(16): p. 8250-8257.
10. Hogg, P.J., *Perspectives - Composites in armor.* Science, 2006. **314**(5802): p. 1100-1101.
11. Gao, G.H., T. Cagin, and W.A. Goddard, *Energetics, structure, mechanical and vibrational properties of single-walled carbon nanotubes.* Nanotechnology, 1998. **9**(3): p. 184-191.
12. Garcia, E.J., et al., *Fabrication of composite microstructures by capillarity-driven wetting of aligned carbon nanotubes with polymers.* Nanotechnology, 2007. **18**(16): p. -.
13. Garcia, E.J., et al., *Fabrication and nanocompression testing of aligned carbon-nanotube-polymer nanocomposites.* Advanced Materials, 2007. **19**(16): p. 2151-+.
14. Garcia, E.J., Wardle, B.L., Guzman de Villoria, R., Wicks, S., Ishiguro, K., Yamamoto, N., and A.J. Hart, *Aligned Carbon Nanotube Reinforcement of Advanced Composite Ply Interfaces*, in *AIAA-2008-1768, 49th AIAA Structures, Dynamics, and Materials Conference.* 2008: Schaumburg, IL.

High Strength Metal-Carbon Nanotube Composites

A. Goyal[*], D.A. Wiegand[**], F. J. Owens[**] and Z. Iqbal[*]

[*]New Jersey Institute of Technology, University Heights,
Newark, New Jersey 07102, USA, ag28@njit.edu, iqbal@njit.edu
[**]US Army Research, Development and Engineering Center, Picatinny,
New Jersey 07806, USA, donald.wiegand@us.army.mil, frank.owens1@us.army.mil

ABSTRACT

Uniform dispersion of pre-synthesized nanotubes in metal matrices is difficult to achieve, and there is substantial damage to the nanotubes during subsequent composite fabrication. The demonstration of an *in-situ* process involving chemical vapor deposition of single-wall and multi-wall carbon nanotubes into iron matrices without the concomitant formation of iron carbides is reported here. It was found that the yield strength of carbide-free iron-carbon nanotube composites increased up to 45% with about 1 wt % of infiltrated single wall carbon nanotubes (SWNTs), and 36% with ~1 wt % multiwall carbon nanotubes (MWNTs), relative to that of similarly treated pure iron matrices of the same piece density. Vickers hardness coefficients were also substantially enhanced - 74% and 96%, respectively, for composites with SWNTs and MWNTs relative to the metal matrices without nanotubes.

Keywords: metal composites, nanocomposites, chemical vapor deposition, carbon nanotubes, yield strength

1 INTRODUCTION

Carbon nanotubes have been extensively investigated for developing polymer matrix nanocomposites and a number of such nanocomposites are already being used in various applications [1]. Metallic composites, particularly those of iron and aluminum, containing carbon nanotubes would offer distinct advantages over polymeric composites, but the development of metal matrix composites remains in its infancy [2] in spite of its great potential, primarily because of high fabrication costs and difficulty in scaling up. In one study [3], 5% to 10 % by weight of pre-synthesized arc-grown multiwall carbon nanotubes were dispersed in aluminum matrices resulting in an increase of hardness, but no tensile strength measurements were performed. In another study [4], mechanical dampening characteristics of MWNT/magnesium composites formed by high pressure infiltration were studied and compared with that for the pristine metal matrix, but no significant improvement was observed. A detailed study was performed by Flahaut et al [5] using pre-synthesized nanotubes, which were then hot pressed with iron and Al_2O_3. The nanotubes suffered significant damage due to

the high temperatures of around 1500-1600 °C used in this process and no substantial increase in mechanical properties was reported. Goh et al. [6] reinforced magnesium matrices with carbon nanotubes using powder metallurgy techniques and reported an increase of only 0.2% in yield strength and 0.18 % in ductility. Some of the issues which preclude mechanical property enhancement in metal-nanotube composites arise due to the high processing temperatures used in infiltrating pre-synthesized nanotubes into the metal matrix. Moreover, the methods used do not lead to good dispersion of the nanotubes in the metal matrix and efficient nanotube-metal contact and pinning. We have developed a scaleable chemical vapor infiltration technique [7, 8] which forms metal matrix nanocomposites with enhanced yield strength. Specifically, catalytic chemical vapor deposition was used to infiltrate metal matrices with single and multiwall carbon nanotubes using carbon monoxide and acetylene as carbon sources. Vickers hardness numbers and stress-strain curves were measured to characterize the mechanical properties of the composites formed.

2 EXPERIMENTAL

The catalyst and promoter precursors, iron acetate, cobalt acetate and molybdenum acetate, each 0.01 weight % of total solution, respectively, were dissolved in ethanol. Typically 3-5 gms of micron-sized iron powder was soaked in this solution, dried, and pressed into thin cylindrical pellets under an applied load of 5000 Kg. The pellets were 13 mm in diameter and between 4 and 5 mm in thickness with piece densities of 5.7 to 6.10 gm/cc. The average porosities of all the samples used in the mechanical measurements were the same. The catalyst loading in our samples was very low. The amount of catalyst metal in the iron matrix was estimated to be about 0.003 weight % of the total weight of the pellet.

The pellets were placed in a quartz boat in a horizontal quartz tube high temperature furnace. The quartz tube was pumped down to about 10^{-3} torr. In case of SWNTs, the reactor was back-filled with flowing pure hydrogen for 30 minutes to an hour to reduce the oxides to metals at 500ºC. In second step the carbon source, carbon monoxide (CO), was introduced at 700ºC into the reactor at a flow rate of 100 standard cubic centimeters per minute (sccm) for 30 minutes-1 hour to deposit SWNTs within the matrix. In case of MWNTs a single step protocol involving heating to 800°C under flowing argon followed by switching the gas flow to a mixed carbon source of acetylene, CO and argon with flow rates of 6, 100 and 300 sccm, respectively, at atmospheric pressure, was used. Reference pellets were prepared with the same weight of iron powder and applied load, followed by same thermal cycles used to grow SWNTs and MWNTs. After completion of the deposition, the system was allowed to cool to room temperature under flowing argon.

The characterization of the composites was carried out by micro-Raman spectroscopy, x-ray diffraction (XRD), and field-emission scanning electron microscopy (FE-SEM). Vickers hardness measurements were conducted on using a LECO micro-hardness tester (LM 700, LECO Corp.). A load of 10 Kgf (kilogram force) at ambient temperature with a dwell time of 5 seconds was selected, average values are reported. Stress-strain data were obtained in compression with a MTS servo hydraulic system operated at a constant displacement rate so as to give a strain rate of about 0.00004/sec.

3 RESULTS AND DISCUSSIONS

The representative micro-Raman spectra obtained using 632.8 nm laser excitation from the surface of the nanocomposites are shown in Figs, 1a-c. In Fig 1a clear evidence for the formation of SWNTs is provided by the appearance of the characteristic SWNT lines associated with the carbon-carbon bond tangential modes near frequencies of 1591 cm^{-1} and 1552 cm^{-1} (the latter appearing as a well-defined shoulder in the spectrum), and the lines at 190, 247, 259 and 279 cm^{-1} due to the radial breathing modes (RBMs) of individual tubes of different diameters. The line observed at 293 cm^{-1} is likely to be associated with Fe_2O_3 present as an impurity in the iron matrix. The diameters (d, nm) of the individual SWNTs can be determined from the RBM frequency ω. For bundled SWNTs: d = $(238/\omega)^{1.075}$ [9]. The broad line at 1327 cm^{-1} assigned to defects and amorphous carbon was found to be relatively weak. This indicated that rather defect-free SWNTs are formed with relatively little amorphous carbon present. In Figure 1 b-c the Raman spectrum does not show the relatively sharp radial breathing mode (RBM) lines that are indicative of SWNTs . The lines at 1323 cm^{-1} and 1581 cm^{-1} observed for samples prepared with acetylene mixed with carbon monoxide can be assigned to the disordered (D) and graphitic (G) modes of MWNTs [10]. Concomitant growth of iron carbide also occurs when acetylene alone is used as the carbon feed – this aspect will be covered in more detail later in the discussion of the XRD results.

In Figures 2 a-b the diameters of 10 to 20 nm estimated from the FE-SEM images are an order of magnitude larger than the individual SWNT diameters obtained from the RBM Raman frequencies, indicating that the SWNTs formed are bundled into "ropes". The individual iron particles are likely to be anchored in place by the nanotube bridges. It is estimated from the measured increase in

weight after nanotube deposition that typically 1 weight % or 2.2 volume % of SWNTs are incorporated into the

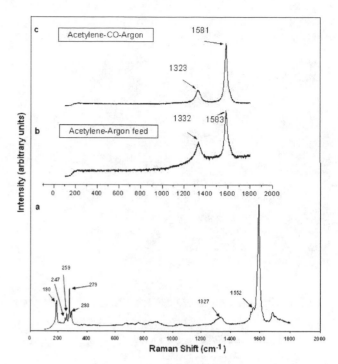

Figure 1: Raman spectra excited at 632.8 nm for: a) iron-SWNT, b) Iron-MWNT with acetylene feed, c) iron-MWNT with CO-acetylene feed.

starting iron matrix. FE-SEM images are consistent with largely MWNT formation, and the images shown in Fig. 2 c and d indicate somewhat denser growth of MWNTs compared to that of SWNTs. Fig. 2c depicts a low magnification image showing sizable MWNT penetration to a depth of 150 to 160 µm. A lower concentration of nanotubes is evident below 160 µm and through the approximately 0.5 mm thickness of the piece. A high magnification image taken from a region about 160 µm inside the top surface of the composite showing dense growth of MWNTs is displayed in Fig. 2d. Measured weight changes indicate a MWNT loading of up to 1 weight % ~ 4.48 volume % in the optimized iron-MWNT composites, which is similar to that obtained for the iron-SWNT composites.

XRD measurements (Fig. 3a) showed the presence of pure iron as indicated by reflections at 2θ values of 45, 65, 83, and 99 degrees. XRD reflections from SWNTs are not detected due to the relatively low weight % loading of SWNTs in the nanocomposites. Reflections associated with the iron-rich cementite Fe_3C phase were clearly absent. An XRD line at a 2θ value of 26 degrees was observed near the expected (001) reflection of graphite, but its intensity is too high for it to be attributed to a carbon phase. We tentatively assign this reflection to an iron sub-oxide formed in the iron matrix under our fabrication conditions. XRD patterns for the composites obtained using acetylene and argon, and a mixture of acetylene, CO and argon, respectively, are shown in Fig. 3b and c. The XRD pattern for the composite prepared using acetylene and argon shows sharp reflections due to iron carbide, Fe_3C, while a mixture of CO, acetylene and argon is carbide free.

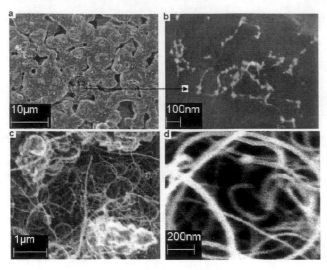

Figure 2: SEM images a) low magnification iron-SWNT, b) higher resolution SWNTs, c-d) MWNTs dispersed in iron matrix.

In order to understand why iron carbides are not formed when acetylene is mixed with CO, we propose the following sequence of reactions during *in-situ* growth with acetylene. The Fe_3C impurity phase is formed by reaction 1 below:

$$3Fe_2O_3 + 8H_2 + C_2H_2 \rightarrow 2Fe_3C + 9H_2O \qquad (1)$$

It involves the reduction of Fe_2O_3 which is typically present in the iron matrix as an impurity phase, by hydrogen (formed by the initial dissociation of acetylene) followed by the adsorption of carbon from acetylene decomposition. The dissociation of acetylene is further enhanced by the presence of iron as catalyst. Iron is supersaturated with carbon and leads to the formation of iron carbide. Introducing CO initiates the occurrence of concurrent reactions 2 and 3 below. The presence of CO results in the formation of carbon nanotubes and CO_2 following the disproportionation reaction 2 in the presence of catalysts and reaction 3. In addition to that, reaction 3 scavenges hydrogen to form carbon nanotubes and prevents the reduction of Fe_2O_3 to Fe_3C via reaction 1.

$$2CO \rightarrow C + CO_2 \qquad (2)$$
$$CO + H_2 \rightarrow C + H_2O \qquad (3)$$

The above reaction sequence is consistent with the XRD data, which shows no evidence for the formation of the Fe_3C phase when CO is introduced into the carbon precursor feed.

Compressive stress-strain curves from two representative samples, with and without nanotubes show significant differences. The flow stress, the stress for the

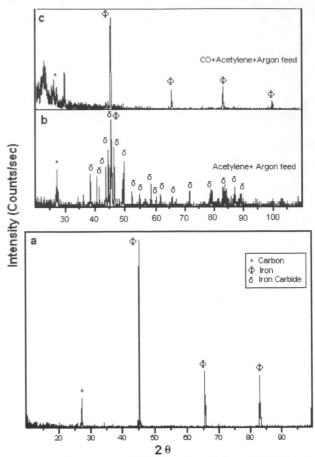

Figure 3: XRD a) iron-SWNT pellet, b) iron-MWNT pellet with acetylene feed, c) iron-MWNT with CO-acetylene.

significant plastic flow, is taken at the intersection of a straight line passing through the initial linear part of the stress-strain curve at low strains and a straight line passing through the linear work hardening part of the curve at larger strains. For the data in Fig. 4a, the flow stress is about 45% higher for the sample containing the SWNTs compared to the value for the reference iron sample. This flow stress is numerically equivalent to a yield stress obtained by a 0.4% strain offset technique. In addition, the work hardening coefficient, the slope in the latter linear part of the stress-strain curve at larger strains, is greater by a factor of about 3.4 (a 240% increase) for the sample containing the SWNTs relative to the reference sample. Thus, the mechanical strength of the sample containing SWNTs is significantly increased and much greater work is required to deform this sample plastically in the work hardening range. Since mechanical properties depend on porosity, it is important to emphasize that the porosity of the reference sample and the sample with SWNTs is the same and therefore differences in porosity cannot account for the enhanced mechanical strength. A sample was grounded by 170 µm on both sides and compression tests were repeated.

A stress-strain curve taken after grinding again indicates that the flow stress increased by an amount consistent with the work hardening during the initial compressions. Thus, the removal of the thin surface layers containing the highest densities of SWNTs had a minimal effect on the flow stress. These results indicate that the higher flow stresses observed in the samples containing the SWNTs are due to bulk effects and further that the SWNTs are distributed throughout the samples of thicknesses of the order of 0.6 cm. In case of MWNTs, the increase in upper and lower yield strength is 36 and 43% respectively as shown in Fig. 4 b. The lower yield point for the reference sample is at 179 MPa and for the iron-MWNT sample it is at 253.3 MPa. The upper yield point for the reference sample is at 200 MPa and at 276 MPa for the iron-MWNT composite.

The observed increase in strength of the nanotube-iron composites can be attributed to the mechanical support provided within the cavities by the nanotubes. The SWNTs are largely concentrated in cavities in the matrix as indicated by the FE-SEM images shown in Fig. 2. It is well-known that iron is an excellent catalyst towards nanotube growth [11] and can partially dissolve and bond to carbon to provide supporting bridges at the cavities. Bonding might also occur at nano-sized catalysts embedded in the larger iron particles of the matrix. Because of this support additional dislocation pinning may not be necessary. High porosity decreases mechanical strength because the average stress inside the material is greater than the average applied stress [12-13]. Providing support in the pores will lower the average stress in the material, which determines dislocation motion and yield. Thus yield will occur at higher values of the applied stress.

Overall this support at the cavities will offset in part the effect of the cavities in weakening the iron matrix, resulting in higher mechanical strength. The very large increase in the work hardening coefficient or slope suggests that the mechanism of work hardening may be different in the samples containing SWNTs than in the reference samples. For example, if the work hardening of the reference sample of Fig. 4 is due to long range dislocation-dislocation interactions, then the much higher work hardening slope of the sample containing SWNTs may be due to dislocation pile up at barriers introduced by the treatment which this sample received. When these barriers are SWNTs, it is expected that they are located not only in the cavities, but also distributed throughout the iron matrix.

Vickers hardness indices correlate with the tensile strength and fatigue resistance [14]. Typical average values of the Vickers hardness (HV) indices HV/10 (with 10 Kgf) of 95.2 for the reference sample and 135.7 for iron-SWNT sample showed an increase in the hardness index by 74%. For the carbide free iron-MWNT composites Vickers numbers showed enhancement in average hardness by 97.5%, which is substantially higher than that of an iron-SWNT composite with a similar concentration of nanotubes.

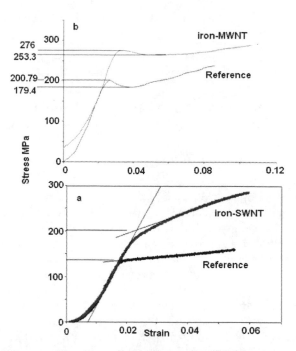

Figure 4: Compressive stress-strain curves for a) iron-SWNT and b) iron-MWNT composite.

REFERENCES

[1] A. Morgan, Material Matters 2, 20 (2007).

[2] P.K. Rohatgi and B. Schultz, Material Matters 2, 16 (2007).

[3] R. Zhong, H. Cong and P, Hou, Carbon, 41, 848, 2002

[4] J. Yang and R. Schaller, Mater. Sci. Eng. A. 370, 512, 2004

[5] E. Flahaut, A. Peigney, Ch. Laurent, Ch. Marlière, F. Chastel, A. Rousset, Acta Mater. 48, 3803, 2000.

[6] C.S. Goh, J. Wei, L.C. Lee, M. Gupta, Nanotechnology. 17, 7, 2006.

[7] A. Goyal, D. A. Wiegand, F. J. Owens and Z. Iqbal, J. of Mat. Res. 21, 522, 2006.

[8] A. Goyal, D. A. Wiegand, F. J. Owens and Z. Iqbal, Chem. Phys. Lett., 442, 365, 2007.

[9] L. Alvarez, A. Righi, S. Rols, A. Anglaret, J.L. Suavajol, E. Muñoz., W.K. Maser, A.M. Benito, M.T. Martínez, and G.F. de la Faunte, Phys. Rev. B. 63, 153401, 2001.

[10] C. Thomsen, S. Reich, H. Jantoljak, I. Loa, K. Syassen, M. Burghard, G.S. Duesberg, S.Roth, Appl. Phys. A. 69, 309, 1999.

[11] A.M. Cassell, N.R. Franklin, T.W. Tombler, E.M. Chan, J. Han and H. Dai, J. Am. Chem. Soc. 121, 7975, 1999.

[12] J.C. Wang, J. Mat. Sci. 19, 801, 1984.

[13] F.P. Knudsen, J. Am. Chem. Soc. 42, 376 (1959).

[14] W.D. Callister Jr.: Materials Science and Engineering: An Introduction, 6th ed. (John Wiley and Sons: New York, 2003), pp. 111-134.

Development of Stab Resistant Body Armor Using Silated SiO2 Nanoparticles Dispersed into Glutaraldehyde

H. Mahfuz, V. Lambert, P. Bordner and V. Rangari

Nanocomposites Laboratory, Department of Ocean Engineering,
Florida Atlantic University, Boca Raton, FL 33431, vlamber4@fau.edu

ABSTRACT

The development of flexible body armor has first been based on Shear Thickening Fluid (STF) which is a mixture of polyethylene glycol (PEG), water and silica at a specific weight ratio. The making process has been improved in incorporating through sonicating the silica particles to the mixture. Although the performances obtained were improved the stab resistance was not completely optimized to stop a penetration more than $45J/g/cm^2$. The intention in this investigation was to study the actual components of the armor composite and to improve its stab performances. The outcome is a completely new approach regarding the fabrication route which differs from the previous STF making process. PEG has been removed and an extra component has been incorporated into the mixture. It is believed that this is the addition of this new chemical: Glutaraldehyde which provides the phenomenal stab performances.

Keywords: shear thickening fluid, glutaraldehyde, stab resistance

1 INTRODUCTION

The idea of making flexible body armor has been around for some time [2/3]. A crucial need has emerged from the battlefield to reduce the weight of the body armor in keeping its resistance. A lot of researches [2/3/11/12] have demonstrated the high potential of the STF regarding an impact but experiences conducted with glutaraldehyde has shown better results. This paper will then concentrate on the properties of the new Kevlar composite based on the addition of Glutaraldehyde.

Utilizing the different improvements that have been made for the STF making process, the new composite follows nearly the same fabrication route. It stays highly flexible and as presented in this paper the performances are extremely interesting.

2 EXPERIMENTATION

2.1 Materials and Synthesis

Previous fabrication procedure included 30nm size silica particles dispersed, by sonication, directly into a mixture of Ethanol, PEG and Silane at a ratio of 55:45 by weight of Silica to PEG. The new fabrication route differs from this making process by substituting Glutaraldehyde for PEG.

The Glutaraldehyde is incorporated according to the amount of silane agent present in the solution. Different types of silane can be used but the key component within the silane, which determines the amount of glutaraldehyde to introduce, is the amino-groups. The amino-groups are functional groups that contain a basic nitrogen atom. In our case, the silane possesses di-amino-groups and it is expected that the Glutaraldehyde creates "bridges" between the amino-groups.

Glutaraldehyde is not new; it is usually employed in medical and dental environment to disinfect equipments.
Also used in biochemistry applications [13] as a fixative, the glutaraldehyde kills cells by cross-linking their proteins. This cross-linker is an aldehyde such as formaldehyde a well known cross-linker and each of these cross-linkers induces subsequent covalent bonding. The covalent "bridge" created is stable mechanically and thermally making it hard to break.

Once the mixture is completed with all the different components, the solution is homogenized and mixed through sonic cavitation. After sonication for about three hours, the mixture was used to soak 12 layers of Kevlar fabric cut in dimensions of 12 in x 12 in. To impregnate the fabric, the layers were placed in a Ziploc along with the sonicated mixture. After a curing process of about 24h, the Kevlar composite is then dried out in a pre-heated furnace at 110°C for approximately thirty minutes. After each set of fabric is baked all the ethanol has evaporated. The 12 layers of Kevlar impregnated with the silated-silica-gluta mixture resulted in an areal density of $0.220g/cm^2$. This fabrication procedure completely differs from the regular STF making process and mechanical tests show the performances of this new Kevlar composite.

2.2 Testing

Different sets of this Kevlar-composite have been fabricated using the above procedure, and using a drop tower constructed based on the Stab Resistance of Personal Body Armor, NIJ Standard-0115.0 (NIJ115) results from several experiments have been recorded. According to the NIJ115 standard a backing material, a nylon drop mass with a weight of about 2000 grams and a NIJ115 engineered spike were used to impact the target (fabric) at drop heights ranging from 0.05m to 1.0m. The velocities just prior to impact were recorded through a laser speed trap. Using the measured impact velocity, the total mass of the spike and drop mass; the actual impact energy was calculated. An interesting way to compare the different fabric is the depth of penetration recorded and measured by damaged witness papers placed underneath each layers of backing material. Doing this test gives excellent information to determine the stab resistance performances of each Kevlar composite.

3 RESULTS AND DISCUSSION

3.1 Stab Test

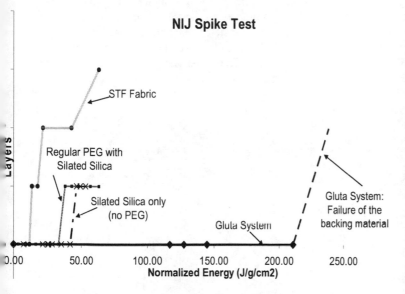

Figure 1: *Results for different Kevlar composite normalized with aerial density after spike impacts*

The impact energies were normalized by dividing them by the areal density of the respective fabric composite and plotted against the penetration depth. The results, shown in Figure 1, are significant. Indeed the gluta-system clearly shows high improvements compare to the STF fabric that has been already improved by removing the PEG. The '0-level penetration' increases by 4.5 times the No PEG and by 10 times than that of STF system.

An important point has to be made regarding the gluta-system result. Indeed the fabric showed so good resistance to the impact that at a certain energy level the failure is not occurring from the fabric but from the backing material. At high energies the Kevlar composite is not penetrated but is pushed through the backing material by the spike. To a certain extent the fabric is going to be penetrated. The penetration is however very small and no bigger than a dot which is another crucial difference with the previous composites fabricated.

The gluta-system is consequently a major improvement in the researches leading to the manufacture of a flexible body armor. The increase of the stab resistance to about 4.5 times more than the Silated-Silica without PEG can only be

explained by the addition of Glutaraldehyde into the solution.

3.2 Microstructure

In an attempt to explain the improvements in the stab resistance of the Gluta-System-Kevlar composites over the different Kevlar composites, SEM and FTIR studies have been performed. A thin coating of Silated-Silica with Glutaraldehyde formed over the surface of the Kevlar fabric is shown in Figure 2. This coating is encompassing every single yarn of the fabric. Previous studies have shown that this coating offers the resistance during the spike penetration. The coating is consisted of agglomerated silica particles embedded in the body of the matrix as seen in Figure. 3.

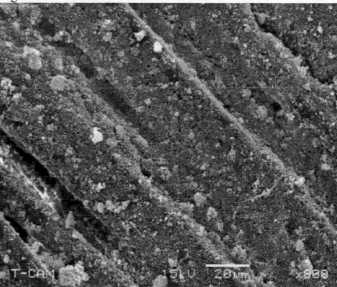

Figure 2: *A thin coating of the Silated silica particles with glutaraldehyde on the surface of the Kevlar fabric*

Figure 3: *Agglomerated Silated SiO_2 particles and Glutaraldehyde mixture*

The mechanism of this particular resistance property to the spike penetration can be explained by the analysis of the operation mode of the glutaraldehyde. Indeed extensive researches have been conducted in the bio-chemistry field regarding the fixative property of the chemical. Usually used to fix cells for electron microscopy, the glutaraldehyde attacks cells by crosslinking their proteins. It specifically works as an amine-reactive homobifunctional crosslinker. Our interest in this chemical stays in its property to create strong covalent bonds that "bridges" one silated-silica particles to another. This specific reaction is made possible by using monomeric glutaraldehyde which polymerizes by aldol condensation reaction. This reaction usually occurs at alkaline pH values.

In a simple way the aldol condensation is a basic organic reaction in which an enolate ion reacts with a carbonyl compound to form strong covalent bonds between two carbons. This reaction requires an alcoholic environment. In our case the aldehyde group of glutaraldehyde is linking with the silated silica by creating bonds with amino-groups as well as other glutaraldehyde groups as shown on Figure.4. As a result it creates a network between the different compounds present in the solution.

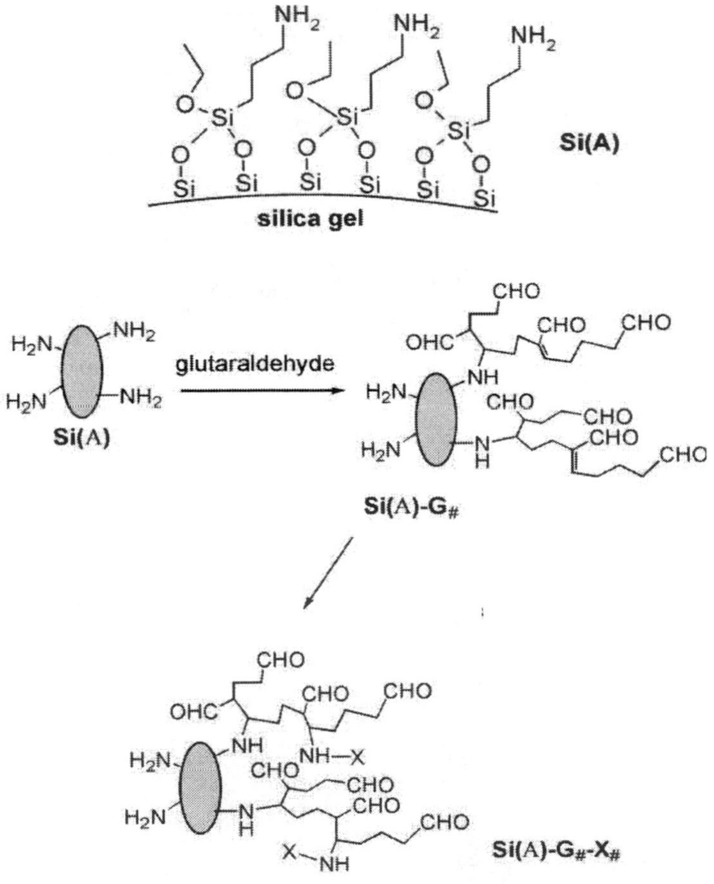

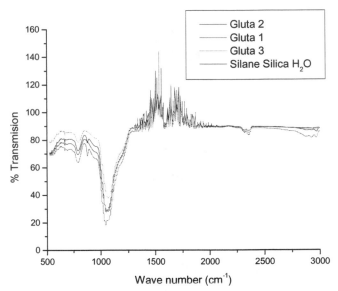

Figure:4. Glutaraldehyde cross-linking reaction

To further understand exactly the microstructure of the Glutaraldehyde Kevlar Composite FTIR (Fourrier Transform InfraRed spectroscopy) has been performed on the glutaraldehyde system as well as on the silated silica particles. The FTIR approach reveals the type of bonds and their frequencies within the solution by measuring the absorption or the transmission of infrared waves. In our case the spectrum is recorded with the transmission against the wave number. Four sets have been tested such as three different Gluta-system based in different ratio and the silated silica. Results shown in the Figure 5, demonstrate that the gluta-system performs better than the silated silica with 30 nm particles. The chemical bond revealed by the analysis is the siloxane (Si-O-Si) bond at around 1100 cm^{-1}. The peak of the gluta1 is a relative improvement compared to the silated silica particles. The peak observed and the area covered by the gluta1 curve shows quantitatively that the addition of glutaraldehyde has increased the concentration of Si-O-Si bond strength and number.

Figure: 5 FTIR spectrum of different gluta-systems and silated- SiO$_2$

4 SUMMARY

Although this new Kevlar composite stands apart from the usual STF fabrication route the stab resistant performance of the Gluta-system-Kevlar composite shows a remarkable increase compared to the previous Kevlar composite fabricated. Origins of these improvements are believed to be, for one, by the replacement of the PEG to Silane and Glutaraldehyde, and secondly, the size advantage of employing nanometer size particles which increases the number of links between the different components. The addition of glutaraldehyde as a linker between the silated silica particles is consequently believed to be the key component in increasing the stab resistance property.

REFERENCES

1. Y.S. Lee and N. J. Wagner, *Rheol. Acta* **42** (2003) 199.
2. Y.S, Lee, E.D. Wetzel and N.J. Wagner, *Journal of Materials Science* **38** (2003) 2825-2833.
3. E. D. Wetzel, Y. S. Lee, R. G. Egres, K. M. Kirkwood, J. E. Kirkwood, and N. J. Wagner, *Proceedings of NUMIFORM*, 2004.
4. Egres, R. G., Lee, Y. S., Kirkwood, J. E., Kirkwood, K. M., Wetzel, E. D. and Wagner, N. J. *Proceedings of the 14th International Conference on Composite Materials*, July 2003.
5. Nathaniel C., Mahfuz, H., Rangari, V., Ashfaq, A. and Jeelani, S., "Fabrication and Mechanical Characterization of Carbon/Epoxy Nanocomposites," Composite Structures, 67 (2005) 115-124.
6. Mahfuz, H., Adnan, A., Rangari, V.K., Hasan, M.M., Jeelani, S., Wright, W.J. and DeTeresa, S.J., "Enhancement of Strength and Stiffness of Nylon 6 Filaments through Carbon Nanotubes Reinforcements," *Applied Physics Letters*, 88, 1 (2006).
7. Rodgers, R., Mahfuz, H., Rangari, V., Chisholm, N., and Jeelani, S., "Infusion of Nanoparticle into SC-15 Epoxy; an Investigation of Thermal and Mechanical Response," *Macromolecular Materials & Engineering*, 2005, 290, 423-429.
8. R. Birringer and H. Gleiter, in "Encyclopedia of materials science and engineering", R. W. Cahn, (Ed.), Pergamon Press, 1988, Suppl. Vol. 1, p. 339.
9. H. Gleiter, "Nanocrystalline Materials," in Progress in Materials Sci., 33 (1989) 223-315.
10. R. Dagani, "Nanostructural materials promise to advance range of technologies," Chem. Eng. & News, (Nov. 1992) 18-24.
11. H. Mahfuz, F. Clements and J. Stewart "Development of Stab Resistant Body Armor Using Fumed SiO2 Nanoparticles Dispersed into Polyethylene Glycol (PEG) through Sonic Cavitation" in NSTI 2006 paper 1143.
12. Hassan Mahfuz, Floria E. Clements "Enhancing the Stab Resistance of Flexible Body Armor using Functionalized SIO_2 Nanoparticles" in ICCM-16-2007
13. Hyunsook Kim,a Myoung-soon Kim,a Hyesun Paik,a Yeon-Sook Chung,a In Seok Hong and Junghun Suha,"Effective Artificial Proteases Synthesized by Covering Silica Gel with Aldehyde and Various Other Organic Groups" in Bioorganic & Medicinal Chemistry Letters 12 (2002) 3247–3250

Synthesis, Reliability and Applications of Nanocrystalline CVD-grown Diamond and Micro Device Fabrication

M. Wiora[*], K. Brühne[*], A. Caron[*], A. Flöter[**], P.Gluche[**] and H.-J. Fecht[*,***]

[*]Ulm University, Institute of Micro and Nanomaterials, Albert-Einstein-Allee 47, 89081 Ulm, Germany, hans.fecht@uni-ulm.de
[**]Gesellschaft für Diamantprodukte GFD, Lise-Meitner-Str. 13, 89081 Ulm, andre.floeter@gfd-diamond.com
[***]Research Centre Karlsruhe, Insitute of Nanotechnology, 76021 Karlsruhe, Germany

ABSTRACT

Nanocrystalline diamond layers have been deposited using hot-filament chemical vapor deposition. By variation of the deposition conditions we are able to adjust the grain size over a wide range from 300 nm up to values lower than 10 nm resulting in very smooth surfaces. Investigations of the mechanical properties show a decrease of the Young's modulus with decreasing grain size. Finally, potential industrial applications of nanocrystalline layers like diamond coatings of cutting tools and diamond tooth wheels are presented.

Keywords: nanocrystalline diamond, hot filament chemical vapor deposition, micro mechanical parts, cutting tools

1 INTRODUCTION

The correlation between the micro- and nanostructure of a material and its physical and chemical properties is the key issue in materials development. Considerable progress has been achieved recently by the development of new processing technologies (hot filament chemical vapor deposition – HFCVD [1-3]) and new materials in a nanocrystalline state with superior mechanical strength and tribological properties [4-6].

Further processes based on lithographic techniques known from silicon technology allow further microstructuring of CVD-diamond. So far, the microstructuring of highly oriented columnar diamond [7] has been hampered by the fact that the internal microstructure is being reproduced by plasma etching yielding rather rough surfaces. This problem now can be overcome by the production of nanocrystalline diamond, i.e. when the grain size is smaller than the acceptable surface roughness. It can be expected that microparts (microtoothed wheels, atomically sharp cutting edges, functionalized diamond surfaces etc.) can be produced on a reliable basis in the near future.

2 EXPERIMENTAL

Nanocrystalline diamond (NCD) films were grown in a CemeCon CC800/Dia hot-filament CVD reactor [8]. Both silicon (100) wafers with a thickness of 350 μm and a diameter of 75 mm and WC-Co hard metal components were used as substrates. In order to achieve a nanocrystalline structure, a new process has been developed to achieve high nucleation densities and controlled growth conditions using a mixture of H_2, CH_4, N_2, O_2 and Ar as a feedstock gas. Only this allows to obtain extremely smooth surfaces which are not available using other techniques, for example by microwave plasma enhanced CVD. The substrates were heated only by thermal radiation of the filaments to a temperature of ~ 750 °C. The gas pressure was fixed to a constant value during deposition in the range between 3 and 25 bar.

Scanning electron microscopy (SEM) was performed using a Zeiss Leo 1540. The roughness of the diamond surface was measured by an atomic force microscopy (AFM) (Dimension 3100, Digital Instruments). A Laser surface acoustic wave system (LSAW) by Fraunhofer IWS Dresden carried out the determination of the Young's modulus. A Philips X'Pert instrument was used for X-Ray diffraction (XRD) measurement. For the analysis presented here, we used a standard Bragg-Brentano configuration. An estimation of the average grain size in the films was obtained by application of the well-known Scherrer formula [9]

$$L = \frac{K\lambda}{\beta_{FWHM}\cos\theta} \tag{1}$$

where θ and β_{FWHM} are the angle and the full width at half maximum of the considered XRD peak, respectively, λ the wavelength of the used $Cu_{K\alpha}$ X-Rays and K a form factor, which is assumed to be equal to one for our estimation.

3 SYNTHESIS

For comparing the morphology of nanocrystalline diamond it is of great importance that all films are of the same thickness, as both the grain size and the roughness of a thin film usually increases with increasing film thickness. Thus, all films shown in this section are ~ 1.5 µm thick. Figure 1a, 1c and 1e show three representative SEM micrographs of diamond thin films deposited on Si wafers. The morphology changes from polycrystalline structure (Fig. 1a) with a grain size of 300 nm over a nanocrystalline (Fig. 1c) with a grain size of 60 nm to a fine-grained nanocrystalline morphology (Fig. 1c) with a grain size of 8 nm. By slight changes of the deposition conditions all grain sizes within this range can be deposited.

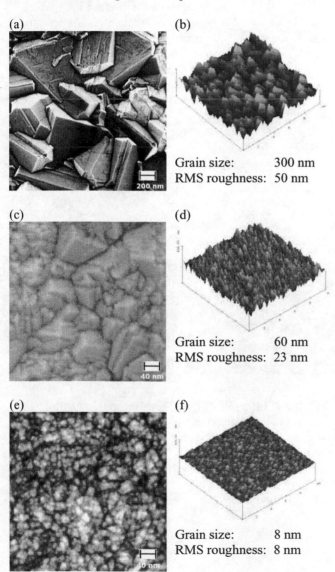

Figure 1: SEM (a, c, e) and AFM (b, d, f) micrographs of diamond samples with grain sizes of 300 nm (a, b), 60 nm (c, d) and 8 nm (e, f).

The AFM measurements (Fig. 1b, 1d, 1e) reveal that the RMS roughness of the diamond surface decreases with decreasing grain size leading to a very smooth surface with an RMS roughness of ~ 8 nm for the fine-grained sample. Samples with such a low roughness will show low friction which is important for applications shown in the next section.

The Young's modulus is an important material property which affects the stress – strain behavior, bending, propagation of cracks and delamination from a substrate. Figure 2 illustrates the dependence of the Young's modulus on the grain size. A decreasing grain size leads to a decrease of the Young's modulus. This is caused by a higher ratio of the grain boundary regions at the fine-grained samples leading to a higher fraction of non-sp^3-bonds which reduces the Young's modulus.

The nanocrystalline samples almost reach the calculated value for randomly oriented polycrystalline diamond of 1143 GPa [10]. However, a Young's modulus of ~ 750 GPa for the fine-grained samples with a grain size lower than 10 nm is still a remarkable result compared with other materials like tungsten carbide which exhibits a Young's modulus of ~ 500 GPa [11].

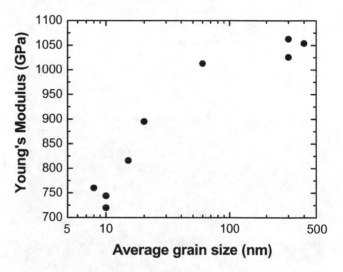

Figure 2: Young's modulus depending on the grain size.

4 APPLICATIONS

Because of the fact that the NCD show a low residual stress, a high hardness and Young's modulus this material is suitable for industrial applications. In this chapter we present the use of diamond for toothed wheels as an example for micro mechanical parts and the diamond coating of cuttings tools.

4.1 Diamond micro mechanical parts

The main advantage can here be taken from the excellent mechanical properties and the low coefficient of friction. Thus, a lubricant-free operation of diamond micro gears at high revolution speed, low wear, low moment of inertia, high efficiency and high reliability is achievable. To evaluate the potential of this application, diamond micro gears have been designed, fabricated and characterized. Figure 3 shows a Scanning Electron Micrograph (SEM) of a diamond micro gear.

Regarding the small size of the diamond toothed wheels, mounting and adjusting them one to another, leads to problems not known in standard gear production. To enable a sufficient vertical overlap for precise movement, their thickness was chosen to be 150 μm. Even with very high aspect ratio etching processes, a lateral under etch of the mask cannot be entirely suppressed. However, the optimization of the plasma processes allowed the realization of micro structured diamond toothed wheels having sidewall angles of 90° ± 2° at a thickness of 150 μm.

From the toothed wheels, a micro gear was assembled using an aluminum base plate with bores of 0.2 mm diameter for the center axis of each wheel. A nickel-wire of 200 μm diameter was used as axis material. The vertical adjustment of the wheels was realized gluing copings on top of the axis. For actuation, a Faulhaber micro motor series 0206H was used, which was specified to max. 100,000 rpm with a maximum torque of 7.5 μNm. The original output-wheel was replaced by a diamond wheel.

Figure 3: Scanning electron micrograph of a micro mechanical diamond toothed wheel.

The diamond gear could be successfully driven in this test setup with a maximum revolution speed of 35,000 rpm. After more than 200 min at 20,000 rpm, the gear showed a parasitic destructive breakdown due to a failure of the motor, which was caused by metal particles from the base plate and the axes. A strong wear can be observed after the operation, which is caused by the still too rough diamond sidewall surfaces. The blocking of the gear caused by

particles, resulted not in a breakdown of the wheels, but of the motor itself. Wear of the diamond wheels could not be observed.

4.2 Diamond cutting tools

After a chemical pre-treating in order to decrease the cobalt concentration of our tungsten carbide (WC) substrates, we deposited a 10 to 15 μm-thick NCD film onto the substrate. The radius of curvature of the cutting edge, however, increases from an initial value of ~ 1 μm to about 15 μm after deposition. Although the hardness of the tool surface has been significantly increased by the diamond film, the tool is no longer usable since the radius of curvature is too large and the tool is not cutting any more. This is the reason why we introduced a plasma sharpening procedure in order to reduce the procedure. Figure 4 shows a schematic view of this process. By etching the diamond layer using Reactive Ion Etching (RIE) from both the top side and the bottom side the radius of curvature of the cutting edge is decreased to values of smaller than 0.2 μm, which is noticeable smaller than that of the original work piece.

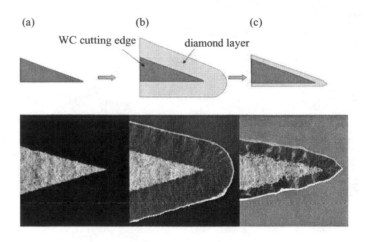

Figure 4: Principle of the plasma sharpening. After depositing the uncoated tungsten carbide (WC) cutting tool (a) with a NCD layer of the thickness ~ 15 μm the radius of curvature is increased. After etching the diamond layer by RIE from the top side and the bottom side (c) the radius of curvature of the cutting edge is decreased to values even smaller than the original work piece.

Several blades were tested in order to compare the life span of the diamond coated, sharpened, cutting edges with those of uncoated carbide or ceramic blades. The experiment was performed using cutting edges having a cutting angle of 15° by cutting a plastic foil with Titanium oxide filling material. The thickness of the foil was approx.

NSTI-Nanotech 2008, www.nsti.org, ISBN 978-1-4200-8503-7 Vol. 1

0.2 mm. The number of tested blades was 150. The carbide and the ceramic blades had comparable life spans of 1.5 and 2.0 days respectively. The sharpened, diamond coated, blades showed life spans of approx. 36 days. The life span increase of the sharpened, diamond coated, blade compared to the carbideblades was therefore 24 times (see also Fig. 5).

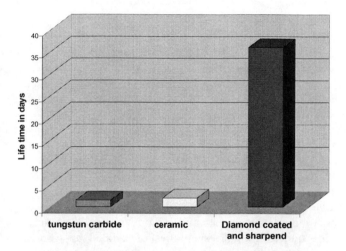

Fig. 5: Comparison of the life span of tungsten carbide, ceramic and sharpened, diamond coated, cutting blades.

5 CONCLUSION

Nanocrystalline diamond layers on silicon and on tungsten carbide were prepared using hot-filament vapor deposition. By slight changes of the source gases we are able to adjust the crystalline size over a wide range from 300 nm to less than 10 nm. The latter sample exhibit a very smooth surface with an RMS roughness of 8 nm.

We presented the use of diamond for toothed wheels as an example for micro mechanical parts and the diamond coating of cuttings tools. For the latter we showed a new technique to reduce the radius of curvature of the cutting edge improving the cutting properties and extending the lifetime of the tool by a factor of 10.

ACKNOWLEDGEMENT

Financial support by the German BMBF under contract no. 13N8998 is gratefully acknowledged.

REFERENCES

[1] S. Matsumoto, Y. Sato, M Tsutsami and N Setaka, J. Mater. Sci. **17** (1982) 3106.

[2] A. Afzal, C. A. Rego, W. Ahmed and R. I. Cherry, Diamond Relat. Mater. **7** (1998) 1033.

[3] S. Pecoraro, J. C. Arnault and J. Werckmann, Diamond Relat. Mater. **13** (2004) 342.

[4] M. Roy, V. C. George, A. K. Dua, P. Raj, S. Schulze, D. A. Tenne, G. Salvan and D. R. T. Zahn, Diamond Relat. Mater. **11** (2002) 1858.

[5] C. Z. Gu and X. Jiang, J. Appl. Phys. **88** (2000) 1788.

[6] S. O. Kucheyev, J. Biener, J. W. Tringe, Y. M. Wang, P. B. Mirkarimi, T. van Buuren, S. L. Baker, A. V. Hamza, K. Brühne and H.-J. Fecht, Appl. Phys. Lett. **86** (2005) 221914.

[7] A. Flöter, H. Güttler, G. Schulz, D. Steinbach, C. Lutz-Elsner, R. Zachai, A. Bergmaier and G. Dollinger, Diamond Relat. Mater. **7** (1998) 283.

[8] S. Schwarz, S.M. Rosiwal, M. Frank, D. Breidt and R.F. Singer, Diamond Relat. Mater. **11** (2002) 589.

[9] P. Scherrer, Gött. Nachr. **2** (1918) 98.

[10] F. Szuecs, M. Werner, R.S. Sussmann, C.J. Pickles, H.J. Fecht, J. Appl. Phys. **86** (1999) 6010.

[11] ASM Engineered Materials Reference Book, ASM International (1989), pp.182.

Residual Stress Measurements of High Spatial Resolution

D. Vogel[*], F. Luczak[**], B. Michel[*]

[*]Fraunhofer IZM, Micro Materials Center, Volmerstrasse 9b, 12489 Berlin, Germany
dietmar.vogel@izm.fraunhofer.de
[**]Berliner Nanotest und Design GmbH, Volmerstrasse 7b, D-12489 Berlin, Germany
info@nanotest.org

ABSTRACT

Residual stresses in semiconductor and MEMS devices superposing functional and environmental loading are one of the crucial reliability issues. Most of routinely used stress measurement methods today allow stress measurements with only limited spatial resolution. The authors present two approaches under development, which are aiming at stress measurements on micro- and nanoscale devices with proven spatial resolution. The first method bases on stress release due to local material removal by a focused ion beam. SEM images before and after ion milling captured in a FIB equipment are compared to each other by cross correlation algorithms. As a result, residual stresses are evaluated from measured stress relief deformation. The second method refers to EBSD techniques. Displacements of Kikuchi diffraction patterns within a mono-crystalline material area are used to map incremental values of residual stress. Application capability of both stress measurement methods are demonstrated by experiments examples.

Keywords: residual stress, MEMS, semiconductor, stress relief, EBSD

1 INTRODUCTION

Development of new semiconductor and MEMS devices gives rise to several reliability issues to be analyzed and solved. Among them mechanical stresses between and in structural components covering some dimensional magnitudes are one of the concerns. They appear as residual stresses resulting from a multitude of different component processing steps manufacturing micro and nano components.

Unfortunately, stress determination in MEMS/NEMS, semiconductor devices and their packaging is everything else as a simple task. Finite element simulations should include modeling of complete production steps, having impact on final stress profiles. Realistically, this approach cannot be realized for complex devices. Experimentally, the choice of available stress measurement methods with micro- or even nanoscopic spatial resolution is rather limited. The authors developed a new promising stress measurement technique allowing access to stresses in microscopic and nanoscopic system areas. This method bases on the specific testing feasibilities provided by focused ion beam (FIB) equipment. Ion milling is utilized to release very locally residual stresses on components of interest. Generated this way surface deformations around the milled area are measured by digital image correlation (DIC) algorithms. As a result originally existing residual stresses are computed from measured stress release deformations. After a brief introduction to local stress measurement methods this method is described in more detail in chapter 3.

However, measurement of residual stresses with high spatial resolution keeps being a difficult task and is until now a topic under investigation and development. Because of the quite different capabilities of tackled approaches comparison and application of more than one method to a particular problem seems to be desirable. In that sense the authors try to set up different techniques to be combined with each other. A second tool reported here, applies the measurement of slight changes in electron diffraction pattern due to lattice stresses. The application of the method to thin stressed layers is illustrated in chapter 4.

2 LOCAL STRESS MEASUREMENT METHODS

Most available today measurement methods for residual stresses suffer from the fact that they average over larger object areas and cannot be applied very locally, i.e. within areas of a micron or of even substantially smaller size. This is the case for standard equipment used to measure wafer or die bow and conclude from curvatures on thin layer stresses [1]. Only a couple of known tools really allow very local measurement, for instance CBED [2,3], EBSD based approaches [4,5] and microRaman [3] stress techniques. Currently effort is made to extend their scope of measurement towards nanoscopically sized object. Nevertheless these methods also possess specific restrictions for application, in the case of microRaman, e.g., only a limited group of materials of interest can be accessed, which provide suitable Raman lines for Raman shift measurements. However, Raman technique currently is under development to extend spatial measurement resolution significantly beyond the limits of diffraction limited optical resolution, i.e. to address areas of less than 100 nm lateral size [6,7]. EBSD stress investigations demand a reference stress state as incremental strain and stress values can be determined only. Advantageous with respect to EBSD measurements is that different elements of the strain / stress tensor can be derived in a direct way and independently from each other.

The developed stress release method of the authors [8-10] intends to meet the goal of high spatial resolution, at the same time considering the method suitable for a larger variety of materials. This technique concludes from the stress release deformation field on stress states utilizing a priori knowledge of the kind of stress field. Analytical derivation or finite element simulation is used to describe stress relaxation due to ion milling and to determine stresses quantitatively by the amount of stress relief displacements. If uncertain, implemented stress hypotheses have to be examined by experimental modifications of stress relief pattern in order to make sure that the right theoretical treatment of the mechanical problem is applied. Stress measurements with the stress release methods, established by the authors as the fibDAC technique, are described in more detail in the following section.

3 STRESS RELIEF TECHNIQUE

3.1 Theory

In contrast to the classical hole drilling method [11] stress relief in semiconductor or MEMS structures, e.g. in thin layers on substrates, is produced by trench milling. Commonly, this choice is made for two reasons. Firstly, trench type geometry provides higher sensitivity with respect to stresses under measurement. Because stress relaxation at milling patterns on thin layers around 100 nm thickness is in the scope of some nanometers even for higher stresses, application of most efficient stress relief structures is necessary. Secondly, trench milling causes a distinguished stress sensitivity in the direction perpendicular to the trench. With some simplifications, the mentioned kind of mechanical problem can be described by an elastic plane strain approach, and was published in [12].

Following [12], the displacements generated when a surface trench is milled into a thin layer are normal to the trench plane and dependent on trench depth, the total film thickness and the ratio of the Young's modulus of the film to that of the substrate. The analytical solution bases on the assumption that the trench is infinitely long and narrow compared with its depth. The trench extends not further than the first interface between the thin film and the substrate. The displacement component perpendicular to the trench direction equals [13]

$$u_x = \frac{2.243}{E'_f} \sigma_o \int_0^a f_1^2 \cdot f_2 \, da \qquad (1)$$

Introducing an angle θ, defined by

$$\theta = arctg\left(\frac{x}{a}\right) \qquad (2)$$

the integrated functions f_1 and f_2 can be written as

$$f_1\left(\frac{a}{h}\right) = \left(1 - \frac{a}{h}\right)^{1/2-S}\left(1 + \lambda\frac{a}{h}\right)$$

$$f_2(\theta) = \cos\theta\left(1 + \frac{\sin^2\theta}{2(1-\nu_f)}\right)* \qquad (3)$$

$$*(1.12 + 0.18\sec h(\tan\theta))$$

x is the independent coordinate perpendicular to the trench, a is the trench depth, h is the thin film thickness. E'_f accounts for the plane strain Young's modulus $E'_f = E_f/(1-\nu_f^2)$ of the film, ν_f for the Poisson ratio and σ_0 for the residual stress. Elastic mismatch between substrate and thin film is taken into account in $f_1(a/h)$ through parameters S and λ. Both of them are related to Dundur's parameters α, β that measure the mismatch in the plane tensile modulus across the interface and in the in-plane bulk modulus, respectively. The parameter S can take values between 0 and 1, being 0.5 in the absence of elastic mismatch, whereas λ is typically small in magnitude and equal to zero in the case of identical film and substrate materials. A detailed description of the matter can be found in [14]. A common way of stress extraction using formulas (1) to (2) is given in the next section.

Besides the analytical solution depicted above, the stress release fields can be computed more accurately by finite element simulations. The numerically obtained x-displacement field is compared to the experimental data. A best fit procedure incorporating a stress scaling factor allows to get the elastically stored residual stress released by ion milling.

3.2 Measurement Example

In order to illustrate capabilities of the stress release method based on solution (1) to (3), a measurement example of a 350 nm thick low pressure chemical vapor deposited (LPCVD) silicon nitride film is given below. The deposition was performed onto a 300 μm thick silicon substrate at 800 ºC under a residual pressure in the range of 190-210 mTorr.

SEM images before and after ion milling were taken at a magnification of approximately 40,000. Under these circumstances accuracy of every single measurement point of displacements equals approximately 0.3 nm. The monitored region was placed at the center of the trench. A displacement field with the displacement component perpendicular to the trench plane has been obtained. As expected, the layer under tensile residual stress experienced an opening under stress release. Fig. 1 shows that the slot depth is asymmetrical; depths of 212 and 272 nm have been milled at left and right sides of the trench, respectively.

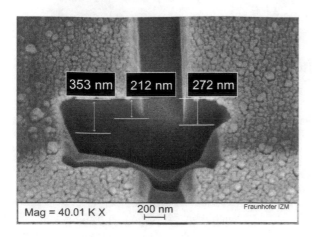

Figure 1: Cross-section of a milled trench in the Si_3N_4 layer on top of the Si substrate

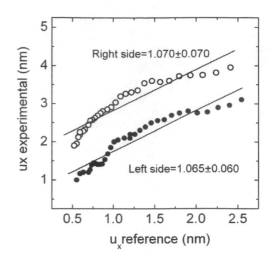

Figure 2: Residual stress determination by trench opening measurement. Fitting of experimental displacement data versus data obtained from the analytical expression (1) at 1 GPa of pre-selected residual stress. Values of the displacement component perpendicular to the trench direction have been averaged over a certain length of the whole trench size.

For stress evaluation from the displacement field formulas (1) to (3) have been applied. An approach similar to that described in [12] has been used to extract the isotropic in-plane stress. To make the analysis easier and to improve accuracy, displacement field data points have been averaged along the trench direction. I.e., an average of all displacements at the same distance from the slot has been computed. This implies that all data points within the measurement window on the specimen are assumed to originate from the middle of the trench. The slow drop down of displacements along the trench length justifies this approach. First, a theoretical displacement curve u_x^{ref} is computed from (1) by introducing experimental parameters (i.e. material properties, trench depth and thin film thickness) in the analytical expression, as well as an arbitrary reference stress value. In this way, the analytical expression yields a reference displacement $u_x^{ref}(x)$ at each distance from the trench. Respective measured and averaged displacement values $u_x^{meas}(x)$ are then plotted versus the computed theoretical values taken for the reference stress. Consequently, the slope of the curve yields the ratio between the real stress of the sample and the one computed as reference. Here, a presumed linear behavior of the mechanical response due to ion milling has been taken into account. The intercept of the linear curve with the abscissa axis accounts for rigid body displacement occurred between image captures (before and after ion milling).

Fig. 2 shows the mentioned plot of reference vs. measured displacements for the sample. Data is fitted for the left as well as for the right side of the trench, independently. The presented procedure is advantageous because of the separate treatment of both left and right side displacement fields. In this way the correct trench depth value can be used for either of the trench sides. So, slight differences in trench depths on both sides of the trench lead not to additional errors of stress evaluation. In this method the amount of absolute displacement and the displacement gradient, moving away from the trench rim, is taken as a measure for the released stress, instead of the trench opening / closing amount.

Material parameters used in the analytical expression of displacements in (1) to (3) and the values of parameters α, β, S and λ have been summarized in Table 1. The Young's modulus of the silicon nitride under study has been obtained from independent nano-indentation measurements.

Parameter	Material	
	silicon	LPCVD Si_3N_4
Young's modulus (GPa)	169 [17]	215 [13]
Poisson ratio	0.27 [17]	0.32 [16]
α	0.120	
β	0.023	
S	0.524	
λ	0.0095	

Table 1: Summary of the material properties and mismatch parameters used in the analytical formulas (1) to (3)

In Fig. 2 the measured displacements $u_x^{meas}(x)$ are plotted against those estimated using the analytical expression $u_x^{ref}(x)$ and assuming a reference tensile stress value of 1 GPa for the right and the left side of the trench, respectively. Stress extraction results obtained on the right and left sides of the trench agree quite well and despite the relatively high data dispersion (R^2 yields a value of 0.89 and 0.91 for left and right side of the trench, respectively). Consequently, the stress value of the layer is determined within an accuracy of 6%. Average stress between both sides yields a value of 1070 ± 70 MPa, which is in very good agreement (4%) with a value obtained from the wafer bending test.

4 EBSD BASED STRESS MEASUREMENT

The presented above fibDAC stress relief method is a versatile tool, largely independent on the kind of material to be investigated. However, some restrictions are set by the need of an appropriate residual stress model in order to connect stress relief with stress state. Another method measuring directly crystal lattice deformations relies on tracking of electron back scattered diffraction (EBSD) pattern. It doesn't exhibit the mentioned disadvantage and generally allows stress field measurements with high spatial resolution. Some first test measurements with a commercially available equipment carried out on stressed silicon are reported in the following.

The method makes use of the slight movements of Kikuchi diffraction patterns captured by an EBSD detector [4, 5], which occurs if crystal lattices are deformed by internal stresses. As a result of normal and shear strains as well as crystal rotations, positions of zone axes displace on the phosphorus screen. Combining at least four spatially separated measurements of zone axes displacements within one Kikuchi pattern allows to calculate the deformation tensor values $\frac{\partial u_i}{\partial x_j} (i \neq j)$ and $\frac{\partial u_i}{\partial x_i} - \frac{\partial u_j}{\partial x_j} (i \neq j)$. If the elastic crystal constants are known, all strain and stress tensor components $\varepsilon_{i,j}$ and $\sigma_{i,j}$ can be computed as well. Measurement values are spatially incremental data, i.e strains and stresses have to be adjusted to a known reference state.

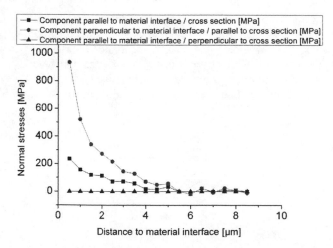

Figure 3: EBSD based stress measurement on Si (stressed by a Si_3N_4 layer), measurement direction perpendicular to layer/substrate interface, $\sigma_{33} = 0$ by definition

Fig. 3 shows measured stresses on a cross sectioned specimen of Si with a highly stressed Si_3N_4 layer on top. It has to be mentioned, that a severe stress re-distribution takes place at the free surface of the cross section. So far measured stresses are not the same as in the inner part, far from the surface. The inner stresses can be computed with the help of finite element simulations on base of the measurement data.

5 CONCLUSIONS

Two advanced methods of very local stress measurement on MEMS and semiconductor structures have been presented. The first technique uses stress relief deformations generated by trench milling with a focused ion beam equipment. Stresses are determined by trench opening / closing displacements. The second technique bases on the analysis of displacements of local EBSD Kikuchi pattern. The first tool allows measurement for a wide range of materials. The latter one is restricted to single crystalline materials, but allows to compute moderate fields of strain and stress tensor components.

REFERENCES

[1] L.B. Freund, S. Suresh, "Thin Film Materials", Cambridge University Press, 2003.
[2] J.A. Nucci, R.R. Keller, St. Kraemer, C.A. Volkert, M.E. Gross, Proc. of Mat. Res. Soc. Symp., Vol. 612, D8.5.1 –D8.5.6.
[3] V. Senez, A. Armigliato, I. de Wolf, G. Carnevale, R. Balboni, S. Frabboni and A. Benedetti, J. Appl. Phys., 94, 5574, 2003.
[4] A.J. Wilkinson, Ultramicroscopy, 62, 237-247, 1996.
[5] A.J. Wilkinson, G. Meaden, D.J. Dingley, Ultramicroscopy, 106, 307-313, 2006.
[6] .C. Georgi, M. Hecker, E. Zschech, Appl. Phys. Letters, 90, 171102, 2007.
[7] R. Ossikovski, Qu. Nguyen, G. Picardi, Physical Review, B 75, 1, 2007.
[8] D. Vogel, et al., Proc. of SPIE, Vol. 5766 (2005).
[9] N. Sabaté, D. Vogel, A. Gollhardt, J. Keller, B. Michel, C. Cané, I. Gràcia, J. R. Morante, Appl. Physics Letters 88, 071910, 2006
[10] D. Vogel, A. Gollhardt, N. Sabate, J. Keller, B. Michel, H. Reichl, Proc. of ECTC 2007, Reno 1490-1497, 2007.
[11] ASTM Standard "Test Method for Determining Residual Stresses by the Hole-Drilling Strain-Gage Method", E837-01e1, 2001.
[12] Ki-Ju Kang, et al, Journal of Engineering Materials and Technology, 126, October, 457-464, (2004).
[13] J.W. Hutchinson, Z. Suo, Advances in Applied Mechanics, 29, 63-191, 1992.
[14] J. Dundurs, D.B. Bogy, Journal of Applied Mechanics, 36, No. 3, 650, 1969.

Metallic submicron wires and nanolawn for microelectronic packaging. Concept and first evaluation.

S. Fiedler, M. Zwanzig, R. Schmidt, W. Scheel

Fraunhofer Institute Reliability and Microintegration (IZM), Berlin, MDI-BIT,
Gustav-Meyer-Allee 25, 13355 Berlin, Germany,
stefan.fiedler@izm.fraunhofer.de

ABSTRACT

Starting with a brief review of submicron wire production techniques, promising applications of submicron and nano wires are discussed. Observed attractive properties of metal rod decorated metal surfaces, so called nano lawn, for low energy joining and flip chip bonding are presented. The potential of these materials for microelectronic packaging by decoration of joining parts is shown: Due to close proximity of different grain sizes in intercalated nanolawn pieces (joining partners) recrystallization phenomena can be used for fluxless joining. Thus, new strategies for electrical interconnection of functional components can be designed by choice of morphology and composition of galvanically cast metal rods and wires. The experimental proof of principle has been performed with gold nanolawn. Based on original work new applications are presented. Related use of structures decorated with submicron wires and nanowires is discussed for microsystem technology.

Keywords: nanolawn, nano wires, nano interconnects, flip chip

1 SUBMICRON WIRE PRODUCTION TECHNIQUES

The functions of nanowires (NWs) which can be developed by suitable microelectronic packaging lay mainly in interconnect formation, sensorics, and photonics. With the common understanding of a wire, NWs would be expected to be cylindrical conductive strands with diameters < 100 nm of infinite length. Whereas common wires are drawn from one single metal, those NWs cannot be drawn and do not necessarily consist of one single material only. As different the materials as different the techniques to produce corresponding wires. Although rod-like colloidal particles [1] are often mentioned as NWs in the literature, they should rather be regarded as rods or crystal needles. Electrically conductive carbon nanotubes also have been presented for wiring purposes [2]. We will focus there exclusively on metallic wires and wire-like structures, e.g. pillars standing on a substrate or lawn like structures generated by template techniques and rods, "mown" from those "lawns". A simplified comparison of NW generation principles is shown in table 1.

wires	,usual' w.	nano wires
PROCEDURE	drawing	wire growth and writing
STARTING MATERIAL	bulk rods	single building units
BUILDING UNITS	n.a.	cations, atoms, clusters (NC's, NP's, rods)
ENERGY BALANCE ?	positive	negative chem. reduction & bonding, assembly
PHASE TRANSITIONS	none (~crystallinity)	cryst. growth from ions or atoms VLS / LS / SLS
DRIVING FORCE	mechanical	external force gradients DC: ELPHO / Plating AC: DEP i.e. inhomog. E-fields
TEMPLATE	drawing plate	w/o external internal
TEMPLATE MATERIAL		molecules hard soft polymer

Table 1: Wire generation: Classical wires vs. NWs.

Even "template free" techniques are characterized by a certain soft barrier: Specific crystal growth directions are shielded by low molecular weight adsorbates. We follow the depicted in Fig. 1 scheme to produce wires and lawn, using "hard" polymer exotemplates. Practically important nanoporous exotemplates are AAO and TEM (See below).

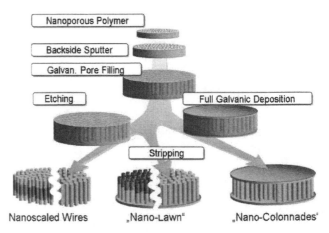

Figure 1: The nanolawn pedigree.

1.1 Anodically oxidised aluminum (AAO)

Electrochemically tunable nanopores in a rigid matrix like aluminum are the ideally hard exotemplate. Pore filling with metals by galvanic plating [3] or with pressurized molten metals seems to be suitable for mass fabrication [4].

1.2 Etched particle track polymer membranes

Since these materials are commercially available as size selective filter membranes, they represent a cheap and easily accessible template. Etched particle tracks in polymers [5] have been galvanically filled at first for analytical purposes [6] and later on explored for different applications [7, 8, and others]. As pore size and shape can be varied [9], and polymer composition can be tuned, they became popular for different applications in nanotechnology. Adhesivless basic materials for flexible PCBs have been presented recently [10]. Other industrial applications have been reviewed by P. Apel [11]. Galvanic metal deposition in such pores has been in the focus of R. Neumanns group at GSI Darmstadt [12, 13].

2 MORPHOLOGY OF NANOLAWN

Depending on pore diameter, orientation and density, different nano lawn morphologies can be generated. Fig. 2 shows single metal structures, fabricated with polycarbonate filter foils (Pore diameter 600 nm). If the pores are oriented perpendicular to the membrane plane, a high amount of free standing single wires can be prepared.

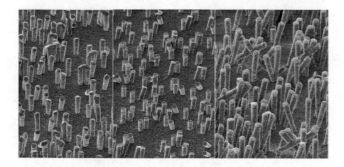

Figure 2: Single metal lawn. From left to right: Au, Pt, and Ag rods, 600 nm diameter. Note different crystallinity.

We used both membrane types with different pore density for the generation of bimetallic wires (See Fig. 3). Under certain conditions single crystal (grain) growth occurs, paving the way towards promising new applications.

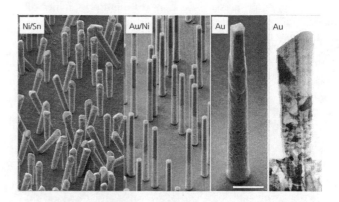

Figure 3: Bimetallic and single crystalline wires (FIB-SEM)

3 APPLICATION POTENTIAL

Especially monodisperse nanowire preparations may be attractive for advanced packaging by generation of
• active and passive photonic [14,15] or HF structures [16], stop-band filters, or antenna structures [17-19],
• for nanowire arrays and stacks to built-up logic and memory circuitry and devices [20-22], for applications in molecular electronics [23] and electronic olfaction or magnetic sensing [24],
• for nanowire arrays as nanoelectrode ensembles (NEE) or nanodisc electrodes with enhanced sensitivity in wet electrochemical and gas analysis [25,26], or as field emitter [27-31]. We have been proposing the use of nanolawn for low temperature joining [32]. Now extend the concept towards a generalized approach and technological principle.

4 LOW TEMPERATURE JOINING BY NANO INTERCALATION BONDING

If galvanically cast nanolawn foils (Comp. Fig. 1) are joined by intercalation face-to-face, grains of different size along the wires of both joining partners are brought into close proximity. During a subsequent annealing step, recrystallization effects cause growth of common grains. Fig. 4 shows the NIB concept for nano lawn, other nanostructures can be intercalated as well.

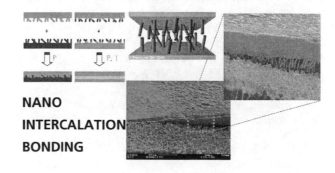

NANO INTERCALATION BONDING

Figure 4: Principle of nano intercalation bonding (NIB).

Depending on pressure applied, different proximities of intercalating structures combined with plastic deformation at sliding surfaces have been observed (Fig. 5).

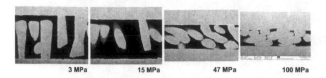

| 3 MPa | 15 MPa | 47 MPa | 100 MPa |

Figure 5: Applied bonding pressure for gold nanolawn. FIB-SEM analysis

With rising annealing temperature or/and duration we observed complete fusion at the contact zone, resulting in a fluxless soldering like bonding process (Fig. 6).

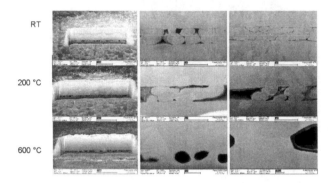

RT

200 °C

600 °C

Figure 5: Grain growth with annealing. Joining partners (top and bottom) have been bonded at 100 MPa. Samples have been cut (FIB-SEM).

5 CONCLUSION

If the joining, i.e. bonding pads of flip-chip parts can be fractally structured or otherwise nano-sculptured, recrystallization processes after or even during joining can be used to produce single metal or metal alloy interconnections in a fluxless process.

Acknowledgments
The authors want to thank Ellen Auerswald and Mathias Minkus for fruitful discussion and practical support. Intramural funding (IZM-Eigenforschung) is greatly acknowledged.

REFERENCES

[1] Xia YN, Yang PD, Sun YG, Wu YY, Mayers B, Gates B, Yin YD, Kim F, Yan YQ (2003) One-dimensional nanostructures: chemistry, physics & applications. Adv Mater 15:353–389.

[2] T.J. McDonald, D. Svedruzic, Y.-H. Kim, J.L. Blackburn, S. B. Zhang, P.W. King, and M.J. Heben (2007) Wiring-Up Hydrogenase with Single-Walled Carbon Nanotubes. Nano Lett., 7 (11), 3528 -3534.

[3] Gelves GA, Murakami ZTM, Krantz MJ, Haber JA (2006) Multigram synthesis of copper nanowires using ac electrodeposition into porous alu-minium oxide templates. J Materials Chemistry 16:3075-3083.

[4] Dresselhaus MS, Lin Y-M, Rabin O, Black MR, Dresselhaus G (2004) Nanowires. In: Springer Handbook of Nanotechnology (Bhushan B- ed.) 99-145, Springer ISBN 3-540-01218-4.

[5] Akapiev, G.N.; BarashenkovV.S., Samoilova, L.I., Tretiakova, L.I., Shegolev, V.A. (1974) K metodike izgotovlenia yadernykh filtrov. Dep. Publ. OIYaI, Dubna, Б1-148214.

[6] Spohr R (1990) Ion tracks and microtechnology. Principles and applications. Braunschweig: Vieweg ISBN 3-528-06330-0.

[7] C.R. Martin: "Nanomaterials: A membrane based synthetic approach", Science 266 (1994) 1961–1966.

[8] P. Apel, R. Spohr: "Introduction to ion track etching in polymers", http://www.iontracktechnology.de.

[9] Ferain E, Legras R (2001) Pore shape control in nanoporous particle track etched membrane. Nucl Instr Meth in Phys Res 174:116-122.

[6] Spohr R (1990) Ion tracks and microtechnology. Principles and applications. Braunschweig: Vieweg ISBN 3-528-06330-0.

[10] M. Danzinger, 4. BAIKEM-Kooperationsforum „Leiterplattentechnologie" 2008 Zukunftsmarkt: Hochfrequenztechnik. Nürnberg. 22. Januar 2008.

[11] Apel, P. (2003) Swift ion effects in polymers: industrial applications. Nuclear Instruments and Methods in Physics Research B 208 11–20.

[12] Dobrev D, Vetter J, Angert N, Neumann (1999) Electrochemical growth of copper single crystals in pores of polymer ion-track membranes. R APPLIED PHYSICS A- MATERIALS SCIENCE & PROCESSING 69 233-237.

[13] Enculescu I, Toimil-Molares ME, Zet C, Daub M, Westerburg L, Neumann R, Spohr R (2007) Current perpendicular to plane single-nanowire GMR sensor. APPLIED PHYSICS A 86 43–48.

[14] Y.T. Pang, G.W. Meng, Q. Fang, L.D. Zhang: "Silver nanowire array infrared polarizers", Nanotechnology 14 (2003) 20-24.

[15] R.B. Wehrspohn, J. Schilling: "Electrochemically prepared pore arrays for photonic-crystal applications", MRS Bulletin 26, August (2001) 623-626.

[16] A. Saib, D. Vanhoechenacker-Janvier, J.-P. Raskin, A. Crahay I. Huynen: "Microwave tunable filters and nonreciprocal devices using magnetic nanowires", Proceedings of the 1st IEEE Conference on Nanotechnology, IEEE-NANO 2001 (Maui, Hawaii, USA Oct 28-30, 2001) 260-265.

[17] J. Choi, G. Sauer, K. Nielsch, R.B. Wehrspohn, U. Gösele: "Hexagonally Arranged Monodisperse Silver Nanowires with Adjustable Diameter and High Aspect Ratio", Chem. Mater. 15 (2003) 776-779

[18] P. Mühlschlegel H.-J. Eisler, O.J.F. Martin, B. Hecht, D.W. Pohl: "Resonant Optical Antennas", Science 308 (2005) 1607-1609

[19] G. Schider, J.R. Krenn, A. Hohenau, H. Ditlbacher, A. Leitner, F.R. Aussenegg, W.L. Schaich, I. Puscasu, B. Monacelli, G. Boremann: "Plasmon dispersion relation of Au and Ag nanowires", Physical Review B68 (2003) 1555427/1-155427/4

[20] N.I. Kovtyukhova, T.E. Mallouk: "Nanowires as building blocks for self-assembling logic and memory circuits", Chem. Eur. J. 8(19) (2002), 4355-4363.

[21] Z. Zhong, D. Wang, Y. Cui, M.W. Bockrath, C.M. Lieber: "Nanowire Crossbar Arrays as Address Decoders for Integrated Nanosystems", Science 302 (2003) 1377-1379.

[22] S. Valizadeh, L. Hultman, J.-M. George, P. Leisner: "Template synthesis of Au/Co multilayered nanowires by electrochemical deposition", Adv. Funct. Mater. 12(11-12) (2002) 766-772.

[23] O. Reynes, S. Demoustier-Champagne: "Template electrochemical growth of polypyrrol and gold-polypyrrole-gold nanowire arrays", J. Electrochem. Soc. 152(9) (2005) D130-D135.

[24] M. Lindeberg, K. Hjort: "Interconnected nanowire clusters in polyimide for flexible circuits and magnetic sensing applications", Sensors Actuators A105 (2003) 150-161

[25] C.R. Martin: "Nanomaterials: A membrane based synthetic approach", Science 266 (1994) 1961–1966.

[26] V.P. Menon, C.R. Martin: "Fabrication and evaluation of nanoelectrode ensembles", Anal. Chem. 67 (1995) 1920-1928.

[27] D. Dobrev, J. Vetter, R. Neumann, N. Angert: "Conical etching and electrochemical metal replication of heavy-ion tracks in polymer foils", J. Vac. Science Technology B: Microelectronics and Nanometer Structures 19(4) (2001) 1385-1387

[28] S.H. Jo, J.Y. Lao, Z.F. Ren, R.A. Farrer, T. Baldacchini, J.T. Fourkas: "Field-emission studies on thin films of zinc oxide nanowires", Appl. Phys. Lett. 83(23) (2003) 4821-4823.

[29] J. Liu, Z. Zhang, Y. Zhao, X. Su, S. Liu, E. Wang: "Tuning the Field- Emission Properties of Tungsten Oxide Nanorods", small 1(3) (2005) 310-313.

[30] C.R. Sides, C.R. Martin: "Nanostructured electrodes and the lowtemperature performance of Li-ion batteries", Advanced Materials 17(1) (2005) 125-128.

[31] L. Vila, P. Vincent, L. Dauginet-De Pra, G. Pirio, E. Minoux, L. Gangloff, S. Demoustier-Champagne, N. Sarazin, E. Ferain, R. Legras, L. Piraux, P. Legagneux: Growth and Field-Emission Properties of Vertically Aligned Cobalt NanowireArrays, NanoLetters 4(3) (2004) 521-524.

[32] Fiedler S, Zwanzig M, Schmidt R, Auerswald E, Klein M, Scheel W, Reichl H (2006) Evaluation of Metallic Nano-Lawn Structures for Application in Microelectronic Packaging. In: 1st Electronics Systemintegration Technology Conference, Dresden, Sept 2006 2 886-891 (ISBN: 1-4244-0553-x).

Thermal Cycling of Buried Damascene Copper Interconnect Lines by Joule Heating*

D. T. Read and R. H. Geiss

Materials Reliability Division, Materials Science and Engineering Laboratory
National Institute of Standards and Technology, Boulder, CO, USA, read@boulder.nist.gov

ABSTRACT

We report tests to failure of 300-nm-wide damascene copper interconnect lines in silicon oxide dielectric, under high amplitude, low frequency alternating current. The cyclic minimum and maximum resistances were obtained from the measured voltage and current waveforms, and remained essentially constant over the lifetime tests. In the lines tested to failure under voltage control at the highest current levels, observed features of the remaining copper deposits seemed to indicate repeated melting over multiple cycles of current. The lifetimes, plotted against temperature, formed a nearly straight line on a semi-log plot, even though the failures, particularly those run under voltage control, became considerably less catastrophic for the longer lifetimes. Understanding the individual and combined effects of the temperature, current, and thermomechanical stresses will open up the possibilities for utilizing these electrical tests in systematic assessments of interconnect reliability and quality control.

Keywords: alternating current, fatigue, stress, temperature, voids

1 INTRODUCTION

Measurements of the lifetime-to-failure of nanoscale structures under relevant stresses allow the reliability of such structures to be quantified. Although extraction and testing of well established specimens, such as microtensile or instrumented indentation, is extremely challenging for nanoscale material features, nanoscale structures that are accessible electrically can often be stressed thermomechanically, by Joule heating, as well as by the "electron wind" that occurs at high current densities and is responsible for electromigration [1]. Stressing by electrical means may offer advantages over other means of reliability assessment, such as mechanical probing, for certain types of nanoscale structures, such as buried features.

Interconnect structures in advanced ULSI (ultra large scale integration) devices are leading-edge nanoscale structures. The smallest line widths are now less than 100 nm, and are buried as many as ten layers deep in the damascene structure. Reliability challenges for the near term include accounting for the effect of surrounding the copper lines with mechanically soft dielectric layers, or perhaps even air, and for the possibility of voids in the copper lines.

*Contribution of the U.S. National Institute of Standards and Technology; not subject to copyright in the U.S. Work supported by the NIST Office of Microelectronics Programs.

Several recent papers describe stressing of films or lines of metal conductors by use of thermomechanical stresses from Joule heating, *e.g.*, [2]. For certain cases of material and geometry, microstructural evidence of mechanical deformation, in particular, dislocations, has been reported [3]. Again for particular cases, the plot of cyclic temperature *vs* lifetime has been found to be similar to the S-N curve for mechanical fatigue, which allowed the extraction from the electrical test data of a fatigue strength that approximated the ultimate tensile strength [4].

A previous report compared the general failure behavior of 300 nm lines and vias in oxide and low-k dielectrics under high amplitude alternating current [5]. Here, additional details of the electrical behavior and SEM (scanning electron microscopy) images of the failure sites are presented. The main benefit of the SEM analyses presented here is to show the type of observations that can be made. Many more experiments will be needed before the sort of observations shown here can be related with a specifiable level of confidence to particular failure modes.

2 EXPERIMENTS

We have applied high amplitude, low frequency (100 Hz here) alternating current to 300 nm damascene copper interconnect lines. A substantial set of lifetime data under current control has been accumulated [5]. In these tests the voltage waveform contains higher harmonics that increase with the imposed current, and the failures are usually so catastrophic that failure investigation by microscopy is uninformative because the details have been obliterated.

Recently we have implemented some new experimental capabilities: operation under voltage control, with adjustment of the voltage waveform to (nearly) eliminate higher harmonics in the power waveform, so that the power becomes a sinusoidal function of time; and recording of both the cyclic maximum and the cyclic minimum electrical resistance during the course of the lifetime tests. The dc resistance is obtained as previously by interrupting the high amplitude alternating current and measuring the resistance at low, static values of current. We have also begun to examine this set of failed lines using SEM and EBSD (electron backscatter diffraction).

2.1 Specimens

Two-level damascene structures were obtained from a commercial source. Here we report tests of M1, the first copper level above the silicon substrate. The lines tested were electrodeposited copper, typical of recent electronics industry practice; their nominal dimensions were 300 nm wide, 500 nm deep, and 400 µm long. "Shield lines", not electrically connected to the test lines, ran parallel to the test lines on both sides. The dielectric was SiO_2.

2.2 Electrical Measurements

The cyclic minimum and maximum resistances were obtained from the measured voltage and current waveforms by fitting each using a discrete fourier transform and then applying a low-pass Weiner filter [6]. The maximum resistance was obtained directly from the fitted waveforms, at the maxima of current and voltage. L'Hôpital's rule was applied to obtain the minimum resistance, at the zero-crossing of the current and the voltage. The cyclic extrema of temperature were deduced from these resistance values [7].

The power waveform was also analyzed using the discrete fourier transform. The instantaneous power was available as the product of the instantaneous current and voltage values. Under voltage control at frequency f_v, the fundamental frequency of the power ias $2f_v$, and the lowest harmonic was at $4f_v$. Under current control without waveform adjustment, the power amplitude at $4f_v$ sometimes reached 20 % of the fundamental amplitude. The straightforward waveform adjustment that we have implemented reduced the second harmonic of the power to less than 1 % of the fundamental. The purpose of maintaining a single frequency in the power waveform was to impose heat generation at a single frequency. This would greatly simplify the use of closed form solutions for heat conduction, such as those found in Carslaw and Jaeger [8]. The prospective use of the present ac measurements for thermal properties is beyond the scope of this paper.

2.3 SEM Examination

Removal of the upper dielectric layers was necessary in order to examine the failed lines by SEM. This was accomplished by etching the tested specimens in a dilute hydrofluoric acid solution for 4 to 20 minutes, depending on the thickness of the dielectric. Both imaging, for general appearance and topography, and EBSD, to locate grain boundaries and determine grain orientations, were performed in the SEM.

3 RESULTS

The maximum cyclic resistance depended linearly on the maximum cyclic power. By deducing the maximum cyclic temperature from the maximum cyclic resistance [7], we found that the thermal resistance of these lines to ambient was about 700 °C/W. Typically the three recorded resistances, namely, the cyclic extrema and the dc value, remained nearly constant throughout the lifetime of these lines (Fig. 1). As indicated in the figure, the cyclic minimum resistance in this test is approximately equal to

the dc resistance, indicating that the copper line cooled almost to ambient temperature during each power cycle. This behavior was typical for imposed currents at 100 Hz. Occasionally the cyclic maximum resistance increased by a few ohms just before failure, but this was difficult to detect consistently since the resistance measurements were made only every 30 seconds. The lifetimes, plotted against temperature, formed a nearly straight line on a semi-log plot (Fig. 2), even though the failures, particularly those run under voltage control, become considerably less catastrophic for the long lifetimes. The current-control and voltage-control failure trends cannot be distinguished in the presence of the scatter in the data, even though the power waveforms are different in these two situations.

Our attempts to examine the failure sites of tests run under current control have been frustrated by the catastrophic nature of the failures: we find only what looks like the aftermath of a small explosion. These failures occur predominantly near, but not at, the ends of the test lines (Fig. 3). The failure sites of tests run under voltage control are not as catastrophic (Fig. 4).

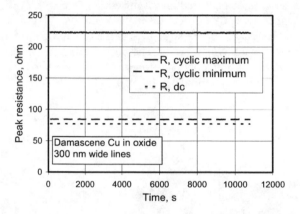

Figure 1. Cyclic maximum, cyclic minimum, and dc resistances plotted against time during a test.

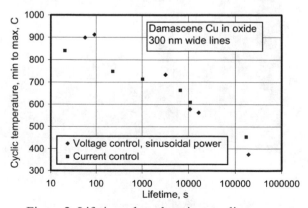

Figure 2. Lifetime plotted against cyclic temperature excursion for a series of tests.

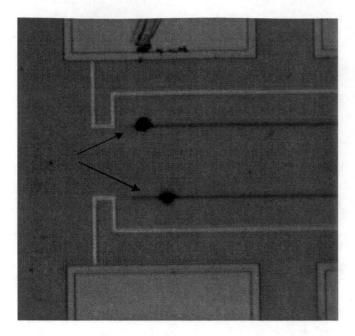

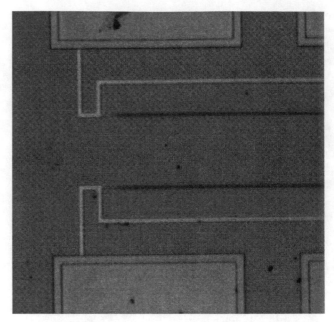

Figure 3. Optical micrograph showing two failed lines. These failures occurred under current control with cyclic temperatures of 455 ºC (lower line) and 555 ºC (upper line). The large copper structures at the top and bottom of the image are contact pads. The buried test line appears dark, rather than copper-colored. The failure sites are indicated by arrows.

SEM and EBSD observations of lines failed under different conditions are shown in Figs. 5-8.

Figure 4. Failures under voltage control with cyclic temperatures of 545 ºC(lower line) and 618 ºC (upper line). Again, the large copper structures at the top and bottom of the image are contact pads, and the buried test line appears dark, rather than copper-colored. The failure regions are not visible in this optical micrograph.

In a set of lines tested under current control at a cyclic temperature amplitude of 786 ºC, we observed a slight increase in topography (Fig. 5a). Observation of multiple EBSD images, such as shown in Fig. 5b, indicates a slight increase in grain length along the lines. Averages of samples of over 40 grains gave an increase in grain length from 0.67 to 0.91 µm. A tendency for the grain boundaries to become more perpendicular to the line axis was also seen.

300 nm

a)   b)

Figure 5. a) SEM images comparing untested line (left) and line tested to failure (734 s at $\Delta T = 786$ ºC) (right); specimen tilted 70 º to enhance topography. b) EBSD images from the same lines; colors represent crystal orientation, used to distinguish grains.

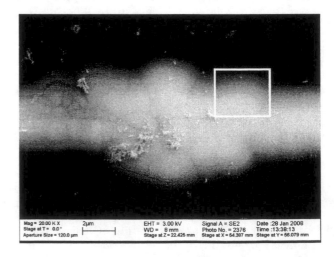

Figure 6. SEM image of failure site showing deposit of melted copper (large bright region in center). The region in the white box is shown in Fig. 7.

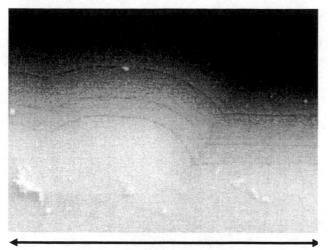

4 μm

Figure 7. Section of SEM micrograph of failed line shown in part a of this figure, expanded to show features indicative of repeated melting.

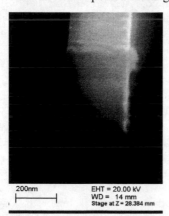

Figure 8. SEM image of 300 nm wide damascene copper line tested under ac to failure under voltage control with a cyclic temperature of 733 °C, and a lifetime of 3220 s. Note the very "square" failure surface, nearly perpendicular to the line axis.

In the lines tested under voltage control at the highest current levels, copper deposits near the center of the length of the lines and wider than the original damascene lines are found (Fig. 6,7). These deposits appear to be the result of melting, and show features that seem to indicate that the copper was melted repeatedly over multiple cycles of current. Lines tested under voltage control at lesser current values showed very abrupt rectangular failures, almost as if the line had been cut in a controlled fashion (Fig. 8).

4 DISCUSSION

The smooth trend of lifetimes with cyclic temperature shown in Fig. 2 suggests similar failure mechanisms at both the longest lifetimes tested, over 100,000 s, meaning over 20 million heating-cooling cycles of 400 °C, and at the shortest lifetimes tested, around 10 s, producing only a few thousand temperature cycles that reach 900 °C on average.

These initial microscopic observations of electrically-driven open-circuit failures suggest that failure occurs when a grain boundary or a void perpendicular to the line gradually opens up to become crack-like. This would reduce the local cross-section of copper carrying the current, leading to a local thermal runaway under current control, or to an eventual, less catastrophic open-circuit failure under voltage control. Signs of local melting at failure sites have been observed for failures under both current control and voltage control.

5 REFERENCES

[1]   Ogawa, E. T.; Lee, K. D.; Blaschke, V. A.; Ho, P. S. Electromigration reliability issues in dual-damascene Cu interconnections, *IEEE Transactions on Reliability* **51** (4), 403-419, 2002.

[2]   Monig, R.; Keller, R. R.; Volkert, C. A. Thermal fatigue testing of thin metal films, *Review of Scientific Instruments* **75** (11), 4997-5004, 2004.

[3]   Geiss, R. H.; Read, D. T. Defect behavior in aluminum interconnect lines deformed thermomechanically by cyclic joule heating, *Acta Materialia* **56** (2), 274-281, 2008.

[4]   Barbosa III, N.; Keller, R. R.; Read, D. T.; Geiss, R. H.; Vinci, R. P. Comparison of Electrical and Microtensile Evaluations of Mechanical Properties of an Aluminum Film, *Metals and Materials Transactions* **38A**, 2160-2167, 2007.

[5]   Read, D. T.; Geiss, R. H.; Alers, G. A. Study of Fatigue Behavior of 300 nm Damascene interconnect Using High Amplitude AC Tests, in *Materials, Process, Integration and Reliability in Advanced Interconnects for Micro- And Nanoelectronics. Mater. Res. Soc. Symp. Proc. Vol. 990;* edited by Lin, Q.; Ryan, E. T.; Wu, W.; Yoon, D. Y., editors; Materials Research Society: Warrendale, PA, 2007; pp. 121-126.

[6]   Press, W. H.; Teukolsky, S. A.; Vetterling, W. T.; Flannery, B. P. *Numerical Recipes in C, The Art of Scientific Computing;* Second ed.; Cambridge University Press: Melbourne, 1992.

[7]   Carslaw, H. S.; Jaeger, J. C. *Conduction of Heat in Solids;* Second ed.; Clarendon Press: Oxford, 1959.

[8]   Schuster, C. E.; Vangel, M. G.; Schafft, H. A. Improved estimation of the resistivity of pure copper and electrical determination of thin copper film dimensions, *Microelectronics Reliability* **41** (2), 239-252, 2001.

Nanoscale Deformation Measurements – Concepts for Failure and Reliability Assessment at the Nanoscale

B. Michel[*], A. Gollhardt[*] and J. Keller[* **]

[*]Fraunhofer Institute for Reliability and Microintegration, Dept. Micro Materials Center Berlin,
Gustav-Meyer-Allee 25, 13355 Berlin, Germany, bernd.michel@izm.fraunhofer.de
[**]AMIC Angewandte Micro-Messtechnik GmbH, Volmerstraße 9B, 12489 Berlin, Germany

ABSTRACT

The paper presents two methods for deformation measurement at the nanoscale level. The first method is based on Scanning Probe Microscopy (SPM) in combination with Digital Image Correlation (DIC). The technique serves as the basis for the development of the nanoDAC method (nano Deformation Analysis by Correlation), which allows the determination and evaluation of 2D displacement fields based on SPM data. The second approach for nanoscale deformation measurements is the so-called fibDAC (FIB, Focused Ion Beam) method. It provides the classical hole drilling method for residual stress measurement for the nanoscale region. The ion beam of the FIB station is used as a milling tool which causes the stress release. With the combination of fibDAC and finite element analysis stresses of silicon microstructures of MEMS devices or at other pre-stressed materials or surface coatings can be determined. Both presented methods can be applied for experimental reliability evaluation in microelectronics packaging, MEMS and NEMS. In addition residual stress determination at ultrathin layers and at microstructural features of bulk materials can be approached.

Keywords: nanoDAC, fibDAC, nanodeformation, residual stress measurement

1 INTRODUCTION

"Nanoreliability" is a name to describe a branch of research activities taking into account effects arising from the nanoscale new for reliability analysis and lifetime estimation. An advanced nanoreliability approach requires new tools for local stress and stain evaluation taking into account structural effects and physics of failure concepts combined with those continuum-based calculations which are suited to describe the essential procedures of the dominant failure mechanisms sufficiently well. Strain fields are coupled with thermal effects, diffusion phenomena, vibrations and various kinds of local changes of structure (defects, defect interactions etc.). Most of these phenomena have been shown to be related with internal stresses. That's why an advanced reliability analysis based on local deformation field characterization also requires an incorporation of local intrinsic stress analysis. By means of X-ray diffraction some information will be obtained, but very often this is not sufficient enough because local strains on a submicrometer scale have to be taken into account.

The increasing interface-to-volume ratio in highly integrated systems and nanoparticle filled materials and unsolved questions of size effect of nanomaterials are challenges for experimental reliability evaluation. To fulfill this needs the authors developed the nanoDAC method (nano Deformation Analysis by Correlation), which allows the determination and evaluation of 2D displacement fields based on scanning probe microscopy (SPM) data. In-situ SPM scans of the analyzed object are carried out at different thermo-mechanical load states. The obtained topography-, phase- or error-images are compared utilizing grayscale cross correlation algorithms. This allows the tracking of local image patterns of the analyzed surface structure. The measurement results of the nanoDAC method are full-field displacement and strain fields. Due to the application of SPM equipment deformations in the micro-, nanometer range can be easily detected. The method can be performed on bulk materials, thin films and on devices i.e microelectronic components, sensors or MEMS/NEMS. Furthermore, the characterization and evaluation of micro- and nanocracks or defects in bulk materials, thin layers and at material interfaces can be carried out.

2 NANODAC PRINCIPLE

Digital image correlation methods on gray scale images were established by several research groups. In previous research the authors developed and refined different tools and equipment in order to apply scanning electron microscopy (SEM) images for deformation analysis on thermo-mechanically loaded electronics packages. The respective technique was established as microDAC, which means micro Deformation Analysis by means of Correlation algorithms [5]. The microDAC technique is a method of digital image processing. Digitized micrographs of the analyzed objects in at least two or more different states (e.g. before and during/after mechanical or thermal loading) have to be obtained by means of an appropriate imaging technique. Generally, images extracted from a variety of sources such as SEM or laser scanning microscopy (LSM) can be utilized for the application of digital cross correlation. The basic idea of the underlying mathematical algorithms follows from the fact that images commonly allow to record local and unique object patterns, within the more global object shape and structure. These patterns are maintained, if the objects are stressed by thermal or mechanical loading. In the case of atomic force microscopy (AFM) topography images structures (patterns) are obtained by the roughness of the analyzed object surface. Figure 1 shows examples of AFM topography images taken at a crack tip of a polymeric material.

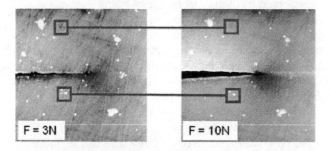

Figure 1: AFM topography scans [15 µm × 15 µm] at a crack tip of a polymer CT (compact tension) specimen; the scans are carried out at different load states.

Markers indicate typical local patterns (i.e. topographic features) of the images. In most cases, these patterns are of stable appearance, even if severe load is applied to the specimens so that they can function as a local digital marker for the correlation algorithm. The cross correlation approach is the basis of the DIC technique. A scheme of the correlation principle is illustrated by Fig. 2.

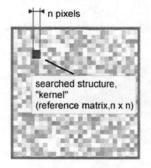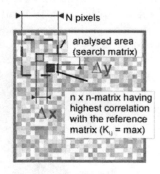

Figure 2: Displacement evaluation by cross correlation algorithm; (left) detail of a reference image at load state 1; (right) detail of an image at load state 2 [6].

Images of the object are obtained at the reference load state 1 and at a different second load state 2. Both images are compared with each other using a cross correlation algorithm. In the image of load state 1 (reference) rectangular search structures (kernels) are defined around predefined grid nodes (Fig. 2, left). These grid nodes represent the coordinates of the center of the kernels. The kernels themselves act as gray scale pattern from load state image 1 that have to be tracked, recognized and determined

by its position in the load state image 2. In the calculation step the kernel window (n × n submatrix) is displaced inside the surrounding search window (search matrix) of the load state image 2 to find the position of best matching (Fig. 2, right). This position is determined by the maximum cross correlation coefficient which can be obtained for all possible kernel displacements within the search matrix. The described search algorithm leads to a two-dimensional discrete field of correlation coefficients defined at integer pixel coordinates. The discrete field maximum is interpreted as the location, where the reference matrix has to be shifted from the first to the second image, to find the best matching pattern. For enhancement of resolution a so-called subpixel analysis is implemented in the utilized software [6]. The two-dimensional cross correlation and subpixel analysis in the surroundings of a measuring point primarily gives the two components of the displacement vector. Applied to a set of measuring points (e.g. to a rectangular grid of points with a user defined pitch), this method allows to extract the complete in-plane displacement field. Commonly, graphical representations such as vector plots, superimposed virtual deformation grids or color scale coded displacement plots are implemented in commercially available or in in-house software packages [7, 8]. Finally, taking numerically derivatives of the obtained displacement fields $u_x(x,y)$ and $u_y(x,y)$ the in-plane strain components ε and the local rotation angle ρ are determined.

For images originating from scanning probe microscopy (SPM) techniques the described approach has been established as so-called nanoDAC method (nano Deformation Analysis by Correlation) [1]. This method is particularly suited for measurement of displacement fields with highest resolution focused on MEMS/NEMS devices and micro and nano-structural features of typical microelectronics materials.

3 CRACK EVALUATION

In a typical nanodeformation measurement session in-situ AFM scans of the analyzed object are carried out at different thermo-mechanical load states as shown in Fig. 1. In the illustrated case an AFM topography signal serves as the image source. It is also possible to use other SPM imaging signals such as Phase Detection Microscopy or Ultrasonic Force Microscopy. The AFM scans are taken at the vicinity of a crack at a compact tension (CT) crack test specimen, Fig. 3. The CT-specimen is loaded with the force F by a special tension/compression testing module so that a Mode I (opening) loading of the crack tip is enabled. Figure 3 shows the CT-specimen and parts of the loading device under the AFM.

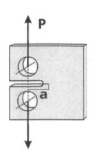 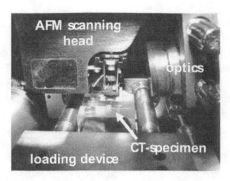

Figure 3: (left) Compact tension (CT) specimen; (right) In-situ loading under the AFM.

For images of the discussed loading of a thermoset polymer CT-specimen as given in Fig. 1 an extracted vertical (crack opening) displacement field is illustrated in Fig. 4.

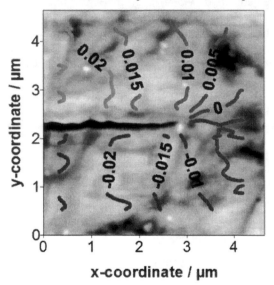

Figure 4: Crack opening displacement field in vertical (y)-direction [µm] determined by means of nanoDAC; in the background of the contour lines an AFM topography scan is illustrated.

Due to the application of SPM equipment deformations in the micro-, nanometer range can be easily detected. Currently the accuracy of the nanoDAC method for displacement field measurement is 1 nm for scan sizes of 2 µm, where the accuracy is determined by the thermo-mechanical stability of the SPM system. Details on the effect of thermal drifts and typical SPM related stability issues are discussed in [9]. In addition this reference shows compensation strategies for such error sources. The measurement technique can be performed on bulk materials, thin films and on devices i.e. microelectronic components, sensors or MEMS/NEMS. Furthermore, the characterization and evaluation of micro- and nano-cracks or defects in bulk materials, thin layers and at material interfaces can be carried out. An example of the determination of crack parameters based on nanoDAC displacement fields is shown in the following section.

4 FIBDAC PRINCIPLE

Measurement of residual stresses is an important demand for MEMS and sensor development. Loading of devices can produce stresses, which superpose with inherent residual stresses. Because internal stresses cannot be measured directly as forces, indirect approaches have to be looked for. One classical method is the release of residual stresses by material removal and subsequent measurement of induced respective strains at the object surface. The method became established as hole-drilling strain gage method [10], where through or blind holes are processed mechanically into the material. Released strains are commonly measured by strain gages attached to the object surface. Unfortunately, mechanical or laser based material removal is restricted in size. Also strain gages can not be placed easily on the object surface of sensors or MEMS. For these reasons, FIB milling seems to be an effective tool to extend the hole-drilling approach to submicron or even nano scale. Accompanying FIB material removal with spatially high resolution deformation measurement methods like DIC or Moiré is another prerequisite to downscale the classical method to the micro and nano region.

Therefore the fibDAC (focused ion beam based Deformation Analysis by Correlation) was developed. In the presented example the ion beam of the FIB station is used as a milling tool which causes the stress release at silicon microstructures of a MEMS device. Figure 5 shows an overview of the device (gas sensor) and the FIB-milled hole for stress release measurement.

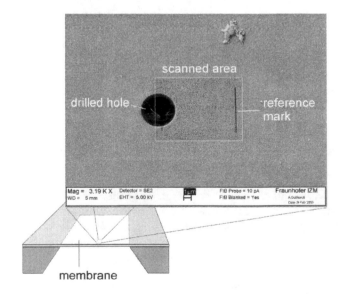

Fig 5: Scheme of micromachined membrane and overview of the scanned area after hole-milling process

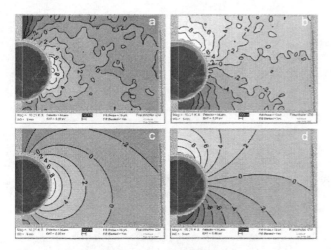

Figure 6: a) Displacement in horizontal direction, u_x
b) Displacement field in vertical direction u_y
c) u_x analytical fit d) u_y analytical fit; all data in [nm]

The analysis of the stress release is achieved by DIC applied to load state SEM images captured in cross beam equipment (combination of SEM and FIB). The results of the DIC analysis are displacement fields $u_x(x,y)$ and $u_y(x,y)$ with particular displacement of the analyzed image. Figure 6(a) and (b) shows the experimental contour lines for u_x and u_y respectively. The experimental results can now be fitted to displacement fields calculated by finite element analysis or analytical solutions, Fig. 6 (c) and (d). Fitting has to be performed for the whole 2-D displacement field with respect to the rigid body translation and rotation. With the knowledge of other material parameters (Young's modulus, Poisson's ratio for isotropic materials) residual stresses can be evaluated.

In another step which is not discussed in this paper the resolution of the method has been improved by the application of trench milling instead of milling of holes [11]. Thereby the accuracy of the method has been improved.

5 CONCLUSIONS

A new approach for nanodeformation measurement based on SPM images has been established. With the nanoDAC method research of nanomechanical effects can be addressed directly with the combination of nanoanalytical SPM techniques and digital image correlation (DIC) software codes. In addition the fibDAC method gives access to high resolution residual stress measurements. Stresses are determined by trench opening/closing displacement measurement based on DIC results in combination with analytical solutions or Finite-Element modelling. Both methods are suited for fundamental and applied research at MEMS/NEMS and micro- or nanostructural building blocks of bulk materials.

ACKNOWLEDGEMENTS

The authors wish to acknowledge the collaboration with Neus Sabate and Carles Cane from Departament d'Electronica, Universitat de Barcelona (EME-UB), Spain.

REFERENCES

[1] Keller, J.; Vogel, D.; Schubert, A. and Michel, B.: Displacement and strain field measurements from SPM images, in Applied Scanning Probe Methods, B. Bhushan, H. Fuchs, and S. Hosaka, eds., pp. 253–276, Springer, 2004.

[2] Vogel, D. and Michel, B.: Microcrack evaluation for electronics components by AFM nanoDAC deformation measurement, in Proceedings of the 2001 1st IEEE Conference on Nanotechnology. IEEE-NANO 2001, pp. 309–312, 2001.

[3] Vogel, D.; Keller, J.; Gollhardt, A. and Michel, B.: Evaluating microdefect structures by AFM based deformation measurement, in Proc. of SPIE Vol. 5045, Testing Reliability, and Application of Micro- and Nano-Materials Systems., pp. 1-12, 2003.

[4] Puigcorbe, J.; Vogel, D.; Michel, B.; Vila, A. Gracia, I.; Cane, C. and Morante, J., Journal of Micromechanics and Microengineering 13(5), pp. 548-556, 2003.

[5] D. Vogel, A. Schubert, W. Faust, R. Dudek, and B. Michel, Microelectronics Reliability 36(11-12), pp. 1939–1942, 1996.

[6] Dost, M.; Kieselstein, E. and Erb, R., Micromaterials and Nanomaterials (1), pp. 30-35, 2002.

[7] UNIDAC, Chemnitzer Werkstoffmechanik GmbH, www.cwm-chemnitz.de

[8] ADASIM, Image Instruments GmbH, www.image-instruments.de/ADASIM/

[9] Michel, B. and Keller, J.: Nanodeformation Analysis Near Small Cracks by Means of nanoDAC Technique in: G. Wilkening and L. Koenders (eds.) Nanoscale Calibration Standards and Methods, Wiley-VCH, Weinheim, pages 481-489, 2005.

[10] Standard Test Method for Determining Residual Stresses by the Hole-Drilling Strain-Gage Method (E837-01e1), ASTM 2001.

[11] N. Sabate; D. Vogel; A. Gollhardt; J. Keller; C. Cane; I. Gracia; J.R. Morante and B. Michel; J. Micromech. Microeng. 16 (2006) 254–259

Surface functionalized magnetic nanoparticles as switchable building blocks for soft nanotechnology

H. Dietsch[*], M. Reufer[*], V. Malik[*] and P. Schurtenberger[*]

[*]Adolphe Merkle Institut and Fribourg Center for Nanomaterials, 1700 Fribourg, Switzerland

ABSTRACT

α-Fe$_2$O$_3$ (hematite) particles can be synthesized by forced hydrolysis of iron salts. Playing with physical and chemical means allows the control of size and shape of the obtained ferromagnetic particles. Here we will summarize our recent work on the synthesis of hematite cubes and spindle type particles, which can be obtained with large aspect ratios such as 7. These particles can then be coated with a silica layer, allowing the creation of a specific surface functionality for their incorporation in polymer matrices or simply for their stabilization in a given solvent. We demonstrate that a moderate magnetic field (2-12mT) is sufficient to orient our particles, and how we can use them as nanoprobes to investigate local rheological properties of complex fluids.

Keywords: α-Fe$_2$O$_3$, spindle particles, cubic particles, active microrheology, nanocomposite.

1 INTRODUCTION

Hematite particles have gained interest in the last years as interesting magnetic nanoparticle model systems. This is due to the existing possibilities to tailor their shape and size by playing with various physical and chemical means such as the precursor concentration [1] and type [2], the variation of the reaction time [3, 4], the pH [5], the temperature, or the ionic strength [5, 6]. More recently, we demonstrated that the intermediate crystalline structure (akaganeite) formed during the synthesis of hematite can also be used as a starting point to control the monocrystallinity of the subsequently obtained particles [7].

Depending on the solvent or the medium in which the particles have to be dispersed, the particle-particle and particle-medium interactions play an important role in any attempt to avoid aggregation. This is particularly important when using hematite particles as magnetic building blocks for nanocomposites or as tracer particles in more complex environments. Here we demonstrate how one can coat hematite particles with a silica layer as a convenient basis for further surface modifications through an additional silane coupling agent [8]. The method used follows the general scheme proposed by Graf *et. al.* [9]. It consists in adsorbing a non-ionic amphiphilic polymer polyvinylpyrrolidone (PVP) on the surface of our colloidal particles. A modified Stöber approach [10] then leads to a homogeneous coating with a silica layer of controlled thickness. Subsequent additional synthesis steps then allow us to prepare magnetic core-shell particles with a specific surface functionalization that can then be used in various applications.

Here we concentrate in particular on microrheological applications of these particles. Microrheology uses the fact that the motion of colloidal tracer particles in a complex medium reflects the underlying rheological properties of the suspending matrix. Most studies published so far have primarily concentrated on passive microrheology, where particle tracking or diffusing wave spectroscopy combined with the application of a generalized Stokes-Einstein relation has been used to measure the viscoelastic properties of complex fluids [11, 12]. Very few groups have extended the microrheology approach to active measurements where tracer particles are moved via optical tweezers or other external fields through a complex fluid [13]. While such experiments are much more demanding experimentally and theoretically, they also offer in principle an extension of microrheology to the non-linear regime that is not accessible for classical microrheology based on thermal fluctuations only.

Here, we focus on two aspects of our hematite-based nanoparticle research. We first describe the synthesis of functionalized hematite particles with controlled shape and size. We then describe the lay-out of a simple set-up based on [14] that allows us to control the orientation of the hematite nanoparticles in a medium. This then allows us for example to create interesting nanocomposite systems with magnetic properties via in-situ polymerization. However, in this presentation we will primarily focus on a demonstration of the feasibility of active microrheology experiments where we use the magnetic nanoparticles as a nanorheometer.

2 EXPERIMENTAL SECTION

2.1 Materials and methods

Spindle shaped hematite particles were synthesized using iron (III) perchlorate (Alfa Aesar) or iron III chloride (Sigma-Aldrich), urea (Fluka) and NaH$_2$PO$_4$, 2H$_2$O (Fluka). For the coating of spindle hematite with silica, Tetraethylorthosilicate (TEOS), Tetramethylammonium-hydroxide (TMAH, 25% in methanol), and ethanol were purchased by VWR, Polyvinylpyrrolidone (PVP 90K,

Sigma-Aldrich) is dissolved in ethanol before being added to the particle to be coated. All chemicals were used without further purification. MilliQ water (Resistivity-18.2 MΩ-cm) was used to for the synthesis.

Carbon TEM grids (300 Mesh) were prepared 4 hours before the measurement; we typically dry one droplet of colloidal solution (~1%) at ambient temperature on the grid and store it in a dust free box.

2.2 Synthesis of hematite particles with controlled size, aspect ratio and surface functionality.

Here, we present as an example, the synthesis of spindle type hematite with an aspect ratio of 7 and its silica coating for surface functionalization.

Spindle hematite cores used for the coating were synthesized using forced hydrolysis of iron perchlorate precursor in the presence of urea and phosphate salt based on Ocana's method [2]. In a typical reaction 115.5g of iron perchlorate are dissolved with 1.62g phosphate salt and 15g of urea in 2.5L of MilliQ water into a tightly closed glas bottle. The solution is kept for 24 hours in an oven at 98°C. After cooling, the so-obtained solution is centrifuged and repeatly washed with water.

The thus-obtained spindle type hematite particles are then coated with silica based on Graf's method [9]. In a typical reaction, 4 g of PVP are dissolved in 60 mL of MilliQ water using mechanical stirring and ultrasonication. 222 mg spindle hematite particles are then added to this polymer solution. The solution is stirred for 24 hrs at ambient temperature. The dispersion is repeatly centrifuged to wash the particles from the non adsorbed polymer. Hematites with adsorbed PVP are then finally dispersed into 210g of ethanol in which 27 g H_2O and 150 mg of 25% tetramethylammoniumhydroxide (TMAH) in methanol are added. A mixture of 9 ml TEOS and 6 ml of ethanol are injected into the stirred particle suspension. This coating step can be repeated and permits a perfect control of the thickness of the silica shell combined with a decreasing aspect ratio of the initial spindle form which becomes ellipsoidal.

2.3 Cross polarizer set-up

A temperature controlled sample holder is located between two pairs of Helmholz coils as shown in the cartoon on figure 1. Two sinusoidal currents with phase shift of 90° drive the coils and create a magnetic field with constant amplitude that rotates in the plane perpendicular to the optical axis with a variable rotation frequency. The subsequent orientation of the magnetic particles is probed with a laser beam in cross polarization geometry. This design allows us to determine the phase lag α of the particles when compared to the driving magnetic field as function of the frequency.

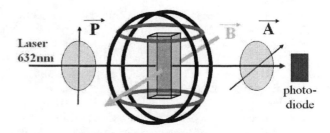

Figure 1: Helmholz coil set-up that allows to create a rotating external magnetic field and probe the induced orientation of the magnetic nanoparticles using a combination of crossed polarizers and a photodiode.

3 RESULTS AND DISCUSSION

3.1 Functionalized core-shell hematite particles: Overview and characterization

The combination of hematite-based nanoparticles with a subsequent silica coating provides us with a series of building blocks that offer the possibility to create functionalized magnetic nanoparticles with a wide range of different architectures and properties. Figure 2 presents a schematic overview of the degree of complexity that is accessible with this system. Depending upon the choice of the synthesis parameters, we can tailor the particle size, the degree of anisotropy as well as create core-shell particles with an internal structure provided by several layers and subsequent surface functionalization. This is also demonstrated by electron microscope images of hematite particles illustrating the range of sizes, aspect ratios and functionalities which can be obtained form the different synthesis pathways of hematite.

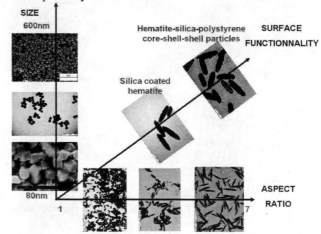

Figure 2: TEM and SEM images of hematite particles (a) with an aspect ratio of 7, coated with 25nm (b), 55nm (c) and 95nm (d) of silica.

The enormous variability in size offered by the system is shown with the example of isotropic magnetic hematite particles. As illustrated by the different electron micrographs, we can create single crystal cubes [7], round shaped or pseudocubical magnetic particles with sizes ranging from 80nm to 600nm. These different systems can be obtained through a variation of the iron III chloride concentration from 0.01M to 0.06M in water. Three different mechanisms of formation were reported in the past [7, 15].

We are not restricted to cubic or globular shape, but can create rod-like shapes with a varying degree of particle anisotropy with axial ratios between 1 and 7. This can be achieved through a variation of the concentration of NaH_2PO_4 used. The largest aspect ratio obtained is 7, which is the synthesis example described in the experimental section.

Finally, surface functionality can then be obtained using a core-shell structure where surfactants, adsorbed polymers or a layer of silica are attached to the hematite particles. The size of the silica layer can conveniently be controlled and allows for a further surface modification through more than 2000 silane coupling agents [16]. In figure 2, we illustrate this with two examples of silica-coated spindle type hematite and hematite-silica-polystyrene core-shell-shell particles, respectively.

The size, shape and internal structure of the colloidal particles can be characterized in detail and on a large range of different length scales using a combination of electron microscopy (SEM and TEM) and small-angle x-ray scattering (SAXS). Figure 3 shows the characteristic pattern evolution observed as we increase the size of the silica layer on the hematite spindle cores.

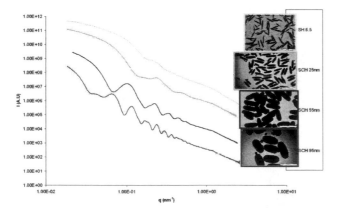

Figure 3: Small Angle X-Ray Scattering (SAXS) data and EM pictures obtained from (a) bare hematite cores; and core-shell structures coated with (b) 25nm, (c) 55nm and (d) 95nm of silica.

3.2 Hematite-(silica) nanomagnets and their applications

The small permanent magnetic moment of the hematite particles provides us with a number of different opportunities where we can use an external magnetic field to either induce an orientation of the particles and thus create a nanostructured material with anisotropic magnetic, mechanical or optical properties, or use the particles as a "nanorheometer" to locally probe the viscoelastic properties of a complex fluid. The feasibility of the latter application is demonstrated in Figure 4. Here we use the Helmholz coil set-up described in Figure 1 in order to orient and turn the particles at different rotation frequencies f and magnetic field strength B and measure their orientation. Figure 4 then shows the phase lag expressed by the phase angle α between the driving field and the orientation of uncoated hematite particles as a function of B and f, where B varies between 2mT and 6mT. Figure 4 demonstrates that for the simple fluid chosen the phase lag $\sin\alpha$ is inversely proportional to the strength of the applied field and increases linearly with a rescaled frequency f/B. A full analysis of the particle dynamics in the external driving field then shows that we can relate the relationship between phase lag and magnetic torque generated to the frequency-dependent viscoelastic properties of the suspending medium.

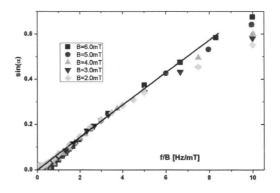

Figure 4: Phase lag between external magnetic field and induced particle orientation as a function of the magnetic field rotation frequency scaled by the applied magnetic field strength.

4 CONCLUSIONS

Hematite nanoparticles represent a very interesting model system that provides us with full control of the final size, shape and internal structure through a series of synthesis, coating and functionalization steps. The final coating step with a functional layer based on silica and silane coupling agents then allows us to disperse the particles in virtually any medium which can be either a simple liquid or a polymer solution or melt. Due to their magnetic properties with a weak permanent magnetic moment we can orient the particles using an external field but avoid aggregation or chaining usually observed in ferrofluids based on magnetic particles with stronger moments. This provides us with a wealth of interesting applications where we either explore their magnetic properties to create anisotropic materials with tunable properties or use them to investigate the viscoelastic properties of the suspending medium.

REFERENCES

[1] E. Matijevic and P. Scheiner, Journal of Colloid and Interface Sci., 63, 509, 1978.

[2] M. Ocana, M. P. Morales and C. J. Serna, Journal of Colloid and Interface Sci., 212, 317, 1999.

[3] T. Sugimoto, S. Waki, H. Itoh and A. Maramatsu, Journal of Colloid and Interface Sci., 109, 155, 1996.

[4] T. Sugimoto, Chem. Eng. Technol. 26, 313, 2003.

[5] T. Sugimoto and Y. Wang, Journal of Colloid and Interface Sci., 207, 137, 1998.

[6] F. Jones, M. I. Ogden, A. Oliveira, G. M. Parkinson and W. R. Richmond, CrysEngComm, 5, 159, 2003.

[7] V. Malik, P. Schurtenberger, V. Trappe, B. Grobety and H. Dietsch, in preparation.

[8] F. D. Osterholtz and E. R. Pohl, Journal of Adhesion Science and Technology, 6, 127, 1992.

[9] C. Graf, D. L. J. Vossen, A. Imhof and A. van Blaaderen, Langmuir, 19, 6693, 2003.

[10] W. Stöber, A. Fink and E. Bohn, Journal of Colloid and Interface Sci., 26, 62, 1968

[11] T. G. Mason and D. Weitz, Phys. Rev. Let., 74, 1250, 1995.

[12] T. G. Mason, K. Ganesan, J. H. van Zanten, D. Wirtz and S. C. Kuo, Phys. Rev. Let., 79, 3282, 1997.

[13] E.M. Furst, Soft Materials, 1, 167, 2003.

[14] O.Sandre, J. Browaeys, R. Perzynski, J. C. Bacri, V. Cabuil and E. Rosensweig, Phys. Rev. Let. E, 59, 1736, 1999.

[15] J. K. Bailey, C. J. Brinker, M. L. Mecartney, Journal of Colloid and Interface Sci., 157, 1, 1993.

[16] Silane agent can be purchased on www.abcr.de

Dispersion and incorporation of optical nanotracers

J. Samuel[*], O. Raccurt[*], O. Poncelet[*], F Tardif[*] and O. Tillement[**]

[*]Commissariat à l'Energie Atomique, Laboratory of Tracer Technologies, 38054 Grenoble, France
[**]Laboratory of Physico-Chemistry of Luminescent Materials, 69622 Villeurbanne, France

ABSTRACT

Fluorescent silica nanoparticles are used as nanotracers to qualify the level of dispersion of nanoparticles into hydrophilic and hydrophobic polymers. In this context, a perfect control of their surface is researched. Reverse micelle is a well known sol-gel process to synthesize silica nanoparticles. But it is observed that using this method, there is an effect of the dye nature on the surface. This effect is erased by growing a thin silica layer on the surface of the nanoparticles. Then, two organosilane molecules have been successfully grafted, making the nanoparticles hydrophilic or hydrophobic and so well suited to fill homogeneously hydrophilic or hydrophobic polymers. Each adapted nanoparticle is indeed well dispersed into the corresponding polymer matrix. This dispersion is characterized thanks to the fluorescence properties of the nanoparticles.

Keywords: nanotracer, dispersion, polymer, nanoparticle, surface functionalization.

1 INTRODUCTION

The nanotechnologies offer new perspectives in numerous industrial and research domains today. New applications, thanks to the enhancement of the mechanical [1,2] or optical [3] properties of some polymers have been shown by the incorporation of nanoparticles into these polymers. Polysiloxane based nanoparticles encapsulating one or several dyes (organic or inorganic) offer several interesting properties : protection of the dyes against photo bleaching and chemical alterations; exaltation of the fluorescence; good dispersion in water and easy surface functionalization because of the reactive silanol groups. Thanks to these properties, the fluorescent silica nanoparticles can be used as nanotracers, to monitor the behavior of the nanoparticles into the polymer matrix

A control of the surface is thus being researched. To achieve a good dispersion in an aqueous medium, two parameters have to be considered: the hydrophilic/hydrophobic character of the surface and its Zeta Potential. It is well known that the highest the absolute value of the Zeta Potential is, the most stable the colloidal suspension is [4]. The Zeta Potential also gives an idea of the hydrophilic character of a surface. An hydrophobic surface has a Zeta Potential value close to zero [5]. Many studies have already been done especially on the grafting of an aminopropyl group onto the silica surface (APTES,

APMS, N-2-(aminoethyl)-3-APTES) [6,7]. It appears that the isoelectric point is shifted toward higher pH values, the Zeta Potential is higher at low pH and these particles are hydrophilic because they can form Hydrogen bounds with water molecules. [8] also studied the effect of another molecule (PhTES) which makes the particles hydrophobic and so well suited to fill hydrophobic polymers.

The dispersion of 40 nm colloidal nanoparticles encapsulating organic dyes synthesized by a reverse micro-emulsion sol-gel method and the quality of the dispersions obtained are studied here.

2 MATERIALS AND METHODS

2.1 Materials

Triton X-100 is purchased from Sigma-Aldrich; Hexanol, Cyclohexane, Aqueous Ammonia (30%wt in water) were also purchased from Sigma Aldrich; TEOS (Tetraethylorthosilicate) was obtained from Sigma Aldrich; N-2-(amino ethyl)3-aminopropyltriethoxysilane 99% (Di-Apts) and Trichloro(1H,1H,2H,2H-perfluorooctyl)silane 97% (FDTS) were purchased from Sigma Aldrich. The dyes studied are Rhodamine B and Fluorescein. The polymer tested are PolyMethylMethAcrylate (PMMA) (average Mw=350) and Polyvinyl Alcohol (PVA) (average Mw = 31000) purchased from Sigma Aldrich.

2.2 Preparation of the silica nanoparticles

The Silica nanoparticles are synthesized by a reverse microemulsion process proposed by R. P. Bagwe[9]. The idea is to create a Water in Oil microemulsion by mixing Cyclohexane and Water. The microemulsion is stabilized by a non ionic surfactant (Triton X-100) and a co surfactant (Hexanol). The incorporation of the hydrophilic dye into the system is followed by the hydrolysis-condensation of the silica precursor (TEOS) catalyzed by aqueous Ammonia into the nanoreactors. The mixture is reacted for 24h under magnetic stirring. The nanoparticles are released from the microemulsion by the addition of Ethanol in excess. They are separated from the reaction mixture by centrifugation at 4000 rpm for 15-30 min and washed three times with anhydrous Ethanol or with Water. The nanoparticles redispersed into anhydrous Ethanol can be encapsulated by a next addition of silica precursor. The encapsulation is done in a basic medium during 6h at 70°C. The nanoparticles can also be functionalized by the addition of a coupling agent such as FDTS or Di-Apts. The mixture is stirred overnight and then washed one time with anhydrous Ethanol and two times with Water or Acetone depending on the dispersion medium researched.

2.3 Preparation of the loaded polymers

The materials studied are all made by evaporation in an appropriate solvent : Chloroform for the PMMA and Water for the PVA.

PMMA: 2g of PMMA pallets are added to a 50 mL Chloroform solution and stirred for 3h. An adapted quantity of silica nanoparticles is added to the mixture when all the PMMA pallets have been dissolved. The colloidal suspension is then stirred 1h and deposed on a thin glass under controlled atmosphere. The evaporation of the Chloroform gives a thin PMMA loaded layer on a glass film.

PVA: 5g of PVA pallets are added to a 100 mL Water bath heated at 80°C. An aqueous solution containing di-Sodium Tetra Borate and a certain amount of silica nanoparticles is prepared and mixed with the first bath. The colloidal suspension is then deposed on a thin glass under controlled atmosphere. The evaporation of the Water gives a thin PVA loaded layer on a glass film.

2.4 Methods

The size and the dispersion of the silica nanoparticles are monitored by Transmission Electronic Microscopy. Zeta Potential measurements are carried out with a ZetaSizer Nano associated with an automatic Titrator MPT-2 supplied by Malvern Instruments to characterize surface charge and hydrophilicity. The hydrophilic/hydrophobic character is also measured by an adapted Dichloromethane method [5]. This method consists on mixing an aqueous phase containing the nanoparticles and a small amount of Hexadecane. The two solvents are stirred and let 30 min. Then 1mL of the aqueous phase is removed and the fluorescence intensity is measured by spectroscopic fluorescence. The hydrophilic/hydrophobic character is given by the ratio between the fluorescence intensity initially measured into the aqueous phase and the fluorescence intensity finally measured. The fluorescent properties of the nanoparticles are observed with a FS920 fluorimeter supplied by Edinburgh Instruments. The homogeneity of the incorporation into the polymer matrix is performed using confocal imaging.

3 RESULTS AND DISCUSSION

This section is divided into three main parts: the first part is the study of primary fluorescent silica nanoparticles; some characterizations are performed to see the effect of the fluorescent dye on the surface of the nanoparticles; the second part is an attempt to a perfect surface control; a control of the Zeta Potential and of the hydrophilic/hydrophobic character is thus researched; the

final part is the incorporation of the nanoparticles within the polymer matrix characterized by confocal microscopy.

3.1 Fluorescent silica nanoparticles

The Fluorescent silica nanoparticles synthesized are spherical, relatively monodispersed and of 40 nm diameter.

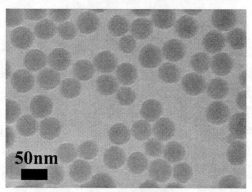

Figure 1: TEM image of the silica nanoparticles synthesized.

It is possible to control the particle size by modifying the ratio [surfactant]/[water] and by this way sizes of 12 nm can be obtained [10]. Other advantages of the reverse micellar synthesis are that nanoparticles are rather monodispersed and that several hydrophilic dye molecules can be incorporated into the silica matrix in contrast with the Stöber method which needs a modification of the dye to ensure its incorporation into the silica matrix.

The Zeta Potential is shown Fig.2 as a function of the pH value for two different dyes and for non fluorescent silica.

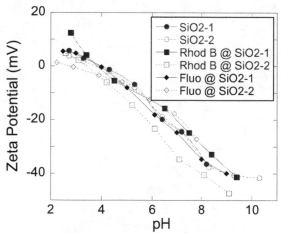

Figure 2: Zeta Potential as a function of the pH for silica nanoparticles without dye or with different dyes.

It appears that the nature of the dye has an effect on the Zeta Potential. Since the synthesis process is the same regardless of the dye molecule incorporated, the surfaces should be identical. Moreover two different measurements with the same dye (Rhodamine B) or with different dyes

(Rhodamine B and Fluorescein) give two different Zeta Potential behaviors. The only values constant are the values obtained for the silica nanoparticles without dye. This difference could therefore be explained by the presence of dye molecules at the surface of the nanoparticles.

3.2 Attempts to control the surface

An attempt to erase the effect of the dye on the surface has consisted on the growth of a non fluorescent silica shell around the primary nanoparticles. The Fig.3 shows that this encapsulation has been done without agglomeration of several primary nanoparticles into a same silica shell.

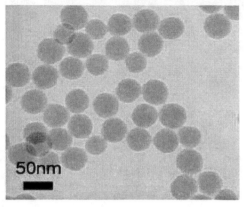

Figure 3: TEM image of the silica nanoparticles encapsulated.

A silica shell of 5 nm in diameter has been grown around the primary nanoparticles. This silica shell has an effect on the Zeta Potential behavior of the fluorescent silica nanoparticles. The Fig.4 shows that encapsulated silica nanoparticles containing Rhodamine B have now the same surface, so there might be no more dye at the surface.

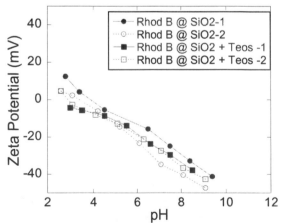

Figure 4: : Zeta Potential as a function of the pH for the silica nanoparticles with and without shell.

This method is therefore an efficient way to control the surface of fluorescent nanoparticles synthesized by reverse microemulsion and so to make efficient nanotracers.

In order to fill homogeneously these nanoparticles into hydrophilic or hydrophobic polymers, it is now essential to modify their surface with adapted chemical molecules. Two molecules have been tested: Di-Apts and FDTS. The first characterization performed is the behavior of the zeta potential of the chemically modified nanoparticles as a function of the pH.

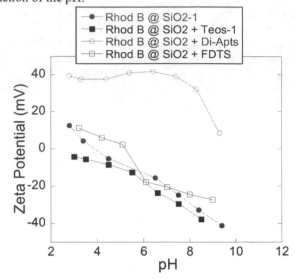

Figure 5: Zeta Potential as a function of the pH for the functionalized silica nanoparticles.

The Zeta Potential becomes highly positive with the Diamino group at acidic pH values and the isoelectric point is shifted toward higher pH values. As a consequence, amino modified nanoparticles will be highly dispersed at low pH values. The fluoric function also modified the surface of the nanoparticles because the Zeta Potential becomes relatively lower in absolute value and the isoelectric point is shifted. Therefore the nanoparticles have been successfully modified. The most hydrophobic seems to be the fluoric functionalized ones.

To control the hydrophobicity of the nanoparticles a Dichloromethane test is done (Fig.6).

	Di-Apts	FDTS	Teos
Ratio	**0,967**	**0,509**	**0,915**

Figure 6: Ratio Hydrophilic/Hydrophobic character for chemically modified silica nanoparticles.

This test confirms that the fluoric compound confers an hydrophobic character to the fluorescent silica nanoparticles and that the diamino molecule gives the most hydrophilic nanoparticles. These two kinds of modified nanoparticles seem to be well suited to fill hydrophilic and hydrophobic polymers.

It is now important to verify the effect of the encapsulation and the different graftings on the intrinsic fluorescence properties of the silica nanoparticles. Indeed the idea is to provide fluorescent nanoparticles whose properties are not dependant on the surface, especially for the emission wavelength. The spectroscopic measures performed Fig.7 show that there is no difference between the dye fluorescence, the fluorescent silica nanoparticles and the functionalized silica nanoparticles.

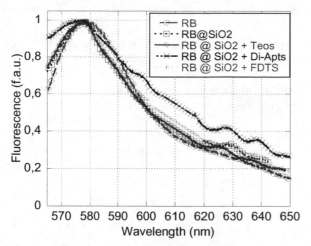

Figure 7: Fluorescence spectroscopy of fluorescent silica nanoparticles doped with Rhodamine B with different surfaces and compared to free Rhodamine B for an excitation at 545 nm.

3.3 Incorporation into a polymer matrix

The nanoparticles being characterized, the aim of the study is now to qualify the level of dispersion obtained into an hydrophobic polymer (PMMA) and an hydrophilic one (PVA). The Fig.8 shows that hydrophobic nanoparticles are more adapted than hydrophilic ones to be homogeneously incorporated into a PMMA matrix. Identically, hydrophilic nanoparticles are more suited for an hydrophilic PVA matrix. So the surface functionalization previously described is essential to achieve a better dispersion of the nanoparticles into a polymer matrix.

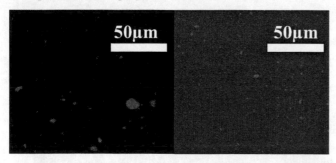

Figure 8: Confocal imaging of fluorescent silica nanoparticles doped with Rhodamine B with different surfaces (left: hydrophilic; right: hydrophobic) into a PMMA matrix.

The use of fluorescent silica nanoparticles is therefore an interesting way to characterize the homogeneity of the incorporation of nanoparticles into polymers.

The Fig.9 shows moreover that homogeneous dispersions can easily be obtained, combining cavitation forces provided by ultrasonication and an adapted surface functionalization.

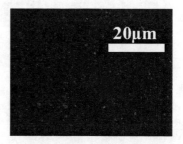

Figure 9: Confocal imaging of hydrophilic fluorescent silica nanoparticles doped with Rhodamine B into a PVA matrix.

4 CONCLUSION

Fluorescent silica nanoparticles are efficient nanotracers to monitor the incorporation of nanoparticles into polymers. Their fluorescence and surface properties are versatile and can be easily tailored. The homogeneous incorporation of hydrophobic nanoparticles into an hydrophobic polymer has been shown, so as the heterogeneous incorporation of hydrophilic nanotracers into an hydrophobic polymer.

REFERENCES

[1] C. L. Wu, M. Q. Zhang, M. Z. Rong and K. Friedrich, Composites Science and Technology, 65, 635-645, 2005.

[2] P. Rosso, L. Ye, K. Friedrich and S. Sprenger, Journal of Applied Polymer Science, 100, 1849-1855, 2006.

[3] F. Goubard, F. Vidal, R. Bazzi, O. Tillement, C. Chevrot and D. Teyssié, Journal of Luminescence, 126, 289-296, 2007.

[4] E. B. Souto, S. A. Wissing, C. M. Barbosa and R. H. Müller, European Journal of Pharmaceutics and Biopharmaceutics, 58, 83-90, 2004.

[5] S. Klitzke and F. Lang, Colloids and Surfaces : Physicochem. Eng. Aspects, 303, 249-252, 2007.

[6] K. N. Phar, D. Fullston and K. Sagoe-Crentsil, Journal of Colloid and Interface Science, 315, 123-127, 2007.

[7] T. Jesionowski, Colloids and Surfaces A : Physicochem. Eng. Spects, 222, 87-94, 2003.

[8] T. Jesionowski and A. Krysztafkiewicz, Colloids and Surfaces A : Physicochemical and Engineering Aspects, 207, 49-58, 2002.

[9] R. P. Bagwe, L. R. Hilliard and W. Tan, Langmuir, 22, 4357-4362, 2006.

[10] T. Jesionowski, Journal of Materials Science, 37, 5275-5281, 2002.

Total Flexibility in Thin Film Design Using Polymer Nanocomposites

T. Druffel[*], O. Buazza, M. Lattis and S. Farmer

Optical Dynamics, 10106 Bluegrass Parkway, Louisville, KY 40299, USA
[*]tdruffel@opticaldynamics.com

ABSTRACT

Optically clear, mechanically flexible nanocomposites are created by doping host polymers with nanoparticles having higher or lower refractive indices. The composite materials preserve the viscoelastic properties of the binding polymers and exhibit infinite tunability between the refractive index limits of the system. These properties provide unique benefits for multi-layer thin film optical filters. Additionally, the nanoparticles are dispersed in a fluid or bound in a polymer matrix in use, thus reducing toxic risks that may be associated with raw particles. A simple and safe method of producing highly engineerable optical filters is presented.

Keywords: nanocomposite, multilayer filter, refractive index, nanoparticle

1 INTRODUCTION

Metal oxide coatings of varying refractive index have been employed in electromagnetic filters ranging from the UV through to the IR regions and have applications as broad band and narrow band pass filters. Filters in the visible region are generally used both to reduce the reflections from the surface of a lens to aid sight and as reflective filters for fashion eyewear. Reflective filters in the IR region serve as heat rejecters in broad band applications and as sensors in narrow band applications.

Thin-film optical filters have been around for over a century and chemical vapor deposition techniques have been predominately the manufacturing choice. The technique generally includes the deposition of metal-oxide ¼ wavelength thin film layers of varying refractive index to get a change in the optical response from the surface of a substrate. These can include broad band antireflective and reflective coatings as well as edge and band gap filters.

Traditional vacuum deposited anti-reflective coatings have been around since the 1930's and actually performed well when coated on a glass ophthalmic lens since the coatings themselves were ceramic. During the 1970's, manufacturing improvements allowed for polymer lenses to gain general acceptance as an alternative for glass; however, anti-reflective coatings did not fair well on the plastic substrates due to the major differences in the strain behavior of the coating and the lens. Significant progress has been made in this technology, but the disparity in the strain domains continues to be an issue.

Deposition of these layers onto a polymer substrate have typically been accomplished using vapor deposition techniques such as CVD, and sol-gel methods. In these cases the deposited metal oxides (ceramics) have mechanical properties that have strain domains that differ significantly from the polymer substrate. (figure 1) This has caused problems with crazing and cracking of the thin film filters during high strain phenomena such as thermal cycling and mechanical deformations.[1, 2]

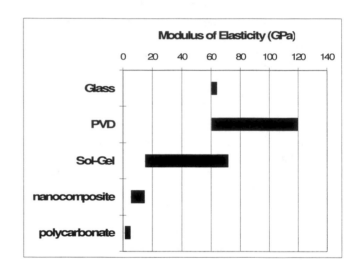

Figure 1: Comparison of the modulus of elasticity of coating materials and a polymer substrate.

More recently researchers have proposed and successfully employed nanoparticles in polymers to build a nanocomposite having an engineered refractive index. These films typically employ spherical particles that have dimensions that are under 100 nm in a polymer. Similar to the artisans of cathedral glass, some have turned to metal chlorides.[3] Yu et. al. produced thin films on the order of several microns using a colloidal silica and acrylic monomer cured in the presence of heat.[4] Others have developed flexible polymer and nanoparticle multilayers into an antireflective coating.[5]

A multilayer nanocomposite coating can be used as a narrow-band reflector whose peak wavelength, bandwidth, and reflectance are determined by the thicknesses and refractive indices of the layers.[6] Traditional deposition techniques using pure dielectric layers are limited to discrete refractive indices and the resulting coatings are extremely brittle.[7]

We have developed a method to deposit a nanocomposite thin film in which the refractive index of the film is tunable over a large range.[8] The method relies on evaporative spin coating in which a solvent born, liquid dispersion of nanoparticles and UV curable monomers is deposited onto a spinning substrate and is subsequently cured. Dispersions of nanoparticles in a carrier solvent are readily available; however, we have synthesized the titanium dioxide dispersion. and in this case titanium dioxide was produced in a solvent.[9] Nanocomposite coatings of up to 15 layers have been prepared for the visible and NIR regions with very discrete layers (as shown by transmission electron microscopy).[10] The resulting stacks exhibit very good mechanical flexibility on polymer substrates as shown using nanoindentation techniques.[11]

Here we demonstrate the abilities of nanocomposites to control filter response through their engineerable refractive indices. With layer counts exceeding 30 (more than twice the layer count of our previous efforts), very little haze is observed. Three formulations of titanium dioxide and UV curable monomers were prepared to produce the following characteristics in the finished films: (1) 1.65 refractive index with a thickness of 73 nm (2) 1.75 refractive index with a thickness of 69 nm (3) 1.85 refractive index with a thickness of 65 nm.

2   BACKGROUND

Our group has been investigating the use of nanoparticles in a UV curable monomer. This requires stabilization methods of nanoparticles favoring the use of solvents that are compatible with the monomers and work well for the spin on process. The spin coating method has been widely studied and allows for a simple low cost deposition of thin films. This simple technique deposits nanocomposite films with high visible light transparence while maintaining the modified refractive index and mechanical strength.

The nanocomposite thin films consist of metal-oxide nanoparticles and a UV-cured acrylate polymer which acts as a binder. The nanoparticles are used to both engineer the refractive index of the individual layers and to improve the mechanical properties of the film. The nanoparticles are initially suspended in a solvent along with an acrylate monomer and a photoinitiator. To insure transparency and avoid light scattering, it is very important that the nanoparticles are not agglomerated and that the primary particle size be much less than 100 nm .

The refractive index of the layers is controlled by adjusting the volume ratio of nanoparticles and monomer, with the refractive index bounded by the pure monomer at the minimum and at approximately 60 volume percent nanoparticles at the maximum. This corresponds well with the theoretical close packing of spheres, and we have noted that the refractive index and modulus of the films reaches a maximum at this point.

The thin film filter design utilizes a simple stack of ¼ wave thickness layers of alternating high and low refractive index materials. The reflectivity off a planer surface from a wave perpendicular to it is

$$R = \left(\frac{n_0 - n_{sub}}{n_0 + n_{sub}} \right)^2 \qquad (1)$$

where

R = reflection
n_o = index of refraction of air (1.00)
n_{sub} = index of refraction of the substrate

If polycarbonate is the substrate (n=1.586 at 550 nm) then the reflection of an incident light wave perpendicular to a surface is 5.1 percent. In this case the reflectance is a function of the refractive index which varies with the wavelength. A coating of multiple thin films of different refractive indices on the substrate can be used to interfere with the reflected waves and for a film having a ¼ wave thickness the reflectance can be computed as

$$R = \left(\frac{(n_0 - Y)}{(n_0 + Y)} \right)^2 \qquad (2)$$

In this case the refractive index of the substrate has been replaced by the admittance of the surface (Y) and is a ratio of the total tangential magnetic and electric fields as described by Macleod.[6] The admittance for a system of i alternating high and low index films is

$$Y = \left(\frac{n_{high}^{(i+1)}}{n_{sub} n_{low}^{(i-1)}} \right) \qquad (3)$$

The first three equations can be used to model the magnitude of reflectance for a multi-layered filter. Continuing with the case of a dual refractive index system, the width of the notch filter is a function of the ratio of refractive indices of the layers

$$\Delta g = \frac{2}{\pi} \sin^{-1} \left(\frac{n_{high} - n_{low}}{n_{high} + n_{low}} \right) \lambda \qquad (4)$$

where

Δg is the half width of the notch

From this equation it can be seen that as the ratio of refractive indices increases, the width of the notch filter will do the same.

This simple relationship allows for the determination of the filter response at a specified wavelength (λ) for which the layers thickness is equal to $\lambda/4d$, where d is the optical thickness of the film. The optical distance is the product of the physical distance and the refractive index of the medium which relates to the phase shift of light traveling through a vacuum at this physical distance. In the case that the layer thickness is not a quarter wave thickness, then the computation of the reflectance is more rigorous and generally requires the use of a computer

3 EXPERIMENTAL

The individual layers in the stack were spin-coated onto a substrate using a machine by Optical Dynamics. This well understood technique controls the layer thickness by balancing the centrifugal forces of a developing thin film to the viscous forces that increase as evaporation takes place. The repeatability of this method is extremely high as long as the coating environment is controlled such that the evaporation rate stays constant. This method can also be extended to coat surfaces with roughness on the order of several microns.

After the solvent is evaporated 50-150 nm film of a UV-curable monomer and nanoparticles remain. In this case the monomer is a trimethylolpropane triacrylate (TMPTA). The film is then cured using a pulse xenon UV source lamp, leaving a polymer nanoparticle composite. Subsequent layers are then added on top of the previous layer to build the filter.

The thickness of the layers was determined by applying and curing the individual film onto a hard substrate such as glass. Part of the film was scratched from the surface of the glass and the step height determined using a profilometer (model number XP-1) by Ambios Corporation. The accuracy of the method was confirmed by measuring thickness in several locations and on multiple layers.

The optical response of the final article is measured using a contact probe spectrophotometer model F20 by Filmetrics. The refractive index of the two layers was determined using equations (2) and (3) and the physical thickness as measured by profilometry. The software package TFCalc by Software Spectra, Inc. was used to confirm the calculations of the equations.

The titanium dioxide dispersion was synthesized in our laboratory using a hydrothermal process. The titania nanoparticles have a mean particle size diameter of approximately 20 nm and are functionalized to improve the stability in an alcohol. The functional groups on the particles also aid in the adhesion to the polymer matrix, which keeps the films from cracking under the large strains.

4 RESULTS

To confirm the infinite refractive index control of these nanocomposites we choose to build optically reflective filters centered at 480 nm. The refractive idecis of each of the layers was controlled by varying the ratio of nanoparticles to polymer. The intent was to show that the increase in layers produced the correct response in reflectance and that the width of the notches followed the ratio of nanoparticle loadings.

In this study we have used an anatase form of titanium dioxide nanoparticle to engineer the refractive index. The refractive index of the cured TMPTA is approximately 1.48 and the anatase titanium dioxide is approximately 2.2. At a volume percentage of 60 percent the theoretical refractive index would be 1.91, however the surface modifications to the nanoparticles has dropped the refractive index.

Three refractive indices of 1.65, 1.75 and 1.85 were chosen to demonstrate the refractive index engineering capabilities. Figures (2) and (3) show the response of multi-layer of two refractive indices and indicate that the reflectance magnitude increases as the layers are added as expected. In addition the width of the filter is also directly controlled by varying the ratio of nanoparticles in each film which is directly related to the refractive index. The calculated reflectance magnitude is reported in table 1.

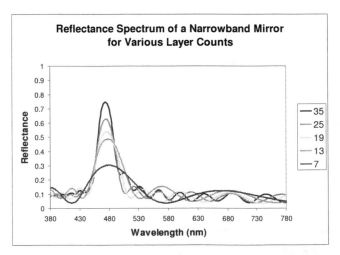

Figure 2: Narrow reflectance filter using refractive indices of 1.75 and 1.85 with layer counts of 7, 13, 19, 25 and 35.

NSTI-Nanotech 2008, www.nsti.org, ISBN 978-1-4200-8503-7 Vol. 1

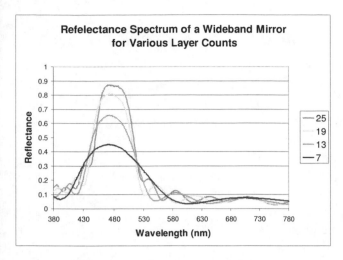

Figure 3: Wide reflectance filter using refractive indices of 1.65 and 1.85 with layer counts of 7, 13, 19 and 25 layers.

Layers	Narrow	Broad
7	27.3	40.8
13	40.0	64.0
19	52.2	79.9
25	62.9	89.4
35	76.7	

Table 1: Calculated reflectance magnitudes of the narrow and broad filters.

Finally to test the mechanical flexibility of the system we simply applied the films to a very thin substrate and bent it around an extremely small radius as shown in figure 4. The results is that there was no damage to the films.

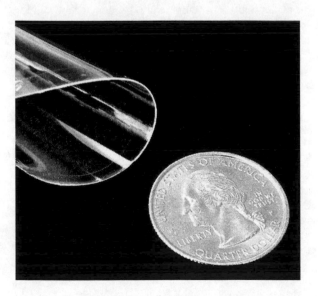

Figure 4: Cast, flexible substrate having a nanocomposite anti-reflective coating demonstrating the mechanical flexibility of the finished coating

5 CONCLUSION

The results clearly show that mechanically flexible thin film filters can be assembled using nanocomposites of inorganic nanoparticles embedded in an organic polymer. Furthermore, the use of a composite allows for an infinite tunability of refractive index between the boundaries of the materials used. The measured results of the final films closely resemble their theoretical models. The goal of this work was to show the capabilities of this simple process in building multilayer thin films that are both optically and mechanically flexible.

REFERENCES

[1]   C. Charitidis, M. Gioti, S. Logothetidis, S. Kassavetis, A. Laskarakis, I. Varsano, *Surface and Coatings Technology* **2004**, *180-181*, 357.
[2]   Y.-W. Rhee, H.-W. Kim, Y. Den, B. R. Lawn, *Journal of American Ceramics Society* **2002**, *83*, 1066.
[3]   H. M. Zidan, M. Abu-Elnader, *Physica B* **2005**, *355*, 308.
[4]   Y.-Y. Yu, W.-C. Chen, *Materials Chemistry and Physics* **2003**, *82*, 388.
[5]   Z. Wu, J. Walish, A. Nolte, L. Zhai, R. E. Cohen, M. F. Rubner, *Advanced Materials* **2006**, *18*, 2699.
[6]   H. A. Macleod, *Thin Film Optical Filters*, Institute of Physics, London **2001**.
[7]   S. Grego, J. Lewis, E. Vick, D. Temple, *Thin Solid Films* **2007**, *515*, 4745.
[8]   K. Krogman, T. Druffel, M. Sunkara, *Nanotechnology* **2005**, *16*, S338.
[9]   N. Mandzy, E. A. Grulke, T. Druffel, *United States Patent Application*, **2005**.
[10]   T. Druffel, N. Mandzy, M. Sunkara, E. Grulke, *Small* **2007**, *accepted*.
[11]   T. Druffel, K. Geng, E. Grulke, *Nanotechnology* **2006**, *17*, 3584.

Structural and Magneto-Optical Properties of Co-doped ZnO Thin Films Prepared by Sol-Gel Method

J. Neamtu[*1], C.M. Teodorescu[**], G. Georgescu[*], J. Ferré[***], T. Malaeru[*], I. Jitaru[****]

[*]National Institute of Electrical Engineering, Splaiul Unirii 313, 074204 Bucharest, Romania, jenica.neamtu@gmail.com
[**]National Institute of Materials Physics, Atomistilor 105b, 077125 Magurele, Romania, teodorescu@infim.ro
[***]Laboratoire de Physique des Solides , UMR CNRS 8502, Bât. 510, 91405 Orsay CEDEX, France, ferre@lps.u-psud.fr
[****]"Politechnica" University Bucharest, Splaiul Independentei 313, 060042 Bucharest, Romania, ioanajitaru2002@yahoo.com

ABSTRACT

Recent experiments evidenced room temperature ferromagnetism in ZnO:Co diluted magnetic semiconductors, however the diluted semiconductor nature and the ferromagnetism strongly depends on the preparation method employed. We have investigated the ability of the sol-gel route for the synthesis of cobalt doped ZnO films grown on Si (100) and glass substrates. A homogeneous and stable $Zn_{1-x}Co_xO$ sol was prepared by dissolving zinc nitrate hexahydrate and cobalt acetate tetrahydrate in a PVP (polyvinylpyrrolidone) solution, followed by annealing at 800 ºC. Local structure studies of the $Zn_{1-x}Co_xO$ thin films by XANES (X-ray absorption near-edge structure) and EXAFS (extended X-ray absorption fine structure) proved the co-existence of a few amount of small metal cobalt aggregates with Co in non-stoichiometric Co_nO_m aggregates, with $n > m$, and $m \sim 4$. Low temperature (1.5 K) Kerr magnetometry give a direct proof of a superparamagnetic behavior of the magnetic aggregates and allowed an estimate of the number of magnetically active atoms in Co aggregates, which is close to the number of Co excess atoms inferred from XANES and EXAFS considerations: $n - m \sim 3$ atoms per aggregate. To this main superparamagnetic phase, a weak ferromagnetic phase with coercitive field of $\sim$ 200 Oe is superimposed, most probably duc to mctal cobalt nanoclusters.

Keywords: zinc oxide; cobalt doping; sol-gel growth; XAFS; magnetic properties

1 INTRODUCTION

Diluted magnetic semiconductors (DMS) are structures with magnetic ions inserted into semiconductor lattices, and ferromagnetic ordering is mediated by charge carriers [1]. A few years ago, it turned out that most of incomplete $3d$ shell metal ions can be used to produce ZnO-based DMS with room temperature ferromagnetism [2]. However, the preparation conditions and levels of doping are crucial for the achievement of this goal. Of the $3d$ elements, Co is of special interest, since it provides a good compromise between the ionic radius matching that of zinc (0.58 vs. 0.6 Å in tetrahedral coordination; 0.75 vs 0.74 Å in octahedral coordination), allowing easy substitution of Zn in the ZnO wurtzite structure. Measurable ferromagnetism at room temperature was reported in cobalt-doped zinc oxide [3].

However, the origin of this ferromagnetism is controversial. First studies assessed an extrinsic origin of this ferromagnetism [4] connected to non-substitutional positions of Co ions in the ZnO lattice. Further studies evidenced the origin of ferromagnetism as being Co aggregates or clusters embedded in ZnO [5]. Subsequent structural and magnetic determinations revealed substitutional placement of Co ions into the ZnO lattice [6].

The sol-gel method was successfully utilized in the past to synthesize room temperature Co-doped zinc oxide [6]. We adopted this technique since it is relatively cheap and provides control of the doping concentrations. As will be seen, the method did not provide diluted magnetic semiconductor properties, but instead the synthesis of very small cobalt oxide non-stoichiometric aggregates with a weak ferromagnetic phase superimposed.

Kittilstved *et al.* [7] investigated both the pre-edge peak in the X-ray absorption near-edge structure (XANES) and the extended X-ray absorption fine structure (EXAFS) in cobalt doped ZnO prepared by chemical vapor deposition. Striking similarities are reported between the Fourier transforms of the Co K-edge and the Zn K-edge EXAFS signals, while all statements about the pre-edge peak in XANES (generated by $1s$? $3d$ dipole forbidden transitions) rely on the similarity of the Co K-edge spectrum irrespective on the sample processing.

In this Communication we report joint magnetometry (magneto-optical Kerr effect, MOKE) and local structure (XANES and EXAFS) studies of sol-gel synthesized ZnO:Co thin films, in order to emphasize the

---

[1] corresponding author

interplay between structure, stoichiometry, and magnetic properties. The analysis of the pre-edge peak in XANES is used in order to quantify the number of $3d$ vacancies per absorbing atom and thus to infer the Co ionization state [8]. EXAFS analysis determines the local atomic environment and the average aggregate size and composition. Magnetic measurements, which evidenced superparamagnetic behavior at low temperatures (1.5 K), are used in order to derive the spin configuration of Co ions.

2 EXPERIMENTAL

$Zn_{1-x}Co_xO$ ($x = 0.2 – 0.3$) films were synthesized by using spin-coating method; the films were grown by sol-gel process either on Si (100) or glass substrates. Zinc nitrate hexahydrate $Zn(NO_3)_2 \cdot 6H_2O$ and cobalt acetate tetrahydrate $Co(CH_3COO)_2 \cdot 4H_2O$ were used as precursors for preparation, and $(1 - x)$ $Zn(NO_3)_2 – (x)$ $Co(CH_3COO)_2 –$ PVP – H_2O (where PVP are polyvinyl-pyrrolidone) as gelling agents. A mixture of $Zn(NO_3)_2$, $Co(CH_3COO)_2$ and PVP in aqueous solution was refluxed 30 hours at 65 °C under continuous stiring. The polycrystalline Co-doped ZnO thin films have been deposited from the above constituents by spin coating method (1500 RPM 30 seconds) on Si/SiO₂ and Crown glass substrates. The pre-heated temperature for film stabilization after each layer deposition was 230 °C. The final films have been crystallized at 800 °C, during 1 – 4 hours.

Room and low temperature ZnO:Co (1.5 K) magneto-optical (MO) measurements were performed using green He-Ne laser light. X-ray absorption fine structure (XAFS: EXAFS and XANES) measurements were performed at the Doris storage ring facility in Hasylab, Hamburg, Germany (A1 beamline), by using a Si(111) double crystal monochromator. The measurements on Co-doped zinc oxide thin layers were performed in fluorescence mode by using a 7-pixel SiLi detector and selecting the channels corresponding to the Co fluorescent radiation. For consistent data interpretation, standards of metal Co and of Co(II)O were measured in the same run, in transmission mode on a 5 μm Co foil for the metal and of a pressed pellet of CoO powder mixed with cellulose.

3 RESULTS AND DISCUSSIONS

Irrespective of preparation conditions, the XANES and EXAFS spectra of cobalt doped ZnO looked similar, both at the Co K-edge and at the Zn K-edge. This allowed the determination of Co concentration into ZnO, by comparison between the Co and Zn fluorescence signals. This concentration (atomic ratio Co:Zn) ranges from 3.6 ± 0.2 % to 10.8 ± 0.3 %. Therefore, a first result of this study would be that, in this concentration range, the local order about Co and Zn atoms is sensibly the same. Therefore, in the following discussion all spectra obtained at the Co K-edge were summed in order to improve the statistics.

3.1. X-ray absorption near-edge structure spectroscopy (XANES)

Figure 1 presents XANES spectra obtained at the Co K-edge on metal Co, on Co-doped ZnO, on Co(II)O. The first remark is the striking similarity between the XANES of the Co (II) oxide and that of the Co in ZnO, with the exception of a small chemical shift of about 0.5 eV. Moreover, both Co K-edge XANES spectra exhibited the pre-edge peak, which is a sign of dipole forbidden transitions [19,24] $1s$? $3d$. Assuming that the integral amplitude of this peak is proportional to the number of $3d$ holes and the $4s^0 3d^7$ electronic configuration of Co^{2+} in CoO, one could infer the number of $3d$ vacancies as being $n_h(3d) = 2.0 \pm 0.1$ in ZnO:Co. From here, with a good approximation, a $3d^8$ configuration can be derived. The question is whether this configuration belongs to Co^+ ($4s^0 3d^8$) or to neutral Co atoms ($4s^1 3d^8$). The general aspect of XANES spectra and also the derived energy of the Co K edge suggest that the hypothesis of neutral Co can be ruled out, at least in what concerns the pre-edge peak: the XANES spectrum of metallic Co has completely different XANES resonances and also the pre-edge aspect is different. Thus, one obtains the value of the ionization state of Co in the ZnO:Co samples, which is (in average) $+1.0 \pm 0.05$.

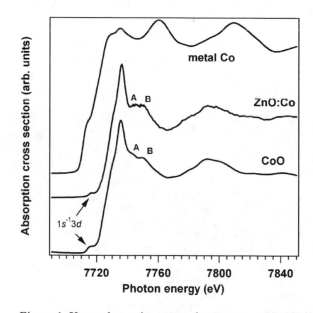

Figure 1: X-ray absorption near-edge structure (XANES) at the cobalt K-edge of metal cobalt, cobalt doped zinc oxide, and cobalt (II) oxide.

3.2. Extended X-ray absorption fine structure spectroscopy (EXAFS)

To gain more insight in the interplay between Co substituting Zn in ZnO and Co forming cobalt (II) oxide, we proceed with the analysis of the EXAFS data. The EXAFS function is defined as the relative variation of the absorption coefficient with respect to the atomiclike case $[\mu_0(k)]$: $\chi(k) = [\mu(k) - \mu_0(k)] / \mu_0(k)$. k is the photoelectron wave vector, $hk^2 / (2m) = h\nu - E_0$, where E_0 is the energy of the absorption threshold and $h\nu$ is the photon energy ($h = h/(2\pi)$ the Planck constant, m the electron mass). Figure 2 presents the k^3-weighted EXAFS spectrum of the three situations discussed previously. Figure 3 presents the moduli of the Fourier transforms of the k^3-weighted EXAFS function. The EXAFS data analysis followed the standard procedure discussed in Ref. [9]. Again, from Fig. 2 the presence of Co nanoclusters can be ruled out, since the EXAFS spectrum of metal particles should not deviate considerably from the EXAFS of bulk metal, which is quite different from the Co K-edge EXAFS of ZnO:Co. The EXAFS spectrum which is the most similar to that of Co in ZnO is again that of cobalt (II) oxide. This result is in line with the XANES observation.

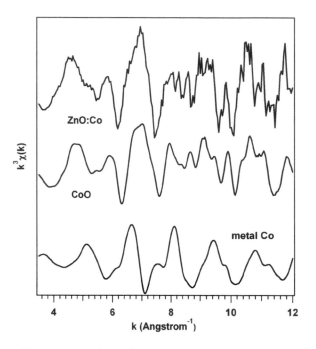

Figure 2: Cobalt K-edge extended X-ray absorption fine structure (EXAFS), weighted by the cube of the photoelectron wave vector k of metal cobalt, cobalt doped zinc oxide, and cobalt (II) oxide.

More precise analysis can be done with the Fourier transforms (FT) of the EXAFS function (Fig. 3). Again, the Fourier transform of the EXAFS of Co metal is completely different. The Co EXAFS spectrum of Co in ZnO is close to the Co EXAFS spectrum in Co(II)O in the cubic rocksalt structure [10], at least in what concerns the first three coordination shells (with interatomic distances of 1.64, 2.58, and 3.13 Å). This is a first considerable difference from the EXAFS data presented in Ref. [7], where the FT of the Co K-edge EXAFS in ZnO:Co are similar to that of the Zn K-edge EXAFS in ZnO. More details about this

comparison are discussed elsewhere [11]. The cobalt (II) oxide rocksalt (NaCl) structure presents around each Co atom a first coordination shell at distance $a/2 = 2.27$ Å [27] constituted by 6 oxygen atoms, followed by a second coordination shell at distance $a/2^{1/2} = 3.21$ Å with 12 Co atoms and a third shell at $3^{1/2}a/2 = 3.94$ Å with 8 oxygens. The difference of 0.6-0.8 Å with respect to the observed interatomic distances in the Fourier transform are due to the k dependence of phase shifts involved in the EXAFS function, namely the backscattering phase shift and the central atom phase shift [9].

However, the amplitudes of the FT peaks of the Co EXAFS in ZnO:Co are different from that of bulk Co(II)O. These aspects are detailed in Ref. [11]; here we will present just the conclusions. The most plausible explanation of the FT amplitude variation is that the largest amount (~ 87 %) of cobalt forms non-stoichiometric cobalt (II) oxide aggregates with stoichiometry close to Co_7O_3, with a smaller amount (~ 13 %) which forms cobalt (Co^0) nanoclusters.

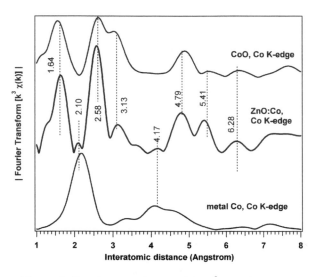

Figure 3: Fourier transforms of the k^3-weighted EXAFS functions from Figure 2.

3.3. Magneto-optical Kerr effect (MOKE) measurements

Figure 4 presents MCD- magneto-optical Kerr effect (MOKE) measurements obtained at low temperature (1.5 K) on the film with Co atomic concentration of 11 %. The MOKE data suggest the coexistence of the superparamagnetic phase with a weak ferromagnetic phase with coercitive field of about 50 Oersted. The experimental data were fitted a constant times the Brillouin function $B_J[g\mu_B B_a/(k_B T)]$, where g is the gyromagnetic factor of the electron (~ 2), μ_B the Bohr magneton, B_a the applied magnetic field (in Teslas) and $k_B T$ the Boltzmann factor. The fit was performed in order to derive the value of the

NSTI-Nanotech 2008, www.nsti.org, ISBN 978-1-4200-8503-7 Vol. 1

total moments participating to the paramagnetism J. For values of J exceeding the individual atomic moments, we may speak about superparamagnetism. The net result of the fit was $J = 3.0 \pm 0.1$, while the resulting temperature was 1.53 ± 0.7 K. Good quality fits were obtained also with smaller values of J, but the resulting temperature was unphysical (below 1 K). Setting values of J larger than 3 in the Brillouin formula visibly degradates the fit. Consequently, it may be inferred that the magnetic properties of these samples are governed by small nanoparticles with moments of about 6 Bohr magnetons, i.e. with three magnetically active metal-like Co atoms ($3d^8$) or two active Co^{2+} ions ($3d^8$).

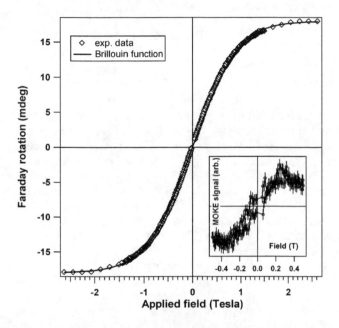

Figure 4: Magneto-optical Kerr effect (MOKE) signals for the ZnO:Co sample with 11 % cobalt atomic percent. Insert: signal in the small field region obtained by subtracting from the experimental data the superparamagnetic contribution.

By subtracting the fit with the Brillouin function from the raw experimental data, one may estimate that the "residual" ferromagnetic phase (insert in Fig. 4) accounts for about 3 % of the observed magnetic signal. This ferromagnetic phase presents a coercive field, estimated to around 0.02 T, i.e. 200 Oe.

4   CONCLUSION

At present, the sol-gel route chosen seems to be inappropriate to synthesize Co-doped ZnO materials with diluted magnetic semiconductor behavior. Nevertheless, the fact that a weak ferromagnetic phase was observed at very low temperature (1.5 K) is encouraging.

Instead, the proposed method seems to be appropriate to synthesize materials where very small Co_nO_m non-stoichiometric aggregates (a few atoms, most probable stoichiometry Co_7O_4) are formed, coexisting with structures where Co is in a chemical state and atomic environment very close to Co metal. It is also inferred that these metal nanoclusters are not very large, being estimated at few more than ten Co atoms. The utility of the present study stems in being amongst the first studies to combine (i) magnetometry, including fitting of the superparamagnetic behavior with: (ii) pre-edge structures in XAFS giving insight about the ionization state and the number of $3d$ vacancies; (iii) analysis of XANES resonances; (iv) analysis of EXAFS spectra. The joint analysis of magnetism and X-ray absorption spectroscopic data yielded a consistent picture about the intimate structure of the synthesized materials.

This work was supported by the Romanian National Authority of Scientific Research under contracts CEEX- No. 69/2005, CEx05-D11-32 No. 32/2005, PN II 71-063/2007.

REFERENCES

[1] T. Dietl, H. Ohno, F. Matsukura, J. Cibert, and D. Ferrand,, Science 287, 1019 (2000); H. Ohno, D. Chiba, F. Matsukura, T. Omiya, E. Abe, T. Dietl, Y. Ohno, and K. Ohtani, Nature 408, 944 (2000);
[2] M. Yamanouchi, D. Chiba, F. Matsukura, and H. Ohno, Nature 428, 539 (2004).
[3] K. Ueda, H. Tabata, and T. Kawai, Appl. Phys. Lett. 79, 988, (2001); M. Kobayashi, Y. Ishida, J.I. Hwang, T. Mizokawa, A. Fujimori, K. Mamiya, J. Okamoto, Y. Takeda, T. Okane, Y. Saitoh, Y. Muramatsu, A. Tanaka, H. Saeki, H. Tabata, and T. Kawai, Phys. Rev. B 72, 201201, (2005).
[4] S.C. Wi, J.S. Kang, J.H. Kim, S.B. Cho, B.J. Kim, S. Yoon, B.J. Suh, S.W. Han, K.H. Kim, K.J. Kim, H.J. Song, H.J. Shin, J.H. Shim, and B.I. Min, Appl. Phys. Lett. 84, 4233, (2004).
[5] X.H. Han, G.Z. Wang, J.S. Jie, X.L. Zhu, and J.G. Hou, Thin Solid Films 491, 249, (2005).
[6] H.J. Lee, S.Y. Jeong, C.R. Cho, and C.H. Park, Appl. Phys. Lett. 81, 4020, (2002).
[7] K.R. Kittilstved, D.A. Schwartz, A.C. Tuan, S.M. Heald, S.A. Chambers, and D.R. Gamelin, Phys. Rev. Lett. 97, 037203, (2006).
[8] D. Mardare, V. Nica, C.M. Teodorescu, and D. Macovei, Surf. Science 601, 4479 (2007).
[9] B.K. Teo, EXAFS: Basic Principles and Data Analysis, Springer, Berlin, 1983.
[10] J. Dicarlo and A. Navrotsky, J. Am. Ceram. Soc. 76, 2465 (1993).
[11] J. Neamtu, J. Ferré, G. Georgescu, T. Malaeru, I. Jitaru, D. Macovei, and C.M. Teodorescu, Thin Solid Films, *submitted*.

Magnetotransport Properties and Tunnel Effect of Thin Film Nano-Structures

Jenica Neamtu[*], M. Volmer[**], R.V. Medianu[***]

[*] National Institute for Research& Development in Electrical Engineering, Splaiul Unirii 313, Bucharest 030138, Romania, jenica.neamtu@gmail.com
[**] Transilvania University Brasov, B-dul Eroilor 29, Brasov Romania, volmerm@unitbv.ro
[***] National Institute for Laser, Plasma&Radiation Physics, Magurele-Bucharest, Romania, rares.medianu@inflpr.ro

ABSTRACT

Magnetotransport and tunnel effect measurements were made on nanostructured thin films. Nano-structures as FM/NM/FM/AFM were deposited by magnetron-sputtering on the oxidized Si wafers, as continuous films and cross-stripe structures. FM (Ferro-Magnetic) denotes NiFe (Py), Co or a combination using Py-Co layers. NM (Non-Magnetic) denotes Cu or Al_2O_3 layers. AFM (Anti-Ferro-Magnetic) denotes FeMn layers. These nano-structures present anisotropic magnetoresistance (AMR) and tunnel magnetoresistance (TMR) effects. On the unpatterned films we made magnetoresistance (MR) and Hall effect measurements in order to study the magnetotransport properties of the samples and surface quality. The effective thicknesses of the oxide layers were estimated by tunnel effect measurements and compared with in situ quartz microbalance measurements.

Keywords: thin films, anisotropic magnetoresistance effect, tunneling effect, Hall effect, microstructure

1 INTRODUCTION

The investigation of the electrical conductivity and magnetoresistance of magnetic multilayers and granular films plays an important role in the design and fabrication process of magnetic micro and nanosystems for spintronics. The aim of this work is to study some multilayered nanosystems and to present a fast and reliable method to characterize the electrical and magnetic properties of these structures. Using magnetoresistance (MR), Hall Effect and Tunneling Effect we can estimate the quality and other properties of the deposited films. Due to low pressure of the Ar gas and the presence of a small amount of O_2 in the deposition chamber, our samples obtained by sputtering method present intermixed regions of FM and NM layers. These intermixed regions are responsible for the increasing of the Hall resistivity which can be very attractive for magnetic sensors. The saturation fields obtained when the magnetic field is applied perpendicular to the film plane are less than the values predicted from the shape anisotropy and give us information regarding the films roughness. The extraordinary Hall Effect can be used as a simple and sensitive tool for the study of the magnetic properties of thin films and can lead to a new generation of magnetic sensors and memory devices.

2 EXPERIMENTAL

The samples investigated were prepared by r.f. magnetron sputtering into a ER 3119 CRYO VARIAN equipment. The base pressure was 5×10^{-7} torr. The Ar pressure during the deposition process was 5×10^{-3} torr and the rate of deposition was 0,01 nm/s. The samples, deposited onto oxidized Si wafers processed at laser quality, were denoted by S_1, S_2, S_3 and S_4. The structure of S_1 is Si/SiO$_2$/Py(17.1nm)/Co(3.3 nm)/Al$_2$O$_3$(1.3 nm)/Co(3.3 nm)/Py(17.1 nm)/FeMn(45 nm)/Py(8.6 nm), the structure of S_2 is Si/SiO$_2$/Py(17.1nm)/Co(3.3 nm)/Cu(2 nm)/Co(3.3 nm)/Py(17.1 nm)/FeMn(45 nm)/Py(8.6 nm) while for S_3 the structure is Si/SiO$_2$/Py(17.1 nm)/Al$_2$O$_3$(1.3 nm)/Py(17.1 nm)/FeMn(45 nm)/Py(8.6 nm) and for S_4 the structure is Si/SiO$_2$/Py(21 nm)/Al$_2$O$_3$(2 nm)/Py(21 nm).

The structures were obtained both as continuous (unpatterned) films and as tunnel junctions (samples $S_{1,3,4}$) by using shadow masks to create a cross-geometry structure of area $2.25\cdot10^{-6}$ m^2 as shown in figure 2a. The insulating barrier Al$_2$O$_3$ was grown by natural oxidization in air of the Al layer at room temperature for 36 h. For electrical conductivity, Hall Effect and tunnel effect measurements we used the dc four point technique. All these configurations are connected through a switching box to a Keithley system composed by a current source 6221 and a nanovoltmeter 2182A. Additionally, two instrumentation amplifiers (EI-1040) and an external DAQ board (Lab Jack U12) were used for test measurements.

3 RESULTS AND DISCUSSION

Figure 1 shows the AFM topography of the sample S_3. The maximum roughness of the film surface is about 5.69 nm. We can see sharp columnar grains that are growing perpendicular to the film surface. This value of the roughness and the surface aspect suggest the existence of intermixing effects between the adjacent FM and NM layers which will produce nonmagnetic layers at the interfaces. These factors will affect the effective thickness of the Al$_2$O$_3$ layers, the barrier height and the amplitude of the tunnel magnetoresistance effect (TMR). The thinner intermixing layer we have, the greater GMR and TMR value we can expect.

Figure 1: The AFM image of the surface for sample S_3

Figure 2 shows the experimental setup and the results of tunnel effect measurements made on samples $S_{1, 3, 4}$. The current-voltage data at room temperature were fitted by Simmons' theory of tunneling [1] to obtain values of tunnel barrier height (F) and the effective thickness (s) of the oxide layer. Excepting sample S_1 the experimental data and the fitting curves show typical behavior for a tunnel junction with metallic electrodes. The estimated thickness of the oxide layer, measured with the quartz microbalance during the deposition process, is smaller, for each sample, than the value obtained from tunnel experiments. For sample S_1 we obtained, as shown in Figure 2a, an almost linear characteristic with a high value, F =7.33 eV, of the tunnel barrier height and an effective thickness of about 7.41 nm. The values for F that we obtained for samples S_3 and S_4 are slight smaller then the values reported for FM/Al_2O_3/FM junctions. On the other hand, the thicknesses that we obtained from tunnel measurements for the oxide layers, s=4.74 nm compared with 1.3 nm for S_3 and s=7.18 nm compared with 2 nm for S_4 as presented in Figure 2b and Figure 2c respectively, show us that the adjacent permalloy layers are partially oxidized [2]. Good insulators like Al_2O_3 lead to F in the eV range whereas the magnetic oxides have F of fractions of eV [3] and lead to a decreases of the measured barrier height. The value for F that we obtained for sample S_3 (about 1.16 eV) is slightly smaller than the value reported for FM/Al_2O_3/FM which is about 1.6 eV. The barrier height for sample S_4 was found to be F =0.54 eV. The value of resistance of the sample S_1 is about 19.7 kO and suggests a high degree of oxidation of the metallic layers. It was shown in [2] that at least 2 nm from the metallic layer deposited on to substrate is oxidized due to SiO_2 layer. On the other hand we can have collision intermixing between the NiFe layer and the Co layer. In addition, during the growth of the Al_2O_3 layer, the metallic layer (Co) is being oxidized. In the same way, we can assume that at least 2 nm from the top layer is also

oxidized. Because of these reasons the structure S_1 presents a high resistance. From the behavior illustrated in Figure 2a, we can assume that the conduction in this structure is made by electron tunneling between metallic grains which are surrounded by thin oxide barriers [4]. These facts explain the high values obtained for the barrier height and width. In other words, our measurements reveal an oxidized granular structure, slightly bellow the percolation threshold. We obtained an activation energy for conduction of about 1.6 eV.

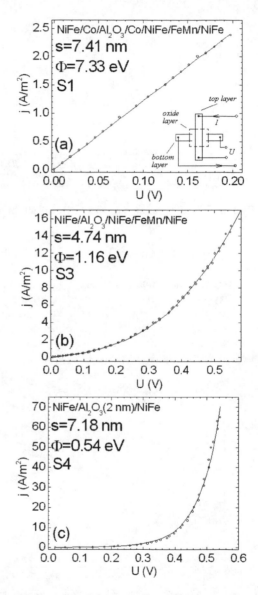

Figure 2: Results of the tunneling effect measurements made on (a) sample S_1, (b) sample S_3 and (c) sample S_4.

The replacement of the Al_2O_3 layer with a Cu layer, in sample S_2, leads to a decrease of about 7 times of the film resistance. The I-U characteristic of the sample is linear. However, the resistance of the sample is still high and we believe that the first two layers (NiFe/Co) are oxidized. The top layers are partially oxidized too. The temperature dependence of the resistance has the same aspect like for S_1 but the activation energy is much lower, about 0.27 eV.

For the sample S_3 we obtained a tunnel magnetoresistance effect (TMR) of about 0.5% with the magnetic field applied perpendicular to the top layer but in the film plane. The field dependence of the TMR effect for this sample is presented in Figure 3.

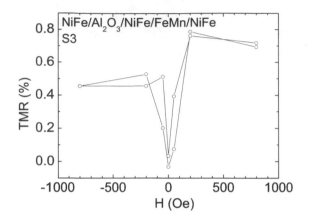

Figure 3: The field dependence of the TMR effect for the sample S_3.

However, because the bottom layer is almost oxidized, we believe that the observed effect is in fact due to anisotropic magnetoresistance effect (AMR) that take place in the top layer. The Hall Effect measurements were performed on the unpatterned films. We obtained a response in magnetic field only for samples S_2 and S_3. Figure 4 shows the field dependencies of the extraordinary and planar Hall Effect (PHE) in samples S_2 and S_3.

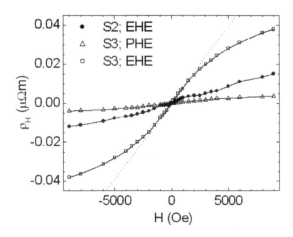

Figure 4: The field dependencies of the Hall resistivity for samples S_2 and S_3; EHE means the extraordinary Hall Effect and PHE the planar Hall Effect.

The high value of the extraordinary Hall resistivity, ρ_H, for sample S_3 reveals a structure which behaves like a ferromagnet-insulator mixture in the vicinity of the percolation threshold [5, 6]. This result is confirmed by the value of the activation energy for conduction E_a=0.91 eV which is lower than the activation energy for sample S_1, that is almost completely oxidized, and higher than the activation energy for sample S_2, which has Cu as interlayer. The unpatterned structure, S_3, presents the giant Hall Effect which is very attractive for magnetic field sensors. The planar Hall Effect is very small for sample S_3 and virtually absent for sample S_2. The absence of the Co layer for sample S_3 permits to saturate more easily. The Cu layer has a shunting effect and lowers the value of the Hall resistivity.

Now, we can compare these values of the Hall resistivity with other measurements made on NiFe (10 nm) and NiFe (2 nm)/Al_2O_3 (1 nm)/NiFe (2 nm) prepared by thermal evaporation and presented elsewhere [7]. For the NiFe sample the magnetic phase is well defined and the PHE effect is relatively high. The sample saturates at a magnetic field which is less than the value predicted from the shape anisotropy and presents hysteresis at low fields [7]. However, the values of the Hall resistivity are about 50 times smaller than the values obtained for sample S_3. Figure 5 shows, for comparison, the field dependencies of the Hall resistivity for NiFe (10 nm) and NiFe (2 nm)/Al_2O_3 (1 nm)/NiFe (2 nm) thin films.

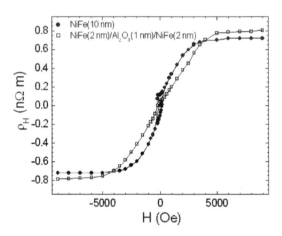

Figure 5: Field dependencies of ρ_H for Py(10 nm) and Py(2 nm)/Al_2O_3(1 nm)/Py(2 nm) thin films (Py denotes NiFe alloy) deposited by thermal evaporation on oxidized Si substrates.

The decrease of the saturation fields is related with the surface quality. This is the roughness effect [8]. For a 10 nm thin film of permalloy obtained by thermal deposition the saturation field decreases to about 4000 Oe. This leads to a roughness effect about 8 nm, which is in good agreement with the AFM measurements. For NiFe (2 nm)/Al_2O_3 (1 nm)/NiFe (2 nm) structure the roughness effect is about 2.5 nm. We expect to have permalloy bridges through the oxide layer. The structure behaves like a granular ferromagnet above the percolation limit which is about 1.5-2 nm for permalloy layers [8]. On the other hand the films deposited by sputtering method at low Ar pressure are smooth. From Hall Effect measurements we have for sample S_3 a saturation field of about 7000 Oe. This leads to a roughness effect of about 7 nm which is very low compared with the film roughness. These electrical measurements show the importance of the deposition rate related with the Ar pressure. At low deposition rate, as we had in this work, the rate of the gaze incorporation in the layers is high and affects the conduction properties [9]. Even the sputtering method becomes in this case a not very clean one. The conduction mechanism will be mainly by multiple tunneling effects between metallic grains separated by non conducting regions due to oxides and gazes incorporated on the substrate. This effect is illustrated in Figure 6, for sample S_1. The I-V characteristic is nonlinear and confirms the presence of isolated regions between the metallic grains.

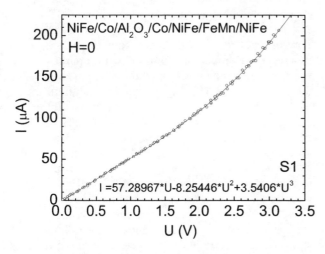

Figure 6. The conduction measurements made on the unpatterned sample, S_1.

4 CONCLUSIONS

We used electrical conductivity, tunnel effect, Hall effect and MR effect measurements to investigate the magneto-electric properties and the quality of some samples deposited by RF magnetron sputtering. At low Ar pressure we obtained structures that are rather granular films than well defined multilayer structures. Using tunneling experiments we can estimate the thickness and the quality of the oxide interlayer. By controlling the oxidation process these samples can exhibit giant Hall Effect which is very attractive for practical applications.

REFERENCES

[1] J. G. Simmons, J. Appl. Phys. **34**, 1793, 1963.

[2] T. S. Plasket, P. P. Freitas, N. P. Barradas, M. F. da Silva and J. C. Soares, J. Appl. Phys. **76**, 6104, 1994.

[3] J. S. Moodera, G. Mathon, J. Magn. Magn. Mater. **200**, 524, 1999.

[4] Y. Suezawa, Y. Gondo, J. Magn. Magn. Mater. **126**, 524, 1993.

[5] A. Gerber, A. Milner, J. Magn. Magn. Mater. **242-245**, 90, 2002.

[6] J.C. Denardin, M. Knobel, X.X. Zhang, A.B. Pakhomov, J. Magn. Magn. Mater. **262**, 15, 2003.

[7] M. Volmer, J. Neamtu, J. Magn. Magn. Mater. **272-276**, 1881, 2004.

[8] T. Lucinski, G. Reiss, N. Mattern, L. van Loyen, J. Magn. Magn. Mater. **189**, 39, 1998.

[9] Milton Ohring, The Materials Science of Thin Films, Academic Press, 95-96, 1992, ISBN 0-12-524990-X

Gold nanorods: Synthesis and modulation of optical properties

N. R. Tiwari[*], S. A. Kalele[*] and S. K. Kulkarni[*]
[*]DST Unit on Nanoscience, Dept of Physics, University of Pune, Pune-411007.
India
Tel: +91-20-25692678-319
Fax: +91-20-25691684
e-mail: neha@physics.unipune.ernet.in

Abstract

Gold nanorods are increasingly becoming popular due to their potential in the field of biotechnology, photonics and optoelectronics. They display optical extinction which can be tuned from visible to infra red region of the electromagnetic spectrum just by changing the aspect ratio (ratio of length to width). However, synthesis of gold nanorods of desired aspect ratio by chemical method is a challenging task. Numerous parameters involved in the synthesis makes the control of the aspect ratio rather difficult. It is also important to reduce the percentage of spherical nanoparticles formed as a byproduct. We have investigated the effect of all the parameters involved in the synthesis of nanorods on the longitudinal plasmon band and its tunability. Also, transformation of gold nanorods to peanut like shape was observed when they were subjected to different pH environments. The mechanism of this shape transformation is explained in detail.

Keywords: gold nanorods, optical properties, seed mediated growth, pH, shape transformation

Metal nanoparticles offer a wide area of application owing to their size and shape dependent optical properties [1]. Recently, there has been significant interest in gold nanorods due to their potential application in the field of biomedical imaging, photo thermal therapy and optoelectronics [2]. Gold nanorods exhibit two different absorption bands namely transverse and longitudinal plasmon band corresponding to oscillation of electrons along the width and length of nanorods when electromagnetic radiation is incident on them. The position of longitudinal plasmon band can be shifted towards infra red region by changing the aspect ratio of nanorods. Gold nanorods of different aspect ratios have been synthesized by various techniques with considerable success [3, 4]. Aspect ratio of gold nanorods is found to be extremely sensitive to nucleation conditions which are further dependent on large number of experimental parameters involved in the seed mediated synthesis.

We have synthesized gold nanorods in aqueous solution by seed mediated growth method and studied the dependence of longitudinal plasmon band on various experimental parameters like temperature, pH, amount of seed, ageing of seed, and other experimental conditions. Apart from these, we observe a strong modulation of optical properties of pristine gold nanorods on addition of KOH in different amounts [5]. Significant blue shift in the absorption spectra was observed on increasing the amount of KOH with the effect being maximum for the rod having highest aspect ratio. We attribute the blue shift in the absorption spectra to the pH dependent shape transformation of gold nanorods and the mechanism is explained in detail. A slight change in the shape of gold nanorod can lead to significant changes in their optical absorption spectrum. Therefore, there is a need to study the shape transformations of gold nanorods by various methods so as to ensure their optical and mechanical stability for applications in sensors or optoelectronic devices.

Experimental

Gold nanorods of different aspect ratios were synthesized by seed mediated growth method. The seed mediated growth method [3] consists of two steps: 1) Formation of spherical nanoparticles of size 3-4 nm and 2) growth of these particles in a growth solution which contains a surfactant, week reducing agent, silver nitrate.

Spherical gold nanoparticles (seed): Five milliliters of 0.2 M CTAB solution was mixed with 5 ml 5×10^{-4} M $HAuCl_4$. This solution was stirred, and 0.6 ml of 0.01 M ice-cold $NaBH_4$ was added to it all at once. Vigorous stirring of the solution was continued for 2 min.

Growth solution: Three samples, A, B, and C, were prepared by addition of 0.2, 0.25, and 0.3 ml of 0.004 M $AgNO_3$ solution, respectively, to a mixture of 0.2 M CTAB and 0.15 M benzyl dimethyl hexadecyl ammonium chloride. To these solutions, 5 ml of 1×10^{-3} M $HAuCl_4$ was added followed by a gentle mixing of the solution and addition of 70 µl of 7.88×10^{-2} M ascorbic acid in each case. Addition of ascorbic acid changed the color of the growth solutions from dark yellow to colorless. Finally, 12 µl of seed solution was added and the solutions were left undisturbed for 24 hours.

Characterization

Absorption measurements were carried out with a Perkin Elmer (Lambda 950) UV-Vis-NIR spectrometer. pH of the solutions were recorded with MFRS Toshniwal Inst CL 54 pH meter. Transmission Electron microscopy images were obtained using a JEOL 1200 EX instrument.

Results and discussion

Synthesis of gold nanorods by various methods is well documented [3]. These methods include electrochemical method, seed mediated growth method, synthesis in alumina membranes etc. Out of these, seed mediated growth method is most popular because maximum tunability of longitudinal plasmon band can be achieved.

As mentioned in the experimental section, seed mediated growth method comprises of two steps viz. formation of seed nanoparticles and growth of these seeds in a growth solution. Seed solution is usually aged for 5 minutes after the preparation and then added to the growth solution. The growth solution consists of more precursors of gold, CTAB and sometimes a cosurfactant BDAC, silver nitrate and a week reducing agent ascorbic acid.

When the seed nanoparticles are put into the growth solution, they start growing. CTAB has more affinity towards {110} facets of gold nanoparticles. As they grow, CTAB starts preferentially adsorbing on side facets (i.e. {110}) of particles due to which growth is restricted along that direction. This leads to development of rod like structures.

Effect of ageing of seed solution: Absorption spectrum of seed solution taken immediately after preparation showed slight hump at 530 nm indicating the formation of very small (3-4 nm). On keeping this solution for several hours, increase in the intensity of peak was observed along with the narrowing of size distribution which indicated the formation of bigger sized particles by Oswald ripening. As the particle grows, it develops certain crystal facets, which determines the final shape of the nanoparticles. Some of these crystal facets can be selectively

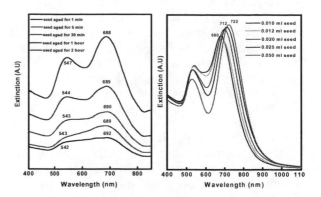

Fig. 1: Effect of (a) seed ageing and (b) amount of seed volume on longitudinal plasmon band

capped so that the particle acquires a particular shape. Hence it is very important to add seed solution after certain period of time in order to avoid the side effects which include formation of large sized particles with arbitrary shape. Fig. 1a shows absorption spectrum of different solutions of gold nanorods synthesized by varying the ageing period of seed. Seed solutions aged for 1, 5, 30, 60 and 120 minutes were added to different growth solutions. Absorption spectra of these growth solutions were recorded after 24 hours. It is clearly observed from the figure that the transverse and longitudinal plasmon band were more distinct and resolved in case of growth solution with seed aged for 2 hours. Intensity of longitudinal plasmon bad increased on increasing the ageing period of seed solution without much change in the peak position of longitudinal plasmon band. However, if the seeds were aged for longer time (not shown in the fig), only one peak was observed at 530 nm indicating the formation of spherical nanoparticles. This can be attributed to the fact that on increasing the ageing period beyond 2 hours, the particle no more has relatively unstable {110} facets, which are responsible for rod formation. Also, for very small ageing period viz. 1 or 2 minutes two peaks are not well resolved. The reason for this could be the incomplete decomposition of $NaBH_4$ within one or two minutes and this ultimately can lead to new nucleation of gold particles when added into the growth solution forming spherical particles along with rod like shapes.

Amount of seed solution: Amount of seed solution was varied from 10 to 50 µl. Fig. 1b shows the effect of increasing the amount of seed solution on the longitudinal plasmon band. In this case seed solution was aged for 5 minutes. The longitudinal plasmon band was found to shift towards higher wavelength on increasing the amount of seed solution. The effect of amount of seed solution on aspect ratio of nanorods is rather controversial. It is expected that increasing the amount of seed solution should lead to shift of the plasmon band towards lower wavelengths due to a decrease in the aspect ratio. Since more number of seed particles is available for growth to form nanorods, the gold from the precursor will reduce on different seed particles rather than already growing rod like particles leading to decrease in the length of the nanorod and hence decrease in the aspect ratio. However, this is not observed in our case. On increasing the amount of seed solution from 10 to 50 µl longitudinal plasmon band shifted from 680 to 722 nm i.e red shift is observed. This indicates an increase in the aspect ratio. The reason could be that on increasing the seed solution, although length of the particle decreases, width decreases further, ultimately causing increase in the aspect ratio and hence the longitudinal plasmon band shifts towards higher wavelength.

Amount of silver nitrate: This factor was found to be most crucial in tuning the longitudinal plasmon band. El-Sayed et al [3] have achieved wide tunability in the absorption band by using silver nitrate. Silver nitrate assists in binding of CTAB to {110} facets of gold nanorods. It is suggested that Ag precipitates on {110} facet of gold nanorods thereby restricting the growth

along that direction and formation of rod like structure. The Ag ion decreases the charge density on bromide ion. This leads to decrease between the neighboring CTAB molecules and hence template elongation leading to an increase in the aspect ratio and red shift in the longitudinal plasmon band. We could tune the longitudinal plasmon band from 720 to 900 nm (fig. 2) by varying the amount of silver nitrate from 0.25 to 0.5 ml.

pH and temperature: pH and temperature of growth solution was varied in order to observe the effect on longitudinal plasmon band (Fig. 3a). Rod formation was observed only in acidic pH. At basic pH, only one peak at 530 nm was observed indicating the formation of spherical nanoparticle. On increasing the pH, reducing power of ascorbic acid increases. This causes reduction of Ag ions to metallic silver which no longer assists in formation of rods.

On increasing the temperature of growth solution, stability of CTAB bilayer reduces causing gradual decrease in the intensity of longitudinal plasmon band (Fig. 3b). This was accompanied by an increase in the intensity of transverse plasmon band. Formation of gold nanorods requires quiescent growth conditions i.e the process is kinetically controlled. Increase in the temperature leads to the enhancement of the rate of reaction and hence does not favor the formation of nanorod of desired aspect ratio.

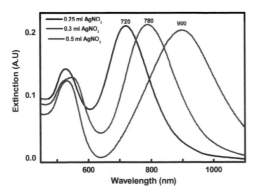

Fig. 2: Effect of amount of silver nitrate on longitudinal plasmon band

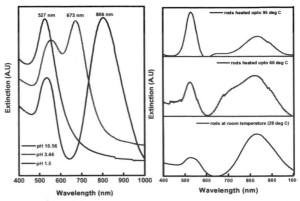

Fig 3: Effect of a) pH and b) temperature on longitudinal plasmon band

NSTI-Nanotech 2008, www.nsti.org, ISBN 978-1-4200-8503-7 Vol. 1

Modulation of optical properties of gold nanrods on addition of KOH:

Once the condition for formation of nanorods of desired aspect ratio was optimized, their behavior in different pH environments was studied [5]. Nanorods having longitudinal plasmon band at 896 nm was chosen (Fig. 4). The nanorods were not separated from growth solution in this case. Molarity of $HAuCl_4$ and $AgNO_3$ in the growth solution is such that it contains unreacted gold and silver ions in the solution. On adding different amounts of KOH, strong shift in longitudinal plasmon band towards lower wavelength along with an increase in the intensity of transverse plasmon band was observed. TEM images (Fig 4) of gold nanorods before and after addition of KOH is shown. A transformation in the shape of gold nanoparticles from rods to peanut like structure is observed.

The shape transformation of gold nanorods in the presence of growth solution has been explained in detail [5]. The side facets of gold nanorods are completely covered with a bilayer of CTAB. The amine group in the outer layer is protonated. Addition of KOH changes the pH of the solution. With increase in pH up to 7 there is an increased probability of ion pair formation between OH^- ions and CTA^+ ions leading to less coverage of gold nanorods by CTAB. At basic pH, reducing power of ascorbic acid increases. It has been reported [6] that during the synthesis of gold nanorods, all the gold and silver atoms in the growth solution are not reduced. Therefore, at high pH (above 7), unreacted gold and silver ions in the growth solution get reduced and deposit on the gold nanorod surface which is already less protected by CTAB. Since ends of the rods are less protected as compared to side faces (preferential binding of CTAB towards {110} facets), faster deposition of unreacted gold ions takes place on {111} facets at the ends at high pH. This could lead to a shape transition of gold nanorods into peanut or dumbbell like structure, ultimately causing decrease in the aspect ratio and blue shift in the absorption spectra. Blue shift in the absorption spectra is also accompanied by an increase in the intensities of transverse plasmon band as compared to the longitudinal plasmon band possibly indicating an increase in the width. Increase in the width of the particle leads to decrease in the aspect ratio of rods. It has been well established that absorption maximum of longitudinal plasmon band decreases linearly with the decrease in aspect ratio [7]. For very high pH values (~ 10-12), again a red shift in the spectra are observed which could be due to complete removal of CTAB, leading to aggregation of gold nanorods as irregular structures. In this case, longitudinal absorption maximum of gold nanorods can again slightly red shift.

Since response of gold nanorods to factors like changes in refractive index, pH, aggregation etc are being used for sensing various analytes, it becomes important to note that pH dependent shape transformations of gold nanorods can be a very effective tool in this regard. This study can be very helpful in monitoring the bimolecular or enzymatic reactions which exhibit pH changes, especially acidic to basic transition.

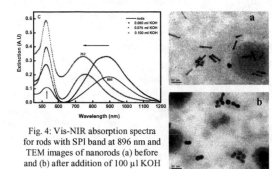

Fig. 4: Vis-NIR absorption spectra for rods with SPl band at 896 nm and TEM images of nanorods (a) before and (b) after addition of 100 μl KOH

Conclusions

Dependence of plasmon band of gold nanorods on various parameters was studied using UV-Vis spectrometer. Plasmon absorption maximum of gold nanorods was found to depend on the temperature, pH, amount of seed solution, ageing period of seed solution and amount of silver nitrate. Modulation in optical properties of gold nanorods was done on addition of different amounts of KOH. The effect was attributed to the shape transformation of gold nanrods due to change in pH.

References

1. C. Burda, X. Chen, R Narayanan, and M. A. El-Sayed, *Chem. Rev.* (2005)**,** 105, 1025.
2. Jorge Pérez-Juste, Isabel Pastoriza-Santos, Luis M. Liz-Marzán and Paul Mulvaney, *Coordination chemistry reviews* (2005), 249, 1870.
3. B. Nikoobakht and M. A. El-Sayed, *chem. Mater* (2003), 15, 1957.
4. N. R Jana, L. Gearheart and C. J Murphy, *J. phy chem. B* (2001), 105, 4065.
5. N. R. Tiwari, S. A. Kalele and S. K. Kulkarni, *Plasmonics* (2007), 2, 231-236.
6. Oredorff CJ, Murphy C, J Phy Chem B (2006) 110, 3990.
7. Link S, Mohamed M B, El-Sayed M A, J Phys Chem B (1999), 103, 3073–3077.

An FPGA Architecture Using Vertical Silicon Nanowire Transistors

A. Bindal*, D. Wickramaratne*, S. Hamedi-Hagh* and T. Ogura**

*San Jose State University, San Jose, CA, USA, ahmet.bindal@sjsu.edu
**Halo LSI, Hillsboro, OR, USA, togura@halolsi.com

ABSTRACT

This study presents an FPGA architecture using vertical silicon nanowire transistors. Cylindrical surrounding gate MOS devices of 2nm in radius and 10nm in length are used to produce ultra-low power CMOS circuits for the particular FPGA architecture. Each FPGA cluster consists of three 4-input Look-Up-Tables (4LUT) and a highly reconfigurable bus composed of eight interconnecting wires for cluster-to-cluster connectivity. Post-layout simulation results indicate that the worst-case propagation delays of a 4-LUT are 62ps during read and 68ps during write operations; the worst-case propagation delay of a cluster increases to 72ps for a fan-out of 1 and 97ps for a fan-out of 3 identical clusters. The worst-case power dissipation is approximately 3.1μW for a 4-LUT and 10.2μW for a cluster at 10GHz. The cluster layout which contains three 4-LUTs occupies approximately 8.0μm².

Keywords: nanowire, silicon, fpga, architecture, low-power.

1 SGFET DESIGN

Both NMOS and PMOS transistors are enhancement-type with undoped, cylindrical silicon bodies perpendicular to SOI substrate used mainly for latch-up prevention. Each transistor has 2nm thick gate oxide and a metal gate tailored to produce 300mV threshold voltage to provide sufficient noise immunity for a 1V power supply operation. Detailed cross section and layout of a single transistor is shown in Figure 1. 3D device simulations including quantum mechanical effects [1] are performed to obtain NMOS and PMOS I-V characteristics and circuit models.

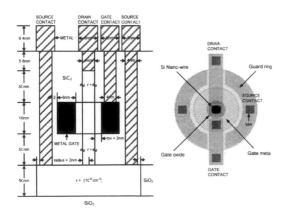

Figure 1: A nanowire transistor cross-section and layout.

The first task of the transistor design is to determine metal work function values for each NMOS and PMOS transistor to produce a threshold voltage of 300mV. This design process is described in detail in [2, 3] for device radii between 2nm and 20nm and shown below in Figure 2.

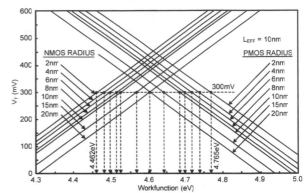

Figure 2: Threshold voltage versus metal work function.

Channel length of each transistor is varied until 1pA or smaller static OFF current is obtained at each radius. Intrinsic transient time and intrinsic energy of each NMOS and PMOS transistor are subsequently measured and plotted against each other to select the best device geometry as shown in Figure 3.

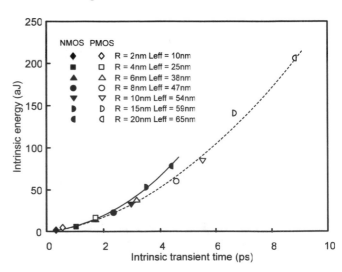

Figure 3: Intrinsic energy versus intrinsic transient time of the transistors with $I_{OFF} \leq 1pA$.

In Figure 3, intrinsic transient time determines the time interval for a transistor to charge/discharge the gate capacitance of an identical transistor when it is fully on and is a quick way of measuring the ON current characteristics of a transistor. Intrinsic energy, on the other hand, corresponds to the integration of instantaneous power delivered (received) to (from) the gate capacitance of an identical transistor as a function of time and measures the dynamic power dissipation of a transistor. The most desirable transistor geometry is found to have 2nm radius and 10nm effective channel length configuration which produces minimal gate capacitance as expected.

2 FPGA ARCHITECTURE

The FPGA architecture consists of an array of clusters interconnected in a network as shown in Figure 4.

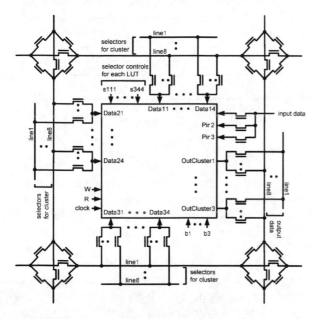

Figure 4: The FPGA architecture.

In this architecture, there are eight interconnecting wires between clusters; each interconnecting wire uses 6-transistor "traffic pole" switch (TPS) at the corners of a cluster to make the overall inter-cluster wiring highly configurable. Each inter-cluster wire is connected to cluster input by an 8-1 pass-gate MUX. Each cluster produces three outputs, each of which is connected to adjacent clusters with interconnecting wires and TPS. Each cluster contains three 4-Look-Up-Tables (4-LUT) as suggested by [4]. Each LUT has four data inputs; its output can be registered, routed to the neighboring LUTs or other clusters via bypass paths as shown in Figure 5. This flexible configuration produces complex gate-level implementations as well as state machines.

The 4-LUT is composed of 16 memory cells, each of which is connected to a single output using an array of pass-gate transistors in the form of a large 16-1 MUX as shown in Figure 6. The same pass-gate configuration also serves as 1-16 DEMUX connecting a single write input to any one of 16 memory cells.

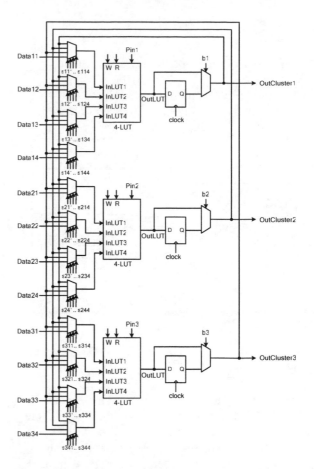

Figure 5: The cluster architecture containing three 4-LUTs.

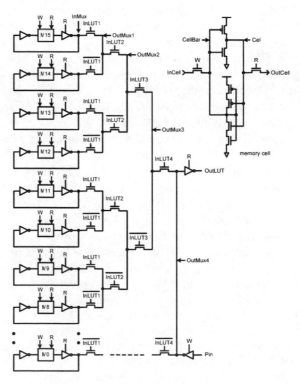

Figure 6: The 4-LUT circuit (M0-M15 are memory elements shown in the inset).

3 CIRCUIT DESIGN

3.1 Circuit operation and waveforms

The waveforms obtained during write operation where logic 1 is written into a memory cell is shown in Figure 7.

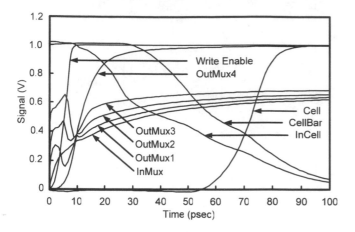

Figure 7: Waveforms of a 4-LUT during write operation.

The node names in Figure 7 are indicated in Figure 6 for cross reference. To write data into a memory cell, the rising signal at the "Pin" input propagates through the write-enabled (W-enabled) tri-state inverter and four pass-gate transistors in series (selected by InLUT1 through InLUT4) prior to arriving the "InMux" node. Figure 7 shows the successive deterioration of this signal by displaying the waveforms at each node from "OutMux4" to "InMux" every time the signal passes through a pass-gate transistor. The waveform at the "InMux" node shows both voltage level and slow rise time problems. The inverter at this node re-establishes the voltage levels of this signal and eliminates the slow rise time before the signal is sent to the memory cell. The re-conditioned signal at the output of the inverter (the "InCell" node) propagates through a pass-gate transistor to the "CellBar" node of the memory cell and is stored at the "Cell" node. In order to reduce the contention between the PMOS transistor of the "re-conditioning" inverter and the NMOS transistor of the "memory cell" inverter (or the NMOS transistor of the "re-conditioning" inverter and the PMOS transistor of the "memory cell" inverter), two extra transistors are added to the cell inverter in the feedback loop as shown in the inset of Figure 6.

The waveforms obtained during read operation where logic 1 is read form a memory cell is shown in Figure 8. The node names in Figure 8 are also indicated in Figure 6 for cross reference. To read data from a memory cell, the tri-state inverter next to the memory cell and the one at the output of 4-LUT must both be read-enabled. Similar to the write operation, the falling signal at the "InMux" node propagates through four pass-gate transistors in series and arrives at the "OutMux4" node. Even though Figure 8

shows some signal degradation in terms of slow fall times at the nodes "OutMux1" through "OutMux4", this deterioration is not as severe as losing part of the signal level due to a threshold voltage drop; NMOS pass-gate transistors do not allow any threshold voltage drop when transmitting logic 0. The slow rising signal at the "OutLUT" node is due to a large capacitive load at this terminal.

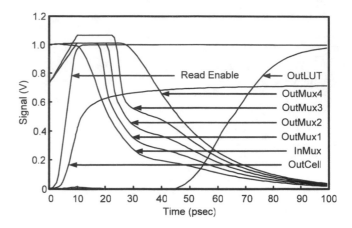

Figure 8: Waveforms of a 4-LUT during read operation.

The inter-cluster bus architecture composed of eight interconnecting wires may also be exposed to signal degradation when transmitting logic 1 through a series of NMOS pass-gate transistors. Connecting two clusters diagonally spaced for more than four cluster lengths is not recommended due to threshold voltage drop in logic 1 level and progressively longer transition times (slow nodes) as the number of pass-gate transistors increase. Figure 9 shows inter-cluster waveforms for transmitting logic 1 from one cluster to the next diagonally placed three cluster lengths away. The stored bit at the cluster of origin shifts from 1V to 0.7V when it goes through the pass-gate transistor of the memory cell and arrives at the "OutCell" node. However, the voltage level is re-established by the R-enabled tri-state inverter at the "InMux" node despite a long fall time. The rising signal at the "OutLUT" propagates through 2-1 MUX placed at the output stage of the cluster as shown in Figure 5 and arrives at the "OutCluster" node. The signal that departs from the "OutCluster" node propagates through seven pass-gate NMOS transistors (one at the 1-8 output DEMUX, five at 6-transistor switch boxes and one at the 8-1 input MUX) and arrives one of the "Data" terminals (Data11 through Data34) of the destination cluster; it suffers from threshold voltage drop and exhibits a very slow rise time as expected. Both of these issues are resolved by 4-1 MUX inverters at the cluster input as shown in Figure 5; progressively better waveforms are generated at the "DataBar" and "InLUT" nodes of the 4-1 MUX before the signal is allowed to propagate through the rest of the destination cluster as shown in Figure 9.

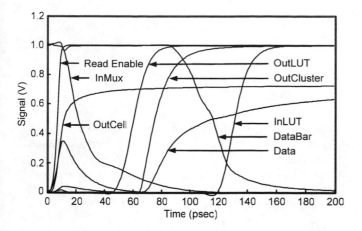

Figure 9: Typical waveforms obtained in a 4-LUT, a cluster and on inter-cluster wires connecting two clusters diagonally three cluster lengths apart.

3.2 Delays and power dissipation

Post-layout, worst-case propagation delays between Read Enable port of a cluster and its output are measured as a function of fan-out (the total number of clusters connected to a cluster output). The rise delay changes with $T_R = 9FO + 61$ in ps where FO corresponds to fan-out; the fall delay similarly changes with $T_F = 12.5FO + 59.5$ in ps.

Post-layout, worst-case wire delays are also measured between a cluster output and a neighboring cluster input as a function of fan-out and diagonal inter-cluster distance. The worst-case interconnecting wire delay is $T_W = 4FO + 23$ in ps for 1 diagonal cluster spacing and $T_W = 14.5FO + 54.5$ in ps for 4 diagonal cluster spacing.

The worst-case, average dynamic power dissipation is measured at 10GHz for a 4-LUT and a cluster. Power dissipation of a 4-LUT is 2.15μW during write and 3.08μW during read. Similarly, power dissipation of a cluster is 7.09μW during write and 10.16μW during read. Both 4-LUT and cluster power dissipations during read cycle are approximately 40% higher compared to write.

3.3 Cluster layout

The layout of a single cluster is shown in Figure 10. All three 4-LUTs in the cluster are stacked on top of each other; each 4-LUT occupies in the neighborhood of 2.6μm^2 layout area. The total layout area of a cluster is approximately 8.0μm^2.

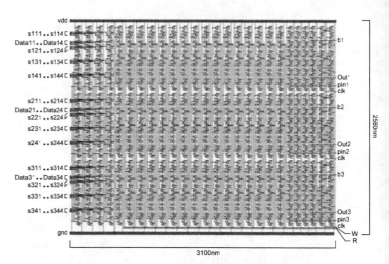

Figure 10: The layout of a FPGA cluster containing three 4-LUTs.

REFERENCES

[1] A. Wettstein, A. Schenk, W. Fichtner, "Quantum Device-Simulation with Density Gradient Model on Unstructured Grids", IEEE Elec. Dev., Vol. 48, No. 2, p. 279-283, 2001.

[2] A. Bindal, S. Hamedi-Hagh, "The Design of a New Spiking Neuron Using Silicon Nano-Wire Technology", Nanotechnology (Institute of Physics), Vol. 18, 095201, 2007.

[3] A. Bindal, A. Naresh, P. Yuan, K. K. Nguyen, S. Hamedi-Hagh, "The Design of Dual Work Function CMOS Transistors and Circuits Using Silicon Nano-Wire Technology", Trans. IEEE Nano., Vol. 6, No. 3, p. 291-302, 2007.

[4] E. Ahmed, J. Rose "The Effect of LUT and Cluster Size on Deep-Submicron FPGA Performance and Density", IEEE Trans. VLSI, Vol. 12, No. 3, p. 288-298, 2004.

Reliability of Nano Devices and Systems: Nuclear Spins of Stable Magnetic Isotopes as the Reliability Factors of Biomolecular Nanoreactors

V.K. Koltover

Institute of Problems of Chemical Physics, RAS, Chernogolovka, Moscow Region, Russia
Fax: +7(496)55223507, E-mail: koltover@icp.ac.ru

ABSTRACT

The trend of nanoscaling brings engineering down to the dimensions of molecular structures. However, it also poses the problem of how to create a reliable system from molecular components which experience permanent thermal and environmental fluctuations. Reliability ("robustness") is defined as the ability of a device to perform its function for a given time under given conditions. Fortunately, engineers may learn wisdom by the examples of Nature ("bionics"), to achieve the high systems reliability while dealing with unreliable components. There are several lines of creating reliable devices from unreliable functional elements in nanoengineering, the main of which is a preventive maintenance of components and prophylaxis of failures. The goal of this paper is to demonstrate how magnesium-25 and other kinds of stable magnetic isotopes are promising, on the spin-chemistry basis, for optoelectronics and nanophotonics to improve efficiency and reliability of the relevant molecular devices.

Key words: nanotechnology, nanoengineering, reliability, nanoreactors, isotopes

1 THE PROBLEM OF RELIABILITY

Nanotechnology – a catchall term for engineering materials sized between 1 and 100 billionths of a meter – is widely seen as having enormous scientific and commercial potential. The idea that engineering can be extended towards the molecular scale, to electronic devices of high efficiency based on single molecules and so on, has become currency over the past decade. This trend of nanoscaling brings engineering down to the dimensions of molecular structures. At the same time, it brings technical devices down to nanoengineering and, as a consequence, intrinsic instability of functional parameters due to thermal, mechanic and other fluctuations. This aggravates the problem of reliability in nanoengineering (see, for examples, references [1] and [2]).

Reliability ("robustness") is defined as the ability of a device to perform its function for a given time under given conditions [3]. In mathematical theory of reliability, a time-dependent m-dimensional vector $Y(t)$ is brought into use, each k-element of which ($k = 1, 2, ..., m$) corresponds to the relevant functional parameter of the device, and the relevant admissible limits of the parameters are introduced. If values of Y_k are in the limits, then the state of the device is defined as normal operation during the given time. If any of Y_k occurs beyond the limits, then we say that the device is in the state of failure. Inasmuch as the device performs its functions in the presence of random factors all Y_k are random values. Moreover, the admissible limits of the functional parameters may also be time-dependent and stochastic. Accordingly, the quantitative characteristic of reliability is the probability R of the non-failure operation in the given interval of time, $R(t) = P(\tau > t)$, where τ is a random value of time of the failure-free operation. Another important characteristic of reliability is the so-called hazard or failure-rate function, $\lambda(t) = -R'/R$, which has the meaning of the conditional probability of failure per unit time provided the device operated failure-free up to the given moment. With time, when wear-out of functional elements comes into play, the failure rate of the device begins to grow up (period of wear-out failures) [3].

How to create a reliable system from unreliable components which experience permanent fluctuations? Of course, thermal fluctuations become essentially lower if a device can operate at helium temperatures. The alternative way of achieving the high systems reliability while dealing with unreliable components is to learn wisdom by the examples of Nature.

2 MALFUNCTIONS OF BIOLOGICAL CONSTRUCTIONS

Similarly to technical devices, biological devices are not perfectly reliable in operation: malfunctions happen alternating with normal operation functions [4-6]. The simplest biological device is enzyme, the function of which consists in catalyzing a specific biochemical reaction. From the point of view of engineering, an enzyme is actually a "robot" specialized in accelerated assembly and disassembly of specific molecules. At present, there is a conception regarding the enzyme catalytic mechanism according to which the catalysis is performed as a process of step-by-step electron-conformation interactions, which are induced by attachment of the substrate molecule to the enzyme molecule [7]. The sequences of electron-conformation interactions represent the displacements and reorientations in enormous quantities of atomic groups. Therefore, temperature and other fluctuations bring about random accidents in the sequences of electron-conformation interactions, i.e. the possibility of conformational relaxation changes in an "incorrect" direction. Such accidents develop in one or another functional violation of the enzyme (inactivation, violation of selectivity, etc.), i.e. in failure. Thus, the conformational fluctuations set limits to reliability of enzymes.

Energetic demands of every operation in living systems are met by molecules of adenosine triphosphate (ATP) which are synthesized from adenosine 5'-diphosphate (ADP) and inorganic phosphate (P_i). Most of ATP is produced in the special organelles of cells, the so-called mitochondria. Therefore, a mitochondrion exemplifies the biomolecular nanoreactor of primary importance. There are specific enzymes organized in the respiratory electron transport chains (ETC) in the inner membrane of the mitochondrion. Each ETC consists of four lipoprotein complexes (I–IV), coenzyme Q which operates as an electron shuttle between complexes I/II and III, and cytochrome C as the electron shuttle between complexes III and IV [8].

Normal functioning of the ETC's enzymes lies in the transport of electrons, one by one, from the electron donor molecules ("oxidative substrates", NADH/succinate) on the chain to the end enzyme, cytochrome oxidase of complex IV, from which the electrons are transferred to molecules of oxygen with two-electron reduction of oxygen into water. Free energy released during the electron transport is used in complex V for synthesis of ATP from ADP and P_i by oxidative phosphorylation [8].

The mitochondrial enzymes have very ancient evolutionary origin and, hence, seem to be ones of the most reliable molecular machines. But yet their reliability characteristics are not perfect because these molecular machines experience conformational fluctuations. In consequence, normal elementary acts of the electron transport alternate with random malfunctions at complexes I and III when an electron, rather than waits for transport to the next enzyme of ETC, goes directly to an adjacent oxygen molecule resulting in production of superoxide free radical ($O_2^{\bullet-}$). Meanwhile, chemical products of $O_2^{\bullet-}$, the so-called reactive oxygen species (ROS), are toxic and initiate free-radical damages in the biomolecular nanoreactors. Therefore, the malfunctions in ETC followed by production of $O_2^{\bullet-}$ are considered as the failures of vital importance for living cells [4-6].

The chloroplast of green plants exemplifies another case of malfunctions. The main functional elements of the chloroplast nanoreactor are molecules of chlorophyll, the derivatives of Mg-protoporphyrin complexes. Chlorophyll molecules (Chl) are specifically arranged in pigment protein complexes which are embedded in the lipoprotein thylakoid membranes of chloroplasts. The function of the vast majority of Chl (antenna Chl) is to absorb light and transfer that light energy by resonance energy mechanism to specific energy sinks, the reaction centers of the chloroplast nanoreactor photosystem I and photosystem II.

Eventually, the light energy is used by the enzyme components of the thylakoid membrane electron transport chains for synthesis of ATP and NADPH which are thereafter employed for synthesis of carbohydrates and so on [8]. Similarly to the mitochondrial nanoreactor, normal electron transport alternates with random malfunctions resulting in generation of $O_2^{\bullet-}$ [6, 8].

While performing their function, the light-exited antenna Chl molecules are in the singlet state ^{1}Chl (electron spin $S = 0$). However, there is the probability of order 10^{-4} of the radiationless relaxation into the triplet state ^{3}Chl ($S = 1$) followed by the chemical reaction with oxygen

$$^3\text{Chl} + {}^3\text{O}_2 \quad \rightarrow \quad \text{Chl} + {}^1\text{O}_2$$

in which singlet oxygen $^1\text{O}_2$ evolves [8]. Thus, normal functioning of the antenna chlorophyll is attended by production of $^1\text{O}_2$. The $^1\text{O}_2$ molecules are essentially more reactive than usual triplet O_2 molecules and initiate photodynamic damages followed by inactivation of the biomolecular nanoreactor. Therefore, the "erroneous" relaxation of ^{1}Chl attended by production of $^1\text{O}_2$ should be considered as the malfunction (failure) of the antenna chlorophyll device.

3 RELIABLE SYSTEMS FROM UNRELIABLE COMPONENTS

There are several lines of creating reliable biological devices from unreliable components. One of them is redundancy, when redundant components of the same type are introduced in the system to fulfill one and the same function. There are different kinds of redundancy, among them - structural (insertion of superfluous amount of functional elements), functional (exploitation of the elements capable to carry out additional functions besides their basic), informational (surplus information), temporal (superfluous time of functioning). For example, all essential biomolecular constructions are present in cells in superfluous amounts. The redundant amounts of enzymes, mitochondria, chloroplasts, and other organelles represent the examples of the structural and functional reservation.

Another line of enhancement of reliability is to supply the possibilities of repair of damaged components. However, when a functional element is replaced once its failure or damage takes place, the failure rate may become intolerably high. In order to decrease the failure rate of the system its elements should be replaced before they become wearout or damaged. Indeed, renewal processes proceed in all complex biological systems, starting from the cell level. The DNA/RNA template synthesis, the so-called turnover, provides the preventive maintenance replacement of cell components. This preventive replacement of functional elements, which follows the pattern preset in the cell genome, is the main line of providing the high reliability of biological systems despite low reliability of cell functional elements, enzymes and so on [4-6].

In mitochondria, there is a specific antioxidant enzyme, the so-called superoxide dismutase (mtSOD), the function of which is to trap $O_2^{\bullet-}$. This enzyme catalyzes the reaction of dismutation of $O_2^{\bullet-}$ into hydrogen peroxide (H_2O_2) and oxygen (O_2) thus protecting cell structures from $O_2^{\bullet-}$ and its toxic chemical products. It makes its defense "job" in cooperation with two other specific enzymes, catalase and glutathione peroxidase, which catalyze decomposition of H_2O_2 into H_2O and O_2. Thus, the antioxidant enzyme system provides the preventive maintenance against the active forms of oxygen.

The calculations, based on the experimental data and the "Birth and Death" model, often used in the mathematical theory of reliability, show that the rate parameter for the "free-radical failures" of ETC in normal mitochondria $\lambda \approx 0.25$ s^{-1} (about 1 radical every 4 s). The rate parameter for elimination of $O_2^{\bullet-}$ by SOD (probability of elimination of one $O_2^{\bullet-}$ per unit time):

$$\mu \approx -\Delta n(t)/n(t)\Delta t = k_e[E] \approx 1.3 \cdot 10^4 \text{ s}^{-1}.$$

Here $n(t)$ is the number of $O_2^{\bullet-}$ in a mitochondrion, $\Delta n(t)$ is the number of the radicals eliminated during the time interval Δt, k_e and $[E]$ are the reaction rate constant and the concentration of mtSOD, respectively. Then the probability of slipping of $O_2^{\bullet-}$ through the mtSOD defense has been estimated to be:

$$z = (\lambda/\mu)/(1+\lambda/\mu) \approx 1.9 \cdot 10^{-5}.$$

Thus, only 2 radicals from every 100,000 may penetrate the defense system. Nonetheless, it was estimated that the longevity of human brain could reach 250 years, should the reliability of the mitochondrial nanoreactors and the reliability of the antioxidant enzyme defense be absolutely perfect [4-6].

In the photosynthetic nanoreactors of chloroplasts, apart from supcroxide dismutase which trap $O_2^{\bullet-}$, there is another special antioxidant system, the so-called carotenoids ("xanthophylls"), which trap $^1O_2$. The reactivity of carotenoids, such as β-carotene, lycopene, zeaxanthin, with singlet oxygen is characterized by the rate constants of the diffusion order, of $3 \cdot 10^{10} \cdot L \cdot M^{-1} \cdot s^{-1}$. The carotenoids effectively catch the $^1O_2$ molecules thereby preventing photodynamic damages of biomolecular components in the chloroplast nanoreactor.

4 PREVENTIVE MAINTENANCE VIA SPIN-CATALYSIS

Beyond the energy control, any chemical reaction as electron-nuclear rearrangement of reactants into products is controlled by angular momentum (spin) of reactants. The reaction is rigorously forbidden if it requires a change in total electron spin, a spin evolution, of the reactants. To lift the ban, spins of the reactants must be changed. The acceleration of the spin evolution and the relevant acceleration of the reaction may be performed via spin-catalysis, be it an external magnetic field ("magnetic field effect"), interaction with unpaired electrons of external paramagnetic ions and free radicals ("electron spin catalysis"), or hyperfine coupling with magnetic nuclei ("magnetic isotope effect") [9].

The general theory of spin catalysis forecasts that spin moments of magnetic nucleons can afford preventive antioxidant effects in nanoreactors. The currently available data suggest that it takes place with regard to the nuclear spin moment of magnesium-25.

Amongst three stable magnesium isotopes, ^{24}Mg, ^{25}Mg and ^{26}Mg, only ^{25}Mg has the nuclear spin (I). It has been recently discovered that the rate of oxidative phosphorylation of ADP in cell mitochondria with magnetic nuclei ^{25}Mg ($I = 5/2$) is about twice higher than that with the spinless, nonmagnetic nuclei ^{24}Mg or ^{26}Mg ($I = 0$). There was no difference between ^{24}Mg^{2+} and ^{26}Mg^{2+} effects in oxidative phosphorylation [10]. It is well-known that oxidative phosphorylation is a complex multi-step reaction [8]. The observations made by Buchachenko and his co-authors demonstrate that a rate-limiting step in oxidative phosphorylation is a spin-selective process, a certain intermediate ion-radical pair, the spin evolution of which is catalyzed by the magnetic isotope. It may be the adenine anion-radical of ADP, bound in the ATP-synthase active center and coupled with the ETC's cation ("hole"), as it was first suggested in our paper [7], or it may be the intermediate ion-radical pair of [ADP]$^{\bullet-}$ with Mg$^+$ cation, as it was proposed in [10, 11].

No matter what the spin evolution, the catalytic effect of the magnetic ^{25}Mg isotope transpires that phosphorylation of ADP should proceed essentially slower when nucleus of Mg^{2+} has no spin magnetic moment than in the case when nucleus of Mg^{2+} has the spin magnetic moment. The retardation of oxidative phosphorylation causes the relevant delay in transport of electrons through the ETC enzymes. While the input of the mitochondrial nanoreactor is overflowed with electrons, it increases probability of the malfunction in the nanoreactor with one-electron reduction of oxygen and formation of radical $O_2^{\bullet-}$. As a result, the yield of $O_2^{\bullet-}$ as the by-product of electron transport in the mitochondrial nanoreactor is bound to be much lower in the presence of ^{25}Mg-ADP by comparison with ^{24}Mg- or ^{26}Mg-ADP. In essence, magnesium-25 exerts the preventive antioxidant effect. Hence, the magnetic isotope can perform the preventive maintenance against the active forms of oxygen, thereby providing the higher system reliability [12].

Quite the reverse, prooxidant, effect of magnetic ^{25}Mg is to be exerted in the case of antenna chlorophyll of green plants. Firstly, the nuclear spin of ^{25}Mg can catalyze the radiationless conversion of photoexited chlorophyll molecules into the triplet state ^{3}Chl (electron spin $S = 1$). Secondly, it can significantly accelerate the above mentioned reaction of the triplet chlorophyll with oxygen in which the singlet $^1O_2$ oxygen evolves. Indeed, the pair of

NSTI-Nanotech 2008, www.nsti.org, ISBN 978-1-4200-8503-7 Vol. 1

initial reactants, ^{3}Chl and ^{3}O$_2$, has total electron spin $S = 2$ and five spin states which correspond to spin projections $S_z = \pm 2, \pm 1, 0$. On the other hand, the pair of products, Chl and ^{1}O$_2$, has total spin $S = 0$. Hence, out of five possible spin state of the reactants there is a single state from which the reaction is open (permissible). Other four "channels" of the reaction are spin-forbidden since the total electron spin of the system must have been changed in the course of the reaction. Consequently, there is little likelihood of significant rate even though this reaction may be catalyzed by spin-orbital or spin-rotational coupling.

However, slow or no spin evolution of the reactants can be accelerated by the nuclear spin of magnesium-25. As a result, one can expect for the essentially higher yield of ^{1}O$_2$ and, thereafter, essentially more the ^{1}O$_2$ induced photodynamic damages of biomolecular components in photosynthetic nanoreactors when chlorophyll molecules contain ^{25}Mg instead of the spinless ^{24}Mg or ^{26}Mg. Thus, in the chloroplast nanoreactor, the nuclear spin catalyst should decrease the system reliability [12].

5 SUMMARY

Functional elements of molecular dimensions in nanoengineering devices are susceptible to environmental fluctuations. That raises the question as to whether operation of the nanoscaling devices can be reliable. One can lower temperature up to helium values at which thermal fluctuations are less. The alternative way is to follow the general theory of reliability and bionics. Insertion of redundant amount of the functional nanocomponents, which fulfill one and the same function, magnifies the system reliability. Another line of creating the highly reliable systems from unreliable components is preventive maintenance or prophylaxis of failures.

Keeping in mind the possible free-radical damages, researchers who fabricate new composite materials embedded with organic dye molecules, metal atoms, and nanoparticles should provide them with appropriate antioxidants as well.

In design of reliable spin-dependent molecular devices for optical communications, quantum information processing, computational schemes and the like, stable magnetic isotopes hold much promise. Moreover, along this nanoengineering line the general principles of spin-chemistry, amongst them – control of chemical reactivity by selective isotope modification and relevant microwave variations, can be employed in optoelectronic and nanophotonic devices.

ACKNOWLEDGMENT

This work was sponsored by the Russian Foundation for Basic Research, initiative project No. 07-03-00897.

References

[1] D. P. Vallett, "Failure analysis requirements for nanoelectronics," IEEE Trans. Nanotechnology, 1, 117–121, 2002.

[2] B. Wunderle and B. Michel, "Progress in reliability research in the micro and nano region", Microelectronic Reliab., 46, 1685-1694, 2006.

[3] D.K. Lloyd and M. Lipov. "Reliability: Management, Methods and Mathematics", New Jersey, Prentice Hall, 1977.

[4] V.K. Koltover, "Reliability concept as a trend in biophysics of aging", J. Theor. Biol., 184, 157-163, 1997.

[5] V.K. Koltover, "Reliability of electron transport in biological systems and the role of the oxygen free radicals in aging", Control Sciences, No. 4, 40-45, 2004.

[6] V.K. Koltover, "Reliability of biological systems: Terminology and methodology", Longevity, Aging and Degradation Models (eds. V. Antonov, C. Huber, M. Nikulin, and V. Polyschook), St-Petersburg, SPbSPU, 1, 98-113, 2004.

[7] L.A. Blumenfeld and V.K. Koltover, "Energy transformation and conformational transitions in mitochondrial membranes as relaxation processes", Mol. Biol. (Moscow), 6, 161-166, 1972.

[8] A. White, Ph. Handler, E.L. Smith, R.L. Hill, and I.R. Lehman, "Principles of Biochemistry", New York et al., McGraw Hill, 1981.

[9] A.L. Buchachenko and V.L. Berdinsky, "Electron spin catalysis", Chem. Rev., 102, 603-612, 2002.

[10] A.L. Buchachenko, D.A. Kouznetsov, S.E. Arkhangelsky, M.A. Orlova, and A.A. Markarian, "Spin biochemistry: Magnetic ^{24}Mg–^{25}Mg–^{26}Mg isotope effect in mitochondrial ADP phosphorylation", Cell Biochem. and Biophys., 43, 243-252, 2005.

[11] A.L. Buchachenko, "New Isotopy in Chemistry and Biochemistry", Moscow, Nauka, 2007.

[12] V.K. Koltover, "Antioxidant and prooxidant effects of magnetic isotopes in biomolecular nanoreactors", Free Radical Biol. and Med., 43, S. 69, 2007.

High Speed Resonant Tunneling Diode Based on GaN & GaAs: A Modeling & Simulation Approach

V. Arjun, D.Sikdar and V.K.Chaubey

Electronics & Instrumentation Engineering Group, Birla Institute of Technology and Science, Pilani, India.
arjun.vijaykumar@gmail.com

Abstract

Double barrier Quantum well structure has been considered to model the Resonant Tunneling Diode (RTD) in terms of structural parameters to compute the electrical transmission characteristics. The transmission matrix for a RTD structure having any arbitrary barrier has been developed to simulate the resonant tunneling current with an applied voltage across GaAs and GaN based devices. The RTD structures have been extended as Resonant Tunneling Transistor (RTT) by introducing an additional optical control over the base quantum well. A significant tunneling improvement has been reported in case of optically activated RTT structures.

Keywords: RTD, Barrier, Laser, Double Barrier Quantum Well (DBQW), Refractive Index, Bandgap, Tunneling Current.

1. Introduction:

Quantum effect devices (QED) are based on quantum mechanical tunneling for controlled carrier transport through multi layer semiconductor structures when sub-band energies of adjacent quantum well coincides [1-3]. A smaller parasitic and compatibility of integration of such resonant structures with CMOS technology, has attracted theoretical and experimental researchers to develop appropriate models and application circuits based on Resonant Tunneling Diode (RTD) application in quantum electronics and optoelectronic sub-systems[4-5]. The RTD is perhaps the most promising candidate for digital circuit applications due to its negative differential resistance (NDR) characteristic, structural simplicity, ease of fabrication, inherent high speed, flexible design freedom, and versatile circuit functionality [6-8].

Fast speeds, low noise, reduced parasitic capacitances and high data rate have been reported in Transmitter module of GaAs HBTs [9] for Gigabit applications. Similar characteristics can be modeled using RTD structure, exploiting the tunneling current for say GaAs/GaAlAs employing SPICE [10]. These investigations have mainly focused on GaAs based materials and are modeled under numerical approximations like Breit-Wigner formula [11].

In this present work, a general model has been developed to model tunneling behavior of RTD using Transmission Matrix [12]. Any arbitrary potential can be discritised in small steps where tunneling can be approximated using transmission matrix concept. All these transmission probabilities are multiplied to obtain the tunneling co-efficient for the entire barrier. The analysis has been applied to two promising material GaAs and GaN based RTD structures. The details of this are given in the section on simulation and results. I-V characteristics for GAN and GaAs are simulated using the concept mentioned above and an approach to increase the peaks current in these I-V graphs is the main focus of our work.

Apart from studying the tunneling as an electrical phenomenon [13], we have investigated how light of different energies affect the tunneling process [14] in the two compound semiconductor materials chosen. Tunneling Current variation with different energies of barriers, barrier heights and electron energies has been studied. The change in refractive index of a semiconductor on interaction with photon is exploited to derive the relation between the band gap of a semiconductor and incident light [15-17]. A Bandgap variation ultimately leads to re-alignment of quantum states hence impacting the tunneling mechanism. All these theoretical explanations have been verified through MATLAB simulations using the transmission matrix model developed in case of GaAs and GaN based RTD structures.

2. Resonant Tunneling Diode (RTD) Structures and Model:

A RTD relies on quantum phenomenon such as tunneling and energy quantization to give an unusual current-voltage characteristic. For this quantum phenomenon to occur the dimensions of the active layer have to be on the order of the De-Broglie wavelength of electron. A typical RTD will have barriers of 2 nm thickness sandwiched with a lower Bandgap material of 5 nm thickness.

A Double Barrier Structure consisting of two Barriers and a quantum well structure is depicted in figure 1(a) for the unbiased case and figure 1(b) for the biased case.

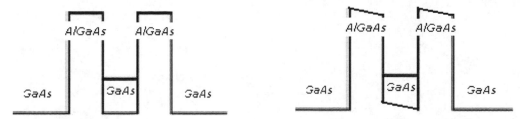

Fig 1a: Typical RTD structure with no Voltage Bias   Fig 1b: Typical RTD structure with Voltage Bias

Evidently as the states in the quantum well align with the ones in the barrier, the quantum mechanical tunneling probability of electrons increases appreciably and determines the operating conditions and speed of such devices

High speed THz optical applications using compact solid state devices are possible through development of Integrated RTD Laser Diodes [18]. The front-end design of an optical receiver circuit can incorporate the RTD for High speed applications because of its insignificant parasitic as compared to conventional photodetectors used in the receivers. Further, in logical applications, the device can be biased with such a threshold potential that the overall structure can be made to go into tunneling mode to ease transmission for a logical high. For logic 0 it goes to switch off mode due to absence of resonant tunneling in the device and causing high switching speed. In Oscillator Circuits consisting of R, L, C and RTD, the Negative Differential Resistance (NDR) exhibited by RTDs can be used to compensate for unavoidable ohmic losses in such circuits [4]. RTD with a light controlled terminal can be used to increase tunneling current hence making it realizable for optoelectronics quantum devices.

The basic idea behind the simulation of the characteristics of a Resonant Tunneling Diode (RTD) is *transmission matrix*. Given any barrier height we have first found out what is the co-efficient of transmission across the barrier through quantum mechanical tunneling as the particle energy here we are dealing with is lesser than the energy of the potential barrier.

The transmission co-efficient across any arbitrary potential, as shown in figure 3, has been computed based on the known potentials.

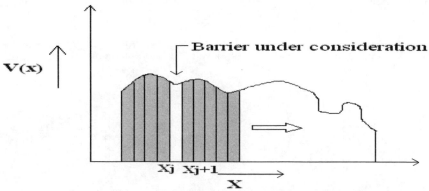

Fig 2: Discretization of the arbitrary potential barrier

Eigen functions can be found by discretizing the potential in small interval size and writing the wave vectors (kj) for respective barrier as given below

$$kj = (2mE - eVj)^{0.5} / (h/2\pi) \tag{1}$$
where m is the effective mass of the electron, E is the energy of the incident electron, Vj is the applied potential across the barrier and h is the Plank's constant.

Now the coefficients of Eigen functions at the interface can be written as

$$\begin{bmatrix} Aj \\ Bj \end{bmatrix} = Pjstep \begin{bmatrix} Cj \\ Dj \end{bmatrix}$$

Where Pjstep can be written as : (2)

$$
\text{Pjstep} \quad = \quad 1/2 \begin{bmatrix} 1 + kj+1/kj & 1 - kj+1/kj \\ \\ 1 - kj+1/kj & 1 - kj+1/kj \end{bmatrix}
$$

Where A, B, C, D are the co-efficients as shown in the figure4.

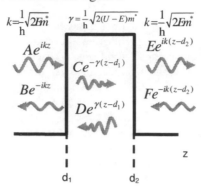

Fig 3: Details of transmission and reflection co-efficients of a barrier.

Obviously boundary conditions yield

$$
\begin{bmatrix} A \\ B \end{bmatrix} = \frac{1}{2} \begin{bmatrix} \left(1 - \dfrac{\gamma}{ik}\right)e^{-ikd_1} & \left(1 + \dfrac{\gamma}{ik}\right)e^{-ikd_1} \\ \left(1 + \dfrac{\gamma}{ik}\right)e^{ikd_1} & \left(1 - \dfrac{\gamma}{ik}\right)e^{ikd_1} \end{bmatrix} \begin{bmatrix} C \\ D \end{bmatrix}
$$

(3)

The matrix representation for the second boundary can also be developed in the same procedure to relate coefficients C, D with E, F to provide an equation as

$$
\begin{bmatrix} C \\ D \end{bmatrix} = \frac{1}{2} \begin{bmatrix} \left(1 - \dfrac{ik}{\gamma}\right)e^{\gamma(d_2-d_1)} & \left(1 + \dfrac{ik}{\gamma}\right)e^{\gamma(d_2-d_1)} \\ \left(1 + \dfrac{ik}{\gamma}\right)e^{-\gamma(d_2-d_1)} & \left(1 - \dfrac{ik}{\gamma}\right)e^{-\gamma(d_2-d_1)} \end{bmatrix} \begin{bmatrix} E \\ F \end{bmatrix}
$$

(4)

Using this matrix formulation, the reflection and transmission from the boundaries can be generalized as:

$$
\begin{bmatrix} a_l \\ b_l \end{bmatrix} = \overline{\overline{M}}_{l,l+1} \begin{bmatrix} a_{l+1} \\ b_{l+1} \end{bmatrix}
$$

(5)

$$
\overline{\overline{M}}_{l,l+1} = \frac{1}{2} \begin{bmatrix} \left(1 + \dfrac{k_{l+1}}{k_l}\right)e^{-ik_l(d_{l+1}-d_l)} & \left(1 - \dfrac{k_{l+1}}{k_l}\right)e^{-ik_l(d_{l+1}-d_l)} \\ \left(1 - \dfrac{k_{l+1}}{k_l}\right)e^{ik_l(d_{l+1}-d_l)} & \left(1 + \dfrac{k_{l+1}}{k_l}\right)e^{ik_l(d_{l+1}-d_l)} \end{bmatrix}
$$

(6)

The wave functions within the barrier can be modeled using a diagonal matrix Pjfree taking care of the phase transformation and free propagation inside the barrier and can be expressed as:

NSTI-Nanotech 2008, www.nsti.org, ISBN 978-1-4200-8503-7 Vol. 1

$$Pjfree = \begin{bmatrix} \exp(-ikjLj) & o \\ \\ o & \exp(ikjLj) \end{bmatrix}$$

(7)

Hence, the complete relation between the wave function on one side to another side is given by propagation matrix of jth barrier and that is given by **Pj = Pjstep*Pjfree**

Finally for n discrete steps $P_{complete}$ is written as

$$P = \prod_{j=1}^{j=N} Pj$$

(8)

This expression can be evaluated for the given structural parameters under different bias conditions and thus provides the transmission probability across the device for an arbitrary potential barrier. The present simulation has taken a large number of discrete step functions to emulate the band bending of the barrier in case of applied potential.

3. Device Characterization and Discussion:

MATLAB simulations are performed to study the behavior of RTD structure based on GaAs/AlGaAs and GaN/AlGaN to estimate the transmission co-efficient through asymmetric double barrier at fixed voltage bias. Figure 4 shows the structural cross section and transmission characteristics for the considered structure.

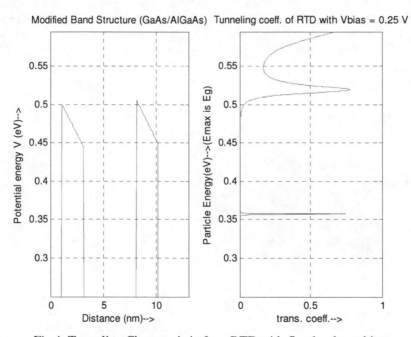

Fig 4: Tunneling Characteristic for a RTD with fixed voltage bias

The graph between tunneling current and. the voltage bias allows us to find the bias voltage at which resonant tunneling current peaks for the RTD device. The structural parameters considered for the simulation for the RTD structure based on GaAs/AlGaAs (shown in figure 1a) are: first barrier ($Al_{.4}GaAs_{.6}$ having Eg as 1.93 eV) width = 2 nm, height = 0.5eV; second barrier ($Al_{.4}GaAs_{.6}$ having Eg as 2.13 eV) width=2nm, height =0.7eV; Width of the well (GaAs having Eg 1.43 eV) in between =5nm. Under the assumption of a fixed current of 1 μA in the emitter side of the device The I-V characteristic has been simulated and presented in the figure 5.

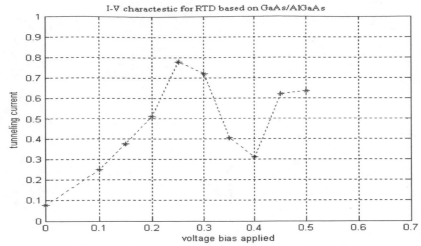

Fig 5: I-V Characteristic for RTD based on GaAs/AlGaAs

The I-V characteristics obtained is in close proximity to the experimentally reported results in [7]. The above figure clearly shows that peak tunneling current (0.78µA) occurs at voltage bias of 0.25V and valley current (0.31µA) at a voltage bias of 0.4V. Obviously, the I-V characteristic graph shows a NDR region to make it suitable for switching and oscillating circuits. The slope of NDR region is calculated as $[(V_{peak}-V_{valley})/ (I_{peak}-I_{valley})]$ and comes out to be -0.319 for the considered typical structural parameters. The simulation has been extended for another promising material like GaN, to evaluate its suitability in quantum tunneling device applications and to establish its application in optoelectronics device design.

The I-V characteristics for GaN/AlGaN RTD structure have been simulated and the graph is reported in the figure 6. The simulation parameters are: first barrier ($Al_{.35}Ga_{.65}N$ having Eg as 4.194 eV) width = 2 nm, height = 0.694eV; second barrier ($Al_{.5}Ga_{.5}N$ having Eg as 4.6 eV) width=2nm, height =1.1eV; Width of the well (GaN having Eg 3.5 eV) in between =5nm.

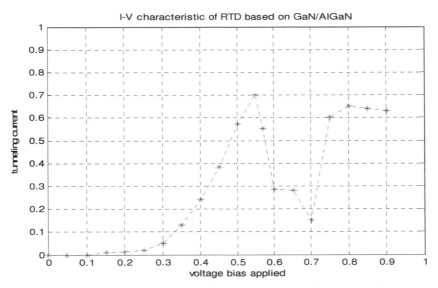

Fig 6: I-V Characteristic for RTD based on GaN/AlGaN

The I-V characteristics obtained is qualitatively similar to the case of GaAs based structure but with an quantitative difference. The above figure clearly shows that peak tunneling current (0.7µA) occurs at voltage bias of 0.55V and valley current (0.15µA) at a voltage bias of 0.7V and shows a NDR of -0.272. This characteristics show a scope to manipulate the peak tunneling current and the NDR value for the structures by the enhancement of the tunneling current.

4. Enhancement in Resonant Tunneling: Proposed Method

The present investigation proposes optical enhancement of tunneling current using laser light shone on the quantum well of the RTD structure. This can be viewed as Resonant Tunneling Transistor (RTT) with the base (quantum well) being controlled optically. The physics under lying the phenomena is generation of excess carrier through optical means and alteration of the optical density of the medium by the laser light. There are a few circumstances, where a normally non-transparent medium can be made transparent through coherent interaction with laser light of particular frequencies, which have been extensively studied in atomic and molecular systems [14]. Due to the laser-matter interaction there is a laser induced change in the material properties. It has been reported that the refractive index varies quite dramatically as the energy of the incident photon changes [15]. Change in refractive index in tern changes the effective band gap of that material [16-17]. The lowering of band gap energy implies increase in the density of states in the well of RTT, in other words it brings quantum energy states of that semiconductor material more close to each other. This ultimately increases the probability of resonant tunneling. This is because; more closely spaced quantum states in the well have higher probability of being aligned with the energy states of the emitter and collector sides. This enhancement in resonant tunneling current was expected theoretically and the same has also been verified through the simulations.

The change in refractive index of a material with photon energy is well established in literature. So the values of refractive indices for different photon energy, for GaAs and GaN are used in the simulations. From [16-17] two mathematical relations reported for Bandgap energy-Refractive Index relation are used. The relation for GaAs is determined using the Ravindra Formula and for GaN using the Moss Formula. This is because; Bandgap of GaAs falls in the region where Ravindra Relation gives results closest to the experimental values reported [15]. Similarly Moss relation is found to be best suit for GaN. These two relations are given below.

Ravindra Formula: $E_g n^4 = 108$ eV $\qquad$ (8)

Moss Formula: $E_g n^4 = 95$ eV $\qquad$ (9)

The simulation results and tabulations for the GaAs/AlGaAs RTT structure under considerations is shown below:

Table 1: Photon energy and their corresponding n values to give different tunneling currents

Photon Energy(eV)	Refractive Index (n)	Eg from (Ravindra Model)	Tunneling Current For bias 0.25V
0	3.3	1.35	.78
0.42	3.08	1.2	0..8476
.49	3.147	1.1	0.85
0.55	3.223	1	0.8625
0.62	3.3	0.9	0.8977
0.7	3.408	0.8	0.898
1.25	3.524	0.7	0.9085
1.3	3.546	0.6	0.9085
1.33	3.55	0.5	0.9085

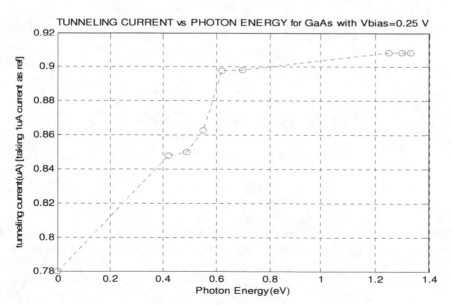

Fig 7: Tunneling current vs. Photon energy for GaAs/AlGaAs RTT

The above simulation is performed at voltage bias of 0.25V that gave maximum resonant tunneling current (figure 5). It is clear from the above graph that without any light shown on the well the tunneling current obtained is 0.78μA. We report that there can be an increase in resonant tunneling current upto 16.47% in the above system with light of different energies. For photon energies closer to the band gap of the material the tunneling current saturates.

The simulation results and tabulations for the GaN/AlGaN RTT structure under considerations is shown below:

Table 2: Photon energy and their corresponding n values to give different tunneling currents

Photon energy (hv)(eV)	Refractive Index(n)	Modified Band Gap Eg (eV)(Moss Relation)	Tunneling Co-efficient at 0.55 V Bias
0	2.28	3.5	0.7
1.0	2.33	3.48	0.7
1.5	2.35	3.46	0.7
1.6	2.364	3.445	0.7
1.7	2.371	3.43	0.7
1.8	2.378	3.38	0.705
1.9	2.385	3.345	0.705
2.0	2.39	3.34	0.7051
2.25	2.41	3.2	0.79
2.5	2.44	2.68	0.827
2.75	2.48	2.511	0.86
3	2.54	2.282	0.86
3.25	2.6	2.07	0.86

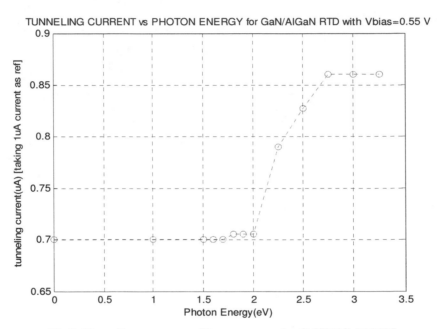

Fig 8: Tunneling current vs. Photon energy for GaN/AlGaN RTT

The above simulation is performed at voltage bias of 0.55V that gave maximum resonant tunneling current (figure 6). It is clear from the above graph that without any light shown on the well the tunneling current obtained is 0.7µA.We report that there can be an increase in resonant tunneling current upto 22.85% in the above system with light of different energies. GaN is a wide Bandgap semiconductor .Hence photon energies upto 2 eV have no influence on the tunneling current. The tunneling current increases sharply for photon energies ranging from 2-2.75 eV and after which the current saturates.

Conclusion:

Double barrier quantum well structures have been used to model the resonant tunneling devices. A simple transmission matrix has been developed to evaluate the quantum mechanical characteristics of the electron wave across the structure. The developed model has been used to simulate the resonant tunneling current with an applied voltage across GaAs and GaN based RTD structures. It is observed through simulations that the tunneling currents for these structures peak at 0.25V and 0.55V respectively. The present investigation reveals the NDR behavior for both the structures but the former with a higher value to confirm its suitability for oscillator applications. It is also observed that optical exposure in the well leads to a significant improvement in the tunneling behavior. It is observed that GaN based RTT shows nearly 22.85 % increase in the tunneling current as against a 16.47 % increase in the case of GaAs based RTT, establishing the superiority of GaN based quantum structures for optoelectronic applications.

References:

[1] M Tsuchiya, H.Sakaki and J. Yaschino "Room temperature of differential negative resistance in AlAs/GaAs/AlAs Resonant tunneling diode" Jpn. J. Appl. Phys Vol 24 1985

[2] P.Capasso, S.Sen, A.C.Gossard, A.L.Hutchinson and J.H.English "IEEE Electron Device Letters" (1986) Pg 573-575

[3] T.C.L.A.Sollner, E.R.Brown and H.Q.Le "Microwave and Millimeter wave resonant tunneling devices" Physics of Quantum Devices, Ed. F.Capaso, Springer Verlag, 1989 pp 276-279

[4] Mayukh Bhattacharya and Pinaki Mazumder, "Augmentation of SPICE for Simulation of Circuits Containing Resonant Tunneling Diodes", IEEE Transactions on Computer-Aided Design of Integrated Circuits and systems, Vol. 20, No. 1, Jan 2001, pp 67-73

[5] Jian Ping Sun et. al., "Resonant Tunneling Diodes: Models and Properties" PROCEEDINGS OF THE IEEE, VOL. 86, No. 4, APRIL 1998, pp 56-62

[6] David A. B. Miller, Fellow, IEEE, Kai Ma, Student Member, IEEE, and James S. Harris, Jr., Fellow, IEEE "MSM-Based Integrated CMOS Wavelength-Tunable Optical Receiver" IEEE Photonics Technology Letters, Vol. 17, No. 6, Jun 2005 pp 101-104

[7] "Analysis of hetereojunction bipolar transistor/resonant tunneling diode logic for low power and high-speed digital applications", IEEE Trans. Electron Devices, 40, (4), 1993 pp. 685-691

[8] Behzad Razavi "Design of Integrated Circuits for Optical Communications", ISBN 0-07-282258-9, 2003

[9] M.Menouni et. al "14 Gbit/s digital optical transmitter module using GaAs HBTs and DFB laser, Electronic letters , 1st Feb 1996, Vol. 32, No. 3

[10] E.R.Brown, O.B.McMahon et.al. "SPICE model of the resonant-tunneling diode, Electronic letters, 9th may 1996, vol. 32 No. 10 pp 97-103

[11] Gerjuoy E and Coon D.D, "Analytic S-matrix considerations and time delay in resonant tunneling", Superlattice Microstruct., 1989, 5, (3), pp. 305-315

[12] John H Davies, "The Physics of Low Dimensional Semiconductors -An introduction", Cambridge University Press, 1996

[13] Brown E.R et.al. , "Oscillation up to 712GHz in InAs/AlSb resonant –tunneling diodes Appl. Phys. Lett., Vol.50, No. 20, 1991, pp 11-15

[14] Ajit Srivastava et al "Laser-Induced Above-Band-Gap Transparency in GaAs", Physical Review Letters, Vol 93, No 16, Oct 2004 , pp 123-126

[15] P. Herb and L. K. J. Vandamme , "General Relation between Refractive Index and Energy Gap in Semiconductors ", Infared Phys. Technol. Vol. 35, No. 4, pp. 609-615, 1994

[16] N.M. Ravindra et al "Energy gap–refractive index relations in semiconductors – An overview"., Infrared Physics & Technology 50 (2007) pp. 21–29

[17] N. M. Ravindra and V. K. Srivastav, "Variation Research note of Refractive Index with Energy gap in Semiconductors", Infrared Physics Vol 19, pp 603-604

[18] Thomas J .Slight and Charles N. Ironside, "Investigation into the Integration of a Resonant Tunneling Diode and an Optical Communications Laser: Model and Experiment", IEEE Journal of Quantum Electronics, Vol 43, No.7, July 2007, pp 145-148

[19] Liang Tang, David A.B.Miller, Krishna C Saraswat, Joseph Mattio 'C – Shaped nanoaperture-enhanced germanium photodetector'; Optic Letters Vol 31, No.10, May 2006, pp 123-129

Fluidic Molecular Processing and Interfacing Devices

Marina Alexandra Lyshevski[*] and Sergey Edward Lyshevski[**]

[*]Microsystems and Nanotechnologies, Webster, NY 14580-4400, USA
[**]Department of Electrical Engineering, Rochester Institute of Technology, Rochester, NY 14623, USA
E-mail: E.Lyshevski@rit.edu and Sergey.Lyshevski@mail.rit.edu

ABSTRACT

We study *fluidic* processing devices and interfacing modules to perform data processing, interfacing and other related tasks utilizing molecular electrochemomechanical transitions and interactions. We consider various *fluidic* devices as *modular* primitives in order to: (i) Develop and examine alternative processing paradigms typifying *natural* processing platforms; (ii) Interface microapparatuses with biomolecular and *natural* systems establishing a sound micro-bio interface. Data processing and memory storage by *fluidic* devices cannot be equated to information processing exhibited by living systems. However, the proposed solution contributes to the *natural* and molecular data processing, memories, communication and interfacing. The *fluidic* devices are examined studying the synthesis issues, electrochemomechanical transitions, control of *microscopic* particles, etc. The device physics, expected performance and functionality are reported and discussed.

Keywords: biomolecule, *fluidic*, interfacing, processing

1. INTRODUCTION

The activity of brain neurons has been extensively studied using single microelectrodes as well as microelectrode arrays to probe and attempt to affect the activity of a single neuron or assembly of neurons in brain and neural culture. Unfortunately, these attempts have been only partially successful due to enormous fundamental, experimental and technological problems. For example, a great deal of effort has been applied attempting to integrate neurons and microelectronics with a very modest progress [1, 2]. It is unlikely that the micro-bio interface can be achieved using conventional *solid*-centered microelectronic solutions. This paper contributes to the aforementioned developments by examining alternative solutions in design of interfacing modules.

We also focus on a sound solution in processing by researching computing and memory utilizing molecular electrochemomechanical transitions and interactions. From processing viewpoints, our solution may have a limited practicality due to technological difficulties and existence of outstanding solutions, e.g., microelectronics and integrated circuits (ICs). However, the developments undertaken are very important and contribute to long-standing problems of biophysics. It may be expected that *fluidic* processing devices and interfacing modules may

affect or empower biotechnology, science, engineering and medicine by contributing to:
1. Biophysics fundamentals;
2. Basic research in *natural* processing (*natural* computing);
3. Sound technological developments;
4. Interfacing and integration of implantable biocompatible microelectronics, microelectrodes, microsensors and other apparatuses.

These problems are forefronts of science, engineering and technology. We propose a *fluidic* molecular processing device which mimics, to some extent, a neuronal processing-and-memory primitive or a brain neuron. In general, data processing and memory storage can be accomplished by various electrochemomechanically-induced transitions, interactions and events. For example, release, propagation and binding/unbinding of movable molecules result in state transitions to be utilized. Due to unsolved fundamental problems, complexity and technological limits, one may not coherently comprehend, mimic and prototype information processing in living systems. We typify 3D topologies/organizations of biosystems, utilize molecular hardware, and employ molecular transitions. These innovations imply novel synthesis, design, aggregation, utilization, functionalization and other features. We proposed to utilize specific electromechanical transitions (bond formation/braking, electron exchange, electron flow and other) between stationary biomolecules and
- *information/routing/executing* carriers for processing devices,
- *interfacing* and *routing* carriers in interfacing modules.

Various molecules and ions, utilized in neurons, are examined. Examples are reported.

2. *FLUIDIC* PROCESSING AND INTERFACING DEVICES

Utilizing 3D topology/organization of biomolecular assemblies, observed in *natural* systems, the *engineered fluidic* molecular processing device (^{MP}device) is illustrated in Figure 1. The inner enclosure can be made of proteins, porous silicon or polymers to form membranes with fluidic channels with the binding sites which ensure the functionality and selectivity. The *information/routing/ executing* or *interfacing* carriers are encapsulated in the outer enclosures or cavities. The release and steering of different carriers are controlled by the control apparatus utilizing *passive* and *active* mechanisms. The electrochemomechanical transitions, caused by the *information* carriers, result in the logic and memory events. Multiple-valued state transitions imply that computing, processing and memory storage can be performed on the high radix. Using *routing* carriers, persistent and robust morphological reconfigurable networking can be achieved. This ensures a reconfigurable networking processing-and-memory organization.

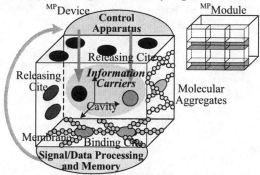

Figure 1. *Synthetic fluidic* ^{MP}*device and* ^{MP}*module with 3×2×2 devices*

The released *routing* carriers are steered in the fluidic cavity to the binding sites resulting in the binding/unbinding of *routers* to the stationary molecules. The binding/unbinding events lead to reconfigurable routing ensuring reconfiguration.

The proposed *fluidic* ^{MP}device mimics, to some extent, a neuron with synapses, membrane, channels, cytoplasm, microtubules, organelles, etc. Specific ions, molecules and enzymes propagate in the synaptic cleft and can pass through the membrane channels. In the proposed ^{MP}device, the carriers pass the porous membrane and propagate in the cavity. The molecules (*information* and *routing* carriers) bind to the specific receptor sites, while enzymes free molecules from binding sites. Binding and unbinding of molecules result in the electrochemomechanical transitions. The released carriers are controlled by changing the electrostatic potential or thermal gradient. Control of carriers cannot be accomplished through preassigned steady-state conditional logics, synchronization, timing protocols and other conventional concepts.

The motion and dynamics of the carrier release, propagation, binding/unbinding and other events are examined. The goal is to achieve a controlled motion and functionality of *microscopic* carriers. Distinct control mechanisms (electrostatic, electromagnetic, thermal, hydrodynamic, etc.) allow one to uniquely utilize selective control ensuring high performance and enabling functionality. The controlled *active* transport of molecules and ions in the fluidic cavity and channels are examined. Having proposed a hardware solution, the carrier displacement r_i is controlled by the control vector **u**. Released carriers propagate in the fluidic cavity. These carriers are controlled by a control apparatus varying $F_n(t,\mathbf{r},\mathbf{u})$ and $V_k(\mathbf{r},\mathbf{u})$ [2]. This apparatus is comprised of molecular assemblies which change the temperature gradient or the electric field intensity. The state transitions occur in the anchored processing polypeptide or organic

molecular complexes as *information/routing/executing* or *interfacing* carriers bind/unbind. For example, conformational *switching*, charge changes, electron transport and other phenomena can be utilized. The transition time of electronic, photoelectric and electrochemomechanical state transitions ranges from psec to μsec. The molecular hardware predefines the phenomena (effects) utilized, as well as device functionality. For example, *electromechanical switching* can be accomplished by biomolecules. It is feasible to design and potentially synthesize aggregated networks of reconfigurable *fluidic* ^{MP}devices. These devices and modules are characterized in terms of input-output functionality, activity, performance, etc. To overcome the fabrication deficiencies, one may utilize the cultured neurons as modules.

To relate neurons to the proposed concept, one recalls the so-called *fluid mosaic model* of the biological plasma membrane. In the membrane, most of the lipids and membrane proteins are held together by hydrophobic attractions which are weaker then covalent bonds. The certain biological conditions (temperature, pH, etc.) must be guaranteed. The lipid bilayer is the main fabric of the membrane, while proteins affect the membrane functionality. There are more then 50 different plasma proteins. For example, glycoproteins (proteins with covalently bonded carbohydrates) are important in intracellular recognition. The selective permeability of membrane is defined by the lipid bilayer specificity, as well as by transport proteins and ion channel plasma membrane proteins.

There are two types of transport across the biological membrane depending on the particles being transported, e.g., passive and active. The channel proteins form hydrophilic pores across membrane. They are narrow, highly selective and referred as *ion channels*. The transport efficiency and its rate were extensively studied. It was found that ion channel transport, as compared to transport mediated by carrier proteins, ensures rate ~1000:1. Control of the transport through the ion channel is essential to maintain basic cell functions.

In most cases, receptors are transmembrane proteins on the target-cell surface. When they bind a released extracellular molecule (ligand), receptors become activated and undergo transitions which define cell activity. Receptors also can reside inside the cell, and the ligand has to enter the cell to activate it. Neurons possess various complex processing-related transitions and mechanisms. The functionality depends on the biomolecule-protein recognition, binding, unbinding and other transitions.

The release and suppression of neurotransmitters, and their role on action potential and other transitions were extensively studied in the literature. For example, some results are reported in [3, 4].

3. ENERGETICS ESTIMATES

The energetics of neurotransmitters, biomolecules and *information carriers* is analyzed. In neurons, synaptic changes proceed through the release of neurotransmitters, activation/deactivation of synaptic receptors, protein conformational changes, etc. The activation of the acetylcholine receptors is the most researched. We estimate the binding energy of the low-molecular-weight neurotransmitters. It is reported that binding is stabilized by noncovalent bonds because they can be rapidly formed and broken. The neurotransmitters bind to the membrane's receptors by means of hydrogen bonds. We consider the intermolecular attraction between hydrogen atoms in a polar bond and unshared electron pair on a nearby electronegative florin, oxygen and nitrogen atoms of other molecule. That is, we study the polar H–F, H–O and H–N bonds. The dipole-dipole attractions between electronegative F, O and N with H, result in a very strong bond. Typically hydrogen bonds are formed with two or more other atoms. The energies of *symmetric* and *asymmetric* hydrogen bonds vary from 4 kJ/mol to 25 kJ/mol. The bond energies and geometry of neurotransmitters and receptors vary. For low-molecular-weight neurotransmitters (acetylcholine, γ-aminobutyric acid, catecholamines, 5-hydroxytryptamine, etc.), we found that the energy of the hydrogen-ligand complex is ~ -70 kJ/mol. Enzymatic activity, stereochemistry, geometry of coordination, interatomic distances, electrostatics, electronegativity of hydrogen-bonded systems, are the factors which significantly affect the energetics. One obtains the estimate for energy to be $\sim -2 \times 10^{-17}$ J per each neurotransmitter ($\sim -1.7 \times 10^{-17}$ J for acetylcholine).

Neurotransmitters are inactivated (removed from the synaptic cleft either by specific hydrolytic enzymes or by specific membrane transport proteins which uptake neurotransmitter back into either the nerve terminal or neighboring glial cells) and re-processed. Acetylcholinesterase is the enzyme which breaks acetylcholine (ACh) into choline and acetate by means of hydrolysis. It controls the rate of the reaction, and by favoring certain geometries in the transition state, lovers activation energy. We derived that the upper ehthalpy estimate for the hydrolysis reaction is $\Delta H = +28$ kJ/mol at pH7 and 25°C. Hence, one finds 7×10^{-18} J for each hydrolized ACh molecule. The $\Delta H = +1.17 \pm 0.1$ kJ/mol, reported in [5], results in 2.8×10^{-19} J per a single molecule.

The derived energetics of biomolecular electrochemo-mechanical transitions is with the projected solid-state microelectronic devise energetics. In particular, the energy of switching is expected to be reduced to $\sim 1 \times 10^{-16}$ J.

4. BROWNIAN DYNAMICS: MOLECULAR MOTION AND TRANSPORT

The Brownian dynamics of N particles is usually described by the second-order stochastic differential equations

$$m_i \frac{d^2 \mathbf{r}_i}{dt^2} = F_v \left(\frac{d \mathbf{r}_i}{dt} \right) + \sum_{i,j}^{N} F(t, \mathbf{r}_{ij}) - \frac{\partial V(\mathbf{r}_i)}{\partial \mathbf{r}_i} - \sum_{i,j}^{N} \frac{\partial V(\mathbf{r}_{ij})}{\partial \mathbf{r}_{ij}} + \xi_i, (1)$$

where m is the mass of particles; F_v is the viscous friction force, $F_v = \eta v$; η is the viscous friction coefficient; $\xi(t)$ is the Gaussian white noise, $\langle \xi(t) \rangle = 0$ and $\langle \xi(t) \xi(t') \rangle = 2 \eta k_B T \delta(t - t')$; k_B is the Boltzmann constant; T is the absolute temperature.

We examine the stochastic particle dynamics integrating stochastic mechanical, thermal, electromagnetic, hydrodynamic, noise-induced and bistable phenomena. In three-dimensional space, the resulting translational equations of motion are derived using the displacement vector $\mathbf{r}$, velocity vector $\mathbf{v}$ ($\mathbf{v} = d\mathbf{r}/dt$), and extended state vector $\mathbf{q}$. Those $\mathbf{r}$, $\mathbf{v}$ and $\mathbf{q}$ are the state variables. One has

$$m_i \frac{d^2 \mathbf{r}_i}{dt^2} = F_v \left(\frac{d \mathbf{r}_i}{dt} \right) + \sum_{i,j,n} F_n(t, \mathbf{r}_{ij}) - \sum_k \frac{\partial V_k(\mathbf{r}_i)}{\partial \mathbf{r}_i}$$
$$- \sum_{i,j,k} \frac{\partial V_k(\mathbf{r}_{ij})}{\partial \mathbf{r}_{ij}} + f_r(t, \mathbf{r}, \mathbf{q}) + \xi_{ri},$$
$$\frac{d \mathbf{q}_i}{dt} = f_q(t, \mathbf{r}, \mathbf{q}) + \xi_{qi}, i = 1, 2, ..., N, \quad (2)$$

where $f_r(t, \mathbf{r}, \mathbf{q})$ and $f_q(t, \mathbf{r}, \mathbf{q})$ are the nonlinear maps.

In (2), forces $F_n(t, \mathbf{r})$ and potentials $V_k(\mathbf{r})$ of electromagnetic, hydrodynamic, thermal and other origin can be varied by changing distinct physical variables (voltage, current, temperature gradient, viscosity, etc.). Therefore, notations $V_k(\mathbf{r}, \mathbf{u})$ and $F_n(t, \mathbf{r}, \mathbf{u})$ are used to define the varied asymmetric potentials and forces.

For controlled particles, the equations of motion, that provides a high-fidelity translational model, are

$$m_i \frac{d^2 \mathbf{r}_i}{dt^2} = F_v \left(\frac{d \mathbf{r}_i}{dt} \right) + \sum_{i,j,n} F_n(t, \mathbf{r}_{ij}, \mathbf{u}) - \sum_k \frac{\partial V_k(\mathbf{r}_i, \mathbf{u})}{\partial \mathbf{r}_i}$$
$$- \sum_{i,j,k} \frac{\partial V_k(\mathbf{r}_{ij}, \mathbf{u})}{\partial \mathbf{r}_{ij}} + f_r(t, \mathbf{r}, \mathbf{q}) + \xi_{ri},$$
$$\frac{d \mathbf{q}_i}{dt} = f_q(t, \mathbf{r}, \mathbf{q}) + \xi_{qi}, i = 1, 2, ..., N. \quad (3)$$

The rotational motion is described by

$$J_i \frac{d^2 \boldsymbol{\theta}_i}{dt^2} = T_v \left(\frac{d \boldsymbol{\theta}_i}{dt} \right) + \sum_{i,j,n} T_n(t, \boldsymbol{\theta}_{ij}) - \sum_k \frac{\partial V_{Tk}(\boldsymbol{\theta}_i)}{\partial \boldsymbol{\theta}_i}$$
$$- \sum_{i,j,k} \frac{\partial V_{Tk}(\boldsymbol{\theta}_{ij})}{\partial \boldsymbol{\theta}_{ij}} + f_\theta(t, \boldsymbol{\theta}, \mathbf{q}_\theta) + \xi_{\theta i},$$
$$\frac{d \mathbf{q}_{\theta i}}{dt} = f_\theta(t, \boldsymbol{\theta}, \mathbf{q}_\theta) + \xi_{\theta i}, i = 1, 2, ..., N, \quad (4)$$

where $\boldsymbol{\theta}_i$ is the angular displacement, and $d\boldsymbol{\theta}_i/dt = \omega_i$; ω_i is the angular velocity; J_i is the moment of inertia. Other notations are similar to the translational motion.

These stochastic differential equations should be numerically solved. The linearization cannot be performed.

The particle motion is due to the time-varying applied force $F_n(t, \mathbf{r})$, potential $V_k(\mathbf{r})$, and noise $\xi(t)$. Forces can be varied and controlled, and the control vector $\mathbf{u}$ is used. The variations of $V_k(\mathbf{r}, \mathbf{u})$ result in changes of $\frac{\partial V_k(\mathbf{r}, \mathbf{u})}{\partial \mathbf{r}}$ which provide the force terms. Hence, the motion of Brownian

NSTI-Nanotech 2008, www.nsti.org, ISBN 978-1-4200-8503-7 Vol. 1

particles is controlled by the time-varying applied force $F_n(t,\mathbf{r},\mathbf{u})$ and potential $V_k(\mathbf{r},\mathbf{u})$.

The van der Waals force can be found using distinct methods. For example, using the interaction energy between spherical particles, as given as $-\dfrac{H}{6d_{ij}}\dfrac{r_i r_j}{r_i + r_j}$, one

finds the force as [6-8] $F_{wij} = \dfrac{H}{6d_{ij}^2}\dfrac{r_i r_j}{r_i + r_j}$, where r_i and r_j

are the radii; d_{ij} is the intersurface distance between particles; H is the Hamaker constant [9], and if $H_i \neq H_j$, the equivalent Hamaker constant is $H=(H_i H_j)^{1/2}$.

The force between a single molecule and sphere (membrane) is $F_{wmi} = -\dfrac{8\pi\rho_m B R_1^3 d_{mi}}{(d_{mi}-R_1)^4 (d_{mi}+R_1)^4}$, where d_{mi} is

the nearest distance between the center of sphere and molecule; ρ_m is density of sphere; B is the van der Waals constant.

The van der Waals interaction energy between a particle of radius r_i and a composite membrane with an outer layer with thickness h is [10]

$$E_{Wmi} = -\frac{H_i}{6}\left[r_i\left(\frac{1}{d_{mi}} - \frac{1}{d_{mi}+h} \right) - \ln\left(\frac{d_{mi}}{d_{mi}+h} \right) \right].$$

Hence, the force is found to be

$$F_{Wmi} = \frac{H_i}{6}\left[r_i\left(\frac{1}{d_{mi}^2} - \frac{1}{(d_{mi}+h)^2} \right) - \frac{h}{d_{mi}(d_{mi}+h)} \right].$$

We study the *activating information* carriers. Ions, diffused into the neuron through the membrane ionic channels, are considered to be the *regulating information* carriers. The number of ions in the synaptic cleft is defined by their concentration. Letting the ionic concentration for Na^+, Cl^-, K^+ and Ca^{2+} to be 140, 100, 5 and 2 mM, in the synaptic cleft with 25 nm separation between membranes (L=25 nm), we study the Brownian motion of 1 neurotransmitter as well as 19, 14, 1 and 1 Na^+, Cl^-, K^+ and Ca^{2+} ions.

The ions interact with a polar neurotransmitter which propagates in the synaptic cleft. We study the motion of 36 particles in three-dimensional space. This results in 324 first-order stochastic differential equations for the translational motion (3). The electric dipole moment for GABA is 4.8×10^{-29} C-m. The length of GABA is 0.91 nm. The neurotransmitter mass and diffusion coefficient are m_{GABA}=1.71×10^{-25} kg and D_{GABA}=4×10^{-11} m²/s.

The masses, diffusion coefficients (at 37°C) and ionic radii of Na^+, Cl^-, K^+ and Ca^{2+} ions used in the analysis are [2]: m_{Na}=3.81×10^{-26} kg, m_{Cl}=5.89×10^{-26} kg, m_K=6.49×10^{-26} kg, m_{Ca}=6.66×10^{-26} kg, r_{Na}=0.95×10^{-10} m, r_{Cl}=1.81×10^{-10} m, r_K=1.33×10^{-10} m, r_{Ca}=1×10^{-10} m, D_{Na}=1.33×10^{-9} m²/s, D_{Cl}=2×10^{-9} m²/s, D_K=1.96×10^{-9} m²/s and D_{Ca}=0.71×10^{-9} m²/s. The relative permittivities of presynaptic and postsynaptic membranes are ε_{rp}=2.3 and ε_{rP}=2.1.

The controlled motion of a neurotransmitter from the origin (0, 0, 0) to the binding cite (0, 0, 25 nm) is reported in Figure 2 for different simulation runs [2]. Here, L=25 nm.

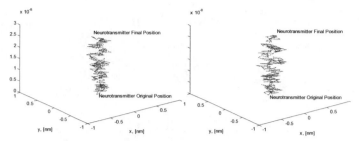

Figure 2. Controlled neurotransmitter displacement in the synaptic cleft

5. CONCLUSIONS

We reported the *fluidic* processing and interfacing devices. These developments are envisioned to be utilized in foreseen processing systems and for micro-bio interfacing. The synthesis aspects are covered. The use of biomolecules as *information* and *interfacing* carriers was introduced. It was demonstrated that the controlled motion of carriers in the fluidic cavity can be accomplished. The feasibility studies for interfacing of *fluidic* devices with *natural* cells are performed.

REFERENCES

1. F. Frantherz, *Neuroelectronics Interfacing: Semiconductor Chips Wth Ion Channels, Nerve Cells, and Brain, Handbook of Nanoelectronics and Information Technology*, Ed. R. Waser, pp. 781-810, Wiley-VCH, Darmstadt, Germany, 2005.

2. S. E. Lyshevski, *Molecular Electronics, Circuits, and Processing Platforms*, CRC Press, Boca Raton, FL, 2007.

3. B. Katz and R. Miledi, "A study of synaptic transmission in the absence of nerve impulses," *Journ. Physiology*, vol. 192, issue 2, pp. 407-436, 1967.

4. S. Thomas and R, Robitaille, "Differential frequency-dependent regulation of transmitter release by endogenous nitric oxide at the amphibian neuromuscular synapse," *The Journal of Neuroscience*, vol. 21, issue4, pp. 1087-1095, 2001.

5. J. M. Sturevant, "The enthalpy of hydrolysis of acetylcholine," *Journ. Biol. Chem.*, vol. 247, pp. 968-969, 1972.

6. B. V. Derjaguin, L. Landau, "Theory of the stability of strongly charged lyophobic sols and of the adhesion of strongly charged particles in solutions of electrolytes," *Acta Phys. Chem. USSR*, vol. 14, pp. 633-622, 1941.

7. E. J. Verwey and J. T. G. Overbeek, *Theory of the Stability of Lyophobic Colloids*, Elsevier, Amsterdam, 1948.

8. J. N. Israelachvili, *Intermolecular and Surface Force*, Academic Press, London, 1992.

9. R. Tadmor, "The London - van der Waals interaction energy between objects of various geometries," *J. Phys.: Condens. Matter*, vol. 13, pp. 195-202, 2001.

10. H. C. Hamaker, "The London - van der Waals attraction between spherical particles," *Physica*, no. 10, pp. 1058-1072, 1937.

Chemical Energy to Mechanical Motion Through Catalysis

Ayusman Sen

Department of Chemistry
The Pennsylvania State University
University Park, PA 16802, USA
Email: asen@psu.edu

Abstract

We have demonstrated that one can build nanomotors "from scratch" that mimic biological motors by using catalytic reactions to create forces based on chemical gradients. These motors are autonomous in that they do not require external electric, magnetic, or optical fields as energy sources. Instead, the input energy is supplied locally and chemically. Depending on the shape of the object and the placement of the catalyst, different kinds of motion can be achieved. Additionally, an object that moves by generating a continuous surface force in a fluid can, in principle, be used to pump the fluid by the same catalytic mechanism. Thus, by immobilizing these nanomotors, it is possible to developed micro/nanofluidic pumps that transduce energy catalytically.

Introduction

Nano and microscale moving systems are currently the subject of intense interest due in part to their potential applications in nanomachinery, nanoscale assembly, robotics, tribology, fluidics, and chemical/biochemical sensing. Most of the research in this area has focused on using biological motor proteins in artificial systems, or using "molecular motors" such as rotaxanes that are powered externally. These have serious limitations with regards to fuel source, stability, and adaptability. As described below, we have demonstrated that one can build nanomotors "from scratch" that mimic biological motors by using catalytic reactions to create forces based on chemical gradients.

Results

In designing catalytic nanomotors and devices derived from them, it is important to understand the principles that govern their motion in the micrometer and nanometer regimes. We have established that electrokinetics is the dominant mechanism for the spontaneous motion of bimetallic nanorods and microgears suspended in hydrogen peroxide (H₂O₂) solutions (see Figs. 1 and 2) [1].

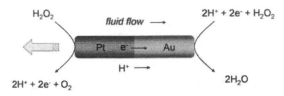

Figure 1. A schematic illustrating self-electrophoresis. Hydrogen peroxide is oxidized to generate protons in solution and electrons in the wire on the platinum end. The protons and electrons are then consumed with the reduction of H₂O₂ on the gold end. The resulting ion flux induces motion of the particle relative to the fluid, propelling the particle towards the platinum end with respect to the stationary fluid [1].

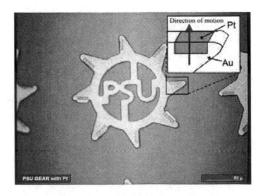

Figure 2. 100 μm diameter gold "microgears" with platinum "teeth" can rotate ~360°/sec in aqueous hydrogen peroxide systems [1].

In addition to controlling the motion of the hydrogen peroxide-driven systems by geometric design of moving structures, motion may also be controlled magnetically [2]. For example, hydrogen peroxide-powered nanorods that contain magnetic segments could be aligned when an external magnetic field is applied (Fig. 3). Nickel segments with a length shorter than the segment radius were used to ensure the easy axis of magnetization was orthogonal to the direction of motion. The rod motion was propelled by the hydrogen peroxide decomposition (and not by attraction to the magnet) and their direction could be remotely controlled using relatively weak magnetic fields. This motion is analogous to the behavior of motile bacteria that align themselves with Earth's weak magnetic field, and in fact these magnetotactic bacteria have magnetic moments on the same order as the nickel-containing rods.

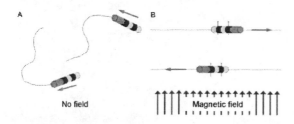

Figure 3. Schematic illustrating movement of Pt/Au/Ni/Au/Ni/Au rods without applied magnetic field (**A**), and with applied field (**B**) in aqueous H_2O_2. Note that rods align *perpendicular* to the applied magnetic field resulting in rod motion ~90 degrees to the applied field [2].

Most recently, we have discovered that the platinum/gold nanorods also exhibit chemotaxis (Figure 4) [3], traditionally defined as the movement of "organisms" toward or away from a chemical attractant or toxin by a biased random walk process. Again, this is the first example outside living systems. Our work also reveals that chemotaxis does not require "temporal sensing" mechanism commonly attributed to bacteria. This behavior provides a novel way to direct particle movement towards specific targets. Chemotaxis also offers a new method of sorting and separating particles of similar mass and size. Only those that are catalytically active move in response to the chemical gradient. From a fundamental standpoint, our work can be a starting point for the design of new motors for collective functions, such as catalytically driven swarming and pattern formation.

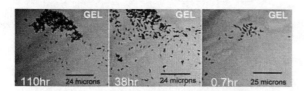

Fig. 4. The changing distribution of platinum-gold rods in a hydrogen proxide concentration gradient. The gel (soaked in 30% hydrogen peroxide) appears

in the upper left part. The images were taken at 0.7 hour, 38 hours, and 110 hours [3].

One noteworthy aspect of the work involving hydrogen peroxide and metallic structures relates to immobilizing the catalyst onto a fixed surface. While freely suspended metal nanorods move with respect to the bulk solution, by Galilean invariance an immobilized metal structure in the presence of hydrogen peroxide will induce fluid flows at the interface between the structure and the fluid. We have demonstrated this fluid pumping effect on a gold surface patterned with silver, another known hydrogen peroxide decomposition catalyst (Fig. 5) [4]. When aqueous hydrogen peroxide containing colloidal tracer particles was deposited onto these bimetallic surfaces, the tracers either followed a convection-type fluid flow towards the micron-sized silver surface or formed patterns as they were pushed away from the catalyst, depending on the zeta potential of the tracer (Figure 10).

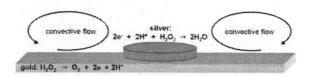

Figure 5. A catalytic micropump consisting of a silver disk on a gold substrate. The electrochemical decomposition of hydrogen peroxide establishes a weak electric field. This field causes tracer particles to migrate *towards* or *away* from the silver depending on their *surface charge*. Particles migrating towards the silver follow *electroosmotic convection* (arrows) near the catalyst surface [4].

Future Directions

The small scale and local conversion of chemical to mechanical energy may find use, among other applications, as micropumps and motors for micro/nanomachinery. The catalytically induced electrokinetic phenomenon could find use in microfluidics and lab-on-a-chip applications where the use of pressure gradients or externally generated electric fields may be difficult or undesirable.

In addition to micropump applications, the ability of particles that are both catalytic and asymmetric to locally convert chemical to mechanical energy offers the possibility of designing and controlling micro- and nanoscale machines that can interact with biological systems, such as individual cells. The idea of designing and developing devices that can interact intimately with biological systems at the cellular level is an exciting one, and will benefit from addressing two aspects of biocompatibility: one that allows the device to work properly in biological systems, and one that will not interfere in an undesirable way with the biological host. This will require developing new approaches (or adapting current ones) to allow for biocompatible and bioavailable materials and fuels.

The examples in this paper demonstrate the fulfillment of the minimum requirements for nanoscale machines, namely initiating motion and, to a lesser degree, controlling the direction of that motion at the micron and nanometer scales. If, in addition to a motor and a director, one could incorporate a device to perform some useful operation, such as carrying cargo, entirely new classes of micro and nanoscale devices become possible. By incorporating materials that selectively adsorb or desorb a substance upon action of some stimulus, from either the immediate environment or a remote source, such devices could be engineered to deliver small-scale amounts of therapeutic agents, or other payload, to specific regions in the human body. In charting future directions, two major areas will need to be addressed. First, while we know that catalysts can be used to impart motion to otherwise inanimate objects, we need to more fully understand and develop the mechanisms of chemical to mechanical energy transduction.

References

[1] Paxton, W.F.; Sundararajan, S.; Mallouk, T.E.; Sen, A. *Angew. Chem. Int. Ed.* **45**, 5420-5429 (2006).
[2] T. R. Kline, W. F. Paxton, T. E. Mallouk, A. Sen, *Angew. Chem. Int. Ed.* **44**, 744-746 (2005).
[3] Hong, Y.; Blackman, N.M.K.; Kopp, N.D.; Sen, A.; Velegol, D. *Phys. Rev. Lett.* **99**, 178103-178106 (2007).
[4] T. R. Kline, W. F. Paxton, Y. Wang, D. Velegol, T. E. Mallouk, A. Sen, *J. Am. Chem. Soc.* **127**, 17150-17151 (2005).

Tensile Stress Effect on Indium Diffusion in Silicon Substrate: *Ab-initio* Study

Young-Kyu Kim, Soon-Yeol Park, and Taeyoung Won

Department of Electrical Engineering, School of Information Technology Engineering,
Inha University, Incheon, Korea 402-751
twon@hsel.inha.ac.kr

ABSTRACT

In this paper, we present our *ab-initio* study on energy configurations, minimum energy path (MEP), and migration energy for neutral indium diffusion in a uniaxial and biaxial tensile strained {100} silicon layer. Our *ab-initio* calculation of the electronic structure allowed us to figure out transient atomistic configurations during the indium diffusion in strained silicon. We found that the lowest-energy structure (In_S - Si_i^{Td}) consists of indium sitting on a substitutional site while stabilizing a silicon self-interstitial in a nearby tetrahedral position. Our *ab-initio* calculation also revealed that the next lowest energy structure is In_i^{Td}, the interstitial indium at the tetrahedral position. We employed the nudged elastic band (NEB) method for estimating the MEP between the two structural states. The NEB method implies that the diffusion pathway of neutral indium is kept unchanged while the migration energy of indium fluctuates in strained silicon.

Keywords: *Ab-initio* calculation, strained silicon, indium, diffusion pathway.

1 INTRODUCTION

Recently, a heavy ion implantation has attracted a great of attention as a promising method for achieving a very shallow junction in silicon microelectronic devices. Due to its heavier mass, indium is considered to be one of the alternatives to boron atom as a p-dopant [1,2]. Now, strain engineering has also received a lot of attention due to the enhanced mobility of carriers [3].

Consequently, studies on the properties of indium diffusion in strained silicon would help us to develop successful process integration. That is because modeling of diffusion profiles of the dopant impurities is imperative for design of the devices due to dopant diffusion in silicon crystal plays a central role[4]. In this work, we investigated the diffusion pathway as well as the energy barrier of

indium in an effort to obtain the physical parameters of indium in silicon through *ab-initio* calculations.

2 COMPUTATIONAL DETAILS
2.1 Diffusion Pathway of Indium

The *ab-initio* calculations were implemented within density functional theory (DFT) with Vienna *Ab-initio* Simulation Package (VASP) which combines ultrasoft pseudopotentials[5] and generalized gradient approximation (GGA) in the Purdew and Wang formulation. The simulation condition was set as the followings: a cutoff energy E_c = 150.62 eV, $2 \times 2 \times 2$ grid for the *k*-points mesh of Monkhorst-Pack[6], and a $3 \times 3 \times 3$ simple cubic super-cell (216 atoms). The optimized silicon lattice constant for GGA in our system is 5.461Å. An energy landscape for In-Si complex can be provided by employing VASP. The lowest-energy structure was found to be In_s - Si_i^{Td}, as shown in Fig. 1. The second lowest-energy structure is found to be In_i^{Td}, as shown in Fig. 2.

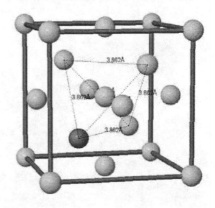

Fig. 1 A plot illustrating In_s-Si_i^{Td} configuration wherein the indium atom (red-colored) sits on a substitutional site and stabilizes self-interstitial silicon (yellow-colored) in a tetrahedral position.

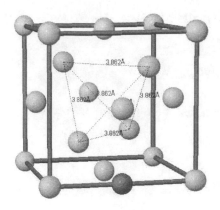

Fig. 2 In_i^{Td} structure wherein the interstitial indium atom(red-colored) is in the tetrahedral position.

The energy difference between the first and the second lowest energy configuration has been calculated to be 0.492 eV. The energy landscape allows us to figure out the diffusion pathway of neutral indium as the following:

$$In_S - Si_i^{Td} \rightarrow In_i^{Td} \rightarrow In_S - Si_t^{Td} \quad (1)$$

2.2 Correlation between Stress and Strain

Fig. 3 is a schematic diagram illustrating a Metal-Oxide Semiconductor Field Effect Transistor (MOSFET) structure with a strained silicon layer on silicon-germanium buffer. To introduce uniaxial and biaxial strain in silicon, we applied the lattice constant of relaxed $Si_{1-x}Ge_x$ to the two crystallographic directions on the (100) plane since the strained silicon grown on relaxed SiGe buffer layer has the same lattice constant as SiGe.

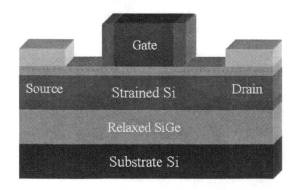

Fig. 3 Schematic diagram showing the strained silicon MOSFET structure employed in this work.

We investigated the change of $Si_{1-x}Ge_x$ lattice constant with respect to x (Ge mole fraction) wherein the lattice constant of $Si_{1-x}Ge_x$ can be expressed as:

$$a_{SiGe} = (1-x)a_{Si} + xa_{Ge} \quad (2)$$

where a_{Si} and a_{Ge} are lattice constants of Si and Ge, respectively. The lattice constant in the perpendicular direction was calculated from the elastic theory while "In-plane" and "Out-of-plane" strain can be calculated from the following three equations:

$$\varepsilon_\| = (a_{SiGe} - a_{Si})/a_{Si} \quad (3)$$

$$\varepsilon_\perp = (a_\perp - a_{Si})/a_{Si} \quad (4)$$

$$\varepsilon_\perp / \varepsilon_\| = -2(C_{12}/C_{11}) \quad (5)$$

where C_{11} and C_{12} are elastic constants of Si.

For our *ab-initio* calculations, we assumed C_{11} = 156Gpa and C_{12} = 56Gpa with basis on density functional theory results[7]. The results of elastic constant of silicon allow us to derive stresses in strained Si from Hooke's laws[8] with the following equations (6):

$$
\begin{bmatrix} \sigma_X \\ \sigma_y \\ \sigma_z \\ \sigma_{yz} \\ \sigma_{xz} \\ \sigma_{xy} \end{bmatrix} =
\begin{bmatrix}
C_{11} & C_{12} & C_{13} & C_{14} & C_{15} & C_{16} \\
C_{21} & C_{22} & C_{23} & C_{24} & C_{25} & C_{26} \\
C_{31} & C_{32} & C_{33} & C_{34} & C_{35} & C_{36} \\
C_{41} & C_{42} & C_{43} & C_{44} & C_{45} & C_{46} \\
C_{51} & C_{52} & C_{53} & C_{54} & C_{55} & C_{56} \\
C_{61} & C_{62} & C_{63} & C_{64} & C_{65} & C_{66}
\end{bmatrix}
\begin{bmatrix} \varepsilon_x \\ \varepsilon_y \\ \varepsilon_z \\ \varepsilon_{yz} \\ \varepsilon_{zx} \\ \varepsilon_{xy} \end{bmatrix} \quad (6)
$$

where C is the elastic compliance, ε is the strain, and σ is the stress. We formulated a correlation between the stress and strain as a function of germanium mole fraction, as shown in Tables I and II, with energy parameters of In-Si complex in uniaxial and biaxial strained silicon.

2.3 Minimum Energy Path

In order to search for the minimum energy path (MEP), we performed the nudged elastic band (NEB) calculation[9] which stems from a transition state theory (TST)[10]. The calculation approach in our work is based on the optimization of a number of intermediate images along the reaction path. Each initial image finds the possible lowest energy while maintaining equal spacing to the neighboring images.

The NEB is an efficient method for finding saddle points and minimum energy paths between the given initial and final states of diffusion. In our previous section, we searched the initial and final states of indium diffusion. The initial state is the lowest energy configuration (In_s - Si_i^{Td}) and the final state the second lowest energy configuration (In_i^{Td}). Repeating the transitions between the two states, neutral indium diffuses in silicon. Therefore, if we investigate the MEP from the initial state to the final state, we can now obtain the energy barrier for indium migration.

Fig. 4 shows the minimum energy path of indium in uniaxial strained silicon calculated by the NEB method wherein stress induced from 0, 1 and 2 GPa, respectively. Column-axis is the migration energy along the MEP of Si: In from In_s - Si_i^{Td} to In_i^{Td}.

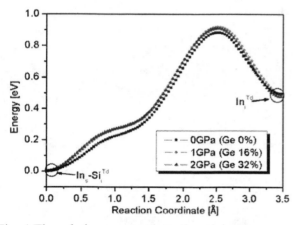

Fig. 4 The relative energy along the minimum energy path from In_s-Si_i^{Td} to In_i^{Td} in uniaxial strained silicon.

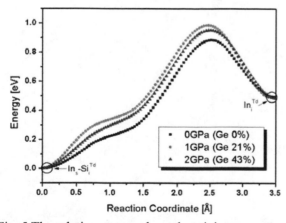

Fig. 5 The relative energy along the minimum energy path from In_s-Si_i^{Td} to In_i^{Td} in biaxial strained silicon.

Fig. 5 also shows the minimum energy path and migration energy of indium in biaxial strained silicon wherein 0, 1 and 2 GPa stresses are induced to silicon layer. Finally, we

listed the migration energy in Table I and II with total energies for the initial (In_s - Si_i^{Td}) and final (In_i^{Td}) configurations in 216 atom super cell.

Stress (GPa)	Strain(ratio)	Total energy (eV)		Migration energy (eV)
		In_s-Si_i^{Td}	In_i^{Td}	
0	0	-1167.628	-1167.136	0.884
1	0.006	-1167.643	-1167.162	0.920
2	0.013	-1167.483	-1167.009	0.910

Table I. Correlation among the strain ratio, Ge concentration (%), total energy (eV) of initial and final configurations and migration energy (eV) as a function of stress in uniaxial strained silicon.

Stress (GPa)	Strain(ratio)		Total energy (eV)		Migration energy (eV)
	X axis	-Y axis	In_s-Si_i^{Td}	In_i^{Td}	
0	0	0	-1167.628	-1167.136	0.884
1	0.009	0.006	-1167.552	-1167.065	0.984
2	0.017	0.012	-1167.111	-1166.631	0.952

Table II. Correlation among the strain ratio, Ge concentration (%), total energy (eV) of initial and final configurations, and migration energy (eV) as a function of stress in biaxial strained silicon.

Referring to Table I, the migration energy from the initial (In_s - Si_i^{Td}) to final (In_i^{Td}) configuration is 0.884, 0920, and 0.910 eV for stress-free, 1Gpa stress, and 2Gpa stress, respectively. We can observe that the migration energy for uniaxial strain increases by 4 % with the 1 GPa stress while the migration energy then decreases with further increase of stress. Table II reveals that the migration energy under biaxial strain increases by 11% with the 1 GPa stress while the migration energy then returns to decrease with further increase of stress. If we consider the simulation region (217 atoms: 216 Si atoms + 1 In atom), the change in the total and migration energy are significantly noticeable which will retard the diffusion. We can see that the strain does not affect the migration pathway, however, it changes the diffusivity of indium based on changing total and migration energies.

3 RESULTS AND DISCUSSION

We investigated the effect of strain on indium diffusion for differently stressed In-Si complexes. In uniaxial tensile strained silicon, total energies for initial (In_s - Si_i^{Td}) and final (In_i^{Td}) configurations (Si: 216 atoms + In: 1 atom) were found to be decreasing until the threshold of 0.5GPa stress. In other words, the super cell which composed 216 Silicon atoms plus an indium atom becomes a more stable state than unstrained structure for super cell. This is the reason why In-Si complex has the highest barrier energy when 0.5GPa stress is induced for super cell. In the meanwhile, our *ab-initio* calculation revealed that the lowest total energy in biaxial tensile strained silicon occurs when 0.25GPa stress is induced. We also found that the barrier energy of In-Si complex is highest when we introduce a stress of 1GPa. That is because the lattice constant along the Y-axis decreases continuously while the lattice distance along the X-axis increases.

4 CONCLUSION

Atomic diffusion is an important phenomenon in condensed matters. Dopant diffusion plays a central role in fabricating the electronic devices in Si crystal. On this account, modeling of diffusion profiles of the dopant impurities is imperative for design of the devices. In this paper, we reported our *ab-initio* calculation on the tensile uniaxial and biaxial strain effects on indium atom migration for the research on ULSI devices. Our theoretical study enlightens us how the stress imposed on silicon substrate affect the total and migration energy of In-Si complex, while migration pathway of In-Si complex does not affect strain on silicon layer.

ACKNOWLEDGMENT

This research was supported by the Ministry of Knowledge Economy, Korea, under the ITRC (Information Technology Research Center) support program supervised by the IITA (Institute of Information Technology Advancement) (IITA-2008-C109008010030) and the authors would like to acknowledge the support from KISTI Supercomputing Center(KSC-2007-S00-1013).

REFERENCES

[1] I. C. Kizilyalli, T. L. Rich, F. A. Stevie, and C. S. Rafferty, *J. Appl. Phys.* 80, 4944 (1996).

[2] P. B. Griffin, M. Cao, P. Vande Voorde, Y. L. Chang, and W. G. Greene, *Appl. Phys. Lett.* 73, 2986 (1998).

[3] L. Lin, T. Kirichenko, B. R. Sahu, G. S. Hwang, and S. K. Banerjee, *Phys. Rev. B* 72, 205206 (2005).

[4] J. W. Jeong and A. Oshiyama, *Phys. Rev. B* 64, 235204 (2001).

[5] G. Kresse and J. Hafner, *J. Phys. Condens. Matt.* 6, 8245 (1994).

[6] H. J. Monkhorst and J. D. Pack, *Phys. Rev. B* 13, 5188 (1976).

[7] W. A. Brantley, *J. Appl. Phys.* 44, 534 (1973).

[8] H. Noma, H. Takahashi, H. Fujioka, M. Oshima, Y. Baba, K. Hirose, M. Niwa, K. Usuda, and N. Hirashita, *J. Appl. Phys.* 90, 5434 (2001).

[9] H. Jónsson, G. Mills and K. W. Jacobsen, Nudged Elastic Band Method for Finding Minimum Energy Paths of Transitions, ed. B. J. Berne, G. Ciccotti and D. F. Coker, World Scientific, 385 (1998).

[10] Vasp Tools: http:// theory.cm.utexas.edu/vtsttools/.

Numerical Stress Analysis on Thermal Nano-Imprint Lithography

Bum-Goo Cho, Soon-Yeol Park and Taeyoung Won

Department of Electrical Engineering, School of Information Technology Engineering,
Inha University, 253, Yonghyun-Dong, Nam-Gu, Incheon, 402-751, Korea
E-mail: cbg@hsel.inha.ac.kr

ABSTRACT

In this paper, we investigate the stress distribution of polymer film in thermal nano-imprint lithography (NIL). In order to simulate this process, commercially available software was employed for NIL process simulation. The proposed model is imprinting a rigid SiO_2 stamp with a rectangular line pattern into a viscoelastic Polymethyl methacrylene (PMMA) film. The distribution of stress in the polymer film is calculated for the detail analysis of deformation behavior. These calculated results represent asymmetric von Mises stress distribution of the polymer around the external line caused by the squeezing flow under flat space.

Keywords: stress distribution, thermal nano-imprint lithography, viscoelastic, Polymethyl methacrylene film, von Mises stress

1 INTRODUCTION

Nano-imprint lithography (NIL) is one of the most promising techniques. Traditional photolithographic approaches make pattern transfer through the use of photons or electrons for modifying the chemical and physical properties of the resist. NIL relies on a direct mechanical deformation of the resist material and can therefore achieve resolutions beyond the limitations which are set by a light diffraction or a beam scattering. It enables both high resolution up to sub-25 nm which is not accomplishable in photolithography and fast throughput compared to serial processing methods such as electron beam or scanning probe microscopy (SPM) lithography, so it is expected as an alternative lithography method to conventional ones [1]. In addition, NIL is expected to realize a low cost and high-throughput production. In the thermal NIL process, a high speed imprinting was reported such as a role-to-role imprint [2]. In order to improve the NIL technology, it is essential to understand the deformation behavior of polymer during the imprinting process. Success of the pattern transfer in NIL depends on exact deformation of a polymer film according to stamp patterns and clear separation of a stamp from a polymer film [3]. Despite the prospective potential for nano-scale pattern transfer, there are few literature publications on the numerical modeling on NIL processes [4-6]. In this work, we modeled the NIL process and employed commercially available software, COMSOL Multiphysics, for the implementation of our model. In this paper, we report the stress distribution of the polymer deformation process on the imprinting pressure.

2 SIMULATION AND RESULTS

COMSOL Multiphysics bases its implementation of the structural mechanics application modes on the equilibrium

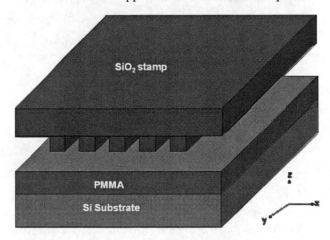

Figure 1: Schematic diagram illustrating the physical model.

equations expressed in the global stress components. For static analysis, substitution of the stress-strain relationship and the strain-displacement relationship into the static equilibrium equation produces Navier's equation of equilibrium expressed in the displacements. It is possible to completely describe the strain conditions at a point with the deformation components and their derivatives.

Figure 1 shows the schematic diagram of physical model about imprinting process. The mold has five line patterns with 1um whose width is 50nm and height is 50nm. Each line patterns are at intervals of 50nm. The polymer materials we used in our simulation were Polymethyl methacrylene (PMMA) and the imprint temperature was 140℃. The SiO$_2$ mold was modeled as a rigid body, and the PMMA resist was modeled as a viscoelastic material. Table 1 shows the material constants used for numerical analyses. 10N/m^2 constant pressure was applied on the top of the mold and kept for various time periods during pressing process. After the pressing process, PMMA specimens were quickly cooled down so that the specimens almost kept the shape at the end of the pressing process.

Table 1: Material constants of PMMA resist.

Material constant	Value
Young's modulus	3 GPa
Poisson's ratio	0.4
Heat capacity	1420 J/K
Thermal expansion coefficient	$7.0*10^7$ K^{-1}
Thermal conductivity	0.19 W/(m*K)
Density	1190 kg/m^3

The stress distributions of the polymer are shown in Figure 2. The method is based on an equation that specifies the 2D pressure distribution for a given polymer thickness and a stamp velocity. The equation is derived from 3D Navier-Stokes equations with the understanding that the polymer motion is largely directed along the substrate surface. The contour of simulated results represents the residual von Mises stress. The major cause of the residual von Mises stress is that the thermal shrinkage arisen in cooling process uniformly. The nonuniformity of the

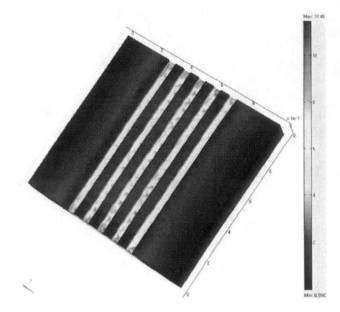

Figure 2: Stress distribution of the polymer.

residual von Mises stress is caused by the distribution of viscous strain at the end of the pressing process.

Figure 3(a), 3(b) and 3(c) show cross-sectional profiles of von Mises stress distribution of the polymer by x-y, y-z and z-x plane, respectively. These calculated results represent asymmetric von Mises stress distribution of the polymer around the external line caused by the squeezing flow under flat space. Similarly, von Mises stress of the polymer around the outside of the line is higher than the inside of the line.

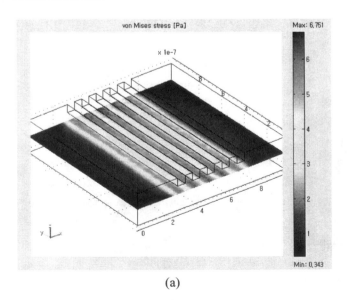

(a)

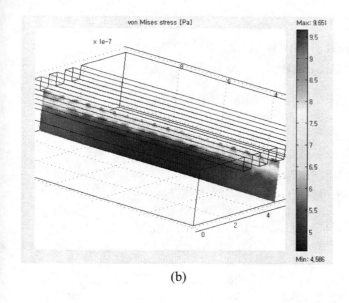

(b)

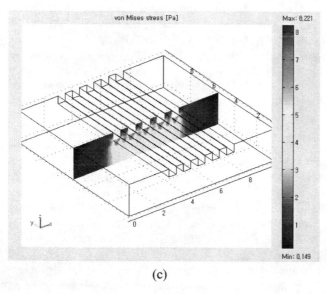

(c)

Figure 3: Cross-sectional profiles of von Mises stress distribution. (a) x-y plane, (b) y-z plane, (c) z-x plane.

CONCLUSION

A numerical simulation model imprinting a rigid SiO2 stamp into a viscoelastic polymer film is suggested in order to analyze the pattern transfer and its related phenomena in thermal NIL. The behavior of polymer deformation is investigated in detail by means of stress distribution analysis. In the imprint process, significant compressive stress is generated under the stamp pattern due to the compression of the polymer film, which causes the polymer to flow and to fill the cavity of the stamp for both patterns.

Numerical results explained characteristic phenomena which never appear in continuous line-and-space patterning cases. We believe that numerical simulations with more accurate material properties have possibilities to agree with experiments better enough for practical use. Our simulation approach could further be extended to accommodate deformation of the stamp and the sub-polymer platen.

ACKNOWLEDGEMENT

This research was supported by the Ministry of Knowledge Economy, Korea, under the ITRC (Information Technology Research Center) support program supervised by the IITA (Institute of Information Technology Advancement) (IITA-2008-C109008010030).

REFERENCES

[1] Stephen Y. Chou, Peter R. Krauss, and Preston J. Renstrom, Science 272, 85 (1996)

[2] T. Makcla, T. Haataninen, P. Majander, and J. Ahopelto, International Conference on Electron, Ion, Photon Beams and Nanotechnology, Paper No. PI-5 (2007)

[3] L. J. Guo, J. Phys. D: Appl. Phys. 37 (2004) R123.

[4] H. D. Rowland, A. C. Sum, P. R. Schunk, W. P. King, J. Micromech. Microeng., 15(2005).

[5] T. Eriksson, H. K. Rasmussen, J. Non-Newtonian Fluid. Mech., 127(2005).

[6] Y. Hirai, S. Yoshida, T. Kanakugi and T. Konishi, J. Vac. Sci. Technol., B22(2204)

Fabrication and Characterization of Nanoscale Heating Sources ("Nanoheaters") for Nanomanufacturing

Harshawardhan Jogdand,[1] Gokce Gulsoy,[2] Teiichi Ando,[2] Julie Chen,[1] Charalabos C. Doumanidis,[3] Zhiyong Gu,[4,*] Claus Rebholz[3] and Peter Wong[5]

1. Dept of Mechanical Engineering, University of Massachusetts Lowell, Lowell, MA
2. Dept of Mechanical and Industrial Engineering, Northeastern University, Boston, MA
3. Dept of Mechanical and Manufacturing Engineering, University of Cyprus, Nicosia, Cyprus
4*. Dept of Chemical Engineering, University of Massachusetts Lowell
One University Ave., Lowell, MA 01854; Zhiyong_Gu@uml.edu
5. Dept of Mechanical Engineering, Tufts University, Medford, MA

ABSTRACT

Nanoheater systems based on the exothermic reaction between aluminum and nickel were developed. Multilayered and powder-based heterostructures were fabricated and characterized, before and after ignition experiments, for their geometries and structures as well for the ignition and exothermic transformation kinetics. A novel ultrasonic powder consolidation (UPC) technique[1] was employed to produce ignitable Al/Ni compacts from elemental powders. Oxide free, fully dense Al matrix and intimate Al/Ni interfaces were obtained. Consolidates were successfully ignited and the reaction self-propagated throughout the compact resulting in the formation of NiAl phase. The ignition characteristics of multilayered structures and sequential thermal generation and conduction were investigated by IR thermal camera measurements.

Keywords: thermal manufacturing, heat sources, nanostructures

1. INTRODUCTION

Active use of heat to alter the geometry, structure and properties of solids is central to material removal, deposition, joining, shaping, and transformation processes in macroscale manufacturing. However as the nanoscale is approached, traditional heat sources are not always compatible because of fundamental technical limits. Spatial and temporal dimensions of nanoscale structures and phenomena pose a difficulty in application of traditional heat sources. The characteristic lengths and times of heat transfer approaches in macroscale are incompatible with the process requirements at nanoscale. Also there are difficulties in controlling temperature distribution and duration. Therefore, there is an outstanding need for new disruptive heat sources, enabling fine local selectivity and time exposure control in nanoscale thermal processing. Such sources could revolutionize manufacturing as well as on-board thermal actuation and autonomous powering of miniature devices and systems.

This paper addresses research in manufacture and operation of nanoheater systems based on the exothermic reaction[2] between aluminum and nickel. Multilayered nanoheaters comprised of alternating thin layers of Al and Ni (ranging around 20-100 nm each) are fabricated with physical vapor deposition, while powder-based Al-Ni composites are developed using ultrasonic joining. These nanoheaters, which can be either embedded within the bulk or on the surface of another material, can be ignited by an external ignition source. Reaction enthalpies released as a result, act as localized heat fluxes transferred to desired areas, primarily by conduction. The spatial and temporal profile of this heat source can be tailored by altering the structures of the nanoheaters. Nanoheaters with sufficiently fine-scale distributions of Al-Ni interface may ignite spontaneously when moderate bulk heating is applied to the material that contains the nanoheaters. Such nanoheaters can be ignited at low temperatures, at which no harm occurs to the substrate or bulk material itself. Applications for this new technology may include, but are not limited to, biomedical therapies[3] (e.g., heating cells and tissues), clean-energy enablers (e.g., pre-heating fuel cell membranes), sustainable products (e.g., reconfiguring polymer microstructures), MEMS and lab-on-a-chip applications.

2. NANOHEATER FABRICATION AND CHARACTERIZATION

To achieve controlled heat generation and distribution, one must have (1) full understanding of the effect of process parameters on the fabricated structure and (2) validated models of the heat output. These require information and validation from characterization of the composition, geometry and configuration of the fabricated nanoheaters

2.1 Multilayered Thin Films

Multilayered Al-Ni thin films are fabricated by a sputtering system at the University of Cyprus. The samples were characterized using JEOL 7401 field-emission scanning electron microscopy (FE-SEM), **Figure 1**. For this specimen the alternating layers of Al and Ni were in the 70-100 nm range. Additional SEM results were used to help establish sputtering process parameters. Since the specimens were fragile, they were sandwiched between two layers of backing material, e.g., silicon wafers, and glued by thin layers of epoxy. Fine polishing with 0.3 µm diamond suspension produced artifact-free smooth surfaces essential for microscopy. All the samples were observed for layer thickness, uniformity, degree of intermixing, defects, chemical composition, and phase identification after reaction.

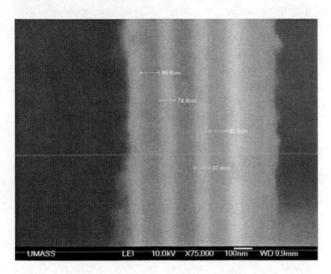

Figure 1 FE-SEM image of alternating thin films of Al and Ni deposited by sputtering at the University of Cyprus.

2.2 Powder-based Composites

Ultrasonically consolidated compacts were produced from Al and Ni powders. Al (99.5% pure) and Ni (99.8% pure) powders, 7-15 and 45-150 µm in size, respectively, were mixed in 1:1 molar ratio for 45 minutes in a cylindrical powder mixer revolving at 500 rpm. The powder mixture was subjected to ultrasonic consolidation at 300 °C in air using a punch and die arrangement as schematically shown in **Figure 2A**. The die is made of a 0.78 mm thick nickel plate with a die hole of 3.4 mm in diameter. Ultrasonic vibration was applied parallel to the die surface at a frequency of 20 kHz with vibration amplitude of 10 µm under uniaxial pressures of 70 to 250 MPa to study the effect of the stress on the microstructure, ductility and ignition behavior of the samples produced. The consolidation time was set at 1 s, which was followed by an additional 0.02 s of an after burst time. Compacts 3.3-3.8 mm in diameter were produced.

Figure 2B shows an SEM image of a compact ultrasonically consolidated under a uniaxial loading of 210 MPa. The compact shows a high degree of densification and a composite microstructure where Ni particles (lighter grey) are well dispersed in the Al matrix (darker grey). Thus, the Al powder particles took most of the deformation needed to produce the consolidated composite while the Ni particles

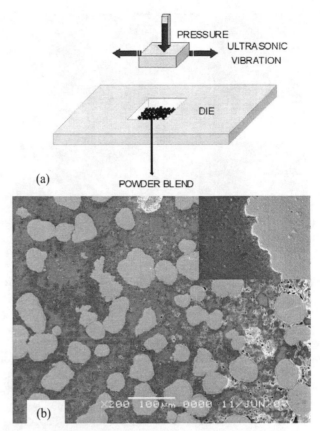

Figure 2 (A) Schematic of ultrasonic powder consolidation (UPC). **(B)** FE-SEM image of an Al/Ni compact consolidated at 300 °C and 210 MPa for 1 s. Inset reveals intimate contact between the Al matrix and the Ni particles.

being harder remained virtually undeformed. In most of the specimen volume, densification was complete and the contact between the Al matrix and the Ni particles was well-established as seen in the inset. Some porosity, however, was also noticed at the Al/Ni interface in the edge regions of the specimen where insufficient supply of aluminum around the Ni particles resulted. EDX results of the consolidated specimen showed no evidence of oxide formation in the matrix as well as in the Al/Ni interface.

Preliminary experiments were also performed with nano-flakes, 200 - 300 nm thick and 50 - 100 µm across, produced by a proprietary hammer milling technique. Fully dense consolidates with no delamination were obtained at 300 °C under 160 and 200 MPa.

3. THERMAL CHARACTERIZATION OF NANOHEATERS

A commercial multilayered nanoheater sample (RNT, Inc., Hunt Valley, MD) was ignited and the electromagnetic radiation (infrared band) was captured using an infrared pyrometry camera (SC4000, FLIR Systems, Billerica, MA). Thermal images reveal that the self-propagating reaction reached a peak temperature of 1181 °C in 46 ms. **Figure 3** shows 4 instances at time intervals as the reaction progressed.

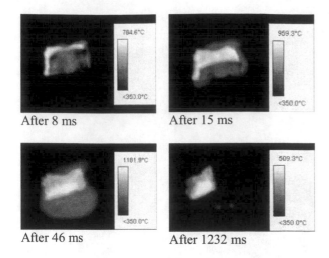

After 8 ms   After 15 ms

After 46 ms   After 1232 ms

Figure 3 Infrared (IR) images of the exothermic reaction of an ignited multilayered Al-Ni film at different times.

Ignition tests were also performed on the ultrasonically consolidated specimens by heating them with a torch in open air (**Figure 4A**). Compacts with sufficient densification (consolidated at 160 MPa or above) ignited after a very brief exposure to the propane torch flame. **Figure 4B** shows the microstructures of the specimen after the ignition. The prior Ni particles are hardly seen in the ignited specimen, indicating that the exothermic reaction has propagated throughout the compact. EDX analysis indicated that the resultant phase produced was NiAl and that slight oxidation occurred during ignition due to the ambient conditions.

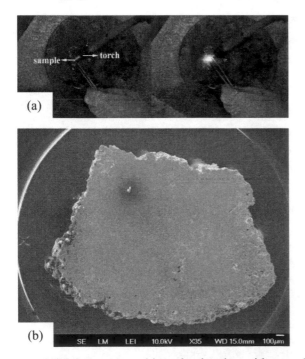

Figure 4 (A) A compact subjected to heating with a torch ignites in 150 ms. **(B)** SEM image of the sample after the self-propagation of the exothermic reaction upon ignition.

Reaction did not self-propagate in powder-based compacts with insufficient densification. Consolidates made from the nanoflakes were ignited even more readily.

4. THERMAL MEASUREMENT AND MODELING

One key advantage of nanoheaters is the ability to provide controlled, intense, and rapid heat to a very localized region, thus reducing energy, processing time, and collateral damage. To design the nanoheaters properly, it is important to predict the heat output and the affected adjacent area and time-temperature profile developed in that area after the nanoheater ignition. A series of experimental studies were initiated that can be used to validate the models for (1) 1-D infinite plane heat conduction and (2) 2-D sequential ignition.

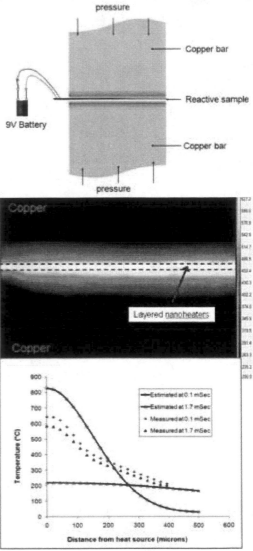

Figure 5 (A) Schematic of 1-D heat conduction setup **(B)** Temperature profile from FLIR IR measurements. **(C)** Comparison of model (lines w/dots) and experimental (dots) measurements of 1-D temperature distribution in infinite length copper rod after instantaneous heat input of 53 kJ/m^2.

4.1 1-D Infinite Plane Heat Conduction

The 1-D transient heat conduction generated by a multilayered nanoheater film (80 μm) was estimated and measured. The nanoheater was in intimate contact with the entire bottom surface of a copper rod (k=383-391 W/m·K), which is used as a bulk material. Air was used for insulation purposes. An estimate was made by applying transient heat conduction equations for 1-D conduction in bulk media.[4] It was assumed that nanoheater heat release was instantaneous and no temperature gradient existed along the thickness of the nanoheaters. If the diameter and length of copper are much larger than the thickness of the nanoheater film (with insulated sides), the temperature distribution can be predicted. **Figure 5A** shows a schematic of this 1-D setup. **Figure 5B** shows the thermal imaging measurements of temperature distribution with the FLIR camera. **Figure 5C** shows estimated and measured temperature profile along the copper rod. For estimated values (lines), the increase in temperature is initially localized to within 500 μm of the heat source at 0.1 ms, and the material beyond that distance is still at room temperature. Within 2 ms, the temperature has dropped rapidly, as it is a decaying function of time, and the heat flow has traveled beyond the 1000 μm distance. However, there is large difference between estimated and experimental values, probably due to the assumption of instantaneous heat generation. Further study is currently under investigation.

4.1 2-D Sequential Nanoheater Ignition

In addition to localized heating, the nanoheaters can be designed to ignite in a sequential manner. Initial experiments have been conducted to quantify the control of the sequential ignition. **Figure 6A** shows a schematic of three nanoheater "islands" connected by a narrow nanoheater strip (all materials are multilayered films). The first island can be ignited and if sufficient heat is transmitted by the inter-connecting strips to the second one, that island will ignite too. In the experimental setup, rectangular films were fixed to a silicon wafer and then narrower strips of films were attached between each pair of films. The nanoheaters were ignited in open air, and the reaction propagated rapidly from island to island until all of the material had been reacted (**Figure 6B**). Further studies are planned to quantify this process.

5. CONCLUSIONS

Fabrication of ignitable Al/Ni heterostructures has been successfully accomplished by sputtering of multilayer films and powder-based deformation bonding (UPC). Al and Ni powders as well as nanoflakes were consolidated under different processing conditions. An oxide and pore free Al matrix and intimate interface between the Al matrix and Ni particles were achieved. Torch heating ignited the powder consolidates, causing a self-propagating reaction that produced NiAl with only slight oxidation due to external

effects. The UPC route will be further explored to establish consolidation conditions for nanoflakes.

The ability to capture and quantify ignition and temperature distribution with microscale temporal and spatial resolution was demonstrated. Improved quantification of heat output and ignition conditions will be conducted using differential scanning calorimetry (DSC) and infrared cameras for the model cases and the sequential ignition studies. Models of both the exothermal reaction kinetics and the macroscale heat conduction are in progress and results will be compared with the experimental results.

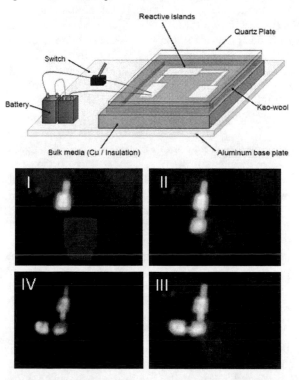

Figure 6 (A) Schematic of 2-D sequential ignition set-up. **(B)** Sequential ignition recorded through a quartz plate by the FLIR camera (red color indicates ignited nanoheaters).

ACKNOWLEDGEMENTS

The authors acknowledge the financial support for the research by the National Science Foundation (DMI 0531127 and 0738253), the European Commission through Marie Curie Chair "UltraNanoMan" (MEXC-CT-2004-006680) and Marie Curie Excellence Team Project "NanoHeaters" (MEXT-CT-2005-0023899). They would also like to thank FLIR Systems (Billerica, MA) for their assistance.

REFERENCES

[1] I. Gunduz, Ph.D. Thesis, Northeastern University, Boston, MA, 2006.
[2] P. Zhu, J. C. M. Li, and C. T. Liu, Mater. Sci. Eng. A239-240, 532, 1997.
[3] S. Mornet et al., J. Mater. Chem. 14, 2161, 2004.
[4] D. Poulikakos, "Conduction Heat Transfer", Prentice-Hall, Englewood Cliffs, NJ, 1994.

Multiple Gate Approach - Solution of Scaling & Nano-MOSFETs

Lamba, V.K[*], Engles, D. [**], Malik, S.S. [***]

[*]Deptt of ECE, HCTM Kaithal, (India), lamba_vj@hotmail.com;
[**] Deptt of ECE, GNDU Amritsar, (India);
[***] Deptt of Physics, GNDU Amritsar, (India).

ABSTRACT

In the nm scale, the aggressive scaling of MOSFETs is expected to culminate in dual-gate (DG) architectures on SOI substrates. DG MOSFETs are widely accepted to be the ultimate design that silicon can deliver in terms of on and off currents. So far, the design efforts on these novel structures have concentrated on ideal geometries and doping profiles. The paper describes the evolution of the SOI MOSFET from single-gate structures to multiple gate (double-gate, tri-gate, π-gate, and gate-all-around) structures. Increasing the "effective number of gates" improves the electrostatic control of the channel by the gate and, hence, reduces short-channel effects. Due to the very small dimensions of the devices, one-and two-dimensional confinement effects are observed, which results in the need of developing quantum modeling tools for accurate prediction of the electrical characteristics of the devices. It also includes the effect of silicon thickness, when we are using multiple gates, the effects of different gate potentials on multiple gate MOSFETs, Doping concentration of Source/Drain/Channel region, thickness of SiO_2 layer, and more.

Keywords: MuGFETs, Mobility, Doping.

1 INTRODUCTION

Due to the continuous and aggressive scaling, silicon MOS devices have rushed into the sub-100 nm regime. In the nano-scale regime, main challenges in device design are the suppression of short channel effects and the management of quantum effects. The short channel effects and quantum effects would severely degrade the device performance in the nano-scale MOSFETs. For scaling as projected in the ITRS to take place, the industry must be capable of cost-effectively fabricating MOSFETs with the required characteristics, which includes meeting the $I_{sd,leak}$ and transistor performance, control of SCE, acceptable control of the statistical variability of the MOSFETs parameters, acceptable reliability, etc.

The paper is organized as follows. Section II consist of scaling challenges and its potential solutions, Section III gives details of model used, Section IV, & V includes device simulation and conclusion respectively.

II MOSFET SCALING CHALLENGES & POTENTIAL SOLUTIONS

It turns out that there are numerous difficult challenges that arise as the technology is scaled with succeeding years and that significant technological innovations will need to be implemented in relatively rapid succession to deal with these challenges are:

2.1 Mobility enhancement:

The first difficult challenge with scaling is increasing the I_{dsat} as much as required while holding $I_{sd,leak}$ to acceptable values. The preferred solution is to enhance the mobility beyond that attainable with standard silicon channels by using strained Si, possibly strained SiGe channels. Strained Si channels with enhanced mobility were implemented [1]. Mobility enhancement up to 1.8 times be obtainable for strained Si channels has been reported for nMOSFETs, and even higher for pMOSFETs [2].

2.2 Gate leakage reduction:

The gate leakage is due to direct tunneling of electrons through the gate dielectric, which increases sharply as the gate dielectric equivalent oxide thickness (EOT) is reduced. The ITRS specified maximum allowable gate leakage current density ($J_{g,limit}$) is closely related to the $I_{sd,leak}$. The potential solution being actively pursued by the industry is to use high-k gate material [3]. As a result, for the same EOT, the physical thickness is larger for the high-k dielectric than for silicon dioxide, so the direct tunneling and thus the gate leakage current should be lower for high-k dielectrics.

2.3 Performance using metal gate:

As Polysilicon gate depletion increases the equivalent EOT, and hence reduces the value of I_{dsat} that can be attained. As the EOT scales with succeeding years, the influence of polysilicon depletion on EOT becomes proportionately greater, and according to the ITRS forecasts that by 2008, polysilicon depletion needs to be reduced below that attainable with polysilicon electrodes if the performance requirements on I_{dsat} and τ are to be met. Metal gate electrodes, which have virtually no depletion, are being developed as the most likely potential solution [4]. To be able to set V_t to the appropriate value for pMOSFETs, the work function of the metal gate electrode must be near the silicon valence band. For nMOSFETs, the work function must be near the silicon conduction band. It is likely that different metals with appropriate work function for the pMOSFET and nMOSFET, respectively, will be used [5].

2.4 MuGFETs structures:

Even with use of the above technology innovations, effective scaling of classical bulk MOSFETs is expected to become increasingly challenging for 2008 and beyond. Achieving adequate control of SCE for such devices will be difficult. Exceedingly high values of channel doping will be needed to control these effects, and high doping will lead both to reduced mobility and increased band-to-band tunneling leakage current. Furthermore, the total number of dopant atoms in the channel for such small MOSFETs is relatively small, which leads to large and irreducible statistical variation in the number and placement of the atoms, and hence to unacceptable statistical variation in V_t, and degrades the short channel characteristics and sub-threshold slope through an increase of penetration of the drain electric field lines in the channel region [6-7]. A potential solution is the use of ultra-thin-body fully depleted silicon-on-insulator [8] technology. These have lightly doped channels, and V_t is set by the work function of the gate electrode not by dopant atoms in the channel, so the variation in the number of dopant atoms does not affect V_t [9]. Scalability and control of SCE are significantly enhanced, although reliably controlling the thickness will presumably be a major challenge. To prevent the electric field lines originating at the drain from terminating under the channel region, special multiple-gate structure devices have been reported. MuGFETs have attracted much attention owing to their ability to enhance the electrostatic control of the gate over the channel Thus MuGFETs are very attractive, because of threshold characteristics, higher drive current, better short channel effect, effective control, less drain induced barrier lowering etc.

III MODEL

Fig.1 shows the existing gate configuration for thin-film SOI MOSFETs: 1) single gate; 2) double gate; 3) triple gate; 4) pi gate; 5) omega gate; and 6) GAA configuration. The most familiar is the double gate where the gate is formed around the channel, or the two gates are electrically connected. Other form is to have the two gates be electrically independent and be controlled separately; here the second gate is used to shift threshold voltage of the first gate. The sub-group of triple gates is π-gate, and Ω gate. In case of π-gate, structure gate electrodes extends to some depth in the buried oxide on both sides of device, and gate is in shape of Greek letter π; and in case of Ω gate, it closely resembles to GAA as its gate almost wraps around the body, having top gate like the conventional UTB-SOI, side walls like FinFETs, and special gate extension under the silicon body. Here, a full quantum-mechanical treatment of electron transfer, in particular of the electron confinement in the direction across the ultra thin channel, and the source -to-drain tunneling in the direction along the channel is must. Indeed the lateral effect provides one of the major limitations for transistor scaling [6-7]. Another major effect is a gradual loss of electrostatic control of the channel potential as the gate length L is decreased. So we address these problems by the self consistent solution of 1D Schrödinger equation and 2D Poisson equation. The simulated structures have a uniform doping concentration in the channel and source/drain regions. Abrupt source and drain junction are used. The simulations are carried out using single charge carrier.

IV DEVICE SIMULATION

Electrical characteristics of devices were simulated (using Multi-gate Nanowire FET on http://nanohub.org [10-11]) for the silicon island having width and thickness of 25 nm,

while the gate oxide thickness is 1.5nm. Tungsten is used as gate (work function = 4.63 eV), doping concentration in channel is uniform and equal to 1×10^{15} cm^3. Simulation is performed for gate lengths L, of 10, 20, 30, 40, 50 & 60 nm. We have used a width of 25 nm, to prevent the electrical field lines from the drain from terminating on the back of the channel region, because of the bottom part of the π, Ω and GAA

4.1 Natural length model

It gives a measure of the short channel effect inherent to a particular device structure. The concept of "natural length", λ represents the extension of the electric field lines from the drain in the channel region [12]. A device is said to be free of short channel effects if the gate length is at least 5-8 nm larger than λ. Suzuki et al [13] has given expression for λ and given by

$$\lambda = \sqrt{\frac{\varepsilon_{Si}}{2\varepsilon_{ox}}\left(1 + \frac{\varepsilon_{ox}t_{Si}}{4\varepsilon_{Si}t_{ox}}\right)t_{Si}t_{ox}} \qquad (1)$$

Where t_{si}, t_{ox} are the silicon and oxide thickness and ε_{Si}, ε_{ox} are the permittives of Si, and SiO$_2$. In 2007, Lee et.al. Calculated the expression for generalizing the λ concept to all MuGFETs by writing

$$\lambda_n = \sqrt{\frac{\varepsilon_{Si}}{n\varepsilon_{ox}}\left(1 + \frac{\varepsilon_{ox}t_{Si}}{4\varepsilon_{Si}t_{ox}}\right)t_{Si}t_{ox}} \qquad (2)$$

Where n is the effective no of gates, whose value is calculated from the dependence of threshold voltage on silicon film thickness, and its value decrease as the no of gates increases from single gate MOSFETs to GAA MOSFETs because of the increasing influence of the gate over the potential in the channel region.

4.2 Back Channel conduction & DIBL model.

As the vertical potential profile underneath a thin-film SOI device is controlled by potential lines originating from the source and drain rather than by the front gate, the potential inside the buried oxide increases linearly with drain bias. This lateral potential coupling from the source and drain causes back-channel conduction and DIBL. Fig. 2(a) & (b) shows the drain-induced barrier lowering effect and the threshold voltage roll-off in fully depleted SOI MOSFETs with different gate structures and different effective gate lengths. The DIBL is defined as the difference in threshold voltage when the drain voltage is increased from 0.1 to 1 V. The threshold voltage roll-off, ΔV_{th}, is defined as the threshold voltage measured at $V_{DS} = 0.1V$ at any gate length minus the threshold voltage at L = 50nm. DIBL is most effectively suppressed by the quadruple-gate structure, but the Ω and π -gate device comes a close second and third. The reason for less short channel effects in Ω & π -gate device is that the lower part of the gate sidewalls effectively acts as a back gate through lateral field effect in the buried oxide. Similarly, it can be observed that the threshold voltage roll-off is minimized by the use of the quadruple-gate structure, but the Ω & π -gate device shows an excellent behavior as well.

4.3 Subthreshold swing Model

From the linear relationship between threshold voltage and film thickness, one can derive the general threshold voltage law for multi-gate devices:

$$V_{thN} = V_{Fb} + 2\phi_f + \frac{qN_A}{C_{ox}} \cdot \frac{t_{Si}}{N} \qquad (3)$$

We have used N = 2.0 for a double-gate, N = 2.3 for electrically independent double gate, N = 3.0 for a triple-gate, N = 3.14 for a π-gate, N = 3.4 for an Ω-gate, and n = 4.0 for a GAA. The equivalent gate number N depends on the gate extension depth for π-gate devices and the lateral extension depth of the side gate in Ω-gate devices. Thus we conclude from Fig. 3, that subthreshold swing degradation is smallest in the GAA, Ω & π gate.

4.4 Model based on doping concentration

To reduce short channel effects, the substrate doping concentration should be increased; Fig. 4 shows the threshold voltage roll-off, ΔV_{th}, as a function of effective gate length for different gate structures and different doping concentrations. As the doping concentration increases ΔV_{th}, is reduced due to the reduction of sharing charge between the gate and source/drain junction. It is also worth noticing that the sensitivity of the ΔV_{th} to the doping concentration is minimized in the GAA structure compared to the others.

4.5 Model based on Channel Width & Silicon Film Thickness

In the present simulation, the silicon island is assumed to have a rectangular cross section and the gate oxide has a uniform thickness in all devices (It is assumed that there is no gate oxide thinning at the edges of the silicon island.) In thin-film single, double and quadruple-gate devices operating in the subthreshold region most carriers flow through the middle of the film due to the volume inversion. In a triple-gate device, the electric field, from source and drain encroaches on the channel region and tends to induce more inversion charge at the bottom of silicon body as device width is increased. When the back channel subthreshold current becomes important, the subthreshold swing degrades significantly. A triple-gate device behaves like a single-gate device when the channel width becomes large. Fig. 5 shows that the threshold voltage of double-gate and quadruple-gate devices increases when the channel width is increased, while the threshold voltage of the triple, Ω & π-gate device decreases. Figs. 6 show the DIBL and subthreshold slope as a function of the channel width for different gate structures, L and T_{Si} are both equal to 25 nm and N_A is equal to 1 x 10^{15}cm^{-3}. Both the DIBL and subthreshold slope increase slightly as the channel width increases in double-gate and quadruple-gate devices, but they increase abruptly in the triple, Ω & π-gate device. The DIBL increases with channel width in the triple, π & Ω-gate device but still its value is lower than that of the double-gate device for channel widths below 45 nm. The subthreshold slope of the Triple, Pi and Omega -gate device increases with channel width but is lower than the double-gate device. The larger DIBL and subthreshold swing in triple, Ω & π -gate devices is due to the encroachment of the electric field lines from drain on the back of the channel region. This effect increases with device width. The degradation in the quadruple-gate structure is the smallest and, followed by Ω & π -gate once again. Even though the use of thin silicon film increases the source and drain resistance [12-14], a small silicon film thickness is required to improve the SCE immunity and subthreshold slope. If the silicon film is ultra thin, energy quantization effects start to appear, which influence the threshold voltage and the I-V characteristics of SOI MOS devices [15-19] Fig. 7 shows the threshold voltage as a function of film thickness with different gate structures. The threshold voltage is almost constant for the quadruple-gate MOSFET but it decreases slightly for the Ω, π & triple-gate structures. Surprisingly the threshold voltage of the double-gate device decreases when the film thickness is increased.

IV Conclusion

The paper establish the guideline for minimum gate length to avoid short channel effects and excess subthreshold swing degradation for different gate structures and different gate lengths W = T = 25 nm, N_A = 1 x 10^{15} cm^{-3}. In the case of a double-gate device the minimum gate length is 40 nm. In the case of Triple gate is 35nm. Pi-gate 30 nm, while for omega and GAA it is 25nm.

V References

1. T. Ghani et al, Int. Elect. Device Meeting Technical Digest, Washington, DC, pp. 978-980, Dec. 2003.
2. K. Uchida, et al., Int. Elect. Device Meeting Technical Digest, Washington, DC, pp. 135-138, Dec. 2005.
3. H. R. Huff, et al., Proce. Of Int. Workshop on Gate Insulators, Tokyo, Japan, Nov. 2001.
4. Q. Lu, et al., VLSI Symp. Dig. Of Tech. Papers, pp. 45-46, June 2001.
5. Z. B. Zhang, et al., VLSI Symposium Dig. of Tech. Papers, Kyoto, Japan, pp. 50-51, June 2005.
6. T. Hiramoto, Jpn. J. Appl. Phy., vol. 42, p.1975, 2003.
7. Park JT, IEEE Trans Elect. Dev 2002;49 (12): p. 2222
8. J.T Park, et al., IEEE Trans. Elec. Dev 2002 49, p.2222
9. P. M. Zeitzoff, et al., Nano & Giga Electronics (ed. by J. Greer), Elsevier Press, Amsterdam, 2003.
10. M. Shin, J. Appl. Phys. **101**, 024510 (2007).
11. Simulations are performed at http://nanohub.org
12. H. Majima et. al. IEDM Tech. Dig., 2001, pp. 733–36.
13. B. Majkusiak et. al IEEE Trans. Electron Devices, vol. 45, pp. 1127–1134, May 1998.
14. Y. Omura et. al IEEEElectron Device Lett., vol. 18, pp. 190–193, May 1997
15. Yan RH, et. al. IEEE Trans Electron Dev 1992;39(7): 1704–1710.
16. Suzuki K, et. al. IEEE Trans Electron Dev 1993; 40(12): 2326–2329.
17. J. P. Colinge, Silicon-on-Insulator Technology: Materials to VLSI, 2nd ed. Boston, MA: Kluwer, 1997.
18. B. Majkusiak Solid-State Electron., vol. 45, pp. 607–611, 2001.
19. G. Baccarani et. al. IEEE Trans. Electron Devices, vol. 46, pp. 1127–1134, Aug. 1999

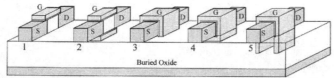

Fig. 1: Different gate configurations for SOI devices: 1) Single gate; 2) Double gate; 3) Triple gate; 4) GAA gate; 5) Pi-gate MOSFET.

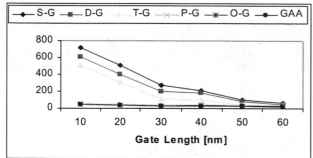

(a)

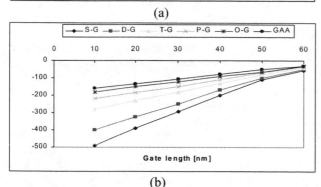

(b)

Fig. 2; DIBL and threshold channel roll-off in fully depleted SOI MOSFETs with different gate structures and different gate lengths W = T_{Si} = 25 nm, N_A = 1 x 10^{15} cm^3.

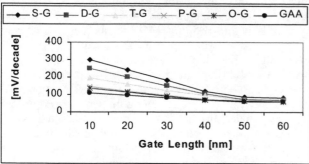

Fig. 3: Subthreshold swing in fully depleted SOI MOSFETs with different gate structures and different gate lengths W = T_{Si} = 25 nm, N_A = 1 x 10^{15} cm^3.

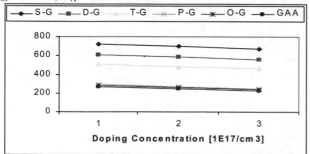

Fig. 4: DIBL in fully depleted SOI MOSFETs with different gate structures and different doping concentrations as a function of gate lengths W = T_{Si} = 25 nm, N_A = 1 x 10^{15} cm^3, 2 x 10^{15} cm^3, and 3 x 10^{15} cm^3

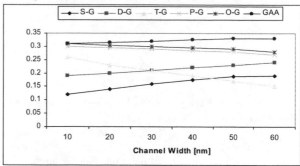

Fig. 5. Threshold voltage in fully depleted SOI MOSFETs with different gate structures and different channel widths W = T_{Si} = 25 nm, N_A = 1 x 10^{15} cm^3

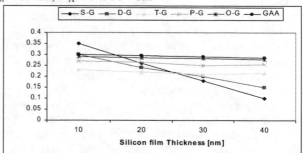

Fig. 6. DIBL in fully depleted SOI MOSFETs with different gate structures and different channel widths W = T_{Si} = 25 nm, N_A = 1 x 10^{15} cm^3

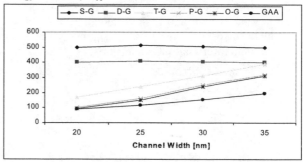

Fig. 7. Threshold voltage in fully depleted SOI MOSFETs with different gate structures and different silicon film thickness L = W = 25 nm, N_A = 1 x 10^{15} cm^3

Remark:

S-G stands for Single gate;
D-G stands for Double gate;
T-G stands for Triple gate;
P-G stands for Pi Gate
O-G stands for Omega Gate
GAA stands for Gate all around MOSFETs.

New Type of High Sensitive Detection of Particles Based on DeFET

Mohamed F. Ibrahim, Fahmi Elsayed, Yehya H. Ghalab, Wael Badawy

Department of Electrical and Computer Engineering
University of Calgary, Calgary, Canada. T2N1N4
mibrahim@ucalgary.ca, f.elsayed@ucalgary.ca, yghallab@ucalgary.ca, badawy@ucalgary.ca

ABSTRACT

This research work presents a new implementation of the CMOS electric-field sensor, titled as Differential Electric Field Sensitive Field Effect Transistor (DeFET). The DeFET is successfully used in detection of small partials and it is a suitable candidate to be used in biochemical and environmental applications. The developed implementation of the DeFET provides an improved sensitivity, thus it can be used to detect tiny particles in the range of nano meter scale. The direct application of the proposed sensor is the air pollution and environmental detection. The chip is implemented using 180 nm CMOS technology; it contains both the actuation and sensing parts. The sensors are distributed in two different arrays (uniform and nonuniform arrays).

Keywords: DeFET, electrode, micro, nano, sensing, environmental.

1. INTRODUCTION

Movement of particles caused by polarization effects in a nonuniform electric field is called Dielectrophoresis (DEP) phenomenon. In 1951, Pohl was the leader of describing this Dielectrophoresis (DEP) phenomenon [1, 2, 7]. Either direct (DC) or alternating (AC) electric fields can cause DEP [3]. A finite separation between equal amounts of positive and negative charges is found in any dipole (induced or permanent). The electric field will align with the dipole, but one side of the dipole will be in a region with lower field intensity than the other, if the electric field is nonuniform. This will produce an uneven charge density in the particle; i.e., uneven charge alignment in the particle; and it will be forced to move toward the regions of greater field strength. If the effective polarizability of the medium is lower than that of the particle then the particle will exhibit positive dielectrophoretic behavior since it will be attracted to regions of greater field intensity [4, 5]. Negative dielectrophoretic behavior or repulsion from regions of greater electric field intensity, is observed in medium with greater polarizability than that of the particles [4, 5].

In the early studies utilizing DEP to manipulate cells, electrodes of different shapes were employed in order to produce nonuniform electric fields. Pin-plate and pin-pin electrodes were used to separate live and dead yeast cells and achieved collection of yeast cells on the electrodes [4, 6]. Recently, arrays of microelectrodes and AC electric fields are used to carry out the DEP applications, due to the availability of micro-fabrication techniques. The DEP forces and heating effects are improved due to the scaling down of microelectrodes' dimensions which generate higher electric field intensity due to the decreasing of space between electrodes [5, 6]. A lot of cell manipulate devices (based on programmable DEP-force cages by using 3D structures of electrodes) were used for cells trapping, levitating and dragging for many applications [6].

Recently, one of the hottest areas of research is Lab-on-a-chip based on DEP phenomenon. It has many applications in the pollution, medical, pharmaceutical, and biological fields. To increase the effectiveness of Lab-on-a-chip it needs to integrate functions such as actuation, sensing, and processing. Lab-on-a-chip is faster, better and cheaper biological analysis. There are two different Lab-on-a-chip approaches are proposed [9, 10, 12]. The first one [9] is the 1st lab-on-a-chip approach for detection of microorganism and electronic manipulation. It combines impedance measurement with Dielectrophoresis to trap and move particles, while monitoring their location and quantity into the device. The prototype has been realized using standard printed circuit board (PCB) technology. The sensing part in this approach can be performed for any electrode by switching it from the electrical stimulus to a transimpedance amplifier, while all the other electrodes are connected to ground. In 2003, the second Lab-on-a-chip was proposed [10]. It is a microsystem for cell manipulation and detection based on standard 0.35μm CMOS technology. This lab-on-a-chip consists of two main units, the actuation unit, and the sensing unit. The chip surface implements a 2D array of micro sites, each consisting of a superficial electrodes and embedded photodiode sensor and logic. The actuation part is based on the DEP technique. While the sensing part depends on that particles in the sample can be detected by the changes in optical radiation impinging on the photodiode associated with each micro-site. During the sensing, the actuation voltages are halted, to avoid coupling with the pixel readout. However, due to inertia, the cells keep their position in the liquid.

There are three main disadvantages of the recent Lab-on-a-chips. First: there is no real time detection of the cell response under the effect of the non-uniform electric field, as we halted the actuation part and activate the sensing part. Second: the sensing part in these two lab-on-a-chips depends on the inertia of the levitated cells. Thus, only cells with higher inertia can be sensed and detected by using these two lab-on-a-chips. Third: Although the detection of the position of the levitated cells can be done, based on these two systems, the actual intensity of the non-uniform electric field that produces the DEP force can't be detected.

In this paper, a new configuration of an electric-field sensor with different geometry is presented, thus, a modification of the differential electric-field sensitive field-effect transistor (DeFET) is presented and discussed. Also, this paper presents the DeFET's theory of operation and presents test results which validate the geometry modifications effect of the proposed DeFET. We are planning to design and prototype a new device that senses and characterizes the response for difference nano and micro size particles. This prototype will have applications in the area of environmental monitoring where fine dust matters can be detected and characterized for pollution and occupancy air quality. It can be used in other applications which deal with small size particles, such as biocells and DNA. The fabricated chip is based on a standard 0.18-μm Taiwan Semiconductor Manufacturing Company (TSMC) CMOS technology, and it includes the sensors which are designed with different dimensions in order to study the effect of changing the dimension on the output voltage which consequently affects the sensitivity of the sensor. In addition, a 16x1 multiplexer is implemented on chip to multiplex 16 sensors; the multiplexer reduces the chip area and provides accurate scanning and monitoring of different sensors.

2. THE PROPOSED CMOS LAB-ON-A-CHIP

The proposed lab-on-a-chip is implemented in a standard CMOS 0.18μm TSMC technology. Fig. 1 shows the die photo, the total die area is 1.5mm x 1mm. The proposed lab-on-a-chip consists of three main parts. (1) The actuation part, which is a quadrupole electrode configuration to produce the required non-uniform electric field profile, and consequently a DEP force to levitate the cell that we want to characterize, (2) the sensing part, which is two sets of 8-array of the Differential Electric Field Sensitive Field Effect Transistor (DeFET), one set is uniform sensors, i.e. with the same dimension, and the other set is non-uniform sensors with different dimensions. (3) Multiplexing part: a 16x1 multiplexer is implemented on chip to multiplex 16 sensors, hence, to reduce the chip area and provides accurate scanning and monitoring of different sensors.

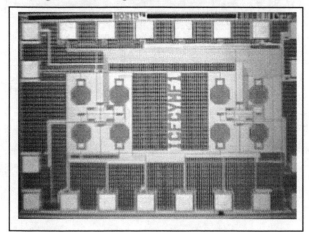

Figure 1.   Photo of the layout of the fabricated proposed DeFET circuit.

The Actuation Process: A quadrupole configuration of electrodes are used to perform the actuation process, see Fig.2, using this configuration we can control the profile of the non-uniform electric field by connecting the whole four electrodes or some of them. Also, the quadrupole levitator comprises an azimuthally symmetric electrode arrangement capable of sustaining passive stable particle levitation. For these reasons, we selected the quadrupole electrode configuration as an actuation part in our design [5].

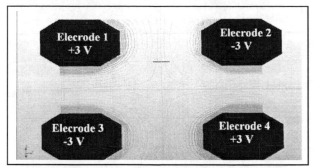

Figure 2.   The quadrupole configuration

The Sensing Process: The sensing part is composed of a two sets, each set consists of 8 DeFET sensor array. This array located around the central point, where the cell will be passed (as shown in Figure 1). In sensing process, the manipulating electric field is a non-uniform electric field (i.e. the electric field is a function of the distance). Thus, we can detect the electric field by using the Electric Field Sensitive Field Effect Transistor [eFET], as a new electric field sensor [11, 12]. Fig. 3 shows the physical structure of the eFET. It consists of two adjacent drains, one source, and two adjacent floating gates with distance "d" between them. For the eFET, it is equivalent to two identical enhancement MOSFET devices. The current flow of the two drain currents occur, under the influence of the non-uniform electric field over the gates. As the drain current dependence on the gate voltage, the eFET can sense the difference between the two gates voltage, which reflects the intensity of the applied non-uniform electric field. To increase the measurement range of the eFET, we can use the CMOS concept to implement the Differential Electric Field Sensitive Field Effect Transistor (DeFET) sensor, and this sensor is the basic sensing block in the sensing part of the proposed lab-on-a-chip [11, 12].

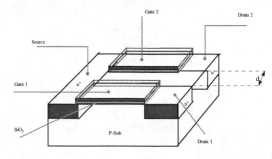

Figure 3.   Physical structure of an eFET [12].

3. DeFET'S THEORY OF OPERATION

The DeFET consists of two complementary eFETs, one of them is a P eFET type and the second is an N eFET type. The equivalent circuit of the DeFET is shown in Fig. 4. From Fig. 4, the two gates of P eFET and N eFET are connected with each other, and there is a cross coupling between the two drains of the P eFET and the N eFET. I_p and I_n are functions of the two applied gate voltages V_{in1} and V_{in2}, respectively. The output current I_{Out} is equal to the difference between these two drain currents I_p-I_n (i.e. I_{Out} = I_p-I_n). The DeFET is designed to achieve that I_{Out} is directly related to the difference between the two applied gate voltages $(V_{in1}$-$V_{in2})$, and V_{in1}-V_{in2} is equal to the applied electric field above the two gates (E) multiplied by the distance (d) between them (i.e. $(V_{in1}$-$V_{in2})/d = E$), where d is the distance between the two split gates. So, I_{Out} is related directly to the intensity of the applied non-uniform electric field if the distance "d" is constant. Thus by measuring I_{Out} we can detect the intensity of the non-uniform electric field [12].

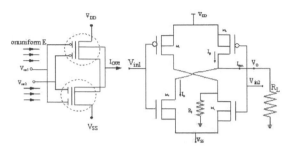

Figure 4. An equivalent circuit of a DeFET [12]

Using simple analysis, the expression relates I_{Out} and E can be driven [11]. From Fig. 4, the output current (I_{out}) is:

$$I_{out} = I_p - I_n \qquad (1)$$

The sensitivity S_1 is given by

$$S_1 = \frac{dI_{out}}{dE} \qquad (2)$$

As a linear equation, we can express the sensitivity (I_{out}) in terms of the output current and the electric filed as follows:

$$S_1 = \frac{d(I_p - I_n)}{dE}, \quad S_1 = \frac{d(I_p - I_n)}{d(\Delta V)} \cdot \frac{d(\Delta V)}{dE} \qquad (3)$$

$$(I_p - I_n) = -g_m \Delta V, \quad \frac{d(I_p - I_n)}{d(\Delta V)} = -g_m, \qquad$$

$$\frac{d(\Delta V)}{dE} = -d \qquad (4)$$

From (3) and (4), the sensitivity can be given as:

$$S_1 = g_m d \qquad (5)$$

$$V_{out} = I_{out} R_L = S_1 R_L E + \text{Constant}, \quad S = S_1 R_L \qquad (6)$$

From (5) into (6) $\qquad S = g_m \cdot d \cdot R_L \qquad (7)$

$$V_{out} = g_m \cdot d \cdot R_L \cdot E \qquad (8)$$

Where; gm: transconductances of the Transistors, d : the distance between the two split gates, and R_L: load resistance

Equation (7) shows a liner relationship between the DeFET's sensitivity and the distance between the two split gates. Thus, if we have an array of DeFET sensors with different distance (d), then we can obtain higher sensitivity.

4. SIMULATION RESULTS

The proposed CMOS lab-on-a-chip has been tested for both DC and AC response. Electrodes 1 and 3 are connected together with the same voltage and Electrodes 2 and 4 have out of phase voltage with the same amplitude.

For the DC response, the measured results are shown in Table 1 and Figs. 4 and 5. These results clarify the relationship between the output voltages of each sensor with respect to the variation of its geometry (the distance between the two split gates (d)) at different electrodes' voltages (see equation 8). The circuit is tested at (d) equals 0.5, 1, 1.25, 1.5 and 2 um, respectively.

The ratio between the sensors output voltages and the output of (d=2 um) sensor is shown to clarify the effect of changing the geometry of the sensors on the output voltage. From Table 1, Figs. 4 and Fig.5, we can observe the following:

- The sensitivity changes with (d) (The slope of the curves), see equation 7.

- The change of the output voltage is observed with all geometries; therefore, if we are aiming low power consumption, we can use lower electrode voltages with acceptable sensitivity. However, it depends on the properties of the particle that we are targeting.

Table 1: The effect of changing the distance between gates (dB)

Electrodes 1&3 Voltage	Output Voltage w.r.t. Vout at d= 2um (dB)				
	Sensor with d=0.5um	Sensor with d=1um	Sensor with d=1.25um	Sensor with d=1.5um	Sensor with d=2um
5 V	5.3852	5.3014	5.3014	5.3014	0
4 V	4.7948	4.7948	4.72095	4.6464	0
3 V	4.44365	4.44365	4.4102	4.376632	0

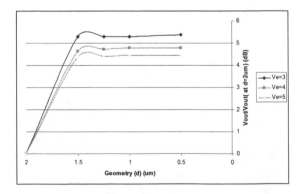

Figure 5. The measured output voltage for different DeFET geometry sensors at different electrodes voltage (Electrode 1 and 3 voltage = - Electrode 2 and 4 voltage).

The electric field and its sensation can be simultaneously actuated in a real time; this is a significant advantage over the proposed in [13, 14, 15, 16].

NSTI-Nanotech 2008, www.nsti.org, ISBN 978-1-4200-8503-7 Vol. 1

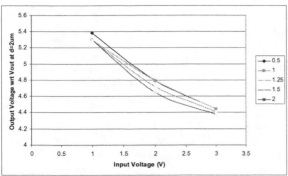

Figure 6. The output voltage vs. the input voltage for different sensor geometries.

5. APPLICATIONS OF THE PROPOSED LAB-ON-A-CHIP

The proposed lab-on-a-chip can be used in many applications, such as; extract properties of the media, to analyze biocells, e.g. the membrane capacitance. It has also application in the DNA analysis, to detect the radius of the DNA molecule. In cancer research area, it detects the distribution of the charges on the surface of the cancer cell and consequently a suitable antibody can be selected to be used in cancer treatment [12]. Moreover, it can be used as an impedance sensor, to measure the impedance of the cell above the sensor using transimpedance amplifier as a read-out circuit, this application is important in cell detection [9]. This sensor is based on CMOS standard technology; thus, we can easily integrate the read-out circuitry in addition to the actuation and sensing parts, which already integrated.

Finally, The DeFET, as it is high sensitive electric field sensor, it can be used within a printed circuit board (PCB) prototype to trap, concentrate, and quantify different types of micro and nano particles. This sensor can also be used in the DNA analysis, to detect the radius of the DNA molecule. In cancer research area, it can be used to detect the distribution of the charges on the surface of the cancer cell and consequently a suitable antibody can be selected to be used in cancer treatment [17].

6. CONCLUSION

A new integrated lab-on-a-chip has been proposed. It's based on CMOS 0.18µm TSMC technology. It consists of two main parts; which are the actuation and the sensing parts. The simulation results verify the theory of its operation. This research work will open a new area of research field which is to use the electric field as a diagnosis tool that can be used in different applications. Also, it can be used as an impedance sensor, to measure the impedance of the cell above the sensor using transimpedance amplifier as a read-out circuit, this application is important in cell detection and characterization. Moreover, this sensor is based on CMOS standard technology; thus, we can easily integrate other read-out and conditioning circuitry in addition to the actuation and sensing parts, which already integrated.

7. ACKNOWLEDGEMENT

The authors want to acknowledge, Canadian Microelectronics Corporation (CMC), Dr. Graham Jullien from Electrical and computer Department at University of Calgary, for his advice and academic help.

REFERENCES

[1] Pohl, H., The Motion and Precipitation of Suspensoids in Divergent Electric Fields. Applied Physics, 1951. 22: p. 869-871.

[2] Jones, T.B., Electromechanics of Particles. 1995, USA: Cambridge University Press. 265.

[3] Pohl, H., Some Effects of Nonuniform Fields on Dielectrics. Applied Physics, 1958. 29: p. 1182-1188.

[4] Crane, J. and H. Pohl, A study of living and dead yeast cells using dielectrophoresis. Journal of the Electrochemical Society, 1968. 115(6): p. 584-586.

[5] Betts, W.B., The potential of dielectrophoresis for the real-time detection of microorganisms in foods. Trends in Food Science and Technology, 1995. 6(2): p. 51-58.

[6] Blake A. Simmons, Eric B. Cummings, et al. "Separation and Concentration of Water-Borne Contaminants Utilizing Insulator-Based Dielectrophoresis" SANDIA REPORT, January 2006.

[7] H.A. Pohl, Dielectrophoresis, Cambridge University Press, Cambridge, 1978.

[8] M. Washizu, and O. Kurosawa "Electrostatic manipulation of DNA in microfabricated structures ", IEEE Transactions on Industry Applications, vol. 26, no.6, pp. 1165-1172, 1990.

[9] Gianni Medoro, Nicoló Manaresi, Andrea Leonardi, Luigi Altomare, Marco Tartagni, and Roberto Guerrieri," A Lab-on-a-Chip for Cell Detection and Manipulation," IEEE Sensors Journal, vol. 3, no. 3, pp. 317-325, June 2003.

[10] N. Manaresi, A. Romani, G. Medoro, L. Altomare, A. Leonardi, M. Tartagni, and R. Guerrieri," A CMOC Chip for Individual Manipulation and Detection", ISSCC 03, pp. 486-488. 2003.

[11] Yehya H. Ghallab, Wael Badawy "A Novel Cmos Lab-On-A-Chip For Biomedical Applications" IEEE Circuits and Systems, 2005. ISCAS 2005. IEEE International Symposium on 23-26 May 2005 Page(s):1346 - 1349 Vol. 2

[12] Yehya H. Ghallab, and Wael Badawy" DeFET, A Novel Electric field Sensor for Lab-on-a-Chip and Biomedical Applications ", IEEE Journal of Sensor, vol.6, no.4, pp. 1027-1037, August 2006.

[13] Yehya Ghallab, and W. Badawy "Sensing methods of Dielectrophorieses from Bulky instruments to Lab-on-a-chip", IEEE Circuit and Systems Magazine, Q3 issue, vol. 4, pp.5-15, 2004.

[14] P. Fortina, S. Surrey, and Lj. Kricka, "Molecular diagnostics: hurdles for clinical implementation," Trends Mol. Med., vol. 8, pp. 264–266, 2002.

[15] L. E. Hartley, K.V.I.S. Kaler, and R. Paul, "Quadrupole Levitation Of Microscopic Dielectric Particles", Journal of Electrostatics, vol.46, pp. 233-246, 1999.

[16] S. Liu, J. Wei, and G. Sung, "SPICE Macro Model for MAGFET and Its Applications", IEEE Transactions on Circuits and Systems- II: Ana log and Digital Signal Processing, vol.46, no.4, 1999.

[17] Lee Hartley, Karan V. I. S. Kaler, and Orly Yadid-Pecht "Hybrid Integration of an Active Pixel Sensor and Microfluidics for Cytometry on a Chip" IEEE Transactions On Circuits And Systems, VOL. 54, NO. 1, JANUARY 2007

Quantum Transport in Nano Mosfets

Lamba, V.J[1], Derick Engles[2] , Malik,S.S [3], Gupta, A[4].

[1]Deptt of ECE, HCTM, Kaithal, India. lamba_vj@hotmail.com
[2]Deptt of ECE, GND University Amritsar, India.
[3]Deptt of Physics, GND University Amritsar, India
[4]Deptt of EEE, HCTM, Kaithal, India.

ABSTRACT

In this work a general method for describing quantum electron transport will be introduced. These formulations are based on Schrödinger equation, Pauli master equation, density matrix, Wigner function, and the Green function. Here, we describe the microscopic quantum theory of electron transport in silicon devices based on the NEGF formalism. We review the NEGF formalism and derive its key equations. We include the electron-phonon interactions and other scattering mechanisms such as the impurity scattering and the surface roughness scattering. For the electron-electron interactions, we used the assumption that each electron moves independently and sees only the average field generated by all the other electrons.

Keywords: Length scale, Green function, Scattering, Broadening, Mobility.

I INTRODUCTION

Various quantum mechanical formulations have been used for the modeling of the carrier transport in semiconductor devices. During the 1980s advances in the semiconductor technology made it possible to fabricate conductors so small that the quantum nature of electrons could be directly observed in transport experiments. A typical systems used for these studies include quantum dots [1], point contacts [2], which are formed by manipulating the electrostatic potential in a two-dimensional electron gas by means of gate electrodes. The size of these semiconductor structures is of the order μm, which characterizes them as mesoscopic systems intermediate between the microscopic and macroscopic. Several important discoveries such as the quantum Hall effect and quantization of conductance are all results of the intense study of electron transport [3].

The paper is organized as follows. In section II, we start with an introduction of length scales, used to characterize electron transport. In section III, describes the Hamiltonian of a system. Section IV, includes green function and study of properties of kinetic equations. The electron-phonon interactions, other scattering mechanisms (including surface roughness scattering) will be included in section V. A simple nano-mosfet will be discussed in section VI.

II LENGTH SCALES

A key quantity in the description of electron transport is the electrical conductance relating the current through a conductor to the applied voltage. For a macroscopic sample the conductance is given by

$$G = \sigma \frac{A}{L} \tag{1}$$

Where, A and L are the cross-sectional area and the length

of the sample, respectively, and σ is the material dependent conductivity. When the dimensions of the sample become sufficiently small the quantum nature of the charge carriers becomes important, and the classical relation (1) breaks down. Below we introduce three length scales which are important in microscopic description of electron transport.

Fermi wave-length: The wave function of an electron in a crystalline metal contains a Bloch factor, exp(ik.**r**), which determines the complex phase of the wave function throughout the crystal [4]. The wavelength of this oscillating phase associated with a conduction electron at the Fermi level is called the Fermi wavelength, $\lambda_F = 2\pi/k_F$. In semiconductors the density of conduction electrons is very low, it can be several nanometers.

Phase coherence length: The phase coherence time, τ_φ, is the time during which the phase of an electron is completely lost due to interactions with the dynamic environment. The dynamic environment consists of all quantum degrees of freedom which interact with the electron such as phonons, impurities, and the other conduction electrons. The phase coherence length, l_φ, is defined as the distance a conduction electron moves before its phase is lost, i.e. $l_\varphi = \upsilon_F \tau_\varphi$,

Mean free path: The mean free path is defined from the momentum relaxation time, τ_m, which is the time it takes for a conduction electron to lose its initial momentum through scattering events. In a perfect crystal the electron moves unhindered and thus $\tau_m = \infty$. In a real system, however, the presence of scatterers will reduce τ_m to a finite value. It should be noted that the scattering events relaxing the momentum can be elastic as well as inelastic and thus there is no a priori relation between τ_m and τ_φ.

The relative size of the length scales introduced above defines, in combination with the characteristic system dimension, L, various different transport regimes. The systems considered in the present work is assumed to be in the phase-coherent regime where $l_\varphi > L$. These systems are large compared to the atomic scale, and the large λ_F characteristic for these systems, renders the electrons insensitive to the detailed atomic structure of the host material. This, of course, simplifies some aspects of the theoretical modeling of transport in semiconductor systems.

III HAMILTONIAN OF SYSTEM

Let us consider an isolated device and its energy levels are described using a Hamiltonian H, a Hartree potential U and energy eigen-states of the electron ε_α,

$$\left(H + U \right) \psi_\alpha \left(\vec{r} \right) = \varepsilon_\alpha \psi_\alpha \left(\vec{r} \right) \tag{2}$$

The Hamiltonian of the considered system, H, is composed

of four different components as

$$H = H_e + H_p + H_{e-p} + H_{is} \qquad (3)$$

where H_e is the Hamiltonian of non-interacting electrons, H_p is the Hamiltonian of free phonons, H_{e-p} is the electron-phonon interaction Hamiltonian, & H_{is} is the Hamiltonian for impurity scattering. In this section, we introduce these components one by one.

Hamiltonian of Non-interacting Electrons

As we use the Hartree approximation to simplify the electron-electron interaction. Therefore, the Hamiltonian of non-interacting electrons can be written as [5]

$$H_e = \int dr \psi^\dagger(\vec{r})\left[T(\vec{r}) + U(\vec{r})\right]\psi(\vec{r}) \qquad (4)$$

where, $T(\vec{r})$ is one-electron K.E., & $U(\vec{r})$ is self-consistent electrostatic P.E. operator. The expression for the one-electron kinetic energy operator is obtained from the effective mass approximation. Since we consider the transport of electrons in the conduction band of silicon, we use the usual parabolic, ellipsoidal energy band structure, which is found to be a reasonable approximation even in the nanoscale devices [6].

Hamiltonian of Free Phonons

The Hamiltonian of free phonons can be written as [7]

$$H_p = \sum_{q\lambda} \hbar\omega_{q\lambda}\left(a_{q\lambda}^\dagger a_{q\lambda} + \frac{1}{2}\right) \qquad (6)$$

where $\omega_{q\lambda}, a_{q\lambda}^\dagger$, and $a_{q\lambda}$ are the angular frequency, the creation operator, and the annihilation operator for mode λ, and wave-vector **q**, respectively.

Electron-Phonon Interaction Hamiltonian

In the first quantization picture, the electron-phonon interaction Hamiltonian felt by the electrons in the conduction band can be obtained from the deformation potential theory as [8]. As in the continuous medium approximation, the ion displacement field y can be written in terms of the phonon creation and annihilation operators. Thus, we obtain the single-electron interaction Hamiltonian in terms of the phonon creation and annihilation operators as

$$\varphi(\vec{r}) = \frac{1}{\sqrt{V}}\sum_{q\lambda} M_{q\lambda}\left[a_{q\lambda} + a_{-q\lambda}^\dagger\right]e^{iq.\vec{r}} \qquad (7)$$

where the electron-phonon matrix element $M_{q\lambda}$, is

$$M_{q\lambda}(\vec{r}) = i\left(\frac{\hbar}{2\rho V \omega_{q\lambda}}\right)^{1/2}\sum_{i=1}^{3}\sum_{j=1}^{3}\Xi_{lj}q_l\xi_{j\lambda} \qquad (8)$$

Which, has the following property: $M_{-q\lambda} = M_{q\lambda}^*$. From the obtained single-electron operator for the electron-phonon interaction Hamiltonian (7), we obtain the second quantized electron-phonon interaction Hamiltonian as

$$H_{e-p} = \int dr \psi^\dagger(\vec{r})\psi(\vec{r})\varphi(\vec{r}) \qquad (9)$$

Hamiltonian for Impurity Scattering [7]

Impurity scattering is an interaction between a moving carrier and a fix ionized atom. Its Hamiltonian is given by

$$H_{is} = \sum_{i}\int dr \psi^\dagger(\vec{r})V(\vec{r} - \vec{R}_i)\hat{\psi}(\vec{r}) \qquad (10)$$

where $\hat{\psi}^\dagger$ and $\hat{\psi}$ are creation and annihilation operators. and $V(\vec{r} - \vec{R}_i)$ is the potential describing the interaction between a carrier at site **r** and an impurity at site $\mathbf{R}_i$.

IV GREEN FUNCTION

Here, the main equations governing the behavior of non equilibrium Green's functions are presented as well as the derivation of two important physical quantities resulting from their solution, the carrier density and the current density. The total Hamiltonian H(t) of system is given by:

$$H(t) = H + H_{ext}(t) = H_0 + V + H_{ext}(t) \qquad (11)$$

Where, H_0 is the non-interacting part of the Hamiltonian, V contains all the interactions (carrier-carrier, carrier-phonon, impurity scattering ...) and $H_{ext}(t)$ is an external perturbation driving the system out of equilibrium. Thus

$$H_{ext}(t) = \int dx \psi^\dagger(x)U(x,t)\psi(x) \qquad (12)$$

Here, U(x,t) is the external potential. In general an isolated device, and its energy levels are described using a Hamiltonian H, a Hartree potential U and energy eigen-states of the electron, ε_α by (2) and it's Hamiltonian by (3). In general, the electron density matrix in real space is given by

$$\left[\rho(\vec{r},\vec{r}';E)\right] = \int_{-\infty}^{+\infty} f(E - E_f)\delta\left(\left[EI - H\right]\right)dE \qquad (13)$$

Here $\delta(EI-H)$ is the local density of states, rewriting further using the standard expansion form equation (13) become

$$\delta(EI - H) = \frac{i}{2\pi}\left(\left[G(E) - G^+(E)\right] \text{ where } G(E) = \left[\left(E - i0^+\right)I - H\right]^{-1}\right)$$

G(E) is the retarded Green's function while $G^+(E)$, its conjugate complex transpose, is called the advanced Green's function. In the time domain, the Green's function can be interpreted as the impulse response of the Schrödinger equation where in the present scenario the impulse is essentially an incoming electron at a particular energy. In the energy domain the Green's function gives the energy eigen-values for the eigen-states that are occupied in response to the applied impulse. As the electron density in the channel is the product of the Fermi function and the available density of states and for an isolated device is written as

$$\rho(\vec{r},\vec{r}';E) = \int_{-\infty}^{+\infty} f(E - E_f)\left[A(E)\right]\frac{dE}{2\pi} \qquad (14)$$

The real portion of the diagonal elements of the density matrix, represent the electron density distribution in the channel. To understand the process of current flow, consider an isolated device having a single energy level ε. The source and drain contacts have an infinite distribution of electronic energy states. When the isolated device with single energy level ε is connected to the source and drain contacts, some of the density of states around this energy level ε will spill over from the contacts into the channel. This process is known as energy level *broadening*. If the

Fermi levels in the source and drain are equal, the amount of broadening will be equal on both sides and hence the net current flow in the channel will be zero. When a positive bias is applied on the drain side, the Fermi level on the drain side is lowered according to equation (15), opening up states below the channel energy level ε in the drain.

$$E_{f_2} = E_{f_1} - qV_D \tag{15}$$

The electrons entering the channel with energy ε now have states with lower energies available in the drain to escape to. This causes the channel current to become non-zero. As the applied bias is increased linearly, more and more states between ε and ε_{f2} become available to remove electrons from the channel causing the source to increase its supply of electrons into the channel. This phenomenon results in a linear increase of current in the channel. Eventually, the difference in the channel energy level ε and the drain Fermi level ε_{f2} is so great that there are no additional states around the energy ε in the drain for the channel electrons to escape. The current reaches saturation such that the number of electrons leaving the drain will equal the number of electrons entering from the source. This process also explains why experimental measurements [9] have shown that the maximum measured conductance of a one-energy level channel approaches a limiting value $G_0 = 2q^2/\hbar = 51.6(K\Omega)^{-1}$. The above analogy of a one-energy level system is applicable to nanoscale thin films and wires where available energy levels along the confined dimension are very limited in addition to being spaced far apart from adjacent energy levels.

In the NEGF formalism the coupling of the device to the source and drain contacts is described using self-energy matrices Σ_1, Σ_2. The self-energy term can be viewed as a modification to the Hamiltonian to incorporate the boundary conditions. Accordingly, equation (2) and (13) can be rewritten as

$$\left(H + U + \Sigma_1 + \Sigma_2\right)\psi_\alpha\left(\vec{r}\right) = \varepsilon_\alpha \psi_\alpha\left(\vec{r}\right) \tag{16}$$

The self-energy terms Σ_1 and Σ_2 originate from the solution of the contact Hamiltonian. In this semi-infinite system, which is connected to the channel, there will be an incident wave from the channel as well as a reflected wave from the contact. The wave function at the interface is matched to conserve energy resulting in the boundary condition,

$$\Sigma_j = -t \exp\left(ik_j a\right) \tag{17}$$

where t, the inter-unit coupling energy resulting from the discretization is given by

$$t = \frac{\hbar^2}{2m^* a^2} \tag{18}$$

Here k_j corresponds to wave vector of the electron entering from the channel while a corresponds to the grid spacing. The broadening of the energy levels introduced by connecting the device to the source and drain contacts is incorporated through, Gamma functions Γ_1 and Γ_2 given by

$$\Gamma_1 = i\left(\Sigma_1 - \Sigma_1^+\right) \text{ and } \Gamma_2 = i\left(\Sigma_2 - \Sigma_2^+\right) \tag{19}$$

The self-energy terms affect the Hamiltonian in two ways.

The real part of the self-energy term shifts the device eigenstates or energy level while the imaginary part of Σ causes the density of states to broaden while giving the eigenstates a finite lifetime. The electron density for the open system is now given by

$$[\rho] = \int_{-\infty}^{+\infty} \left[G^n\left(E\right)\right]\left(\frac{dE}{2\pi}\right) \tag{20}$$

$G^n(E)$ represents the electron density per unit energy and is given by

$$G^n\left(E\right) = G\left(E\right)\Sigma^{in}\left(E\right)G^+\left(E\right) \text{ where } \left[\Sigma^{in}\left(E\right)\right] = \left[\Gamma_1\left(E\right)\right]f_1 + \left[\Gamma_2\left(E\right)\right]f_2$$

For plane wave basis functions, the current through the channel is calculated as the difference between the inflow and the outflow at any given contact.

$$I_j = -\frac{q}{\hbar}\int_{-\infty}^{+\infty} trace\left[\Gamma_j A\right]f_1 - trace\left[\Gamma_j G^n\right] \tag{21}$$

where the subscript j indexes the contacts. For a two-terminal device $I_1 = -I_2$. The devices are examined in the present work is a Double gate MOSFET.

Phenomenological parameter η in NEGF formalism

The NEGF treatment of the junction side branches can be visualized by dealing with the floating probes in a similar way as the left and right contacts. Each probe can be viewed as connected with a reservoir with chemical potential $\mu_s(i)$, with i denotes the ith probe in the mesoscopic system. In general, the self-energy function Σ_s depend on the system density matrix and the Green's function, which requires an iterative solution of the problem. Principally, the NEGF formalism provides clear descriptions on calculating Σ_s for each kind of scattering mechanism. However, we will not go into any of these models here. Instead we will present results obtained from a phenomenological model that captures some of the important features of dissipative transport by a phenomenological parameter η, just like the phenomenological Büttiker probe coupling parameter ε, which is used for describing coupling strength of the scatterer and is related with the inelastic scattering rate. As the imaginary potential method gives the other way to describe the energy dissipation and the partially coherent transport, and because it modifies the Hamiltonian directly, it is more convenient to connect this method to the NEGF mechanisms. Based on the imaginary potential method [4], if we put one Büttiker probe in each mesh point, the scattering self-energy term will be:

$$\Sigma_s = -i\begin{bmatrix} \eta_1 & 0 & 0 & \cdots \\ 0 & \eta_2 & 0 & \cdots \\ 0 & 0 & \eta_3 & \cdots \\ \cdots & \cdots & \cdots & \cdots \end{bmatrix} \tag{22}$$

where, η is phenomenological parameter, which is related with the inelastic scattering energy relaxation time by:

$$\eta = \hbar / 2\tau \tag{23}$$

The device Green's function then has the form:

$$G = [EI - H - \Sigma_1 - \Sigma_2 - \Sigma_s] \tag{24}$$

where, Σ_l, Σ_2, and Σ_s denote the coupling with the left contact, right contact, and the reservoir connected to Büttiker probe respectively. As we have discussed about the imaginary potential method, the equation (24) gives the proper dynamic equation to study the dissipation processes inside the device.

Kinetic equations

The central part of the NEGF formalism is the kinetic equations, which relates the correlation function G^n and G^p with the scattering function Σ_{in} and Σ_{out}. The in-scattering function Σ_{in} tells the rate at which electrons are scattered in the floating reservoir, and Σ_{out} tells the rate at which holes are scattered in the floating reservoir or electrons are scattered out of it. They can be related with the scattering self-energy term Σ_s by the equation:

$$\Sigma_{in} = [F_s(\mu_s)] \cdot i \cdot [\Sigma_s - \Sigma_s^+] = [F_s(\mu_s)] \cdot \Gamma(E)$$
$$\Sigma_{out} = [1 - F_s(\mu_s)] \cdot i \cdot [\Sigma_s - \Sigma_s^+] = [1 - F_s(\mu_s)] \cdot \Gamma(E) \quad (25)$$

Here, we let each lattice site float to a thermal reservoir with a chemical potential μ_s different from each other, and $F_s(\mu_s)$ are the carrier distribution function with chemical potential μ_s. As we know that the Green's function $G^R(r, r1)$ is solution of above equation with the source term equal to a delta function. For an arbitrary source function, the solution of equation (25) has the form:

$$\psi(r) = \int G^R(r, r1) S(r1) dr1 \quad (26)$$

Multiplying equation (26) by the complex conjugate, we can have:

$$\psi(r)\psi(r')^* = \int G^R(r, r1) G^R(r, r1')^* S(r)S(r1')^* dr1 dr1' \quad (27)$$

Noting that G^n represents the correlation between wave functions, and Σ_{in} represents the correlation between sources,

$$G^n(r, r') \sim \psi(r)\psi(r')^*$$
$$\Sigma_{in}(r1, r1') \sim S^R(r1, r1')^* \quad (28)$$

we can write:

$$G^n(r, r') = \int G^R(r, r1) \Sigma_{in}(r1, r1') G^A(r, r1') dr1 dr1' \quad (29)$$

In matrix notation it will be $G^n = G^R \cdot \sum_{in} \cdot G^A$ which thus proves equation (25).

The density matrix of the system can then be written as:

$$2\pi[\rho(E)] = G^R \cdot (\Gamma_l + \Gamma_r) \cdot G^A + G^R \cdot \sum_{in} \cdot G^A = F_l \cdot A_l + F_r \cdot A_r + F_S \cdot A_S$$

The carrier density and the current can then be found from the density matrix in above equation, where $F_{l,r,s}$ and $A_{l,r,s}$ are the Fermi distribution and spectral function for source, drain, and each Büttiker probe. Thus NEGF formalism gives us a sophisticated way to study the scattering behavior inside the ultra-scaled transistors.

V SIMULATION OF DOUBLE GATE Si MOSFETs

The central problem for the simulation study to include the scattering processes is to determine the coupling strength of Büttiker probe or the phenomenological parameter η in the scattering self energy term. η is the adjusting parameter, which means it can be found by comparing the simulation results with experimental observations. It should be pointed out that there are extra scattering processes for Si MOSFETs, comparing with a simple Si slab. For Si double

gate MOSFET, it will be helpful to get reasonable values of these parameters by studying the scattering processes considering the device configurations.

A lot of attention has been paid to two major scattering processes for Si MOSFETs scaled to ~10nm quasi-ballistic region. One is the surface roughness scattering of electrons at the Si/SiO2 interface, and the other is the electron-electron scattering. Here we will not have a detailed discussion on the scattering mechanisms. For a phenomenological modeling of the device behavior in the inversion region, we will use the mobility mapping technique discussed in previous works [10-14]. The imaginary potential or the de-phasing scattering self-energy term η can be related to the energy relaxation time by $\eta = \hbar/2\tau$ as shown in equation (28). The energy relaxation time can be related with mobility by the equation [11]:

$$\mu = q\tau/m^* \quad (30)$$

So we can approximately have:

$$\eta \sim \hbar q / (2\mu m^*) \sim 0.003 * \left[\frac{\mu}{200} \cdot \frac{m^*}{m_e}\right]^{-1} eV \quad (31)$$

Table 1, shows the relation between effective mobility and effective channel length for the measurement of real devices [10]. Although it is not accurate to get information from this table on the effective mobility value we are to use in our simulations due to the difference in fabrication details and device configurations, for a simple evaluation of the scattering effects on carrier transport in the ~10nm double gate MOSFET, this table can give us some clue of the range of the mobility value we can put in our simulations. Here we used the $\mu \sim 165 cm^2/(V.s)$ for the transport dephasing calculations.

Table 1: Measured μ_{eff} vs. L_{eff} relation for different E_{eff} [10].

S.No	μ_{eff} (cm²/Vs)		Effective channel length (nm)
	E_{eff}(MV/cm) = 0.8	E_{eff}(MV/cm) = 1.1	
1	205	185	20
2	210	200	30
3	220	205	40
4	240	220	45
5	245	235	50
6	260	245	65
7	285	255	85

In figure 1, we show our simulation results by the mobility mapping technique [12-14]. It is found that due to the scattering processes, the on current is reduced to ~54% of the ballistic current. It is just what is expected to see based on the increase of the on-state channel resistance. For generations, transistors are operating close to 50% of the ballistic current. The calculation results here show that the phenomenological model can give a reasonable evaluation of Si MOSFETs scaled to ~10nm region.

VI CONCLUSION

Here, we reviewed, and derived the quantum kinetic equations starting from the Hamiltonian of the electron-phonon system, and we also derived the self-energy functions for the electron-phonon interactions, which is found to be spatially local. As a result, we obtained the simplified quantum kinetic equations, which will constitute the basis of our study on the quantum transport in the semiconductor devices. We also introduced two kinds of expressions for the current density, and the linear response of the current in the near equilibrium condition. Finally, we introduced a simple Double gate MOSFET example, and applied the NEGF formalism to obtain the analytic expressions for the retarded Green function and the other physical quantities

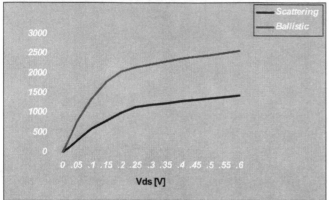

Figure 1: The simulation result includes scattering (Blue), compared with the ballistic (Pink) results. The device is the double-gate MOSFET with effective channel length of 10nm, and the channel thickness is 7nm, and oxide thickness for both top and bottom sides are 1.5nm. The source-drain doping is 10^{20} cm^{-3}. The mapping mobility used in the simulation is $165 cm2/(V.s)$.

VII REFERENCES

1. R. C. Ashoori., *Nature,* 379: 413-419, 1996.
2. L. N. Pfeifier, et. al., *Nature,* 411:51-54, 2001.
3. D. K. Ferry, et. al., Cambridge University Press, 1997.
4. G. Grosso, et. al., *Solid State Physics*, Academic Press, 2000.
5. A. L. Fetter , et. al., *Quantum theory of many-particle system*, New York: McGRAW-HILL, 1971.
6. C. Jacoboni , et. al., *Reviews of Modern Physics*, vol. 55, no. 3, pp. 645-705, July 1983.
7. G. D. Mahan, "*Many-Particle Physic*", 2nd ed. Newyork: Plenum Press, 1990.
8. J. Bardeen , et. al.,*Phys. Rev.*, vol. 80, no. 1, pp. 72-80, Oct. 1950
9. Szafer, A., et. al.,*Physical Review Letters*, 62, 300-303.
10. S. Takagi, A. , et. al.,IEEE, Trans. Electron Dev., Vol. 41(12), pp2357-2362, 1994.
11. A. Rahman, , et. al.,IEEE Trans. Electron Dev., Vol. 49(3), pp481-489, 2002.
12. W. Hansch, , et. al., *IEEE, Solid State Electronics,* vol. 32, No. 10, pp. 839-849, 1989.
13. A. Svizhenko et al., J. of Appl. Phys., 91:2343–2354, 2002.
14. U. Landman, et al., *Phys. Rev. Lett.*, vol. 85, no. 9, pp.1958–1961, 2000.

Development of New High Performance Nanocrystalline Hard Metals

N. Jalabadze, A. Mikeladze, R. Chedia, T. Kukava, L. Nadaraia and L. Khundadze

Technical University of Georgia, 77, Kostava St., Tbilisi 0175, Georgia, jalabadze@gtu.ge

ABSTRACT

It is known that performance of nanocrystalline hard metals suppresses that of ordinary materials. Lots of techniques for manufacturing nanopowders were developed and successfully used but preparation of nanocrystalline bulk pieces still faced certain problems. The resulted materials sintered by conventional technologies were nanocrystalline pieces. However physicomechanical testing of the pieces showed that their performance was not improved to an expected level, a possible reason considered to be an excessive free carbon: remaining in the material during the synthesis and playing an important role in preserving nanocrystallinity, at the same time it makes obstacles to obtaining hard metal pieces with high performance. Problems of elimination of excessive free carbon during the process of manufacturing of new high performance nanocrystalline hard metal pieces were solved by using a method and a device developed in the RCSR on the basis of Spark Plasma Synthesis (SPS) method.

Keywords: nanotechnology, SPS, hard metals

Carbides of transition hard metals, in particular, titanium carbide, are materials with unique properties they are good subjects for numerous investigations [1, 2]. A fact of broadening of the production of titanium carbide and its wide application in various areas, has transferred it to the ranks of the materials widely used in the industry and not only in the area of solid state physics as it was used earlier as a convenient and model material for research studies [2]. High demands are made of the materials with titanium carbide content used in rocket production, aircraft, nuclear power and microelectronics industry. A probability of using titanium carbide to that or other production is defined by a complex variety of properties [3], one of them and the most important, being structural condition of the material. Most promising is using of titanium carbide in microcircuitry in the electronic industry [4]. Main disadvantage of using hard metals based on titanium carbide if compared with those based on tungsten carbide is lack of elasticity of the material though we think that this kind of disadvantage can be removed in case if the alloys are fabricated on nanocrystalline level. Physical-mechanical properties of nanocrystalline materials significantly differ from those with crystalline structure. Nanocrystalline carbides are characterized with excellent catalytic properties [5].

Fabrication of hard metals by powder metallurgy technique embraces the following steps: production of the powders of hard metal components, preparation of the charge of hard metals, forming, sintering and control. We have developed a nanotechnology for manufacturing of powders of hard metals. After compaction and sintering of the obtained nanopowders with standard technology the structure of alloys remains nanocrystalline. Investigation of physicomechanical properties showed that their characteristics are a little better than those of the obtained by standard technology. However the hard metals with nanocrystalline structure were to have much higher performance. A reason preventing to achieve much better results appeared to be an excessive free carbon which remains there during the synthesis of nanopowders. Namely the excessive free carbon was a reason for the preserved nanostructural state at synthesising the samples by an ordinary technology: due to being plated on carbide particles an excessive free carbon makes obstacles to their growth. The same excessive free carbon was an obstacle on the way of obtaining hard metal samples with high performance since the structure was nanocrystalline. All details of the technological cycle of fabricating nanopowders will be studied within the scope of the project. Mechanism of formation of carbon will be established during the process of synthesizing and an excessive carbon will be eliminated. Reduction of the amount of free carbon in hard metal nanopowders will make possible to prepare pieces with high physicomechanical and operation properties by using SPS device.

New technology is based on realization of the methods elaborated by group of Georgian Technical University. The new technology provides formation of nanocrystalline material. This method is based on thermo-chemical synthesis. Carbides - hard components of hard metals –are obtained by a high-temperature synthesis, and therefore they are coarse-grained. Despite the fact that they are disintegrated to $\sim 0.1 \mu m$ upon grinding, during the process of sintering they grow up to 1-$10 \mu m$. Bonding components of hard alloys are not disintegrating. Upon grinding they undergo plastic deformation and after heating their size increases up to ten and even hundreds of microns. Issued from the above mentioned it is clear that for the obtaining of dispersed system it is necessary to conduct a low temperature synthesis. Such a synthesis may be provided by obtaining of chemical compounds (chemical synthesis) and their subsequent pyrolysis (thermal synthesis). The subsequent operations are similar to the technologies common for making hard alloys, but with one difference - in such case the plasticization of powder is not necessary. Nanocrystallinity of bonding material of hard alloys promotes to good ability of compressing the powders. It should be noted that upon heating, grains of the components of hard alloys are not coarsening when the process is thermo-chemical synthesis and a final product remains nanocrystalline.

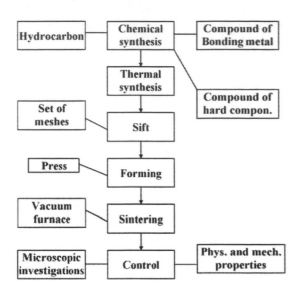

Fig.1 Schematic drawing of a technological cycle for the production of nanocrystalline hard metals

The thermo-chemical method (Fig.1.) is based on formation of nanocrystalline hard metal charge because of high temperature (~900oC) chemical interaction of titanium hydride and bonding metal salts with hydrocarbon compounds and the resulting product is a hard alloy charge where all components are in nanocrystalline (of an order of 200-300 nm) state. Sintering of such material will possibly lead to coarsening of grains and certainly the material will not remain nanocrystalline. But if nanocrystalline carbide particles will be coated with nano particles of bonding material or demarcated by them, then they will not grow upon sintering and hard metal will preserve nanocrystallinity.

A number of new techniques for powder consolidation aimed at fully dense bulk nanocrystalline materials have been proposed in recent years [6-11]. Preparation of bulk pieces requires compaction and sintering of the obtained nanocrystalline powders. This process is connected with lots of problems, namely, it is very difficult to preserve nanocrystalline structure of powders in the bulk. Standard methods for manufacturing of bulk material are: cold compaction with further sintering, hot pressing, sintering under high pressure, electric discharge synthesis, shock-

wave sintering and gasostate sintering. Basing on these methods by different companies there were designed and built various installations: for hot pressing, max temperature 2400o C, max pressure 40MPa (1000A, Thermal Technology INC., USA); High temperature graphite furnace, max temperature 2400o C, equipped with quenching facilities (1000-4560-FP20, Thermal Technology INC., USA); Spark Plasma Sintering unit (SPS), max temperature 2000o C, max. load 20000 kN (Dr Sinter 2050, Sumitomo Coal Mining Co., Ltd, Japan); Microwave sintering unit, 6 kW operating at 2.45 GHz (S6G Cober Electronics Inc., USA) and etc Using of these methods and of an appropriate installation is not effective because of intensive growth of grains which stimulates formation of an ordinary structure instead of the desired nanostructure.

One way to prevent the processes of grain growth is adding of inhibitors to the powders. However this route is not the best one because composition of the material will possibly be changed due to contaminations brought in the powders: nanocrystallinity may improve properties of the material, however adding of inhibitors may reduce these properties. But very often such additions have negative effect on the alloys. It seems that an excessive free carbon appearing in our experiments at thermochemical synthesis works as a natural inhibitor. Pressing of hard metal in the presence of excessive free carbon is difficult. Only some parts of hard alloy can be consolidated and the structure becomes porous. Nanocrystalline structure in agglomerates is preserved. In such cases, due to porosity, strength characteristics of the structure of the alloys are not as high as required of nanocrystalline structure. Therefore improving of characteristics induced by preservation of nanocrystallinity is not as efficient as expected. Another way to prevent the processes of grain growth is: guiding of sintering processes in the time limited to a certain extent. This route is realizable in the installation based on using SPS method which is considerably new and it can be used to conduct in situ preparation and synthesis of composites with superfine microstructures. In spite of the fact that there are already designed and constructed the SPS method-based industrial installations, physical essence of the provided processes are not yet clarified to final extent.

As notified earlier nanocrystalline hard metals can be fabricated by an ordinary sintering technology. Presence of excessive free carbon impedes processes of grain growth but does not promote significant increase of physicomechanical performance of the alloys. We have developed a new device for sintering of nanocrystalline hard metals. Principle of working of the device was based on plasma-sparkling sintering method. Fig.2 shows press-form for forming and sintering cylindrical hard metal samples. Powder is isolated from matrix and high-ampere pulse current passes through puncheon to powder.

1. Matrix
2. Upper Plug
3. Lower Plug
4. Insulator
5. Graphite

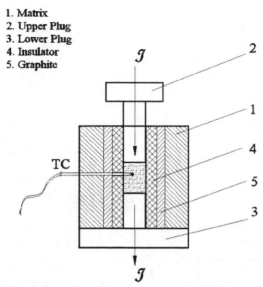

Fig.2. Press-form for the synthesis of nanocrystalline hard metal.

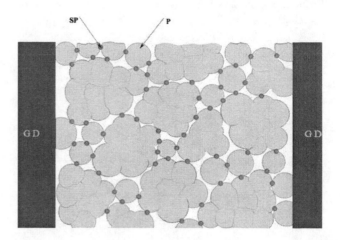

Fig.3. Scheme of hard metal structure-formation at sintering by using ark-plasma method.

Principle of working of the SPS-based device is the following: Passing of a pulsed DC of low voltage and high magnitude through a powder material creates high temperature mainly in the area of particle contact. High temperatures created in microseconds are not sufficient for spreading through the whole mass of a powder and hence, temperature of powder mass remains rather low and the processes of grain growth do not proceed. However the temperatures created between the surfaces of powder particles are quite sufficient for providing the processes of synthesizing and therefore the obtained material remains nanocrystalline. Devices of the SPS type available worldwide are intended for using only of conductive powders, or the materials capable to gain conductivity after heating. Otherwise there can not be created any spark between powder particles and subsequently, plasma can not be created. Therefore using of such devices for dielectric materials is not appropriate. Another problem while applying such devices is using of nanocrystalline powders with high rate of aggregation.

For measuring the temperature chromel-alumel thermocouple is introduced in the powder. Passing of pulse current through the powder provides the process of sintering due to creating a sparkle and the followed up plasma (Fig.3) between contact points of hard metal particles. Heat is released only in the contact points between grains. Duration of pulsing can be varied from one to several tens of milliseconds. During such a little period of time only a surface is heated and the heat can not be spread through the grains. Therefore the temperature of a sample is much less than that at the contact points of grains. High interfacial contact temperature promotes sintering of the sample and due to low integral temperature - prevents the process of grain growing thus providing for maintenance of nanocrystalline structure.

The developed device also solves a problem of excessive free carbon creating at the fabrication of nanocrystalline hard metals based on titanium carbide. The solution is analogous: in this case it is also necessary to bring an excessive amount of free carbon into the charge for fabricating nanopowders with normal structure. This excessive free carbon furtherly makes obstacles to the alloys to be normally sintered. If carbon introduced into the charge is of fewer amounts, then there is detected presence of new structure compounds besides the main phase in nanopowders. Fig.4 shows X-ray diffraction pattern of the nanopowder with an excessive amount of the introduced free carbon: there are observed molybdenum, tungsten and their carbides (W_2C and Mo_2C) (Fig. 4a). It is easy to overcome this disadvantage if the powders are sintered on the developed device (Fig. 4b)

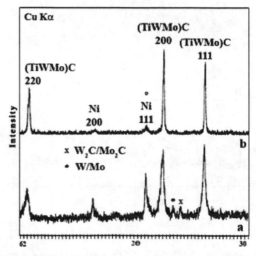

Fig.4. X-ray diffraction patterns of hard metal TiC-Ni-Mo-W: a- nanopowder with molybdenum, tungsten and their carbides, b- sintered alloys with normal structure.

From the nanopowders with rather defective structure (Fig.5a) can be easily fabricated alloys with normal structure (Fig. 5b) if the SPS method and the developed device are applied.

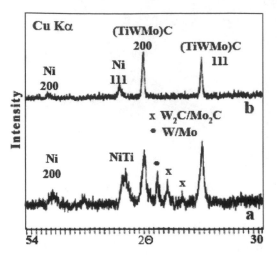

Fig.5. X-ray diffraction patterns of hard metal TiC-Ni-Mo-W: a- nanopowder with W/Mo, W_2C/Mo_2C, NiTi, b-sintered alloys with normal structure

Unique experiments were conducted on fabrication of nanocrystalline hard metal with the help of the developed device immediately from the alloy components omitting the procedure of preparing initial nanopowders. Fully sintered nanocrystalline hard metal of the (TiW,Mo)C-Ni system was obtained from the charge comprised of titanium hydride, nickel chloride, molybdenum- and tungsten oxides and soot (Fig. 6).

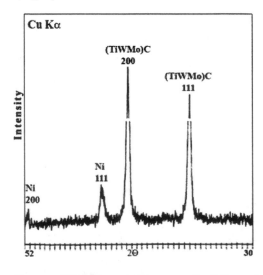

Fig.6. X-ray diffraction patterns of fully sintered nanocrystalline hard metal of the (TiW,Mo)C-Ni system obtained from the charge comprised of TiH_2, $NiCl_2$, MoO_3, WO_3 and soot.

REFERENCES

[1] Samsonov G.V., Upadhaya G.M., Neshpor V.S. – Physical Materials Science of Carbides. –Kiev: "Naukova Dumka", 456, 1974.

[2] Kyparisov S.S., Levinski Ju.V., Petrov A.P. Titanium Carbide: Preparation, Structure, Application.- M. "Mettallurgia", 216, 1987

[3] Alekseev S.A., Andrievski R.A., Dzodziev G.T., Kalkov A.A. Tungsten-free sintered hard metals based on titanium carbide and titanium carbonitride–M: "TSNIITSVETMET" (Russ), 44, 1979

[4] Wendell S. Williams, "Transition metal carbides, nitrides, and borides for electronic applications", J. of Materials, 38, 1997.

[5] Taeghwan Hyoen, Mingming Fang, and Kenneth S. Suslick, "Nanostructured Molybdenum Carbide: Sonochemical Synthesis and Catalytic Properties", J. Am. Chem. Soc., 118, 5492-5493, 1996.

[6] Y. V. Bykov, K. I. Rybakov, and V. E. Semenov, "High–temperature Microwave Processing of Materials," Journal of Physics D: Applied Physics, 34, R55-R75, 2001.

[7] V. Mamedov, "Spark Plasma Sintering as Advanced PM Sintering Methods," Powder Metallurgy, 45 [4], 323-328, 2002.

[8] J. R. Groza, "Nanocrystalline Powder Consolidation Methods," Nanostructured Materials, Noyes Publications, Williams Andrew Publishers, NY, , 115-178, 2002.

[9] K. C. Cho, R.H. Woodman, B. R. Klotz, and R. J. Dowding, "Plasma Pressure Compaction of Tungsten Powders," Materials and Manufacturing Processes, 19 (4), 619-630, 2004.

[10] D. Jia, K. T. Ramesh, and E. Ma, "Effects of Nanocrystalline and Ultrafine Grain Sizes on Constitutive Behavior and Shears Bands in Iron," Acta Materialia 51, 3495-3509. 2003,

[11] E. Lassner and W. Schubert, "Tungsten: Propertics, Chemistry, Technology of Element, Alloys, and Chemical Compounds", Kluwer Academic/Plenum Publishers, NY 1998.

Mechanical Characterization of Nickel Nanoparticles Elastomer Composites

Heather Denver[1], Timothy Heiman[1*], Elizabeth Martin[2*]
Amit Gupta[2] and Diana-Andra Borca-Tasciuc[1]

[1]Mechanical, Aerospace and Nuclear Engineering Department,
[2]Chemical and Biological Engineering Department,
Rensselaer Polytechnic Institute
110 8th St, Troy, NY 12180, Phone: (518) 276-3385, Fax: (518) 276-6025
{denveh, heimat,martie2, guptaa6, borcad}@rpi.edu

ABSTRACT

Polymer composites exhibiting magnetic properties are becoming increasingly important for a variety of applications. This work presents fabrication and mechanical characterization of magnetic polymer nanocomposites based on polydimethylsiloxane (PDMS) with nickel nanoparticles as fillers. Allyltrimethoxysilane (ATS) was used as surfactant to enhance nanoparticle dispersion in the polymer matrix. Composites with nanoparticles concentrations of 5, 10 and 15 vol. % respectively as well as pure PDMS specimens were fabricated and their mechanical properties were tested. Up to 70% increase in the elastic modulus is reported for composites with particle concentration of 15 vol. % and cured at 100°C.

Keywords: polydimethylsiloxane, nickel nanoparticles, nanocomposites

1. INTRODUCTION

Magnetic polymer composites consisting of polymer matrix and magnetic fillers are emerging as a new class of multi-functional materials. They have tunable elastic modulus (via application of an external magnetic field), tunable electromagnetic properties (such as index of refraction and RF absorption), may be heat-activated to self heal (via hysteresis or magnetic relaxation losses in alternating external magnetic field) and can be used for selective, remote thermal and magnetic actuation. Potential applications include low-loss cores in power transformers [1], electromagnetic shielding [2,3] adjustable vibration dampers [4] as well as microelectromechanical systems (MEMS) magnetic sensors or actuators [5-10] and medical devices [11]. Polydimethylsiloxane (PDMS) polymer is of particular interest to the microsystems and medical field, being employed in a wide range of applications in the areas of microfluidics and biotechnology [12,13]. In this context, this work presents fabrication and mechanical property characterization of PDMS composites with 100 nm diameter nickel nanoparticles as filler. Studies on particle dispersion in the polymer matrix are carried out via scanning electron microscopy, while mechanical properties are measured through standard tensile test.

2. FABRICATION

Nickel nanoparticles, 100 nm in diameter were procured from Argonide Corporation (Sanford, FL). Prior to mixing with PDMS, the nanoparticles were coated with allyltrimethoxysilane surfactant. The ATS solution was prepared by adding 2 wt. % ATS to a mix of 95% ethanol and 5% deionized water. After the pH of the solution was adjusted to 5 by the addition of acetic acid, it was stirred with a magnetic stir bar for 5 minutes. Then it was added to the nickel nanoparticles in a ratio of 100 mL per 25 grams of nickel. The nickel was ultrasonicated for 3 minutes and then was rinsed twice in ethanol to remove the ATS excess. The nanoparticles were then left for 24 hours at 60°C to cure the ATS coating and to evaporate any residual ethanol.

PDMS polymer (Sylgrad 184) was acquired from Dow Corning Corporation (Midland, MI). It consists of a siloxane base and a cross-linking agent, which must be mixed in 10:1 volumetric ratio. The ATS coated nickel nanoparticles were first dispersed in PDMS base elastomer by utilizing an ultrasonicator for 4 hours. Next, a Hauschild, FlackTek Speed Mixer was used for 30 minutes to enhance nanoparticle dispersion. During the last 3 minutes in the Speed Mixer, the curing agent was added to the base. A vacuum oven was then used to degas the sample for 30 minutes to remove the air bubbles. Finally the samples were poured into a mold and heated to 100°C for 2 hours to cure. This process was followed to fabricate composites with different concentrations of nickel nanoparticles of 5, 10 and 15 vol. % respectively. In addition, similar procedure was followed to prepare pure PDMS samples for comparison of mechanical properties.

However, in many applications it is desirable to have the flexibility of curing the composite at room temperature. Hence, for comparison, samples with similar nanoparticle concentrations were prepared curing the composite at

---

*These authors have contributed equally to this work.

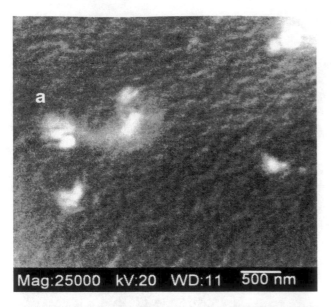

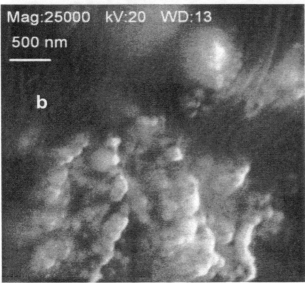

Figure 1: SEM of composites with 1 vol. % nickel nanoparticle: a) ATS coated and b) bare nanoparticles.

ambient temperature for 24 hours. To synthesize these specimens, the same steps were followed as for the samples cured at 100°C, except for vacuum degassing.

2. MECHANICAL PROPERTIES TESTING

The tensile test specimens had a gage length of 1 inch and a width of 0.25 inches. The mold was made from aluminum to prevent interactions between the ligands in the polymer. Tests were carried out with an Instron 5480 employing the ASTM (E8-04) Standard Test Method. After the sample was loaded into the testing apparatus, the video strain gage was activated, all variables were cleared and the speed of the grips was set to 10 mm/minute. Each specimen was pulled until failure while data was collected by a computer acquisition system for further analysis.

3. RESULTS AND DISCUSSION

3.1 Structural Characterization

The quality of the dispersion was evaluated in the preliminary stage of the fabrication process in order to ensure that particle uniformity is acceptable. Figure 1a) is a scanning electron micrograph (SEM) of a sample with 1 vol. % nickel nanoparticles coated with ATS. This sample showed uniform particle dispersion without noticeable agglomeration. In contrast, PDMS composites with bare nickel nanoparticles filler of same concentration were observed to present significant particle agglomeration, as shown in Fig. 1b. It was concluded that addition of ATS is critical for ensuring uniform nanoparticle dispersion.

3.2 Mechanical Properties

Figure 2 shows the measured tensile stress as function of the applied strain for samples with different concentrations of ATS coated nickel fillers that were cured at 100°C. As seen from this figure, at higher nanoparticle concentration the slope of the stress curve increases which indicates an increase in the elastic modulus. Figure 3 shows the measured elastic modulus as function of nanoparticle concentration for the composite samples with ATS coated nanoparticles that were cured at 100°C and room temperature respectively. The measured elastic modulus of plain PDMS, 2.57 MPa and 1.73 MPa respectively, are comparable with literature values (1.7-3.7MPa) [13,14]. The variation observed is may be due to the difference in curing temperature[14] and possible voids in the sample cured at room temperature which was not degassed. The elastic modulus of the specimens cured at 100°C increases with particle concentration. For example, the elastic modulus of 15 vol. % nickel nanoparticle- PDMS is

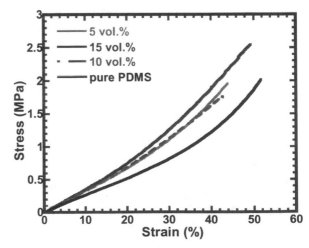

Figure 2: Stress versus strain obtained from tensile experiments carried out on pore PDMS and Ni nanoparticle PDMS composites cured at 100°C.

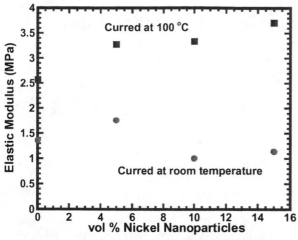

Figure 3: Elastic modulus as function of nanopartiucle concentration for specimens curred at 100°C (squares) and room temperature (circles).

c

3.71 MPa, which represent ~ 70% increase over that of pure PDMS cured at the same temperature. Although predicting mechanical properties of polymer nanocomposites often requires extensive modeling that needs to take into account inclusion/matrix interfacial characteristics [15], the elastic modulus of the PDMS nickel composite is expected to increase with an increase in nanoparticle concentration, since nickel has a much higher Young's modulus of 200 GPa. However, samples cured at room temperature show an inconsistent trend of elastic modulus, which apparently decreases as particle concentration increases. Since samples cured at room temperature were not degassed during preparation, the decrease in elastic modulus may be partially due to presence of the air bubbles within the specimens. In addition, differential scanning calorimetry studies were carried out on specimens with similar compositions. These studies revealed that for temperatures below 60°C the cross-linking process is not consistent (the peak in heat flow when this was plotted as function of time varied from sample to sample), suggesting possible problems in sample curing. These indicate that both, degassing and higher curing temperature, are critical in order to produce composites with good mechanical properties.

In addition to mechanical characterization, simple qualitative tests where samples were subjected to the magnetic field from a strong permanent magnet, indicated that these samples have acquired magnetic properties. Quantitative magnetic characterization of the composites is currently under way.

4. SUMMARY

This work presents fabrication and mechanical property characterization of magnetic polymer composites based on PDMS and nickel nanoparticle as fillers. It is found that ATS addition to nickel nanoparticles greatly enhances their dispersions in the polymer matrix. Mechanical property characterization shows that composites with 15 vol. % of nickel particles have an elastic modulus 70% higher than pure polymer samples when samples are cured at 100°C.

Acknowledgements. The authors would like to thank to Professor Linda Schadler for permission to use the FlackTek Speed Mixer TM and useful discussions.

REFERENCES

[1]K. W. E. Cheng, C. Y. Tang, D. K. W. Cheng, H. Wu, Y. L. Ho., and Y. Lu, Proceeding of 33rd Annual IEEE Power Electronics Specialists Conference, 3, 1254, 2002.
[2]F. Tsuda, H. Ono, S. Shinohara, and R. Sato, IEEE International Symposium on Electromagnetic Compatibility. Symposium Record, 2, 867, 2000.
[3]L. K. Lagorce, and M. G. Allen, Proceedings. 1996 International Symposium on Microelectronics (SPIE), 2920,176,1996.
[4]S. Abramchuk, E. Kramarenko, D. Grishin, G. Stepanov, L. V. Nikitin, G. Filipcsei, A. R. Khokhlov, and M. Zrinyi, Polymers for Advanced Technologies, 18, 513, 2007.
[5]H. S. Gokturk, T. J. Fiske, and D. M. Kalyon, IEEE Transactions on Magnetics, 29, 4170, 1993.
[6]L. K. Lagorce and M. G. Allen, Proceedings. IEEE, The Ninth Annual International Workshop on Micro Electro Mechanical Systems. An Investigation of Micro Structures, Sensors, Actuators, Machines and Systemsp, 85, 1996.
[7]L. Lagorce, D. Kercher, J. English, O. Brand, A. Glezer, and M. Allen, Proceedings. 1997 International Symposium on Microelectronics, 3235, 494, 1997.
[8] L. K. Lagorce and M. G. Allen, Journal of Microelectromechanical Systems, 6, 307, 1997.
[9] Lagorce, L.K., Brand, O. and Allen, M.G. "Magnetic microactuators based on polymer magnets," Journal of Microelectromechanical Systems, Vol. 8, p. 2 (1999).
[10]J. G. Boyd, D. C. Lagoudas, and S. Cheong-Soo Proceedings of the SPIE - The International Society for Optical Engineering, 5055, 268, 2003.
[11]P. R. Buckley, G. H. McKinley, T. S. Wilson, W. Small, W. J. Benett, J. P. Bearinger, M. W. McElfresh, and D. J. Maitland, IEEE Transactions on Biomedical Engineering, 53, 2075, 2006.
[12]S. K. Sia, and G. M. Whitesides, Electrophoresis, 24, 3563, 2003.
[13]X. Q. Brown, K. Ookawa, and J. Y. Wong, Biomaterials, 26, 3123, 2005.
[14] K. L. Mills, X. Zhu, S. Takayama, M. D. Thouless, Journal of Materials Research, 23, 37, 2008.
[15]S. Saber-Samandari, and A. Afaghi-Khatibi, Polymer Composites, 28, 405, 2007.

Mechanical and Electrical Properties of CNT/Inorganic Nanocomposites Fabricated by Molecular Level Mixing Process

Soon Hyung Hong and Chan Bin Mo

Dept. of Materials Science and Engineering, Korea Advanced Institute of Science and Technology, Daejeon, Korea, shhong@kaist.ac.kr

ABSTRACT

CNT/Inorganic nanocomposites with homogeneously distributed CNTs in inorganic matrix with strong interfacial strength are fabricated by a novel fabrication process, i.e. molecular level mixing process. Molecular level mixing process enables CNTs to be mixed and react with ions of inorganic matrix in molecular level overcoming agglomeration problem of CNTs. The fabricated CNT/Inorganic nanocomposites show outstanding multifunctional behavior for various applications. For the structural applications of CNT nanocomposites, CNT/Cu nanocomposite could be fabricated with excellent strength, increased by 3 times, and high elastic modulus, increases by 1.6 times, compared to those of Cu. For functional applications of CNT nanocomposites, the CNT/Co nanocomposite can be applied to high-efficiency field emitter for applications as back light unit, field emission display. CNT/Inorganic nanocomposites can be applied to various applications EMI shielding materials and electrode materials for energy storage and conversion.

Keywords: carbon nanotube, nanocomposites, fabrication process, molecular level mixing process

1 INTRODUCTION

Several researchers have attempted to fabricate CNT reinforced metal or ceramic matrix nanocomposite materials by means of traditional powder metallurgy process,[1-3] which consists of mixing CNTs with matrix powders followed by sintering or hot pressing. However, these attempts were not successful to fabricate CNT/Inorganic nanocomposites with homogeneously dispersed CNTs in the matrix. This is mainly due to strong agglomeration of CNTs in powder forms: the van der Waals forces between CNTs cause them to mutually attract each other rather than homogeneously disperse. Furthermore, if CNT/Inorganic nanocomposites are manufactured by the conventional process, most of the CNTs are preferentially located on the surfaces of the metal or ceramic powders after mixing.[1-3] The conventional process inhibits the diffusion of matrix materials across or along the powder surfaces; hence, sintering cannot proceed without damaging the CNTs or removing them from the powder surfaces. Even if sintering is successful, CNTs are mostly located at grain boundaries of the matrix and are insignificant in improving material performance. At the same time, the most important processing issue is how to obtain good interfacial bonding between carbon nanotubes and matrix. In the case of CNT/polymer nanocomposites, the interfacial strength between the CNTs and the polymer matrix is strong because they interact at molecular level.[4] In the case of CNT/metal or CNT/ceramic nanocomposites, however, the interfacial strength cannot be expected to be strong because the CNTs and the matrix are merely blended.

2 EXPERIMENTAL PROCEDURES

The strategy for developing a novel fabrication process for CNT/Inorganic nanocomposite basically involves molecular level mixing of the reinforcement(CNT) and the matrix material in a solution instead of the conventional powder mixing. This new process produces CNT/Inorganic

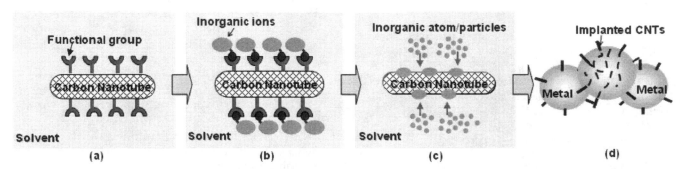

Fig. 1. Schematics depicting strategies and procedures for the molecular level mixing process, (a) functionalization of carbon nanotube, (b) reaction between the ions and the functional group on the carbon nanotube surface, (c) nucleation and growth of inorganic particles by reduction or solvent evaporation, (d) carbon nanotube/metal nanocomposite powders in which carbon nanotubes are homogeneously implanted.

composite powders where the CNTs are mainly located within the inorganic powders rather than on surfaces of them; the chemical bonding between the CNTs and the inorganic matrix ions provide homogeneous distribution of CNTs as well as high interfacial strength between CNT and inorganic matrix.

The molecular level mixing process for fabricating CNT/Inorganic composite powders consists of 4 steps. First, CNTs are dispersed in a solution, to make a stable suspension by attaching functional groups on the CNT surfaces. (Fig. 1a) There are several chemical methods for attaching functional groups on the CNT surfaces.[5] Once the functional groups are attached on the CNTs, the electrostatic repulsive force between the CNTs could overcome the Van der Waals force to form a stable suspension within the solvent. Second, a soluble salt containing matrix ions is dissolved in the CNT suspension. Sonication treatment is introduced to disperse the inorganic matrix ions among the suspended CNTs and to promote chemical reaction between the ions and the functional groups on CNT surfaces. The third step is to dry the solution consisted of CNTs and ions. During this process, the solvent and ligands are removed and the inorganic matrix ions on CNTs are oxidized to form powders. The fourth step is calcination and reduction processes to obtain chemically stable crystalline nanocomposite powders. The nanocomposite powders obtained in the third step are generally existed as oxides. These powders are changed into CNT/Metal nanocomposite powders by a reduction process. The reduced CNT/Inorganic nanocomposite powders show that the CNTs are homogeneously implanted in inorganic matrix as shown in Fig. 1d.

3 RESULTS AND DISCUSSION

3.1 Microstructure of CNT/Inorganic Nano-composites

The CNT/Cu nanocomposite powders fabricated by molecular level mixing process consisted of homogeneously dispersed CNTs within Cu powders as shown in Fig. 2. The most important feature of this process is that CNTs and matrix ions are mixed each other at molecular level. That is, the CNTs are located within the powders rather than on the powder surfaces. The morphologies of the CNT/Cu and CNT/Co powders show an ideal composite microstructure, which displays that CNTs are homogeneously implanted in the powders (Fig. 2).

The CNT/Metal nanocomposite powders fabricated by the molecular level mixing process was consolidated into bulk CNT/Metal nanocomposite by spark plasma sintering process, which can produce a high heating rate of 100°C/min and rapid consolidation through high joule heating and generation of spark plasma between powder-to-powder contacts. The consolidated CNT/Metal nanocomposite shows homogeneous distribution of carbon nanotubes within the matrix (Fig. 3).

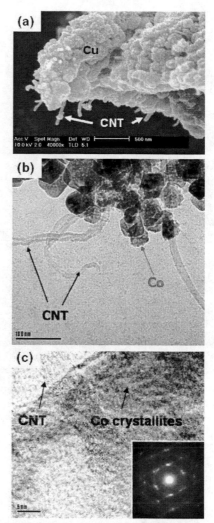

Fig. 2. Microstructures of carbon nanotube/metal nanocomposite powders, (a) SEM image of implanted type carbon nanotube/Cu nanocomposite powders [6], (b) TEM image of necklace type carbon nanotube/Co nanocomposite powders, (c) HRTEM Image of necklace type carbon nanotube/Co nanocomposite powders.[8]

3.2 Mechanical Properties of CNT/Cu Nano-composites

The mechanical properties of CNT/Cu nanocomposite were characterized by compressive test. As shown in Fig. 4a, the compressive yield strengths of CNT/Cu nanocomposites were much higher than that of Cu matrix, which is fabricated by the same process without adding CNTs. 5 volume percent CNT reinforced Cu matrix nanocomposite shows yield strength of 360MPa, which is 2.3 times higher than that of Cu. In the case of 10 volume percent CNT reinforced Cu, the yield strength is 485MPa, which is more than 3 times higher than that of Cu. Moreover, Young's modulus of CNT/Cu nanocomposite increases as the volume fraction of carbon nanotubes is increased, as shown in Fig. 4b[6].

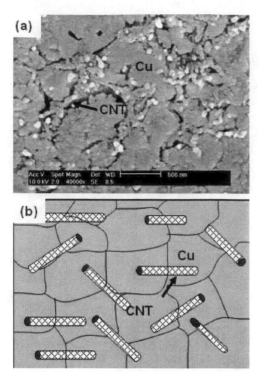

Fig. 3. Microstructure of sintered CNT/Cu nanocomosite (a) SEM microstructure of CNT/Cu nanocomposite showing homogeneous distribution of CNTs within Cu matrix, (b) schematics microstructure of CNT/Cu nanocomposites.

The remarkable strengthening effect of CNTs in CNT/Cu nanocomposite was due to a high load-transfer efficiency caused by strong interfacial strength between CNTs and Cu, which originated from strong chemical bonds formed during the molecular level mixing process. The strengthening efficiency (R) of reinforcement can be expressed as

$$R = (\sigma_c - \sigma_m) / V f \sigma_m$$

R : Strengthening efficiency of reinforcement
σ_c: yield strength of composite
σ_m: yield strength of matrix
V_f: volume percent of reinforcement

The strengthening efficiency, defined as the strengthening effect of a given volume percentage of reinforcement on the matrix, for carbon nanotubes is much higher than those of SiC particles or SiC whiskers, which are the most widely used reinforcements for metal matrices.[6] This indicates that carbon nanotubes are the most effective reinforcements. Such a high strengthening effect of CNTs has been confirmed in other CNT/Inorganic nanocomposites such as CNT/Co nanocomposite[7] and CNT/Alumina nanocomposite[9] fabricated by molecular level mixing process.

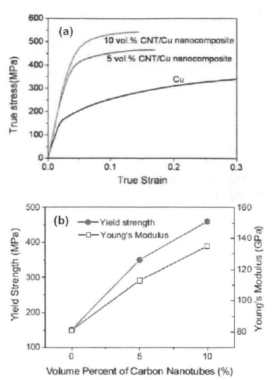

Fig. 4. Mechanical properties of CNT/Cu nanocomposites. (a) the stress-strain curves of CNT/Cu nanocomposites obtained by compressive test, (b) yield strength and Young's modulus of CNT/Cu nanocomposites according to the volume percentage of CNTs[6].

3.3 Electrical Properties of CNT/Co Nanocomposites

CNT/Metal nanocomposites can be applied to not only structural material but also functional materials such as field emitter. The CNT-implanted Co nanocomposite field emitter is fabricated from necklace type CNT/Co nanocomposite powders by a screen-printing process followed by a sintering process as shown in Fig 5a. Before sintering, the necklace type CNT/Co nanocomposite powders are buried in organic binders (Fig. 5b), which are thermally decomposed during the sintering process (Fig. 5c). During the sintering process, the Co nanoparticles are sintered together and form a dense metallic layer in which the CNTs are implanted. CNTs are straightened and aligned perpendicular to the substrate so that they stand upright on the surface of metallic layer as shown in Fig. 5d. The CNTs tend to be aligned perpendicular to the substrate because the base of the CNT is implanted in the Co metal layer during the sintering process.[8]

CNT/Co nanocomposite field emitter, fabricated by sintering of CNT/Co nanocomposite powders, shows good field emission properties with low turn-on field of 1.28V/μ m, high current density of 4.5mA/cm² at 3V/μ m and homogeneous field emission as shown in Fig. 6b. Good field emission properties were due to low electrical resistivity by strong interfacial bonding between CNTs and Co and homogeneous dispersion of CNTs in Co matrix.

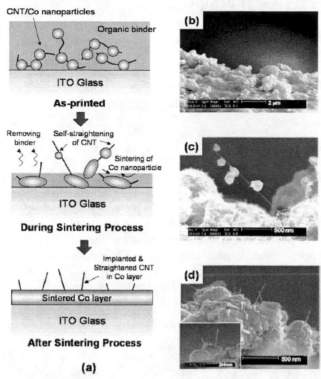

Fig. 5. Fabrication process of CNT/Co nanocomposite field emitters. (a) Schematic depiction of the fabrication process and formation mechanism for CNT-implanted Co nanocomposite emitters. (b) Cross-sectional SEM image of screen-printed necklace type CNT/Co powders with an organic binder, (c) SEM image showing Co nanoparticles threaded by a straight CNT after sintering, and (d) SEM images of the CNT-implanted Co nanocomposite emitter after sintering.[8]

4 CONCLUSIONS

The critical issues on fabrication of CNT/Inorganic nanocomposites are homogeneous dispersion of CNTs and strong interfacial bonding between CNTs and matrix. Molecular level mixing process is a novel process to solve critical issues. CNT/Inorganic nanocomposites with homogeneously dispersed CNTs within the matrix were fabricated and showed highly enhanced mechanical properties by effective load transfer and excellent field emission properties by low electrical resistivity due to strong interfacial bonding between CNTs and the matrix. It is expected that molecular level mixing process can contribute not only for development of high strength/modulus structural components but also for development of various functional materials such as field emitters, EMI shielding materials and electrode materials for energy storage and conversion applications.

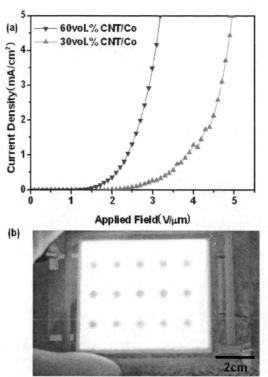

Fig. 6. Fied emission properties of CNT/Co nanocomposites. (a) Field emission curves of 60vol.% CNT/Co and 30vol.% CNT/Co nanocomposite field emitters, (b) field emission image of CNT/Co nanocomposite field emitters.

ACKNOWLEDGEMENT

This research was supported by a grant (code #: 07K1501-00500) from the 'Center for Nanostructured Materials Technology' under the '21st Century Frontier R&D Programs' of the Ministry of Science and Technology, Korea.

REFERENCES

[1] G. D. Zhan, et al., Nature Materials, 2003, 2, 38.
[2] E. Flahout, et al. Acta Materialia, 2000, 48, 3803.
[3] X. Wang et al., Nature Materials, 2004, 3, 539.
[4] A. Hirsch, et al., Nature Materials, 2002, 1, 190.
[5] J. Liu, et al., Science, 1998, 280, 1253.
[6] Cha S. I. et al, Adv. Mater., 2005, 17, 1377.
[7] Jeong, Y. J. et al., Small, 2007, 3, 840.
[8] Cha S. I. et al., Adv. Mater., 2006, 18, 553.
[9] S. I. Cha, et al., Scripta Mater., 2005, 53, 793.

Mechanical properties under nanoindents in gold-reinforced poly(vinyl alcohol) nanocomposite of free standing films

P. Tripathy[1]*, A. Mishra[1], S. Ram[1] and H. -J. Fecht[2]

[1]Materials Science Centre, Indian Institute of Technology, Kharagpur-721 302, India
[2]Werkstoffe der Elektrotechnik, Universität Ulm, Albert Einstein Allee-47, Ulm, D-89081, and Forschungszentrum Karlsruhe, Institute of Nanotechnology, D-76021, Karlsruhe, Germany
*e-mail: tripathypuspanjali@gmail.com

ABSTRACT

Free-standing films of gold nanoparticles reinforced poly(vinyl alcohol) (PVA) nanocomposites, with Au-contents varied from 0.1 to 2.0 wt%, are synthesized by using a novel in-situ chemical method. Microstructure in these films reveal the presence of spheroid or cuboid shaped Au-nanoparticles (NPs) in 0.1 wt% Au-content sample to thin platelets of triangular, square, rectangular, or hexagonal shaped NPs in sample with 2.0 wt% Au-content. The structural anisotropic NPs are formed as a result of the preferential adsorption of PVA on selective Au-crystals facets. In correlation to the microstructures the measurement of reduced elastic modulus (E_r) and hardness (H) under nanoindentation technique show the influence of Au-reinforcement in the virgin PVA film. The H-value changes from an initial value 0.29 GPa in the virgin PVA to a value 0.23 GPa in a 0.5 wt%, or 0.31 GPa in a 2.0 wt% Au-content. Similarly, the E_r-value decreases from 8.19 GPa in the PVA film to that of 6.48 GPa in the 0.5 wt% Au-PVA films and then results in an increase of the final value up to 7.3 GPa in the 2.0 wt% Au-PVA films.

Keywords: Gold nanomaterials, Polyvinyl alcohol, Nanocomposites, Mechanical properties, Nanoindentation.

INTRODUCTION

In recent years, the physical and chemical properties of metal-polymer nanocomposites have attracted much attention for their promising applications in catalysis, electronics, and photonics technologies [1-3]. Such nanocomposites not only recombine the advantageous properties of the metals and polymers but also offer many new characters that are often missing in a single phase material. The intrinsic properties of a metal nanoparticle are mainly determined by its size, shape, composition, crystalinity, and structure. Under certain conditions of morphological anisotropy, gold nanoparticles exhibit anisotropic optical absorption properties associated with the collective oscillation of conduction electrons known as the surface plasmon resonance. The uniqueness of Au-NPs and the unique inorganic-organic interactions of them with certain polymer molecules have motivated us to synthesize Au-NPs of different sizes and shapes in polymer nanofluids and Au-polymer nanocomposite films of selective compositions. In part of our extensive studies of the optical, electrical, and mechanical properties in the Au-PVA free-standing nanocomposite films here in this investigation, we report the preliminary results of the mechanical properties such as hardness (H) and reduced elastic modulus (E_r) under nanoindentation technique.

One of the key areas of interest in the study of nanocomposites is the characterization of their mechanical properties at the macro-, micro- or nano-levels. The mechanical property measurement is very much essential for proper design and device fabrication using the Au-PVA optical films. Traditional mechanical test methods are currently not desirable options since they require a larger size specimen (i.e., more material of the nanocomposite) than is normally available due to their limited or one of a kind fabrication method. Hence, a desirable method such as nanoindentation is one which would require a small size specimen or a limited quantity of material for characterizing mechanical behavior from a small quantity of available material [4,5]. Nanoindentation offers a unique method by which the in situ properties of a nanocomposite may be probed. Using this technique, mechanical properties such as elastic modulus, hardness, and fracture toughness can be determined from the simple indentation load–displacement relationship at the micro or nanoscale without imaging the indentation.

Small metal particles as filler support a large interfacial area in the composite system. The surface-interface controls the degree of interaction between two components and thus controls the final properties of the nanocomposite. Geometry of the sample as films, with a molecularly oriented structure of polymer, governs the microstructure and other properties of interest. In this work of Au-PVA nanocomposite films, the changes in the E_r and H-values with Au-content indicate the role of the Au-reinforcement in modifying such properties in virgin PVA. The results useful for fabricating stable optical films for possible applications are analyzed in correlation to the microstructure.

EXPERIMENTAL DETAILS

PVA (average molecular weight 72000, and fractional hydrolysis 98%) solution was prepared by heating and stirring a 3 g/dl PVA in distilled water at 50-60°C. An aqueous $HAuCl_4 \cdot 3H_2O$ (0.05 M) was added to the PVA solution drop by drop at the same temperature as used in preparing a polymer precursor solution of reactive PVA molecules with refreshed surfaces. The Au-contents were varied from 0.1 to 2.0 wt%, in the Au-PVA composition series in an attempt to exploring mechanically reliable, stable, and practically useful optical films in a semiconductor of hybrid polymer nanocomposite structure. Changing of an achromatic color in the virgin PVA solution in the beginning to a light blue, a bluish violet, a purplish red or to the equilibrium faint yellowish according to the final Au/PVA ratio ensured a caloric $Au^{3+} \rightarrow Au$ reaction. The sample was kept for 20-30 h of aging and then excess water was evaporated at bit high temperatures 60-70°C to cast films in specific moulds of a silicate glass. The films could be pilled off easily without any surface roughness and used for the microstructure and mechanical properties measurements.

A scanning electron microscope (SEM) (Oxford model Leo 1550 of accelerating voltage 10 kV) was used to study size and shape in nanocrystals in the films. The emission spectra of Au-PVA films were studied using a Perkin Elmer Model-LS 55 luminescence spectrometer, with a pulsed xenon lamp of excitation source (20 kW power). The mechanical properties of these free standing films were measured using a nanoindenter (Nanoindenter XP with Berkovich triangular diamond indenter). The force required to press the sharp a sharp diamond indenter into tested material was recorded as a function of indentation depth. During an indentation, corresponding values of load and displacement of the diamond tip (indentation depth) were recorded, and from the resulting curve the H and the E_r-values were calculated.

RESULTS AND DISCUSSION

A mechanochemical stretching in stirring under heating conditions dispersed PVA molecules get as thin as molecular layer [6,7]. PVA molecules of such enlarged surfaces (involve refreshed OH groups free from the H-bonding) serve as a reducer to induce a surface enhanced $Au^{3+} \rightarrow$ Au reaction [8-10]. It occurs in templates of such PVA molecules. As the Au atoms mainly confine to the vicinity of the template surface, once the concentration of them reaches a critical value, they nucleate and grow as nanocrystals (NCs) in a core-shell structure with the surface oxidized PVA molecules.

Fig. 1 shows the SEM images in (a) 1.0 and (b) 2.0 wt% Au-PVA free standing films. In 1.0 wt% Au-PVA film, Au NCs of near cuboids or spheroids (80-100 nm diameters d) are observed. A magnified region of the film for visual clarity is given in the inset. Whereas thin platelets

(20-30 nm thickness δ) of platelets triangular, pentagonal and hexagonal shapes, with average width β = 300-700 nm are observed in 2.0 wt% Au-PVA sample as shown in Fig. 1b. Also there are few cuboids with some typical rods, bowels, or tea-cup shaped particles. Multi-shaped particulates as observed here are useful for designing reinforced composites of superior optical and other useful properties.

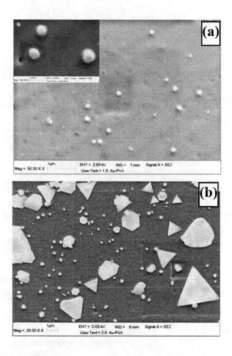

Fig. 1. SEM images in (a) 1.0 and (b) 2.0 wt% Au-PVA films. A magnified image of a selected area in Fig. 1a is given in the inset.

The emission in 2.0 wt% Au-PVA film occurs over shorter wavelengths 400-600 nm to the absorption region, with three to four distinct bands as given in the Fig. 2. It compares 440 nm emission observed in Au-nanocolloids (~ 5 nm size) in water [11]. We measured the sample by irradiating under identical conditions at excitation wavelength λ_{ex} = 372 nm by a xenon lamp. The λ_{ex}-value 372 nm is chosen according to the maximal excitation of the emission as demonstrated with a typical excitation spectrum in the inset of Fig. 2, which corresponds to an average 475 nm emission value in a 2.0 wt% Au-sample. A multiplet band structure of excitation spectrum (of four bands 266, 342, 372, or 400 nm) is very similar to the emission bands.

The four bands, which for example have 412, 437, 468 and 500 nm values in 2.0 wt% Au-PVA sample, include part of the surface-enhanced vibronic PVA transitions in a complex composite system. The first two bands ascribe the $5d^{10}6s^1 \longrightarrow 5d^96s^1p^1$ interband transition (IBT), viz., IBT-I (core) and IBT-II (shell) bands, while the other two ones refer to a vibronic band 2905 or 2885 cm^{-1}, i.e., the C-H stretching vibration ν (CH) of 2917 cm^{-1} in the IR spectrum [6]. It confirms that the PVA molecules extend a short of chemical bonding to the Au-metal surface of a metal-polymer complex (shell). A vibronic band thus occurs in the electron-phonon coupling in a core-shell structure. Thin platelets, which share large interfaces in such specific structure, favor an intense emission in such multiplet bands.

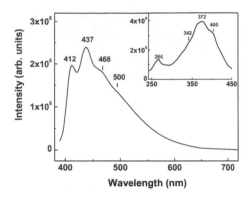

Fig. 2. An emission spectrum in 2.0 wt% Au-PVA film with λ_{exc} = 372. The inset represents a typical excitation spectrum for average emission band 475 nm in the sample.

To observe the mechanical reliability of these optical Au-PVA films, the load-displacement curves were studied under nanoindenter. The load-displacement curves for 0.1 to 2.0 wt% Au-PVA samples along with virgin PVA film are shown in Fig. 3. Quantitative analyses of the each load-displacement indent cycle of mechanical property variation determine the H and E_r of the Au-PVA films. Indentations were performed in load-control mode to a load as high as 1000 μN. Evidently, no fracture is observed during loading.

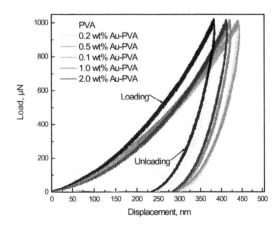

Fig. 3. Load–displacement curves for PVA and Au-PVA films. The film compositions are given in the plot as inset.

Several models have been proposed to provide the mechanical properties of materials using the data from nanoindentation tests by Doerner and Nix [12], Cheng and Cheng [13], Oliver and Phar [5] etc. Liu et al. in 2006

proposed a model based on the Burgers viscoelastic concept to describe the nanoindentation behaviors of polymeric materials [14]. In our case also the viscoelasticity behaviors of the Au-PVA nanocomposite films can be confirmed from Fig. 3. A nose like shape is observed during the unloading in all the samples indicates the decrease in the viscosity parameter as proposed by Liu et al.

Both E_r and H can be readily extracted directly from the nanoindentation curve [5,12-14]. E_r is determined based on the knowledge of the tip shape function (A) and the load-displacement curve (load P and displacement h) [5,14]. The E_r value accounts for the fact that the measured displacement includes contributions from both the specimen and the indenter.

$$E_r = \frac{\sqrt{\pi}}{2} \cdot \frac{dP}{dh} \cdot \frac{1}{\sqrt{A}} \qquad (1)$$

Here, dP/dh is the slope of the unloading curve (Fig. 3). The H-value was calculated using the relation between indention load and projected contact area as follows.

$$H = \frac{P_{max}}{A_{prjected}} \qquad (2)$$

The variation of H and E_r -value as a function of selective Au-contents was plotted in Fig. 4.

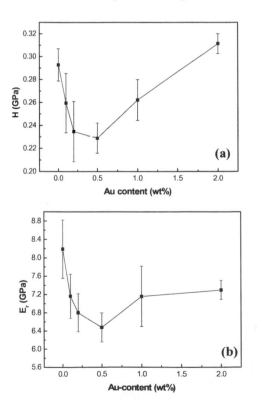

Fig. 4. Variation of (a) hardness H and (b) reduced modulus E_r as a function of selective Au-contents.

The Au-reinforcement is found to influence very sensitively both the Er and H values. A nonlinear variation is observed in the H-value while increasing Au-content in the film. The H-value changes from an initial value 0.29 GPa in the virgin PVA to a value 0.23 GPa in a 0.5 wt%, or 0.31 GPa in a 2 wt% Au-content (Fig. 4a). The H attains a value as low as 0.23 GPa in a 0.5 wt% Au-PVA. The morphology of the Au in PVA for this particular Au-content leads to the minimum value of the H-value in the film. After 0.5 wt% of Au in Au-PVA, the H-value instantaneously increases and attains a maximum in 2.0 wt% Au-content. Similarly, the E_r -value decreases from 8.19 GPa in the PVA film to that of 6.48 GPa in the 0.5 wt% Au-PVA films and then results in an increase of the final value up to 7.3 GPa in the 2 wt% Au-PVA films (Fig. 4b). These results can be correlated to the microstructures obtained in the composite films.

Interestingly, both the H and E_r -values decrease up to 0.5 wt% Au-content with the formation of nearly spherical shaped Au-NCs, but above that there is an increase in the hardness with the formation of polygonal Au-platelets in the nanocomposite structure. Also beyond 0.5 wt% Au-content the E_r -value of the film increases but it is lesser than that of the virgin PVA film. These changes in the E_r and H-values indicate the role of the Au-reinforcement in. modifying such properties in PVA and similar linear polymers. The Au-polymer surface interfaces in these examples seem to be one of the sensitive chemical parameters, which are responsible for the variation in the physical properties as a function of the Au-content.

CONCLUSIONS

Au-nanoparticles doped poly(vinyl alcohol) (PVA) composite films are synthesized with selective Au-contents from 0.1 to 2.0 wt% and their mechanical properties are studied under nanoindentation technique. The changes in the E_r and H-values indicate the role of the Au-reinforcement in modifying such properties in PVA and similar linear polymers. These results are useful for fabricating stable optical films for possible applications.

ACKNOWLEDGEMENTS

University Grant Commission (UGC), Government of India, is acknowledged for providing the research fellowship.

References

[1] R. C. Hayward, D. A. Saville and I. A. Aksay, Nature 404, 56, 1999.

[2] J. Zhong, W. Y. Wen, and A. A. Jones, Macromolecules 36, 6430, 2003.

[3] M. -C. Daniel and D. Astruc, Chem. Rev. 104, 293, 2004.

[4] A. C. Fischer-Cripps, Nanoindentation. New York: Springer-Verlag; 2002.

[5] W.C. Oliver and G.M. Pharr, J. Mater. Res. 19, 3, 2004.

[6] S. Ram and T. K. Mandal, Chem. Phys. 303, 121, 2004.

[7] A. Gautam, P. Tripathy, and S. Ram, J. Mater. Sci. 41, 3007, 2006.

[8] P. Tripathy, S. Ram, and H. -J. Fecht, Plasmonics 1, 121, 2006.

[9] S. Ram, P. Tripathy, and H. -J. Fecht, J. Nanosci. Nanotech. 7, 3200, 2007.

[10] P. Tripathy, A. Mishra, and S. Ram, Mater. Chem. Phys. 106, 379, 2007.

[11] J. P. Wilcoxon, J. E. Martin, F. Parsapour, B. Wiedenman and D. F. Kelley, J. Chem. Phys. 108, 9137, 1998.

[12] M. Doerner and W. D. Nix, J. Mater. Res. 1, 601, 1986.

[12] Y.-T. Cheng and C.-M. Cheng, Appl. Phys. Lett. 73, 614, 1998.

[13] C.-K. Liu, S. Lee, L.-P. Sung, and T. Nguyen, J. Appl. Phys. 100, 035503, 2006.

[14] Q. K. Jiang, X. P. Nie, J. Z. Jiang, N. Deyneka-Dupriez, and H. -J. Fecht, Scripta Materialia 57, 149, 2007.

A comparison of boron hydride- and hydrocarbon-based thiol derivatives assembled on gold surfaces.

T. Baše[*], Z. Bastl[**], M.G.S. Londesborough[*] and J. Macháček[*]

[*]Institute of Inorganic Chemistry of the Academy of Sciences of the Czech Republic, v.v.i. 250 68 Husinec-Řež, č.p. 1001, Czech Republic, tbase@iic.cas.cz, michaell@iic.cas.cz, jmach@iic.cas.cz
[**]J. Heyrovský Institute of Physical Chemistry of the Academy of Sciences of the Czech Republic, v.v.i.182 23 Prague 8, Dolejškova 2155/3, zdenek.bastl@jh-inst.cas.cz

ABSTRACT

The elements carbon and boron uniquely form extensive series of compounds with hydrogen; hydrocarbons and boron hydrides. In comparison to carbon, which forms various chain- or ring-like structures, boron, an electron deficient element, forms *quasi*-aromatic clusters architecturally derivable from a twelve-vertex icosahedral unit. Herein, we focus on a comparison of hydrocarbon and boron hydride based systems anchored to a gold surface via one or more thiol groups. This study includes the aspects of bonding, electrochemistry, and thermal, radiation and oxidation stabilities.

Keywords: gold, boron hydride, carborane, hydrocarbon, thiol

1 INTRODUCTION

Hydrocarbon chemistry plays an important role in our everyday life. It is a chapter of chemistry that is well understood, and many books, articles and reports have been addressed to its principles and applications [1]. Boron hydrides do not occur naturally and therefore its chemistry is a product of human curiosity and scientific investigation. The structures formed by boron hydrides are dominated by the fact that atoms of boron have one electron less than do atoms of carbon whilst having the same number of electronic orbitals. This fact has led to boron being sometimes referred to as electron deficient [2]. The electron-precise nature of carbon and the electron-deficient character of boron is responsible for most of the structural differences between their respective hydride series. For example, the bonds between sp3-hybridised carbon atoms in an alkane chain can rotate freely and thus give the molecule a relatively high degree of flexibility for adopting various conformations. In an attempt to share electron density boron hydrides form clusters, based on triangular 3-centre, 2-electron bonds, which lead to generally more rigid structures. Such clusters can be assigned to three basic classes according to the rules of Wade and Williams [3]. The comparison of the selected thiolated representatives of the BH and CH series of compounds assembled on gold surfaces brings several interesting revelations. The thiol derivatives of both can be easily assembled onto gold surfaces either from a solution or gas phase to form a densely packed monolayer. Despite the fact that alkanes can take up various conformations, they are reported to preferably form a straight-line arrangement when anchored via thiol groups to a gold flat surface [4]. However, disorders can be found in the form of various ball-like structures generated spontaneously by some of the alkyl chains. The straight-line assembly of alkanethiols, sometimes compared to a crystalline phase, is most likely favored due to attractive van der Waals forces between individual chains. Disorders similar to those in the monolayers of alkanethiolates are not likely to occur with the thiolated boron hydride clusters because of their steric demands. This feature typical of clusters allows us to precisely orient various functional groups at a gold surface, and construct more sophisticated assemblies without any predictable imperfections on very large areas of the surface. Figure 1 shows molecules of $1,12-(HS)_2-1,12-C_2B_{10}H_{10}$, and octanethiol ($C_8H_{17}SH$) on a (111) gold flat surface.

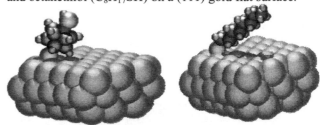

Figure 1. Space filling models of $1,12-(HS)_2-1,12-C_2B_{10}H_{10}$ and octanethiol ($C_8H_{17}SH$) on a gold (111) surface.

2 RESULTS AND DISCUSSION

On the basis of investigations on thiolated hydrocarbons that have been described in the last three decades [5], and information that we have collected on thiolated BH clusters [6], we would like to report on a comparison of several aspects associated with these two classes of compounds. After brief consideration of the relevant differences between BH and CH chemistries, we would like to focus on the aspects of bonding and the character of the thiol groups attached to either hydrocarbons or boron hydrides, thermal stabilities and desorption experiments from gold nanoparticle surfaces, oxidation and radiation stabilities, and electrochemistry.

2.1 The aspects of bonding

It is recognized that alkanethiols bind to a gold flat surface as alkanethiolate units [7]. The value of binding energy for the S2p electrons is ~ 162.0-162.2 eV. To better understand this aspect, however, it is worthwhile to take a deeper look into the character of a thiol group and, in particular, its acidity. The acidity of a thiol group, respectively the value of pK_a, attached to an alkyl chain is ~ 10-11. The value of pK_a of a single thiol group attached to aromatic hydrocarbons is ~ 7-8. This decrease in basicity seen for thiolated aromatic compounds is due to the stabilization of the conjugate base by the electronically delocalized aromatic system. This means, that thiolated hydrocarbons are generally considered as relatively weak acids, and, within this group of compounds, the stronger acids are the aromatic thiols. Quite differently, if we look at the carborane skeleton, the value of pK_a depends on the vertex to which a thiol group is attached, if it is either to a boron or carbon vertex. Figure 2 shows two dithiol derivatives of $1,2-C_2B_{10}H_{12}$. The thiol groups are attached to the carbon atoms, indicated with grey arrows, in $1,2-(HS)_2-1,2-C_2B_{10}H_{10}$, and to boron atoms, marked with green arrows, in $9,12-(HS)_2-1,2-C_2B_{10}H_{10}$.

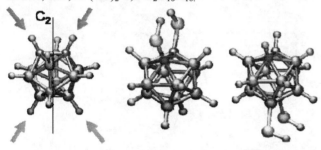

Figure 2. A cluster molecule of $1,2-C_2B_{10}H_{12}$ and its two dithiol derivatives: (from left) $1,2-(HS)_2-1,2-C_2B_{10}H_{10}$ and $9,12-(HS)_2-1,2-C_2B_{10}H_{10}$.

Let us take this icosahedral carborane skeleton as a suitable system for the evaluation of BH vertices, and therefore, the respective B-SH groups. The molecule of $1,2-C_2B_{10}H_{12}$ cluster is a relatively strong dipole with the value of 4.3 D. The selective thiolation of this carborane derivative on either the 1 and 2 carbon vertices, or the 9 and 12 boron positions has been described in the literature [8]. Both B-SH and C-SH dithiol derivatives of the $1,2-C_2B_{10}H_{12}$ have also been used for the purposes of a gold surface modification [6]. Table 1 lists the values of pK_a, and the binding energies of S2p electrons. The sulphur atoms in $9,12-(HS)_2-1,2-C_2B_{10}H_{10}$, that are attached to a gold film, bear a higher negative charge in comparison to both $1,2-(HS)_2-1,2-C_2B_{10}H_{10}$, and alkanethiols attached to a gold surface. This can potentially have an impact on the stability of this particular derivative.

Table 1. The pK_a values and the binding energies of S2p electrons (eV) for the selected representatives of thiolated hydrocarbons and boron hydrides.

	pK_1 , pK_2	S2p (BE)
$1,2-(HS)_2-1,2-C_2B_{10}H_{10}$	4.47 , 8.87	162.2[6]
$9,12-(HS)_2-1,2-C_2B_{10}H_{10}$	5.5 , 10.45	161.7[6]
$1-(HS)-1,2-C_2B_{10}H_{11}$	3.30[8]	-
$9-(HS)-1,2-C_2B_{10}H_{11}$	3.30[8]	-
1-octanethiol ($C_8H_{17}SH$)	~ 10-11	162.2[9]
1-butanethiol (C_4H_9SH)	~ 10-11	162.0[10]
Benzenethiol (C_6H_5SH)	~ 7-8	161.9

2.2 Thermal stability

Using the technique of electron ionization in a heated inlet of a mass spectrometer, we were able to make some qualitative assessments about the desorption of both the carboranethiolate and alkanethiolate species from the surface of gold nanoparticles. The samples suitable for these experiments were prepared using the two-phase method [11]. The average size of the nanoparticles is ~ 2-5 nm. The solid products, as prepared, had black waxy appearances. However, the carboranethiol-stabilized particles were dissolvable in acetone, and can be purified by chromatography on a silica-gel column, using acetone as the eluting agent. To understand the material before and after the chromatography purification, we carried out desorption experiments for both of these samples, and a typical desorption curve with mass spectra at two points is shown in Figure 3. The desorbing species were identified as $C_2B_{10}H_{10}$ clusters and, as shown in the second mass spectrum, tetraoctylammonium with its characteristic fragmentation.

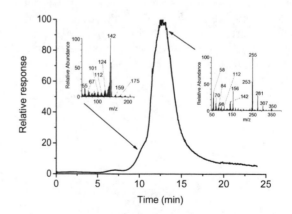

Figure 3. Total ion current profile of the desorbing molecules from the surface of gold nanoparticles before chromatography purification, obtained with a temperature program starting at 35 °C (0 min) followed by 20 °C/min increase to 280 °C (12.25 min), where it was held to the end of analysis.

After chromatography samples did not exhibit the waxy character as before, and the mass spectra did not show any fragments characteristic of tertaoctylammonium. It is worthy to note that only $C_2B_{10}H_{10}$ clusters were observed to escape from the surface, and, from this view, gold nanoparticles were stabilized purely by carboranethiol derivatives. Further details of the desorption experiments of the chromatography purified carboranethiol-stabilized nanoparticles were described recently [6]. Indeed, all attempts to purify the alkanethiol-stabilized gold nanoparticles were not successful, and therefore, the fragments characteristic of tertaoctylammonium ions are observed in all desorption curves apart from the desorption products originating from the molecules of alkanethiolate units. These latter species escape from the surface as dialkydisulfide molecules, and leave the surface of gold nanoparticles bare. As a consequence of the desorption process, the aggregation of 'naked' nanoparticles is observed, and the desorption product is not dissolvable in any solvent. At this point, it is worthy to note another difference between the hydrocarbon- and boron hydride-based thiol derivatives. The carboranethiolate units escape from the surface as carboryne $C_2B_{10}H_{10}$ clusters, and leave the atoms of sulphur atoms on the surface. This means that the carborane-S-Au bond is cleaved at between the sulphur and carborane cluster. In comparison to alkanethiolates on gold, the escaping 'carboryne' clusters can be stabilized by their *quasi*-aromatic character. Also the formation of a disulphide molecule during the desorption is less probable because of steric demands of the vicinal dithiolcarborane These sulphur atoms can henceforth stabilize the surface of gold nanoparticles, and the desorption product is still nicely dissolvable to a colloidal solution. Methanethiol stabilized gold particles were prepared for the purposes of a comparison. Figure 4 shows that the methanethiolate species escape from the surface as dimethyldisulphide molecules. Therefore, these molecules do not leave sulphur atoms on the surface as carboranethiol derivatives. Also the desorbing temperature is lower by approximately 35 °C than with carboranethiol derivatized gold nanoparticles.

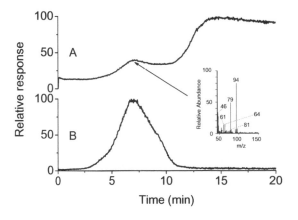

Figure 4. (A) Total ion current profile of the desorbing molecules from the nanoparticle surface stabilized by methanethiol. The mass spectrum indicates species that escape from the surface at 185 °C. (B) Extracted profile for the ion-radical m/z = 94.

2.3 Oxidation and radiation stability

The carboranethiol derivatives are white solids, which can be purified by sublimation, and which are soluble in various organic solvents including for example hexane, chloroform, acetone, or ethanol. To dissolve these derivatives in water, however, it is necessary to prepare the respective thiolate salts with for example sodium (Na$^+$) counter ion. The sodium carboranethiolate salts can oxidize to disulphide molecules in an aqueous solution because water usually contains traces of dissolved oxygen. Figure 5A shows the XP spectrum of S2p electrons in a solid disodium salt of $1,12-(S^-)_2-1,12-C_2B_{10}H_{10}$. The spectrum shows two components. The major one corresponds to thiolate sulphur atoms with a binding energy of 161.2 eV, and the minor component has a binding energy of 163.0 eV, which can be attributed to sulphur atoms typical of a disulphide molecule. The product of this partial oxidation is shown in the inset of Fig. 5. This oxidation occurs also on a gold surface at ambient conditions, and Figure 5B shows that the disulfide component becomes major.

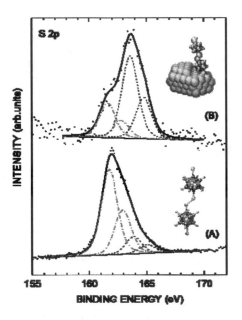

Figure 5. XP spectra of S2p electrons for a (A) solid disodium salt of $1,12-(S^-)_2-1,12-C_2B_{10}H_{10}$ and (B) $1,12-(HS)_2-1,12-C_2B_{10}H_{10}$ assembled on a gold surface. There are two components with binding energies of 161.2 eV, 163.0 eV (A), and 161.6 eV, 163.6 eV (B).

The carborane cluster species are extremely stable toward heating. They do not decompose at temperatures of up to 700 °C. This behavior has been discussed previously in the literature. Also, these species do not exhibit any decomposition upon their X-ray [6] or radioactive irradiation [12]. Compounds based on boron hydrides have been studied and successfully applied for the retreatment of nuclear waste.

2.4   Electrochemistry and surface pasivation

The electrochemical investigation provides us useful information about the interface between gold and the environment of an electrolyte, usually an aqueous solution. A redox system based on $[Fe(CN)_6]^{3+}$ and $[Fe(CN)_6]^{4+}$ complex ions has used to investigate how effectively the alkanethiol or carboranethiol self-assembled monolayers block the access to the surface. Alkanethiol derivatives assembled on a gold surface have been reported to block the surface as a function of the alkyl chain length [4]. The longer the alkyl chain is, the more effective is the passivation of the surface. Thus, the oxidation and reduction Fe^{2+}/Fe^{3+} can easily occur on a surface modified with methanethiol because a molecule of methanethiol has only one carbon atom. The molecules of octanethiol derivative block the surface more effectively. Carboranethiol derivatives can also block the surface of a gold surface. However, in this case, the passivation of a gold film depends on the orientation of the $1,2\text{-}C_2B_{10}H_{12}$ icosahedral cluster [6]. The precise orientation of the cluster changes also the value of work function from the gold surface. The same effect can be achieved only if perfluorinated, or partially fluorinated alkanes are used for the modification of a gold surface.

3   EXPERIMENTAL SECTION

The experimental data including the methods and instrumentation are in detail described in the literature [6]. Desorption experiments were monitored by MS analysis of the gaseous phase evolving from the samples. The time dependency of the desorption experiments, presented in Figures 3 and 4, was obtained by using the following experimental setup: 35 °C (0 min), linear increase of temperature by 20 °C/min up to 280 °C (12.25 min), 280 °C kept till the end of analysis.

4   CONCLUSION

In this report, we compared selected aspects of hydrocarbon- and boron hydride-based thiol derivatives assembled on gold surfaces. Both systems have some features in common, they easily anchor to gold surfaces, and to some extent, block it for further reactions that usually occur on the interface. However, several interesting and specific phenomena were observed with the carboranethiol derivatives. These molecules are more stable toward heating than their alkanethiol counterparts, and exhibit different behavior during heating. Also, they not exhibit any decomposition upon X-ay irradiation. There was not any decomposition observed in the XP spectra even after heating to the temperatures of 400 °C. In comparison

with most of the organic compounds, they have rigid structures derived from an icosahedron. The precise thiolation of its vertices allows us to attach these molecules to gold flat film via either one or two vertices. Therefore, these rigid architectures can be used as potential building blocks for further molecular assemblies.

5   ACKNOWLEDGEMENT

We thank to Dr. Zbyněk Plzák for MS measurements. For financial support, we thank the Ministry of Education, Youth and Sports of the Czech Republic (grant no. LC06041) and the Grant Agency of the Academy of Sciences of the Czech Republic (grant nos. KAN400480701, 1ET400400413, and project nos. AVOZ40320502, AV0Z40400503).

REFERENCES

[1]   G.A. Olah, Hydrocarbon Chemistry, Wiley-Interscience, 2003, Second Ed., New Jersey.

[2]   See for example E.L. Muetterties, Boron Hydride Chemistry, Academic Press 1975, New York.

[3]   See, for example: R. E. Williams, Inorg. Chem., 10, 210-214, 1971. K. Wade, J. Chem. Soc., Chem. Commun. 792-793, 1971. K. Wade, Adv. Inorg. Chem. Radiochem., 18, 1-66, 1976. R. W. Rudolph, Acc. Chem. Res., 9, 446-452, 1976. K. Wade, M. E. O'Neill, Compr. Organomet. Chem., 1, 25-35, 1987.

[4]   M.D. Porter, T.B. Bright, D.L. Allara, C.E.D. Chidsey, J. Am. Chem. Soc., 109, 3559-3568, 1987.

[5]   See for example A. Ulman, Chem. Rev. 96, 1533-1554, 1996 and references therein.

[6]   T. Baše, Z. Bastl, Z. Plzák, et al. Langmuir, 21 (17), 7776-7785, 2005.

[7]   M. Wirde, U. Gelius, T. Dunbar, D.L. Allara, Nuclear Instruments and Methods in Physics Research B, 131, 245, 1997.

[8]   J. Plešek, S. Heřmánek, Collect. Czech. Chem. Commun., 44, 24, 1979. J. Plešek, Z. Janoušek, S. Heřmánek, Collect. Czech. Chem. Commun., 45, 1775, 1980. J. Plešek, S. Heřmánek, Collect. Czech. Chem. Commun., 46, 687, 1981.

[9]   H. Rieley, G.K. Kendall, F.W. Zemicael, T.L. Smith, S. Yang, Lanmuir, 14 (18), 5147-5153, 1998.

[10]   M.-C. Bourg, A. Badia, R.B. Lennox, J. Phys. Chem. C 104, 6562-6567, 2000.

[11]   M. Brust, M. Walker, D. Bethell, D.J. Schiffrin, R. Whyman, J. Chem. Soc., Chem. Commun., 801, 1994.

[12]   Y. Marcus, A.K. Sengupta, Ion Exchange and Solvent Extraction, Vol. 17, J. Rais, B. Grüner, Chapter 5, Marcel Dekker Inc. 2004, New York, Basel.

Effects of the Building Block on the Morphology and Properties of Porous CdSe Nanostructured Framework

Hongtao Yu*, Robert Bellair**, Rangaramanujam M. Kannan** and Stephanie L. Brock*

*Department of Chemistry and **Department of Chemical Engineering and Material Science, Wayne State University, Detroit, USA, sbrock@chem.wayne.edu

ABSTRACT

A major challenge in designing and making potentially valuable nanostructures is to find suitable ways to effectively tune their properties and function. Recent work in our lab has demonstrated a powerful strategy to engineer the morphology and properties of metal chalcogenide 3-D nanostructured assemblies prepared by sol-gel methods. Here we show that by altering the shape of CdSe building blocks from dot to rod, the morphology of the gel can be altered from colloidal to polymeric. Notably, the polymeric (rod) aerogel has twice the surface area of the colloidal (dot) aerogel. Rheological studies of aerogel-PDMS (polydimethylsiloxane) composites indicate that the rod aerogel structure is stronger than the dot aerogel, and the reinforcement is due to the structure of the gel, not the particle geometry. Finally, in addition to dots and rods, the morphological consequences of CdSe branched and hyperbranched nanoparticles on resultant aerogels will also be discussed.

Keywords: CdSe nanodots, CdSe nanorods, morphology, gel stength

INTRODUCTION

Quantum dots (semiconducting nanoparticles) as a novel class of materials, have the potential to reform many current technologies and accordingly have received a great deal of attention from materials scientists. In the past several years, the research priority on semiconducting nanoparticles has changed from studies of size and shape dependent electronic and optical properties to designing more complex nanostructures[1,2] and organizing simple nanometer scale building blocks into functional architectures.[3,4] Recently, a general methodology has been developed in our lab to assemble quantum dots into 3-D porous networks without the presence of intervening surface ligands that can moderate dot-dot interactions.[5] Experimental data suggest that the resultant architectures maintain the crystalline phase and quantum confinement of the building blocks and exhibit a colloidal morphology similar to that of a base-catalyzed silica aerogel.[6] While the quantum dot aerogels exhibit relatively high surface areas, the structures are fragile and their surface areas are considerably lower than traditional silica aerogels. In order to alter and enhance the inherent properties of the metal chalcogenide aerogel framework, including the morphology, gel strength, surface area and porosity, here we employed a new strategy to engineer these radical properties in semiconducting metal chalcogenide aerogels by changing the shape of the building block.[7]

Among the quantum dot systems, CdSe materials have been extensively studied by the scientific community due to the ability to precisely control the size and shape of CdSe nanoparticles and the fact that the optical properties of the particles can be tuned throughout the visible spectrum by varying the size and shape.[8,9] There are two major shapes associated with CdSe nanoparticles: dot and rod. The rod shape demonstrates an anisotropic geometry and is expected to be a more rigid building block, compared with an isotropic dot. On the other hand, compared with an identical-diameter dot, the rod is less quantum-confined. Recent research suggests that the surface free energy of the apexes of CdSe nanorods are more chemically reactive than the facets along the axis due to the presence of more dangling bonds on the surface Cd atoms of the end facets.[10,11] Indeed, it is this difference in reactivity that drives the formation of the nanorod in solution growth.[12] We postulate that when assembling rod-shaped CdSe nanoparticles into an aerogel network, this effect can significantly influence the way these nanorods interconnect and therefore lead to different inherent morphology, surface characteristics and porosity of resultant aerogels relative to analogs composed from spherical particles. We find that the CdSe aerogels assembled from rod-shaped building blocks exhibit a totally different morphology, much higher surface areas and better gel strength when compared with the corresponding dot aerogel.[7]

EXPERIMENTAL

CdSe dot nanoparticles. CdO powder (0.050 g, 0.37 mmole) was added to a mixture of TDPA (n-tetradecylphosphonic acid, 0.20 g, 0.72 mmole) and distilled TOPO (trioctylphosphine oxide, 4.0 g, 10.3 mmole), heated at 150 °C and kept under Ar flow for 30 min to remove residual water. Then the temperature was set to 320 °C and the brownish mixture was left under Ar flow for 6-7 hours, resulting in a colorless solution. The temperature was reduced to 150 °C. A solution containing 0.032 g selenium (0.30 mmole) in 2.5 mL TOP was rapidly injected at 150 °C. The temperature was then raised at a rate of 10 °C per 10 min up to 230 °C, and the solution was kept at this temperature for 4 hours before cooling down to 80 °C. 4 mL of toluene was injected and the particles precipitated with an excess of ethyl alcohol. To purify the particles, the solution was centrifuged and the sediment was re-dispersed in toluene. After a second precipitation with ethyl alcohol, the sediment was kept for further analysis.

CdSe rod nanoparticles. CdO powder (0.15 g, 1.1 mmole) was added to a mixture of TDPA (0.70 g, 2.5 mmole) and distilled TOPO (4.0 g, 10.3 mmole), heated at 150 °C and kept under Ar flow for 30 min to remove residual water. Then the temperature was set to 320 °C and the brownish mixture was left under Ar flow for 4 hours. This step resulted in a colorless solution. The temperature was then reduced to 270 °C. A solution of 0.095 g selenium (1.1 mmole) in 1.5 mL TOP was rapidly injected at 270 °C. Then the temperature was decreased to 230 °C, and kept for 3 hours before cooling to 80 °C. Isolation and purification were conducted as described for dot-shaped CdSe nanoparticles.

MUA capping of CdSe nanoparticles, gelation and aerogel formation. For dot and rod-shaped CdSe nanoparticles, the 11-mercaptoundecanoic acid (MUA) solution was prepared by dissolving 0.8 g MUA (3.3 mmole) in 10 mL methanol solution with tetramethylammonium hydroxide pentahydrate (TMAH) added to achieve a system pH of 10.5-11. The solid precipitate of CdSe nanoparticles was then dispersed in the MUA solution and left stirring in a static Ar environment at 30 °C overnight. At room temperature, an excess amount of ethyl acetate was added to precipitate MUA-capped nanoparticles. The solution was centrifuged and the sediment was re-dispersed in methanol. After a second precipitation with ethyl acetate, the sediment was dispersed in 10 mL methanol for gelation.

Gelation was achieved by adding 10 μL of 3% tetranitromethane (TNM) to 2 mL aliquots of CdSe sols. The mixture was shaken vigorously and subsequently allowed to sit undisturbed for gelation. The resulting wet gels were aged for 7 days under ambient conditions. Aged gels were exchanged with acetone 6-7 times over 3 days and then transferred to a SPI-DRY model critical point dryer where they were subsequently washed and immersed in liquid CO_2 over 6 hours. The CO_2 exchanged gels were dried under supercritical conditions by raising the drier temperature to 39 °C, maintaining that temperature for 30 min, followed by venting of CO_2 gas to obtain CdSe aerogel.

Composite preparation. CdSe/polydimethylsiloxane (PDMS) composites were prepared first by mixing 0.1 g of CdSe aerogel or nanoparticles to excess cyclohexane and sonicating the mixture for 2 minutes. Then a solution of 4.9 g of PDMS in cyclohexane was added into the mixture and stirred for 24 hours, followed by being frozen at -40 °C and drying under vacuum to remove all solvents.

X-ray Powder Diffraction. Powder X-ray diffraction (PXRD) analysis was employed to study the structure, phase and crystallinity of the nanoparticles and resultant aerogels. A Rigaku RU 200B X-ray diffractometer (40 kV, 150 mW, Cu-Kα radiation) with rotating anode was used

for X-ray diffraction measurements. Powdered samples were deposited on a low background quartz (0001) holder coated with a thin layer of grease. X-ray diffraction patterns were identified by comparison to phases in the International Centre for Diffraction Data (ICDD) powder diffraction file (PDF) database (release 2000).

Transmission electron microscopy. Transmission electron microscopy (TEM) was employed to study the three CdSe different building blocks and the morphology of the resultant aerogels. The TEM analyses were conducted in the bright field mode using a JOEL FasTEM 2010 HR TEM analytical electron microscope operating at an accelerating voltage of 200 kV. Particle samples were prepared by depositing a drop of a dilute toluene dispersion of CdSe nanocrystals on carbon-coated copper grids and subsequently evaporating the solvent. Aerogel samples were prepared on carbon-coated copper grids by first grinding the aerogel to fine powders and then pressing the TEM grid onto the dried powder.

Surface area and porosimetry. The surface areas of CdSe aerogels were obtained by applying the BET model to nitrogen adsorption/desorption isotherms. A Micromeritics ASAP 2010 surface area analyzer was used to produce nitrogen physisorption isotherms at 77 K on powdered CdSe aerogel samples. The data were fit by using a Brunauer-Emmett-Teller (BET) model to determine the surface areas of the aerogels. The average pore diameter and cumulative pore volume were calculated by using the Barrett-Joyner-Halenda (BJH) model. Samples were degassed under vacuum at 100 °C for 48 hours prior to the analysis and employed a 30 s equilibrium interval and a 5 cc dose for a total running time of about 13 hours. Three independently prepared samples were analyzed for polymer-type aerogels prepared from rod building blocks, and compared with previously studied colloid type aerogels prepared from spherical building blocks.

Optical absorption and photoluminescence measurements. Optical absorption measurements of MUA capped CdSe nanoparticles in methanol were obtained on a Hewlett-Packard (HP) 8453 spectrophotometer. A dilute CdSe nanoparticle solution was analyzed against a methanol blank in the region from 400 nm to 700 nm. A Jasco V-570 UV/VIS/NIR spectrophotometer equipped with an integrating sphere was used to measure the optical diffuse reflectance of CdSe aerogels. Powdered aerogel samples were evenly spread on a sample holder pre-loaded with a reflectance standard and measured from 200 nm to 1500 nm. The band gaps of the samples were estimated from the onset of absorption in data converted from reflectance.

Emission properties of the CdSe nanoparticle precursors and corresponding aerogels were investigated using photoluminescence spectroscopy. A Cary Eclipse (Varian, Inc.) fluorescence spectrometer with 5 nm excitation and emission slits was used for photoluminescence studies. A dilute MUA-capped CdSe nanoparticle solution in methanol was placed in a 10 mm quartz optical cell and analyses were done under ambient conditions. Powdered aerogel samples were sealed in evacuated quartz tubes and analyses were done at liquid nitrogen temperature. Monolithic aerogel samples were analyzed at room temperature.

RESULTS AND DISCUSSION

CdSe semiconducting nanoparticles were synthesized by typical high temperature colloidal methods to yield highly monodisperse CdSe nanodots and nanorods.[8] These nanoparticles were then surface modified with thiolate ligands by treatment with 11-mercaptoundecanoic acid in the presence of tetramethylammonium hydroxide pentahydrate. Finally, controlled oxidative removal of the surface thiolate ligands led to gel formation, concomitant with the production of disulfide as a byproduct, and supercritical CO_2 drying yielded aerogel monoliths. TEM images in Figure 1 demonstrate the different building blocks and the morphology of the resultant aerogels. The CdSe rod aerogel exhibits a totally different morphology than the colloidal type dot aerogel and is similar to an acid-catalyzed silica aerogel with a polymer-type framework. This confirmed the hypothesis that altering the shape of the building block can be alternative way to change the morphology of resultant aerogels, instead of changing the gelation mechanism as is done with silica.[13] The high resolution TEM images reveal the crystallinity of the building blocks for both types of aerogel (Figure 2). The lattice fringes can be attributed to hexagonal CdSe in all cases, confirmed by powder X-ray diffraction measurements on bulk samples. We postulate that the mechanism for the formation of the polymer morphology involves preferential connectivity of rods at the ends due to the higher surface free energies of the apical facets. This leads to a more penetrating aerogel framework than for the dots.

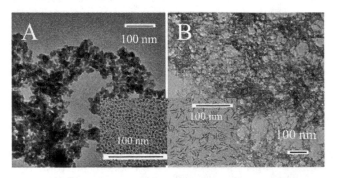

Figure 1: TEM images of A) CdSe dot aerogel, B) CdSe rod aerogel. Insets are different building blocks (dots and rods). The data were reproduced with permission from reference 7. Copyright 2008 American Chemical Society.

The surface area and porosity analyses were conducted with a Micromeritics model ASAP 2010 surface area analyzer using nitrogen physisorption isotherms performed at 77 K on powder aerogel samples. The data in Table 1

show that the CdSe rod aerogels exhibit double the surface area value of the dot aerogels based on the BET model, and over twice the cumulative pore volume using BJH model (based on cylinder pore). This strongly proves that altering the building blocks is an effective way to engineering the critical parameters of quantum dot aerogels. More interestingly, in silica aerogels, the polymer-type aerogel (acid-catalyzed) exhibits lower surface areas than the colloid-type aerogel (base-catalyzed);[14] conversely in CdSe aerogels, the polymer-type rod aerogel shows a much higher surface areas than colloid-type dot aerogel.

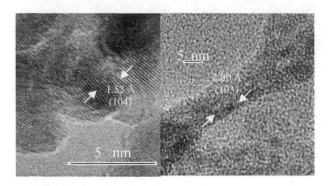

Figure 2: High resolution TEM images of CdSe dot (left) and rod aerogels (right). The data were reproduced with permission from reference 7. Copyright 2008 American Chemical Society.

Table 1: Brunauer-Emmett-Teller (BET) surface areas, Barrett-Joyner-Halenda (BJH) absorption average pore diameters and BJH cumulative pore volumes of CdSe colloidal and polymeric aerogels (average of three independent samples); primary particle size and optical bandgap of the building block and aerogel. The data were reproduced with permission from reference 7. Copyright 2008 American Chemical Society.

CdSe aerogel	colloidal	polymeric
BET surface area (m^2/g)	107 ± 10	239 ± 7
average pore diameter (nm)	35.4 ± 2.4	24.5 ± 1.5
Cumulative pore volume (cm^3/g)	0.59 ± 0.31	1.64 ± 0.11
size of building block (nm)	3.17 ± 0.31 (dot)	$3.4 \pm 0.33 \times 22.7 \pm 2.2$ (rod)
Absorption onset value Primary particle / aerogel (eV)	2.15 / 2.12	2.06 / 2.02

The band gap and emission measurements reveal that both types of CdSe aerogel maintain a nearly identical degree of quantum confinement as seen in the respective building blocks from which they were assembled. This is manifested in a nearly identical band gap value (Table 1) and emission peak energy as is observed in the particle components. More impressively, the CdSe rod aerogel monolith shows much stronger band edge emission peak than the dot aerogel monolith with the same mass and under the identical experimental conditions (Figure 3). The emission intensity ratio for the maxima between the rod monolith and dot monolith is 25:1.

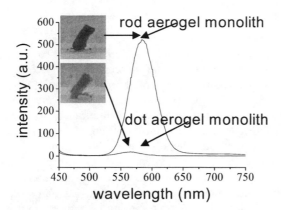

Figure 3: Graph of room temperature emission spectra (λ_{ex} = 440 nm) of CdSe dot and rod aerogel monoliths (pictured in insects) with the same mass (dot: 0.0183g/ rod: 0.0181g). The data were reproduced with permission from reference 7. Copyright 2008 American Chemical Society.

Another distinct difference between the CdSe dot gel and rod gel is that the rod gel exhibits a much smaller tendency to shrinkage of the gel body-i.e., the syneresis effect in which a wet gel expels solvent is less pronounced in the rod gel relative to the dot gel. Thus rod gels can be inverted in contrast to dot gels (Figure 4 inset). This phenomenon indicates that the CdSe polymer-type rod aerogel might be much stronger than colloid-type dot aerogel. To quantify this gel strength property, rheological studies were conducted on aerogel-PDMS (polydimethylsiloxane) composites for a weight percentage of 5% CdSe. Rheology data show that rod aerogel-PDMS composites exhibit a much higher complex viscosity than dot aerogel-PDMS composites and rod aerogels result in an enhanced system modulus over the dot aerogel. (Figure 4) This enhancement is not due to the identity of the building block, since composites formed from dot and rod primary particles show identical (and considerably lower) viscosities relative to the aerogels. This strongly suggests that the enhanced strength is a consequence of the morphology of the interconnected networks.

Altering the building block appears to be an effective way to engineer the basic properties of metal chalcogenide semiconducting aerogels. This novel strategy might offer a powerful manner to tune specific properties of the nanostructure for targeted applications. In addition to the comparison between CdSe dot and rod aerogels, the effects of other shapes, including CdSe branched and hyperbranched nanoparticles, on the morphology and properties of resultant aerogels will be also discussed.

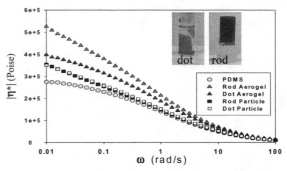

Figure 4: Complex viscosity data for PDMS and PDMS composites of dots, rods, and corresponding aerogels (5 wt. % CdSe). The insets show the effect of syneresis on shrinkage of dot gels relative to rod gels. The data were reproduced with permission from reference 7. Copyright 2008 American Chemical Society.

ACKNOWLEDGEMENT

We thank Dr. Yi Liu for assistance with TEM measurements. This work was supported in part by the Donors of the Petroleum Research Fund (AC-43550), NSF (DMR-0701161, DMR-0216084), and the Institute of Manufacturing Research (WSU).

REFERENCES

[1] D. J. Milliron, S. M. Hughes, Y. Cui, L. Manna, J. Li, L.-W. Wang and P. A. Alivisatos, Nature, 430, 190-195, 2004.
[2] R. D. Robinson and P. A. Alivisatos, Science, 317, 355-358, 2007.
[3] Z. Tang, Z. Zhang, Y. Wang, S. C. Glotzer and N. A. Kotov, Science, 314, 274-278, 2006.
[4] E. V. Shevchenko, D. V. Talapin, N.A. Kotov, S. O'Brien and C. B. Murray, Nature, 439, 55-59, 2006.
[5] J. L. Mohanan, I. U. Arachchige and S. L. Brock, Science, 307, 397-400, 2005.
[6] I. U. Arachchige and S. L. Brock, J. Am. Chem. Soc., 128, 7964-7971, 2006.
[7] H. Yu, R. Bellair, R. M. Kannan and S. L. Brock, J. Am. Chem. Soc., 10.1021/ja801212e, 2008.
[8] Z. A. Peng and X. Peng, J. Am. Chem. Soc., 123, 183-184, 2001.
[9] X. Peng, Adv. Mater. 15, 459-463, 2003.
[10] T. Mokari, E. Rothenberg, I. Popov, R. Costi and U. Banin, Science, 304, 1787-1790, 2004.
[11] J. E. Halpert, V. J. Porter, J. P. Zimmer and M. G. Bawendi, J. Am. Chem. Soc., 128, 12590-12591, 2006.
[12] Z. A. Peng and X. Peng, J. Am. Chem. Soc., 123, 1389-1395, 2001.
[13] N. Hüsing and U. Schubert, Angew. Chem. Int. Ed., 37, 22-45, 1998.
[14] M. L. Anderson, C. A. Morrris, Stroud, R. M.; C. I. Merzbacher and D. R. Rolison, Langmiur, 15, 674-681, 1999.

Nanostructuring Plastic Displays within the Flexible Electronics Paradigm

M. Andrews[*], P. Frechette*, T. Li*, B. Fong*, W. Fong, I. Shih**

[*]Plastic Knowledge, 424 Guy St, Montreal, Quebec, H3J 1S6, Canada mandrews@plastic-k.com
[**]McGill University, Montreal, Quebec, H3A 2K6, Canada

ABSTRACT

This presentation describes our advances in fabricating a fully functional color, plastic liquid crystal display that uses an all-inorganic thin film transistor array based on a nanocrystalline II-VI semiconductor. The display itself represents an interesting context in which to explore the ways in which the key attributes of a flexible electronics platform are enhanced with nanostructured materials and processes, including inorganic nano-printing inks, nanocrystalline materials and nanocomposites, and nanostructured interfaces.

Keywords: flexible displays, flexible electronics, smart plastic, plastic LCD, nanocomposite, nanocrystalline semiconductor

1 BACKGROUND

It is likely that the transformations anticipated by flexible electronics will be achieved by merging not only large-scale electronic platforms with traditional materials and industries, but also by recruiting nanotechnologies, and hybrid combinations of organic and inorganic materials into the service of device fabrication. There is widespread consensus that the transformational potential of flexible electronics is enormous. In applications like RFID tags, sensors, lighting, logic and memory, authentication, anti-tampering, energy and displays, a simple basis set of drivers can be identified that gives major impetus to market applications estimated to be on the order of $136 B [1]. Plastic Knowledge has been engaged in developing a variety of electronic devices on plastics with a strong focus on manufacturing potential. Our long term perspective on flexible electronics emphasizes manufacturable embedded functionality (embedded intelligence) in plastics.

The plastic film manufacturing industry is mature. Plastics and plastic composites have long been printed on (inked), electroplated, metallized by evaporative coating, and supported all manner of dielectrics, in discrete panel, roll-to-roll processes, and in 3D geometries. Nevertheless, plastic substrates for flexible electronics create new manufacturing challenges not only in creating dense arrays of electronic devices, but in elaborating those devices through volume manufacturing into hyphenated plastic structures like smart cards and displays. Successful implementations of flexible electronics require judicious recruitment of interdisciplinary solutions. These are likely to include nanotechnologies, and hybrid combinations of organic and inorganic materials in the service of flexible electronics. Currently, the nascent industry of flexible electronics is attracting a plethora of potentially scalable solutions based on printing technologies and re-tooled versions of standard patterning technologies. Solutions will also emerge from resolving conflicting demands of scale that on the one hand might, for example, demand conductors on the scale of microns to perhaps kilometers. These conflicting demands declare themselves when flexible electronics is called upon to do more than just create devices. The flexible display is an interesting context in which to examine some of the requirements that must be met to realize functionality. Indeed, nano-, micro-, meso- and macro-scale structures must all converge properly to make a functional plastic display. The presentation below outlines some of the features that are important in this context.

2 EXPERIMENTAL

As in previous experiments [2] we used 200 μm thick films of a custom-formulated polyarylester compound. The plastic film was modified so that micron scale devices could be fabricated with multilevel mask registrations despite the fact that films were heated to temperatures on the order of 220 °C. The plastic has a thermal expansion coefficient of ~50 ppm/°C and exhibits optical transparency better than 90% across the visible wavelength range. The polymer was hard-coated with a propriety silica nanoparticle composite. Processing was conducted on samples up to 5 inch diagonal. Plastic substrates were first degassed in a preparatory chamber to remove primarily water, residual organic solvents and unreacted monomer by a programmed slow ramp to 250 °C, followed by baking for 6 h at that temperature, and then slow cooling under vacuum to room temperature. The plastic is quite hydrophobic, absorbing a maximum 0.4% water, which is easily removed during the baking cycle. After this treatment, films were then rapidly transferred in a clean room environment. The transistor consists of an aluminum gate, aluminum oxide dielectric layer, custom silicon oxynitride dielectric (PECVD), a II-VI semiconductor and aluminum source and drain electrodes. Processing was done under class 100 conditions. Aluminum was deposited on the plasma roughened substrate by sputter deposition. Films on the order of 0.75 μm thickness were subsequently patterned by photolithography and wet etched to create the gate lines.

The gate region was quantitatively anodized at neutral pH. This gate dielectric was subsequently annealed to give the required isolation layers. Leakage currents were on the order of 10^{-10} A or smaller. For aluminum surface quality studies, aluminum was deposited by argon ion sputter deposition (4 mTorr, 1000W) onto substrates that were uncooled or cooled to ~ 6 °C. Foils were characterized by reflectivity at 632.8 nm (HeNe laser). CdS was chosen to prototype the initial experiments, despite the lower mobilities compared with CdSe. CdS films were formed at 70 °C by reaction of thiourea, ammonia (complexant) and $CdCl_2$. Pixel electrodes were fabricated from gold films or from ITO. Liquid crystal material was vertically aligned with a polyimide alignment layer. The LC was sandwiched between the lower (TFT equipped) plastic substrate and the upper plastic layer coated with ITO. Spacers maintained a uniform 3 μm cell gap. Color switching under the control of off-board electronics was achieved by flashing side illuminating red, blue and green light emitting diodes. Atomic force microscopy, focused ion beam (FIB), and SEM experiments were conducted at the Thin Films Group facilities of the Ecole Polytechnique in Montreal. Ink jet printing was carried out in collaboration with Optomec using the M^3D Aerosol Jet® printer system. Organic TFTs were printed according to designs and formulations developed by Frisbie and coworkers [5].

3 RESULTS AND DISCUSSION

One of the first length scales we encounter in fabricaging a plastic LCD is associated with free volume and structural relaxations in the plastic substrate. In polymer films, glassy-state structural relaxation, often referred to as physical aging, can produce a time dependence of the modulus, brittleness, permeability, and dimensional stability. Structural relaxations are often associated with redistributions of free (excluded) volume in the polymer, whose dimensions are frequently on the nanoscale. Summed over the dimensions of a polymer film, nanoscale relaxations can interfere with photolithography, where mask-to-mask registration can be crucial to device fabrication. As others have found, we observe that it is important to impart dimensional stability to polymer substrates by annealing out short and longer term relaxation mechanisms. Algorithms have been developed for optical steppers for partial compensation of dimensional changes in plastic substrates, but it is not clear yet to what extent or how robust these compensation schemes are, or how dependent they are on the type of polymer film or individual film properties. Dimensional stability in our hands has been worked into our program of developing Smart Plastic[TM] substrates with a kind of "shape memory" that permits the substrate to return to its original dimensions after each processing step, independent of whether the process involves heating, exposure to aqueous acid, base, organic solvents or photoresist and strippers. The Smart Plastic[TM] strategy reduces manufacturing constraints by having the polymer itself resolve some of the conflicting demands placed on it through processing, rather than requiring that the manufacturing method do so. In our hands, multilayer polymer films offer wider latitude for solving problems of dimensional stability during device fabrication.

Forces on a TFT plastic backplane can be divided broadly into those that are internal and those that are external. External forces are those associated with macroscopic deformations like bending or shaping. Internal forces are those that are introduced by microlithography, including photoresist deposition and patterning, metal film deposition, wet and dry etching, solvent development, stripping (lift-off) and thermal annealing. These forces, which are difficult to control, result in differential thermal expansion and contraction and dimensional changes caused by water and organic solvent absorption and evaporation, and also by differences in thermo-mechanical properties of deposited inorganic and organic devices. Stiff device films

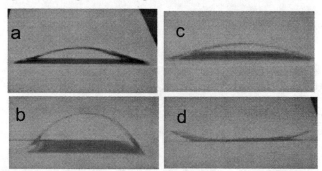

Figure 1. Changes in curvature due to differential stresses after different processing steps in improperly conditioned plastic substrates (TFT side is on top). a) after deposition of aluminum data lines; b) after lift-off (acetone solvent exposure); c) after 5 min standing in ambient atmosphere; d) after 250°C heat treatment in air, 20 min.

on compliant polymer substrates therefore place significant demands on the TFT backplane in terms of stress control. .Figure 1 above shows a sequence of deformations that result when there is stress mismatch during strain following several TFT fabrication steps on a plastic film that has not been properly stressed compensated. Significant effort has gone into developing stress management schemes [3]. These include applications of stress compensating dielectric layers and the fabrication of devices at island points of zero stress on films. Another compensation strategy that has not be much discussed in the literature is to build up multilayer

Figure 2. Active matrix backplane developed on stress compensated plastic substrate.

composite plastic films with built-in stress compensation. When this is done, as in our approach, the deformations observed in Figure 1 disappear to give the result shown in Figure 2.

Interfaces, like those between polymer and semiconductor materials, dielectrics, metal and metal oxide conductors, and even photoresists can modify polymer mechanical motions over nanometer scales at device interfaces. For example, there is substantial evidence that the effects of nano-confinement on polymer chain dynamics depend on the surfaces and interfaces to which polymers are attached. Interfacial interactions are now known to modify dynamics in the interior regions of glass formers [4], but the implications for surface adhesion and stress response of overlaid electronic components extending over the nanometer scale are not yet understood. Interfaces can affect adhesion and film quality in other areas. Differences in thermo-mechanical properties of polymers and inorganic electronic components – including gate and data lines - deposited on them can lead to mechanical failure through cracking and delamination. Polycrystalline aluminum films in the range 0.1-10 µm belong to the so-called mesoscale, and are highly inhomogeneous, so that local stress and structural variations may play important roles in lithographic patterning, electronic transport properties and mechanical performance. Specular reflectance R_s gives a measure of aluminum film quality through the rms surface roughness

$$\frac{R_s}{R_0} = e^{-(4\pi\delta\cos\vartheta_0 / \lambda)^2} \quad (1)$$

Where R_0 is the total reflectance and θ_0 is the angle of incidence of the light on the surface. Figure 3 shows that reflectance of Al from the uncooled films is about 20% lower than that of the mirror-finish obtained on cooled plastic substrates. Plastic films that were not cooled during aluminization emerged from the sample chamber with a

milky appearance, despite the fact that deposition on neighboring pristine silicon wafer test pieces gave mirror-like finishes. We attribute the milky appearance to poor thermal conductivity of the plastic, differential thermal expansion of the polymer/hard coat surface, and enhanced physical and chemical reactions between high energy projectiles with the composite hardcoat. Adventitious contaminants released from the substrate polymer tend to generate large numbers of large grain boundaries. Surface scattering shows up more clearly in the right hand plots in Figure 3 where we observe significant anisotropic scattering around the central reflectivity peak for the uncooled sample. It is known that aluminum films can suffer from hillock formation and significant degradation in

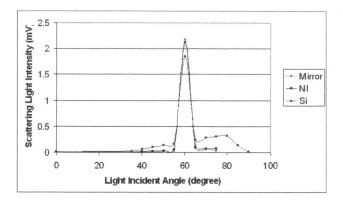

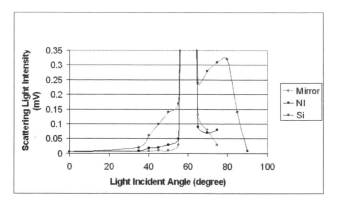

Figure 3. Reflectivity measurements for plasma-deposited aluminum on uncooled (Si) and cooled (Nl) polymer substrates as compared with a high quality optical mirror (Mirror). Top figure shows complete reflectance spectra. Bottom figure shows expanded zone at the base of the peaks in the maxima of the reflectance spectra.

reflectivity [4]. Hillock formation leading to diffuse light scattering can start at temperatures on the order of 70-80 °C. Moreover, films incorporating impurities from interfacial reactions may be brittle, fracturing and delaminating more easily. During deposition, the optical quality and physical properties of the resulting aluminum film on the plastic depend on the degree of outgassing from the polymer, substrate temperature, microstructural responses of the plastic during thermal cycling in and out of

NSTI-Nanotech 2008, www.nsti.org, ISBN 978-1-4200-8503-7 Vol. 1

the vacuum chamber, aluminum deposition rate and uniformity, and thickness of the metal layer. The organosilica nanoparticle hardcoat can exhibit surface topography that differs with etch conditions due to differential etch rates of the silica and polymer binder matrix. Roughness at the hardcoat surface due to the plasma etch will be transferred to the Al film, hence it is important to minimize etch-related roughness. AFM analysis revealed that the hardcoat overlayer exhibits an rms surface roughness < 2 nm. The AFM measurements indicate good surface uniformity with no contamination by larger silica particles that might act as significant Al nucleation sites. Al overcoatings show approximately 15 nm rms roughness and good film uniformity, but we note that there are occasional hillocks and imperfections on the order of 150 nm in height by ~ 500 nm in diametrer. The polycrystalline Al films are typically nanograin materials. Thicker Al films have larger grains and are also somewhat more elastic. Differences in mechanical properties at the mesoscale depend on intergrain and intragrain (size dependent) deformations and interactions, and these in turn affect adhesion, ductility and the quality of surface anodization during formation of the gate dielectric of the TFT. With mirror-finish films, Al gate regions (conductance ~3.5 x 10^{-6} Ωcm) could be quantitatively anodized at neutral pH. Annealing of the gate dielectric gave isolation layers with breakdown voltages on the order of 8 MV/cm. Capacitance measurements yielded a dielectric constant of ~9 for the annealed anodized oxide. Focused ion beam analysis showed that there was no diffusion of aluminum across the hardcoat barrier.

Thin film transistors that we have developed from II-VI semiconductors also have features that scale from nanometers to microns. Our method uses an ion-by-ion chemical bath technique [1] that was developed as a low temperature alternative to vapor deposition. As discussed previously [1], the ion growth mechanism yields CdS (or CdSe) films that can exhibit lattice orientation perpendicular to the substrate. CdS, for example, is deposited as a mixture of hexagonal close packing cubic phases. Not unexpectedly, the interfaces between the nano-grains are metastable so that brief periods of annealing at 250 °C improved the I-V characteristics of the transistors. Figure 4 is an optical micrograph showing the source and drain electrodes with the II-VI nanocrystalline semiconductors (yellow band) interposed. These TFTs were then incorporated into a 3 inch diagonal QVGA all plastic display. The display uses RGB LEDs and off-board field sequential color (FSC) drive electronics to achieve color mixing. The FSC architecture requires 2/3 less transistors than conventional LCDs, and there is no need for a color filter. Without optimization of the liquid crystal, switching speeds of 120 subframes/sec over a 3 μm cell gap have been achieved. In a similar manner, we have fabricated a plastic display that embodies the key features of FSC color mixing in a display that has been shaped into

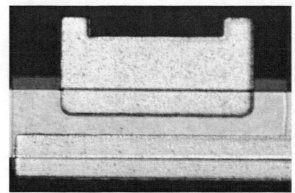

Figure 4. Optical micrograph of the source/drain region of a II-VI semiconductor (yellow band) TFT created by low temperature ion-by-ion chemical bath deposition

a convex curvature. While the display is not intended to be reversibly flexed once the shape has been defined, the achievement advances a basic principle of flexible electronics, namely that devices embedded in plastic can be elaborated into sophisticated value-added systems like functional curved (conformal) displays.

In an effort to develop Smart Plastic™ substrates that are agnostic to the choice device fabrication process, we fabricated arrays of low and high performance devices by hybridizing vapor deposition techniques with inkjet printing. For example, nanoparticle silver and dielectric inks were used to create capacitors on the plastic substrate (Figure 5).

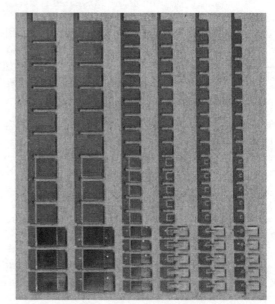

Figure 5. Capacitors created by inkjet printing silver and dielectric nanoparticle inks on bottom electrode lithographically patterned aluminum electrodes.

After sintering at ~80°C in a convection oven, the silver electrodes emerged some 5x rougher (80 nm rms) by AFM than the aluminum electrodes. Surface plasma treatment of the plastic gave strongly adherent devices (tape test). No attempt was made to optimize the dielectric used to make the capacitors, since the objective was to determine that functioning multilevel structures could be obtained by precision overprinting of dielectric and conductor.

A primary objective of organic electronics is to develop inexpensive materials and methods to make high performance devices on flexible substrates. Accordingly, functional organic (ion gel) TFTs (Figure 6) were successfully printed by inkjet onto the Smart Plastic™ substrate. The material compositions and performance characteristics of these OTFTs have been published elsewhere [5].

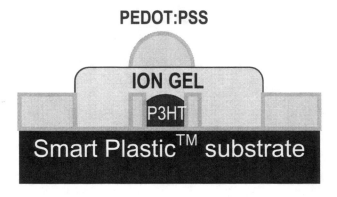

Figure 6. Schematic of an organic TFT deposited by inkjet printing onto a Smart Plastic™ substrate.

In summary, examination of selected materials and processes required to create a functional color plastic liquid crystal display identifies ways that nanostructures feature in the flexible electronics paradigm. In the display context, nanostructures are recruited to scale across micron, meso- and macro-dimensions. For displays of the type discussed in this paper, relevant nanostructured materials and processes include inorganic nano-printing inks, nanocrystalline materials and nanocomposites, and nanostructured bulk polymer films and their structure determining interfaces. In this perspective, the elaboration of flexible electronics into commodity products will benefit from ongoing research in nanoscience and nanotechnology. Finally, since there no common polymer platform for flexible electronics, a useful strategy we are pursuing is to create (smart) polymer composite substrates that are more or less agnostic to device fabrication processes. This was illustrated in simplified form by building TFTs using both standard lithographic processing and inkjet printing techniques on the same substrate.

ACKNOWLEDGMENTS

The authors gratefully acknowledge the collaboration and assistance of U. Berger, M. Renn and J. Paulsen of Optomec for use of the M3D Aerosol Jet® printing system. The generous assistance of Professor C.D. Frisbie, (Departments of Chemistry and Chemical Engineering and Materials Science, University of Minnesota), for materials and design of the OTFTs is greatly appreciated. For partial financial support for this work we thank the Industrial Research Assistance Program (IRAP) of the National Research Council (NRC) of Canada.

REFERENCES

[1] iNEMI Organic and Printed Electronics Roadmap, 2007.
[2] F. MacNab, P. Frechette, B. Fong, T. Li, I. Shih and M.P. Andrews, NSTI Nanotech 1 301, 2007.
[3] *Flexible Flat Panel Displays*, Edited by Gregory P. Crawford, Ch 12, 2005.
[4] C. Kylner and L. Mattsson, Thin Solid Films, 348, 222 (1999).
[5] J. Lee, M. J. Panzer, Y. He, T. P. Lodge, and C. D. Frisbie, J. Amer. Chem. Soc., 129, 4532, 2007.

Towards Automation in the Characterization of Nano-Structured Materials and Devices

K. Weishaupt, U. Schmidt*, T. Dieing, and O. Hollricher

WITec GmbH, Hoervelsingerweg 6, 89081 Ulm, Germany, www.witec.de
*e-mail address corresponding author: ute.schmidt@witec.de

ABSTRACT

The combination of a confocal Raman microscope and an atomic force microscope (AFM) in an automated microscope system is used to study large samples in terms of chemical composition and morphological conformation on the nanometer scale. Recent developments in CCD detectors and ultra-high throughput spectrometer enable the acquisition of Raman spectra in microseconds, thus enabling the acquisition of 2D spectral arrays consisting of ten-thousands spectra in less than 5 minutes. The evaluation of spectral features, such as peak intensity, peak position etc, lead to Raman images revealing either chemical or stress distributions within the analyzed materials. The automated sample positioner, allows the automated execution of pre-defined measurement sequences on any user defined selection of measurement points on the sample. By turning the microscope turret, the confocal Raman microscope is transformed into an AFM, allowing to perform high resolution topographical images on the same pre-selected positions of the sample.

Keywords: Confocal Raman Microscopy, AFM, automated system for large sample analysis

1 INTRODUCTION

The characterization of nanostructured materials implies knowledge about their chemical and structural properties, leading to a growing demand for characterization methods for heterogeneous materials on the nanometer scale. However, certain properties are difficult to study with conventional characterization techniques due to either limited resolution or the inability to chemically differentiate materials without inflicting damage or using invasive techniques such as staining. By combining various analytical techniques such as Raman spectroscopy, confocal microscopy and AFM in one instrument, the same sample area can be analyzed with all implemented methods, leading to a better understanding of nanostructured materials.

Raman spectroscopy, a chemical analysis technique, combined with confocal microscopy enables the unique Raman imaging of heterogeneous materials [1-8]. The power of Raman imaging stems from the high chemical information content of molecular vibrational spectra. In the Raman spectral imaging mode, a complete Raman spectrum is recorded at every image pixel, leading to a two-dimensional array consisting of ten-thousands of complete Raman spectra. From this array images are extracted by analyzing various spectral features (sum, peak position, peak width, etc). Differences in chemical composition, although completely invisible in optical images, will be apparent in the Raman image and can be analyzed with a resolution down to 200 nm [3,6,9]. The recently implemented high speed EMCCD camera and ultra-high throughput spectrometer (UHTS 300) allow the reduction in Raman data acquisition time by another factor of 10 [10, 11]. If higher resolution is required, by simply turning the microscope turret, the confocal Raman microscope can be transformed into an AFM. Using this imaging technique, structures below the diffraction limit can be visualized from the same sample area [9].

This article describes further instrumental improvements which allow the automated analysis of large samples. Beside the acquisition of large area scans [12], through special scripting functions it is possible to predefine measurement sequences on any user defined selection of measurement points on the sample, guaranteeing the most comprehensive surface analysis tool for systematic and routine research tasks.

2 INSTRUMENTATION

The new developed alpha500 RA microscope from WITec (www.witec.de) is a highly modular and flexible microscopy system. It combines a high throughput confocal Raman microscope for 3D chemical imaging and an AFM for high resolution morphological imaging in an automated system for large samples. An image of the instrument is shown in Fig. 1. The piezo scanner, with a scan range of 200x200 μm^2, is mounted on top of a motorized x-y-z stage. This stage is driven by stepper motors and allows an expansion of the scanning range in x-y direction to 150x100 mm^2 with a step size of 100 nm. Besides expanding the scanning range, this stage can be used to perform multi-area/multi-point measurements on any user-defined number of measurement points. An example of such a point raster is shown in Fig. 2. The table on the right side lists the coordinates of the points of interest, whereas the graph on the left side displays the points of interest and the travel path for automated point measurements. Through scripting functions several consecutive tasks can be executed automatically, without any online process control by an operator during the measurements.

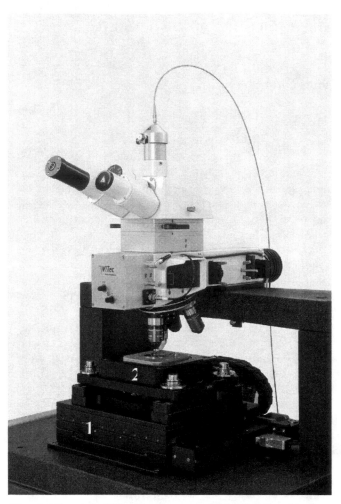

Fig. 1: The alpha500 AR microscope, highlighting the motorized xy-stage (1) and the high accuracy piezo scanner (2)

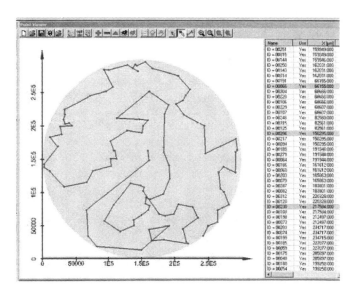

Fig. 2: Example of point raster for automated multi-point /multi-area measurements.

3 EXAMPLE MEASUREMENTS AND DISCUSSIONS

A series of AFM and Raman measurements were performed on a DRAM (dynamic random access memory) chip (Infineon Technologies), to prove the automation capabilities of the alpha500 RA. In a first step white light video images were recorded on three different positions of the sample (Fig. 3). For this procedure the following scripts were used:

a) auto-illumination for optimum white light illumination of the sample,

b) auto-focus performs an auto-focus based on the video images,

c) the snapshot acquires a video image in each pre-selected point.

The video images from Fig. 3 were recorded with an Olympus 100x (NA = 0.95) objective and reveal the different pattern selected from the DRAM chip.

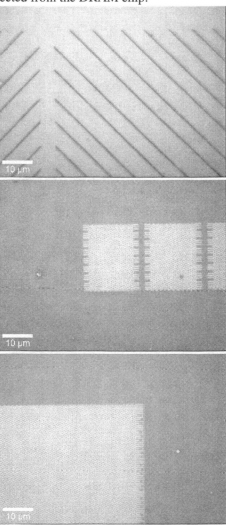

Fig. 3: Automated recorded video images from three different areas of the DRAM.

NSTI-Nanotech 2008, www.nsti.org, ISBN 978-1-4200-8503-7 Vol. 1

Parts of the three selected areas shown in Fig. 3 were then imaged in Raman spectral imaging. In this imaging mode, a complete Raman spectrum is recorded in every image pixel, leading to a 2D array of 150x150 Raman spectra. The integration time for each individual Raman spectrum was 0.01 s, thus the array of 22500 Raman spectra was recorded in less than 4 minutes. By evaluating the position of the first order Si Raman band at 520/cm, stress images of the selected areas can be obtained. As shown in a previous paper [13], indents in Si produce a stress field which expands around the indent as a function of load force. Fig. 4 shows the stress images obtained from the three pre-selected areas.

By simply turning the microscope turret, the confocal Raman microscope was transformed to an AFM. By using the same positioning raster as used before for the acquisition of the Raman images, and the additional auto-tip-approach scripting function, AFM images were recorded from the same sample areas. Fig. 5 shows the AFM images recorded from the three selected sample areas. The line structure in area one consists of parallel grooves which are 80 nm deep. The pits from areas two and three have a diameter of 300 nm and a depth of 300 nm.

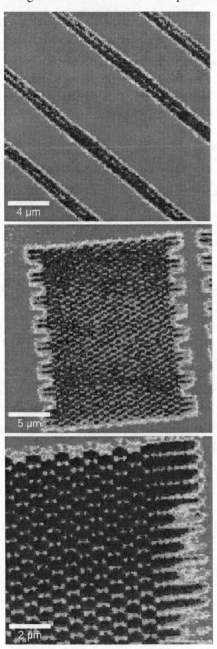

Fig. 4: Raman stress images recorded on three different areas of the DRAM.

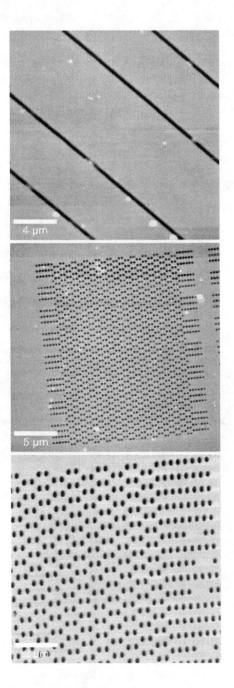

Fig. 5: AFM topography images measured from the three different areas of the DRAM.

4 CONCLUDING REMARKS

The capabilities of an automated confocal Raman AFM system are demonstrated. By combining two nondestructive sample analysis methods in one automated system, one and the same sample area can be characterized by the implemented measuring methods. Multi-point/multi-area measurements can be routinely performed on various areas of the sample.

The motorized sample stage can be used to perform large overview scans, whereas the additional piezo-scan-stage allows the acquisition of high resolution images from selected areas of the overview image [12].

The implemented scripting routines such as auto-focus, auto-AFM-tip-approach, etc. guarantee standardized routine measurements procedures without any online process control by an operator during measurement.

REFERENCES

[1] O. Hollricher, OE Magazine, Nov., 2003.

[2] U. Schmidt, A. Jauss, W. Ibach, and O. Hollricher, Microscopy Today, 13, 30, 2005.

[3] U. Schmidt, S. Hild, W. Ibach, and O. Hollricher, Macormol. Symp. 230, 133, 2005.

[4] A. Jauss and H. Fischer, Imaging and Microscopy 4, 24, 2005

[5] U. Schmidt, Imaging and Microscopy 2, 34, 2005.

[6] A. Jauss, H. Fischer, and O. Hollricher, Imaging and Microscopy 1, 17, 2006.

[7] O. Hollricher, W. Ibach, A. Jauss, and U. Schmidt, FutureFab Intl, 21 (2006).

[8] U. Schmidt, W. Ibach, J. Mueller, and O. Hollricher, SPIEProc. 6616 Pt. 1, 66160E-1, 2007.

[9] U. Schmidt, F. Vargas, T. Dieng, K. Weishaupt and O. Hollricher, Nanotech, 4, 48, 2007.

[10] T. Dieing and O. Hollricher, Vibrational Spectroscopy, in press.

[11] Ultra-fast Confocal Raman Imaging, Application Note , www.witec.de 2008.

[12] alpha500 – Large area Scans of Tablets, WITec Application Note, www.witec.de 2008.

[13] U. Schmidt, W. Ibach, J. Mueller, K. Weishaupt, and O. Hollricher, Virational Spectroscopy, 42, 93 2006..

Design and Parameters of Cellulose Filter Media with Polymer Nanofiber Layer

Stanislav PETRIK*, Miroslav MALY* and Lukas Rubacek*

*Elmarco Ltd., Liberec, Czech Republic
petrik@elmarco.com, maly@elmarco.com, rubacek@elmarco.com

ABSTRACT

Nanofiber polymer layer deposited on regular cellulose filter media usually shifts air filtration parameters of this material several classes higher. Very fine nanofiber web of basis weight of 0.01 gsm to 0.1 gsm improves filtration efficiency for submicron particles by hundreds of percent, while lowering air permeability by only tens of percent. This evident benefit becomes to be widely commercially accessible due to the availability of productive industrial-scale Nanospider™ technology.

Influence of morphology of the nanofiber layer (fiber diameter, porosity, fiber shape and orientation) on filtration parameters of the air filtration media has been studied. Preparation method of the filtration media samples and their analysis will be described in the paper. It will be shown that the morphology of nanofiber layer plays usually more important role than its basis weight. Some new physical/geometrical values will be introduced to reflect these relations.

Keywords: nanofiber, air filtration, cellulose media

1 INTRODUCTION

Influence of morphology of the nanofiber layer (fiber diameter, porosity) on filtration parameters of the air filtration media has been studied. All samples analyzed in this paper have been prepared by using an industrial production NanospiderTM machine. Preparation method of the filtration media samples and their analysis are described in the paper. It has been shown that the morphology of nanofiber layer plays usually more important role than its basis weight. Some new physical/geometrical values are introduced to reflect these relations.

Nanofiber polymer layer deposited on regular cellulose filter media usually shifts air filtration parameters of this material several classes higher. A very fine nanofiber web of basis weight of 0.01 gsm to 0.1 gsm improves the filtration efficiency for submicron particles by hundreds of percent, while lowering the air permeability by only tens of percent. This evident benefit becomes to be widely commercially accessible due to the availability of productive industrial-scale NanospiderTM technology [1].

Usual parameter discussed (and required) by nanofiber filtration media users is the basis weight of the nanofiber layer. However, experience with final (product) parameters of filters shows, that there is not a very strong correlation between basis weight, filtration efficiency and permeability (or pressure drop), at least in the case of cellulose substrate combined with relatively low-weight polymer nanofiber layer.

Mechanism of filtration process on fibrous media has been discussed in several publications [2-6]. For example, for cellulose filtration media of solidity ß in the range of 0.11-0.33 and an average pore diameter of 12-84 µm, it has been shown that pressure drop depends on fiber diameter df according to relation [2]:

$$\Delta p = \frac{\mu \cdot v \cdot w_b \cdot h}{d_f^2 \cdot \rho_f \cdot \left(-0,984 \cdot \ln \beta - 0,47\right)} \qquad (1)$$

where μ is air dynamic viscosity, w_b - media basis weight, h - media thickness, ρ_f - fiber density, β - filter solidity (or packing density) - volume of fibers/volume of filter.

However, in the case of low basis weight nanofiber layer, where molecular (or transition) flow regime takes place, following equation is considered to be valid [5]:

$$\Delta p = \frac{\mu \cdot v \cdot w_b \cdot h}{r_f \cdot \lambda} \qquad (2)$$

where r_f is radius of nanofiber and λ means free path of molecules.

Hence, the combination of cellulose medium with polymer nanofiber layer will require more complex models to describe behaviour of the final filtration material. Systematic experimental study of correlations between final product parameters (filtration efficiency, pressure drop) and morphology of cellulose/nanofiber media can provide useful data for both theoretical understanding of filtration mechanisms and practical design of air filters.

2 EXPERIMENTAL

For this study, we prepared 42 samples of controlled basis weight and fiber diameters (6 series, 7 samples in each). Nanospider production machine has been used, as it provides good long-term consistency of filtration media parameters (16 hours run with repeatability in the range of ±5%).

To obtain various basis weights, substrate speed had been varied from 0.2 m/min to 4 m/min for each series of samples, while polymer solution parameters (concentration, etc.) together with electric field intensity determined the range of nanofiber diameter. Nanofiber diameter distribution has been measured by using scanning electron microscope (SEM). Basis weights were obtained either directly by using analytical balances Mettler (higher values) or by extrapolation from its known dependence on substrate velocity (lower ones). Pressure drop and initial gravimetric filtration efficiency have been chosen as representatives of product parameters. They were measured according to EN 779 using NaCl aerosol at following settings: air flow speed: 5 m/min, sample area 100 cm2, flow rate 50 l/min.

3 RESULTS AND DISCUSSION

Fig. 1 illustrates very good correlation between initial gravimetric filtration efficiency and pressure drop, regardless of nanofiber layer parameters (df, wb). As pressure drop monitoring can be relatively easily incorporated into the nanofiber production line, it can be used as a very good parameter for the on-line quality control of final filtration media.

Samples in Fig. 2 show how different nanofiber layers can exhibit similar filtration properties. The first sample is made of thin nanofibers (around 80 nm) and lower basis weight (0.03 gm-2), while the second one is characterized by almost 2-times thicker fibers (144 nm) and more than 3-times higher basis weight (0.10 gm-2).

To consider the influence of both basis weight and fiber diameter on final product filtration parameters, we can define a simple value (Relative Fiber Length Lf) expressing the total length of nanofibers (in kilometers) deposited on unit surface of filtration medium (in square meters):

$$L_f = \frac{w_b}{\pi \cdot \rho \cdot \left(\frac{d_f}{2}\right)^2} \tag{3}$$

where ρ is density of material of nanofibers.

As it can be seen from the graphs in Fig. 3, both filtration efficiency and pressure drop are in much better correlation with Relative Fiber Length Lf than with nanofiber layer basis weight (wb).

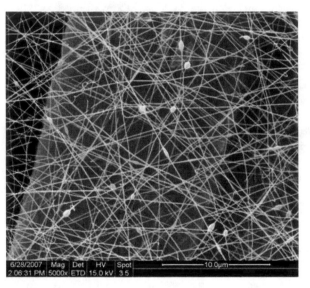

$IGE = 73 \pm 1$ % $\qquad L_f = 6114$ km.m^{-2}

$\Delta p = 168 \pm 5$ Pa $\qquad w_b = 0.03$ gm^{-2}

$d_f = 83 \pm 22$ nm

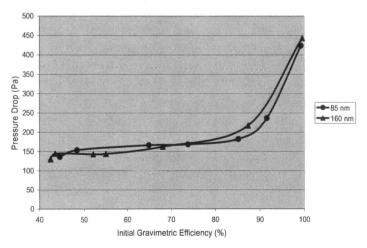

Fig. 1

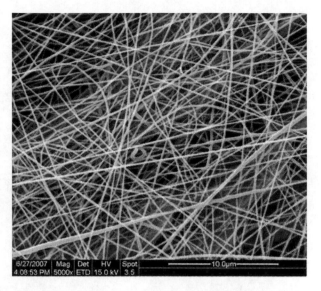

$IGE = 68 \pm 1\ \%$ $L_f = 6128\ \mathrm{km.m^{-2}}$

$\Delta p = 163 \pm 4\ \mathrm{Pa}$ $w_b = 0.10\ \mathrm{gm^{-2}}$

$d_f = 144 \pm 36\ \mathrm{nm}$

Fig. 2

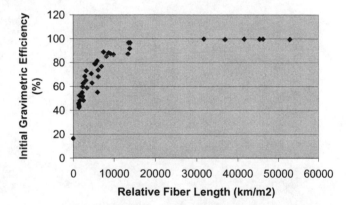

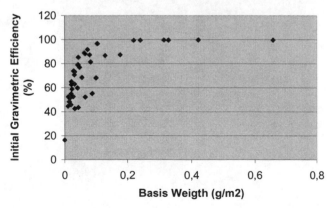

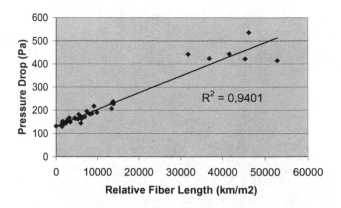

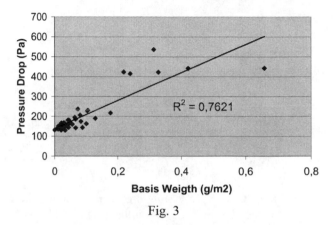

Fig. 3

4 CONCLUSIONS

Experimental results obtained in this study showed that:

- Basis weight of nanofiber layer does not predict filtration parameters well enough. The same initial gravimetric efficiency can be obtained using nanofibers of very different mean diameters, or in other words, of very different basis weights.
- Filtration properties of cellulose media with low-weight polymer nanofibrous layer depend mostly on the total length of nanofibers per media surface unit ("Relative Fiber Length").
- "Relative Fiber Length" is in good correlation with both pressure drop and filtration efficiency.
- Image analysis of SEM pictures can be used for predicting filtration properties of the media.

For more detailed understanding and predicting filtration properties of the media studied here, it will be useful to investigate similar relations using:

- Measurements of fractional filtration efficiency of the samples.
- Observations of the role of dust cake formed on media surface during filter lifetime or in field tests.
- Investigations of the effect of final filter design, pleating of the media, etc.

5 REFERENCES

[1] Jirsak, O., Sanetrnik, F., Lukas, D., Kotek, V., Martinova, L., Chaloupek, J., "A method of nanofibers production from a polymer solution using electrostatic spinning and a device for carrying out the metod", The Patent Cooperation Treaty WO 2005/024101, 2005.

[2] Myedvyedyev, V. N., Nyevolin, V. F., and Kagan, M. R., 1984, "A Method of Pressure Drop Calculation in Cellulose-Type Filters for Dust Free Air and Under Dust Loading," Engine Construction, No. 5, pp. 25-27.

[3] Jaroszczyk, T., Fallon, S. L., Liu, Z. G., Schwartz, S. W., Holm, Ch. E., Badeau, K. M., and Janikowski, E., "Direct Flow Air Filters – A New Approach to High Performance Engine Filtration", Filtration, Vol. 6, No. 4, 2006, pp. 280-286.

[4] Cheng, Y. S., Allen, M. D., Gallegos, D. P., and H. C. Yeh, "Drag Force and Slip Correction of Aggregate Aerosols," Aerosol Science and Technology, 8, 1988, pp. 199-214.

[5] Pich, J., "The pressure drop in fabric filters in molecular flow," Staub-Reinhalt. Luft, Vol. 29, No 10, October 1969, pp. 10-11,

[6] Pich, J., "Pressure characteristics of fibrous aerosol filters," J. of Colloid and Interface Science, Vol. 37, No 4, December 1971, pp.912-917.

Permeability and Protein Separations: Functional Studies of Porous Nanocrystalline Silicon Membranes

Jessica L. Snyder[*], Maryna Kavalenka[**], David Z. Fang[**], Christopher C. Streimer[**], Philippe M. Fauchet[**], and James L. McGrath[***]

[*]Department of Biochemistry and Biophysics
[**]Department of Electrical and Computer Engineering
[***]Department of Biomedical Engineering
jessica_snyder@urmc.rochester.edu
jmcgrath@bme.rochester.edu

ABSTRACT

We have developed an ultrathin (15nm) freestanding nanoporous silicon based membrane. This new material, termed porous nanocrystalline silicon (pnc-Si), is fabricated using standard photolithography techniques. The intrinsic characteristics of this new material and its relatively facile fabrication make it superior to commercial membranes and other currently studied nanomembranes. We have assembled these membranes into simple diffusion and centrifugal apparatuses. We show that we can achieve sharp protein separations of concentrated complex protein mixtures. We additionally show that air and water permeabilities through pnc-Si membranes are greater than both polycarbonate track etched membranes and carbon nanotube membranes.

Keywords: silicon, membrane, ultrathin, separation, protein

1 INTRODUCTION

Clinical, laboratory, and industrial research frequently make use of nanoporous membranes, and the demand for such a membrane continues to grow. These membranes are used to perform molecular separations by size and charge [1]. An ultrathin membrane with well-defined monodisperse pores would be the ideal instrument for molecular separations. Unfortunately, there are no commercially available membranes that are nanoporous and have a nanoscale thickness.

Polymeric membranes are the most common nanoporous membranes used for molecular separations, and these are made from materials such as cellulose and polyether sulfone [2]. These membranes are employed to perform procedures such as hemodialysis and are a common material in laboratory spin columns. They are referred to as tortuous path membranes since they have a mesh-like morphology and lack discrete pores. Polymeric membranes have thicknesses on the order of hundreds of micrometers to millimeters, and the large surface area of the winding pathways causes sample loss and biofouling due to protein adsorption [2]. The molecular weight cutoff for these membranes is approximate due to the lack of well-defined cylindrical pores. Polymeric membranes function best by separating molecules that differ in weight by whole magnitudes. These membranes, though, are easily and inexpensively made, and perform well as an initial rough purification step.

While nanomembranes can be engineered to have controlled pore sizes and are able to perform sharper separations, those that are commercially available are still thick and have long cylindrical pores. Track etched membranes (Whatman, Sterlitech) are manufactured by bombarding a thin plane (~6 μm) of polycarbonate or polyester with high energy ions and subsequently etching pores where bonds were severed. This creates membranes with pore sizes between .01 and 20 μm [2]. The pores are generally well defined, but porosity must be kept low to reduce the chances of pores overlapping and creating larger holes in the membrane. Pores self assemble by anodization in alumina membranes (Anopore, Whatman), and by changing the voltage of anodization, the pore size of these membranes can be tuned between 20 and 200 nm [3]. The

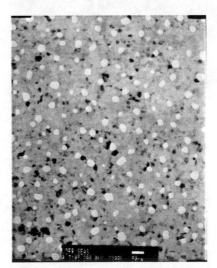

Figure 1 - TEM image of 15 nm thick porous nanocrystalline silicon membrane. The film is composed of silicon nanocrystals oriented in different directions; those that satisfy the Bragg conditions appear as black spots upon the gray film. Pores in the membrane appear as white circles. Pore diameter and porosity are measured directly from TEM micrographs.

porosities are higher than track etched membranes, and the pores are very homogeneous. Both track etched and alumina membranes are convenient to manufacture, but their relative thickness means increased resistance to fluid flow [4] and a larger surface area for protein adsorption. Porous silicon nanomembranes and carbon nanotube membranes [5] are currently being researched, but both of these membranes also have cylindrical pores and face the same problems as commercially available engineered nanomembranes.

We have developed a novel ultrathin nanomembrane with well-defined disc shaped, rather than cylindrical, pores [6]. The material, called porous nanocrystalline silicon, is fabricated using standard photolithography techniques. Pore sizes can be tuned in a thermodynamically driven self-assembly step during pnc-Si manufacture. Because of the thinness of the material, we can readily measure porosities and the diameters of the well-defined pores by TEM (Figure 1). The intrinsic characteristics of this new material and its relatively facile fabrication make it superior to commercial membranes and other currently studied nanomembranes. We have built apparatuses to test the membrane in both diffusion and pressurized flow modalities. We have studied the permeability and performed separations of concentrated complex protein mixtures using this novel nanomembrane.

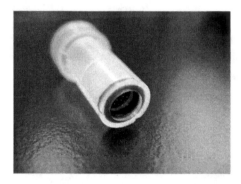

Figure 2 – Centrifuge tube insert. Circular silicon chips with 2 pnc-Si silts are sandwiched between and o-ring and a polypropoline retention ring, sealing it inside the polypropoline insert. These inserts are placed in a 1.5 mL centrifuge tube and filled with the solution to be tested and spun in a centrifuge.

2 METHODS

2.1 Membrane Fabrication

Membranes were fabricated as discussed in [6]. We used a photolithography mask that allowed us to create a wafer with 64 .6 cm chips, each containing two .5 mm by 2 mm slits spanned with pnc-Si membrane material.

2.2 Diffusion Apparatus and Protein Separations

We built four ~55 uL reservoirs in a block of erylite. We filled each reservoir with 60 uL of phosphate buffered saline (PBS) so that the fluid curved slightly above the reservoir. We placed a membrane chip above each reservoir so that the membrane came in contact with the PBS. A 2 ul drop of 10mg/mL bovine brain extract was pipetted into the well of the silicon chip. Diffusion of the protein solution was allowed to occur for 24 hours.

After 24 hours we recovered the retentate, or the solution remaining above the membrane, and the filtrate, or the solution in the reservoir. Both retentate and filtrate were analyzed using SDS polyacrilimide gel electrophoresis (SDS-PAGE) and were compared to a control solution that did not see the membrane.

2.3 Centrifuge Tube Apparatus

Harbec Plastics (Ontario, NY) built a polypropoline centrifuge tube insert that would enable us to perform centrifugation using our pnc-Si membrane (Figure 2). The membrane is sealed in the tube with an o-ring and plastic retention ring. Solutions to be tested can be pipetted into the insert and will pass through the membrane during centrifugation.

2.4 Permeability Testing

Water permeability was tested using the centrifuge tube insert. 500 uL of water was pipetted into the device and the underside of the membrane was wetted using 200 uL water. The device was spun in a 15 mL conical at 1000 rpm for 2 hours. The volume of water to pass through the membrane was measured and the permeability (volume per time per active membrane per psi) was calculated.

Air testing was performed using the setup as illustrated in Figure 3. The volume of nitrogen to pass through the membrane was measured using the scale on a horizontal u-tube, and the permeability was calculated.

3 RESULTS AND CONCLUSIONS

In Figure 4 we show that we obtained a sharp separation of the 10 mg/mL bovine brain extract. Proteins under 90 kD begin to appear in the filtrate, and the lowest molecular weight species are hardly detected in the retentate. 30kD proteins are sharply separated from 90kD proteins in this gel, and this represents only a 3x difference in size. Protein size effects and absorption will effect the cutoffs that we see for different protein mixtures, and this is an area of ongoing research.

We plot the permeability values we obtained against two engineered nanomembranes, polycarbonate track etched membranes and carbon nanotube

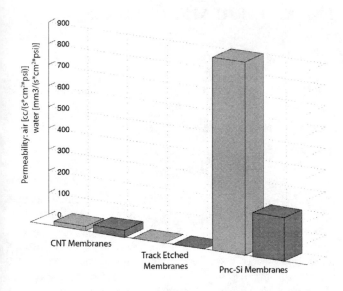

Figure 3 – Air and water permeability of nanomembranes. Here we contrast the permeabilities of CNT membranes and track etched (TE) membranes from [5] with the permeabilities of pnc-Si membranes. Pnc-Si membranes are an order of magnitude thinner than CNT and TE membranes, and we show here that there is less resistance to air and water flow through pnc-Si membranes.

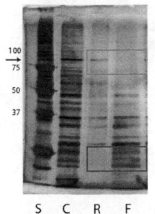

Figure 4 – Brain Extract Separation. S = Molecular Weight Standards, C = Control (protein that has not seen the membrane), R = Retenetate, F = Filtrate. Here we show a fractionation of 10 mg/mL bovine brain extract. The red box outlines a ~90kD protein that appears in the retenetate, but not the filtrate. The blue box outlines a series of low molecular weight proteins that appear more heavily in the filtrate. We consider our cutoff to be around 75k in this separation.

membranes [5]. Permeability of air and water is a magnitude higher though pnc-Si membranes, as expected due to the relative thickness of track etched and nanotube membranes.

In conclusion, we show that the characteristics of pnc-Si membranes enable them to perform better than other current nanomembranes. Sharp molecular separations are achieved because of the well-defined pores and lower surface area for adsorption. The air and water permeaility of pnc-Si membranes is much greater than cylindrical pore membranes, and presumably so will be the molecular transport. In the future we plan to study the effects of pore distribution on molecular weight cutoffs and the efficiency of filtration through pnc-Si membranes.

REFERENCES

[1] M. Menon and A. Zydney. J. Mem. Sci, 181, 179-184, 2001.

[2] C. van Rijn. "Nano and micro engineered membrane technology." Volume 10. Elsevier, 2004.

[3] R. Furneaux, W. Rigby, and A. Davidson. Nature, 337, 147-149, 1989.

[4] C. van Rijn, M. van der Wekken, W. Nijdam, and M. Elwenspoek. J. Microelectromechanical Sys, 6, 48-54, 1997.

[5] J. Holt et al., Science, 312, 1034-1037

[6] C. Striemer, T. Gaborski, J. McGrath, and P. Fauchet. Nature, 445, 749-753, 2007.

INTERACTION OF POLYAMIDOAMINE (PAMAM) DENDRIMERS WITH GLASSY CARBON SUPPORTED BILAYER LIPID MEMBRANES

Sachin R. Jadhav* and R. Mark Worden**

*Department of Chemical Engineering and Materials Science, Michigan State University, East Lansing, MI-48824, jadhavsa@msu.edu
** Department of Chemical Engineering and Materials Science, Michigan State University, East Lansing, MI-48824, worden@egr.msu.edu

ABSTRACT

Electrochemical techniques were used to monitor the formation of a 1,2-dioleoyl-*sn*-glycero-3-phosphocholine supported bilayer lipid membrane (sBLM) on a glassy carbon electrode and to investigate interactions of polyamidoamine (PAMAM) dendrimers with the sBLM. Dendrimers of generation 2 through generation 4 (G2-G4) did not cause significant damage to the sBLM, whereas G5-G7 dendrimers created large defects in sBLM. In cyclic voltammetry studies, the peak current obtained after dendrimer addition was used to estimate the defect area. Impedance spectroscopy showed an increase in capacitance values after addition of G5-G7 dendrimers, consistent with a decrease in sBLM surface coverage. Nanoparticle mass, diameter, and surface charge density all increase significantly with generation number, suggesting that these variables may influence nanoparticles' ability to rupture sBLM. The method presented here provides a convenient and sensitive way to detect interactions between nanoparticles and lipid bilayers, and may be useful in evaluating nanoparticles' safety profile or suitability as drug or gene delivery vehicles.

Keywords: Glassy carbon electrode, supported bilayer lipid membrane, PAMAM, dendrimer, nanoparticles, impedance spectroscopy, cyclic voltammetry, drug delivery.

1 INTRODUCTION

Polyamidoamine (PAMAM) dendrimers are hyperbranched polymeric nanoparticles that are formed in concentric layers. As each subsequent layer (generation) is added, the molecular size, molecular weight, and number of functional groups increases by a known amount. This property makes dendrimers well suited for use as monodisperse nanoparticles having a desired size, shape, and surface charge [1]. Recently, PAMAM dendrimers have also been investigated as non-viral vectors for the gene delivery [2, 3] that offer the potential for low immunogenicity [4, 5] and may protect the gene from DNase activity. The mechanism of gene delivery involves internalization of the charged polymer by host cells, through a mechanism that may involve disruption of the host cell membrane [6]. Dendrimers' ability to cross cell membranes also makes them an alternative vehicle for drug delivery [7, 8]. Atomic force microscopy studies have shown that PAMAM dendrimers form 15-40 nm holes in 1,2-di-myristoyl-*sn*-glycerophosphocholine (DMPC) supported bilayer lipid membrane (sBLM) on mica [9-12]. These studies concluded that the high surface charge density of amino groups found in higher generation dendrimers was responsible for disruption of bilayer membranes.

sBLMs are more robust and stable than unsupported planar bilayer membranes [13] and can be deposited on a variety of hydrophilic surfaces, such as oxidized glassy carbon, silica, and mica [13]. Even though a sBLM is only about 5 nm thick, it is an excellent electrical insulator. Thus, disruption of the sBLM by nanoparticles could, in principle, be measured electrochemically. This paper describes a rapid, sensitive, and convenient electrochemical method to detect disruption of sBLM by nanoparticles. The approach involves fabricating a sBLM on a glassy carbon electrode (GCE), and then using electrochemical methods to monitor interactions between nanoparticles and the sBLM. Disruption of a 1,2-dioleoyl-*sn*-glycero-3-phosphocholine (DOPC) sBLM formed on a GCE was investigated using cyclic voltammetry (CV) and electrochemical impedance spectroscopy (EIS).

2 EXPERIMENTAL

2.1 Methods

Each GCE (Bioanalytical Systems, West Lafayette, IN) was sequentially polished with 1000, 300 and 50 nm alumina slurry, followed by washing with de-ionized (DI) water and methanol. The electrode was ultrasonicated for 2 min in DI water to remove physically adsorbed alumina, given final rinses with methanol and DI water, and then dried under a nitrogen stream. The GCE was placed in 100 mM NaCl solution and oxidized at a potential of 1500mV for 3 min [14]. It was then washed with DI water and dried under a nitrogen stream. Five µL of dioleoyl-*sn*-glycero-phosphocholine (DOPC) lipid solution (5 mg/mL in

NSTI-Nanotech 2008, www.nsti.org, ISBN 978-1-4200-8503-7 Vol. 1

chloroform) was applied to the electrode surface, and the electrode was immediately immersed in an electrolyte solution (100 mM sodium phosphate buffer, pH 7.4, containing 1 mM of potassium ferrocyanide and 1 mM of potassium ferricyanide). After 20 min, PAMAM nanoparticles (G2 - G7, provided by Dr. Steve Kaganove, Michigan Molecular Institute, Midland, MI) were added to obtain a final concentration of 20 μM (based on the surface amino groups). CV was then used to monitor changes in the sBLM's electrochemical properties until a steady-state situation was observed.

2.2 Instrumentation

Electrochemical measurements were performed using a CHI660B electrochemical workstation (CH Instruments Inc., Austin, TX). EIS was performed in 100 mM sodium phosphate buffer, pH 7.4, containing 1 mM potassium ferrocyanide and 1 mM potassium ferricyanide. The dc potential was $E^o = 230$ mV vs a Ag/AgCl reference electrode, and a 5 mV sinusoidal potential was applied across the frequency range of 0.1 to 10,000 Hz. A modified Randle's equivalent circuit model [15] was fit to the impedance data using Z-view software (Scribner Associates, Southern Pines, NC). CV was performed in same electrolyte solution as used for EIS studies. The potential was cycled in a range of 500 mV to -200 mV at a scan rate of 50 mV/s.

3 RESULTS AND DISCUSSION

sBLM have been deposited on hydrophilic supports such as mica and silica and on the conductive materials such as platinum. sBLM formation on metal surfaces requires cutting the metal while it is immersed in the lipid solution, to ensure that the lipids contact a fresh metal surface [13, 16]. GCE provide a convenient and cost-effective alternative to metal electrodes such as platinum, offering good conductivity, a wide potential window, and the possibility of chemical functionalization [17]. To increase the hydrophilicity of the surface, we oxidized the GCE at 1.5 V for 3 min in a 100 mM NaCl solution. The resulting formation of carboxylate ions imparts a net negative charge to the surface of electrode, which associates with the positively charged choline moiety of the DOPC lipids.

Upon cooling, lipid bilayers undergo a phase transition from a highly fluid, liquid-crystalline phase (L_α) to a more viscous, gel phase (L_β). Mecke et al [18] reported that polycationic polymers selectively interact with the lipid bilayers in the liquid-crystalline phase (i.e., above the transition temperature). To ensure that the sBLM was fully fluid during the room-temperature studies, the unsaturated phospholipid DOPC (phase transition temperature = -4°C) was chosen. In contrast, saturated phospholipids have a transition temperature of about 28°C.

sBLM formation was monitored using CV and EIS. Figure 1A shows the CV curves for bare GCE and DOPC sBLM. The bare GCE curve exhibited the classic duck shaped profile, with well-defined peaks for ferrocyanide oxidation and ferricyanide reduction. The cathodic and anodic peak currents were 12.3 and 12.5 μA, respectively. The peak splitting of 70 mV was indicative of a reversible redox reaction. The sBLM curve was essentially flat, with no redox peaks, indicating that the sBLM forms a effective dielectric barrier that prevents redox species from reaching the electrode surface.

Figure 1B shows EIS curves of the bare GCE and the DOPC sBLM. The bare GCE showed a very small charge transfer resistance (6 Ωcm^2), and most of the impedance spectrum was representative of Warburg impedance, indicating mass transfer resistance. The bare GCE capacitance (C_{GCE}) was 35 μF/cm^2. The modified Randle's equivalent circuit shown in Figure 1B (inset) was fit to the sBLM impedance data, giving a charge transfer resistance (R_{ct}) of 85 KΩcm^2. R_{ct} values greater than 1 MΩcm^2 have been reported for tethered lipid bilayer systems, suggesting that the sBLM formed on the GCE has pin-hole defects. Nevertheless, the sBLM still effectively shields the electrode from the ferricyanide.

The Randle's equivalent circuit assumed for the sBLM involves a parallel arrangement of capacitors: one corresponding to the membrane (C_m), and the other corresponding to membrane defects (C_{defect}). The overall capacitance is the sum of both capacitances. At the defects, where the electrolyte can contact the electrode surface, the capacitance per area of defect should be similar to C_{GCE}. Before dendrimer addition (Figure 1B, squares), the overall capacitance was determined to be 0.75 μF/cm^2, a value typical of those reported for sBLM-coated electrodes [19]. Under these conditions, the fraction of the total area occupied by defects, and thus the C_{defect} term, is negligibly small.

After sBLM formation, PAMAM nanoparticles were added to the electrolyte solution, and their interaction with the sBLM was monitored electrochemically. The system typically reached steady state 15 min after the addition of dendrimer solution, but the electrochemical parameters were monitored for 30 min to ensure a constant value. The dendrimer concentration was calculated based on surface amino groups to ensure that the same number of charged groups was used in all experiments.

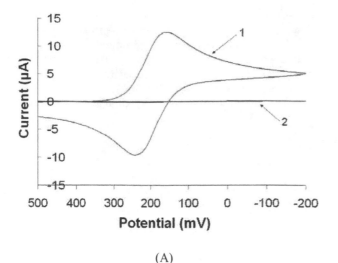

(A)

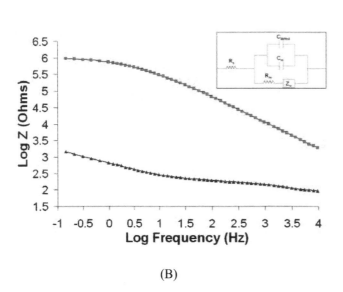

(B)

Figure 1: (A) Cyclic voltammogram for bare GCE (curve 1) and sBLM on GCE (curve 2) and, (B) impedance spectrum of sBLM (squares) compared with bare GCE (triangles). The assumed equivalent circuit is shown in the inset.

The CV curves obtained after addition of G2 and G3 dendrimers are shown in Figure 2A. No redox peaks were obtained for G2, and only a small peak current of 0.12 μA was obtained for G3. These results indicate that lower generation dendrimers cause minimal defects in the sBLM, and are consistent with the AFM study by Mecke *et al* [9], which found that G2 and G3 primarily adsorbed on the sBLM, without creating obvious holes. The CV curves obtained after addition of G4-G7 dendrimers are shown in Figure 2B. The peak current increased roughly linearly with generation number for G5 – G7, suggesting that larger dendrimers more effectively create defects in the bilayer, through which the redox species can reach the electrode surface. The corresponding EIS curves (data not shown) indicated that the overall capacitance increased significantly with generation number, presumably due to a larger fraction

of electrode area being exposed to the electrolyte solution, and hence a significant C_{defect} value.

The peak current values for G2-G7 dendrimers are summarized in Table 1. These values were used with the following equation to calculate the approximate sBLM defect area value for each generation [15]:

$$i_p = (2.69 \times 10^5) \, n^{3/2} A \, C_{bulk} (Dv)^{1/2}$$

where i_p is the peak current (μA), n is the number of electrons transferred (1), A is the area of electrode (0.0706 cm^2), C_{bulk} is the bulk concentration of redox species (1×10^{-6} mol/cm^3), D is the diffusion coefficient (cm^2/s) and v is the scan rate (0.05 V/s). The diffusion coefficient of the redox species (8.4×10^{-6} cm^2/s) was calculated from the slope of the plot of peak current *vs* scan rate for a bare GCE surface.

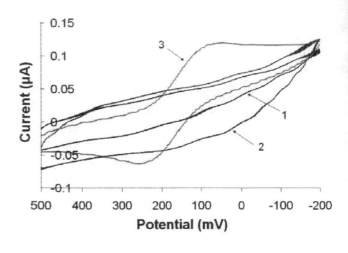

(A)

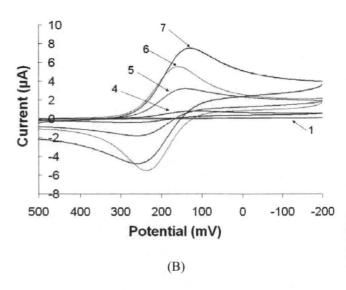

(B)

Figure 2: Cyclic voltammograms showing the effect of (A) lower generation PAMAM dendrimers: G2 (curve 2) and G3 (curve 3) and, (B) higher generation PAMAM dendrimers: G4 (curve 4), G5 (curve 5), G6 (curve 6) and G7 (curve 7). Curve 1 represents the sBLM before dendrimer addition.

Generation	Mol. weight	Surface amino groups/ molecule	Diameter (nm)	Peak current (µA)	Area of electrode exposed (mm²)
Bilayer	N/A	N/A	2.2	0.07	0.0402
2	3256	16	2.9	0.1	0.0574
3	6909	32	3.6	0.12	0.0689
4	14215	64	4.5	0.69	0.3962
5	28826	128	5.3	3	1.7229
6	58048	256	6.7	5.24	3.0093
7	116493	512	7.6	7.09	4.0718
Bare GCE	N/A	N/A	N/A	12.3	7.0639

Table 1: Characteristics of PAMAM dendrimers, along with the parameters showing their effect on SBLM

Based on the data provided in Table 1, G7 dendrimers have a 36 times greater mass, 2.6 times greater diameter, and a 4.6 times greater surface charge density than G2 dendrimers. The relative contributions of these properties on the efficacy of sBLM disruption by PAMAM dendrimers is not yet understood and requires further study.

4 CONCLUSION

A biomimetic interface consisting of a sBLM deposited on a GCE was used to study the interaction of G2-G7 PAMAM dendrimers with lipid bilayers. CV and EIS studies showed that dendrimer addition dramatically decreased impedance and increased ferricyanide peak currents, presumably by creating defects in the sBLM. Higher generation dendrimers had a much stronger effect on the sBLM than did lower generation dendrimers. CV data were used to estimate the defect area as a function of generation. EIS results showed an increase in overall capacitance values following dendrimer addition, consistent with a decrease in sBLM surface coverage. The molecular mass, diameter, and surface charge density of PAMAM dendrimers all increase with generation number, although the relative roles of these nanoparticle properties in efficacy of lipid bilayer disruption is not understood. The method described in this paper provides a rapid and convenient way to screen nanoparticles for strong interactions with lipid bilayers and may have utility in assessing nanoparticles' toxicity or suitability for intracellular genes or drug delivery.

Acknowledgement

We would like to thank Dr. Steve Kaganove and Tracy Zhang from Michigan Molecular Institute, Midland, MI, for providing the samples of PAMAM dendrimers. Financial support from the National Science Foundation under award number DBI-0649847 is gratefully acknowledged.

REFERENCES

[1] F.W. Zeng, S.C. Zimmerman, Chemical Reviews, 97, 1681, 1997.

[2] C. Dufes, I.F. Uchegbu, A.G. Schatzlein, Advanced Drug Delivery Reviews, 57, 2177, 2005.

[3] J.H. Zhou, J.Y. Wu, N. Hafdi, J.P. Behr, P. Erbacher, L. Peng, Chemical Communications, 2362, 2006.

[4] R. Duncan, L. Izzo, Advanced Drug Delivery Reviews, 57, 2215, 2005.

[5] N. Malik, R. Wiwattanapatapee, R. Klopsch, K. Lorenz, H. Frey, J.W. Weener, E.W. Meijer, W. Paulus, R. Duncan, Journal of Controlled Release, 65, 133, 2000.

[6] D.M. Domanski, B. Klajnert, M. Bryszewska, Bioelectrochemistry, 63, 189, 2004.

[7] H.L. Crampton, E.E. Simanek, Polymer International, 56, 489, 2007.

[8] R. Esfand, D.A. Tomalia, Drug Discovery Today, 6, 427, 2001.

[9] A. Mecke, I.J. Majoros, A.K. Patri, J.R. Baker, M.M.B. Holl, B.G. Orr, Langmuir, 21, 10348, 2005.

[10] S.P. Hong, A.U. Bielinska, A. Mecke, B. Keszler, J.L. Beals, X.Y. Shi, L. Balogh, B.G. Orr, J.R. Baker, M.M.B. Holl, Bioconjugate Chemistry, 15, 774, 2004.

[11] S.P. Hong, P.R. Leroueil, E.K. Janus, J.L. Peters, M.M. Kober, M.T. Islam, B.G. Orr, J.R. Baker, M.M.B. Holl, Bioconjugate Chemistry, 17, 728, 2006.

[12] A. Mecke, S. Uppuluri, T.M. Sassanella, D.K. Lee, A. Ramamoorthy, J.R. Baker, B.G. Orr, M.M.B. Holl, Chemistry and Physics of Lipids, 132, 3, 2004.

[13] E. Sackmann, Science, 271, 43, 1996.

[14] L.W. Du, X.H. Liu, W.M. Huang, E.K. Wang, Electrochimica Acta, 51, 5754, 2006.

[15] a.L.F. A. Bard, Electrochemical Methods, 2nd edition ed., John Wiley and Sons, New York, 2001.

[16] H.T.T.a.A.L. Ottova, Membrane Biophysics as viewed from experimental bilayer lipid membranes, 1 ed. ed., Elsevier, Amsterdam; New York, 2000.

[17] O. Niwa, H. Tabei, Analytical Chemistry, 66, 285, 1994.

[18] A. Mecke, D.K. Lee, A. Ramamoorthy, B.G. Orr, M.M.B. Holl, Langmuir, 21, 8588, 2005.

[19] H. Gao, J. Feng, G.A. Luo, A.L. Ottova, H.T. Tien, Electroanalysis, 13, 49, 2001.

Lithium selective adsorption on low-dimensional TiO₂ nanoribbons

Qin-Hui Zhang, Shao-Peng Li, Shu-Ying Sun, Xian-Sheng Yin, Jian-Guo Yu

State Key Lab of Chemical Engineering, East China University of Science and Technology,

Shanghai, P. R. China, qhzhang@ecust.edu.cn

ABSTRACT

Mesoporous titania nanoribbons were synthesized via an optimized soft hydrothermal process and the derived titania ion-sieves with lithium selective adsorption property were accordingly prepared via a simple solid-state reaction between Li_2CO_3 and TiO_2 nanomaterials followed by the acid treatment process to extract lithium from the Li_2TiO_3 ternary oxide precursors. First, mesoporous titania nanoribbons were prepared and the formation mechanism was discussed; Second, the physical chemistry structure and texture were characterized by powder X-ray diffraction (XRD), (high-resolution) transmission electron microscopy (TEM/HRTEM), selected-area electron diffraction (SAED) and N_2 adsorption-desorption analysis (BET); Third, the lithium selective adsorption properties were studied by the adsorption isotherm, adsorption kinetics measurement and demonstrated with the distribution coefficient of a series of alkaline and alkaline-earth metal ions.

Keywords: adsorption; ion-sieve; lithium; low-dimensional; nanoribbons; titania

1 INTRODUCTION

Low-dimensional TiO_2-related materials with high morphological specificity, such as nanotubes, nanowires, nanosheets, nanofibers and nanoribbons had attracted particular interests in advanced application as catalyst [1], semiconductors for solarenergy conversion [2], gas sensors [3] and electrochemical materials [4], as it was discovered by Kasuga [5]. However, doubts exist concerning the formation and construction of these low-dimensional TiO_2-related materials [6, 7], and the lithium selective adsorption into titania ion-sieves from aqueous resources including brine or sea water has been rarely studied.

In this paper, the optimized synthesis, characteristic and consequent lithium ion-sieve property of titania nanoribbon were systematically studied. First, the optimized synthesis, with a direct soft chemical hydrolysis from TiO_2 nanoparticles to large quantities of quite uniform titania nanoribbons, was realized and the titania ion-sieves were accordingly prepared via a solid-state reaction followed by the acid treatment process to extract lithium from the Li_2TiO_3 precursor. Second, the structure and texture characteristics were characterized by XRD, TEM/HRTEM, SAED and BET analysis; Third, the lithium selective adsorption properties were studied by the batchwise adsorption isotherm and demonstrated with a series of

alkaline and alkaline-earth metal ions, including Li^+, Na^+, K^+, Ca^{2+} and Mg^{2+}.

2 EXPERIMENTAL

2.1 Synthesis of titania nanoribbons and lithium ion-sieves

All chemicals used in this work were AR reagents, except where otherwise indicated. P25 (5 g, from Degussa Co. Ltd.) and an aqueous solution of NaOH (10 mol·l⁻¹, 500 ml) were mixed for 0.5 h in an ultrasonic bath, then transferred into a Teflon-lined stainless steel autoclave, sealed, and maintained at 448 K for 48 h; after the reaction was completed, the resulted white precipitate was separated by filtration and washed with 0.1 mol·l⁻¹ hydrochloric acid solution and deionized water until the conductance of the supernatant lucid solution reached the same level with the deionized water (the pH value now is 7), followed by ultrasonic assisted dispersion in anhydrous ethanol for 0.5 h and dried at 333 K for 8 h. By optimizing the NaOH concentration and the reaction conditions of hydrothermal process, titania nanoribbons could be produced by using different particle size (from several nanometers to several hundreds of micrometers) of the starting anatase or rutile TiO_2 powder.

The as-synthesized titania nanoribbons (denoted as TO) were mixed with Li_2CO_3 with a molar ratio of 1:1, and calcined at 1173 K for 24 h in static air to insert the lithium into the Ti-O lattice and finally to form the Li_2TiO_3 tenary oxide precursors (denoted as LTO); the lithium extraction from Li_2TiO_3 precursors was carried out in 0.1 mol·l⁻¹ HCl solution at 333 K for 72 h until the lattice lithium was completely extracted (in situ Measured by Metrohm 861 IC with Mrtrosep C2 100/4.0 column); the acid-treated materials were filtered, washed with deionized water and dried at 333 K for 8 h to obtain the final titania ion-sieves (denoted as STO).

2.2 Characterization of samples

The bulky phase of the samples were examined by powder XRD analysis on a Rigaku D/max 2550 X-ray diffractometer with monochromatized CuKα radiation (λ = 1.54056 Å), operating at 40 kV, 100 mA and scanning rate of 10° · min-1; The pore structure of the derived aggregates was characterized by N_2 adsorption at 77 K using an adsorption apparatus (Micromeritics, ASAP 2010 V5.02); surface area of the samples was determined from the

Brunauer-Emmett-Teller (BET) equation and pore volume, from the total amount of nitrogen adsorbed at relative pressures of ca. 0.96; The microstructure and morphology of the samples were analyzed using a low-magnification JEOL JEM-1200EX TEM (60 kV) and the layered structure of the nanoribbons was revealed by HRTEM and SAED on a JEOL JEM-2100F TEM (200 kV) after the samples were dispersed by ultrasonic in anhydrous ethanol for 10 min and then placed onto the cuprum grid for observations.

2.3 Lithium selective adsorption measurement

The lithium ion adsorption isotherm was carried out by stirring (130 r·min^{-1}) 100.0 mg titania ion-sieves in 50.0 ml LiCl solution (pH = 9.19, adjusted by buffer solution comprised of 0.1 mol·l^{-1} NH$_4$Cl and 0.1 mol·l^{-1} NH$_3$·H$_2$O, the molar ratio equal to 2.0) with different initial Li$^+$ concentration for about 144 h at 303 K till the attainment of equilibrium and the Li$^+$ content in supernatant lucid solution was determined in situ by IC; the exchange capacity or the amount of Li$^+$ adsorbed per gram of TiO$_2$ ion-sieve was calculated according to equation (1):

$$Q = (C_0 - C_e) \cdot V / W \qquad (1)$$

in which, Q is the amount of metal ion adsorbed per gram adsorbent, mmol·g^{-1}; C_0, the initial concentration of metal ions, mol·l^{-1}; C_e, equilibrium concentration of metal ions, mol·l^{-1}; V, solution volume, ml; W, adsorbent weight, g.

The uptake behaviors of lithium ions compared with other ions in brine or seawater was carried out by stirring 100.0 mg of the ion-sieves in 10.0 ml solution (pH = 9.19) containing Li$^+$, Na$^+$, K$^+$, Ca^{2+} and Mg^{2+} of about 10 mmol·l^{-1}, respectively, for 144 h at 303 K till the attainment of equilibrium and the metal ions in supernatant lucid solution were also determined in situ by IC; The distribution coefficient (K_d) and separation factor (α^{Li}_{Me}) were calculated according to equation (2) and (3):

$$K_d = (C_0 - C) \cdot V / (C \cdot W) \qquad (2)$$

$$\alpha^{Li}_{Me} = K_d (Li) / K_d (Me) \quad Me: Li, Na, K, Ca \text{ and } Mg \qquad (3)$$

3 RESULTS AND DISCUSSION

The Powder XRD diffraction patterns of the nanoribbon TiO$_2$ oxide, Li$_2$TiO$_3$ ternary oxide precursor and TiO$_2$ ion-sieve are presented in Figure 1.

Figure 1 TO give the pattern of the as-synthesized titania nanoribbons. The relatively broad Bragg peaks, due to the finite size of the particles, could not be assigned to the anatase, rutile or any known phase of titania in JCPDS card, nevertheless, the strong reflection peak at low degree of 2θ = 11.06 ° might correspond to the interlayer (or

tunnel-tunnel) spacing of titania with a long-range ordered structure; combined with TEM/HRTEM analysis shown in Figure 2 (TO) and 3, it was indexed as a kind of layered or tunnel-constructed titania units.

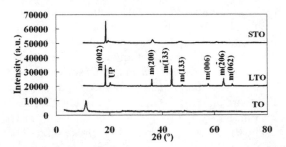

Figure 1: Powder XRD patterns of the nanoribbon TiO$_2$ oxide, Li$_2$TiO$_3$ precursor and derived TiO$_2$ ion-sieve. m (hkl): monoclinic Li$_2$TiO$_3$ crystal face; UP: unknown phase.

The XRD patterns of Figure 1 LTO were indexed to almost pure monoclinic phase [S.G.: C2/c (15)] of Li$_2$TiO$_3$ (JCPDS 33-831, a = 5.069Å, b = 8.799 Å, c = 9.759 Å), although there existed a faint unknown peak at 2θ = 20.24°. All the peaks became sharper than those of TO, indicating larger size distribution than the tiania nanoribbons as shown in Figure 2, which was resulted from the high temperature calcinations during the lithium insertion into Ti-O lattice procedure.

Figure 1 STO was of the derived titania ion-sieves. The reflection patterns of STO ion-sieve were somewhat similar to those Li$_2$TiO$_3$ precursor in Figure 1 LTO with broader and weaker peaks, however, they should not be indexed as monoclinic Li$_2$TiO$_3$ since nearly 96.5% of the lithium in Li$_2$TiO$_3$ had been extracted during the acid treatment process, quantitatively measured by in situ IC; and any known titania in JCPDS card was not matched. It implied that, although the Li$^+$ extracted from the titania nanoribbon derived Li$_2$TiO$_3$ precursor resulted in smaller lattice dimension (with the reflection peaks shifting a little to lower degrees), the Ti-O lattice in LTO precursor was very stable during the Li$^+$ extraction process and the locations of titanium in the crystal structure were well maintained. The specific characteristic arised partly from their rigid structure with little swelling or shrinking in aqueous phases and brought about a strong steric effect or an ion-sieve effect for various ions depending on the hydrated or dehydrated size of the adsorbed ions.

We carefully compared the synthesis process with those in the related literature, and confirmed the point of view of Li et al. [8] and Kasuga et al. [5, 6] that the washing and dispersing process was key to obtain the layered or tubular structure. In such earlier reports on one-dimensional titania nanostructures, the washing process merely with water would result mainly vesicles, shrinking under irradiation with an electron beam and finally changing to aggregate-like depositions. In this case, anhydrous ethanol was used to disperse the white precipitate after washing with HCl solution and deionized water, and typical low-magnification TEM image was shown in Figure 2 TO. Samples dispersed in ethanol yielded mainly monodisperse nanoribbons with 20-100 nm in width and several micrometers in length, implying that ethanol was a good dispersant and aided the formation of low-dimensional nanostructure. Additionally, the use of ethanol enabled the formation of aggregation-free titania nanoribbons with quite uniform geometry, evidenced by the pore-size distribution obtained from the Barrett-Joyner-Halenda (BJH) deposition curve (inset of Figure 2 TO), indicating that the ethanol washed and dispersed nanoribbons maintained a pore size distribution centered around 43.7 nm, with no other peaks in the range from 1 to 100 nm. The pore-size distribution, unlike the TEM observations, was obtained statistically, thus the appearance of the single broad peak was attributed to capillary condensation in the slit-shaped mesopores with parallel walls and implied the separation of most nanoribbons from one another although some of the nanostructures looked thicker than others.

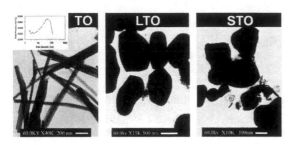

Figure 2: TEM images of TiO_2 oxides, Li_2TiO_3 ternary oxide precursors, and final TiO_2 ion-sieves.

HRTEM image (Figure 3-a) showed an individual nanoribbon with a rolled end; from the rolled region, it was clearly seen that the ribbon was very thin (<5 nm). Due to the limit of our current reaction apparatus, the effect of a temperature higher than 453 K would be further investigated; Figure 3-b was a typical HRTEM image of a nanoribbon with well-defined structure. Three sets of lattice fringes could be observed in the lattice resolved image with the face intervals of 0.79, 0.37 and 0.36 nm, respectively, exactly corresponding to the stripe images of the (200), (110) and (202) lattice plane (inset SAED pattern of Figure 3-b) of the monoclinic $H_2Ti_3O_7$. The fringes parallel to the (200) plane corresponded to an interplannar distance of about 0.79 nm and this fringe spacing was also comparable to the shell spacing (0.75-0.8 nm) of titania nanotubes reported recently [8, 9]. The other two sets of fringes, with smaller spacing of 0.36 and 0.37 nm, could be correspond to the (202) and (110) plane of the $H_2Ti_3O_7$ crystal structure; In many related reports to analyze the structure of the low dimension nanostructures, similar phenomena were discovered and described as $H_2Ti_3O_7 \cdot xH_2O$ [9, 10, 11], $Na_xH_{2-x}Ti_3O_7$ [8], $H_2Ti_4O_9 \cdot H_2O$ [12], and $H_2Ti_2O_4(OH)_2$ [13]. The SAED pattern also confirmed that the titania nanobibbons were single crystal.

After titania nanoribbons were calcinated with Li_2CO_3 to form Li_2TiO_3 at 1173 K for 24 h and treated with HCl solution for about 72 h to extract the lattice lithium, the obtained ion-sieve samples all changed to irregular particles with much larger size as shown in Figure 2 LTO and STO. This was resulted from the agglomeration during the high temperature solid reaction process and the lattice protons exchanged with lithium ions might also cause the destruction of hydrogen bonding, and as a result the low dimensional nanomaterials were destroyed to irregular particles, which was also evidenced in our related work focused on the effect of different polymorphs and the size effect of MnO_2 nanocrystals on the lithium ion extraction process from spinel $LiMn_2O_4$ [14].

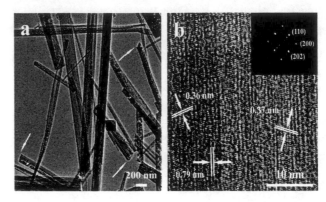

Figure 3: HRTEM images of one straight nanoribbon (a) and HRTEM (inset with SAED) image of a single titania nanoribbon growing along (200) (b).

The porosity of the ion-sieves should be closely related to their aggregation framework and Figure 4 showed the N2 adsorption-desorption isotherm of the derived ion-sieves. The nanoribbon derived ion-sieve STO had a much more alleviated desorption delay branch, an indication of the loosening of the aggregate structure, which might be resulted from the layered structure from the titania nanoribbons. Pore size distribution analysis via the DFT method, applicable for a complete range of pore size, was inserted in Figure 4, also indicating that the derived ion-sieves afforded a bimodal mesoporeous size distribution the peak pore size (D_p) centered around 3.9 and 11 nm of STO.

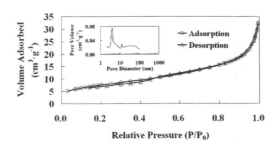

Figure 4: N_2 adsorption-desorption isotherms and pore distribution curves (insets) of the STO ion-sieves.

Figure 5 shows the Li$^+$ adsorption isotherm of STO ion-sieve and simulation according to Freundlich equation. The data present linearity with the congruence of R^2 = 0.8765, indicating that the Li$^+$ exchange process in the experiment is accorded with the Freundlich adsorption isotherm with the maximum Li$^+$ adsorption quantity to be 3.69 mmol·g^{-1}. The adsorption constants are calculated to be n = 12.56, K$_f$ = 3.47 via the linear slope and intercept with the y-axis.

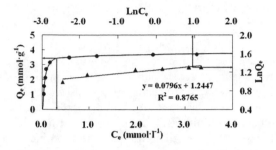

Figure 5: Li$^+$ adsorption isotherm of STO ion-sieve and simulation according to Freundlich equation. T = 303 K, pH = 9.19, V = 50 ml, W = 100 mg.

Metal Ion	Q$_e$ mmol·g^{-1}	K$_d$ ml·g^{-1}	α^{Li}_{Me}	CF 10^{-3}·l·g^{-1}
Li$^+$	0.997	13837.70	1.0	99.48
Na$^+$	0.044	20.83	664.3	17.25
K$^+$	0.003	0.31	45352.1	0.30
Mg^{2+}	0.209	17.00	814.0	14.53
Ca^{2+}	0.024	4.20	3294.0	4.03

Table 1: Li$^+$ adsorption selectivity on STO ion-sieves. T = 333 K, pH = 9.19, V = 10 ml, W = 0.10 g

Table 1 showed the selectivity of lithium ions compared with uptake behaviors for other typical alkaline and alkaline-earth metal ions in brine or sea water including Na$^+$, K$^+$, Ca^{2+} and Mg^{2+}. The equilibrium distribution coefficients (K$_d$) of lithium ion were 13837.7 of STO, and the selectivity of all metal cations was in the order of Li$^+$>> Na$^+$> Mg^{2+}> Ca^{2+}> K$^+$, indicating much higher selectivity for Li$^+$ than that of Na$^+$, K$^+$, Ca^{2+}, and Mg^{2+}; Additionally, the remarkable discrimination of the separation factor (α^{Li}_{Me}) between Li$^+$ and other cations (with the lowest multiple of 664.3) also meant that there existed no cross effect during the competitive adsorption process in the complex solution.

4 CONCLUSION

Titania nanoribbons with 20-100 nm in width and several micrometers in length were synthesized by a simple soft hydrothermal method (448 K for 48 h) followed by washing with HCl and deionized water and dispersing in ethanol. The equipment required was simple and alkali solutions were reusable, implying that this method had potential utilization for large-scale industrial production. Li$_2$TiO$_3$ precursor and final TiO$_2$ ion-sieves were prepared through the process of calcinations at 1173 K for 24 h, acid treatment with HCl for 72 h and dryness at 333 K for 8 h. According to XRD, TEM/HRTEM, SAED and BET characterization, the as-synthesized nanoribbons were mainly mesoporous monoclinic H$_2$Ti$_3$O$_7$ and the final TiO$_2$ ion-sieves were proved to have a remarkable lithium selective adsorption capacity, implying the promising application in lithium extraction from aqueous resources including brine or seawater with low lithium content, a novel utilization aspect for low-dimensional titania up to date.

5 ACKNOWLEDGEMENT

This work was supported by NSFC (No. 20576031 and 20576014), Risingstar Project (No. 05QMX1414) and National 863 Project (2008AA06Z111).

REFERENCES

[1] D. W. Bahnemann, S. N. Kholuiskaya, R. Dillert, A. I. Kulak and A. I. Kokorin, Appl. Catal., B: Environmental, 36, 161, 2002.
[2] A. Hagfeldt, M. Grätzel, Chem. Rev., 95, 49, 1995.
[3] Y. F. Zhu, J. J. Shi, Z. Y. Zhang, C. Zhang and X. R. Zhang, Anal. Chem., 74, 120, 2002.
[4] I. Moriguchi, R. Hidaka, H. Yamada, T. Kudo, H. Murakami and N. Nakashima, Adv. Mater., 18, 69, 2006.
[5] T. Kasuga, M. Hiramatsu, A. Hoson, T. Sekino and K. Niihara, Langmuir, 14, 3160, 1998.
[6] T. Kasuga, M. Hiramatsu, A. Hoson, T. Sekino and K. Niihara, Adv. Mater., 11, 1307, 1999.
[7] Y. C. Zhu, H. L. Li, Y. R. Koltypin, Y. R. Hacohen and A. Gedanken, Chem. Commun., 24, 2616, 2001.
[8] X. M. Sun and Y. D. Li, Chem. Eur. J., 9, 2229, 2003.
[9] G. H. Du, Q. Chen, R. C. Che, Z. Y. Yuan and L. M. Peng, Appl. Phys. Lett., 79, 3702, 2001.
[10] Q. Chen, W. Z. Zhou, G. H. Du and L.-M. Peng, Adv. Mater., 14, 1208, 2002.
[11] Y. Suzuki and S. Yoshikawa, J. Mater. Res., 19, 982, 2004.
[12] A. Nakahira, W. Kato, M. Tamai, T. Isshiki, K. Nishio and H. Aritani, J. Mater. Sci., 39, 4239, 2004.
[13] M. Zhang, Z. S. Jin, J. W. Zhang, X. Y. Guo, J. J. Yang, W. Li, X. D. Wang and Z. J. Zhang, J. Molec. Catal. A: Chem., 217, 203, 2004.
[14] Q. -H. Zhang, S. Y. Sun, S. P. Li and J. G. Yu, Chem. Eng. Sci., 62, 4869, 2007.

Nanoporous Alumina Membranes Based Microdevices for Ultrasensitive Protein Detection

M. G. Bothara*, R. K. Reddy*, T. Barrett**, J. Carruthers*** and S. Prasad*

*Electrical and Computer Engineering Department, Portland State University,
160-11 Fourth Avenue Building, 1900 SW Fourth Avenue, Portland, OR, USA, sprasad@pdx.edu
**Department of Veteran Affairs, Oregon Health Sciences University Portland OR, barretth@ohsu.edu
***Department of Physics, Portland State University, Portland, OR, USA,

ABSTRACT

Electrical and chemical properties of nanostructured alumina membranes have been used to improve sensitivity and performance of Si-based microdevices for protein sensing. Trapping of proteins within the nanoscale well-like structures is an experimental demonstration of "molecular crowding" phenomenon retaining the functionality of proteins due to confinement in small spaces i.e. nanowells. Protein conjugation in nanowells causes charge perturbation in the electrical double layer at the interface between biomolecule and gold electrode, measured as capacitance change. We have demonstrated the detection of cardiovascular biomarkers, C-reactive protein (CRP) and Myeloperoxidase (MPO), from purified samples and human serum. The device performance metrics - sensitivity, selectivity, speed and dynamic range of detection - have been measured to quantify the efficacy of these nanomonitors and compared to standard immunoassays.

Keywords: alumina, nanopore, biomolecule detection, macromolecular crowding, cardiovascular

1 INTRODUCTION

Every year about 10% of the population of United States undergoes cardiovascular surgery. One million of these patients have adverse effects after the surgery It has been projected that within the next two decades surgical patients would increase by 25%, costs by 50%, and complications by 100% as the population ages [1-4]. The current surgical burden is burgeoning as a surgical crisis. Our best allay is to improve the outcomes after surgery to reduce the after effects. Currently, this is being addressed by using medicines, which were developed to address the perioperative ischemia. Perioperative ischemia is the best know predictor for postoperative cardiovascular morbidity and mortality [5-8]. Perioperative ischemia and infarction are thought to occur from plaque rupture of an unstable or vulnerable plaque in most cases.

Due to the complex nature of the problem at hand, one single biomarker is not sufficient to predict the risk factor of a patient (Fig. 1). The heterogeneity of the condition stems from the fact that many markers have to be considered as potential factors for this condition.

Monitoring multiple biomarkers at the same time would provide a holistic image of the patient's condition. To perform such a test there are currently no tests that could do these proteins at the same time, which would require a patient to make multiple visits and also trained technicians to run the tests. We wanted to address these problems by creating a multiplexed test that is easy to perform in a clinical environment (doctor's office), which would require very low quantities of patient's samples. The various proteins identified as potential markers are as follows:

1. CRP
2. Myeloperoxidase (MPO)
3. Ox-LDL
4. Tissue factor
5. CD40 ligand.

In this current work, we use the first two proteins CRP and MPO as a proof of concept demonstration of the technology. This paper describes the development of a biomimetic electrochemical device for protein sensing that works on the principle of excluded volume associated with the phenomenon of macromolecular crowding [9]. The macromolecular crowding phenomenon refers to the influence of mutual volume exclusion upon the energetics and transport properties of protein molecules within a crowded, or highly volume-occupied, medium. Because of steric repulsion, no part of any two macromolecules can be in the same place at the same time. We have developed a nanoarray membrane based system that has utilized this phenomenon towards building an electrical immunoassay with selective protein biomarker localization in confined spaces.

2 MATERIALS AND METHODS

2.1 Materials

We have developed an electrical immunoassay technology that works on the principle of creation of the electrical double layer at the liquid/metal interface and modulation of the double layer due to the addition of charged species such as proteins. Integration of both detection and the measurement makes the device label free and easier to use. The Nanomonitor comprises of two parts. They are the micro-fabricated platform fabricated by overlaying the Au electrodes on the Si substrate using standard lithographic techniques; and the nanoporous membrane implanted on the platform as represented in Fig. 1(A).

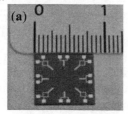 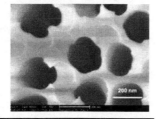

Figure 1: (a) Microfabricated platform with 8 sensing sites. (b) SEM micrograph showing the nanopores in the membrane.

Alumina was chosen as the membrane materials because the chemistry to make them is well known and the pore diameter could be well controlled. Alumina membranes are helpful:

1. Adsorption of protein onto their surfaces due to simple van der Waals bonds. When sample liquid flows on top of the membrane due to capillary effect the molecules in the solution flow to the transducer element

Figure 2: Charge distribution across liquid-electrode interface forming the equivalent circuit.

(microfabricated platform) at the bottom of the well. Due to the flow of charge on the electrodes, this causes the accumulation of the ions in solution to adsorb on the surface creating the double layer. The electrical double layer consists of an array of charged species and/or oriented dipoles existing at electrode interface. A simple equivalent circuit is given in Fig. 2.

2. Micro encapsulation of the biomarkers in each well facilitates ease of the transfer of electrons and ions exclusive to that well, providing very little cross talk electrically across wells.

3. Entrapment of the biomarker very close to the transducer element (microfabricated platform of gold) improves the signal transduction and sensitivity of the device.

A 250nm thick layer of aluminum film is thermally evaporated onto the microelectrode platform. This acts as the anode for the two-step anodization process required for fabricating the nanopores. The cathode is a platinum wire. Both the anode and cathode were immersed in an electrochemical bath. A constant voltage of 45 V and a constant current density of 20 mA/cm^2 are applied across the electrodes. A mixture of 0.1 M sulfuric acid and 0.4 M oxalic acid is used as the electrolyte to obtain a uniform structure of the pores. The conditions for the electrochemical bath were optimized to get a uniform pore size of 200 nm. The pore size is modified to match the protein dimensions suitable for protein entrapment. This

results in the formation of nanowells. Hence on an individual sensing site there area approximately quarter million nanowells and the electrical conductance signal is cumulated and averaged out over all the nanowells on the sensing site. The pore dimensions were tailored to 200 nm for the two cardiac markers that were evaluated: C-reactive protein and Myeloperoxidase.

2.2 Methods

The first step in detection of the proteins was to saturate the nanoporous membrane with the antibody and then inoculate the region with sample for detection. Two inflammatory biomarkers, C-reactive protein (CRP) and Myeloperoxidase (MPO), were identified as proteins of study for the development of the device and to demonstrate the detection. The purified samples were obtained from EMD Biosciences, San Diego, CA. The baseline measurements were recorded with 0.1X PBS because the dilutions for both antibody and the samples were made in 0.1X PBS. The interfacial voltage and the frequency of voltage signals are recorded. Experimental protocol included the cleaning of the chip before the experiment and then embedding the nanoporous membrane on the chip. Baseline measurements with 0.1X PBS were done, and then the NM was functionalized with the chemical cross-linker dithiobis (succinimidyl propionate) (DSP), a homo-bifunctional, amine-reactive cross-linker. The antibodies that were then inoculated on the surface are immobilized by the covalent linking with DSP. The chip was incubated for a period of 15 min at room temperature after measurements were made. The antibody saturation for both CRP and MPO are shown in Figure 3.

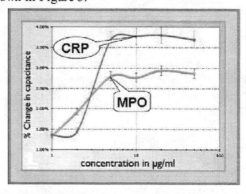

Figure 3: Antibody saturation measurements for proteins CRP and MPO.

Next, the proteins were inoculated followed by 15 min incubation and measurements. The multiplexed data with both the antibodies of CRP and MPO immobilized on the surface are tested with a solution of both the proteins. Further experiments were conducted with the human serum (Fischer Scientific) to check for the nonspecific binding issues associated with most immunoassay setups.

3 RESULTS

The results for both CRP and MPO detection in both pure and serum samples are presented here (Fig. 4). There were

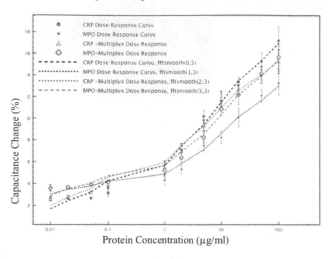

Figure 4: Dose response curves for proteins CRP and MPO when detected in pure and serum

several data points in the dosee response curve from 100pg/ml to 100µg/mL. Each data point was evaluated in triplicate. There was linear correlation between the % change in capacitance and the concentration.

The values in this plot were obtained at 4 kHz, where the change in capacitance was found to be the highest. The detection limit was found to be ~200 pg/ml for CRP and ~500 pg/ml for MPO. The time taken for the detection of proteins was about 120 – 180 seconds.

4 DISCUSSION

In the studies conducted, the device has been tested with standard synthesized samples obtained from the vendor. The samples have also been validated with ELISA Vendor validation datasheets.

4.1 Validation Experiments

Optical detection through fluorescence markers and ELISA are two commonly used protein detection techniques. Thus in order to validate the NM results we designed two experiments to qualitatively and quantitatively demonstrate antigen-antibody conjugation on the NM.

Fluorescence Validation

Three NM chips were prepared such that the first chip (negative control) was coated with only the DSP linkers, while the remaining two NMs were functionalized with DSP and the anti-CRP with the concentrations of anti-CRP being 10 µg/ml and 100 µg/ml according to the procedure

described before. Fluorescence marker kit Alexa 594 (Invitrogen Corporation, OR) was used to tag the CRP-protein according to the specified protocol. This tagged protein was reconstituted to a concentration of 80 µg/ml in 1X PBS and introduced on the control and the two experimental NMs and incubated for 30 minutes. All the NMs were washed under 1 ml of flowing 1X PBS. The NMs were observed under fluorescence microscope activated by red light (~594 nm wavelength) before and after the washing step. From the images shown in Figure 5 it is evident that with the increase in anti-CRP concentration, the amount of CRP localized on the NM increased. Also there was no binding of the CRP on the NM without the anti-CRP in the control experiment.

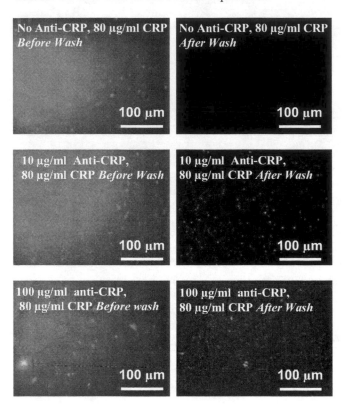

Figure 5: Sequence of images taken before and after wash with varying concentrations of anti-CRP showing increasing fluorescence, thereby indicating higher amount of anti-CRP and CRP conjugation on

Gold ELISA Validation

The second validation experiment was conducted by placing small pieces of Au coated Si wafer in an ELISA plate and treating the pieces with the DSP linker and antibody solutions in the ELISA plate. Then a HRP (horseradish peroxidase) conjugated detection antibody was added followed by the TMB (3,3′,5,5′-tetramethylbenzidine) substrate to give the change in color as seen in Figure 8. The plate was washed three times after each step with 3% BSA/PBS solution. A total of twelve samples consisting of four controls, six test samples and two blank samples were tested in duplicates. Table 1 gives the details of the testing conditions. Rows A-D in columns 1 and 2 were set as four negative controls. The next three samples (cells E1, E2, F1, F2, A3 and A4) were test samples with increasing concentrations of anti-CRP. Similarly, three different concentrations of a second antibody, anti-HAPT (anti- haptoglobin) were tested in columns 3 and 4 of rows B-D. The last four cells (E3, E4, F3 and F4) servd as blanks without any Au coated Si-pieces.

	1&2	3&4
A	Au, HRP-Antibody, TMB (**No DSP, No Anti-CRP**)	Au, DSP, **Anti-CRP (100 µg/ml)**, HRP-Antibody, TMB
B	Au, DSP, HRP-Antibody, TMB (**No Anti-CRP**)	Au, DSP, **Anti-HAPT (0.4 µg/ml)**, HRP-Antibody, TMB
C	Au, Anti-CRP, HRP-Antibody, TMB (**No DSP**)	Au, DSP, **Anti- HAPT (2 µg/ml)**, HRP-Antibody, TMB
D	Au, DSP, Anti-CRP (50 µg/ml), TMB (**No HRP-Antibody**)	Au, DSP, **Anti- HAPT (10 µg/ml)**, HAPT, HRP-Antibody, TMB
E	Au, DSP, **Anti-CRP (10 µg/ml)**, HRP-Antibody, TMB	Blank
F	Au, DSP, **Anti-CRP (50 µg/ml)**, HRP-Antibody, TMB	Blank

Table 1: Details of the solutions added for the twelve samples in Gold ELISA experiment

Figure 6 shows a digital photograph of the Gold ELISA plate. The cells in which the TMB substrate solution turned blue indicated that HRP-conjugated antibody had been captured by the antibody attached to the Au in that cell. This in turn validated the binding of the antibody on Au coated Si piece which forms the metallized platform for NM electrical detection.

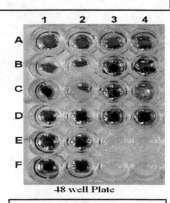

Figure 6: Digital image of the Gold ELISA plate

5 CONCLUSIONS AND FUTURE WORK

In this article, we have demonstrated the design, fabrication, and operation of electrical biosensor for protein biomarker detection. We have selected two inflammatory proteins, CRP and MPO, as the study proteins to demonstrate the operation of the device prototype. These protein biomarkers were chosen as they are thought to be biomarkers of the vulnerable coronary plaque rupture state. The basis of the electrical biosensor functioning is the perturbation to the Helmholtz layer due to the binding of the relevant proteins from a test sample. This perturbation results in a modulation to the electro-ionic distribution of the interfacial electrical double layer formed in each of the nanowells/pores). These electrical signals can be measured as change in capacitance of the electrical double layer and be correlated to the concentration of the protein. It is clear from the current work that it is possible to selectively identify surface-charged proteins by measuring the variations to the Helmholtz layer in nanoscale confined spaces. The current work is a feasibility study indicating that changes at the nanoscale can be captured and correlated to protein-binding events thus indicting a promise in the development of a technique that is rapid and label free for the detection of protein molecules.

REFERENCES

[1] Mangano, D. T., J. Cardiothorac. Vasc. Anesth. 2004, 18(1), 1-6.
[2] Mangano, D. T.; Browner, W. S.; Hollenberg, M.; London, M. J.; Tubau, J. F.; Tateo, I. M. N. Engl. J. Med. 1990, 323(26), 1781-1788.
[3] Mangano, D. T.; Hollenberg, M.; Fegert, G.; et al. J. Am. Coll. Cardiol. 1991, 17(4), 843-850.
[4] Mangano, D. T.; Wong, M. G.; London, M. J.; Tubau, J. F.; Rapp, J. A., J. Am. Coll. Cardiol. 1991, 17(4), 851-857.
[5] Mangano, D. T.; Layug, E. L.; Wallace, A.; Tateo, I. N. Engl. J. Med. 1996, 335(23), 1713-1720.
[6] Wallace, A.; Layug, B.; Tateo, I.; et al. McSPI Research Group. Anesthesiology 1998, 88(1), 7-17.
[7]. Brady, A. R.; Gibbs, J. S.; Greenhalgh, R. M.; Powell, J. T.; Sydes, M. R., J. Vasc. Surg. 2005, 41(4), 602-609.
[8] Juul, A. B.; Wetterslev, J.; Kofoed-Enevoldsen, A.; Callesen, T.; Jensen, G.; Gluud, C., Am. Heart J. 2004, 147(4), 677-683.
[9] Li, Feiyue; Zhang, Lan; Metzger, R. M., Chem. Mater. 1998, 10, 2470-2480.

Conformational Changes of Acetylcholine During Spontaneous Diffusion Through a Nano/Microporous Gel

E. Vaganova[*], H. Ovadia[**], S. E. Lyshevski[***], V. Khodorkovsky[****] I. F. Pierola[*****], S. Yitzchaik[*]

[*]The Chemistry Institute and the Farkas Center for Light-Induced Processes
The Hebrew University of Jerusalem, Israel, gv@cc.huji.ac.il, sy@cc.huji.ac.il
[**]Hadassah University Hospital, Jerusalem, Israel, OVADIA@hadassah.org.il
[***]Department of Electrical Engineering, Rochester Institute of Technology
Rochester, New York 14623, Sergey.Lyshevski@mail.rit.edu
[****]Université de la Méditerranée, CNRS UMR6114, Marseille, France, khodor@luminy.uni-mrs.fr
[*****]Facultad de Ciencias, Universidad a Distancia (UNED), 28040 Madrid, Spain, ipierola@ccia.uned.es

ABSTRACT

This paper reports results which aim to enhance the understanding of mechanisms and transitions taking place in biological systems [1]. In order to mimic these low energy processes, as well as to perform characterization and data acquisition on these processes, we have developed a new methodology. We study the conformational changes of acetylcholine (ACh) during propagation through a nano/microporous polymer system by measuring the fluorescence lifetime of the label molecule – fluorescein.

Keywords: acetylcholine, fluorescein, porous polymer gel, diffusion, fluorescence lifetime

1 INTRODUCTION

Acetylcholine (ACh) attracts a lot of attention due to its modulatory role in the central nervous system. However, ACh arrived within the evolutionary scheme long before the design of the nervous system and functional synapses [2].
[2]. The scheme of ACh is presented below.

Torsional rotation in the ACh molecule can occur around the bonds θ_1, θ_2 and θ_3. Since the methyl groups are symmetrically disposed around θ_3 and constraints may be placed on θ_1 by the planar acetoxy group, the most important torsion angle determing ACh conformation in solution is θ_2. The *gauche* conformation is predominant in solution. However, the *trans* conformation may also be an active conformation. It was shown that the conformations of this flexible molecule differ substantially depending on the type of the adjacent reacting molecule [2]. Structural modifications change molecular activity [2].

We analyze the effect of the diffusion of ACh through the nano/microporous poly (*N*-vinylimidazole) (PVI) gel on its conformational changes. The synthesized porous chemically crosslinked polymer PVI gel has micropores with average diameter $\sim 1.5\times10^{-9}$ m, mesopores with diameters varying from 2×10^{-8} to 2×10^{-7} m, and interconnected macropores with an average diameter $\sim 3\times10^{-6}$ m [3]. Figure 1 shows a TEM image of the pores on the nanometer scale.

To investigate the conformational changes of ACh during the spontaneous diffusion through a nano/microporous gel, the fluorescence lifetime of the label molecule (fluorescein) was studied. Fluorescein is one of the best known fluorophores. It has two deprotonated forms, e.g., monoanion and dianion (scheme of the dianion is presented as),

which have lifetimes 3-4 ns and 4-5 ns [4]. The pKa of the monoanion-dianion transition is 6.3 in water. However, it substantially is increased at the lipid-water interface of micelles and bilayers [5]. Papers [4, 6] demonstrated that the monoanion and dianion behave as independent, non-interacting species [4, 6].

2 EXPERIMENTS

Fluorescence lifetime measurements were made by means of FluoTime 200 (PicoQuant GmbH) with a pulsed diode laser PDL 800-B, $\lambda = 410$ nm, pulse FWHM 54 psec, repetition frequency 40 MHz, peak power 219 mW. The Data Analysis Software FluoFit (PicoQuant GmbH) was used for lifetime measurements. For pulse control a solution of Ludox was used. The average fluorescence lifetime is evaluated as

$$\tau_{av} = \alpha_i \tau_i + \alpha_j \tau_j, \qquad (1)$$

where τ_i and τ_j are the fluorescence lifetimes; α_i and α_j are the fractional coefficients describing the contribution of the mono- and di-anion, respectively.

For the experiment, 0.045 g of the gel was swollen in water (the final weight is 0.548g) and placed in a solution of fluorescein for five days at the stabilized temperature of 22.0±0.1°C. To remove the fluorescein from the surface of the gel, the gel was washed at least three times in triple distilled water TDW (18 MΩ/cm). The removal of the free fluorescein was monitored by UV-VIS spectroscopy. Subsequent to washing, the porous gel, loaded with the fluorescein molecules, was cut into two pieces of equal weight. One piece was placed in a water solution of ACh and the other one in TDW. The diffusion of fluorescein out of the gel into the bath was monitored by fluorescence lifetime and spectral absorption measurements. Measurements were made in the gel and also as a function of distance from the gel surface.

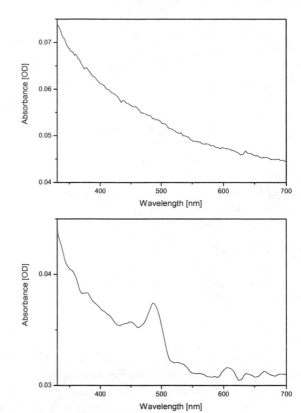

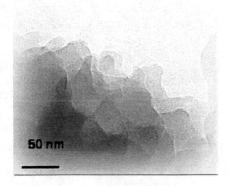

Figure 1. HR TEM image of the porous poly (*N*-vinyl imidazole) PVI gel.

3   RESULTS

There was no indication of diffusion of individual fluorescein molecules out of the porous PVI gel into the TDW bath in which it was immersed. However, the absorption spectrum of the PVI gel/ACh/TDW bath demonstrated the absorption at 488 nm. This indicates the diffusion of a fluorescein/ACh complex as reported in Figure 2.

The investigation of the fluorescence lifetime of the fluorescein provides an evidence that the fluorescence lifetime depends on:
1. Duration of the diffusion process;
2. Distance (propagation length) above the PVI gel as documented in Figure 3.

Further control experiments show that fluorescein/ACh/TDW at different ratios of fluorescein/ACh do not show any fluorescence lifetime changes during 200 hours.

Figure 2. Absorption spectra of the bath containing both the aqueous solution of ACh and the porous gel.
(a) Immediately after immersion of the porous gel, loaded with fluorescein, in the aqueous solution;
(b) The same sample after 50 hours of storage in the dark. The number of fluorescein molecules expelled from the gel after 50 hours storage is $\sim2.4\times10^{13}$.

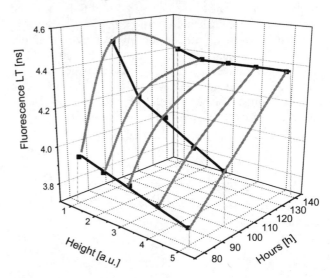

Figure 3. Fluorescein fluorescence lifetime dependence on the diffusion duration and the distance above the PVI gel. The fluorescence lifetime $\tau_{average}$, diffusion duration and distance are plotted on Z-, Y- and X-axes, respectively. The distance is plotted in the a.u, and a.u. is 0.8 cm.

The experimental data illustrates the role of diffusion of ACh into the gel, and the resulting conformational changes, on the formation of the different complexes, such as PVI/fluorescein, PVI/Ach and ACh/fluorescein.

4 DISCUSSION

The data reported in Figures 2 and 3 show that the ACh penetration in the gel induces fluorescein diffusion out of the gel in such a way that the dynamic and steady state behavior is described as

$$-\frac{dC_{ACh}}{dt} = \frac{dC_{Fl}}{dt}, \qquad (2)$$

where C_{ACh} and C_{Fl} are the acetylcholine and fluorescein molar concentrations.

In the initial phase (**I**) (0 - 80 hours) of diffusion of the fluorescein molecules out of the gel, a fluorescence lifetime of 3.6-3.8 ns was recorded around a whole volume of the sample (practically with monoexponential decay). That means that the fluorescein molecules are present as the monoanion throughout an entire PVI/Fl/ACh/water solution. The fluorescein molecules exhibit the same fluorescence lifetime (3.6-3.7 ns) in aqueous solution with the concentration 1×10^{-5} M^{-1}.

During the second phase (**II**) of the diffusion (80-100 hours) in the area located near the gel, as shown in Figure 3, fluorescein's fluorescence lifetime is significantly increased to ~4.5 ns (the long fluorescence lifetime component constitutes ~65%) in addition to the prolongation of the diffusion duration, and an increase in the concentration of the fluorescein molecules in solution. The absorption spectra are not shown in Figure 3. The value of the fluorescence lifetime is evidence for the presence of the dianion near the surface of the gel. With increasing propagation length, the fluorescence lifetime is reduced to 3.7 ns which is typical for the monoanion.

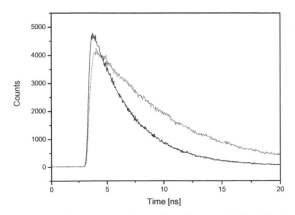

Figure 4. Fluorescence intensity decay of the fluorescein inside the gel (red curve) and in the solution (black curve) after solution storage during 80 hours.

The third phase (**III**) (100-140 hours) is a saturation (steady state) behavior of the concentration of the fluorescein molecules. The average fluorescence lifetime of the fluorescein was 4.35 nsec, which corresponds to a double exponential decay. This value was consistently measured over the whole volume of the sample.

We summarize the diffusion phases as:

(**I**) (0-80 hours) This stage demonstrates a homogeneous distribution of the fluorescein monoanions over almost the whole volume of the sample, and the lifetime is 3.7 ns, e.g., mainly the monoexponential decay;
(**II**) (80-100 hours) This phase demonstrates that dianion fluorescein molecules are expelled near the gel surface, and the lifetime becomes 4.5 ns, corresponding to the monoexponential decay;
(**III**) In the saturated steady state, the distribution of lifetimes is almost constant throughout the sample volume with an average value of ~4.35 ns. We calculated the differential absorption spectra of the solutions as the difference of the absorption of the solution during [80 100] hours and [100 140] hours diffusion time. The results are documented in Figure 5. The measurements were made at the distance from 0.8 to 2.4 cm above the gel.

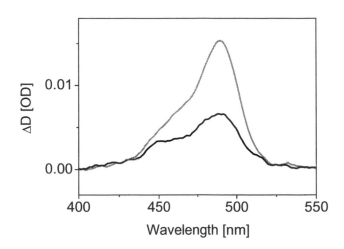

Figure 5. Differential absorption spectra of the solution (distance is from 0.8 to 2.4 cm above the gel) during the period [80 100) hours diffusion time (red curve) and during [100 140] hours diffusion time (black curve)

In agreement with the fluorescence lifetime measurements, these spectra show a decrease in the rate of the diffusion of fluorescein molecules in solution as the sample approaches saturation due to a redistribution of the mono- and dianions during the spontaneous diffusion. The maximum absorption of the fluorescein monoanion is at

455 nm, while the maximum absorption dianion is at 488 nm [4].

It seems that during the initial phase, ACh molecules expel the free fluorescein molecules located in the micropores. The expelled molecules in the monoanion form diffuse into solution through the nanopores. The value of the fluorescence lifetime shows that the monoanion form of the fluorescein is present throughout practically a whole volume of the sample. This means that the ACh molecules in the solution exist in the non-reactive, "closed" *gauche* form, which does not form complexes with fluorescein.

During the second phase, ACh molecules interact with fluorescein molecules which are bound to PVI at the pore surface. In this case, the fluorescein exhibits a fluorescence lifetime ~4.5 ns as reported in Figures 3 and 4. The fluorescence lifetime, which is ~ 4.5 ns, is similar to that of the free fluorescein dianion [3].

During the equilibrium phase (saturation), the collective effect of the combination of the two different forms of the fluorescein, results in an average fluorescence lifetime from 4.3 to 4.4 ns.

5 CONCLUSIONS

We conclude that:

1. ACh molecules expel fluorescein molecules from the micro/nanoporous PVI gel.
2. Intermolecular interactions of ACh with fluorescein molecules which are bound to PVI result in different forms of the fluorescein molecules being expelled from the gel.
3. Our data illustrate the role of diffusion of ACh into the gel and the resulting conformational changes, i.e. from the non-reactive, *gauche* closed form in solution to the active *trans* form inside the gel pores. These changes are in turn due to competition among the three different molecular complexes-PVI/fluorescein, PVI/Ach and Ach/fluorescein. Our experiments confirm that the flexible Ach can play the role of a regulator in the process of molecular transport.

6 ACKNOWLEDGEMENT

Evgenia Vaganova gratefully acknowledges financial support from the Israel Ministry for Immigrant Absorption.

REFERENCES

[1] S. E. Lyshevski, *Molecular Electronics, Circuits, and Processing Platforms*, CRC Press, Boca Raton, FL, 2007.
[2] P. Taylor and J. H. Brown, *Acetylcholine*, in *Basic Neurochemistry*, Ed. G. J. Siegel, B. W. Agranoff, R. W. Allers and P. B. Molinoff, Raven Press, NY, pp. 232-248, 1995.
[3] V. Calvino-Casilda, A. J. López-Peinado, E. Vaganova, S. Yitzchaik, I. E. Pacios and I. F. Pierola "Porosity inherent to chemically crosslinked polymers. Poly (*N*-vinylimidazole) hydrogels", *JPC B*, in press.
[4] N. Klonis, W.H. Sawyer "The thiorea group modulates the fluorescence emission decay of fluorescein-labeled molecules", *Photochem. Photobiol.*, 77, pp. 502-509, 2003.
[5] N. O. Mchedlov-Petrossyan, V. N. Kleshchevnikiva, "Influence of the cetyltrimethylammonium chloride micellar pseudophase on the protolytic equilibria of oxyxanthene dyes at high bulk phase ionic strength", *J. Chem. Soc. Faraday Trans.*, 90, pp. 629-640, 1994.
[6] N. Klonis andW. H. Sawyer, "Spectral properties of the prototropic forms of fluorescein in aqueous solution", *J. Fluoresence*, 6, pp. 147-157, 1996.

Raman study and DFT calculations of amino acids

V. Sonois[*,***], A. Estève[**], A. Zwick[***], P. Faller[*], W Bacsa[***]

[*] LCC CNRS UPR, 205 route de Narbonne, Univ. Toulouse, 31077 Toulouse, France, faller@lcc-tlse.fr
[**] LAAS CNRS UPR, 7 av du Colonel Roche, Univ. Toulouse, 31077 Toulouse, France asteve@laas.fr
[***] CEMES CNRS UPR, 29 rue Jeanne Marvig, Univ. Toulouse, 31055 Toulouse France, bacsa@cemes.fr

ABSTRACT

Intense Raman signals were observed of nano gram quantities of three amino acids (histidine, valine, glycine) using visible laser excitation (488nm) at relatively low laser power (5mW) and short acquisition time (2min). Considerable variations in the Raman signal were observed when changing the pH of the amino acid solution. Scanning electron microscopy reveals considerable differences in the crystallite structure of histidine, valine and glycine and for different pH. Narrow vibrational bands are correlated with formation of crystallites. H/D substitution and pH dependence experiments were used to identify vibrational bands associated with the functional groups able to exchange protons. The observed spectral bands are directly compared with density functional calculations to assign the vibrational bands in histidine, valine and glycine. The effect of hydration is studied at neutral pH in the theoretical calculations.

Keywords: Raman spectroscopy, amino acids, density functional calculations, scanning electron microscopy, micro-crystals

1 INTRODUCTION

Raman spectroscopy is a powerful non-invasive tool to obtain information on structure, function and reactivity of biological molecules such as proteins. [1] The vibrational spectra of amino acids in peptides and proteins depend sensitively on organization and interaction with its environment and give hence important information on conformation and function. Amino acids are the basic building blocs of proteins and have been studied by Raman spectroscopy and theoretical simulations have been used to assign the observed spectra. [2-4] The native 20 amino acids have no chromophores in the visible spectral range, no electronic resonances can be used to enhance the Raman signal in the visible spectral range and relatively large samples volumes (powders, crystals) or concentrations are necessary to study amino acids. To take advantage of electronic resonances of chromophores of amino acids such aromatic residue side chains and peptide bonds one can use UV Raman spectroscopy. UV Raman has been increasingly applied to study proteins and their secondary structure. [5,6]

We have recently observed an enhancement of the Raman signal of the amino acid histidine using visible excitation at low laser power (<5mW). This was achieved by drying droplets of mM solutions of histidine on SiO_2/Si surfaces. [7] The drying process formed spontaneously needle shaped micro crystals (diameter 80nm, several micrometers long). We have applied the same method to other amino acids and investigated the influence of pH and H/D substitution on the vibrational spectra. Scanning electron microscopy of the micro crystals is used to clarify if the shape of the crystals influences the observed enhancement. A strong influence of the pH of the solution on the vibrational spectra is found. Amino group were studied by proton/deuterium substitution experiments. We find that intense Raman signals can also be observed for amino acids such as valine and glycine. We compare experimental Raman spectra to *ab-initio* calculations and propose an assignment of the observed vibrational Raman bands.

2 EXPERIMENTAL

Histidine, glycine and valine (Sigma Aldrich) in their zwitterionic form (($^+H_3$NCHRCOO$^-$) were first dissolved in 1ml of de-ionized water (or D_2O) at a concentration of 30mM and then single droplets (15µl) were deposited on SiO_2/Si plates. The droplet was dried on a heating plate at 80°C. For the liquid or powder sample spectra, a 15µl droplet or 0.3-0.4 mg respectively was deposited on a glass plate (estimated thickness 50 micrometers).

Raman spectra were recorded (Dilor XY) using 488nm excitation. The incident laser beam was focused on the sample through a microscope with an x100 objective. The resulting spot size is less than 1µm². The power on the sample has been 4mW.

3 THEORETICAL CALULATIONS

All calculations were performed using state of the art density functional theory (Gaussian-03 Package). [8] The considered geometries and vibrational frequency were calculated within the Becke's three parameters exchange hybrid functional B3LYP associated with the Generalized Gradient Approximation (GGA) of Lee Yang and Parr. [9,10] The electronic wave functions were described by the 6-311++G** basis set. The polarization and diffuse functions are crucial to the treatment of both electrostatic interactions and hydrogen bonds. [3] This approach has proven to be efficient in describing structural, electronic

NSTI-Nanotech 2008, www.nsti.org, ISBN 978-1-4200-8503-7 Vol. 1

and vibrational properties of many molecular systems such involving amino acids.

4 RESULTS AND DISCUSSION

4.1 Raman intensity and scanning electron microscopy

Glycine is the simplest amino acid and valine has a hydrophobic side chain in contrast to the more hydrophilic side chain in histidine. No Raman signal was observed from a 30mM solution of glycine using the 488nm excitation line at 4mW. But after drying a single droplet of solution we recorded intense Raman signals (Figure 1) which are similar to what is observed for spectra recorded from powder but corresponding to a much larger quantity.

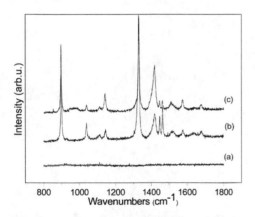

Figure 1: Raman spectra of glycine (a) 30mM in H_2O at pH 7.8 (b) powder (c) solution 30mM in H_2O at pH 7.8 dried on silica

Figure 1 compares the Raman signal of glycine recorded on the SiO_2/Si surface with glycine in powder form. No signal is detected at comparable experimental conditions for the solution and the spectrum is comparable to the spectrum recorded from powder of glycine after crystallization. We find the same behavior for valine. Amino acids can have different protonation states depending on pH which might influence the molecular packing in the micro crystals.

pH values between different pKa values were selected to have the most uniform protonation state. The Raman spectra of dried glycine from a solution at pH 1.2, 7.8 and 12 are shown in Figure 2. The intensities of the Raman signal are comparable at pH 1.2 and pH 7.8. At pH 12, the intensity is significantly lower (Figure 2, 10x longer acquisition time at pH 12). The observed vibrational bands depend clearly on the pH of the solution. Figure 3 shows the pH dependent Raman spectra of valine. We have found similarly intense signals at pH 1, but significant lower signals at pH 12 (acquisition time x10 longer). We recorded spectra for histidine at four different pH values (pH 1.6, 3.8, 7.8, and 12).

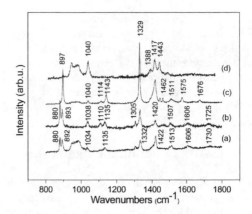

Figure 2: Raman spectra after evaporation of glycine dissolved in H_2O de-ionized at different pH (a) pH 1.2 (b) pH 2.4 (c) pH 7.8 and (d) pH12 on silica.

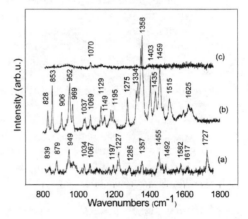

Figure 3: Raman spectra after evaporation of valine dissolved in H_2O de-ionized at different pH (a) pH 1 (b) pH 7.8 and (c) pH 12 on silica.

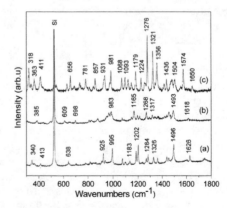

Figure 4: Raman spectra of aqueous histidine at different pH evaporated on silica (a) pH 1.6 (b) pH 3.8 (c) pH 7.8. The intense vibrational band at 521cm[-1] corresponds to the SiO_2/Si substrate.

Figure 4 shows an intense spectrum for pH 1.6, 3.8 and 7.8. At pH 12 the spectrum disappeared nearly completely.

For all three amino acids micro crystals, no vibrational bands due to the presence of water (OH: 3400cm⁻¹) were observed, in contrast of previous study on larger crystals. [11] This does not exclude the presence of water in the crystalline structure but shows that the amount of water in the crystals must be small.

4.2   Density Functional Calculations

We used density functional calculations to get a better insight on the organization of amino acids and their ionic nature in condensed form. We find through structural optimization that a totally dehydrated zwitterionic form of a single histidine molecule is unstable, giving rise to a proton transfer from the NH_3^+ to the COO^- group upon minimization. To explore the zwitterionic form of the amino acids, we focus on the polar regions and investigate how they couple through intermolecular interactions or with residual water molecules. For this purpose, we investigate the vibrational response of the amino acids as a function of the degree of saturation of their polar regions by water. We build different hydration models for the glycine with up to six surrounding water molecules (table 1). For valine and histidine, we only consider the simplest cases using two water molecules. We use the information derived from glycine for the assignment of the main vibrational bands.

Glycine ($^+H_3NCH_2COO^-$) is the simplest amino acid and is abundant in various proteins and enzymes (up to 7.5%). It is the only amino acid which has no stereo isomer (not optically active). Several studies of Raman bands of glycine are reported in the literature. [3,12,13] The observed Raman bands of glycine at pH 7.8 are listed in table 1.

In order to test the assignment, in particular of the vibrational modes that correspond to the NH_3^+ group, we made H/D exchange experiments (Figure 5). The only exchangeable protons of glycine at neutral pH are the three protons of the NH_3^+ group (CH_2 does not exchange and COO^- is de-protonated).

When H was replaced by D in NH_3^+ several vibrational bands appeared at 832cm⁻¹, 971cm⁻¹, 992cm⁻¹, 1006cm⁻¹, 1278cm⁻¹, and some bands disappeared or are attenuated at 1575cm⁻¹, 1511cm⁻¹, 1143cm⁻¹ and 897cm⁻¹. The band at 1511cm⁻¹ disappears in the D_2O spectrum and corresponds accordingly to the assignment to the NH_3^+ deformation vibration. The bands at 1114cm⁻¹ and 1143cm⁻¹ disappear in the D spectrum and they correspond to the vibrations related to N (N-H, N-C) according to the assignment. The band at 897cm⁻¹ is attenuated in the D spectrum and is attributed to C-C vibration indicating that the D influences the C-C vibration. Some of the vibrations are shifted in energy by incorporating D. The vibrational band at 1329cm⁻¹ is shifted to 1320cm⁻¹ in 100% D_2O. This is in line with the theoretical calculations which show the contribution of NH_3 to this band. The same is found for the 1420cm⁻¹ band which shifts to 1410cm⁻¹.

Wavenumbers (cm⁻¹)	Calculation (cm⁻¹)	Assignment
360	318	δ(CCN)
597	574	δ(CCN)
696	681	δ(CCN)
897	853 - **891**	ν(C-C), δ(COO)$_{sc}$
1040	954 - **1017**	ν(C-N)$_{st}$
1114	1088 - **1153**	δ(CH₂)$_{tw}$, δ(NH₃)$_{tw}$
1143	1122 - **1194**	δ(NH₃)$_r$, δ(CH₂)$_w$
1329	1273 - **1371**	ν(COO)$_s$, δ(CH₂), δ(NH₃)
1420	1306 - **1357**	δ(CH₂)$_{tw}$, δ(NH₃)$_{tw}$
1445	**1416**	ν⁸, δ(CH₂)$_w$, δ(NH₃)$_r$
1462	1475 - **1482**	δ(CH₂)$_{sc}$
1511	1618 - **1575**	δ(NH₃)$_{umb}$
1575		
1676	1776 - **1694**	ν(COO)$_{as}$
2984	2957	ν(CH₂)$_s$
3019	3015	ν(CH₂)$_{as}$

Table 1: List of experimental and calculated vibrational frequencies for glycine (pH 7.8). Bold numbers in the calculation column account for glycine surrounded by six water molecules (normal text accounts for two water molecules around the amino acid).

The vibrational band at 1278cm⁻¹ which is not present in the protonated spectrum, can be attributed to a N-D deformation vibration. In agreement with the assignment, we find the bands of CH_2 at 2984cm⁻¹ and 3019cm⁻¹ which are not influenced by D substitution (i.e. 2986cm⁻¹ and 3020cm⁻¹). It is not clear why the band at 1676cm⁻¹, assigned to ν(COO) disappears in the D spectrum. A possible explanation is that the COO is engaged in a hydrogen bond which is modified as a result of H/D substitution. As the signal to noise ratio is the same in 100% H_2O and 100% D_2O we can conclude that the effect of signal increase is active in both cases. The band at 897cm⁻¹ is shifted to lower and the band at 1676cm⁻¹ is shifted to higher energies when the pH is reduced to pH 1.2. This is consistent with the assignment of the two bands to COO vibrations (see Table 1).

Valine is an amino acid and has a branched-chain amino acid like leucine and isoleucine. Valine contains a single hydrocarbon side chain making it hydrophobic and is usually found in the interior of proteins. Its hydrophobicity contributes with other hydrophobic amino acids to the tertiary and quaternary structures of proteins. The spectrum of valine contains a larger number of vibrational bands than glycine due to the larger side chain (iosopropyl group).

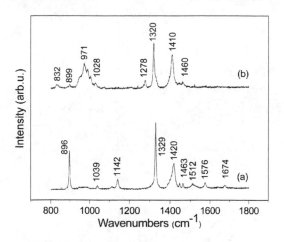

Figure 9: Raman spectra after evaporation on silica of glycine in solution (a) 100% H_2O, (b) 100% D_2O.

Only the Raman spectra at low and neutral pH have shown intense bands. At low pH the COO^- is protonated and forms a COOH. As a result the bands assigned to COO^- are expected to shift most upon reducing the pH. The band of COO^- at $1323cm^{-1}$, $906cm^{-1}$ and $853cm^{-1}$ are not found at low pH. The bands which shift less ($<5cm^{-1}$) are in the high frequency range at $2979cm^{-1}$, $2921cm^{-1}$, $2890cm^{-1}$ and in the intermediate frequency range at $1358cm^{-1}$, $1334cm^{-1}$, $1195cm^{-1}$, $1069cm^{-1}$ and $969cm^{-1}$. All these bands are assigned to vibrations including C-H of the isopropyl side or C alpha chain. At high pH we observe only one narrow band at $1069cm^{-1}$. Its origin is not clear at this point.

We observe that vibrational bands are influenced by crystallization. The bands are narrower, shifted and new bands appear in microcrystalline form. [7] By varying the pH in histidine we find that almost all bands are shifted. This shows that the pH variation affects not only bonds directly related to groups of atoms that are de/protonated (NH_3^+, COO^- and imidazole group), but the entire molecule. We propose that the presence of the imidazole group with its possible delocalized electrons and the formation of hydrogen bonds with the carboxylate group influence the electronic structure of the microcrystal.

In the case of the H/D exchange using D_2O, most bands shift in the $800-1800cm^{-1}$ region. The bands above $2900cm^{-1}$, assigned to localized CH_2 vibrations are not affected by the D/H exchange as expected.

5 CONCLUSION

We show that the method to crystallize histidine to observe intense Raman signals using visible laser excitation can be extended to other amino acids. We find that the pH influences the spectrum considerably which shows that the crystalline order depends on the protonation state of the molecule. This simple method allows detecting intense Raman signals of glycine, valine and histidine at neutral and acidic pH at the nano gram level. We expect that other amino acids dried on surfaces the same way exhibit the same enhancment on the Raman signal. We note, however, that high pH resulted in reduction of the Raman signal which could be related to the overall negative charge or the presence of Na^+ ions in the solution.

The observed narrow Raman bands provide an interesting basis for the study of the crystalline structure using state of the art DFT calculations. The results of the theoretical modeling show the importance of the degree of saturation by water of each polar region which affects strongly the calculated vibrational modes. We find that the zwitterionic form of the amino acids give better agreement with observed modes compared to the neutral form where the polar regions appear to be saturated. It is not clear at this point how the saturation is taking place.

ACKNOWLEDGMENT

We thank Pascal Puech for helpful discussions. We thank CALMIP and IDRIS supercomputer centers and financial support of the following projects: ACI-INTERFACE-PCB, ITAV-ALMA, ANR-NANOBIO-M.

REFERENCES

[1] Maiti, N. C.; Apetri, M. M.; Zagorski, M. G.; Carey, P. R.; Anderson, V. E. *JACS* **2004**, *126*, 2399.

[2] Pandiarajan, S.; Umadevi, M.; Rajaram, R. K.; Ramakrishnan, V. *SpecChim. Acta A:* **2005**, *62*, 630.

[3] Derbel, N.; Hernandez, B.; Pfluger, F.; Liquier, J.; Geinguenaud, F.; Jaidane, N.; BenLakhdar, Z.; Ghomi, M. *JPC, B* **2007**, *111*, 1470.

[4] Dammak, T.; Fourati, N.; Abid, Y.; Boughzala, H.; Mlayah, A.; Minot, C. *SpecChim Acta A:* **2007**, *66*, 1097.

[5] Chi, Z.; Chen, X. G.; Holtz, J. S. W.; Asher, S. A. *Biochemistry* **1998**, *37*, 2854.

[6] Caswell, D. S.; Spiro, T. G. *JACS* **1986**, *108*, 6470.

[7] V. Sonois, P. Faller, W.S. Bacsa, N. Fazouan and A. Estève, Tech. Proceed. of the Nanotechnology Conf, **Vol 2**, p 37 – 40 (2007), Sonois, V.; Faller, P.; Bacsa, W.; Fazouan, N.; Esteve, A. *Chem Phys Lett* **2007**, *439*, 360.

[8] Frisch, M. J. T. et al *Gaussian 03*, Revision C.02; Gaussian, Inc.: Wallingford CT, 2004.

[9] Lee, C.; Yang, W.; Parr, R. G. *PR B* **1988**, *37*, 785.

[10] Becke, A. D. *The J. of Chem. Phys.* **1993**, *98*, 5648.

[11] Faria, J. L. B.; Almeida, F. M.; Pilla, O.; Rossi, F.; Sasaki, J. M.; Melo, F. E. A.; Mendes Filho, J.; Freire, P. T. C. *J of Raman Spec.* **2004**, *35*, 242.

[12] Kumar, S.; Rai, A. K.; Singh, V. B.; Rai, S. B. *SpecChimica Acta A:* **2005**, *61*, 2741.

[13] Furic, K.; Mohacek, V.; Bonifacic, M.; Stefanic, I. *J of Mol Struc* **1992**, *267*, 39.

Effect of surfactants on the size-distribution of starch nanoparticles during wet grinding

N. I. Bukhari[*], Y. B. Kang[*], Y. K. Hay[**], A. B. A. Majeed[***], M. Nadeem[****], S. H. Bai[**]

[*]School of Pharmacy, International Medical University, Kuala Lumpur
[**]School of Pharmaceutical Sciences, Universiti Sains Malaysia, Penang
[***]Faculty of Pharmacy, Universiti Technologi Mara, Kuala Lumpur
[****]Chemical Engineering Department, University Technology PETRONAS, Perak, Malaysia

ABSTRACT

Conventional and non-conventional surfactants were utilized to fabricate corn-starch nanoparticles. The surfactants used were pluronic F-68, span 20, Aerosol TR and Aerosol OT. Starch, in premix with either of the surfactants with concentration 0.5% to 2% were grinded in Netsch MiniCer® laboratory circulation machine (Netsch, Germany) at 25 Hz. Zicronium beads of 1mm diameter were used as the grinding media. At 0.5 hr grinding time, 0.5% pluronic reduced corn starch to the lowest size of 335 ± 5.57 nm. The other surfactants yielded particle size equal to or above 382 ± 9.5 nm. However, with 1 hr grinding, 1% pluronic F-68 reduced the particle size of corn starch to the minimum level of 353 ± 14.50 nm. At 1 hr grinding, the other surfactants reduced the size equal to or above 399 ± 9.50 nm. The lowest PDI was 0.393 ± 0.06 and 0.417 ± 0.02, respectively with 1% TR-OT at 0.5 hr and 2% TR-OT at 1 hr grinding. Pluronic demonstrated the highest efficiency in size reduction followed by ATR-AOT (1:4).

Keywords: starch, nanoparticles, surfactants

1   INTRODUCTION

Starch has unique molecular properties, which are the causes of its enhanced and limited applications, simultaneously. Starch it self as well as its end products are biocompatible, non-toxic and non-immunogenic. Thus, starch has been reported to be well tolerated in animals and human [1-3]. On the other side, the processing of starch is not straightforward due to the involvement of several parameters in its properties [4-5], therefore, the above factors make development of starch nanoparticles intricate too.

With the increasing demand of nano scale materials, the fine grinding has found wide applications in pharmaceutical and nonpharmaceutical industries [6 -9]. Grinding is achieved by high-speed stirred mills, also called nano or media mill. In the present study, wet grinding using Netsch MiniCer® laboratory circulation machine (Netsch, Germany) (Figure 1) along with the processing aid was used to prepare starch nanoparticles of desired size and the poly dispersity index (PDI).

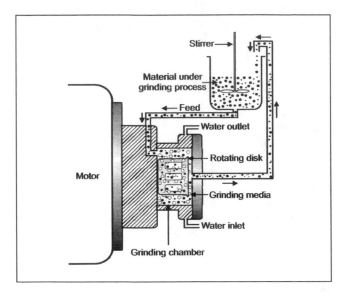

Figure 1: Schematic of Netsch MiniCer® laboratory circulation machine

2   EXPERIMENTAL

2.1 Materials

Corn starch pluronic F-68 (Sigma), span 20 (R&M Chemicals), Aerosol TR (ATR) (Cytec) and Aerosol (AOT)(Cytec) were used as such without any processing or modifications. Zirconium oxide beads of 1 mm diameter were used as the grinding media for a metal-free fine grinding.

2.2 Preparation of Starch Premixes

Added 3.75 gm of starch in about 400 ml of water and heated the mixture at about 76°C until the starch dissolves in water. The mixtures were allowed to cool down naturally to the room temperature and adjusted the pH at 5.5. To the mixtures, added the 4 ml of 0.25% of cross linker, heated to 50°C for 2 hours. The dispersion was allowed to cool down to the room temperature before addition of the required amount of processing aids and stirred well before going to next step. The required amount of surfactants (0.5 to 2%, BOS) was added and stirred well before loading into the nanomill.

2.3 Nanomilling of Starch Premixes

The premixex were treated in Netsch MiniCer® laboratory circulation machine (Netsch, Germany) (Figure 1). The maximum pressure in grinding tank was fixed as 2 bar (200 kPa). To the grinding chamber charged 140 ml of beads, loaded the premix into the collecting vessel and operated the nanomill according to the instructions in the manufacturer's manual [10].

2.4 Determination of Particle Size and PDI

Zetasizer Nano SZ (Malvern Instruments Ltd., UK), integrated with data processing software was used for the measurement of particle size and PDI.

3   RESULTS

The present work is part of an ongoing project on the study of critical factors for the preparation of starch nanoparticles of varying properties by using wet grinding. This paper reports the effect of different surfactants on the particle size and PDI of corn starch.

The effect of different surfactants on the particle size of corn starch is given in Figure 2. The surfactants with different concentrations yielded varied sizes of the corn starch. At grinding time of 0.5 hr, the particles ranged from 335 ± 5.57 nm to 946 ± 105.20 nm. Pluronic at concentration of 0.5% produced the minimum particle size of 335 ± 5.57 nm. At the same grinding time, TR-OT combination (1:4) at concentration of 1% yielded particle size of 382 ± 9.5 nm. At 0.5 hr grinding, the rank of efficiencies of other surfactants in size reduction in the decreasing order was; 1% pluronic (418 ± 3.0 nm) > 2% TR-OT (426.67 ± 16.01 nm) > 0.5% span 20 (478.67 ± 2.31 nm) > 0.5% TR-OT (578.67 ± 18.04 nm) > 1% Span 20 (762.33 ± 24.04 nm) > 2% span 20 (940 ± 105.20 nm). The minimum standard deviation was observed to be 2.31 with 0.5% span 20.

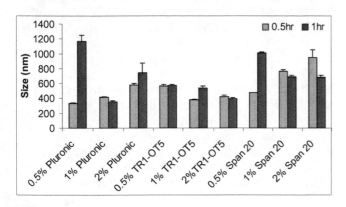

Figure 2: Effect of different surfactants on the particle size of corn starch.

When the material under study was exposed to 1 hr in the grinding zone of nanomill, the particle size range was

observed to be 353.33±14.50 nm to 1166.67 ± 80.83 nm. The minimum particle size of 353 ± 14.50 was noted with 1% pluronic. The second lowest particle size, 399 ± 9.54 was produced with 2% TR-OT. The efficiency of other surfactants with decreasing order was: 1% TR-OT (542.0 ± 27.22 nm) > 0.5% TR-OT (575 ± 15.40 nm) > 2% span 20 (631 ± 27.62 nm) > 1% span 20 (689 ± 19.98 > 2% pluronic (744 ± 130.98) > 5% pluronic (1167.7 ± 80.83). As indicated by the standard deviation, the minimum variation was observed in case of 2% TR-OT blend. In the majority of the cases above, it seemed that the increased grinding time was ineffective in reducing the particle size particularly in the presence of 0.5% and 2% pluronic, 1% TR-OT and 0.5% span 20.

Figure 3 illustrates the effect of surfactants on the PDI of the corn starch.

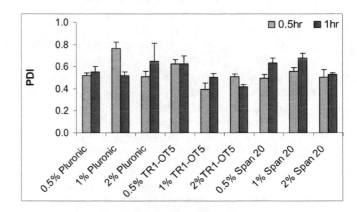

Figure 3: Effect of different surfactants on the PDI of corn starch nanoparticles

After 0.5 hr grinding, at 25 Hz, the PDI ranged from 0.393 ± 0.06, produced by 1% TR-OT to 0.764 ± 0.06 yielded by 1% pluronic. In case of 1 hr grinding, the PDI range was found to be 0.417 ± 0.02, resulted from grinding in presence of 2% TR-OT combination to 0.676 ± 0.05. The time of grinding seemed to have negative effects in presence of 1% TR-OT, 0.5% span 20 and 2% span 20 and otherwise, with rest of the surfactants used in this study.

4   DISCUSSION

A couple of studies using physical treatment of starch have been cited in literature yet, studies on the development of starch particles employing the wet grinding technique are scarce. The use of nanomill offers certain advantages. In the nanomills, the size can be reduced to micron or submicron sizes without use of the organic solvents with manipulation of the operational and formulative variables [11]. In the present study, the particle size and PDI was reduced using different surfactants. The minimum particle size of 382 ± 9.54 nm with the lowest PDI of 0.393 ± 0.06 was achieved with 1% TR-OT blend after 0.5 hr of wet grinding. After 1 hr of grinding, the lowest particle size, 399 ± 9.54 nm with minimum PDI of 0.417 ± 0.020, was achieved with 2% TR-OT blend. Size reduction in a media mill is predominantly due to shear, compressional and torsional stresses produced by high speed stirring and moving grinding media [12-14]. In a study on starch composite, Yilmaz et al [15] observed reduction of average droplet size of starch composite significantly under the influence of HLB ranging from 0-10. Another study supported the size reeuction in presence of the surfactant action [16]. In this study, the surfactants shown to had influenced the reduction in the particle size and PDI of corn starch.

Stenger and coworkers reported that nanomilling, beside size reduction simultaneously causes mechanochemcial modification in the materials [16]. Stress-led mechanical activation creates highly energetic particle surface that may induces reactivity potentials with environment, phase transformations and amorphsisation. Thus, the resultant particles may have totally different properties such as higher solubility, reactivity or better catalytic performance [18-19]. Keeping the above in view, the nanomill-grinded corn starch in the present study is expected to have varied characteristics. In the present study, the corn starch nanoparticles having size less than 500 nm with low PDI were achieved. The particles of size range 50-3,000 nm are taken up by the Peyer's patches and through the lymphatics, subsequently translocated to blood [20-22]. It has also been reported that the particles of size, 0.5–1.0 μm are phagocytozed by lymphocytes which, are migrated to the draining lymph nodes [20-21]. Thus, the corn starch nanoparticles may be used, as nano-platform for oral drug delivery, targeting carriers to lymphatics and or for other applications.

The PDI values are bit higher and need to be reduced further. To accomplish this, further study is underway and in upcoming reports elsewhere, the critical factors for particle size of different starches will be presented.

CONCLUSION

Surfactants has shown varying effects on the particle size and PDI of corn starch nanoparticles. Surfactants, particularly the pluronic and TR-OT blend along with other processing and ingredient aids are helpful in developing the corn starch nanoparticles.

REFERENCES

1. Wesslén & Wesslén, 2002. Synthesis of amphiphilic amylase and starch derivatives. *Carbohydrate Polymers, 47, 303-311.*
2. Artursson P, Berg A & Edman P. 1989. Biochemical and cellular effects of degraded starch microspheres on macrophages. *International Journal of Pharmaceutics. Vol 52,183-190.*
3. Bjork E,Bjurstrom and Edman P. (1991). Morphologic examination of rabbit nasal mucosa after nasal administration of degradable starch microspheres.
International Journal of Pharmaceutics. 75: 73-79.
4. Donovan, J.W. (1979). Phase transition of starch-water system. *Biopolym.* 18: 263-275.
5. Martinze, I, Partal, P., Munoz, J., Gallegos, C. (2003). Influence of thermal treatment on the flow of starch-based food emulsions. *Eur. Food Res. Technol.* 10: 676-685.
6. Bordes, C., Garcia, F., Snabre, P., Frances, C. (2002). On-line characterization of particle size during an ultrafine wet grinding process. Powder Technol. 128: 218-228.
7. Gracia, F., Le-Bolay, N., Frances, C. (2002). Changes of surface and volume properties of calcite during a batch wet grinding process. Chem. Eng. J. 85: 177-187.
8. Stenger, F., Mende, S., Schwedes, J., Peukert, W. (2005a). The influence of suspension properties on the grinding behavior of alumina particles in submicrone size range in stirred media mills. Poweder. Technol. 156: 103-110.
9. Stenger, F., Mende, S., Schwedes, J., Peukert, W. (2005b). Nanomilling in stirred media mills. Chem. Eng. Sci. 60: 4556-4565.
10. Netsch MiniCer® Manufacturer's manual, Netsch, Germany, 11-05-05.
11. Varinot, C., Berthiaux, H., Dodds, J. (1999). Prediction of product size distribution in association of stirred mills. Poweder Technol. 105: 228-236.
12. Blecher, L., Kwade, A., Schwedes, J. (1996). Motion and stress intensity of grinding beads in a stirred medial Mill. Part I. Energy density distribution and motion of single grinding beads. Powder Technol. 86(1):59-68.
13. Theuerkauf. J. Schwedes. J. (1999). Theroetical and experimental investigation on particle and fluid motion in stirred media mills. Powder Technol. 105(3): 406-412.
14. Gao, M.W., Forssberg, E. (1995). Prediction of product size distribution for a stirred ball mill. Powder Technol. 84(2): 101-106.
15. Yilmaz, G., Jongboom, R.O.J., Van-Soest, J.J.G., Feil, H. (1999). Effet of glycerol on the morphology of starch-sunflower oil composites. *Carbohyd. Polym. 38*: 33-39.
16. Ubrich, N., Bouillot, P., Pellerin, C., Hoffman, M. and Maincent, P. (2004). Preparation and characterization of propranolol hydrochloride nanoparticles: a comparative study. *J. Contr. Rel.* 97(2), 291-300.
17. Stenger, F., Gotzinger, M., Jakob, P., Peukert, W. (2004). Mechanochemical changes in nanozided α-Al2O3 during wet dispersing in stirred media mills. Part. Part. Syst. Charact. 21: 31-38.
18. Rougier, A., Soiron, S., Haihal, I, Aymard, I., Taouk, B., Tarascon, J. K. (2002). Influence of grinding on the catalytic properties of oxides. Powder Technol. 128: 139-147.
19. Kostic, E., Kiss, S., Boskovic, S., Zec., S. (1997). Mechanical acticavion of the gamma to alpha transition in Al2S3. Poweder. Technol. 91: 49-54.

20. Jani, P. McCarthy, D.E., Florence A.T. (1992). Nanosphere and microsphere uptake via Peyer's patches: observation of the rate of uptake in the rate after a single oral dose. *Int. J. Pharmaceut.* 86: 239-246.

21. Randolph, G.J., Inaba, K., Robbiani, D.F., Steinman, R.M. and Muller, W.A (1999). Differention of phagocytic monocytes into lymph node dendritic cells. *Immunity.* 11(6): 753-761.

22. Jung, T., Kamm, W., Breitenbach, A., Kaiserling, E., Xiao, J. X., Kissel, T. (2000). Biodeegradable nanopartciles for oral delivery of peptides. Is there a role for polymers to affect mucosal uptake? *Eur. J. Pharm Biopharm.* 50(1): 147-160.

Molecular Dynamics of Self-Assembled Monolayer Formation in Soft Nanolithgoraphy

D. Heo[*], M. Yang[*], and J. Jang[**]

[*]Department of Chemistry, Chungbuk National University, Cheongju 361-763, Korea
[**]Department of Nanomaterials Engineering, Pusan National University, Miryang 627-706, Korea,
jkjang@pusan.ac.kr

ABSTRACT

Molecular dynamics simulation is performed to study the growth mechanism of self-assembled monolayer in the AFM tip-assisted soft nanolithography such as in dip-pen nanolithography. We investigate how the droplet created around the tip spreads out to become a monolayer on the substrate. The previous diffusion model assumes that molecules diffuse on top of molecules already adsorbed on the substrate. In contrast, our molecular simulation shows that a molecule on top pushes out a molecule below it and the molecule just pushed out in turn pushes out a molecule next to it. The monolayer grows through such a serial pushing. The present large scale (40 nm diameter) simulation reveals new features. For a relatively weak adsorbate-substrate binding, the monolayer has irregular branches. As the adsorbate-substrate binding strengthens, the monolayer becomes compact, and reflects the rotational symmetry of substrate. A substrate with a hexagonal symmetry results in a hexagonal monolayer. An extremely strong molecule-substrate binding removes such an effect of the substrate anisotropy, giving rise to a circular monolayer. The monolayer periphery shows an initial diffusional growth in its time dependence followed by a slow expansion. The rates of self-assembled monolayer growth exhibit a turn-over behavior with increase in the attractive force between the adsorbate and substrate.

Keywords: dip-pen nanolithography, molecular dynamics simulation, monolayer, growth dynamics

1 INTRODUCTION

An atomic force microscope (AFM) tip serves as a useful tool for the deposition of monolayer on various substrates [1]. Due to its sharp asperity, this nanoscale tip serves as a point source of molecules which are usually designed to bind to a substrate. Currently, we poorly understand the mechanism of the monolayer growth at the molecular level. Due to the continuous downward flow of molecules from the tip, a droplet forms around the tip. This multilayered droplet subsequently spreads out to form a monolayer. As molecules in the upper layers step down to the substrate, the monolayer periphery broadens on the substrate. Exactly how this growth occurs? Elucidating this point will advance our understanding of the monolayer growth utilizing a nanoscale tip.

In our prior molecular dynamics (MD) simulation [2], we found that a molecule in the upper layer pushes a molecule below it out of its place, and the molecule just pushed out in turn pushes molecules next to it, and so on. Our MD however has been performed for a small-sized monolayer with a diameter of about 9 nm. Hence, it is not clear whether the above pushing mechanism should hold for a large monolayer. Herein, we investigate the growth dynamics of monolayer with a size comparable to typical dip-pen nanolithography experiments. For monolayer diameters up to 24 nm, we run MD simulations with trajectory lengths up to 1.5 ns. We investigate whether a novel mechanism emerges for such a large monolayer. We study how the monolayer shape depends on the molecule-substrate binding energy by systematically varying this energy in simulation.

2 SIMULATION DETAILS

We consider the deposition of a nonpolar, spherical molecule on gold (111) substrate. The molecular mass is set identical to that of 1-octadecanethiol ($CH_3(CH_2)_{17}SH$, ODT). We have also performed a simulation that explicitly takes into account the alkyl chain of ODT (by using a united atom model), and such a realistic simulation agrees with our coarse grained simulation.

The AFM tip is modeled as a hemisphere made of silicon atoms and ODT molecules are coated on the surface of the tip [figure 1]. The every interaction (molecule-molecule, molecule-tip atom, and molecule-gold atom interactions) is assumed to be a pairwise Lennard-Jones (LJ) potential [3], $U(r) = 4\varepsilon\left[\left(\sigma/r\right)^{12} - \left(\sigma/r\right)^{6}\right]$. LJ parameters, ε and σ, for the tip atom (silicon) and molecule are 0.4184 kJ/mol and 0.4 nm [4] and 5.24 kJ/mol and 0.497 nm, respectively. ε of our molecule is taken from that of stearic acid ethyl ester which is similar to ODT in mass. σ (= 0.497 nm) of our molecule is chosen to reproduce the experimental structure of the ODT monolayer on Au (111). σ for gold is 0.2655 nm [7] but ε value for gold has been systematically varied in order to examine the effects of molecule-substrate binding energy. The Lorentz-Berthelot combination rule [3] has been used for the interactions between unlike atomic or molecular species. The dissociation energy of ODT and gold has been estimated as 3.182 kcal/mol [8]. We have set the lowest value of ε for molecule-substrate interaction as 3.182 kcal/mol. To inspect the effects of the molecule-substrate binding strength, we have considered additional values of ε.

The radius our hemispherical tip is 3.4 nm. Before starting the simulation of molecular deposition, we coat 2097 molecules on the tip by running a separate MD simulation. To do so, we positioned the molecules at the cubic lattice points near the tip. Then we artificially increased ε of tip 100 times its original value and ran MD simulation for 300 ps. Because of the artificially strong tip attraction, molecules spontaneously stick to the tip surface. The coated tip is used as the initial condition of MD [figure 1].

Figure 1: The initial configuration of molecular dynamics simulation. Total of 2097 molecules are coated on a spherical tip (made of 914 silicon atoms) above Au (111) surface (a single layer consisting of 17503 atoms).

The total number of molecules is 2097. The vertical distance from the tip end to the substrate is 1.3 nm. We include only a single layer of Au (111) substrate in simulation. The horizontal boundary of the gold substrate is a circle with a lateral diameter of 80 nm, and the substrate consists of 17503 gold atoms. The tip and gold atoms are frozen during simulation but they interact with molecules through LJ potentials. We propagated the molecular trajectory by using the velocity Verlet algorithm [3]. We used a time step of 3 fs, and the total time length of simulation was 1.5 ns. The temperature of our system was fixed to 300 K by using the themostat proposed by Berendsen et al. [9]

3 RESULTS

In all the cases, we found the pushing mechanism described in the Introduction prevails. We observed molecules sitting on top of other molecules for short times, but such molecules soon pushed molecules below to make their ways down to the substrate. No molecule stayed long enough on top and made it to the periphery to hop down to the substrate. The above pushing mechanism persisted even for a monolayer as large as 24 nm in diameter and for the strongest molecule-substrate binding energy considered in this work.

In figure 2, we present the final (t=1.5 ns) MD snapshot of the monolayer (top view) for four different molecule-substrate binding strengths. The tip is not drawn for visual clarity. For a relatively weak molecule-substrate binding (figure 2(a)), molecules easily move between the 3-fold hollow binding sites of the substrate. The periphery of the monolayer has many branches which fluctuate significantly in shape. This is in qualitative agreement with the recent DPN experiment using 1-dodecylamine on mica [10]. As the binding strength increases (ε =6.2 kcal/mol, figure 2(b)), the irregular branches of the monolayer are missing and molecules aggregate to form a more compact pattern. The monolayer however is not perfectly compact but has some holes in it. Intriguingly, the monolayer is non-circular and looks like a hexagon. Due to the substrate anisotropy (6-fold rotational symmetry), the molecular motion on the substrate depends on its direction. The monolayer grows faster in the direction from the center to one of 6 vertices of the hexagon. Along these 6 directions, a molecule sitting at one of the hollow binding sites can move to an adjacent hollow site easily. That is, a molecule actually does not have to move on top of a gold atom in going from one hollow site to adjacent one. It can pass through the valley between two gold atoms. The direction from the center to one of 6 vertices is significantly more favored than other directions.

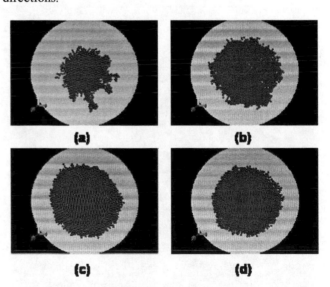

Figure 2: Final monolayer shapes for various molecule-substrate binding energies ε s. We have drawn snapshots taken at t=1.5 ns for 4 different binding energies, 3.1 kcal/mol (a), 6.2 kcal/mol (b), 12.4 kcal/mol (c), and 24.8 kcal/mol (d).

The rotational anisotropy of the substrate again manifests itself in the case of a stronger binding energy, ε =12.4

kcal/mol (figure 2(c)). A hexagonal shape of the monolayer boundary still exists. For the extremely strong molecule substrate binding (ε =24.8 kcal/mol, figure 2(d)), however, the monolayer periphery assumes a compact circular shape. Due to an extremely strong molecule-substrate binding, the molecular motion on the substrate is slow. The movement from one binding site to another takes more energy than in the previous cases. The difference in the activation energy depending on direction however becomes relatively small compared to the activation energy itself. As a result, the molecular motion becomes isotropic and the monolayer periphery becomes circular.

We quantitatively study the growth of the monolayer radius. We kept track of the number of molecules which constitute the monolayer at a given time t, $N(t)$. To do so, we chose molecules whose vertical distances from the substrate are within 0.45 nm. Among such molecules, we checked the intermolecular distance of every possible pair and declared the pairs with intermolecular distances below 0.95 nm as neighbors. A molecule is treated as a part of the monolayer if it is a neighbor of any molecule that forms the monolayer. Then the monolayer radius at time t, $R(t)$, is defined as $R(t)^2 = N(t)/(\pi\rho)$, where ρ is the surface density of the perfect monolayer (4.64 nm^{-2}). In Figure 3, we draw the radial growth of the monolayer for various binding energies. For all the cases, the radial growth is fast initially and then becomes slow at later times.

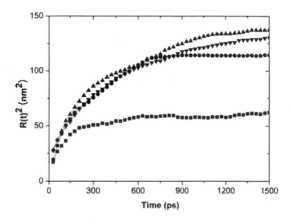

Figure 3: Radial growth of the monolayer for various molecule-substrate binding energies. We plot the radius squared, $R(t)^2$, vs. time for ε = 3.1 kcal/mol (squares), 6.2 kcal/mol (circles), 12.4 kcal/mol (upper triangles), and 24.8 kcal/mol (lower triangles).

There are two distinct phases in the monolayer growth, launching and expansion phases. During the initial launching phase, molecules flow down fast from the tip and move rapidly on the substrate. This launching phase persists until the area directly under the tip is all covered with molecules. Molecules are strongly pulled down from the tip, and the molecular spreading looks nearly inertial. This launching phase is followed by an expansion phase where the nascent monolayer around the tip expands slowly. For the expansion of the monolayer, it takes a series of pushing that needs to propagate to reach the periphery. Sometimes, many molecules move collectively toward the periphery to expand the monolayer area. The molecular motion and the monolayer growth in the expansion phase are significantly slower than in the initial launching phase.

Figure 4 shows how the final (t=1.5 ns) monolayer size depends on the molecule-substrate binding energy. The figure illustrates a turn-over behavior of the monolayer size with respect to the molecule-substrate binding energy. Up to the binding energy of 8.7 kcal/mol, increasing the binding strength raises the monolayer size. This reflects an enhanced attractive force of the substrate pulls down molecules from the tip more strongly, making the downward molecular flow from the tip faster. A further increase in the binding energy however makes the growth rate smaller. Due to a very strong binding to the substrate, molecules are now less mobile than for a smaller binding energy. The pushing of molecules from the center toward the periphery is resisted by molecules strongly sticking to the substrate.

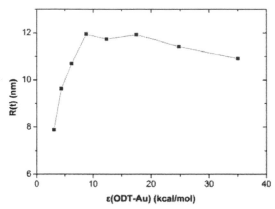

Figure 4: Final monolayer size vs. the molecule-substrate binding energy, ε. The radius $R(t)$ of the monolayer at t=1.5 ns is plotted as a function of the Lennard-Jones ε for the molecule-substrate interaction.

We also considered a cylindrical tip which contains ODT molecules in it, mimicking the "fountain-pen tip" used in DPN [11]. The radius and height of our cylindrical tip are 8.0 nm and 24.1 nm, respectively. The total number of molecules is 5602. The vertical distance from the tip end to the substrate is 1.3 nm. Figure 5 shows the final monolayer shapes for both the cylindrical tip and the hemispherical tip. Regardless of the molecule-substrate binding energy, the monolayer shape is similar for both tips. Therefore, one can expect that the qualitative conclusions obtained in this work will remain intact for different tip shapes.

NSTI-Nanotech 2008, www.nsti.org, ISBN 978-1-4200-8503-7 Vol. 1

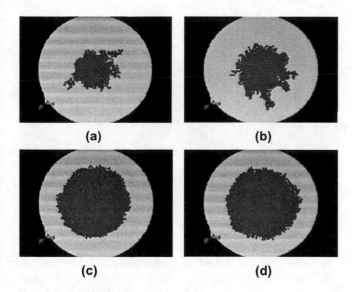

(a)　　　　　　　(b)

(c)　　　　　　　(d)

Figure 5: Dependence of the monolayer structure on the tip shape. In figures (a) and (b), we draw the final monolayer shape for the cylindrical tip and for the hemispherical tip, respectively. The molecule-substrate binding energy ε is 3.2 kcal/mol in both cases. Figures (c) and (d) show the final monolayer shapes of the cylindrical tip and hemispherical tip, respectively. In both caes, ε is 25.5 kcal/mol.

4 CONCLUSIONS

In contrast to its wide applications of dip-pen nanolithography, our understanding of the molecular mechanism of this technique and the timescale of the monolayer growth is at its infancy. We thus have performed molecular dynamics simulations to study the growth mechanism and rate, and the shape of the monolayer deposited from a nanoscale tip. Using the coarse grained molecular model which captures the essential features of alkanethiol, we have examined the monolayer growth. The pushing mechanism found in our previous study of a small sized monolayer holds for a monolayer with a diameter up to 24 nm. That is, molecules deposited from the tip push out molecules already on the substrate, and molecules pushed out in turn push other molecules nearby. When such a pushing propagates to the periphery, sometimes in a collective manner, the monolayer size grows. We have investigated how the monolayer structure is affected by the molecule-substrate binding energy. For a weak binding energy, the monolayer pattern is limited in size and has irregular branches. As the binding strength increases, the monolayer becomes compact and dense, consistent with experimental observations in dip-pen nanolithography.

We found the monolayer becomes hexagonal due to the substrate anisotropy for a moderate binding strength. An extremely strong molecule-substrate binding erases this anisotropy effect, giving a circular periphery. The monolayer growth occurs in two phases, an initial fast launching phase and a slow expansion phase. Interestingly, the speed of

monolayer growth shows a turn-over behavior with respect to the increase in the binding strength. The growth speed initially increases with raising the molecular binding energy, reflecting the enhanced attraction from the substrate. A further rise in the binding energy however slows down the growth. This means that an extremely strong binding strength can make molecules immobile and block the propagation of molecular pushing toward the periphery.

We also simulated the monolayer growth by using a cylindrical tip. The monolayer does not change much in shape by changing the tip. This implies our current conclusions are quite general regardless of the tip shape.

REFERENCES

[1] C. A. Mirkin, ACS Nano 1, 79, 2007.

[2] Y. Ahn, S. Hong, and J. Jang J. Phys. Chem. B 110, 4270, 2006.

[3] M. P. Allen and D. J. Tildesley, Computer Simulation of Liquids, Clarendon Press, 1987.

[4] L. Zhang and S. Jiang, J. Chem. Phys. 117, 1804, 2002.

[5] J. Zhou, X. Lu, Y. Wang, and J. Shi, Fluid Phase Equlib. 172, 279, 2000.

[6] C. A. Alves, E. L. Smith, and M. D. Porter, J. Am. Chem. Soc. 114, 1222, 1992.

[7] L. Zhang, R. Balasundaram, S. H. Gehrke, and H. Jiang, J. Chem. Phys. 114, 6869, 2001.

[8] L. Zhang, W. A. Goddard III, and S. Jiang, J. Chem. Phys. 117, 7342, 2002.

[9] H. J. C. Berendsen, J. P. M. Postma, W. F. van Gunsteren, A. DiNola, and J. R. Haak, J. Chem. Phys. 81, 3684, 1984.

[10] P. Manandhar, J. Jang, G. C. Schatz, M. A. Ratner, and S. Hong, Phys. Rev. Lett. 90, 115505, 2003.

[11] S. Deladi, N. R. Tas, J. W. Berenschot, G. J. M. Krijnen, M. J. de Boer, J. H. de Boer, M. Peter, and M. C. Elwenspoek, Appl. Phys. Lett. 85, 5361, 2004.

Nanotechnology for Building Security

G. Elvin

Green Technology Forum
9801 Fall Creek Rd. #402, Indianapolis, IN, USA, 46256, elvin@greentechforum.net

ABSTRACT

What more can be done to protect building occupants from the threat of terrorist attack? One answer may be nanotechnology, which promises to heighten building security through stronger materials, more powerful sensors, and improved air and water filtration. Access barriers such as walls and windows will benefit from nanocomposite reinforcement, and access control features including biometric devices will be enhanced by advances in micro-electro mechanical systems (MEMS). Nanocomposite reinforcement will also improve building response to explosive blasts and chemical/biological/radiation (CBR) threats. Weapon detection systems are already benefiting from nanosensors capable of detecting some explosives and toxins down to the parts-per-trillion range. Finally, nanofiltration and nanoporous materials will enhance air and water filtration.

Keywords: building security, homeland security, defense, construction, nanosensors

1 NEW THREATS, NEW RESPONSES

Despite increased vigilance and federal anti-terrorism spending exceeding $55 billion per year, building security remains a major concern for most Americans. For those working in large governmental or institutional buildings, fears are particularly acute. What more can be done to protect building occupants from the threat of terrorist attack? What potentially effective strategies, tactics or technologies remain untried? One answer may be nanotechnology. Nanotechnology, the manipulation of matter at the molecular scale, promises to heighten building security and homeland security through the introduction of stronger materials and more powerful sensors. Many of the nanotech innovations now available or in development will have applications for building security. The primary areas of application will be in the strengthening of materials, the advent of nanosensors, and improved air and water filtration.

Aggressors have a variety of weapons and tactics at their disposal, with new ones appearing with alarming frequency. Potential weapons include explosives and incendiaries, stand-off weapons, small arms, and airborne and aqueous toxins. Car bombs, mail bombs, handguns, and chemical/biological/radiation (CBR) threats such as the anthrax used in attacks on the U.S. Congress in 2001,

are just some of the current weapons of choice. New technologies mean new threats, including dirty bombs whose nuclear capabilities threaten almost unimaginable destruction and biological bombs like those unleashed in the Tokyo subway attacks of 1995.

Providing physical and psychological security is one of the primary functions of architecture. In order to provide security, buildings must prevent unauthorized entry and harm to occupants. In recent years, the threat of terrorist attack has become the greatest fear of occupants in governmental and institutional buildings. While almost no amount of physical barriers, technology, or protective service can thwart a determined suicide bomber, measures can be taken to protect building occupants, detect intruders, and react effectively in the event of attack. Improving building security can also reduce the threat of attack by making buildings less attractive targets.

2 EMERGING NANOTECHNOLOGIES

Nanotechnology, the understanding and control of matter at the molecular level, is bringing remarkable changes to industries ranging from electronics and medicine to automotive and apparel. Today there are over 700 products on the market using nanotechnology with a value of approximately $13 billion, with sales expected to top $1 trillion by 2015.

Nanotechnology is bringing remarkable changes to the materials and processes of building. Already this new science has brought architectural advances to market including self-cleaning windows, smog-eating concrete, and wi-fi blocking paint.

In the near future, nanotechnology will contribute to building security through stronger materials, nanosensors, and air and water filtration. Access barriers such as walls and windows could incorporate stronger materials benefiting from nano-reinforcement utilizing carbon nanotubes and nanocomposites. Carbon nanotubes can be designed up to 250 times stronger than steel, yet 10 times lighter. Nanotubes are the building blocks for hundreds of applications across many industries, from sports equipment to medication delivery. Nanotubes for reinforcing glass, concrete, masonry and plastics are all in development. The cost of carbon nanotubes is declining and will eventually make nano-reinforced building materials a reality.

As nano-reinforced building materials gain market share, a complementary technology, nanosensors, will also emerge to help improve building security. Nanosensors integrated throughout our buildings and public spaces will collect and transmit a vast quantity of information about our environment and its users. Already, nanosensors smaller than a penny are marketed for detecting airborne toxins such as carbon monoxide and anthrax in and around buildings. These nanosensors will play a growing role in building security and antiterrorism efforts.

Weapon detection systems are also benefiting from commercially available nanosensors capable of detecting some explosives and toxins down to the parts-per-trillion range. Reaction technologies enabling a building to respond to explosive blasts and chemical/biological/radiation (CBR) attack will improve, as structural hardening and shear resistance in building components are enhanced with nanocomposites and nano-reinforcement, leading to blast-resistant building envelopes, sacrificial exterior walls, and shatter-resistant exterior walls and window glass.

Finally, air and water filtration systems are already one of the primary markets for the rapidly developing technologies of nanofiltration and nanoporous materials.

3 DETECTION NANOTECHNOLOGIES

Detection systems are typically deployed at building entry control points to screen individuals and their belongings for hidden firearms, explosives, and other potentially harmful materials. Current detection technologies include imaging devices and sensors for detecting explosives, pathogens and chemicals. The X-ray machines, metal detectors, and explosive detectors found in airport terminals are examples of today's detection technology.

Nanotech-based chemical sensors can provide high sensitivity, low power and low cost portable tools like the "Carbon Nanotube Sensors for Gas Detection" available for licensing from NASA's Ames Research Center [1]. Similarly, the Lawrence Berkeley National Laboratory has "Miniature Airborne Particle Mass Monitors" for monitoring ventilation systems available for licensing [2]. A prototype nanotech-engineered biosensor from Michigan State University could help detect multiple pathogens faster and more accurately than current devices [3]. And a wristwatch-sized device developed at the University of Michigan can detect toxic gasses at the level of 100 parts per trillion, recognizing mustard gas in a building's air supply in just 4 seconds [4].

In the marketplace, Nano-Proprietary is developing a simplified photo acoustic sensor (PAS) platform capable of identifying trace amounts of most gases and vapors,

miniaturizing and improving the sensitivity of the PAS sensor down to parts per billion levels [5]. ND Life Sciences, a subsidiary of NanoDynamics, Inc., has received a $738,653 Small Business Innovation Research (SBIR) Award to work with chemical detection specialists ICx-Agentase to speed development of a nano-enabled biocatalytic air monitor capable of detecting hazardous nerve agents at extremely low concentrations in air.

Bioident Technologies Inc., a co-winner in the semiconductor category for the seventh annual *Wall Street Journal* contest for Technology Innovation, produces the PhotonicLab Platform, which enables rapid in-vitro diagnostics, chemical and biological threat detection. NanoSensors, Inc. also markets sensors to detect explosive, chemical and biological agents.

Intrusion detection systems use sensors to detect unauthorized entry or attempted entry by monitoring motion, vibrations, heat, or sound. Closed circuit television (CCTV) is an example of current intrusion detection technology. Applied Nanotech has been awarded a $750,000 contract for a "Dual Sensor Module for Human Detection" from the Homeland Security Advanced Research Project Agency. The company will design, develop and demonstrate a high reliability, low cost, low power chemical sensor with the ability to operate in harsh environments to detect humans [6].

4 NANOTECHNOLOGY FOR REACTION

Reaction technologies include those that improve a building's resistance to terrorist attack, as well as the ability of the building to react and respond. They can be incorporated into structural materials, components, and systems including mechanical and electrical.

4.1 Nanocoatings

Surfaces in buildings can be treated with antimicrobial nanoparticles to kill microbes and bacteria that come in contact with them. These antimicrobial treatments, whether integrated during the manufacturing process or applied to existing surfaces, could reduce the threat of biological attack. Researchers at North Carolina State University have developed a nanocoating that can kill most visuses and bacteria when exposed to visible light. Early tests have shown that it kills 99.9 percent of influenza viruses and 99.99 percent of vaccinia virus [7].

Elsewhere, Yale researchers have discovered that single-walled carbon nanotubes can kill bacteria like the common pathogen E. coli by severely damaging their cell walls, offering the first direct evidence that carbon nanotubes have such powerful antimicrobial activity [8].

NanoViricides is creating special purpose nanomaterials designed to attack viruses and dismantle them. The company is developing nanoviricides to fight bird flu, influenza, HIV, hepatitis C, rabies and dengue fever [9].

4.2 Fire retardants

Both surfaces and structural components like columns and beams can be made more resistant to terrorist attack through improved fire resistance. Plastics are particularly combustible unless they incorporate flame-retardant chemicals. Alternative fire retardants are now being developed in the manufacture of nanocomposite plastics. In nanocomposite, nanoparticles (clay, metal, carbon nanotubes) act as fillers in a matrix [10].

Other research is exploring the synergistic effect of carbon nanotubes and clay for improved fire resistance. Researchers at the University of Warwick have found a way of replacing the soap used to stabilize latex emulsion paints with nanotech sized clay armor that can create a much more hard wearing and fire resistant paint [11].

4.3 Self-healing materials

Eventually, building materials based on nanotechnology may even become self-healing. Researchers at Rensselaer Polytechnic Institute are at work on an epoxy material infused with a wire grid and carbon nanotubes that can detect and repair structural problems in airplanes. When a crack is detected, voltage is increased to the carbon nanotubes, generating heat which melts the epoxy that fills the crack [12].

Leeds NanoManufacturing Institute is designing an experimental house with walls containing nanopolymer particles that will turn into a liquid when squeezed under pressure, flow into the cracks, and then harden to form a solid material. The house walls will also contain wireless, battery-less sensors and radio frequency identity tags that collect data about the building over time [13].

4.4 Nano-reinforced Glass and Concrete

Glass may be strengthened as well in the nano-enabled future. University of Texas at Dallas nanotechnologists have produced transparent carbon nanotube sheets that are stronger than the same-weight steel sheets [14]. And engineers at the Air Force Research Laboratory are testing a new kind of transparent armor made from aluminum oxynitride that could stop armor-piercing weapons from penetrating vehicle windows [15].

In the marketplace, Solutia Inc., the world's largest producer of polyvinyl butyral (PVB) protective interlayer used to manufacture laminated glass, has launched Vanceva Secure, a nanotech-based product [16]. And 3M has created Prestige Ultra Safety & Security transparent window film using polyester nanomaterials [17].

Traditional structural materials like concrete will also eventually benefit from nano-reinforcement. Vanderbilt University assistant professor Florence Sanchez recently won a CAREER Award from the National Science Foundation (NSF) for her research into more durable nano-structured cement. Sanchez is investigating how nanofibers made of carbon could be added to a concrete bridge, allowing it to monitor itself for cracks [18].

4.5 Air Purification Technologies

Building systems, particularly air and water supply, are also vulnerable to attack. A building's heating, ventilating, and air-conditioning (HVAC) systems can become an entry points and distribution system for many hazardous contaminants, including chemical/biological/radiation (CBR) agents like arsine, nitrogen mustard gas, anthrax, and radiation from a dirty bomb. Air filtration systems can protect a building and its occupants from the effects of a CBR attack, and nanotech has already entered the air filtration market.

Commercially available products include the Nano e-HEPA (High Efficiency Particulate Arrest) filtration system from Samsung, which uses a dust filter coated with 8-nanometer silver particles to kill airborne health threats, including 99.7 percent of influenza viruses [19]. Other nanotech-based air filtration products include the Ultra-Web nanofiber media from Donaldson Filtration Systems [20], ConsERV brand energy recovery ventilator products by Dais Analytic Corporation [21], And NanoBreeze room air purifiers [22].

4.6 Water Purification Technologies

Finally, water purification can also benefit from emerging nanotechnologies. Some nanoparticles have a high surface area and reactivity, and can be used to render heavy metals like lead and mercury insoluble, reducing their contamination. Dendrimers, with their sponge-like molecular structure, can clean up heavy metals by trapping metal ions in their pores [23].

Photocatalytic nanomaterials enable ultraviolet light to destroy pesticides, industrial solvents and germs. Titanium dioxide, for example, can be used to decontaminate bacteria-ridden water. When exposed to light, it breaks down bacterial cell membranes, killing bacteria like E. coli [24]. Purification and filtration of water can also be achieved through nanoscale membranes or nanoscale polymer "brushes" coated with molecules that can remove poisonous metals, proteins and germs.

Seldon Laboratories has delivered prototype portable water purification systems to the Air Force for testing

NSTI-Nanotech 2008, www.nsti.org, ISBN 978-1-4200-8503-7 Vol. 1

[25]. Water purification nanotechnologies available for licensing include "Biofunctional Magnetic Nanoparticles for Pathogen Detection," from Hong Kong University of Science and Technology [26].

The complete 100+ page report, "Nanotechnology for Building Security," is available from Green Technology Forum at greentechforum.net/security.

REFERENCES

[1] S. Venkatesh, "Carbon Nanotube Sensors for Gas Detection," http://128.102.216.35/factsheets/view.php?id=125

[2] L. Gundel, "Miniature Airborne Particle Mass Monitors," http://www.lbl.gov/tt/techs/lbnl1850.html

[3] A. El Amin, "Nanotech biosensor developed for multipathogen detection," foodproductiondaily.com, Oct. 8, 2007, http://www.foodproductiondaily.com/news/ng.asp?n=80381&m=1FPDO08&c=vlnrklrxctxggzl

[4] G. Elvin, "Wireless Sensors Sniff Out Toxins," smallplans.com, May 11, 2006, http://www.smallplans.blogspot.com/2006/05/wireless-sensors-sniff-out-toxins.html

[5] "Photo Acoustic Sensors (PAS)," Nano-Proprietary, http://www.nano-proprietary.com/TechnologyPlatforms/PAS.asp

[6] "Sensor For Cargo Ships Contract Awarded By Homeland Security To Nano-Proprietary," azonano.com, August 17, 2007, http://www.azonano.com/news.asp?newsID=4748

[7] "Novel Nano-Coating Kills Viruses and Bacteria when Exposed to Light," azonano.com, November 2, 2006, http://www.azonano.com/news.asp?newsID=3278

[8] "Yale scientists use nanotechnology to fight E. coli," nanotechnology.com, August 31, 2007, http://www.nanotechnology.com/news/?id=11335

[9] "THE NANOTECHNOLOGY REVOLUTION IN BIOPHARMACEUTICS," nanoviricides.com, http://www.nanoviricides.com/

[10] R. Delobel, "Fire-retardant plastics," Cordis Technology Marketplace, December 18, 2006, http://cordis.europa.eu/fetch?CALLER=OFFR_TM_EN&ACTION=D&RCN=3112

[11] "Nanotech Clay Armour Creates Fire Resistant Hard Wearing Latex Emulsion Paints," sciencedaily.com, July 27, 2007, http://www.sciencedaily.com/releases/2007/07/070726104821.htm

[12] G. Elvin, "Self-healing Nanomaterial Detects and Repairs Structural Damage," October 12, 2007, http://www.greentechforum.net/category/news/2007/10/12/self-healing-nanomaterial-detects-and-repairs-structural-damage/

[13] "Special House Walls Containing Nano Polymer Particles," azonano.com, April 4, 2007, http://www.azonano.com/news.asp?newsID=3930

[14] G. Elvin, "Transparent nanotube sheets stronger than steel," smallplans.com, October 17, 2005, http://smallplans.blogspot.com/2005/10/transparent-nanotube-sheets-stronger.html

[15] L. Lundin, "Air Force testing new transparent armor," Air force Link, October 17, 2005, http://www.af.mil/news/story.asp?id=123012131

[16] "Laminated Glass Protective Interlayer," azom.com, October 16, 2001, http://www.azom.com/details.asp?ArticleID=954

[17] T. Feran, "New film for windows blocks heat, not view," Columbus Dispatch, August 26, 2007, http://www.dispatch.com/live/content/home_garden/stories/2007/08/26/WINDOW.ART_ART_08-26-07_H1_5F7L7DJ.html?sid=101

[18] "Prestigious Award for Nano-Fiber Reinforced Concrete," azonano.com, December 8, 2005, http://www.azonano.com/news.asp?newsID=1718

[19] Azonano.com, "Samsung Launches Nano e-HEPA Air Purifier System," February 27, 2004,

[20] Donaldson, "Nanofiber Technology Is Cleaner," http://www.ultrawebisalwaysbetter.com.au/cleaner.htm

[21] Dais Analytic Corporation, "Welcome to ConsERV," 2005, http://www.conserv.com/

[22] K & W Products Inc., "NanoBreeze", 2006, http://www.nanobreeze.com/index.html

[23] S. Mann, "Nanotechnology and Construction," Institute of Nanotechnology, 2006

[24] S. Gray, "Nanotechnology Applications in Water Management," 2005, http://www.nanovic.com.au/downloads/water_management.pdf

[25] B. Edwards, "Windsor firm expects to add nearly100 jobs," Rutland Herald, December 14, 2006, http://www.rutlandherald.com/apps/pbcs.dll/article?AID=/20061214/NEWS/612140343/1003/NEWS02

[26] Bing XU, Biofunctional Magnetic Nanoparticles for Pathogen Detection," Hong Kong University of Science and Technology and RandD Corporation Ltd, 2005, http://www.ttc.ust.hk/new_selected/doc/patent%20186S.pdf

Refined Coarse-Grain Modeling of Stamp Deformation in Nanoimprint Lithography

S. Merino[*], A. Retolaza[*], A. Juarros[*], H. Schift[**], V. Sirotkin[***], A. Svintsov[***] and S. Zaitsev[***]

[*]Fundación Tekniker. Avda. Otaola, 20, 20600 Eibar, Guipúzcoa, Spain
[**]Laboratory for Micro- and Nanotechnology, Paul Scherrer Institute, 5232 Villigen PSI, Switzerland
[***]Institute of Microelectronics Technology, RAS, Chernogolovka, Moscow district, 142432, Russia

ABSTRACT

A refined version of the IMPRINT software is applied for simultaneous calculation of the resist viscous flow in thermal nanoimprint lithography and the stamp/substrate deformation. This version applies a modified coarse-grain method as well as takes into account the composition and elastic properties of the imprint setup (the stamp/substrate + "pressure buffer layers"). The presented comparison of calculated and experimental results confirms the potential of the IMPRINT software as an efficient tool for the reduction of the stamp bending.

Keywords: nanoimprint lithography, stamp and substrate deformation, computer simulation

1 INTRODUCTION

The inhomogeneous distribution of the residual layer thickness is a vital issue in thermal nanoimprint lithography (NIL). Using the simulation of NIL, this problem can be alleviated by optimizing the stamp geometry and by choosing process parameters.

In [1, 2] the IMPRINT software for modeling of NIL process has been presented. The software takes into account the stamp/substrate bending during squeeze flow and is able to predict the distribution of the residual resist thickness with an accuracy better than 10% [2, 3]. It should be noted that the above-mentioned results have been obtained using the deformation model in which the stamp/substrate are represented as semi-infinite regions (an elastic medium bounded by a plane).

In [3] a dramatic effect of the stamp thickness on the distribution of the residual resist thickness has been described. The experiments were performed for a typical R&D case where the grating is surrounded by a large unstructured area. For the simulation of these experiments the IMPRINT software has been modified. The software has been adapted for the calculation of the extensive deformation for the imprint setup (the stamp/substrate + "pressure buffer layers"). Consequently, the latest version of the IMPRINT software has been supplemented by: a multilayer model of the stamp/substrate deformation; and an adaptive multi-grid realization of the coarse-grain method for the simulation of structures having regard to surroundings.

Figs. 1-3 demonstrate the implementation of the modified IMPRINT software.

2 EXPERIMENTS

The experiments were performed for 4" silicon stamps with thickness of 400 and 1000 μm. On the stamps nine different arrays (gratings with 12 μm period) are placed. They are comprised of different areas (1×1 mm^2, 2×2 mm^2 and 4×4 mm^2) and different fill factors (0.25, 0.5 and 0.75). The stamps were imprinted on 300 nm thick of mr-I 7030 (Micro Resist Technology GmbH) coated on silicon substrates 500 μm in thickness. The stamp cavities depth was 170 nm. The imprint temperature was 140°C. The imprinting process lasted 1200 s.

In the simulation, the resist dynamic viscosity was taken to be 5×10^3 Pa·s. This value gave the best fit of calculated residual thickness distribution to the experimental one. For the calculation of the stamp and substrate deformation, elastic properties of single-crystalline silicon were used: modulus of elasticity – 10^{11} Pa, Poisson's ratio – 0.2.

By the coarse-grain simulation, two embedded grids were applied: the first (fine) grid was 128×128 pixel; the second (coarse) grid was 256×256 pixel.

In Fig. 4 experimental and simulated results are compared. It is evident that for the 1000 μm stamp the experimental and simulated results agree very closely. Slightly worse agreement is observed for the 400 μm stamp. The reason is the lack of information about the exact elastic properties of the "pressure buffer layer".

ACKNOWLEDGEMENT

The partial support of the EC-funded project NaPa (Contract no. NMP4-CT-2003-500120) is gratefully acknowledged. The content of this work is the sole responsibility of the authors.

REFERENCES

[1] V. Sirotkin, A. Svintsov, H. Schift and S. Zaitsev, Microelectron. Eng. 84, 868, 2007.

[2] V. Sirotkin, A. Svintsov, S. Zaitsev and H. Schift, J. Vac.Sci. Technol. B 25, 2379, 2007.

[3] N. Kehagias, V. Reboud, C.M. Sotomayor Torres, V. Sirotkin, A. Svintsov and S. Zaitsev, Microelectron. Eng. doi:10.1016/j.mee.2007.12.041

[4] S. Merino, A. Retolaza, H. Schift and V. Trabadelo, Microelectronic Eng. doi:10.1016/j.mee.2008.01.045

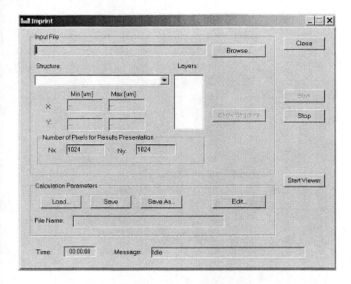

Figure 1: Main window of the IMPRINT software.

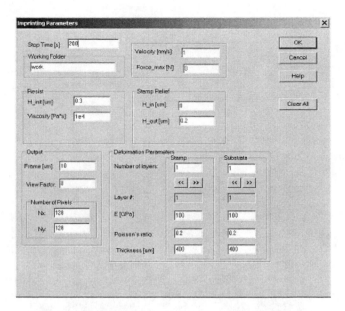

Figure 2: The IMPRINT software: list of simulating parameters (the default values).

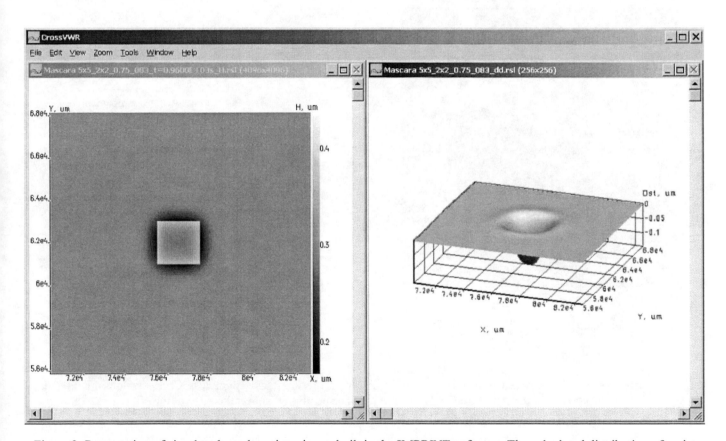

Figure 3: Presentation of simulated results using viewer built in the IMPRINT software. The calculated distribution of resist thickness (left window) and the calculated distribution of the stamp deformation (right window).

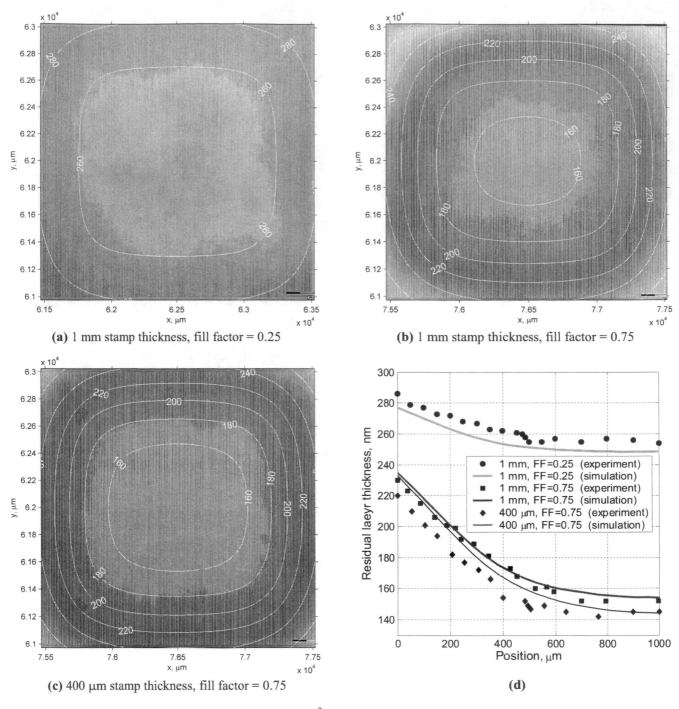

(a) 1 mm stamp thickness, fill factor = 0.25

(b) 1 mm stamp thickness, fill factor = 0.75

(c) 400 µm stamp thickness, fill factor = 0.75

(d)

Figure 4: **(a)-(c)** Optical microscopy images for a 2×2 mm² array size imprinted from different silicon stamp thickness and at different fill factors. White isolines indicate the calculated distribution of the residual layer thickness (numbers signify the thickness in nanometers). **(d)** Measured and simulated distributions of the residual layer thickness. Cross-sections are directed along the y-axis and through the centre of the grating. The position 0 shows the edge of the grating.

NSTI-Nanotech 2008, www.nsti.org, ISBN 978-1-4200-8503-7 Vol. 1

Plasma Treatment of Nanofillers for Polymer Nanocomposites

Andrew C. Ritts[1], Qingsong Yu[1,2], Hao Li[2]

[1] Center for Surface Science and Plasma Technology, Department of Chemical Engineering,
University of Missouri, Columbia, MO 65211, USA, yuq@missouri.edu
[2] Department of Mechanical and Aerospace Engineering,
University of Missouri, Columbia, MO 65211, USA, liha@missouri.edu

ABSTRACT

Plasma nanocoatings can create a variety of desirable surface functionalities and thus to tailor the surface characteristics of multiwalled carbon nanotubes (MWNTs) and silicon carbide nanofibers (SiCNFs) for improved dispersion capability in polymer matrices. Ultrasonication was used to disperse MWNTs and SiCNFs into epoxy Epon 815c resin. The amine groups on the coating surface are believed to strengthen the interfacial interactions through chemical bond formation between the fillers and matrix. Plasma nanocoated MWNTs and SiCNFs were characterized using SEM, XRD, FTIR, surface contact angle and pH value measurements. Mechanical testing results showed that all SiCNF reinforced nanocomposites were found to be stronger than the MWNT reinforced nanocomposites. Plasma coated (PC) MWNTs better dispersed in the Epon 815C resin than the uncoated nanotubes and enhanced the mechanical properties significantly. Plasma treated SiCNFs increased the tensile strength of the epoxy by 40% with only 1.0wt% loading.

Keywords: low-temperature plasma nanocoatings, amine, interface, polymer nanocomposite, and SiC nanofibers

1 INTRODUCTION

Because of their lightweight and significantly improved properties, polymer nanocomposites reinforced with nanosize fillers make up a new class of materials. Polymer nanocomposites have low percolation thresholds (~0.1 to 2vol%) requiring only minute quantities of nanofillers to significantly enhance the properties of the composites.[1] Nanotubes and nanofibers are excellent choices for the reinforcement of polymer nanocomposites due to their high aspect ratio, and outstanding mechanical, thermal, and electrical properties.[2] The comparative experimental strengths of carbon nanotubes (CNTs) and SiCNFs are reported to be 100 GPa[3] and 50 GPa[4] respectively, much higher than their microscale counterparts. Controlling the strength transfer of these nanofillers can form a new class of high-strength polymeric materials never before seen. To date, two major challenges exist in developing novel polymer nanocomposites. First, a homogeneous dispersion of nanofillers in their host polymer matrices must be achieved.[5] Second, an enhanced interfacial adhesion must be attained in order to provide effective load transfer between polymer matrices and the reinforcing nanofillers.[3]

Previous work has applied various techniques to functionalize CNTs, however this requires defects in the CNTs lowering the theoretical strength of the tubes significantly.[2] SiCNFs are stable compounds which can be functionalized without producing defects making SiCNFs a better candidate as a filler in polymer nanocomposites. Plasma technologies have been utilized as an environmentally friendly way to enhance the surface properties of fibrous materials for composite materials since the 1960's.[6] However plasma coating of nanopowders introduces unique challenges due to increased surface energies at the nano-level which produces agglomerations. Powder-plasma coating reactors must be designed to prevent agglomerations during its operation and allow the plasma coating to prevent agglomerations when introduced to the polymer matrix.

In this proceeding, a low-temperature nanocoating process is proposed for treating nanosized fillers for use in composite material applications. Specifically amine rich surface functionalities are deposited on MWNTs and SiCNFs which are used in epoxy nanocomposites and are compared to their non-coated counterparts. The amine functional groups on the nanofiller are believed to chemically bond the filler to the epoxy matrix enhancing the interfacial tension.

2 EXPERIMENTAL

MWNTs with 10-30nm in diameter, 20-40µm in length, and 95% purity were purchased from Helix Material Solutions. SiCNFs produced from various carbon sources were made and characterized in lab as previously reported.[7] The following abbreviations will be used to compare different types of SiC nanofibers: SMG is 99.9% submicron scale graphite with average particle size of 500nm purchased from Nanostructured and Amorphous Materials, MG is 99.995% micron scale graphite with 2-10µm purchased from Alfa Aesar, and LG is 99.99% 44µm average particle size graphite purchased from Sigma Aldrich. MWNTs were also used to synthesize SiC NFs.

2.1 Amine Plasma Treatment of MWNTs and SiCNFs

Plasma conditions were chosen by determining deposition rates of the amine coating on Si wafers prior to the filler treatment. One gram of nanofiller was plasma treated with 50% Ar, 50% allylamine at 100mtorr with 6 or 10 watts of power, and 40 or 60 minutes of treatment. Figure 1 shows

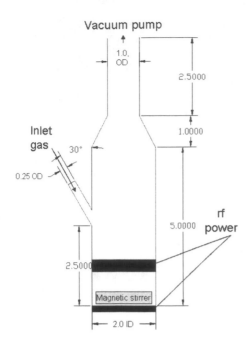

Figure 1. Low-temperature plasma powder reactor (dimensions in inches)

the plasma reactor setup where the electrodes were placed one inch apart. The powder was stirred at a rate of 300 rotations per minute, pushing the powder into a direct contact with the amine plasma. Approximately 700mg of plasma treated nanofiller was used in the epoxy composite, and the excess was characterized by SEM, XRD, FTIR, surface contact angle, and pH value measurements.

2.2 Epoxy Nanocomposite Fabrication

Fillers were incorporated into the Epon 815c epoxy resin (Miller-Stephenson) via Ultra Sonic horn (Branson Ultrasonics Corporation) for 4 minutes with 8 microtip limit and 50% duty cycle (~40W). The Epon 815c resin was cured using 12 parts of Epicure 3223 (Miller-Stephenson) for 100 parts of epoxy resin by weight. Resin, filler, and curing agent were mixed with a magnetic stirrer for 5 minutes at 300 rotations per minute. Five samples were cast in an aluminum dog bone shaped mold and cured for 1 hour at 100°C. Dog bone samples were tested in a tensile testing machine (MTS).

3 RESULTS AND DISCUSSION

Before manufacturing the composite, a thorough characterization of the filler and matrix material should help determine the treatment method of the filler. Epon 815c resin contains 86.4% bisphenol-A-(epichlorhydrin) and 13.6% N-butyl glycidyl ether, both of which contain at least one epoxide group. Epicure 3223 curing agent mainly consists of diethylenetriamine, which allows the matrix to crosslink. Coating the nanofiller with amine groups can allow the filler to be chemically incorporated into the matrix as seen in

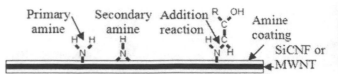

Figure 2. Depiction of an epoxide reaction with a primary amine functionality on the surface of a nanofiller.

figure 2. Epoxides react with amines by adding each individual part into a single molecule. Primary amines have a higher probability to react compared to secondary amines, but secondary amines can also react in this manner.

3.1 Characterization of PC MWNTs and PC SiCNFs

Nanofillers characterized by SEM seen in figure 3 compare strait SiCNFs (40-100nm diameter) with tortuous MWNTs (10-30nm in diameter). Tortuous nanotubes should affect the composite more as whiskers than fibers due to their relative aspect ratio in a single direction; in contrast the straight SiCNFs should give better load transfer to the epoxy matrix due to their significantly larger relative aspect ratio. However, the 50-70% yield of SiCNFs is significantly lower than the 95% yield of MWNTs. When 1wt% of SiC is added to the epoxy only 0.5-0.7wt% of the SiC is nanofibers. CNTs also have significantly higher strengths than SiCNFs, however these strengths may be misleading when looking at

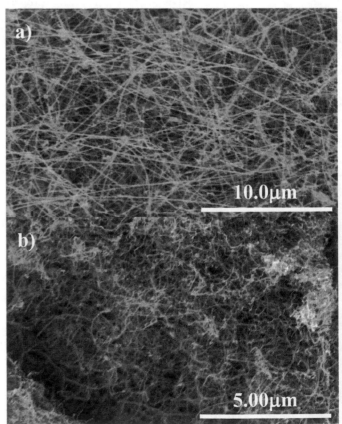

Figure 3. SEM images of SiCNFs synthesized in our lab (a) and MWNTs purchased (b)

Table1. ΔpH from plasma treated and untreated nanofillers

Material	mass (mg)	ΔpH
PC SiCNF from MWNT	100	0.40
SiCNF from MWNT	100	-0.55
PC SiCNF from SMG	300	0.86
SiCNF from SMG	300	-0.77
PC MWNT	100	0.68
MWNT	100	0.1

Table 2. Tensile testing results of epoxy and untreated MWNTs and SiCNFs nanocomposites.

Filler	Tensile Strength (MPa)	STD (MPa)
None	34.32	1.31
MWNTs	22.70	8.23
SiCNFs1*	31.12	3.26
SiCNFs2**	32.97	2.65
SiCNFs3***	31.40	5.48

*SiCNFs synthesized from MWNT
**SiCNFs synthesized from MG
***SiCNFs synthesized from LG

these materials as fillers for composite materials as mentioned previously. XRD determined that SiCNFs are β-SiC with some stacking faults. Previous TEM analysis has shown a thin amorphous SiO_x and C coatings on the surface of the fibers.

After plasma treatment approximately ~100 mg of PC SiCNFs from MWNT precursor was able to increase the pH of 50 mL of distilled water by ~0.40 where ~100 mg of untreated SiCNFs lowered the pH by ~0.55 as seen in table 1. The PC nanofillers were extremely hydrophobic so they were dispersed into distilled water using the ultra sonic horn. All PC nanofillers increased the pH of distilled water significantly due to the amine groups that can be found on the surface. Amines become protonated in water producing hydroxyl groups in solution, thus increasing the pH. Untreated MWNTs have little effect on the pH; however SiCNFs decrease the pH of the significantly. This is believed to be due to the SiOH groups produced insitu during the nanofiber fabrication. These SiOH will deprotonate in water producing H^+ which in turn decreases the pH. It is very difficult to determine amine plasma coatings on nanopowders with FTIR, but small NH peaks on PC SiCNF from 2560 to 2700 can be seen when compared with untreated SiCNFs. When a silicon wafer is PC under the same amine plasma conditions for 10 minutes, contact angle measurements can show a change in surface energy. At different heights in the reactor there are different deposition rates of amine coating which has an effect on the surface energy, however there is a significant change between the coated and plain wafers. When Epon 815c resin droplets came in contact with amine plasma coatings they spread across the wafer until reaching the edge, where untreated samples had contact angles around 30°.

Our results indicated that PC MWNTs and PC SiCNFs had better dispersed in the Epon 815C resin than the uncoated fillers which should increase the tensile strength of their composites significantly. When uncoated samples were tested the nanofiller would accumulate at the bottom of the beaker, even after sonication. Therefore the epoxy composites formed with out PC had less filler than recorded, and the probability of aggregates in the samples increased significantly.

3.2 Mechanical Testing of Epoxy Nanocomposites

Mechanical testing results showed that all SiCNF reinforced nanocomposites were found to be stronger than the MWNT reinforced nanocomposites. The SiO_x coatings found on the SiC are believed to enhance the interface better than the MWNT surface. The slight polarity of the SiOH groups should also enhance the dispersion of the SiCNFs in the epoxy resin. The SiCNFs samples made from larger carbon precursors had lower strengths than their smaller analogues. SiCNFs made from MWNTs had the highest strengths, closely followed by SiCNFs from MG. This is believed to be due to the size of the SiC by-products produced and their role as defects in the composite. However, it was found that all nanocomposites prepared from untreated MWNTs and SiCNFs had inferior mechanical strength as compared with the controls of the pure epoxy samples as seen in table 2. SiCNF epoxy nanocomposites made from MG and MWNT have statistically the same strengths, which are very close to pure epoxy. When the SiCNFs are plasma coated the improved dispersion and interfacial tension should increase the strengths significantly.

PC nanofillers in contrast showed vast improvement compared to their untreated counterparts. Table 3 shows tensile strength tests with PC and pure epoxy. PC SiCNFs from SMG precursor out perform all other samples with a tensile strength of 54.95 MPa, with more than a 40% increase from the highest plain epoxy sample. The Young's Modulus of a majority of the samples were close to 21.5 GPa; in contrast PC MWNTs increased the Young's Modulus by 12.8%. Stress strain curves of PC fillers compared with the best plain epoxy samples are found in Figure 4. Certain PC MWNT and PC SiCNF samples failed at the clamp during testing, which implies that the test section is stronger than the recorded values. Both PC nanofillers had the largest strain measurements around 1.0029, but MWNTs also have the

Table 3. Tensile testing results epoxy and PC MWNTs and PC SiCNFs nanocomposites.

Filler	Tensile Strength (MPa)	STD (MPa)	Young's Modulus (GPa)	Max Strain
plain1	38.84	3.85	21.47	1.0018
plain2	34.23	5.71	21.41	1.0016
MWN1	52.64	16.29	24.24	1.0029
SiC1*	54.95	8.60	21.56	1.0028

*SiCNFs synthesized from SMG

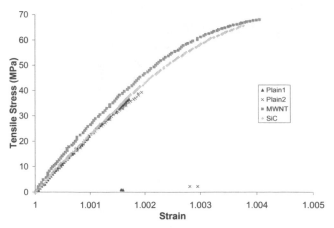

Figure 4. Tensile stress strain curves of epoxy composites with PC MWNTs, PC SiCNFs, and plain epoxy.

largest standard deviation.

The fractured interface was examined using HR-SEM to study the fracture mechanisms. Figure 5 shows SEM images at various magnifications showing PC MWNT pullout in the epoxy nanocomposite fracture surface. MWNT loading in the epoxy was relatively small and provided for infrequent pullout. The average diameter of the MWNTs found on the surface was around 50nm; much larger than the original 10-30nm implying the MWNTs were strongly bonded with the epoxy. Similar fiber pullout was seen on the PC SiCNF epoxy fracture surface. All non PC epoxy nanocomposites fracture surfaces had similar topologies and no fiber pullout

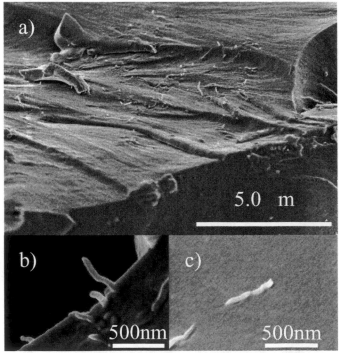

Figure 5. SEM images of PC MWNT pullout of epoxy nanocomposite fracture surface

was found using HR-SEM. A few agglomerations could be found in MWNT composites that were not PC.

4 CONCLUSIONS

Low-temperature plasma coatings for nanopowder applications have promise in improving the mechanical properties of polymer nanocomposites by enhancing dispersion of the nano-filler and interfacial tension between the matrix and composite. PC SiCNFs in particular have increased the tensile strength of epoxy nanocomposites by 40% with only 1wt% loading. PC MWNTs have increased the Young's modulus of the composite by 12.8%. Future work involves enhancing the properties of other polymer composite materials with plasma coating technologies.

ACKNOWLEDGEMENT

This work has been supported in part by University of Missouri Research Council and the Department of Education's GAANN (Graduate Assistance in Areas of National Need) Program. Andrew C. Ritts is a GAANN fellow.

REFERENCES

[1] Vaia, R.A. and H.D. Wagner, *Framework for nanocomposites.* Materials Today, 2004. **7**(11): p. 32-37.

[2] Breuer, O. and U. Sundararaj, *Big Returns From Small Fibers: A Review of Polymer/Carbon Nanotube Composites.* Polymer Composites, 2004. **25**(6): p. 630-645.

[3] Wagner, H.D. and R.A. Vaia, *Nanocomposites: issues at the interface.* Materials Today, 2004. **7**(11): p. 38-42.

[4] Wong, E.W. and P.E. Sheehan, *Nanobeam mechanics: Elasticity, strength, and toughness of nanorods and nonotubes.* Science, 1997. **277**(5334): p. 1971.

[5] Graff, R.A., et al., *Achieving Individual-Nanotube Dispersion at High Loading in Single-Walled Carbon Nanotube Composites.* Advanced Materials, 2005. **17**(8): p. 980-984.

[6] Tatoulian, M., et al., *Plasma Deposition of Allylamine on Polymer Powders in a Fluidized Bed Reactor.* Plasma Processes and Polymers, 2005. **2**(1): p. 38-44.

[7] Ritts, A.C., H. Li, and Q. Yu, *Synthesis of SiC Nanofibers with Graphite Powders.* Proceedings Coco Beach J. Am. Ceramics Soc., 2007.

Thermal and Mechanical Response of POSS (Epoxy Cyclohexyl) coated Nanophased expandable foam core materials

Wanda D. Jones, Vijaya K. Rangari,*and Shaik Jeelani.
Center for Advanced Materials (T-CAM), Tuskegee University
Tuskegee, AL 36088

ABSTRACT

Epoxy Cyclohexyl Polyhedral Oligomeric Silsesquioxanes (POSS) nanoparticles were sonochemically coated on expandable thermoplastic (Expancel®) microspheres. The foam core materials are then fabricated using a compression molding technique. Morphological investigation of the neat and nanophased coated microspheres have been carried out using scanning electron microscopy (SEM). Thermal characterizations have also been carried out using Thermogravimetric Analysis (TGA), and Differential Scanning Calorimetry (DSC) to determine decomposition temperature and glass transition temperature of these materials. Mechanical analysis such as Quasi-static compression tests have been carried out for both nanocomposite and neat foam systems. Details of the synthesis, thermal and mechanical characterization are presented in this paper.

Keywords: Foam, Sonochemical, POSS, Expandable microspheres.

1 INTRODUCTION

Sandwich composites materials are increasingly being used in various industries because of their superior bending-stiffness-to-weight ratio. Sandwich composite consists of three main parts; two thin, stiff and strong faces separated by a thick, light and weaker core. The core plays an important role in enhancing the energy absorption capability during an impact and determines the extent of damage in the structure. [1] If the properties of the core can improved, the overall performance of the sandwich structure will be enhanced as well. Recently researchers have shown interests in improving polymeric materials physical, mechanical, thermal and chemical properties using nanoparticles as filler materials. Nanoparticles embedded in polymer matrix have attracted increasing interest because of the unique properties displayed by nanoparticles. Due to nanometer size of these particles, their physicochemical characteristics differ significantly from those of molecular and bulk materials [2-3]. Nanoparticle-polymer nanocomposites synergistically combine the properties of both the host polymer matrix and the discrete nanoparticles there in. Such nanocomposite materials are expected to have novel thermal and mechanical properties [4-5]. Recently we have shown that the nanoparticles can enhanced the mechanical performance of

core materials [6-7]. One distinct group of nanoparticles that have gained researchers' attention are Polyhedral Oligomeric Silsesquioxanes (POSS). POSS are organic silica compounds with the general formula being $(RSiO_{1.5})$. Advantages to using these nanoparticles are 1) they are analogues to the smallest possible particles of silica with diameters from one to three nanometers 2) they can be functionalized in various ways. By changing the functional groups (R) on a POSS molecule, the characteristics and surface activity can be changed. This variety in fuctionalization is the source of one of the main differences between POSS molecules and customary fillers [8]. These nanoparticles can have a significant affect on the thermal and mechanical properties of core materials. High-performance structural foam materials are fabricated using a blowing agent (surfactants, hydrocarbons) in liquid polymers to expand and form rigid, low-density foams. Some of the leading thermoplastic foams made in this way are polymethacrylimide (PMI) and partly cross-linked polyvinyl chloride (PVC), with trade names Rohacell [9], Divinycell [10] and Expancel [11]. The hollow thermoplastic microspheres produced by Expancel, Inc., under the trade name Expancel® these microspheres are small, spherical plastic particles consisting of a polymer shell encapsulating a hydrocarbon gas. When the gas inside the shell is heated, it increase in pressure and the thermoplastic shell softens, resulting in a dramatic increase in the volume of the microspheres. Researchers used these microspheres for various applications such as car protection (corrosion resistance, acoustic insulation, gap fillers, underbody coatings) [12], Young-wook and his coworkers developed a closed-cell silicon oxycarbide foams with cell densities greater than 10^9 cells/cm^3 and cells smaller than 30 μm were obtained from a preceramic polymer using expandable microspheres [13]. Lev et al studied the reinforcement of microspheres in PVC with the aramid fibers and reported the improved mechanical properties [14].

In the present manuscript we study the thermal and mechanical properties of thermoplastic polymeric foam materials using POSS nanoparticles as fillers materials.

2 EXPERIMENTAL

Expancel-092-DU-120 is unexpanded thermoplastic polymer (particles sizes 28-38µm) was received from Expancel Inc, Samples for this study were prepared as follows. Expancel polymeric powder and known percentages (2%, 4%, and 6% by weight) of Epoxy Cyclohexyl POSS nanoparticles was dispersed in n-hexane using a high intensity ultrasonic horn (Ti-horn, 20 kHz, 100 W/cm^2) at 5°C for 30 minutes. The mixture was then dried in a vacuum for 12 hours and remaining n-hexane was removed by heating the sample to 60°C for 1 hour. The free falling dry mixture transferred to a compression mold and heated to 190°C held for 15 minutes. The mold is cooled to room temperature. The samples are taken out and cut to the required dimensions for thermal and mechanical characterization.

Thermogravimetric analysis (TGA) of various specimens was carried out under nitrogen gas atmosphere on a Mettler Toledo TGA/SDTA 851^e apparatuses. The samples were cut into small pieces 10-20 mg using a surgical blade. The TGA measurements were carried out from 30°C to 800°C at a heating rate of 10°C/minutes. Differential scanning calorimetry (DSC) experiments were carried out using a Mettler Toledo DSC 822^e from 30°C to 200°C at a heating rate of 10°C/min under nitrogen atmosphere. The morphological analysis was carried out using JEOL JSM 5800 Scanning electron microscopy (SEM). The sample were precisely cut into small pieces and placed on a double sided carbon tape and coated with gold/palladium to prevent charge buildup by the electron absorption by the specimen.

In order to investigate the compression response, the specimens were tested in the thickness direction using Zwick/Roell Material Testing Machine. ASTM C365-57 was followed for this quasi-static compression test and the size of the specimen is 12.7 mm x 25 mm x 25 mm respectively. The load cell used on the Zwick/Roell machine is approximately 2.5 k N. The test is carried out in displacement control mode and the cross-head speed was 1.27 mm/min. In order to maintain evenly distributed compressive loading, each specimen was sanded and polished with high a high accuracy, where the opposite faces were parallel to one another. TestXpert software was used to analyze the load-deflection data recorded by the data acquisition system.

3 RESULTS AND DISCUSSION

Thermogravimetric analysis (TGA) measurements were carried out to obtain information on the thermal stability of neat and nanophased foam. The TGA results are presented in Table 1. TGA results clearly shown that the foam disintegrates in three steps: first step is corresponds to the loss of organic vapor and the second major weight loss is corresponds to the rupture of the microspheres and finally the third weight loss is corresponds to the decomposition of the polymer itself. The first weight loss corresponds to the weight loss of organic vapor delayed by infusion of POSS nanoparticles on the surface of polymeric microspheres. It continued up to 6 wt% loading of POSS. The expansion of the microspheres increases with the amount of POSS added. The thermal studies of higher loading of POSS are under investigation.

The glass transition temperature (*Tg*) of the expancel was obtained from the DSC curves, and the scans were carried out at a heating rate of 10°C/min in a nitrogen atmosphere. The Tg results were presented in table 1. The Tg measured for neat expancel is ~ 93°C and the nanophased expancel is Tg increased to 105°C. The reason for increase may be due to the restricted polymer chain movement in presence of the POSS.

Table 1: TGA and DSC results of neat and nanophased expancel foam and expancel foam/epoxy composites

Material	First	Second	Third	Weight Retain (%)	DSC results Tg$^\circ$C
Neat expancel (a)	147	299	406	34	93.18
2% POSS expancel (b)	201	299	413	40	105.23 ± 1
4%- POSS - expancel (c)	200	297	412	42	105.04 ± 1
6%- POSS - expancel (d)	194	302	417	34	105.18 ± 1

To understand the mechanical behavior of POSS coating on the polymeric foam the compression tests were carried for all samples. Stress-strain curves for tested samples are shown in figure 1 and the results are presented in table 2.

Table 2: Compression properties of neat and nanophased expancel foam

Material	Compressive Strength (kPa)	Compression Modulus (kPa)
(a) Neat Expancel foam	400	312
(b) 2% EpoxyCyclohexyl POSS	448	471
(c) 4% EpoxyCyclohexyl POSS	517	952
(d) 6% EpoxyCyclohexyl POSS	491	954

It is observed from figure 2 that the compressive strength of the POSS/expancel for the system is higher than the neat uncoated expancel sample. This improvement may be the result of increasing the interfacial bonding on the surface of the nanoparticles.

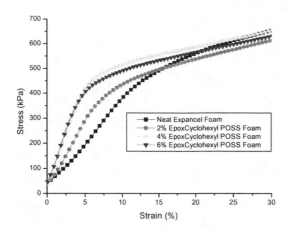

Figure 1: Compressive stress-strain curves of neat and nanophased expancel foam

SEM analysis has been carried out to understand the morphology of the POSS coating on the expancel foam. Figure 2(a) show that all the microspheres are unexpanded and typical sizes measured are about 40-100μm. Figure 2(b) shows the microspheres in open expansion Figure 2(3) shows the microspheres in compression molding expansion.

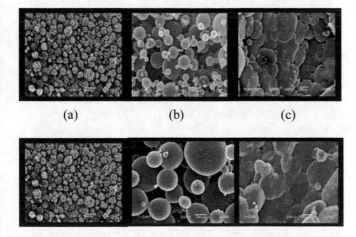

(a)　　　　(b)　　　　(c)

Figures 2 and 3　SEM micrographs of (a) unexpanded, (b) open expansion and (c) compression molding expansion of neat Expancel foam core materials

4 CONCLUSIONS

- Sonochemical method has been developed to coat POSS nanoparticles on expandable polymeric microspheres.
- Improved thermal and mechanical properties were observed for nanocomposites with 4% POSS loading

5 ACKNOWLEDGEMENTS

The authors would like to thank the NSF-IGERT, NSF-PREM, Alabama EPSCoR, and Alabama Commission on Higher Education, for financial support and Expancel Inc for providing polymeric sample.

REFERENCES

[1]　Mallick, P.K., 1993, Fiber-Reinforced Composites, Marcel Dekker Inc., US, 2nd Edition.

[2]　Antonietti M, Goltner C. Superstructures of functional colloids: chemistry on the nanometer scale. Angew Chem Int Ed Engl 1997; 36 910.

[3]　Schmid G. Clusters and colloids. VCH Weinheim. 1994.

[4]　Schmid G. Large clusters and colloids metals in the embryonic state. Chem Rev, 1992; 92, 1709.

[5]　Hirai H, Wakabayashi H, Komiyama M. Preparation of polymer protected colloidal dispersions of copper. Bull Chem Soc Jpn 1986; 59,367.

[6]　Renee´ R, Mahfuz. H, Rangari V K, Nathanie C, Jeelani S J, Infusion of SiC Nanoparticles Into SC-15 Epoxy: An Investigation of Thermal and Mechanical Response. Macromol. Mater. Eng. 2005, 290, 423–429.

[7]　Hassan Mahfuz, Vijaya K. Rangari, Mohammad S. Islam, Shaik Jeelani Fabrication, synthesis and mechanical characterization of nanoparticles infused polyurethane foams, Composites: Part A 35 2004, 453–460.

[8]　K. Pielichowski, et al., Advanced Polymer Science, 201, 225–296 (2006).

[9]　Rohm A G. Technical Information for ROHACELL Foam. 2002.

[10]　DIAB Inc. Technical Information for Divinycell Foam; 2002.

[11]　Elfving K and Soderberg B. Versatile microspheres offer dual benefits Reinforced Plastics June 1994; 64.

[12]　Tomalino M, Bianchini G, Hear-expandable microspheres for car protection production Progress in Organic Coatings. 1997; 32,17-24

[13]　Kim Y W, Kim S H, Kim H D, Park C B. Processing of closed-cell silicon oxycarbide foams from a preceramic polymer. Journal of Material Science 2004;39, 5647– 5652

[14]　Vaikhanski L, Nutt S R. Fiber-reinforced composite foam from expandable PVC microspheres. Composites: Part A 2003 ; 34,1245–125

Silica and silica-alumina composite monoliths with a hierarchical pore structure containing the MFI-type zeolitic films

W. Pudlo[1], V. Pashkova[2], M. Derewinski[2], A.B. Jarzebski[1,3] *

[1]Faculty of Chemistry, Silesian University of Technology, Strzody 7, 44-100 Gliwice, Poland
[2]Institute of Catalysis and Surface Chemistry, Polish Academy of Sciences,
Niezapominajek 8, 30-239 Krakow, Poland
[3]Institute of Chemical Engineering, Polish Academy of Sciences, Baltycka 5,
44-100 Gliwice, Poland, Andrzej.Jarzebski@polsl.pl

ABSTRACT

Pre-shaped amorphous silica and silica/alumina monoliths containing an extensive system of meso-/macropores were used as parent solids for the preparation of a new type of hierarchical porous systems. Impregnation of the monolith pore walls with a template solution and subsequent hydrothermal recrystallization resulted in the formation of materials consisting of amorphous silica or silica/alumina carcass covered with a thin layer of the MFI crystals. The differences in the porous structure of the parent monoliths appeared to have critical effect on the porosity and deposition of the zeolitic phase in the prepared composites. The results of characterisation of both parent and modified monoliths carried out by XRD, N₂ adsorption and SEM are discussed.

Keywords: monoliths, hierarchical pore structure, zeolitic films

1 INTRODUCTION

The size of micropores in zeolites may significantly hinder or even restrain transport of bigger reactants inside the crystals. Synthesis of new catalysts containing an active phase in the form of fine and dispersed zeolite crystals and a system of meso- and macropores (serving as arteries for unrestricted transport of the molecules) could overcome such limitations.

Previous research on the preparation of the multimodal porous materials containing zeolitic phase showed the possibility of introducing nanodomains of the desirable microporous phase (FAU, BEA, MFI) into the meso/macroporous aluminosilicate structure via partial recrystallization of amorphous walls [1, 2]. The present work aimed to apply this method to the preparation of a new type of porous solids using pre-shaped silica and silica/alumina monoliths, with very well developed system of the meso/macropores [3-5]. Application of these materials, which retain the size and shape after recrystallization, should eliminate costly and time-consuming shaping step in the preparation of more efficient zeolite-based catalyst.

2 EXPERIMENTAL

The synthesis of the silica monolith was carried out according to the procedure reported previously [6]. The same procedure was applied to obtain silica/alumina monolith.

Figure 1. Parent silica (left) and silica-alumina (right) monoliths with a hierarchical pore structure.

In this case, 0.95 g of aluminium nitrate nonahydrate (Al(NO₃)₃×9H₂0) was used as alumina precursor (Si/Al molar ratio of the product equal to 40). After dissolving the aluminium source in 1M nitric acid, the polyethylene glycol (PEG), tetraethoxysilane (TEOS) and cetyltrimethyl-ammonium bromide (CTABr) were subsequently added. The gels obtained were aged at 40°C for 6 days and then dried at 60°C for four days and finally calcined in air at 550°C for 5h. The synthesis procedure used for the preparation of the silica and silica/alumina monoliths enabled to tailor their size and shape (Figure 1). The applied process of recrystallization consisted of: impregnation of a dry monolithic sample with 40 wt% tetrapropylammonium hydroxide solution (TPAOH), ageing the impregnated monolith for up to 1.5 h, and its heating at 175°C for the period ranging from 12 to 44 h. The products i.e., partially recrystallized monoliths (composites) were calcined for 12 h at 550°C in dry air.

The composites obtained were structurally characterized by XRD. Their porosity was determined from

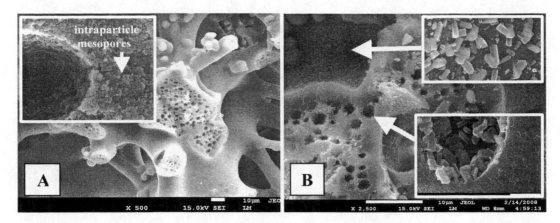

Figure 2. A) Parent silica monolith: "corridor-like" macropores and spherical voids; intra-particle mesopores. B) Recrystallized silica monolith: deposit of silicalite-1 crystals on the walls and inside the spherical void. The bar size is 10 μm.

the N_2 adsorption/desorption isotherms. Scanning electron microscopy (SEM) was applied to study the morphology and architecture of materials and to verify the presence of the zeolite phase.

3 RESULTS AND DISCUSSION

3.1 Recrystallization of the silica monolith

The SEM images of the obtained silica monolith showed a developed system of macropores, with interconnected large empty spaces inside a monolith and numerous smaller, spherical macrovoids located in its skeleton (Figure 2A). The adsorption measurements revealed a considerable amount of mesopores, as indicated by the presence of a hysteresis loop (Figure 3A). A careful analysis of SEM images showed, that the mesopores were mainly of an interparticles origin and related to the free spaces between fine silica grains forming the monolith structure (Figure 2A).

The microscopic analysis revealed that the applied process of the controlled recrystallization did not change the initial shape and size of the monolith. The structural analysis (XRD, data not presented) showed, that the MFI was the sole microporous phase present in the material. Recrystallization time of about 12 h proved quite sufficient to obtain the composite product containing a considerable amount of the zeolitic phase. Longer treatment, up to 44 h, did not produce other crystalline phases. The SEM investigations (Figure 2B) showed that the macroporous character of the parent silica monolith was retained in the recrystallized solids. The newly developed microporous phase had a form of fine silicalite-1 crystals (MFI), deposited on the walls of macropores. Similar MFI crystals were also found inside the spherical macrovoids (cf. Figure 2B), which were accessible for the template

molecules through the mesopores present in the pristine material (intraparticle mesopores).

The recrystallization process significantly changed adsorption properties of the modified silica monoliths (Figure 3). The isotherm of composite (B) was shifted upward for low values of the relative pressure (p/p_o), in comparison to the isotherm from parent material – A. This signifies the presence of micropores originating from the zeolitic phase. Additionally, the changes in the shape and position of the hysteresis loop registered for the recrystallized material and seen in the pore sizes distributions (results not given) revealed a considerable presence of newly formed smaller mesopores, at the expense of large mesopores, the amount of which considerably decreased. Such modification of porous structure could be explained by the changes in the system of intraparticle mesopores, caused by interaction of the template molecules with silica particles, resulting in their partial dissolution during the hydrothermal process. This idea was supported by the closer inspection of the silica skeleton images in the parent and recrystallized preparations (Figure 2B)

3.2 Recrystallization of the silica/alumina monolith

The SEM analysis of the silica/alumina monolith revealed that introduction of Al and small changes in the synthesis procedure resulted in the generation of less developed pore system with some large macropores. This observation was supported by the adsorption data. The same data (results not given) showed, however, the presence considerable number of irregular micropores. Spherical voids similar to those detected in purely siliceous material were also found in the silica/alumina system. Similarly to the silica monolith, the pure MFI phase was the only microcrystalline phase formed during

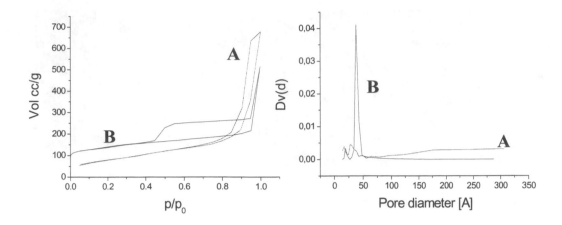

Figure 3. Nitrogen adsorption/desorption isotherms and pore size distribution (calculated on the basis of the desorption branch) of parent (A) and recrystallized (B) silica monolith.

the partial recrystallization. Yet, the synthesis time necessary to obtain a similar level of transformation into the zeolite phase was longer due to nore difficult formation of aluminium-containing zeolite phase. SEM analysis of the recrystallized silica/alumina monolith proved, that the zeolite crystals were formed exclusively on the outer surface of macroscopic flow-through "corridors" (Figure 4). No MFI crystals were found inside the isolated spherical voids located in the walls. The lack of mesopores and limited transportation of the template molecules through micropores is the most likely explanation of the lack of MFI crystals inside these empty spaces.

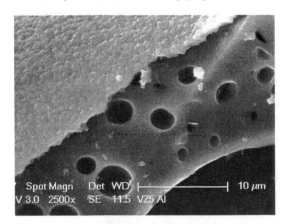

Figure 4. Zcolite MFI laser created on the wall of silica/alumina monolith.

4   CONCLUSIONS

The method of controlled hydrothermal recrystallization appears suitable to transform partially the macroporous skeletons of pre-shaped parent silica and silica/alumina monoliths into a zeolite phase. As a result composite materials containing the layer of fine MFI crystals, deposited in the macroporous structure are obtained. The shape of the parent materials is very well retained. Clear differences in texture of the parent silica and silica/alumina monoliths have a critical effect on the porosity and deposition of the zeolite phase in the prepared composites.

Our future studies will mainly focus on the recrystallization of silica monoliths impregnated with aluminum compounds. This will aim to exploit their vast advantages such as high thermal stability and very well developed meso/macroporosity. We believe that by combining high acidity and shape-selectivity of the zeolites with the highly open structure of silica monoliths we should significantly boost the catalytic performance of these materials.

REFERENCES

[1] J.H. Koegler, C.Y. Yeh, P.J. Angevine, 2003, U.S. Patent No 0147805.
[2] V.O. Pashkova, P. Sarv, M. Derewiński, 2007, Stud. Surf. Sci. Catal., 170A, 289-296.
[3] J.-H. Smatt, S. Schunk, M. Linden, 2003, Chem. Mater., 15, 2354-2361.
[4] T. Amatani, K. Nakanishi, K. Hirao, T. Kodaira, 2005, Chem. Mater., 17, 2114-2119.
[5] W. Pudło, W. Gawlik, J. Mrowiec-Białoń, T. Buczek, J.J. Malinowski, A.B. Jarzębski, 2005, Inż. Chem. Proc., 27, 177-185.
[6] V. Pashkova, W. Pudło, E. Bielańska, Z. Czuła, A. B. Jarzębski, M. Derewinski, 2007, Proceedings of the XIVth Zeolite Forum, Kocierz, ISBN: 978-83-60514-05-4, 155-160.

TiO$_2$-Ag Porous Nanocomposites for Advanced Photocatalytic Processes

L. Baia[*], M. Baia[*], F. Vasiliu[**], L. Diamandescu[**], A. Peter[***], V. Cosoveanu[***], V. Danciu[***]

[*]Faculty of Physics, Babes-Bolyai University, 400084, Cluj-Napoca, Romania, lucb@phys.ubbcluj.ro
[**]National Institute of Materials Physics, P. O. Box MG-7, 77125 Bucharest, Romania, fvasiliu@infim.ro
[***]Faculty of Chemistry and Chemical Engineering, Babes-Bolyai University, 400028, Cluj-Napoca, Romania, vdanciu@chem.ubbluj.ro

ABSTRACT

Porous nanocomposites based on TiO$_2$ aerogels and silver colloidal particles were obtained by sol-gel method. The samples were heat-treated at 500 ^{0}C for 2h with the aim to obtain a preponderant TiO$_2$ anatase phase. The photocatalytic performances of the composites were monitored by using salicylic acid as test molecule. The promising results obtained in comparison with the commercial product Degussa P25 directed our attention towards the understanding of the morphological and structural particularities of the composite samples. The morphological data revealed the existence of high specific surface area for the as-prepared samples and of a relatively small one for the heat-treated samples. The influence of the anatase content was found to be insignificant, while that of TiO$_2$ nanoparticles size was found to be an important parameter besides the presence of silver particles and aggregates inside the TiO$_2$ porous network.

Keywords: Nanocomposites, TiO$_2$ aerogels, silver particles, photocatalysis.

1 INTRODUCTION

Titanium dioxide based aerogels are porous solid materials with high surface area, ultra-low density and high homogeneity that make them attractive for diverse applications in solar energy conversion, pigments, electronic devices, etc. [1-4]. Since TiO$_2$ has a high photosensitivity, non-toxic nature, large band-gap and stability the photo-assisted catalytic decomposition of organic pollutants in water and air, employing titanium dioxide as photocatalysts, is a very promising method. Its photocatalytic performances can be improved by retarding the electron-hole recombination process that is in direct competition with space-charge separation of the electron and the hole. A possibility is the loading of electron accepting species on the TiO$_2$ surface. It was shown that the contact of Ag nanoparticles with those of the TiO$_2$ grains influences in a favourable way the energetic and interfacial charge transfer processes and leads to an enhancement of the photocatalytic activity [5]. Thus, the synthesis and characterization of novel heterostructures based on TiO$_2$-Ag nanoparticles from the structural and morphological point of view with the purpose of improving the photocatalytic activity still remains an imperative challenge.

In the present work, porous nanocomposites based on TiO$_2$ aerogels and Ag colloidal particles were synthesized by sol-gel method and using the preparation chemical route of the silver colloidal particles. In order to enable the formation of the TiO$_2$ anatase phase, which exhibits improved photocatalytic performances, the nanoarchitectures were subjected to a thermal treatment and were comparatively characterized from the structural and morphological point of view. In this respect, several complementary investigation techniques like Brunauer–Emmet–Teller (BET), transmission electron microscopy (TEM), Raman spectroscopy and X-ray diffraction were used. The photocatalytic activity tests carried out on the obtained composites revealed exceptional results in comparison with the commercial product Degussa P25. In order to explain the nanocomposites photocatalytic performances the structural parameters have been correlated with the morphological ones.

The heat treatment parameters were settled as those established in a study previously carried out in our laboratory, in which a morphological and structural optimization process was performed with the aim of improving the photocatalytic efficiency of the TiO$_2$ aerogel [6].

2 EXPERIMENTAL

2.1 Samples preparation

TiO$_2$ gels prepared by sol-gel method using titanium isopropoxide (TIP), HNO$_3$, EtOH and H$_2$O (1/0.08/21/3.675 molar ratio) were allowed to age for a few weeks. The TiO$_2$ aerogel was obtained after a supercritical drying with CO$_2$ (T = 40 ^{0}C and p = 1400 psi) was applied [6, 7].

The silver colloidal suspension was prepared in the TiO$_2$ aerogel presence according to the following procedure [8]: 300 ml 2x10^{-3} M NaBH$_4$ kept at ice temperature was added to 100 ml 5x10^{-3} M AgNO$_3$ solution under vigorous stirring. The reaction mixture was brought to a boil with vigorous stirring on a magnetic stirring hot plate. The boiling was continued for 60 minutes. The aerogels were filtered from the solution and dried at 100 ^{0}C.

The obtained composites were subjected to a thermal treatment at 500 ^{0}C for 2 hours in order to obtain the desired TiO$_2$ anatase phase. The as-prepared and heat-treated composite samples will be further denoted as **TiO2@Ag** and **TiO2@Ag-ht**, respectively.

2.2 Samples measurements

The photocatalytic activity of the composites was established from the degradation rate of salicylic acid. The decrease in time of the salicylic acid concentration ($C_0 = 5 \times 10^{-4}$ M for all investigated samples) was monitored by using a Jasco V-530 UV-Vis spectrophotometer ($\lambda = 297$ nm). The samples immersed in salicylic acid solution were irradiated with UV light from a medium pressure Hg lamp HBO Osram (500 W). The working temperature was 20-22^{0}C and the solution pH 5.3. Before UV irradiation as well as before UV-Vis measurements, the cell with the sample was kept in dark for 15 minutes in order to achieve the equilibrium of the adsorption-desorption process.

The surface area of the samples was determined by the BET method, in a partial pressure range of $0.05 < P/P_0 < 0.3$ by using a home made equipment. The krypton adsorption was carried out at 77 K. Before each measurement the samples were heat cleaned at 333 K for 2 hours.

A Nd-YAG laser ($\lambda = 1064$ nm) was employed for the recording of Raman spectra of the prepared composites. The FT-Raman spectra were recorded using a Bruker Equinox 55 spectrometer with an integrated FRA 106 Raman module, a power of 30 mW incident on sample and a resolution of 4 cm^{-1}.

The phase content and the particle dimensions were determined using a DRON X-ray powder diffractometer linked to a data acquisition and processing facility; CuK$_\alpha$ radiation ($\lambda = 1.540598$ Å) and a graphite monochromator were used.

A JEOL 200 CX TEM operating at an accelerating voltage of 200 kV was employed to obtain bright (BF) and dark (DF) field images as well as the electron diffraction patterns of the nanoparticles system.

3 RESULTS AND DISSCUSSION

The disappearance of salicylic acid from the solution during the photocatalytic tests was monitored versus time (see Fig. 1); the photodecomposition reaction follows a pseudo-first order kinetics. The apparent rate constant was calculated by plotting $\ln(C_0/C)$ vs time. The slope of the plot after applying a linear fit represents the apparent rate constant [9, 10].

The determined apparent rate constants (k_{app}) for the photodegradation of the salicylic acid are presented in Table 1 together with other morphological and structural parameters. One can observe a considerable increase of the

apparent rate constant of the **TiO2@Ag** composite in comparison with both Degussa and **TiO2@Ag-ht** samples. However, the improvement of the photocatalytic performances of the heat-treated composite is two times higher than those of the commercial product.

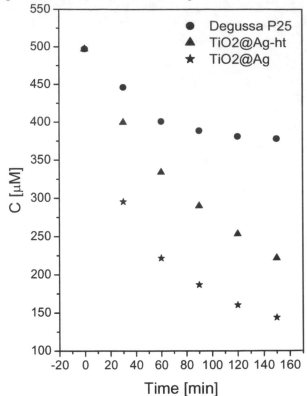

Figure 1: Photodegradation profiles of salicylic acid on nonannealed (**TiO2@Ag**) and annealed (**TiO2@Ag-ht**) nanocomposites as compared to the commercial powder (**Degussa P25**).

Sample	k_{app} x 10^3 [min^{-1}]	BET surface area [m^2/g]	Crystalline phase composition (%)	Average particles size [nm]
TiO2@ Ag	13.46	487	22.3 – A 77.7 – B	10 – A 9 – B 5 ÷ 15 – Ag
TiO2@ Ag-ht	6.62	70	88.5 – A 11.5 – B	13 – A 9 – B 10 ÷ 25 – Ag
Degussa P25	2.5	50	80 – A 20 – R	20 – A 30 – R

Table 1: The apparent rate constant, BET surface area, crystalline phases composition, and average particles size for the investigated samples. Abbreviations: A - anatase, B - brookite, R - rutile and Ag - silver

By looking at the data obtained from the BET analyses of the synthesized composites in comparison with the corresponding value reported [11] for the commercial product (Table 1) one can infer that the specific surface area decisively influence the photocatalytic performances, but only in the situation when it is extremely high, e.g. **TiO2@Ag** composite in comparison with Degussa P25.

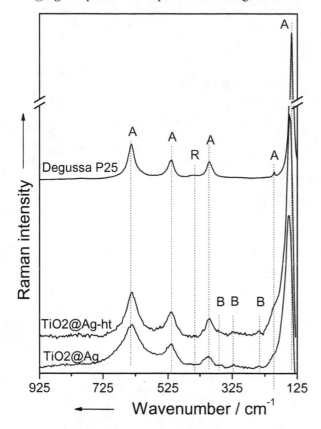

Figure 2: Raman spectra of the synthesized composite samples (**TiO2@Ag** and **TiO2@Ag-ht**) together with the spectrum of the commercial powder (**Degussa P25**). Abbreviations: A - anatase, B - brookite and R – rutile.

The Raman spectra of synthesized composites are displayed in Fig. 2 together with that of the commercial sample. Well-defined bands at 144, 197, 399, 517 and 639 cm^{-1} that correspond to the well-known five fundamental vibrational modes of TiO_2 anatase (denoted with A) with the symmetries of E_g, E_g, B_{1g}, A_{1g} and B_{1g}, respectively [12] can be seen in all recorded spectra. A close analysis of the spectra reveals the presence of other three very small Raman bands around 245, 320 and 365 cm^{-1}, especially in the spectra of the composite samples. Such spectral features were previously observed, when ultrafine TiO_2 powders with grain sizes around 10 nm were investigated, and were attributed to the existence of a little amount of brookite [12-14]. Therefore, the weak intense features from the Raman spectra of the composites can be associated with the presence of brookite phase and are denoted with B. The rutile phase presence can be seen as a band located at 440 cm^{-1}, only in the Raman spectrum of Degussa P25. Unfortunately, no quantitative analysis can be performed on the investigated samples by using Raman spectroscopy.

Thus, in order to get further insights into the samples structure X-ray diffraction measurements were done (data not shown). X-ray diffraction patterns confirmed the presence of the anatase phase accompanied by a rather significant fraction of brookite phase. Rietveld refinement reveals an increase of anatase phase content (from 22.3 % to 88.5 %) as a cost of the brookite phase amount (from 77.7 % to 11.5 %) after the heat treatment was applied (see Table 1). As shown by particle dimension calculation (using Scherrer equation) the mean particle diameter of anatase phase is about 10 nm for the as-prepared composite and 13 nm for the heat-treated sample, while the mean size of the brookite nanoparticles are preserved even after annealing (~ 9 nm). By comparing both the crystalline phase composition and average particles size obtained for **TiO2@Ag-ht** sample (88.5 %-anatase, 11.5 %-brookite and 13 nm-anatase, 9 nm-brookite, respectively) with that of the commercial powder (80 %-anatase, 20 %-brookite and 20 nm-anatase, 30 nm-rutile, respectively) one can conclude that the size of the TiO_2 nanoparticles has an important influence on the photocatalytic performances, besides that represented by the presence of TiO_2-Ag nanoparticles contact.

TEM results show that, as compared to the microstructure of **TiO2@Ag**, the **TiO2@Ag-ht** sample presents higher sizes (see Table 1) and more non-homogeneous distribution of silver particles, which can also affect the photocatalytic behavior. Many metallic particles are amorphous in the both samples, but some big particles become monocrystalline.

4   CONCLUSIONS

Porous nanocomposites based on TiO_2 aerogels and silver particles were successfully synthesized by sol-gel method. A major amount of TiO_2 anatase was obtained by applying a heat treatment at 500 ^{0}C for 2h. The photocatalytic performances of the composites were tested by using salicylic acid as test molecule and it was found that both as-prepared and heat-treated composites show a considerable improvement of the photocatalytic activity in comparison with the commercial product Degussa P25. The positive photocatalytic results were completed with information derived from morphological and structural investigations. The morphological data revealed the existence of a huge pores surface area for the as-prepared samples and a relatively small one for the heat-treated samples. The anatase content was found to not decisively influence the photocatalytic performances, while the size of the TiO_2 nanoparticles was found to be an important parameter besides the presence of silver particles and aggregates inside the TiO_2 porous network. The size and distribution of silver particles can also modify the photocatalytic behavior of these nanocomposites.

REFERENCES

[1] N. Hüsing, U. Schubert, Angew. Chem. Int. Ed. 37, 22, 1998.

[2] Z. Zhu, M. Lin, G. Dagan, M. Tomkiewicz, J. Phys. Chem. 99, 15945, 1995.

[3] S. Kelly, F. H. Pollak, M. Tomkiewicz, J. Phys. Chem. B 101, 2730, 1997.

[4] G. Dagan, M. Tomkiewicz, J. Phys. Chem. 97, 12651, 1993.

[5] P. D. Cozzoli, E. Fanizza, R. Comparelli, M. L. Curri, A. Agostano, J. Phys. Chem. B 108, 9623, 2004.

[6] L. Baia, A. Peter, V. Cosoveanu, E. Indrea, M. Baia, J. Popp, V. Danciu, Thin Solid Films 511-512, 512, 2006.

[7] L. Baia, M. Baia, A. Peter, V. Cosoveanu, V. Danciu, J. Optoelect. Adv. Mater. 9(3), 668, 2007.

[8] P. C. Lee, D. Meisel, J. Phys. Chem. 86, 3391, 1982.

[9] A. Barau, M. Crisan, M. Gartner, A. Jitianu, M. Zaharescu, A. Ghita, V. Danciu, V. Cosoveanu and I. Marian, J. Sol-Gel Sci. Techn. 37, 175, 2006.

[10] V. Danciu, L. Baia, V. Cosoveanu, M. Baia, F. Vasiliu, L. Diamandescu, C. M. Teodorescu, M. Feder, J. Popp, J. Optoelect. Adv. Mater.-Rapid Comm. 2(2), 76, 2008.

[11] S. Bakardjieva, J. Šubrta, V. Štengla, M. J. Dianezb, M. J. Sayaguesb, App. Catalysis B: Environmental 58, 193, 2005.

[12] Y.H. Zhang, C.K. Chan, J.F. Porter, W. Guo, J. Mater. Res. 13, 2602, 1998.

[13] G. Busca, G. Ramis, J.M. G. Amores, V.S. Escribano, P. Piaggio, J. Chem. Soc. Faraday Trans. 90, 3181, 1994.

[14] M. Gotic, M. Ivanda, A. Sekulic, S. Music, S. Popovic, A. Turkovic, K. Furic, Mater. Lett. 28, 225, 1996.

Influence of the Crystallite Size of BaTiO$_3$ on the Dielectric Properties of Polyester Reactive Resin Composite Materials

B. Schumacher* **, H. Geßwein*, T. Hanemann*, J. Haußelt*

* Forschungszentrum Karlsruhe GmbH, Institut für Materialforschung III
Hermann-von-Helmholtz-Platz 1, 76344 Eggenstein-Leopoldshafen, Germany
** corresponding author: benedikt.schumacher@imf.fzk.de

ABSTRACT

Commercially available barium titanate powder was temperature treated to induce grain growth at temperatures between $500°C$ and $1200°C$. The permittivities of polyester reactive resin composite materials with a filler load of $60m\%$ where examined in dependence of the heat treatment. A permittivity of 25 at $150Hz$ was achieved using a treatment temperature of about $1000°C$ while a composite containing the original powder exhibits a permittivity of 11. Temperature treatment of commercial BaTiO$_3$ powder is one crucial step in optimizing high d_k composite materials without having direct influence over the powder production process.

Keywords: barium titanate, crystallite size, permittivity

1 INTRODUCTION

The industry needs for highly integrated and compact electronics are steadily increasing. The use of integrated passives technologies is one way to reduce the number of components on top of a printed circuit board (PCB) and therefore reduce precious surface area [1].

To produce embedded capacitors polymeric materials with high dielectric constant, processing temperatures below $250°C$ and good mechanical stability are necessary [2]. In order to be successful on the market the developed materials need to comply with traditional PCB production processes. To reach high capacities with single layer assemblies the produced high permittivity layers have to be as thin as possible.

This can be achieved through inorganic filler loadings within a polymeric matrix. Barium titanate is a widely used material for high permittivity applications. It's dielectric properties depend strongly on the crystallite size and lattice [3]. To reduce the layer thickness - and therefore increase the capacitance - small particles are needed. In sintered barium titanate a maximum permittivity has been found at grain sizes around $1\mu m$.

The permittivity of BaTiO$_3$ particles also highly depends on the powder production process [4]. But powder production processes are often not in hand of the composite material manufacturer. Commercially available BaTiO$_3$ powder has been optimized by temperature treatment for high permittivity and low particle size.

2 EXPERIMENTAL

A nano scale barium titanate powder (Nanoamor #1150XW, 99.6%, cubic, $85 - 128nm$ average particle size, $8 - 12\frac{m^2}{g}$ specific surface area) was used as a basic raw material for all experiments. To initiate crystallite growth $100g$ samples of the powder where temperature treated (Carbolite, RHF 17/3E) at different temperatures between $500°C$ and $1200°C$ with a heating rate of $2\frac{K}{min}$, a holding time of one hour and a cooling rate of $5\frac{K}{min}$. For temperature treatment the powder was loosely given into a ZrO$_2$ crucible (Frialit-Degussit, FZY-Material) and covered with a ZrO$_2$ disc. The treated samples where characterized using BET, He-pycnometrie, REM and XRD (Siemens, D505).

X-ray diffraction patterns were collected at room temperature from 20° to 80° 2θ with a step size of 0.02 and a fixed counting time of $20s$ using Cu K_α radiation with a graphite monochromator in reflection geometry. The full-profile-fitting refinements were carried out by the Rietveld method using the *FullProf* program [5]. To determine the crystallite size, phase contents and the tetragonality ($\frac{c}{a}$ ratio) of the powders. A pseudo-Voigt function was used for the peak profile shape. The instrumental contribution to peak broadening was determined with annealed BaF$_2$ powder as a standard material.

Using a spherical model a mean particle radius r_a was calculated using (1); V being the volume of a sphere, A being it's surface area, $\rho[\frac{g}{m^3}]$ being the density derived from He-pycnometrie and $s_s[\frac{m^2}{g}]$ being the specific surface area derived from BET measurements.

$$r_a = \frac{3 \cdot V}{A} = 3 \cdot (\rho \cdot s_s)^{-1} \qquad (1)$$

The temperature treated barium titanate powders where then incorporated into a polyester reactive resin (Carl Roth). To reduce the viscosity $20m\%$ of styrene where added to the reactive resin. As release agent $2m\%$ of INT-54 (Würtz) where added. Three mass percent of MEKP where used as cold hardener. The composite was stirred at $800\frac{rev}{min}$ (IKA dissolver stirrer $29mm$ diameter) for 30 minutes before the MEKP was added. Discs where produced by pouring the composite into cavities made from silicone (Wacker, Elastosil M 4370 A). The composite discs where hardened at $50°C$ for three hours.

The discs where abraded to get plane surfaces and its dimensions (thickness and diameter) where determined. For dielectric measurement electrodes where attached to both sides of each disc using conductive silver paint (RS Components, Silver Conductive Paint). The complex capacity of the discs was measured between 150Hz and 10MHz using an impedance analyzer (Agilent, HP4194A) and the permittivity was calculated using (2) with measured capacity of the disc C, average thickness of the disc t and radius of the disc r.

$$\epsilon_r = \frac{C \cdot t}{\pi \cdot r^2 \cdot \epsilon_0} \qquad (2)$$

Unless specified otherwise measurements where conducted at room temperature.

3   RESULTS AND DISCUSSION

The optical morphology of the temperature treated powders starts to change at about $1000°C$. This can be observed from the REM images shown in figure 1. At this point grain growth across particle boundaries starts to take place and is visually recognizable.

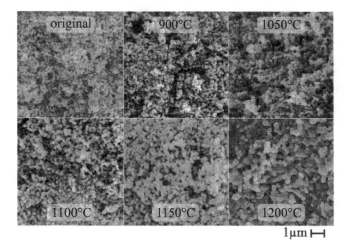

Figure 1: REM images of heat treated BaTiO$_3$ nanopowder (maximum Temperature given).

The density of the powder samples increases steadily from $5.53\frac{g}{cm^3}$ to $5.94\frac{g}{cm^3}$ (see. fig. 2), coming close to the literature value of $6.08\frac{g}{cm^3}$ [6]. The BET surface area slightly increases from $58.0\frac{m^2}{cm^3}$ to $71.5\frac{m^2}{cm^3}$ at $650°C$ temperature treatment. Above $650°C$ treatment temperature the BET surface area decreases steadily to $6.5\frac{m^2}{cm^3}$ due to grain and particle growth of the powder.

The average particle radius derived from BET and He-pycnometrie measurements increases slowly up to a

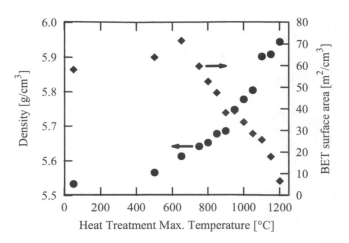

Figure 2: Density and BET surface area of heat treated BaTiO$_3$ nano scale powder.

temperature treatment of $1100°C$. Above that temperature the particles start to grow rapidly as can be observed from figure 3. This is in good agreement with the REM images (see fig. 1).

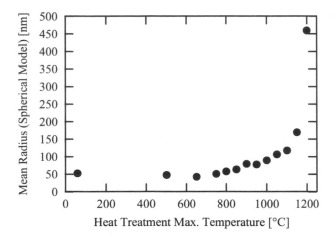

Figure 3: Average particle radius derived from BET and density using a spherical particle model (see (1))

The original BaTiO$_3$ powder consists of two different phases. A tetragonal fraction t and a cubic fraction c in equal parts (see tab. 1, compare [7]). With growing crystallites the tetragonal fraction increases and at a treatment temperature above $1000°C$ the cubic fraction vanishes. The lattice parameter of the cubic phase a_c changes rapidly when heating the powder to $500°C$ (see tab. 1 and fig. 4 top two plots peak shift) and stays constant for any higher heat treatment. The $\frac{c}{a}$ ratio of the tetragonal phase undergoes similar changes than the cubic phase, but only stays constant up to $800°C$ and rises with higher temperatures. This is in good agreement with the particle diameter (see fig. 3) as larger crystallites show a higher $\frac{c}{a}$ ratio [7].

The x-ray density ρ_{xrd} undergoes less rapid changes

NSTI-Nanotech 2008, www.nsti.org, ISBN 978-1-4200-8503-7 Vol. 1

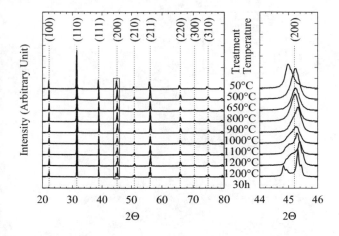

Figure 4: XRD patterns of temperature treated BaTiO$_3$ powders with varying treatment temperature.

Table 1: Physical properties of temperature treated BaTiO$_3$ powders.

$T[°C]$	$t[\%]$	$c[\%]$	$\frac{c}{a}$	$a_c[\mathring{A}]$	ρ_{he}	ρ_{xrd}
50	47	53	1.0071	4.0274	5.53	5.95
500	65	35	1.0053	4.0085	5.56	6.01
650	73	27	1.0051	4.0093	5.61	6.01
800	74	26	1.0051	4.0082	5.65	6.01
900	79	21	1.0055	4.0086	5.68	6.01
1000	87	13	1.0058	4.0097	5.78	6.01
1100	100	0	1.0062	---	5.90	6.01
1200	100	0	1.0066	---	5.94	6.02

than the He-pycnometrie density ρ_{he} (see tab. 1). It rises constantly from $5.95\frac{g}{cm^3}$ to $6.02\frac{g}{cm^3}$, coming close to the BaTiO$_3$ density of $6.08\frac{g}{cm^3}$.

Figure 5 shows the composite viscosity at three different temperatures, fixed shear rate of $100s^{-1}$ and constant solid load of $60m\%$ as function of the BaTiO$_3$ heat treatment conditions. With increasing heat aging temperature up to 800°C a slight viscosity increase can be observed (section I). Between 800°C and 1100°C filler heat treatment temperature the viscosity remains almost constant (section II), followed by a pronounced drop at larger temperatures (section III).

A clear correlation of the viscosity with the filler preprocessing cannot be detected, hence the influence of several filler characteristics on the flow behavior have to be considered. The thermal treatment changes the specific surface area, the average particle size, the particle size distribution as well as the particles morphology resulting in a superimposed impact on the composite viscosity.

The impact of the particle heat treatment on the flow behavior can be depicted from viscosity curves representing the three different sections. Figure 6 presents typical flow curves of composites containing BaTiO$_3$ treated at 500°C, 1150°C and 1200°C measured at a

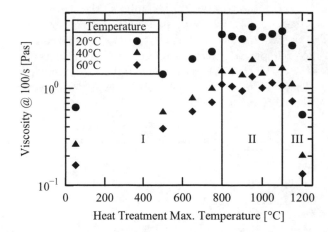

Figure 5: Viscosity of BaTiO$_3$ composite material at $60m\%$ and a shear rate of $100s^{-1}$

composite temperature of 60°C. While the composite with the filler aged at the lowest temperature show a slight viscosity increase in whole shear rate range, the mixture with the 1150°C treated filler exhibit a pronounced viscosity rise between 10 and $20s^{-1}$. Finally the composite with the filler aged at 1200°C possesses an almost Newtonian flow.

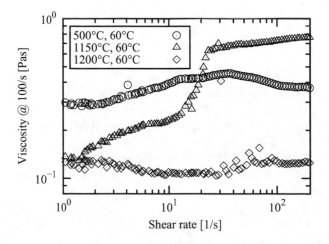

Figure 6: Viscosity of BaTiO$_3$ composite material with $60m\%$ filler load at 60°C. Heat treatment temperatures of the filler are given.

The permittivity of polyester reactive resin composite material with a filler load of $60m\%$ highly depends on the temperature treatment of the filler as can be seen in figure 7. Up to 900°C treatment temperature the permittivity rises steadily from 11 to 25. Between 900°C and 1150°C a plateau can be observed with an absolute maximum of 27 at 1050°C temperature treatment. At temperature treatments of the filler above 1150°C the permittivity of the composite drops rapidly.

The highest permittivity is reached in the same temperature region, where the tetragonal shoulder of the

(200) peak starts to develop visually in the XRD diagram (see fig. 4). Crystal structure and size of the BaTiO$_3$ powder show as expected a large influence on the dielectric properties of the composite material. Barium titanate crystallites with small but mostly tetragonal phases offer the best dielectric behavior (compare tab. 1 cubic fraction c and fig. 7).

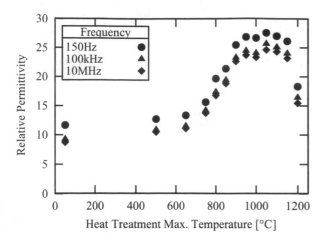

Figure 7: Permittivity at room temperature of BaTiO$_3$ filled polyester resin ($60m\%$) with varying heat treatment of the BaTiO$_3$ filler.

The permittivity of such composite materials is - as well as the permittivity of sintered barium titanate - dependent on the temperature of the material. A temperature dependent measurement of reactive resin composite with $60m\%$ of temperature treated BaTiO$_3$ nanopowder filler (treated at $1050°C$) is shown in figure 8. The permittivity rises nearly linear from 23.4 at $-20°C$ to 29.9 at $60°C$. A peak around room temperature as seen in [3] that results from a crystal phase transition (orthorombic to tetragonal) can not be observed in the composite material.

For minimal grain and particle size in combination with optimal dielectric performance a temperature treatment of $950°C$ to $1050°C$ has to be enforced on the filler before incorporating it into the composite material.

Recent publications about the permittivity of epoxy / BaTiO$_3$ composite materials show values of $27 - 35$ at $60Vol\%$ filler load [8] and 38 at $50Vol\%$ [9]. Choi et al. published a permittivity of 12 for a PI / BaTiO$_3$ composite at $30Vol\%$ filler load [10]. With the method described in this paper values of $25-27$ could be reached at much lower filler loads of about $22Vol\%$ permitting improvements concerning the filler load or lower viscosity of the composite material.

Crystal modification by temperature treating commercially available BaTiO$_3$ seems to be a useful and cheep route for the optimization of high d_k composite materials.

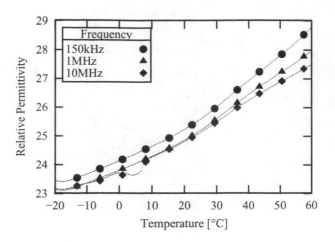

Figure 8: Temperature dependent permittivity of BaTiO$_3$ composite material ($60m\%$). The filler is temperature treated at $1050°C$ for one hour.

4 CONCLUSION

The dielectric properties of commercially available barium titanate powder have been optimized by temperature induced grain growth. The permittivity of a polyester reactive resin composite material can double by use of a temperature modified barium titanate filler. The tetragonal phase is the crucial factor for maximizing the permittivity. The transition point where all cubic phases are transformed to tetragonal phases is the point of maximum permittivity.

For further development, the temperature treated barium titanate powder needs to be sieved and conditioned. A dispersing agent is needed to improve the stability of the composite while hardening. To maximize the permittivity the filler load of the composite needs to be increased.

ACKNOWLEDGMENTS

The authors would like to thank Mrs. M. Offermann for the BET and He-pycnometrie measurements.

REFERENCES

[1] S.K. Bhattacharya, R.R. Tummala, J. Mat. Sci.-Materials in Electronics, 11, 253-268, 2000.
[2] Xu, Wong, Composites A, 38, 13-19, 2007
[3] Kinoshita et al., J. App. Phy., 47, 1, 371-373, 1976
[4] Wada et. al., Key Eng. Mat., 301, 27-30, 2006
[5] Rodriguez-Caraval, Physica B, 192, 55-69, 1993
[6] Sigma Aldrich MSDS Barium Titanate Powder 338842 Version 1.4, 2006
[7] Aoyagi et al., J. Phys. Soc. Jpn., 71, 5, 1218-1221, 2002
[8] Dang et al., Comp. Sci. & Techn., 68, 171, 2008
[9] Ramajo et al., Comp. Part A, 38, 1852, 2007
[10] Choi et al., Mat. Lett., 61, 2478, 2007

Preparation of Transparent Electrodes Based on CNT-s Doped Metal Oxides

M. Paalo*, T. Tätte*, A. Juur *, A. Lõhmus*, U. Mäeorg**, I. Kink*

* Institute of Physics, University of Tartu, and Estonian Nanotechnology
Competence Center, 142 Riia St, 51014 Tartu, Estonia, madis.paalo@fi.tartu.ee
** Institute of Organic and Bioorganic Chemistry, University of Tartu,
2 Jakobi St., 51014 Tartu, Estonia

ABSTRACT

In present work we demonstrate preparation of transparent and electrically conductive electrodes suitable for utilization in solar cells and as different displays. Materials as fibers and films are readily obtained by using inexpensive chemical method where processing transition metal alkoxides together with single-wall carbon nanotubes (SWCNT-s) leads to metal oxide/SWCNT composite materials. Since transition metal oxide/CNT hybrid materials have good characteristics of both substances, it would be stimulating subject for industry.

Keywords: transparent electrodes, SWCNT composites, sol-gel, alkoxides

1 INTODUCTION

In recent years electrically conductive and transparent electrodes have been among the most important issues in optoelectronics. The field of applications of these electrodes varies from solar cells to different displays. As increased need for transparent electrode material indium tin oxide (ITO) has raised the price of indium around 10 times since the year 2000 [1], it has set the search for alternative materials into focus.

Carbon nanotubes (CNT-s) are widely known for extremely good mechanical and electrical properties and are applicable in a wide optical range [2]. For that reason, CNT-s have been used as dopants to improve electrical and mechanical characteristics of different transparent materials in order to create new electrode materials [3]. Being the cheapest, the main materials tested as matrixes are different organic polymers [3]. Still, these materials have some drawbacks compared to ITO, like lower chemical and mechanical stability.

To overcome these problems we have started studies to design electrodes based on high refractive index transparent metal oxides like TiO_2 and SnO_2, which are among the most stable compounds. As these materials also have high hardness, they are often used as protective layers [4].

We have shown in our earlier works that it is possible to prepare self-standing oxide fibers and films from high viscosity transition metal alkoxides [5]. These oxide structures are obtained by well-known, relatively simple and cost-effective sol-gel procedure. Method is based on oligomeric concentrates, which are obtained from neat metal alkoxides by addition of water, followed by removal of the solvent. The concentrates are highly viscous materials that can be transformed to desired shapes (fibers, films, powders etc.) at room temperature by appropriate combination of mechanical and chemical treatments (Fig. 1.).

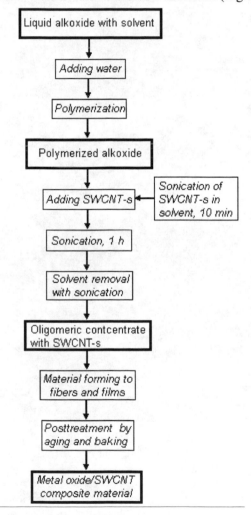

Figure 1. Preparation of CNT-e composite materials by sol-gel method.

In current work we demonstrate the preparation of transparent electrodes as a result of abovementioned treatment steps. Due to the simplicity of the method such electrodes could easily be introduced into the industrial production.

2 ELECTRICALLY CONDUCTIVE AND TRANSPARENT ELECTRODES

As mentioned above, preparation of electrodes is based on sol-gel method [5], [6]. Initially neat liquid transition metal alkoxide (transition metal: Ti, Sn) is polymerized by adding water in suitable solvent. Presonicated single-walled carbon nanotubes (SWCNT-s) in solvent are then introduced into the polymerized alkoxide. Sonication is needed to attain homogeneous dispersion of nanotubes in solution and afterwards in polymerized alkoxide [7]. Before extraction of solvent and alcohol, material is processed in ultrasonic bath for approximately one hour to assure uniform medium. Obtained oligomeric concentrate with SWCNT-s continues to polymerize via cross-linking when introduced to humid air and can only be stored in dry atmosphere. Material will polymerize in humid environment until it becomes completely solid, containing still some organics and water.

Figure 2. SEM images of a CNT-doped SnO_2 fiber. The fiber was mechanically broken prior to imaging in order to reveal the doped CNT-s. It can be seen that CNT-s are oriented along the axis of the fiber.

Electrodes as fibers are formed by pulling oligomeric mass in humid air [5], [6]. Dimensions of the fibers depend greatly on viscosity of the precursor material, humidity of the surrounding atmosphere, pulling speed and on nanotube aggregates remained in the matrix after sonication. Films are prepared by spin-coating or dip-coating.

SEM analyzes revealed that orientation of the nanotubes was anisotropic with a maximum alongside axis of the fiber (Figure 2). This could be explained by axial forces acting on nanotubes when polymerized alkoxide is drawn (Figure 3.).

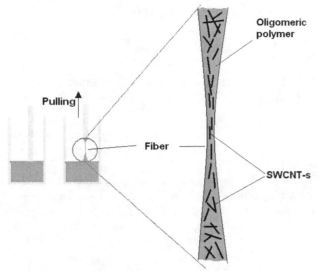

Figure 3. Fiber pulling process is orienting CNT-s.

After baking to 340 °C in air, dense oxide ceramics, doped by aligned nanotubes, were achieved in shape of fibers or films. Heating was needed to remove organics and water still in material after polymerization. Baking temperature was selected so that carbon nanotubes would not oxidize [8] and other organics could be removed as much as possible.

Fiber electrodes with diameters greater than 1,5 mm turned opaque as a result of heating, because organic remnants could not emerge from inside. Smaller than 1,5 mm strands remained on visual observation as transparent and were up to 25 mm long (Figure 4).

Figure 4. Electrically conductive and transparent tin(IV)oxide electrodes doped by SWCNT-s and with a diameter from 40 to 50 μm.

Electrodes electrical conductivities were measured by 4-point method. Maximum conductivity of 500 S/m was measured on fibres doped by 0,1 wt% SWCNT. It was also noted that fibers heated up to 400 °C, experienced a drop of conductivity down to 0,00015 S/m. This was in accordance

with prior knowledge from thermogravimetric analysis [8], which states that all SWCNT-s should be oxidized at this temperature.

3 CONCLUSIONS

Current study proves that oxide/CNT composites films and fibers are potential materials for transparent electrodes applications. As transition oxides preserve good chemical and mechanical properties, these materials are promising in utilization as protective layers. Our results prove that SWCNT-doping significantly improves the conductivity of the fiber electrodes (up to 500 S/m).

ACKNOWLEDGEMENTS

This work was supported by the Estonian Nanotechnology Competence Centre and Estonian Science Foundation (Grants ETF 7612, 6537, 6163, 6660).

REFERENCES

[1] "Indium Price Supported by LCD Demand and New Uses for the Metal", http://geology.com/articles/indium.shtml
[2] Collins P.G., Avouris P., "Nanotubes for electronics", Scientific American, 283 (6), 62-9, 2000.
[3] Breuer, O., Sundararaj, U., „Big returns from small fibers: a review of polymer/carbon nanotube composites", Polymer Composites, 25 (6), 630-645, 2004.
[4] Zywitzki O., Modes T., Sahm H., Frach P., Goedicke K, Glöβ D., "Structure and properties of crystalline titanium oxide layers deposited by reactive pulse magnetron sputtering", Surface and Coatings Technology, 180-181, 538-543, 2004.
[5] Tätte, T., Paalo, M., Kisand, V., Reedo, V., Kartushinsky, A., Saal, K., Mäeorg, U., Lõhmus, A., Kink, I., "Pinching of alkoxide jets – a route for preparing nanometre level sharp oxide fibres." Nanotechnology 531-535, 18, 2007.
[6] Tätte, T., Talviste, R., Vorobjov, A., Part, M., Paalo, M., Kiisk, V., Saal, K., Lõhmus, A., Kink, I., „Preparation and applications of transition metal oxide nanofibres and nanolines", Technical Proceedings of the 2008 Nanotechnology Conference and Trade Show, Nanotech 2008
[7] Ham, H. T., Choi , Y. S., Chung, I. J., „An explanation of dispersion states of single-walled carbon nanotubes in solvents and aqueous surfactant solutions using solubility parameters", Journal of Colloid and Interface Science, 286, 216-223, 2005.
[8] Musumeci, A. W., Silva, G. G., Martens, W. N., Waclawik, E. R., Frost, R. L., "Thermal decomposition and electron microscopy studies of single-walled carbon nanotubes", Journal of Thermal Analysis and Calorimetry, 88 (3), 885-891, 2007.

Novel Hybrid Materials Of Cellulose Fibres And Nanoparticles

A. C. Small[*], J. H. Johnston[*]

[*]Victoria University of Wellington and The MacDiarmid Institute for Advanced Materials and Nanotechnology, Wellington, New Zealand, jim.johnston@vuw.ac.nz

ABSTRACT

Hybrid materials are of interest due to the potential synergistic properties that may arise from the combination of two or more precursors. Such precursors are paper fibres (cellulose) and magnetic or photoluminescent nanoparticles. These materials exhibit the inherent properties of the substrate, in particular flexibility and strength, and also the properties of the surface bonded nanoparticles.

Keywords: cellulose, nanoparticles, magnetic, photoluminescent.

1 INTRODUCTION

Cellulose, $(C_6H_{10}O_5)_n$, is a long-chain polymeric carbohydrate of β-glucose. It is the fundamental structural component of green plants, in which the primary cell wall of the plant is predominantly cellulose, and the secondary wall contains cellulose with variable amounts of lignin. Lignin and cellulose, considered together, are termed lignocellulose, which (as wood) is the most common biopolymer on earth. Cellulose monomers (β-glucose) are linked together through 1,4 glycosidic bonds by condensation. Each monomer is oriented 180° to the next, as seen in figure 1 below, and the chain is built up two units at a time.

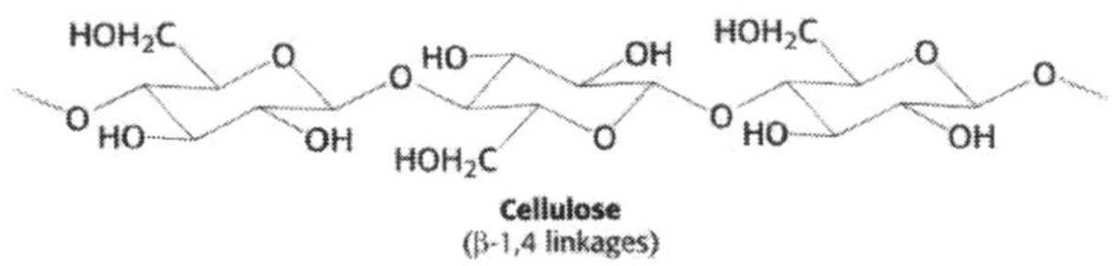

Figure 1: Structure of cellulose $(C_6H_{10}O_5)_n$.

The cellulose chains are formed into micro fibrils that constitute the basic framework of the plants cell. In micro fibrils, the multiple hydroxyl groups on the glucose units hydrogen bond with each other, holding the chains firmly together and contributing to their high tensile strength [1]. Cellulose and its derivatives (paper, nitrocellulose, cellulose acetate, etc.) are principal materials generated by industry and see a considerable economic investment. Magnetically responsive cellulose fibres allow the investigation of new concepts in papermaking and packaging, security paper, and information storage. Potential applications are in electromagnetic shielding,

magneto-graphic printing and magnetic filtering. Doped ZnS nanoparticles have attracted large amounts of interest since 1994 when Bhargava et al [2]. reported for the first time that Mn^{2+} doped ZnS could yield high quantum luminescence efficiency. As well as ZnS:Mn nanoparticles, a large amount of work has been dedicated to investigating the synthesis and properties of copper doped ZnS nanoparticles. The combination of doped ZnS nanoparticles and cellulose fibres could lead to the development of new security papers and cost effective display technology.

2 EXPERIMENTAL

Magnetite (Fe_3O_4) nanoparticles were synthesized by adding aqueous ammonia drop wise from a burette to solutions containing dissolved $FeCl_2.4H_2O$. Black precipitates were formed immediately [3]. As particle size is dependant on both the concentration of the precursor solution and the rate precipitation, various concentrations of precursor solution were used and the precipitation rate was kept constant. These are summarized in table 1.

In a typical synthesis of Mn^{2+} or Cu^{2+} doped ZnS particles, $ZnCl_2$, $MnCl_2$ or $CuCl_2$ and sodium citrate solution were mixed and stirred at constant speed for 10 minutes. From a burette, Na_2S was added drop wise. A white precipitate was formed immediately. The resulting suspension was centrifuged and then washed with distilled water.

A colloidal suspension of these nanoparticles was then added to an approx. 2 wt. % suspension of bleached *pinus radiata* Kraft paper fibres and stirred vigorously for approximately 2 hours, after which they were filtered and washed with H_2O. The resultant coated fibres were then sonicated for 20 minutes in order to remove any loosely bound nanoparticles.

3 RESULTS AND DISCUSSION

3.1 Magnetic Nanoparticles

Pinus radiata Kraft fibres coated with Fe_3O_4 nanoparticles retain the inherent properties of the fibre; tensile strength, flexibility, and the ability to be made into a sheet. The fibres also gain the magnetic properties of the surface bound nanoparticles. Consistent with the literature [8], particle size decreases as the concentration of $FeCl_2.4H_2O$ decreases. X-ray line broadening indicates that samples have an average particle size of between 12 and 26 nm, depending on solution concentration. By varying the concentration of Fe^{2+} in the precursor solution, different particle sizes were obtained (table 1).

Solution conc. (%)	Average particle size (nm)	Saturation magnetisation (emu g^{-1})	Coercive field (Oe)
3.00	26	62	112
0.60	24	62	122
0.05	15	70	42
0.025	12	63	19

Table 1: Effect of precursor solution concentration on particle size, saturation magnetisation and coercive filed of Fe_3O_4 nanoparticles.

Magnetisation versus applied field curves are presented in figures 2 and 3.

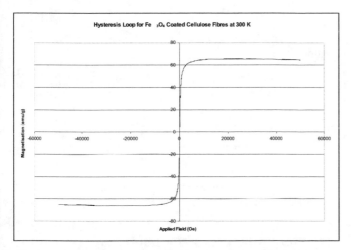

Figure 2: Magnetic hysteresis loop for Fe_3O_4-coated cellulose fibres at 300 K. Sample has an Fe loading of 2.88 %.

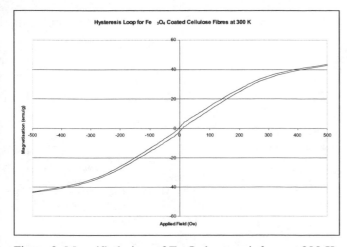

Figure 3: Magnified view of Fe_3O_4 hysteresis loop at 300 K. A small remnant magnetization is visible.

Saturation magnetisation is independent of particle size but dependent on temperature. Fe_3O_4-coated cellulose fibre samples show a saturation magnetisation of ~ 65 emu g^{-1}. This value is comparable to the values of saturation magnetisation of bulk Fe_3O_4, indicating that the combination of cellulose fibres and magnetic nanoparticles does not alter the magnetic properties of the nanoparticles. The magnetic moment is also comparable to materials with similar Fe_3O_4 loadings [4]. Lowering the concentration of Fe^{2+} in the precursor solution results in the formation of smaller particles, and changes the coercive field accordingly. Particle sizes calculated using the Debye-Scherrer approximation for XRD are only average particle sizes, hence a smaller average particle size means a larger number of particles will be in the superparamagnetic range, and as a consequence, the coercive field is lowered. Coercive fields at 300 K for these materials are very low; H_c = ~ 20-120 Oe. A coercive field as low as this is ideal for application in electromagnetic shielding. The materials also possess small remnant fields of ~ 3-11 emu g^{-1} at 300 K, as shown in the magnetic hysteresis loops in figures 2 and 3 above. Optical microscope images show that the coated fibres retain the same morphology as the precursor Kraft fibres. The brown colour of the fibres (a change from the original white) indicates they are coated with the Fe_3O_4 nanoparticles. From the SEM image shown in figure 4, the completeness of the coating can be seen. The Fe_3O_4 nanoparticles completely encapsulate the fibre and follow its morphology, similar to previous research involving cellulose fibres coated with conducting polymers [5].

Figure 4: SEM images of Fe_3O_4 cellulose fibres.

This is also confirmed by examining the EDS map for Fe (figure 5), which shows full coverage of the fibre surface. Nanoparticles are present on the surface of the cellulose fibres in agglomerations of ~ 100 nm, as no surfactant was used in the synthesis procedure. This differs from the particle size calculated using the Debye-Scherrer approximation, where the individual crystallite size was calculated as being between 12-26 nm. XRF analysis shows the samples to be 0.5-2.8 % Fe. The Fe 2p XPS spectrum of the Fe_3O_4 nanoparticles (Figure 6) shows the Fe_3O_4 and α-FeOOH phases to be present [6], consistent with XRD results. A comparison between the Fe 2p XPS spectrum of Fe_3O_4 coated cellulose fibres and Fe_3O_4 nanoparticles (figure 7) shows a considerable shift to lower binding energy of up to 1.5 eV for both the 1/2 and 3/2 spin multiplicities of the coated samples. This shift confirms that there is chemical bonding between the Fe_3O_4 nanoparticles and the cellulose fibre surface, presumably through hydrogen bonding between the O in the Fe_3O_4 and α-FeOOH nanoparticles and the H of the OH groups present in cellulose. The shape of the Fe 2p XPS spectrum of Fe_3O_4-coated cellulose fibres does not change compared to that of the Fe_3O_4 nanoparticles by themselves. This indicates that the chemical bonding between the Fe_3O_4 nanoparticles and the cellulose fibre does not alter the chemistry of the nanoparticles. Even after repeated washing and sonication steps, the particles remain bound to the surface. This is contrary to other reports in the literature which state that particles on the fibre surface are removed in the washing step

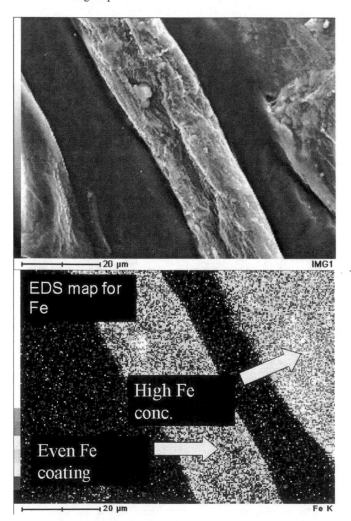

Figure 5: EDS images of Fe_3O_4 coated cellulose fibres, SEI image (above) and elemental Fe map (below).

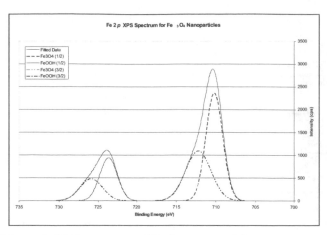

Figure 6: XPS spectrum of Fe_3O_4 nanoparticles.

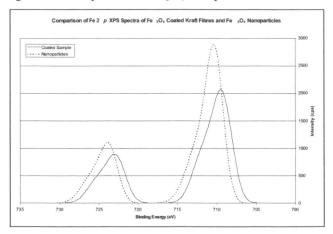

Figure 7: Comparison between Fe 2p XPS spectra of Fe_3O_4-coated Kraft fibres and Fe_3O_4

3.2 Photoluminescent Nanoparticles

Figure 8 shows the photoluminescence emission spectrum of ZnS:Mn powders synthesized by the aforementioned methods.

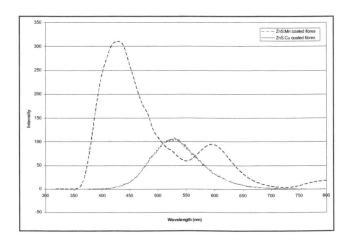

Figure 8: Photoluminescence emission spectra of doped ZnS nanoparticles.

An emission peak can be seen at approximately 590 nm, which can be attributed to the ^1A_g-^2T_g transition in Mn^{2+}. The larger peak at 480 nm can be attributed to an emission due to S^{2-} vacancies in the ZnS lattice [7]. In the case of ZnS:Cu, a broad blue-green emission can be seen at approximately 520 nm. The source of this emission has been fiercely debated. A recent theory suggests the emission arises from the transition of an excited electron from an S^{2-} vacancy in the ZnS lattice to the t_{2g} excited state of the Cu^{2+} [8]. Again, on bonding to cellulose fibres, the emissions are unaffected. X-ray photoelectron spectroscopy (XPS) shows that bonding is occurring between the doped ZnS nanoparticles and the cellulose surface, due to shifts in the 3/2 and 1/2 spin multiplicities of the 2p electrons of both Zn and S. In some cases the shift is large up to 1.4 eV (figure 10).

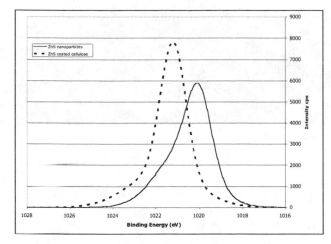

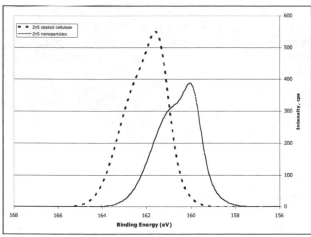

Figure 10: XPS spectra for $2p$ electrons in ZnS:Mn nanoparticles and ZnS:Mn coated cellulose fibres. Zn $2p$ above, and S $2p$ below.

On examining electron microscope images (figure 11), it can be seen that spherical nanoparticles of doped ZnS have been formed. It is difficult to deduce particle size from these images as the particles have agglomerated due to the absence of a surfactant during the synthesis procedure. From x-ray line broadening analysis, the average particle size is ~15 nm for both ZnS:Mn and ZnS:Cu nanoparticles. Electron dispersive spectroscopy (EDS) indicates that the particles are indeed ZnS, and in both cases, the dopant can be mapped as well.

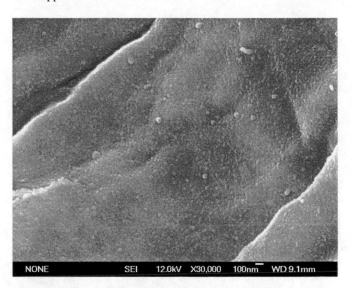

Figure 11: SEM micrograph showing the surface of the cellulose fibre coated with ZnS:Mn nanoparticles

4 CONCLUSIONS

Fe_3O_4 and Cu^{2+} and Mn^{2+} doped ZnS nanoparticles were synthesized and hybrid materials of these and cellulose fibres were prepared and characterized using a number of techniques. It has been shown that the nanoparticles coat the surface evenly and retain the magnetic and photoluminescent properties of their standalone powders.

REFERENCES

[1] Wallenberrger and Weston, "*Natural fibers, polymers and composites.*", Kluwer Academic Publishers, 2004.
[2] R. N. Bhargava, D. Gallagher, X. Hong, A. Nurmikko, *Phys. Rev. Lett., 72*, 416, 1994.
[3] P. Berger, N. B. Adelman, K. J. Beckman, D. J. Campbell, A. B. Ellis, G. C. Lisensky, *J. Chem. Ed.*, 76, 943, 1999.
[4] R. H. Marchessault, P. Rioux, L. Raymond, *Polymer,* 33, 4024, 1992.
[5] M. J. Richardson, J. H. Johnston, T. Borrmann, *Curr. Appl. Phys.* 6, 462, 2006.
[6] J. F. Anderson, M. Kuhn, U. Diebold, *Surf. Sci. Spectra*, 4, 266, 1998.
[7] L. Sun, C. Liu, C. Liao, C. Yan, J. Mater. Chem. 9, 1655, 1999.
[8] P. Yang, M. Lu, D. Xu, G. Zhou, Appl. Phys. A, 73, 455, 2001.

Novel Polymeric Technologies for Pulp and Paper Products

G. Jogikalmath, L. Reis, P. Kincaid, M. Berg and D. Soane

NanoPaper LLC, 35 Spinelli Pl., Cambridge, MA, USA, gjogikalmath@soanelabs.com

ABSTRACT

NanoPaper LLC develops solutions for the paper industry by integrating self-assembly of polymers and nanoparticles into pulp and paper products. By using surface modification of polymeric and particle additives, superior properties can be achieved in existing technologies and completely new product categories can be created. Surface modification also increases interactions between additives and cellulose fibers to promote retention and make better use of materials for a more economically and environmentally friendly product. NanoPaper has developed technologies using these principles to create high bending stiffness laminates with reduced pulp, grease resistant coatings that eliminate fluorinated compounds, and pigments that more effectively interact with fibers to reduce waste in colored paper grades. Nanotechnology and surface modification are crucial in all of these technologies to produce enhanced eco-friendly and economical products for the pulp and paper industry.

Keywords: paper, packaging, pigments, grease coating, surface modification

1 HIGH BENDING STIFFNESS PAPERBOARD LAMINATES

High bending stiffness is desirable in many paper and paperboard applications. This property is often achieved by manufacturing dense, high caliper sheets or boards. In addition to consuming large amounts of pulp, such paperboards or sheets must be subsequently fluted and glued to prepare high stiffness laminates. There exists a need for a single-step process to yield laminates of high stiffness with reduced weight, using inexpensive materials. Moreover, traditional papermaking, whether to produce paper or paperboard, results in stresses on the environment due to energy and resource usage. We are attempting to address these issues by producing high bending stiffness laminates that are made with minimal use of pulp while using renewable materials to impart bending stiffness to the laminate.

Our process uses inexpensive expandable fillers and other expansion agents that react in situ during laminate manufacturing. The core imparts high bending stiffness to these specially engineered laminates. The density and pore size are easily tunable, leading to precise control over stiffness and cost. We have also demonstrated different structural compositions as seen in Figure 1. Furthermore the expandable fillers are surface modified to impart different surface properties such as hydrophobicity and oleophobicity to the paperboard. The structure also offers significant advantages in applications where sound damping and thermal insulation are required. Finally, certain grades of our laminates are made using materials approved for food contact. Initial results show a 2x improvement in weight by volume compared to commercially available paperboards of comparable stiffness.

Figure 1: Images of expandable laminate core.

2 GREASE RESISTANT COATINGS

Grease resistant and oil-resistant coatings are used in a variety of applications including paper and board used in food packaging. Many of these treatments or coatings use fluorinated materials, and others use high amounts of polyolefins or other plastics. An alternative coating is needed to promote consumer safety in products as well as to reduce environmental impact. Our grease resistant coating has eliminated the need for these compounds. It is a durable thin coating that is resistant to folding and creasing while maintaining its oleophobic properties. The grease resistance can be seen in Figure 2 where a half-treated sample shows the coating's effectiveness as a barrier. Additionally, it performs well in a wide range of temperatures including the higher temperatures sometimes required in food packaging.

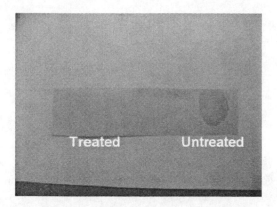

Figure 2: Grease resistance of paper both treated (left side) and untreated (right side) with grease resistant coating.

4 CONCLUSIONS AND FUTURE DIRECTIONS

NanoPaper LLC has exhibited use of surface modification and nanotechnology to solve problems in the pulp and paper industry. We have been able to apply these technologies in a wide range of applications. High-stiffness paperboard, grease resistant coatings, and high-retention coloring systems are a few examples of applications in which nanotechnology can be utilized in the pulp and paper industry. Ongoing experimentation is underway to give additional desirable properties to the traditional and novel products by means of surface modification.

3 HIGH RETENTION PIGMENT COLORING SYSTEM

In producing colored paper, it is difficult to obtain consistent colors because small changes in dye concentrations significantly affect the end product. Also, dyes impart color to processing equipment requiring elaborate and time-consuming cleaning protocols. We have developed a high retention coloring system that can be surface modified to bond more effectively with the cellulose fibers. Common filler materials can be functionalized with dyes and encapsulated with special polymers to increase fiber interaction and retention. This allows for a decreased amount of waste in drainage waters as well as efficiency improvements in cleaning. Figure 3 shows the clear filtrate and improved retention from paper made with surface modified pigments as compared to the filtrate from paper made with traditional dyeing techniques.

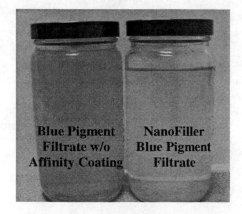

Figure 3: On the left is filtrate from traditional dyeing and on the right is filtrate from using surface modified pigments.

Functionalization of multiwalled carbon nanotubes with β-lactoglobulin

Jinjin Zhou, Qiang Yang, Xuejun Pan and Sundaram Gunasekaran[*]

[*]Department of Biological Systems Engineering, University of Wisconsin, Madison
460 Henry Mall, Madison,WI, 53706
guna@wisc.edu

ABSTRACT

β-Lactoglobulin (β-LG) can be successfully immobilized on the multiwalled carbon nanotubes (MWNTs) to form nanohybrid via nonconvalent bonding with 1-pyrenebutuanoic acid succinimidyl ester (PSE) as connector. PSE was first attached to the MWNTs by the π-π stacking interaction. Then, β-LG was immobilized on the MWNTs in an aqueous solution by a reaction between amine group of β-LG and N-hydroxysuccinimide of the PSE. The resultant functionalized MWNTs were characterized by UV-vis, fluorescence spectroscopy, confocal microscopy and scanned electron microscopy (SEM). Results showed that the functionalized MWNTs were quite soluble in aqueous medium. The effects of conjugation of carbon nanotubes to protein backbones on the rheological properties of protein-carbon nanotube hybrids were also determined using a dynamic rheometer. The core-shell structure of nanohybrid can be observed from SEM images.

Keywords: MWNTs, β-LG, nanohybrid, core-shell structure, encapsulation and delivery

1 INTRODUCTION

Carbon nanotubes have been researched as potential carriers for the delivery of biological molecules into cells. Noncovalent functionalization of the carbon nanotubes (CNTs) has proven to be a promising tool because of undamaged sp2 nanotube structure of the CNTs[1]. Pyrene-containing compounds have been widely used to functionalize the CNTs[1-4].

The immobilization of proteins on CNTs is attracting much attention because the resultant hybrid nanomaterials are potentially useful in many aspects such as biomolecular nanosensors and novel encapsulation and delivery materials. Many proteins have been immobilized on single-walled carbon nanotubes by the nonvalent sidewall functionalization methodology[5].

β-LG, the major whey protein component in the milk of ruminants, is a small globular protein widely used as food ingredient because of its nutritional value and functionalities. β-LG exhibits ligand binding properties and appearss to be resistant to degradation of the pepsin in the stomach in its native structure[6, 7].

The major goal of the present paper is to immobilize β-LG on MWNTs using the 1-pyrenebutuanoic acid succinimidyl ester (PSE) as a connector to synthesize the novel β-LG/MWNTs hybrids. The resulted hybrids may possess the advantages of both protein and carbon nanotube, exhibiting unique structure and encapsulation properties for targeted delivery of bioactive and pharmaceutical compounds.

2 EXPERIMENTAL SECTION

2.1 Materials

MWNTs, PSE and β-LG were obtained from Sigma-Aldrich (St. Louis, MO). All other reagents were of analytical grade. Milli-Q water (resistivity 18.2 MΩ cm) and N, N-dimethylformamide (DMF) were used in the experiments for suspension preparation.

2.2 Preparation of MWNTs/PSE

10 mg MWNTs, 20 mg PSE and 5 mL DMF were added into a 10 mL three-necked tube. After sonication for 30 min, the suspension reacted overnight at room temperature. Then, the product was centrifugally separated and was washed by DMF three times. The resultant solid was dried overnight under vacuum at 40 °C.

2.3 Functionalization of MWNTs

2 mg of previously modified MWNTs and 20 mg β-LG were dispersed in 10 mL water in a test tube under sonication for 30 min. The dispersion was kept stirred overnight at room temperature to ensure complete reaction. After reaction, dispersion was separated centrifugally and was washed by water three times. The resulting β-LG/MWNTs hybrids were isolated and dried overnight under vacuum at 40 °C.

2.4 Conjugation characterization

UV-vis spectrophotometer (UV-1601PC, Shimadzu Corporation) was used to investigate the π-π stacking between MWNTs and modified β-LG. Fluorescence spectroscopy was carried out on a fluorescence spectrometer (LS 55, Perkin Elmer). Microstructure of the β-LG/MWNTs conjugates suspension were studied using a confocal scanning laser microscope (CLSM, Zeiss Axiovert

NSTI-Nanotech 2008, www.nsti.org, ISBN 978-1-4200-8503-7 Vol. 1

200M) with 100x oil-immersed objective lens. Scanning electronic microscopy (SEM, Hitachi S900) was used to observe the surface structure of β-LG/MWNTs nanohybrid.

2.5 Rheological characterization

The apparent viscosity of the β-LG/MWNTs hybrid dispersions was measured by steady shear viscometry in a shear rate range of 0.1 to 10 s^{-1}. 0.1g β-LG was dispersed into 2 mL solvent containing 1 mL deionized water and 1 mL DMF. Certain amount of MWNTs was then added to β-LG dispersion to synthesis β-LG/MWNTs hybrids with relative concentration of 1 wt% and 2 wt% β-LG/MWNTs hybrids according to the method described above. The viscoelastic behaviors of β-LG/MWNTs hybrids were evaluated by dynamic small-amplitude oscillatory shear (SAOS) tests via frequency sweep and time sweep. Steady shear and SAOS tests were performed using a controlled stress dynamic rheometer with cone and plate measurement cell (Bohlin CVO-R, Malvern Instruments Inc.). Rheological properties of β-LG, β-LG/MWNTs mixture and β-LG/MWNTs hybrids were compared.

3 RESULTS & DISCUSSION

Dai et al. [1] reported PSE could combine carbon nanotubes and all kinds of protein together. Here, we adopted the PSE as a connector preparing the β-LG/MWNTs hybrids, as illustrated in **scheme 1**.

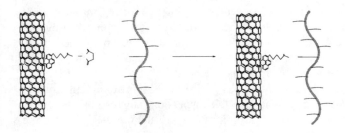

Scheme 1 Functionalization of multiwalled carbon nanotubes with β-LG

UV-vis spectrum was used to characterize the β-LG/MWNTs hybrids. **Fig. 1** shows the UV-vis spectra of MWNTs/PSE, β-LG and β-LG/MWNTs hybrids. From **Fig. 1**, β-LG has an adsorption at 293 nm. For the β-LG/MWNTs hybrids, a similar adsorption at 293 nm. This result verified the hybrid structure of β-LG/MWNTs.

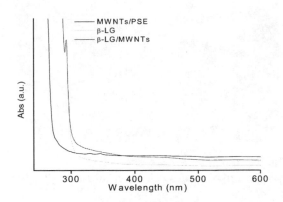

Fig. 1 UV-vis spectra of MWNTs/PSE, β-LG and β-LG/MWNTs hybrids

Fluorescence spectroscopy is one of the powerful techniques to investigate the structure of protein/carbon nanotube hybrids. It is well known that carbon nanotubes can quench the fluorescence efficiency of many proteins. In our experiments, the fluorescence quenching phenomena of the modified MWNTs on β-LG was also observed obviously, as shown in **Fig. 2**. β-LG exhibited strong characteristic adsorption at 330 nm. After forming hybrid, fluorescence efficiency of β-LG has been completely quenched by MWNTs.

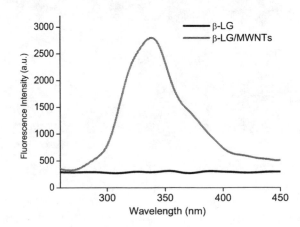

Fig. 2 Emission spectra of β-LG and β-LG/MWNTs in D$_2$O (λ excitation =235nm)

Suspensions of synthesized **β-LG/MWNTs** hybrids showed clear conjugation between MWNTs and protein molecules as observed by CLSM from **Fig. 3**. Some aggregates of hybrids are also observed, which is in good agreement with the rheological study described below.

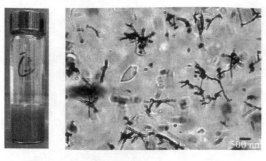

Fig. 3 Photograph and Confocal microscopy of β-LG/MWNTs hybrids in suspension (Scale bar stands for 500 nm).

The core-shell structure of β-LG/MWNTs hybrids (Fig. 4) can be observed from SEM images, where protein molecules (A) surround the carbon nanotube (B) surfaces. From Fig. 4-2, partically coated β-LG/MWNTs hybrids (C) and fully coated hybrid (D) can be observed. When the β-LG concentration reach the gelation critical point, the incorporation of carbon nanotube may reinforce the gel structure by offering higher mechanical strength.

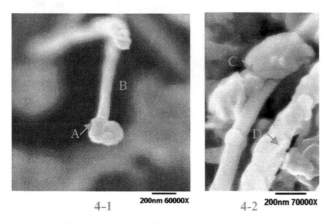

Fig. 4 SEM images of β-LG/MWNTs nanohybrids

Shear-thinning behavior of all the suspensions (**Fig. 5**) were identified across the experimental shear rate ranges, which may account for the breakdown of aggregates in the suspension due to the hydrodynamic forces generated during shear. The existence of MWNTs tends to raise the viscosity at higher shear rate compared to that of protein suspension, indicating the mechanical strength of MWNTs brought to the protein molecules resulting in the less protein molecular rearrangement under hydrodynamic forces. It therefore showed higher resistance to flow exhibiting the higher viscosity.

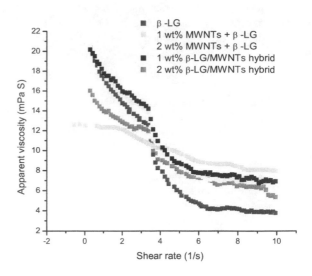

Fig. 5 Steady shear viscosity of β-LG, β-LG/MWNTs mixture and β-LG/MWNTs hybrid suspensions

The elastic modulus, G', from **Fig. 6** showed strong frequency-dependence behavior across the full experimental frequency range, indicating the existence of relaxation processes occurring even at short time scales. The moduli showed an approximately power-law scaling, $G' \sim \omega^n$, with n=0.86 for β-LG/MWNTs hybrids dispersion. Since the concentration of protein in this study is low, there is less opportunity to form cross-linking among protein chains. Although MWNTs have been reported to exhibit strong elastic properties[8], the magnitude of G' is significantly dependant on concentration and network/aggregates structure. The low concentration of MWNTs (0.05~0.1%) in our samples may not form crosslinking structure but some aggregates, which in turn is significantly frequency-dependent. The G' of MWNTs added β-LG and MWNTs- β-LG hybrids are one magnitude lower than that of β-LG dispersion, suggesting that β-LG make more contributions on G' than MWNTs and the decrease of the moduli may attribute to the steric effect of MWNTs in diluted protein dispersions. It can also be seen that the higher concentration of MWNTs showed higher G' under the low frequency whereas the difference diminished as the frequency increased, indicating the loose aggregates structure formed by MWNTs exist.

NSTI-Nanotech 2008, www.nsti.org, ISBN 978-1-4200-8503-7 Vol. 1

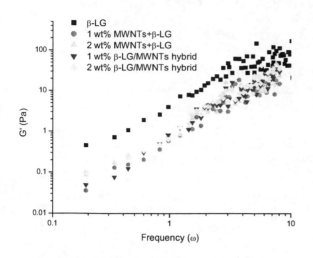

Fig. 6 Dynamic shear storage moduly of β-LG/MWNTs hybrids suspension

4 CONCLUSIONS & FUTURE WORK

β-LG can be successfully immobilized on the MWNTs to form nanohybrid via nonconvalent bonding with PSE as connector. Core-shell structure of the hybrids was identified and shear-thinning behavior was found. Conjugating MNNTs with β-LG affects the rheological properties and the hybrids in suspension showed strong frequency dependency. Further studies on concentrated suspension of β-LG/MWNTs need to be carried out and the effects of carbon nanotube incorporation on β-LG functionalities, such as emulsification and gelation are worth to be investigated in future. The introduction of carbon nanotube to functional proteins may enhance their encapsulation properties for targeted delivery of bioactive and pharmaceutical compounds.

REFERENCES

[1]   Chen, R.J., et al., *Noncovalent Sidewall Functionalization of Single-Walled Carbon Nanotubes for Protein Immobilization*. J. Am. Chem. Soc., 2001. **123**(16): p. 3838-3839.

[2]   Ou, Y.Y. and M.H. Huang, *High-Density Assembly of Gold Nanoparticles on Multiwalled Carbon Nanotubes Using 1-Pyrenemethylamine as Interlinker*. J. Phys. Chem. B, 2006. **110**(5): p. 2031-2036.

[3]   Guldi, D.M., et al., *Functional Single-Wall Carbon Nanotube Nanohybrids-Associating SWNTs with Water-Soluble Enzyme Model Systems*. J. Am. Chem. Soc., 2005. **127**(27): p. 9830-9838.

[4]   L. Hu, Y.L.Z., K. Ryu, C. Zhou, J.F.Stoddart, G.Gruner *Light-Induced Charge Transfer in Pyrene/CdSe-SWNT Hybrids*. Advanced Materials, 2008. **20**(5): p. 939-946.

[5]   Patrick G. Holder and M.B. Francis, *Integration of a Self-Assembling Protein Scaffold with Water-Soluble Single-Walled Carbon Nanotubes*. Angewandte Chemie International Edition, 2007. **46**(23): p. 4370-4373.

[6]   Morr, C.V. and E.Y.W. Ha, *Whey-Protein Concentrates and Isolates - Processing and Functional-Properties*. Critical Reviews in Food Science and Nutrition, 1993. **33**(6): p. 431-476.

[7]   Kontopidis, G., C. Holt, and L. Sawyer, *Invited Review: beta-lactoglobulin: Binding properties, structure, and function*. Journal of Dairy Science, 2004. **87**(4): p. 785-796.

[8]   Fan, Z.H. and S.G. Advani, *Rheology of multiwall carbon nanotube suspensions*. Journal of Rheology, 2007. **51**(4): p. 585-604.

Förster transfer in coupled colloidal type-II and type-I quantum dots

Lev G. Mourokh[1,2,3], Igor L. Kuskovsky[1], Anatoly Yu. Smirnov[3], and Hiroshi Matsui[4]

[1]Department of Physics, Queens College (CUNY), Flushing, NY 11367, USA,
levmurokh@physics.qc.edu
[2]Department of Enginerring Science and Physics, College of Staten Island (CUNY), Staten Island, New York 10314, USA
[3]Quantum Cat Analytics, 1751 67 Street #E11, Brooklyn, NY 11204, USA
[4]Department of Chemistry, Hunter College (CUNY), New York, NY 10021, USA

ABSTRACT

We examine theoretically resonant non-radiative exciton (Förster) transfer in the system of colloidal core-shell quantum dots (QDs) of different types. We show that when the type-I QD (acceptor) is coupled with the type-II QD (donor), the electron-hole pair optically excited in the type-II QD can be transferred to the type-I QD non-radiatively, if the recombination time of the former is larger than the time of the exciton transfer. Correspondingly, the time-resolved photoluminescence signal from the type-I QDs would be modified both quantitatively and qualitatively. We also discuss the feasibility to use this effect in a *pathogen detection platform* aimed at a single molecule level. With the QDs functionalized with antibodies to target pathogens, their presence will manifest itself by allowing or disallowing the Förster transfer by either bringing QDs closer to each other or changing electrostatic environmental properties, respectively. This "On-or-Off" scheme can work in both liquids and gases.

Keywords: type-II colloidal core-shell quantum dots, time-resolved photoluminescence, Förster transfer

Recently, efficient chemical and biological sensors become extremely important because of their applications in homeland security and environmental control. To protect public and soldiers from chemical/biological attack, food poisoning, or hazardous biological leakage, we need to establish improved diagnostic methods and sampling strategies in order to identify pathogens more rapidly and precisely. In order to satisfy this requirement, improved diagnostic methods are necessary to combine *high specificity*, *sensitivity*, *low power consumption*, and *low unit cost* with ultrahigh throughput and ultrafast detection. One of the most promising directions to build such detectors is the use of the artificial nanoscale structures known as quantum dots (QDs). They have discrete energetic spectra but in the contrast with usual atoms the level spacing can be readily and precisely engineered just through variation of their size [1]. The solubility of colloidal QDs in solution further increased the range of potential applications [2]. For example, their conjugation to biological molecules that can recognize and bind specific types of pathogens represents a great opportunity to perform ultrasensitive biomolecular detections in a readily controllable way. Although those types of bioconjugated QDs have been applied for biomolecular sensors since binding events between biological molecules and QDs can be probed optically due to their distinct change in the emission property, the optical detection by QDs systems in single biological molecule level has not been achieved yet. This ultimate resolution limit will have a significant impact for the improved future sensors. In the present paper, we discuss the physical principles of bioconjugated QDs and the strategy to develop new pathogen sensors with high sensitivity.

The vast majority of investigated QDs are type-I heterojunctioned QDs whose narrower bandgap material is a potential well for both electrons and holes. There exists, however, another group of semiconductor heterojunctions, so-called *type-II systems*, whose band alignment has a staggered character; i.e. the lower potential energy for electrons and the higher energy for holes or vice versa. Thus, electrons and holes are separated in real space, which gives rise to a *longer lifetime of the emission* [3,4]. Moreover, the lifetime can be controlled by optimizing the intensity of excitation, external electric and magnetic fields (see discussions and device examples in Refs. [5-7]). Furthermore, typc-II heterostrutures suppress Auger recombination [8,9] that shortens the lifetime of the electron-hall separation and becomes a significant obstacle for successful implementation of nanocrystal-based electronic devices [10,11].

Among various models for sensitive biodetectors, the use of the Förster energy transfer [12] (i.e., the fluorescence resonant energy transfer (FRET) based on the Förster process) between QDs and biomolecules (e.g. Refs. [13,14]) is of special interest. For example, when maltose-binding proteins with dye conjugates were attached to the type-I CdSe-ZnS core-shell QDs, the photoluminescence (PL) signal from the QDs was quenched due to the Förster energy transfer of optical excitation to these proteins. However, once maltose in solution replaced these dye conjugates, PL signal from the QDs was restored and the PL signal amplitude was dependent on the maltose concentration. Method of DNA sensing using Förster energy transfer was proposed in Ref. [15] where the event of coupling of DNA with dye conjugates to QDs can cause the PL signal quenching. It should be noted that in both these approaches, the signal from FRET donor has been measured.

Here we propose a novel Förster transfer-based sensing system where the detection is not made by the Förster transfer between the QD and an attached biomolecule but rather between *two colloidal QDs* coupled by the pathogen. Since those QDs conjugate antibodies to target the pathogen, the pathogen binding to the antibodies bridges two QDs, which induces the Förster transfer between these QDs. The major advantage over the previous schemes [13,14] is that our approach is not limited to detect molecules which must work as the quenching centers for the QD PL. Our sensing system allows one to detect *any molecule* in the "ON-or-OFF" manner because the Förster transfer signal between two QDs is observed only when the target pathogen exists in solution and bridges these QDs. Another advantage over approaches proposed in Refs. [13-15] is that we plan to study the PL signal from FRET *donors* whose life-time is larger and Förster energy transfer leads not only to quantitative changes of the signal, but to the modified *shape* of the time-resolved PL as well.

The physical origin of the Förster transfer is the electron-hole Coulomb interaction and the matrix element of such transfer (in resonant conditions) is given by

$$V = -\langle 2|U_c|1 \rangle \sim \frac{1}{\varepsilon \cdot r_{12}^3} \qquad (1)$$

where |1> and |2> are the exciton wavefunctions in the first and second dots, respectively; r_{12} is the distance between the dots, and ε is the environmental dielectric constant. The corresponding transfer time, τ_{FRET}, is inversely proportional to the matrix element squared, and thus is linear function of the dielectric constant squared and the sixth power of the interdot distance.

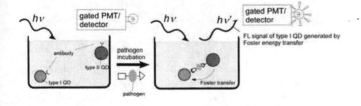

Figure 1: Proposed pathogen sensing platform in solution; Antibody-conjugated quantum dots are coupled in solution via the pathogen bridging under detection.

In our scheme, when the type-I QD (acceptor) is coupled with the type-II QD (donor), the electron-hole pair optically excited in the type-II QD can be transferred to the type-I QD non-radiatively if the recombination time of the former is larger than τ_{FRET}. Due to the long PL decay time of type-II QDs, the binding event between targeted agents and QDs is detected by the characteristically delayed PL signals from the type-I QDs after the Förster transfer occurs (i.e., "On

state"). If there is no target agent in the solution, the dots are far from each other and the Förster transfer, which is extremely sensitive to the distance between the donor and the acceptor, is impossible (i.e., "Off state"). When the dots are functionalized to the specific antibodies for the target pathogen, these QDs are coupled by this pathogen into QD-pathogen-QD system, in which these QDs are close enough to facilitate the exciton transfer (see Fig. 1).

Indeed, assuming for the simplicity [16] an exponential decay for donors and acceptors, the rate equation for the acceptors is

$$\frac{dN_A}{dt} = -\frac{N_A}{\tau_A} + \frac{N_D^0}{\tau_{FRET}} \exp\left[-\left(\tau_{FRET}^{-1} + \tau_D^{-1}\right) \cdot t\right]$$
(2)

with the solution

$$N_A(t) = N_A^0 \exp\left[-\frac{t}{\tau_A}\right] + \frac{1}{\tau_A^{-1} - (\tau_{FRET}^{-1} + \tau_D^{-1})}$$
$$\times \frac{N_D^0}{\tau_{FRET}} \left(\exp\left[-(\tau_{FRET}^{-1} + \tau_D^{-1}) \cdot t\right] - \exp\left[-\frac{t}{\tau_A}\right]\right) \qquad (3)$$

Here, $N_A(t)$ is the number of excited acceptors at the time moment, t, (this number determines acceptor PL intensity); N_A^0 and N_D^0 are the numbers of acceptors and donors at the maximum of the laser pulse ($t = 0$), respectively; τ_D and τ_A are the PL lifetime of donors and acceptors, respectively. In Eq (3), in addition to usual exponential decay expressed by the first term, there is the second term that could increase the number of excitons in the acceptors due to the Förster transfer of excitations from the donors. It should be noted that this term is more pronounced if the number of donors is larger than the number of acceptors.

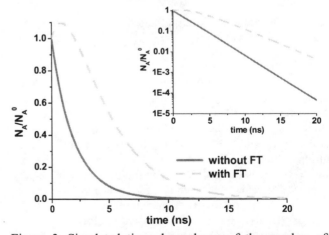

Figure 2: Simulated time dependence of the number of excitons in acceptors. The main panel (linear scale) demonstrates the characteristic increasing of the number of excitons in acceptors when QDs are coupled with the Förster transfer, whereas the inset (in logarithmic scale)

shows the increase of the decay time by orders of magnitude by the Förster transfer.

The time dependence of the number of excited acceptors is shown in Fig. 2 for $\tau_A = 2$ ns, $\tau_D = 10$ ns, $\tau_{FRET} = 4$ ns, and $N_D^0/N_A^0 = 3$. It is evident that the time-resolved PL signal changes both qualitatively and quantitatively when the Förster transfer occurs. As shown in the inset of Figure 2, the characteristic increase of PL intensity is observed in the initial phase under the influence of the Förster transfer and the acceptor is drastically increased as a function of PL decay time due to the extension of the excitonic lifetime of the type II QDs. This result indicates that the Förster transfer-based spectral change by binding the donor QD and the acceptor QD via the pathogen-antibody binding can be applied to sense and assay the targeted pathogens.

The scheme described above makes it possible to detect pathogen agents in solutions. Alternatively, the sensing platform in Fig. 3 can also be used to detect pathogens, which could be convenient for the gaseous environmental sensors. Arrays of type-II and type-I QDs are immobilized on a surface and then QDs are spaced with the distance allowing the Förster transfer. When the pathogen attached to these QDs, the environmental dielectric constant increases and the Förster transfer is strongly suppressed (see Eq. (1)). This binding event can be monitored by the emission from the type I QDs as shown in Figure 2.

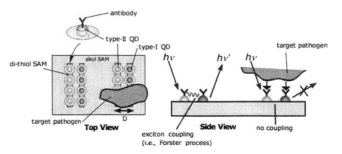

Figure 3: Illustration of the pathogen-sensing platform in gaseous environment.

In conclusion, we proposed a novel method suitable for detection of *any* biomolecules by coupling two heterogeneous QDs, type II and type I, in solution. Since the Förster transfer from the type II QD to the type I QD produces the delayed emission from the type I QDs, this characteristic emission can be applied as the signal for the pathogen sensing. The detection of the Förster transfer induced by the pathogen binding in the time domain makes the proposed scheme inherently more sensitive than other techniques detecting by the quenching of the donor or acceptor PL intensity. In the alternative platform, the array of the acceptor QDs and the donor QDs, already electrostatically coupled, can detect the pathogen by disallowing the Förster transfer via the pathogen binding on these QDs and switching the emission spectra to the non-Förster characteristics.

The work of L.G.M. is partially supported by NSF NIRT, grant ECS-0609146. H.M. acknowledges gratefully partial supports from U.S. Department of Energy (DE-FG-02-01ER45935), the National Science Foundation CARRER Award (ECS-0103430), and the National Institute of Health (2-S06-GM60654). Hunter College infrastructure is supported by the National Institutes of Health and the RCMI program (G12-RR-03037).

REFERENCES

1. A. P. Alivisatos, Nat. Biotechnol. 22, 47 (2004).
2. A. P. Alivisatos, Science 271, 933 (1996).
3. S. Kim, et al., J. Am.Chem. Soc. 125, 11466 (2003).
4. Y. Gu, et al., Phys. Rev. B 71, 045340 (2005); Y. Gu, et al., Solid State Commun. 134, 677 (2005).
5. A. Raghavachari, *Near-Infrared Applications in Biotechnology*, Marcel Dekker (2001).
6. E. H. Sargent, Adv. Mat. 17, 1 (2005).
7. M. P. Mikhailova and A. N. Titkov, Semicond. Sci. Technol. 9, 1279 (1994).
8. G. G. Zegrya and A. D. Andreev, Appl. Phys. Lett. 67, 2681 (1995).
9. S. A. Ivanov, et al., J. Phys. Chem. B 108, 10625 (2004).
10. V. I. Klimov, et al., Phys. Rev. B 60, 13740 (1999).
11. V. I. Klimov, et al., Science 287, 1011 (2000).
12. T. Forster, in *Modern Quantum Chemistry*, ed. O. Sinanogly, Academic Publishers, New York (1965), p. 73.
13. P. T. Tran, et al., Physica Status Solidi B 229, 427 (2002).
14. I. L. Medintz, et al., Nat. Materials 2, 630 (2003).
15. C.-Y. Zhang, et al., Nat. Materials **4**, 826 (2005).
16. It is known that time-resolved PL of type-II QDs is usually non-exponential (e.g., Refs. 3, 4 and references therein), whereas that of type-I QDs is often fitted by two exponentials (e.g. Klimov, et al., Phys. Rev. B **60**, R2177 (1999)); however, after a relatively long time, the decay for both QD types can be looked at as a single exponent, and the characteristic times are usually selected after such a delay. The only requirement we have here is that the decay of a type-II QDs be slower than that of type-I QDs.

Adsorption of Colloid Gold Nanoparticles on Charged Surface

Shien-Der Tzeng,[*] Chih-Shin Luo,[**] Shangjr Gwo,[***] and Kuan-Jiuh Lin[**]

[*] Department of Electrical Engineering, National Chung Hsing University
250 Kuo Kuang Rd., Taichung 402, Taiwan R.O.C., sdtzeng@dragon.nchu.edu.tw
[**] Department of Chemistry, National Chung Hsing University, kjlin@dragon.nchu.edu.tw
[***] Department of Physics, National Tsing-Hua University, gwo@phys.nthu.edu.tw

ABSTRACT

Colloidal gold nanoparticles have many important applications, such as bio-molecular sensors and electronic devices. To fabricate devices, one of the most significant processes is to assemble gold nanoparticles on the substrate surface. In this work, we use the advanced random sequential adsorption (ARSA) model to simulate the adsorption properties of gold nanoparticles on a surface with positively charged self-assembled monolayer (SAM). By comparing the results of experiment and ARSA simulation, we can deduce some properties of the colloid system. Besides, we demonstrate that the saturation density and the radial distribution function of adsorbed nanoparticles can be well predicted from the ARSA simulation. The influence of Hamaker constants used in the simulation is also discussed. These results will be advantageous on the understanding and control of the self-assembly of nanoparticles on charged surface.

Keywords: nanoparticle, self-assembly, colloid

1 INTRODUCTION

Colloidal particles have many applications and attractive properties for the fabrication of nano-structures and nano-devices. Their ability of self-assembly can be used to form not only ordered structures, but also semi-disordered structures on the surface. The formation properties of self-assembled nanoparticle structures allow us to study the particle-particle interactions or particle-surface interactions with a variety of conditions. Such study is important for the understanding of many physical and biological phenomenon. In this work, we introduce the advanced random sequential adsorption (ARSA) simulation method to study the interactions of gold nanoparticles in colloidal solution. When a strong attractive interactions present between particle and substrate, the adsorption of particle on surface will be highly irreversible, and the adsorbed particles have low lateral mobility, resulting in a random spatial distribution. Their adsorption curves, densities, and radial distribution functions can be calculated and compared to the experiment results. These compares could give useful information about the properties of the colloid system.

2 MATERIALS

2.1 Colloidal Gold Solution

A commercial colloidal gold solution (Sigma, G1527) is used as the source of nanoparticles. The diameter of particles is 8.8 ± 0.6 nm. The solution is first diluted by deionized water, and then the ionic strength and pH value are adjusted by adding NaCl and HCl aqueous solutions, respectively. The ionic strength I_C is 2.91 mM, and pH value is 6. The density of nanoparticles in the solution is estimated to be 1.1×10^{12} particles/cm^3.

2.2 APTMS Substrates

Silicon substrates functionalized with 3-aminopropyl-trimethoxysilane (APTMS) are used as the substrate for the assembly of colloidal gold nanoparticles. Silicon substrates are cleaned by water, ethanol, acetone, and then treated in an oxygen plasma cleaner to activate the silicon oxide surfaces. After plasma treated, the substrates are immersed into APTMS for about 1 day to form an amine-terminated self-assembled monolayer on the surface. The excess APTMS molecules on the surface are washed away by water. Then, the substrates are placed into prior colloidal gold solution for about 24 hours to adsorb gold nanoparticles.

3 ADVANCED RANDOM SEQUENTIAL ADSORPTION (ARSA) MODEL

A 3-D RSA simulation method used for simulate the adsorption of colloidal particles has been reported by Lenhoff et al.[1]. Particles with radius a are attempted to placed on the surface at random position sequentially. The probability of a successful adsorption is $\exp(-U_b/k_BT)$, where U_b is the maximum of the total interacting energy U, which includes the electrostatic and van der Waals potential energies. The electrostatic potential between the substrate and the approaching particle is $U_{PS}^{el}(h) = B_{PS}\, e^{-\kappa a h}$. The van der Waals potential between the substrate and the approaching particle is

$$U_{PS}^{vdW}(h) = -\frac{A_{132}}{6 k_B T}\left[\frac{1}{h} + \frac{1}{h+2} + \ln\left(\frac{h}{h+2}\right)\right].$$

The electrostatic potential between the approaching particle and the particle already at the surface is

$$U_{PP}^{el}(r) = \frac{B_{PP}}{r} e^{-\kappa a(r-2)} .$$ The van der Waals potential

between the approaching particle and the particle already at the surface is

$$U_{PP}^{vdW}(r) = -\frac{A_{131}}{6 k_B T}\left[\frac{2}{r^2-4} + \frac{2}{r^2} + \ln\left(1 - \frac{4}{r^2}\right)\right].$$

These potentials have been scaled by thermal energy $k_B T$. Where r is the dimensionless distance between the approaching particle and the particle already at the surface, h is the dimensionless distance between the surface of particle and the planar surface (scaled by the particle radius a), $\kappa^{-1} \sim 5.6$ nm is the Debye length corresponding to 2.91 mM ionic strength. A_{131} is the Hamaker constant for two particles interacting through the solution, and A_{132} is for the particle interacting with the substrate through the solution. A_{131} and A_{132} are assumed to be 11.5×10^{-20} J and 2.5×10^{-20} J. B_{PS} and B_{PP} are the characteristic energies given by [2]

$$B_{PS} = \left(\frac{4\pi\varepsilon\varepsilon_0 k_B T a}{e^2}\right)\left(\frac{y_P + 4\gamma\Omega\kappa a}{1 + \Omega\kappa a}\right)\left(4\tanh\left(\frac{y_S}{4}\right)\right) ,$$

and $B_{PP} = \left(\frac{4\pi\varepsilon\varepsilon_0 k_B T a}{e^2}\right)\left(\frac{y_P + 4\gamma\Omega\kappa a}{1 + \Omega\kappa a}\right)^2$, where y_P

~ -3.7 is the dimensionless potential (scaled by $k_B T/e$) of the particle. $y_S \sim +0.04$ is the dimensionless potential of the APTMS substrate surface at pH6. γ and Ω are defined by

$$\gamma = \tanh\left(\frac{y_P}{4}\right) \text{ and } \Omega = \left(\frac{y_P - 4\gamma}{2\gamma^3}\right).$$ The area used for the

RSA simulation is 1 μm × 1 μm with a periodic boundary condition.

Figure 1(a) shows a typical result of RSA simulation. The density of particles is only about 1350 μm⁻² and there is no aggregation of these particles. However, the actual adsorption result with experimental conditions corresponding to the simulation, as shown in Figure 1(b), has much higher particle density (~1750 μm⁻²), and there is some aggregation of particles. More obvious difference between RSA simulation and experiment can be seen on their corresponding radial distribution functions $g(r)$, as shown in Figure 1(c). The peak value of $g(r)$ at $r \sim 24$ nm corresponds to simulation is much larger than that to experiment. Besides, the latter is broader. Moreover, there is a peak appear at distance $r \sim 2a = 8.8$ nm, which is related to the aggregation of particles.

To improve the 3-D RSA simulation results, we introduce an advanced RSA (ARSA) simulation [3]. Some additional modifications are applied in the 3-D RSA simulation. They are related to the consideration of the size distribution of nanoparticles, the potential distribution of nanoparticles, and the treatment of the overlap of adsorbed particles.

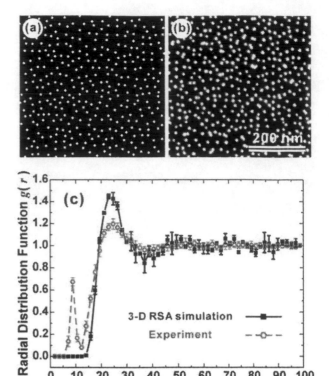

Figure 1: (a) RSA simulation result, (b) SEM image of gold nanoparticles on APTMS/Si substrate. (c) Radial distribution functions $g(r)$ corresponding to the results of RSA simulation and experiment.

Figure 2(a) shows the result of ARSA simulation. Although the density of particles (~1520 μm⁻²) is still quite lower than the result of experiment, the radial distribution function is very similar to that of experiment, except the peak of $g(r)$ at $r \sim 2a$. We think the additional value of $g(r)$ at $r \sim 2a$ is related to the additional aggregation of particles, which is caused by the contamination of some particles.

Some parameters used in the prior ARSA simulation are very critical, such as the potential of particles, the potential of substrate, and the ionic strength of the solution. For example, the adsorption density decreases about 50% when the ionic strength changes from 2.91 mM to 0.58 mM, as shown in Figure 3. By contrast, some parameters such as the Hamaker constants A_{131} and A_{132} have larger permissible range for simulation. As A_{132} changes from 1×10^{-20} J to 10×10^{-20} J, the adsorption density increases less than 10%. The influence of A_{132} is even less than A_{132}. This means that we don't have to know the exact value of Hamaker constants for the simulation.

Since the adsorption properties strongly depend on the electric properties of the system, we can well control the nanoparticle assembly by adjusting the electric-related conditions, and predict by the ARSA simulation. Figure 4 shows that the experiment results are well predicted by the ARSA simulation with different ionic strength.

NSTI-Nanotech 2008, www.nsti.org, ISBN 978-1-4200-8503-7 Vol. 1

4 SUMMARY

In summary, we have shown that the advanced random sequential adsorption simulation (ARSA) method can be used to well simulate the adsorption properties of nanoparticles on a charged surface. Some electric-related parameters, such as the ionic strength of the solution, obviously influence the adsorption properties. By contrast, the Hamaker constants show relatively small influence.

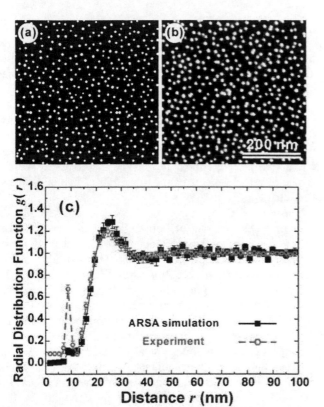

Figure 2: (a) ARSA simulation result, (b) SEM image of gold nanoparticles on APTMS/Si substrate. (c) Radial distribution functions $g(r)$ corresponding to the results of ARSA simulation and experiment.

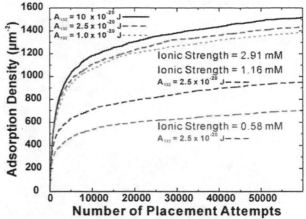

Figure 3: Some adsorption curves calculated by ARSA simulation with different values of ionic strength or Hamaker constant.

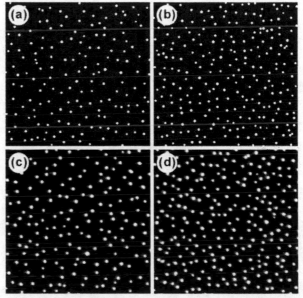

Figure 4: (a)-(b) ARSA simulation result, (c)-(d) SEM images of gold nanoparticles on APTMS/Si substrate. The ionic strength for (a) and (c) is 0.58 mM. The ionic strength for (b) and (d) is 1.16 mM. Potential of particles used for simulation is $y_P = -3.7 \pm 1.3$.

REFERENCES

[1] M. R. Oberholzer, J. M. Stankovich, S. L. Carnie, D. Y. C. Chan, and A. M. Lenhoff, J. Colloid Interface Sci. 194, 138, 1997.

[2] Y. Yuan, M. R. Oberholzer, and A. M. Lenhoff, Colloids Surf., A 165, 125, 2000.

[3] S.-D. Tzeng et al., unpublished, 2008.

Generalized model for the aggregation rate of colloidal nanoparticles and clusters induced by shear in the presence of repulsive interactions

M. Lattuada and M. Morbidelli[*]

ETH Zurich, DCAB-Institute of Chemical and Bio-Engineering
Wolfgang Pauli Strasse 10, 8093 Zurich, Switzerland.
E-mail: massimo.morbidelli@chem.ethz.ch

ABSTRACT

Aggregation of colloidal nanoparticles in the presence of shear is a key step in processing of many particulate materials, polymeric nanoparticles, food products etc. Predicting the rate of aggregation is importance in order to optimize the operating conditions under which coagulation takes place. In this work, we have performed detailed simulations of the aggregation rate of colloidal nanoparticles in the presence of both linear shear flow and repulsive interactions, by numerical solution of the convection-diffusion equation for the pair probability function, from which aggregation rates are computed. Using a simplified model we have interpreted the results of the rigorous calculation and provided a simple criterion to determine the relative importance of all mechanisms involved in the aggregation. The main result is that sufficiently high shear rates can cancel the effect of repulsive interactions and lead to the same aggregation rates found in fully destabilized suspensions.

Keywords: aggregation, colloidal nanoparticles, DLVO, rate constant, pair probability function

1 INTRODUCTION

Coagulation of colloidal nanoparticles in the presence of shear is a key step in processing of many particulate materials, polymeric nanoparticles, food products etc. In the case of charged nanoparticles, coagulation is usually induced out by adding sufficient amount of electrolytes to completely screen all electrostatic interactions. In addition, the suspension is usually sheared to accelerate the coagulation process. However, for certain applications, the addition of large amounts of electrolytes is not beneficial for subsequent processing of the material and removal of electrolytes might be necessary. Therefore, operating with lower amounts of electrolytes, and in particular lower than the critical coagulation concentration, might be desirable. The major obstacle is given by the high sensitivity of the particle stability to the electrolyte concentration, and by the poor understanding of the mechanism of shear induced aggregation in the presence of a repulsive barrier.

For systems that are produced in particulate form, such as polymers and ceramic materials, and then recovered though coagulation, it is well known that quantities such are the broadness of the cluster mass distribution and the cluster structure obtained during coagulation strongly affect physical and mechanical properties of the final product. Therefore, the determination of the time and in some cases of the space evolution of the entire cluster mass distribution during the coagulation process is often desirable. The kinetic approach based on Population Balance Equations (PBE) is a particularly convenient method to accomplish this task. However, the determination of the aggregation rate constants that appear in the PBEs, and which contain all the physical aspects of the aggregation mechanism, is a very challenging task.

In this work, we have performed detailed simulations of the aggregation rate of both colloidal nanoparticles and fractal clusters in the presence of both shear flow and repulsive interactions, by numerical solution of the convection-diffusion equation for the pair probability function, *i.e.* the probability of finding two particles (respectively two clusters) at a given relative position. The equation has been solved for the most important linear flow fields, *i.e.* simple shear and elongational flow.

In order to gain better physical insight of the interplay between the various mechanisms affecting the aggregation, a simple but effective model is presented and used to interpolate the results of the rigorous calculations. This approach provides a simple expression for the aggregation rate that can be used in population balance equation calculations. Furthermore, the simplified model provides a simple criterion to estimate the relative contributions to the aggregation rate due to shear and to repulsive interactions.

2 DESCRIPTION OF THE MODEL

In order to obtain quantitative information about the aggregation rate of two particles or clusters exposed to a flow field gradient and in the presence of repulsive interactions, we have solved the convection diffusion equation for the pair probability function of finding two particles having a given relative position. This approach is not now, having being used by von Smoluchowski almost a century ago, to derive the famous aggregation rate constant K_{11} for particles in a simple shear flow field [1]:

$$K_{11} = \frac{4}{3} G \left(R_1 + R_2 \right)^3 \tag{1}$$

NSTI-Nanotech 2008, www.nsti.org, ISBN 978-1-4200-8503-7 Vol. 1

where G is the shear rate and R_i is the radius of the ith particle.

More recently, the convection-diffusion equation for the pair probability function has been also used to calculate the rate of aggregation of droplets [2], as well as the rate of aggregation of small colloidal particles [3]. The same approach has been also used to describe the microrheology of stable dispersions of monodisperse spherical particles [4].

The convection diffusion equations accounts for all mechanisms of interactions of particles, *i.e.* convection, diffusion, hydrodynamic interactions and colloidal interactions. The colloidal interactions have been described using classical DLVO theory. Hydrodynamic interactions of particles have been described using the conventional approach described by Kim et al. [5], while hydrodynamic interactions of clusters have been described using the approach described by Baebler et al. [6]. No coupling between hydrodynamic and DLVO interactions should be considered [1].

The convection-diffusion equation for the pair probability function P has the following form:

$$\nabla \cdot \left(D(r)\nabla P + \frac{D(r)\nabla \Psi}{kT}C - \mathbf{v}C \right) = 0 \qquad (2)$$

where D is the diffusion coefficient, which is a function of the separation distance r between the particles due to hydrodynamic interactions, k is Boltzmann constant, T is the absolute temperature, Ψ is the total DLVO interaction potential, given by the sum of the attractive Van der Waal contribution, which for equal size particles can be written as [1]:

$$\Psi_a = -\frac{A}{6}\left(\frac{2R_1^2}{r^2 - 4R_1^2} + \frac{2R_1^2}{r^2} + \log\left(\frac{r^2 - 4R_1^2}{r^2} \right) \right) \qquad (3)$$

were A is the Hamaker constant, and a repulsive electrostatic contribution given by:

$$\Psi_r = \frac{4\pi\varepsilon}{e_l^2}\left(4e^{\frac{\kappa(r-2R_1)}{2}} \operatorname{arctanh}\left(e^{\frac{\kappa(r-2R_1)}{2}} \tanh\left(\frac{\zeta e_l}{kT} \right) \right) \right)^2 \frac{R_1^2}{r}\log\left(1 + e^{-\kappa(r-2R_p)}\right) \qquad (4)$$

where e_l is the electron charge, ε is the dielectric permittivity of water, ζ is the surface potential and κ the Debye-Hueckel parameter. The expressions for the velocity profile in the case of elongational flow are reported in reference mmm, while for simple shear they are reported in Van de Venn. The expressions for the hydrodynamic interactions can be also found in the literature, and are omitted here for brevity.

The following boundary conditions are applied:

$$\begin{cases} r = 2R_1 \Rightarrow P = 0 \\ r \to \infty \Rightarrow P = 1 \end{cases} \qquad (5)$$

which express the fact that the pair probability function is zero when two particles (or clusters) touch, while at infinite separation P equals unity.

The solution of the equation has been achieved by using the free Package MUDPACK [7]. After having solved the equation, the rate of aggregation has been calculated by integrating over an arbitrary spherical surface surrounding the reference particle the total flux of particles entering the surface [1]:

$$K_{11} = \iint\limits_S \left(D(r)\nabla P + \frac{D(r)\nabla \Psi}{kT}C - \mathbf{v}C \right) \cdot \mathbf{n}dS \qquad (6)$$

3 RESULTS AND DISCUSSION

The values derived from the solution of equation (2) using equation (6) are plotted in Figures 1, 2 and 3.

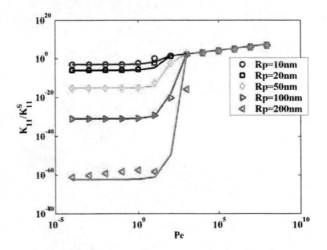

Figure 1: Normalized aggregation rate constant of two particles as a function of Pe number, for various particle sizes.

The quantities on the ordinate are the ratios between the actual rates of aggregation and the Smoluchowski fast diffusion-limited aggregation rate, given by:

$$K_{11}^{S} = \frac{8kT}{3\eta} \qquad (7)$$

where η is the viscosity of the medium.

All the calculations have been performed assuming that the only monovalent symmetric electrolytes are present in the solution, and that the physical properties of particles are those of polystyrene. In Figure 1, for a given electrostatic surface potential of 40mV and a given electrolyte concentration of 10mM, the normalized aggregation rates are shown as a function of Peclet (Pe) number, defined as:

$$Pe = \frac{3\pi\eta G R_1^3}{kT} \tag{8}$$

are shown for different particle sizes. The Pe number represents the ratio between shear contribution and diffusive contribution. It can be seen that, for low values of the Pe number, the rate of aggregation is independent of shear and dominated by diffusion and by DLVO interactions. In fact, it can be proved that the normalized rate constant equals the inverse Fuchs stability ratio for $Pe=0$. However, as the shear contribution becomes more important, a critical shear rate (*i.e.* a critical Pe number) value is reached where an abrupt increase in the rate of aggregation is obtained. For large enough shear rates, the rate of aggregation becomes equal to the one obtained in the absence of any repulsive barrier. By increasing the particle size, it is observed that the DLVO interactions are more and more important, but the general trend observed above is retained. At high enough shear rates, no repulsive barrier is strong enough to resist the effect of shear, and its contribution is cancelled.

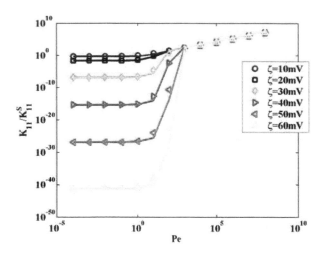

Figure 2: Normalized aggregation rate constant of two particles as a function of Pe number, for various electrostatic potentials.

In Figure 2, the same trend is observed when for a given particle size (50nm), the surface potential is changed. An increase in the surface potential just requires an increase in the critical Pe number necessary to overcome the energy barrier. An identical trend is observed in Figure 3, where the salt concentration is changed, and keeping constant all other parameters. At higher salt concentrations the electrostatic barrier is progressively lowered due to the screening of the electrostatic interaction, reducing the value of shear required to induce fast aggregation.

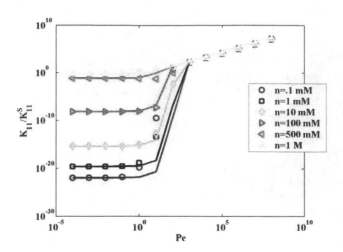

Figure 3: Normalized aggregation rate constant of two particles as a function of Pe number, for various electrolyte concentrations.

In order to rationalize these results, and condense them in a simple equation, a semi-empirical kernel is proposed. This model assumes that the effect of hydrodynamic interactions can be described, within a prefactor, by assuming that the there is a flow field converging towards the reference particle. The advantage of using this unrealistic flow field is that the convection diffusion equation (2) can be solved analytically, and the following expression for the rate of aggregation can be derived:

$$K_{11} = \frac{8\pi D R_1}{2R_1 \int\limits_{2R_p}^{\infty} \frac{e^{-\alpha Pe\frac{R_1}{r} + \frac{\Psi(r)}{kT}}}{r^2} dr} \tag{9}$$

The above equation, when the value of the parameter α is equal to 0.3, can fit very well the results of the rigorous simulations, as shown by the continuous lines in all of the three figures. What equation (9) show in an amazingly simple manner is the fact that the contribution of shear forces directly competes with DLVO interactions. For high enough Pe numbers, where Pe dominates over the maximum of the repulsive barrier, the effect of electrostatic interactions becomes negligible. The other interesting consequence of equation (9) is that it can be applied even in the case of more complex interparticle interactions. Similar conclusions can be drawn in the case of fractal clusters, which are treated at the level of porous particles.

4 CONCLUSIONS

The calculations performed in this work have demonstrated a general feature of aggregation rate constants in the presence of shear and repulsive interactions between particles. At low Peclet numbers, i.e. at low shear rate, the aggregation rate of two particles (or clusters) is controlled by repulsive interactions and not affected by shear. However, as Peclet number is increased, a critical value is reached after which nanoparticles (or clusters) aggregation rate becomes quickly shear controlled. For large enough Peclet number values the aggregation rate becomes the same as if no repulsive barrier was present. This behavior can be rationalized by using the proposed simplified kernel (equation (9)), which clearly shows how the rate of aggregation is determined direct by the competition between shear (expressed by the Peclet number) and the height of the repulsive barrier.

REFERENCES

[1] T.G.M. Van de Venn, "Colloidal Hydrodynamics," Academic Press, 1989.

[2] A.Z. Zinchenko, and R.H. Davis, Phys. Fluids, 10, 2310-2327, 1995.

[3] S. Melis, M. Verduyn, G. Storti and M. Morbidelli, AIChE J., 45, 1383-1393, 1999.

[4] J. Bergenholtz, J.F. Brady, and M. Vicic, J. Fluid Mech., 456, 239-275, 2002.

[5] S. Kim, S.J. Carrila, "Microhydrodynamics", Dover, 2005.

[6] M.U. Babler, J. Sefcik, M. Morbidelli and J. Baldyga, Phys. Fluids, 18, 013302, 2006.

[7] J. Adams, "Multigrid Software for Elliptic Partial Differential Equations: MUDPACK," NCAR Technical Note-357+STR, 1991.

Organic versus Hybrid Coacervate Complexes :
Co-Assembly and Adsorption Properties

J. Fresnais[*], J.-F. Berret[*], L. Qi[**], J.-P. Chapel[**], J.-C. Castaing[**]

[*] : Matière et Systèmes Complexes, UMR 7057 CNRS Université Denis Diderot Paris-VII, Bâtiment Condorcet, 10 rue Alice Domon et Léonie Duquet, 75205 Paris, France
[**] Complex Fluid Laboratory, UMR CNRS/Rhodia 166, Rhodia North-America, R&D Headquarters CRTB, 350 George Patterson Blvd., Bristol, PA 19007 USA

ABSTRACT

We report the co-assembly and adsorption properties of coacervate complexes made from polyelectrolyte-neutral block copolymers and oppositely charged cerium oxide nanoparticles. We first show that the electrostatic complexation resulted in the formation of stable core-shell aggregates in the 100 nm range. The microstructure of the CeO_2-based complexes was resolved using cryogenic transmission electronic microscopy, which revealed that the cores were clusters made from densely packed nanoparticles. The adsorption properties of the same complexes were investigated by stagnation point adsorption reflectometry. The adsorbed amount was measured as a function of time for polymers and complexes using anionically charged silica and hydrophobic poly(styrene) substrates. It was found that all complexes adsorbed readily on both types of substrates up to a level of $1 - 2$ mg m^{-2} at stationary state. Upon rinsing however, the adsorbed layer was not removed. Combining the efficient adsorption and strong stability of the CeO_2–based core-shell hybrids on various substrates, it is suggested that these systems could be used appropriately for coating and anti-biofouling applications.

Keywords: *nanoceria – colloidal complexes – block copolymers - SPAR*

1 INTRODUCTION

In the context of electrostatic complexation, it has been shown that polyelectrolyte-neutral copolymers exhibited interesting clustering and coating properties. These hydrosoluble macromolecules were found to co-assemble with oppositely charged systems, e.g. with surfactants [1-3], polymers [4-8] and proteins [9,10], yielding "supermicellar" aggregates with core-shell structures. As a result, the core of the aggregates was described as a dense coacervate microphase comprising the oppositely charged species. Made from neutral blocks, the corona was identified as surrounding the cores and related to the stability of the whole colloid. In the present paper, the same type of formulations were conducted 7 nm cerium oxide nanoparticles (CeO_2, nanoceria).

Nanoceria were considered because of their potential applications in catalysis [11,12], polishing and coating technologies [13] as well as in biology [14,15]. The main goal of the present survey was to search for synergistic complexation effects in the context of coating and anti-biofouling applications. Cohen Stuart and coworkers have shown that electrostatic complexes made from oppositely charged polyelectrolytes could be efficiently adsorbed on charged hydrophilic and on hydrophobic surfaces, such as silica and poly(styrene) respectively [6,7,16]. These authors have also demonstrated the stability of the deposited layer upon rinsing, and its remarkable anti-fouling properties with addition of proteins [6,7,16]. The approach followed here has consisted to extend the adsorption measurements to new types of electrostatic complexes, namely to complexes built from anionically-coated cerium oxide nanoparticles and cationic-neutral block copolymers [17,18].

2 EXPERIMENTAL

For the electrostatic complexation, we have used poly(trimethylammonium ethylacrylate methylsulfate-b-poly(acrylamide) block copolymer, hereafter noted PTEA$_{11K}$-b-PAM$_{30K}$ and cerium oxide nanocrystals. Light scattering performed in the dilute regime of concentrations have revealed hydrodynamic diameter $D_H = 11$ nm and 8.6 nm, respectively [19]. Cryo-TEM images of single nanoparticles are displayed in Fig. 1a. An image analysis has allowed us to determine the size distribution function which was log-normal with median diameter $D = 6.9 \pm 0.3$ nm and polydispersity $s = 0.15$. Nanoceria were coated by PAA$_{2K}$ using the precipitation-redispersion pathway recently described in the literature. Quantitative determinations using small-angle scattering experiments and chemical analysis have disclosed a brush comprising ~ 50 PAA$_{2K}$ chains adsorbed at the interfaces [20]. A cryo-TEM image of the PAA$_{2K}$-coated nanoceria is displayed in Fig. 1b. This technique revealed isolated particles with size and size distribution identical to those obtained for the bare acidic particles. Mixed solutions of nanocolloids and copolymers were prepared by simple one-shot mixing of dilute solutions prepared at the same concentration c (c ~ 1 wt. %) and same pH .

The amount of adsorbed polymers and coacervates deposited onto silica or poly(styrene) (PS) surfaces was monitored using Stagnation Point Adsorption Reflectometry (SPAR) [21]. Fixed angle reflectometry measured the reflectance at the Brewster angle on the flat substrate. A linearly polarized light beam was reflected by the surface and subsequently splitted into a parallel and a perpendicular component using a polarizing beam splitter. As material adsorbed at the substrate-solution interface, the ratio S between the parallel and perpendicular components of the reflected light varied. The system was analyzed in terms of Fresnel reflectivities for a multi-layer system (substrate, coating, adsorbed layer and solvent), where each layer was described by a complex refractive index and a layer thickness. At the stagnation point the hydrodynamic flow was zero leading to diffusion-limited exchanges between the injected solution and the collecting surface.

Hydrophilic silica substrates were modeled using smooth silicon wafers covered with a layer of ~ 100 nm SiO_2. Hydrophobic poly(styrene) substrate was modeled by a poly(styrene) thin layer of ~ 100 nm deposited on top of an HMDS (hexamethyldisilizane) functionalized silicon wafer by spin-coating a toluene solution (2.5 wt. %) at 5000 rpm.

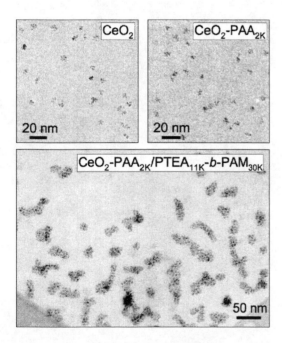

Figure 1 : Cryogenic transmission electron microscopy (cryo-TEM) images of bare and coated cerium oxide nanoparticles. Upper left panel : bare CeO_2 in acidic conditions (pH 1.4); Upper right panel : CeO_2 coated with poly(acrylic acid) with molecular weight 2000 g mol^{-1} (pH 8). Lower panel : CeO_2-PAA_{2K} complexed with the oppositely charged diblock copolymers $PTEA_{11K}$-b-PAM_{30K}.

3 RESULTS

3.1 Cryo-TEM

Cryo-TEM was also performed on CeO_2-PAA_{2K}/$PTEA_{11K}$-b-PAM_{30K} dilute sample, and an illustration of the hybrid aggregates is provided in Fig. 1c (lower panel). The photograph covers spatial field that is 0.36×0.50 μm^2 and displays clusters of nanoparticles [19]. A large visual field was shown here in order to stress that the clusters were dispersed in solutions, a result which was consistent with the visual observations of the solutions. A closer inspection revealed that the aggregates were slightly anisotropic, with a fixed diameter around 20 nm and slight polydispersity in length and morphology. Elongated and branched aggregates were observed too in Fig. 1c, with length comprised between 20 and 100 nm [18].

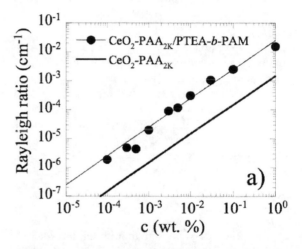

Figure 2 : Concentration dependences of the Rayleigh ratios obtained by light scattering (in the 90°-configuration) for CeO_2-PAA_{2K}/$PTEA_{11K}$-b-PAM_{30K}. The thick straight lines depict the intensities calculated assuming that nanoparticles and polymers were not associated.

3.2 Critical Association Concentration

Fig. 2 show the concentration dependences of the Rayleigh ratio for CeO_2-PAA_{2K}/$PTEA_{11K}$-b-PAM_{30K} formulated at c = 1 wt. %. A linear variation was found down to a concentration of 10^{-4} wt. %. Contrary to the data of the organic complexes [17,18], Fig. 2 did not exhibit a drop of the intensity upon dilution. For this sample, dynamic light scattering has revealed a slightly polydisperse diffusive relaxation mode associated with hydrodynamic diameters D_H = 100 ± 10 nm. In Fig. 2, the scattering intensities obtained from unassociated nanoparticles and polymers were shown for comparison (thick line). The observation of a linear dependence down to the lowest concentration available, and well above the unassociated state supports the assumption that the hybrids are stable toward dilution, and that their critical association concentration is below 10^{-4} wt. %. For the organic complexes discussed in Refs. [17,18], the critical association concentration was of the order of 10^{-2} wt. %.

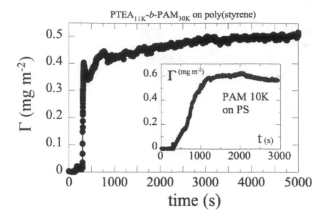

Figure 3 : Adsorption kinetics for $PTEA_{11K}$-b-PAM_{30K} block copolymers on poly(styrene) substrate as received from stagnation point adsorption reflectometry. Measurements were performed with dilute solutions at c = 0.1 wt. %. Inset : adsorption kinetics for poly(acrylamide) molecular weight 10 000 g mol^{-1} operated in the same conditions. In both experiments, rinsing with de-ionized water did not remove the polymer layer.

3.3 Adsorption kinetics for polymers

Adsorption kinetics was first monitored on silica and poly(styrene) model substrates using $PTEA_{11K}$-b-PAM_{30K} polymeric solutions. In dilute solutions, the copolymers are considered as dispersed and non-aggregated. On anionically charged silica surfaces, $PTEA_{11K}$-b-PAM_{30K} were found to adsorb up to a value of Γ_{ST} = 0.8 mg m^{-2}. This adsorption was most likely due to the electrostatic attractive interactions between the opposite charges coming from the cationic polyelectrolyte blocks and the surface. A subsequent rinsing with de-ionized water did not remove the copolymer layer. In the case of a hydrophobic PS surfaces, adsorption was found at a lower level however than on silica surfaces (Γ_{ST} = 0.4 - 0.6 mg m^{-2}). This is a surprising result since in the case a strong polyelectrolyte on an uncharged surface, electrostatics is not expected to contribute to the adsorption energy. In order to understand this phenomenon, reflectometry was also carried out in the same conditions on the different homopolymers with comparable molecular weights, namely PTEA and PAM [18]. On hydrophobic PS surfaces, poly(acrylamide) with molecular weight 10 000 g mol^{-1} was the only polymer to adsorb spontaneously at a low but significant level (Γ_{ST} = 0.5 mg m^{-2}, inset in Fig. 3). Again, in this particular case no desorption occurred on rinsing. One possible explanation is

the release of water molecules trapped in a frozen-like structure close to the hydrophobic surface [7].

3.4 Adsorption kinetics for Hybrid Coacervates

Figs. 4 show the adsorption properties of the hybrid coacervates CeO_2-PAA_{2K}/$PTEA_{11K}$-b-PAM_{30K} on silica and on PS substrates. The two sets of data exhibit the same behavior : after an adsorption process at a level 0.8 − 1.4 mg m^{-2}, no desorption occurred upon rinsing. We explain the absence of desorption observed here as a consequence of the strong stability of the polymer-particle hybrids upon dilution, a result that was already emphasized in a previous report, and also by light scattering. Clearly, the existence of a very low critical aggregation concentration for the particle-polymer hybrids (below 10^{-4} wt. %) guarantees the integrity of the deposited layer on the two substrates [17,18].

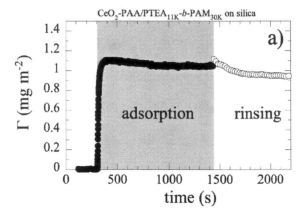

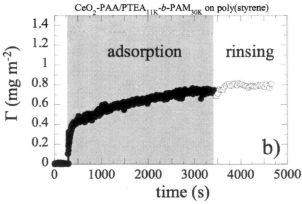

Figure 4 : Adsorption kinetics for organic core-shell SDS/$PTEA_{11K}$-b-PAM_{30K} coacervates complexes on a) anionically charged silica and b) hydrophobic poly(styrene) substrate. Upon rinsing, the reflectometry signal dropped down to a level comprised between 0.3 − 0.5 mg m^{-2}. The decrease of the reflectometry signal suggests that the adsorbed layer has been partially removed during this second phase.

4 CONCLUSION

We have investigated the bulk and surface properties of mixed aggregates resulting from the co-assembly between cerium oxide nanoparticles and block copolymers. Cryo-TEM observations have disclosed unambiguously the cluster-like microstructure of the hybrid aggregates. These results compare very well to those found recently on oxide nanoparticles coated with citrates counterions [19]. The SPAR measurements have shown that organic and hybrid aggregates adsorbed readily on model substrates. At steady state, the adsorbed amount attained values of the order of 1 – 1.5 mg m^2, which correspond to a single and densely packed monolayer of colloids. SPAR experiments were also performed with the copolymers alone, as well as with the homopolymers from which these copolymers were composed. From these tests, it was recognized that poly(acrylamide) with molecular weight 10 K adsorbed spontaneously on PS, a result that was not reported for this substrate. This latter findings was important since it allowed us to conclude that the adsorption of the coacervates was primarily driven by that of the PAM corona.

REFERENCES

[1] T.K. Bronich; T. Cherry; S. Vinogradov; A. Eisenberg; V.A. Kabanov; A.V. Kabanov. Langmuir 14 (1998) 6101.

[2] P. Hervé; M. Destarac; J.-F. Berret; J. Lal; J. Oberdisse; I. Grillo. Europhys. Lett. 58 (2002) 912 .

[3] J.-F. Berret; P. Hervé; O. Aguerre-Chariol; J. Oberdisse. J. Phys. Chem. B 107 (2003) 8111 .

[4] M.A.C. Stuart; N.A.M. Besseling; R.G. Fokkink. Langmuir 14 (1998) 6846 .

[5] A. Harada; K. Kataoka. Science 283 (1999) 65 .

[6] S.v.d. Burgh; A.d. Keizer; M.A.C. Stuart. Langmuir 20 (2004) 1073 .

[7] S.v.d. Burgh; R. Fokkink; A.d. Keizer; M.A.C. Stuart. Colloids and Surfaces A - physicochemical and Engineering Aspects 242 (2004) 167 .

[8] B. Hofs; A. deKeizer; M.A. CohenStuart. J. Phys. Chem. B 111 (2007) 5621 .

[9] F. Cousin; J. Gummel; D. Ung; F. Boue. Langmuir 21 (2005) 9675 .

[10] M. Danial; H.-A. Klok; W. Norde; M.A. CohenStuart. Langmuir 23 (2007) 8003 .

[11] M. Nabavi; O. Spalla; B. Cabane. J. Colloid Interface Sci. 160 (1993) 459 .

[12] M. Das; S. Patil; N. Bhargava; J.-F. Kang; L.M. Riedel; S. Seal; J.J. Hickman. Biomaterials 28 (2007) 1918 .

[13] X. Feng; D.C. Sayle; Z.L. Wang; M.S. Paras; B. Santora; A.C. Sutorik; T.X.T. Sayle; Y. Yang; Y. Ding; X. Wang; Y.-S. Her. Science 312 (2006) 1504 .

[14] R.W. Tarnuzzer; J. Colon; S. Patil; S. Seal. Nano Lett. 5 (2005) 2573 .

[15] J. Chen; S. Patil; S. Seal; J.F. McGinnis. Nature Nanotechnology 1 (2006) 142 .

[16] M.A.C. Stuart; B. Hofs; I.K. Voets; A.d. Keizer. Curr. Opin. Colloid Interface Sci. 10 (2005) 30 .

[17] L. Qi; J.P. Chapel; J.C. Castaing; J. Fresnais; J.-F. Berret. Langmuir 23 (2007) 11996.

[18] L. Qi; J.P. Chapel; J.C. Castaing; J. Fresnais; J.-F. Berret. Soft Matter 4 (2008) 577 – 585.

[19] J.-F. Berret; A. Sehgal; M. Morvan; O. Sandre; A. Vacher; M. Airiau. J. Colloid Interface Sci. 303 (2006) 315 .

[20] A. Sehgal; Y. Lalatonne; J.-F. Berret; M. Morvan. Langmuir 21 (2005) 9359 .

[21] G.J. Fleer; M.A. Cohen Stuart; J.M.H.M. Scheutjens; T. Cosgrove; B. Vincent. Polymers at Interfaces; Chapman & Hall: London, 1993.

Solubilization of Capsaicin and Its Nanoemulsion Formation in the Sonication and Self-Assembly Methods

A. J. Choi[*], C. J. Kim[*], Y. J. Cho[*], J. K. Hwang[**] and C. T. Kim[*]

[*]Nano-Bio Research Group, Korea Food Research Institute, Kyeonggido, Korea, ctkim@kfri.re.kr

[**]Department of Biotechnology, Yonsei University, Seoul, Korea, jkhwang@yonsei.ac.kr

ABSTRACT

In this research, we investigated the optimum condition for the preparation of O/W nanoemulsion containing surfactants and oleoresin capsicum(OC) in four-component systems, and characterized the structure and stability of nanoemulsions by using the ternary phase diagrams of systems. Various types of nanoemulsion including single-layer, double-layers and triple-layers nanoemulsions could be produced depending on the polyelectrolytes such as alginate and chitosan. O/W nanoemulsions of OC could be prepared by the ultra-sonication process at the ration of mixture of OC:Tween 80(1:0.7) and by self-assembly method at the ratio of mixture of OC:Tween 80(1:3) with a particle size of 20-100nm and having a good stability during storage. The ultra-sonication method may be more powerful tool to prepare the nanoemulsion than the self-assembly process, it might be due to the formation capacity of nanoemulsion phase.

Keywords: nanoemulsion, capsaicin, solubilization, bioactive ingredient, self-assembly, stability

INTRODUCTION

Recently, a bioactive lipids such as phytosterols, carotenoids, and ω-3 fatty acids have generated interest in the functional food as they providing specific health benefits to humans. However, these bioactive lipids still have many problems to use as nutraceutical or functional foods because of their physicochemical and physiological properties, including solubility, stability, and bioavailability [1]. Nanoemulsions have excellent longterm thermodynamic stability, and they are capable of solubilizing considerable amounts of water-soluble and oil-soluble compounds. Nanoemulsions have also potential advantages over macroemulsions offering sustained controlled release, improved bioavailability and high stability for bioactive ingredients [2]. Capsaicin has been used for many years in food additives to prevent the development of arteriosclerosis, reduce high blood pressure and to improve prothrombin, thrombin and partial thromboplastin times [3]. It is effective to the growth of human leukemic cells, gastric, and hepatic carcinoma cells in vitro [4]. And also recently it has a profound antiproliferative effect on prostate cancer cells, inducing the apoptosis of both androgen receptor positive and -negative prostate cancer cell lines [5]. However, up to now, known bioactive substances imparting functionality to human body lack in solubilization, stability, and bioavailability, which results in unsuccessful quality enhancement and commercialization.

MATERIALS AND METHODS

1. Materials

Oleoresin capsicum (OC, SHU 100,000) was purchased from G&F Co., Seoul, Korea. Chitosan (Mw 330,000, 93% of deacetylation) solution was prepared by dispersing 0.05 wt% in distilled water and stirring for 2 hr at room temperature.

2. Preparation of nanoemulsion

The composition of three-component nanoemulsion system included oleoresin capsicum(OC) , Tween 80 and water. An oil-in-water emulsion was prepared by sonication or self-assembly method. Chitsan and alginate solution were used to form the double- or triple-layers nanoemulsions bsed on the single nanoemulsions [6].

3. Ternary phase diagram

Existence of a clear, one-phase nanoemulsion region in a three-component mixture was determined by the construction of ternary phase diagrams. Ternary mixtures with varying compositions of OC, Tween 80 and water were prepared. OC concentration was varied from 1.9 to 55.6% (w/w), Tween 80 concentration was varied from 1.3 to 42.9% (w/w) and water concentration was varied from 27.8 to 42.9% (w/w).

4. Particle size measurements

The mean droplet size and size distribution were determined by laser light scattering (Nanotrac [TM]250, Microtrac Inc., PA, U.S.A) at 25°C.

5. ζ-potential measurements

3 ml of nano-emulsions were injected into the measurement chamber of of a particle electrophoresis instrument (Zetasizer Nanoseries ZS, Malvern Instrument,

UK), and the ζ-potential was determined by measuring the direction and velocity that the droplets moved in the applied electric field.

RESULTS AND DISCUSSIONS

1. Phase diagram of nanoemulsions

The phase diagrams indicating the behavior of the systems composed of OC, Tween 80 and water, respectively, and area of nanoemulsion existence are shown Fig. 1. Area enclosed within the solid line represents the extent of nanoemulsion formation, which was a clear and one-phase region. In place of an extensive nanoemulsion region, 2-phase and large liquid crystal regions were formed at high surfactant (60-84 wt %), low OC concentrations, together with very large, cloudy multiphase regions. It is clear that the nanoemulsions could be obtained from OC (1.92-27 %, w/w), Tween 80 (1.93-27.3%, w/w) and water (54.1-96.2 %, w/w) within range of 15-150 nm.

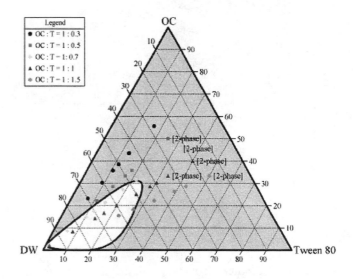

Figure 1 : Ternary phase diagram of nanoemulsion formed by the ultra-sonication on the system of OC/Tween 80/water.

2. Electrical charge of nanoemulsions

The electrical charge (ζ-potential) on the emulsion droplets of double- and triple layers nanoemulsions which layered with chitosan, alginate and the combination of both them by self-assembly, exhibited a positive or negative as the pH increased (Fig. 2). The ζ-potential reached a maximum value −50.3, −39.6, and −31.3 mV at pH 8 for AN, CN, and CAN respectively. The fact that the AN had more higher negative ζ-potential than CA can be attributed to the fact that the anionic charge of sodium alginate. Since most foods or biopolymers can be digested easily in the colon whereas pH value is in the range of 6.5–7.0, it is expected that this type of nanoemulsions seems to be a good candidate for effective functional food delivery system. It is reported that the stable multilayer emulsions could be produced by adsorbing chitosan or alginate onto oil droplets stabilized by a nonionic surfactant [7].

3. AFM observation of nanoemulsion

Atomic force microscpy (AFM) observations of the double-layers nanoemulsion with alginate (AN) was made immediately after the emulsion was prepared and after storage for four weeks at 25°C, and the resultant micrographs are presented in Fig. 3. In general, the particles were slightly smaller than 80 nm in diameter, which supported the results of particle sizing by the dynamic light scattering method. Most of particles appear spherical or irregular in shape, which were likely due to the agglomeration of smaller particles. The images showed little differences between the samples analyzed immediately after the emulsion was prepared and after four weeks storage, suggesting that the nanoemulsion had a good stability during storage.

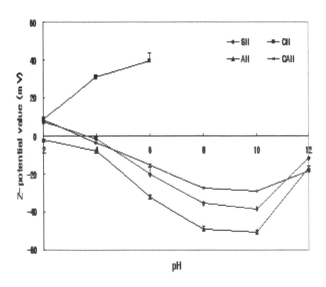

Figure 2 : ζ-potential value of nanoemulsions. SN: single-layer nanoemulsion, CN:double-layers nanoemulsion with chitosan, AN:double-layers nanoemulsion with alginate, CAN:triple-layers nanoemulsion with alginate/chitosan.

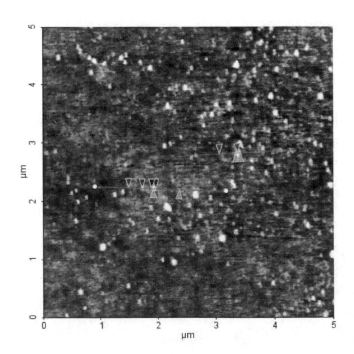

Figure 3 : AFM image of double-layers nanoemulsion with alginate (AN).

CONCLUSIONS

1. O/W nanoemulsions containing bioactive ingredient including oleresin capsicum and surfactant such as Tween 80, glycol and sucrose fatty acid ester could be produced by using ultra-sonication and self-assembly method.

2. Optimal conditions for the preparation of food nanoemulsions by ultra-sonication were confirmed to be mixture of OC : Tween 80 : water = 1 : 0.7 : 2 on the ternary phase system, and the droplet diameter was in the range of 50 to 100 nm.

3. The nanoemulsion could be formed by self-assembly when the mixture ratio of OC : Tween 80 : water = 1 : 3 : 5 on the ternary phase system, and the droplet diameter was in the range of 10 to 20 nm.

4. Double- and triple-layers nanoemulsions prepared with biopolymers such as alginate and/or chitosan by self-assembly, showed a particle size of below 20nm.

5. In conclusion, the double-layers nanoemulsions incorporated with alginate and chitosan can be expected to improve the stability and bioavailability of bioactive ingredient, therefore it may be apply for the production of functional foods containing nutraceutical ingredients.

ACKNOWLEDGMENT

This work supported by the Food Nanotechnology Development Project (2008), Ministry of Education, Science Technology, Korea.

REFERENCES

[1] D. J. McClements, E. A. Decker and J. Weiss, Emulsion-based delivery systems for lipophilic bioactive components, J. Food Sci., 72, R109, 2007.

[2] J. Flangan and H. Singh, Microemulsions: A potential system for bioactives in food, Crit. Rev. Food Sci. & Nutr., 46, 221, 2006.

[3] R. K. Kempaiah, H. Manjunatha and K. Srinivasan, Protective effect of dietary capsaicin on induced oxidation of low-density lipoprotein in rats, Molecular and Cellular Biochem., 275, 7, 2005.

[4] K. Ito, T. Nakazato and K. Yamato, Induction of apoptosis in leukemic cells by homovanillic acid derivative, capsaicin, through oxidative stress: Implication of phosphorylation of p53 at ser-15 residue by reactive oxygen species. Cancer Res., 64, 1071, 2004.

[5] A. Mori, S. Lehmann, J. O'Kelly, T. Kumagai, J. C. Desmond, M. Pervan, W. H. McBride, M. Kizaki and H. P. Koeffler, Capsaicin, a component of red peppers, Inhibits the growth of androgen-independent, p53 mutant prostate cancer cells, Cancer Res., 66, 3222, 2006.

[6] C. T. Kim, Y. J. Cho, C. J. Kim, I. H. Kim, Y. H. Kim and A. J. Choi, Research report of Korea Food Research Institute, No. E071000-07085. Development of process technology for functional food nanocomposite. In: Report of the Development of Food Nanotechnology, December 31, Kyeonggi, Korea, 123-266, 2007.

[7] S. Mun, E. A. Decker and D. J. McClements, Influence of droplet characteristics on the formation of oil-in-waater emulsions stabilized by surfactant-chitosan layers, Langmuir, 21, 6228, 2005.

Modeling evaporation and ring-shaped particle deposition of a colloidal microdroplet, with applications in biological assays

Rajneesh Bhardwaj and Daniel Attinger[*]

Laboratory for Microscale Transport Phenomena, Columbia University, New York, NY 10027,
*Corresponding author. Tel: +1-212-854-2841; fax: +1-212-854-3304;
e-mail: da2203@columbia.edu (D. Attinger)

ABSTRACT

Biological microarrays can be manufactured by spotting colloidal drops and evaporating them. The remaining deposits, however are not always homogeneous and often exhibit ring-like patterns. In this work, a numerical modelling is made for the evaporation of a colloidal nanoliter drop on a solid, non-isothermal substrate. The equations governing fluid, heat and mass transport are expressed in a Lagrangian framework. The boundary conditions at the drop free surface are the mass and energy jump conditions. The diffusion of vapor in the gas surrounding the drop is solved numerically, determined by the drop-substrate geometry and thermodynamic conditions. The formation of a peripheral ring is observed during the evaporation of the drop. The proposed modeling will allow the determination of optimum processing conditions to make spot arrays from drying colloidal drops.

Keywords: evaporation, colloidal drop, ring formation, particle deposition

1 INTRODUCTION

Colloidal drops evaporating on a solid substrate can be used to deposit or organize small particles. The pattern left after the evaporation often looks like the ring-like structure of the dried DNA drop [1] in Figure 1a. Transport phenomena during the evaporation of a colloidal drop evaporation are complex and coupled: the fluid dynamics is transient and severely influenced by the shrinking free surface and wetting conditions; heat transfer occurs by evaporation, convection and conduction to the substrate; mass transfer takes place by the diffusion of liquid vapor in the atmosphere and diffusion of particles in the bulk liquid. Deegan and co-workers made a significant contribution to the understanding of the deposit formation [2]. They explained the formation of a peripheral ring-pattern or "coffee ring" by showing that the diffusion equation predicts the evaporative flux to be highest near the wetting line. This phenomenon, together with a pinned wetting line, results in a radially outward flow inside the drop so that most particles are convected towards the wetting line. Recently, Maenosono et al [3] studied the growth of a nanoparticle-ring during the evaporation of pyridine and

water drops. They observed two stages of ring evolution, a ring buildup followed by a receding of the wetting line. They predicted ring growth dynamics and found a reasonable agreement with experiments for both the ring growth and the final width. Researchers have also looked the formation of multiple rings during the drying of the colloidal drop. Adachi et al. [4] suggested that the "stick slip motion" of the wetting line caused the formation of stripe patterns (since particles accumulate at the wetting line when it sticks) and they proposed a model for the stick slip motion. Due to the complex, coupled physics involved, most theoretical models reported are based on relatively crude assumptions such as fluid flow with negligible inertia [5-8], negligible gravity [5-8], small wetting angle [2, 8], a spherical cap shape of the free surface [2, 5-10], pinned wetting line throughout the evaporation [5-8, 10-14] and negligible heat transfer between the drop and the substrate [2, 6, 8, 14, 15]. However, the ring formation kinetics depends upon the evaporation rate, flow and thermal fields. Thus a full numerical solution for the fluid flow, temperature field and particle diffusion is highly desirable to predict particle deposition and pattern growth. In this paper, we present a two-dimensional axisymmetric finite element model that solves the mass conservation, Navier-Stokes, energy equations, diffusion equation for particle transport and diffusion equation for the vapor concentration outside the drop for the evaporation of a nanoliter colloidal drop.

2 NUMERICAL MODEL

We extend a mathematical modeling for drop deposition [16, 17] to account for the evaporation and particle transport in the colloidal drops. The 2D axisymmetric finite element model solves the unsteady equations of mass, momentum and heat transfer. All governing equations are expressed in a Lagrangian framework, which provides accurate modeling of the free surface motion. The boundary conditions at the free surface are the mass and energy jump conditions. While the equations for mass conservation, Navier-Stokes and heat transfer equations are given in [16, 17], we describe here the mathematical model for particle transport and evaporation: The value of evaporative mass flux j at the free surface is obtained by solving the quasi-steady diffusion equation for the water vapor concentration

c: $\nabla^2 c = 0$. The boundary conditions for the above governing equation are:

(a) at $r > R_{cap}, z = 0$; $\partial c / \partial z = 0$

(b) at $r = 0$; $\partial c / \partial r = 0$

(c) at $r = \infty, z = \infty$; $c = Hc_\infty$

(d) at free surface $c = c_{int}$

where R_{cap} is the wetted radius, c_{int} is the saturated density (kg/m^3) of water vapor near the interface and c_∞ is density of water vapor in the far-field (ambient). These densities are calculated using data in [18]. At the free surface of the drop, we assume 100% of relative humidity H, while in the far-field H is measured. In these calculations, the far field ($r = \infty, z = \infty$) is considered at $r = 20R_{cap}, z = 20R_{cap}$. The evaporative mass flux j at the free surface can be expressed as:

$$\mathbf{j}(r,T) = D_{LG}(T) \left[\frac{\partial c}{\partial r}\mathbf{n}_r + \frac{\partial c}{\partial z}\mathbf{n}_z \right]_{free\_surface}$$

where D_{LG} is the diffusion coefficient of liquid in surrounding gas. At the free surface of the drop, the hydrodynamic and thermodynamic vapor-liquid jump conditions [10, 19] are applied:

$$\mathbf{j}.\mathbf{n} = \rho(1 - X_f)(\mathbf{v} - \mathbf{v}_f).\mathbf{n} \text{ and } jL = -k\nabla T.\mathbf{n}$$

where, $\mathbf{v}$ is velocity of the liquid at the free surface, $\mathbf{v}_f$ denotes the velocity of the free surface, $\mathbf{j}$ is the evaporative mass flux at the free surface (kg/m^2-s), $\mathbf{n}$ is the outward normal unit vector at the liquid-air interface ρ is the density of the liquid. X_f is the volume concentration of particles at the free surface and L is the latent heat of evaporation of the liquid (J/kg). We assume that the particles have same density as water. The governing equation for particles transport is given by [20-22] : $DX / Dt = D_{PL}\nabla^2 X$ where D_{PL} is the diffusion coefficient of particles in the fluid. Boundary conditions are given by:

@ $r = 0$, $\partial X / \partial r = 0$ (Axisymmetry).

@ $z = 0$, one of the following two cases is possible:

    1. Non-sticky substrate, like silicon: $\partial X / \partial z = 0$

    2. Sticky substrate, like silicone: $X = 0$ (Perfect sink boundary condition).

Physically, the perfect sink boundary condition means that all particles approaching the surface deposit instantaneously [23]. The mathematical model presented is solved using the Galerkin finite element method [16]. The computational domain (Figure 1b) is discretized with a mesh of linear, triangular elements. The unsteady solution for the fluid dynamics is based on the scheme given by Bach and Hassager [24]. The time step is constrained to the smallest time scale of the problem (free surface oscillations, about 15 ns) so that a whole evaporation simulation (2 minutes for a 20nL drop) would involve 8 billion time steps, implying a prohibitive computing time. To resolve this issue, a two-step temporal integration scheme is developed for this problem. This scheme is described as follows: first, a converged, instantaneous solution of the

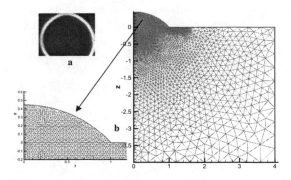

Figure 1: (a) Ring-like pattern formation during the drying of a DNA drop. Drop wetted radius is about 500 micrometer [1] (b) Typical mesh in the computational domain.

fluid flow and the evaporation flux are obtained using a short time step ($\Delta\tau_{short}$, order of 15 ns) using scheme in [24]. Second, assuming that the fluid flow remains steady, a new location of the free surface due to the evaporation is calculated using a time step 4 orders of magnitude larger ($\Delta\tau_{long}$, order of 150 µs). Then, the first scheme is applied again and so on. In order to avoid losing particles during the shrinkage of the free surface, a source term is added to the diffusion equation for the free surface elements. Increase in the concentration of the particles with time due to the evaporation is given by XS / ρ where ρ is the density of the liquid and the source term is defined as: $S = j\delta A / \delta V$ where j is the evaporative mass flux [kg of water m^{-2}s^{-1}], δA is the area from which liquid evaporates [m^2] and δV is the volume of the liquid in which particles accumulate [m^3]. This two-step approach is justified as long as the long time step is small with regard to the evaporation of the drop (quasi-steady approach), so that the fluid flow and the evaporation flux can be considered as steady for the duration of the long time step.

3 RESULTS

Results are presented in this section for the evaporation of a 20 nL water drop with 298 micrometer initial wetted radius and 49° initial contact angle on a flat substrate for four different cases: (a) Drop contains 1 micron particle diameter on sticky or non-sticky substrate (b) Drop contains 10 nm particle diameter on sticky or non-sticky substrate. The value of the diffusion coefficient for 1 micron and 10 nm particles are calculated as 4e-13 and 4e-11 m^2/s respectively, assuming spherical particles by Stokes-Einstein equation [25]. The initial temperature of the drop is taken as 25°C (ambient temperature). A relatively high substrate temperature (125°C) was selected to enhance evaporation. A perfect thermal contact is assumed at the drop-substrate interface. The ambient temperature and relative humidity are recorded as 25°C and 40% respectively. The values of dimensionless numbers, Reynolds, Weber, Froude number are 330, 4.1 and 342 respectively. The variation of viscosity with temperature is taken into account. Initially, the drop contains a uniform 0.5% concentration of particles. The thermophysical properties for water and fused silica used in simulations are presented in [26]. Figure 2 shows streamlines and isotherms (left) during the evaporation of colloidal 20 nL water drop containing 1 micron particles on a non-sticky fused silica substrate heated at 125°C.

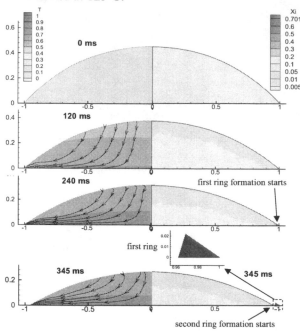

Figure 2: Particle concentrations (right), streamlines and isotherms (left) for the evaporation of a 20 nL water drop for 1 micron particles.

During the evaporation (t = 0 to 345 ms), the fluid flow inside the drop is radially outward. This outward flow is due to the fact that the pinning of the wetting line ensures that the liquid evaporating from the edge is replenished by the liquid from the interior [2]. The value of the evaporative flux is also higher near the wetting point. This corresponds to a higher mass loss near the wetting line. This outward flow pattern has maximum velocities on the order of 10 micrometer/sec. The heat transfer is mainly governed by three mechanisms: conduction of heat from the substrate, evaporative cooling along the free surface and convective effects inside the drop. Flat isotherms in Figure 2 show that the conduction heat transfer dominates over convection and evaporative cooling along the free surface. Figure 2 shows the concentration of particles (right) at different times during the evaporation of the drop. The evaporation of the liquid results in build up of concentration along the free surface. The particles accumulate faster near the wetting line than on the top of the drop since the evaporation is highest near the wetting line. As time evolves, more particles are convected to the edge of the drop where they accumulate and start forming a ring (at t = 240 ms). A concentration value of 0.7 is considered as the criterion of the ring formation because it corresponds to maximum packing of the particles. When height of the ring formed reaches to 8 micrometer, wetting line is allowed to recede

and pinned to a different location near the first ring. At this time, width of the ring formed is 12 micrometer. The formation of second ring starts at t = 345 ms. The mass distribution of the particles in the drop vs. radial distance is shown in Figure 3 for different times during the evaporation of the drop for 10 nm particles.

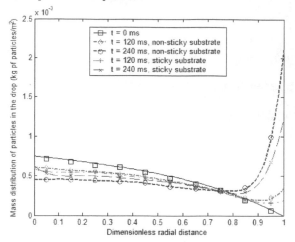

Figure 3: Mass distribution of particles at different radial locations for different times during the evaporation of the drop for the non-sticky and the sticky substrate for 10 nm particles.

Two cases are compared: (a) non-sticky substrate (b) sticky substrate. In case of non-sticky substrate, profile of mass distribution of particles, which is approximately spherical cap initially, develops a spike near the contact line at t = 240 ms. This spike is representative of the ring of the particles. At t = 240 ms, mass distribution for the sticky substrate shows that the particles have deposited along the drop-substrate interface. Figure 4 shows the mass distribution of the particles in the drop vs. radial distance for 1-micron particles. Two cases are compared for the stick and the non-sticky substrate as explained above. In this case, mass distribution profiles shows that the particles deposition along the drop-substrate interface is negligible.

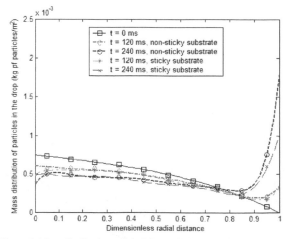

Figure 4: Mass distribution of particles at different radial locations for different times during the evaporation of the drop for non-sticky and sticky substrate for 1-micron particles.

4 CONCLUSIONS

A numerical model for the evaporation of a nanoliter evaporating colloidal drop is presented. Results are presented for the evaporation of colloidal 20 nL water drop on a fused silica substrate heated at 125°C. The flow, thermal and particles concentration fields are shown in the drop. Ring growth and deposition of particles onto the substrate are predicted at different times during the evaporation of the drop.

ACKNOWLEDGEMENTS

The authors gratefully acknowledge financial support for this work from the Chemical Transport Systems Division of the US National Science Foundation through grant 0622849.

REFERENCES

[1]   I. I. Smalyukh, O. V. Zribi, J. C. Butler, O. D. Lavrentovich, and G. C. L. Wong, "Structure and dynamics of liquid crystalline pattern formation in drying droplets of DNA," *Physical Review Letters*, vol. 96, pp. 177801, 2006.

[2]   R. D. Deegan, O. Bakajin, T. F. Dupont, G. Huber, S. R. Nagel, and T. A. Witten, "Capillary flow as the cause of ring stains from dried liquid drops," *Nature*, vol. 389, pp. 827-829, 1997.

[3]   S. Maenosono, C. D. Dushkin, S. Saita, and Y. Yamaguchi, "Growth of a Semiconductor Nanoparticle Ring during the Drying of a Suspension Droplet," *Langmuir*, vol. 15, pp. 957-965, 1999.

[4]   E. Adachi, A. S. Dimitrov, and K. Nagayama, "Stripe Patterns Formed on a Glass Surface during Droplet Evaporation," *Langmuir*, vol. 11, pp. 1057-1060, 1995.

[5]   H. Hu and R. G. Larson, "Analysis of Microfluid flow in an Evaporating sessile droplet," *Langmuir*, vol. 21, pp. 3963-3971, 2005.

[6]   H. Hu and R. G. Larson, "Evaporation of a sessile droplet on a substarte," *Journal of Physical Chemistry B*, vol. 106, pp. 1334-1344, 2002.

[7]   Y. Popov, "Evaporative Deposition Patterns Revisited: Spatial Dimensions of the Deposit," *Physics Review E*, vol. 71, pp. 36313, 2005.

[8]   R. D. Deegan, O. Bakajin, T. F. Dupont, G. Huber, S. R. Nagel, and T. A. Witten, "Contact line deposits in an evaporating drop," *Physics Review E*, vol. 62, pp. 756, 2000.

[9]   R. G. Picknett and R. Bexon, "The Evaporation of Sessile or Pendant Drops in Still Air," *Journal of Colloid and Interface Science*, vol. 61, pp. 336-350, 1977.

[10]   R. Mollaret, K. Sefiane, J. R. E. Christy, and D. Veyret, "Experimental and Numerical Investigation of the Evaporation into Air of a Drop on a Heated Substrate," *Chemical Engineering Research and Design*, vol. 82, pp. 471-480, 2004.

[11]   H. Hu and R. G. Larson, "Analysis of the Effects of Marangoni stresses on the Microflow in an Evaporating Sessile Droplet," *Langmuir*, vol. 21, pp. 3972-3980, 2005.

[12]   O. E. Ruiz and W. Z. Black, "Evaporation of Water droplets Placed on a Heated Horizontal surface," *Journal of Heat Transfer*, vol. 124, pp. 854, 2002.

[13]   F. Girard, M. Antoni, S. Faure, and A. Steinchen, "Evaporation and Marangoni Driven Convection in Small Heated Water Droplets," *Langmuir*, vol. 22, pp. 11085-11091, 2006.

[14]   Widjaja E, N. Liu, M. Li, R. T. Collins, O. A. Basaran., and M. T. Harris, "Dynamics of sessile drop evaporation: A comparison of the spine and the elliptic mesh generation methods," *Computers and Chemical Engineering*, vol. 31, pp. 219-232, 2007.

[15]   R. D. Deegan, "Pattern Formation in Drying Drops," *Physics Review E*, vol. 61, pp. 475-485, 2000.

[16]   J. M. Waldvogel and D. Poulikakos, "Solidification Phenomena in Picoliter Size Solder Droplet Deposition on a Composite Substrate," *International Journal of Heat and Mass transfer*, vol. 40, pp. 295-309, 1997.

[17]   R. Bhardwaj, J. P. Longtin, and D. Attinger, "A numerical investigation on the influence of liquid properties and interfacial heat transfer during microdroplet deposition onto a glass substrate," *International Journal of Heat and Mass Transfer*, vol. 50, pp. 2912-2923, 2007.

[18]   G. J. VanWylen, R. E. Sonntag, and C. Borgnakke, *Fundamentals of Classical Thermodynamics*, 4th ed: John Wiley, 1994.

[19]   J. P. Burelbach, S. G. Bankoff, and S. H. Davis, "Nonlinear stability of evaporating/condensing liquid films," *Journal of Fluid Mechanics*, vol. 195, pp. 463-494, 1988.

[20]   R. Blossey and A. Bosio, "Contact line deposits on cDNA microarrays: A 'Twin-spot effect'," *Langmuir*, vol. 18, pp. 2952-2954, 2002.

[21]   Y. Y. Tarasevich and D. M. Pravoslavnova, "Drying of a multicomponent solution drop on a solid substrate: Qualitative analysis," *Technical Physics*, vol. 52, pp. 159-163, 2007.

[22]   T. Heim, S. Preuss, B. Gerstmayer, A. Bosio, and R. Blossey, "Deposition from a drop: morphologies of unspecifically bound DNA," *Journal of Physics: Condensed Matter*, vol. 17, pp. S703-S716, 2005.

[23]   L. Song and M. Elimelech, "Calculation of particle deposition rate under unfavourable particle-surface interactions," *Journal of the Chemical Society, Faraday Transactions*, vol. 89, pp. 3443-3452, 1993.

[24]   P. Bach and O. Hassager, "An Algorithm for the Use of the Lagrangian Specification in Newtonian Fluid Mechanics and Applications to Free-Surface Flow," *Journal of Fluid Mechanics*, vol. 152, pp. 173-190, 1985.

[25]   R. B. Bird, W. E. Stewart, and E. N. Lightfoot, *Transport Phenomena*. New York: John Wiley & Sons, 2007.

[26]   R. Bhardwaj and D. Attinger, "A Numerical Model for the Evaporation of a Nanoliter Droplet on a Solid, Heated Substrate," in *ASME International Mechanical Engineering Congress and Exposition*. Seattle, 2007.

[27]   D. R. Lide, *CRC Handbook of Chemistry and Physics on CD-ROM*, 81st ed: Chapman and Hall/CRC, 2001.

Synthesis of Cu Nanoparticles for Preparation of Nanofluids

Velasco Abreo A.[1], **Perales-Perez O.**[2] and **Gutiérrez G.**[1]

[1]University of Puerto Rico, Mechanical Engineering Department, Mayagüez, Puerto Rico 00681
[2]University of Puerto Rico, Department of Engineering Science & Materials, Mayagüez, Puerto Rico 00681-9044

ABSTRACT

Copper nanoparticles have been prepared by reduction of Cu(II) ions in aqueous and non-aqueous solutions. The time at which elemental Cu was formed from starting 0.016M Cu solutions decreased from 10 hours to only 6 minutes when the concentration of hydrazine was increased from 0.059M up to 0.7M, respectively. The corresponding average crystallite size decreased from 30nm down to 16nm. The complete reduction of in ethylene glycol was achieved after 30 seconds when a NaOH/Cu mole ratio of 50 was used. The average crystallite size was estimated at 25 nm. In both approaches, the Cu-forming reaction involved the formation of intermediate Cu_2O, which underwent dissolution and subsequent reduction into elemental Cu. UV-vis measurements evidenced the formation of nanosize Cu crystals. The surface plasmon resonance band was blue-shifted after intensive sonication suggesting a decrease in crystal size during the preparation of the nanofluid in ethylene glycol.

Keywords: copper nanocrystals, aqueous reduction, polyol reduction, nanofluids.

1. INTRODUCTION

Stable suspensions of nanometer solid particles in suitable solvents, so-called nanofluids, have shown enhanced thermal conductivity when compared with the fluid base. This feature enables these suspensions to be considered a promising material for efficient and effective thermal management in different systems, [1-4]. Nanofluids bearing metal nanoparticles, e. g. Ag or Cu, exhibit in general more remarkable thermal conductivity than those containing oxide nanoparticles [5-7]. The preparations of metal bearing nanofluids have been attempted by direct condensation of copper vapor in ethylene glycol [8]. Although the thermal conductivity measurements suggested an increase in the thermal conductivity of the nanofluid, the selected preparation route did not allow any possibility to restrict particle size at the nanoscale. The reasons behind the increase in thermal conductivity of nanofluids are still unclear; however, the big surface area in nanosize particles, the Brownian motion of suspended cystals and the establishment of fluid layers onto nanoparticles should be involved with this phenomenon. These enhanced thermal conductive nanofluids will find inmediate applications in cooling systems and in the design of smaller machines and and components [8,9].

Despite of the technological importance of such a fluid, there is still a lack of systematic research on other possibilities to synthesize copper nanocrystals and the corresponding nanofluids where the tuning on crystal size could become possible. On this basis, the present work was focused on the optimization of the synthesis conditions of Cu nanoparticles and their subsequent stabilization in ethylene glycol. Copper nanoparticles have been synthesized through the reduction of Cu ions by hydrazine in water as well as by taking advantage of the reducing power of polyol solutions in excess of hydroxide ions. The preparation of the copper-bearing stable nanofluids has also been attempted by dispersing synthesized nanoparticles through intensive ultrasonication in ethylene glycol.

2. EXPERIMENTAL

2.1 Materials

All reagents were of analytical grade and were used without further purification. Required weights of Cu(II) sulfate pentahydrate salt, $CuSO_4.5H_2O$, with 98.0 – 102.0 % in purity (Alfa Aesar) and Cu(II) acetate salt, $Cu(CH_3CO_2)_2$, (99%, Strem Chemicals), were dissolved in high purity water and ethylene glycol, respectively. Hydrazine, (N_2H_4), (+98%, Alfa Aesar) and ethyelen glycol, $(HOCH_2CH_2OH)$, (99%, Alfa Aesar) were used as reductant agents in aqueous and non-aqueous media, respectively. NaOH (+98%, Sigma - Aldrich) was also used to accelerate the reduction reaction in ethylene glycol.

2.2 Synthesis of Copper Nanoparticles

A. Aqueous reduction of Cu(II) ions: Required amounts of hydrazine and Cu(II) sulfate solutions were contacted with N_2-purged water at room temperature conditions. Possible re-oxidation of Cu precipitates was avoided by conducting the reduction reaction in tightly closed reaction vessels. At the end of the contact time, the suspension was treated by centrifugation. Recovered solids were washed twice and storaged in ethanol.

B. Polyol route: A Cu(II) acetate solution in ethylene glycol was added to a boiling ethylene glycol solution containing NaOH . The experimental set-up considered the use of a condenser to recover volatilized solvent. The experiments were carried out at 500 rpm of stirring intensity using a mechanical stirrer. The reacting solutions were heated at different heating rates provided by a suitable control of the power of the mantle heater device. At the end of the reaction time, the polyol solution was allowed to cool down. Copper nanoparticles were recovered by centrifugation and washed in ethanol two times prior to their storage and characterization.

2.3 Nanofluid Preparation.

Cu nanoparticles produced by the above described methods were dispersed in ethylene glycol under intensive and prolonged ultrasonication cycles (up to 7 hours). Obtained suspensions were characterized by Uv spectroscopy as a function of time to determine the suspension stability against sedimentation. The estimated volumetric fraction of Cu nanoparticles in ethylene glycol was 0.05%.

2.4   Characterization Techniques

Structural analysis of the powders was carried out by x-ray diffraction (XRD) using the Cu-Kα radiation.   The average crystallite size of produced powders was estimated by using the Scherrer's equation for the (111) and (200) peaks. UV-vis spectroscopy was used to determine optical properties of the Cu nanoparticles in different suspending media. Particle morphology was examined by high-resolution transmission electron microscopy (HRTEM) and scanning electron microscopy (SEM).

3.   RESULTS AND DISCUSSION

3.1   Cu(II) Reduction in Hydrazine.

Figure 1 shows the XRD patterns for the solids produced after reduction of 0.0016M Cu(II) in 0.059M aqueous hydrazine solutions after different reaction times. The solids formed at earlier times consisted of partially reduced cuprous oxide (Cu_2O). The drop in the intensity of the Cu_2O peaks and the simultaneous increase and sharpening of the peaks corresponding to elemental copper, suggested that the final formation of elemental Cu from starting Cu(II) solutions took place through the formation of partially reduced Cu(I) oxide as an intermediate phase. This intermediate should have been redissolved at prolonged reaction times and finally reduced to the zero state. The average crystallite size for Cu nanoparticles varied from 25 nm to 30 nm when the reaction time was prolonged from 4 hours to 24 hours, respectively. In order to evaluate the effect of the initial concentration of Cu(II) species on the average crystallite of produced Cu nanoparticles, a 0.059M hydrazine solution was contacted with Cu(II) solutions in the range of concentrations between 0.016M and 0.064M. The reaction was interrupted when the complete reduction of Cu was realized. Figure 2 shows the XRD patterns for the solids produced at different initial Cu(II) concentration. Only peaks corresponding to elemental Cu were detected, suggesting the complete reduction of Cu(II) species. It was observed that the lowering of the Cu(II) concentration made the time required to form elemental Cu shorter. The reaction time was shortened from 34 hours to 12 hours when the Cu(II) concentration varied from 0.016M to 0.064M. The corresponding average crystallite sizes varied between 25nm and 27 nm.

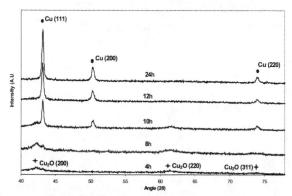

Figure 1: XRD patterns of powders synthesized at different reaction times. Reactants concentrations were 0.0016M Cu and 0.059M N_2H_4

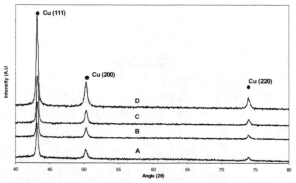

Figure 2: XRD patterns of powders synthesized at 0.059M N_2H_4 and different Cu concentrations. A: 0.016M Cu; B: 0.032M Cu; C: 0.048M Cu and D: 0.064M Cu.

The effect of the hydrazine concentration on the kinetic of Cu(II) reduction and the corresponding crystal size was also evaluated. Figure 3 shows the XRD patterns of the solids produced from a 0.0016M Cu(II) solution and different concentrations of hydrazine. Once again, the reduction reaction was interrupted once the formation of elemental Cu was completed. As evident, the reduction of Cu(II) species required shorter times when the concentration of hydrazine was increased from 0.059M up to 0.70M. The corresponding reaction times and average crystallite size varied from 12hours to 30 minutes and from 30nm to 17nm, respectively. The nanoparticles were suspended in ethanol and characterized by UV-vis. The spectra, shown in figure 4, displayed a surface plasmon (SP) resonance band at around 280nm that can be attributed to Cu nanoparticles [9].

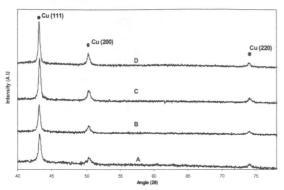

Figure 3: XRD patterns of Cu powders synthesized at different concentrations of hydrazine. A: 0.7M N_2H_4; B: 0.24M N_2H_4: C: 0.12M N_2H_4 and D: 0.06M N_2H_4. The Cu concentration was 0.0016M.

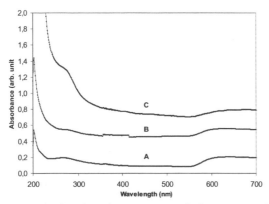

Figure 4: UV-vis absorption spectra of Cu nanoparticles suspended in ethanol. Copper was sinthetized from 0.016M Cu_2SO_4 and using 0.12M (A), 030M (B) and 0.65M (C) hydrazine solutions.

TEM and SEM images of figure 5 correspond to the Cu particles synthesized using 0.064M Cu and N_2H_4 0.013 M . after 30 hours and 24 hours of reaction, respectively. Although strongly aggregated, the images evidenced the nanocrystalline nature of the particles.

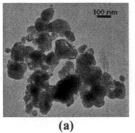 

(a)   **(b)**

Figure 3: TEM (a) and SEM images of Cu nanoparticles synthesized using 0.013M N_2H_4 and 0.064M $CuSO_4$.

3.2   Cu(II) Reduction in Ethylene glycol

XRD patterns in figure 6 correspond to the solids produced in ethylene glycol at different NaOH/Cu mole ratios, R'. The initial Cu(II) concentration was 0.0014M. The presence of NaOH accelerated the formation of elemental Cu; the reaction time was shortened from 1 hour to 30 s when the 'R' values varied from 0, i.e. no NaOH, to 50. The corresponding average crystallite size decreased from 39nm to 21nm. The excess of NaOH would have enhanced the reducing conditions in the polyol media and/or accelerated the dissolution of any intermediate phase under strongly alkaline conditions. Under these conditions the enhancement of the nucleation rate, and hence the reduction in crystal size, could be expected.

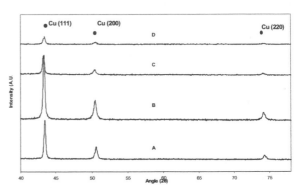

Figure 6: XRD patterns of Cu particles synthesized in ethylene glycol at NaOH/Cu mole ratio, 'R', of 0 (B); 1 (C), 10 (D) and 50 (E). The initial Cu concentration was 0.0014M.

Figure 7 shows the UV-vis spectra of the Cu nanoparticles synthesized at different 'R' values and suspended in ethanol.

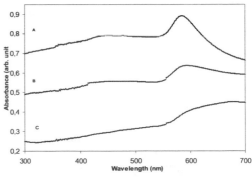

Figure 7: UV/VIS absortion spectrum of copper nanoparticles synthetized in ethylene glycol and suspended in ethanol. The 'R'values were 0 (A), 1 (B) and 50 (C). Initial Cu concentration was 0.014M.

The spectra displayed a SP resonance band at around 590nm, which is also typical of Cu particles. Mott et al. observed the SP band at ~600nm [10] when ~100nm Cu nanoparticles were suspended in hexane. As known, the exact position of the SP may depend on several factors including particle size, shape, type of solvent and capping agent, if any. Furthermore, the observed narrowing of the SP band in the solid produced at R= 50 suggests a less size polydispersity.

3.3 Nanofluid Stability

Copper nanoparticles synthesized in aqueous (hydrazine reduction) and non-aqueous (polyol reduction) media were suspended in ethylene glycol. Copper nanoparticles reduced by hydrazine were treated by 0.001M HCl solution in ethanol prior to their sonication. The volumetric % of the nanocrystals in the fluid was estimated at 0.05 %. In order to promote the nanofluid stability against sedimentation, the suspension was intensively ultrasonicated for 7 hours. The final suspension exhibited an orange tone and showed no evidence of settling of particles even after 96 hours. The UV-vis spectra were collected during 25 minutes at 5 minutes interval to determine the stability of the suspension.

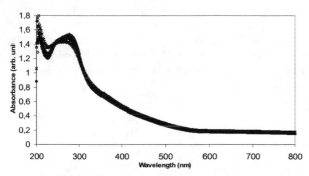

Figure 8: UV/VIS absortion spectra of Cu nanoparticles synthesized and suspended in ethylene glycol. The initial Cu concentration and 'R' were 0.0014M and 10, respectively.

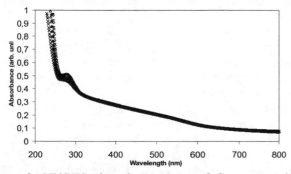

Figure 9: UV/VIS absortion spectra of Cu nanoparticles synthesized in 0.70M of hydrazine and suspended in ethylene glycol. The initial Cu concentration was 0.0016M.

As figures 8 and 9 show, the SP resonance band was centered on 280nm and no remarkable shifting was observed in the measurement period. This fact may suggest the stability of the suspension at least during the 25 minutes-period of data collection. Nevertheless, the position of the SP band was blue-shifted with respect to the SP position shown in figure 7. On the consideration that no dissolution of Cu nanoparticles could have taken place in

ethylene glycol media, this blue shift can be a consequence of the reduction in the nanoparticles size (fragmentation) by intensive ultrasonication. The position of the SP resonance band is in agreement with other reports where the size of Cu nanoparticles was in the 2-4nm range [11,12].

4. CONCLUDING REMARKS

We have successfully synthesized copper nanoparticles in aqueous and non-aqueous phase. The rate of the reduction reaction was strongly dependent on the concentration of hydrazine (aqueous route) and the NaOH/Cu mole ratio, 'R', in ethylene glycol media. Suitable hydrazine concentration and 'R' values were conducive to a dramatic shortening of the reaction time and, consequently, to a remarkable decrease in crystal size of Cu nanoparticles. Long-term stable nanofluids were prepared by intensive ultrasonication of Cu nanoparticles in ethylene glycol. Ultrasonication may have fragmented Cu crystals as suggested by blue-shifted SP resonance band.

ACKNOWLEDGEMENTS

This material is based upon work supported by the National Science Foundation under Grant No. 0351449.

REFERENCES

1. J. A. Eastman, U. S. Choi, S. Li, G. Soyez, L. J. Thompson, R. J. DiMelfi. *Mat. Sci. Forum*, vols. 312 – 314, 629 – 634, 1999.
2. A.-R.A. Khaled, K. Vafai. Intern. J. *Heat Mass Transf.*, 48, 2172–2185, 2005.
3. S. A. Putnam. D. G. Cahill, P. V. Braun. *J. Appl. Phys.* 99, 2006.
4. Eden Mamut. Romanian J. Phys., Volume 51, No 1-2, 5-12, 2006.
5. Putnam, S. A., Cahill, D. G., Braun, P. V., Ge, Z. and Shimmin, R. G., *J. Appl. Phys.*, 2006, **99**, 084308.
6. M-S. Liu, M. C-C. Lin, C.Y. Tsay, C-C. Wang. *Internat. J. Heat Mass Transfer*, 49, 3028-3033, 2006.
7. C. Wu, B. P. Mosher, T. Zeng. *J. Nanoparticle Res.*, 2006.
8. M. Aslam, G. Gopakumar, T.L Shoba, et al. *J. Coll. Interface Sci.* 255 (2002), p. 79 – 90
9. J.A. Eastman, S.R. Phillpot, S.U.S. Choi, P. Keblinski. *Annu. Rev. Mater. Res.* 2004. 34:219–46
10. S. P. Jang, U. S. Choi. *Appl. Phys. Lett.*, 2004. 84, p 21
11. D. Mott, J. Galkowski, L. Wang, J. Luo, C. Zhong. *Langmuir*, 23 (10), 5740 -5745, 2007
12. L. Balogh and D. A. Tomalia. *J. Am. Chem. Soc.* 120 (1998), p. 7355.

Biofunctional core-shell nanoparticle deposition for biochip creation by printing processes

Jolafin Plankalayil[1], Sandra Genov[2], Kirsten Borchers[1], Andrzej Grzesiak[3], Thomas Hirth[1,2], Achim Weber[1], Günter E. M. Tovar[1,2]

[1] Fraunhofer-Institute for Interfacial Engineering and Biotechnology IGB
[2] Institute for Interfacial Engineering, University of Stuttgart
[3] Fraunhofer-Institute for Manufacturing Engineering and Automation IPA
Nobelstr. 12, 70569 Stuttgart, Germany, guenter.tovar@igb.fraunhofer.de

ABSTRACT

In order to set up an automated production-line for nanoparticle-based bio-chips, a piezo driven ink-jet printing process was developed which is biocompatible and allows for high flexibility as well as accuracy in nanoparticle deposition.

Highly dense and homogenous 3-dimensional amino-modified core-shell nanoparticle surfaces were achieved via inkjet printing with 2-methyl-1-pentanol based ink. Nanoparticles dispersed in a poly ethylene glycol (PEG) based ink with adjusted physical properties for jetting such as viscosity and surface tension were printed onto glass substrates which resulted in loosely packed nanoparticle monolayer coatings after washing procedures to remove low volatile PEG 200. 23-mer oligonucleotide spotted onto inkjet printed amino-modified particle layers showed homogenous and sharply defined DNA spots. Proteins (streptavidin) were suspended in glycerol-based ink and printed on epoxy-silanized glass and kept their bioactivity after this printing procedure.

Keywords: inkjet printing, bio-ink, piezo-driven, micro-array, ink formulation

1 INTRODUCTION

Nanostructured core-shell particles with tailor-made affinity surfaces are utilized to generate micro-structured affinity surfaces by micro-spotting the particles to form densely packed amorphous nanoparticle layers [1]. These layers provide an enormous surface enlargement and thus attract notice to chip-based analytical or diagnostic technologies [2]. Core-shell nanospheres, constituted from a silica core and an organic shell, can be used to couple specific capture-proteins or DNA-probes by application of state-of-the-art bioconjugate chemistry [3].

Further to a spotting of biofunctional nanoparticles in a micro arrayed format, the generation of a covering nanoparticle-coat in a user-defined shape [4] allows for flexible design of microstructures and for fabrication of large bioconjugative interfaces on various substrates, eg. glass and polymer.

In view of the advantages that have been experienced in the past decade using biochip-based approaches to screening type experiments, biochip-production should be made a cheap, flexible and reproducible process [5]. Ink-jet printing is a relatively straightforward fabrication process in order to assemble reproducible nanoparticle-coatings.

This process enables economical coating of substrates by integrating the molecular functionalization of substrates in a fully automated process. In contrast to classical microarray printing, many formats of 2-D drawings can be rasterized into X- and Y-coordinates to deposit materials in a corresponding printed pattern. Inkjet printing technology has recently been used to fabricate electronic, medical, optical and polymeric devices [6, 7].

Not only biomolecules but also viable mammalian cells can be printed for different applications such as tissue engineering [8]. Ink-jet printers can dispense fluid droplets with volumes in the picoliter to microliter range. In contrast to thermal ink-jet printing piezoelectric ink-jet printing is a thermally constant process that can be carried out at room temperature. The chemical properties of inks determine their jettability: surface tension and viscosity are two primary material properties that determine the success of a printing process.

To ensure simple adaptability to current bio-chip technology, we aim to develop ink formulations containing functional components, which can be used to feed ink-jet printers to generate automated production lines for nanoparticle-based bio-chips.

We will present results concerning activation of the substrates, preparation of PEG-based nanoparticle inks and 2-methyl-1-pentanol based inks. Viscosity and surface tensions are physical parameters affecting the characteristics of printability. An important criteria for not only wetting the printed surface properly but also for the formation of homogenous particle layers, is the adjustment of the surface energy of the ink to the needs of the substrates.

Ink formulation considering the maintenance of the biological function during the printing process was additionally observed both in form of micro-spotting Cy5 labeled oligonucleotide on the jetted organo functionalized core-shell particle layers as well as by jetting streptavidin and analyzing the bioactivity of the micro-structures in a fluorescent assay followed by an optical scan.

NSTI-Nanotech 2008, www.nsti.org, ISBN 978-1-4200-8503-7 Vol. 1

2 MATERIAL AND METHODS

Chemicals

The NH_3 solution (p.a. 25 wt. %), H_2O_2 (30%), ethanol (HPLC grade), poly-(diallyldimethylammoniumchloride) (PDADMAC, MW ~100000, 20 wt. % in dH_2O) and poly-(sodium 4-styrenesulfonate) (SPS, MW~70000), 2-methy-l-1 pentanol (99%) were purchased from SAF (Taufkirchen, Germany). tetraethoxysilane and 3-aminopropyl-2-triethoxysilane were purchased from ABCR (Karlsruhe, Germany). Polyethylenglycol 200 (PEG 200) was purchased from Th. Geyer.

Functional nanoparticles

Monodisperse suspensions of spherical core-shell nanoparticles were generated via the method of Stoeber [9]. For detailed information concerning synthesis of the particles and their surface modification, see elsewhere [3]. Briefly: nanoparticulate silica spheres were synthesized from a mixture of 13.7 ml NH_3 solution (25 wt. %) 20 ml H_2O, and 5 g tetraethoxysilane in 400 ml ethanol in a sol-gel reaction. The average diameter of the particles used in this work was 117 ± 16 nm. The organic particle shell was prepared by employing 3-aminopropyltriethoxysilane, thus introducing amino-functions to the particle surface.

Pre-treatment of the substrates

Cleaning microscope glass slides (Menzel GmbH & Co KG, purchased via SAF, Taufkirchen, Germany) was carried out in 2% HELLMANEX solution (Hellma, Müllheim, Germany). The glass-surface was subsequently hydroxylated by incubation at 3:1 (v/v) NH_3 solution (25wt. %) and H_2O_2 (30%) for 40 min at 70°C and rinsed in ultrapure water. The surface activation was achieved by coating of glass surfaces with polyelectrolytes (PE) using layer-by-layer technique by Decher [10]. The coating of the substrate cyclic olefin copolymer (COC) (TOPAS 8007) was processed the same way.

Microspotting and Fluorescent read-out

DNA-micro-arrays were generated by using a GMS 417 pin-ring micro-arrayer. Fluorescence was detected using an Arrayworksx Biochip Reader (Applied Precision, Issaquah, WA, USA).

Organo-functional nanoparticle inks

The nanoparticles with a mean diameter of 117 ± 16 nm were dispersed in two different solvents. To obtain alcohol based inks the previously prepared nanoparticle suspensions were directly dispersed into 2-methyl-1-pentanol. Concerning polyethylenglycol (PEG) based ink the nanoparticle suspensions were dispersed in a premixed composition of poly ethylene glycol 200 (PEG 200) with solvent. The mixing-ratio of premixed solutions was adjusted in according to the jettability of ink-jet printer Dimatix Materials Printer 2800 (Dimatix Inc. Santa Clara, USA). Dimatix Materials Printer 2800 requires viscosities of $10 \leq \eta \leq 12$ mPa*s and surface tensions of $28 \leq \sigma \leq 33$ mN*m[1]. Surface tensions were determined by using a SITA Dyno Tester tensiometer (SITA Messtechnik, Dresden, Germany) and viscosity by using a Brookfield rheometer LV DV – III (Brookfield E.L.V., Middleboro, Mass., USA) at contant temperature of 25°C. Printing was proceeded at room temperature with 40°C plate temperature The formulated ink was filled in a printing cartridge after filtering through 450 nm Filter.

3 RESULTS

Considering the needed ink properties from piezo-driven Dimatix Materials Printer, several mixing-ratios were established. To obtain a basic ink in which nanoparticle suspension can be dispersed, PEG 200 was mixed with the solvents. Viscosity was adjusted to approximately 12 mPa s after this. Table 1 shows the physical properties of PEG 200 basic ink. First printing experiments were done with PEG 200 with Solvent 1 to check how stable the process is.

Table 1: Ink properties such as viscosity and surface tension for selected basic ink measured at 25°C to achieve optimum printing performance.

Basic ink	Viscosity [mPa*s]	Surface tension [mN*m⁻¹]
PEG 200 + Solvent 1	12.16	31.95
PEG 200 + Solvent 2	11.93	29.6
2-Methyl-1-Pentanol	5.49	24.04
Glycerol + Additives	11.43	36.7

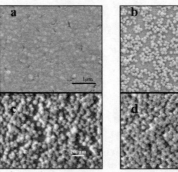

Figure 1 Scanning electron micrographs (SEM) of Amino-modified core-shell nanoparticle layers printed by a piezo-driven ink-jet printer. a) Printed particle monolayer covered by PEG, b) Loosely packed particle coating after washing step in ultrapure water to remove water soluble PEG, c) 3-fold particle layer printed onto pre-coated COC with 1-methyl-2-pentanol based ink, 20 μm drop-spacing led to loosely packed coating and d) 15 μm drop-spacing led to a highly densely packed 3-D particle layer.

Due to pre-processed glass substrates with a two-fold PDADMAC-SPS layer as described above, the amino-modified nanoparticle arranged after printing as a monolayer. PEG 200 is low-volatile and the deposited ink layer did not evaporate. Post processing, by washing the glass substrate in ultrapure water, removed PEG 200 but resulted in loosely packed monolayer particle coating as shown in Figures 1 a and b. Ultrasonication for 30 min prevented aggregation of particles at the nozzles

The printability of PEG 200 based ink on COC depends on its surface energy. With an 8-fold coating with PDADMAC-SPS which reduced the contact angle from 90° to approximately 60°, the programmed pattern could be printed (data not shown).

An alcohol-based ink is able to provide appropriate properties for printing without adding surfactants to reduce surface tension. Solvent-based nanoparticles were dispersed directly into 2-methyl-1-pentanol and physical properties were determined. Not only the printing process was stable without particle aggregation for at least 3 hours but also 3-dimensional patterning could be achieved. In addition to adjusting ink properties for an optimal printing process, printing parameters such as drop-spacing resulted in different particle coatings as shown in Figure 1 c and d. Drop-spacing is the distance in X and Y, center to center, of the drops that the in-jet printer will deposit to create the pattern. This parameter is most useful for altering the amount of jetted ink per unit area (fill density). The amount of jetted ink with 15 µm drop-spacing higher than that with 20 µm which resulted in loosely packed particle coatings with 20 µm.

Biofunctionality of the printed particles was tested by micro-spotting Cy5 labeled oligonucleotides (23-mer) followed by fluorescent scans. The oligonucleotides were spotted in 10 µM, 5 µM, 2.5 µM and 1.25 µM dilutions onto the COC surface which was coated with 4-fold PE and printed 3-fold with nanoparticles and resulted in homogenous and sharply defined DNA Spots (Figure 2).

10 µM   5 µM   2..5 µM 1.25 µM

Figure 2 Bioactivity proven by fluorescent micrographs. Left hand: Cy5 labeled Oligonucleotide (23-mer) spotted onto ink-jet printed amino-modified core shell nanoparticle layer. Right hand: Streptavidin suspended in glycerol-based ink were printed onto epoxy-silanized glass substrate with a piezo-based ink-jet printer.

The protein streptavidin was dissolved in a glycerol-based ink and printed onto epoxy-silanized glass substrates. The microstructered patterns bioactivity was shown by fluorescence micrographs (Figure 2) thus showing that the bioactivity was maintained during the printing process.

4   CONCLUSIONS

Biochip fabrication needs processes for micro-structuring the surface which can easily be automated. This study shows that the process of nanoparticle printing to obtain bioactive 3-D surfaces which was previously established by contact printing techniques [1, 2, 4] can be successfully transferred to corresponding ink-jet printing processes. Creating homogenous and controllably packed particle surfaces is important for DNA micro-array technology as well as for protein micro-array technology to provide surface enlargement and hence maximize the observable flourescence signal intensities after biomolecular assay interactions [1]. The 2-methyl-1-pentanol-based ink showed optimal printing performance for bioactive core-shell nanoparticles. The alcohol enabled for a controlled printing process and evaporated excellently from the slightly heated substrates. The alcohol-based ink-jet printing process led to homogenous particle layers. For highest resolution of the printed patterns, surface energy of inks and substrates must be harmonized. The method shown here enables for micro-structure fabrication eg. for multiplex DNA hybridization experiments or protein-protein interaction screening.

Acknowledgements
The authors thank Monika Riedl for the SEM images, Marion Herz for nanoparticle synthesis and Markus Knaupp for experimental assistance (all Fraunhofer IGB, Stuttgart), Stefan Güttler (Fraunhofer IPA, Stuttgart) for help with the viscosimetry, and Ingo Wirth and Mathias Müller (both Fraunhofer IFAM, Bremen) for helpful discussions. This work was supported by the Fraunhofer-Gesellschaft under Grant No. MAVO 815098.

REFERENCES

[1] Borchers K., Weber A., Brunner H., Tovar G.E.M., *Anal. Bioanal. Chem.*, 383, 738, 2005
[2] Tovar G.E.M., Weber A., *Ency. Nanosci. Nanotechnol.*, 1, 277, 2004
[3] Schiestel T., Brunner H., Tovar G.E.M., *J. Nanosci. Nanotechnol.*, 4, 504, 2004
[4] Weber A., Knecht S., Brunner H., Tovar G.E.M., *Eng. Life Sci.*, 4, 93, 2004
[5] Sumerel J. et al., *Biotechnol. J.*, 1, 976, 2006
[6] Song J.H., Nur H.M., *Mater. Proc. Technol.*,155–156, 1286, 2004
[7] Lemmo A.V., Rose D.J., Tisone T.C., *Curr. Opin. Biotechnol.*, 9, 615, 1998
[8] Xu T. et al., *J.Biomater.*, 26, 93, 2005
[9] Stöber W., Fink A., *J Colloid Interf. Sci.*, 8, 62, 1968,
[10] Decher G., *Science,* 277, 1232, 1997

Surface Nanostructuration: from Coatings to MEMS Fabrication

R. Pugin, N. Blondiaux, A. M. Popa, E. Scolan, A. Hoogerwerf, T. Overstolz, M. Liley, M. Giazzon,
G. Weder, H. Heinzelmann

Swiss Center for Eletronics and Microtechnology (CSEM SA)
Nanotechnology and Life Sciences Division
Jaquet-Droz 1, Case Postale CH-2002 Neuchâtel, Switzerland, raphael.pugin@csem.ch

ABSTRACT

We have developed a number of complementary methods using polymer self-assembly, sol-gel texturation and reactive ion etching techniques on a mix-and-match basis. The resulting nanostructured surfaces can have striking properties: super-hydrophobic, anti-fogging and anti-reflective to name but a few. Moreover, polymeric surface nanostructures are also very suitable as nanoscale etch masks for the transfer of the structure into the underlying material through reactive ion etching. This process leads to more durable nanostructures with aspect ratios of 10:1 in hard materials such as silicon, silicon nitride or quartz. Potential applications of these techniques include the fabrication of nanoporous membranes as well as the design of nanostructured surfaces for the analysis of the influence of nanotopographies on cell adhesion, orientation and growth properties.

Keywords: nanostructures, nanofabrication, self-assembly, polymer, membranes

INTRODUCTION

The systematic creation of nanoengineered surfaces with controlled nanoscale topographies (containing at least one dimensional feature below 100nm) is seen as an essential step for obtaining surfaces with radically new and unique size-dependent chemical or physical properties.

The continued development of lithography technologies allows the production of surface patterns with features sizes well below 100 nm. However, the increasing cost and complexity of standard top-down nanolithography techniques (e.g. deep UV, e-beam, Focused Ion beam) makes difficult their integration for mass production in an industrial environment. Only few examples, where the expensive nanofabrication processes are compensated by replication techniques (e.g. hot embossing, nanoimprint lithography) could find industrial applications. A prominent replication process that allows one to produce highly controlled surface topographies with structure sizes below one micrometer is the production of compact discs (CDs). The next generation of ultra high-density optical storage, Blu-Ray Discs is currently reaching patterning in the sub-micron region with track pitch and pit length of 320 and 140 nm respectively, but again the master fabrication is expensive and time consuming.

Other ways of creating surface structures in the ~10-100 nm range may represent alternatives if they offer advantages in reduced production cost, smaller feature sizes, or more flexibility regarding the morphology and size of nanometric structures. Systems that show ordering and pattern formation through self-assembly may offer some of these advantages. Among these nanopatterning approaches, surface structuring using polymer, block-copolymer and sol-gel self-assembly are currently attracting great attention.

POLYMER SELF-ASSEMBLY

Polymer demixing uses the phase separation properties of polymers. The resulting structures are generally larger than the length of the polymer chain (micrometre, sub-micrometre range) and have peculiar morphologies (worm-like or "dots") depending on the processing conditions. They are stochastic with respect to shape and order but present a well-defined length-scale. Various parameters can be used to tune the size and shape of the structures (Figure 1). The length-scale of the structures can easily be controlled by modifying the solution-concentration or of the rotational speed during spin coating.[1] To modify the morphology of the features, the ratio between both polymers can be adjusted.[2] For symmetrical blends, the system phase-separates via spinodal decomposition, leading to the typical worm-like structures. In the case of very asymmetrical blends, the phase separation follows a nucleation and growth process, leading to the "dots in a matrix" pattern.

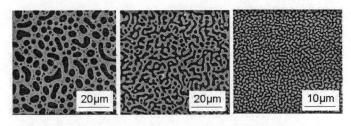

Figure 1: AFM images of phase-separated PS/PMMA thin films for different solution-concentrations (CSEM).

In order to create even smaller structures, we also investigated self-organization of block-copolymers (BC). BC is a special class of polymer with two or more polymer chains (or blocks) chemically bound to each other covalently. When deposited on a surface phase separation occurs intermolecularly and the two blocks can only separate to a distance compatible with the size of the chains. This constraint leads BC to separate into periodic microphases, i.e. into domains that are each rich in one of the constituent blocks. The size of the domains is on the order of the size of the macromolecules, i.e. ~10-100 nm. The properties of the constituent polymers, the number of monomeric units in each block, along with the relative proportion of the polymers within the BC determine the resulting equilibrium morphologies (gyroid, cylindrical or spherical domains). The characteristics of the pattern are significantly influenced by parameters such as the film thickness, solvent-selectivity and substrate influence, in relation to the BC molecular characteristics. The primary means of realizing this has been through (i) changing the relative lengths of the blocks, and (ii) the total BC molecular weight.[3-4] Tunability of micellar nanostructures obtained by dissolving BC in selective solvents has been shown for their size and spacing in a systematic manner (Figure 2).[5]

Figure 2: AFM images of arrays of BC micelles spin-coated onto silicon substrates. The spacing of the features can be tuned by using solutions of micelles of different concentrations. All the images are of 50 nm z scale and 1 μm x 1 μm scan size (CSEM).

This very flexible process of polymer self-assembly was already used for various applications where surface structuring plays a major role. Wahleim *et al.* for instance showed that it can be used to fabricate antireflective coatings.[6] Among potential applications of BC surface nanopatterns already identified are isoporous membranes.[7] However, a direct use of polymer nanostructured thin films is not always possible since the structures produced are easily damaged through mechanical wear and present often small aspect ratios. To overcome this problem, we used nanostructured polymer thin films as etch-masks for the transfer of the polymeric structures into the underlying substrate (hard material such as Si or quartz) using standard etching processes (Figure 3). This leads to more durable nanostructures in the form of arrays of evenly spaced pillars with aspect ratios of up to 1:10 in hard materials of technical importance such as silicon, silicon nitride or quartz.

Figure 3 : SEM images of polymer/BC structures transferred into Si by DRIE (CSEM).

The main advantages of this approach are the fabrication of durable surface structures with lateral sizes controlled via the polymer self-assembly, while the depth of the structures is defined by the etching process (from few tens of nm to few μm). Three potential applications of such nanostructured surfaces have been identified.

First, subsequent surface functionalisation with perfluoro silane leads to a substantial reduction in surface wettability. The water contact angle increases from 111° on flat surfaces to 150° in the case of the pillar substrates thus demonstrating the surface roughness induced enhancement in hydrophobicity (Figure 5).

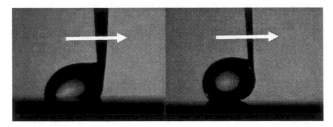

Figure 4 : Photographs of a 10μL water droplet moved on a flat (left) and nanostructured (right) silicon surface silanized with a perfluorosilane. The arrows indicate the direction of the movement (CSEM).

Based on previous work [8] a second application currently under investigation at CSEM is the control of biological cell growth on structured surfaces. For instance, Gallagher et al. have shown that cell growth can be completely inhibited by nanostructuring the surface. [9]

Third, CSEM has recently developed a silicon based freestanding nanoporous membrane in which the pore diameter can be tuned between 10 and 20 nm. The membrane is 60 nm thick and has been fabricated by combining standard microfabrication processes and block-copolymer lithography (a nanostructured block-copolymer thin film is used as etch mask for the transfer of the self-assembled structure into the underlying material by deep reactive ion etching). With this technology, the formation of pores with aspect ratios up to 1:5 could be demonstrated (Figure 5). [10] The nanoporous membranes are fabricated with a support structure in order to facilitate their manipulation and integration in macroscopic devices.

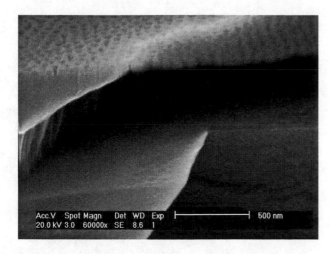

Figure 5 : SEM image of reinforced nanoporous membranes (CSEM).

Applications including ultrafiltration and prevention of particle emission are currently evaluated. Extension of this fabrication processes to the development of nanostructured metal thin films for plasmonics applications is currently under investigations. Present technical developments are concerning the fabrication of thicker suspended membranes with higher aspect ratio pores, narrow pore size distribution and improved mechanical robustness.

SOL-GEL NANOSTRUCTURATION

Complementary to polymer self-assembly, sol-gel processes are very versatile and well-suited for the formation of layers with controlled homogeneity, thickness, porosity and associated surface structures. Thin films with highly ordered structures can even be obtained by templating the gelation process by selected surfactants or block-copolymers. The control of surface properties such as the morphology and crystallinity of sol-gel derived thin films is of great importance for practical applications such as antireflection, photocatalysis, abrasion resistance, water-repellency, electric conductivity, and gas barrier properties for various substrates. A simple scalable method has been developed for the preparation of controlled 'grass-like' nanostructured surfaces with promising antireflective and wettability properties.

The preparation of precipitated "grass-like" thin films with a roughness of around 100 nm has been reported[11] for Al_2O_3-based coatings on a large variety of substrates (glass, ceramics, metals, plastics). Our approach uses the design of sol-gel surface nanostructures tailored to give outstanding properties for a specific application. The first experimental step consists of synthesizing an alcoholic sol of alumina nanoparticles by the controlled hydrolysis of alkoxide precursors stabilized by strong chelating agents. The sol can then be deposited on glass, silicon or polymeric substrates, typically using spin- or dip-coating processes. After annealing, transparent and crack-free alumina thin films are

obtained. Subsequent immersion of the dried alumina films in hot water, at a temperature between 50 to 100°C, leads to the formation of "grass-like" surface nanostructures that are caused by successive dissolution-precipitation steps of the initial alumina-based film (Figure 6). The depth of the nanostructures can be easily controlled by varying the immersion time.

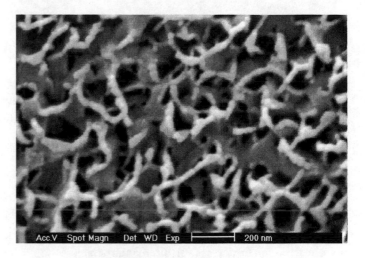

Figure 6: Scanning Electron Microscopy (SEM) picture of an alumina-based coating after hot water treatment (CSEM).

The optical properties of the nanostructured sol-gel layers have been intensively investigated. For most structures, the reflection spectrum at normal incidence shows a strong reduction of intensity for wavelengths in the range 400-1200 nm. The lowest reflectivity is 5.5% for a glass slide coated on one side. Moreover no significant variation in reflectance was observed on varying the incidence angle: a slightly increase from 5.5% at normal incidence up to 7% at 45° was obtained. In addition, colorimetric characterizations of both transmitted and reflected light demonstrate a very color-neutral behavior on changes of incidence angles when compared with bare glass sample.

While intrinsically super-hydrophilic (water contact angle smaller than 5°), such highly nanotextured metal oxides surfaces become super-hydrophobic and self-cleaning upon silanisation with perfluorinated silane (water contact angle 160°, hysteris <10°, roll off angle <10°; as evidenced in Figure 7).

Figure 7: Picture of water droplet on a superhydrophobic nanostructured alumina surface (CSEM).

CONCLUSION

The self-assembly of polymer systems and controlled sol-gel texturation processes are promising strategies to create structures on the nanometer scale. As we gain a deeper understanding of the structruring processes, we will achieve better control of feature size/morphology and resulting properties. The simplicity of this very versatile toolbox, along with its potential low-cost for generating small features over large areas, makes it very attractive for applications in various different fields.

ACKNOWLEDGMENT

The work reviewed in this paper was partly funded by the State Secretariat for Education and Research and by European Union research framework programme FP6. The authors thank them for their financial support.

REFERENCES

[1] DalnokiVeress K., Forrest J.A., Stevens J.R., Dutcher J.R., *Physica A*, 239, 87, **1997**

[2] DalnokiVeress K., Forrest J. A., Stevens J.R., Dutcher J.R., *Journal of Polymer Science Part B-Polymer Physics*, 34, 3017, **1996**

[3] Wu C., Gao J., *Macromolecules*, 33, 645, **2000**

[4] Milchev, A. Bhattacharya, Binder K., *Macromolecules*, 34, 1881, **2001**

[5] Krishnamoorthy S., Pugin R., Liley M., Heinzelmann H., Brugger J., Hinderling C., *Adv. Func. Mater.*, 16, 1469, **2006**

[6] Walheim S., Schaffer E., Mlynek J., Steiner U., *Science*, 283, 520, **1999**

[7] Peinemann, K.V., Abetz V., Simon, P.F.W. *Science*, 6, 992, **2007**

[8] Dalby M.J., Giannaras D., Riehle M.O., Gadegaard N., Affrossman S., Curtis A.S.G., *Biomaterials*, 25, 77, **2004**

[9] Gallagher et al, *IEEE Trans Nanobio*, Vol1, **2002**

[10] Hoogerwerf A., Hinderling C., Krishnamoorty S., Overstolz T., *CSEM Scientific reports* , **2006**

[11] Tadanaga K., Katata N., Minami T., *J. Am. Ceram. Soc.*, 80, 1040, **1997**

Nanostructured thin layers of vanadium oxides doped with cobalt, prepared by pulsed laser ablation: structure, chemistry, morphology, and magnetism

C.M. Teodorescu [*1], G. Socol[**], C. Negrila[*], D. Luca[***], D. Macovei[*]

[*] National Institute of Materials Physics, Atomistilor 105b, 077125 Magurele, Romania, teodorescu@infim.ro
[**] National Institute of Lasers, Plasma and Radiation Physics, Atomistilor 409, 077125 Magurele, Romania, gabriel.socol@inflpr.ro
[***] "Alexandru Iona Cuza" University, Carol I Blvd. No. 11, 700506 Iasi, Romania, dumitru.luca@uaic.ro

ABSTRACT

Cobalt-doped vanadium oxide thin layers prepared by pulsed laser ablation are investigated by (i) X-ray photoelectron spectroscopy, (ii) the local atomic order by X-ray absorption near-edge structure (XANES), (iii) the morphology of the films was investigated by atomic force microscopy (AFM), and (iv) magnetic properties were quantified by magneto-optical Kerr effect (MOKE). In most cases, the chemical composition of the host matrix was found to be the vanadium (5+) oxide V_2O_5 at the sample surface and lower ionization states (+2 and +3) in the bulk. Co ions are found either in high ionization state Co^{5+} (for samples synthesized in high vacuum condition, denoted by VO1), or with Co in lower ionization states Co^{4+} (for samples synthesized in a mixture of argon and oxygen atmosphere, denoted by VO2). Consistent information was obtained from chemical shifts from individual core level scans in XPS, compared with existent data in litterature, and the amplitude of the pre-edge peak in XANES, which is a sign of quadrupole $1s \to 3d$ dipole-forbidden transition and whose amplitude is proportional to the number of $3d$ vacancies per atom. AFM revealed big particles with sizes > 100 μm for VO1 samples, whereas smaller nanoparticles with sizes ranging between 20 and 30 μm were observed for VO2 samples. VO1 samples presented very high coercive fields with a relatively low saturation magnetisation at room temperature, whereas VO2 samples presented double-loop hysteresis curves, indicating the coexistence of exchange bias between two kinds of magnetic moieties with strong anisotropy.

Keywords: vanadium oxide; pulsed laser deposition; XPS; XAFS; magnetic properties

1 INTRODUCTION

The synthesis of Diluted Magnetic Semiconductors (DMS) is amongst the outstanding results of the last two decades in materials physics [1,2]. The last years (2003-2007), diluted magnetic semiconductors were succesfully synthesized based on (nonmagnetic) transition metal oxides (ZnO, TiO_2) doped with magnetic ions (Cr-Ni). Recently, also our group succeded PLD synthesis of cobalt doped ZnO which has shown definite room temperature ferromagnetism (manuscript in preparation).

In DMS, magnetic ordering is produced between isolated magnetic ions via the doube exchange or RKKY interaction [3], where the magnetic ions interaction is mediated by the charge carriers from the matrix. Hence, possibility of control of magnetic ordering via the density of the charge carriers is provided by these systems. A wide range of applications may be foreseen starting with this phenomenon, such as induced ferromagnetism by light irradiation [1], DMS based spintronic devices [2], ferromagnetism control *via* applied electric fields, particularly the possibility of domain wall displacement by varying the charge density. A novel application was demonstrated in the possibility of ferromagnetic ordering *via* molecular adsorption at surfaces [4], by using carriers donated by molecular species for mediating ferromagnetism by double exchange.

This Contribution deals with investigating the possibility of synthesis of materials where magnetic ions are isolated in matrices which exhibit Mott-Hubbard metal-insulator transition (MIT) [5]. The chosen host materials are in this study vanadium oxides, where MIT can be achieved at temperatures interesting for applications, in the range -50 ºC up to +60 ºC [6,7]. Indeed, first explorations revealed interesting magnetic properties, such as uniaxial magnetic anisotropy or persistent orbital currents on materials starting with doped vanadium oxide [8]. The practical realization of such materials would open controlling possibilities of the ferromagnetic ordering occurence through the strong variation of charge density following the Mott-Hubbard transition from the insulating to the metal state.

2 EXPERIMENTAL

The materials were synthesized by pulsed laser deposition (PLD) in NILPRP Magurele. The targets were synthesized by sintering a mixture of vanadium oxide (VO_2) and cobalt oxide (CoO), in various proportions.

---

[1] corresponding author

Pulsed laser deposition was performed by using an excimer laser (KrF) with fluence of 9 J/cm², in controlled atmosphere: (i) in vacuum (10^{-5} mbar = 10^{-3} Pa); (ii) in oxygen atmosphere (10^{-3} mbar = 10^{-1} Pa). Depositions were realized on single crystal Si(001) and quartz SiO_2(001). Substrates were heated during deposition, in order to improve crystalinity and to avoid the formation of macroscopic droplets.

X-ray photoelectron spectroscopy (XPS) was performed by using a VG-ESCA MK II installation of NIMP Magurele with Al K_α radiation and a hemispherical electron analyzer with 100 mm radius, operating at pass energy of 100 eV for overview spectra, and at 50 eV for individual core level spectra. Room temperature magneto-optical Kerr effect (MOKE) measurements were performed in NIMP Magurele by using a MOKE system with He-Ne laser radiation and a pizoceramic modulator. X-ray absorption fine structure (XAFS: EXAFS and XANES) measurements were performed at the Doris storage ring facility in Hasylab, Hamburg, Germany (A1 beamline), by using a Si(111) double crystal monochromator. The measurements on Co-doped vanadium oxide thin layers were performed in fluorescence mode by using a Si:Li detector and selecting the channels corresponding to the fluorescent radiation of interest (Co or V K_α). For consistent data interpretation, standards of metal Co and of Co(II)O were measured in the same run, in transmission mode on a 5 μm Co foil for the metal and of a pressed pellet of CoO powder mixed with cellulose. Atomic force microscopy (AFM) was performed in the University of Iasi by using a home-made equipment.

3 RESULTS AND DISCUSSIONS

Although several investigations were performed, in the following we discuss mainly on the differences presented by samples prepared in high vacuum conditions (10^{-5} mbar) and in oxygen atmosphere (10^{-3} mbar).

3.1. X-ray photoelectron spectroscopy (XPS)

Figure 1 presents overview XPS spectra on the two samples of interest (VO1, synthesized at 10^{-5} mbar and VO2, prepared at 10^{-5} mbar). One notices the presence of V, O and Co photoelectron and Auger electron signals, and of carbon contamination from the inherent contamination layer. A first observation is that the carbon contamination visibly decreases when the synthesis is performed in oxygen atmosphere; at the same time, in this case, the Co 2p, V 2p, and O 1s signals show noticeable increases.

One observes also the presence of Cu and, for the VO1 sample, of Zn - but this is connected to the copper or brass plates on which the samples were mounted. The most intriguishing result here is the presence of the N 1s level in both samples. Usually, this level does not come from the contamination layer. Therefore, it may be infered that in this case nitrogen is present in the sample itself. The integral amplitude of this level is 4,5 ± 0,5 % from the oxygen level. Therefore, it may be infered that the synthesized samples are nitrogen doped from the very preparation moment.

Figure 2 presents the V 2p and O 1s core level electron distribution curves (EDC). One notices that there is no observable chemical shift between both samples. The resulting binding energies (BE) are 17,64 eV for V $2p_{3/2}$, 525,16 eV for V $2p_{3/2}$, and 531,0 eV for O 1s. Also, the O 1s spectrum contains a "shoulder" due to the oxygen from the contamination layer, at BE of around 533 eV, corresponding to oxygen in ionocovalent bindings C=O, C-OH, etc. The position of the principal O 1s component from the EDC is close to the reported value in literature for O^{2-}. With the levels of vanadium, we have performed a complete analysis of the reported data in literature.

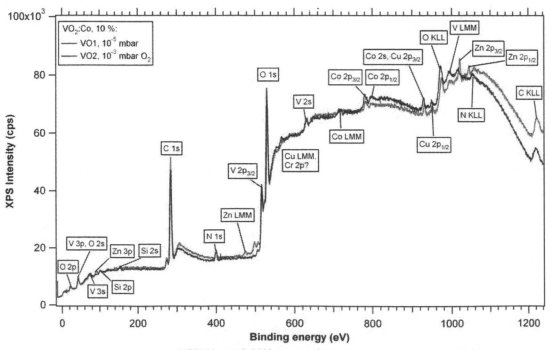

Figure 1: overview XPS scan of the two samples of interest VO1 and VO2, synthesized at different oxygen pressures.

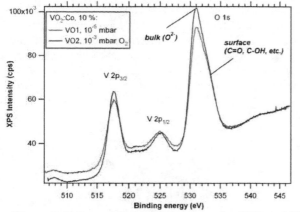

Figure 2: V 2p and O 1s EDC on samples VO1 and VO2

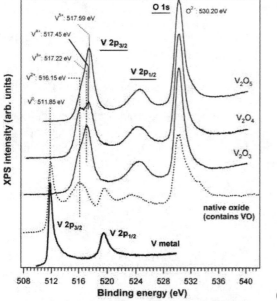

Figure 3: V 2p and O 1s EDC on several vanadium oxides

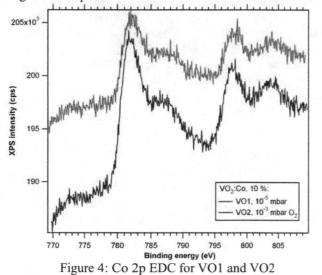

Figure 4: Co 2p EDC for VO1 and VO2

Figure 3 shows the analysis of the V 2p data from different vanadium oxides, as vanadium exist in several oxidation states, from (II) to (V). The data are extracted from Ref.

[9]. By comparing these data with the experimental BE derived from Fig. 2, one may infer the presence of vanadium in a (+5) ionization state, therefore by forming the compound V_2O_5. This result is a bit deceiving, since this compound does not present Mott-Hubbard transition.

Figure 4 presents the Co 2p EDC for both samples. Comparing with data from literature [9], cobalt here is found in a (IV-V) ionization state; the ionization state seems to be a bit *lower* when the sample is prepared in oxygen atmosphere (+4.7 for VO1, +4.3 for VO2).

3.2. X-ray absorption near-edge spectroscopy

Figure 5 presents the V K-edge XANES on both samples. From the analysis of the pre-edge region, where dipole forbidden transitions $1s \rightarrow 3d$ manifest [10], one derives the average number of $3d$ vacancies and hence the vanadium oxidation state. From Figure 5, different ionization states may be infered for vanadium, namely +5 for VO2 and around +2 (+3) for VO1, which corresponds to vanadium (II) or VO, and to vanadium (III) or V_2O_3. A similar compound is actually identified in the vanadium native oxide on surfaces in vacuum (see Fig. 2) and was recently evidenced for thin films prepared by using a plasma atomization source [11].

This result seems to contredict XPS results, but one has to take into account that the investigated depth is around 1-2 nm in XPS owing to the photoelectron escape depth [12], whereas in X-ray absorption recorded in fluorescence mode, the investigated depth is several hundreeds of nm (up to micrometer range). Therefore, although at the surface vanadium pentoxide is formed in both cases, by varying the oxygen pressure one may succeed to synthesize vanadium oxides with V in lower ionization states, such as VO or V_2O_3, which exhibit Mott-Hubbard transition.

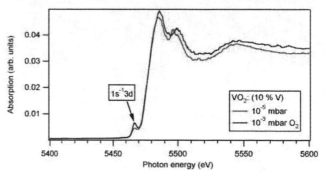

Figure 5: Vanadium K-edge X-ray absorption near-edge structure (XANES)

3.3. Magneto-optical Kerr effect investigations

Figure 6 presents MOKE hysteresis loops obtained on both samples. Sample VO1 exhibits a strong (super) paramagnetic behaviour, with superposed a weak ferromagnetic component with strong coercitive field, suggesting the presence of two magnetic phases: one which is superparamagnetic (might originate from vanadium oxide aggregates), the other being a kind of DMS (cobalt ions interacting by double exchange). The hysteresis cycle obtained for sample VO2 suggest the occurence of a strong anisotropy, superposing a ferro- and an antiferromagnetic phase, possibly interacting by exchange bias [8,13]. Due to the lack of space, we shall not discuss here AFM data; we just mention that these images support the co-existence of two distinct phases in both kind of samples (isolated nanoparticles in VO1, percolated nanoparticles in VO2).

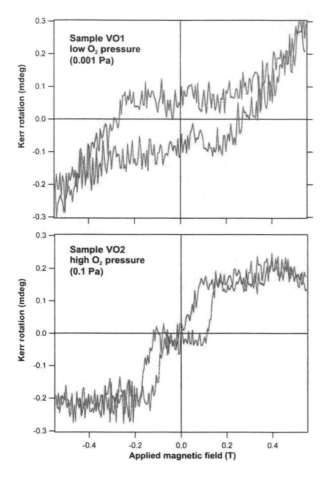

Figure 6: Magneto-optical Kerr effect

4   CONCLUSION

Very first results have been presented, by correlated reactivity and chemical composition measurements (XPS) with electronic structure (XANES), magnetism (MOKE) and morphology (AFM) for samples formed by magnetic ions embeded in a Mott-Hubbard material (vanadium oxide). We identified compostion variations between surface and bulk of materials; the bulk composition seems to be tunable through preparation parameters. Promising ferromagnetic behaviour is observed, also dependent on the preparation conditions. It

is suggested that the VO2 hysteresis loops might be connected to the percolation threshold of nanoparticles evidenced by AFM, whereas the ferromagnetic phase from VO1 might be connected to the presence of vanadium suboxides (VO, V_2O_3) which provide carriers for intermediating ferromagnetism and hence is a true DMS behaviour.

This work was supported by the Romanian National Authority of Scientific Research under contracts CEx05-D11-32 No. 32/2005 and PN II 71-063/2007.

REFERENCES

[1] H. Munekata, H. Ohno, S. Von Molnar, A. Segmuller, L.L. Chang, and L. Esaki, Phys. Rev. Lett. **63**, 1849 (1989); H. Ohno, Science **281**, 951 (1998).

[2] T. Dietl, H. Ohno, F. Matsukura, J. Cibert, and D. Ferrand, Science **287**, 1019 (2000).

[3] M.A. Ruderman and C. Kittel, Phys. Rev. **96**, 99 (1954); T. Kasuya, Progr. Theor. Phys. **16**, 45 (1956); K. Yosida, Phys. Rev. **106**, 893 (1957); E.V. Macocian and S. Filip, J. Optoel. Adv. Mater. **8**, 1098 (2006); D.J. Priour, Jr. and S. Das Sarma, Phys. Rev. Lett. **97**, 127201 (2006); G. Tang and W. Nolting, Phys. Rev. B **75**, 024426 (2007); E.Z. Meilikhov, Phys. Rev. B **75**, 045204 (2007).

[4] K.R. Kittilstved and D.R. Gamelin, J. Am. Chem. Soc. **127**, 5292 (2005); K.R. Kittilstved, N.S. Norberg, and D.R. Gamelin, Phys. Rev. Lett. **94**, 147209 (2005).

[5] R.J.O. Mossanek and M. Abbate, Phys. Rev. B **75**, 115110 (2007).

[6] S. Ilani, A. Yacoby, D. Mahalu, and H. Shtrikman, Science **292**, 1354 (2001); V.Y. Butko and P.W. Adams, Nature **409**, 161 (2001).

[7] M.W. Haverkort, Z. Hu, A. Tanaka, W. Reichelt, S.V. Streltsov, M.A. Korotin, V.I. Anisimov, H.H. Hsieh, H.J. Lin, C.T. Chen, D.I. Khomskii, L.H. Tjeng, Phys. Rev. Lett. **95**, 196404 (2005); M. Demeter, M. Neumann, and W. Reichelt, Surf. Science **454**, 41 (2000); K. Nagashima, T. Yanagida, H. Tanaka, and T. Kawai, Phys. Rev. B **74**, 172106 (2006); *ibid*. J. Appl. Phys. 101, 026103 (2006); H.T. Kim, Y.W. Lee, B.J. Kim, B.G. Chae, S.J. Yun, K.Y. Kang, K.J. Han, K.J. Yee, and Y.S. Lim, Phys. Rev. Lett. **97**, 266401 (2006).

[8] J. Zhang, R. Skomski, Y.F. Lu, and D.J. Sellmyer, Phys. Rev. B **75**, 214417 (2007).

[9] B.V. Crist, *Handbook of Monochromatic XPS Spectra, Vol. 2: Commercially pure binary oxides*, XPS International Inc. (2005).

[10] D. Mardare, V. Nica, C.M. Teodorescu, and D. Macovei, Surf. Science **601**, 4479 (2007).

[11] B. Sass, C. Tusche, W. Felsch, F. Bertran, F. Fortuna, P. Ohresser, and G. Krill, Phys. Rev. B **71**, 014415 (2005).

[12] S. Hüfner, *Photoelectron Spectroscopy: Principles and Applications*, Springer, Berlin (2003).

[13] G.A. Prinz, Science **250**, 1092 (1990).

Engineering Superhydrophobic and Superoleophobic Surfaces

Anish Tuteja[1], Wonjae Choi[2], Joseph M. Mabry[3], Gareth H. McKinley[4], Robert E. Cohen[5]

[1]*Department of Chemical Engineering, Massachusetts Institute of Technology, atuteja@mit.edu*
[2]*Department of Mechanical Engineering, Massachusetts Institute of Technology, wonjaec@mit.edu*
[3]*Air Force Research Laboratory, Edwards Air Force Base, Joseph.Mabry@edwards.af.mil*
[4]*Department of Mechanical Engineering, Massachusetts Institute of Technology, gareth@mit.edu*
[5]*Department of Chemical Engineering, Massachusetts Institute of Technology, recohen@mit.edu*

ABSTRACT

The combination of *surface chemistry* and *roughness* on multiple scales imbues enhanced repellency to the lotus leaf surface when in contact with a high surface tension liquid such as water. This understanding has led to the creation of a number of biomimetic superhydrophobic surfaces (i.e. apparent contact angles (θ^*) with water greater than 150° and low contact angle hysteresis). However, surfaces that display contact angles of $\theta^* > 150°$ with organic liquids having appreciably lower surface tensions (i.e. superoleophobic surfaces) are extremely rare. Calculations suggest that creating such a surface would require a surface energy lower than any known material. In our recent work (*Science, 318, 1618, 2007*) we demonstrated how a third factor, *re-entrant surface curvature,* in conjunction with chemical composition and roughened texture can be used to design surfaces that display extreme resistance to wetting from alkanes such as decane and octane. Here, we extend that work by designing a number of different nano-fiber surfaces (composed of a hydrophilic polymer (PMMA) and extremely low surface energy fluorinated molecules, fluoroPOSS) through electrospinning, which incorporate re-entrant curvature. By systematically changing the various design parameters we are able to assess the effects of surface geometry on both the apparent contact angle and hysteresis. Further, we perform a Zisman analysis on various PMMA - fluoroPOSS blends (using a series of alkanes) to estimate their surface energy. Rather surprisingly, it is observed that the Zisman analysis yields a negative value of surface energy for certain blends. These results are explained and reconciled with existing literature.

Keywords: superoleophobic, superhydrophobic, oil-repellent, self-cleaning, Zisman analysis.

1 INTRODUCTION

The most widely-known example of a natural superhydrophobic surface is the surface of the lotus leaf (*Nelumbo nucifera*). It is textured with small 10-20 μm protruding nubs which are further covered with nanometer size epicuticular wax crystalloids (see inset of Fig. 1a) [1]. Numerous studies have suggested that it is this combination of surface chemistry plus roughness on multiple scales [2-5] on the lotus leaf's surface that allows for the trapping of air underneath a water droplet (γ_{lv} = 72.1 mN/m), thereby imbuing the leaf with its characteristic superhydrophobicity (see Fig. 1a). However, a liquid with a markedly lower surface tension like hexadecane (γ_{lv} = 27.5 mN/m) rapidly wets the lotus surface leading to a contact angle of ~ 0° (see Fig. 1b), clearly demonstrating the leaf's oleophilicity. Indeed, in spite of the plethora of superhydrophobic surfaces now available, there are no naturally occurring superoleophobic surfaces [6-11] (i.e. surfaces that display contact angles greater than 150° with organic liquids such as alkanes having appreciably lower surface tensions than water).

2 RESULTS AND DISCUSSION

2.1 Electrospun superoleophobic fibers

Recent work in our laboratories [6] has led to the development of a new class of hydrophobic POSS molecules (radius 1-2 nm; see Fig. 2a) in which the rigid silsesquioxane cage is surrounded by fluoroalkyl groups. A number of different molecules with different organic groups (including 1H,1H,2H,2H-heptadecafluorodecyl (referred to as fluorodecyl POSS); 1H,1H,2H,2H-tridecafluorooctyl (fluorooctyl POSS) have now been synthesized, and this class of materials is denoted generically as fluoroPOSS. The fluoroPOSS molecules contain a very high surface concentration of fluorine containing groups, including $-CF_2$ and $-CF_3$ moieties. The high surface concentration and surface mobility of these groups, as well as the relatively high ratio of $-CF_3$ groups with respect to the $-CF_2$ groups results in one of the most hydrophobic and lowest surface energy materials available today [12]. By varying the mass fraction of fluoroPOSS dispersed in various polymers, we can systematically change the surface energy of the polymer-fluoroPOSS blend (see Fig. 3).

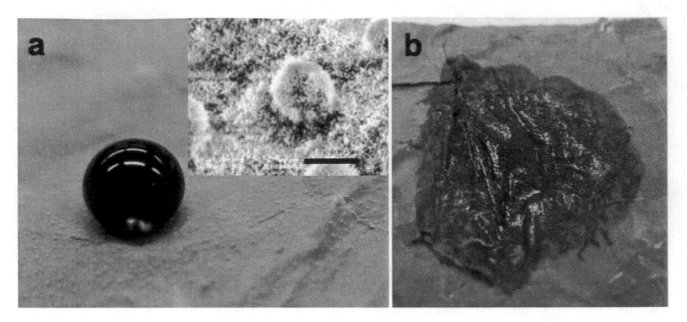

Figure 1. a. A droplet of water (colored with methylene blue) on a lotus leaf surface. The inset shows an SEM micrograph of the lotus leaf surface; the scale bar is 5 μm. b. The wetted surface of the lotus leaf after contact with a droplet of hexadecane. (Some images adapted from previous work [6].)

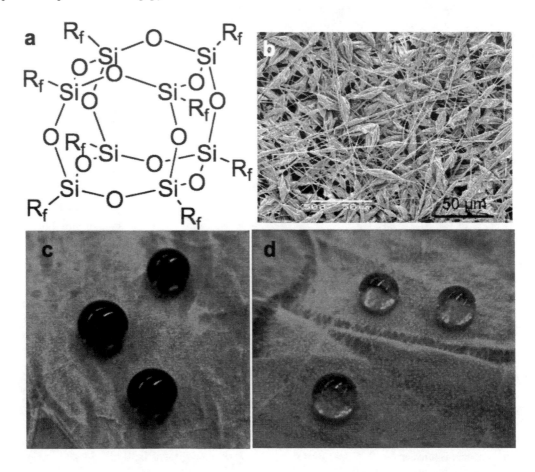

Figure 2. a. The general molecular structure of fluoroPOSS molecules. The alkyl chains (R_f) have the general molecular formula $-CH_2CH_2(CF_2)_nCF_3$, where n = 0, 3, 5 or 7. b. A scanning electron microscope (SEM) micrograph for an electrospun surface containing PMMA + 9.1 wt% fluorodecyl POSS. c and d. Droplets of water and hexadecane (colored with 'oil red O')

on a lotus leaf surface covered with electrospun fibers of PMMA + 44 wt% fluorodecyl POSS. A reflective surface is visible underneath the droplets in both pictures, indicating the presence of microscopic pockets of air. (Some images adapted from previous work [6].)

In the present work, we study blends of a moderately hydrophilic polymer, poly(methyl methacrylate) (PMMA, M_w = 540 kDa, PDI ~ 2.2) and fluorodecyl POSS as a model system. Fig. 2b shows the beads on a string morphology of these fluorodecyl POSS-PMMA blends created by electrospinning [13-16]. The complex re-entrant [4, 6, 11, 17, 18] topology allows us to support a composite (solid-liquid-air) interface with various liquids including water and alkanes such as hexadecane (see Fig. 2c and 2d), leading to extremely high apparent contact angles, even though the constituent nano-fibers themselves may be hydrophilic or oleophilic respectively. In contrast to many lithographic or vapor-deposition techniques, electrospinning is a benign single step process and the fluoroPOSS-PMMA blends can be deposited on a lotus leaf to confer it with oleophobicity, in addition to superhydrophobicity, as shown in Figs. 2c and 2d.

In our recent work [6] we also established the various design parameters that affect the robustness of the composite interface, allowing for the creation of extremely non-wetting rough surfaces, even though their corresponding smooth surfaces may be easily wetted by a given liquid. By systematically changing the various design parameters for the electrospun nano-fiber surfaces, we can evaluate the effects of surface geometry on both the apparent contact angle and hysteresis for various liquids. This enables us to develop surfaces that repel practically any liquid as evidenced by apparent contact angles greater than 150° with methanol (γ_{lv} = 22.7 mN/m), decane (γ_{lv} = 23.9 mN/m) and octane (γ_{lv} = 21.6 mN/m), on electrospun surfaces containing 56 wt% fluorodecyl POSS.

2.2 Estimation of solid surface energy (γ_{sv})

Previous work by Shibuichi et al. argued that for a chemically homogeneous, smooth surface to exhibit $\theta > 90°$ with any liquid, its solid surface energy (γ_{sv}) must be less than one-fourth the liquid surface tension, (γ_{lv})/4 [8, 9]. Careful studies of monolayer films by Zisman et al. [7] show that the contributions to the overall magnitude of surface energy of a flat surface decreased in the order -CH$_2$ > -CH$_3$ > -CF$_2$ > -CF$_2$H> -CF$_3$, and based on this analysis, the lowest solid surface energy is estimated to be ~ 6.7 mN/m (for a hexagonally closed packed monolayer of –CF$_3$ groups on a surface) [7, 19]. Taken in conjunction, these studies explain the absence of non-wetting surfaces displaying equilibrium contact angles > 90° with decane and octane [7-10, 20], as a solid surface would need to have a surface energy of ~ 5 mN/m to display $\theta > 90°$ with these liquids [6].

However, recently a few groups have reported extremely low γ_{sv} values; for example Coulson et al. [20, 21] report surface energy values as low as 1.5 mN/m for coatings created by pulsed plasma polymerization of 1H,1H,2H-perfluoro-1-dodecene.

Thus, the issue of the minimum surface energy seems to be a bit controversial and unresolved in the literature. The problem stems from the fact that measurement of equilibrium contact angles only provides an indirect estimate of the surface energy, and typically involves extrapolation or assuming an additive decomposition of γ_{sv} into dispersive and H-bonding / polar contributions. The most accurate determination of surface energies requires the measurement of the work of adhesion, and this is not often done [12].

There are several different methods of using contact angles to estimate the surface energy of a material (e.g. the Zisman analysis [7], the Owens-Wendt analysis [22], and Girifalco-Good-Fowkes-Young [23, 24] analysis), and each of these methods typically yields a different value for the computed surface energy, depending on the surface under study. Thus, previous studies have noted that these methods should only be used to obtain an estimate of the actual surface energy, which can be useful in comparing and ranking different surfaces (say with different degree of fluorination) as long as the same method is used for each surface [12].

Indeed, Coulson et al. also report two different measures of surface energy. They obtain values of γ_{sv} =1.5 mN/m (on a smooth glass substrate coated by pulsed plasma polymerization of 1H,1H,2H-perfluoro-1-dodecene [21]) and 4.3 mN/m (on a smooth glass substrate coated by pulsed plasma polymerization of 1H,1H,2H,2H-heptadecafluorodecyl acrylate [20]) using the Zisman analysis, or γ_{sv} =8.3 mN/m [21] and 10 mN/m [20] using the Owens-Wendt method for the same two surfaces. It is therefore unclear as to which method provides a more accurate value for γ_{sv}. An indication that the Zisman analysis might be providing a γ_{sv} value lower than the actual value for their surface comes from the values of octane contact angles obtained by Coulson et al. As mentioned above, if $\gamma_{sv} < \gamma_{lv}/4$, the equilibrium contact angle θ measured experimentally should be greater than 90°. In contrast, Coulson et al. report values of advancing contact angle, θ_{adv} = 74° and receding contact angle, θ_{rec} = 35° respectively on their coatings of 1H,1H,2H-perfluoro-1-dodecene when using octane (γ_{lv} = 21.7 mN/m).

We have also computed the surface energy of the various spincoated PMMA + fluoroPOSS surfaces (r.m.s

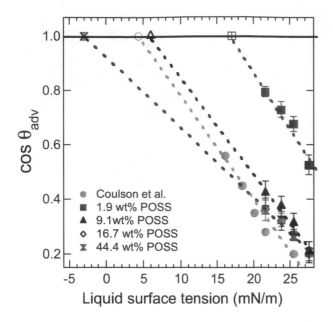

Figure 3. Zisman plot for various spincoated PMMA+fluoroPOSS films. The data from the Zisman analysis performed by Coulson et al. [20] for surfaces prepared by pulsed plasma polymerization of 1*H*,1*H*,2*H*,2*H*-heptadecafluorodecyl acrylate are also included for comparison. (Some data adapted from previous work [6].)

roughness for all spincoated surfaces was less than 4 nm) using the Zisman and the Owens-Wendt methods. For a spincoated surface containing 44.4 wt% POSS we obtain values of γ_{sv} = -3 mN/m and γ_{sv} = 7.8 mN/m (with the dispersive component of surface energy, γ_d = 6.6 mN/m and the polar component, γ_p = 1.2 mN/m) using the Zisman and the Owens-Wendt method respectively. Fig. 3 shows the Zisman analysis for four different spincoated PMMA + fluoroPOSS films, as well as, the data for the Zisman analysis done by Coulson et al. [20].

Clearly, the negative value of the surface energy obtained from the Zisman analysis of our surfaces are spurious (and arise solely form the extrapolation process employed), however, these calculations again point out the limitations of the various methods that use measurements of equilibrium contact angles to compute γ_{sv}. It is however clear from the data in Fig. 3 that, as would be expected, the surface energy of our PMMA + fluoroPOSS blends decreases with increasing POSS concentration and for high fluoroPOSS concentrations, the calculated interfacial energy seems to approach values consistent with those obtained by Coulson et al.

3   CONCLUSIONS

We have shown that in spite of the absence of any naturally oleophobic material, it is possible to engineer superoleophobic surfaces through the incorporation of re-entrant curvature within the surface texture. This enables us to create surfaces that repel practically any liquid as evidenced by apparent contact angles greater than 150° with methanol (γ_{lv} = 22.7 mN/m), decane (γ_{lv} = 23.9 mN/m) and octane (γ_{lv} = 21.6 mN/m), on certain electrospun surfaces. Further, we have also computed the surface energies of various PMMA-fluoroPOSS blends and found that the extrapolation involved is the Zisman analysis may lead to spurious values of the estimated solid surface energy.

4   REFERENCES

[1]   W. Barthlott, and C. Neinhuis, Planta **202**, 1 (1997).
[2]   Y. Yu, Z. H. Zhao, and Q. S. Zheng, Langmuir **23**, 8212 (2007).
[3]   M. Callies, and D. Quere, Soft Mat. **1**, 55 (2005).
[4]   L. Cao, H. H. Hu, and D. Gao, Langmuir **23**, 4310 (2007).
[5]   A. Otten, and S. Herminghaus, Langmuir **20**, 2405 (2004).
[6]   A. Tuteja *et al.*, Science **318**, 1618 (2007).
[7]   W. A. Zisman, *Relation of the equilibrium contact angle to liquid and solid construction. In Contact Angle, Wettability and Adhesion, ACS Advances in Chemistry Series.* (American Chemical Society, Washington, DC. , 1964), Vol. 43, pp. 1.
[8]   K. Tsujii *et al.*, Angew. Chem. Int. Ed. Engl. **36**, 1011 (1997).
[9]   S. Shibuichi *et al.*, J. Colloid Interface Sci. **208**, 287 (1998).
[10]   W. Chen *et al.*, Langmuir **15**, 3395 (1999).
[11]   A. Ahuja *et al.*, Langmuir (2007).
[12]   M. J. Owen, and H. Kobayashi, Macromol. Symp. **82**, 115 (1994).
[13]   D. H. Reneker *et al.*, J. Appl. Phys. **87**, 4531 (2000).
[14]   M. Ma *et al.*, Langmuir **21**, 5549 (2005).
[15]   M. Ma *et al.*, Macromolecules **38**, 9742 (2005).
[16]   M. Ma *et al.*, Adv. Mater. **19**, 255 (2007).
[17]   S. Herminghaus, Europhys. Lett. **52**, 165 (2000).
[18]   J.-L. Liu *et al.*, J. Phys.: Cond. Matt. **19**, 356002 (2007).
[19]   T. Nishino *et al.*, Langmuir **15**, 4321 (1999).
[20]   S. R. Coulson *et al.*, Chem. Mater. **12**, 2031 (2000).
[21]   S. R. Coulson *et al.*, Langmuir **16**, 6287 (2000).
[22]   D. K. Owens, and R. C. Wendt, J. Appl. Poly. Sci. **13**, 1741 (1969).
[23]   L. A. Girifalco, and R. J. Good, J. Phys. Chem. **61**, 904 (1957).
[24]   F. M. Fowkes, Ind. Eng. Chem. **56**, 40 (1964).

Hypertransparent Nanostructured Superhydrophobic Self-Cleaning Coatings on Glass Substrates

Y. D. Jiang[*], Y. Smalley[**], H. Harris[***], and A. T. Hunt[****]

*n*Gimat Co., Atlanta, GA, USA, [*]yjiang@*n*Gimat.com, [**]ysmalley@*n*Gimat.com,
[***]hharris@*n*Gimat.com, [****]ahunt@*n*Gimat.com

ABSTRACT

In this paper, high quality SiO_2 based hypertransparent superhydrophobic coatings with double-roughness microstructure were successfully deposited onto glass substrates by the Combustion Chemical Vapor Deposition (CCVD) technique. A contact angle of higher than 165°, a rolling angle of <5°, a haze of <0.5%, and an increased transmittance by 2% higher and a reflectance of 2 % lower than bare glass have been achieved.

Keywords: nanostructure, superhydrophobic, self-cleaning, coating, CCVD

1 INTRODUCTION

Studies of superhydrophobic self-cleaning surfaces have been attracting increasing interest in recent years as a result of numerous new prospects for both fundamental research and practical applications. The applications of self-cleaning surfaces include architectural glass for homes and commercial buildings, automotive glass, shower doors, solar panel glass covers, nanochips, and microfluidic systems [1-3]. With wide usage of self-cleaning surfaces, over $100 million a year in energy savings will be generated as a result of removing the need for washing, scrubbing, and chemical polishing of windows, ceramics, and other surfaces. In addition to a reduction in cleaning requirements, these superhydrophobic surfaces have additional benefits, such as improved safety when driving in severe rain and snow, and improved efficiency in solar cells (anti-reflective and self-cleaning properties).

The development of superhydrophobic self-cleaning surfaces was first inspired by the observation of natural cleanness of lotus leaves [4] and other plant leaves [5]. The typical superhydrophobic self-cleaning effect in nature is found from lotus leaves, which is revered as a symbol of purity for its ability to maintain clean leaves even in heavily contaminated waters. In studies of lotus leaves by scanning electron microcopy (SEM) [5,6], it was revealed that the key features of the lotus leaf are a microscopically rough surface consisting of an array of randomly distributed micropapillae with diameters ranging from 5 to 10 μm. These micropapillae are covered with waxy hierarchical structures in the form of branch-like nanostructures with an average diameter of about 125 nm. Motivated by the nature and the superhydrophobic self-cleaning performance of the lotus leaf, many techniques have been being developed to create nanostructures mimicking the lotus effect from many materials, both in organic and inorganic [7,8]. However,

existing hydrophobic coatings either have low transmittance which is not suitable for windows and solar panels or not as hydrophobic as lotus leaves.

In this work, an open atmosphere CCVD technique was employed to deposited SiO_2 based superhydrophobic coatings onto glass substrates that actually increase light transmission and are thus hypertransparent. The coatings morphological, hydrophobic, and other physical properties are presented.

2 EXPERIMENTAL PRECEDURES

2.1 Superhydrophobic coatings by the CCVD Process

In the CCVD process [9], as shown in Figure 1, precursors, which are the metal-bearing chemicals used to coat an object, are dissolved in a solvent, which typically also acts as the combustible fuel. This solution is atomized to form submicron droplets by means of the proprietary Nanomiser™ device. These droplets are then convected by an oxygen containing stream to the flame where they are combusted. A substrate (the material to be coated) is coated by simply drawing it over the flame plasma. The heat from the flame provides the energy required to evaporate the ultrafine droplets and for the precursors to react and to vapor deposit on the substrates. The CCVD technique uses a wide range of inexpensive, soluble precursors that do not need to have a high vapor pressure. The key advantages of the CCVD technique include:

- Open-atmosphere processing
- High quality at low cost
- Wide choice of substrates and
- Continuous production capability

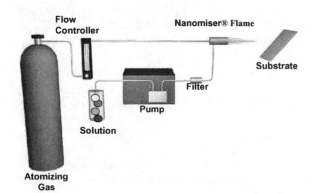

Figure 1. Schematic representation of the CCVD system

In this work, all the coatings were fabricated by the CCVD technique addressed previously. Prior to depositions, the glass substrates were ultrasonically cleaned in organic solvents such as isopropanol, rinsed in deionized water, and blown dry using nitrogen. The substrate was then mounted on a metal chunk or the top of a back heater. Key process parameters include deposition temperature, solution concentration, motion speed, coating thickness, and coating composition.

2.2 Analytical techniques

To measure water contact angle (CA), the coated specimens were surface treated with a fluorinated silane by immersing the specimens in a hexane solution of the fluorinated silane (with a volume ratio of silane to hexane of 1:50) for 10 min. The specimens were then rinsed by hexane and deionized water and blown dry by nitrogen gas. Equilibrium, receding, and advancing CAs were measured by a CA measuring system (G10, Kruss USA). Equilibrium CAs were measured using deionized water droplets of approximately 1 – 2 mm in diameter. If not indicated specifically, all the CAs in the following sessions are equilibrium water CAs. To measure advancing CA, a water droplet of about 1 – 2 mm in diameter was first placed on the film surface. The cursor line was placed in front of the water droplet. The droplet was then enlarged by pushing more water through the needle. While the droplet volume increases, the contact line between water and the solid surface moved forward. The advancing CA was measured once the droplet front reached the cursor line. To measure receding CA, a large water droplet of about 5 mm in diameter was placed on the solid surface first. The cursor line was placed between the syringe needle and the droplet front. The droplet volume was decreased by sucking water back into the syringe. When the contact line began moving and the droplet front reached the cursor line, the receding CA was recorded. During the measurements of advancing and receding CAs, the syringe needle was always in the water droplet. Three data points were tested on all samples.

The coating's morphology was observed by SEM (Hitachi s-800). Transmittance and reflectance in the visible range were measured by a spectrometer (PERKIN-ELMER Lambda 900 UV/VIS/NIR spectrometer). Surface roughness (as root mean square, RMS) was evaluated by an optical profilometer (Burleigh Instruments, Inc.). Haze in the visible range was characterized by a haze meter (BYK Gardner Haze Meter).

3 RESULTS AND DISCUSSION

Superhydrophobic coatings have been grown by many techniques such as CVD [7] and sol-gel [8]. These techniques require costly starting materials, and/or are time consuming and have low throughput. The low cost CCVD technology offers an attractive alternative to grow nanostructured superhydrophobic coatings on glass and plastic substrates with good yield and high throughput potential. SiO_2 is chosen as primary coating material because of its low cost, ease to make, and refractive index match between the coating and the substrate, which reduces reflectivity of the coated specimens.

To achieve low haze and high transmittance, the feature size of the coating must be much smaller than the visible wavelength to reduce large light scattering. Figure 2 shows the SEM image of a SiO_2 coating on glass. The coating has a rough surface and double surface roughness, in which coarse features are composed of nanostructures of 30 to 200 nm. The double roughness morphology is similar to the topology of lotus leaf. The nanometer sized hierarchical structure is essential to simultaneously achieve low haze, increased transparency, and superhydrophobicity.

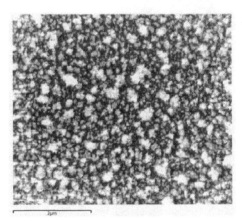

Figure 2. SEM image of a typical CCVD SiO_2 based superhydrophobic coating on glass substrate

The relationship between CA and surface roughness of the SiO_2 coatings is present in Figure 3. It is clear that CA increases rapidly with the increase of surface roughness up to about 1.5 nm. When surface roughness reaches 1.5 nm or higher, CA remains almost constant. A CA of over 165° was achieved with surface roughness ranging from 1 to 20 nm. For reference, the bare glass substrate coated with the same fluorinate silane has a CA of approximately 110° with a surface roughness of about 0.70 nm. For comparison, photographs of water droplets sitting on bare and silane/SiO_2-CCVD coated glass are shown in Figure 4. Water droplets spread out on bare glass with a CA of less than 20°, showing the nature of hydrophilicity of glass (Figure 4 (a)). After coated with the nanostructured SiO_2 film and surface treated with the fluorinated silane, water droplets bead up to form spheres, showing superhydrophobicity (Figure 4 (b)).

Haze of the superhydrophobic coatings is another important property for many applications. Haze of the bare glass is about 0.2%. As shown in Figure 5, when surface roughness is less than 2.0 nm haze increases slightly with increasing roughness. Beyond a roughness of 3 nm haze increases almost exponentially with increasing roughness. For samples with surface roughness less than 2.0 nm, a haze of less than 0.5% was obtained with a contact angle of over 165° and a CA hysteresis of less than 5°.

Figure 3. Contact angle of the SiO₂ based coatings as a function of surface roughness

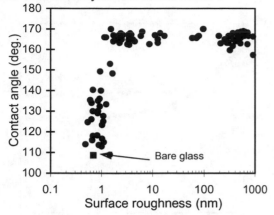

Figure 4. Optical images of water droplets sitting on (a) bare glass substrate and (b) fluorinated silane coated CCVD SiO₂ coating on glass substrate

Figure 6 shows typical transmittance and reflectance spectra of one side CCVD SiO₂ coated and bare glass substrates in the visible range. The transmittance and reflectance of bare glass are in the range of 91.8 to 92.6% and 7.4 to 8.5%, respectively while those the CCVD SiO₂ coated samples are in the range of 93.9 to 94.5% and 5.6 to 6.2, respectively. The SiO₂ coated glass is hypertransparent since it has increase transmission by about 2% higher. Reflectance is about 2% lower than bare glass substrate, suggesting the CCVD SiO₂ coatings reduce reflection in the visible range and are of anti-reflection, which will benefit

many applications, especially solar cells, lighting and imaging. For imaging applications the *n*Vision™ coating line is being launched based on the properties of these coatings.

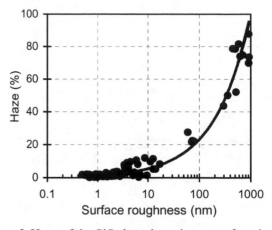

Figure 5. Haze of the SiO₂ based coatings as a function of surface roughness

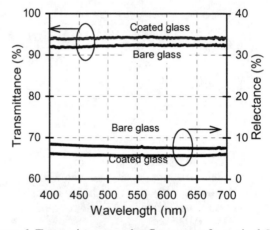

Figure 6. Transmittance and reflectance of a typical SiO₂ based superhydrophobic coating on glass

In this work, process parameters such as deposition temperature, deposition time/lap, and flame motion speed were investigated to optimize the hydrophobicity. Figure 7 shows the CA and CA hysteresis of SiO₂ coatings deposited at a certain solution concentration and different processing conditions as a function of motion speed. As shown in Figure 7 (a), all other samples maintain almost the same CAs with the increase of motion speed. CA hysteresis, which is directly related to rolling angle, is another important factor to evaluate superhydrophobic self-cleaning surfaces. The smaller the CA hysteresis is, the easier the water droplets roll off the surface, namely the higher the self-cleaning performance. Figure 7 (b) shows the effects of motion speed on CA hysteresis as a function of deposition variables including one or two laps. All samples but one show nominal change in angle hysteresis with the increase of motion speed.

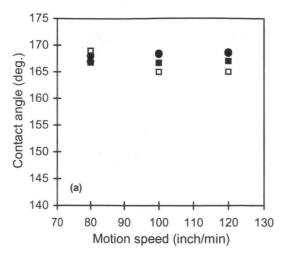

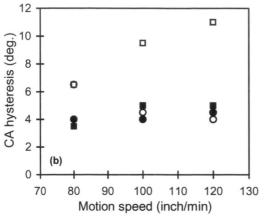

Figure 7. (a) CA and (b) CA hysteresis of CCVD SiO$_2$ coatings deposited at different processing conditions as a function of processing motion speed. Each symbol represents a change in processing condition.

Abrasion resistance is an important factor for many practical applications of self-cleaning surfaces. Abrasion tests were conducted at *n*Gimat by moving the samples across a defined distance on a polishing cloth surface. Force was determined by the weight of the samples themselves. CA was measured after each two passes across the abrasion surface. In total, twenty passes were completed for each sample. Figure 8 shows abrasion test results of two differently processed CCVD SiO$_2$ coatings. The initial CA of the A-processed coating was 170°. It decreased rapidly in the first ten passes of abrasion. After ten passes the contact angle remained almost constant at 150°. It was noticed that after the abrasion test, the rolling angle increased significantly. To increase the strength of the superhydrophobic surfaces, a sample was differently B-processed then coated with silane. As shown in Figure 8, its initial CA was 168°, about 2° lower than that of the A-processed one. In the first 4 passes, the CA decreased rapidly to 159°. After 4 passes, the CA decreased minimally. After 20 passes the final CA was 157°, which is

8° higher than that of the A-deposited sample, suggesting the B-processing increased the coating's strength.

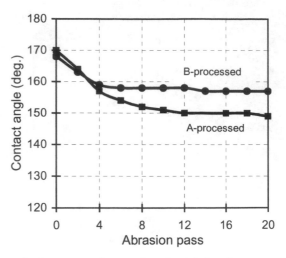

Figure 8. Contact angle as a function of abrasion pass of A and B processed CCVD coatings on glass

4 SUMMARY

As a summary, hypertransparent superhydrophobic surfaces with high performance have been successfully prepared on glass substrates by the CCVD technique. A contact angle of 170°, a rolling angle of <5°, a haze of <0.5%, and a transmittance 2% higher and a reflectance of 2 % lower than bare glass have been achieved.

This work was supported by the Deportment of Energy (DOE) through Grant No. DE-FG02-04ER84007. The authors would like to thank George E. Sakoske at Ferro Corporation for rolling angle measurements.

REFERENCES

[1] A. Nakajima, K. Hashimoto, and T. Watanabe, Chem. Monthly 132, 31, 2001.
[2] N. A. Patankar, Langmuir 20, 8209, 2004.
[3] D. Quere, A. Lafuma, and J. Bico, Nanotechnology 14, 1109, 2003.
[4] W. Barthlott and C. Neinhuis, Planta 202, 1 (1997)
[5] L. Feng, S. H. Li, Y. s. Li, H. J. Li, L. J. Zhang, J. Zhai, Y. L. Song, B. Q. Liu, L. Jiang, and D. B. Zhu, Adv. Mater. 14, 1857, 2002.
[6] B. He, J. Lee, and N. A. Patankar, Colloids and Surfaces A: Physicochem. Eng. Aspects 248, 101, 2004.
[7] Y. Y. Wu, H. Sugimura, Y. Inoue, and O. Takai, Chem. Vap. Deposition 8, 47, 2002.
[8] H. M. Shang, Y. Wang, S. J. Limmer, T. P. Chou, K. Takahashi, and G. Z. Cao, Thin Solid Film 472, 37, 2005.
[9] A. T. Hunt, W. B. Carter, and J. K. Cochran, Jr., App. Phys. Lett. 63, 266, 1993.

A Markedly Controllable Adhesion of Superhydrophobic Spongelike Nanostructure TiO$_2$ films

Y. K. Lai[*], C. J. Lin[*], J. Y. Huang[*], H. F. Zhuang[*], L. Sun[*] and T. Nguyen[**]

[*]State Key Laboratory for Physical Chemistry of Solid Surfaces
College of Chemistry and Chemical Engineering, Xiamen University, Xiamen 361005, China
[**]National Institute of Standards and Technology, Gaithersburg, MD 20899, USA

ABSTRACT

A simple electrochemical and self-assembled method was adopted for the fabrication of superhydrophobic sponge-like nanostructured TiO$_2$ surface with markedly controllable adhesions. Water adhesion ranged from ultra-low (5.0 µN) to very high (76.6 µN) can be tuned through adjusting the nitro cellulose dosage concentrations. The detailed experiments and analyses have indicated that the significant increase of adhesion by the introducing of nitro cellulose is ascribed to the combination of hydrogen bonding between the nitro groups and the hydroxyl groups at the solid/liquid interfaces and the disruption of the dense packed hydrophobic PTES molecule. A mechanism has been proposed to explain the formation of superhydrophobic TiO$_2$ films with distinct adhesion.

Keywords: superhydrophobic, titanium dioxide, nanostructures, adhesion, hydrogen bonding

1 INTRODUCTION

In recent years, superhydrophobic surfaces, with a water contact angle (CA) greater than 150°, have attracted much interest due to their importance in theoretical research and practical application.[1-3] It is well-known that the preparation of superhydrophobic surfaces is the combination of rough structure and the coating of low surface energy materials.[4-9] There are two extremely superhydrophobic cases in nature, that is, superhydrophobic lotus leaves with a sliding angle (SA) lower than 10° and superhydrophobic gecko feet with highly adhesive force. These findings have inspired the creation of functional materials with self-cleaning and novel adhesive by mimicking their special structures.[10-12] However, the research on superhydrophobic surfaces with ultra-low or highly adhesive force to water droplet was always neglected and only had been theoretically or experimentally discussed to a very limited extent.[13-15] It is expected that superhydrophobic surface with ultra-low or highly adhesion to water will have many potential applications, such as microanalysis and liquid transportation without loss.

In this work, we fabricated sponge-like structure TiO$_2$ films by a simple electrochemical method, and then modified the as-prepared samples with 1H, 1H, 2H, 2H-perfluorooctyltriethoxysilane (PTES) and certain concentration of nitro cellulose (NC) to obtain superhydrophobicity with highly distinct adhesive forces by self-assembled technique. The sponge-like film with superhydrophobic and ultra-low adhesive force was achieved by sole PTES modification. However, when a certain NC was added into the PTES modified solution, the films not only retain superhydrophobicity but also have highly adhesive forces to water droplet. Extensive experimental results indicate that NC concentration has a great effect on both the contact angles and adhesive force for water droplet. This remarkable phenomenon and further mechanism analysis offer us an insight into how to control the surface chemical compositions to regulate the superhydrophobicity with a markedly controllable adhesive force that varies a wide range (5.0-76.6 µN). This simple method on the control of water adhesion under ambient condition without resorting to other special coating processes may be conveniently applied to other superhydrophobic surfaces.

The sponge-like nanostructure TiO$_2$ films were fabricated by electrochemical anodizing purity titanium sheets in 0.5 wt % HF electrolyte with Pt counter electrode. The anodizing was carried out at 50 V for 20 min. The TiO$_2$ film was then rinsed with deionized water and dried with dry N$_2$. The as-prepared TiO$_2$ films were then treated with a methanolic solution of hydrolyzed 1 wt % PTES (Degussa Co.) for 3 h and subsequently baked at 140 °C for 1 h.

2 RESULTS AND DISCUSSION

Figure 1a, b show the typical top and cross-sectional SEM images of the as-anodized samples, respectively. It can be seen that the surface has a uniformly-distributed sponge-like nanostructure, and the thickness of nanostructured TiO$_2$ film is approximately 500 nm. Water rapidly spreads and wets this as-anodized sponge-like film without PTES modification due to side penetration of the liquid by capillary forces, indicating such sample is superhydrophilic. However, it is observed that the droplets with spherical shapes slide spontaneously, and hardly come to rest even when it is placed gently onto the PTES modified sponge-like TiO$_2$ surface. The water CA on such sponge-like structure film is as high as 160° (shown in the inset of Figure 1a), while that on a regular flat TiO$_2$ surface is only 115°, which is in good agreement with the results of coating fluorocarbon hydrophobic layer on smooth surface by self-assembly.

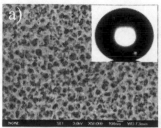

Figure 1. A typical top (a) and cross-sectional (b) SEM images of the as-prepared TiO_2 sponge-like structure film by 50 V. The inset shows the shape of water droplet on the corresponding PTES modified film.

To understand the superhydrophobicity of the sponge-like TiO_2 surface, we describe the contact angle in terms of the Cassie equation:[16] $\cos\theta_r = f_1 \cos\theta - f_2$. Here, f_1 and f_2 are the interfacial area fractions of the sponge-like TiO_2 surface and of the air in the interspaces surrounding the TiO_2 material, respectively (i.e., $f_1 + f_2 = 1$); θ_r (160°) and θ (115°) are the contact angles on the rough sponge-like TiO_2 surface and on the self-assembled monolayer of PTES on a flat TiO_2 surface, respectively. It is easy to deduce from this equation that contact angle of the rough surface (θ_r) increases with increasing the air fraction (f_2). According to the equation, the f_2 value of the rough, sponge-like TiO_2 surface is estimated to be 0.90. Therefore, we can realize that the large fraction of air trapped in the rough surface (i.e., to prevent water from penetrating) is important to the superhydrophobicity. These results demonstrate that the combination of a uniquely rough structure with large air fraction and a low surface energy coating modification are definitively vital for the superhydrophobicity.

The roll-off behavior was recorded with a high-speed photography as shown in Figure 2a. The white arrow heads at the bottom of the figures show the sliding direction of water droplet on the sponge-like TiO_2 thin film. The sliding behavior indicates that this film has an exceptionally low resistance to water droplet rolling. It is also interesting to note that the water droplet bounces off this sponge-like structure film when it is dropped from a certain height above the surface. The mean advancing and receding angles of water droplet on the surface are 160.1° and 159.3°, respectively, indicating that this unique structure film has an ultra-low CA hysteresis (only about 0.8°). Up to now, only limited information on the ultra-low water hysteresis phenomenon has been reported.[17] The reason for such ultra-low hysteresis can be attributed to the large fraction of air trapped in the rough sponge-like nanostructure TiO_2 surface, which significantly decreases the contact area. On the other hand, it has been suggested that the unstable discontinuous three-phase (solid-liquid-air) contact line of the rough TiO_2 surface can lead to a smaller hysteresis than the well-ordered two-dimensional surface.[18]

In contrast to the rolling behavior of water droplet on the sole PTES-modified sponge-like structure film, the behavior of water droplet on the surface of the same film that was modified by a mixture of PTES and NC is totally different. Figure 2b shows the shape of a water droplet on the PTES and 0.2 mg/mL NC (PTES-NC) modified sponge-like structure film with different tilt angles. The static water CA on the horizontal surface of this material is also very high, approximately 153.6°, indicating that these films retain their superhydrophobicity after the incorporation of NC. For these films, the water droplet adheres firmly to the surface and can resist against its gravitational forces when the sample is tilted vertically (90°) or even turned upside down (180°), indicating a strong adhesive effect exists between the water droplet and the "sticky" TiO_2 sponge-like structure surface.

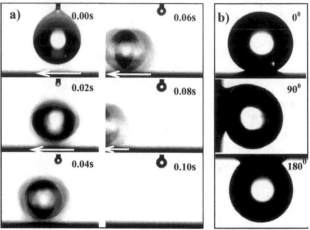

Figure 2. (a) Behaviors of water droplet on the PTES-modified 50 V sponge-like structure surface. In right top of images show the time sequence. (b) Shapes of water droplet on the PTES and 0.2 mg/mL NC modified 50 V sponge-like structure surface with different tilt angles: 0°, 90° and 180°.

Figure 3 shows the typical results of X-ray photoelectron spectroscopy (XPS, VG ESCALAB MK II) for sponge-like TiO_2 films before and after NC modification. It is indicated that the strong F 1s peak along with a C 1s peak due to C-F in addition to that of C-H and the attenuation of the Ti 2p and Ti 2s peaks comfirm the presence of PTES monolayer. This is in good agreement with the report by Wang et al.[19] However, the further addition of NC, which damages the uniform arrangement of PTES molecule and leads to the decrease of the C-F on surface, result in the slightly CA decrease and highly adhesive force of the superhydrophobic nanostructure surface with water droplet. The detailed XPS spectra for the corresponding N 1s region of the sponge-like TiO_2 films with and without adsorbed NC are shown in the inset of Figure 3. Without NC modification, there is only one peak in the N 1s spectrum at a binding energy (BE) about 400.5 eV, which could be assigned to adsorbed nitrogen gas. However, another peak attributed to nitrate is observed at a BE about 407.2 eV with PTES-NC mixed modification. This indicates the presence of NC on the sponge-like TiO_2 surface.[20]

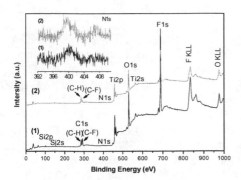

Figure 3. XPS spectra of the sponge-like surface with PTES modification (1) and PTES and 0.2 mg/mL NC mixed modification (2). The inset shows the coresponding N 1s XPS detail spectra.

Figure 4a shows a typical force-distance curve of the sole PTES-modified sponge-like structure film when it approaches and retracts from the water droplet. From the linear force-distance relationship during the contacting and retracting processes, it reveals that the water droplet always maintains a spherical shape without any noticeable distortion. The adhesive force the water droplet is subjected to pull down by the sponge-like structure surface is only approximately 5.0 µN. The low value is attributed to the large fraction of air trapped beneath the water droplet. This effect greatly reduces the solid-liquid interface and the rolling resistance and, thus, cannot create a high adhesive force to water droplet. The force-distance curve shows a marked change when the sponge-like structure TiO₂ nanostructure was co-assembled with PTES and 0.2 mg/mL NC under identical experimental conditions (Figure 4b). It is obvious that an attractive force is created as soon as this sticky, superhydrophobic surface makes contact with the water droplet. When the sample was withdrawn from the droplet, the adhesive force gradually increases and the droplet shape changes from spherical to elliptical. Then, just before the detachment, it yields the highest adhesive force of approximately of 76.6 µN. To our surprise, no any detectable water remained on the PTES-NC modified film surface is observed. The final balance force comes back to nearly zero also confirms that there is no water abrupt loss, except a faint evaporation of the water droplet during the measurement process under ambient environment. This high adhesive force is fifteen times stronger than that required to remove a 3 mg water droplet away from the sole PTES-modified sponge-like TiO₂ nanostructure surface.

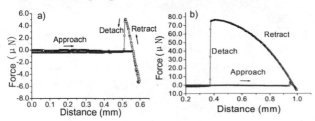

Figure 4. Typical force-distance curve of the sponge-like structure TiO₂ film approaches and retracts a 3 mg water droplet: (a) PTES modification; (b) PTES and 0.2 mg/mL NC mixed modification. The arrows represent the sample's moving direction to water droplet.

To determine the effects of NC concentration on the adhesion of PTES-NC modified TiO₂ nanostructure surfaces, other samples of superhydrophobic, sponge-like TiO₂ nanostructure modified with different NC concentrations were fabricated. The experimental results indicate that NC concentration has a drastic effect on adhesive force change. Figure 5 shows the influence of NC concentration in the sample on water CA and adhesive force. When the amount of NC increases from nothing to 0.04 mg/mL and 0.067 mg/mL, the adhesive forces between the water droplet and the sponge-like structure film increase sharply from 5.0 µN to 19.8 µN and 70.7 µN, respectively. It is noted that the largest deviation of the adhesion values measured in this study is ± 4.2 µN, indicating that the superhydrophobicity of the modified sponge-like surface is uniform and highly reproducible. In contrast, the water CA decreases gradually and then reaches a minimum near 153.5° with an increase of the NC concentration. This difference in the behavior between CA and adhesion suggests that other factors besides the wettability may be involved in the adhesion with water of these PTES-NC modified samples.

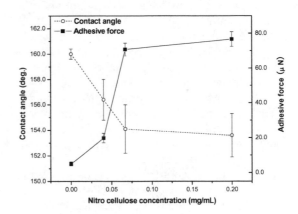

Figure 5. The inter-relationships between water CA and adhesive force with the NC concentration on the sponge-like structure TiO₂ surface.

A possible model for surface chemical composition change during the self-assembled process of mixed PTES-NC on hydroxylated sponge-like TiO₂ surfaces and the formation of hydrogen bonding between NC nitro group and water hydroxyl is shown in Figure 6. When NC is introduced into the PTES system, a competition occurs between NC and PTES molecules for the hydroxyl on hydroxylated TiO₂ surfaces (Figure 6a). This would lead to a disruption of the dense packed hydrophobic PTES molecule and thus a decrease of CA to some extent.[21] The hydrophilic nitro groups on the self-assembled layer surface can readily form hydrogen bonding with water (Figure 6b), which directly provides a good adhesion between the PTES-NC modified, sponge-like TiO$_2$ layer and water. Therefore, a combination of the hydrogen bonding offered by the nitro groups on the surface and the disruption of the dense packed hydrophobic PTES molecule is the primary factor responsible for the significant increase of the adhesion between PTES-NC modified, sponge-like TiO$_2$ film and water

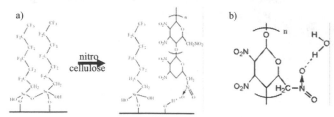

hydroxylated TiO$_2$ surface   hydrogen bonding

Figure 6. A model of surface chemical composition change during the PTES-NC mixed self-assembled layer on the hydroxylated sponge-like TiO$_2$ surfaces (a) and the conformation of hydrogen bonding association between nitro group and hydroxyl group of the water droplet (b).

3   CONCULSIONS

By using a facile electrochemical oxidation and mixed self-assembled processes with proper control of the nitrocellulose concentration, we have fabricated a sponge-like nanostructure of superhydrophobic TiO$_2$ thin films that have a wide range of the adhesive force. The results provide new insights into how to vary the wettability and adhesion of superhydropbobic surfaces. These stable TiO$_2$ nanostructure superhydrophobic films with exceptional adhesion effect can potentially be used in many industries, and be further extended to control the adhesion of a wide variety of superhydrophobic functional materials.

Acknowledgment. The authors thank the National Nature Science Foundation of China (Grants Nos: 50571085, 20773100, 20620130427) and the Key Scientific Project of Fujian Provincc, China (Grant No: 2005HZ01-3). The authors would like to thank Prof. Jiang Lei and Dr. Gao Xuefeng of the National Center for Nanoscience and Nanotechnology Center for Molecular Science for providing the access to facilities for water contact angle and adhesion measurement.

REFERENCES

[1] a) R. Wang, K. Hashimoto, A. Fujishima, M. Chikuni, E. Kojima, A. Kitamura, M. Shimohigoshi, T. Watanabe, Nature 388, 431, 1997. b) A. Nakajima, A. Fujishima, K. Hashimoto, T. Watanable, Adv. Mater. 11, 1365, 1999.

[2] K. Ichimura, S. K. Oh, M. Nakagawa, Science 288, 1624, 2000.

[3] a) Y. Liu, L. Ma, B. H. Liu, J. L. Kong, Chem. Eur. J. 11, 2622, 2005. b) R. Blossey, Nat. Mater. 2, 301, 2003.

[4] a) X. F. Gao, L. Jiang, Nature 432, 36, 2004. b) T. L. Sun, L. Feng, X. F. Gao, L, Jiang, Acc. Chem. Res. 38, 644, 2005. c) L. Feng, S. Li, Y. Li, H. Li, L. Zhang, J. Zhai, Y. Song, B. Liu, L. Jiang, B. Zhu, Adv. Mater. 14, 1857, 2002.

[5] a) J. Y. Shiu, C. W. Kuo, P. Chen, C. Y. Mou, Chem. Mater. 16, 561, 2004. b) P. S. Swain, R. Lipowsky, Langmuir 14, 6772, 1998. c) S. Herminghaus, Europhys. Lett. 52, 165, 2000.

[6] P. Aussillous, D. Quéré, Nature 411, 924, 2001.

[7] X. Y. Lu, C. C. Zhang, Y. C. Han, Macromol. Rapid Commun. 25, 1606, 2004.

[8] K. K. S. Lau, J. Bico, K. B. K. Teo, M. Chhowalla, G. A. J. Amaratunga, W. I. Milne, G. H. Mckinley, K. K. Gleason, Nano Lett. 3, 1701, 2003.

[9] Y. K. Lai, C. J. Lin, H. Wang, J. Y. Huang, H. F. Zhuang, L. Sun, Electrochem. Commun. 10, 387, 2008.

[10] R. Wang, K. Hashimoto, A. Fujishima, M. Chikuni, E. Kojima, A. Kitamura, M. Shimohigoshi, T. Watanabe, Adv. Mater. 10, 135, 1998.

[11] a) K. Autumn, Y. A. Liang, S. T. Hsieh, W. Zesch, W. P. Chan, T. W. Kenny, R. Fearing, R. J. Full, Nature 405, 681, 2000. b) A. K. Geim, S. V. Dubonos, I. V. Grigorieva, K. S. Novoselov, A. A. Zhukov, S. Y. Shapoval, Nat. Mater. 2, 461, 2003.

[12] M. H. Jin, X. J. Feng, L. Feng, T. L. Sun, J. Zhai, T. J. Li, L. Jiang, Adv. Mater. 17, 1977, 2005.

[13] Z. Yoshimitsu, A. Nakajima, T. Watanabe, K. Hashimoto, Langmuir 18, 5818, 2002.

[14] a) W. Chen, A. Y. Fadeev, M. C. Hsieh, D. Oner, J. F. Youngblood, T. J. McCarthy, Langmuir 15, 3395, 1999. b) D. Oner, T. J. McCarthy, Langmuir 16, 7777, 2000.

[15] M. Miwa, A. Nakajima, A. Fujishima, K. Hashimoto, T. Watanabe, Langmuir 16, 5754, 2000.

[16] A. B. D. Cassie, S. Baxter, Discuss. Faraday Soc. 40, 546, 1944.

[17] L. Gao, T. J. McCarthy, Langmuir 23, 9125, 2007.

[18] a) W. Chen, A. Y. Fadeev, M. C. Hsieh, D. Oner, J. F. Youngblood, T. J. McCarthy, Langmuir 15, 3395, 1999. b) A. Marmur, Langmuir 19, 8343, 2003.

[19] A. F. Wang, T. Cao, H. Y. Tang, X. M. Liang, C. Black, S. O. Sally, J. P. McAllister, G. W. Auner, H. Y. S. Ng, Colloid Surf. B-Biointerfaces 47, 57, 2006.

[20] J. Chastain, Handbook of X-ray Photoelectron Spectroscopy; Perkin-Elmer: Minnesota, 1992.

[21] D. Schondelmaier, S. Cramm, R. Klingeler, J. Morenzin, Ch. Zilkens, W. Eberhardt, Langmuir 18, 6242, 2002.

Silane-based water repellent and easy-to-clean surfaces for concrete structure improvement

M. A. Kargol[*]

[*]Danish Technological Institute, Concrete Center, Gregersvej, DK-2630 Taastrup, Denmark, marta.kargol@teknologisk.dk

ABSTRACT

Long-term performance of concrete surfaces is one of the most important problems in the construction film. Modifications of concrete surfaces by targeted deposition of surface functional additives that make the surface hydrophobic allowed to improve the aesthetic durability of the concrete. Silane-based water repellent materials using modified silicon precursor were synthesized in a laboratory and applied on concrete substrates. The initial chemical and physical properties of the treated and untreated surface were tested at two different curing times of concrete specimens with water-to-cement ratios of 0.38 and 0.5, and using different silane-based protective systems. The longer curing time of the concrete specimens the slightly better hydrophobic effect was observed after treatment. The interactions between the applied chemical protective treatments and the concrete surface were examined. We presented a few preliminary experiments and results.

Keywords: silanes, water repellent surfaces, concrete treatment

1 INTRODUCTION

The durability of concrete structures decreases over time due to the deterioration of its constituents that are exposed to the outside environment. Most of the damage can be caused by water since aggressive agents such as chloride, sulphate, water-dissolved carbon dioxide and sulphur dioxide penetrate into concrete and react harmfully with the cement paste [1].

In addition to its civil engineering properties, concrete also has an architectural and aesthetic function. When the concrete structure ages, the visible appearance of the surface gradually changes as it is exposed to the environment, i.e. weather, pollution and biological growth. Often this aging does not take place homogeneously over the surface which combined with dirt and biological deposition of the surface reduces the aesthetic appearance of concrete. Modifications of concrete surfaces by targeted deposition of surface functional additives, e.g. with agents that make the surface hydrophobic, allow to minimize the visual aging and thus improve the aesthetic durability of the concrete. The long-lasting hybrid sol-gel systems developed and applied for plastic and metal surfaces are not always applicable to more complex concrete surfaces due to its heterogeneous, porous and rough nature.

Water repellent and easy-to-clean coatings for inorganic substrates gained strong attention during the last few years and various formulations, based on silicones or alkylpolysiloxanes besides others were developed [2]. The advantages of easy-to-clean treated materials are that dirt is easier removed. Moreover, aggressive cleaning agents are not necessary to be used for the created low energy surfaces. Water repellent effect of the substrate treated with silicones is caused by the formation of a hydrophobic organic polysiloxane thin film on the substrate, which changes the contact angle of water with the capillary surface and therefore reduces the capillary absorption. The hydrophobic polysiloxane is mainly chemisorbed on the substrate trough siloxane bonds [3].

So far, there has been few works published concerning the effectiveness of surface treatments in improving concrete durability. The effect of the protection treatment has been evaluated by testing water absorption and water evaporation [4-8], chloride permeability [6-7, 9-10] and chloride diffusion [7, 8] or water and water vapor permeability [4-5, 10]. However, there is deficiency of published literature on the performance of surface treatments for concrete with relation to the microstructure of treated concrete surface and coating-substrate interactions.

The aim of this work was to gain knowledge on the interactions between the applied chemical protective treatments and concrete surface, thereby to obtain a more fundamental understanding of the physical and chemical properties responsible for the aging of the concrete surface. The experiments were carried out on high performance concrete specimens of different composition and history, and treated with silane-based material developed in a laboratory. Moreover, the concrete specimens were treated with free different available coating systems for comparison purpose.

The initial chemical and physical properties of the treated and untreated surface were tested, i.e. hydrophobicity, surface morphology, penetration depth, adhesion. Here we presented a few preliminary experiments and results.

2 EXPERIMENTAL

Two concrete mixes with different water-to-cement (w/c) ratios were manufactured and examined at two different curing times. Old concrete with w/c ratio of 0.38 corresponds to 150 days of curing (O *I*) and young concrete with w/c ratios of 0.38 and 0.5 corresponds to 14 days (Y *I* and Y *II* respectively). Concrete mix proportions and compressive strength are reported in Table 1.

Silane-based water repellent materials using modified silicon precursor were synthesized and applied on concrete substrates. The silicon precursor was obtained during the hydrolysis and condensation reactions of tetraethoxysilane in the presence of hydrochloric acid (0.2 M). A homogeneous solution was stirred under reflux at 323 K for 5 h. The excess of ethanol was removed under reduced pressure. A stable sol was modified with alkylalkoxysilane to improve its hydrophobic properties and then applied with a spray pistol on the concrete surfaces.

Additionally, three different commercial products were applied on concrete substrate: solvent-free hydrophobizing alkylalkoxysilane, aqueous 'easy-to-clean' silane-based solution and aqueous 'anti-graffiti' silane-based solution. The experiments were performed on cubic concrete specimens with a size of 70×100×20 mm and 50×70×25 mm. All the specimens were treated at the same time in order to maintain the chemical and physical characteristic of the material.

The untreated and treated samples were dried at room temperature and kept in the plastic box with low humidity. The hydrophobic effect was first judged by water dropping on the treated concrete surface and the appearance of beading-effect.

The static water contact angles of various treated concrete surfaces were measured by a Drop Shape Analysis System DSA 10 Mk2 apparatus (Krüse GmbH Germany). A water drop of 5 μl from a syringe was placed on the treated sample. After the tip of the needle was separated from the drop, the CCD camera captured the side view of water drop and the contact angle was measured using a profile of the drop.

	O *I*	Y *I*	Y *II*
w/c ratio	0.38	0.38	0.5
Curing time (days)	150	14	14
Cement CEM II/LL (kg/m^3)	493.2	493.2	421.3
Aggregate 0/4 NCC RN E 0/4 (kg/m^3)	669.2	669.2	669.2
Aggregate 1/4 NCC Vesterhavsral (kg/m^3)	1013.6	1013.6	1013.6
Superplasticizer BASF Glenium 540 (kg/m^3)	5.15	5.15	2.25
Water (kg/m^3)	187.4	187.4	210.6
Density (kg/m^3)	2278	2278	2225
Compressive strength (N/mm^2)	70	70	46.7

Table 1: Concrete mix proportions and compressive strength.

For each sample, at least three measurements from different surface locations were averaged.

Microstructural features of untreated and treated samples were investigated by scanning electron microscopy (Philips XL 30) equipped with an energy-dispersive X-ray (EDX) unit.

3 RESULTS AND DISCUSSION

In order to have a preliminary surface characterization in terms of hydrophobicity of the treated surface, static contact angles were determined by using water as a probe liquid. The treated concrete surfaces were highly hydrophobic without any correlation with two different water-to-cement ratios of substrate.

Data presented in Table 2 show that water contact angles were ranged between 101 ° and 133 ° indicating a very low wettability with respect to probe liquid. The low values of standard deviation may be considered as evidence of high surface homogeneity and relatively smooth surfaces after treatment [11].

	O *I*		Y *I*		Y *II*	
	0.38		0.38		0.5	
Coating system	$\theta(°)$	$\delta(°)$	$\theta(°)$	$\delta(°)$	$\theta(°)$	$\delta(°)$
Developed silane	120	2	116	3	117	3
Hydrophobizing silane	105	4	101	2	102	2
Easy-to-clean silane	130	2	132	1	133	1
Anti-graffiti silane	117	3	111	2	114	5

Table 2: Water static contact angles, θ and standard deviations, δ measured for different coating systems.

Penetration depth of applied silane-based systems was preliminarily studied by scanning electron microscopy. The concrete specimens were impregnated with silver nitrate solution for 24 hours before microscopic analysis due to similarity in the composition of applied coating systems and substrates. SEM micrographs for concrete specimen treated with developed silane-based material are presented in Figure 1. We clearly observed the brighter layer of the carbonated paste in the concrete surface. The presence of dark thin layer on the top can indicate the silane-based solution penetrating the concrete specimen. However, this statement requires further investigation.

Since the aim in porous materials is mostly to prevent water penetration, the decrease in water uptake should be the criterion for evaluation of hydrophobic treatments' effectiveness. Measuring the contact angle after treatment probably can be right in a theoretical sense. If a surface is exposed to contaminants like oil or hydrophilic particles, the surface free energy is changed, so is the contact angle. However, the contamination might be only superficial, within the pores the applied treatment can be still active and the water repellency is intact. For this purpose, the treated concrete specimens were exposed to deionised water for several months.

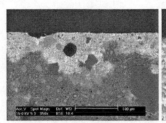

Figure 1: SEM micrographs for concrete surface treated with developed silane-based material and impregnated with silver nitrate solution.

Only the treated concrete surface was in contact with water while the remaining five sides of cubic concrete specimens were sealed with an epoxy resin. The water uptake will be investigated in our further studies.

4 CONLUSIONS

The concrete specimens with different history and composition were manufactured and examined. We obtained a stable water repellent solution using modified silicon precursor for concrete structure improvement. Hydrophobic effect of the treated concrete surfaces with different curing times was slightly better for older concrete specimens. The initial SEM experiments allowed to distinguish the penetrated part of the treated concrete specimen. Effect of surface treatments on aesthetic durability of the concrete surface when exposed to physical/chemical aging will be investigated. Moreover, studies of the microstructural and porosity changes of untreated, treated and leached specimens using SEM and SEM-EDX will be performed. Principle research scientist

ACKNOWLEDGEMENT

Financial support for this research is provided by the European Research Network NANOCEM. This support is highly appreciated.

REFERENCES

[1] T. P. Lees, in Durability of Concrete Structures, edited by G. Mays (E&FN Spon, London), 15, 1992.

[2] S. Giessler, E. Just, R. Störger, Easy-to-clean Properties-Just a Temporary Appearance?, Thin Solid Films, 502, 252-256, 2006.

[3] K. Ren and D. A. Kagi, Study of Water Repellent Effect of Earth Substrates Impregnated with Water-Based Silicones, J. Chem. Tech. Biotechnol., 63, 237-246, 1995.

[4] A. Barbucci, M. Delucchi, G. Cerisola, Organic Coatings for Concrete Protection: Liquid Water and Water Vapor Permeabilities, Progress in Organic Coatings, 30, 293-297, 1997.

[5] M. Delucchi, A. Barbucci, G. Cerisola, Study of the Physico-Chemical Properties of Organic Coatings for Concrete Degradation Control, Construction and Building Materials, 11, 365-371, 1997.

[6] M. Levi, C. Ferro, D. Regazzoli, G. Dotelli, A. Lo Presti, Comparative Evaluation Method of Polymer Surface Treatments Applied on High Performance Concrete, Journal of Materials Science, 37, 4881 – 4888, 2002.

[7] A. A. Almusallam, F. M. Khan, S. U. Dulaijan, O. S. B. Al-Amoudi, Effectiveness of Surface Coatings in Improving Concrete Durability, Cement and Concrete Composites, 25, 473–481, 2003.

[8] S. Assié, G. Escadeillas, V.Waller, Estimates of Self-compacting Concrete 'Potential' Durability, Construction and Building Materials, 21, 1909– 1917, 2007.

[9] L. Basheer, D. J. Cleland, A. E. Long, Protection Provided by Surface Treatments against Chloride Induced Corrosion, Materials and Structures, 31, 459-464, 1998.

[10] M. P. M. C. Rodrigues, M. R. N. Costa, A. M. Mendes, M. I. E. Marques, Effectiveness of Surface Coatings to Protect Reinforced Concrete in Marine Environments, Materials and Structures, 33, 618-626, 2000.

[11] P. S. Swain and R. Lipowsky, Contact Angles on Heterogeneous Surfaces: A New Look at Cassie's and Wenzel's Laws, Langmuir, 14, 6772-6780, 1998.

Nanotube Formation and Surface Study of New Ternary Titanium Alloys

*H.C. Choe, Y.M. Ko, ¹W. A. Brantley

College of Dentistry, Chosun University, 2ⁿᵈ stage of Brain Korea 21 for College of Dentistry, Gwangju, Korea
¹College of Dentistry, Ohio State University, Columbus, OH, USA
*hcchoe@chosun.ac.kr

ABSTRACT

New titanium alloys with improved biocompatibility have been developed for dental and biomedical applications. The osseointegration of implants depends on surface properties of these alloys, and the nature of the oxide film at the metal-oxide-bone. Also, stress acting on the interface between bone and implant can influence the bone resorption, due to stress-shielding effects. So we have tried to improve this effect by manufacture of new ternary alloy containing Ta, Nb, and Zr alloying elements. In this study, nanotube formation and surfaces of new ternary titanium alloys have been investigated using various electrochemical methods after anodizing on the Ti-30Ta-XZr and Ti-30Nb-XZr alloy surfaces, where X = 3 and 15 (wt %) The polarization resistance of nanotube-formed new ternary alloys was lower than that of the corresponding non-nanotube-formed alloy. The diameter and depth of the nanotubes could be controlled, depending upon the composition and titanium alloy phased for osseointegration of bio-implant.

Keywords: Nanotube, Corrosion, Polarization resistance, Ti-Ta-Zr, Ti-Nb-Zr alloy, Implant

1 INTRODUCTION

Pure titanium and its alloys are drastically used in implant materials due to their excellent mechanical properties, high corrosion resistance and good biocompatibility [1]. However, the widely used Ti-6Al-4V is found to release toxic ions (Al and V) into the body, leading to undesirable long-term effects. Ti-6Al-4V has much higher elastic modulus (100-120GPa) than cortical bone (10-30 GPa) [1]. Therefore, titanium alloys with low elastic modulus have been developed as biomaterials to minimize stress shielding [2].

Recently, Ti-Nb and Ti-Ta based alloy systems have been studied and found to display both lower elastic moduli

and higher tensile strengths than are common for metals and alloys. Nb, and Ta can be stabilized to β-phase of Ti alloy and β-phase structure exhibits about 60-80 GPa of Young's modulus [1]. Some researcher reported that the Ti-30%Ta alloy with martensite α"-phase has the potential to become a new candidate for biomedical application due to its good combination of low modulus and high strength. To improve bone tissue integration on implant surfaces, various techniques have been used to increase the roughness of the implant surfaces [3]. Cell adhesion and proliferation depend on surface roughness [4] and metal ion dissolution [5]. The electrochemical formation of novel highly ordered oxide nanotube layers has been reported for Ti anodization in fluoride-containing acid electrolytes at moderate voltage[6]. Nanotube formation on the Ti oxide is important to improve the cell adhesion and proliferation in clinical use. It is possible to control nanotube size and morphology for biomedical implant use. Factor is applied voltage, alloying element, current density, time, and electrolytes. Recently, many results of nanotube formation have been reported on effects of these factors without consideration of alloying element in ternary alloys,

In this study, nanotube formation and surface characteristics of Ti-30Nb-xZr and Ti-30Ta-xNb alloy with low elastic modulus have been investigated using various electrochemical methods.

2 MATERIALS AND METHODS

Ternary Ti-30Nb-xZr(x=3,15wt%) and Ti-30Ta-xNb (x=3,15wt%) alloys were prepared by using high purity sponge Ti (G&S TITANIUM, Grade. 4, USA), Ta, Zr and Nb sphere (Kurt J. Lesker Company, 99.95% wt.% in purity). Two kinds of Ti alloys prepared using the vacuum arc melting furnace. The weighed charge materials were prepared by the vacuum arc furnace, under the purified Ar gas into water cooling copper hearth chamber in vacuum atmosphere of 10^{-3}torr, and controlled atmosphere in chamber by method to keep vacuum again. Also, before melting, the constituents were cleaned with methanol to minimize oxygen quantity and surface contaminants in chamber and pure Ti was melted six times with purified argon gas. After that, melting treatment was carried out more than six times by reversing the alloy sponges in order to homogenize, and each time was held in the molted state for 5 min. in a high purity argon atmosphere by arc furnace with water-cooled copper hearth and solution treatment was carried out for 1hr at 1050°C in an argon atmosphere, followed by water quenching to stabilize the β phase

Microstructures of the alloys were examined by optical microscopy (OM, OLYMPUS BM60M, JAPAN) and scanning electron microscopy (SEM, HITACHI-3000, JAPAN). The specimens for the OM and SEM analysis were etched in Keller's solution consisting of 2 mℓ HF, 3 mℓ HCl, 5 mℓ HNO$_3$ and 190 mℓ H$_2$O.

Nanotube formation was carried out with a conventional three-electrode configuration with a platinum counter electrode and a saturated calomel (SCE) reference electrode. Experiments were performed in 1M H$_3$PO$_4$ with small additions of NaF(0.8wt%). All experiments were conducted at room temperature. Electrochemical treatments were performed by using potentiostat (EG&G Co, 362, U.S.A). The electrochemical treatment consisted of a potential ramp from the open-circuit potential (Eocp) to an end potential at 10 V with a scan rate of 500mV/s followed by holding the sample at 10V for 2h.

In order to investigate the corrosion behavior, potentiostatic test was carried out using a potentiostat(EG&G PARSTAT 2273) test at constant potential 300 mV in 0.9% NaCl solution at 36.5±1℃. Electrochemical impedance spectroscopy was performed (10 mHz to 100 kHz) in 0.9% NaCl solution at 36.5±1℃. All electrochemical characteristics were performed in a standard three-electrode cell having specimen as a working electrode and a high carbon as counter electrode. The potential of working electrode was measured against a saturated calomel electrode (SCE) and all given potentials were referred to this electrode. After electrochemical corrosion tests, the surfaces of each specimen were investigated by using FE-SEM.

3 RESULT AND DISCUSSION

Fig. 1 is FE-SEM micrographs showing the microstructure of homogenized Ti alloy surface. Figure 1(a) and (b) show the microstructure of Ti-30Ta-3Zr and Ti-30Ta-15Zr. Figure 1(c) and (d) show the microstructure of Ti-30Nb-3Zr and Ti-30Nb-15Zr alloy, respectively. The microstructures of the Ti-30Ta-xZr alloys had a needle-like appearance. And the thick of needle like -α structure increased as Zr content increased. It is confirmed that Zr content played a role to stabilizer of α phase[7]. The quenched alloys exhibit the lamellar martensite α' structure at below 20wt% Ta, and exhibit the needle-like orthorhombic martensite α" structure from 30 to 50wt% Ta. We confirmed that the microstructure changes from α phase to β-phase through XRD and β phase increases, as Nb content increases[7]. With these results, we could make an analogical inference that Nb and Ta is the β stabilizing element as the β-phase increases according to the addition of Nb and Ta. The literature reports that the amount of needle-like orthorhombic martensite α" structure increases with increasing cooling rate due to the transformation of β phase into α" phase[8].

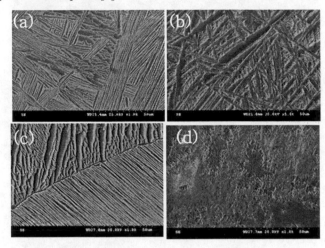

Fig. 1. FE-SEM micrographs of homogenized Ti-30Ta-xZr and Ti-30Nb-xZr alloys.
(a) Ti-30Ta-3Zr   (b) Ti-30Ta-15Zr
(c) Ti-30Nb-3Zr   (d) Ti-30Nb-15Zr

Fig. 2 shows the evolution of TiO$_2$ nanopores grown from Ti films in 1M H$_3$PO$_4$ + 0.8wt% NaF electrolyte solution for 2h at an applied anodic potential of 10V. . Figure 2(a) and (b) show the nanotube morphology of Ti-30Ta-3Zr and Ti-30Ta-15Zr. Figure 2(c) and (d) show the nanotube morphology of Ti-30Nb-3Zr and Ti-30Nb-15Zr alloy, respectively. Nanotube formed in β phase showed the regular array with pore diameters around 150nm and the wall thickness in the range of 20nm and 100nm interspace of nanotube, but nanotube formed in α phase showed irregular array with small(50nm) and large diameter(200nm) of nanotube. It can be concluded that α phase has Zr rich with martensite α" structure and high stress energy in matrix. Consequently, and stable nanotube can be formed in β phase compared to α phase. Small and large nanotube morphplogy showed predominantly in case of increasing Zr content to β phase.

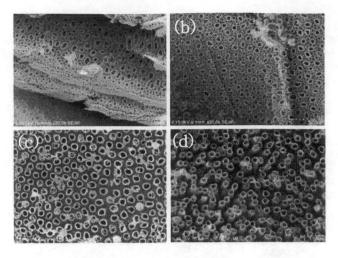

Fig. 2. FE-SEM image of Ti-Nb-Zr alloy nanotube layer formed in 1M H₃PO₄ + NaF(0.8wt%) for 2h at 10V.
(a) Ti-30Ta-3Zr   (b) Ti-30Ta-15Zr   (c) Ti-30Nb-3Zr,
(d) Ti-30Nb-15Zr

structure is easily dissolved in halides ion contained electrolytes [9]. The high value of R_p implies a high corrosion resistance of alloy, that is, a low rate of released metallic ion into the electrolytic solution or nanotube on the surface. The Bode plot results indicated that the corrosion behavior of alloy in solution was under charge-transfer controlled because of the local variation of aggressive ion like Cl⁻, preferentially attacking or damaging the oxide film and nanotube. The Bode plots for all the alloys before nanotube formation showed near capacitive response in the high and middle frequency region. The Bode plots for all the alloys after nanotube formation showed near capacitive response in the lower and middle frequency region which was characterized by slope≈ -1 in the log|z| vs. log(f) curve. It seems that the alloying element has been responsible for the resulting better corrosion resistance of nanotubed Ti-30Ta-xZr alloy and Ti-30Nb-xZr alloy. From potentiostatic test, surface stability (current density vs time) of nanotubed alloy showed the lower than that of non-nanotubed alloy without Ti-30Nb-3Zr alloy.

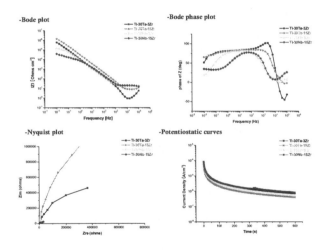

Fig. 3. AC impedance and potentistatic results of non-nanotube formed surface in in 0.9% NaCl solution at 36.5±1℃. (a) Bode plot, (b) Bode-phase plot, (c) Nyquist plot, (d) Potentiostatic curves

The polarization resistance(R_p)of non-nanotube formed alloy is higher than that of nanotube formed alloy. From impedance tests of non-nanotubed alloys, the polarization resistance of Ti-30Ta-3Zr alloy was higer than that of Ti-30Nb-xZr alloy, whereas, in case of nanotubed alloys, the polarization resistance of Ti-30Nb-3Zr alloy was higer than that of other alloy. Zr additions to Ti-30Nb alloy have improved electrochemical corrosion behavior due to nobler characteristics to the alloys [8]. The polarization resistance of nanotubed alloy was lower than that of non-nanotubed alloy as shown in Table 1. It was considered that nanotubed surface has unstable surface due to amorphous structure without heat treatment for crystallinity. Amorphous

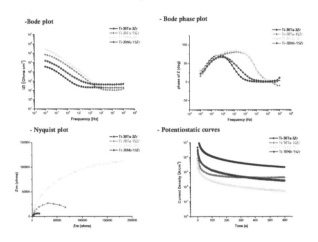

Fig. 4. AC impedance and potentistatic results of nanotube formed surface in in 0.9% NaCl solution at 36.5±1℃. (a) Bode plot, (b) Bode-phase plot, (c) Nyquist plot, (d) Potentiostatic curves

Table 1. EIS parameters of non-nanotubed and nanotubed Ti-Ta-Zr and Ti-Nb-Zr alloys.

Non treatment	Ti-30Ta-3Zr	Ti-30Ta-15Zr	Ti-30Nb-3Zr	Ti-30Nb-15Zr
Rp (Ω cm²)	5.87×10^5	1.35×10^6	1.46×10^5	3.77×10^4
Rs (Ω cm²)	0.873	7.6575	15.225	15.53
Nanotube formed	Ti-30Ta-3Zr	Ti-30Ta-15Zr	Ti-30Nb-3Zr	Ti-30Nb-15Zr
Rp (Ω cm²)	3.49×10^3	6.93×10^4	2.53×10^5	$1.56 \times 10^{4^4}$
Rs (Ω cm²)	19.756	10.103	22.686	41.874

NSTI-Nanotech 2008, www.nsti.org, ISBN 978-1-4200-8503-7 Vol. 1

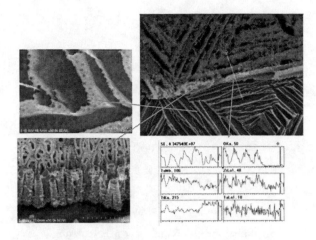

Fig. 5. FE-SEM image and line profiles of Ti-30Nb-3Zr alloy nanotube layer formed in 1M H_3PO_4 + NaF(0.8wt%) for 2h at 10V after potentiostatic corrosion test.

Fig. 5 shows FE-SEM image and line profiles of Ti-30Nb-3Zr alloy nanotube layer formed in 1M H_3PO_4 + NaF(0.8wt%) for 2h at 10V after potentiostatic corrosion test. Some nanotube morphology was changed from circle to parabolic style and tore off from tip of tube compared to non-corroded surface (Fig.2). From cross-section image shows pore tubes with many rings on their wall. This is that the regularity of the rings corresponds to the periodicity of current oscillations of current –time curves [10]. From line profile, Nb content in the nanotube covered region was slightly higher than that of uncovered region due to formation of Nb_2O_5 on the nanotube film.

4   CONCLUSIONS

The corrosion and polarization resistance of nanotube-formed new ternary alloys was lower than that of the corresponding non-nanotube-formed new ternary alloy.
The diameter and depth of the nanotubes could be controlled, depending upon the composition and titanium alloy phased for osseointegration of bio-implant.

ACKNOWLEDGEMENT

"This work was supported by research funds from 2nd stage of Brain Korea 21 for College of Dentistry"

REFERENCES

[1] G. He, M. Hagiwara, "Ti alloy design strategy for biomedical applications", Materials Science and Engineering, C 26, 14-19, 2006.
[2] D. Kuroda, M. Niiomi, "Design and mechanical properties of new β-type titanium alloys for implant materials", Materials Sci and Eng A, 243, 244, 2001.
[3] H.C. Choe, Y.M.Ko, W.A.Brantley, "Nano-Surface Behavior of Osteoblast Cell-Cultured Ti-30(Nb,Ta) with Low Elastic Modulus" NSTI-Nanotech-2007, 2, 744-747, 2007.
[4] K. Anselme, "Osteoblast adhesion on biomaterials", Biomaterials, 21, 667-681, 2000.
[5] T. Hanawa, "Metal ion release from metal implants", Matr. Sci. & Eng., 24, 745-752, 2004.
[6] V. Zwilling, M. Aucouturier, E. Darque-Ceretti, A. Boutry-Forveille, " Anodic oxidation of titanium and TA6V alloy in chromic media, An electrochemical approach," Electrochemi Acta, 45, 921-929, 1999.
[7] J. J. Park, H. C. Choe, Y.M. Ko, "Corrosion Characteristics of TiN and ZrN coated Ti-Nb alloy by RF-sputtering", Materials Science Forum , 539-543, 1270-1275, 2007.
[8] D. Q. Martins, W. R. Osorio, M.E.P.Souza, R. Caram, A. Garcia, Electrochemica Acta, 53, 2809-2817, 2008.
[9] H.C. Choe, Y.M.Ko, S.S.Kim, W.A.Brantley, "Surface Characteristics of Tooth Ash Coatrings on Ti and Ti-6Al-4V Deposited by Pulsed Laser Method", AEPSE2007, NSTI-Nanotech-2007, 41, 2007.
[10] Jan M. Macak, K. Sirotna, P. Schmuki, "Self-organized porous titanium oxide preparaed in Na2SO4/NaF electrolytes", Electrochemical Acta, 50, 3679-3684, 2005.

Phenomena of Nanotube Formation on the Surface of Cp-Ti, Ti-6Al-4V, and Ti-Ta alloys

K. Lee, *H.C. Choe, Y.M. Ko, [1]W. A. Brantley

College of Dentistry, Chosun University, 2nd stage of Brain Korea 21 for College of Dentistry, Gwangju, Korea
[1]College of Dentistry, Ohio State University, Columbus, OH, USA
*hcchoe@chosun.ac.kr

ABSTRACT

In this study, the phenomena of nanotube formation on the surface of Ti alloys by electrochemical method. Nanotube layer formed on Ti alloys in 1 M H_3PO_4 electrolyte with small additions of F^- ions. The pore growth behavior was clearly different on individually phase and was also different for alloys. On the α-phase of Cp-Ti and β-phase of Ti-30Ta alloy, the nanotube showed clearly highly ordered TiO_2 layer. In case Ti-30Ta alloy, the pore size of nanotube was smaller than of Cp-Ti due to β stabilizing Ta element. In this case of Ti-6Al-4V alloy, the α phase showed stable porous structure; the β-phase was dissolved entirely.

Keywords: Nano-structure, Ti-Ta alloy, dental implant

1 INTRODUCTION

Commercial pure titanium(Cp-Ti) and TI-6Al-4V are widely used as a dental root implant material in clinical dentistry and as orthopedic implant material because of it mechanical strength, stability, and good compatibility with bone tissue[1-3]. This high degree of biocompatibility is thought to be due to the titanium surface forming a stable oxide layer(TiO_2 matrix) that presumably aid in the formation of an osteogenic extracellular matrix at the implant-bone tissue interface, which is important for osseointergration[4-5]. Recently published experimental studies seem to indicate that on nanotube formation, a nanotube of the thickness of the native oxide, will result in very strong reinforcement of the bone response. Webster et al.[6] reported that on nanograined ceramics such as alumina oxide and titanium oxide (~100 nm regime), improved bioactivity of implant and enhanced osteobalst adhesion are observed. Since the appearance of the first report on formation of nanoporous anodic oxide film of Ti and Ti-6Al-4V alloy in 1999 by Zwilling et al.[7], a significant number of reports is available on the anodization of Ti in different fluoride solutions [8-10]. Synthesis of homogenously ordered straight forward because of a sing phase microstructure, however, show a dual phase α+β microstructure. The α-phase, which has a hcp structure, is enriched with hcp stabilizing element such as Al, oxygen, nitrogen, etc.: whereas β-phase, with a bcc structure, is enriched with beta-stabilizing elements such as V, Nb, Mo, Ta, etc. because of the difference in chemistries of these phases, the formation of the nanotubluar oxide layer was not uniform in the Ti alloys having dual phase microsture as one phase could get etched preferentially by the electrolyte. Tsuchiya et al. [11] anodized Ti -28Nb-13Ta-4.6Zr alloy in two different microstructural conditions and observed uniform presence of ordered oxide nanotubes when the microstructure was a single β-phase. In the dual phase (α+β) condition, the oxide layer was not uniform and contained nano-scratches and etched pore wall.

In this study, the phenomena of nanotube formation on the surface of Ti alloys by electrochemical method. Nanotube layer formed on Ti alloys in 1 M H_3PO_4 electrolyte with small additions of F^- ions. The pore growth behavior was clearly different on individually phase and was also different for alloys. On the α-phase of Cp-Ti and β-phase of Ti-30Ta alloy, the nanotube showed clearly highly ordered TiO_2 layer.

2 MATERIALS AND METHODS

2.1 Alloy preparation

Titanium and the following titaium alloys were studied : Ti-6Al-4V and Ti-30Ta (the numbers signify the wt.%). They were manufactured by arc melting on a water-sealed copper hearth under an argon gas atmosphere with a non-consummable tunsten elecetrode. Each ingot was melted six times by invering the metal to ensure homogeneous melting. Fig. 1 shows the heating condition for Ti alloys. Cp-Ti and Ti-30Ta alloy were homogenized in argon atmospher at 1000℃(which is above the β transus temperature) for 24h followed by a rapid quenching in ice water. Ti-6Al-4V was carried out at 1000℃ for 12h. After water quenching the Ti-6Al-4V were aged at 560℃ for 2h and water quenching. They cut to a diameter of 10 ㎜ and a thickness of 2 ㎜ followed by diamond cutter. Prior to anodizing, polished

NSTI-Nanotech 2008, www.nsti.org, ISBN 978-1-4200-8503-7 Vol. 1

sample were etched (the mixture of HNO_3 and HF aqucous solution) to remove the surface oxided, followed by ultra sonic cleaning in acetone for approximately 10 min.

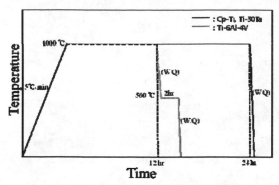

Figure 1: Schematic diagram to illustrate the heating condition for the Ti alloys

2.2 Anodization treatments

Electrochemical experiments were carried out with conventional three-electrode configuration with a platinum counter electrode and a saturated calomel reference electrode. The sample was embedded with epoxy resin, leaving a square surface area of 10 ㎟ exposed to the anodizing electrolyte, 1 M H_3PO_4 containing 0.64 wt.% NaF. Anodization treatments were carried out using a scanning potentiostat (EG&G Co., Model 362, USA). All experiments were conducted at room temperature. The electrochemical treatments consist of potential was first swept from the open-circuit potential to desired final potential with a sweep rate of 500 mV/s, then the potential was held for 2h. After the treatments the anodized sample were rinsed with distilled water and dried with dry air stream.

2.3 Material characterization

The microstructures of the studied alloys were investigated by optical microscope (OM, Olympus BM60M)) and scanning electron microscope (Hitachi S-3000). The samples for the OM and SEM analysis were etched in Keller's solution. In order to identified the phase constitutions of Ti alloys, X-ray diffraction (XRD, X-pert Pro) analysis with a Cu-Kα radiation were performed.

Structural characterization of the anodized samples was carried out with a field emission scanning electron microscope (FE-SEM, Hitachi S-4800). The scanning electron microscope was capable of energy dispersive spectrometer(EDS, Oxford).

3 RESULT AND DISCUSSION

3.1 Microstructural observations

Fig. 1 shows the optical microstructures of Cp-Ti, Ti-6Al-4V and Ti-30Ta alloys. It controlled by heat treatment in Ar atmosphere followed by 0℃ water quenched. It exhibits martensite structure, and Cp-Ti alloy was acicular structure of α-phase, Ti-6Al-4V alloy was lamellar structure of α+β phase and Ti-30Ta alloy was needle-like structure of β-phase.

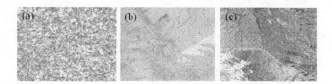

Figure 2 : OM micrographs showing the microstructure of heat treated Ti alloys (× 200) . (a) Cp-Ti, (b) Ti-6Al-4V, (c) Ti-30Ta

Fig. 2 show the SEM image and EDS result of Cp-Ti, Ti-6Al-4V and Ti-30Ta alloys. It can be seen that acicular, lamellar and needle-like structure, respectively. As a result of EDS, chemical composition was in accord with alloys design.

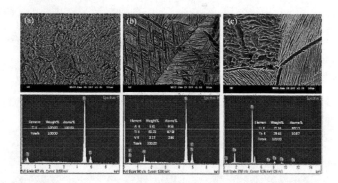

Figure 2 : SEM micrographs and EDS results of heat treated Ti alloys (a) Cp-Ti, (b) Ti-6Al-4V, (c) Ti-30Ta

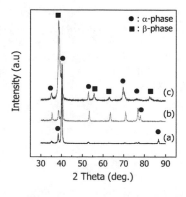

Figure 3: XRD diffraction patterns of heat treated Ti alloys (a) Cp-Ti, (b) Ti-6Al-4V, (c) Ti-30Ta

Fig. 3 shows the X-ray diffraction profiles of Cp-Ti, Ti-6Al-4V and Ti-30Ta alloys. The peaks were indexed using the JCPDS diffraction data of Ti-alloys. Cp-Ti alloy and Ti-30Ta alloy can be seen that only reflections from α and β phase, respectively. And Ti-6Al-4V alloy was identified α+β phase. Ti-30Ta alloy has the beta phase in XRD peaks.

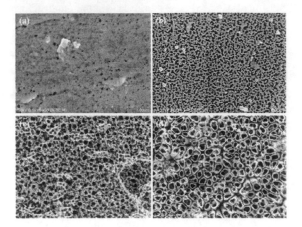

Figure 4: FE-SEM images of the anodic porous oxide layers form at final voltage after periods of potential sweep : (a) 0 min, (b) 30 min, (c) 60 min, (d) 120 min

Fig. 4 shows FE-SEM images of the anodic porous oxide layers formed at final voltage after periods of potential sweep. Nanotube diameter increased predominantly as anodized time increased. The nuclei of nanotube on Cp-Ti surface is very small in case of anodizing time for 30min as shown in Fig. 4(b), whereas, in the case of anodizing time for 120min, nanotube diameter size is about 100nm. It is confirmed that nanotube of diameter and depth depend on anodizing time.

Fig. 5 shows (a) top and (b) cross-section of anodized on Cp-Ti surface. The clear nanotube structures with large diameter have been formed on Cp-Ti compared to other specimen. The nanotubes of Cp-Ti have an inner average diameter of 80~100 nm with a tube-wall thickness of about 20nm. Ti-6Al-4V alloy shows stable porous nanotube formation in the only α-phase structures, whereas, sponge-like structure was formed in β-phase.

The nanotube size formed on β-phase of Ti-30Ta surface has smaller porous than that of nanotube formed Cp-Ti surface on α-phase surface. It can be concluded that it is not easy to dissolute oxide film for nanotube formation due to formation of Ta oxide like Ta_2O_5 on the passive film.(Fig 5a, b).

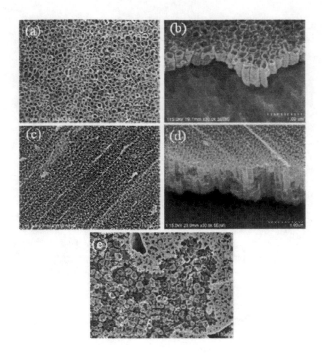

Figure 5: FE-SEM images of nanotube layers formed at different phase of Ti alloys (a) Cp-Ti top-view, (b) Cp-Ti cross-section, (c) Ti-30Ta top-view, (d) Ti-30Ta cross-section, (e) Ti-6Al-4V top-view

4 CONCLUSIONS

1. From the microstructure analysis, Cp-Ti showed acicular structure of α-phase, Ti-6Al-4V showed lamellar structure of α+β phase and Ti-30Ta showed needle-like structure of β-phase.
2. As a result of EDX, chemical composition was in accord with alloys deign. Cp-Ti, Ti-6Al-4V, and Ti-30Ta alloys, individually α, α+β and β were identified by XRD.
3. The nuclei of nanotube on Cp-Ti surface is very small in case of anodizing time for 30min whereas, in the case of anodizing time for 120min, nanotube diameter size is about 100nm. Therefore nanotube of diameter and depth depend on anodizing time.
4. The nanotubes of Cp-Ti have an inner average diameter of 80~100 nm with a tube-wall thickness of about 20nm. Ti-6Al-4V alloy shows stable porous nanotube formation in the only α-phase structures, whereas, sponge-like structure was formed in β-phase. The nanotube size formed on β-phase of Ti-30Ta surface has smaller porous than that of nanotube formed

ACKNOWLEDGEMENT

"This work was supported by research funds from Chosun University (2007) and 2nd stage of Brain Korea 21 for College of Dentistry"

REFERENCES

[1] A. Schoeder, F. Sutter, G. Krekeler, Oral implantolgy. Newyork : Thieme Medical, 35-58, 1991

[2] T. Albrektsson, P. I. Branemark, H. A. Hansson, K. Kasemo, K. Larsson. I. Loundstrom, D.H. McQueen, R. Skalak, "The interface zone of inorganic implants *in vivo* : Titanium implants in bone" Ann Biomed Eng, 11, 21-27, 1983

[3] D.F. Williams, "Titanium and titanium alloys : Biocompatibility of clinical implant materials", Boca Raton, FL:CRC Press, Vol 1, 9-44

[4] B. Kasemo, J. Lausmaa, "Biomaterial and implant surface : a surface scienc approach" Int J Oral Maxillofac Implant, 3, 247-259

[5] C.M. Stanford, J. C. Keller, "Osseoinetergration and matrix production at the implant surface" Crt Rev Oral Biol Med 2, 83-101, 1991

[6] T. J. Webster, C. Ergun, R. H. Doremus, R.W. Siegel, R. Bizios, "Enhanced functions of osteoblasts on nano-phase ceramics" , 21, 1803-1810, 2000

[7] V. Zwilling, E. Darque-Ceretti, A. Bountry-Forveille, D. David, M. Y. Perrin, M. Ancounturier, "Sturucture and Physicochemistry of Anodic oxide Films on Titanium and TA6V Alloy", Surf Interface Anal, 27, 629-637, 1999

[8] G. K. Mor, O. K. Varghes, M. Paulose, K. Shankar, C. A. Grines, "A review on highly ordered, vertically oriented TiO_2 nanotube arrays: Fabrication, material properties, and solar energy applications", Sol. Energy Mater Sol Cells, 90, 2011-2075

[9] S. P. Albu, A. Ghicov, J. M. Macak, P. Schmuki. "250 μm long anodic TiO2 nanotubes with hexagonal self-ordering" Phys Stat Sol, 1, R65-R67, 2007

[10] K. S. Raja, T. Gandhi, M. Misra," Effect of water content of ethylene glycol as electrolyte for synthesis of ordered titania nanotubes", Electrochem Commun, 9, 1069-1076, 2007

[11] H. Tsuchiya, J. M. Macak, A. Chicov, Y. C. Tang, S. Fusimoto, M. Niinomi, T. Noda, P. Schmuki, "Nanotube oxide coating on Ti–29Nb–13Ta–4.6Zr alloy prepared by self-organizing anodization", Electrochim Acta, 52, 94-101, 2006

Electrochemical Characteristics of Nanotube Formed Ti-Zr Alloy

W.G.Kim, *H.C. Choe, Y.M. Ko, [1]W. A. Brantley

College of Dentistry, Chosun University, 2nd stage of Brain Korea 21 for College of Dentistry,
Gwangju, Korea
[1]College of Dentistry, Ohio State University, Columbus, OH, USA
*hcchoe@chosun.ac.kr

ABSTRACT

In this paper, Ti-Zr(10, 20, 30 and 40 wt%) alloys were prepared by arc melting and nano-structure controlled for 24 hr at 1000 °C in argon atmosphere. Formation of oxide nanotubes are conducted by anodizing a Ti-Zr alloy in H_3PO_4 electrolytes with small amounts of fluoride ions at room temperature. The corrosion properties of the specimens were examined through potentiodynamic test (potential range of -1500 ~ 2000 mV) in 0.9% NaCl solution by potentiostat (EG&G Co, PARSTAT 2273. USA). Microstructures of the alloys were examined by optical microscopy (OM), scanning electron microscopy (SEM) and X-ray diffractometer (XRD). Diameter of nanotube was not depended on Zr content, but interspace of nanotube was predominantly depended on Zr content. It is confirmed that, ZrO_2 oxides play a role to formation on the surface.

Keywords: Nano-structure, Corrosion, Ti-Zr alloy, dental implant

1 INTRODUCTION

Titanium alloys are expected to be much more widely used for implant materials in the medical and dental fields because of their superior biocompatibility, corrosion resistance and specific strength compared with other metallic implant materials. The use of titanium and its alloys implant applications has mainly been limited to the alloy Ti-6Al-4V and to CP-Ti [1,2]. For medical application titanium and Ti-6Al-4V have been used since 1960s, with Ti-6Al-4V gradually replacing CP-Ti due to the increased mechanical strength of plates, nails, screws and endoprostheses [3].

Recently, however, much concern has developed over the issue of biocompatibility with respect to the dissolution of aluminum and vanadium ions and the possibility of any toxic effects [4-6]. Consequently, other titanium alloys are currently being considered as alternatives to the Ti-6Al-4V alloy. Therefore, Ti-alloy, Al and V free and composed of non-toxic element such as Nb and Zr as biomaterials have been developed. Especially, Zr element belongs to same family in periodic table as Ti element. Addition of Zr to Ti alloy has an excellent mechanical properties, good corrosion resistance and biocompatibility [7].

The high degree of biocompatibility of Ti alloys is usually ascribed to their ability to form stable and dense oxide layers consisting mainly of TiO_2. The native oxide layers on Ti are usually 2-5nm thick and are spontaneously rebuilt in most environments whenever they are mechanically damaged. It is believed that thicker and more stable TiO_2 based oxide surfaces are generally favorable for surface bioactivity [8,9]. Spark anodization is one of the conventional routes to increase the biocompatibility of titanium and its alloys. This process typically leads to the formation of a disordered oxide structure (irregular pores with lateral features from 1 to 10 μm) several hundreds of nanometers thick [10,11]. In contrast to this approach, the electrochemical formation of novel highly ordered oxide nanotube layers has been reported for Ti anodization in fluoride containing acid electrolytes at moderate voltages [12]. Such TiO_2 structures consist of arrays of nanotubes with diameters in the 100 nm range and thickness up to about 400~500 nm.

In order to investigate the electrochemical characteristics of nanotube formed Ti-xZr alloy for biomaterials have been researched using by electrochemical methods.

2 MATERIALS AND METHODS

2.1 Alloy preparation

Ti (G&S TITANIUM, Grade. 4, USA)alloys containing Zr(Kurt J. Lesker Company, 99.95 % wt% in purity) up to 10,20, 30 and 40 wt% were melted six times to improve chemical homogeneity using the vacuum arc melting furnace. And heat treatment was carried out at 1000℃ for 24h in order to homogenization in argon atmosphere.

The specimens for electrochemical test were prepared by using various grit emery papers and then finally, polished with 0.3 μm Al_2O_3 powder. All of polished specimen was ultrasonically cleaned and degreased in acetone.

2.2 Microstructure analysis

Microstructures of the alloys were examined by optical microscopy (OM, OLYMPUS BM60M, JAPAN) and scanning electron microscopy (SEM, HITACHI-3000, JAPAN). The specimens for the OM and SEM analysis were etched in Keller's solution consisting of 2 ml HF, 3 ml HCl, 5 ml HNO_3 and 190 ml H_2O.

In order to identify the phase constitutions of the Ti-xZr alloys, X-ray diffractometer (XRD, Philips, X'pert Pro MPD) analysis with a Cu-Kα radiation were performed.

2.3 Anodization test

Electrochemical experiments were carried out with conventional three-electrode configuration with a platinum counter electrode and a saturated calomel reference electrode. The sample was embedded with epoxy resin, leaving a square surface area of $10mm^2$ exposed to the anodizing electrolyte, 1M H_3PO_4 containing 0.5wt% NaF. Anodization treatments were carried out using a scanning potentiostat (EG&G Co., Model 362, USA). All experiments were conducted at room temperature. The electrochemical treatments consist of potential was first swept from the open-circuit potential to desired final potential with a sweep rate of 500mV/s, then the potential was held for 2h. After the treatments the anodized sample were rinsed with distilled water and dried with dry air stream.

2.4 Electrochemical test

The corrosion behaviors were investigated using potentiostat (EG&G Co, 2273A) in NaCl solution at 36.5 ± 1℃. A conventional three-electrode cell with a high dense carbon as counter electrode, saturated calomel (SCE) as reference electrode, and specimens as working electrode, connected to a potentiostat, was used to conduct the potentiodynamic test.

3 RESULT AND DISCUSSION

3.1 Microstructural observations

Fig. 1,2 shows the microstructures of Ti-xZr alloys with different Zr contents (10, 20, 30 and 40 wt.%). The microstructures of Ti-10Zr and Ti-20Zr alloy showed lamellar structure and needle-like structure, these phase changed gradually to almost needle-like structure in Ti-40Zr alloy. Consequently, microstructures of Ti-xZr alloys were changed from lamellar structures to needle-like structure as Zr content increased.

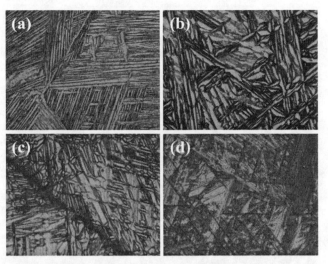

Fig. 1. OM micrographs of homogenized Ti-xZr alloys (a) Ti-10Zr (b) Ti-20Zr (c) Ti-30Zr (d) Ti-40Zr

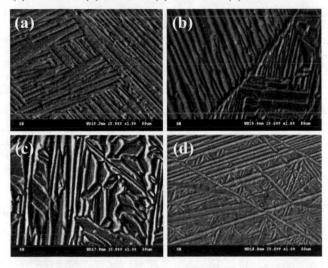

Fig. 2. SEM micrographs of homogenized Ti-xZr alloys (a) Ti-10Zr (b) Ti-20Zr (c) Ti-30Zr (d) Ti-40Zr

As a result of interpretation using software (Nwetown Square. JCPDS win, USA) for each peak, X-ray diffracto- meter (XRD) of the homogenized Ti-xZr alloys are summarized in Fig. 3, which indicate that the phase transformation in the Ti-xZr alloys sensitive to Zr content. It suggested that β → α transformation progressed gradually with increasing Zr content due to Zr displacement [13]. Each diffraction peak shifted to a lower angle with increasing Zr content. The absence of additional peaks is consistent with single-phase.

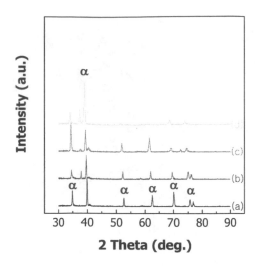

Fig. 3. X-ray diffracto-meter of Ti-xZr alloys
(a) Ti-10Zr (b) Ti-20Zr (c) Ti-30Zr (d) Ti-40Zr

3.2 Nanotube structures

The microstructures shown in Fig. 4 are taken from the Ti-20Zr, which were TiO_2 nanotube layer formed on Ti substrate. In the microstructures that are presented, beta phase appears dark area and the alpha phase shows the light area. The microstructure of Ti-20Zr, provided in Fig. 4, reveals elongated alpha/beta phase interface. Nanotube was formed mainly on the alpha phase with many tube like stacked ring. It is confirmed that number of ring is related to nanotube formation time and to the periodicity of current oscillations. In the alpha phase, shape of nanotube is different from that of beta phase.

Fig. 4. SEM micrographs of TiO_2 nanotube layer formed Ti-20Zr alloy

Fig. 5 shows a typical SEM image of TiO_2 nanotube layer prepared by anodizaiton of titanium at 10V in 1M H_3PO_4 + 0.5wt% NaF for 2h. The nanotubes have an inner

average diameter of 150 ~ 200 nm with a tube-wall thickness of about 20nm.

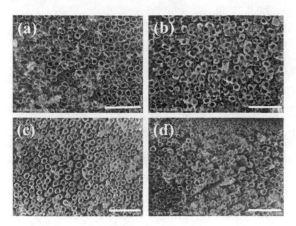

Fig. 5. SEM top-view images of TiO_2 nanotube layer formed on Ti substrate
 (a) Ti-10Zr (b) Ti-20Zr (c) Ti-30Zr (d) Ti-40Zr

But, for Zr content of 10wt% (Fig. 5a), the interspace of TiO_2 nanotubes was very small, 60nm. As the Zr content increased, the interspace of TiO_2 nanotubes increased 70, 100 and 130 nm, respectively (Fig. 6).Diameter of nanotube was not depended on Zr content, but interspace of nanotube was predominantly depended on Zr content. It is confirmed that, ZrO_2 oxides play a role to formation on the surface.

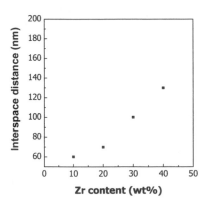

Fig. 6. Relation between the Zr content and the interspace of TiO_2 nanotubes under the same magnifications as for Fig. 5.

3.3 Electrochemcial characteristics

Fig. 7 shows the results of potentiodynamic test (potential range of -1500 ~ 2000mV) in NaCl solution, which was conducted in order to investigate the effect of Zr content on the polarization curve. It can be seen in Fig. 7(a) that the Ti-40Zr alloy has the highest resistance to corrosion. It is thought that increase of corrosion resistance with Zr content is attributed to the a few nm thick passive film such as TiO_2 and ZrO_2 formed rapidly on the specimen surface. A few nm thick passive films could restrict the movement of metal ions from the metal surface to the solution, thus minimizing corrosion (Fig. 7(b)).

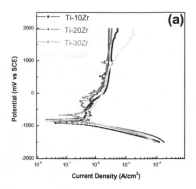

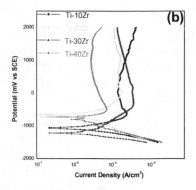

Fig.7. The polarization curves of Ti-xZr alloys after potentiodynamic test in NaCl solution at 36.5±1 ℃
   (a) as-received   (b) nanotube formed

4 CONCLUSIONS

1. Microstructure properties observed by OM and SEM changed from lamellar structure to needle-like structure with increasing Zr content.

2. Microstructure changes from β phase to α phase through XRD and α phase increases according to the amount of Zr added

3. From the results of TiO_2 nanotube layer formed in the Ti-xZr alloy, nanotubes have an inner average diameter of 150 ~ 200 nm with a tube-wall thickness of about 20nm. As the Zr content increased, the interspace of TiO_2 nanotubes increased ~60, ~70, ~100 and ~130 nm, respectively.

4. From the results of polarization behavior in the Ti-xZr alloys, the current density of nanoformed Ti-40Zr in the passive region was higher than that of other alloys.

ACKNOWLEDGEMENT

"This work was supported by research funds from Chosun University (2007) and 2nd stage of Brain Korea 21 for College of Dentistry"

REFERENCES

[1] M. Aziz-Kerrzo, KG. Conroy, AM. Fenelon, ST. Farrell, CB. Breslin, "Electrochemical studies on the stability and corrosion resistance of titanium-based implant materials", biomaterials, 22, 1531-1539, 2001.

[2] LJ. Knob, DL. Olson, "Metals handbook corrosion", 13, 669, 1987.

[3] E. Eisenbarth, D. Velten, M. Muller, R. Thull and J. Breme, "Biocompatibility of β- stabilizing elements of titanium alloys", Biomaterials, 25, 5705-5713, 2004.

[4] M.A. Khan, R.L. Williams, D.F. Williams, "The corrosion behavior of Ti-6Al-4V, Ti-6Al-7Nb and Ti-13Nb-13Zr in protein solutions", Biomaterials, 20, 631-637, 1999.

[5] KL. Wapner, "Implications of metallic corrosion in total knee arthroplasty", Clin Orthop, 271, 12-20, 1991.

[6] J. J. Park, H. C. Choe, Y. M. Ko, "Corrosion Characteristics of TiN and ZrN coated Ti-Nb alloy by RF-sputtering", Materials Science Forum, 539-543, 1270-1275, 2007.

[7] D. Kuroda, M. Ninomi, M. Morigana, Y. Kato, T. Yashiro, "Design and mechanical properties of new β type titanium alloys for implant materials", Materials Science and Engineering A, 243, 244-249, 1998.

[8] Yang BC, Uchida M, Kim HM, Zhang XD, Kokubo T, "Preparation of bioactive titanium metal via anodic oxidation treatment", Biomaterials, 25, 1003-1010, 2004.

[9] Dyer CK, Leach JSL, "Breakdown and efficency of anodic oxide grown on titanium. J. Electrochem. Soc., 125, 1032-1038, 1978

[10] marchenoir JJC, Loup JP, Masson JE, "Etude des couches poreuses formees par oxidation anodique du titane sous fortes tensions", Thin Solid Films, 66, 357-369, 1980.

[11] Zwilling V, Aucouturier M, Darque-Ceretti E, Boutry-Forveille A, "Anodic oxidation of titanium and TA6V alloy in chromic media, An electrochemical approach", Electrochemical Acta, 45, 921-929, 1999.

[12] Gong D, Grimes CA, Varghese OK, Chen Z, Dickey EC, "Titanium oxide nanotube arrays prepared by anodic oxidation, J. Mater. Res., 16, 3331-3334, 2001.

[13] A. V. Dobromyslove, V. A. Elkin, Scripta Materialia, "Martensitic transformation and metastable β-phase in binary titanium alloys with d-metals of 4-6 periods", 44, 905-910, 2001.

Effects of β- Stabilized Alloying Element on the Nanotube Morphology of Ti-Alloy

S.H. Jang, *H.C. Choe, Y.M. Ko, [1]W.A. Brantley

College of Dentistry, Chosun University, 2nd stage of Brain Korea 21 for College of Dentistry, Gwangju, Korea
[1]College of Dentistry, Ohio State University, Columbus, OH, USA
*hcchoe@chosun.ac.kr

ABSTRACT

Titanium alloys show attractive properties for biomedical applications where the most important factor are biocompatibility, corrosion resistance, low modulus of elasticity, very good strength-to-weight ratio, reasonable formability and osseointegration. The aim of this study was to investigate the effects of Nb content(10, 20, 30 and 40wt%) on microstructure and nanotube formation of Ti-xNb alloy samples. Electrochemical anodization of the Ti-xNb alloys were carried out by potentiostatic experiments in 1M H_3PO_4 + 0.8wt% NaF electrolyte at room temperature. Under specific sets of conditions highly self-organized titanium oxide nanotubes are formed with diameters varying from approx. 55 to 220 nm and length from approximately 727 nm to 2 μm.

Keywords: anodization, nanotube, Ti-Nb alloy

1 INTRODUCTION

Titanium and its alloys have been widely used for structural biomaterials such as artificial hip joints and dental implants due to their excellent specific strength and corrosion resistance. Amongst conventional biomaterials, pure Ti as well as Ti-6Al-4V alloy exhibit excellent properties for surgical implant applications. These properties are achieved by a thin titanium oxide based layer that is always present on these alloys surface [1]. By adding alloying elements to titanium, such as Al and V, its mechanical properties are improved [2-3]. However, due to toxicity effects caused by Al and V and high elastic modulus, new alloys that present lower elastic modulus and do not contain these elements are receiving a great deal of attention. Tissue reaction studies have identified Ti, Nb, Zr and Ta as non-toxic elements as they do not cause any adverse reaction in human body. It is well known that recent biomaterials research has focused on β-Titanium alloys due to increased biocompatibility and decreased Young's modulus. Some studies have shown that Ti-Nb alloys ranging from 20 to 50 wt% Nb exhibit a modulus of elasticity of about 60GPa [4-6], which is closer to that of

bone when compared to those of other conventional alloys applied to orthopedic implants. The increase on Nb content tends to decrease the modulus of elasticity and stabilizes the β-titanium phase. It has also been reported that the yield strength increases and the elongation decreases with increasing Nb content [7].

The high degree of biocompatibility of Ti alloys is usually ascribed to their ability to form stable and dense oxide layers consisting mainly of TiO_2. The natural oxide is thin (about 3-8 nm in thickness) and amorphous, stoichiometrically defective. It is known that the protective and stable oxides on titanium surfaces are able to provide favorable osseointegration [8]. The stability of the oxide depends strongly on the composition, structure and thickness of the film. As a consequence, great efforts have been devoted to thickening and stabilizing surface oxide on titanium to achieve desired biological response [9].

Anodic oxidation is efficient to control the thickness, composition and topography of the oxide film on titanium and can be applied for implant surface modification. Using anodic oxidation, TiO_2 is formed with a chemical bond between the oxide and Ti substrate that likely results in enhanced adhesion strength. Indeed, a bone-like apatite layer is formed on TiO_2 in simulated body fluid. Furthermore, researchers have suggested that TiO_2 with a 3D micro/nanoporous structure may enhance apatite formability when compared to dense TiO_2 [10].

In the present study, we chose Nb as a binary alloying element because it has excellent biocompatibility and be expected to act as a β- stabilizer. The purities of the starting raw materials were 99.99% for Ti and 99.9% for Nb. For the purpose of improvement in biocompatibility, the anodic TiO_2 layer on Ti-xNb alloys were fabricated by electrochemical method in 1M H_3PO_4 with small amounts of fluoride ions, and then the effect of Nb content on the nanotube size, the morphology of Ti oxide layer formed by the anodic oxidation method was investigated.

2 EXPERIMENTAL

2.1 Materials and surface preparation

The purities of the starting raw materials were 99.99% for Ti and 99.9% for Nb. A series of Ti-xNb alloys (10, 20, 30 and 40wt%) (in this study, all the percentages are shown in weight %) were prepared. Ti-xNb alloys of 10 ㎜ in diameter were prepared by arc melting on a water-sealed copper hearth under an argon gas atmosphere with a non-consumable tungsten electrode. Each ingot was melted six times by inverting the metal to ensure homogeneous melting. The ingots were heat-treated for 6h at 1000℃ for homogenization after water quenching. They were cut to a diameter of 10 ㎜ and a thickness of 2 ㎜ followed by degreasing with acetone. The as-received group specimens (10, 20, 30 and 40wt%Nb), which were used as the control groups, were prepared by polishing the Ti-xNb disk surfaces with up to #2000 SiC paper and ultrasonically cleaning in ethanol and distilled water.

2.2 Nanotube formation treatment

Electrochemical experiments were carried out with a conventional three-electrode configuration with a platinum counter electrode and a saturated calomel (SCE) reference electrode. Experiments were performed in 1M H_3PO_4 with small additions of NaF(0.8wt%). All experiments were conducted at room temperature. Electrochemical treatments were performed by using a scanning potentiostat (EG&G Co, 362, U.S.A). The electrochemical treatment consisted of a potential ramp from the open-circuit potential (Eocp) to an end potential at 10 V with a scan rate of 500mV/s followed by holding the sample at 10V for 2h.

2.3 Surface characterization

Structural characterization of the nanotube formed samples was carried out with a field emission scanning electron microscope (Hitachi FE-SEM 4800, Japan). The scanning electron microscope was capable of energy dispersive spectrometer (EDS). The cross-sectional and surface of Ti substrate images were obtained from mechanically scratched samples. In order to identify the phase constitutions of the Ti-xNb alloys, x-ray diffractometer (XRD, philips, X'Pert Pro) analysis with a Cu-K α radiation were performed.

3 RESULT AND DISCUSSION

3.1 Microstructural observations

Fig. 1 shows the optical microstructures of Ti-xNb alloys with different Nb contents (10, 20, 30 and 40 wt.%). It was controlled by heat treatment at 1000 °C for 6hr in Ar atmosphere followed by water quenching. Fig. 1(a, b) shows the average microstructure, which is mainly composed of α+β colonies. The microstructure is formed by very fine and almost continuous grain boundary α and α parallel plates, which grew from the grain boundary, dispersed in a β matrix. The clear

needle-like traces of martensite in the β grain were observed in the image of Fig. 1 (c). However, in Ti-40Nb alloys (Fig. 1(d)), no trace of martensite phase is seen compared to Ti-10Nb and ti-20Nb due to complete β-phase formation. The apparent volume fraction of martensite decreased with increasing Nb content in Ti-xNb alloys, as seen in Fig. 1. This may be associated with β-phase stability, i.e., the stability of β-phase would be enhanced with increasing Nb content, since Nb is known to be a β-phase stabilizer [11-12].

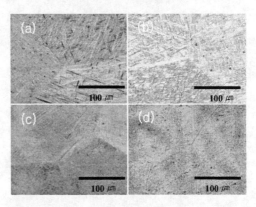

Fig. 1. OM micrographs showing the microstructure of heat-treated (W.Q) Ti-xNb alloys. (a) Ti-10Nb (b) Ti-20 Nb (c) Ti-30 Nb (d) Ti-40 Nb

Fig. 2 Shows the XRD profiles for Ti-xNb alloys in which solution treatment was carried out at 1000℃ for 6h followed water quenching. We prepared samples, varying Nb content, in order to investigate the alloying element effect on phase formation behavior. The structure of alloy was changed from α-phase (100) to β-phase (110) with increase of Nb content. It can be explained that Nb acted as β-phase stabilizer like Nb and Ta[12].

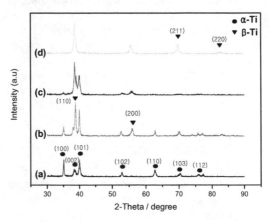

Fig. 2. XRD diffraction patterns of Ti-xNb alloys . (a) Ti-10Nb (b) Ti-20Nb (c) Ti-30Nb (d) Ti-40Nb

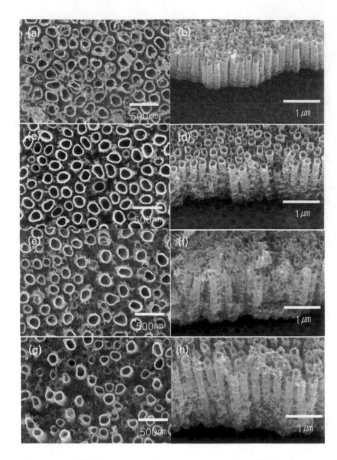

Fig. 3. SEM images of titanium oxide obtained by anodization in 1M H₃PO₄ + 0.8wt% NaF at room temperature. (a,b) Ti-10Nb (c,d) Ti-20Nb (e,f) Ti-30Nb (g,h) Ti-40Nb

Fig. 3 and 4 show SEM images of top, cross-sectional and bottom view. The cross-sectional and bottom views were taken from mechanically scratched samples where some pieces flaked off and were upside down. It is apparent Fig. 3 and Fig. 4 that the nanotubes are open on the top while on the bottom they closed. Fig. 3 shows SEM images of the resulting layer formed at 10V in 1M H₃PO₄ containing 0.8wt% NaF. Under specific sets of conditions highly self-organized titanium oxide nanotubes are formed with diameters varying from approx. 55 to 220 nm and length from approx. 727 nm(Ti-10Nb) to 2 μm(Ti-40Nb) . It is obvious that the amount of Nb strongly influences the pore formation process and morphology of the porous surfaces. As the Nb contents increase to 10 (Fig. 2-a,b), 20 (Fig. 2-c,d), 30 (Fig. 2-e,f) and 40wt% (Fig. 2-g,h), the average nanotube length varies : 727 nm, 944 nm, 1.5 μm and 2 μm respectively. The average pore spacing(un-nanotube formation area, initially) increases from 55 nm to 432 nm, as Nb content increases. According to Fig.3, higher contents of Nb in Ti alloy cause a increase in pore spacing and pore length. Fig. 4(a) shows an SEM image of nanotube bottom with pore size of approx. 225 nm prepared using anodization voltage of 10V for

2h. It can be seen that the nanotube array is uniform over the substrate.

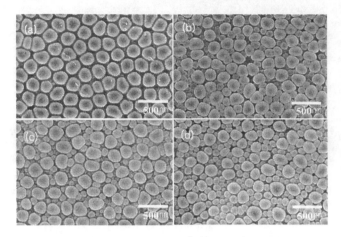

Fig. 4. SEM bottom-view image of TiO₂ nanotubes formed at 10V for 2h in 1M H₃PO₄ + 0.8wt% NaF. (a) Ti-10Nb (b) Ti-20Nb (c) Ti-30Nb (d) Ti-40Nb

However, from a careful look at Fig. 4(b-d), it can be seen that the self-organized nanotube layers consist of arrays with two distinctly different tube diameters of 220 and 55 nm at the bottom. A regular assembly is formed where the 220 nm large nanotubes are surrounded by several 55 nm small tubes. Also at the bottom, smaller nanotubes appear to protrude as Nb content increased. Our results indicate that the composition of the alloys has a great influence on the two-size-scale structure, and two-size-scale structure appeared predominantly as Nb content increased as shown in Fig. 4(d). We think that the site of small size tube was at un-nucleated area(large interspaced tube in Fig. 3-g). This area is necessary for such self-organization in the fluoride-containing electrolyte.

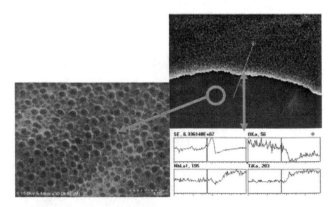

Fig. 5. Line analysis showing the TiO₂ nanotubes formed Ti-30Nb alloy at 10V for 2h in 1M H₃PO₄ + 0.8wt% NaF.

Fig. 5 shows the line profile at nanotube covered and uncovered area after TiO₂ nanotubes formation in 1M H₃PO₄ + 0.8wt% NaF solution. On the nanotubed surface,

NSTI-Nanotech 2008, www.nsti.org, ISBN 978-1-4200-8503-7 Vol. 1

Nb content was lower than that of uncovered nanotube film surface, whereas oxygen content was higher on the nanotubed surface. It is confirmed that nanotubed surface mainly consisted of TiO_2 oxide and Nb_2O_5 oxide film. In the matrix, Ti and Nb content were same before nanotube formation like manufactured alloy.

4 CONCLUSION

1. The structure of alloy was changed from α-phase to β-phase with increase of Nb content. The structure of alloy was changed from α-phase(100) to β-phase(110) with increase of Nb content.
2. Titanium oxide nanotubes were formed with diameters varying from approx. 55(small tube) to 220 nm(large tube) and length from approx. 727 nm(Ti-10Nb)to 2 μm(Ti-40Nb).
3. The composition of the alloys has a great influence on the two-size-scale structure, and two-size-scale structure appeared predominantly as Nb content increased, and small size nanotube was nucleated at pore spacing area un-nucleated, initially.
4. From the line profile analysis, on the nanotubed surface, Nb content was lower than that of uncovered nanotube film surface, whereas oxygen content was higher on the nanotubed surface.

ACKNOWLEDGEMENT

"This work was supported by funds from Chosun University(2007) & 2nd stage of Brain Korea 21 for College of Dentistry"

REFERENCES

[1] J. Lausmaa, B. Kasemo, H. Mattson and H. Odelius. Appl. Surf. Sci. 45, 189 (1990).
[2] R. Boyer, G. Welsch, E.E. Collings. Materials Properties Handbook: Titanium Alloys, ASM, Metals Park, OH, 1994.
[3] G. Lutjering, Mat. Sci. Eng. A 263(1999) 117.
[4] E.B. Taddei, V.A.R. Henriques, C.R.M. Silva, C.A.A. Cairo. Mater. Sci. Eng. 24C (2004) 683.
[5] Y.H. Hon, J.Y. Wang, Y.N. Pan, Metall. Mater. Trans. 44 (2003) 2384.
[6] C.M. Lee, C.P. Ju, J.H.C. Lin, J. Oral Rehabil. 29 (2003) 314.
[7] H.S. Kim, S.H. Lim, I.D. Yeo, W.Y. Kim, Mater. Sci. Eng. 449-451A (2007) 322.
[8] Kieswetter K, Schwartz Z, Dean DD, Boyan BD. Crit Rev Oral Biol Med 7(1996) 329
[9] Xiaolong Zhu, Jun Chen, Lutz Scheideler, Rudolf Reichl, Juergen Geis-Gerstorfer, Biomaterials 25(2004) 4087.
[10] Ketul C. Popat, Lara Leoni, Craig A. Grimes, Tejal A. Desai, Biomaterials 28(2007) 3188.
[11] J. J. Park, H. C. Choe, Y.M. Ko, Materials Science Forum, 539-543, 1270-1275, 2007.
[12] H.C. Choe, Y.M.Ko, W.A.Brantley, NSTI-Nanotech-2007, 2, 744-747, 2007.

Nano-diamond Coatings for Aluminum Substrates

Rodger Blum, Dinesh Kalyana-Sundaram, Rajeev Nair, Pal Molian

Department of Mechanical Engineering
Laboratory for Lasers, MEMS, and Nanotechnology
Iowa State University
Ames, IA 50011-2161

ABSTRACT

Laser-induced phase transition and sintering of nano-diamond powders (4-8 nm) was used to produce a thick (10-50 µm) strongly adherent nano-diamond/diamond-like carbon (nD-DLC) coatings on aluminum allow 319 Nano-diamond powders produced by detonation synthesis were electrostatically sprayed on the aluminum substrate coupons followed by a continuous wave CO_2 laser heating in a controlled fashion to cause a liquid-phase nano-sintering which yielded a strongly adherent coating. . A transition to diamond-like carbon produced a Vickers hardness of 2250 kg/mm^2, and a transition to complete nano-diamond produced a Vickers hardness of 9000 kg/mm^2. The potential application of the work is hard and wear resistant coatings for light-weight engine components to improve their wear resistance.

Keywords: nanocoatings, diamond, tribology, aluminum, laser sintering

1 INTRODUCTION

Diamond is the hardest material known and finds great number of applications because of its unique and attractive physical and chemical properties. In addition to its super hardness, it also exhibits chemical inertness, higher wear resistance, and higher thermal conductivity. Nano-diamond is emerging as a new type of material with even better properties than single and poly crystalline diamonds. Commercially produced nano-diamond powders are used in various applications including abrasives for the optical and semiconductor industries, for durable and hard coatings, polymer reinforcements, lubricant additive for engines and other moving parts, protein absorbent and even medicinal drugs [1-4]. Nano-diamond powders are also being used for making new types of rubber and are important components in reprocessing worn tire rubber [1]. Nano-diamond coatings are generally synthesized by chemical vapor deposition (CVD) and plasma deposition techniques. These techniques are bound by time and cost constraints, and can be hazardous. The gases used during the chemical vapor deposition can be toxic, flammable, corrosive, and even explosive [5]. The high cost of chemical vapor deposition and plasma deposition is because of the need for a vacuum to create the plasma and then the reactor used needs to be more sophisticated to contain the plasma [5]. Other downfall to the plasma deposition process is that its capability of high energy plasma could damage the substrate [5]. The time constraint can be seen in these processes by the fact that the fastest growth rate seen is 900 µm per hour but at this speed nothing was said about the uniformity of the layer, its nucleation density, and the stability of the operation [6]. A more typical growth rate would be .5-5 µm per hour; there are a few instances of 20 µm per hour in the plasma enhanced CVD process [6, 7]. So the CVD process is relatively slow.

In this work, we developed novel techniques to achieve nano-diamond coatings on aluminum A319 by CO_2 laser–sintering process. Aluminum alloy A319 is a widely used material in the production of automotive components due to its excellent mechanical properties and good castability characteristics [8]. Laser sintering technique is the continuous rastering of the laser beam on the sample surface. Laser sintering is a widely used technique for many rapid proto-typing processes including selective laser sintering (SLS) [9, 10], and laser engineered net shaping (LENS). These processes are layered based meaning that they make parts layer by layer [9, 10]. In the selective laser sintering process a layer of powder is spread over the previous layer and sintered to that layer by a raster filling motion of the laser [9, 10]. The layer of powder is only about as thick as the particle size of the powder being used [9, 10]. In our experiments, laser sintering is used to bond a thick (25-35 µm) layer of namo-diamond powder to an aluminum substrate. A continuous wave CO_2 laser (10.6 µm) was used for this process. The goal of the experiment is to strengthen the surface hardness of the aluminum by laser inducing a phase transition of nano-diamond powder and then sintering it to the substrate. This nano-diamond coating should make the aluminum more durable with less friction. The success of the project could help us in developing longer life engine components.

2 EXPERIMENTAL DETAILS

2.1 Preparation of Nano-diamond Powders

Nano-diamond powders of 4-8 nm were produced by shock detonation synthesis and were confirmed from TEM images. In the detonation synthesis process, powerful explosives are mixed with the composition $C_aH_bN_cO_d$ with a negative oxygen balance in a non-oxidizing medium to yield condensed carbon phase that involves nano-diamond particles [1]. The result of this process be 75% nano-diamond powder [1]. A cleaning process is done to the resultant powder to remove almost all of the impurities. The nano-diamond powder used in this experiment had no more than 2% impurities.

2.2 Preparation of Aluminum Coupons

Aluminum ingot of A319 was purchased from Custom Alloy Light Metals. The composition of A319 can be seen in Table 1 and it conforms to the ASTM specifications. Rods of 25 mm (~1") dia x 250 mm long (~10") were prepared from the ingot by die casting. After the aluminum was cast into rods it was heat treated to convert it to 319-T6. T6 has a higher tensile strength. Following the heat treatment, the aluminum was then cut into coupons/discs of approximately 3.2 mm (1/8") thickness. Once in coupon form the disk surfaces needed a good surface finish. To accomplish this a Clemco Industries Dry Blast Cabinet with Dust Collector (Model ACDFM) was used to sandblast the surfaces.

Table 1: Composition of A319

Element	WT %	Element2	Wt %3
Si	5.860	Ti	0.160
Fe	0.678	Sn	0.021
Cu	3.519	Pb	0.032
Mn	0.246	Ni	0.121
Mg	0.056	OET	< .500
Cr	0.104	Sr	0.000
Zn	0.901	Al	87.8

2.3 Electrostatic Spray Coating

After the aluminum A319 coupons were made, nanodiamond were electrostatically spray coated on one of the surfaces of the coupon. The deposition of nanodiamond powder was done by a home-built corona charging based electrostatic spray coating setup. The powder was deposited by first being negatively charged. Once charged the nano-particles follow the electric field lines toward the grounded substrate (aluminum), forming a uniform coating. The coating is loosely adhered to the substrate by electrostatic forces which may cause the coating to decay as the electrostatic forces dissipate. Since the size of the diamond particles are in the order of nanometers, agglomeration of the particles is a potential issue that has to be overcome to achieve uniform deposition. To prevent agglomeration of the particles, a jet mill was integrated into the electrostatic spray coating setup (Fig 1). Besides the physical properties of the powder (particle density, particle shape, average particle size, particle size distribution, and dielectric properties), process parameters including electrode-substrate distance, electrical voltage applied at the electrode, and the main air pressure affect the formation of the coating. A coating of 25-35 µm

was obtained at a voltage of 60 kV with the electrode to substrate distance of 150 mm.

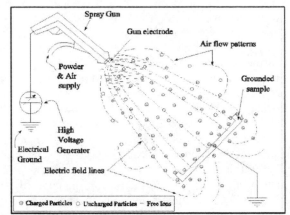

Figure 1: Schematic of the electrostatic spray coating

2.4 Laser Sintering Setup

A 1.5 kW Continuous wave CO_2 laser (Spectra Physics – 820) of 10.6 µm wavelength with CNC controlled worktable was used for experiments as shown in Fig 2. The coupons were held in place by means of a vice clamped on to the table. Sintering was performed using a focused beam of 0.2 mm spot size with an overlap of 5% between passes. A lens of 127 mm (5") focal length helped in achieving the desired spot size. Argon gas at 2.3 x 10^{-2} m^3/s (50ft³/min) was used as an assist gas during the sintering process. Argon is used as an assist gas because it prevents oxidation as it is an inert gas [11]. Since it is an inert gas it would not interact with the liquid melt metal at the laser-substrate interface [11]. Preliminary studies of single pass treatment were made to identify the best set of parameters for laser-sintering. After identification, laser sintering was performed to completely cover the surface of the coupon in a raster-frame motion and it took 70 passes for sintering to be completed.

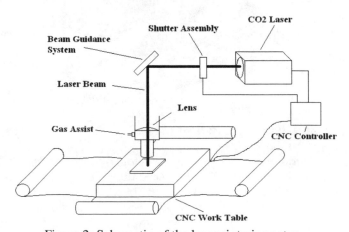

Figure 2: Schematic of the laser sintering setup

2.5 Measurement Techniques

To identify the mechanical and tribological properties of the finished coating the following tests were performed: hardness by Vickers micro-hardness test, identification of the carbon phase by Raman Spectroscopy, friction and wear test, and surface roughness measurement by optical profilometer. The micro-hardness of the heat treated aluminum is known to be around 113 kg/mm^2 but this can vary depending on the temperature it is held at [12]. Again the goal is find the hardness to be between 2000-9000 kg/mm^2. Following the hardness tests, the surface will be characterized using a Renishaw inVia Raman Microscope at a wavelength of 488 nm. With this process we are expecting to get a surface that consists of a mixture of nano-diamond, diamond-like-carbon, and graphite. These three carbon phases will yield three very different Raman peaks. Diamond will give a definite peak at 1332 cm^{-1}, and graphite will yield a definite peak at 1580 cm^{-1}, while diamond like carbon can yield a range of peaks but the main peak is at 1355 cm^{-1} [13]. Thus Raman Spectroscopy will be plotted through the range of 1100-2000 cm^{-1}.

3 RESULTS AND DISCUSSION

3.1 Effect of Laser Parameters

Laser powers in the range 200 W-1000 W were studied. Higher powers than 400 W caused vaporization of the coating. Lower powers than 200 W did not cause melting of the substrate which is necessary for liquid phase sintering. Parameter studies coupled with Raman spectroscopy yielded the most optimum parameters at 200 W and 85 mm/sec (200 in/min). Fig 3 shows the nano-diamond covered coupon.

Figure 3: Coupon sintered with optimal parameters

3.2 Cleaning the Samples

Ultrasonic water cleaning was found to remove the outer layer of carbon that was produced during laser sintering. At first this layer was found to be easily scratched away and was thought to damage the Raman test we were taken. The coupons were cleaned in an ultrasonic bath for 3 minutes using deionized water. The samples were then removed from the bath and left to air dry. It was found that cleaning of the coupon gave better Raman peaks.

3.3 Hardness Tests

A Wilson Tukon Microhardness Tester (Model 200) was used in taking Vickers micro-hardness tests. These tests were done using a 25 gram weight which is better for superficial layers and avoids the risk of puncturing the coating. The Vickers hardness tests were used as a basis for characterizing the process parameters. The uneven spread of the nano-diamond powder after laser-sintering resulted in a rougher surface. The rougher surface profile combined with the higher hardness of the diamond particles makes it difficult to clearly identify the endpoints of the indentation caused by the instrument probe. Along with the micro-hardness test, a macro hardness test was also done. In this test a Leco Hardness Tester (RT-120) with a test weight of 15kg was used. Fig. 4 shows the results of plain aluminum coupon and a coupon with the coating on it. This was done to see the results of the test. As shown by the figure there is not much difference between the two surfaces, which is not what was expected. If there was a difference in the two surfaces, we expected there to be some chip formation or something around the indention of the coated sample. Since we were neither able to get an accurate micro-hardness nor macro hardness reading, we decided to use Raman Spectroscopy for determining what laser parameters yield the best result.

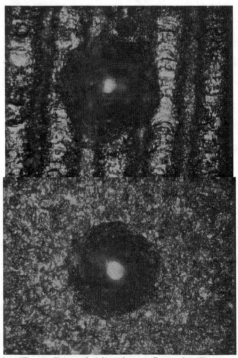

Figure 4: (Top) Coated Aluminum Sample (Bottom) Plain Aluminum Sample

3.4 Raman Spectroscopy

Raman Spectroscopy of the single-pass parameters which were done at 200 W and a 1000 W with a range of laser speeds. These powers were chosen to understand the effect of lower and higher powers. Also 1000W would yield an interface temperature (> 900°C) high enough to melt the aluminum which will enable sintering of the nano-diamond powders to take place. We observed no definite diamond peaks at 1332 cm⁻1 in the Raman analysis of the laser sintered coupons both at 200 W and 1000 W. We also observed rounded peaks of diamond like carbon and graphite at 1350 cm⁻1 and 1550 cm⁻1. We even had a spectrum that carried one big peak from about 1350 to 1550 cm⁻1. A higher graphite count was noticed which might primarily be due to the high heat created at 1000 W except at high speeds of nearly 245 mm/s (600 inch/min). At high speeds the spectra showed some diamond-like carbon (peak at 1350 cm⁻¹) starting to appear but the graphite peak count was still higher. From the tests conducted we narrowed down the laser parameters by comparing the intensity of diamond, diamond-like carbon, and graphite peaks. The parameter that gave us the best result was 200W at 84.7 mm/s (200 in/min), and we used these parameters to laser sintered the whole surface of the coupon. Raman Spectra of both the laser-sintered sample and nano-diamond powder can be seen in Fig 5. The spectrum shows an upward slope, but our focus is on the two peaks in the center of the spectrum. These are the diamond like carbon and graphite peaks. The figure shows that the intensity of the peaks is higher than that found in the original nano-diamond powder.

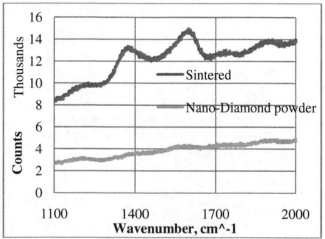

Figure 5: Spectra of the optimal parameters and nano-diamond powder.

3.5 Friction and Wear

Friction was measured using a custom-built reciprocating ball-on-flat microtribometer that can produce a microscale (apparent area ~ 1000 µm²) multi-asperity contact [14]. The details of equipment are found elsewhere [14]. In short, the microtribometer uses a spherical probe placed at the end of a crossed I-beam structure, which is lowered using a linear stage to apply a desired normal load to the sample. The normal and friction (lateral) forces are measured using semiconductor strain gages on the cantilevers. Friction forces can be resolved to approximately ± 5 µN and normal forces to approximately ± 15 µN. All samples were cleaned with DI water in an ultrasonic bath for 3 min and dried using dry air

prior to tests. To obtain the coefficient of friction, ramped load tests were performed in which the load was increased linearly with the sliding distance. An AISI 52100 steel ball of radius 1.2 mm and RMS roughness of 4 nm was used to probe test the coupon. The load was increased from 0.2 to 200 mN as the probe was moved across a stroke distance of 45 mm at 10 mm/s. The average friction coefficient of the bare aluminum and nano-diamond sintered aluminum was 0.13 and 0.22 respectively. The friction tests were performed at room temperature of 24°C and 30% relative humidity.

A 100 cycle reciprocating sliding wear test was performed under the same friction test conditions at a constant load of 100 mN and a stroke speed of 5 mm/s for a total sliding distance of 8 m (80 mm/cycle). Wear track profiling was carried out using an optical profilometer. The wear depth of bare aluminum was around 3 µm and that of nano-diamond coated aluminum was around 8 µm. The wear tests done on the nano-diamond coated aluminum are not accurate due to the ridges on the coupon. The 1.2 mm radius steel ball probe used in the test was not small enough to pass between the ridges. Hence during the test, the probe was sitting between two ridges there by wearing both ridges down before reaching the surface of the coupon. The measurement of wear of 8 µm includes both the ridge height (~5-6 µm) and the depth worn into the surface. Therefore more work needs to be done to improve the surface quality so that accurate tests can be done.

Surface roughness was also measured with the optical profilometer. The measurements for this test were done on the laser path (parallel to the ridges). We found the surface roughness of the nano-diamond coated coupon to be 2.4 µm. This result is a little higher than that of the bare aluminum which was found to be 1.4 µm. This shows that if we can produce a coupon without ridges, the resulting surface roughness would be equivalent to the original substrate.

4 CONCLUSION

We have investigated a laser-based coating technique of nano-diamond particles on aluminum substrate. The coating consisted of nano-diamond, diamond-like carbon and graphite phases with reasonable adherence of the nano-powder to the substrate. However, we were unable to measure hardness equivalent to that of the nano-diamond for the following reasons/constraints. Rastering was done to achieve a complete coverage of the surface of the coupon due to focused spot sizes (~200 µm) used (5% overlap in between passes). This resulted in uneven surfaces vowing to higher surface roughness, higher friction coefficient, and difficulty in measuring surface hardness. This can be avoided if we could use a larger focus spots and a larger spot size will also help to reduce number of rastering passes. Further studies are being conducted to understand the feasibility of using larger spot sizes. We will also explore the use of gold plated sample holder with DI water to get more explicit Raman spectrum.

Acknowledgements

The authors would like to thank the National Science Foundation for the Grant 0738405. We also appreciate Dr. Anatolli Frishman for acquiring the nano-diamond powder for us, Mr. Wenping Jiang for electrostatically spray coating our samples & providing the information on the setup, and Mr. Hal Sailsbury for his help in characterizing the surface after sintering. We would also like to thank Mr. Satyam Bhuyan (Mechanical Engineering, Iowa State University) and Mr David Eisenmann (Scientist, Center for Non-destructive Testing, Iowa State University) in helping us to measure tribological properties and surface roughness.

REFERENCES

[1] V.Y. Dolmatov, "Detonation synthesis of ultradispersed diamonds: properties and applications," Russian Chemical Reviews, 70, 607-626, 2001.

[2] G. Post, V.Y. Domatov, V.A Marchukov, V.G. Sushchev, M.V. Veretennikova, and A.E. Sal'ko, "Industrial synthesis of ultradisperse detonation diamonds and some fields of their use," Russian Journal of Applied Chemistry, 75, 755-760, 2002.

[3] G.P. Bogatyreva, M.A. Marinich, N.A. Oleynik, and G.A. Bazaliy, "Nanodispersed Diamond Adsorbents for Biological Solution Cleaning, in Nanostructructured Materials and Coatings for Biomedical and Sensor Applications," Kluwer Academic Publishers, 111-119, 2002.

[4] G.P. Bogatyreva, M.A. Marinich, and V.L. Gvyazdovskaya, "Diamond- an adsorbent of a new type," Diamond and Related Materials, 2002-2005, 2000.

[5] K.L. Choy, "Chemical vapour deposition of coatings," Progress in Materials Science, 48, 57-170, 2003.

[6] W. A. Yarbrough, and Russell Messier, "Current Issues and Problems in the Chemical Vapor Deposition of Diamond," Science, 247, 688-696, 1990.

[7] J.C. Angus and Cliff C. Hayman, "Low-Pressure, Metastable Growth of Diamond and Diamond-like Phases," Science, 241, 913-921, 1988.

[8] E. Cerri, E. Evangelista, S. Spigarelli, P. Cavaliere and F. DeRiccardis, "Effects of thermal treatments on microstructure and mechanical properties in a thixocast 319 aluminum alloy," Materials Science and Engineering, 284, 254-260, 2000.

[9] J.P Kruth, X. Wang, T. Laoui and L. Froyen, "Lasers and materials in selective laser sintering," Assembly Automation, 23, 357-371, 2003.

[10] D.T. Pham, S. Dimov, F. Lacan, "Selective laser sintering: applications and technological capabilities," Proceedings of the Institution of Mechanical Engineers, 213, 435-449, 1999.

[11] D.K.Y. Low, L. Li and A.G. Corfe, "The influence of assist gas on the mechanism of material ejection and removal during laser percussion drilling," Proceedings of the Institution of Mechanical Engineers, 214, 521-527, 2000.

[12] R. Mahmudi, P. Sepehrband and H.M. Ghasemi, "Improved properties of A319 aluminum casting alloy modified with Zr," Materials Letters, 60, 2606-2610, 2006.

[13] A.C. Ferrari and J. Robertson, "Raman spectroscopy of amorphous, nanostructured, diamond-like carbon, and nanodiamond," Philosophical Transactions of the Royal Society of London, 362, 2477-2512, 2004.

[14] Bhuyan, S., et al., "Boundary lubrication properties of lipid-based compounds evaluated using microtribological methods," Tribology Letters, 22(2), 167-172, 2006.

Oxygen plasma treated super-hydrophobic nozzle for electro-spray device

Youngjong Lee[1], Doyoung Byun[1]*, Si Bui Quang Tran[1], Sanghoon Kim[2], Baeho Park[2], Sukhan Lee[3]

[1] Konkuk University Korea, Aerospace Information Engineering
[2] Konkuk University Korea, Department of Physics
[3] Sungkyunkwan University, Korea, Schools of Information and Communication Engineering

* Corresponding to Phone:82-2-450-4195, E-mail: dybyun@konkuk.ac.kr

ABSTRACT

In this study, electro-spray device with super-hydrophobic nozzle is presented and fabricated instead of protruded nozzle. The super-hydrophobic nozzle is created by roughening the surface of the polyfluorotetraethylene (PTFE) or Teflon coated surface by argon and oxygen plasma treatment. We present a simple process of fabricating super-hydrophobic nozzle which is able to allow itself to enhance the stability of liquid meniscus at the outlet of the nozzle. Through the ion beam treated nozzle, the liquid does not overflow and keeps more stable and repeatable ejection of liquid jet.

Keywords: Ion beam treatment, Super-hydrophobic surface, Electro-spray

1 INTRODUCTION

Electro-hydrodynamic spraying, or electro-spray, has been a subject of intensive research in recent years; for instance, mass spectrometry [1]-[3], printing technology [4]-[6], and biological micro-arrays [7]. As a mechanism that allows for the dispersal of very fine liquid droplets, its potential applications are seen as numerous. When a low, constant flow rate of a liquid is passed through a capillary, and the meniscus of the liquid subject to an electric field beyond certain strength, an electric charge is induced on the meniscus, and a combination of electrostatic, hydrostatic, and capillary forces elongate the liquid into a conical form known as a Taylor-cone [8]-[10].

Depending on the flow rate of the liquid and the electric field applied, these cones emit fine particles in a variety of classified regimes or modes, one of which is known as the "cone-jet" mode. This cone-jet spraying mode, in which a steady jet of charged droplets is emitted from the apex of the Taylor-cone, allows for spray drops in the sub-micron range [10]. This emission of charged droplets is properly termed "electro-spray" [9].

Conventional inkjetting devices based on thermal bubble or piezoelectric pumping, however, have some fundamental limitations including size and density of the nozzle array, and ejection frequency, both primarily due to thermal problems. Mechanical jetting has limits in the density of the nozzle array, while the ejection frequency is limited by physical properties, and jetting reliability limited due to the difficulty of fabrication.

On the other hand, electro-hydrodynamic jet printing [4,6], or electrostatic field induced jetting device [5], based on the direct manipulation of liquid by an electric field, appears more promising. Using a continuously focused colloid jet, Lee, et al. [6] have introduced the electro-hydrodynamic printing of silver nano-particles as a direct writing technology. Park, et al. [4] have been able to use electro-hydrodynamic spraying to print images and electrode structures from gold of line width ~2μm. Such structures may be used in the manufacture of circuitry. Lee, et al. [5] have developed an electro-spray nozzle specifically for drop-on-demand inkjet printing. Their design has been able to provide relatively stable and sustainable droplet ejections under a wide variety of applied voltages. They showed successfully the feasibility of the electrostatic force to eject liquid droplet for the application to industry. However their technology should be extended to multi-nozzle device which can be manufactured massively and reproduced with appropriate yield.

Hyrdophobicity of a given surface is known to enhance the stability of a liquid meniscus in contact with that surface, and hence the stability of a cone-jet [11]. However, even on a hydrophobic surface against which a meniscus has a contact angle of around, a cone-jet is not stable and the meniscus on top of the nozzle may overflow onto the surrounding surface. We need to produce a jetting nozzle that can eliminate these instabilities during jetting. This objective is most readily achieved by fabricating some sort of protruding nozzle; however, a protruding nozzle is difficult to manufacture.

Thus, we realize that a flat nozzle composed of a super-hydrophobic material (that is, with a static contact angle greater than 150°[12]) is the most advantageous, as the resulting high-contact angle of the liquid meniscus at the nozzle's opening diminishes potentially hazardous leaks to the nozzle's surface and thus ensures long term stability and repeatability of the electro-spray process.

A known process for creating a super-hydrophobic surface is ion beam treatment [13]. The efficacy of the ion beam in modifying the topology and wetting characteristics of polymer surfaces has been established [14]. Capps, et al. [15] have reported on the effectiveness of argon and argon-oxygen ion beams on polyfluorotetraethylene (PTFE) in particular.

In this article, in order to fabricate polymer based electro-spray device with super hydrophobic nozzle, we use Teflon AF (amorphous fluoropolymers) 1600 and PTFE (polyfluorotetraethylene) plate with ion beam treatment to fabricate the super-hydrophobic surface.

2 FABRICATION OF SUPER-HYDROPHOBIC SURFACE

2.1 Ion beam treatment experiments

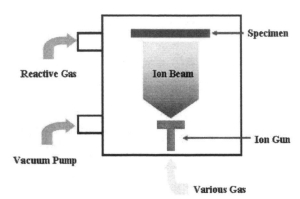

Fig1. Ion beam treatment equipment

To fabricate super-hydrophobic surface we use ion beam treatment process as shown in Fig. 1. Argon and oxygen are supplied to a vacuum chamber and then plasma ion beam is created and bombards polymer surface. In order to obtain the super-hydrophobic surface, the optimal conditions of modification were examined varying argon and oxygen concentration and energy levels. In this study we use Teflon AF 1600 and PTFE. Teflon AF 1600 is coated on a substrate and treated by 1.5 keV ion beam made from argon 3 sccm and oxygen 3 sccm gases. And PTFE is treated under the condition of argon 2 sccm and oxygen 2 sccm, and 1.5 keV.

2.2 Contact angle measurements

Contact angles of 2μl DI water droplet is measured on the treated surface using CCD camera and X-Y stage to investigate the effects of exposure time and energy level.

Figure 2 shows the shows the contact angle results of treated Teflon surface. Because the thickness of the Teflon layer affects much a super-hydrophobic surface characteristic, we tested several Teflon layers varying the spin coating speed from 350 rpm to 1000 rpm. In the case of 350 rpm, Teflon layer thickness is around 5 μm, which is measured by alpha step machine. For 700 rpm and 1000 rpm the thickness is measured to be around 3 μm. Contact angle could be lager than 150° when the exposure time is 5

min for the layer by 350 rpm spin coating, while the others show lower contact angle (nearly 100°) after 5 min. This observation depicts that 3 μm layer of Teflon is not enough to generate super0hydrophobic surface. The Teflon layer should be etched to form microscale structures on the surface by means of ion beam. Therefore, if the thickness of the layer is thin, the micro-structured Teflon surface may be too much.

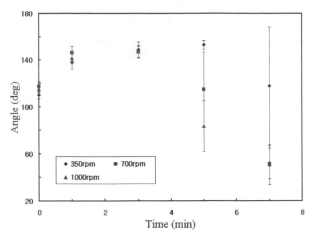

Fig2. Treated Teflon layer contact angle results

Figure 3 shows the contact angle results of treated PTFE surface and degradation characteristics are observed during longer than 2 months. Ion beam treated PTFE surface, hydrophobicity is increasd as expose time and contact angle is increased. First observation contact angle in July was 155°and after two month later, the contact angle did not change and kept same contact angle. With these results, PTEF surface can be treated permanent super-hydrophobic surface and can be applied to super-hydrophobic nozzle for electro-spray device.

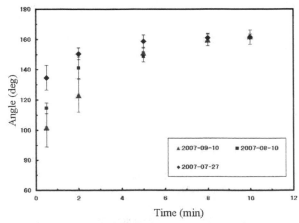

Fig3. Ion beam treated PTEF contact angle results

To try out quantitative data of nano structure after the ion beam treatment on the PTFE, we take AFM (atomic force microscope) and SEM (scanning electron microscope) data. Figure 4 show results of PTFE and treated PTFE surfaces. We see the width of the nano-scale structures is approximately 200nm and the height is 1.2μm, respectively. Due to these nano-scale structures, treated PFTE surface can enhance the liquid meniscus contact angles to super-hydrophobic ranges.

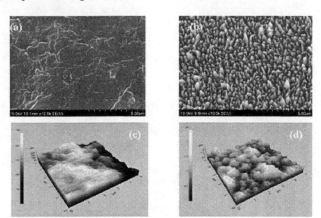

Fig4. SEM and AFM results. (a) and (c) PTFE, (b) and (d) treated PTFE

3 RESULTS AND DISCUSSIONS

3.1 Ion beam treated nozzle

To fabricate PMMA reservoir and PTFE plate nozzle (the nozzle diameter is 700μm), we used CNC machine and laser cutting machine. And this PTFE nozzle treated super-hydrophobic by ion beam for forming super-hydrophobic surfaces. The treated ion beam condition was Ar 2 sccm, O2 2sccm and 1.5kev.

The Al plate (thickness 0.2mm) is used for ground electrode, and using drilling machine fabricate the hole (hole diameter is 2mm). The gap between nozzle and ground electrode set up 3 mm. Using syringe pump delivered mixed solution. For solution, mixed D.I water 50%, methanol (CH3OH) 49%, and acetone (CH3COCH3) 1%.

When high voltage supply, meniscus shape is changed Taylor-cone and tiny jet spread from the apex of the cone shape meniscus on the ion beam treated surface nozzle.

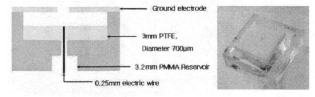

Fig5. The schematic of Ion beam treated surface PTFE nozzle.

3.2 Sequential elector-spray results

Figure 6 and 7 show the sequential images of electro-spray through hydrophobic PTFE nozzle as well as super-

hydrophobic treated PTFE nozzle. For the PTFE nozzle, as the electro-spray operates repeatedly, the meniscus widely spreads around the nozzle and also shape of the meniscus changes to make the spray to be inefficient.

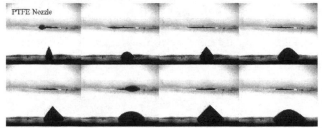

Fig6. The sequential images of Electro-spray on the PTFE surface nozzle. Flow rate is 5μl/min and operating voltage is 4.0kV.

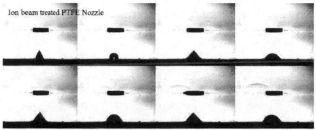

Fig7. The sequential images of Electro-spray on the Ion beam treated PTFE surface nozzle. Flow rate is 5μl/min and operating voltage is 4.0kV.

On the other hand, for the ion beam treated PTFE surface nozzle, as shown in Fig. 7, the liquid doesn't overflow and it keeps the same initial cone shape and position. Because of the super-hydrophobic surface, the liquid can be sustained to form a meniscus with high contact angle.

3.3 Electro-spray on the various nozzle

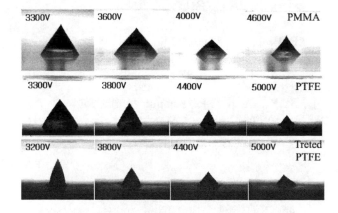

Fig7. Electro-spray through various nozzle

In case of PMMA nozzle, meniscus didn't keep the initial position and shape as increasing operating voltage. The PTFE nozzle case, at the low operating voltage range from 3.3kV to 3.8kV, liquid overflow so the meniscus width is larger than initial cone width.

But the Ion beam treated PTFE nozzle case, even at low operating voltage ranges; the meniscus width is nearly same with initial meniscus width. And as increasing voltage, the position of meniscus did not change.

According to these results, the super-hydrophobic surface nozzle is useful for forming cone jet spray and helpful to keep the initial cone shape and position also.

4 CONCLUSION

In this paper, to fabricate a super-ydrophobic surface, Teflon layer and PTFE (Polytetrafluoroethylene) plate are treated by argon and oxygen ion plasma beam. We vary argon and oxygen flow rate and energy level to investigate the effects of exposure time and energy level.

For the ion beam treated PTFE surface nozzle, the liquid does not overflow and keep the cone shape position even at the low voltage ranges. Because of the super-hydrophobic surface, the liquid can be sustained to form a meniscus with high contact angle. And the width of the meniscus at bottom is same as the nozzle diameter.

Using the ion beam treatment on PTFE surface, super-hydrophobic surface can be fabricated and allow the nozzle to keep the stable cone-jet even at the low voltage ranges.

5 ACKNOWLEDGEMENTS

This work was supported by a grant from the Korea Research Foundation (KRF-2006-005-J03301) and the National Research Laboratory program, Korea Science and Engineering Foundation Grant (R0A-2007-000-20012-0). SBQT acknowledges partial support from the Korea Research Foundation (KRF-2005-D00208).

REFERENCES

[1] J.B Fenn, M. Mann, C.K. Meng, S.K. Wong, C. Whitehouse, Science 246, 64–71 (1989).

[2] C.II. Yuan and J. Shiea, Analytical Chemistry 73, 1080-1083 (2001).

[3] J.S. Kim and D.R. Knapp, Journal of American Society for Mass Spectrometry 12, 463-469 (2001).

[4] J.U. Park, M. Hardy, S.J. Kang, K. Barton, K. Adair, D.K. Mukhopadhyay, C.Y. Lee, M.S. Strano, A.G. Alleyne, J.G. Georgiadis, P.M. Ferreira, and J.A. Rogers, Nature Materials 6, 782-789 (2007).

[5] S. Lee, D. Byun, D. Jung, J. Choi, Y. Kim, J.H. Yang, S.U. Son, S.B.Q. Tran, and H.S. Ko, Sensors and Actuators A: Phys. **141**, 506-514 (2008).

[6] Dae-Young Lee, Yun-Soo Shin, Sung-Eun Park, Tae-U Yu, and Jungho Hwang, Applied Physics Letters 90, (2007)

[7] M.D. Paine, M.S. Alexander, K.L. Smith, M. Wang, and J.P.W. Stark, Journal of Aerosol Science 38, 315-324 (2007).

[8] J. Fernández de la Mora, Journal of Fluid Mechanics 243, 561-574 (1992).

[9] J. Fernández de la Mora, Annual Review of Fluid Mechanics 39, 217–243 (2007).

[10] M. Cloupeau, M. and B. Prunet-Fuch, Journal of Aerosol Science 25, 1021-1036 (1994).

[11] P. Lozano, P, M. Martínez-Sánchez, and J.M. Lopez-Urdiales, Journal of Colloid and Interface Science 276, 392–399 (2004).

[12] E. S. Yoon, S. H. Yang, H. S. Kong, and K. H. Kim, Tribology Letters 15, No. 2, 145-154 (2003).

[13] L. Zhu, Y. Y. Feng, X. Y. Ye, and Z. Y. Zhou, Sensors and Actuators A 130–131, 595–600, (2006).

[14] J. Lee, Y. Seo, K. Hanseong, K. Taejin, and K. Sehyun, Fukugo Zairyo Shinpojiumu Koen Yoshishu 30, 65-66 (2005).

[15] N. Capps, L. Lou, M. Amann, Advanced Energy Industries Inc. Application Note (website: http://www.advanced-energy.com/upload/File/Sources/SL-IONPOLY-260-01.pdf)

Fluorination effects on tribological characteristics of hydrogenated amorphous carbon thin films

M. Rubio-Roy, S. Portal, M.C. Polo, E. Pascual, E. Bertran, J.L. Andújar

FEMAN Group, Dept. Física Aplicada i Òptica, IN2UB, Universitat de Barcelona
C/Martí i Franquès 1, E08028, Barcelona, Spain, mrubioroy@ub.edu

ABSTRACT

The main effect of the introduction of fluorine to diamond-like carbon (FDLC) thin films is the reduction in surface free energy (SFE). Regarding tribological properties, frequent allusions to low friction are found in the literature. However, studies proving a direct relationship between fluorine content in DLC films and macroscopic friction reduction are hard to find, while others state fluorine does not affect friction significantly. The present study completes a previous one, in which we introduced fluorine leading to a reduction in SFE, and establishes which is the relationship between films' fluorination degree and their tribological characteristics. In order to complete the study, optical emission actinometry measurements were carried out, and plasma composition results are discussed in terms of their relationship with chemical properties of the films.

Keywords: amorphous carbon, fluorine, tribology, actinometry, pulsed-DC

1 INTRODUCTION

Hydrogenated amorphous carbon (a-C:H) films are good candidates for biomedical applications [1-5], requiring smooth surfaces, biocompatibility and singular tribological characteristics. The addition of elements to a-C:H films, such as oxygen, silicon, fluorine or metals, may greatly modify their properties [2,6] (stress, hardness, wear, surface energy, friction, biocompatibility or bulk electrical and optical properties). Therefore, they are of interest for applications like protective and biocompatible coatings (e.g. for biomedical sensors and tools) and also for antisticking[7], low friction [8-9] or self-cleaning surfaces as well as low-κ layers [10-11].

Apart from its characteristic high hardness and low wear rate, diamond-like carbon (DLC) thin films exhibit low friction and low surface free energy (SFE). The introduction of fluorine to these films has proved to reduce even more SFE [12], therefore being useful as anti-sticking layers, and it has also been beneficial reducing its dielectric constant, for microelectronic applications. Furthermore, its antithrombogenic [13-14] properties have widened the fields of application to biomedical devices or instrumentation.

In this study, our goal was to study in detail the effects of the incorporation of fluorine to amorphous carbon (a-C:H:F) thin films on their tribological properties. C-F_x bonds shown to have, as already reported, an important effect on diminishing surface energy. In this work we report the slight increase of friction of a-C:H:F films deposited by pulsed-DC plasma enhanced chemical vapour deposition (PECVD).

2 EXPERIMENTAL DETAILS

Fluorine-containing amorphous carbon thin films were deposited by PECVD in a CH_4 and CHF_3 environment at 10 Pa with a total gas flow of 25 sccm. The power source for the process was an ENI RPG-50 providing 1.3 W/cm^2 at 100 kHz. More details about the deposition process can be found in a previous study [12].

Optical emission actinometry (OEA) measurements were performed in order to determine the relative concentrations of the different species in the plasma. Compared to direct measurement techniques like laser induced fluorescence (LIF), OEA requires less specialized equipment and it is not so experimentally demanding [15]. Moreover, it usually provides a better signal to noise ratio. On the other side, actinometry procedure is based on several requirements, which have to be checked in order to ensure reliable results. Basically, they are: comparable threshold energies for reactive species and actinometers, similar excitation cross sections and comparable de-excitation paths.

Kiss et al [15] extended the validity of OEA for fluorocarbon plasmas despite of the big difference in threshold energies between Ar* (13.5 eV) and CF* (6.1 eV) or CF$_2$* (4.5 eV). The predictable small variation of electron energy distribution (EEDF) for the selected range of variation of process parameters, which is also applicable in the present study, validates the applicability of OEA.

Actinometry requires the addition of a small amount of an inert gas such as argon to the plasma. The intensities of the different peaks corresponding to the species of interest are then referred to that of a selected Ar line. In the present study the line selected as actinometer was Ar 750.4 nm, while F 703.7 nm, CF_2 251.9 nm, CH 430.5 nm and H 656.3 nm were monitored. Spectra were collected for different CHF_3 to CH_4 flow ratios. A StellarNet EPP2000C-UV-VIS spectrometer, using an integration time of 20000 ms and an averaging of 4 to 6 measurements, was used. Precautions in order to reduce the collection of light coming from the plasma sheath, where EEDF is expected to vary during the power signal cycle, were taken.

Regarding friction measurements, a Nanotribometer from CSM Instruments was used. The measurements were performed in ball-on-disk configuration during 10^4 cycles, with a linear speed of 1 mm/s and applying 50 mN of normal load. The selected counterpart was a 3 mm in diameter ball of WC with the ST-117 cantilever (100mN of maximum normal load). Ambient conditions were kept between 30% and 35% of relative humidity (RH), and between 24°C and 27°C of temperature.

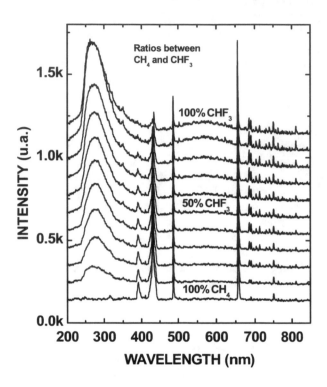

Figure 1: Raw optical emission spectra of CH_4+CHF_3 for different quantities of CHF_3 gas. A constant gas flow of 0.5 sccm of Ar was added to the plasma as actinometer

XPS was used to determine the F to C ratio in the films as well as the bonding structure. A PHI ESCA-5500 Multitechnique System, with a monochromatic X-Ray source (Al K_α line of 1486.6 eV energy and 350 W) placed perpendicular to the analyzer, was used. The chamber pressure remained under 10^{-6} Pa during the analyses. It was calibrated using the $3d_{5/2}$ line of Ag with a full width at half maximum (FWHM) of 0.5 eV. The sample was at 45° from both the detector axis and the X-Ray source. All samples were subjected to a general scan with electron pass energy of 187.85 eV and 0.8 eV/step in order to check the presence of any other elements apart from C, H and F. Moreover, high resolution spectra at 11.75 eV of electron pass energy and 0.1 eV/step were collected on C 1s and F 1s regions of the spectrum. Neither sputtering nor neutralization of the sample was carried out in order not to modify the bonding

structure or composition of the surface. MULTIPAK software package was used to quantify the relative content of F and C in the sample. The percentages of F contents referred in the text are relative to the total F and C content.

3   RESULTS AND DISCUSSION

In order to establish the possible relation between tribological properties of films and their chemical characteristics, we have performed an actinometry study of the plasma conditions during the film's growth as a function of the relative concentration of the precursor gases.

Having considered the previously exposed assumptions, actinometry provides the evolution of each species with respect to themselves, by referring the emission intensity to Ar, whose concentration is kept constant throughout the different experiments.

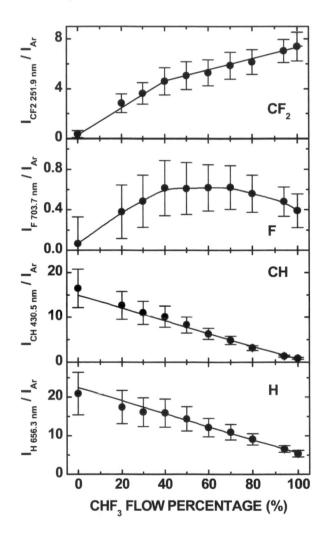

Figure 2: Concentrations of F and CF_2, relative to Ar. Each series of dots is only comparable to itself. All lines are to guide the eyes, only.

As shown in Figure 2, actinometry results have evidenced linear evolutions of CH radical, and atomic H. CH radical actinometry shows this specie is only produced by dissociation of CH_4 gas.

Regarding fluorinated species, both CF_2 and atomic F show a non linear behaviour, with a change of trend around 40%. On the other hand atomic F shows a plateau region where F concentration is saturated, and further addition of CHF_3 does not affect it. For pure CHF_3, F emission decreases again. The simultaneous changes of trends of CF_2 and F are probably connected with the observed decrease of H content. Dissociated H is a source of electrons, which therefore promotes a higher electron and ion density of the plasma. Moreover, as hydrogen is removed from the plasma, film would receive weaker bombardment, which could explain the higher measured deposition rate [12].

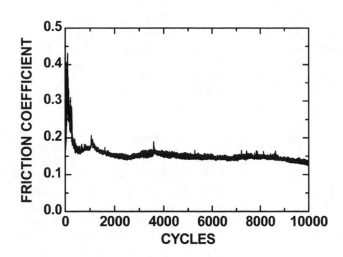

Figure 4: Friction coefficient evolution over 10^4 cycles, for 10% of CHF_3 flow ratio.

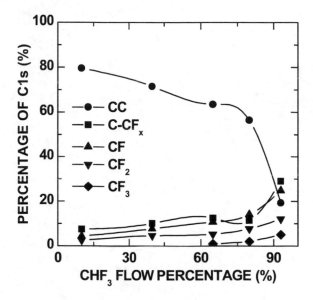

Figure 3: Contribution to C1s band of each fluorinated peak area depending on the F incorporation in the films ([F]/[F+C]). All lines are to guide the eyes, only.

The main specie responsible of C-C bonds (cross-linking) is CH. Its emission is reduced to almost zero as CHF_3 is introduced in the chamber. Looking at XPS results shown at figure 3, it can be seen there is an inverse relationship between CH emission and C-C bonds in the film.

With respect to friction, a first period of stabilization has been observed in all measurements (in figure 4, first 500 cycles). During this period, friction coefficient fluctuates, achieving high values. After a settling time, friction achieves a stable low value, sometimes getting it with an exponential decay.

This first period could correspond to the removal time of ultra thin superficial layers as neither optical microscopy observation nor profilometry measurement showed any wear track on the samples after the test. These layers could be due to several reasons: surface roughness (typically a few nanometers); a top nanometric layer of amorphous carbon with different properties than bulk, due to the subplantation [16-17] process which rules DLC growth; contamination left by cleaning process (carried out with a small cotton wetted with isopropanol) prior to measurements; or water adsorption from air.

This last hypothesis would be supported by the fact that samples are introduced in the measurement chamber from different ambient conditions (relative humidity and temperature) and a certain time is required in order to achieve an equilibrium state. The exponential decay, previously mentioned would be in agreement with this possibility.

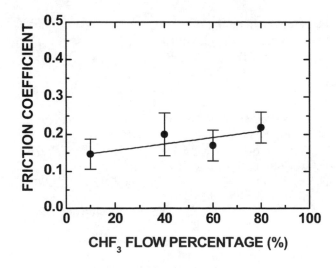

Figure 5: Friction coefficients for different fluorine contents

When comparing the final friction values of each film, we have not observed important differences in their coefficients. However, as shown in figure 5, there is a slight trend towards higher friction when fluorine is added. This could be attributed to a less cross-linked structure of the films when F is added, which is coherent with XPS results shown in figure 3.

4   CONCLUSIONS

Fluorine incorporation in DLC thin films only shows a slight increase of friction coefficient values, from 0.15±0.05 to 0.23±0.05. However, our experiments have evidenced a strong dependence on the ambient conditions during the friction measurements. A fine humidity and temperature control is advisable during measurements to avoid friction derives.

In addition, the strong fluctuations observed on each friction measurement, due to the initial top layers wear process, masked nominal friction values, forcing us to carry out long measurements in order to get a stable friction values.

ACKNOWLEDGEMENTS

This study was partially supported by the Generalitat de Catalunya (Project No. 2005SGR00666) and the MEC of Spain (Project SURFaC: DPI2007-61349). One of the authors (M.R.) acknowledges financial support from CSIC of Spain (Grant No: UAC-2005-0021) and the helpful discussions with Prof. J. Esteve.

REFERENCES

[1] F.Z. Cui, D.J. Li, Surf. Coat. Technol., 131, 481-487, 2000

[2] R. Hauert, Diamond Rclat. Mater., 12, 583-589, 2003

[3] A. Grill, Diamond Relat. Mater., 12 , 166-170, 2003

[4] A.H. Lettington, Carbon, 36, 555-560, 1998

[5] M. Allen, B. Myer, N. Rushton, J. Biomed. Mater. Res., 58 , 319-328, 2001

[6] M. Grischke, A. Hieke, F. Morgenweck, et al., Diamond Relat. Mater., 7, 454-458, 1998

[7] K. Nakamatsu, N. Yamada, K. Kanda, et al., Jpn. J. Appl. Phys., 45, L954-L956, 2006

[8] A. Grill, Surf. Coat. Technol., 94-5, 507-513, 1997

[9] C. Donnet, Surf. Coat. Technol., 100, 180-186, 1998

[10] J.W. Yi, Y.H. Lee, B. Farouk, Thin Solid Films, 374, 103-108, 2000

[11] A. Grill, V. Patel, C. Jahnes, J. Electrochem. Soc., 145, 1649-1653, 1998

[12] M. Rubio-Roy, E. Bertran, E. Pascual, M.C. Polo, J.L. Andújar, Diamond Relat. Mater., (in press)

[13] T. Saito, T. Hasebe, S. Yohena, et al., Diamond Relat. Mater., 14, 1116-1119, 2005

[14] T. Hasebe, A. Shimada, T. Suzuki, et al., J. Biomed. Mater. Res. A, 76A , 86-94, 2006

[15] L.D.B. Kiss, J.P. Nicolai, W.T. Conner, et al., J. Appl. Phys., 71 , 3186-3192, 1992

[16] Y. Lifshitz, Diamond Relat. Mater., 8, 1659-1676, 1999

[17] P.J. Fallon, V.S. Veerasamy, C.A. Davis, et al., Phys. Rev. B, 48, 4777-4782, 1993

Au Nanoparticle Coupled- Surface Plasmon Resonance Immuno-sensor for Sensitivity Enhancement

Hyo-Sop Kim*, Jin-Ho Kim*, Sung-Ho Ko**, Yong-Jin Cho**, and Jae-Ho Kim*

*Department of Molecular Science and Technology, Ajou University
Suwon, 443-749, Republic of Korea
**Korea Food Research Institute, Sungnam, 463-746, Republic Korea

ikari06@ajou.ac.kr

ABSTRACT

In order to overcome the sensitivity limitation of SPR, nanoparticle-coupled SPR biosensors have explored since nanoparticles may significantly enhance the sensitivity by 1-2 orders of magnitude. Although the nanoparticle conjugated method contributes to the enhancement of the sensitivity, it does not take an advantage of SPR from the viewpoint of label-free detection. Herein we demonstrate the sensitivity enhancement with Au nanoparticles coupled-SPR immuno-sensor chip on which the specific size and surface density of the particles controlled as a label-free detection system. Au nanopariticles were synthesized and selected with a specific size. Aminoethanethiol was used as a coupling layer on bare Au substrate for immobilization of Au particles. Surface density of immobilized 30 nm Au nanoparticles on bare Au film was estimated as 1×10^9 ea/cm^2 using atomic force microscopy. With systematic control of the size of the Au nanoparticle and thickness of the bare Au film, it was found that 30 nm Au particles on 50 nm thick Au film demonstrated the largest resonance angle shift for surface reaction on the chip. This enhancement may result from both the effective surface area increase and the amplification of the resonance resonance effect. For human immunoglobulin G (hIgG) antibody-antigen analysis, this Au nanoparticle immobilized Au substrate shows highest SPR sensitivity comparing to the conventional bare Au film chip.

Keywords: SPR, colloidal Au particle, chemical sensor

1 INTRODUCTION

Surface plasmon resonance (SPR) has been widely employed for protein-protein interaction studies in which changes in refractive index close to a thin metal surface are monitored. [1-3] The need for significant surface refractive index changes for detection, however, has limited its applications for ultra sensitive analysis. To address this limitation, several signal-amplification approaches have been developed.[4-5] Of those, the use of colloidal Au nanoparticles to enhance SPR signals associated with biochemical binding events has attracted great attention.[6-8] In this method, molecular-recognition probes are pre-conjugated to metallic nanoparticles. The binding events between the probes and their partners on the surface lead to subsequent immobilization of particles on SPR substrates. As a result, remarkably large changes in both the position of the SPR angle and the magnitude of film reflectivity are observed. The successful applications of this method have been demonstrated for sandwich immunoassays and DNA hybridization detection by various groups.[4, 6]

In addition, the effect of nanoparticle binding on SPR response has been investigated as a function of particle composition, size, surface coverage, and substrate metals.[9-10]

Though this colloid conjugated method has overcome the sensitivity, it does not take an advantage of SPR from the viewpoint of label-free detection. Herein we investigated the sensitivity enhancement in SPR immuno-sensor coupling colloidal Au nanoparticles which have the specific size and surface density on sensor chips as label-free detection system.

2 EXPERIMENTAL

Reagents. In this research, we used reagents as received without further purification. All solvents used were HPLC grade. Deionized water was purified to > 18.3 MΩ with a Millipore system. Cystamine dichloride, nitric acid, hydrochloric acid, hydrogen tetrachloroaurate, sodium citrate, ethanol, methanol were purchased from Aldrich. Glass substrates (BK7, n = 1.517, 18 mm x 18 mm x 0.15 mm) was purchased from Matsunami glass.

Au Colloid Synthesis. Hydrogen tetrachloroaurate 0.08 g was dissolved in 170 ml D.I. water. Hydrogen tetrachloroaurate solution was heated with stirring up to 100 ℃. Added 4 mM ~ 15 mM sodium citrate aqueous solution. After 2 hr, solution was slowly cooled at room temperature. [11]

Au Film Preparation. Glass substrates were cleaned by sonicating in ethanol, D.I. water and then treated by ozone for 10 min each. Au films were deposited on the cleaned glass substrates by Au shot (99.999 %, Eiko) in Eiko IB-3 ion coater system. Au substrates (thickness 50 nm) were deposited at a pressure of 0.05 torr for 210 s and used freshly. Au films were modified by treating with 10 mM aqueous solution of cystamine dichloride for a period of 24 hr.

Fabrication of Au colloid immobilized substrate. Cystamine modified Au film on a glass substrate was exposed to Au colloids solution for 30 min at room temperature for fabrication of Au nanoparticle coupled substrate for SPR measurement (Scheme. 1).

Immobilization of antibody. The Au colloid coated substrate was immersed in MHDA 1 mM ethanol solution for 2 hr. Then the substrate was activated in EDC/NHS (0.4 M/0.1 M) aqueous solution for 10 min. The activated substrate was quickly immersed in 0.1 mg/ml Human IgG solution (pH 7.4 PBS) for 1 hr.

SPR Instrument. Excitation of the surface plasmon is accomplished using a 1 cm diameter triangle prism (BK 7 glass, Sigma product) which is index-matched via a microscopy immersion oil (n = 1.516) to a BK 7 substrate onto previously deposited Au film. This assembly is then affixed to a home-built batch cell (volume ~1 ml) with the Au film exposed solution. The SPR excitation source is a cylindrical 5 mW HeNe laser (674.5 nm, Melles Griot) which is further polarized by 500 : 1 visible-optimized linear polarizer (Newport, 10-LP VIS). Stage rotation and data collection are controlled through the computer interface that was programmed by K-MAC Instrument. A typical SPR scan was run 0.04° resolution and a stage rotation rate of $0.2°$ s^{-1}.

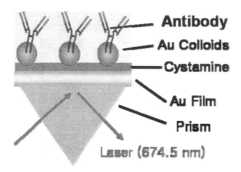

Scheme 1. Particle-enhanced SPR sensing architectures

Atomic Force Microscopy (AFM). AFM images acquired were obtained using a PSIA Instruments XE-100 system, operated in non-contact mode with an acquisition frequency of 312 kHz and line density of 512. Standard 200 μm etched silicon probes were used.

Field Emission Scanning Electron Microscope (FE-SEM). The SEM images were obtained using a JEOL 6340F, under typical working conditions of 5 kV.

3   RESULTS AND DISCUSSTION

The dimension of the synthesized Au colloid size was measured using FE-SEM (Figure 1). It was found that the diameter of Au colloid was varied according to the change of citrate concentration in the Au solution (Table 1).

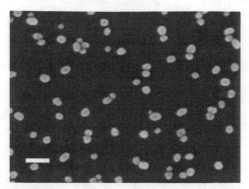

Figure 1. FE-SEM image of synthesized Au colloid of 9 mM sodium citrate concentration. Scale bar was 100 nm.

Table 1. The diameter of synthesized Au colloid of different sodium citrate concentration.

Citrate conc.	4 mM	9 mM	15 mM
Au colloid diameter (nm)	60 ± 5	30 ± 3	20 ± 3

Au colloids were deposited on the Au film with various surface densities. The surface morphology of Au colloids immobilized on Au films was evaluated using AFM (Figure 2). Surface density of Au colloids was calculated using PSIA image analysis program. The surface densities of Au colloids immobilized substrates were differ approximately 100 times by 10 times diluted Au colloids solution (Table 2).

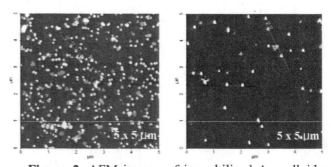

Figure 2. AFM images of immobilized Au colloid on Au film. Scan size was 5 x 5 μm.

Table 2. The surface densities of immobilized Au colloid on substrate at immobilize conditions. (ea/cm^2)

Diameter	Undiluted	Diluted (x 10)
20 nm	5 x 10^{11}	3 x 10^{9}
30 nm	1 x 10^{11}	1 x 10^{9}
60 nm	0.2 x 10^{11}	0.5 x 10^{9}

By immobilization of Au colloid, SPR curve was moved right direction and was broaden (Figure 3). Resonance angle shifts were showed linear correlation with alcohol

NSTI-Nanotech 2008, www.nsti.org, ISBN 978-1-4200-8503-7 Vol. 1

concentration (Figure 4). SPR signal enhancement was different by changes of diameter and surface density. Because deposited Au thickness on glass substrates was maintained constant, SPR resonance shift was originated from the effect of immobilized Au colloid.

The relative signal enhancement efficiency by Au colloid diameter and surface density was summarized in Table 3. The SPR response of substrate of 20 nm Au colloid immobilized was enhanced at high surface density, but low surface density was not enhanced. The 30 nm Au colloid substrate was enhanced higher at low surface density. The SPR response was very low for 60 nm Au colloid immobilized on substrate. The maximum SPR response was observed for Au colloid of 30 nm diameter, immobilized on substrate at low surface density.

Au colloid immobilized substrate with 30 nm diameter and 1×10^9 ea/cm^2 surface density was used to detect SPR signal by incremental concentration changes for around 20 % ethanol solution. In this is the typical alcoholic concentration of the consumable liquors (Figure 5). Confirmed SPR signal by alcohol concentration was enhanced using Au colloid.

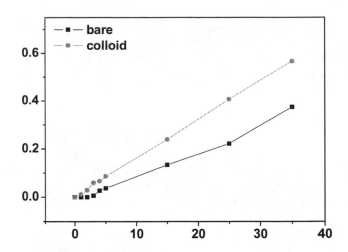

Figure 5. SPR signal changes by small changes in concentration for 20 % ethanol solution. Au colloid immobilized on substrate was with 30 nm diameter and 1×10^9 ea/cm^2 surface density.

Table 3. SPR signal enhancement efficiency (%) by Au colloid immobilized condition at 25 % (v/v) MeOH/H$_2$O solution.

	20 nm	30 nm	60 nm
Undiluted	21.4	14.3	- 25
Diluted (x 10)	0	33	- 7.7

In summary, we found that of the maximum SPR response was observed at 30 nm Au particle size and 1×10^9 ea/cm^2 surface density on the SPR substrate. The existence of electromagnetic interactions between the metallic nanoparticles and the surface, and light scattering are also thought to contribute signal enhancement. Especially for relative small separation and large particles, the electromagnetic coupling between surface and the metallic nanoparticles is expected. While quantification of the contribution from each factor was not possible in current work, our results highlight the importance of the dielectric of the particle layer on Au film.[12-14]

4 RELATION

In conclusion, to enhance the SPR signal through changed conc. of alcohol aqueous solution, we immobilized Au colloid on substrate. Enhancement of SPR signal efficiency was depended on the size and the surface density of immobilized Au colloid on substrate. A quantified equation for describing relationship between surface density and size of the Au colloids is needed further investigation.

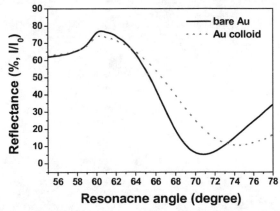

Figure 3. SPR curves obtained from a bare Au film, Au colloid immobilized surface. SPR curve were moved right direction and broaden. Au colloid diameter was 30 nm, surface density was 1×10^9 ea/cm^2.

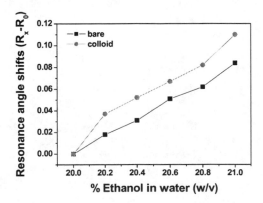

Figure 4. SPR signal was enhanced to detect concentration changes of methanol solution. Au colloid immobilized on the substrate was 30 nm diameter.

5 ACKNOWLEDGEMENT

This study is supported by the research project of 'Development of Food Nanotechnology', Korea Food Research Institute and BK21 Program of the Ministry of Education and Human Resources Development for the Molecular Science and Technology in Ajou University, Republic of Korea.

6 REFERENCE

[1] G. Wegner, Anal. Chem. 75, 4740, 2003

[2] S. Lofas, Mod. Drug Discovery, 6, 47, 2003

[3] J. M. McDonnell, Curr. Opin. Chem. Biol. 5, 572, 2001

[4] H. Q. Zaho, L. Jiang, J. Nanopart. Res. 3, 321, 2001

[5] R. J. Pei, E. K. Wang, Anal. Chem. Acta. 453, 173, 2002

[6] L. A. Lyon, M. J. Natan, J. Anal. Chem. 70, 5177, 1998

[7] P. –T. Leung, M. F. Finlan, Sens. Actuators B 22, 175, 1994

[8] J. C. Riboh, C. R. Yonzon, J. Phys. Chem. B, 107, 1772, 2003

[9] L. A. Lyon, M. J. Natan, J. Phys. Chem. B, 103, 5826, 1999

[10] E. Hutter, J. H. Fendler, Chem. Phys. B, 105, 11159, 2001

[11] G. Frens, Nature: Phys. Sci 241, 20, 1973

[12] L. A. Lyon, M. J. Natan, Anal. Chem. 70, 5177, 1998

[13] H. Lin, M. J. Natan, J. Am. Chem. Soc. 122, 9071, 2000

[14] H. Lin, M. J. Natan, J. Phys. Chem. B, 108, 10973, 2004

Synthesis and X-ray Photoelectron Spectroscopy Studies of Poly(vinyl pyrrolidone) Capped Gold Nanoparticles in a Hybrid Nanocomposite

A. Mishra[1], P. Tripathy[1], S. Ram[1*], and H. -J. Fecht[2]

[1]Materials Science Centre, Indian Institute of Technology, Kharagpur-721 302, India
[2]Werkstoffe der Elektrotechnik, Universität Ulm, Albert Einstein Allee-47, Ulm, D-89081, and
Forschungszentrum Karlsruhe, Institute of Nanotechnology, D-76021, Karlsruhe, Germany

*e-mail: akhileshiitkgp@gmail.com

ABSTRACT

A polymer composite of Au-metal reinforced N-polyvinyl pyrrolidone (PVP) is synthesized in a shape of thin laminates. The process involves a direct reduction of the $Au^{3+} \rightarrow Au^0$ in presence of active PVP molecules under hot conditions in water at 60-70°C. A stabilized Au-PVP surface interface of Au-nanoparticles capping in stable PVP polymer molecules is formed. The surface regulating PVP polymer insulates the particles from Au^{3+} and further aggregation. A viscous sample, after evaporating the water at 60-70°C, is casted as a film or in a shape of a thin laminate. Varying the Au-content 0.05 -1.0 wt% in Au-PVP, films are studied in terms of optical absorption, X-ray photoelectron spectroscopy (XPS), and X-ray diffraction. The Au-metal reflects in two characteristic $4f_{7/2}$ and $4f_{5/2}$ XPS bands of 83.1 and 86.8 eV respectively. The shifting of the binding energy E_b in the Au 4f peaks to a lower value relative to the bulk value confirms the microscopic interaction between the Au nanoparticles and the PVP polymer in the hybrid composites.

Keywords: Gold nanoparticles, Nnanocomposites, PVP polymer, XPS.

1 INTRODUCTION

In the past decades, there has been a strong interest in the research and development of the Au-polymer nanocomposites for several applications. It has been shown that the optical and other properties of Au or Ag-metal nanoparticles in a hybrid nanocomposite highly depend on the size, shape and distribution in a matrix [1-4]. Dispersion, stabilization, and immobilization of metal particles are a big challenge. As filler small particles support a large interfacial area in composites. The interface controls the degree of interaction between two components and thus controls the final properties. Geometry of the sample as films, with a molecularly oriented structure of polymer, governs the microstructure and other properties of the composites. A suitable polymeric addition during the synthesis allows dispersion of the metal particles, prevents the oxidation and coalescence, and templates a long-time stability of dispersed particles [1, 4-7]. Commonly used additives for the purpose, especially in the case of Au-

nanoparticles, are poly(vinyl pyrrolidone) (PVP) [5-7], poly(ethylene glycol) [8], and poly(vinyl alcohol) (PVA) [9-13]. Particularly, pyrrolidone groups of PVP molecules of refreshed surfaces extend strong covalent bonding (through the donation of the N lone pair of electrons) to nascent Au-surfaces. As a result of the lowered effective surface energy, the concerned particles no longer tend to separate or aggregate so easily. Our group [9-13] developed Ag- and Au-reinforced PVA nanocomposite colloids and films. A single step in-situ process involved no additional stabilizer or immobilizer. Here, we report synthesis of the Au-PVP films (0-1 wt% Au) and their characterization in terms of optical absorption, X-ray photoelectron spectroscopy (XPS), and X-ray diffraction.

2 EXPERIMENTAL DETAILS

A highly pure PVP (weight average molecular weight ~ 40,000 and polymerization number = 360) from Aldrich chemicals was used to prepare an aqueous PVP solution (10 g/dl) by magnetic stirring in a 100 ml batch at 60-70°C temperatures. 10-20 min stirring ensures a colorless transparent solution. Then, an aqueous $HAuCl_4 \cdot 3H_2O$ (0.05 M) of an analytical grade (99.99%) (Merck chemicals) was added drop by drop to it during the stirring at this temperature. As we reported in reaction with PVA [9, 12], an instantaneous $Au^{3+} \rightarrow Au$ reaction follows via a displacive Au^{3+} ligand reaction with PVP in a polymer complex. The PVP molecules act as a reductant as well as a stabilizer to resulting Au of nanoparticles. Average color changes from a beautiful wine-red via a light blue to a brown, depending on the Au^{3+}-content (0.05-1.0 wt% Au). Thus obtained nanocolloid of Au-nanoparticles embedded in PVP molecules and of those dispersed in the aqueous PVP solution has been cooled down and aged for 24 h. The viscous Au-PVP sample obtained by evaporating part of water at 60-70°C temperature has been casted in form of thin laminates in a mould of glass or plastic.

The optical absorption of the Au-PVP films was studied by a uv-visible spectrometer (Ocean Optics, Inc. Model-SD 2000). A scanning electron microscope of Oxford model Leo1550 was used to study microstructure in these samples. The phase analysis of the samples was carried out with X-ray diffractograms, which were measured by PW 1710 X-ray diffractometer with 0.15405 nm CuKα radiation as the

X-ray source at a scanning rate 0.05°/s over the diffraction angle 2θ in the 10-100° range. A VG ESCALB MK-II X-ray photoelectron spectrophotometer was used to measure the typical XPS spectra from the Au-PVP polymer films. This involved exciting the sample, under a reduced pressure ~10^{-8} Pa, with Mg $K\alpha_{1,2}$ radiation of hν = 1253.6 eV energy operating at 12 kV and 20 mA.

3 RESULTS AND DISCUSSION

3.1 UV-visible Spectra in Au-PVP Nanocomposites

The photographs in Fig. 1a compare the developments in the apparent visible colors in forming the Au-PVP nanocomposite films of 0, 0.05, 0.10, 0.20, and 1.00 wt% Au-contents. An intense pink color occurs in the 0.05 wt% Au-content film. The color turns into a blue when raising the Au-content to 0.1 wt%. A relatively faint and brown color follows in a 0.2 wt% Au-content. The color appears dark brown when the Au-content is raised as high as 1.0 wt% in the Au-PVP films. The observation demonstrates that the color is intrinsic of the optical properties of the Au-reinforced PVP molecules of hybrid nanocomposite particles. The volume fraction, shape, and size in the Au-PVP nanocomposite complex particles govern the optical properties Occurring of selective intense colors of sample infers primarily a preponderance of the involvements of the electronic transitions in the visible region. Scattering from the composite particles is another source of their apparent colors. Zhou et al. [7] synthesized purple color films of icosahedral Au-nanocrystals on the glass slide by a dip-coating method.

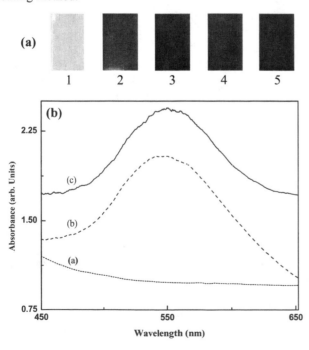

Fig. 1: (a) Apparent colors in Au-PVP nanocomposites films; 1 to 5 having 0, 0.05, 0.1, 0.2, and 1.0, wt% Au, respectively, and (b) optical absorption spectra in the Au-PVP films of (a) 0, (b) 0.05, and (c) 0.1, wt% Au-contents.

Fig. 1b shows the absorption spectra in the 450-650 nm range for three selective samples of Au-PVP nanocomposite films of (a) 0, (b) 0.05, and (c) 0.1 wt% Au-content, respectively. This specific region of the spectrum is characteristic of the surface plasmon resonance (SPR) absorption in Au-nanoparticles. No absorption occurs over this region in a pure PVP film. The 0.05 wt% Au-PVP film, which reflects in a bright pink characteristic color in Fig. 1a, exhibits a strong absorption band, with maximum absorption wavelength λ_{max} = 545 nm. It shifts to red at 550 nm upon increasing the Au-content to 0.1 wt% in the sample. Two bands at about 570 and 640 nm in a purple color film of icosahedral Au-nanocrystals were reported by Zhou et al. [7] Enhanced particle-particle interactions in larger Au-contents in such Au-PVP nanocomposite share the energy loss at the expense of the SPR absorption process, i.e., a red-shift of the λ_{max} value in one kind of the particles. The final shape and size, which determine the interfacing in Au-PVP composite particles, more effectively tune the SPR band in terms of both the λ_{max} value.

3.2 Microstructure and X-Ray Diffraction in Au-PVP Nanocomposites

The crystalline nature of Au-PVP films are analyzed with X-ray diffractograms. Fig. 2 shows X-ray diffractogram in a typical 1.0 wt% Au-PVP film, ~ 3 mm thickness. Such a thick sample of sufficiently large Au-content is chosen in order to meet a reasonably resolved diffractogram of lattice reflections from small Au-particles, which are covered in a thin PVP surface layer. Six peaks of diffractogram (from the Au-crystal lattice) occur, in the 10-100° range of the diffraction angle 2θ, in superposition of a broad scattering background in part of the noncrystalline PVP polymer matrix. As marked therein, it involves a broad diffraction halo of wavevector q = 15.9 nm^{-1}. The Au-metal surface supports occurrence of the polymer counterpart surface layer of this structure. This is feasible if the polymer deposits in thin layers, especially over the Au-particles, in a hybrid Au-PVP composite structure.

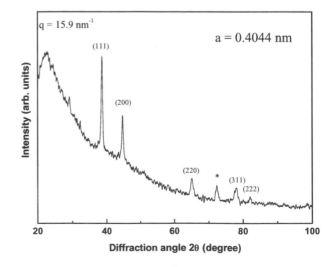

Fig. 2: X-ray diffractogram in 1.0 wt% Au-PVP nanocomposite films.

NSTI-Nanotech 2008, www.nsti.org, ISBN 978-1-4200-8503-7 Vol. 1

As in bulk Au-metal [14], the sharp diffraction peaks in Fig. 2 are assigned to (111), (200), (220), (311), and (222) lattice reflections in an Fm3m fcc crystal structure. The intensities (I_p) as well as the interplanar spacings (d_{hkl}) in these peaks differ from the values in bulk Au-metal. The d_{hkl} values determine a lattice parameter $a = 0.4044$ nm (density $\rho = 19.77$ g/cm^3) against the bulk value $a = 0.4079$ nm ($\rho = 19.28$ g/cm^3) [14]. Using the fwhm-values (fwhm: full width at half-maximum I_p-value) in these peaks in the Debye-Scherer formula [15] yields ~ 30 nm Au-crystallite size.

Fig. 3 portrays SEM images from the surfaces in (a) 0.1 and (b) 0.5 wt% Au-PVP films. The whitish contrasts of sharp images ascribe the Au-particles. Such Au-particles, which, in fact, are capping in thin PVP polymer films, are dispersed in a polymer matrix of less whitish (diffuse) contrasts in a cloud type of the features. Occurrence of the polymer in the Au-surface layers and the matrix obscures virginal shapes in the Au-particles.

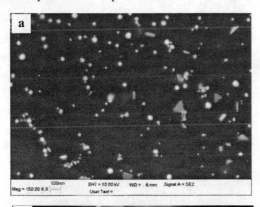

Fig. 3: SEM images in (a) 0.1 and (b) 0.5 Au-PVP films.

At low magnification, the particles appear in nearly spherical shapes, 40-50 nm diameters, distributing uniformly on a polymer surface. Few are triangular prisms. The polymer supported Au-particles have a narrow size distribution, with 70-80 % particles of the average value. Average width (D) in such images is 40-50 nm while the height varies from 40 to 100 nm. The D-value ascribes the size in such Au-crystallites along with the polymer surface layer (5 -10 nm in the thickness) in correlation to the 30 nm value derived by the fwhm values. Bigger particles are clusters of such crystallites. This is feasible in an early

stage of the growth process by a recombination reaction of the crystallites.

3.2 XPS in Au-PVP nanocomposite films

The Au-metal nanoparticles are embedded in the PVP polymer matrix via chemical linkage with PVP molecules. Studies on the nature of the microscopic interaction between Au particles and PVP polymer chains are important in understanding the Au-PVP hybrid composite system. The XPS studies are performed for the free-standing films of pure and reinforced PVP polymer molecules in form of Au-PVP nanocomposites. Fig. 4a shows a typical XPS spectrum measured from the pure PVP films. A prominent C1s band with normalized intensity value $I_p = 100$ occurs of binding energy E_b-value of 281.6 eV confirming the carbon occurring in the PVP molecular in layers. Furthermore, the O1s peak (Fig. 4b) appears as the second most intense peak with a value $I_p = 40$ at the average peak value at ~ 530.7 eV. As shown this band has a doublet structure. A deconvoluted spectrum consists of two band components of 529.9 and 531.6 eV of the E_b-values. The O1s band component of 531.6 eV is assigned to the O^{2-} bonded to the C atom (in form of the C=O group) in the PVP molecules.

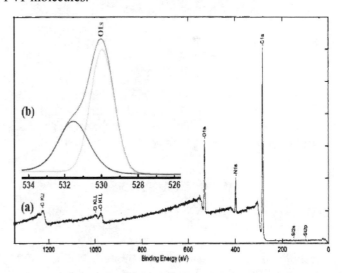

Fig. 4: (a) XPS spectrum in the virgin PVP films, with (b) a close-up of the O 1s band.

In order to study the valence state of gold and its interaction with matrix, in 0.1 wt% Au-PVP nanocomposite films, we analyzed the characteristic Au $4f_{7/2}$ and $4f_{5/2}$ XPS bands. As portrayed in Fig. 5, a well-resolved double structure of the concerned XPS signal occurs with the two components of 83.1 and 86.8 eV E_b-values, respectively, with a separation ΔE_d between the two bands of 3.7 eV. The bands are assigned to the spin-orbit spitted components of the Au-4f level in the pure Au-metal [16]. In comparison to the bulk, the Au nanoparticles have relatively lower E_b-values in the $4f_{7/2}$ and $4f_{5/2}$ Au-bands in the Au-PVP nanocomposites. The shifting in E_b-values confirms the microscopic interaction between the Au nanoparticles and the PVP polymer in the hybrid composites. A surface polymer coating of Au-particles, or a reinforcing of Au-particles in a polymer matrix in a composite structure, causes a modification in E_b values according to interactions between the two phases. The shift in the E_b-value can be attributed to chemical bonding with the surroundings and interparticle interactions. In general, a coating of the Au nanoparticle by a thin surface layer PVP in nanocomposite structure is of interest in this work causes a shift in E_b-values in the characteristic bands according to the prominent interactions between the two phases.

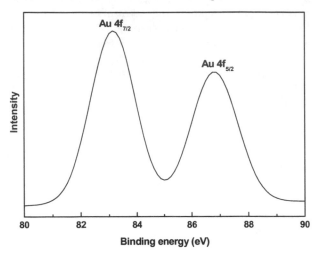

Fig. 5: The Au 4f7/2 and 4f5/2 XPS bands in the 0.05 wt % Au-PVP nanocomposites film.

The XPS confirms that PVP molecules exist on the surface of the Au cores, which play a very important role in the formation and evolution of the different shape and size of the of the Au nanoparticles, e. g, twinned Au plates, polyhedral plates, truncated triangular shaped particles as observed in the SEM (Fig. 3).

4   CONCLUSIONS

The colored Au-PVP nanocomposite films are synthesized by a simple in-situ $Au^{3+} \rightarrow Au$ conversion reaction in an aqueous medium of PVP molecules followed by casting of the Au-PVP nanocolloids. The Au-SPR absorption band, which occurs over 450-650 nm, varies sensitively in its position as well as the intensity when varying the Au-content 0-1 wt%. An intense bluish-pink color of a nanocomposite occurs in the 0.05 wt% Au, with the absorption maximum λ_{max} at 545 nm which shifts to 550 nm upon raising the Au-content to 0.1 wt%. The XRD peaks in Au-PVP nanocomposite confirm the lattice reflections in an Fm3m fcc crystal structure. The polymer supported Au-particles have a narrow size distribution, with 70-80 % particles of the average size value 40-50 nm. The shifting in E_b-values in XPS bands confirms the microscopic interaction between the Au nanoparticles and the PVP polymer in the hybrid composites. The results are useful for designing and fabricating novel metal-polymer nanocomposites in specific shapes of thin films, sheets, plates, wires, or cylinders useful for electronic and optical devices.

ACKNOWLEDGEMENTS

University Grant Commission (UGC), Government of India, is acknowledged for providing the research fellowship.

REFERENCES

[1]   Y. Sun and Y. Xia, Science 298, 2176, 2002.

[2]   D. Ibano, Y. Yakoyo, and T. Tominaga, Chem. Lett. 32 574, 2003.

[3]   M. Tsuji, M. Hashimoto, Y. Nishizawa, M. Kubokawa, and T. Tsuji, Chem. Eur. J. 11, 440, 2005.

[4]   C. Kan, X. Zhu, and G. Wang, J. Phys. Chem. B 110, 4651, 2006.

[5]   M. Yamamoto, Y. Kasiwagi, T. Sakata, H. Mori, and M. Nakamoto, Chem. Mater. 17, 5391, 2005.

[6]   C. E. Hoppe, M. Lazzari, I. P. Blanco, and M. A. L. Quintela, Langmuir 22, 7027, 2006.

[7]   M. Zhou, S. Chen, and S. Zhao, J. Phys. Chem. B. 110, 4510, 2006.

[8]   C. S. Ah, Y. J. Yun, H. J. Park, W. J. Kim, D. H. Ha, and W. S. Yun, Chem. Mater. 17, 5558, 2005.

[9]   A. Gautam, P. Tripathy, and S. Ram, J. Mat. Sci. 41, 3007, 2006.

[10] A. Gautam, G. P. Singh, and S. Ram, Synth. Met. 157, 5, 2007.

[11] S. Ram, A. Gautam, H.-J. Fecht, J. Cai, J. Bansmann, and R. J. Behm, Phil. Mag. Lett. 87, 361, 2007.

[12] S. Ram, P. Tripathy, and H.-J. Fecht, J. Nanosci. Nanotech. 7, 3200, 2007.

[13] P. Tripathy, A. Mishra, and S. Ram, Mater. Chem. Phys. 106, 379, 2007.

[14] W.F. McClume, Powder Diffraction File JCPDS (Joint Committee on Powder Diffraction Standards), Swarthmore, Pennsylvania: (International Centre for Diffraction Data) 04-0784, 1979.

[15] B. D. Cullity, "Elements of X-ray diffraction", Addison-Wesley, Reading, Massachusetts, 1978.

[16] A. Patnaik and C. Li, J. Appl. Phys. 83, 3049, 1998.

Replication of micro/nano combined structure using micro/nano combined aluminum stamp

Kyoung Je Cha* and Tai Hun Kwon*

* Department of Mechanical Engineering, POSTECH, KOREA

ABSTRACT

This paper presents an easy and efficient fabrication method of plastic replicas of micro/nano combined structure (MNCS) using an aluminum stamp with micro/nano combined structure (ASMNCS) on its surface. ASMNCS could be fabricated by two steps: i) the first step is to introduce micro patterns on an electro-polished aluminum plate by pressing it by micro structured stamp; ii) the second one is to form nano dimple array onto the micro structured aluminum plate making use of the anodic aluminum oxide (AAO) technique. MNCS is successfully replicated on the polymer films via hot embossing process using ASMNCS as a mold. Experimental measurement of contact angle indicates that MNCS in so fabricated film increases the hydrophobicity over the surface which has just microstructure only.

Keywords: Micro/nano combined structure (MNCS), aluminum stamp with micro/nano combined structure (ASMNCS), anodic aluminum oxide (AAO), hot embossing, contact angle

1 INTRODUCTION

In recent years, the need of micro/nano combined patterns draws many researchers' attention. These micro/nano combined patterns have many interesting phenomena, i.e. self cleaning, water-repellency, adhesion-enhancement, optical effect, drag reduction and so on [1]. These micro/nano combined patterns could be applied to micro fluidic devices, micro optical devices and molecular diagnosis [2].

Fabrications of micro/nano combined pattern require not only MEMS but also NEMS technologies [3]. In general, patterning techniques can be categorized into two methods: top-down approach and bottom-up approach [4]. The top-down approaches often use the traditional micro fabrication methods. Photolithography and ink-jet printing belong to this category. Bottom-up approaches, in contrast, use the chemical properties of single molecules to cause single-molecule components to automatically arrange themselves into some useful conformation. These approaches utilize the concepts of molecular self-assembly and molecular recognition [4]. The precision of top-down approaches has its limit. On the other hand, bottom-up approaches could easily make structures below several tens nano meter, but with a weakness in its productivity.

In this regard, we present an easy and efficient fabrication method of plastic replicas of MNCS. The proposed method merges the top-down and bottom-up approaches in fabricating ASMNCS. Micro patterns were formed on an aluminum plate by pressing it by micro structured stamp. And nano dimple array was subsequently formed onto the micro structured Al plate by AAO technique [5]. Then the micro/nano combined structure is replicated on the polymer film via hot embossing process using ASMNCS as a mold.

We have investigated the effect of hot embossing pressure on the transcriptability in hot embossing process and characterized the hydrophobicity of so fabricated MNCS film by measuring the contact angle.

2 FABRICATION METHOD

Figure 1 schematically shows the overall fabrication method of ASMNCS and polymeric replica.

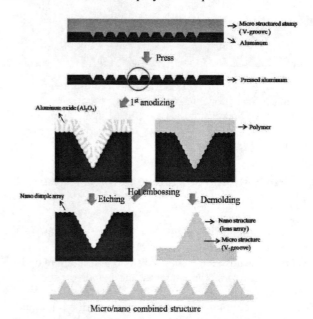

Figure1. Schematic diagram representing the overall fabrication procedure of micro/nano combined aluminum stamp and the replication of polymeric film.

3 FABRICATION AND EXPERIMENT

3.1 Fabrication of micro stamp

In the present work, as an illustration of the proposed method, a microscale V-groove patterned nickel stamp (2 × 2cm²) is prepared by nickel electroforming as the micro structured stamp. Figure 2(a) shows the fabrication procedure for the nickel stamp. First, the microscale V-groove patterns were fabricated by well-known Si anisotropic wet etching [6]. The Ni layer (100nm thick) is deposited on the etched Si wafer using E-beam evaporator as a seed layer for the electroforming. Figure 2(b) shows the nickel stamp of which the SEM image is shown in Fig.2(c).

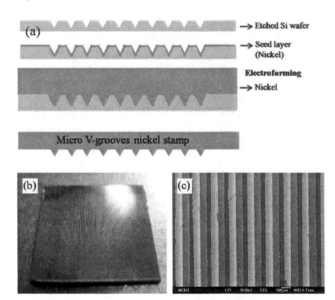

Figure 2. Images of (a) schematic representation of fabrication process for micro scale nickel stamp, (b) actual micro V groove nickel stamp and (c) SEM image of a micro V grooves nickel stamp's surface (width: 164µm, space: 39.6µm).

3.2 Fabrication of micro/nano structured stamp

A pure aluminum plate (99.999%) was electropolished in a mixture of ethanol and perchloric acid to remove surface irregularities. An electro polished aluminum plate is pressed by a micro stamp such that the negative pattern is replicated onto the surface of the aluminum plate.

The anodic aluminum oxide (AAO) technique is applied on the micro structured aluminum plate in the following manner. Micro V-groove patterned aluminum plate was anodized at 180V in 0.1M phosphoric acid at -5°C for 12hours. Figure 3 shows cross-section image of the anodized aluminum template. The anodic porous alumina was formed on inclined surface as well as flat surface.

After the anodization was complete, the formed aluminum oxide is removed in a mixture of 1.8% chromic acid and 6wt% phosphoric acid at 65°C for 8hours. Figure 4(a) ~ (d) show the SEM images of the aluminum stamp's surface. Figure 4(a), (b), (c) and (d) show that ASMNCS was successfully fabricated. It may be noted that nano dimple arrays (inter pore size: 400nm) are uniformly patterned on the microscale V-grooves surface and edge portion as well as on the flat surface.

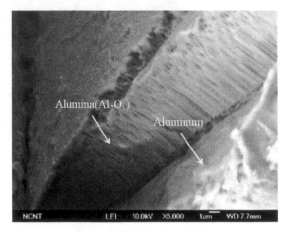

Figure 3. SEM image of anodic porous alumina on inclined surface.

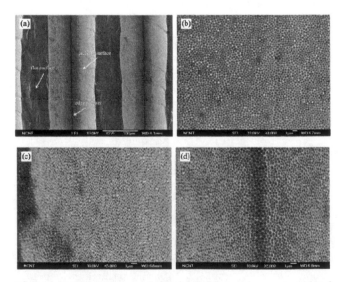

Figure 4. SEM image of fabricated micro/nano combined stamp: (a) stamp's several portions, (b) flat portion, (c) inclined portion and (d) edge portion.

3.3 Replication of micro/nano combined structure via hot embossing

The primary aims of this study are to fabricate an Al stamp with MNCS and to replicate MNCS on polymeric film using Al stamp as a mold via hot embossing. For this study, our group designed and constructed the hot embossing equipment. PMMA films were embossed by different embossing pressures to investigate its effect on the transcriptability. And COC films were used to check the hydrophobicity by means of contact angle. Processing conditions of hot embossing process are summarized in Table 1.

Table 1. Processing condition of hot embossing process

	PMMA		COC
Embossing temp. (°C)	130		140
Holding time (min)	10		10
Pressure (MPa)	22	36.6	21.6

3.4 Measurement of contact angle

Contact angle could be measured to characterize property of micro/nano combined structure. So we measured contact angles of COC films of three kinds: the first with MNCS, the second with only micro V-groove structure and the third with no structure (i.e. just a flat surface). In the case of V-grooves, contact angle was measured in two directions: one along the V-groove direction (say, longitudinal direction), the other transversal direction. Contact angle was measured by a commercial contact meter (DSA100, Krüss Drop shape analysis system). The volume of water droplet used for the measurements was 5μl. Contact angle was measured at three different positions for each sample and average values were calculated.

4 RESULT

4.1 Hot embossing

Figure 5 and 6 show SEM images of the replicated PMMA films at several locations. It has been found that micro V-grooves pattern was well replicated not only at 36MPa, but also at 22MPa. However, a nano lens array was not so well reproduced on replicated PMMA film at 22MPa. In particular, the edge portion of replicated film was scarcely filled with polymer. In this case, relatively high pressure is required for good transcriptability of MNCS film.

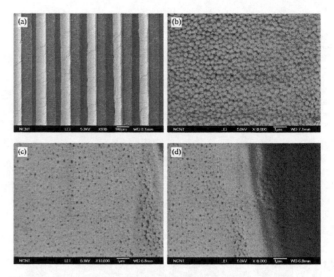

Figure 5. SEM images of the replicated PMMA film (22MPa): (a) film's several portions, (b) flat portion, (c) inclined portion, and (e) edge portion.

Figure 6. SEM images of the replicated PMMA film (36.6MPa): (a) film's several portions, (b) flat portion, (c) inclined portion, and (e) edge portion.

4.2 Contact angle characteristic

We have measured contact angles of replicated COC films, a flat surface and film with microstructure only as mentioned above. It may be mentioned that the contact angle of a flat COC film is 91.5° Figure 7 summarizes the measured contact angles for comparison between MNCS film and only micro structured film. Contact angle on MNCS COC film is higher than that on only micro V-grooves film by 5° in the transverse direction and by about 10° in the longitudinal direction. It has been confirmed that nano structure added on top of microstructure certainly increases the hydrophobicity.

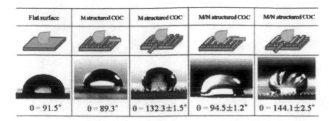

Flat surface	M structured COC	M structured COC	M/N structured COC	M/N structured COC
θ = 91.5°	θ = 89.3°	θ = 132.3±1.5°	θ = 94.5±1.2°	θ = 144.1±2.5°

Figure 7. Photos of a water drop on COC films and comparison of contact angle between MNCS film and only micro V-grooves film.

5 CONCLUSION

In this paper, we present an easy and efficient fabrication method of plastic replicas of MNCS. The MNCS films of PMMA and COC have been successfully replicated using an aluminum stamp with micro/nano combined structure (ASMNCS) via hot embossing. Relatively high pressure was required for good transcriptability of MNCS in the hot embossing process. And MNCS have a favorable effect on the hydrophobicity of COC film.

ACKNOWLEDGEMENTS

The authors would like to thank Defense Acquisition Program Administration and Agency for Defense Development (UD060049AD) and the Korea Science and Engineering Foundation(KOSEF) grant funded by the Korea government(MOST) (R01-2005-000-10917-0)

REFERENCES

[1] Metin Sitti, "Synthetic Gecko Foot-Hair Micro/Nano -Structures as Dry Adhesives", Appeared in Journal of Adhesion Science and Technology, VOL.18, pp.1055-1074, 2003

[2] Dong Sung Kim, Han Ul Lee, Nam Hyo Kim, Kun-Hong Lee, Dong-Woo Cho and Tai Hun Kwon, "Fabrication of microchannel with nanopillar array using micromachined AAO", Micro- and Nano-Engineering, In: Proceedings of 32nd international conference on micro- and nano-engineering: P-NSC45

[3] Jang Min Park, Nam Hyo Kim, Bong-Kee Lee, Kun-Hong Lee, Tai Hun Kwon, "Nickel stamp fabrication and hot embossing for mass-production of micro/nano combined structures using anodic aluminum oxide", Microsystem Technology, online first, 2008

[4] D. Mijatovic, J. C. T. Eijkel and A. van den Berg, "Technologies for nanofluidic systems: top-down vs. bottom-up—a review", Lab Chip, 5, 492-500, 2005

[5] Ralf B. Wehrpohn, "Ordered Porous Nanostructures and Applications: chapter3.Highly Ordered Nanohole Arrays in Anodic Porous Alumina", ISBN: 0-387-23541-8

[6] Chang Liu, "Foundations of MEMS: Chapter 10. Bulk Micromachining and Silicon Anisotropic Etching", ISBN: 0-13-147286-0

Molecular Dynamics Study of Ballistic Rearrangement of Surface Atoms during High Energy Ion Bombardment on Pd (001) Surface

Sang-Pil Kim and Kwang-Ryeol Lee

Computational Science Center, Korea Institute of Science and Technology, Seoul 139-791, Korea,
spkim@kist.re.kr (S.-P. Kim), krlee@kist.re.kr (K.-R. Lee)

ABSTRACT

Atomic behavior during ion bombardment was investigated by using three dimensional classical molecular dynamics (MD) simulation. It was observed that significant amount of surface atoms were rearranged when Ar ions bombarded the Pd (001) surface in addition to the erosion of surface atoms. Quantitative analysis showed that the rearranged atoms are three times as many as sputtered atoms regardless to the energy and angle of incidence ions. Contrary to the conventional concepts which describe the surface structure evolution based on the erosion theory, the rearranged atoms were turned out to play a significant role in forming the surface morphology as shown in the simulated surface morphology by the bombardment of many Ar ions.

Keywords: molecular dynamics, ion beam sputtering, surface structure

1 INTRODUCTION

High energy ion bombardment on solid surface has attracted much attention owing to its capability to fabricate ordered nanoscale structures such as self aligned quantum dots [1]. Moreover, it is worth while considering a peculiar technology beyond the conventional approach such as 'Top-down' and 'Bottom-up'. The simplest application of ion bombardment is to collect the sputtered atoms on the substrate to form a thin film. Most researchers have utilized sputtering as deposition tool and they focused on the reaction phenomena between recoiled ions from the target and substrate atoms to deposit. However, since a suggestion by Facsko *et al.* was introduced, it has taken an enormous attention to researchers whose aim was to manufacture the nano patterns on the surface [1]. Such a peculiar process resulted in enhancing the possibility for designing nano sized patterning by cheap and simple method.

Using this technique, it can be applied to various aspects such as the ordered adsorption of large molecules [2], optoelectronic devices [1], molding templates [3], manipulating magnetism [4,5], tuning the chemical reactivity of catalytically active surface [6] and for manipulating film texture [7] *etc*. Also there are many results for obtaining various nano patterns with respect to the incident energy, angle of ion source, temperature, and substrate materials [8-11].

Theoretical studies on the ion bombardment have been developed based on the Sigmund's theory [12] and Bradley-Harper (BH) instability model [13]. It is assumed in these theoretical works that the surface roughening or structure evolution by ion bombardment was proportional to the sputter yield. However, some results based on the atomistic simulation were reported that ion bombardments result in not only erosion of target atoms but also rearrangement on the surface [14, 15]. As an example, Fig. 1 shows the result of the atomic configuration when Ar ion with 10 keV energy bombarded on Au (001) surface by using MD simulation. This dynamic simulation evidently shows that some atoms were sputtered from the surface and some of the surface atoms form a rim around the crater by ballistic rearrangement.

In this work, we focused on the rearrangement of the surface atoms to reveal the mechanism of surface structure evolution during the high energy Ar bombardment on Pd (001) surface. To obtain quantitative data, statistical analysis was performed for 1,000 events of Ar bombardment on randomly chosen surface position. We also simulate the surface structure evolution by sequential Ar bombardment. Autocorrelation function of the bombarded surface was employed to characterize the representative surface pattern.

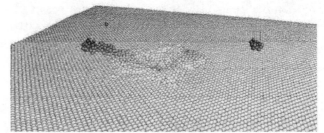

Figure 1: Snapshot of atomic configuration during Ar bombardment on Au (001) with 10 keV incident energy.

2 CALCULATION PROCEDURE

The accuracy of MD is largely dependent on the reliability of interatomic potential in use. In this work, two kinds of interatomic potentials were combined and utilized for simulating ion bombardment. To describe the thermal and mechanical behaviors of the materials, we used embedded atom method (EAM) potential which is well-known and reliable for the metallic system such as Pd [16].

EAM potential is expressed as follows:

$$E_{tot} = \frac{1}{2} \sum_{i,j(j\neq i)} \phi(r_{ij}) + \sum_i F_i\left(\sum_{j\neq i} \rho_j(r_{ij})\right) \quad (1)$$

Here, $\varphi(r_{ij})$ is a pairwise interaction between atoms at a distance between i and j atoms, $\rho(r_{ij})$ is electron density contributed by atom j. With this approximation for the electron density, the effect of electrons on individual ions can be simply but efficiently described. Using embedding function $F(\rho)$, energy can be obtained from the electron density. Coupling with the electron density functions, many-body properties such as crystal structure, stacking fault energy, and phase stability can be described.

For describing highly repulsion behavior at short distance, we used Ziegler, Biersack and Littmark (ZBL) potential [17] which is composed of Coulombic potential and screening function. ZBL potential is the most commonly used in ion bombardment simulations and it is related to the atomic number Z_1 and Z_2. This function is expressed as the product of the Coulombic potential and a screening function as follows:

$$V(R) = \frac{Z_1 Z_2 e^2}{4\pi\varepsilon_0 R} \sum_{k=1}^{N} c_k \exp(-b_k R / a) \quad (2)$$

where e is the electron charge, ε_0 is the permittivity of vacuum, R is the interatomic distance between two atoms. In the screening function, c_k and b_k are the coefficients defined differently for potentials and each value listed in table 1. a is the screening length which is given by

$$a = \frac{0.8856 \times a_0}{Z_1^{0.23} + Z_2^{0.23}}, \quad (3)$$

where $a_0 = 0.529\text{Å}$ is the Bohr length.

When Ar atom collides with Pd atoms with high energy, atomic states far from their equilibrium position should be simulated. EAM potential is generated based on the data of near equilibrium state. Therefore, to describe the high energy ion collisions more accurately, pairwise potential, $\varphi(r_{ij})$, should be modified by joining smoothly to the ZBL potential at small separations. To avoid splining errors and derivative discontinuity problem, switching function should be utilized for interpolating between EAM-pair and ZBL potential. Switching function is generally formed as follows:

$$V(r) = V_{ZBL}(r)S(r) + V_{EAM}(r)(1-S(r)) \quad (4)$$

which is applied in a region $r_1 < r < r_2$, and has the following properties: $S(r_1)=1$, $S(r_2)=0$ and $S'(r_1)=S'(r_2)=0$. We used cosine function as follows:

$$S(r) = \frac{1}{2}\cos\left(\frac{r-r_1}{r_2-r_1}\pi\right) + \frac{1}{2} \quad (5)$$

The potentials used in this study were rigorously benchmarked by using the calculated or experimental observed physical properties of Pd.

In this study, we investigated atomic behavior around the surface when highly accelerated Ar ion impacts on the Pd (001) surface with various energies and incident angles.

To elucidate the atomic reaction on the surface, one thousand individual trials were performed for statistics and quantitative parameters such as sputtering yield (Y_{spt}), rearrangement yield (Y_{rear}) were extracted. Moreover, spatial distribution of these behaviors also obtained after normalizing process. To show the effect of the incident energy of Ar ion, we selected 0.5, 1.0 and 2.0 keV energy and to show the effect of the incident angle, we selected 0 (normal), 30, 45, 60 and 75 degrees of incident angle with projecting toward [011] direction. Considering Ar incident energy, the substrates were prepared 3-types which scale is large enough for Ar and substrate atoms to react without simulation errors. The size of the substrate for 0.5, 1.0, and 2.0 keV was $20\times20\times20$ (a_0^3), $24\times24\times20$ (a_0^3) and $30\times30\times30$ (a_0^3), where a_0 is equilibrium lattice constant for each metal and total number of the substrate atoms was 32,000, 46,080 and 108,000, respectively. All calculations were performed by using LAMMPS code [18].

i	1	2	3	4
c_i	0.1818	0.5099	0.2802	0.02817
b_i	3.2	0.9423	0.4029	0.2016

Table 1: Parameters used in screening function.

3 RESULTS AND DISCUSSION

We computed Y_{spt} from the total amount of sputtered atoms for various simulation conditions. Figure 2(a) shows the change of Y_{spt} as a function of incident energy and angle. The unit of Y_{spt} is the number of atoms per incident ion. Y_{spt} varies with the incident angles and the highest yield was obtained at the incident angle of 60 degrees. Such a tendency becomes more obvious as the incident energy increased. However, Y_{spt} were decreased rapidly after 60 degrees and near zero sputter yield was obtained at 75 degrees. We also computed Y_{rear} by counting the total atoms which remained above the original surface after the bombardment. Figure 2(b) shows the change of Y_{rear} at same conditions of Y_{spt} (Fig. 2(a)). The tendency for the incident angle was similar to that of Y_{spt}. However, it must be noted that the value of Y_{rear} was much larger than that of Y_{spt}.

The ratio between Y_{spt} and Y_{rear} was in the range of 2.8 ± 0.4 regardless of the incidence energy and incidence angle. This means that when Ar ion bombards on the Pd (001) surface, Y_{rear} was always shown to be about 3 times larger than Y_{spt}. It was reported that when Xe atom bombards on the Pt (111) surface, the ratio was almost constant in the range from 3 to 5 [19,20]. Similar results have been reported in various systems [21-23]. Our simulation results are in good agreement with the previous report on the ratio between erosion and rearrangement. The present results show that the surface morphology in nano meter scale should be investigated based on the new concept or model that includes the effect of the rearranged atoms.

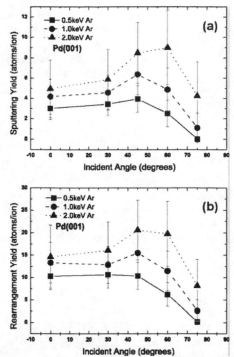

Figure 2: (a) Sputtering and (b) rearrangement yields with respect to the incident energy and angle of Ar on Pd (001). Error bars denote the standard deviation of each case.

To confirm the rearrangement effect on the formation of surface patterns, many Ar atoms were bombarded on the Pd (001). The position of bombarding Ar was randomly selected in the x and y direction. The Ar atoms of 0.5 keV of incident energy bombarded the surface in surface normal. Fig. 3 (a) shows the surface morphology after 4,200 bombardments of Ar atoms. The colors of each atom denote the relative height of the atoms. Orange color denotes the initial surface height. Blue one is an atom which locates lower than initial surface, while the red one is an atom which locates higher than initial surface due to the rearrangement of surface atoms. From the data shown in Fig. 2 (a), 4,200 Ar bombarding would sputter 12,600 (3.9 ML) Pd atoms. Even though such a large number of Pd atoms were eroded, some areas on the surface are still remained higher than the initial surface. It is evident in Fig. 3 (a) that the rearranged atoms play an important role in evolving the surface patterns.

In order to investigate the qualitative analysis of the surface pattern, two dimensional autocorrelation function of Fig. 3 (a) surface was calculated. Fig. 3 (b) is the autocorrelation function, which reveals that the correlation image is a little distorted diamond shape along <100> and <010> directions. This result is in consistent with an experimental result [24] that shows the diamond-shape autocorrelation function. It must be noted that this calculation did not consider the long range surface diffusion but the ion induced ballistic rearrangement. It can be thus said that the long range diffusion would be negligible in the surface morphology evolution.

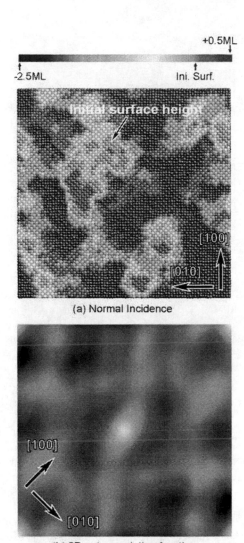

(a) Normal Incidence

(b) 2D autocorrelation function

Figure 3: (a) Atomic configuration after 4,200 (1.73×10^{15} ions/cm^2) Ar bombardments on Pd(001) surface (15.56×15.56 nm^2). Color depth denotes the height. (b) 2D autocorrelation function image of (a).

The 4 fold symmetry in the surface morphology is closely related to the distribution of the rearranged atoms. The lateral distribution of the rearranged atoms was computed by analyzing statistically the molecular dynamic simulation results. The lateral positions of rearrangement atom were normalized by the impact point. Fig. 4 shows the lateral distribution of the rearranged atoms when Ar atoms of 0.5 KeV bombarded the Pd (001) surface in normal direction. The intensity of the color denotes the degree of probability of rearranged atoms. The shape of the distribution has the 4 fold symmetric features where the edges are parallel to the <110> direction. It can be said that the 4 fold anisotropic surface pattern of Fig. 3 results from the accumulation of the rearranged atoms of the anisotropic lateral distribution as shown in Fig. 4.

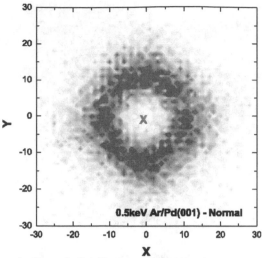

Figure 4: Lateral distribution of rearrangement atoms on Pd(001) with 0.5 keV Ar ion bombardments for normal incidence. x axis coincides with <100> direction and y axis with <010> direction of (001) surface. 'X' means the normalized impact point.

4 CONCLUSION

The most significant result of the present work is to show that significant rearrangement of the surface atom occurs during high energy ion bombardment. Because the number of rearranged atoms is much larger than that of sputtered atoms, the surface structure evolution during ion bombardment can be dominated by the distribution of the rearranged atoms. This result suggests that the kinetic theories of the surface structure evolution include the concept of the rearranged atoms in addition to the negative deposition concept.

The present simulation of Ar ion bombardment on Pd (001) surface also demonstrates the formation of 4 fold symmetric autocorrelation of the surface. By quantitative analysis of lateral distribution of rearranged atoms, we could conclude that ballistic rearrangement of the surface atom plays a significant role in the surface structure evolution. These results are in good agreement with the previous experimental observation.

ACKNOWLDEGEMENT

This work was financially supported by the KIST core capability enhancement program (2E20330).

REFERENCES

[1] S. Facsko, T. Dekorsy, C. Koerdt, C. Trappe, H. Kurz, A. Vogt, and H. L. Hartnagel, Science 285, 1551, 1999.

[2] P. Chaudhari, J. Lacey, J. Doyle, E. Galligan, and S. C. A. Lien, Nature 411, 56, 2001.

[3] O. Azzaroni, P. L. Schilardi, R. C. Salvarezza, R. Gago, and L. Vazquez, Appl. Phys. Lett. 82, 457, 2003.

[4] Y. J. Chen, J. P. Wang, E. W. Soo, L. Wu, and T. C. Chong, J. Appl. Phys. 91, 7323, 2002.

[5] R. Moroni, D. Sekiba, B. Mongeot, G. Gonella, C. Boragno, L. Mattera, and U. Valbusa, Phys. Rev. Lett. 91, 167207, 2003.

[6] L. Vattuone, U. Burghaus, L. Savio, M. Rocca, G. Costantini, F. Mongeot, C. Boragno, S. Rusponi, and U. Valbusa, J. Chem. Phys. 115, 3346, 2001.

[7] K. C. Ruthe, and S. A. Barnett, Surf. Sci. 538, L460, 2003.

[8] W. L. Chan and E. Chason, J. Appl. Phys. 101, 121301, 2007.

[9] M. J. Aziz, Ion Beam Science: Solved and Unsolved Problems 52, 187, 2006.

[10] S. Vogel and S. J. Linz, Europhys. Lett. 76(5), 884, 2006.

[11] U. Valbusa, C. Boragno, and B. Mongeot, Mater. Sci. and Eng. C 23, 201, 2003.

[12] P. Sigmund, J. Mater. Sci. 8, 1545, 1973.

[13] R. M. Bradley, and J. M. E. Harper, J. Vac. Sci. Technol. A 6(4), 2390, 1988.

[14] A. Friedrich, H. M. Urbassek, Surf. Sci. 547, 315, 2003.

[15] E. M. Bringa, K. Nordlund, and J. Keinonen, Phys. Rev. B 64, 235426, 2001.

[16] S. M. Foiles, M. I. Baskes and M. S. Daw, Phys. Rev. B 33, 7983, 1986.

[17] J. F. Ziegler, J. P. Biersack, U. Littmark, The Stopping and Range of Ions in Solids (Pergamon, New York 1985).

[18] S. J. Plimpton, J. Comp. Phys. 117, 1, 1995.

[19] H. Gades, H. Urbassek, Phys. Rev. B 50, 11167, 1994.

[20] P. Mishra, and D. Ghose, Phys. Rev. B 74, 155427, 2006.

[21] H. Hansen, A. Redinger, S. Messlinger, G. Stoian, Y. Rosandi, H. M. Urbassek, U. Linke, and T. Michely, Phys. Rev. B 73, 235414, 2006.

[22] G. Costantini, B. Mongeot, C. Boragno, and U. Valbusa, Phys. Rev. Lett. 86, 838, 2001.

[23] C. Busse, H. Hansen, U. Linke, and T. Michely, Phys. Rev. Lett. 85, 326, 2000.

[24] T. C. Kim, C. M. Ghim, H. J. Kim, D. Y. Noh, N. D. Kim, J. W. Chung, J. S. Yang, Y. J. Chang, T. W. Noh, B. Kahng, and J. S. Kim, Phys. Rev. Lett. 92, 235414, 2006.

Templated virus deposition: from molecular-scale force measurements to kinetic Monte Carlo simulations

S. Elhadj[*], R. Friddle[*], G. Gilmer[*], A. Noy[*], J.J. De Yoreo[*]

[*] Lawrence Livermore National Laboratory, Livermore, CA, USA, elhadj2@llnl.gov

ABSTRACT

The use of macromolecular scaffolds for hierarchical organization of molecules and materials is a common strategy in living systems that leads to emergent behavior. Here we describe an effort to relate interaction force measurements between viruses and modified substrates to the energy landscape during virus assembly on surfaces. Potentials and binding energies are then used in kinetic Monte Carlo simulations to predict assembly morphology under controlled conditions replicated experimentally. We use atomic force microscope (AFM) tips functionalized with specific chemical species to measure interactions in the assembly system, which includes Cow Pea Mosaic Virus (CPMV). CPMV virus particles were engineered to express specific functional groups to modulate the strength and kinetics of interactions and assembly morphology. We show that the CPMV morphological evolution predicted by the simulations correlates with AFM observations.

Keywords: virus, force, monte carlo, energy, assembly

1 RESULTS AND DISCUSSION

One advantage of using macromolecular scaffolds for hierarchical organization is that it generates micron-scale structures from nm-scale building blocks possessing high-density functionality defined at Å-scales by active sites, typically on proteins complexes such as viral capsids. Figure 1a shows a model representation of CPMV self assembled virus on SAM modified gold substrate (Figure 1b).

Using interaction force measurements between viruses and modified substrates to derive the energy landscape during virus assembly provides a means to explore the nature of the governing interactions and to investigate the role of solvent interactions on inter-viral potentials. The virus are specifically engineered to express various functional groups on its surface that can be used for attachment and/or patterning on templated substrates to guide the virus assembly (Figure 2).

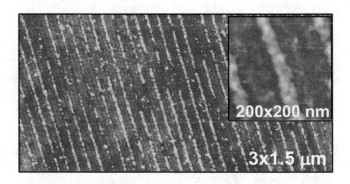

Figure 2: AFM image of a chemically modified trenches on a SAM background to form an array of line template for bonding to CPMV cys modified virus.

The experimental configuration to measure force curves and image surfaces is schematically shown in Figure 3, along with a typical force curve derived between –SH modified and AFM tips and cys-modified-CPMV attached to the surface.

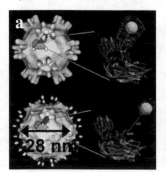 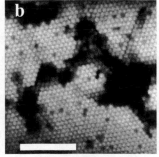

Figure 1: Model of CPMV virus (http://johnsonlab.scripps.edu/research/images/) and AFM image of self assembled CPMV virus on a gold substrate (scale bar = 500 nm).

28 nm

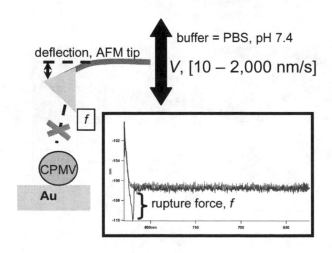

buffer = PBS, pH 7.4

deflection, AFM tip

V, [10 – 2,000 nm/s]

rupture force, f

CPMV

Au

Figure 3: Schematic of force measurement setup indicating AFM tip pull off rates and bond breaking for a given rupture force.

The results of repeated force measurements describe a stochastic process [1] that depends on the pull-off rates of the tip (rate = $k_s \times v$ N/s), which, in turn, depends on the spring constant of the cantilever/tip, k_s, and the linear velocity of the tip, v, as it approaches and retracts from the CPMV virus attached to the surface. Preliminary measurements made between dithiolbezene (DTB) modified surfaces on gold covered tips and surfaces are shown in Figure 4.

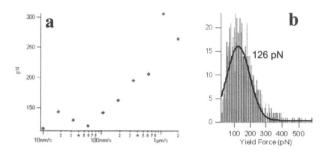

Figure 4: Force spectrum (a) and distribution (b) measurements for DTB modified surfaces.

Analysis of the force spectrum in Figure 4 indicates a distance to the transition state along the reaction coordinate of 0.12 nm during bond breaking and a binding energy on the order of $\Delta G = 6 k_B T$.

Monte Carlo simulations of the binding of viruses to modified surfaces are shown in Figure 5, along with the AFM images of viruses deposited on templated substrates (Figure 2). These simulations relate binding strength (Figure 4) to specific patterning features which correlate with observed CPMV virus assembly patterns. This model description provides thus a means to tune the assembly of CPMV virus by modifying attachment parameters through chemical and biochemical modification and achieve arbitrary patterns on the nanoscale.

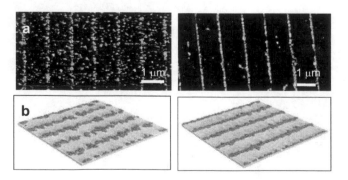

Figure 5: (a) AFM images of patterned CPMV virus and (b) corresponding Monte Carlo simulation of the virus assembly on templated surfaces.

2   TEXT FORMAT

The title should be in boldface letters centered across the top of the first page using 14-point type. First letter capitals only for the title. Insert a blank line after the title, followed by Author Name(s) and Affiliation(s), centered and in 12 point non-bold type. The paper begins with the abstract and keywords followed by the main text. It ends with a list of references.

REFERENCES
[1] Evans E and Ritchie K 1997 *Biophys J* **72** 1541-55

Scanning Probe-Based fabrication of 3D Nanostructures via Affinity Templates, Functional RNA and Meniscus-Mediated Surface Remodeling

Sung-Wook Chung[1*], Andrew D. Presley[1,2], Selim Elhadj[1], Saphon Hok[1], Sang Soo Hah[1], Alex A. Chernov[1], Matthew B. Francis[2], Bruce E. Eaton[3], Daniel L. Feldheim[3], James J. DeYoreo[1,4*]

[1]Chemistry, Materials, Earth and Life Sciences Directorate,
Lawrence Livermore National Laboratory, 7000 East Ave., Livermore, CA 94551
[2]Department of Chemistry, University of California, Berkeley, Berkeley, CA 94720-1469
[3]Department of Chemistry & Biochemistry, University of Colorado, Boulder, Boulder, CO80309-0215
[4]Molecular Foundry, Lawrence Berkeley National Laboratory, 1 Cyclotron Road, Berkeley, CA 94720
*E-mail: chung20@llnl.gov, jjdeyoreo@lbl.gov

ABSTRACT

Developing generic platforms to organize discrete molecular elements and nanostructures into deterministic patterns at surfaces is one of the central challenges in the field of nanotechnology. Here we review three applications of the atomic force microscope (AFM) that address this challenge. In the first, we use two-step nanografting to create patterns of self-assembled monolayers (SAMs) to drive the organization of virus particles that have been either genetically or chemically modified to bind to the SAMs. Virus-SAM chemistries are described that provide irreversible and reversible binding, respectively. In the second, we use similar SAM patterns as affinity templates that have been designed to covalently bind oligonucleotides engineered to bind to the SAMs and selected for their ability to mediate the subsequent growth of metallic nanocrystals. In the final application, the liquid meniscus that condenses at the AFM tip-substrate contact is used as a physical tool to both modulate the surface topography of a water-soluble substrate and guide the hierarchical assembly of Au nanoparticles into nanowires. All three approaches can be generalized to meet the requirements of a wide variety of materials systems and thus provide a potential route towards development of a generic platform for molecular and materials organization.

Keywords: scanning probe microscopy, scanning probe nanolithography, nanostructures, affinity templates, virus, RNA

1   INTRODUCTION

The ability to organize macromolecular complexes, nanoparticles, and wires into predetermined patterns at surfaces has enormous potential in the fields of electronics, photonics, biology, and medicine. For example, assembly of quantum dots or light-harvesting dendrimers [1] into ordered arrays could lead to high-efficiency solar cells [2-9] and photonic solids with tunable optical properties [10]. Organization of semiconductor nanorods and wires into complex patterns of interconnects could form the basis of ultra-high density logic circuits [11-13]. An ability to deposit biomolecules such as proteins and viruses onto specific sites or into ordered arrays would facilitate the emerging revolution in biological imaging that will be made possible by Free Electron Laser (FEL) photon sources [14, 15] and potentially lead to high-throughput protein structure determination. Perhaps most intriguing is the idea of combining biological molecules such as proteins, peptides, and oligonucleotides with inorganic nanostructures to construct hierarchical biological-inorganic architectures. This has the potential to lead to integration of fundamental biomolecular motifs into nanoelectronic devices, nanoelectronic components into *in vivo* medical devices, and perhaps even biomolecule driven assembly of circuits or devices [16].

To realize these possibilities, robust methods for organizing discrete molecular elements and nanostructures into deterministic patterns are needed. Indeed, developing a generic platform for doing so has emerged as one of the central challenges in the field of nanotechnology. Approaches that rely on self-assembly into 2D film or 3D solids [17] do not provide a route to deterministic patterns incorporating multiple elements or functionalities into predetermined geometries. Printing methods [18, 19], which have the potential for organizing nanostructures through creation of templates that control the location of attachment or deposition at surfaces, can rarely achieve patterning below 50 – 100 nm and then controlling orientation of individual nanostructures is difficult. One obvious approach is to use a patterning technique with much greater spatial resolution to define these templates. Electron beam and ion beam lithography, nano-imprinting, and scanning probe nanolithography (SPN) have all been used for this purpose [20-33]. SPN offers an excellent combination of resolution (< 20 nm) and registration (< 100 nm), feature architecture, chemical flexibility, low cost and potential for high-throughput fabrication. Dip-Pen Nanolithography (DPN) [23-26, 29, 32, 34-36] and Scanning Probe Nanografting (SPNG) [30-33] are two powerful SPN techniques suitable for creating biochemical templates. DPN is a SPN technique where an atomic force microscope (AFM) tip is used to transfer functional molecules to a reactive surface, usually via a solvent meniscus that forms at the tip-substrate contact [23, 24, 29,

35, 37, 38]. This meniscus mediates the transfer of molecules from the tip to the substrate, can be composed of water or organic solvents, and enables surface patterning on length scales as small as 10 nm.

SPNG uses an AFM probe to displace the "resist" molecules of a self-assembled monolayer (SAM) initially deposited uniformly on a solid substrate such as Au or SiO$_x$. Application of the probe to the surface under high load force and high scan speed creates "trenches" of SAM-free surface surrounded by the background resist. These trenches are then subsequently re-functionalized with molecular linkers to form a functionalized pattern. Here we report using these SPN techniques to: 1) create biochemical affinity templates for organization of macromolecular complexes such as Cowpea Mosaic Virus (CPMV) and Tobacco Mosaic Virus (TMV), 2) pattern functional oligonucleotides that mediate the subsequent formation of Pd nanocrystals, and 3) use the meniscus as a physical tool to both modulate the surface topography of the substrate and guide the hierarchical assembly of Au nanoparticles into nanowires.

2 RESULTS AND DISCUSSION

2.1 Use of Scanning Probe Microscopy to Create Affinity Templates Designed to Bind Sites on Modified Macromolecular Complexes

We have used CPMV and TMV as model macromolecular complexes. The CPMV capsid is a sphere with a diameter of 28 nm and is composed of 60 copies each of large and small coat proteins arranged with pseudo three-fold symmetry [39-42]. A high resolution AFM image of CPMV particles showing 5-fold symmetry of the virus is shown in Figure 1. CPMVs were generically engineered to present either cysteine (Cys) [42] or histidine (His) [41] residues at specific sites on the capsid surface. These were used for chemoselective immobilization.

The TMV capsid is a 300 nm long rod, with an outer diameter of 18 nm and an inner diameter of 4 nm. 2130 identical copies of the coat protein self assemble into the rod-like helical structure around genomic RNA [43-45]. In the absence of RNA, the aggregation state of TMV coat protein monomers can be controlled by adjusting the pH and ionic strength of buffered solution. Aggregates, including ca. 4 nm thick double disks and micron-length rods can be assembled [44, 46, 47]. TEM and AFM images of Tobacco Mosaic Virus coat Protein (TMVP) disk aggregates showing a doughnut shape are given in Figure 1. A methodology for modification of native [48] and engineered [47, 49] residues on the capsid surface has been developed. In this study, chemoselective immobilization via engineered cystein residues is utilized.

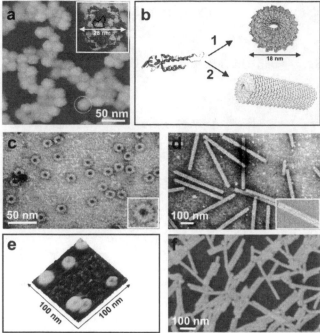

Figure 1. (a) High-resolution AFM topographic (height) image of typical CPMV particles on mica surface. (b) Schematic of assembly of TMVP monomers into disk aggregates (**1** at pH = 7; 400 mM potassium phosphate buffer) and rod aggregates (**2** a pH = 5.5; 100 mM sodium acetate buffer). (c) TEM image of TMVP disk aggregates. (d) TEM image of rod-shaped TMV capsids. (e) AFM topographic image of TMVP disk aggregates on atomically flat Au surface. (f) AFM topographic image of rod-shaped TMV capsids on atomically flat Au surface.

Starting with the methoxy-terminated poly(ethylene glycol) (PEG) alkanethiol (MeO-PEG-SH) functionalized atomically flat gold substrates described above, patterns of various alkanethiol linkers were deposited by either DPN or SPNG to produce affinity templates. Typically, these templates were comprised of lines or dots having a minimum dimension ranging from 10 – 100 nm, separated by 100 – 1000 nm (Figure 2-a, b, c, and d). These features provided chemoselective attachment sites for CPMV and TMVP disks.

For CPMV, nickel(II)-chelating nitrilotriacetic acid (Ni-NTA) terminated alkanethiol (Ni-NTA-PEG-SH) linkers were used to reversibly bind His-CPMVs through a metal coordination complex (Scheme 1-a) and maleimide-terminated functional groups were used to covalently attach Cys-CPMVs. Figures 2-e and 2-f show AFM images of CPMVs assembled on linear templates demonstrating two different morphologies depending on the choice of immobilization chemistry. His-CPMV particles bind almost exclusively along Ni-NTA patterned lines showing a fully populated single line of His-CPMV particles [33] with virtually no inter-viral binding. However, Cys-CPMV particles assembled on maleimide-functionalized lines exhibit dendritic patterns extending away from the lines reminiscent of epitaxial growth in a regime of diffusion limited aggregation [32]. This lack of order and control over assembly of Cys-CPMV on maleimide templates is likely due to two factors: The first is a strong irreversible attractive interaction between Cys-CPMV particles and the template, which prevents any virus reorganization from providing a means for achieving well ordered packing along the lines. The second is the formation of covalent disulfide bonds between neighboring virus particles, which leads to uncontrolled lateral aggregation.

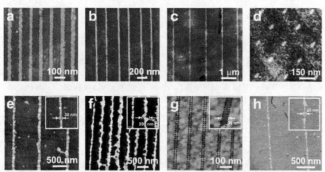

Figure 2. AFM topographic images of typical affinity templates of various alkanethiol linker and dot patterns (line width = ~ s5 nm) separated by (a) 100 nm, (b) 200 nm, (c) 1 μm, and (d) 150 nm. (e) AFM topographic image of His-CPMV chemoselectively bound on Ni(II)-NTA-terminated line patterns. Each line is populated with a single layer of His-CPMV. The measured height (~28 nm) closely matches with that previously reported in the literature. (f) AFM topographic image of Cys-CPMV chemoselectively bound on maleimide-terminated line patterns. (g) AFM topographic image of TMVP disk aggregates nonspecifically bound on methoxy-terminated "resist" SAM (dotted line represent patterns initially intended to TMVP disk immobilization). (h) AFM topographic image of TMVP disk aggregates where Cys-residues were chemoselectively bound on maleimide-terminated line patterns.

Initially, a similar MeO-PEG-SH SAM was employed to resist non-specific adsorption of TMVP on affinity templates. However, this SAM did not sufficiently resist attachment of TMVP disks. Most of the disks bound the MeO-PEG-SH SAM, resulting in the "inverted" pattern shown in Figure 2-g. Preferred attachment of TMVP disks on the resist area is presumably due to relatively significant hydrophobic interactions with this SAM [50, 51]. It is to be expected that TMVP disks, which represent an intermediate assembly state, are subject to greater hydrophobic interactions than fully assembled, negatively-charged capsids. Screening of various functional SAMs led to the development of a more hydrophilic OH-PEG-SH SAM. This monolayer is significantly more hydrophilic, as indicated by water contact angle (ca. 30° vs ca. 50° for MeO-PEG-SAM), and sufficiently resists protein adsorption. With the improved resist SAM in hand, the cysteine-based immobilization strategy similar to that employed for CPMV also proved successful for TMVP

disks. Figure 2-h shows AFM images of TMVP disks assembled only on maleimide-terminated linear templates.

These results show that SPM-based fabrication of biochemical affinity templates provides a means for creating chemoselective features ranging from tens of nanometers to microns, a size scale that is relevant for viruses and other functional macro-biomolecular complexes. SAM functionality can be chosen to selectively complex or react with modified viral particles, and organization can be directed by altering the surface chemistry of the complexes by chemical or genetic means. Changes to capsid surface chemistry can also modulate virus-virus interaction, as can alterations to solution environment such as pH and ionic strength. We are now using these affinity templates to interrogate the role of viral interaction on directed assembly kinetics and morphology.

2.2 Use of Scanning Probe Microscopy to Pattern Functional Molecules that Drive the Formation of Materials

Biomolecule *in vitro* selection methods [52-57] such as phage display, ribosome display, and SELEX have been enormously successful at the discovery of peptides and oligonucleotides with high binding affinities to proteins and catalytic activities toward organic reactions [58-72]. These methods have also recently been adapted to the discovery of biomolecules that possess high binding affinities to inorganic solids or that can mediate or catalyze solid-state reactions. (Note the differences in "mediation" and "catalysis" have recently discussed [73].) For example, RNA sequences have been selected from a large random sequence RNA library (ca. 10^{14} unique sequences) that mediate the formation of metal crystals when incubated with solutions containing the metal complex [$Pd_2(DBA)_3$] [74, 75]. Moreover, one sequence isolated from the initial library mediated the formation of crystals with hexagonal morphology while another sequence yielded crystals with cubic morphology [74, 75]. While the detailed atomic level structure, composition, and formation mechanism of these crystals are still being investigated, the selected RNA sequences can be thought of as biomolecule codes for materials. As such, we found them to be attractive candidates for creating spatially well-defined patterns of materials on surfaces. We have used one of these RNA sequences (referred to as sequence Pd017) to mediate the formation of Pd nanocrystals on surfaces. Atomically flat gold substrates were first functionalized with MeO-PEG-SH to resist non-specific attachment of excess RNA and Pd precursor. SPN was utilized to pattern maleimide linkers (Mal-PEG-TA) for the covalent immobilization of Pd017 that was modified on the 5' end with guanosine monophosphorothioate (GMPS).

Figure 3. AFM topographic images of typical affinity templates of maleimide-terminated alkanethiol grid patterns (line width = ~ 25 nm) separated by (a) 100 nm, (b) 250 nm, (c) 500 nm. (d) AFM topographic image of hexagonal PD nanocrystals on Pd017 RAN molecule chemoselectively immobilized on linear affinity template where bottom left inset shows phase image. Inset (top left) shows TEM image of hexagonal Pd nanocrystals formed in solution, mediated by Pd017 RAN, and (top right) the AFM image of surface grown hexagonal Pd nanocrystals, mediated by Pd017 RNA molecules immobilized on the surface.

Figure 3-a, b and c show AFM images of maleimide-functionalized templates with grid patterns. Pd017 was then covalently immobilized onto the patterns via coupling of the maleimide and GMPS moieties. Subsequent bathing of the chip in an aqueous THF solution containing the $[Pd_2(DBA)_3]$ precursor generated nanocrystals on the substrate. Figure 3-d shows AFM images of hexagonal nanocrystals formed on a patterned Pd017 surface. AFM height measurements showed that Pd nanocrystals were ca. 25 nm thick and 250 nm wide, as previously found when nanocrystals were mediated by Pd017 in solution [74, 75]. Most of the hexagonal nanocrystals grew exclusively from the patterned area of the template, indicating the presence of functionally active immobilized RNA. However, it is currently unclear why the coverage of RNA-mediated nanocrystals synthesized on patterned surface is not high. Because it is unknown both how many RNA molecules are immobilized on the substrate (i.e. the covalent conjugation reaction efficiency between the maleimide functional group and 5'-GMPS moiety of Pd017 RNA) and how many of the immobilized RNA molecules are indeed functionally active, it is difficult to define the overall yield of Pd nanocrystals on the surface. As these molecular-level details are understood and optimized, we anticipate that greater control over the position and efficiency of inorganic nanostructure synthesis on patterned surfaces will be afforded. The demonstration that RNA sequences patterned on surfaces can mediate the formation of nanocrystals suggests a new method for integrating nanoscale materials on surfaces. As more RNA sequences coding for different materials are discovered, it should be possible to affect the orthogonal synthesis of nanoscale materials in ways that could lead to functional device architectures.

2.3 Use of Scanning Probe Microscopy and its Meniscus as a Tool for Patterning and Modifying Surfaces

Our initial experiments with meniscus-based fabrication of nanostructures focused on the repair of topographic defects in crystal surfaces. As Figure 4 (a1 – a5) shows, the grooves made by mechanical etching, as described above, are eliminated as the meniscus is rastered over the area. We analyzed the time dependence of this effect and have derived a general expression to describe it [76]. The underlying physical driver for this process is the relaxation of curvature driven by the Gibbs-Thomson effect. The presence of the meniscus provides a solution environment for rapid transport of material from regions of high curvature to those of low curvature, but the rate is determined largely by an effective diffusivity related to the step stiffness and the step kinetic coefficient [76].

Following our observation of groove-filling, we explored the effect of scanning on a groove-free KDP surface. We discovered that, when the meniscus is allowed to remain stationary, a convex "dot" forms beneath it. If the tip is rastered back and forth along a line, a convex "line" is formed. Figures 4-d and e show examples of a KDP "dot" and "line" fabricated via this method. The driving force and physical principles underlying this process are not clear. However, given that material is transported to the meniscus from surrounding flat regions, that the method only works at certain humidities high enough to ensure that the entire KDP surface is covered by a one-to-two monolayer thick film of water, and that the width is defined by the diameter of the meniscus, we conclude that the chemical potential of the solution with respect to solid KDP is different in the bulk fluid of the meniscus than it is within the thin water layer due to its comparatively water-poor volume [77]. The resulting difference in local KDP solubility may be the main driver behind the formation of these nanostructures, although some process involving continuous evaporation may also be important [78]. Whatever the mechanism, this process provides a facile method to form 3D structures that can be used for forensic purposes since these features disappear progressively, to minimize surface energy, when exposed to humid air, (i.e., when the surface has been tampered with) [76]. In principle, this method and process can be applied to many organic and inorganic materials and surfaces that have a finite solubility in water or other solvents with sufficient vapor pressure near room temperature.

NSTI-Nanotech 2008, www.nsti.org, ISBN 978-1-4200-8503-7 Vol. 1

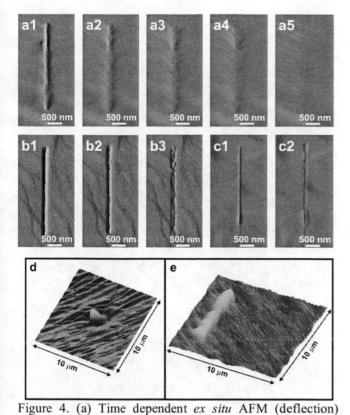

Figure 4. (a) Time dependent *ex situ* AFM (deflection) images of repair of topographic groove defects in KDP crystal surfaces. Initially the grooves were fabricated via SPNG at time = 0 (a1). Then the AFM tip was continuously scanned along the long axis of the groove back and forth at 2 Hz (tip speed: 24 μm/s) at 85% R.H. and the grooves were refilled wit hKDP dissolved in the surrounding water meniscus. Each AFM image was captured at time = 4 min (a2), 9 min (a3), 13 min (a4), and 71 min (a5). (b) Time –dependent *ex situ* AFM (deflection) images of mechanically formed KDP trenches as templates to guide and fill in 2 nm diameter Au nanoparticles. At time = 0 (b1), Au nanoparticles start to spontaneously diffuse and assemble in the trench on a KDP crystal surface. Au nanoparticles show coalescence at time = 42 min (b2) and eventually produce a Au wire at time = 76 min (b3) at room temperature under ambient condition R.H. = 85%). (c) Another example of Au wire formation nat time = 0 (c1) and time = 400 min (c2) under the same experimental conditions as (b). (d) AFM (deflection) image of a KDP "dot" overgrown on KPD crystal surface at 85% R.H. when the water meniscus at the AFM tip-surface interface was allowed to remain stationary for ~ 15 min. The height of the dot is 900 nm but can be made as small as a few nanometers by reducing the contact time, size of the tip-meniscus, and humidity [76]. (e) AFM (deflection) image of a KDP "line" overgrown on KDP crystal surface at 85% R.H. when the water meniscus was scanned along the long axis of the wire for ~ 10 min. The width and height of the line are 800 and 10 nm, respectively.

Finally, we found that we could use the mechanically formed trenches as templates to guide and fill in foreign materials such as Au nanoparticles. Figure 4 (b1–b3 and c1–c2) shows AFM images of 2 nm diameter Au nanoparticles selectively assembled in a groove on a KDP crystal surface. When an aqueous solution of Au nanoparticles was deposited on a KDP surface by solution casting, the majority of Au nanoparticles were observed to spontaneously localize in pits or trenches. Depending on the environmental conditions, these Au nanoparticles were sometimes observed to coarsen into a solid Au wire. Figure 4 (b1–b3 and c1–c2) shows this process within a groove on a KDP surface. Our preliminary explanation why the gold nanoparticles accumulate in the trench to form a nanowire is as follows. Nanoparticles deposited on the surface can experience Brownian motion facilitated by the molecularly thin liquid condensate layer. The gold nanoparticles reach the trench as a result of this motion where they became physically trapped. The wetting angle of gold by bulk water in air is ca. 70° [79] suggesting a moderate, but still positive, wetting. This hydrophilicity probably makes removal of a nanoparticle from the bulk aqueous solution unfavorable. Therefore, the nanoparticles should accumulate in a trench filled by the bulk solution and remain there by such capillary trapping. In the trench, the nanoparticles sinter to minimize surface energy although the details of the structure in the metallization process of the nanoparticle aggregates (Figure 4 b and c) are still being investigated. The geomctry of the physical templates dictates the resulting organization of the desired materials, which, when combined in orthogonal steps, can lead to hierarchical structures whose resulting functions depend on the combined properties of the individual nanomaterials.

ACKNOWLEDGEMENTS

We acknowledge professors John E. Johnson, Tianwei Lin, and Dr. Anju Chatterji from the Scripps Research Institute for preparing and supplying genetically engineered Cys- and His-CPMV particles. We acknowledge the contributions of Drs. Ted Tarasow, Lina Gugliotti and Julie Perkins to their initial helps on development of synthetic methods and optimization of RNA mediated reactions. S.W.C. and J.D.Y. acknowledge U.S. Department of Energy (DOE), Office of Basic Energy Science (BES), Division of Materials Science and Engineering for support of this research. S.W.C. acknowledges Lawrence Livermore National Laboratory (LLNL) Laboratory Directed Research and Development Program (LDRD Laboratory-Wide Funding, 06-LW-051) for support of this research. A.D.P. and M.B.F. acknowledge Laboratory Directed Research and Development Program of Lawrence Berkeley National Laboratory (LBNL) under the Department of Energy (DOE) Contract DE-AC02-05CH11231. A.D.P. acknowledges LLNL Student Employee Graduate Research Fellowship Program (SEGRF). B.E.E. and D.L.F. acknowledge NSF, the W. M.

Keck Foundation, and U.S. Department of Energy (DOE). This work performed under the auspices of the U.S. Department of Energy by Lawrence Livermore National Laboratory under Contract DE-AC52-07NA27344.

REFERENCES

[1] S. Bahatyrova, R. N. Frese, C. A. Siebert, J. D. Olsen, K. O. van der Werf, R. van Grondelle, R. A. Niederman, P. A. Bullough, C. Otto, C. N. Hunter, "The native architecture of a photosynthetic membrane," Nature, 430, 1058, 2004.

[2] I. Gur, N. A. Fromer, C. P. Chen, A. G. Kanaras, A. P. Alivisatos, "Hybrid solar cells with prescribed nanoscale morphologies based on hyperbranched semiconductor nanocrystals," Nano Lett., 7, 409, 2007.

[3] I. Gur, N. A. Fromer, M. L. Geier, A. P. Alivisatos, "Air-stable all-inorganic nanocrystal solar cells processed from solution," Science, 310, 462, 2005.

[4] W. U. Huynh, J. J. Dittmer, A. P. Alivisatos, "Hybrid nanorod-polymer solar cells," Science, 295, 2425, 2002.

[5] E. Klimov, W. Li, X. Yang, G. G. Hoffmann, J. Loos, "Scanning near-field and confocal Raman microscopic investigation of P3HT-PCBM systems for solar cell applications," Macromolecules, 39, 4493, 2006.

[6] A. J. Nozik, "Quantum dot solar cells," Physica E, 14, 115, 2002.

[7] R. D. Schaller, V. M. Agranovich, V. I. Klimov, "High-efficiency carrier multiplication through direct photogeneration of multi-excitons via virtual single-exciton states," Nat. Phys., 1, 189, 2005.

[8] R. D. Schaller, V. I. Klimov, "High efficiency carrier multiplication in PbSe nanocrystals: Implications for solar energy conversion," Phys. Rev. Lett., 92, 2004.

[9] P. R. Yu, K. Zhu, A. G. Norman, S. Ferrere, A. J. Frank, A. J. Nozik, "Nanocrystalline TiO2 solar cells sensitized with InAs quantum dots," J. Phys. Chem. B, 110, 25451, 2006.

[10] Y. A. Vlasov, N. Yao, D. J. Norris, "Synthesis of photonic crystals for optical wavelengths from semiconductor quantum dots," Adv. Mater., 11, 165, 1999.

[11] Y. Huang, X. F. Duan, Y. Cui, L. J. Lauhon, K. H. Kim, C. M. Lieber, "Logic gates and computation from assembled nanowire building blocks," Science, 294, 1313, 2001.

[12] N. A. Melosh, A. Boukai, F. Diana, B. Gerardot, A. Badolato, P. M. Petroff, J. R. Heath, "Ultrahigh-density nanowire lattices and circuits," Science, 300, 112, 2003.

[13] P. D. Yang, F. Kim, "Langmuir-Blodgett assembly of one-dimensional nanostructures," Chemphyschem, 3, 503, 2002.

[14] H. N. Chapman, A. Barty, S. Marchesini, A. Noy, S. R. Hau-Riege, C. Cui, M. R. Howells, R. Rosen, H. He, J. C. H. Spence, U. Weierstall, T. Beetz, C. Jacobsen, D. Shapiro, "High-resolution ab initio three-dimensional x-ray diffraction microscopy," J. Opt. Soc. Am. A, 23, 1179, 2006.

[15] H. N. Chapman, S. P. Hau-Riege, M. J. Bogan, S. Bajt, A. Barty, S. Boutet, S. Marchesini, M. Frank, B. W. Woods, W. H. Benner, R. A. London, U. Rohner, A. Szoke, E. Spiller, T. Moller, C. Bostedt, D. A. Shapiro, M. Kuhlmann, R. Treusch, E. Plonjes, F. Burmeister, M. Bergh, C. Caleman, G. Huldt, M. M. Seibert, J. Hajdu, "Femtosecond time-delay X-ray holography," Nature, 448, 676, 2007.

[16] S. W. Chung, D. S. Ginger, M. W. Morales, Z. F. Zhang, V. Chandrasekhar, M. A. Ratner, C. A. Mirkin, "Top-down meets bottom-up: Dip-pen nanolithography and DNA-directed assembly of nanoscale electrical circuits," Small, 1, 64, 2005.

[17] C. B. Murray, C. R. Kagan, M. G. Bawendi, "Synthesis and characterization of monodisperse nanocrystals and close-packed nanocrystal assemblies," Annu. Rev. Mater. Sci., 30, 545, 2000.

[18] Y. N. Xia, G. M. Whitesides, "Soft lithography," Annu. Rev. Mater. Sci., 28, 153, 1998.

[19] Y. N. Xia, G. M. Whitesides, "Soft lithography," Angew. Chem. Int. Ed., 37, 551, 1998.

[20] P. Bhatnagar, S. S. Mark, I. Kim, H. Y. Chen, B. Schmidt, M. Lipson, C. A. Batt, "Dendrimer-scaffod-based electron-beam patterning of biomolecules," Adv. Mater., 18, 315, 2006.

[21] D. Falconnet, D. Pasqui, S. Park, R. Eckert, H. Schift, J. Gobrecht, R. Barbucci, M. Textor, "A novel approach to produce protein nanopatterns by combining nanoimprint lithography and molecular self-assembly," Nano Lett., 4, 1909, 2004.

[22] L. J. Guo, "Nanoimprint lithography: Methods and material requirements," Adv. Mater., 19, 495, 2007.

[23] D. Ginger, H. Zhang, C. Mirkin, "The evolution of dip-pen nanolithography," Angew. Chem. Int. Ed., 43, 30, 2004.

[24] R. D. Piner, J. Zhu, F. Xu, S. H. Hong, C. A. Mirkin, ""Dip-Pen" Nanolithography," Science, 283, 661, 1999.

[25] K. B. Lee, S. J. Park, C. A. Mirkin, J. C. Smith, M. Mrksich, "Protein nanoarrays generated by Dip-Pen Nanolithography," Science, 295, 1702, 2002.

[26] L. M. Demers, D. S. Ginger, S. J. Park, Z. Li, S. W. Chung, C. A. Mirkin, "Direct patterning of modified oligonucleotides on metals and insulators by dip-pen nanolithography," Science, 296, 1836, 2002.

[27] K. Salaita, Y. H. Wang, J. Fragala, R. A. Vega, C. Liu, C. A. Mirkin, "Massively parallel dip-pen nanolithography with 55000-pen two-dimensional arrays," Angew. Chem. Int. Ed., 45, 7220, 2006.

[28] R. A. Vega, D. Maspoch, K. Salaita, C. A. Mirkin, "Nanoarrays of single virus particles," Angew. Chem. Int. Ed., 44, 6013, 2005.

[29] B. L. Weeks, A. Noy, A. E. Miller, J. J. De Yoreo, "Effect of dissolution kinetics on feature size in dip-pen nanolithography," Phys. Rev. Lett., 88, 255505, 2002.

[30] K. Wadu-Mesthrige, N. A. Amro, J. C. Garno, S. Xu, G. Y. Liu, "Fabrication of nanometer-sized protein patterns using atomic force microscopy and selective immobilization," Biophys. J., 80, 1891, 2001.

[31] K. Wadu-Mesthrige, S. Xu, N. A. Amro, G. Y. Liu, "Fabrication and imaging of nanometer-sized protein patterns," Langmuir, 15, 8580, 1999.

[32] C. L. Cheung, J. A. Camarero, B. W. Woods, T. W. Lin, J. E. Johnson, J. J. De Yoreo, "Fabrication of assembled virus nanostructures on templates of chemoselective linkers formed by scanning probe nanolithography," J. Am. Chem. Soc., 125, 6848, 2003.

[33] C. L. Cheung, S. W. Chung, A. Chatterji, T. W. Lin, J. E. Johnson, S. Hok, J. Perkins, J. J. De Yoreo, "Physical controls on directed virus assembly at nanoscale chemical templates," J. Am. Chem. Soc., 128, 10801, 2006.

[34] J. H. Lim, D. S. Ginger, K. B. Lee, J. Heo, J. M. Nam, C. A. Mirkin, "Direct-write Dip-Pen Nanolithography of proteins on modified silicon oxide surfaces," Angew. Chem. Int. Ed., 42, 2309, 2003.

[35] C. A. Mirkin, S. H. Hong, L. Demers, "Dip-pen nanolithography: Controlling surface architecture on the sub-100 nanometer length scale," Chemphyschem, 2, 37, 2001.

[36] A. Noy, A. E. Miller, J. E. Klare, B. L. Weeks, B. W. Woods, J. J. DeYoreo, "Fabrication of luminescent nanostructures and polymer nanowires using dip-pen nanolithography," Nano Lett., 2, 109, 2002.

[37] S. Rozhok, P. Sun, R. Piner, M. Lieberman, C. A. Mirkin, "AFM study of water meniscus formation between an AFM tip and NaCl substrate," J. Phys. Chem. B, 108, 7814, 2004.

[38] B. L. Weeks, J. J. DeYoreo, "Dynamic meniscus growth at a scanning probe tip in contact with a gold substrate," J. Phys. Chem. B, 110, 10231, 2006.

[39] A. Chatterji, W. F. Ochoa, T. Ueno, T. W. Lin, J. E. Johnson, "A virus-based nanoblock with tunable electrostatic properties," Nano Lett., 5, 597, 2005.

[40] T. Douglas, M. Young, "Virus particles as templates for materials synthesis," Adv. Mater., 11, 679, 1999.

[41] I. L. Medintz, K. E. Sapsford, J. H. Konnert, A. Chatterji, T. W. Lin, J. E. Johnson, H. Mattoussi, "Decoration of discretely immobilized cowpea mosaic virus with luminescent quantum dots," Langmuir, 21, 5501, 2005.

[42] Q. Wang, T. W. Lin, L. Tang, J. E. Johnson, M. G. Finn, "Icosahedral virus particles as addressable nanoscale building blocks," Angew. Chem. Int. Ed., 41, 459, 2002.

[43] B. Bhyravbhatla, S. J. Watowich, D. L. D. Caspar, "Refined atomic model of the four-layer aggregate of the tobacco mosaic virus coat protein at 2.4-angstrom resolution," Biophys. J., 74, 604, 1998.

[44] A. Klug, "The tobacco mosaic virus particle: structure and assembly," Philos. Trans. R. Soc. London B, 354, 531, 1999.

[45] R. Pattanayek, G. Stubbs, "Structure of the U2 Strain of Tobacco Mosaic-Virus Refined at 3.5 Angstrom Resolution Using X-Ray Fiber Diffraction," J. Mol. Biol., 228, 516, 1992.

[46] P. J. G. Butler, "Self-assembly of tobacco mosaic virus: the role of an intermediate aggregate in generating both specificity and speed," Philos. Trans. R. Soc. London B, 354, 537, 1999.

[47] R. A. Miller, A. D. Presley, M. B. Francis, "Self-assembling light-harvesting systems from synthetically modified tobacco mosaic virus coat proteins," J. Am. Chem. Soc., 129, 3104, 2007.

[48] T. L. Schlick, Z. B. Ding, E. W. Kovacs, M. B. Francis, "Dual-surface modification of the tobacco mosaic virus," J. Am. Chem. Soc., 127, 3718, 2005.

[49] M. Demir, M. H. B. Stowell, "A chemoselective biomolecular template for assembling diverse nanotubular materials," Nanotechnology, 13, 541, 2002.

[50] G. P. Lopez, H. A. Biebuyck, R. Harter, A. Kumar, G. M. Whitesides, "Fabrication and Imaging of 2-Dimensional Patterns of Proteins Adsorbed on Self-Assembled Monolayers by Scanning Electron-Microscopy," J. Am. Chem. Soc., 115, 10774, 1993.

[51] C. Palegrosdemange, E. S. Simon, K. L. Prime, G. M. Whitesides, "Formation of Self-Assembled Monolayers by Chemisorption of Derivatives of Oligo(Ethylene Glycol) of Structure $HS(CH_2)_{11}(OCH_2CH_2)$Meta-OH on Gold," J. Am. Chem. Soc., 113, 12, 1991.

[52] A. D. Ellington, J. W. Szostak, "Invitro Selection of Rna Molecules That Bind Specific Ligands," Nature, 346, 818, 1990.

[53] C. Tuerk, L. Gold, "Systematic Evolution of Ligands by Exponential Enrichment - Rna Ligands to Bacteriophage-T4 DNA-Polymerase," Science, 249, 505, 1990.

[54] S. W. Lee, C. B. Mao, C. E. Flynn, A. M. Belcher, "Ordering of quantum dots using genetically engineered viruses," Science, 296, 892, 2002.

[55] C. B. Mao, C. E. Flynn, A. Hayhurst, R. Sweeney, J. F. Qi, G. Georgiou, B. Iverson, A. M. Belcher, "Viral assembly of oriented quantum dot

nanowires," Proc. Natl. Acad. Sci. U.S.A., 100, 6946, 2003.

[56] B. E. Eaton, "The joys of in vitro selection: chemically dressing oligonucleotides to satiate protein targets.," Curr. Opin. Chem. Biol., 1, 10, 1997.

[57] B. E. Eaton, B. Holley, in *Evolutionary Methods in Biotechnology* S. Brakmann, A. Schwienhorst, Eds. (Wiley-VCH, Weinheim, Germany, 2004) pp. 87-111.

[58] D. Nieuwlandt, M. West, X. Q. Cheng, G. Kirshenheuter, B. E. Eaton, "The first example of an RNA urea synthase: Selection through the enzyme active site of human neutrophile elastase," Chembiochem, 4, 651, 2003.

[59] B. Seelig, A. Jaschke, "A small catalytic RNA motif with Diels-Alderase activity," Chem. Biol., 6, 167, 1999.

[60] G. Sengle, A. Eisenfuhr, P. S. Arora, J. S. Nowick, M. Famulok, "Novel RNA catalysts for the Michael reaction," Chem. Biol., 8, 459, 2001.

[61] J. W. Szostak, D. Bartel, A. Hager, A. Das, P. Lohse, "Isolation of new ribozymes form pools of random sequences.," FASEB J., 10, C1, 1996.

[62] T. M. Tarasow, S. L. Tarasow, B. E. Eaton, "RNA-catalysed carbon-carbon bond formation," Nature, 389, 54, 1997.

[63] T. W. Wiegand, R. C. Janssen, B. E. Eaton, "Selection of RNA amide syntheses," Chem. Biol., 4, 675, 1997.

[64] A. Eisenfuhr, P. S. Arora, G. Sengle, L. R. Takaoka, J. S. Nowick, M. Famulok, "A ribozyme with michaelase activity: Synthesis of the substrate precursors," Bioorg. Med. Chem., 11, 235, 2003.

[65] S. Fusz, A. Eisenfuhr, S. G. Srivatsan, A. Heckel, M. Famulok, "A ribozyme for the aldol reaction," Chem. Biol., 12, 941, 2005.

[66] F. Q. Huang, C. W. Bugg, M. Yarus, "RNA-catalyzed CoA, NAD, and FAD synthesis from phosphopantetheine, NMN, and FMN," Biochemistry, 39, 15548, 2000.

[67] M. Illangasekare, G. Sanchez, T. Nickles, M. Yarus, "Aminoacyl-Rna Synthesis Catalyzed by an Rna," Science, 267, 643, 1995.

[68] P. A. Lohse, J. W. Szostak, "Ribozyme-catalysed amino-acid transfer reactions," Nature, 381, 442, 1996.

[69] S. Tsukiji, S. B. Pattnaik, H. Suga, "An alcohol dehydrogenase ribozyme," Nat. Struct. Biol., 10, 713, 2003.

[70] M. Wecker, D. Smith, L. Gold, "In vitro selection of a novel catalytic RNA: Characterization of a sulfur alkylation reaction and interaction with a small peptide," RNA, 2, 982, 1996.

[71] C. Wilson, J. W. Szostak, "In-Vitro Evolution of a Self-Alkylating Ribozyme," Nature, 374, 777, 1995.

[72] B. L. Zhang, T. R. Cech, "Peptidyl-transferase ribozymes: trans reactions, structural characterization and ribosomal RNA-like features," Chem. Biol., 5, 539, 1998.

[73] D. L. Feldheim, B. E. Eaton, "Selection of biomolecules capable of mediating the formation of nanocrystals," ACS Nano., In Press., 2007.

[74] L. A. Gugliotti, D. L. Feldheim, B. E. Eaton, "RNA-mediated metal-metal bond formation in the synthesis of hexagonal palladium nanoparticles," Science, 304, 850, 2004.

[75] L. A. Gugliotti, D. L. Feldheim, B. E. Eaton, "RNA-mediated control of metal nanoparticle shape," J. Am. Chem. Soc., 127, 17814, 2005.

[76] S. Elhadj, J. J. De Yoreo, A. A. Chernov, Submitted, 2007.

[77] S. Garcia-Manyes, A. Verdaguer, P. Gorostiza, F. Sanz, "Alkali halide nanocrystal growth and etching studied by AFM and modeled by MD simulations," J. Chem. Phys., 120, 2963, 2004.

[78] H. Shindo, M. Ohashi, O. Tateishi, A. Seo, "Atomic force microscopic observation of step movements on NaCl(001) and NaF(001) with the help of adsorbed water," J. Chem. Soc., Faraday Trans., 93, 1169, 1997.

[79] M. E. Abdelsalam, P. N. Bartlett, T. Kelf, J. Baumberg, "Wetting of regularly structured gold surfaces," Langmuir, 21, 1753, 2005.

Fabrication of biosensor arrays by DPN and multiple target detection by triple wavelength fast SERRS mapping

Robert J. Stokes,[*] Jennifer A. Dougan,[*] Ross Stevenson,[*] Eleanore Irvine,[*] Jason Haaheim,[**] Tom Levesque,[**] Karen Faulds,[*] Duncan Graham[*]

[*] Centre for Molecular Nanometrology, Department of Pure and Applied Chemistry, WestCHEM, University of Strathclyde, Thomas Graham Building, 295 Cathedral Street, Glasgow, G1 1XL, United Kingdom. Robert.stokes@strath.ac.uk, Duncan.Graham@strath.ac.uk
[**] NanoInk, Inc., 8025 Lamon Ave, Skokie, IL USA. j.haaheim@nanoink.net

ABSTRACT

We have used Dip-Pen Nanolithography (DPN) to create biosensor arrays on micro- and nano-structured surfaces. Bespoke linker and labelling materials enables detection of biological molecules by surface enhanced resonance Raman scattering (SERRS). Careful DPN-directed placement of the biological species or capture chemistry, within the array, facilitates rapid read out via ultra fast Raman line mapping. Writing DPN features that complement the spectroscopic collection geometry allows the lateral resolution and detection speed to be optimised. Applied in a DPN directed array format, we show that SERRS offers several advantages over conventional fluorescence detection. The information rich nature of the SERRS spectrum allows multiple levels of detection capability to be embedded into each pixel, further increasing the information depth of the array. Effective practical application of multiple target detection by SERRS in a plasmonic array format is demonstrated, using suitable dye labels and resonant wavelengths of excitation.

Keywords: DPN, SERS/SERRS, Plasmonics

1 INTRODUCTION

Surface enhanced [resonance] Raman scattering (SE[R]RS) is a technique whereby signals from suitable molecules are enhanced enormously by close proximity to high electric field gradients at metal surfaces, in some cases to the extent where scattering from single molecules can be observed.[1] The technique is flexible and has been demonstrated to be effective in number of biodiagnostic applications including gene probes,[2,] and DNA detection[3]. A significant part of the overall enhancement in SERRS derives from the additional 'resonance' with the molecular chromophore. The resonance-enhanced spectrum from a reporter dye is often less complex in appearance than would normally be expected from a larger molecule as only selected vibronic states are enhanced (to a lesser or greater extent) when probed with a single wavelength of excitation (λ_{ex}).

Therefore, in many cases only narrow SERRS lines (~0.5 nm) that conform to Raman, resonance and surface selection rules are observed. This is a significant advantage of the technique when applied in real assay a number of characteristic bands within each dye class are enhanced to a greater extent than other materials in the matrix. Two notable examples of this are the distinctive N=N modes that can be observed in the family of SERRS azo dyes,[4] and the unique carbonyl modes of the squarylium type reporter.[5]

A further advantage of SERRS is that the excitation wavelength (λ_{ex}) can be selected anywhere in the optical range and wavelength selectivity can be observed using some combinations of dye reporters.[6]

A number of effective SERS surfaces have been reported in recent years, including those made by nanosphere lithography,[7] silver metal island films [8] and nanostructured gold surfaces.[9, 10] Herin, we demonstrate the effective combination of DPN and fast line scanning spectroscopy to gold SERS surfaces to create effective and efficient biosensor arrays.

2 EXPERIMENTAL

DPN was performed using an NScriptor™ (Nanoink, Skokie, IL) instrument and an environmental chamber to control temperature and humidity. Klarite™ SERS substrates were obtained from D3 Technologies (Glasgow, UK) and modified with self assembled monolayers appropriate to each experiment before use. In a number of experiments solutions for lithography were prepared using "Just Add DNA" solutions (Nanoink), typically 1% B in A solution being used as a start point for optimization in each case. HPLC purified thioctic acid modified oligonucleotide sequences were purchased from ADT Bio (Southampton, UK) and used without further purification. The areas surrounding the DPN-written active features could be modified with materials of the generic form thiol-aklyl-polyethylene glycol, to prevent non specific binding of oligonucleotides. Self-assembly steps were typically performed over 15 minutes in a sealed chamber. Effective hybridizations were achieved at room temperature by immersion slide into solutions containing the target oligonucleotides in PBS buffer within a sealed hybridization chamber.

SERRS spectroscopy was performed using a Renishaw *InVIA* Raman system (Renishaw, UK) coupled to an inverted microscope (Leica, Germany) equipped with a Streamline™ mapping stage. Three laser wavelengths of excitation were used; a 632.8 nm (HeNe, ~30 mW), 785 nm (diode, ~180 mW) and 830 nm (diode, ~170 mW). The laser at the surface was line focused and attenuated as appropriate to the experiment (typically >0.5 mW, using a delivery optic with NA > 0.5).

3 RESULTS AND DISCUSSION

3.1 Surface chemistry and DPN

It has been recently demonstrated that very sensitive SERRS measurements from dye labelled oligonucleotides could be recorded from a commercially available SERRS surface (Klarite™, D3 Technologies, UK).[11] Furthermore, careful control of the surface chemistry (as apposed to drop coating) lead to a dramatic reduction in the relative standard deviation (~40% to ~10%) of the signal obtained from the positive control region of the substrate area. Following a similar methodology, active areas of the SERS surface were coated with a thioctic acid derived linker. 3' Amino modified oligonucleotides with 5' dye modifications could then be immobilized *via* an inbuilt N-hydroxy succinimidyl (NHS) ester *via* facile chemistry. These idealised self-assembled films resulted in strong SERRS signals, typically up to 25,000 counts per second against a background of 200-300 in blank regions that had not been modified with the linker. Surface passivation of designated blank regions was necessary when using amino modified oligonucleotides, as the interaction between the primary amine and the gold surface is strong enough to result in SERRS signals even after 10 washes (10 ml 0.5 M PBS). 5' Dye labelled oligonucleotides could be deposited directly by this method in order to study the efficiency of the surface modification. Alternatively, a stand could be immobilized that could be used to capture by hybridized, a second dye labelled oligonucleotide or a target strand and a dye labelled complement in a three strand sandwich approach.

All three schemes described above were transferred to direct DPN writing of thioctic acid (and thiol) modified oligonucleotide sequences, resulting in similar strengths of signals from areas as small as 10 x 10 microwells. In an indirect approach the thioctic acid NHS linker could be written directly by DPN (with similar flexibility to

mercaptohexadecanoic acid) and deposition of the primary capture strand could be achieved by NHS coupling to the 3' primary amine. However in this case it was necessary to process the written features rapidly as the exposure to water vapour in the DPN process could lead to hydrolysis of the material rendering it ineffective. A number of alternative bespoke surface chemistries, suitable for DPN, are currently under investigation and will be reported shortly.

DPN writing was performed in single surface microwells (~1.3 μm diameter) to establish the whether effective SERRS, suitable in magnitude and spectral resolution, could be obtained the smallest features on the surface (shown below in figure 1). DPN writing of DNA arrays prepared in this manner could also scaled up using multipen arrays to cover larger areas (up to 800 x 800 μm).

Figure 1 A-frame tip interaction with (Klarite™) SERS surface demonstrating how single plasmonic microwell can be selectively functionalized by DPN.

3.2 Spectroscopic detection by SERRS

The high sensitivity of the SERRS method allowed detection of dye-labeled oligonucleotide capture from single plasmonic array "pixels" ~1 μm² in area (examples of which are shown unprocessed in figure 2). Additionally, the information rich nature of the SERRS spectrum allows multiple levels of detection capability to be embedded into each pixel, further increasing the information depth of the array. Individual spectral components from dyes similar in structure (such as Cy3.5™ and Cy5™) could be easily identified.

In idealized conditions, a large number of dye reporters resonant at the red and near IR could be written by DPN and identified using SERRS. These included IR700, Bodipy

650, Cy3.5™, Cy5™, Cy5.5™, ROX, TAMRA and a number of modified squarylium materials. Using suitable combinations of dye labels wavelength selectivity could be observed e.g. Cy3.5™ (λ_{ex} 632.8 nm) and Cy5.5 (λ_{ex} 785 nm). Therefore, effective practical application of multiple target detection by SERRS in a plasmonic array format could be demonstrated. Although standard Klarite surface were studied in this case the versatility of the surface plasmon derived SERRS response can be further optimized by tuning the aspect ratio of the surface features. Spectral differentiation could between features could be achieved either by curve fitting of multi-component analysis of each spectra.

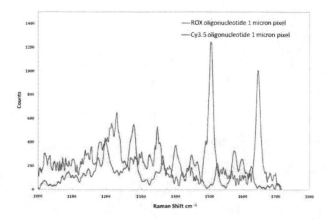

Figure 3 Unprocessed SERRS spectra obtained from ROX and Cy3.5™ dye-labelled oligonucleotide sequences written into single plasmonic microwell.

To demonstrate that the massively scalable potential of DPN written arrays[12] is not fundamentally incompatible with SERRS rapid multi wavelength SERRS readout of array features and arrays was performed. Fast SERRS mapping on the surfaces was achieved using a Streamline™ Raman system (Renishaw, UK). Optimization of the system and procedure allowed large areas of the slide to be analyzed quickly. The whole surface of a Klarite™ slide (4 mm x 4 mm) could be examined in under 10 minutes using 200 µm steps. This allowed the location of positive control areas to be quickly identified. Areas of interest could be examined at the maximum resolution for the surface (i.e. 1.3 µm steps) using higher power magnification (although the minimum step size of the could be as small as 100 nm). At higher magnifications a typical 25 x 25 µm array area could be scanned in high detail within a few minutes. The use of controlled surface chemistry results in highly efficient SERRS, which in turn, allows the spectrometer to operate at the shortest effective accumulation time per pixel. This is key to rapid scanning times and the use of low laser powers to avoid sample damage.

The Streamline™ mapping system works by line focusing the laser incident on the surface and rapidly scanning simultaneously collecting spectra from the whole area of the charge coupled device, collecting ~1000 spectra per minute. The method is naturally suited to the Klarite surface as plasmon derived enhancement is obtained from areas adjacent to each individual pixel. The strongest enhancement occurs where a standing wave are set up in the base of each microwell. This means that the use of DPN to write in the base of each well results in an efficient deposition technique in this case as little material is wasted coating surrounding areas.

The fact that the surface plasmons are able to travel some distance across the gold surface means that in order to achieve effective lateral resolution from a single dye reporter it is necessary to leave a spacing of one or two 'blank' microwells between active sites (~5 µm maximum). However, the fact that SERRS provides such distinct spectra from different dye reporters, that can be overlaid whilst retaining a quantitative response, means that array features can be placed by DPN in closer proximity than the optical method would normally allow. Writing DPN features that complement the spectroscopic collection geometry allows the lateral resolution and detection speed to be optimized. Massively parallel DPN has been recently demonstrated, capable of covering areas > 1 cm^2, and the geometry of these MEMS-generated DPN arrays can be tailored to the form factor of detecting substrates. This flexibility is key to enhancing the throughput of this combined technique by many orders of magnitude.

4 REFERENCES

[1] a) S. Nie, S. Emory, *Science*, **1997**, *275*, 1102-1106 b) K. Kneipp, Y. Wang, H. Kneipp, L.T. Perelman, I. Itzkan, R. R. Dasari, M.S. Feld, *Phys. Rev. Lett.* **1997**, *78(9)*, 1667-1670.

[2] T. Vo-Dinh, K. Houck, D.L. Stokes, *Anal. Chem.*, **1994**, *66(20)*, 3379-3383.

[3] Y.C. Cao, R. Jin, C.A. Mirkin, *Science*, **2002**, *297(5586)*, 1536-1540.

[4] (a) D. Graham, C McLaughlin, G. McAnally, J. C. Jones, P. C. White, W.E. Smith, *Chem Commun.* 1998, 1187. (b) G.M. McAnally, C. McLaughlin, R. Brown, D. Robson, K. Faulds, D.R. Tackley, W.E. Smith, D. Graham, *Analyst*, 2002, **127**, 838.

[5] R.J. Stokes, A. Ingram, J. Gallagher, D. Armstrong, W.E. Smith, D. Graham, *Chem. Commun.* **2008**, 567.

[6] K. Faulds, F. McKenzie, D. Graham, *Angew. Chem. Int. Ed.Engl.* **2007**, *46*,1829-1831.

[7] A. J. Haes, C. L. Haynes, A. D. McFarland, S. Zou, G. C. Schatz, and R. P. Van Duyne, *MRS Bulletin*, **2005**, 30, 368.

[8] L.R. Allain, T. Vo-Dinh, *Anal. Chim. Acta.*, **2002**, *469*, 149-154.

[9] Daniel M. Kuncicky, Brian G. Prevo and Orlin D. Velev, *J. Mater. Chem.*, **2006**, 16, 1207.

[10] N.M.B. Perney, J.J. Baumberg, M.E. Zoorob, M.D.B. Charlton, S. Mahnkopf, C.M.Netti, *Optics Express*, **2006** *14 (2)*,847-857.

[11] R.J. Stokes, A. Macaskill, J.A. Dougan, P.G. Hargreaves, H.M. Stanford, W.E. Smith, K. Faulds, D. Graham, *Chem. Commun.* **2007**, 2811-2813.

[12] K. Salaita, Y. Wang , J. Fragala , R. A. Vega , C. Liu, C. A. Mirkin, *Angew. Chem. Int. Ed. Engl.* **2006**, 45, 7220.

Lipid Dip-Pen Nanolithography for functional biomimetic membrane systems

S. Lenhert[*], H. Fuchs[**]

[*]Forschungszentrum Karlsruhe GmbH, Institut für NanoTechnologie
76344 Eggenstein-Leopoldshafen, Germany, steven.lenhert@kit.edu
[**]Physikalisches Institut, Universität Münster, D-48149 Münster,
Germany, fuchsh@uni-muenster.de

ABSTRACT

Dip Pen Nanolithography (DPN) is uniquely capable of integrating of multiple materials (or inks) with both high resolution and high throughput. Phospholipid-based inks are particularly well suited for this purpose for several reasons. First, their lyotropic liquid crystalline nature enables reproducible tip-coating using multiplexed inkwells for humidity controlled delivery of different inks to different tips in an array, for multiplexed patterning. Second, their chemical diversity allows a variety of different functional groups to be directly integrated into the ink. Third, since the ink transport from the tip to the substrate is based on self-organization and adhesion just about any surface can patterned in this way, including insulating glass or hydrophobic polymer surfaces such as polystyrene. Finally, being a major structural and functional component of biological membranes phospholipids are compatible with the vast amount of molecular resources provided by nature, for instance lipophilic materials and membrane bound proteins.

Keywords: dip pen nanolithography, phospholipid, membrane, liquid crystal, interface

1 INTRODUCTION

Dip-Pen Nanolithography (DPN) is a versatile tool for nanotechnology by using the tip of an Atomic Force Microscope (AFM) as an ultrasharp pen.[1-3] Being a constructive method of scanning probe lithography, DPN can be carried out in a massively parallel fashion, in principle enabling the resolution of electron beam lithography, integration capabilities typical of inkjet printing, and a high throughput comparable to microcontact printing.[3-15] In particular, DPN makes it possible to integrate different materials on scales (both in size and complexity) that appear impossible to reach by any other direct-write fabrication method. Such a method is especially desirable for the fabrication of biomolecular arrays, and opens entirely new possibilities in the study and development of nanobiotechnology.[16-21]

Phospholipids are ubiquitous biological molecules that self-assemble under physiological conditions to form the bilayer structure of biological membranes. Established methods for generating supported lipid bilayer membranes include as vesicle fusion, Langmuir-Blodgett transfer or self-spreading.[22] Self-spreading from dehydrated lipid multi-layer stacks is particular applicable here as it will be shown that it is directly compatable with lipid DPN.

Several methods have been used for the patterning of supported lipid bilayers, for instance vesicle fusion onto pre-patterned substrates,[23] direct photolithography,[24] microcontact printing,[25] microarraying,[26] stensiling,[27] and molecular editing by Atomic Force Microscopy (AFM).[28, 29] These and other methods for generating heterogeneous phospholipid arrays, are severely limited either in their lateral resolution or in the ability to integrate multiple lipids on a single surface, which is desireable for creating surfaces that mimic biological membranes.[30-32]

2 PHOSPHOLIPID-BASED DPN

Here an approach to DPN is used that is based on non-covalent adhesion and humidity control of the liquid crystalline phase of phospholipid inks. This makes it possible to pattern phospholipids on a variety of substrates. The chemical structure of a typical phospholipids ink (DOPC) is shown in Figure 1 along with a schematic illustration of the lipid DPN process. In contrast to the transport behavior of most other inks, where the ink transport can be controlled by the tip-contact time and scan speed, respectively, as well as humidity, phospholipid inks tend to stack into multilayer structures where the thickness of the film can be controlled by those same parameters.[4]

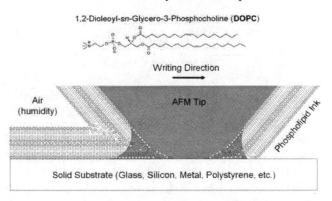

Figure 1: Chemical structure of a typical phospholipid ink (DOPC), and schematic illustration of the writing process.

A crucial step in the DPN process is to coat the tip with the ink to be patterned. Typically, AFM tips are coated for DPN either by thermal evaporation onto the tips or by dip-coating, i.e. immersing the tip into a solution containing the material to be patterned and then allowing the solvent to evaporate from the coated tip. Phospholipids (and related materials) provide a third alternative for tip coating because their fluidity depends on their degree of hydration. In the case of DOPC at room temperature, the bulk material behaves like a solid below 40% humidity, but becomes viscous fluid above that humidity. At 75% humidity, the pure (hydrated) phospholipids material readily flows onto tip. The tip-coating process using microfluidic inkwells (commercially available from the company NanoInk) is drawn in Figure 2.[33]

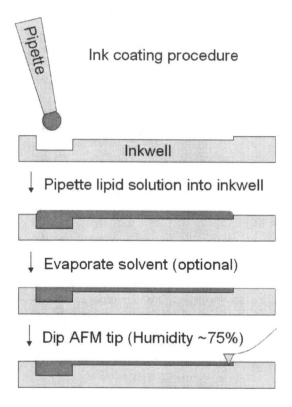

Figure 2: Schematic illustration of the method for coating the AFM tips using inkwells.

The ability to adjust the fluidity (or viscosity) of the ink in this manner facilitates the delivery of different ink materials to different tips in an array for multiplexed patterning. In comparison, it is difficult to get a thick coating of solution based inks onto two or more neighboring tips of an AFM tip array primarily for two reasons. First, the tip is coated mostly by solvent, and upon evaporation of the solvent much of the ink volume disappears. Second, solution based inks that have low viscosity and surface tension can wick out of inkwells when the AFM tip contacts them, thus coating the entire chip and contaminating the tips. Phospholipid based inks provide a

general solution to these problems and an example is shown in Figure 3. In this case, small amounts (1 Mol%) of fluorescently labeled lipids were dispersed in the DOPC ink.

Figure 3: Top: optical micrograph of a one dimensional tip array having a phospholipid ink selectively delivered to every second tip in an array. Bottom: Fluorescence image of different fluorescently doped lipids (Rhodamine DOPE in red and FITC-DOPE in green) selectively applied to different tips on a parallel array. The 2 channel fluorescence signal is overlayed with the brightfield image of the cantilever array.

The rate of ink transport during the writing process can then be precisely controlled by lowering the humidity within a range of 40-75% for DOPC. In contrast to the transport behavior of most other DPN inks, where the dot and line width could be controlled by the tip-contact time and scan speed, respectively, as well as humidity, phospholipid inks tend to stack into multilayer structures where the thickness of the film can be controlled by those same parameters. Careful adjustment of the humidity and scan speed has enabled line widths down to 100 nm and thickness as low as a single bilayer.[4]

3   SUBSTRATE DEPENDANCE

In order to generate biomimetic lipid membranes from the lipid multilayer patterns it is necessary to immerse the patterns into aqueous solution. As biological membranes are fluid, it is generally undesirable to covalently stabilize the patterns for this purpose. However, depending on which substrate is used (and which type of biomembrane one wishes to mimic), the multilayer patterns can be immersed into water while retaining their multilayer structure, or they can be spread to form supported lipid bilayer membranes.

3.1 Supported Lipid Multilayers

An example of a heterogeneous lipid multilayer pattern that was written in parallel and immersed in water without loss of lateral resolution is shown in Figure 4, and a hypothetical structure of the multilayer patterns is drawn. From a physical chemistry perspective, the stability of the multilayer patterns can be understood in terms of substrate wettability. That is, the interfacial tensions involved (solid/water, solid/lipid, water/lipid) must be such that spreading is thermodynamically unfavorable, thus resulting in a non-zero contact angle.[34] The multilayer structures under water provide the possibility of reducing the friction of membrane bound materials with the substrate, as typically the goal of polymer supported lipid membranes.[35] Furthermore, the entirely new possibility of encapsulating materials within the multilayers becomes accessible, further characterization is necessary in order to test these capabilities as well as to determine the precise supramolecular structure of the multilayer patterns under water.

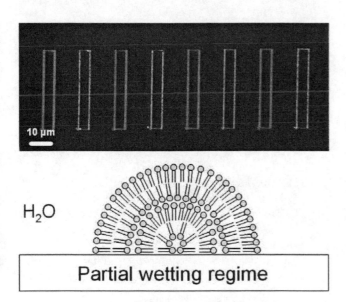

Partial wetting regime

Figure 4: Top: 2 channel fluorescence image of 2 different fluorescently labeled lipids patterned in parallel on a glass surface (used as received without further treatment) while immersed in water. The fine lateral structure of the patterns (visible down to ~200 nm by fluorescence) is preserved.
Bottom: hypothetical structure of the phospholipid multilayers on a partially wetting surface under water.

3.2 Spread Supported Lipid Bilayers

A better characterized example of an artificial biomimetic surface is the supported lipid bilayer formed by self-spreading. When a dehydrated multilayer stack is printed onto a hydrophilic surface, and immersed into water, it can spread to form a homogenous and fluid lipid bilayer.[36-42] The formation of a supported lipid bilayer in this way can be confirmed by fluorescence microscopy. The spreading of the small multilayer spots can be watched in real time, providing insights into dynamic membrane organization processes. Upon equilibration, the resulting thin films then show a homogeneous fluorescence intensity indicating the bilayer has been formed. A final test of the function of the bilayer is through fluorescence recovery after photobleaching experiments (FRAP), in order to determine the fluidity of the membranes as shown in Fig 5 (A-E). Once the bilayers have spread, the spreading stops and the bilayer patterns remain stable in water for at least several weeks. As a variety of functional lipids are readily available both from biological and synthetic sources, the combination of DPN with self-spreading therefore provides a reliable method for generating multi-component biomimetic membrane systems.

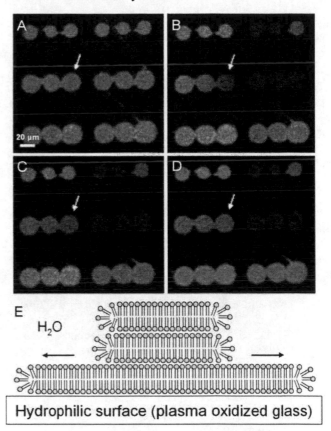

Hydrophilic surface (plasma oxidized glass)

Figure 5: Supported lipid bilayer patterns formed by spreading DPN deposited multilayers onto a hydrophilic surface. Top: shows fluorescence recovery after photobleaching. A) image before bleaching. B) image after bleaching. C) image after 10 minutes, and D) image after 20 minutes. The arrow shows the bleached and recovered part of the patterns. E) shows a hypothetical structure of the phospholipid multilayers spreading on a hydrophilic surface under water.

4 CONCLUSIONS

The use of phospholipids as an ink for DPN opens several new possibilities both for materials integration as well as the fabrication of biomimetic surfaces. The non-covalent interactions with the substrate make the ink generally applicable to a variety of surfaces. The humidity dependant nature of the inks fluidity makes it ideal for both multiplexed tip coating as well as writing. Furthermore, the supramolecular structure of the resulting micro and nanostructures can be dynamically modulated by the underlying substrate. The ubiquitous nature of phospholipids in biological systems provides a wide repertoire of functional materials that can be purified or synthesized, and now integrated on the appropriate micro and nanoscopic length scales.

REFERENCES

[1] K. Salaita, Y. H. Wang, C. A. Mirkin, *Nat Nanotechnol* **2007**, *2*, 145.

[2] D. S. Ginger, H. Zhang, C. A. Mirkin, *Angew Chem Int Edit* **2004**, *43*, 30.

[3] R. D. Piner, J. Zhu, F. Xu, S. H. Hong, C. A. Mirkin, *Science* **1999**, *283*, 661.

[4] S. Lenhert, P. Sun, Y. H. Wang, H. Fuchs, C. A. Mirkin, *Small* **2007**, *3*, 71.

[5] K. B. Lee, S. J. Park, C. A. Mirkin, J. C. Smith, M. Mrksich, *Science* **2002**, *295*, 1702.

[6] K. B. Lee, E. Y. Kim, C. A. Mirkin, S. M. Wolinsky, *Nano Lett* **2004**, *4*, 1869.

[7] L. M. Demers, C. A. Mirkin, *Angew Chem Int Edit* **2001**, *40*, 3069.

[8] J. Haaheim, R. Eby, M. Nelson, J. Fragala, B. Rosner, H. Zhang, G. Athas, *Ultramicroscopy* **2005**, *103*, 117.

[9] S. H. Hong, J. Zhu, C. A. Mirkin, *Science* **1999**, *286*, 523.

[10] A. Ivanisevic, C. A. Mirkin, *J Am Chem Soc* **2001**, *123*, 7887.

[11] K. H. Kim, N. Moldovan, H. D. Espinosa, *Small* **2005**, *1*, 632.

[12] N. Moldovan, K. H. Kim, H. D. Espinosa, *J Micromech Microeng* **2006**, *16*, 1935.

[13] K. S. Ryu, X. F. Wang, K. Shaikh, D. Bullen, E. Goluch, J. Zou, C. Liu, C. A. Mirkin, *Appl Phys Lett* **2004**, *85*, 136.

[14] K. Salaita, S. W. Lee, X. F. Wang, L. Huang, T. M. Dellinger, C. Liu, C. A. Mirkin, *Small* **2005**, *1*, 940.

[15] K. Salaita, Y. H. Wang, J. Fragala, R. A. Vega, C. Liu, C. A. Mirkin, *Angew Chem Int Edit* **2006**, *45*, 7220.

[16] D. L. Wilson, R. Martin, S. Hong, M. Cronin-Golomb, C. A. Mirkin, D. L. Kaplan, *P Natl Acad Sci USA* **2001**, *98*, 13660.

[17] K. B. Lee, J. H. Lim, C. A. Mirkin, *J Am Chem Soc* **2003**, *125*, 5588.

[18] L. M. Demers, D. S. Ginger, S. J. Park, Z. Li, S. W. Chung, C. A. Mirkin, *Science* **2002**, *296*, 1836.

[19] X. G. Liu, Y. Zhang, D. K. Goswami, J. S. Okasinski, K. Salaita, P. Sun, M. J. Bedzyk, C. A. Mirkin, *Science* **2005**, *307*, 1763.

[20] R. A. Vega, D. Maspoch, K. Salaita, C. A. Mirkin, *Angew Chem Int Edit* **2005**, *44*, 6013.

[21] Rafael A. Vega, C. K. F. Shen, D. Maspoch, Jessica G. Robach, Robert A. Lamb, Chad A. Mirkin, *Small* **2007**, *3*, 1482.

[22] A. N. Parikh, J. T. Groves, *Mrs Bull* **2006**, *31*, 507.

[23] J. T. Groves, N. Ulman, S. G. Boxer, *Science* **1997**, *275*, 651.

[24] C. K. Yee, M. L. Amweg, A. N. Parikh, *Adv Mater* **2004**, *16*, 1184.

[25] S. Majd, M. Mayer, *Angew Chem Int Edit* **2005**, *44*, 6697.

[26] V. Yamazaki, O. Sirenko, R. J. Schafer, L. Nguyen, T. Gutsmann, L. Brade, J. T. Groves, *Bmc Biotechnology* **2005**, *5*.

[27] P. Sharma, R. Varma, R. C. Sarasij, Ira, K. Gousset, G. Krishnamoorthy, M. Rao, S. Mayor, *Cell* **2004**, *116*, 577.

[28] B. L. Jackson, J. T. Groves, *J Am Chem Soc* **2004**, *126*, 13878.

[29] J. J. Shi, J. X. Chen, P. S. Cremer, *J Am Chem Soc* **2008**, *130*, 2718.

[30] K. D. Mossman, G. Campi, J. T. Groves, M. L. Dustin, *Science* **2005**, *310*, 1191.

[31] D. Falconnet, G. Csucs, H. M. Grandin, M. Textor, *Biomaterials* **2006**, *27*, 3044.

[32] D. J. Irvine, J. Doh, B. Huang, *Curr Opin Immunol* **2007**, *19*, 463.

[33] D. Banerjee, N. A. Amro, S. Disawal, J. Fragala, *J Microlith Microfab* **2005**, *4*, 230.

[34] F. Brochard-Wyart, in *Soft Matter Physics* (Ed.: C. E. W. M. Daoud), Springer-Verlag, Berlin Heidelberg New York, **1995**, pp. 1.

[35] M. Tanaka, E. Sackmann, *Nature* **2005**, *437*, 656.

[36] J. Radler, H. Strey, E. Sackmann, *Langmuir* **1995**, *11*, 4539.

[37] J. Nissen, S. Gritsch, G. Wiegand, J. O. Radler, *Eur Phys J B* **1999**, *10*, 335.

[38] J. Nissen, K. Jacobs, J. O. Radler, *Phys Rev Lett* **2001**, *86*, 1904.

[39] H. Nabika, A. Fukasawa, K. Murakoshi, *Langmuir* **2006**, *22*, 10927.

[40] K. Furukawa, K. Sumitomo, H. Nakashima, Y. Kashimura, K. Torimitsu, *Langmuir* **2007**, *23*, 367.

[41] M. C. Howland, A. W. Szmodis, B. Sanii, A. N. Parikh, *Biophys J* **2007**, *92*, 1306.

[42] B. Sanii, A. N. Parikh, *Soft Matter* **2007**, *3*, 974.

NSTI-Nanotech 2008, www.nsti.org, ISBN 978-1-4200-8503-7 Vol. 1

Commercially Available High-Throughput Dip Pen Nanolithography®

J. R. Haaheim*, O. A. Nafday*, V. Val*, J. Fragala**, R. Shile**

*NanoInk, Inc., Skokie, IL, USA, jhaaheim@nanoink.net, onafday@nanoink.net, vval@nanoink.net
**NanoInk, Inc., Campbell, CA, USA, jfragala@nanoink.net, rshile@nanoink.net

ABSTRACT

Dip Pen Nanolithography® (DPN®) is an inherently additive SPM-based technique which operates under ambient conditions, making it suitable to deposit a wide range of biological and inorganic materials. Massively parallel two-dimensional nanopatterning with DPN is now commercially available via NanoInk's 2D nano PrintArray™, making DPN a high-throughput, flexible and versatile method for precision nanoscale pattern formation. By fabricating 55,000 tip-cantilevers across a 1 cm² chip, we leverage the inherent versatility of DPN and demonstrate large area surface coverage, routinely achieving throughputs of 3×10^7 µm² per hour. Further, we have engineered the device to be easy to use, wire-free, and fully integrated with the NSCRIPTOR's scanner, stage, and sophisticated lithography routines. We herein discuss the methods of operating this commercially available device, subsequent results showing sub-100 nm feature sizes and excellent uniformity (standard deviation < 16%), and our continuing development work. Simultaneous multiplexed deposition of a variety of molecules is a fundamental goal of massively parallel 2D nanopatterning, and we will discuss our progress on this front, including ink delivery methods, tip coating, and patterning techniques to generate combinatorial libraries of nanoscale patterns. Another fundamental challenge includes planar leveling of the 2D nano PrintArray, and herein we describe our successful implementation of device viewports and integrated software leveling routines that monitor cantilever deflection to achieve planarity and uniform surface contact. Finally, we will discuss the results of 2D nanopatterning applications such as: 1) rapidly and flexibly generating nanostructures; 2) chemically directed assembly and 3) directly writing biological materials. We will demonstrate flexibly generated nanostructures that are useful in that they complement and exceed the capabilities of existing techniques (e.g., nano imprint lithography and e-beam lithography); we will also demonstrate nanostructures that are valuable for plasmonics surface studies, particularly SERS enhancement.

Keywords: Dip Pen Nanolithography, DPN, Scanning Probe Lithography, SPL, Scanning Probe Microscopy, SPM, AFM, nanoscale lithography, nanoscale deposition, direct deposition, nanofabrication

1 INTRODUCTION

Dip Pen Nanolithography is NanoInk's patented process for deposition of nanoscale materials onto a substrate. The DPN process uses a coated scanning probe tip (the "pen") to directly deposit a material ("ink") with nanometer-scale precision onto a substrate [1] (Fig. 1a). Fig. 1b demonstrates this concept scaled to portray two-dimensional (2D) arrays of tips. The vehicle for deposition can include pyramidal scanning probe microscope tips, hollow tips, and even tips on thermally actuated cantilevers. It is an amazingly robust and versatile technique, and can deposit a variety of organic and inorganic molecules onto a variety of substrates [2] under ambient conditions (Fig. 1). Further, thermal DPN (tDPN) grants access to an even wider range of ink materials by enabling solid ink deposition via a heated tip [3].

Table 1 provides an instructive look at DPN's place among nanopatterning techniques: it is highly scalable with the use of multi-pen arrays; it is a technique that enables both bottom up nanofabrication (self-assembly,

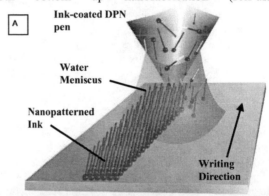

Fig. 1: (a) Schematic of the Dip Pen Nanolithography (DPN) process. A molecule-coated AFM tip deposits ink via a water meniscus onto a substrate. (b) Schematic representation of the DPN process scaled up for massively parallel nanopatterning. The graphic depicts the ultimate aim of rapidly creating a variety of structures on the fly, with different inks on each tip.

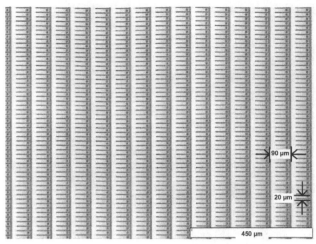

Fig. 2: Optical microscope image of the 2D nano PrintArray (tips facing up) showing the pitch, spacing, and high yield. 832 individual tips are shown, roughly 1.5% of the entire array.

templating) [4] and top down fabrication via etch resist-based "inks" [5]; and it is high resolution (14 nm line widths, 20 nm pitches) [6]. DPN is a direct-write technique, so materials of interest can be placed exactly (and only) where desired. Among sub-50 nm techniques – such as e-beam lithography – DPN is the only one that can directly deposit molecules under ambient conditions [1, 7, 8]. Further, NanoInk's platform system, the NSCRIPTOR™ (Fig. 2), is an instrument and software package enabling nanoscale registry and alignment, sophisticated CAD design, and high quality AFM imaging.

Since DPN's inception in 1999 [1], a great deal of research territory has been explored; we will not attempt to chronicle that body of work here, as several excellent reviews have already done so [9-14]. Rather, we will focus on the development of enabling technology – namely, NanoInk's 2D nano PrintArray. From that point we will elaborate several applications enabled by this fundamental capability of massively parallel nanopatterning.

2 MASSIVELY PARALLEL NANOPATTERNING INNOVATIONS

While the DPN process has been successfully demonstrated with a variety of capabilities in academic and government research labs, critics had naturally pointed out its initially serial nature, and as such have argued against its commercial viability. Recently, with a view to overcome the serial nature of the DPN process, we initiated efforts to perform massively parallel nanopatterning with cantilever arrays. The resulting collaboration with the Mirkin group at Northwestern University produced a vital proof-of-principle: massively parallel DPN patterning over cm^2 areas retains essentially all of the critical attributes of single pen DPN [15]. With throughput exceeding $1 \times 10^7 \ \mu m^2/hr$, and a dot size standard deviation of only 16%, they demonstrated sub-100 nm massively parallel nanoscale deposition with a 2D array of 55,000 pens on a centimeter square probe chip (Figs. 2-3). In that work [15] an image

Fig. 3: SEM image showing multiple rows of cantilevers attached to silicon ridges. The inset shows individual cantilevers, while also highlighting the 7.5 um tall sharpened tips and inherent cantilever curvature (~ 14 μm bow).

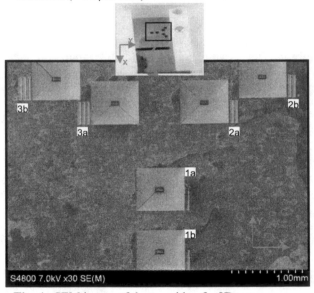

Fig. 4: SEM image of the top side of a 2D nano PrintArray, showing all six viewports as oriented when looking through the NSCRIPTOR optics. An optical image of the device with the wedge is shown for reference and orientation.

of the Jefferson nickel was imported into InkCAD™ software (the NSCRIPTOR system interface), transformed to a map of dots, and then 55,000 identical patterns were generated with ODT ink. The patterned ODT later served as an etch resist on the gold layer. This massively parallel approach to DPN works because DPN is effectively force-independent, and thus generally forgiving with respect to probe array leveling with low-spring-constant silicon nitride cantilevers. To date, there is no other way to flexibly pattern a variety of materials at this unprecedented resolution (80 nm). Additionally, 55,000 tips in 1 cm^2 is the highest cantilever density ever reported. Fundamentally, this enables flexible direct-writing with a variety of molecules and simultaneously generating 55,000 duplicates at the resolution of single-pen DPN.

In spite of the impressive results detailed above, several prominent engineering hurdles stood between the devices these groups used and a robust commercial offering. Significant challenges included facile device mounting, routinely and accurately leveling the array with respect to the substrate, ensuring uniform contact of all of the tips when the array is meant to write, and making sure no tips are touching when the array is retracted. To overcome these challenges, we introduced etched viewports (Fig. 4), a precisely machined magnetic attachment wedge (Fig. 4 inset), and semi-automated leveling routines. We have engineered the device to be easy to use, wire-free, and fully integrated with the NSCRIPTOR scanner, stage, and sophisticated lithography routines. Unlike previous prototypes, it is now possible to view the substrate through the handle wafer, align to pre-existing surface features (such as inkwells), and in principle even align a laser to a cantilever for imaging. Massively parallel two-dimensional nanopatterning with DPN is now commercially available via NanoInk's 2D nano PrintArray™ (Figs. 2-3), making DPN a high-throughput, flexible and versatile method for precision nanoscale pattern formation.

Fig. 5 shows the results of this type of nanofabrication with the 2D nano PrintArray. Leveling is one of the most important aspects for successfully using the 2D nano PrintArray. The 2D nano PrintArray exists as a flat 1 cm^2 square chip whose goal is to be brought into uniform contact with a substrate. The sophisticated software control (Fig. 6a) easily enables the degree of planarity required to level the device, and to fine-tune the z-positioning prior to lithography. Leveling is accomplished by examining cantilever deflection through these viewports (Fig. 6b) at three different points, noting the z-height differences, and then entering these numbers into the software to calculate the planarity corrections of the three z-motors. Contrasted with earlier methods, the viewport leveling takes only a few minutes; the NanoInk logo data shown in Fig. 5 was generated in under 30 minutes, from mounting the probes to

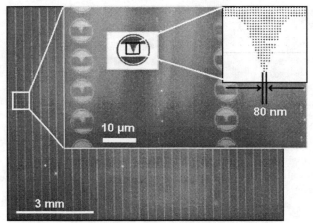

Fig. 5: Optical images of the NanoInk logo following the successful process of mounting, leveling, and printing ODT with the 2D nano PrintArray. The design includes 2250 dots of 80 nm diameters. Patterns were generated using ODT as the etch-resist. Only a fraction of the 55,000 printed logos are shown.

finishing the etch-resist process.

3 APPLYING 2D DPN

The 2D nano PrintArray's capabilities are constantly evolving, and the above represent only a sampling of what is possible. 2D nanopatterning currently falls into three broad categories: 1) rapidly and flexibly generating nanostructures (e.g., Au, Si) via etch resist techniques; 2) chemically directed assembly and patterning templates for either biological molecules (e.g., proteins, viruses, cell adhesion complexes), or inorganics (e.g., carbon nanotubes, quantum dots); and 3) directly writing biological materials.

Using established templating techniques, these advances enable screening for biological interactions at the level of a

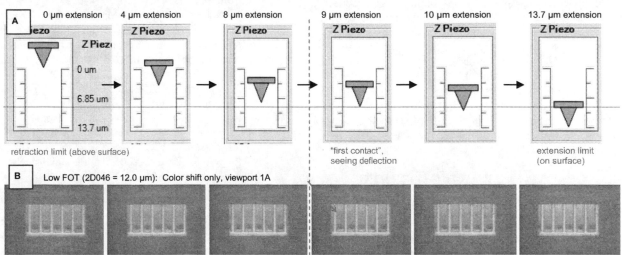

Fig. 6: Mapping the visual progression of cantilever deflection for a viewport; (a) The sequence of positions of the Z-piezo tool used to bring the cantilevers of a given viewport into contact with the surface. (b) These cantilevers are less curled (FOT = 12.0 μm), but display a dramatic color shift across the whole cantilever. At the point of first contact, note the subtle color and shade change at the base of the cantilever (shown inset). These subtle shifts become obvious when the Z-piezo is repeatedly extended to 9.0 μm and retracted. The shift becomes dramatic at an extension of 13.7 μm.

few molecules, or even single molecules. This in turn can enable engineering the cell-substrate interface at sub-cellular resolution. This technology allows users to routinely pattern libraries of small molecules over very large areas, and realistically practice single cell experimentation. Using 2D nanopatterning, the process is scalable and can cover large areas for statistically significant investigations of these individual bioprocesses. For example, DPN-generated arrays have been demonstrated to monitor single-cell infectivity from virus-particle nanoarrays [16]. In this work, Vega and coworkers immobilized antibodies on DPN-patterned MHA-Zn_2^+ regions. These nanoarrays were later incubated with fluorescent SV5 viral solution. The resulting virus nanoarrays were used for CV1 cell infectivity studies.

Further, because these inks can be used as etch-resist materials [17],[18], we can perform maskless rapid prototyping across large areas, forming combinatorial arrays of metallic or solid-state features varying in size, spacing, and shape. There are a variety of useful things one can do once one has the capability to rapidly generate arbitrary gold nanostructures on silicon oxide across a cm^2 area. Numerous researchers in the field of SERS would benefit from a method of quickly generating arrays of noble metal nanostructures. The most important requirement for a SERS substrate is its ability to increase the electromagnetic field at the surface; nanoscale noble metal structures (e.g., Ag, Au) have this ability through their interaction with light that has been tuned to the resonance frequency of the conductive electrons surrounding the metal. The field enhancing properties of these structures are particularly sensitive to the structures' size, shape, environment, inter-structure distance, and inter-particle distance. As such, it is extremely desirable to have the ability to fabricate such arbitrary metallic patterns – a clear strength of massively parallel DPN. Such patterns are typically fabricated by slow and often costly serial e-beam lithography. More crude methods, such as nanosphere lithography (i.e., "polystyrene drop-coating"), have reduced costs considerably, but only by sacrificing reproducibility. Further, this 2D etch-resist technique can generate very small metallic structures next to very large ones – something Nano Imprint Lithography (NIL) has a difficult time accomplishing. Finally, this 2D patterning method is not limited to any particular shape: notably, we can generate closely spaced arcs and circles, which is a weakness of e-beam lithography. All of these approaches are maskless with a quick-turn time, flexible, inexpensive, and require little to no chemistry expertise.

4 SUMMARY AND OUTLOOK

With the 2D nano PrintArray, we are advancing DPN as a technique for high-throughput nanopatterning. With such technology now proven and in practice, desirable future developments could include laser feedback on viewable cantilevers for immediate imaging, and automated step-and-repeat lithographic routines. But simultaneous multiplexed deposition of a variety of molecules remains a fundamental goal of massively parallel 2D nanopatterning. We are currently involved in research to demonstrate this application, although significant engineering challenges are involved in getting several hundred or thousand different molecules onto different tips. Massively parallel multiplexed DPN, enabled by multiplexed selective ink delivery, is a fundamental requirement of a variety of biological applications, and a direction of important development. In this regard, we anticipate the need for universal inks of nearly identical properties to ensure even fluidic control, tip loading, and ink transport from the tip. We are approaching this ink delivery challenge through a variety of methods, including InkTrough™ channels and different vapor coating techniques. Such a capability would enable multiplexed combinatorial libraries of nanoscale patterns across large areas.

REFERENCES

1. R.D. Piner, et al., Science, 1999. **283**(5402): p. 661-663.
2. Ginger, D.S., H. Zhang, and C.A. Mirkin, Angew. Chem. Int. Ed., 2004. **43**: p. 30-45.
3. Sheehan, P.E., et al., Appl. Phys. Lett., 2004. **85**(9): p. 1589-1591.
4. Demers, L.M., et al., Angew. Chem. Int. Ed., 2001. **40**(16): p. 3071-3073.
5. Zhang, H., S.W. Chung, and C.A. Mirkin, Nano Lett., 2003. **3**(1): p. 43-45.
6. Haaheim, J., et al., Ultramicroscopy, 2005. **103**: p. 117-132.
7. Hong, S.H., J. Zhu, and C.A. Mirkin, Science, 1999. **286**(5439): p. 523-525.
8. Haaheim J., et al., Ultramicroscopy, 2005. **103**(2): p. 117-132.
9. D.S. Ginger, H. Zhang, and C.A. Mirkin, Angewandte Chemie-International Edition, 2004. **43**(1): p. 30-45.
10. Huck W.T.S., Angewandte Chemie-International Edition, 2007. **46**(16): p. 2754-2757.
11. J. Haaheim and O.A. Nafday, Scanning, 2008. **In press**.
12. K. Salaita , Y. H. Wang, and C. A. Mirkin Nature Nanotechnology, 2007. **2**(3): p. 145-155.
13. Lenhert S., in *Nanotechnology, Vol. 2 - Nanoprobes*, H. F., Editor. 2008, WILEY-VCH Weinheim: Berlin. p. In print.
14. Rosner B.T. and Demers L.M., in *Dekker Encyclopedia of Nanoscience and Nanotechnology*, Schwarz J.A., Contescu C.I., and Putyera K., Editors. 2005, Taylor and Francis Group: New York. p. 1-14.
15. Salaita K.S., et al., Angewandte Chemie-International Edition, 2006. **45**(43): p. 7220-7223.
16. Vega R.A., et al., Small, 2007. **3**(9): p. 1482-1485.
17. Zhang H., Chung S. W., and Mirkin C. A., Nano Letters, 2003. **3**(1): p. 43-45.
18. K.S. Salaita, et al., Nano Letters, 2006. **6**(11): p. 2493-2498.

Nanolithographic Patterning of Catalysts for Synthesis of Carbon Nanotubes

Rohit V. Gargate *, Debjyoti Banerjee **

* Mechanical Engineering Department, Texas A&M University, College Station, TX 77843-3123

ABSTRACT

A novel process was developed for synthesis of carbon nanotubes (CNT) which does not require the traditional Chemical Vapor Deposition (CVD) synthesis techniques. Catalyst precursors were deposited in bulk using a "wet process" on the MEMS enabled micro-heater elements. The microheaters were coated with a layer of fullerene apriori (of 150 nm thickness) using Physical Vapor Deposition (PVD). The chip was then heated using integrated microheaters and external heaters in an inert atmosphere to obtain CNT. Thus, in this process we obviate the Chemical Vapor Deposition (CVD) process for synthesis of CNT (SWCNT and MWCNT). This work also proves the feasibility for a portable hand held instrument for synthesis of CNT "on demand".

Keywords: Carbon nanotubes, Dip Pen Nanolithography, MEMS, Nanotechnology, Raman spectroscopy.

1 INTRODUCTION

Carbon nanotubes (CNT) were discovered by Ijima using the arc-discharge process in 1991 [1], and since then, various techniques have been devised to synthesize carbon nanotubes. CNT are considered for various applications owing to their novel mechanical [2] and electronic properties [3]. Processes like Chemical Vapor Deposition or "CVD" [4], and the HiPCo process [5] were developed and optimized subsequently. Mass production of nanotubes would allow their commercial use [6]. Furthermore if the synthesis process can be controlled so as to have a desired type of nanotubes with controlled diameters, properties of nanotubes could be even more exploited.

This work is an extension of our earlier investigation for using Dip Pen Nanolithography (DPN) techniques [7-9] for the synthesis of CNT and controlling the chirality of the synthesized CNT [10-12]. In this paper, we try to present a refined process for the synthesis, which overcomes some of the hurdles mentioned in the commercial processes described above.

2 EXPERIMENTAL APPARATUS

Custom designed MEMS (Micro Electro Mechanical Systems) platforms were used for the synthesis process for carbon nanotubes (the MEMS substrates were fabricated by MEMS Exchange Inc.). VA. These platforms were in the form of square chips. The dimension of the substrate was 3.5 x 3.5 x 0.5 mm. Three types of serpentine heaters were designed for the substrate. Titanium was used as the material for the heater. Three different types of heaters with

variation in length and orientation were used on the substrate. The lengths of the microheaters are 18 mm, 19 mm and 42 mm. The thickness of the heater elements is 100nm.

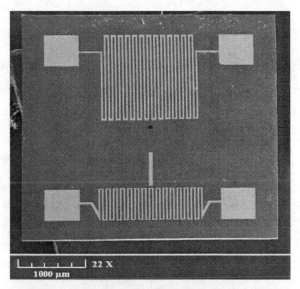

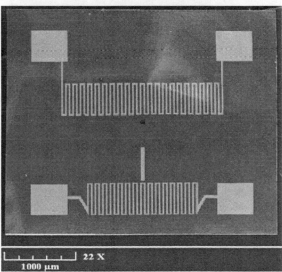

Figure 1. Scanning Electron Microscopy (SEM) images of the two types of substrates used in the synthesis of Carbon Nanotubes (CNT). There are three different types of micro-heaters microfabricated insitu on the silicon nitride membrane, a backside square opening etched in silicon for exposing the membrane at the center, and a vertical alignment marker (gold) perpendicular to the heater aligned with the membrane.

Two types of MEMS platforms (substrates) were fabricated for the synthesis of the CNT, with each substrate having two heaters on either edge of the top side. A silicon nitride membrane was deposited on the top side in the middle portion of the silcon chip. The membrane had a back side opening of 60 µm x 60 µm square. A metal pad was deposited near the edge of the square membrane and perpendicular to the heaters - using "lift-off" process to serve as an 'alignment mark'. The chip structure is shown in **Figure 1**.

The Mbraun Labmaster double glove box system with gas purification (located at the Center for Nanomanufacturing, University of Texas, Austin); was used for the vapor deposition of fullerene layer on this chips. This instrument consists of a two chamber glove box system. A spin coater (specialty coating systems model P6700) is available inside the smaller glove box chamber. An Edwards's thermal deposition chamber with four sources is available inside the larger glove box. A vacuum oven is mounted at one end of the glove box. Multiple chips are mounted on a glass slide with a double sided tape, and this glass slide is placed inverted on the roof of in the thermal deposition chamber. Fullerene powder (Manufactured by Nano-C Inc., Westwood, MA) was placed in the target container on the chambers base. The container was electrically heated at a pressure of 1 mTorr to a temperature of 550 °C to enable sublimation of the fullerene powder. The deposition rate by sublimation was estimated to be 20 nm/s for a total deposition time of 10 s. The fullerene thickness was estimated to be 150 nm at the end of the vapor deposition process using a profilometer.

Metal catalysts (Nickel chloride, Cobalt Chloride, Palladium Chloride) were prepared by mixing the solid powder into water with minimal heating for uniformity. Sodium Hydroxide was used in the preparation of the $PdCl_2$, since it doesn't form a direct solution with water.

A digital DC power supply with enough power rating is used, which has displays for monitoring the current and voltage. A wire thermocouple is used to the measure the temperature, with the welded end on the chip and the open end connected to a thermocouple reader.

MINCO mica external heaters were used to raise the temperature of the coated substrates in the CNT synthesis experiments. Suspended in mid-air Individual heaters can be heated to temperatures of around 200 °C, So three heaters were mounted together , stacked above each other and connected electrically in parallel. The heater assembly was used to heat the chips to 550 °C, required for the synthesis process.

A dessicator is used as the chamber for the synthesis process. The dessicator has two openings. One of this is connected to a Nitrogen tank and the other serves as an inlet for the electrical connections. The external heaters, Chips, and the thermocouples are placed in the dessicator, 2 alligator pin clamps are used for suspending the heaters and positioning the thermocouple. Also a humidity sensor is positioned inside the dessicator.

3 EXPERIMENTAL PROCEDURE

It has been demonstrated experimentally that Carbon Nanotubes (CNT) can be synthesized with Fullerene as the source for carbon, where the synthesis temperature is ~500°C [13]. Metal catalysts also play a significant role in the formation of carbon nanotubes. The catalyst particles serve as nucleation sites for formation of CNT from fullerene [14]. Nickel Chloride, Cobalt Chloride, and Palladium chloride were used as catalyst precursors in this experiment. These precursors were prepared in an aqueous solution as mentioned above. Each catalyst was deposited as a bulk layer on separate fullerene coated chips with the aid of a syringe. The catalyst precursor solution allowed to dry at room temperature after dispensing it on the fullerene coated chips.

In an initial attempt, heater connections were made across the bond-pads of one of the heater per chip. The connections were possible with normal electrical wires available commercially with very small diameters, in accordance with the bond-pad areas. The apparatus was placed in a dessicator and connected to nitrogen supply from a pressurized cylinder. The experiments were performed in an inert atmosphere by passing the nitrogen into the dessicator. A humidity sensor was also placed inside the dessicator. The relative humidity in the air was 38.2% at the beginning before passing nitrogen gas. After a duration of 20 minutes, the humidity was found to be 3.8% inside the dessicator. At this stage, current was passed through the microheaters in the chip. However, the chips were damaged when they reached a temperature of ~300 °C due to stress concentration. Since this temperature was not suitable enough for CNT synthesis, external MINCO heaters were therefore used in this process. The external heaters connected in a parallel assembly as discussed above. The chips were then placed on the MINCO heater assembly for direct contact heating. The temperature of the heaters was then raised slowly in steps of 10 °C per minute until the temperature reached 510 °C. It has been reported that lower synthesis temperatures yield a tighter distribution of the diameters (and correspondingly the chirality) of the synthesized CNT [15]. The heater assembly was maintained at this temperature for 15 minutes and the power supply was then cutoff, along with the flow of nitrogen. The characterization of these samples was then carried out using Raman Spectroscopy and Scanning Electron Microscopy.

4 RESULTS AND DISCUSSION

The synthesized CNT can be characterized to determine the material properties using several technqiues which include SEM/ TEM, Raman spectroscopy, absorption spectroscopy, etc. SEM/TEM and Raman spectroscopy are widely used for their convenience and ease. Imaging techniques (which includes SEM and TEM) provides a direct visual observation. Raman spectroscopy provides distinct information of the vibrational modes

NSTI-Nanotech 2008, www.nsti.org, ISBN 978-1-4200-8503-7 Vol. 1

which are detected using peaks in known Raman shifts in wavenumber. These peaks are seen at three locations, first being the Graphite Peak (G-Peak) around 1600 cm⁻¹, second, Defect Peak (D-Peak) around 1350 cm⁻¹, and third, the Radial Breathing Modes (100-300 cm⁻¹). RBM peaks are caused by the vibration along the radial direction of the individual CNT, and are widely used to obtain information about the distribution of the diameters of the synthesized single walled CNT (SWCNT) [16]. When the samples were excited by 633 µm laser, after the experiment (Model: LabRam IR Confocal Raman Spectroscopy - Microscopy Instrument, Manufacturer: JY Horiba, Japan), the above mentioned peaks in RBM, G and D bands were distinctly visible for the experiments using Palladium catalyst. **Figure 2** shows the obtained Raman spectra.

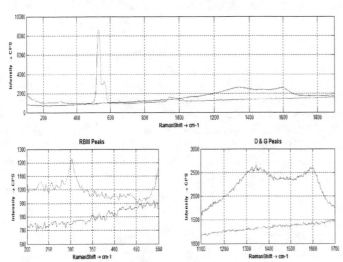

Figure 2: The Raman spectra of the sample before (in Red) and after (in Blue) the synthesis of CNT from Fullerene coated chips.

Figure 3: Image of the PdCl2 bulk deposited sample. Conductive copper tape used for mounting the sample is observed at the top. The conductive tape is used to minimize the effects of charge trapping.

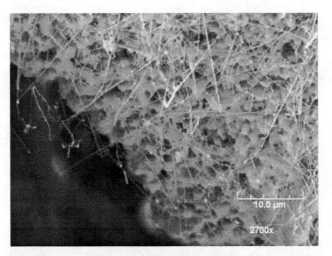

Figure 4: Top: The synthesized Carbon Nanotubes (CNT) are observed on the edge of the MEMS chip. Bottom: Image obtained at a higher magnification shown the synthesis experiment yields carbon nanotube bundles (SWCNT and MWCNT) and carbon nano-fibers which are estimated to be 20-200 nm in diameter.

As seen from **Figure 2** the peaks at 1340 & 1600 cm⁻¹ are clearly visible. Since the D band is more intense, it indicates presence of multi-walled carbon nanotubes [11, 12]. The peak at 300 cm⁻¹ seen in the RBM region and the peak at 520 cm⁻¹ in indicative of silicon, which is the substrate material. For confirmation of the data from Raman spectra, the Palladium samples were observed using Scanning Electron Microscopy. **Figure 3** shows the over all view of the bulk deposited sample after the experiment. **Figure 4** shows that the diameters of carbon nanotubes obtained in this synthesis experiment ranged from 20 nm to 200nm. This variation of the diameter of the carbon nanotubes and the haphazard location mixed with carbon fibers was expected owing to the bulk deposition of the metal catalyst. Energy dispersive Spectroscopy (EDS) techniques were used to analyze the elemental composition of the sample. Data obtained by EDS unit (inbuilt into the SEM apparatus) is shown below in Table 1.

Table 1

Element	Weight(%)
Silicon (Si)	13.11
Palladium (Pd)	3.67
Carbon (C)	83.22
Total	**100.00**

5 CONCLUSION & FUTURE DIRECTION

In this work we have developed a novel process for synthesis of carbon nanotubes (CNT) without requiring the conventional synthesis techniques such as Chemical Vapor Deposition (CVD) process or the process gasses that are used in conventional synthesis techniques. CNT can be synthesized at temperatures of around 500°C. The bulk deposition of the catalyst over vapor deposited fullerene layer yields a mixture of carbon nanotubes with different diameters. By optimizing this process for catalyst deposition, with the help of techniques like Dip Pen Nanolithography, it would be possible to synthesize carbon nanotubes of a desired diameter and hence a 'specific-chirality' at desired locations.

Acknowledgement: The authors gratefully acknowledge the guidance and financial support by Dr. Amit Lal from the Micro Technology Office (MTO) of the Defense Advanced Research Projects Agency (DARPA). The authors also gratefully acknowledge the financial support and monitoring by Dr. Ryan Lu, Mr. Richard Nguyen and Mr. Al Navasca from the Space and Naval Warfare Center (SPAWAR). Support from the Office of Naval Research (ONR) is also acknowledged. Support from the Mechanical Engineering Department at Texas A&M University and the Texas Engineering Experimentation Station (TEES) is also gratefully acknowledged. The FE-SEM acquisition was supported by the National Science Foundation under Grant No. DBI-0116835. During the course of execution of this study, the authors and their research activities were also supported partially through various other research programs and are duly acknowledged here: National Science Foundation (CBET Grant No. 0630703), Micro/Nano-Fluidics Fundamental Focus Center (DARPA-MF3) through the University of California at Irvine, New Investigations Program (NIP 2005) of the Texas Space Grants Consortium (TSGC), Office of Naval Research (ONR), Air Force Office of Scientific Research (AFOSR) through the American Society for Engineering Education (ASEE), the Summer Faculty Fellowship (SFFP) Program at the Air Force Research Labs. (AFRL), Army Research Office (ARO) SBIR Phase II subcontract through Lyntech Inc. and the National Science Foundation (NSF) SBIR Phase I through NanoMEMS Research LLC. We also thank Dr. S. Sinha from the Department of Physics at the University of New Haven for helpful discussions on catalysis and CVD of CNT.

6 REFERENCES

[1] Iijima S.: Helical Microtubules of Graphitic carbon, Nature(1991) 354 56-58.
[2] Salvetat J.: Mechanical properties of carbon nanotubes, Appl. Phys. (1999) 69 255-260.
[3] P.G. Collins., et al: Nanotubes for electronics, Scientific American (2000) 283, 62-69.
[4] Ebbesen T, and Ajayan, P: Large-Scale Synthesis of Carbon Nanotubes. Nature(1992) 358 220–222.
[5] Ouellette J: Building the Nanofuture with Carbon Tubes. The Industiral Physicist,(2002) December 18-21.
[6] Robertson J: Growth of nanotubes for electronics. Materials Today 10, (2007) No 1-2, 36-45.
[7] D. Banerjee, N. Amro, and J. Fragala, "Optimization of microfluidic ink-delivery apparatus for Dip Pen Nanolithography™", *SPIE Journal of Microlithography, Microfabrication and Microsystems ("JM³")*, (2005), vol. 4, pp. 023014-023021.
[8] B. Rosner, T. Duenas, D. Banerjee, R. Shile, and N. Amro, "Functional extensions of Dip Pen Nanolithography™: active probes and microfluidic ink delivery", *IEEE Journal of Smart Materials and Structures*, (2005), Vol. 15, page S124-S130.
[9] J.A. Rivas-Cordona and D. Banerjee, "Microfluidic device for delivery of multiple inks for Dip Pen Nano-lithography", *SPIE Journal of Micro/Nanolithography, MEMS and MOEMS, ("JM³")*, (2007), Vol. 06, No. 03, pp. 033004-12.
[10] D. R. Huitink, D. Banerjee, and S.K. Sinha, "Precise control of carbon nanotube synthesis of a single chirality", *Paper No. IMECE2007-42588, Proceedings of ASME-IMECE, 2007*, Nov. 11-15, Seattle, WA.
[11] R. Gargate and D. Banerjee, "In-Situ synthesis of carbon nanotubes on heated scanning probes using dip pen techniques", *Scanning: The Journal of Scanning Probe Microscopies,* 2008 (in print).
[12] R. Gargate and D. Banerjee, SPIE DSS
[13] R.E. Morjan, O.A. Nerushev, M. Sveningsson, F. Rohmund, L.K.L. Falk, E.E.B. Campbell: Growth of carbon nanotubes from C60, Appl. Phys. A 78, 253–261 (2004).
[14] [Yiming Li, et al,: Growth of Single-Walled Carbon Nanotubes from Discrete Catalytic Nanoparticles of Various Sizes,J. Phys. Chem. B (2001) 105, 11424-11431].
[15] Robertson J: Growth of nanotubes for electronics. Materials Today 10, (2007) No 1-2, 36-45.
[16] Lefrant S., et al: RAMAN & SERS studies of Carbon notubes,Spectroscopy of Emerging Materials (2004), 127-138
[17] Jorio A, Saito R, Dresselhaus G, Dresselhaus M, "Determination of nanotubes properties by Raman spectroscopy", Philosophical Transactions on the Royal Society of London Part A, 362, 2311-2336 (2004).
[18] Thomsen C, Reich S, Maultzsch J, "Resonant Raman spectroscopy of nanotubes", Philosophical Transactions o the Royal Society of London Part A, 362, 2337-2359 (2004).

A Brief History of Thiols: An Assembly of Self-assembly

D. Graham[*]

[*]Asemblon Inc, 15340 NE 92nd Street; Suite B,
Redmond WA, USA, dgraham@asemblon.com

ABSTRACT

For decades, thiols have been used to control surface chemistry and study surface interactions in a wide range of fields. The elegantly simple method of forming thiol self-assembled monolayers has enabled researchers to unleash their creative minds and begin to tap into true surface engineering designs. This brief overview will highlight the history of the use of thiols in self-assembly from simple solution deposition of homogeneous monolayers to the creation of patterned surfaces with UV photopatterning, stamps, and dip pen lithography. The versatility and simplicity of thiol self-assembly has kept the method in the forefront as one of the most common surface modification methods, and has lead to new developments and opportunities in nanotechnology.

Keywords: self-assembly, surface chemistry, patterning, surface functionalization

1 EARLY THIOL HISTORY

The purpose of this document is not to provide a comprehensive summary of alkanethiols and their uses, but to give a sampling of the history of alkanethiol self-assembly and highlight some of the accomplishments and discoveries found from their use. Well written papers already exist in the literature that provide a comprehensive review of alkanethiols and their uses[1-3].

The history of self assembly is often started by discussing the early work of Zisman [4]. Zisman truly was a pioneer in the area of self-assembly and made very significant and important contributions to the literature. His earliest publication in 1946 on self-assembly introduced the assembly of a long chain alcohol, amine and carboxylic acid on glass and clean metal surfaces. From Zisman's work it is typical to jump to the 1980's when alkyltrichlorosilanes on silicon oxide were introduced [5] and then to the introduction of alkanethiols on gold by Nuzzo and Allara [6]. However it is interesting to note that alkanethiols were used for self-assembly long before Nuzzo and Allara published their first paper on the subject in 1983. The earliest mention of 'self-assembly' of thiols that I could find is U.S. patent 2,841,501 titled 'Silver Polish' published in 1958 [7]. While this patent is far from a scientific publication, the author had keen insight into the assembly mechanism of alkanethiols, in this case on silver. In the patent the author describes, though excusing himself from presenting proof, the assembly mechanism, chemical reaction, stabilizing factors, and minimum chain length to form a stable self-assembled monolayer (SAM) on silver. It is interesting to note the accuracy of the author's hypotheses with regards to these ideas. The author describes the formation of a silver thiol bond (Ag-S-R) once the oxide has been removed by the abrasives in the polish. It is known that sulfur forms a semi-covalent bond with gold with a bonding energy around 45 kcal/mol [8]. The author then describes that after rinsing off the polish the resulting coating consists of thiol molecules that are aligned with their alkyl tails standing up away from the surface. This ordered arrangement of the thiol chains has been established by surface analytical methods such as infrared spectroscopy [9]. The author then gives insight into the factors that affect stability of the layers stating that straight chain alkanethiols work best since they can pack closer together. Finally the author states that alkanethiols with chains of 12 carbons or longer work best and form the most stable layers. Interestingly this corresponds well with current research that shows that in order to get an ordered monolayer one should use alkanethiols with a chain length of at least 10 carbons [10]. It is not know whether the author of the patent was aware of the work of Zisman, however he did show good insight into what was happening on the surface.

Though the use of alkanethiols on metals was potentially being used by silver owners around the country, their use in the scientific literature did not surface for many years after the 'Silver Polish' patent was issued. Fortunately the pioneering work of Nuzzo and Allara kick started a literal explosion of publications with their seminal work in 1983 demonstrating the assembly of disulfides onto gold substrates [6].

2 MODEL SURFACES AND CHARACTERIZATION

After the discovery of alkanethiol assembly on gold, there was a rush of papers characterizing the surfaces of alkanethiol monolayers. Due to the relative ease in controlling surface chemistry using alkanethiols, they have been used as controlled surfaces for studies using practically every surface analytical method imaginable. Alkanethiol SAMs have been characterized by RAIRS, Raman, XPS, FTIR, HREELS, NEXAFS, helium atom scattering, X-ray diffraction, contact angle goniometry, ellipsometry, SPR, ToF-SIMS, AFM, STM, electron diffraction, SFG, and electrochemistry, just to name a few. Alkanethiols have been used to form monolayers on a wide variety of substrates including gold, silver, copper, nickel, platinum, palladium, mercury, zinc selenide, and cadmium selenide. This flexibility and wealth of surface analytical data has been a great advantage for users of alkanethiol SAMs as it provides a plethora of data and experimental details with which to compare one's own surfaces.

Most early studies of thiol SAMs focused on the structure and assembly mechanism of single and dual component monolayers on gold[10-14]. These and other studies led to the well recognized structure of alkanethiol SAMs shown in Figures 1 and 2 [15].

Understanding the structure of SAMs and discovering the ease and versatility of their use supplied researchers with a tool that could enable their creativity and has resulted in a rapid growth of unique applications of alkanethiol SAMs.

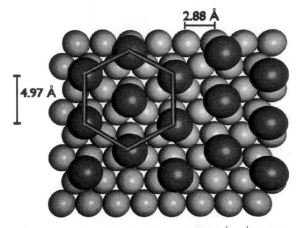

Figure 1. Schematic showing tyipcal ($\sqrt{3}$x$\sqrt{3}$)R30° hexagonal close pack arrangement of alkanethiols on gold.

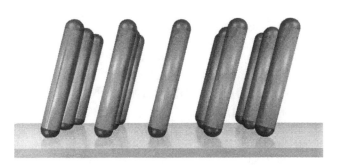

Figure 2. Schematic showing an ordered monolayer on gold.

3   PATTERNING SAMS

Though patterning SAMs is not the only expression of the creativity demonstrated with alkanethiols over the years, patterning has enabled and stimulated many unique applications and ideas about how to use alkanethiol SAMs.

One of the first patterning methods introduced was referred to as micromachining. This method involved forming a monolayer of one alkanethiol and then removing the thiol from designated regions by carefully removing the metal from those areas with a surgical scalpel blade or cut carbon fiber[16]. The sample was then introduced to a second thiol solution which allowed deposition of the second thiol in the areas of freshly exposed metal. Though this method was able to create micron sized patterns, it did not provide very precise control and resulted in the deformation of the metal surface in the local area of the pattern. However, within the same decade several more robust and flexible methods emerged that allowed controlled patterning of alkanethiols. These include, but are not limited to, UV photopatterning, microcontact printing, and dip pen lithography.

3.1 UV Photopatterning

In 1993 Tarlov *et. al.* combined observations from two previous studies, one showing that alkanethiols on gold can be oxidized by exposure to UV light in air [17], and another showing that oxidized thiols can be displaced by fresh thiols by dipping the sample in a thiol solution [18], and showed that the combination of these two discoveries creates patterned surfaces [19].

UV Photopatterning enabled patterning of alkanethiols without touching the surface. It also used established technology in the fabrication of photomasks, and only required the use of a suitable UV light source in order to create the pattern.

Early examples of UV photopatterning showed feature replication down to around 6 μm [20]. Recently Sun *et. al.* have shown line widths below 50 nm using a UV laser with a near field scanning optical microscope [21].

3.2 Microcontact Printing

In 1994 the Whitesides group introduced a new method of patterning that eliminated the need for expensive photolithography equipment and the use of clean rooms. This new method was microcontact printing and worked by the same simple process that is used to transfer ink onto paper using a rubber stamp [22]. The main process limitation of microcontact printing is creating the masters that are used to form the stamps. The masters are typically created using standard lithography processes on a silicon wafer. Features of desired sizes and shapes are created on the master. Then an elastomeric polymer, typically polydimethyl siloxane, is poured over the master and cured.

The stamp is then removed from the master, cut to size, and rinsed. The stamps can then be inked with an alkane thiol solution and placed on a gold substrate. All the areas of contact with the stamp get coated with a monolayer, while the unexposed areas remain uncoated. The sample can then be placed in a second thiol solution to fill in the uncoated regions. Features as small as 30 nm have been reported using microcontact printing [23]. It is interesting to note that though it has not received much attention in the literature, it is known that low molecular weight silicones are transferred along with the thiols during the stamping process [24, 25].

Microcontact printing has been used to create a wide range of surfaces with different feature sizes and shapes. The simplicity of the method and accessibility of the materials makes it an ideal platform for development and exploration of surface chemistry.

3.3 Dip-pen Lithography

In 1999 the Mirkin group developed a method for patterning that enabled precise control of the placement of alkanethiols on a surface by using an atomic force microscope tip to write using thiols as ink [26]. The discovery of dip pen lithography came out of an investigation of the capillary that forms between an AFM tip and the surface when working in air. The Mirkin group was able to use this meniscus of water that forms in air to pull thiols off the AFM tip and deposit them onto the surface. Initial work with dip-pen lithography was limited to writing small features over limited areas due to the write time limitations of a single AFM tip. However, recently NanoInk, a company specializing in dip-pen nanolithography, has introduced a system that is capable of writing with 55,000 tips simultaneously enabling parallel writing over large areas in short times.

4   ALKANETHIOLS AND NANOTECHNOLOGY

The vast knowledge of alkanethiol self assembly, their versatility in control of surface chemistry, and the wide range of methods that can be used to deposit them, has kept alkanethiols in the forefront of surface science over the last 30 years. This is still true today where alkanethiols are playing a critical role in the field of nanotechnology.

Nanomaterials are either created from the top down, by removing material until one gets nanoscale features, or from the bottom up, where nanoscale features are created using self-assembly. Alkanethiol SAMs are nanoscale surfaces themselves with typical layer thicknesses of 1 to 3 nm, and have been used successfully to create nanoscale patterns using various deposition methods. Alkanethiols are not only useful in the creation of nano-thick layers or nanoscale patterns, but they are also used extensively in the formation of nanoparticles and nanorods. Nanoparticles are often modified with alkanethiols containing specific chemical functionalities that enable the researcher to tailor the surface properties of the particles and engineer them for specific applications.

It is only a matter of time until researchers devise ways to control alkanethiol chemistry on nanoscale features in the same way they pattern and manipulate the chemistry of larger scale substrates. The flexibility and control provided by alkanethiols is sure to inspire future scientists to develop new and innovative ways of engineering surface chemistry and solving important problems through scientific discovery.

REFERENCES

[1]   Love, J.C., Estroff, L.A., Kriebel, J.K., Nuzzo, R.G., and Whitesides, G.M., Chemical Reviews, 105, 1103-1169, 2005

[2]   Schreiber, F., Journal Of Physics: Confensed Mater, 16, R881-R900, 2004

[3]   Ulman, A., Chemical Reviews, 96, 1533-1554, 1996

[4]   Bigelow, W.C., Pickett, D.L., and Zisman, W.A., Journal Of Colloid and Interface Science, 1, 513, 1946

[5]   Sagiv, J., Journal of the American Chemical Society, 102, 92, 1980

[6]   Nuzzo, R.G. and Allara, D.L., Journal of the American Chemical Society, 105, 4481-4483, 1983

[7]   Murphy, J.G., *Silver Polish*, U.S.P. Office, Editor. 1958, James G. Murphy: United States of America.

[8]   Dubois, L.H. and Nuzzo, R.G., Annual Reviews in Physical Chemistry, 43, 437-463, 1992

[9]   Nuzzo, R.G., Dubios, L.H., and Allara, D.L., Journal Of The American Chemical Society, 112, 558-569, 1990

[10]   Bain, C.D., Troughton, E.B., Tao, Y.-T., Evall, J., Whitesides, G.M., and Nuzzo, R.G., Journal Of The American Chemical Society, 111, 321-335, 1989

[11]   Nuzzo, R.G., Fusco, F.A., and Allara, D.L., Journal Of The American Chemical Society, 109, 2358-2368, 1987

[12]   Nuzzo, R.G., Zegarski, B.R., and Dubois, L.H., Journal Of The American Chemical Society, 109, 733-740, 1987

[13]   Bain, C.D., Evall, J., and Whitesides, G.M., Journal Of The American Chemical Society, 111, 7155-7164, 1989

[14]   Bain, C.D. and Whitesides, G.M., Journal Of The American Chemical Society, 111, 7164-7175, 1989

[15] Poirier, G.E. and Tarlov, M.J., Langmuir, 10, 2859, 1994

[16] Abbott, N.L., Folkers, J.P., and Whitesides, G.M., Science 257, 1380-1382, 1992

[17] Huang, J. and Hemminger, J.C., Journal Of The American Chemical Society, 115, 3342-3343, 1993

[18] Tarlov, M.J. and Newman, J.G., Langmuir, 8, 1398-1405, 1992

[19] Tarlov, M.J., Donald R. F. Burgess, J., and Gillen, G., Journal Of The American Chemical Society, 115, 5305-5306, 1993

[20] Huang, J., Dahlgren, D.A., and Hemminger, J.C., Langmuir, 10, 626-628, 1994

[21] Sun, S., Chong, K.S.L., and Leggett, G.J., Nanotechnology, 16, 1798-1808, 2005

[22] Kumar, A. and Whitesides, G.M., Applied Physics Letters, 63, 2002-2004, 1993

[23] Odom, T.W., Thalladi, V.R., Love, J.C., and Whitesides, G.M., Journal of the American Chemical Society, 124, 12112-12113, 2002

[24] Graham, D.J., Price, D.D., and Ratner, B.D., Langmuir, 18, 1518-1527, 2002

[25] Glasmastar, K., Gold, J., Andersson, A.-S., Sutherland, D.S., and Kasemo, B., Langmuir, 19, 5475-5483, 2003

[26] Piner, R.D., Zhu, J., Xu, F., Hong, S., and Mirkin, C.A., Science 283, 661-663, 1999

Low Cost Fabrication of Micro- and Nanopores in Free-Standing Polymer Membranes for Study of Lipid Adsorption

Junseo Choi, Anish Roychowdhury and Sunggook Park[*]

[*]Department of Mechanical Engineering and Center for Bio-Modular Multiscale Systems
Louisiana State University, Baton Rouge, LA70803, USA, sunggook@me.lsu.edu

ABSTRACT

This study presents low cost fabrication of free-standing membranes in polymer with perforated pores down to sub-μm diameter, which provides platforms for fundamental studies of many biosystems. For the fabrication, a combination of imprint lithography and a sacrificial layer technique was employed in order to obtain a clean, fully released, and mechanically stable membrane with perforated pores. Lift-off resist (LOR) was used as a sacrificial layer first while SU-8 resist spin-coated on the LOR layer was used as the active membrane layer in which micro- and nanopores patterns are formed via a combined thermal- and UV-imprint process. With this method, we could achieve a large area, free-standing SU-8 membrane with micropores up to 4 inch diameter.

As a demonstration of the use of the membrane in the study of a biosystem, the membrane was exposed to a solution with lipid vesicles. Lipid vesicles preferentially adsorb at the pore sites in the membrane, the extent of which depends on a surface treatment with poly(L-lysine)-graft-poly(ethylene glycol) (PLL-g-PEG) performed prior to the lipid adsorption. We will also show integration of the polymer membrane into microfluidic devices made of polydimethysiloxan, which allows for in-situ study of lipid adsorption.

Keywords: thermal- and UV-imprint process, perforated micro- and nanopores, polymer membranes, lipid adsorption, conductivity measurements

1 INTRODUCTION

The ability to mimic micro- and nanostructures existing in biosystems is important because it provides tools and platforms to answer many relevant fundamental questions without necessity of performing time-consuming and costly in-vivo experiments. One interesting structure is micro- and nanoscale pores. As an example, a cell membrane consists of a lipid bilayer and different proteins, both sides of which are exposed to different chemical and biological environments. Many cell functions are mediated via transportation of materials and signals through numerous nanopores existing in the membrane. Naturally occurring nanopores in a cell membrane, however, suffer from a number of inherent limitations for use in the study of biosystems. This includes poor thermal, mechanical and chemical stabilities, and poor controllability over the pore sizes and locations. Thus, an artificial micro- and nanopores in a synthetic membrane will allow for overcoming such limitations as well as controlling environments of both sides of the membrane.

Artificial membranes have been used in modeling cell membranes to study specific biological phenomena such as the formation and structure of lipid micro-domains of rafts [1], peptide/lipid interactions [2], cell-adhesions [3]. Lipid bilayers in artificially fabricated micro- and nanopores are also useful in understanding chemical and mechanical properties of the lipid bilayers, such as surface tension effects on the mechanical stability of the lipid bilayers. Furthermore, nanopores can be instantaneously formed upon introduction of α-haemolysin, into the lipid biolayer. Such naturally-formed nanopores have been used to study DNA transport through the nanopores [4].

Membrane structures with perforated micro- and nanoscale pores have been produced by a number of methods. Synthetic polymer membranes with random sizes and pore locations are most widely used in separation. Also commercially available are membranes fabricated by ion track etching in polycarbonate [5] and anodization of aluminum [6], which produce randomly distributed and a hexagonal array of nanopores, respectively, with pore diameter as small as 10 nm. Perforated micro- and nanopores at designated locations with controlled pore sizes can be produced using high end nanofabrication techniques. Lo et al. demonstrated fabrication of sub-5 nm pores using focused ion beam and electron beam techniques [7]. Electron beam in the transmission electron microscope followed by a size shrinkage by laser heating was also used to produce nanoscale pores in SiO_2 [8]. However, those methods do not allow for control over both the size and location of pores and the high yield of production. Therefore, it is needed to develop a flexible method to fabricate such membrane structures with micro- and nanoscale pores at designated locations with high throughput.

We have previously developed a fast and high-throughput process to produce free-standing polymer membranes down to sub-μm pore diameter by using modified imprint lithography and a sacrificial layer technique [9, 10]. The self-supporting mechanical stability was achieved by using a UV-curable polymer SU-8. In this work, we extend the process to the fabrication of large area membranes up to 4 inch diameter and a systematic study on the adsorption behavior of lipid vesicles at the membrane

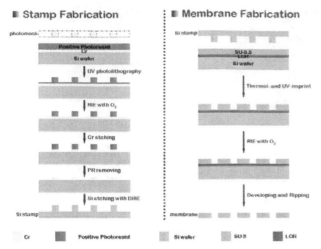

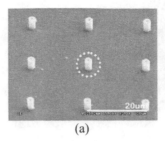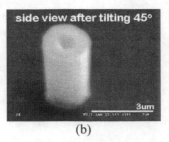

Figure 2. SEM Images for Si posts with 2μm produced by photolithography and DRIE.

Figure 1. Process schematics for fabricating imprint stamps (left) and SU-8 membranes with perforated micro- and nanopores (right).

surface upon exposure to a lipid solution. In addition, we will demonstrate a concept of how to integrate the membrane with a fluidic device that can be used for in-situ study of lipid adsorption.

2 EXPERIMENTAL

Figure 1 shows the process schemes for fabricating imprint stamps and SU-8 membranes with perforated micro- and nanopores. The process involves a number of sub-processes: photolithography and deep reactive ion etching (DRIE) for stamp fabrication, and modified imprinting process using a double resist layer and lift-off for the membrane fabrication.

2.1 Stamp Fabrication

Si stamps were fabricated using a combination of photolithography and semiconductor micromachining techniques. First, photolithography was done with a custom designed photomask in a 'Quintel' UV exposure station at the Center for Advanced Microstructures and Devices, Louisiana State University. The resist patterns were transferred down to Si substrate using a DRIE process which was performed at the Micro Electronics Research Center, Georgia Institute of Technology. The DRIE process involved alternating etch and sidewall passivation cycles so as to maintain a high degree of anisotropy, resulting in almost vertical sidewall profiles. Figure 2 shows scanning electron microscopy (SEM) images for a Si stamp containing micro-posts of 1.5 μm diameter. Nearly vertical sidewalls with scallop-like features resulting from the DRIE process are seen.

Prior to imprinting, the stamp surface was treated with a fluorinated silane in the vapor phase in a home-made chemical vapor deposition chamber in order to reduce the adhesion to the resist.

2.2 Fabrication of SU-8 Membranes

A double resist layer was used for imprinting: Lift-off Resist (LOR, MicroChem) as a sacrificial layer and SU-8 as the active membrane layer. First, a 4 inch quartz wafer was sequentially spin-coated with 1 μm and 5 μm thick LOR and SU-8 layers, respectively. Imprint lithography was performed with a commercial nanoimprinter (Obducat 6") which allows for both thermal and UV imprinting. In order to define pore structures in SU-8 which is a UV-curable resist, a modified imprinting process combining thermal and UV imprint was employed. Imprinting was carried out at 65°C, which is close to the glass transition temperature (T_g) of 55°C for uncured SU-8. An imprint pressure of 5 MPa was used. These conditions are comparable to those used in a reversal imprint process (40-85°C and 1-5 MPa) reported by Hu et al. [11]. After the temperature was reduced to 50°C, which is slightly lower than T_g, the sample was exposed to UV light for 10 sec. For a UV imprint process, UV light is usually shone through a transparent stamp to harden a UV-curable resist. However, in our process, we could avoid the requirement of fabricating a transparent stamp by using a transparent quartz wafer as substrate. Then, the quartz substrate was separated from the Si stamp at 40°C and baked at 95°C for 5 min to complete the cross-linking of SU-8.

Finally, a free-standing SU-8 membrane with perforated micropores was lifted-off by dissolving the LOR sacrificial layer with a MF319 solution. It took ~ 3 hours to complete the lift-off process.

2.3 Selective Immobilization of Lipid Vesicles

The feasibility of using the perforated micro- and nanopores in the SU-8 membrane to mimic a cell membrane was examined by studying the adsorption behavior of lipid vesicles to the membrane surface. Prior to the lipid adsorption, some of the SU-8 membrane samples were treated with a poly-L-lysine-grafted-(polyethylene glycol) (PLL-g-PEG) solution in order to prevent non-specific adsorption of lipid vesicles at the membrane surface. For synthesis and treatment of PLL-g-PEG, we followed a synthetic path described in [10].

A di-palmitoylphosphatidylcholine lipid solution (Avanti Polar Lipids) was mixed with chloroform (Purity=99.9%, Fisher Scientific) at different volume ratios. After staining with a fluorescent dye (5'-Cy5-oligonucleotide-amine-3'), the solution was dispensed on the membranes. Excessive lipid vesicles were then washed off with DI water and the membrane was observed under a fluorescence microscope (Leica).

2.4 Integration of the Membrane into a PDMS Microfluidic Devices

The SU-8 membrane in a free-standing form is very versatile in many applications. One example is a fluidic interconnecting component in modular microfluidic devices. As a demonstration of this concept, we have fabricated a modular microfluidic device containing the SU-8 membrane with micro- and nanopores, which allows for in-situ study of lipid layer formation at the pore sites. In order to fabricate simple microfluidic channels, SU-8 negative photoresist was patterned on silicon via photolithography, which was replicated by casting PDMS (10:1 mass ratio of silicone elastomer to curing agent). After curing overnight at room temperature, the PDMS replica was peeled off from the master. The SU-8 membrane was then sandwiched between the two PDMS microchannels. The PDMS devices were bonded at 100°C for 2 hr on a hot plate and connected with inlet and outlet ports.

3 RESULTS AND DISCUSSIONS

The free-standing SU-8 membranes with micropores were fabricated by a combination of a modified imprinting process and a sacrificial layer technique. The most important requirement for the fabrication involved selection of polymer materials that allow for imprinting of high aspect ratio structures with good replication fidelity and at the same time had enough mechanical stability to be free-standing. Although thermoplastic polymers such as PMMA and PC are widely used for imprinting, it does not provide enough mechanical stability. Therefore, SU-8 was chosen as the membrane layer because it is the resist widely used for high aspect ratio microstructures in the LIGA process and thus high mechanical stability was expected. The Young's modulus of PMMA and SU-8 is reported to be in the range of 1.8-3.1 GPa and 3.5-7.5 GPa, respectively [12].

We have previously reported an optimal condition for thermal imprinting into SU-8 layers [9, 10]. Best imprint results were obtained when imprinting was performed at 135°C. However, high thermal stress and adhesion generated during molding and cooling often resulted in a peeling of imprinted SU-8 from the substrate at the demolding step. This problem becomes more significant for large area imprinting and we hardly obtained good imprint results when a 4 inch stamp fully covered with micropillars

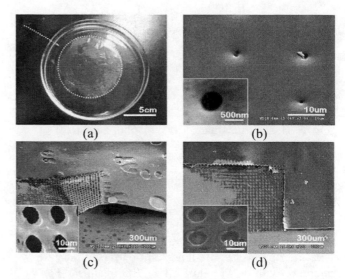

Figure 3. (a) Photograph for a SU-8 membrane after release from substrate and (b), (c) and (d) are SEM images from the top surface of membrane with nanopores, top and bottom surfaces of the released membrane with micropores, respectively.

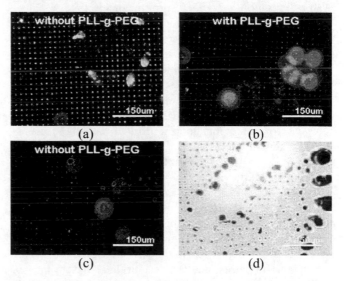

Figure 4. Fluorescence images after exposure to a lipid solution for SU-8 membranes with or without PLL-g-PEG treatment at a different volume ratio of lipid to chloroform. The volume ratio is 1:100 (a), (b) and 1:10 (c), (d).

was used. In this experiment, the thermal imprint process was combined with UV-imprint in order to reduce the imprinting temperature. We were able to achieve high aspect ratio micropores over 4 inch diameter at a low imprint temperature close to T_g of SU-8. Figure 3(a) shows a photograph for a 4 inch diameter, free-standing SU-8 membrane after release from substrate. Figure 3(b), (c) and (d) show SEM images from the top surface of membrane with sub-micrometer pores, top and bottom surfaces of the released membrane with micropores, respectively.

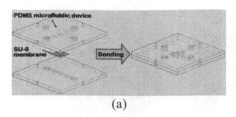 

(a)	(b)

Figure 5. Schematic image (a) and photograph (b) of a fluidic device integrated with the membrane.

The free-standing membranes with micro- and nanoscale pores having access from both sides can be used as a platform to mimic a cell membrane. The first step to mimic a cell membrane is to selectively immobilize lipid bilayer at the pore sites. Figure 4 shows that fluorescence images after exposure to a lipid solution for different SU-8 membranes with or without PLL-g-PEG treatment. For the SU-8 membrane with well-defined micropores (Figure 4(a) and (b)), lipid vesicles preferentially adsorb at the pore sites in the membrane. However, when the membrane surface was treated with PLL-g-PEG prior to the lipid adsorption, the fluorescence signal becomes weaker. Given that PLL-g-PEG prevents non-specific adsorption of the lipid vesicles at the membrane surface, it would be surface tension force by which the lipid vesicles can remain at the pore sites stably. This in turn gives an insight to a bilayer formation at the pore sites. The adsorption behavior is dramatically changed where a strong fluorescence signal emanates from the corrugated SU-8 surface created due to failed imprints, as shown in Figure 4(c) and (d). This result indicates the importance of achieving good imprinted patterns in using the free-standing SU-8 membranes to selectively immobilize the lipid vesicles at the pore sites, more broadly to mimic the cell membrane.

Figure 5(a) shows a schematic diagram of how to integrate the free-standing SU-8 membrane into a modular microfluidic device. The membrane with perforated micropores was sandwiched by two PDMS microfluidic devices. The microchannels in the two PDMS devices were so aligned to be perpendicular to each other. One advantage of such a crossed orientation is that only micropores in the membrane which are located within the overlapped area between the two microchannels will be active in the transportation of substances through the pores. Therefore, the number of active pores can be controlled simply by using microchannels of different widths. This will also alleviate the requirement of high accuracy in aligning two PDMS devices for bonding. The fabricated PDMS device integrated with a SU-8 membrane is shown in Figure 5(b).

4 CONCLUSIONS

A low-cost and flexible method was developed using all parallel processes to produce large area, free-standing polymer membranes up to 4 inch diameter which contain perforated micropores. For the fabrication, a combination of imprint lithography and a sacrificial layer technique were used. Selective adsorption of lipids at the pore sites in the SU-8 membrane was achieved using lipid vesicles stained with a fluorescent dye and PLL-g-PEG chemistry, demonstrating that the membrane can be used as platforms to study transport behavior through lipid layers. In addition, we have developed a concept of how to integrate the membrane structure into microfluidic devices.

This technique developed in this study can be extended to produce a membrane with smaller pore size. Of course, this requires high aspect ratio imprinting of nanostructures in order to achieve enough mechanical strength for free-standing, which is still challenging. However, an improved imprinting process combined with high quality stamp fabrication will enable fabrication of free-standing membranes with nanosized pores at low cost and with high throughput.

ACKNOWLEDGEMENT

This research was supported by National Science Foundation CAREER Award (CMMI-0643455) and by the Louisiana Board of Regents–RCS (Contract No. LEQSF (2006-09)–RD–A–09).

REFERENCES

[1] C. Yuan, J. Furlong, P. Burgos, and L. Johnston, Biophys. J, 82, 2526, 2002
[2] D. Takamoto, M. Lipp, A. Nahmen, K. Lee, A. Waring, and J. Zasadzinski, Biophys. J., 81, 153, 2001
[3] S. Sivasankar, W. Brieher, N. Lavrik, B. Gumbiner, D. Leckband, Proc. Natl. Acad. Sci. U.S.A., 96, 11820, 1999
[4] C. Dekker, Nature Nanotechnology, 2, 209, 2007
[5] Y. Komaki and S. Tsujimura, Science, 199(4327), 421, 1978
[6] J. O'Sullivan and G. Wood, Proceedings of the Royal society of London Series A, Mathematical and Physical Sciences, 317(1531), 511, 1970
[7] C. Lo, T. Aref, and A. Bezryadin, Nanotechnology, 17(13), 3264, 2006
[8] S. Wu, S. Park, and X. Ling, Nano Lett., 6(11), 2571, 2006
[9] A. Roychowdhury and S. Park, Proceedings for NSTI Nanotech Conference 07, 453, 2007
[10] A. Roychowdhury, J. Choi, A. Yi, F. Xu, and S. Park, Proceedings for ASME International Mechanical Engineering Congress & Exposition (IMECE) 07, IMECE2007-43813, 2007
[11] W. Hu, B. Yang, C. Peng, and S. Pang, J. Vac. Sci. Technol. B, 24(5), 2225, 2006
[12] H. Khoo, K. Liu, and F. Tseng, J. Micromech. Microeng., 13, 822, 2003

Surface Plasmon Polariton Assisted Organic Solar Cells

Changsoon Kim*, Jung-Yong Lee**, Peter Peumans**, and Jungsang Kim*

* Department of Electrical and Computer Engineering, Duke University
Durham, North Carolina 27708
** Department of Electrical Engineering, Stanford University
Stanford, California 94305

ABSTRACT

We propose a lateral tandem cell system consisting of organic thin-film photovoltaic devices. The crucial element of the system is a surface plasmon polariton (SPP) assisted organic solar cell employing a metallic grating electrode. In the SPP-assisted solar cell, the incident light resonantly excites an SPP mode to increase the optical field intensity in the absorption layer. As a result, a high absorption efficiency is maintained when the thickness of the absorption layer is decreased below the exciton diffusion length, thereby overcoming the 'exciton diffusion bottleneck' present in conventional organic solar cells. For a model structure, where an organic multilayer is sandwiched by a planar cathode and a grating anode, both consisting of Ag, we show, using the finite element method, that the absorption efficiency of a 10-nm-thick absorption layer with the absorption coefficient of 10^5 cm^{-1} exceeds 80% for TM-polarized incident light with a wavelength of 765 nm. We show that the resonance can be tuned by varying the grating period. We also discuss design guidelines for the lateral tandem cell system, and estimate its performance.

Keywords: organic solar cell, surface plasmon, grating

1 INTRODUCTION

Unlike conventional inorganic solar cells, absorption of a photon in an organic solar cell creates a tightly bound exciton, which must be dissociated into an electron and a hole to generate electrical power [1], [2]. Since the efficient dissociation of the excitons occurs at a donor-acceptor (DA) interface, 'useful' excitons that eventually contribute to electrical power are those generated within a diffusion length from the DA interface. The diffusion length ($L_D < \sim 10$ nm) of excitons in organic semiconductors, both in polymeric and small molecular materials, is in general smaller than the optical absorption length ($1/\alpha \sim 100$ nm, where α is the absorption coefficient), presenting an inherent trade-off between the efficiencies of exciton dissociation and photon absorption (so-called 'exciton diffusion bottleneck') [3]. Optical resonances can be utilized to enhance the light absorption without increasing the thickness of the ac-

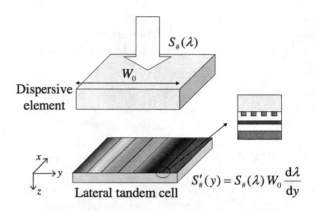

Figure 1: Schematic configuration of a lateral tandem cell system.

tive layer beyond L_D. However, there exists an inherent trade-off between the magnitude of the enhancement and the bandwidth over which it occurs. Here, we propose a lateral tandem cell (LTC) system capable of overcoming this trade-off. The LTC consists of many sub-cells that are connected in series. The resonance we exploit is that of a surface plasmon polariton (SPP) mode present in an organic solar cell employing a metallic grating electrode, and the tunability of the resonance is demonstrated by numerical calculations based on the finite-element method. The LTC system offers a straightforward platform for both optical and electrical optimization. We discuss design guidelines for the LTC system, and estimate its performance.

2 LATERAL TANDEM CELL - CONCEPT

Figure 1 shows a basic concept of a lateral tandem cell (LTC). The dispersive element spatially separates the incoming solar photons according to their wavelengths (λ) to achieve a mapping between λ and the position (y) on the LTC, and the local photonic and electronic properties of the LTC are optimally tuned at all y. Specifically, the structure of the device is continuously varied along the y-direction to enable the optical fields incident on y to resonantly excite the cell at that location. Furthermore, organic materials are chosen so that (1) the absorption spectrum of the optically active

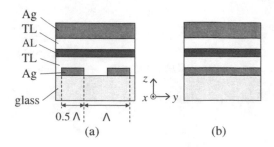

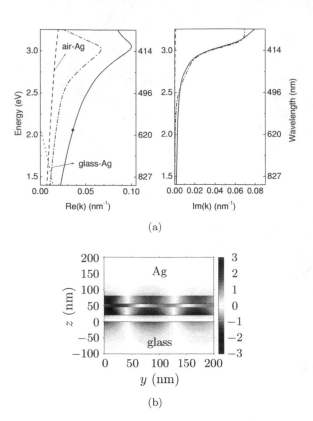

(a)

Figure 2: (a) Layer structure of a model device. TL: charge transport layer, AL: absorption layer, Λ: grating period. (b) Layer structure of a planar device, where the grating electrode in (a) is replaced by a planar layer.

material matches well with $\lambda(y)$ and (2) energy loss associated with exciton dissociation and charge transport processes is minimized to obtain a large open-circuit voltage (V_{OC}) and fill factor (FF). The LTC is divided into sub-cells, which are connected in series with the adjacent sub-cells. The short-circuit current that each sub-cell generates can be tuned by varying the width (along the y-axis) of the cell, providing a straightforward method for balancing the short-circuit currents, as required for a tandem cell connected in series [4].

3 SURFACE PLASMON POLARITON ASSISTED ORGANIC SOLAR CELLS

A model device that we consider is schematically shown in Fig. 2(a). It has a structure comprising: glass substrate / 20-nm-thick Ag grating electrode / 45-nm-thick charge transport layer / 10-nm-thick absorption layer / 25-nm-thick charge transport layer / 150-nm-thick Ag electrode. The charge transport layers have a refractive index of $\tilde{n} = 2$. The absorption layer has a refractive index of $\tilde{n} = 2 + in'$, with n' varying with wavelengths to yield $\alpha = 10^5$ cm^{-1} in all spectral regions. First, we calculate the photonic band structure of a planar device shown in Fig. 2(b), where the grating electrode in the model device is replaced with a planar Ag layer. By considering the grating electrode as perturbation of the planar electrode, we semi-quantitatively understand the dependence of resonant energy on the grating period. The rigorous calculations for the model device are achieved by performing electromagnetic and excitonic analyses using the finite-element method. In this paper, we limited our analysis to the TM-polarization (the magnetic field $\parallel x$).

3.1 Planar structure

Figure 3(a) shows the complex photonic band structure, i.e. energy (E) vs. complex in-plane wavevector (k), of the planar device. The mode drawn with

(b)

Figure 3: (a) Complex photonic band diagram of the planar device for TM-polarization. Dashed line: surface plasmon polariton (SPP) mode propagating along the air-Ag interface. Dash-dot line: SPP mode along the glass-Ag interface. Solid line: SPP mode that has an appreciable field intensity in the organic layers. Dotted line in the left panel: Solid line folded at $\Re(k) = 0.018$ nm^{-1}. (b) z-component of the electric field of the mode at $(k, E) = (0.036 + i0.0028$ nm$^{-1}, 2.07$ eV) indicated with a dot in (a).

a dashed line corresponds to a surface plasmon mode propagating along the air-Ag interface, while the mode drawn with a dash-dot line corresponds to a surface plasmon mode propagating along the glass-Ag interface. The third mode represented as a solid line is our main interest. As shown in Fig. 3(b), it is a surface plasmon polariton (SPP) mode with an appreciable field intensity in the organic layers. The SPP mode lies below the light line corresponding to the glass layer, and hence cannot be excited by external fields. However, the introduction of the grating electrode 'folds' the band diagram, allowing the external fields to couple with the SPP mode. For a given grating period (Λ), the intersection of the folded branch at the E axis can be considered a zeroth order estimation of the photon energy that resonantly couples with the device. For example, to get a coupling at $\lambda = 600$ nm ($E = 2.07$ eV), the boundary of the 1st Brillouin zone needs to be at $\pi/\Lambda = 0.018$ nm^{-1},

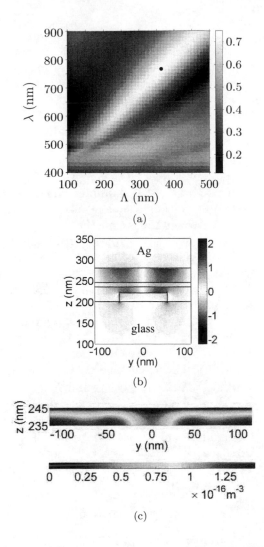

(a)

(b)

(c)

Figure 4: (a) External quantum efficiency (η_{ext}) of a model device as a function of the grating period (Λ) and illumination wavelength (λ). A black dot represents (Λ, λ) that maximizes η_{ext}. (b) z-component of the electric field corresponding to (Λ, λ) = (232 nm, 600 nm). (c) Exciton density profile for (Λ, λ) =(232 nm, 600 nm).

indicating that $\Lambda = 175$ nm. We expect that the resonance can be tuned by varying the grating period; as Λ increases, the coupling occurs at a lower energy. Furthermore, Fig. 3(a) shows that the SPP mode becomes lossy as $\lambda < 500$ nm, which is due to the absorption in Ag. Therefore, we expect the absorption efficiency of the organic layer in the model device to decrease as λ decreases below 500 nm, limiting the device efficiency.

3.2 Grating structure

We calculate the external quantum efficiency (η_{ext}, defined as the probability that an incident photon creates a charge carrier collected at the electrodes) spectrum of the model device as Λ is varied. First, we calculate the absorption of electromagnetic energy in the

absorption layer for the case of normal incidence. Subsequently, steady-state exciton density profile was calculated by solving a diffusion equation under the following assumptions: the diffusion length (L_D) of the excitons is 20 nm; the exciton generation rate is proportional to the time-averaged absorbed power; the exciton dissociation velocity at the interface between the absorption and upper charge transport layers is infinite, while that at the interface between the absorption and lower charge transport layers is zero. The short-circuit current is then calculated from the steady-state exciton density profile, from which we obtain η_{ext}. The similar analysis for planar multilayer structures was outlined in detail in [1]. Both electromagnetic and diffusion simulations were done using a finite-element method software package [5].

Figure 4(a) shows the calculated η_{ext} as a function of (Λ, λ). The expected dependency (resonant energy decreases with increasing Λ) is clearly visible. The absorption efficiency of the 10-nm-thick absorption layer (η_{abs}, defined as the probability that an incident photon is absorbed in the absorption layer) peaks at 0.81 when $\lambda = 765$ nm ($\Lambda = 364$ nm). On this condition, $\eta_{ext} = 0.75$ as marked by a black dot in Fig. 4(a), indicating that 92% of the photo-generated excitons ($L_D = 20$ nm) reach the interface bewteen the absorption layer and the upper charge transport layer. By varying Λ along the bright white region in Fig. 4(a), we can achieve $\eta_{abs} > 0.7$ over a large spectral region ranging from $\lambda = 550$ nm to 900 nm. Λ that maximizes η_{ext} for $\lambda = 600$ nm is 232 nm, which is larger than our zeroth order estimation (175 nm) in the previous section. However, the z-component of the corresponding electric field shown in Fig. 4(b) resembles that shown in Fig. 3(b), confirming that our perturbative approach is valid.

4 LATERAL TANDEM CELL - DESIGN

To demonstrate that an LTC can enhance the device performance compared to conventional devices, we perform an analysis under the following assumptions; a dispersive element in the LTC is capable of achieving a spectral distribution, $\lambda(y)$, at the surface of the LTC from the incoming solar photon flux; the LTC consists of N_{sc} sub-cells that are connected to the adjacent sub-cells in series; in all spectral regions where there is non-negligible amount of photon energy, an absorption material with $\alpha \simeq 10^5$ cm^{-1} is available. Under solar illumination with a photon number flux per wavelength, $S_\#(\lambda)$, a region of the LTC ranging from $y = y_i$ to y_j generates a short-circuit current per unit length (in the

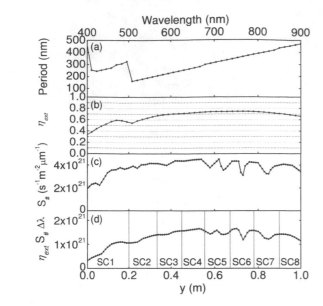

Figure 5: (a) Grating period maximizing the external quantum efficiency at all locations on a lateral tandem cell. We assume that the wavelength of photons incident on y linearly varies from 400 nm to 900 nm. (b) External quantum efficiency as a function of position. (c) Photon number flux per wavelength with an integrated power of 100 mW/cm^2. (d) Integrand in Eq. (1). $\Delta\lambda = 500$ nm. The vertical lines represent the partitioning of the lateral tandem cell into 8 sub-cells. SCi: i^{th} sub-cell.

x-direction)

$$I'_{SC} = e \int_{y_i}^{y_j} \eta_{ext}(y) \, S'_{\#}(y) \, \mathrm{d}y$$
$$= e \int_{y_i}^{y_j} \eta_{ext}(y) \, S_{\#}(\lambda) \, W_0 \, \frac{\mathrm{d}\lambda}{\mathrm{d}y} \, \mathrm{d}y, \qquad (1)$$

where $S'_{\#}(y)$ is the photon number flux on the LTC surface at y, W_0 is the width of the dispersive element, and e is the electron charge. For given $\eta_{ext}(y)$ and $\lambda(y)$, evaluation of Eq. (1) over the entire LTC length is a conserved quantity, and is defined to be $I'_{SC,tot}$. To maximize the power-conversion efficiency (η_P), the LTC is partitioned into N_{sc} sub-cells, each generating the identical short-circuit current, $I'_{SC,tot}/N_{sc}$. This partitioning is achieved in a straightforward manner by successively determining the widths of the sub-cells. When $V_{OC}^{(i)}$ and $FF^{(i)}$ denote, respectively, the open-circuit voltage and the fill factor of the ith sub-cell, η_P of the LTC is given by

$$\eta_P = \frac{I'_{SC,tot} \left\{ \frac{1}{N_{sc}} \sum_i V_{OC}^{(i)} \, FF^{(i)} \right\}}{W_0 \int S_{\#}(\lambda) \, \frac{hc}{\lambda} \, \mathrm{d}\lambda}, \qquad (2)$$

where h is Planck's constant, and c is the speed of light in vacuum. The advantage of the LTC approach is apparent in Eq. (2). Since the local photonic property of the sub-cells is optimized to allow a very thin ($\sim L_D$) active material to efficiently absorb the incident photons, the LTC maximizes $I'_{SC,tot}$. Furthermore, the energetics and charge transport property of the organic materials in each sub-cell are optimized to increase the terms in the curly brackets.

We estimate η_P for an LTC where the sub-cells are the SPP-assisted cells described in the previous section. First, from the result shown in Fig. 4(a), we calculate the grating period at each location along the y-direction [$\Lambda_{max}(y)$] that maximizes η_{ext}, assuming that $\lambda(y)$ increases linearly from 400 nm to 900 nm over the width of the LTC (W_{LTC}). We further assume that $W_{LTC} = W_0$. $\Lambda_{max}(y)$ varies almost linearly with y over a large spectral region (λ from 500 nm to 900 nm). By continuously varying Λ along the y-direction, we can achieve $\eta_{ext} > 0.7$ for $\lambda = 595$ nm to 860 nm. $I'_{SC,tot}$ under TM-polarized AM1.5 solar illumination with an integrated intensity of 100 mW/cm^2 is proportional to the area under the curve shown in Fig. 5 (d), and is found to be 210 A/m. Figure 5 (d) also shows the width of each sub-cell with the identical short-circuit current. Assuming that $V_{OC}^{(i)} = (h \, c) / (e \, \lambda(y^{(i)})) - 0.8 \, V$, and $FF^{(i)} = 0.7$, where $y^{(i)}$ is the location of the right boundary of the ith sub-cell, $\eta_{P,TM} = 15\%$. Since we consider the TM-polarized illumination only, the lower bound of η_P for the unpolarized case is 7.5%. However, we note that η_P can be substantially increased if the device structure is optimized for both polarizations.

5 CONCLUSION

We proposed an LTC system consisting of SPP-assisted organic solar cells. Our analysis based on the finite-element method showed that a SPP resonance present in such devices enables a 10-nm-thick absorption layer to absorb 81% of the incident power at $\lambda = 765$ nm. The tunability of the resonance makes the SPP-assisted cell a good candidate to be used in a system where photons in the solar spectrum are delivered to their corresponding matched cells. We provided design guidelines for the LTC system, and estimated its performance.

REFERENCES

[1] P. Peumans, A. Yakimov, and S. R. Forrest, *J. Appl. Phys.*, **93**, 3693 (2003).

[2] H. Hoppe and N. S. Sariciftci, *J. Mater. Res.*, **19**, 1924 (2004).

[3] S. R. Forrest, *MRS Bull.*, **30**, 28 (2005).

[4] A. Yakimov and S. R. Forrest, *Appl. Phys. Lett.*, **80**, 1667 (2002).

[5] COMSOL, Inc., Burlington, MA 01803.

Synthesis of one-dimensional titanium dioxide nanostructures

M. Shaffer*, B. Cottam*, S. Chyla*, R. Menzel*, A. Bismarck**

*Department of. Chemistry, Imperial College, London SW7 2AZ, UK, m.shaffer@imperial.ac.uk
**Department of. Chemical Engineering, Imperial College, London SW7 2AZ, UK

ABSTRACT

Finely structured titanium dioxide is a technological material of long-standing importance for many applications including pigments and catalysis. There is growing interest in smaller, truly nano-sized titanium dioxide particles with well-defined crystallinity and a range of geometries from spheres to rods and tubes, that are relevant to applications in composites, photovoltaics, sensors, and catalysis. High aspect ratios, in particular, introduce high surface to volume ratios, network forming abilities, and opportunities to control anisotropic properties. Here we report a number of different synthetic strategies for producing high aspect ratio titanium dioxide nanostructures.

Keywords: titanium dioxide, nanorods, microfluidics, templating, sol-gel synthesis

1 SOL-GEL ROUTES

Titanium dioxide is commonly obtained via hydrolysis of metal alkoxides or halides; however, enhanced control over the reaction can be achieved in non-hydrous conditions. In the absence of water, the surface does not become hydroxylated, leaving the organic ligands bound tightly to the surface. In addition, the reaction temperature can be raised to improve the crystallinity of the products without leading to ripening or sintering.

A comparison will be made between nanorods synthesised via hydrolytic and non-hydrolytic routes, using different structure directing agents. Typical products are small, single crystal nanorods of anatase (~ 3 × 25 nm), although aging reactions under suitable conditions yield single crystal rutile nanorods (15 x 135 nm). The hydrolytic synthesis can be dramatically accelerated when performed on a microfluidic chip, as compared to a conventional bulk reaction. The acceleration is attributed to improved contact between the immiscible reagent phases [1].

2 TEMPLATING

A second strategy is based on high temperature templating reactions on carbon nanotubes. In this case, titanium iodide is used as a reagent to convert aligned arrays of carbon nanotubes into titanium carbide. Subsequent oxidation reactions yield aligned arrays of anatase or rutile nanorods, depending on the reaction temperature. This route allows the versatility of carbon nanotube synthesis to be leveraged to create titania structures of different dimensions [2].

3 HYDROTHERMAL

Hydrothermal treatment in strongly basic conditions has been used to convert titanium containing starting materials into 'nanotubes' or 'nanoribbons'. Early work identified these structures as anatase, but more recently it has become clear that layered titanate phases are more likely. The influence of changing the synthesis starting material, and washing conditions, on the nature of the product, casts light on the growth mechanisms involved. [3]

4 CONCLUSIONS

A variety of routes to titania nanorods have been explored, yielding products with different dimensions, orientation, crystallinity, and phase. These materials are relevant to a range of applications, particularly in photovoltaics [4] and composite materials; preliminary application data will be provided, time permitting.

REFERENCES

[1] B F Cottam, S Krishnadasan, A J. deMello, J C. deMello and M S. P. Shaffer, Accelerated synthesis of titanium oxide nanostructures using microfluidic chips, Lab on a chip, 7, 167-169, DOI: 10.1039/b616068a, 2007

[2] B Cottam, M Shaffer, Synthesis of aligned arrays of TiO₂ nanowires via a high temperature conversion of carbon nanotubes, Chem Comm, 4378 - 4380, 2007

[3] R Menzel, A Peiro, J Durrant, M Shaffer, Impact of Hydrothermal Processing Conditions on High Aspect Ratio Titanate Nanostructures, Chem Mat, 18, 6059-6068, DOI 10.1021/cm061721, 2006

[4] J Bouclé, S Chyla, M Shaffer, J Durrant, D Bradley, J Nelson, Hybrid Solar Cells from the Blend of Poly(3-hexylthiophene) and ligand-capped TiO₂ Nanorods, Adv. Func. Mat., In Press, DOI: 10.1002/adfm.200700280, 2008

Fabrication and properties of nanoscale metallic arrays in polymers

O.P Valmikanathan*, S. Bhowmik**, O. Ostroverkhova*, B. Shanker***, and S.V. Atre*

* Oregon State University, Corvallis, OR 97331, sundar.atre@oregonstate.edu
** Hewlett-Packard, Corvallis, OR 97331
*** Michigan State University, East Lansing, MI 48224

Abstract

Bottom up and top down approaches were explored for creating random and ordered metallic arrays in polymer matrix, respectively. For the bottom up method, the synthesis of Au, Pd and Ag nanoparticles stabilized by reverse micelles from an amphiphilic copolymer was performed. TEM images revealed partially ordered structures of Au nanoparticles with an average particle size less than 15 nm. In contrast, Ag and Pd were mostly disordered structures. Top down fabrication of Au features in PMMA were fabricated by milling with a focused ion beam. The features were about 50 nm in width with a 20 nm periodic spacing. The two approaches provide model systems for understanding the optical and electrical properties of nanostructured materials at 2 limiting extremes of order.

Keywords: nanoparticles, nanocomposites, reverse micelles, self-assembly.

1. INTRODUCTION

In the recent years, metals/polymers nanocomposites emerged as an important field in the nanotechnology and nanodevices due to their enhanced optical, thermal, electrical and catalytic properties. Our prior studies on the palladium/polycarbonate nanocomposites [1] showed that morphology of the nanoparticles determines the properties of the resulting nanocomposites. Similar structure-property relationships were noted by Chatterjee *et al* [2], Liu *et al* [3] and Wang *et al* [4] in the metal/polymer nanocomposites. In order to avoid agglomeration in nanocomposites, functionalized organic ligands [1, 5] or polymers [3, 4, 6, 7] have been employed as dispersing agents. Polymers have often been preferred over organic ligands due to the convenience in handling, reduced post-synthesis treatments and more direct applications.

Dispersions of metal salts in homopolymers have not been successful in achieving periodic nanostructures. However, a few recent studies [8] explored copolymer systems in the synthesis of periodic structures. In the present work, we explored a simplified route to synthesize self-assembled Au, Ag and Pd nanoparticles with an amphiphilic copolymer like PS_b_PEO as a stabilizing agent. PS_b_PEO was selected due to the presence of both hydrophilic and hydrophobic blocks which helps have been reported to aid the formation of periodic structures with nanoparticles. Focused ion beam milling was used to create Au nanoscale periodic arrays supported on silicon wafers. Polymethyl methacrylate was spun coated from solution to create nanocomposites from these patterned arrays. The two approaches are considered in turn.

2. NANOPARTICLE DISPERSIONS

All analytical grade chemicals used in the synthesis were purchased from Sigma-Aldrich USA. Deionized water with a resistivity of 18×10^6 Ω-cm was obtained from a Millipore unit. Lab synthesized diblock copolymers PS_b_PEO was used as such. The stepwise procedure followed in the atom transfer radical polymerization of PS-b-PEO polymerization is given elsewhere [Journal of applied polymer science, 2006. Vol. 102, pp. 4304-4313]. In the synthesis, 100 mg of powdered PS_b_PEO is dispersed into the acetone/water mixture. 1ml of 0.1M metal salt solution is slowly injected into the polymeric solution through micro-syringe. The metal salt solution is then reduced with a slow injection of 1M NaBH$_4$ to the reaction mixture. The reduction is accompanied by a drastic color change indicating the nanoparticles formation as shown in Figure 1. The copolymer stabilized metal nanoparticles were allowed to settle down. The acetone/water mixture is decanted and the nanocomposites sediments were washed thoroughly with deionized water. The nanocomposites were then dried and dispersed in toluene for further characterization. All steps were carried out under normal room conditions.

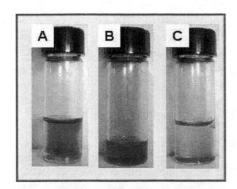

Figure 1: Nanocomposites dispersed in toluene; a. Au/PS_b_PEO b. Ag/PS_b_PEO c. Pd/PS_b_PEO

2.1 Optical Properties

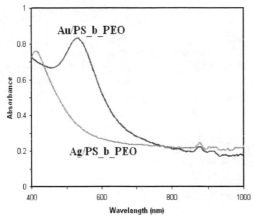

Figure 2: UV-Vis absorption spectra of Au and Ag nanoparticles.

As shown in the Figure 2, appearance of new absorbance peak at 530 nm and 420nm confirms the Au and Ag nanoparticles formation respectively. Similar results were noted by Liu *et al* [3]. Pd nanoparticles however failed to show any such characteristic peak as supported by the early studies on Pd/PMMA nanocomposites by Aymonier *et al* [6]. In the absence of PS_b_PEO all the reduced metal particles tend to settle down and form macro-sized powders that are found not dispersable in toluene. In the presence of PS_b_PEO, the metal nanoparticles tend to remain completely dispersed in toluene.

2.2 Morphology

Figure 3 shows the TEM images of metal/polymer nanocomposites. Au/PS_b_PEO and Pd/PS_b_PEO nanocomposites exhibited self-assembled nanoparticles of 15nm size. There have been very few prior studies reporting self-

assembled metal nanoparticles when capped by a copolymer via wet method. Prior work by Moller *et al* [8] on Au/ PS_b_PEO, involved the drying of the metal salt-polymer mixture followed by an electronic reduction of metal salt. In the present work, the stability might be attained due to the reverse micelles from PS_b_PEO with PEO blocks (head) being the hydrophilic part while the PS (tail) is the hydrophobic part.

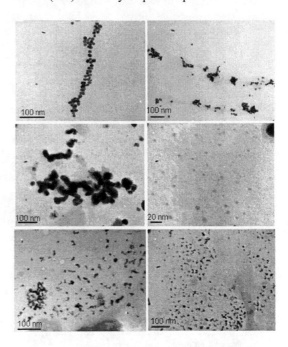

Figure 3: TEM images of (top) Au/PS_b_PEO (~15nm), (middle) Ag/PS_b_PEO (~5nm), and (bottom) Pd/PS_b_PEO (~10nm).

A schematic representation of metal nanoparticles adhering to the hydrophilic PEO blocks is shown in the Figure 4. Further FTIR and X-ray diffraction studies need to be performed to confirm the possible chemical interactions between the metal nanoparticles and the PEO blocks.

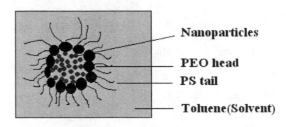

Figure 4: Schematic representation of capping of metal nanoparticles by the inverse micelles from PS_b_PEO

3. PERIODIC METALLIC ARRAYS

Controlling light at sub-wavelength scales is of considerable interest due to applications in broadband filters, sensors, substrates for miniature antennas, and sub-wavelength cavities. Metal-dielectric nanostructures have received attention due to their highly structure-sensitive optical properties and due to possibility to create various distributions of enhanced localized electric fields that allow for sub-diffraction-limit optical imaging. For example, a periodic arrangement of features (such as perforations in Figure 5A) yields frequency-dependent reflection and transmission properties (Figure 5B), which can be manipulated by changing the size, shape, and distance between the features. However, the challenges to produce such nanostructures remain.

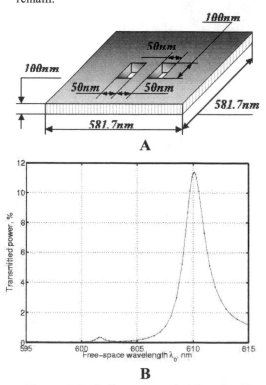

A

B

Figure 5: Periodic nano-perforations in silver film and enhanced transmission predicted theoretically

Our principal objective is to design periodic nanostructures with requisite optical properties and develop the means by which they can be reproducibly fabricated with relative ease and adapted for various applications. To this end, we will use a recently developed, rigorous theoretical/numerical tool set by which we can systematically understand the physics of electromagnetic interactions of materials at nanoscale regimes. This approach will be used to develop design rules for nanocomposites with specified periodicity and nanoscale features with: (a) predictable frequency-selective responses to electromagnetic radiation and (b) predictable distribution of enhanced electric fields (antenna-like effects). Obtaining such material systems hinges on the availability of nanofeatures with accurately controlled size and shape attributes, and methods for arranging them into two and three-dimensional structures of uniform periodicity. The nanoscale periodic structures will be fabricated using two approaches: self-assembly and nanofabrication. Structure-property-processing relationships of these material systems and architectures will be developed to refine and validate the design rules.

In this study, we explored the focused ion beam milling technique for fabricating such structures. Furthermore, by systematic varying of the nanostructure parameters, we will be able to explore dependence of the optical properties on the nanoscale features of our structures and compare experimental findings with results of calculations.

3.1 Nanofabrication

An attractive proposition would be to have a possible device on Si for applications in integrating with opto-electronic devices. Our approach was to create polymer nanocomposites with periodic structures the focused ion beam machining of metal nanofeatures and embedding them in polymer monolayers using self-assembly. Ion beam patterning involves patterning based on Coulombic interactions between positively charged ions and charged electrons and nuclei of sample atoms. Initial structures were performed on transparent SiO_2 membranes on supporting Si skeletal structure (Figure 6).

Figure 6: Schematic representation of periodic Au nanostructures on a SiO_2 membrane fabricated on a Si wafer

NSTI-Nanotech 2008, www.nsti.org, ISBN 978-1-4200-8503-7 Vol. 1

An optical image of the membrane substrate for the nanofabricated structure is shown in Figure 7.

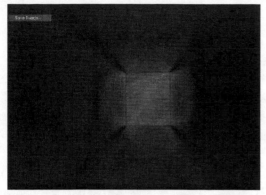

Figure 7: Top down view of a 100 μm square membrane formed on Si by etching

An optical micrograph of the patterned Au nanostructures on the membrane is shown in Figure 8.

Figure 8: Optical image of a nano-patterned Au structure on oxide membrane.

The SEM image of the nanostructured patterns is shown in Figure 9.

Figure 9: SEM view of nano-patterned grid showing a 61.7nm structure with 16.7nm spacing on an Au patterned sample.

The samples were characterized further by AFM to better understand the topographical details of the periodic structures (Figure 10).

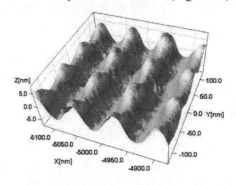

Figure 10: AFM characterization of patterns in the tapping mode rendered in 3D

In summary, a bottom-up approach based on nanoparticle synthesis in a block copolymer solution provided partially ordered to disordered structures in the present work. In contrast, highly periodic structures were obtained by top-down fabrication methodologies using ion beam milling as an example.

REFERENCES

1. S.V. Atre, O.P.Valmikanathan, V.K. Pillai, I.S. Mulla, and O. Ostroverkhova. *Nanotech 2007*, 1, 158.

2. S.K. Jewrajka, and U. Chatterjee. *Journal of Polymer Science Part A: Polymer Chemistry*, 44, 1841, 2006.

3. F.K. Liu, S.Y. Hsieh, F.H. Ko, and T.C. Chu, *Colloids and Surfaces A: Physiochemical and Engineering Aspects*, 231, 31, 2002.

4. H. Wang, X. Qiao, J. Chen, and S. Ding. *Colloids and Surfaces A: Physicochemical and Engineering Aspects*, 256, 111, 2005.

5. M. Brust, M. Walker, M. Bethell, D. Schiffrin, and D.J. Whyman. *Journal of Chemical Society and Chemical Communication*, 20, 801, 1994.

6. Y.Ne, X. Ge, Z. Zhang and Q. Ye. *Materials Letters*, 55, 171, 2002.

7. C. Aymonier, D. Bortzmeyer, R. Thomann, and R. Mulhaupt. *Chemistry of Materials*, 15, 4874, 2003.

8. M. Moller, J.P. Spatz, and A. Roescher, *Advanced Materials*, 8, 337, 1995.

9. J. B. Pendry, A. J. Holden, D. J. Robbins and W. Stewart. *IEEE Trans, Microwave Theory Tech.*, 47, 2075, 1999.

Dynamic study on Nanometer size square Permalloy (NiFe) antidot arrays; use as Monolithic Microwave localize band-pass filter

Bijoy K. Kuanr[*], Leszek M. Malkinski[**], Minghui Yu[**], Donald Scherer II[**], R. E. Camley[*] & Z. Celinski[*]

[*]Department of Physics, University of Colorado, Colorado Springs, CO 80918, USA
[**]Advanced Materials Research Institute, University of New Orleans, New Orleans, Louisiana 70148, USA

ABSTRACT

Nanometer sized Permalloy antidot arrays with different square hole sizes (1200 X 1200, 800 X 800, and 400 X 400 nm^2) have been fabricated by means of electron-beam lithography and lift-off techniques. We report here the dynamic properties of antidot arrays at GHz frequencies by using a flip-chip geometry and Network Analyzer based Ferromagnetic Resonance (NA-FMR) techniques. The dynamic excitations in the antidot arrays exhibit multiple resonance modes for the magnetic field applied in the plane of the array. Two distinct effective anisotropy field patterns split the uniform resonance into double resonance modes. The double resonance modes show uniaxial in-plane anisotropy and the easy axes are orthogonal. The magnitude of the induced effective anisotropy field decreases with the increase of square-hole size. The higher order resonance mode peaks move to low frequency with increase of the square-hole size at a constant applied field. We have demonstrated the possible application of antidot arrays as a magnetically tunable localized band-pass filter.

Keywords: Nano-holes, NA-FMR, spin dynamics, band-pass filter.

1 INTRODUCTION

The dynamic properties of magnetic nanostructures are drawing extensive attention due to their potential application for ultrahigh density data storage [1, 2]. It was proposed that a memory bit could be trapped between consecutive holes along the intrinsic hard axis of the antidot nanostructure [2]. One advantage of antidots over dots is that they can overcome the superparamagnetism limitation of isolated magnetic dots while preserving the properties of the magnetic film. This makes antidots a promising candidate for ultrahigh density data storage. One of the interesting dynamic phenomenona is the spin wave excitation, which arises as a result of quantization in small structures when the structure dimensions become comparable to the wavelength of the spin waves. Spin wave modes have been observed in magnetic dots and wires arrays with Brillouin light scattering (BLS) and ferromagnetic resonance (FMR)

experiments. Though FMR is a powerful tool in investigating spin wave spectra very few FMR experiments have been conducted on such nano-magnets. Yu et al. [6, 7] studied the ferromagnetic resonances (FMR) in micron-sized square and rectangular Permalloy antidot arrays with circular hole size around 1.5 μm in diameter and separation from 3 to 7 μm. All the square and rectangular antidot arrays show double resonances with uniaxial in-plane anisotropy, which are the consequence of a dipolar field distribution producing two regions with different demagnetization field patterns [6]. In addition the main uniform mode, lateral spin waves were observed in antidot arrays which were attributed to lateral confinement from the vacant holes [7]. In this work, we have fabricated lithographically patterned antidot arrays with different nano-scale sizes of square holes on top of flat Si substrates. We performed a detailed Network Analyzer based FMR investigation in the frequency domain on the spin wave excitations in permalloy antidot arrays. A detailed study on the size dependences of the dynamic magnetic properties of these lithographically patterned nano-scale antidot arrays is presented.

2 EXPERIMENT

A Permalloy film with nominal composition of 81% Ni and 19% Fe and thickness of 100 nm was deposited [8] onto a resin by magnetron sputtering at the rate of 0.2nm/s at an Ar pressure of 3 mTorr. The thickness of the films was measured using a crystal growth monitor. Three square Permalloy antidot arrays were fabricated having a constant thickness of 100 nm using electron-beam lithography and lift-off techniques. First, a thin layer of polymethyl methacrylate (PMM) resist was spun onto Si(100) substrates, and then patterned with an LEO 1530 VP field emission scanning electron microscopy (FESEM) system operating at 30 kV. After development, a Permalloy film with thickness of 100 nm was deposited using a magnetron sputtering system with research S-gun and base pressure of 2×10^{-7} Torr. The argon pressure during deposition was 3.0 mTorr. The deposition rate was controlled by an INFICON IC 6000 quartz monitor and was kept at 0.2 nm/s. The Permalloy film was coated with a 5 nm Cu layer to prevent oxidation. Ultrasonic assisted lift-off in acetone was used to obtain the patterned antidot arrrays. The whole antidot pattern consists of an array of 15 X 15 patches; the lateral size of each patterned patch is 100 X 100 μm². The spacing between each patch is around 10 μm, and the size of the whole pattern is 1.65 X 1.65 mm². The SEM images of the square arrays show well defined structures as seen in Fig. 1

In our earlier work [8], the static magnetic properties of these arrays were studied using a Quantum Design MPMS XL superconducting quantum interference device (SQUID) magnetometer at 300 K.

In the present work, a Cu-coplanar waveguide structure on a GaAs substrate was used as a transmission line to propagate an electromagnetic wave from a Network Analyzer. The width of the signal lines was 12 μm and the length of the device was 6 mm. The coplanar waveguides were designed for a 50 Ω characteristic impedance. The antidot array was flipped on top of the transmission line with the array length parallel to an external dc magnetic field. The frequency was swept from 0.05 to 40 GHz at zero or a fixed external magnetic field (H). The device characterization was done using a vector network analyzer along with a micro-probe station. Noise, delay due to uncompensated transmission lines connectors, its frequency dependence, and crosstalk which occurred in measurement data, have been taken into account by performing through-open-line (TOL) calibration using NIST Multical® software [14]. The exact resonance frequency (f_{res}) and frequency linewidth (Δf) [15] were obtained from Lorentzian fits to the experimental data.

3. RESULTS & DISCUSSION

The SEM images of the three antidot arrays with different square-hole sizes are shown in Fig. 1. Well defined square holes were achieved in the antidot arrays with hole widths of 1200, 800 and 400 nm. Decreasing the hole width to 400 nm causes the corner of the hole to be a little rounded. The spacing between the vacant holes is fixed at 400 nm for all the three antidot patterns.

Fig.2 shows the transmission response of the 800X800 nm² antidot structures. One sees two resonance modes of almost equal intensity. The first mode is at a lower frequency and the second is at a higher frequency in comparison to the single uniform resonance mode observed for a continuous Permalloy film of same thickness. For the antidot arrays, the uniform mode splits into two distinct resonance modes. In between these two resonance modes we observed a band-pass region (shown by a double arrow line in Fig.2), which can be tuned by an external magnetic field. It is also observed that the bandwidth of the pass-band increases with the increase in the square hole-size. This is because at a constant magnetic field (say 4 kOe); the higher resonance mode for all the three antidot arrays (S1-S3) occurs at the same frequency of 21 GHz,

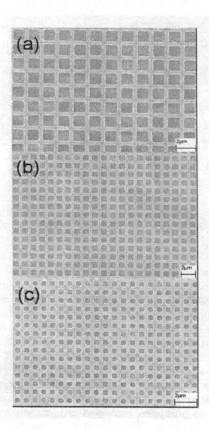

Fig. 1: SEM images of the antidot arrays with different square hole sizes: [(a) 1200 X 1200, (b) 800 X 800, and (c) 400 X 400 nm²]. The separation between holes is 400 nm.

whereas, the lower resonance occurs at 16, 15 and 14 GHz for sample S1 to S3, respectively.

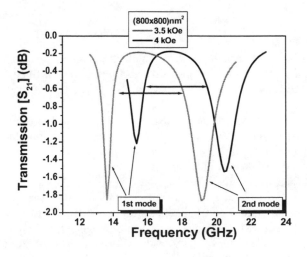

Fig. 2: The transmission response of the Permalloy antidot sample 800X800 nm² exhibiting double resonances at 3.5 and 4 kOe fields. The localized band-pass region is due to double notches.

In contrast, the continuous Py film resonance occurs at 18.5 GHz, i.e. in between the two modes of the antidot array. We also observed that the higher mode is broader than the lower mode. This may be due to multiple additional modes around that frequency, which are the outcome of the localized edge modes due to the sharp drop of the effective field near the hole edges.

Fig.3 shows the in-plane magnetic field dependence of both the modes for the 800 x 800 nm² (S2) antidot array along with the resonance frequency of the continuous Py film. When the magnetic field was along the plane, the sharp uniform precessional resonance mode was observed at lower frequency, and multiple resonance peaks also showed up at higher frequency. The multiple resonance peaks are associated with the excitation of quantized standing spin wave modes due to both the perpendicular and the lateral confinements [7, 9]. The double uniform resonance modes are originating from two regions experiencing different demagnetization field distributions.

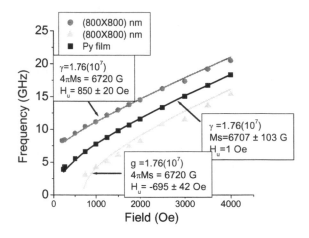

Fig. 3: Magnetic field dependence of double mode frequencies for sample S2 along with the frequency for a Py continuous film. The solid line is the theoretical fit to the FMR relation.

The inhomogeneous demagnetizing field has both static and dynamic components. When the field is applied along one side of a hole, the static magnetic poles are mainly distributed around the side perpendicular to the field. Therefore, the effective static dipole field has the opposite orientation of the external field in this case. In contrast, the long line of magnetic material parallel to the applied field will have some dynamic demagnetizing fields along its side edges. The distinct orientations of the effective fields in these two regions cause the split of the uniform mode. The theoretical fitting in Fig. 3 to these modes includes uniaxial in-plane anisotropy which is the consequence of effective fields with orthogonal orientation. In a microwave experiment, when the magnetic field is applied along the length of the hole, the spins precess in the uniform mode at FMR and they generate a demagnetizing field with its origin in the "magnetic charges" present on the surfaces. These local demagnetizing fields are responsible for the up-shifting of the second mode resonance frequency of the hole in comparison [5] to a continuous film. The absorption spectra of all the hole arrays show comparable line-shape.

One can use these ideas to obtain a simple, yet reasonable, estimate for the frequencies. We consider two fundamental elements: 1) a finite rectangular bar with its long axis oriented perpendicular to the applied field, and 2) an infinitely long bar of finite width with the long axis oriented along the external field. The Kittel formula [16]

$$f = \gamma \sqrt{\left(H + 4\pi(N_y - N_z)M_s\right)\left(H + 4\pi(N_x - N_z)M_S\right)}$$

gives the frequency as a function of the applied field H, the saturation magnetization M_s, and the demagnetizing factors N. For case (1) and using sample S2 above, the finite bar has a length of about 900 nm, a width of 400 nm and a thickness of 100 nm. This leads to demagnetizing factors $N_x = 0.712$, $N_y = 0.086$, and $N_z = 0.201$. For case (2) the length is infinite, and the width and thickness are the same as in case (1). This gives demagnetizing factors $N_x = .770$, $N_y = .230$, and $N_z = 0$. In Fig. 4 we present the resultant frequencies as a function of applied field.

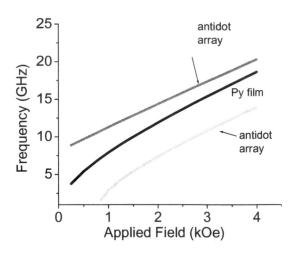

Fig.4: Simple estimate for the magnetic field dependence of double mode frequencies for sample S2 along with the frequency for a Py continuous film.

Considering the simplicity of the model, the result is surprisingly close to the experimental data of Fig. 3. We emphasize that the static magnetization patterns and dynamics can be much more complicated than the simple picture presented here as illustrated in Ref [17] Furthermore, it is not obvious whether one should use a finite or infinite length structure in case (1). Using the infinite length lowers the frequency of the lower mode by about 2 GHz. Nonetheless, Fig. 4 clearly shows the effective anisotropy is due to the structure of the antidot array.

NSTI-Nanotech 2008, www.nsti.org, ISBN 978-1-4200-8503-7 Vol. 1

Fig.5 shows the effective anisotropy field arising from the dipolar field distribution of the different antidot arrays as a function of the size of the antidot. The anisotropy field is observed to decrease with the increase of hole size. It is also observed that the frequency linewidth of the notches increases with the decrease in hole size.

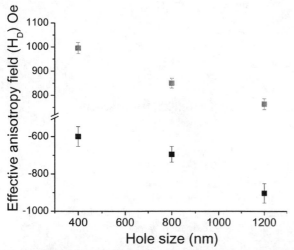

Fig. 5. Observed effective anisotropy field as a function of square antidot size.

4.   CONCLUSION

Gyromagnetic resonance is observed using NA-FMR for Py antidot arrays. For an external field applied in the plane of the sample and along one side of the square, two resonances are observed. The frequencies of the resonances are controlled by the hole-size as well as by the applied magnetic field. We also observed that the frequency linewidths of the antidot arrays are dependent on the size of the hole. The two resonances can be modeled by introducing two distinct effective anisotropy fields. The magnitude of the induced anisotropy decreases with increase in the square hole size. The frequency linewidth of the higher order mode was observed to be larger than the lower one. The possible application of antidot arrays as a magnetically tunable local band-pass filter is shown.

The work at UCCS was supported by DOA Grant No. W911NF-04-1-0247

REFERENCES

[1]. L. Torres, L. Lopez-Diaz, and O. Alejos, J. Appl. Phys. **87**, 5645 (2000).

[2]. C. C. Wang, A. O. Adeyeye, and N. Singh, Nanotechnology **17**, 1629, (2006).

[3]. A. O. Adeyeye, J. A. C. Bland, and C. Daboo, Appl. Phys. Lett. **70**, 3164, (1997).

[4]. C. C. Wang, A. O. Adeyeye, N. Singh, Y. S. Huang, and Y. H. Wu, Phys. Rev. B **72**, 174426 (2005).

[5]. L. J. Heyderman *et al.*, Phys. Rev. B **73**, 214429 (2006).

[6]. C. T. Yu, M. J. Pechan, and G. J. Mankey, Appl. Phys. Lett. **83**, 3948, (2003).

[7]. C. T. Yu, M. J. Pechan, W. A. Burgei, and G. J. Mankey, J. Appl. Phys. **95**, 6648 (2004).

[8]. Minghui Yu, Leszek Malkinski, Leonard Spinu, Weilie Zhou, and Scott Whittenburg, J. App.Phy. **101**, 09F501, (2007).

[9]. M. Nisenoff and R. W. Terhune, J. Appl. Phys. **36**, 732 (1965).

[10]. P. E. Wigen, C. F. Kooi, and M. R. Shanabarger, J. Appl. Phys. **35**, 3302, (1964).

[11]. M. J. Pechan, C. T. Yu, R. L. Compton, J. P. Park, and P. A. Crowell, J. Appl. Phys. **97**, 10J903 (2005).

[12]. G. N. Kakazei, P. E. Wigen, K. Yu Guslienko, V. Novosad, A. N. Slavin, V. O. Golub, N. A. Lesnik, Y. Otani, Appl. Phys. Letter, v.85, p. 443 (2004)

[13]. G. Gubbiotti, Phys. Rev. B, v.72, p.224413-1 (2005)

[14]. R. B. Marks, IEEE Trans. Microwave Theory Tech., vol. 39, p. 1205, (1991).

[15]. Bijoy K. Kuanr, R. E. Camley and Z. Celinski, Applied Physics Letters, **87**, 012502 (2005).

[16] C. Kittel, Phys Rev., **73**, 155 (1948).

[17] I. Guedes, M. Grimsditch, V. Metlushko, P. Vavassori, R. Camley, B. Ilic, P. Neuzil, and R. Kumar, Phys. Rev. B **67** 024428 (2003)

Micro-patterning layers by flame spray aerosol deposition

A. Tricoli[1], M. Graf[2], F. Mayer[2], S. Kühne[3], A. Hierlemann[3] and S. E. Pratsinis[1]

[1]Particle Technology Laboratory, [3]Physical Electronics Laboratory,
ETH Zurich, CH-8092 Zurich, Switzerland
[2]Sensirion AG, CH-8712 Stäfa Zürich, Switzerland

Keywords: patterning, sintering, nanoparticles, sensors, solid-oxide fuel cells.

ABSTRACT

Here we present a CMOS-compatible, two-step method for deposition and *in-situ* mechanical stabilization of gas sensitive, metal-oxide microlayers on wafer-level. Lace-like highly porous, Pt-doped SnO_2 nanostructured layers are deposited at wafer-level on 69 microsensors. Second, these layers are converted in well-adhered, cauliflower-like structures (figure 1b, inset). The resulting sensor layer performance is characterized using the analytes CO and EtOH on microsensor devices.

1 INTRODUCTION

The development of low-cost, portable, metal-oxide gas sensors with high sensitivity, selectivity and material stability bears considerable scientific and commercial potential (Eranna *et. al.*, 2004). Highly sensitive nano-material synthesis by direct, aerosol-based methods offer unique advantages in comparison to wet-routes including crack-free, highly pure deposits, and the fact that only few process steps are required (Madler *et al.*, 2006). Sputtering, spray pyrolysis, cluster beam deposition, spray pulverization, combustion chemical vapor deposition (Liu *et al.*, 2005) and, recently, flame spray pyrolysis (FSP) have been applied to yield nanostructured sensing layers. The FSP freshly-deposited layers, in particular, consist of highly-porous (98%), loosely interconnected, soft nanostructures (Madler *et al.*, 2006). These, however, can be easily destroyed under mechanical stress and require stabilization. As we have shown lately it is possible to restructure the morphologies of these layer by in-situ annealing reaching higher mechanical stability (Tricoli et al., 2008).

2 RESULTS AND DISCUSSION

Figure 1 shows TEM images of powder samples that were collected on a filter placed downstream of the sensor deposition area from flames A (Fig. 1a, b) and D (Fig. 1c, d). Mostly polyhedral particles are made which are similar to previously FSP-made and vapor-fed, flame-made SnO_2 particles. The number of small particles ($d_{TEM} < 10$ nm) in the TEM images decreases with increasing FSP flame enthalpy density, in agreement with the decrease in average grain size (SSA) and crystal sizes.

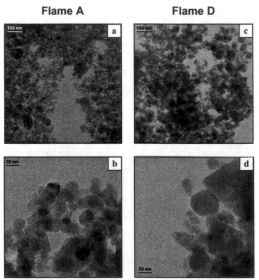

Flame A **Flame D**

Figure 1: TEM images of SnO2 particles produced by FSP. The particle size increases with increasing flame enthalpy.

Figure 2 shows the resistance of a sensor with a nanostructured, transparent SnO_2 layer ($d_{XRD} = 12$ nm) at different ethanol concentrations by heating the substrate at 220 °C. The response of the sensor response was in the range of seconds and a stable resistance was reached promptly. The sensor response to ethanol was always large in comparison to previous studies.

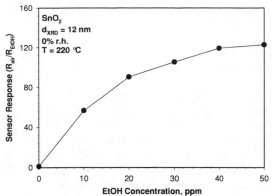

Figure 2: Sensor response to increasing ethanol concentrations.

3 CONCLUSIONS

Uniform, regular, macroporous Pt/SnO$_2$ layers have been patterned simultaneously on microsensors on wafer-level down to a diameter of 100 μm at 20 μm resolution. Gas microsensors showed a detection limit to CO of 1 ppm and fast response and recovery times. The layers had a large response also to EtOH ranging from 60 to 120 for concentrations varying from 10 to 50 ppm at 220 °C. Recent studies have reduced this to 100 ppb.

Financial support was provided from the Swiss Commission for Technology and Innovation KTI under grant 7745.1 and the ETH CCMX-NANCER program.

Eranna, G., Joshi, B.C., Runthala, D.P., & Gupta, R.P. (2004). *Crit. Rev. Solid State Mat. Sci.*, 29, 111-188.

Madler, L., Roessler, A., Pratsinis, S.E., Sahm, T., Gurlo, A., Barsan, N., & Weimar, U. (2006). *Sens. Actuators, B*, 114, 283-295.

Liu, Y., Koep, E., & Liu, M.L. (2005). *Chem. Mat.*, 17, 3997-4000.

Tricoli, A., Graf, M., Kühne, S., Mayer, F., Hierlemann, A., & S.E. Pratsinis (2008). Adv. Mater, In Press

DNA-Templated Assembly of Protein Complexes at Nanoscale

G. E. Sroga

Rensselaer Polytechnic Institute

Ricketts 101, 110 8th Street, Troy, NY 12180-3590, USA, srogag@rpi.edu

ABSTRACT

Directed, biologically-driven self-assembly has the potential to yield hybrid multicomponent architectures with applications ranging from sensors and diagnostics to nanoelectronic devices. Critical to these applications is to gain control over the precise orientation and geometry of biomolecules interacting with one-another and with surfaces. Such control has thus far been difficult to achieve in even the simplest biomolecular designs. Using a novel strategy for generation of multicomponent biological nanoarchitectures, the DNA-templated assembly of multiprotein complexes recognizing methylated DNA was achieved. The reassembly of two fragments of TEM-1 β-lactamase, each one fused with a specific DNA recognition factor, into a catalytically active protein was achieved by using the cognate DNA elements of these factors. This strategy could potentially become a useful tool in studies of genomic DNA methylation in the context of cellular epigenetic processes.

Keywords: fragmented enzyme reporter, reassembly driven by methylated dsDNA, nanoscale biomolecular architectures, catalytic function

1 INTRODUCTION

Methylation of genomic DNA is a common characteristic of living organisms such as bacteria, plants or animals. However, its role varies widely between different organisms. Even within the animal genomes, methylation patterns differ substantially from undetectable in nematodes to global methylation in vertebrate genomes. In vertebrates, DNA methylation occurs predominantly at position 5 of cytosines when followed by guanosine (CpG). CpG islands are GC-rich regions of DNA, stretching for an average of about 1 kb, and for example in humans, they coincide with the promoters of approximately 60% of genes transcribed by RNA polymerase II [1]. DNA methylation is known to play an essential role in gene silencing [2, 3] and mammalian development [4]. Thus, the promoters having CpG islands could be subjected to regulation by methylation. However in the case of genomic DNA methylation, there could be subtle differences between identical genomes that escape detection by strategies available nowadays, for example, by current microarray technologies. Therfore, there is a clear need for experimental strategies with capabilities to address emerging issues of epigenetics.

One mechanism by which DNA methylation can cause transcriptional repression is by direct interference with the binding of sequence-specific transcription factors to DNA. Indeed, some transcription factors have been shown to be unable to bind to their target sequences, which became methylated [5, 6]. More indirect mechanism of repression also exists and is supported by observations that DNA methylation can repress transcription at some distance [7, 8] as well as only after chromatin assembly [9]. Several proteins have been identified, which bind specifically to methylated DNA in any sequence context [10]. These proteins have similar structures and make specific contacts in the major grove of methylated DNA, and therefore, became the focus of this study aiming to develop a simple experimental approach to investigate changes in genomic DNA methylations during some developmental processes.

Specifically, two well characterized proteins were selected, human MBD2 protein that binds to methylated CpG islands with K_d = 2.7 nM [11] and Zinf268 that binds to its cognate dsDNA sequence motif with K_d = 6 nM [12]. As MBD2 binding affinity for nonmethylated-CpG sites is at least 70-fold lower then for methylated CpGs, it was inferred that the difference in MBD2 the binding affinities for methylated and nonmethylated DNA should allow selective targeting of the methylated CpG ($_m$CpG) sites. Conversely, it was expected that the exquisite specificity of Zif268 transcription factor for 5'-GCGGGTGC-3' sequence motif would facilitate control over targeting of the engineered protein complexes to the desired DNA sites. In order to be able to detect attachment of the respective DNA-binding proteins to DNA, a reporter was needed that would permit monitoring of the assembly process.

The enzyme-based assays are simple and relatively inexpensive. *E. coli* TEM-1 β-lactamase was selected as a reporter as this monomeric enzyme appears to meet all the essential criteria of a desired protein reporter. It can be easily expressed in *E. coli* and is not toxic to both prokaryotic and eukaryotic cells. As eukaryotic cells do not contain endogenous β-lactamase activity, potential problems caused by an unspecific background would not be encountered. Structure and function of *E. coli* TEM-1 β-lactamase are well characterized [13] and its activity can be simply assayed by the hydrolysis of nitrocefin using a colorimetric assay. It was also shown that the enzyme reassembles from its fragments into a functional protein [14] and this was particularly relevant to the discussed work as one of its main goals has been to explore the structural complementation of the fragmented β-lactamase in conjunction with the presence of investigated DNA.

To this end, a potentially general approach for the design and synthesis of structured and functional protein assemblies reporting the degree of double-strand DNA (dsDNA) methylation was developed. Exchanging MBD2 for other natural or synthetic proteins binding methylated dsDNA and Zif268 for other transcription factors, respectively, may yield more complete set of tools to study global changes of DNA methylation during different biological processes.

2 EXPERIMENTAL PROCEDURES

2.1 Cloning, Expression, and Purification of the Proteins

Unless otherwise stated, all molecular biology methods used in the presented study were performed according to the standard protocols described by Sambrook et al. [15] and Sambrook and Russell [16]. The *E. coli* gene of TEM-1 β-lactamase originated from pCRII-TOPO vector (Invitrogen). Full-length *E. coli* TEM-1 β-lactamase, two fragments of the enzyme, Bla1 and Bla2 [17], and the investigated fusion proteins were generated using the PCR method. The M182T mutation was introduced into the Bla1 sequence to enhance the stability of the protein [18]. The Bla1 fragment of TEM-1 β-lactamase (H26 to G196) was fused with Zif268 and the Bla2 fragment (L198 to W290) with MBD2, respectively. In each case, a 15 amino acid linker separated the Bla domain from the fused protein.

Periplasmic expression of the respective proteins was achieved after recloning of the gene constructs into the pBAD/gIII-D (Kana^R) vector (Invitrogen). Protein expression was conducted according to the manufacturer's instruction. Cell fractionation and extracts' analysis was performed as described by Sroga and Dordick [19].

2.2 Methylation of DNA *in vitro*

Biotinylated DNA fragments (B-DNA) contained Zif268 sequence motif (5'-GCGGGTGC-3') separated from the methyl-CpG ($_m$CG) site (i.e., the binding target for MBD2 domain) by 10 bp DNA spacer. Cognate nonmethylated DNA and random DNA sequence were used in the control experiments. All DNA fragments contained a 16 bp linker sequence at the 3'-end. The DNA fragments were generated using a method published previously [20] and attached to the streptavidin-coated paramagnetic particles (SA-PMPs) [20] after their methylation.

Metylation of dsDNA was conducted using *Sss*I methylase (New England Biolabs) according to the manufacturer's protocol.

2.3 Assembly of Engineered Proteins onto DNA Attached to SA-PMPs

E. coli periplasmic fractions served as the direct source of Bla1-Zif268 and Bla2-MBD2 fusion proteins. To perform the DNA-driven reassembly of TEM-1 β-lactamase, the appropriate periplasmic fractions were mixed in various combinations with PMPs carrying different DNAs. After washings, the PMPs with the attached complexes were directly used in the nitrocefin assay.

2.4 Nitrocefin Assay

The assay was performed using standard 96-well micro-titer plates (MT-plates). To the wells of MT-plate containing 20 or 40 μL of the appropriate DNA-PMPs in the nitrocefin assay buffer (100 mM phosphate, pH 7.0), 20μL of a given protein sample(s) were added to the final volume of 200 μL, mixed and placed immediately in the HTS 7000 Plus Bio Assay Reader (Perkin Elmer, Norwalk, CT) with the HTS 2.0 software. Absorbance readings were recorded at 486 nm at one min intervals for 20 min. The SA-PMPs in 200 μL of the nitrocefin assay buffer served as a blank. Initial reaction rate data were collected and calculated using the HTS 2.0 software taking the molar extinction coefficient for hydrolyzed nitrocefin ($\varepsilon_{486\ hydrolyzed\ nitrocefin}$ = 20.5 x10^3 M^{-1} cm^{-1}) into account.

3 RESULTS AND DISCUSSION

To meet the requirements of a simple assay for *in vitro* detection and quantitation of genomic DNA methylations, two engineered fragments of *E. coli* TEM β-lactamase were tested for their ability to reassemble onto a methylated-dsDNA template. The assembly occurred through the appropriate DNA recognition factor that was fused in frame with a given fragment of β-lactamase (Figure 1).

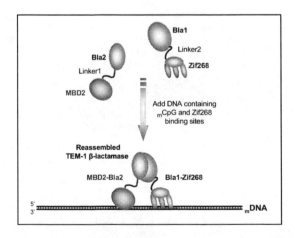

Figure 1: Schematic showing the principle of the methylated-dsDNA detection through the assembly of the split TEM-1 β-lactamase.

Each protein fragment of β-lactamase was designed with a 15 amino acid linker that separated it from the DNA-binding factor in order to prevent potential steric clashes. Conversely, selection of the 10 bp spacing between the binding sites was based on the typical distance observed between natural DNA binding motifs (also named boxes) within promoters for various transcription factors as well as the results from the previous work [20].

Catalytic activity of TEM-1 β-lactamase was determined using the colorimetric substrate nitrocefin that changes color from yellow to red (peak absorbance at 486 nm) upon hydrolysis. The activity of the enzyme was restored only after addition of methylated-DNA-PMPs to the protein samples containing a mixture of the Zif268-Bla1 and MBD2-Bla2 fusion proteins (Figure 2). The observed initial reaction rate, v_0 [Hydrolyzed nitrocefin], for hydrolytic activity of reassembled TEM-1 β-lactamase was 0.56 ± 0.04 µM/40 µL periplasmic fraction min^{-1} (Table 1). The specificity of the binding was further confirmed by showing the lack of enzymatic activity in the samples containing only one type of the assembly component and methylated cognate DNA or both protein constructs and either nonmethylated-cognate or non-cognate DNA (Figure 2). In summary, this part of the investigation demonstrated that the Bla1 and Bla2 fragments can be brought together to form an active enzyme only in the presence of methylated cognate dsDNA.

% of DNA on SA-PMPs A: $_m$CpG-(N)$_{10}$-TFBS$_{Zif268}$ B: CpG-(N)$_{10}$-TFBS$_{Zif26}$	V_0 [Hydrolyzed nitrocefin] [µM/40 µL periplasmic fraction min^{-1}]
100% **A**	0.56 ± 0.04
50% **A** + 50% of **B**	0.29 ± 0.03
100% **B**	ND[1]

[1] ND = Not Detected. Values are the average of four parallel assays. Average amount of DNA attached to 75 µl particles was 600 ng (21 pmoles). To ensure similar binding-space availability, the length of all tested DNA fragments was ca. 75 bp. Thus, molar concentrations and masses of the DNAs used were comparable.

Table 1: Initial reaction rates (v_0 [Hydrolyzed nitrocefin]) for hydrolytic activity of reassembled TEM-1 β-lactamase investigated using different percentage of methylated and nonmethylated cognate dsDNA fragments attached to SA-PMPs.

Catalytic activity of the reassembled TEM-1 β-lactamase can be observed as an increasing intensity of the red color product (Figure 2). To test that proteins of *E. coli* periplasm do not cause background hydrolysis of nitrocefin, periplasmic fraction of *E. coli* host was incubated with nitrocefin under the standard assay conditions. No observable hydrolysis was detected over the assay period as well as within 1 hr of its completion. The rationale for

testing periplasmic fractions directly in the assay instead of the purified proteins was to develop more robust and faster to perform experimental procedure. Notably, similar results were obtained when the assay was performed using purified fusion proteins.

In a separate set of experiments, two- and three-fold higher concentration of engineered proteins was used for attachment and this resulted in approximately 1.8-fold and 2.8-fold increase in v_0, respectively. These data imply that in addition to the identification of DNA methylation, a concentration of the reassembled TEM-1 β-lactamase from its fragments can be simply calculated, as under the usual *in vitro* enzyme assay conditions reaction rate is directly proportional to the enzyme concentration, and hence, to the level of recombinant DNA-binding factors. Thus, it was calculated that on average periplasmic fractions contained about 18 – 19 % of a given fusion protein.

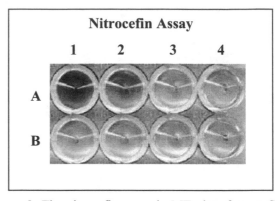

Figure 2: The nitrocefin assay in MT-plate format. Sample **A1**: 20 µL $_m$CpG-TFBS$_{Zif268}$ dsDNA-PMPs with 160 ng (6 pmoles) of bound DNA plus Bla1-Zif268 and MBD2-Bla2; **A2**: 10 µL $_m$CpG-TFBS$_{Zif268}$ dsDNA-PMPs with 80 ng (3 pmoles) of bound DNA plus Bla1-Zif268 and MBD2-Bla2; **A3**: 20 µL $_m$CpG-TFBS$_{Zif268}$ dsDNA-PMPs and Bla1-Zif268; **A4**: 20 µL $_m$CpG-TFBS$_{Zif268}$ dsDNA-PMPs and Bla2-MBD2; **B1**: 20 µL CpG-TFBS$_{Zif268}$ dsDNA-PMPs (i.e., non-methylated) plus Bla1-Zif268 and MBD2-Bla2; **B2**: 20 µL random-dsDNA-PMPs plus Bla1-Zif268 and MBD2-Bla2; **B3**: 20 µL $_m$CpG-TFBS$_{Zif268}$ dsDNA-PMPs and the periplasmic fraction of the *E.coli* host; **B4**: 20 µL SA-PMPs, Bla1-Zif268 and MBD2-Bla2. The amount of DNA attached to the paramagnetic particles used in the samples A3 to B4 was 160 ng (6 pmoles).

The major challenge in developing methods to study diverse biological processes *in vitro* is to be able to extrapolate the results to the *in vivo* cellular events. This issue is especially relevant to the study of methylation-dependent modulation of gene expression. The use of SA-PMPs permits generation of the particles with various percentage of cognate methylated-dsDNA on the "background" of cognate nonmethylated-dsDNA and this could mimic the variation in the levels of biological genome methylation. One may expect that more engineered proteins would assemble onto the larger number of methylated-dsDNA sites as compared to the particles carrying less of these sites and this would correspond to quantitative differences in the observed enzyme activity. Indeed, the measured v_0 values for the reassembled TEM-1 β-lactamase confirmed the aforementioned hypothesis (Table 1).

The availability of rapid and robust methods for detection of DNA methylation is important for elucidating roles DNA methylation-based gene regulation and related epigenetic processes. The strategy presented here should be useful for selective, qualitative and quantitative analysis of methylated-DNA - protein interactions through the enzymatic activity of the reassembled tag. Because under the usual *in vitro* assay conditions reaction rate is directly proportional to the enzyme concentration – and in this study also to the concentration of the fused DNA-binding proteins – the degree DNA methylation can be simply evaluated in the terms of the enzyme arbitrary units. The work presented here relies on the concept of the enzyme/protein-fragment complementation (PFC) that was developed for the quantitative detection of dynamic protein-protein interactions *in vitro* and *in vivo* [21]. The crucial feature of PFC is that the fragments are designed not to fold spontaneously without being brought into close proximity by the interaction of the proteins to which they are fused. Without a spontaneous folding, there is no a false positive signal and this is in contrast to the assay systems that rely on naturally occurring and spontaneously associating subunits of the enzymes fused to interacting proteins. The central problem of those assays is that subunits, even if weakly associating, are always able to do this to some extend, meaning that there is a constant background of spontaneous assembly.

Although a number of cell biological applications use GFP and other fluorescent proteins as tags [22], the split-enzyme reporter was selected instead of GFP for a number of reasons. For example, the somewhat disappointing sensitivity of the GFP tag is probably inherent result of its lack of signal amplification. Unlike enzymes, GFP can not catalytically process an indefinite number of substrate molecules. Each GFP molecule produces at most one fluorophore. It has been estimated that 1 μM well-folded wild-type GFP molecules is required to equal the endogenous autofluorescence of a typical mammalian cell [23], that is, to double the fluorescence over the background. Mutant GFPs with improved extinction coefficients might improve this detection limit, but still 0.1 μM GFP is approx. 10^5 copies per typical cell of 1 – 2 pL volume. Moreover, this estimate already assumes perfect GFP maturation; any maturation/folding problems would raise the threshold copy number even further.

In conclusion, an approach facilitating detection of methylated dsDNA was developed. One of the most attractive aspects of this strategy would be to extend some of its concepts to develop an assay functioning in the *in vivo* cellular setting, and this is currently undergoing investigation.

This work was supported by the Chemical Engineering Instructor Program, School of Engineering, Rensselaer Polytechnic Institute.

REFERENCES

[1] F. Antequera and A. Bird, Proc. Natl. Acad. Sci. USA 90, 11995 – 11999, 1994.

[2] B. Neumann and D. P. Barlow, Curr. Opin. Genet. Dev. 6, 159 – 163, 1996.

[3] A. Razin and H. Cedar, Cell 77, 473 – 476, 1994.

[4] E. Li, T. H. Bestor and R. Jaenisch, Cell 69, 915 – 926, 1992.

[5] M. Comb and H. M. Goodman, Nucleic Acid Res. 18, 3975 – 3982, 1990.

[6] S. M. M. Iguchi-Ariga and W. Schaffner, Genes Dev. 3, 612 – 619, 1989.

[7] H. Cedar, Cell 53, 3 – 4, 1988.

[8] S. U. Kass, J. P. Goddard and R. L. P. Adams, Mol. Cell. Biol. 13, 7372 – 7379, 1993.

[9] G. Buschhausen, B. Wittig, M. Graessmann and A. Graessmann, Proc. Natl. Acad. Sci. USA 84, 1177 – 1181, 1987.

[10] B. Hendrich and A. Bird, Mol. Cell. Biol. 18, 6538 – 6547, 1998.

[11] M. F. Fraga, E. Ballestar, G. Montoya, P. Taysavang and P. A. Wade, Nucleic Acid Res. 31, 1765 – 1774, 2003.

[12] N. P. Pavletich and C. O. Pabo, Science 252, 809 – 817, 1991.

[13] L. Maveyraud, R. F. Pratt and J.-P. Samama, Biochemistry 37, 2622 – 2628, 1998.

[14] A. Galarneau, M. Primeau, L.-E. Trudeau and S. W. Michnick, Nature Biotechnol. 20, 619 – 622, 2002.

[15] J. Sambrook, E. F. Fritsch and T. Maniatis, "Molecular Cloning: A Laboratory Manual," 2nd ed., CSH Laboratory Press, Cold Spring Harbor, NY, 1989.

[16] J. Sambrook and J. D. W. Russell, "Molecular Cloning: A Laboratory Manual," 3rd ed., CSH Laboratory Press, Cold Spring Harbor, NY, 2001.

[17] A. Galarneau, M. Primeau, L.-E. Trudeau and S. W. Michnick, Nature Biotechnol. 20, 619 – 622, 2002.

[18] V. Siedarki, W. Huang, T. Palzkill and H. F. Gilbert, Proc. Natl. Acad. Sci. USA 98, 283 – 288, 2001.

[19] G. E. Sroga and J. S. Dordick, Biotechnol. Bioeng. 78, 761 – 769, 2002.

[20] G. E. Sroga and J. S. Dordick, Biotechnol. Bioeng. 94, 312 – 321, 2006.

[21] I. Remy and S. W. Michnick, BioTechniques 42, 137 – 144, 2007.

[22] I. Ghosh, C. I. Stains, A. T. Ooi and D. J. Segal, Mol. BioSyst. 2, 551 – 560, 2006.

[23] K. D. Niswender, S. M. Blackman, L. Rhode, M. A. Magnuson and D. W. Piston, J. Microsc. 180, 109 – 116, 1995.

Bimodal Macro- Mesoporous Silica Network

G. Abellán, A.I. Carrillo, N. Linares, J. García Martínez*

Inorganic Chemistry Department. University of Alicante, Carretera San Vicente s/n,
E-03690, Alicante, Spain.
e-mail: j.garcia@ua.es URL: www.ua.es/grupo/nanolab

ABSTRACT

Bimodal macro-mesoporous silica networks have been prepared in a simple one-pot synthesis using an inexpensive non-toxic surfactant and tetraethoxysilane as a silica precursor. These novel materials show high pore volumes with interparticle macropores and templated mesopores (average pore size 3.0 nm) in 20 nm thick walls. The key properties of these materials can be tailored by controlling the experimental conditions (temperature, pH, stirring and surfactant properties and concentration) to obtain surface areas over 1000 m^2/g and total pore volumes of 2.0 cm^3/g.

Keywords: bimodal porous silica, non-toxic surfactants, mesoporous

1 INTRODUCTION

Mesostructured solids have attracted much attention in the last years due to their wide range of applications in catalysis, controlled delivery, separation techniques, optical devices and sensors. [1-6] Surfactant-templated synthesis is a convenient widely used technique for the preparation of mesostructured materials.[2,3] This strategy allows for the precise control of porosity by adjusting key synthesis parameters. Surfactant-templated materials show some remarkable properties, such as high surface area and narrow pore size distribution [4-6]. Especially relevant is the case of ordered mesoporous silica, such as MCM-41 and related materials, firstly reported in the early 90´s [8]. The principal feature of the MCM-41 is its periodic and ordered unimodal mesoporous structure. Although these properties have been suggested to be useful in catalysis, the poor acidity and stability of these materials have limited their applications [7-8]. Multimodal materials combine the benefits of high surface area micro- and mesoporosity with the accessible diffusion pathways of macroporous networks [9]. There are several reports on materials with various pore combinations: micro-mesopores, micro-macropores, meso-macropores, or trimodal micro-meso-macropores.[10].

In order to induce porosity, directing agents like tetraalkylammonium ions are typically used to produce microporous materials, like zeolites, block copolymers or long chain surfactants have been extensively used to prepare a wide variety of mesoporous solids, and hard templates, like polystyrene latexes, nanoparticles, or carbon black to make macroporous materials [6,10].

S. Mann and co-workers have synthesized macroporous sponge-like monoliths and mesoporous thin films using starch gels and sponges in combination with preformed silicalite nanoparticles.[10] Zhang et al. have described the synthesis of bimodal nanoparticulated silicas using a single non-ionic templating surfactant.[11] Cationic surfactants, like cetyltrimethy-lammonium bromide (CTAB), have been widely used to produce MCM-41 and other mesoporous materials. [12,13] In order to prepare materials with bimodal porosity two or more surfactants have been used simultaneously. [14,15] In many cases, the precise control of the pH is an important variable to finely tune the porous properties of these templated structures.

Using a modified synthesis of the so called "atrane route", L. Huerta and co-workers have synthesized a novel bimodal mesoporous material with pore volumes in excess of 2.0 cm^3/g. The strategy is based on the use of complexes like atranes or silatranes as hydrolytic inorganic precursors and surfactants as template [7]. L. Huerta and co-workers also shows that the pore and particle sizes in hierarchic porous silicas are highly dependent on the surfactant nature [7,9]. These materials, although prepared using a different strategy, show a similar structure to the ones that we have produced.

Herein we described the synthesis of a bimodal porous silica network prepared at room temperature in the presence of a cationic surfactant, namely the quaternary amine tallow tetramine. These materials have large surface area, high pore volume, and interpaticlc porcs (20-100 nm) formed by mesoporous (3 nm) walls (20 nm thick). The key properties of these materials can be tailored controlling the experimental conditions (temperature, pH, stirring and surfactant properties and concentration) to produce materials with more than 1000 m^2/g and total pore volumes of 2.00 cm^3/g in a one-pot synthesis.

2 EXPERIMENTAL SECTION

2.1 Materials

The surfactant used as template was tallow tetramine, kindly supplied by Tomah3 Products Inc.

All reagents, tallow tetramine, tetraethoxysilane (98%, Aldrich, denoted as TEOS) and ammonium hydroxide (30%, Sigma-Aldrich, NH4OH) were used as received without further purification.

2.2 Synthesis

In a typical synthesis, 0.275 g of tallow tetramine was magnetically stirred (400 rpm) overnight in 25 ml of deionised water. Alternatively, orbital stirring (60 rpm) was used also overnight. Then, the pH was adjusted to 9.75 with NH_4OH (30%). At this value a viscous solution was obtained. Then, the copolymerization of 1.36 g of TEOS was carried out by base-catalyzed hydrolysis in the presence of tallow tetramine. After a 15h of orbital stirring at room temperature, a white suspension was obtained and the resulting solid was washed first with water and then with ethanol, filtered out, and air dried. Finally, the surfactant was removed by calcination at 550 ºC for 8h.

The two different stirring methods (magnetic 400 rpm versus orbital 60 rpm) play a key role in the properties of the final material. More severe stirring (magnetic) produces a clear solution, which suggests that the surfactant was completely dissolved after overnight stirring. On the contrary, milder orbital stirring (60 rpm) does not allow for a complete dissolution of the surfactant after the same amount of time, producing a turbid suspension. The bimodal silica networks have been denoted as BSN1 and BSN2, where 1 and 2 refers to magnetic and orbital stirring of the surfactant, respectively.

2.3 Characterization

The morphology of the bimodal silica networks was investigated by transmission electron microscopy (TEM) and scanning electron microscopy (SEM). TEM analysis was performed using a JEM-2010 electron microscope (JEOL, Japan). The instrument, operated at 200 kV, has a resolution of 0.14 nm. Samples for TEM studies were prepared by dipping a sonicated suspension of the sample in ethanol on a carbon-coated copper grid. Scanning Electron Microscopy (SEM) analysis of all the samples was carried out using a JEOL JSM-840 microscope.

N_2 adsorption/desorption isotherms were obtained at 77 K in a Quantacrome Autosorb-6 volumetric adsorption analyzer. Before the adsorption measurements, all samples were outgassed at 523 K for 4 h at $5*10^{-5}$ bars. The pore size distribution was calculated from desorption branch of isotherms using the BJH method.

3 RESULTS AND DISCUSSION

3.1 Sample Morphology: TEM and SEM Analysis

Representative TEM micrographs of BSN1 (magnetic) and BSN2 (orbital) silica materials are shown in Figures 1a and 1b, respectively. The insets in Figures 1a and 1b are a detail of BSN1 and BSN2 obtained at higher magnification. BSN1 and BSN2 materials show a tortuous open structure which form irregularly-shaped intraparticle pores. At higher magnification, non-ordered mesopores are observed inside

the network walls which are 20 nm thick (see inset in Figure 1a and 1b). The intraparticle mesopores are only a few nanometers in diameter whereas the interparticle pores are significantly larger. In all cases, the samples were highly homogeneous. No dense phases were observed.

a)

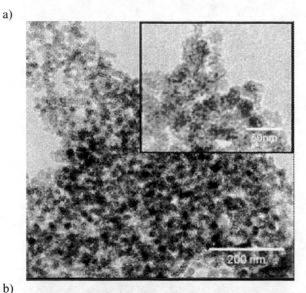

b)

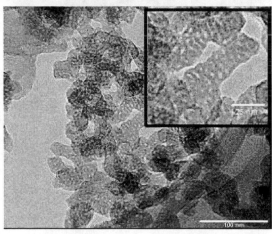

Figure 1: Representative TEM micrographs of calcined bimodal silica materials showing a very open network with both interparticle macro- and intraparticle mesopores: (a) BSN1 (magnetic) and (b) BSN2 (orbital). The detail shown as an inset in (b) was obtained at higher magnification for BSN2 (scale bar represents 25 nm).

The SEM micrographs of BSN materials shown in Figure 2 are typical of an open and irregular structure formed by nanosized particles. These materials are highly homogeneous independently of the stirring method used (see Figure 2a and b).

a)

b)

Figure 2: Representative SEM images of calcined silica bimodal material BSN2 (orbital) at two different magnifications.

3.2 Nitrogen physisorption isotherms

a)

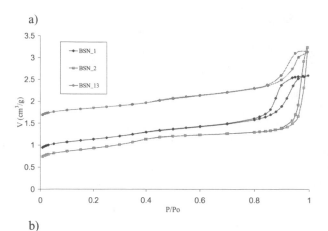

b)

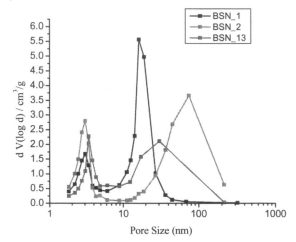

Figure 3: (a) Nitrogen adsorption isotherms of various mesoporous silicas prepared using tallow tetraamine as template (curves are shifted 0.5 cc/g for clarity) and b) Pore size distribution plot, obtained using the BJH method for the desorption branch.

Sample	Surface area[a] $(m^2 \, g^{-1})$	Pore size[b] (nm) (small pore)	Pore size[b] (nm) (large pore)	Pore Volume[c] $(cm^3 \, g^{-1})$ (small pore)	Pore Volume[c] $(cm^3 \, g^{-1})$ (large pore)	Total Pore[c] Volume $(cm^3 \, g^{-1})$
1- BSN1[d]	1015.26	3.05	16.2	0.72	1.18	1.90
2- BSN2[e]	1034.95	3.04	74.7	0.74	2.01	2.74
3- BSN13[f]	804.02	3.41	30.4	0.63	1.00	1.64

a The surface area was obtained using the BET method.

b Pore size was obtained using the BJH on the desorption branch of the isotherm.

c Pore volume was obtained from the isotherm. For the small mesopores the pore volume was measured at P/Po = 0.6 and for the total pore volume at P/P0 = 0.99. The pore volume of the large mesopore was estimated subtracting both values.

d Magnetic stirring t = 15 h.

e Orbital stirring. t = 15 h.

f Magnetic stirring t = 48 h.

Table 1: Textural parameters of different bimodal macro- mesoporous silica networks.

Nitrogen physisorption at 77K was used to study the porous properties of our materials. Figure 3a shows the isotherms of calcined BSN1, BSN2, BSN13. These isotherms have two distinctive gas uptakes. The first one, which is related to nitrogen condensation in small mesopores, appears at intermediate relative pressures (P/Po = 0.2 − 0.4). The second noticeable gas uptake occurs at higher relative pressures (P/Po = 0.8 − 1.0) and it is due to the filling of much larger mesopores or even macropores. In all cases, only the second uptake shows a hysteresis loop, indicating an irregularly shaped interparticle porosity. Their pore size distributions, obtained using the BJH method, show two distinct peaks at 3 nm (surfactant templated mesopores) and at 20-80 nm (interparticle porosity) as shown in Figure 3b. Table 1 contains some relevant textural parameters of these samples. These data is consistent with the electron micrographs, and with a very open bimodal network.

The formation of these bimodal materials has been related to the competition between kinetic and thermodynamic parameters. While intraparticle mesopores are due to the template effect of the surfactant aggregates, the large mesopores and the macropores appears as a consequence of a nucleation and growth of the primary mesoporous nanoparticles, forming an open and disordered network. Moreover, the proposed macropores formation mechanism involves the collision and aggregation of primary nanoparticles, which produces a disordered open network [7,9,16,17].

According to Amorós et al. [7,9], these structures seem to be formed by nanoparticle aggregates in the form of large clusters with fractal structure. In our case, the dimensions of the silica forming nanoparticles are between 40-70 nm, depending on the sample and the mesopores are around 6-7 nm in both BSN1 and BSN2-like network silica.

The combination of templated mesoporosity and interparticle meso- and macroporosity in these materials is expected to increase the accessibility of bulky molecules to the interior of this very open structure.

4 CONCLUSIONS

A simple one-pot synthesis of novel bimodal porous silica networks has been described using tallow tetramine surfactants. The judicious control of the pH conditions allows for the preparation of materials with very high pore volumes and open structures. Two different stirring methods have been studied at different reaction times obtaining the large pore volumes for orbital stirring and the sharp pore size dispersions results for magnetic stirring, during 15 hours.

The main advantages of the preparation herein described are its simplicity, the low cost and toxicity of the surfactant used (tallow tetramine), and the use of the pH as structure directing parameter. The materials produced are macro- mesoporous bimodal, show a large surface area and pore volumes and a very open structure.

ACKNOWLEDGEMENT

This research was supported by the Spanish MCyT (CTQ2005 − 09385 - C03 - 02).). J.G.M. is grateful for financial support under the Ramón y Cajal Program.

REFERENCES

[1] G.J.A.A. Soler-Illia, C. Sanchez, B. Lebeau, J. Patarin, Chem. Rev. 102, 4093, 2002.

[2] L. Wang et al., Micropor. Mater. 86, 81-88, 2005.

[3] T. Linssen, K. Cassiers, P. Cool, E.F. Vansant, Adv. Colloid Interf. Sci. 103, 121, 2003.

[4] D. Trong On, et al., Appl. Catal. 253, 545, 2003.

[5] M.E. Davis, Nature, 417, 813, 2002.

[6] J. García-Martínez in Highlights of Chemistry. Nanostructured Porous Materials. Building matter from the bottom-up. Wiley-VCH (Ed. Bruno Pignataro), 2007.

[7] L. Huerta, C. Guillem, J. Latorre, A. Beltrán, R. Martínez-Máñez, M. Marcos, D. Beltrán, P. Amorós, Solid State Sciences 8, 940–951, 2006.

[8] C.T. Kresge, M.E. Leonowicz, W.J. Roth, J.C. Vartuli, J.S. Beck, Nature 359, 710, 1992.

[9] J.M. Morales et al. Solid State Sciences 7 415–421, 2005, and references therein.

[10] B. Zhang, S.A. Davis, S. Mann, Chem. Mater. 14, 1369, 2002, and references therein.

[11] W. Zhang, T.R. Pauly, T.J. Pinnavaia, Chem. Mater. 9, 2491, 1997.

[12] P. Ågren, M. Lindén, P. Trens, S. Karlsson, Stud. Surf. Sci. Catal. 128 297, 2000.

[13] J. El Haskouri, D. Ortiz de Zárate, C. Guillem, J. Latorre, M. Caldés, A. Beltrán, D. Beltrán, A.B. Descalzo, G. Rodríguez, R. Martínez, M.D. Marcos, P. Amorós, Chem. Commun 330, 2002.

[14] J. Sun, Z. Shan, T. Maschmeyer, J.A. Moulijn, M.-O. Coppens, Chem. Commun. 2670, 2001.

[15] J. Sun, Z. Shan, T. Maschmeyer, M.-O. Coppens, Langmuir 19 8395, 2003.

[16] R.K. Iler, The Chemistry of Silica. Solubility, Polymerization, Colloid and Surface Properties, and Biochemistry, John Wiley & Sons, New York, 1979.

[17] C.J. Brinker, G.W. Scherer, Sol–Gel Science. The Physics and Chemistry of Sol–Gel Processing, Academic Press, New York, 1990.

SCIL – A New Method for Large Area NIL

Johann Weixlberger[*], Jan van Eekelen[**] and Marc Verschuuren[**]

[*]Suss MicroTec Lithography GmbH
Schleißheimerstr. 90, 85748 Garching, johann.weixlberger@suss.com
[**] Philips Research, High Tech Campus Eindhoven, 5656AE

ABSTRACT

A new imprint technology for sub-50nm patterning will be introduced, bridging the gap between small rigid stamp application for best resolution and large area soft stamp usage with usual limited printing resolution below 200nm – SCIL Substrate Conformal Imprint Lithography is an enabling technology offering best of two worlds – large area soft stamps with repeatable sub-50nm printing capability, avoiding stamp deformation as no contact force applied, non-UV based curing at room temperature and allowing high aspect ratios even up to 1:10

The technology will be introduced, results for various applications be shown.

Keywords: NIL, large area imprint, non-UV based room temperature NIL-process

1 INTRODUCTION

NIL Nano Imprint Technology has gained an important status as patterning technology when it comes to applications that require sub-100 nm features at reasonable costs, such as optical gratings and other periodical patterns.

Basically two different stamp materials are in use, selected corresponding to the imprint area: rigid quartz stamps for best printing resolution, demonstrated printed features down to sub-10 nm [1] but on rather small areas like 10x10mm, definitely smaller than 1″ due to physical limits of substrate/stamp planarization issues.

Soft PDMS stamp allow larger imprint areas up to substrate size (even up to 8″ is applicable), but provide a trade-off in terms of printing resolution – small features may not be printed in a repeatable stability as the stamp allows and even requires a certain distortion in order to adapt to substrate topography. Even in a very few micron range of substrate waviness a stamp distortion takes place, what can distort patterns in the sub-100 nm range. For applications in microlenses or similar the dimensions are in the sub-mm down to 1 micron range, so not impacted by this effect.

Applications in the optical field like gratings for light tracking (planar/vertical light coupling/de-coupling) in LEDs, VCSELs etc. the pitch dimensions get into a sub-200nm range, thus being close to the limits of soft imprint stamps.

A method or material to overcome this trade-off is needed in order to further drive the use of NIL as an applicable technology.

2 SCIL SUBSTRATE CONFORMAL IMPRINT LITHOGRAPHY

In order to address a.m. issues an imprint method based on soft PDMS stamps was considered to provide best results. An approach that does not apply mechanical imprint forces, a rigid stamp backplane avoiding lateral deformation but still allowing bending for planarisation and de-molding was considered as applicable approach.

2.1 Material Selection

The material combination chosen is a rather thin PDMS layer of 400 – 700 μm thickness, produced by casting from a (rigid) master stamp. Additionally this PDMS-layer is attached to a 200μm glass backplane – enabling soft bending but avoiding lateral deformation of PDMS.

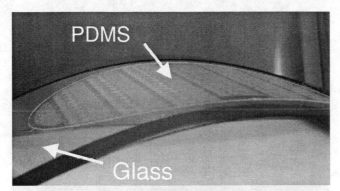

PDMS stamp (6") on 200 μm AF-45

In order to avoid imprint forces an imprint material with excellent wetting properties to PDMS was chosen – allowing pattern filling by means of capillary forces. A slightly chemically modified Sol-Gel material is the choice providing the required process properties. Following sketch explains the physics of the actual SCIL imprint process:

step 1

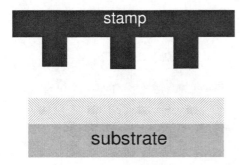

Wet sol-gel layer ~ 70nm

Step 2

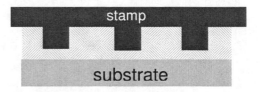

Capillary force pulls stamp into wet coating

Step 3

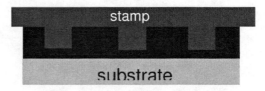

Solvent diffuses into PDMS

Step 4

Remove stamp from rigid sol-gel layer

2.2 Sol-Gel Imprint Process

The sol-gel imprint resist contains SiO_2 precursors; upon application onto the substrate by spin coating the low boiling solvent evaporates.
After application of the stamp the remaining high boiling solvent diffuses in the PDMS.
Solidification can be done within 5 minutes and depends on solvent as well as resist preparation.

2.3 Process results

Based on the fact of solvent diffusion into the PDMS stamp the question of saturation effects arises – but due to the low amount of Sol-Gel (less than 100 nm layer thickness) compared to the PDMS stamp mass (400 – 700 μm thickness) the amount of solvent going into the stamp is neglect able, especially when considering that the time window between two imprints allow the solvent to vanish completely.

The effect of solvent transfer is expected to cause a certain shrinkage; this effect happens with a predictable amount of 7%, but only in z-direction [2].

Additional shrinkage only applies when curing the Sol-Gel above 200°C

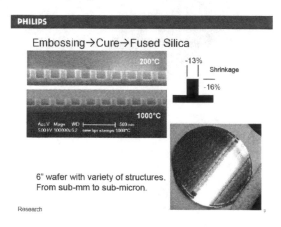

On the other hand this material allows a high temperature treatment compared to other imprint polymers that don't.

Furthermore the imprinted pattern provides excellent residual layer thickness uniformity, what can be removed with a RIE etch process and subsequently the pattern can act as etch mask into substrate or metal layers.

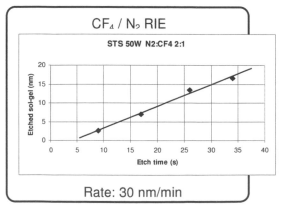

2.4 Substrate Conformal Imprints

Particles should not be the rule – but show the conformal printing capability in this images.

3 APPLICATIONS

The excellent performance in respect to substrate conformity and pattern fidelity over large areas (< 0,03 % pattern deviation measured over a 15x15 mm2 area [3]) makes this imprint technology to a powerful tool, especially for applications like high brightness LEDs, VCSELs and other optical devices but also for hard disc drives - e.g. next generation "patterned media" require sub-50nm concentric features printed onto various disc sizes with extreme pattern fidelity requirements. Both applications have been addressed already and feasibility is proven

4 REFERENCES

[1] Tomi Haatainen, Jouni Ahopelto, Gabi Grützner:" step&stamp imprint litho", SPIE3997-103
[2] Marc Verschuuren: " 3D photonic structure by sol-gel imprinting", MRS2008 San Jose
[3] Marc Verschuuren: " 3D photonic structure by sol-gel imprinting", MRS2008 San Jose

UV-NIL with optimal droplets

Vadim Sirotkin, Alexander Svintsov, Sergey Zaitsev

Institute of Microelectronics Technology, RAS, 142432 Chernogolovka, Moscow district, Russia

ABSTRACT

A homogeneous residual layer thickness in nanoimprint lithography (NIL) is a serious problem in step and flash (UV-)NIL. Improvement of thickness homogeneity could be expected from optimized size of droplets at resist dispensing in case of UV-NIL. The optimization in droplet size must exclude the stamp geometry involving areas in which the resist has to flow laterally over large distances so stamp geometry should be considered at optimization. The paper is devoted to development and critical analysis of an optimizing algorithm, which take into account only filling factor (geometry) of a stamp and does not consider the following resist flow.

In current realization of the approach a specially developed algorithm transfers stamp geometry defined in standard GDSII (or ACAD) format into rectangular (square) cells and calculates the filling factors taking into consideration stamp depth and desirable residual resist thickness. Then depending of the jet model continuous or discreet volume is calculated and saved for further use by control system of a UV-NIL machine.

Keywords: UV NIL, step and flash NIL, optimization, droplet dispensing

1 INTRODUCTION

A homogeneous residual layer thickness in nanoimprint lithography (NIL) is a serious problem in both thermal NIL and in step and flash (UV-)NIL. Improvement of thickness homogeneity could be expected from optimized size of droplets at resist dispensing in case of UV-NIL. The optimization in droplet size must exclude the stamp geometry involving areas in which the resist has to flow laterally over large distances so stamp geometry should be considered at optimization. Also the optimization should consider process of resist wetting and spreading at imprint analyzing resist viscous flow. The paper is devoted to development of an optimizing algorithm, which take into account *only* filling factor (geometry) of a stamp and does not consider the following resist flow.

2 OPTIMAZING APPROACH

In current realization of the approach a specially developed algorithm transfers stamp geometry defined in standard GDSII (or ACAD) format into rectangular (square) cells (Figure 1) and calculates the filling factors taking into consideration stamp depth and desirable residual

resist thickness (Figure 2). Then depending of the jet model continuous or discreet volume is calculated and saved for further use by control system of a UV-NIL machine

Several alternatives could be adopted as jet work model. Two extreme models of jet dispensing are of main interest here (Figure 3):

-a "continuous" model when the jet is able to provide a drop with infinitesimal accuracy and

-"discreet" model when final drop consists of several droplets of some minimal volume.

Other jet models like "threshold" model or "nonlinear in time" model can be easy incorporated in the approach

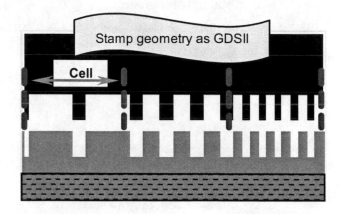

Figure 1 Schematic presentation of stamp defined in GDSII format with different density divided into square cells

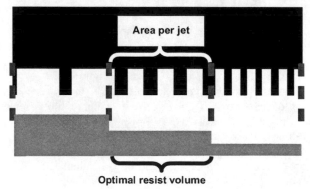

Figure 2 Calculation of ideal resist volume considering stamp depth and residual resist thickness.

Final results as a matrix of volume per elementary area is transferred to a system controlling jet dispensing (Figure 4).

Jet model

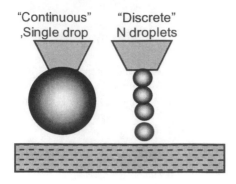

Figure 3 Two extrime jet dispensing model "continuous" and "discrete".

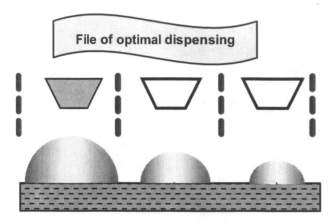

Figure 4 ".Final data presenting resist volume shoud be deposite by jet in the center of elementary area calculated from GDSII file.

2.1 Continuous model

The first simulation was performed with a test structure, which contains areas with variety of filling factors. The calculations were performed in "continuous" jet model. Schematic 3D presentation is shown on Figure 6.

Figure 5 shows geometry of a binary stamp as downloaded from GDSII format where black area corresponds to protrusions in <u>imprinted</u> structure. Figure 6 illustrates work of the algorithm where 24x24 cells covering the whole test structure are filled with different drops with volumes corresponding to calculated filling factor. Cell filling factor is schematically illustrated in bottom of Figure 6 in form of 3D presentation as drops spreading on a substrate. In reality dispensed resist has a shape of spherical segment diameter of which is dependent on wetting angle.

Figure 5 Original binary stamp in GDSII format, black area corresponds to protrusions after imprint

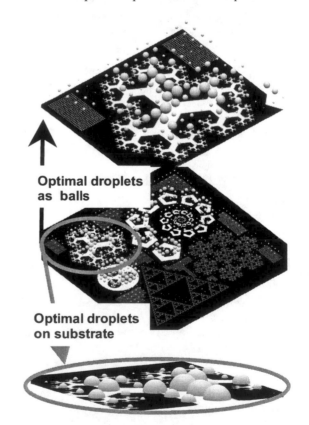

Figure 6 Schematic 3D presentation of optimal resist volume should be dispensed in an each cell (within square area marked in Fig.2). Each droplet comprises a spherical segment corresponding to wetting angle about 90^0. Diameter is calculated in "continuous" model

2.2 Qualitative analysis

Let us make a quantitative analysis of dispensing. Consider a drop of some minimal size V_{min} and find residual resist thickness h if the drop is pressed with an absolutely flat stamp to fill an elementary area of size A. A is equal to $d*d$ where d is distance between jets in line. Then

$$h = \frac{V_{min}}{A}$$

Estimation for typical values A=0.5mm*0.5mm V_{min}=60pl, gives h=240nm. This value is too high to perform optimization of drop value because typical expected value of residual thickness is 25-30nm and expected stamp depth is 150nm

One of solutions of the problem is to increase distance between jets in jet-line, for example one can use not all jets in line but only a half of them. This increases d in two times and results to residual thickness 60nm, acceptable but still larger than desirable value.

Another solution (and maybe the best) is to decrease minimal drop value. It is known that there is ink-jet with V_{min}=3pl [1].

2.3 Discrete model

To simulate variation of jet distance, influence of other like residual thickness and stamp depth a practical test stamp submitted by Dr. Holger Schmidt (Lehrstuhl fuer Elektronische Bauelemente Universitaet Erlangen-Nuernberg) in GDSII was used (see Figure 7). Black areas represent places where residual thickness should high so the stamp represents

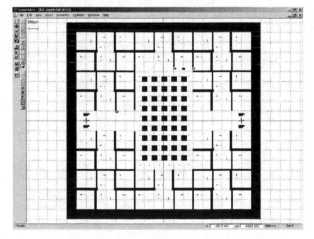

Figure 7 "Erlangen" practical test structure.

Firstly optimal dispensing was calculated for the following parameter set (H_residual=20nm; Stamp_depth=**150**; V_drop=60pl; d=1mm) and result is shown in Figure 8. Analysis of optimal volumes shows that the optimal volume of most cells smaller than 60pl. Then stamp depth was increased to value 250nm and optimization in discrete model was performed (Figure 9).

Due to large drop volume optimal drops number was 1 or 2 also there are two cells where resist should be dispensed at all.

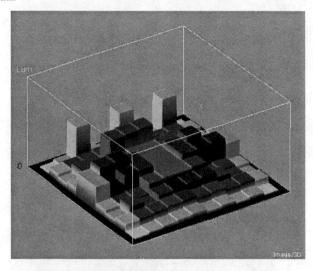

Figure 8 Optimal drop volumes for continues dispensing

Figure 9 Optimal drop volume for Hstamp=250nm, disrete dispensing (V_{min}=60pl)

The next Figure 10 shows optimal drops calculated for jet pitch d equal to 2mm. It is seen that number of optimal drops belongs to range 1-5 what is better than previous case. But still there are cell with zero dispensing. Nevertheless stamp depth was decreased up to normal value used in practical imprinting.

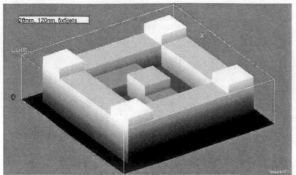

Figure 10 Optimal drop volume for Hstamp=150nm, discrete model

3   DISCUSSION AND CONCLUSION

Realization of optimal dispensing performed in the paper showed serious difficulties on this way. It turned that minimal drop volume like 60pl is too large to provide flexible volume tuning to stamp structure. This could be overcome with increasing of inter-jet distance but this leads to a situation when single jet will be more suitable and advantage of jet-line would be lost.

Another solution could be decreasing minimal drop volume.

Also important to note that the simulations performed clearly show that some model of drop spreading definitely should be considered. Only such model can help in consideration of temporal picture of drop spreading and can be a useful tool to understand whether current distance between jets is acceptable or not.

ACKNOLEDGEMENT

Dr. Holger Schmidt (Lehrstuhl fuer Elektronische Bauelemente Universitaet Erlangen-Nuernberg) is greatly appreciated for collaboration. The partial support of the EC-funded project NaPa (Contract no. NMP4-CT-2003-500120) is gratefully acknowledged. The content of this work is the sole responsibility of the authors.

REFERENCES

[1]   http://www.epson.com/cgi-bin/Store/consumer/consDetail.jsp?BV_UseBVCookie=yes&infoType=Specs&oid=63070381&category=Products

Self-assembly of Functional Groups Inside High Aspect Ratio Silicon Nanopores

S. Moghaddam[*,**], E. Pengwang[*], R. Masel[**], and M. Shannon[*]

[*]Department of Mechanical Science and Engineering at University of Illinois at Urbana-Champaign,
Urbana, IL, USA, saeedmog@uiuc.edu
[**]Department of Chemical and Biomolecular Engineering at University of Illinois at Urbana-Champaign
Urbana, IL, USA

ABSTRACT

In this study, development of a technique for self-assembly of molecules with functional end groups inside high aspect ratio silicon nanopores is reported. A 20-µm thick porous silicon membrane with pore sizes of 5-7 nm was fabricated. The structure was then hydroxylated to enable silane-based self-assembly. A setup was fabricated to hold the membrane die between two chambers charged with dry helium. A 1 mM solution of 3-mercaptopropyl-trimethoxysilane (MPTMS) in benzene was supplied to the top side of the membrane. A constant flow of helium over the bottom of the membrane allowed continuous evaporation of the solvent from the bottom of the pores, so that the fresh solution could get into the pores from top. The pores were charged with solvent approximately 10000 times, to supply enough MPTMS molecules to fill the estimated OH sites within the pores. Penetration of the MPTMS molecules down to the bottom of the pores was verified by measuring sulfur variation through the membrane thickness using Time of Flight-Secondary Ion Mass Spectroscopy (ToF-SIMS) technique.

Keywords: self-assembly, nanoporous silicon, functionalization, biosensing, proton exchange membrane

1 INTRODUCTION

In recent years, functionalization of porous structures has garnered intense interest. In biosensing technology, interaction between biomolecules and materials through an intermediate layer commonly with amines, carboxylic acids, and thiol functional groups is required. Porous structures provide a great advantage over flat surfaces because they provide significantly higher binding capacity (i.e. number of sites for capturing molecules) due to their high specific surface area [1]. Functionalized porous membranes can also greatly advance the fuel cell technology. Development of mechanically and thermally stable proton exchange membranes (PEMs) through assembly of functional groups within a porous solid structure enables development of the next generation membrane electrode assembly (MEAs) with enhanced performance, lifetime, and reliability.

Functionalizing porous materials has been actively researched. For example, sulfonic-functionalized porous structures were made through co-condensation of tetraethoxysilane (TEOS) and 3-mercaptopropyl-trimethoxysilane (MPTMS) in presence of tri-block copolymer Pluronic 123 [2] or porous particles (Si-MCM-41 or Si-SBA-15 powder) were functionalized with MPTMS [3]. The common approach in self-assembly of monolayers (SAMs) inside pores of a membrane (e.g. porous silicon membrane in [4]) has been to soak the membrane in a solution containing SAMs, but poor results has been achieved since diffusion is relied upon to deliver molecules through torturous nanopores with high aspect ratios.

A simple estimation suggests that one pore volume of a 1 mM solution contains 3-4 orders of magnitude fewer molecules than necessary for complete coverage of all hydroxylated sites on the wall of a long (several microns) nanopore. Therefore, high aspect ratio nanopores should be filled with solution thousands of times to supply enough solute molecules to the pores. Increasing the solution concentration to reduce the number of filling times is not an option, since it leads to self-polymerization of the solute molecules. The approach taken in this study has been to extract the depleted solvent from the bottom of the membrane pores continuously while the solute-rich solvent is supplied to the top of the membrane.

An experimental apparatus has been fabricated for this purpose and a set of recipes has been developed to assemble MPTMS molecules fairy uniformly throughout a 20-nm thick silicon membrane. Details of the developed processes are described in the following sections.

2 FABRICATION PROCESS

A 20-µm thick silicon membrane was fabricated through KOH etching of a p-doped silicon wafer. The membrane was anodized in HF electrolyte in a standard two-chamber bath. Details about the apparatus and the process could be found in [5]. Since the primary goal for this effort was to develop a PEM in which a pore size of several nanometers is necessary to maintain a low fuel crossover level, 5-7 nm in diameter pores (cf. Fig. 1) were produced inside the membrane. This was achieved by varying several parameters including concentration of the HF electrolyte, anodization current density, and silicon doping level. The pores grew through the entire membrane thickness, except through an approximately 50-nm thick layer at the backside of the membrane that was later etched using reactive ion etching process. The anodization process and subsequent back etching of the membrane with SF_6 plasma rendered the porous membrane surface hydrophobic due to high concentration of fluorine on the surface and inside the pores. The surface of the pores was cleaned and hydroxylated. This process left the pores slightly wider (7-10 nm) than their original size (cf. Fig. 2).

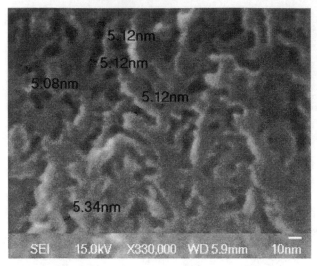

Figure 1: SEM view of the cross section of the middle of the anodized membrane with 5-7 nm in diameter pores. Anodization was conducted in 25% HF electrolyte at a current density of about 20 mA/cm2. The silicon wafer used was p-doped with a resistivity of 0.01-0.02 Ω-cm.

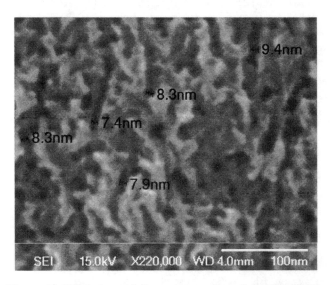

Figure 2: SEM view of the cross section of the membrane after hydroxylation. The cleaning/hydroxylation process was conducted in $NH_4OH/H_2O_2/H_2O$ solution. Continuous oxidation of the silicon backbone attacked by OH^- and its subsequent dissolution in aqueous ammonia (NH_3) resulted in under-etching the fluorine terminated Si as well as fluorine compounds on the surface of the pores.

In a preliminary test, self-assembly of MPTMS molecules on a hydroxylated silicon surface was conducted. Results of X-ray Photoelectron Spectroscopy (XPS) confirmed presence of thiol groups on the surface. In addition, Angular Resolved XPS (ARXPS) was conducted to ensure that the MPTMS reacted with the surface through its silane end (i.e. the outermost layer on the surface is thiol).

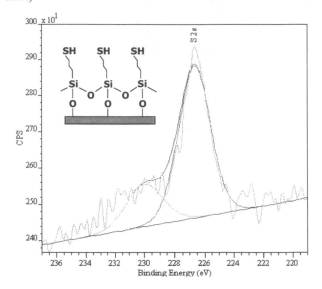

Figure 3: Sulfur 2s binding energy of thiol (SH) terminated surface. Peak at 226.7 eV corresponds to SH. Peak at 230.5 eV corresponds to sulfonate (SO_3H) group. Some of the thiol groups were oxidized in ambient.

2.1 Functionalization Apparatus

As mentioned earlier, one pore volume of the solvent contains orders of magnitude less solute molecules necessary for all OH sites inside a long pore. For example, one volume of a 5-nm diameter, 20-μm long pore only contains 232 MPTMS molecules, when filled with a 1 mM solution, while the number of potential OH sites on the pore wall is close to 2 million. A relatively dilute solution of 1 mM was selected to avoid self-polymerization of the MPTMS molecules. Therefore, the pores should be filled thousands of times to supply the number of molecules needed in the pores. In order to achieve this goal, a setup was fabricated to allow holding the porous membrane die between top and bottom compartments of a chamber and flow the solution through the pores. Figure 4 shows a top view of the setup during operation.

Immediately after the hydroxylation process, the membrane die was placed within a fixture designed for holding the membrane between the two compartments of the setup main chamber. The chamber was put under vacuum and purged with helium multiple times to minimize water vapor presence inside the chamber. Then, the MPTMS in benzene solution was supplied to the solution reservoir on top of the

Solution supply line

Main chamber of the setup

Solution reservoir filled with liquid

Figure 4: Top view of the main chamber of the nanopore self-assembly setup and close view of the sample holder and solution supply line during operation.

membrane. While the top chamber was charged with helium and the vacuum and helium lines connected to it were closed, the lines connected to the bottom compartment were slightly open to maintain a slow flow of dry helium. The process was continued for up to 2 days to complete (i.e. supply approximately 10000 pore volumes of solution to the pores). The estimated liquid velocity inside the pore during the process was on the order of 1 μm/s.

3   VERIFICATION

Penetration depth of the functional group inside the membrane was verified using Time of Flight-Secondary Ion Mass Spectroscopy (ToF-SIMS) with depth profiling. As can be seen in Figure 5, the sulfur penetrated through the entire membrane thickness. Results show a higher count for all elements at both top and bottom of the membrane. This is counterintuitive, considering the fact that the anodization process left the pores larger at the top of the membrane than its middle. Also, the hydroxylation process enlarged the pores at the bottom of the membrane and made it more porous than its middle. Thus, silicon count should have been the lowest at both sides of the membrane. We speculate that higher count at both sides of the membrane

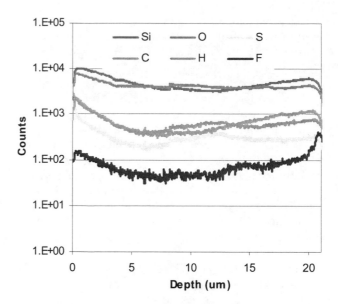

Figure 5: Time of Flight-Secondary Ion Mass Spectroscopy (ToF-SIMS) results (phased depth profile using a 22 kV Au+ analysis beam and a 2 kV Cs+ sputtering beam) showing composition of the 20-μm thick membrane. A spot on the membrane was carved in gradually until reached to the bottom of the membrane. Note that the sudden spike in fluorine content at the bottom of the membrane is due the back-etching process used to open up the pores.

is an artifact of the technique employed. Generally, more surface area near the beam results in generation of more secondary ions. Hence, a higher number of counts could result from a more porous surface. If this argument is true, variation of the elements count could also represent a change in the pore size. Interestingly, it appears that the silicon, oxygen, and fluorine count curves, which represent the initial construct of the porous membrane (i.e. before functionalization), have a relatively similar slope.

Also, the variations in the count curves for sulfur and hydrogen (and to some extent carbon at the first half of the curve) that represent the MPTMS presence in the pores follow a similar slope. These curves have a noticeably steeper slope than that of silicon. This suggests that the higher sulfur count up to a certain depth might not be due only to the surface effect described above, but perhaps suggests sulfur concentration is higher close to the top surface. It is not clear, however, whether this higher concentration is due to the higher surface area of the more porous structure at the top of the membrane or the presence of physisorbed molecules on the monolayer.

4   CONCLUSIONS

A nanoporous silicon membrane with pore aspect ratio of several thousands was successfully functionalized. This was achieved through development of a set of recipes for fabrication of a porous silicon membrane, its hydroxylation, and development of a new technique for supplying the necessary number of molecules deep inside the nanopores. The results were verified using ToF-SIMS with depth profiling.

ACKNOWLEDGMENTS

The authors would like to acknowledge and appreciate the support provided by DARPA. This work was carried out in part in the Frederick Seitz Materials Research Laboratory Central Facilities, University of Illinois, which are partially supported by the U.S. Department of Energy under grants DE-FG02-07ER46453 and DE-FG02-07ER46471. The authors would like to thank Drs. Tim Spila and Rick Haasch for their assistance with the ToF-SIMS and XPS analysis.

REFERENCES

[1] M. Arroyo-Hernandez, R.J. Martin-Palma, J. Perez-Rigueiro, J.P. Garcia-Ruiz, J.L. Garcia-Fierro, and J.M. Martinez-Duart, "Biofunctionlaization of surfaces of nanostructured porous silicon," Mater. Sci. and Eng., 23, 697-701, 2003.

[2] S. Mikhailenko, D. desplantier-Giscard, C. Danumah, and S. Kaliaguine, "Soild electrolyte properties of sulfonic acid functionalized mesostructured porous silica," Micropor. Mesopor. Mater., 52, 29-37, 2002.

[3] R. Marschall, I. Bannat, J. Caro, M. Wark, "Proton conductivity of sulfonic acid functionalized mesoporous materials," Micropor. Mesopor, Mater. 99, 190-196, 2007.

[4] T. Pichonat and B. Gauthier-Manuel, "Realization of porous silicon base miniature fuel cells," J. Power Sources, 154, 198-201, 2006.

[5] Volker Lehmann, "Electrochemistry of Silicon," Wiley-vch, 2002.

Electric Field Assisted-Assembly of Perpendicular Oriented ZnO Nanorods on Si Substrate

O. Lupan[*,**], L. Chow[**], G. Chai[***], S. Park[**] and A. Schulte[**]

[*]Technical University of Moldova, Blvd. Stefan cel Mare 168, Chisinau MD-2004, Moldova,
lupanoleg@yahoo.com
[**]Department of Physics, University of Central Florida, Orlando, PO Box 162385, FL, USA,
chow@mail.ucf.edu   lupan@physics.ucf.edu
[***]Apollo Technologies, Inc., 205 Waymont Court, 111, Lake Mary, FL 32746, USA,
guangyuchai@yahoo.com

ABSTRACT

In this work, we report a new electric-field assisted assembly technique used to vertically-align ZnO nanorods on the Si substrate during their growth. In the assembly process, the forces that induce the alignment are a result of the polarization of the electric field. The synthesis was carried out at near-room temperature (90° C). Novel measurements on these structures show encouraging characteristics for future applications.

The phase purity, composition and morphology of the synthesized products by the assembly method were examined by XRD, TEM, and SEM. Room-temperature micro-Raman spectroscopy was performed to examine the properties of the self-assembly ZnO nanorods on Si substrate structures. Such highly oriented and ordered ZnO nanorods could be beneficial for field emission, solar cells, LEDs and spintronic applications.

Keywords: ZnO nanorod, self-assembly, nanorod arrays, nanowires, silicon

1 INTRODUCTION

ZnO is of importance for fundamental research as well as relevant to industrial and high-technology applications. ZnO can also be considered as an alternative wide band gap semiconductor for photonic devices [1]. ZnO nanorods/nanowires are becoming common building blocks for the next generation electronic devices [2].

In this context, self - assembly and assisted - assembly of nano-architectures has attracted high interest driven by the demands of technology and engineering. The large interest is motivated by necessity in development of new fabrication tools for novel electronic, optoelectronic and magnetic properties for versatile applications in nanotechnology. Chemical and electric field assisted-assembly offer a new opportunity to create heterostructures in multicomponent systems and to manufacture of nanodevices. They include nanorods-based ultra-violet lasers [3], nanosensors [2] and light emitting diodes [4, 5]

with higher performances and significantly lower cost in comparison to the traditional lithographic fabrication. Thus, it is important to study the novel fabrication techniques of semiconducting nanorods-nanowires for future applications.

At the same time, from a fundamental point of view, it is crucial to study the structure and assembly of such novel materials to enable tailoring their properties for novel and improved nanodevices fabrication. Efficient manipulation, positioning and alignment of one-dimensional (1-D) ZnO nanowires present key challenges toward the integration of nanostructures with larger scale systems.

The presence of a direct current (DC) electric field during synthesis yields a better organized growth, because nanorods will align with electric-field lines. This behavior can be attributed to the polarizability of 1-D nanorods and the electrophoretic effect [6].

ZnO nanoarchitectures can be assembled on different types of substrates (e.g. glass, silicon, sapphire) by patterning the seed or catalyst layer, and usually the ZnO nanorods are well-aligned with their c-axis perpendicular to the substrate surface [3]. In this respect, typically gold is used to catalyze the growth of nanorods from vapor–liquid–solid (VLS) method and silver is used to facilitate the assembly of ZnO nanorods from solution [7, 8].

In this paper, we demonstrate electric field assisted-assembly of perpendicularly oriented ZnO nanorods on a Si substrate during growth from aqueous solution. We demonstrate the flexibility to order ZnO nanorods on substrates through the strength of the electric field.

2 EXPERIMENTAL

The starting materials, zinc sulfate ($Zn(SO_4)\cdot7H_2O$, 99.9 % purity) and ammonia NH_4OH (29.6%) were used as received without further purification.

The hydrothermal synthesis of vertically-aligned ZnO nanorod arrays was carried out by dissolving $Zn(SO_4)\cdot7H_2O$ and ammonia NH_4OH (29.6%) in 20 mL of deionized water (18.2 MΩ·cm). The complex solution was then transferred to a reactor [9]. A DC electric field was applied during growth. The substrates, Si or glass slides were placed in vertical and horizontal positions in order to study the effect of the electric field direction on the synthesized samples. The reactor was heated at 90 °C for 20 min and then cooled to 40 °C.

After synthesis, substrates were washed with distilled water several times and then dried in a hot air flux at 150 °C. ZnO nanorods with different architectures were also prepared with the same procedure.

X-ray diffraction (XRD) pattern were obtained with a Rigaku 'D/B max' diffractometer, CuK$_\alpha$ radiation (λ=1.54178 Å) were used, operating at 40 kV and 30 mA. All samples were measured in a continuous scan mode at 20 – 90 ° (2θ) with a scanning range of 0.01°/s). Peak positions and relative intensities of synthesized nanorod arrays were compared to values from Joint Committee on Powder Diffraction Standards (JCPDS) card for ZnO (JCPDS 036-1451) [10].

The morphology of the products was obtained using a scanning electron microscope (SEM, JEOL and a Hitachi S800) and high resolution transmission electron microscopy (TEM) (FEI Tecnai F30 TEM). Room-temperature micro-Raman spectroscopy experiments were performed with a Horiba Jobin Yvon LabRam IR system at a spatial resolution of 2 μm. Raman scattering was excited with the 633 nm line of a He-Ne laser with less than 4 mW of power at the sample.

3   RESULTS AND DISCUSSION

The phase composition and phase purity of ZnO nanorods were identified by XRD analysis. Typical patterns are depicted in Fig. 1. All diffraction peaks of the products are well indexed as the hexagonal phase of wurtzite zinc oxide (space group $P6_3mc$(186); a = 0.32498 nm, c = 0.52066 nm, JCPDS card #036-1451). The cell parameters of a and c of obtained samples are 3.2502 Å and 5.2096 Å calculated by using the following equation [9]

$$\frac{1}{d_{(hkl)}^2} = \frac{4}{3}\left(\frac{h^2 + hk + k^2}{a^2}\right) + \frac{l^2}{c^2} \qquad (1)$$

using the d_{hkl} measured on the XRD pattern. From Fig. 1(a) for ZnO nanorods on Si grown under electric field assistance, only diffractive peaks in the pattern which belong to ZnO wurtzite structure were observed. The full width of half maximum (FWHM) of (002) diffraction peak is only 0.51°. Thus, Fig. 1(a) shows the X-ray diffraction patterns determined by the (002) plane, indicating that zinc oxide has a single ZnO phase.

Fig. 1(b) shows the X-ray diffraction pattern for ZnO nanorods on Si grown without electric field assistance. It can be observed that the sample has a good crystalline phase, but dominated by the (101) plane. From the Fig. 1 one can conclude that all intensities indicate the good crystalline ZnO material.

The typical morphology of the ZnO nanorods on glass prepared by hydrothermal synthesis after the growth time of 20 min is shown in Fig. 2(a). All products are hexagonal nanorods with closed ends (see inset Fig 2(a)). The radius of nanorods is around 400 nm and the length is about 2-3 μm. These ZnO nanorods are randomly distributed on the substrate.

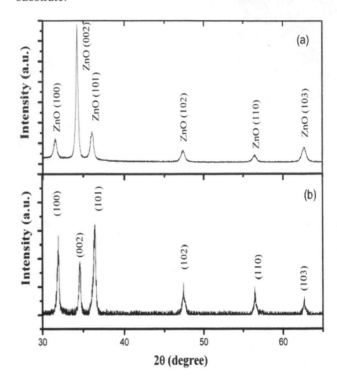

Figure 1: XRD patterns of (a) quasi-aligned ZnO nanorods on Si grown under electric field assistance and (b) ZnO nanorods on Si grown without electric field assistance.

An enlarged SEM view in the inset of Fig. 2(b) illustrates hexagonal symmetry of ZnO nanorods on Si grown under relatively weak electric field (~5·10³ V/m) assistance. The alignment of ZnO nanorods on Si(111) substrate as shown in Fig. 2(c) increases under stronger electric field (~5·10⁴ V/m) assistance. The diameter and length of the nanorods (Fig 2c) are smaller than those grown on glass substrates (Fig 2a). The radius is about 100 nm and can be observed that all have the same dimensions.

The perpendicular orientation of the ZnO nanorods is controlled by the electric field, but lattice match between the two materials may play a role. The lattice mismatch between ZnO and Si (111) is ~3.5% [11]. Based on the observed nanorod growth from SEM image (Fig 2c), it is concluded that stronger fields will contribute to better alignment. At the same time we notice that some ZnO nanorods grow at a slightly inclined angle. Comparing with glass substrates the mosaic distribution is narrower for Si. But slightly tilted ZnO nanorods can be observed on the Si substrate.

NSTI-Nanotech 2008, www.nsti.org, ISBN 978-1-4200-8503-7 Vol. 1

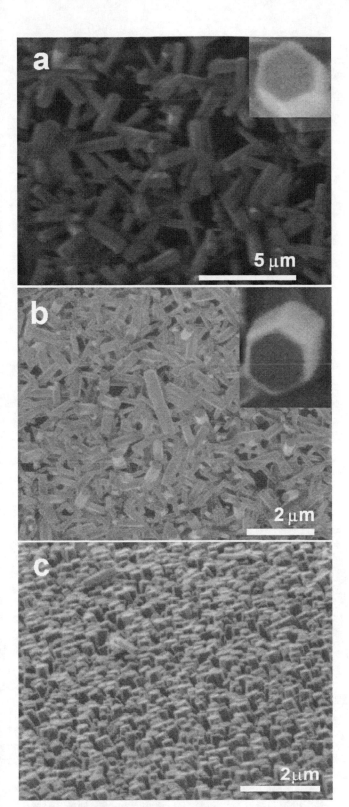

Figure 2: SEM images of (a) ZnO nanorods on glass substrate, (b) ZnO nanorods on Si substrate grown under relatively weak electric field assistance and (c) quasi-aligned ZnO nanorods on Si wafers grown under stronger electric field assisted-assembly.

Therefore, the ZnO nanorods reflect more uniformity and vertical alignment on Si compared to glass substrate. In order to study the vibrational properties of the ZnO nanorod arrays room-temperature Raman measurements were performed. The spectral shape and the peak position of the Raman shift are related to the crystallinity, structural disorder, residual stress, and defects in the investigated samples. Figure 3 shows the micro-Raman spectrum of ZnO nanorods on Si substrates with growth time of 20 min on glass without assistance and on the Si under stronger electric field assisted-assembly. The wurtzite phase ZnO belongs to C_{6v}^4 or $6mm$ symmetry group. All observed peaks can be assigned to phonon modes which correspond to those of ZnO [13].

The optical phonons at the Γ point in the Brillouin zone belong to the representation [14]:

$$\Gamma_{opt} = 1A_1 + 2B_1 + 1E_1 + 2E_2. \qquad (1)$$

A_1 and B_1 modes are polar and split into transverse optical (TO) and longitudinal optical (LO) phonons with different frequencies due to the macroscopic electric fields associated with the LO phonons. The interatomic forces can cause anisotropy, that's why A_1 and E_1 modes have different frequencies. Due to the fact that the electrostatic forces dominate the anisotropy in the short–range forces, the TO–LO splitting is larger than the A_1 E_1 splitting. In the case of lattice vibrations with A_1 and E_1 symmetry, the atoms move perpendicular and parallel to the c–axis, respectively. The A_1 and E_1 modes are Raman and infrared (IR) active. The non-polar IR inactive E_2 (E_2(low), E_2(high)) modes are Raman active. The B_1 modes are IR and Raman inactive (silent modes). From group theory the $A_1 + E_1 + 2E_2$ modes are Raman active.

The peak at 438 cm^{-1} (Fig. 3 a and b) is attributed to E_2(high) mode of non-polar optical phonons, which is a characteristic Raman active branch of hexagonal ZnO. The strong E_2(high) mode demonstrates that the ZnO nanorods are of good crystalline hexagonal structure which corroborates with XRD data (Fig 1a).

Compared to bulk material the E_2(high) mode show a 1cm^{-1} blue-shift. One of the reasons is that nanorods possess piezoelectric effect which causes the shift of Raman modes. Another one can be attributed to the optical-phonon confinement (which is anisotropic and has different effect on different phonon modes). Also shift can be attributed to the strain variation.

As E_2(high) mode is close to that of bulk ZnO (437-439 cm^{-1}), the sample with oriented nanorod arrays can be regarded as unstrained. The peak at 332 cm^{-1} corresponds the E_{2H}-E_{2L} mode and shows broadening (about 70 cm^{-1}) (Fig. 3b) due to multiphonon processes.

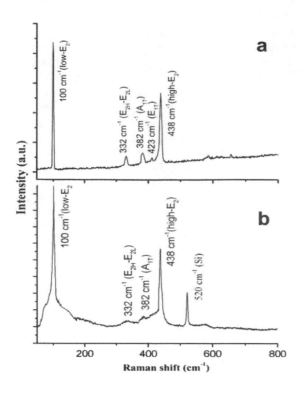

Figure 3: Micro-Raman scattering spectrum of (a) ZnO nanorods on glass substrate and (b) quasi-aligned ZnO nanorods on Si wafers grown under electric field assisted-assembly.

The multiphonon processes occur when phonon wave vectors are shifted away from the center of the Brillouin zone [15]. Another dominant peak at 100 cm^{-1} is commonly observed in the wurtzite structure ZnO [16] and is attributed to E_2(low) mode of non-polar optical phonons.

The weak peak is observed at 580 cm^{-1} of E_1(low) mode which show that our nanorod arrays have low quantity of impurities and structural defects (oxygen vacancies and Zn interstitials [17]).

4 CONCLUSIONS

This is a new report on electric-field assisted assembly technique used to vertically-align ZnO nanowires on the Si substrate during their growth in only 20 min. Well aligned ZnO nanorods were synthesized without employing any seeds, metal catalysts on glass and Si(111) substrates without any pre-coated buffer layers. The alignments of the ZnO nanorods on the different substrates depend on the growth conditions, substrate type and the strength of the applied electric-field. Similarly, the XRD, SEM and micro-Raman measurements showed the distinct appearance of ZnO nanorods on two types of substrates. Consequently, the crystal structure of ZnO nanorods is related to the type of the substrate and direction of electric field used.

To summarize, a technique for localized growth and alignment of ZnO nanorods on Si substrates to yield a assisted-assembled system has been demonstrated. This technique allows to ZnO nanorods synthesis and alignment using a DC electric-field. Enhanced control over ZnO nanorod alignment and organization is evident at higher field strengths. This process with its improved control yields a simple and reproducible fabrication method, which could be used in fabrication of nanodevices.

Acknowledgments The research described in this publication was made possible in part by Award No. MTFP-1014B Follow-on of the Moldovan Research and Development Association (MRDA) and the U.S. Civilian Research and Development Foundation (CRDF). Dr. L. Chow acknowledges partial financial support from Apollo Technologies Inc and Florida High Tech Corridor Program.

REFERENCES

[1] S. Choopun, H. Tabata, T. Kawai, J. Cryst. Growth 274, 167, 2005.

[2] O. Lupan, G. Chai, L. Chow, Microelectron. J. 38, 1211, 2007.

[3] M. Huang, S. Mao, H. Feick, H. Yan, Y. Wu, H. Kind, E. Weber, R. Russo, P. Yang, Science 292, 1897, 2001.

[4] X. F. Duan, Y. Huang, Y. Cui, J. F. Wang, C. M. Lieber, Nature 409, 66, 2001.

[5] R. Hauschild, H. Kalt, Appl. Phys. Lett. 89, 123107, 2006.

[6] M. Yan, H. T. Zhang, E. J. Widjaja, R. P. H. Chang, J. Appl. Phys. 94, 5240, 2003.

[7] J. Henzie, J. Barton, C. Stender, T. Odom, Accounts of Chemical Research 39, #4, 2006, p.249

[8] J. W. P. Hsu, Z. R. Tian, N. C. Simmons, C. M. Matzke, J. A. Voigt, J. Liu, Nano Letters 5, 83, 2005.

[9] O. Lupan, L. Chow, G.Chai, B. Roldan, A.Naitabdi, H. Heinrich, A. Schulte, Mater. Sci. Eng. B 145, 57, 2007.

[10] Joint Committee on Powder Diffraction Standards, Powder Diffraction File No 36-1451.

[11] A. Nahhas, H. K. Kim, J. Blachere, Appl. Phys. Lett. 78, 1511, 2001.

[12] J. S. Suh, K. S. Jeong, J. S. Lee, I. Han, Appl. Phys. Lett. 80 2392, 2002.

[13] (a) K. A. Alim, V. A. Fonoberov, M. Shamsa, A. A. Balandin, J. Appl. Phys. 97, 124313, 2005.

(b) M. Rajalakshmi, A. K. Arora, B. S. Bendre, S. Mahamuni, J. Appl. Phys. 87, 2445, 2000.

[14] C. Bundesmann, N. Ashkenov, M. Schubert, D. Spemann, T. Butz, E. M. Kaidashev, M. Lorenz, M. Grundmann, Appl. Phys. Lett. 83, 1974, 2003.

[15] T. C. Damen, Phys. Rev. 142, 570, 1966.

[16] Y. J. Xing, Z. H. Xi, Z. Q. Xue, X. D. Zhang, J. H. Song, R. M. Wang, J. Xu, Y. Song, S. L. Zhang, D. P. Yu, Appl. Phys. Lett. 83, 1689, 2003.

[17] K. Vanheusden, W. Warren, C. Seager, D. Tallant, J. Voigt, B. Gnade, J. Appl. Phys. 79, 7583, 1996.

DNA-guided Assembly of Organized Nano-Architectures

Oleg Gang[1], Mathew M. Maye[1], Dmytro Nykypanchuk[1], Huiming Xiong[1], Daniel van der Lelie[2]

[1]Center for Functional Nanomaterials, [2]Biology Department,
Brookhaven National Laboratory, Upton, NY 11973, USA

ogang@bnl.gov

ABSTRACT

Incorporation of DNA into nano-object design provides a unique opportunity to establish highly selective and reversible interactions between the components of nanosystems. Assembly approaches based on the nano-object's addressability promise powerful routes for creation of rationally designed nano-systems for the development of novel magnetic, photonic and plasmonic metamaterials. DNA provides a powerful platform due its unique recognition capabilities, mechanical and physicochemical stability, and synthetic accessibility of practically any desired nucleotide sequences. Recently, strategies based on DNA programmability for a pre-designed placement of nanoparticles in one- and two-dimensions using scaffolds have been demonstrated [1, 2]. However, in three dimensions, where theory predicted a rich phase behavior[3, 4], experimental realization has remained elusive[5], with nanoscale systems forming amorphous aggregates.

Keywords: DNA, self-assembly, hybrid systems, nanoparticles, superlattices

1 DNA-GUIDED ASSEMBLIES

Two approaches are widely used for the DNA-guided assembly of 3D nanoparticle systems [6, 7] (Scheme 1): (i) direct hybridization of two types of complementary single-stranded (ss) DNAs attached to the particle's surface and (ii) particle hybridization with linker DNAs, whose two ends are complementary to the mutually non-complementary DNA attached to particles. Each approach has its own advantages: a direct hybridization allows for tailoring interparticle interaction via DNA shell design, while a linker strategy is

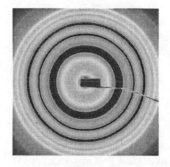

Scheme 1. Approaches for DNA-guided assembly 3D structures from nanoparticles.

attractive due to its potential of building various architectures from a given set of nanoparticle by changing a linker design. Herein we report a structural study of DNA-guided nanoparticle systems in which we have observed formation of crystalline ordered assemblies using both approaches described above.

1.1 Direct hybridization

We have systematically studied [8] the structure of self-assembled aggregates in a binary system of nanoparticle coated with complementary single stranded (ss) DNA for various DNA lengths. In each assembly system, a set of DNA-capped gold nanoparticles, ~11 nm with different DNA shells were allowed to assemble via DNA hybridization into meso-scale aggregates. The complementary outer recognition sequences of the DNA-capping provided the driving force for A and B particle assembly. The length of the recognition sequence, N_a=15

Figure 1. (Left) SAXS scattering pattern from crystalline DNA-guided nanoparticles assembly (50 bp DNA). (Right) Corresponding structure factor indicates bcc lattice.[8]

base pairs (bp), sets the scale of adhesion (per hybridized linker), thus attraction energy $E_a \sim N_a$, from ~30 kT at room temperature, to ~0 kT at DNA melting temperature. The length, N(15, 30, 50, 75), of DNA and the flexibility of the non-complementary internal spacer allowed for tuning the range, $d_r \sim N^{3/5}$, of repulsive interaction and its strength $E_r \sim (N^{3/5}/(N^{3/5}-cN_a))$,[9, 10] where c is defined by persistence length and molecule surface density and is constant for all studied systems and N is a length of ssDNA. Thus the use of multiple systems with constant E_a (i.e., recognition sequence), and varied d_r (i.e., spacer lengths), allowed for effective interparticle potential tuning. We have modulated the interparticle potential via length of ssDNA and for sufficiently soft interaction potentials (long ssDNA) discovered the formation of 3D nanoparticle assemblies with crystalline long range order (~ micron) using synchrotron small angle x-ray scattering (SAXS) measurements (beamline x21 at NSLS, BNL). The SAXS patterns in Fig. 1 reveals multiple orders of resolution limited Bragg's peaks for a system containing DNA shells of 50 base pairs, demonstrating its crystalline 3D structure, remarkable degree of long-range ordering, and crystallite sizes of ~ micron, as estimated from scattering correlation length. The system with shorter or rigid DNA form only disordered structures. The measured lattice parameters for the observed bcc structure are ~35 nm at 30 °C and ~42.4 nm at 28°C for system with 50 bp and 75 bp. Once formed, near DNA melting temperature, these crystalline structures were reversible, as confirmed by multiple assembly-disassembly cycles, without a noticeable loss of ordering quality or changes in system behavior.

1.2 Hybridization via DNA linkers

To explore an assembly strategy using DNA linkers we generated a binary set of gold nanoparticles [11], which were covered with ssDNA containing a 15 bp outer recognition part and a 15 bp poly-dT, which serves as a spacer separating the recognition sequence from the particle surface [12]. The ends of the linker ssDNAs are complementary to the respective ends of ssDNAs on nanoparticles, and are separated by a central flexible poly-dT fragment which contains n nucleotides, having a length of 0, 15, 30 or 70 bp (denoted accordingly as Sys-L0, Sys-L15, Sys-L30 and Sys-L70). Each system was formed by mixing an equal mole of the two types of ssDNAs capped nanoparticles and a linker ssDNA (DNA/particle mole ratio 36:1) in 0.3 M PBS buffer.

Temperature dependent synchrotron SAXS is utilized to characterize the structure of assembled aggregates. Figure 2 illustrates SAXS patterns obtained after annealing samples at few degrees below assembly melting temperature for various studied linkers. At this temperature, a well-defined crystalline order is manifested by an increased number of diffraction rings and their decreased widths. Analysis of the detected seven orders of Bragg's peaks positions reveals a ratio $q/q_1 = 1:2^{1/2}:3^{1/2}:4^{1/2}:5^{1/2}:6^{1/2}:7^{1/2}$, which corresponds to bcc lattice. The relative intensities of the diffraction peaks are also in accordance with the predictions for bcc lattice. The first peak thus arises from diffraction of {110} planes. For Sys L15, the lattice constant of the unit cell (a) is ~39 nm and the nearest-neighbor distance d_{nn} is ~34 nm at 56 °C. As shown in Figure 2b, Sys-L70 with the longest linker is the most prone to order improvement during annealing. For Sys-L15, although there was little perfection during initial heating, the order increased tremendously after annealing. However, for Sys-L0 with no flexible part in the linker, the

structure of nanoparticles assembly remains disordered after the annealing process.

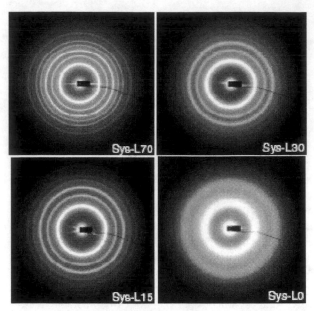

Figure 2. SAXS patterns for systems with different linkers length 0 bp (Sys-L0), 15 bp (Sys-L15), 30 bp bp (Sys-L30) and 70 bp (Sys-L70) at 25°C after annealing at 55°C for 2 hours [12].

SUMMARY

In summary, we have demonstrated that DNA mediated interaction between nanoparticles can result in formation of assemblies with 3D crystalline order (bcc lattice). For tailored DNA designs, ordered nano-architectures are obtained using both discussed assembly strategies, i.e. using a direct hybridization of particle DNA shells or DNA linkers. The flexibility of the DNA is determined to play an important role in the formation of ordered phase. The described approach offers a unique way for building 3D ordered materials from a broad range of nanoscale objects compatible with DNA functionalization. Phase behavior of these systems and formation more complex multi-component architectures are still to be explored.

Research was supported by the U.S. DOE Office of Science and Office of Basic Energy Sciences, under Contract No. DE-AC-02-98CH10886.

REFERENCES

[1]   J. W. Zheng, P. E. Constantinou, C. Micheel, A. P. Alivisatos, R. A. Kiehl, and N. C. Seeman, "Two-dimensional nanoparticle arrays show the organizational power of robust DNA motifs" Nano Letters **6**, 1502 (2006).
[2]   J. D. Le, Y. Pinto, N. C. Seeman, K. Musier-Forsyth, T. A. Taton, and R. A. Kiehl, "DNA-templated self-assembly of metallic nanocomponent arrays on a surface" Nano Letters **4**, 2343 (2004).

[3]   A. V. Tkachenko, "Morphological diversity of DNA-colloidal self-assembly" Physical Review Letters **89**, 148303 (2002).

[4]   D. B. Lukatsky, and D. Frenkel, "Phase behavior and selectivity of DNA-linked nanoparticle assemblies" Physical Review Letters **92**, Art. No. 068302 (2004).

[5]   S. Y. Park, J. S. Lee, D. Georganopoulou, C. A. Mirkin, and G. C. Schatz, "Structures of DNA-linked nanoparticle aggregates" Journal of Physical Chemistry B **110**, 12673 (2006).

[6]   S. J. Park, A. A. Lazarides, C. A. Mirkin, and R. L. Letsinger, "Directed assembly of periodic materials from protein and oligonucleotide-modified nanoparticle building blocks" Angew. Chem.-Int. Edit. **40**, 2909 (2001).

[7]   M. M. Maye, D. Nykypanchuk, D. van der Lelie, and O. Gang, "DNA-Regulated micro- and nanoparticle assembly" Small **3**, 1678 (2007).

[8]   D. Nykypanchuk, M. M. Maye, D. van der Lelie, and O. Gang, "DNA-guided crystallization of colloidal nanoparticles" Nature **451**, 549 (2008).

[9]   S. T. Milner, and T. A. Witten, "Bending moduli of polymeric surfactant interfaces" J. Phys. **49**, 1951 (1988).

[10]   D. Nykypanchuk, M. M. Maye, D. van der Lelie, and O. Gang, "DNA-Based Approach for Interparticle Interaction Control" Langmuir **23**, 6305 (2007).

[11]   M. M. Maye, D. Nykypanchuk, D. van der Lelie, and O. Gang, "A Simple method for kinetic control of DNA-induced nanoparticle assembly" J. Am. Chem. Soc. **128**, 14020 (2006).

[12]   H. Xiong, D. Van der Lelie, and O. Gang, "DNA Linker-Mediated Crystallization of Nanocolloids" J. Am. Chem. Soc. **130**, 2442 (2008).

A novel self-assembled and maskless technique for highly uniform arrays of nano-holes and nano-pillars

W. Wu, D. Dey, O. Memis, A. Katsnelson, H. Mohseni[*]

[*]EECS department, Northwestern University, Evanston, IL, USA, hmohseni@ece.northwestern.edu

ABSTRACT

We present a low-cost and high-throughput process for realization of two-dimensional arrays of deep sub-wavelength features using a self-assembled monolayer of HCP silica and polystyrene microspheres and well-developed photoresist lithography. This method utilizes the microspheres as super-lenses to fabricate highly uniform nano-holes and nano-pillars arrays over large areas on conventional positive and negative photoresists, and with a high aspect ratio. The period and diameter of the holes and pillars formed with this technique can be controlled precisely and independently. We also present our 3-D FDTD modeling, which shows a good agreement with the experimental results. The technique is simple, fast, economical, and is a convergence of bottom-up and top-down lithography.

Keywords: lithography, microspheres, nanoholes and nanopillars

1 INTRODUCTION

Nowadays, nanostructures and nanopatterns can be formed by top-down and bottom-up methods. Top-down methods mostly make use of lithography techniques to generate small patterns/structures in polymers of high sensitivity and then transfer them into other materials. Bottom-up methods assemble the identical building blocks into the desired structures. Top-down methods have been used in industry for a very long time and they are still the mainstream methods. However, they are expensive or inefficient to be applied into certain areas. For example, to generate a large area of periodic uniform nanoholes and nanopillars, using photolithography method requires expensive exposure instruments and mask sets [1] while e-beam lithography and focal ion beam milling methods will be extremely time-consuming [2]. With bottom-up methods being developed more and more widely, they have shown many advantages compared with the top-down methods, such as lower-cost, higher-efficiency, and easier-operation. However, it is still not as mature-developed as the top-down methods. The convergence of both methods has become an important research direction of nanoscale lithography [3][4].

Here we present an unconventional lithography method using the colloidal microspheres as nano-lenses to focus UV light in photoresist. We simulated that microspheres of silica have a strong focus ability to get a small focused beam size with high intensity and they can be self-assembled into a large hexagonally closed packed (HCP) monolayer [5]. So, using the technique, a large area of nanoscale patterns or structures can be generated in photoresist. Our simulations show that the focused beam waist is a very weak function of the sphere diameters and in hence extremely uniform pattern size can be achieved. By changing the exposure energy and development time of the photoresist we are also able to control the sizes of the patterns as well as changing the periodicity of the patterns by using spheres of different diameters. This method is a low-cost and high-throughput technique for realization of two-dimensional arrays of deep sub-wavelength features with high uniformity. The arrays of nanoholes and nanopillars generated by the method can be potentially applied in many areas, such as photonic crystals [6], storage devices [7], solar devices [8] and ion pumps [9].

2 SIMULATIONS

We used 3D-Finite-difference time-domain (3D-FDTD) method to simulate how the UV light was propagating through silica microspheres and photoresist. We tested four different sizes of silica microspheres and the simulation results were shown in figure 1 (a). By comparing the full-width at half-maximum (FWHM) of the focused beams by four different spheres, we found that the focused beam sizes have a very weak function with the sizes of spheres. As shown in figure 1(b), the cross-section distribution of light intensity after being focused showed that the FWHM of the focused beam almost didn't change with the spheres' sizes. Besides silica spheres, we found that polystyrene spheres (PS) also had a similar focus property as the silica ones because of the close refractive indices. In fact, FWHM of the focused light intensity is a good indication of the photoresist exposure, since the developing rate usually changes by almost an order of magnitude for a 50% optical intensity change around the photoresist threshold dose [10].

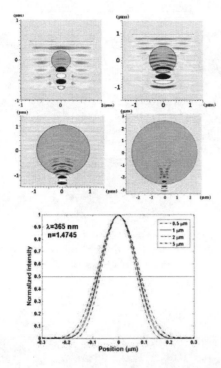

Figure 1: (a) 3D-FDTD simulations of light's electrical field profile of being focused by silica microspheres with different sizes; (b) normalized light intensity distribution after being focused by silica microspheres.

3 EXPERIMENT

All experiments are done in a class-100 clean room. Two kinds of photoresist, positive photoresist AZ 5214-E and negative photoresist ma-N 405, were used. 10 wt.% aqueous suspensions of transparent silica microspheres were bought from Bangs Lab Inc. for formation of HCP monolayer. We spin photoresist on GaAs substrate at 3000 rpm for 60 seconds and prebaked the sample at 95 degrees for 1 minute. We used the convective self-assemble method to form a large area of HCP monolayer of silica spheres on top of the photoresist. The samples with photoresist covered with a single microspheres layer were exposed by conventional photolithography instrument (Quintel Q-4000) under low exposure energy with a broad wavelength centered at about 400 nm. Before development, the spheres were removed by either HF acid solution or ultrasonication in D.I. water. The photoresist was developed using AZ-300 MIF developer. We deposited metal films in high vacuum using Edwards Electron Beam Evaporator. Figure 2 is the process chart for the technique.

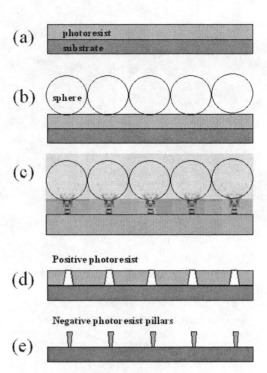

Figure 2: processing chart for our technique: (a) spin-on deposition of photoresist on a substrate, (b) a monolayer of microspheres formed on top of the photoresist, (c) UV light exposure of the photoresist covered with the particle lenses, (d) nanoholes formed after development of positive photoresist, (d) nanopillars of negative photoresist formed after being developed.

4 RESULTS

Figure 3(a) shows the SEM image of a typical monolayer of silica spheres with diameter of ~0.97 um formed on top of the negative photoresist. A monolayer of HCP microspheres is easy to form under optimized conditions of the temperature, humidity and the concentration of spheres. Figure 3(b) shows the top view of SEM images of the developed positive photoresist. The diameter of the holes is about 250 nm. The periodicity of these holes is 0.97 um: almost identical to the diameter of the spheres. The ratio of the feature size to the wavelength used is about 0.625. As shown in figure 3(c), we used another negative photoresist, ma-N 405 to fabricate the photoresist nanopillars. Figure 3 (d) is the enlarged image of the negative pillars. The diameter of the nanopillars is about 250 nm and the thickness of the photoresist is about 500 nm. The nanopillars have a high aspect ratio and it is good for the lift-off process using these nanopillars.

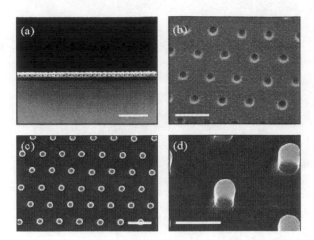

Figure 3: (a) SEM images of a single layer of microspheres (~0.97 μm diameter) on top of photoresist, the scale bar is 5 μm; (b) AZ5214 photoresist nanoholes after microspheres removal and photoresist development, the scale bar is 1 μm; (c) ma-N 405 photoresist used as negative photoresist to form nanopillars of photoresist, the scale bar is 1 μm; (d) high aspect ratio of nanopillars of negative photoresist, the scale bar is 500 nm.

We are also able to change the size of the holes and the lattice periods of the photoresist nanoholes array precisely and independently. The diameters of these nanoholes have been controlled with different exposure and development time, while the lattice periods were changed using different sizes of spheres. Figure 4 shows a uniform HCP arrays of photoresist holes with the diameters of about 300, 500, and 700 nm and lattice periods of about 500, 1000, 2000, and 4000 nanometers. In each line of the figure, the hole arrays have almost the same sizes with different periods. In each column, the array of holes has the same periods with different diameters.

Using the nanoholes and nanopillars of photoresist, we successfully produced a large area of highly uniform hexagonally packed metal nanoposts and nanoholes in multi-layers of metals by lift-off process, as shown in figure 5. As shown in figure 5 (a), we produced an array of gold posts with a thickness of 70 nm and 5 nm Cr as the adhesion layer. Figure 5 (b) is the enlarged view of the posts, which clearly shows the hexagonal distribution and the surface is very smooth. Figure 5 (c) and (d) are the nanoholes with the 100 nm gold and 5 nm Ti as the adhesion layer. The holes are perfectly circular and the edge of these holes is very smooth. These metal nanoposts and nanoholes can be potentially applied into photonic crystals, and also for further processing as metal masks.

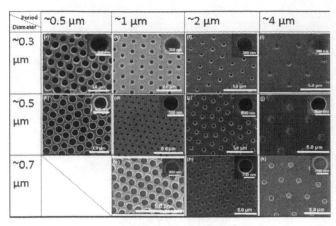

Figure 4: SEM images of uniform HCP arrays of nanoholes with controllable diameters and periods in the positive photoresist.

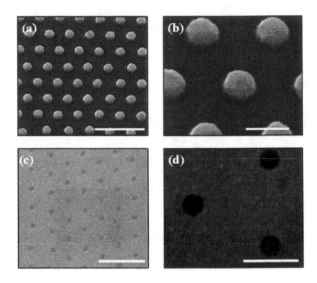

Figure 5: SEM images of (a) an array of gold nanoposts with a thickness of 70nm by lift-off process with 5nm Cr as the adhesion layer, the scale bar is 2 μm; (b) the enlarged view of the gold nanoposts, the scale bar is 500 nm; (c) gold nanoholes with a thickness of 100nm by lift-off process with 5nm Ti as the adhesion layer, the scale bar is 2 μm; (d) the enlarged view of the nanoholes in gold films, the scale bar is 500 nm.

NSTI-Nanotech 2008, www.nsti.org, ISBN 978-1-4200-8503-7 Vol. 1

5 CONCLUSIONS

We have successfully developed a novel mask-less and self-assembled sub-wavelength photolithography technique for forming highly uniform nano-holes and nano-pillars. The technique utilizes the self-assembled and super-focusing property of silica microspheres and applies them into the maturely developed photolithography system. It is a convergence of the bottom-up and top-down methods with the advantages of being simple, fast, economical, and compatible with current photolithography sources and photoresist. In hence it can be alternatively applied into some areas.

REFERENCES

[1] Peters, J. H., "Status of EUVL mask development in Europe," Proc. SPIE Int. Soc. Opt. Eng., 5853, 297-307, 2005

[2] Gates B. D. et al, "New Approaches to Nanofabrication: Molding, Printing, and Other Techniques", Chem. Rev., 105, 1171-1196, 2005

[3] Cheng J. Y. et al, "nanostructures engineering by templated self-assembly of block polymers", Nature Materials, 3, 823-828, 2004

[4] Park M. et al, "Block copolymer lithography: Periodic arrays of similar to 10(11) holes in 1 square centimeter," Science, 276, 1401-4, 2005

[5] Denkov N. D., Nagayama K. et al, "2-Dimensional crystallization," Nature, 361, 26-26, 1993

[6] Masuda H. et al., "Photonic band gap in anodic porous alumina with extremely high aspect ratio formed in phosphoric acid solution," J. Appl. Phys., 39, 1039-1041, 2000

[7] Weekes S. M., Ogrin F. Y. and Murray W. A., "Fabrication of Large-Area Ferromagnetic Arrays Using Etched Nanosphere Lithography," Langmuir, 20, 11208-11212, 2004

[8] Chiu W. L. et al., "Sub-wavelength Texturing for Solar Cells using Interferometric Lithography," Advances in Science and Technology, 51, 115-120, 2006

[9] Siwy Z., Fulinski, A., "A nanodevice for rectification and pumping ions" Am. J. Phys. 72, 567-574, 2004

[10] See for example Shipley 1800 series photoresist development curves at http://cmi.epfl.ch/materials/Data_S1800.pdf

Templated Self-Assembly of 5 nm Gold Nanoparticles

L. E. Ocola, X.-M. Lin

Center for Nanoscale Materials, Argonne National Laboratory, Argonne, IL

ABSTRACT

Combining top-down electron beam lithography and bottom-up colloidal chemical assembly can potentially lead to large scale patterning of nanocrystals on arbitrary substrates. We report on the templated self-assembly of 5 nm gold nanoparticles in both trenches and in geometries that lend themselves to large area self-assembly. Preliminary results show that templates that retain the self-assembly symmetry can be used to induce self-assembly at distances significantly larger than trench geometry constraints.

Further progress in templated self-assembly will lead to novel nanophotonic and chemical sensing devices, along with the means to extend top-down nanofabrication below the 10 nm barrier.

1. INTRODUCTION

Large-area self-assembly from colloidal solutions has been a topic of ever increasing research. Initial attempts used biomaterials such as DNA to act as locators for the dispersed nanostructures [1-3]. Langmuir Blodgett and other kinetically driven self-assembly of nanostructures [4-7] have also been used as bottom-up approaches for large area self-assembly. Most of these methods though have been unsuccessful to cover large areas without defects or grain boundary formation.

Solvent dewetting, combined with top-down lithography, has also been used to align particles in topographically defined surfaces [8, 9] with single particle resolution. However, this has only been achieved in very narrow area templating using particles larger than those used here. Combining top-down electron beam lithography and bottom-up colloidal chemical assembly can potentially lead to large-scale patterning of nanocrystals on arbitrary substrates if the constraint of large amount of patterning requirement can be avoided. In this paper we provide initial results on attempting to induce large-area self-assembly by fabricating template constraints that mimic the natural self-assembly symmetry of the nanostructures. In this particular case the nanostructures are 5-nm gold nanoparticles, the symmetry is hexagonal closed-pack, and the template constraints used are triangles.

2. EXPERIMENTAL

Line trench and equilateral triangle pillar templates were fabricated on silicon wafer substrates of about 20 mm x 10 mm per side. The substrates were pre-treated with a mild Piranha etch with a ratio of sulfuric acid to peroxide of 1:1, for about 5 minutes. This made the surface hydrophilic. The samples were spin coated with about 80 nm to 100 nm of hydrogen silsesquioxane (HSQ) negative resist, without a subsequent bake. The resist was then exposed using a 100 KV JEOL 9300 FS electron beam lithography system with doses between 1100 $\mu C/cm^2$ and 1800 $\mu C/cm^2$. The patterns used were a variety of linear gratings and arrays of equilateral triangles of different sizes, ranging from 10 nm to 200 nm per side in the design layout. After exposure the sample was developed in a TMAH solution 0.26 N (MF-CD 26, Microposit) for 45 seconds and rinsed in an IPA:DI water solution 1:5, both at 55 $^\circ$C. Finally the sample was dried with a nitrogen gun. A second set of sample substrates were treated to be hydrophobic by triple surface priming with hexamethyldisilazane (HMDS) after exposure and development.

A colloid dispersion of monodisperse (<5%) dodecanethiol-ligated gold nanocrystals in toluene [6], with a typical size of 5 nm to 6 nm, was applied to the template surface via pipette. Toluene wetted the hydrophilic substrate surface very well, two to three drops were sufficient to cover the entire surface. Meanwhile on a hydrophobic surface the contact angle was about 70 degrees and only the center of the sample was covered.

Two methods of colloid deposition were used. The first one, called "drop-and-flow", consists of applying drops of solution, waiting for about 30 s, and then tilting the sample so the excess liquid would flow off the substrate. The second method, called "intermittent-spin", used a spin coater. The sample was placed in the spin coater before the solution drops were applied. After waiting about 60 s, the sample was spun at 300 rpm for 15 sec, then stopped for another 15s, for at least 3 cycles, and then a final fast spin at 1000 rpm to remove excess liquid was applied. This method was efficient in providing uniform thin films on the entire substrate for both hydrophobic and hydrophilic samples. The drop-and-flow method had problems with the hydrophobic surface because the solution did not wet the surface well. Results of both methods are discussed in the next section.

3. RESULTS AND DISCUSSION

Using the drop-and-flow technique, assembly of 5 nm nanoparticles in sub-50 nm trenches was achieved (Fig 1). The gratings were patterned with variable spaces ranging from 30 nm to 110 nm. The purpose of the design was to

NSTI-Nanotech 2008, www.nsti.org, ISBN 978-1-4200-8503-7 Vol. 1

determine what was the maximum trench opening at which the gold nanoparticles could self assemble without grain defects appearing over large distances. In figure 1, it is clear that the 30-nm trench at the right is narrow enough that the nanoparticles pack in with no defects. The tilt technique did not yield good results for the large-area templates. We observed evidence that fluid flow was dominant and interactions with the triangles were not significant, as illustrated in Figure 2.

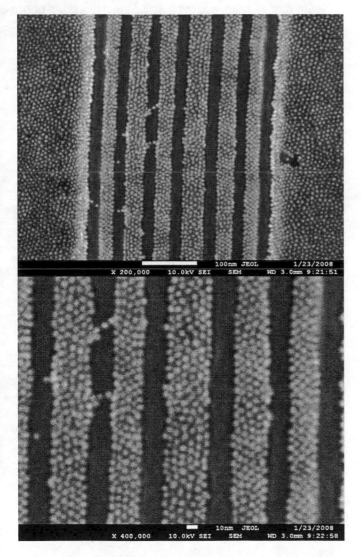

Figure 1. Gold nanoparticles arranged inside trenches defined by a grating of seven lines of HSQ using drop-and-flow deposition. The gold nanoparticles outside the grating show traditional grain boundary structures. (top) SEM micrograph at 200 KX magnification. (bottom) SEM micrograph at 400 KX magnification.

When the dispersion was allowed to sit longer on the sample surface, as in the intermittent spin method, better results for large area templating were obtained for both hydrophilic and hydrophobic surfaces. The results are shown in Figure 3 for a hydrophobic surface sample and Figure 4 for a hydrophilic surface sample.

Figure 2. Gold nanoparticles after drop and flow application on a pattern on a hydrophilic surface with 50 nm equilateral triangles.

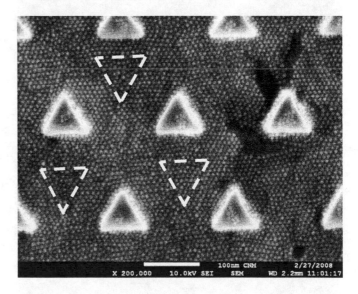

Figure 3. Gold nanoparticles arranged on around 100 nm (top) triangular pillars of HSQ using intermittent-spin deposition on a hydrophobic surface. Dashed triangles are placed for visual assistance to highlight template influence of pillars.

Although similar self-assembly results were obtained on both the hydrophobic and hydrophilic surfaces, the film quality was superior on the hydrophilic surface. The hydrophobic surface exhibited significant amount of coverage defects as seen in the gaps in Figure 3.

Both Figures 3 and 4 show that nanoparticle ordering influenced by templates that reflect the hexagonal closed pack symmetry that is inherent to nanoparticle spheres can be achieved at distances larger than the size of the templates themselves. Improvement on line edge definition and deposition technique should be able to permit then registered templated self-assembly covering wafer size areas, instead of the small gaps found in the trench experiments, due to the high precision pattern placement capabilities of current lithography tools over large areas. In addition, the symmetry based template approach can be extrapolated to nanoparticles with different symmetries, such as cubes or nanorods. In those cases, squares or rectangular templates may be used.

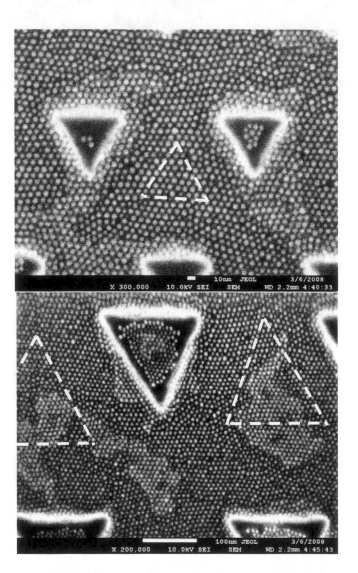

Figure 3. Gold nanoparticles arranged on around 80 nm (top) and 120 nm (bottom) triangular pillars of HSQ using intermittent spin deposition on a hydrophilic surface. Dashed triangles are placed for visual assistance to highlight template influence of pillars. Bottom image is skewed due to sample charging.

4. SUMMARRY AND CONCLUSIONS

We have reported on the templated self-assembly of 5 nm gold nanoparticles in both trenches and in geometries that lend themselves to large-area self-assembly. Preliminary results show that templates that retain the self-assembly symmetry can be used to induce ordered self-assembly at distances significantly larger than trench geometry constraints and larger than the size of the template size. This method may be applied to nanoparticles of other shapes, such as cubes and nanorods.

5. ACKNOWLEDGEMENTS

This work was supported by UChicago Argonne, LLC and the Department of Energy under Contract No. DE-AC02-06CH11357.

Use of the Center for Nanoscale Materials was supported by the U. S. Department of Energy, Office of Science, Office of Basic Energy Sciences, under Contract No. DE-AC02-06CH11357.

REFERENCES

1. N. Kossovsky, D. Millett, A. Gelman, E. Sponsler, H. J. Hnatyszyn, Nature Biotechnology 11, 1534, 1993
2. C. A. Mirkin, R. L. Letsinger, R. C. Mucic, J. J. Storhoff, Nature 382, 607, 1996
3. E. Braun, Y.Eichen, U. Sivan, G. Ben-Yoseph, Nature 391, 775, 1998
4. F. Caruso, R. A. Caruso, and H. Möhwald, Science, 282, 1111, 1998
5. A. Stein, Microporous and Mesoporous Materials, 44-45, 227, 2001
6. T. P. Bigioni, X.-M. Lin, T. T. Nguyen, E. I. Corwin, T. A. Witten and H. M. Jaeger, Nature Materials, 5, 265, 2006
7. X.-M. Lin, A. C.S. Samia, Journal of Magnetism and Magnetic Materials 305, 100, 2006
8. J. A. Liddle, Y. Cui, and P. Alivisatos, J. Vac. Sci. Technol. B, 22, 3409, 2004
9. T. Kraus, L. Malaquin, H. Schmid, W. Riess, N. D. Spencer, H. Wolf, Nature Nanotechnology 2, 570, 2007

Precision Placement and Integration of Individual Carbon Nanotubes in Nanodevices by Fountain-Pen controlled Dielectrophoresis

T. Schwamb[*], N. C. Schirmer[*] and D. Poulikakos[*]

[*] Laboratory of Thermodynamics in Emerging Technologies, Institute of Energy Technology, Department of Mechanical and Process Engineering, ETH Zurich, CH-8092 Zurich, Switzerland

ABSTRACT

We demonstrate the feasibility of a novel method to deposit individual carbon nanotubes (CNTs) at desired locations. The concept integrates the deposition of CNTs by dielectrophoresis onto a fountain-pen setup. Herein, we show the successful deposition of individual CNTs on two-point electrode gaps. Our approach goes in the direction of realizing fully automated manufacturing processes of CNT-based nanodevices. Furthermore, an integrated post-processing soldering step lowering the electrical contact resistance between the deposited CNT and the electrodes is discussed.

Keywords: carbon nanotube deposition, dielectrophoresis, fountain-pen

1 INTRODUCTION

The rapid progress in the research of nanomaterials drives the creation of novel methods and devices in which the advantages of the established silicon chip technology are exploited in combination with the potential of special nanoparticles such as nanowires and carbon nanotubes (CNTs). To this end, several research groups demonstrated the feasibility of CNT-integrated field-effect transistors [1,2]. An obstacle for the realization of CNT-integrated devices is the precise, reliable and high yield integration of CNTs in electric circuits and other MEMS. Current research in nanotechnology is concerned with the precise handling of individual CNTs [3,4].

Our goal herein is to contribute a simple, mass production capable method to assemble CNT-integrated electric circuits and subsequent precise soldering of the electrical contacts of these circuits. This is done by combining the method of dielectrophoresis (DEP) with a fountain-pen setup.

2 DIELECTROPHORESIS

An electric dipole exposed to an electric field aligns parallel to the electric field lines. As long as the electric field is homogenous, no net force acts on the electric dipole. In an inhomogeneous electric field, the non-uniformity of the electric field exerts a net force on the dipole. The force on an electric dipole exposed to an electric field can be described by

$$\mathbf{F} = (\mathbf{p} \cdot \nabla)\mathbf{E} \qquad (1)$$

Therefore the net force $\mathbf{F}$ on the particle is a linear function of the electric field gradient $\nabla\mathbf{E}$ and the dipole moment $\mathbf{p}$ [5].

An electric dipole is induced to an uncharged polarizable particle subjected to an electric field. The resulting net force on the polarized particle in a non-uniform electric field is the dielectrophoretic force $\mathbf{F}_{DEP}$.

The induced dipole moment $\mathbf{p}$ of a particle dissolved in a liquid dielectric depends on the geometry of the particle, the external electric field and the solvent. The geometry of a CNT can be assumed to be a long prolate spheroid. This gives an induced dipole moment which is described by [6,7]

$$\mathbf{p} = \frac{1}{2}\pi r^2 l \varepsilon_m \, \mathrm{Re}(K)\mathbf{E} \qquad (2)$$

In equation 2, r and l are the radius and the length of the CNT, respectively, ε_m the absolute permittivity of the solvent, $\mathbf{E}$ the external electric field and $\mathrm{Re}(K)$ the real part of the complex polarization factor.

The complex polarization factor K, which is known as Clausius-Mossotti factor for spherical particles, depends on the complex permittivity of the medium and of the particle. Equation 3 describes K for an elongated particle aligned to the electric field [8].

$$K = \frac{\varepsilon_p^* - \varepsilon_m^*}{\varepsilon_m^*} \qquad (3)$$

Here the absolute complex permittivity of the medium ε_m^* and the particle ε_p^*, respectively, can be expressed as follows [9]:

$$\varepsilon^* = \varepsilon - i\frac{\sigma}{f} \qquad (4)$$

Where $i = \sqrt{-1}$, σ is the electric conductivity and f is the frequency of the external electric field. For a perfect dielectric ($\varepsilon \gg \sigma$), the absolute complex permittivity ε^* approximately equals the absolute permittivity ε.

The direction of the dielectrophoretic force relative to the electric field is denoted by K. A positive K means that the particle is moved to the region of high electric field strength. A negative K means that the particle is forced to the region of low electric field strength.

Due to the frequency dependency of ε^*, the complex polarization factor K is a function of the frequency. From equation 3 one can derive two extrema:

$$f \to 0 : \operatorname{Re}(K) = \frac{\sigma_p - \sigma_m}{\sigma_m}$$

$$f \to \infty : \operatorname{Re}(K) = \frac{\varepsilon_p - \varepsilon_m}{\varepsilon_m} \qquad (5)$$

If the electric conductivity of the solvent can be neglected, the frequency dependence of K is only governed by the frequency dependence of ε_p^*.

Applying a biased ac field, the time-averaged dielectrophoretic force comes out to be [6]:

$$\left\langle F_{DEP} \right\rangle = \frac{1}{2} \pi r^2 l \varepsilon_m \operatorname{Re}(K) |E|^2 \qquad (6)$$

3 EXPERIMENTS

For the CNT deposition, a low concentrated CNT-water suspension was introduced into the micropipette of the fountain-pen setup. The micropipette was manufactured from borosilicate capillary tubes in a pipette puller (Zeitz DMZ Puller). The puller was programmed to deliver micropipettes with tip openings in the range of some micrometers. With x-, y- and z-axis translation stages the micropipette tip opening was placed precisely above an electrode gap on a silicon chip where an individual CNT was to be deposited. Applying pressure to the micropipette, a microdroplet of CNT-suspension was squeezed out of the opening. Subsequently, the pipette was moved down in the z-axis direction until the droplet touches the substrate surface. Due to the fountain-pen principle [10], the microdroplet on the substrate surface is maintained despite evaporation, by a constant flow of CNT-suspension from the inside of the pipette. At the moment a microdroplet has formed on the substrate surface above the electrode gap (figure 1), the ac voltage was switched on and the DEP was initiated. The voltage was applied for a time interval of 60 seconds.

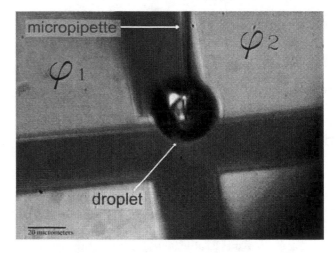

Figure 1: The picture, taken with a CCD camera, shows the pipette placed over the electrode gap which is covered by a spherical droplet of CNT solution. At the same moment the ac voltage is applied to the electrodes for up to 60 seconds. The electrical configuration is also shown, depicting the potential configuration on the electrodes in blue (φ_1) and green color (φ_2).

The fountain-pen DEP process described above is schematically depicted in Figure 2. The electric field inside the microdroplet attracts an individual CNT through the micropipette opening to the desired position above the electrode gap.

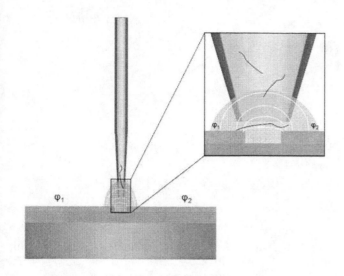

Figure 2: Schematic of the presented method. The micropipette is placed above an electrode gap as depicted. The electrical field between the electrodes (light gray color) results from an applied potential difference ($\Delta\varphi 12$). This field is indicated by the narrow white lines within the CNT (black lines) in the suspension (blue color).

After the voltage was switched off, the micropipette was retracted from the droplet. Instantly, the cut off of CNT-suspension supply initiates the shrinking of the droplet sitting on the substrate. The evaporation driven shrinking extinguishes the droplet within a few seconds.

4   RESULTS & DISCUSSION

The fountain-pen controlled DEP method was proven to be able to deposit CNTs precisely. Figure 3 depicts the successful deposition of an individual, multi-walled CNT over an electrode gap of approx. 400 nm.

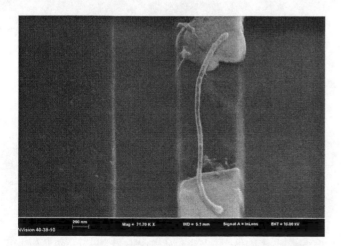

Figure 3: A CNT of 45 nm diameter was successfully deposited by the fountain-pen method. The electrode gap is approx. 400 nm wide.

Due to the precise positioning of the micropipette opening over the electrode gap with translation stages, the confinement of the micropipette walls and the small droplet volume in which the DEP was carried out, CNTs are placed precisely and individually at the desired location. At the same time, most of the possible spaghetti-like-bundled CNTs are held back in the micropipette.

In combination with the soldering process described in ref. [11], a precise soldering of the fountain-pen deposited CNT is possible within the same fountain-pen setup. The integration of the soldering process allows the manufacturing of complete CNT-based nanodevices within one setup and process chain. Furthermore, the process has the potential to yield computerized manufacturing by automating the positioning of the silicon chip and the micropipette. This results in the possibility for a larger scale, automated assembly. Further research will investigate the possibility to combine the fountain-pen based deposition with four-point electrodes which have been described in ref. [12].

In summary, we presented a simple novel method capable of aiding the assemblage future CNT-integrated chips in a fully automated larger scale process.

The financial support of the ETH Forschungskommission is kindly acknowledged.

REFERENCES

[1] Martel, R.; Schmidt, T.; Shea, H. R.; Hertel, T.; Avouris, P. *Appl. Phys. Lett.* **1998**, 73, (17)

[2] Dockendorf, C.P.R.; G. Hwang; B.J. Nelson; C.P. Grigoropoulos; D. Poulikakos *Appl. Phys. Lett.* **2007**, 91, (24)

[3] Banerjee, S.; White, B. E.; Huang, L. M.; Rego, B. J.; O'Brien, S.; Herman, I. P. *J. of Vac. Sc. & Tech. B* **2006**, 24, (6)

[4] Abrams, Z.R.; Ioffe, Z.; Tsukernik, A.; Cheshnovsky, O.; Hanein, Y. *Nano Lett.* **2007**, 7, (9)

[5] Pohl, H. A. *J. Appl. Phys.* **1951**, 22, (7)

[6] Kim, J. E.; Han, C. S. *Nanotechnology* **2005**, 16, (10)

[7] Jones, T.B. *J. of Electrostatics* **1979**, 6

[8] Dimaki, M.; Boggild, P. *Nanotechnology* **2004**, 15, (8)

[9] Kadaksham, J.; Singh, P.; Aubry, N. *Mech. Res. Commun.* **2006**, 33, (1)

[10] Choi, T.Y.; Poulikakos, D.; Grigoropoulos, C. P. *Appl. Phys. Lett.* **2004**, 85, (1)

[11] Dockendorf, C.P.R.; Steinlin, M.; Poulikakos, D.; Choi, T. Y. *Appl. Phys. Lett.* **2007**, 90, (19)

[12] Schwamb, T.; Choi, T.Y.; Schirmer, N.; Bieri, N.R.; Burg, B.; Tharian, J.; Sennhauser, U.; Poulikakos, D. *Nano Lett.* **2007**, 7, (12)

A Novel Technique for Segregation
Of HiPCO CNT'S Using INKJET Technology

Dr. Tulin Mangir, Juan Chaves, Mukul Khairatkar, Sindy Chaves.

California State University Long Beach
Long Beach, California, 90840-8303
temangir@csulb.edu, jchaves@csulb.edu, mkhairat@csulb.edu, schaves@csulb.edu

ABSTRACT

In paper we describe a technique to purify and segregate carbon nano tubes (CNTs) from the iron impurities to use the CNTs for thin films. After extensive research into a proper and repeatable cleaning process we developed a new method of using bacteria to remove impurities from CNTs. This gives us the ability to manipulate the CNT. We also developed an apparatus based on inkjet technology that deposit the bacteria containing CNT on conductive surfaces such as silicon, plastic, or any other conductive surface. In this way we create a simple way to manipulate the CNTs for the creation of organic LED, Thin Films, or even electronic devices such as transistors using a simple process. Our methods specifically apply to CNTs obtained by the HiPCO process. However, we assume the method can be extended to other cases. We expect this approach will lead to finding novel conductors, structures and thin films for future nano-engineered devices that could be investigated further for use in nano-scale electronics, implants and bio-materials.

Keywords: HiPCO CNT, Inkjet, bacteria, purification, segregation.

1 INTRODUTION

Today, the race to find materials with stronger and faster carrier mobility and smaller thickness for the fabrication of thin films that will yield smaller geometries and flexible electronic devices led to the experimentation with CNTs. CNTs have attracted a lot of interest because their unique physical properties and because they show a lot of potential for their use in microelectronic and nanoelectronic applications. Hence, by developing an industrial process, they can be use in areas such as display technology, as thin film transistors in liquid crystal displays or Organic Light Emitting Diodes (OLED), as well as thin films for the manufacturing of CMOS devices and Field Effect Transistors (FETs). One of the physical properties of CNTs that made them very attractive to the manufacturing of FET is their carrier transport property. It is been estimated that the carrier mobility of a single CNT will be in the range of 1000 cm^2/Vs [1]. Although, the carrier mobility of CNT seems to be a great advantage in the manufacturing of thin films, in reality it gets compromise somehow by the introduction of carbonaceous impurities as well as metal impurities due to the manufacturing process of CNTs. As an example a FET made out of a single CNT can carry limited currents generally in the range of nA or μA, but by using CNT thin film mesh the range can be significantly increase[1]. Hence, for CNT to be compatible with today's device fabrication at low cost and high throughput it needed to be in more aqueous stage e.g., a liquid material or solution, since the process of CNT growth requires high temperatures of 900 °C.

In our research we utilize SWCNTs (Single Walled Carbon Nano Tubes) obtain by the HiPco (High-pressure CO dissociation) process. One of the major drawbacks using this type of material was the amount of iron impurities contain in the CNTs, due to the use of iron as a catalyst in the process. Hence, a pioneering cleaning method was developed in order to obtain pristine SWCNT that could be use for the creation of high quality thin films, and the creation of the liquid material required to the manufacturing process of thin films.

2 EXPERIMENTAL

The main purpose of our research was to create thin films for use in electronic applications. Consequently, we needed a simple way to purify the SWCNT from the iron contaminants induce thru the HiPco process at the time of their growth. After experimenting with several methods already being researched to clean the same type of material from their impurities, we concluded that these methods caused structural damage of the SWCNT which will also led to a change in the electrical properties. Hence, after trying different methods we created a more passive process for the purification that will prevent damage in the SWCNT structure, but at the same time eliminating over 90% of the contaminants such carbonaceous and iron impurities, leaving a pristine material to be use in the creation of thin films. For such reason we devise the use of a biological agent for the purification process. After, research several ways of processing the SWCNT dust and exposing it to a variety of biological agents, we found that by using bacteria we achieve the cleaning task without affecting the condition and/or structure of the CNTs; thus eliminating most of the mineral contaminants in the process*. The biological properties of the bacteria and the ubiquitousness of it, was so appealing for their use in this cleaning process.

(a)

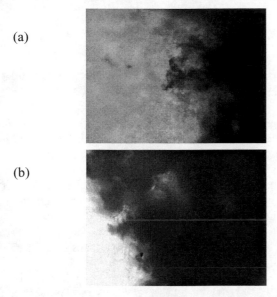

(b)

Figure 1: (a) Optical image of CNT dust when bacteria are just applied; (b) Optical Image of CNT dust after 4h once the bacteria start the cleaning process. The red area is the Iron oxide from the Iron being pulled out from the inside of the CNTs.

After several attempts to formulate a procedure to expose the CNT dust to biological cleaning agents, we finally create a working protocol to generate a high yield of material with pristine SWCNT for the use in the manufacturing process of thin films*.

Thru this biological cleaning process, the purified SWCNTs have been found to have higher surface areas retaining also their physical characteristics of conductivity, compare to the acid treatments or microwave treatments which were demonstrated to have good results in the elimination of metallic catalytic impurities but with a rise in structural and electrical changes in the end material [2]. Also this biological cleaning process was demonstrated to be superior and more cost effective that the gas oxidation process, in which the material has to be exposed to Argon or Oxygen gas flow for several hours, annealed and then clean by HCl acid exposing the material to structural damage [3-4]. In the biological process, the SWCNT were exposed to the right amount of biological agents with the right amount of CNT dust and then water rinse to eliminate the untreated material, making the cleaning process cost effective by eliminating the use of expensive acids or gases thru the process. Consequently, the process achieves high yields of the clean material reducing the cost and reducing the waste of the CNT dust during the process.

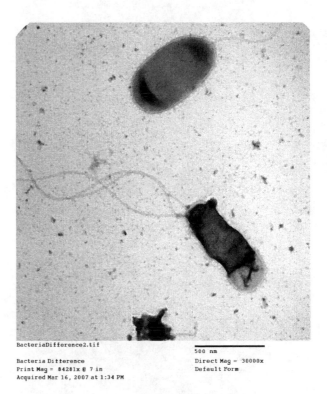

BacteriaDifference2.tif

Bacteria Difference
Print Mag = 84281x @ 7 in
Acquired Mar 16, 2007 at 1:34 PM

500 nm

Direct Mag = 30000x
Default Form

Figure 2: Bacteria Difference one without CNT and one with SWCNT inside.

In the biological cleaning process one of the bacterium main nutrients is Iron which is needed for the production of nitrogen. Using this to our advantage the bacterium is grow in a starving environment (lack of Iron nutrients) and then is released in an environment containing only the raw CNT. Once the bacteria colony reaches the raw material the absorption process starts. During this process each bacterium works as a "biological cleaning machine" absorbing the raw CNT to extract the Iron from the SWCNT. In figure 2 a TEM micrograph shows the difference between a normal bacterium and one after "ingesting" the raw CNT material. After the CNT are absorbed the bacterium breaks down the carbonaceous cocoons were the iron is enclosed leaving only clean SWCNT.

This absorption process is very significant in our research since it accomplishes two tasks at once. First, the cleaning of the material and segregation of individual SWCNT form their tangled bundles was achieved. Second each bacterium performs as a SWCNT "container or vessel", were the clean pristine material is enclosed after the cleaning process. Hence, allowing an easy manipulation of a SWCNT thin film mesh without disturbing the structural characteristics of the thin film as well as the conductivity properties of it.

Figure 3 shows an SEM micrograph of the SWCNT mesh contained inside of an individual bacterium.

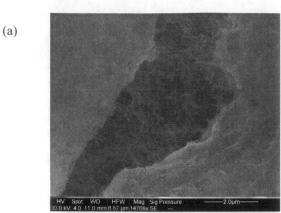

(a)

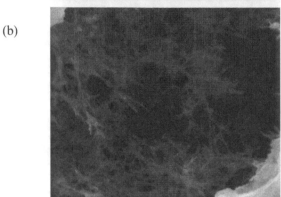

(b)

Figure 3: (a) SEM micrograph of the SWCNT thin film inside the bacterium; (b) a closer image of the SWCNT mesh thin film inside the bacterium.

Once the absorption and cleaning process is done the bacteria gets selected and clean from any leftover raw material by a rinse process. After the rinse process is complete the bacteria vessels are ready to be deposited into the surface to create the thin film.

2.1 Inkjet Dispersion of the SWCNT

The next step is the segregation of the individual SWCNT and the separation from the bacterium. This is achieved by using Inkjet technology developed by Hewett Packard [4]. A special ink was created with the bacteria containing the SWCNT thin film mesh inside*. The ink was developed with a specific ratio that with yield a consistent number of bacteria population with SWCNT in them for every 50 ml of solution, making a very dense bacteria ink.

An inkjet cartridge was modifying to host the special ink and a printer apparatus was developed to use the special cartridge, for the process of deposited the SWCNT mesh virtually in any surface*.

The principle of inkjet technology provides the last step of converting the material back from the wet environment to a dry environment. Hence, the method allows the use of the clean SWCNT in the manufacturing process of electronic devices*.

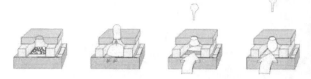

Figure 4: Inkjet process to release droplets of ink in a surface [4].

By using the thermal inkjet technology the solution is superheated to about 1,000,000 °C/second, the solution were the bacteria is contain is ejected without boiling but disintegrating the bacteria in the process virtually depositing the mesh of SWCNT in the surface. Each droplet released is about 150pl making this process very accurate to deposit the material virtually in any space or surface. Once the material is deposit in the surface the surface exposed to a quick heat treatment to eliminate any residual of the biological agent.

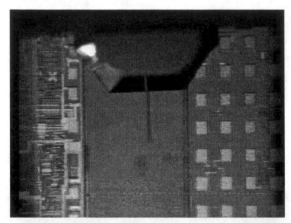

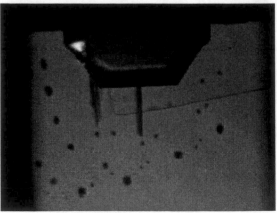

Figure 5: Pictures of the special ink droplet taking by the onboard camera of the AFM

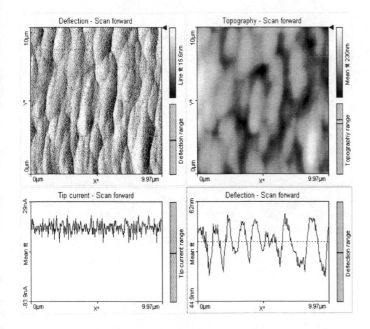

Figure 6: AFM image of the SWCNT mesh after deposited thru the inkjet nozzle on a silicon surface.

3 DISCUSSIONS AND CONCLUSION

Our results has shown that our biological cleaning and segregation process of HiPco SWCNT is less invasive, less damaging and more cost effective than processes that involved the use of HCl, Oxygen or Argon gas, as well processes that uses microwaves to clean the impurities in the raw material. Also, with the use of inkjet technology in conjunction with our biological cleaning process allow us to achieve the segregation of individual clean SWCNT and the manipulation of the thin film mesh form by them. Hence, with the creation of the cleaning protocol, the creation of the special bacteria ink, and the creation of the ink depositing apparatus we achieve an industrial process in the manufacturing of thin films for electronic applications. Further discussions of the major topics of this work and most of our data results are going to be presented at the conference.

REFERENCES

[1]   Xuliang Han, Daniel C. Janzen, Jarrod Vaillancour, and Xuejun Lu. 2006. A flexible **thin-film** transistor with high field-effect mobility by using **carbon nanotubes**. *Nanotechnology Materials and Devices Conference, 2006. NMDC 2006. IEEE* 1 (2007-11-27) : 296-297.

[2]   Ester Vázquez, Vasilios Georgakilas, and Maurizio Prato. 2002. Microwave-assisted purification of HIPCO carbon nanotubes. *Chemical Communications* , no. 20: 2308-2309.

[3]   Chiang, I. W., B. E. Brinson, A. Y. Huang, P. A. Willis, M. J. Bronikowski, J. L. Margrave, R. E. Smalley, and R. H. Hauge. 2001. Purification and Characterization of Single-Wall Carbon Nanotubes (SWNTs) Obtained from the Gas-Phase Decomposition of CO (HiPco Process). *Journal of Physical Chemistry B* 105, no. 35: 8297-8301.

[4]   KIM Yeji, MINAMI Nobutsugu, ZHU Weihong, KAZAOUI Said, AZUMI Reiko, and MATSUMOTO Mutsuyoshi. 2003. Langmuir-Blodgett films of single-wall carbon nanotubes: Layer-by-layer deposition and in-plane orientation of tubes. *Japanese journal of applied physics* 42, no. 12 (December 10, 2003) : 7629-7634.

[5]   Rob Beeson. Thermal Inkjet: Meeting the Applications Challenge.

*   **This new approach is under a patent pending process, reason why some details were omitted from this paper.**

Special Thanks to the Army Research Office for their support in this investigation thru their Grant "Assessing the Integrity and Integration Issues of Nano- Sensors/Nanostructures-Equipment".

Periodic Nanowell Array using Template-Assisted Nanosphere Lithography

Seungwon Jung and Junghoon Lee

Department of Mechanical and Aerospace Engineering, and Institute of Advanced Machinery and Design
Seoul National University, Seoul, Korea
jleenano@snu.ac.kr

ABSTRACT

Nanosphere lithography (NSL) has been studied as the effective method of fabricating nano-scale array. This paper presents template-assisted NSL to obtain high-quality crystal in regularity and coverage. A periodic array of 100 nm deep nano trenches with two different widths, 400 and 450 nm was fabricated by nano imprint lithography (NIL), and used as the template. When polystyrene nanospheres with the diameter of 100 nm were spin-coated on the substrate with the template morphology, they formed a crystal array with improved periodicity in both lateral and longitudinal directions. We also demonstrated nanowell array through subsequent deposition and etching steps.

Keywords: nanosphere lithography, template-assisted NSL, nano imprint lithography, nano fabrication

1 INTRODUCTION

Nano-scale fabrication has been developed for scaling down in microelectronics, sensors, and many other applications. The conventional lithographic methods have been used for nano patterns. Photolithography, most widely used in modern microelectronic fabrication, is parallel, simple, quick, and inexpensive process. Photolithographic resolution is, however, limited by the diffraction of the UV light. Many approaches such as electron beam lithography, X-ray lithography, extreme UV lithography, and scanning probe lithography have been studied in order to overcome this problem [1-4]. However, high cost, complexity, and long process time still prevent the wide use of above approaches to reduce the features.

Among non-optical lithographic methods studied as alternatives in recent days, nanosphere lithography (NSL) has been proved useful when a regular pattern on a large area was needed [5-8]. Advantages include low cost, quick process, and the freedom in choosing various materials on diverse substrates [5, 6]. In NSL, many approaches have been developed such as drop-coating, spin-coating, Langmuir-Blodgett method, and electrical method [9-12].

Many studies have been reported for nanosphere array above 300 nm in diameter using the above methods. No previous method, however, claims successful results with well-defined 100 nm features on a reasonably large area. The large variation of bead sizes and strong Van der Waals force between beads prevent the formation of a single domain, leading to multiple, fragmented domains.

This paper introduces template-assisted NSL that is the new method combined the morphology-driven arrangement with spin coating. We show 100 nm patterning over a large area through the optimization of spin coating and the use of nano trench on a substrate fabricated by nano imprint lithography (NIL). Also demonstrated are the fabrication of nanowell array by the shrinkage of nanosphere, metal deposition, and lift-off process.

2 EXPERIMENT

2.1 Preparation of Substrate

The nano-trench on a substrate was prepared using NIL (NANOSYS 420, NND co., Korea) [13, 14]. The mold was fabricated by electron beam lithography and inductive-coupled plasma (ICP) etching of silicon. Completed mold was coated with self assembled monolayer (SAM, heptadecafluoro-1,1,2,2-tetrahydrodecyl trichlorosilane) for an easy removal of the mold from the substrate after imprint. A silicon wafer with 100 nm-thick SiO_2 was used as the substrate. Thermoplastic polymer (mr-I 8010, Micro Resist Technology, Germany) was coated on the substrate. The NIL was performed at 190 °C and 40 bars. The NIL was followed by the removal of residual polymer. The SiO_2 layer was etched with reactive ion etching (RIE) until the silicon surface was exposed. Finally polymer was stripped with piranha treatment (4:1, $H_2SO_4{:}H_2O_2$).

After diced into 1 mm ×1 mm pieces, the substrate with the nano-trench was pretreated with piranha for 3 h @ 120 °C to clean the substrate and grow the native oxide on the silicon surface. These treatments changed the silicon surface into a hydrophilic one. The surface was further treated with RCA treatment (5:1:1 H_2O: NH_4OH: H_2O_2) with sonication for 30 min to increase the hydrophilicity of the substrate. To maintain cleanliness and wettability, the completely treated substrate was stored in deionized water until next continues.

2.2 Assembly of Nanospheres

Polystyrene nanospheres (Duke Scientific Co., USA) with the diameter of 100 nm were spin-coated on the substrate. In our approach, spin coating was completed in three steps as shown in Figure 1. First, 60 μg nanosphere solution was spin-coated on a substrate at 500 rpm for 10 s to spread the solution. Second, the substrate was rotated at a critical spin rate for 2 min to form and arrange the crystal of nanospheres. Various spin rates from 100 rpm to 1500 rpm were used to find the optimal condition for crystal formation. Finally, to remove the nanosphere solution from the substrate except for the nanospheres assembled in the trench, the spin rate was ramped up to 2000 rpm at the acceleration of 100 rpm/s.

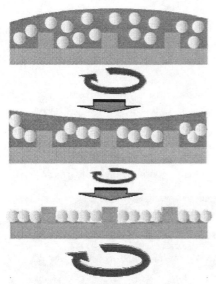

Figure 1. Schematic of template-assisted NSL that consists of spin coating with three steps: 1) spreading, 2) arrangement, 3) removal

2.3 Fabrication of Nanowell Array

The nanosphere array produced through template assisted NSL was shrunk by O$_2$ plasma to reduce the diameter of the nanospheres. The ashing with the plasma continued for 20 s at flow rate of 50 sccm, RF power 50 W, and pressure 50 mTorr. After the shrinkage, 10 nm-thick chromium was deposited on the spaced nanosphere array by e-gun evaporator (ZZS550-2/D, Maestech Co., Korea). The deposition rate was kept very low to obtain a uniform metal layer. Then the nanospheres were lifted off using toluene with sonication for 30 min.

3 RESULTS & DISCUSSION

The quality of the template-assisted NSL method heavily depends on spin coating. To obtain optimal spin rate for the best formation of nanosphere array with 100 nm in diameter, nanospheres were coated on the substrate without the nano template at various spin rates; 100, 200, 500, 800, 1000, 1200, and 1500 rpm.

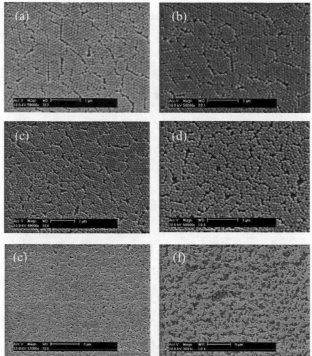

Figure 2. The SEM images of nanosphere array without nano template using spin rate of (a) 200, (b) 500, (c) 800, (d) 1000, (e) 1200, and (f) 1500 rpm. As the spin rate increased, the coverage was poorer and more domains were generated.

The relationship between the spin rate and the formation of nanosphere crystal is shown in Figure 2. The increase of the spin rate raised the number of the cracks, while reducing the domain size and the coverage on a substrate. Lower spin rate enabled better coverage and larger single domain. However, spin rate lower than 200 rpm deteriorated the formation of nanosphere crystal. This trend means that there exists the critical spin rate for the best crystal formation of nanospheres. For current process 200 rpm was determined as an optimum spin rate.

The perfect nanosphere crystal, however, could not be achieved using spin coating only. Even the best result at the spin rate of 200 rpm among the above ones had many cracks. It is likely that there was no force to rearrange the imperfect nanospheres without cracks. The breaking of the domains and reformation into the single domain were needed to produce the large scale patterning. We introduced a nano-template to enforce the periodic, crack-free arrangement of the nanospheres. Dimensions of the trench fabricated by NIL were 100 nm in depth with two different widths, 400 and 450 nm as shown in Figure 3.

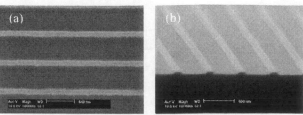

Figure 3. The SEM images of the nano template used in this experiment: (a) top view and (b) side view of nano template. The nano-template was 100 nm in depth with two different widths, 400 and 450 nm.

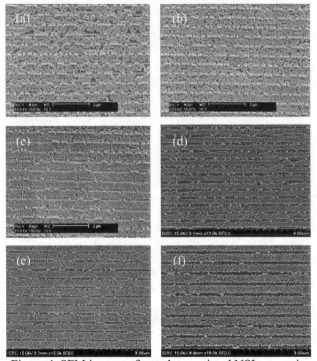

Figure 4. SEM images of template-assisted NSL at a spin rate of (a) 1500, (b) 1200, (c) 800, (d) 500, (e) 200, and (f) 100 rpm.

Figure 4 shows the results of template-assisted NSL at various spin rates. Like previous results of spin coating, crystal at 200 rpm produced the best coverage. There are additional advantages due to the use of nano-template. The nano-template helped the rearrangement of crystals as a passive guidance. The crystals of nanospheres were formed from several spots throughout the substrate at the early stage of arrangement process. Without any guidance, the crystals combined by Van der Waals interaction were never broken and rearranged, resulting in multiple domains. However, when guided by the template, the crystals could be relocated in the trench. The nano-template broke and rearranged the crystals by fitting them in the trench. Also, the cracks in the crystal array were reduced by the nano template. The cracks were almost always formed during the crystal formation of nanospheres on the substrate. Generally the crack is produced by (1) the distribution of

nanoshpere diameters, (2) the collision between domains, and (3) the spots missing nanospheres in the array. In the template assisted NSL, however, the cracks were not continued on in the array and the effect of the crack, if any, was kept small. The template allows the freedom in one dimension, and the crack stops when bumping against the wall.

The gap size of the nano-templates was also a key factor for the crystal formation as the gap size is determined by

$$l_g = [(n-1)\sin 60° + 1]d \qquad (1)$$

, where n is the number of nanosphere rows per gap and d the average diameter of nanospheres. We used nano trench with two types of widths. One was 450 nm which was calculated with 100 nm in diameter and 5 nanosphere rows per gap. The other was 400 nm which was not suited with the current dimension of the nanosphere.

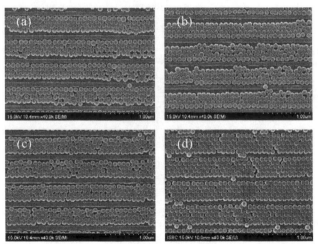

Figure 5. The crystals in 400 nm in width were placed (a) at the top or (b) at the bottom or (c) randomly. (d) 450 nm trench was packed perfectly. These results show the importance of the correct dimension of width.

Figure 5 (d) clearly shows that the nano template with precisely calculated width (450 nm) resulted in the formation of highly periodic nanosphere crystal. In contrast, the crystals were placed on the either side (Figure 5 (a), (b)) or randomly (Figure 5 (c)) when the width was 400 nm. In case of (a) and (b), biased side was determined by the direction of the centrifugal force.

Figure 6 shows the results of the fabrication process of nanowells. Shrunk nanospheres with O_2 plasma were shown in Figure 6 (a) and (b). The diameter of nanospheres after the plasma was reduced by 30 % in size. The periodic nanowell array was produced by the metallization and lift-off process as clearly shown in Figure 6 (c). We obtained the nanowell array extended in lateral direction through template assisted NSL.

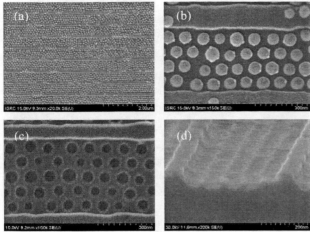

Figure 6. (a) and (b) shrunk nanospheres with O_2 plasma. (c) and (d) the nanowell array after metallization and lift-off process.

4 CONCLUSION

In this paper, we developed the template-assisted NSL which combines the active coating method (spin coating) and the passive guidance (nano-template). The nano-template fabricated by NIL helped the periodicity of nanosphere array. The nanowell array was fabricated through the subsequent steps of shrinkage, metallization, and lift-off process. We believe our methods will enable many applications that need a precise periodic nano pattern with simple process.

ACKNOWEDGEMENT

This work was supported by the National Core Research Center Program of the Ministry of Science and Technology (R15-2003-032-01002-0) and in part by the Micro Thermal System Research Center of the Korea Research Foundation (ERC, F0001022-2006-31).

REFERENCES

[1] C. Vieu, F. Carcenac, A. Pépin, Y. Chen, M. Mejias, A. Lebib, L. Manin-Ferlazzo, L. Couraud and H. Launois, "Electron beam lithography: resolution limits and applications", Applied Surface Science, 164, 111-117, 2000

[2] G. Simon, A. M. Haghiri-Gosnet, J. Bourneix, D. Decanini, Y. Chen, F. Rousseaux, H. Launois, "Sub-20 nm x-ray nanolithography using conventional mask technologies on monochromatized synchrotron radiation", J. Vac. Sci. Technol. B., 15(6), 2489-2494, 1997

[3] G. F. Cardinale, C. C. Henderson, J. E. M. Goldsmith, P. J. S. Mangat, J. Cobb, S. D. Hector, "Demonstration of pattern transfer into sub-100nm polysilicon line/space features patterned with extreme ultraviolet lithography", J. Vac. Sci. Technol. B., vol. 17(6), 2970-2974, 1999

[4] A. Majumdar, P. I. Oden, J. P. Carrejo, L. A. Nagahara, J.J. Graham, J. Alexander, "Nanometer-scale lithography using the atomic force microscope", Appl. Phys. Lett., 61, 2293, 1992

[5] K. Kempa, B. Kimball, J. Rybczynski, J. Y. Lao, W. Z. Li, Z. F. Ren et al, "Photonic Crystals Based on Periodic Arrays of Aligned Carbon Nanotubes", Nano Letters, 3(1), 13, 2003

[6] A. L. Thangawng, J. Lee, "Fabrication of a Micro/Nano Integrated Roughened Structure using Nanosphere Lithography(NSL)", Proceeding of IMECE04, IMECE2004, 62248, 2004

[7] Ferenc Jarai-Szabo, Simion Astilean, Zoltan Neda, "Understanding self-assembled nanosphere patterns", Chemical Physics Letters, 408, 241-246, 2005

[8] E. Kumacheva, R. K. Golding, M. Allard and E. H. Sargent, "Colloid Crystal Growth on Mesoscopically Patterned Surfaces: Effect of Confinement", Advanced Materials, 14(3), 221, 2002

[9] S. Astilean, "Fabrication of periodic metallic nanostructures by using nanosphere lithography", Romanian Reports in Physics, 56, 340-345, 2004

[10] J. C. Hulteen, R. P. Van Duyne, "Nanosphere lithography: A materials general fabrication process for periodic particle array surfaces", J. Vac. Sci. Technol. A, 13, 1553, 1995

[11] Karl-Ulrich Fulda, Bernd Tieke, "Langmuir films of monodisperse 0.5 µ m spherical polymer particles with a hydrophobic core and a hydrophilic shell", Advanced Materials, 6, 288, 1994

[12] S. O. Lumsdon, E. W. Kaler, O. D. Velev, "Two-Dimensional Crystallization of Microspheres by a Coplanar AC Electric Field", Langmuir, 20(6), 2108-2116, 2004

[13] Stephen Y. Chou, Peter R. Krauss, and Preston J. Renstrom, "Imprint of sub-25nm vias and trenches in polymers", Appl. Phys. Lett., 67(21), 3114-3116, 1995

[14] Stephen Y. Chou, Peter R. Krauss and Preston J. Renstrom, "Nanoimprint Lithography", J. Vac. Sci. Technol. B., 14(6), 4129-4133, 1996

Nanomaching of Silicon nanoporous Structures by Colloidal Gold Nanoparticle

J. Zhu[*], H. Bart-Smith[**], M. R. Begley[**], R. G. Kelly[***], G. Zangari[***], and M. L. Reed[*]

[*]Department of Electrical and Computer Engineering,
[**]Department of Mechanical and Aerospace Engineering,
[***]Department of Materials Science and Engineering,
University of Virginia, Charlottesville VA 22904

ABSTRACT

We report an efficient method for fabricating nanoporous structures in crystalline silicon, using colloidal gold nanoparticles in HF/H_2O_2 etchant. Colloidal gold in HF/H_2O_2 solutions accelerates the etching rate by two orders of magnitude and simultaneously introduces anisotropy in the etching process, with preferential penetration along the <100> directions. Random nanoporous structure, tailored nanochannel and nanopore were developed with this method.

Keywords: gold nanoparticle, nanoporous silicon

1 INTRODUCTION

Porous silicon has drawn a great deal of attention in applications such as optoelectronics [1], photonic crystals [2], optical filters [3], membranes and molecular sieves [4, 5]. Porous silicon is usually synthesized by electrochemical anodization in HF-containing aqueous or organic solutions [6-8]; its morphology is determined by the anodization conditions, the electrolyte chemistry, as well as the Si doping type and the dopant concentration. Certain morphologies, such as straight pores with narrow pore size distributions, can be obtained only under strictly controlled synthesis conditions [9]. Surface modification of silicon by the deposition of discontinuous metal films has been found to accelerate the etching process [10] and in some cases to generate straight pores [11]; however, the control of pore size by this method is limited and the lowest achievable dimensions are of the order of 50-100 nm. Here we demonstrate control of the nanostructure of porous silicon by nanoparticle-assisted etching of silicon in electrolytes containing HF and H_2O_2, which accelerates the etching rate by two orders of magnitude and simultaneously introduces anisotropy in the etching process, with preferential penetration along the <100> directions. The process does not require an external electrochemical cell or potentiostat and can thus be achieved without electrical leads and contacts.

2 POROUS SILICON FORMED WITH GOLD NANOPARTICLE ASSISTED ETCHING

Etching of silicon by immersion in HF/ H_2O_2 solutions occurs very slowly, at a rate of about 1 nm/min [12]. Inclusion of H_2O_2 in an HF solution decreases the surface roughness by rendering the etch process isotropic [13]. Addition of colloidal gold nanoparticles (AuNPs) to the HF/ H_2O_2 solution (AuNPs-HF/ H_2O_2), as we report here, results in a considerable enhancement of the silicon etching rate, up to approximately 100 nm/min. Additionally, the etching process in presence of the AuNPs is no longer isotropic. As shown later, this procedure can be used to synthesize porous silicon structures from starting material with different crystal orientations, as well as different doping types and concentrations.

The AuNPs used in this study were MesoGold from Purest Colloids, Inc. (Westampton, NJ), with a purity of 99.99% and a nominal diameter of 3.2 ± 0.3 nm. The etching solution (AuNPs-HF/H_2O_2) was a 10:5:1 mixture (by volume) of 10 ppm colloidal AuNPs in deionized water, 30% H_2O_2, and 49% HF (the latter two supplied by Mallinckrodt Baker, Inc., Mallinckrodt, NJ). The prepared etchant is stable at room temperature for several months.

Etching experiments were carried out on 2 inch diameter silicon wafers with various doping types, with a doping concentration ranging from 2×10^{14} to 10^{20} cm^{-3}, and different orientations. After sequential spin cleaning with trichloroethylene, isopropanol, and methanol, the samples were lowered into a polypropylene beaker containing the AuNPs-HF/ H_2O_2, at room temperature, and a magnetic stirrer. After a timed etch, the samples were rinsed with deionized water and dried with nitrogen. Porous silicon regions formed on all silicon samples, as evidenced by scanning electron microscopy (SEM) images.

The electron micrographs in Fig.1 a-c (30, 60, 300 s etching, respectively) show the temporal evolution of the porous silicon region formed by this method on a P-type (001) Si substrate. During the early etching phases we observe pit formation at discrete locations. The etching paths increase in density and length until they cover the whole surface; a corresponding increase in surface roughness and depth of the porous silicon region results. In Fig. 1d, a tilted (10°) cross section reveals that the silicon has been converted to a porous region to a depth of approximately 150 to 200 nm. Control experiments, using

NSTI-Nanotech 2008, www.nsti.org, ISBN 978-1-4200-8503-7 Vol. 1

samples immersed in the HF/ H_2O_2 etchant without AuNPs, do not result in formation of any porous structures.

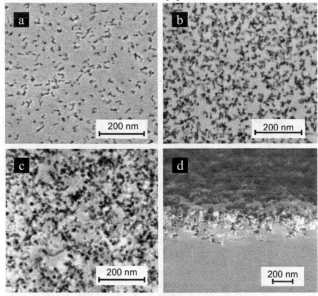

Figure 1: Scanning electron micrographs (SEM) of nanoporous network structures in p-type (001) silicon induced by colloidal gold nanoparticles (AuNPs) in an HF/H_2O_2 etchant. In a-c, the temporal evolution of the porous silicon formation is evident. a, 30 sec etching. b, 60 sec etching. c, 5 min etching. d. Cross sectional view [10° tilt] of the sample etched for 5 min, showing the porous silicon layer approximately 150 - 200 nm thick.

The SEM images in Fig. 1 suggest that the etching process results in the formation of channels with approximately uniform diameter into the Si. To clarify this observation, we fabricated arrays of porous silicon regions by patterning a Si surface using optical lithography. A hexagonal pattern of exposed Si regions approximately 1.2 μm in diameter was defined on a 600 nm thick thermally grown SiO_2 film. After a 300 s etch in the HF/H_2O_2 solution containing AuNPs, the oxide layer was stripped using a buffered HF solution. Examination of the surface, Fig. 2, shows the resulting porous silicon regions. As depicted in Fig. 2 a, for P-type (001) Si substrate, the edges of the etched circles are not sharp, but instead reveal a network of channels extending outwards, each about 20 nm wide, which are exclusively aligned along the [010] and [100] directions, both of which are in the plane of the (001) silicon surface.

We also observe that the nanometer sized channels in Fig. 2 a, which extend away from the edges of the porous silicon arrays, usually terminate with one of two different features. Bright end regions are indicated by black arrowheads, while dark end regions are indicated by white arrowheads. We hypothesize the former are AuNPs that have localized the dissolution along a <100>-oriented channel into the Si, while the later are vias through which the AuNPs escaped from or into the silicon surface by turning 90° into another <100> direction. Only one bright

or one dark dot is observed along any single nanochannel, suggesting that the nanochannels are the result of penetration of AuNPs into the bulk Si.

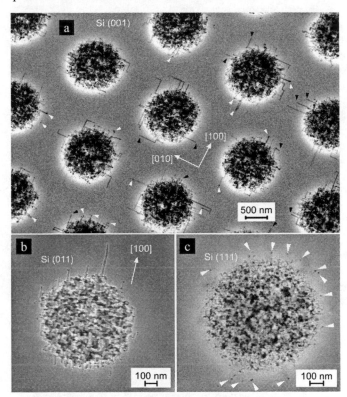

Figure 2: SEM of nanoporous Si arrays with different Si substrate orientations. Porous silicon patterns approximately 1.2 μm in diameter were fabricated by etching lithographically defined regions with HF/H_2O_2 solution containing AuNPs. a, For a (001) Si substrate, at the periphery of the etched areas are numerous nanochannels extending away from the porous regions, always along <100> directions. Two features are commonly observed at the ends of the nanochannels: bright areas, indicated by black arrowheads, and white areas indicated by white arrowheads. b, Nanochannels arising from penetration of AuNPs are in one direction only, the single <100> direction lying in the (011) surface plane of the silicon. c, No nanochannels extend from the periphery, consistent with the absence of <100> directions lying in the (111) surface plane. However we observe numerous nanometer-scale openings some distance away from the periphery; we interpret these as exit holes of nanochannels formed by localized dissolution along <100> directions by AuNPs that have penetrated into the Si bulk.

Confirmation of the preferred penetration along <100> was obtained by conducting experiments on Si wafers with other orientations. Fig. 2 b shows the result of etching an (011) oriented silicon substrate in the AuNPs-HF/ H_2O_2 solution for 5 min; nanochannels resulting from the AuNP penetration are oriented only in one direction, that of the single <100> direction that lies in the (011) surface plane.

Fig. 2 c shows the result of an etch experiment using a (111) oriented Si substrate; there are no nanochannels apparent on the wafer surface, consistent with the fact that no <100> directions lie in the (111) plane. However, we observe numerous small openings about 20 nm in diameter (indicated by arrowheads in Fig. 2 c) near the periphery of the masked region, but separated from the porous Si area. We interpret these as exit points for nanochannels formed by localized dissolution by the AuNPs in the bulk Si which have reemerged after making a 90° turn into another <100> direction.

3 DISCUSSION AND CONCLUSION

To further understand the influence of the AuNPs on the etching process, we prepared solutions of AuNPs in HF/H$_2$O$_2$ electrolytes using two other gold colloids, one synthesized in our laboratory following the method of Turkevich et al14, with average nanoparticle diameters of about 50 nm, and the other commercially available from Ted Pella Inc. (Redding CA), with nanoparticles having an average diameter of 25 nm. After 10 min immersion of silicon wafers into either of these solutions, no nanoporous structures formed.

We believe this lack of porous Si formation is due to the different adsorption behavior of the NPs onto the silicon surface. Gold nanoparticles in water or in weakly acidic solutions are usually negatively charged due to the presence of a citrate/acrylate stabilizing coat remaining after the manufacturing process [15]. This negative charge results in Coulombic repulsion from a silicon surface, which is negatively charged at these pH values [16]. When the pH is decreased, the citrate anions stabilizing the AuNPs are transformed into neutral citric acid, thus allowing adsorption to the silicon surface, while at the same time favoring agglomeration of the NPs. We tested the adsorption properties of the three types of AuNPs on silicon surface in 0.1% HF solutions. Our results show that, after 2 min of immersion, MesoGold NPs were adsorbed and uniformly dispersed on the silicon surface, Fig. 3a. In contrast, the other two types of AuNPs, with diameters of 50 and 25 nm, tended to aggregate when adsorbed on the silicon surface, Fig.3b.

We also investigated the influence of the HF concentration (from 0.1% to 2%) on the adsorption behavior of the MesoGold. We found that there is no significant change in the density of AuNPs adsorption in various HF: H$_2$O solutions (no H$_2$O$_2$) for concentrations exceeding 0.1%, which indicates that a wide range of HF concentration will favor the AuNPs adsorption. The areal density of adsorbed AuNPs varied from 124 to 171 m^{-2}, much higher than those recently reported by Woodruff et al [16] (3 to 4 counts/m^2). The high density of adsorbed nanoparticles and the lack of aggregation may perhaps be explained in terms of the extremely small size of the

MesoGold nanoparticles, which renders them more susceptible to Brownian scattering than to aggregation interactions.

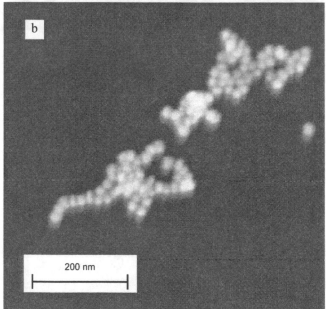

Figure 3: Colloidal gold nanoparticle adsorption on silicon in 1% HF solution. a, High density adsorption of 3.2 nm diameter AuNPs was achieved with the MesoGold colloids. b, 25 nm AuNPs from another source (Ted Pella Inc.) aggregated and covered the silicon surface at considerably lower density. Both SEM images were taken after 2 min of immersion, followed by rinsing with deionized water and drying with nitrogen.

We hypothesize that the etching process may be the result of the formation of local electrochemical cells at the silicon surface, induced by the presence of adsorbed AuNPs and of hydrogen peroxide. In this scenario, the H_2O_2 is reduced to water at the adsorbed AuNPs, by the following reaction [17] in acidic solution:

$$H_2O_2 + 2H^+ + 2e^- \rightarrow 2H_2O$$

The work function of AuNPs is about 5.4 eV within a wide range of particle diameter [18], which is higher than the electron affinity of silicon (4.05 eV). When AuNPs contact with crystalline silicon, it is energetically favorable for electrons to transfer from silicon to the nanoparticles, resulting in a more efficient reduction of hydrogen peroxide in the vicinity of the AuNPs. The migration of electrons at the AuNPs-Si contact also leaves behind compensating holes on silicon surface. The holes thus made available would promote silicon oxidation through the two possible reaction paths [19]:

$$Si + 2h^+ \rightarrow Si^{2+} \quad \text{followed by} \quad Si^{2+} + 2H^+ \rightarrow Si^{4+} + H_2$$

or the direct oxidation:

$$Si + 4h^+ \rightarrow Si^{4+}$$

In this way, the silicon atoms are etched and disappear into the etchant, as a soluble product such as $HSiF(OH)_2$ [20].

The overall process would be made possible by adsorbed AuNPs and would thus proceed in their proximity; the formation of elongated channels may consequently be explained by the fact that the AuNPs tend to remain localized at the regions that are being etched away. The preference for etching in the <001> direction can be explained in terms of the Current Burst Model by Föll's group [21], whereby {111} pore surfaces are passivated at a higher rate, thus favoring etching along the <100> directions. Fluctuations in the kinetics of pore wall passivation would cause the observed occasional changes in the direction of the etching channel.

In summary, we have demonstrated an efficient method for producing porous silicon, capable of achieving very small and uniform pore sizes, without the need for an external electrochemical cell or potentiostat. This method may open up fruitful possibilities for developing controlled nanochannel networks in crystalline silicon.

REFERENCES

[1] V. Lehmann, U. Gosele, Appl. Phys. Lett. 58, 856-858, 1991.
[2] S. R. Nicewarner-Pena, R. G. Freeman, B. D. Reiss, L. He, D. J. Pena, I. D. Walton, R. Cromer, C. D. Keating, M. J. Natan, Science 294, 137-141, 2001.
[3] V. Lehmann, S. Ronnebeck, Sens. Actuators, A, 95, 202-207, 2001.
[4] J. Fu, J. Yoo, J. Han, Phys. Rev. Lett. 97, 018103 , 2006.
[5] C. C. Striemer, T. R. Gaborski, J. L. McGrath, P. M. Fauchet, Nature 445, 749-753, 2007.
[6] D. R Turner, J. Electrochem. Soc. 105, 402-408 , 1958.
[7] A. Uhlir, Jr., Bell Syst. Tech. J. 35, 333-347, 1956.
[8] R. L. Smith, S. D. Collinsa, J. Appl. Phys. 71, R1-R22, 1992.
[9] H. Föll, M. Christophersen, J. Carstensen, G. Hasse, Mater. Sci. Eng. R 39, 93–141, 2002.
[10] X. Li, P. W. Bohna, Appl. Phys. Lett. 77, 2572-2574, 2000.
[11] K. Tsujino, M. Matsumura, Adv. Mater. 17, 1045-1047, 2005.
[12] S. Koynov, M. S. Brandt, M. Stutzmann, Appl. Phys. Lett. 88, 203107, 2006.
[13] N. Imou, T. Ishiyama, Y. Omuraa, J. Electrochem. Soc. 153, G59-66, 2006.
[14] J. Turkevich, P. C. Stevenson, J. A Hillier, Discuss. Faraday Soc. 11, 55-75, 1951.
[15] S. Diegoli, P. M. Mendes, E. R. Baguley, S. J. Leigh, P. Iqbal, Y. R. Garcia Diaz, S. Begum, K. Critchley, G. D. Hammond, S. D. Evans, D. Attwood, I. P. Jones, J. A. Preece, J. Exp. Nanosci. 3, 333-353, 2006.
[16] J. H. Woodruff, J. B. Ratchford, I. A. Goldthorpe, P. C. McIntyre, C. E. D. Chidsey, Nano Lett. 7, 1637-1642, 2007.
[17] A. K. Shukla, R.K. Raman, Annu. Rev. Mater. Res. 33, 155–68, 2003.
[18] M. Schnippering, M. Carrara, A. Foelske, R. Kötz, D. J. Fermín. Phys. Chem. Chem. Phys. 9, 725-730, 2007.
[19] X. G. Zhang, J. Electrochem. Soc. 151, C69-C80, 2004.
[20] P. Allongue, V. Kielinc, H. Gerischer, Electrochimca Acta. 40, 1353-1360, 1995.
[21] C. Jager, B. Finkenberger, W. Jager, M. Christophersen, J. Carstensen, H. Foll, Mat. Sci. Eng. B 69, 199-204, 2000.

Fabrication and Integration of Nanobolometer Sensors on a MEMs Process

S.F. Gilmartin[*], K. Arshak[**], D. Collins[*], D. Bain[*], W.A. Lane[*], O. Korostynska[**],
A. Arshak[**], B. McCarthy[***] and S.B. Newcomb[****]

[*]Wafer Fabrication Dept., Analog Devices Inc., Limerick, Ireland, stephen.gilmartin@analog.com
[**]University of Limerick, Limerick, Ireland, khalil.arshak@ul.ie
[***]Tyndall National Institute, Cork, Ireland, brendan.mccarthy@tyndall.ie
[****]Glebe Scientific Ltd., Newport, Tipperary, Ireland, simon@sonsam.ie

ABSTRACT

In this work, we combine electron beam lithography (EBL) with conventional microscale metal deposition and etch process technologies, to create bolometer devices with nanoscale feature critical dimensions (CDs). We report the creation of titanium (Ti) bolometer devices with 70 nm minimum feature CDs, and total bolometer film thicknesses ranging between 60 nm and 150 nm. Our new nanobolometer devices are integrated with conventional CMOS/MEMs fabrication technologies, creating thermally isolated sensors with nanoscale feature sizes on a 0.5 μm CMOS base process. We also present temperature coefficient of resistance (TCR) data for the new devices, and show a nanobolometer TCR performance of 0.22%/K at 70 nm CDs, comparable to microscale bolometer devices.

Keywords: nanobolometer, nanolithography, sensor, etch, MEMs

1 INTRODUCTION

Uncooled bolometer devices have advantages such as lower cost, better reliability performance and portability, when compared with cryogenically cooled sensors. However, uncooled bolometer detectors have drawbacks, such as low detectivity due to self-heating under constant bias [1], thermal noise, and sensitivity of the sensor material to the fabrication process [2]. To produce highly sensitive uncooled infrared (IR) bolometers, the development of devices that use sensor elements made from high-TCR and low-noise materials is important. In the production of uncooled IR bolometer devices, trade-offs are required between sensor design criteria, choice of bolometric sensor material, and the best device performance deliverable by the wafer fabrication process.

Materials commonly used as bolometer sensor films include vanadium oxides [3]-[6], metals and semiconductors. Vanadium oxide is used primarily to achieve a sufficiently high sensor TCR performance for use in IR sensing applications. However, the deposition and processing conditions for vanadium oxide films are narrowly defined, and they generally require annealing at moderate to high temperatures. Vanadium oxide is also difficult to integrate with mainstream CMOS/mixed-signal wafer fabrication, due to contamination concerns.

Semiconductor films, such as amorphous polysilicon, are compatible with CMOS and mixed-signal fabrication technologies [7], but usually require high temperature annealing. Metal films, such as platinum (Pt) and Ti, are readily compatible with mainstream fabrication processes, [1], [8]-[11]. However, Pt and Ti are susceptible to changes in material and electrical characteristics (including TCR) through the bolometer fabrication flow.

Serpentine sensor element designs, with microscale feature CDs, have been used to increase the effective thermal resistance of the bolometer for a given pixel area. The serpentine structure raises the length-to-width ratio of the sensor resistor, and device sensitivity can be increased through the increase of this ratio. Bolometer IR detectors are also typically designed for high thermal isolation, as this makes them more receptive to small differences in environmental temperature. Surface micromachining [12]-[16], or bulk micromachining [17]-[18] techniques have been used in microscale bolometer fabrication, to create thermal-isolating cavities beneath the sensor areas.

In our work, we explored the trade-offs between sensor design, bolometeric material, and the impact of the fabrication process flow on bolometer TCR performance, at nanoscale sensor element CDs. We used Ti bolometric sensor films ranging in thickness from 60 nm to 150 nm, and MEMs-based silicon bulk micromachining (SBM), to create thermal-isolating cavities beneath the sensor pixels. Ti and SBM process options were chosen because of their compatibility with CMOS, mixed-signal and MEMs wafer fabrication processing.

Fig. 1 shows a schematic of our CMOS/MEMs-integrated nanobolometer process scheme. We incorporated serpentine bolometer sensor elements within our pixel design, to increase the length-to-width ratio of the sensor resistors. Using EBL, we further maximised the length-to-width ratio of the resistor, by creating bolometer devices with sensor element minimum CDs ranging between 350 nm and 70 nm. TCR performance was maintained at nanoscale CDs, and we report TCR values of 0.22%/K at 70 nm CDs. We inserted our nanobolometer fabrication module within the backend of a 0.5 μm CMOS fabrication flow, preventing exposure of the bolometeric layer to medium or high temperature process steps.

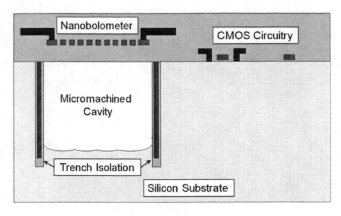

Fig. 1: CMOS/MEMs-integrated nanobolometer process scheme.

2 EXPERIMENTAL METHOD

We used 200 mm substrates running on a 0.5 μm CMOS base process. Prior to CMOS and nanobolometer processing, we pre-defined 50 μm deep silicon trenches in our wafers, and then filled the trenches with a SiO_2 sidewall liner layer and a polysilicon fill layer.

2.1 Nanoscale Resist Features

We used a JEOL 6000FS EBL tool (spot-beam, vector scan), ZEP-520 positive EBL resist, and ZED-N50 resist developer to create our nanoscale bolometer resist features. Resist thickness, exposure and develop processing parameter set points were optimised through a designed experimental matrix. We used post-develop resist thickness, profile and CD data, and post-etch profile and CD data, across local and global topographies, to site our final nanolithography process. We set our final resist spin thickness at 120 nm (as spun on planar substrates), EBL exposure at 100 μC, and resist develop time at 30 seconds. Fig. 2 shows a planar CDSEM image of a nanobolometer serpentine resist feature patterned using our final nanolithography process. The resist features shown have 60 nm minimum CDs.

2.2 Nanoscale Etched Features

Ti bolometer layers ranging from between 60 nm and 150 nm in thickness were deposited on a 1.5 μm thick SiO_2 layer. After EBL resist patterning, the Ti layers were then etched in a LAM Research 9600 plasma etch chamber using Cl_2/BCl_3-based etch chemistries [19]. The plasma nanoscale etch set points were determined through a designed experimental matrix. Our final Ti etch process was determined from analysing responses of Ti, resist and SiO_2 etch rates, and feature profile and CDs, to variations in etch process parameter set points. Following Ti etch, residual resist was then removed using H_2O/O_2 plasma etch chemistries, and polymer residues were removed using a solvent wet strip process. Fig. 3 shows a planar CDSEM image of a Ti nanobolometer serpentine etched feature, defined using our final nanoscale etch process. The etched features shown have 70 nm minimum CDs. The inset in Fig. 3 shows a TEM micrograph cross-section through a Ti nanobolometer meander. The image shows well-defined Ti features, 65nm layer thickness, with 70 nm minimum CDs and 60 nm separation between dense features.

The meander structure layouts were modified (positively biased) to create wider features at the extremes of the serpentine layouts on the nanoscale devices. The pattern biasing was carried out to prevent etch microloading adversely affecting isolated and semi-isolated structure definition [20]. The biasing of the meander layouts can be seen at the top part of the CDSEM images shown in Fig. 2 and Fig. 3.

Fig. 2: CDSEM images, of an EBL-defined nanobolometer serpentine resist feature with 60 nm minimum CDs.

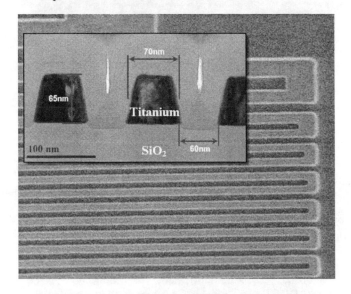

Fig. 3: CDSEM image of an etched nanobolometer serpentine Ti feature with 70 nm minimum CDs. Inset shows a TEM micrograph section through the 70 nm lines.

2.3 Bulk Silicon Cavity Release

The etch-defined nanobolometer structures were capped with passivating dielectric layers and then processed through an SBM fabrication module, to release the final sensor structures from the bulk silicon substrates. The cavity silicon release etch was done on a Xactix xenon diflouride ($XeFe_2$) etch tool. The cavity release process creates a 40 μm deep cavity beneath the sensor platform, providing thermal isolation from the bulk substrate. The pre-defined trenches contain a SiO_2 sidewall liner. The $XeFe_2$ etch process has a high selectivity to the SiO_2 trench layer, and the liner serves as a lateral etch-stop for the silicon cavity etch. The trenches therefore define the lateral boundary for the silicon cavities beneath the sensor pixels. Fig. 4 shows an optical image, planar view, of a fully fabricated nanobolometer pixel, with 70 nm minimum CDs, post-silicon cavity release processing, with the cavity boundary trench visible around the pixel perimeter.

Fig. 4: Optical image, planar view, of a sensor pixel, with 70 nm minimum CDs, after silicon cavity release.

3 RESULTS

3.1 TCR MEASUREMENTS

We used 4-point probe measurements to investigate the TCR performance of the released nanobolometer pixels. Forced current was limited during measurement, to prevent Joule/self- heating in the devices [21]. The device TCR was measured by controlling temperature of the substrate, and measuring resistances at 298K and 373K. The impact of sensor element line width on TCR was also investigated. A range of devices were fabricated, with minimum sensor element CDs ranging from 350 nm to 70 nm, while maintaining the same pixel size and sensor element layout area across all devices.

Fig. 5 shows a graph of the dependence of nanobolometer device TCR on minimum sensor element CD, using the same sensor element layout area on all devices. Fig. 5 shows the TCR reducing as the minimum sensor element CD shrinks. A similar TCR dependence on CD has been previously reported for small-geometry metal interconnects [22]. In the absence of significant self-heating in the devices, the observed TCR and CD dependence can be attributed to surface and grain boundaries scattering processes, as the mean free path of the charge carriers become comparable to the dimensions of the etched nanobolometer features.

The TCR performance of our nanobolometer devices was maintained at 0.22 %/K, at 70 nm sensor element CDs. The sensor TCR values we report are comparable to fabricated microscale bolometer devices referenced earlier in this paper. Our new nanobolometer devices may be suitable for use in high-resolution IR sensing and gas detection applications.

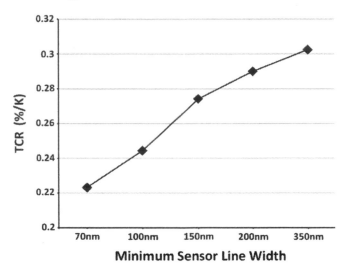

Fig. 5: Graph of 80 nm thick Ti bolometer TCR vs. minimum sensor line width. Temperature delta was taken between 293K and 373K.

4 CONCLUSIONS

We have shown experimentally that EBL and microscale plasma etch technologies can be used to create Ti bolometer sensor elements with nanoscale CDs. We have fabricated devices with Ti sensor layer thicknesses ranging from between 60 nm and 150 nm, and with minimum sensor element CDs down to 70 nm.

Our nanobolometer devices were integrated with a 0.5 μm CMOS process node and a MEMs-based SBM process module, illustrating the flexibility of the new sensor process. Our new devices delivered a TCR performance of 0.22 %/K at nanoscale sensor element CDs, comparable to many microscale bolometer devices. Possible uses for the nanobolometer devices we report include high-resolution IR sensing and gas detection applications.

ACKNOWLEDGEMENTS

The authors acknowledge Christine Tsau, Analog Devices, Cambridge, MA, USA, for facilitating the SBM wafer processing. S.F. Gilmartin acknowledges Science Foundation Ireland (SFI), for supporting EBL processing used in this work, through the National Access Programme (NAP).

REFERENCES

[1] M. Ramakrishna, et al., "Highly sensitive infrared temperature sensor using self-heating compensated microbolometers," Sensors and Actuators A: Physical, vol. 79, no. 2, 2000, pp. 122-127.

[2] S. Chen, et al., "Fabrication and performance of microbolometer arrays based on nanostructured vanadium oxide thin films," Smart Materials and Structures, vol. 16, no. 3, 2007, pp. 696-700.

[3] D. Murphy, et al., "Multi-spectral uncooled microbolometer sensor for the Mars 2001 orbiter THEMIS instrument," Proc. Aerospace Conference Proceedings, 2000 IEEE, 2000, pp. 151-163 vol.153.

[4] R. Kumar, et al., "Room temperature deposited vanadium oxide thin films for uncooled infrared detectors," Materials Research Bulletin, vol. 38, no. 7, 2003, pp. 1235-1240.

[5] J. Dai, et al., "Low temperature fabrication of VOx thin films for uncooled IR detectors by direct current reactive magnetron sputtering method," Infrared Physics & Technology, vol. 51, no. 4, 2007, pp. 287-291.

[6] B. Cole, et al., "High performance infrared detector arrays using thin film microstructures," Proc. Applications of Ferroelectrics, 1994.ISAF '94., Proceedings of the Ninth IEEE International Symposium on, 1994, pp. 653-656.

[7] E. Socher, et al., "Modeling, design and characterization of surface micromachined polysilicon microbolometers," Proc. Electrical and Electronics Engineers in Israel, 2002. The 22nd Convention of, 2002, pp. 56-57.

[8] M. Gonzalez and E.R. Hodgson, "Electrical and mechanical behaviour of improved platinum on ceramic bolometers," Fusion Engineering and Design, vol. 82, no. 5-14, 2007, pp. 1277-1281.

[9] K. Tsutsumi, et al., "The experimental study of high TCR Pt thin films for thermal sensors," Proc. Sensors, 2002. Proceedings of IEEE, 2002, pp. 1002-1005 vol.1002.

[10] J.S. Shie, et al., "Characterization and modeling of metal-film microbolometer," Journal of Microelectromechanical Systems, vol. 5, no. 4, 1996, pp. 298-306.

[11] H. Esch, et al., "The stability of Pt heater and temperature sensing elements for silicon integrated tin oxide gas sensors," Sensors and Actuators B: Chemical, vol. 65, no. 1-3, 2000, pp. 190-192.

[12] A. Tanaka, et al., "Infrared focal plane array incorporating silicon IC process compatible bolometer," IEEE Transactions on Electron Devices, vol. 43, no. 11, 1996, pp. 1844-1850.

[13] F. Niklaus, et al., "Uncooled infrared bolometer arrays operating in a low to medium vacuum atmosphere: Performance model and tradeoffs," SPIE, Bellingham WA, WA 98227-0010, United States, 2007, pp. 65421.

[14] K.M. Chang, et al., "Design of low-temperature CMOS-process compatible membrane fabricated with sacrificial aluminum layer for thermally isolated applications," Sensors and Actuators, A: Physical, vol. 134, no. 2, 2007, pp. 660-667.

[15] W.B. Song and J.J. Talghader, "Design and characterization of adaptive microbolometers," Journal of Micromechanics and Microengineering, vol. 16, no. 5, 2006, pp. 1073-1079.

[16] M. Almasri, et al., "Uncooled multimirror broadband infrared microbolometers," Journal of Microelectromechanical Systems, vol. 11, no. 5, 2002, pp. 528-535.

[17] L. Dong, et al., "Design and fabrication of single-chip a-Si TFT-based uncooled infrared sensors," Sensors and Actuators A: Physical, vol. 116, no. 2, 2004, pp. 257-263.

[18] D.S. Tezcan, et al., "A low-cost uncooled infrared microbolometer detector in standard CMOS technology," IEEE Transactions on Electron Devices, vol. 50, no. 2, 2003, pp. 494-502.

[19] S.F. Gilmartin, et al., "Fabricating nanoscale device features using the 2-step NERIME nanolithography process," Microelectronic Engineering, vol. 84, no. 5-8, 2007, pp. 833-836.

[20] R.A. Gottscho, et al., "Microscopic uniformity in plasma etching," Journal of Vacuum Science & Technology B: Microelectronics Processing and Phenomena, vol. 10, no. 5, 1992, pp. 2133-2147.

[21] A. Scorzoni, et al., "On the relationship between the temperature coefficient of resistance and the thermal conductance of integrated metal resistors," Sensors and Actuators A: Physical, vol. 116, no. 1, 2004, pp. 137-144.

[22] J.F. Guillaumond, et al., "Analysis of resistivity in nano-interconnect: full range (4.2-300 K) temperature characterization," Proc. Interconnect Technology Conference, 2003. Proceedings of the IEEE 2003 International, 2003, pp. 132-134.

Silylation Hardening for Mesoporous Silica Zeolite Film

Takamaro Kikkawa

Research Center for Nanodevices and Systems, Hiroshima University
1-4-2 Kagamiyama, Higashi-hiroshima, Hiroshima, 739-8527 Japan
kikkawat@hiroshima-u.ac.jp

ABSTRACT

Pure silica zeolite films were formed for ultra-low-k interlayer dielectrics in ultra-large integrated circuits. Hydrothermal crystallization and vapor phase transport synthesis methods were developed. Self-assembled mesoporous silica was formed by use of a surfactant. Silylation hardening was achieved by 1,3, 5, 7 tetramethylcyclotetrasiloxiane (TMCTS) vapor treatment. The zeolite formation by hydrothermal crystallization and vapor phase transport can reduce O-H bond in the silica films, resulting in the reduction of leakage current by a factor of 1/10. The TMCTS vapor treatment can reduce the leakage current by 4 orders of magnitude due to the decrease of Si-OH and O-H bonds. The elastic modulus of 5.18 GPa and the dielectric constant of 1.96 were achieved simultaneously by TMCTS silylation hardening process for the mesoporous silica zeolite thin film.

Keywords: silylation, zeolite, surfactant, low-k, mesoporous silica

1 INTORDUCTION

In order to overcome problems of signal delays in interconnects in ultra-large scale integrated circuits (ULSI), low-dielectric-constant (low-k) interlayer dielectric films arc nccdcd. Mcsoporous silica whose pore sizes are ranging from 2 nm to 10 nm has been studied as a potential candidate for ultra low-k materials[1]. However, mechanical strength of the low-k film is degraded when mesosized pores are introduced into skeletal materials to reduce the film density. Zeolite is a promising candidate as an advanced low-k material[2-4], which has microporous crystalline structure so that the Young's modulus (100 GPa) is larger than that of dense silica (70 GPa) and the density (e.g., 1.76 g/cm^3 for silicalite) is lower than that of dense silica (2.1-2.3 g/cm^3).

In this paper, pure-silica zeolite films are prepared by hydrothermal synthesis and vapor phase transport (VPT) methods, and the effect of silylation on electrical and mechanical properties are investigated.

2 ZEOLITE FROMATION

2.1 Hydrothermal Crystallization

The precursor solution of tetra-butyl ammonium hydroxide (TBAOH), tetraethyl orthosilicate (TEOS) and ethyl alcohol (EtOH) were mixed and stirred at the room temperature. Hydrolysis of TEOS was caused by EtOH. The precursor was heated in an autoclave for 110 hours at 100ºC, then cooled down to the room temperature, and heated up again for 10 hours at 100ºC. The suspension prepared by the hydrothermal crystallization method contained zeolite nanoparticles.

To reduce the k-value, butanol and a surfactant of ethylene oxide propylene oxide ethylene oxide triblock co-polymer, $(EO)_{13}(PO)_{20}(EO)_{13}$, were added to the suspension while stirring so that mesoscopic size pores of several nm in diameter were formed. The zeolite silica film was formed on a Si wafer by spin coating. After the film was pre-baked on a hot plate for one hour at 90ºC, it was calcined at 400ºC in air and annealed at 400ºC for 5 h. TMCTS vapor treatment was carried out at 400ºC and annealed in the N_2 atmosphere so that the inner pore wall surface was silylated by TMCTS molecules, forming polymer cross-linking network.

2.2 Vaporphase Transport

A precursor solution was prepared by use of TEOS, catalyst nitric acid, water and EtOH. Surfactant template Brij78® (Polyethylene glycol stearyl ether $C_{18}H_{37}(OCH_2CH_2)_{20}OH$) and the precursor solution were mixed. The final molar ratio of the solution was TEOS: Brij78: H_2O: HNO_3 = 1:0.01:10:0.01. The resulting solution was deposited on a Si (100) wafer by a spin-coating method to form a homogeneous thin layer and cured at 150°C for 1 min under nitrogen.

The VPT was carried out in an autoclave in which a wafer with spin-coated silica film was set on a perforated plate in the middle of the vessel. A liquid phase mixture of ethylenediamine (EDA), triethylamine (Et_3N) and water. filled the bottom of the autoclave. The autoclave was heated up to 200°C and kept at that temperature during synthesis. They were exposed to the vapor mixture of EDA, Et_3N and water under an autogenous steam pressure. As-synthesized products were rinsed with acetone and blow-dried with nitrogen and were calcined at 400°C for 4 hours in dry air.

3 RESULTS AND DISCUSSION

Figure 1 shows the framework view of ZSM-48 zeolite, which has a one-dimensional 10-membered channel with a diameter of 5.6 Å× 5.3 Å. Figure 2(a) shows the results of the XRD patterns. The diffraction peaks of Sample No. 1 and 2 matched those of ZSM-48 zeolite, which has two groups of the polytypes: monoclinic unit cell and orthorhombic unit cell. The measured XRD patterns showed only two peaks at 7.54° and 8.74° at low angle so that the zeolite crystal was the orthorhombic unit cell having parameters of a = 24.66 Å, b = 8.4 Å, c = 24.66 Å and β = 109.47°. Figure 2(b) shows the XRD patterns of the samples synthesized by the VPT method (Sample No. 1, 5, 6 and 7).

Figure 3(a) shows the FTIR spectra of the samples synthesized by the VPT method (Sample No. 1, 4, 5, 6, 7 and 8). The spectrum of sample No. 8 is formed by the sol-gel technique and spin-coating method. For sample No. 2, 6 and 7, the absorption peak was not observed around 550 cm^{-1}, which is assigned to the presence of the 5-membered ring of tetrahedral SiO_2 in the framework and characteristic structure of ZSM-48 zeolite. Another peak around 470 cm^{-1} was attributed to the Si-O symmetric bend and was not structure sensitive. In Fig. 3(b), the broad absorption band related to O-H stretching bonds between 3000 cm^{-1} and 3800 cm^{-1}, and the absorption peak at 3740 cm^{-1} related to the isolated Si-OH bond were observed. The absorption peaks of O-H bonds were suppressed effectively for samples 2, 6 and 7, so that ZSM-48 zeolite was intrinsically hydrophobic.

Figure 4 shows the film thickness and refractive index as a function of the VPT time for the sample 4, 5 and 8. The VPT time equal zero denotes the sample formed by the sol-gel technique and spin-coating method. This result showed that the film shrinkage was suppressed by the VPT method.

Figure 5 shows the dielectric constant at 1 MHz and the leakage current density at 1 MV/cm as a function of the VPT time for the sample 4, 5 and 8, which were obtained from IV and CV measurements. The dielectric constant decreased to 2.7 with increasing vapor phase transport time and the leakage current was suppressed to an order of 10^{-8} (A/cm^2). Reductions of the dielectric constant and the leakage current were caused by the hydrophobicity obtained from the contribution of EDA, Et_3N and water.

MEL type zeolite which has 10 membered ring whose pore diameters are 0.53 and 0.54 nm was made by a hydrothermal crystallization method as shown in Fig. 6. The existence of zeolite was confirmed by the absorption peak around 560 cm^{-1} which is associated with an asymmetric stretching mode of five-membered ring blocks in MEL-type zeolite skeletal [13] as shown in Fig. 7.

The pore wall surface was silylated by TMCTS molecules that reacted with Si-OH groups on the pore wall surface as shown in Fig. 8 (a), and formed the polymer cross-linking network as shown in Fig. 8 (b).

FT-IR absorbance spectra of zeolite and porous silica with and without silylation were compared as shown in Fig. 9. A broad absorption in zeolite around 3513 cm^{-1}, which is associated with the O-H group, was less than that of porous silica film. Both absorptions by isolated Si-OH at 3745 cm^{-1}

and the O-H group were dramatically decreased by the silylation treatment. Furthermore, the existence of C-H, $SiHCH_3O_2$ and $SiHO_3$ bonds were exhibited in Fig. 9, indicating the film hydrophobic. Pore size distribution was calculated from the measurement of SAXS as shown in Fig. 10. The pore distribution of about 4 nm due to the surfactant was confirmed, but zeolite micropores of 0.5 nm could not be identified by SAXS. The distribution of meso-sized pores slightly decreased from 4.512 nm to 4.399 nm by the TMCTS treatment because the pore surface was covered with cross-linked TMCTS molecules.

The dependences of porosity on the surfactant concentration of the zeolite film and porous silica film were shown in Fig. 11(a). Figure 11(b) shows the dielectric constant versus surfactant concentration. Zeolite shows lower dielectric constants than porous silica and its dielectric constant decreased from 2.206 to 1.962 by silylation.

Elastic modulus of zeolite increased from 3.312 to 5.180 GPa by the TMCTS treatment as shown in Fig. 12. The dielectric constant of 1.96 was achieved with the elastic modulus of 5.18 GPa for TMCTS silylated porous zeolite film. Figure 13 shows Weibull plot for the life time of time dependent dielectric breakdown (TDDB). The life time of zeolite with silylation was longer than the porous silica with silylation by a factor of approximately 2.5.

4 CONCLUSION

Zeolite formation by hydrothermal crystallization and the VPT can reduce O-H bond in the silica films, resulting in the reduction of leakage current by a factor of 1/10. The TMCTS vapor treatment can reduce the leakage current by 4 orders of magnitude due to the decrease of Si-OH and O-H bonds. The elastic modulus of 5.18 GPa and the dielectric constant of 1.96 were achieved simultaneously by TMCTS silylation hardening process.

REFERENCES
[1] T. Kikkawa, S. Chikaki, R. Yagi, M. Shimoyama, Y. Shishida, N. Fujii, K. Kohmura, H. Tanaka, T. Nakayama, S. Hishiya, T. Ono, T. Yamanishi, A. Ishikawa, H. Matsuo, Y. Seino, N. Hata, T. Yoshino, S. Takada, J. Kawahara, K. Kinoshita, Tech. Dig. IEEE International Electron Devices Meeting, pp 99-102, 2005.
[2] T. Yoshino, G. Guan, N. Hata, N. Fujii and T. Kikkawa, Extended Abstracts of Solid-State Devices and Materials, p. 58, Japan Society of Applied Physics, Kobe (2005).
[3] T. Seo, T. Yoshino, N. Ohnuki, Y. Seino, N. Hata, T. Kikkawa, Ext. Abst. of Inter. Conf. on Solid State Devices and Materials, pp. 928-929, Tukuba, Sept. 18-21, 2007.
[4] Y. Cho, T. Seo, K. Kohmura and T. Kikkawa, Ext. Abst. of Inter. Conf. on Solid State Devices and Materials, pp. 382-383, Tukuba, Sept. 18-21, 2007.

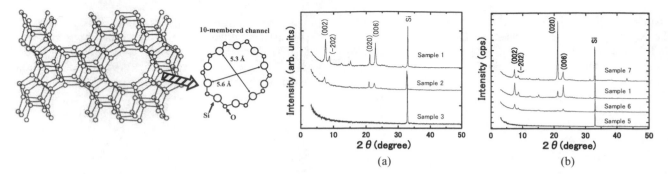

Fig. 1. A schematic diagram of ZSM-48 zeolite framework.

Fig. 2. XRD patterns for ZSM-48 zeolite films formed by vapor phase transport method. (a) Sample No. 1, 2, and 3. The mixture ratio of EDA: Et3N: H2O are 1: 2: 1, 2: 1: 1 and 1: 1: 2, respectively. (b) Sample No. 1, 5, 6, and 7. The mixture ratio of EDA: Et3N: H2O is 1: 2: 1. Synthesis times are 6days, 2hours, 4days, and 8 days, respectively.

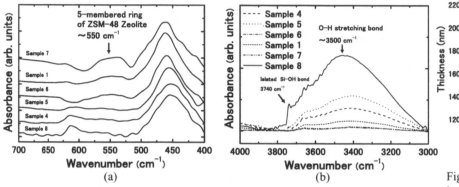

Fig. 3. FTIR spectra of the samples synthesized by the VPT method (Sample No. 1, 4, 5, 6, 7 and 8). (a) The absorption peak around 550 cm^{-1} related to the 5-membered ring of ZSM-48 zeolite. (b) The absorption peak related to O-H stretching bonds and isolated Si-OH bond.

Fig. 4. Film thickness and refractive index of the sample synthesized by the VPT method as a function of vapor phase transport time for the sample 4, 5 and 8. Sample No. 8 was formed by the sol-gel technique and spin-coating method without VPT.

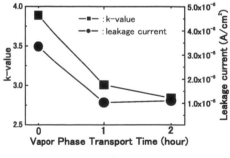

Fig. 5. k-value and leakage current as a function of vapor phase transport time for the sample 4, 5 and 8. Sample Sample No. 8 was formed by the sol-gel technique and spin-coating method without VPT.

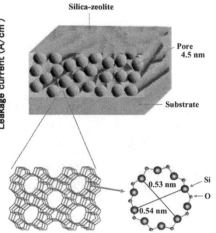

Fig. 6. A schematic diagram of mesoporous silica low-k film and MEL-type zeolite cristal framework.

Fig. 7. Fourier transform FT-IR spectra of MEL type pure silica zeolite which was formed by hydrothermal crystallization method.

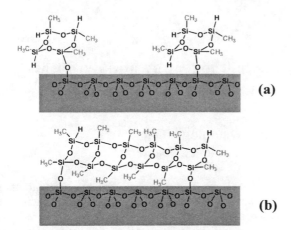

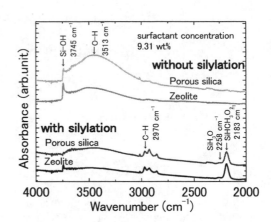

Fig. 8. Schematic illustration of silylation for inner surfaces of porous silica. (a) Adsorption of tetramethylcyclotetrasiloxane (TMCTS). (b) After cross linking of TMCTS.

Fig. 9. FT-IR spectra of zeolite and porous silica films with and without TMCTS silylation.

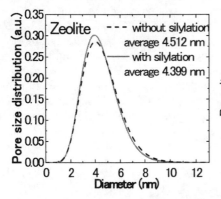

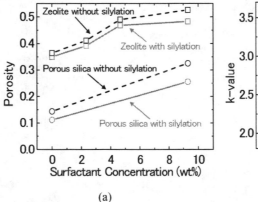

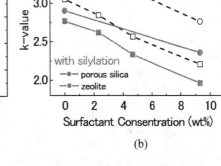

Fig. 10. Pore size distribution of zeolite films with and without TMCTS silylation.

(a)

(b)

Fig. 11. Dependence of surfactant concentration for zeolite and porous silica films with and without TMCTS silylation. (a) Porosity versus surfactant concentration. (b) k-values versus surfactant concentration.

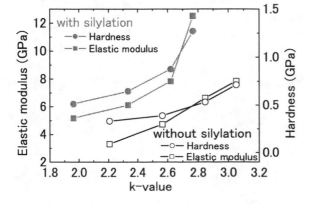

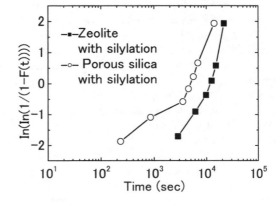

Fig. 12. Elastic modulus and hardness versus dielectric constant for zeolite and porous silica films with and without TMCTS silylation.

Fig. 13. Weibull plot of time dependent dielectric breakdown life time at 3.6 MV/cm in 200°C for zeolite and porous silica films with TMCTS silylation.

Vertically aligned carbon nanowires (CNWs)

: Top-down approach using photolithography and pyrolysis

Seok Woo Lee[*], Jung A Lee[**] and Seung S. Lee[*]

[*]Korea Advanced Institute of Science and Technology, Daejeon, Korea
[**]Korea Research Institute of Standards and Science, Daejeon, Korea

ABSTRACT

Top-down approach for vertically aligned CNWs has not reported yet although it is necessary for immediate integration of CNWs with micro and nano systems.

In this paper, vertically aligned pyrolyzed carbon nanowires (CNWs) is fabricated by photolithography with modification and pyrolysis as a top-down approach which the control of shape and position of CNWs is reliable. Sub-μm high aspect ratio (HAR) SU-8, negative tone photoresist, tip array fabricated using modified photolithography is transformed to carbon structure with shrinkage which reduces the dimension of CNWs to a few hundreds of nanometer or less. EDS analysis shows that fabricated CNWs is consist of pure carbon. The atomic structure of fabricated CNWs is amorphous as shown in XRD analysis.

We expect that the new fabrication method for vertically aligned CNWs as top down approach has a huge potential to expand an application to various micro and nano systems for their controllable positioning, array, and shape.

Keywords: carbon nanowires, vertically aligned, pyrolysis, SU-8 nano tip, high aspect ratio.

1 INTRODUCTION

A vertically aligned carbon nanowires (CNWs) offers a lot of opportunity to break through the ordinary micro and nano systems electrically, mechanically, and chemically [1]. The fabricating methods can be characterized as bottom up having the advantage of reducing scale of the structures rather than positioning at several specific points and well controlled shapes of the structures. Top-down approach for vertically aligned CNWs has not reported yet although it is necessary for immediate integration of CNWs with micro and nano systems.

In this paper, vertically aligned pyrolyzed carbon nanowires (CNWs) is fabricated by photolithography with modification and pyrolysis as a top-down approach which the control of shape and position of CNWs is reliable. Sub-μm high aspect ratio (HAR) SU-8, negative tone photoresist, tip array fabricated using modified photolithography is transformed to carbon structure with shrinkage which reduces the dimension of CNWs to a few hundreds of nanometer or less as shown in Figure 1.

Pyrolyzed carbon fabricated by photolithography and photoresist thermal decomposition at high temperatures in inert atmosphere has been widely studied [ref]. Pyrolyzed carbon derived from photoresist has many advantages such as batch fabrication, fine resolution, and reproducibility. For that reasons, various applications such as microbatteries, image sensors, and biochemical sensors utilizing pyrolyzed carbon have been reported [2-4].

One of popular precursor of pyrolysis is SU-8, a negative tone thick photoresist. Its high sensitivity and low absorbance for UV light offer strength of high aspect ratio micro structure fabrication [5-7]. However, the realization of high aspect ratio sharp SU-8 tip still remain as a considerable challenge for a dimensional limitation or a complexity of process as the structure is reduced to micro or nano scale. In our recent research, this limitation has been solved by modification of photolithography to apply high aspect ratio SU-8 nano tip as a middle step of a fabrication of pyrolyzed carbon nanowire array and shape estimation method has been provided using numerical analysis of Huygens Fresnel diffraction principle [8].

2 FABRICATION

The fabrication process of HAR SU-8 tip and pyrolysis for CNWs is shown in Figure 2. The fabrication process of HAR SU-8 tip on fused silica substrate is based on one step UV photolithography with exposed dose control [8]. Chromium thin film circular apertures on a fused silica substrate are fabricated by conventional photolithography and etching processes (a). These apertures define the pattern of SU-8 (SU-8 100, Microchem, Co.) as a surface mask of backside exposure process (b). In the UV exposure process, UV band pass filter (λ=365 nm, bandwidth=10 nm, 079-0550 band-pass filter, OptoSigma Co.) is used for i-line exposure into SU-8 media. An exposed dose control

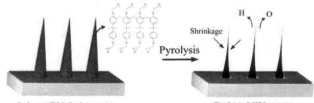

Sub μm SU-8 tip array   Carbon NW array

Figure 1 Schematic view of Top-down approach for CNWs fabrication.

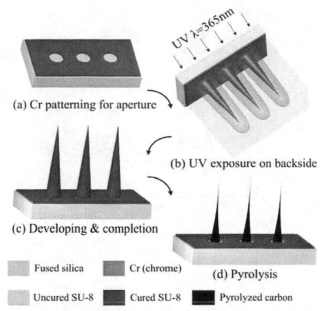

(a) Cr patterning for aperture

UV λ=365nm

(b) UV exposure on backside

(c) Developing & completion

Fused silica   Cr (chrome)
Uncured SU-8   Cured SU-8   Pyrolyzed carbon

(d) Pyrolysis

Figure 2 Fabrication process based on photolithography and pyrolysis.

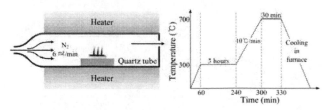

Figure 3 Pyrolysis process and temperature condition.

can produce tip shape of SU-8 whose radius is decreased. After development process, SU-8 tip array is completed (c).

The fabricated SU-8 tip is carbonized by pyrolysis process and vertically aligned CNWs following the shape of the SU-8 tip is fabricated (d). Figure 3 shows the schematic view of quartz tube furnace for pyrolysis and its temperature profile. Nitrogen gas flow of 6 mℓ/min provides an inert environment and prevent an oxidization of SU-8 tip. The SU-8 tip is dehydrated and other volatile chemicals are evaporated, in 300℃ for 3 hours and carbonized when temperature is increased to 700℃. The increasing ratio of temperature is 10℃/min. The temperature is maintained at 700℃ for 30min and cooled to ambient temperature naturally in the furnace with nitrogen atmosphere.

3 RESULTS

Figure 4 shows SEM images of the fabricated HAR SU-8 tips with φ1.0 and 1.4 μm diameter of aperture on fused silica wafer for various exposed dose in photolithography. When the exposed doses are varied from 100 to 200 mJ/cm², the fabricated tip is sharpened to sub-μm width by

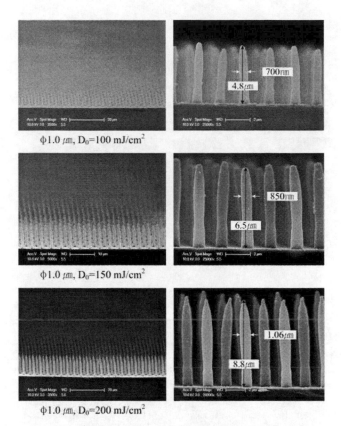

φ1.0 μm, D_0=100 mJ/cm²

φ1.0 μm, D_0=150 mJ/cm²

φ1.0 μm, D_0=200 mJ/cm²

Figure 4 Fabricate results of HAR SU-8 tip array for various aperture diameter and exposed dose.

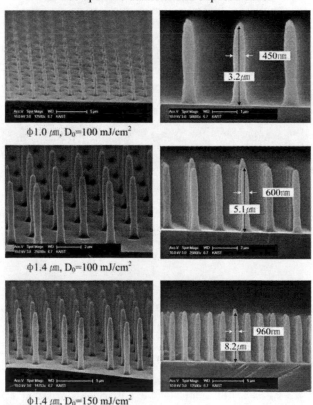

φ1.0 μm, D_0=100 mJ/cm²

φ1.4 μm, D_0=100 mJ/cm²

φ1.4 μm, D_0=150 mJ/cm²

Figure 4 Fabricate results of Carbon nanowires.

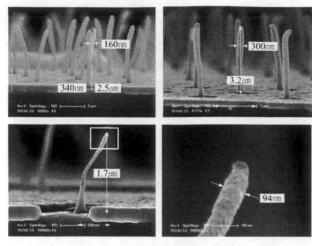

Figure 6 Occasional results of CNWs.

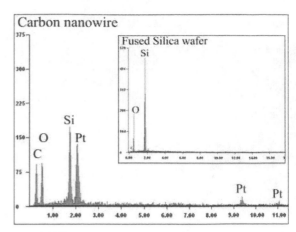

Figure 7 EDS analysis of fabricated CNW compared with fused silica wafer.

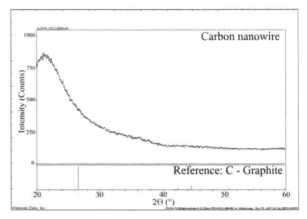

Figure 8 XRD anlaysis shows that the atomic structure of fabricated CNW is amorphous.

diffraction of exposed UV. The width of the SU-8 tips on 1 μm diameter circular aperture are 700 nm, 850 nm, and 1.06 μm and their aspect ratios are 6.9, 7.6, and 8.3 when the exposed doses are 100, 150, and 200mJ/cm², respectively. It is found that the aspect ratio of the tip is increased more than 10 on a larger circular aperture. This tendency is also shown in micro scale SU-8 tip whose aspect ratio is almost 30 when the diameter of circular aperture varied from 3 to 10 μm [2]. Figure 5 shows SEM images of the fabricated vertically aligned CNWs pyrolyzed from SU-8 tip array shown in previous figures. It is found that vertically aligned CNWs on fused silica wafer are well fabricated. When the SU-8 tip is pyrolyzed to CNW, a decomposition of hydrogen and oxygen induces shrinkage of dimension of the CNW. The SU-8 tip on 1 μm diameter circular aperture whose height and width are 4.8 μm and 700 nm is carbonized to CNW whose height and width are 3.2 μm and 450 nm. The widths are 600 and 960 nm and aspect ratios are 8.5 and 8.6 when the exposed doses of SU-8 tips, the precursor of CNW, are 100 and 150 mJ/cm² and its diameter of circular aperture is 1.4 μm.

The fabricated CNWs occasionally shows narrow and high aspect ratio features which width is lower than 100 nm and aspect ratio is more than 10 in Figure 6.

4   ANLYSIS

EDS analysis (EDAX, Ametek Inc.) shows the chemical composition of fabricated CNWs. It is found that fabricated CNW consists of pure carbon following the significant peak of carbon in Figure 7. The peaks of Si and O represent a fused silica wafer and the peak of Pt represents Pt coating on CNWs surface for SEM analysis. In XRD analysis using thin film X-ray diffractometer (D/MAX-RC(12kW), Rigaku Co.) a significant peak of intensity compared with reference of graphite was not found in Figure 8. It does

means that the atomic structure of fabricated CNWs is amorphous.

5   CONCLUSION

In this paper, we found that a pyrolysis process and modified photolithography can be used as top-down method of vertically aligned carbon nanowires (CNWs). Width and aspect ratio of the fabricated CNWs are several hundred nanometer width and around 8. In occasional results, it is found that width of fabricated CNWs are reduced to below 100 nm and aspect ratio is almost 20. These results mean that under 100 nm width CNWs can be fabricated by the proposed process with optimal pyrolysis condition. Hence, optimization of pyrolysis condition is needed to reduce dimension of CNWs as a future works. Furthermore, the mechanical and electric properties of CNWs can be controlled by the conditions.

We expect that the new fabrication method for vertically aligned CNWs as top down approach has a huge potential to expand an application to various micro and nano systems for their controllable positioning, array, and shape.

ACKNOWLEGMENT

This study is supported by brain korea 21 of korea research foundation and kaist institute for information technology convergence.

REFERENCES

[1] A. V. Melechko, et al., "Vertically aligned carbon nanofibers and related structures: Controlled synthesis and directed assembly," J. Appl. Phys. 97, 041301 ,2005.

[2] C. Wang, G. Jia, L. H. Taherabadi, and M. J. Madou, "A novel method for the fabrication of high-aspect ratio C-MEMS structures," J. Microelectromech. Syst. 14, 348, 2005.

[3] O. J. A. Schueller, S. T. Brittain, and G. M. Whitesides, "Fabrication of glassy carbon microstructures by soft lithography," Sens. Actuat. A, 72, 125, 1999.

[4] J. A. Lee, et al., "Biosensor utilizing resist-derived carbon nanostructures," Appl. Phys. Lett., 90, 264103, 2007.

[5] J. A. Lee, et al., "Fabrication and Characterization of freestanding 3D carbon microstructures using multi-exposures and resist pyrolysis," J. Micromech. Microeng. 18, 03512, 2008.

[6] A. d. Campo and C Greiner, "SU-8: a photoresist for high-aspect-ratio and 3D submicron lithography," J. Micromech. Micoeng. 17, R81, 2007.

[7] M. Han, W. Lee, S.-K. Lee, and S. S. Lee, "3D microfabrication with inclined/rotated UV lithography," Sensor Actuat. A, 111, 14, 2004.

[8] S. W. Lee and S. S. Lee, "Application of Huygens-Fresnel diffraction principle for high aspect ratio SU-8 micro/nanotip array," Opt. Lett. 33, 40, 2008.

InGaAs/GaAs quantum dot with material mixing

I. Filikhin, V. M. Suslov and B. Vlahovic

Department of Physics, North Carolina Central University,
Durham, NC 27707, USA, vlahovic@nccu.edu;

ABSTRACT

An effective model for description of the electronic structure of the InGaAs/GaAs quantum dots (QDs) is presented. The model includes a single sub-band approach with an energy dependent electron effective mass and an effective potential which simulates the total effect of the strain and piezoelectricity. Based on our previous calculations for pure InAs/GaAs quantum objects which reproduce both the capacitance-gate-voltage (CV) experimental data and the ab initio calculations, we expand the model for the InGaAs quantum dots with significant Ga fraction. It is found that our model accurately describes the CV and photoluminescence (PL) from QDs with Ga fractions data up to 15%. We illustrate the accuracy of our model by comparing the calculated results and those obtained with the atomic pseudopotential model. We found that considerable difference of these models, appeared in the calculations, related to strength of the electron confinements.

Keywords: quantum dots, single carrier levels, optical properties

1 INTRODUCTION

The fabrication process of nano-sized self-assembled InAs/GaAs quantum dots (QD) and quantum rings (QR) may give the strained quantum structures with controlled geometrical properties [1,2]. There are experimental indicating [3,4,5] that a significant amount of Ga is incorporated into the quantum dots. This material mixing in the initially pure InAs quantum dots occurs during the growth process due to interdiffusions of the QD/substrate materials and can not be carefully controlled. The spatial distribution of the Ga fraction is not definitely known. Theoretical analysis [6-8], based on the effective mass kp-theory calculations, taking into account inter-band interactions, strain and piezoelectric effects by an ab initio manner, shows good ability to match experimental data. However, at present time the ab initio description is limited to the case of pure InAs QDs. In this work, we propose an effective model for strained InAs/GaAs quantum dots to study the character and magnitude of the changes to the energy spectrum of a single carrier arising from presence of the Ga fraction in QD. The model is based on a single sub-band approach for InAs/GaAs QDs with a energy dependent electron effective mass [9-11]. An additional potential V_s is included in the model to simulate the total

effect of interband interactions, the strain and piezoelectricity. We have shown [12] that these effects may be taken into account in an effective manner through this approach. Using the model we have reproduced the results of the ab initio calculations [6-7] and the experimental capacitance-gate-voltage (CV) data [1]. We apply the effective model to study the InGaAs quantum dots with significant Ga fraction. Calculated results for QDs with 15% Ga fraction match both CV and photoluminescence (PL) data simultaneously. We illustrate the accuracy of our model by comparison of the calculated results with those obtained within the framework of the atomic pseudopotential model [13-16]. We found that the considerable difference of these models, displayed in the calculations, related to strength of the electron confinements.

2 EFFECTIVE MODEL

The described 3D heterostructure is modeled utilizing the kp-perturbation single subband approach with the energy dependent quasi-particle effective mass [8-11]. The energies and wave functions of a single carrier in a semiconductor structure are the solutions of the nonlinear Schrödinger equation:

$$\left(-(\nabla \frac{\hbar^2}{2m^*(r,E)} \nabla) + V(r) - E \right) \psi = 0 , \qquad (1)$$

where $V(r)$ is the band gap potential, proportional to the energy misalignment of the conduction (valence) band edges of InAs QD (index 1) and GaAs substrate (index 2). $V(r) = V_c$ inside the substrate, and $V(r) = 0$ inside the quantum dot. The electron effective mass $m^* = m^*(x,y,z,E)$ is linearly dependent on energy for $0 < E < V_c$ and varies within the limits of the QD/substrate bulk effective mass values. The magnitude of V_c is defined as $V_c = \kappa \left(E_{g,2} - E_{g,1} \right)$, where E_g is the band gap and the coefficient $\kappa < 1$ is different for the conduction and valence bands. We use values for κ from Ref. [12]: $\kappa^{CB} = 0.54$ and $\kappa^{VB} = 0.46$. With experimental values $E_{g,1} = 0.42$ eV, $E_{g,2} = 1.52$ eV the band gap potential for the conduction band (valence band) is $V_c = 0.594$ eV ($V_c = 0.506$ eV). Bulk effective masses of InAs and GaAs are $m^*_{0,1} = 0.024 \, m_0$ and $m^*_{0,2} = 0.067 \, m_0$, respectively. Here m_0 is free electron

NSTI-Nanotech 2008, www.nsti.org, ISBN 978-1-4200-8503-7 Vol. 1

mass. We use value of $m^* = 0.4\,m_0$ as the effective mass of heavy hole for both QD and the substrate. The presented band structure model is applied to "unstrained" InAs/GaAs structures. A realistic (ab initio) 3D model for QD has to take into account the band-gap deformation potential, the strain-induced potential, and the piezoelectric potential,

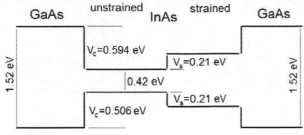

Figure 1. Band structure model for InAs/GaAs quantum dot.

in addition to the band-gap potential [6-8]. Alternatively one can take these potentials into account in an effective manner. In [17] we have introduced the potential V_s, which simulates the integrated effect of all QD potentials aforementioned. The effective potential V_s has an attractive character and acts inside the volume of the QD. The magnitude of the potential can be chosen to reproduce experimental data. This effective model was tested in [12] by comparison of the results of the calculations for QDs and quantum rings (QR) with the ab initio solution of the works [6-7] in which 8-band **kp**-Hamiltonian has been used. It was shown that there is a good agreement of our results with those of the realistic calculations. The effective potential obtained from the comparison has value of 0.21 eV for the pure InAs/GaAs quantum dot. The band structure of our model is shown in Fig. 1. In the figure the effective model with potential V_s is denoted as "strained".

3 QD/SUBSTRATE MATERIAL MIXING

An InAs quantum dot having a semi-ellipsoidal shape embedded into the GaAs substrate is considered. Geometrical parameters of the QD are the height H, and circular base $b = 2R$. The previously reported geometry for experimental fabricated QD is defined by $b \cong 20$ nm, $H \cong 7$ nm [1, 18]. The cross section of QD considered in this section is shown in Figure 2.

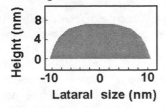

Figure 2. Cross section of semi-ellipsoidal shaped InAs quantum dot embedded in GaAs substrate

The effective potential $V_s = 0.31$ eV was chosen for the interpretation of the capacitance-gate-voltage (CV) experiments of Ref. [1]. In the experiments the spectra of few electrons tunneling into QDs ware observed. There are two s and four p-shell electron levels below the GaAs conduction band edge [1, 18]. The first level of the d-shell is located above this threshold, but it can be observed by the Zeeman effect in a magnetic field.

	Ga, 10%	Ga, 15%	Ga, 20%	0.31 eV	Exp.
m_{QD}^*/m_0	0.055	0.056	0.058	0.057	0.057 $\pm$ 0.007 [1]
$\Delta E(e)$	222	200	177	185	
$\Delta E(h)$	211	194	167	176	
$e_1 - e_0$	47	46	45	46	44*[1]; 49*[18]
$e_2 - e_1$	55	54	51	52	
$h_0 - h_1$	10	9	10	9	
$h_1 - h_2$	11	11	11	11	
E_{e0e0}^c	21.1	20.9	20.8	20.8	21.5[1]; 18.9[18]
E_{e0e1}^c	18.3	18.1	18.0	18.0	24[14]; 13.0[18]
E_{e1e1}^c	17.3	17.1	16.9	17.0	~18[14]
E_{h0h0}^c	25.2	25.0	24.8	24.9	
E_{e0h0}^c	22.8	22.7	22.5	22.6	
E_{e0h0}^{ex}	1060	1104	1154	1137	1098[14]
d_{00}	0.08	0.13	0.08	0.15	0.4$\pm$0.1 [4]

Table 1. Calculated single electron (hole) energy-level spacing e (h), electron (hole) binding energy $\Delta E(e)$ ($\Delta E(h)$), electron-electron, electron-hole and hole-hole Coulomb energies $E_{\alpha\beta}^c$ (α,β =e,h), excitonic band gap E^{ex} (in meV), exciton dipole moment d_{00} (in nm) and effective mass of the QD material for semi-ellipsoidal shaped InGaAs QDs (Ga fraction in %) embedded in GaAs.

With this potential the electron spectra calculation results in localization of the s-shell electron level with respect to the conduction band edge of the GaAs substrate, similar to that can be derived from the CV measurements (~180 meV). First order perturbation theory calculations [17] give a spectral picture that is in an agreement with the CV data. The effect of non-parabolicity, taken into account in our model by means of the energy dependent effective mass approximation, leads to a change in the electron effective mass of QD with respect to the bulk value. For the QD considered, the effective mass in InAs increases from the initial bulk value of $0.024 m_0$ to $0.057 m_0$ (for p-wave electron), which is agrees with the experimental value of $0.057 m_0 \pm 0.007 m_0$, obtained in CV measurements by the Zeeman splitting of p-shell levels [1, 18]. From Table 1 one can see that the agreement between our results (the column "0.31eV") and the experiment data is satisfactory. The small numerical disagreement for few-electron spectrum obtained in [17] we have explained by the uncertainty of the QD geometry. In Ref. [12] we found that the small variations of the QD cross section can lead to quite significant changes of the electron level structure. In that work we did not consider the uncertainty which is related with material mixing of the quantum dot and substrate. From the experiments [3-5] it is evident that a significant quantity of Ga diffuses into the initially pure InAs quantum dots during the fabrication process. A more realistic model must take into account the material mixing. We assume that the effective model describes pure InAs/GaAs QD with the magnitude of the effective potential V_s =0.21 eV. This assumption is justified from comparison our model results with *ab initio* calculations [6-7]. Differences between the effective potential strengths, obtained for the experimental data (V_s =0.31 eV) and the *ab initio* calculation (V_s =0.21 eV), may be related to existence of the Ga fraction in the experimentally fabricated quantum dot. For V_s =0.21 eV potential we have performed calculations varying the Ga fraction in QD. Note that the effective electron mass and the band gap for $In_x Ga_{1-x} As$ material changes linearly with respect to the value of Ga fraction, assuming a homogenous distribution of Ga in the QD volume. The results of the calculations are listed in Table 1. The most reliable results we have gotten for the case of the 15% Ga fraction in QD. For this case we have also matched the experimental value of the transition energy for recombination of exciton pair (E_{e0h0}^{ex}).

4 COMPARISON WITH ATOMISTIC PSEUDOPOTENTIAL MODEL

The theoretical analysis of the electron state in the quantum dot with different Ga fraction distributions was also performed by A. Zunger and coworkers in Refs. [13-16] applying the atomistic pseudopotential approach. We have compared predictions of our effective model with the realistic calculations [6-7, 13] in Ref. [12] for the case of the pure InAs/GaAs quantum dots. As a result we concluded that the pseudopotential approach has strong electron confinement which does not allow us to reproduce the capacitance-gate-voltage measurement results simultaneously for QDs and QRs. In the present work, we continue the comparison of our approach and the atomistic pseudopotential treatment for the quantum dots having the material mixing with a substrate. We consider lens shaped InGaAs quantum dot embedded into GaAs substrate. The circular base of QD has a radius of 25.2 nm and a height varying from 2 nm to 10 nm. Calculations for a single

electron and heavy hole energy levels are listed in Table 2 along with corresponding results of [14]. The Ga fraction in QD was chosen to be 15%. The height of QD was equal 7 nm and 3.5 nm for our calculations and the atomistic calculations, respectively. In spite of the quite different heights, one can see good agreement between the two calculations. The results of calculations are also in an agreement with experimental data given in Table 2. Coincidence of the calculations can be explained by the strong electron confinement of the atomic model. The decreasing of the vertical size of the quantum object can be compensated by the stronger confinement.

	[14]	Effective Model	Exp.
m_{QD}^* / m_0	--	0.055	0.057 ± 0.007 [1]
$\Delta E(e)$ $\Delta E(h)$	204 201	209 196	
$e_1 - e_0$	52	42	44*[1]; 49*[18]
$e_2 - e_1$	60	43	
$h_0 - h_1$	11	8	
$h_1 - h_2$	9	10	
E_{e0e0}^c	28	19.7	21.5[1]; 18.9[18]
E_{e0e1}^c	24	16.7	24[14]; 13.0[18]
E_{e1e1}^c	26	15.7	~18[14]
E_{h0h0}^c	30	23.8	
E_{e0h0}^c	29	21.2	
E_{e0h0}^{ex}	1083	1093	1098[14]
d_{00}	0.05	0.14	0.4 ± 0.1 [4]

*the quantization energy.

Table 2. Calculated result for lens-shaped InGaAs quantum dots with the Ga fraction of 15%. The notations identical to the those in Table 1. The QD has H =3.5 nm in [14] and H =7 nm in our calculations.

Note that the calculated exciton dipole moment d_{00} does not reproduce the experimental value as reported in Ref. [4]. This may indicate the existence of a non-homogenous distribution of the Ga fraction in vertical direction [4]. The agreement between other physical observables calculated by our model and the experimental data may be explained by the effective nature of our description. We also compared the effective model results along with the results obtained within the atomic pseudo-potential approach [14-16]. In these calculations the dependence of the transition energy of electron – heavy hole recombination for the InGaAs/GaAs quantum dots on the Ga fraction and vertical

NSTI-Nanotech 2008, www.nsti.org, ISBN 978-1-4200-8503-7 Vol. 1

size were obtained. The transition energy as a function of the Ga fraction value shown in Fig. 3 was calculated for two case of the QD height $H =3.5$ nm (open marks) and $H =7$ nm (solid marks). An interpretation of the results can be also based upon the stronger electron confinement in pseudopotential model compared with our model as well as in the *ab initio* model [12]. The compensation of such strong confinement can be obtained by increasing the Ga mixture in QD and/or decreasing the vertical size of QD. Note that the available experimental data for LP spectra rifer to the energy region from 1.0 eV to 14 eV. (see a review in [15]).

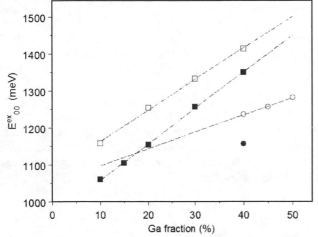

Figure 3. Transition energy of electron – heavy hole recombination (between the ground states) for various Ga fractions and the height H of the InGaAs/GaAs QD. The circles correspond to results of [15,16]. Our results are shown by the squares. The case $H =3.5$ nm ($H =7$ nm) correspond open (solid) marks. The straight lines are drawn only to lead the eye.

For the effective model calculations with $b \cong 20$ nm, this range approximately corresponds to quantity of the Ga fraction from 15% to 40%.

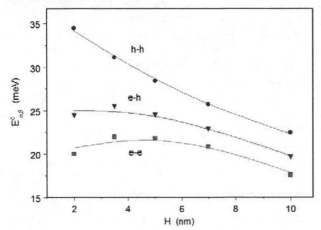

Figure 4. The height dependence of electron-electron, electron-hole and hole-hole Coulomb energies $E^c_{\alpha\beta}$ (α, β =e,h). The QD has a Ga fraction of 20%.

Continuing the comparison the effective model with molecular pseudopotential model, we present in Fig. 4 the result of our calculation of the Coulomb matrix elements for electron-electron, electron-hole and hole-hole interaction. In our calculations the lines corresponding to results for the matrix elements of different interaction do not cross, in contrast to results of the atomic pseudopotential model [16] for QD height of ~5nm. Note that the value of ~20 meV for the Coulomb energy of two electrons tunneling into s-shell of a QD was obtained from the CV experiments reported in [1,18]. Our calculations reproduce this value well, as one can see in Fig. 4.

5 CONCLUSION

In the framework of a effective model we have obtained satisfactory description of the CV and PL experimental data [1] for InGaAs/GaAs quantum dot with the Ga fraction about 15%. From the experimental PL data we found that available concentration of Ga in QD (with $b \cong 20$ nm) cannot be more than 40%. Analyzing the results calculated within the effective model and the atomic pseudopotential model we found considerable difference of both models related to strength of the electron confinements.

This work was partially supported by the Department of Defense and NASA through Grants, W911NF-05-1-0502 and NAG3-804, respectively.

REFERENCES

[1] B. T. Miller et al., Phys. Rev. B 56, 7664-6769, 1997.
[2] J.M. Garsia et al., Appl. Phys. Lett. 71, 2014, 1997.
[3] P. B. Joyce, et al. Phys. Rev. B 58, R15 981-984, 1998
[4] P. V. Fry, et al., Phys. Rev. Lett. 84, 733-736, 2000.
[5] I. Kegel, et al., Phys. Rev. Lett. 85, 1694, 2000.
[6] C. Pryor, Phys. Rev. B 57, 7190, 1998.
[7] M. Grundmann, et al., Phys. Rev. B 52, 11969, 1995; O. Stier, M.Grundmann, D.Bimberg, Phys. Rev. B 59, 5688, 1999.
[8] J. I. Climente, J. Planelles, F. Rajadell, J. Phys.: Condens. Matter 17, 1573-1582, 2005
[9] S. S. Li and J. B. Xia, J. Appl. Phys. 89, 3434, 2001.
[10] Y. Li, et al., J. Appl. Phys. 90, 6416-6420, 2002.
[11] I. Filikhin, at al., Molecular Simulation 31, 779, 2005.
[12] I. Filikhin, V. M. Suslov, M. Wu and B. Vlahovic, Physica E 40, 715-723, 2008.
[13] L.-W. Wang, J. Kim and A. Zunger, Phys. Rev. B 59, 5678-5687, 1999.
[14] A. J. Williamson, L. W. Wang and A. Zunger, Phys. Rev. B 62, 12963, 2000.
[15] J. Shumway et al., Phys. Rev. B 64, 125302, 2001.
[16] G. A. Narvaez, G. Bester, A. Zunger, Phys. Rev. B 72, 245318-245328, 2005.
[17] I. Filikhin, V. M. Suslov and B. Vlahovic, Phys. Rev. B 73, 205332-205335, 2006; I. Filikhin, E. Deyneka and B. Vlahovic, Sol. Stat. Comm. 140, 483-486, 2006.
[18] R. J. Warburton, et al., Phys. Rev. B 56, 6764, 1997.

Large Scale Synthesis of SiC Nanofibers from Various Carbon Precursors

Andrew C. Ritts[1], Hao Li[2], Qingsong Yu[1,2]

[1] Center for Surface Science and Plasma Technology, Department of Chemical Engineering,
University of Missouri, Columbia, MO 65211, USA, yuq@missouri.edu
[2] Department of Mechanical and Aerospace Engineering,
University of Missouri, Columbia, MO 65211, USA, liha@missouri.edu

ABSTRACT

Submicron and micron sized graphitic particles were compared with multiwalled carbon nanotubes (MWNTs) as precursors for producing several grams of SiC nanofibers in a 3'' diameter hotwalled chemical vapor deposition reactor. Nickel nitrate and the carbon precursor were dispersed in tetrahydro furan and dried to disperse the nickel used as a catalyst. Samples were characterized by SEM, TEM, EDS, and XRD. Carbon precursor size and shape, and nickel catalyst concentration are compared. While all carbon precursors produced single crystalline SiC nanofibers, large sized carbon precursors also produced large sized SiC particles by direct conversion. Small amounts of SiC fibers with low aspect ratios and wide ranging diameters and lengths were produced when little or no catalyst was used with carbon precursors. Surface area and volume of the carbon precursor is believed to have a large effect on the quality and quantity of SiC nanofibers produced.

Keywords: chemical vapor deposition, vapor-liquid-solid mechanism, and SiC nanofibers

1 INTRODUCTION

Nanotubes and nanofibers have great promise as reinforcing material in composites due to their outstanding mechanical, thermal, and electrical properties.[1] There is increased interest in hetero-atomic nanomaterials due to their unique electronic properties which can be controlled by changing their chemistry.[2] SiC fibers and whiskers have been widely used and studied as a reinforcing material for composites for their extraordinary properties, such as high thermal conductivity, high thermal stability, excellent mechanical strength, and chemical inertness.[3] However, most industrial applications use micron sized whiskers with low aspect ratios, which have much lower strength compared to their nanosized counterparts. Single crystalline SiC nanofibers also have great promise in producing high toughness and strength composite materials.

Many methods are currently used to produce SiC nanofibers including chemical vapor deposition, thermal plasma synthesis, carbon nanotube confined reactions, and template catalyst free methods; however it is a challenge to economically produce bulk amounts of high quality SiC nanofibers. Processes which produce high quality SiC nanofibers often limit the quantity of fibers synthesized by limiting the growth to two dimensions on a wafer.[4] Other processes use carbon nanotubes confined reactions to fabricate SiC nanofibers which can limit them as fillers in composite materials due to their high cost. For many methods it is difficult to consistently grow large quantities of high quality SiC nanofibers. Despite the various growth methods, only three known mechanisms can fabricate SiC nanofibers: vapor-liquid-solid (VLS), vapor-solid (VS), and direct conversion (DC) of carbon nanotubes. Our previous work used graphite particles as an inexpensive carbon precursor to make SiC nanofibers in bundles along with SiC particles with similar sizes to their graphitic precursor. In this proceeding submicron and micron sized graphitic particles were compared with MWNTs as precursors for producing several grams of SiC nanofibers.

2 EXPERIMENTAL

A horizontal hot-walled chemical vapor deposition reactor was assembled from an MTI GSL1600-80X Bench-top High Temperature Tube Furnace and MKS mass flow controllers. Silica, with 0.5-10 mm particle size and 99% purity (Sigma Aldrich, WI) and silicon with 325 mesh (less than 44 microns) and 99% purity (Sigma Aldrich), were mixed with a 2:1 molar ratio and then put into high purity alumina crucible and leveled, as previously described.[5] Then carbon precursors were added to the crucible and leveled. The crucible was put inside a hot wall CVD chamber with 100 sccm Ar/H_2 (5%H_2) gas. Certain trials utilized 10 sccm of methane, hydrocarbons, or CO_{2g}. The temperature was raised to 1500 $^\circ$C at 150 $^\circ$C/hour and held for 15 hours.

Various carbon precursors shown in table 1 were used such as such as MWNT, submicron graphite (SMG), micron sized graphite (MG), and large graphite (LG). Nickel (II) nitrate hexahydrate (Sigma Aldrich) was used to prepare nickel nitrate ($Ni(NO_3)_2$) solutions. Catalyst quantities chosen were 10μmol, 25 μmol, 50 μmol, or 200 μmol moles of Ni^{+2} per gram of carbon precursor. When catalysts are used, carbon source materials were soaked and mixed in nickel nitrate tetrahydro furan solutions at room temperature, then dried over night. The dried carbon source was placed on top of the Si/SiO_2 mixture for ease of separation after the reaction. When nickel nitrate was used, the H_2 in Ar/H_2 gas was used to reduce the nickel nitrate to nickel nanoparticles. The as-prepared samples were characterized with SEM (Scanning Electron Microscopy), EDS (Energy Dispersive Spectrometer), TEM (Transmission electron spectroscopy) and XRD (X-Ray Diffraction).

NSTI-Nanotech 2008, www.nsti.org, ISBN 978-1-4200-8503-7 Vol. 1

Table 1. Carbon precursor size, purity, and volume to surface area ratio

Name	Company	Purity	Size	Volume/Surface Area
MWNT	Helix Material Solutions	95%	10-30 nm diameter; 0.5-20 μm length	2.48-7.49 nm
SMG	Nanostructured and Amorphous Materials	99.9%	450 nm diameter	150 nm
MG	Alfa Aesar	99.995%	1-15 μm diameter	333-5000 nm
LG	Sigma Aldrich	99.99%	44 μm diameter	14700 nm

3 RESULTS AND DISCUSSION

Several grams of β-SiC were produced from graphitic precursors and compared to SiC nanofiber samples produced from MWNTs. Various factors impacted the size and quantity of the SiC nanofibers produced.

3.1 SiC Nanofibers From Graphite

Previously our process was unable to synthesize high quality nanofibers using MG[5]; however a new catalyst mixing method has produced similar SiC nanofibers from MG as from LG, seen in figure 1, using 25 μmol of Ni^{+2} per

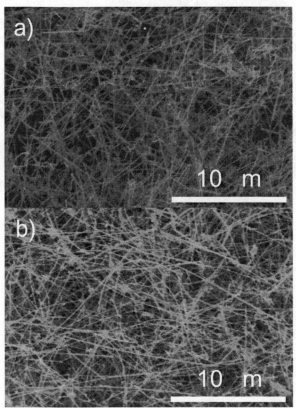

Figure 1. SEM images of SiC nanofibers from LG (a) and MG (b) with Ni catalyst

gram of carbon. The believed primary growth mechanism for SiC nanofibers in all graphitic samples is VLS due to the presence of nickel nanoparticles found and characterized by EDS. Also, when little or no Ni catalyst was present few nanofibers were formed. If excess Ni catalyst was provided for the synthesis of SiC nanofibers, large Ni particles were found on the ends of the nanofibers as seen in figure 2. The size of the Ni particle depends directly on the concentration of catalyst provided for the reaction; however the nanofiber diameter tended to be less than 100nm despite the larger size of the Ni particle, excluding the 200μmol sample. This is believed to be due to the increased rate of reaction of the crystal growth in the liquid Ni particle in the <111> direction. TEM images indicate that the growth direction of the VLS formed SiC nanofibers are in the <111> direction. The limited quantities of CO_g and SiO_g in the liquid Ni are also believed to affect the diameters of the SiC nanofibers synthesized. Dispersion of catalyst into the carbon precursor is essential for consistent fabrication of high quality SiC nanofibers. If the samples do not evenly disperse the Ni catalyst throughout the carbon source, bundles of nanofibers will be produced similar to our previous work.[5] Poorly dispersed Ni produces various yields of SiC nanofibers with wide ranging Ni particle sizes under similar CVD conditions.

Direct conversion of the graphite to SiC particles decreases the potential yield of SiC nanofibers in the sample. However, previous studies have shown CO_{2g} is produced during the nanofiber growth process, which then reacts with more carbon source to produce more carbon monoxide.[6,7] This increases the theoretical yield over 50%; however losses

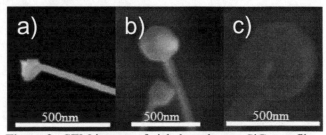

Figure 2. SEM images of nickel catalyst on SiC nanofibers synthesized from MWNT with 25 μmol (a), 50 μmol (b), and 200 μmol (c) of Ni^{+2} catalyst

of CO_g diffusion from the crucible can lower the yield significantly.

3.2 SiC nanofiber Reaction Kinetics and Diffusion

Reaction kinetics and diffusion affect the SiC nanofiber synthesis in every process; however either kinetics or diffusion can be the rate determining step of the process at different times of the fabrication. The duration of the kinetically limited step is believed to be controlled by the size and shape of the carbon precursor.

Initially the silica and silicon react bringing forth SiO_g which diffuses and comes into contact with the surface of the carbon source as seen in equation 1. This produces the CO_g which is required to produce the high quality SiC nanofibers by either VS or VLS mechanism as seen in equation 2. Equation 1 also directly converts the surface of the carbon source into SiC. This is believed to be a kinetically limited step which should produce a spike in localized CO_g concentration. This spike should increase in magnitude and duration as the volume to surface area ratio decreases.

$$SiO_g + 2C(carbon\_source) \rightarrow SiC(DC) + CO_g \qquad (1)$$
$$SiO_g + 3CO_g \rightarrow SiC(nanofibers) + 2CO_{2g} \qquad (2)$$

After the surface of the carbon source has been completely converted to SiC the SiO_g must diffuse through the SiC shell to produce the CO_g essential for the continuation of the SiC nanofiber growth process. Previously reported results show a time-length effect believed to occur in part from the increased time necessary for the gases to diffuse through the SiC layer of the carbon precursor.[5] When the volume to surface area ratio is large, more time is spent in the diffusion limiting step. This produces a relatively constant source of CO_g increasing the yield of SiC nanofibers. Le Chatelier's principle dictates that lower localized concentrations of CO_g and CO_{2g} will limit the diffusion of the gases from the crucible, in turn increasing the yield.

3.3 MWNT and Graphite For SiC Nanofiber Growth

Different shaped carbon precursors such as MG and MWNT can produce very similar SiC nanofibers as seen in figure 3, however the yields and by-products of each process varies. The shape of directly converted SiC from MWNTs is very similar to their MWNT precursors as seen in figure 4. After the lower quality, anfractuous, directly converted SiC nanofibers are sonicated some fibers fracture into whiskers or particles as seen in figure 4c, while the fibers believed to be from VLS have similar high aspect ratios after sonication. The anfractuous nanofibers act more like whiskers or particles and have less promise in nanocomposites that require enhanced bending and tensile strengths. This renders the SiC formed by direct synthesis an unwanted by-product regardless of the carbon source.

High quality SiC nanofiber yields are highest, with sufficient time, from graphitic precursors with large particles

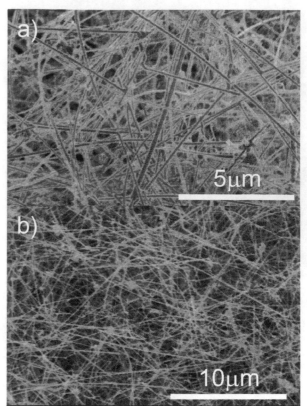

Figure 3. SEM images of SiC nanofibers synthesized from MWNT (a) and MG (b) with Ni catalyst

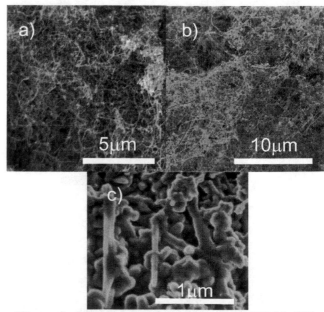

Figure 4. SEM image comparison of MWNT (a), SiC nanofibers created from DC of MWNTs (b), and SiC nanowhiskers and nanoparticles observed after sonication (c)

providing the system with a steady source of CO_g. Yields of 50-70% were achieved with ~2.0 g of MG per trial, resulting in ~3.5 g of SiC. Higher yields can be produced by modifying the growth process. SMG does not provide as high of yields

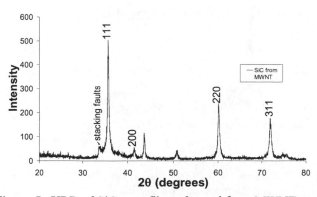

Figure 5. XRD of SiC nanofibers formed from MWNT

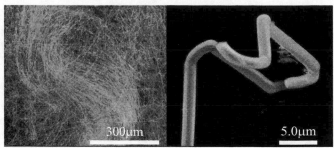

Figure 6. SiC fibers synthesized with hydrocarbon addition

believed to be caused by the small volume to surface area ratio.

The quality of SiC nanofiber powders from MWNT and MG was confirmed using XRD. Figure 5 shows a typical XRD spectrum taken from a SiC nanofiber powder fabricated from MWNTs. Previous TEM investigations show SiC nanofibers synthesized are single crystalline and have thin SiO_x and/or amorphous carbon coatings. These coatings can be tailored by changing the SiO_2/Si molar ratio and CO_g concentration respectively. When the nanofibers are used in composite materials the thin coatings can provide a better interface and dispersion in the matrix material.

3.4 Reactive Gases

Reactive gases can also be used to help increase yield of the SiC nanofibers. When CO_{2g} was introduced into the system the yield of SiC nanofibers increased. The average production of SiC powder from 1.0 g of carbon source is 1.6-1.7 g. The highest yield of SiC nanofibers from that powder is between 50-70% from MG. With CO_{2g} introduced with the 1.0g of carbon the average production of SiC increases to 2.0g. When CO_{2g} is introduced into the system there are limited ways for it to incorporate into the SiC process. First the CO_{2g} can react with the carbon source producing CO_g. This removes impurities from the SiC and helps to increase the yield. The CO_g produced can then incorporate itself into the SiC nanofiber through VLS or VS mechanisms. The CO_{2g} can also incorporate itself into the wall of the SiC nanofiber producing an amorphous coating which increases the diameter of the fiber; however this is unlikely and unsubstantiated from SEM images. An added theoretical benefit is the CO_{2g} limits the diffusion of the CO_{2g} produced from SiC nanofiber by Le Chatelier's principle. This helps to make the SiC nanofiber production yields more consistent, allowing the process to become one step closer to large scale industrial manufacturing.

Methane additions did little to enhance the yield and quality of SiC nanofibers; however some unique larger submicron/micron nanofibers were produced when certain hydrocarbons were added to the system. These large white fibers were located on the outside of the powder and were easily distinguished from the grey, green, blue, and tan colors normally produced and can be seen in figure 6. Colors of the SiC produced depended on the type of precursor and catalyst used.

4 CONCLUSIONS

Currently MG produces the highest yields of SiC nanofibers under scale up conditions. By tailoring other experimental conditions, such as reactive gases, higher yields can be achieved. All carbon sources produce unwanted by-products by DC; however limiting the size of the by-product can limit the effect seen by those by-products. This provides great promise for using relatively low cost SiC nanofibers as fillers for composite materials. Future work will investigate the mechanisms and tailor experimental conditions to produce higher yields of bulk SiC nanofibers in an economic manner.

ACKNOWLEDGEMENT

This work has been supported in part by the Department of Education's GAANN (Graduate Assistance in Areas of National Need) Program. Andrew C. Ritts is a GAANN fellow.

REFERENCES

[1] Breuer, O. and U. Sundararaj, *Big Returns From Small Fibers: A Review of Polymer/Carbon Nanotube Composites.* **Polymer Composites**, 2004. 25(6): p. 630-645.

[2] Zhang, Y., et al., *Coaxial Nanocable: Silicon Carbide and Silicon Oxide Sheathed with Boron Nitride and Carbon.* **Science**, 1998. 281(5379): p. 973-975.

[3] Koumoto, K., et al., *High-Resolution Electron Microscopy Observations of Stacking Faults in B-SiC.* **J. Am. Ceram. Soc.**, 1989. 72(10): p. 1985-87.

[4] Wang, C.S., et al., *Large-scale synthesis of b-SiC/SiOx coaxial nanocables by chemical vapor reaction approach.* **Physica E**, 2007. 39: p. 128-132.

[5] Ritts, A.C., H. Li, and Q. Yu, *Synthesis of SiC Nanofibers with Graphite Powders.* **Proceedings Coco Beach J. Am. Ceramics Soc.**, 2007.

[6] Chiu, S.C., C.W. Huang, and Y.Y. Li, *Synthesis of High-Purity Silicon Carbide Nanowires by a Catalyst-Free Arc-Discharge Method.* **J. Phys. Chem. C**, 2007. 111(28): p. 10294-10297.

[7] Tang, C.C., et al., *Growth of SiC nanorods prepared by carbon nanotubes-confined reaction.* **Journal of Crystal Growth**, 2000. 210(4): p. 595-599.

Identification of Endohedral Metallofullerenes by Method of UV-VIS Spectroscopy

V. Dobrovolsky, N. Anikina, O. Krivuschenko, S. Chuprov, O. Mil'to and A. Zolotarenko

Institute for Problems of Materials Science of NAS of Ukraine,
Kiev-150, P.O.Box 195, 03150 Ukraine, lab67@materials.kiev.ua

ABSTRACT

The endofullerenes attract a close attention of theorists and practicians and especially after appearance of possibility to produce them in macroscopic amounts. This has allowed the investigation of their atomic-molecular architecture, electron structure, the determination of levels for their possible applications. The search for the optimum composition of metal-carbon composites has been done. Analytical control over the EMF content in soot has been performed according to the specially developed procedure. At first the products of arc synthesis have been extracted from fullerene-containing soot by toluene using fractional extraction. After complete extraction of hollow fullerenes, extraction from "poor" soot has been continued in dimethylformamide at the boiling point. identification The fullerenes identification in the extracts has been preformed by the method of UV-VIS spectroscopy.

Keywords: endohedral metallofullerene, soot, arc synthesis, evaporation, absorption, spectroscopy

1 INTRODUCTION

As the class of new synthesized compounds endofullerenes attract a close attention of theorists and practicians and especially after appearance of possibility to produce them in macroscopic amounts. This has allowed the investigation of their atomic-molecular architecture, electron structure, the determination of levels for their possible applications.

2 EXPERIMENTAL

The simplest method of producing endohedral metallofullerenes (EMF) is the electric arc method. Although there exist many theoretical works focused on the formation of fullerenes in carbon-containing plasma, the mechanism of EMF formation in the arc discharge plasma is still unclear [1]. Therefore the search for optimum parameters of arc EMF synthesis is a topical problem. Efficiency of the arc method for EMF producing depends on the technological parameters of synthesis, the composition of metal-carbon composites, the preliminary treatment of graphite anodes and a number of other reasons.

The search for the optimum composition (that increases EMF yield) of metal-carbon composites has been done in the present work. Analytical control over the EMF content in soot has been performed according to the specially developed procedure. At first the products of arc synthesis have been extracted from fullerene-containing soot by toluene using fractional extraction. After complete extraction of hollow fullerenes, extraction from "poor" soot has been continued in dimethylformamide at the boiling point. Identification of fullerenes in the extracts has been preformed by the method of UV-VIS spectroscopy.

3 RESULTS AND DISCUSSION

Fig.1 shows UV-VIS spectra measured in dimethylformamide extracts from the "poor" soot produced in arc evaporation of metal-carbon composites for five different compositions. We believe that the spectra are absorption spectra of EMF. The form of the spectra confirms this supposition. As a rule, absorption spectra of EMF have a long "tail" directed to the long wave region [2, 3]. The maximums in characteristic absorption bands for EMF are usually observed in the range of 500-1000 nm. Fig.2 shows UV-VIS-NIR spectra recorded in the present work and taken from [4].

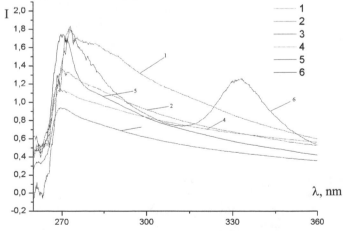

Figure 1: UV-VIS spectra of M@C$_n$ measured in dimethylformamide extracts: spectrum 1 – (M=Ni, treatment 1), spectrum 2 – (M=Ni, treatment 2), spectrum 3 – (M=Y), spectrum 4 – (M=Cu), spectrum 5 – (M=Co), spectrum 6 – C$_{60}$ and C$_{70}$ mixture in dimethylformamide solution.

As Fig. 2 shows, the positions of maximums in absorption bands in the spectra of EMF that contain different metal coincide practically. The observed identity of the spectra is within the framework of the empiric rule that concerns the relation between absorption characteristics and the isomer structure of EMF. UV-VIS-NIR spectra of one metal are analogous to those of the other metal independently on the type of an encapsulated atom when the cage structure and the charge of atoms are identical [2, 3]. When we proceed from the opposite, EMF prepared in this work (their spectra are given in Fig.2, a, b) have $Y_2 @ C_{82}$ and $Ni_2 @ C_{82}$ formulae, respectively. In spectrum 6 (Fig.1) measured in the dimethylformamide solution of C_{60} and C_{70} fullerenes mixture the absorption band with the maximum at 335 nm is characteristic of C_{60} and C_{70} fullerenes. The absence of this absorption band in spectra 1-5 indicates that dimethylformamide extracts contain only EMF. Hence we can judge about the quantitative content of EMF in the extract and consequently in soot by the value of optical density of the EMF spectrum at the chosen wavelength. Based on the comparison of intensities of absorption spectra given in Fig.1, we can conclude that the largest EMF yield has been achieved during evaporation of modified anode 1 (Fig.1, spectrum 1).

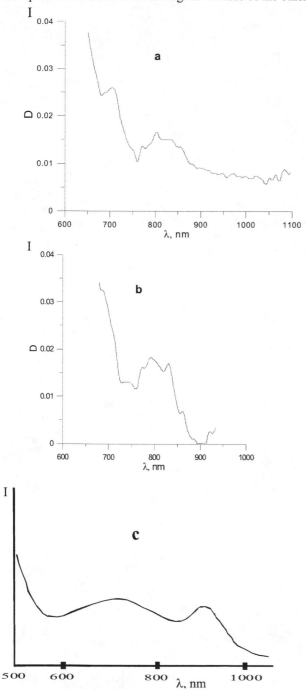

Figure 2: UV-VIS spectra: a) spectrum for $\underline{Y_2@C_{82}}$ in toluene, b) spectrum for $\underline{Ni_2@C_{82}}$ in toluene, c) spectrum for $\underline{Y_2@C_{82}}$ in CS_2 [3].

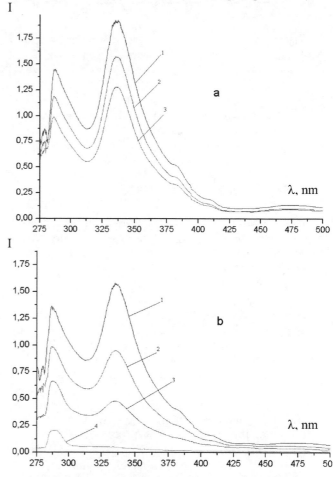

Figure 3: a) UV-VIS spectrum 1, 2, 3 measured in fractions 1, 2, 3, respectively, of toluene extraction from the soot containing only "hollow" fullerenes; b) UV-VIS spectra 1, 2, 3, 4 measured in fractions 1, 2, 3 and in the last fraction of toluene extraction from the soot containing C_{60}, C_{70} and $\underline{Ni@C_n}$, respectively.

Fig.3 shows the spectra measured in toluene extracts of the products of arc synthesis. The soot has formed during arc evaporation of a graphite anode (Fig.3a). Fig.3b shows spectra for the soot containing EMF. Intensity of spectra (Fig.3a) in each subsequent fraction decreases. In this case the value of A_1/A_2 ratio remains constant (A_1 and A_2 – optical densities at 335 and 287 nm, respectively) (Fig.4, curve 2). The value of the ratio in spectra (Fig.3b) decreases (Fig.4, curve 2).

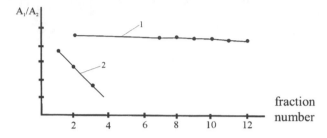

Figure 4: Curves 1 and 2 illustrate how the values of A_1/A_2 ratio change in UV-VIS spectra given in Fig.3a and 3b, respectively.

The decrease in the value of the A_1/A_2 ratio occurs due to the decrease in the concentration of hollow fullerenes in each subsequent fraction of the extract. Intensity of the band at 287 nm decreases slower than that at 335 nm.

4 CONCLUSIONS

In parallel with C_{60} and C_{70}, EMF is also washed out with toluene, and EMF absorption intensity is very high in the UV region. Hence we can conclude about the presence or the absence of EMF in the soot produced by the spectra for only toluene fractions.

REFERENCES

[1] A.S. Fedorov, P.V. Novikov and G.N. Churilov, "Influence of electron concentration and temperature on endohedral metallofullerene $Me@C_{84}$ formation in a carbon plasma", Chemical Physics, V. 293, N 2, 253-261, 2003.

[2] H. Shinohara, "Endohedral metallofullerenes", Rep. Prog. Phys., V. 63, 843-892, 2000.

[3] T. Inoue, T. Tomiyama, T. Sugai and H.Shinohara, "Spectroscopic and structural study of Y_2C_2 carbide encapsulating endohedral metallofullerene: $(Y_2C_2)@C_{82}$", Chem. Phys. Letters, V. 382, N3-4, 226-231, 2003.

[4] T. Inoue, T. Tomiyama, T. Sugai, et.al., "Trapping a C_2 radical in endohedral metallofullerenes: synthesis and structures of $(Y_2C_2)@C_{82}$ (isomers I, II and III)", J. Phys. Chem. B, V. 108, N 23, 7573-7579, 2004.

Vapor Deposited Nanolaminates

B. Kobrin, M. Grimes and N. Dangaria
Applied Microstructures, Inc.
1020 Rincon Circle, San Jose, CA 95131 USA
Tel: 1-(408) 907-2885; Fax: 1-(408) 907-2886
e-mail: boris_kobrin@appliedmst.com

ABSTRACT

We present the latest developments in nanocoatings deposition technology, realized using Molecular Vapor Deposition (MVD®) method. Multi-layer nanolaminates of metal oxides (alumina, titania, silicon oxide, …etc.) and self-assembled monolayers (SAMs) sequentially deposited using an automated system in-situ without breaking a vacuum between layers show improved lifetime and environmental stability (mechanical impact, immersion in liquids, …etc.) compared to just SAM layer alone. We have found that different metal oxides are preferable as an adhesion layers to various materials and as seed layers for organic coatings deposition, thus previously implemented dual layers should be complemented with additional intermediate layer. We show results of trilayer MVD nanolaminates improvement in thermal and catalytic stability, discuss recently developed novel MEMS lubrication schemes, and promising results in moisture barrier applications.

Keywords: MVD, molecular vapor deposition, metal oxides, organic films, nano-laminates, adhesion layers

1. INTRODUCTION

MVD® surface engineering has already proved to be enabling technique for MEMS yield and lifetime improvement, Inkjet nozzles passivation, molds pre-treatment for nanoimprint Lithography, and many other applications [1-3]. The most important advantage of this method is it's applicability to wide varietry of substrate materials. Lack of bonding sites in non-oxide-based material surfaces, like polymers and some metals, is compensated with "artificial" oxide surface, which is grown in-situ from a vapor phase. Such an adhesion layers are based on metal-oxides, which can be bound with most materials using non-covalent-type linkage. For example, alumina have an ability to incorporate into polymer pores and thus create quite strong bond with the surface. At the same time, we have found that metal-oxide which forms strong bond with substrate material is not the best (or the most stable) surface for functional organic coating. Another intermediate layer is necessary to create strong adhesion and provide maximum density of functional coatings.

2. EXPERIMENTAL

Alumina, titania, and heptadecafluoro-1,1,2,2-tetrahydrodecyltrichlorosilane (FDTS) coatings were vapor deposited from liquid precursors (obtained from Gelest Inc. and Sigma) using the MVD100 vapor deposition system manufactured by Applied MicroStructures, Inc. Surface cleaning and hydroxylation of the substrates were performed *in-situ* using a remote RF oxygen plasma source. The metal oxide adhesion layers and FDTS films were grown sequentially at temperatures between 50-80ºC without exposure of the substrate to ambient conditions during the cycling process.

Water contact angles were taken using a Rame-Hart Goniometer. Water vapor transmission rate (WVTR) is measured using MOCON Permatran 3/33 water permeation measurement tool at 37 C. and 85% RH.

3. RESULTS AND DISCUSSION

Fig 1. represents a simple nanolaminate comprised of 2 layers, the first- adhesion metal-oxide layer, and, the second – functional Self-assembled monolayer (SAM). Total thickness of this nanolaminate can be adjusted by controlling thickness of metal-oxide within a practical range of 50-500 A. We were able to apply SAM to many polymers, metals, semiconductors and glasses using this scheme. Recently we found that stability of this coating can be further improved by addition of another coating (intermediate) between metal-oxide and SAM.

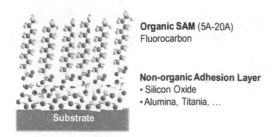

Organic SAM (5A-20A)
Fluorocarbon

Non-organic Adhesion Layer
• Silicon Oxide
• Alumina, Titania, …

Substrate

Figure 1: Schematics of dual-layer of metal-oxide and SAM

Fig. 2 demonstrates advantages of such 3-layer stack (especially for A1/B1/F1 materials) for thermal stability on Si wafer. Self-assemble coating F1 degrades quite rapidly with temperature; adhesion layer A1 helps with stability, especially up to 250 C; different tri-layer schemes allow to extend usability of MVD coatings up to 500 C.

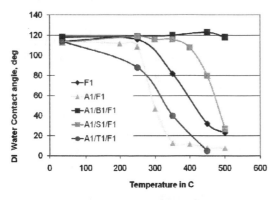

Fig. 2 Temperature stability for single, double and triple-layer stacks on Si

This approach has also allowed forming quite durable hydrophobic layer on Ni material, as shown on Fig. 3. Triple-layer approach gives performance on Ni similar to Si material.

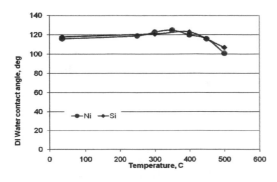

Fig. 3 Triple-layer stacks temperature stability on Si and Ni

Triple-layers also provide good catalytic stability. Fig. 4 and 5 demonstrate performance of dual and triple coatings deposited on Si and SU8 photoresist materials in DI water immersion. Here again, the best stability is achieved with triple-layer configuration.

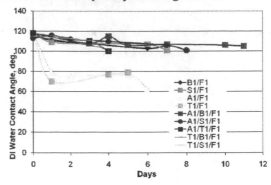

Fig. 4 DI water immersion stability of double and triple-layer stacks deposited on Si

Similar stability improvement has been obtained on polymer materials. For example, Fig. 5 represents immersion stability of dual and triple-layers deposited on SU8 photoresist material.

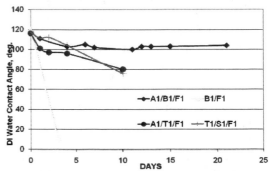

Fig. 5 DI water immersion stability of double and triple-layer stacks deposited on SU8 polymer

Recently we have discovered that nanolaminates deposited using MVD technology can provide quite impressive moisture permeation properties. Fig. 6 shows results of moisture transmission rate measurements on Polyethylene naphtalite (PEN) film (5 mil) with and without MVBD alumina coating. We observed 3 orders of magnitude drop in WVTR for <50 A thick coating deposited on both sides of the film. We plan to continue this development and complement MBD alumina with one or number of organic MVD coatings to improve WVTR beyond 10^{-3} g/m2*day region.

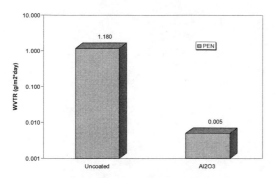

Fig. 6 Water vapor transmission rate (WVTR) for MVD metal-oxide deposited on PEN

Another interesting application of MVD nanolaminates are wear resistant lubrication coatings for MEMS and release coatings for Nanoimprint Lithography. The test structures, based on a double comb drive design, were fabricated in the Sandia SUMMiT V process. Their operation has been extensively discussed in the literature [4]. Brielfy, one set of combs is used to apply a static load on a beam, which is pulled against a fixed post; the other set is used for tangential actuation to rub the beam against the fixed post (Fig. 1).

Fig.7 represents a polysilicon wear test structure we used for testing this coating performance. (crossing beams only; the driving combs are outside of the field of view). The fixed post is shown blown up in the inset for clarity. Apparent contact pressure was about 80 MPa, sliding frequency – 3 Hz.

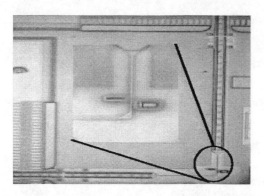

Fig. 7 Optical micrograph of wear tester device.

The results of this tests are presented on The carbon-doped alumina coatings were applied to the released microstructures in an MVD100 system by molecular vapor deposition, a technique similar to atomic layer deposition. Trimethylaluminum and water were used as precursors and nitrogen as the purge gas. The chamber temperature was kept below 100° C. A multi-layer process resulted in an alumina coating, 5.8 nm thin and a very smooth surface

Coating	Cycles-to-Failure
Native oxide	8×10^3
Vapor SAM	4×10^5
SiC LPCVD	Testing halted after 10^6
Alumina MVD®	Testing halted after 2×10^7

Table 1. Lifetime of various coatings

We have performed wear tests by applying a tangential actuation voltage (sine wave plus DC offset, 80 V peak-to-peak, 3 Hz). An 80 V DC load was applied to the load actuator, corresponding to a mean contact pressure of 80 MPa. Device failure (i.e., the inability of the tangential actuation voltage to slide the beam across the fixed post, with the static load applied) usually occurs within a few thousand cycles of operations for uncoated devices [3] (Table I). In contrast, the microstructures coated with carbon-doped alumina have so far undergone testing without failure for over 7×10^6 cycles. This is approximately seven times longer than the SiC-coated structures tested in a similar way.

SUMMARY

We demonstrated enhancement of MVD® deposition technology using multi-layer nanolaminate schemes. Triple layers nanolaminates, combined from different types of metal-oxides and organic self-assembled monolayers provide higher thermal and immersion stability. MVD metal oxides also show very impressive performance as lubrication coatings for MEMS and Nanoimprint Lithography. Nanolaminates deposited by MVD technique hold a good promise for moisture barrier applications.

REFERENCES

[1] B. Kobrin, M. Grimes, "Surface Engineering for Microfabrication", Solid State Technology, Jan 2008
[2] B. Kobrin, "Better MEMS, NIL with Surface Engineering", Small Times, Nov/Dec 2007

[3] B. Kobrin, T. Zhang, M. Grimes, K. Chong, R. Nowak, J. Chinn 2006 Reliability of Self-Assembled Hydrophobic Coatings: Chemical Resistance and Mechanical Durability *NSTI Nanotech* May 8-12 Boston MA

[4] W.R. Ashurst, C. Carraro, and R. Maboudian, Tribological impact of SiC encapsulation of released polycrystalline silicon microstructures, Tribology Lett. **17**(2) (2004) 195-198.

PECVD synthesis of Si nanowires.

I.I. Kravchenko

Nano-Research Facility, University of Florida, P.O.Box 116621,
Gainesville, FL32611, USA, kravch@phys.ufl.edu

ABSTRACT

Silicon nanowires 42 nm in diameter and up to 70 micrometer long have been obtained by utilizing a common (for an average nanofabrication facility) plasma equipment set that did not require special modifications. Further studies are needed to improve the synthesis process parameters and gold catalyst particle distribution.

Keywords: Si nano-wire, PECVD, gold catalyst, nano-particle

1   INTRODUCTION

Since the discovery of carbon nanotubes, other one-dimensional as well as differently nanostructured systems have been attracting increasing attention due to potentially unlimited technological applications. These materials are expected to help overcome existing technological limits of the semiconductor industry [1] and provide new venues to obtain higher device density and to develop new devices (e.g. single-electron [2]) that utilize new unexplored principles. Silicon nanowires (SiNWs) have already been employed to create test prototypes of field effect transistors [3,4], nanosensors [5,6] and thermoelectric devices [3]. Besides, due to quantum confinement SiNWs manifest properties of a direct band-gap semiconductor [7,8] that, unlike bulk Si, may be of interest for optoelectronics applications.

A number of Si nanowire growth methods such as laser ablation [9], thermal evaporation [10], CVD [11], plasma enhanced CVD (PECVD) [12,13], have already been reported. The CVD methods require assistance of heavy metal catalysts such as Au, Cu, Ag, Ni, Ga, In [11-14]. Catalytic growth gives the benefit of lower synthesis temperatures and flexibility in choosing the sample areas where the wires are needed. If the PECVD method is chosen the precursor gas molecules are already dissociated in the plasma discharge, so the temperature required is only slightly above of Si-catalytic metal eutectic point.

2   EXPERIMENTAL

It is an unfortunate but established fact that conducting nano-scale research requires expensive and complex instrumentation that an average research laboratory rarely can afford.   Therefore, there is trend to concentrate equipment, expertise and highly skilled dedicated staff in user facilities that allow uninterrupted operations and experience transfer. On the other hand, continuous flow of diverse ideas, tasks, and users compounded by competition among research groups often exclude time consuming or highly specialized modifications to existing equipment that put additional strain on researchers and stuff members to find ways to fully utilize and/or expand available capabilities. This work is a result of such explorations.

A commercial STS310PC plasma enhanced CVD system (PECVD) initially designed to deposit silicon oxides and nitrides was utilized to synthesize SiNWs. The precursor gas was 2% silane diluted with nitrogen. The hot flat bottom electrode where the substrates are located during deposition is grounded. The other electrode which also serves as a gas shower head is located about 2 cm above.

Silane/nitrogen mixture flow, gas process pressure and the process temperature were variable. The 13.56 MHz RF power was kept constant at 30W. The deposition time was usually 2 minutes. The process temperature was set at 380°C.

A KJL CMS-18 multi-target sputtering deposition system along with a Unaxis RIE-ICP etcher was needed to create catalyst particles on the sample surface. Process pressures in both systems were 5 mtorr.

Gold was selected as the catalytic material. The Au-Si eutectic temperature is 363°C which is within working temperature range of the PECVD system. To prevent gold diffusion into the sample bulk the substrates (20x20 mm pieces of Si(100) wafer) were coated with a 1 micrometer thick PECVD silicon oxide. The substrates were deliberately made larger to compare SiNWs growth condition on the sample flat surface and near its edge where the plasma electrical field was disturbed.

It has been reported recently that in co-sputtered Au/SiO_2 films, gold will precipitate in bulk even at room temperatures and create nanometer-size particles in the SiO_2 matrix if Au/Si atomic ratio exceeds a critical value [15].

In this work, a 500 nm composite film was deposited by simultaneous sputtering from SiO_2 and Au targets. The sputtering rates were adjusted in such a way that the Au/Si atomic ratio would be close to 1. The resulting film was etched back 60 nm in the RIE-ICP system by using a CHF3-based chemistry. The process parameters were tuned to obtain isotropic SiO_2 matrix etching. SEM inspection of the resulting surface revealed exposed clusters averaging 15 nm in size (Fig.1).

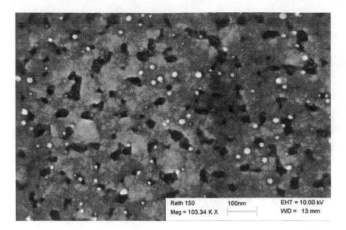

Figure 1. Catalyst particles after plasma surface preparations.

3   RESULTS AND DISCUSSION

First attempts to grow silicon nanowires were not successful due to a very high deposition rate. Resulting films were unstructured masses of amorphous silicon. Only after gas flow and process pressure were reduced to the lowest possible values when the plasma was still stable, were the SiNWs obtained (Fig.2).   As it can be seen from the picture the catalytic particle is at the tip of the nanowire indicating the tip-growth mechanism which in this case may be described by the VLS model [16], according to which Si diffuses into a particle before solidifying in the shape of wire.

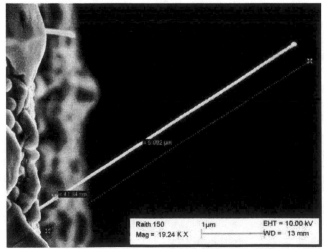

Figure 1. A 5 micrometer long silicon nanowire. The size of the wire at the root and at the tip is the same, 42 nm across. The catalyst particle is visible at the top.

It should be noted that the wires were observed only at the edges of the substrate. SEM inspections of the sample areas away from the edge revealed interwoven worm-like structures (Fig.3a) occasionally dotted with protruding short stubs of nanowires (Fig.3b).   This indicates that the nanowire growth was hindered by the incoming flux of ions and other radicals to the surface, causing a random movement of the catalyst particles over the sample surface and resulting in the worm-like shapes of crystallized silicon.

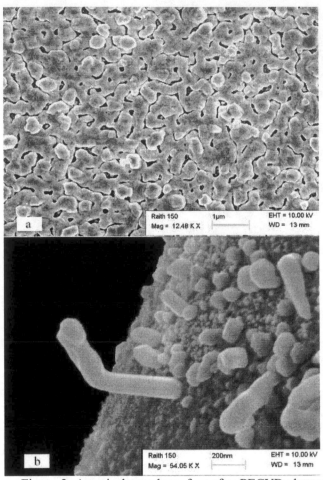

Figure 3. A typical sample surface after PECVD plasma process.

At the same time, different plasma conditions at the sample edges were more favorable for the SiNWs growth. To test these assumptions a grounded cage made of 100x100 tungsten mesh was placed over the substrates during deposition. Although it did not lead to the nanowire growth on the flat areas of the samples, the density and length of SiNWs were visibly bigger at the edges (Fig.4). A nanowire as long as 70 micrometers long was discovered that amounts to a 35microns/min growth rate. It was also observed that the wires were not tapered, i.e. the thickness at the root and at the tip was the same (around 40 nm) indicating a negligible mass loss by the gold particle as it was moving along the wire during the growth.

To complete this paper a different attempt of growing Si nano-structures should be mentioned. A piece of Si(100) wafer with a native silicon oxide intact was sputter coated with a thin (7.5 nm) gold film and processed in the PECVD system with parameters identical to the experiments when the SiNWs shown in Fig.4 were obtained. SEM inspections of the sample did not discover any nanotubes but, instead, revealed a self-organized array of interconnected 270 nm diameter nanoballs (Fig.5). It is too early to speculate on origin and the growth mechanism of these Si-based nanostructures, but one can already come to the conclusion that possibilities of obtaining a variety of structures through process parameters changing and catalyst tailoring look promising.

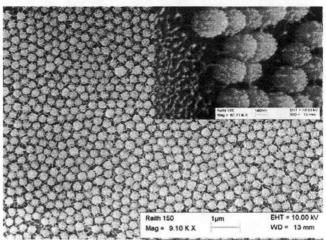

Figure 5. An array of 270 nm diameter nanoballs grown on Si surface.

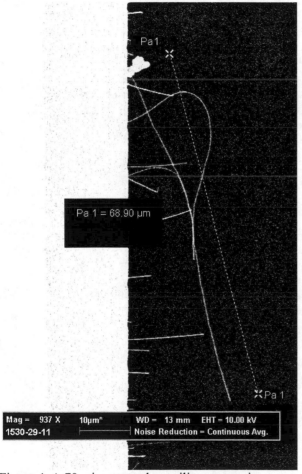

Figure 4. A 70 micrometer long silicon nanowire.

4 CONCLUSIONS

This paper shows that with a proper catalyst conditioning it is possible to grow silicon nanowires and orher Si-based nanostructures in a typical nanofabriacation facility. Tens-of-micrometers-long wires and a high growth rate are achievable with help of commercially available PECVD equipment.

REFERENCES

[1] L.E.Brus, Appl.Phys.A 53, 465, 1991

[2] A.Tilke, R.H.Blick, H.Lorenz, J.P.Kotthaus, J.Appl.Phys. 89, 8159, 2001.

[3] Y.Cui, C.M.Lieber, Science, 291, 851, 2001

[4] M.C.McAlpine, R.S.Friedman, S.Jin, K.-H.Lin, W.U.Wang, C.M.Lieber, Nano Lett, 3, 1531, 2003.

[5] Y.Cui, Q.Wei, H.Park, C.M.Lieber, Science, 293, 1289, 2001

[6] J.H.Hahm, C.M.Lieber, Nano Lett, 4, 51, 2004.

[7] L.T.Canham, Appl.Phys.Lett., 57, 1046, 1990.

[8] G.D.Sanders, C.Yia-Chung, Phys.Rev.B, 45, 9202, 1992.

[9] A.M.Morales, C.M.Lieber, Science, 279, 208, 1998.

[10] H.Pan, S.Lim, C.Poh, H.Sun, X.Wu, Y.Feng, J.Lin, Nanotech., 16, 417, 2005.

[11] J.Westwater, D.P.Gosain, S.Tomiya, S.Usui, H.Ruda, J.Vac.Sci.Tech.B, 15, 554, 1997.

[12] S.Hoffman, C.Ducati, R.J.Neill, S.Piscanec, A.C.Ferrari, J.Geng, R.E.Dunin-Borkowski, J.Robertson, J.Appl.Phys., 94, 6005, 2003.

[13] H.Griffiths, C.Xu, T.Barrass, M.Cooke, F.Iacopi, P.Vereecken, S.Esconjauregui, Surf.&Coat.Tech., 201, 9215, 2007.

[14] G.A.Bootsma, H.J.Gassen, J.Cryst.Growth, 10, 223, 1971.

[15] K.-H. Jung, J.-W.Yoon, N.Koshizaki, Y.-S.Kwon, Current Appl.Phys. in press, 2007

[16] R.S.Wagner, W.C.Ellis, Appl.Phys.Lett., 4, 89, 1964.

Quantum Gates Simulator Based on DSP TI6711

Victor H. Tellez[1, 3], Antonio Campero[2], Cristina Iuga[2], Gonzalo I. Duchen[3]

[1]Electrical Engineering Departament, [2]Chemical Departament, DCBI, Universidad Autonoma Metropolitana Iztapalapa, Av. San Rafael Atlixco 186, Col. Vicentina, Iztapalapa 09340 D.F. Mexico, email: vict@xanum.uam.mx
[3]SEPI, ESIME Culhuacan, Av. Santa Ana 1000, Col. San Francisco Culhuacan, C.P. 04430, D.F.

ABSTRACT

Quantum theory has found a new field of application in the information and computation fields during recent years. We developed a Quantum Gate Simulator based on the Digital Signal Processor (DSP) DSP TI6711 using the Hamiltonian in the time- dependent Schrödinger equation. The Hamiltonian describes the Quantum System by manipulating a Quantum Bit (QuBit) using unitary matrices. Gates simulated are conditional NOT operation, Controlled-NOT Gate, Multi-bit Controlled-NOT Gate or Toffoli gate, Rotation Gate or Hadamard transform and *twiddle* gate, all useful in quantum computation due to their inherently reversible characteristic. With the simulation process, we have obtained approximately 95% fidelity action of the gate on an arbitrary two and three QuBit input state. We have determined an average error probability bounded above by 0.07 ± 0.01.

1 INTRODUCTION

The basic unit of storage in a quantum computer is the *qubit*. A qubit is like a classical bit in that it can be in two states, zero or one. The qubit differs from the classical bit in that, because of the properties of quantum mechanics, it can be in both these states simultaneously [1, 2].A convenient method for representing a qubit state is the ket notation defined by Dirac [3]. In this notation the ket $|0\rangle$ denotes the zero state and the ket $|1\rangle$ represents the one state. This notation is convenient because it labels the qubit state, and therefore only those states with non zero amplitude need to be explicitly written.

A quantum computer performs operations on *qubits*, whose value can be one or zero or any *superposition* of one and zero. A quantum computer performs transformations on these qubits to implement logic gates. These quantum logic gates create correlations between qubits, referred to as *entanglement*, which allows the representation of an exponential number of states using a polynomial number of qubits. Combinations of these logic gates define quantum circuits.

All operations in a quantum computation are achieved by means of transformations on the qubits contained in quantum registers. A transformation takes an input quantum state and produces a modified output quantum state.

Typically transformations are defined at the gate level, i.e. transformations which perform logic functions. Transformations that correspond to physical processes can also be defined. These lower level transformations are then composed so that they implement gate operations.

Because of the laws of quantum mechanics each transformation of the quantum state space, other than a measurement, must leave the quantum superposition of the state intact. More specifically each transformation must be unitary.

2 METHODOLOGY

For the development of the project, we use the DSP, like the TI TMS6711, with architectural optimizations to speed up processing. This DSP can be connected on classical personal computer for transfer the data between them, and the architectural features is the next:

Program flow:
- Floating-point unit integrated directly into the data-path.
- Pipelined architecture
- Highly parallel accumulator and multiplier
- Special looping hardware. Low-overhead or Zerooverhead looping capability

Memory architecture:
- DSPs often use special memory architectures that are able to fetch multiple data and/or instructions at the same time:
- Harvard architecture
- Use of direct memory access
- Memory-address calculation unit

A quantum logic gate is a transformation which performs a logic function on the input state and produces a new output state[6]. Circuits are constructed as sequences of these gates in the same manner as is used in conventional digital circuits. The gates that perform a conditional "not operation" are useful in quantum computation because they are inherently reversible. A single bit gate performs an unconditional not operation, and multibit-gates negate the resultant bit conditionally based on the input bits.

2.1 Controlled NOT gate

A reversible version of a conventional exclusive or gate is constructed by retaining the value of one of the inputs. This gate, called the controlled-not gate [5], is defined by the truth table shown in Table 1.

Table 1: truth table for controlled not gate

Input Qubits		Output Qubits	
A	B	A'	B'
0	0	0	0
0	1	0	1
1	0	1	1
1	1	1	0

The controlled-not gate leaves the qubit unchanged and flips the value of only if is set. The controlled-not gate can be reversed by performing another controlled-not gate. The logic symbol for this gate is shown in figure 1.

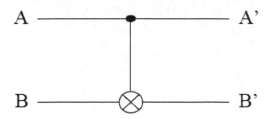

Figure 1: Logic symbol for the controlled not gate

$$\begin{bmatrix} 1 & 0 & 0 & 0 \\ 0 & 1 & 0 & 0 \\ 0 & 0 & 0 & 1 \\ 0 & 0 & 1 & 0 \end{bmatrix} X \begin{bmatrix} a_{00} \\ a_{01} \\ a_{10} \\ a_{11} \end{bmatrix} = \begin{bmatrix} a_{00} \\ a_{01} \\ a_{11} \\ a_{10} \end{bmatrix}$$

(1)

The matrix 1 show the transformation of the logic controlled not gate.

2.2 Multi-bit Controlled-not Gates

Multi-bit controlled-not gates are defined by adding additional controlled inputs. The resultant bit is only flipped if the logical AND of all the input qubits is one. These multi-bit gates are useful in the construction of logic circuits because of this AND property. Figure 2 shows the logic symbol for the three bit controlled-controlled not gate. This gate is also called the Toffoli gate after its designer [6]. It transforms the state $|1\rangle|1\rangle|0\rangle$ to $|1\rangle|1\rangle|1\rangle$ and the state $|1\rangle|1\rangle|1\rangle$ to $|1\rangle|1\rangle|0\rangle$ leaving all other states unchanged.

Figure 2: Logic symbol for the multi controlled not gate.

2.3 Rotation Gates

The *Hadamard* transform is a single bit rotation gate. The matrix 2 shows its definition. The logic symbol used in circuit diagrams is shown in Figure 3. The Hadamard transform applied to a qubit that is in the state $|0\rangle$ creates a state that is in the equal superposition of the $|0\rangle$ and $|1\rangle$ states. The Hadamard transform is also used in the encoding and error correction circuits.

$$H = \frac{1}{\sqrt{2}} \begin{bmatrix} 1 & 1 \\ 1 & -1 \end{bmatrix}$$

(2)

Figure 3. The Hadamard transformation

2.4 Twiddle Gates

The quantum FFT circuit requires an additional gate, the *twiddle* gate shown in the matrix 3. A twiddle is performed between two bits, denoted by the bit positions and as part of an FFT performed across L qubits.

$$T = \begin{bmatrix} 1 & 0 & 0 & 0 \\ 0 & 1 & 0 & 0 \\ 0 & 0 & 1 & 0 \\ 0 & 0 & 0 & w^m \end{bmatrix}$$

(3)

3 RESULTS AND DISCUSSION

In the figure 3, we show the simulation for the entropy on gate C-Not.

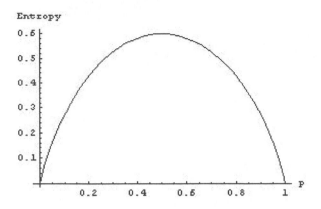

Figure 3: Entropy for C-Not gate

In the figure 4, we presents the fidelity for C-Not gate. In the matrix 4, we show the density matrix for C-Not gate, while in the matrix 5 we represent the matrix density for Hadamard gate.

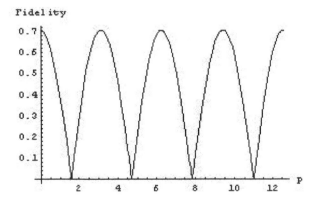

Figure 3: Fidelity for C-Not gate

Classical computation theory began for the most part when Church and Turing independently published their inquiries into the nature of computability in 1936 [7]. For our purposes, it will suffice to take as our model for classical discrete computation. So we are able to simulate the possible quantum processor using different gates and approach to the development of the Von Neuman Model[8].

The next step of this project is to integrate different gates and emulate as the classical components of the VLSI that integrated the classical semiconductor.

$$
\begin{pmatrix}
\frac{1}{8} & 0 & 0 & \frac{1}{8} & \frac{1}{8} & 0 & 0 & \frac{1}{8} & \frac{1}{8} & 0 & 0 & \frac{1}{8} & \frac{1}{8} & 0 & 0 & \frac{1}{8} \\
0 & 0 & 0 & 0 & 0 & 0 & 0 & 0 & 0 & 0 & 0 & 0 & 0 & 0 & 0 & 0 \\
0 & 0 & 0 & 0 & 0 & 0 & 0 & 0 & 0 & 0 & 0 & 0 & 0 & 0 & 0 & 0 \\
\frac{1}{8} & 0 & 0 & \frac{1}{8} & \frac{1}{8} & 0 & 0 & \frac{1}{8} & \frac{1}{8} & 0 & 0 & \frac{1}{8} & \frac{1}{8} & 0 & 0 & \frac{1}{8} \\
\frac{1}{8} & 0 & 0 & \frac{1}{8} & \frac{1}{8} & 0 & 0 & \frac{1}{8} & \frac{1}{8} & 0 & 0 & \frac{1}{8} & \frac{1}{8} & 0 & 0 & \frac{1}{8} \\
0 & 0 & 0 & 0 & 0 & 0 & 0 & 0 & 0 & 0 & 0 & 0 & 0 & 0 & 0 & 0 \\
0 & 0 & 0 & 0 & 0 & 0 & 0 & 0 & 0 & 0 & 0 & 0 & 0 & 0 & 0 & 0 \\
\frac{1}{8} & 0 & 0 & \frac{1}{8} & \frac{1}{8} & 0 & 0 & \frac{1}{8} & \frac{1}{8} & 0 & 0 & \frac{1}{8} & \frac{1}{8} & 0 & 0 & \frac{1}{8} \\
\frac{1}{8} & 0 & 0 & \frac{1}{8} & \frac{1}{8} & 0 & 0 & \frac{1}{8} & \frac{1}{8} & 0 & 0 & \frac{1}{8} & \frac{1}{8} & 0 & 0 & \frac{1}{8} \\
0 & 0 & 0 & 0 & 0 & 0 & 0 & 0 & 0 & 0 & 0 & 0 & 0 & 0 & 0 & 0 \\
0 & 0 & 0 & 0 & 0 & 0 & 0 & 0 & 0 & 0 & 0 & 0 & 0 & 0 & 0 & 0 \\
\frac{1}{8} & 0 & 0 & \frac{1}{8} & \frac{1}{8} & 0 & 0 & \frac{1}{8} & \frac{1}{8} & 0 & 0 & \frac{1}{8} & \frac{1}{8} & 0 & 0 & \frac{1}{8} \\
\frac{1}{8} & 0 & 0 & \frac{1}{8} & \frac{1}{8} & 0 & 0 & \frac{1}{8} & \frac{1}{8} & 0 & 0 & \frac{1}{8} & \frac{1}{8} & 0 & 0 & \frac{1}{8} \\
0 & 0 & 0 & 0 & 0 & 0 & 0 & 0 & 0 & 0 & 0 & 0 & 0 & 0 & 0 & 0 \\
0 & 0 & 0 & 0 & 0 & 0 & 0 & 0 & 0 & 0 & 0 & 0 & 0 & 0 & 0 & 0 \\
\frac{1}{8} & 0 & 0 & \frac{1}{8} & \frac{1}{8} & 0 & 0 & \frac{1}{8} & \frac{1}{8} & 0 & 0 & \frac{1}{8} & \frac{1}{8} & 0 & 0 & \frac{1}{8}
\end{pmatrix}
$$

(4)

The central problem that we will concern ourselves with repeatedly in these notes is the problem of universality. That is, given an arbitrarily large function f, is it possible to identify a universal set of simple functions that can be used repeatedly in sequence to simulate f on its inputs.

$$
\begin{pmatrix}
\frac{1}{16}(1+(-1)^{2a})(1+(-1)^{b})^{2} & 0 & \frac{1}{16}(1+(-1)^{2a})(1+(-1)^{b})^{2} & 0 \\
0 & \frac{1}{16}(1+(-1)^{2a})(-1+(-1)^{b})^{2} & 0 & \frac{1}{16}(1+(-1)^{2a})(-1+(-1)^{b})^{2} \\
\frac{1}{16}(1+(-1)^{2a})(1+(-1)^{b})^{2} & 0 & \frac{1}{16}(1+(-1)^{2a})(1+(-1)^{b})^{2} & 0 \\
0 & \frac{1}{16}(1+(-1)^{2a})(-1+(-1)^{b})^{2} & 0 & \frac{1}{16}(1+(-1)^{2a})(-1+(-1)^{b})^{2}
\end{pmatrix}
$$

(5)

4 CONCLUSIONS

Hamiltonian (21) is sufficiently generic to represent most models for candidates of physical realizations of quantum computer hardware. The spin-spin term in Eq. (3) is sufficiently general to describe the most common types of interactions such as Ising, anisotropic Heisenberg, and dipolar coupling between the spins. Furthermore, if we also use spin-1/2 degrees of freedom to represent the environment then, on this level of description, the interaction between the quantum computer and its environment is included in model. In other words, the Hamiltonian (3) is sufficiently generic to cover most cases of current interest.

5 REFERENCES

[1] D. Deutsch. "Quantum Theory, the Church-Turing Principle and the Universal Quantum Computer" *Proceedings Royal Society London A* **400**, pp 97-117. 1985.

[2] R. Feynman. "Simulating Physics with Computers" *International Journal of Theoretical Physics* **21**, p 467. 1982.

[3] P.A.M. Dirac. *The Principles of Quantum Mechanics*. Oxford, Clarendon Press, 4th Ed. 1958.

[4] T. Toffoli. "Bicontinuous Extensions of Invertible Combinatorial Functions" *Mathematical Systems Theory*. **14**, p 13-23. 1981.

[5] D. Divincenzo. "Two-bit Gates are Universal for Quantum Computation" *Physical Review A*, **51**, Number 2. February 1995.

[6] A. Ekert. "Quantum Computation" *Proceedings of the ICAP meeting, Boulder, CO*. 1994. 179

[7] D. Atkins et al. "Advances in Cryptology" *ASIACRYPT'94*. Springer-Verlag. 1994.

[8] A. Barenco, C. Bennett, R. Cleve et al. "Elementary Gates for Quantum Computation" *Physical Review A*. March 1995.

[9] A. Barenco, A. Ekert, K. Suominen, and P. Torma. "Approximate Quantum Fourier Transform and Decoherence" *Physical Review A*. January, 1996.

Bismuth Triiodide Sheets Assisted Solution Synthesis of Cadmium Sulfide Binary Crystals and Branched Nanowires

Haiyan Li, Jun Jiao*

Department of Physics, Portland State University
P.O. Box 751, Portland, Oregon 97207
*Corresponding author. Tel. 503-725-4228, Fax. 503-725-2815. E-mail: jiaoj@pdx.edu

ABSTRACT

We report here on a solution-phase procedure for the synthesis of CdS binary crystals and branched nanowires (NW). These branches are perpendicularly aligned on both sides of the trunk along a 2D plane. This is attributed to the adoption of the BiI_3 sheet as a template, which was used to confine the growth of CdS NWs. The perpendicular alignment is dominated by both the electrostatic force of the Bi-I layer and the lattice match between the (001) plane and the (110) plane of wurtzite CdS. This 2D feature growth of branched NWs shows great potential in assembling NWs.

Keywords: Cadmium Sulfide, binary crystals nanowire, branched nanowire, Bismuth Triiodide sheets, synthesis

1 INTRODUCTION

Assembling nanowires (NW) and tailoring nanostructures into branched NWs are two approaches to increase the structural complexity of NWs and enable greater functionality [1-3]. As one of the most important II-VI group semiconductors, CdS NWs have been widely synthesized and demonstrate desirable potential in optoelectronics [4-8]. It is crucial to fabricate CdS branched NWs to further tune their properties and exploit their applications [9]. CdS multi-pod nanostructures have been synthesized through a thermal evaporation process [10,11]; and CdS tetra-pod nanostructures with rigidly geometrical shapes were obtained by a solution method [12-14]. Note that using those methods all the branches of each multi-pod nanostructure only grew on the same crystal nucleus. The alignment of the branches along the trunk is imperative to the effective assembly of NW devices. Jung, et al. reported the synthesis of CdS branched NW heterostructures via sequential seeding of gold nanocluster catalysts in a metal-organic chemical vapor deposition process [15]. Yao et al. synthesized the pine-like branched NWs via a solvothermal approach [16]. However, the random distribution and poor alignment of the branches on the trunk limit their further application in nanodevice fabrication and integration.

In this report, we demonstrate a low-temperature solution-phase approach for synthesis of CdS binary crystal and branched NWs. These branches are parallel to each other and perpendicular to the NW trunk. The planar confinement of growth was realized by introducing bismuth triiodide (BiI_3) sheets as 2D templates. BiI_3 was selected due to its layered structure as well as its orderly aligned polar Bi-I bonds exposed on the surface [17]. To lower the reaction temperature, we adopted a solution reaction system which employs an organic solvent with a boiling point higher than 300°C. Cadmium acetate dihydrate [Cd$(CH_3COO)_2 \cdot 2H_2O$] and S powder were used as the Cd source and S source correspondingly.

2 EXPERIMENTAL SECTION

BiI_3 (99.999%, Alfa Aesar)/ethanol solution was coated on the Si substrate and dried naturally. The mixture of Cadmium acetate dihydrate [Cd$(CH_3COO)_2 \cdot 2H_2O$, 0.2 g, 99.999%, Alfa Aesar] and Dotriacontane ($C_{32}H_{66}$, 5 g, 97%, Alfa Aesar) was treated at 160 °C for 2 h before the substrate covered with BiI_3 was introduced. The reaction lasted for 24 h at 185 °C after adding sulfur powder (S, 99.9%, 0.15 g, Alfa Aesar); the substrate was pulled out and coated with BiI_3 sheets. Then the substrate was put into the mixture again. The products were pulled out in 24 h and washed in isopropanol at 160 °C for 30 min, and dried at 200 °C for 0.5 h.

The morphology and internal structure of the CdS binary crystals and branched NWs were analyzed using an FEI Sirion XL30 field emission scanning electron microscope (SEM) equipped with an energy-dispersive X-ray (EDX) spectrometer, and a FEI Tecnai F-20 transmission electron microscope (TEM).

3 RESULTS AND DISCUSSION

SEM analyses indicate that only a few CdS branched NWs were formed without introducing BiI_3 sheets during the synthesis. The number of branched NWs increased after introducing BiI_3 sheets which suggests that the formation of branched NWs is related to the BiI_3 sheets (Figure 1a). Each of the branched NWs has thin branches on both sides of a thick NW trunk. All the branches on the same NW trunk aligned themselves along a 2D plane (Figure 1b). Interestingly, there is a 90° angle between the trunk and the branches (Figure 1c). The energy-dispersive X-ray (EDX) spectrum on the as-synthesized sample reveals that these branched NWs consist of Cd and S with an atomic percent proportion of 1:0.98, respectively (Figure 1d). Note that Bi and I elements were not detected. The morphology of CdS branched NWs is distinct from the CdS multi-pod nanostructure with randomly aligned branches, as well as the CdS tetra-pod nanostructure with an angle of 109.5° between two fingers [10-14]. This indicates that the formation of 2D branched NWs is not only different from the radial growth achieved by a thermal evaporation process but also from the zinc blende/wurtzite phases twinning occurred in the solution-based synthesis of CdS NWs. Due to the low reaction temperature (185 °C), our method for growing branched NWs differs from the seeded growth which is based on the SLS mechanism [18, 19]. The former is affected by the BiI$_3$ sheets.

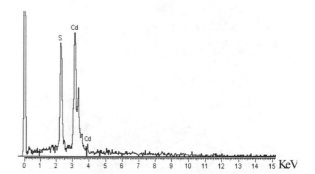

Figure 1: (a)-(c) are three representative SEM images of the branched NWs synthesized with the reaction time of 24 h after introducing the BiI$_3$ sheets. (d)) is an EDX spectrum of branched CdS NWs.

To understand the growth mechanism of branched nanostructures, we carried out a synthesis experiment with a reaction time of 12 h after introducing the BiI$_3$ sheets. Tiny residual BiI$_3$ sheets were found on the as-synthesized NWs (Figure 2a). As the reaction time was increased up to 24 h, the NW branches formed in the locations of the tiny BiI$_3$ sheets. These NW branches aligned perpendicularly on the side of a straight NW trunk are shown in Figure 2b. Further TEM characterization demonstrated that some of the NW trunks and individual NWs have a binary crystal structure as shown in Figure 2c and 2d. A selected area electron diffraction (SAED) pattern taken from the trunk confirms that [0001] is the preferred growth direction for the wurtzite CdS NW (inset of Figure 2c) [20]. The twin diffraction points marked by the circle reveal that the SAED pattern was formed by superimposing two sets of diffraction points. This further confirms that the trunk was composed of two parallel aligned NWs with the same growth orientation. Also, in Figure 2e, some NWs demonstrated a parallel growth phenomenon which may ultimately lead to the formation of binary crystalline NWs or the binary crystal trunk.

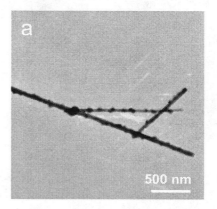

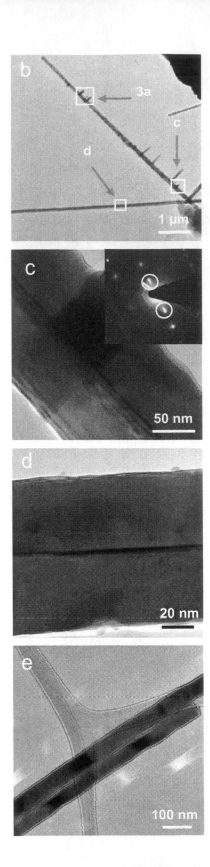

Figure 2: (a) TEM image of the CdS NWs synthesized with the reaction time of 12 h after introducing the BiI₃ sheets. (b) TEM overview of the 2D branched CdS NWs synthesized with the reaction time of 24 h after introducing

the BiI₃ sheets. (c), and (d) are high resolution TEM images of the areas marked on image (b). Inset of (c) is an electron diffraction pattern recorded from the trunk as shown in (c). (e) is a TEM image of two parallel aligned NWs grown for 24 h after introducing the BiI₃ sheets.

We found that the part near the tip of the trunk does not have the binary crystalline nanostructures (Figure 3a). This suggests that the binary crystal nanostructure of the trunk was converted into a single crystalline structure during its elongation and the branches were formed on the trunk with or without the binary crystals nanostructure. Interestingly, the branch that formed on the single crystalline trunk has a polyhedral shape at both tips. A high resolution transmission TEM (HRTEM) image reveals that the CdS branch is single crystalline. The spacing of the lattice fringes along the growth orientation was 0.67 nm (Figure 3b). This confirms that the growth is along the [0001] direction of wurtzite CdS [14]. We found no Bi attached to the body or the tip of any NW. This suggests that the formation of the trunk or the branches is attributed to the epitaxial growth of CdS crystal.

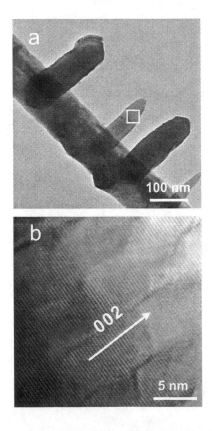

Figure 3: (a) is high resolution TEM images of the areas marked on Figure 2b. (b) is a HRTEM image taken from the area marked by a rectangle on (a).

It is obvious that in our reaction system, BiI₃ sheets provided the induced force between the 2D plane (the surface of BiI₃ sheet) and the reactants to assist the formation of a binary crystals nanostructure and branched NWs. Based on the structural phenomena that the size of the branches is much smaller than that of the trunk, we propose that the formation of branched NWs is attributed to a two-step growth. First, as shown in Figure 4, the CdS NWs result from anisotropic growth along the c-axis of wurtzite structures. Second, when BiI_3 sheets were added, some CdS NWs were covered by the BiI_3 sheet. As Cd^{2+} cations were absorbed on both sides of the BiI_3 sheet, CdS branches prefer to be formed on the sides of the trunk after alternative deposition of S and Cd^{2+} cations. During the growth of CdS branches along the [0001] direction, there is a lattice mismatch between the (001) plane of the branches and the (110) plane of the trunk. However the effect of this mismatch is indistinctive. This is largely attributed to the pre-alignment of Cd^{2+} cations and S by the electrostatic force of the Bi-I layer.

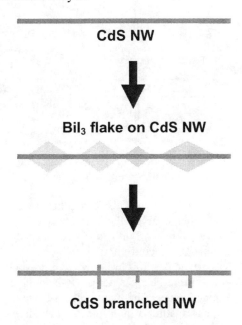

CdS NW

BiI₃ flake on CdS NW

CdS branched NW

Figure 4: Schematic diagrams of the growth of branched CdS NW assisted by the BiI_3 sheets on the CdS NW.

4 CONCLUSION

In this study, 2D confined growth was demonstrated by the synthesis of CdS NWs with binary crystals and branched features. This growth is attributed to the coordination between the CdS crystal anisotropic growth and the electrostatic force of the Bi-I layer. Further study is under way to synthesize orderly aligned complex nanostructures with various chemical compositions using the 2D template. This may ultimately lead to the fabrication and assembly of nanodevices by the method introduced here.

Acknowledgment: This work is supported by the National Science Foundation under the grants of ECCS-0217061, ECCS-0348277, and ECCS-0520891.

REFERENCES

[1] Lieber, C. M.; Wang, Z. L. *MRS Bull.* **32**, 99, 2007.
[2] Wang, D.; Qian, F.; Yang, C.; Zhong, Z.; Lieber, C. M. *Nano Lett.* **4**, 871, 2004.
[3] Yan, H.; He, R.; Johnson, J.; Law, M.; Saykally, R. J.; Yang, P. *J. Am. Chem. Soc.* **125**, 4728, 2003.
[4] Duan, X. F.; Huang, Y.; Agarwal, R.; Lieber, C. M. *Nature.* **421**, 241, 2003.
[5] Duan, X.; Niu, C.; Sahi, V.; Chen, J.; Parce, J. W.; Empedocles, S.; Goldman, J. L. *Nature* **425**, 274, 2003.
[6] Ma, R. M.; Dail, L, Qin, G. G. *Nano Lett.* **7**, 868, 2007.
[7] Jie, J. S.; Zhang, W. J.; Jiang, Y.; Meng, X. M.; Li, Y. Q.; Li, S. T. *Nano Lett.* **6**, 1887, 2006.
[8] Huang, Y.; Duan, X.; Lieber, C. M. *Small.* **1**, 142, 2005.
[9] Barrelet, C. J.; Greytak, A. B.; Lieber, C. M. *Nano Lett.* **4**, 1981, 2004.
[10] Dong, L.F.; Gushtyuk, T.; Jiao, J. *J. Phys. Chem. B* **108**, 1617, 2004.
[11] Shen, G.; Lee, C. *Crystal Growth&Design.* **5**, 1085 (2005).
[12] Jun, Y.; Lee J.; Choi J.; Cheon, J. *J. Phys. Chem. B* **109**, 14795, 2005.
[13] Carbone, L.; Kudera, S.; Carlino, E.; Parak, W. J.; Giannini, C.; Cingolani, R.; Manna, L. *J. Am. Chem. Soc.* **128**, 748, 2006.
[14] Chu, H.; Li, X.; Chen, G.; Zhou, W.; Zhang, Y.; Jin, Z.; Xu, J.; Li, Y. *Crystal Growth&Design.* **5**, 1801, 2005.
[15] Jung,Y.;Ko, D.; Agarwal, R. *Nano Lett.* **7**, 264, 2007.
[16] Yao, W.; Yu, S.; Liu, S.; Chen, J.; Liu, X.; Li, F. *J. Phys. Chem. B* **110**, 11704, 2006.
[17] Nason, D.; Keller, L. *J. Cryst. Growth.* **156**, 221, 1995.
[18] Fanfair, D. D.; Korgel, B. A. *Chem. Mater.* **19**, 4943, 2007.
[19] Dong, A.; Tang, R.; Buhro, W. E. *J. Am. Chem. Soc.* **129**, 12254, 2007.

3D Nanostructured Silicon Relying on Hard Mask Engineering for High Temperature Annealing (HME-HTA) Processes for Electronic Devices

M. Bopp[*/**], P. Coronel[*], F. Judong[*], K. Jouannic[*], A. Talbot[*], D. Ristoiu[*], C. Pribat[*], N. Bardos[*], F. Pico[*], M.P. Samson[*], P. Dainesi[**], A.M. Ionescu[**] and T. Skotnicki[*]

[*]ST Microelectronics, 850 rue J.Monnet, BP.16, 38926 Crolles, France
[**]LEG2, Ecole Polytechnique Fédérale de Lausanne, 1015 Lausanne, Suisse

ABSTRACT

A 3D nanostructuration of silicon through hard mask engineering and high temperature annealing (HME-HTA) in hydrogen ambiance is reported The use of a nitride/oxide double hard mask stack on silicon during the etching of bulk structures allows for leaving a patterned nitride thin film on the structures surface during the high temperature annealing, after having removed the top oxide layer. This solution will be referred as the *nitride-capped approach*, which is an alternative to the use of a single sacrificial oxide hard mask for a free Si surface annealing (referred as the *mask-less approach*). The nitride-capped approach opens new technological and design possibilities when using 2D arrays of various geometry trenches. Implications and potential device applications are discussed, such as the role played by the silicon-nitride interface during the annealing process, the role of the remaining nitride layer, and the possibility to explore this 3D technique to solve the planar independent double gate transistor challenge.

Keywords: Silicon reflow, hydrogen annealing, 3D nanostructures, hard mask, autoaligned devices

1 SILICON REFLOW

Annealing silicon at high temperatures in a hydrogen ambiance has been reported to induce surface diffusion of silicon [1]. The result of such process at low pressure and high temperature on an etched cylindrical trench is the formation of a buried cavity within the bulk substrate (Figure 1a). By arranging these trenches in patterns like in a row or in a 2D matrix, these cavities can connect to each others if the initial trenches are close enough and will form a buried pipe or a buried planar cavity (Figure 1b). respectively

The use of a nitride hard mask for such a HTA process has been mentioned in [2] to remain stable during the HTA step. Moreover, it minimizes the surface diffusion at its interface with silicon. Indeed, it has been observed that the mechanism of cavity formation is different from nitride-capped structures compared to the case of a mask-less HTA. This observation has been confirmed by the mean of simulations in the frame of surface diffusion studies [3].

The deformation of the trench starts from its bottom, the nitride layer inhibiting the contribution of the upper part of the trench during the closing process, which normally plays

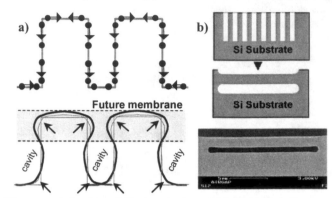

Figure 1: a) Surface diffusion during mask-less annealing: b) Example of a silicon membrane obtained with a H2 HTA

an important role in the cavity's formation mechanism (Fig 2). This is what enables the formation of multiple cavities for such trenches with a lower aspect ratio compared to the mask-less configuration.

Nevertheless, to the best of our knowledge, the behavior at the nitride-silicon interface has not really been taken into account yet. In cited papers, the main object of interest was the possibility for formation of multiple cavities allowed by the use of the nitride layer. In simulations, in order to give morphological predictions of the shape of an annealed structure with a nitride hard mask, the nitride-silicon interface is defined by a blocked surface condition.

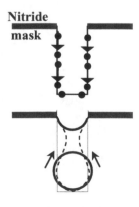

Figure 2: perfectly blocked nitride-silicon interface assumed during surface diffusion simulations.

The work presented here intends to investigate the effects of a nitride hard mask on the hydrogen HTA of periodical structures, what has to our knowledge not been performed yet. In particular, we will see the role of the nitride-silicon interface on arrays of trenches and show what happens at this interface strongly influences the morphological results after annealing of the structures, and opens interesting possibilities for electronic devices fabrication.

2 EXPERIMENTS

2.1 Test wafers preparation

For this study, we compare two approaches, with: (i) a reference wafer annealed in hydrogen ambiance with a hard mask free surface (mask-less approach) and (ii) wafers annealed with a patterned nitride layer on the silicon surface (nitride capped approach). We used bulk wafers with two different types of hard masks for the silicon etching step:

- 3000Å TEOS hard Mask (mask-less annealing)
- 1600 Å Nitride hard mask deposited on Silicon 3000 Å TEOS additional layer. (capped annealing)

Wafers for mask-less and nitride-capped approach have been through the same optical lithography and silicon etching steps. Before the annealing, the TEOS layer has been removed on both wafers. After an annealing at 1100°C during 10min with a pressure of 10Torr, SEM observations have been performed.

Observed patterns consist of 225µm² areas of circular trenches arranged periodically in matrices. Diameter (CD) of the circular trenches varies from 600nm to 1µm. As the largest trenches would take more time to eventually form a cavity compared to the smallest structures, we are able to make qualitative observations on the silicon diffusion dynamics by observing the different matrices.

2.2 Observations

Observations on the reference wafer obtained with mask-less approach show that after annealing, adapted designs formed membranes as expected (Figure 3). Typical membrane and buried cavity thicknesses obtained are 1µm and 600nm, respectively. It should be noted that these dimensions are scalable by modifying the design of the matrices and by further processing of the membrane, like an oxidation/oxide etching step.

By comparing some annealed mask-less structures and nitride-capped structures, we observed that the silicon structure resulting of the annealing was almost equivalent: In Figure 4, we see single cavities about to contact each others to form the plate cavity. Moreover, the surface of the membrane can be considered planar in both structures, what was not supposed to be the case with the nitride-capped

structures (Fig 4b). Notice that the dashed line traced in Fig 4a corresponds to the substrate surface (line out of observation plane), whereas the arrow markers are placed along the line corresponding to the remaining nitride surface (line in observation plane)

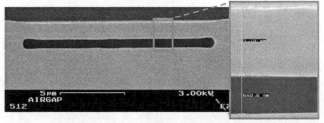

Figure 3: Buried cavity obtained after annealing of a trench matrix on the reference wafer (thicknesses: 1.10µm/669nm)

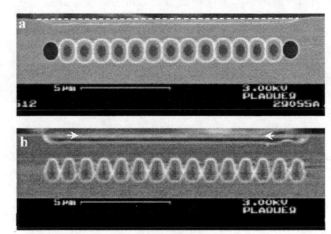

Figure 4: Same trench matrix processed with a) mask-less annealing b) nitride-capped annealing

We observed indeed on matrices with low Aspect Ratio (AR) and low density (AR=4.5, CD/Trench Spacing = 1), (Figure 5a), a profile close to the ones predicted by simulations in Figure 2, however, for higher densities and higher AR a different situation can occur.

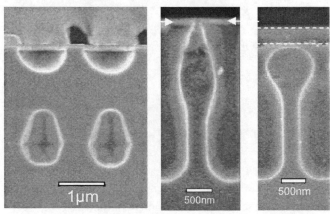

Figure 5: a) semi isolated trenches with hard mask after annealing b) annealed capped silicon trench wall in denser trenches array c) annealed mask-less silicon trench wall in denser trenches array, with 'match shape'

We can see on Figure 5 b) that the trench wall between two trenches is about to detach from the nitride layer, indicated by the arrow markers. Indeed, at these temperatures, even if silicon tends to stay attached to the nitride layer, the surface diffusion of silicon occurring at the edges of the trench wall reduces more and more the surface of silicon in contact with the nitride layer. The contact area with the hard mask layer shrinks until the silicon detaches at some point. The trench wall is then supposed to be free to behave like the one in mask-less structures shown in Figure 5.c) In that mechanism, silicon in the area of the trenches matrix (lower dashed line) levels down compared to the initial substrate level (upper dashed line) and the walls between the trenches are shaping their profile in a 'match shape'

The non-spherical shape of the single cavities in Fig 5 b is explained by the interaction between the hard mask and silicon. Their shape is likely to evolve and to change to spheres like in the mask-less case, but after some additional time.

A suspended membrane can be obtained with both approaches, the interest in using a nitride mask is motivated by new geometrical features brought compared to the mask-less annealing process. Indeed, thanks to its stability and its structure, maintaining the nitride hard mask allows obtaining virtually two cavities (Fig. 6): one bottom closed cavity and one upper cavity delimited by the mask with built-in access (hard mask apertures). We have therefore the possibility to work on the membrane without etching new access.

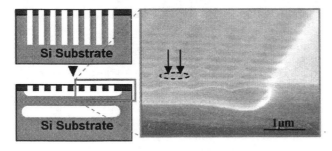

Figure 6: Annealing of nitride-capped structures allows obtaining two cavities -top and bottom-, considering the one delimited by the Si membrane surface and nitride layer.

3 PERSPECTIVES

3.1 3D structuration by HME-HTA as a novel engineering tool on bulk silicon

Based on the reported results, one can observe that the nitride capped approach offers a new tool for nanostructuration of bulk silicon and can be part of an interesting 3D Front End development. Combining design of cavities and etching, one can obtain a wide range of cavities heights and membranes thicknesses (still to be explored experimentally). Moreover, adjustments by processing are still possible, like the thinning of the silicon

membrane by oxidation / oxide etching step mentioned in the mask-less approach and still available in this new configuration.

The interest of the remaining nitride layer is multiple: not only does it delimit the upper cavity, but it contains build-in access apertures for filling the upper cavity with a conformal material, and can as well be considered as a protective layer during the CMP process to planarize the deposited material. Nitride hard mask can as well simply be reused after the annealing step for its original function, ie a hard mask. Many trenches apertures geometries can lead to the same result we obtained during the silicon reflow step (like lines, but with a very high AR required, see [3]). A smart design of the mask apertures can be imagined so that the membrane and cavities formation is not disturbed during annealing, and yet the mask patterns can be used for a photolithographic step later in the process.

3.2 Application to the planar IDG transistor

The main challenges in realizing a planar independent double gate transistor are the alignment between the back gate and the front gate, and allowing contacting the two gates. Concerning the gate alignment issue, process with wafer bonding has shown good results with 10nm gate-length planar IDG transistors [5]. Some complex processes like in [6] relying as well on SOI substrates have been experimented

The HME-HTA nanostructuration allows obtaining two separated, superposed, autoaligned cavities. Each one of them can be processed separately to create a gate, being filled with different gate materials if necessary, as shown on Fig 7a). With such a technique, one can avoid wafer bonding. Of course, there is a main problem remaining: the bottom gate may be not isolated easily if working on a bulk substrate. However, simulations of mask-less structures annealing on SOI have been performed in [3] and the results show that one of the possible morphological configurations after annealing is to have the bottom cavity in contact with the oxide, like on Fig 7b) what would isolate the bottom gate.

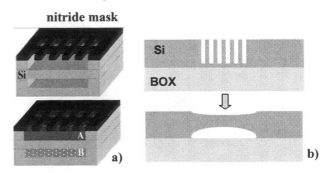

Figure 7: a) Filling of independent autoaligned cavities with different materials (A and B). b) One of the morphologies obtained for a mask-less approach simulated on SOI [3].

4 CONCLUSION

3D nanostructuration of silicon through hard mask engineering and high temperature annealing (HME-HTA) in hydrogen ambiance with a nitride-capped approach was reported and compared to a mask-less approach. Some of the main technological characteristics and the future potential of this 3D technique have been discussed.

REFERENCES

[1] T. Sato et al, IEDM Tech.Dig., p29, 1999.

[2] T. Sato et al. Jpn J. Appl.Phys. Vol 39 (2000) pp. 5033-5038

[3] E. Dornel, *Evolution morphologique par diffusion de surface et application à l'étude du démouillage de filmes minces solides*, PhD thesis, Université Joseph Fourier, Grenoble, Nov 2007.

[4] M. Vinet et al, *Bonded Planar Double-Metal-Gate NMOS Transistors Down to 10 nm*, IEEE Electron Device Letters, Vol. 26, No. 5, May 2005, pp. 317-319.

[5] K. W. Guarini et al, *Triple-Self-Aligned, Planar Double-Gate MOSFETs: Devices and Circuits*, IEDM 2001, pp 425-427.

Introduction of Mesopority in Zeolites Using Nanoparticles as Hard Templates

A.I. Carrillo, N. Linares, J. García Martínez[*]

Department of Inorganic Chemistry. University of Alicante, Carretera San Vicente s/n,
E-03690, Alicante, Spain.
e-mail: j.garcia@ua.es URL: www.ua.es/grupo/nanolab

ABSTRACT

Silicalite-1 zeolite with intracrystalline mesoporosity has been synthesized by using nanosized MgO as a hard template. The inorganic nanoparticles have been embedded inside the silicalite-1 crystals during the crystallization step and then easily removed by simple acid wash. All the mesoporous materials have been obtained by hydrothermal treatment using tetrapropylammonium as a structure directing agent. All samples were characterized by X-ray diffraction (XRD), scanning electron microscopy (SEM), transmission electron microscopy (TEM), and N_2 physisorption at 77K. The results suggest that MgO nanoparticles are suitable hard templates for the synthesis of hierarchical porous materials.

Keywords: Silicalite-1; Mesoporous; Hard template, Nanoparticles.

1 INTRODUCTION

The development of ordered, thermal and hydrothermal stable mesoporous materials with zeolite-like crystalline frameworks is highly desirable to reduce the diffusion limitations of these microporous materials.[1] Zeolites are widely used in catalysis, purification, adsorption, separation and other industrial applications because their stability, high internal surface, and controllable chemistry.[2] However, their small channels and cavities (typically below 1 nm) significantly reduce some of their applications. Different strategies have been tried to overcome this limitation. Among them, the direct incorporation of mesoporosity (intracrystalline mesoporosity) is especially desirable. Mesopores have been introduced into zeolites mostly by dealumination,[3] recrystallitation,[4] or by using various templates such as surfactants,[5] cationic polymers,[6] carbon black,[7] and nanoparticles.[8]

The use of templates is especially convenient because it allows for an easy pore size control. The templates used so far include tree types, namely soft, hard, and hierarchical templates. The soft templates, which can be subsequently removed by heat treatment, are usually organic-based molecules, such as polymers and surfactant micelles. The hard templates, which can be leached away by using alkali or acid washes or calcination, actually refer to nanosized solids, for example carbon black, or polymer microspheres.

In some cases, nanocasting was used to produce inverse replicas of a porous structure. The pores are filled with a secondary material, followed by a post-treatment, and the removal of the template. The hard template method has been recently used to fabricate various novel porous structures, which are difficult to synthesize using the soft template method. For example, zeolites, ordered mesoporous silicas (*e.g.*, MCM-48, SBA-15, and SBA-16), and inverse opals have been used to fabricate microporous, mesoporous and macroporous carbons, respectively. Finally, hierarchical templates, which combines different templates with controlled structures at different length scales, have been used to prepare bi-modal and tri-modal porous structures with the primary objective of minimizing diffusion resistance.[9]

A wide variety of nanoparticles have been used as hard templates to produce mesoporous zeolites. H. Kato et al. describe the synthesis of mesoporous zeolites by using platinum nanoparticles as mesopore-forming templates.[8] A.H. Janssen et al. present an exploratory study on the generation of mesopores by templating with carbon during zeolite synthesis.[7] H. Zhu et al. have recently reported the preparation of silicalite-1 single crystal with intracrystal pores in the range of 50-100 nm by using the nanosized $CaCO_3$ as a hard template. In this case, the $CaCO_3$ nanoparticles embedded in the silicalite-1 crystals are removed by acid wash, which give rise to intracrystal mesopores within the zeolite crystal.[2]

Herein, we describe the use of MgO nanoparticles as hard templates to obtain mesoporous zeolites [2,8]. These nanoparticles can be conveniently removed by chemical treatment (usually acid wash) without damaging the zeolite. The final mesoporous materials prepared using this strategy shows a bimodal pore size distribution, i.e. the pores due to their crystalline framework (micropores) and the mesopores resulting from the removal of the nanoparticles

2 EXPERIMENTAL SECTION

2.1 Materials.

MgO nanoparticles used as a hard template were obtained from Sigma-Aldrich. According to the vendor, the particle size distribution of MgO is in the range of 25-50 nm.

NSTI-Nanotech 2008, www.nsti.org, ISBN 978-1-4200-8503-7 Vol. 1

3-mercaptopropyltriethoxysilane (MPTES, 80%, Fluka), NaOH (99%), tetrapropylammonium bromide (98%, Aldrich, denoted as TPABr), tetraethoxysilane (98%, Aldrich, denoted as TEOS) were commercial available and used in the synthesis without further purification.

Finally, sulfuric acid (96%) from Panreac was used to remove the MgO nanoparticles after convenient dilution.

2.2 Synthesis of Silicalite-1 with Intracrystal Mesopores.

In a typical synthesis of mesoporous silicalite-1, 0.10 g of MgO nanoparticles was added to a clear solution of tetrapropylammonium bromide, water, NaOH, TEOS as silica precursor and MPTES to functionalize MgO nanoparticles. The resulting solution underwent ultrasonic agitation for 15 min in order to produce a homogeneous dispersion of MgO nanoparticles in the silica gel. The final mixture has a molar composition of about $0.05TPABr:0.15NaOH: 1TEOS:43.2H_2O$, the ratio molar of MPTES:MgO nanoparticles is 1:1. The resulting gel was transferred into a Teflon-lined stainless steel autoclave, where it was heated under autogeneous pressure for 72 h in an oven at 150 ºC. After cooling to room temperature, the products were recovered by filtration, washed with water and dried at room temperature for 24 h. The organic structure-directing agent trapped in the pores was thoroughly removed by calcination at 550 ºC for 8 h. Finally, the inorganic MgO nanoparticles were removed by chemical treatment with sulfuric acid using a 0.01 M H_2SO_4 solution during 10 min or citric acid to pH =3 during 30 min. The resulting zeolite was water washed, filtered out, and dried at room temperature for 24 h.

For comparison purposes, silicalite-1 without MgO nanoparticles was synthesized using the same experimental conditions and procedure.

2.3 Characterization Methods.

X-ray powder diffraction patterns (XRD) were obtained with a Bruker D8-Advance diffractometer using CuKα radiation; the scanning velocity was 0.1º/min over the range of 1.7-50º (2θ). Scanning electron microscopy (SEM) features of all the samples were observed by SEM (JEOL JSM-840). Transmission electron microscopy (TEM) experiments were performed on a JEM-2010 electron microscope (JEOL, Japan) having 0.14 nm instrumental resolution equipment operated at 200 kV. Samples for TEM studies were prepared by dipping a carbon-coated copper grid into a suspension of samples in ethanol that was presonicated. N_2 adsorption/desorption isotherms were measured at 77 K on a Quantacrome Autosorb-6 volumetric adsorption analyzer. Before the adsorption measurements, all samples were outgassed at 523 K for 4 h. The pore size distribution was calculated from adsorption branches of isotherms using the BJH method, with the software attached to the equipment.

3 RESULTS AND DISCUSSION

3.1 X-ray diffraction (XRD).

Figure 1 shows the XRD patterns of a) silicalite, b) silicalite with 10 % of MgO nanoparticles functionalized with MPTES, c) silicalite with 25 % of hard template and functionalizate with MPTES, and d) silicalite with 50 % of MgO nanoparticles and functionalizate with MPTES. Sample b show high crystallinity and all the peaks are located at the same angles, which correspond to the pure silica MFI structure. No peaks due to the MgO nanoparticles were observed, as expected because their small size. Therefore, the incorporation of MgO nanoparticles does not hinder the crystallization of silicalite.

Figures 1c and 1d show than silicalites with high quantity of hard template lose the structure crystalline of zeolite. TEM image of silicalite-1b, with 25 % of MgO nanoparticles, show an opening structure of the crystal that corresponds with the adsorption branch in nitrogen physisorption when we can see that these samples have more mesoporosity than silicalite-1a.

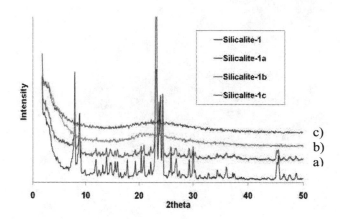

Figure 1: XRD patterns of silicalites with a different quantity of the hard template MgO: a) silicalite without MgO b) silicalite with 10 % of nanoparticles, c) silicalite with 50 % of hard template and d) silicalite with 50 % of MgO nanoparticles.

3.2 Transmission electron microscopy (TEM)

The transmission electron microscopy is a powerful technique for direct observation of secondary pores in zeolites.[6]

Figure 2 shows a TEM image of mesoporous silicalite-1b. The zeolite has lost the crystallinity and we can see an opening structure.

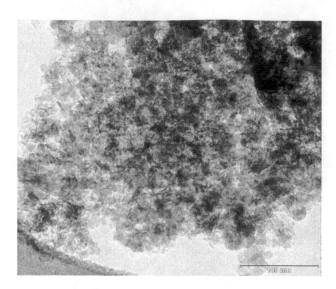

Figure 2: TEM image of silicalite-1b after it has been washed by sulfuric acid.

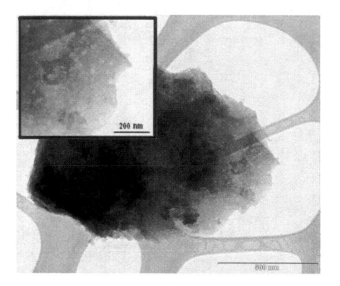

Figure 3: Low magnification TEM micrograph of mesoporous silicalite-1a. In this case we can see the crystallinity of silicalite-1.

The crystals are not well facetted, as some irreguraly shaped intracrystralline mesoporosity is observed throughout the sample. The TEM micrographs show that the intracrystalline pores are not ordered as the regular mesopores observed in the more conventional amorphous mesoporous materials, like MCM-41 silica.

3.3 Scanning electron microscopy (SEM)

The SEM images in Figures 4-5 give details on silicalite-1 morphology. The low magnification image shows that particles are uniform.

Figure 4: SEM image of the silicalite-1a, with 10 % of MgO nanoparticles.

Figure 5: SEM image of mesoporous silicalite-1a.

The results of scanning electron microscopy for the hard-templated-directed silicalite-1 show that samples are highly crystalline. Estimated from the SEM overview, the average crystal size of silicalite-1a/MgO composite and the mesoporous silicalite-1 are in the range of 10 microns. So the morphology and size of crystals has been maintained after MgO removal.

3.4 Nitrogen physisorption

Figure 6 shows the isotherms of silicalite with 10 wt% of MgO washed with citric acid during 30 min (circles and dotted line), silicalite with 10 wt% MgO washed with sulfuric acid 2 M during 10 min (crosses and solid line), silicalite with 25 wt% MgO washed with sulfuric acid 2 M for 10 min (squares), and silicalite with 50 wt% MgO washed with sulfuric acid 2 M for 10 min (triangles). The materials with higher amount of mesoporosity are silicalite 1b and 1c, i.e. those that were prepared using 25 wt% and 50% wt MgO. These materials, however do not show any crystallinity, (see Figure 1), which suggest that there is a limit in the amount of template that can be used.

Figure 7 shows the pore size distribution of mesoporous silicalite-1a-1 washed with citric acid 30 min and silicalite-1a-2 washed with sulfuric acid 10 min. In both cases, there is a peak around of 20 nm that it corresponds with the size of nanoparticles. This is an evidence of the template effect of the MgO.

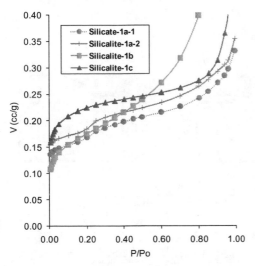

Figure 6: N_2 adsorption isotherm of mesoporous silicalite obtained by using different concentrations of hard template. Curve of silicalite-1a-1 is shifted +0.02 cc/g for clarity.

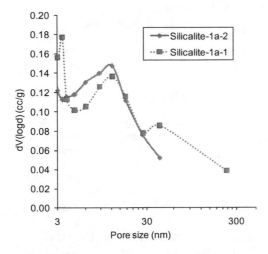

Figure 7: Mesopore size distributions of silicalite-1a-1, dot line, and silicalite-1a-2, solid line, calculated from the adsorption branch by the BJH method.

4 CONCLUSIONS

Silicalite-1 with intracrystalline mesopores (average pore size aprox. 20 nm) was prepared using different concentrations of MgO nanoparticles as hard template. The MgO nanoparticles were embedded in the silicalite-1 crystal during the crystallization process. Afterwards, they were removed by controlled acid wash without damaging the silicalite crystals besides producing intracrystalline mesoporosity. We have used MPTES to functionalize the MgO nanoparticles. The pore size distribution obtained by using BJH method exhibits a pore size distribution centered at20 nm that it corresponds with size of MgO nanoparticles. The combined use of X-ray diffraction, TEM and SEM images analysis show that the crystallinity is preserved even after the acid wash step when the concentration in MgO nanoparticles is small.

Mesoporous zeolites with both microporosity and intracrystalline mesoporosity, like the ones herein described, are desirable because their enhanced diffusion properties and therefore potential application in catalysis, separation, and adsorption of bulky molecules.[10,11]

ACKNOWLEDGEMENT

This research has been funded by the Ministerio de Educación y Ciencia (CTQ2005 – 09385 - C03 - 02). J.G.M. is grateful for financial support under the Ramón y Cajal program and N.L. for a FPI fellowship.

REFERENCES

[1] H. Li et al., Micropor. Mater., 106, 174-179, 2007.
[2] H. Zhu et al., Chem. Mater., 20, 1134-1139, 2008.
[3] A.H. Janssen, A.J. Koster, K.P. de Jong, Angew. Chem.Int. Ed., 40, 1102, 2001.
[4] I. Ivanova et al., Stud. Sruf. Sci. Catal., 158, 121, 2005.
[5] F.-S. Xiao, L. Wang, C. Yin, Y. Di, J. Li, R. Xu, D. Su, R. Schlogl, T. Yokoi, T. Tatsumi, Angew. Chem., Int. Ed. 45, 3090, 2006.
[6] M. Choi, H.S. Cho, R. Srivastava, C. Venkatesan, D.-H. Choi, R. Ryoo, Nature Mater., 5, 718, 2006.
[7] A.H. Janssen, I, Schmidt, C.J.H. Jacobsen, A.J. Koster, K.P. de Jong, Micropor. Mesopor. Mater., 65, 59-75, 2003.
[8] H. Kato, T. Minami, T. Kanazawa, Y. Sasaki, Angew. Chem., Int. Ed. 43, 1271-1274, 2004.
[9] X.S. Zhao, J. Mater. Chem., 16, 623-625, 2006.
[10] C. H. Christensen, I. Schmidt, A. Carlsson, K. Johannsen, K. Herbst, J. Am. Chem., 127, 8098, 2005.
[11] I. Schmidt, A. Krogh, K. Wienberg, A. Carlsson, M. Brorson, C. J. H. Jacobsen, Chem. Commun., 2157, 2157.

Proposal of Porous Chromium Film Fabrication Method for an IR Absorber

Mitsuteru Kimura and Masatoshi Hobara

Faculty of Eng., Tohoku-Gakuin University, Tagajo, Japan
Phone:+81-22-368-7162, Fax:+81-22-368-8535, e-mail: kimura@tjcc.tohoku-gakuin.ac.jp

ABSTRACT

Novel fabrication method of porous Chromium (Cr) film for an infrared (IR) absorber is proposed and total absorption factor above 96.2% at about 10 μm of IR wavelength, which is corresponding to the peak wavelength of our body temperature of 37 °C, is achieved. The porous Cr film is formed on the silicon substrate. Our infrared absorber is composed of porous Cr film remained after chemical etching removal of co-sputtered copper (Cu) with Cr, and the composition rate of Cu and Cr is varied so as to be the graded composition film: Cr:Cu (20:80)/ Cr:Cu (50:50) / Ti (as an underlying metal) on the thermally grown SiO_2 thin film formed on the Si substrate .

Keywords: IR absorber, porous Chromium, Silicon, graded material, needle-like crystals

1 INTRODUCTION

Many kind of IR absorbers have been used for thermal IR sensors, such as gold black, infrared absorbing dyes, and silicon nitride film, however, they have problems for micromachining fabrication process using the alkaline anisotropy etchant, such as hydrazine [1], [2]. The stable and high absorption film in infrared wavelength of 8-14 μm are strongly demanded, especially for the human body temperature detection.

2 CONCEPT OF THE IR ABSORBER AND FABRICATION METHOD OF THE POROUS CHROMIUM FILM

In this paper following essential matters for the infrared absorber（IR absorber）are considered. 1. The infrared absorber should have many free electrons in it, because their motions due to their acceleration by the induced electric field originated from the incident infrared light make energy loss by collisions at grain (particle) boundaries. This fact leads to that the infrared absorber should be composed of high conductivity materials such as metals and be less rusty metals (only surface oxidation is allowed). 2. Most of free electrons in the metal particle should lose the kinetic energy within one period of the infrared light. This fact leads to the suitable particle size of the infrared absorber. 3. Reflection from the infrared absorber should be less. This fact leads to that the surface region of the infrared absorber should be less density. 4. The effective optical thickness of the infrared absorber will be better to be about 1/4 of peak wavelength (10 μm) at 37 °C because of interference effects between the incident IR and the reflected one.

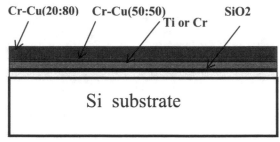

Fig.1 Multilayer of Cr-Cu with different composition deposited on the Si substrate.

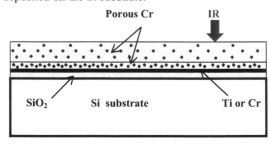

Fig.2 Porous Cr film after etching removal of Cu component shown in Fig.1 as an IR absorber.

Our infrared absorber is composed of porous Cr film remained after chemical etching removal of co-sputtered copper (Cu) with Cr, and the composition rate of Cu and Cr is varied (graded composition film: Cr:Cu (20:80)/ Cr:Cu (50:50) / Ti or Cr (as an underlying metal)) according to the essential matters described above. In Fig. 1 multilayer of Cr-Cu with different composition formed on the Si substrate is shown. In Fig. 2 a sketch of the porous Cr film after etching removal of Cu shown in Fig.1 as an IR absorber is shown.

3 EXPERIMENTAL RESULTS

3.1 Morphology Dependences of Crystallized Cr on the Underlying Metals and the Substrate Temperature Ts

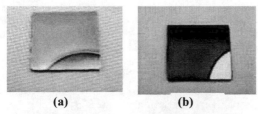

(a)　　　　**(b)**

Fig.3 Sputtering deposited Cr-Cu film on the Si substrate shown in Fig.1 (a), and porous Cr film shown in Fig.2 (b) for Ti as the underlying metal.

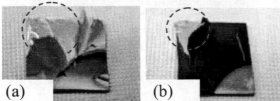

(a)　　　　**(b)**

Fig. 4 Sputtering deposited Cr-Cu film come off and clacked inside the Si substrate before Cu etching removal (a) and after that (b) for the Cr underlying metal at 300 °C.

Sputtering deposited Cr-Cu film (Cr:Cu (20:80)/ Cr:Cu (50:50) / Ti on the SiO_2 thermally formed on the Si substrate) at the substrate temperature Ts of 400 °C before Cu etching removal as shown in Fig.1 (a) and the porous Cr film after the chemical etching of Cu component using the nitric acid as shown in Fig.2 (b) are shown in Fig.3 (a) and (b), respectively. Conditions of the sputtering depositions as a series experiments are as follows: Ti: 3min.(200W); Cr-Cu (50:50):8min. (500W);　Cr-Cu (20:80): 1hr .(500W).

We can see that there are no clacks on the deposited Cr-

Fig. 5 SEM images of sputtering deposited Cr-Cu films after Cu etching removal for different Ts for the Ti underlying metal.

Fig. 6 SEM images of sputtering deposited Cr-Cu (20:80) films after Cu etching removal for different Ts for the Cr underlying metal.

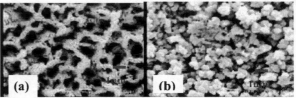

(a)　　　　**(b)**

Fig. 7 SEM images of Cr-Cu (20:80) (a), and Cr-Cu (50:50) (b) deposited at Ts=400 °C after Cu etching removal.

(a)　　　　**(b)**

Fig. 8 SEM images of Cr-Cu (20:80) deposited at Ts=400 °C before Cu etching out for different underlying metals of Cr (a) and Ti (b)

Cu film for this Ti underlying metal, however, most deposited Cr-Cu films formed above Ts>200°C for the underlying metal of Cr film are come off and cleaved inside the silicon substrate but not at the boundary of the deposited Cr-Cu film and the SiO_2 film as shown in Fig.4 (a) and (b). The thermally grown SiO_2 film layer of about 0.3 μm thick on the Si substrate is lost in this case of Cr underlying metal.

In Fig.5 SEM images of porous Cr film deposited at different substrate temperature Ts of 100 °C and 400 °C and deposited on the Ti film as the underlying metal, and in Fig.6 that on the Cr film are shown.

We have found that morphology of crystallized Cr in the co-sputtered Cu depends strongly on the underlying metal of Cr and Ti, and also on the substrate temperature Ts, and that as to the Cr film, mesh-like crystals composed of

Fig. 9 Cross sectional view of the porous Cr film deposited on the Ti film at Ts=400°C.

many needle-like crystals are grown.

In Fig.7 comparison of SEM images of Cr-Cu (20:80) (a), and Cr-Cu (50:50) (b) deposited at Ts=400 °C after Cu etching removal are shown. From these Fig.7 (a),(b) we can see that larger Cu content in the sputtering deposited films from different Cu : Cr ratio targets leads to the mesh structured film.

In Fig.8 SEM images of Cr-Cu (20:80) deposited at Ts=400°C before Cu etching removal for different underlying metals of Cr (a) and Ti (b) are shown. We can see that porous Cr-Cu (20:80) films composed of many layers are grown at this time before the Cu etching removal. We have confirmed that these porous Cr-Cu (20:80) film composed of many layers are already formed at the temperature below Ts=100 °C although their pore sizes become less than that formed at higher Ts. We can find that the porous Cr-Cu (20:80) composed of rich Cu film material as shown in Fig.8 has needle-like Cr crystals inside these porous but planar materials judging from the appearance of needle-like Cr crystals after Cu etching removal as shown in Fig. 5 (Ts=400 °C) and Fig. 7 (a).

In Fig.9 cross sectional view of the porous Cr film deposited on the Ti film as the underlying metal at Ts=400°C is shown. We can see that the porous Cr film is composed of the perpendicular nano size needle-like crystals.

3.2 IR Spectra for Transmissivity T and Reflectivity R of the Porous Cr

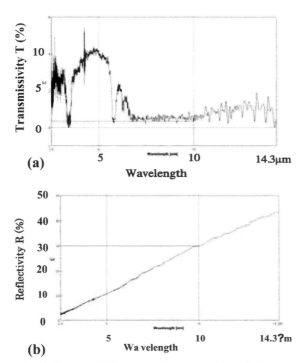

(a)

(b)

Fig.10 Spectra of the transmissivity T (a) and the reflectivity R (b) of the porous Cr film as an IR absorber at T_s=100 °C.

The spectra of the transmissivity T and the reflectivity R of the porous Cr film as an IR absorber are measured using the FT-IR equipment. In the reflectivity R measurements the sputtering deposited gold film on the SiO_2/mirror polished Si substrate is used as a reference. The incident angle of the focused IR light measured from the surface plane of the sample of the porous Cr film is about 45°. In the transmissivity T measurements the SiO_2/mirror polished Si substrate is used as a reference, however in this case the incident IR light is irradiated from the backside rough surface of the reference Si substrate in order to prevent the larger total reflection at the mirror polished Si surface.

In Fig.10 spectra of the transmissivity T (a) and the reflectivity R (b) of the porous Cr film of that in Fig.5 (Ts=100°C) as an IR absorber formed at T_s=100 °C and underlying metal of Ti as shown in Fig. 2 are shown. The transmissivity T and the reflectivity R in the sample formed at this Ts = 100 °C are 0.4% and 30% at wavelength λ =10 μm, respectively. The total absorption factor at about 10 μm of IR wavelength is about 69.6% because of having very large reflectivity R in this wavelength.

In Fig.11 the spectra of the transmissivity T (a) and the reflectivity R (b) of the porous Cr film as the IR absorber at T_s=400 °C and underlying metal of Ti are shown, which is the same sample as that shown in Fig.9.

In this case the deposition conditions are different from

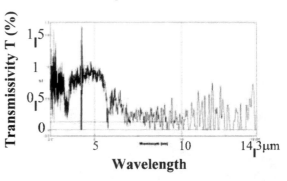

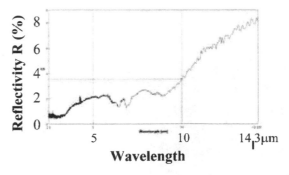

Fig.11 Spectra of the transmissivity T (a) and the reflectivity R (b) of the porous Cr film as an IR absorber at T_s=400 °C.

that of normal series experiments as shown in Fig 7-10 and next Fig.12 to get the good IR absorption. In this experiment in Fig.11 the Ti deposition time is enlarged to be 15 min. to reduce the transmissivity T, and the film deposition time of Cr-Cu (50:50) layer is also a little enlarged to be 10 min. to increase the IR absorption. The thickness of the Cu removed film of the deposited Cr-Cu (20:80) layer is about 5μm, judging from the SEM image shown in Fig. 9 and this will be the suitable thickness for the IR absorber in the 8-14 μm wavelength.

In this porous Cr film formed at Ts=400°C, the transmissivity T and the reflectivity R have been only 0.2 % and 3.6 % at the wavelength of λ =10 μm, respectively. Therefore very large total absorption factor of about 96.2%, at near 10 μm of IR wavelength is achieved.

In Fig.12 substrate temperature Ts dependences of the reflectivity R of the porous Cr film for various IR wavelengths are shown. In these sample-preparation conditions such as the sputtering deposition time and the sputtering RF power for various sputtering materials are unified to be Ti:3 min. (200W); Cr-Cu (50:50):8 min. (500W); Cr-Cu (20:80):1hr. (500W) as mentioned above.

We can see from these results that as increasing the substrate temperature Ts, the reflectivity R is deceased for every IR wavelength, which directly reflects and leads to the larger total IR absorption factor.

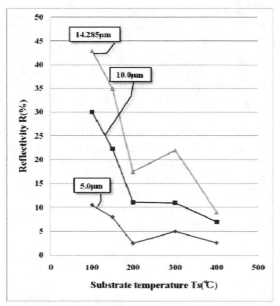

Fig.12 Substrate temperature Ts dependence of the reflectivity R of the porous Cr film (Parameter: IR wavelength).

4 DISCUSSIONS

In Fig.13 an example of enlarged SEM images shown in Fig.7(a) of the mesh structure composed of many needle-

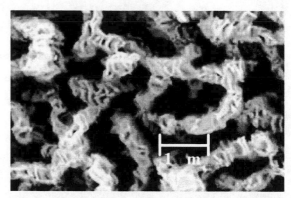

Fig.13 SEM image of the needle-like Cr crystals revealed after Cu etching out from the Cu-Cr (80:20) deposited film formed at Ts=400 °C.

like Cr crystals revealed after Cu etching removal from the Cu-Cr (80:20) deposited film at Ts=400 °C is shown. We have found that these structures of mesh like and needle-like crystals are already observed in the Cu rich samples sputtering deposited film at Ts=100 °C, however lower temperature depositions leads to smaller pore size and larger reflectivity R in the wide range of the IR spectra.

As shown in Fig.9 and in Fig.11 the sample with better characteristics for the IR spectra of 8-14μm wavelength has relatively larger thickness of about 5μm for the Cr-Cu (20:80) film than that of expected one, however, the effective optical thickness must be less than this 5μm because of porous one.

5 CONCLUSIONS

Novel fabrication method of porous Chromium (Cr) film for an infrared (IR) absorber is proposed and total absorption factor above 96.2% at about 10 m, being the peak IR wavelength radiated from near room temperature substance such as the human body temperature, is achieved.

This fabrication method can be applied to the catalytic materials such as the catalytic Pt film because of having nano size pores and easy fabrication.

ACKNOWLEDGEMENT
Authors would like to thank to Mr. Masuji Satoh for his experimental help and advices. This work is supported by funds of Japan Science and Technology Agency (JST), and is carried out in the High Tech-Research Center of Tohoku-Gakuin University.

REFERENCES
[1] Neli, R.R., Doi, I.,Diniz, J.A. Swart, J.W. , "Development of process for far infrared sensor fabrication", A. Physical, 132 (1), pp.400-406, 2006.
[2] Parsons, A. D.; Pedder, D. J., "Thin-film infrared absorber structures for advanced thermal detectors", Journal of Vacuum Science and Technology A, vol. 6, pt. 2, pp. 1686-1689, 1988.

Research on Preparation of Nano-Barium Titanate and Dielectric Property of Barium Titanate Ceramic Capacitor

Feng Xiuli Liu Xiaojing Cui Yixiu

Institute of Electronic Engineering ,China Academy of
Engineering Physics ,P.O.Box919-516 Mianyang,621900, China

ABSTRACT

A mild alcohol-thermal synthesis of nano-$BaTiO_3$ from $H2TiO_3$ and $Ba(OH)_2$ at an Atmospheric Press was systematically studied and the optimized preparation conditions were obtained. When the reaction is carried out at a temperature of 79℃,an atmospheric press for 6h, the moral ratio of $Ba(OH)_2$: H_2TiO_3=1.0, pH=11, the cubic crystal structure of nano-$BaTiO_3$ was obtained. The dielectric ceramic was fabricated with nano-$BaTiO_3$. By comparison with ordinary ceramic, the ceramic with nano-$BaTiO_3$ from alcohol-thermal method could lower sintering temperature from 1300℃ to 1150℃, change Courier point to 70℃,make dielectric constant 20000.

Keywords: nanometer powder, Barium titanate, solvent-thermal method, dielectric property

1   INTRODUCTION

Barium titanate is increasing importance in the preparation of dielectric capacitors or transducers due to its high dielectric constants, ferroelectric properties, piezoelectric properties and positive temperature coefficient (PTC)effect [1-5]. In particular, nano-$BaTiO_3$ exhibits unique features that strongly differ from those of bulk phase, such as switch of Curie temperature and increase in dielectric constant [6-7]. Many authors emphasized the importance of their synthesis process on the properties of BaTiO3 powder ,thus many methods have been developed to synthesize $BaTiO_3$ with specific size, such as solid state method, sol-gel process, precipitation and hydrothermal. Solid-state method[8] did not satisfy the needs of electronic ceramic industries due to lack of uniform nanocrystalline $BaTiO_3$.Sol-gel process[9] needs high temperature over 700℃ and expensive alkoxides, which goes against practical application. Precipitation[10] using inorganic salts has shown success in producing $BaTiO_3$, but it also need high temperature during sintering. Hydrothermal methods[11-15] with inorganic salts offers a promising approach to prepare $BaTiO_3$, whereas it needs high temperature(160-300℃) and high pressure(4-8MPa). Michael Z-C[16] reported that hydrothermal proceeded at temperature 100℃,but the reaction time is as long as 24h-72h,the particle size is 0.2-1um.The reports concerning low temperature alcohol-thermal synthesis of nano-$BaTiO_3$ with low-priced H_2TiO_3 at ambient pressure to prepare nano-$BaTiO_3$ has not been published.

Qinghua University reported in 2006 that ceramic based nano-$Ba(Zr,Ti)O_3$ powder exhibits 23000 of dielectric constant, but the high dielectric constant ceramic based pure nano-$BaTiO_3$ powder have not been reported heretofore.

In this paper we report a mild alcohol –thermal route to $BaTiO_3$ nanoparticles;the mole ratio of Ba to Ti in precursor and reaction time are used as variable.The formation process of Barium titanate ,the sinterring temperature of ceramic and the dielectric property are discussed.

2   INTRODUCTION

H_2TiO_3 was scattered in alcohol at room temperature (marked A); $Ba(OH)_2$ was scattered in alcohol in a closed bottle at 79℃ (marked B).Aqueous ammonia and A were gradually added to B respectively, which were stirred and refluxed for some time at 79℃. The suspension were cooled to room temperature, separated by centrifugation, washed with the acid, distilled water and alcohol, dried at ambient temperature. Thus the $BaTiO_3$ powders were gained.

$BaTiO_3$ powders were characterized with TEM (JEM-2010 model, JEOL Corporation, Japan) and X-ray diffraction(XRD) model D/MAX-ⅡP, Tokyo, Japan.. $BaTiO_3$ capacitor were characterized with SEM (SEM-5900LV,JEOL Corparation) and the capacitance at different temperature was tested by LCR apparatus (Tianjin No.6 radio Factory,China).

3   RESULTS AND DISCUSSION

3.1   The Effect of Ba/Ti Molar Ratio

When the reaction is carried out at pH=1, 79℃,6h,the effect of Ba/Ti molar ratio in the precursors on the products is shown in Fig.1.

It can be seen from Fig.1 that with the increase Ba/Ti moral ratio in the precursors, the Ba/Ti moral ratio in the powders all the while increased.The XRD patterns of the sample prepared by the different moral ratio of Ba/Ti in reactants at 79℃ for 6h were showed in Fig.2.

NSTI-Nanotech 2008, www.nsti.org, ISBN 978-1-4200-8503-7 Vol. 1

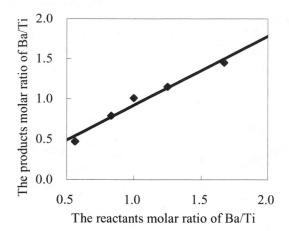

Figure 1 :The curves of molar ratio of Ti/ Ba

（reaction conditions：pH:11, reaction time:6h,temperature:79℃）

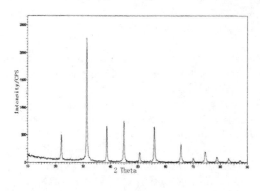

a

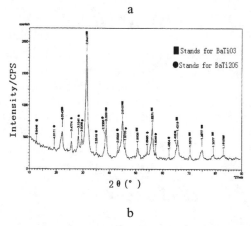

b

Figure 2: XRD patterns of Barium Titanate

a:Ba/Ti moral ratio 1.0;b Ba/Ti moral ratio 0.8;

Fig.2.a is the XRD patterns of sample from reactants moral ratio of Ba to Ti：1.0, it is clear that all diffraction peaks can be indexed to the cubic structure of $BaTiO_3$, Fig.2.b is the XRD patterns of sample from reactants moral

ratio of Ba to Ti：0.8,there are peaks of $BaTi_2O_5$ (moniclinic) and $BaTiO_3$ (cubic) ,which suggest the reactants moral ratio of Ba to Ti affected the constitute of products. To obtain pure $BaTiO_3$, it is necessary to control reactants moral ratio of Ba to Ti., that is the optimum Ba/Ti moral ratio in the precursors is 1.0.

The reaction process abides by the following steps:

$$H_2TiO_3 + Ba(OH)_2 \longrightarrow BaTiO_3 + 2 H_2O \qquad (1)$$

when H_2TiO_3 is excess,the reactiom is as follows

$$H_2TiO_3 + BaTiO_3 \longrightarrow BaTi_2O_5 + H_2O \qquad (2)$$

3.2 The Effect of Reaction Time on Products

Table 1 showed the effect of reaction time on products. It is clearly seen that Ba/Ti molar ratio and the content of $BaTiO_3$ in the products increased with the increase of reaction time from table 1.The increase extent was not large after 6h, which indicated the reaction can complete when the reaction is 6h .

Time(h)	Ti/Ba in product	$BaTiO_3$ （%）
1	0.32	-
3	0.48	30.8
5	0.98	80.4
6	1.01	92.1
10	1.02	93.5

（ reactants moral ratio of Ba / Ti:1.0, temperature:79℃）

Table 1: The effect of reaction time on powder

3.3 TEM Analysis

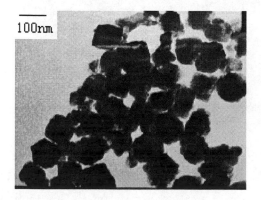

Figure 3 : TEM micrograph of $BaTiO_3$

TEM photos showed that the BaTiO₃ nanoparticles were square in shape, the size was about 50nm .Particle size distribution was in narrow range, and no agglomerates were observed.

3.4 SEM Analysis

The micrograph of BaTiO₃ ceramic at different sinterring temperature are showed in figure 4.

1150℃

1200℃

1250℃

Figure 4: SEM micrograph of BaTiO₃ ceramic

It can be seen that the diameter of BaTiO₃ in ceramic increased with increaing temperature,which conform with the relationship between sinterring temperature and the diameter.

SEM photos showed that the particle size in BaTiO₃ ceramic sinterred at 1150℃ was about 200nm,Particle size distribution was in narrow range.

3.5 DielectricConstant of BaTiO₃ Capacitor

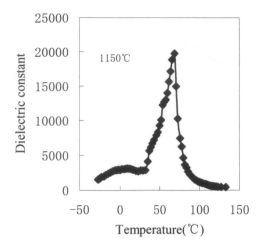

Figure. 5 Temperature dependence of dielectric constant of BaTiO₃ capacitors

The relationship between temperature and dielectric constant is nonline,which indicated the ceramic sinterred at 1150℃ from nano-BaTiO₃ is ferroeletric ceramic. By comparison with ordinary BaTiO₃ ceramic capacitor, the ceramic capacitor fabricated by nano-BaTiO₃ could not only lower sintering temperature from1300℃ to1150℃, but also change Courier point from 130 to 70℃, improve dielectric constant from 6000 to 20000,which can satisfy the development of microdevies.

4 CONCLUSION

When Ba/Ti moral ratio in the precursors is 1.0 and the reaction time is 6h, the reaction of synthesis pure BaTiO₃ has completed. one step synthesis BaTiO₃ particles without sintering are cubic structure and the size is 50nm. The ceramic capacitor fabricated by this kind of BaTiO₃ is ferroeletric ceramic , sintering temperature is 1150℃, Courier point is 70℃, dielectric constant at Curie point is 20000.

REFERENCE

[1] Dang-Hyok Yoon,Burtrand I. Lee , J. Europ.Ceram. Soc. 24,739,2004
[2] M.Rajendran,M.S.Rao,J.Solid state Chem. 113, 239,1994
[3] B. D. Stojanovic, C. R. Foschini, V. Z. Pejovic,etc., J.Europ.Ceram. Soc.24,1467,2004
[4] Y. C. Chan, Y. Wang, D. P. Webb,etc., Materials Science and Engineering B.47,197,1997

[5] P. Padmini, N. S. Hari ,T. R. N. Kutty Sensors and
      Actuators A: Physical. 50,34,1995
[6] B.R.Li,S.G.Lu,Electronics Components and
      Materials, 23,2004
[7] K.Uchino,E.Sdanaga,T.Hirose,J.Am.Ceram.Soc.
      34, 2319,1999
[8] F.Valdivieso,M. Pijolat, M. Soustelle. Chemical
      Engineering Science, 51,2535,1996
[9] K. Kusakabe, K. Ichiki and S. Morooka,Journal of
      Membrane Science, 95,171,1994
[10] S. van der Gijp, M. H. J. Emond, A. J. A.
      Winnubst etc. J. Europ.Ceram. Soc. 19,683,1999
[11] Ying Ma,Elizabeth Vileno,Steven L,
      etc. ,Chem.Mater.. 9, 3203, 1997
[12] Reza Asiaie,WeidongZhu,Sheikh A.Akbar,etc.,
      Chem. Mater.,8,226,1996
[13] Xu Huarui,Gao lian,Guo Jingkun.Functions
      materials, 5,558,2001
[14]Prabir K.Dutta,J.R.Gregg.Chem.Mater.,4,843,1992
[15] D. Völtzke, H. P. Abicht, J. Woltersdorf ,etc.,
      Mater. Chem. . Phys . 73,274,2002
[16] Michael Z.-E. Hu,Vino Kurian,E.Andrew Payzant,
      etc.,Powdew Technology,110,2,2000Wongduan

*Corresponding author: Tel:(086) 816-2485505,

Fax:(086) 816-2487594

E-mail：fengxiuli345@sina.com

A Novel Technique for Purification and Segregation Of HiPCO CNT'S Using INKJET Technology

Dr. Tulin Mangir, Juan Chaves, Mukul Khairatkar, Sindy Chaves.

California State University Long Beach
Long Beach, California, 90840-8303
temangir@csulb.edu, jchaves@csulb.edu, mkhairat@csulb.edu, schaves@csulb.edu

ABSTRACT

In paper we describe a technique to purify and segregate carbon nano tubes (CNTs) from the iron impurities to use the CNTs for thin films. After extensive research into a proper and repeatable cleaning process we developed a new method of using bacteria to remove impurities from CNTs. This gives us the ability to manipulate the CNT. We also developed an apparatus based on inkjet technology that deposit the bacteria containing CNT on conductive surfaces such as silicon, plastic, or any other conductive surface. In this way we create a simple way to manipulate the CNTs for the creation of organic LED, Thin Films, or even electronic devices such as transistors using a simple process. Our methods specifically apply to CNTs obtained by the HiPCO process. However, we assume the method can be extended to other cases. We expect this approach will lead to finding novel conductors, structures and thin films for future nano-engineered devices that could be investigated further for use in nano-scale electronics, implants and bio-materials.

Keywords: HiPCO CNT, Inkjet, bacteria, purification, segregation.

1 INTRODUTION

Today, the race to find materials with stronger and faster carrier mobility and smaller thickness for the fabrication of thin films that will yield smaller geometries and flexible electronic devices led to the experimentation with CNTs.
CNTs have attracted a lot of interest because their unique physical properties and because they show a lot of potential for their use in microelectronic and nanoelectronic applications. Hence, by developing an industrial process, they can be use in areas such as display technology, as thin film transistors in liquid crystal displays or Organic Light Emitting Diodes (OLED), as well as thin films for the manufacturing of CMOS devices and Field Effect Transistors (FETs). One of the physical properties of CNTs that made them very attractive to the manufacturing of FET is their carrier transport property. It is been estimated that the carrier mobility of a single CNT will be in the range of 1000 cm^2/Vs [1]. Although, the carrier mobility of CNT seems to be a great advantage in the manufacturing of thin films, in reality it gets compromise somehow by the introduction of carbonaceous impurities as well as metal impurities due to the manufacturing process of CNTs. As an example a FET made out of a single CNT can carry limited currents generally in the range of nA or μA, but by using CNT thin film mesh the range can be significantly increase[1]. Hence, for CNT to be compatible with today's device fabrication at low cost and high throughput it needed to be in more aqueous stage e.g., a liquid material or solution, since the process of CNT growth requires high temperatures of 900 °C.

In our research we utilize SWCNTs (Single Walled Carbon Nano Tubes) obtain by the HiPco (High-pressure CO dissociation) process. One of the major drawbacks using this type of material was the amount of iron impurities contain in the CNTs, due to the use of iron as a catalyst in the process. Hence, a pioneering cleaning method was developed in order to obtain pristine SWCNT that could be use for the creation of high quality thin films, and the creation of the liquid material required to the manufacturing process of thin films.

2 EXPERIMENTAL

The main purpose of our research was to create thin films for use in electronic applications. Consequently, we needed a simple way to purify the SWCNT from the iron contaminants induce thru the HiPco process at the time of their growth. After experimenting with several methods already being researched to clean the same type of material from their impurities, we concluded that these methods caused structural damage of the SWCNT which will also led to a change in the electrical properties. Hence, after trying different methods we created a more passive process for the purification that will prevent damage in the SWCNT structure, but at the same time eliminating over 90% of the contaminants such carbonaceous and iron impurities, leaving a pristine material to be use in the creation of thin films. For such reason we devise the use of a biological agent for the purification process. After, research several ways of processing the SWCNT dust and exposing it to a variety of biological agents, we found that by using bacteria we achieve the cleaning task without affecting the condition and/or structure of the CNTs; thus eliminating most of the mineral contaminants in the process*. The biological properties of the bacteria and the ubiquitousness of it, was so appealing for their use in this cleaning process.

(a)

(b)

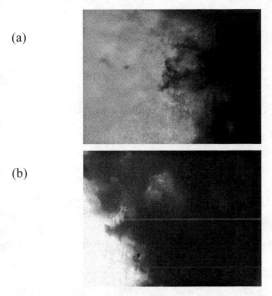

Figure 1: (a) Optical image of CNT dust when bacteria are just applied; (b) Optical Image of CNT dust after 4h once the bacteria start the cleaning process. The red area is the Iron oxide from the Iron being pulled out from the inside of the CNTs.

After several attempts to formulate a procedure to expose the CNT dust to biological cleaning agents, we finally create a working protocol to generate a high yield of material with pristine SWCNT for the use in the manufacturing process of thin films*.

Thru this biological cleaning process, the purified SWCNTs have been found to have higher surface areas retaining also their physical characteristics of conductivity, compare to the acid treatments or microwave treatments which were demonstrated to have good results in the elimination of metallic catalytic impurities but with a rise in structural and electrical changes in the end material [2]. Also this biological cleaning process was demonstrated to be superior and more cost effective that the gas oxidation process, in which the material has to be exposed to Argon or Oxygen gas flow for several hours, annealed and then clean by HCl acid exposing the material to structural damage [3-4]. In the biological process, the SWCNT were exposed to the right amount of biological agents with the right amount of CNT dust and then water rinse to eliminate the untreated material, making the cleaning process cost effective by eliminating the use of expensive acids or gases thru the process. Consequently, the process achieves high yields of the clean material reducing the cost and reducing the waste of the CNT dust during the process.

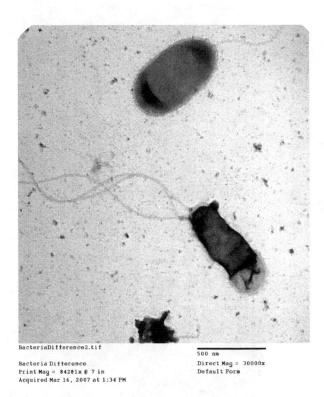

BacteriaDifference2.tif

Bacteria Difference
Print Mag = 84281x @ 7 in
Acquired Mar 16, 2007 at 1:34 PM

500 nm

Direct Mag = 30000x
Default Form

Figure 2: Bacteria Difference one without CNT and one with SWCNT inside.

In the biological cleaning process one of the bacterium main nutrients is Iron which is needed for the production of nitrogen. Using this to our advantage the bacterium is grow in a starving environment (lack of Iron nutrients) and then is released in an environment containing only the raw CNT. Once the bacteria colony reaches the raw material the absorption process starts. During this process each bacterium works as a "biological cleaning machine" absorbing the raw CNT to extract the Iron from the SWCNT. In figure 2 a TEM micrograph shows the difference between a normal bacterium and one after "ingesting" the raw CNT material. After the CNT are absorbed the bacterium breaks down the carbonaceous cocoons were the iron is enclosed leaving only clean SWCNT.

This absorption process is very significant in our research since it accomplishes two tasks at once. First, the cleaning of the material and segregation of individual SWCNT form their tangled bundles was achieved. Second each bacterium performs as a SWCNT "container or vessel", were the clean pristine material is enclosed after the cleaning process. Hence, allowing an easy manipulation of a SWCNT thin film mesh without disturbing the structural characteristics of the thin film as well as the conductivity properties of it. Figure 3 shows an SEM micrograph of the SWCNT mesh contained inside of an individual bacterium.

(a)

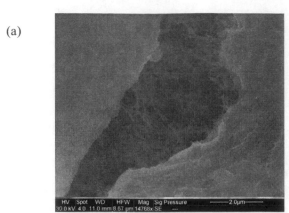

(b)

Figure 3: (a) SEM micrograph of the SWCNT thin film inside the bacterium; (b) a closer image of the SWCNT mesh thin film inside the bacterium.

Once the absorption and cleaning process is done the bacteria gets selected and clean from any leftover raw material by a rinse process. After the rinse process is complete the bacteria vessels are ready to be deposited into the surface to create the thin film.

2.1 Inkjet Dispersion of the SWCNT

The next step is the segregation of the individual SWCNT and the separation from the bacterium. This is achieved by using Inkjet technology developed by Hewett Packard [4]. A special ink was created with the bacteria containing the SWCNT thin film mesh inside*. The ink was developed with a specific ratio that with yield a consistent number of bacteria population with SWCNT in them for every 50 ml of solution, making a very dense bacteria ink.

An inkjet cartridge was modifying to host the special ink and a printer apparatus was developed to use the special cartridge, for the process of deposited the SWCNT mesh virtually in any surface*.

The principle of inkjet technology provides the last step of converting the material back from the wet environment to a dry environment. Hence, the method allows the use of the clean SWCNT in the manufacturing process of electronic devices*.

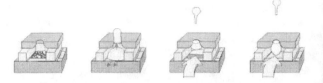

Figure 4: Inkjet process to release droplets of ink in a surface [4].

By using the thermal inkjet technology the solution is superheated to about 1,000,000 °C/second, the solution were the bacteria is contain is ejected without boiling but disintegrating the bacteria in the process virtually depositing the mesh of SWCNT in the surface. Each droplet released is about 150pl making this process very accurate to deposit the material virtually in any space or surface. Once the material is deposit in the surface the surface exposed to a quick heat treatment to eliminate any residual of the biological agent.

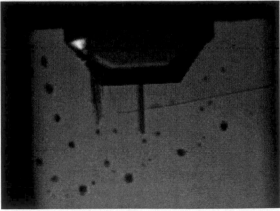

Figure 5: Pictures of the special ink droplet taking by the onboard camera of the AFM

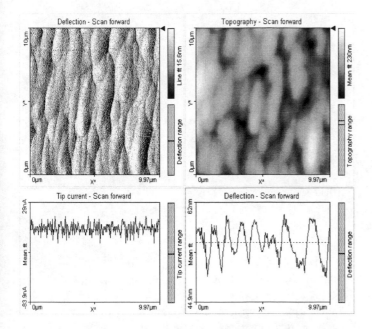

Figure 6: AFM image of the SWCNT mesh after deposited thru the inkjet nozzle on a silicon surface.

3 DISCUSSIONS AND CONCLUSION

Our results has shown that our biological cleaning and segregation process of HiPco SWCNT is less invasive, less damaging and more cost effective than processes that involved the use of HCl, Oxygen or Argon gas, as well processes that uses microwaves to clean the impurities in the raw material. Also, with the use of inkjet technology in conjunction with our biological cleaning process allow us to achieve the segregation of individual clean SWCNT and the manipulation of the thin film mesh form by them. Hence, with the creation of the cleaning protocol, the creation of the special bacteria ink, and the creation of the ink depositing apparatus we achieve an industrial process in the manufacturing of thin films for electronic applications. Further discussions of the major topics of this work and most of our data results are going to be presented at the conference.

REFERENCES

[1] Xuliang Han, Daniel C. Janzen, Jarrod Vaillancour, and Xuejun Lu. 2006. A flexible **thin-film** transistor with high field-effect mobility by using **carbon nanotubes**. *Nanotechnology Materials and Devices Conference, 2006. NMDC 2006. IEEE* 1 (2007-11-27) : 296-297.

[2] Ester Vázquez, Vasilios Georgakilas, and Maurizio Prato. 2002. Microwave-assisted purification of HIPCO carbon nanotubes. *Chemical Communications* , no. 20: 2308-2309.

[3] Chiang, I. W., B. E. Brinson, A. Y. Huang, P. A. Willis, M. J. Bronikowski, J. L. Margrave, R. E. Smalley, and R. H. Hauge. 2001. Purification and Characterization of Single-Wall Carbon Nanotubes (SWNTs) Obtained from the Gas-Phase Decomposition of CO (HiPco Process). *Journal of Physical Chemistry B* 105, no. 35: 8297-8301.

[4] KIM Yeji, MINAMI Nobutsugu, ZHU Weihong, KAZAOUI Said, AZUMI Reiko, and MATSUMOTO Mutsuyoshi. 2003. Langmuir-Blodgett films of single-wall carbon nanotubes: Layer-by-layer deposition and in-plane orientation of tubes. *Japanese journal of applied physics* 42, no. 12 (December 10, 2003) : 7629-7634.

[5] Rob Beeson. Thermal Inkjet: Meeting the Applications Challenge.

*   **This new approach is under a patent pending process, reason why some details were omitted from this paper.**

Special Thanks to the Army Research Office for their support in this investigation thru their Grant "Assessing the Integrity and Integration Issues of Nano- Sensors/Nanostructures-Equipment".

Dip Pen Nanolithography of Silver Nanoparticle-based Inks for Printed Electronics

Mohammed Parpia, Emma Tevaarwerk, Nabil A. Amro,*
Raymond Sanedrin, Hung-Ta Wang, Jason Haaheim

NanoInk, Inc
8025 Lamon Ave
Skokie, IL 60077

Corresponding author, emma@nanoink.net

ABSTRACT

Here, we present liquid-ink based nanoscale deposition of commercially available silver nanoparticles onto defined locations on untreated silicon dioxide substrates. We monitor the flow of the ink from the cantilevers to the substrate using specially designed transparent cantilevers with a high spring constant (0.5 N/m) in an open geometry which prevents tip clogging that plagues both inkjet printing[10] and nanofountain probes[6, 11]. We discuss the deposition mechanism for the silver NP inks, including methods for creating nanoscale arrays, lines, as well as varying the size and shape nanostructures deposited with the silver NP based inks.

Keywords: dip pen nanolithography, nanoparticle, direct-write, nanoscale, ink

1 INTRODUCTION

Nanoparticles (NPs) offer a wide range of unique properties when compared with their bulk counterparts, including optical, electrical, catalytic and quantum properties. Metal nanoparticles' unique properties have lead to new applications in electronic circuits and waveguides[1]. The most common metals used in electronic devices are gold, copper and silver. A requirement for many metal nanoparticle applications is the controlled placement of the nanoparticles in the desired locations on a surface. Recently developed nanoparticle lithography approaches include the chemically directed assembly of nanoparticles on to locally modified surfaces [2, 3], the directed assembly of NPs using a meniscus and mask [4], direct printing of nanoparticle inks onto heated substrates by inkjet printing[5] and the direct deposition of liquid gold NP solutions onto a functionalized surface[6-8]. These additive approaches offer advantages over standard subtractive microfabrication techniques which require masks, larger volumes of material, expensive facilities and extreme processing conditions.

Of the additive techniques, dip pen nanolithography (DPN) offers several advantages including computer driven, maskfree lithography, ambient working conditions, inexpensive equipment, and compatibility of a variety of inks with various substrates.

Here, we present liquid-ink based nanoscale deposition of commercially available silver nanoparticles onto defined locations on untreated silicon dioxide and gold substrates. We monitor the flow of the ink from the cantilevers to the substrate using specially designed transparent cantilevers with a high spring constant (0.5 N/m) in an open geometry which prevents tip clogging that plagues both inkjet printing[10] and nanofountain pen probes[6, 11]. We discuss the deposition mechanism for the silver NP inks, including methods for creating nanoscale arrays, lines, as well as varying the size of dots deposited with the silver NP based inks.

2 LIQUID NANOPARTICLE INKS

Liquid inks involve the use of a transport medium (a "carrier" solvent) in which the nanoparticles are suspended and stabilized. Combined, this carrier solvent and nanoparticle source form the nanoparticle ink that is coated onto the outer surface of the cantilever. Many conditions must be met in the development of liquid inks DPN, because of the nature of the deposition mechanism. These factors result from the open tip geometry, ambient conditions, and nature of the ink and substrate, and may include:

1) a slow evaporation rate of the ink from the cantilever,
2) a homogeneous ink solution with a long shelf life,
3) uniform wetting and loading of the ink on the cantilever,
4) an affinity of the ink for the substrate, and
5) controlled transfer of ink from the cantilever to the substrate while writing.

Meeting all of these conditions often requires and multiple trials optimization, in particular if the ink is to perform in the same manner over many repetitions. Here, we present two inks optimized for DPN of silver NPs, one based on a glycerol carrier solvent and silver nanoparticles acquired from

P-Chem Associates, another based on heptadecane/α-terpineol mixture and a silver nanoparticle paste from Harima Chemicals.

Like in a macroscale fountain pen, the loading and wetting of the ink on the cantilever is key in determining written feature size, in addition to the nature of the ink itself. Controlled volumes of these inks are loaded onto the cantilevers using a microfluidic-based solvent delivery system, Universal Inkwells, and our newly developed transparent cantilevers.

3 ACCESSORIES FOR LIQUID INKS

In general, with contact mode cantilevers that are metal coated, investigation of the loading and wetting of the ink on the cantilever is accomplished by flipping the cantilever over and looking at it under an optical microscope. A thin, uniform & continuous film across the surface area of the cantilever indicates that it will write uniform feature shapes and sizes. The ability to investigate ink loading and wetting on the cantilever *in situ*, both immediately after loading and while writing, is of key importance in controlling feature size from experiment to experiment and tip to tip. To aid in the study of ink loading, wetting and flow, we use transparent cantilevers shown in Figure 1. The optical micrographs (a) show inking (inset), and resulting ink loading and wetting of transparent cantilevers with AgNP ink. Patterning from two different transparent cantilevers (b) shows a clear difference between properly & improperly inked cantilevers. The left has ideal ink loading and wetting resulting in good flow and small, uniform arrays, while the right has poor ink loading and wetting, and hence non-uniform flow from the cantilever and unreliable spot sizes.

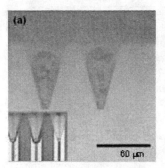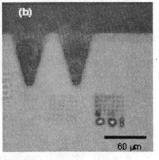

Figure 1: (a) Proper Ink loading and wetting on two cantilevers, (b) resulting pattern from properly loaded cantilever (left) and poorly loaded and wetted cantilever (right).

4 Patterning with Inks & Cantilevers

Once loaded with ink, the cantilevers are brought into contact with the surface, and then withdrawn to leave behind a droplet of ink. It is not necessary to align the laser on the back of the cantilever during this process because this process can be done without feedback. Indeed, this is in contrast to the diffusive inks, where feedback is needed during ink deposition. In fact, we observed that if we attempted to draw a feature without multiple withdrawals from the surface, the carrier solvent would deposit but the nanoparticles would remain on the cantilever; thus when curing the deposited droplets the carrier solvent evaporated and no particles remain behind. In this manner a dot-matrix like printing method was used to create both arrays and lines. After ink deposition, the substrates were annealed at between 300 – 500 °C to evaporate off the carrier solvent as well as to sinter the nanoparticles together into a continuous structure. The nanoscale alignment feature of InkCAD enabled AFM imaging of the deposited pattern after the substrate was removed from the system for annealing.

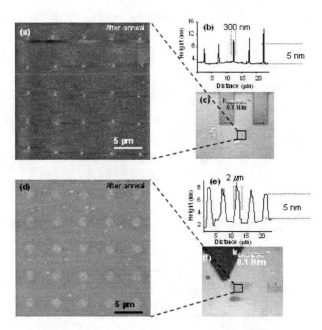

Figure 2: Patterning of dots with aqueous (*a*, *b* and *c*) and organic (*d*, *e* and *f*) silver nanoparticle inks.

Figure 2 shows the differences between features obtained by deposition of aqueous (*a*, *b* and *c*) and organic (*d*, *e* and *f*) silver nanoparticle inks. After deposition, an anneal at relatively low temperatures (300 – 500°C) boils off the carrier solvent and sinters the metal nanoparticles into a solid feature. For the same tip-sample contact time of 0.01 ms, deposition of the organic silver nanoparticle ink resulted in ~2 micron sized features (*d*), whereas the aqueous silver nanoparticle ink yielded ~300 nm features (*a*). We postulate that this difference is primarily due to differences in surface tension between the two inks. Since feature size is controlled by the size of the droplet deposited, a smaller droplet results in smaller features. Surface tension is an interfacial property that depends on the strength of intermolecular forces in a liquid. We postulate that our water based ink has more hydrogen bonding interactions between individual molecule than our organic ink. In addition to smaller droplet sizes, surface tension helps reduce feature sizes by drawing the nanoparticles towards the center of the feature. This pulling in of the nanoparticles takes place at the moving edge of the droplet as it boils off. Feature sizes are relatively uniform for the nanoscale silver features formed in this manner. Across 10 randomly selected dots in five separately written arrays, the average dot diameter is 287 nm with a standard deviation of about 35 nm.

Since ink is deposited when the cantilever withdraws from the surface, lines cannot be easily obtained by dragging the cantilever across the surface. Lines are realized, however, by close deposition of a line of spots of silver nanoparticle ink, as shown in Figure 3. Post annealing, the lines are between 300 and 800 nm wide and 5 nm tall. Continuous features are obtained as the nanoparticles are carried to the center of the feature by the evaporating solvent, before they sinter together.

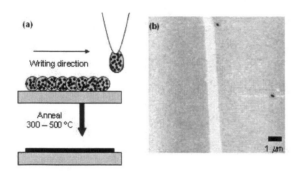

Figure 3: (a) Schematic showing fabrication of lines of silver by close deposition of AgNP ink and (b) topography image revealing continuous silver line on SiO2 after annealing at 300 – 500°C.

We can control the dimensions of the nanoscale features deposited with the silver NP ink. With the liquid silver NP inks this is accomplished by repeatedly striking the tip in the same location on the sample, in order to deposit more ink to the surface. The results of such an experiment are shown in Figure 4, post anneal. The first spot (x1) was formed by repeatedly spotting at the same location 10 times, and is both taller and broader than the second (x2) which was formed by spotting five times, and the third (x3) spot formed by striking three times.

The ink remains on the surface with each subsequent cantilever strike because the adhesive forces between the sample surface and the NP suspension are stronger than the adhesive forces between the tip and the suspended NP solution. In all likelihood a dry cantilever would remove previously deposited ink while a wet cantilever allows the deposition of additional ink to the substrate.

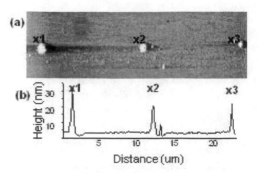

Figure 4: Variation in the size of the deposited material is accomplished by spotting

5 CONCLUSIONS

In conclusion, two types of silver nanoparticle ink (aqueous and organic) were developed to consistently obtain solid features on untreated SiO$_2$. Inks resulted in features that were about 2 μm in diameter, whereas nanoscale features were achieved using aqueous inks due to higher surface tension. Transparent cantilevers were used to investigate ink loading and wetting as well as the flow of ink from the cantilever to the substrate in real time. Methods for controlling the size and shape of the deposited nanostructures are discussed, both for dots and lines.

REFERENCES

1. Zhang, X.P., et al., *Metallic photonic crystals based on solution-processible gold nanoparticles.* Nano Letters, 2006. **6**(4): p. 651-655.
2. Ginger, D.S., H. Zhang, and C.A. Mirkin, *The evolution of dip-pen nanolithography.* Angewandte Chemie-International Edition, 2004. **43**(1): p. 30-45.
3. Piner, R.D., et al., *"Dip-pen" nanolithography.* Science, 1999. **283**(5402): p. 661-663.
4. Kraus, T., et al., *Nanoparticle printing with single-particle resolution.* Nature Nanotechnology, 2007. **2**(9): p. 570-576.
5. Fuller, S.B., E.J. Wilhelm, and J.M. Jacobson, *Ink-Jet Printed Nanoparticle Microelectromechanical Systems.* Journal of the Microelectromechanical Systems, 2002. **11**(1): p. 54-60.
6. Wu, B., et al., *Direct deposition and assembly of gold colloidal particles using a nanofountain probe.* Langmuir, 2007. **23**(17): p. 9120-9123.
7. Ben Ali, M., et al., *Atomic force microscope tip nanoprinting of gold nanoclusters.* Langmuir, 2002. **18**(3): p. 872-876.
8. Prime, D., et al., *Nanoscale patterning of gold nanoparticles using an atomic force microscope.* Materials Science & Engineering C-Biomimetic and Supramolecular Systems, 2005. **25**(1): p. 33-38.

9.   Thomas, P.J., G.U. Kulkarni, and C.N.R. Rao, *Dippen lithography using aqueous metal nanocrystal dispersions.* Journal of Materials Chemistry, 2004. **14**(4): p. 625-628.

10.   Sirringhaus, H. and T. Shimoda, *Inkjet Printing of Functional Materials.* MRS Bulletin, 2003: p. 802-806.

11.   Kim, K.H., N. Moldovan, and H.D. Espinosa, *A nanofountain probe with sub-100 nm molecular writing resolution.* Small, 2005. **1**(6): p. 632-635.

Nanolithographic Patterning of Catalysts for Synthesis of Carbon Nanotubes

Rohit V. Gargate *, Debjyoti Banerjee **

* Mechanical Engineering Department, Texas A&M University, College Station, TX 77843-3123

ABSTRACT

A novel process was developed for synthesis of carbon nanotubes (CNT) which does not require the traditional Chemical Vapor Deposition (CVD) synthesis techniques. Catalyst precursors were deposited in bulk using a "wet process" on the MEMS enabled micro-heater elements. The microheaters were coated with a layer of fullerene apriori (of 150 nm thickness) using Physical Vapor Deposition (PVD). The chip was then heated using integrated microheaters and external heaters in an inert atmosphere to obtain CNT. Thus, in this process we obviate the Chemical Vapor Deposition (CVD) process for synthesis of CNT (SWCNT and MWCNT). This work also proves the feasibility for a portable hand held instrument for synthesis of CNT "on demand".

Keywords: Carbon nanotubes, Dip Pen Nanolithography, MEMS, Nanotechnology, Raman spectroscopy.

1 INTRODUCTION

Carbon nanotubes (CNT) were discovered by Ijima using the arc-discharge process in 1991 [1], and since then, various techniques have been devised to synthesize carbon nanotubes. CNT are considered for various applications owing to their novel mechanical [2] and electronic properties [3]. Processes like Chemical Vapor Deposition or "CVD" [4], and the HiPCo process [5] were developed and optimized subsequently. Mass production of nanotubes would allow their commercial use [6]. Furthermore if the synthesis process can be controlled so as to have a desired type of nanotubes with controlled diameters, properties of nanotubes could be even more exploited.

This work is an extension of our earlier investigation for using Dip Pen Nanolithography (DPN) techniques [7-9] for the synthesis of CNT and controlling the chirality of the synthesized CNT [10-12]. In this paper, we try to present a refined process for the synthesis, which overcomes some of the hurdles mentioned in the commercial processes described above.

2 EXPERIMENTAL APPARATUS

Custom designed MEMS (Micro Electro Mechanical Systems) platforms were used for the synthesis process for carbon nanotubes (the MEMS substrates were fabricated by MEMS Exchange Inc.). VA. These platforms were in the form of square chips. The dimension of the substrate was 3.5 x 3.5 x 0.5 mm. Three types of serpentine heaters were designed for the substrate. Titanium was used as the material for the heater. Three different types of heaters with

variation in length and orientation were used on the substrate. The lengths of the microheaters are 18 mm, 19 mm and 42 mm. The thickness of the heater elements is 100nm.

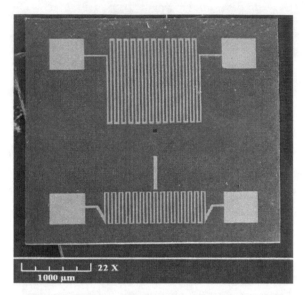

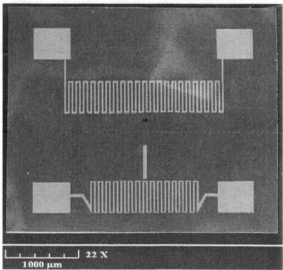

Figure 1. Scanning Electron Microscopy (SEM) images of the two types of substrates used in the synthesis of Carbon Nanotubes (CNT). There are three different types of micro-heaters microfabricated insitu on the silicon nitride membrane, a backside square opening etched in silicon for exposing the membrane at the center, and a vertical alignment marker (gold) perpendicular to the heater aligned with the membrane.

Two types of MEMS platforms (substrates) were fabricated for the synthesis of the CNT, with each substrate

NSTI-Nanotech 2008, www.nsti.org, ISBN 978-1-4200-8503-7 Vol. 1

having two heaters on either edge of the top side. A silicon nitride membrane was deposited on the top side in the middle portion of the silcon chip. The membrane had a back side opening of 60 μm x 60 μm square. A metal pad was deposited near the edge of the square membrane and perpendicular to the heaters - using "lift-off" process to serve as an 'alignment mark'. The chip structure is shown in **Figure 1.**

The Mbraun Labmaster double glove box system with gas purification (located at the Center for Nanomanufacturing, University of Texas, Austin); was used for the vapor deposition of fullerene layer on this chips. This instrument consists of a two chamber glove box system. A spin coater (specialty coating systems model P6700) is available inside the smaller glove box chamber. An Edwards's thermal deposition chamber with four sources is available inside the larger glove box. A vacuum oven is mounted at one end of the glove box. Multiple chips are mounted on a glass slide with a double sided tape, and this glass slide is placed inverted on the roof of in the thermal deposition chamber. Fullerene powder (Manufactured by Nano-C Inc., Westwood, MA) was placed in the target container on the chambers base. The container was electrically heated at a pressure of 1 mTorr to a temperature of 550 °C to enable sublimation of the fullerene powder. The deposition rate by sublimation was estimated to be 20 nm/s for a total deposition time of 10 s. The fullerene thickness was estimated to be 150 nm at the end of the vapor deposition process using a profilometer.

Metal catalysts (Nickel chloride, Cobalt Chloride, Palladium Chloride) were prepared by mixing the solid powder into water with minimal heating for uniformity. Sodium Hydroxide was used in the preparation of the $PdCl_2$, since it doesn't form a direct solution with water.
A digital DC power supply with enough power rating is used, which has displays for monitoring the current and voltage. A wire thermocouple is used to the measure the temperature, with the welded end on the chip and the open end connected to a thermocouple reader.

MINCO mica external heaters were used to raise the temperature of the coated substrates in the CNT synthesis experiments. Suspended in mid-air Individual heaters can be heated to temperatures of around 200 °C, So three heaters were mounted together , stacked above each other and connected electrically in parallel. The heater assembly was used to heat the chips to 550 °C, required for the synthesis process.

A dessicator is used as the chamber for the synthesis process. The dessicator has two openings. One of this is connected to a Nitrogen tank and the other serves as an inlet for the electrical connections. The external heaters, Chips, and the thermocouples are placed in the dessicator, 2 alligator pin clamps are used for suspending the heaters and positioning the thermocouple. Also a humidity sensor is positioned inside the dessicator.

3 EXPERIMENTAL PROCEDURE

It has been demonstrated experimentally that Carbon Nanotubes (CNT) can be synthesized with Fullerene as the source for carbon, where the synthesis temperature is ~500°C [13]. Metal catalysts also play a significant role in the formation of carbon nanotubes. The catalyst particles serve as nucleation sites for formation of CNT from fullerene [14]. Nickel Chloride, Cobalt Chloride, and Palladium chloride were used as catalyst precursors in this experiment. These precursors were prepared in an aqueous solution as mentioned above. Each catalyst was deposited as a bulk layer on separate fullerene coated chips with the aid of a syringe. The catalyst precursor solution allowed to dry at room temperature after dispensing it on the fullerene coated chips.

In an initial attempt, heater connections were made across the bond-pads of one of the heater per chip. The connections were possible with normal electrical wires available commercially with very small diameters, in accordance with the bond-pad areas. The apparatus was placed in a dessicator and connected to nitrogen supply from a pressurized cylinder. The experiments were performed in an inert atmosphere by passing the nitrogen into the dessicator. A humidity sensor was also placed inside the dessicator. The relative humidity in the air was 38.2% at the beginning before passing nitrogen gas. After a duration of 20 minutes, the humidity was found to be 3.8% inside the dessicator. At this stage, current was passed through the microheaters in the chip. However, the chips were damaged when they reached a temperature of ~300 °C due to stress concentration. Since this temperature was not suitable enough for CNT synthesis, external MINCO heaters were therefore used in this process. The external heaters connected in a parallel assembly as discussed above. The chips were then placed on the MINCO heater assembly for direct contact heating. The temperature of the heaters was then raised slowly in steps of 10 °C per minute until the temperature reached 510 °C. It has been reported that lower synthesis temperatures yield a tighter distribution of the diameters (and correspondingly the chirality) of the synthesized CNT [15]. The heater assembly was maintained at this temperature for 15 minutes and the power supply was then cutoff, along with the flow of nitrogen. The characterization of these samples was then carried out using Raman Spectroscopy and Scanning Electron Microscopy.

4 RESULTS AND DISCUSSION

The synthesized CNT can be characterized to determine the material properties using several technqiues which include SEM/ TEM, Raman spectroscopy, absorption spectroscopy, etc. SEM/TEM and Raman spectroscopy are widely used for their convenience and ease. Imaging techniques (which includes SEM and TEM) provides a direct visual observation. Raman spectroscopy provides distinct information of the vibrational modes which are detected using peaks in known Raman shifts in wavenumber. These peaks are seen at three locations, first being the Graphite Peak (G-Peak) around 1600 cm⁻¹, second, Defect Peak (D-Peak) around 1350 cm⁻¹, and third, the Radial Breathing Modes (100-300 cm⁻¹). RBM peaks are caused by the vibration along the radial direction of the individual CNT, and are widely used to obtain information about the distribution of the diameters of the synthesized single walled CNT (SWCNT) [16]. When the samples were excited by 633 µm laser, after the experiment (Model: LabRam IR Confocal Raman Spectroscopy - Microscopy Instrument, Manufacturer: JY Horiba, Japan), the above mentioned peaks in RBM, G and D bands were distinctly visible for the experiments using Palladium catalyst. **Figure 2** shows the obtained Raman spectra.

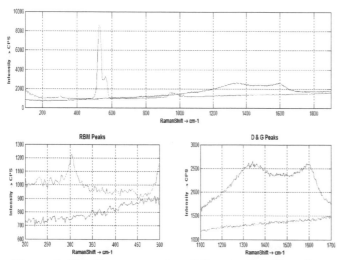

Figure 2: The Raman spectra of the sample before (in Red) and after (in Blue) the synthesis of CNT from Fullerene coated chips.

Figure 3: Image of the PdCl2 bulk deposited sample. Conductive copper tape used for mounting the sample is observed at the top. The conductive tape is used to minimize the effects of charge trapping.

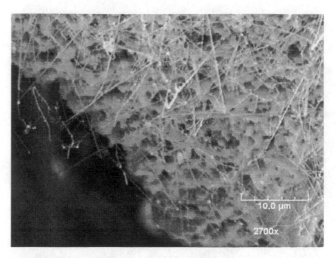

Figure 4: Top: The synthesized Carbon Nanotubes (CNT) are observed on the edge of the MEMS chip. Bottom: Image obtained at a higher magnification shown the synthesis experiment yields carbon nanotube bundles (SWCNT and MWCNT) and carbon nano-fibers which are estimated to be 20-200 nm in diameter.

As seen from **Figure 2** the peaks at 1340 & 1600 cm⁻¹ are clearly visible. Since the D band is more intense, it indicates presence of multi-walled carbon nanotubes [11, 12]. The peak at 300 cm⁻¹ seen in the RBM region and the peak at 520 cm⁻¹ in indicative of silicon, which is the substrate material. For confirmation of the data from Raman spectra, the Palladium samples were observed using Scanning Electron Microscopy. **Figure 3** shows the over all view of the bulk deposited sample after the experiment. **Figure 4** shows that the diameters of carbon nanotubes obtained in this synthesis experiment ranged from 20 nm to 200nm. This variation of the diameter of the carbon nanotubes and the haphazard location mixed with carbon fibers was expected owing to the bulk deposition of the metal catalyst. Energy dispersive Spectroscopy (EDS) techniques were used to analyze the elemental composition of the sample. Data obtained by EDS unit (inbuilt into the SEM apparatus) is shown below in Table 1.

Table 1

Element	Weight(%)
Silicon (Si)	13.11
Palladium (Pd)	3.67
Carbon (C)	83.22
Total	**100.00**

5 CONCLUSION & FUTURE DIRECTION

In this work we have developed a novel process for synthesis of carbon nanotubes (CNT) without requiring the conventional synthesis techniques such as Chemical Vapor Deposition (CVD) process or the process gasses that are used in conventional synthesis techniques. CNT can be synthesized at temperatures of around 500°C. The bulk deposition of the catalyst over vapor deposited fullerene layer yields a mixture of carbon nanotubes with different diameters. By optimizing this process for catalyst deposition, with the help of techniques like Dip Pen Nanolithography, it would be possible to synthesize carbon nanotubes of a desired diameter and hence a 'specific-chirality' at desired locations.

Acknowledgement: The authors gratefully acknowledge the guidance and financial support by Dr. Amit Lal from the Micro Technology Office (MTO) of the Defense Advanced Research Projects Agency (DARPA). The authors also gratefully acknowledge the financial support and monitoring by Dr. Ryan Lu, Mr. Richard Nguyen and Mr. Al Navasca from the Space and Naval Warfare Center (SPAWAR). Support from the Office of Naval Research (ONR) is also acknowledged. Support from the Mechanical Engineering Department at Texas A&M University and the Texas Engineering Experimentation Station (TEES) is also gratefully acknowledged. The FE-SEM acquisition was supported by the National Science Foundation under Grant No. DBI-0116835. During the course of execution of this study, the authors and their research activities were also supported partially through various other research programs and are duly acknowledged here: National Science Foundation (CBET Grant No. 0630703), Micro/Nano-Fluidics Fundamental Focus Center (DARPA-MF3) through the University of California at Irvine, New Investigations Program (NIP 2005) of the Texas Space Grants Consortium (TSGC), Office of Naval Research (ONR), Air Force Office of Scientific Research (AFOSR) through the American Society for Engineering Education (ASEE), the Summer Faculty Fellowship (SFFP) Program at the Air Force Research Labs. (AFRL), Army Research Office (ARO) SBIR Phase II subcontract through Lyntech Inc. and the National Science Foundation (NSF) SBIR Phase I through NanoMEMS Research LLC. We also thank Dr. S. Sinha from the Department of Physics at the University of New Haven for helpful discussions on catalysis and CVD of CNT.

6 REFERENCES

[1] Iijima S.: Helical Microtubules of Graphitic carbon, Nature(1991) 354 56-58.

[2] Salvetat J.: Mechanical properties of carbon nanotubes, Appl. Phys. (1999) 69 255-260.

[3] P.G. Collins., et al: Nanotubes for electronics, Scientific American (2000) 283, 62-69.

[4] Ebbesen T, and Ajayan, P: Large-Scale Synthesis of Carbon Nanotubes. Nature(1992) 358 220–222.

[5] Ouellette J: Building the Nanofuture with Carbon Tubes. The Industiral Physicist,(2002) December 18-21.

[6] Robertson J: Growth of nanotubes for electronics. Materials Today 10, (2007) No 1-2, 36-45.

[7] D. Banerjee, N. Amro, and J. Fragala, "Optimization of microfluidic ink-delivery apparatus for Dip Pen Nanolithography™", *SPIE Journal of Microlithography, Microfabrication and Microsystems ("JM³")*, (2005), vol. 4, pp. 023014-023021.

[8] B. Rosner, T. Duenas, D. Banerjee, R. Shile, and N. Amro, "Functional extensions of Dip Pen Nanolithography™: active probes and microfluidic ink delivery", *IEEE Journal of Smart Materials and Structures*, (2005), Vol. 15, page S124-S130.

[9] J.A. Rivas-Cordona and D. Banerjee, "Microfluidic device for delivery of multiple inks for Dip Pen Nano-lithography", *SPIE Journal of Micro/Nanolithography, MEMS and MOEMS, ("JM³")*, (2007), Vol. 06, No. 03, pp. 033004-12.

[10] D. R. Huitink, D. Banerjee, and S.K. Sinha, "Precise control of carbon nanotube synthesis of a single chirality", *Paper No. IMECE2007-42588, Proceedings of ASME-IMECE, 2007*, Nov. 11-15, Seattle, WA.

[11] R. Gargate and D. Banerjee, "In-Situ synthesis of carbon nanotubes on heated scanning probes using dip pen techniques", *Scanning: The Journal of Scanning Probe Microscopies,* 2008 (in print).

[12] R. Gargate and D. Banerjee, SPIE DSS

[13] R.E. Morjan, O.A. Nerushev, M. Sveningsson, F. Rohmund, L.K.L. Falk, E.E.B. Campbell: Growth of carbon nanotubes from C60, Appl. Phys. A 78, 253–261 (2004).

[14] [Yiming Li, et al,: Growth of Single-Walled Carbon Nanotubes from Discrete Catalytic Nanoparticles of Various Sizes,J. Phys. Chem. B (2001) 105, 11424-11431].

[15] Robertson J: Growth of nanotubes for electronics. Materials Today 10, (2007) No 1-2, 36-45.

[16] Lefrant S., et al: RAMAN & SERS studies of Carbon notubes,Spectroscopy of Emerging Materials (2004), 127-138

[17] Jorio A, Saito R, Dresselhaus G, Dresselhaus M, "Determination of nanotubes properties by Raman spectroscopy", Philosophical Transactions on the Royal Society of London Part A, 362, 2311-2336 (2004).

[18] Thomsen C, Reich S, Maultzsch J, "Resonant Raman spectroscopy of nanotubes", Philosophical Transactions o the Royal Society of London Part A, 362, 2337-2359 (2004).

A Novel Method for the Production of a Vast Array of Metal, Metal Oxide, and Mixed-Metal Oxide Nanoparticles

B.F. Woodfield,[*] S. Liu,[*] J. Boerio-Goates,[*] and L. Astle[**]

[*]Brigham Young University, Provo, UT, USA. Brian_Woodfield@chem.byu.edu
[**]Cosmas, Inc., Provo, UT, USA. Lynn_Astle@byu.edu

ABSTRACT

In this paper we discuss the materials that can now be produced using a novel method for producing high-purity metal, metal oxide, and mixed-metal oxide particles. The particles formed are uniform in size with dimensions that can range from 1 nm to greater than 10 μm. The metal oxides produced by this method include (but are not limited to) the transition metals, rare earth metals, and the Groups I, II, and III metals of the Periodic Table. Mixed-metal oxides of any combination of the aforementioned metals with any stoichiometry can also be produced. Additionally, the final oxidation state of the metal can be controlled. The present method is unique and advantageous because it uses a simple process to prepare large quantities of a vast array of metal oxides, mixed metal oxides, and metals with purity levels as high as 99.999+% with tight control of the particle size (±10%). This method is also much lower in cost and more environmentally benign than other techniques currently available.

Keywords: synthesis, nanoparticles, metals, metal oxides, mixed metal oxides

1 INTRODUCTION

Nanomaterials possess unique and useful chemical, physical, and mechanical properties, and they can be used in a wide range of industrial, biomedical, and electronic applications [1-3]. Exploration of the properties and uses for nanoscale metal oxides, metals, and alloys is underway in a variety of disciplines such as chemistry, physics, materials science, and engineering. The importance of such new interdisciplinary investigations may be realized in the design and characterization of advanced materials. Studies of nanometer-sized metal oxides, mixed-metal oxides, metals, and alloys provide powerful examples of how material performance can be optimized by controlling particle size.

The metal oxides, including mixed-metal oxides, especially those containing transition metals, encompass a wide range of technologically important materials ranging from gas sensors, catalytic supports, optical coatings, electrodes in batteries, and catalysts to gate dielectrics in the semiconductor industry [4-5]. In all of these roles, the surface or interfaces of the oxide dominates the activity. Because of their greater percentage of surface area, nano-crystalline metal oxides are often much more reactive than their conventional or bulk metal oxide counterparts. Considerable effort has already been focused on the synthesis and use of transition metal oxides at the nanoscale [6-8]. For example, zinc oxide nanoparticles are useful in the rubber industry in the areas of activation, acceleration, biochemical activity, dielectric strength, heat stabilization, latex gelatin, light stabilization, pigmentation, reinforcement, rubber-metal bonding, and tack retention. Also, ZnO nanoparticles can be greatly modified in their optical properties to increase the absorption of light in the visible region, a process known as sensitization, which is generally carried out by addition of certain dyes that are absorbed on the surface of ZnO nanoparticles [9-11].

Nanocrystalline metal oxides, including mixed-metal oxides, of elements in the rare earth block also play a vital role in many areas of chemistry and physics. Ceria (cerium oxide), for instance, has been recently used for applications requiring ionic conductivity at low temperatures, such as the manufacture of solid oxide fuel cells and other electrochemical applications [12]. An important property required to use these materials in such applications is nano-crystallinity, that is, to have crystalline particles whose dimensions are in the range of 5 to 100 nm.

Nanometer metals and alloys [13-16] have been extensively used as structural materials in such high performance components as medical implants and aerospace components; they can also be used as catalysts, corrosion-resistant coatings, and permanent magnets. The economic impact of these materials is as unappreciated as the diversity of their applications. For example, the alloys constitute a very important class of materials for electromagnetic applications, such as: (1) magnetic, magnetoresistive, and magnetostrictive applications; (2) superconductor applications, the most powerful superconducting magnets are built with Nb_3Sn; (3) semiconductor and optical applications, alloys are indispensable for optoelectronic devices such as optical switches, diffraction gratings, filters, solar cells, photodetectors, light emitting diodes, and lasers; (4)

magneto-optical applications; and (5) thermoelectric applications.

Techniques for producing nanocrystalline materials fall into one of three general categories [17-19], namely, mechanical processing (milling), chemical processing (precipitation), or thermal processing (evaporation and condensation). It is beyond the scope of this paper to provide a thorough review of existing synthetic techniques. However, conventional techniques for nanoparticle production have been plagued by drawbacks in (a) the formation of uniformly sized and high-purity nanoparticle products, (b) easily producing particles sizes of less than 10 nm, (c) controlling surface morphologies, (d) producing large or even industrial scale quantities, and (e) the high cost (time, specialized equipment, and materials) of production. Additionally, synthetic techniques developed for one metal or mixed-metal oxide often do not transfer to other important nano-oxides. New methods are needed that overcome these problems, enabling the synthesis of a wide range of nano-oxide materials in large quantities using a single transferable technique. Also needed are new and reliable metal oxide nanoparticle products with improved characteristics and properties for use in diverse applications.

2 EXPERIMENTAL

We have recently developed a truly unique and transferable method for producing large quantities of nanometer metal or mixed-metal oxide powders with ultra-high purities and with tight control of the particle size from amorphous to bulk, as well as for producing nanometer metal and metal alloy powders. In general terms, the technique consists of mixing metal-containing starting materials in appropriate stoichiometric ratios to form a novel precursor material, which is later heated to form the nanomaterial product. Details on the actual procedures and reactants will be reported elsewhere. The purpose of this paper is to provide a description of the range of materials that can now be produced and a demonstration of some of their unique properties. Knowledge of the availability of these types of materials is important for those developing new applications. To this end, a summary of the materials that can be produced, the advantages, and unique features of these new nanoparticles and the synthetic method are given below:

1. Practically *all* metal oxides in the transition, rare earth, and group III, IV, etc. blocks can be produced with well controlled particle sizes from 1 nm to bulk.
2. An innumerable array of mixed-metal oxides of any combination of metals (including group I and group II) and stoichiometric ratios with controlled particle sizes from 1 nm to bulk can also be produced. Not only does this open up the ability to produce, for the first time, a vast assortment of nano-sized mixed metal oxides, but this also provides a new synthetic pathway to produce novel metal oxide combinations.
3. The size distribution of the particles is very tight with typical distributions of less than 10%.
4. The chemical purity can be as high as 99.999+% and is only dependent on the purity of the starting materials. Surface morphologies are uniform, surfaces are contaminant-free, the surface water is minimized, and the particles are crystalline. Indeed, the powders produced with this method are free-flowing and minimally agglomerated.
5. The oxidation state of the product can also be controlled to produce high oxidation state oxides (such as Co_3O_4), low oxidation state oxides (such as CoO), and even nano metals and metal alloys.
6. The process is green and produces no solid or liquid waste products.
7. Industrial scale quantities of high purity product can be produced in less than 3 hours. By hand a hundred grams of product can be made in that 3-hour window. With automation, it will be a simple matter to produce kilograms of product.
8. The cost of the product is essentially the cost of the starting materials. No exotic or expensive equipment (plasmas, lasers, vacuums, etc.) are required.

3 RESULTS

Shown in Table 1 is a representative list of materials and their resulting particle sizes that we have synthesized using our new synthetic method.

Material	Size (nm)
CoO	8 ± 1
Co_3O_4	8 ± 1
NiO	$3 \pm 0.5, 9 \pm 1$
CuO	8 ± 1
ZnO	$8 \pm 1, 16 \pm 1$
Fe_2O_3	12 ± 1
Fe_3O_4	10 ± 1
$LiCoO_2$	10 ± 1
In_2O_3	8 ± 1
SnO_2	12 ± 1
Al_2O_3	$2 \pm 0.5, 8 \pm 1$
$NiFe_2O_4$	7 ± 1
$Zn_{0.4}Co_{0.6}Fe_2O_4$	8 ± 1
$Li_{0.15}Zn_{0.3}Ni_{0.4}Fe_{2.15}O_4$	8 ± 1
Y_2O_3	$1 \pm 0.5, 13 \pm 1$
Nd_2O_3	9 ± 1
Ag_2O	65
Ni	40
TiO_2	7
ZrO_2	5
$(Y)ZrO_2$	7
Bi_2O_3	10
Ni, Co alloys	40 - 70

Table 1: Representative examples of materials and the resulting particle sizes synthesized with our new generalized methodology.

As is clearly evident, the range of materials that can be produced using a single methodology is remarkable including transition metal oxides, semi-metal oxides, rare earth oxides, mixed-metal oxides, Li-doped oxides, and nano metals.

We have performed careful characterizations on many of these samples and can show that these materials are phase pure, chemically pure, have a uniform particle size, and maintain their physical properties. For example, shown in Figure 1a is the powder x-ray diffraction (XRD) for CoO collected with a scan rate of 0.2 2θ/min. As can be seen, the CoO product is phase pure even using the slowest scan rate of the diffractometer. The average particle size of 8 nm was calculated using the Scherrer formula from the peak width at half maximum for the principle XRD peak. Shown in Figure 2a is a transmission electron micrograph (TEM) of the CoO sample where the size uniformity is evident and the particle size is consistent with the average size calculated from the XRD data.

Further characterizations of the CoO sample were performed using inductively coupled plasma (ICP) to check for chemical impurities. The ICP results showed the chemical impurities to be less than 10 ppm. The water content of the CoO sample was measured by thermogravimetric analysis (TGA). A sample of CoO was heated under an atmosphere of He gas and the mass change was measured. From these measurements, the water content was calculated to be 2.23%. This same sample of CoO was heated again but in a reducing atmosphere (5% H_2 gas) and to high temperatures. The resulting mass change was due solely to the reduction of CoO to Co metal from which the oxygen content could be calculated. The combined oxygen and water content measurements allowed us to determine the molecular formula of the CoO sample to be $CoO_{1.006} \cdot 0.095 H_2O$. Notice that the water content for these materials is extremely low.

In order to further characterize the physical properties of the CoO sample, we have also measured the specific heat of the sample in the region of the antiferromagnetic transition and compared the results to previous measurements on a high-purity single crystal, a 60 μm powder sample, and a previous 7 nm sample that had typical levels of phase and chemical impurities for a sample produced by precipitation. Shown in Figure 3 is a comparison of the specific heat results for the various samples. Note for the single crystal, the antiferromagnetic transition is sharp and high, the 60 μm sample still has a sharp transition that is not as high as the single crystal transition (as expected), and the 7 nm sample has an identifiable transition that is broad and at a lower temperature. What is surprising is that the specific heat results for our 8 nm CoO sample, while lower than the 60 μm powder data, are still sharp and well defined and are similar in shape to the 60 μm powder data. These results show that the CoO sample produced using our technique must be highly crystalline and, consequently, maintains its magnetic properties even at the nanoscale.

Shown in Figures 1b, 1c, and 1d are further examples of XRD data for other samples produced by our technique, and shown in Figure 2b, 2c, and 2d are further examples of TEM images. Important points that should be noted include (a) the phase purity of the samples as shown in the XRD data, (b) the uniform particle sizes as shown by the TEM images, (c) the well-defined and small size of the NiO sample, and (d) the Y_2O_3 particles shaped as rods and with clearly defined fringe patterns that provide further support for the highly crystalline nature of our particles.

Figure 1: X-ray powder diffraction data for (a) CoO, (b) NiO, (c) $NiFe_2O_4$, and $Li_{0.15}Zn_{0.3}Ni_{0.4}Fe_{2.15}O_4$. Particle sizes were calculated using the Scherrer formula using the most intense peak.

Figure 1a: XRD of CoO

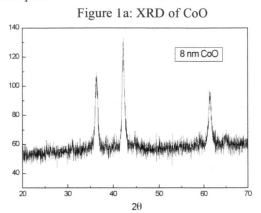

Figure 1b: XRD of NiO

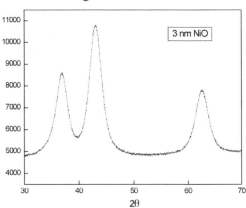

Figure 1c: XRD of $NiFe_2O_4$

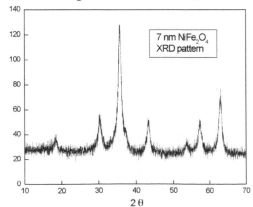

NSTI-Nanotech 2008, www.nsti.org, ISBN 978-1-4200-8503-7 Vol. 1

Figure 1d: XRD of $Li_{0.15}Zn_{0.3}Ni_{0.4}Fe_{2.15}O_4$

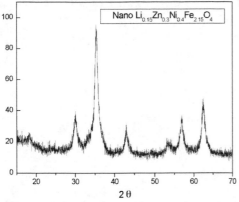

Nano $Li_{0.15}Zn_{0.3}Ni_{0.4}Fe_{2.15}O_4$

2 θ

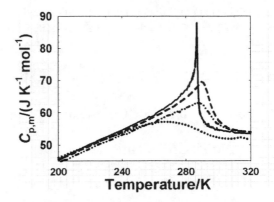

Figure 2: Transmission electron microscopy images of for (a) CoO, (b) NiO, (c) $NiFe_2O_4$, and (d) Y_2O_3. Notice that the Y_2O_3 forms as rods and highly crystalline nature of the particles as evidenced by lattice patterns.

Figure 3: Specific heat of various CoO samples including a high-purity annealed single crystal (□), a 60 □m powder (- - -), a 7 nm CoO of poor quality (•••), and an 8 nm CoO sample produced using our technique (••).

(a)

(b)   (c)

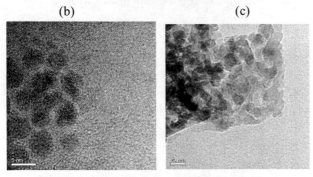

(d)

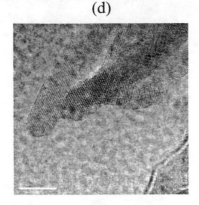

REFERENCES

[1] M. Gell, *Mater. Sci. Eng.* A204 1-2, 246 (1995).
[2] C.C. Koch, *Nanostruct. Mater.* 2(2), 109 (1993).
[3] H. Gleiter, *Nanostruct. Mater.* 1(1), 1-20 (1992).
[4] F.T. Quinlan, R. Vidu, L. Predoana, M. Zaharescu, M. Gartner, J. Griza, P. Stroeve, *Ind. Eng. Chem. Res.* 43, 2468 (2004).
[5] A. Szatvanyi, M. Crisan, D. Crisan, A. Jitianu, L. Stanciu, M. Zaharescu, *Rev. Rom. Chim.* 47(12), 1255 (2002).
[6] G.F. Gaertner, P.F. Miquel, *Nanostruct. Mater.* 4(3), 559-568 (1993).
[7] J.L. Katz, P.F. Miquel, *Nanostruct. Mater.* 4(5), 551-557 (1994).
[8] B. Gunther, A. Kumpmann, *Nanostruct. Mater.* 1(1), 27-30 (1992).
[9] S. Pillai, J. Kelly, D. McCormack, R. Ramesh, *J. of Mater. Chem.* 14, 1572 (2004).
[10] D.W. Bahnemann, C. Kormann, M.R. Hoffmann, *J. Phys. Chem.* 91, 789 (1987).
[11] E. A. Meulenkamp, *J. Phys. Chem. B*, 102, 5566 (1998).
[12] G. Balazs, R. Glass, *Solid State Ionics* 76, 155 (1995).
[13] K.S. Suslick, S.B. Choe, A.A. Cichoulus, M.W. Grinstaff, *Nature* 353, 414 (1991).
[14] K.S. Suslick, T. Hyeon, M. Fang, *Chem. Mater.* 8, 2172 (1996).
[15] X. Cao, Y. Koltypin, R. Prozorov, G. Kataby, A. Gedanken, *J. Mater. Res.* 12, 402 (1997).
[16] K.V.P.M. Shafi, A. Gedanken, R. Goldfarb, I. Felner, Y. Koltypin, *J. Appl. Phys.* 81, 6901 (1997).
[17] C.C. Koch, *Nanostruct. Mater.* 2(2), 109 (1993).
[18] B. Gunther, A. Kumpmann, *Nanostruct. Mater.* 1(1), 27 (1992).
[19] Y. Mizoguchi, M. Kagawa, M. Suzuki, Y. Syono, T. Hirai, *Nanostruct. Mater.* 4(5), 591 (1994).

Structure-Property Correlations in ZnO Tetrapods and Spheres prepared by Chemical Vapour Synthesis

Revathi Bacsa[*], Yolande Kihn[**], Marc Verelst[**], Jeannette Dexpert[**], Wolfgang Bacsa[**] Philippe Serp[*]

[*] Laboratoire de Chimie de Coordination, Composante ENSIACET, 118 Route de Narbonne Toulouse 31077, France, Revathi.Bacsa@ensiacet.fr
[**]CEMES CNRS UPR, 29 rue Jeanne Marvig, Univ. Toulouse, 31055 Toulouse France

ABSTRACT

We present experimental results on the synthesis and characterization of ZnO nanorods, tetrapods and nanospheres prepared by the gas phase oxidation of zinc vapour followed by the collection of the oxide particles in an aerosol. Spherical nanoparticles or fine nanorods of ZnO consisting principally of tetra pod like structures are obtained selectively depending on the process parameters. Intermediate structures can also be synthesized. Transmission electron microscopy of the particles and rods shows that these are single crystals with diameters in the 5-50nm range. We find that the tetrapods have high absorption in the UV and nearly 100% reflectivity in the visible. UV absorption coefficients for suspensions of tetrapods are found to be substantially higher (x10) than for commercial nanocrystalline powders.

Keywords: ZnO, nanoparticlea, nanorods, tetrapos, nanospheres, TEM, UV absorption, Raman spectroscopy

1 INTRODUCTION

Due to its tunable electrical properties and stable chemical structure, zinc oxide has been used in varistor and sensor applications. High purity nano-crystalline ZnO with controlled particle size is used in medicine and cosmetic industry for transparent UV screens since it absorbs both UV A and UV B and has a low photo catalytic activity when compared to TiO_2 [1-3]. To ensure high absorption coefficients and transparency, the development of scaleable synthetic procedures that allow the large scale preparation of size and shape controlled ZnO is necessary. While solution methods suffer from the disadvantages of being laborious and susceptible to contamination by anions or organic additives, the gas phase chemistry has the advantage of producing high purity crystalline products in a single step reaction and is often preferred due to its lower cost [4,5]. Aerosol processes are increasingly being used for the large scale production of metal and oxide nanoparticles [6]. Various precursors have been used for the gas phase synthesis of ZnO that include Zn metal, diethyl zinc and more recently, organometallic precursors such as heterocubane [7-9]. In this com-munication we report the gas phase synthesis of high purity, nanocrystalline and non agglomerated spheres and rods of ZnO produced via a one step gas to particle conversion in an aerosol using zinc metal as the source. We find that spherical or rod shaped particles can be produced selectively in the aerosol by controlling the temperature of zinc vapor and that of oxidation.

2 EXPERIMENTAL

The reactor used was a 60 cm long tubular quartz gas flow reactor (Figure 1). All experiments were performed under atmospheric pressure. Zinc metal powder (325 mesh size) contained in an alumina boat was introduced in the central zone of the three zone furnace maintained under flowing Ar/N_2.

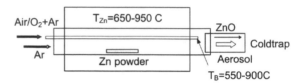

Figure 1. Schematic diagram of the reactor used for the synthesis.

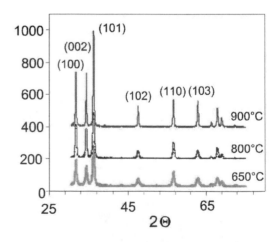

Figure 2. X-ray powder diffractogram of the nano-particles produced at Zn temperatures of 650, 800 and 900°C.

A mixture of air/oxygen +Ar (in the ratio 1:1) in parallel or cross flow is introduced so as to react with Zn vapour at the point B. Zn temperature was varies from 650-900°C and

T_B was varied from 550-900C. The residence time of the zinc vapour was also varied by moving the point B. The Zinc vapor oxidized giving rise thick white fumes or fluffy particles that were collected at the cold end using liquid nitrogen traps. The rate of reaction depended principally on the oxidation temperature. The yield varied from 10 to 50% increasing with increasing oxidation temperatures from 550-900C. The fluffy white material was collected and characterized without further treatment.

3 STRUCTURE AND OPTICAL PROPERTIES

X-ray powder diffractograms of as produced material showed the crystalline phase of wurtzite structure for all temperatures. No preferred orientation was observed (Fig. 2). BET surface areas measured from nitrogen adsorption isotherms showed values ranging from 14 to 28m^2/g depending on temperature and morphology with tetrapods showing the highest value and are comparable to those for commercial samples (Degussa ADNANO20) (Fig. 3).

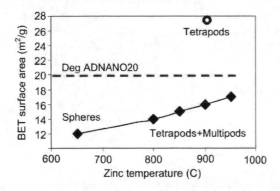

Figure 3. BET surface area as a function of synthesis temperature compared to the SSA for Degussa ADNANO 20.

Figure 4 shows electron micrographs of 4a), 4b) nanospheres and 4c), 4d) nanorods produced at T_{Zn} =650C and 900°C respectively. At 650°C, the ZnO sample consists of nanoparticles with an average size of 26 nm. The average size was measured from over a hundred randomly selected particles. The smaller particles are spherical whereas the larger ones are faceted. At T=900°C, nanorods with length over 1 μm and with maximum diameter of 50 nm are obtained. Intermediate structures can also be synthesized. In contrast, the CVD ZnO collected from the reactor walls shows a mixture of morphologies for all the temperatures showing that the gas phase nucleation permits a better shape control. The atomic ratios of zinc and oxygen are calculated from EDX spectra to be approximately equal to 1. High resolution TEM showed that the spheres and rods are single crystals. The rods consist mostly of tetrapod type structures but also contain multipod type structures consisting of rods growing outward from plates of ZnO and

there is no abrupt change in growth direction. For the tetrapods, there is no correlation between the growth direction of the four legs. The aspect ratios for the rods vary between 10 and 500 and the narrowest of them are around 5-8 nm in width.

The optical absorption spectra for colloidal solutions of ZnO particles and rods in ethanol are shown in Figure 5. All the dispersions are found to be stable for 24 h after sonication for 5 minutes in an ultrasonic bath. The absorption maximum varies from 366 nm for nanoparticles to 371 nm for the rods. In all cases, a shoulder is observed at 358 nm.

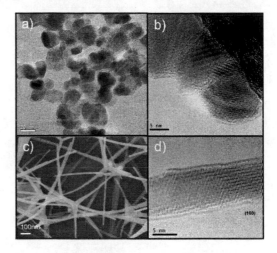

Figure 4. Electron microscopy images of ZnO nanospheres and tertrapods synthesized by CVS at two different temperatures: a) and b) T_{Zn}=650C, T_B=550C, c) and d) T_{Zn}=900C,T_B= 900C.

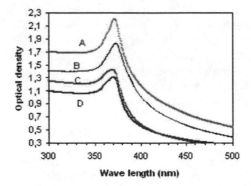

Figure 5. Optical absorption spectra of ZnO colloids A) nanorods, B) intermediate structures, C) nanoparticles and D) Degussa AD Nano 20 in ethanol (0.1mg/mL).

This shoulder is more pronounced for the nanoparticles and for Degussa ZnO than for the nanorods. The shoulder at 358nm is attributed to exciton absorption and is blue shifted with respect to bulk ZnO (380 nm) due to the small size of the colloids. Spanhel and Anderson [10] have measured the exciton spectra of ZnO colloids as a function of the aggregate size. According to this study, absorption at 358nm corresponds to a size of around 5.8 nm that corresponds to the lowest observed aggregate size of our nanoparticles. The band at 275 nm observed by Spanhel et al is absent in our case showing the absence of smaller aggregates. The absorption at 370 nm corresponds to the band edge absorption. For the same concentration, the absorption coefficients are higher for the nanorods by one order of magnitude when compared to spheres showing the influence of shape on the UV absorption efficiency of nanocrystalline ZnO.

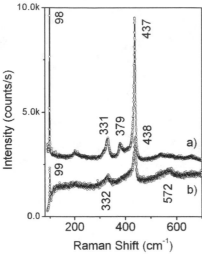

Figure 6. Raman spectra recorded at 632 nm of Zn nanorods (a) and Zn nanospheres (b).

The Raman spectrum of the ZnO nanorods in figure 6.a shows an intense peak at 437cm^{-1} corresponding to the high energy non-polar E_2 mode involving the oxygen sublattice. This band is at the same spectral position as in bulk ZnO. A less intense band is observed at 331cm^{-1} which is attributed to disorder induced second order scattering as has been reported for quantum dots of ZnO [11]. The band at 379 cm^{-1} can be assigned to the polar A_1 (TO) mode and we find in the spectral region of the A_1 (LO) mode at 572 cm^{-1}, a broad band. The intense and narrow (2 cm^{-1}) low energy non-polar E_2 mode of the Zn sublattice is observed at 98 cm^{-1}, slightly down shifted in energy from bulk ZnO (102 cm^{-1}). Figure 6.b shows spectra for nanospheres. The spectral bands are less intense with spectra differences in the region between the A_1 (TO) band at 379 cm^{-1} and E_2 band at 437cm^{-1} and A_1 (LO) band at 572 cm^{-1}. It is not clear at this stage whether a new band emerges between the A_1 (TO) and E_2 band or the E_2 band is asymmetric due to electronic coupling due to mobile charges. This indicates structural differences between nanorods and nanospheres. Surface induced structural changes for nanospheres in the small size range are more important which can lead to dynamic fluctuations between different crystal orientations. The Raman spectrum shows clearly the formation of the wurzite structure for nano spheres and nanorods.

4 CONCLUSIONS

Our results show that a simple substrate-free gas phase oxidation permits the formation of nanoparticles in the form of spheres and rods even at high temperatures without the use of organic stabilizer or organometallic precursors. We assume that Zn vapor forms metal clusters that react instantaneously with oxygen in the mixing zone of oxygen and metal vapor to produce primary particles of zinc oxide in the gas phase that coalesce and aggregate to form larger particles. These particles are directed towards the collector by thermophoresis. Large agglomerates are not observed in this synthesis. Rods or spheres are formed depending on the temperature of oxidation. Intermediate forms are also observed. BET surface areas comparable to commercial nanocrystalline ZnO have been observed for these nanoobjects. The nanorods show ten times higher absorption for UV radiation when compared to commercial nanoscale ZnO powders and can thus function as efficient UV absorbers.

ACKNOWLEDGMENT
The authors thank L. Datas and Y. Thibaud for SEM and EDS analysis. Financial support from Agence Nationale de recherche (ANR Projet PRONANOX, ANR-05-RNMP-002) France is gratefully acknowledged.

REFERENCES
[1] A.C. Dodd, A.J. McKinley, M. Saunders, T. Tsuzuki, J. Nanopart. Res. 8 (2006) 43.
[2] Z.W. Pan, Z.R. Dai, Z.L. Wang, Science, 291 (2001) 1947.
[3] L. Vayssieres, N. Beerman, S. Lindquist, A. Hagfeldt, Chem. Mater. 13 (2001) 4365.
[4] K. Nakaso, K. Okuyama, M. Shimada, S.E. Pratsinis, Chem. Eng. Sci., 98 (2003) 3327.
[5] R. Bacsa, Y. Kihn, M. Verelst, J. Dexpert, W. Bacsa, P. Serp, Surf. Coat. Tech., 201 (2007) 9200.
[6] M. Height, L. Maedler, S. Pratsinis, Chem. Mater. 18 (2006) 572.
[7] S. Polarz, A. Roy, M. Merz, S. Halm, D. Schroder, L. Schneider, G. Bacher and M. Driess, Small, 2005, 5, 540-552.
[8] N. Ramgir, D. Late, A. Bhise, M. More, I. Mulla, D. Joag, K. Vijayamohanan, J. Phys. Chem. B, 110 (2006) 18236.
[9] G. Shen, Y. Bando, D. Chen, B. Liu, C. Zhi, D. Golberg, J. Phys. Chem. B 110 (2006) 3973.
[10] L. Spanhel, M. Anderson, J. Am. Chem. Soc. 113 (1991) 2826.
[11] K.A. Alim, V.A. Fonoberov, A.A. Balandin, Appl. Phys. Lett. 86 (2005) 053103

Two-nozzle flame synthesis of Pt/Ba/Al$_2$O$_3$ for NO$_x$ storage

R. Büchel[a+b], R. Strobel[a], S.E. Pratsinis[a], A. Baiker[b]

[a]Department of Mechanical and Process Engineering, ETH Zurich, Switzerland
[b]Department of Chemistry and Applied Biosciences, ETH Zurich, Switzerland.

ABSTRACT

NO$_x$ storage-reduction (NSR) is applied for exhaust gas treatment of lean fuel engines The proximity between Pt and BaO in Pt/Ba/Al$_2$O$_3$ NSR catalysts affect the storage-reduction due the different spillover length the different interaction with the support. Here, we applied a two-nozzle flame synthesis way to control the location of the deposition of platinum with Pt either on the alumina support or on the Ba storage component. Differently deposited Pt were elucidated by electron microscopy techniques. NSR behavior of these catalysts was investigated in a microreactor by switching between lean and rich conditions. Our studies on the effect of the preferential deposition of platinum on the alumina support or Ba storage component corroborate that the remote control of the different constituents is a crucial factor for high performance of NSR catalysts. Flame synthesis based on a two-nozzle system was shown to be a suitable tool for controlling the proximity of storage and reduction components in NSR catalysts.

Keywords: NOx storage-reduction catalysts, NSR, Flame spray pyrolysis, lean NOx trap

1 INTRODUCTION

NO$_x$ storage-reduction (NSR) is applied for exhaust gas treatment of lean fuel engines [1]. The proximity between Pt and BaO in Pt/Ba/Al$_2$O$_3$ NSR catalysts has been found to affect the storage-reduction behavior due to its effect on spillover of NO$_x$ species[2]. Here we applied flame synthesis to control the location of deposition of platinum either on the alumina support (Figure 1) or on the Ba storage component (Figure 2). Pt on Al$_2$O$_3$ is known for its good NO to NO$_2$ conversion. And NO$_2$ was shown to be stored best on the BaCO$_3$. However, due the longer spillover this system has a disadvantage compared to Pt directly deposited on BaCO$_3$.

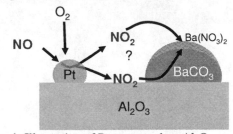

Figure 1: Illustration of Pt supported on Al$_2$O$_3$.

For Pt in close contact with Ba (see Figure 2) the spillover distance is minimal. On the other hand the NO to NO$_2$ oxidation rate is expected to be lower and at temperatures around 600°C BaPtO$_3$ crystals can form reducing the catalytic activity [3].

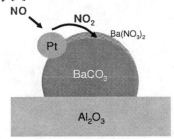

Figure 2: Illustration of Pt in close contact to BaCO$_3$.

2 EXPERIMENTAL

The Pt/Ba/Al$_2$O$_3$ catalysts were prepared using a two-nozzle FSP set up as illustrated in Figure 3 with an angle φ of 160° between the two nozzles that results in mixing of the two flame plumes after 34 cm. A detailed description of the setup can be found in [4].

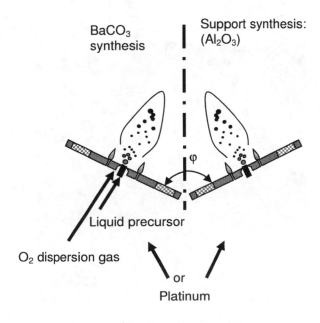

Figure 3: Schematic figure of the used 2 FSP setup

The specific surface area (SSA) of the as-prepared powders was measured by BET (Micrometrics Tristar). X-ray diffraction (XRD) patterns were recorded with a Bruker D8 Advance instrument. The Pt dispersion was measured by CO-pulse chemisorption at 40°C on a Micromeritics Autochem II 2920 unit [5].

For scanning transmission electron microscopy (STEM), the catalyst material was dispersed in ethanol and deposited onto a perforated carbon foil supported on a copper grid. The STEM images were obtained with a high-angle annular dark-field (HAADF) detector attached to a Tecnai 30F microscope (FEI; field emission cathode, operated at 300 kV), showing the metal particles with bright contrast (Z contrast).

The NO_x storage-reduction (NSR) measurements were performed in a fixed-bed reactor. The reactor was allowing rapid switching between oxidizing and reducing conditions. The NO and NO_2 concentrations in the effluent gas were monitored using a chemiluminescence detector (ECO Physics, CLD 822S), and other gases were analyzed by means of a mass spectrometer (Thermostar, Pfeiffer Vacuum). NO_x conversion for a full cycle (one storage and one reduction) was derived from following equation:

$$NO_x \text{conversion} = \frac{NO_{x,in} - NO_{x,out}}{NO_{x,in}} \times 100\% \qquad (1)$$

The NSR was measured at 300°C by switching 10 times between oxidizing (3 min in 667 ppm NO and 3.3% O_2 in He) and reducing atmospheres (1 min in 667 ppm NO, 1333 ppm C_3H_6 in He). The total gas flow rate for all experiments was 60 mL/min.

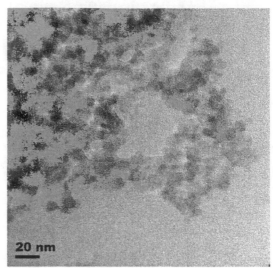

Figure 4: Overlap of TEM and Ba-4M mapping. The green aras indicate the presence of Ba.

3 RESULTS AND DISCUSSION

Preferential deposition of Pt on Al_2O_3 (PtAl-Ba) or $BaCO_3$ (Al-BaPt) was made with the two-nozzle FSP setup as illustrated in Figure 3. The powders resulted in well mixed $Pt/Ba/Al_2O_3$ mixtures as can be seen in Figure 4. The presence of Pt did not measurably influence the materials characteristics of γ-Al_2O_3 or monoclinic $BaCO_3$ in all samples. After 2-3 weeks stored at ambient conditions the monoclinic $BaCO_3$ gradually transformed into thermodynamically more stable orthorhombic $BaCO_3$ [6]. The catalysts exhibited a specific surface area of 140 m^2/g and were non-porous. CO chemisorption showed CO/Pt molar ratios of 0.3 and 0.25 for PtAl-Ba resepectively Al-BaPt. LA-ICP-MS measurements showed that the measured

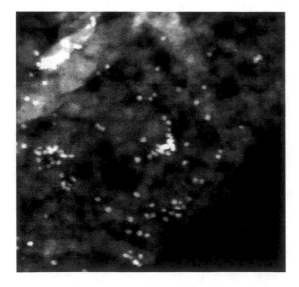

Figure 5: Pt on top of Al_2O_3 in a $Pt/BaAl_2O_3$ matrix

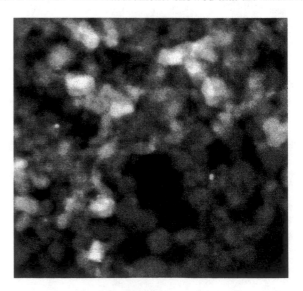

Figure 6: Pt on top of $BaCO_3$ in a $Pt/Ba/Al_2O_3$ matrix

NSTI-Nanotech 2008, www.nsti.org, ISBN 978-1-4200-8503-7 Vol. 1

weight ratios Pt:Ba:Al$_2$O$_3$ was 1:21.3:100 corresponding very well to the prepared ratios of 1:20:100. It has been shown before that FSP made powders conserve the prepared weight ratios rather well [7].

Preferential deposition was confirmed by STEM analysis where Pt particles appeared as bright, spherical dots. In Figure 5 PtAl-Ba can be see how Pt particles are distributed on the alumina. In areas with high Ba content also some Pt particles can be seen there fore the separation is not complete as a small fraction of Pt particles eventually do not follow the main production stream or nucleate after the Al-Ba mixture. In Figure 6 the case with Pt on BaCO$_3$ (Al-BaPt) is shown. Not many Pt particles can be seen as Pt as well a as Ba appear wit similar brightness on STEM pictures.

In Figure 7 the NO$_x$ exhaust gas for a constant NO is plotted for the 4th cycle. The storage-reduction cycles of the three catalysts showed that the performance of PtAl-Ba is best in the beginning, whereas for the reduction (regeneration) Al-BaPt is superior and after 6 cycles it performs better than PtAl-Ba. PtAl-BaPt catalyst perform in between the two curves suggesting that the described effects are independent and can be seen as an other indicator for successful separation of Pt on BaCO$_3$ or Al$_2$O$_3$, respectively.

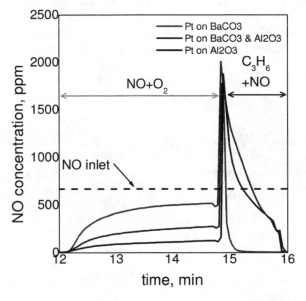

Figure 7: NO storage for different position Pt/Ba/Al$_2$O$_3$ during the 4th cycle.

4   CONCLUSIONS

Two nozzle flame spray pyrolysis is able to separate the position of Pt and produce Pt/Ba/Al$_2$O$_3$ catalysts with preferentially deposition of Pt on BaCO$_3$ and/or Al$_2$O$_3$. The storage-reduction cycling. The location of Pt deposition affected the stability and deactivation behavior of the catalysts as emerged from experiments where the catalysts where exposed to multiple storage-reduction cycling.

PtAl-Ba showed the highest NO$_x$ storage activity, however, NO$_x$-conversion decreased for higher cycle numbers due to the insufficient regeneration of the Ba-nitrates at 300°C. Al-BaPt, on the other hand, maintained its original storage reduction performance under the given conditions.

REFERENCES

[1]   Epling, W.S., et al., "Overview of the fundamental reactions and degradation mechanisms of NOx storage/reduction catalysts," Catalysis Reviews-Science and Engineering.46.163-245, 2004.

[2]   Cant, N.W., I.O.Y. Liu, and M.J. Patterson, "The effect of proximity between Pt and BaO on uptake, release, and reduction of NO$_x$ on storage catalysts," Journal of Catalysis.243.309-317, 2006.

[3]   Casapu, M., et al., "The fate of platinum in Pt/Ba/CeO$_2$ and Pt/Ba/Al$_2$O$_3$ catalysts during thermal aging," Journal of Catalysis.251.28-38, 2007.

[4]   Strobel, R., et al., "Two-nozzle flame synthesis of Pt/Ba/Al$_2$O$_3$ for NO$_x$ storage," Chemistry of Materials.18.2532-2537, 2006.

[5]   Piacentini, M., et al., "Flame-made Pt-Ba/Al$_2$O$_3$ catalysts: Structural properties and behavior in lean-NO$_x$ storage-reduction," Journal of Catalysis.243.43-56, 2006.

[6]   Strobel, R., et al., "Unprecedented formation of metastable monoclinic BaCO$_3$ nanoparticles," Thermochimica Acta.445.23-26, 2006.

[7]   Hannemann, S., et al., "Combination of flame synthesis and high-throughput experimentation: The preparation of alumina-supported noble metal particles and their application in the partial oxidation of methane," Applied Catalysis a-General.316.226-239, 2007.

Nanopowders Synthesis at Industrial-Scale Production Using the Inductively-Coupled Plasma Technology

R. Dolbec, M. Bolduc, X. Fan, J. Guo, J. Jurewicz, T. Labrot, S. Xue and M. Boulos

Tekna Plasma Systems Inc., 2935 Boul. Industriel, Sherbrooke (QC) J1L 2T9 CANADA
richard.dolbec@tekna.com

ABSTRACT

The increasing demand for nanopowders (particles size <100 nm) having very specific properties calls for the development of new technologies that could bring nanopowders synthesis at the industrial scale. The production of rather large volumes of nanopowders involves processing equipments that can provide a complete control of the synthesis conditions in a continuous producing mode with strong reliability, as well as low processing costs. More importantly, such equipments must be designed to ensure the safe recovery and handling of the ultra-fine constituents. Inductively-coupled plasma (ICP) is one of the most promising approach in the production of a wide range of nanopowders with tailored properties, either at laboratory or industrial scales. The ICP technology developed by Tekna Plasma Systems Inc. will be briefly described, while highlighting specific characteristics that make this technology particularly attractive for the synthesis of various types of nanopowder.

Keywords: Inductively-coupled plasma, nanopowder, production.

1. INTRODUCTION TO ICP TECHNOLOGY

At a sufficient high energy, solids can be melted to liquids and vaporized to form gases, which are ionized to generate a plasma. Plasmas are partially ionized gases containing ions, electrons, atoms and molecules, all in local electrical neutrality. Plasmas can be generated by different means, including inductively coupled plasma (ICP, or induction plasma). ICP are generated through the electromagnetic coupling of the input electrical energy into the discharge medium. More specifically, radio frequency (RF) AC currents in a coil (Figure 1a) generate an oscillating magnetic field that couples to the partially ionized gas flowing through the coil (the discharge cavity), generating thereby a stable discharge [1]. Under typical low power conditions (torch power < 100 kW; oscillator frequency of ~3 MHz), the discharge is found to present a diameter of ~20 – 30 mm (Figure 1b), while for high power industrial installation (torch power > 100 kW; oscillator frequency of 200 – 400 kHz), the discharge volume can reach 50 – 100 mm in diameter by 200 – 600 mm long [2]. These physical parameters represent the basic criterion for the design of Tekna's ICP torches [3].

A selection of ICP torches designed and manufactured by Tekna Plasma Systems Inc. is shown in Figure 1c. The ICP torch consists in a water-cooled ceramic confinement tube surrounded by an induction coil of 3 to 7 turns, which is connected to the RF power supply through the tank circuit. The gas distributor head located on the upstream part of the torch is used for the introduction of different types of gas into the discharge cavity. This specific design allows a particular flow pattern that insures a stable discharge in the center of the coil. A water-cooled stainless steel probe, which is inserted through the torch head, is used to inject the reactants (*i.e.* gaseous species, powders, or liquids) coaxially into the center of the discharge. The temperature at the injection point is typically > 10000 K (see modelling results in Figure 2), even though the plasma is generated at atmospheric pressure or under soft vacuum conditions (*i.e.* down to ~1 psi). The downstream end of the torch is a water-cooled exit nozzle that acts essentially as an interface between the plasma torch and the processing chamber. The shape of the nozzle can be either convergent or divergent, depending on the needs of materials processing.

Because of the unique plasma torch design, the ICP technology arises to be a versatile and reliable processing method that became highly attractive in the synthesis and surface treatments of advanced materials over the last decades. Such growing interest for the ICP technology is mainly due to unique features summarized as follow:
- No electrodes (consumable);
- High purity environment (absence of electrode erosion);
- Axial injection of feedstock in the highest temperature zone of the plasma;
- Rather long residence time within the hot gas stream (up to ~500 ms, depending on the reactor design, in comparison to typically < 1 ms in DC plasma unit);
- Large-volume plasma;
- Discharge in various types of atmospheres, namely inert, reducing, corrosive or oxidizing;
- Rather high throughput.

One of the main advantage of the ICP technology is the processing flexibility regarding the chemistry of the plasma gas. Indeed, the absence of electrodes (as found in conventional DC plasma torches) allows plasma generation not only under inert or reducing environments, but also under oxidizing atmosphere [4]. Depending on the nature of

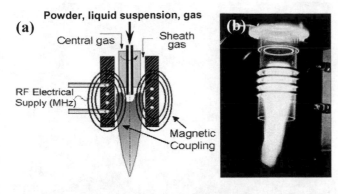

(a) Operation principle of the induction plasma torch. (b) Low power plasma discharge in air. (c) Induction plasma torches manufactured by Tekna Plasma Systems Inc.

(a)

Powder, liquid suspension, gas

Central gas   Sheath gas

RF Electrical Supply (MHz)

Magnetic Coupling

(b)

Figure 1: (a) Operation principle of the induction plasma torch. (b) Low power plasma discharge in air. (c) Induction plasma torches manufactured by Tekna Plasma Systems Inc. [3]. From left to right: PL-70, PL-50 and PL-35 (PL-100 not shown). The number refers to the internal diameter (in mm) of the ceramic confinement tube.

the gas mixture injected in the discharge cavity and, more importantly, on the ionization potential of these gases, various torch performances may be obtained. The gas selection is thus found to depend essentially on chemical reactions to be promoted or avoided in the reactor.

2. NANOPOWDERS SYNTHESIS

The versatility of the ICP technology offers the possibility to modify or produce advanced powders at the nanometer scale. Indeed, the powder spheroidization or the nanopowder synthesis can be performed on the same plasma unit, which needs only minor design modifications. Also, the ability of ICP technology allows the processing of a wide variety of advanced materials of specific properties at a relatively high yield and affordable production cost [4]. This convenient technology is well suitable to face the growing interest that arises not only from academic institutions and research centers, but more recently from industries in their search for a reliable and high capacity manufacturing technology. The two fundamental key features that make the ICP technology attractive are the very high temperature processing and the high quench rate. Since the temperature prevailing in the center of the discharge can reach more than 10000 K, reaction rates are much faster than those found in conventional methods. The local temperature remains fairly high over rather long axial distance, corresponding thereby to residence times sufficiently long to enable the evaporation of most materials. As an illustration, modeling works (for instance, see Figure 2) suggest that the on-axis temperature in PL-50 and PL-70 torches rises above 2000 K for typically 500 ms, and is found to exceed 4000 K over the first ~50 ms. On the other hand, the high quench rate at the exit of the reactor, which is typically ~10^5 K/s, prevents products dissociation

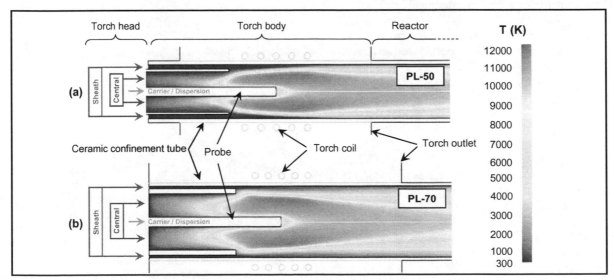

Figure 2: Modeling work showing temperature profiles in Tekna plasma torches operated under typical processing conditions. (a) PL-50 torch operating at 70 kW; reactor pressure: 4 psi; probe: 7.92 mm diameter; Torch gas composition: sheath: 120 slpm Ar + 10 slpm H₂; central: 20 slpm Ar; carrier / dispersion: none. (b) PL-70 torch operating at 100 kW; reactor pressure: 15 psi; Probe: 9.52 mm diameter; torch gas composition: sheath: 180 slpm Ar + 10 slpm H₂; central: 40 slpm Ar; carrier / dispersion: none. A temperature scale (in Kelvin) is also provided on the right hand side of the figure.

and is responsible for particles condensation as ultrafine powders with a typical mean particle size in the nanometer range (e.g. 10 – 100 nm). Moreover, the supersaturation of vapour species provides the driving force for particles condensation, leading to homogeneous nucleation in the gas phase. This permits the production of nanoparticles that have a rather narrow particles size distribution [4].

A wide range of nanopowders have been produced so far using the ICP technology developed by Tekna and many notable works can be found in the literature (see [5 – 9] as a non-exhaustive list). The various types of nanomaterial produced in these studies were synthesized by combining appropriately the plasma gas with the reactant injected into the torch, which can be a gas, a liquid, a solid or a liquid suspension. Other examples of nanopowders prepared with the ICP technology are presented in Figure 3. These materials have been produced at Tekna's facility as part of our R&D program using in-house plasma units dedicated to the development of new processes and new nanomaterials synthesis approaches. While R&D efforts may be required to develop nanomaterials that exhibit very specific properties, general trends can be followed to produce the required material at either small or large production rates. For instance, reducing plasmas (e.g. Ar/H$_2$) are commonly used to produce pure metallic nanopowders with oxygen-free surfaces, such as aluminum nanoparticles (Figure 3a) that are known to be extremely reactive in their non-passivated state. On the other hand, oxygen-rich plasmas (like Ar/O$_2$, Ar/Air and Ar/CO$_2$) are suitable in the synthesis of oxide nanomaterials such as TiO$_2$ (Figure 3b), while inert plasma are more appropriate for maintaining precursor stoichiometry. For example, B$_4$C nanopowders can be successfully produced under an Ar/He inert plasma using B$_4$C micrometric powders as reactant (Figure 3c).

The physico-chemical properties of the nanopowders can be modified through various experimental parameters specific to the ICP technology. One of the most important parameter in the nanoparticle synthesis process is the quench gas flow [7]. The quench gas flow controls the temperature of the gas stream, which in turn influences the degree of vapor super-saturation (S_v) in a gas mixture. S_v is defined as the ratio of the vapor pressure over the saturation vapor pressure at a given temperature. It is also inversely proportional to the gas temperature. In other words, when the temperature of the gas stream carrying the vapor species is sufficiently lowered, S_v exceeds a threshold value beyond which an homogeneous nucleation can occur followed by the vapor condensation in order to form solid nanoparticles. The more rapid is the change in temperature, the finer are the nanoparticles. The quenching effect on the gas stream temperature is illustrated through the modeling work presented in Figure 4. In this figure, a hot gas stream generated by an ICP torch (not shown) is flowing freely in a reactor (Figure 4a). In this case, gas cooling occurs rather slowly, essentially due to natural heat transfer phenomena.

Figure 3: Typical scanning electron microscopy (SEM) images of nanoparticles prepared using Tekna's ICP technology: (a) Pure Al with tailored mean particle size (from left: 100 nm, 60 nm, 25 nm); (b) TiO$_2$ (mean particle size of 100 nm); (c) B$_4$C (mean particle size of 80 nm).

When a quench gas flow is injected radially (yellow arrows on Figure 4b), the gas temperature drops suddenly upstream. The graph in Figure 4c shows the variation of the gas temperature as a function of the axial distance along the dashed line from Figure 4b. It shows that the gas stream exits the ICP torch at a very high temperature (about 9000 K). However, as soon as the hot gas stream encounters the quench gas flow, a cold front is generated and the temperature drops drastically down to ~1000 K within an on-axis distance of less than 2 cm. In this cold front zone, the nanoparticles can nucleate and coalesce up to their final dimension. The diameter of the particles can be determined based on the following equation:

$$d_p = \frac{4\sigma_0}{\rho\left(\dfrac{\kappa}{m}\right)T \ln S_V} \qquad (1)$$

where σ_0 and ρ are surface tension and density of the liquid, respectively, while κ is the Boltzmann constant, m the mass of the vapor molecule and T the local temperature. Obviously, the particle size is predominantly determined by the degree of super-saturation S_v, which is inversely proportional to d_p. Such theoretical concept is supported experimentally. As an example, the influence of the quench gas flow on the mean nanoparticle size is illustrated in Figure 3a. The SEM image shows that under low quench flow conditions (Figure 3a, left), particles are relatively large in size (~100 nm). In contrast, high quench flow conditions (Figure 3a, right) limit particles coalescence after the nucleation stage, leading to nanoparticles that have a much smaller mean diameter (~25 nm).

Figure 5: *Typical 60 kW plasma unit commercialized by Tekna for laboratory-scale nanopowders synthesis. The nanopowder produced in the reactor is carried up to the secondary filter mounted on a glove-box, allowing safe nanopowder collection and handling and packaging under inert environment without interrupting nanopowder production.*

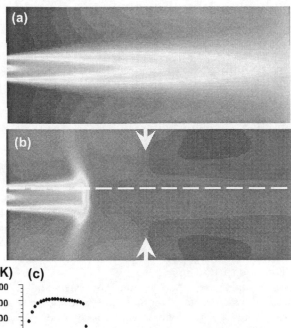

Figure 4: *Influence of the quench gas flow on the gas stream temperature. (a) Temperature profile of the gas stream without quench. (b) Same conditions as in (a), but with a quench gas flow injected radially (arrows). (c) Gas temperature as a function of the axial distance identified by the dashed line in (b).*

The turn-key plasma units developed and manufactured by Tekna are characterized by a high versatility and a very good reliability with appropriate robustness (allowing operation in harsh environment) as well as many fully automated functions for ease of operation. Moreover, the systems are designed to permit nanopowder synthesis in a continuous production mode. These features make the ICP technology particularly attractive in the synthesis of a wide range of nanopowders, at laboratory or industrial scales. Laboratory-scale nanopowders production requires unit with a torch power of typically 30 or 60 kW (two standard power levels commercialized by Tekna), which is normally sufficient to develop new nanomaterials and produce small batches. A typical integrated laboratory-scale (*i.e.* 60 kW) plasma unit is presented in Figure 5. On the other hand, an industrial-scale production requires higher torch power (*i.e.* 100 and 200 kW), essentially to achieve higher production rates.

1 CONCLUSION

The ICP technology commercialized by Tekna Plasma Systems Inc. is perfectly adapted to face the growing needs in new nanopowders development at the laboratory scale or industrial-scale production. The worldwide-recognized expertise in ICP technology developed by Tekna, since its incorporation in 1990, can be utilised to design specific processes or to customize the equipments according to particular needs. By combining versatility and capability, the nanopowder synthesis units proposed by Tekna can produce safely and affordably high-quality advanced nanomaterials.

REFERENCES

[1] M.I. Boulos, P. Fauchais and E. Pfender, "*Thermal Plasmas – Fundamentals and Applications*", Plenum Press, Vol. 1, 1994.

[2] M.I. Boulos, High Temp. Chem. Processes 1, 401, 1992.

[3] U.S. Patent # 5 200 595 and International PCT/CA92/00156.

[4] M.I. Boulos and E. Pfender, MRS Bulletin 8, 65, 1996.

[5] I.A. Castillo and R.J. Munz, Plasma Chem. Plasma Proc. 25, 87, 2005.

[6] Y.-L. Li and T. Ishigaki, J. Phys. Chem. B, 108, 15536, 2004.

[7] B.M. Goortani, N. Mendoza and P. Proulx, Int. J. Chem. Reactor Eng., 4, A33, 2005.

[8] J.W. Shin et al., Plasma Sources Sci. Tech., 15, 441, 2006.

[9] X.H. Wang et al., J. Am. Chem. Soc. 127, 10982, 2005.

Synthesis of Porous Nanostructured Silica Particles by an Aerosol Templating Method

H.D. Jang[*], H. Chang[*], J.H. Park[**], T.O. Kim[***], and K. Okuyama[****]

[*]Korea Institute of Geoscience & Mineral Resources, Daejeon, Korea, hdjang@kigam.re.kr
[**]Sogang University, Seoul, Korea, smartian@sogang.ac.kr
[***]Kumoh National Institute of Technology, Gumi, Korea, tokim@kumoh.ac.kr
Hiroshima University, Higashi-Hiroshima, Japan, Okuyama@hiroshima-u.ac.jp

ABSTRACT

Nanostructured spherical silica (SiO_2) particles having both mesopores and macropores were prepared by using an aerosol templating method with colloidal mixtures of polystyrene latex (PSL) particles and silica nanoparticles. The as-prepared particles showed bimodal size distribution consisting of mesopores ranged 2 to 20 nm and macropores ranged 60 to 160 nm. As the PSL size decreased, mesopores size increased due to a reduction in packing rates of primary SiO_2 nanoparticles composing the walls of nanostructured porous particles. With an increment of the weight ratio of PSL/SiO_2, mesopores size increased but mesopores volume decreased due to the broken structure of particles and the reduction in the packing rates. Mesopores disappeared when the furnace temperature was 900 °C. The residuals of organic template were detected when furnace temperature and flow rate of carrier gas were below 600 °C and above 3 l/min, respectively.

Keywords: macro-mesoporous silica; nanostructures; aerosol templating; catalytic supporter

1 INTRODUCTION

Silica nano-materials with meso- and macroporosity are of great interest due to their variety of potential applications in catalysts, separations, coatings, chromatography, low dielectric constant fillers, pigments, microelectronics, and electro-optics. Especially, for the catalytic application of the nanostructured porous materials, controlled pore size as well as high surface area is required. Therefore, important progress has been made in the controlled synthesis of porous structures [1-4].

Sphere templating methods were able to fabricate macroporous (pore size > 50 nm) materials as well as mesoporous (2 nm < pore size < 50 nm) ones. The methods have several common steps; packing of sphere templates, filling of the interstitial space, and sphere remove [1]. The conventional templating method is time-consuming and somewhat complicate. Recently, Iskandar *et al.* reported in situ production of spherical silica particles containing self-organized mesopores [5] and the controllability of pore size and porosity [6]. However, their works were limited at the structural investigation of the mesoporous silica (SiO_2)

particles. Although the optical properties of nanostructured porous SiO_2 particles having macroporous structure were reported by Chang and Okuyama [7], the characteristics of porous particles for the catalytic application have not been reported in detail, for example, specific surface area, pore size distribution and so on.

In this study, we report on the pore size-controlled synthesis of nanostructured porous SiO_2 particles having both mesopores and macropores via an aerosol templating method. Both of the pores are expected to play an important role in the application as a catalytic supporter, i.e., mesopores increase the activity of the catalysts and macropores permit reactant to approach to the inside of the porous materials. We investigated the effects of the size and mixing ratio of organic template spheres (polystyrene latex particles) in a colloidal mixture, precursor concentration, furnace temperature, and flow rate of carrier gas on the properties of as-prepared particles such as pore size distribution, particle size, particle morphology, and residual of the organic template.

2 EXPERIMENTS

The polystyrene latex (PSL) particles used as an organic template were prepared by following procedure [8]. The mixture of styrene monomer (6.0 g) and *n*-hexadecane (0.25 g) was added to the solution of sodium dodecyl sulfate (SDS) in 24 g of water. The mixture was stirred for 1 h, followed by ultra-sonication for 1 min to form miniemulsion. The miniemulsion was then heated to 70 °C, followed by the addition of 0.12 g of potassium persulfate for polymerization, under a nitrogen atmosphere, and the reaction proceeded for 2 h.

Colloidal suspension of silica nanoparticles (SS-SOL 30, Shin Heung Silicate Co., Korea) was used as a silica source for the preparation of nanostructured porous SiO_2 particles. The nominal average particle diameter of SiO_2 particles was 15 nm. Precursor suspension was prepared by mixing the colloidal suspensions of the PSL and the silica nanoparticles.

Our strategy for the controlled synthesis of nano structured porous SiO_2 particles having spherical shape is the same method with previous research [5-7]. Fig. 1 shows an experimental apparatus used in this work, which consists of an ultrasonic atomizer, an electric furnace, and a filter

NSTI-Nanotech 2008, www.nsti.org, ISBN 978-1-4200-8503-7 Vol. 1

sampler. The ultrasonic atomizer was used to generate micron-sized droplets of the precursor suspension. The droplets were then carried by 1.0, 3.0 or 5.0 l/min of air into the low temperature zone of the electric furnace, in which submicron-sized and spherical SiO_2-PSL mixed particles were prepared by drying of solvent. And then, when the mixture particles were introduced into a high temperature zone, the evaporation of the PSL particles and the restructuring of SiO_2 nanoparticles resulted in nano structured porous SiO_2 particles. The length and diameter of heating zone were 54 cm and 25 mm, respectively. As-prepared nanostructured porous SiO_2 particles were finally collected by the filter sampler.

The morphology of PSL particles and nanostructured porous SiO_2 particles was observed by a transmission electron microscope (TEM; CM12, Philips), a field emission scanning electron microscopy (FE-SEM; Sirion, FEI) and a scanning electron microscope (SEM; JSM-6380LA, JEOL; S-4800, Hitachi), respectively. The size distribution of PSL particles was determined from TEM micrographs by measuring at least 100 particles. N_2 adsorption-desorption isotherms at 77 K were obtained using ASAP 2400 equipment from Micromeritics. The Brunauer, Emmett, and Teller (BET) equation was used for specific surface calculations. The pore size distributions were calculated using the Barrett, Joyner, and Halenda (BJH) method [9]. Fourier transform infrared (FT-IR) spectra were measured by a Nicolet 380 spectrometer (Thermo Electron Co.) from samples prepared as KBr pellets.

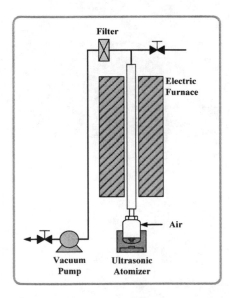

Fig. 1. A schematic diagram of an experimental apparatus for the synthesis of nanostructured porous SiO_2 particles by an aerosol templating method.

3 RESULTS AND DISCUSSION

TEM micrographs and size distributions of the as prepared PSL particles are shown in Fig. 2. The morphology of the PSL particles was spherical and particles size changed with the variation of amount of SDS at the fixed process conditions. The average particle diameter of PSL increased from 62 nm to 161 nm as the amount of SDS decreased from 40 mM to 50 mM. The particle size was very uniform. It was also found that the surface area of PSL was inversely proportional to the adding amount of SDS.

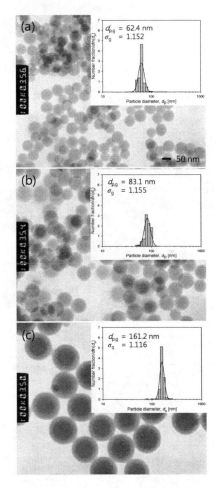

Fig. 2. TEM micrographs and particle size distributions of size-controlled PSL particles prepared with different concentrations of SDS by utilizing a miniemulsion method: (a) 40 mM, (b) 20 mM, (c) 5 mM.

Fig. 3 shows FE-SEM micrographs ((a)-(d)) and desorption pore size distribution (e) of nanostructured porous SiO_2 particles prepared with different sizes of PSL and furnace temperatures ((a) 161 nm, 600 °C; (b) 161 nm, 900 °C; (c) 83 nm, 600 °C; (d) 62 nm, 600 °C) while keeping the weight ratio of PSL/SiO_2, precursor concentration and flow rate of carrier air at 1.43, 0.1 M and 1 l/min, respectively. As-prepared nanostructured porous particles were nearly spherical and their porosity could be easily controlled by changing PSL sizes. The average diameter of macropores on the particle surfaces observed by SEM was corresponded with PSL sizes. Desorption pore size distributions (Fig. 3(e)) and specific surface areas revealed the particles contained mesopores as well as macropores except the particles prepared at 900 °C. The desorption pore size distributions showed broad distributions of mesopores with diameter of 2-10 nm. With a decrement of PSL size, the mesopores size slightly increased. From this result, we could presume that the packing rate of primary SiO_2 nanoparticles composing the walls of nanostructured porous particle decreased when the smaller PSL was used as templates. Besides, we also found that the porous particles prepared at 900 °C could not contain mesopore due to sintering of intra-particles. It means that the sintering should be suppressed by keeping the furnace temperature below 900 °C for the application as a catalytic supporter.

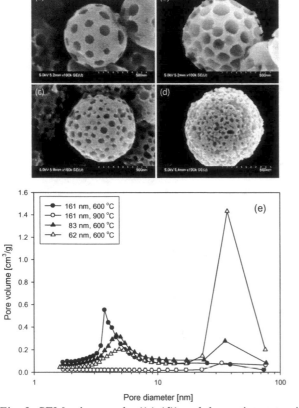

Fig. 3. SEM micrographs ((a)-(d)) and desorption pore size distributions (e) of nanostructured porous SiO_2 particles prepared with different sizes of PSL and furnace temperatures ((a) 161 nm, 600 °C; (b) 161 nm, 900 °C; (c) 83 nm, 600 °C; (d) 62 nm, 600 °C) while keeping the weight ratio of PSL/SiO_2, precursor concentration and flow rate of carrier air at 1.43, 0.1 M and 1 l/min, respectively.

Fig. 4((a)-(d)) shows SEM micrographs of nano structured porous SiO_2 particles prepared from various precursors having different weight ratios of PSL/SiO_2 ((a) 0.48, (b) 1.43, (c) 2.39, (d) 4.77) while keeping the other experimental conditions at constant. The spherical shape

and porous structure were observed regardless of the weight ratios. With an increment of the weight ratio of PSL/SiO_2, the number of macropores increased. However, further increase in the PSL ratio led to the brittle and broken structure. For three contacting spheres (PSL particles) of diameter D (here, D = 161 nm) with a hexagonal packing arrangement, the maximum diameter, d, of a smaller sphere (SiO_2 nanoparticle) that can fit within the void space between sphere is d ≈ 0.155D ≈ 25 nm [6]. The increase in PSL/SiO_2 ratio led to a hexagonal packing arrangement of PSL in a PSL/SiO_2 mixed particle (Fig. 4(c) and 4(d)) and reduced the void space for SiO_2 nanoparticles to be packed. Considering the nominal diameter (15 nm) of SiO_2 nanoparticles, the maximum diameter (d ≈ 25 nm) is too small for SiO_2 nanoparticles to be packed and composed of the internal structure. Therefore, when the PSL packed hexagonally was removed, the brittle and broken structure was obtained. The weight ratio for the stable formation of macroporous particles was below 2.39 in our experiment. The pore size distribution showed broad distributions of mesopores with diameter of 3-6 nm (Fig. 4(e)). With an increment of the weight ratio of PSL/SiO_2, the mesopores size increased but the mesopores volume decreased gradually because the increment of PSL/SiO_2 ratio led to the brittle and broken structure and a reduction in the packing rates of primary SiO_2 nanoparticles.

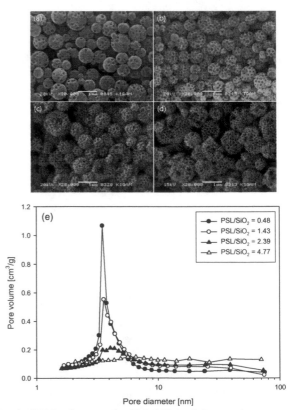

Fig. 4. SEM micrographs ((a)-(d)) and desorption pore size distributions (e) of nanostructured porous SiO_2 particles prepared from various precursors having different weight

ratios of PSL/SiO$_2$ ((a) 0.48, (b) 1.43, (c) 2.39, (d) 4.77; PSL size: 161 nm).

In order to investigate the effect of precursor concentration on the particle size, nanostructured porous SiO$_2$ particles were prepared from different concentrations of SiO$_2$ nanoparticles. The furnace temperature, PSL diameter, weight ratio of PSL/SiO$_2$, and flow rate of carrier air were kept at 600 ^{o}C, 161 nm, 2.39 and 1 l/min, respectively. The geometric mean diameters of the as-prepared particles according to the increment of precursor concentration were 0.56 μm (0.05 M), 0.68 μm (0.1 M), 0.91 μm (0.5 M) and 1.07 μm (1.0 M).

The furnace temperature and the flow rate of carrier gas were manipulated to find process conditions for the removal of residual of the organic template. Fig. 5 shows the FT-IR spectra of nanostructured porous SiO$_2$ particles prepared at different furnace temperatures and flow rates of carrier air: (a) 400 ^{o}C, 1 l/min; (b) 500 ^{o}C, 1 l/min; (c) 600 ^{o}C, 1 l/min; (d) 600 ^{o}C, 3 l/min; (e) 600 ^{o}C, 5 l/min. The FT-IR spectra revealed the C-H stretching vibrations at 3000-2850 cm^{-1}, the C=O stretching at 1715 cm^{-1}, C-H bending at 1375 cm^{-1}, C-O stretching at 1230 cm^{-1}, and Si-O-Si stretching at 1100 cm^{-1}. The vibrations by carbon content such as methyl and carbonyl groups continuously disappeared with an increment of furnace temperature and a decrement of flow rate of carrier gas. It was required to remove the residual of the PSL in our experiments at the furnace temperature above 500 ^{o}C and flow rate of carrier gas below 5 l/min.

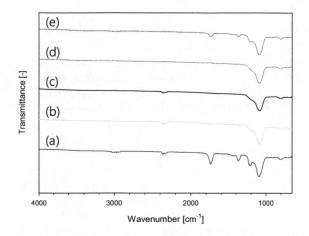

Fig. 5. FT-IR spectra of nanostructured porous SiO$_2$ particles prepared at different furnace temperatures and flow rates of carrier air: (a) 400 ^{o}C, 1 l/min; (b) 500 ^{o}C, 1 l/min; (c) 600 ^{o}C, 1 l/min; (d) 600 ^{o}C, 3 l/min; (e) 600 ^{o}C, 5 l/min.

4 CONCLUSIONS

Pore size-controlled synthesis of nanostructured porous silica particles containing both mesopores and macropores by the aerosol templating method was successfully conducted for the catalytic application. The smaller PSL template led to the decrement of the packing rate of primary SiO$_2$ nanoparticles composing the walls of nanostructured porous particles, resulting in the larger volume of mesopores. High furnace temperature reduced mesopores volume due to the sintering of intraparticles. Therefore, for the application as a catalytic supporter, the sintering should be suppressed by keeping the furnace temperature below 900 ^{o}C. The increment of PSL/SiO$_2$ ratio resulted in a brittle and broken structure of particles and a reduction in the packing rates of primary SiO$_2$ nanoparticles. It was required to remove the residual of the PSL in our experiments at the furnace temperature above 500 ^{o}C and flow rate of carrier gas below 5 l/min.

ACKNOWLEDGEMENTS

This research was supported by the Basic Research Project of the Korea Institute of Geoscience and Mineral Resources (KIGAM) funded by the Ministry of Science and Technology of Korea.

REFERENCES

[1] A. Stein, Microporous Mesoporous Mater. 44-45 (2001) 227.
[2] O.D. Velev, T.A. Jede, R.F. Lobo, A.M. Lenhoff, Nature 389 (1997) 447.
[3] B.T. Holland, C.F. Blanford, A. Stein, Science 281 (1998) 538.
[4] J.E.G.J. Wijnhoven, W.L. Vos, Science 281 (1998) 802.
[5] F. Iskandar, Mikrajuddin, K. Okuyama, Nano Lett. 1 (2001) 231.
[6] F. Iskandar, Mikrajuddin, K. Okuyama, Nano Lett. 2 (2002) 389.
[7] H. Chang, K. Okuyama, J. Aerosol Sci. 33 (2002) 1701.
[8] K. Landfester, N. Bechthold, F. Tiarks, M. Antonietti, Macromolecules 32 (1999) 5222.
[9] E. Barret, L. G. Joyner, P. P. Halenda, J. Am. Chem. Soc. 73 (1951) 373.

^{1} Nanomaterials group, Korea Institute of Geoscience and Mineral resources, 30 Gajeondong, Yuseonggu, Daejeon, 305-350, Korea, Ph: (42) 868-3612, Fax: (42) 868-3418, hdjang@kigam.re.kr

Photocatalytic Activity of Nitrogen Doped Nano-titanias and Titanium Nitride towards Methylene Blue Decolouration

Z. Zhang [a,b], X. Weng [a,b], K. Gong [a] and J. A. Darr [a]

[a] Department of Chemistry, University College London, Christopher Ingold Laboratories, 20 Gordon Street, London WC1H 0AJ, UK

[b] School of Engineering and Materials Science, Queen Mary, University of London, Mile End Road, London, E1 4NS, UK, zhice.zhang@qmul.ac.uk

ABSTRACT

The potential for synthesis nitrogen-doped TiO_2 system with various nitrogen loadings was conducted by nitridating "as prepared" nano-TiO_2 made by a continuous hydrothermal flow synthesis (CHFS) process and exposing it to ammonia/argon atmosphere at range of temperatures (from 400 to 1100 °C). This process has proved to be simple, and nitridation reaction was easy to control. Products ranging from N-doped TiO_2, and phase pure titanium nitride (TiN) were obtained at increasing heat treatment temperatures. Our results suggested that the TiN phase started forming at 800 °C. Pure phase TiN was obtained at 1000 °C after 5 h nitridation. A shift of the band gap to a lower energy and increasing absorption in the visible light region were observed with increasing calcination temperature from 400 to 700 °C. The results of photodegradation of methylene blue (MB) solution using simulated light irradiation suggested that at lower N-doping levels, the photocatalytic activity of TiO_2 after nitrogen doping was greatly improved.

Keywords: nitrogen doping, nitridation, TiO_2, TiN, photocatalyst

1 INTRODUCTION

Recently, visible-light activated titanium dioxide photocatalysts have been obtained by doping anionic elements into the host lattice. These anionic dopant species were found to be better than transition-metal dopants, with respect to stability of the catalyst. The red shift of the band-gap into the visible-range makes it possible to carry out a wide range of photocatalytic investigations under visible light. Asahi et al.[1] reported that band-gap narrowing by nitrogen doping in TiO_2, yields high photocatalytic activities under visible light. They prepared N-doped TiO_2 films by means of reactive magnetron sputtering in N_2 (40%)/ Ar gas mixture and then annealed in N_2. Meanwhile, they synthesized N-TiO_2 powders by treating anatase TiO_2 (BET surface area, 270 m^2/g) in the NH_3 (67%)/Ar atmosphere at 600 °C for 3 hours. Samples included the total N contents of ca. 1 at.%. Asahi et al.[1] declared that the active sites for photocatalyst, under visible light, were due to the substitutional nitrogen into the oxide lattice.

Nitrogen doped TiO_2 was made by sputtering TiO_2 in N_2 reactive atmosphere,[1] annealing TiO_2 or titanium compounds at elevated temperature under the NH_3 gas flow,[2;3] using N-containing precursors in a sol-gel process,[4] oxidising the TiN by annealing in air,[5] pulsed laser deposition,[6] chemical vapour deposition (CVD) using titanium metal-organic precursors in ammonia gas,[7] ion-beam-assisted deposition (IBAD) technique using TiO_2 vapour and nitrogen ions.[8] The range of doping concentrations reported recently for photocatalysts has been very narrow, generally lower than 1 at.%. However, TiO_2 with higher nitrogen load, or even titanium nitride (TiN), are also of interest in other applications. The common methods of nitride or oxynitride thin film creation are physical vapor deposition (PVD), reactive sputtering, cathodic Arc Deposition or electron beam heating and chemical vapor deposition(CVD), which are required to react with nitrogen under high-energy, vacuum conditions. However, these synthesis routes always involve in expensive apparatus and complicated synthesis process.

The aim of the current work was to obtain light active N-doped photocatalysts by calcinations of continuous hydrothermal flow synthesis (CIIFS) nano-TiO_2 (ca. 4.8 nm in diameter) in ammonia/argon atmosphere. This method was also used to synthesize nitride nanopowders with different nitrogen loadings from nano-TiO_2 by simply elevating the heating temperatures. The photocatalytic activity of the obtained catalysts was estimated on a basis of decomposition of MB solution under the light irradiation.

2 EXPERIMENTAL

2.1 Materials

Titanium (IV) bis(ammonium lactato) dihydroxide (also called TiBALD, 50 wt% in water) was obtained from Sigma-Aldrich (Dorset, UK). Mixed ammonia /argon (60:40 vol%) gas was obtained from BOC (UK), with purity quoted higher than 99%. Methylene blue (high purity, ≥99.9%) was obtained from Acros Organics UK. All chemicals were used as received.

2.2 Experimental methods

The "as prepared" virgin nano-powders used for heat treatments were made using a continuous hydrothermal flow synthesis (CHFS) system (see Figure 1). For the synthesis of nano-TiO_2, an aqueous solution of TiBALD (0.4 M) was pumped to meet a flow of room temperature water at a "Tee" junction and this mixture was brought to mix with a stream of superheated water at the reaction mixing point (a countercurrent mixer), whereupon rapid precipitation of crystalline anatase occurred. The slurry was then collected HPLC pump rates of 20, 10 and 10 mL/min were used for superheated water (via Pump 1), titania precursor (TiBALD) solution (Pump 2) and the cold water feed (Pump 3), respectively. The third feed (Pump 3) was used in order to give inline dilution. Slurries were then centrifuged (5000 rpm for 60 min), washed twice, and freeze dried at *ca.* 1.33×10^{-4} MPa overnight (22 h). According to the mass of the dry powder, the yield was calculated as *ca.* 90 %.

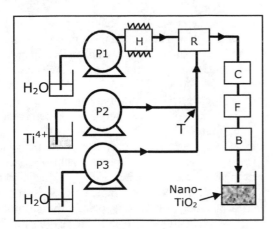

Figure 1 Schematic representation of the three-pump (P1, P2 and P3) continuous hydrothermal flow synthesis system that was used to prepare nano-TiO_2. Key: P = pump, C = cooling, F = filter, B = back-pressure regulator, R = reactor

Freeze-dried nano-TiO_2 (ca. 0.5 g), synthesized by CHFS, was placed on a platinum plate before being transferred into the tube furnace. The furnace was heated from room temperature to a set point (this ranged from 400 to 1100 °C over all the experiments) at rate of 10 °C/min, and maintained for up to 5 h. Then the furnace was allowed to cool down at 20 °C /min, under flowing NH_3 /Ar. The flow rate of feeding gas was kept at *ca.* 200 mL/min.

A 50 mL aliquot of standard MB solution was added into each glass beakers along with 0.01 g of the particular nitridated sample. The mixtures were mixed using a magnetic stirrer bead for 30 min in the dark to allow

adsorption-desorption equilibrium of dye on the catalyst surface to be established. The photocatalytic reactors containing the catalyst were irradiated with light for 60 min periods for a total of four hours. After each period of illumination the absorbance of the MB was measured. The calculated concentrations of the MB with irradiation time are given for the $TiO_{2-x}N_x$ samples.

2.3 Characterizations

Freeze-drying was performed using a Vitris Advantage Freeze Dryer, Model 2.0 ES, supplied by BioPharma; the solids were frozen in liquid nitrogen and then freeze dried for 22.5 h at 1.33×10^{-4} Mpa. X-ray powder diffraction (XRD) data were collected on a Siemens D5000 X-Ray diffractometer using Cu-Kα radiation (λ= 0.15418 nm). Data were collected over the 2θ range 20-80 ° with a step size of 0.02° and a count time of two seconds. Crystallite size was calculated from XRD pattern peak half-widths using the Scherrer equation. BET surface area measurements (multipoint) were performed on a Micromeritics Gemini analyzer. All powders were degassed at 80 °C for 2 h prior to BET analysis. UV-Vis absorption spectra of methylene blue (MB) solutions and colours of the samples were recorded using a PerkinElmer (Lambda 950) UV-Vis spectrophotometer.

3 RESULTS AND DISCUSSIONS

The "as prepared" TiO_2 was pure phase anatase, with surface area ca. 290 m^2/g. Identical powdered nanosized TiO_2 samples were heat treated in NH_3 /Ar gas under different temperatures each and then cooled to give the final powders. The colours of powders ranged from "off white" for "as prepared" nanosized TiO_2, to "dark brown" for pure TiN that was nitridated at 1000 °C for 5 hrs at high gas flow rate (200 mL/min); see Figure 2.

Figure 2 Photo of samples nitridation under different conditions: CHFS synthesized pure TiO_2 power (off white) and TiO_2 powders nitridated at different temperatures (ranging 400 to 1100 °C) for 5 h.

3.1 Phase and morphology

In the XRD data for nitridated samples (Figure 3), single anatase-like phases were observed on all samples nitridated at 600 °C and below. Rutile peaks were first observed at the sample nitridated at 700 °C, which coincided with findings reported in our previous study on the heat treatment of nano-TiO_2 powders in air. It can be assumed that the presence of the nitrogen does not have any significant influence on the transformation temperature of anatase to rutile. Distinct peaks assigned to TiN appeared in the XRD plot of the sample heat treated at 800°C in NH₃/Ar, indicating that the cubic-phase TiN began to form. A small amount of rutile (TiO₂) was also observed. Phase pure TiN was obtained at 1000°C after 5 h heating in NH₃/Ar.

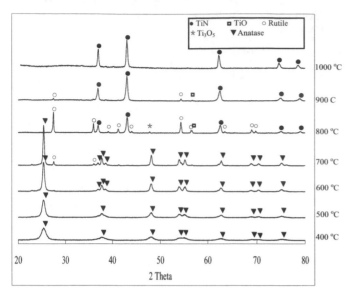

Figure 3 XRD patterns take at room temperature of nanocrystalline TiO_2 powders nitrided at different temperatures, in the range 400 to 1000 °C in NH₃/Ar (flow rate of 200 mL/ min) for 5 h.

The crystallite size of nitrided TiO_2 samples was calculated from XRD pattern peak half width by applying the Scherrer equation and is illustrated in Figure 4. It is suggested that increasing the temperature generally, increased crystallite sizes during nitridation. Surface areas of all nitrided samples generally decreased with nitridation temperature and dwelling time in the furnace, from 98 m²/g (400 °C, 5 h) to 9 m²/g (1100 °C, 5 h). Figure 4 shows the relationship between the particle size (calculated by BET results, based on the assumption of a perfect sphere), crystallite size (determined from XRD half peak width by application of the Scherrer equation), and the nitridation temperature (same dwelling time of 5 h). A critical temperature is observed at 900 °C; above this temperature, the particle size increased rapidly from 62.2 nm (at 900 °C) to 105.7 nm (at 1100 °C). It was also found that powders annealed at lower gas flow rate (100 mL/min) at 900 °C/ 5 h, had noticeably lower surface area (15.0 nm vs. 16.5 nm) than those made at high rate (200 mL/min), which suggested the greater presence of ammonia gas could hinder the crystal growth of titania at some degree. This was also observed by Chen *et al* [9] and may be due to the adsorption of NH₃ molecules which prevent active surfaces from contacting and agglomerating. However, NH₃ molecules will decompose when electron transfer occurs from surface O^{2-} to NH₃ at high temperature. Such hindrance is expected to be weaker (due to the

decomposition of NH₃) especially for longer dwelling times and higher temperatures [9]. It typically gave an average crystallite size of 79 nm and particle size (calculated by BET surface area assuming a sphere) of 90 nm, respectively,

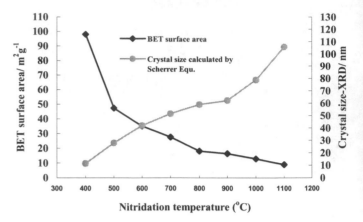

Figure 4 BET surface area and crystallite size (calculated from XRD by Scherrer Equation) dependence of nitrided powders prepared at different temperatures.

UV-Vis spectra measured by diffuse reflectance of the samples. Applying the Kubelka-Munk model, the absorbances were determined and band gap edge position was calculated for the samples. Samples nitrided at 400, 500, 600, and 700 °C showed the band gap absorption onset at 419.6, 451, 485, and 526 nm, corresponding to energy band gaps of 2.97, 2.75, 2.56, and 2.36 eV, respectively. Consequently, the optical absorption edges of these N - doped samples shift to the lower energy region compared to "as prepared" pure nano-TiO_2 (E_g= 3.12 eV), and the absorptions after nitrogen doping are stronger in the visible range. It is expected that nitrogen doping contributed to the red shift because of the narrowing of the band gap. In the case of catalysts nitrided at a temperature of 800 °C (at which phase transition start taking place), the absorption edge in the visible range is at 430 nm (2.89 eV). The samples heated at 900 °C and 1000 °C revealed broad absorption bands across the spectral range, that's due to the dark colour of the high nitrogen loaded samples.

3.2 Photocatalytic decolorization of MB

The photocatalytic reactors containing catalysts were irradiated with visible light after 30 min stirring in the dark to allow the dye adsorption to reach equilibrium onto the catalyst's surface. At 60 min intervals (during the irradiation), the light was switched off and the intensity of the 662 nm UV-Vis band absorbance was quickly measured for each sample.

Samples nitrided at low temperatures (eg. 500 and 600 °C) generally exhibited better photocatalytic activities over the 4 h timeframe of the experiment. (Figure 5) The photocatalytic activities of these yellowish catalysts are

NSTI-Nanotech 2008, www.nsti.org, ISBN 978-1-4200-8503-7 Vol. 1

even better than the high surface area starting material "as prepared" nano-TiO$_2$. After partial N-doping into TiO$_2$, the local structures of titanium dioxides species are altered, which not only improves the absorption of titania in the visible light region but also decreases the recombination of photoinduced electrons and holes. Therefore, it is reasonable that the photocatalytic activities of nitrogen doped titania are improved accordingly. By comparison the powder treated at 400 °C showed comparably lower photocatalytic degradation activity even though its surface

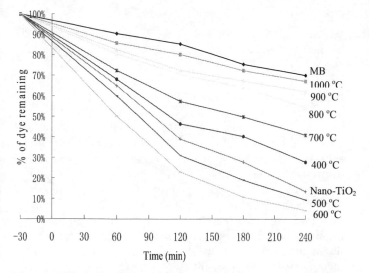

Figure 5 Photocatalytic degradation of the methylene blue dye for TiO$_2$ powders nitrided at different temperatures for 5 h dwelling time, as a function of irradiation time under a visible-light source. Experiment was undertaken in dark for the first 30 min, and then irradiated in visible light.

area was higher than others. There was no improvement in photodegradation efficiency for samples nitrided at the other temperatures used. For samples heated at higher temperatures, the decrease in surface area of samples resulted in less active. This is despite the fact that some of these samples have the highest displacement of their absorption radiation edges in the visible range.

4 CONCLUSIONS

A simple method for the synthesis of nanocrystalline TiO$_2$, nitrogen-doped TiO$_2$, and titanium nitride powder has been developed. In this process, TiN nanocrystalline powder can be obtained by the direct nitridation of nanocrystalline TiO$_2$ powder for 5 h in an NH$_3$/Ar gas flow of 200 mL/min. Particle size of TiN nanocrystalline powder obtained at 1000 °C/5h is ca. 89.5 nm.

Nitrogen-doped TiO$_2$ (i.e. titanium oxynitride) photocatalysts with only the anatase phase and a relatively small crystallite size (11.4 to 51.5 nm) were obtained under the nitridation temperatures in the range 400 to 700 °C. Powders prepared at 600 °C had the best photocatalytic

activity towards decolorization of MB under visible-light irradiation, even better than the "as prepared" starting TiO$_2$ nanopowder. The higher activity of the N-doped samples could be correlated with the modification in their absorption spectra.

ACKNOWLEGEMENTS

EPSRC is thanked for funding the "High Throughput Inorganic Nanomaterials Discovery" project for developing better photocatalysts [EPSRC Grant Reference: EP/D038499/1] (JAD, KG). QMUL is thanked for the college scholarship (ZZ, XW). M. Phillips, V. Ford, J. Caulfield, M. Willis, Z. Luklinska, and N. Mordan are thanked for technical assistance.

REFERENCES

(1) Asahi, R.; Morikawa, T.; Ohwaki, T.; Aoki, K.; Taga, Y. *Science* **2001**, *293*, 269-271.
(2) Ramanathan, S.; Oyama, S. T. *Journal of Physical Chemistry* **1995**, *99*, 16365-16372.
(3) Li, J. G.; Gao, L.; Sun, J.; Zhang, Q. H.; Guo, J. K.; Yan, D. S. *Journal of the American Ceramic Society* **2001**, *84*, 3045-3047.
(4) Ihara, T.; Miyoshi, M.; Iriyama, Y.; Matsumoto, O.; Sugihara, S. *Applied Catalysis B-Environmental* **2003**, *42*, 403-409.
(5) Morikawa, T.; Asahi, R.; Ohwaki, T.; Aoki, K.; Taga, Y. *Japanese Journal of Applied Physics Part 2-Letters* **2001**, *40*, L561-L563.
(6) Suda, Y.; Kawasaki, H.; Ueda, T.; Ohshima, T. *Thin Solid Films* **2004**, *453-54*, 162-166.
(7) Pradhan, S. K.; Reucroft, P. J.; Yang, F. Q.; Dozier, A. *Journal of Crystal Growth* **2003**, *256*, 83-88.
(8) Wu, P. G.; Ma, C. H.; Shang, J. K. *Applied Physics A-Materials Science & Processing* **2005**, *81*, 1411-1417.
(9) Chen, H. Y.; Nambu, A.; Wen, W.; Graciani, J.; Zhong, Z.; Hanson, J. C.; Fujita, E.; Rodriguez, J. A. *Journal of Physical Chemistry C* **2007**, *111*, 1366-1372.

Controlled Encapsulation of a Hydrophilic Drug Simulant in Nano-Liposomes Using Continuous Flow Microfluidics

A. Jahn[1,3], J.E. Reiner[1], W.N. Vreeland[2], D. DeVoe[3], L.E. Locascio[2] and M. Gaitan[1]

[1]NIST, Semiconductor Electronics Division, EEEL, Gaithersburg, MD 20899 USA
[2]NIST, Biochemical Science Division, CSTL, Gaithersburg, MD 20899 USA
[3]University of Maryland, Biomedical Engineering, College Park, MD 20742 USA

ABSTRACT

A new method to tailor the size and size distribution of nanometer scale liposomes and control loading of liposomes with a model drug in a continuous-flow microfluidic design is presented. Size and size dispersion are determined with tandem Asymmetric Flow Field Flow Fractionation (AF[4]) and Multi-Angle Laser Light Scattering (MALLS). Fluorescence Correlation Spectroscopy (FCS) combined with Fluorescence Cumulant Analyis (FCA) allow for determining the number of encapsulated molecules [1]. Results show that this system allows for control of loading efficiency as well as minimization of encapsulant consumption.

Keywords: liposomes, encapsulation, fluorescence spectroscopy, hydrodynamic focusing, microfluidics

1. INTRODUCTION

Liposomes are a promising example of nanoparticles for a wide variety of medical applications including quantized reagent packets for the delivery of drugs, therapeutic agents and proteins [2-4], the encapsulation of contrast agents for enhanced magnetic resonance imaging [5], and model systems for the study of biological membranes. Traditional liposome preparation methods are mostly conducted through mixing of bulk phases, leading to inhomogeneous chemical and/or mechanical conditions during the formation process, hence producing liposomes that are often polydisperse in size and lamellarity. Traditional methods are often accompanied with additional process steps such as membrane extrusion or sonication to yield the desired homogenous liposome populations. A previously reported microfluidic method to produce homogeneous liposome populations omitting size altering post processing steps is applied to determine its feasibility for controlled encapsulation of a hydrophilic drug simulant [6,7].

A stream of lipids dissolved in isopropyl alcohol is hydrodynamically focused and sheathed between two oblique aqueous buffer streams in a microfluidic channel as shown in Figure 1. The laminar flow in the microfluidic channel enables controlled mixing at the liquid-liquid diffusive interface causing the lipids to be insoluble and self-assemble into vesicles. While in traditional bulk mixing techniques (e.g., test tubes, macroscale fluidic containers, and mixers) individual fluid elements experience uncontrolled mass transfer profiles and mechanical stresses; microfluidics enable precise and reproducible control of the flow conditions and hence reproducible fluidic mixing on the micrometer length scale. This reproducibility allows for facile adjustment of the lipid self-assembly and the resultant liposome size and size dispersion. Using this technique, the liposome size is tunable over a mean diameter of 50 nm to 150 nm by adjusting the aqueous to solvent volumetric flow rate ratio.

For drug delivery applications, the encapsulation efficiency needs to be controlled as well; further, it is desirable to minimize the waste of the encapsulant not incorporated into the interior aqueous space of the liposome. In contrast to conventional mixing techniques, where liposomes are formed in a bulk aqueous solution containing a homogenous concentration of the water-soluble encapsulant, the microfluidic method allows confining the encapsulant to the immediate vicinity where lipids self-assemble into liposomes and concomitant encapsulation is expected to occur. Here we report on the ability to control the loading efficiency, which is the concentration of molecules encapsulated in liposomes with respect to a starting concentration, by adjusting the flow parameters in the microfluidic channel network.

2. EXPERIMENTAL SECTION[1]

2.1 Device Fabrication

Microfluidic channels were fabricated in a silicon wafer as previously described [7]. Briefly, a microchannel network was transferred with photolithographic techniques onto a silicon wafer (76.2 mm (3 in.) diameter, 305 μm to 355 μm thick, Nova Electronics Materials, Inc., Carollton, TX), subsequently etched with deep reactive ion etching and sealed by anodic bonding to a borosilicate glass wafer.

---

[1] Certain commercial materials and equipment are identified in order to adequately specify the experimental procedures. In no case does such identification imply recommendation or endorsement by the National Institute of Standards and Technology, nor does it imply that the items identified are necessarily the best available for the purpose.

The resulting channels have a rectangular cross section with a depth of 100 µm and a width of 42 µm or 64 µm. Commercially available Nanoports (F-124S, Upchurch Scientific, Oak Harbor, WA) were bonded to the backside of the silicon wafer. Polyetherketone (PEEK) tubing connected the Nanoports to a syringe. All fluidic reagents were injected through a 0.2 µm syringe filter (Anatop, Whatman, NJ) by syringe pumps (model PHD 2000, Harvard Apparatus Inc., Holliston, MA).

2.2 Preparation of Lipid Mixture and Model Drug

Dimyristoylphosphatidylcholine (DMPC), cholesterol (both Avanti Polar Lipids Inc., Alabaster, AL), and dihexadecyl phosphate (DCP) (Sigma Aldrich) in a molar ratio of 5:4:1 were dissolved in chloroform (Mallinckrodt Baker Inc., Phillipsburg, NJ). The chloroform solvent was evaporated under a stream of nitrogen to form a dry lipid film on the bottom of a scintillation vial. The vials were subsequently stored in a desiccator for at least 24 h to ensure complete solvent evaporation. The dried lipid mixture was redissolved in 4 mL dry isopropyl alcohol (IPA) at a 5 mmol/L concentration of total lipid. Phosphate buffered saline (PBS) solution (10 mmol/L phosphate, 2.7 mmol/L potassium chloride, 130 mmol/L sodium chloride, 3 mmol/L sodium azide, pH 7.4) was used as a hydration buffer. Sulforhodamine B (SRB) dye was dissolved at 0.5 mmol/L concentration in phosphate buffered saline.

2.3 Liposome Preparation

Nanoscale liposomes were prepared with a flow focusing technique described earlier [6,7]. Briefly, a lipid mixture dissolved in IPA is injected into the left center channel of the microfluidic network shown in Figure 1. SRB dye dissolved in PBS is injected into the two inner side channels and PBS without SRB dye into the two outermost side channels all intersecting with the center channel. Liposome size and size distribution studies were performed at a constant volumetric flow rate (VFR) of 100 µL/min and an aqueous to solvent flow rate ratio (FRR$_{A/S}$) varying from 14 to 39, defined as the ratio of the overall buffer VFR$_A$, Q$_A$, to IPA VFR$_S$, Q$_S$. The VFRs of PBS in the inner and outer side channels were identical. Sulforhodamine B was present only in the PBS of the inner side channels at a concentration of 0.5 mmol/L. Encapsulation studies were performed with a constant FRR$_{A/S}$ of 35 and a total VFR of 200 µL/min. The total SRB dye content in the system was varied between 5 %v/v to 40 %v/v of the total liquid volume by adjusting the VFRs of the outer and the inner side channels, respectively. A volume of 500 µL of the liposome effluent was collected from the outlet channels of the microfluidic network and subsequently filtered through a polyacrylamide column with a MWCO of 6 kDa to remove non-encapsulated SRB.

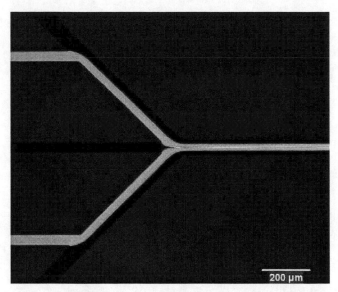

Figure 1: False color confocal fluorescent micrograph of the microchannel network showing the fluorescent intensity of 0.5 mmol/L SRB in PBS injected into the left two inner side channels. PBS without SRB dye is injected into the left two outer side channels, and lipid mixture is injected into the left center channel.

2.4 Light Scattering and Asymmetric Flow Field Flow Fractionation (AF4)

High resolution separation of the liposome population according to their hydrodynamic radius and subsequent size distribution analysis was performed using AF^4 with multiangle laser light scattering (MALLS) and quasi-elastic light scattering (QELS) (model DAWN EOS and QELS, Wyatt Technology, Santa Barbara, CA) as previously described [7]. The liposome radii of the eluted fractions were monitored using MALLS and a QELS detector with data processing using vendor supplied software (ASTRA, Wyatt Technology, Santa Barbara, CA).

2.5 Fluorescent Spectroscopy

Fluorescence measurements were performed on an inverted microscope operating *via* confocal detection as described earlier [1]. A frequency doubled Nd:YAG laser operating in continuous wave mode at 532 nm was sent through the back aperture of a water-immersion objective to excite SRB dye molecules. Liposome and free dye samples were prepared by transferring 100 µL of solution into a well formed in a microscope slide. An avalanche photodiode detects the photons emitted from the SRB dye. The detected photons create transistor-transistor logic (TTL) pulses that are recorded with a data acquisition card and subsequently analyzed to create a photon counting histogram and fluorescence autocorrelation for 10 second time intervals. For each sample, we collect 20 to 25 ten-second intervals of data in order to perform ensemble averaging and to extract standard deviations for each average. All results are reported with standard error bars at the 95 % confidence interval. The free dye brightness, Q_{SRB}, is calibrated with 7 to 10 separate measurements of approximately 1 nmol/L SRB in PBS. The autocorrelation of each liposome sample is used to extract the average number of background dye molecules, N_d, contained within the laser excitation volume. It is assumed that a liposome containing n SRB molecules will be n times brighter than a free SRB molecule as long as the concentrations are within the non-self quenching regime of SRB. For an ensemble of liposomes, each having an integer number of SRB molecules, the first two cumulants of the fluorescent signal are given by

$$\langle P \rangle = N_d Q_{SRB} + N_{lip} Q_{SRB} \langle n \rangle \qquad \text{, and} \qquad (1)$$

$$\langle \Delta P^2 \rangle - \langle P \rangle = N_d Q_{SRB}^2 + N_{lip} Q_{SRB}^2 \langle n^2 \rangle \quad , \qquad (2)$$

where $\langle n \rangle$ is the average number of molecules contained within a liposome, $\langle P \rangle$ is the average number of photons detected per unit time, Q_{SRB} is the free dye molecular brightness measured in the number of detected photons per molecule per unit time, N_{lip} is the average number of liposomes regardless of how many molecules they contain within the laser excitation volume, and $\Delta P = P - \langle P \rangle$. Equations (1) and (2) can be manipulated to arrive at the following expression,

$$J = \frac{\langle n^2 \rangle}{\langle n \rangle} = \frac{1}{Q_{SRB}} \frac{\langle \Delta P^2 \rangle - \langle P \rangle - N_d Q_{SRB}^2}{\langle P \rangle - N_d Q_{SRB}} \quad . \qquad (3)$$

In this work we assume that all liposomes in a sample are equally bright so that $\langle n^2 \rangle = \langle n \rangle^2$ and therefore J is equal to the average number of molecules contained within the liposomes. A more realistic approach assumes a non-zero variance for the distribution of encapsulated molecules, but this is beyond the scope of this work.

3 Results and discussion

Hydrodynamic focusing allows for fast and controlled mixing in a microfluidic format with the benefit of reduced sample consumption. The flow conditions in the micro channels are laminar with Reynolds numbers Re of about 20, which allows for mixing based entirely on molecular diffusion in a direction normal to the liquid flow streamlines. Increasing the FRR$_{A/S}$ reduces the focused lipid solvent stream (center stream) width and thereby the diffusion length for mixing of the lipid solvent stream and the aqueous side streams, thereby reducing the distance in the center channel distal from the mixing intersection to

reach the critical alcohol concentration [7]. At a critical alcohol-to-water ratio the lipid monomers become insoluble and spontaneously self-assemble into closed spherical structures sequestering the surrounding fluid. The in this way formed liposomes resemble kinetic equilibrium structures that vary in radii according to the force field and chemical conditions in the microchannel. As shown in Figure 2, it is possible to control the mean geometric diameter from about 68 nm to 34 nm by reducing the FRR$_{A/S}$ from 14 to 39 at a VFR of 100 µL/min.

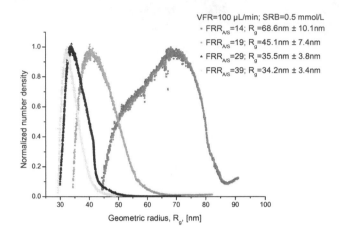

Figure 2: Liposome size and size distribution for different FRR$_{A/S}$ at a constant total VFR of 100 µL/min. Increasing FRR$_{A/S}$ from 14 to 39 results in a decrease of liposome mean geometric radius from about 68 nm ± 10.1 nm to 34 nm ±3.4 nm (one standard deviation).

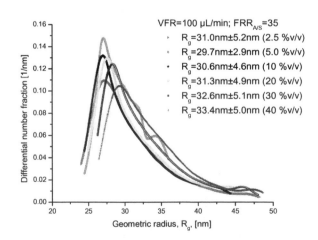

Figure 3: Reproducible liposome size and size distribution at constant VFR of 100 µL/min and FRR$_{A/S}$ of 35 for 6 different volume fractions of SRB in the total sample volume.

Control of the loading efficiency of SRB into liposomes was studied by injecting different volume fractions of SRB (2.5 %v/v, 5 %v/v, 10 %v/v, 20 %v/v, 30 %v/v, 40 %v/v) with a total VFR of 200 µL/min. In

NSTI-Nanotech 2008, www.nsti.org, ISBN 978-1-4200-8503-7 Vol. 1

addition to changing the number of molecules loaded into the liposomes, a reduced volume fraction of SRB was determined that allowed encapsulation without adversely affecting the loading efficiency of liposomes. Figure 3 shows the liposome size and size distribution obtained during the encapsulation studies. It further demonstrates the high reproducibility of liposome formation achievable in a microfluidic format.

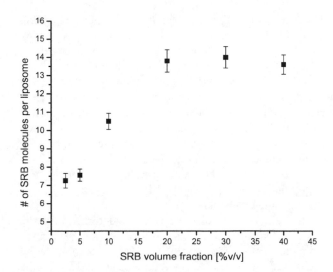

Figure 4: Data shows the number of SRB molecules per liposome along with the 95 % confidence interval. We assume that the average number of encapsulated molecules is equal to the parameter J. Reducing the SRB volume fraction reduces the number of encapsulated molecules from about 14 to 7.

Figure 4 shows that the volume fraction of SRB can be reduced significantly before a moderate change in the amount of encapsulated dye is detected. A 40 fold reduction in SRB content reduces the total number of encapsulated SRB molecules by about a factor of 2. In contrast to common batch fabrication methods, microfluidics has the ability to spatially localize the encapsulant to the immediate vicinity where encapsulation is expected and thereby reduce the encapsulant waste substantially without adversely affecting the liposome's loading efficiency. Control over loading efficiency of SRB into liposomes can be achieved below 20 %v/v of SRB of the total sample volume. This allows for tuning the loading efficiency beginning from an initial SRB concentration. The control over loading efficiency is again facilitated by the laminar flow conditions in the microchannel which enables controlled mixing at the liquid-liquid diffusive interface prior to mixing with the center stream. In macro scale batch processing this can only be achieved by replacing the entire buffer volume. Filling all side channels with PBS containing SRB is representative of common batch processing.

In addition to reduced sample consumption, which is desired from an economical standpoint, the continuous flow microfluidic approach allows the control of the concentration of the substance to be encapsulated from an initial starting concentration *via* controlled diffusive mixing. This enables fine control over the loading efficiency of a drug simulant into liposomes.

4 Conclusion

The formation of liposomes and encapsulation of a hydrophilic drug simulant (SRB) has been demonstrated using a microfluidic technique. It was shown that microfluidics enables reproducible and fine control over liposome size and size distribution, tunable loading efficiency of liposomes, and much reduced encapsulant consumption without adversely affecting the loading efficiency by confining the encapsulant to the region of interest. The simplicity of this liposome formation and drug encapsulation strategy could allow for implementation in point-of-care drug encapsulation, eliminating shelf life limitations of liposome preparation and reducing encapsulant consumption. Fluorescence correlation spectroscopy for encapsulation efficiency measurements has the potential for integration in future lab-on-a-chip applications for online liposome characterization.

REFERENCES

[1] Reiner, J.E. et al., Proc. SPIE noise and fluctuations in biological, biophysical, and biomedical systems, 6602, pp. 66020I-66021-66012, 2007

[2] Ramachandran, S. et al., Langmuir, 22, pp. 8156-8162, 2006

[3] Andresen, T.L. et al., Progress in Lipid Research, 44 (1), pp. 68-97, 2005

[4] Abraham, S.A. et al., Methods in Enzymology, 391, pp. 71-97, 2005

[5] Martina, M.S. et al., Journal of American Chemical Society, 127 (30), pp. 10676-10685, 2005

[6] Jahn, A. et al., Journal of American Chemical Society, 126 (9), pp. 2674-2675, 2004

[7] Jahn, A. et al., Langmuir, 23 (11), pp. 6289-6293, 2007

Production of Polymer Nanosuspensions Using Microfluidizer® Processor Based Technologies

T. Panagiotou*, S.V. Mesite*, J. M. Bernard*, K. J. Chomistek* and R. J. Fisher**

*Microfluidics Corporation, Newton, MA, USA mimip@mfics.com
**Massachusetts Institute of Technology, Cambridge, MA, USA

ABSTRACT

Polymer nanoparticles are often used for controlled drug delivery of active pharmaceutical ingredients (APIs). Microfluidizer® Processor based technologies offer two options for production of polymer nanoparticles. The first is an emulsion method, which involves dissolving the polymer and API in the oil phase of an emulsion and then subsequent removal of the oil. The second is a precipitation method, in which the polymer and API are dissolved in a solvent and then forced to precipitate inside the high shear mixing zone when mixed with an antisolvent. These methods are compatible with a wide variety of polymer/API systems. The focus of this work is to identify the effects of varying key parameters such as process pressure, relative flow rates of the streams, and formulation on the particle size distribution.

This article showcases polymer nanosuspensions in the range of 50-500 nm that were prepared with two different polymers, using both techniques. Furthermore, these tests indicate that an API was successfully encapsulated within the nanoparticles.

Keywords: polymer, nanoparticles, nanoemulsion, drug encapsulation, Microfluidizer®

1 INTRODUCTION

The creation and use of chaperone systems in drug delivery and diagnostic imaging has greatly broadened the applications, and thus needs, for polymer nanosuspensions.[2,3] The enhancement of surface to volume ratios obtained when these nanosuspensions are created provides unique capabilities for functionalization of the surface required for specificity. Encapsulation of APIs and contrast agents within these biocompatible polymers is readily accomplished using versatile Microfluidizer based technologies for processes that are reproducible and scalable. Furthermore, the probability of physical property changes due to processing is reduced. Especially when compared to sonication, the most commonly used laboratory scale technique, where cavitation and sono-chemistry issues may arise.

Two techniques are reported here that can create nanosuspensions of many different polymers types with varying particle sizes by controlling the formulation and process variables. Microfluidics Reaction Technology (MRT) was used with the solvent/anti-solvent precipitation method. Particle size distribution can be controlled by varying parameters such as processing pressure, degree of supersaturation and the ratio of solvent and anti-solvent streams.[4] This process has the advantage of producing nanosuspensions in a single step which is ideal for process intensification. The emulsion evaporation method was implemented using a Microfluidizer Processor; i.e., dissolving a polymer in a solvent, creating a nanoemulsion with an immiscible continuous phase, then removal of the solvent to produce the nanosuspension. Particle size distributions can be controlled by varying process parameters and/or formulation.[5]

Both systems control the amount and form of energy dissipation that occurs at specific locations in the system, i.e., directed toward maximizing the useful work in forming surfaces and interfaces. Narrow flow channels convert the energy input to high fluid velocities. These jet streams impinge upon each other in precision fabricated micro-liter sized interaction chambers (Figure 1). Various degrees of mixing intensity (i.e., macro-, meso-, or micro-mixing) and associated level of turbulence intensity (i.e., eddy sizes) are obtained depending upon the energy dissipation rate. The size of the smallest eddies formed, and thus the Kolmogorov scale for the desired diffusion and reaction coordinates, are in the 50-200 nanometer range. This platform can achieve processing pressures of up to 276 MPa (40,000 psi), generate fluid velocities of over 400 m/s and achieve energy dissipation values exceeding 10^7 W/kg.[6].

2 EXPERIMENTAL APPARATUS

The core of this technology is a continuous microreactor (reaction chamber) based on impinging jet design, described earlier, see Figure 1. Two opposing jets form as fluids flow through two microchannels within the chamber. The jets collide inside a microliter volume where the fluids mix at the nanometer scale. Average fluid velocities inside the channels may exceed 400 m/s, which is orders of magnitude higher than existing impinging jet reactors.[7] A planar array of opposed pairs of such channels ensures effective scaling up of the technology.

High velocities through the channels are achieved by applying high pressures to the fluid upstream of the channels. Pressures up to 207 MPa (30,000 psi) are required for such velocities generated using a hydraulically or pneumatically driven pressure multiplier referred to as an intensifier.

Depending on the application, a variety of feed systems can be used. For simple "top-down" processing, which includes particle size reduction, a single feed system is used. For "bottom-up" processes which involve particle generation as a result of chemical or physical processes, a multiple stream feed system is used.[7] This system delivers multiple, separate streams to the processor at controlled rates. Mixing of the streams is minimized prior to the chamber, and maximized inside the chamber. Therefore particle formation is suppressed prior to the reaction chamber. Uniform mixing at the nanometer scale inside the reaction chamber ensures uniform particle production conditions in addition to nanoparticle formation.

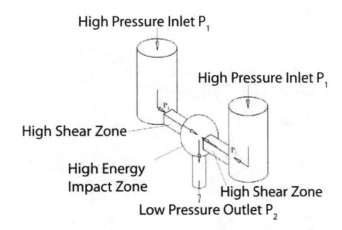

Figure 1. Schematic of the flow path of the reaction chamber. Copyright by Microfluidics, 2007.

3 PROCEDURE

Nanosized polymer particles were generated using two methods: (a) emulsion-evaporation, and (b) precipitation.

3.1 Emulsion Method

This method is a "top down" approach that involves the formation of a stable emulsion of a polymer/solvent solution with an immiscible non-solvent liquid and the subsequent removal of the solvent. For some of these tests, an API, carbamazepine, was added to the solvent phase to be incorporated inside the polymer particles.

Microfluidizer® Processors are the industry standard in forming nanoemulsions. For this method, a nanoemulsion was formed by first preparing a coarse emulsion with the solvent and aqueous streams using the IKA T-25 high shear mixer, then processing with a Microfluidizer® processor. The nanoemulsion size was controlled by varying the processing pressure, the number of passes and the concentration of the oil phase.

The solvent was then removed from the emulsion leaving only the polymer particles suspended in the water phase. There are many different ways of removing the solvent such as evaporation and co-solvent extraction. When the API was involved, the goal was to incorporate it within the polymer particle.

This method can be used for any polymer/API/solvent/non-solvent system. For illustration, this method was used to make poly(lactic-co-glycolic acid) (PLGA) particles. The polymer was dissolved in dichloromethane (DCM) at concentrations between 10 and 80 mg/ml. It was then mixed at concentrations of 1-10% dichloromethane with water that contained poly(vinyl alcohol) (PVA) to form a coarse emulsion. The processing pressure was varied between 70 and 140 Mpa and multiple passes were performed with some of the material. All these tests were performed on the M-110EH Microfluidizer® Processor with the F20Y (75 μm)–H30Z (200 μm) chamber configuration.

The solvent was then removed using several different methods that have different driving forces to obtain particles with different sizes. The evaporation method was performed in a rotovap at 25 kPa absolute for 10-20 minutes depending on the concentration of the dichloromethane. The temperature of the sample was maintained at room temperature using a water bath. The co-solvent extraction process was performed by mixing the emulsion with a co-solvent immediately after processing. The co-solvent does not dissolve the polymer, but is miscible with both the water and organic phases.

3.2 Precipitation Method

This method is a "bottom up" process that involves the precipitation of the polymer from a solution, by adding a polymer/solvent/API solution to a miscible anti-solvent. The addition of the anti-solvent results in a supersaturated condition with subsequent polymer dissolution. These streams were mixed inside the interaction chamber at various shear rates by controlling the orifice size and processing pressure.

A surfactant was added to the anti-solvent (water) in order to: (a) to stabilize the nanoparticles and limit their growth, and (b) to minimize agglomeration of the particles and thereby to create a stable suspension. A non-ionic surfactant was used, Solutol® HS 15 (polyoxyethylene esters of 12-hydroxystearic acid) from Bayer.

Nanosuspensions of two different polymers, Poly(epsilon-caprolactone) (PCL) and poly(D,L-lactide-co-glycolide) (PLGA) were produced using the precipitation method. These polymers were dissolved in acetone at concentrations ranging from 10mg/ml to 40 mg/ml. These solutions were mixed with water that contains a surfactant with flow ratios in the range of 1:2 – 1:10. Process pressures were varied between 35 – 140 mPa.

3.3 Drug Encapsulation

To date, there have only been qualitative measurements of the amount of drug that was encapsulated within the polymer nanoparticles during these tests. These were obtained by performing two replicate tests, both with the same concentrations of API; one with the and one without polymer. These samples were analyzed using optical microscopy to identify any large drug particulates.

4 ANALYSIS

4.1 Particle Size Analysis

The particle size distribution of these samples was measured using the Malvern Zetasizer® which uses dynamic light scattering. The samples were measured at 25°C with water as the continuous phase and PLGA as the particle phase. The results given are the Z-Average, which is a volume weighted average.

4.2 Electron Microscopy

Two different electron microscopy techniques were used for this process. The emulsion evaporation samples were analyzed using a transmission electron microscope (TEM); model JEOL, JEM 1010 TEM, operated at 60 kV. A staining material was used to increase the contrast of the particles. The samples that were prepared using the precipitation technique were analyzed using a scanning electron microscope (SEM); Hitachi S-4800 FESEM.

4.3 Light Microscopy

To determine if the API was encapsulated within the polymer particles, the samples were analyzed using a light microscope. Although it is unable to achieve resolution at the nanoparticle scale, it is powerful enough to see preliminarily whether or not the API has been encapsulated.

5 RESULTS

5.1 Emulsion method

Table 1. Results from the processing of the polymer nanoparticles using the emulsion method.

#	C PLGA (mg/ml)	% DCM	Pres. (MPa)	# of Passes	Evap. (nm)	Co-Solv. (nm)
1	10	5	70	1	223	129
2	10	5	70	2	100	76
3	10	5	70	3	114	99
4	10	1	70	1	3254	124
5	10	10	70	1	168	116
6	10	5	105	1	127	84
7	10	5	140	1	140	149
8	40	5	70	1	193	146
9	80	5	70	1	168	119

The results from the emulsion tests are shown in Table 1. All of these tests were performed with 1% PVA dissolved in the water phase to stabilize the emulsion. The concentration of the PLGA in the DCM is given as "C PLGA"; the amount of oil phase that is mixed with the water phase as "% DCM"; the process pressure as "Pres" and the Z-average particle size for the two different solvent removal techniques as "Evap." for the solvent evaporation technique and "Co-Solv." for the co-solvent extraction technique.

Figure 3 is a TEM image of particles formed using the emulsion method; sample #9. The black specks that are present in the picture are identified as the contrast agent, phosphotungstic acid, which was used to enhance imaging.

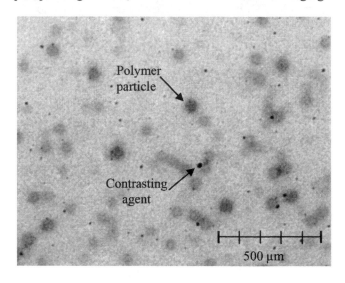

Figure 3. TEM image of the polymer nanoparticles generated after test #9 of the emulsion method.

5.2 Precipitation Method

Table 2. Results from the processing of the polymer nanoparticles using the precipitation method.

#	Polymer	% Acetone	C Poly. (mg/ml)	Shear (s⁻¹ X 10⁶)	Z-ave. (nm)
1	PCL	10	20	1.2	445
2	PCL	10	20	1.2	517
3	PCL	10	20	6.0	281
4	PCL	10	40	6.0	341
5	PCL	10	10	6.0	258
6	PCL	10	20	6.8	280
7	PLGA	10	20	6.0	230
8	PLGA	9	10	6.8	184
9	PLGA	17	10	6.8	173
10	PLGA	25	10	6.8	177
11	PLGA	33	10	6.8	212

The results from the precipitation tests are presented in Table 2. All of these tests were performed with 1% Solutol dissolved in the water phase to stabilize the dispersion. The type of polymer used is given as "Polymer"; the concentration of the polymer in the acetone as "C Poly"; the amount of acetone that is mixed with the water phase as "% acetone"; the shear rate, which is a function of pressure and orifice size, as "Shear"; and the Z-average particle size as "Z-ave." Figure 4 is a SEM image of particles formed via the precipitation method; sample #3.

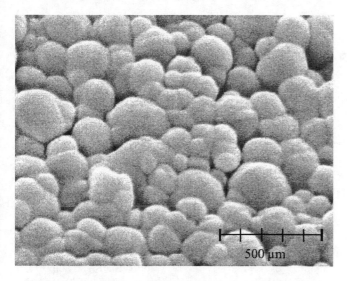

Figure 4. SEM image of the polmer nanoparticles generated after test #3 of the precipitation method.

5.3 Drug Encapsulation

Pictures taken with the optical microscope from the drug encapsulation tests can be found in Figure 5. These samples were prepared with the conditions from test #10 of the precipitation method. The absence of an API particles in 5a indicate that the drug is encapsulated within the polymer as opposed to the free crystalline form seen in 5b.

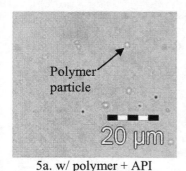 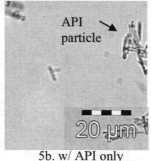

| 5a. w/ polymer + API | 5b. w/ API only |

Figure 5. Optical microscope images from test # 10 of the precipitation tests with polymer and API (5a) and with API only (5b).

6 CONCLUSIONS

Polymer Nanosuspensions in the range 50-500 nm, with two different polymers, have been created successfully using both the emulsion and the precipitation methods. By controlling the processing parameters, nanosuspensions with various polymer sizes and densities were created.

For the emulsion method, the dispersions that were prepared by co-solvent extraction were, in general, smaller than those prepared by the evaporation method. This may be due to the stability of the emulsion after processing or the agglomeration of the particles either during or after drying process. The co-solvent extraction step was performed immediately after processing. Some time (5-30 min.) elapsed before the evaporation technique was performed which may have enabled the emulsions to ripen.

By varying the process pressure (70-140 MPa) and number of passes (1-3, the size of the polymer particles varied in the range of 75-250 nm. Given a desired formulation, it is likely that the particle size of the dispersion can be controlled by selecting the appropriate processing conditions.

It appears that the API was encapsulated within the polymer nanoparticles during these tests. Future work will involve quantifying both the amount of API encapsulated and the release rate as a function of process parameters.

REFERENCES

[1] J. Vandervoort, K. Yoncheva, A. Ludwig. "Influence of the Homogenization Procedure on the Physicochemical Properties of PLGA Particles." *Chem. Pharm. Bull.* **52**(11) 1273-1279 2004.

[2] S. Kummar, M. Gutierrez, J. Doroshow, A. Murgo. "Drug development in oncology: classical cytotoxics and molecularly targeted agents." *British Journal of Clinical Pharmacology* **62** (1) 15–26, 2006.

[3] S. McNeil, "Nanotechnology for the biologist." *Journal of Leukocyte Biology* (78) 585-594 2005.

[4] U. Edlund, A. Albertsson, "Degradable Polymer Microspheres for Controlled Drug Delivery." *Advances in Polymer Science: Degradable Aliphatic Polyesters.* 67-105 2002.

[5] T.Crowley, S. Meadows, E. Kostoulas, F. Doyle III, "Control of particle size distribution described by a population balance model of semibatch emulsion polymerization." *Journal of Process Control* **10** (5T) 419-432 2000.

[6] Panagiotou, Thomai et. al. *Production of Stable Drug Nanosuspension Using Microfluidics Reaction Technology.* Nanotech 2007 Technical Proceedings of the 2007 Nanotechnology Conference and Trade show.

[7] B. Johnson, R. Prud'homme "Chemical processing and micromixing in confined impinging jets." *AiChe Journal,* **49** (9) 2264-2282 2003.

Formation of Pyrophoric α-Fe Nanoparticles from Fe(II)-Oxalate

Rajesh Shende,[*] Alok Vats,[*] Zac Doorenbos,[*] Deepak Kapoor,[**] Christopher Haines,[**] Darold Martin,[**] and Jan Puszynski[*]

[*]Department of Chemical and Biological Engineering,
South Dakota School of Mines & Technology,
Rapid City, SD 57701 USA, Jan.Puszynski@sdsmt.edu
[**]Armament Research, Development, and Engineering Center
Picatinny Arsenal, NJ, 07806 USA

ABSTRACT

The formation of pyrophoric Fe-nanoparticles from Fe(II)-oxalate decomposition under H_2 atmosphere was investigated. The oxalate was synthesized by controlled nucleation process involving addition of oxalic acid in the $FeCl_2.2H_2O$ solution followed by separation and drying. As-synthesized oxalate was decomposed in a quartz tubular reactor at 450-520°C and reduced with a gases mixture containing 5 vol% of H_2 in N_2. After exposing it to air at room temperature, the fine Fe particles generated a temperature of about 800°C in less than 1 sec. In other approach, Fe-oxalate nanorods were synthesized by solid state synthesis method using polyethylene glycol surfactant. TEM images indicated nanorod like morphology with the diameter of about 20 nm and length of 70-80 nm. The fine oxalate particles were embedded in a porous Y_2O_3 matrix coated on a metallic substrate and reduced at the experimental condition stated above. Upon exposure to air, these substrates were found to be extremely reactive.

Keywords: Fe-oxalate, synthesis, nanorods, α-Fe, reactive substrate

1 INTRODUCTION

Iron (Fe), the most ubiquitous transition metal abundantly available in the Earth's crust, is the structural backbone of our modern infrastructure. It is therefore paradoxical that as a nanoparticle, Fe has been somewhat neglected in favor of other metals such as Au, Ag, Pt, Ni, Cu etc. This is understandable because Fe is extremely reactive. Fe as fine particles have long been known to be pyrophoric [1], which might be the major reason that Fe nanoparticles have not been fully investigated. Fe nanoparticles have several important applications in catalysis [2], magnetic resonance imaging [3], magnetic data storage [4], coatings [5], synthesis of carbon nanotubes [6], synthesis of highly oriented and ordered nanostructures for field emission devices [7] etc. Further research is currently being pursued to explore electrical, dielectric, magnetic, optical, imaging, catalytic, biomedical and bioscience properties of Fe nanoparticles.

Although fine Fe particles have been long known as pyrophoric, only few studies [8-10] investigated the pyrophoric behavior of Fe nanoparticles. These reactive nanoparticles can be produced by thermal decomposition of Fe precursor [11], selective leaching of a metal from an alloy [9], e.g. Fe-Al, and reduction of Fe compounds such as oxides or oxalates in H_2 environment. Recently, Lawrence Livermore National Laboratory (LLNL) has developed the sol-gel methodology to generate high surface area porous Fe [8]. These investigators however, found out that the porous Fe prepared by sol-gel route is non-pyrophoric but ignitable by thermal source. They believe that the addition of W or Sn particles to the sol-gel derived Fe nanoparticles can improve their pyrophoric action [8]. Other investigators [9] have studied the reduction of Fe-oxalate in H_2 atmosphere to produce α-Fe nanoparticles. These nanoparticles were non-pyrophoric but they were very effective for scavenging O_2 from Ar atmosphere [12]. α-Fe, however, has been reported to be the most active form of Fe that is pyrophoric [8-10].

Very recently, we reported the synthesis of pyrophoric particles from micron size Fe-oxalate particles [10]. These particles were produced by thermal decomposition followed by reduction of Fe-oxalate in the H_2 environment. In the present investigation, we report synthesis of Fe-oxalate nanoparticles using $FeCl_2.2H_2O$ and oxalic acid in presence of polyethylene glycol surfactant. In using a surfactant the nucleation will occur inside the micelles, which will produce well-defined nanostructures such as nanorods. Although the synthesis of oxalate nanorods using the reverse micellar route employing cetyltrimethyl ammonium bromide surfactant has been reported [11], the synthesis of pyrophoric nanoparticles from the oxalate nanorods has not been reported so far. This study further includes the embodiment of oxalate particles inside Y-OOH gel to fabricate the reactive substrates. The results presented here exhibit the pyrophoric response of the active Fe produced by thermal decomposed followed by reduction in H_2 and N_2 environment.

2 EXPERIMENTAL

2.1 Materials

Ferrous chloride (FeCl$_2$.2H$_2$O) and oxalic acid (HOOCCOOH) used for the synthesis of Fe-oxalate were procured from Sigma-Aldrich, WI. The PEG-600 surfactant used to prepare nanoscale oxalate powder was obtained from Alfa-Aeser. Yttrium chloride (YCl$_3$.7H$_2$O), ethanol, Brij-76 surfactant and propylene oxide required to prepare mesoporous gel were procured from Sigma Aldrich (WI). Stainless steel foils were purchased from McMaster Carr.

2.2 Synthesis of Fe(II)-oxalate

In one synthesis approach, FeCl$_2$.2H$_2$O was dissolved in de-ionized water and the solution was heated to 60-70°C under magnetic stirring. Oxalic acid was dissolved in de-ionized water and the solution was added drop wise to the FeCl$_2$.2H$_2$O solution. After addition, the contents were maintained for 2 h under constant stirring. The solution mixture was allowed to cool down to the room temperature and the contents were centrifuged to recover the precipitate. Synthesis of Fe-oxalate nanorods was attempted using polyethylene glycol (PEG-600) surfactant. In this, oxalic acid, FeCl$_2$.2H$_2$O and PEG-600 were homogeneously mixed together for about 30 min. The contents were diluted with de-ionized water and ethanol, sonicated for 4 h and finally centrifuged to recover the precipitate.

2.3 Synthesis of Y-OOH gel embedded with Fe-oxalate particles

To synthesize the Y-OOH gel, Solution-A containing YCl$_3$.7H$_2$O in ethanol and Solution-B of Brij-76 (20-30 wt%) in ethanol were prepared by ultrasonic mixing. Solution-B was added drop wise to the Solution-A and after mixing these two solutions at room temperature, propylene oxide was added, which resulted in a transparent gel in 5 min. The gel was aged for 1-2 hrs and Fe-oxalate powder was added to it. This synthesis route is designated as route-I. The oxalate loading was varied between 30-80 wt% based on the weight of YCl$_3$ initially taken. In another synthesis route, designated as route-II, the oxalate powder was first dispersed in the solution mixture of Solution-A and Solution-B and later, propylene oxide was added to achieve the gel formation. The time required for the gel formation was 5-6 min.

2.4 Fabrication of reactive substrate

Fabrication involves dip coating a metallic foil with a Y(yttrium)-sol solution followed by coating a layer of the oxalate and finally, dip-coating this foil in Y-sol solution. To synthesize the Y-sol solution, YCl$_3$.7H$_2$O and Brij-76 (20-30 wt%) were dissolved in ethanol by ultrasonic mixing and propylene oxide was added. A metallic foil was dip coated with this Y-sol solution. The films were annealed at 100°C for 5 min. Next, Fe(II)-oxalate slurry in acetone containing 20wt% polycarbonate binder was prepared and applied on the dip-coated Y-OOH sol solution using a doctor blade casting technique. After drying the film for few minutes, the metallic foil previously coated with the oxalate was dip coated in Y-sol solution and dried at 90°C.

2.5 Thermal decomposition, reduction, and oxidation

The decomposition experiments were performed in an isothermal flow reactor where Fe(II)-oxalate powder or foil was placed inside a quartz tube supported in a split furnace. The quartz tube was connected with two mass flow controllers to monitor the flow rates of H$_2$ and N$_2$ gases. Oxalate sample was heated to 450-500°C under the constant flow of N$_2$ and reduced with H$_2$ (5 vol%) for 5 min. The sample was cool down to the room temperature under the flow of N$_2$. The α-Fe samples removed from the reactor and exposed to air at ambient conditions to allow it to undergo oxidation. The temperature was continuously recorded using a pyrometer mounted on the tubular flow reactor.

2.6 Characterization

As-synthesized oxalate powder, nanorods, and commercial oxalates were analyzed by Fourier transform infrared (FTIR) spectroscopy. Microstructural characterization of various samples was performed using scanning and transmission electron microscopy.

3 RESULTS AND DISCUSSION

The FTIR spectra of as-synthesized Fe-oxalate showed the –CO and –OCO vibrations of the oxalate ligand at 1636 and 800 cm^{-1}, respectively whereas the bridging vibrations of oxalate were found in the region of 1370-1300 cm^{-1}. Absorption corresponding to Fe was observed in the region of 480-520 cm^{-1}. The other characteristics signatures of the as-synthesized oxalate were similar to the commercial oxalate samples.

The as-synthesized oxalate was thermally decomposed, reduced in a tubular quartz reactor, and finally, exposed to ambient air allowing α-Fe to undergo oxidation as per the procedure outlined in the Experimental section. The effect of reducing environment on the peak temperature (T$_{peak}$) generated during oxidation of α-Fe is shown in Figure 1. It can be observed that when 100vol% H$_2$ was used at the decomposition temperature of 500°C, the peak temperature produced during oxidation was around 600°C. However, while decomposing the oxalate at the same temperature but reducing with 5vol% H$_2$, the T$_{peak}$ was increased to 735°C.

At 450°C decomposition temperature and using 5vol% H₂, the T_{peak} decreased to about 400°C. Thus, we selected H₂ of 5 vol% with 95 vol% N₂ as the most effective reducing environment.

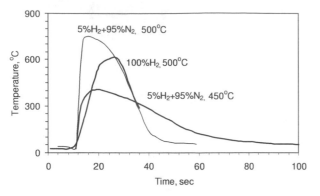

Figure 1: Transient temperature profile of the obtained from oxidizing the a-Fe produced from reducing the Oxalate at various processing conditions.

Mechanism of Fe-oxalate decomposition and reduction is rather complex involving several reactions. Thermal decomposition of oxalate dihydrate leads to dehydration at low temperatures up to about 200°C [10]. In presence of H₂, oxalate reduces to FeO (unstable), which further reduces to Fe₃O₄ and finally it leads to α-Fe. The gaseous products produced concomitantly with FeO further reacts with Fe to form Fe₃C that reduces to Fe and CH₄. We have also performed thermal decomposition, reduction, and oxidation of α-Fe obtained from Fe₂O₃ and Fe₃O₄, and the T_{peak} observed are given in Table 1, which also summarizes T_{peak} for the commercial oxalates. We can notice that the T_{peak} generated by oxidation of the α-Fe produced from as-synthesized oxalate is around 800°C, which is the maximum among those reported in the table.

Compound	Source	H₂, vol%	T_{peak}, °C
Fe_2O_3	Alfa-Aesar	5	700 ± 25
Fe_3O_4	Alfa-Aesar	5	650 ± 20
$F_eC_2O_4.2H_2O$	Baker, NJ	5	620 ± 10
$F_eC_2O_4.2H_2O$	Sigma Aldrich	10	560 ± 05
$F_eC_2O_4.2H_2O$	As-synthesized	5	800 ± 10

Table 1 Peak temperature (T_{peak}) observed during oxidation of pyrophoric α-Fe

Scanning electron microscopy (SEM) images of as-synthesized Fe-oxalate and after reduction and oxidation are shown in Figure 2a and b. The microstructure reveals plate and rod-like particle morphology. These plates and rods are few μm in size and they are stacked together. TEM image of as-synthesized nanoscale oxalate is shown in Figure 3, which shows nanorod morphology. Under higher

resolution the majority of nanorods were found to have an aspect ratio of approx. 3.0. Oxidation of α-Fe produced from the decomposition of oxalate nanorods was found to be pyrophoric as well.

Figure 2: SEM images of the oxalate a) as-synthesized, and b) after reduction and oxidation.

Figure 3: SEM images of the oxalate nanorods a) as-synthesized, and b) after dispersion in ethanol.

Y-OOH gels containing the oxalate were prepared by using the two synthesis routes as elaborated in the Experimental section. In route-I, the oxalate was added to the Y-OOH gel whereas, in route-II, the oxalate was initially dispersed in the Y-sol-solution and later the gelation was achieved using propylene oxide. Fe-oxalate was added to the gel in the amount of 30-80% (w/w) based on the YCl₃ weight originally taken. The gels prepared by synthesis route-I were decomposed to 500°C in N₂ atmosphere and reduced in presence of 5 Vol% H₂. The transient temperature profiles obtained after exposing the gel containing the reduced oxalate to ambient air are shown in Figure 4.

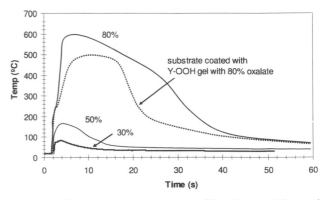

Figure 4: Transient temperature profiles obtained from air oxidation of α-Fe produced from reducing the Y-OOH gels mixed with the oxalate in the amount of 30-80% (w/w).

NSTI-Nanotech 2008, www.nsti.org, ISBN 978-1-4200-8503-7 Vol. 1

We can observe that the T_{peak} for the gels mixed with the oxalate increases as the loading of the oxalate increases from 30 to 80% (w/w). At the level of 30 wt% oxalate mixed with Y-OOH gel, the temperature barely reach to 100°C whereas for 80 wt% oxalate, the temperature increased to 600°C. Another important observation is that the temperature sustained for longer time at the higher levels of the oxalate, which is obvious because the active Fe content produced from higher loadings of the oxalate is higher.

In other experiments, the gels prepared with the synthesis route-II (where propylene oxide addition was performed after dispersing the oxalate in Y-sol solution) were heated to 500°C in N_2 atmosphere and reduced in presence of 5 Vol% H_2. After exposing α-Fe to air, the T_{peak} obtained by this route were lower than route-I where oxalate was added to Y-OOH gel previously made with propylene oxide. It is possible that the propylene oxide may have degrading effect on the oxalate converting it at least partly into its oxide and α-Fe produced from the oxide is less pyrophoric than the one generated from the oxalate (Table 1).

The pyrophoric response of Y-OOH film loaded with 80%oxalate and coated on a metallic substrate is shown in also shown in Figure 4. The T_{peak} generated during oxidation of reduced film is around 550°C, which is slightly lower than the T_{peak} achieved for the Y-OOH gel mixed with the oxalate at 80% concentration level (Figure 1). In the green state, the film was free of any cracks; however, during decomposition and reduction, several cracks were developed. The thickness of the films prepared by doctor blade type of casting technique was approx. 300 μm. We believe that the longer path length for the diffusion of gas species produced during decomposition might be the cause for the crack formation. Currently, we are making efforts to develop defect free reactive substrates containing Fe-nanoparticles.

4 CONCLUSIONS

Synthesis of Fe-oxalate in the form of nanorods has been accomplished using a controlled nucleation process using $FeCl_2$ and oxalic acid with the use of PEG-600 surfactant. The aspect ratio of the as-synthesized nanorods was about 3.0. The α-Fe produced after decomposition and reduction of the micron and nano size oxalates at 500°C using 5vol%H_2 was found to be reactive in air as the pyrophoric temperature as high as 800°C was achieved. The films fabricated by coating the Y-OOH gel embedded with Fe-oxalate on a metallic substrate were found to be reactive as well.

ACKNOWLEDGEMENT

The authors gratefully acknowledge the financial support by the Army Research Development Engineering Center, Picatinny Arsenal, NJ (Contract no. WI5QKN-06-D-0006).

REFERENCES

[1] E.A. Shafranovsky,Yu. I. Petrov . "Aerosol Fe Nanoparticles with the Passivating Oxide Shell", Journal of Nanoparticle Research. 6, 1, 71-90, 2004.

[2] Xuanke Li, Zhongxing Lei, Rongcui Ren, Jing Liu, Xiaohua Zuo, Zhijun Dong, Houzhi Wang, Jianbo Wang, "Characterization of carbon nanohorn encapsulated Fe particles", Carbon, 41, 15, 3068-3072, 2003.

[3] Stefan G. Ruehm, Claire Corot; Peter Vogt, Stefan Kolb, Jörg F. Debatin, "Magnetic Resonance Imaging of Atherosclerotic Plaque With Ultrasmall Superparamagnetic Particles of Iron Oxide in Hyperlipidemic Rabbits" Circulation, 103 , 415-422, 2001.

[4] S. A. Majetich, Y. Jin, "Magnetization Directions of Individual Nanoparticles" Science, 284, 5413, 470 – 473, 1999.

[5] Satoshi Tomita, Masahiro Hikita, Minoru Fujii, Shinji Hayashi, Keiichi Yamamoto, "A new and simple method for thin graphitic coating of magnetic-metal nanoparticles", Chemical Physics Letters, 316(5-6),361-364, 2000

[6] Yoshikazu Homma, Takayuki Yamashita, Paul Finnie, Masato Tomita, Toshio Ogino, "Single-Walled Carbon Nanotube Growth on Silicon Substrates Using Nanoparticle Catalysts", Jpn. J. Appl. Phys., 41, L89-L91, 2002

[7] C. H. Liang, G. W. Meng, L. D. Zhang, Y. C. Wu, Z. Cui, "Large-scale synthesis of β-SiC nanowires by using mesoporous silica embedded with Fe nanoparticles", Chemical Physics Letters, 329, 3-4, 323-328, 2000.

[8] A.E. Gash, J.H. Satcher Jr., R. L. Simpson, "Preparation of porous pyrophoric iron using sol-gel method", US Patent 20060042417, Appl. No. 11/165734.

[9] US Patent 4435381, "Pyrophoric foil and article, and pyrophoric technique", Alloy Surfaces Company Inc., March 6, 1984.

[10] R. V. Shende, A. Vats, Z.D. Doorenbos, D. Kapoor, D. Martin, J.A. Puszynski, "Formation of pyrophoric iron particles by H_2 reduction of oxalates and oxides", 7th International Symposium on Special Topics in Chemical Propulsion, Kyoto, Japan, 2007.

[11] A.K. Ganguli, T. Ahmed, "Nanorods of iron oxalate synthesized using reverse micelles: facile route for alpha Fe_2O_3 and Fe_3O_4 nanoparticles", J. Nanoscience Nanotechnology, 7, 6, 2029, 2007.

[12] P. Pranda, V. Hlavacek, M.L. Markowski (2002) "Ultrafine iron powder as an oxygen scavenger for argon purification", Ind. Eng. Chem. Res., 41, 4837, 2002.

Rhodamine B Isothiocyanate-Modified Ag Nanoaggregates on Dielectric Beads: A Novel Surface-Enhanced-Raman-Scattering and Fluorescent Imaging Material

Kwan Kim*, Hyang Bong Lee, Hee Jin Jang, and Ji Won Lee

Department of Chemistry, Seoul National University, Seoul 151-742, Korea
kwankim@snu.ac.kr

ABSTRACT

Rhodamine B isothiocyanate (RhBITC) is a prototype dye molecule that is widely used as a fluorescent tag in a variety of biological applications. We report in this work that once RhBITC adsorbs on Ag on silica beads, it exhibits not only a strong surface-enhanced-Raman-scattering (SERS) signal but also a fairly intense fluorescence, contrary to the usual expectation that the fluorescence will be quenched by metal nanoaggregates. The RhBITC-modified silica beads are well dispersed in ethanol, and they are also readily coated in water with polyelectrolytes for their further derivatization with biological molecules of interest that can bind to target molecules. The application prospects of these materials are thus expected to be very high especially in the areas of biological sensing and recognition that rely heavily on optical and spectroscopic means.

Keywords: surface-enhanced Raman scattering, SERS, fluorescence, silver, silica bead, rhodamine B isothiocyanate, molecular sensing/recognition, biotin-streptavidin interaction

1 INTRODUCTION

Noble metallic nanostructures exhibit a phenomenon known a surface-enhanced Raman scattering (SERS) in which the Raman scattering cross-sections are dramatically enhanced for the molecules adsorbed onto them [1]. When the adsorbed molecules are subjected to resonance Raman scattering, it is called a surface-enhanced resonance Raman scattering (SERRS) by which even single molecule detection is known to be possible, suggesting that the enhancement factor can reach as much as 10^{14}-10^{15} [2,3]; the effective Raman cross sections are then comparable to the usual fluorescence cross sections. In recent years, there has also been considerable interest in the application of SERS/SERRS in biomolecular detection [4,5]. Although fluorescence is currently the principal detection method in bioassays, it has inherent drawbacks such as photobleaching, narrow excitation with broad emission profiles, and peak overlapping in multiplexed experiments. The latter limitations can be overcome by means of SERS/SERRS. Several groups have thus developed various types of SERS-active tagging materials that can be used in diagnostic bioassays [4-7].

Noble metals such as Au and Ag can support nanoparticle plasmon resonances in the ultraviolet, visible, and near-infrared regions of the spectrum that can be modified by varying the nanoparticle size and shape. The usual, solid metallic nanospheres and nanorods exhibit relatively weak plasmonic tunability compared with metallic nanoshells composed of a dielectric core and a concentric metal shell. This is because nanoshells exhibit plasmon resonances that are critically dependent on the inner and outer shell dimensions due to the hybridization of the two fixed-frequency plasmon modes supported by the inner cavity and outer surface of the nanoshell [8]. Moreover, the plasmon resonance of solid silver nanoparticles appears at shorter wavelength than that of gold, along with stronger and sharper resonance strength [9]. The plasmon tunability of a silver nanoshell is thus greater than that of a gold nanoshell. However, it has not been routine to fabricate a silver nanoshell on dielectric beads.

We demonstrate in this work that Ag can be deposited onto the silica beads simply by soaking them in ethanolic solutions of $AgNO_3$ and butylamine. The extent of silvering could be adjusted by varying the relative concentrations of butylamine and $AgNO_3$. Upon the deposition of silver, the UV/vis absorption peak at ~420 nm gradually red-shifts, finally showing a very broad feature extending from near-UV to near-infrared regions. In accordance with the electromagnetic enhancement mechanism in SERS, the Ag-deposited silica beads are efficient SERS substrates that can be used as core materials of SERS/SERRS-based biosensors. Specifically, we report that once rhodamine B isothiocyanate (RhBITC) adsorbs on Ag on silica beads, it exhibits not only a strong SERS/SERRS signal but also a fairly intense fluorescence, contrary to the usual expectation that the fluorescence will be quenched by metal nanoaggregates. To our knowledge, this is the first report informing the simultaneous observation of fluorescence and SERS/SERRS for dye molecules assembled on metal nanoaggregates.

2 EXPERIMENTAL

Tetraethyl orthosilicate (TEOS; 99%), silver nitrate (99%), butylamine (99%), poly(allylamine hydrochloride) (PAH, MW~70,000), poly(acrylic acid) (PAA, MW~450,000), and RhBITC (97%) were purchased from Aldrich and used as received. Other chemicals, unless

NSTI-Nanotech 2008, www.nsti.org, ISBN 978-1-4200-8503-7 Vol. 1

specified, were of regent grade. Highly pure water (Millipore Milli-Q system), of resistivity greater than 18.0 MΩ·cm, was used throughout. The synthesis of PLL-g-poly(ethylene glycol) (PEG) and biotinylated-PLL-g-PEG was based on the protocols described by Huang et al. [10] and the detailed processes were reported in our recent publication [11].

Monodisperse silica particles were prepared using the Stöber-Fink-Bohn method [12], comprising the base-catalyzed hydrolysis of TEOS in water-ethanol mixtures. The silica particles with a mean diameter of 250 nm thus prepared were cleaned by centrifugation (10000 rpm for 15 min) and redispersion in absolute ethanol repeated 10 times. When silver was deposited onto silica particles, a polypropylene container was used as the reaction vessel to avoid nonspecific silvering of the reaction vessel. Specifically, the cleaned silica in ethanol was added into the silvering medium to a final concentration of 0.11 mg/mL (w/v, dried silica mass/ethanol) and then incubated for 50 min at 50 ± 1°C with vigorous shaking. As a silvering mixture, the concentrations of $AgNO_3$ and butylamine were maintained at 1 mM. The silver-coated silica particles were finally rinsed and redispersed in ethanol.

For the self-assembly of RhBITC on silver, 0.1 mg of silver-coated silica beads were placed in a small vial into which 2 mL of 0.1 mM ethanolic RhBITC solution was subsequently added. After 12 h, the solution phase was decanted and then rinsed with highly pure water. The remaining solid particles were left to dry in a vacuum for 2 h. Subsequently, polyelectrolyte layers were formed by the sequential dipping of the RhBITC-modified silver-coated silica beads into the PAA and PAH solutions (0.1 mg mL^{-1}) for 10 min at room temperature. In the interim, to change the polyelectrolyte solution, silver-coated silica beads were intensively rinsed with water. At the final stage, the PAA-derivatized silver-coated silica beads were electrostatically reacted with biotinylated-PLL-g-PEG.

To construct a dose–response curve for streptavidin, a glass slide was initially soaked in a piranha solution to assume negatively charged surfaces. The glass slide was subsequently dipped in PLL-g-PEG (20 μg mL^{-1}) solutions for 10 min. After washing the slide with water and then drying it in a nitrogen atmosphere, 1.5 μL of biotinylated-PLL-g-PEG (100 μg mL^{-1}) was pipetted onto it to obtain a 5-mm domain. The glass slides coated with biotinylated-PLL-g-PEG were soaked in a streptavidin solution at various concentrations for 10 min, followed by extensive rinsing with a phosphate-buffered saline (PBS) solution and drying by a nitrogen stream. The streptavidin-attached slides were subsequently immersed in a solution containing silver-coated silica beads (1 mg mL^{-1}) derivatized consecutively with RhBITC, PAA–PAH, and biotinylated-PLL-g-PEG. After 10 min, the glass slides were washed with water, and the dried slides were finally subjected to Raman spectroscopy measurements.

UV/vis absorption spectra were obtained using a SCINCO S-2130 spectrometer. Field emission scanning electron microscopy (FESEM) images were obtained with a JSM-6700F field emission scanning electron microscope operated at 5.0 kV. Transmission electron microscopy (TEM) images were obtained on a JEM-200CX transmission electron microscope at 200 kV. Confocal Laser Scanning Microscope images were obtained with a MRC-1024 Confocal Laser Scanning Microscope. The 568-nm line from an Ar-Kr laser was used at the excitation source. X-ray diffraction (XRD) patterns were obtained on a Bruker D5005 powder diffractometer for a 2θ range of $30°$ to $80°$ at an angular resolution of $0.05°$ using Cu$K\alpha$ (1.5406 Å) radiation. IR spectra were measured using a Bruker IFS 113v Fourier transform IR spectrometer equipped with a globar light source and a liquid nitrogen cooled wide-band mercury cadmium telluride detector. Raman spectra were obtained using a Renishaw Raman spectrometer (model 2000) equipped with an integral microscope (Olympus BH2-UMA). The 514.5-nm line from a 20-mW Ar$^+$ laser (Melles-Griot model 351MA520) or the 632.8-nm line from a 17-mW He–Ne laser (Spectra Physics model 127) was used as the excitation source. The Raman peak intensities of RhBITC were normalized with respect to that of a silicon wafer at 520 cm^{-1}.

3   RESULTS AND DISCUSSION

Monodisperse silica particles can be readily prepared by the well-known Stöber-Fink-Bohn method [12] since the particle size depends largely on the relative concentration of reactants. The silica particles synthesized were maintained stable without agglomeration in ethanol. The dried silica powder exhibited a very broad O-H stretching band in the region of 3200-3600 cm^{-1} in the transmission infrared spectrum. This indicates that the surfaces of silica particles are terminated with OH groups; these OH groups must be in a H-bonded state. Owing to these characteristics, the silica powder can be dispersed in water and ethanol. To deposit silver onto the silica particles, the colloidal silica was dispersed in a reaction mixture consisting of ethanolic $AgNO_3$ and butylamine. As described in the Experimental Section, the reaction mixture was incubated for 50 min at 50 ± 1°C. The concentration of $AgNO_3$ and butylamine were maintained at 1 mM. Figure 1 shows a typical FE-SEM image of silica particles taken after the deposition of silver.

Figure 1: FE-SEM image of Ag-coated 250-nm silica beads.

The deposition of silver can also be confirmed from the XRD data. As Ag is deposited onto the silica particles, four distinct XRD peaks are clearly observed at 2θ values of 38.1°, 44.3°, 64.4°, and 77.3° (data not shown), corresponding to the reflections of (111), (200), (220), and (311) crystalline planes of cubic Ag, respectively. It has to be mentioned that we could not identify any colloidal silver formed in the bulk. This indicates that no nucleation center existed in the solution phase. The reduction of silver must have occurred only on the surfaces of silica particles. Butylamine is a very weak reductant, so nucleation centers hardly seem to form in the solution. However, once silver ions are bound to the anionic oxygen sites of the silica particles, silver nitrate will be reduced by butylamine, anchoring onto silica surfaces. The weak reductant characteristics of butylamine can also be confirmed from the observation that the silvering of silica particles hardly takes place when the reaction vessel is maintained at room temperature. The reduction of silver nitrate is facilitated upon increasing the temperature. The formation of a silver shell can be readily controlled at 50°C. However, silver particles may form even in the bulk at temperatures much higher than 50°C.

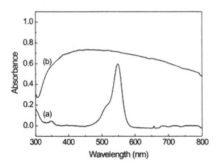

Figure 2: UV/vis spectra of (a) RhBITC and (b) Ag-coated silica beads.

Figure 2a shows the UV/vis absorption spectrum of 0.1 mM RhBITC in ethanol. The absorption maximum is located at 548 nm, so the resonance Raman scattering would be expected to occur when the excitation wavelength is below 548 nm. Figure 2b shows the UV/vis extinction spectra of the Ag coated silica particles. The spectrum measured for bare silica powder was featureless. As the particles become coalesced into a network-like structure, a distinct peak is no longer observed. Instead, a very broad band appears, extending from near-UV to near-infrared regions (see Figure 2b). Much the same UV/vis extinction spectrum was observed after the adsorption of RhBITC. Since the absorption band of RhBITC extends between 450-600 nm, we may well expect to observe a distinct Raman spectrum for RhBITC adsorbed on silver-coated silica beads not only by 514.5-nm radiation but also by 632.8-nm radiation. As shown in Fig. 3, distinct Raman spectra are indeed observed by using both the 514.5- and 632.8-nm radiation. The former must then be a SERRS

spectrum, while the latter is largely a SERS spectrum. Due to the resonance Raman scattering effect, the Raman signal of RhBITC in Fig. 3a is about one and half times more intense than that in Fig. 3b.

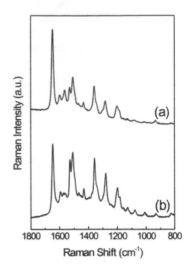

Figure 3: SERS/SERRS spectra of RhBITC on Ag/250-nm silica taken at (a) 514.5 nm and (b) 632.8-nm excitation.

As mentioned in the Introduction, we hope to use the RhBITC-modified silver-coated silica beads as a core material of SERS-based biosensors (see Scheme 1). The SERS-marker molecules like RhBITC on Ag have then to be stabilized in one way or the other for the modified beads to be used in buffer solutions. We have reported recently that the layer-by-layer (LbL) deposition of cationic and anionic polyelectrolytes is a useful strategy to protect SERS marker molecules assembled on micrometer-sized Ag particles [13]. We confirmed that PAA and PAH were consecutively deposited onto the RhBITC-modified silver-coated silica beads.

Scheme 1: Fabrication of Ag-coated silica beads usable as a template of biosensor operating via SERS/SERRS.

Positively charged poly(L-lysine) may then adsorb fairly well onto the PAA layer. Accordingly, we have subsequently deposited biotinylated PLL-g-PEG onto the outermost PAA layer of RhBITC-adsorbed Ag on silica beads and then confirmed their interaction with streptavidin molecules by monitoring the SERS/SERRS peaks of the RhBITC. We evaluated further the sensitivity of the biotinylated RhBITC Ag/silica beads in recognizing streptavidin molecules by constructing a dose–response

NSTI-Nanotech 2008, www.nsti.org, ISBN 978-1-4200-8503-7 Vol. 1

curve. After the fabrication of biotinylated glass slides, SERS/SERRS spectra were obtained as a function of the streptavidin concentration, ranging from 10^{-6} to 10^{-13} gmL^{-1}, as shown in Fig. 4a. Fig. 4b shows the normalized SERS/SERRS intensity of the characteristic band of RhBITC at 1647 cm^{-1}; all the SERS/SERRS peaks were normalized with respect to the peak intensity of a silicon wafer at 520 cm^{-1}. A very intense SERS/SERRS spectrum is obtained as long as the concentration of streptavidin is above 10^{-9} gmL^{-1}. At 10^{-10} gmL^{-1}, the number of polystyrene particles adsorbed on the biotinylated glass substrates decreases, resulting in a lowering of the SERS/SERRS intensity. The SERS/SERRS peak becomes very weak at 10^{-11} gmL^{-1} and is hardly detected at 10^{-12} gmL^{-1}, however. Considering the simplicity of the present method, the detection sensitivity indicated by these results is indeed remarkable.

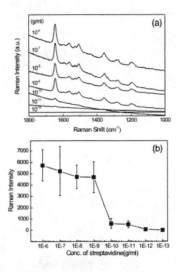

Figure 4: (a) A typical SERS/SERRS spectrum of RhBITC measured after biotinylated RhBITC/Ag/silica particles were allowed to interact via streptavidin with other biotinylated layers on glass. (b) SERS/SERRS intensity of the characteristic band of RhBITC at 1647 cm^{-1} measured as a function of streptavidin concentration; the SERS/SERRS intensities were the average of 10 different measurements with the error bars denoting their standard deviation.

Considering the appearance of SERS/SERRS peaks, the fluorescence of RhBITC is expected to be quenched. Surprisingly, however, the remaining fluorescence is strong enough to be detected by fluorescence microscopy. Figure 5 shows a typical confocal laser scanning microscope image of RhBITC-modified Ag on 250-nm SiO$_2$, taken simply after spreading them on a glass slide. To our knowledge, this is the first report informing the simultaneous observation of fluorescence and SERS for dye molecules assembled on metal nanoaggregates. The RhBITC-modified

silica beads are thus expected to be invaluable in the multiple analyses of biomolecules.

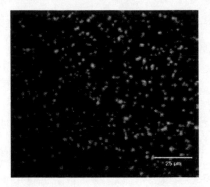

Figure 5: Confocal laser scanning microscope image of RhBITC-modified Ag on 250-nm SiO$_2$ on a glass slide; excitation wavelength was 568-nm.

ACKNOWLEDGEMENT

This work was supported by the Korea Science and Engineering Foundation (Grants R01-2006-000-10017-0 and R11-2007-012-02002-0).

REFERENCES

[1] M. Moskovits, Rev. Mod. Phys. 57, 783, 1985.

[2] S. Nie and S.R. Emroy, Science 275, 1102, 1997.

[3] K. Kneipp, Y. Wang, H. Kneipp, L.T. Perelman, I. Itzkan, R.R. Dasari and M.S. Feld, Chem. Phys. Lett. 78, 1667, 1997.

[4] Y.C. Cao, R. Jin and C.A. Mirkin, Science 297, 1536, 2002.

[5] D.S. Grubisha, R.J. Lipert, H.Y. Park, J. Driskell and M.D. Porter, Anal. Chem. 75, 5936, 2003.

[6] S.P. Mulvaney, M.D. Musick, C.D. Keating and M.J. Natan, Langmuir 19, 4784, 2003.

[7] A.F. McCabe, C. Eliasson, R.A. Prasath, A. Hernandez-Santana, L. Stevenson, I. Apple, P.A.G. Cormack, D. Graham, W.E. Smith, P. Corish, S.J. Lipscomb, E.R. Holland and P.D. Prince, Faraday Discuss. 132, 303, 2006.

[8] E. Prodan, C. Radloff, N.J. Halas and P. Nordlander, Science 302, 419, 2003.

[9] U. Kribig and M. Vollmer, Optical Properties of Metal Cluster, Springer, New York, 1995.

[10] N.P. Huang, J. Vörös, S.M. De Paul, M. Textor and N.D. Spencer, Langmuir 18, 220, 2002.

[11] K. Kim, H.K. Park and N.H. Kim, Langmuir 22, 3421, 2006.

[12] W. Stöber, A. Fink and E. Bohn, J. Colloid Interface Sci. 26, 62, 1968.

[13] K. Kim, H.S. Lee, H.D. Yu, H.K. Park and N.H. Kim, Colloid Surf. A, 316, 1, 2008.

Polarimetric and Photonic Properties of 2D Submicroparticle Arrays

S. Portal*, M. A. Vallvé**, O. Arteaga*, J. Ignés-Mullol**, A. Canillas* and E. Bertran*

* FEMAN group, IN2UB, University of Barcelona, Martí i Franquès, 1, 08028 Barcelona, Spain
** SOC&SAM group, IN2UB, University of Barcelona, Martí i Franquès, 1, 08028 Barcelona, Spain
sabineportal@hotmail.com

ABSTRACT

Self-assemblies of submicrometric sized particles in large area compact monolayers constitute a promising field with applications as new optical submicrostructured devices of macroscopic dimensions. These structured surfaces present a photonic band, directly associated to the self-assembled structure and to the nature of the used material. In this work, monodisperse silica submicrospheres with diameter sizes of about 300 nm and about 380 nm were synthesized by sol-gel process. They were arranged in monolayers showing hexagonal structure on 2.5 cm diameter glass disks by means of the Langmuir-Blodgett technique. Morphology and structural arrangements were determined by electron microscopy. The optical properties of the submicroparticle arrays were studied by optical transmittance measurements giving several photonic bands in the 340-440 nm region. Further measurements by phase-modulated transmission spectroscopic ellipsometry revealed optical anisotropy in the samples. This result suggests that the detected amount of birefringence and dichroism must be related to the periodical structure and to the presence of defects in the arrangement.

Keywords: silica particle, 2D crystal, Langmuir-Blodgett, birefringence.

1 INTRODUCTION

The field of photonic crystals (PC) has rapidly expanded in the last decades. These materials, due to the periodicity of their dielectric constant, are characterized by a photonic bandgap that forbids light propagation [1]. A complete photonic bandgap can only be theoretically achieved in 3D PC [2]. However, due to the simplicity of their fabrication, 1D and 2D PC are more developed [3]. 2D monolayer crystals which derivate from 2D PC [4] possess interesting properties that allow a wide range of applications: from micrometric sized lenses [5], catalytic devices [6] to biological applications [7].

Concerning optical anisotropy in general, although linear birefringence is usually considered as an undesirable attribute in crystallography, the possibility of controlling and modifying this property is of great interest for photonic applications from reflection coatings to light confinement [8]. Linear birefringence is characterized by a difference in the speed of light for two orthogonal linear polarisation of light passing through a sample. Birefringence can be evidenced by placing the sample between two crossed polarizers. Recently, photoelastic modulators have been used to measure low levels of retardation and when used with a suitable optical setup, are also able to accurately measure the direction of the anisotropy fast axis [9,10].

In this work, we have synthesized silica particles by sol-gel method and used the Langmuir-Blodgett technique to organize them in compact monolayers on glass substrates. Our aim was focused on the optical properties of these 2D crystal monolayers or particle monolayer (PM), and more specifically on the photonic and optical properties related to the size and arrangement of the particles.

2 EXPERIMENTAL

2.1 Sample Production

The silica particles were synthesized by the Stöber method [11] from tetraethylorthosilicate (TEOS, high purity ≥99.0%, Fluka), absolute ethanol (Aldrich- 98%) and ammonia (NH_4OH, 25%, Merck). The solution was stirred for 24 h at ambient temperature to let the hydrolysis and condensation reactions of the alkoxide precursor to be completed. The excess reactant and by-products were eliminated by centrifugation before collection of the sol-gel particles and redispersion in fresh ethanol. The centrifugation cycle was repeated 5 times.

Glass substrates were cleaned in a Piranha solution, rinsed with Milli-Q water and immersed vertically inside the Langmuir-Blodgett trough filled with Milli-Q water. Meantime, the synthesized particles were diluted in a mixture of alcohol (ethanol or methanol) and chloroform (1:3) and sonicated for 5 min before spreading them on the water surface. The trough barriers were brought closer (with a constant rate of 10 mm/min) until a pressure of 5 mN/m was reached. The substrate was lifted from the trough with a speed of 2 mm/min at constant pressure, dragging on its surface a monolayer of silica particles.

2.2 Characterization techniques

The morphology of silica particles, their size and the assembly structure were determined by Environmental

NSTI-Nanotech 2008, www.nsti.org, ISBN 978-1-4200-8503-7 Vol. 1

SEM (ESEM Quanta 200 FEI). This apparatus allows non-destructive observations. An accelerating voltage of 15 kV, an emission current between 90-100 μA and a water vapour pressure of 0.5-0.9 Torr were used in this study.

Transmittance measurements were performed with a UV-2101 PC UV-Vis Scanning Spectrophotometer (Shimadzu), between 300 nm and 800 nm in normal incidence using unpolarized light. The reference of each sample was the bare glass substrate.

We used a standard phase modulated ellipsometry scheme (polarizer, photoelastic modulator, sample, and analyzer) in transmission mode (TSE), working in the UV-visible range, to measure the linear birefringence (LB) and the linear dichroism (LD) of the sample. In our setup, the sample could be rotated in its plane (azimuthal angle), describing a complete circle, with a minimum step of 1°.

The anisotropy axes of the samples were evaluated by performing several spectral measurements of LD and LB for different azimuthal angular positions of the sample (ϕ); thus, by rotating the sample, we performed a scan that allowed us to find LB_{max} and LD_{max} for every wavelength. More details of the transmission ellipsometer setup can be found in reference [12]. The Mueller matrix for a non-depolarizing linearly anisotropic uniaxial sample at normal incidence with its principal axis $\phi=0°$ is given by [12]:

$$M = \begin{pmatrix} \cosh LD & -\sinh LD & 0 & 0 \\ -\sinh LD & \cosh LD & 0 & 0 \\ 0 & 0 & \cos LB & -\sin LB \\ 0 & 0 & \sin LB & \cos LB \end{pmatrix} \quad (1)$$

where $LB = \dfrac{2\pi d(n_o - n_e)}{\lambda}$ and $LD = \dfrac{2\pi d(\kappa_o - \kappa_e)}{\lambda}$ (2).

and d is the thickness of the sample, λ is the wavelength of light, n_O and n_E are the ordinary and extraordinary refractive indexes, and k_O and k_E are the ordinary and extraordinary extinction coefficients. When LB and LD are small (as those of the samples we have studied where they are not higher than 0.02 rad) equation (1) can be approximated as follows [12]:

$$M = \begin{pmatrix} 1 & -LD & 0 & 0 \\ -LD & 1 & 0 & 0 \\ 0 & 0 & 1 & -LB \\ 0 & 0 & LB & 1 \end{pmatrix} \quad (2)$$

If the orientations of the polarizer, modulator and analyzer of the sample are strategically chosen (A = 45°, P-M = 45°, M = -45°) the Mueller matrix of the sample can be completely determined with a single phase-modulated ellipsometry measurement. In this configuration, the matrix elements m_{01} and m_{23} (that respectively are related to LD and LB) can be directly obtained from the fundamental and second harmonic amplitudes and phases of the detected signal. LB and LD were measured between 280 and 750 nm.

3 RESULTS AND DISCUSSION

Two kinds of samples were produced here: before Langmuir-Blodgett deposition a mixture of ethanol and chloroform was used for PM sample 1 (PM1), whereas ethanol was replaced by methanol for PM sample 2 (PM2). Moreover, slight differences introduced in the chemical synthesis made the particle size used for PM1 a bit smaller than the one used for PM2.

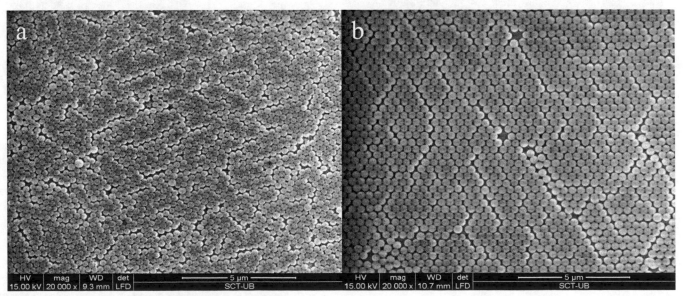

Figure 1: PM1 (a) and PM2 (b) observed by SEM

On figure 1, we observe that the particles are arranged in hexagonal structured domains. Many dislocations are present in the monolayer and could be the result of monolayer breaking due to the competition between gravitational and capillary forces during the particle arrangement on the substrate by the Langmuir-Blodgett process. Particle size calculation performed with ImageJ software gave D = 297 ± 5 nm for PM1 and D = 380± 5 for PM2.

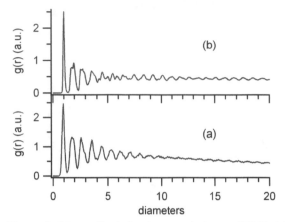

Figure 2: Normalized density correlations of PM1 (a) and PM2 (b). Positional order is greatly enhanced in (b) with respect to (a). The interparticle distance is represented on the horizontal axis.

In figure 2, peaks in the positional correlations correspond to the normalized probability of interparticle distances. The analysis has been performed in a region of size 50 x 50 particles for PM1 (a) and 200 x 200 particles for PM2 (b). Position correlations decrease rapidly with distance for PM1 as indicated by the amplitude of oscillations (a). This fast decay is related to the existence of dislocations that divide the monolayer in numerous ordered particle domains. As for PM2 (b), they decay significantly more slowly than PM1, and characteristic peaks are clearly resolved. For instance, the next-nearest neighbour ($\sqrt{3}$ D) and the second neighbour (2 D) peaks can be distinguished only in PM2. This structural analysis illustrates the higher quality of PM2. Moreover this study shows that methanol is the best solvent to obtain large films of well ordered particles. It is actually well known that the bigger is the particle the better is the arrangement [13].

On transmittance spectra (figure 3), PM1 presents a gap at 340 nm while PM2 presents a gap at 435 nm and small non-well resolved gaps are noticeable at 340 and 360 nm. Light that propagates through the crystal is scattered by the individual particles of the crystal [14]. Photonic gaps are due to the coupling of the grazing scattered light, whose in-plane component k_D is equal to the reciprocal network vector G with the eigenmodes $k(\omega)$ of the 2D crystals [15, 16]:

$$k(\omega) = k_0 n_{eff} = k_d = \frac{2\pi}{a} \qquad (3)$$

where k_0 is the wave vector in vacuum, n_{eff} is the effective refractive index of the 2D crystal and a the interlinear distance between particles. The position of the gap contains information about the size of the particles. For instance if we consider the interlinear distance $a = \frac{\sqrt{3}}{2} \cdot D$, typical of triangular lattices, with the assumption that the crystal is isotropic and we can apply the effective medium theory through a constant refractive index, with n_{eff}=1.287, we found a particle size of 305 nm for PM1 and 390 nm for PM2, which are closed to the values measured on the SEM pictures. The transmittance of fig.3 depends on the polarization of incident light according to the activation of the resonance associated to particular directions (hexagonal symmetry) in our 2D photonic crystal.

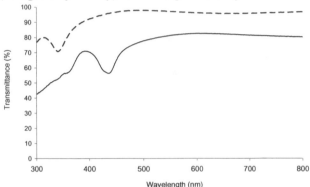

Figure 3: Transmittance spectra of PM1 (dashed line) and PM2 (bold line)

The isotropy of a perfect 2D crystal having hexagonal symmetry implies null values for LB and LD. However peaks are found in the LB and LD spectra detected by ellipsometry (figure 4). Although their values are small, their presence and their evolution with the sample azimuthal rotation indicate that the monolayer is uniaxially anisotropic and that the anisotropy axis is on the plane of sample. The detected peak positions for PM1 and PM2 appear in the regions that coincide with the photonic gap locations found in the transmittance spectra (figure 3). The analogous positions are due to the fact that LD (eq. 2) is proportional to one polarization component value parallel to the anisotropy direction through k, which undergoes a variation at the wavelength resonance of the crystal.

A correspondence between every birefringent and dichroic peak is evident and the experimental dispersion relation for the real (LB) and imaginary (LD) parts of the anisotropy shows Lorentz oscillator characteristics [17]. The imaginary part of the anisotropy reaches its extreme values when the real part vanishes at energies slightly lower and higher than the absorption.

The presence of extra peaks (340 and 360 nm of PM2) found in TSE spectra confirms the existence of second-order photonic bands. Whispering gallery modes or Mie resonance of individual spheres may also be involved in

NSTI-Nanotech 2008, www.nsti.org, ISBN 978-1-4200-8503-7 Vol. 1

this phenomenon [18,19]. However, no significant change in the amplitude of anisotropy was found between PM1 and PM2 despite the differences of particle size and quality of the arrangement (PM1 on fig. 1 appears more disordered than PM2).

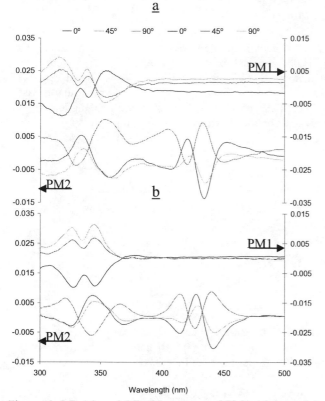

Figure 4: LB (a) and LD (b) spectra of PM1 (right scale) and PM2 (left scale) at 0°, 45° and 90° from bottom to top. Vertical axis scales are in radians.

The deposition technique could be responsible of the anisotropy found in the PM. Effectively the uniaxial birefringence can be related to the stretching of the monolayer in the extraction during deposition due to the competitive effect of capillary and gravitational forces, resulting in the formation of oriented phases in the 2D crystal. Within the lattice, the mismatches of the sample are not completely isotropically distributed (figure 1), and dislocations could be more frequent in one direction than in the others. We are taking further measurements here to prove whether the direction of the anisotropy is macroscopically introduced during the film transfer in the Langmuir process, so that the axis of the anisotropy is related with the dipping direction [20] or whether uniaxial birefringence is an intrinsic property of hexagonal 2D crystals.

4 CONCLUSION

Photonic gaps associated to uniaxial anisotropy were revealed in 2D hexagonal crystals of silica submicrosphere monolayer. This study showed the effect of order on the peaks appearance and the maintenance of anisotropy despite a high quality 2D crystal.

Possible applications of these anisotropic nanoparticle monolayer films are opal effect for application in jewellery, narrow-band polarizing filters and 2D-gratings in the UV-visible range for photonic applications.

ACKNOWLEDGEMENTS
This study was supported by the Generalitat de Catalunya (Project 2005SGR00666) and the MEC of Spain (Project DPI2006-03070). The authors thank Serveis Cientifico-Tècnics of the Universitat de Barcelona (SCT-UB) for measurement facilities. O. A. and M. A. V. acknowledge the financial support from MEC of Spain through the grant FPU AP2006-00193 and project BQU2003-05042-C02-01 respectively

REFERENCES
[1] E. Yablonovitch, Phys. Rev. Lett., 58, 2059, 1989.
[2] M. Maldovan and E.L.Thomas, Nat. Mater., 3, 593, 2004.
[3] T. Stomeo, A. Passaseo, R. Cingolani, and M. De Vittorio, Superlattices Microstruct, 36, 265, 2004.
[4] Y. Kurokawa, H. Miyazaki, Y. Jimba, Phys. Rev. B: Condens. Matter, 65, 20, 2002.
[5] Hirai T. and Hayashi S., Colloids Surf. A, 153, 503, 1999.
[6] Y. Li., W Cai, B. Cao, G. Duan, F. Sun, C. Li and L. Jia, Nanotechnology, 17, 238, 2006
[7] C. Huwiler, M. Halter, K. Rezwan, D. Falconnet, M. Textor and J. Vörös, Nanotechnology, 16, 3045, 2005.
[8] V. Ksianzou, R.K. Velagapudi, B. Grimm, and S. Schrader, J. Appl. Phys., 100, 6, 2006.
[9] T.C. Oakberg, SPIE Proceedings, 17, 2873, 1996.
[10] B.L. Wang and T.C. Oakberg, Rev. Sci. Instr ,70, 3847, 1999.
[11] W. Stöber, A. Fink, and E. Bohn, J. Colloid Interface Sci., 26, 62, 1968.
[12] O. Arteaga, Z. El-Hachemi, and A. Canillas, Phys. Stat. Sol. (a) Accepted
[13] M. Shishido, D. Kitagawa, Coll. Surf. A, 311, 32 2007.
[14] C.F. Bohren and D.R. Huffman, Absorption and Scattering of Light by Small Particles,New York: Wiley, 1983.
[15] H. Miyazaki and K. Ohtaka, Phys. Rev. B, 58, 6920, 1998.
[16] M. Lester and R.A. Depine, Opt. Commun., 127, 189, 1996.
[17] H.G. Tompkins.and E.A. Irene, "Handbook of ellipsometry", Springer 159, 2005.
[18] H.T. Miyazaki, H. Miyazaki, K. Ohtaka and T. Sato, J. Appl. Phys., 87, 7152, 2000.
[19] L. Shi, X. Jiang and C. Li, J. Phys.: Condens. Matter, 19, 176214, 2007.
[20] D.K. Schwartz, Surf. Sci. Rep., 27, 241, 1997.

Advanced fluorescent nanotracers: a broad field of application

O. Raccurt, J. Samuel, O. Poncelet, S. Szenknect, F. Tardif

Commissariat à l'Energie Atomique, Department of Nano-Materials, Laboratory of Tracer Technologies, 38054 Grenoble, France

ABSTRACT

Optical and magnetic nanoparticles are both attractive fields of investigation which are gaining in importance due to their potential markets e.g. polymer nano-composite, colloidal dispersion, environmental monitoring or biomedical diagnostics, etc. A very large set of fluorescent particles can be bought or synthesized thanks to recent progress in nanotechnologies. Their diameter, their surface functionalization, their spectral signature can be tailored for specific applications. This article describes the interesting potentiality for tracing nanoparticles into three different applications such as the monitoring of the distribution of nanocharges in polymer composites, the characterization of colloidal stability of sols in liquids of various polarity and finally, the colloidal transport in porous media. In all the case, nanotracers may be constituted by engineered fluorescent particles chemically and structurally similar to the colloids of interest.

Keywords: nanotracers; nanoparticles; fluorescence; surface functionalization, colloidal particles as transport agents

1 INTRODUCTION

Optical and magnetic nanoparticles are both attractive fields of investigation which are gaining in importance due to their potential markets e.g. polymer nano-composite [1,2], colloidal dispersion, environmental monitoring or biomedical diagnostics [3], etc. Among all the spectroscopic mechanisms used in order to monitor tracers, fluorescence is the most attractive because of its simplicity to use and inherently low detection limits. A very large set of fluorescent particles can be bought or designed. Their diameter, their surface functionalization, their spectral signature can be tailored for specific applications. For instance, these fluorescent nanotracers with controlled size and surface properties offer an interesting potential for different application such as nanofiller homogeneity monitoring in polymer composite [4], characterization of colloidal stability in liquids and colloidal transport in porous media.

2 NANOTRACERS

First, how define what a nanotracer is? First at all, a nanotracer has to be similar in size, density, surface charges to the species which have to be traced. So it can be reasonably expected that the nanotracer particles will physically behave like the nanocharges which have to be traced. Moreover the nanotracer must be detectable using a particular physical property for example luminescence or magnetism. In each case of application presented here, we have chosen to use fluorescent nanotracers because it is easy to detect them by classical fluorescence imaging and to set analytical apparatus around process equipment.

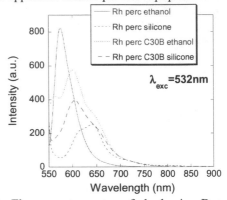

Figure 1: Fluorescent spectra of rhodamine B perchlorate and nanotracers closite 30 B nanoclay with intercaled rhodamine B perchlorate in ethanol and silicone oil. All tracers are excited at 532 nm.

Typically there are two approaches for designing a nanotracer, the first one is the direct incorporation of a fluorescent organic dye in the nanoparticle, this is a convenient way for clays wherein some ionic exchanges can be carried out. The second one is to encapsulate a fluorescent organic dye during the synthesis of nanoparticles, this can be done by using the sol-gel chemistry of silica.

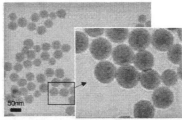

Figure 2: TEM image of silica nanotracers including organic fluorophores embedded into a silica matrix covered by a nanoshell, whose zeta potential can be tailored for each application.

Montmorillonite nanoclay material is a good example to illustrate the first approach. We incorporated in a Montmorillonite nanoclay (South Clay Co.) by ionic exchange an organic fluorescent dye. To ensure the exchange, the dye has to be positively charged, so we have chosen Rhodamine B perchlorate. Intercalation of fluorescent dye is characterized by ATG, RX and fluorescent measurement. Figure 1 shows the fluorescent spectral data for Rhodamine B perchlorate only and intercalated into clay (MMT Closite 30B from South Clay Co.) in non polar solvent. The split of the peak from one to tree is representative from the position of organic dye compared to the clay material: free in polymers, adsorbed on the external surface of nanoclay particles and intercalated between the layers of the nanoclay (see figure 1). So in this case, the fluorescence measurements will monitor both the distribution of the nanocharges but also the affinity of the particles versus the organic part of the composite.

The second approach can be illustrated by the synthesis of fluorescent silica nanoparticle by reverse micelle sol-gel route. More detail of the synthesis protocol is given in another abstract of this conference [1]. The process controlled the size actually from 40 nm to 200 nm. The surface properties were modified by addition of silane coupling agents to obtain the appropriate surface potential or hydrophobic or hydrophilic properties. Two example of application of this family of nanotracer will be discussed for the measurements of the dispersion of nanoparticles in an organic binder and for the tracing of the mobility of natural colloids into soils column laboratory experiments

3 EXEMPLES OF APPLICATIONS

3.1 Luminescent clays nanotracers in polymers[1]

The quality of the mixing of nanocharge in polymer matrix is a key point to design in a reproducible way high-tech composite. For the first application, dispersion of nanoparticles such as luminescent nano-clays into polymer has been studied by measurement of the fluorescence. For nano-clays, the intercalation of the fluorescent organic dyes into natural clay is studied. These fluorescent nano-clay are used to monitoring the mixing and the exfoliation process during extrusion [3].

The objective of this experiment is to observe the dispersion of nanoclay during the mixing into the extrusion of polymers. The experimental setup is show in figure 3. A transparent tube with a standard screw extrusion is installed in closed circuit. The PDMS polymer is used for this transparence at the wavelength scale for fluorescence measurement. The viscosity of PDMS is chosen to simulate the viscosity of a real molten polymer in the screw. Nanotracers (Closite 30B with rhodamine B perchlorate intercaled) are first dispersed into 20 mL of PDMS at 1% in mass concentration. This "labeled" volume is introduced at the beginning of the screw with a syringe. Four cameras

[1] the authors would like to thank A. Esposito, J Balcaen and J. Duchet from LMM/IMP UMR CNRS for their collaborative work on the study of mixing nanoclay

with adapted long band pass filters shoot the distribution of nanotracer during extrusion (see fig 4).

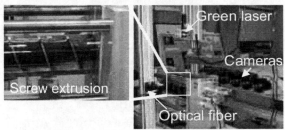

Figure 3: Experimental display for tracing luminescent nanoclay mixing into the polymer during extrusion processes, right. Zoom view of screw extrusion at the end of mixing show laser plan excitation and nanotracers into polymers (red), left.

Figure 4: Imaging the fluorescent nanotracer into the screw extrusion during mixing

A green laser (532 nm, 10 mW) with lense is used to excited nanotracers by a laser plane (see figure 3 left). An optical fiber record at a portable spectrophotometer (USB2000 from Ocean Optics) is placed at the end of the screw to measure the fluorescence in function of time during the experiment. The time acquisition is synchronized to the rotation rate of the screw. Evolution of spectral response of the nanotracer in function of time during the mixing is showing in figure 5.

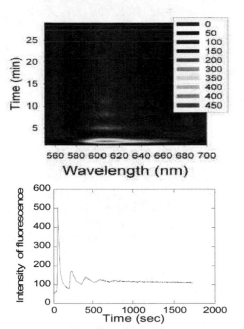

Figure 5: Evolution in function of time of the spectral fluorescence of tracer during the mixing (top complete spectral data, bottom @ 610 nm)

The results show an amortized oscillator function characteristic of a mixing process. First, immediately after the injection of tracer, the maximum of fluorescence become after 2.5 minutes corresponding to the time of tracers become from the detector. After that the maximum of fluorescence @ 610 nm decrease with oscillation to tend with a constant value. The mixing is completed after 5 cycles.

3.2 Luminescent silica nanotracers to measure the dispersion of nanocharges into polymers

The second application concerns the characterization of the dispersion and the colloidal stability in different solvents. In this part we will show the impact of the surface properties on the dispersion of fluorescent silica nanoparticle into polymeric material. Fluorescent silica nanoparticles synthesized by sol-gel micro-emulsion process showing in [5]. The hydrophobic or hydrophilic surface properties are tuned by the covalent bounding of silane coupling agents.

Figure 6: Confocal microscopy image of silica nanotracer with fluoro/hydrophobic (left) and amino/hydrophylic (right) surface function into PVA polymer

The fluorescent silica nanoparticles (diameter: 40nm) are easy to disperse in water or in acetone (Fig. 2). By modifying the protocol, other sizes can be easily obtained [6,7]. Two chemical functions have been successfully covalently linked to the nanoparticles' surface: a hydrophilic amino based molecule and a hydrophobic/oleophobic one containing a fluoro carbon moiety. These molecules have an effect of the physical behavior of the nanoparticles' surface, illustrated by the zeta potential value in aqueous medium (Fig. 9). Confocal microscopy imaging (Fig. 6) shows the distribution of the luminescent silica nanoparticles in PVA films. Unambiguously, the surface treatments of the silica particles greatly impact the distribution of particle in PVA.

3.3 Luminescent silica nanotracers for tracing colloidal transport in soil

The silica fluorescent nanoparticle are using for tracing the colloidal transport into the soil. Recent field and laboratory experiments have identified colloid-facilitated transport of

contaminants as an important mechanism of contaminant migration through groundwater. Groundwater is an important receiving environment for bulk chemicals of daily use and is therefore expected to be a sink for nanoparticles as well. Indeed, the mobility of a particle in a porous medium is strongly limited by its tendency to deposit on the surface of the grains, the so-called "filtration-effect". The process of nanoparticle deposition during flow through porous media is commonly assumed to take place in two rate-limiting steps [8]: first transport to the surface of the collector by Brownian diffusion, interception, or sedimentation (for the larger ones), then, attachment to the surface. The kinetics of the transport step depends primarily on physical factors such as size, shape, density of the nanoparticles, flow velocity and pore geometry. The kinetics of the attachment step is controlled by interparticle forces: van der Waals, electrostatic, steric repulsion and hydrophobic forces between nanoparticles and grain surfaces. Due to the large number of factors influencing nanoparticles mobility in porous media, we propose here to use versatile model colloidal tracers, called "nanotracers" to gain insight into the impact of the mechanisms that drive the transport and the deposition of nanoparticles in a natural porous medium.

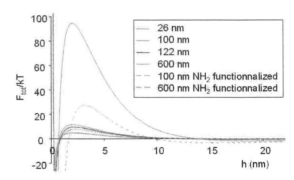

Figure 7: Comparison of total energy potential of particles (DLVO theory) as a function of interaction distance with collector surface between nanotracers (100 and 600 nm) and natural colloids (26 and 122 nm).

To demonstrate the existence of colloid-facilitated transport, three criteria must be fulfilled: (1) soil colloidal particles must be mobilized; (2) pollutants must associate with the colloids and (3) colloids must be transported through the porous media. Our aim is to determine the physicochemical conditions that lead to soil colloid generation and transport in a heterogeneous porous media and compare them with the classical mobilization theory. As this theory hardly fits to natural heterogeneous porous media, we are used new colloidal tracers to simulate the transport of natural particles. Monodispersed silica fluorescent nanoparticles were obtained from a modified Stöber synthesis [9,1]. The use of nanotracers with controlled properties allows quantification of the impact of size and the zeta potential on the mobility of the particles individually. Therefore, after production of functionalized nanotracers using amino silane coupling agents [10,1], and characterization by Dynamic Light Scattering technique (Fig. 8 and 9), leaching experiments were performed in repacked columns under conditions representative of groundwater flow. The soil used is a calcareous alluvial deposit from a fluvial shallow aquifer. The produced nanotracer suspensions have been injected into columns of 25-30 cm in length and fluorescence followed in the outflow to determine the breakthrough curves.

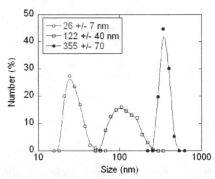

Figure 8: Size distribution of three different silica fluorescent nanotraceurs measured by Dynamic Light Scattering (Malvern Nanosizer)

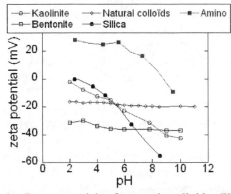

Figure 9: Zeta potential of natural colloids (Kaolinite, bentonite and unknown natural colloids) and nanotracer with and without amino functionnalization

These results have been described using the classical convection-dispersion equation, with a first-order irreversible process for particle deposition. Correlations between transport and deposition parameters using size and zeta potential of the nanotracers have been established under environmentally relevant conditions. The deposition rates have also been compared with predictions made using the DLVO theory and the classical filtration theory [11], that it is known to be hardly applicable for particles transport through natural porous media. Experimental measurement of the irreversible deposition rate (Kirr) of colloids is performing into a column of sol. The measurement of the fluorescence of tracers before and after the soil column give an information of the colloid mobility and the quantity of nanoparticle fixing into the soil. Figure 10 show the irreversible deposition rate of nanoparticles in

function of size and the flow rate of water through the column. The results obtained indicate that the use of designed colloidal tracers is a promising tool for the estimation of nanoparticles transport parameters.

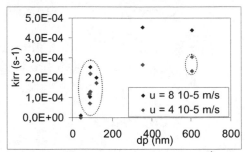

Figure 10: *Kirr* irreversible deposition rate [s^{-1}] in function of particle size measured in a laboratory column experiment

4   CONCLUSION

Throw four application we demonstrated the interest of nanotracers into the measurement of nanoparticles dispersion or movement into a media. The fluorescence techniques offer a good sensibility in the case of the media is transparent (like polymers) for in-situe observation. Fluorescent nanotracers offer a very interesting panel of possibility to observe and measure the comportement of nanoparticles into a media with a wild range of application.

REFERENCES

[1]   J. Samuel, O. Raccurt, O. Poncelet, O. Tillement, F. Tardif, Nanotech 2008, 2008.

[2]   J.L. Chavez, J.L. Wong, A.V. Jovanovic, E.K. Sinner and R.S. Duran, IEE Proc.-Nanobiotechnol., Vol. 152, No. 2, April 2005.

[3]   A.Burns, H. Ow and U. Wiesner, Chem. Soc. Rev., 35, 1028–1042., 2006.

[4]   P. H. Maupin, J. W. Gilman, R. H. Harris, Jr., S. Bellayer, A. J. Bur, S. C. Roth, M. Murariu, A. B. Morgan, J. D. Harris, Macromol. Rapid Commun, 25, 788–792. 2004.

[5]   R. P. Bagwe ; Langmuir ; N°22 ; 4357 – 4362 ; 2006

[6]   T. Jesionowski ; Journal of Materials Science ; N°37 ; 2002 ; 5275 – 5281.

[7]   J. W. Goodwin; Colloid. Polym. Sci ; N°268 ; 1990 ; 766 – 777.

[8]   Kretzschmar, R., Borkovec, M., Grolimund, D., Elimelech, M., 1999. Advances in Agronomy, 66: 121-193.

[9]   Ow H., D.R. Larson, M. Srivastava, B.A. Baird, W.W. Webb and U. Wiesner, 2005. Nanoletters, 5 (1): 113-117.

[10] Jesionowski T. and A. Krysztafkiewicz, 2002. Colloids and Surfaces a-Physicochemical and Engineering Aspects, 207 (1-3): 49-58.

[11] Ryan, J.N. and Elimelech, M., 1996. Colloids Surfaces A: Physicochem. Eng. Aspects 107: 1-56.

Development of Nanoparticle-Based Gold Contrast Agent for Photoacoustic Tomography

Yi-Shan Yang[1], Srikant Vaithilingam[2], Herb Te-Jen Ma[2], Saeid Salehi-Had[1]
Ömer Oralkan[2], Butrus (Pierre) T. Khuri-Yakub[2], Samira Guccione[1]

[1]Department of Radiology and [2]Department of Electrical Engineering,
Stanford University, Stanford, CA, USA
ysyang@stanford.edu

ABSTRACT

Photoacoustic tomography (PAT) utilizes non-ionizing energy to obtain structural/functional information with high spatial resolution and sensitivity. Heterogeneous absorption of optical energy in biological tissues results in differentially expressed acoustic signals yielding spatial/temporal information. A critically underdeveloped area in PAT is the development of contrast agents. Gold-containing nanoparticles can be a good choice due to their high bio-compatibility, low toxicity, and high feasibility of surface modification/conjugation. Here we present a broad evaluation of gold nanosphere with diameters 2-60nm for PAT contrast agents at ~550nm. The acoustic signal enhances with increasing size. Encapsulation of gold nanospheres using liposomes shifted optical absorption to > 600nm with increased enhancement. The highest contrast enhancement was obtained for 5nm-Au and 30nm-Au nanosphere encapsulated in PDA. Future work focuses on modifying the liposome for in-vivo targeted delivery.

Keywords: photoacoustic tomography, gold nanosphere, gold nanoparticle, contrast agent, molecular imaging

1 INTRODUCTION

Photoacoustic tomography (PAT) is an emerging imaging technique. The object of interest is heated with laser pulses. The laser power is absorbed by the object and converted to heat. An acoustic wave is generated based on the specific thermo-elasticity property of the object. This acoustic wave is detected using ultrasound transducers [1]. This technology combines the advantages from both optical and ultrasound imaging, while overcoming several of their disadvantages [2]. It utilizes non-ionizing energy to obtain both structural and functional information with high spatial resolution, high sensitivity, and low scattering. Several biological molecules have intrinsic PAT signal, such as hemoglobin and melanin. Applications have been reported using these endogenous contrast agents including tumor angiogenesis monitoring, functional brain mapping, blood oxygenation mapping and cancer detection [3-5].

In order to extend applications to molecular imaging, the development of exogenous contrast agents for this modality becomes critical. Contrast agents with proper size, shape, composition and functionality will be able to increase the signal intensity, penetration depth and specificity for this imaging modality. Typically, metal-containing particles provide excellent efficiency in light absorption [6]. One of the most promising candidates for *in vivo* applications is gold-containing nanoparticles. They have high *in vivo* bio-compatibility and low toxicity. In addition, these particles are amenable to surface modification and conjugation for increased contrast enhancement or targeting capability. To date, gold-based nanoparticles that have been explored and developed include nanoshells [7,8], nanorods [9-11], and nanocages [12,13] with feasibility of bio-conjugation.

In the present study we characterize a broad array of gold-containing nanosphere suitable for PAT contrast agents with different size and optical absorption properties. Potential application and future directions for *in vivo* PAT imaging is discussed.

2 MATERIALS AND METHODS

Gold nanospheres with diameter 2-60nm were purchase from Corpuscular, Inc. (Cold Spring, NY). Phosphatidylcholine (PC) was purchased from Avanti Polar Lipids, Inc. (Alabaster, AL). 10,12-pentacosadiynoic acid (PDA) was purchased from Alfa Aesar (Ward Hill, MA). PC and PDA-based liposomes were prepared based on procedures published previously [14]. Characterizations of the nanoparticles were performed using dynamic light scattering with multi angle particle sizer, scanning electron microscopy, UV-Vis absorption spectroscopy, and inductively coupled plasma atomic emission spectroscopy.

Four groups of nanoparticles were prepared. Group A: gold nanospheres with diameters ranging from 2-60nm. Group B: gold nanospheres encapsulated in PC-based liposomes. Group C: gold nanospheres encapsulated in PDA-based liposomes. Group D: control substances including pure vegetable oil, deionized water and dye. Vessel-mimicking phantom samples were prepared by injecting nanoparticles into the polyethylene tubes with I.D. of 0.045". Both tube ends were heat-sealed and secured using clamps to maintain a fixed distance to the acoustic transducer.

NSTI-Nanotech 2008, www.nsti.org, ISBN 978-1-4200-8503-7 Vol. 1

Q–switched Nd:YAG laser or OPO tunable laser was used to generate pulse energy. The Nd:YAG laser features a fixed wavelength at 532nm with pulse duration of 10ns and frequency of 20Hz. OPO tunable laser features tunable wavelength 680-2500nm with 3-5ns pulse width and 20Hz frequency. Acoustic waves were detected using either a 2-D array of capacitive micromachined ultrasonic transducers (CMUTs) or piezoelectric ultrasound transducer. A thorough design and fabrication of the 2-D CMUT has been published elsewhere [15]. Piezoelectric ultrasound transducer had a center frequency of 25 MHz, focal depth of 8 mm. The acoustic wave was received and the output was amplified, digitized and recorded by a digital oscilloscope synchronized to the laser.

3 RESULTS

Figure 1 shows the normalized molar absorbance of the nanospheres. Group A: gold nanospheres with diameters of 2-60nm have maximum absorption wavelength around 520-540nm. Their molar absorbance increased with size. Group B: encapsulation of the gold nanosphere with PC-based liposomes. They have size average of ~85nm and shelf life of at least 2 months. Encapsulation of the nanospheres did not significantly alter their maximum absorption wavelength. The bandwidth of the absorption increased and the molar absorbance increased as compared with the gold nanospheres alone. PC-encapsulated 30nm Au had the highest molar absorbance. In contrast, encapsulation of the gold nanosphere with the PDA-based liposomes (Group C) significantly altered the optical absorption profile. These particles have an averaged size of 120nm and a shelf life of at least 5 months. PDA alone has a wide absorption band from 500-700nm. Thus when the gold nanosphere is encapsulated in PDA liposomes, the composite nanoparticle has a wide absorption band extended to 700nm due to the additive optical absorption from the two components. PDA-5nmAu and PDA-30nmAu had the highest molar absorbance.

Photoacoustic signals were generated by applying laser pulse to the nanoparticle samples at approximately the wavelength where the nanoparticle had maximum optical

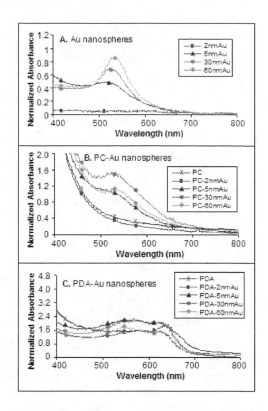

Figure 1. Optical absorption properties (A) Gold nanospheres, (B) PC-encapsulated gold nanospheres and (C) PDA-encapsulated gold nanospheres.

absorption. Setup of samples were described earlier. Four groups of samples were measured and the individual photoacoustic signals are presented in Figure 2. Gold nanospheres (Group A) showed moderate photoacoustic signals that peaked at 30nm. Encapsulating of the nanosphere with PC-based liposome did not significantly enhance the photoacoustic signal (Group B). In contrast, PDA-based encapsulation provided a significant increase in contrast enhancement by approximately 5 folds (Group C). Upon encapsulation, 5nm and 30nm gold nanosphere had the highest photoacoustic signal. Pure oil and deionized water were used as negative controls (Group D).

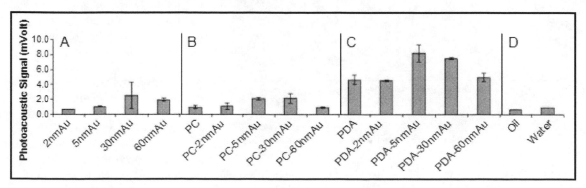

Figure 2. PA signal intensity of gold nanoparticles obtained using CMUT transducers with laser pulses at 532nm

Since PDA encapsulated gold nanosphere (Group C in Figure 2) showed a board optical absorption profile from 500-700nm (Figure 1C), we have further investigated the photoacoustic properties of this group of molecules using a tunable laser, allowing us to obtain photoacoustic signals at a higher wavelength that is more suitable for future clinical application. Signal intensities for PDA-encapsulated gold nanosphere from either 532nm laser pulses received by CMUT transducer, or 685nm laser pulses received by piezoelectric single transducer are comparable (Figure 3, dashed vs. solid bars).

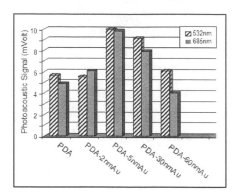

Figure 3. Comparison of the PA signal intensity of PDA-encapsulated gold nanosphere using CMUT at 532nm (dashed) and piezoelectric transducer at 685nm (solid).

Photoacoustic image of gold nanoparticles was obtained using piezoelectric transducer coupled to the OPO laser tuned at 685nm. Sample setup is shown in Figure 4A. Gold nanoparticles were sealed in tubes mimicking blood vessels. The tubes were aligned and secured on the holder and immersed in water for imaging. Images were acquired by slewing the transducer-laser unit across all the tubes in x direction (90mm) and a small area along the tube in the y direction (4mm) (Figure 4A, the

dashed area). The laser was focused on the center of the tube for maximize signal coverage. Figure 4B presents the macro view of the photoacoustic contrast enhancement from the top across all the sample tubes at the same orientation as shown on Figure 4A. Note that the dimension on the x (90mm) and the y axis (4mm) is not to scale. The y axis is elongated for better visualization.

The orientation of the image slice relative to the sample tubing is shown on the top of Figure 4C. Dashed area represents the area where the images were acquired. The contrast enhancement was along the vessel phantom walls, as shown on the bottom of Figure 4C. Since acoustic signals are likely to arise at interfaces between different materials, it is reasonable to question whether the contrast signal is from the interfaces or from the contrast agent. If solely from the interfaces, the signal intensity should remain the same across all the tubes regardless of the type of contrast agents kept in the tubes. Figure 4D shows the images for all the samples (zoomed-in of Figure 4B). The differential signal intensities along the vessel-mimic tube walls indicate that the laser was the best absorbed and the contrast was further enhanced along the vessel-mimic tube wall. For PDA-encapsulated gold nanoparticles, the highest contrast enhancement was observed when the gold nanospheres are 5nm and 30nm in diameter. For gold nanosphere along, the contrast enhancement is fairly weak regardless of the particle size. Since gold nanospheres alone have a rather narrow optical absorption band in the region of 550nm (Figure 1A), at 685nm the gold nanospheres should not generate significant contrast enhancement. Thus the more intense signals observed for the PDA-encapsulated gold nanospheres suggests that this class of nanoparticles should be further evaluated and characterized for use as a photoacoustic contrast agent. In particular the PDA-5nm and PDA-30nmAu gives the best PAT contrast enhancement.

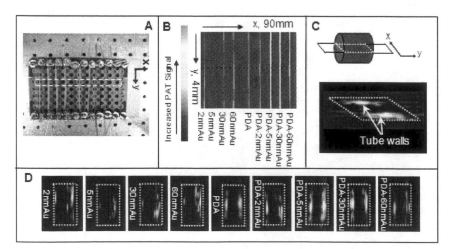

Figure 4. Photoacoustic images of gold nanoparticles PAT contrast agent using piezoelectric transducer with laser wavelength at 685nm. Images of gold nanospheres were also shown. (A) Sample setup with a scanned area of 90mm x 4mm marked with the dashed square. (B) Top view of the images in color scale acquired by scanning through the dashed area marked in (A). Note that for better visualization, the y axis has been elongated and not in scale with the x axis. (C) Top: schematic orientation of the plane, marked with a dashed box, where the image was acquired in relative to the sample tube. Bottom: representative photoacoustic image shows high contrast enhancement along the tube walls. (D) Serial photoacoustic images for PDA-encapsulated gold nanospheres and gold nanospheres alone with different sizes as labeled. Dashed boxes represent image slices corresponding to what on top of (C). Images are shown in the XY plane as indicated in (A) with a dynamic range of 40 dB.

4 DISCUSSION AND SUMMARY

In the present study we have characterized different gold-based PAT contrast agents. We had hypothesized that optimum PAT contrast agents would have a metallic core the would act as a heat sink, coated with and external material that would have an enhanced expansion in response to the core heating and expansion of the core material. We demonstrated that although gold nanospheres alone generate PAT contrast enhancement, composite materials made from a metallic (gold core) in combination with and external coating produced and enhanced PAT signal. Encapsulation of gold nanospheres using polymerized liposomes altered the photoacoustic properties of these nanoparticles. Phosphotidylcholine (PC)-based liposomes slightly increased the contrast enhancement, while pentacosadiynoic acid (PDA)-based liposomes extended the optical absorption energy to around 700nm with significant increase of the contrast enhancement. For these liposome-encapsulated contrast agents, the PAT signal can be generated from both the liposome and the gold nanosphere, and the signal can be modulated by the interaction between the two components. Liposome encapsulation provides an environment where the liposome, depending on the composition, may generate additional photoacoustic wave at the same or different absorption energy region. PC-based liposomes do not have significant optical absorption above 500nm. Therefore the photoacoustic signals for PC-encapsulated gold nanospheres are mainly from the gold nanospheres only and remained similar upon encapsulation. Alternatively, PDA-based liposomes is chromatic with a wide bandwidth of optical absorption up to 700nm, thus generating additional photoacoustic signal when used with gold nanosphere. However, the increased PAT signal in PDA-based liposome became less significant when the size of the gold nanospheres increased. The highest contrast enhancement was obtained for the PDA-encapsulated 5nm and 30nm Au nanospheres. Since the size of the PDA-based liposome is ~120nm for optimal thermodynamic stability, the liposome accommodates fewer particles with increased gold nanoparticle size, resulting in decreased PAT signal.

Since PAT signals were measured using two different setups: Laser pulses at 532nm with CMUT 2-D array transducer, or laser pulse at 685nm with piezoelectric ultrasound transducer, the reliability and feasibility of either system is demonstrated, providing flexibility for future development in instrumentation. PDA-encapsulated 5nm and 30nm gold nanosphere generated the best contrast enhancement. When using vessel phantoms, the best enhancement occurred along the wall. This behavior suggests that the contrast enhancement may be further increased if the agents can be targeted to vessels, which holds potential application for monitoring and therapy in tumor angiogenesis.

In summary, we have shown the capability of preparing the optimal contrast agent for photoacoustic tomography by fine-tuning the size and the composition of the nanoparticles. We believe nanoparticles with a core that is an optimum heat sink, coated with material that readily expand upon heating and can provide functionalization and surface modification would be ideal PAT contrast agents. Here, among the contrast agents we tested, the small size (5nm or 30nm) gold nanosphere encapsulated in the PDA-based polymerized liposome produced the strongest PAT signal. It is advantageous to have liposome encapsulation since polymerized liposomes can readily be modified to carry other molecules such as fluorophores and targeting agents, which have been well-characterized with proven applications for molecular targeting and therapeutic delivery [16]. We have synthesized gold nanorods with optical absorption in the near-IR region (900nm). They can be encapsulated in the liposomes in a similar manner. Future work focuses on further development of the gold nanorods with optical absorption at near-IR region, and the modification of the liposome to make these agents suitable for in-vivo vascular targeted imaging.

REFERENCES

[1] M. H. Xu and L. H. V. Wang, Review of Scientific Instruments, 77, 041101, 2006.

[2] V. Ntziachristos, J. Ripoll, L. H. V. Wang and R. Weissleder, Nature Biotechnology, 23, 313, 2005.

[3] J. Laufer, D. Delpy, C. Elwell and P. Beard, Physics in Medicine and Biology, 52, 141, 2007.

[4] H. F. Zhang, K. Maslov, M. Sivaramakrishnan, et al., Applied Physics Letters, 90, 053901, 2007.

[5] J. T. Oh, M. L. Li, H. F. Zhang, K. Maslov, et al., Journal of Biomedical Optics, 11, 034032, 2006.

[6] J. A. Copland, M. Eghtedari, V. L. Popov, N. Kotov, et al., Molecular Imaging and Biology, 6, 341, 2004.

[7] M.-N. Li, J. A. Schwartz, J. Wang, G. Stoica, et al., Progress in Biomedical Optics and Imaging - Proceedings of SPIE, 6437, 2007.

[8] L. Xiang and F. Zhou, Key Engineering Materials, 364-366 II, 1100, 2008.

[9] A. Agarwal, S. W. Huang, M. O'Donnell, K. C. Day, et al., Journal of Applied Physics, 102, 2007.

[10] K. Kang, H. Sheng-Wen, S. Ashkenazi, M. O'Donnell, et al., Applied Physics Letters, 90, 223901, 2007.

[11] V. P. Zharov, E. I. Galanzha, E. V. Shashkov, et al., Journal of Biomedical Optics, 12, Article, 2007.

[12] S. E. Skrabalak, J. Chen, L. Au, X. Lu, et al., Advanced Materials, 19, 3177, 2007.

[13] X. M. Yang, S. E. Skrabalak, Z. Y. Li, Y. N. Xia, et al., Nano Letters, 7, 3798, 2007.

[14] R. W. Storrs, F. D. Tropper, H. Y. Li, et al., Journal of the American Chemical Society, 117, 7301, 1995.

[15] S. Vaithilingam, I. O. Wygant, P. S. Kuo, X. Zhuang, et al., Proceedings of SPIE - The International Society for Optical Engineering, 6086, 2006.

[16] J. D. Hood, M. Bednarski, R. Frausto, S. Guccione, et al., Science (Washington D C), 296, 2404, 2002.

Gold Nanoparticles as Colourants in High Fashion Fabrics and Textiles

James H. Johnston*, Michael J. Richardson* and Kerstin A. Burridge*

*School of Chemical and Physical Sciences, Victoria University of Wellington,
PO Box 600, Wellington 6140, New Zealand, jim.johnston@vuw.ac.nz

ABSTRACT

This paper presents the novel use of gold nanoparticles with different particle sizes and hence colours, as stable colourfast colourants on wool and cotton fibres for use in high quality fabrics and textiles for high end fashions. The synthesis of gold nanoparticles using monomeric or polymeric amines that serve both as a reductant to facilitate the reduction of Au^{3+} to Au^0, and as a linker to attach the nanoparticles to the amino acids in wool keratin proteins is discussed. X-ray photoelectron spectroscopy supports the bonding of the gold nanoparticles to the wool through the amine linker molecules. Electronmicroscopy shows that the nanoparticles are present on the cuticles of the fibre surface and are concentrated at the edges of these cuticles. A range of coloured fibres and fabrics have been produced.

Keywords: gold, nanoparticles, wool, textile, fashion

1 INTRODUCTION

Wool fibres have been used since historical times in fabrics and textiles for garments, furnishings and floor coverings. The wool fibres are typically about 10-40 microns in diameter and several tens of centimeters in length. They are woven into yarn that is used to make a very wide range of fabrics and textiles. The yarn or the fabric and textiles can be dyed using conventional technology to produce many different colours. *Merino* wool with its smaller fibre diameter of about 10-20 microns is highly valued for fine fabrics and commands a price premium. Wool belongs to the group of fibrous proteins consisting of α-keratins and comprises long polyhedral inner cortical cells surrounded by flattened external cuticle cells and partially by an external fatty acid monolayer, mainly 18-methyleicosanic acid. The helix arrangement of the proteins gives wool its flexibility, elasticity and crimp properties. The fatty acid layer provides some hydrophobicity to the surface which can influence further processability such as dyeing, enhancing shrink resistance and anti-static properties.

Colouration of wool fibres is achieved by dyeing with either natural dyes such as *madder (red), indigo (blue)* or synthetic dyes which are typically anthraquinone-based (the first example *alizarin* was synthesised in 1869) or azo-based compounds, typically under mild acidic conditions of pH=3-5. Colour changes resulting from exposure to UV light, repeated washing and surface abrasion from wearing are issues that adversely affect colorfastness and hence the overall quality of the fabrics and textiles. Colourfastness is particularly important in high quality fabrics and textiles for the high fashion market.

Nobel metals such as gold and silver have long been known to form stable colloids of nanosize particles. Such nanoparticles exhibit different colours due to surface plasmon resonance effects which result from the interaction of incoming electromagnetic radiation in the visible region with the collective plasmon oscillations at the metal surface [1]. The colour is dependent on the particle size and shape. Gold colloids with spherical particles of about 7-20 nm particles appear as a deep red wine colour, which progressively changes through to a blue colour with increasing particle size up to about 80-100 nm. Such colours are stable towards UV light and do not change providing there is no change in particle size, through the growth or reduction of discrete nanoparticles or agglomeration. Interestingly, colloidal gold nanoparticles have been used to colour glass dating back to about the 17th century. Also, the Lycurgus Cup dating back to around 400AD used gold and silver nanoparticles as colourants. The science was not understood until 1869 when Michael Faraday recognized and explained the role of gold as a colourant in general terms. In 1908 Mie [2] provided a theoretical explanation by solving Maxwells equations for the absorption and scattering of electromagnetic radiation by very small metallic particles – nanoparticles.

In this paper we present the novel development and use of gold nanoparticles of different sizes and colours as stable colourfast colourants on *merino* wool and cotton for high value fashion fabrics and textiles. This utilizes proprietary science and technology developed by us [3,5,6] which relates to the controlled reduction of Au^{3+} to Au^0 and the incorporation of them into a variety of wool, cotton and synthetic fibres and their corresponding fabrics and textiles. Control of the nanosize of the Au^0 particles and their binding onto the fibres are critical and govern their use in this interesting application.

2 NANOGOLD – WOOL

Gold nanoparticles in the range of about 2 -100 nm can be readily synthesised by treatment of a Au^{3+} containing solution, usually as the complex $AuCl_4^-$ ion with a reducing agent such a tri-sodium citrate [4], or monomeric and polymeric amines which act as both a reductant and a

NSTI-Nanotech 2008, www.nsti.org, ISBN 978-1-4200-8503-7 Vol. 1

stabilizer [3]. The citrate can also act as a stabilizer. The colour of the resulting gold colloid depends on the size of the gold nanoparticles. Examples of these colours are shown in Fig. 1.

Figure 1: Colloids of gold nanoparticles of different particle size.

The UV-Visible spectra of the red and blue coloured colloids of gold nanoparticles prepared using amines are shown in Fig. 2. This shows that the absorbance due to the transverse plasmon band [1] shifts from about 510 nm (red colloid) to about 575 nm (blue colloid) with increasing particle size. The longitudinal plasmon band at about 750 nm is also shown (green colloid). A transmission electronmicroscope image of typical gold nanoparticles is shown in Fig. 3 [3].

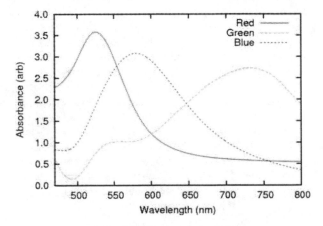

Figure 2: UV-Visible spectra of red and blue colloids of gold.

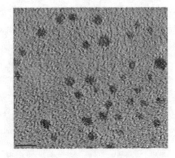

Figure 3: Transmission electronmicroscope image of gold nanoparticles. Marker bar = 5 nm.

The gold nanoparticles present in a colloid can be readily attached to wool, cotton or synthetic fibres using a linker, which can typically be an amine. When monomeric or polyamines are used as the reductant, they also conveniently function as linker molecules [3]. As gold has a strong affinity for S, it is likely that the gold nanoparticles would bind directly to the S of the cystine amino acids in the keratin proteins. However, when a stabilizer or linker is used, the gold nanoparticles are effectively isolated by the surrounding molecules and as such it is likely that the linker attaches to the N, S, O or OH groups of the amino acids in the protein chain through hydrogen bonding.

Photoelectron spectra for the Au 4f electrons of gold nanoparticles show the typical 5/2 and 7/2 peaks for gold metal. However these have been shifted to lower binding energies by about 0.6 eV, probably as a result of the amino groups bound to the gold increasing the negative charge at the gold surface (Fig. 4a) [5]. When the gold nanoparticles are bound onto *merino* wool these 5/2 and 7/2 peaks shift to higher binding energies indicating chemical bonding of the nanoparticles to the fibre through the amine (Fig. 4b) [5]. The peaks also broaden asymmetrically which probably reflect a range of possible binding sites with a slightly different chemical environment, or a component of partially reduced Au^+ [5]. The former is the preferred interpretation.

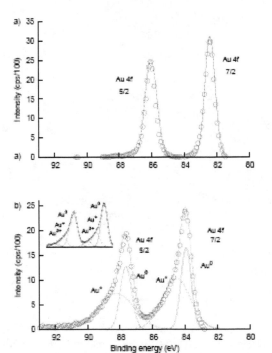

Figure 4: Au 4f X-ray photoelectron spectra for: (a) gold nanoparticles synthesised with polyamine; (b) gold nanoparticles synthesised with polyamine and bound to a *merino* wool fibre [5].

The photoelectron spectrum for the N 1s electrons for gold nanoparticles synthesised using polyamine [3] has been resolved into two peaks relating to N-Au bonds from the gold bonding to the amine linker and N-C bonds from the amine itself (Fig. 5a) [5].

a)

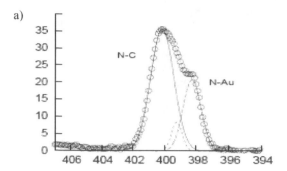

b)

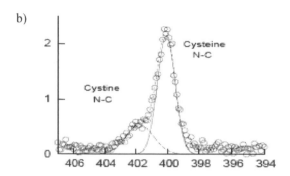

c)

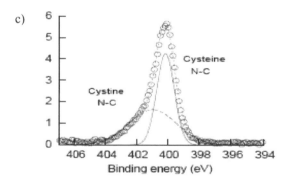

Figure 5: N 1s X-ray photoelectron spectra for: (a) gold nanoparticles synthesised using a polyamine; (b) *merino* wool fibre; (c) gold nanoparticles bound to *merino* wool fibre [5].

The N 1s spectrum for *merino* wool fibre shows N-C bonds consistent with the cystine and cysteine amino acids in the keratin protein (Fig. 5b). For gold nanoparticles bound to the *merino* wool fibre, the N 1s spectra show the N-C cystine and N-C cysteine bonds. The N-C cystine peak is broadened and shifted to lower binding energies suggesting that the gold nanoparticles are predominately associated with the cystine amino acids (Fig. 5c). Also, the disappearance of the N-Au peak observed for gold

nanoparticles when they are bound to the wool fibre, further suggests electron interaction between the gold and wool fibre through the amine nitrogen atoms [5].

Fig. 6a shows the scanning electronmicroscopy images of a *merino* wool fibre coated with gold nanoparticles where polyethyleneimmine was used as the reductant and linker. The characteristic morphology and surface cuticles of a wool fibre are shown in Fig. 6a. At higher magnification it can be seen that the cuticles are covered with an essentially uniform layer of gold nanoparticles of about 10 nm in size (Fig. 6b). Some agglomeration is present as shown by the larger particles of about 50 nm in size. The overall fibre is a pink-red colour.

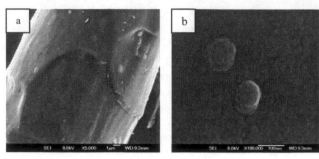

Figure 6: SEM images at different magnifications of a *merino* wool fibre coated with gold nanoparticles.

When a proprietary new methodology developed by us [6] is used to produce nanogold wool fibres at lower gold concentrations, it can be seen from backscattered electronmicroscope images that the gold nanoparticles adhere mainly to the edges of the cuticles (Fig. 7). This is due to the higher surface free energy associated with an edge rather than a planar surface. As the density of surface nanoparticles increases, there is a corresponding increase in surface coverage, although the edge coverage still dominates (Fig. 7). This is further illustrated by energy dispersive X-ray analysis for Au, which shows the highest concentration of gold at the cuticle edges (Fig. 8).

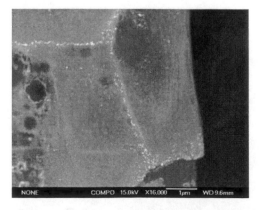

Figure 7: Backscattered electronmicroscope image of a *merino* wool fibre with gold nanoparticles on the surface and mainly at the cuticle edges.

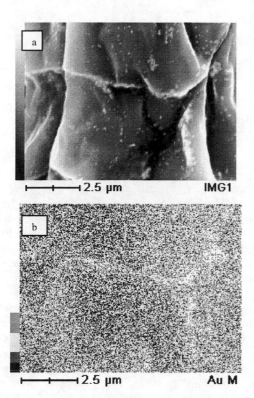

Figure 8: (a) backscattered electronmicroscope image of image of a *merino* wool fibre with gold nanoparticles on the surface; (b) energy dispersive X-ray analysis pattern for Au showing a concentration at the cuticle edges

3 NANOGOLD – WOOL AND COTTON FABRIC

Using the methodology discussed here [3,5] and more particularly new proprietary science and technology developed recently by us [6], we have coated raw top *merino* wool, *merino* wool yarn, *merino* wool fabric and cotton thread and fabric with gold nanoparticles of different sizes and hence colours, to provide new and potentially high value fabrics and textiles for the high end fashion markets. These products are colourfast since the gold nanoparticles are not degraded by UV light as many traditional dyes are. However with the current high cost of gold, the resulting nanogold products are expensive and is therefore suited to the high end market sector. In particular, our new proprietary technology enables a range of "softer" and more desirable colours to be produced. Hence we have used fine *merino* wool and added further value by the novel use of gold nanoparticles as colourants. In addition, as gold exhibits mild antimicrobial properties, these fabrics and textiles have some inherent anti-microbial resistance. Examples of wool and cotton coated with gold nanoparticles are shown in Fig. 9.

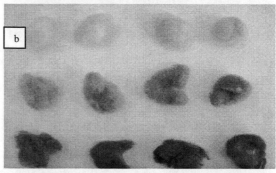

Figure 9: gold nanoparticle coloured samples of (a) wool yarn; (b) top (loose) wool; (c) cotton fabric (top) and wool yarn (bottom).

REFERENCES

[1] T.J. Norman Jr., C.D. Grant, D. Magana and J.Z. Zhang, J. Phys. Chem. B, 106, 7005-7012, 2002.

[2] G. Mie, Ann. Physik, 25, 377, 1908.

[3] S. Link and M.A. El-Sayed, J. Phys. Chem. B, 103, 4212-4217, 1999.

[4] M.J. Richardson, J.H. Johnston and T. Borrmann, Eur. J. Inorg. Chem., 13, 2618-2623, 2006.

[5] M.J. Richardson and J.H. Johnston, J. Coll. Int. Sci., 310, 425-430, 2007.

[6] J.H. Johnston, F.M. Kelly, K.A. Burridge, M.J. Richardson and T. Borrmann, NZ Patent Application, 2008.

ACKNOWLEDGEMENTS

We gratefully acknowledge support from AgResearch Ltd (NZ) and the World Gold Council (London).

Synthesis of highly concentrated silver nanoparticles

G. Klein, E-M. Meyer, L. Si-Ahmed*

Metalor Technologies SA, Av. du Vignoble 2, 2009 Neuchâtel, Switzerland
* lynda.si-ahmed@metalor.com

ABSTRACT

Two new preparation routes for the synthesis of highly concentrated Silver nanoparticles are presented here first. Silver nanoparticles were successfully synthesized in mild conditions using a high loading of silver acetate and water-soluble polymers as Poly-(vinylpyrolidone) PVP, which was found here to act as both reducing and protecting agent. At the end of the reaction, after supernatant removal, highly concentrated silver nanoparticles (up to 40wt%) are obtained and characterized by TEM where 3-50 nm spherical silver nanoparticles can be observed. The second preparation route reported concerns the synthesis of highly concentrated silver nanoparticles (30wt%) in presence of polyethylene glycol and tert-butanol in mild conditions. The as-collected silver pastes were characterised by TEM and revealed spherical silver nanoparticles with 50nm particle sizes.

Keywords: wet-synthesis, emulsions, polymers, silver nanoparticles, high silver nanoparticles concentration.

1 INTRODUCTION

Silver nanoparticles are nowadays of great interest; either for their antibacterial properties or as a low temperature sintering conductive filler in electronic applications. In both cases, silver nanoparticles have to be optimally dispersed in a polymeric matrix while meeting stringent cost targets customary to such large volume markets. Current vapor phase processes are seen as highly investment-intensive leading to small chances of meeting cost targets, while recently published wet processes for silver nanoparticles suffer from drawbacks regarding cost effective large scale production. In some cases [1], the starting silver concentrations in the reactive medium are so low (0.06wt%) that large capacity reactors and time-consuming concentrations would have to be implemented. In other cases [2], the starting concentrations are satisfactory, but the proposed synthetic routes involve non-friendly organic solvents and reactants as toluene, phenylhydrazine and alkylamine. Tert-butanol is more of interest since it is a non-toxic product and could be used as solvent.

Compared to main reducing agents, as hydrazine, sodium borohydride and aldehyde, reported for the preparation of silver nanoparticles, polyethylene glycol and polyvinylpyrolidone are environmentally friendly products. For example, polymers of ethylene glycol are widely used in pharmaceutical and biomedical industries. polyvinylpyrolidone (PVP) has been widely used as a surface passivation and stabilising agent in the synthesis of silver nanoparticles. [3,4] Although Silvert et al. observed the promoting effect of nucleation of PVP in the formation of silver particles using ethylene glycol as the reducing agent [5] there are few reports about the reducing properties of these 2 polymers for the fabrication of nanoparticles.[6] Recently, Luo et al. reported, for the first time, that polyethylene glycol is able to act as both reducing agent and stabilizer when used in the synthesis of silver nanoparticles in the absence of other chemicals. [1] In this case the maximum silver loading achieved is 0.12wt%.

Compared to the classical stabilising agents usually used, PEG and PVP are expected to show a weaker surface coordination to the silver nanoparticles, allowing a more active silver surface in device applications.

Herein, we report a new route to synthesize highly concentrated silver nanoparticles (30-40wt%) in mild conditions using aqueous solution containing PVP, as reducing and protecting agent, and silver acetate. We will also show that the addition of a reducing agent such as ascorbic acid, at the end of the reaction, doesn't seem necessary. We will also present a new synthetic route of highly concentrated silver nanoparticles using tert-butanol and PEG.

2 EXPERIMENTAL

Synthesis of silver nanoparticles via the PVP process.

For a typical synthesis using PVP, 5 g of Polyvinylpyrolidone (PVP) (MW 10000) was dissolved in 200 ml ultra-pure water at 50 °C under magnetic stirring. Then, 5g of Silver acetate (64.6 wt% Ag) was added to the solution at 50 °C. The mixture was kept during 45 minutes at 80 °C. Then, the mixture was left to cooling. At room temperature, a 20 mM ascorbic acid solution was added to the mixing under magnetic stirring. After 30 minutes reaction, the solution is kept without magnetic stirring, leading to a decantation. The supernatant is easily removed and the obtained silver paste is weighted. Ultra-centrifugation was also performed on the mixing to extract the silver paste.

Synthesis of silver nanoparticles via the PEG process.

For a typical synthesis using PEG, 1 g of polyethylene glycol (PEG) (MW 1500) was dissolved in 80 ml *tert*-butanol at 50 °C under magnetic stirring. Then, 10 g of silver acetate (64.6 wt% Ag) was added to the solution at 50 °C. The mixing was kept during 45 minutes at 80 °C.

Transmission Electron Microscopy (TEM) analyses of the particles were undertaken using a Philips CM200. Samples for TEM were prepared by placing a drop of the solution on carbon coated copper grid and dried at room temperature

3 RESULTS AND DISCUSSION

During the synthesis, color changes are observed showing the reduction process, and indicating the formation of silver nanoparticles. As soon as silver acetate is added into the yellowish PVP-water solution, the color turns from yellow to brown. At the end of the reaction, the color turns to dark green. After elimination of the supernatant, the silver paste has a content of 40 wt% silver. When the mixing is concentrated by ultra-centrifugation (2'000 rpm) the obtained silver content is higher than 45 wt%. The obtained dark green product was characterized by Transmission Electron Microscopy (TEM), as shown in Figure 1, where round shaped silver nanoparticles with particles size up to 50 nm can be observed. Observed colors changes clearly imply that PVP plays both the role of stabilizer and reducing agent in these synthesis conditions, whereas, in the literature it is mainly reported that PVP acts as a protecting agent in presence of a reducing agent [3,4].

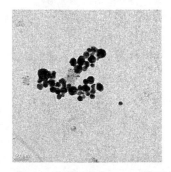

Figure 1: TEM pictures of 50nm Silver nanoparticles

The ascorbic acid (AA) is also used here as a reducing agent with the aim of finalising the reduction reaction. AA has a coordination affinity with Ag^+, while having a limited reduction potential preventing the agglomeration of the reduced silver nanoparticles. Thus, ascorbic acid can in a first step be linked to Ag^+ ions in a stable way allowing the

electron transfert to occur on a second step. But in this case the addition of ascorbic acid didn't show a clear color modification. The color of the solution remained dark-green. But we could see as a first result that the TEM pictures shows particles of 2 to 25 nm. (figure 2A, 2B)

The same reaction results were obtained when replacing water by other solvents as *tert*-butanol and the reducing agent by PEG. The translucid PEG/*tert*-butanol solution turns from white to green color when silver acetate is added. At the end of the reduction process, particles of up to 50 nm are measured by TEM. To increase the silver concentration, a modified synthesis was realised with *tert*-butanol and PEG. Similarly to [1], we carried out experiments using PEG as both solvent and reducing agent. A silver loading of 9wt% maximum was achieved by this method, since the mixture became too viscous for easy manipulation. With the choice of *tert*-butanol as solvent, the silver loading achieved, at the beginning of the reaction, was of 16 wt% in the presence of PEG as reducing and protecting agent. A green color was observed at the end of the reaction showing that the reduction occurred. After elimination of the supernatant, a 30 wt% silver content in the silver paste is obtained. Round shaped particles are observed by TEM with sizes going up to 50 nm. (same as figure 1)

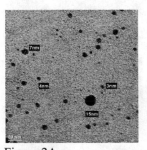

Figure 2A   Figure 2B

Figure 2A an 2B: TEM pictures of 2 to 25 nm Silver nanoparticles

A study of the effect of silver concentration on the particles size synthesis was also carried out. As reported in table 1, we modified the silver concentration added at the beginning of the reaction from 1.5 wt% to 9 wt% silver, keeping the concentrations of PVP and PEG identical. The solvents used are water, and *tert*-butanol. As summarized in table 1, there is no color difference when a higher silver quantity is loaded for the same quantity of reducing agent probably indicating incomplete reduction. The silver nanoparticles sizes observed by TEM for the four products are all between 25 to 50 nm.

Products	1	2	3	4
Solvent	water	water	tert-butanol	tert-butanol
Reducing Agent	PVP	PVP	PEG	PEG
% Ag	1.5	9	7	9
Color of the product	dark green	dark green	green	green
Final wt % Ag	40	40	30	30
TEM (nm)	25-50	25-50	25-50	25-50

Table 1. Conditions and Results for the synthesis of silver nanoparticles.

For all the experiments, careful TEM examination did not reveal particles larger than 50 nm. They were not agglomerated, but effectively stabilised by the polymer. However, there are strong self-organising forces within the polymer-nanoparticle system. These interactions are demonstrated by TEM pictures performed on samples of silver paste which were diluted with 20 % of ethylene glycol/water (pictures 3A and 3B).

Figure 3A.   Figure 3B.
Figure 3A: Silver nanoparticles into the polymer matrix.
Figure 3B: is a magnification of Figure 3A.

4 CONCLUSION

Highly stable concentrated silver nanoparticles (above 40 wt%) were successfully prepared using mild synthetic conditions, under atmospheric pressure and friendly solvent without concentration step. In this procedure, PVP and PEG acted as both reducing agent and stabilizer for silver nanoparticles. Particles in the 2 to 50 nm size range can easily be produced in a one-step synthesis. Stability tests should be carried out in the future. Considering their ease of fabrication and active surface, such nanoparticles are expected to have wide applications in electronics and other related fields.

REFERENCES

[1] C. LuO, Y. Zhang, X. Zeng, Y. Zeng, Y. Wang J. Colloid Interface Sci., *288*, 444, **2005**

[2] Y. Li, Y. Wu, B. S. Ong J. Am.Chem. Soc, *127,* 3266, **2005**

[3] H. P. Choo, K. Y. Liew, H.F. Liu, C.E. Seng, J. Mol. Catal. A Chem. 165, 127, **2001**

[4] H. S. Shin, H. J. Yang, S. B. Kim, M. S. Lee, J.Colloid Interface Sci. 274, 89, **2004**

[5] P.-Y. Silvert, R.H. Urbina, N. Duchauvelle, V. Vrjayakrishnam, K. Tekaia-Elhsissen, J. Mater. Chem. 6, 573, **1996**

[6] P. Raveendran, J. Fu, S.L. Wallen, J. Am. Chem. Soc. 125, 13, 940, **2003**

Emulsions as Templates for Nanostructured Materials

C. Solans[*], J. Esquena, J. Nolla

Institut d'Investigacions Químiques i Ambientals de Barcelona, (IIQAB)
Consejo Superior de Investigaciones Científicas (CSIC)
CIBER de Bioingeniería, Biomateriales y Nanomedicina (CIBER-BBN)
Jordi Girona, 18-26, 08034, Barcelona, Spain

[*]Corresponding author: csmqci@cid.csic.es

ABSTRACT

The preparation of nanostructured materials by emulsion templating is of particular interest among the soft nanotechnology processes. The use of emulsions allows good control of size and morphology, mild reaction conditions and technological feasibility. This work focuses on the use of nano-emulsions and highly concentrated emulsions as templates for the preparation of nanoparticles and dual meso / macroporous materials, respectively. Nano-emulsions, characterized by droplet size in the range 20-200 nm, appear transparent or translucent and possess stability against sedimentation or creaming. It will be shown that nanoparticles with controlled size and low polydispersity have been obtained by our group from nano-emulsions prepared by low-energy methods. Highly concentrated emulsions, characterized by an internal phase volume fraction higher than 0.74, consist of deformed (polyhedrical) droplets separated by a thin film of continuous phase, a structure resembling gas-liquid foams. We have obtained dual meso / macroporous materials from highly concentrated emulsions. The resulting materials combine the advantages of high specific surface area (due to the presence of mesopores) with the accessible diffusion pathways (associated with macropores). These properties make the meso/macroporous materials very important in many industrial applications.

Keywords: Nano-emulsion, highly concentrated emulsion, nanoparticle, meso/macroporous, template.

1 TEMPLATING IN NANO-EMULSIONS

Nano-emulsions are emulsions with droplet size in the range 20-200 nm. They are also referred to in the literature as miniemulsions, submicrometer-sized emulsions, finely dispersed emulsions, ultra fine emulsions, etc. [1].Due to their characteristic size, nano-emulsions possess stability against sedimentation or creaming and appear transparent or translucent (Figure 1).

Fig. 1: Visual aspect of a nano-emulsion (droplet size 80 nm).

These properties make them interesting systems as reaction media (e.g. to obtain nanoparticles). Nano-emulsions are generally prepared by dispersion or high-energy emulsification methods in which emulsification is achieved by high shear stirring, high-pressure homogenizers ultrasound generators, etc. However, nano-emulsions can also be prepared by condensation or low-energy methods, making use of the phase transitions that take place in the system during the emulsification process. Nano-emulsions with droplet sizes as low as 20 nm and high kinetic stability have been obtained either by the phase inversion temperature (PIT) or the phase inversion composition (PIC) methods [2-4]. Phase behavior studies have shown that similar phase transitions occur in both methods[2,3] and that the size of the nano-emulsion droplets is governed by the surfactant structure (bicontinuous or lamellar) present at the inversion point.

The use of nano-emulsions for the preparation of polymeric nanoparticles (the so-called miniemulsion polymerization method) is well known [1, 5-7]. However, in the published work related to this subject, the nano-emulsions used are obtained by high-energy methods (ultra-sonication, high-energy homogenization, etc.).In this paper, the use of nano-emulsions obtained by low-energy methods as templates for the preparation of nanoparticles is described

The nano-emulsions of the O/W type were prepared by the phase inversion temperature (PIT) method by cooling the samples from a temperature slightly higher than the corresponding PIT of the system (e.g. 50 ºC) to 25 ºC. The oil component of the nano-emulsions consisted of a monomer (e.g. butyl acrylate), the surfactants were of the polyoxyethylene type. Phase behavior studies as a function of surfactant concentration and temperature at constant water/oil weight ratio allowed to determine the phase transitions taking place during the emulsification process and to rationalize the formation of nano-emulsions with minimum droplet size. Nano-emulsions with droplet size of the order of 50 nm and low polydispersities were obtained, as determined by dynamic light scattering. The polymerization reactions were initiated by addition of an initiator (e. g. a redox pair) to the nano-emulsions. An example of nanoparticles obtained is showed in Figure 2.

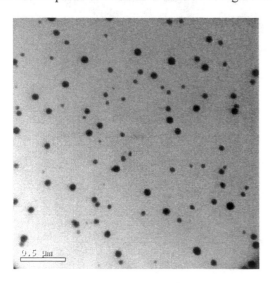

Fig. 2: Transmission Electron Microscopy (TEM) image of polymeric nanoparticles obtained in nano-emulsions prepared by a low-energy method.

The results obtained showed that nanoparticle size is of the same order as that of nano-emulsion droplets (e.g. 50 nm).

2 TEMPLATING IN HIGHLY CONCENTRATED EMULSIONS

Highly concentrated emulsions are an interesting class of emulsions characterized by an internal phase volume fraction exceeding 0.74, the critical value of the most compact arrangement of uniform, undistorted spherical droplets [8,9]. Consequently, their structure consists of deformed (polyhedrical) and/or polydisperse droplets separated by a thin film of continuous phase, a structure resembling gas-liquid foams. They are high-internal-phase (HIPE) emulsions which are also referred to in the literature as gel emulsions [10-14], hydrocarbon gels [15], biliquid foams [16], etc. A typical example of a highly concentrated emulsion is shown in Fig. 3.

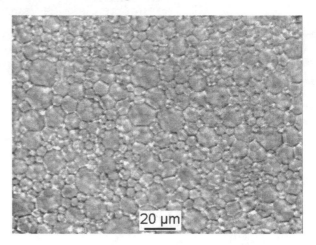

Fig. 3: Image obtained by optical microscopy of a highly concentrated emulsion.

Highly concentrated emulsions are very appropriate systems for the preparation of low-density macroporous or meso/macroporous materials, which have very large pore volume and very low bulk density. These materials can be obtained by reactions in the continuous phase of highly concentrated emulsions followed by the removal of the dispersed phase components [17-25]. The solid foams produced by this method have interconnected sponge-like macropores and possess low bulk densities. Different monomers can be used to obtain macroporous solid foams, and the results show that a very wide variety of low-density materials can be obtained [17-25]. An example of such materials is shown in Fig. 4.

Fig. 4: Scanning electron microscopy (SEM) image of a meso/macroporous dual silica, obtained by templating in a highly concentrated emulsion.

The meso/macroporous dual structure consists of polydispersed macropores and monodispersed mesopores, which are well interconnected.

CONCLUSIONS

Emulsions can be used as templates for the preparation of a very wide variety of nanostructured materials, achieving a good control of size and morphology. Nanoparticles and dual meso / macroporous materials can be obtained by the use of nano-emulsions and highly concentrated emulsions, respectively, as reaction media. Nano-emulsions, which can be prepared by low-energy methods, have droplet sizes in the range 20-200 nm depending on composition and preparation methods. Consequently they allow to control the size of nanoparticles. Highly concentrated emulsions are characterized by an internal phase volume fraction higher than 0.74 and consist of deformed (polyhedrical) droplets separated by a thin film of continuous phase. Dual meso / macroporous materials that consists of polydispersed macropores and monodispersed mesopores have been obtained in highly concentrated emulsions.

ACKNOWLEDGEMENTS

The authors acknowledge financial support by the Spanish Ministry of Science and Education DGI (Grants CTQ2005-09063-C03-02 and CTQ2005-08241-C03-01/PPQ) as well as "Generalitat de Catalunya", DURSI (Grant 2005SGR-0018) and CIDEM (Xarxa IT).

REFERENCES

[1] C. Solans, P. Izquierdo, J. Nolla, N. Azemar, M. J. Garcia-Celma Curr Opi. Colloid Interface Sci 10, 102-110 (2005)

[2] D. Morales, J. M. Gutiérrez, M. J. Garcia-Celma, C. Solans. Langmuir 19, 7196-7200 (2003).

[3] P. Izquierdo, J. Feng, ‚J. Esquena Th. F. Tadros, J.C. Dederen, M. J. Garcia, N. Azemar C. Solans J. Colloid Interface Sci. 285, 388-394 (2005).

[4] N. Sadurni, C. Solans, N. Azemar, M. J. Garcia-Eur. J. Pharm. Sci, 26, 438445 (2005).

[5] E.D. Sudol, M.S. El-Aasser: Miniemulsion Polymerization. In: Emulsion Polymerization and Emulsion Polymers, edited by P.A. Lovell and M.S. El-Aasser. John Willey and Sons, 1997, 699-722.

[6] M. Antonietti, Landfester K: Polyreactions in miniemulsions. Prog Polym Sci 27 (2002), 689-757.

[7] Asua J.M.: Miniemulsion Polymerization. Prog Polym Sci 27 (2002), 1283-1346.

[8] K.J. Lissant. J. Colloid. Interface Sci., 22:462, 1966.

[9] H.M. Princen. J. Colloid. Interface Sci., 71:55, 1979.

[10] H. Kunieda, C. Solans, N. Shida and J.L. Parra. Colloids Surf., A 24:225, 1987.

[11] C. Solans, J.G. Domínguez, J.L. Parra, J. Heuser and S.E. Friberg, Colloid Polym. Sci., 266:570, 1988.

[12] C. Solans, R. Pons and H. Kunieda. In: B.B. Binks (ed.), Modern Aspects of Emulsion Science. Cambridge: Royal Society of Chemistry, 1998, pp 367-394.

[13] C. Solans, J. Esquena, N. Azemar, C. Rodríguez and H. Kunieda. In: D.N. Petsev (Ed.), Emulsions: Structure, Stability and Interactions. Amsterdam: Elsevier, 2004, Pp 367-394.

[14] J.C. Ravey, M.J. Stébé, S. Sauvage. Colloids Surf. A, 91:237, 1994.

[15] G. Ebert, G. Platz And H. Rehage. Berichte Der Bunsengesellschaft, 92:1158, 1988.

[16] O. Sonneville-Aubrun, V. Bergeron, T. Gulik-Krzywicki, B. Jönsson, H. Wennerström, P. Lindner and B. Cabane. Langmuir, 16:1566, 2000.

[17] D. Barby, Z. Haq. European Patent 0060138 (Unilever), 1982.

[18] J.M. Williams, D.A. Wrobleski. Langmuir, 4:656-662, 1988.

[19] N.R. Cameron, D.C. Sherington, L. Albiston And D.P. Gregory Colloid Polym. Sci., 274:592-595, 1996.

[20] E. Ruckenstein and J.S. Park. Polymer, 33:405-417, 1992.

[21] E. Ruckenstein. Adv. Polym. Sci., 127:3-58, 1997.

[22] J Esquena, Gsrr Sankar, C Solans. Langmuir, 19:2983-2988, 2003.

[23] H. Maekawa, J. Esquena, S. Bishop, C. Solans, B.F. Chmelka. Adv. Materials, 15:591-596, 2003.

[24] C. Solans, J. Esquena, N. Azemar. Curr. Opinion Colloid Interface Sci., 8:156-163, 2003.

[25] J. Esquena, C. Solans. "Highly Concentrated Emulsions As Templates For Solid Foams", En: "Emulsions And Emulsion Stability" (Ed. J. Sjöblom) Francis & Taylor (Surfactant Science Series), New York (2006).

Synthesis, Extraction and Surface Modification of (Zn,Mn)Se Nanocrystals Using Microemulsions as Templates

T. Heckler, Q. Qiu, J. Wang, B. Mei and T.J. Mountziaris[*]

Department of Chemical Engineering, University of Massachusetts, Amherst, MA 01003
[*] Corresponding author: tjm@ecs.umass.edu

ABSTRACT

We report the synthesis of ZnSe:Mn quantum dots (QDs) in the dispersed phase of a microemulsion template consisting of p-xylene as the continuous phase, water as the dispersed phase, and an amphiphilic block copolymer (Pluronic P105) as the surfactant. The Zn and Mn precursors were Zn-acetate and Mn-acetate, respectively, dissolved in the aqueous dispersed phase. Hydrogen selenide gas was bubbled through the microemulsion. A single ZnSe QD forms in each droplet of the microemulsion via an irreversible reaction between the precursors and coalescence of the resulting nuclei. Mn^{2+} ions slowly diffuse into the ZnSe lattice. The emission peak attributed to Mn^{2+} ions at 585 nm increases substantially with time eventually dominating over the ZnSe peak and reaching a plateau. The QDs were extracted from the template by adding water and inducing phase separation. They were subsequently capped with mercapto-undecanoic acid or dihydro-lipoic acid and dispersed in water.

Keywords: quantum dots, diluted magnetic semiconductors, synthesis, microemulsion template, zinc manganese selenide.

1 INTRODUCTION

Semiconductor nanocrystals or quantum dots (QDs) have size-tunable optical properties that make them excellent candidates for applications in clinical diagnostics, photovoltaics, optoelectronics, and spintronics[1, 2]. CdSe QDs that emit in the visible part of the spectrum have been the most widely studied system. ZnSe QDs emit in the blue and violet parts of the spectrum and are a less toxic alternative to CdSe for *in vivo* applications[3]. ZnSe QDs have a quantum confinement threshold of 9nm and tunable emission between 380 nm and 460 nm. Hines et al. reported a method for synthesizing highly luminescent ZnSe QDs using injection of organometallic precursors in a hot coordinating solvent[4].

Doping of II-VI semiconductors with transition metals yields diluted magnetic semiconductors (DMS), with materials having a chemical composition $Zn^{II}_{1-x}Mn_xB^{VI}$ being the most widely studied compounds[5]. DMS-based nanocrystals have potential applications in spintronics and can be used to form QDs emitting at wavelengths that are longer than the wavelength corresponding to the gap emission of the original bulk semiconductor. Typical examples of such materials are ZnS:Mn[6], and ZnSe:Mn[7-10]. The introduction of the Mn^{2+} ion in the ZnSe lattice results in an emission peak at 585nm. In general, the doping of ZnSe nanocrystals with transition metals expands their emission range into the visible part of the spectrum_and has additional advantages, such as the reduction of self-quenching and the elimination of the need to control the particle size for tuning the emission wavelength.[10]

In this paper, we report the synthesis of ZnSe:Mn nanocrystals using a microemulsion template. The technique is based on the templated growth of ZnSe nanocrystals reported by Karanikolos et al.[11-13] The microemulsion template consists of p-xylene as the continuous phase, water as the dispersed phase, and a poly(ethylene oxide)-poly(propylene oxide)-poly(ethylene oxide) (PEO-PPO-PEO) block copolymer (Pluronic P105) as the surfactant. This template is very stable and has very slow droplet-droplet coalescence kinetics. Because of that, it minimizes particle aggregation and allows precise control of particle size by manipulating the concentration of the precursor in the dispersed phase. The technique is easy to scale up and employs less expensive chemicals compared to the hot injection method.

2 EXPERIMENTAL SECTION

Microemulsion templates were prepared by mixing 82.1 wt% p-xylene, 2.4 wt% aqueous solution of Zn- and Mn-acetate, and 15.5 wt% Pluronic P105 surfactant.[13] The concentration of Zn-acetate in the aqueous solution was used to control the final size of the QDs. The typical Mn:Zn molar ratio in the solution was 1:9. Hydrogen selenide gas diluted in hydrogen was bubbled through the micro-emulsion to enable the formation of ZnSe QDs. The hydrogen selenide gas diffuses through the surfactant layer and reacts with Zn-acetate to form ZnSe. The Mn^{2+} ion is incorporated into the lattice though a slow diffusion process. The hydrogen selenide flow was kept on for 10 minutes to convert Zn-acetate to ZnSe. The microemulsion was subsequently purged with nitrogen and was analyzed using fluorescence spectroscopy. The extraction of particles from the microemulsion template was accomplished by adding water and inducing phase separation. Surface modification of the QDs with mercapto-undecanoic acid and dihydro-lipoic acid was performed. The surface modified QDs were dispersed in water.

3 RESULTS AND DISCUSSION

Figure 1 shows the evolution of fluorescence emission spectra from a microemulsion containing ZnSe:Mn QDs with increasing time in storage. For this case the initial concentration of Zn-acetate precursor in the dispersed phase was 0.1 M and the Mn:Zn molar ratio in that solution was set at 0.1. The initial spectrum exhibits a single peak at 393nm that corresponds to ZnSe QDs. The spectra obtained after 3, 4 and 5 days in storage exhibit two major peaks, one at 393 nm that corresponds to ZnSe and one at 585 nm that corresponds to the Mn^{2+} ion. The relative intensity of the Mn^{2+} peak with respect to the ZnSe peak increases with time in storage, most probably due to the diffusion of the Mn^{2+} ion into the ZnSe lattice, until equilibrium with the surrounding solution is established. Initially, the ZnSe emission peak is dominant, but the Mn^{2+} peak eventually dominates. These observation suggest that ZnSe QDs are formed first, and Mn^{2+} diffuses slowly into them to replace Zn^{2+} in the crystal lattice.[5] It has been reported that for a single ZnSe:Mn QD, the emission intensity ratio should be directly proportional to the number of Mn^{2+} ions incorporated into the nanocrystal.[8] In Figure 2, the intensity ratio of the Mn^{2+} and ZnSe peaks is plotted as a function of storage time. The emission intensity ratio increases with time in storage at room temperature and reaches a plateau after about 20 days.

The effects of varying the Mn to Zn ratio in the precursor solution, while keeping the initial Zn-acetate concentration constant at 0.3M, were also studied. The observed emission intensity that corresponds to the Mn^{2+} peak is plotted in Figure 3 as

function of time in storage. The data shows that when the Mn^{2+} concentration in the precursor solution is increased, the evolution of the Mn^{2+} peak emission intensity from the ZnSe:Mn QDs is slower. This is counter-intuitive and suggests that the rate of incorporation of Mn^{2+} incorporation into the ZnSe QD is not simply proportional to the Mn^{2+} concentration in the surrounding solution. An investigation of the underlying Mn^{2+} incorporation mechanism is currently underway in our laboratory.

4 CONCLUSIONS

We report the synthesis of ZnSe:Mn QDs in microemulsion templates using aqueous solutions of Zn-acetate and Mn-acetate, and hydrogen selenide gas. The reaction takes place at room temperature and the size of the particles is controlled by the concentration of Zn-acetate in the dispersed phase. The ratio of the emission peak intensity at 585nm attributed to the Mn^{2+} ion to the emission peak attributed to ZnSe increases with time in storage until a plateau is reached. This indicates that ZnSe QDs are formed first and the Mn^{2+} ion subsequently diffuses into the ZnSe lattice.

This work was supported by the UMass President's Office S&T Funds, an NSF-IGERT grant (TH, BM), a Clare Boothe Luce Fellowship (TH), and an NSF-MRSEC grant. The authors thank Prof. Pablo E. Visconti for access to a Horiba Fluorog3 fluorometer.

6 FIGURES

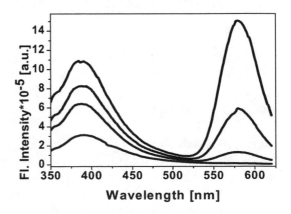

Figure 1: Fluorescence spectra of ZnSe:Mn QDs after 2, 3, 4, and 5 days in storage in the microemulsion. The emission intensity increases with time.

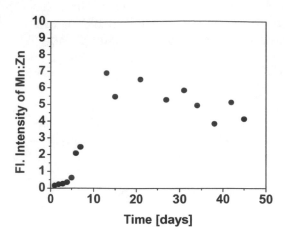

Figure 2: Evolution of Mn^{2+} to ZnSe fluorescence emission intensity ratio in ZnSe:Mn QDs with time in storage in the microemulsion.

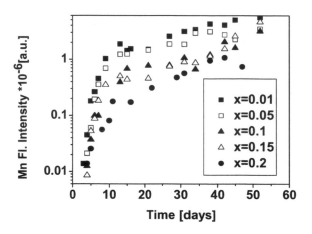

Figure 3: Evolution of Mn^{2+} peak fluorescence emission intensity from ZnSe:Mn QDs for various Mn to Zn molar ratios in the initial precursor solution.

REFERENCES

[1] Alivisatos, A., "Semiconductor clusters, nanocrystals, and quantum dots," *Science*, 271, 933-937, 1996.

[2] Murray, C.; Kagan, C.; Bawendi, M., "Synthesis and characterization of mono-disperse nanocrystals and close-packed nanocrystal assemblies," *Annual Review of Materials Science*, 30, 545-610, 2000.

[3] Derfus, A.; Chan, W.; Bhatia, S., "Probing the cytotoxicity of semiconductor quantum dots," *Nano Letters*, 4, 11-18, 2004.

[4] Hines, M.; Guyot-Sionnest, P., "Bright UV-blue luminescent colloidal ZnSe nanocrystals," *Journal of Physical Chemistry B*, 102, 3655-3657, 1998.

[5] Furdyna, J., "Diluted Magnetic Semiconductors," *Journal of Applied Physics*, 64, R29-R64, 1988.

[6] Bhargava, R. N.; Gallagher, D.; Hong, X.; Nurmikko, A., "Optical properties of manganese-doped nanocrystals of ZnS," *Physical Review Letters*, 72, 416-419, 1994.

[7] Norris, D.; Yao, N.; Charnock, F.; Kennedy, T., "High-quality manganese-doped ZnSe nanocrystals," *Nano Letters,* 1, 3-7, 2001.

[8] Zu, L.; Norris, D. J.; Kennedy, T. A.; Erwin, S. C.; Efros, A. L., "Impact of ripening on manganese-doped ZnSe nanocrystals," *Nano Letters*, 6, 334-40, 2006.

[9] Pradhan, N.; Goorskey, D.; Thessing, J.; Peng, X., "An alternative of CdSe nanocrystal emitters: pure and tunable impurity emissions in ZnSe nanocrystals," *Journal of the American Chemical Society*, 127, 17586-7, 2005.

[10] Pradhan, N.; Peng, X., "Efficient and color-tunable Mn-doped ZnSe nanocrystal emitters: control of optical performance via greener synthetic chemistry," *Journal of the American Chemical Society*, 129, 3339-47, 2007.

[11] Karanikolos, G. N.; Alexandridis, P.; Itskos, G.; Petrou, A.; Mountziaris, T. J., "Synthesis and size control of luminescent ZnSe nano-crystals by a microemulsion-gas contacting technique," *Langmuir,* 20, 550-3, 2004.

[12] Karanikolos, G.; Alexandridis, P.; Mallory, R.; Petrou, A.; Mountziaris, T.J., "Templated synthesis of ZnSe nanostructures using lyotropic liquid crystals," *Nanotechnology*, 16, 2372-2380, 2005.

[13] Karanikolos, G.; Law, N.; Mallory, R.; Petrou, A.; Alexandridis, P.; Mountziaris, T.J., "Water-based synthesis of ZnSe nanostructures using amphiphilic block copolymer stabilized lyotropic liquid crystals as templates," *Nanotechnology*, 17, 3121-3128, 2006.

Synthesis of nanocrystalline CoFe$_2$O$_4$ using citrate-urea assisted combustion process for lithium battery anode

I. Prakash*, N. Nallamuthu*, P. Muralidharan*, M. Venkateswarlu**, N. Satyanarayana*

* Department of Physics, Pondicherry University, Puducherry 605 014, India.
** HBL Power systems Ltd., Hyderabad- 500 078, India

ABSTRACT

Nanocrystalline CoFe$_2$O$_4$ particles are prepared using citric acid urea assisted combustion process by varying the metal ions, citric acid and urea ratio as 1:1:0.5, 1:1:1 and 1:1:2. The prepared sample is characterized using XRD, FTIR, DSC and SEM-EDS techniques. The impurity free, nano size (~15 nm) crystalline CoFe$_2$O$_4$ is obtained for M:CA: urea ratio 1:1:0.5 sample.

Keywords: Nanocrystalline CoFe$_2$O$_4$, SEM-EDS, DSC, XRD, FTIR

1. INTRODUCTION

CoFe$_2$O$_4$ magnetic nanoparticles with high coercivity, moderate magnetization and very high magneto crystalline anisotropy having wide range of applications in the field of ferro fluids, magneto optics, spintronics, biomedical applications, data storage devices, catalysts, sensors, etc. Recently, CoFe$_2$O$_4$ nanostructures have also used as anode material in lithium ion battery and showed an improved electrochemical performance [1-3]. CoFe$_2$O$_4$ form normal, inverse spinel or mixed structure, according to the preparation temperature. CoFe$_2$O$_4$ nanoparticles exhibit variety of unusual magnetic properties when compared to the its bulk, which in turn will have wide rage of commercial applications. Magnetization, hysteresis and phase transition temperature depend on the size of the magnetic nanoparticles. Magnetic nanoparticles can be prepared using sol-gel, co precipitation, polyol, combustion process, etc, [4-6]. Among the available wet chemical processes, combustion route is capable of producing nanocrystalline powder at a lower calcination temperature in a short time. In our present study, CoFe$_2$O$_4$ nanocrystals are prepared using combustion process by varying the metal ions, citric acid and urea fuel ratio as 1:1:0.5, 1:1:1 and 1:1:2. The synthesized sample is calcined at different temperatures and characterized using XRD, FTIR, DSC, SEM-EDS techniques to confirm the formation of CoFe$_2$O$_4$ nanocrystals.

Corresponding author E-mail: nallanis2007@gmail.com
Phone: +91-413-2654404

2. PREPARATION OF CoFe$_2$O$_4$ USING CITRIC ACID - UREA ASSISTED COMBUSTION PROCESS

Nanocrystalline CoFe$_2$O$_4$ powder sample was prepared using citric acid urea assisted combustion process by varying the metal ions, citric acid and urea ratio as 1:1:0.5, 1:1:1 and 1:1:2. All the analar grade precursor chemicals of Cobaltus nitrate, Ferric nitrate nanohydrate, citric acid and urea were used for the synthesis of Nanocrystalline CoFe$_2$O$_4$ powder. The required amount of metal nitrate solutions were taken and sonicated for 10 minutes. The mixture of solution was continuously stirred at 353 K for an hour. The required amount of citric acid and urea were added to the above solution on continuous stirring. The solution was evaporated at 353 K for 8 hours under constant stirring, which results the polymeric resin. The resins were further dried in an oven at 333 K for 24 hours to remove the excess water and the drying process caused the formation of dried foamy polymeric intermediate. Nanocrystalline CoFe$_2$O$_4$ powders were obtained by calcining the polymeric intermediates at 448 K and above. The complete process was investigated through XRD, FTIR, DSC and SEM techniques.

3. CHARACTERIZATION

The powder X-ray diffraction patterns were recorded on a PANalytical XPertPro X-ray diffractometer with Cu K$_\alpha$ radiation of wavelength λ=1.5418 A$^\circ$ and scanned from 80-10^0. FTIR spectra were recorded on pellet samples using Shimadzu FTIR/8300/8700 spectrophotometer in the range of 4000 – 400 cm^{-1} with 2 cm^{-1} resolution for 20 scans. Thin transparent pellet samples were prepared using the dried gel sample, heated from room temperature to higher temperatures (623 K), and grounded well with spectra pure KBr powder taken in 1:20 ratio. The DSC measurements were made on Mettler Toledo Star e System; module DSC 821e/500/575/414183/5278 under ambient air atmosphere. The polymeric intermediate sample, 3 mg, was taken in aluminium crucible and heated from 300 K to 773 K at a heating rate of 10K per minute. The microstructure of polymeric intermediate as well as the final size of the CoFe$_2$O$_4$ particles and their elemental distribution were obtained through SEM – EDS measurements using Hitachi Scanning Electron microscope S 3400N.

4. RESULT AND DISCUSSION

4.1 XRD

Fig 2 shows the powder X-ray diffraction patterns of $CoFe_2O_4$ samples of three different compositions; 1:1:0.5, 1:1:1 and 1:1:2. From Fig 2, it was observed that the XRD patterns of as synthesized $CoFe_2O_4$ polymeric intermediates of composition 1:1:1 and 1:1:2 show the

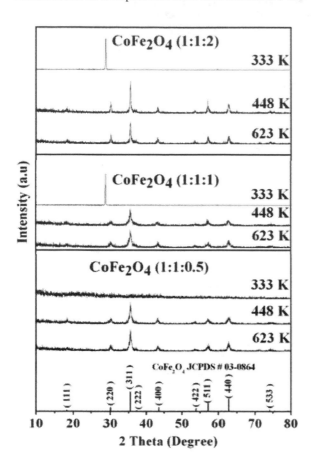

Fig. 1 XRD patterns of $CoFe_2O_4$ samples of three different compositions (M:CA:Urea = 1:1:0.5, 1:1:1 and 1:1:2).

characteristic diffraction patterns of ammonium nitrate. This indicates that the addition of excess urea to the sample is not participated in the formation of polymeric matrix, rather it forms ammonium nitrate. The XRD patterns of the as synthesized $CoFe_2O_4$ sample of composition 1:1:0.5 shows complete amorphous nature, which indicate that the added urea and citric acid are completely take part in the polymeric matrix formation and hence, showed amorphous nature. The XRD patterns of the calcined polymeric intermediate powders of

compositions 1:1:0.5, 1:1:1 and 1:1:2 show the characteristic diffraction patterns of the $CoFe_2O_4$ phase and further calcination increase the crystallite size of the sample. The formation of $CoFe_2O_4$ crystalline phase was confirmed by comparing the diffraction patterns with JCPDS # 03-0864 data. The crystallite size of the $CoFe_2O_4$ sample was calculated by using Scherer's formula and it is found to be 15 nm for 1:1:0.5, 12 nm for 1:1:1 and 30 nm for 1:1:2 samples, which confirm that the prepared crystalline $CoFe_2O_4$ samples are in nano size. The lowest crystallite size is observed for 1:1:1 composition of the $CoFe_2O_4$ sample.

4.2 FTIR

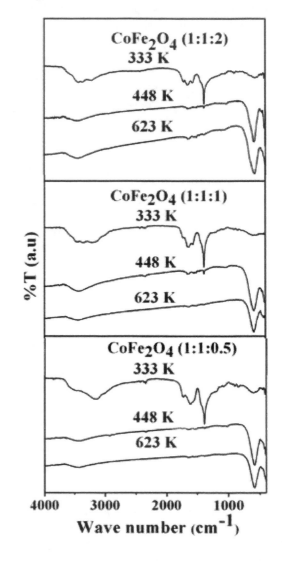

Fig. 2 FTIR spectra of $CoFe_2O_4$ samples of three different compositions (M: CA: Urea = 1:1:0.5, 1:1:1 and 1:1:2).

Fig. 2 shows the FTIR spectra of $CoFe_2O_4$ samples of three different compositions. From fig. 2, all the three different compositions of the synthesized samples showed the characteristic vibrational bands of organic groups and nitrates at 3430, 3173, 1725, 1616, 1384, 1139, 1078, 901, 834 cm^{-1}. The above observed IR confirm the presence of citric acid and urea in the sample. The FTIR spectra of calcined samples of three different compositions showed the characteristic IR bands at 600 and 430 cm^{-1}, which correspond to the formation of tetrahedral and octahedral structured spinel $CoFe_2O_4$ and also confirms the removal of organic residuals [7-9]. Nanocrystalline $CoFe_2O_4$ sample with high purity was observed for 1:1:0.5 composition. The presence of organic impurities decreases the performance of lithium battery when this material is used as lithium battery anode. Hence organic impurity free $CoFe_2O_4$ sample is taken for further studies.

4.3 DSC

Fig 3 shows the DSC thermogram of $CoFe_2O_4$ sample of all compositions. From fig. 3, it was observed that the heat generation is more for 1:1:0.5 composition sample and all the organic fuels are completely burned around 473 K. There is no more exothermic peak observed in the sample after 620 K.

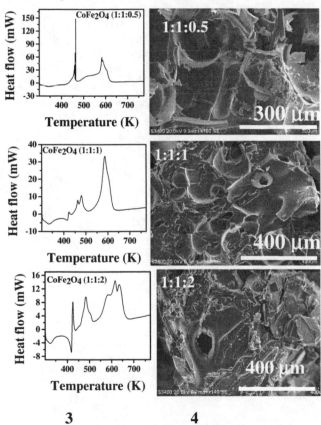

3   **4**

Fig. 3 DSC thermograms, Fig. 4 SEM images of foamy polymeric intermediates of all compositions (M:CA:Urea = 1:1:0.5, 1:1:1 and 1:1:2) dried at 333 K.

The sample of composition 1:1:1 and 1:1:2 showed poor heat generation during combustion reaction. The sample showed two endothermic peaks one corresponds to

evaporation of water and other endotherm may be due to the melting of the sample before volume expansion. The poor exothermic reaction continues for a longer time and resulted unburnt carbon retain in the sample calcined around 623 K. The DSC result confirmed that the $CoFe_2O_4$ sample of composition 1:1:0.5 generate more heat during combustion reaction and resulted high pure fine powdered sample calcined at 623 K. For other composition, the prolonged exothermic reaction results in impurities. This result is also reflected in FTIR analysis

4.4 SEM

Fig 4 shows the SEM images of as synthesized polymeric intermediate of CoFe2O4 samples with different citric acid urea ratios. $CoFe_2O_4$ sample of composition 1:1:0.5 showed better polymerization, displayed very soft polymeric intermediate with high volume expansion and large pores. The $CoFe_2O_4$ samples of compositions 1:1:1 and 1:1:2 showed poor polymerization with less volume expansion and hard nature. The same is reflected as a crystalline form in XRD results. The SEM images of the above compositions $CoFe_2O_4$ samples showed the porous nature of as well as flakes. The porous nature will help for storing oxygen during combustion and better combustion was observed for the better polymerized sample with large pores. The same result is reflected in DSC results.

4.5 SEM-EDS

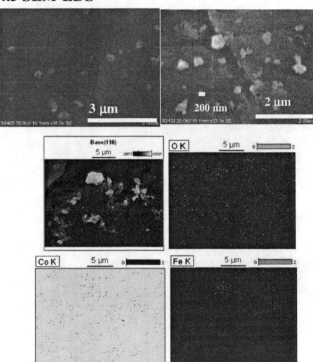

Fig. 4 SEM-EDS image of $CoFe_2O_4$ sample prepared with 1:1:0.5 composition

The fine powdered $CoFe_2O_4$ sample of composition 1:1:0.5 calcined at 623 K is dispersed in acetone and a little drop is dispersed over the aluminium stub and dried for SEM-EDS measurements.

Element Line	Net Counts	Weight %	Atom %	Formula
O K	0	27.52	57.41	
Fe K	84	49.58	29.63	Fe_2O_3
Fe L	0	---	---	
Co K	33	22.90	12.97	CoO
Co L	11	---	---	
Total		100.00	100.00	

Table 1 SEM-EDS results of $CoFe_2O_4$ sample prepared with 1:1:0.5 composition

Fig. 5 shows the SEM and EDS elemental mappings of $CoFe_2O_4$ sample of composition 1:1:0.5. From SEM image, the $CoFe_2O_4$ particles are dispersed uniformly and the particle size was varied form 150 to 250 nm. The elemental analysis confirmed the presence and uniform distribution of Co, Fe and O elements in the $CoFe_2O_4$ sample.

5. CONCLUSION

The nanocrystalline $CoFe_2O_4$ sample was prepared using citric acid and urea assisted combustion process. The metal ions, citric acid and urea fuel ratio was varied as 1:1:0.5, 1:1:1 and 1:1:2. The crystalline phase formation was confirmed from XRD results. The crystallite size was calculated using Scherer's formula and it is found to be 15 nm for 1:1:0.5, 12 nm for 1:1:1 and 30 nm for 1:1:2 composition sample. The FTIR results confirm the bond and structural formation of citric acid, urea and $CoFe_2O_4$. The complete removal of impurity at 623 K for 1:1:0.5 composition is also confirmed from FTIR results. $CoFe_2O_4$ sample of composition 1:1:0.5 showed high pure final product at 623 K compared to other compositions. SEM image of as synthesized polymeric intermediate sample showed the porous nature of the sample. The SEM-EDS analysis of the sample of composition 1:1:0.5 showed the uniform elemental distribution. The particle size varied form 150 nm to 250 nm. The elemental result showed the presence and uniform distribution of Co, Fe and O elements in the $CoFe_2O_4$ sample.

ACKNOWLEDGMENT

NS is gratefully acknowledged CSIR, DRDO and DST, Govt. of India, for receiving the financial support in the form of major research projects. IP acknowledge Jawaharlal Nehru Memorial Fund (JNMF) for receiving fellowship for doing Ph.D.

REFERENCES

[1] T. Hyeon, Y. Chung, J. Park, S. S. Lee, Y-W Kim, and B. H. Park, J. Phys. Chem. 106, 6831, 2002.
[2] J. M. Wesselinowa and I Apostolova, J. Phys.: Condens. Matter 19, 1, 2007.
[3] M. George, Swapna S Nair, K. A Malini, P. A. Joy and M. R. Anantharaman, J. Phys. D: Appl. Phys. 40, 1593, 2007.
[4] S-Y Zhao, R. Qiao, X. L. Zhang, and Y. S. Kang, J. Phys. Chem. C 111, 7875, 2007.
[5] S Chkoundali, S. Ammar, N. Jouini, F. Fievet, P. Molinie, M. Danot, F. Villain and J-M Greneche, J. Phys.: Condens. Matter 16, 4357, 2004.
[6] C. Cannas, A. Musinu, D. Peddis, and G. Piccaluga, Chem. Mater. 18, 3835, 2006.
[7] S.HXiao, W. F. Jiang, L. Y. Li, X. J. Li, Materials Chemistry and Physics, 106, 82, 2007.
[8] N. Kasapoglu, A. Baykal, Y. Koseoglu and M. S. Toprak, Scripta Materialia, 57, 441, 2007.
[9] G. Socrates, "Infrared and Raman Characteristic Group Frequencies", John Wiley and Sons, New York, 2001.

In-situ Incorporation of Pd-MPTES Complex in Zeolite and MCM-41 type Silica

N. Linares, A.I. Carrillo, J. García Martínez[*]

Department of Inorganic Chemistry. University of Alicante, Carretera San Vicente s/n,
E-03690, Alicante, Spain.
e-mail: j.garcia@ua.es URL: www.ua.es/grupo/nanolab

ABSTRACT

Monodispersed Pd nanoparticles have been successfully synthesized by controlled reduction of Pd(II) complexes. In the first step, the Pd(II) complexes were obtained by using mercaptopropyltriethoxysilane (MPTES) or aminopropyltriethoxysilane (APTES) as ligands. In a subsequent step, these Pd-complexes were *in-situ* incorporated to different nanostructured silica materials. Finally, the Pd complex was thermally decomposed to produce Pd nanoparticles homogeneously dispersed in nanostructured silica. The resulting materials can be used as catalysts for a variety of reactions. In addition, we have studied the possibility to use this method with others metals.

The method herein described yields novel nanostructured silica materials having highly dispersed and strongly anchored Pd nanoparticles into their structures. Consequently, these materials are expected to be highly active and stable catalysts.

Keywords: Pd complex, zeolites, nanostructured materials, MCM-41, catalysis, nanoparticles

1 INTRODUCTION

The widespread use of Pd and other noble metal catalysts in many industrial processes, especially in the fine-chemical industry, has lead to several efforts devoted to improve their activity and stability.

In recent years, most studies have been focused on the use of soluble Pd complexes with various types of ligands with the aim of increasing the activity and selectivity of the catalysts. [1] These homogeneous catalysts have been successfully used in numerous processes including C–C bond formation and oxidation reactions. The main limitations are the catalyst recovery and recycling, and the catalyst deactivation due to the aggregation of the nanoparticles formed *in situ* during reaction.

To overcome these problems, heterogeneous catalysts are used because they limit the loss and agglomeration of metal and to make their handling and recovery easier. [2]

Nowadays, two different approaches are used to prepare heterogeneous catalysts. First, many Pd complexes have been heterogeneized by covalently bounding their ligands to various solid supports. Most of the novel heterogenized catalysts are based on silica. [3] Nanostructured silica materials are widely used as catalyst support because their large surface area, controllable surface chemistry and porosity, excellent stability (chemical and thermal) and good accessibility.[4]

On the other hand, Pd nanoparticles are typically loaded on porous supports. The high surface area-to-volume ratio of noble metal nanoparticles makes them highly attractive for catalysis. Usually, supported Pd catalysts have lower activity than the homogeneous ones and require more drastic reaction conditions. However, they are often more stable and can be recovered and reused. These advantages overcome their lower activity in comparison with the homogeneous catalysts.

Conventional synthesis techniques of supported metal nanoparticles such as wet or dry impregnation, deposition-precipitation, deposition-reduction, and ion-exchange methods, still have considerable limitations, mainly in terms of homogeneity, dispersion and reproducibility. It is highly desirable to develop heterogeneous Pd catalysts more homogeneous and stable for industrial applications.

In last years, many examples of new Pd heterogeneous catalysts have been described, as Kobayashi's incarcerated catalysts [5], ionic gels [6], Pd-containing perovskites [7], and physically immobilized catalysts using a zeolite shell [8].

In 2007, J.D. Webb et al. synthesized mesoporous silica functionalized by mercaptopropyltrimethoxysilane [9]. After Pd loading, the resulting catalyst showed optimal characteristics when used for the Suzuki–Miyaura coupling reaction.

Recently, we described a simple method to incorporate Pd nanopaticles in porous materials by functionalization of the nanoparticles with trialkoxysilanes. This strategy is based on three steps: first, the metal nanoparticles were produced by controlled reduction in toluene and phase transfer to water, then they were functionallized with mercapto-terminal trialkoxysilanes, and finally they were incorporated into the porous material by co-polymerization with a silica alkoxyde (tetraethoxysilane).[10]

Herein, a novel method to obtaining nanostructured silica materials supported Pd catalyst with high loading and dispersion by means *in situ* Pd complex incorporation is brought forward. This approach does not require the preparation of metal nanoparticles because the Pd precursor is directly incorporated in the silica support using a Pd complex. Moreover, as herein described, this method allows for *in situ* incorporation the Pd complex in several nanostructured silica materials i.e. MCM-41 type silica, zeolite A and zeolite Y during their synthesis.

2 EXPERIMENTAL SECTION

2.1 Materials

Reagents were purchased from the following suppliers: mercaptopropyltriethoxysilane (MPTES), aminopropyltriethoxysilane (APTES) and sodium metasilicate were obtained from Fluka. Tetraethylorthosilicate (TEOS), cetyltrimethylammonium bromide (CTAB), sodium silicate solution (27% SiO_2), tetrapropylammonium bromide (TPABr), sodium hydroxide, and sodium tetrachloropalladate(II) were supplied by Aldrich. Sodium aluminate was obtained from Riedel-de Haën. All chemicals were used as received without further purification.

2.2 Synthesis of Pd-MCM-41 type silica

Parent silica-based MCM-41 was synthesized as described elsewhere [11] taking into account that Pd-complexes are unstable in ammonia solutions. NH_4OH was therefore avoided as a base to prevent the formation of Pd ammonia complexes. Pd/MCM-41 2 wt% was typically prepared dissolving 0.24 g NaOH, 0.099 g Na_2PdCl_4 and 0.164 ml MPTES in 54 ml of distilled water under orbital stirring. A brown-red solution was obtained after a few minutes. Then, 2.19 g CTAB was dissolved. When a clear solution was obtained, 6.25 g TEOS was added dropwise under vigorous stirring to ensure its complete hydrolysis. After that, the solution was transferred to a 100 ml Teflon-lined stainless steel autoclave and heated at 100 °C for 24 h. After cooling to room temperature, the solid product was water washed, filtered out, and air dried overnight. The surfactant was removed by calcination at 550°C for 8 h.

2.3 Synthesis of Pd-zeolite A

Zeolite Type A was prepared according to the procedure reported by R.W. Thompson and M.J. Huber [12]. For 1 g Pd/zeolite A 2 wt%, 0.072 g sodium hydroxide, 0.055 g Na_2PdCl_4 and 0.164 ml MPTES were dissolved in 8 ml of distilled water. The solution, which took a brown-red, was divided into two equal volumes in polypropylene bottles. 0.826 g sodium aluminate was dissolved in one of these parts and 15.48 g sodium metasilicate in the other. The silicate solution was added to the aluminate solution quickly under vigorous stirring. A thick orange gel was obtained after 6 hours of orbital stirring. This product was transferred to a Teflon-lined stainless steel autoclave and heated at 99 °C for 4 h. The solid product was water washed, filtered out, and air dried overnight.

2.4 Synthesis of Pd-zeolite Y

Zeolite Type Y was synthesized using a gelling process described by D. M. Ginter, A. T. Bell and C. J. Radke [13]. Pd incorporated in zeolite Y was obtained using a three-step synthesis procedure. A typical synthesis solution for zeolite Y seeds was prepared by dissolving in a plastic bottle 0.204 g sodium hydroxide, 0.02 ml MPTES, 0.012 g Na_2PdCl_4 and 0.105 g sodium aluminate in 0.997 g distilled water. Finally, 1.136 g sodium silicate was added. After 10 min of mild stirring a orange gel was obtained. Finally, the bottle was capped and the solution let age at room temperature for 1 day.

The second step is the preparation of a feedstock gel. 0.007 g sodium hydroxide, 0.125 ml MPTES, 0.077 g Na_2PdCl_4 and 0.655 g sodium aluminate were dissolved in 6.55 ml of distilled water. Then, 7.122 g sodium silicate solution was added under vigorous stirring.

Finally, 0.165 g seed gel was slowly added to the feedstock gel under vigorous hand stirring, the composition molar of the mixture is $10SiO_2:1Al_2O_3:4.62Na_2O:180H_2O$. The beaker was energetically shaken to improve the mixing.

The crystallization was carried out in a Teflon-lined stainless steel autoclave at 100°C for 5 h. An orange-brown solid and a clear supernatant were obtained. This solid was water washed, filtrated out, and air dried overnight.

2.5 Instrumentation

Transmission Electron Microscopy (TEM) studies were carried out using a JEOL JEM-2010 instrument operated at 200 kV. The Pd-nanoparticles-containing silica was suspended in ethanol, sonicated and a few drops placed over a carbon-coated copper grid. In combination with TEM studies, chemical analysis was carried out using a INCAEnergyTEM – EDS Microanalysis system The porous texture of the samples was characterized by N_2 adsorption at 77 and 273 K in an AUTOSORB-6 apparatus. The samples were previously degassed for 4 h at 523 K at $5*10^{-5}$ bars. The nanostructure of these materials were characterised by X-ray powder diffraction (Bruker D8-Advance) using a CuKα radiation. The scanning velocity was 0.1°/min from 1.7 to 50 2theta (°).

3 RESULTS AND DISCUSSION

3.1 TEM images of the Pd containing zeolite and MCM-41 type silica

Two TEM micrographs at of a Pd/zeolite A are presented in Figure 1. More specifically, Figure 1.b) shows a magnification of the area indicated in Figure 1.a) with a red square. It is possible to distinguish Pd nanoparticles in this zoomed-in image as dark spots. The analysis of this image indicates that these nanoparticles are a very small size (around 1.5 nm) and high dispersion in the support.

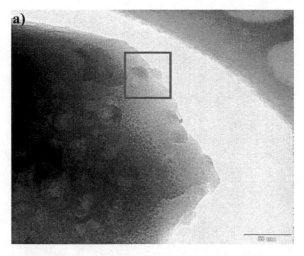

Figure 1: Two transmission electron micrographs at two different magnifications of the same sample of Pd-zeolite A.

The TEM micrographs shown in the Figure 2 and 3 of the Pd/MCM-41 type silica materials confirm their mesoporosity. These images show a sample containing 2 wt% Pd, before (Figure 2) and after calcination (Figure 3). The sample before calcination, Figure 2, has the Pd incorporated in atomic form as Pd-MPTES complex, therefore, it is impossible to recognize at the image. Nevertheless, the Pd presence was confirmed by microanalysis with EDS system. Figure 3 shows a micrograph of the same sample after calcinations. In the case, highly disperse Pd nanoparticles formed during the thermal treatment were observed, the size of these particles change between 4 and 8 nm. It is likely that the Pd was partially oxidised producing a incomplete PdO.

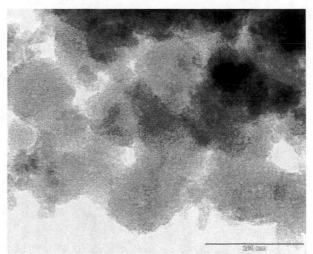

Figure 2: TEM micrograph of a 2 wt% Pd/MCM-41 type silica sample, before surfactant removal.

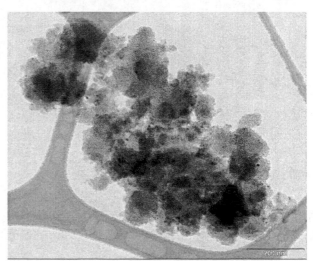

Figure 3: TEM micrograph of a 2 wt% Pd/MCM-41 type silica sample, after calcination treatment to remove the surfactant.

3.2 XRD of Pd containing nanostructured silicas

The X-ray diffraction (XRD) patterns of the Pd-MPTES incorporated zeolite Y an A are shown in Figure 4. For comparison purposes, the XRD patterns conventional zeolites have been added. The XRD pattens of all the Pd-MPTES incorporated zeolites present well-defined peaks with similar intensities that the original zeolites. No shift was observed in the position of the Pd containing zeolite, suggesting that the unit cell of the zeolites (U.C.S zeolite Y = 24.794 Å, zeolite A = 12.283 Å) was not change due to the presence of Pd (U.C.S Pd/zeolite Y = 24.757 Å, Pd/zeolite A = 12.283 Å). No other phases were identified as no spurious peaks were observed, neither broad peaks nor change in the baseline due to amorphous material. Also, no peaks that could be indexed to Pd metal were observed which is in agreement with the very small size of the Pd nanoparticles and Pd content in the sample. All these observations are consistent with the TEM analysis. These results suggest that Pd nanoparticles were successful *in situ* incorporated in zeolites without damaging their structural integrity.

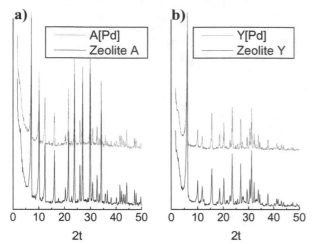

Figure 4: XRD patterns of a) Pd-MPTES zeolite A (2 wt% Pd) (top) compared to a sample of zeolite A (down) and b) Pd-MPTES zeolite Y (2 wt% Pd) (top) compared to a sample of zeolite Y (down).

3.3 Nitrogen isotherms at 77K of Pd containing MCM-41 type silica

The mesoporous nature of the calcined materials was confirmed by nitrogen adsorption at 77K, which produces a type IV isotherms (see Figure 5) with a sharp uptake at 0.3 – 0.4 P/P_0.

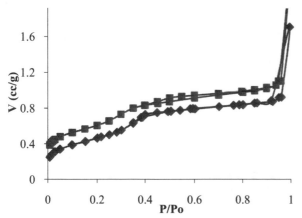

Figure 5: N_2 isotherms at 77K of a sample prepared with Pd:MPTES molar ratio of: 4 (squares, top) and 1 (rhombus, bottom). The curve at the top is shifted +0.15 cc/g for clarity.

In the hybrid MCM41-type materials so obtained both the surfactant and the functional group can be removed by calcination, producing a composite material with Pd nanoparticles incorporated in a mesoporous support.

4 CONCLUSIONS

The incorporation of Pd nanoparticles in zeolite A and Y and in MCM-41 type silica has been described using a two step process. The condensation of a Pd-MPTES complex with TEOS under hydrothermal conditions (zeolites) or in a surfactant solution (MCM-41) and complex thermal decomposition. The final materials have highly dispersed fairly small (1-3 nm) Pd nanoparticles.

In this work we have developed an approach to obtain highly dispersed and strongly anchored Pd nanoparticles incorporated in different silica matrices. This method is based on the use of a Pd complex as building blocks for the construction of silica materials, producing composite materials with Pd nanoparticles homogeneously dispersed in nanostructured supports. The cavity, in which the metal nanoparticle is located after the functional groups are removed, inhibits the migration of the nanoparticle and thus the agglomeration/sintering. This method is general and should also serve to incorporate other metal nanoparticles in silica matrices.

ACKNOWLEDGMENTS
This research has been funded by the Ministerio de Educación y Ciencia (CTQ2005-09385-C03-02). J.G.M. is grateful for financial support under the Ramón y Cajal program and N.L. for a FPI fellowship. Cristina Almansa Carrascosa is kindly thanked for assistance with TEM.

REFERENCES
[1] S. Brase et al., "Metal-Catalysed Cross-Coupling Reactions", Wiley-VCH, 99, 1998.
[2] L. Yin, J. Liebscher, Chem. Rev., 107, 133, 2007.
[3] V. Polshettiwara, A. Molnár, Tetrahedron, 63, 6949, 2007.
[4] I. Yuranov et al, J. Mol. Catal. A: Chem., 192, 239, 2003.
[5] R. Akiyama, S. Kobayashi, J. Am. Chem. Soc., 125, 3412, 2003.
[6] W. Solodenko et al., Eur. J. Org. Chem., 3601, 2004.
[7] S.P. Andrews, A.F. Stepan, H. Tanaka, S.V. Ley, M.D. Smith, Adv. Synth. Catal., 347, 647, 2005.
[8] N. Ren el al., J. Catal., 246, 215, 2007.
[9] J. D. Webb, S. MacQuarrie, K. McEleney, C. M. Crudden, Journal of Catalysis 252, 97, 2007.
[10] J. Garcia-Martinez et al., Micropor. Mesopor. Mater. (submitted)
[11] Y. Zheng, Z. Li, Y. Zheng, X. Shen, L. Lin, Materials Letters 60, 3221, 2006.
[12] R.W. Thompson, M.J. Huber, J. Cryst. Gr. 56, 711, 1982.
[13] D.M. Ginter, A.T. Bell, C.J. Radke, "Synthesis of Microporous Materials", 1, 6, 1992.

Production of Nanoparticles Sizes of Active Pharmaceutical Ingredients (API) by Wet Comminution with the DYNO®-MILL

Stanley Goldberg*, Frank Long**, and Norbert Roskosch*
*GLEN MILLS INC., Clifton, NJ, USA
**WILLY A. BACHOFEN AG Maschinenfabrik, Muttenz, Switzerland

KEYWORDS

bead milling, DYNO®-MILL, nanoparticle production, comminution

ABSTRACT

The ability to produce nanoparticles by the comminution of suspended solids is improved when using minute ceramic beads with diameters of 0.05-0.10mm in a newly developed bead mill the DYNO®-MILL Model RESEARCH LAB RL. A detailed description of the equipment is included. Example of an inorganic pigment ground at various energy levels and feed rates is presented. Another important example is presented where an Active Pharmaceutical Ingredients (API) had been comminuted in the DYNO®-MILL Model MULTI LAB to a size in the two-digit nanometer range.

INTRODUCTION

Taking full use of small grinding media (beads) of 0.05-0.10mm and the new bead mill DYNO®-MILL Model RESEARCH LAB, experiments were run to demonstrate the ability to produce nanometer sized particles. Whereas larger beads can take up more kinetic energy, which is important for the breakage of particles and true grinding, a packing of smaller beads contains much more beads in the same volume. A high number of beads is especially an advantage for the dispersion of fine particles, where the number of interactions between the particles and the beads play a major role. In stirred media mills, the handling of smaller beads is quite different than in the case of larger beads, because the beads are carried much stronger by the product suspension, which means that the viscosity plays a major role.

EXPERIMENTAL SETUP

The DYNO®-MILL Model RESEARCH LAB (W. A. BACHOFEN AG; GLEN MILLS INC.) that was used for all experiments is shown in Figure 1 and the setup of the grinding chamber is shown in Figure 2. The product enters through a funnel into the inlet and onto the conveying screw (auger), which pumps the product into the grinding chamber, which has a cooling mantle. Since the material is transported by the conveying screw, no external pump is required. The grinding chamber contains the patented DYNO®-Accelerator to throw the beads, which is mounted on the same shaft as the conveying screw. At the outlet of the grinding chamber, a chrome screen is mounted to hold back the grinding beads. If the mill is operated in the circulation mode, the material goes through a steel pipeline and back into the funnel. All grinding beads that were used are made of yttrium stabilized zirconium oxide (specific gravity 6.0).

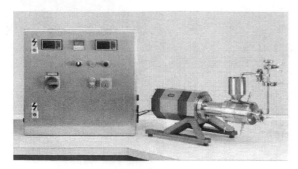

Figure 1: DYNO®-MILL
Model RESEARCH LAB

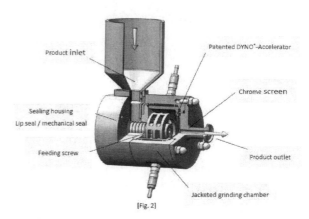

Figure 2: Mill internals showing feed
hopper, feed auger, Accelerator®, outlet

BOTTOM-UP or TOP-DOWN

The two ways to produce particles in the nanometer size range are (1) by building up from starting atoms/molecules or (2) by comminution of larger starting particles (Figure 3). All solid materials start with individual atoms or molecules that build up atom by atom, or molecule by molecule, to form a nano-structure. If the size growth process can be arrested at the nanometer scale, the attainment of small particles is achieved.

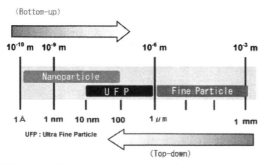

Figure 3: Paths to Nanoparticles

NUMBER OF BEADS vs. DIAMETER

Since the count of beads (grinding media) greatly increases as their diameter decrease, there are more contact points, more collisions, and more active surfaces involved with the

However, in wet processes, such as crystallization and precipitation, or dry processes such as chemical vapour deposition (CVD) the growth may not stop until very large particles are produced. The larger particles, well over tens or hundreds of microns in size, can be either individual large crystals or an agglomeration of minute particles strongly bound together by surface forces.

Producing nanomaterials or structures when starting with a large sized mater will require a high energy comminution technique. This paper describes using such a method, the wet bead milling DYNO®-MILL.

comminution. The problem of how to retain such small beads is solved by use of outlet screens with small openings or employing a centrifugal slinger to repel the beads while allowing product to escape the milling zone.

GRINDING OF AN INORGANIC PIGMENT

A suspension of an inorganic pigment, with a particle size D90 of 5.227 μm, in water was milled with 0.1 mm beads at a filling degree of 65% in the circulation mode. The solid concentration was 30 mass percent. The operating parameters are listed in Table 1.

Time (Minutes)	Agitator Tip Speed (m/sec)	Throughput flow rate (kg/hr)
10	10	7
20	12	7
30	14	6
40	14	5
50	14	5

Table 1: Operating Parameters When Grinding an Organic Pigment

The speed had to be increased, due to a rise in the viscosity. In the end, the minimal shear rate that had to be applied to keep the suspension flowing rose to about 6 Pa, despite the relatively low solid concentration. In the end, however, a particle size D90 of 736 nm was achieved. The frequency curves of the product before grinding (red), after 30 minutes (green), and after 50 minutes (blue) are shown in Figure 8.

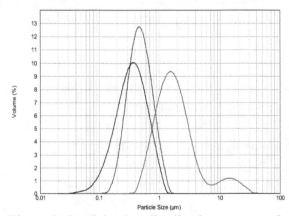

Figure 8: Particle size distribution curves of raw pigment, after 30 minutes, and after 50 minutes

At first the reduction of the particle size is high, because the particles in the range around 10 μm are being "truly grounded". As soon as the particle size is below 1 μm, the dispersion plays a larger role, which can also be seen in the rise of the viscosity. In order to grind further, dispersant would have to be added, because a reduction of the particle size from over 5 μm to less than 1 μm leads to an enormous increase in the solid surface area. Even though no agglomerates were measured, there is practically no difference in the amount of particles in the range above 1 μm between 30 minutes and 50 minutes. It has to be mentioned that 100 nm is at the limit of the measuring range of the instrument. Figure 9 shows the change in the values of D90 over time

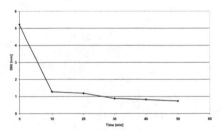

Figure 9: Particle size D90 of an inorganic pigment, as a function of the circulation time

NANOPARTICLES OF API

By employing small diameter beads, long milling times, and high energy inputs, an API that started with D50 of 5.6μ was driven to a D50 of 13nm (Figure 10). Eventually, there was some re-agglomeration to the 100 to 200nm range, but that was acceptable for the pharmacological efficacy was not impeded.

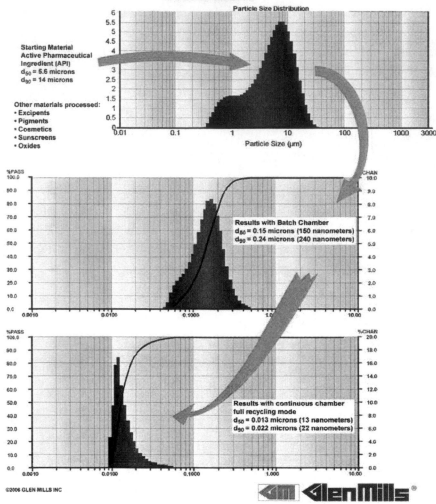

Figure 10: API Comminuted to 13nm with DYNO®-MILL ML

CONCLUSIONS

Using grinding beads with diameters from 0.80mm down to 0.10mm, stable grinding processes can be achieved if one takes into account the higher amount of liquid and dispersant that is required when the particle size is below 1 μm. As with larger beads the addition of more power does not lead to a lower particle size when an optimum has been surpassed. At lower particle sizes, the increased interactions between particles and the liquid lead to a strong increase in the viscosity, and this is especially important if the product is pumped in circulation or more than one pass through the mill. Depending on the product, the viscosity can already increase at solid concentrations of 30 mass percent after a certain grinding time.

Successful comminution of suspended API particles to the nanometer size has been demonstrated. This allows normally insoluble drug compounds to penetrate the cells. The bead mill method (DYNO®-MILL) is demonstrated to be a good platform for this work.

Silver nanoparticles synthesized on titanium dioxide fine particles

N Niño-Martínez[1, 2], G A Martínez-Castañón[3], A Aragón-Piña[2], F Martínez-Gutierrez[4], J R Martínez-Mendoza[1] and Facundo Ruiz[1].

[1]Facultad de Ciencias, UASLP,
[2]Instituto de Metalurgia, Facultad de Ingeniería, UASLP,
[3]Maestria en Ciencias Odontológicas, Facultad de Estomatología, UASLP
[4]Facultad de Ciencias Químicas, UASLP,
Álvaro Obregón 64, C. P. 78000, San Luis Potosí, S. L. P., México.

ABSTRACT

Silver nanoparticles with a narrow size distribution (20 nm mean size) were synthesized over the surface of two different commercial TiO_2 particles using a simple aqueous reduction method. The reducing agent used was $NaBH_4$; different molar ratios TiO_2:Ag were also used. The nanocomposites thus prepared were characterized using TEM, STEM, SEM, EDS, XPS, XRD, DLS and UV-VIS absorption spectroscopy; the antibacterial activity was assessed using the standard microdilution method, determining the minimum inhibitory concentration (MIC) according to the National Committee for Clinical Laboratory Standards. From the microscopy studies (TEM and STEM) we observed that silver nanoparticles are homogeneously distributed over the surface of TiO_2 particles and TiO_2:Ag molar ratios play an important role. The size of silver nanoparticles was controlled in the range of 10 - 30 nm. It was found that the antibacterial activity of the nanocomposites increases considerably comparing with separated silver nanoparticles and TiO_2 particles.

Keywords: metal nanoparticles, silver and titanium oxide composites, and antibacterial activity.

1 INTRODUCTION

Titanium dioxide (TiO2) is one of the most popular semiconductor materials, it can be found commercially available and can be used in many catalytic applications [1-8]. TiO2 has a wide band gap (3 eV and 3.23 eV for anatase and rutile respectively) which makes this material transparent to visible light, it is, no photon absorption occurs at wavelengths beyond 380 nm and catalytic reactions using pure TiO2 must be carried out using ultraviolet photons. Titanium dioxide is a material that also presents antibacterial activity [9-12]; this antibacterial activity has been studied over E. coli and B.

Megaterium using environmental light [13]. Few studies have investigated the application of TiO2 in life science [14]. It has been reported that catalytic and bactericide properties of TiO2 can be improved by growing particles of a noble metal (Ag, Au or Cu) over its surface [13]. In this work, silver nanoparticles were synthesized on the surface of TiO2 fine particles using a simple aqueous reduction method and the composites thus obtained (TiO2@Ag) were characterized using TEM, STEM, SEM, EDS, XRD, UV-VIS spectroscopy, DLS and XPS. An antibacterial activity test (NCCLS M7-A4, 1997) was conducted in order to confirm the improved bactericide properties of the composites obtained.

2 EXPERIMENTAL SECTION

2.1 Materials

TiO_2 particles (DuPont™ Ti-Pure® R-902 and Degussa P25), $AgNO_3$ (Sigma Aldrich, ACS Reagent), $NaBH_4$ (Sigma Aldrich, ACS Reagent) and NH_4OH (30 % w/w aqueous solution, Sigma Aldrich, ACS Reagent) were used as received without further purification.

2.2 Synthesis Method

For a typical procedure, 0.2000 g (2.5 mmol) of commercial TiO_2 particles were dispersed in 100 mL of deionized water by using ultrasonic by approximately five minutes, immediately 0.0169 g (0.1 mmol) or 0.0425 g (0.25 mmol) of $AgNO_3$ were added. The solution was magnetically stirred for about 30 minutes at pH = 7. After this, sodium borohydride, which was previously dissolved in 10 mL of deionized water, was added as reducing agent. The pH of the reaction media was adjusted to 10 by dropping NH_4OH, finally the solution was magnetically stirred for 30 minutes. After this time, the products obtained (TiO2@Ag) were filtered, washed and dried for further characterization. Hereafter, DuPont™ particles will be named as TiO_2_1 and Degussa particles will be named as TiO_2_2. Three different samples were synthesized; the samples obtained using TiO$_2$_1 and molar ratios of 25:1 and 10:1 (TiO$_2$:Ag) will be named as TiO$_2$_1@Ag25 and TiO$_2$_1@Ag10 respectively. The sample prepared using TiO$_2$_2 and a molar ratio of 10:1 (TiO$_2$:Ag) will be named as TiO$_2$_2@Ag10.

2.3 Characterization

The produced composites were characterized by UV-VIS spectroscopy using a S2000-UV-VIS spectrometer from OceanOptics Inc. Dynamic Light Scattering analysis was performed in a Malvern Zetasizer Nano ZS. X-Ray Diffraction pattern were obtained on a GBC-Difftech MMA model, with Cu K$_\alpha$ irradiation at λ= 1.54 Å. Transmission Electron Microscopy (TEM) analysis was performed on a JEOL JEM-1230 at an accelerating voltage of 100 kV, the STEM images were obtained on a JEOL 2010F. Scanning Electron Microscopy (SEM) analysis was performed on a Phillips XL-30 SEM equipped with an EDS spectrometer EDAX DX-4 Model. XPS analysis of the powder samples was carried out using a Kratos AXIS ULTRA XPS system fitted with a monochromated Al K$_\alpha$ X-ray source and a hemispherical analyser with eight channeltrons. The source was operated at 10 mA and 15 kV. UV-VIS spectroscopy, SEM, EDS, XRD and XPS analysis were made using dried powders and TEM, STEM and DLS analysis were made using aqueous dispersions of the TiO$_2$@Ag composites.

2.4 Antibacterial test

The antimicrobial activity of the synthesized composites was tested using the standard microdilution method, which determines the minimum inhibitory concentration (MIC) leading to inhibition of bacterial growth (NCCLS M7-A4, 1997). Disposable microtitration plates were used for the tests. The composites in dispersion form were diluted 2-128 times with 100 µL of Mueller-Hinton broth inoculated with the tested bacteria at a concentration of 10^5 CFU/mL. The minimum inhibitory concentration (MIC) was read after 24 h of incubation at 37 °C as the MIC of the tested substance that inhibited the growth of the bacterial strain. The dispersions were used in the form in which they had been prepared. Therefore, control bactericidal tests of solutions were performed containing all the reaction components.

3 RESULTS AND DISCUSSION

3.1 Synthesis

Silver ions (Ag$^+$) can be deposited over the surface of TiO$_2$ particles by a cationic adsorption. TiO$_2$ is an anphoteric oxide with an isoelectronic point IEP = 6 [15]. When the pH value of a TiO$_2$ dispersion is lower than 6 the main surface specie is —OH$_2$$^+$, when the pH value of a TiO$_2$ dispersion is bigger than 6 the main surface specie is —O$^-$, in the latter case the surface of TiO$_2$ particles is negatively charged and silver ions can be deposited over its surface [13]. In this work, in order to ensure a complete adsorption of the silver ions, a mixture of TiO$_2$ particles and silver ions (added as silver nitrate) was magnetically stirred for about 30 minutes at pH = 7. After that, the reduction reaction proceeded on the surface of TiO$_2$ particles.

3.2 TEM and STEM

Using TEM we can confirm the size of the TiO$_2$_1 particles and, the most important information extracted is the irregular thin layer observed on the surface of TiO$_2$ particles, this could be a layer made of SiO$_2$ and Al$_2$O$_3$ (according with the results obtained in EDS analysis). TiO$_2$_2 particles have a spherical morphology and a particle size ranging from 15 nm to 70 nm. Using TiO$_2$_1 particles and increasing the amount of silver nitrate in reaction we can not produce more Ag nanoparticles as we expected, instead silver nanoparticles already formed on the surface of TiO$_2$_1 grow. The reason of this unexpected behavior could be the presence of the irregular SiO$_2$-Al$_2$O$_3$ thin layer on the surface of TiO$_2$_1 particles. SiO$_2$ and Al$_2$O$_3$ have no reactivity if they are not activated [16], so Ag nanoparticles are formed only on the spots where there is no SiO$_2$-Al$_2$O$_3$ thin layer; in a moment, these spots are replete and the remaining silver ions are deposited over the first silver nanoparticles formed and finally they grow. If we use TiO$_2$_2 instead of TiO$_2$_1 and if we maintain the concentration of Ag$^+$ as a constant, the amount of silver nanoparticles over the surface of TiO$_2$ particles increases considerably, the reason of this could be, again, the presence of the SiO$_2$-Al$_2$O$_3$ thin layer on the surface of TiO$_2$_1 and the fact that Degussa P25 is reported as the most reactive phase of TiO$_2$ [17].

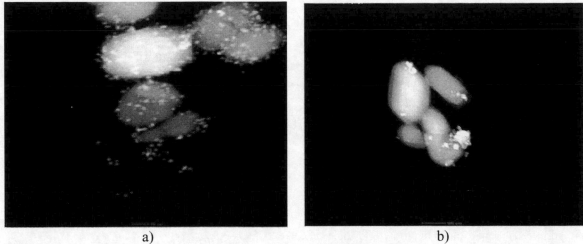

a)   b)

Figure 1. STEM images of the samples a) TiO$_2$_1@Ag25 and b) TiO$_2$_1@Ag10. These images show that silver nanoparticles are bigger in sample TiO$_2$_1@Ag10 than in sample TiO$_2$_1@Ag25.

3.3 Antibacterial results

Minimum inhibitory concentration values were obtained for the synthesized composites tested against *E. coli* (Gram negative bacteria, ATCC 25922) and *S. aureus* (Gram positive bacteria, ATCC 25923). The results are presented as average values on table 1 (the Kruskal-Wallis test was applied). Control sample containing all the initial reaction components showed no antibacterial activity.

Material	Minimum Inhibition Concentration[c] (μg/mL)	
	Bacteria	
	E. coli	*S. aureus*
TiO$_2$_1	-[a]	-[a]
TiO$_2$_2	-[a]	-[a]
TiO$_2$_1 @Ag25	130.2 (0.651)	250 (1.25)
TiO$_2$_2 @Ag10	358.5 (35.94)	333.3 (33.3)
TiO$_2$_2 @Ag10	190.1 (19.01)	208.3 (20.67)
Ag nanoparticles[b]	13.02	16.67

[a] No antibacterial activity was found with the concentrations tested in this work.
[b] 20 nm Ag nanoparticles were synthesized under the same conditions as the composites but without the presence of TiO$_2$ particles.
Values on parentheses represent the calculated content of silver in the composites.

Table 2. Minimum Inhibition Concentrations of TiO$_2$ particles, Ag nanoparticles and TiO$_2$@Ag composites.

TiO$_2$_1@Ag25 sample has higher antibacterial activity than the other composites and present higher antibacterial activity than TiO$_2$ particles. If we compare the MIC of TiO$_2$_1@Ag25 with that of silver nanoparticles we can see that the MIC of the latter is lower but the silver content in TiO$_2$_1@Ag25 sample is much lower (almost 20 times) than the MIC of silver nanoparticles, then, we can say that there is a real synergetic antibacterial activity in these composites. The fact that TiO$_2$_1 and TiO$_2$_2 particles showed no antibacterial activity is due to the test conditions, the test was performed on dark. It is reported [8, 14] that bactericide activity of TiO$_2$ is directly related to ultraviolet light absorption and formation of free radicals, so in dark conditions TiO$_2$ particles present no bactericide activity which is consistent with our results. By the other hand, all the TiO$_2$@Ag composites show antibacterial activity even though no light is present. The antibacterial mechanism of these composites is under investigation by our group.

4 CONCLUSIONS

Silver nanoparticles were synthesized over the surface of two different commercial TiO_2 particles. The composites thus obtained were characterized, using XRD, XPS and UV-VIS analysis it was demonstrated that the nature of the nanoparticles prepared is elemental silver; these silver nanoparticles are well distributed over the surface of TiO_2 particles and their average size is about 20 nm. The antibacterial activity of TiO_2 nanoparticles was improved and is dependent on the sort of TiO_2 particles. In this work, the best results were achieved using TiO_2 Dupont^{TM} particles, a TiO_2:Ag molar ratio of 1:25.

REFERENCES

[1] Kominami H, Murakami S, Kato J, Kera Y and Ohtani B 2002 Correlation between some physical properties of titanium dioxide particles and their photocatalytic activity for some probe reactions in aqueous systems *J. Phys. Chem. B.* **106** 10501-10507.

[2] Egerton T A, Tooley I R 2004 Effect of changes in TiO_2 dispersion on its measured photocatalytic activity *J. Phys. Chem. B.* **108** 5066-5072.

[3] Noguchi T, Fujishima A, Sawunyama P and Hashimoto K 1998 Photocatalytic degradation of gaseous formaldehyde using TiO_2 film *Environmental Science and Technology* **32** 3831-3833.

[4] Karvinen S M 2003 The effects of trace element doping on the optical properties and photocatalytic activity of nanostructured titanium dioxide *Ind. Eng. Chem. Res.* **42** 1035-1043.

[5] Aarthi T and Madras G 2007 Photocatalytic degradation of rhodamine dyes with nano-TiO_2 *Ind. Eng. Chem. Res.* **46** 7-14.

[6] Vinodgopal K, Wynkoop D E and Kamat P V 1996 Environmental photochemistry on semiconductor surfaces: photosensitized degradation of a textile azo dye, acid orange 7, on TiO_2 particles using visible light *Environmental Science and Technology* **30** 1660-1666.

[7] Xu N, Shi Z, Fan Y, Dong J, Shi J and Hu M 1999 Effects of particle size of TiO_2 on photocatalytic degradation of methylene blue in aqueous suspensions *Ind. Eng. Chem. Res.* **38** 373-379.

[8] Wolfrum E J, Huang J, Blake D M, Maness P, Huang Z, Fiest J and Jacoby W A 2002 Photocatalytic oxidation of bacteria, bacterial and fungal spores, and model biofilm components to carbon dioxide on titanium dioxide-coated surfaces *Environ. Sci. Techno.* **36** 3412-3419.

[9] Sunada K, Kikuchi Y, Hashimoto K and Fujishima A 1998 Bactericidal and detoxification effects of TiO_2 thin film photocatalysts *Environ. Sci. Technol.* **32** 726-728.

[10] Sunada K, Watanabe T and Hashimoto K 2003 Bactericidal activity of copper-deposited TiO_2 thin film under weak uv light illumination *Environ. Sci. Technol.* **37** 4785-4789.

[11] Jacoby W A, Maness P C, Wolfrum E J, Blake D M and Fennell J A 2000 Mineralization of bacterial cell mass on a photocatalytic surface in air *Environ. Sci. Technol.* **32** 2650-2653.

[12] Linkous C A, Carter G J, Locuson D B, Ouellette A J, Slattery D K and Smitha L A 2000 Photocatalytic inhibition of algae growth using TiO_2, WO_3, and cocatalyst modifications *Environ. Sci. Technol.* **34** 4754-4758.

[13] Fu G, Vary P S and Lin C 2005 Anatase TiO_2 nanocomposites for antimicrobial coatings *J. Phys. Chem. B* **109** 8889-8898.

[14] Zhang A and Sun Y 2004 Photocatalytic killing effect of TiO_2 nanoparticles on Ls-174-t human colon carcinoma cells *World J. Gastroenterol.* **10** 3191-3193.

[15] Zanella R, Giorgio S, Henry C R and Louis C 2002 Alternative methods for the preparation of gold nanoparticles supported on TiO_2 *J. Phys. Chem. B* **106** 7634-7642.

[16] Pastoriza-Santos I, Gómez D, Pérez-Juste J, Liz-Marzán L M and Mulvaney P 2004 Optical properties of metal nanoparticle coated silica spheres: a simple effective medium approach *Phys. Chem. Chem. Phys.* **6** 5056–5060.

[17] Singh H K, Muneer M and Bahnemann D 2003 Photocatalysed degradation of a herbicide derivative, bromacil, in aqueous suspensions of titanium dioxide *Photochem. Photobiol. Sci.* **2** 151-156.

[18] Bowering N, Croston D, Harrison P G and Walker G S 2007 Silver modified degussa P25 for the photocatalytic removal of nitric oxide *International Journal of Photoenergy* **2007** 1-8.

Free-standing SnO$_2$ nanoparticles synthesized by hydrothermal route

A.L. Fernández-Osorio* , A. Vázquez-Olmos**, R. Sato-Berru**, D.Casas-Gutiérrez*

*Facultad de Estudios Superiores Cuautitlán, ana8485@servidor.unam.mx
**Centro de Ciencias Aplicadas y Desarrollo Tecnológico, Universidad Nacional Autónoma de México,
México, D.F., 04510, México, america.vazquez@ccadet.unam.mx

ABSTRACT

In this contribution we present the obtaining of SnO$_2$ nanoparticles with average diameters of 5 nm by hydrothermal route from SnCl$_2$ and temperature of 150 C. The nanostructures were studied by UV-visible electronic absorption and Raman spectroscopy [1], their crystal structure were determined from XRD patterns and by HRTEM images.
XRD showed the presence of single phase of cassiterite structure, as found from XRD line broadening the crystallite sizes of all powders were in the nanometric range.

Keywords: tin oxide, nanoparticles, semiconductors, chemical preparation

1 INTRODUCTION

SnO$_2$ with a rutile type crystalline structure is an n-type wide band gap (3.5 eV) semiconductor that presents a proper combination of chemical, electronic and optical properties that make it advantageous in several applications. Due to its physical properties, such as transparency and semiconductivity, it is an oxide of great interest from the technological point of view for gas sensors, white pigments for conducting coatings for furnaces and electrodes, ultraviolet optical fibers, dye based solar cells, optoelectronic devices, and catalysts, an increasing interest in the use of anodes of SnO$_2$ in lithium batteries has been recently noticed [1-4]
One area of primary importance is the field of solid state gas sensors for environmental monitoring, where SnO$_2$ has been established as the predominant sensing materials.
Tin oxide (SnO$_2$) has been the material of choice for semiconductor gas sensors, which detect reducing gases in air from a change in electrical resistance. According to research on the sensing mechanism, a reduction in grain size leads to an increase in sensitivity.
The microstructure of SnO$_2$ could be controlled by temperatures treatment, doping and method of preparation.
Many methods had been developed to synthesize SnO$_2$ nanoparticles such as homogeneous precipitation, sonochemical, hydrothermal, microemulsion, sol-gel, and polymeric precursor [4-10] methods among others. Even though the development of agglomerates is to be avoieded, their growing is somehow inevitable due to the small

diameter of the oxide particles and to the presence of the compounds involved in the mentioned procedures, mainly solvents.
In the present study, had been synthesized free-standing SnO$_2$ nanoparticles by hydrothermal route using SnCl$_2$ like precursor.

2 EXPERIMENTAL

2.1 Materials

Tin chloride, SnCl$_2$.2H$_2$O (98% Aldrich) and Ammonium hydroxide NH$_4$OH (98% Aldrich) were purchased and used as received, without further purification. Ultra pure water (18 MΩcm^{-1}) was obtained from a Barnstead E-pure deionization system.

2.2 Preparation of nano-SnO$_2$

An aqueous solution of 0.163 M SnCl$_2$.2H$_2$O was prepared by mixing 2.03g of tin chloride with 30 ml of distilled water, NH$_4$Cl was added with stirring, a white precipitate was immediately formed, which was then separated from the aqueous solution. To remove the ammonium and chloride ions, the precipitate was washed 4 times with distilled water. The obtained precipitate was dried at room temperature, the dried powder was fired at 150°C for 2 h in air to finally obtain a pale yellow powder of tin dioxide nanoparticles..
The chemical reaction proceeds according to:

$$SnCl_2.2H_2O + NH_4OH \rightarrow Sn(OH)_2 + 2NH_4^+ + 2Cl^-$$

$$Sn(OH)_2 \xrightarrow[2h]{150°} SnO_2 + H_2$$

3 CHARACTERIZATION

The UV-visible electronic absorption spectra of the powdered samples were obtained by diffuse reflectance technique, with an Ocean Optics HR4000 miniature fiber optic spectrometer. The Raman spectra, from 100 to 900 cm^{-1}, were evaluated using a Nicolet Almega XR Dispersive Raman Spectrometer and detected by a CCD camera, at 25 seconds and a resolution of ~4 cm^{-1}. The excitation beam was a Nd:YVO$_4$ 532 nm laser and the incident power on the sample was ~3 mW. The X-ray diffraction patterns were performed at room temperature with Cu Kα radiation (λ = 1.5406Å) in a D5000 Siemens diffractometer; diffraction intensity was measured between 2.5° and 70°, with 2θ step of 0.02° for 0.8 s per point. High-resolution transmission electron microphotographs (HR-TEM) were obtained in a JEOL 2010 FasTEM analytical microscope, operating at 200 kV, by deposition of a drop of the powdered transition metal oxide dispersed in N,N'-dimethylformamide (DMF) onto 300 mesh Cu grids coated with a carbon layer.

4   RESULTS AND DISCUSSION

The powders were analysed by X-ray powder diffraction. XRD pattern (Fig.1) reveals the formation of a single nanocrystallyne product, which was identified as cassiterite, all diffractions peaks can be perfectly indexed to the rutile type structure (JCPDS card 21-1250), with a unit cell described by the space group P42/mnm and lattice parameters a = 4.738 and c = 3.188 Å.

These results are consistent with those of the bulk. In order to determine the average crystallite size, a peak broadening method was applied using the classical equation over all reflections, finding out to be of 5.74 nm

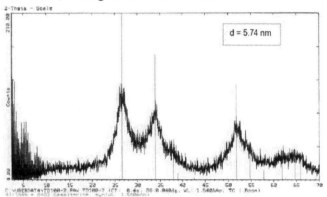

Figure 1 XRD pattern of SnO_2 nanoparticles. All peaks can be indexed to SnO_2 cassiterite, card 21-1250

The optical response of SnO_2 Nps was evaluated by UV-Visible electronic absorption spectra, obtained by DRS. As shown in figure 2, the spectra exhibits one broad absorption band, centered at 630 nm for the SnO_2 Nps.

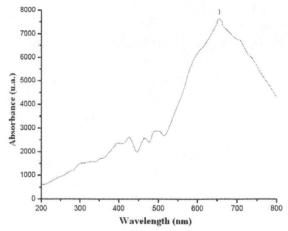

Figure 2. UV-visible spectra of SnO_2 Nps

Moreover, since the Raman spectroscopy is a nondestructive technique which in the last years has been extensively used in nanostructure characterization, we obtained the corresponding SnO_2 Nps Raman spectra, as shown in figure 3. These spectra clearly exhibit five well-defined peaks at 230, 472, 630, and 773 cm^{-1}, assigned to the Raman-active modes of the SnO_2 with B_{2g}, E_g and A_{1g} symmetries, respectively. In the SnO_2 Nps Raman spectrum, the peaks appear broadened and additionally a little red shift is observed. This behavior has been observed by other authors, and related to size effect.

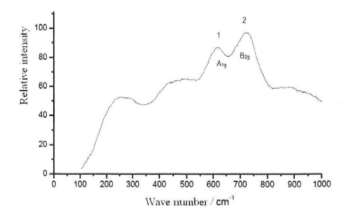

Fig 3 Raman spectra SnO_2 Nps

The HR-TEM micrographs corroborate the formation of small nanocrystals with dimensions close to those determined by X-ray diffraction patterns (figure 4). These nanocrystals have dimensions of 4.7 nm X 4.3 nm (the smallest one) and 6.2 nm X 5 nm (the largest one). While figure 5 shows a nanocrystal with dimensions of 5.4 nm X 4.7 nm;

NSTI-Nanotech 2008, www.nsti.org, ISBN 978-1-4200-8503-7 Vol. 1

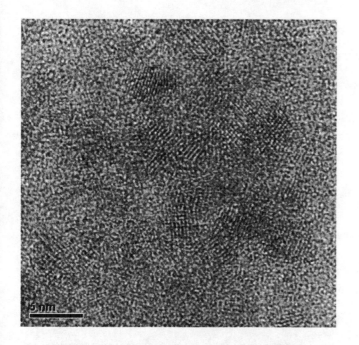

Figure 4 HR-TEM micrograph of SnO₂ Nps

The interplanar distances determined from their corresponding electron diffraction patterns confirm that the nanocrystals are composed of SnO₂. (Fig.5 and Fig.6)

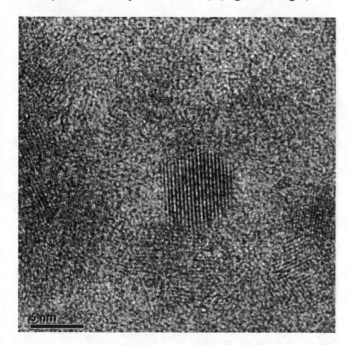

Figure 5. HR-TEM Micrograph of SnO₂ Nps

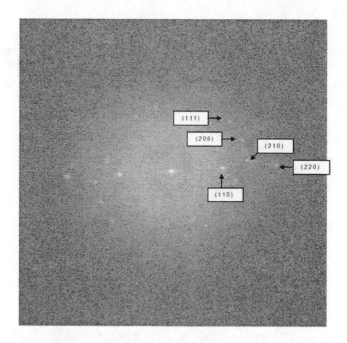

Figure 6. Electron diffraction pattern of SnO₂ Nps

5 CONCLUSION

Free-standing single nanocrystalline SnO₂ have been synthesized by the coprecipitation method and posterior thermal treatment, at temperature of 150°C, from the tin chloride and ammonium hidroxide. The average crystal size of the nanoparticles varies from 4.7 to 5.3 nm. The approach described in this study can be readily scaled up to fabricate large quantities of these nanocrystals.

Acknowledgments

The support of PAPIIT project IN106007

REFERENCES

[1] Noriya Izu, Norimitsu Murayama, Woosuck, Toshio Itoh, Ichiro Matsubara, Mat. Lett. 62, 313-316, 2008

[2] Yi Chun Chen, Jin-Ming, Yue-Hao Huang, Yu-Run Lee, Han C.Shih., Surf. & Coat. Tech. 202, 1313-1318, 2007

[3] G.Brambilla, V.Pruneri, L.Reeckie,, Appl.Phys.Lett. 76, 807, 2000

[4] A.M. Mazzone, Solid State Comm.143, 481-486, 2007

[5] C.Ararat Ibarguen, A.Mosquera, R.Parra, M.S.Castro, J.E.Rodriguez-Paez, Mat.Chem. and Phys. 101, 433-440, 2007

[6] Mira Ristic, Mile Ivanda, Stanko Popovik, Svetozar Music, J. Non.Cryst.Solids 303, 270-280, 2002

[7] R.Y.Sato-Berru, A.Vázquez-Olmos, A.L.Fernández-Osorio "Micro-Raman Investigation of transition metal doped ZnO nanoparticles" J.Raman Spectrosc.

Glycerol Assisted Polymeric Precursor Route for the Synthesis of Nanocrystalline LiCoO$_2$ Powders

S. Vivekanandhan[*], M. Venkateswarlu[**], N. Satyanarayana[*]

[*]Department of Physics, Pondicherry University, Pondicherry- 605 014, India
[**]HBL Power Systems Ltd, Hyderabad- 500 078, India
[*] Corresponding author: E-mail: nallanis2007@gmail.com, Phone: +91-413-2654404

ABSTRACT

Glycerol assisted polymeric precursor route was investigated for the synthesis of nanocrystalline layered LiCoO$_2$ powders using metal nitrates as metal ion sources, citric acid and glycerol as fuels as well as polymerizing agents. The synthesis process was investigated through TG/DTA, FTIR, XRD, and SEM analysis. It is observed that the addition of glycerol to citric acid caused the formation of foamy porous intermediate. From FTIR and XRD analysis, it was found that the ethylene glycol assisted polymeric precursor process lead to the formation of organic free single phase crystalline LiCoO$_2$ powder at 600 $^{\circ}$C for 12 hours and the crystallite size is found to be 45 nm.

Keywords: nanocrystalline LiCoO$_2$, citric acid, glycerol, XRD, FTIR, TG/DTA, SEM.

1 INTRODUCTION

LiCoO$_2$ with Layer structure has been widely used as a cathode material in lithium batteries [1, 2]. Recently, it is found that the nanostructured LiCoO$_2$ powders exhibit an enhanced electrochemical performance compared to their bulk. Physiochemical properties of the nanocrystalline LiCoO$_2$ powders like crystallite size, homogeneity, chemical stability etc., are mainly depending on the synthesis process [3-5]. A number of wet chemical routes such as sol-gel, combustion, polyol, hydrothermal, etc., were investigated for the synthesis of nanocrystalline metal oxides including LiCoO$_2$ powders [6-9].

Among them, polymeric precursor route is found to be a simple and effective soft combustion process for the synthesis of nanocrystalline multi-component oxide powders relatively at lower temperatures [10]. Selection of suitable fuel is an important issue in combustion process. Carboxylic acids like citric acid, tartaric acid, etc., have been extensively investigated for these propose [10, 11]. Apart from that, poly hydroxyl alcohols such as ethylene glycol, polyvinyl alcohol, etc., are used as polymerizing agent (Pechini process) in order to enhance the uniform distribution of metal ions and avoid their precipitation during evaporation [12]. Recently, we have investigated a novel process based on glycerol as polymerizing agent along with citric acid for the successful synthesis of nanocrystalline LiNiVO$_4$ powders relatively at lower temperature [13].

Hence, in the present work, glycerol assisted polymeric precursor route was investigated for the synthesis of nanocrystalline layered LiCoO$_2$ powders using citric acid and glycerol as polymerizing and chelating agents. The complete process was investigated through TG/DTA, FTIR, XRD and SEM techniques.

2 EXPERIMENTAL TECHNIQUES

Fig. 1 shows the schematic of polymeric precursor route for the synthesis of nanocrystalline LiCoO$_2$ powders. Required stoichiometric amounts of lithium nitrate (99.8 %, S. d. Fine-Chem. Ltd.) and cobalt (GR grade, Merck, India) nitrate solutions were made and mixed with citric acid (GR grade, Merck) and glycerol (SQ grade, Qualigens) solutions. Resulting transparent clear pink colour solution was evaporated at 80 $^{\circ}$C under constant stirring. Continuous evaporation lead to the formation of polymeric resin and it was further dried at 150 $^{\circ}$C for 12 hours for the removal of excess water. During the drying process the resin has expanded upto 30 times and caused the formation of high foamy nature. Further, the intermediate was calcined at 600 $^{\circ}$C in order to get nanocrystalline LiCoO$_2$ powders.

FTIR spectra were recorded using FTIR - 8000 spectrometer of Shimadzu, Japan. The measurement was carried out between 400 and 4000 cm^{-1} with KBr diluter. X-ray diffraction (XRD) experiments were carried out using a Panalytical, X-ray powder diffractometer, with Cu Kα radiation. The average crystallite size was calculated from the Scherrer's formula employing line – broadening technique. NBS standard silicon was used for estimation of instrumental broadening [14]. Thermal behavior of the polymeric intermediates was investigated by simultaneous TG/DTA using Labsys thermal analyzer, Setaram, France. Approximately, 3mg of polymeric intermediate was heated at a rate of 10 $^{\circ}$C min^{-1} between 30 – 600 $^{\circ}$C. All thermal studies were performed in flowing oxygen. Microstructure of the polymeric intermediates was taken using scanning electron microscope, Jeol, Japan.

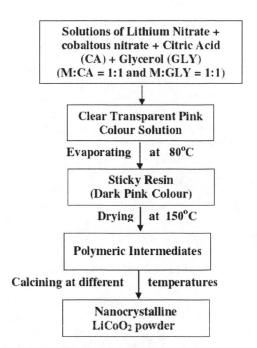

Solutions of Lithium Nitrate +
cobaltous nitrate + Citric Acid
(CA) + Glycerol (GLY)
(M:CA = 1:1 and M:GLY = 1:1)

↓

Clear Transparent Pink
Colour Solution

Evaporating | at 80°C

↓

Sticky Resin
(Dark Pink Colour)

Drying | at 150°C

↓

Polymeric Intermediates

Calcining at different | temperatures

↓

Nanocrystalline
LiCoO₂ powder

Fig. 1 Flow chart for the synthesis of nanocrystalline
LiCoO₂ powder by Polymeric precursor route

3 RESULTS AND DISCUAAION

Fig 2a shows the photograph of synthesized foamy polymeric intermediate and the respective microstructure obtained from SEM is shown in fig. 2b. SEM image shows the porous structure with large size of voids. Fig. 3a and 3b are the FTIR and XRD results of polymeric intermediate respectively. From fig. 3a the observed peaks at 3417-3444 cm^{-1} and 2926-2974 cm^{-1} are respectively due to the stretching frequencies of OH (due to the adsorbed moisture) and aliphatic CH groups from citric acid and glycerol [15]. The shoulder band at 1725 cm^{-1} observed in citric acid derived intermediate is corresponds to the asymmetric vibration of bridging COO^- group and the respective symmetric vibration is observed at 1314 cm^{-1}. From fig. 3a, the IR peaks observed at 1608-1632 cm^{-1} and 1400-1415 cm^{-1} region for all the intermediates are respectively due to the asymmetric and symmetric stretching vibration of COO^- groups, which confirm the chelation of metal ions by citric acid [15]. The observed peak free XRD patterns for the intermediates confirm their amorphous nature of the polymeric structure.

The TG/DTA thermogram of the polymeric intermediate is shown in fig. 4. Initial weight loss about 3%, observed between 75 and 100 °C is due to the removal of absorbed moisture. The major weight loss observed in TG curve between 250 and 415 °C is responds to the decomposition of organic derivatives and the respective exothermic peak was observed in DTA curve. Though the combustion reaction completes at 410 °C, there is a gradual weight loss in TG curve up to 500 °C, which indicates the presence of un-decomposed organic residuals even after the completion of combustion reaction. Hence, 600 °C is optimized for the calcinations of polymeric intermediate in order to get organic free LiCoO₂ powder.

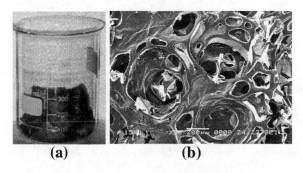

Fig. 2. Photograph (a) as well as scanning electron micrograph (b) of polymeric intermediate

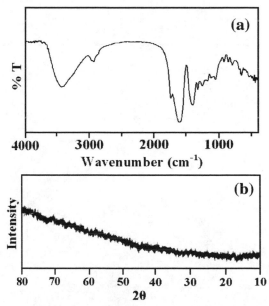

Fig. 3 FTIR (a) and XRD (b) analysis of polymeric intermediate

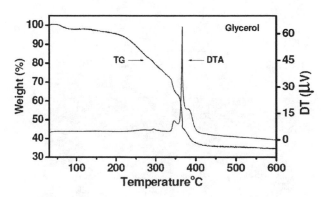

Fig. 4 TG/DTA thermograms of polymeric intermediate

FTIR, XRD and SEM analysis of the synthesized $LiCoO_2$ powders are respectively shown in fig 5a, 5b and 5c. In FTIR spectra, absence of peaks, which are belongs to organic derivatives, indicate their complete decomposition during the calcinations process at 600 °C. The observed new peaks at 510- 520 cm^{-1} and 580-610 cm^{-1} region are attributed to the asymmetric stretching modes of $[CoO_6]$ group, which confirms the formation of $LiCoO_2$ structure [16, 17]. From fig. 5b, the observed major peaks at 45°, 38° and 19° respectively for (104), (101) and (003) planes confirm the formation of $LiCoO_2$ phase. Further, it was confirmed by comparing their XRD pattern with JCPDS data and the crystallite size calculated using XRD data is found to be 45nm. The cell parameters obtained for the $LiCoO_2$ powders prepared by citric acid assisted process are a = 2.807 Å, c = 14.028 Å and c/a = 4.998, which are very much comparable with reported values of hexagonal cell in the literature [18, 19]. SEM image of $LiCoO_2$ powder exhibit an agglomeration of fine $LiCoO_2$ particles.

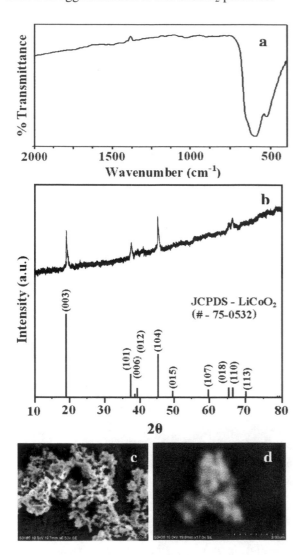

Fig. 5 FTIR (a), XRD (b) and SEM (c & d) analysis of Synthesized $LiCoO_2$ powder

4 CONCLUSIONS

Glycerol assisted polymeric precursor route has successfully investigated for the synthesis of nanocrystalline $LiCoO_2$ powders. Addition of glycerol to citric acid in the precursor solution caused formation of the porous foamy polymeric intermediate. From, TG/DTA, FTIR and XRD analysis, it is confirmed that the decomposition of synthesized foamy intermediate at 600 °C results ultra fine nanocrystalline $LiCoO_2$ powders. Crystallite size of the $LiCoO_2$ powders prepared by glycerol assisted polymeric precursor process is found to be 45 nm. Scanning electron micrograph of synthesized $LiCoO_2$ powders exhibits the agglomeration of fine particles.

5 ACKNOWLEDGEMENTS

Dr. N. Satyanarayana gratefully acknowledges DST, CSIR, and DRDO, government of India for financial support through major research project grants. SV acknowledges the CSIR, Government of India, for the award of Senior Research Fellowship.

6 REFERENCES

[1] M. S. Whittingham, Chem. Rev., 104 (2004) 4271-81.

[2] J. M. Tarascon, M. Armand, Nature, 414 (2001) 359-367.

[3] A. S. Arico, P. Bruce, B. Scrosati, J. M. Tarascon and W. V. Schalkwijk, Nature Mater., 4 (2005) 366-376.

[4] J. Jamink and J. Maier, Phys. Chem. Chem. Phys., 5 (2003) 5215-5219.

[5] E. Stura, C. Nicolini, Anal. Chim. Acta, 568 (2006) 57.

[6] L.J. Fu, H. Liu, C. Li, Y.P. Wu, E. Rahm, R. Holze, II.Q. Wu, Prog. Mater. Sci., 50 881–928 (2005).

[7] S. Vivekanandhan, M. Venkateswarlu, N. Satyanarayana, Mater. Chem. Phy., 91 (2005) 54–59.

[8] L. W. Tai, P. A. Lessing, J. Mater. Res., 7 (2) (1992) 502-510.

[9] S. W. Kwon, S. B. Park, G. Seo, S. T. Hwang, J. Nucl. Mater., 257 (1998) 172-179.

[10] R. Ganesan, S. Vivekanandhan, T. Gnanasekaran, G. Periaswami and S.S. Raman, J. Nucl. Mater. 325 (2004) 134–140.

[11] W. Liu, G. C. Farrington, F. Chaput, B. Dunn, J. Electrochem. Soc., 143 (3) (1996) 879-884.

[12] L. W. Tai, P. A. Lessing, J. Mater. Res., 7 (2) (1992) 511-519.

[13] S. Vivekanandhan, M. Venkateswarlu, N. Satyanarayana, Mater., Lett. 58 (2004) 1218– 1222.

[14] H. P, Klug, L. E. Alexander, X-ray Diffraction Procedures for Polycrystalline and Amorphous Materials, Wiley, New York, 1954.

[15] G. Socrates, Infrared and Raman Characteristic Group Frequencies, John Wiley and Sons, New York, 2001.

[16] K.J. Rao, H. Benqlilou-Moudden, B. Desnat, P. Vinatier, A. Levasseur, J. Solid State Chem. 165 (2002) 42–47.

[17] C. Julien, M.A. Camacho-Lopez, T. Mohan, S. Chitra, P. Kalyani, S. Gopukumar, Solid State Ionics 135 (2000). 241–248.

[18] Y. Gu, D. Chen, X. Jiao, J. Phys. Chem. B 109 (2005) 17901–17906.

[19] G.T.K. Fey, C.Z. Lu, T. Prem Kumar, Y.C. Chang, Surf. Coat. Technol. 199 (2005) 22.

Structure of Ni nanoparticles/TiO$_2$ films prepared by sol-gel dip-coating

A. García-Murillo[*], E. Ramírez-Meneses[*], J. Ramírez-Salgado[**],
G. Sandoval-Robles[***], V. Montiel-Palma[****], H. Dorantes-Rosales[*****],
P. Del Angel-Vicente[**]

[*] Centro de Investigación en Ciencia Aplicada y Tecnología Avanzada-IPN
Unidad Altamira, Km. 14.5 Carretera Tampico-Puerto Industrial, C.P. 89600,
Altamira, Tamaulipas. México., angarciam@ipn.mx, esramirez@ipn.mx
[**] Programa de Ingeniería Molecular, Instituto Mexicano del Petróleo, Eje Lázaro Cárdenas No. 152,
C.P. 07730, México, D.F., ramirezj@imp.mx, pangel@imp.mx
[***] Tecnológico de Cd. Madero Instituto Tecnológico de Ciudad Madero
Av. 1° de Mayo esq. Sor Juana Inés de la Cruz s/n Col. Los Mangos C.P.89440
Cd. Madero Tamaulipas, México., jgsandor@hotmail.com
[****] Centro de Investigaciones Químicas, Universidad Autónoma del Estado de Morelos,
Av. Universidad 1001, Colonia Chamilpa C.P. 62201 Cuernavaca, Morelos, México.,
vmontielp@ciq.uaem.mx
[*****] Departamento de Metalurgia, ESIQIE - IPN, C.P. 07300 México, D.F., hdorantes@ipn.mx

ABSTRACT

Titania thin films were synthesized by sol-gel dip-coating method with metallic Ni nanoparticles (~10 nm) from an organometallic precursor. Titania matrix and Ni nanoparticles were prepared separately. The precursors employed to prepare Titania were titanium isopropoxide, isopropanol, methanol and acetic acid. Nickel nanostructures were synthesized from Ni(COD)$_2$ (COD=cycloocta-1,5-diene) under H$_2$ atmosphere in THF with 1,3-diaminopropane as stabilizer.

Colloidal solution of Ni nanoparticles was added to sol. Ni/TiO$_2$ sol system was used to coat glass spheres substrates and further heat treatment at 400 °C.

The photocatalytic activity of the Ni/TiO$_2$ films was evaluated in H$_2$ evolution from decomposition of ethanol using a mercury lamp for UV light irradiation. The prepared Ni/TiO$_2$ was characterized using AFM, UV-Vis, TEM and HR-TEM.

Keywords: sol-gel, thin films, nickel, nanoparticles, organometallic.

1 INTRODUCTION

Metallic nanoparticles as catalyst have been extensively studied due to their particular properties which are quite different from those of bulk materials, for example their large surface area.

Titanium oxide has a great importance due to its excellent photocatalytic properties[1] as well as their industrial applications related to photo-splitting of water[2]

photocatalyst[3], photovoltaic devices[4]. However, TiO$_2$ exhibits a relative high energy bandgap (~ 3.2 eV) and can only be excited efficiently by high energy UV irradiation that constrains the practical usage of the TiO$_2$. Efforts have been made to extend the energy absorption range of TiO$_2$ from UV to visible light or to further improve the photocatalytic activity of TiO$_2$ by adding foreign metallic elements[5,6]. Ag can serve as electron trap aiding electron-hole separation, and can also facilitate electron excitation by creating a local electric field. Nanocrystalline TiO$_2$/Ag composite thin films were evaluated by degrading methylene blue UV exhibing good photocatalytic efficiency[7]. Another element of interest to improve photocatalytic performances is nickel, titania thin film with metallic Ni nanoparticles on its surfaces showed high efficiency in photocatalysis of hydrogen evolution from decomposition of ethanol[8].

On the other hand, On the other hand, the synthesis of metal nanoparticles has received considerable attention due to their unusual properties and potential applications. A number of methods such as hydrazine reduction of nickel chloride[9], electrochemical[10], thermal evaporation [11], laser-assisted gas phase photonucleation [12], hydrogen plasma metal reaction[13], physical vacuum deposition (PVD) process[14] and organometallic method[15] have been employed to obtain nickel nanoparticles. Additionally, many researchers have reported various synthetic methods for nickel nanoparticles supported on the surfaces of others materials.

Nickel nanoparticles with a diameter of 1-10 nm can be supported on the surface of titania particles[16,17]. A large variety of methods have been developed to prepare TiO$_2$

films by sol-gel process in which dip-coating[18], chemical vapor reductive deposition (CVRD)[8], MOCVD[19], aerosol-assisted process[20], etc. Comparing to these methods the sol-gel process possesses many merits. Metallic particles could be included homogeneously into the sol allowing coating onto various materials even pure metals.

Therefore, in this paper we investigated the feasibility of preparing Ni nanoparticles/TiO_2 thin films grown onto commercial glass spheres by sol-gel dip-coating technique. The as-deposited Ni/TiO_2 films were characterized by TEM, HR-TEM, XRD and AFM. The photocatalytic properties were evaluated by degrading ethanol under UV-irradiation.

Additionally, we report that Nickel nanoparticles deposited onto the surface of the glass spheres improve the separation of photo-produce electron-hole pairs, resultantly enhance the photocatalytic activity of TiO_2[21].

2 EXPERIMENTAL

2.1 Synthesis of metallic nanoparticles

The reaction was carried out, as a standard procedure in a Fischer–Porter bottle. In a typical procedure, a solution of THF (50 ml) including 500 mg of the precursor Ni(COD)$_2$ (Aldrich) and 1,3-diaminopropane (99%, Aldrich) 5 equivalents per Ni atom was reacted at 70 °C for 20 h, under dihydrogen pressure (3 bar), in Fischer-Porter bottle leads to the formation to dark gray colloids. The reaction is slow at lower temperatures[15]. The obtained solutions were purified by hexane washings. Finally, the resulting solution was evaporated in vacuum until the residue was completely dry.

2.2 Synthesis of of titanium oxide "sol" with Ni nanoparticles

Figure 1 shows the schematic flowchart of the experimental procedure. Titanium tetraisopropoxide (Ti(OiPr)$_4$ (99.9 %, Aldrich) was mixed with isopropanol (iPrOH-Fermont) and stabilized in presence of acetic acid acetic (AcOH-Fermont), with a molar ratio AcOH/Ti=5.84. The obtained solution was diluted with methanol. The solution was then stirred at room temperature for 1 h. After the reaction time a stable yellowish "sol" is obtained. Then, the metallic nanoparticles were dispersed into the titanium "sol" forming a gray colloidal stable solution.

2.3 Elaboration of Ni/TiO_2 thin films by "dip-coating" technique

Commercial glass spheres used as substrates (3 different diameter sizes: 6 mm, 4 mm and 3 mm) were carefully cleaned with methanol and were wet uniformly with the

Ni/TiO_2 sol using a withdrawal speed of 2.5 cm min^{-1}. After film deposition stage the sample was heat-treated in an oven at 100 °C for 15 min with the aim of remove the most volatile organic compounds and a second coating was done. Finally, the sample was thermally treated at 400 °C for 3 h in order to densify and promote the crystallization of the film.

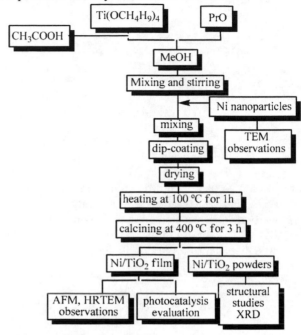

Figure 1. Flowchart of the experimental procedures

3 RESULTS AND DISCUSSIONS

3.1 Phase composition

The phase of the sample was determined by XRD (D8 Advance Bruker, Germany) using a Cu Kα radiation.
Figure 2 shows the XRD patterns of heat-treated TiO_2 and Ni/TiO_2 films. It can be found that the TiO_2 film pattern (a) and Ni/TiO_2 film (b), exhibited nanocrystalline anatase-TiO_2 phase. It is also notice that not additional XRD peaks corresponding to Ni addition can be revealed. This may be attributed to the well dispersed nanocrystalline Ni particles in the TiO_2 matrix, which are not large enough for XRD detection. The XRD also revealed the absence of any other characteristic peaks of a mixed oxide phase being formed due to an interaction between Ni and TiO_2 support.

3.2 AFM studies

Atomic force microscopy AFM observations of TiO_2 and Ni/TiO_2 films were carried out at room temperature using a Nanoscope IV (Veeco D3100). Figure 3 (a-b) shows the AFM top-view image of TiO_2 and Ni/TiO_2 films respectively. Figure 3(a) shows AFM image corresponding to TiO_2 film heat treated at 400 °C for 3 h which is characterized for mean roughness (Ra) of 1.747 nm and a RMS (Rq) of 2.219 nm. The TiO$_2$ nanosized grains formed on the glass surface was found to be 18.1 nm.

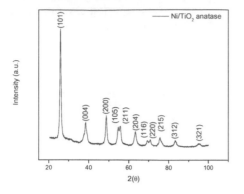

Figure 2. XRD powders patterns of the TiO$_2$ (a) and Ni/TiO$_2$ films (b).

Compared to Ni nanoparticles obtained on TiO$_2$ (Figure 3b) have very fine size with good uniformity in structure and composition. The mean roughness (Ra) was 0.347 nm and the RMS (Rq) was 0.440 nm. The average grain size of Ni/TiO$_2$ has a diameter of 5 nm (Figure 3b).

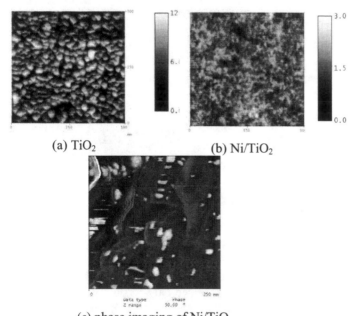

(a) TiO$_2$ (b) Ni/TiO$_2$

(c) phase imaging of Ni/TiO$_2$

Figure 3. a) top-view of AFM topography micrographs of TiO$_2$, b) AFM topography micrographs of Ni/TiO$_2$, and c) AFM phase imaging of Ni/TiO$_2$.

Phase Imaging is a powerful extension of TappingMode™ AFM. It provides nanometer-scale information about surface structure and properties often not revealed by other SPM techniques. By mapping the phase of the cantilever oscillation during the TappingMode scan, phase imaging goes beyond simple topographical mapping to detect variations in composition, adhesion, friction, viscoelasticity, and numerous other properties. Phase

imaging was used in both sample TiO$_2$ and Ni/TiO$_2$, in the first one no phase contrast was found that means the presence of one compound in the second, phase contrast become visible as is shown in the Figure 3c.

Mapping of different components in composite materials, as Ni/TiO$_2$ sample, is show in the Figure 3c. In this figure Ni nanoparticles appears as light areas in the phase image. The size of this light areas or Ni nanoparticles is of an average of 14 nm. The particle distribution of Ni in the TiO$_2$ is almost homogeneous.

3.3 TEM and HRTEM studies

Specimens for TEM analysis were prepared by the slow evaporation of a drop of the colloidal solution after the purification process deposited onto a holey carbon covered copper grid. TEM experiments were performed on a JEOL-1200 EX electron microscope, operating at 120 KV. TEM micrographs of Ni nanoparticles before mixing with TiO$_2$ solution, revealed the formation of nickel nanostructures with an average size of 10 nm (Figure 4).

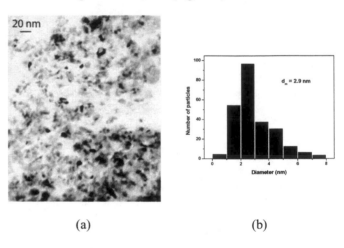

(a) (b)

Figure 4. TEM images of Ni nanoparticles prepared from Ni(COD)$_2$ in THF in presence of 5 equiv. 1,3-diaminopropane.

The microstructure of the Ni/TiO$_2$ film was observed using a JEOL 2200FS field emission gun (FEG) TEM equipped with in-column energy filter (Omega Filter because of shape of the electronic trajectories) to produce a high-end, optimally configured TEM for energy filtered imagery. Figure 5 shows HRTEM micrographs of Ni/TiO$_2$ film.

Figure 5(a) shows Ni/TiO$_2$ sample with particle size less than 15 nm. This parameter was measured by digital micrograph software of GATAN Company. The grain size is very similar of that obtained by AFM. The Figure 5(b) shows HRTEM micrograph at higher magnification clearly shows the lattice planes of monocrystal TiO$_2$. Interplanar spacings d were determined by FFT, the space between the lattice plane was 0.352 nm which corresponds well to d

value of (101) plane for anatase, which is in agreement with the theoretical data.

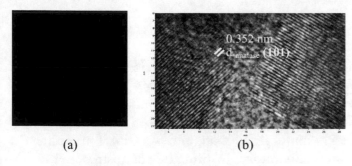

(a)   (b)

Figure 5. TEM (a) and HRTEM (b) of Ni/TiO$_2$ film micrographs calcined at 400 °C for 3 h at two different magnifications.

3.4 Photocatalytic Evaluation

In order to evaluate the photocatalytic activity of TiO$_2$ and Ni/TiO$_2$ films, the photodegradation of ethanol was performed at 25 °C under UV illumination.

For the visible light experiment, the sample was illuminated with a 25 W mercury 2 lamps (Trojans) at a distance of 10 cm from the glass reactor.

100 g of glass spheres films were placed in a 30 cm long and 5 cm diameter cylindrical air-filled static glass reactor with a pyrex window by degrading a mixture composed by ethanol (1.05x10^{-4} gmol/min), water (molar ratio EtOH/H$_2$O= 0.3) and N$_2$ (20 ml/min, Praxair) pretreated at 100 °C. The obtained products (hydrogen and acetaldehyde) were analyzed using gas chromatography using a Varian Star 3400-TCD chromatograph equipped with Porapak Q column (3 m).

The photocatalytic properties with the presence of Ni nanoparticles reached values of 90, 550 and 586 μmol after 12, 25 and 40 min the exposure time. TiO$_2$ and Ni/TiO$_2$ films exhibited 23 and 40 % degradation of ethanol respectively after the same time of UV exposure.

Conclusions

In this study, we prepared nanocrystalline TiO$_2$ and Ni/TiO$_2$ composite thin films by sol-gel dip-coating technique. XRD results showed that these films exhibed anatase phase.

The photocatalytic properties of the prepared thin films were evaluated by degrading ethanol under UV. It was observed that Ni/TiO$_2$ system calcined at 400 °C for 3 h show high catalytic activity on the photodegradation of ethanol.

In the absence of Ni nanoparticles, the degradation of ethanol was dramatically lower reaching a value of about 23 % about 12 min after the lighting of the light source. Pure Ni/TiO$_2$ film showed 44 % degradation of ethanol after 40 min of UV exposure.

In order to extend the utility of Ni/TiO$_2$ material further studies related to TiO$_2$ and Ni/TiO$_2$-based nanocomposites at different calcinations temperatures are now in progress.

The hydrogen production rate observed in the Ni/TiO$_2$ system, reveal very efficient photocatalytic ethanol degradation.

References

[1] C. He, Y. Yu, X. Hu, A. Larbot, Appl. Surf. Sci. 200, 239-247, 2002.

[2] A. Fujishima, K. Honda, Nature, 238, 37-38, 1972.

[3] B. O'Regan, M. Gratzel, A low-cost, Nature, 353, 737-740, 1991.

[4] A. L. Linsebigler, G. Lu, J. T. Yates, Chem. Rev. 95, 735-758, 1995.

[5] S. Castillo, T. Lopez, Appl. Catal., B. Environ. 15, 203-209, 1998.

[6] I. M. Arabatzis, T. Stergiopoulos, M. C. Bernard, D. Labou, S. G. Neophytides, P. Falaras, Appl. Catal., B. Environ, 42, 187-201, 2003.

[7] C. C. Chang, J. Y. Chen, T. L. Hsu, C. K. Lin, C. C. Chan. Thin Solid Films, 516, 1743-1747, 2008.

[8] M. Yoshinaga, H. Takahashi, K. Yamamoto, A. Muramatsu, T. Morikawa, J. Colloid and Interface Science, 309, 149-154, 2007.

[9] D.E. Zhang, X.M. Ni, H.G. ZhengT, Y. Li, X.J. Zhang, Z.P. Yang, Mater. Lett., 59, 2011– 2014, 2005.

[10] Y. Zhao, E. Yifeng, L. Fan, Y. Qiu, S. Yang., Electrochim. Acta 52, 5873–5878, 2007.

[11] C. Bittencourt, A. Felten, J. Ghijsen, J.J. Pireaux, W. Drube, R. Erni, G. Van Tendeloo, Chemical Physics Letters, 436, 368–372, 2007.

[12] H. He, R.H. Heist, B.L. McIntyre, T.N. Blanton, Nanostructured Materials, 8, 7, 879-888, 1997.

[13] X.G. Li, T. Murai, A. Chiva, S. Takahashi, J. Appl. Phys. 86, 1867-1873, 1999.

[14] E. Verrelli, D.Tsoukalas, K. Giannakopoulos, D. Kouvatsos, P. Normand , D. E. Ioannou. Microelectronic Engineering, 84, 1994–1997, 2007.

[15] N. Cordente, M. Respaud, F. Senocq, M. J. Casonove, C. Amiens and B. Chaudret, *Nano Lett.* 1, 565-568, 2001.

[16] H. Takahashi, Y. Sunagawa, S. Myagmarjav, K. Yamamoto, N. Sato, A. Muramatsu, Mater. Trans. 44, 2414-2416, 2003.

[17] S. Myagmarjav, H. Takahashi, Y. Sunagawa, K. Yamamoto, N. Sato, E. Matsubara, A. Muramatsu, Mater. Trans., 45, 2035-2038, 2004.

[18] Z. Yuan, J. Zhang, B. Li, J. Li., Thin Solid Films, 515, 7091-7095, 2007.

[19] F. D. Duminica, F. Maury and R. Hausbrand , Surface and Coatings Technology, 201, 9349-9353, 2007.

[20] Petr O. Vasiliev, Bertrand Faure, Jovice Boon Sing Ng and Lennart Bergström, Journal of Colloid and Interface Science, 319, 144-151, 2008.

[21] P.V. Kamat, J. Chem. Rev., 93, 267-300, 1993.

Biphenyl-functionalized Ethane-silica Hybrid Materials with Ordered Hexagonal Mesoporous Structure

R. Rivera Virtudazo[*], E. Magdaluyo, Jr.[**], L. dela Cruz[***], E. Castriciones[****] and H. Mendoza[*****]

[*]Ceramic Engineering Department, Mariano Marcos State University, Batac, Ilocos Norte, Philippines, rvrv26@yahoo.com
[**]Department of Mining, Metallurgical and Materials Engineering, University of the Philippines, Diliman, Quezon City, Philippines, edmagdaluyo@gmail.com
[***]National Science Research Institute, University of the Philippines, Diliman, Quezon City, Philippines
[****]Inorganic Synthesis and Computational research Laboratory, Institute of Chemistry, University of the Philippines, Diliman, Quezon City, Philippines, ecastriciones@yahoo.com
[*****]Department of Mining, Metallurgical and Materials Engineering, University of the Philippines, Diliman, Quezon City, Philippines, herman.mendoza@up.edu.ph

ABSTRACT

An ordered mesoporous hybrid xerogel has been prepared using a sol-gel route involving the co-condensation of 4,4'-bis(triethoxysilyl)biphenyl and 1,2-bis(triethoxysilyl)ethane. Chemical treatment was performed under basic conditions, using cetyltrimethylammonium bromide as a structure-directing agent. The obtained mesoporous organosilica materials were subsequently characterized. FTIR spectra confirmed that the biphenyl moiety is covalently linked in the ethane-silica framework. The interplanar spacing was observed between 8 to 9 nm, surface area from 1196.99 to 1578.57 m^2/g, total volume from 1.2 to 0.75 cm^3/g and primary pore diameter from 3.54 to 3.56 nm. Particle morphology of the biphenyl-functionalized organosilica produced rod-shaped, wormlike and spherical particles. The TEM images revealed hexagonal array of mesopores and lattice fringes along and perpendicular to the pore axis.

Keywords: biphenyl, organosilica, surfactant-templated

1 INTRODUCTION

Many new developments in the field of hybrid materials have been reported [1-3]. The synthesis to generate ordered hexagonal mesoporous structure is made of inorganic silica incorporating organic moieties in the framework [4]. The functionalization represents a useful tool to fine-tune hybrid materials for specific physical, chemical and surface properties, as well as better thermal and hydrothermal stabilities for a wider range of applications in catalysis, adsorption, separation, sensing technology and nanoelectronics [5-6]. It has been shown that the properties and structure of mesoporous materials can be varied based on the synthesis process and conditions, such as base concentrations, temperatures, and the nature of the organic moiety to be incorporated [7-10].

In this paper, we report the synthesis of novel periodic mesoporous biphenyl-functionalized ethane-silica. The preparation involves 4,4'-bis(triethoxysilyl)biphenyl and 1,2-bis(triethoxysilyl)ethane co-condensation. Investigation was extended on the influence of the surfactant ratio and heat treatment on the properties of the bifucntional hybrid material.

2 METHODS

The 4,4'-bis(triethoxysilyl) biphenyl organosilane precursor was synthesized based on the general procedure described by Shea et al. For the surfactant-templated polymerization of organosilica, the procees was done by employing 1:1:8 mole ratio of the organosilica precursor monomer, bis(triethoxysilyl)ethane and tetraethoxysilane (TEOS) respectively. The mixture under basic media (62.0{H_2O}:0.64{CTAB}: 0.25{NaOH} ratio) was stirred for twenty-four hours and placed in an oven at 95 –100 ° C for three hours. The products were then filtered, washed and dried. Removal of excess surfactant from the product was done by reflux, and the white xerogel again filtered out, washed with ethanol to neutral pH and dried in a vacuum oven. The surfactant ratio and heat treatment were subsequently varied to investigate the effect on the properties of the obtained organosilica materials. All the resulting products were characterized using Fourier Transform Infrared Spectroscopy (FTIR), Powder X-ray Diffraction (XRD), Transmission Electron Microscopy (TEM), and Scanning Electron Microscopy (SEM).

3 RESULTS AND DISCUSSIONS

The success of polymerization and incorporation of biphenyl moiety in the organosilica framework was analyzed using Fourier Transform Infrared Spectroscopy (FTIR). Table 1 summarizes the functional groups present in the obtained mesoporous hybrid material.

Table 1: FTIR peak assignment

Functionality		Wavenumber (cm^{-1})
C – H stretch	Phenyl	2930
	Aliphatic	2895
C=C vibration	Phenyl	1633
C–H bend	Aliphatic	1409
Si–O–Si	Asymmetric	1081
C–H	Aromatic	964
Si–O–Si	Symmetric	974

The intense absorption bands at 1081 and 794 cm^{-1} are characteristics for the asymmetric and symmetric stretching respectively of the Si–O–Si bonds. The C–H deformation vibrations at 2895 and 1409 cm^{-1} were observed due to the aliphatic part of the bridging silsesquioxanes incorporated into the structure. The peaks at 2930, 1633 which is assigned to the C=C vibration of the aromatic structure, and 964 cm^{-1} were indicative of the biphenyl moiety. This confirms that the synthesized biphenyl-bridged organosilane monomer was covalently linked in the channels of the hybrid organosilica material.

Formation of enhanced rope- and rod-based morphologies, as well as observation of larger particles with gyroid and granular characteristics was observed under SEM when surfactant was varied. With longer heat treatment, spherical particles morphology were greatly observed.

Transmission electron microscopy images reveal many lattice fringes with hexagonal array of mesopores. The surface structure of the organosilica material may result from alternating hydrophilic and hydrophobic layers, composed of silica-biphenyl-silica-ethane functionality respectively. This material is formed as a result of structure directing interactions between the biphenyl-bridged organosilane monomer to the other silsesquioxane molecules, and between the precursor molecules and surfactant. The hexagonal structure was confirmed by XRD analysis and the pattern showed a two-dimensional *p6mm* space group with interpalnar spacing ranges from 8 to 9 nm.

Samples of solvent-extracted mesoporous organosilica under varying surfactant ratio and heat treatment were tested for N$_2$ adsorption/desorption isotherms as shown in Figure 2. Analyses showed that the materials exhibit type IV isotherms. This is typical for the well ordered mesoporous material with narrow pore size distribution. All the formulations exhibit an H1 hysteresis loop at the range of 0.55-0.80 relative pressure (P/P$_o$).

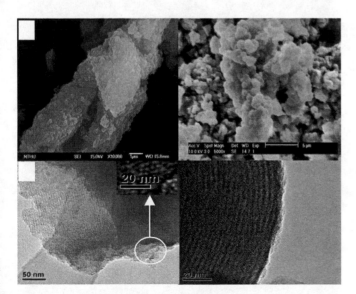

Figure 1: SEM (A) and TEM (B) images of the organosilica

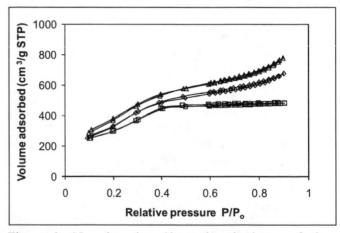

Figure 2: N$_2$ adsorption /desorption isotherm of the solvent-extracted mesoporous organosilica: $\triangle$ surfactant ratio (0.32), heat treatment (3 hrs); $\diamond$ surfactant ratio (0.64), heat treatment (3 hrs); $\square$ surfactant ratio (0.64), heat treatment (9 hrs)

Table 2: Effect of surfactant ratio and heat treatment on the structure of organosilica derived from N$_2$ isotherms

Surfactant ratio	Heat treatment (hrs)	Surface area (m^2/g)	Pore volume (cm^3/g)	Pore diameter (nm)
0.32	3	1196.99	0.750	3.56
0.64	3	1578.57	1.205	3.55
0.64	9	1423.07	1.049	3.54

The specific surface area ranges from 1196.99 to 1578.57 m^2/g, and the total pore volume from 1.2 to 0.75 cm^3/g. Using the BJH desorption method, the primary pore diameter from ranges from 3.54 to 3.56 nm. This indicates that no greater effect on the increasing surfactant concentration and heat treatment in terms of pore diameter.

4   CONCLUSION

Biphenyl-containing mesoporous hybrid organosilica materials have been synthesized and characterized. The organic group was covalently bonded in the framework and formed a two-dimensional hexagonal crystal structure.

5   REFERENCES

[1]   B. Hatton, K. Landskron, W. Whitnall, D. Perovic, & G Ozin.   Accounts of Chemical Research 38:305.2005.

[2]   K. Yamamoto, Y. Sakata, Y. Nohara, Y. Takahashi, & T. Tasumi. Science 300:470.2003.

[3]   M. Davis. "Ordered Porous Materials for Emerging Applications." Nature 417: 813-821. 2000.

[4]   T. Asefa, M. Machlachlan, N. Coombs, & G. Ozin "Periodic Mesoporous Organosilica with Organic Group Inside the Channel Walls." Nature 402:867-871. 1999.

[5]   C. Baleizao, B. Gigante, D. Das, M. Alvaro, H. Garcia & C. Corma. "Synthesis and Catalytic Activity of a Chiral Periodic Mesoporous Organosilica. "Chemical Communications 15:1860-1861. 2003.

[6]   M. Burleigh, M. Markowitz, S. Jayasundera, S. Spector, C. Thomas, & B. Gaber. Journal of Physical Chemistry B 107:12628. 2003.

[7]   A. Dolye, B. Hodnett, Micropor. Mesopor. Mater. 58 255. 2003.

[8]   T. Asefa, M. Kruk, M. Machlachlan, N. Coombs, H. Grondey, M. Jaroniec, & G. Ozin. "Novel Bifunctional Periodic Mesoporous Organosilica: Synthesis, Charcaterization, Properties and In-situ Selective Hydrocarbon-alcolysis Reactions of Functional Groups." Journal of American Chemical Society 123: 8520-8530. 2001.

[9]   G. Temtsin, T. Asefa, S. Bittner, & G. Ozin. "Aromatic PMOs: Tolyl, xylyl and Dimethoxyphenyl Groups Integrated within the Channel Walls of Hexagonal Mesoporous Silica." Journal of Materials Chemistry 11:3202-3206. 2001.

[10]   S. Inagaki, S.Guan, T Ohsuna & O. Terasaki. "An Ordered Mesoporous Organosilica Hybrid Material with a Crystal-like Wall Structure." Nature 416:304-307. 2002.

Free Energies of Low-Indexed Surfaces of Anatase and Rutile TiO$_2$ Terminated by Non-metals

C.H. Sun[*,**], H.G. Yang[*], S. Smith[*,**], Q.S. Zhang[*], G. Liu[*], J. Zou[***], H.M. Cheng[****], G.Q. Lu[*]

[*]ARC Centre of Excellence for Functional Nanomaterials, School of Engineering and Australian Institute of Bioengineering and Nanotechnology, The University of Queensland, QLD 4072, Australia
maxlu@uq.edu.au
[**]Centre for Computational Molecular Science, Biological and Chemical Sciences
Faculty and Australian Institute for Bioengineering and Nanotechnology, The University of Queensland, QLD 4072, Australia, s.smith@uq.edu.au
[***]Centre for Microscopy and Microanalysis, The University of Queensland, QLD 4072, Australia,
[****] Shenyang National Laboratory for Materials Science, Institute of Metal Research, Chinese Academy of Sciences, 72 Wenhua Road, Shenyang 110016, China

ABSTRACT

The structures and free energies (γ) of low index surfaces of anatase and rutile TiO2 terminated by nonmetals (H, B, C, N, O, F, Si, P, S, Cl, Br, and I) have been studied with the frameworks of density functional theory. It is found that, (i) the surface chemistry and stabilities can be adjusted using different nonmetal adsorbates; and (ii) the relative stabilities of these low-indexed surfaces of both anatase and rutile TiO$_2$ vary with the adsorbates, indicating that specific surfaces (such as highly reactive surfaces) are obtainable with use of different precursor and crystallographic controlling agents. The above calculated results may serve as a guideline for the synthesis of various metal oxides.

Keywords: TiO$_2$, anatase, rutile, surface free energy, surface stability

1 INTRODUCTION

The current interest in titanium dioxide (TiO$_2$) for advanced photochemical applications has prompted a number of studies to analyze its surface stability and reactivity with an aim to synthesize highly active surfaces [1-3]. Unfortunately, surfaces with high reactivity usually diminish rapidly during the crystal growth process as a result of the minimization of surface energy. So it is an open challenge to synthesize highly active surfaces through controlling their surface chemistry. For instance, in the case of anatase TiO$_2$, (101) is the most stable and frequently observed surface amongst various low-index surfaces, due to its relatively low surface free energy (γ=0.44 J/m^2 [4]). However, the high stability/low reactivity of the majority (101) surface makes it often difficult to understand the observed reactivity of anatase. A typical example is that recent sum frequency generation studies of the coadsorption of water and methanol on anatase nanoparticles show the presence of hydroxyls and methoxy groups, with the latter

being strongly bound to the surface and capable of replacing surface hydroxyls [5-6]. This cannot be result from (101) surfaces - on which both water and methanol can only be molecularly adsorbed [7-9]. Gong et al. believed that the minority (001) surface exhibits a high reactivity and can account for the above results based on detailed density functional theory (DFT) calculations [10]. Thereby, anatase TiO$_2$ single crystals with a large percentage of {001} facets are predicted to have high reactivity, which underlines the importance of developing effective approaches to control the stabilities of different surfaces.

Typically, the PH value is a decisive factor for controlling the final sample size and shape [11-12]. However, both hydrated (acid) and oxygenated (basic) surfaces show higher surface free energies than clean conditions [13-14], which is not helpful for the stabilization of highly reactive surfaces. High γ for H- and O-terminated surfaces are mainly caused by the high bonding energies (D_0) of H-H (436.0 kJ/mol) and O-O (498.4 kJ/mol) [15]. Therefore, to find a low D_0 element with high bonding to Ti might be a solution for stabilizing the faceted surfaces. In this paper, we carried out systematic investigation of 12 non-metallic atoms X (X= H, B, C, N, O, F, Si, P, S, Cl, Br, I) using first-principle quantum chemical calculations, in which clean surfaces of (001), (101) and (100) were used as references. Based on such calculations, it is found that adsorbate atoms can effectively change the relative stabilities of different crystal facet, which may serve as a guideline for the selecting of morphology controlling agent for experimental synthesis.

2 COMPUTATIONAL METHODS

In each case, stoichiometric slab models (1×1), consisting at least 6 atomic layers, are employed. In clean situations, all low-indexed surface contain fivefold Ti on two slides of each slab, which is saturated by X (X=H, B,

C, N, O, F, Si, P, S, Cl, Br, I). During the structural optimization, all atoms are relaxed without any constraint before the total energies are calculated. All calculations have been carried out using density functional theory (DFT) within the generalized-gradient approximation (GGA) [16], with the exchange-correlation functional of Perdew-Burke-Ernzerhof (PBE) [17-18]. This has been implemented in the Vienna ab initio simulation package (VASP) [19-20], which spans reciprocal space with a plane-wave basis, in this case up to a kinetic energy cutoff of 450 eV.

3 RESULTS AND DISCUSSION

3.1 Anatase TiO$_2$

According to early experimental studies, the surface of anatase TiO$_2$ with the equilibrium morphology consists of two facets, {001} and {101}, whose surface free energies are calculated. The calculated values of surface free energies of (001) and (101) of anatase TiO$_2$ terminated by different adsorbates are listed in Table 1. Based on these results, it is found that: (i) among 12 non-metal-terminated surfaces and the clean surfaces, the surfaces terminated by different nonmetals present different surface free energies, suggesting that the surface chemistry and stabilities can be adjusted using different nonmetal adsortates; and (ii) the relative stabilities of (001) and (101) vary with the adsorbates. Although in most cases, (101) is more stable than (001), however, with the termination of Si and F, (001) is more stable, indicating that the percentages of these surfaces may be controllable using suitable morphology controlling agents.

X	(001)	(101)
Clean	0.92	0.39
H	3.76	3.11
B	1.47	-0.21
C	4.18	1.85
N	8.12	6.31
O	2.88	2.46
F	-0.55	-0.22
Si	0.93	1.31
P	3.73	3.25
S	2.84	2.56
Cl	1.61	1.40
Br	1.98	1.92
I	1.18	0.83

Table 1: Calculated surface free energies (in unit of J/m^2) of (001) and (101) of anatase TiO$_2$.

3.2 Rutile TiO$_2$

Based on the equilibrium morphology of rutile TiO$_2$, four surfaces, including (001), (100), (011) and (110), are studied here. The calculated values of surface free energies for rutile TiO$_2$ terminated by different adsorbates are listed in Table 2, with clean surfaces being the reference. Similar with anatase TiO$_2$, the surface free energies of rutile TiO$_2$ strongly depend on the surface adsorbates. For instance, the surface free energies of (101) may change from -3.30 J/m^2 (B-terminated) to 9.06 J/m^2 (N-terminated), suggesting that the controlling of the surface adsorbates is an effective way to adjust the surface stability and reactivity in a wide range. Among 12 nonmetals, F and Si can improve the stabilities of all four surfaces investigated. Another interesting feature is the change of the relative stabilities of these surfaces. For clean surfaces, the stability is in the sequence of (110) > (100) > (101) > (001). However, with the coverage of nonmetals, such sequence has been totally changed. Importantly, the highly active surfaces of (001) and (101) present high stabilities when they are terminated by F and B. Following these calculations, large areas of (001) and (101) can be obtainable if the surface of rutile samples is covered by these atoms, which can serve as a guideline for experimentalists focusing on the synthesis of highly active rutile samples.

X	(001)	(100)	(110)	(101)
Clean	1.49	0.60	0.27	1.10
H	1.46	3.71	2.56	1.81
B	0.41	-0.97	0.61	-3.30
C	0.44	2.06	2.80	1.77
N	9.80	9.10	5.87	9.06
O	1.14	3.21	2.26	2.85
F	-0.76	-0.33	-0.24	-0.69
Si	-1.26	-0.22	0.20	-1.51
P	1.39	2.99	2.11	1.01
S	0.85	2.16	1.42	1.87
Cl	1.49	1.35	0.80	1.18
Br	2.54	1.35	0.82	1.71
I	1.75	0.56	0.34	0.71

Table 2: Calculated surface free energies (in unit of J/m^2) of (001), (100), (110) and (101) of rutile TiO$_2$.

4 CONCLUSIONS

The surface free energies of anatase and rutile TiO$_2$ have been calculated with the frameworks of density functional theory with the exchange-correlation functional of GGA. Two conclusions can be summarized from our calculations: (i) the surface chemistry and stabilities can be adjusted using different nonmetal adsortates; and (ii) the relative stabilities of these low-indexed surfaces of both anatase and rutile TiO$_2$ vary with the adsorbates. The above conclusions indicate that the controlling of the surface

NSTI-Nanotech 2008, www.nsti.org, ISBN 978-1-4200-8503-7 Vol. 1

adsorbates is an effective way to adjust the surface stability and reactivity in a wide range, which is of fundamental importance for the control synthesis of TiO_2.

5 ACKNOWLEDGEMENT

The financial support from the ARC Centre of Excellence for Functional Nanomaterials Australia and from the University of Queensland to this work is acknowledged. We also acknowledge Shenyang National Laboratory for Materials Science, Institute of Metal Research, for allowing the access to the high performance computing clusters.

REFERENCES

[1] K.I. Hadjiivanov and D.G. Klissurski, Chem. Sov. Rev. 25, 61, 1996.

[2] A.L. Linsebigler, G. Lu, J.T. Yates, Chem. Rev. 95, 735, 1995.

[3] A. Hagfelt and M. Grätzel, Chem. Rev. 95 (1995) 49.

[4] M. Lazzeri, A. Vittadini, A. Selloni, Phys. Rev. B 65 119901, 2002.

[5] C.Y. Wang, H. Groenzin, M.J. Shultz, J. Am. Chem. Soc. 126, 8094, 2004.

[6] C.Y. Wang, H. Groenzin, M.J. Shultz, J. Phys. Chem. B 108, 265, 2004.

[7] A. Vittadini, A. Selloni, F.P. Rotzinger, M. Grätzel, Phys. Rev. Lett. 81, 2954, 1998.

[8] A. Tilocca and A. Selloni, J. Phys. Chem. B 108, 19314, 2004.

[9] A. Tilocca and A. Selloni, Langmuir 20, 8379, 2004.

[10] X.Q. Gong and A. Selloni, J. Phys. Chem. B 109, 19560, 2005.

[11] A. Zaban, S.T. Aruna, S. Tirosh, B.A. Gregg, Y. Mastai, J. Phys. Chem. B, 104, 4130, 2000.

[12] Y. Gao, S.A. Elder, Mater. Lett. 44, 228, 2000.

[13] A.S. Barnard and P. Zapol, Phys. Rev. B 70, 235403, 2004.

[14] A.S. Barnard, P. Zapol, L.A. Curtiss, Surf. Sci. 582, 173, 2005.

[15] K.P. Huber and G. Herzberg. "Molecular Spectra and Molecular Structure Constants of Diatomic Molecules," Van Nostrand, New York, 1979.

[16] W. Kohn and L.J. Sham. Phys. Rev. B 140, A1133, 1965.

[17] J.P. Perdew, K. Burke, and M. Ernzerhof. Phys. Rev. Lett. 77, 3865, 1996.

[18] G. Kresse and D. Joubert. Phys. Rev. B 59, 1758, 1999.

[19] G. Kresse and J. Furthmüller. Phys. Rev. B 54, 11169, 1996.

[20] G. Kresse and J. Furthmüller. Comput. Mater. Sci. 6, 15, 1996.

Scattering properties of dense clusters of nanoparticles

M. Lattuada, L. Ehrl, M. Soos and M. Morbidelli[*]

ETH Zurich, DCAB-Institute of Chemical and Bio-Engineering
Wolfgang Pauli Strasse 10, 8093 Zurich, Switzerland.
E-mail: massimo.morbidelli@chem.ethz.ch

ABSTRACT

In this work, we present a strategy aimed at improving the utilization of Small Angle Light Scattering (SALS) for the characterization of dense clusters of spherical particles. By making use of a tunable fractal dimension Monte-Carlo algorithm, we generate dense clusters with a desired fractal dimension (d_f). In order to analyze clusters with a d_f larger than 2.5, we introduce a new algorithm which is capable of making clusters progressively denser and reach d_f equal to 3. The cluster structure is than characterized by means of its pair-correlation function, which is used to compute their scattering properties through a mean-field version of the T-Matrix theory. The scattering profiles from mean-field T-matrix theory compensate for the limitations of the more commonly used RDG theory and are effectively used in the analysis of SALS data from coagulation experiments of polymer colloids.

Keywords: small angle light scattering, clusters, Monte-Carlo simulations, pair correlation function, T-Matrix theory

1 INTRODUCTION

In sheared induced coagulation processes of colloidal nanoparticles, which are customarily used in most industrial processes, dense clusters of particles are usually formed. The characterization of their size, structure and distribution is a key factor for the quality control of the final product. However, the challenges encountered in the characterization of suspensions of dense clusters are numerous. While several microscopy-based techniques have been proposed that can access information such as size, shape, and structure of clusters, their long analysis times and difficult preparation do not make them the ideal techniques for routine analysis. On the other hand, small angle light scattering (or laser diffraction) techniques are more commonly used to perform this characterization, because of their speed, excellent statistics and great simplicity. Nevertheless, a major obstacle in the effective utilization of this technique is the lack of realistic structural models for shapes other than spheres or cylinders. In the case of dense clusters of spheres, the commonly used Rayleigh-Debye-Gans (RDG) theory fails because of the intracluster multiple light scattering effects, which are particularly important for particles with a size comparable to that of the wavelength of the incident radiation. These conditions are unfortunately more the rule rather than the exception in industrially relevant coagulation processes. In addition, there is also a lack of structural information on the morphology of dense clusters, which makes the determination of scattering profile even more difficult.

In this work, a methodology is presented that aims at overcoming these limitations. First of all, a tunable fractal dimension Monte-Carlo algorithm, initially proposed by Thouy and Jullien [1], is used to generate dense clusters with a desired fractal dimension. Since it is well known that the tunable fractal dimension algorithm cannot generate clusters with fractal dimension larger than 2.5, a new procedure has been developed to create cluster with fractal dimension up to 3. This procedure starts from clusters with a fractal dimension equal to 2.5 and creates a Voronoi tessellation of the space occupied by the cluster, which is made progressively denser by moving particles initially located on its surface to its interior. In this manner, the entire range of cluster fractal dimension encountered in typical aggregation processes, ranging from 1.8 to 3, is covered. The cluster structure is than characterized by means of its pair correlation function. The pair correlation function is finally used to compute scattering properties of cluster used a mean-field version of the T-Matrix theory, proposed by Botet et al.[2], which can provide reliable scattering behavior of dense clusters with arbitrary primary particle size. The predictions of the mean field T-matrix theory are compared to RDG theory predictions, to show the difference between these two approaches. In addition, the application of these results to the analysis of coagulation experiments of polymer colloids is discussed.

2 MONTE-CARLO GENERATION OF CLUSTERS

The Monte-Carlo tunable fractal dimension algorithm used in the first part of this work is virtually identical to that proposed by Thouy and Jullien [1]. According to the original idea, a given number of particles are sequentially connected to form progressively bigger clusters, making sure that at each step the clusters formed fulfill certain criteria, *i.e.* their fractal dimension and prefactor are assigned, so that the relationship between mass i and size:

$$R_g = R_p \cdot \left(\frac{i}{k} \right)^{\frac{1}{d_f}} \qquad (1)$$

NSTI-Nanotech 2008, www.nsti.org, ISBN 978-1-4200-8503-7 Vol. 1

where R_g is the cluster radius of gyration, R_p is the primary particle size and k is the prefactor, gives directly the target radius of gyration for a given mass. During the formation of a cluster, several possible combinations are considered, until equation (1) is fulfilled within a given precision. The process is repeated by randomly selecting at each step one pair of clusters.

However, it is well documented that this strategy is very effective in generating clusters with a fractal dimension up to 2.5 [1]. Above this value, it is highly unlikely to find a random combination of clusters that can form an extremely dense cluster. However, since in the case of clusters formed during shear aggregation processes it is not unusual to form cluster with a fractal dimension of 2.6 or higher, an alternative algorithm is introduced in order to form highly dense clusters. In real shear aggregation processes, the high fractal dimension values are obtained as a result of a combination of aggregation, breakage and restructuring of each cluster due to the action of the fluid. Since a realistic simulation of the real phenomenon is physically very challenging, once again a totally empirical approach is followed. The densification process starts from a cluster with a d_f =2.5, and first creates a Voronoi tessellation of the space occupied by the cluster. The Voronoi tessellation is then used to find the amount and position of empty spaces inside the cluster. Then, particles located on the outer surface of the cluster are progressively moved to its interior, until all empty spaces are occupied. The procedure is extremely effective, and leads to the generation of clusters up to d_f =3. As an example, Figure 1 shows a typical cluster made of 1000 particle with a d_f =2.5.

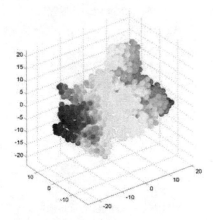

Figure 1. MC cluster with 1000 particles and d_f =2.5.

The structural characterization of the clusters is carried out by determining the average particle-particle correlation function $g(r)$, which gives the probability of finding a particle at a distance r from the average particle. The $g(r)$ function contains all most important structural properties of clusters, such are number of nearest neighbors, fractal

regime and upper size [3]. An example of particle-particle correlation function is given in Figure 2 for a cluster made of 10000 particles and with d_f =2.2.

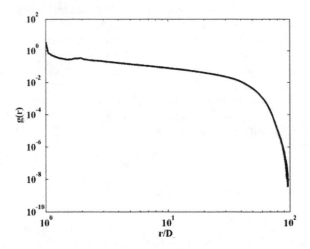

Figure 2. Particle-particle correlation function $g(r)$ of a cluster with 10000 particles and d_f =2.2.

3 LIGHT SCATTERING THEORY

According to the simplified RDG theory [4], which neglects any multiple scattering effects, the intensity of the radiation scattered by one cluster made of identical subunits is proportional to the square of the cluster mass. The angular dependence of the scattered radiation $I(q)$ (where q is the scattering wave vector), contains two contributions: the form factor of one single subunit $P(R_p,q)$, depending upon the subunit size and shape, and the structure factor $S(q)$, which depends upon the relative positions of the subunits. Quantitatively, this reads:

$$I(q) \sim i^2 P(R_p,q)S(q) \qquad (2)$$

where i is the number of particles in the cluster. The consequence of equation (2) is that in a suspension containing a large number of clusters with different masses, the bigger cluster will contribute much more to the total intensity of the scattered radiation compared to the small clusters. The above mentioned description, however, fails to provide the correct description in the case when multiple light scattering is playing a role, as it happens in the case of dense clusters, or in the case of clusters made of large particles. In this case, it is not possible anymore to factorize the form factor from the structure factor, since the intensity of light scattered by any particle depends specifically on its location inside a cluster. The rigorous T-matrix theory [4] accounts for this effect, and uses an extension of Mie theory accounting for the contribution from all the neighbor particles to scattering of any specific particle.

However, in the case of fractal clusters, where there can be several realizations of clusters with the same mass and structural properties, Botet et al [2]. proposed a mean-field version of T-Matrix theory, which used a similar formulation as the RDG one, except that the form factor is corrected to account for the multiple light scattering contribution that the average particles experiences in the cluster. In other words, the form factor is not anymore just a function of size and optical properties of the particle, but is also a function of the cluster structure. The correction factor can be determined if the particle-particle correlation function is known. It is worth noting that the structure factor is not changed from RDG to mean-field T-matrix theory.

4   RESULTS AND DISCUSSION

One of the consequences of the mean-field T-matrix theory is that the intensity of the scattered radiation is not anymore proportional to the square of the mass of the cluster, but to a power of the mass which depends on its fractal dimension and on its primary particle size. This implies that:

$$I(q) \sim i^{\alpha} P(R_p, q) S(q) \qquad (3)$$

where α is a number that can be as low a *2/3* in the case of very large spheres. For clusters with d_f up to 2.5 and particle size of 700nm, the zero angle scattered intensity is plotted in Figure 3. It can be clearly seen that the exponent is decreasing as the fractal dimension increases, going to 1.5 for d_f=2.5. This means that large clusters will contribute much less to the overall scattering in a solution compared to what RDG theory predicts.

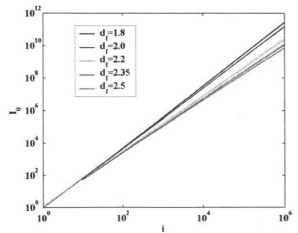

Figure 3. Zero angle scattered as a function of the cluster mass for cluster with different d_f values, made of 700nm polystyrene particles (incident light having a wavelength of 633nm)

In order to give a flavor of what can be done with the broad collection of structural and scattering properties we have available, we have chosen to treat a couple of sets of experimental SALS data of coagulating polymeric colloidal particles. The first case is that of a dispersion of particles undergoing DLCA aggregation. We have used our library of structural properties to invert SALS data collected from a sample of 70nm polystyrene particles in DLCA conditions (fractal dimension 1.8). The inversion of the data, in order to recover the cluster mass distribution, has been achieved using CONTIN algorithm. CONTIN [5] is a well known powerful algorithm for the solution of ill posed integral equations, such as the one arising from the inversion of scattering data. In order to provide a clear idea of the substantial advantage of our approach over conventional structural models, we have treated the same data using both the fractal model we have developed and the conventional one which assumes that the dispersion is made of a population of spheres, instead of a population of clusters. The results are shown in Figure 4. Figure 4a shows the experimental data and all the fittings, while Figure 4b shows the distributions. Clearly, using the scattering profiles of spheres leads to a totally meaningless distribution, whereas the use of fractal scattering profiles give a distribution quite close to the one predicted by Population Balance Equations [3]. Interestingly, all of the fittings of the experimental data are almost equivalent. This results show the importance of using the correct structural model to invert SALS data.

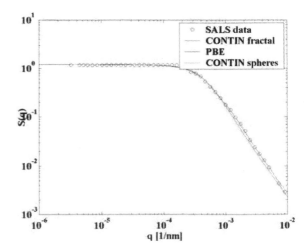

Figure 4a. SALS experimental data from DLCA cluster of 70nm particles (d_f =1.8). Fitting using distributions calculated from PBE, extracted using CONTIN and both fractal model and sphere model are shown.

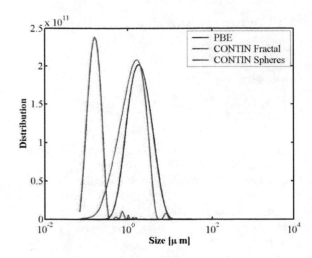

Figure 4b. Comparison of cluster mass distributions calculated and extracted from data shown in Figure 4a. PBE calculations results, and CONTIN results with both fractal model and sphere model.

5   CONCLUSIONS

We have shown in this work a methodology to better treat SALS data from cluster of particles. First of all, a library of cluster structures has been generated using a tunable cluster-cluster aggregation Monte-Carlo algorithm, which generates clusters with a fractal dimension up to d_f =2.5. In order to go beyond that limit, a densification algorithm has been proposed, which using a Voronoi tessellation of space to find the empty space inside the cluster and fill it. The structure of the fractal clusters is then characterized in terms of particle-particles correlation function. The correlation function is also the required information for computing scattering properties of clusters within the framework of mean-field T-matrix algorithm, which provides realistically into account the intra-cluster multiple scattering. These properties have been shown to be effectively used in combination with CONTIN algorithm to invert experimental SALS data of colloidal clusters in order to extract the cluster mass distribution.

REFERENCES

[1] R. Thouy, and R. Jullien, JOURNAL OF PHYSICS A, 27, 2953-2963, 1994.

[2] R. Botet, P. Rannou, and M. Cabane, APPLIED OPTICS, 36, 8791-8797, 1997.

[3] M. Lattuada, H. Wu, and M. Morbidelli, J Colloid Interface Sci, 268, 106-120, 2003.

[4] M. I. Mishchenko, L. D. Travis, and A. A. Lacis, "Scattering, Absorption, and Emission of Light by Small Particles", Cambridge University Press, Cambridge, 2002.

[5] S.W. Provencher, Computer Physics Comm, 27, 229-242, 1982

Synthesis of Janus Nanorods for Assemblies of Thermoresponsive Rod Pairs

Robert I. MacCuspie[*,**], Kyoungweon Park[*,**], Rajesh Naik[*], Richard A. Vaia[*]

[*]Air Force Research Lab, Nanostructured and Biological Materials Branch
2941 Hobson Way, Bldg 654, Wright-Patterson AFB, OH 45433, robert.maccuspie@wpafb.af.mil
[**]National Research Council, Washington, DC

ABSTRACT

Janus particles offer unique functionality; for example micron-scale Janus spheres are used for electronic paper. At the nanoscale, reproducible, scalable fabrication of Janus particles is one of the key enabling technologies for "bottom-up" assembly processes. For example, self-assembly of Au nanorod pairs created from Au Janus rods would provide a bottom-up, self-limiting assembly pathway alternative to current e-beam lithography methods. Commonly, Janus particles are synthesized using one of two masking concepts, either self masking resulting from directional vapor deposition or substrate masking to block non-directional modification chemistry. Using surface masks and non-directional chemistry, the ligand size was found to provide a limited range of tunability of the extent of asymmetric functionalization. This was found to agree well with geometrical-based theoretical predictions. The extent of asymmetry in the nanoparticle coverage was found to affect the protoassemblies of rods, where the greater the attractive coverage the larger the protoassemblies formed. This approach can potentially enable affordable fabrication of tangible quantities of thermoresponsive nanorods pairs for fundamental and application studies of gold nanorods pair spacing parameters and the effect on optical properties such as surface plasmon resonances.

Keywords: Janus, gold nanorods, self-limiting assembly, nanorod pairs

1 INTRODUCTION

The need for precisely engineered nanostructures is universal (electronics, medicine, materials, etc). Protoassemblies of smaller component engineered nanoparticles provides a key step to achieving a wider range of material properties for these applications[1-6]. For example, paired metal nanorods structures can be used to create magnetically inductive loops for novel optical properties[7-9]. To date, most paired structures have been formed using top-down approaches in a 2D fashion, such as those used to make gold rod pairs for negative index of refraction applications. Oil-water interfaces[10, 11] and surface masking are other approaches to perform an asymmetric ligand exchange[12] or asymmetric metal deposition[13] onto a microparticle or nanoparticle. Attempts to scale-up the reaction throughput using parallel

fluid flows have successfully created micron-sized Janus particles[14-16]. Ideally, a surface-masking approach is desired that can utilize a smaller 3D substrate such as colloids to increase the surface area to volume ratio available for surface masking reactions[17] while maintaining a nanoscale size of the particles being functionalized.

This work selects gold nanorods as the target Janus particle, with the goal of creating self-limiting self-assembling rod-pair structures that may potentially have optical properties, such as surface plasmon resonances, that depend on the spacing parameter between the rods. Previous work into nanorods assemblies[18-26] has shown the self-limiting rod-pair assembly to be elusive. The current challenge is to create a bottom-up assembly approach that enables self-limiting assembly of nanorod-pairs. By controlling the "sticky" surface areas of the nanorods, we propose to drive reactions towards increased rod pair yield and decreased random, amorphous aggregation. Creating Janus particles will require several key components to be combined successfully. The first will be the controlled and reversible attachment of the nanorods to a surface-masking substrate. The second will be addressing chemistry specifically to the unmasked nanorod surfaces. Additionally, high throughput reactions, and scalability of the reaction are important factors that should be desired.

2 RESULTS AND DISCUSSION

The reversible deposition and release of the nanorods is accomplished primarily through electrostatic interactions. The CTAB surfactant inherently on the nanorods has a native positive charge in aqueous solutions. Using metal oxide surfaces that have a slight negative charge, or polymer cation exchange resins, the rods are able to bind to the masking surface. The energy for the subsequent release of the nanorods will be provided by sonication.

It must be demonstrated that this process by itself will not damage the particles in any significant fashion. In Figure 1, rods are attached to a masking surface and imaged by SEM. XPS data shows the relative gold content on the masking substrate as increasing quantities of gold nanorods are incubated with the masking substrate.

The surface-masked nanorods can now have chemistry selectively performed only on the exposed surface of the nanorods. The grafting density of the ligand being exchanged onto the unmasked surfaces depends on the ratio of the nanorod diameter to the ligand's van der Walls diameter. For small organic molecules (< 16 carbons), and a 10nm diameter rod, approximately 78 to 89% of the nanorod surface is available for ligand exchange, based on simple geometric and trigonometric calculations.

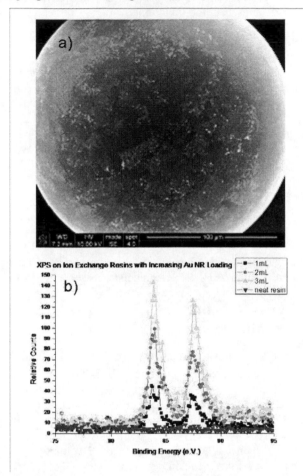

Figure 1: a) SEM of gold nanorods deposited onto a surface-masking colloid; b) XPS showing increasing Au absorbtion on surface as quantity of Au nanorods is increased.

Recent work by Ionita, et. al. using EPR experiments demonstrated that thiol molecules are not able to "walk" around laterally on the surface of a gold nanoparticle[27]. This suggests that once the thiol molecules are attached in a Janus fashion to the nanorods by the surface-masking technique, the nanorods will retain their asymmetric character of their new chemical coating.

A series of thiols with different chain lengths were examined for their grafting density on the surface-masked nanorods to explore this concept. Samples of the rods were then deposited onto fresh silicon wafers for XPS analysis. The atomic compositions were observed as well as the ratios of atoms in various bonding states such as C-C vs. C-S. Comparing these observations with predictions by the molecular formulae of the surfactants and the ligands exchanged, the relative stoichiometries of each on the surface of the nanorod could be estimated.

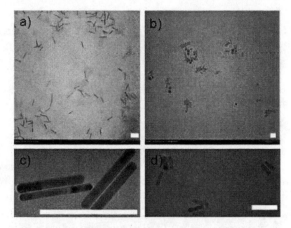

Figure 2. Equimolar rod concentrations of a) single nanorods and b) nanorod pairs, self-assembled from Janus nanorods made by surface masking methods; rod pairs from c) 3C and d)16C linkers; scale bars=100nm.

The series of Janus nanorod samples was then allowed to self-assemble into rod pairs based on either hydrophobic-hydrophilic interactions between the rods and solvent, or chemical cross-linking between terminal functional groups on the ligands. The resulting structures were imaged by TEM. Examples of chemical cross-linked nanorod pairs are shown in Fig 2b, and are in eqiumolar concentration to single nanorods shown in Fig 2a. Rod pairs were typically found to be at least 30% of the structures observed.

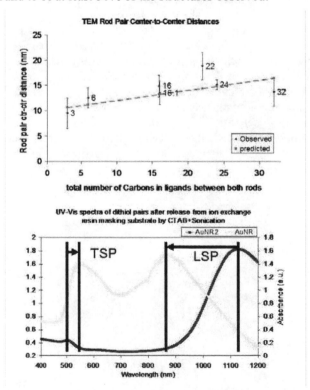

Figure 3. Spacing parameters between nanorods as a function of linker size.

The space between the nanorod pairs was then measured and plotted as a function of the size of the linkers in Figure 3. A prediction line is overlaid in Figure 3 representing the approximate length of the short-chain organic molecules in a purely trans conformation. The observed spacing results agreed well with the predicted values, demonstrating that a static series of various spacings between nanorods is achievable.

Also in Figure 3, the UV-Vis spectra of single rods is compared to the rod-pairs. The yield of rod-pairs released from the surface-masking substrate was found here to be about ten percent compared to the stock single nanorod solution. The peak around 520nm is from the transverse surface plasmon resonance (TSP), and is a function of the diameter of the gold in the structure. One would expect a red-shift in the TSP when the amount of gold in the structure doubles with the formation of rod pairs. The peak around 1100nm is from the longitudinal surface plasmon resonance (LSP), and is a function of the aspect ratio of the gold structure. Since higher aspect ratios lead to red-shifts in the LSP, one would expect a blue-shift in the LSP when a rod pair structure is formed, as the aspect ratio is now cut in half by the doubling of the width of the gold structure.

To expand the series of spacing parameters available, larger ligands were attached using the same EDC coupling chemistry. Amine-terminated molecules such as the coiled-coil peptides shown in Figure 4 were attached to a carboxyl-terminated thiol 90% Janus grafted nanorod. The 3-MPA had been attached to the nanorods in a Janus fashion with approximately 90% grafting density. The predicted size of the coiled-coil peptide is 2.45nm, and the observed coating thickness was 2.36nm. This experiment is a proof-of-concept that larger amine-containing molecules, such as biomolecules or polymers, can be successfully attached to gold nanorods in a Janus fashion for later use in self-limiting self-assembly of rod-pair structures. The use of amine-terminated thermo-responsive polymers could then be employed to create a dynamic array of rod pair spacing parameters.

Figure 4. Coiled-coil peptide coating gold nanorods with a 90% Janus grafting density.

3 CONCLUSIONS

In conclusion, using surface masks and non-directional chemistry, Janus nanorod functionalization was achieved. The ligand size was found to provide a limited range of tunability of the extent of asymmetric functionalization. A limited range of static tunability of rod-spacings was also provided by varying the ligand size. Finally, using a model system of a coiled-coil peptide attached to gold nanorods in a Janus fashion, proof of concept was provided that large linkers, such as thermoresponsive polymers containing amine groups, could be attached to gold nanorods in a Janus fashion, enabling future work into dynamic rod-pair spacing parameters.

MATERIALS AND METHODS

3.1 Au NR synthesis:

Au nanorods were grown by a seed-mediated, bi-surfactant method[28]. All chemicals were ordered from Sigma Aldrich and used as received.

3.2 Janus Rods by Ion Exchange Resin Negative Masking:

Au NR were absorbed onto DOWEX 50W-X8 resin for 24 hrs. The clear supernate was removed, and the resin was washed 3X with milliQ water. Next, the thiol ligand was reacted with the exposed nanorods surface for 24 hrs, at a stoichiometry of 2:1 thiol:surface gold atoms. The resin was washed 2X, then 0.1M CTAB solution was added. The resulting suspension was sonicated for 10 minutes to release the rods from the resin.

3.3 Amine Coupling to Janus Nanorods:

3-mercaptopropionic acid (MPA) was added by negative masking to gold nanorods. Before the Janus MPA rods were released, typically, 267uL of 2.5nM rod solution was added to 0.100g of resin. After removing the supernate and washing with milliQ water, a 1:250 dilution of MPA in water was added to the resin and reacted for 24 hrs. After another wash, 300uL of a 50ng/uL peptide solution was added to the resin with a fresh solution of 100uL of 10mM N-hydroxysuccinimide and 100uL of 75mM N-(3-Dimethylaminopropyl)-N'-ethylcarbodiimide hydrochloride (EDC). After another milliQ water wash, the rods were released using CTAB and sonication.

3.4 Instrumentation:

A Phillips 200CM Lab6 TEM was used at 200kV acceleration voltage and 400 mesh carbon-coated copper grids (Ted Pella). A Cary 500 UV-Vis spectrophotometer was used with a 1mm path length quartz cuvette. SEM images were obtained using a Quanta SEM. A Surface Science XPS was used; detector angle was 45 degrees, and samples were observed in multiple locations and rotated on the stage such that discrepancies arising from observation geometry could be averaged out. For image analysis, NIH Image J was used.

4 REFERENCES

1. Glotzer, S.C., et al., *Self-assembly of anisotropic tethered nanoparticle shape amphiphiles.* Current Opinion in Colloid & Interface Science, 2005. **10**(5): p. 287-295.

2. Horsch, M.A., Z. Zhang, and S.C. Glotzer, *Self-Assembly of Laterally-Tethered Nanorods.* Nano Lett., 2006. **6**(11): p. 2406-2413.

3. Mark, A.H., Z. Zhenli, and C.G. Sharon, *Simulation studies of self-assembly of end-tethered nanorods in solution and role of rod aspect ratio and tether length.* The Journal of Chemical Physics, 2006. **125**(18): p. 184903.

4. Carney, R.P., et al., *Size Limitations for the Formation of Ordered Striped Nanoparticles.* J. Am. Chem. Soc., 2007.

5. X. Gao, L.Y.R.M.H.M., *Controlled Growth of Se Nanoparticles on Ag Nanoparticles in Different Ratios.* Advanced Materials, 2005. **17**(4): p. 426-429.

6. Adeline Perro, et al., *Design and synthesis of Janus micro- and nanoparticles.* Journal of Materials Chemistry, 2005. **15**: p. 3745-3760.

7. Alexander V. Kildishev, et al., *Negative refractive index in optics of metal-dielectric composites.* JOSA B, 2006. **23**(3): p. 423-433.

8. Uday K. Chettiar, A.V.K., Thomas A. Klar†, and Vladimir M. Shalaev, *Negative index metamaterial combining magnetic resonators with metal films.* Optics Express, 2006. **14**(17): p. 7872-7877.

9. Govyadinov, A.A., V.A. Podolskiy, and M.A. Noginov, *Active metamaterials: sign of refraction index and gain-assisted dispersion management.* Physics.optics, 2007.

10. Jiang, S. and S. Granick, *Janus balance of amphiphilic colloidal particles.* The Journal of Chemical Physics, 2007. **127**(16): p. 161102-4.

11. Hong, L., S. Jiang, and S. Granick, *Simple Method to Produce Janus Colloidal Particles in Large Quantity.* Langmuir, 2006. **22**(23): p. 9495-9499.

12. Li, B. and C.Y. Li, *Immobilizing Au Nanoparticles with Polymer Single Crystals, Patterning and Asymmetric Functionalization.* J. Am. Chem. Soc., 2007. **129**(1): p. 12-13.

13. Lu, Y., et al., *Asymmetric Dimers Can Be Formed by Dewetting Half-Shells of Gold Deposited on the Surfaces of Spherical Oxide Colloids.* J. Am. Chem. Soc., 2003. **125**(42): p. 12724-12725.

14. T. Nisisako, T.T.T.T.Y.T., *Synthesis of Monodisperse Bicolored Janus Particles with Electrical Anisotropy Using a Microfluidic Co-Flow System.* Advanced Materials, 2006. **18**(9): p. 1152-1156.

15. Roh, K.-H., D.C. Martin, and J. Lahann, *Biphasic Janus particles with nanoscale anisotropy.* Nat Mater, 2005. **4**(10): p. 759-763.

16. Roh, K.H., M. Yoshida, and J. Lahann, *Water-Stable Biphasic Nanocolloids with Potential Use as Anisotropic Imaging Probes.* Langmuir, 2007. **23**(10): p. 5683-5688.

17. Adeline Perro, et al., *Towards large amounts of Janus nanoparticles through a protection–deprotection route.* Chemical Communications, 2005: p. 5542-5543.

18. C. J. Murphy, C.J.O., *Alignment of Gold Nanorods in Polymer Composites and on Polymer Surfaces.* Advanced Materials, 2005. **17**(18): p. 2173-2177.

19. Gole, A. and C.J. Murphy, *Biotin-Streptavidin-Induced Aggregation of Gold Nanorods: Tuning Rod-Rod Orientation.* Langmuir, 2005. **21**(23): p. 10756-10762.

20. Murphy, C.J., et al., *Anisotropic Metal Nanoparticles: Synthesis, Assembly, and Optical Applications.* J. Phys. Chem. B, 2005. **109**(29): p. 13857-13870.

21. Orendorff, C.J., P.L. Hankins, and C.J. Murphy, *pH-Triggered Assembly of Gold Nanorods.* Langmuir, 2005. **21**(5): p. 2022-2026.

22. Hu, X., et al., *Well-ordered end-to-end linkage of gold nanorods.* Nanotechnology, 2005. **16**(10): p. 2164-2169.

23. Pan, B., et al., *DNA-Templated Ordered Array of Gold Nanorods in One and Two Dimensions.* J. Phys. Chem. C, 2007.

24. Vial, S., et al., *Plasmon Coupling in Layer-by-Layer Assembled Gold Nanorod Films.* Langmuir, 2007. **23**(8): p. 4606-4611.

25. Pierrat, S., et al., *Self-Assembly of Small Gold Colloids with Functionalized Gold Nanorods.* Nano Lett., 2007. **7**(2): p. 259-263.

26. Pramod, P., S.T.S. Joseph, and K.G. Thomas, *Preferential End Functionalization of Au Nanorods through Electrostatic Interactions.* J. Am. Chem. Soc., 2007. **129**(21): p. 6712-6713.

27. Ionita, P., et al., *Lateral Diffusion of Thiol Ligands on the Surface of Au Nanoparticles: An Electron Paramagnetic Resonance Study.* Anal. Chem., 2008. **80**(1): p. 95-106.

28. Park, K., *Synthesis, Characterization, and Self – Assembly of Size Tunable Gold Nanorods,* in *School of Polymer, Textile and Fiber Engineering.* 2006, Georgia Institute of Technology: Atlanta, GA. p. 241.

Synthesis and Characterization of Novel ZnO/Whey Protein Nanocomposite

Liang Shi, Jinjin Zhou and Sundaram Gunasekaran[*]

[*]Department of Biological Systems Engineering
University of Wisconsin-Madison
460 Henry Mall, Madison,WI, 53706
guna@wisc.edu

ABSTRACT

Nanocrystalline zinc oxide (ZnO) particles coated with whey protein (WP) were synthesized in the weak basic aqueous solution condition at room temperature. The X-ray diffraction (XRD) and transmission electron microscopy (TEM) measurements confirmed the ZnO/WP nanoscaled composite structure. The average composite granules size was about 300 nm and the embedded ZnO nanoparticles were uniform and monodisperse with an average diameter of 65 nm

Keywords: ZnO, whey proteins, nanocomposite, synthesis

1 INTRODUCTION

Zinc (Zn) is one of the essential micronutrients and serves important and critical roles in human being growth, development, and well-being. Zn is essential to support child growth, lower the risk of common infections, prevent adverse outcomes of pregnancy, and improve other aspects of human health function. In addition, Zn is also required for the metabolic activity of numerous enzymes in our body and is considered essential for cell division and the synthesis of DNA and protein.

Nonetheless, Zn is deficient in the diet of many segments of the world population. Zn is the number-one nutritional deficiency in U.S. children based on a Tufts University study. More than 50 % of poor children and 30 % of non-poor children, ages 1 to 5, get less than 70 % of the recommended dietary allowance (RDA) of zinc [1-3]. Zn deficiency can be attributed to many reasons including inadequate intake, malabsorption, increased requirements and/or losses, and impaired utilization. Of these, inadequate dietary intake of absorbable Zn is the primary reason. Phytate, which is present in staple foods such as cereals and pulses, has a strong negative effect on Zn absorption from. Therefore, preventive and therapeutic interventions should be implemented in places where Zn deficiency is likely to be prevalent. If suitable Zn fortificants can be developed to successfully fortify staple foods, it will help alleviate Zn deficiency.

Zn is available in different forms that may be used for supplementation and fortification. The US Food and Drug Administration (FDA) has listed Zn chloride, Zn gluconate, Zn oxide, Zn sterate, and Zn sulfate as generally recognized as safe (GRAS). Among these, ZnO has well been used as foodstuff. It will decompose into Zn ions after going into human body, which were proven in medical papers as the indispensable elements for human body due to Zn deficiency syndrome [4]. Further more, ZnO is less expensive than other formation and most commonly used in food industry [5]. Wheat products fortified with ZnO have proven to possess good Zn absorption [6]. Therefore, ZnO appears to be the best choice for dietary Zn fortification.

Recently, nanotechniques have been applied in the food industry [7,8]. Nanoparticles improves the bioavailability of nutraceutical compounds, especially poorly soluble substances originated from their subcellular size.

Nanocrystalline ZnO is a very interesting material due to its excellent properties and promising applications in various fields including food industry. Protein-based nanoparticles are particularly interesting because they are relatively easy to prepare and their size distribution can be controlled [9]. Whey proteins (WPs), by-products of cheese manufacturing, are an excellent choice to serve as the delivery matrix for ZnO owning to their functionalities such as gelling, foaming and emulsifying capacity. Besides enhancing the status of these essential micronutrients, Whey protein encapsulated ZnO may also improve the protein content of the fortified diet.

On the other hand, hybrid inorganic-organic nanocomposite materials have currently attracted great interest of researchers because of their multifunctionality induced by combination of different compounds [10]. ZnO incorporated into nanocomposites, when added to foods, will be protected from interactions with food components that may impair their bioavailability and that may result in detrimental reactions. Since WP delivery and controlled release properties have been addressed [11], WP encapsulated nanocomposites may also survive the gastric environment and become available in the intestine and readily absorbed due to their nanoscale size. The incorporation of nanocrystalline ZnO into WPs to form nanocomposite may possess the unique functionalities and potential applications in material science and food sciences.

Herein we report the synthesis of WP-coated nanocrystalline ZnO particles via a simple method under mild basic condition. The formation mechanism of the ZnO/WP nanocomposite is also discussed.

2 EXPERIMENTAL SECTION

2.1 Materials

ZnO nanoparticle was purchased from Sigma-Aldrich (St. Louis, MO). Whey protein isolate (WPI, 98 wt% protein) was obtained from Davisco Foods International, Inc. (Een Praire, MN). Milli-Q water (resistivity 18.2 MΩ cm) was used for sample preparation. All other reagents are analytical grade and were used without further purification.

2.2 Preparation of ZnO/WP nanocomposite

0.05 M buffer tris(hydroxymethyl) aminomethane at pH 8 was prepared by dissolving 50 mL 0.1 M Trizma base and appropriate 0.1 M HCl aqueous solution, then diluted to 100 mL with distilled water. Appropriate amount of (24 mg) WP was added into 20 mL previously made buffer with stirring. After complete dissolution, WP Solutions was firstly heated at 80 °C for 30 min. After cooling to room temperature, 0.8 mL 0.5 M $Zn(NO_3)_2$ aqueous solution was added drop-wise into WP solutions under constant stirring. The reaction was proceed at 40 °C for 5 h after addition of $Zn(NO_3)_2$. Then, the obtained precipitate was centrifuged at 10,000 rpm for 10 min and collected, washed with deionized water several times to remove the byproducts. After drying in vacuum at 30 °C for 4 h, the final white powder product was obtained.

2.3 ZnO/WP nanocomposite characterization

The overall crystallinity of the product was examined by a powder X-ray diffraction unit (Scintag Pad V with a Ge solid-state detector; Cu Kα radiation) with the solid specimens mounted on a low background quartz holder. Detailed microstructure analysis was carried out using a transmission electron microscopy (TEM, PhilipsCM120). A particle size analyzer (90Plus, Brookhaven Instruments Corporation, New York, USA) was used to determine the granular average diameter distribution of ZnO/WP composite. The differential scanning calorimetry (DSC) was carried out on a TA Instruments Modulated DSC system. Measurements were conducted from 20 to 500 °C, at a heating rate of 10 °C min^{-1} under N_2 atmosphere.

3 RESULTS & DISCUSSION

The evidence for phase structure of the ZnO/WP composite was obtained by X-ray powder diffraction (XRD) pattern, as shown in **Fig. 1**. All the diffraction peaks marked with star can be indexed to those of hexagonal ZnO. After refinement, the lattice constants, $a = 3.251$ Å, $c = 5.210$ Å, are obtained, which was very close to the reported value for ZnO (a = 3.253Å, c = 5.209, JCPDS card, No.80-0075). The broadening of the ZnO XRD peaks suggests that the grain sizes were on nanometer scale. The average particles size was estimated to be 70 nm using the Scherrer equation [12]:

$$D = (K\lambda)/\beta(\cos \theta) \qquad (1)$$

Where, K is the shape factor of the average crystallite, λ is the wavelength for the Kα$_1$ (1.540 56 Å), β is the full width at half-maximum of the diffraction line and θ is the Bragg's angle. The rest XRD diffraction peaks should arise from WP.

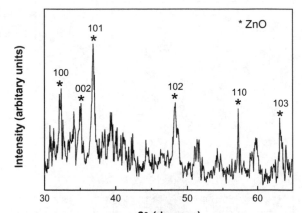

Fig. 1 XRD pattern of the synthesized ZnO/WP composite

The morphology of the ZnO/WP composite was investigated by TEM. **Fig. 2** shows a typical TEM image of the ZnO/WP composite. The dark part is ZnO and the pale part is WP. It clearly shows that we obtained a true composite of ZnO and WP, i.e., a WP granule of about 300 nm embedded with several ZnO nanoparticles. These nanoparticles are uniform and monodisperse with average diamtere of 65 nm, which correlates very well with XRD results (70 nm). The ZnO/WP particles were stable under the electron beam in vacuum used for TEM, suggesting a strong bond between ZnO and WP. The corresponding selected area electron diffraction (SAED) pattern of ZnO nanoparticle is shown in the inset of **Fig. 2**. It reveals bright spots, suggesting well-crystallized diffraction pattern of ZnO

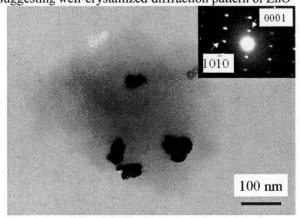

Fig. 2 TEM image of the ZnO nanoparticles embedded in the WP matrix; Inset: selected area electron diffraction pattern of ZnO nanoparticle.

particles. The SAED pattern can be indexed as a hexagonal crystal structure (wurtzite), which is in agreement with the XRD result.

The granular size distribution of ZnO/WP composite examined with a particle size analyzer is shown in **Fig. 3**, which indicates that the composite granules are uniform and the granular size distributes mainly at about 300 nm. This is consistent with the TEM observance result.

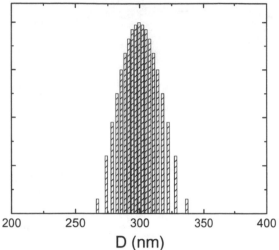

Fig. 3 Histogram of granular size distribution of ZnO/WP nanocomposite

For the UV-Vis absorbance measurement, the as-prepared powder sample was ultrasonically dispersed in distilled water before examination. The room temperature optical absorbance spectrum is shown in **Fig. 4**, which indicates that the sample is transparent in the visible region.

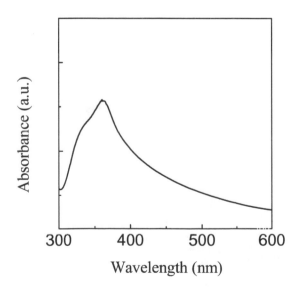

Fig. 4 UV-visible absorbance spectra of the ZnO/WP nanocomposite

A sharp absorbance peak located at about 362 nm correspond to the band gap of 3.42 eV. This is almost in accordance with the value of bulk ZnO [13], suggesting excellent crystal quality of the ZnO nanoparticles. So, no blue shift was observed in the UV-Vis absorbance spectrum, revealing the nanoscaled ZnO particles were not small enough to show quantum confinement related effects. An asymmetric tail can also be found on the peak shoulder side with higher wavelength, induced by light scattering. Since the nanocrystalline ZnO sample is not soluble in water, some incident light may be scattered by the dispersed particles due to inhomogeneous distribution.

To investigate the thermal properties of the samples, differential scanning calorimetry (DSC) of pure WP and as prepared ZnO/WP nanoscaled composite were carried out at nitrogen atmosphere, as shown in **Fig. 5**. The DSC curves indicate that both samples have a broad endothermic peak around 100 °C, which should be attributed to the loss of absorbed water. A small endothermic peak centered at 250 °C can also be observed for the two samples. This peak may arise from the loss of chemical bonging water. WP and ZnO/WP nanoscaled composite both show endothermic peak around 310 °C, which is associated with the decomposition of whey protein. However, the peak for ZnO/WP nanoscaled composite is much stronger and sharper than that of pure WP, revealing the decomposition of whey protein is remitted in some degree. This suggests there may exist some interactions, such as electrostatic attraction, hydrogen bonding or O-Zn-O bonding, between protein molecules of WP and ZnO. The stabilization of ZnO/WP nanoscaled composite under the electron beam for TEM measrurements confirms also the existence of the interaction between WP and ZnO. Similar behavior has been reported for polymer-inorganic composites [14].

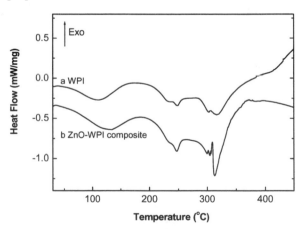

Fig. 5 DSC curves of WPI and Zn/WPI nanocomposite.

The formation mechanism of the ZnO/WP composite may be proposed as follows. During our synthesis, WP solution was first heated at 80 C for 30 min. Under this condition, WPs denature into soluble WP aggregates. This solution is adjusted to weak basic, PH = 8, with buffer. At this PH value, WP solution (the isoelectric point is 5.2 for β-lactoglobulin (BLG), the major component of whey proteins) is negatively charged. Therefore, electrostatic attraction

NSTI-Nanotech 2008, www.nsti.org, ISBN 978-1-4200-8503-7 Vol. 1

between the positively charged Zn^{2+}-water complex and the carboxylic groups of the negatively charged protein forms a composite. The WP may also coordinate Zn^{2+} ions via either primary amino groups or sulfhydryl group. The configuration of the whey proteins in the aggregate solution is that of random coil. These random coils behave as organic matrixes holding many hydrated Zn cations on the negatively charged amino acids. Deprotonation of these Zn aquo-complexes results because binding of water to Zinc increases the acidity of the water. The buffer consume the released protons, keeping the zinc complex hydrolyzing continuously. This deprotonation of water molecules associated with the Zn cations produces Zn hydroxides that presumably then lose additional protons to form ZnO. With the reaction continuing, the formed ZnO nanoparticles bind with whey proteins.

4   CONCLUSIONS

In summary, we report a facile process to synthesize ZnO/WP nanoscaled composite of about 300 nm. The ZnO/WP composite has been characterized by XRD, UV-Vis, DSC and TEM. TEM images show that typical sample is composed of uniform WP coated ZnO nanoparticles. The average size of ZnO nanoparticles is 65 nm. The TEM and DSC results indicate a strong binding between ZnO and WP.

REFERENCES

[1] Fanjiang G, Kleinman RE. 2007. Nutrition and performance in children. *Current Opinion in Clinical Nutrition and Metabolic Care* 10:342-7

[2] Sazawal S, Dhingra U, Deb S, Bhan MK, Menon VP, Black RE. 2007. Effect of zinc added to multi-vitamin supplementation containing low-dose vitamin A on plasma retinol level in children - A double-blind randomized, controlled trial. *Journal of Health Population and Nutrition* 25:62-6

[3] Yu HH, Shan YS, Lin PW. 2007. Zinc deficiency with acrodermatitis enteropathica-like eruption after pancreaticoduodenectomy. *Journal of the Formosan Medical Association* 106:864-8

[4] Smith JW, Tokach MD, Goodband RD, Nelssen JL, Richert BT. 1997. Effects of the interrelationship between zinc oxide and copper sulfate on growth performance of early-weaned pigs. *Journal of Animal Science* 75:1861-6

[5] Gibson RS, Ferguson EL. 1998. Assessment of dietary zinc in a population. *American Journal of Clinical Nutrition* 68:430s-4s

[6] de Romana DL, Salazar M, Hambidge KM, Penny ME, Peerson JM, et al. 2005. Longitudinal measurements of zinc absorption in Peruvian children consuming wheat products fortified with iron only or iron and 1 of 2 amounts of zinc. *American Journal of Clinical Nutrition* 81:637-47

[7] Arbos P, Arangoa MA, Campanero MA, Irache JM. 2002. Quantification of the bioadhesive properties of protein-coated PVM/MA nanoparticles. *International Journal of Pharmaceutics* 242:129-36

[8] Ko S, Gunasekaran S. 2006. Preparation of sub-100-nm beta-lactoglobulin (BLG) nanoparticles. *Journal of Microencapsulation* 23:887-98

[9] MacAdam AB, Shafi ZB, James SL, Marriott C, Martin GP. 1997. Preparation of hydrophobic and hydrophilic albumin microspheres and determination of surface carboxylic acid and amino residues. *International Journal of Pharmaceutics* 151:47-55

[10] Yao KX, Zeng HC. 2007. ZnO/PVP nanocomposite spheres with two hemispheres. *Journal of Physical Chemistry C* 111:13301-8

[11] Gunasekaran S, Xiao L, Eleya MMO. 2006. Whey protein concentrate hydrogels as bioactive carriers. *Journal of Applied Polymer Science* 99:2470-6

[12] Warren BE. 1969. *X-ray diffraction*. Reading: Wesley

[13] Yang PD, Yan HQ, Mao S, Russo R, Johnson J, et al. 2002. Controlled growth of ZnO nanowires and their optical properties. *Advanced Functional Materials* 12:323-31

[14] Sui XM, Shao CL, Liu YC. 2005. White-light emission of polyvinyl alcohol/ZnO hybrid nanofibers prepared by electrospinning. *Applied Physics Letters* 87:-

Ferrite-Silica-Insulin Nanocomposites (FeSINC) for Glucose Reduction

Neelam Dwivedi[*#], Arunagirinathan M.A[*], Somesh Sharma[#], Jayesh Bellare[*]

* Department of Chemical Engineering, Indian Institute of Technology Bombay, Mumbai-400076, India.
# Nicholas Piramal Research Centre, Mumbai-400063, India.
neelam@iitb.ac.in, agnathan@che.iitb.ac.in, someshs@nicholaspiramal.co.in, jb@iitb.ac.in

ABSTRACT

Proteins find more stable environment upon encapsulation in a silica host, because of polymeric silica frame that grows around the macromolecule and protects them from denaturation. Silica-insulin nanocomposite (SINC) and ferrite coated SINC (FeSINC) prepared by polyelectrolytic condensation of silica precursor on insulin were studied for their ability to control glucose levels. SINC was prepared by acid- base catalysed polymerization in presence of insulin at room temperature by modified Stober's process. FeSINC nanoparticles were prepared by co-precipitation of both ferric and ferrous salts on the bovine insulin loaded silica nanoparticle. The presence of ferrite coating in FeSINC was identified using VSM and quantified from XRF study. The intermolecular interactions in these nanocomposites were studied by FTIR and Raman spectroscopy. An *in vivo* study indicated that FeSINC was biologically active in reducing glucose levels as compared to SINC.

Keywords: silica, ferrite, insulin, nanocomposite.

1 INTRODUCTION

Formation of silica nanosphere in the presence of soft template is well known [1]. It is established that silica coating is chemically inert, biocompatible, hydrophilic and inexpensive. Here, we studied this inexpensive way to encapsulate protein in silica and silica-ferrite hosts as silica-insulin nanocomposite (SINC) and ferrite coated SINC (FeSINC) to avoid denaturation of insulin in the gastric environment. Silica precursor which has a negative charge binds to insulin that is polycationic at pH 2, thereby, forming silica insulin nanocomposite (SINC) with gradual increase in pH to neutral [2].

Insulin silica-ferrite nanocomposite was prepared by acid-base catalysed reaction. The maghemite ferrite coating above the silica prevents the extensive growth of polymeric silica framework thereby favoring the formation of FeSINC nanoparticles instead of large micrometer sized particles. The outer maghemite surface present in FeSINC is biocompatible and can be functionalized by transferrin receptor present in gut line of intestine. In vivo activity and physiochemical properties like microstructure, morphology, magnetic nature, encapsulation efficiency, interaction among the constituent of these SINC, FeSINC were studied using microscopy and spectroscopy. Parameters like pH, ionic strength and amount of protein were found to control the formation of ferrite-silica-insulin nanocomposite. In vivo studies were carried out to check bioactivity of formulation.

2 EXPERIMENTAL

2.1 Materials

Tetra ethyl orthosilicate and bovine insulin were obtained from Sigma-Aldrich. Ferrous chloride, ferrous sulphate, HCl, ammonia and ethanol of analytical grade obtained from local suppliers were used without further purification.

2.2 Synthesis of nanocomposites

SINC, FeSINC were synthesized by 2-step acid-base catalysed method. For SINC, acid catalysed hydrolysis of alkoxysilane precursors was carried out for 10min under stirring in ethanol at a mole ratio of 0.9:17:4 (TEOS: Ethanol: H2O). Later, the reaction mixture transformed into gel with increase in silica network formation as the pH was increased to 7 using ammonia. In order to obtain insulin loaded silica, insulin dissolved in HCl (0.001 N) was added before adding the catalyst. The concentration of insulin in silica sol was 0.4 mg/ml. Polymerisation of silica in presence of insulin was controlled by storing at $0^{\circ}C$. In FeSINC system, the silica framework was controlled by addition of aqueous ferrous sulphate and ferric chloride at a mole ratio of 0.04: 0.08 [3]. The magnetite initially formed over the silica network was transformed into maghemite by decreasing the pH from 7 to 2. The reaction mixture was stirred for 2hr, centrifuged and washed with distilled water. The resultant FeSINC material was lyophilized at $-60^{\circ}C$ and stored at $4^{\circ}C$.

2.3 Characterisation of nanocomposites

The synthesized materials were characterized by XRF (Model-PW2404, X-ray tube with Rh target, PANAnalytical, Netherlands), TEM (Technai-12, FEI), FTIR (Magna 550, Nicolet Instruments Corporation, USA), confocal Micro Raman spectrometer (Labram HR 800, Horiba Jobin Yvon). The magnetic nature of FeSINC was analysed by vibrating sample magnetometer VSM (Model 7410, Lakeshore USA). BET measurement (Model: ASAP 2020, Micromeritics) was carried to ascertain the porous nature of the nanocomposites. Insulin content was measured by HPLC (Waters, Photo diode array detector) using a gradient method.

TEM: Room temperature replica of samples were prepared by placing a drop of sample on a cover slip and allowed to dry at room temperature. Replicas of the sample were made with a 2nm Pt/C and 40nm carbon coating over it and the coated specimen was transferred to sodium hypochlorite and allowed to stand for 1hr to detach the replica from the specimen. Detached replica was further washed with D/W, lifted over bare carbon grid and examined under TEM.

XRF: Silica and ferrite present in FeSINC was quantified by XRF. The ferrite-silica and ferrite-silica-insulin nanocomposite were mixed with methyl cellulose and propanol, dried under IR light, made into pellet and observed by XRF spectrometer.

HPLC: Insulin content of the samples was analysed using HPLC–UV at 210 nm. Gradient elution was performed using 30% acetonitrile and 70% TFA (0.1%) at a flow-rate of 1ml/min and injection volume of 20 µl. Insulin was detected at a retention time of 5.7 min with detection limit of 0.01 mg/ ml. Encapsulation of insulin in SINC and FeSINC was around 55% and 20% respectively.

FTIR: Samples were mixed with KBr and compressed into discs at 20 KN force at room temperature. Bovine insulin powder, SINC, ferrite-silica, FeSINC was analysed.

Confocal Micro Raman Spectroscopy: Samples were focused using Olympus optical microscope with 20 X objective connected with Raman spectrometer. Ar laser, 514.5 nm, 20 mW, 6 A was used. Spectra were recorded as an average of 3 scans with a time span of 30 seconds each.

2.4 Biological activity of SINC and FeSINC in Wistar rats

The biological activity of the insulin encapsulated in SINC and FeSINC was tested in a rat model (Wistar rats, 200+/-25g, n=3) by measuring decrease in blood glucose levels. Animals fasted overnight (16hr) were anesthetized with isoflurane. Experimental animals received a subcutaneous injection of bovine insulin (4 IU/Kg) while the control animals with distilled water and test animals with 25 IU/Kg of insulin encapsulated in desired formulations. Blood samples were withdrawn at 0, 15, 30, 60, 120, 240min after subcutaneous injection. Plasma glucose levels were measured by glucose analyzer.

3   RESULTS AND DISCUSSION

SINC and FeSINC nanocomposites studied here were formed by supersaturation of anionic silanol species above the insulin molecule by polyelectrolytic condensation due to polycationic nature of insulin at pH 2. Initiation of silica growth above the surface of insulin occurs through hydrolysis of silica precursor at lower pH (2-5) that promotes colloid formation by release of alcohol followed by silanol condensation leading to gelation at pH 7. Presence of silica coating around insulin prevents it from denaturation due to pH and enzymes. Silica and SINC nanoparticles prepared were 100-200 nm (Fig. 1, 2) as observed under TEM.

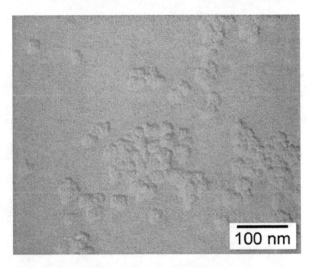

Figure 1: Room temperature replica TEM image of silica nanosphere.

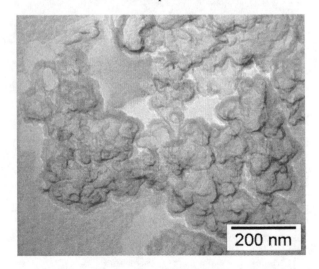

Figure 2: Room temperature replica TEM image of SINC nanocomposite.

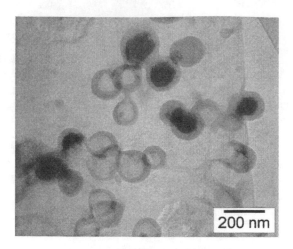

Figure 3: Room temperature replica TEM image of FeSINC nanocomposite.

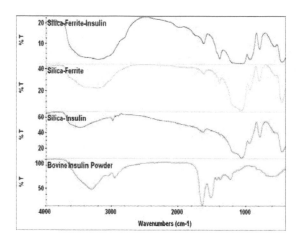

Figure 4: FTIR spectra of bovine insulin powder, SINC, silica-ferrite and FeSINC.

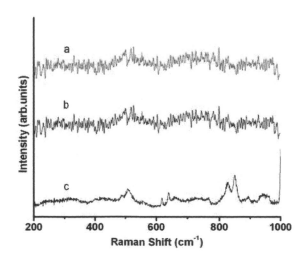

Figure 5: Confocal micro Raman spectra of (a) SINC, (b) FeSINC and (c) bovine insulin powder.

Fused structures of these nanoparticles are due to extended polymeric network of outer silica layer. The maghemite ferrite coating around the silica framework in FeSINC favors well dispersed spherical shape nanoparticles of less than 100 nm by disrupting the silica network with impregnated ferrite particles (Fig. 3). Insulin present in SINC and FeSINC was confirmed from infrared absorption around 3288 cm^{-1} (amide A band), 1644 cm^{-1} (amide I), 1514 cm^{-1} (amide II) and 1236 cm^{-1} (amide III), in comparison with standard bovine insulin by FTIR (Fig. 4). Silica coating in these insulin containing nanocomposite was evident from Si-O stretching at 1080 cm^{-1}, Si-OH at 950 cm^{-1}, Si-O-Si bending at 800 cm^{-1} and Si-O bending at 470 cm^{-1} [4]. The biologically active state of insulin in SINC and FeSINC nanocomposites was identified from Raman shifts due to S-S skeletal bending at 495,505, 517 cm^{-1} and C-S stretching at 668, 680 cm^{-1} by Confocal Micro Raman Spectroscopy in comparison with standard insulin (Fig. 5). Presence of disulphide stretching and amide stretching confirms the chemical stability of insulin encapsulated in SINC and FeSINC [5, 6, and 7].

The XRF results show the maghemite content to be 5% in ferrite silica and 2.5% in FeSINC. Presence of maghemite ferrite in FeSINC was evident from magnetic nature as shown in Fig. 6, 7. Super paramagnetic nature of FeSINC indicated that the ferrite present is less than 50 nm in size. From the magnetization plot, it is apparent that the ferrite content is less in FeSINC which correlates with XRF data. The synthesis protocol involves interaction of insulin and silica at the first stage during which most of the silanol species would have interacted with insulin. In presence of insulin very few ferric and ferrous ion interact with silanol species due to low availability of silanol host, thus leading to a decrease in ferrite content and low magnetic moment of FeSINC.

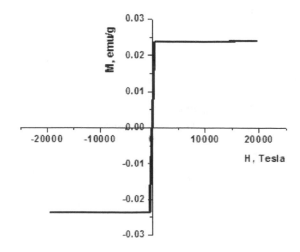

Figure 6: Magnetization of ferrite-silica nanocomposite.

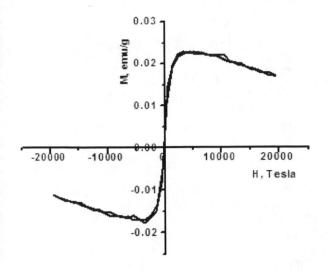

Figure 7: Magnetization of FeSINC nanocomposite.

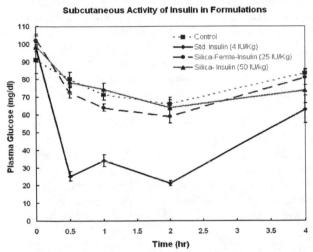

Figure 8: Subcutaneous activity of insulin encapsulated in nanocomposites compared with standard bovine insulin.

BET results indicated that ferrite-silica and FeSINC are porous while SINC is nanocomposites compared with standard bovine insulin.nonporous. Presence of ferrite favored loosely packed silica network by domain segregation thereby forming porous material. Room temperature processed silica xerogels are extensively used as carriers for the controlled release of enzymes, proteins and pharmaceutical substances [8]. Silica coating present in SINC and FESINC prevent it from rapid degradation in the gut lining. Release of insulin from ferrite-silica nanocomposite (FeSINC) may occur through both diffusion and dissolution through pores present in these nanoparticles. Therefore, chemical and structural characteristics of the silica xerogel strongly affect their drug release behavior [9]. The insulin monomer has multiple ionizable groups due to six amino acid residues capable of attaining positive charge and other 10 amino

acid residues capable of attaining negative charge [10]. This polyelectrolytic nature of insulin can be attributed for the entrapment of insulin in FeSINC. The decrease in blood glucose level was observed in test animals administered with FeSINC (25 IU/Kg) in contrast with SINC (50 IU/Kg). Although the insulin content was high in SINC, its ineffectiveness in glucose reduction can be due to its compact nature unlike porous in FeSINC. The porous nature of FeSINC helps to reduce the glucose level in blood by releasing the entrapped insulin (Fig. 8). Native state of insulin in FeSINC formulation is confirmed by its bioactivity in corroboration with spectroscopy. Thus the present formulation may be used for oral delivery of insulin.

4 CONCLUSION

Sol-gel derived silica is biocompatible and biodegradable inorganic carrier material. Its bioresorbability occurs by hydrolysis of siloxane bonds in human body and it is excreted via kidneys [11]. The FeSINC formulation described here was found to be effective in reducing glucose levels. Silica framework is a probable host for biological activity of proteins that is otherwise vulnerable for denaturation. The FeSINC formulation may be used for oral delivery of insulin which could enhance the absorption of nanoparticle in intestine through transferrin receptors.

REFERENCES

[1] Hentze, Srinivasa R. Raghavan, Craig A. McKelvey and Eric W. Kaler, Langmuir, 19, 1069-1074, 2003.

[2] J. Brange, L. Langkjoer, Pharmacol. Biotechnol. 5, 315–350, 1993.

[3] Xianqiao Liu, Jianmin Xing, Yueping Guan, Guobin Shan, Huizhou Liu, Colloids and Surfaces A: Physicochem. Eng. Aspects, 238, 127-131, 2004.

[4] Thibaud Coradin, Jacques Livage, Colloids and surfaces B: Biointerfaces, 21, 329-336, 2001.

[5] Nai-Teng Yu, C. S. Liu and D. C. O'Shea, Mol. Biol., 70, 117-132, 1972.

[6] Nai-Teng Yu, C. S. Liu, J. Culver and D. C. O'Shea, Biochim. Biophys. Acta, 263, 1-6, 1972.

[7] H. Fabian, P. Anzenbacher, Vib. Spectrosc., 4, 125–148, 1993.

[8] Bottcher H, slowik P, Sub W, J Sol Gel Sci Technology 13: 277-281, 1998.

[9] Dahlback B, Lancet, 335, 1627-1632, 2000

[10] Bruno Sarmento a, Domingos Ferreira a, Francisco Veiga b, Anto´nio Ribeiro c, Carbohydrate Polymers, 66, 1–7, 2006.

[11] Kortesuo P, Ahola M, Karlson S, Kangasniemi I, Kiesvaara J, Yli-Urpo AJ, Biomaterals 21: 193-198, 2001.

Particles as Protein Markers: Nanoscale Microscopy Towards Picoscale

B.D Shriniwas[*], R.Sharma[**], A.Sharma[***], C.J.Chen[**]

[*]Department of Zoology, RTM Nagpur University, Nagpur, Maharastra, India, bharuds312@gmail.com
[**]Center of Nanobiotechnology, Florida State University, Tallahassee, FL 32301 US
[***]Nanotechnology Division, Department of Electrical Engineering, MP A&T University,Udaipur, Rajasthan, India

ABSTRACT

We proposed the proteins-nanogel particles as potential nanogel adducts for their commercial value in measurement of enhanced metabolic energy mechanism by muscle contractile protein conformational changes at picoscale. Picoscale (at the down-side level of 10^{-12}) is the minutest limit of molecular detection till date. In muscle, tropomyosin-calcium bound nanogels trigger the conformational changes at picoscale (10^{-12} meters or 10^{-2} °A) and offer a promise of safe and rapid modality to increase the energy holding capacity of muscle tropomyosine-calcium protein assembly. Till date, electron microscopy and biophysical techniques can predict these submolecular physical dimensions without any information of metabolic energy mechanism or dynamicity. The proposed picoscale measurement of in vivo protein concentration and molecular dynamic events in *Heteropneustes fossilis* fish muscle may open a window to predict subphysiological, submolecular conformation to design hyperenergetic marine diets to get high quality fish food. The present report shows the emerging trend of picotechnology in bioengineering, *Heteropneustes fossilis* fish protein characterization by PAGE-gel electrophoresis, 10-100 nm nanoadducts by electron microscopy with possibility of nanogel-peptide(polyacylate-polyethylene-iron oxide) adducts as predictors of fish muscle proteins participating to enhance the capacity of tropomyosin-calcium hyperexcitation to measure it at picoscale in muscle tissue. In conclusion, the nanogel-protein adduct as bio-technology has tremendous commercial potentials to design marine diets, aquaculture, nanogel-drug carriers.

Key words: Tropomyosin-calcium, fish, protein, nanoparticles

1. INTRODUCTION

In fish muscle, the Ca^{2+} microdomains generated around the mouth of open ion channels represent the basic building blocks from which cytosolic Ca^{2+} signals in muscle tropomyosin are constructed. Recent improvements in optical imaging techniques using nanoparticles allow these calcium microdomains as single channel calcium fluorescence transients (SCCaFTs) to speculate channel properties and replace patch-clamp recordings. We report use of protein marker paramagnetic nanoparticles with a review of recent advances in single channel Ca^{2+} and other multimodal imaging by MRI, PET, total internal reflection fluorescence microscopy (TIRFM) as the technique of choice for recording SCCaFTs from voltage- and ligand-gated plasmalemmal ion channels. The old technique of 'optical patch-clamp possibly permits simultaneous imaging of hundreds of channels and provides millisecond resolution of calcium gating kinetics and sub-micron spatial resolution of channel locations; and seems applicable to evaluate the effect of seasonal changes on fish muscle membrane channels that display partial permeability to Ca^{2+} ions [1,2]. Picoscope has given hope of studying further intricacies of these molecules in muscle. To confirm the role of calcium and proteins in fish at different seasonal temperatures, we performed PAGE electrophoresis to distinguish several muscle protein candidates on scan in the electrophoretic mobility range of 0.5-2.5 eV/min.

2 PREPARATION OF NANOPARTICLES

A. Synthesis and Modification of Maghemite Nanoparticles

Iron oxide ($\gamma Fe_{2\alpha}$) particles of size 5-10 nm were synthesized using a three-step process of (i) coprecipitation of iron-oxide by mixing 0.1 M ferrous chloride and 0.1 M ferric chloride (1:2) with 10 M sodium hydroxide with continuous stirring for one hour at 20° C at 90°C, (ii) peptidization with 2M nitric acid, and (iii) sonication for 10 min. at 90° C. at 50% amplitude. For efficient binding with polyethylene, iron-oxide powder was stirred with 30% w/w sodium oleate at 1000 rpm speed for 2 hours. The mixture of iron oxide and sodium oleate was washed and dried under vacuum to yield fine 5-10 nm iron oxide particles.

B. Formation of the Polymer Composite Particles

A batch process was developed for preparing composite particles. A 0.05% w/w; 10 ml solution of polyethylene wax (number average MW of 700 g/mole (Honeywell Corp.) was prepared using decaline or octamethylcyclotetrasiloxane OMCTS (Dow Chemical Company) at 150°C. To this solution, iron oxide powder was added with 30%-50% w/w polyethylene, and sonicated at 50% amplitude for 30 seconds. Then, 10 ml of tetraglyme ("TG") (obtained from Sigma-Aldrich) was added to the iron-oxide polyethylene mixture at 150°C, and sonicated at 50% amplitude for about 30 seconds. Next, the mixture was immediately cooled to about 0°C in ice water held at 0°C. Within three to four minutes, the polyethylene-iron-oxide mixture transformed into an emulsion with microdroplets made of supercooled polyethylene wax solution and iron oxide dispersed in a continuous phase of nonsolvent. The emulsion was warmed to room temperature 27°C for 45 minutes to make polyethylene and maghemite particles suspended in the emulsion. This emulsion was cooled to -10°C for half an hour to form a macrophase separated system made of thin reddish-brown sandwiched between top solvent layer and non-solvent bottom layer. The reddish-brown layer of polyethylene/iron oxide particles was centrifuged to isolate the particles from solvent mixture as shown in Figure 1 and described elsewhere.

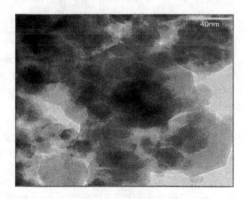

Figure 1: The 35 nm sized poly Ethylene – FeO nanoparticles are shown under Scan Electron Microscope (magnification x 10^5).

3. POLY AGAROSE GEL ELECTROPHORESIS (PAGE) OF FISH PROTEINS

Discontinuous SDS-PAGE analysis was performed under standard Laemmli conditions. The PEG–protein conjugates were loaded with an average of 7.5 mg (15 mL load volume) per well and run at 110 V constant voltage per gel. The gel dimension was $10.167.360.75 \text{ cm}^3$. The running buffer was 0.025 M Tris–0.19 M glycine–0.05% SDS. The gels were silver-stained for the protein portion. Native PAGE characterization of PEG–HAS conjugates and native PAGE analysis of the PEG–protein conjugates was carried out by loading samples with an average of 7.5 mg(15 mL load volume) per well and run at 110 V constant voltage per gel. The stacking gel, separatory gel, and the running buffer were prepared in the same way as the SDS-PAGE, except that no SDS was used. The gels were silver-stained for the protein portion.

4. RESULTS

SDS-PAGE has been a popular method to characterize PEG–proteins. The assumptions made were that the higher molecular weight gel band ladders represented one molar incremental additions of PEG to the protein. Most of the gel bands are smeared or blurred, especially for the PEG 5000. Three standard PEG 5000, 10000, 20000, were used as samples to compare the unknown fish proteins [3, 4]. At 110 eV, HS albumin, 5 kDa, 10 kDa, 20 kDa PEG gave specific spots as shown in Figure 2. However, major proteins were seen at locations of 31 kDa, 45 kDa, 65 kDa, 94 kDa as shown in Figure 2. The *Heteropneustes fossilis* fish muscle protein electrophoresis was used to identify different protein candidates based on their pI values as shown in Figure 2.

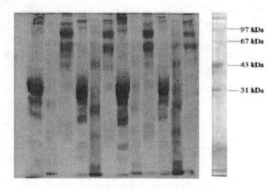

Figure 2: A representative PAGE elctrophoresis scan of fish muscle protein fractions. Notice the different proteins at electrophoretic mobility (pI locations).

5. NANOPROBES FOR MUSCLE PROTEINS AND PICOTECHNOLOGY

Recently, advancements are made in the field of protein markers and enhanced protein detection limits more and more accurate and with high sensitivity upto $1/100^{th}$ fraction (10 picomoles) of nanoscale.

For example the calcium regulatory behavior of calmodulin protein and calcium ion channel in single muscle cell can be observed up to the fraction of nanovolt or less. These outcomes are challenging and depend on the success of high precision and minute size of biosensor probes in use. We focus here on fluoro-sensitive green protein and phosphor sensitive immunospecific protein markers as shown in Table 1. These markers have uniqueness to generate images and measurement capability. The current trend is towards development of multimodal and multifunctional protein markers. Still the art is in infancy. However, the application of proteins and proteomics is expanding in marine biology, fishries, health science, cell biology and molecular biology.

Table 1: Muscle proteins and calcium-Magnesium regulatory markers are shown in muscle by different biochemical, physical, imaging and scanning techniques. Notice that some markers act as picogram level in action(shown by *).

Muscle Protein	Calcium/Mg	Technique used (reference)
Tropomyosin	Mg-Calmodulin	Biochemical (1, 11)
	Conformation	NMR (2)
	MRI Imaging	Nano-FluoroProbe(1)
	Optical Imaging	Nano-FluoroProbe(1)
	FRIT-imaging	Green Protein(1,2)
	Ca^{++} channel imaging	*SCCaFTs (1,2)
	Ca^{++} channel imaging	*TIRFM (1,2)
	Electrophoresis	2D-3D SDS/PAGE(3)
	Myosin conc	Biosensor (1)
Myosin	Mb4-myosin	Nano-Feo-AbMyo(1)
	Electrophoresis	2D-3D SDS/PAGE(3)
Actin	Ca-ATPase	Enzyme/histology(5)
	Filament Protein	*FluoroNanoProbe(2)

Table 2: Protein markers and protein probes in imaging applications of muscle proteins

Muscle Protein	Nanoprobe	Technique used (reference)
Tropomyosin	FETNIM-Imaging	MRI(15)
Myosin-Actin	FTIR Imaging	IR(15)
	CLIO-imaging	Nano-FeO Probe(15)
	MION- imaging	Nano-FeOProbe(15)
	Dendrimer imaging	Poly-nanoprobe(1)
	Cd-S imaging	*Quantum dots (1)
	Multimodal imaging	*Zn-Gd-Fe probe (15)
	Nanogel imaging	Polymer composits
	Mb4-FeO-CNT	Biosensor (15)
	Nanochips	Ion channel sensor(2)

6. EFFECT OF TEMPERATURE AND SEASON ON FISH MUSCLE PROTEIN AND CALCIUM CONTENT

Proteins with all other cellular constituents are in a state of continuous turnover. Protein turnover is offcourse a function of the rate of synthesis and the rate of degradation, both of which are under separate control (5).This phenomenon may significantly enhance the organism's ability to readily adapt to the changes in the environment (6). The energy associated with spawning is derived from liver (18%) somatic tissue (33%) and gonad (48%) for males in adult pacific cod (10). During maturation both protein synthesis (7, 13) and phosphorylation of proteins (5, 10) may occur. Fish muscle Protein content ranged from 15.9 to 16.5 percent during November2000 to March 2001. Maximum Protein content was recorded in summer months when water temperature was 23.5 degree centigrade (March 2001)and minimum proteins was in winter months i.e. December 2000 and January 2001 when the water temperature was 14.8 degrees Centigrade and 10.9 degree centigrade. It was also observed that in winter months the fishes consumed less feed which resulted in poor growth. Therefore, protein content was decreased in winter months. In summer months when the temperature started increasing, fishes also started taking more feed in comparison to winter months resulting in higher protein percentage. When the temperature became less than 15 degree centigrade the protein percentage decreased, while at higher temperatures of 20 degree centigrade and above the protein percentage have also increased. The protein content was significantly affected by the season in rohu. (3). Similar types of results were observed in cyprinus carpio. Variation in calcium content recorded in different months. Calcium content in rohu ranged from 620 to 650(mg/100g) in our experiments, from November 2000 to March 2001. Maximum calcium content was recorded in summer and minimum in winter. Like other contents it also increased and decreased with increasing and decreasing temperature. It may be due to the intake of feed in various months (3).Similar type of results were observed by Missima et al. (11).

7.CONCLUSION

In conclusion, the protein detection methods and current use of nanoprobes in muscle proteins is highlighted. The nanotechnology and its application in biology, health science and diagnostics is expanding as real-time, sensitive, monitoring rapid method. In marine biology, muscle protein characteristics play significant role in fish growth at optimized fish living water temperature conditions.

8. ACKNOWLEDGEMENTS

The authors appreciate the fish experiments at facility provided at RTM University of Nagpur for PAGE electrophoresis data. The nanoparticle preparation was done at Center of Nanobiotechnology, FSU.

9. REFERENCES

[1].Shuai J, Parker I. Optical single-channel recording by imaging Ca^{2+} flux through individual ion channels: theoretical considerations and limits to resolution Cell Calcium, 37, 4:283-299,2005.

[2].Demuro A, Parker I. Imaging single-channel calcium microdomains Cell Calcium 40(5-6):413-422,2006.

[3]. Zheng CY, Ma G, Su Z. Native PAGE eleminates the problem pf PEG-SDS interaction in SDS-PAGE and provides an alternative to HPLC in characterization of protein PEGylation. Electrophoresis 28:2801-2807,2007.

[4].Chun, S. Y., Billig, H., Tilly, J.L., Furuta, I., Tsafriri, A. and Hsueh, A. J. W. Gonadotropin suppression of apoptosis in cultured preovulatory follicles.. mediatory role of endogenous insulin-like growth factor I, Endocrinology, 135. 1845,1994.

[5].Goldberg, A.L. and Odessey, R. Regulation of protein and amino acid degradation in skeletal muscle. Excerpta.Med.Int.Congr. Scr., 333:187-199,1974.

[6]. Goldberg, A.L., and Dice, J.F. Intracellular protein degeneration in mammalian and bacterial cells.Anu.Rev.Biochem.43:835-869,1974.

[7].Gur,G.,Melamed,P.,Levavisivan,B.,Holland, C.,Gissis,A.,Bayer,D.,Elizur,A.,Zohar,Y.,and Yaron,Z.Long term testreone treatment stimulates GTH II synthesis and release in the pituitary of the black carp,Mylopharygodon piceus, In:Proc.5th Int.Symp. Reprod.Physiology.Fish, (eds.Goetz, F. and Thomas, P.), Fish Symp.95, Austin32,1995.

[8]. Laemmli.U.K.Clevage of structural protein during the assembly of head of bacteriophage T4, Nature, 227:680-685,1970.

[9].Lowry O.H.,Roserbrough,N.J., Farr, A.L.and Randall ,R.J.(1951).Protein measurement with the folin phenol reagent.J.Biol.Chem., 193: 265-275.

[10]. Maller, J.L. and Smith, D.S. (1985). Two-dimensional polyacrylamide gel analysis of changes in proteins phosphorylation during maturation of Xenopus oocyte. Develop. Biol., 109: 150-156.

[11]. Missima, T., Yokoyama TL, Yano K and Tsuchimoto K.The influence of rearing water temperature in the properties of Ca2+ and Mg2+ ATP activity on carp myofibril. Nippon Suison Gappaishi, 56: 477-487,1990.

[12] Chandra shekhar, A.P. Rao and A.B.Abidi. Changes in muscle biochemical composition of Labeo rohiota (Ham.) in relation to season, Indian J.Fish., 51(3):319-323,2004

[13]. Smith, R.L., Paul, A.J.and Paul, J.M. Seasonal changes in energy and the energy cost of spawning in gulf of Alaska Pacific Cod.J.Fish biol., 36:307-316, 1990.

[14]. Wasserman, P.M., Richter, J.D.and Smith, L.D.Protein synthesis during maturation promoting factor and progesterone induced maturation of Xenopus oocyte. Develop Biol., 89: 152-158, 1982.

[15]. Santra S, Bagwe RP, Dutta D, Stanley JT, walter GA, tan W, Moudgil BM, Mericle RA. Synthesis and characeterization of fluorescent, radio-opaque, and paramagnetic silica nanoparticles for multimodal bioimaging applications. Advanced Materials 17,2165-2169, 2005.

Convective heat transfer to Ag – and Cu – ethylene glycol nanofluids in the minichannel flow

Grzegorz Dzido[*], Klaudia Chmiel-Kurowska[*], Andrzej Gierczyki[*],
Andrzej B. Jarzębski[*,**]

* Faculty of Chemistry, Silesian University of Technology, Strzody 7, 44-100 Gliwice, Poland,
Andrzej.Jarzebski@polsl.pl
** Institute of Chemical Engineering, Polish Academy of Sciences, Bałtycka 5, 44-100 Gliwice, Poland

ABSTRACT

A paper deal with experimental investigations of The results of experimental studies on convective heat transfer in nanofluids made of Ag or Cu nanoparticles (0.15% and 0.25% vol.) and ethylene glycol are given. Nanoparticles were synthesized in the polyol process. Experiments were conducted at constant heat flux conditions. The 25% increase in the average heat transfer coefficient relative to that for pure ethylene glycol was obtained. Most significant increase was observed in the entrance region.

Keywords: nanofluid, enhanced heat transfer, minichannel flow

1 INTRODUCTION

Rapid progress in miniaturization of elctronic, optoelectronic or electro-mechanical equipment fuels a search for efficient cooling systems. In recent years many concepts were suggested to improve the cooling of miniaturized devices, e.g. using microchannel heat exchangers or microchannel heat sinks. However, the most common coolants such as water, oil or ethylene glycol possess poor thermal properties, i.e. the convective heat transfer and heat conductivity coefficient. It was found that the addition of solid particles with sizes less than 100 nm can significantly improve these properties, and the suspensions obtained behave as typical fluids. In this reason they are usually called nanofluids [1]. Studies on the practical application of nanofluids are still under development [2]. The experiments performed show that heat transfer in nanofluid is quite complex and it is strongly affected by such factors as: particle size, pH, micro-convection and particle-fluid interrelations. To elucidate these phenomena and to obtain their quantitative description further investigations on microchannel heat transfer are necessary [3,4].

There are scarce reports on convective heat transfer performance of nanofluids. Li and Xuan [5,6] investigated the convective heat transfer coefficient of Cu-water nanofluid in tubular laminar and turbulent flows at constant heat flux, to find that nanofluid displayed enhanced heat transfer behavior. For a given Reynolds number, the heat transfer coefficient was about 60% higher for the fluid with 2 vol.% Cu nanoparticles in comparison with pure water. In another experiment they found that the ratio of Nusselt number for nanofluid to that of pure water, under the same conditions, rose from 1.05 to 1.14 with increasing the volume fraction of nanoparticles from 0.5% to 1.2%, respectively.

Wen and Ding [7] studied heat transfer in laminar flow at constant wall heat flux for Al_2O_3-water system. They observed the increase of heat transfer coefficient with increase in Reynolds number and nanoparticles concentration, particularly at the entrance section and found that the thermal developing length was longer for nanofluid than for pure water. The enhanced heat transfer for nanofluids was explained by decrease in the thermal boundary layer thickness, induced by non-uniform distribution of thermal conductivity and viscosity, originating from the Brownian motion of nanoparticles. Opposite results were presented in the papers by Yang et al. [8] and Pak and Cho [9].

This paper presents the results of experimental studies on heat transfer to the Ag - and Cu - ethylene glycol nanofluids in the minichannel laminar flow regime.

2 SET-UP AND EXPERIMENTAL CONDITIONS

The set-up used for heat transfer experiments is shown in Fig. 1. Fluid from a reservoir (1) was pumped by a dosing pump (2), through a dumper (3) into a preliminary heater (4) and delivered to the test section (6), connected to the junction head (5) by means of insulating materials. The construction of both heads enabled to measure temperature of the fluid at both inlet (T_{in}) and outlet (T_{out}) and also pressure drop. A testing section was made of copper tube 500 mm in length and 1.4 mm I.D. with attached K-type thermocouples. Controlled heat flux was generated by means of the resistance wire wrapped on the testing tube and connected to the adjustable DC supply. Such a system provided constant heat flux, q, conditions. To minimize heat losses the whole section was covered with a plastic tube of 80 mm I.D. and heavily insulated using polyurethane foam. Additionally two control thermocouples were inserted radially into the insulating layer to estimate the amount of heat losses. From the testing section liquid was transported to a cooler (4) and finally back to the reservoir. The experiments were conducted in the laminar flow regime in the range of 30<Re<160.

In addition to temperature measurements at both ends of the test section, wall temperatures at five positions along the channel axis and temperature inside the insulation layer were also measured by means of K-type thermocouples. Additionally input heating power and volumetric flow rate of the working fluid were determined to check the thermal balances.

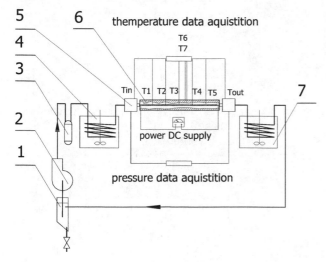

Figure 1: Experimental set-up; 1- reservoir, 2- pump, 3- pressure oscillation dumper, 4 – preliminary heater, 5- junction head for the tested section, 6 – test section, 7 - cooler.

3 PREPARATION OF NANOFLUID

Ag - or Cu - ethylene glycol nanofluid was prepared in the one-step polyol batch process ([10]). The prescribed amounts of Ag or Cu acetates and polyvinylpyrrolidone (PVP) used as capping agent were dissolved in ethylene glycol and then irradiated with a microwave (domestic microwave oven) at high power level. When a boiling point was approached the mixture was cooled for 60 seconds. Such sequence was repeated ten times. The nanofluid obtained remained stable for over a month. A typical particle size distributions (PSD) obtained using DLS method (Malvern Sizer S90) are presented in Figs. 2 and 3. An average diameter (by intensity) was equal to 325 and 136 nm respectively. One can expect that the size of bare particles should be smaller than those indicated by measurements in which sizes of particles (or their aggregates) are taken together with a layer of PVP. Therefore, the obtained diameter may be overestimated.

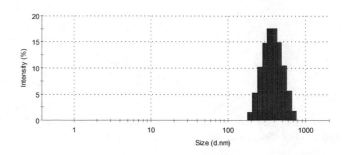

Figure 2: PSD of Cu particles in the nanofluid tested; Cu 0.25% vol. Cu:PVP 1:5 (mole).

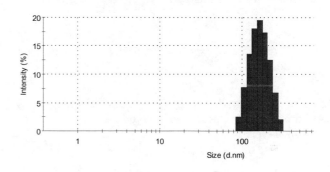

Figure 3: PSD of Ag particles in the nanofluid tested; Ag 0.15% vol. Ag:PVP 1:7 (mole).

4 DATA EVALUATION

Local Local values of heat transfer coefficient h_i at x_i position were calculated according to the Newton law:

$$h_i = \frac{q}{T_{wi} - T_{fi}} . \qquad (1)$$

where T_{wi} was calculated from the Fourier law for conduction in circular systems:

$$T_{wi} = T'_{wi} - \frac{Q}{2 \cdot k_{Cu} \cdot \pi \cdot x_i} \ln\left(\frac{d_o}{d_i}\right). \qquad (2)$$

Local liquid temperature was calculated assuming the linear distribution:

$$T_{fi} = T_{in} + \frac{\pi \cdot d_i \cdot q \cdot x_i}{m_f \cdot C_p} . \qquad (3)$$

The value of heat flux q was calculated on the base of energy conservation and the Ohm law from:

$$q = \frac{Q}{\pi d_o L}, \qquad (4)$$

where the value of heat transferred Q was calculated as a difference of heat generated in the resistive wire (Q_g) and lost to the surroundings (Q_l):

$$Q_g = \frac{U^2}{R} \qquad (5a)$$

$$Q_l = \frac{2 \cdot k_{is} \cdot \pi \cdot s \cdot (T_7 - T_6)}{\ln\left(\dfrac{r_7}{r_6}\right)} \quad . \tag{5b}$$

The global heat transfer coefficient was defined as:

$$h_{av} = \frac{q}{\overline{T_w} - \overline{T_f}} \tag{6}$$

where T_w is the mean temperature of five wall readings and T_f is the average liquid temperature.

5 PROPERTIES OF NANOFLUID

Density ρ_f and specific heat capacity C_f of nanofluid conventional were calculated using the most conventional formulas [8]:

$$\rho_f = (1 - \Phi_v)\rho_l + \Phi_v \rho_p \quad . \tag{7}$$

Similarly, the effective specific heat was calculated using the equation:

$$C_f = (1 - \Phi_v)C_l + \Phi_v C_p \quad . \tag{8}$$

Temperature dependence of nanofluid viscosity was approximated by means of empirical formulas of polynomial type:

$$\eta_f = aT^2 + bT + c \quad . \tag{9}$$

A viscosity values were measured using Brookfield DV II+ viscosymetr. Coefficients a, b and c are presented in table 1.

medium	$a \cdot 10^5$	$b \cdot 10^3$	c
ethylene glycol	0.499	-3.483	0,613
Ag – ethylene glycol, ϕ_v=0.15%	3.400	-23.408	4.052
Cu - ethylene glycol, ϕ_v=0.15%	0.977	-6.941	1.244
Cu – ethylene glycol, ϕ_v=0.25%	10.896	-73.848	12.576

Table 1: Coefficients of empirical equation (9).

6 RESULTS AND DISCUSSION

The experiments showed significant difference between the values of heat transfer coefficient (HTC) determined for the host liquid (ethylene glycol) and the Ag - and Cu - based nanofluids. This can be seen from the values of mean HTCs plotted against the Reynolds number (Figs. 4, 5). An average excess of hav value over that for a host liquid is about 25%. Experiments were conducted at constant heat fluxes in the range of 795 - 3190 W/m². Approximate contents of Ag and Cu particles were 0.15% vol. and 0.15-0.25% vol., respectively. The experiments with nanofluids were performed for the Reynolds numbers lower than those with pure ethylene glycol. This was caused by the presence of capping fluid (PVP) which significantly increased viscosity of nanofluid with respect to the host liquid. Thus

the increase in h_{av} may only be explained by the presence of fine metallic particles.

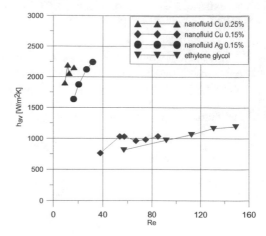

Figure 4: Average heat transfer coefficient vs. Re number for different nanofluids and ethylene glycol at q=1400W/m².

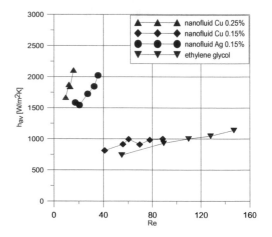

Figure 5: Average heat transfer coefficient vs. Re number for different nanofluids and ethylene glycol at q=1900W/m².

It can be seen from Figs. 6 - 7 that for in the case under study the values of h_{av} depend on moderately the applied heat flux. Distributions of the local h_i values along the test tube are shown in Fig. 8. It can be observed that the main part of excess in the heat transfer coefficient for nanofluid over the value for the host liquid is in the entrance region of the tested section and decreases with the axial distance. It could be caused partially by disturbances at the same inlet from the junction head to the tube. This behavior can be explained by a larger value of heat conduction coefficient that strongly affects the convective heat transfer. But nanoparticles themselves also affect the boundary layer thickness [11]. The exact mechanism of the thermal resistance reduction in this region is still not entirely clear.

NSTI-Nanotech 2008, www.nsti.org, ISBN 978-1-4200-8503-7 Vol. 1

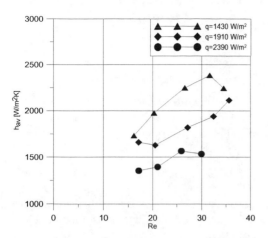

Figure 6: Average heat transfer coefficient vs. Re number for 0.15 % Ag - ethylene glycol nanofluid.

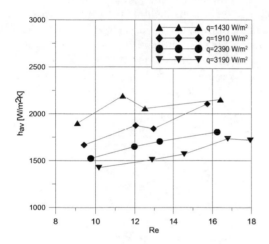

Figure 7: Average heat transfer coefficient vs. Re number for 0.25 % Cu-ethylene glycol nanofluid.

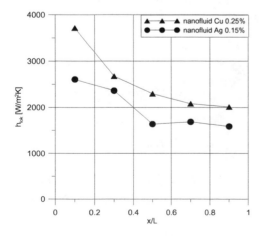

Figure 8: Variation of local heat transfer coefficient (Re=20, q = 1400 W/m^2).

7 CONCLUSIONS

Application of copper or silver nanoparticles dispersed in ethylene glycol can significantly enhance convective heat transport in a laminar flow regime. This effect is particularly notable in the entrance region and decreases with a distance from the inlet. Relatively high viscosities of the current nanofluids may seriously limit their widespread applications. Further investigations should address this issue and be directed to finding alternative capping agents.

Acknowledgments

The support of the Polish Ministry for Science and Education for this work under Grant R14 026 02 is gratefully acknowledged.

REFERENCES

[1] S. U. S. Choi, Two are better than one in nanofluids, presented at the Colloquium on Micro/Nano Thermal Engineering, Seoul National University, Seoul, Korea, February, 17, 2002.

[2] V.Trsaksri, S. Wongwises, Critical review of heat transfer characteristics of nanofluids, Renewable and Sustainable Energy Reviews, 11, 512, 2007.

[3] X-Q Wang, A.S. Mujumdar, Heat transfer characteristic of nanofluids – a review, Int. J. of Thermal Sciences, 46, 1, 2007.

[4] G. Dzido, K. Chmiel-Kurowska, A. Gierczycki, A. Jarzębski, Suspensions of nanoparticles. New properties and applications, Przemysl Chemiczny 86(12), 1217, 2007.

[5] Y. Xuan , Q. Li, Heat transfer enhancement of nanofluids, Int. J. Heat Fluid Flow, 21, 58, 2000.

[6] Y.Xuan, Q.Li Investigation on convective heat transfer and flow features of nanofluids, J. of Heat Transfer 125, 151, 2003.

[7] D.Wen, Y.Ding, Formulation of nanofluids for natural convective heat Transfer applications, Int. J. of Heat and Fluid Flow 26, (6), 855, 2005.

[8] Y. Yang, Z.G. Zhang, E.A. Grulke, W.B. Anderson, G. Wu, Heat transfer properties of nanoparticle-in-fluid dispersions (nanofluids) in laminar flow, Int. J. of Heat and Mass Transfer 48,(6), 1107, 2005.

[9] B.Pak, Y.Cho, Hydrodynamic and heat transfer study of dispersed fluids with submicron metallic oxide particles, Experimental Heat Transfer 11(2), 151, 1998.

[10] M. Figlarz, F. Fievet, J. P. Lagier, US Patent 4539041 USA, Sep. 3, 1985.

[11] D. Wen, Y. Ding, Experimental investigation into convective heat transfer of nanofluid at the entrance region under laminar flow conditions, Int. J. of Heat and Mass Transfer 47, 5181, 2004.

Process Intensification Strategies for the Synthesis of Superparamagnetic Nanoparticles and Fabrication of Nano-Hybrid

Suk Fun Chin[1], *K. Swaminathan Iyer*[1], *Colin L. Raston*[1] and *Martin Saunders*[2]

[1] Center for Strategic Nano-Fabrication, School of Biomedical, Biomolecular and Chemical Sciences,The University of Western Australia, Crawley, 6009 Australia

[2]Center for Microscopy, Characterization and Analysis, The University of Western Australia, Crawley, W.A. 6009   Australia

ABSTRACT

Continuous flow spinning disc processing (SDP), which has extremely rapid mixing under plug flow conditions, effective heat and mass transfer, allowing high throughput with low wastage solvent efficiency, is effective in gaining access to superparamagnetic Fe_3O_4 nanoparticles at room temperature. These are formed by passing ammonia gas over a thin aqueous film of $Fe^{2+/3+}$ which is introduced through a jet feed close to the centre of a rapidly rotating disc (500 to 2500 rpm), the particle size being controlled with a narrow size distribution over the range 5 nm to 10 nm, and the material having very high saturation magnetizations, in the range 68–78 emu g^{-1}. SDP also shown to be effective for fabrication of superparamagnetic carbon nanotubes composite. Ultra fine (2-3 nm) magnetite (Fe_3O_4) nanoparticles were uniformly deposited on single-walled carbon nanotubes (SWCNTs) *in situ* by modified chemical precipitation method using SDP in aqueous media at room temperature under continuous flow condition

Keywords: Spinning disc processor , superparamagnetic, continuous flow technology, magnetic nanoparticles, carbon nanutubes

1   INTRODUCTION

Traditional fluid based synthesis techniques for the production of nanoparticles have inherent limitations such as poor particle size distribution and reproducibility, and difficulties in scalability for commercial production. Process intensification, by means of spinning disc processing (SDP), potentially offers an avenue for the production of monodisperse nanoparticles with tuneable and controllable properties. SDP, Figure 1, is a rapid flash nano-fabrication technique with all reagents being treated in the same way, and is in contrast to traditional batch technology where conditions can vary across the dimensions of the vessel.[1] The reagents are directed towards the centre of the disc, which is rotated rapidly (300 and 3000 rpm) resulting in the generation of a very thin fluid film (1 to 200 µm). The thinness of the fluid layer and the large contact area between it and the disc surface facilitates very effective heat and mass transfer. The drag forces between the moving fluid layer and the disc surface enable very efficient and rapid mixing. The greatest strength of SDP synthesis is the broad range of control possible over all the operating parameters involved in nanoparticle formation, enabling the simultaneous and individual optimization of many interdependent operating mechanisms, with the ultimate goal of achieving very narrow particle size distributions.[2] Another feature of SDP is that it is continuous flow, readily allowing scale-up of the ensuing product formation.

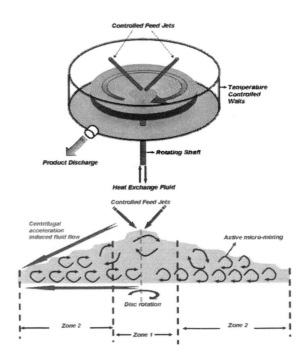

Figure 1. (a) Schematic representation of a SDP, (b) Hydrodynamics of the fluid flow over a spinning surface.

The most common cost effective and convenient way to synthesize Fe_3O_4 nanoparticles is by co-precipitating ferrous and ferric salt solutions with a base, such as

NSTI-Nanotech 2008, www.nsti.org, ISBN 978-1-4200-8503-7 Vol. 1

aqueous NaOH or NH$_4$OH. [3] However, the size distribution of the Fe$_3$O$_4$ nanoparticles produced using this method is normally very broad. Consequently, the downstream purification and isolation process is more expensive and is time and energy intensive. Furthermore, scale-up of this method using conventional reactors can be problematic given the inhomogeneous agitation and areas of localized pH variations, resulting in the precipitation of non-magnetic iron oxides.[4] Herein we demonstrate the successful synthesis of Fe$_3$O$_4$ nanoparticles via co-precipitation using NH$_3$ gas as a base source using spinning disc processing (SDP) under scalable and continuous flow conditions. To our knowledge, this is the first use of NH$_3$ gas as a precipitating agent to make Fe$_3$O$_4$ nanoparticles in a thin fluid film. The technology offers a realistic route towards large scale synthesis of Fe$_3$O$_4$ nanoparticles with precise control within the 10 nm size range.

The efficient SDP capability of fabricating magnetic nanoparticles decorated single wall carbon nano-tubes (SWCNTs) were also demonstrated. A novel yet simple method to coat superparamagnetic Fe$_3$O$_4$ nano-particles of narrow size distribution on SWCNTs *in situ* by modified chemical precipitation method using SDP in aqueous media at room temperature under continuous flow conditions was reported in this paper.

2 EXPERIMENTAL

2.1 Synthesis of Fe$_3$O$_4$ nanoparticles

In a typical synthesis, aqueous solutions of Fe$^{2+/3+}$ precursors were prepared by dissolving FeCl$_2$.4H$_2$O (10 mM) and (20 mM) FeCl$_3$.6H$_2$O (1:2 molar ratios) in deoxygenated ultrapure Mili-Q water. The SDP was a Protensive 100 series with integrated feed pumps to direct the reactants onto the rotating disc. The above solutions were delivered onto the disc surface using one feed jet at 1.0 mls^{-1}, using continuous flow gear pumps (MicroPumps). Grooved and smooth stainless steel discs with 100 mm diameter were used which were manufactured from 316 stainless steel with the grooved disc having 80 concentric engineered grooves equally spaced at 0.6 mm depth.

The reactor chamber was purged with argon gas before the reaction to remove oxygen. Ammonia gas was then fed into the sealed reactor chamber at a constant flow rate. Black suspensions of Fe$_3$O$_4$ nanoparticles were collected from beneath the disc through an exit port.

The samples collected were immobilized with a permanent magnet and supernatant solutions were decanted. Samples were re-dispersed in deoxygenated ultrapure Mili-Q water.

2.2 Fabrication of Fe$_3$O$_4$ decorated Carbon Nanotubes (CNT)

For purification and functionalization of SWCNTs, 10 mg of SWCNTs were dispersed in 5 ml of 1: 1 mixture of 70% HNO$_3$ and 98% H$_2$SO$_4$ aqueous solutions in the reaction chamber. The reaction was carried out in a CEM Focused Microwave Synthesis System, Discover Model. The microwave power was set at 300 W, and the pressure was 12 bar, and the temperature was set at 130 °C for 30 minutes. After the reaction, the SWCNTs were filtered, washed and re-dispersed in 100 mL of ultra pure Milli-Q water and sonicated for 15 minutes. The functionalized SWCNTs were dispersed in water and the container purged with N$_2$ gas to remove oxygen then 10 mM of FeCl$_2$.4H$_2$O and 20 mM of FeCl$_3$.6H$_2$O (1:2 molar ratios) was added and the mixture stirred for 1 hour. After that, the solution was filtered to remove excess Fe$^{2+/3+}$ and the resulting carbon nano-tube and Fe$^{2+/3+}$ complex was re-dispersed in deoxygenated Mili-Q water.

Solutions/suspensions of CNTs and Fe$^{2+/3+}$ were fed from one feed and the deoxygenated NH$_4$OH aqueous solution was fed from the other under an atmosphere of high purity (99.9%, BOC Gasses) argon gas, within the sealed reactor chamber. The disc surface was manufactured from 316 stainless steel. The 10 cm grooved disc was used for the current study, with 80 concentric engineered grooves equally spaced approximately 0.6 mm in height. Samples were collected from beneath the disc through an exit port.

3 RESULTS AND DISCUSSION

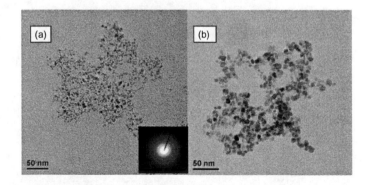

Figure 2. TEM images of Fe$_3$O$_4$ nanoparticles (10 mM Fe^{2+}) synthesized at 25 °C on grooved disc with disc rotating speed: (a) 2500 rpm and (b) 500 rpm

The operating parameters of the SDP were effective in controlling the size, size distribution and shape of the Fe$_3$O$_4$ nanoparticles in the presence of NH$_3$ gas feed as the base rather than NH_4OH aqueous solution. In a typical synthesis, the samples were synthesized using 10 mM of Fe^{2+} and 20 mM of Fe^{3+} aqueous solution and were fed at $1mls^{-1}$ onto the rotating grooved disc. A high rotation speed of 2500 rpm resulted in ultrafine (3-5 nm) particles with asymmetrical shape. A lower speed of 500 rpm, resulted in spherical nanoparticles of around 10 nm in size, Figure 2.

The fluid layer of the SDP has a plug flow characteristic, where the particles produced are constantly being removed from the nucleation and growth zones by radial propagation across the disc. At high rotation speeds the non-linear wave regime is dominant in the fluid film, which in turn ensures greater gas adsorption into the thin film. Consequently the nucleation process becomes dominant, thereby resulting in a number of ultra-small magnetic nanoparticles. At lower spin speeds the wave regime is no longer predominant, and the velocity of the traversing waves are highly reduced, and so is the absorption of the corresponding NH_3 gas into the flowing film. The number of nucleation sites is thereby decreased and the growth process becomes dominant at lower speeds resulting in the formation of larger particles.

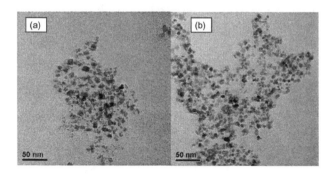

Figure 3. TEM images of Fe_3O_4 nanoparticles (10 mM Fe^{2+}) synthesized on smooth disc with disc rotating speed of 500 rpm, at (a) 25 °C and (b) 120 °C.

At constant speed, samples synthesized using the grooved disc had a narrower particle size distribution compared to the smooth disc. The particle size for the samples synthesized on the grooved disc ranged from 3-5 nm (Figure 2a), whereas the particle size distribution of a sample prepared using the smooth disc ranged from 3 -12 nm (Figure 3a). When using the grooved disc, the shear forces and viscous drag between the moving fluid layer and the periodic grooved surface give rise to more efficient turbulent mixing within the fluid layer and this in turn results in homogeneous reaction conditions in the thin film. The grooves on the disc also result in the formation of waves on the flowing film, thereby amplifying the amount of the NH_3 gas adsorbed. However, on a smooth disc the micro-mixing is not as efficient resulting in a broader size

distribution. Increasing the temperature of the disc offers scope to overcome the aforementioned size distribution issues associated with the smooth disc. At higher temperature the absorption efficiency of the gas is reduced in the flowing fluids, however the growth conditions become highly amplified [5]. As a result, the size distribution of the samples prepared on the smooth disc became very narrow with particle sizes in the range 8-10 nm for the reactions at 120 °C, Figure 3(b).

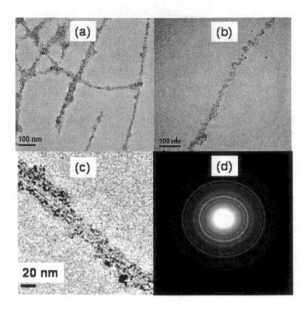

Figure 4. TEMs of Fe_3O_4–SWCNT composite at (a and b) low resolution; (c) high resolution and (d) associated SAED pattern for material synthesized using SDP

For the fabrication of Fe_3O_4-SWCNT composite, ultrafine (2-3nm) of Fe_3O_4 nano-particles with very narrow size distribution were observed to be uniformly coated onto the CNTs surface, Fig 4 (a), (b) and (c). As can be seen from the TEM images, the distribution of Fe_3O_4 nano-particles on the SMWNTs surface is very uniform and no local aggregation is observed with a very high coverage density. The ability to effectively decorate the SWCNTs with Fe_3O_4 nano-particles relates to the functionalisation of the nano-tubes with COO^- moieties that can bind directly to $Fe^{2+/3+}$ ions. [6] The use of this simple yet strategic approach reduces the down-stream processing dramatically. This is associated with purification/separation steps to remove excess nanoparticles that grow independently in solution and are not associated with the CNTs.

4   CONCLUSION

In conclusion, the particle size of Fe_3O_4 nanoparticles

can be controlled by judicious choice of the operating parameters of SDP technology. The capabilities of SDP in controlling particle size and stability have been clearly demonstrated with the ability to produce ultra-small superparamagnetic magnetite nanoparticles with high saturation magnetizations, and with narrow size distributions. We have also developed a novel, simple, rapid, cost effective and scalable method to decorate CNTs with 2-3 nm Fe_3O_4 nano-particles using SDP. The resulting CNTs show high coverage density. The apporach also avoids the need for subsequent purification and separation of the composite material from excess Fe_3O_4 nano-particles formed in solution, unlike in the case of using batch reactions. Futhermore the ability of SDP to fabricate superparamagnetic SWCNT composites under continuous flow in scalable quantities is significant in any down stream applications.

REFERENCES

1. P. Oxley, C. Brechtelsbauer, F. Ricard, N. Lewis and C. Ramshaw, *Ind. Eng. Chem. Res.* 39, 2175, 2000.
2. K. Swaminathan Iyer, C. L. Raston and M. Saunders, *Lab Chip*, 7, 1800, 2007.
3. T. Fried, G. Shemer, G. Markovich, *Adv. Mater.* 13, 1158, 2001
4. L. T. Vatta, R. D. Sanderson, K. R. Koch, *J. Magn. Magn. Mater.* 311, 114, 2007
5. J. H. Wu, S. P. Ko, H. L. Liu, S. S. Kim, J. S. Ju, Y. K. Kim, *Mater. Lett.* 61, 3124, 2007
6. B. He, M. Wang, W. Sun, Z. Shen. *Mater. Chem. Phys.* 95, 289, 2006

Conductance Enhancement of Nano-Particulate Indium Tin Oxide Layers Fabricated by Printing Technique

M. Gross, I. Maksimenko, H. Faber, N. Linse, P. J. Wellmann

University of Erlangen-Nuremberg, Materials Department 6, Martensstr. 7, 91058 Erlangen, Germany

ABSTRACT

To improve the conductance of nano particulate indium tin oxide (In_2O_3:Sn, ITO) layers we have applied a number of treatments: post bake, infiltration by an ITO precursor solution with subsequent sol-gel transformation into ITO, treatment with etching agents as well as annealing in vacuum. We obtained a maximum conductance of 120 $\Omega^{-1}cm^{-1}$ by combining the precursor infiltration technique with vacuum annealing. Considering the layer thickness of 5.2 μm, this value corresponds to a sheet resistance of about 16 $\Omega/\square$ at a maximum transmittance of 83 %. These nano particulate layers should be applicable as electrodes in optoelectronics.

To understand the mechanisms behind the conductance improvement and to fathom the limits we did some investigations: long-term resistance measurements gave an insight in oxygen vacancy balance concentration. From infrared absorption measurements we inferred the free charge carrier density and at present we carry out XPS measurements which yield information about the alteration of oxygen vacancy concentration. Considering these data we discuss two conductance improving mechanisms: The increase of the free charge carrier density by a rise of oxygen vacancy concentration as well as a release of trapped charge carriers from the aggregate surfaces.

Keywords: ITO, nanoparticles, transmittance, conductance

1 INTRODUCTION

Indium tin oxide (ITO, In_2O_3:SnO_2) thin films exhibit a high conductance at coexistent brilliant transparency [66Gro,83Ray,86Ham]. Therefore ITO is widely-used as transparent electrode material in optoelectronic applications as TFT-LCD-s [99Kat,07Gai], touch screens, organic light emitting diodes [07Gär], organic solar cells [04Bra,06Wal] and electro-chromic devices [06Gra]. The state-of-the-art ITO layer fabrication techniques are physical vapour deposition methods (especially sputtering) [83Ray,03Fra], laser deposition [01Suz] or evaporation [93Rau,01Bae]. Typical for these layers are conductance values of more than 10,000 $\Omega^{-1}cm^{-1}$ [93Rau,01Suz]. As drawback, the great values demand cost intensive vacuum processes as well as patterning by an additional etching step [99Kat] which involves a strong material loss. Indium tin oxide layers deposited from nanoparticle dispersions are of emerging interest due to the printability which comes along with large area fabrication potential as well as direct patterning possibility [06Ald,07Gro]. Nano particulate layers exhibit comparatively low conductance [06Ald,07Gro,03Ede] which hitherto impeded their use in modern optoelectronic applications despite their advantages of low production costs and a significant lower absorption coefficient.

In a previous work we found a conductance increase from 0.018 $\Omega^{-1}cm^{-1}$ to 17 $\Omega^{-1}cm^{-1}$ by using a temperature treatment in air at 550°C [07Gro]. Here we present further post annealing treatments: post bake, infiltration by an ITO precursor solution with subsequent sol-gel transformation into ITO, treatment with etching agents as well as annealing in vacuum.

2 EXPERIMENTAL

We fabricated nano particulate ITO layers by spin-coating a commercially available (Evonik Degussa GmbH) ethanolic dispersion with a Headway EC101 spin coater on 20 mm × 20 mm × 0.14 mm glass substrates with subsequent annealing at 550°C for 30 min in air. This leads to approx. 1000 nm thick layers which are the initial samples for all following treatments:

Post bake treatment means a temperature step at 250°C for 5 min in air.

Infiltration treatment was done by dipping the samples for 15 s in an ITO precursor solution which was prepared following the recipe of Daoudi et al. [03Dao]. Afterwards the precursor gelification was carried out by drying the samples at 150°C for 30 min. Transformation into ITO however takes place during subsequent annealing at 550°C for 30 min in air.

Etching treatment was carried out by dipping the samples in aqueous 1 M HCl solution for 15 s and washing with water.

Vacuum annealing means temperature treatment in a fused silica tube at 550°C for 30 min at a pressure of about 10^{-5} mbar.

We characterised the nano particulate ITO layers with a number of techniques: Sheet resistance measurements in a linear four-probe setup using a Keithley SMU 236. Layer thickness was determined with a DekTak IIA mechanical profilometer. High resolution SEM (Hitachi S4800) visualised the layer morphology and porosity which itself was calculated via refractive index calculation using the method of Manifacier et al. [76Man,07Gro]. The required transmittance spectra were measured with a Perkin Elmer

Lambda 19 spectrometer. There from we also deduced the free charge carrier concentration.

3   RESULTS AND DISCUSSION

As deposited layer conductance is only 0.018 $\Omega^{-1}cm^{-1}$, but it can be increased by annealing in air at 550°C to 5.7 $\Omega^{-1}cm^{-1}$. In Fig. 1 we present an overall view how several additional post annealing treatments impact the layer conductance.

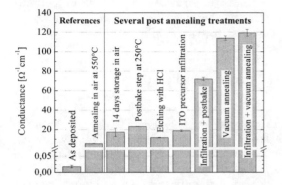

Fig. 1: Impact of several post annealing treatments on conductance of printed nano particulate ITO layers

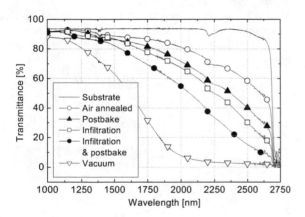

Fig. 2: Near-IR transmittance spectra of various treated ITO layers. The shift of the plasma absorption edge towards shorter wavelength indicates an increase in free charge carrier concentration [86Ham]

3.1   Post bake treatment

We found that the layer conductance increases with storage time after annealing at 550°C in air up to a saturation level (see Fig. 3). A 14 days storage in environment for example leads to a conductance of 17,4 $\Omega^{-1}cm^{-1}$, storage in dry air to 21.1 $\Omega^{-1}cm^{-1}$. This effect was also reported by Ederth et al. [02Ede]. As presented in Table 1 we found 250°C as the optimum post bake temperature. This post bake treatment for only 5 min leads to a conductance increase to

23.2 $\Omega^{-1}cm^{-1}$ which is comparable to the conductance after storage.

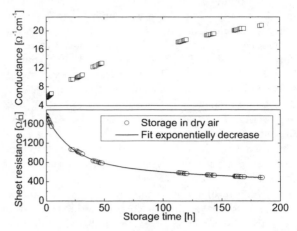

Fig. 3: Conductance and sheet resistance of annealed samples with storage time in dry air

Post bake temperature	20	100	150	200
Sheet resistance [$\Omega/\square$]	2956	2446	1275	967
Post bake temperature	250	300	400	500
Sheet resistance [$\Omega/\square$]	743	856	1272	2010

Table 1: Sheet resistance versus post bake temperature

The mechanism behind the conductance increase after sample storage, as well after post bake treatment is not fully understood. The phenomenon can be attributed either (i) to an acceleration of the kinetic of oxygen diffusion out of the ITO material or (ii) to modification of surface band bending. The oxygen vacancy balance concentration in the ITO particle interior at the annealing temperature of 550°C should not significantly change after heat treatment. Therefore, the conductivity increase after storage of the samples for 14 days or post bake at 250°C is most likely related to a change of surface band bending with a resulting greater surface conductivity due to an increased charge carrier concentration. Beside surface band bending, enhanced oxygen diffusion along grain boundaries in the nano particulate layers may impact oxygen vacancy and, hence, charge carrier concentration, at grain boundaries within the nano-particulate system. The latter is supported by the date in Fig. 2, that indicate a shift of the plasma edge towards shorter wavelength (see also [86Ham, 03Ald]).

3.2   Infiltration treatment

After nanoparticle dispersion deposition and annealing step the ITO layers exhibit a porosity of more than 40 % [07Gro]. The aim of infiltration experiments first was to decrease the porosity and to increase the contact area between the nanoparticle aggregates by bringing in additional ITO material.

Thereby we found that there is no impact if the layers are infiltrated in vacuum or in air. If the samples stand in a puddle of precursor solution the rising infiltration edge can be observed demonstrating that infiltration really takes place. More homogeneous results however are obtained by dipping the samples into the precursor solution for about 15 s and subsequent spinning away the surplus solution.

Precursor infiltration into the layers and subsequent transformation to ITO leads to an increased layer conductance of 19.0 Ω^{-1}cm^{-1} (Fig. 1) due to an increase in free charge carrier concentration (plasma edge shift in Fig. 2).

By the combination of infiltration and post bake treatment a significant higher conductance of 72.1 Ω^{-1}cm^{-1} is achieved (Fig. 1). Thereby the plasma edge shift is larger in comparison to that of both single treatments. This indicates the presence of a second mechanism of charge carrier increase:

At the surface of the nanoparticle aggregates probably exists a depletion layer due to charge carrier trapping in surface defect states and by surface band bending. We now suggest that the precursor solution dissolves this depletion layer and probably partially replaces it with a new generated undepleted sol-gel ITO layer.

This thesis is supported by some results: On the one hand we found no significant conductance increase after multiple infiltration steps and on the other hand we unexpectedly obtained no porosity decrease too. It is rather a slight increase (see Table 2). Multiple infiltration hereby means a repetition of the sequence: precursor infiltration, gelification and precursor transformation.

Infiltration steps	0	1	2	3	4	5
Porosity [%]	42.4	43.2	43.7	43.7	44.0	43.8

Table 2: ITO layer porosity after some infiltration steps

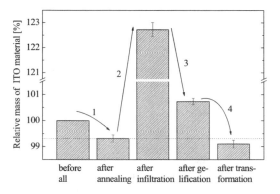

Fig. 4: Relative mass of ITO pellets along the infiltration process sequence

To verify this porosity results we measured the mass of ITO pellets along the infiltration process sequence. We used pellets because of the poor ITO to substrate mass ratio in the case of standard thin film samples. The relative masses are presented in Fig. 4. The mass decrease from green body to annealed pellet (1) presumably dues to evaporation of adsorbed species from surface. The infiltration step itself leads to a significant mass increase (2) due to the

additional mass of precursor solution solvent in the pores. The gelification (3) is connected with the loss of this solvent mass. Only the mass of ITO precursor molecules remains. These molecules (indium chelate complex ions) [03Dao] are disturbed during transformation at 550°C and all organic is burned out (4). The complete infiltration and transformation process is connected with a slight mass loss which verifies the found minor porosity increase.

These results as well as the deed that etching of ITO layers in 1 M HCl acid also leads to a conductance increase (see Fig. 1, details will be presented elsewhere) affirms the hypothesis of dissolving respectively etching of a surface depletion layer by the precursor solution and refutes the first assumption of partially fulfilling of the pores with sol-gel ITO material.

3.3 Vacuum infiltration

It is noteworthy that conductance enhancement by infiltration and post bake is long-term stable and the achieved ITO layers are still in the oxidised yellowish state indicating a relative low charge carrier concentration. It is well known that temperature treatment of ITO layers in vacuum or reducing atmospheres leads to a further conductance increases since oxygen diffuses out which increases oxygen vacancy concentration and with it the charge carrier density [02Ede,03Ald]. Thereby the ITO is transformed to the reduced bluish state.

We observed conductances up to 113.8 Ω^{-1}cm^{-1} (Fig. 1) after annealing at 550°C at a pressure of about 10^{-5} mbar, whereas the achieved conductance strongly depends on the vacuum quality.

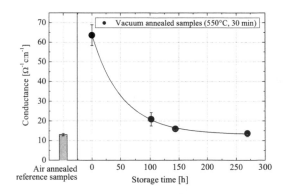

Fig. 5: Time dependent conductance of vacuum annealed nanoparticulate ITO layers

With increasing storage time in air the conductance is declining due to back-diffusion of oxygen (see at Fig. 5). The significant increase in free charge carrier concentration appears in the strong plasma edge shift towards shorter wavelength (Fig. 2) and the widening of the optical band gap due to Burstein-Moss-effect (Fig. 6).

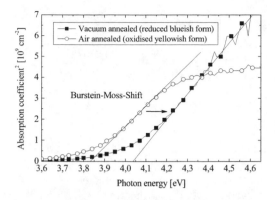

Fig. 6: Widening of the optical bandgap after vacuum annealing due to Burstein-Moss-effect

3.4 Best conductance

By combining infiltration treatment and vacuum annealing we obtained a maximum conductance of 119.3 $\Omega^{-1}cm^{-1}$ which represents a slight increase compared to single vacuum annealing. A possible reason is the additional release of charge carriers from surface trap states as discussed above, whereas the mechanism of charge carrier generation by vacuum annealing does not impact these trapped electrons. But as mentioned above, the achieved conductance by vacuum annealing strongly depends on the experimental conditions. Therefore this slight increase perhaps can have other reasons.

However, infiltrated and vacuum annealed nanoparticulate layers exhibit a conductance high enough to be applicable in modern optoelectronics. For example we achieved layers with a sheet resistance of 17 $\Omega/\square$ at a coexistent transmittance through the complete stack of glass substrate and ITO layer of more than 80 %. The latter may be applied to organic light emitting device processing.

4 OUTLOOK

At present we carry out XPS (ESCA) measurements of variously treated ITO layers to clarify the mechanisms behind the conductance increase: The oxygen O^{2-}:1s ESCA peak of ITO typically has a small shoulder and can be separated into two super-positioning single peaks O^{2-}_{I}:1s at binding energies in the range of 531.2 ... 531.6 eV and O^{2-}_{II}:1s at about 529.9 eV. The first peak is generated by O^{2-}-ions from oxygen-deficient regions, whereas the second one belongs to O^{2-}-ions having neighbouring In atoms with their full complement of six nearest neighbouring O^{2-}-ions [77Fan]. The O^{2-}_{I}:1s / O^{2-}_{II}:1s-ratio therefore should be a direct indicator for the number of oxygen vacancies in the material. By combining these results with optical measurements of plasma edge we hopefully are able to clearly separate both mechanisms: charge carrier density increase by rising oxygen vacancy concentration on the one hand and by release of trapped carriers on the other hand.

ACKNOWLEDGEMENT

This work is financially supported by DFG (contract number GRK 1161) and Evonik Degussa GmbH. Fruitful discussions with Dr. Dieter Adam and Dr. Anna Prodi-Schwab from Evonik Degussa GmbH, Creavis Technologies and Innovation are gratefully acknowledged.

REFERENCES

[66Gro] R. Groth, Phys. Stat. Sol. 14 (1966) 69

[83Ray] S. Ray, R. Banerjee, N. Basu, A.K. Batabyal, A.K. Barua, J. Appl. Phys 54 (1983) 3497

[86Ham] I. Hamberg, C.G. Granqvist, J. Appl. Phys. 60(11) (1986) R123

[99Kat] M. Katayama, Thin Solid Films, 341 (1999) 140

[07Gai] M. Gaillet, L. Yan, E. Teboul, Thin solid films, in press 2007

[07Gär] C. Gärditz, A. Winnacker, F. Schindler, R. Paetzold, Appl. Phys. Lett. 90 (2007) 103506

[04Bra] C.J. Brabec, Sol. Energ. Mat. Sol. C. 83 (2004) 273

[06Wal] C. Waldauf, M.C. Scharber, P. Schilinsky, J.A. Hauch, C.J. Brabec, J. Appl. Phys 99 (2006) 104503

[06Gra] C.G. Granqvist, Nat. Mater. 5 (2006) 89

[03Fra] P. Frach, D. Glöß, K. Goedicke, M. Fahland, W.M. Gnehr, Thin Solid Films 445 (2003) 251

[01Suz] A. Suzuki, T. Matsushita, T. Aoki, Y. Yoneyama, M. Okuda, Jpn. J. Appl. Phys. 40 (2001) L401

[93Rau] I.A. Rauff, Mater. Lett. 18(3) (1993) 123

[01Bae] J.W. Bae, J.S. Kim, G.Y. Yeom, Nucl. Iinstrum. Meth. B 178 (2001) 311

[06Ald] N. Al-Dahoudi, M.A. Aegerter, Thin Solid Films, 502 (2006) 193

[07Gro] M. Gross, A. Winnacker, P.J. Wellmann, Thin Solid Films 515 (2007) 8567

[03Ede] J. Ederth, P. Heszler, A. Hultåker, G.A. Niklasson, C.G. Granqvist, Thin Solid Films, 445 (2003) 199

[03Dao] K. Daoudi, B. Canut, M.G. Blanchin, C.S. Sandu, V.S. Teodorescu, J.A. Roger, Thin Solid Films 445 (2003) 20

[76Man] J.C. Manifacier, J. Gasiot, J.P. Fillard, J Phys E Sci Instrum 9 (1976) 1002

[02Ede] J. Ederth, A. Hultåker, P. Heszler, G.A. Niklasson, C.G. Granqvist, A.v. Doorn, C.v. Haag, M.J. Jongerius, D. Burgard, Smart Mater. Struct. 11 (2002) 675

[03Ald] N. Al-Dahoudi, PhD thesis, university of Saarland and INM (2003)

[77Fan] J.C.C. Fan, J.B. Goodenough, J. Appl. Phys. 48 (1977) 3524

Palladium/Polymer Nanocomposite Based Chemiresistive SO$_2$ Sensor

D. Meka*, V. Onbattuvelli**, S.V. Atre**, S. Prasad*

*Department of Electrical Engineering, Portland State University, Portland, OR 97201
**Department of Industrial and Manufacturing Engineering, Oregon State University, Corvallis, OR 97331

ABSTRACT

We report a highly sensitive palladium/polymer nanocomposite based SO$_2$ gas sensor. The palladium/polycarbonate (Pd/PC) nanocomposite was synthesized by utilizing two different techniques. The ex-situ method involved the homogenous mixing of pre-synthesized Pd nanoparticles with the PC matrix. The in-situ method on the other hand involved the synthesis of Pd nanoparticles in the presence of PC matrix. The semi-conducting behavior of the *in situ* Pd/PC nanocomposites and the chemical affinity of Pd nanoparticles towards the lewis acid gases were efficiently employed to develop the SO$_2$ sensor. A chemiresistive technique was used in the detection of the SO$_2$ gas molecules in which the change in the electrical resistance of the nanocomposite associated with the adsorption of SO$_2$ molecules by the nanocomposite was measured.

Keywords: palladium, polycarbonate nanocomposites, chemiresistive sensor, sulfur dioxide, electrical conductivity.

1. INTRODUCTION

Sulfur dioxide (SO$_2$) is one of the six common air pollutants that is released into the atmosphere which causes a wide variety of health and environmental impacts like respiratory effects, visibility impairments, acid rains, etc., because of the way it reacts with other substances in the air [1-3]. Therefore, it is useful to measure the emission of this gas into the atmosphere by various sources.

Developing a lower ppm level sensitive, reliable, low cost, solid state fluid sensor in a portable format has always been a challenge. Using metal nanoparticles to solve the above problem constitutes an important research area in the field of nanotechnology. Volkening et al. [4] reported the H$_2$ sensing capability of nanocrystalline Pd particles. However, utilizing the metal nanoparticles alone had some limitations in the sensor application domain due to handling constraints and easily oxidizable nature of the metal oxide nanoparticles [5]. As an alternative metal-polymer nanocomposites have been utilized for the same purpose. For example, Methanol sensing by palladium/polyaniline (Pd/PANI) nanocomposites was reported by Athawale et al. [6]. These nanocomposites have demonstrated tunable optical, thermal and electrical properties that are utilized as the parameters used for sensing [4-6]. From the earlier work [7-10], it can be inferred that the morphology of the nanoparticles depend on several factors such as the molecular weight of the protecting agent; metal salt: protecting agent ratio; functional groups in the protecting agents; temperature of the reaction; reducing agent; reduction rate and the mode of synthesis.

In this study, we report the varied electrical behavior during chemiresistive detection of SO$_2$ utilizing the Pd/PC nanocomposite synthesized by two different methods: the *ex-situ* and *in-situ* methods with different Pd content.

2. CHEMIRESISTIVE SO$_2$ SENSOR

The principle of operation of the chemiresistive sensor is based on the measurement of resistance change associated with the adsorption/reaction of gaseous analyte by/with the nanomaterial matrix [11,12]. The prototype gas sensor functions based on the chemiresistive principle, where variations in the resistance of the SO$_2$ gas sensitive Pd/PC nanocomposite is observed and measured due to selective reaction between the nanocomposite and the SO$_2$ gas molecules that in turn decreases the number of free electrons resulting in a concentration dependant resistance increase.

This chemiresistive SO$_2$ sensor consists of a two main components: the Pd/PC nanocomposite film that functions as the active sensing element and the electrical circuitry supporting the nanocomposite film. Each component is discussed in turn.

2.1. Pd/PC Nanocomposites

In the *ex situ* method, C$_{12}$H$_{25}$SH-protected Pd nanoparticles were prepared using the Brust method [13]. The Pd nanoparticles were then homogenously mixed with a solution of 40 mg of PC in 20 ml of CH$_2$Cl$_2$ (1.6 μM) followed by film casting at room temperature. In the case of the *in situ* method, PC (40 mg) was dissolved in CH$_2$Cl$_2$ (20 ml) (1.6 μM). 15 mg of PdCl$_2$ was first dissolved in 2ml of conc. HCl so as to form a complex [PdCl$_4$]$^{2-}$, and was further dissolved in 48 ml water to form a 1mM solution. This biphasic mixture was stirred continuously using a magnetic stirrer for 30 minutes. A freshly prepared solution of NaBH$_4$ in 20 ml water (0.1M) was added drop-wise to the mixture. The color of the reaction mixture changed rapidly from golden yellow to black, indicating the formation of Pd nanoparticles. After stirring for 3 hours, the organic phase was separated, washed with water and was directly cast into film at room temperature. Soon after the reduction nearly all of the reduced Pd nanoparticles get transferred from the aqueous phase to the organic phase. The *ex situ* nanocomposite yielded a dispersed mixture in contrast to the *in situ* nanocomposite which produced an agglomerated mixture. (Figure 1)

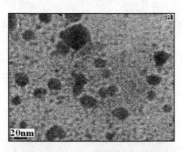
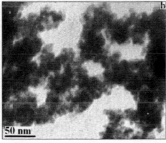

Figure 1. TEM image of the *ex situ* and *in situ* Pd/PC nanocomposites showing (a) Dispersed Pd nanoparticles, and (b) Agglomerated Pd nanoparticles, respectively.

The *in situ* and *ex situ* two nanocomposites were packaged on to a miniature base platform. The base platform comprised of a metallic interdigitated microelectrode structure (Figure 2). Coating the interdigitated electrode (IDE) with the nanocomposite resulted in an active sensing area that was investigated for its sensitivity and selectivity for SO_2 detection. 1wt. % of carbon nanoparticles (CNPs) was added into these nanoparticles in order to improve their electrical conductivity.

2.2. Electrical Circuitry

When the Pd/PC nanocomposite was exposed to SO_2 gas, it was observed that the SO_2 gas molecules resulted in a in the change of the nanocomposite's electrical resistance. In order to measure this change in electrical resistance in a continuous manner, we employed a Wheatstone's bridge circuit, which detected the change in the electrical resistance by measuring the change in resistance between the balanced and imbalanced condition. The amount of imbalance observed in the Wheatstone bridge was observed to be a function of the concentration of SO_2. The measurand obtained was an output voltage proportional to the change

in the electrical resistance of the nanocomposite (Figure 2). In the entire circuit, the IDE pattern functions as an electrical resistor whose resistance changed following reaction between the Pd/PC nanocomposite and the SO_2 gas molecules.

Figure 2 (a) The optical micrograph of the chemiresistive sensor chip with the interdigitated electrodes coated with polymer nanocomposite. Each sensor array comprised of 20 digits. Each digit is ~200µm in width and 2mm in length with 300µm spacing. (b) The optical micrograph of the chemiresistive sensor covered with the Pd/PC nanocomposite forming a homogenous layer on the surface.

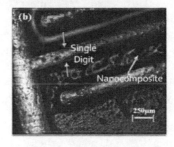
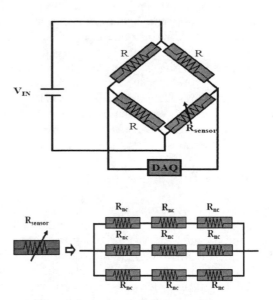

Figure 3. The equivalent circuit of the entire experimental setup and equivalent resistance of the nanocomposite are shown. The resistor R_{sensor} in the Wheatstone bridge circuit is the resultant resistance of the nanocomposite.

3. RESULTS AND DISCUSSION

3.1. Morphology of Pd/PC nanocomposites

The TEM image of the *ex situ* nanocomposite with 2vol.% Pd, revealed dispersed Pd nanoparticles of ~15 nm embedded in PC matrix (Figure 1a). Based on earlier reports on the synthesis and morphology of n-alkanethiol-protected Pd nanoparticles, the presence of dodecanethiol on the surface of the Pd nanoparticles in the present study is likely to ensure the separation of the nanoparticles even after mixing with PC. Although an identical metal salt: thiol ratio and reducing agent were used in the present study, an increase in the size of the nanoparticles was found than the prior work by Brust et al.[13]. This may be due to the absence of the surfactant, in the reaction mixture which helps in phase transfer of reduced Pd nanoparticles. The effect of increased temperature of the reaction mixture from ice-cold condition in the earlier studies in comparison to the reaction room temperature may have also contributed to the increased size of the nanoparticles.

In contrast to the above system, the *in situ* nanocomposites of Pd nanoparticles (2 vol.% on a stoichiometric basis) in PC showed significant agglomeration (Figure 1b). Similar observations on agglomeration were reported by Chen et al using Pd/mercapto-poly(ethylene glycol) [7], and Chatterjee et al with Au/poly(dimethylamino ethyl methacrylate-methyl methacrylate) copolymers [8]. Wang et al have suggested that in order to obtain discrete nanoparticles, the rate of adsorption of organic ligands on the surface of nanoparticles should equal the rate of nanocluster formation [14]. Accordingly, organic ligands with lower molecular weight have generally been found to be more effective in limiting the nanoparticles size. In addition, the nature of interactions between the polymer and the surface of the nanoparticles may also play a role in determining the morphology of the resulting nanocomposites. The following sections examine the consequences of the differences in morphology on the resulting electrical properties and the SO_2 sensing capabilities of these nanocomposites.

3.2. Electrical Properties

PC is electrically insulating in nature with a volume resistivity of about 2×10^{14} Ω-m. No significant difference was observed for the *ex situ* nanocomposites (with 2 vol. % Pd (on a stoichiometric basis)) with resistivity of 7.2×10^{13} Ω-m. However, the *in situ* nanocomposite films having 2 vol. % Pd (on a stoichiometric basis) showed a linear increase in the current with the voltage indicating a constant resistance of about 440 Ω and thus a resistivity of 2.1×10^5 Ω-m.

Similar results were observed by Athawale et al [5] on Pd/polyaniline nanocomposites and by Rao et al [15] on Pd/polypyrrole nanocomposites. However, these results

involved conducting polymers for electrical conductivity studies on nanocomposites. In the present study an electrically insulating polymer (PC) was used instead of a conducting polymer. The reasons for the differences in electrical conductivity between the *in situ* and *ex situ* are not known although there may be a link between the differences in microstructure as seen in Fig 1.

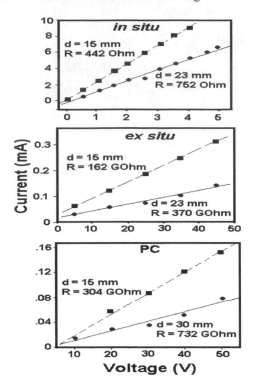

Figure 4. Electircal properties of 2 vol. % Pd/PC nanocomposites compared that of PC.

3.3.Sensitivity towards SO_2

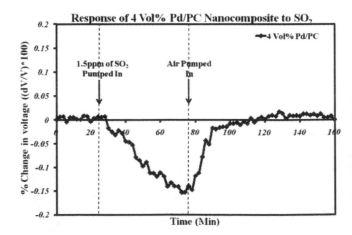

Figure 5. The electrical response of 4 vol. % *in situ* Pd/PC nanocomposite film with 1 wt. % CNPs following exposure to SO_2.

The influence of SO_2 exposure on the Pd/PC nanocomposite was monitored by mesuring the change in voltage as a function of time (Figure 5). We observed that there was a change in the electrical conductivity of the Pd/PC *in situ* nanocomposite when it was exposed to SO_2. This change in electrical conductivity was due to the adsorption of SO_2 molecules by the Pd/PC nanocomposite. However, this change in electical resistance due to the adsorption of the SO_2 gas molecules was not observed in the case of *ex situ* Pd/PC nanocomposite.

In summary, the variations in synthetic procedure affects the morphology and electrical properties of Pd/PC nanocomposites. These differences can be used for designing chemiresistive sensors with tailored performance for detecting chemical species.

4. REFERENCES

1. W. Dab, S. Medina, P. Quénel, Y. L. Moullec, A. L. Tertre, B. Thelot, C. Monteil, P. Lameloise, P. Pirard, I. Momas, R. Ferry and B. Festy, J Epidemiol Community Health **50**, 42 (1996).

2. A. P. d. Leon, H. R. Anderson, J. M. Bland, D. P. Strachan and J. Bower, J Epidemiol Community Health **50**, 63 (1996).

3. A. Pönkä and M. Virtanen, J Epidemiol Community Health **50**, 59 (1996).

4. F. A. Volkening, M. N. Naidoo, G. A. Candela, R. L. Hold. and V. Provenzanoe, Nanostructured Materials **5**, 373 (1995).

5. A. A. Athawale, S. V. Bhagwat and P. P. Katre, Sensors and Actuators, B: Chemical **114**, 263 (2006).

6. M. Chen, J. Falkner, W. H. Guo, J. Y. Zhang, C. Sayes and V. L. Colvin, Journal of Colloid and Interface Science **287**, 146 (2005).

7. S. K. Jewrajka and U. Chatterjee, Journal of Polymer Science Part A: Polymer Chemistry **4**, 1841 (2006).

8. F. K. Liu, S. Y. Hsieh, F. H. Ko and T. C. Chu, Colloids and Surfaces A: Physiochemical and Engineering Aspects **231**, 31 (2002).

9. C. Aymonier, D. Bortzmeyer, R. Thomann and R. Mulhaupt, Chemistry of Materials **15**, 4874 (2003).

10. S. V. Atre, O.P. Valmikanathan, V. K. Pillai, I. S. Mulla and O. Ostroverkhova In *Nanotech 2007* (2007), Vol. 1, p 158.

11. F. J. Ibanez, U. Gowrishetty, M. M. Crain, K. M. Walsh and F. P. Zamborini, J Anal. Chem. **78**, 753 (2006).

12. M. D. Wita, E. Vannestea, H. J. Geisea and L. J. Nagelsb, Sens. and Actua. B: Chem. **50**, 164 (1998).

13. M. Brust, M. Walker, M. Bethell, D. Schiffrin and D. J. Whyman, Journal of Chemical Society and Chemical Communication **20**, 801 (1994).

14. H. Wang, X. Qiao, J.Chen and S. Ding, Colloids and Surfaces A: Physicochemical and Engineering Aspects **256**, 111 (2005).

15. C.R.K.Rao and D. C. Trivedi, Catalysis Communications **7**, 662 (2006).

Structure-property relationships in Pd/PC nanocomposites based chemical sensors

O.P. Valmikanathan*, D. Meka**, S. Prasad** and S.V. Atre*

* Oregon State University, Corvallis, OR 97331, sundar.atre@oregonstate.edu
**Portland State University, Portland, OR 97201, sprasad@pdx.edu

ABSTRACT

In this paper, we study the effect of morphology on the thermal stability of palladium/polycarbonate (Pd/PC) nanocomposites. Pd/PC nanocomposites were synthesized via two different techniques, *ex situ* and *in situ*. Discrete Pd nanoclusters of ~ 15nm size were formed in the absence of PC in the reaction mixture (*ex situ* method) while agglomeration of Pd nanoclusters was noticed in the presence of PC in the reaction mixture (*in situ* method). Variations in thermal stability were noted for the various Pd/PC nanocomposites as a function of their morphology. These results have implications for the operating life-times of chemical sensors containing Pd/PC nanocomposites as active elements.

Keywords: nanocomposites; *in situ* and *ex situ* synthesis, morphology; electrical conductivity

1. INTRODUCTION

The enhanced surface area and the ability to tune the electrical characteristics have made metal nanoparticles a better candidate for the optical, thermal and electrical applications. However, their applications are limited due to the difficulty in handling, quick oxidation of the metal particles [1]. Thus, protecting these nanoparticles by organic ligands [2-3] including polymers [4-6] becomes necessary. Prior studies [4-5] show that the properties of the nanoparticles were controlled by the selection of the protecting agent as well as the distribution of the nanoclusters. Polymers are preferred over the organic ligands due to the convenience in handling, reduced post-synthesis treatment and more direct applications.

Generally, polymer-protected nanoclusters can be prepared by two different synthetic methods. In the *ex situ* method, organic ligand-protected nanoclusters are initially prepared followed by homogenous mixing with a polymer solution [7-8]. In contrast, the *in situ* method, involves the preparation of nanoclusters in the presence of a polymer [5-6]. This method generally does not involve additional organic ligands other than the polymer, as protecting agents. The resulting solutions from either method can subsequently be cast into films.

In this paper, we report an *in situ* method for preparing nanocomposite films by reducing palladium chloride ($PdCl_2$) in the presence of polycarbonate (PC) dissolved in dichloromethane. We also synthesized Pd/PC nanocomposite films by an *ex situ* method involving the

dispersion of dodecanethiol-protected Pd nanoclusters in a solution of PC in dichloromethane. It was found that the synthesis method had a significant impact on the morphological characteristics and the corresponding optical, thermal and electrical properties of the nanocomposite films. The effect of varying Pd content and involving a phase transferring agent (tetraoctylammonium bromide -TOABr) on the properties of the nanocomposites were also studied. The electrical conducting behavior of the *in situ* nanocomposites and the affinity of the Pd metal towards the Lewis acid gases have resulted in the construction of a SO_2 sensor out of the Pd/PC nanocomposites. The SO_2 sensing capabilities of the Pd/PC nanocomposites have been discussed in detail in another paper. In this paper, we present the results from our investigations in the synthesis and thermal stability of Pd/PC nanocomposites. These issues are of importance in determining the time and temperature combinations affecting the use of Pd/PC nanocomposite-based chemical sensors.

2. EXPERIMENTAL SECTION

2.1. Synthesis of Pd/PC nanocomposites

All analytical grade chemicals purchased were used as received. In the *ex situ* method, $C_{12}H_{25}SH$-protected Pd nanoparticles were prepared using Brust method [2]. The Pd nanoparticles were then homogenously mixed with a solution of 40 mg of PC in 20 ml of CH_2Cl_2 (1.6 µM) followed by film casting at room temperature. In the case of the *in situ* method, PC (40 mg) was dissolved in CH_2Cl_2 (20 ml) (1.6 µM). 15 mg of $PdCl_2$ was first dissolved in 2ml of conc. HCl so as to form a complex $[PdCl_4]^{2-}$, and was further dissolved in 48 ml water to form a 1mM solution. This biphasic mixture (along with or without 10ml of 0.5 mM TOABr) was stirred continuously using a magnetic stirrer for 30 minutes. A freshly prepared solution of $NaBH_4$ in 20 ml water (0.1M) was added drop-wise to the mixture. The color of the reaction mixture changed rapidly from golden yellow to black, indicating the formation of Pd nanoparticles. However, in the presence of TOABr the color change was from golden yellow to greenish yellow. After stirring for 3 hours, the organic phase was separated, washed thrice with 250ml of with ethanol/water mixture (60/40 v/v %) and was directly cast into film at room temperature. The schematic of the Pd/PC nanocomposites synthesis is given in Figure 1.

NSTI-Nanotech 2008, www.nsti.org, ISBN 978-1-4200-8503-7 Vol. 1

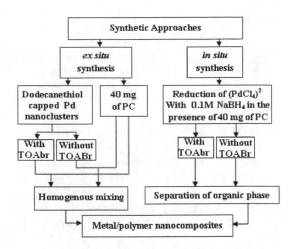

Figure 1: Schematic representation of *in situ and ex situ* Pd/PC nanocomposites synthesis.

3. RESULTS AND DISCUSSION
3.1. Morphology of Pd/PC nanocomposites

The TEM image of the *ex situ* nanocomposite with 2 vol. % Pd (on a stoichiometric basis) revealed dispersed Pd nanoclusters of ~15 nm embedded in PC matrix (Figure 2a). Based on earlier reports on the synthesis and morphology of n-alkanethiol-protected Pd nanoclusters, the presence of dodecanethiol on the surface of the Pd nanoclusters in the present study is likely to ensure the separation of the nanoclusters even after mixing with PC. However, the average particle size of the Pd nanoclusters in previous studies was found to be ~ 5nm size; using the Brust method [2]. Although the identical metal salt: thiol ratio and reducing agent were used in the present study, an increase in the size of the nanoclusters was found. The effect of increased temperature of the reaction mixture and absence of any phase transferring agents may have contributed to the increased size of the nanoclusters [9]. A difference in the concentration of reducing agent may have also contributed to the increase in the average size of nanoclusters.

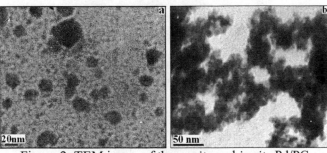

Figure 2: TEM image of the *ex situ* and *in situ* Pd/PC nanocomposites showing (a) dispersed Pd nanoparticles, and (b) agglomerated Pd nanoparticles, respectively .

In contrast to the above system, *in situ* nanocomposite of Pd nanoclusters (2 vol. % on a stoichiometric basis) in PC showed significant agglomeration (Figure 2b). Similar

observations on agglomeration were reported by Liu et al using Au/PMMA [6], and Chatterjee et al with Au/poly(dimethylamino ethyl methacrylate-*b*-methyl methacrylate) copolymers [4]. Thus, morphological changes in nanocomposites appear to be strongly dependent on the specific polymer system and reaction conditions.

However, when synthesized in the presence of phase transferring agents the size of the Pd nanoparticles in 2 vol. % *ex situ* nanocomposites were controlled to 5nm size (Figure 3a). The Pd nanoparticles size was further controlled to 3nm by reducing the Pd content to 1 vol. % (Figure 3b). Similar results were reported by Brust *et al.* [2] on Pd capped with dodecanethiol.

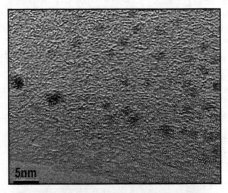

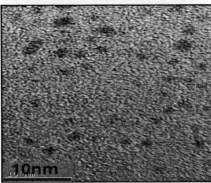

Figure 3: TEM image of the *ex situ* Pd/PC nanocomposites (2 vol.% and 1 vol.%) with phase transferring agents showing (a) dispersed Pd nanoparticles ≤ 5nm size and (b) Pd nanoparticles ≤ 3nm size, respectively.

The possible mechanism involved in the nanoclusters formation is given Figure 4. Now, when a metal salt is reduced into metal, the nanoparticles formed tend to grow into clusters due to their enhanced surface energy. Further growth of such nanoclusters can be terminated by protecting them with organic ligands. Thus, the size of the metal nanoclusters might possibly be controlled by various ways. As per Brust *et al* [2] and Liu *et al* [6] one major route is to expedite the process of shifting nanoclusters from aqueous phase to the organic phase by adding phase transferring agents (TOABr) which was proven in our case. Prior studies also talks about the other factors like metal content [5] and molecular weight of the capping agent.

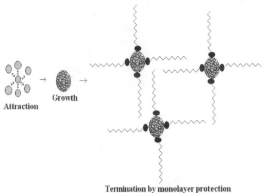

Figure 4: Mechanism involved in the Pd nanoclusters formation.

Wang *et al* have suggested that in order to obtain discrete nanoclusters, the rate of adsorption of organic ligands on the surface of nanoclusters should equal the rate of nanocluster formation [10]. Accordingly, organic ligands with lower molecular weight have generally been found to be more effective in limiting the nanoclusters size. The wide-ranging behavior of agglomeration in nanocomposites prepared by the *in situ* methods may also be due to the differences in the conformations of the polymer chain in the different studies. These conformational differences can arise from variations in molecular weight, solvent, and temperature. Consequently, the mobility of the polymer during adsorption on the nanocluster surface can be affected, thereby limiting the agglomeration of the nanoclusters. In addition, the nature of interactions between the polymer and the surface of the nanoclusters may also play a role in determining the morphology of the resulting nanocomposites. Further studies are needed to better understand the differences in morphologies observed between the *in situ* and *ex situ* nanocomposites when synthesized with varying TOABr concentration and also with PC of different molecular weight. The following sections examine the consequences of the differences in morphology on the resulting properties of the nanocomposites.

3.2. Thermal Properties

The thermogravimetric (TGA) curves for the Pd/PC nanocomposites and PC at different heating rates and atmospheres are shown in the Figure 5. For each sample, the corresponding differential thermogravimetric analysis (DTGA) plots in the Figure 6 revealed the temperature at which the maximum rate of weight loss (T_{max}) occurs. The DTA plots are represented with normalized weight Ψ (%) which can be expressed as,

$$\Psi = \left[\frac{W_i - W_{end}}{W_0 - W_{end}} \right] * 100 \qquad (1)$$

where, W_i is the weight (%) of the sample at the given temperature, W_0 is the weight (%) at the starting of TGA experiment and W_{end} is the weight (%) at the ending of TGA experiment.

The onset temperature of degradation (T_{onset}) and the end temperature of degradation (T_{end}) can be obtained from the TGA. Their difference ΔT thus, represents the temperature range for thermal degradation of a given sample. The ratio of ΔT with the heating rates (r) gives the overall degradation time of the sample. From the data, it was seen that the thermal stability of PC and the nanocomposites increases with increase in the heating rates. The results are supported by the shifts in the values of T_{onset} and T_{max} to the higher range with increase in the heating rates. Also, the overall degradation time ($\Delta T/H$) for the sample decreases with increase in the heating rates. Similar results were obtained by Peng et al [11] for poly (vinyl alcohol)/silica nanocomposites. A possible reason for such behavior may be explained as follows. The rate of degradation depends on the rate of heat absorption by PC which once again depends on the heating rate. Thus, even though when the nanocomposites showed enhanced thermal stability at higher heating rates, their overall degradation time may highly get reduced.

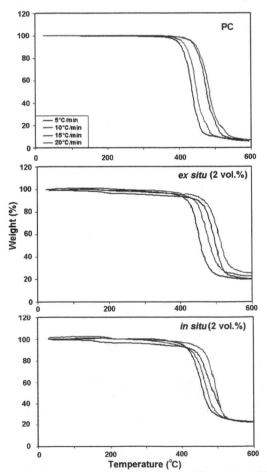

Figure 5: Thermogravimetric analysi of 2 vol. % Pd/PC nanocomposites and PC at different heating rates.

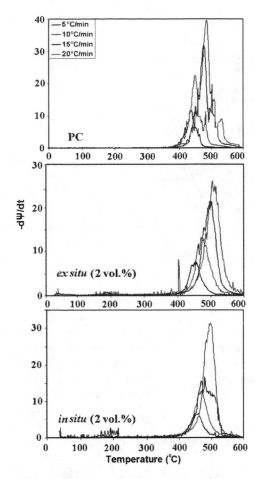

Figure 6: DTGA data for 2 vol. % Pd/PC nanocomposites and PC at different heating rates.

Further analysis of the data revealed that increasing the Pd content in the nanocomposites, increases the T_{onset} and T_{max}. Similar results were obtained by Aymonier et al [12] for Pd/PMMA (Pd content: 0.01 vol.%), Xia et al [13] for Cu/LDPE (Cu content : 17 wt.%) and Hsu et al [14] for Au/polyurethane (Au content: 0.065 wt%). The increment in the thermal stability may possibly be due to the increase in the PC-nanocluster interfacial area. Further research is required to find the existence of a percolation limit of Pd above which the thermal stability of the nanocomposites may remain unaffected.

It could also be determined that for any given Pd concentration, *ex situ* nanocomposite is more thermally stable than the *in situ* ones with equal or less Pd content. A similar increase in thermal stability was noted in earlier reports by Huang et al [7] for Au/poly(methyl styrene) (particle size: 3.5 nm and Au content: 5 wt. %), Aymonier et al [12] for Pd/PMMA (particle size: 2.5 nm) and Hsu et al [13] for Au/polyurethane (particle size: 5 nm). Aymonier et al and Hsu et al also observed that the thermal stability of the nanocomposites increase with increase in metal concentration and with decrease in particle size and agglomeration. These results roughly correlate with changes

in the area of the polymer-nanocluster interface and are in close agreement with the morphology of the nanocomposites.

4. CONCLUSIONS

Pd/PC nanocomposites were synthesized via two different techniques, ex situ and in situ respectively. The morphology was found to control the thermal properties and thus the time-temperature limits on the SO_2 sensing capabilities of the nanocomposites. The particle size and morphologies were found controlled by involving the phase transferring agent (TOABr) in the reaction mixture and by varying the metal content. Further research is required to explore the factors possibly affecting the morphology and properties of the nanocomposites. These studies will be useful for tailoring metal-polymer nanocomposites for chemical sensing applications.

REFERENCES

1. W.P. Wuelfing, S.M. Gross, D.T. Miles, R.W. Murray, *Journal of American Chemical Society*, 120, 12696, 1998.
2. M. Brust, M. Walker, M. Bethell, D. Schiffrin, and D.J. Whyman, *Journal of Chemical Society and Chemical Communication*, 20, 801, 1994.
3. M. Aslam, G. Gopakumar, T.L. Shoba, I.S. Mulla, K. Vijayamohanan, S.K. Kulkarni, J. Urban and W.Vogel , *Journal of Colloid and Interface Science*, 255, 79, 2002.
4. S.K. Jewrajka, and U. Chatterjee, *Journal of Polymer Science Part A: Polymer Chemistry*, 44, 1841, 2006.
5. C. Aymonier, D. Bortzmeyer, R. Thomann, and R. Mulhaupt, *Chemistry of Materials*, 15, 4874, 2003.
6. F.K. Liu, S.Y. Hsieh, F.H. Ko, and T.C. Chu, *Colloids and Surfaces A: Physiochemical and Engineering Aspects*, 231, 31, 2002.
7. H. Huang, Q. Yuan and X. Yang. *Colloids and Surfaces B: Biointerfaces*, 39, 31, 2004.
8. S. Chattopadhyay, and A. Datta, Synthetic *Metals*, 155, 365, 2005.
9. K.Koga, T. Ikeshoji and K.Sugawara, *Physical Review Letters*, 92, 11507-1, 2004.
10. H. Wang, X. Qiao, J. Chen, and S. Ding, *Colloids and Surfaces A: Physicochemical and Engineering Aspects*, 256, 111, 2005.
11. Z. Peng L.X. Kong LX, S.D. Li, and P. Spiridonov. *Journal of Nanoscience and Nanotechnology*. 6, 3934-3938, 2006.
12. C. Aymonier, D. Bortzmeyer, R. Thomann, and R. Mulhaup. *Chemistry of Materials* 15, 4874- 4878, 2003.
13. X. Xia, S. Cai, and C. Xie. *Materials Chemistry and Physics* 95, 122-129, 2006.
14. S.H. Hsu, C.W. Chou, and S.M. Tseng. *Macromolecular Materials and Engineering* 289, 1096-1101, 2004.

Toughening mechanisms of nanoparticle-modified epoxy polymers

R.D. Mohammed [a], B.B. Johnsen [b], A.J. Kinloch [c], A.C. Taylor [c], S. Sprenger [d]

[a] C-IDEAS, The University of Trinidad and Tobago, O'Meara Campus, Arima, Trinidad and Tobago
[b] Norwegian Defence Research Establishment, Postboks 25, 2027 Kjeller, Norway
[c] Department of Mechanical Engineering, Imperial College London, South Kensington Campus, London SW7 2AZ, UK
[d] Nanoresins AG, Charlottenburger Strasse 9, 21502 Geesthacht, Germany

ABSTRACT

An epoxy resin, cured with an anhydride, has been modified by the addition of silica nanoparticles. The particles were introduced via a sole gel technique which gave a very well-dispersed phase of nanosilica particles of about 20 nm in diameter. AFM and TEM showed that the nanoparticles were well-dispersed in the epoxy matrix. T_g was unchanged by the addition of the nanoparticles, but both the modulus and toughness increased. The fracture energy increased from 100J/m for the unmodified epoxy to 460J/m for the epoxy with 13 vol% of nanosilica. The fracture surfaces were inspected using SEM and AFM, and the results were compared to various toughening mechanisms proposed in the literature. The microscopy showed evidence of debonding of the nanoparticles and subsequent plastic void growth. A theoretical model of plastic void growth was used to confirm that this mechanism was indeed most likely to be responsible for the increased toughness that was observed due to the presence of the nanoparticles.

Keywords: epoxy, nanoparticles, fracture

1 INTRODUCTION

Epoxy polymers are widely used for the matrices of fibre reinforced composite materials and as adhesives. When cured, epoxies are amorphous and highly-crosslinked (i.e. thermosetting) polymers. This microstructure results in many useful properties for structural engineering applications, such as a high modulus and failure strength, low creep, and good performance at elevated temperatures.

However, the structure of such thermosetting polymers also leads to a highly undesirable property in that they are relatively brittle materials, with a poor resistance to crack initiation and growth. Nevertheless, it has been well established for many years that the incorporation of a second microphase of a dispersed rubber, e.g. [1], or a thermoplastic polymer, e.g. [2], into the epoxy polymer can increase their toughness. Here the rubber or thermoplastic particles are typically about 1–5μm in diameter with a volume fraction of about 5–20%. However, the presence of the rubbery phase typically increases the viscosity of the epoxy monomer mixture and reduces the modulus of the cured epoxy polymer.

Hence rigid, inorganic particles have also been used, as these can increase the toughness without affecting the glass transition temperature of the epoxy polymer. Here glass beads or ceramic (e.g. silica or alumina) particles with a diameter of between 4 and 100μm are typically used, e.g. [3]. However, these relatively large particles also significantly increase the viscosity of the resin, reducing the ease of processing. In addition, due to the size of these particles they are unsuitable for use with infusion processes for the production of fibre composites as they are strained out by the small gaps between the fibres.

More recently, a new technology has emerged which holds promise for increasing the mechanical performance of such thermosetting polymers. This is via the addition of a nanophase structure in the polymer, where the nanophase consists of small rigid particles of silica [5]. Such nanoparticle modified epoxies have been shown to not only increase further the toughness of the epoxy polymer but also, due to the very small size of the silica particles, not to lead to a significant increase in the viscosity of the epoxy monomer.

The aims of the present work were to investigate the fracture toughness of epoxy polymer modified with silica nanoparticles, and to establish the structure/property relationships. The toughening mechanisms which may be operating will be reviewed, and the mechanism most likely to be responsible will be identified.

2 MATERIALS

The materials were based upon a one-component hot-cured epoxy formulation. The epoxy resin was a standard diglycidyl ether of bis-phenol A (DGEBA) with an epoxy equivalent weight (EEW) of 185 g/mol, 'Bakelite EPR 164' supplied by Hexion Speciality Chemicals, Duisburg, Germany. The silica ($SiO2$) nanoparticles were supplied as a colloidal silica sol in the resin matrix, 'Nanopox F400', by Nanoresins, Geesthacht, Germany. The particles are synthesised from aqueous sodium silicate solution [5]. They then undergo a process of surface modification with organosilane and matrix exchange, to produce a masterbatch of 40 wt% (26 vol%) silica in the epoxy resin. The nanosilica particles had a mean particle size of about 20 nm, with a narrow range of particle-size distribution; laser light scattering shows that almost all particles are between 5 and 35 nm in diameter. The particle size and excellent dispersion of these silica particles remain unchanged during any further mixing and/or blending operations. Further, despite the relatively high silica content of 26 vol%, the nanofilled epoxy resin still has a comparatively low viscosity due to the agglomerate-free colloidal dispersion of the nanoparticles in the resin. The small diameter and good dispersion of the nanoparticles of silica have been previously reported and shown [4]. The curing agent was an accelerated methylhexahydrophthalic acid anhydride, namely 'Albidur HE 600' supplied by Nanoresins, Geesthacht, Germany.

Bulk sheets of unmodified epoxy and nanosilica-modified epoxy polymers were produced to determine the properties of the polymers. Firstly, the simple DGEBA resin was mixed together with given amounts of the nanosilica-containing epoxy resin. The value of the EEW of the blend was then measured via titration. Secondly, the stoichiometric amount of the curing agent was added to the mixture, which was poured into release-coated moulds and pre-cured for 1h at 90°C, followed by a cure of 2h at 160°C.

The densities of the plates were measured. An epoxy density of 1100 kg/m^3 and a silica density of 1800 kg/m^3 were calculated. The volume fraction of silica was calculated from the known weight fractions using the measured densities.

3 EXPERIMENTAL

The glass transition temperature, Tg, of the various polymers was measured using differential scanning calorimetry. The sample was heated to 175°C at a rate of 10°C /min, and then cooled to 0°C. The sample was then heated again to 175°C, and the results quoted are from this second heating run.

Tensile dumbbell specimens were tested at a displacement rate of 1 mm/min and a test temperature of 21 □C, according to the ISO standard test method [6,7]. The strain in the gauge length was measured using a clip-on extensometer, and the Young's modulus, E, was calculated.

The single-edge notch bend (SENB) tests were used to determine the fracture toughness according to the relevant ISO standard [8], using a displacement rate of 1 mm/min and a test temperature of 21°C. Four replicate specimens were tested for each blend composition. The machined notch was sharpened by drawing a razor blade across the notch tip before testing. All the specimens failed by unstable crack growth, and hence only a single initiation value of the fracture toughness was obtained from each specimen.

Thin sections, approximately 60-80 nm thick, of the blends were cryo-microtomed (at -50°C) for subsequent examination using transmission electron microscopy (TEM). For atomic force microscopy (AFM) studies, a smooth surface was first prepared by cutting samples on a cryo-ultramicrotome at temperatures down to -100°C. The AFM scans were performed in tapping mode using silicon probes, and both height and phase images were recorded.

4 RESULTS

4.1 Thermo-mechanical Results

The tensile and fracture results for various nanosilica contents are given in Table 1.

Table 1
Glass transition temperatures, modulus and fracture properties of the anhydride-cured epoxy polymer containing nanosilica particles

Nanosilica content (wt%)	Nanosilica content (vol%)	T_g (°C) DSC	T_g (°C) DMTA	E (GPa)	K_{Ic} (MN m$^{-3/2}$)	G_{Ic} (J/m^2)
0	0	143	153	2.96	0.59	103
4.1	2.5	137	152	3.20	1.03	291
7.8	4.9	136	154	3.42	1.17	352
11.1	7.1	141	151	3.57	1.18	343
14.8	9.6	138	152	3.60	1.29	406
20.2	13.4	138	150	3.85	1.42	461

4.2 Fracture Surfaces

SEM micrographs of the fracture surfaces of selected epoxies are shown in Figure 1. Crack growth occurs from left to right. The fracture surface of the unmodified epoxy polymer is shown in Fig. 1a, where the direction of crack propagation is from left to right. The fracture surface is relatively smooth and glassy, which is typical of a brittle thermosetting polymer [10], and shows that no large-scale plastic deformation has occurred during fracture. These observations agree well with the low measured toughness of the material, where K_{Ic} = 0.59 MN m$^{-3/2}$. In addition, there are apparent steps and changes of the level of the crack which can be observed in Fig. 1a. These features are feather markings, which are caused by the crack forking due to the excess of energy associated with the relatively fast crack growth. This repeated forking and the multi-planar nature of the surface are ways of absorbing this excess energy in a very brittle material [11]. The fracture surface of the nanosilica-modified materials showed similar features to those of the unmodified epoxy polymer, as shown in Fig. 1b and c. Crack forking and feather markings are observed, and the fracture surfaces have a brittle appearance. However, the addition of nanosilica did not give an apparent increase in the roughness observed by scanning electron microscopy, unlike for micrometre-sized particles, e.g. [9].

4.3 Toughening micromechanisms

The results were compared to various toughening mechanisms proposed in the literature, and most explanations were discounted, including crack pinning, immobilized polymer and crack deflection [13].

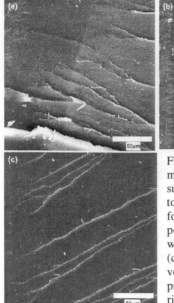

Fig. 1. Scanning electron micrographs of fracture surfaces, showing the precrack towards the left of the image, for (a) unmodified epoxy polymer, (b) epoxy polymer with 2.5 vol% nanosilica, and (c) epoxy polymer with 13.4 vol% nanosilica. (Crack propagation is from left to right.)

4.3.1 Plastic Void Growth

The toughening mechanisms associated with micrometer-sized particles have frequently been shown to be due to debonding of the particles followed by plastic void growth. Indeed, Kinloch and Taylor [10] have also demonstrated that the voids around particles closed-up when the epoxy polymer was heated above its T_g and allowed to relax. The debonding process is generally considered to absorb little energy compared to the plastic deformation of the matrix. However, debonding is essential because this reduces the constraint at the crack tip and hence allows the matrix to deform plastically via a void growth mechanism.

High resolution scanning electron microscopy (FEG-SEM) of a fracture surface of the polymer containing 9.6 vol% nanosilica, see Fig. 2, showed the presence of voids around several of the nanoparticles. This shows that plastic void growth of the epoxy matrix, initiated by debonding of the nanoparticles, has occurred. The diameter of these voids is typically 30 nm. These voids were also observed in the fracture surfaces of samples with different contents of nanosilica. Although the samples are coated to prevent charging in the electron microscope, the voids are not an artefact of the coating as they could not be observed on a coated fracture surface of the pure epoxy polymer, see Fig. 3. Also, the nanosilica modified samples appeared similar whether they were coated with platinum or gold.

In addition, similar voids were observed by AFM of uncoated fracture surfaces, see Fig. 4. However, the apparent diameter of the nanoparticle in the void highlighted in Fig. 4 is 30 nm, as shown in the graph in Fig. 4, whereas transmission electron microscopy has shown that the mean particle size is actually around 20 nm. This discrepancy is due to the tip-broadening effect when the AFM is used to identify such small features. As the tip

radius of the AFM probe is about 10 nm, this makes features that are protruding out of a surface appear larger than their true size in the micrographs. The void diameter in this case is about 70 nm, and it would appear that AFM can only be reliably used to detect the largest voids.

Voids with no nanoparticles were also observed with FEGSEM. Here the particles associated with these voids will be situated in the opposite fracture surface, or have fallen out of the surface completely during fracture, as is commonly observed with micrometre-sized particles [10]. (It should be noted that the diameters of most of these holes are less than those discussed above, as the matrix is unlikely to fail across the widest point of the void. Further, the coating, which is 5 nm thick, will partially fill the voids, and hence the observed size may be smaller than the true (uncoated) diameter.)

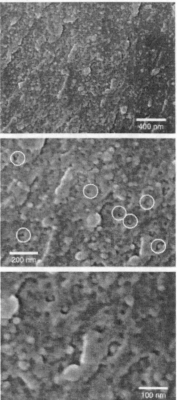

Fig. 2. Scanning electron micrographs (FEG-SEM) of the fracture surface of the epoxy polymer containing 9.6 vol% nanosilica. (Voids with nanoparticles are circled in the central image.)

Fig. 3. Scanning electron micrograph (FEG-SEM) of the fracture surface of the unmodified epoxy polymer.

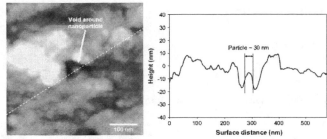

Fig. 4. Atomic force micrograph (height image) of a fracture surface of the epoxy polymer containing 9.6 vol% nanosilica as well as the surface profile of the line drawn across the nanosilica particle and void

4.3.2 Modelling of the contribution from the plastic void growth mechanism

To confirm whether the observed debonding and plastic void growth which occur for the nanoparticle-modified epoxy could be responsible for the toughening effect, the increase in toughness can be compared to a theoretical model. A suitable model for this is by Huang and Kinloch [12] where the contribution to the increase in fracture energy from the plastic void growth mechanism, ΔG_v, is given by:

$$\Delta G_v = (1 - \mu^2_m/3)(V_v - V_f)\sigma_{yc}r_{yu}K^2_{vm}$$

where μ_m mm is a material constant, V_v is the volume fraction of voids, V_f is the volume fraction of particles, σ_{yc} is the compressive yield stress of the unmodified epoxy polymer, r_{yu} is the radius of the plastic zone of the unmodified epoxy polymer, and K_{vm} is the maximum stress concentration factor of the von Mises stress in the plastic matrix. A comparison of the predicted and measured toughening increments is given in Table 2.

Table 2
Measured and predicted toughening increments

Nanosilica content (wt%)	Nanosilica content (vol%)	Toughening increment, Ψ (J/m^2)	
		Measured	Predicted
4.1	2.5	188	107
7.8	4.9	249	209
11.1	7.1	240	297
14.8	9.6	303	394
20.2	13.4	358	540

4.4 Conclusions

An epoxy resin cured with an anhydride has been used. This was modified by the addition of silica nanoparticles, manufactured using a sol-gel process, which were 20 nm in diameter. These particles were well-dispersed through the epoxy matrix with no agglomeration observed using transmission electron and atomic force microscopies. The addition of the nanoparticles did not affect the glass transition temperature; the Tgs of the unmodified and nanoparticle-modified epoxy polymers were measured to be in the range of 140±4°C using differential scanning calorimetry. Dynamic mechanical thermal analysis confirmed this observation. The addition of nanoparticles

increased the modulus of the epoxy polymer as expected. The fracture toughness of the polymers was measured, and a K_{Ic} of 0.59 MN m$^{-3/2}$ was recorded for the unmodified epoxy. Addition of the nanoparticles increased the fracture toughness with a maximum value of 1.42 MN m$^{-3/2}$ being measured for the epoxy polymer with 13.4 vol% of nanoparticles. These values were converted to fracture energies, G_{Ic}, using the measured modulus. The unmodified epoxy polymer gave $G_{Ic} = 103$ J/m^2, and a maximum fracture energy of 460 J/m^2 was calculated. Hence there is a significant toughening effect due to the addition of the silica nanoparticles. Observation of the fracture surfaces using scanning electron and atomic force microscopies showed nanoparticles surrounded by voids, providing evidence of debonding of the nanoparticles and subsequent plastic void growth. An analytical model of plastic void growth was used to confirm whether this mechanism could be responsible for the increased toughness. The mean void diameter was measured from the micrographs, and the model was used to predict the toughening increment (compared to the fracture energy of the unmodified epoxy polymer). The predicted values agreed well with the measured values, indicating that debonding of the nanoparticles and subsequent plastic void growth were most likely to be responsible for the increase in toughness that was observed due to the presence of the nanosilica particles.

4.5 References

[1] Drake RS, Siebert AR. SAMPE Quart 1975;6(4):11e21.
[2] Pascault JP, Williams RJJ. Formulation and characterization of thermoset-thermoplastic blends. In: Paul DR, Bucknall CB, editors. Polymer blends volume 1: formulation. New York, USA: John Wiley & Sons; 1999. p. 379-415.
[3] Young RJ, Beaumont PWR. J Mater Sci 1975;10:1343e50.
[4] Kinloch AJ, Taylor AC, Lee JH, Sprenger S, Eger C, Egan D. J Adhes 2003;79(8-9):867-73.
[5] Sprenger S, Eger C, Kinloch AJ, Taylor AC, Lee JH, Egan D. Adhaesion Kleben Dichten 2003;2003(3):24-8.
[6] ISO-527-1. Plastics e determination of tensile properties - part 1: general principles. Geneva: ISO; 1993.
[7] ISO-527-2. Plastics e determination of tensile properties - part 2: test conditions for moulding and extrusion plastics. Geneva: ISO; 1996.
[8] ISO-13586. Plastics e determination of fracture toughness (G_{Ic} and K_{Ic}) - linear elastic fracture mechanics (LEFM) approach. Geneva: ISO; 2000.
[9] Kinloch AJ, Taylor AC. J Mater Sci 2006;41(11):3271-97.
[10] Kinloch AJ, Taylor AC. J Mater Sci 2002;37(3):433-60.
[11] Andrews EH. Fracture in polymers. 1st ed. Edinburgh: Oliver & Boyd; 1968.
[12] Huang Y, Kinloch AJ. J Mater Sci 1992;27(10):2763e9.
[13] Johnsen, B.B., Kinloch A.J., Mohammed R.D., Taylor A.C., Sprenger S., Polymer 2007;48:530-541.

Self-Assembly of Polymers and Nanoparticles in Personal Care Products

M.C. Berg, N.A. Suddaby and D. Soane

Soane Labs, LLC
35 Spinelli Place, Cambridge, MA 02138 USA mcberg@soanelabs.com

ABSTRACT

Soane Labs develops and applies nanotechnology and advanced polymer chemistry to the personal care industry trough novel colorants and high-affinity polymers. By controlling the spontaneous self-assembly of polymers and nanoparticles at interfaces, the color, look, and feel of skin or hair can be modified. Furthermore, by modifying the surface using self-assembly of polymers, safe and environmentally friendly materials can be used with less toxicity than many materials currently used in personal care products. For example, many of today's commercial coloring products involve reactive precursor blends, which cause staining and other undesirable effects. In contrast, Soane Labs' hybrid pigments can impart both desired physical attributes and enhanced color. The combination of hybrid pigments and self-assembled polymers have been found to produce both hair and skin cosmetics with unique properties. Soane Labs has been successful in developing many hair care treatments using this area of nanotechnology including a system for advanced hair hold, easy-to-apply hair color, and UV-blocking/conditioning agents. Such nanotechnology products can lead to better product performance, high safety for consumers, and better economics for manufacturers.

Keywords: cosmetics, colorants, personal care, self-assembly, hair care

1 NOVEL EFFECTS PIGMENTS

Nacreous pigments are often composed of layered metal oxides that interfere with reflected light to produce a subtle iridescent color. These shimmering particles then have to be mixed with other dyes or pigments to produce a brilliant color in a coating or film. Soane Labs has developed a technology to attach colorants to nacreous pigments in a thin enough layer to sustain the shimmering color, yet thick enough to produce a brilliant scattering color. By providing these effects in one single pigment, many disadvantages to small molecule dyes such as bleeding or migration can be avoided and unique color effects can be produced. Figure 1 shows samples of these novel effects pigments over a range of color dispersed in a lacquer.

Figure 1. Novel effects pigments in lacquer.

2 HAIR COLOR

Traditional hair color treatments often involve harsh chemistry such as peroxides and oxidative couplers that can leave the hair damaged after treatment. The hair color treatment developed by Soane Labs does not chemically modify the hair yet is substantive to friction and shampoo washing after application. The treatment is a nanostructured material with a high affinity for the hair surface. These products can potentially benefit consumers both in the salon and in home applications. Figure 2 provides pictures of (a) a control sample of blonde hair and (b) blonde hair dyed black. A range of colors is being developed for both light-to-dark and dark-to-light transitions.

(a)

(b)

Figure 2. Images of (a) control blonde hair and (b) blonde hair dyed black.

3 HAIR STYLING

Hair styling products are particularly susceptible to humid conditions or other forms of moisture. In addition, many commercial products leave a stiff and/or tacky feel. Our unique hair styling polymeric system holds the desired hair shape under extreme humidity conditions and will withstand limited water submersion. It also has a softening agent built into the polymeric system to leave the hair with a desirable soft texture. Figure 3 depicts hair after 24 hours in 95% humidity conditions with various amount of styling polymer.

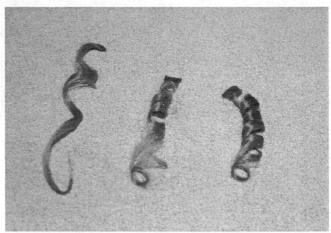

Figure 3. Hair samples after 24 hours in 95% humidity conditions.

4 CONDITIONER

Soane Labs has developed a conditioning polymer designed to have a high affinity for hair and impart a soft feel. Due to its high affinity, the polymer can withstand water and oil residues to leave a soft feel for a longer time compared to many conditioning polymers on the market. In addition, UV blockers have been successfully incorporated into the polymer synthesis to potentially block UV radiation from damaging the hair. In addition to softening and UV protection, the polymer can act as a base material for adding a variety of functionalities due to its attraction to the surface of hair.

5 CONCLUSIONS

Soane Labs has developed multiple active materials for use in the cosmetic industry. Many of these materials provide unique properties that can differentiate products from the competition. The underlying technology behind the Soane Labs products is self-assembly of polymers or polymer/nanoparticle systems onto surfaces.

With New Dispersing Technologies Towards Nanoscale Systems

Dr. L. Fischer[*] and Dr. C. Thomas[**]

[*]AC Serendip Ltd.
Fahrenheitstrasse 1, Bremen DE 28359, Germany, ludger.fischer@ac-serendip.com
[**]Serendip AG, Sihleggstrasse 23, Wollerau CH 8832, Switzerland, chris.thomas@serendip.ag

ABSTRACT

The presentation will give an overview of state-of-the-art technologies and today's possibilities as well as technical and regulatory limitations. The focus will be on the potential of a new generation of homogenizers, which allow for the efficient and energy-saving dispersion of cosmetic formulations. Examples of modern formulations with its specific technological requirements are presented.

Keywords: nanoscaled systems, technologies, emulsion, lipsomes, dispersion, formulations

1 NANOTECHNOLOGY

During the last decades, nanotechnology has become an integral part in research and development. Effort in the field of nanoparticle synthesis and analysis as well as the development of their industrial fabrication processes allowed for the emergence of an interdisciplinary science with new and widespread applications. Simultaneously, the development of suitable analytical techniques such as electron microscopy permitted deeper insights in the structure of materials. As indicated by the name, the main feature in nanotechnology is the particle dimension and it is known that the consequent reduction of the latter one can give rise to significant changes in the particle properties.

Nanoparticle production is described either by the "bottom-up" or the "top-down" approach, e.g., the assembly of particles starting from the molecular level or the subsequent particle downsizing of macro-material. Very often, the pathway for molecular assembly is not accessible as raw products provided by nature exist as macro-sized and "pre-formed" material. Here, new technologies are required that allow for efficient and economical dispersing resp. homogenization of materials towards nanoscale products.

2 WHY ARE SMALL PARTICLE SIZE AND HOMOGENEITY IMPORTANT?

In various application fields, small particle size is essential for an improvement of product quality. For example, colors in paints or lacquers develop an enhanced level of brilliance and may even display special effects, when the pigments in the film are small and homogeneous in size. This holds true also for coatings, where the scratch resistance is based on effects from the small particle dimension. Further, dispersions based on waxes, resins and various other classes are economically important in various industrial sectors and their properties depend considerably on size. When applications in the field of cosmetics are concerned, the search for a suitable type of transport medium (e.g. emulsion, liposomes, etc.) is as important as the particle physico-chemical parameters, such as size and size distribution. This is due to the general observation that small particles are often more stable and more capable of undergoing skin penetration. As a consequence, the delivery effect is more pronounced in this case.

In the following, we would like to give an example illustrating the importance of size and homogeneity. In certain cases, products with pleasing optical properties with high degree of transparency are required. Generally, transparency is enhanced, when the particles in the dispersion are preferably small and homogeneous. The optical aspect of the sample can be visualized by laser light transillumination. In Fig. 1, a dispersion with a broad spectrum of particles sizes in the submicron range is shown before homogenization. At first sight, the sample appears indeed homogeneous but it is perceptible that the opalescent character is limited due to the blearing effect resulting from the fraction of large particles. Dynamic Light Scattering (DLS) analysis reveals that the hydrodynamic mean diameter is about 300 nm and that particle sizes range from 80 nm to 800 nm circa (yellow curve).

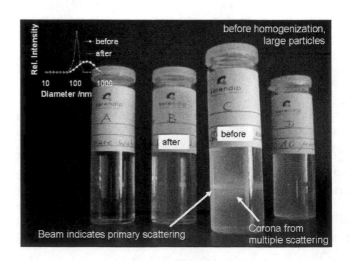

Figure 1: Laser light scattering of non-homogenized sample (mean diameter circa 300 nm). The laser beam is surrounded by a diffuse corona of multiple scattering processes, and indicates a broad spectrum of large particles.

During the homogenization process the mean diameter of the particles is strongly reduced (after few homogenization passages only) and the size distribution clearly shifts to smaller particle sizes (see Fig. 2, red curve). When laser light transilluminates the homogenized sample, the enhanced transparency and the less diffuse character of the scattered light become visible. This is due to the fact that small particles scatter light less efficiently than large ones.

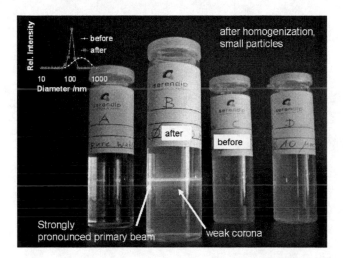

Figure 2: Laser light scattering of homogenized sample (mean diameter circa 120 nm). The laser beam is more pronounced and the corona from multiple scattering is reduced, indicating smaller and more narrowly distributed particles.

However, likewise important is the effect on the shape of the size distribution, which undergoes a significant narrowing upon homogenization. The homogenized sample completely lacks particles above 200 nm circa, which is important as the particle properties are strongly correlated with their dimension. Therefore, a narrow size distribution of particles corresponding to an enhanced level of homogeneity promises products with advantageous properties and a smaller spectrum of side effects as imposed by "odd" particles.

3 LOW PRESSURE HOMOGENIZATION: EFFICIENT AND ENERGY-SAVING DISPERSING OF PARTICLES

Summarizing, the given examples point towards the general need for the development of new technologies that are able to fulfil the market requirements of various products (color brilliance, scratch resistance, etc.).

The Serendip Low Pressure Nanogenizer technology (LPN, see Fig. 3) is based on a well-known high pressure homogenizing technology. The core of our technology is a new pressure cell that is designed according to the latest scientific know how, considering the optimised effects of friction, turbulence and cavitation. The technology combines the advantages of valve and nozzle systems allowing for very low working pressures and thus gentle process conditions. The technology opens the door to high quality dispersions in the field of nanoparticles and aims particularly to supersede conventional technologies such as bead mills, etc.

Figure 3: Serendip LPN 60

As the LPN works at moderate pressures (e.g. 300 or 500 bar) compared to high pressure homogenizers (up to 2500 bar), energy consumption and wear are considerably reduced. Fig. 4 reports about the mean particle decrease as a function of the number of homogenization passages. In the diagram, the effect on the particle mean size upon LPN homogenization at two working pressures (300 and 500 bar) is schematically illustrated and compared to the effect upon conventional bead mill processing. Clearly, the homogenization process with the LPN leads to significantly better results concerning the homogenization rate as well as the final particle size. As one can see, the final homogenization result is obtained after few passages only, corresponding to much shorter production times and thus less energy consumption.

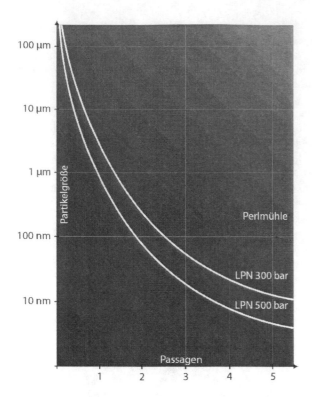

Figure 4: Mean particle size decrease during
homogenization as a function of the number of passages for
the Low Pressure Nanogenizer (LPN) and a conventional
bead mill

NSTI-Nanotech 2008, www.nsti.org, ISBN 978-1-4200-8503-7 Vol. 1

Molecular approach for the synthesis of supported nanoparticles on active carbon using noble metal clusters

C. Willocq[*], D. Vidick[*], S. Hermans[*], A. Delcorte[**], P. Bertrand[**], V. Dubois[***] and M. Devillers[*]

*Université catholique de Louvain, Unité de Chimie des Matériaux Inorganiques et Organiques
Place Louis Pasteur 1/3, B-1348 Louvain-la-Neuve, Belgium
**Unité de Physico-Chimie et de Physique des Matériaux
Place Croix du Sud 1, B-1348 Louvain-la-Neuve, Belgium
***Institut Meurice, Avenue Emile Gryson 1, B-1070 Brussels, Belgium

ABSTRACT

Noble metal clusters were used as precursors for active carbon supported nanoparticles. The synthetic strategy implied first the synthesis of organometallic clusters, i.e. multinuclear complexes with mainly carbonyl ligands. Second, the active carbon surface was treated in order to create at its surface anchoring sites for the clusters in the form of chelating phosphine groups. In a third step, mixed-metal or Pd-based clusters were introduced on the surface of the functionalized support and of the intact carbon for comparison. A fourth step consisting in thermally treating the obtained solids led to activated 'naked' nanoparticles. Characterization at each step of the synthesis relied on XPS, SIMS and TEM mainly. The series of obtained Pd/C materials were tested as hydrogenation catalysts and gave competitive activities.

Keywords: nanoparticle, cluster, carbon, palladium, heterogeneous catalysis

1   INTRODUCTION

Active carbons are widely used in the industry as supports in heterogeneous catalysis because they are cheap, robust and present a high specific surface area. However, they are seldom functionalized [1, 2] and metallic precursors are generally simply deposited on their surface without any particular interaction leading to inhomogeneous solids with particles of variable shape, size and composition.

Clusters are organometallic compounds containing a minimum of three metallic atoms linked by minimum two metal-metal bonds and surrounded by a stabilizing ligand layer. Thus, being well-defined entities, they are known to be ideal precursors for the synthesis of nanoparticles of controlled size and composition [3]. They were already chemically grafted on dendrimers [4] or polymer beads [5] for example but never on carbon.

Here, we report the chemical functionalization of an active carbon support and the grafting of noble metal clusters on this modified support in order to control the nanostructure of supported nanoparticles thanks to a molecular strategy. After a soft thermal activation,

homogeneous supported nanoparticles of 1-10 nm in size were obtained. These particles are active in hydrogenation heterogeneous catalysis.

2   RESULTS AND DISCUSSION

2.1   Cluster Synthesis

A series of palladium clusters of increasing nuclearity were obtained by using a simple strategy based on the reaction of $Pd_2(dba)_3$ with PPh_3 in various amounts. When running the reaction with a 1:1 stoichiometric ratio, the cluster $[Pd_{10}(CO)_{12}(PPh_3)_6]$ was isolated. This cluster was characterized by infrared spectroscopy, ^{31}P-NMR, ^{1}H-NMR and X-ray crystallography. Its molecular structure consists of an octahedron of Pd atoms with four Pd-Pd edges bridged by additional Pd atoms, surrounded by terminal phosphines and carbonyl ligands (Fig. 1).

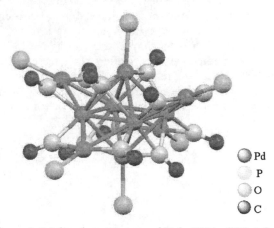

Pd
P
O
C

Figure 1: Molecular structure of $[Pd_{10}(CO)_{12}(PPh_3)_6]$. Phenyl groups have been omitted for clarity.

With 3 equivalents PPh_3, the $[Pd_4(CO)_5(PPh_3)_4]$ cluster was obtained. Again, characterization by IR and NMR confirmed the formulation and X-ray crystallography was used to determine the molecular structure. It consists of a Pd_4 tetrahedron with terminal phosphines on each metal atom and bridging CO's on Pd-Pd edges (Fig. 2). With 0.5 equivalent PPh_3 in the same reaction, a cluster of higher nuclearity is obtained, which corresponds to an approximate formula $[Pd_n(CO)_x(PPh_3)_y]$ where n>10 and y<x.

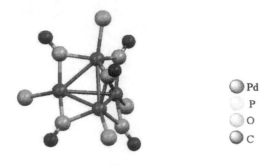

Figure 2: Molecular structure of $[Pd_4(CO)_5(PPh_3)_4]$. Phenyl groups have been omitted for clarity.

Two Pd clusters of the literature were also prepared following the published procedures [6, 7]: namely $[Pd_{12}(CO)_{17}(PBu^n_3)_5]$ and $[Pd_{10}(CO)_{12}(PBu^n_3)_6]$. Characterization by infrared spectroscopy and ^{31}P-NMR confirmed the success of the syntheses. A known mixed-metal species $[Ru_5PtC(CO)_{14}(COD)]$ was finally prepared from $(PPN)_2[Ru_5C(CO)_{14}]$ and $[Pt(COD)Cl_2]$ [8].

2.2 Support Functionalization

The strategy envisaged for the carbon support functionalization is illustrated in Figure 3. First, the number of surface acidic functions were increased by treatment with HNO_3 and quantified by the so-called Boehm's titration method and X-ray Photoelectron Spectroscopy (XPS). Second, the carboxylic groups were activated by treatment with $SOCl_2$ to convert them to acyl chlorides. Third, ethylenediamine was grafted onto these groups, leaving pending amine functions, which were transformed into chelating phosphines by the action of $H_2CO/HPPh_2$ in a final step to give the functionalized support (noted C_{PPh2}).

2.3 Cluster-derived Nanoparticles

In order to study the grafting mechanism of molecular clusters on the phosphine-containing support, the Ru_5Pt mixed-metal cluster was brought in contact with the functionalized carbon; and characterized by Secondary Ion Mass Spectrometry (SIMS), XPS and TEM.

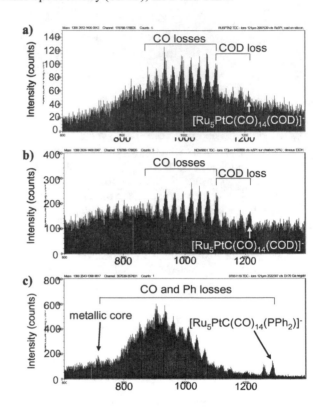

Figure 4: Negative SIMS spectra of the cluster $[Ru_5PtC(CO)_{14}(COD)]$: a) pure unsupported cluster as reference, b) cluster deposited on C_{SX+}, and c) cluster grafted on C_{PPh2}.

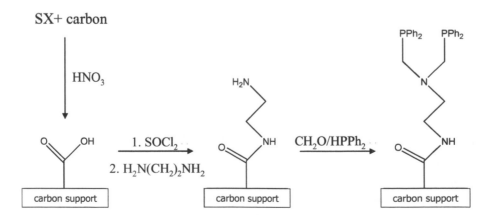

Figure 3: Functionalization of the carbon support.

The mass spectrum obtained for the pure, unsupported cluster (Fig. 4a), and for the cluster deposited on the unmodified support (Fig. 4b) fitted perfectly, showing that the cluster could be deposited in a molecularly intact fashion on the carbon support. However, when the cluster was grafted onto the P-containing support (Fig. 4c), peaks at higher masses were obtained, confirming the occurrence of a surface chemical reaction in this case. In addition, XPS showed that the Ru/Pt ratio present initially in the cluster was retained on the surface. We believe that a ligand exchange mechanism occurred, as illustrated in Figure 5.

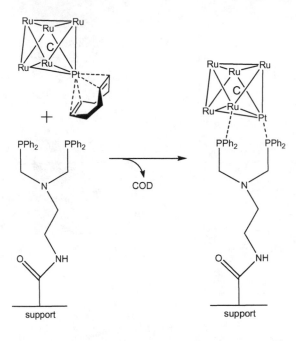

Figure 5: Grafting model for [Ru$_5$PtC(CO)$_{14}$(COD)] on C$_{PPh2}$. Carbonyl ligands are omitted for clarity.

Figure 6: TEM image of the cluster [Ru$_5$PtC(CO)$_{14}$(COD)] grafted on C$_{PPh2}$ and thermally treated.

After gentle thermal treatment, to remove the ligands, nanoparticles of very small sizes are formed on the support. These were evidenced by TEM (Fig. 6). Analysis by EDX spectroscopy within the microscope confirmed the presence of both ruthenium and platinum.

The same strategy was applied to the series of homometallic Pd clusters described above in section 2.1: (i) grafting onto the phosphine-functionalized support (ii) characterization by SIMS, XPS and TEM, (iii) thermal activation and subsequent re-evaluation by the same characterization techniques. In addition, the activated samples were also analysed by elemental analysis and CO chemisorption to determine metal loading and dispersion, respectively (Table 1). TEM showed that this method of preparation allowed to obtain carbon-supported palladium nanoparticles of 1 to 10 nm in size (Figure 7).

Precursor	Pd loading (wt.%)		Dispersion (%)	
	C$_{SX+}$	C$_{PPh2}$	C$_{SX+}$	C$_{PPh2}$
[Pd(dba)$_2$]	2.97	2.44	3	3
Pd$_{10}$ Med	2.35	1.68	29.1	17.4
Pd$_{12}$ Med	2.78	2.11	25.6	7.8
Pd$_{10}$	1.94	1.98	7.4	12.3
Pd$_4$	1.02	1.81	3.3	8.1
Pd$_n$	2.61	2.93	22.1	9.1

Table 1: Results of elemental analysis and CO chemisorption for the activated Pd/C samples (Med = cluster prepared following the method described by Mednikov et al. [6, 7]).

2.4 Application in Catalysis

The obtained Pd/C materials were tested as catalysts in the hydrogenation of nitrobenzene into aniline. It was found that the activity was competitive when compared to a commercial catalyst. Moreover, the samples prepared on the unmodified carbon for comparison gave generally worse results (Figure 8).

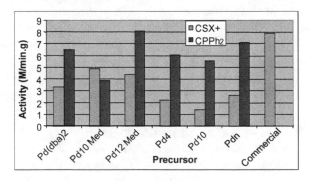

Figure 8: Catalytic activity by accessible g of Pd in the hydrogenation of nitrobenzene.

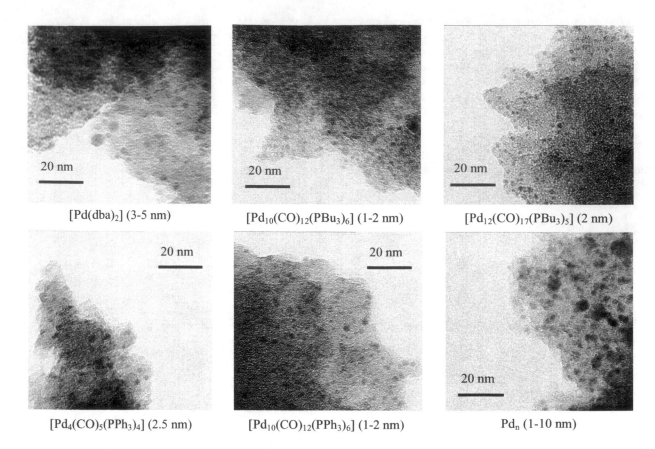

[Pd(dba)$_2$] (3-5 nm) [Pd$_{10}$(CO)$_{12}$(PBu$_3$)$_6$] (1-2 nm) [Pd$_{12}$(CO)$_{17}$(PBu$_3$)$_5$] (2 nm)

[Pd$_4$(CO)$_5$(PPh$_3$)$_4$] (2.5 nm) [Pd$_{10}$(CO)$_{12}$(PPh$_3$)$_6$] (1-2 nm) Pd$_n$ (1-10 nm)

Figure 7: TEM images of the various Pd clusters grafted on C$_{PPh2}$ and thermally treated (mean particle sizes are given in parentheses).

3 CONCLUSION

In this work, we have shown that molecular clusters are ideal precursors for the preparation of nanoparticles of controlled size and composition. Moreover, we have been able to functionalize an active carbon support in order to introduce chelating phosphine groups on its surface. The clusters can then ideally be grafted onto the modified support in order to secure the desired nanoparticles. A series of Pd/C samples were tested in hydrogenation catalysis and shown to be as active as a commercial material.

4 ACKNOWLEDGMENTS

The authors acknowledge the F.N.R.S. and F.R.I.A. for a research fellowship to CW. They also acknowledge Prof. E. M. Gaigneaux for access to XPS, Prof. R. Legras for access to SEM and TEM, the NORIT firm for supplying the carbon support and J.-F. Statsijns for technical support.

This work has been performed within the framework of the Interuniversity Attraction Poles Programme of the Belgian State, Belgian Science Policy, Project INANOMAT, P6/17.

REFERENCES

[1] J. A. Diaz-Aunon et al., Stud. Surf. Sci. Catal. 143, 295, 2002.
[2] A. R. Silva et al., Catal. Today 102, 154, 2005.
[3] J. C. Fierro-Gonzalez et al., J. Phys. Chem. B 110, 13326, 2006.
[4] M. T. Reetz et al., Angew. Chem. Int. Ed. 36, 1526, 1997.
[5] C. M. G. Judkins et al. Chem. Commun. 2624, 2001.
[6] E. G. Mednikov, N. K. Eremenko, S. P. Gubin, Y. L. Slovokhotov, Y. T. Struchkov, J. Organomet. Chem., 239, 401, 1982.
[7] E. G. Mednikov, N. K. Eremenko, S. P. Gubin, J. Organomet. Chem. 202, C102, 1980.
[8] S. Hermans; T. Khimyak; B. F. G. Johnson, J. Chem. Soc., Dalton Trans. 3295, 2001.

Evaluation of Antibacterial Activities of Zinc Oxide-Titanium Dioxide Nanocomposites Prepared by Sol-Gel Method

Chang Hengky*, Leung Henry**

*Biomedical Engineering Group, School of Engineering (Manufacturing), Nanyang Polytechnic, 180 Ang Mo Kio Avenue 8 Singapore 569830, hengky_chang@nyp.gov.sg
**School of Chemical and Life Sciences, Nanyang Polytechnic, 180 Ang Mo Kio Avenue 8 Singapore 569830, henry_leung@nyp.gov.sg

ABSTRACT

The antibacterial activity of zinc oxide nanoparticles embedded at 10, 20 and 30 wt% in titanium dioxide glass matrix synthesized via sol-gel route was studied via conductimetry assay against Staphylococcus aureus and Escherichia coli bacterial. The zinc oxide-titanium dioxide nanocomposite was characterized by Field Emission Scanning Electron Microscope and X-Ray Diffractometer to observe the microstructure morphology and crystallinity phase of the synthesized nanocomposite. The nanocomposite granules microstructure has been analyzed using FESEM and show a very high surface area qualitatively and the x-ray spectrum does not show TiO_2 crystal phase suggesting that TiO_2 is in amorphous glass phase. Nanocomposites with 20wt% and 30wt% of ZnO nanoparticles in TiO_2 solgel matrix (will be refer as TiO_2-ZnO20 and TiO_2-ZnO30 respectively) inhibited 40% to 95% of both antibacterial proliferation from different batch of nanocomposite products. Both nanocomposites selectively inhibit towards *E.Coli* compare with *S. Aureus*. A clear dose-dependent response between TiO2-ZnO20 and TiO2-ZnO30 was recorded in S. *Aureus* assay.

1 INTRODUCTION

ZnO powders applications have been found for varistors and other functional devices and also can be used as enforcement phase in wear resistant phase, anti-sliding phase, and antistatic phase in composites in consequence of its high elastic modulus and strength properties and current characteristic as an n-type semiconductor material [1-3].

Recently, antibacterial activity of ceramic powders has attracted attention as a new technique that can substitute for conventional methods using organic agents. Ceramic powders of zinc oxide (ZnO), calcium oxide (CaO) and magnesium oxide (MgO) were found to show marked antibacterial activity [4–6]

The use of ZnO has the following advantages: It contains mineral elements essential to humans and exhibit strong antibacterial activity in small amounts without the presence of light. It was found that ZnO exhibits antibacterial activity at Ph values in the range from 7 to 8 [7], and these values are suitable for use in water used for washing. The antibacterial activity of ZnO is considered to be due to the generation of hydrogen peroxide (H_2O_2) from its surface [8]. However, the use of ZnO in powder form for antibacterial water treatment is limited as it will cause water to be turbid and the nano particles will also flow with water stream and contaminated to locations where clear water is required. Thus, ZnO powder has to find a way in a form composite for many applications in water purification, antibacterial coatings and etc.

Thin films or nanoscale coating of ZnO nanoparticles on suitable substrates is also potential as substrates for functional coatings, printing, UV inks, e-print, optical communications (securitypapers), protection, barriers, portable energy, sensors, photocatalytic wallpaper with antibacterial activity etc.[9,11–24]. The nanocomposite coating of ZnO nano particles with biomolecules, oil, pigments (calcium carbonate, clay, talc, silicates, TiO2, etc.), polymers, plastics etc. has been reported with the help of suitable binders and cobinders [10]. However, to the best of our knowledge till now no paper has reported on the synthesis of nanocomposites of Zinc Oxide nanoparticles with the TiO_2 amorphous glass as matrix via sol gel method.

In the present work, nanocomposites containing nano particle zinc oxide (average size 20 nm) in titania sol gel matrix has been synthesized with various loading weight percentage of nano zinc oxide powder with respect to titania sol and dried at 40°C for 8 hours. After the nanocomposites were dried into flakes, they were characterized using electron microscopy and x-ray diffractometer to analyse the structure and morphology of the nanocomposites. In this study, the antibacterial activities of the nanocomposites with various nano particles zinc oxide weight percentage were also studied against Staphylococcus aureus and Escherichia coli by conductimetric assay.

2 EXPERIMENTAL

2.1 Chemicals and materials

A commercially available reagents grade alkoxide solution of Titanium (IV) Isopropoxide (TTIP) (purity > 98%, Sinopharm Chemical Reagent Co. Ltd), Isopropyl alcohol (IPA), deionised water with resistivity of 18.3 MΩ and Hydrochloric acid 3 Molar concentration were used to make the TiO_2 solgel. ZnO nanoparticles (purity > 99.7%, Sinopharm Chemical Reagent Co. Ltd) with an average particles size (nm) of 20 and specific surface area of more than 90 m2/g. All the chemicals and materials mentioned were used as purchased without further purification.

2.2 Nanocomposite synthesis technique

Firstly, the TiO2 sol was prepared by mixing the TTIP ($Ti(OC_3H_7)_4$) with solvent, acid catalyst and deionised water that will lead to series reactions of hydrolysis and condensation. The TTIP alkoxide solution was mixed and stirred using magnetic stirrer in isopropyl alcohol at a molar ratio of 1:14.7 for 30 minutes before four drops of hydrochloric acid (3M) was added into this solution and followed by another four drops of deionised water into the above solution. The reaction mixture was keep stirred and aged for 1 hour before various weight percentages of ZnO nano particles (10, 20 and 30 wt %) were added into the TiO_2 sol. The resultant mixture of the ZnO nanoparticles in TiO_2 sol was stirred for 1 hour and followed by sonication process in ultrasonic bath for 10 minutes. After the sonication, the TiO_2 sol containing ZnO nano particles were transfer into petri dish and covered by laboratory film (Parafilm) with few holes for the evaporation of volatile solvent during heat treatment in oven at 40°C for around 6 hours until the nanocomposites solution dried and form granules prior to characterization follow by heat treatment at 400°C for 1 hour.

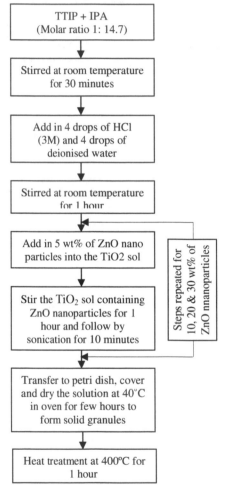

Fig. 1 Flow chart for the preparation of TiO_2 glass – ZnO nanoparticles nanocomposites granules by the sol-gel technique.

2.3 Nanocomposites materials characterization

The ZnO nanoparticles and the prepared nanocomposites granules surface morphology and elemental composition was characterized using FESEM (JEOL-JSM7500F) with built-in EDS. XRD patterns were obtained using PANAlytical X'Pert Pro MPD advanced powder X-ray diffractometer (using Cu Kα = 1.54056 A° radiation) with scanning range of 2 theta from 15° to 85°.

2.4 Antibacterial analysis technique

A collection strain of *E. coli* and *S. aureus* (American Type Culture Collection, Rockville, MD) has been used in this study. Bacteria from frozen stock cultures were grown aerobically to late logarithmic or early stationary phase in LB broth (Oxoid Ltd, Basingstoke, UK) at 37.8C. Cells were harvested by centrifugation and re-suspended in fresh medium. Inocula were prepared by adjusting the cell suspension to predetermined optical densities (OD) corresponding to 10^8 CFU/ml.

The antibacterial analysis has been performed following the method of Weiss et al. with minor modifications. 2 ml of the bacterial inoculums (approximately 10^6 bacteria) were placed at each well on 24 well culture plate against different concentration/weight of test material for 24hr.

Bacterial growth at 24hr was assay by colorimetric method (Drummond et al, 2000) 50ul of 10% resazurin solution (Sigma, St. Louis) was added to each well and the plate was then incubate at 37°C for 1hr. The OD in each well was measured at excitation 485nm and emission 530nm in a microplate reader (Tecan, Switzerland). All experiments, carried out under aseptic conditions, were repeated three times to ensure reproducibility.

3 RESULTS AND DISCUSSION

3.1 Nanocomposites materials characterization

The synthesized ZnO-TiO2 nanocomposite granules microstructure were characterized by FESEM and shown as follows:

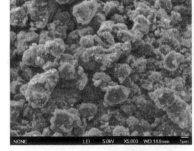

Fig. 2 SEM micrograph of ZnO-TiO210 (10 wt% ZnO in TiO2 sol-gel matrix)

Fig. 3 SEM micrograph of ZnO-TiO20 (20 wt% ZnO in TiO2 sol-gel matrix)

Fig. 4 SEM micrograph of ZnO-TiO30 (30 wt% ZnO in TiO2 sol-gel matrix)

It was observed that the higher the ZnO nanoparticles content, the higher the surface area as shown in figure 2 to 4. The higher surface area size on the surface of the nanocomposite granules is hypothesized to cause higher effectiveness in bacterial killing effect.

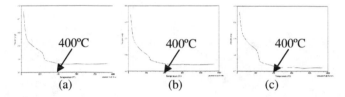

(a)   (b)   (c)

Fig. 5 Thermogravitometry (TGA) analysis of TiO2-ZnO nanocomposites (a) TiO2-ZnO10, (b) TiO2-ZnO20, (c) TiO2-ZnO30

Thermogravitometry analysis was performed on the TiO2-ZnO10, TiO2-ZnO20 and TiO2-ZnO30 nanocomposites and shown that after 400°C, all the organic and volatile components from the alkoxide precursors have been removed totally. Thus, heat treatment temperature parameter was set at this temperature for all the samples

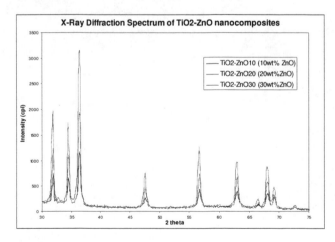

Fig. 6 X-Ray diffractometry spectrum of TiO2-ZnO10, TiO2-ZnO20 and TiO2-ZnO30

As shown in X-Ray Diffractometry spectrum, only ZnO crystals were shown and no TiO2 spectrum indicating that the TiO2 is in amorphous glass matrix. Also shown that the higher the ZnO nanoparticle weight percentage, the intensity peaks was also higher.

3.2 Antibacterial analysis

The antibacterial activity of TiO2-ZnO20 and TiO2-ZnO30 nanocomposites materials was tested using E. coli and S. aureus in comparison with ZnO power as positive controls, the results of which are presented in Fig. 6. TiO2-ZnO10 nanocomposite was not tested due to inability to form granules with required size for the antibacterial test. Data were represented as % survival of control against weight of ZnO powder present after 24hr incubation. Bacterial survival rate drops as the ZnO power concentration increased for both E. coli and S. aureus. The bacteria growth reduced by half at 15mg/ml and 18mg/ml respectively. Similar results were observed in both TiO2-ZnO20 and TiO2-ZnO30 nanocomposites materials. However, ZnO powder present in TiO2-ZnO20 and TiO2-ZnO30 nanocomposites contain 20% and 30% w/w of ZnO powder respectively. Our current result suggested that ZnO nanocomposites materials have better antibacterial activity compared with ZnO powder alone.

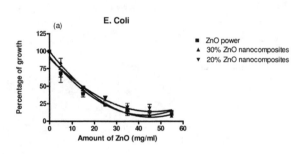

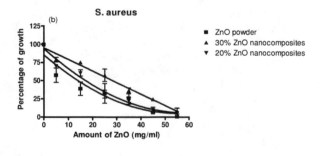

Fig. 7 Effect of ZnO nanocomposites on the growth of E. coli and S. aureus. Plots present mean % of growth of E. Coli (a) and S. aureus (b) exposed to different concentration of ZnO present for 24 hr incubation (n=4).

4 CONCLUSION

This antibacterial activity studies of ZnO nanopowder embedded in TiO2 amourphous matrix approach has shown excellent results against E. Coli and S.Aureus bacteria and proven the concept of creating nanocomposites of ZnO and TiO2 which can be used in granules flakes form as an antibacterial material or can be coated to any surfaces when the nanocomposite is in the sol form. Thus, the problem of turbidity caused by ZnO nanopowder in water or ZnO nanoparticle ashes in dry application such as air purification will be solved by this nanocomposite. The TiO2-ZnO nanocompsoite material has been characterized using FESEM, EDS, XRD. Considering the potential implication of this TiO2-ZnO nanocomposite material, it may potentially prove useful as nanomedicine based antimicrobial agents at selective therapeutic dosing regimes and coated layers can be varied using suitable chemistry for desired applications.

5 ACKNOWLEDGEMENT

The author would like to take this opportunity to express his sincere gratitude to Nanyang Polytechnic Nanomaterials Laboratory, Biomedical Engineering Hub and Biomolecular Laboratory, School of Life Sciences technical staffs (Choy Pei Ye and Tan Soek Soo) in helping him throughout the whole period of the research project. This study was conducted as part of the Nanyang Polytechnic's nanotechnology initiative effort to raise the applied research activity in the nanotechnology field.

6 REFERENCES

[1] M. Singhal, et al., Mater. Res. Bull. 32 (2) (1997) 236.

[2] C.M. Lieber, et al., US Patent 5 897 945, Application No. 606892, April 27, 1999.

[3] D.W. Yuan, et al., J. Mater. Sci. 34 (1999) 1293.

[4] Yamamoto O, Hotta M, Sawai J, Sasamoto T, Kojima H. J Ceram Soc Jpn 1998;106:1007.

[5] Yamamoto O, Sawai J, Sasamoto T. J Inorg Mater 2000;2:451.

[6] Sawai J, Igarashi H, Hashimoto A, Kokugan T, Shimizu M. J Chem Eng Jpn 1995;28:288.

[7] Yamamoto O, Hotta M, Sawai J, Sasamoto T, Kojima H. J CeramSoc Jpn 1998;106:1007.

[8] Sawai J, Kojima H, Igarashi H, Hashimoto A, Shoji S, Kokugan T etal. J Ferment Bioeng 1998;86:521.

[9] U. Ozgur, Y. I. Alivov, C. Liu, A. Teke, M. A. eshchikov, S. Dogan, V. Avrutin, S. J. Cho and H. Morkoc, J. Appl. Phys., 2005, 98, 041301.

[10] C. Kugge, V. S. J. Craig and J. Daicic, Colloids Surf., A, 2004, 238,1.

[11] O. Yamamoto, Int. J. Inorg. Mater., 2001, 3, 643.

[12] O. Yamamoto, M. Komatsu, J. Sawa and Z. E. Nakagawa,
J. Mater. Sci.: Mater. Med., 2004, 15, 847.

[13] J. Sawai, S. Shoji, H. Igarashi, A. Hashimoto, T. Kokugan, M. Shimizu and H. Kojima, J. Ferment. Bioeng., 1998, 86, 521.

[14] S. J. Pearton, D. P. Norton, K. Ip, Y. W. Heo and T. Steiner, J. Vac. Sci. Technol., B, 2004, 22, 932.

[15] R. Brayner, R. Ferrari-Iliou, N. Brivois, S. Djediat, M. F. Benedetti and F. Fievet, Nano Lett., 2006, 6, 866.

[16] L. Q. Jing, X. J. Sun, J. Shang, W. M. Cai, Z. L. Xu, Y. G. Du and H. G. Fu, Sol. Energy Mater. Sol. Cells, 2003, 79, 133.

[17] W. F. Shen, Y. Zhao and C. B. Zhang, Thin Solid Films, 2005, 483, 382.

[18] S. H. Bae, S. Y. Lee, B. J. Jin and S. Im, Appl. Surf. Sci., 2001, 169, 525.

[19] T. P. Niesen and M. R. De Guire, J. Electroceram., 2001, 6, 169.

[20] E. M. Bachari, S. Ben Amor, G. Baud and M. Jacquet, Mater. Sci.Eng., B, 2001, 79, 165.

[21] R. U. Ibanez, J. R. R. Barrado, F. Martin, F. Brucker and D. Leinen, Surf. Coat. Technol., 2004, 188–89, 675.

[22] N. Golego, S. A. Studenikin and M. Cocivera, J. Electrochem. Soc., 2000, 147, 1592.

[23] S. Chaudhuri, D. Bhattacharyya, A. B. Maity and A. K. Pal, Surf. Coat. Adv. Mater., 1997, 246, 181.

[24] D. Bahnemann, Sol. Energy, 2004, 77, 445.

Nanoscaled Mg(OH)₂ Used as Flame Retardant Additive

Zhenhua Zhou, Zhihua Wu, Martin Fransson and Bing Zhou

Headwaters Technology Innovation, LLC
1501 New York Avenue, Lawrenceville, NJ 08648, zzhou@headwaters.com

ABSTRACT

Headwaters Technology Innovation (HTI) has applied aqueous colloidal chemistry with great success in the development of "bottom-up" nanomaterials and nanotechnologies. One such material is a nano-scaled magnesium hydroxide, NxCat® Mg(OH)₂, which is primarily used as a flame retardant additive. In order to maximize the water release and flame extinguishing effect, HTIG's manufacturing process allows for very small crystallites to form (~3 nm) and gives elaborately modified surface properties, leading to both a superior surface area and dispersion in the polymer/composite materials. With usage of NxCat® Mg(OH)₂, the plastic with 20% Mg(OH)₂ loading has better fire extinguish performance than that of conventional Mg(OH)₂.

Keywords: Mg(OH)₂, nanoparticles, flame retardant

1 INTRODUCTION

Various chemical compounds are used to retard ignition and burning in plastics. Supplied in powder, liquid or pellet form, flame retardants are incorporated into polymer formulations either as additives supplied to the polymer formulation during compounding or reacted into the polymer structure during polymerization [1]. The main flame retardants materials are halogens, phosphorus, inorganics and melamine compounds. Efforts are in place to ensure that the environment, health and safety issues for humans are sustained. In the past halogenated flame retardants were commonly used, but has been phased out by tighter restriction on the use of Green-house-gases. The use of halogen-free flame retardants is being adopted and a voluntary shift towards their use is also significantly increasing. The search for better and innovative flame retardant additives has resulted in a heterogeneous mix of compounds. New solutions to flammability are being offered by the $2.3 billion flame retardants market, which making up about 27% of the $8.6 billion 'performance' additives market [2].

Metal Hydroxides are well recognized alternatives to halogens. The most widely applied are aluminum hydroxide and magnesium hydroxide, the later being increasingly sought after in automobile and wire and cable applications. Magnesium hydroxide undergoes endothermic decomposition with water release at 630 °F (332 °C). The endothermic decomposition of Mg(OH)₂ which occurs during combustion is its flame retardant mechanism. For combustion to occur, there must be fuel, oxygen and heat. By absorbing some of the heat, magnesium hydroxide prevents or delays ignition and retards combustion of the polymeric material. The water released during decomposition has the effect of diluting the combustible gases and acting as a barrier, preventing oxygen from supporting the flame.

$$Mg(OH)_2 \rightarrow MgO + H_2O \quad 1316 \text{ J/g} \qquad (1)$$

The smoke suppression properties of magnesium hydroxide are believed to be due to the dilution effect of the water vapor on the combustible gases or due to a char formation effect on the polymer.

However, up to 60% loading has to be used due to the low efficiency of the micro-size Mg(OH)₂, which impair the mechanical performance of plastic. In order to maximize the water release and flame extinguishing effect, HTIG's manufacturing process allows for very small crystallites to form (~3 nm) and gives elaborately modified surface properties, leading to both a superior surface area and dispersion in the polymer/composite materials.

2 EXPERIMENTS

2.1 Preparation

Headwaters Technology Innovation (HTI) proprietary technology has been applied to make nanoscaled magnesium hydroxide. The technology achieves great success in the development of "bottom-up" nanomaterials and nanotechnologies by using aqueous colloidal chemistry.

2.2 Characterization

Specimens were prepared for TEM analysis by dispersing the catalyst powder ultrasonically in isopropanol. A drop of the suspension was then applied onto a Cu grid covered with holy carbon and was dried in air. These samples were then examined by JEOL 2010F TEM/STEM operated at 80-200 KV.

Microscope: Microscope photographs were taken of each burn surface.

X-ray diffraction (XRD) measurements were carried out on a Philips PW1800 using Cu radiation at 40KV/30mA over the range of 20° to 70° with a step size of 0.05° and a counting time of 14 hours. Once the pattern was obtained, the phases were identified with the aid of the Powder Diffraction Files published by the International Centre for Diffraction Data.

2.3 Test

Samples were prepared by fluxing 80 grams of base resin on the Wright mixer. The flame retardant agent was added after flux of the base resin and fluxed into the molten polymer. The loading of flame retardant is maintained in 20 wt%. The mixture is placed in a compression mold and pressed. Observations were made with respect to dispersion, color and removal of the polymer from the mixer. The plaques were then subjected to horizontal burn testing to characterize differences in flammability.

3 RESULTS AND DISCUSSION

The TEM image indicates that the average particle size of nano-scaled magnesium hydroxide is ~3 nm, as showed in Figure 1. There is no apparent aggregation occurs in this sample. The nanoparticles are loosely bonded together to form a stable conglomeration. In Figure 2, the TEM lattice image indicates a Tweed microstructure due to a slight variation in crystal chemistry. It is possible that the material

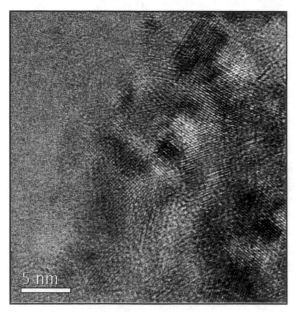

Figure 1: Transmission electron microscopy of NxCat® $Mg(OH)_2$

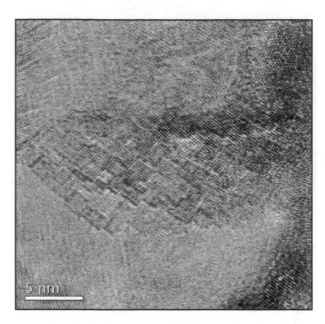

Figure 2: High resolution TEM micrograph showing Tweed microstructure

is slightly substoichiometric and is separating into the stoichiometric $Mg(OH)_2$ and hydrogen deficient MgO phases. These types of structures have been observed in simple binary oxides such as SnO and PbO. In these situations, the decomposition into two phases occurs over very short distances (10 Å to 100 Å) and the microstructure is often periodic. In addition, the decomposition seems to propagate in crystallographically soft directions so that the structure evolves on only 1-2 crystallographic planes. Thus, for X-ray analysis, one set of planes will appear to have the modulated structure (resulting in a very small particle size), while the other planes will not exhibit the modulations (resulting in a larger particle size). For the present sample, the 011 planes of $Mg(OH)_2$ exhibit a particle size of approximately 29Å, while the other planes exhibit a particle size of 117Å, as shown in figure 3. The MgO phase is more difficult to analyze since only one peak is observable and that gives an average size of 15Å. All of the remaining MgO peaks are obscured by overlap from the majority phase.

Two phases seem to be present in the diffraction pattern shown in Figure 3. The majority phase (~93%) is $Mg(OH)2$ (Brucite structure) as expected, but there is also a smaller quantity (~7%) of MgO (Periclase structure). The peak profiles are a bit unusual since some of the $Mg(OH)_2$ peaks are much broader than others. Typically, all of the peaks broaden in a similar way if the particle shape is uniform. But if the particle has a plate-like morphology, then some of the diffraction peaks will broaden more than others. Those peaks which represent planes parallel to the short dimension in the particle will broaden more, while those

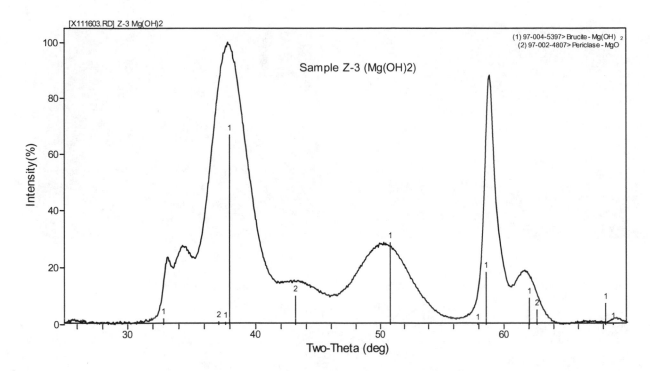

Figure 3: X-ray diffraction pattern for NxCat® Mg(OH)$_2$

planes perpendicular to the short dimension will result in diffraction peaks that broaden less.

The XRD results are consistent with the TEM images and indicate the presence of a Tweed microstructure. In such structures, the mottled structure usually lies in a single plane, which breaks up the structure into smaller domains. This is supported by the XRD results, which exhibit a small dimension for the 011 planes and a much larger dimension for the remaining planes. Thus, it appears that the material is hydrogen deficient and as a result, separates into two phases which differ in the hydrogen and oxygen concentration.

The horizontal burn testing has been used to characterize differences in flammability. The test results are showed in Table 1. Compared to conventional flame retardant, NxCat® Mg(OH)$_2$ give lower burn rate, i.e., 14 mm/min vs. 19.1 mm/min. It has been proven that flammability of flame retardant compounds is improved or optimized when the flame retardant agent particle size is small and the distribution of the flame retardant agent within the polymer matrix is very homogenous. From the previous TEM and XRD results, the smaller particle size has been observed for NxCat® Mg(OH)$_2$. Microscope photographs were taken of each burn surface after horizontal burn testing for further study of the flame retardant distribution in plastic. The distribution NxCat® Mg(OH)$_2$ in resin base at 20% is adequate and surface is glossy, as showed in Figure 4(a). Foaming present only in small localized domains. The surface of plastic with the conventional Mg(OH)$_2$ at 20% is glossy with large domains of Mg(OH)$_2$ with porosity in area surrounding these domains, as showed in Figure 4(b). The image of Figure 4(a) demonstrate better distribution and less aggregation with NxCat® Mg(OH)$_2$ compared to that of conventional Mg(OH)$_2$, showed in Figure 4(b). Apparently, both of smaller particle size and uniform distribution contribute the excellent fire extinguish performance.

For some flame applications, the plastic color has to be maintained as white or transparent. According to the observation, the NxCat® Mg(OH)$_2$ had acceptable white color after mixing, heating and pressing. All these features improve NxCat® Mg(OH)$_2$ application feasibility.

In conclusion, Headwaters Technology Innovation (HTI) proprietary technology can provide NxCat® Mg(OH)$_2$ with very small particle size (~ 3nm) and uniform distribution in plastic, which give improved performance of flame retardant.

Sample	Burn Time (seconds)	Burn Length (mm)	Burn Rate (mm/min)
Conventional Mg(OH)$_2$	235	75	19.1
NxCat® Mg(OH)$_2$	321	75	14.0

Table 1: horizontal burn testing to characterize differences in flammability.

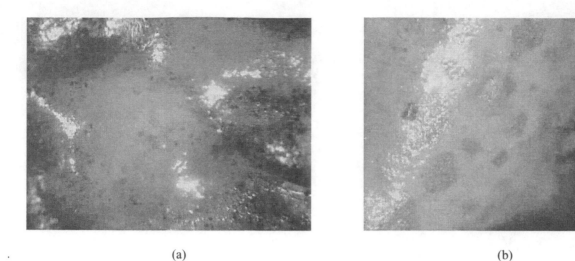

(a)   (b)

Figure 4: Microscope photographs of $Mg(OH)_2$ filled plastic after burn testing

(a) NxCat® $Mg(OH)_2$; (b) Conventional $Mg(OH)_2$

REFERENCES

[1] Ann Innes, Plastics Additives & Compounding, 4, 22-26, 2001

[2] John Murphy, Plastics Additives & Compounding, 4, 16-20, 2001

NanoZeolites – porous nanomaterials for CleanTech, Encapsulation and Triggered-release Applications

Wayne Daniell, Jürgen Sauer, Andreas Kohl

NanoScape AG
Am Klopferspitz 19, Planegg-Martinsried, D-82152, Germany
E-mail: daniell@nanoscape.de; http://www.nanoscape.de

ABSTRACT

Zeolites are well known porous materials established in applications which employ their large surface area, controllable porosity and tuneable surface chemistry. Through a reduction in their particle size by an order of magnitude, significant increases in product performance have been achieved and, more notably, new applications have been accessed which until now were unsuitable for classical zeolite materials.

Keywords: zeolites, porous, encapsulation, triggered release

1 INTRODUCTION

Zeolites were first discovered at the end of the 18th century and came to prominence in the 1960's when synthetically produced zeolites were employed in large industrial applications as molecular sieves for filtration and purification purposes, and as catalysts in the cracking of crude oil. However, recent advances in hydrothermal synthesis techniques [1] and procedures to modify the surfaces of the crystalline particles [2] have led to the development of a new form of these materials – the NanoZeolites, with particle size in the lower sub-micron range (e.g. NanoZeolite-FAU, see Fig.1).

Figure 1: TEM micrograph of crystalline NanoZeolite Faujasite (mean particle size ca. 50nm)

These porous, nanocrystalline materials have breathed new life into an established material class and possess significantly improved properties (such as higher specific surface area), leading to performance enhancement in many known applications.

Furthermore, the availability of nanoscale zeolites has opened up new applications, which were until now unserviceable with microcrystalline particles.

2 NANOZEOLITES

Advancements in synthesis and purification techniques have taken the mean particle size of zeolites down from the few microns range to around 100 nm (an order of magnitude). Furthermore, narrower particle size distributions can be achieved in comparison with classical microcrystalline zeolites which are currently commercially available (see Figure 2). This not only means a more homogeneous material and an increase in the surface area (per gram of particles), but also increases the kinetics of adsorption/desorption processes.

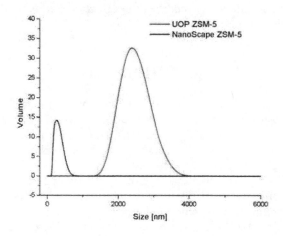

Figure 2: Comparison of particle size distribution of NanoZeolite MFI with a commercial microcrystalline MFI zeolite

This property can be put to good use in the form of thin film adsorber coatings in industrial dehumidification and/or energy recovery systems [3]. The use of nanoscale zeolites means: a reduction in weight (less adsorber material required to coat surfaces effectively); smaller cross-sections (due to thinner films) and thereby lower air stream resistance; enhanced adsorption/desorption kinetics; and a higher overall efficiency of the complete system (through savings in energy). The use of an LTA zeolite structure (with pore diameter 3-5 Å) furthermore imposes a high selectivity towards moisture adsorption, and reduces risk of bacterial contamination. Operational lifetime of the systems is also lengthened due to the better adhesion of the nanoparticles onto the substrate (typically aluminium foil or stainless steel).

3. OPEN-PORE APPLICATIONS

NanoZeolites lend themselves to applications in which the porous material is applied either as a thin-film coating or as an additive dispersed within a matrix. The so-called "open pore" applications include using the zeolite as a filter, adsorber or catalyst. A combination of the pore size and pore-surface chemistry can be employed to effectively and selectively filter out, adsorb, desorb or catalytically convert gas or liquid phase molecules. When coated onto the surfaces of various substrates (aluminium, steel, ceramic, carbon fibres), NanoZeolite thin films have significantly contributed to improving the performance of air purification filters and exhaust-gas catalysts. On fibre mats, these nanoparticle coatings have increased the active surface area without causing a significant loss in porosity (which would reduce airflow rate). For water purification applications NanoZeolites have proven suitable for both polymer and ceramic membranes. In the former case, the inorganic material is used as an additive to an organic polymer matrix; whereas for ceramic membranes a coating is employed, which modifies the effective pore size of the ceramic material.

4. ENCAPSULATION

NanoZeolites can also be employed as hosts for functional molecules such as dyes or catalysts (see Fig. 3). Through a suitable encapsulation process (e.g. impregnation, grafting etc.), the guest molecules are brought into the pore structure of the NanoZeolite and, depending on the size of molecule (in relation to pore dimensions) and the pore wall – guest molecule interactions, reside either in the pore channels or the central cavities of the unit cell. These encapsulated guest molecules exhibit more stability than when solvated and an increased resistance to chemical, mechanical and thermal influences. Furthermore, through encapsulation the application range of the functional molecule can be extended, e.g. organic dyes dispersed in aqueous systems; photochromic dyes in

polymers with high flexural modulus. Through this technique the use of (and reliance upon) organic solvents can be reduced, and the lifetime of certain molecules significantly prolonged.

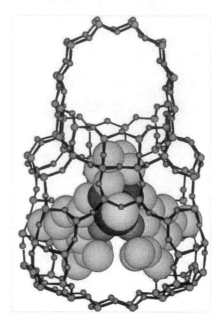

Figure 3: Functional molecule encapsulated within the pore structure of a NanoZeolite host

For certain applications (e.g. catalysts, sensors) it is not the encapsulation, protection or transport of the molecule that is of importance, but its release into a surrounding medium. A slow release can be achieved through careful selection of parameters such as: the strength of host-guest interaction (physical/chemical); and the steric confinements imposed upon the guest through encapsulation. For larger, bulkier molecules a nanoscale mesoporous host is often more appropriate, with pore diameters reaching up to 100Å (see Fig. 4).

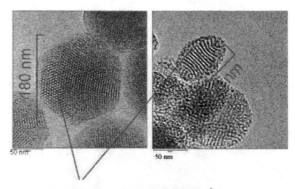

Pores with diameter ca. 50-100 Å available for loading with functional molecules

Figure 4: TEM micrographs of nanometre sized mesoporous materials

NSTI-Nanotech 2008, www.nsti.org, ISBN 978-1-4200-8503-7 Vol. 1

A controlled or triggered release can be achieved through use of any external stimulus ("trigger"). This can be temperature, pressure or a burst of radiation (e.g. UV light).

Thermal triggers typically focus on the increased vibrations which occur within the zeolite framework when heat is applied. The oscillations of the cage structure can cause a pore mouth to expand to a size large enough to allow the guest molecule to exit the pores. A change in pressure can also force a molecule out through a pore opening, though more commonly the effect is that of varying the equilibrium adsorption state of weakly bound molecules. Attachment of guest molecules to the pore channel walls through special linkers, or the fitting of "flaps" over the pore mouth openings, allows the use of UV light as a trigger. There is scope here not only to control the speed of release but also the wavelength range over which the trigger activates. This leads to a highly sensitive and extremely controllable release mechanism.

REFERENCES

[1] T. Bein and V. P. Valtchev, "Synthesis and stabilisation of nanoscale zeolite particles" EP 1334 068 B1

[2] J. Sauer and A. Kohl, "Coated molecular sieve" WO 2008/000457 A2

[3] J. Sauer, T. Westerdorf, and H. Klingenburg "Humidity and/or heat-exchange device for example plate heat-exchanger, sorption rotor, adsorption dehumidifying rotor or the similar" WO 2006/079448 A1

Non-destructive 3D Imaging of Nano-structures with Multi-scale X-ray Microscopy

Steve Wang, S. H. Lau, Andrei Tkachuk, Fred Druewer,
Hauyee Chang, Michael Feser, Wenbing Yun

Xradia, Inc.
5052 Commercial Circle
Concord, CA 94520

ABSTRACT

X-ray microscopy offers a unique combination of large penetration depth, high resolution, and high sensitivity to elemental composition. When combined with computed tomography (CT) techniques, the full three-dimensional structure of a sample can be obtained non-destructively and often with little sample preparation. With Xradia's high-resolution x-ray optics technology, we have pushed the imaging resolution to 40 nm in our nanoXCT™ system. This product has been applied to a wide range of applications including single-cell level biological microscopy, failure analysis of integrated circuits, and fuel cell development. By further integrating this high-resolution system with a projection-type x-ray micro-CT system, we have developed a multi-scale x-ray CT system with variable resolution between 40 nm to 40 μm, and field of view from 20 μm to 40 mm. This system offers unprecedented non-destructive imaging capabilities for structures ranging from tens of nm to tens of μm size scale – a three orders of magnitude "zoom" range.

Keywords: x-ray microscopy, CT, nanoXCT, microXCT.

1 INTRODUCTION

X-ray imaging offers several unique capabilities that are favorable for non-destruction imaging and analysis: (1) very large penetration depth; (2) ability to image exact 3D structure with the use of computed tomography (CT) technique [1]; (3) high elemental sensitivity; and (4) no charging effect and very low radiation damage. For example, Fig.1 shows the $1/e$ attenuation length of several materials commonly used the fabrication of nano-structures as a function of x-ray energy [2]. Multi-keV x-rays can penetrate tens to hundreds of um of light metal and semiconductor materials, and many millimeters of organic materials, such as photo-resist or soft tissue. Furthermore, this length scale is increased by an order of magnitude for x-rays with tens of keV energy. In contrast, the penetration depth of other commonly used imaging techniques based on visible light and electrons is limited to no more than several microns by absorption and multiple scattering. *Therefore, x-ray radiation is fundamentally better suited for probing internal structures in a non-invasive and non-destructive fashion.*

This key advantage of x-ray radiation has been recognized since its discovery and is widely used in medical imaging and industrial inspection applications, typically at mm-scale resolution. However, recent developments in x-ray optics and detector technology have lead imaging resolution to micron to nm scale. As a result, x-ray imaging technology has become increasingly more common used for studying micro-structure.

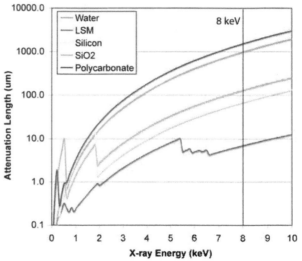

Figure 1. Attenuation length of several materials as a function of x-ray energy.

X-ray imaging is traditionally performed in a direct-projection scheme where a spatially resolved detector is placed behind the subject to record its shadow radiograph. This technique can achieve up sub-micron resolution in 2D imaging and several micron resolution in 3D imaging.

In the past two decades, advances in x-ray optics have lead to the ability to directly magnify x-ray images. Pioneered at synchrotron radiation facilities in Germany and US, x-ray microscopes based on this technology have demonstrated better than 30 nm resolution [3]. Xradia has developed table-top x-ray lens-based imaging system, nanoXCT™, that is able to routinely perform 3D imaging operations at 40 nm resolution in a highly automated fashion [4]. The multi-scale system described in this paper combines the nanoXCT™ system with a direct-projection system to produce a highly versatile x-ray imaging system capable of variable resolution between 40 nm to 40 um, with field of view ranging from 20 um to 40 mm. This large 3 orders of magnitude length scale is unique among imaging technologies and provides unprecedented non-destructive imaging capabilities for a diverse range of applications in biotechnology, nanotechnology, energy research, and advanced materials, *etc*.

2 MULTI-SCALE IMAIGN SYSTEM

The multi-scale x-ray imaging system consists of a nanoXCT™ module for imaging at 40 nm resolution and 20 um field of view, and a microXCT™ direct-projection module for imaging at 1 um or coarser resolution and field of view larger than 0.5 mm. These two modules share a sample positioning system that is able to co-register a sample's region of interest with an accuracy of 1 um.

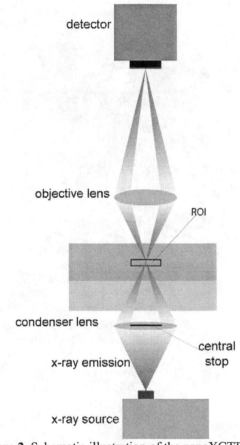

Figure 2. Schematic illustration of the nanoXCT™ system.

The nanoXCT™ is the world's only table-top x-ray CT system with tens of nm resolution. As illustrated schematically in Fig. 2, it resembles a conventional light microscope, consisting of an x-ray source, a condenser lens, an objective lens and a CCD detector. The key features of nanoXCT™ include: (1) penetration depths in the mm range for organic samples and in the 100-um range for light metal and semiconductors, (2) 50 nm resolution that is uniform in 3D, (3) multiple imaging modalities including absorption or Zernike phase contrast to optimize image contrast, and (3) automated data acquisition and analysis.

The nanoXCT™ uses rotating anode x-ray generator and a Fresnel zone plate lens as the objective lens. By using different target material, characteristic x-ray emission at energies between 5 keV and 20 keV can be generated. For example, at a typical 8 keV emission energy from Cu target, organic materials have over 1 mm attenuation length, indicating that tissues or small organs of several mm thickness can be examined without sectioning. For example, single cells within a specimen, such as engineered tissue, can be imaged at 50 nm resolution in 3D. This provides a completely new way to study biological systems near its native state. Similarly, for nanotechnology and materials research applications, the x-ray beam penetrates over 100 um of most solid semiconductor materials such as silicon. This allows fully functional multi-level nano-electronic, MEMS or NEMS devices, or fuel cell devices to be examined with minimal modification, and also possibly while the device is in live operation.

The microXCT™ module uses a direct-project x-ray imaging scheme, but has a unique design that uses a very high-resolution detector system [5]. In a projection system, the geometrical magnification is:

$$M = \frac{L_s + L_d}{L_s}, \qquad (1)$$

where, L_s is the source to sample distance and the L_d is the sample to detector distance. We can then derive from a geometric argument that the achievable resolution of this system is:

$$\delta \geq \max\left(\frac{M-1}{M}s, \frac{\delta_{\text{det}}}{M}\right), \qquad (2)$$

where s is the size of the x-ray source spot and δ_{det} is the detector resolution. From this simple relationship, we see that in order to achieve high resolution, one can make M close to 1. In this case, the sample plane overlaps the detector plane, and we arrive at the geometry of *proximity (or contact-printing) mode*, where the system resolution depends entirely on that of the detector. Traditional film-based radiography operates in this mode to take advantage of high-resolution of the recording medium. An alternative imaging mode is to let $M \gg 1$, that is, increase the magnification so that the features in the sample are magnified sufficiently to be sampled even with a detector with coarse resolution. This lead to the *projection mode*, where the resolution is roughly the source size. A high resolution imaging system can be based on either projection or proximity mode. The trade-offs, or the challenges, are that with the proximity mode, one must have a detector with high resolution while with the projection mode, one must have a source with very small spot size.

Traditional industrial high-resolution imaging systems typically operate in the projection mode. In order to achieve highest resolution, the sample must be placed very close the source, often within millimeters distance. This requirement is acceptable for 2D imaging but presents a severe limitation on the sample size for 3D imaging. Xradia's microXCT™, however, uses a unique combination of moderate x-ray source size and high detector resolution to overcome this restriction and achieve a 1-um resolution for both 2D and 3D imaging with large samples. This is the highest resolution achieved in the industry. At this highest resolution, the nanoXCT™ provides a 0.5 mm field of view, but wider field of view of up to 40 mm can be achieved by different magnification settings in the detector optics. Furthermore, operating at x-ray energy of 100-150 kVp, the microXCT™ is able to image many millimeters to centimeters thick samples of organic and semiconductor material or light metal, such as complete semiconductor packaging, circuit boards, and bones.

3 APPLICATION EXAMPLES

We illustrate the capabilities of the multi-scale system with an application in analyzing integrated circuit devices. Modern microprocessors die contain multiple layers of interconnects with line width ranging from 45 nm to microns and a thickness of up to 10 um. The die is typically mounted on a packaging that contains multi-level structure with micron-sized conductors and a total thickness of several mm. They contain a wide range of feature sizes that is well suited for the multi-scale x-ray imaging system. Furthermore, the analysis can be performed non-destructively in most cases with little sample preparation.

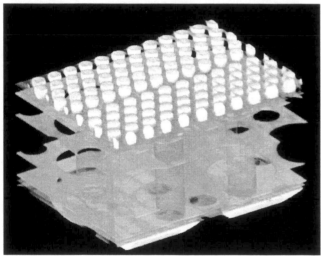

Figure 3. A microprocessor packaging imaged by the microXCT™ module.

Fig. 3 shows a typical flip-chip packaging imaged with the microXCT™ module. The 3D image is obtained by taking a series of projection image at different view angles and then mathematically reconstruct the sample's 3D structure. This 3D image is then generated with iso-surface volume rendering. Key features of the sample such as the μBGA contacts and through holes can be identified from the 3D image, but we can also examine detailed features by studying the cross-sectional images of the 3D volume data,

as shown in Fig. 4. This is commonly referred to as *virtual cross-sectioning*. In this example, a μBGA ball containing foreign particles can be identified from both the "layer plane" and the "cross-sectional plane".

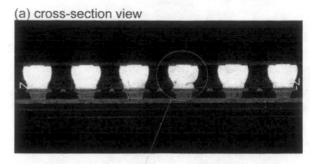

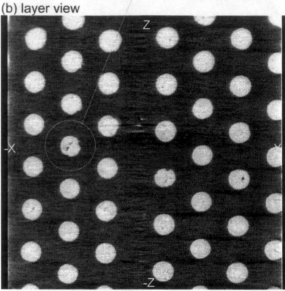

Figure 4. Virtual cross-sectioning images that allows one to view the 3D internal structure without physically sectioning the sample. One μBGA ball containing small trapped particles can be identified in both cross-section views.

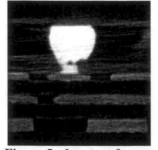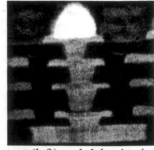

Figure 5. Images of poor contact (left) and delamination (right) defects in a packaging sample.

More difficult micron-sized defect types such as poor wetting and delamination can also be identified from the 3D images as shown in Fig. 5. Traditionally, in order to observe these defects, one must be cut and polish the sample to reach the suspected failure location and use a visible light microscope or scanning electron microscope

(SEM) to identify the failure. This physical cross-sectioning process is destructive, time-consuming and often prone to introducing new defects.

Figure 6. An integrated circuit die imaged with the nanoXCT™ module.

The nanoXCT™ module is able to perform 3D CT imaging at 40 nm resolution. As the microXCT™ is intended for studying micron-scale feature such as packaging or engineered tissues, the nanoXCT™ is well-suited for imaging nanometer-scale features such as semiconductor dies or individual biological cells. Fig. 6 shows an integrated circuit die with 5 metal layers imaged with nanoXCT™ module. The finest lines in this sample with 80 nm width are clearly resolved in the image, but the overlapping features make it difficult to distinguish them. These features, however, can be resolved in depth with the CT technique. Fig. 7 shows several virtual cross-sectional images extracted from the reconstructed 3D volume data. The metal structure of the sample can then be analyzed from these images.

In addition to semiconductor applications, the x-ray CT technique can be applied to a wide range of fields such as in biotechnology to examine osteoporotic bone or engineered tissues samples, composite materials, and NEMS or MEMS devices, *etc*. The non-destructive x-ray micro-CT technique is increasingly more accepted as an alternative to physical cross-sectioning in these applications and may in some cases completely replace destructive testing and evaluation procedures.

4 SUMMARY

Multi-scale x-ray microscopy offers a unique combination of large penetration depth, high resolution, and wide scale range. When combined with the CT technique,

the full three-dimensional structure of a sample can be obtained non-destructively and often with little sample preparation. This system offers unprecedented analytical and imaging capabilities that will significantly improve the productivity of technologies requiring studies at nanometer to micron scale.

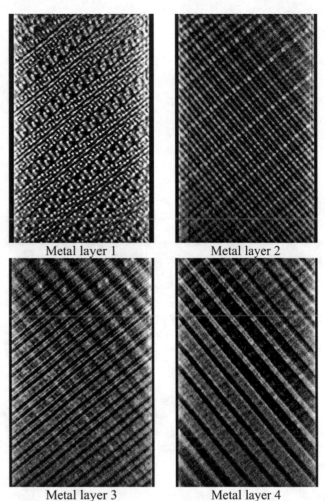

Metal layer 1   Metal layer 2

Metal layer 3   Metal layer 4

Figure 7. Four lower metal layers from the 3D data.

REFERENCES

[1] Haddad, W.S., *et al.*, *Ultra high resolution x-ray tomography.* Science, 1994. **266**: p. 1213--1215.

[2] Henke, B., Gullikson, E., and Davis, J., *Atomic Data and Nuclear Data Tables*, 1993. 54:181-342.

[3] Jacobsen, C. and Kirz, J., "X-ray microscopy with synchrotron radiation," *Nature Structural Biology 5 (supplement)*, pp. 650-653, 1998.

[4] Wang, Y., *et al.*, *Conference Proceeding from International 29th Symposium for Testing and Failure Analysis*, 29 227-233, 2002.

[5] Wang, Y., *et al.*, Conference Proceeding from International 29th Symposium for Testing and Failure Analysis, 29 227-233, 2002.

An Embedded Atom Method for Alloy Nanoparticles

Bin Shan, Jangsuk Hyun, Ligen Wang, Sang Yang, Neeti Kapur, and John B Nicholas

Nanostellar Inc

3696 Haven Ave, Redwood City, CA, USA, jnicholas@nanostellar.com

ABSTRACT

One of the key problems in studying nanoparticles is the characterization of their segregation behavior. We present here the development and application of a high-accuracy embedded atom method (EAM) potential for the simulation of metal alloy nanoparticles. The potential was parameterized by a large set of density functional theory (DFT) calculations of metal nanoparticles in addition to bulk properties. The EAM potential accurately reproduces heats of formation, bulk moduli, chemical potentials and geometries. We used this EAM potential in Monte Carlo simulations of PdAu nanoparticles ranging from 55-atoms ($\sim$ 1nm) to 5083-atom particles ($\sim$ 5nm). The potential can be utilized for computational screening of different binary and ternary alloy particles for catalysis.

Keywords: embedded atom method, PdAu, alloy, segregation, catalyst

1 Introduction

Metal alloy nanoparticles have been widely used in many industries, including electronics and catalysis. Their large surface to bulk ratio, as well as their quantum confinement effects, make nanoparticles attractive for many potential applications [1]. Despite their appealing properties, the average number of atoms in a particle usually goes well beyond a thousand, making direct first-principles density functional theory (DFT) calculations prohibitively expensive [2].

Over the years, many empirical potentials and force fields have emerged for modeling large numbers of atoms and molecules [3]–[6]. The empirical methods ignore the fast movement of the electrons and only deal with the slower movement of atom and ions, thus providing a highly effective means of predicting the energetic and structural properties of nanoparticles. The Embedded Atom Method (EAM) is one of the most successful in describing metallic systems [3]. However, most previous work has focused on the study of bulk behavior and the parameterized EAM potentials generally do not faithfully reproduce nanocluster behavior. Recently, some efforts has been made toward this direction by including additional parameters that reflect the local asymmetric charge distribution [7].

In this paper, we report the development, parameterization, and application of an EAM potential that is benchmarked to state-of-the-art DFT calculations. To parameterize potentials that are accurate for nanoparticles, we calculated standard bulk properties, such as rose curves, sheer distortions, and surface energies, as well as the energies of 3 to 147-atom clusters, all within DFT framework at the Perdew-Burke-Erzernhof functional (PBE) level [11]. The parameterized EAM potentials of the pure metals reproduce faithfully both bulk properties and cluster geometries and energies. The parameterized cross-pair potentials for alloys agree well with heats of formation, and maintain the correct energy ordering of the alloy nanoparticles under different segregation configurations. As an example of the utilization of our EAM potential, we carried out Monte-Carlo (MC) simulations of PdAu alloy nanoparticles ranging from 55 to 5083 atoms. We demonstrate how different factors like temperature, particle size, and the alloy composition ratio influence the surface morphology of the nanoparticle. Our computational study provides a comprehensive dataset of alloy nanoparticle segregation behavior that gives insight into catalyst characterization and identification.

2 Computational Methods

We briefly describe in the following sections the procedures we used in the DFT calculations, the EAM parameterization for pure elements, the cross-pair potential fitting, and the MC simulations of PdAu nanoparticles.

2.1 EAM for Pure Elements

Conceptually, the total energy of a given system within the EAM framework can be expressed as

$$E_{tot}(\rho) = F(\rho) + \sum_{i<j} \phi(r_i, r_j) \tag{1}$$

where the first term is the embedding function representing the energy of embedding an atom in a free electron sea of density ρ, while the second term is a correction term in the form of pair-wise interactions between

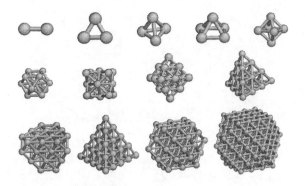

Figure 1: Clusters that are included in the parameterization of EAM. They are Pd_2, Pd_3, Pd_4, Pd_5, Pd_6, Pd_{13}, Pd_{14}, Pd_{19}, Pd_{20}, Pd_{31}, Pd_{35}, Pd_{55}, Pd_{147}, respectively

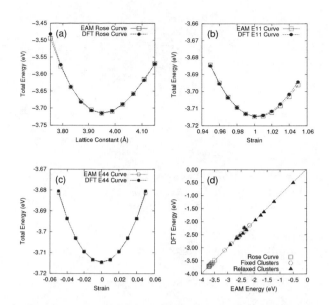

Figure 2: A comparison between DFT and EAM predictions. a) Rose Curve. b) Elastic constant E_{11}. c) Elastic constant E_{44}. d) Binding Energies for clusters.

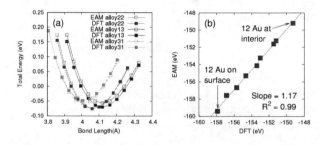

Figure 3: a) The alloy heat of formation for PdAu alloy in 1:3, 2:2, and 3:1 ratio. b) Correlation between DFT and EAM for $Pd_{43}Au_{12}$ nanoparticles with different segregation profiles.

neighboring atoms located at r_i and r_j. Such empirical potentials avoid solving the electronic wavefunction and are capable of handling much larger systems than quantum mechanical calculations. Many different types of embedding functionals and pair-wise interactions have been proposed to give an accurate estimate of the total energy of the system [5], [8]–[10]. We have based our work on Zhou's scheme where spline functions were used to interpolate the embedding functions at different charge density regions [9], [12]. To add additional flexibility to the EAM potential, especially the low electron density range, which usually occurs in coordinately unsaturated atoms, we added additional splines.

In addition to the bulk properties that EAM potentials are commonly parameterized against, such as cohesive energies, rose curves, sheer distortions and surface energies, we have included in our fit a series of nanoparticles to improve the quality of our EAM potential. Figure 1 shows an example of 13 Pd cluster geometries that was included in the parameterization of Pd potential. All geometries of the clusters were optimized by DFT and their binding energies were obtained. Both the relaxed and un-relaxed geometries were used in the parameterization. The relaxation and total energy calculations were done using the Vienna ab initio simulation package [13] with the PBE exchange-correlation functional [11].

Figure 2 shows the comparison between DFT data and EAM results for Pd bulk properties and nanoparticle binding energies. The parameterized potential not only reproduces all the bulk properties, but also accurately maps the binding energies of different-sized clusters. We followed the same protocol in fitting the EAM potential for Au. In both cases, the correlation coefficient (R^2) between DFT data and EAM prediction is larger than 0.98, indicating excellent correlation.

2.2 Cross Pair for Alloys

We used the same functional form as Zhou's pairwise potential for the cross pair potential [9], [12], but have re-parameterized the pair potential to the DFT heat of formation for PdAu alloys with 3:1, 2:2, and 1:3 composition ratios. Figure 3 demonstrates the excellent correlation between the DFT heat of formation curve as a function of lattice constant.

We have also validated the accuracy of the cross pair by comparing the energies of a 55-atom $Pd_{43}Au_{12}$ nanoparticles with different segregation profiles. The top right point on figure 3b corresponds to the $Pd_{43}Au_{12}$ particle where 12 Au atoms are all segregated to the surface, while the left bottom point represents the case where 12 Au atoms are in the interior of the nanoparticle. The points in between represents intermediate states where some Au has segregated to the surface. The

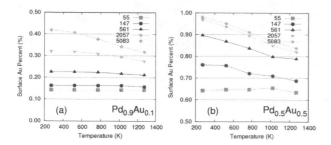

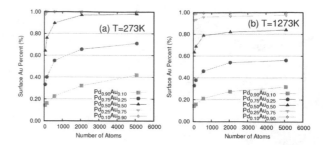

Figure 4: Segregation behavior as a function of temperature for different sized PdAu nanoparticles a) Pd rich nanoparticle. b) 1:1 ratio PdAu nanoparticle.

Figure 5: Segregation behavior as a function of particle size (a) at T=273K (b) at T=1273K

energy ordering between particles with different segregation profiles is in exact agreement with DFT result.

2.3 Monte-Carlo Simulations

In order to probe the equilibrium properties of PdAu nanoparticles, we used MC simulations with the Metropolis algorithm [14]. The simulations were run for PdAu nanoparticles with different composition ratios ($Pd_{0.9}Au_{0.1}$, $Pd_{0.75}Au_{0.25}$, $Pd_{0.5}Au_{0.5}$, $Pd_{0.25}Au_{0.75}$, $Pd_{0.1}Au_{0.9}$), different particle sizes (55, 147, 561, 2057, 5083 atoms) and at different temperatures (273K, 523K, 773K, 1023K, 1273K). A random swap of two atoms fixed on the lattice is attempted and is either accepted or rejected according the the Metropolis rule [14]. This procedure is repeated for over one million steps for each of the cases we studied. The final morphology of the particle is obtained by taking a statistical average of the last 10000 steps, after the particle has reached equilibrium.

3 Results and Discussions

Since the surface energy of Au is smaller than that of Pd, we expect that Au atoms will segregate to the particle surface to minimize the total energy of the system. A compensating effect is the negative heat of formation of the PdAu alloy, which tends to promote PdAu mixing rather than keeping the elements totally separated. These two competing mechanisms determine the final morphology of the nanoparticle. In the following sections, we discuss how different factors, including temperature, particle size and composition ratios, would change the surface concentration of Au in a PdAu alloy.

3.1 Segregation vs Temperature

Figure 4 shows the effect of temperature on the surface morphologies of PdAu nanoparticles. Figure 4a is for a Pd rich alloy ($Pd_{0.9}Au_{0.1}$), and Figure 4b is for PdAu with 1:1 composition ratio. Layer by layer osillation in concentration is generally observed, where the outmost layer is Au rich while the sub-layer is Pd rich. For the current study, we focus on the surface layer only and analyze how the Au concentration changes with

different factors. It can be seen that the the surface concentration of Au is significantly enriched as compared to its bulk composition ratio, and the surface composition of small particles is very resistant to temperature change. This effect is primarily due to the fact that small nanoparticles are dominated by vertex and edge sites, where Au and Pd atoms have significant energy differences. Even the highest temperature we studied (1273K) is unable to invert the surface Au concentration. For larger nanoparticles however, we do see a significant decrease in surface Au concentration with higher temperature. The surface Au concentration change with respect to temperature is most significant for PdAu alloys with Au concentration in the 25% to 50% range. Larger Au concentrations lead to an Au-dominated surface irrespective of temperature, while for smaller Au concentrations, the change is limited by the amount of Au available.

3.2 Segregation vs Particle Size

Figure 5 shows the surface segregation versus particle size at T=273K and T=1273K, respectively. For Au rich nanoparticles, the surface forms a monolayer of Au. In the Pd rich particles, we see an initial sharp increase in the Au concentration followed by an asymptotic plateau. For each series of curves with same color, the PdAu composition ratio is kept fixed. Thus, smaller nanoparticles having larger surface-to-volume ratios develop a smaller Au surface concentration. This is the main reason for the slope change in Au concentration versus particle size. If we make the assumption that the maximum number of surface Au atoms is limited by the number of Au atoms in the alloy, which is true for Pd rich alloys under lower temperature, we expect the surface Au concentration will follow a $N^{1/3}$ curve with respect to the total number of atoms. This qualitatively agrees with what we see in the lower portion of Figure 5.

3.3 Segregation vs Particle Composition

We finally discuss the segregation behavior with respect to the Pd:Au composition ratio (Figure 6). The

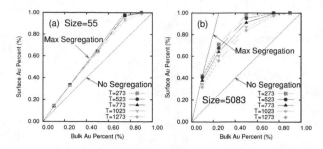

Figure 6: Segregation behavior as a function of Au composition for a) 55-atom particle. b) 5083-atom particle.

diagonal line across the graph indicates the case of no segregation while the other dotted line shows the maximum possible degree of segregation as permitted by the number of available Au atoms in the particle. It is clear from the graph that for small size nanoparticles, the Au atoms segregate under all temperatures. Au continues segregating to the surface until it reaches a full monolayer. The degree of segregation is less pronounced in larger nanoparticles, where the negative heat of formation of PdAu keeps a fraction of Au inside the nanoparticle to maintain the energetically favorable equilibrium structure. It can also be seen that the segregation of Au is less pronounced at higher temperatures.

4   Conclusions

We have developed and parameterized a set of EAM potentials for the modeling of alloy nanoparticles. These potentials yield accurate results for both bulk properties and nanoparticle geometries and energies. We used the EAM potential to study the segregation behavior of PdAu nanoparticles with different sizes, composition ratios, and at different temperatures. Our MC simulation indicates an almost full segregation of Au to the surface in small nanoparticles. In large nanoparticles, especially for lower Au concentrations, a temperature dependent Au surface segregation is observed. The influence of different factors on segregation that has been presented here gives insight into PdAu nanoparticle morphology.

REFERENCES

[1] C. Burda, X. Chen, R. Narayanan, and M.A. El-Sayed, Chem. Rev. 105, 1025 (2005).

[2] W. Kohn, Rev. Mod. Phys. 71, 1253 (1999).

[3] M.S. Daw and M. I. Baskes, Phys. Rev. B 29, 6443 (1984).

[4] A.P. Sutton and J. Chen, Phil. Mag. Lett. 61, 139 (1990).

[5] R.A. Johnson, Phys. Rev. B 39, 12554 (1989).

[6] D.W. Brenner, Phys. Rev. B 42, 9458 (1990).

[7] B. Lee and K. Cho, Surf. Sci 15, 1982 (2006).

[8] J. Cai and Y.Y. Ye, Phys. Rev. B 54, 8398 (1996).

[9] X.W. Zhou, R. A. Johnson, and H.N.G. Wadley, Phys. Rev. B 69, 144113 (2004).

[10] H.R. Gong, L.T. Kong, W.S. Lai, and B.X. Liu, Phys. Rev. B 66, 104204 (2002).

[11] J.P. Perdew, K. Burke, and M. Ernzerhof, Phys. Rev. Lett. 77, 3865 (1996).

[12] H.N.G Wadley, X. Zhou, R.A. Johnson, and M. Neurock, Prog. Mat. Sci. 46, 329 (2001).

[13] G. Kresse, J. Furthemuller, Comp. Mat. Sci. 6, 15 (1996).

[14] N. Metropolis, A.W. Rosenbluth, M.N. Rosenbluth, A.H. Teller, and E. Teller, J. Chem. Phys. 21, 1087 (1953).

Ultrasonic Pulsed Doppler (USPD): A Novel Ultrasonic Method for Characterizing Suspensions of Nanoparticles

Steven A. Africk

Prodyne Corp. 30 Fenwick Road, Waban, MA, USA and
Massachusetts Institute of Technology, Cambridge, MA, USA, safrick@att.net

ABSTRACT

The Ultrasonic Pulsed Doppler (USPD) method is a novel, inexpensive and rapid means to characterize nanoparticle suspensions. Using only a single transducer, and applicable with almost arbitrary sample volumes, USPD measures the ultrasonic backscatter in the 15 Mhz range from moving particles as small as sub-ten nanometers. Particle motion spreads the backscatter over a range of frequencies and the shape of a high resolution power spectrum (bin widths of several Hz) is analyzed to determine the backscattered power that is the measure of the suspension concentration and/or particle mechanical properties. The spectral shape can also characterize particle size. Changes in particle properties (e.g. compressibility change due to functionalization) can be detected. Measurements can be made on batch samples and for particle streams which can support online process monitoring. To date, backscatter measurements have been made with particles as large as Islets of Langerhans (120 μm) and as small as sub ten-nanometer particles such as dendrimers, titanium dioxide colloids and SDS micelles.

Keywords: nanoparticles, ultrasonics, concentration measurement, particle size distribution, quality control, quality assurance, process monitoring

1 INTRODUCTION

The Ultrasonic Pulsed Doppler Methodology was originally developed as a simple, inexpensive and rapid means to measure the concentration of Islets of Langerhans in preparations for transplantation for treatment of Type I diabetes. Subsequently, we have endeavored to extend its capabilities downward in particle size and have found that it can measure properties of single-digit nanometer-sized particles.

The technique is based on the measurement of ultrasonic backscatter from moving particles. In the 15 Mhz frequency range, the wavelength of the interrogating signal is much larger than the diameter of the particles and acoustic Rayleigh scattering occurs. The outgoing scattered wave $p_s(r)$ at a distance r from a particle modeled as a fluid sphere of radius a in response to an incident plane wave $p_i(r)$ can be expressed in terms of an angular distribution factor Φ:

$$\Phi = \frac{rp_s(r)}{p_i(r)} = \frac{1}{3}k_0^2 a^3 \left[\frac{\kappa_1 - \kappa_0}{\kappa_0} - \frac{3(\rho_1 - \rho_0)}{2\rho_1 + \rho_0} \right] \quad (1)$$

where k_0 is the acoustic wavenumber ω/c. The terms in brackets represent the contrasts between the particles' properties and the suspending medium. The first represents compressibility where κ_0 is the compressibility of the medium and κ_1 is that of the particle and the second density where ρ_0 is that of the medium and ρ_1 is that of the particle. The former term denotes spherical monopole reradiation while the latter denotes dipole radiation. Compressibility contrast dominates backscatter for many particles.

The second important feature of USPD is the measurement of Doppler shifted backscatter due to motion of the particles. By measuring backscatter at frequencies differing from that of the interrogating signal, clutter due to reflections of the interrogating signal is eliminated. Consequently, the noise floor of the USPD measurements is set by the electronics and not the acoustical conditions in the measurement volumes. Doppler shifts can be generated by three sources of particle motion. First, the fluid can be stirred, leading to motions both toward and away from the transducer. Second, if the fluid is flowing through a conduit (e.g. exiting a manufacturing or other processing step) this motion can supply the needed Doppler shift. Finally, if it is of sufficient power, the interrogating signal itself can give rise to acoustical streaming, whereby a velocity away from the transducer is generated by the nonlinear deposition of momentum into the fluid.

2 SYSTEM ARCHITECTURE

A sketch of the USPD system in Figure 1 shows a batch processing chamber as it is presently configured.

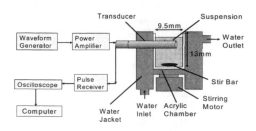

Fig. 1: Schematic of USPD Batch System

In this system the sample suspension is confined in a thermally-jacketed vessel containing a stirrer bar. A single 1/8" OD transducer extends into the chamber horizontally. In the figure it is shown extended into the fluid. In this position, the system can be self calibrated by measuring the reflectivity of the back wall of the chamber. For backscatter measurements the transducer face is flush with the inside wall of the chamber.

The transducer is focused so that there is a small volume about 1.7 mm from the transducer face in which the ultrasonic fields are concentrated and strongest. It is in this volume that most of the backscatter is generated. Consequently, the size and shape of the container is not a critical element and as long as a uniform flow through the focal volume is maintained the measurement can be made. This suggests that USPD can be used in almost arbitrary fluid volumes.

A conceptual configuration that can support real-time monitoring of a particle manufacturing process is shown in Figure 2 below. Here the flow of the particles gives rise to the requisite Doppler shifts. The transducer is shown outside the flow region, connected acoustically to it by a window.

then, if necessary, amplified and sent to the transducer. The backscattered signal is detected by the same transducer and the total electronic signal containing both interrogating and backscattered signals is sent to a pulse receiver for additional amplification and signal conditioning.

The deep-memory oscilloscope (presently a LeCroy 6030 series) has been the heart of the system. It performs the A/D function and a two-million point FFT of the entire signal to produce a very high resolution power spectrum with a several-Hz bin width. The shape of these high resolution power spectra contains the information required to compute the concentration and other properties of the suspensions. Individual power spectra are then averaged to make a single measurement. At present, 250 power spectra can be analyzed in approximately 8 minutes to form a useful average power spectrum. With increasing processing speed the time required may be reduced.

Evolution of the system now includes the incorporation of all the electronics into a 10 in^3 package with improved A/D capabilities (14 vs. 8 bits) and lower noise floor. This package could readily be incorporated into existing analytical instruments and process equipment for online monitoring.

3 SIGNAL PROCESSING

As indicated above, the key signal characteristic is the power spectrum of the backscattered energy. Figure 3 illustrates the power spectra for water (no particles) and a 10% wt/wt solution of 40 nm carboxylated polystyrene beads.

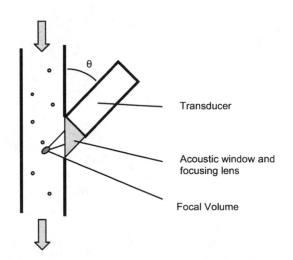

Fig. 2: In-Line Monitoring of Particle Streams

The electronics suite includes a signal generator that generates the interrogation signal, a series of tone bursts separated by quiet periods during which the backscattered signal is detected. The signal from the signal generator is

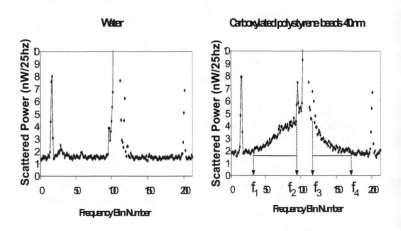

Fig. 3: Examples of Power Spectra with and without particles. Frequency bins are 25 Hz wide.

On the left side of Figure 3, the power spectrum for water shows a main peak associated with the interrogating signal centered at frequency bin 106. On either side of this peak are flat regions representing the noise floor of the measurement and containing two spurious peaks around bins 10 and 200 associated with the electronics. The right side of Figure 3 shows a spectrum in the presence of the particles. Here, there is significant power in triangular regions of the spectrum below and above the main peak, with the limits of this power denoted by frequencies f_1 and f_2 below the peak and f_3 and f_4 above it. The total backscattered power is proportional to the area of the two triangles adjacent to the main peak. The concentration measured is proportional to this backscattered power and to the square of the angular distribution function in Eq. (1).

4   APPLICATIONS

Ultrasonic backscatter by the USPD system has been demonstrated over a broad range of particle systems, including:

Islets of Langerhans (~ 120 μm)
Stem Cell Spheroids (~ 120 μm)
Cells (~ 50 μm)
Polymer Beads (40 nm - 150 μm)
Perfluorocarbon Emulsions (~ 100 nm)
Carbon Nanotubes (10 nm diameter)
Dendrimers (4 nm)
Titanium Dioxide Colloids (4 nm)
SDS micelles (5nm)
BSA protein
Nanoparticle inks (50- 120 nm)

An example of a high-precision concentration calibration curve, that for human islets, is shown below in Figure 4. Currently, work is progressing on development of concentration curves for particles at 50 nm and smaller.

Other potential applications of USPD include monitoring of physical properties of particles by making use of the dependence of their backscatter on particle size, compressibility and density. Measurements of these properties may allow real-time tracking of modifications to particles as they are functionalized for specific applications.

Two examples of this capability have been the monitoring of changes in backscatter due to the incorporation of palmitic acid molecules within SDS micelles (Figure 5), and the observation of backscatter increases during the expansion of BSA molecules due to a decrease in pH (Figure 6).

USPD has also demonstrated the capability of measuring particle size. In the absence of stirring, the interrogating signal alone can induce particle velocity away from the transducer. As this velocity is correlated with particle size,

size can be measured by the induced Doppler shift. Early indications are that this size can be measured with precision approaching 10% in the nanometer range. Thus, a binomial particle size distribution with peaks varying by this amount can be resolved by USPD.

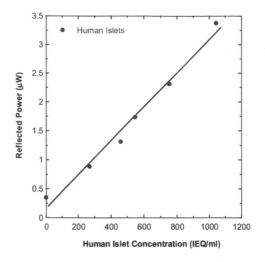

Fig. 4: Concentration Calibration Curve for Human Islets

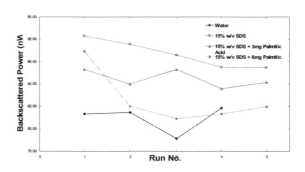

Fig. 5: Decrease in backscatter observed with uptake of palmitic acid into SDS micelles

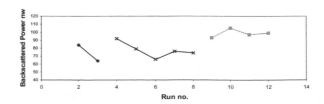

Fig. 6:   Increase in backscatter from BSA with pH change (runs 9-12)

5 CONCLUSIONS

The USPD system has shown promise as a tool for rapid measurements of concentration, particle size and compressibility for submicron particles using batches of materials or with streams of particles for manufacturing process control. This system promises to be simpler and less expensive than competing methods, and it may provide information related to the mechanical properties of the particles that cannot be easily obtained in any other way and high resolution particle sizing.

USPD technology is protected by pending U. S. patents.

Nanoparticle Measurement by Spectroscopic Mie Scattering

R. Trutna, M. Liu, D. Chamberlin and J. Hadley

Agilent Technologies, Inc., 1400 Fountaingrove Parkway, Santa Rosa, CA, USA,
judith_hadley@agilent.com

ABSTRACT

In this paper we present a spectroscopic technique for measuring particle size. The technique utilizes a UV-Visible spectrophotometer (in the range from 190 to 1100nm) to measure the attenuation spectrum of a particle dispersion due to scattering. Using the Mie scattering theory, we compute the particle size distribution and particle concentration that best matches the measured scattering spectrum. While the operating range is material dependent, we have found that the instrument can measure particles sizes from 10 nm to 15 microns. The technique involves minimal sample preparation and clean up, and is fast (less than 10 seconds). Our results show that the UV-Visible spectroscopic technique can be used to resolve complex multimodal dispersions. Particle size populations can be resolved as close as 1:1.5 size ratio. The technique also allows a sensitive determination of as low as 5% small particles in a large-sized particle dispersion.

Keywords: nanoparticle size, UV-Visible, light scattering, dispersions

1 INTRODUCTION

The size and shape of nanoparticles are their most important characteristic because they determine many other features of the behavior of nano suspensions. The rate of settling, the ease with which they can be filtered, their flow properties when poured or pumped and their capacity to absorb surface molecules all depend on particle size. The optical properties of latexes (gloss and sheen) depend on particle size, as does drug solubility. Colloidal suspensions often exhibit a wide range of particle sizes. For some purposes it is enough to know the minimum, maximum and average size. But in many cases it is necessary to have a complete knowledge of the details of the distribution, width and the presence of multiple modes. Accurate determination of the particle size and characterization of the distribution of nanoscale objects will improve our understanding of their functioning and behavior. However current techniques capable of nanoscale resolution have many limitations. In this paper we report experimental work on identifying nanosize particle size populations in complex samples with more than one size population.

2 EXPERIMENTAL

2.1 Materials

Two types of commercial particles were used for the experiments reported in this paper. All polystyrene particles were monodisperse polymer microsphere suspensions in water, NIST traceable size standards supplied by Duke Scientific Corporation. All colloidal gold particles were monodisperse gold nanoparticles dispersed in water manufactured by British BioCell International and available through Ted Pella Inc.

2.2 Instrumentation

Based on it's expertise in UV-Vis spectrophotometry, Agilent has developed an accurate, simple to use and maintain particle size instrument. The optical system shown in Figure 1 produces high dynamic range, low stray light and a highly collimated beam. It uses two light sources, a deuterium and a tungsten lamp. The shutter consists of 2 parts; one serves simply as an optical on/off switch while the second is a filter used to reduce stray light in the UV range. The 1024 element diode array is a series of silicon photodiodes. The diode array combined with the wide spectrum grating allows multiple wavelengths to be collected simultaneously for fast data collection. The measurement time is on the order of 1 second, and particle size distribution calculation can take up to approximately 5 seconds. This results in a total data acquisition time of less than 10 seconds. This enables measurement of samples where dispersion stability is an issue or where particles are dense and the sedimentation rate makes traditional measurement techniques challenging. Since the technique does not rely on observing the Brownian motion of particles, the samples may be stirred or flowed during the measurement via the use of a stir bar accessory or utilizing flow through cells and a peristaltic pump.

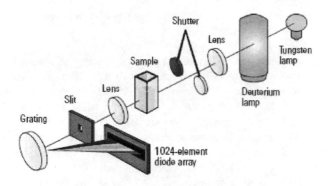

Figure 1: Optical path of particle sizing
spectrophotometer

When solutions contain particles, light passing through the solution is attenuated due to scattering. The wavelength dependence was calculated by Gustav Mie in 1908[1]. Mie developed a rigorous method to calculate the intensity of light scattered by uniform spheres. The shapes of light scattering spectra are distinguishable for particles that are much larger than the wavelength of light λ but differ in size by as small as $\lambda/10$, and the particle size information can be obtained by analyzing these spectral shapes. Mie theory is an exact solution of Maxwell equations on light scattering by a spherical scatterer of arbitrary size. Mie solution requires only three parameters: the size and the refractive indicies of the scatterer and the medium. Using Mie's mathematical theory, the particle size distribution and volume concentration which produce the best agreement with the measured spectra is determined.

3 RESULTS AND DISCUSSION

Multimodal distributions are usually thought of and represented as being the sum of two or more normal (or log normal) distributions. In some industrial situations it is important to be able to accurately distinguish the presence of a true bimodal distribution. For example, the presence of a larger population might interfere with the main process. The particle size methods that first separate the different sizes and then measure them are intrinsically better able to detect the presence of a bimodal distribution. The time required to first separate sizes and then measure however can be on the order of hours. The UV-Vis particle sizing spectrophotometer configuration described above is capable of resolving bi- and multi modal particle size distributions. We measured colloidal gold of nominal 80 nm diameter and nominal 150 nm diameter. The SEM micrograph in Figure 3 captures the mix of these two size populations. First, we measured these colloids separately. The 80 nm colloidal gold was measured to be 78 nm in diameter at a concentration of 0.000322% by volume (red line, Figure 2) and the 150 nm diameter colloid was determined to be 156 nm in diameter at a concentration of 0.000344% (blue line, Figure 2). These colloids were then mixed in a 1:1 ratio by volume. The black line in Figure 2 shows the particle size distribution of the mixture of gold standard dispersions. The spectral fitting algorithm clearly broadens the peaks in the case of the mixture as compared to the monodisperse measurements. However, The particle size measurement correctly shows the nearly 1:1 relationship in concentration of the two peaks. In fact, integration of the peaks reveals that the relative concentration of the two modes is 48% 78nm and 52% 156 nm Au – exactly the expected ratio. The concentration of the mixed sample was measured as 0.000331% - a difference of <1% from the expected value for the mixture. In addition to the detection of these modes, the instrument correctly determines the relative amounts of the two modes and the correct total particle concentration on a %v/v basis.

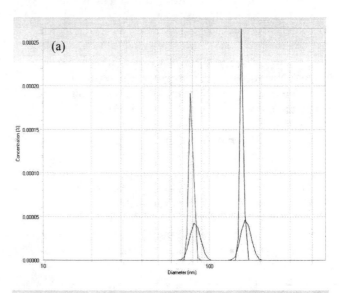

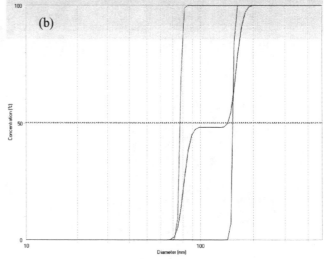

Figure 2: Absolute (a) and cumulative (b) particle size distribution of pure 78 nm Au colloid (red), pure 156 nm Au colloid (blue), and a mix of 50% 78 nm gold and 50% 156 nm gold (black). (Ted Pella Inc.)

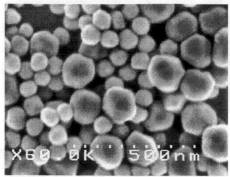

Figure 3: SEM micrograph of 78 nm and 156 nm gold particles (Ted Pella Inc.)

To test the limit of the size resolution capability of this technique, 50/50 mixtures of different size populations of polystyrene particle standards were combined. Particle size measurements were taken as the size ratio was decreased. Figure 4 demonstrates the resolution capability of the UV-Visible particle sizing instrument with a sample of equal concentrations of 2μm and 3μm polystyrene latex (in blue). The UV-Vis spectroscopic technique allows us to resolve bimodal distributions as close as a size ratio of 1:1.5. The particle size distribution of the same 2μm polystyrene sample alone is also shown in figure 4. Comparing these distributions, it appears that the broadening of the 2μm peak in the mixture is an artifact of the fitting algorithm as seen in Figure 2. There appears to be a trend in close bimodal samples of the smaller peak becoming artificially broadened. Figure 4 compares this bimodal sample with a 3μm and 5μm bimodal polystyrene sample (in green). Such 1:1.5 ratio size resolution has also been demonstrated for metal nanoparticles as well.

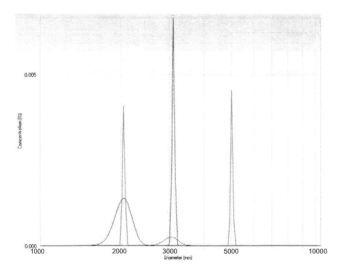

Figure 4: Particle size distribution of a mix of 50% 2μm and 50% 3μm polystyrene standard (Duke Scientific Corp.) (blue) demonstrating 1:1.5 size ratio resolution capability. Red line: monodisperse 2 μm polystyrene standard alone;

Green line: mixture of 50% 3 μm polystyrene and 5 μm polystyrene standards.

A common challenge for light scattering techniques is the measurement of small particles in a population of large particles. Typically the presence of a few large particles can scatter too much light, hiding the presence of small particles. For industrial processes where dispersions are unstable or are designed around 2 or more size populations, the ability to accurately detect the individual populations is critical. The spectral dispersion of the UV-Vis particle sizing spectrophotometer allows accurate measurement of mixtures of small and large particles in suspension. The blue graph is figure 5 demonstrates the ability to detect a mixture of 9% 91nm polystyrene particles in a background of 91% 3μm polystyrene particles. As well the converse situation is also shown by the red graph in figure 5, 9% of 3μm polystyrene in a background of 91% 91nm polystyrene. The limit of the small particle sensitivity has been tested and determined to be 5%. We have been able to detect 5% v/v 91 nm polystyrene latex particles in a background of 95% v/v 3 μm particles.

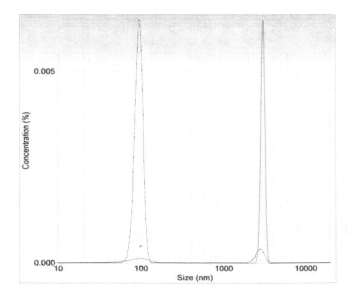

Figure 5: Mixture of 9% 3μm polystyrene (Duke Scientific Inc.) in a background of 91% 91nm polystyrene (in red) and 9% 92nm polystyrene in a background of 91% 3μm polystyrene (in blue).

4 CONLCUSION

Agilent has developed an accurate, simple to use and maintain particle size analyzer built around the robust spectrometric hardware provided by the Agilent UV-Vis platform in the wavelength range 190nm to 1100nm. The spectral attenuation of the colloidal particles is accurately measured in this range and the software calculates the particle size distribution. Bimodal particle size distributions can be measured with size difference as close as 1:1.5 while maintaining accuracy in both the size of the particles *and* the relative concentrations of the modes. Using Mie theory, the particle size distribution and volume concentration which produce the best agreement with the measured spectra is determined. We believe this combination of fast measurement time, high resolution, and ease of sample preparation offers distinct advantages over existing techniques and enables new measurements.

REFERENCES

[1] G. Mie, "Beitrage zur Optik truberModien, speciel Kollaidaler Metallosungen", Annalen der Physik, vol 25, no. 3, pp.337-447, 1908.

Real-time Crystallization of Organoclay Nanoparticle Filled Natural Rubber under Stretching

J. Carretero-González[*], R. Verdejo[*], E. P. Giannelis[**], S. Toki[***], B. S. Hsiao[***], M. A. López-Manchado[*]

[*]Institute of Polymer Science and Technology, CSIC, Madrid, Spain, jcarretero@ictp.csic.es
[**]Cornell University, Ithaca, New York, USA
[***]Stony Brook University, New York, USA

ABSTRACT

The inclusion of clay nanoparticles (nanoclay) in cross-linked natural rubber (NR) induces an early promotion and enhancement on crystallization under uniaxial deformation. We evaluate and monitor the molecular structure of the polymer network and its changes during elongation by using in-situ synchrotron radiation. Nanoclay introduces an increase of the crystalline content and a dual crystallization mechanism related to the alignment of the nanoparticles during stretching. This study could be the starting point on the designing of a new generation of elastomers materials without the trade-off exhibited in conventional rubber composites.

Keywords: natural rubber, nanoclay, crystallization.

1   INTRODUCTION.

Layered-silicate nanofillers can improve the physical, mechanical, and thermal properties of polymeric matrices.[1-3] This behavior has been explained by the formation of a reinforcing nanofiller network (exfoliated or/and intercalated), within which the polymer chains are confined. However, this explanation is not sufficient to paint a complete picture of the reinforcing mechanism taking place in many systems.

The inclusion of highly anisotropic clay nanoparticles (nanoclay) in NR decreases the network defects and induces an early promotion and enhancement on crystallization under uniaxial deformation. We evaluate the molecular structure of the polymer network and its changes during uniaxial deformation by using dielectric spectroscopy and in situ synchrotron X-ray diffraction, respectively.

2   MATERIALS AND METHODS.

The dispersion of the layered silicate on NR was studied by X-ray diffraction (XRD) and Transmission Electron Microscopy (TEM). The viscoelastic properties at low deformation of the materials were studied by Dynamic-Mechanical Analysis (DMA). The distribution of the NR network components was determined by Dielectric Spectroscopy (DS) and the crystallization coupled to mechanical elongation was monitored by in-situ Wide-Angle X-ray (WAXD) synchrotron radiation.

3   RESULTS AND DISCUSSION.

By XRD (Figure 1) and TEM studies (Figure 2) we have evidenced that rubber chains easily penetrates into the layered silicate leading intercalate structures and partial exfoliation.

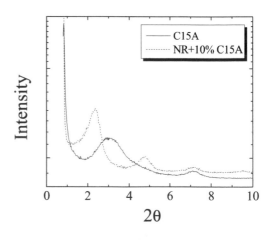

Figure 1. XRD patterns for organoclay (C15A) and for the nanocomposite (NR+10% C15A).

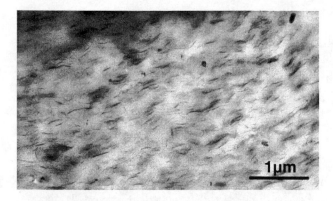

Figure 2. TEM image of NR/C15A nanocomposite.

Dielectric relaxation experiments (Figure 3) of dry and swollen specimens provided a qualitative evaluation of the degree of homogeneity on the network components through the distribution of relaxation times corresponding to the segmental and normal mode.

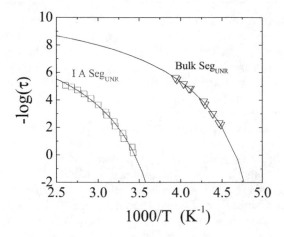

Figure 3. Arhenious plot of bulk NR´s segmental mode and interfacially adsorbed (I. A.) rubber chains.

The results from the study of the induced-crystallinity in the NR materials during uniaxial elongation by synchrotron radiation were rationalized, among others parameters, in terms of the degree of order imposed by the nanoclay on the distribution of the topological constrains of the polymer network as a consequence of a better distribution of the curatives on the NR matrix during the mixing procedure.

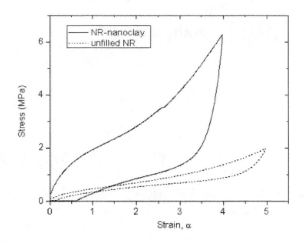

Figure 4. Stress-strain curves for the NR nanocomposites and for the unfilled sample.

In order to understand the improvement of the mechanical properties (Figure 4) we investigated the possibility of any bound rubber formation on the outer surface of fillers like in carbon black or silica bids composites. Evidences have been found in the nanocomposites by dielectric spectroscopy and the existence of a possible relaxation mode suggests a strong adhesion with the fillers (interfacially adsorbed polymer IA) that corresponds to a glass transition substantially higher than the bulk glass transition (Figure 3). Further investigation of the meaning and properties of that mode is underway.

REFERENCES

[1] E. P. Giannelis, Adv. Mater. 8, 29-35, 1996.
[2] M. Arroyo, M. A. López-Manchado, B. Herrero, Polymer, 44, 2447-2453, 2003.
[3] Q. H Zhang., A. B.Yu, G. Q. Lu (Max), D. R. Paul, J. Nanosci. Nanotech. 10, 1574-1592, 2005.

Acknowledgements

The authors gratefully acknowledge support from the Cornell Center for Materials Research (CCMR). JCG is grateful to Spanish Ministry of Education (MEC) for a FPI fellowship concession.

Nano cellulose crystallites: optical, photonic and electro-magnetic properties

D. Simon, Y. Kadiri and G. Picard

Ahuntsic College, 9155, rue St-Hubert, Montreal (Quebec) Canada H2M 1Y8,
dominique.simon@collegeahuntsic.qc.ca

ABSTRACT

Nano cellulose crystallites (NCC) are nano tubes, about 200 nm long and 10 nm in diameter. This is a natural product, found in trees. With unmodified NCCs, optical properties like iridescence and polarization were found either in aqueous suspension or in a dry state. To better understand and exploit the optical properties of NCCs, a mathematical model to describe the nano-crystalline cellulose NCC is presented. We use a periodic potential and allow the electrons to interact with each other by a weak coulombian potential. Thus, our model of the NCC is a finite cylinder; r and L are respectively, its radius and length. The carbon atoms are on the surface of the cylinder. They are arranged periodically. The model represents correctly the NCCs and is therefore fit for exploitation in the field of wood-based nano devices.

Keywords: **cellulose, crystallite, optics, photonic, magnetic**

1 INTRODUCTION

The nano crystalline cellulose (NCC) is one of the structuring components of trees. This is a natural product that was long time ago been identified by the pulp and paper industry. The NCCs were found together with the long fibers, essential elements for the preparation of paper. Typically, during the paper process, the fine fibers were separated from the long ones, and were simply rejected in massive volumes. As a consequence, the NCCs remained byproducts for nearly a century, its inherent qualities remaining largely unused.

Due to the present strong nanotechnological dynamism, the properties of the NCCs were revisited, and examined under a fresh perspective. Indeed, NCCs are 2D crystals, rolled up in a tubular geometry, about 200 nm long and 5 nm in diameter. The geometrical and chemical similarity with CNT is interesting, both containing carbon atoms. NCCs present mechanical properties comparable to fiber glass. Combined with the thinness of the nano tubes, an unusual flexibility is obtained, one of the best. For the fabrication of nano devices, this mechanical characteristics make the NCCs good candidates for commercial applications.

The NCC inner and outer surfaces are also chemically active: grafting of other chemical groups is therefore possible, providing the NCCs various potential fields of applicaitons. NCCs could also be deposited to coat surfaces. The permeation of gases though a fibril coat can be tailored to any value. This has an immediate application in solar cells, fuel cell or else, packaging in food industries. Indeed, the new trend is green, and the NCCs are fit for this challenge for those reasons. Together, the material is good for making composite structures, foams, and stratified layers for a large variety of applications.

The mechanical properties of the NCCs were also simulated with success. The modelisation by computer of the nano crystalline cellulose mechanical properties provides an excellent tool for nano engineering the product is goods for consumers. In this work, another interesting aspect of the NCCs that was not extensively investigated is presented. The optical and photonic properties of the nano crystals are approached from the electro-magnetic point of view. Indeed, with the unmodified NCCs, optical properties like iridescence and polarization were found either in aqueous suspension or in a dry state. As a matter of fact, NCC's naturally tends to self align under specific physico-chemical conditions. The interaction of the visible electro-magnetic waves with the monocrystal tubes apparently generates some rotational in the polarization plane. Moreover, various materials can be introduced in the tube inner volume, the lumen, providing them with electro-magnetic properties suitable for various specific applications.

2 MATHEMATICAL MODEL

We introduce a mathematical model to describe the nano-crystalline cellulose NCC. We use a periodic potential and assume the electrons to interact with each other by a weak coulombian potential.

Then we analyze the effects of light on the NCC described by a simple one-dimensional model. This external perturbation is a classical time function electromagnetic field polarized in the direction of the NCC. It is processed using the theory of linear response. The disruption is seen as a small perturbation of the system.

The NCC is a crystalline structure cylindrically shaped. Its length is about 200 nm and its diameter is about 10 nm. Thus, our model of the NCC is a finite cylinder; r and L are respectively, its radius and length. The carbon atoms are on the surface of the cylinder. They are arranged periodically. Each atom has six electrons. We assume that only one electron can relocate and be seen as a "free" particle. The two heart electrons are localized around the nucleus. The other three electrons participate in the inter-atomic bonds. The NCC is modeled as N-electrons system on the surface of a cylinder. These electrons are subjected to a periodical potential. It is assumed that the atoms are immobile. We adopt the Born-Oppenheimer approximation. We neglect the interaction between electrons and phonons. We use the periodic boundary conditions for the Hamiltonian describing the system.

One point of the cylinder is located by $x \in \left[-\dfrac{L}{2}, \dfrac{L}{2}\right]$ along the axis cylinder and by $y \in \left[-\pi r, \pi r\right]$ following the circumference. We use a periodic potential in x and y.

We model the potential by the V_{at} function. a and b are, respectively, the longitudinal and the transversal periods of atoms. So, for each fixed x and y,

$$V_{at}(x+a, y) = V_{at}(x, y) = V_{at}(x, y+b) \qquad (1)$$

The rectangle $a \times b$ contains in most configurations over 4 atoms. L, N, a and b are linked because the length of the cylinder as well as the perimeter are multiples of an entire number of atoms. We assume that it has n atoms in the $a \times b$ rectangle area. We choose N pairs and we obtained:

$$\frac{L}{a} \times \frac{r}{b} \times n = N \qquad (2)$$

where n is the electrons number in $a \times b$.

We suppose the potential attractive. First, we suppose that we have a single electron on the cylinder where N ions are present. Its Hamiltonian is:

$$\mathbf{H}_1 = -\frac{\hbar}{2m_e}\frac{\partial^2}{\partial x^2} \otimes 1 + 1 \otimes -\frac{\hbar}{2m_e}\frac{\partial^2}{\partial y^2} + V_{at}(x, y) \qquad (3)$$

with m_e the electron mass.

V_{at} depends on the physical parameters ε, the cylinder permittivity and e, the electron load, so:

$$V_{at}(x, y) = \frac{e^2}{\varepsilon}\tilde{V}_{at}(x, y) \qquad (4)$$

where $\tilde{V}_{at}$ is a none dependent on physical constants function.

The same goes for the potential for interaction between electrons. We make the change of variables: $\tilde{x} = x/a_0$, $\tilde{y} = y/a_0$ where:

$$a_0 = \frac{\hbar^2 \varepsilon}{m_e e^2} \qquad (5)$$

Energy will be a multiple of the Rydberg constant of the material:

$$R_y = \frac{\hbar^2}{2m_e a_0} \qquad (6)$$

Thus the physical constants will not appear explicitly in the Hamiltonian.

We want to study some optical properties of the NCC which heavily depend of the interactions between the electrons. So we will introduce in the Hamiltonian a potential describing this coupling. We choose naturally a "periodical" Coulomb potential usually used when we work in a finite box, with periodic boundary conditions.

We note that (x, y), $x \in \mathbf{R}$, $y \in \left[-\pi r, \pi r\right]$, the coordinates of a point on the infinite cylinder. We introduce the distance d_{12} between two particles $(x_1; y_1)$ and $(x_2; y_2)$ on the surface of the cylinder:

$$d_{12} = d_{12}(x_1 - x_2, y_1 - y_2) \qquad (7)$$

$$d_{12} = \sqrt{(x_1 - x_2)^2 + 4r^2 \sin\frac{y_1 - y_2}{2r}} \qquad (8)$$

$\left|2r\sin\dfrac{y_1 - y_2}{2r}\right|$ is the length of the rope connecting two points y_1 and y_2 on the circle of radius r.

The function

$$V_c^r(x,y) = \frac{\lambda}{\sqrt{x_2 + 4r^2 \sin^2 \dfrac{y}{2r}}} \qquad (9)$$

for $\quad x \in \mathbb{R}, x \in \left[-\pi r, \pi r \right]$

with $\lambda > 0$. The expression above with $\lambda = 1$ is the Coulomb potential on the infinite cylinder.

We choose to treat the case of weak interactions between electrons. The parameter λ is then very small.

The Hamiltonian associated with the N electrons system is formally defined by

$$H_C = \sum_{j=1}^{N} \left(-\frac{1}{2}\frac{\partial^2}{\partial x_j^2} - \frac{1}{2}\frac{\partial^2}{\partial y_j^2} + V_{al}(x_j, y_j) \right)$$
$$+ \sum_{j \neq k=1}^{N} V_C^{r,L}(x_j - x_k, y_j - y_k) \qquad (10)$$

The first simplification of the problem is that the radius of the NCCs is very small. This characteristic property of the NCCs led to the strengthening of interaction between electrons.

We established that the potential between electrons is:

$$v_L(x_1 - x_2) =$$

$$\frac{1}{(2\pi r)^2} \int_{-\pi r}^{\pi r} \int_{-\pi r}^{\pi r} V_C^{r,L}(x_1 - x_2, y_1 - y_2) dy_1 dy_2 \qquad (11)$$

$$= \frac{1}{2\pi r} \int_{-\pi r}^{\pi r} V_C^{r,L}(x_1 - x_2, y) dy \qquad (12)$$

And the atomic potential projected on the circle is:

$$v_{al}(x) = \frac{1}{2\pi r} \int_{-\pi r}^{\pi r} V_{al}(x, y) dy \qquad (13)$$

$$= \frac{1}{2\pi} \int_{-\pi}^{\pi} V_{al}(x, ry) dy \qquad (14)$$

3 PHYSICAL MODEL

After introducing a mathematical model for the NCCs, summarized in the Hamiltonian $\mathbf{H}_c$, we used one of the properties of the NCC, their small diameter, to reduce the problem to a one-dimension Hamiltonian $\mathbf{H}_0$. This better reflects the low dimensional nature of the nano objects, in the shape of hollow nano cylinders.

A low amplitude and frequency variable monochromatic light is passed through the NCCs. Part of this light is absorbed by the electrons and allows the system cylinder + N electrons moving in an excited state. The other part, not absorbed, travels through the cylinder and is measured. Thus we obtained the absorption spectrum, depending on the frequency of light.

We apply an electromagnetic field classic, time-dependent, modeling light. We want to know the linear answer of the system to this disturbance.

To model the incident monochromatic light, we introduce the electric field $E(t)$, $E_0 > 0$ is the field amplitude and $\omega > 0$ its frequency. The adiabatic lighting is controlled by the parameter $\eta > 0$:

$$E(t) = E_0 \cos(\omega t) \exp(\eta t) \qquad (15)$$

The Space is one-dimensional, and we are on a circle. It is impossible to define the disruption caused by the light from the electric field as we do normally in the Coulomb gauge, because the potential partner is not periodic. We use a different gauge:

$$a(t) = \mathrm{Re}\left(\frac{\exp(i\omega t + \eta t)}{i\omega + \eta} \right)$$

$$= \frac{\eta \cos(\omega t) + \omega \sin(\omega t)}{\omega^2 + \eta^2} \exp(\eta t) \qquad (16)$$

$$a'(t) = \frac{da(t)}{dt} = \mathrm{Re}(\exp(i\omega t + \eta t))$$

$$= \cos(\omega t) \exp(\eta t) \qquad (17)$$

so: $\qquad E(t) = E_0 a'(t) \qquad (18)$

The N electrons system with the light disturbance is described by the Hamiltonian:

$$H(t) = \frac{1}{2} \sum_{j=1}^{N} \left\{ \left(\frac{1}{i} \frac{\partial}{\partial x_j} - a(t)E_0 \right)^2 + v_{at}(x_j) \right\}$$

$$+ \frac{1}{2} \sum_{j \neq k=1}^{N} v_L(x_j - x_k) = H_0 + W(t) \qquad (19)$$

$$W(t) = -a(t)E_0 \sum_{j=1}^{N} \frac{1}{i} \frac{\partial}{\partial x_j} + \frac{1}{2} Na^2(t)E_0^2 \qquad (20)$$

4   CONCLUSION

In this work, we study the linear response of the NCC to the light excitation. We use two approaches: numerical simulation using *ab initio* methods and the theory of perturbation to get the absorption spectrum of the NCC. Then we graft metal atoms on the cylindrical wall of the NCC to change the conduction properties of the NCC and measure its influence on the optical response.

REFERENCES

M.K. Kostov, M. W. Cole and G.D. Mahan, Phys. Rev. B, 66, 075407, 2002.

Wannier G. H., Phys. Rev., 52, 1937.

Elliot R. J., Phys. Rev., 108, 6, 1384-1389, 1957

Epidermal Growth Factor Receptor-Targeted Engineered Gelatin Nanovectors for Gene Delivery and Transfection in Pancreatic Cancer Cells

Padmaja Magadala and **Mansoor Amiji**

Department of Pharmaceutical Sciences, School of Pharmacy, Northeastern University, Boston, MA 02115

ABSTRACT

Pancreatic cancer is a leading cause of cancer-death, mostly owing to lack of efficient therapeutic options. Upwards of 90% pancreatic tumors over express epidermal growth factor receptor (EGFR) and this largely influences the disease aggressiveness and poor clinical outcome in patients. Non-viral gene therapy has achieved increased attention over the last two decades. In this study type B gelatin-based nanoparticles were characterized and evaluated for gene delivery in pancreatic cancer. Further, these nanoparticles were surface-modified to actively target EGFR on the surface of Panc-1 cells.

1. INTRODUCTION

Pancreatic cancer is the fourth leading cause of cancer-related deaths in the United States (1). The poorly understood etiology, lack of early diagnostic tools, and effective treatment options of pancreatic cancer contribute to its increasing incidence and mortality rates. Epidermal growth factor receptor (EGFR) family includes EGFR (or *erb*B-1), HER-2 (or *erb*B-2 or neu), HER-3, and HER-4. Over 90% of human pancreatic cancer cases over express EGFR family members (2). Upon ligand binding, activated EGFR plays an important role in cell growth, differentiation, migration, and metastasis. Its positive signaling was found to cause increased proliferation, decreased apoptosis, enhanced tumor cell motility and angiogenesis. Also, EGFR is expressed in different non-transformed cell types in the tumor microenvironment that are involved in tumor growth and progression, including endothelial cells on the neovasculature. Targeting the EGFR receptors for therapeutic purpose has become possible with the recent introduction of chimeric and humanized monoclonal antibodies (3).

Gene therapy comprises transfer of genetic constructs intended to alter the neoplastic potential of the cancer cell. Various viral and non-viral delivery systems are currently being examined pre-clinically and clinically in different models of pancreatic cancer. In the development process of a nanoparticulate drug delivery system for *in vivo* gene therapy application, several design criteria are necessary. First, the nanovector must efficiently and stably encapsulate DNA and protect against degradation in the systemic circulation and upon cellular uptake. Second, the delivery system must be able to overcome biological barriers and accumulate at the target site and be internalized in cells by non-specific or receptor-mediated endocytosis. Third, the internalized nanovector must release plasmid DNA and allow for nuclear import and transfection. Lastly, the nanovector matrix must be degradable and non-toxic for chronic administration. Various nano-particulate systems based on natural and synthetic polymers like gelatin, chitosan, and poly(D,L-lactide-co-glycolide) (PLGA) have been extensively investigated (4).

Gelatin-based delivery systems are known for their biocompatibility and biodegradability. Over the last several years, our lab has engaged in studies of gelatin-based nanoparticulate systems for drug and gene delivery. Kaul and Amiji (5) have shown that gelatin and poly(ethylene glycol) (PEG)-modified gelatin nanoparticles are efficiently endocytosed by different types of cells and accumulate in the perinuclear region. Additionally, Kommareddy and Amiji (6) have demonstrated that, PEG-modified thiolated gelatin nanoparticles can be successfully used for *in vivo* delivery of therapeutic gene encoding for anti-angiogenesis sFlt-1 factor in an orthotopic human breast cancer xenograft model.

PEG is a biocompatible water soluble polymer that are increasingly chosen for shielding drug delivery systems, particularly for proteins, peptides and antibody fragments. Over the last two decades, PEG surface modification has become an attractive choice of drug delivery, largely due to its reduced renal clearance, reduced proteolysis and immunogenicity. Therefore, the use of PEG spacers helps in potential reduction of steric interference and long circulating nanoparticles.

For surface modification of gelatin nanoparticles for EGFR targeting, we have utilized a peptide ligand. Recently, various researchers have reported the successful identification of peptide ligands by screening phage display libraries (7). In earlier gene delivery studies, EGFR-targeting-peptide-conjugated polyethylenimine (PEI) vectors were shown to be less mitogenic, but highly efficient at transfecting genes into EGFR overexpressing cells and tumor xenografts (8). EGFR-targeted gelatin-based engineered nanovectors (GENS) were synthesized by conjugating the EGFR-targeting peptide ligand via a heterobifunctional PEG spacer. Reporter plasmid DNA expressing enhanced green fluorescent protein (GFP) was encapsulated and the transfection was studied in EGFR over-expressing human pancreatic adenocarcinoma cells (Panc-1).

2. EXPERIMENTAL METHODS

2.1. Synthesis and Characterization of EGFR-Targeted Gelatin Nanoparticles

Type B gelatin (225 bloom strength) was used to make the nanoparticles (GENS) by the solvent displacement method (5). EGFP-N1 plasmid DNA, expressing GFP, was encapsulated in the nanoparticles and the surface was modified with a heterobifunctional PEG linker for covalent attachment of peptide. To enhance particle uptake by EGFR targeting, we used the 11-amino acid peptide -YHWYGYTPQNVI- (8). Mean particle size and size distribution as well as surface charge (zeta potential) values were determined for the blank and DNA-loaded nanoparticles. The spherical shape and surface morphology of the particles was confirmed by scanning electron microscopy (SEM).

The plasmid DNA loading efficiency in the control and EGFR peptide-modified gelatin nanoparticles was quantified dissolving the nanoparticles in protease-containing buffer and quantifying the released DNA using PicoGreen® dsDNA fluorescence assay. Stability of the encapsulated plasmid DNA was examined by agarose gel electrophoresis.

2.2 EGFR Expression in Panc-1 Cells

Baseline EGFR expression in human pancreatic adenocarcinoma (Panc-1) cells was analyzed by Western blot and immunocytochemistry techniques. Other pancreatic cancer (Capan-1) cell line, along with SKOV3 (human ovarian cancer) and NIH3T3 (murine fibroblast) cells were used as positive and negative controls of EGFR over-expression. Briefly, the different cells were grown in fetal bovine serum supplemented DMEM and RPMI media and lysed at 80% confluence. The cell lysate was analysed for total protein concentration using the NanoOrange® protein quantitation kit. Briefly, 50 µg of total protein was subjected to western blot analysis using SDS page gel, nitrocellulose membrane and anti-EGFR antibody, to confirm the receptor expression in Panc1 cells. Immunocytochemistry analysis was performed by incubating the cells with a primary EGFR antibody and then with secondary antibody conjugated with horse-radish peroxidase. Addition of the peroxidase substrate diaminobenzidine results in the formation of brown precipitate.

2.3 Cellular Uptake and GFP Transfection Studies

Panc1 cells, cultured in DMEM at 37°C and 5% CO_2 atmosphere, were seeded at 3×10^5 cells per well in 6-well cell culture plates. Cover slips were added some culture plates to facilitate microscopic analysis. Plasmid EGFP encapsulated, surface modified and functionalized gel nanoparticles were prepared as described above. Panc1 cells were incubated with control and modified nanoparticles, with final DNA concentration of 30 µg/well, in serum-free medium. At the end of 4 hours, treatments were removed, and growth medium was replaced in the culture wells following a wash with PBS. At 24, 48, 72 and 96 hour time points, transfected cells were harvested, washed and analyzed for green fluorescence intensity by fluorescence activated cell sorting (FACS). Transfected cells on cover slips were mounted on glass slides for fluorescence microscopy imaging.

3. RESULTS AND DISCUSSION

Under controlled solvent displacement, DNA-encapsulated gelatin nanoparticles of less than 200 nm in diameter were reproducibly prepared (Figure 1). These were further crosslinked with glyoxal. Further modification of the gelatin nanoparticles with bifunctional PEG and functionalization with the EGFR-targeting peptide showed that the modification process had no significant influence on particle size (<200 nm) and zeta potential. The similarity of zeta potential values to those of the blank nanoparticles indicates the encapsulation of the plasmid DNA as opposed to adsorption on the surface. The SEM images reveal the smooth surface morphology and spherical shape of these nanoparticles. The percent cell viability as a function of the polymer and EGFR-peptide concentrations was evaluated. These studies indicate that gelatin is highly biocompatible showing 99% cell viability. However, for the EGFR-peptide, we found that the relative cell viability at the highest concentration of 200 µM was 88%. This could most likely be due to the inhibitory effect of the peptide on the cell proliferation than actual cell kill. In comparison, the positive control, poly(ethyleneimine), has shown a relative cell viability of only 40%.

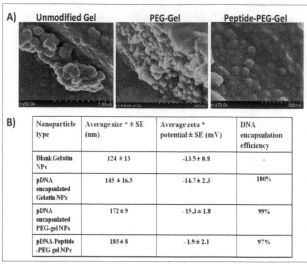

Figure 1. Characterization of the control and EGFR peptide-modified gelatin nanoparticles. (A) Scanning electron microscope images of unmodified, PEG-modified and peptide-modified gelatin nanoparticles and (B) size, surface charge, and percent DNA encapsulation in the control and peptide-modified gelatin nanoparticles.

An understanding of the EGFR expression pattern in different pancreatic cell lines was pertinent at this point. Past literature suggests that large variation in EGFR expression among the pancreatic adenocarcinoma cell lines was common. In fact, this is suspected to be a major challenge in the development of receptor targeting drug delivery system for pancreatic cancer treatment. Using Western blot and immunocytochemistry techniques with standard protocol (Figure 2), we found that Panc-1 cells showed increasing levels of EGFR over expression. However, in comparison, Capan-1 (another pancreatic cancer) cells showed significantly lower EGFR expression levels, which was comparable to the human ovarian cancer (SKOV3) cells. Lastly, the NIH3T3, murine fibroblast cells showed no detectable levels of EGFR over expression. This is most likely due to the efficient endocytotic sequestration and subsequent degradation of the EGF receptor in normal cells.

On the other hand, competition binding studies confirmed that over 70% of the binding and uptake mediated by the EGFR-targeted GENS could be competed, while the control nanoparticles showed no such competition. By the *in vitro* particle uptake

studies, EGFR peptide-modified GENS resulted in higher uptake by the Panc-1 cells than normal murine fibroblasts (NIH3T3) due to EGFR over-expression. Additionally, we found that uptake of peptide-modified nanoparticles in the Panc-1 cells was time and dose dependent. These results confirm the specificity of the EGFR-targeting peptide in cell lines and, therefore, suggest possible receptor-mediated uptake of the functionalized nanoparticles.

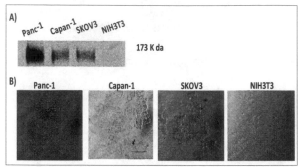

Figure 2. Epidermal growth factor receptor over-expression analysis by (A) Western blot and (B) immunocytochemistry methods in Panc-1 and Capan-1 pancreatic adenocarcinoma cell lines, SKOV3 ovarian adenocarcinoma cells, and NIH3T3 murine fibroblasts.

The DNA loading efficiency of control and peptide-modified GENS was >95%. Surface modification and functionalization of nanoparticles has insignificant effect on the encapsulation of pDNA itself. In the presence of the reducing agent glutathione, release of the encapsulated DNA was found to be 100% in five hours.

With unmodified GENS, the GFP transfection efficiency (Figure 3) in Panc-1 cells was 11%. Upon PEG-modification, the transfection efficiency increased to 30%. EGFR-targeted peptide conjugation to PEG modified nanoparticles led to further increase (~50%) in the transfection efficiency of GFP. This is predominantly attributed to the effective targeting of the functionalized nanoparticles to the EGFR receptor expressed on the tumor cell surface. In addition, the enhanced transfection levels are also due to the prolonged stability of the nanoparticles upon PEG modification. Lipofectin®-complexed plasmid DNA, used as a positive control, showed lower transfection efficiency and higher toxicity over time.

However, previous reports (9, 10) have shown that transduction levels in tumor cell lines were not significantly different from non-targeted transfection levels. Therefore, it is hypothesized that this insignificant increase may be due to the lack of external force *in vitro* to increase vector-to-cell interaction as can be seen *in vivo* post intravenous administration. This emphasizes the popular belief that a wider range of factors controling the tumor microenvironment *in vivo* cannot be completely and successfully mimicked in the *in vitro* models.

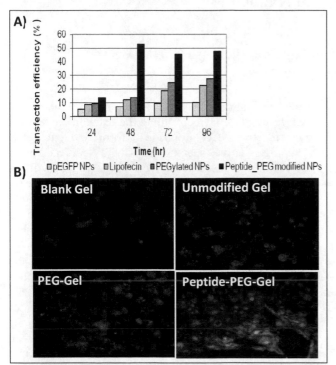

Figure 3. (A) Quantitative and (B) qualitative green fluorescent protein expression efficiency following administration of EGFP-N1 plasmid in control and EGFR peptide-modified gelatin nanoparticles in Panc-1 human pancreatic cancer cells. Quantitative and qualitative transfection was evaluated by flow cytometry and fluorescence microscopy, respectively.

4. CONCLUSIONS

The results of this study showed that type B gelatin nanoparticles, when conjugated with EGFR-targeting peptide ligand, can mediate safe and specific gene transfer to EGFR over-expressing tumor cells in vitro. Furthermore, these results demonstrate the potential to active targeting *in vivo* while achieving efficient gene therapy in pancreatic cancer.

5. ACKNOWLEDGEMENTS

Padmaja Magadala is an IGERT Fellow in Nanomedicine Science and Technology training program. This program is supported by the National Cancer Institute (NCI) and the National Science Foundation (NSF). SEM was performed at the Electron Microscopy Facility of Northeastern University (Boston, MA).

REFERENCES

1. American Cancer Society: Cancer Facts and Figures 2007. Atlanta, GA. American Cancer Society, 2007. Accessed on January 3, 2008 from http://www.cancer.org/downloads/STT/CAFF2007PWSecured.pdf

2. Woodburn, J. R. (1999). The epidermal growth factor receptor and its inhibition in cancer therapy. *Pharmacol. Ther.* **82:** 241–250.

3. Abbruzzese JL, Rosenberg A, Xiong Q, *et al.*, (2001). Phase II study of anti-epidermal growth factor receptor (EGFR) antibody cetuximab (IMC-C225) in combination with gemcitabine in patients with advanced pancreatic cancer. *Proc Annu Meet Am Soc Clin Oncol.* pp. 518.

4. Amiji, M.M. *Polymeric Gene Delivery: Principles and Applications* (2005). CRC Press, New York, NY.

5. Kaul G. and Amiji M. Cellular interactions and in vitro DNA transfection studies with poly(ethylene glycol)-modified gelatin nanoparticles. *J Pharm Sci.,* **94(1):** 184-198 (2004).

6. Kommareddy S and Amiji M. (2007). Antiangiogenic gene therapy with systemically administered sFlt-1 plasmid DNA in engineered gelatin-based nanovectors. *Cancer Gene Ther.,,* **14(5):** 488-98.

7. Brissette, R and Goldstein, NI. (2007). The use of phage display peptide libraries for basic and translational research. *Methods Mol. Biol.,* **383:** 203-13.

8. Li Z, Zhao R, Wu X, et al. (2005). Identification and characterization of a novel peptide ligand of epidermal growth factor receptor for targeted delivery of therapeutics. *FASEB Journal,* **19(14):**1978-85.

Polyurethane Nanocomposites Containing NCO-Functionalized Carbon Nanotubes and NCO-Functionalized Nanoclays

T. Nguyen, B. Pellegrin, A. Granier, and J. Chin

National Institute of Standards and Technology, Gaithersburg, MD, USA

Email: tinh.nguyen @nist.gov

ABSTRACT

We have developed an effective method to covalently functionalize CNTs and nanoclays that carry free isocyanate (NCO) groups, which readily form covalent bonds with polymer matrices containing hydrogen–active groups. The functionalization employs a diisocyanate molecule in which the two end NCO groups have different reactivities. By controlling the reaction conditions, the more reactive NCO groups form covalent bonds with hydrogen-active species on the nanoparticles surface, while the less reactive NCO groups are available for reacting with the polymers. Mechanical properties and photodegradation under UV radiation of an acrylic polyurethane (PU) containing NCO-functionalized (f) (MWCNTs and nanoclays) have been evaluated. Tensile moduli of composites containing NCO-f nanopartciles are much higher than those without or with unfunctionalized nanomaterials. The incorporation of functionalzed or unfunctionalized MWCNTs at 1 % mass fraction appears to stabilize the degradation of PU under UV light, but clay nanoparticles even at 5 % mass fraction have no effect on the UV resistance of this polymer. The results show that NCO-f(MWCNTs and nanoclays) greatly enhance the stress transfer efficiency of acrylic polyurethanes.

Keywords: nanocomposites, NCO, functionalization, polyurethane, nanotubes, nanoclays

1 INTRODUCTION

Carbon nanotubes (CNTs) and nanoclays are being studied intensively as the ultimate reinforcing materials for polymeric composites because of their exceptional mechanical properties and high aspect ratio. For this application, the nanofillers must be well dispersed in the polymer matrix and/or form strong interfacial bonds to effectively transfer load from the matrix to the nanofillers. To meet these requirements, considerable effort has been devoted to chemically functionalize these materials (1-4). We have developed a method to covalently attach an organic molecule to the MWCNTs and nanoclays surfaces that carry free isocyanate (NCO) groups. Such functionalized nanoparticles are highly desirable because NCO is very reactive towards hydrogen-active species such as OH, NH, and COOH, which are common in synthetic and natural polymers. The uniqueness of this functionalization method is the use of a diisocyanate

molecule containing two end NCO groups that have different reactivities. By controlling the reaction conditions, the more reactive NCO groups will form covalent bonds with hydrogen-active species on CNTs and nanoclays surfaces, while the less reactive NCO groups will be available for reacting with the polymer matrices of interest. Further, since the diisocyanate molecule contains a large hydrocarbon chain, this functionalization would lead to better dispersion of nanoparticles in polymers. For polyurethane (PU) nanocomposites, it is expected that the NCO-terminated nanofillers will react readily with the polyol component during curing to form covalent bonds (urethane linkages) between the matrix and the nanofillers. This should increase the load transfer efficiency of PU/CNTs and PU/nanoclays composites. This paper reports on the mechanical and photodegradation performance of an acrylic urethane containing NCO-f(MWCNTs and nanoclays).

2 EXPERIMENTAL

2.1 Materials

Multi-walled nanotubes (purity > 95%, diameter 10-30 nm, length 0.5-40 μm, and synthetic nanoclays (Laponite, chemical compositions in mass fraction: 66.2 % SiO_2, 30.2 % MgO, 2.9 % Na_2O, and 0.7 Li_2O) were obtained from commercial sources. Isophrone diisocyanate (IPDI) was used for the NCO functionalization. One particular characteristics of this diisocyanate is one NCO group is substantially more reactive than the other (5). Dibutyltin dilaurate was used as the catalyst. The polymer matrix was an acrylic polyurethane typically used in automobile coatings, consisting of an acrylic polyol and a biuret hexamethylene tri-isocyanate. After curing, this PU forms a crosslink network. Anhydrous acetone and ethylene glycol dimethyl ether (DME) were used as the solvents for the functionalization of CNTs and nanoclays, respectively. Anhydrous reagent grade acetone was used for composite processing.

2.2 NCO Functionalization of MWCNTs and Nanoclays

The procedure used to functionalize MWCNTs and nanoclays that carry terminal NCO groups is summarized in the schemes shown in Figs. 1a and 1b. More complete descriptions of the functionalization are given elsewhere

(6,7). Since the covalent functionalization requires that hydrogen-active species must be present in the nanoparticles, MWCNTs were first oxidized to generate COOH groups on their surface (COOH-fMWCNT) by sonicating and refluxing as-received MWCNTs (AR-MWCNT) in 9.5 mol/l HNO_3 solution for 1 h, followed by 24 h reflux with constant stirring at 125 °C. The suspended CNTs were then filtered, washed thoroughly with water and dried in vacuum. This procedure produced about 4.5 % mass fraction of COOH groups on the surface of CNTs. For nanoclays, surface SiOH species serve as the reactive sites for the functionalization. Because of its pellet structure, nanoclays were first exfoliated to increase the number of reactive sites before functionalization. This was accomplished by sonicating clay nanoparticles in DME for 30 minutes, followed by refluxing at 85 °C for 4 h under constant stirring.

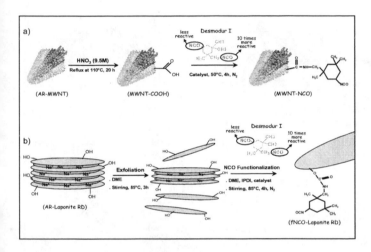

Figure 1. Steps and conditions used for NCO functionalization of MWCNTs (a) and nanoclays (b).

The NCO functionalization was carried out by placing COOH-f CNTs or DME-containing exfoliated nanoclays in a three-neck flask containing appropriate amounts of dibutyltin dilaurate catalyst and solvents (acetone for CNTs and DME for nanoclays). The suspension was heated for an period of time (50 °C and 7 h for CNTs and 85 °C and 4 h for nanoclays), under constant stirring and N_2 atmosphere. As shown in the schemes, amide linkages are generated between the diisocyanate (IPDI) molecule and the COOH-fCNTs, and these bonds may be formed on both the sidewalls and ends of the carbon nanofillers, because the COOH groups could be generated on both locations (8). On the other hand, urethane bonds are formed between the IPDI and clay nanoparticles. After cooling to room temperature, NCO-f(MWCNTs and nanoclays) were filtered, sonicated in acetone for 1h, and washed with acetone. These steps were repeated at least three times to ensure that most of the physically-sorbed and excess IPDI molecules have been removed from the functionalized

materials. After drying in a vacuum oven at 70 °C for 12 h, the functionalized materials were stored in N_2 at 4 °C until use.

2.3   Preparation   of   Polyurethane Nanocomposites

PU nanomposites were fabricated at 0.5, 1, and 2 % mass fraction loadings for MWCNTs and at 0.5, 1 and 5% loadings for nanoclays. The amounts of nanofillers added to the matrix were based on the effective mass of the nanoparticles, which were obtained by thermogravimetric analysis (TGA). In addition to NCO-f(MWCNT and nanoclays), composites containing AR (MWCNT and nanoclays) were also made for comparison. PU/MWCNT and PU/nanoclay composites were prepared using NCO:OH ratios of 1.3:1 and 1:1, respectively.

Figure 2 illustrates the steps and conditions used for preparation of PU/MWCNT and PU/nanoclay composites. Nanofillers were first sonicated in large amount of anhydrous acetone for 2 h. After adding polyol, the nanofiller suspensions were stirred and sonicated for an additional 1 h using an 80 KHz tip sonicator. The suspension was then subjected to high shear mixing at 210 rad/s (2000 rpm) for 30 min at room temperature (24 °C) to further break up nanofillers aggregations. The triisocyanate component was then added to the suspension and stirred at 315 rad/s (3000 rpm) for another hour. After degassing for 1 h at room temperature, the mixture was placed on a release paper and drawn down using an applicator. All coated films were cured at ambient conditions (24 °C and 45 % RH) for 4 days, followed by post-curing for 4 h at 130 °C in an air-circulating oven. FTIR analysis showed no further decrease of the NCO band intensity at longer curing times. All cured films had a thickness between 75 µm and 100 µm.

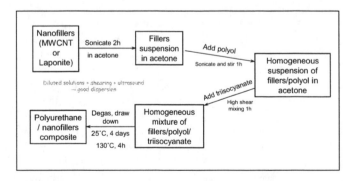

Figure 2. Steps and conditions used for preparation of PU/MWCNT (a) and PU/nanoclay (b) composites.

2.3 Characterization of NCO-f(MWCNTs and Nanoclays) and Polyurethane Nanocomposites

NCO-functionalized nanoparticles were characterized by transmission Fourier transform infrared (FTIR) spectroscopy and TGA; the latter can provide the amounts of IPDI attached to the nanofillers. Tensile modulus, tensile strength, elongation at break, and glass transition temperature (T_g) of PU nanocomposites were measured by Dynamic Mechanical Thermal Analysis (DMTA). The mechanical properties were measured on 45 mm x 5 mm samples (thickness was measured for each sample) with an extension rate of 0.02 mm s^{-1}. T_g was obtained on 30 mm x 10 mm samples at 5 Hz and 0.1 % strain, from 30 °C to 170 °C. T_g was taken as the maximum of the tan δ curve. At least five samples were analyzed for each composite. The standard deviations of the mechanical properties and T_g were ± 10 % and ± 1 %, respectively. The UV resistance of PU and its nanocomposites were investigated using a 2 m integrating sphere-based weathering device, referred to as SPHERE (Simulated Photodegradation via High Energy Radiant Exposure) (9). This device utilizes a mercury arc lamp system that produces a collimated and highly uniform UV flux of approximately 480 W/m^2 in the 290 nm to 450 nm range. This device can also precisely control the relative humidity and temperature. In this study, film samples of 2 cm x 2 cm μm were exposed in the SPHERE UV chamber at 50 °C and 70 % relative humidity. They were removed at specified time intervals for photodegradation measurement using FTIR in the attenuated total reflection mode (FTIR-ATR).

3 RESULTS AND DISCUSSION

3.1 Mechanical Properties of PU Composites Containing NCO-*f*(MWCNTs and Nanoclays)

FTIR and TGA results clearly showed that this approach is effective for covalently attaching IPDI monomer to the CNTs (both SWCNTs and MWCNTs) and nanoclays (6,7). We have also demonstrated that if the two or three end NCO groups in the di or triisocyanate molecules have the same reactivity such as diphenylmethane 4,4'-diisocyanate (MDI), no free NCO groups are observed (6). Therefore, such isocyanate molecules are not suitable for the NCO functionalization. NCO-f(MWCNTs and nanoclays) have been prepared, and the performance of PU composites containing these functionalized materials has been evaluated. Figures 3a and 3b show the effects of 1 % mass fraction loading of NCO-*f*(MWCNTs and nanoclays) on mechanical properties and T_g of PU. Results for unfilled PU and composites containing AR(MWCNT and nanoclays) are included for comparison.

The addition 1 % mass fraction of both unfunctionalzied and functionalized (MWCNTs and nanoclays) substantially increases the modulus of PU. However, NCO-functionalized materials have a greater influence. For example, the modulus increases nearly 40 % and 65 % after incorporating 1 % NCO-*f*MWCNTs and NCO-*f*nanoclays, respectively, but only 23 % and 42 % for the corresponding unmodified nanoparticles. On the other

hand, the addition of NCO-*f*nanoparticles decreases the elongation at break of both types of nanocomposites, with NCO-*f*MWCNTs appear having a greater effect. Figure 3 shows little effect of unfunctionalized or functionalized (MWCNT or nanoclays) on T_g of PU. The modulus increase with NCO-*f*(MWCNTs and nanoclays) suggests that this functionalization is effective in improving the stress transfer in these PU nanocomposites.

a

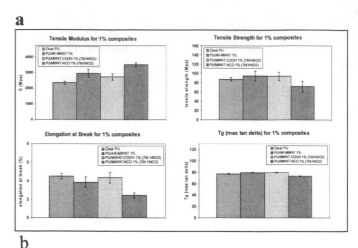

b

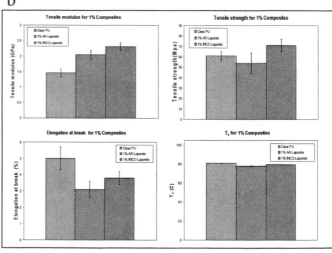

Figure 3. Tg and mechanical properties of NCO-*f*MWCNT (a) and NCO-*f*nanoclay (b) composites. Results of unfilled and PU filled with unfunctionalized nanoparticles are included for comparison.

3.2 Photodegradation of PU Composites Containing NCO-*f*(MWCNT and Nanoclay)

The effects of NCO-*f*(MWCNTs and nanoclays) on UV resistance of PU are displayed in Figs. 4a and 4b. This figure presents relative changes of the 1520 cm^{-1} band (NH group) of the PU film with and without 1 % mass fraction of NCO-functionalized and unfunctionalized (MWCNTs and nanoclays) as a function of time exposed in the SPHERE UV chamber. A decrease of the 1520 cm^{-1} band

NSTI-Nanotech 2008, www.nsti.org, ISBN 978-1-4200-8503-7 Vol. 1

intensity is attributed to an increase in the chain scission of the PU polymer. The intensity changes have been normalized to both the initial absorbance and that of the least-changed band (1172 cm⁻¹, due to C-O) to minimize the effects due to thickness differences between samples and contact variations by the ATR probe on the sample. The identification of the least-changed band was provided by a transmission FTIR analysis on a 7 μm thick film on a CaF₂ substrate exposed to the same UV/RH/temperature conditions. As seen in this figure, adding 1% mass fraction of unfunctionalized or NCO-*f*MWCNT in PU greatly enhances the UV resistance of this polymer. The reason for the adverse effect on photodegradation by the AR-MWCNTs at 2 % mass fraction loading is unknown and being investigated. The addition of nanoclays, functionalized or unfunctionalzied, even at 5% loading, has little effect on the UV resistance of this PU.

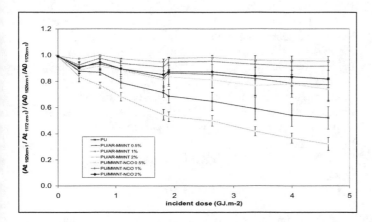

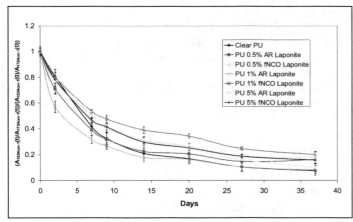

Figure 4. FTIR intensity changes as a function of time exposed to UV light for PU/NCO-*f*MWCNT (a) and PU/NCO-*f*nanoclay (b) composites. Results of unfilled and PU filled with unfunctionalized nanoparticles are included for comparison.

4 CONCLUSION

We have developed an effective method to covalently functionalize (CNTs and nanoclays) that carry terminal NCO groups, which can readily form covalent bonds with polymer matrices that contain hydrogen active species. Iincorporation of functionalized or unfunctionalized MWCNTs at 1 % mass fraction appears to stabilize the degradation of PU under UV light, but nanoclays even at 5 % mass fraction loading have no effect on the UV resistance of this polymer. The results show that NCO-*f*(MWCNTs and nanoclays) substantially increase the modulus and enhance the stress transfer efficiency of acrylic polyurethanes.

5 REFERENCES

1. S. B. Sinnott, Chemical Functionalization of Carbon Nanotubes, *J. Nanosci. Nantech.*, 2 (2002) 113-123.
2. K. Balasubramanian and M. Burghard, Chemically Functionalized Carbon Nanotubes, *Small,* 1 (2005) 180-192.
3. M. Park, I.K. Shim, E.Y. Jung, J.H. Choy, Modification of external surface of Laponite by silane grafting, *J. of Phys. & Chem. Solids,* 65 (2004):499-501.
4. N.N. Herrara, J.M. Letoffe, J.P. Reymond, E. Bourgeat-Lami, Silylation of Laponite clay particules with monofunctional and trifunctional vinyl alkoxysilanes, *J. Mater. Chem.,* ,15 (2005) 863-871.
5. O. Lorentz, H. Decker, G. Rose, *Angew Makcromol. Chem.,* 122 (1984) 83.
6. A. Granier, T. Nguyen, N. Eidelman, J. Martin, *Proc, Adhesion Society Meeting,* 2007, p. 346.
7. B. Pellegrin, A. Granier, and T. Nguyen, *Proc Adhesion Society Meeting,* 2008, p.199.
8. B. Zhao, H. Hu, E. Bekyarova, M. Itkis, S. Niyogi, R. Haddon, Carbon Nanotubes: Chemistry, *Dekker Encyclopedia of Nanoscience and Nanotechnology,* 2004, pp 493-505.
9. J. Chin, E. Byrd, N. Embree, J. Garver, B. Dickens, T. Fin, and J. W. Martin, Accelerated UV Weathering Device Based on Integrating Sphere Technology, *Review Scientific Instruments,* 75 (2004) 4951-4959.

Nanostructured LZO Films: Synthesis and Characterizations

H. S. Chen[*], R. V. Kumar[**] and B. A. Glowachi[***]

Department of Materials Science and Metallurgy, University of Cambridge, Pembroke Street
Cambridge CB2 3QZ, U.K.
[*]hsc28@cam.ac.uk
[**]rvk10@cam.ac.uk
[***]bag10@cam.ac.uk

ABSTRACT

Nanocrystalline lanthanum zirconium oxide (LZO) films have been successfully synthesized by a wet-chemical route, using dried lanthanum acetate, zirconium propoxide, propoionic acid, acetic acid glacial, and anhydrous methanol. X-ray 2θ-scan diffraction patterns of the LZO film deposited on the Ni/W substrate displayed only the (400) peak. X-ray ω-scan diffraction data showed that the LZO films exhibited out-of-plane alignment along <400>. Field emission gun scanning electron microscope images have indicated that LZO films are composed of nanoparticles with an average size of about 25 nm. The Taguchi Design has been employed to examine the effects of composition and solution chemistry of the precursors in the synthesis of LZO. The optimum experimental conditions have been evaluated as: 0.7 ± 0.5 for the ratio of the lanthanum acetate to the propionic acid, 0.08 ± 0.02 mol/kg for the precursor concentration, and 850 ± 10 °C for the heating process using 5 ± 1 °C per minute.

Keywords: LZO, Nanocrystalline, film

1 INTRODUCTION

Development of new synthesis methods for nanostructured materials and demonstration of improved and/or unusual properties of these materials have increased rapidly within recent years. Nanostructured materials are available in a variety of compositions of metal oxides and as metals supported on metal oxides or vice versa, which have led to many investigations of their exceptional chemical, physical and electric properties.

The attractiveness of the sol-gel method of synthesis arises from the fact that virtually any metal oxide system can be examined, and no special equipment as such is required. Sol-gel methods comprise of precipitation of nano-scaled colloid suspension of particles from a solution formed by dissolving a salt in a suitable liquid, usually water, which is then modified by addition of an acid or a base to obtain a gel network. Alternatively, it is also common to use polymeric alkoxide route, where a metal oxide is dissolved or supported in a suitable liquid usually alcohol, which is then activated by an acid or a base resulting in a gel network.

Lanthanum zirconium oxide ($La_2Zr_2O_7$ or LZO) is a well known material with a pyrochlore.structure[1] In fact, LZO has two different crystal structures, one is cubic fluorite type, and the other is cubic pyrochlore type. The LZO is thought to be a potential candidate for applications as high dielectric constant materials in the Si-based industry, thermal barriers, radiation resistant layer and others.[2] In the past decade, the number of the scientific reports about the synthesis and the properties of LZO has been growing rapidly. This study reports a sol-gel synthesis method to obtain a nanocrystalline LZO film and Taguchi Design statistical method has been used to optimise the film.

2 EXPERIEMNTAL

2.1 Preparation of LZO Precursor

Lanthanum acetate hydrate powder (99.9%, Aldrich) was first dried at 170°C for an hour. Zirconium (IV) propoxide (70wt%, Aldrich), propoionic acid (99.5%, Fluka-Garantie), acetic acid glacial (analytical reagent grade, Fisher), and methanol anhydrous (99+%, Aldrich), all as liquids were used as received. LZO precursor sol was prepared in a stoichiometric ratio of La and Zr. Lanthanum solution was first prepared by dissolving dried lanthanum acetate in propionic acid at a temperature near 80 °C. The solution was then cooled to room temperature after it transformed to a clear liquid. Zirconium solution was prepared by rapidly adding glacial acetic acid into zirconium propoxide and then was mixed by a magnetic stirrer for 15 minutes. The zirconium solution was then added into the lanthanum solution and mixed by the stirrer for another 30 minutes. The LZO precursor sol was obtained eventually by diluting the above mixture with methanol to a concentration of 0.2 mol/kg.

2.2 Preparation of LZO Film

For the coating process to deposit a LZO film, the precursor sol was first diluted to a desired concentration and then loaded into the ink-jet printing system. A substrate, which was cleaned in an ultrasonic system sequentially by ethanol and acetone and thermally treated at 850 °C to remove all organics, was placed 30 mm below the nozzle. The precursor solution was loaded into a reservoir in the ink-jet printer, and pressure was applied to the reservoir by a mechanical pump.[3] A computer-controlled electronic system was used to allow the nozzle to open for a fixed time period so that a set amount of an ink could be deposited. LZO films were printed as a 5x25 mm² track on textured Ni substrates of 5x40 mm² at room temperature under ambient atmosphere. The sample was then moved to a pre-purged furnace heated at 10 °C/min to 900°C for 1 hour under Ar-5%-H₂ atmosphere and then cooled in the furnace which was switched off.

2.3 Characterization

Weight change as a function of temperature of the lanthanum acetate, LZO precursor solutions, and dried LZO precursor solution films, were examined by thermogravimetry analysis (TGA, TA Instruments Q500). Spectroscopy of the lanthanum acetate, the zirconium propoxide, the as-prepared LZO precursor solution and both the dried and the thermally treated LZO precursor films, were carried out by Fourier transform infrared (FTIR, Bruker Optics Tensor 27 FT-IR) spectrometer. X-ray diffractometer (Phillips PW 1830/00 with Cu K radiation) was employed to analyze the crystal structure and also the out-of-plane texture of LZO films. The 2-θ scan was used to examine the crystalline phase of the LZO films. Out-of-plane orientation was judged by estimating the FWHM (expand this) values of the diffraction peaks obtained from the ω-scan mode at a fixed 2θ value with 1/12° of diversion slit. The better the out-of-plane texture, the narrower the FWHM of the diffraction peak from the ω-scan. The surface morphology of the LZO films was investigated by field emission electron scanning microscope (FEGSEM, JEOL 6340F) with a working distance of 6 mm and an acceleration voltage of 5 keV. In the current study, surface integrity was defined as a ratio of Vs to Vf, which are the area of a solid part and the total area of the film, respectively, calculated by ImageJ version 1.37.

3   RESULTS AND DISCUSSIONS

Figure 1 gives the FTIR spectra of the LZO precursor mixture and LZO precursor film (dried overnight at room temperature) along with the zirconium propoxide and the lanthanum acetate hydrate data. Not only the LZO precursor mixture but also the dried LZO precursor films exhibited a similar characteristic to a combination of the patterns of the zirconium propoxide and the lanthanum acetate hydrate, implying there was no significant change in bonding at the molecular level among elements after LZO precursor was formed, though some of the peaks were slightly shifted and several extra peaks had appeared that can be ascribed to the formation of complexes.

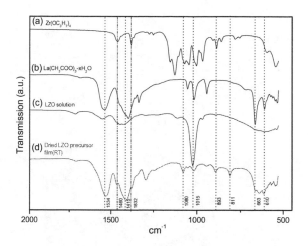

Figure 1: FTIR spectra of chemicals and LZO precursors

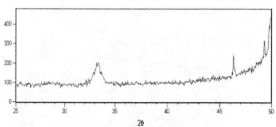

Figure 2: XRD of LZO film.

Figure 3: FEGSEM images of a LZO film.

Figure 2 shows a XRD pattern of a LZO film on the Ni substrate after heating at 900 °C for an hour. The pattern has lost some diffraction peaks such as (111) and displays only a (400) peak. This indicates that the LZO film have a lattice alignment along <400> directions. FEGSEM images of the LZO film after the heating process are shown in Figure 3. The LZO precursors formed a dense film containing nanocrystallites with a size of about 25 nm, but generated some pine holes on the surface. Figure 4 (a) and (b) provide a display of the AFM images of a LZO film. The images clearly show that the LZO films are composed of nanoparticles, which are conical in shape with the cone being perpendicular to the substrate surface. As can be observed, the nanocrystlaline LZO have grown as individual cones along the c-axis, giving visual evidence for the XRD data.

(a)

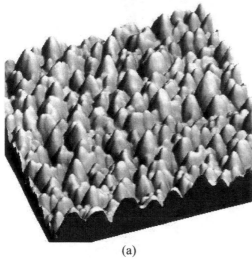

(a)

Figure 4: AFM image of a LZO film.

Control Factors / Levels	Cn (mol/kg)	Ratio of La(Ac)₃ to Propionic Acid	Heating Rate (°C/min)	Heating Temp. (°C)
	A	B	C	D
1 (Low)	0.05	0.7	1	850
2 (Medium)	0.08	1.0	5	900
3 (High)	0.1	1.3	10	1000

Table 1: Chosen control factors and levels.

Figure 5: FEGSEM images of a LZO film (for Taguchi Desgin).

The optimum experimental conditions given from Taguchi design are: 0.7±0.5 for the ratio of lanthanum acetate to propionic acid, 0.08±0.02 mol/kg for the precursor concentration, and 850±10 °C for the heating process at a heating rate of 5±1 °C per minute. As shown in Figure 5, the LZO film, corresponding to the optimum parameters is of a good quality and free of pin hole with a surface integrity value of 0.99.

5 CONCLUSIONS

A dense, smooth, pin-hole-free lanthanum zirconium oxide films have been successfully synthesized by using ink-jet printing from a liquid ink prepared by dried lanthanum acetate, zirconium propoxide, propoionic acid, acetic acid glacial, in anhydrous methanol. The Taguchi Design has successfully been employed to find the optimum experimental conditions, which are 0.7±0.5 for the ratio of the lanthanum acetate to the propionic acid, 0.08±0.02 mol/kg for the precursor concentration, and 850±10 °C for the heating process using 5±1 °C per minute. The LZO particles are conical in shape and are textured along the c-axis.

REFERENCES

[1] J. W. Seo, J. Fompeyrine, A. Guiller, G. Norga, C. Marchiori, H. Siegwart, and J.-P. Locquet, Appl. Phys. Lett. 83, 5211, 2003.
[2] A. Chartier and C. Meis, Phys. Rev. B 65, 134116 2002.
[3] T. Mouganie, M. A. Moram, J. Sumner, B. A. Glowachi, B. Schoofs, I. V. Driessche, and S. Hoste, J. Sol-Gel Sci. Tech. 36, 87, 2005.

4 TAGUCHI DESIGN

The LZO precursor preparation contains at least eight components in the system. Given the number of steps in the subsequent heating process there were at least ten factors. So it was difficult to clarify the role of each factor in the solution, and also difficult to adjust experimental conditions to gain better results by trial and error. To investigate experimental variables, Taguchi Design was employed for the current study.

The following major factors have been identified in the synthesis of LZO: ratio of lanthanum acetate to propionic acid; concentration of diluted solution used in the coating system; heating temperature; and the heating rate. Thus for four control factors and three factor levels, the $L^9(3^4)$ orthogonal array was employed in our study. Other variables including mixing conditions of La solution and Zr solution were fixed to reduce the overall complexity in this analysis. Detail discussion will be published elsewhere.

Quartz Crystal Microbalance with Dissipation Monitoring (QCM-D): Real-Time Characterization of Nano-Scale Interactions at Surfaces

Archana Jaiswal[*], Stoyan Smoukov[**], Mark Poggi[*] and Bartosz Grzybowski[**]

*Q-Sense, Inc. 808 Landmark Drive, Suite 124,
Glen Burnie, MD 21061, U.S.A, archana.jaiswal@q-sense.com
**Northwestern University. Department of Chemical and Biological Engineering.
2145 Sheridan Rd., Evanston, IL 60208-3120, U.S.A.

ABSTRACT

There has been an increasing demand for analytical tools to quantify the interactions and/or reactions of nano-scale particles, polymers and bio-molecules, with a variety of surfaces, in real time. Understanding the behavior of such molecules at the nano-scale enables researchers to optimize the conditions for desired results at macro scale.

Quartz Crystal Microbalance with Dissipation Monitoring (QCM-D), a nanomechanical acoustic-based analytical technique, provides in-situ, real-time characterization of materials at interfaces. The present report is focused on applications of the QCM-D technique in real-time characterization of Au/Ag nanoparticle adsorption, and the interaction of various biomolecules such as proteins, DNA, and viruses with other biological materials, polymers or inorganic surfaces.

Keywords: QCM-D, Nanoparticles, Polymers, Protein, DNA.

INTRODUCTION

Quartz crystal microbalances (QCMs) have been used for decades to monitor the growth of thin films in vacuum. More recently, invention of QCMs that operate in liquid has opened up numerous possibilities of applying this technique in the fields of nanotechnology, biotechnology, biology, drug delivery, etc., where real time, in-situ analysis of interactions of nano-scale molecules/particles at various surfaces, in a liquid medium, is of critical importance.

A QCM consists of a thin quartz disc sandwiched between a pair of electrodes. Due to the piezoelectric properties of quartz, these crystals can be excited to oscillate at their resonance frequency by applying an AC voltage across the electrodes. In the event of any adsorption onto the surface of the oscillating quartz crystal, the frequency decreases. The resonance frequency (F) of the crystal depends on the total oscillating mass, including water coupled to the oscillation.

For a thin and rigid film, the decrease in frequency is linearly proportional to adsorbed mass, defined by the Sauerbrey relation[1]:

$$\Delta m = -\frac{C \cdot \Delta f}{n}$$

$C = 17.7 \text{ ng Hz}^{-1} \text{ cm}^{-2}$
$n = 1,3,5$, etc., is the overtone number.

When the adsorbed film is not rigid, the Sauerbrey relation becomes invalid. By measuring multiple frequencies and the energy dissipation of the quartz sensor it is possible to determine if the adsorbed film is rigid or soft. In contrast, conventional QCMs only measure changes in frequency and hence fail to fully characterize a soft material.

With QCM-D, simultaneous measurements of changes in resonance frequency (ΔF) and energy dissipation (ΔD) are performed by periodically switching off the driving power to the sensor crystal and recording the decay of damped oscillation as the adsorption and/or structural changes take place (Figure 1). While changes in frequency (ΔF) provide information about mass changes, changes in dissipation (ΔD) provide structural information about the viscoelastic properties of adsorbed films in real time.

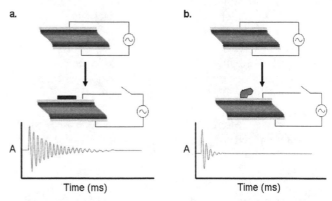

Figure 1. Cartoon depicting the viscoelastic differences between (a) a rigid material and (b) a soft viscoelastic material. In each case, both the change in frequency (change in mass) and change in dissipation (change in viscoelasticity) are monitored simultaneously

Figure 2 represents typical frequency (ΔF) and energy dissipation (ΔD) changes vs. time plot for a protein-protein interaction. By measuring ΔFs and ΔDs at multiple frequencies and applying a viscoelastic (Voigt) model, the adhering film can be characterized in detail and viscosity, elasticity and the correct thickness of the adsorbed films may be extracted.

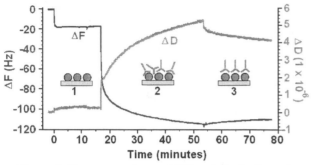

Figure 2. Frequency change (ΔF) and dissipation change (ΔD) vs. time plot for the reaction of a globular protein e.g. antigen with elongated protein e.g. antibody. Step 1: adsorption of antigen on to the crystals surface, Step 2: reaction of antibody with the adsorbed antigen molecules, Step 3: rinsing of loosely bound molecules and reorganization

MONITORING NANO-SCALE INTERACTIONS AND REACTIONS BY QCM-D

Applications of QCM-D can be divided into two broad categories:

1-Surface interactions: Interactions between nano-scale molecules and surfaces as well as interactions between various molecules on a surface are studied. Examples include: nanoparticle-substrate interactions, biomolecular interactions (protein, DNA, antibody), polyelectrolyte multilayer buildup, enzyme-substrate interactions, etc.

2- Surface reactions: Reactions within the adsorbed layer on the surface are studied. Examples include: cross-linking of proteins, polymers, conformation changes, hydration/swelling of polymers and other thin films, etc.

This article describes the results of a recent QCM-D experiment carried out to study the adsorption of Au/Ag nanoparticles, both in static and dynamic modes, onto SiO₂ surfaces.

We also present other examples of QCM-D applications that have been reported elsewhere in detail. Specifically the study of Ti-specific Phage display viruses, buildup of polyelectrolyte multilayers and DNA-PNA hybridization are briefly described here.

1 STUDY OF CHARGED NANO-PARTICLE ADSORPTION

In an earlier study it has been established that surface adsorption of oppositely charged nanoparticles is driven by cooperative electrostatic interactions and does not require chemical ligation or layer-by-layer schemes[2]. The composition and quality of the coatings can be regulated by the types, the charges, the relative concentrations of nanoparticles and the pH.

In the present study, we have investigated the effect of static and dynamic motion of charged nanoparticles solutions on their adsorption on silica surfaces.

1.1 Experimental

Materials: Silicon dioxide-coated 5-MHz quartz sensor crystals were cleaned using the following protocol: UV/ozone treatment for 10 minutes followed by cleaning in a 2% SDS solution for 30 minutes and rinsing with DI water and thereafter UV/ozone treatment for another 10 minutes.

Three concentrations of 1:1 Au/Ag nanoparticle slurries; 50 µg/ml and 150 µg/ml and 400 µg /ml in DI water were used. A mercaptoundecanoic acid coated Au nanoparticles (AuMUA) (2 mM) slurry was used as a non-coating control solution.

Methods: The experiment was carried out on a Q-Sense E4 instrument. An E4 sample chamber has four removable flow modules, each holding one sensor. In the present study, all four flow modules containing silicon dioxide coated crystals were used in a parallel configuration. In the first three modules, solutions of coating nanoparticles (1:1 Au/Ag); 50 µg/ml, 150 µg/ml and 400 µg/ml respectively, were pumped. The fourth module was used as a control, where solution of non-coating nanoparticles (AuMUA) was pumped. Such parallel configuration of flow modules enabled four parallel, real time and simultaneous measurements of frequency (F) and dissipation (D) changes for different concentrations of nanoparticles. All the measurements were performed at 25°C, and under both static and flow modes of nanoparticles solution. A peristaltic pump was used to give a flow of 50 µl /min.

ΔF and ΔD plots were recorded at 5 overtones (3rd 5th 7th 9th and 11th) and fundamental frequency (5 MHz) of crystal sensor. For clarity reason, data for only the third overtone (15MHz) are shown in the present report.

1.2 Results and discussions:

Deposition of 1:1 Au/Ag nanoparticles on the silica surface began as soon as the solutions were introduced into

NSTI-Nanotech 2008, www.nsti.org, ISBN 978-1-4200-8503-7 Vol. 1

the flow modules. Continuous increase in the adsorbed mass on the crystal due to deposited nanoparticles induced a continuous decrease in frequency as shown in figure 3.

Dissipation change of these crystals showed an interesting behavior; a small but sharp increase in ΔD was observed as the nanoparticle layer begins to form (Inset Fig 3), which then approaches its original value relatively slowly. This result indicates that initially, the nanoparticle layers are soft (represented by increase in dissipation) with some water entrapped within clusters. As more particles are added, the rigidity of these films increases which is represented by a decrease in dissipation

Addition of non coating AuMUA nanoparticles did not show any shift either in frequency or dissipation indicating that these nanoparticles did not adsorb onto silica surface. Each sensor was subsequently evaluated in a scanning electron microscope (SEM) and all micrographs correlated with the QCM-D results (data not shown).

Adsorption in flow mode and static mode: effect of concentration:

Figure 3 represents a comparison between the frequency and dissipation responses of three crystals that will be referred to as 1, 2, and 3, on exposure to 50µg/ml, 150 µg/ml and 400 µg/ml respectively.

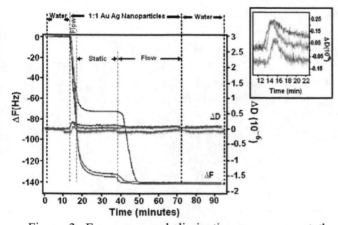

Figure 3. Frequency and dissipation responses, at the 3rd overtone of crystal 1, 2, and 3 on exposure to 50 µg/ml (blue lines), 150 µg/ml (green lines) and 400 µg/ml (red lines) respectively. (Inset: Close up mage of "ΔD vs Time" plot at the point of nanoparticle solution injection)

Flow was continued for 4 minutes in order to fill the flow modules, and then adsorption under static conditions was monitored. Changes in frequency and dissipation of the silica crystals were recorded both during initial flow and static modes. ΔF of crystal 1 that was exposed to the lowest concentration (50µg/ml) reached a constant value of -72 Hz in about 14.5 minutes, indicating no further adsorption of nanoparticles. Whereas frequency changes of crystals 2 and

3, exposed to higher concentrations (150 µg/ml and 400 µg/ml respectively), took a longer time (~23 min) to reach their saturation values. Final ΔFs of crystal 2 and 3 at the saturation point in static mode were higher (~-130 Hz) than that of crystal 1 (-72 Hz) indicating that the particles coverage on crystal 1 is significantly less than coverage on crystals 2 and 3 in the *static mode*.

Once the frequency and dissipation changes stabilized, flow of the corresponding nanoparticle solutions was resumed over the initially-adsorbed layers. This resulted in further significant decreases in ΔF of crystal 1 and relatively smaller decrease in ΔFs of crystals 2 and 3. These results indicate that the lower concentrations probably did not have enough nanoparticles accessible to the silica surface in *static mode* therefore resulting in only partial coverage (49% on crystal 1). Whereas higher concentrations had sufficient amount of nanoparticles to largely cover the silica surface (91% and 93% on crystal 2 and 3 respectively).

Further, continuous flow of the nanoparticle solution enabled additional nanoparticles to immobilize onto the crystal surface, leading to a fully saturated layer. It is interesting to note that the final value of ΔF change (Figure 3) is similar in all the three cases indicating that concentration of nanoparticles does not affect the surface coverage as long as the solution is in *flow mode*. In the *static mode* however, better coverage was obtained with higher concentrations.

Washing the adsorbed nanoparticles layer with DI water did not remove any significant mass which indicates that these particles are stably adsorbed on the silica surfaces and there are no detectable loosely bound nanoparticles present on the surface.

1.3 Conclusion:

The Au/Ag nanoparticles form very rigid layers on silica surfaces that can be defined by the Sauerbrey equation where frequency change is linearly proportional to adsorbed mass. Attachment of these nanoparticles to the silica surface is quite stable and not affected after several hours rinsing in DI water.

At lower concentrations, the particles adsorbed at a slower rate as compared to the higher concentrations. In *static mode*, lower concentration reached saturation much more rapidly than higher concentrations but with a much lower value of frequency change. This indicates a partial coverage of ~50% whereas the coverage at higher concentration was close to 92%. This result indicates that in *static mode*, better surface coverage is obtained with higher concentrations.

A coverage close to one hundred percent, was achieved in the flow mode of the nanoparticle solutions. The final frequency change was the same in all three cases indicating that the concentration of the solution does not have any influence on the extent of surface coverage in *Flow mode*. However, rates of adsorption would be affected by concentrations.

2 OTHER EXAMPLES OF QCM-D APPLICATIONS

2.1 Phage Display – Titanium Specific Peptides[3]

Peptides that bind specifically to various metals and metal oxides have great potential in the nanotechnology/ semiconductor industry to probe nano-scale structures.

Sano et al., have developed phage viruses with specific peptide sequences that specifically bind titanium surfaces. The peptides were isolated by several rounds of biopanning and their specificity towards titanium (Ti) was verified by QCM-D. In the biopanning process, a diverse ($\sim 2.7 \times 10^9$ sequences) phage display library was exposed to titanium oxide beads. The phages with affinity towards Ti bound to the Ti surface and the rest remained in solution. The non-bound phages were then removed from the solution and bound phases were separated from the beads. The phages specific for Ti were then magnified by infection of E. coli.

After several rounds of biopanning, when sufficient specificity was reached, the phages are separated and exposed to Ti crystal sensor mounted in a QCM-D chamber. As a control experiment, phage library of mixture of random DNA sequences was exposed to a control Ti surface. QCM-D results showed comparable shifts in ΔF in both control and experimental cases whereas a huge increase in ΔD was observed for adsorption of Ti specific phages. Change in ΔD for control phage was minimal. Significant increase in ΔD suggests that only very end of the øTi-12-3-1 phages interacted with the Ti surface and therefore take on a vertical orientation. This orientation promotes a greater amount of trapped water therefore increasing the viscoelastic properties of the film (high dissipation value). Control phages were randomly oriented on the Ti surface and were rigid (low dissipation value).

2.2 Build-up of polyelectrolyte multilayers[4]

Halthur et al., have reported in-situ characterization of a multilayer build-up of poly(L-glutamic acid) (PGA) and poly(L-lysine) (PLL) on silica and titanium surfaces, with and without an initial layer of polyethyleneimine (PEI), by means of ellipsometry and QCM-D.

A biphasic buildup was found in all systems, where the length of the first slow-growing phase is dependent on the structure of the initial layers. In the second, fast-growing phase, the film thickness grows linearly while the mass increases in a nearly exponential fashion with the number of deposited layers. Combination of ELP (measuring dry mass) and QCM-D (measuring wet mass) enables calculation of water content that show that the film density increases in the later phase as the multilayer film builds up.

2.3 Characterization of DNA-PNA Hybridization[5]

Höök et al., have used QCM-D to characterize the bound state of single-stranded peptide nucleic acid (PNA) and deoxyribose nucleic acid (DNA) in relation to their ability to function as selective probes for fully complementary and single-mismatch DNA. Two different immobilization strategies of single-stranded PNA and DNA have been explored (thiol coupling to gold and biotin coupling to streptavidin).

The QCM-D technique has been used to characterize the successive steps of various immobilization strategies of single stranded PNA and DNA and the subsequent hybridization with fully complementary and various mismatched sequences of DNA. Both thiol-PNA and thiol-DNA bound to gold surfaces were rigid (low ΔD) and did not have affinity towards fully complementary DNA. Biotin-PNA and biotin-DNA coupled to a 2-D streptavidin layer, were bound in a flexible state (high ΔD), and had high ability to hybridize with both single mismatch and fully complementary DNA.

SUMMARY

QCM-D is a powerful tool for the study of nano-scale interactions and/or reactions on various surfaces. By measuring both changes in frequency and energy dissipation of the oscillating quartz, at various overtones, useful information regarding the viscoelastic properties of materials can be obtained in real time. High sensitivity of this technique allows in-situ detection of structural changes such as conformational changes, cross linking, swelling, etc. during and/or after the adsorption process.

REFERENCES

[1] G.Z. Sauerbrey, Phys. 155, 206, 1959.
[2] S.K.Smoukov, K.J.M.Bishop, B.Kowalczyk, A.M. Kalsin and B.A.Grzybowski, .JACS., 129(50), 15623, 2007
[3] Ken-Inchi Sano and K Shiba, JACS, 125, 14234, 2003.
[4] T. Halthur, and U Elofsson, Langmuir 20, 1739,2004
[5] F.Höök,, A. Ray, B.Norden and B.Kasemo, Langmuir, 17, 8305,2001.

Characterization of Different Surfaces Morphology in Heterogeneous Catalyst

Rajib Mukherjee[*], Ahmet Palazoglu[**] and Jose A. Romagnoli[*]

[*] Cain Department of Chemical Engineering, Louisiana State University,
Baton Rouge, LA 70803, USA, jose@lsu.edu
[**]Dept. of Chemical Engineering and Materials Science, UC Davis,
Davis, CA 95616, USA, anpalazoglu@ucdavis.edu

ABSTRACT

Structures and property of surfaces are very important in different chemical, physical and biological processes. In the case of heterogeneous catalysis, reaction takes place at the surface of the active metal crystals (Pt, Rh, Ni etc) supported primarily on metal oxides. The interaction between the reactant molecules (adsorbate) and the surface are important in the performance of the catalyst. Surface has defects in terms of vacancies, dislocations and grain boundaries. The property of the surface depends on its crystal lattice structures and defects on them. This surface property influences the structure of the adsorbate on the surface. A first principle density functional theoretical (DFT) calculation is generally used to find the activation energy of different surface reaction on different catalysts [1]. In this paper we have simulated the energy of Ni (111) surface with and without defects. A fractal dimension of the crystal surface is found using wavelet transformation.

Keywords: crystal planes, surface energy, surface defects, wavelet transform

1 INTRODUCTION

The performance of various chemical, mechanical, biochemical and semiconductor products is governed by the characteristics of their surfaces. For example, the performance of heterogeneous catalysts is controlled by the characteristics of the molecular interfaces and nanostructure of their crystal surfaces. The surface structure of these crystals due to surface kinetics and surface diffusion governs the function of the material. The interaction of the reactant with the active centers and the support determine the performance of a catalyst. It is thus important to thoroughly characterize its surface properties to optimize its fabrication process. Understanding the surface characteristics at the microscopic level is essential in order to relate the surface characteristics to the performance of the product. To create an application of desired features, we first have to understand the surface characteristics in the microscopic level and then find the relation of product performance with the surface characteristic. This eventually will allow us to improve the product performance through optimizing the manufacturing process towards novel tailored nanostructures. Thus, the key to control fabrication of applications whose performance matrix is governed by its surface features is to characterize the surface. In this way we can optimize product performance.

The characterization of these nanostructured materials often involves surface imaging using Scanning Electron Microscopy (SEM), Transmission Electron Microscopy (TEM), high resolution electron microscopy (HREM) and several Scanning Probe Microscopy (SPM) methods such as Scanning Tunneling Microscopy (STM) and Atomic Force Microscopy (AFM). Those techniques can also reveal additional information about a catalyst (e.g., elemental surface composition, oxidation state, dispersion, acid/base properties). A high resolution electron microscopy (HREM) image of the catalyst surface can be used to obtain the crystal lattice structure of the active metal. The defects on the crystal surface of the crystal lattice structure can be found. The different crystal surface with or without defects has different property associated with them. While microscopy images can provide detailed, qualitative information about the features of a surface, for the applications mentioned above it is critical to obtain a quantitative description of the surface morphology.

In this paper, an estimation of the surface energy of Ni (111) with and without defects in terms of vacancies is made using CPMD [2] program. This surface energy of the catalyst is related to the absorption energy of the reactant molecules. Furthermore, the fractal dimension of the surface changes under the presence of defects. A regular surface will have a fractal dimension of two while a surface with defects in terms of vacancies will have fractal dimension between two and three. In this work, a fractal dimension of the crystal surface is found with and without defects using wavelet transformation.

2 SURFACE FREE ENERGY

The surface free energy σ at a given temperature and pressure is given by the difference in Gibbs free energy of the surface and sum of the potential of each species in bulk.

$$\sigma(T,P) = \frac{1}{A}\left(G_s - \sum_i \mu_i N_i\right) \qquad (1)$$

where G_s is the Gibbs free energy of the surface and μ_i is the chemical potential of each species at the reference temperature and pressure and N_i is the number of atoms of specie i and A is the surface area [3]. The surface energy is thus a function of composition and configuration. For solid and single species say nickel and zero temperature the equation reduces to

$$\sigma = \frac{1}{A}(E - e_{Ni}N_{Ni}) \qquad (2)$$

where E is the internal energy of the surface and e_{Ni} is the energy per atom in the bulk nickel phase.

3   SURFACE FRACTALS FROM WAVELET VARIANCE

A surface is generally characterized by its fractal dimension. Fractal dimension can be obtained from the variance of wavelet coefficients. In wavelet transformation, a wavelike function called wavelets $(\psi_{a,b})$ is used to transform a function $f(x)$ in space or time into another form by convolution. The normalized unidirectional wavelet function is often written as:

$$\psi_{a,b}(x) = \frac{1}{\sqrt{a}}\psi\left(\frac{x-b}{a}\right) \qquad (3)$$

where a is the dilation parameter (also called scale, inverse of resolution) and b is the translation parameter [4]. The transformation integral is written as:

$$T(a,b) = \int_{-\infty}^{\infty} f(x)\psi_{a,b}^*(x)dx \qquad (4)$$

The asterisk is because the complex conjugate of the wavelet function is used for transformation.

Self affinity is defined as: when x changes to λx then the function $f(x)$ changes to $\lambda^H f(x)$ where H is the Hurst exponent also known as roughness exponent. Thus we can write:

$$f(x) = \lambda^{-H}f(\lambda x) \qquad (5)$$

Taking wavelet transformation on both sides with dilation parameter a and translation parameter b we get

$$T[f(x)](a,b) = T[\lambda^{-H}f(\lambda x)](a,b) \qquad (6)$$

When $f(\overline{x})$ happens to be a function of two variables (a plane, $\overline{x}$ is a vector x_1, x_2) then the location parameter $\overline{b}$ must also be a vector with two variables $\overline{b}\ (b_1, b_2)$. The wavelet transformation is given by

$$= \frac{1}{a}\int_{-\infty}^{\infty}\int_{-\infty}^{\infty}\lambda^{-H}f(\lambda\overline{x})\psi^*\left(\frac{\overline{x}-\overline{b}}{a}\right)dx_1 dx_2 \qquad (7)$$

Taking $\lambda\overline{x} = \overline{x}'$ i.e. $\lambda x_1 = x_1'$ and $\lambda x_2 = x_2'$ we can write

$$= \lambda^{-H-1}\frac{1}{\lambda a}\int_{-\infty}^{\infty}\int_{-\infty}^{\infty}f(\overline{x}')\psi^*\left(\frac{\overline{x}'-\lambda\overline{b}}{\lambda a}\right)dx_1'dx_2' \qquad (8)$$

$$= \lambda^{-H-1}T[f(\overline{x})](\lambda a, \lambda\overline{b}) \qquad (9)$$

And thus we can write wavelet transformation as

$$T[f(\overline{x})](\lambda a, \lambda\overline{b}) = \lambda^{H+1}T[f(\overline{x})](a,\overline{b}) \qquad (10)$$

This for a dyadic grid $\left(\lambda = 2^m\right)$ and taking square becomes:

$$T^2[f(\overline{x})](\lambda a, \lambda\overline{b}) = 2^{m(2H+2)}T^2[f(\overline{x})](a,\overline{b}) \qquad (11)$$

When we plot log base 2 of the wavelet coefficient variance at different scale index m for a rough 2D plane, the slope s is related to the Hurst exponent as

$$H = \frac{s-2}{2} \qquad (12)$$

For a surface the roughness or Hurst exponent is related to the fractal dimension (D) through

$$D = 3 - H \qquad (13)$$

4   COMPUTATIONAL DETAILS

In this paper we have worked with a Ni(111) surface. A DFT estimation of the surface energy is made for both regular surface as well as surface with defects using CPMD program. The wave function was expanded on a basis set of plane wave with cutoff of 70 Ry. We have used a norm conserving pseudo potential of Trouiller-Martins with Becke-Lee-Yang-Parr functional and non linear core correction (NLCC). Geometry optimization is done for the nickel face centered cubic (fcc) structure and Ni(111) surfaces. The fcc structure is simulated with periodic boundary condition (PBC) and different surfaces are simulated with PBC in two dimensions (x-y plane). In order to obtain the energy per atom in the bulk, the number of unit cell is increased until a constant energy per atom is obtained. The surface free energy is obtained from the difference of energies of surface structure and similar number of atoms from the bulk fcc structure. In this paper we have calculated the surface free energy of a Ni(111) plane using equation 2. A change in average surface energy is calculated for surface with defects like vacancies.

5 RESULTS AND DISCUSSIONS

An fcc structure of nickel is created using CPMD program to obtain the energy of bulk nickel atom. A periodic boundary condition (PBC) is used. The number of crystal unit cell is increased until a constant energy per atom is obtained. The lattice parameter of nickel is found to be 6.66 au. or 3.52 angstrom. This is close to experimental value of 6.659 au [3]. Its structure is shown in figure 1.

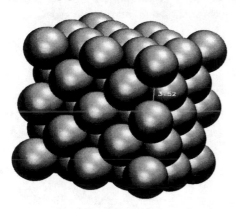

Figure 1: FCC Nickel crystal structure.

A nickel (111) surface is created using CPMD program to obtain the energy of the nickel atom forming a surface. A periodic boundary condition is applied in two dimensions (x and y). Due to lattice relaxation at the metal surface the distance between the (111) planes at the surface is taken to be 4.014 au instead of 3.845 as found in the bulk. It is calculated from the 4.4% relaxation of the nickel (111) surface [5]. The number of atoms in a layer is increased until a constant energy per atom is obtained. It is shown in figure 2. The number of layers is increased till a constant surface energy is obtained. The energy is found to be same for a two and three layer slab. So the modeling is done with a two layer slab to represent the surface.

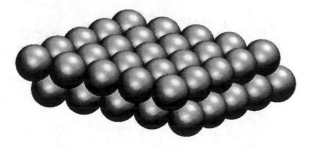

Figure 2: Ni (111) surface without defects.

The (111) plane of fcc structure is modeled with a surface area of 0.994e-18 m^2. Equation 2 is used to calculate the surface energy. It is found to be 2.19 J/m^2. This is in agreement with the values of 2.01 J/m^2 to 2.45 J/m^2 reported in different literatures [6, 7].

When we introduce defects in the surface in terms of vacancies, a significant change in the surface energy occurs. A nickel (111) surface is created with few vacancies as shown in figure 3.

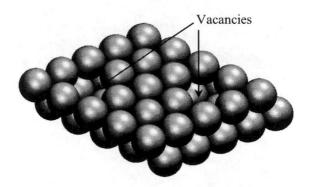

Figure 3: Ni (111) surface with defects in terms of vacancies (1).

The average energy per atom is found to be higher in a surface with vacancies. The surface energy is found to be 2.49 J/m^2. This higher energy gives an enhanced reactivity of the surface. Vacancies created in another pattern are shown in figure 4.

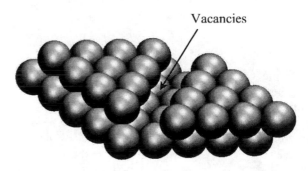

Figure 4: Ni (111) surface with defects in terms of vacancies (2).

The surface energy in this case is found to be 2.74 J/m^2. This indicates that along with the number of vacancies, the location of the vacant sites influences the energy of the surface.

In order to obtain a fractal dimension of the surface we need to create a surface with dyadic $\left(2^m\right)$ number of atoms. A fractal dimension of the surface is created by simulating the same surface in a periodic manner. This will keep the fractal dimension same as it is based on the self affine nature of the surface. A simulated surface with fractal dimension estimation using wavelet variance is shown in figure 5 and 6 respectively.

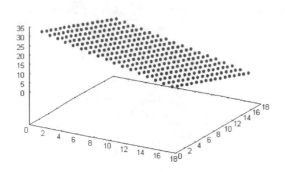

Figure 5: Simulated (111) lattice plane without defects.

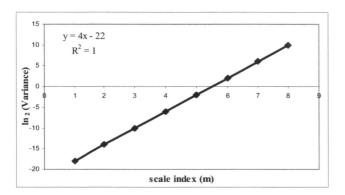

Figure 6: log of wavelet variance for a regular (111) lattice plane shown in figure 5 with scale index.

The fractal dimension is calculated from the slope of the line in figure 6 using equation 12 and 13. For a regular surface the fractal dimension is found to be 2.

Surface with similar vacancies as shown in figure 3 are simulated (shown in figure 7) and the fractal dimension are found from the slope of the wavelet variances shown in figure 8 as before. It was found to be 2.417. Similarly, the fractal dimension of the surface shown in figure 4 is found to be 2.035. Because of the regular nature of defect in the surface of figure 4 it has a lower fractal dimension.

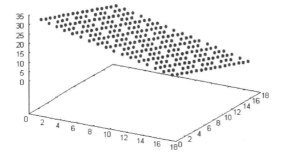

Figure 7: Simulated (111) lattice plane with vacancies as defects.

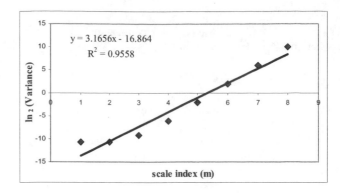

Figure 8: log of wavelet variance for a (111) lattice plane with defects as shown in figure 7 with scale index.

The higher fractal dimension of the surface shows the roughness of the surface in atomic length scale. A higher reactivity of the rough surface can be related with the fractal dimension.

6 CONCLUSIONS

Surfaces with defects have higher energies than regular surfaces and they are more reactive in nature. In this paper, using the pseudo potentials of the elements, the average surface properties for regular surface as well as for surface with defects were estimated using ab initio density functional theory simulation. We have shown how defects in the surface in terms of vacancies increased the surface energy. Furthermore, it was shown that a fractal dimension of the surface can be estimated using wavelet transform and it can be used to characterize the surface properties. Similar studies can be made with different surface planes of different active metals. Defects in terms of dislocations and grain boundaries can also be studied.

REFERENCES

[1] R.A.van. Santen, M. Neurock, Wiley-VCH Verlag GmbH & Co. KGaA, Weinheim, 2006.
[2] CPMD V3.11, copyright INTERNATIONAL BUSINESS MACHINES CORPORATION (1990-2006) and MAX PLANCK INSTTUTE FUER FESTKOERPERFORSCHUNG STUTTGART (1995-2001).
[3] G. Kalibaeva, R. Vuilleumier, S. Meloni, A. Alavi, G. Ciccotti, R. Rosei, J. Phys. Chem. B, 110, 3638-3646, 2006.
[4] P.S. Addison, Institute of Physics Publishing, Bristal and Philadephia, 2002.
[5] R.P. Gupta, Physical Review B, 23, 12, 1981.
[6] S. Hong, Y.-H. Shin, J. Ihm, Jpn. J. Appl. Phys., 41, 6142, 2002.
[7] F.R. de Boer, R. Boom, W.C.M. Mattens, A.R. Miedema, A.K. Niessen, Cohesion in metals, North-Holland: Amsterdam, The Netherlands, 1988.

Nanoparticle Characterization Using Light Scattering Technologies

Ren Xu

Particle Characterization, Beckman Coulter, Inc.
11800 SW 147 Ave., Miami, FL 33116-9015, USA, ren.xu@coulter.com

ABSTRACT

Characterization of various nanoparticles is on the center stage in nanotechnology development. The subjects for nanoparticle characterization are focused on particle size and particle surface charge determination. The latest development in particle size analysis using dynamic light scattering and surface charge determination using electrophoretic light scattering for nano or even sub-nano particles in concentrated suspension is summarized.

Keywords: particle sizing, zeta potential, concentrated suspension, light scattering, PCS

1 SIZE ANALYSIS

For size analysis of dry nanoparticles transmission electron microscopy (TEM) or scanning electron microscopy (SEM) and small angle X-ray scattering (SAXS) are the only viable choices. Microscopic methods are accurate with high resolution, can provide shape information, and are often the final judgment for standard reference materials. In most cases, they only yield information from 2-D projected areas of particles. Particle orientation in the prepared sample can alter the result significantly. The biggest drawback is that in spite of modern image analysis means the number of particles in focus that can be inspected in any field of view is limited. Thus, for a polydisperse sample, an adequate statistical representation of entire sample can be an exhaustive, if not impossible, task. In addition, the analysis process is slow and expensive.

Dynamic light scattering (DLS) that utilizes time variation of scattered light from suspended particles in liquid under Brownian motion to obtain their hydrodynamic size distribution is the most popular technology in sizing nanoparticles. DLS has been used to measure macromolecules and small particles in dilute suspension since coherent light sources, i.e., commercial lasers, became available in the 1970's [1]. Lately, demand arises for measuring smaller particles in nanometer range that requires instruments being more sensitive in picking up weak scattering signals from small particles or molecules, and for measuring particles in concentrated suspension that requires instruments being able to avoid multiple scattering but still extract correct particle motion information. To measure particles in concentrated suspension, three techniques have been developed, i.e., frequency analysis, photon cross correlation function, and back scattering. Using these techniques, particle size measurement in suspension up to concentration of 40% or higher can be realized. Among them, the back scattering arrangement has been proven being the best approach in effectively avoiding multiple scattering and maximizing signal strength.

However, in any case, even though multiple scattering can be avoided or filtered, particle-particle interaction always exists so particles' motions are not pure Brownian but affected or constrained by interaction. The affection of interaction to measured size cannot be simply predicted or calculated. Therefore, the size and its distribution yielded from measurement are only apparent and not the real hydrodynamic size.

For measuring subnanometer particles or molecules, because of their fast motion and weak scattering, an instrument has to have a high power light source and a detector with fast photoelectron response and high sensitivity. Small and rugged coherent laser diodes having high powers with low costs have mostly replaced traditional He-Ne laser as the industrial standard for DLS instrumentation. Even though some avalanche photodiodes now can be used in DLS, for strict photon counting, thanks to advancement and reduced cost in photoelectron detectors, high grade photomultiplier tube is still a better choice. In addition, a different type of correlator that correlates photon arrival time interval instead of time domain photon counting often has to be used in situations when during a sampling time there are few or no photons. For example, if the photon count rate is 10^6/sec, at a sampling time of 50 ns, there will be in average 0.05 photons. In this instance, photo counting will be inefficient and a time-of-arrival correlation has to be used in obtaining a good autocorrelation function. With the combination of high sensitive photo detector, efficient and precision optics and fast time-of-arrival correlator, size measurement of fullerene molecules or even thiamin molecules (Mw=337 Dalton) can be achieved.

2 ZETA POTENTIAL DETERMINATION

For small particles in liquid, there is no satisfactory technique to determine surface charge of particles. The common practice is to determine electric potential of a particle at a location away from the particle surface, somewhere in the diffuse layer. This location, related to particle movement in liquid, is called slipping plane or shear plane. The potential measured at this plane is called zeta potential. Zeta potential is a very important parameter to colloidal or nanoparticles in suspension. Its value is closely related to suspension stability and particle surface morphology, therefore it is widely used in product stability study and surface adsorption research.

In zeta potential determination of suspended particles, even though there exist three types of technologies, i.e., electrophoretic light scattering method, acoustic methods and electroacoustic method, due to its measurement sensitivity, accuracy, and versatility, electrophoretic light scattering is by far the best technique that has been widely used in many applications [2]. However, classical electrophoretic light scattering using a small scattering angle typically between 8 to 30 degrees cannot be used for concentrated samples. Since zeta potential, unlike particle size or molecular weight, not only is the property of particles but also the environment surrounding particles, e.g., pH, ionic strength and even the type of ions in the suspension. Therefore, in many instances, even though zeta potential of suspended particles is measured after diluted by, typically, DI water and high resolution and accurate result is produced, the value has little or even opposite relation with the true value in the original environment. Therefore, the measured value may have no practical usefulness or sometimes even misleads the user.

To measure zeta potential distribution in concentrated suspension is a technological challenge. Acoustic methods only yield average value with low sensitivity provided that the solid concentration is known. The back scattering approach used in size measurement cannot be adopted due to interference of Brownian motion to oriented electrophoretic motion at large scattering angles. For example, for a 250 nm particle with zeta potential of 60 mV, when it is subjected to a 30 V field, the electrophoretic motion will produce a Doppler shift of 55 Hz at a scattering angle of 10 degrees and the Brownian motion of the particle will cause a 3.5 Hz peak broadening. If the measurement is performed at a scattering angle of 160 degrees, the Doppler shift and peak broadening will be 108 Hz and 430 Hz, respectively, making accurate determination of zeta potential impossible.

Lately, a unique optical arrangement for measuring zeta potential in concentrated suspension has been invented [3]. In this invention, a unique electrode that is conductive but transparent to both illuminating and scattering lights is used in electrophoretic light scattering measurement. The incident light entering from one side of a thick window is refracted by the window and exits the window to the sample cell through a surface, which is perpendicular to the first surface, on which a thin metal coating serves as the transparent electrode. Particles are moving electrophoretically in the field created by this electrode and another ordinary electrode. Light scattered from particles near the window surface are refracted twice before exiting from the other side of the window. This configuration

enables a similar arrangement as in back scattering PCS but at a much smaller scattering angle (~30 degrees). The additional advantage of this technique is that because of the scattering volume location, there is no electroosmotic flow typically caused by surface charge of cell side walls. Therefore, the Doppler shift measured is only from particles motion without interference from liquid motion. Particle size measurement in concentrated samples can also be performed using this sample cell when electric field is not applied.

This invention has been utilized in an instrument (Delsa^{TM}Nano from Beckman Coulter, Inc.), which is capable of performing nanoparticle size measurement using forward scattering and back scattering PCS at multiple scattering angles and zeta potential measurement for nanoparticles in either low concentration or turbid samples, in addition to its capability of measuring zeta potential of solid surface or film. Nanoparticles or molecules in various concentrations can be characterized with the combination of precision optics, high sensitive and fast response photo detector, and the transparent electrode technology. For example, particle size increment from 2.8 nm to 6.2 nm and zeta potential increment from 7.2 mV to 7.6 mV when the generation of dendrimers increased from 3 to 5 have been successfully captured.

3 COUNCLUSION

Ensemble particle characterization techniques that characterize particulate systems, not a few single particles, will continue to play a very important role in nanotechnology, especially when more and more nonmaterial are being transformed from academic and laboratory research to production of different scales. In quality control environment, sample analysis often has to be performed quickly and easily without complicated sample preparation procedure and with minimum alteration to the sample. The technologies introduced in this article, i.e., back scattering and electrophoretic motion detection through a transparent electrode, can meet such demands: both particle size and zeta potential measurements can be performed in concentrated suspension without dilution. On the other hand, with particle dimension decreasing to a few nanometers or smaller, requirements for high sensitivity and fast photoelectric response become important for any light scattering instrument.

REFERENCES

[1] R. Xu, "Particle Characterization: Light Scattering Methods." Kluwer Academic Publishers, Dordrecht, the Netherlands, 223-288, 2000.
[2] R. Xu, "Particle Characterization: Light Scattering Methods." Kluwer Academic Publishers, Dordrecht, the Netherlands, 289-343, 2000.
[3] M. Sekiwa, K. Tsutsui, K. Morisawa, T. Fujimoto and A. Toyoshima, "Electrophoretic Mobility

Measuring Apparatus," US Patent Pub. App.
20040251134, 2004.

The real-time, simultaneous analysis of nanoparticle size, zeta potential, count, asymmetry and fluorescence

R. Carr, P. Hole, A. Malloy, J. Smith, A. Weld and J. Warren

NanoSight Ltd., Old Sarum Park, Salisbury, Wiltshire, UK
bob.carr@nanosight.co.uk

ABSTRACT

A new nanoparticle sizing and characterization technique is described which allows nanoparticles in a suspension to be individually but simultaneously detected and analysed in real time using a low cost instrument. Nanoparticles of all types and in any solvent can be detected, sized and counted through video-based tracking of their Brownian motion when illuminated by a specially configured laser beam. Depending on particle type, nanoparticles as small as 10nm can be visualized (though not imaged). Advances in the technique are described which allow each particle to be simultaneously analysed not just in terms of its size but also light scattering power (mass/refractive index) and electrophoretic mobility in an applied electric field. Finally, an unprecedented ability to quantify nanoparticle asymmetry in real time, again on a particle-by-particle basis is outlined

Keywords: nanoparticle, analysis, multiparameter, size, zeta

1   INTRODUCTION

The analysis of nanoparticle size is now a central requirement in the development of a wide range of particulate-based materials and substances in which their properties and behaviour are enhanced when produced at nanoscale dimensions. Besides the well established use of electron microscopy and scanning probe microscopy methods, optical ensemble techniques such as Dynamic Light Scattering (DLS), also known as Photon Correlation Spectroscopy (PCS), have long been used routinely in the analysis of nanoparticle dispersions. A number of commercial manufacturers supply instrumentation capable of rapidly and accurately determining particle size from a wide range of sample types.

However, it is well recognised that dynamic light scattering methods become unreliable when presented with samples which contain a wide range of particle sizes, i.e. are polydisperse, and that the basic information obtained, the intensity weighted mean size (the 'z-average'), does not always reflect the sample composition accurately. Furthermore, successful analysis of the correlation function by classical deconvolution algorithms to extract, for instance, bimodal distributions are realistically limited to sample types containing only two or at best three sized particle types, each needing to differ from each other by a size factor of, in practice, >3:1. Finally, DLS is limited in its ability to allow the user to recognise when the sample is unsuitable for analysis by that method and that the data (i.e. the particle size distribution profile) obtained should accordingly be treated with some suspicion.

We report here on a new light scattering technique, Nanoparticle Tracking Analysis (NTA) for determining nanoparticle size through tracking and analysing the trajectories described by individual nanoparticles undergoing Brownian motion in a fluid and which has recently been made commercially available [1-6].

2   THE OPTICAL SYSTEM

The technique is based on laser light scattering in which a finely focussed laser beam (of arbitrary wavelength but which is commercially available at 534nm, 25mW [2]) is passed through a sample of liquid containing a dilute suspension of nanoparticles. The beam is caused to refract at the interface between the liquid sample and the optical element through which it is passed such that it describes a path which is close to parallel to the glass-sample interface. Particles resident in the beam (which is approximately 100μm wide by 25μm deep), are visualised by a conventional optical microscope aligned normally to the beam axis and which collects light scattered from each and every particle in the field of view.

Given NTA is not an imaging technique <u>per se</u>, the total magnification of the system is quite modest (x100 via a 20x 0.4 NA long working distance microscope objective) and for a suitably diluted sample the particles are visualised as light scattering centres moving under Brownian motion in the field of view (approx. 100x80μm, i.e. matched to the beam width) of a camera (640x480 pixels each of 9x9μm) located on a C-mount on the microscope assembly.

The sample chamber is approximately 250μl in volume and 500μm deep and sample introduced by syringe via a Luer port. The sample is allowed to thermally equilibrate for 20 seconds prior to analysis

3   ANALYSIS OF BROWNIAN MOTION.

A video of typically 20-60 seconds duration is taken (30 frames per second) of the moving particles. The video is analysed by a proprietary analysis programme [3] on a frame by frame basis, each particle being identified and located automatically and its movement tracked and measured from frame to frame. The thresholds for particle identification can be user adjusted, as can the gain and shutter speed settings of the camera, thus allowing the user to optimise the image for a particular sample type. Particles diffusing into the scattering volume are identified and followed for the duration of the particle presence in the beam or until they diffuse to within a certain distance of an adjacent particle at which point tracking is ceased. This eliminates the possibility of analysing particle trajectories which cross behind each other and which might lead to erroneous results. Particles must be detected by the camera in order to be tracked. The limits to particle sizes which can be analysed by NTA are determined primarily by particle size and refractive index.

3.1 Visibility of particles and detection limits

The amount of light scattered by a particle in any given direction is a function of many variables including incident illumination power, wavelength, angle and polarisation; particle size, refractive index (real and imaginary) and shape, as well as solvent refractive index. Similarly, the amount of light falling on a detector and strength of the resultant signal is dependent, of course, on the efficiency of the collection optics (e.g. Numerical Aperture) and the spectral response and sensitivity of the camera.

The theory of light scattering is well established and the formula for Rayleigh scattering of small particles of radius a, refractive index n_1 in a liquid of refractive index n_2 is given below.

$$\frac{I}{I_{in}} = \frac{16\pi^4 a^6}{r^2 \lambda^4}\left(\frac{n^2 - 1}{n^2 + 2}\right)\sin^2\psi \tag{1}$$

Where λ is the wavelength of the incident light beam, n relative ref index (n_2/n_1), I_{in} is incident power per unit area, I the scattered power per unit area a distance r from the scattering region and ψ is the angle between the input polarisation and the scattering direction. The total scattering into an aperture of collection angle θ (numerical aperture $NA = \sin\theta$) is then:

$$P_{scat} = \frac{64\pi^4 a^6}{\lambda^4}\left(\frac{n^2 - 1}{n^2 + 2}\right)\eta I_{in}$$

$$\eta = \frac{(1 - \cos\theta)}{4} + \frac{(1 - \cos^3\theta)}{12} \tag{2}$$

For a single mode 25mW laser diode at $\lambda = 534$nm and using the camera available on the commercially available instrument discussed in this paper (Marlin, Allied Vision Ltd.), the limits of detection depend primarily on the size and refractive index (R_i) of the nanoparticle. In practice,

high R_i materials such as colloidal silver can be visualised down to approx 10nm diameter. For nanoparticles of slightly lower Ri, such as metal oxides (e.g. TiO_2), detection limits may be only >20nm. For weakly scattering materials (e.g. polymer, biological, liposome), the smallest particle visible might only be >50-75nm in diameter.

3.2 Diffusion of nanoparticles under Brownian motion..

Nanoparticles in a liquid are under continuous solvent bombardment and move randomly over length scales related to their size. For spherical particles, the diffusion coefficient (Dt), is related to particle diameter by the Stokes-Einstein relationship

$$Dt = \frac{K_B T}{6\pi\eta r_h} \tag{3}$$

where K_B is the Boltzmann constant, T temperature, η is viscosity and r_h the hydrodynamic radius.

For non-spherical particles, Dt is a complex function of particle shape (e.g. prolate ellipsoid, oblate spheroid, etc..) [3] but in the technique described here, Dt is expressed only as a sphere equivalent, hydrodynamic radius. For the majority of nanoparticle samples, particle shape is sufficiently spherical (aspect ratios of <2) for the Stokes-Einstein sphere assumption to sufficiently valid to generate useful information about particle size and size distribution.

4 RESULTS.

4.1 Determination of particle diameter through analysis of Brownian motion

The following results show the ability of the technique to resolve different particles sizes at resolutions that exceed those achievable by DLS [10-13]. Fig 1 shows a well resolved bimodal from a mixture of 100 and 300nm polystyrene calibration microspheres. It is important to note that the plot shows number of particles as a function of size, allowing particle concentrations (numbers of particles/ml) for each (user selectable) size class.

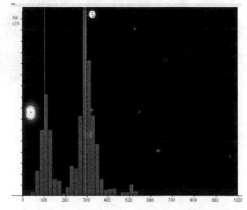

Fig 1. A mixture of 100 and 300nm polystyrene particles as analysed for 166 seconds.

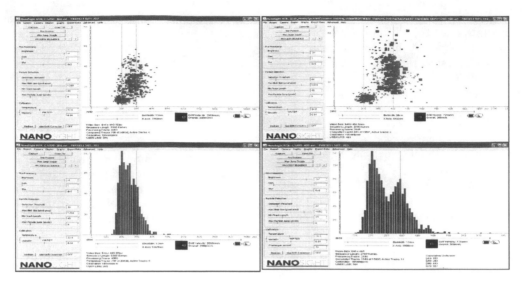

Fig 2. Size plots (a; histograms) and two dimensional plots of size v. intensity (b; scatter plots) for mixtures of 200 and 300nm particles (left) and 200 and 400nm particles (right).

4.2 Simultaneous analysis of particle size and intensity

The analysis program is capable of measuring the intensity of light scattered by the particle at the same time as tracking its Brownian motion from which its size is calculated. This offers the ability to generate two dimensional plots of intensity v. size. As can be seen from Fig 2, the 2D plots of a mixture of titanium dioxide nanoparticles of 200, 300 and 400nm diameter can more easily be resolved by 2D plotting even though the samples of each were not particularly monodisperse. Fig 2a shows plots from a mixture of 200 and 300nm particles, Fig2b is for a mixture of 200 and 400nm particles.

4.3 Analysis of changes in particle refractive index with simultaneous size determination.

In the following example, suspensions of a virus particle were incubated over a 2 hour period in a metal salt solution which resulted in the formation of highly scattering metallic coating on many of the virus particles. While there was, except for the formation of a few aggregates, no significant increase in particle diameter through the formation of a molecularly thin metallic film, and hence no significant change in particle size, the increase in particle refractive index was clearly seen in the 2D, intensity v. size plot Fig. 3

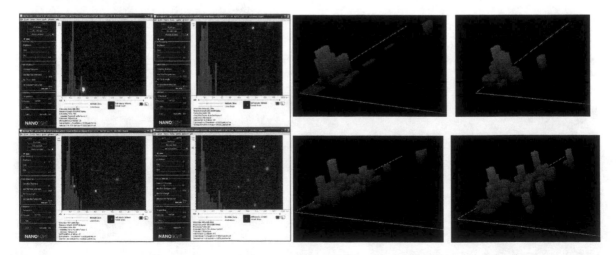

Fig 3. Size profiles (left) obtained for virus particles over a 4 hour incubation period in the presence of a metal salt solution which resulted in the reduction of the metal onto the virus particles causing their refractive index to increase without significant increase in virus size (right). Note that the aggregates can be clearly differentiated by virtue of the increase in both size and scattered intensity

4.4 Simultaneous measurement of particle size and electrophoretic mobility.

By the introduction of electrodes into the sample cell, it is possible to apply an electric field across the liquid region through which the laser beam passes [8,9]. Applying an electric field causes charged particles to move in this field. We have shown, Fig 4, that electrophoretic mobility is linearly dependent on applied voltage

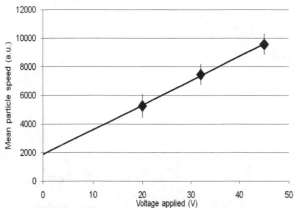

Fig 4. Electrophoretic mobility as a function of applied voltage

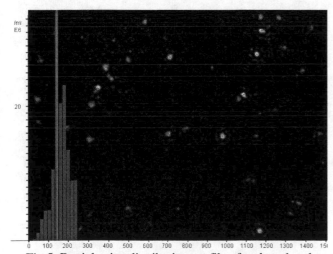

Fig 5. Particle size distribution profile of carboxylated particles with no applied electric field.

It is possible to determine the size of a negatively charged 150nm particle (carboxylated), through analyzing its Brownian motion prior to applying an electric field, Fig 5

On the application of an electric field (30V), the particles are seen to move towards the anode and their electrophoretic tracks can be visualised and analysed by the analysis program (Fig 6). The Brownian motion from which the size is obtained can be simultaneously extracted from the overall electrophoretic motion.

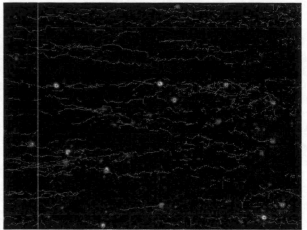

Fig 6. Tracks described by particles moving under an applied electric field.

In Fig 7 can be seen a 2 Dimensional plot of size of particle in a) the absence of an applied electric field and b)_when the filed is switched on. This shows that particle sizing is unaffected by the electrophoretic motion of the particles.

Shown in Fig 8 is a three dimensional plot of Fig 7 in which the vertical axis represents numbers of particles of any given class over the analysis time.

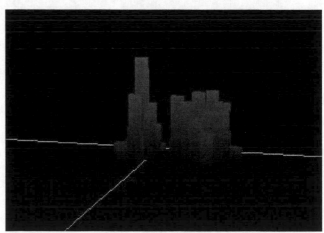

Fig 8 A three dimensional plot of Fig 7 in which the vertical axis represents numbers of particles of any given class over the analysis time.

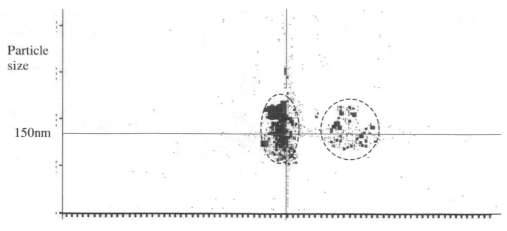

Electrophoretic mobility

Fig 7 shows a 2 Dimensional plot of size of particle in a) the absence of an applied electric field and b)_when the field is switched on. This shows that particle sizing is unaffected by the electrophoretic motion of the particles.

5 CONCLUSION

The technique outlined above allows for the multi-parameter analysis of nanoparticle suspension on a particle-by-particle basis [10-12]. Particles which are known to be asymmetric with an aspect ratio of >2:1 also exhibit a specific optical signature in which the light scattered by the particle varies in intensity on a timescale related to the major axis length through interference effects associated with the rotation of the particle under Brownian motion.

Accordingly, it is possible that complex nanoparticle suspensions comprising many different particle types can be resolved into different populations through the application of two or more of the above measureands allowing higher resolution analyses than have hitherto been possible.

REFERENCES

1. Carr, R (2005), 27th International Fine Particles Research Institute (IFPRI) Meeting and Workshop June 25-29th, 2006, SANTA BARBARA, USA
2. Company literature at www.nanosight.co.uk
3. Malloy, A and Carr, R (2006), Particle & Particle Systems Characterization (Special Issue: Particulate Systems Analysis) 23, (2), p.197 – 204.
4. Kevin Kendall and Maria R. Kosseva (2006) Colloids and Surfaces A: Physicochemical and Engineering Aspects Volume 286, Issues 1-3, 1 September 2006, Pages 112-116
5. Bob Carr, Patrick Hole and Andrew Malloy (2007). Abs. 8th International Congress on Optical Particle Characterisation, 9-13 July 2007 Karl-Franzens University Graz, Austria, p25
6. Bob Carr, Andrew Malloy and Patrick Hole (2007). EuroNanoForum 2007 - Nanotechnology in Industrial Applications, June 19 - 21, 2007, CCD Düsseldorf, Germany
7. Carr, R (2007) NTNE2007 - NanoTechnology Northern Europe (Congress and Exhibition), 27-29th March, Helsinki, Finland
8. Stephan Barcikowski, Ana Menéndez-Manjón, Boris Chichkov, Marijus Brikas and Gediminas Račiukaitis (2007) Appl. Phys. Lett., Volume 91, Issue 8,
9. Barcikowski, S., Hahn, A. and Ostendorf, A. (2007): EuroNanoForum 2007 - Nanotechnology in Industrial Applications, June 19 - 21, 2007, CCD Düsseldorf, Germany
10. Hassan M. Ghonaim, Shi Li, Charareh Pourzand, and Ian S. Blagbrough (2007) British Pharmaceutical Conference BPC2007, Manchester, 10th Sept
11. Iker Montes-Burgos, Anna Salvati, Iseult Lynch, Kenneth Dawson (2007) European Science Foundation (ESF) Research Conference on Probing Interactions between Nanoparticles/Biomaterials and Biological Systems, Sant Feliu de Guixols, Spain, 3 - 8 November 2007
12. Carr, R. et al (2008), European Journal of Parenteral and Pharmaceutical Sciences, April 2008, in prep
13. Hans Saveyn, Bernard De Baets, Patrick Hole, Jonathan Smith and Paul Van der Meeren (2008) PSA2008, Stratford on Avon, September, In Prep

Applications Developments with the Helium Ion Microscope

L. Scipioni*, L. A. Stern, and J. Notte, B. Griffin**

Carl Zeiss SMT, Inc
ALIS Business Unit
1 Corporation Way, Peabody, MA, USA 01960
*l.scipioni@smt.zeiss.com
**Center for Microscopy, University of Western Australia, CRAWLEY, WA

ABSTRACT

The helium ion microscope is offering new windows into nano-scale imaging. This is due to a combination of high source brightness and unique sample interaction dynamics. These dynamics allow new types of sample information to be gathered. Some fundamental advantages conferred by probing a surface with a helium ion beam are the high contrast due to the sensitive material dependence of the secondary electron (SE) yield, the surface sensitivity arising from the low energy of the SE's, and the ability to image low atomic weight materials, such as carbon, due to the greater interaction cross-section of ions with these materials. The low mass of helium confers also a new image acquisition mode available via the collection of backscattered ions. There are also secondary benefits to helium ion microscopy. The Orion™ tool allows imaging of non-conducting samples through the use of charge neutralization via a low energy flood gun. The low mass of the helium ion minimizes beam induced damage. The much longer depth of focus of the microscope allows imaging of three dimensional objects or tilted cross-sections while maintaining focus over the entire field of view. These capabilities become even more important as the size of the features to be imaged becomes smaller than the beam-sample interaction volume found with SEM imaging.

Keywords: helium, ion microscopy, high resolution, ALIS, FIM

1 THE HELIUM ION MICROSCOPE

The helium ion microscope from Carl Zeiss SMT represents a completely new entry into the field of charged particle optics. Several workers have created and characterized gas field ion emitters in various embodiments [1]. The benefits of such an emitter are the atomic sized source ("Atomic Level Ion Source", ALIS) afforded by the principles of field ion emission and the ability to run noble gases, avoiding the use of contaminating metal ions as in the liquid metal ion source. In addition, the ability to use low mass helium ions in particular dramatically reduces the sample damage created by sputtering. Diffraction effects, which limit the ultimate spot size of an electron beam, become irrelevant since the helium ion is 7332× heavier.

The ALIS source is based upon the principles of the field ion microscope, wherein a sharpened metallic wire is cooled to cryogenic temperature and put at a high positive potential. Gas introduced into the region of the tip, after being adsorbed onto its surface by the combined action of low temperature and high field, is ionized at the electric field maximum which is found over each atomic center. The resulting ion beam from each atomic sized emitter is separated in space from all others, giving rise to a multiplicity of possible ion sources.

The key to the ALIS technology is a formation process which produces a special structure on the end of the tip, reducing the emission sites from the hundreds normally visible in the field of view in Figure 1 to a very small number. The microscope's ion gun allows for one emitter from this group to be selected for transmission through the electrostatic ion optics, yielding a single, stable ion beam with high brightness (> 3.4×10^9 A-cm^2-sr). Calculations show that an 0.25nm probe size is possible with the current optics; so far measurements have achieved < 0.6nm. The interested reader may see reference [2] for further information.

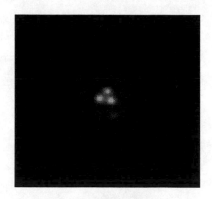

Figure 1: field emission pattern from the ALIS ion source.

2 INTERACTIONS AND IMAGES

The sample information available from any scanning probe microscope depends only partially on the probe size.

There are several factors to consider, and the full range of beam-sample interactions determine which applications are best suited to a given technology. We present here a discussion of several of these, illustrated by applications that benefit from them.

2.1 Beam-Sample Interaction Volume

Instrument considerations notwithstanding, the image resolution from a scanning microscope depends not on the probe size alone but also on the volume in the sample from which the detected signal originates. For an electron beam, a significant portion of the SE signal can consist of type-II electrons – those which arise from backscattered electrons as they exit the sample [3]. The results in a delocalized contribution to the detected signal, which reduces lateral resolution and convolutes surface topology with sub-surface signal. For an incident helium ion beam, the secondary electron energy spectrum peaks below about 5eV and then drops off quickly. Since the escape depth of electrons of this energy is on the order of a couple of nanometers, there is no SE contribution from deeper in the sample. This can be seen more readily in high resolution images, such as the example shown in Figure 2. The sample is evaporated gold on a carbon substrate. In the helium ion micrograph on the left, we see excellent edge definition and also great topological information, including the fine surface detail on the gold particles. This arises from the combination of the sub-nanometer probe size and the true surface-only nature of the SE signal. The SEM image in Figure 1b, while able to image the small particles, suffers from a diffuse contribution to the signal everywhere. Thus the edges of the particles become slightly blurred, the smaller particles have low contrast, and most notably the surfaces of the particles look flat. Thus the combination of small probe size and small signal origination volume provides a unique benefit for high resolution imaging tasks.

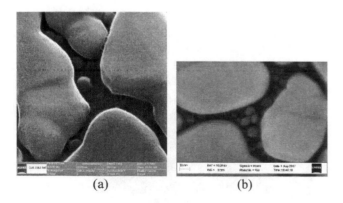

Figure 2: High resolution images of gold particles; 500nm field of view. (a) Orion™ image, (b) SEM image.

2.2 Surface Sensitivity

Related to the concept of the reduced signal generation volume is an improved ability to image thin surface layers and materials. Due to the different energy loss mechanism for a helium ion beam, more signal is generated from the top monolayers of a given sample. This effect can overcome the electron transparency that challenges SEM imaging of very thin layers and objects. This is highlighted in Figure 3, which shows an nitro-biphenyl-thiol (NBPT) monolayer which has been lifted off its substrate and then laid over a gold grid. It is very difficult to see in SEM, requiring the image to be highly saturated in order to get enough contrast to detect it (Figure 3b). This reduces resolution, however, and introduces noise. The helium ion microscope is able to image this free-standing monolayer much more readily. It is obvious from Figure 3a that even such a thin layer of this low atomic weight material is easily detected, as can be judged from the opacity of the grid windows that are covered. This ability can be leveraged into many applications, such as carbon nanotubes (cf. Figure 8).

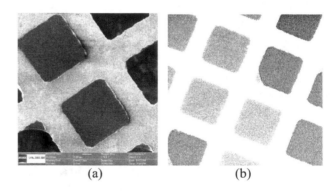

Figure 3: NBPT monolayer stretched over a gold grid. (a) Orion™ image, (b) SEM image.

2.3 Material Contrast

The ability to distinguish different materials in a sample arising from changes in the secondary particle emission from one material as compared to another. For a primary electron beam the SE yield does not change as much as a function of material as for a primary ion beam [2]. This, combined with the greater physical localization of the SE signal from a helium ion probe, allows nano-structured devices to be imaged with greater ease. This can be seen especially in cross sections of layered devices from semiconductor or magnetic storage technologies, where there are stacks of thin layers of several materials. An example of this is shown in Figure 4, which shows a cross section of a tunneling magneto-resistance device. The SEM image of the middle "H2" layer appears to be a single material with a thickness of 31nm. In the Orion™, however, this is revealed to be a triple layer. There is additionally a thin layer just above the substrate which is not captured at all in the SEM. This demonstrates the ability to find new information from lithographically defined structures. We have also made excellent progress in imaging 45nm technology semiconductor devices and will be publishing results in the near future.

A further test of material contrast is the ability to distinguish different materials on a surface. For many applications, very thin layers may be nano-patterned via deposition, selective etching, or by beam induced chemical modification. The need exists to be able to image even single atomic monolayers that have been patterned across a surface. The helium ion microscope shows the ability to do this, as highlighted in Figure 5.

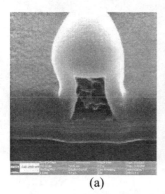 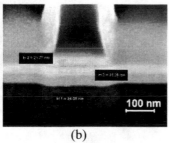

(a) (b)

Figure 4: Tunneling magneto-resistance layer stack, viewed in cross section. (a) Orion™ image, (b) SEM image

The sample imaged in Figure 5 is a NBPT doubled monolayer, the same as in Figure 4, but now over a gold film. It has been patterned by e-beam exposure via a contact mask. The exposure converted the nitrate terminal group to amine. Thus the only contrast mechanism available for imaging these patterns are the substitutions of hydrogen for oxygen on the top monolayer of the surface. The SEM image in Figure 5b shows that it is just possible to start seeing the 40 micron window where exposure had occurred. With the helium ion microscope, however, there is very clear chemical contrast that allows the full pattern to be seen easily. This makes the Orion™ to be a powerful tool for investigations of surface modifications and patterns.

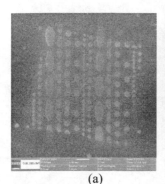 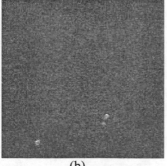

(a) (b)

Figure 5: NBPT double layer over a gold film, patterned with an e-beam mask. (a) Orion™ image, (b) SEM image.

2.4 Backscattered Ion Imaging

A new imaging mode available when imaging with helium ions comes from the collection of Rutherford backscattered ions (RBI mode). Due to the low mass of helium it will backscatter from every element except hydrogen. Since the backscatter probability rises with the atomic number of the target, there is an inherent strong material contrast provided. We have collected RBI data on a set of pure elemental test samples for all the conducting elements from boron to bismuth, as presented in Figure 6. The scatter points show the experimental data, while the solid line shows TRIM simulations for the same beam energy (25keV). There is a good fit of the overall shape, allowing for a qualitative mass mapping in the images. There are peaks in the experimental data, occurring in each period of the elements, centered approximately at the noble metals Cu, Ag, and Au. The origin of these peaks is not explained and not predicted by TRIM. This effect enhances chemical contrast in RBI mode images but also makes quantitative determination of species uncertain. We are working to combine the RBI information with energy selective backscatter detection to provide true elemental identification at deep sub-micron resolution.

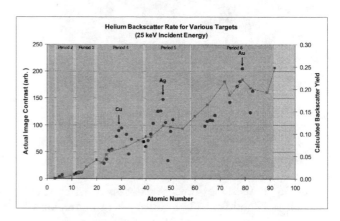

Figure 6: A comparison of experimental and simulation data on the backscattering of 25keV helium ions from a wide variety of chemical elements.

3 ADDITIONAL BENEFITS

We touch briefly here upon two of the additional features of helium ion microscopy and their benefit to specific applications.

3.1 Charge Control

Imaging for samples which charge under a beam will suffer from distortions, noise, loss of contrast, and other image artifacts. Low energy SEM, with the beam energy near the unity SE emission point is often used to deal with this – but at a loss of resolution due to decreased beam brightness. Ion beams allow the neutralization of charge through the use of a low energy electron flood gun to dissipate the build-up of positive charge. The Orion™ microscope is equipped with a flood gun for handling insulating samples. An example of this capability is highlighted in Figure 7. The sample is a chrome on quartz photomask. The electron flood, applied in multiplex with the ion beam, allows for very stable imaging which brings out the shape of the chrome line, the details of the chrome grain structure and the interface between the metal and the quartz.

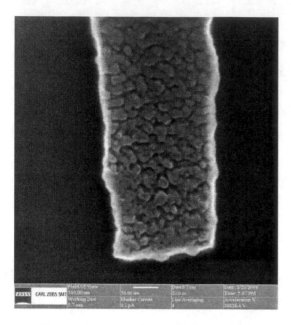

Figure 7: Chrome line on a photomask imaged using charge neutralization.

3.2 Depth of Field

When imaging three dimensional structures or tilted surfaces, it is beneficial to keep as much of the region of interest as possible in focus. This becomes difficult to achieve as the image magnification increases, for the beam convergence angle becomes higher. The low angular divergence of the beam in the ALIS source, combined with the small source size (requiring less de-magnification at the image plane), allows for a depth of field about 5 times greater than an SEM under similar conditions. This is evidenced in Figure 8, where the investigation centered on single walled carbon nanotubes grown on SiGe catalysts. It was found that the nanotubes preferred to grow in a network connected amongst the catalyst particles and suspended in free space over the silicon substrate. For this reason the image in the Figure was taken at a sample tilt of 70°, in order to see beneath the network. Even at this high viewing angle the nanotubes remain in focus, both in the foreground and the background.

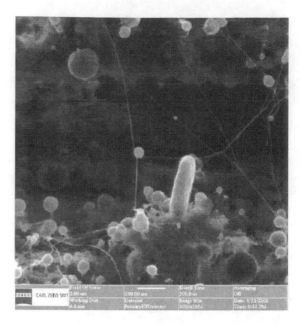

Figure 8: Single wall carbon nanotubes suspended above a silicon surface on SiGe catalyst particles. 70° tilt of sample from beam axis.

4 CONCLUSION

This has been a brief survey of a subset of the capabilities being developed for the helium ion microscope. We hope to have demonstrated that this technology can provide a wealth of information complementary to, and in some cases even beyond, SEM. We continue to research the capabilities for new applications, as we drive to further establish this technology in the microscopy community.

5 ACKNOWLEDGEMENTS

The authors kindly thank Karsten Rott and Andre Beyer of the University of Bielefeld, Germany, Mike Postek of the National Institutes of Standards and Technology, and Prof. Harvey Rutt of the University of Southampton, UK, for several of the samples shown in this paper.

REFERENCES

[1] V. N. Tondare, J. Vac. Sci. Technol. A, 23 (6), p.1498 (2005).
[2] J. Morgan et. al., Microscopy Today, 14 (4), p.24 (2006).
[3] D. C. Joy et. al., SPIE Proceedings: Metrology, Inspection, and Process Control for Microlithography XXI 6518, (2007).

Guided Self-Assembly of Block-Copolymer Nanostructures

A. Karim, B. Berry, S. Kim, A. Bosse, J.F. Douglas, R.L. Jones*[†], R. M. Briber**, H.C. Kim***

* Polymers Division, MS8541, National Institute of Standards and Technology,
Gaithersburg, MD, USA, karim@nist.gov
**Dept. of Materials Science, University of Maryland, College Park, MD, USA
*** IBM, Almaden Research Center, Almaden, San Jose, CA, USA

ABSTRACT

Block copolymers are an important class of soft-materials with significant potential for a variety of potential applications ranging from high-density data storage devices to nanowires and photonics. These potential applications rely on the ability to not only manipulate and control their structure at the nanoscale, but also to accurately measure the structure and morphology of the film in all dimensions. As is well known, diblock copolymer molecules are comprised of two dissimilar polymers covalently bonded at one end. Typically, the individual blocks are thermodynamically incompatible and the system will microphase separate upon heating. It is also desirable to control order, orientation and defects over large areas in most thin film applications of these materials.[1-5] This paper focuses on the dynamics of directed self-assembly of block copolymers in confining geometries such as thin films modulated by processing parameters such as applied fields which can be surface chemical or thermal gradient. We illustrate the surface chemical gradient approach below. The presentation will discuss some of the "tomographic" neutron scattering measurement methods being developed to investigate the structure and morphology of block copolymer thin films in 3-dimensions as a result of such gradient fields and simulation methods to capture aspects of kinetics and dynamics involved in the evolution of thin film structure.

Keywords: block-copolymer, thin-film, directed self-assembly, surface chemical control, scattering, zone annealing

EXPERIMENTAL[6]

Chemical modification of the substrates is obtained using chlorosilane chemistry for covalent self-assembled monolayers (SAM) on silicon wafers followed by ultra-violet ozonolysis (UVO) to generate stable gradient energy test substrates. The substrate is thoroughly washed with toluene and dried. The resulting SAM was then exposed to a gradient of UVO using a modulated UVO exposure device.[7] The graded oxidative process results in surface chemical modification by the generation of a gradient in surface chemical moieties primarily carboxyl functionalities across the sample. This gradient was characterized using water contact angle measurements. Thin films of the poly(styrene-block-methylmethacrylate) or PS-b-PMMA block copolymer were cast on these gradient wettability substrates and annealed above their glass transition temperature and characterized by atomic force microscopy, neutron reflectivity and small angle neutron scattering.

RESULTS

After the block copolymer film (library) was been formed, it was be characterized with atomic force microscopy (AFM), a high resolution technique. An example of this morphology change is shown in Figure 1 where AFM micrographs of an M = 51 k PS-b-PMMA film acquired at h = 75 nm are presented. We observe a progression of surface topography from labyrinthine pattern to smooth surface to islands to co-continuous regions to holes as a function of changing substrate surface contact angle. Essentially, the morphology is observed to change from symmetric to asymmetric wetting of block component on the substrate surface as illustrated in cartoon Figure 2. The nature of the surface patterns formed depended on local film thickness as well as expected due to the quantized nature of the block copolymer films. The evolution dynamics of these film structures offers much insights into the mechanisms behind the ordering processes in these films.

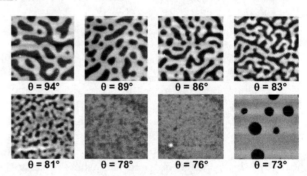

θ = 94° θ = 89° θ = 86° θ = 83°
θ = 81° θ = 78° θ = 76° θ = 73°

Figure 1. 10x10 μm AFM images indicating the progression from holes through a smooth region and back to holes. Substrate water contact angle on the corresponding position are indicated.

[†] Official contribution of the National Institute of Standards and Technology; not subject to copyright in the United States.

Figure 2. Illustration of the effect of surface energy on block copolymer thin films for hydrophobic, neutral, and hydrophilic surface energies.

Neutron Reflectivity (NR) as well as Small Angle Neutron Scattering (SANS) were employed to investigate the ordering kinetics and orientation within deuterated thin films of the block-copolymer. Currently tomographic SANS is being developed to investigate the degree of orientation within these films by collecting scattering data at varying incident angles.

CONCLUSIONS

The morphology of block copolymer thin films as a function of film thickness and substrate surface energy has been characterized using gradient combinatorial mapping techniques. A switch from symmetric to asymmetric film morphology was observed as a function of substrate surface energy variation. At the center of this surface energy range is a "neutral" surface region that can be exploited to obtain vertically oriented lamellae or cylindrical morphology in thin films. Simulations involving self-consistent field theory (SCFT) invoking ordering dynamics with a mobility gradient are being developed to understand some of the observed phenomena.

REFERENCES

[1] Black, C. T.; Guarini, K. W.; Russell, T. P.; Tuominen, M. T. Appl. Phys. Lett. 2001, 79, 409.

[2] Cheng, J.Y.; Ross, C.A.; Smith, H.I.; Thomas, E.L., Adv. Mater. 2006, 18, 2505.

[3] Kim, S. H.; Misner, M. J.; Xu, T.; Kimura, M.; Russell, T. P. Adv. Mater. 2003, 15, 226

[4] Bodycomb, J.; Funaki, Y.; Kimishima, K.; Hashimoto, T. Macromolecules 1999, 32 (6), 2075-2077.

[5] Berry, B.C.; Bosse, A.W.; Douglas, J.F.; Jones, R.L.; Karim, A. Nanoletters 2007, 7 (9), 2789.

[6] Certain equipment, instruments or materials are identified in this paper in order to adequately specify the experimental details. Such identification does not imply recommendation by the National Institute of Standards and Technology nor does it imply the materials are necessarily the best available for the purpose.

[7] Berry, B.C.; Stafford, C.M.; Pandya, M.; Lucas, L.; Karim, A.; Fasolka, M.J. Rev. Sci. Instrum. 2007, 78(7), 072202.

Raman Spectroscopy resolution limits overcome with nanoscale thermal analysis for complete Polymer Blend Characterization

Authors: J. Ye; N. Kojima; M.Reading; K. Kjoller, R.Shetty

Introduction

Characterizing polymer blends can sometimes present a significant challenge for a single technique and the best way is a combination of techniques. In this application note, we discuss an PA6-PET blend which was initially characterized via Raman Spectroscopy at a 500nm spatial resolution. Subsequent thermal characterization at a sub-100nm spatial resolution via the nano-TA revealed a lot of interesting details and more complex sub-structures that were not available from the Raman study and helped in obtaining a detailed understanding of the blend.

nano-TA is a local thermal analysis technique which combines the high spatial resolution imaging capabilities of atomic force microscopy with the ability to obtain an understanding of the thermal behaviour of materials with a spatial resolution of sub-100nm. (a breakthrough in spatial resolution ~50x better than the state of the art). The conventional AFM tip is replaced by a special nano-TA probe that has an embedded miniature heater and is controlled by the specially designed nano-TA hardware and software. This nano-TA probe enables a surface to be visualised at nanoscale resolution with the AFM's routine imaging modes which enables the user to select the spatial locations at which they would like to investigate the thermal properties of the surface. The user then obtains this information by applying heat locally via the probe tip and measuring the thermomechanical response.

Experimental Setup

Raman microscopy was performed with a Confocal Raman system with spatial resolution of 500nm (details are proprietary to Nissan ARC). The nanoscale thermal results were obtained using a Veeco Dimension 3100 AFM equipped with an Anasys Instruments (AI) nano-thermal analysis (nano-TA) accessory and AI micro-machined thermal probe. Imaging and localised thermal analysis (LTA) spatially accurate at the 100nm scale was performed. The contact and tapping modes were used to acquire surface images and LTA employed to determine the glass transition temperature (Tg) of the different domains at a 10°C /s heating rate.

The nano-TA data presented are the cantilever deflection (while the probe is in contact with the sample surface and the feedback turned off) plotted against the probe temperature. This measurement is analogous to the well established technique of thermo-mechanical analysis (TMA) and is known as nano-TA. Events such as melting or glass transitions that result in the softening of the material beneath the tip, produce a downward deflection of the cantilever. Further information on this technique can be obtained at www.anasysinstruments.com.

Results and discussion

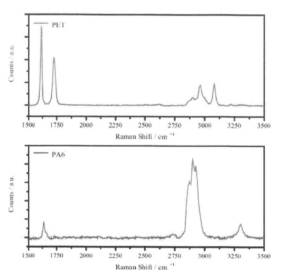

Fig 1. Raman spectra for PET and PA6

The blend of PA6 and PET investigated here was a ratio of 1:3. It was first characterized with Raman spectroscopy and nano-TA was used to further characterize the sample in regions where the spatial resolution of the Raman technique was insufficient and to obtain additional information. In figure 1, the Raman spectra for PET and PA6 are given.

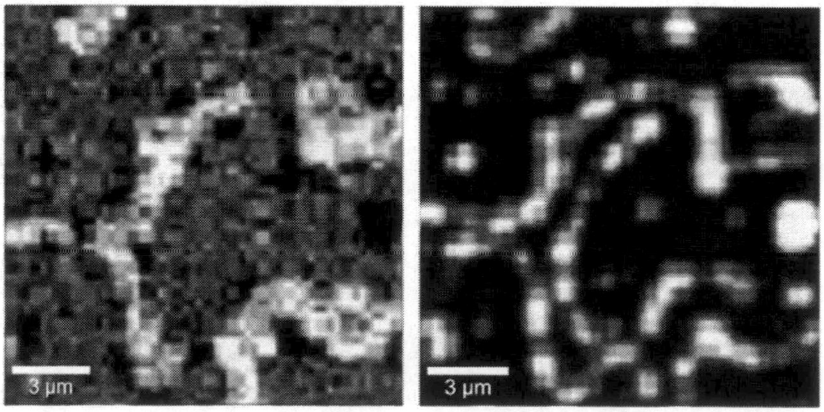

Figure 2 shows Raman images of the blend at different wave numbers; on the left is the Raman image at 2830-2940 cm-1 which highlights the PA6 given its dominant peak at 2900 cm-1. On the right is the Raman image at 1590-1640 cm-1 which highlights the PET regions due to one of the dominant peaks in its spectra occurring at 1600 cm-1. On the basis of this, we can identify occluded

2830 ~ 2940 cm⁻¹   "PA6"   1590 ~ 1640 cm⁻¹   "PET"

Fig 2. Raman images of the blend at the indicated wave numbers

islands of PA6 surrounded by PET with some occluded domains of PET within a matrix whose composition is not well defined.

As stated above, based on the Raman images shown in figure 2, we can see that the polymer blend has occluded regions whose central portion is composed of PA6 surrounded by a layer of

PET. There are also domains of PET which correspond to dark areas in the 2830-2940 cm-1 image and so are not associated with PA6.

Information from nano-TA:

Fig 3 shows an AFM phase image of the edge of the PET region that is in contact with the PA6 as well as the inside PA6. It can be seen that there is structure in the PET regions that surround occluded PA6 domains that is of the order of tens of nanometres and thus much smaller than can be resolved with Raman microscopy. There is a border around the region which is about 100nm wide and beyond this there is another type of structure in the more central region. The

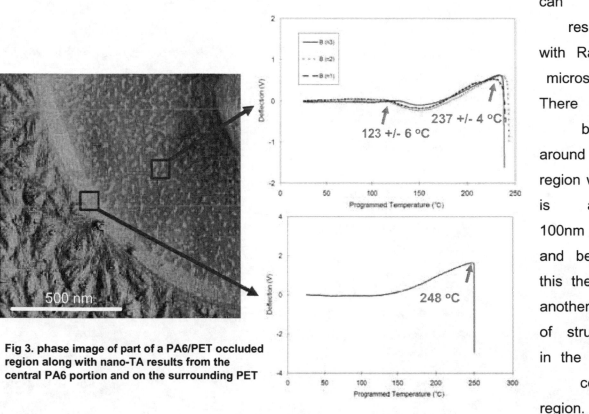

Fig 3. phase image of part of a PA6/PET occluded region along with nano-TA results from the central PA6 portion and on the surrounding PET

border shows a transition at 248°C which can clearly be ascribed to PET which then agrees with the Raman data. The more central PA6 region, that the Raman data shows to be PA6, exhibits a melting transition at 237 °C which is much higher than 220 °C which is the melting temperature of PA6. The curves for this region also show a transition at 123 °C which has all the characteristics of a glass transition. The Tg of PET is around 70 °C while for PA6 it is 50 °C. The rapid heating rates used in these experiments, around 20°C/sec., will give rise to a higher Tg than is seen in more conventional thermal analysis experiments but this cannot explain the greatly elevated temperature that is seen here. The most probable explanation is that this arises from the so-called rigid amorphous fraction. It is well known that amorphous regions located between crystalline domains can be made rigid by the constraints on mobility imposed by the crystalline material and so the Tg of this material is increased. As for the higher melting temperature exhibited by the central region of PA6, we think ultra high heating rate and ultra small melting area leads, in this case, to suppression of the relief of the molecular orientation by the border around the melting area. Thus polymer chains of PA6 crystals are still in an oriented state in the melt, thereby reducing melting entropy and resulting in a higher melting point. This suggests that the melting temperature as measured by nano-TA may give us more information about polymer orientation if we apply different heating rates and this will be the subject of future work.

Figure 4 shows the AFM topography image of the matrix material between the PA6/PET domains has a clear structure with circular domains that are approximately 1-3 microns. The Raman image at 1590-1640 cm-1 (which highlights the PET) does show some features of this size that appear to be PET domains within a matrix that has a composition that must be some combination of the two materials. In figure 8 we can see that the melting temperature within the circular domains

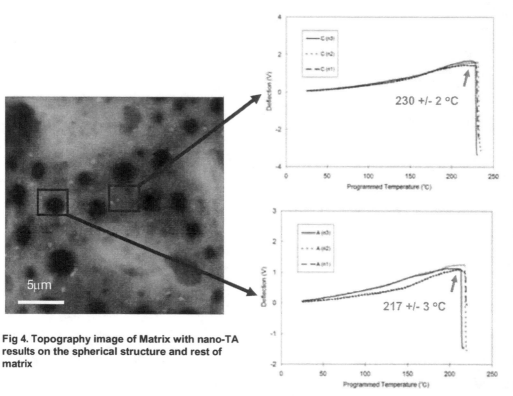

Fig 4. Topography image of Matrix with nano-TA results on the spherical structure and rest of matrix

is 217 +/- 3oC which is the melting temperature for the PA6 (within experimental error). The material between these domains has a higher melting temperature but one that is lower than the melting temperature of pure PET. High resolution AFM images of these also show clear structural differences. It can be concluded that the circular domains seen in the AFM image in figure 4 are not those seen in the Raman image that highlights PET in figure 2 both because of the low melting temperature and because they can be seen to be present at a higher density than the PET domains figure 6 reveals. Consequently there are roughly circular domains on a scale of microns that are PET-rich and also ones that are PA6-rich. Closer inspection of the 2830-2940 cm-1 image (that highlights PA6) in figure 2 does suggest that there are poorly resolved micron-scale features not associated with PA6 and so this lends support to the nano-TA data.

In summary; the PA6/PET blend is a complex system and characterising its structure presents a significant challenge. There are PA6 or PA6-rich domains that are surrounded by PET or PET-rich material. The PET in these features has a substructure; there is a 100nm outer casing that is highly crystalline PET. Within the central PA6 domains there is a region that has a significant amount of amorphous material that has a very high Tg. This elevated glass transition arises because the amorphous domains have become rigid probably because of their intimate association with the crystalline phase. There is the interesting observation that the occluded PA6 domains have a higher-than-expected melting temperature and the explanation for this is the subject of on-going work. In addition to these two-phase domains, there are single phase domains of 1-3 microns of both PET-rich and PA6-rich material. Surrounding all of these is a matrix that is a mixture of PET and PA6 which has a melting temperature intermediate between these two materials.

Conclusions

AFM and nano TA can be used without other techniques to assess the structure of materials difficult to characterise by other means such as Raman microscopy. In the case investigated here excellent agreement and complementarity between the Raman and nano-TA data was achieved in determining the composition of the different domains. The different techniques provide different information, the Raman spectra can give unequivocal information on chemical composition while the AFM and nano-TA provide higher spatial resolution, and information on transition temperature that can be using to identify materials and differentiate easily between crystalline and amorphous phases. These investigations can be seen as generic examples of the general problem of characterizing the structure and composition of materials on a nano-scale and as such the range of applications is potentially vast.

Attaching Biological Molecules to AFM Probes for Nanoscale Molecular Recognition Studies

W. Travis Johnson

Agilent Technologies, Nanotechnology Measurements Division
4330 W Chandler Blvd, Chandler Arizona 85226

ABSTRACT

Atomic Force Microscopy (AFM) is an important tool to study nanoscale molecular interactions. A strong suit of AFM is its ability to measure ligand-receptor interactions with picoNewton sensitivity. These biomolecular interactions are critical factors in a variety of physiological processes; such as the initiation, modulation and termination of DNA replication, transcription, enzyme activity, infection, immune responses, tissue generation, wound healing, cell differentiation, apotopsis, and physiological responses from drugs, hormones and toxic agents. Using AFM, scientists can probe and quantify these interactions in their native environments or perform dynamic experiments *in situ* by removing or adding ions, solutes or other reagents to the sample environment. Ligand-receptor unbinding forces and several kinetic parameters can be calculated and used to infer structural information about the molecular interactions. Nanoscale bioconjugation chemistry and surface chemistry are often required in these studies because, for the AFM to resolve the mechanical forces required to separate the ligands from their targets, the ligands must be immobilized on the AFM probe and the receptors need to be immobilized on stationary substrates.

Keywords: SPM, AFM, molecular recognition, single molecule, biosensor

1 INTRODUCTION

In MRFM (molecular recognition force microscopy) studies, ligand molecules are often covalently attached to the tip of an AFM probe which transforms the AFM probe into a highly specific, single molecule biosensor [1]. In MRFM, single molecule unbinding interactions are observed and quantified individually as the ligand-functionalized AFM probe approaches and then is subsequently withdrawn away from the substrate which contains the immobilized receptor molecules. Therefore, MRFM relies heavily on surface chemistry and nanoscale bioconjugation chemistry. AFM force spectroscopy experiments such as these, can give valuable information about the structure and dynamics of single molecule unbinding events with nanoNewton precision [2]. These techniques have also been applied to gain an understanding of the intramolecular forces involved in protein unfolding

[3]. Topography and recognition (TREC) imaging is another single molecule technique that is based on AFM. TREC also utilizes probe-bound ligands and immobilized receptors [4]. However, unlike force spectroscopy, TREC is a dynamic technique. In TREC, the ligand-functionalized AFM probe is scanned and oscillated over the surface that is covered with immobilized receptors. Specific interactions between the ligand molecules on the AFM probe and receptor molecules on the substrate are resolved as small changes in the probe's oscillation amplitude [5]. TREC imaging allows molecular interactions to be identified on compositionally complex samples such as biological materials. Using an AFM equipped with PicoTREC (Agilent Technologies), which resolves the signals that arise from molecular recognition events from the topography signals, the lateral positions of receptors on cells or other biological surfaces can be resolved with nanometer resolution [6, 7, 8, 9].

2 ATTACHING BIOLOGICAL MOLECULRES TO AFM PROBES

In MRFM and TREC studies it is often advantageous to attach biological molecules to a short tether that is in turn attached to the tip of the AFM probe because the tether permits the molecules to diffuse within a defined volume of space [10]. The tether also imparts upon a ligand the ability to reorient its position as it approaches the target in order that they may bind efficiently.

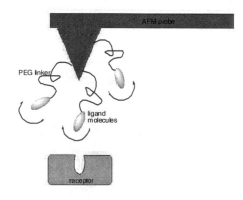

Figure 1. Ligands immobilized to an AFM probe by PEG tethers. The PEG tethers are flexible and allow the ligands to diffuse and bind with the receptor in an optimal manner.

Short polyethylene glycol (PEG) linkers are used in many force spectroscopy applications. PEG is a water soluble, nonadhesive polymer, so, nonspecific interactions can be minimized [1, 11]. Many PEG tethers can be purchased functionalized with variety of useful end groups to permit the attachment of a wide range of molecules. The number of PEG molecules on the tip of the probe can often be controlled. When the number of ligand-receptor binding interactions are minimized, single molecule interactions can be resolved [1, 12]. PEG linkers permit probe-bound molecules to diffuse within in a well defined volume of space, so that the ligands may be more likely to encounter and bind to receptors on the substrate. The steps involved in biological molecule immobilization to AFM probes are: 1. Cleaning: Gentle cleaning methods are preferred (organic solvents, UV-ozone); 2. AFM Probe Activation (amination): Silicon and silicon nitride AFM probes should be silanized with an amniosilane (APTES) or esterified (ethanolamine); 3. PEGylation: Bifunctional PEG linkers are attached to the activated AFM probes.; 4. Bioconjugation: The biological molecules are attached to the PEG tethers and 5. Characterization: The number of molecules on the surface of the AFM probe must be determined.

2.1 AFM Probe Activation

When both silicon and silicon nitride surfaces are exposed to air or water they will naturally oxidize so that the surfaces of silicon and silicon nitride materials are covered in a thin oxide layer that contains numerous reactive SiOH groups. The pKa of SiOH is approximately 6.8 [13], so it is slightly acidic and various reactions have been developed to cover surface oxides in a thin layer of amine groups. Silicon and silicon nitride AFM probes can be esterified using ethanolamine in dry DMSO to add a layer of amine groups [1, 12, 14, 15].

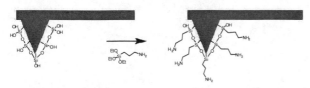

Figure 2. Activation (amination) of AFM probes with aminopropyltriethoxysilane (APTES)

Alkoxy aminosilanes may also be used to prepare aminated AFM probes. Traditional reactions of in solvents such as toluene ethanol or acetone [16, 17, 18, 19, 20] should be avoided because alkoxy aminosilanes have a propensity to form polymers in solution. This can add bulk and roughness to the tip of the AFM probe [21]. However, by performing AFM probe amination reactions in the vapor phase using freshly distilled APTES (aminopropyltriethoxysilane), tip smoothness and tip sharpness can be generally be maintained [1, 7, 9, 22, 23, 24, 25].

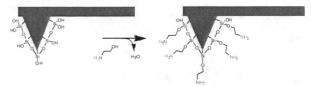

Figure 3. Amination of AFM probes with ethanolamine

2.2 PEGylation and Bioconjugation

PEG tethers can be synthesized or purchased with various end groups so that they can be attached to activated AFM probes and anchor many biological entities, including antibodies, peptides and nucleic acids. Table 1 lists some useful, commonly used, relatively short X-PEG-Y linkers that have been effectively utilized in MRFM and/or TREC studies. References that describe their synthesis or commercial sources, where they can be purchased, are also included in the table.

The NHS ester group is very reactive towards the amine groups on activated AFM probes so NHS-PEG linkers can be attached directly to these probes [11, 26, 27]. Both PEG-PDP and PEG-maleimide are reactive towards sulfhydryl groups. Consequently, cystine terminated peptides and proteins [23] or sulfhydryl modified oligonucleotides [28] can be linked directly to PEG-PDP and PEG-maleimide tethers on AFM probes. In addition, many proteins can be modified with sulfhydryl reactive reagents to facilitate their attachment to immobilized PEG-PDP and PEG-maleimide tethers [8, 9, 11, 22, 23, 29, 30, 31, 32, 33]. The synthesis of a PEG tether possessing an amine reactive aldehyde group at one terminus was recently described [27]. The aldehyde group of this tether can be reacted directly with protein molecules that contain lysine groups available near the protein's surface [6, 27, 34] or even with virus particles [27]. Antibodies or other biological molecules that have free amine groups may be anchored to the PEG-aldehyde tether without laborious protein preactivation and purification steps. Various other PEG molecules with assorted of functional groups can either be synthesized by those skilled in the art of organic chemistry or by contracting with a vendor that will perform custom synthesis.

2.3 Characterization

After the AFM probes have been functionalized with the PEG linkers and the biological molecules, it is imperative that steps be taken to determine the density of the ligand molecules on the AFM probe. The methods described above generally will result in approximately one molecule per effective tip area [12], so just one ligand on the AFM probe may have access to the receptors on the substrate at any given time. The Hinterdorfer lab in Linz

Austria has developed several methods to calculate ligand density on AFM probes [33]. The methods utilize relatively larger silicon or silicon nitride substrates which are treated in parallel under identical reaction conditions along with the AFM probes. The methods are based on (a) direct fluorescence, (b) fluorescently labeled secondary antibodies and (c) horse radish peroxidase (HRP) [12, 32].

3 CONCLUSION

Biomolecular interactions are critical to most biological phenomena. AFM offers unique advantages over many other tools for the study of biological process at the nanometer scale so it has become an important tool. For example, AFM allows scientists to visualize, probe, and analyze biological interactions in their native environments with unprecedented resolution and without the need for extraneous labels or tags. As described above, bioconjugation chemistry and surface chemistry can enhance the power and utility of AFM by allowing specific biomolecular interactions to be quantified and/or located on biological surfaces.

REFERENCES

[1] Riener et al *Analytica Chimica Acta* 479 (2003) 59-75

[2] Noy et al *Annu. Rev. Mater. Sci.* 27 (1997). 381–421

[3] Allison et al *Curr. Opinion Biotechn.* 13, (2002) 47–51

[4] Hinterdorfer, P *Handbook of Nanotechnology* (Ed.: B. Bushan), Springer Verlag, Heidelberg (2004) 475–494

[5] Kienberger et al *Biol. Proc. Online* 6, (2004) 120–128

[6] Chtcheglova et al *Biophysical Journal* 93 (2007) L11-L13

[7] Lee et al *PNAS USA* 104(23) (2007) 9609–9614

[8] Stroh et al *Biophysical Journal* 87 (2004a) 1981-1990

[9] Stroh et al *PNAS* 101, 34 (2004) 2503–12507

[10] Kienberger et al *BIOforum Europe* 06 (2004a) 66-68

[11] Reiner et al *Recent Research Developments in Bioconjugate Chemistry* 1 (2002) 133-149

[12] Hinterdorfer et al *PNAS USA* 93 (1996) 3477–3481

[13] Hubbard Ed. *Encyclopedia of Surface and Colloid Science* Vol 1 (2002) CRC Press. P. 469

[14] Raab et al *Nature Biotechnology* 17 (1999) 902-905

[15] Ray et al *J. Phys. Chem. B* 111 (2007) 1963-1974

[16] Lin et al *Emerging Information Technology Conference Proceedings*, IEEE Aug. (2005)

[17] Ros et al *PNAS USA* 95 (1998) 7402–7405

[18] Schumakovitch et al *Biophysical Journal* 82 (2002) 517–521

[19] Schwisinger et al *PNAS USA* 97(18) (2000) 9972–9977

[20] Vinckier et al *Biophysical Journal* 74 (1998) 3256-3263

[21] Lee et al *Science* (1996) 266, 771-773

[22] Lohr et al *Methods* 41 (2007) 333–341

[23] Kamruzzahan et al *Bioconjugate Chemistry* 17, 6 (2006) 1473-1481

[24] Ratto et al *Biophysical Journal* 86 (2004) 2430–2437

[25] Wang et al *Biophysical Journal* 83 (2002) 3619–3625

[26] Kienberger et al *Single Mol.* 1 (2000) 59–65

[27] Ebner et al *Bioconjugate Chem.* 18 (2007) 1176-1184

[28] Lin et al *Biophysical Journal* 90 (2006) 4236-4238

[29] Bash et al *FEBS Letters* 580 (2006) 4757–4761

[30] Baumgartner et al *PNAS USA* 97(8) (2000) 4005–4010

[31] Baumgartner et al *Single Mol.* 1, 2 (2000) 119-122

[32] Hinterdorfer et al *Nanobiology* 4 (1998) 177-188

[33] Hinterdorfer et al *Colloids and Surfaces B: Biointerfaces* 23 (2002) 115–123

[34] Bonanni et al *Biophysical Journal* 89 (2005) 2783-2791

[35] Kienberger et al *Single Mol.* 1 (2000) 123–128

Mechanical Characterization of Electrostatic MEMS Switches

[*]F. Souchon[1], A. Koszewski[1] D. Levy[1], P.L. Charvet[1]

[1]CEA-LETI-MINATEC, Grenoble, FRANCE, f.souchon@cea.fr

ABSTRACT

This paper presents a methodology developed to characterize the mechanical properties of MEMS switches. Mechanical experiments have been performed to explain the electrostatic behavior of an ohmic electrostatic series switch made by CEA-LETI.

The mechanical properties of switches are characterized by nanoindentation experiments : membrane stiffness, gap heights and contact load. These results have been compared to the results obtained by simple analytical models : the mechanical model shows a good correlation with a membrane stiffness currently around 50-100 N/m, and the electrostatic model gives capacitance values in accordance with measurements.

Based on these results, the electrostatic behavior of switches has been analysed : the influence of the mechanical properties on the ohmic and capacitive responses is pointed out.

Keywords: MEMS, switch, nanoindentation, stiffness, model

1 INTRODUCTION

Over the past years, CEA-LETI has been developing electrostatic MEMS switches. This paper is focused on the mechanical study that has been performed in order to explain the electrostatic behavior of an ohmic electrostatic series switch. Experimental and theoretical aspects are compared through two types of switches which are different from geometric and process flow standpoints.

2 ANALYTICAL MODEL

2.1 Mechanical model

The switch is modeled as a fixed-fixed beam, with two vertical symmetric loads corresponding to the actuating electrostatic force, and two axial stretching forces corresponding to the effect of the residual stress (see Fig. 1).

S : axial force / F_1-F_2 : vertical force

Figure 1 : Mechanical model of the switch

The analytical model is derived using the classic formula as described in reference [1].

(1) is the final equation used to calculate the stiffness of the MEMS switch.

$$k = \frac{F_1 + F_2}{z(d_1) + z(d_2)} \tag{1}$$

where

$$z(d_1) = -\frac{F_1 \sinh pd_1}{Sp \sinh pl} \sinh px + \frac{F_1 d_1}{Sl} x + \frac{M_{01}}{S}\left[1 - \frac{\cosh p\left(\frac{l}{2} - x\right)}{\cosh \frac{pl}{2}}\right] \tag{2}$$

and

$$M_{01} = F_1\left(\frac{2EIu}{l \tanh u}\right)\left(\frac{\sinh pd_1}{S \sinh pl} - \frac{d_1}{Sl}\right) \tag{3}$$

and $p^2 = \dfrac{S}{EI}$, $u = \dfrac{pl}{2}$, $I = \dfrac{wt^3}{12}$, k is the membrane stiffness, F_n is the nth vertical load, z is the deflection of the beam, d_n is the distance from the anchor to the nth vertical load, M_0 is the force moment in the anchor, S is the axial force due to residual stresses, w is the width of the beam, t is the thickness of the beam, l is the length of the beam, E is the Young's modulus.

2.2 Electrostatic model

The actuator's equivalent capacitor is modeled by 2 capacitors in parallel. Each capacitor between symmetric electrodes and coplanar wave guide can be considered as a parallel plate capacitor [2].

Based on these hypothesis, the capacitance of the switch is simply expressed by the following formula (4) :

$$C = \frac{2\varepsilon_0 A}{y_e + \dfrac{t_d}{\varepsilon_r}} \tag{4}$$

where C is the actuator capacitance, ε_o is the vacuum permittivity, ε_r is the relative dielectric constant, A is the electrode surface, t_d is the dielectric thickness, y_e is the air gap between electrodes.

The capacitances for up-state position and down-state position can be easily calculated from equation (4).

3 SWITCH DESCRIPTION

3.1 Design and process flow

The ohmic electrostatic switch was designed for RF applications and implemented on a coplanar wave guide using full wave analysis [3].

The series switch (see Fig. 2) is made of a silicon nitride fixed-fixed membrane with patterned metallic contacts: 2 symmetrical electrodes located inside the membrane actuate the membrane while a center metallic contact with 2 dimples short-circuits the transmission line.

When a biasing voltage is applied between the electrodes and the coplanar wave guide ground plans, the membrane is pulled down and the transmission line is short-circuited by the metallic contact.

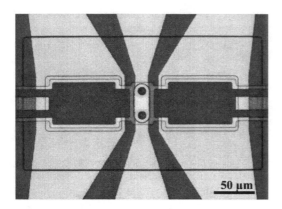

Figure 2 : Top view of a fabricated ohmic electrostatic series switch

The main steps of the switch fabrication process are as follows (see Fig. 3) :

- After a thermal oxidation on a silicon wafer, 2 etching steps are required to create a cavity with bumps,
- A gold layer is deposited and patterned to define coplanar wave guide and RF lines,
- A thick photo resist sacrificial layer is deposited by means of spin coating and then patterned,
- A first silicon nitride layer is deposited, followed by a TiN layer for electrodes. This last layer is patterned before being covered with a second silicon nitride layer,
- The switch contact is then realized by a nitride etching and a gold layer deposition. A last silicon nitride layer is deposited,
- The process ends with pads and membrane opening, and with removal of the sacrificial layer in dry oxygen plasma.

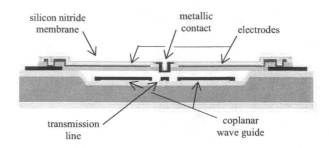

Figure 3 : Schematic stack of the ohmic electrostatic series switch

3.2 Type description

Various types of switches have been designed and manufactured. Each one is different from geometric and process flow standpoints.

This paper presents the results achieved for 2 types of switches among these : A-type and B-type. The 2 types of switches come from 2 runs of fabrication with geometrical design and process flow variations. The main differences are listed below :

- A longer beam for B-type switches,
- A lower residual stress in the membrane for B-type switches,
- Different etching depths to get a bigger electrode gap and a smaller contact gap for A-type switches
- 2 different process flows to realize the switch contact, in particular to etch the silicon nitride and the sacrificial layer before the gold layer deposition.

4 EXPERIMENTAL SET-UP FOR MECHANICAL CHARACTERIZATION

The mechanical properties of the switches have been characterized by nanoindentation experiments. The experiments consist in applying a vertical concentrated load to the indenter tip and measuring the force and the displacement. More precisely, the stiffness of the contact between the indenter tip and the beam is continuously measured by superimposing simultaneously an oscillating force, and the test is automaticaly stopped as soon as the stiffness rises above a limit value in order to keep the switch working.

In practice, nanoindentation experiments have been performed at the membrane center and at the electrode center in order to measure the membrane stiffness and two gap heights: the contact gap between dimples and transmission line, and the electrode gap between the electrodes. The contact load required for getting the ohmic switching have been also extracted from experiments made at the membrane center.

5 RESULTS AND DISCUSSION

5.1 Mechanical behavior

The A-type switch has a usual mechanical behavior as shown in Figure 4 (effective stiffness versus the indent displacement). The membrane center stiffness presents usually a step which corresponds to the membrane free stiffness before the dimples of the metallic contact and the transmission line get into contact. The electrode center stiffness presents 2 steps: one before and another one after contact at the membrane center. The stiffness is respectively around 110N/m at the membrane center, 130 N/m at the electrode center before the contact at the membrane center and 200 N/m after contact. As wanted, the contact gap height is very small compared to the electrode gap height : 130 nm against 650 nm.

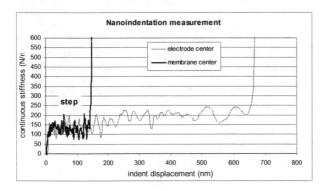

Figure 4 : Nanoindentation measurement A-type switch

The B-type switch does not have the same behavior (see Fig. 5). The membrane center stiffness presents an additional step. The first step corresponds to the membrane free stiffness, the additional step could correspond to a 2 phase contact between the dimples of the metallic contact and the transmission line.

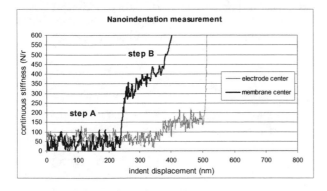

Figure 5 : Nanoindentation measurement B-type switch

These hypothesis have been confirmed by a failure analysis. Various investigations show that the dimples surface is significantly different between the 2 types of switches. Figures 6 and 7 present SEM observations for the 2 switches. The dimples for the A-type have a rough and quite homogeneous surface. Comparatively, B-type switches are less homogeneous with higher peaks which can explain the stiffness trend : after the first contact of the highest peak of 2 dimples on the transmission line, the membrane could twist to put the second dimple in contact. The process flows used to make the switch contact explain these differences between the dimples of the 2 switches.

Figure 6: SEM observation of membrane dimple surface _ A-type switch

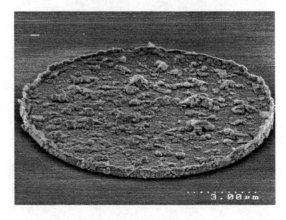

Figure 7: SEM observation of membrane dimple surface _ B-type switch

The free stiffness of B-type switches at the membrane center is around 50 N/m, the stiffness after initial contact is very high between 300 and 500N/m. The classical behavior at the electrode center is similar to the A-type switch with 2 steps : one before and another one after contact at the membrane center, the value are respectively 60 N/m and 160 N/m. The contact gap is a little bit smaller than the electrode gap, respectively 430 and 480 nm.

The contact load required to get the ohmic switching is significantly higher for the B-type switch. This is due to a higher contact gap height and the additional step with a higher stiffness compared to A-type switches.

Table 1 summarizes the mechanical experimental results obtained on the 2 types of switches where K1 is the free stiffness at the membrane center, K2a is the free stiffness at the electrode center, K2b the stiffness at the electrode center after contact at the membrane center, Gc the contact gap height, Ge the electrode gap height, Fc the necessary load for contact switching.

	Stiffness (N/m)			Gap Height (nm)		Load (μN)
	K1	K2a	K2b	Gc	Ge	Fc
A-type	110	130	200	130	650	14
B-type	50	60	150	420	480	58

Table 1 : Mechanical experimental results

The mechanical model detailed in 3.1 has been adapted to fit the nanoindentation experiments (one vertical concentrated load instead of 2). Table 2 presents the experimental and analytical results, and show that the stiffness calculated with the model correlates quite well with the nanoindentation data when the switch behavior is usual.

	Experimental stiffness (N/m)			Analytical stiffness (N/m)		
	K1	K2a	K2b	K1	K2a	K2b
A-type	110	130	200	101	135	214
B-type	50	60	150	48	64	103

Table 2 : Experimental and analytical results
for the membrane stiffness

5.2 Electrostatic behavior

The behavior of the 2 switches (see Fig. 8 and 9) have been identified using a classic single sweeping test and illustrates how the 2 gap heights and the contact load manage the electrostatic behavior.

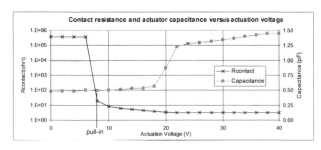

Figure 8 : Single sweeping test
A-type switch

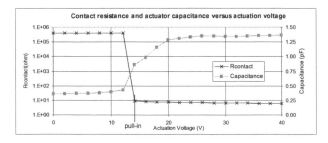

Figure 9 : Single sweeping test
B-type switch

The A-type switch has a significant shift between the ohmic and capacitive responses due to the different gap heights.

The B-type switch has no significant shift between the 2 responses due to similar gap heights. And, the ohmic pull-in voltage is bigger for B-type switches due to the contact load and the relative gap heights.

Tables 3 and 4 summarize the electrostatic results obtained on the 2 types of switches where Cup is the capacitance of the actuator in the up-state position, Cdn the capacitance of the actuator in the down-state position for contact switching, $\Delta C = Cdn - Cup$, Cfull_dn the capacitance of the actuator for electrodes in contact, Vpull-in the actuator voltage for contact switching.

The capacitances have been calculated by using the gap heights given by the nanoindentation experiments; the results given by the analytical model are close to the experimental data for the 2 types of switches.

	Experimental capacitance (pF)				Voltage (V)
	Cup	Cdn	ΔC	Cfulldn	Vpull-in
A-type	0,45	0,48	0,03	1,30	8
B-type	0,33	1,00	0,67	1,38	14

Table 3 : Experimental electrostatic results

	Analytical capacitance (pF)			
	Cup	Cdn	ΔC	Cfulldn
A-type	0,16	0,19	0,03	1,45
B-type	0,21	0,95	0,73	1,45

Table 4 : Analytical electrostatic results

6 CONCLUSION

The paper presents nanoindentation experiments in order to characterize the mechanical properties of MEMS switches : membrane stiffness, gap heights and contact load. These results have been compared to the results given by simple analytical models : mechanical and electrostatic models show a good correlation with experimental data. Finally, the electrostatic behavior of switches are well-understood thanks to these mechanical characterizations.

REFERENCES

[1] S. Timoshenko, "Strength of Materials, Part I and Part II", 3rd edition, Krieger Publishing Company, 1983, 1010 pages
[2] G. Rebeiz, "RF MEMS: theory, design and technology", Willey Inter Science Publisher, New Jersey, 2003, 512 pages
[3] D. Mercier et al., "A DC to 100 GHz high performance ohmic shunt switch", Microwave Symposium Digest, IEEE MTT-S International Volume 3, 6-11 June 2004 pp. 1931-1934

Measurement of nanometric deformations of thin membranes under controlled mechanical loads

Marc Jobin, Vladimir Sidorenko and Raphael Foschia

Ecole d'Ingénieurs de Genève (EIG), University of Applied Sciences
4 rue de la Prairie, 1202 Genève, Switzerland, marc.jobin@hesge.ch

ABSTRACT

We have measured the mechanical deformation of a 500nm thick Si_3N_4 membrane under normal loads between 10uN and 5mN. The loads were applied with a nanoindentor whose tip location onto the membrane surface can be positioned within 5 um. The deformation was measured at nanometer vertical resolution with a home-made interferometric microscope. Deformations were completely reversible without any residual strain for loads up to 0.1 mN.

Keywords: membrane, MEMS, interference microscope, phase shift, nanomechanics, silicon nitride

1 MOTIVATION

The accurate mechanical characterisation of thin membranes is recognized to be of prime importance for the development of membrane-based MEMS. To address this issue, we have built a measurement system which uses a nanoindentor to apply a well controlled force on one side of the membrane, while the deformation is recorded simultaneously on the other side with a home-made interferometric microscope [1]. Such a combination to study the mechanical properties of membrane has been pioneered by Espinosa [2]. The use of a nanoindentor ensures both a very well controlled force and a highly localised load on the membrane, while the use of the interferometric microscope allows us to measure deformations down to nanometric scale. The available forces of our nanoindentor [3] enable deformations from nanometers range to membrane rupture.

2 SAMPLES AND INSTRUMENTATION

2.1 Membrane

A SEM image of the membranes we have used is shown in the inset of Fig. 1. It is a 250um x 250um silicon nitride membrane, with a thickness of 500nm [4]. As seen on the binocular picture, the membrane is optically transparent, which greatly simplify the alignment of the nanoindenter tip with the membrane : as soon as the tip is correctly above the membrane window, the light coming from the interferometric objectives scatter onto the tip and is easily perceived.

Figure 1: Si_3N_4 membrane used in this paper, showing its optical transparency and its geometry

2.2 Integrated Interferometric Microscope/Nanoindentor

Fig 2 is a picture of the actual instruments. The nanoindenteor is placed on top of the setup and the inverted interferometric microscope is below the membrane holder. Three micrometer screws are used to tilt the sample and to approach it up to contact to the nanoindentor tip (diamond cube corner). Once the tip is approached up to contact to the membrane, a slight force (10uN) is applied on the membrane. The resulting deformation is easily seen on the interferogram, which allows for a precise (5um) lateral positioning of the tip on the membrane.

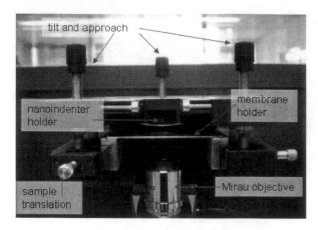

Figure 2 : Picture of the combined nanoindentor and interference microscope

A set of successive loads is applied on the membrane by the nanoindentor tip. During the 3 seconds when the force was held constant (plateau), the topography of the membrane is monitored in phase shift mode (details can be found in Ref [5]) with the interferometric microscope.

A typical interferogram of a deformed membrane is shown in Fig 3, for a load of 70uN in the center of the membrane. In case of accidental rupture of the membrane, this can be unambiguously seen on the interferogram (inset), preventing then erroneous measurement on damaged membranes.

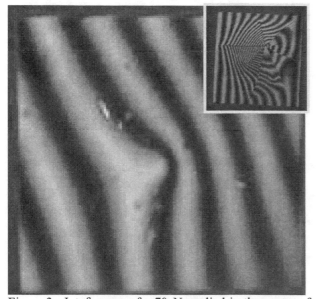

Figure 3 : Inteferogram for 70uN applied in the center of the membrane. Inset : interferogram immediately after the rupture

3 MEMBRANE DEFORMATION

3.1 Force curve

To investigate the mechanical properties of the membrane, we first performed force-distance curves with the nanoindentor on two membrane's locations: in the center and close (20um) from a corner. The ramp loads we used are shown in the inset of Fig. 4. This type of ramp is standard in nanoindentation. In our case, it allows to investigate the possible permanent deformation and creep of the membrane. We have verified by Atomic Force Microscopy that during such experiment, we do not actually indent the membrane with the nanoindentor tip.

The F-d curves are shown in Fig.4. We see that the deformation is totally reversible without any measurable residual strain for loads up to 100 uN. We measured rupture for loads ranging from 1600uN to 1700uN.

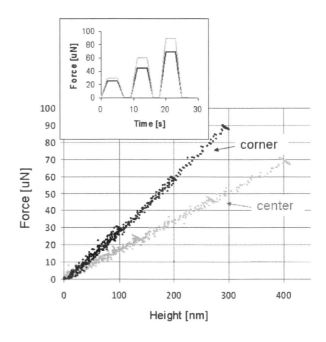

Fig 4 : Force curve in the center and close to a corner of the membrane. Inset : ramp loads.

3.2 Membrane deformation

The actual 3D deformation of the membrane has been measured with the interferometric microscope in Phase Shift Mode (PSM) and a 10x Mirau objective. Five successive $\pi/2$ phase shifted interferograms were acquired during the 3 seconds of constant force (see ramp in inset of Fig 4). The phase map was then constructed with the Hariharan [6] algorithm and the final topography is computed after standard unwrapping procedure.

Fig 5 is the membrane deformation for a 70uN load applied in the center, as measured with the interferometric microscope in Phase Shift Mode.

Fig 5 : Unwrapped topography of deformed membrane under a central load of 70 uN (Field of view : 260um x 260um)

The cross section (Fig 6) is not completely symmetric and shows a maximum deflection of 420nm. The measured membrane spring constant at the center of the membrane is 169 N/m, close to the expected 140 N/m given by the formula :

$$k = \frac{66Et^3}{a^2(1 - v^2)},$$

where E=250 [MPa] is the Young modulus of Si_3N_4, v=0.24 is the Poisson coefficient of Si_3N_4, t is the membrane thickness (500nm) and a its lateral dimension (250um). The measured spring constant increases up to 301 N/m when the tip is brought close to a corner. The inset of Fig 6b shows the nanometric deformation of the membrane, as a flattened portion of the cross section,. As our interferometric microscopes allows for atomic step imaging [7], it can be used to monitor efficiently the nanotopographies of such deformed membranes.

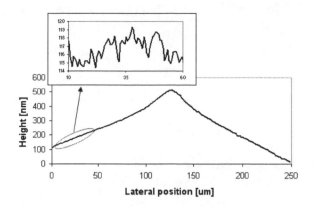

Fig 6 : Corresponding cross section of Fig 5.

4   CONCLUSIONS

We have setup an combined nanoindentor / interferometric microscopes devoted to the investigation at nanometer scale of the deformation of membranes under controlled loads in the range of 10-100uN.

REFERENCES

[1] M. Jobin et al., *NSTI Nanotech* **2**, *695 (2005)*
[2] H.D. Espinosa et al., *J. Appl. Phys.* **94**, 6076 (2003)
[3] Hysitron, model Triboscope
[4] SPI Supplies, http://www.2spi.com
[5] M. Jobin et al., *Proc. SPIE* **6188**, *61880T (2006)*
[6] P. Hariharan et al., *Appl. Opt* **26**, 2504 (1987)
[7] M. Jobin, R. Foschia, *Measurement* (2008), doi:10.1016/j.measurement.2007.12.006

Flow of Viscous Surface Layer during Electropolishing of Tungsten Wire

M. Kulakov, I. Luzinov and K. G. Kornev

School of Materials Science and Engineering, Clemson University,
264 Sirrine Hall, Clemson, SC, 29634, kkornev@clemson.edu

ABSTRACT

We investigated the mechanism of mass transfer during DC electropolishing of a tungsten wire in 2M KOH electrolyte solution. It appears that a viscous layer formed by the products of electrochemical reaction on anode surface has distinguishable optical and physicochemical properties. The viscous fluid flow is initiated by formation and detaching of a small droplet at the wire tip. Therefore, according to Tate's law, effects of surface tension and gravitational forces exerted on the film are comparable. Using high speed flow imaging, we estimated the density of this layer as $\rho \approx 1075 kg/m^3$ and its interfacial tension as $\gamma \approx 5.82 \cdot 10^{-6} N/m$.

Keywords: electropolishing, tungsten, surface tension, density, viscous fluid

1 INTRODUCTION

Metal wires with nanotips are widely used as atom probes [1], charge sources in scanning tunneling microscopes [2-4] and nanoindentation tips [5]. These ultrasharp wires are produced by conventional cutting [6], DC and AC electropolishing [2, 4, 7, 8] occasionally combined with ion milling [3], chemical etching [9] and other methods [10]. Among the listed methods, electropolishing is considered as the most inexpensive, easiest to implement, and most reliable technique for obtaining sharp tips with nanoscale radii of curvature.

During DC electropolishing, necking of a tungsten wire anodically connected to a power supply takes place in a vicinity of electrolyte-air interface, this being accompanied by continuous decrease of a current in the circuit. When the neck breaks under weight of the lower wire part, the current disappears [2]. The resulting tip shapes are dependent on meniscus surrounding wire submerged into electrolyte solution [2, 7, 10]. The products of electrochemical reaction on anode in a form of viscous fluid flow down continuously during tungsten electropolishing and shield the lower part of immersed wire, while lower rates of polishing of the upper wire part are ascribed to insufficient concentration of hydroxyls at the volume enclosed by meniscus [2, 7].

Electrochemical dissolution of tungsten involves anodic reaction of oxidation of metal atoms to their ions. Reaction on cathode corresponds to reduction of water with hydrogen and hydroxyls as products. These reactions follow Equations (1) and (2), respectively [2, 11]:

$$W(s) + 8OH^- \rightarrow WO_4^{2-} + 4H_2O + 6e^- \tag{1}$$

$$6H_2O + 6e^- \rightarrow 3H_2(g) + 6OH^- \tag{2}$$

And overall reaction in the electrolytic cell can be written as:

$$W(s) + 2OH^- + 2H_2O \rightarrow WO_4^{2-} + 3H_2 \tag{3}$$

Equation (1) is rather oversimplified, yet reflects the mechanism of electrochemical reaction. A sequence of complementary reactions were suggested in the References [11, 12]. For alkaline solutions with pH greater than 12, the reaction on anode results first in creation of a tungsten oxide WO_3 followed by formation of WO_4^{2-} ions which are considered as the reaction products. Diffusion of hydroxyls to anode surface was suggested as a rate-controlling factor for tungsten dissolution [12]. The concentration of WO_4^{2-} ions might approach the saturation limit, hence the surface film is expected to have distinguishable physico-chemical properties [13]. While importance of the surface layer on electropolishing process has been discussed for a long time, and some hypotheses were put forward [7, 13, 14], the mechanism of mass transfer has never been studied.

In this paper we investigate the mechanisms of fluid flow observed during electropolishing as the thin film formed at the tungsten wire due to electrochemical reaction is crucial for the process of the tip sharpening. Using high speed flow imaging and measuring the flow parameters, we conclude that the film flow is caused by capillary action of a thin film and free convection of the heavy ions.

2 EXPERIMENTAL PROCEDURES

The materials used in this study were 99.95% commercial pure tungsten wires (Type 1A ASTM F288-96) supplied by Small Parts Inc., Miramar, FL with diameters of 0.3, 0.5 and 0.7 mm. The wires were cut to necessary lengths using ordinary metal cutter.

Schematic illustration of electropolishing equipment is shown in Figure 1. The tungsten wire was annodically connected to DC Instek PSS-2005 programmable power supply obtained from Tequipment.NET, Long Branch, NJ.

For the positively charged electrode, cathode, austenitic non-hardenable chromium-nickel stainless steel 314 was used. All experiments were conducted in 2M water solution of potassium hydroxide (KOH). Container for electrolyte was an ordinary laboratory 150 ml glass beaker. Vertical position of the wire was precisely controlled by a calibrated microscope stage not shown in Figure 1.

We collected the data on wire diameter, depth of immersion into electrolyte, its concentration, voltage, current, and time of electroetching. Electroetching was performed in a constant-voltage mode with a current over time change being recorded by a specifically designed Labview application. At a moment when the current reached zero, the applied voltage was automatically shut off to prevent further polishing of the upper wire part. The electropolishing process was recorded using DALSA 4M-60 camera pointed at the angle of ≈14° to the electrolyte surface with a focus on the wire-electrolyte interface and surrounding meniscus.

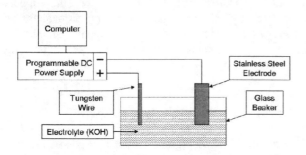

Figure 1 Schematic Illustration of Electroetching Process

Viscous liquid film flowing downward along the anode surface is created during electrochemical dissolving of tungsten. To assess the properties of this fluid, the flow was recorded at 100 fps using Motion ProX3 high speed camera (Princeton Instruments). During electropolishing process, we observed a sporadically appearing droplets at the wire tip, Figure 2a. When the drop detached from the wire tip, the fluid front was tracked and the traveled distance and time were recorded.

The fluid density in the drop ρ was assessed from a force balance for a drop falling in a viscous fluid under its own weight:

$$\rho \times a \times V = F_{fr} - \Delta\rho \times g \times V \qquad (5)$$

where a is acceleration of the viscous fluid droplet, $\Delta\rho = \rho - \rho_{KOH}$ is a density difference between fluid in the drop and surrounding 2M KOH solution, $\rho_{KOH} = 1068\,{kg}/{m^3}$, and V is the drop volume.

Assuming for order-of-magnitude estimates that the drop is spherical, the friction force can be assessed using the Rybczynski-Hadamard equation:

$$F_{Fr} = 4\pi \times r \times \eta \times v \qquad (6)$$

where $r = \sqrt[3]{3V/4\pi}$ is the drop radius, v is the drop velocity, and $\eta = 0.001 P \cdot s$ is the viscosity of surrounding liquid.

Image of the droplet before detachment from the wire was used to estimate the droplet volume, Figure 2b, by rotating a half of the droplet image using AutoCAD. Surface tension at the viscous film-electrolyte interface was estimated using Tate's law representing a force balance between capillary forces and the droplet weight:

$$\gamma = \frac{W}{\pi \times D} \qquad (7)$$

where $W = \Delta\rho \times g \times V$ is the weight of the droplet opposed by buoyancy, and D is the wire diameter.

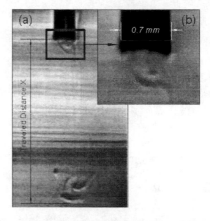

Figure 2 a) Droplet Fall in KOH Solution with Another Droplet Being about to fall from the Tip, b) Magnified Image of the Droplet Before Detachment from the Wire.

3 RESULTS AND DISCUSSIONS

Typically, the upper part of tungsten wire after electropolishing had conical shape, Figure 3a. For example, in Figure 3b we show the tip with a radius of curvature of 115 nm. The lower part of the wire is shown in Figure 3c. An image sequence for a wire dissolution process is presented in Figure 4. The maximum rate of dissolution is observed under meniscus. The part which is below the flat electrolyte-air interface is dissolved at lower rates. Creation of upward pointed meniscus at wire-electrolyte interface indicates hydrophilic properties of the tungsten surface in contact with KOH solution. Equilibrium position of the meniscus is achieved as a balance between capillary and gravitational forces, the former being proportional to the perimeter of contact and, therefore, wire diameter [15]. Bubbles of the emitting hydrogen are occasionally attached to the tungsten wire, Figures 4b, with their subsequent slow motion along the wire surface toward electrolyte-air interface and presumably cause retardation of anodic reaction.

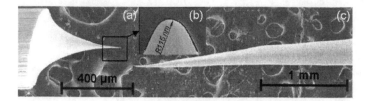

Figure 3 The Upper Part of the Wire at a) Low and b) High Magnification, c) The Lower Part of the Wire

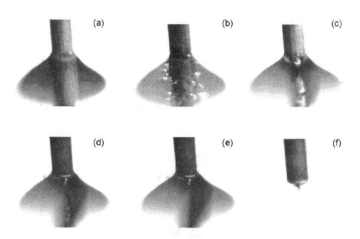

Figure 4 Image Sequence Showing Different Stages of Electropolishing

The distance-time relation for the droplet front was found to be linear, Figure 5, with the velocity being equal to $\approx 3.1 \times 10^{-3} m/s$. After approximately 1.5 s, the front started to smear and we cannot apply Eqs. (5)-(6) anymore. Using Eqs. (5)-(7), we estimated the average droplet volume as V~ $1.63 \cdot 10^{-10} m^3$, the corresponding droplet radius as r~$3.4 \cdot 10^{-4} m$, the density difference between liquids in the drop and in surrounding solution as $\Delta\rho \sim 8 \dfrac{kg}{m^3}$. Consequently, the fluid density in the film formed due to electrochemical reaction is equal to $1075 \dfrac{kg}{m^3}$ not differing significantly from the density of surrounding solution. Surface tension of the viscous fluid in contact with 2M KOH solution was estimated as $5.82 \times 10^{-6} \dfrac{N}{m}$. As seen from above, the studied fluid has distinguishable physicochemical as well as optical properties. Moreover, creation of the droplets at the initial stage of electropolishing suggests that the surface tension and gravitational forces causing motion of the film are comparable.

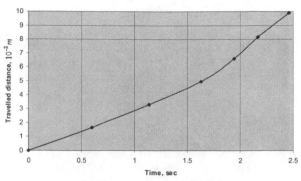

Figure 5 Traveled Distance of the Droplet front vs. Time

4 CONCLUSIONS

In summary, we have presented an experimental method for non-contact measurements of density and surface tension of electrochemical reaction products segregating on the anode surface in a form of thin viscous film. Observations on droplet formation at the wire tip suggest that the effects of surface tension and gravitational forces are comparable. Both of these components should be included into a physical model describing the necking phenomenon during electropolishing.

ACKNOLEDGEMENTS

The authors are very grateful to Dr. Ruslan Burtovyy and Dr. Bogdan Zdyrko for helpful discussions concerning the electrochemistry, Graduate Student Taras Andrukh for extensive technical assistance.

REFERENCES

[1] Miller, M. K. & Smith, G. D. W. *Atom probe microanalysis : principles and applications to materials problems* (Materials Research Society, Pittsburgh, Pa.) (1989).

[2] Ibe, J. P., Bey, P. P., Brandow, S. L., Brizzolara, R. A., Burnham, N. A., Dilella, D. P., Lee, K. P., Marrian, C. R. K. & Colton, R. J., *Journal of Vacuum Science & Technology a-Vacuum Surfaces and Films*, **8**, 3570 (1990).

[3] Zhang, R. & Ivey, D. G., *Journal of Vacuum Science & Technology B*, **14**, 1 (1996).

[4] Klein, M. & Schwitzgebel, G., *Review of Scientific Instruments*, **68**, 3099 (1997).

[5] Grunlan, J. C., Xia, X. Y., Rowenhorst, D. & Gerberich, W. W., *Review of Scientific Instruments*, **72**, 2804 (2001).

[6] Packard, W. E., Liang, Y., Dai, N., Dow, J. D., Nicolaides, R., Jaklevic, R. C. & Kaiser, W. J., *Journal of Microscopy-Oxford*, **152**, 715 (1988).

[7] Xu, D. W., Liechti, K. M. & Ravi-Chandar, K., *Review of Scientific Instruments*, **78,** 073707 (2007).

[8] Fotino, M., *Review of Scientific Instruments*, **64,** 159 (1993).

[9] Bico, J., Vierling, K., Vigano, A. & Quere, D., *Journal of Colloid and Interface Science*, **270,** 247 (2004).

[10] Melmed, A. J., *Journal of Vacuum Science & Technology B*, **9,** 601 (1991).

[11] Kelsey, G. S., *Journal of the Electrochemical Society*, **124,** 814 (1977).

[12] Anik, M. & Osseo-Asare, K., *Journal of the Electrochemical Society*, **149,** B224 (2002).

[13] Landolt, D., *Electrochimica Acta*, **32,** 1 (1987).

[14] Davydov, A. D., Grigin, A. P., Shaldaev, V. S. & Malofeeva, A. N., *Journal of the Electrochemical Society*, **149,** E6 (2002).

[15] Gennes, P.-G. d., Brochard-Wyart, F. & Quéré, D. *Capillarity and wetting phenomena : drops, bubbles, pearls, waves* (Springer, New York) (2004).

Room Temperature Growth of Single Intermetallic Nanostructures on Nanoprobes

Mehdi M. Yazdanapanah, Vladimir V. Dobrokhotov,
Abdelilah Safir, Santosh Pabba, David Rojas and Robert W. Cohn

ElectroOptics Research Institute & Nanotechnology Center
University of Louisville, Louisville, KY, USA, 40292, rwcohn@uofl.edu

ABSTRACT

Many metals readily alloy with gallium at room temperature to form intermetallic crystals. The process is directed to grow single crystals directly on nanoprobes, such as atomic force microscope (AFM) probes, tungsten needles and nanotapered glass pipettes. To date we have grown single Ag_2Ga nanoneedles and $PtGa_6$ nanoplates/blades from supersaturated melts of Ga on AFM tips. Several other alloy crystallites have been identified in bulk, but not yet grown individually on probe tips. For selective growth of the nanostrucutres, metal films from 30-150 nm thick are sputtered onto AFM probes (that are first flashed with a 10 nm Cr adhesion layer). The probes are placed in a nanomanipulator and inserted into a Ga droplet while observing the crystal formation under a scanning electron microscope. So far, Ag_2Ga wires are the most easily grown with a success rate of around 80 %. Single nanowires up to 70 microns long have been grown in this way. Ga_6Pt plates have been grown in this same way, but with a much lower success rate. In mechanical tests we have observed that the nanostructures are very strongly attached to the AFM probe. The Ag_2Ga needles are very flexible and difficult to break or detach from AFM probes.

Keywords: Self-assembly, nanostructures, nanomanipulation, AFM, NEMS

1 INTRODUCTION

In the past decade many efforts have been devoted to the synthesis of metallic nanostructured materials in the forms of nanowires, nanorods and nanobelts for their preferred electrical, optical, mechanical, magnetic and catalytic properties. In the majority of the previously published fabrication methods, synthesis of nanomaterials usually requires expensive precursors and bulky gas fixturing. Also, the particular fabrication method normally can only be used for synthesis of nanostuctures with specific morphologies. Alternatively, lithographic fabrication of freestanding nanostructures requires the use of electron-beam lithography, which has high cost and low throughput.

In this paper we demonstrate a fast and simple method of fabricating metal nanostructures at room temperature. Our method is based on direct self-assembly of gallium-metal (Ga-M) alloys, where gallium is used as a solvent that is incorporated into the final product of synthesis. Depending on the chemical nature of precursors, the nanostructure shapes can be 1-D (nanorods, nanobars), 2-D (nanoplates), or 3-D (cubooctahedrons and icosahedrons). The method of growing a single nanostructure on a single probe appears capable of being performed in parallel, making it suitable for mass production of unique probes, as well as integrated nano-electro-mechanical systems (NEMS).

The general method for developing a metal-tipped probe consists of a first exploratory phase in which the nanocrystals from Ga-M alloys are grown in bulk, followed by a second directed self-assembly phase in which the growth of a single individual crystal is localized to the tip of an AFM cantilever, or similar probe; e.g., a tipless cantilever, a tapered tungsten STM probe, a tapered glass pipette, or any other suitably prepared solid surface.

2 CRYSTALLIZATION AND GROWTH OF NANOSTRUCTURES ON THIN FILMS

Nanostructures readily form at room temperature at the interface between liquid Ga and a solid metal. This is demonstrated by placing 2 μm to 1 mm diameter drops of Ga (99.9 % purity from Alfa Aesar), in contact with sputter deposited metal thin films or foils of Pd (99.9 %), Ag (99.998 %), Pt (99.99 %), Fe (99.95 %), and Co (99.95 % purity) all from Alfa Aesar. The thin films are between 20 nm and 350 nm and the foils are between 25 μm and 125 μm in thickness. Ga is left in contact with the foil from 10 minutes to 24 hours. In some cases, the Ga is not completely reacted with the film. The excess Ga is removed by etching the sample in 1 N HCL at 60 °C for 5 to 10 minutes.

Figure 1 shows the crystals that result from the reaction of Ga with different metal foils. Each material combination produces unique crystal morphologies and in each experiment, structures of nanoscale dimensions are formed. As shown in Figure 1 these structures are hexagonal rods (Ag_2Ga, $CoGa_3$), square bars ($FeGa_3$), cubooctahedrons and icosahedrons (Ga_5Pd) and thin rectangular plates (Ga_6Pt, $FeGa_3$). The plates are frequently less than 10 nm thick, as well as being electron transparent. Note also the occasional tubular needle (Figure 1e) which we believe is templated on the growth of a cluster of nanoneedles, which then grows together to form plate-like sidewalls.

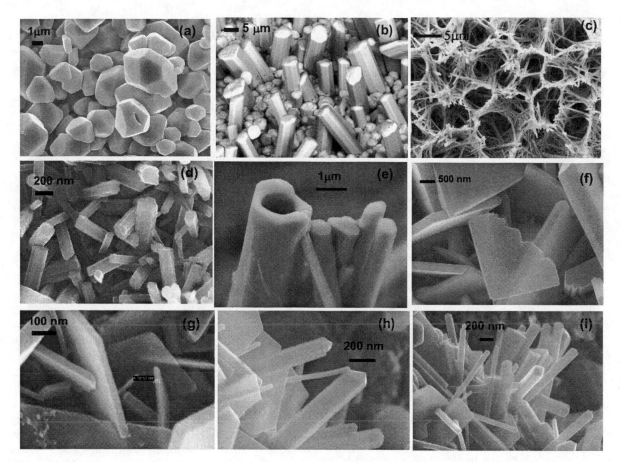

Figure 1. Ga-M crystals that form at the interface between a Ga droplet and a metal foil for foils of M= (a) Pd (b, c, e) Ag, (d) Co, (f, g) Pt, and (h, i) Fe. The reactions were performed at 25° C, except in (b) at 200° C and in (f) at 160° C.

The Ga-M crystallites are associated with *ordered phases* (i.e., Ga and M display a near-periodic atomic arrangement with a high degree of short and long range order) on their Ga-M binary phase diagrams [1,2]. We have confirmed by X-ray diffraction (XRD), energy dispersive spectroscopy (EDS), selective area diffraction (SAD) and high resolution transmission electron microscopy (HRTEM) that most of crystals are ordered phases. For instance, EDS shows that the rods in Figure 1b and c are 32-33 at % Ga which corresponds to the ordered Ag_2Ga phase.

Figures 1b and c show SEM images of Ag_2Ga crystalline needles that grew at 200 °C and room temperature respectively. The larger diameter of the needles in Figure 1b than in Figure 1c corresponds to the increased value of the diffusion constant at the higher temperature [1]. Again, Ga-Pt at 160 °C (Figure 1f) produces larger plates of Pt_6Ga that at room temperature (Figure 1g).

The temperature dependence suggests that growth is nucleated. For the growth of Ga-Pd (Figure 1a) the growth is roughly isotropic, with differences related to the surface free energy of the lower energy low order crystal planes. More precisely, the grains take on the shapes of cubo-octahedrons and icosahedrons [3-9]. The surface energies corresponding to different crystallographic facets typically increase in order of {111}<{100}<{110}. Therefore, the metallic crystals nucleate and grow into quasi-spheres enclosed by mix of both {111} and {100} facets to minimize the total surface free energy.

However, for different metals, Ga-M growth can become highly directional. Probably nuclei grow similar to those found in Ga-Pd followed by strongly preferential growth upon reaching a critical size. Initially, a seed of Ag_2Ga takes the shape of the plate bound by {111} planes from the top and the bottom, and the six side facets by {111} and {100} planes. These lower energy surfaces become much less wetting to Ga (contact angle 90 degrees or more), resulting in strongly enhanced growth at the end of the needle, where there are a great number of incomplete planes and higher surface free energy. Therefore growth becomes highly directional with the incoherent liquid-solid interface growing and continuing to extend the low-energy coherent crystalline facets in its wake, without increasing the needle diameter further. Formation of nanostructures in the form of hexagonal nanorods is also observed in the case of reaction of gallium with silver and cobalt (Figure 1b-d).

3 DIRECTED SELF-ASSEMBLY OF NANOSTRUCTURES ON AFM PROBES

The most interesting application of the crystallization of Ga-M alloys is the possibility for directed self-assembly of individual nanostructures at selected locations and orientations on probes (Figure 2). The method of selective growth presented here, not only can be performed at room temperature, but it produces strongly adhered nanostructures that are positioned both at arbitrarily desired locations and with specified orientations.

First the fabrication process of individual nanostructure on an AFM tip will be described for a Ag_2Ga nanorod [1]. The AFM tips are sputter-coated with ~ 10 nm Cr film followed by a 50 to 200 nm Ag film with a preferred thickness of 100 nm. Small Ga spherical droplets are formed on a Si substrate. First, a small amount of Ga (less than 1 mm diameter) is placed on the Si surface using a tungsten tip. Then the tip is scratched on the Si substrate until several micron wide lines of Ga are formed. Next the sample is dipped in 1 N HCl at 60 °C for 1 minute. This operation removes any surface gallium oxide and causes the gallium lines to dewet into spheres that are usually less than 20 μm in diameter. The sample is then blown dry with nitrogen and immediately transferred into a SEM chamber.

In the SEM the AFM tips are manipulated using a Zyvex nanomanipulator. Coarse mode manipulation is used to position the cantilever close to the Ga droplet and fine mode manipulation (5 nm resolution) is used to dip the tip into and withdraw it from the droplet.

The silver coated AFM tip is dipped into the liquid Ga droplet and partially retracted from the droplet forming a meniscus between the cantilever and the droplet. Ga reacts with and dissolves the silver film, followed by nanoneedle growth. Before the needle formation is complete, the cantilever is pulled further to narrow the meniscus. The needle continues to grow within the meniscus and towards the center of the Ga droplet. As the needle grows the contact angle of the meniscus increases to at least 90 degrees, greatly reducing the force required to withdraw the needle from the Ga. The total time of needle formation on the AFM tip ranges from 5 seconds to 10 minutes, depending largely on the time required for the onset of adequate supersaturation needed to overcome the barrier to nucleated growth, as well as the length of the needle grown.

Another example of directed self-assembly is a Ga_6Pt nanoplate, synthesized on the AFM probe (Figure 2a, b). This fabrication technique is very similar to the fabrication of nanoplates, and includes the same phases of interaction between the AFM-tip and gallium droplet. However, Pt-Ga reaction is significantly slower than Ag-Ga reaction. Initially, the surface of Pt shows a contact angle greater than 90 degrees in liquid gallium. Instant removal of the tip from the Ga droplet shows no plates synthesized. However, in 30-40 minutes after the immersion of the tip into Ga the reaction increases the wetting of the Pt and the contact angle reduces to less than 90 degrees. When the tip is removed from the droplet after 40 minutes, the synthesized nanoplates from 0.2-5 μm in size and ~ 10 nm thick are found extending past the tip. This slow reaction rate was also observed with the non-selective growth of Ga-Pt on foils of Pt.

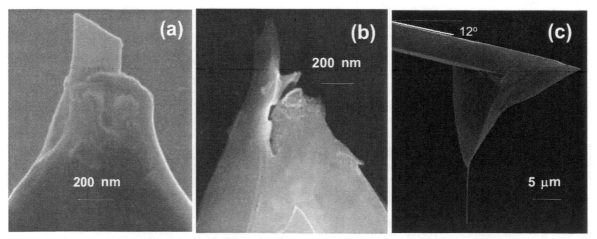

Figure 2. Ga-M nanostructures synthesized on the tips of AFM-probes. (a, b) front and side-view of Ga_6Pt nanoplate, (c) Ag_2Ga nanoneedle.

In contrast, the Ag coated AFM tips initially wet easily and become less wetting on the sidewalls of the Ag_2Ga which causes the needles to grow mostly in length over width. The end of the needles often are rounded incoherent growth fronts, which can more easily accommodate growth than the coherent sidewalls. These features seem to account for the needles being of constant diameter over tens of microns in length (Figure 2c).

4 CONCLUSIONS

A two step procedure for growing metal nanostructures at room temperature on AFM and other nanoprobes was presented. The first step identifies the possible structures for a specific metal in liquid Ga, and the second step uses a Ga meniscus to localize the growth and orientation of the nanostructure with respect to the supporting probe. It was shown that Ga-M alloys can grow in several different crystal morphologies. Prior to this report only Ag_2Ga nanoneedles had been demonstrated [1]. This report shows that Ga_6Pt can be similarly grown. While the other nanostructures containing Fe and Co have not yet been selectively grown, issues related to wetting and reactivity (as observed with Pt) certainly need to be considered further in refining this technique. The unusual shapes and metal nature of these modified AFM-tips recommend them for assorted applications, e.g. as we have recently reported [1,10].

REFERENCES

[1] M. M. Yazdanpanah, S. A. Harfenist, A. Safir, R. W. Cohn, J. Applied Physics, 98, 073510, 2005. Also see S. Saxena A. S. Warner, *"The Science and Design of Engineering Material"*, 2nd ed., McGraw-Hill, 1999.

[2] B. Predel, *"Phase equilibria of binary alloys"*, CD-ROM, edited by O. Madelung, Springer, 2003.

[3] Z.L. Wang, Journal of Physical Chemistry B, 104, 1153-1175, 2000.

[4] S.M. Lee, S.N. Cho, J. Cheon, Advanced Materials, 15, 441-444, 2003.

[5] Y. Xiong and Y. Xia, Advanced Materials, 19, 3385-3391, 2007.

[6] Y. Xiong, H. Cai, B.J. Wiley, J. Wang, M.J. Kim, Y. Xia, Journal of American Chemical Society, 129, 3665-3675, 2007.

[7] B. Wiley, Y. Sun, B. Mayers, Y. Xia, Chemical European Journal, 11, 454-463, 2005.

[8] M.M. Yazdanpanah, *Near room temperature self-assembly of nanostructures by reaction of gallium with metal thin films,* Ph.D. Dissertation, (University of Louisville), 2006.

[9] Y. Xiong, J.M. McLellan, J. Chen, Y. Yin, Z.Y. Li, Y. Xia, Journal of American Chemical Society, 127, 17118-17127, 2005.

[10] V. V. Dobrokhotov, M. M. Yazdanpanah, S. Pabba, A. Safir and R. W. Cohn, "Visual force sensing with flexible nanowire buckling springs," Nanotechnology 19, 035502, 2008.

Use of Lateral Force Microscopy to Elucidate Cooperativity and Molecular Mobility in Amorphous Polymers and Self-Assembling Molecular Glasses

D. B. Knorr, Jr., J. P. Killgore, T.O. Gray and R.M. Overney

University of Washington, Seattle, WA, USA, roverney@u.washington.edu

ABSTRACT

Intrinsic friction analysis (IFA), a novel methodology based on lateral force microscopy is presented wherein isothermal kinetic friction measurements, obtained as a function of velocity, are used to deduce apparent Arrhenius-type activation energies (E_{ac}) of molecular mobilities. If cooperativity exists between molecular mobilities involved, the dissipation energy can carry a significant entropic energy contribution, accounting for the majority of E_{ac} depending on the coupling strength between molecular actuators involved. IFA also provides a means of directly separating enthalpic contributions to E_{ac} from entropic contributions by employing a combination of a relaxation model based on absolute rate theory with thermal activation of plastic deformation. As such, the degree of cooperativity in the system can be discerned. This methodology is illustrated with nanoscale tribological experiments on two systems, (1) monodisperse, atactic polystyrene and (2) self assembling molecular glassy chromophores.

Keywords: tribology, mobility, cooperativity, lateral force microscopy.

1 INTRODUCTION

In many nanoscale technologies the ability to obtain information regarding molecular mobility and cooperativity is critical to understanding processes as well as directing molecular engineering to obtain desired results. Many experimental techniques have been developed to provide insight into molecular relaxations such as dielectric spectroscopy[1], dynamic light scattering[2] and multidimensional NMR[3] to name just a few. In a more recent development, Sills et al. showed that, due to nanoscale contact between the tip and the sample in lateral force microscopy[4], insight into molecular dissipation mechanisms can be obtained for molecular systems providing relaxation information analogous to dielectric spectroscopy[5]. However, the ability to distinguish between entropic (cooperative) and enthalpic (non-cooperative) contributions to the resulting apparent Arrhenius activation energy of a molecular process was not achieved. In this work, we develop a methodology that allows access to such information via nanoscale friction experiments and we apply this methodology to a well understood model system (atactic polystyrene) and a system of practical interest (self-assembling glassy chromophores).

2 METHODOLOGY

2.1 Experimental

The experimental setup for IFA is provided below in Fig. 1. As shown, an atomic force microscope (AFM) is used to scan the surface of a sample isothermally at various velocities (0.1µm/s to 10µm/s). Scan areas were chosen to be 1µm², and the line spacing was set to be 20 nm, which exceeds the contact diameter. The left to right (torsional) photodiode signal is sent to an oscilloscope so that friction loops[6] can be obtained. Experiments were conducted using a Topometrix Explorer AFM using Nanosensor PPP-CONT uncoated cantilevers ($C_N \sim 0.2$N/m). Cantilevers were calibrated on a silicon surface with the native oxide layer present as described in the literature[7]. Subsequently, the tips were scanned on the sample material at temperatures above the glass transition temperature of the material for approximately 1 hour. According to our experience, this process passivates the tip and produces better friction signals.

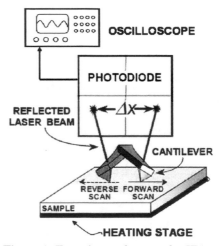

Figure 1: Experimental set-up for IFA.

Samples themselves were prepared as follows. Polystyrene samples were prepared by spin casting a 0.28wt% solution of 56k MW polystyrene in toluene onto a silicon surface, where the native oxide had been removed by HF treatment. Self assembling glassy chromophores were spin-cast onto indium tin oxide (ITO) coated glass substrates from a filtered 5 % chromophore solution dissolved in 1,1,2-trichloroethane. Film thicknesses were on the order of 300nm, which is sufficiently thick to avoid confinement effects.

2.2 Data Analysis

Data obtained are friction force as a function of scanning velocity at various isothermal temperatures $(F_F(v)|_T)$. As such, these data can be analyzed in a manner similar to that used for dielectric spectroscopy measurements. Friction isotherms are first plotted as a function of $\ln(v)$ and then, employing the time-temperature superposition principle[8, 9], are shifted horizontally to form a master response curve, with each temperature having a unique shift factor a_T. An apparent Arrhenius activation energy, E_{ac}, is available by plotting $\ln(a_T)$ vs. $1/T$ according to the following equation:

$$\ln(a_T) = \frac{E_{ac}}{R}\left(\frac{1}{T} - \frac{1}{T_R}\right) \tag{1}$$

As will be discussed below, in some cases vertical shifting is actually necessary in some cases for certain temperatures, thereby introducing a vertical shifting term, $\Delta F_F(T)$. Justification for this, as well as substantial insight into the processes involved in the system can be obtained by comparing Starkweather's understanding of relaxation phenomena in organic systems[10, 11] with Briscoe and Evan's model for activated shear processes[12].

Starkweather employs absolute reaction rate theory to develop the following expression for an apparent Arrhenius activation energy in terms of the relaxation temperature T_R, the peak frequency (f_R), and the activation entropy of the relaxation ΔS^* as[10, 11]:

$$E_{ac} = kT\left[1 + \ln\left(\frac{kT}{2\pi h f}\right)\right] + T\Delta S^* \tag{2}$$

Starkweather also noted that for polymer systems where relaxations were isolated phenomena (e.g., side chain rotation) $\Delta S^*=0$. These relaxations he dubbed "non-cooperative."

Briscoe and Evans[12] developed an activation model for shear processes with apparent activation energy E'_{ac} that involves a slider exerting pressure P over a pressure activation volume, Ω, resulting in shear stress τ acting over stress activation volume, ϕ, being resisted by potential barrier of height Q'.

$$E'_{ac} = Q' + P\Omega - \tau\phi \tag{3}$$

By equating E'_{ac} and E_{ac}, setting $\tau = (F_F + \Delta F_F)/A$, where A is the true contact area, introducing a term $\phi' = \phi/A$, called the apparent stress activation length[13], and employing thermodynamic identities[12] one may show that:

$$\Delta F_F = -\frac{kT}{\phi'}\left[1 + \ln\left(\frac{kTv}{2\pi f v_o}\right)\right] - \frac{T\Delta S^*}{\phi'} \tag{4}$$

where v_o is a characteristic velocity of the system. The key point in this discussion is that equation (4) establishes a clear connection between vertical shifting of $F_F(v)$ data and the cooperativity of the system. As such, cooperativity in the system is indicated by the need for vertical shifting to form the master response curve. We now turn our attention to the first system of interest, a well understood amorphous polymer, atactic polystyrene.

3 POLYSTYRENE

Friction measurements on polystyrene following the IFA method are provided in Fig. 2. As shown, initial raw data over the temperature range 37°C to 123°C (Fig. 2 inset) can be shifted using the procedure described in Section 2 to provide a master curve containing the peak of the α-relaxation (low values of $\ln(a_T^*v)$) and the shoulder of the β-relaxation (high values of $\ln(a_T^*v)$).

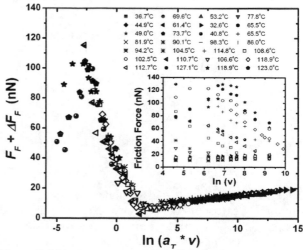

Figure 2: Master curve established for polystyrene from IFA data. (Inset) Raw friction data before horizontal and vertical shifting.

Figure 3 provides plots of $\ln(a_T)$ as a function of $1/T$ (inset) as well as a plot of vertical shifting (ΔF_F) as a function of temperature. Apparent Arrhenius activation energies obtained are 22 kcal/mol and 90 kcal/mol for regions below and above the glass transition temperature, respectively. These values are in reasonable agreement with literature values for the β-relaxation (17.4 kcal/mol [14], 21.5kcal/mol [15] and 24 kcal/mol [16]) and the α-relaxation (80-90kcal/mol[17]), respectively. To determine the degree of cooperativity ($T\Delta S^*$) with Eq. (2), a peak frequency, f, is required, which can be obtained for the α-relaxation from measured peak velocities (Fig. 2) and dissipation lengths, ξ, available from the literature for polystyrene as $f = v/\xi$ [5]. Using this process for data above T_g (α-relaxation) reveals an entropic contribution of ~70 kcal/mol to E_{ac}. The entropic contribution was determined from $v_R = 0.8$ μm/s at $T_R = T_\alpha = 396.2$ K with the previously determined[5] dissipation length of 0.4 nm. Such a large (~80 %) energy contribution is expected for the highly cooperative process of the α-relaxation in polymeric glass formers.[18].

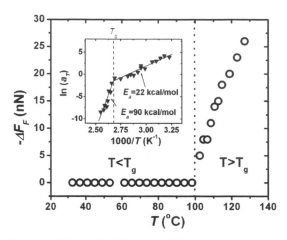

Figure 3: Vertical shifting ($-\Delta F_F$) necessary to produce master curve. (Inset) Plot of $\ln(a_T)$ as a function of $1/T$ resulting in apparent Arrhenius activation energies of 22kcal/mol below T_g and 90 kcal/mol above T_g.

4 SELF ASSEMBLING GLASSY CHROMOPHORES

Having established that the methodology is successful for determining extent of cooperativity in a well understood model system we now turn to a more interesting system of practical interest. For self assembling glassy chromophores which show optical non-linearity [19], the challenge is to acentrically align the chromophores while preventing undesirable aggregation due to dipole-dipole interactions. One such strategy is to add dendrons to the chromophores that are capable of self assembly by face to face phenyl-perfluorophenyl interactions due to complementary quadrupolar moments. Such interactions provide 4-7kcal/mol[20-22] of additional stability to the system per interaction and allow the molecules to create physically linked chains, with each molecule acting as a link in that chain. One such system, dubbed HDFD (Fig. 4 inset), has been shown to have two transition temperatures, at T_1=59°C and T_2=70°C[19].

Understanding the dynamics of such systems is critical to designing molecules with improved electro-optical activity (EO activity). As such, the degree of intermolecular cooperativity ($T\Delta S^*$) was deduced using the above methodology (Fig. 4). For temperatures below T_1, no cooperative motion was expected due to the fact that $\Delta F_F = 0$; however, between T_1 and T_2, vertical shifting was required as well as for $T>T_2$ (Fig. 4 inset). In this case no dissipation length data was available, so a reasonable range was assumed (0.1-20 nm, Fig. 4) and it was found that the entropic contribution to the apparent Arrhenius activation energy was ~0 for $T<T_1$, but increased to be the majority contribution for $T_1<T<T_2$ and $T_2<T$ (Table 1).

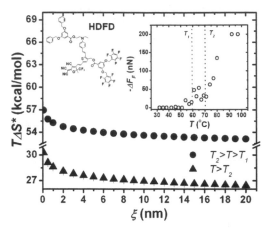

Figure 4: Plot of $T\Delta S^*$ as a function of dissipation length. (Right Inset) Vertical shifting required to generate master curve. (Left Inset) Depiction of HDFD.

The molecular dynamics of the system can be understood as follows. Below T_1 there is only isolated motion of HDFD molecules, as they associate and dissociate with each other locally. Above T_1 there is enough mobility in the system that groups of linked molecules can act cooperatively in rotation or translation. Above T_2, the system exhibits a high degree of cooperativity, although the linkages between molecules have broken down. This is similar to a polymer melt close to the glass transition temperature.[18]

	$T<T_1$	$T_1<T<T_2$	$T_2<T$
E_{ac} (kcal/mol)	23	44	71
$T\Delta S^*$ (kcal/mol)	-	26-30	52-56
% entropic	~0%	59-68%	73-79%

Table 1. E_{ac} and $T\Delta S^*$ contributions for HDFD.

5 CONCLUSIONS

As shown in the examples of polystyrene and self-assembling glassy chromophore HDFD, the intrinsic friction analysis technique provides substantial insight into molecular relaxations in a system. Apart from determination of an apparent Arrhenius activation energy, this technique is capable of providing information relating to the cooperativity, with the presence and extent of vertical shifting necessary to develop a master response curve. This technique promises to provide substantial insight for other organic systems in the future.

REFERENCES

1. Hedvig, P., *Dielectric spectroscopy of polymers.* 1st ed. 1977, New York: John Wiley & Sons.

2. Fischer, E.W., et al., J. Non-Crystalline Solids, 2002. **307-310**: p. 584-601.

3. Tracht, U., et al., Phys. Rev. Lett., 1998. **81**(13): p. 2727-2730.

4. Overney, R.M., et al., *Interfacially confined polymeric systems studied by atomic force microscopy.* 1997: Material Research Society.

5. Sills, S., T. Gray, and R. Overney, J. Chem. Phys., 2005. **123**(13): p. 134902.

6. Overney, R.M., et al., Phys. Rev. Lett., 1994. **72**: p. 3546-49.

7. Buenviaje, C.K., et al., Mat. Res. Soc. Symp., 1998. **522**: p. 187-192.

8. Williams, M.L., R.F. Landel, and J.D. Ferry, JACS, 1955. **77**: p. 3701-3707.

9. Sills, S. and R.M. Overney, Phys. Rev. Lett., 2003. **91**(9): p. 095501(1-4).

10. Starkweather, H.W., Jr., Macromolecules, 1988. **21**(6): p. 1798-802.

11. Starkweather, H.W., Jr., Macromolecules, 1981. **14**(5): p. 1277-81.

12. Briscoe, B.J. and D.C.B. Evans, Proc. R. Lond. A, 1982. **380**: p. 389-407.

13. He, M., et al., Phys. Rev. Lett., 2002. **88**(15): p. 154302-(1 - 4).

14. Wypych, A., et al., Journal of Non-Crystalline Solids, 2005. **351**(33-36): p. 2593-2598.

15. Vyazovkin, S. and I. Dranca, Journal of Physical Chemistry B, 2004. **108**(32): p. 11981-11987.

16. Bershtein, V.A. and V.M. Egorov, *Differential Scanning Calorimetery of Polymers.* 1994, Chichester, West Sussex: Ellis Horwood Ltd.

17. Patterson, G.D. and C.P. Lindsey, Journal of Chemical Physics, 1979. **70**(2): p. 643-5.

18. Donth, E., J. Non-Crys. Sol., 2002. **307-310**: p. 364-375.

19. Kim, T.-D., et al., Journal of the American Chemical Society, 2007. **129**(3): p. 488-489.

20. Reichenbaecher, K., H.I. Suess, and J. Hulliger, Chemical Society Reviews, 2005. **34**(1): p. 22-30.

21. Dunitz, J.D., A. Gavezzotti, and W.B. Schweizer, Helvetica Chimica Acta, 2003. **86**(12): p. 4073-4092.

22. West, A.P., Jr., S. Mecozzi, and D.A. Dougherty, Journal of Physical Organic Chemistry, 1997. **10**(5): p. 347-350.

Nanoscale Resolution Deformation Measurements at Crack Tips of Nanostructured Materials and Interface Cracks

J. Keller[*][**], D. Vogel[*], A. Gollhardt[*] and B. Michel[*]

[*]Fraunhofer Institute for Reliability and Microintegration, Dept. Micro Materials Center Berlin, Gustav-Meyer-Allee 25, 13355 Berlin, Germany, juergen.keller@izm.fraunhofer.de
[**]AMIC Angewandte Micro-Messtechnik GmbH, Volmerstraße 9B, 12489 Berlin, Germany

ABSTRACT

The trend towards the application of nanoparticle filled materials in the aerospace and automotive electronics sectors have led to a strong need in material characterization on the micro and nano scale. Another challenging task is the development and evaluation of interface concepts of biological structures to microelectronic materials such as polymers, metals, ceramics and semi-conducting materials. To fulfil these needs new strategies for reliability assessment on the submicron scale are essential. Under this prerequisite Scanning Probe Microscopy (SPM) serves as the basis for the development of the nanoDAC method (nano Deformation Analysis by Correlation), which allows the determination and evaluation of 2D displacement fields based on SPM data. In-situ SPM scans of the analyzed object are carried out at different thermo-mechanical load states. The images are compared utilizing grayscale cross correlation algorithms. This allows the tracking of local image patterns of the analyzed surface structure. The derived results are full-field displacement and strain fields. Due to the application of SPM equipment deformations in nanometer range can be easily detected. The method can be performed on bulk materials, thin films and on devices i.c microelectronic components, sensors or MEMS/NEMS. Furthermore, the mechanical characterization of material interfaces can be carried out with highest precision.

Keywords: nanoDAC, nanodeformation, AFM, SPM, deformation measurement

1 INTRODUCTION

With the demands for low cost electronics stacking and packaging of ICs is essential for the design of microelectronic systems. In addition to the IC and module level the integration on system level is achieved by embedded active and passive components. Regardless of the integration level thermo-mechanical challenges have to be solved due to the fact that material interfaces become even more important. Another challenge in electronic packaging is the integration of multiple device technologies such as digital, RF and MEMS, optoelectronics on the same packaging platform. Loading such structures thermally and/or mechanically means to stress the structure within submicron and nano-scale volumes caused by severe material mismatch. Therefore, actual loading causes local stresses and strains due to different material properties such as coefficient of thermal expansion (CTE), Young's modulus or time depended viscoelastic or creep properties. The smallest existing material imperfections or initial micro/nano-scale defects can grow under stress and strain and can finally cause failure of the device. Due to these facts efforts have to be made to gain a better understanding of the material responses at submicron and nano scale and at material interfaces. The way to achieve this aim is the combination of displacement and strain measurements on the micro-and nano-scale with modeling techniques such as finite element analysis or molecular modeling. Under this prerequisite Scanning Probe Microscopy (SPM) serves as the basis for the development of the nanoDAC method (nano Deformation Analysis by Correlation), which allows the determination and evaluation of 2D displacement fields based on SPM data.

From the experimental point of view measurements of thermo-mechanically induced deformations and strains at the nano scale can be carried out by state of the art nanotechnological microscopy. Research on the combination of atomic force microscope (AFM) images and digital image correlation (DIC) algorithms proofs the ability to determine nanodisplacements at microelectronic components and MEMS. The authors of the paper made use of AFM equipment for deformation field measurement [1-4]. In this paper the underlying basic principles of the digital image correlation (DIC) method will be presented. The application of nanoscale displacement measurement technique on micro- and nanomaterials will be shown by a crack analysis of a thermoset polymer (cyanate ester thermoset) material.

2 NANODAC PRINCIPLE

Digital image correlation methods on gray scale images were established by several research groups. In previous research the authors developed and refined different tools and equipment in order to apply scanning electron microscopy (SEM) images for deformation analysis on thermo-mechanically loaded electronics packages. The respective technique was established as microDAC, which means micro Deformation Analysis by means of Correlation algorithms [5]. The microDAC technique is a method of digital image processing. Digitized micrographs of the analyzed objects in at least two or more different states (e.g. before and during/after mechanical or thermal loading) have to be obtained by means of an appropriate imaging technique. Generally, images extracted from a variety of sources such as SEM or laser scanning microscopy (LSM) can be utilized for the application of digital cross correlation. The basic idea of the underlying mathematical algorithms follows from the fact that images commonly allow to record local and unique object patterns, within the more global object shape and structure. These patterns are maintained, if the objects are stressed by thermal or mechanical loading. In the case of atomic force microscopy (AFM) topography images structures (patterns) are obtained by the roughness of the analyzed object surface. Figure 1 shows examples of AFM topography images taken at a crack tip of a polymeric material.

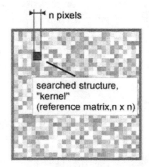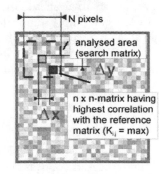

Figure 1: AFM topography scans [15 μm × 15 μm] at a crack tip of a polymer CT (compact tension) specimen; the scans are carried out at different load states.

Markers indicate typical local patterns (i.e. topographic features) of the images. In most cases, these patterns are of stable appearance, even if severe load is applied to the specimens so that they can function as a local digital marker for the correlation algorithm. The cross correlation approach is the basis of the DIC technique. A scheme of the correlation principle is illustrated by Fig. 2.

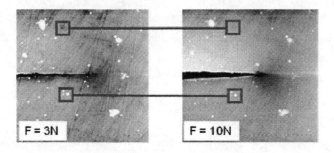

Figure 2: Displacement evaluation by cross correlation algorithm; (left) detail of a reference image at load state 1; (right) detail of an image at load state 2 [6].

Images of the object are obtained at the reference load state 1 and at a different second load state 2. Both images are compared with each other using a cross correlation algorithm. In the image of load state 1 (reference) rectangular search structures (kernels) are defined around predefined grid nodes (Fig. 2, left). These grid nodes represent the coordinates of the center of the kernels. The kernels themselves act as gray scale pattern from load state image 1 that have to be tracked, recognized and determined by its position in the load state image 2. In the calculation step the kernel window (n × n submatrix) is displaced inside the surrounding search window (search matrix) of the load state image 2 to find the position of best matching (Fig. 2, right). This position is determined by the maximum cross correlation coefficient which can be obtained for all possible kernel displacements within the search matrix. The described search algorithm leads to a two-dimensional discrete field of correlation coefficients defined at integer pixel coordinates. The discrete field maximum is interpreted as the location, where the reference matrix has to be shifted from the first to the second image, to find the best matching pattern. For enhancement of resolution a so-called subpixel analysis is implemented in the utilized software [6]. The two-dimensional cross correlation and subpixel analysis in the surroundings of a measuring point primarily gives the two components of the displacement vector. Applied to a set of measuring points (e.g. to a rectangular grid of points with a user defined pitch), this method allows to extract the complete in-plane displacement field. Commonly, graphical representations such as vector plots, superimposed virtual deformation grids or color scale coded displacement plots are implemented in commercially available or in in-house software packages [7, 8]. Finally, taking numerically derivatives of the obtained displacement fields $u_x(x,y)$ and $u_y(x,y)$ the in-plane strain components ε and the local rotation angle ρ are determined.

For images originating from scanning probe microscopy (SPM) techniques the described approach has been established as so-called nanoDAC method (nano Deformation Analysis by Correlation) [1]. This method is particularly suited for measurement of displacement fields with highest resolution focused on MEMS/NEMS devices and micro and nano-structural features of typical microelectronics materials.

3 CRACK EVALUATION

3.1 Experimental set-up

In a typical nanodeformation measurement session in-situ AFM scans of the analyzed object are carried out at different thermo-mechanical load states as shown in Fig. 1. In the illustrated case an AFM topography signal serves as the image source. It is also possible to use other SPM imaging signals such as Phase Detection Microscopy or Ultrasonic Force Microscopy. The AFM scans are taken at the vicinity of a crack at a compact tension (CT) crack test specimen, Fig. 3. The CT-specimen is loaded with the force F by a special tension/compression testing module so that a Mode I (opening) loading of the crack tip is enabled. Figure 3 shows the CT-specimen and parts of the loading device under the AFM.

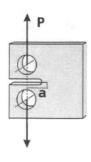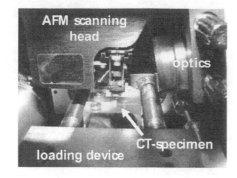

Figure 3: (left) Compact tension (CT) specimen; (right) In-situ loading under the AFM.

For images of the discussed loading of a thermoset polymer CT-specimen as given in Fig. 1 an extracted vertical (crack opening) displacement field is illustrated in Fig. 4.

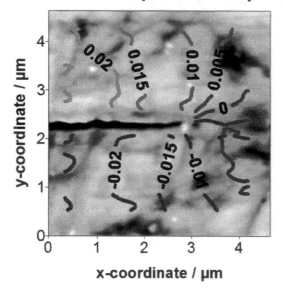

Figure 4: Crack opening displacement field in vertical (y)-direction [µm] determined by means of nanoDAC; in the background of the contour lines an AFM topography scan is illustrated.

Due to the application of SPM equipment deformations in the micro-, nanometer range can be easily detected. Currently the accuracy of the nanoDAC method for displacement field measurement is 1 nm for scan sizes of 2 µm, where the accuracy is determined by the thermo-

mechanical stability of the SPM system. Details on the effect of thermal drifts and typical SPM related stability issues are discussed in [9]. In addition this reference shows compensation strategies for such error sources. The measurement technique can be performed on bulk materials, thin films and on devices i.e. microelectronic components, sensors or MEMS/NEMS. Furthermore, the characterization and evaluation of micro- and nano-cracks or defects in bulk materials, thin layers and at material interfaces can be carried out. An example of the determination of crack parameters based on nanoDAC displacement fields is shown in the following section.

3.2 Crack opening displacement analysis

A straightforward approach for crack evaluation in the AFM is the technique of crack opening displacement (COD) determination. In order to extract the stress intensity factor K_I, crack opening displacements, u_y^u and u_y^l, are measured along both the upper and lower crack boundaries. If determined by linear elastic fracture mechanics they must equal to:

$$u_y^{u,l} = \pm \frac{K_I}{2\mu} \sqrt{\frac{x}{2\pi}} (k+1) \quad \text{for} \quad x \le 0 \quad (1)$$

$$u_y^u = u_y^l = 0 \quad \text{for} \quad x > 0 \quad (2)$$

where μ, is the shear modulus and k is a function of Poisson's ratio, ν; $k = (3-4\nu)$ for plane strain and k = (3-ν)/(1+ν) for plane stress. Taking the square of the difference of upper and lower displacements, we obtain a linear function of the x-coordinate or 0, depending on the position relative to the crack tip:

$$\left(\frac{u_y^u - u_y^l}{2}\right)^2 = Cx \quad \text{for} \quad x \le 0 \quad (3)$$

$$= 0 \quad \text{for} \quad x = 0$$

For the equation above, the crack tip is set at location x = 0. The crack tip location on the real specimen can be found at the interception of a linear fit of the curve Cx with the x-axis. The slope C allows estimating the stress intensity factor K_I, which is a measure of the crack tip load. It is given by:

$$K_I = \frac{E}{1+\nu} \frac{1}{k+1} \sqrt{2\pi C} \quad (4)$$

where E is the Young's modulus.

The discussed analysis is applied to the displacement field measurements presented in Fig. 5. The results of the linear fit according to Eqn. 3 are shown in Fig. 6.

The determined value for K_I with the application of Eqn. 4 equals to 0.056 MPam$^{1/2}$ which is about 1/10 of the critical stress intensity factor for the cyanate ester resin.

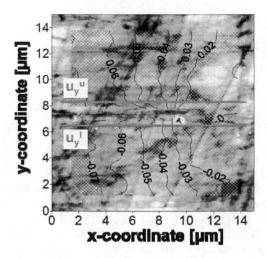

Figure 5: AFM image of crack tip area (15μm × 15μm) with overlaid displacement results in y-direction u_y; lines for the upper and lower crack face are included

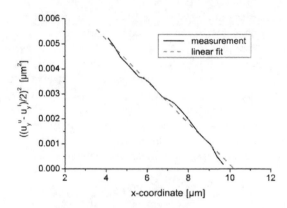

Figure 6: Evaluation of slope C for the calculation of K_I

4 CONCLUSIONS

The principle of digital image correlation based displacement measurements at in-situ loaded structures under the AFM is successfully applied to crack tip evaluation of a polymer material. The measurements were carried out at a commercially available SPM equipped with specially designed loading stages. The presented nanoDAC method is suited for in-situ thermomechanical measurements of MEMS and sensor components. Material data such as fracture properties, Young's modulus, coefficient of thermal expansion, Poisson's ratio can be determined.

REFERENCES

[1] Keller, J.; Vogel, D.; Schubert, A. and Michel, B.: Displacement and strain field measurements from SPM images, in Applied Scanning Probe Methods, B. Bhushan, H. Fuchs, and S. Hosaka, eds., pp. 253–276, Springer, 2004.

[2] Vogel, D. and Michel, B.: Microcrack evaluation for electronics components by AFM nanoDAC deformation measurement, in Proceedings of the 2001 1st IEEE Conference on Nanotechnology. IEEE-NANO 2001, pp. 309–312, 2001.

[3] Vogel, D.; Keller, J.; Gollhardt, A. and Michel, B.: Evaluating microdefect structures by AFM based deformation measurement, in Proc. of SPIE Vol. 5045, Testing Reliability, and Application of Micro- and Nano-Materials Systems., pp. 1-12, 2003.

[4] Puigcorbe, J.; Vogel, D.; Michel, B.; Vila, A. Gracia, I.; Cane, C. and Morante, J., Journal of Micromechanics and Microengineering 13(5), pp. 548-556, 2003.

[5] D. Vogel, A. Schubert, W. Faust, R. Dudek, and B. Michel, Microelectronics Reliability 36(11-12), pp. 1939–1942, 1996.

[6] Dost, M.; Kieselstein, E. and Erb, R., Micromaterials and Nanomaterials (1), pp. 30-35, 2002.

[7] UNIDAC, Chemnitzer Werkstoffmechanik GmbH, www.cwm-chemnitz.de

[8] ADASIM, Image Instruments GmbH, www.image-instruments.de/ADASIM/

[9] Michel, B. and Keller, J.: Nanodeformation Analysis Near Small Cracks by Means of nanoDAC Technique in: G. Wilkening and L. Koenders (eds.) Nanoscale Calibration Standards and Methods, Wiley-VCH, Weinheim, pages 481-489, 2005.

Interfacial Nanocomposite Characterization by Nanoparticle Debonding

J. Killgore and R. Overney*

*University of Washington, Seattle, WA, USA, roverney@u.washington.edu

ABSTRACT

Heated tip AFM (HT-AFM) is an emerging scanning probe technique that can provide local transition behavior in polymer materials. We demonstrate here, that the tool also provides direct access to the interfacial interaction (debonding) strength in reverse selective nanocomposites of poly(trimethyl silyl propyne) (PTMSP) and silica. Local nano thermomechanical analysis identified a transition at 330°C, which subsequent imaging confirmed as degradation. By thermally inducing debonding of the silica particles, interfacial polymer mobility and debonding energy were simultaneously probed. Using a torsional stiffness calibrated thermal AFM probe, probe-particle impact force was continuously monitored to reveal a debonding force of 450 nN and an impact force transition 30°C below the degradation temperature in the neat polymer, confirming the presence of enhanced polymer mobility at the PTMSP-silica interface.

Keywords: HT-AFM, interface, debonding, nanocomposite, poly(trimethylsilyl propyne)

1 INTRODUCTION

Poly(trimethyl silyl propyne) (PTMSP) blended with nano-scale silica particles is a composite system which has garnered significant attention for it's unique properties.[1, 2] Specifically, when used as a gas membrane, PTMSP exhibits reverse selectivity, thereby showing higher permeability to large soluble gases than for small permanent gases. Typically it is expected that the introduction of an impermeable filler material will result in a corresponding volumetric decrease in permeability,[3] which is further amplified by an increase in diffusing path length.[4] However, in the case of PTMSP and silica, enhanced permeability with increased filler concentration is observed. Furthermore, large molecule to small molecule selectivity is also enhanced.[5, 6] The significant permeability enhancement, even with a large concentration of an impermeable phase, suggests that physical or chemical alteration of the polymer has occurred at the composite interface.

Molecular dynamics simulations have demonstrated that Si-SiO2 induced electrostatic repulsion between the polymer and particle results in enhanced mobility of the interfacial PTMSP.[7] Enhanced mobility creates lower density regions around the interface, allowing for faster gas molecule diffusion. This calculation has been supported experimentally by bulk positron annihilation lifetime spectroscopy (PALS), which showed an increase in free volume domain size from 1.4 nm at 0% SiO2 to 1.44 nm at 25% SiO2 loading.[8] Assuming these subdomains remain discrete, permeation obeys solution-diffusion behavior similar to the neat polymer; however, if increased particle loading results in a series of interconnected subdomains, Knudsen type flow becomes significant, and reverse selective behavior begins to break down.

In an effort to better understand the interfacial properties of PTMSP-silica composites, heated tip atomic force microscopy (HT-AFM) is used here to study the adhesive strength between PTMSP and silica nanoparticles. Although initially developed for local nano thermomechanical analysis (nTMA),[9-12] it has recently been shown that resistively heated nanoscopically sharp cantilever probes provide the opportunity to study adhesive interactions between the polymer matrix and a single silica particle.[13] Here, discrete measurements and observations on individual particles allow for comparison of particle size effects and measurement of debonding energy. Furthermore, mobility of the interface is explored by comparison of debonding temperatures with local neat polymer transition temperature measurements.

2 EXPERIMENTAL

2.1 Materials

PTMSP was obtained from Gelest, Inc. and dissolved into a 2 % solution in toluene. Silica particles of 190 nm diameter were prepared by mixing 200 ml anhydrous ethanol, 6 ml tetraethyl orthosilicate and 12 ml of 30 wt% ammonia in water in a glass beaker for 12 hours. The solution was then dried to produce a silica powder. Trace amounts of the dry silica powder were ultrasonically mixed into the polymer solution, which was then added drop wise to a clean glass slide and dried for 7 days in a fume hood.

2.2 Characterization

HT-AFM measurements were performed using a Topometrix Explorer SPM with heated probes provided by Anasys Inc., with tip radii of ~30 nm and normal spring constants of ~1 N/m. The input voltage was generated from a 10 V National Instruments data acquisition card, and custom electronics allowed for real time monitoring of tip voltage and current.

NSTI-Nanotech 2008, www.nsti.org, ISBN 978-1-4200-8503-7 Vol. 1

Calibration of the HT-AFM probe was completed as follows. After performing a topography scan of the sample, the probe was positioned on a suitable surface location and the SPM feedback controller disabled. The power to the heater was ramped at a rate of 1.2 V/min until a transition in the cantilever normal deflection signal was observed. Below the melt temperature, the signal increases steadily as a result of the coupled thermal expansion of the tip and the sample. At the transition temperature, the tip penetrates the sample, as indicated by a rapid decrease in the Z-signal (or change in slope as in PI). The melt/glass transition temperature values for polycaprolactone (PC) (T_m = 60 °C), polyethylene (PE) (T_m = 130 °C), polyethylene terephthalate (PET) (T_m = 238 °C) and polyimide (PI) (T_g = 333 °C) were correlated to HT-AFM input voltage with a sigmoidal fit. By comparing to SM-FM, a well-established surface T_g probing technique that allows the use of SPM levers,[14, 15] we found the HT-AFM calibration in good correspondence on polystyrene.

Heated scans were performed with a constant heater temperature setting. Coarse 200 line 10 μm scans were performed at a scan speed of 20 μm/s with a normal force of ~23nN. Temperatures were ramped in 10 °C increments until significant surface deformation was observed. Subsequently, in a new area, 200 line, 1.5 μm scans were used to provide a detailed analysis of specific particles. For each scan, the normal Z-piezo voltage, the normal lever deflection signal, and the forward and reverse lateral force signals were recorded.

2.3 Debonding Analysis

Figure 1 shows a representative forward direction line trace for the lateral deflection signal. The deflection, as the cantilever impacts an embedded particle, is described by magnitude, ΔF_{max} and duration, α. ΔF_{max} and α are recorded at each temperature from the forward and reverse lateral force signals and compared to the baseline polymer-particle signal, β. Lateral deflection was converted to units of force using the blind calibration technique of Buenviaje et al. on a surface of known friction coefficient (Si <1,1,1>).[16]

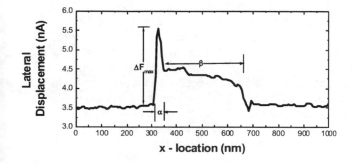

Figure 1. Lateral force line trace of polymer embedded silica nanoparticle.

3 RESULTS AND DISCUSSION

3.1 Transition Analysis

In contrast to most other glassy polymers that exhibit glass transitions at temperatures significantly below their degradation temperature, PTMSP exhibits a complex thermal relaxation behavior wherein degradation precedes the glass transition.[17] Thus, a true glass to rubber transition has not been recorded in the literature, but it is generally accepted to be in excess of 250°C.[18] Figure 2 shows a HT-AFM thermal transition plot on a PTMSP film, revealing a transition temperature, T* of 330 °C, which exceeds the degradation onset.[17]

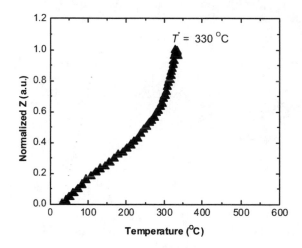

Figure 2: HT-AFM thermal analysis plot on PTMSP. Transition temperature is indicated by the apex of the tip displacement curve.

Subsequent to indentation, by scanning the probed area, HT-AFM also provides information about thermally-induced mechanical indentation properties, and thus provides an opportunity to contrast degradation with a glass transition process. The hole pattern in PTMSP exhibited no plastic rim formation, suggesting that the very rigid conjugated backbone of PTMSP is still highly resistant to deformation at T^* = 330 °C. This suggests that the critical transition in PTMSP is due to decomposition, which consequently preceded the glass transition.

3.2 Particle Debonding – Coarse Scanning

Figure 3(a) and (b) provide 10 μm HT-AFM topography scans at room temperature on PTMSP, recorded before (a) and after (b) a series of scans at probe temperatures up to 336 °C that caused particles to debond from the surface at critical conditions (temperature, pressure and scan rate). Comparing the surface particle density in the two images, there are considerably fewer particles in the post heated scanning image. The debonding occurred at a range of temperatures from ~5 °C below to 5 °C above T*. It should be noted that upon impact of the HT-AFM probe with the silica particle, the probe temperature is expected to be lower than the actual tip temperature as heat is removed by the particle. The magnitude is however to be considered small and within the provided debonding range around T*.

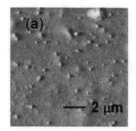 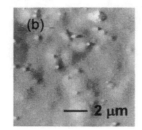

Figure 3. (a) and (b) Room temperature 10 μm HT-AFM topography scans on PTMSP before and after heated scanning respectively.

3.3 Scan Size Effect.

At the same scan frequency, compared to the 10 μm scans discussed above, a 1.5 μm scan results in a larger number of particle impacts per scan (same scan resolution over a smaller area) and an increased equilibrium contact temperature (slower scan speed). Figure 4 shows a silica aggregate in PTMSP before and after heated scanning. A particle height of ~100 nm was measured prior to debonding, while a 40 nm cavity with an additional 20 nm rim was observed after debonding. The particle satisfies the earlier specified conditions of shallow embedding and high topography, allowing debonding to occur at a temperature of 326 °C, consistent with the lower bound debonding temperatures from the 10 μm scans.

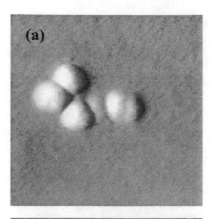

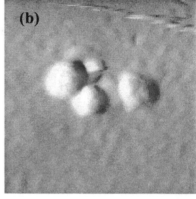

Figure 4. 1.5 μm debonding scans on PTMSP before, (a), and after, (b), heated scanning at 326 °C.

3.4 Debonding Forces.

Investigation of lateral impact forces, i.e., the lateral peak forces ΔF_{max} (see Fig. 1), on silica in PMMA and PTMSP reveal dramatically different thermal responses in the vicinity of the transition temperature. As shown in Figure 5, the impacting force ΔF_{max} for PTMSP decreases slowly from 200 nN at room temperature to 50 nN at ~300 °C, before it sharply increases to ~450 nN prior to debonding at T_{DB}, 326 °C. Softening creates a feedback delay as the probe impacts the particle, and ultimately induces greater torsional deflection of the lever. The onset of the force increase is attributed to the local transition behavior of PTMSP at the particle interface. The onset, T', occurs ~30 °C below T*. The critical temperature T', above which adhesive failure is noticeable, provides insight into the mobility of the polymer phase in close vicinity to the particle. As adhesive failure under such slow sliding conditions and fast normal force feedback control can be expected to be dominated by transition properties, and as $T' < T^* < T_g$, we can conclude that the polymer matrix in the interfacial region possesses increased mobility. This confirms current models, based on global free volume measurements, that showed an increase in free volume on average over the bulk composite.[8] Thus, the experimentally observed increase in polymer matrix

NSTI-Nanotech 2008, www.nsti.org, ISBN 978-1-4200-8503-7 Vol. 1

mobility in the interfacial region allows for the formation of larger free volume cavities which contribute to increased local sorption and diffusion. It can also be concluded, based on the direction of the interfacial transition temperature change, that a repulsive interface is present.[19]

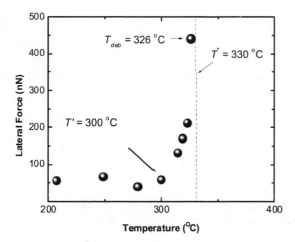

Figure 5. Debonding force ΔF_{max} to remove silica particles embedded in PTMSP as function of the probe temperature.

4 CONCLUSION

In addition to local transition property analysis, HT-AFM provides direct access to manipulation of isolated particles in nanocomposite thin films. Local thermal analysis with subsequent imaging yielded a transition of 330 °C that was found to be caused by thermal degradation. The study and manipulation of embedded nanoparticles in PTMSP revealed an impact force transition temperature ~30°C below the measured degradation transition, suggesting the presence of more mobile polymer chains at the particle interface. This is an interesting finding as the properties of the interfacial region are responsible for the enhanced mass transfer properties of the PTMSP-silica composite over virgin PTMSP. The debonding strength of single particles with the polymer matrix, which offers a simple measure of the interface, was determined to be 450 nN.

REFERENCES

1. Pinnau, I. and L.G. Toy, Journal of Membrane Science, 1996. **116**(2): p. 199-209.
2. Nagai, K., et al., Progress in Polymer Science, 2001. **26**(5): p. 721-798.
3. Barrer, R.M., *Diffusion and Permeation in Heterogeneous Media*, in *Diffusion in Polymers*, A. press, Editor. 1968: London.
4. Bharadwaj, R.K., Macromolecules, 2001. **34**(26): p. 9189-9192.
5. Merkel, T.C., et al., Macromolecules, 2003. **36**(18): p. 6844-6855.
6. De Sitter, K., et al., Journal of Membrane Science, 2006. **278**(1+2): p. 83-91.
7. Zhou, J.-H., et al., Polymer, 2006. **47**(14): p. 5206-5212.
8. Hill, A.J., et al., Journal of Molecular Structure, 2005. **739**(1-3): p. 173-178.
9. Reading, M., et al., American Laboratory (Shelton, Connecticut), 1998. **30**(1): p. 13-17.
10. Reading, M., et al., American Laboratory (Shelton, Connecticut), 1999. **31**(1): p. 13-16.
11. Grandy, D. and K. Kjoller, Microscopy Today, 2006. **14**(4): p. 58,60.
12. Pollock, H.M. and A. Hammiche, Journal of Physics D: Applied Physics, 2001. **34**(9): p. R23-R53.
13. Gray, T., et al., Nanotechnology, 2007. **18**.
14. Ge, S., et al., Physical Review Letters, 2000. **85**(11): p. 2340-2343.
15. Sills, S., et al., J. Chem. Phys., 2004. **120**(11): p. 5334-5338.
16. Buenviaje, C.K., et al., Materials Research Society Symposium Proceedings, 1998. **522**(Fundamentals of Nanoindentation and Nanotribology): p. 187-192.
17. G. Consolati, I.G., M. Pegoraro, L. Zanderighi,, Journal of Polymer Science Part B: Polymer Physics, 1996. **34**(2): p. 357-367.
18. Hill, A.J., et al., Journal of Membrane Science, 2004. **243**(1-2): p. 37-44.
19. Fryer, D.S., et al., Macromolecules, 2001. **34**(16): p. 5627-5634.

Structural fingerprinting of a cubic iron-oxide nanocrystal mixture: A case study

Peter Moeck[1], Sergei Rouvimov[1], Stavros Nicolopoulos[2] and Peter Oleynikov[3]

[1] Nano-Crystallography Group, Department of Physics, Portland State University, P.O. Box 751, Portland, OR 97207-0751, USA; [2] NanoMEGAS SPRL, Boulevard Edmond Machterns No 79, Sint Jean Molenbeek, Brussels, B-1080, Belgium http://nanomegas.com; [3] AnaliTEX, http://www.analitex.com

ABSTRACT

Two novel strategies for the structurally identification of a nanocrystal from either a single high resolution (HR) transmission electron microscopy (TEM) image or a single precession electron diffractogram (PED) are proposed and their advantages discussed in comparison to structural fingerprinting from powder X-ray diffraction pattern. Simulations for cubic maghemite and magnetite nanocrystals are used as case study examples.

Keywords: structure, nanocrystals, transmission electron microscopy, structural fingerprinting

INTRODUCTION

Nanocrystals possess size [1] and morphology [2] dependent properties that are frequently superior to those of the same materials in their bulk form. Any future large-scale commercial "nanocrystal powder-based industry" will need to be supported by structural assessment methods [3]. The quite ubiquitous method of identifying crystal structures is (Cu-tube based) powder X-ray diffraction (XRD) [4], e.g. Fig. 1. That method works best for micrometer-sized crystals and becomes due to peak broadening and (isotropic or anisotropic) shifting less useful to useless for crystals in the nanometer range [5, 6]. XRD patterns of nanocrystals are also made significantly less characteristic by surface relaxation effects [7].

Two novel strategies for the structural identification of nanocrystals in the TEM are, therefore, proposed. Both of these methods are applicable to nanocrystal thicknesses for which the scattering of fast electron can be considered as essentially (quasi-)kinematic. This thickness range is for HRTEM imaging 1 to about 10 nm and for PEDs 10 to 50 nm. In the dynamic scattering limit, these methods become analogous to the well known structural identification methods for single crystals in the TEM that only use information on the projected reciprocal lattice geometry. For a recent review of those methods and more information on the two novel strategies, see ref. [8]. Because cubic maghemite and magnetite possess almost the same lattice constant and "rather similar" atomic arrangements (i. e. nearly cubic densest packings of oxygen with differences in the iron occupancies of the intersites), the XRD patterns are very similar, Fig. 1. Allowing for peak broadening, peak shifting, and surface relaxation, nanometer sized crystals of these two cubic iron-oxide minerals can hardly be told apart and their mixtures can not be quantified by XRD.

Quite independent on the nanocrystal size, there is, however, structure information at the atomic level in a (single) HRTEM image and a (single) precession* electron diffractogram (PED) of a (single) nanocrystal that can be advantageously employed for its structural identification [8-13].

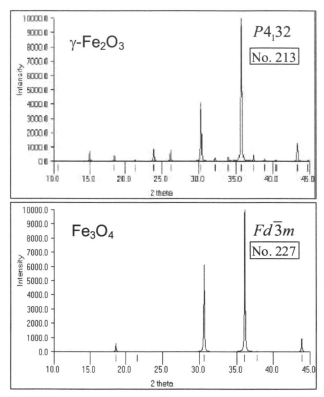

Figure 1: Calculated powder X-ray (Cu-Ka) diffraction patterns for micrometer-sized cubic maghemite, γ-Fe_2O_3, and magnetite, Fe_3O_4, out to the 400 reflections. The space group symbols and their numbers are also given.

This atomic-structure-level structure information is in the case of TEM images after crystallographic image processing [8, 14, 15] structure factor amplitudes and phases out to the point resolution of the microscope, e.g. at least out to 5 nm^{-1} for dedicated (but non-aberration corrected) HRTEMs. In the case of PEDs, this atomic-structure-level information is the structure factor amplitudes and extends to at least twice as far in reciprocal space.

Extracting this kind of information for unknowns, combining it with the extractable projected reciprocal lattice geometry, and comparing it to structural information that is contained for a range of candidate structures in a crystallographic database is the basis of our two novel methods for structural fingerprinting in the TEM [8]. The approximately 20,000 entry mainly inorganic subset [16] of the more than 50,000 entry Crystallography Open Database (COD) [17] may be employed for this purpose.

This paper illustrates that for nanocrystals which scatter fast electrons quasi-kinematically much more structural information can be extracted from either HRTEM images or PEDs than is accessible from powder XRD. Simulations for nanocrystals of cubic maghemite, γ-Fe_2O_3, and magnetite, Fe_3O_4, are used as case study examples.

STRUCTURAL INFORMATION FROM HRTEM IMAGES OR PRECESSION ELECTRON DIFFRACTOGRAMS

Table 1 lists theoretical structure factor amplitudes and phase angles for cubic maghemite and magnetite nanocrystals. Their experimental counterparts can be extracted from HRTEM images that were recorded at a microscope with 0.19 nm point resolution. Figure 2 shows a so called "lattice-fringe fingerprint plot" for magnetite for the same point resolution. This plot was calculated over the Internet (on the fly) from data of the mainly inorganic subset of the COD [16]. We call these plots "lattice-fringe fingerprint plots" because the idea to plot two reciprocal spacings and their acute intersecting angle, (i.e. 3 independent entities), into a two-dimensional (2D) plot originated in connection with Fourier transforms of HRTEM images that showed crossing lattice fringes [13].

{hkl}	γ-Fe_2O_3 $|F|$	γ-Fe_2O_3 α	Fe_3O_4 $|F|$	Fe_3O_4 α
011	0.78	90	-	-
111	0.55	135	1.55	0
012	0.90	90	-	-
112	0.60	0	-	-
022	3.25	0	3.29	180
013	0.50	270	-	-
113	4.41	45	4.85	180
222	0.15	90	1.11	0
023	0.63	0	-	-
123	0.43	180	-	-
004	5.65	180	6.47	0
033	0.38	270	-	-
114	0.38	270	-	-
133	0.28	135	0.37	180

Table 1: Theoretical** structure factor amplitudes (|F| in nm) and phase angles (a in degree) for cubic maghemite, γ-Fe_2O_3, and magnetite, Fe_3O_4. The experimental counterparts to these structure factors can be extracted from Fourier transformed HRTEM images that were taken at a microscope with 0.19 nm point resolution. (There are also tetragonal maghemites with similar stoichiometries and variations in the occupancy of the iron intersites, which we do not consider here).

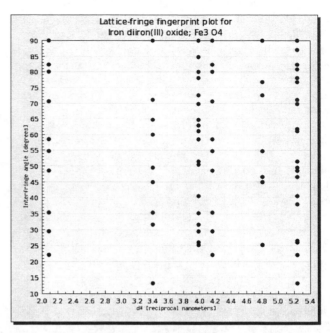

Figure 2: Lattice-fringe fingerprint plot of magnetite for a HRTEM with 0.19 nm point resolution.

The so called "interfringe angle", i.e. the acute angle under which lattice fringes intersect in HRTEM images, is plotted in a lattice-fringe fingerprint plot against the reciprocal lattice vector magnitude. While there are two data points in lattice fringe fingerprint plots for crossed fringes with different spacings, the crossing of two symmetrically related fringes results in just one data point (because the latter possess by symmetry the same spacing). These plots may extend in reciprocal space out to either the point or the instrumental resolution of the microscope. All of the resolvable lattice fringes up to this resolution will be included for a certain crystal structure into these plots.

If derived from PED data, the counterpart to a lattice-fringe fingerprint plot will extend in reciprocal space out to the diffraction limit of the structure. As far as the projected reciprocal lattice geometry is concerned, there is no essential difference between lattice-fringe fingerprint plots that originated from Fourier transformed HRTEM images and their counterparts that originated from PED data.

An initial search in a database of theoretical lattice-fringe fingerprints that is only based on the 2D positions of lattice-fringe data points, Fig. 2, may result in several candidate structures. In the following step, the search can be made more discriminatory by trying to match crystallographic indices to the 2D positions. Because one will always image along one zone axis, all of the indices of the reflections must be consistent with a certain family of zone axes. (As far as the lattice-fringe fingerprint plots are concerned, this follow up search is equivalent to assigning crystallographic indices to the 2D data points.)

Similarly to the classical Hanawalt search strategy of powder X-ray diffraction databases [4], one can divide lattice-fringe fingerprint plots, such as the ones shown in Fig. 2, into 2D geometric data sectors of experimental condition specific average precisions and accuracies and also allow for some overlap between the sectors. Larger reciprocal spacings and interfringe angles can be measured inherently more accurately than smaller reciprocal spacings and interfringe angles. The location of the respectively more accurate data points will be in the upper right hand corners of lattice-fringe fingerprint plots.

The accuracy and precision of the extracted structure factors will depend on how accurately the contrast transfer function of the objective lens can be determined at every point of interest by crystallographic image processing [14, 15]. The accuracy of theoretical structure factors is not precisely known as it depends on the (not precisely known) accuracy of the atomic scattering factors. Nevertheless, the accuracy of theoretical structure factors is likely to be similar for all structure factors because each of them represents the scattering in a certain direction by all of the atoms in the unit cell.

COMPARISON BETWEEN TEM AND XRD DATA FOR STRUCTURAL IDENTIFICATION OF NANOCRYSTALS

If one takes the peak position and peak height in an XRD diffractogram as two pieces of information, there are just 12 such pieces for magnetite (including those from the very weak 222 peak next to the strong 400 peak and the weak 133 peak which falls just outside the angular range of Fig. 1), which can be used for the structural identification of this mineral. In Fig. 2, there are, however, 74 data points for magnetite out to the family of {133} reciprocal lattice vectors. In addition, each of the 6 families of lattice planes in Fig. 2 possesses both structure factor amplitude and structure factor phase angle, see last two columns of Table 1.

If the counterpart of a lattice-fringe fingerprint is for maghemite constructed from PED data, there will be many more data points in the plot as the resolution of such data is not restricted to the point or information limit resolution of the HRTEM. There will, however, be for each family of lattice planes only the structure factor amplitude available for structural fingerprinting in the TEM.

Due to the primitive cubic space group symmetry of maghemite, its theoretical lattice-fringe fingerprint plot counterpart contains about five times more data points (with distinctively different 2D coordinates in the plot) for the same 0.19 nm point resolution of the HRTEM. In addition, due to this space group being not centrosymmetric, the structure factor phase angles can have any value, see third column in Table 1, while they are restricted to be either 0° or 180° for magnetite, see last column of Table 1. Cubic maghemite and magnetite nanocrystals can, therefore, be reliable distinguished on the basis of HRTEM images when they are part of a mixture, as experimentally demonstrated in refs. [9-11].

Since the indices of the three strongest peaks in XRD patterns out to the 400 reflection, Fig. 1, are for magnetite

and maghemite identical, these two iron-oxides can even for micrometer sized crystals not easily be distinguished by the classical Hanawalt [4] approach. Due to XRD peak broadening, peak shifting, and surface relaxation effects, both a distinction between these two minerals and quantification in case of a mixture of these two iron-oxides become for nanocrystals quite impossible.

SUMMARY OF THE KIND OF INFORMATION THAT IS OBTAINABLE FROM TEM FOR STRUCTURAL IDENTIFICATION OF NANOCRYSTALS

The structural information that can be extracted from a HRTEM image is the projected reciprocal lattice geometry, the plane symmetry group, and a few structure factor amplitudes and phases. Except for the structure factor phases, the same kind of information can be extracted from a single PED, but the information that can be used for structural fingerprinting is in this case is not limited by to the point or instrumental resolution of the TEM. PEDs show frequently higher order Laue zones that enable the extraction of structural information in 3D.

More elaborate lattice-fringe fingerprint plots may contain in the third and forth dimension information on structure factor phases and amplitudes. Possibly in a fifth dimension, histograms of the probability of seeing crossed lattice fringes in an ensemble of nanocrystals may be added to lattice fringe fingerprints and may facilitate the structural fingerprinting of a multitude of nanocrystals. The equations for calculating such probabilities for an ensemble of randomly oriented nanocrystals are given in ref. [13]. Instead of employing higher dimensional spaces, one may stick to two-dimensional displays such as Fig. 2 and simply add to selected data points sets of numbers that represent additional information, e.g. structure factor phases and amplitudes with their respective error bars. Because all interfringe angles between identically indexed reflections are the same in the cubic system, space group information can be extracted straightforwardly from lattice-fringe fingerprint plots of cubic crystals even without indexing.

Searching for these kinds of extractable structural information in comprehensive databases and matching it with high figures of merit to that of candidate structures allows for highly discriminatory identifications of nanocrystals, even without additional chemical information as obtainable in analytical TEMs. Structural identification of nanocrystal within the quasi-kinematic electron diffraction limit will be after automation [8] superior to structural fingerprinting from XRD data.

ACKNOWLEDGMENTS

This research was supported by congressional earmark funds for the development of nanometrology strategies to the Oregon Nanoscience and Microtechnologies Institute.

REFERENCES

[1] E. Roduner, *Nanoscopic Materials, Size-dependent Phenomena*, Cambridge (U.K.), Royal Society of Chemistry, 2006

[2] N. Tian et al., "Synthesis of Tetrahexahedral Platinum Nanocrystals with High-Index Facets and High Electro-Oxidation Activity", *Science* **316** (2007) 732-735

[3] C. Saltiel and H. Giesche, "Needs and opportunities for nanoparticle characterization", *J. Nanoparticle Res.* **2** (2000) 325-326

[4] J. Faber and T. Fawcett, "The Powder Diffraction File: present and future", *Acta Cryst. B* **58** (2002) 325-332

[5] J. A. López Pérez et al., "Advances in the preparation of magnetic nanoparticles by the Microemulsion Method", *J. Physic. Chem. B* **101** (1997) 8045-8047

[6] N. Pinna, "X-Ray diffraction from nanocrystals", *Progr. Colloid. Polym. Sci.* **130** (2005) 29-32

[7] W. Lojkowski et al. (editors), *Eighth Nanoforum Report: Nanometrology,* http://www.nanoforum.org, (2006) 79-80

[8] P. Moeck and P. Fraundorf, "Structural fingerprinting in the transmission electron microscope: Overview and opportunities to implement enhanced strategies for nanocrystal identification", *Zeitschrift für Kristallographie* **222** (2007) 634-645; expanded version in **open access:** arXiv:0706.2021

[9] R. Bjorge, "Lattice-Fringe Fingerprinting: Structural Identification of Nanocrystals Employing High-Resolution Transmission Electron Microscopy", *Master of Science Thesis*, Portland State University, May 9, 2007; Journal of Dissertation Vol. **1** (2007); **open access:** http://www.scientificjournals.org/journals2007/j_of_dissertation.htm

[10] P. Moeck and R. Bjorge, "Lattice-Fringe Fingerprinting: Structural Identification of Nanocrystals by HRTEM", in: *Quantitative Electron Microscopy for Materials Science,* eds. E. Snoeck, R. Dunin-Borkowski, J. Verbeeck, and U. Dahmen, Mater. Res. Soc. Symp. Proc. Volume **1026E** (2007), paper 1026-C17-10

[11] P. Moeck et al., "Lattice-fringe fingerprinting of an iron-oxide nanocrystal supported by an open-access database", *Proc. NSTI-Nanotech* Vol. **4** (2007) 93-96; ISBN 1-4200637-6-6

[12] P. Moeck et al., "Lattice fringe fingerprinting in two dimensions with database support", *Proc. NSTI-Nanotech* Vol. **1** (2006) 741-744; ISBN 0-9767985-6-5

[13] P. Fraundorf et al., "Making sense of nanocrystal lattice fringes", *J. Appl. Phys.* **98** (2005) 14308-1-14308-1-10; **open access:** arXiv:cond-mat/0212281v2; *Virtual Journal of Nanoscale Science and Technology* Vol. **12** (2005) Issue 25

[14] S. Hovmöller, "CRISP: crystallographic image processing on a personal computer", *Ultramicroscopy* **41** (1992) 121-135; http://www.fos.su.se/~svenh/index.html

[15] X. D. Zou and S. Hovmöller, "Electron crystallography: imaging and single-crystal diffraction from powders", *Acta Cryst. A* **64** (2008) 149-160; **open-access:** http://journals.iucr.org/a/issues/2008/01/00/issconts.html

[16] http://nanocrystallography.research.pdx.edu/CIF-searchable

[17] http://crystallography.net

---

* The electron precession method is formally analogous to the well known X-ray precession technique (Buerger, M. J.: Contemporary Crystallography. McGraw-Hill, 1970, pages 149-185), but utilizes a precession movement of the electron beam around the microscope's optical axis rather than that of the specimen goniometer around a fixed beam direction. The diffracted beams are de-scanned in such a manner that stationary spot diffraction patterns are obtained. The illuminating beam can be either parallel or focused.

Precession electron add-ons (to newer and older) TEMs have been developed by Dr. S. Nicolopoulos (Tel.: +34 649 810 619, info@nanomegas.com) and coworkers and can be purchased from NanoMEGAS. Prof. P. Moeck's (pmoeck@pdx.edu, Tel.: USA 503 725 4227) Laboratory for "Structural Fingerprinting and Electron Crystallography" at Portland State University's Physics Department is the first demonstration site for this company in the Americas. There is currently only one other commercial precession electron system from NanoMEGAS installed in the USA (at ExxonMobile Research & Engineering Co. Inc, Annandale, NJ), while there are already 26 installations in Europe alone. Profs. S. Hovmöller and X. D. Zou of the Swedish company Calidris, http://www.calidris-em.com, offer IBM-PC compatible software that supports the extraction of structural information from both HRTEM images and PEDs. (This software can also be demonstrated at P. Moeck's lab in Portland, OR.) While Dr. P. Oleynikov is developing dedicated structural-electron-fingerprinting software at the AnaliTEX company, Dr. S. Rouvimov will develop structural-electron-fingerprinting protocols for industrial partners.

Northwestern University (Evanston, IL) possesses a user-built precession electron system. Copies of that system have been installed at the University of Illinois at Urbana-Champaign, Arizona State University, and the National Center for Electron Microscopy at the Lawrence Berkeley National Laboratory.

It is advantageous that the so called "structure-defining" reflections fulfill the quasi-kinematic diffraction approximations sufficiently well even for thicknesses on the order of 20 to 40 nm for crystals that are otherwise known to scatter fast electrons dynamically; C. S. Own, W. Sinkler, and L. D. Marks, *Ultramicroscopy* **106** (2006) 114-122 as well as P. Oleynikov, S. Hovmöller, and X. D. Zou, *Ultramicroscopy* **107** (2007) 523-533. **Precession electron diffraction is, thus, bound to become the "quasi-kinematic electron diffraction fingerprinter's and crystallographer's" preferred operation mode for nanocrystals in the thickness range from approximately 10 to 50 nm.**

** Averaged structure factor amplitude values, calculated with AnaliTEX's program Emap & Simulator from **open-access** data at: http://rruff.geo.arizona.edu/AMS/amcsd.php.

Interaction between alternating current and cathodic protection over nano sized/structured surface metals

V. Bueno*, L. Lazzari *, M. Ormellese *, P. Spinelli **

* Politecnico di Milano. Department of Chemistry, Materials and Chemical Engineering. Via Mancinelli 7 - 20131 Milan, Italy, vbueno@chem.polimi.it
** Politecnico di Torino. Department of Materials Science and Chemical Engineering. Corso Duca degli Abruzzi 24 - 10129 Torino, Italy, p.spinelli@polito.it

ABSTRACT

Serious cases of corrosion of metallic structures under both alternating current (ac) and cathodic protection (cp) have been reported. The objective of this research is to propose a mechanism by which ac affects the nano sized/structured surface films formed on metals under cp. A model will be proposed and experimentally validated, in order to relate the metal corrosiveness to its nano sized/structured surface, under the simultaneous effect of ac and cp. This study intends to show how observations made by the ac-Voltammetry can be used to obtain an insight into the nature and mechanism of electrode processes. Once reactions are identified, the intensity and morphology of the surface damage will be related to the amplitude and frequency of the sinusoidal perturbation. Analytical and kinetic studies will be combined with Atomic Force Microscopy (AFM), performed firstly ex-situ and later in the electrolytic cell while the ac-corrosion is simulated, in order to obtain the nano-surface characterization.

Keywords: nano sized oxides film, ac-voltammetry, cathodic protection, ac-corrosion, AFM.

1 INTRODUCTION

It has been reported that metallic structures present corrosion damages while, under cathodic protection, are interfered with alternating currents. Different authors [1-3] have studied the phenomenon mathematically, but the involved mechanism is still not fully understood. It has been suggested that sinusoidal voltages of small amplitudes do not corrode the metal surface; in fact, the application of alternating signals to measure corrosion rates is a common practice today [4]. The analysis of the corrosion of mild steel under cathodic protection and ac-perturbations (e.g., in soil with basic pH at the surface, due to the high cathodic protection levels) calls for the necessity of an improvement of the study from both, the electrochemical and the physical points of view and its evaluation not only at the micro, but also at the nano-scale. The surface oxides film formed on mild steel in a basic NaOH 1M media was analyzed by two groups of researches: one intended to explain the corrosion of the mild steel when it is under ac perturbation, even being in the range of thermodynamic immunity gained by the cathodic protection level. They suggested the formation of a non protective porous oxide, due to the ac perturbation, that was responsible for the ac-corrosion of the metal [5]. The second group analyzed the oxide film without ac perturbations. Then, based on dc Cyclic Voltagramms and other experiments, a more complete mechanism was proposed, coinciding with the first group, in the presence of that compound responsible of the poor corrosion protection of mild steels under certain conditions [6]. This research, attempts to characterize the surface oxides combining two approaches, i.e. studying the response of the system to dc signals, and analyzing its reversibility to ac perturbations. AFM analysis can complement ac-Voltammetry [7] in the study of the formation and stability of the surface oxides, under ac effects.

2 AC CORROSION. MECHANISM, MODELS AND CHARACTERIZATION

The more important phenomenon to be taken into account in the study of reactions occurring at the metal surface, under the combined effect of ac and dc perturbations, is the presence of the rectifying properties of the electron transfer reactions [7]. These properties are the consequence of changes in reaction rates with differences of potential across the interface. Specifically, when the alternating sinusoidal voltage is applied, periodic concentration changes of the reacting species are produced at that interface, being accompanied by periodical diffusion processes and the flow of a Faradic alternating current, which contains, at least, the harmonic component of the fundamental frequency. Effects of the non-linearity of the redox reactions are evidenced by the partial rectification of the alternating voltage, producing a direct potential and by the appearance of higher harmonics in the alternating current.

2.1 Corrosion mechanism by which ac affects the nano surface film of metals under cp

To propose the ac-corrosion mechanism of the surface metal under cathodic protection and ac-perturbations, this innovative research applies the ac-Voltammetry, used in analytical chemistry with non polarizable working electrodes, onto a reactive one, to identify and characterize the sequence of reactions occurring at the metal surface, in the presence of ac and dc signals. This characterization is made through the study of the fundamental electrochemistry of the double layer and of the Pourbaix Fe diagram. AC-V, based on Polarography, applies an alternating sinusoidal voltage onto a solid electrode, while it is swept in a direct voltage range and the amplitude of the fundamental faradic current is recorded as a function of the dc potential. Phase angle variations and impedance data are recorded too. The reactions to be tested could behave reversible to dc and to ac polarizations, quasi-reversible with nernstian dc reversibility, or quasi-reversible without it. Applying AC-V, electron-transfer reactions, adsorption/desorption processes can be identified through the ac waves, while changes in the base-current could give information about the double layer. Once reactions are identified, the intensity and morphology of the surface damage will be related to the amplitude and frequency of the sinusoidal perturbation.

2.1 Modeling the ac-current vs. dc-voltage characteristic of the ac-corrosion process

Several mathematical models have been developed to simulate the current voltage characteristic of the electrode-electrolyte interface, when an alternating voltage is applied, with or without the presence of a direct potential [1-3]. Some of them have assumed a linear current-voltage relationship, when small amplitudes are applied; others have worked with larger amplitudes. It has been established that the total current will have a direct component coming from the faradic rectification effect and a series of harmonic components of the current due to non linear behavior of the electrode [1]. As a first approach, it is intended to model the amplitude of the first harmonic component of the current, $I(\omega t)$, through the equations established by polarographyc theories for non polarizable working electrodes [7], but taking into account that, for this specific case of a reactive and stationary solid electrode, without renewal of the diffusion layer, the mean concentrations could be affected by the concentration profiles throughout the diffusion layer.

2.2 Nano Surface Characterization by AFM

It is interesting to relate the ac-perturbation parameters, to the damages occurring at the metal surface and, consequently, to its structure and corrosiveness. The study will be done at the micro and nano scales, because one specific nanostructure could be the responsible of the corrosion susceptibility of mild steel to ac and dc effects, or probably, the presence of alternating voltages could modify the nucleation during the formation of the new faces. AFM will be applied to study the corrosion morphology at the metal surface due to ac and dc polarizations, in this paper ex-situ and latter on, in the electrochemical cell.

3 EXPERIMENTAL

3.1 Reactions occurring on the mild steel surface, under ac and dc polarizations

The AC-V technique, allows identifying the reactions and their reversibility to ac and to dc polarizations. A stationary working mild steel electrode, a Platinum counter of large area and a SCE reference, with Luggin capillary, were used. An ASTM electrochemical cell contained the aerated and non stirred solution of NaOH 1M, pH 13-14 approx. An Autolab potentiostat was used with the following experimental setup: amplitude of the sinusoidal signal of 10 mV vs. SCE, phase angle of zero grades, to better filter the charging currents, sweep range of dc potential for a first set was between -1.5 and 1.5 V vs. SCE at Frequencies: 25, 50, 75, 100, 125 and 250 Hz, and for a second set (the analyzed by AFM) it was between -2.5 and -0.7 V vs. SCE, at 50 Hz.

3.2 Morphology of corrosion on the mild steel surface, under ac and dc polarizations

The Atomic Force Microscope (AFM) has been applied, ex-situ and in tapping mode, to study the corrosion morphology occurring at the electrode surface, due to the effect of dc and ac polarizations. The study of the surface of the polished mild steel sample was performed before and after 10 reversible cycles of AC-V, which applied a sinusoidal voltage of 10 mV vs. SCE of amplitude, at 50 Hz of frequency onto a range of direct scanning potential between -2.5 and -0.7 V vs. SCE.

4 RESULTS AND DISCUSION

Figure 1 shows the forward ac-Voltagramms of a mild steel working electrode perturbed with an ac-sinusoidal voltage of 10 mV of amplitude, at six different frequencies.

Four peaks can be observed in Figure 1 in the forward direction of the scan. These peaks correspond to the reactions proposed by previous researchers [6] as follows: I and II are associated to the reduction of some adsorbed hydrogen and to the reaction of Fe^0 and Fe^{+2} forming FeO. Peak III, quasi imperceptible to other techniques, could be the reaction of Fe^0 and Fe^{+2} forming $Fe(OH)_2$. Peak IV corresponds to the formation of magnetite, increasing with cycles and frequencies.

What still remains under verification is the presence of other two peaks (V and VI) that, according to [6] would correspond to the formation of two Fe^{+3} species obtained from the oxidation of magnetite: $3\gamma-Fe2O3$ and $3\alpha-FeOOH$, the last one supposed to be responsible for the corrosion of the metal under certain conditions.

The next test, to be presented in another paper, will be the 2nd harmonic Voltammetry, to verify the presence of those peaks, and to calculate the corrosion rate.

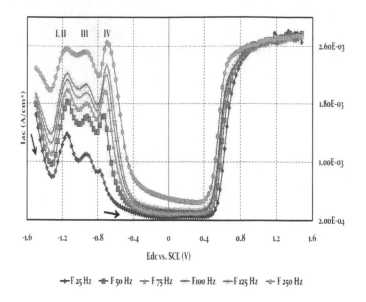

Figure 1: AC-V Dependence of Iac on Edc and Frequency

Figure 2 shows the zoom of experimental forward and backward ac-Voltagramms, for 10mV of sinusoidal amplitude, at six frequencies. The hysteresis observed indicate that dc reversibility does not hold; because the currents tend to depend on the sense of the direct sweep potential. Four peaks can be identified in the forward sense while only two in the backward. There are the crossover potentials where both forward and backward scans yield the same response, being those points the better to calculate transfer coefficients, rate constants and other kinetic parameters needed for the simulation of the ac-waves. The system is not reversible to dc neither to ac polarizations.

The Figure 3 shows the simulation of I and II peaks of the 25 Hz ac-Voltagramm, showed in figure 1.

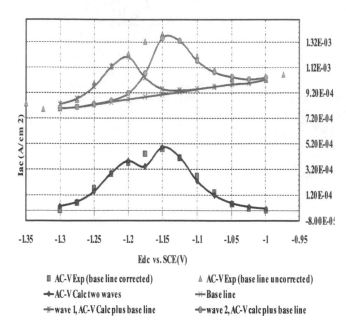

Figure 3: Simulation of peaks I and II of the AC-V at 25 Hz

The upper part of Figure 3 shows one experimental peak composed by two simulated small ones. Those peaks correspond to reactions of the cathodic protection system. At the bottom, the sum of the two simulated peaks is shown.

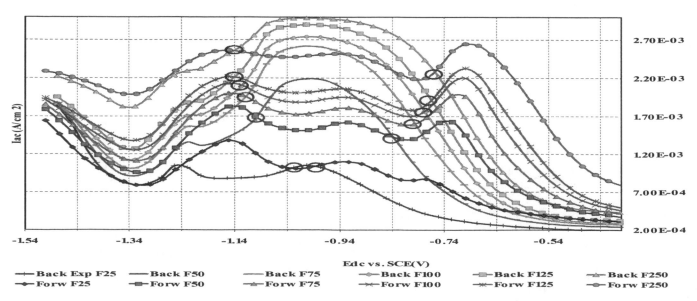

Figure 2: AC-V Dependence of Iac on Edc and Frequency

The analysis by AFM, ex-situ and in tapping mode, of the surface oxides film of the mild steel, after ten cycles of AC-V in the NaOH 1M solution, are shown in Figures 4 and 5. The consecutive cycles of dc plus ac polarization are responsible for the corrosion of the sample in one environment where, in the absence of ac perturbation, it would not occur. Pictures shown are 20 microns wide. The darker color represents a cavity with a maximum depth of 2 microns.

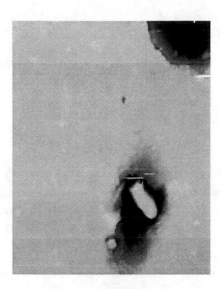

Figure 4: AFM after ten cycles of ACV with 10 mV of amplitude and 50 Hz.

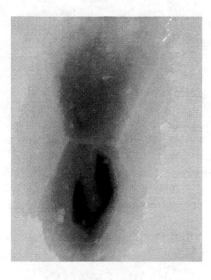

Figure 5: AFM after ten cycles of ACV with 10 mV of amplitude and 50 Hz.

In Figures 4 and 5, the morphology of corrosion can be described as coalescence of pits, some deeper and larger than the others but all like cavities with the corrosion products inside. All found pits were between 5 and 25 microns of surface area and not deeper than 2 microns.

5 CONCLUSION

The AC-V can be useful for the study of redox processes occurring at the surface of a reactive working electrode under small amplitude signals, and for the study of the ac corrosion produced at the surface when bigger amplitudes are potentiostatically applied. The electrochemistry of the double layer has demonstrated that in the NaOH 1M media, and in the range of cathodic protection of mild steel, there are 3 main reactions that are not reversible to dc nor to ac perturbations and it explains the formation of unstable and porous oxides responsible for the ac-corrosion. The dc processes are not nernstian, and potentials reached depend on the sweep sense. An increase in the charge transfer resistance, due to ac perturbations, is observed, that modifies the faradic impedance and capacitative properties of the system. The ac corrosion has been found in form of pits up to 2 microns deep and from 5 to 25 microns long, produced by the application of cycles of a sinusoidal voltage on the direct potential in the basic environment where normally, the mild steel should be cathodically protected in the absence of ac signals. In the future, this work will be complemented with the AFM in the electrolytic cell and with other techniques to identify the components of the oxides film and to characterize the crystalline structure of the metal in the studied conditions.

REFERENCES

[1] U. Bertocci, "AC induced corrosion. The effect of an alternating voltage on electrodes under charge transfer control", CORROSION NACE, 415-4215, Vol. 35, No. 5, May 1979.
[2] J. Devay and L. Meszaros. Study of the rate of corrosion of metals by a Faradaic distortion method, I. Acta Chimica Academiae Scientarium Hungaricae, Tomus 100 (1/4),183-2002, 1979.
[3] H.P. Argawal, Modern aspects of electrochemistry, Plenum, New York, Vol. 20, 1989, ed. J. O'M. Bockris, R.E. White, and B.E. Conway,177-263.
[4] R. W. Bosh, J. Hubrecht, W.F. Bogaerts and B. C. Syret, Electrochemical Frequency Modulation: A New Electrochemical Technique for Online Corrosion Monitoring, CORROSION, pp. 60-70, Jannuary 2001
[5] M. Buchler, P. Schmuki, H. Bohni. J. Electrochim Soc, 145-609, 1998.
[6] S. Joiret, M. Keddam, X.R. Novoa, M.C. Perez, C. Rangel, H. Takenoutti. Use of EIS, ring-disk electrode, EQCM and Raman spectroscopy to study the film of oxides formed on iron in 1M NaOH. Cement & Concrete Composites 24, 7-15, 2002.
[7] A.J. Bard, L.R. Faulkner. Electrochemical methods, Ch. 09, John Wiley & Sons, New York, 1980.

Increased Piezoresistive Effect in Crystalline and Polycrystalline Si Nanowires

K. Reck, J. Richter, O. Hansen and E.V. Thomsen

Department of Micro- and Nanotechnology, DTU Nanotech
Technical University of Denmark, Kgs. Lyngby, Denmark, kasper.reck@mic.dtu.dk

ABSTRACT

Recently, a giant piezoresistance effect has been observed in bottom-up fabricated silicon nanowires [1]. Here we present the results of piezoresistive measurements on boron doped crystalline silicon nanowires at two different doping levels, as well as polycrystalline boron doped silicon nanowires, fabricated using a traditional top-down approach compatible with industrial mass-production techniques. In crystalline silicon nanowires, at a doping level of 10^{17} cm^{-3}, we find an increase in the piezoresistive effect of up to 633% that of bulk silicon. While at a doping level of 10^{20} cm^{-3}, we find a constant or decreasing piezoresistive response, relative to bulk measurements. In polysilicon nanowires we find an increase in the piezoresistive response of up to 39% that of bulk polysilicon. Finally, an increase in the temperature sensitivity is observed in low doped crystalline silicon nanowires.

Keywords: piezoresistivity, silicon nanowires, top-down

1 INTRODUCTION

The piezoresistive effect is much larger in silicon than in most other materials, and silicon piezoresistors have therefore been used for force sensing in several MEMS devices in the past three decades [2, 3]. While there has been a large interest in the temperature and doping dependence of the piezoresistive effect, the dimensional dependency has not been investigated until recently. In self-assembled silicon nanowires, a piezocoefficient of up to $\pi_{[111]} = 3550 \cdot 10^{-11}$ Pa^{-1} has been measured [1]. A large number of micro and nano sensors can thus be greatly improved if such highly piezoresistive silicon nanowires can be integrated with conventional mass-produced sensor designs. Self-assembly of silicon nanowires is still a relatively new fabrication method, and is not yet fully compatible with standard top-down micro- and nanoscale fabrication technologies. Furthermore, self-assembly has several disadvantages including the tendency to prefer growth along specific crystal directions, as well as the difficulty of positioning the catalyst particle and creating electrical contact to the nanowire, e.g. for four-point measurements. While top-down fabrication of silicon nanowires can not be achieved using standard MEMS fabrication methods, such as UV lithography (UVL), one can easily reach the sub-100 nm range using technologies such as e-beam lithography (EBL), nano imprint lithography (NIL) and oxidation thinning. Top-down fabricated highly piezoresistive silicon nanowires thus show great promise of a new generation of highly sensitive piezoresistive sensors.

2 DESIGN

In order to measure the piezoresistive effect in silicon nanowires, a test chip has been designed. The chip is 4 cm long, 5.3 mm wide and approximately 350 μm thick. Each test chip carries 6 dielectrically isolated p-type piezoresistors, as seen in Figure 1. The piezoresistors are located at the center of the chip and oriented along the chip, i.e. in the [110]-direction. Using a four-point bending (4PB) fixture for bending the chip will thus cause a uniaxial stress in the piezoresistors along the length direction. The 6 piezoresistors includes 50, 100, 150, 250, 350 nm wide nanowires as well as a 25 μm wide bulk type reference piezoresistor. All piezoresistors have a width:length ratio of 1:20. This design gives a single chip analysis of the dimension dependence of the piezoresistive effect. The contact surface between the nanowires and the electrical contacts will usually be very small, and contact resistances can therefore be large compared to the resistance of the nanowires. To minimize voltage drops due to contact resistances each piezoresistor is fabricated with integrated four-point probes, see Figure 2. It is known from bulk measurements, that the largest piezocoefficient in p-type silicon is π_{44}. If an uniaxial stress is applied in the [110]-direction, the effective piezocoefficient of the piezoresistors is

$$\pi_{\text{eff}} = \frac{1}{2}(\pi_{11} + \pi_{12} + \pi_{44}) \approx \frac{1}{2}\pi_{44} \qquad (1)$$

3 FABRICATION

The main fabrication steps are illustrated in Figure 3. Crystalline (100)-oriented SOI wafers with a device layer of 340 nm are implanted with boron to a final concentration of either 10^{17} cm^{-3} or 10^{20} cm^{-3}. The polycrystalline device layers are deposited using LPCVD at 620 °C and in-situ doped. Macroscale structures, such

NSTI-Nanotech 2008, www.nsti.org, ISBN 978-1-4200-8503-7 Vol. 1

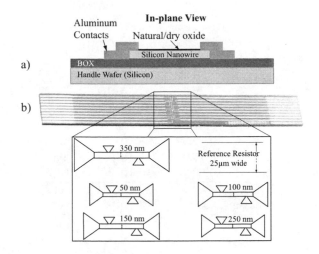

Figure 1: (a) The piezoresistors are lying on top of the buried oxide and thus dielectrically isolated. (b) The test chip consists of 6 piezoresistors of which five are nanowires of different dimensions and one is a 25 μm wide reference resistor. The length to width ratio of the piezoresistors is 20. The illustration is not to scale.

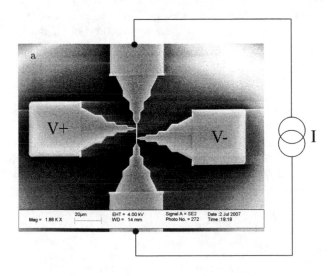

Figure 2: SEM image of a silicon nanowire with integrated four-point probe lying on SiO₂. A constant current is applied using the upper and lower contacts, and the resistance of the nanowire is then found by measuring the resulting voltage drop.

as contacts and test structures are made by standard UVL and transferred to a gold mask by liftoff. Nanowires and four-point probes are made by EBL (JEOL JBX9300FS) in the positive resist ZEP520A and likewise transferred to the gold mask. Using a reactive ion etch (RIE) the structures are defined in the silicon device layer. The gold mask is removed using potassium iodide, and aluminum wires for electrical contact are made by e-beam evaporation.

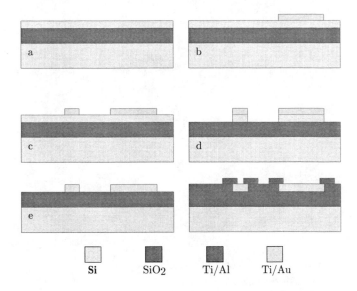

Figure 3: (a) A SOI wafer with crystalline or polycrystalline device layer is implanted with boron. (b) Macroscopic structures are made in 60 nm thick gold by UVL and liftoff. (c) Nanoscopic structures are made in 60 nm thick gold by EBL and liftoff. (d) The gold structures are used as masking in a RIE, thereby defining the structures in the device layer. (e) The gold mask is removed in potassium iodide. (f) Electrical contacts to the piezoresistors are made by e-beam evaporation of aluminum.

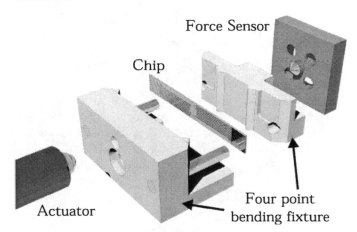

Figure 4: A four-point bending fixture is used to apply an uniaxial stress in the center of the test chip, where the piezoresistors are located. The two parts of the fixture are forced together using a microstep actuator, and the resulting force is measured using a force sensor.

4 EXPERIMENTAL SETUP

The stress dependence of the resistance in the piezoresistors, i.e. the piezoresistive effect, is measured using a four-point bending (4PB) fixture, see Figure 4. The test chip is placed between the two pair of blades, which are then forced together using a microstep actuator. The

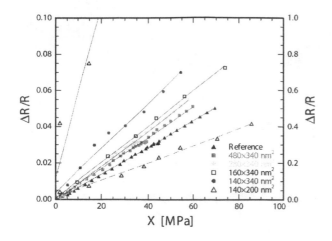

Figure 5: The relative change in resistance for five different dimensioned crystalline silicon nanowires as function of compressive stress. All data except the one fitted by the dashed line are plotted on the left hand side axis.

resulting force on the test chip is measured using a conventional force sensor. Assuming the test chips behaves as a beam subjected to pure bending, the stress in the test chip at the location of the piezoresistors, X, is calculated from the measured force, F, as

$$X = -\frac{6 \cdot F \cdot a \cdot z}{w \cdot t^3}, \tag{2}$$

where a is the distance between the inner and outer blades of the 4PB fixture, z is the distance from the surface of the test chip to the neutral plane (i.e. half the thickness of the test chip), w is the width of the test chip and t is the thickness of the test chip.

The 4PB fixture is situated in an aluminum box with built-in resistive heating and Peltier cooling. Current is applied to the piezoresistors using a HP4145A parameter analyzer, and the resulting voltage drop is measured using a Keithley 2182A nanovoltmeter. Electrical contact to the test chip is obtained using zero insertion force flat flexible cable (FFC) connectors.

5   Results

The relative change in resistance as function of compressive stress for crystalline silicon nanowires at a boron concentration of 10^{17} cm^{-3} is seen in Figure 5. It is observed that the relative change increases as dimensions are reduced. Using the approximation in Equation 1, the π_{44} coefficient for these nanowires has been listed in Table 1. Considering the relatively large dimensions of the nanowires, the increased piezoresistive effect can not be contributed to quantum effects. However, the increase in the number of surface states compared to the number of carriers as dimensions are decreased, might explain the observed increase piezoresistive effect.

Table 1: Approximations of π_{44} for different sized crystalline silicon nanowires at a boron concentration of 10^{17} cm^{-3}.

Dimensions	ρ [$\Omega\cdot$cm]	π_{44} [Pa^{-1}]	$\Delta\pi_{44}/\pi_{44,ref}$
Reference	0.17	$124\cdot10^{-11}$	-
480×340 nm^2	0.17	$140\cdot10^{-11}$	13%
280×340 nm^2	0.17	$165\cdot10^{-11}$	33%
160×340 nm^2	0.12	$198\cdot10^{-11}$	60%
140×340 nm^2	0.17	$245\cdot10^{-11}$	98%
140×200 nm^2	0.09	$910\cdot10^{-11}$	633%
Smith [2]	7.8	$138\cdot10^{-11}$	11%
Tufte et al. [3]	0.02	$113\cdot10^{-11}$	-9%

The relative change in π_{44} for crystalline silicon nanowires at a boron concentration of 10^{20} cm^{-3} as function of nanowire width is seen in Figure 6. At widths down to 150 nm, the piezoresistive effect remains approximately constant. At widths below 150 nm the piezoresistive effect decreases rapidly approaching zero at 50 nm. This rapid decrease can be contributed to an increase in the resistance of the nanowires due to surface scattering. Assuming the mean free path is generally smaller than the dimensions of the nanowire, the conduction through the nanowire will be dominated by bulk-like conductance. At the surface regions, however, the conductance is reduced due to an extra surface scattering term. Due to the lowered conductance, the change in resistance due to the piezoresistive effect is vanishing in the surface regions. When the width of the nanowire is comparable to the width of the surface scattering dominated regions, the piezoresistive response will decrease rapidly, and approach zero as surface scattering becomes dominant. Based on these assumptions the effective piezocoefficient can be found as

$$\pi_{eff} \approx \frac{G_{bulk}}{G_{bulk} + G_{surface}}\pi_{bulk}, \tag{3}$$

where G is the conductance. Using this equation and the theory of Richardson and Nori [4], the obtained data has been fitted as seen in Figure 6. Deviations from the fit are primarily contributed to variations in surface roughness and uncertainties in the nanowire dimensions.

The relative change in resistance for boron doped polycrystalline silicon nanowires under compressive stress is seen in Figure 7. Again the piezoresistive effect increases as dimensions are decreased, with a maximum increase of 39% compared to the reference resistor. Assuming that it is changes in the number of surface states that is responsible for the increased piezoresistive effect, it is not expected that the increase in the piezoresistve effect will be as large in the polysilicon nanowires as in the low doped crystalline silicon nanowires, since the polysilicon nanowires already have a large number of surface states at the grain boundaries.

NSTI-Nanotech 2008, www.nsti.org, ISBN 978-1-4200-8503-7 Vol. 1

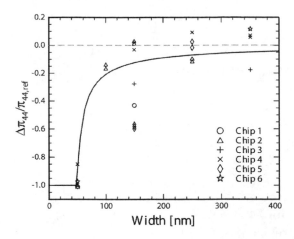

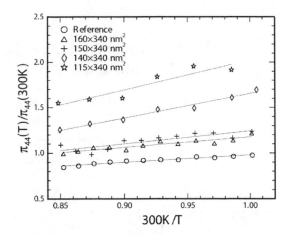

Figure 6: The piezoresistive effect in top-down fabricated high doped crystalline silicon nanowires decreases as the widths are reduced. The thickness of all piezoresistors is 340 nm. The fit is based on the assumption that the decrease is due to surface scattering.

Figure 8: The π_{44} piezocoefficient for low doped crystalline silicon nanowires normalized with respect to the results of Smith [2], as function of inverse temperature.

6 Conclusion

Crystalline and polycrystalline p-type silicon nanowires have been fabricated using a top-down approach. In low doped crystalline silicon nanowires an increased of up to 633% that of the bulk reference resistor was found. This increase is contributed to an increase in the surface states to carrier ratio. In highly doped crystalline silicon nanowires the piezoresistive effect is approximately constant down to widths of 150 nm, whereafter it decreases rapidly due to surface scattering. In polysilicon nanowires an increase of up to 39% were found as dimensions were decreased. The temperature sensitivity of the piezoresistive effect has been found to increase in low doped crystalline silicon nanowires as dimensions are reduced.

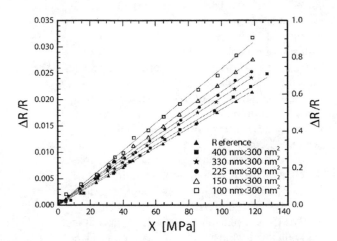

Figure 7: As for low doped crystalline silicon nanowires, an increase in the piezoresistive effect is observed in polysilicon nanowires, as dimensions are decreased. Compared to the reference resistor an increase of up to 39% is measured.

Measurements of the temperature dependence of the piezoresistive effect in the low doped crystalline silicon nanowires are shown in Figure 8. As expected from theory and bulk measurements [5], the piezoresistive effect shows a $1/T$ dependence. However, it is also noted that the temperature sensitivity of the piezoresistive effect increases as dimensions are decreased. Comparing the smallest nanowire with the reference, one finds an approximately 4 times larger temperature sensitivity. It is thus critical that piezoresistive silicon nanowire sensors incorporate temperature compensation.

References

[1] R. He and P. Yang. Giant piezoresistance effect in silicon nanowires. *Nature nanotechnology*, 1(1):42–46, October 2006.

[2] C. S. Smith. Piezoresistance effect in silicon and germanium. *Physical Review*, 94(1):42–49, April 1954.

[3] O. N. Tufte and E. L. Stelzer. Piezoresistive properties of heavily doped n-type silicon. *Physical review*, 133(6A):A1705–A1716, March 1963.

[4] R. A. Richardson and F. Nori. Transport and boundary scattering in confined geometries: Analytical results. *Physical Review B*, 48(20):15209–15217, November 1993.

[5] J. Richter, J. Pedersen, M. Brandbyge, E. V. Thomsen, and O. Hansen. Piezoresistance in p-type silicon revisted. Not yet published, 2008.

Synthesis and Magnetic Characterizations of Manganite-Based Composite Nanoparticles for Biomedical Applications

R. Bah, K.Zhang, T. Holloway, R. B. Konda, R. Mundle, H. Mustafa,
R. R. Rakhimov, and A. K. Pradhan
Center for Materials Research, Norfolk State University, 700 Park Avenue,
Norfolk, Virginia 23504, USA

Xiaohui Wei and D. J. Sellmyer
Department of Physics and Astronomy and Center for Materials Research and Analysis,
University of Nebraska, Lincoln, Nebraska 68588-0113, USA

ABSTRACT

Chemically synthesized highly crystalline lanthanum strontium manganite $LaSrMnO_3$ and Eu-doped Y_2O_3 and their composites yielding nanoparticles of size 30–40 nm are reported in this paper. Magnetic measurements performed on the synthesized nanoparticles and composites showed a magnetic transition at about 370 K with a superparamagnetic behavior at room temperature. The ferromagnetic resonance studies of the nanoparticles showed large linewidth due to surface strains. The composite nanoparticles also displayed luminescent behavior when irradiated with ultraviolet light. The manganites as well as their composite with the luminescent nanoparticles may be very useful for biomedical applications. For this possible application, the nature and effects of the nanoparticles in a biological environment have been studied as such particles are known to be toxic.

Keywords: Magnetic nanoparticles, Toxicity studies, Biomedical applications

I. INTRODUCTION

Magnetic nanoparticles with a Curie temperature above room temperature are needed for most biomedical and magnetofluidic applications. Rare-earth based group of half-metallic ferromagnetic materials, such as manganites with a typical composition $La_{0.7}Sr_{0.3}MnO_3$ (LSMO), are of interest in this context due to its high T_C of 380 K and a large magnetic moment at room temperature.[1–5] The half-metallic manganites are fairly metallic and can have large microwave absorption with the possibility of its use in hyperthermia applications and the large moment can also allow its use in marker experiments in biodetection[6] as well. On the other hand, due to significant magnetoresistance effects in the immediate vicinity of the Curie point of manganites, eddy currents can be utilized as part of the heating mechanism. Hence, one can achieve the selective warming of the given areas of an organism by T_c-limiting production of controlled heat effects by means of an alternating external magnetic field. In general, particles that switch certain inherent properties "on/off" in relation to the relatively simple parameter, such as temperature, will provide possibilities for many potential designs of biomedical applications.

Another interesting material is Eu^{3+} ions, which show luminescent properties due to $^5D_0 \rightarrow ^7F_2$ transitions within and emits red light with a wavelength of 611 nm, and can be used as the red phosphor. Y_2O_3:Eu^{3+} has a lumen equivalent brightness of 70% relative to 611 nm light and radiant efficiency of about 8.7% with better saturation without any detrimental effects.[7,8] With the rapid advancement of biotechnology, nanoparticles of doped lanthanide oxides can be used as promising labels because of their above mentioned optical properties, lack of photobleaching, and long luminescence lifetime (~1 ms).

Coating the nanoparticles with a suitable material offers the possibility of attaching them to antibodies, proteins, medical drugs, etc. Therefore, studies on surface adsorption, the possibility of functionalizing and/or conjugating the particle coating with bioactive components, is also a crucial issue. There are only a few reports available for the synthesis of stochiometric manganite nanoparticles.[9,10] Here, we report the synthesis and characterization of macromolecule encapsulated LSMO and Eu:Y₂O₃ composite nanoparticles.

2. EXPERIMENTS

La$_{0.7}$Sr$_{0.3}$MnO$_3$ (hence called as LSMO) nanoparticles were synthesized by a sol-gel method from their acetate hydrate precursors, which were dissolved in water. This solution was mixed with citric acid solution in 1:1 volume ratio ultrasonically for about 30 min. The mixture was heated in a water bath at 80 °C until all water is evaporated, yielding a yellowish transparent gel. The gel was further heated in an oven at 100 °C which formed a foamy precursor. This precursor decomposed to give black-colored flakes of extremely fine particle size on further heating at 400 °C for 4 h. The flakes were ground and sintered at 800 °C for duration of 2 h. Further heating in O₂ ambient removed the carbon content. The nanopowders were coated with equal quantity ofoctadecyl amine (ODA) by magnetically stirring at 120 °C.

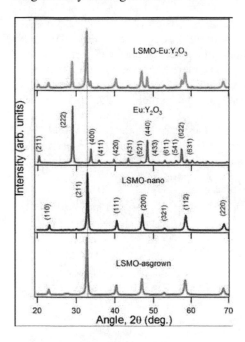

FIG. 1. X-ray diffraction patterns of LSMO, EYO, and composite nanopowders.

A solution of ODA-coated LSMO and chloroform was prepared with a molarity of 10^{-2} M and sonicated for 1 h for dispersion. The solution was rested for 24 h settling of the heavier LSMO particles. The solution containing suspended LSMO-ODA nanoparticles was decanted and purified using methanol several times in order to remove excess ODA from the surface. The final purification was done using magnetic separation in order to remove any carbon content in the solution. The purified powder was naturally dried.

The nanocrystalline Eu $^{3+}$:Y₂O₃ (EYO) powders were synthesized using a combustion technique[11,12] from their respective nitrate solutions. LSMO and EYO nanopowders were thoroughly mixed and heated at 800 °C for 30 min to form a composite. These powders were coated with ODA and purified as described above.

3. RESULTS AND DISCUSSION

Figure 1 shows the powder x-ray diffraction (XRD) patterns of nanoparticle ensembles of manganites, doped rareearth oxides, and their composites. The purified manganite nanoparticles are highly crystalline and share the pseudocubic perovskite structure. However, the broadening of the XRD peaks is influenced by surface strain. Similar effects due to surface strains were observed in EYO nanoparticles. The XRD patterns of composite nanoparticles exhibit the characteristics of both LSMO and EYO nanoparticles.

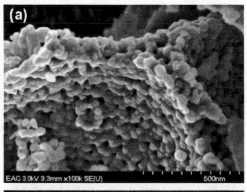

FIG. (2.a) FE-SEM image of LSMO naoparticles with and without ODA coating.

Figures (2a) and (2b) show the representative field emission scanning electron microscope (FE-SEM) images of LSMO nanoparticles with and without ODA coating. The size distribution is rather narrow, and the crystallite size is in the range of 30–40 nm. Analysis of electron diffraction from nanoparticles and XRD patterns, such as those shown in Fig. 1, indicates that both manganite and EYO nanoparticles are highly crystalline and share the pseudocubic perovskite and cubic structures, respectively.

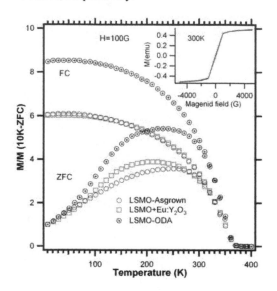

FIG. 3. Temperature dependent field-cooled (FC) and ZFC magnetization of as-synthesized LSMO, ODA-coated LSMO and composite (LSMO+EYO) nanopowders. The inset shows the magnetic hysteresis of LSMO nanoparticles.

Figure 3 shows the temperature (T) dependent magnetization (M) of LSMO and composite nanoparticles in an applied magnetic field (H) of 100 G. All samples show the onset of the magnetic transition at 370 K with a Curie temperature ~360 K. The as grown LSMO and LSMO+EYO composites show very similar magnetization behavior. The samples were cooled from 400 K down to 10 K under a magnetic field and then M was recorded as the sample was warmed to 400 K under the same magnetic field [field cooled (FC)]. Similarly, the samples were cooled from 400 K down to 10 K with zero magnetic field and then M was recorded as the sample was warmed to 400 K under the same magnetic field applied at 10K [zero-field cooled (ZFC)]. The results clearly illustrate the occurrence of

ferromagnetic-to-paramagnetic phase transitions in all three samples as the temperature is increased. On the other hand, the coated nanoparticles showed reduced magnetization compared to LSMO due to reduction in the volume fraction caused by ODA coating, as expected in both FC and ZFC curves. The magnitude of the low temperature saturation magnetization also decreases with ODA coating, although hardly any reduction in the transition temperature was observed. Interestingly, the absolute magnetization increases in composite nanoparticles compared to that of LSMO. One of the reasons for such increase may be due to the local field enhancement at Eu site. However, further studies are necessary to resolve this issue. The slight reduction in transition temperature in these nanoparticles compared to their bulk counterpart (~375 K) may be attributed to the surface-to-volume ratio in the nanoparticles, spin disorders at the surface,[13] as well as due to the presence of superparamagnetic behavior of the nanoparticles. For nanoparticles of the order of tens of nanometers or less, one can see superparamagnetism, where the magnetic moment of the particle as a whole is free to fluctuate in response to thermal energy, while the individual atomic moments maintain their ordered state relative to each other, yielding anhysteretic, but sigmoidal M-H curve. In fact, the superparamagnetic behavior is observed in these nanoparticles, and a magnetic field dependent magnetization curve is presented in the inset of Fig. 3 for LSMO nanoparticles.

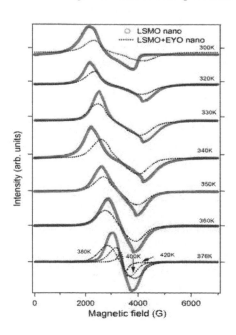

NSTI-Nanotech 2008, www.nsti.org, ISBN 978-1-4200-8503-7 Vol. 1

FIG. 4. Ferromagnetic resonance spectra of LSMO and composite (LSMO+EYO) nanopowders.

Figure 4 shows the ferromagnetic resonance (FMR) spectra of LSMO and composite nanoparticles at various temperatures. FMR studies are probably the most sensitive method for detecting ferromagnetic order as well as the possible existence of other magnetic species. The narrower linewidth is generally taken as a signature of a homogeneous sample. However, a large FMR linewidth, which is in the range of ~2 kG, indicates the effects of surface strains due to the size effect of the nanoparticles. No significant difference in the FMR curves was noticed, except a slight increase in resonance field in EYO nanoparticles. However, a striking difference is the shape of the FMR resonance curve, which loses the ferromagnetic feature as the temperature is increased beyond 350K. This can be attributed due the interface effects in the composite nanoparticles.

4. CONCLUSION

We synthesized LSMO and macromolecule encapsulated LSMO and Eu:Y_2O_3 composite nanoparticles by the chemical routes. Both nanopowders show very good crystalline quality. The composite nanopowders of LSMO and Eu:Y_2O_3 were synthesized by mixing and controlled heat treatment. The FE-SEM and TEM images of the individual nanoparticles and composite nanopowders show that the particles are in the range of 30–40 nm. The LSMO as grown, ODA-coated, and composite nanoparticles show magnetic transition around 370 K. The composite nanoparticles show superparamagnetic behavior at 300 K. The FMR studies of the nanoparticles show large linewidth due to the surface strains. Our results are useful for possible applications in the field of biomedicals.

5. ACKNOWLEDGMENTS

This work is supported by NSF (RISE) project. Research at the University of Nebraska is supported by NSF-MRSEC, ONR, and CMRA.

6. REFERENCES

1. S. S. Davis, Trends Biotechnol. **15**, 217, 1997.
2. H. Y. Hwang, S. W. Cheong, N. P. Ong, and B. Batlogg, Phys. Rev. Lett. **77**, 2041 (1996).
3. A. Chainani, M. Mathew, and D. D. Sarma, Phys. Rev. B **47**, 15397 (1993).
4. Z. Trajanovic, C. Kwon, M. C. Robson, K. C. Kim, M. Rajaswari, S. E. Lofland, S. M. Bhagat, and D. Fork, Appl. Phys. Lett. **69**, 1007, 1996.
5. J. M. De Teresa, C. Marquina, D. Serrate, R. Fernandez-Pacheco, L. Morellon, P. A. Algarabel, and M. R. Ibarra, Int. J. Nanotechnol. **2**, 3 (2005).
6. A. Pankhurst, J. Connolly, S. K. Jones, and J. Dobson, J. Phys. D **36**, R167 (2003).
7. G. Blasse and B. C. Grabmaier, *Luminescent Materials* !Springer, Berlin, (1994).
8. T. Hase, T. Kano, E. Nakazawa, and H. Yamamoto, in *Advances in Electronics and Electron Physics*, edited by P. W. Hawkes Academic, NewYork, 1990, Vol. 79, p. 135.
9. J. J. Urban, L. Ouyang, M.-H. Jo, D. S. Wang, and H. Park, Nano Lett. **4**, 1547, 2004.
10. V. Uskokovic, A. Kocak, M. Drofenik, and M. Drofenik, Int. J. Appl. Ceram. Technol. **3**, 134, 2006; R. Rajagopal, J. Mona, S. N. Kale, T. Bala, R. Pasricha, P. Poddar, M. Sastry, B. L. V. Prasad, D. C. Kundaliya, and S. B. Ogale, Appl. Phys. Lett. **89**, 023107, 2006.
11. E. Zych, Opt. Mater. Amsterdam, Neth. **16**, 445, 2001.
12. K. Zhang, A. K. Pradhan, G. B. Loutts, U. N. Roy, Y. Cui, and A. Burger, J. Opt. Soc. Am. B **21**, 1804, 2004.
13. A. J. Millis, Nature London **392**, 147,1998.

Thin NIL films characterization (viscosity, adhesion) with rheological nano–probe

A.A. Svintsov, O.V. Trofimov, S.I. Zaitsev

Institute of Microelectronics Technology, RAS, Chernogolovka, 142432 Russia.

ABSTRACT

The approach based on the nano-indenter developed for viscosity characterization allows temperature measurement of thermoplastic, thermo- and photo-curable polymers (materials used in NILand UV-NIL) with sub-micron resolution. It could be used in design, tailoring and optimisation of soft matter for different application. In particular:

- Recommended imprint temperature can be deduced from nano-probe experiments.
- Optimal curing time as function of temperature can be established.
- Prediction on mechanical properties at imprint temperature easily can made.
- Surface adhesivity as function of *temperature* can be measured

Keywords: viscosity of thin films, surface adhesivity, AFM

1 INTRODUCTION

Polymers of different kinds are widely used in NanoImprint Lithography (NIL). First of all it is a usage of thermoplasts in thermal NIL. In step and flash imprint lithography (UV-NIL) photo-cured polymers are used. Thermocurable polymers could be used in step and flash approach when photo-exposure is changed with short heating.

Demands of high throughput in nanoimprint lithography (NIL) dictate to make glass transition temperature of polymer matter used in NIL rather low therefore an approach for measurement of viscous properties of tailored polymers should be developed. The method should allow measuring the properties at the statc (films of 100-500nm) in which the polymer is intended to be used. On the other hand it would be highly desirable the method would allow characterizing polymer curing and mechanical properties of the *cured* polymer, which planned to use as cost-effective stamps. One more additional usage of the method could be characterization on adhesion properties of polymer surfaces before and after special treatment predicting release properties of stamps as function of temperature.

2 SETUP AND EXPERIMENTAL METHOD

To meet the requirements a special tool called *rheological nano-probe* was designed and fabricated consisting of several components:

a) generic AFM device (see Figure 1),
b) original stage (made of *invar* to decrease thermal expansion) with heater for *local* heating of a sample at place of indentation,
c) original control software to provide complex tip trajectory and flexibility in data acquisition and treatment.

Small heater placed under a wafer provides heating of the wafer and temperature is controlled with a thermocouple. Resist film is on the opposite side of the wafer and special measurement showed the difference of temperature of wafer both sides does not exceed 2^0C. Rate of heating could be changed with current of the heater. Maximal rate of cooling is defined with heat scattering in environment and normally it is relatively high so the cooling rate could be controlled with weak heating current. Heating-cooling cycle is controllable and normally belongs to range 100-500 seconds. Usual temperature range is RT-200^0C.

Temperature rate is sufficiently small to perform indentation because indentation cycle takes less than 0.1 second. During a heating cycle up to ten thousands probing could be made.

Main measurement action is measuring so-called "loading curve" shown in Figure2 for two temperatures (see also Figure9 and Figure10).

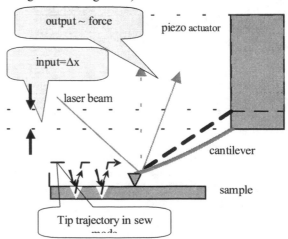

Figure 1: Principal schema of generic AFM and sew-mode of the prob. Elongation (displacement Δx) of an actuator is an input signal, force measured as cantilever bending is output signal resulting to "loading curve".

Principal parameters acquired from "loading curves" are slopes of loading curve (compliance, DZ/DF), residual deformation and viscosity as function of temperature. Specially developed control software realises rather complicated measuring procedures. For example Figure 2 shows a mode when the tip is moving with low velocity then velocity can be increased in controllable manner so resistance to penetration (compliance, viscosity) could be measured at different velocities of indentation.

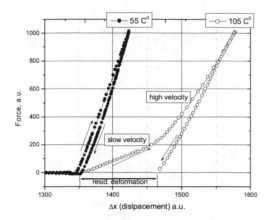

Figure 2 Examples of loading curves as function of temperature. At low temperature (55^0C) the curve shows elastic behaviour (compliance is very small and does not depend on velocity of the tip), at higher temperature compliance becomes higher demonstrating inelastic behaviour (depending on tip velocity).

Concerning compliance and residual thickness acquisition it should be noted that control software not only control tip trajectory but it contains also some postprocessing part which transfers original data considering creep and hysteresys of piezo actuators. This allows measuring real slopes of the loading curves.

3 MODEL AND MEASUREMNT APPROACH

For direct viscosity measurements a special approach was developed based on theoretical analysis.

Considering Navier-Stokes equation penetration of a paraboloid tip (with curvature R) in viscous thin film (of thickness h_0) was considered. Main approximations follow Reynolds approach [1,2] from them it follows that

$$h_0 / R \gg 1 \qquad (1)$$

Details of the theoretical consideration published in [2] and main formula relating viscosity η to measurable quantities is

$$\frac{1}{\eta} \approx \frac{8\pi R^2 (z/h_0)^3}{\int\limits_0^t F(t')dt'} \qquad (2)$$

z is a residual deformation (or penetration depth). The integral over time $\int F(t')dt'$ is simply an area under the loading curve (see Figure2) at its uploading part. The controlling software was modified to allow automatic acquisition of accumulated force value during loading to use formula (2)

To fulfill condition (1) tip with large radius (like 1um) should be used. Therefore a special technological approach was developed. It consists of growing a desirable tip with controllable contamination, in other words decomposition of vacuum oil residuals under intensive e-beam exposure up to pure carbon was used. The exposure was performed with e-beam lithograph under control of NanoMaker [3]. SEM image of the carbon tip is shown in Figure3.

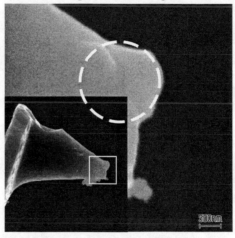

Figure3. Carbon tip of R=700nm grown with controllable contamination

4 EXPERIMENTAL RESULTS
4.1 Thermoplast polymers

Temperature dependence the *compliance* for positive resist (950K PMMA) exposed at different doses is shown as an example in Figure 4. Common feature is constant value of the compliance at low temperatures what corresponds to high viscosity (and here the compliance is equal to inverse stiffness of a cantilever) then the compliance increases what corresponds to viscosity decreasing and to plastic deformation of polymer matter. It is expected that higher exposure results to decrease of molecular weight lower molecular weight what is known results to lower viscosity. It is clearly seen the viscous behavior (and plastic deformation) in accordance to expectations starts at lower temperatures corresponding to lower molecular weight.

These measurements demonstrate that due local character of measurement with setup presented electron exposure dose could be measured and characterized with sub-micron (and even nanometer with finer tip) spatial resolution without any real development.

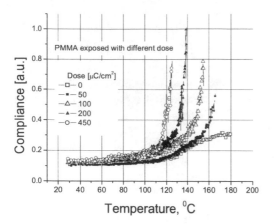

Figure 4 Compliance of 950K PMMA film subjected to different exposures as function of temperature, higher exposures result to more plastic behaviour. In principal distribution of exposure could be measured with nanometer accuracy.

4.2 Thermocurable polymers

The next example illustrates investigation of curing kinetics (Figure 5), thermo-curable polymer (submitted by *micro resist technology GmbH* [4]) was backed at 170^0C for different times.

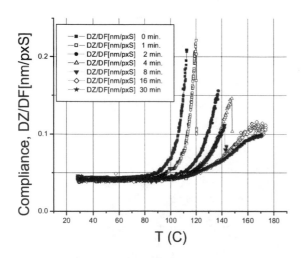

Figure 5 Compliance of a thermocurable resist (from mrt [4]) baked for different times at 170^0C.

Figure 6 shows residual deformation (penetration depth) for backing kinetics. And finally Figure 7 demonstrates viscosity measurements.

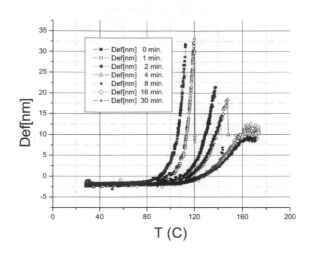

Figure 6 Residual deformation (penetration depth) of a thermocurable resist (from mrt [4]) baked for different times at 170^0C.

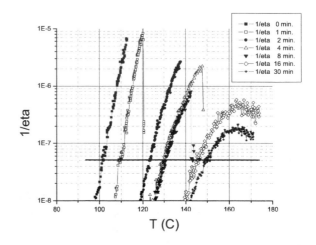

Figure 7 Viscosity of a thermo-curable resist (from mrt [4]) baked for different times at 170^0C..

4.3 Photocurable polymers

As example the developed approach was applied to photo-curable resist (mrJ9030M, N43). The samples were exposed with Hg lamp for different times (0min, 1min, 10min, 30min, 60min). Power of the lamp is 20.6Wt irradiated at range 240-320nm, length of lighting body is 6cm with 19mm of diameter and distance to samples from the lighting body center is 30cm at inclination angle 45°.

Viscosity and residual deformation as function of temperature are shown in Figure8. In spite of long exposure time (1hour) it is seen that curing kinetics is still not accomplished.

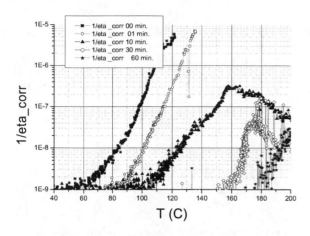

Figure 8 Viscosity of a photocurable resist (from mrt [4]) exposed for different times.

4.4 Comparison of thermo-curable and photo-curable polymers

Comparing curing processes of thermo-curable and photo-curable polymers one can see that they result to quit different results. Accomplished curing can be characterized with temperature shift equal to about 30C. As to photo-curable resist the temperature shit is more than 100C.

This could be explained as if cross-linking of photo-curable polymer results to more dense net with significantly smaller distances between cross-links.

One more conclusion can be done activation energy of viscosity is approximately two times larger for thermo-curable than for photo-curable resists.

4.5 Surface adhesivity

Up to now for investigation some averaged parameters acquired from loading curves were exploited. Example of usefulness of loading curves themselves is illustrated below. One of mrt thermoplastic resist was plasma treated to investigate whether plasma treatment (by Dr. Isabel Obieta Vilallonga from INASMET, Spain) can change and improve surface adhesivity to be used for easier detachment in thermal NIL. Figure 9 and Figure 10 shows loading curves for non-treated and treated samples. Quantitative characterization should include calculation of work done during detaching. But for quantitative judging it is sufficient to pay attention to maximal detaching force. After plasma treatment the maximal force becomes three times less than before. Very important to outline that the approach presented allows measuring surface properties as function of temperature what is not so easy with conventional technique based on measurement of wetting angle. Additionally the approach is very local in comparison to the conventional methods.

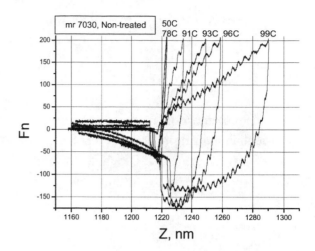

Figure 9 Loading curves as function of temperature before plasma treatment.

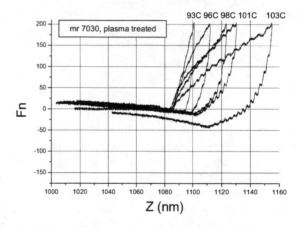

Figure 10 Loading curves as function of temperature after plasma treatment. Surface adhesivity could be easily investigated as function a temperature

ACKNOLEDGEMENT

NaPa partners from mrt (Germany) and Inasmet (Spain are greatly appreciated for collaboration. The partial support of the EC-funded project NaPa (Contract no. NMP4-CT-2003-500120) is gratefully acknowledged. The content of this work is the sole responsibility of the authors.

REFERENCES

[1] L. Landau and E. Lifshitz, Fluid Mechanics (2nd ed., Pergamon, Oxford UK 1987).
[2] A. A. Svintsov, O. V. Trofimov, S. I. Zaitsev, J. Vac. Sci. Technol. B 25(6), Nov/Dec 2007
[3] www.nanomaker.ru
[4] www.microresist.de

Fluorescence tag-based inspection of barrier coatings for organic light emitting diodes and polymer packages

Yadong Zhang[1,2], Yu-Zhong Zhang[5], David C. Miller[1,2], Jacob A. Bertrand[1,3], Ronggui Yang[1,2], Martin L. Dunn[1,2], Steven M. George[1,3,4] and Y. C. Lee[1,2]

[1]DARPA Center for Integrated Micro/Nano-Electromechanical Transducers (iMINT), [2]Department of Mechanical Engineering, [3]Department of Chemistry and Biochemistry, [4]Department of Chemical Engineering and Biochemical Engineering, University of Colorado, Boulder, CO 80309
[5]Invitrogen/Molecular Probes, Inc. Eugene, OR 97402

1. Introduction

Barrier coatings are essential to protect organic light emitting diodes (OLED) and other polymer packaged components from moisture- and oxygen-aided deterioration. One such coating technology is atomic layer deposition (ALD), which can be used to grow a nano-thickness, pin-hole free, and uniform alumina layer on a polymer substrate even at temperatures as low as 33 °C [1]. The water vapor permeability for ALD alumina can be lower than 0.0001 gram/m^2/day, which is 10,000X better than that of a typical polymer [2, 3]. Such improvement may enable flexible polymer packaging for flexible displays, OLED, and other chemically sensitive electronics.

To limit defects and cracks in the barrier coating, which allow the leakage of reactive species, is very important to barrier quality. Gas (oxygen or helium) or water vapor transmission are often used to characterize the barrier coating [4]. The "Mocon" test is commonly used for measuring the oxygen and water vapor transmission rate (WVTR). But, as its sensitivity cannot satisfy barrier transmission specification for the OLED application, the Calcium [5], and HTO tests [6] have been recently used to evaluate ALD barrier coating quality. However, these tests are time-consuming and not easy to implement, and therefore not suitable for rapid quality evaluation expected in a manufacturing environment. It is well known that the measurable permeability is attributed to the defects in the coating, such as pinholes or cracks, which allow the leakage of reactive species [7, 8]. The quality of the barrier coating is directly related to the defect density in the coating. Table1 compares the defect densities between various state-of-the-art barrier coatings. The defect densities are estimated based on the WVTR and oxygen transmission rate (OTR) data using the model described in [7]. For easy comparison, we used the nominal defect size of 0.6 μm as demonstrated in [7]. The ALD barrier coating is of much better quality than the conventional coatings, attributed to the lack of pin-hole defects.

As no barrier coating technology is perfect, coating-related defects have to be inspected and controlled to assure high-yield manufacturing. However, the characteristics of ALD barrier coatings, such as nanoscale thickness, transparency, and smooth surface, make it formidable to inspect the defects directly using common microscopy-enabled methods. In the past, researchers have used O_2 plasma etching or aching to undercut and enlarge defects or cracks [7, 9], which can require extensive sample preparation. In support of fast defect inspection, we have developed novel fluorescent tags which are capable of identifying nanometer-scaled defects. This approach allows rapid quality inspection of ALD alumina barrier coating on polymer substrates used for OLED encapsulation.

The fluorescent tag molecule has a particularly designed lipophilic moiety that facilitates selective binding to defect sites, based on the surface adhesion characteristics of the material system. As shown in Fig. 1, the tag binds solely to the polymer substrate, based on its greater hydrophobicity. The tag molecules bear a fluorescent moiety, allowing for their identification at specific wavelengths. In addition, the tag has been designed to bear a relatively small molecule size, i.e. molecular weight of ~300, allowing the tag molecule to easily enter into nanometer-scale defects.

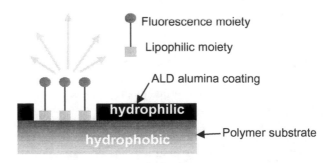

Figure 1. Fluorescent tag labeling mechanism, in which the tag binds solely to the polymer substrate, based on its greater hydrophobicity

2. Mechanical cracks visualization

To demonstrate the tagging application, 25 nm thick ALD alumina barrier films were deposited onto Polyethylene Naphthalate (PEN) substrates. The coated specimens, some of which were mechanically manipulated in order to intentionally generate defects, were then soaked in a fluorescent tag solution for 5 min. Then solvent solution containing 70% ethanol and 30% water was used to wash away the excess tag that was not specifically attached to the film. The sample was then dried using clean dry air and maintained in an ultraviolet-safe environment. The tag can be excited by many commonly used excitation sources, yielding a bright fluorescent signal. A LSM 510 confocal microscope (Carl Zeiss, Inc.) was specifically used for the tag inspection. A 488nm laser source was used to excite the tag and the fluorescent emission (maximum at 515 nm) was measured with a 505-530 nm band pass filter.

We tested a PEN substrate with an ALD alumina coating, and an identically coated PEN substrate bearing intentionally-made scratches, respectively. As shown in Figure2 (a), the fluorescent tag does not attach to the ALD alumina, which consists of an all dark field. For the scratched ALD alumina coating, the tag attached only to the exposed PEN, as shown in Figure2 (b). Figure 2 demonstrates that the tag will selectively attach to the polymer substrate, where it is exposed in the PEN/alumina materials system.

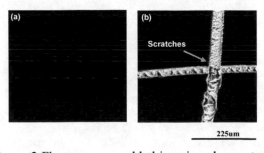

Figure 2 Fluorescence enabled imaging, demonstrating the tag molecules do not adhere to PEN substrate coated with ALD alumina (left); but adhere selectively to scratches introduced to ALD alumina coating (right).

3. Tag molecules-enabled visualization of mechanical "channel-cracks"

The failure mode of "channel cracking" is commonly encountered when a brittle inorganic coating is subjected to mechanical strain or thermal cycling. However, a series of such cracks is not readily observed in transparent films. To demonstrate the use of the fluorescent tag, an external tensile loading was applied to PEN substrates coated with 25 nm of ALD alumina. The fluorescent tag was then applied to these specimens according the previously described procedure. Figure 3 (a) shows cracks identified across the gage section of a specimen that was elongated to 5% strain. Such cracks, which propagated in the direction orthogonal to the applied load, are common when the stress in brittle films exceeds their critical threshold limit.

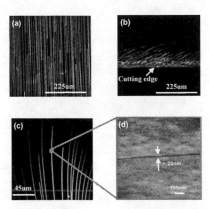

Figure 3 Cracks in the ALD alumina coating rendered visible by the fluorescent tag: (a) series of channel cracks generated at the specimen's interior after a 5% externally applied strain, (b) cracks at edge of specimen resulting from shearing during sample preparation, (c) the fully-developed region of the shear cracks, (d) FESEM image demonstrating the true size of a single shear crack

The cracks in Figure 3 (a) may be distinguished from those at the edges of the specimen, shown in Figure 3 (b), which were generated during the sample preparation. Specifically, these cracks were generated in shear when the specimen was cut to size prior to testing. Figure 3 (b) identifies the unique characteristics of the shear cracks, which quickly arrest near the edge of the specimen. Excellent image contrast was obtained in all of the confocal measurements, allowing cracks to be readily identified despite minimal sample preparation. At the fully formed region of the shear cracks Figure 3(c), the crack opening is observed to be ~20nm using a JSM-7401F field emission scanning electron microscope (FESEM, JEOL Limited), Figure 3(d). In comparison, the standard fracture mechanics solution predicts the crack opening displacement of 28 nm for the alumina coating [10]. Importantly, the true minimum feature size is identified using the FESEM. Disparity in the size (width) of the cracks between the fluorescence and FESEM measurements may be explained according to the point spread distribution of the combined specimen/microscope system. Compared with SEM observation, however, the tagging offers the advantage of continuous observation at low magnification, allowing a large field size to be examined.

4. Visualization of individual defects and particles

In contrast to mechanical cracks, individual defects are generally caused by particulate contamination and/or the substrate surface roughness. In the absence of mechanical cracks, tiny individual defects become the critical features limiting barrier permeation. Fig. 4(b), obtained from the fluorescent tag, shows a defect rich region in 25 nm thick ALD alumina coating deposited on PEN. Prescribed marker features were made on the ALD coating as the white arrows indicated in Fig. 4(b) to facilitate defect location during imaging. Sites #1 and #2, shown in Fig. 4(a) and (c), respectively, were subsequently observed using the FESEM in order to more accurately determine the size of the individual defects. For sites #1 and #2, the diameters of ~200nm and ~1.2um were determined, as indicated in Fig. 4(a) and (c). In addition to site #1, defects smaller than 200nm were also rendered by the tag during continued inspection. This indicates that the minimum detectable defect size for the tag is below 200nm. Fig. 4 also provides information about the morphology of the individual defects. From Fig. 4(a), the oval shaped defect bears a tiny crack at its top end. Concerning the origin of the defect, it could have been generated by a particle contained within on impressed into the specimen. The tag, however, does not necessarily distinguish between contamination occurring before or after coating and the sources of contamination are currently being investigated. Regardless of the origin of the defect, its interaction with alumina coating resulted in crack formation, perhaps based on residual stress or alternately through abrasion. Fig. 4(a) identifies that individual defects serve as the crack initiation sources during the coating processing and subsequent handling. The defect in Fig. 4(c) is explained as a contaminated region, to which alumina was not able to form the chemical bond required for ALD. Information, including the size and morphology of defects, is essential to the improvement of barrier quality. However, defect inspection becomes very cumbersome for transparent and nanoscale-thick coatings using SEM or atomic force microscopy (AFM), because the general defect-locations as well as the defect density cannot be determined at low magnification, whereas examination at high magnification (subject to a small field size) is not very efficient.

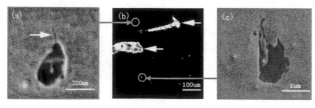

Fig. 4: Individual defects in/on the ALD alumina coating rendered visible by the fluorescent tag: (b) the defect density and location is rendered at low magnification using confocal microscopy, allowing the details of size and morphology to be indentified in high magnification FESEM images, insets (a) and (c).

5. Conclusion and application

A fluorescent tag molecules-enabled visualization technique is developed for rapid inspection and evaluation of the quality of barrier coatings. Excellent selectivity was demonstrated for the material system consisting of a PEN substrate coated with ALD alumina. The tag readily identified channel cracks ~20 nanometers in width and individual defects as small as 200 nanometers in diameter. The excellent image contrast for the taggant advantageously allows the defect location and density to be determined at low magnification, whereupon additional details can be determined at high magnification. This novel method can be applied to barrier quality inspection in research and manufacturing of OLED, polymer packages and other organic components or systems.

Acknowledgement

The authors are grateful to their colleague Dr. Virginia Yong for valuable contribution on FESEM. The studies conducted by the authors from the University of Colorado -Boulder are supported by the DARPA Center on Nanoscale Science and Technology for Integrated Micro / Nano-Electromechanical Transducers (iMINT) funded by DARPA N/MEMS S&T Fundamentals Program(HR0011-06-1-0048).

References

[1] J.W. Elam, and S.M. George, "Low Temperature Al2O3 Atomic Layer Deposition", Chem. Mat., 16 (4), pp. 635-645, 2004

[2] M. D. Groner, S. M. George, R. S. McLean, and P. F. Carcia, "Gas diffusion barriers on polymers using Al2O3 atomic layer deposition", Appl. Phys. Lett. 88, 051907, 2006.

[3]P. F. Carcia, R. S. McLean, M. H. Reilly, M. D. Groner, and S. M. George, "Ca test of Al2O3 gas diffusion barriers grown by atomic layer deposition on polymers", App. Phys. Letters 89, 031915 , 2006.

[4] G. Crawford, Flexible Flat Panel Displays-Chapter 4 (Wiley - SID Series in Display Technology, 2001).

[5] G. Nisato, P. Bouten, P. J. Slikkerveer, W. Bennett, G. Graff, N. Rutherford, and L. Wiese, "Evaluating high performance diffusion barriers: The calcium test", 21st International Asia Display/8th International Display Workshop, Nagoya, Japan, 2001 (Society for Information Display, San Jose, CA, 2001) pp. 1465.

[6] R.Dunkel, R. Bujas, A.Klein, V. Horndt, "Method of measuring ultralow water vapor permeation for OLED displays", Proceedings of the IEEE, Vol. 93, pp 1478-82, 2005.

[7] A. S. da Silva Sobrinho, G. Czeremuszkin, M. Latreche, and M. R. Wertheimer, "Defect-permeation correlation for ultrathin transparent barrier coatings on polymers", J.Vac. Sci. Technol. A, Vol. 18, pp.149–157, 2000.

[8] G. L. Graff and R. E. Williford, "Mechanisms of vapor permeation through multilayer barrier films: Lag time versus equilibrium permeation", J. of Appl. Phys. V96, No. 4 15 pp1840-1849, Aug. 2004.

[9] Sonia Grego, Jay Lewis, Erik Vick, Dorota Temple, "A method to evaluate mechanical performance of thin transparent films for flexible displays", Thin Solid Films, Vol. 515, pp.4745-4752, 2007.

[10] J.L. Beuth Jr., "Cracking of thin bonded films in residual tension", International Journal of. Solids and Structures, Vol. 29, pp.1657-75, 1992.

Nanomechanical characterization of UV curable hyperbranched polymers

J. Kim, Z. Chen, B. Chisholm and S. Patel

Center for Nanoscale Science and Engineering, North Dakota State University, 1805 NDSU Research Park Drive, Fargo, ND 58102 USA

ABSTRACT

Hyperbranched polymers were explored to develop UV-curable, protective coatings for aluminum 2024-T3 and their viscoelastic and mechanical properties were characterized using nanoindentation techniques. The coating system consists of a hyperbranched polyester acrylate (CN2300, Sartomer Inc), a modified hyperbranched polyols, and a photoinitiator (Irgacure 2022). Two different hyperbranched polyols (DPP130 and P1000, Perstorp Polyols Inc.) were partially modified with acryloyl chloride to give acrylate functionality for UV curing. The quasi-static indentation results showed that the reduced modulus and hardness of the UV-cured coatings were significantly affected by the type and amount of the modified polyols.

Keywords: hyperbranched polymers, nanoindentation, UV curable

1 INTRODUCTION

Hyperbranched polymers (HBPs) possess unique physical and chemical properties such as low viscosity and high density of terminal reactive groups compared to their linear analogs [1,2]. The rheological characteristics and high functionalities also make HBPs useful in UV-curable coating applications, showing fast cure, low shrinkage, formulation flexibility, and scratch resistance [3,4]. Even though the blends mixed with HBPs were extensively investigated, few studies on mechanical characterization of the UV curable HBPs are found [4,5]. Klang [4] and Schmidt et al [5] measured Young's modulus of HBPs using a tensile tester. However, the UV curable HBPs result in hard, brittle, and thin films, so that instrumented nanoindentation technique provides a good tool to characterize their mechanical properties. The indentation normally is operated in quasi-static mode, controlled by a feedback loop (load or displacement control). Reduced modulus and hardness are calculated using the contact area and the slope (stiffness) of the unloading segment of the resulting force-penetration depth curve [6]. The quasi-static indenting was conducted on the UV-cured thin films (~100 µm) in a displacement feedback control. The films were prepared from a hyperbranched polyester acrylate, a polyol partially modified with acryloyl chloride, and a photoinitiator, which was initially designed to develop UV-curable, corrosion-resistant coatings. The modified polyol was added to the polyester acrylate in an effort to enhance adhesion of the coating to an aluminum surface (2024-T3). In this work, mechanical properties of the UV-cured hyperbranched polymers were characterized by the nanoindentation technique and the effect of the modified polyols on structure-properties of the final coating was investigated.

2 EXPERIMENTS

2.1 Materials

Hyperbranched acrylate polyester (CN2300, MW=1304 g/mole, 8 acrylates groups per molecule, acrylate equivalent weight=163) and photoinitiator (Irgacure 2022) were purchased from Sartomer Company Inc. and Ciba Specialty Chemicals Inc., respectively. Two different hyperbranched polyols, DPP130 and P1000, were supplied by Perstorp Polyols Inc. DPP130 is a six hydroxyl functional ethoxylated hyperbranched polyether polyol (MW=830 g/mole, hydroxyl equivalent weight=138) and P1000 is a fifteen hydroxyl functional hyperbranched polyester polyol (MW=1500 g/mole, -OH EW=100). Tertiary amine, HPLC grade acetone, and THF were purchased from Sigma-Aldrich Company. All the chemicals were used as received. The acrylated polyols were synthesized as follows: 0.005 mole of P1000 (7.5g) and 0.05 mole of acryloyl chloride (4.53g) were mixed with 15 g of acetone/THF (50:50 wt%) co-solvent for 10 min in a conical reactor equipped with an ice-water bath. 0.1 mole of tertiary amine (~ 10 g) was then slowly charged into the reaction vessel. White precipitate was formed with the addition of tertiary amine. The reaction mixture was allowed to further mix for 30 minutes before filtration. Solvents were evaporated from the filtered clear solution to obtain the final product. The modification procedure of DPP130 with acryloyl chloride was the same as P-1000, in which 0.01 mole of DPP-130 (8.3g), 0.03 mole of acryloyl chloride (2.71g), 0.05 mole of tertiary amine (5 g) and 10 g of acetone were used. ADPP and AP denote the DPP130 and P1000 partially modified with acryloyl chloride, respectively. The chemicals used in this study were summarized in Table 1.

2.2 Preparation of UV-curable coatings

Ten different coating formulations were prepared by adding ADPP and AP to CN2300, increasing loading amount by 10 wt% up to 50 wt% (Table 2). The control was prepared using CN2300 alone. A constant amount of the photoinitiator (4 wt%) was added to the each solution and mixed thoroughly in a 20 ml vial using a magnetic stir bar. The eleven coating solutions were then deposited onto a 4" × 8" Al 2024 panel with the aid of a 3 × 4 stamped Teflon mask to obtain dry film thickness of ~ 100 µm. The coatings were cured in air by passing two times through Fusion LC6B Benchtop Conveyor system with F300 lamp at belt speed 20 ft/min (Fusion UV systems, Inc). The UV light intensity measured using a NIST Traceable Radiometer (International Light model IL1400A) was approximately 2200 mJ/cm^2.

Table 1: Descriptions on chemicals used in the study.

Abbreviation	Description
CN2300	An eight acrylate functional hyperbranched polyester (MW=1304 g/mole, acrylate equivalent weight=163)
DPP130	A six hydroxyl functional ethoxylated hyperbranched polyester polyol (MW=830 g/mole, hydroxyl equivalent weight=138)
P1000	A fifteen hydroxyl functional hyperbranched polyester polyol (MW=1500 g/mole, hydroxyl equivalent weight =100).
ADPP	DPP partially modified by acryloyl chloride(MW=990 g/mole, three hydroxyl and three acrylate functional groups in a molecule, Acrylate equivalent weight=330)
AP	P partially modified by acryloyl chloride (MW=2040 g/mole, five hydroxyl and ten acrylate functional groups in a molecule, Acrylate equivalent weight=204)

Table 2: Coating formulations.

Chemicals	Control (wt %)	ADPP10 to ADPP50 (wt %)	AP10 to AP50 (wt %)
CN2300	96	From 86 to 4	From 86 to 46
ADPP	0	From 10 to 50	0
AP	0	0	From 10 to 50
Photonitiator	4	4	4

2.3 Indentation methods

Quasistatic indentation was performed using Hysitron Triboindenter (Hysitron Incorporated, Minneapolis, MN) mounted with a 5 μm conospherical diamond tip. The contact area (A) was calculated using the following formula.

$$A = -\pi h_c + 2\pi r h_c^2 \qquad (1)$$

Where, r is a radius of the tip and h_c is contact depth.

2.3.1 Quasi-static measurement

Quasistatic nanoindentation was implemented in a displacement control to obtain modulus and hardness of the test sample from a resultant force-penetration depth curve with the following feedback controlled load function - a loading rate of 100 uN/s, a hold time of 30 s at the maximum force to allow viscoelastic and plastic dissipation, and an unloading rate of 100 uN/s. The tip moved towards the surface at a lift height of ~ 500 nm away from the surface and started indenting in the surface until it reached a maximum displacement (~1000nm). The preload at which the indenter will start its indentation on the surface was set to

2 uN. The contact depth and area were calculated using equations 1 and 2.

$$h_c = h_{max} - 0.75 \frac{P_{max}}{S} \qquad (2)$$

Where, h_c is contact depth, h_{max} maximum indenting depth, P_{max} force at the maximum indenting depth, and S stiffness.

The reduced modulus of the specimen (E_r) was then calculated using equation 3 with the known values of contact area (A_c) and the slope (stiffness, S) of the initial portion of the unloading curve. The hardness (H) was calculated using equation 4.

$$E_r = \frac{S\sqrt{\pi}}{2\sqrt{A_c}} \qquad (3)$$

$$H = \frac{P_{max}}{A_c} \qquad (4)$$

The reduced modulus, E_r is defined as $\frac{1}{E_r} = \frac{(1-\upsilon_i^2)}{E_i} + \frac{(1-\upsilon_s^2)}{E_s}$. υ is Possion's ratio and the subscript i and s represent the indenter tip and the indented specimen, respectively.

Hardness and modulus reported here are averaged values of six indentations. Multiple indentations on the 11 coatings prepared on the aluminum substrate were conducted automatically.

3 RESULTS AND DISCUSSION

3.1 Quasi-static indentation
3.1.1 Load-displacement behavior

Figure 1 shows the typical loading-hold-unloading profiles of the UV cured hyperbranched polymers (~100nm). The indentation started at a lifted height of ~ 500 nm away from the surface. The tip was approaching toward the surface and slightly jumped in when the tip sensed the surface. The tip indented into the surface to the predefined maximum displacement (~1000 nm). As the tip was withdrawing from the maximum depth, the coatings recovered elastic portion of the deformation, leaving a residual impression in the surface of the UV-cured coatings. The plasticity index Ψ, defined by the ratio of plastic work to total work during the loading and unloading indentation [7], is ~ 0.74 for all the samples, indicating that the UV-cured films showed considerable plastic deformation. It was also noticed that force (P_{max}) at the maximum penetration depth was relaxed during the hold time. As the amount of ADPP increased to 50 wt%, the P_{max} significantly decreased from 3570 μN to 1010 μN, showing a more significant force relaxation at the maximum depth. On the other hand, the P_{max} was slightly decreased by the incorporation of AP to

CN2300. For AP50 coating, 2740 μN of force is required to penetrate the pre-assigned maximum displacement, which is much higher than the corresponding ADPP50 coating.

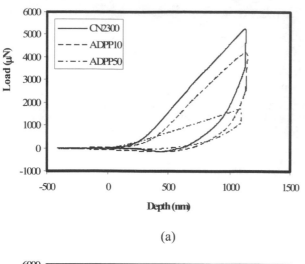

(a)

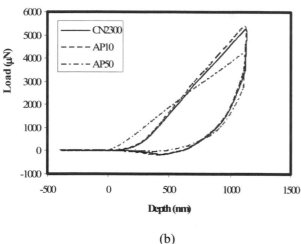

(b)

Figure 1: Load-displacement data for the hyperbranched polymers (CN2300) blended with the modified polyols; (a): ADPP, (b): AP.

3.1.2 Modulus and Hardness

Oliver-Pharr method calculates material stiffness on the assumption that only the elastic deformation is recovered during the unloading. Thus, the reduced modulus and hardness were obtained from the initial unloading segment of the load-displacement response evolved after the force was relaxed at the maximum displacement. As the weight percent of the modified polyols (ADPP) in the coating matrix increased up to 50, the reduced modulus of the UV-cured coatings dramatically decreased from 2.5±0.02 GPa to 0.7±0.02 GPa (Figure 2). This is primarily due to the flexible ether linkages in ADPP backbone and low crosslink density[3]. As the modulus decreases, hardness (H), which represents a material ability to resist indentation, linearly decreases. This linear correlation between hardness and modulus was also observed in other studies [8]. On the

other hand, the coatings blended with AP gradually decreased their mechanical properties. The reduced modulus at the loading of 50 wt% of AP decreased to 1.8±0.06 GPa. As shown in Table 1, ADPP has less acrylate functionality than AP and contains flexible ether groups, so the mechanical properties of the UV-cured hyperbranched films were more significantly affected by the incorporation of ADPP to CN2300. A similar trend was observed in the conventional DMTA measurement (TA instrument, Q800), in which glass transition temperature decreased as the amount of ADPP and AP increased. However, the indentation moduli differ from the Young's moduli measured using a tensile tester (MTS systems corporation, Insight 5 EL). The tensile Young's moduli are lower than the indentation Young's moduli calculated with υ=0.3. The tensile moduli of the coatings blended with ADPP more sharply decrease, as also illustrated in the static indentation results.

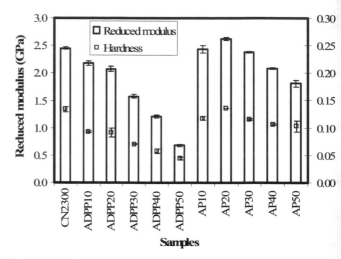

Figure 2: Indentation modulus and hardness of the samples

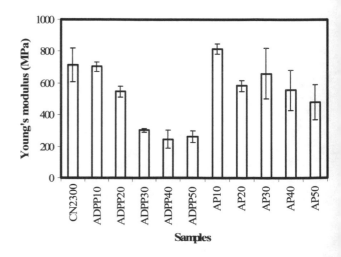

Figure 3: Young's moduli measured by the tensile test. The test speed is 1 mm/min and the modulus was calculated at 0.5% strain.

4 SUMARY

Mechanical and viscoelastic properties of hyperbranched polymers were measured using nanoindentation technique. The UV cured thin films exhibit an elasto-plastic behavior and considerable force relaxation. The moduli of the films calculated using Oliver-Pharr method was higher than those obtained by tensile tests. As the amount of the ADPP increased, the film becomes flexible due to the ether linkage, leading to a low hardness and modulus. On the other hand, incorporation of AP to CN2300 up to 30 wt % did not show any significant impact on mechanical properties. As the amount of AP further increased, the modulus and hardness of the coatings decreased probably due to decrease in crosslink density of the matrix.

REFERENCES

1. A. Hult, M. Johansson and E. Malmstrom, "Hyperbranched polymers", Advances in Polymer Science, 143, 1, 1999.
2. Y.H. Kim and O.W. Webster, "Hyperbranched Polyphenylenes", Macromolecules, 21, 5561, 1992.
3. Z. Chen and D.C. Webster, "Study of the effect of hyperbranched polyols on cationic UV curable coating properties", Polym. Int. 56, 754, 2007.
4. J.A. Klang, "Radiation curable hyperbranched polyester acrylates", Sartomer Inc. technical report.
5. L.E. Schmidt, D. Schmah and Y. Leterrier, J.E. Manson, "Time-intensity transformation and internal stress in UV-curable hyperbranched acrylates", Rheol. Acta. 46, 693, 2007.
6. W.C. Oliver and G.M. Pharr, "An improved technique for determining hardness and elastic modulus using load and displacement sensing indentation experiments, J. Mater. Res. 7, 1564, 1992.
7. B.J. Briscoe, L. Fiori and E. Pelillo, "Nano-indentation of polymeric surfaces", J. Phys. D: Appl. Phys. 31, 2395, 1998.
8. M.L. Oyen, "Nanoindentation hardness of mineralized tissures", Journal of biomechanics, 39, 2699, 2006.

Assessment of the Impacts of Packaging, Long-Term Storage, and Transportation on the Military MEMS

J. Zunino III[*] and D.R. Skelton[*]

[*]U.S. Army RDE Command, AMSRD-AAE-MEE-M

Bldg 60 Picatinny Arsenal, NJ, james.zunino@us.army.mil, donald.skelton@us.army.mil

ABSTRACT

The Army is transforming into a more lethal, lighter and agile force. Enabling technologies that support this transition must decrease in size while increasing in intelligence. Micro-electromechanical systems (MEMS) are one such technology that the Army and DoD will rely on heavily to accomplish these objectives. Conditions for utilization of MEMS by the military are unique. Operational and storage environments for the military are significantly different than those found in the commercial sector. Issues unique to the military include: high G-forces during gun launch, extreme temperature and humidity ranges, extended periods of inactivity (20 years plus) and interaction with explosives and propellants. The military operational environments in which MEMS will be stored or required to function are extreme and far surpass any commercial operating conditions. The impact of these environments on the functionality of MEMS has not been assessed. Furthermore, a standardized methodology for conducting these analyses does not exist.

To facilitate the insertion of MEMS technologies in weapon systems, U.S. Army ARDEC is addressing the information gaps delineated above. This will benefit the MEMS user community by providing data on failure modes induced by packaging, transportation and storage for selected devices. These data will then be used to develop a impacts in the early stages of development.

Keywords: MEMS, Reliability, Quality, Failure Modes, Storage

1 INTRODUCTION

Recent conflicts in the Global War on Terror (GWOT) have demonstrated a need for a lighter, more agile and more lethal force. Therefore, the technologies that support these systems must decrease in size while increasing in intelligence. Micro-electromechanical systems (MEMS) are an enabling technology that will allow the Army and Department of Defense (DoD) to meet these objectives.

The conditions military MEMS must operate in are unique. Operational and storage environments for the military are significantly different than those found in the commercial sector. Issues impacting the utilization of MEMS by the DoD include; high G-forces and spin associated with gun launch, extreme temperature and

standardized methodology for conducting these assessments. Specific activities of the program include:

- Assessment of the operational environments in which the military MEMS device may be utilized
- Determining & developing methods to preserve MEMS devices during long-term storage before they are designed into products
- Independent assessment of reliability that cannot be obtained from private industry
- Establishment of reliability data that will be fed back into development & design to improve MEMS devices

Recent accomplishments of the program include an assessment of the MEMS & NEMS technologies currently of interest to the Department of Defense, Test Guidelines for Environmental Stress Screening (ESS) of MEMS devices and components, Long-Term Storage Test Guidelines, Assessment of the Barriers to Implementation, and other tools and methodologies to facilitate the transition of MEMS & NEMS to the Department of Defense.

The MEMS assessment generated under this program will benefit the MEMS user community by filling the information gap that currently exists for reliability. With the rapid growth of the MEMS industry it is crucial to consider the reliability of this emerging technology and its

humidity ranges, extended periods of inactivity (20 years plus), interaction with explosives and propellants and stringent requirements for safety and reliability. The military operational environments in which MEMS will be stored or required to function are extreme and in general greatly surpass any commercial operating conditions.

The Army needs to rapidly respond and adapt to varying missions and operational environments. This requires the Army to transform into a lighter yet more lethal "objective force," that must be deployable, 70% lighter and 50% smaller than current systems while maintaining equivalent lethality and survivability[1]. Accordingly, the technologies embedded in military vehicles and weapon systems must be decreased in size and weight while providing improved reliability, capability, and intelligence. To meet these requirements, Army scientists and engineers need to capitalize on new technologies and breakthroughs in emerging technologies such as MEMS.

2 MILITARY UTILIZATION

The transition the Army is facing has led to increased interest the utilization of MEMS. The rapid increase in potential MEMS applications have resulted in a greater need for an accurate assessment of quality, reliability, and survivability. The Army is not alone in facing these challenges. The U.S. DoD (U.S. Army, Navy, Air Force, NSA, OSD, DARPA, ONR, Homeland Defense, etc.), NASA, Homeland Security, Health and Human Services Department, National Sciences Foundation (NSF), and Department of Energy (DoE) are implementing MEMS based technologies. With increased utilization, come increased requirements, qualifications, and safety considerations.

Besides being implemented as an enabling technology for new systems, MEMS technology is also being investigated for replacement of obsolete or unprocurable electronics, switches, sensors, storage devices, mechanical systems, and other components in current weapon systems.

2.1 MEMS for the Department of Defense

MEMS technologies involve complex, systematic fabrication approaches utilizing advantages of wafer-based lithography for the miniaturization of multi-component systems and microelectronics to create these advanced devices. MEMS technology enables the manufacture of small systems with increased functionality that will lead to performance enhancement for both current systems and entirely new systems.

The DoD has identified several key benefits of MEMS utilization [2]. These include:

- Cost Savings
- Life Savings
- Weight Reduction
- Space Reduction
- Miniaturization
- Reduce Integration Risk
- Combination / Integration of Components
- Increase Plug and Play Capabilities
- Potential Reduction of System Risk
- Power Reduction
- Advanced Capabilities

The DoD is focusing on four primary areas of interest for MEMS utilization: 1) Inertial Measurement, 2) Distributed Sensing, 3) Power Systems, and 4) Information Technology [3]. While the majority of the DoD's MEMS efforts are focused on these areas, although there are ongoing efforts in other areas such as Safety & Arming, Fuzing technologies, and Energetics. Some examples of advanced MEMS technology applications in DoD Systems include [2]:

- Acoustic Microsensors
- Velocimeters
- Magnetometers
- Micromechanical-Based Spectrometers
- Low Power Wireless Integrated Microsensors
- Surveillance Systems
- Multifunctional Electro-optical Sensors
- MEMS Infrared Remote Micro Sensors
- Magnetic Resonance Imaging Systems (MRI)
- MEMS Un-cooled Infrared (IR) Imaging Arrays
- Gyroscopes
- Embedded Sensor Systems
- Remote Weapon Systems
- Unmanned Vehicles (Land, Air, Space, Sea, etc.)
- Battle Space Awareness
- S&A Devices
- BITs (Built In Test)

The four primary areas are facilitated through cooperation with commercial industries, while the emergence of the secondary area is rapidly increasing. Unfortunately, commercial expertise to assist in MEMS fuzing and energetics is very limited.

2.2 Storage, Transportation, & Operational Environments

There are vast differences between military and commercial operating and storage environments. Commercial MEMS devices are usually designed, fabricated, shipped, and used within a relatively short period of time. Conversely, the DoD expects MEMS systems to function and survive through extended periods of inactivity, which may be greater than 20 years, and potentially extreme transportation profiles. Corrosion, materiel migration, device anomalies, and other potential failure mechanisms, may directly result from the conditions military devices are exposed to. Operational environments for military MEMS are far more severe than those encountered in the commercial sector. Mission profiles for Army weapon systems often include operation in harsh environments. These operating profiles may include any or all of the following [4]:

- Continuous operations in high humidity & moisture
- Continuous operations in areas of high wind
- Significant off-road driving
- Continuous operations in areas of high heat
- Continuous operations in sub-zero temperatures
- Extreme high & low pressured environments
- Continuous operations in sandy regions, where sand ingestion, infiltration, & contamination are commonplace
- Large variations and rapid changes in temperature due to diurnal cycles and deployment from aircraft
- Extreme G-forces due to airdrop delivery & gun launch

Another issue related to the extreme environments is that Military Standards currently used for assessing the functionality and survivability of Systems and Ammunition in potential operating and storage environments were developed to cover 99% of global operational environments [5]. Unfortunately the conditions in the Middle East are often out of this range. The operational, transportation, and storage environments to which military MEMS may be exposed play a critical role in determining reliability of those devices. An assessment of the environment in Iraq [6] recorded temperatures as high as $115F^{\circ}-125F^{\circ}$ with 1.0 kW/m^2 of insolation. Average daytime temperatures ranged from $95F^{\circ}-105F^{\circ}$. Large diurnal cycles occur in the summer because of the inability of the dry air to retain heat. Besides the temperature variations, missions in the Middle East have brought another environment condition to the forefront: sand. Sand exposure and intrusion is a significant issue with all military equipment that is deployed in current operations.

As previously stated, long-term storage is expected to be a major factor affecting the reliability of MEMS devices. Long-term storage assessments must include material compatibility. This assessment should include the materials within and adjacent to the device, hermeticity of packaging, creep of materials, and stresses on interfaces caused by temperature, shock, vibration, and other variables.

Material compatibility may be the most critical issue with regard to long-term storage. The MEMS industry is employing numerous new material combinations in emerging devices, with minimal data on their compatibility. Materials such as metallics, polymers, composites, ceramics, and numerous fluidics are all being incorporated into MEMS devices. With reliability directly affected by material compatibility, understanding all the potential interactions and problems is critical. Material interfaces and their affects are currently under investigation.

Because of the vastly different types of devices, DARPA & Sandia National Labs divide MEMS into four classes based on functionality. ARDEC has added a fifth class which is unique to the military in that the device contains energetic materials. Descriptions of each are detailed below [7]:

Class I - These devices posses no moving parts. Some common examples include RFID, flexible electronics, pressure sensors, thermal indicators, etc.

Class II - Devices have moving parts or components, but no contact surfaces: Gyros, accelerometers, resonators, etc.

Class III - Devices with moving parts that possess impacting or contact surfaces and often include relays, valves, pumps, and fluidics.

Class IV- Devices have moving and rubbing surfaces. Common Class devices types include switches, scanners, and discriminators.

Class V - Unique to the military. Devices contain or are packaged with energetic materials. Some examples are IMUs, Safe and Arming (S&As) devices, fuzing, etc.

In all, the environments that MEMS will be exposed to during deployment play a critical role in determining reliability of these devices.

2.3 Barriers to Implementation

Besides the physical demands, there are other barriers to implementation military MEMS must overcome. Another barrier impeding both military and commercial adoption of MEMS is the lack of documented reliability data for MEMS. This lack of data has adversely affected the "trust" in MEMS as an enabling technology. Furthermore, recent MEMS research has primarily focused on development of new devices, but little consideration is paid to reliability, interfaces or packaging. Vast array of devices types have been developed with little knowledge or consideration of the failure mechanisms, quality, reliability, or safety of these devices. Of the limited data that does exist from the commercial sector, it is nearly impossible to track, or is not readily accessible due to its propriety nature [8]. Also most MEMS manufacturers will not allow unpacked devices "out" until they are packaged, limiting the ability to collect valuable reliability data on fabrication and assembly process employed. As a result, failure mechanisms and failure rates for many devices are not well characterized.

The lack of trust in MEMS as an enabling technology has left many end users reticent to adopt this technology. System developers and program managers are unwilling to utilize MEMS since they are unwilling to assume the potential risks of an "unknown technology". Successful utilization of MEMS in military systems will be largely dependent upon their long-term performance within these systems. There are many aspects of MEMS performance in military applications that have yet to be explored. The unknown responses of MEMS to their storage, transportation, and operational environments must be determined if MEMS are to be widely implemented into DoD systems.

2.4 MEMS Reliability Assessment Program

These barriers must be addressed for successful integration of MEMS into weapon systems. In response to this need, researchers at ARDEC's Materials, Manufacturing, and Prototyping Technology Division have a comprehensive program to reduce and eliminate these barriers.

As part of the team's efforts, Dr. Ivars Gutmanis, of Hobe Corporation, has written a report entitled, "*MEMS*

Standards, Tests, and Applications in U.S. Department of Defense Activities," which provided a snapshot of the current state of MEMS.

Several members preformed a Lean Six Sigma Project, "*MEMS and NEMS Assessment Using Six Sigma,*" to develop risk mitigation tools to facilitate the transition to and implementation of MEMS for the U.S. Army. Some results included [9]:

- MEMS materials of construction selection hierarchy
- Assessment of the impact of corrosion/material degradation on failure mechanisms
- Physics of failure for MEMS devices
- Environmental Stress Screening (ESS) for ARDEC devices
- Test Guidelines
- Barrier to Implementation Matrix
- Cause and Effect
- Assessment and Summary of Current MEMS practices
- Preliminary identification of DoD MEMS utilization
- Benefits of MEMS Utilization

To address the lack of standards and common test methods, test guidelines have been developed with the intent that they become standards for the DoD, and potentially the MEMS industry. The first set of test guidance documents developed were Environmental Stress Screening (ESS) test plans for several Army MEMS devices including the ARDEC S&A and the Common Guidance IMU.

Along with the guidelines, more specific Joint Test Protocols (JTPs) are under development. The intent is to develop JTPs based on MEMS device classes as described previously. Work to date indicates that failure modes are common among device class. Once approved, these documents will form the basis of assessing reliability of potential MEMS for the DoD.

3 CONCLUSIONS

Due to the benefits associated with miniaturization, MEMS, is an enabling technology the DoD plans to utilize to meet their current and future objectives. The DoD continues investigating MEMS and related technologies to react to emerging threats, enhance weapon systems performance, reduce life cycles costs, and improve system reliability. Before the military begins to employ MEMS into weapon systems, MEMS must be highly reliable in extreme environments, and furthermore the reliability must be demonstrated. To be integrated into weapon systems, reliability of MEMS devices must be demonstrated. The current relative inexperience using the emerging MEMS technologies in military applications is a barrier to its implementation. Members of ARDEC and their partners are providing tools to assess and improve the reliability of military MEMS. Through the MEMS Reliability Assessment Program and other programs the US Army is being proactive in meeting these objectives. As

detailed above, ARDEC is continuing to identify the devices, operational conditions, and applications of MEMS in military systems. The development of test protocols and testing of representative devices continues. Work towards the identification or adoption of standards has begun. The information gained from these activities, and the data gained from failure analyses will aid in the transition of MEMS technologies from the labs to the field where it is needed.

REFERENCES

[1] *Future Combat System (FCS)*: Article, www.globalsecurity.org.
[2] J. Zunino, D. Skelton, & R. Mason, *Micro-electromechanical Systems (MEMS) Reliability Assessment Program for Department of Defense Activities,* NSTI / Nanotech May, 2005.
[3] Gutmanis et al, *MEMS Standards, Tests, and Applications in U.S. Department of Defense Activities,* Report for AMSRD-AAR-AEE-P, Picatinny Arsenal, NJ, 2004
[4] I. Gutmanis, *Long-term Storage Performance and Standards of Micro-Electromechanical Systems (MEMS),* Report for Army Corrosion Office, Picatinny Arsenal, NJ, 2006.
[5] Department of Defense Test Method Standard, "Environmental Engineering Considerations and Laboratory Tests," MIL-STD-810F, January 1, 2000.
[6] Skelton et al. *Assessment of the Impact of Iraq Environment on Army Materiel.* Report for Army Corrosion Office. June 2005.
[7] J.L. Zunino and D. Skelton, "U.S. Army Corrosion Office's Storage and Quality Requirements for Military MEMS Program," SPIE Volume 6528 Nano-, Micro-, Bio-Sensors and Systems, 2007 [6528-27].
[8] J. Zunino and D. Skelton, "Department of Defense Need for a Micro-electromechanical Systems (MEMS) Reliability Assessment Program," *Reliability, Packaging, Testing, and Characterization of MEMS/MOEMS IV*, Proceedings of the SPIE, Vol. 5716, pp. 122–130, January 2005.
[9] J. Zunino, D. Skelton, A. Coscia Jr., R. Zanowicz, "*MEMS and NEMS Assessment Using Six Sigma,*" Lean Six Sigma Project for U.S. Army ARDEC, 2006.
[10] Microelectromechanical Systems Opportunities *A Department of Defense Dual-Use Technology Industrial Assessment*, United States DoD, 2000

Advances in electrolytic plating technologies for plating nano & micro structures

L. Derrig & Dr. A. Kuzmin:

Digital Matrix Corp. 92 Madison Avenue, Hempstead, NY 11550 USA +1-516-481-7990

leslie@digitalmatrix.us   Anatoly@digitalmatrix.us www.digitalmatrix.us

Abstract

Electroless and Electrolytic plating has had to keep up with the demands in developing new technologies to meet the needs of filling nano and micro structures. Requirements have been for plating angstrom thin layers with an equal deposition over an entire 300 mm wafer, filling high aspect ratio structures with deep side walls, with out voids and plating up varying width structures at the same time Digital Matrix primary function as a company is to manufacture high-speed, high-precision electrolytic plating equipment. With the demands and requests presented to them, they have been frequently asked to solve the plating tribulations created by various aspects of the MEMS/NEMS industries. A few of the more significant technologies performed are: • Digital Matrix process and system can control the plated deposition over the entire wafer surface with a consistent TTV. (Total Thickness Variation) • The process developed by DM allows the plating up of varying width and height structures to grow at the same rate, eliminating the need to slurry away the excess and lose valuable time and precious metals. • Digital Matrix has developed a bump plating process that can over-plate posts/pillars straight up over the Photo Resist or mask, 20 µms with out the mushroom or muffin affect. Larger bumps can be created with a slight mushrooming formation. This technology is being used to replace the need for solder bumps. • Deep posts and side-walls can be plated, filling in the corners and with no trapped voids. • Exact height plating can be achieved, eliminating the need to slurry excess over-plating. • Angles created in Photo Resist can be plated into the deep corners and acute angled side walls without voids.

Keywords: electrolytic, plating, high-aspect-ratio, vias

The demands put on the MEMS and Nano industries have overwhelmed all aspects of research and development, especially electroplating. When we feel we have made it the smallest possible, we have to make it even smaller, and we have to manufacture trillions of these items in record time, as cheap as possible. Every phase of producing a MEMS or Nano part needs to comply with numerous other demands imposed in manufacturing the substrate and each must meld perfectly to complete the final product…within angstroms.

There are many phases that go into the preparation of a wafer before plating and if every phase is performed within spec, advanced plating techniques have the capabilities of out performing previously impossible tasks.

Wafer plating has generally relied on Electroless plating (Illustration 1) to layer several metals uniformly, but this requires slurry stages to remove excess and over-plated areas. There is limited control of the TTV (total thickness variation) and deposition. It is a slow and arduous process with a limit of 100 micron thickness and real concern of stress. The chemicals weaken over a period of time and are tossed for recycling, a very expensive and environmentally wasteful procedure. As the chemical composition changes, the properties of the metal deposition layer drift. This leads to a non uniform and out of specification result. Properties such as hardness can change dramatically.

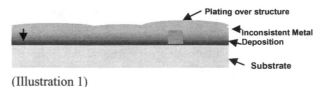

(Illustration 1)

Electrolytic plating, (Illustration 2) also known as High-Speed Plating, has recently become the preferred method because of its capabilities of out performing the Electroless process. The absolute control of plating parameters can be within 1 to 5% of the specifications and can be controlled in angstroms with an excellent thickness uniformity (low TTV). There is no need to slurry excess areas and there is virtually no limit to the height of the plated area. Electrolytic plating can be up to ten times faster than the Electroless process and chemicals are NOT disposed of but maintained recycled within the system. Hardness and other properties are much more consistent and easily maintained. Thickness can range from <0.001 micron to 10,000 microns.

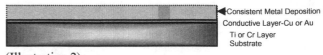

(Illustration 2)

Most metals used in the Electroless process can also be deposited electrolytically, and various formulas are available to meet a wide range of requirements. Digital Matrix has developed their own special formulations to maximize results and have focused on the elimination of organics that generally cause stress.

Nickel Cobalt

NiCo, one of the preferred for its high hardness. Typically NiCo high-speed NiCo plating is plagued with such as internal stress, tensile and brittle structures. It also generally requires special organics additives and constant metered replenishments of Cobalt Sulfamate.

Digital Matrix has reduced the stress from 150,000 psi to 0 psi and created a hardness of 680 HV by maintaining the process parameters such as pH (3.8 to 4.1), Stag Factor (1.6-1.9) and Boric concentration (25 to 35%), temperature (60ºC to 65 ºC) and current density from 0.1A/dm2 to 5.5A/dm2. Co-ion concentration is 5-6 g/L and Ni-ion concentration is 130-140 g/L. DM programs 3 to 9 gradual plating ramp steps from low DC and RPP (reverse pulse plating) current density to high DC and RPP current density.

Copper & Copper Alloys

In the familiar Cu process, organics such as Cuprax, Acid Cu Sulphate, Cyanide and Pyrophosphate solutions are used. Cuprax and Cyanide are used for heavy Cu deposits that are smooth and fine grained. Cuprax Cu used 160 to 120 g/L of High-Efficiency Cuprax Salts and Cyanide Cu uses 5 g/L Sodium & Potassium-Cyanide and Sodium-Sulphite to keep the anodes clean.

The Digital Matrix Cu Pyrophosphate formula requires high-speed, rotating head, electrolytic plating. This very stable process uses Cu-Phosphate 70-90 g/L, Potassium-Pyrophosphate 220-240 g/L and Ammonia Liquor 3-5 ml/L.

The CuAlloy solution has a pH range from 1.8 to 3.8 and the average current density is 0.1A/dm2 to 3.0A/dm2. Temperatures can range from 20ºC to 50ºC. Plating times can vary from 2 to 4 hours for 100 µms and 1 to 2 hours for 50 µms achieving a +/- 1 µm TTV. DM recommends Copper Sulphate 200-280 g/L, Sulphuric Acid 20-50 g/L and additives; Brightener and Refiner.

The results are stress free with pre and post plating, and TTV within +/-1 to 5% and a perfect adhesion to Au and other Cu layers. With an achievable hardness of 250 Vicors, the mechanical strength of a Cu film becomes non-destructive and creates a perfect electrical resistivity of the plated area.

Digital Matrix standard Cu Acid solution can withstand very high plating speeds and meet hardness requirements of approximately 250 Vicors. It is a more stable process than a cyanide based solution. DM's formula uses a Cu-ion concentration of 80 to 90 g/L, Cu Sulfamate concentration of 260 to 270 g/L, Sulfuric Acid (10 to 30 ml/L), Specific gravity: 1.167 to 1.17, with a stag factor of 1.4 to 1.9 and a temperature of 23ºC to 25ºC. Current density is 0.001A/dm2 to 0.008 A/dm2

Other Benefits of Electrolytic plating with DM Equipment

Other benefits of Electrolytic High-Speed Plating are

- Plate to exact height of the structure +/- 1 to 3µm
- Plate without voids and defects
- Plate in corners and sidewalls of high aspect posts
- Smooth back finish with a mirror finish
- No tensile or brittle structures
- Plate 20µm above photo-resist -straight up
- Plate exact thickness thru diameter: +/-1 to 3µm
- Adjust the distance from the Cathode to Anode
- Plate in a fraction of the time of your previous cycle

Plating On Aluminum:

As Silicon wafers become harder to purchase, labs are looking at alternative substrate materials to plate to, such as Aluminum. Digital Matrix has formulated a process that makes plating to Al easier and within the required specifications. With the use of special adhesion layers, such as Ti, Cr & Zn, additional layers, such as Ni, Cu, Au, can be plated to the Al. Unlike traditional Electroless processes, DM has a process that can plate in excess of over 300 microns

Plating Structures:

Digital Matrix has used their High-Speed Electrolytic plating systems to achieve many various results that have helped to revolutionize the Nano and MEMS industry. With the ability to plate with exact and precise properties, the plating equipment and metal plating formulas have opened up new techniques, allowing a more eco friendly and less expensive method to achieve results once thought un-obtainable.

Illustration 3 demonstrates an even more unique Digital Matrix proprietary process that offers multiple opportunities when using an electrolytic process that most would not have considered in the past. The resulting structure can be a 3-D part that can also be removed from the wafer as a micro part. The structure etched into the photo-resist can be prismatic or octagonal in shape and carry the shape up to 20 microns above the photo-resist, or it can create a bump over the PR without a mushroom affect.

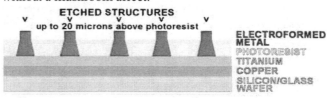

(Illustration 3)

The Bump affect has been used in various applications, but where it has received the most attention is in the Semiconductor packaging industry. With the environmental concerns to replace lead/tin, it is now possible to plate this in any metal, allowing for a harder bump. DM is capable of plating 20 µms over the top of the PR or mask without mushrooming and can grow 45 to 50 microns over with an ever so slight mushroom affect. These can also be done in Nano bumps and with a closer placement. It is much faster and more reliable than the standard bump process and the adhesion is excellent when the preparation is performed properly.

Side-wall deposition and Vias have also created challenges to most facilities. As soon as you let someone know you can plate a side-wall or a via, they immediately push to a smaller structure and higher aspect ratio. The challenges have been to remove the hydrogen bubbles that get trapped in the bottom corners and walls. (Illustration 4)

Then
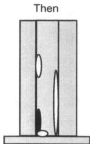

Now
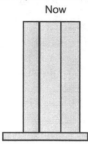

Digital Matrix has a 45° angled, rotating, cathode workholder that spins at an adjustable 60 rpms that will spin out the trapped bubbles, giving strength to the structures for mass production.

Filling structures of varying widths and height on the same wafer would ordinarily cause over-plating on the smaller or narrower structures, while the wider and deeper structures would take longer to fill, often causing a dip or uneven fill. Digital Matrix has developed a program and process to plate at the same rate so that the narrow structures would plate at the same rate as the wider areas. (Illustration 5)

Photo Resist
Metal
Plating

Plating requirements of exact thickness are easily achieved on a DM plating system. With a structure height of 292 microns high it is desirable to plate 292 microns consistently across the entire wafer, we can do this within +/- 1%. This eliminates the need to slurry to remove high edges or center. This potential has been increasingly interesting to those who plate Au or any of the other precious metals. The need to remove and then recover expensive metals is eliminated. The tolerances are also controlled by the vacuum sealed cathode backplate that draws the air from behind the wafer, allowing for more precise results. (Illustration 6)

(Illustration 6) 300 mm vacuum rotating cathode backplate

Plating requirements will always be put to the challenge. It is not enough to work with older processes and technologies to achieve the results required today. What was once the answer, may no longer apply to today's demanding specifications.

Influence of nano size on various properties of $Sn_{1-x}Co_xO_2$ nanocrystalline Diluted Magnetic Semiconductors

K. Srinivas and P. Venugopal Reddy*
Department of Physics, Osmania University,
Hyderabad, A.P, India -500007, * pvreddy@osmania.ac.in

ABSTRACT

With a view to understand the influence of nano size on various structural, electronic and magnetic properties of nanostructured $Sn_{1-x}Co_xO_2$, systematic Investigations have been carried out. After synthesizing the samples by sol-gel route, structural characterization was undertaken by XRD, TEM, FT-IR and Raman techniques. The samples are having the tetragonal rutile structure with single phase without any detectable impurity. The average particle sizes obtained from TEM are in the range of 18-48 nm. The reduction of intensities of Raman peaks, shift in their positions, their shape and size distribution are found to be influenced predominantly by the nano size of the materials. The optical band gap of the materials clearly indicates a red shift with increasing particle sizes. From XPS data, it has been observed that the oxidation state of Co is +2 and changes from low spin state to high spin state as the nano size of the materials increase. Further, the phonon confinement effect and electronic structure have been analyzed. The Magnetization studies clearly indicated the presence of room temperature ferromagnetism in these samples. Further, the analysis of the saturation magnetization, Coercive field clearly indicated that they critically depend on nano size of the materials.

Keywords: tartarate gel, diluted magnetic semiconductors, nano particles, electronic structure, optical properties,

1 INTRODUCTION

Diluted magnetic semiconductors (DMS), such as transition metal ions (Co, Mn, Ni, Fe, Cr etc.,) doped with wide band gap semiconductors such as TiO_2 ZnO, SnO_2 and HfO_2 etc., have been attracting considerable attention in recent years due to their great potential for novel applications in high-density magnetic storage, spintronics devices, bio-medical, magneto fluid and magneto-optical applications etc.,. However, it is of particular importance to understand the role of dimensionality in shaping the spin-polarized electronic structure of nanocrystalline oxide based DMSs because quantum confinement may result in intriguing ferromagnetic properties as they will be potentially useful. Apart from this, the synthesis of stable nanocrystalline DMSs with controllable grain size and surface morphology are of great fundamental and technological interest and has attracted much experimental efforts. Very few reports [1-4] on nanocrystalline Co-doped

SnO_2 powders rather than thin films have been reported with the conflicting results regarding the existence of the room temperature ferromagnetism. In the present work, systematic investigations have been made to understand the nano size dependent structural, electronic and magnetic properties and the results of such a study are presented here

2 EXPERIMENTAL

Nanocrystalline Co doped SnO_2 ($Sn_{0.95}Co_{0.05}O_2$) samples were prepared by the tartaric acid assisted sol-gel method. In this method, starting precursors were converted to tartarates and P^H was adjusted to a value between 5 - 6 by adding ammonia and tartaric acid. Later, ethylene glycol was added as a promoter of tartaric polymerization and also prevents the formation of Cobalt segregations during gelation. The solution was concentrated by slow evaporation (~16 hours) at 65 °C. After slow heating between, 140-160 °C the gel formed as a fluffy porous precursor resin. The prepared precursor was calcined at 200 °C for about 8 hrs and finally the powders were sintered at four different temperatures between 300 – 600°C for about 4 hrs. For comparison purposes, the undoped SnO_2 sample was also prepared in similar experimental conditions.

To understand the structural and morphological properties XRD, TEM, FT-IR and Laser Raman studies were undertaken. The electronic properties were studied using the optical absorption and XPS techniques. Finally, the magnetization measurements were undertaken using a Vibrating Sample Magnetometer (Model: DMSADE-1660 MRS) at room temperature by applying 15 kOe magnetic field.

3 RESULTS AND DISCUSSION

3.1 Structural properties

The XRD patterns of $Sn_{1-x}Co_xO$ (x=0.05,0) powder samples sintered at different temperature were shown in figure 1. It can be seen from these patterns that all the samples are found to exhibit tetragonal rutile structure and are also comparable with JCPDS data [Card No: 41-1445]. The lattice parameters a and c were obtained from (110) and (101) peaks using Scherer's formula and are presented in Table-1. It is clear from the table that there is no systematic variation of lattice parameter.

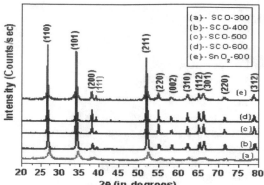

Figure 1: XRD patterns of $Sn_{0.95}Co_{0.05}O_2$ samples

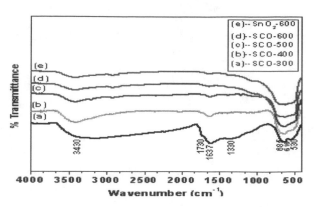

Figure 2: FTIR spectra of Co doped SnO_2 samples (x= 0, 0.05), sintered at different temperatures.

Further, the average crystallite size of the samples (*D*) is determined from the diffraction peaks of (110) and (101), using peak broadening method and are in the range 15 nm-57 nm.

Table 1: XRD and TEM data for the $Sn_{1-x}Co_xO_2$ (X = 0, 0.05) samples.

Sample Name	Sintering Temp (°C)	<D> (nm)	Average Particle Size (nm)	<a> (°A)	<c> (°A)
SCO-300	300	15	18	4.741	3.155
SCO-400	400	24	25	4.726	3.177
SCO-500	500	32	35	4.711	3.163
SCO-600	600	40	45	4.719	3.173
SO-600	600	57	60	4.683	3.155

To investigate the chemical functional groups, FTIR spectra of all the samples were also recorded and are shown in Fig. 2.. Several bands due to fundamentals, overtones and combinations of OH, Sn–O and Sn–O–Sn entities appear in 700 –400 cm^{-1} range. The low wave number region exhibits a strong vibration around 618 cm^{-1} which corresponds to antisymmetric Sn–O–Sn mode of tin oxide. The widening of SnO_2 related bonds with increasing sintering temperature indicate, improvement in the crystallinity by strengthening O-Sn-O bonds with removal of organic impurities involved in the synthesis process.

Figure 3. represents the room temperature Raman spectra of all the samples. The peaks appeared around at 574 and 593 cm^{-1} undoped and Co doped samples are clear indication of nano crystalline behaviour. The most intense As the crystallite size increases, A_{1g} mode position is found to shift gradually from 616 to 631 cm^{-1} and the observed behavior clearly indicates the phonon confinement effect in the lattice. Raman bands around at 475 and 776 cm^{-1} are the vibrational modes E_g and B_{2g}, respectively. In fact, all these

peaks are comparable with those reported in the literature [5-8]. Further, two new bands around at 306 and 695 cm^{-1} with less intensity, whose origin is controversial, were also observed As a matter of fact, both these peaks were also observed in the case of undoped SnO_2 sample also. These peaks are neither due to parasitic phases nor due to the insertion of Co in the SnO_2 lattice in our case and may be originated from the defects such as oxygen vacancies. It is interesting to note from the Raman spectra that the peak intensities of doped samples are found to decrease continuously when compared with the undoped one. As the peak intensity is directly correlated with the insertion of Co in the SnO_2 matrix, the systematic changes in the peak positions and shapes clearly indicate the influence of nanosize on local structure and distribution of magnetic ions.

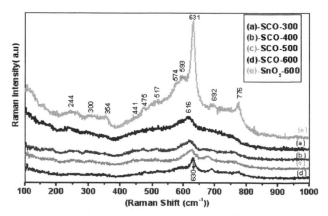

Figure 3: Raman spectra for nanocrystalline Co doped SnO_2 samples (x= 0, 0.05), sintered different temperatures.

Figure 4. illustrates the morphology and the electron diffraction patterns (shown inset) of the samples. It has been observed that the average particle sizes obtained from TEM are found to be in the range 18-48 nm [Table-1]. Incidentally, the crystallite size values obtained from Scherer's formula using XRD data are in agreement with those obtained from TEM studies. A close observation of TEM morphology images with increasing sintering temperature clearly indicate that the nano particles might

NSTI-Nanotech 2008, www.nsti.org, ISBN 978-1-4200-8503-7 Vol. 1

have aggregated and the observed behavior may be attributed to the redistribution of surface atoms due to change in particle's size, shape and surface defects .

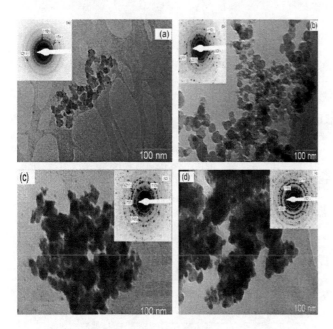

Figure 4: TEM morphology and SAD patterns (shown inset) for the Co doped SnO_2 samples.

3.2. Electronic properties

The electronic properties of all the samples were investigated using XPS and optical absorption measurements. The core level binding energy values of Co $2p_{3/2}$ and Co $2p_{1/2}$ peaks are shown in figure 5. and the values are given in Table-2. It can be seen that B.E values are found to shift towards higher side with decreasing particle size of the samples. One may therefore conclude that the nano size of the samples clearly influence the electronic structure and this may be attributed due to local structural changes associated with variation of particle size and non uniform strain. By comparing the binding energies of the cobalt core level XPS peaks with those in the literature and [9], it has been concluded that the oxidation state of Co in $Sn_{0.95}Co_{0.05}O_2$ samples is +2. Further, two shake-up satellite peaks, as specific evidence and it changing from low spin Co $^{2+}$ state to high spin Co $^{2+}$ with increasing sintering temperature. There is no possibility of forming Co Clusters since the B.E of Co metal cluster is at 778.3 eV and the energy difference between Co $2p_{3/2}$ and $2p_{1/2}$ core levels for metallic Co is 14.97 eV which are not observed in the present investigation.

The room temperature absorption spectra of Co doped SnO_2 samples, sintered at different temperatures along with undoped SnO_2 shown in figure 6. The energy band gap values were estimated using second derivative approach and are given in Table-2 along with the particle size values of all the samples. In the Co doped SnO_2 samples upon Co doping, red shift in band gap values was exhibited where

as with decreasing particle size blue shift in the band gap values was demonstrated.

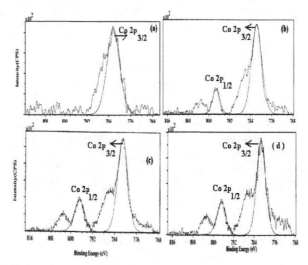

Figure 5: XPS core level deconvoluted spectra of Co 2p for the different particle sizes

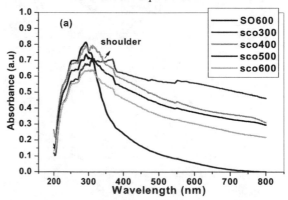

Figure 6: Optical absorbance DR spectra for $Sn_{1-x}Co_xO_2$ (X = 0.05, 0) powder samples

Table 2: Co XPS data Co element and optical band gaps for the $Sn_{1-x}Co_xO_2$ (X = 0.05, 0) powder samples with varying particle size.

Sample Code	Particle size (nm)	Binding Energy (e.V)		ΔB.E (eV)	E_g (eV)
		Co $2p_{3/2}$	Co $2p_{1/2}$		
SCO300	18	782.33	-	-	3.53
SCO400	25	781.66	797.25	15.99	3.47
SCO500	35	781.14	796.93	15.79	3.43
SCO600	45	780.47	794.76	15.39	3.32
SO-600	60	-	-	-	3.70

A shoulder is observed for photon energies right above the direct energy i.e.~3.6–3.9 eV and found to decrease its width with increasing sintering temperature. This appearance of shoulder is due to transitions, from the O(p)— states in the valence band to Sn(s)— states in the conduction band, occurring at points located near the Brillion Zone.

3.2. Magnetic properties

The magnetization measurements were performed using a vibrating sample magnetometer (VSM) at 300K and shown in figure 7. The obtained hysteresis curves exhibited that the $Sn_{1-x}Co_xO_2$ samples were all (except SCO300 sample) ferromagnetic at room temperature. The SCO300 sample might have impurities which are not removed completely as evidenced by FTIR and XPS. It is interesting to note that this is the first report which is reporting room temperature ferromagnetic samples of 5% Co doped SnO_2 nanocrystalline samples. The Ms and Hc values with varying particle size are given Table-3. The results revealed that the precise control of dopant distribution in the lattice and the particle size of the samples enormously influencing the local structure there by magnetic properties.

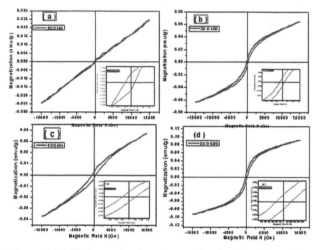

Figure 7: M-H Magnetization Curves for the $Sn_{0.95}Co_{0.05}O_2$ samples sintered at different temperatures.

In the present investigation the week ferromagnetism obtained as a result of bound magnetic polarons (BMPs) coupled with adjacent randomly distributed Co^{2+}. The maximum obtained saturation magnetization value is 0.093 emu/g for the SCO600sample. This rather small value can be explained by the fact that only a small portion of Co spins could be coupled ferromagnetically and the remaining Co are coupled either paramagnetic or antiferromagnetic when are coupled through oxygen vacancies with superexchange interaction.

The spontaneous magnetization (Ms) is decreases with D, suggesting the presence of a surface region (shell) with a reduced magnetization (M). The same behavior is reported earlier for soft magnetic nanoparticles [10]. The non-systematic changes indicating that the saturation magnetization of nano-particles depends not only on the magnitude of the individual atom or spin moment, but also on the particle size, shape or the complicated surface condition of the particles.

4 CONCLUSIONS

In summery, nanocrystalline $Sn_{0.95}Co_{0.05}O_2$ materials were synthesized with the average particle sizes in the range of 18 – 45 nm by tartaric gel route without any additional impurity phases as evidenced by XRD, FTIR, Laser Raman and TEM studies. The gradual shift in A_{1g} mode position from 630 to 616 cm^{-1} with decreasing crystallite size, confirms the phonon confinement effect in the lattice.

The electronic studies using XPS and Optical absorbance measurements revealed that the nano size of the materials influences the binding energy values and optical band gap values predominantly.

Magnetic studies revealed that all the samples above sintered above 300 °C exhibited clear room temperature ferromagnetic behaviour. The values of saturation magnetization (M_s), and other related parameters depend on surface effects generated by changing the particle size and their distribution.

Table 3: Magnetization values for the $Sn_{1-x}Co_xO_2(X=0.05)$ samples sintered at different temperatures

Sample Code	Ms (emu/g)	Hc (Oe)	Average Particle size (nm)
SCO300	0.02456	150.091	18
SCO400	0.06430	215.170	25
SCO500	0.03714	478.047	35
SCO600	0.09292	209.232	45

REFERENCES

[1] A. Bouaine and N. Brihi, G. Schmerber, C. Ulhaq Bouillet, S. Colis, and A. Dinia, J. Phys.Chem. C, 111, 2924,2007.

[2] C.M. Liu, X.T.Zu and W.L. Zhour, J.Phys: Condens. Matter 18, 6001, 2006.

[3] A. Punnose, M.H. Engelhard and J.Hays, Solid. State. Communi. 139, 434, 2006 and other Punnose et. al group references.

[4] C.B. Fitzgerald, M.Venkatesan, A.P.Douvalis, S.Huber, J.M.D.Coey and T.Bakas , J.Appl.Phys.95, 7390, 2004.

[5] S.H.Sun, G.W. Meng, G.X. Zhang, T.Gao, B.Y.Geng, L.D. Zhang and J.Zuo, Chem.Phys. letters 376, 103 2003.

[6] Cunyi Xie, Lide Zhang, and Chimei Mo, Phys. Status Solidi *A* 141, K59 1994.

[7] Soumen Das, Soumitra Kar and Subhadra Chaudhuri, J. Appl. Phys 99, 114303, 2006.

[8] L. Abello, B. Bochu, A. Gaskov, S. Koudryavtseva, G. Lucazeau, and M. Roumyantseva, J. Solid State Chem. 135, 78, 1998.

[9] J.F. Moulder, W.F. Stickle, P. E. Sobol, K. D.Bomben, "Handbook of X-ray Photoelectron Spectroscopy" (Perkin-Elmer Co., Eden Praine) 1992.

[10]. R. Skomski, H. Zeng, M. Zheng, and D.J. Sellmyer, Phys. Rev. B 62, 3900, 2000.

Synthesis and characterization of Dy_2O_3 nano crystalline power by using a combustion Urea process

M. Vijayakumar[1], N. Nallamuthu[1], M. Venkateswarlu[2], N. Satyanarayana[1]*

1. Department of Physics, Pondicherry University, Pondicherry – 605 014, India
2. HBL Power systems Ltd., Hyderabad- 500 078, India
Tel: +91 413 2654404: fax: +91 413 2655260/11
E-mail address: (N.Satyanarayana) nallanis2000@yahoo.com

ABSTRACT.

Urea assisted combustion process was used for the synthesis Dy_2O_3 nano crystalline powder. The formation of crystalline phase and structure of the Dy_2O_3 compound were identified by XRD, FTIR and SEM - EDX measurements.

1. INTRODUCTION

Ceramic materials with high chemical durability, high melting point, high hardness, wear resistance are attractive candidates for use as engineering components in extreme environmental conditions]. The potential of engineering ceramics for high temperature applications is not fully exploited because of the brittle nature of the ceramics [5-6]. One of the established methods of reducing brittleness of ceramics is toughening the ceramics by the controlled application of martensitic phase transformation. Rare earth oxides are the better potential candidates for development as transformation tougheners as they undergo martensitic transformation between 2000 - 500°C accompanied by a volume change of ~9%, compared to tetragonally stabilized ZrO_2. Among the trivalent rare earth oxides, Dy_2O_3 exhibits the highest B→C martensitic transformation temperature at ~ 1900 °C. Metal oxides are usually produced by the thermal decomposition of inorganic precursors such as metal acetate, metal oxalate, metal nitrate, metal hydroxide, etc [1-3]. The step of decomposition often causes physical and chemical changes, which result in a variation in the stoichiometry, crystal structure, surface morphology and activity of the product oxide [4]. In Many wet chemical methods, like, sol-gel, combustion, polyol, etc., have been developed in order to overcome the above mentioned difficulties. Among the available, combustion process is found to simple and capable of producing nanocrystalline powder at a lower temperature in a short time. Therefore, in the present work, the Urea assisted combustion process have been investigated to prepare dysprosium oxide nano crystalline powder and characterized by XRD, FTIR and SEM - EDX.

2. EXPERIMENTAL

Analytical grade chemicals of dysprosium nitrate, citric acid (qualigens fine chemicals) and urea (qualigens fine chemicals) are used as the starting materials. The metal ions, citric acid and urea ratio was kept 1:1:1. The required amount of each precursor chemical was separately dissolved in distilled water. Citric acid, urea and dysprosium solutions were mixed under stirring at 90°C up to 5 hours. The formed clear mixed solution was kept in oven to form the gel. The obtained gel was heated at 473K and 523K for 4 hours. The phase and the structural evaluation of the heat treated gel were monitoring respectively by XRD and FTIR techniques.

3. RESULTS AND DISCUSSION

X' Pert PRO MPD, PANalytical (Philips), X-ray diffractometer with Cu Kα radiation

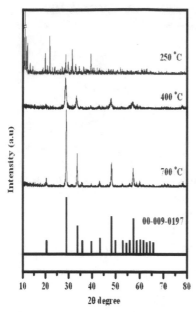

Fig 1. XRD patterns for Dy_2O_3 sample

of wavelength $\lambda = 1.5418$ Å was used to record the XRD patterns for Dy_2O_3 sample between 80° and 10°, 2θ values at a scan rate of 2° per minute. NBS silicon standard was used for the estimation of instrumental broadening. FTIR spectra were recorded using a Shimadzu FTIR-8300/8700

spectrometer in the range 4000–400 cm^{-1} at 40 scans for the pellets of Dy_2O_3 sample. All pellets were made with KBr powder and Dy_2O_3 sample using a KBr mini press. The shape and size of the prepared samples were investigated using a scanning electron microscope (SEM), JEOL-JSM6400 scanning electron microscope with an accelerating voltage of 20 keV.

Fig 1 shows the XRD patterns of the heat treated gel at 250^{0}C, 400^{0}C and 700^{0}C. From Fig 1, the observed peak free XRD patterns of gel heated at 250^{0}C confirm the amorphous nature. For the gel heated at 400^{0}C and 700^{0}C, the observed XRD peaks and its analysis revealed the formation of nanocrystalline Dy_2O_3 phase was confirmed on comparison with the JCPDS. File number #09-0197[9]. The crystallite size was calculated using the Scherer equation[10],

$$D = \frac{k\lambda}{\beta \cos \theta}$$

where D is the crystallite size, k is a constant (=0.9 assuming that the particles are spherical), λ is the wavelength of the X-ray radiation, β is the line width (obtained after correction for the instrumental broadening) and θ is the angle of diffraction. The average crystallize size obtained from XRD data is found to be ~28 nm.

Fig. 2 shows the FTIR spectra of the heat-treated gel at 250^{0}C, 400^{0}C and 700^{0}C. In Fig. 2, the observed variation of IR band positions and also band intensities revealed that the structure of sample changes when it was heated at different temperatures. For the sample heated at 700^{0}C, the observed IR band positions revealed the formation of Dy_2O_3 structure and also the removal of water molecules [11].

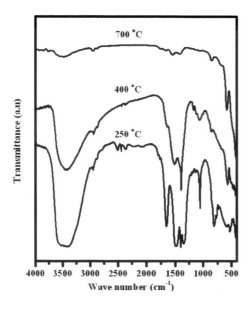

Fig 2. FTIR spectra for Dy_2O_3 sample

Fig 3 SEM EDX Image

are respectively confirm the existence and uniform distribution of O and Dy in the Dy_2O_3 . SEM- EDX results confirm the formation of the Dy_2O_3, shown in table.1, and it is free from organic contamination.

Element	Net Counts	Weight %	Atom %	Formula
O	339	11.45	56.77	O
Dy	2524	88.55	43.23	Dy
Dy	1367	---	---	
Total		100.00	100.00	

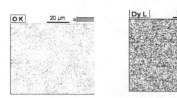

Fig 4. SEM–EDX mappings

The microstructure and elemental analysis of the Dy_2O_3 formation was conform using SEM - EDX measurements. SEM micrograph of fig.4 showed an agglomerated

spherical particles of Dy_2O_3 and their particle size is ~200nm. SEM–EDX spectrum and mappings shown in fig.5 & 6

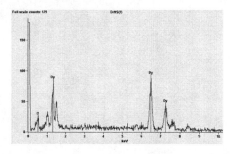

Fig 5. SEM–EDX spectrum

CONCLUSION

SEM–EDX spectrum and mappings SEM–EDX spectrum and mappings Nano crystalline Dy_2O_3 power was Synthesized by using combustion process. The formation of nanocrystalline phase and structure of the Dy_2O_3 compound were identified by monitoring the heat treated dried gel, at various temperatures, from amorphous to crystalline growth through XRD and FTIR measurements. Crystallite size of the Dy_2O_3 powder, calculated using its XRD data and the Scherer equation, $D = k\lambda/\beta cos\theta$, is found to be ~28 nm. SEM micrograph of fig.4 showed an agglomerated spherical particles of Dy_2O_3 and their particle size is ~200nm. SEM–EDX spectrum and mappings, shown in fig.5 & 6, are respectively confirm the existence and uniform distribution of O and Dy in the Dy_2O_3 . SEM- EDX results confirm the formation of the Dy_2O_3, shown in table.1, and it is free from organic contamination

ACKNOWLEDGMENTS

N.S. is gratefully acknowledged **DRDO**, Government of India for granting the funds in the form of major research Project.

REFERENCE

[1] V.D. Alfred, S.R. Buxton, J.P. McBride, J. Phys. Chem. 61 (1957) 117.

[2] B.H. Davis, J. Catal. 52 (1978) 176.

[3] A.K. Galwey, S.G. McKee, T.R.B. Mitchell, M.E. Brown, A.F. Bean, React. Solids 6 (1988) 173.

[4] K. Okabe, K. Sayama, H. Kusama, H. Arakawa, Bull. Chem. Soc. Jpn. 67 (1994) 2894.

[5] A.K. Nohman, H.M. Ismail, G.A.M. Hussein, J. Anal. Appl. Pyrolysis 34 (1995) 265.

[6] M. LeVan, G. Perinet, P. Bianco, Bull. Chem. Soc. Fr. 15 (1968) 6483.

[7] G.A.M. Hussein, H.M. Ismail, Powder Technol. 84 (1995) 185.

[8] Joint Committee on Powder Diffraction Standards, Diffraction Data File, No. 48-0008.

[9] H.P. Klug and L.E. Alexander, X-ray Diffraction Procedures for Polycrystalline and Amorphous Material, John Wiley and Sons, New York (1954).

[10]G. Socrates, Infrared and Raman Characteristic Group Frequencies, John Wiley and Sons, New York (2001).

SYNTHESIS AND CHARACTERIZATION OF NANOCRYSTALLINE TiN POWDER BY REACTIVE MILLING

Joaquina Orea Lara[a], Heberto Balmori Ramírez[b]
Depto. de Ingeniería Metalúrgica
Escuela Superior de Ingeniería Química e Industrias Extractivas
Instituto Politécnico Nacional
Edificio 7, Unidad Profesional Adolfo López Mateos, Col. Lindavista
C. P. 07300, México D.F., México.
(a) jorea9@yahoo.com.mx, (b) hbalmori@ipn.mx

Abstract.

TiN powder was obtained by reaction milling of a titanium powder in air. The Ti powder was milled in an attrition mill or a Spex mill. Characterization of the powders was carried out by means of chemical analysis, X-ray diffraction (XRD) and scanning electron microscopy (SEM). The reaction took place in 12 h in the Spex mill. It took 96 h in the attrition mill to complete the reaction. The XRD results indicate that a cubic TiN-like phase crystalline structure was produced in the two mills, with a lattice parameter of 4.38 Å and 4.23 Å, respectively. The morphology of the obtained powders was nodular, with particle size within the nanometric size range.

Introduction.

The reactive milling is a process where chemical reactions are activated by the mechanical energy of the milling and downsizing of particles takes place [1]. For that reason it is called mechanical synthesis. The method has been strongly indicated as a potential method for the production of new materials, particularly advanced ceramic under controlable conditions [2].

Because titanium is a very reactive element, it can easily form compounds with oxygen, carbon and nitrogen. The objective of the present study has been to study the formation of TiN by the method of reactive milling, starting from elementary Titanium powders milled in air atmosphere. The method used involves a mechanical activation of the reactions of elementary titanium with nitrogen of the environment, by means of the milling of powders. In order to study the effect of the milling intensity, Titanium powders were milled in an attricionator mill and also in a Spex mill.

Experimental Procedure.

Material used. Materials used for the production of the new phase of TiN were Titanium powders with a nominal purity of 99,99 % (Aldrich), and reactive degree methanol that was used as control agent in Spex.

Reactive milling to obtain TiN in a Spex mill. A Spex mill was used equipped with a stainless steel container in which 5g Ti powders were loaded in each test in air atmosphere, using 6 stainless steel small balls of 5 mm diameter as milling elements. In order to avoid that the milled powder stuck in the walls of the containers, first the necessary amount of control agent was determined, making millings with different methanol content, varying from 0 to 0,7 milliliter, in intervals of 0,1 milliliter. Once the optimal amount of methanol was determined, millings were made for different time intervals of 3, 6, 9 and 12h at 400 r.p.m. In all cases, the containers atmosphere was refreshed every 3 hs. Each milling process was followed by X-rays characterization of powders based on the milling time. [3].

Reactive milling to obtain TiN in a atricionador mill.- Another lot of Titanium powders was milled in an Process Union attricionator mill in a 3,25 l stainless steel container, using 50g of powders of Ti and 7 milliliter of methanol control agent in each test. As milling element 3 kg of stainless steel balls of 3 mm diameter were used. This mill is not hermetic, so that air atmosphere is renewed constantly during the milling. The times of milling were 24, 48, 72 and 96 h, at 400 r.p.m. Before extracting powders of the milling container, it is let cooling down for 2hs. In all cases a small sample (~ 0,1 g) was taken from powders every 24 h. The milling process was followed by X-rays characterization of powders based on the milling time.

Characterization of powders.- The evolution of phases of milled powders. was followed by X- ray diffraction (XRD) using a Siemens D 500 diffractometer with k-α radiation of Cu and collimator of 1 mm, doing a scanning from 10 to 100° in 2θ at 2°θ/min with increments of 0.03° 2θ. The lattice parameter of TiN powders was determined by the method of least square. The particle size was measured in a Malvern Zetasizer IV equipment.

Results.

Effect of the milling time.- The time of milling to obtain a new phase depends on the type of mill that is used, of the agent of milling and the speed or intensity of the milling. Figure 1 shows the XRD patterns of milled powders in the attricionator for 0, 24, 48, 72 and 96 h. After 24h milling, Titanium diffraction peaks are broadened and their intensity is considerably reduced, which indicates that particle size diminishes and powders tend to lose chrystallinity. This pattern also shows a very wide and poor defined peak of TiN. After 48h milling, the pattern is flat, which possibly indicates that the Titanium powders amorphize after this milling time. After 72h milling, Titanium signals disappear completely and wide and low intensity TiN peaks appear. This process continues up to 96h milling .

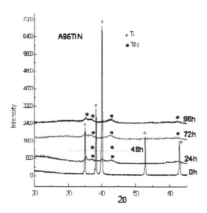

Figure 1. X-rays diffraction patterns of Titanium powders milled in the attricionator for 0, 24, 48, 72 and 96 h.

XRD patterns of Titanium powders obtained in the Spex by different milling times are shown in Figure 2. Table 1 summarizes the phases identified for every time of milling. After 3 h milling, the Titanium phase continues to appear but the peaks are wider and of smaller intensity than in the original powder, which indicates that the size of Titanium particles diminishes and lose chrystallinity, of similar way to which happens in the attricionator. After 6h milling,

wide TiN peaks appear, that are better defined by increasing milling time. Well defined peaks of TiN appear in powders milled for 9 and 12 h. XRD patterns of powders milled for 9 and 12h have two very small peaks that were identified as of Fe. In Table 2 the content of milled Fe powders is reported. Very likely, this element is introduced in the powders as a product of the wearing down of the containers and of the milling elements.

Previous results demonstrate clearly that the formation of TiN in the Spex is much faster than in the attricionator, because the milling is more intense. However, this process has the disadvantages that only small amount of powders can be processed, in addition to which the contamination with Fe is much greater.

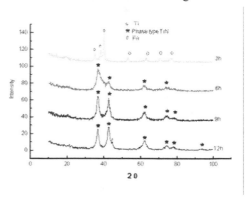

Figure 2. – XRD patterns of Titanium powders milled in the Spex for 3, 6, 9 and 12 h.

Table 2. Identified phases and Iron content of milled powders in the Spex.

Time of milling	Identified phase.	Content of Fe (% weight)
3 h	Ti	0.61
6 h	Ti, TiN	5.62
9 h	Ti, TiN, Fe	----
12 h	TiN, Fe	5.8

Lattice parameters of milled powders for 96 h in the attricionator and 12 h in the Spex were determined by the method of least square. The graph used for the powders milled in the Spex is shown in Figure 3. The powder of TiN obtained by milling in the Spex during 12 h has a lattice parameter of 4,234 Å. The lattice parameter of the powders milled in the attritionator during 96 h was of 4,38 Å. Both values are different from the lattice parameter of pure TiN, which is of 4,242 Å [3].

NSTI-Nanotech 2008, www.nsti.org, ISBN 978-1-4200-8503-7 Vol. 1

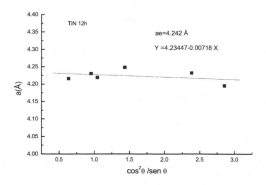

Figure 3. Illustration of the method of square minimums to determine the reticular parameter of the TiN obtained by milling of 12 h in the Spex.

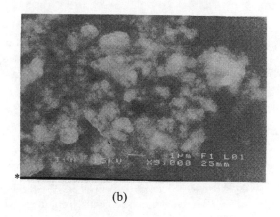

(b)

Figure 4. Micrograph of: (a) powders of original Ti and (b) powders milled in the Spex by 9 h.

Morphology and particle size of milled powders.

In Figure 4 the morphology of powders appears before and after the process of 9h reactive milling of in the Spex. The powders of TiN has an irregular morphology with sizes highcr than 20 μm (Figurc 4á). Milled powders for 9h in the Spex is shown in Figure 4b. This powder is formed by particles with spherical tendency of less than 1μm of diameter, that form clusters.

The distributions of sizes of powders of TiN obtained by milling in the attricionator for 96 h and in the Spex by 12 h appear on Figure 5. Both powders show a Gaussian particle size distribution. Powders milled in the attricionator reaches an average size of around 500 nm, but it has a very wide size distribution from 20 nm up to 5,000. nm (5 μm). On the other hand, the powder milled in the Spex is much finer, of about 178 nm, and has a much narrower size distribution that goes from 45 nm to 515 nm.

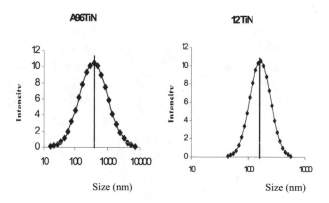

Figure 5. Distribution of sise of milled powders. (a)mMilled powders in attricionator by 96 h, (b) Milled powders in Spex by 12 h.

Discussion of results.

The formation of the TiN phase by reactive milling depends on several factors, such as: Type of mill, atmosphere of milling, balls load to Ti powder lod ratio, milling temperature and balls size In this work, the main difference studied was the type of mill.

XRD results indicate, of general way, that in both mills the reaction between powders and the nitrogen of the air is carried out, appearing a phase identified as TiN type, of cubical structure. The powders milled in the attricionator during 96 h has a lattice parameter 4,38 Å that would correspond to a Ti(C type, N)compound. The powders milled for 9 h in the Spex has a lattice parameter of 4,23 Å, so that it would be a compound of Ti(O, N) type. The greater speed of formation at the Spex is because this mill is a more powerful one than the attricionator.

(a)

The morphology of powders obtained is nodular, with a particle size of the nanométrico order as considered by the analysis made by scanning electron microscopy and from the distribution size measurement of particle of powders. This difference also could be explained by the fact that milling in the Spex is much more intense.

The previous discussion shows the advantages to make the reactive milling in a mill Spex type, although the disadvantages of this method must be taken into account, i.e. a greater contamination with iron, and that only a limited amount of powders can be milled.

Conclusions.

Spherical powders with a TiN type phase can be synthesized very easily by reactive milling from elementary powders of Ti, using a attricionator mill or a Spex, in an air atmosphere. The formation of TiN is much faster in the Spex than in the attricionator. In the attricionator, in 96h milling obtains a type Ti(C, N) compound is obtained, with a lattice parameter of 4,38 Å and an average particle size of 500 nm. In the Spex, in 9h milling a Ti(O type, N) powder is obtained with a lattice parameter of 4,242 Å, and an average particle size of 178 nm. This powder build clusters due to its small size.

References

1. K.K.Chawla: Composite *Materials Science and Engineering* (Springer-Verlag Inc., USA, 1987).
2. P. A. Thornton and V. J. Colangelo: *Ciencia de los Materiales para Ingeniería* (Prentice Hall Hispanoamericana, S.A, México)
3. S. Quintana: *Obtención de Nitruro de Titanio por Molienda Reactiva*. (Tesis de
4. K. Isozaki: Fine Ceramics Report 8 (1990) p. 264.
5. M. Moriyama, H. Auki, Y. Kogayashi and K. Kamata: J. Ceram. Soc. Jpn.101 (1993) p. 279.
6. K. Niihara: J. Ceram. Soc. Japan, 99 (1991) p. 974.
7. K. Niihara y A.Nakahira: Ann.Chim.Fr. 16 (1991) p. 479.
8. Licenciatura en Ing. Metalúrgica, E.S.I.Q.I.E., I.P.N, México, D.F. 1996).
9. T Tsuchida, T, Hasegawa and M. Inagaki: J.Am. Ceram. Soc. 77 (1994) p. 3227.

Y_2O_3 Nanophosphors Synthized by Combustion and Thermal Decomposition Techniques

D. Tatar[***], H. Kaygusuz[*,**], F. Tezcan[**], B. Erim[*], ML Ovecoglu[***] and G. Ozen[**]

Istanbul Technical University, Maslak-Istanbul, Turkey 34469
[*]Department of Chemistry, Faculty of Science & Letters,
[**]Department of Physics, Faculty of Science & Letters
[***]Department of Metallurgical & Materials Engineering,
Faculty of Chemical & Metallurgical Engineering,

ABSTRACT

The combustion and chemical decomposition techniques were used to synthesize Y_2O_3 nanophosphors that are high quality powders as promising materials for the next generation of the display technology. Combustion technique focused on an exothermic reaction between Yttrium nitrate and glycine as fuel. The thermal decomposition technique (to our knowledge, a new technique used to synthesize Y_2O_3 nanocrystalline phosphors) is based on the thermal decomposition of Yttrium alginate gels. The gels in this technique were produced in the form of beads by ionic gellation between a yttrium solution and sodium alginate. Both the wet beads and the Y_2O_3 powders obtained by using the former technique were annealed at 600, 800 and 1000 °C for various annealing times. The products were characterized by X-ray diffraction (XRD) and the crystal size distribution of each product was measured by BET technique. The results explicitly illustrate that the size of the nanocrystalline Y_2O_3 phosphors is influenced by the technique, annealing temperature and the duration of the annealing process. We found that the size of the nanocrystalline Y_2O_3 phosphor varies from a 9nm to about 200nm nanometers.

Keywords: Y_2O_3 nanophosphors, chemical decomposition, combustion, XRD.

1 INTRODUCTION

Significantly improved performance of displays and light emitting devices demands high-quality phosphors having sufficient brightness and long term stability[1]. Lanthanide activated rare earth oxides such as Eu^{3+} and/or Tm^{3+} doped Y_2O_3, remain as promising materials for the next generation display technology because of the following important properties:

1. They are stable in vacuum,
2. They give corrosion-free gas emission under electron bombardment compared with red phosphors used in current field-emission displays and,
3. They are damage resistant and high reflection materials when used in the light-emitting diodes and/or high power UV lasers.

The size of the nanocrystals is significantly influenced with the technique used for preparation as well as the annealing temperature and duration of the annealing process [2]. In the present project we report that the size of the nanocrystalline phosphor material was varied from a few nanometers to several hundred nanometers by annealing the samples at varied temperatures and using two different preparation techniques.

2 EXPERIMENTAL

The combustion and chemical decomposition techniques were used to synthesize Y_2O_3 nanophosphors. Yttrium nitrate, glycine and low viscosity (250 cps % 2 solution in water) acid sodium salt purchased from Sigma-Aldrich. 0,2 M Yttrium nitrate solution and %1 (w/v) alginate solution were prepared by dissolving in ultra pure water. Yttrium-alginate beads were produced by drop wise addition of 30 ml alginate solution in 60 ml of yttrium nitrate solution by means of stainless steel needle. Prepared yttrium-alginate beads were shaken for 30 min with the rate of 150 rpm. After this process, beads were separated from the solution and placed in a silica crucible and heated at 450 ^{0}C for 24 hour. Finally beads were heat treated at 600-800-1000 ^{0}C and at each temperature for 2-4-8-16 hours. X-ray diffraction (XRD) measurements were performed using a Bruker^{TM} D8 Advanced Series powder diffractometer to confirm the formation of Y_2O_3 crystalline phase. All traces were recorded using $CuK\alpha$ and the diffractometer setting in the 2θ range from 20° to 60° by changing the 2θ with a step size of 0.02^o. All samples were ground to fine powder for investigation and Eva Software was used to label peaks in the sample.

3   RESULTS AND DISCUSSION

Fig.1 (a) and (b) show the x-ray diffraction patterns (XRD) of the Y_2O_3 powder samples which were synthesized using the combustion and the chemical decomposition techniques. The diffraction patterns were compared with the data from ICDD and a good agreement was found between the diffractogram obtained and the cubic Y_2O_3 (File # 71-5970). The XRD patterns were obtained by annealing the Y_2O_3 powders at 600°C for the different duration of annealing times. As it can be seen from both figures, no appreciable effect neither of the method used to synthesize the powders nor the time duration of the heat treatment exists on the peaks position and the full-width at the half of the maximum intensity (FWHM).

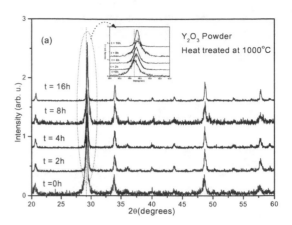

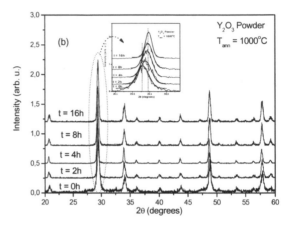

Figure 2:XRD patterns of the Y_2O_3 powders annealed at 1000°C for 16-hours synthesized with (a):combustion (b): chemical decomposition techniques.

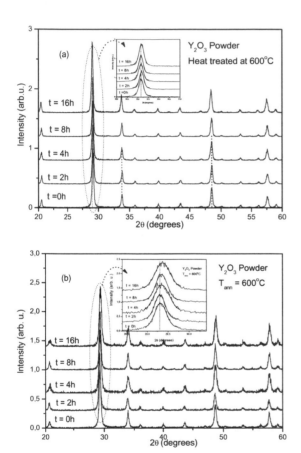

Figure 1: XRD patterns of the Y_2O_3 powders annealed at 600°C for 16-hours synthesized with (a):combustion (b): chemical decomposition techniques.

Fig.2 (a) and (b) show the x-ray diffraction patterns (XRD) of the Y_2O_3 powder samples which were synthesized using two techniques of the samples annealed at 1000°C for 16-hours. As it can be seen from the figures, both of the method used to synthesize the powders and the time duration of the heat treatment has an observable effect on the peaks position and the FWHM.

The crystalline size of the powder samples were determined from the XRD peaks using the Scherrer equation [3] which is given as follows;

$$L = \frac{K\lambda}{\beta Cos(\theta)}$$

where K and λ are the Scherrer shape factor taken as 0.89 and the X-ray wavelength of the CuKα taken as 1.54°A, respectively; θ is the Bragg angle and the β is the pure line broadening. The crystallite size of the Y_2O_3 powders is influenced by both the annealing temperature and the method used to synthesize the powders. The crystalline is determined to be 21±2nm and 35±2nm when the annealing temperature was below 800°C for the powders produced by chemical deposition and the combustion methods as can be seen in Fig. 3. The crystalline size grows much faster above this temperature when the combustion method was used.

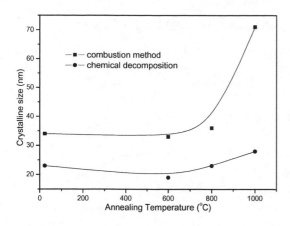

Figure 3:Variation of the crystalline size with the annealing temperature (■):combustion (●): chemical decomposition techniques.

4 CONCLUSIONS

Y_2O_3 nanophosphors powders were synthesized using the combustion and chemical decomposition techniques and heat treated at 600, 800 and 1000oC for 16-hours. The variation of the crystallite sizes were determined using the Scherrer equation to the peaks observed in the XRD patterns. According the experimental results the following conclusions can be made;

1. The crystalline sizes are smaller at each annealing temperature when the powders were prepared using the chemical decomposition technique than those obtained with the combustion method.
2. The crystalline sizes of the powders prepared by both methods do not vary appreciably with the annealing temperature below 800°C.
3. The crystalline size grows much faster above 800°C when the combustion method was used.

REFERENCES

[1] H. Huang, G. Qin-Xu, W.S. Chin, L.M. Gan and C.h. Chew " Synthesis and characterization of Eu:Y2O3 nanoparticles" Nanotehnology 13 318 – 323 (2002).
[2] C.A. Kodaira, R. Stefani, A.S. Maia, M.C.F.C. Felinto, H.F. Brito, J. Lumin.., 127, 616-622, 2007.
[3] C. Suryanarayana, M.G. Norton, "X-Ray Diffraction: A Practical Approach", Plenum Press, 207-220, 1998.

The Effect of Oxygen Beam (O^{+7},100 MeV) and Gamma Irradiation On Polypyrrole Thin Film.

Subhash Chandra [*], S Annapoorni [**], R G Sonkawade [***], J M S Rana [*] and R C Ramola [*]

[*]. Department of Physics, HNB Garhwal University Badshahi Thaul Campus, Tehri Garhwal - 249 199, India
[**]. Department of Physics and Astrophysics, University of Delhi, Delhi - 110 007, India
[***]. Inter University Accelerator Center, Aruna Asaf Ali Marg, New Delhi - 110 067, India

ABSTRACT

Polypyrrole thin films doped with Para-toluene Sulphonic acid were prepared by electrochemical process and a comparative study of the effect of Swift Heavy Ions and Gamma rays irradiation on the structural and optical properties of Polypyrrole (PPY) has been carried out. For this study Oxygen ions (energy 100 MeV, charge state O^{+7}) fluence varies from 1×10^{10} to 3×10^{12} ions/cm^2 and the gamma dose varies from 6.8 to 67 gray. The polymer thin films were characterized by X-Ray diffraction (XRD), UV- Visible spectroscopy and Scanning Electron Microscopy (SEM). XRD pattern shows that after the irradiation the crystallinity was improved with increasing fluence, which could be attributed to cross-linking mechanism. The UV-visible spectra shows a shift in the absorbance edge towards the higher wavelength and significant decrease in the bandgap was found after irradiation. SEM study shows a systematic change in the surface of the polymer. A similar pattern was found with the Gamma irradiation.

Key Words: Ppy, SHI, XRD, UV-Visible, SEM.

1   INTRODUCTION

Polypyrrole is an especially promising Inherently Conducting Polymers, as it is highly conducting, environmentally stable and relatively easy to synthesise. It has recently found applications in a wide range of fields, including chemical and biological sensors [1,2] lighting emitting diodes [3], electromagnetic interference shielding [4] and advanced battery systems [5,6]. Now research has been focused on the continuous monitoring for hazardous chemical vapors present in specified levels. Conducting polymers are good candidates for the sensors because of ease of fabrication, low cost and due to possibility of using the same polymer with the different modifications. The sensing properties of these conducting polymers are associated with detection of some hazardous gases like ammonia and chorine and organic solvent vapors [7,8,9,10]. The physical property of PPY films largely depends on the nature of dopant and methods of sample preparation and also low temperature of polymerization [11] that leads good electrical conductivity.

Various dopants are used for synthesis of Ppy by electrochemical process [12,13]. Due to commercial uses polymers become a subject of scientific and commercial interest. The use of ion irradiation of these materials is of great importance to modify properties of these materials. Any modification of material depends on the structure and the ion beam parameters (ion mass energy and fluence) and the nature of target material itself. By using high-energy heavy ions, dramatic modifications in the polymer material have been observed. In general most of these modifications can be traced back to changes taking place in chemical structure of polymer. High energy ions by electronic excitation and ionization create the tracks in polymers. The latent tracks in the polymers can cause creation of triple bonds, unsaturated bonds and loss of volatile fragments [14]. Some other workers [15] studied the Polyvinyl alcohol (PVA) exposed by 16 MeV electrons and found that for the same transferred energy density heavy ions were more efficient for the damage in polymers than the low energetic ions. [16] studied the Polypyrrole irradiated by 160 MeV Ni^{+12} ions and found that a sharp increase in the degree of crystallinity of polymer and also found a shift in the absorbance peak towards the higher wavelength.

2 EXPERIMENTAL METHOD

0.1 M Pyrrole monomer (Aldrich) and 0.1 M Paratolune sulphonic acid (Lancaster, UK) were distilled in double distilled water. The platinum plate used as counter electrode and electrochemical polymerization of the pyrrole was carried out on ITO, during the polymerization anodic potential was kept 0.8 V. Varying the deposition time controlled the thickness of the thin films. Self standing films of Ppy of size 1cm^2 were irradiated in Material Science Beam Line under high vacuum 5×10^{-6} torr by using the 100 MeV O^{+6} ion with a beam current of 1 pnA available from 15 UD Pelletron at Inter University Accelerator Centre, New Delhi using various fluences ranging from 1×10^{10} to 3×10^{12} ions/cm^2. The thickness in the present work was selected so as to be thin enough to allow the 100 MeV Oxygen ions to completely pass through it. XRD of the polypyrrole thin films were carried out by a Bruker AXS, X-ray diffractometer with Cu-Kα radiation (1.54 A^0) for a wide range of Bragg's angle 2θ (15<2θ<40). UV-Visible Spectra were obtained using a U-3300 Spectrophotometer. SEM images were obtained using a JEOL JSM 840 Scanning Electron Microscope.

3 RESULTS AND DISCUSSION

3.1 X-RAY DIFFRACTION

The XRD pattern of pristine and irradiated PPY irradiated by Oxygen beam is shown in Fig.1. The pristine polypyrrole shows the semi crystalline nature. After irradiation with SHI at low fluence, intensity of the peaks increases, which show the polymer crystallinity increases. This may be due to cross linking of the polymer chains or by the formation of single or multiple helices which produces more crystalline regions in the polymer film which show the crystalline behavior of the polymers with SHI. The degree of crystallinity for the polymers calculated by following formula

$$C = \frac{A}{A}\,X100\%$$

Where A is the total area of the peaks (area of crystalline and amorphous peaks) and A' is the total area under the diffractogram [17]. The % crystallinity calculated by above relation is shown in Table 1.

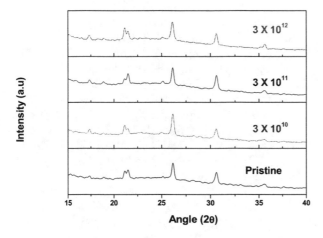

Fig1. XRD of Poypyrrole irradiated with O^{+7} beam.

O^{+7} ion fluences (Ions/cm^2)	% K of Ppy	Gamma rays dose (gray)	% K of Ppy
Pristine	23.02	pristine	23.02
3X10^{10}	26.91	6.8	23.45
3X10^{11}	30.39	12	23.90
1X10^{12}	--	30	25.69
3X10^{12}	32.21	68	26.60

Table 1 : Calculated %crystallinity in Polypyrrole.

The crystallinity of the polymers arises due to the formation of single or multiple helices along their length [18]. [16] Also studied the Polypyrrole irradiated with Ni +^{12} ion and found that the crystallinity of PPY increases after irradiation.

3.2.1 UV-VISIBLE SPECTROSCOPY

The electronic structure and the carrier type in the polymers can be visualized by UV-visible spectra. The UV-visible spectra recorded for the O^{+7} ion beam irradiated polypyrrole Fig 2 & Fig 3. The absorption peak around 450 nm is polaron absorption peak of the conducting polypyrrole [19]. A shift in the absorption peak towards higher wavelength was found indicating a decrease in the energy bandgap of the polymer after SHI irradiation which gives rise increase in the dc conductivity of polymers.

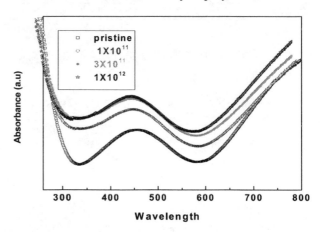

Fig. 2: Absorbance of Ppy irradiated by O^{+} ^{7} ion beam

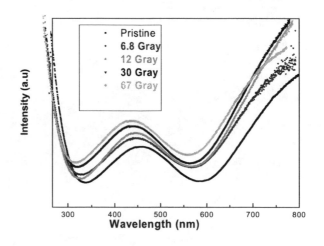

Fig. 3: Absorbance of Ppy irradiated by gamma rays

This shift in the absorption may be produced in creation of free radicals or ions and thus have a capability of increasing the conductivity of the polymers [20]. From the absorption the direct band gap of the polymers was calculated by linear part of the Tauc's plot [21]. The band gap was found 3.4 eV for the pristine PPY, which decrease upto 3.0 eV after irradiation, by oxygen beam.

It may be higher rate of electronic energy loss of oxygen ions, which affect the optical properties of polymers to greater extent. The calculated values of band gap are shown in Table 2.

O^{+7} ion fluences (Ions/cm^2)	band gap (eV)	Gamma rays dose (gray)	band gap (eV)
Pristine	3.4	pristine	3.4
3×10^{10}	----	6.8	3.4
1×10^{11}	3.3	12	3.4
3×10^{11}	3.2	30	3.3
1×10^{12}	3.0	67	3.3

Table 2: Band gap calculation in Polypyrrole.

3. SCANNING ELECTRON MICROSCOPY

The dopant and the solvent used during the synthesis affect the morphology of the polymer and also their physical, chemical and the electrochemical properties [22].

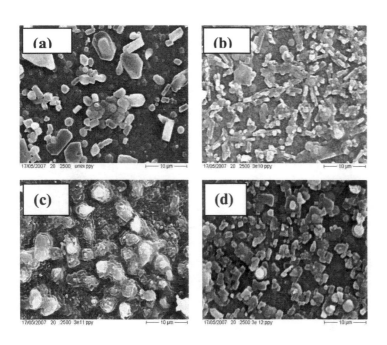

Fig 4. SEM of Poypyrrole irradiated with O^{+7} beam (a) unexposed, (b) 1×10^{11} ions/cm^2, (c) 3×10^{11} ions/cm^2 and (d) 3×10^{12} ions/cm^2. Scales in the images are 10 µm.

Some other workers stated that Ppy prepared by using different dopants shows different morphological structures [23]. The

SEM images recorded for the O^{+7} ion beam irradiated polypyrrole is shown in Fig 4. Grain like structure was observed in the unexposed polypyrrole thin film. After irradiation with SHI microcrystalline structure can be seen at a fluence of 3×10^{10} ions cm^{-2}. But higher fluence of 3×10^{11} ions cm^{-2} & 3×10^{12} ions cm^{-2} exhibit growth of Grain like structure after irradiation. The Grain like structure growth upon SHI irradiation may be due to huge energy deposition by heavy ions by the process of electronic energy loss.

REFERENCES

[1] Ramanaviciene, A & Ramanavicius, A., Crit Rev Anal Chem 32 (2002) 245-252.
[2] Bhat, N. V., Gadre A. P and Bambole V.A., J Appl Polym Sci 88, 22-29, 2003.
[3] Shen, Y and Wan, M., Synth Met 98, 147-152, (1998).
[4] Kim, S.H., Jang, S.H., Byun, S.W., Lee, J.Y., Joo, J.S., Jeong, S.H and Park, M.J., J Appl Polym Sci 87, 1669-1674, 2003.
[5] Levi, M.D., Gofer, Y and Aurbach, D., Polym. Adv. Technol 13, 697-713, 2002.
[6] Kuwabata, S., Masui, S., Tomiyori, H and Yoneyama, H., Electrochim. Acta. 46 91-97, 2000.
[7] Partridge, A.C., Jansen, M.L and Arnold, W.M. Mater. Sci. Eng. C12, 37-42, 2000.
[8] Ampuero, S and Bosset, J.O. Actuators B94, 1-12, 20003.
[9] Bartlett, P.N and Ling-Chung, S.K. Sens. & Actuators 20, 287-292, 1989.
[10] Cabala, R., Meister, V and Potje-Kamloth, K. J. Chem. Soc. Faraday Trans. 93, 131-137, 1997.
[11] Yoon, C. O., Sung, H. K., Kim, J. H., Bassoukov .E., Kim J.H and Lee H., Syn. Met. 99, 201, 1999.
[12] Ramanathan, K., Annapoorni, S., Kumar A and Malhotra, B.D., J Matter. Sci. Lett 15 , 124, 1996.
[13] Rodriguez, J., Grande H. J., Otero T. F., In Handbook of Organic Conductive Molecules and Polymers, H S Nalwa, John Wilely and Sons, Vol 2 Chapter 10 ,1997.
[14] Lee, E. H., Nucl. Instru. And Meth. B 151, 29, 1999.
[15] Bauffard, S, Gervais, B., and Leray, C., Nucl. Instru. And Meth. B 105, 1, 1995.
[16] Fink, D., Chung, W.H., and Wilhelm, M., Rad. Eff. And Deff. In Solids. 133, 209, 1995.
[17] Hussain, A. M. P., Kumar, A., Singh, F., and Avasthi, D.K., J.Phys. D: Appl. Phys. 39 750-755, 2006.
[18] Bruno Scrosati (Ed.), Appliations of Electroactive Polymers, Chapman & Hall, London, 1993.
[19] Chen, J., Too, C.O., Wallace, G.G., and Swierers, G.F., Electrochm. Acta 49, 691, 2004.
[20] Singh Surinder., and Prasher Sangeeta., Nucl. Instru. And Meth. B 222, issue 3-4 518-524, 2004.
[21] Tauc, J and Grigorovici, R, A. Vancu. Phy. Stat. Soli. B. 15, 627, 1966.
[22] Chung, K.M., and Bloor, D., *Polymer* 29, 1709, 1988.
[23] Price, C.J, J. Phys. Chem. 17, 231, 1993.

The Effect of Inclusion of Nanoparticles on the Rheological and Morphological Properties of Triblock Copolymer Gels

M. A. Paglicawan[**] and J. K. Kim[*]

[**]Industrial Technology Development Institute, Philippines, mapaglicawan@yahoo.com
[*]Gyeongsang National University, South Korea, rubber@gnu.ac.kr

ABSTRACT

Nanocomposite materials were prepared by embedding nanosized particles into triblock copolymer gels. The properties related to morphology, viscoelasticity and thermal stability were explored and discussed. Dynamic rheological measurements of the resultant NCTPE gels showed that at temperature between 30 °C to 40 °C below the gel point, the nanocomposite thermoplastic elastomer gels (NCTPEGs) have dynamic storage modulus greater than loss modulus (G' and G''), thereby indicating that at ambient temperature a physical network was still present despite the addition of nanoparticles. Storage modulus slightly increases as the nanoparticles increases. The morphology revealed that nanoparticles used to generate nanocomposite triblock copolymer gels are dispersed generally within the swollen copolymer and or solvent. Thermal degradation was improved with addition of nanoparticles. This research hopefully gives new advancement in the field of nanocomposite polymer gels with wider application.

Keywords: nanocomposite, triblock copolymer, nanoparticles, MWCNT, nanographite platelets

1 INTRODUCTION

Most studies of polymer nanocomposites utilize a homopolymer matrix to which nanofiller is added. In recent years, the colloidal properties of nanocomposite polymer clay gels and solutions have also received considerable attention in the literature. Unusual properties are induced by the physical presence of the nanoparticle and by the interaction of the polymer with the particle and the state of dispersion. Silva et al. [1] have examined the rheological properties of intercalated nanocomposites based on a poly(styrene-*b*-isoprene) (SI) diblock copolymer and a dimethyl/dioctadecyl-substituted montmorillonite (silicate). While numerous fundamental studies on the rheological and morphological properties of molecular networks composed of a microphase-separated multiblock copolymer swollen to a large extent by a low-volatility midblock-selective solvent, [2-6] few comparable efforts have extended studies to thermoplastic elastomer gels in the presence of nanoparticles. The aim of this paper is to determine the effect of these nanoparticles on the rheological, thermal, and morphological properties of nanocomposite thermoplastic elastomer gels (NCTPEGs). The results obtained here further add to the insight gleaned from previous experiment studies of triblock copolymer [7-9] and hopefully gives new advancement in the field of nanocomposite polymer gels with wider application.

2 EXPERIMENTAL PROCEDURE

2.1 Materials

The poly[styrene-b-(ethylene-co-butylene)-b-styrene] (SEBS) triblock copolymers with number molecular weight of 79,000 and 29 wt% S (Kraton G1652) was used as received. Two types of exfoliated Graphite Nanoplatelets (xGnP) were used, xGnP1 which has thickness of ~10nm and an average diameter of 1μm and and xGnP-15 with the same thickness and diameter of 15 μm. Details on the exfoliation process as well as on the morphology of xGnP can be found elsewhere [10]. The MWCNTs was purchased from Iljin Nanotech Co., Korea, synthesized by the chemical vapor deposition (CVD) process with average diameter of 13 nm and length of 10 μm. Purity of the prestine MWCNTs, as received, was 97%. The paraffin oil with molecular weight of 480 g/mol and density of 0.88 g/cm^3 was used, which was supplied by Michang Oil Industrial Co., South Korea.

2.2 Preparation of Nanocomposite Triblock copolymer Gels

Nanoparticles were sonicated in ethanol bath for 8 hours, dried and dispersed in chloroform before it was added to the mixture. Nanocomposite triblock copolymer gels composed of 20% SEBS, 1% Irganox antioxidant, 80 % Paraffin Oil , and 0.5 to 5 wt% nanoparticles were prepared in similar in our previous manner except the temperature is reduced to 130°C [5-6]. The resultant hybrid gels were compression-molded without applying much pressure for 5 minutes at 130 °C to yield sample measuring 2.0 to 2.5 mm thickness.

2.3 Characterization

The elastic storage modulus, G′ and loss modulus, G″ were measured at a temperature range 30-140 °C with increments of 10 °C/min with constant strain (γ_o) of 1% using a strain-controlled Rheometrics Mechanical Spectrometer (RMS800, USA). Parallel plate geometry with 25 mm and with 1.5 or 2.5 mm gap heights was used for measurement.

The transmission electron microscopy (TEM) measurements were carried out with a Bio-TEM transmission electron microscope applying an acceleration voltage of 120 kV. The specimens were cut at -100 °C by an ultra microtome (Ultra cut E, Reichert & Jung) equipped with a diamond knife. Ultra thin sections of approximately 50 nm thickness were stained with the vapor of 0.5% RuO_4 (aq) gas phase for 5 min, i.e. the PS blocks are stained selectively.

To discern the effect of SEBS/and or hydrocarbon oil on the morphology of the platelets, the spatial atomic arrangements of the neat MWCNTs and NCTPEGs were characterized using a Bruker AXS X-ray Diffractometer (Type D8 Advance), Germany and a Siemens generator (Kristalloflex 760). X-ray diffraction (XRD) profiles were recorded using CoK_α radiation in the angular range from 10°-60° (2θ) at an operating voltage of 40kV and a current of 20mA with a wavelength of 1.7902 Å.

The thermal stability of NCTPE gels was evaluated by TG-DTA. TG measurements were carried out on 10 mg sample in a DuPont TA2100 TGA in the temperature range of 30-700 °C at a heating rate of 20 °C/min and a nitrogen flow of 50 ml/min. For determining the degradation temperature of TPE and NCTPE gels, thermogravimetric analysis (TGA) was carried out.

3 RESULTS AND DISCUSSION

3.1 Rheological Properties

All NCTPEGs with hybrid of nanoparticles have been subjected to dynamic mechanical studies to discern the effect of nanoparticles on the rheological behavior of NCTPEGs. The oscillatory shear measurements were focused on the variations of elastic (in-phase) G′ and viscous (out-of-phase) G″, as a function of temperature and frequency. Figure 1(a,b & c) show the change in G′ with temperature heated throughout the melting range for NCTPEGs containing different amounts of nanoparticles. At a temperature between 30 °C to 40°C, an initial plateau over which G′ remains relatively constant or slightly increases with increasing temperature is observed, which means that the rubbery PS domains become glassy below the gel point. Also at this temperature, the NCTPEGs have the property of elastic moduli, where G′ > G″. This indicates that at ambient temperature physical network is still present despite with addition of nanoparticles. As the

temperature increases, an abrupt reduction in G′, which attributed which an indicative of limited network alteration. Eventually plummeting of G′ was observed as the network ultimately collapses. The most common feature observed for all NCTPEGs at low concentration between 0.5 to 1 wt%, G′ values are lesser and drops at lower temperature. It was observed that the magnitude of G′ slightly increases at low concentration. The minimal increase in G′ is probably due poor dispersion that tend the particles to flocculate into large-scale aggregates and thus may not diffuse within the swollen polymer network. Also, the nanoparticles have less connectivity to the underlying copolymer network and weak bonding between the nanoparticles and swollen SEBS/oil system. However, with further addition of 3 to 5wt % concentrations to SEBS/oil gel likewise promotes a modest increase in G′, which is due to high aspect ratio nature of the material.

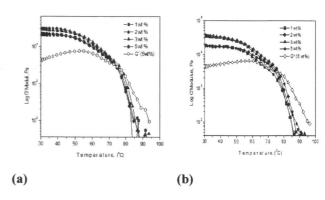

(a)   (b)

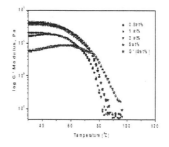

Figure 1. Storage modulus, G' of NCTPEGs (a) xGnP-1, (b) xGnP-15 and (c) MWCNT. Loss modulus (G") (open symbol) curves are also included for 5 wt% concentration.

3.2 Morphological Properties

Figure 2 displays a pair of TEM images collected from NCTPEGs modified with 5 wt % of, (a) xGnP1, (b) xGnP15 and MWCNT. The images of NCTPEGs exhibit morphology composed of a micellar of SEBS in hydrocarbon oil and nanographite. The irregularity shaped, dark features identify the nanoparticles. It is clear that these nanoparticles are flocculated into large-scale aggregates due to high solvent content and hinder to a lesser extent the bridging efficacy of individual copolymer molecules which possibly dictated by poor dispersion. It was observed by Kalaitzidou [11] that xGnP15 has a tendency to roll together and to form some agglomeration during mixing due the intrinsic van der Waals attractions between the individual nanoplatelets. The relatively high magnification

NSTI-Nanotech 2008, www.nsti.org, ISBN 978-1-4200-8503-7 Vol. 1

image in all the figures of NCTPEGs confirm that the matrix consists of SEBS micelles measuring ca.20 nm in core diameter uniformly dispersed throughout the hydrocarbon oil, which agrees well with our previous studies [12]. It is clear from these related images that the micelles do not exhibit discernible indication of long-range order a face- or, more likely, body-centered cubic lattice. This indicates that addition of nanoparticles didn't reduce the intermicellar distance. Moreover, the nanoparticles are formed within the swollen polymer network and not on PS styrene endblock. This finding was also observed in expandable graphite (EG) with the same parent TPE gels [12].

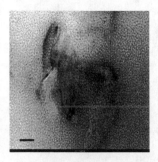
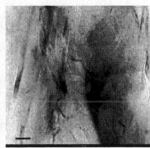
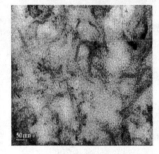

Figure 2. Storage modulus, G' of NCTPEGs (a) xGnP-1, (b) xGnP-15 and (c) MWCNT. Bar size: 50 nm

X-ray diffraction is valuable tool in discerning the extent to which the SEBS/oil matrix alters the layering of the MWCNTs and possibly modify their intrinsic morphology. Figure 3(a & b) represent the X-ray data for NCTPEGs in comparison with parent TPE gels and their neat nanoparticles. From X-ray diffraction (XRD) the 002 peak was observed at $2\theta = 25.95°$ for xGnP (Figure 3a). This xGnP exhibited a clearly discernible basal reflection corresponding to layer spacing 3.40Å. The d-spacing for NCTPEGs containing 1 wt% and 5 wt% is 26.31°, while xGnP15° is 26.05° for 1 wt % and 26.69° for 5 wt %. The shift in xGnP15 and xGnP1 to higher angles is similar to NCTPEGs with EG [12]. This phenomenon was also seen in PP-g-MA layered EG [13]. Cheng reported that the shift to higher angles indicates tighter packing in the crystal unit cell in directions perpendicular to the chain direction [14]. This can be concluded that the xGnP did not intercalate into the gallery of the carbon layers of graphite during mixing. The NCTPEGs with xGnP is a typical conventional microcomposite. This can be explained that even after the expansion or exfoliation process of the intercalated graphite flakes, it is practically not possible to obtain ideally or completely exfoliated graphite layers. The inner layers of the exfoliated platelets may have a graphene nanostructure consisting of multiple graphene sheets [15-16]. As a result, the NCTPEGs consist of multi-layered xGnP which is dispersed within the swollen triblock network on the nanoscale, as seen in TEM (Figure 2 (a & b)). On the other hand, the structure of MWCNTs is similar to the hexagonal close-packed lattice of graphite with the interplanar spacing of the diffraction peak 002 at $2\theta = 25.55°$, which corresponds to interlayer spacing 3.45 Å. Due to extensive layer separation (beyond the resolution of Bragg-Brentano geometry), it is not possible to observe basal reflection peak, leading to intensity loss and disappearance of the unintercalated basal reflection. This behavior was observed for NCTPEGs with MWCNTs (Figure 3b), indicating that TPE gels is successfully exfoliated within the nanoparticles of MWCNTs, which is well supported in TEM images. The XRD patterns also show a broad, but distinct peak at low 2θ, confirming the existence of the intrinsic micellar network of TPE gels. The broadened peak is attributed to the heterogeneous distribution of styrene micelles.

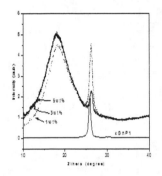
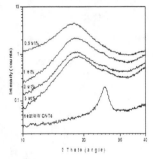

Figure 3. XRD patterns of NCTPEGs (a) xGnP-1 (b) MWCNTs.

3.3 Thermal Stability

The thermograms of NCTPEGs shift towards higher temperature as the heating increases. This shift of thermograms to higher temperature depends primarily on the type of nanoparticles. For each sample, the thermogram revealed that the DTG plot shows a maximum rate of weight loss, so the peak temperatures of degradation (T_p) can be determined. The onset temperatures of degradation (T_{onset}) can be calculated from the TG curves by extrapolating from the curve at the peak of degradation to the initial weight of the polymer. Similarly, the end temperature of degradation can be calculated from the TGA curves by extrapolating from the curve at the peak of degradation forward to the final weight of the polymer. The difference of the peak temperatures and the onset temperatures of degradation is D_T, which represents the temperature range of thermal degradation. These characteristic temperatures are listed in Table 1. The bonding state of EB middle block which dissolves by high content of oil in SEBS/oil system and microstructure features of the matrix and nanoparticles may be play an important role in determining the degradation temperature of the nanocomposites. It appears that the particles reside in the region of EB block swollen by high content of oil thereby increasing the distinct region of oil degradation temperature. However, the second peak that appears at higher temperature remains constant with addition of nanographites. Due to this, the weight loss for swollen EB containing oil and SEBS matrix remains constant with inclusion of nanoparticles. However, the residual yields increase with increasing nanoparticle content, indicating that thermal decomposition of the polymer matrix was retarded in the NCTPEGs/nanoparticles with higher residual yield. This result may be attributed to a physical barrier effect due to the fact that nanoparticles would prevent the transport of decomposition products in the polymer nanocomposites. Comparing the residue of TPE gels without MWCNTs, there is a little residue because the component of the gels is consists only of carbon and hydrogen element.

Table 1. Characteristics of thermal stability.

Code NCTPE gels	Onset temperature, °C T_{onset}		Peak temperature, °C T_{peak}		Residue at 600°C, %
	1st	2nd	1st	2nd	
Parent TPEG	141	413	231	455	0.269
xGnP1-1	163	403	256	454	1.74
xGnP1-5	172	412	273	463	4.95
xGnP15-1	160	410	268	450	0.98
xGnP15-3	167	412	274	453	2.14
xGnP15-5	174	421	291	459	3.20
MWCNT-1	169	410	270	452	0.95
MWCNT-5	174	412	279	464	3.1

4   CONCLUSIONS

The nanocomposite triblock copolymer gels can be generated by addition of nanoparticles such as xGnP and MWCNT into a triblock copolymer of poly(styrene-b-(ethylene-co-butylene)-b-styrene) (SEBS) which is selectively swollen in a midblock-selective solvent. The nanoscale additive can be used to modify the dynamic properties of nanocomposite thermoplastic elastomer gels. Although xGnP did not intercalate nor exfoliated, while MWCNT exfoliated, both dynamic modulus increase to some extent. Thermal degradation was improved with addition of nanoparticles. The area of polymer gels with nanoparticles composites is still in infancy stage and much more needs to be done to fully appreciate the systems.

5   REFERENCES

1. Krishnamoorti, J. Ren and A.S. Silva, J. Chem. Phys. 11, 4968, 2001.
2. N. Mischenko, K. Reynders, M.H.J. Koch, K. Mortensen, J.S. Pedersen, F. Fontaine, R. Graulus and H. Reynaers, Macromolecules, 28, 2054, 1995.
3. J.H. Laurer, J.F. Mulling, S.A.,Khan, R.J. Spontak, J.S. Lin, and R.J. Bukovnik, Polym. Sci: Part B: Polym. Phys., 36, 2513,1998.
4. R.Kleppinger, N. Mischenko, H.L. Reynaers, and M.H.J. Koch, J. Polym. Sci. Part B: Polym. Phys., 37, 1833, 1999.
5. J.K. Kim, M.A. Paglicawan, and B. Maridass, Macromol. Res., 14, 365, 2006.
6. J. K. Kim, M.A. Paglicawan, S.H. Lee, and B. Maridass, J. Elas. Plast., 39, 133, 2007.
7. E. Theunissen, N. Overbergh, H. Reynaers, S. Antoun, R. Jerome, K. Mortensen, Polymer, 45, 1857, 2004.
8. G. J. van Maanen, S. L. Seeley, M. D. Capracotta, S.A. White, R.R. Bukovnik, J. Hartmann, J. D. Martin and R. J. Spontak, Langmuir, 21, 3106, 2005.
9. M.A. Paglicawan, B. Maridass, and J.K. Kim, Macromol Symp.,249-50, 601, 2007.
10. H. Fukushima, Graphite nanoreinforcements in polymer nanocomposites, PhD thesis, Michigan State University, Department of Chemical Engineering and Materials Science, 2003.
11. K. Kalaitzidou, H. Fukushima, L.T. Drzal, Carbon 2007, 45, 1446.
12. M.A. Paglicawan, Properties and Structure of Multicomponent Nanocomposite Thermoplastic Elastomer Gels, Ph.D Thesis, Gyeongsang National University, Department of Polymer Science and Engineering. 2007.
13. F..T. Cerezo, C.M.L. Prestong, R.A. Shanks, Comp Sci Tech, 2007, 292, 155.
14. S. Z. C. Cheng, J.J. Janimak, J. Rodriques, Poly(propylene), Structure and Morphology, J. Karger-Kocsis, Ed., Chapman and Hall: London, 1995.
15. D. Cho,L.T. Drzal, J Appl Polym Sci.,75, 1278, 2000.
16. D. H. Cho, S.H. Lee, G.M. Yang, H. Fukushima, L.T. Drzal, Macromol Mater Eng., 290, 179, 2005.

Replication of nano dimple structures by injection molding with rapid thermal control system

Moonwoo Rha, Jang Min Park, and Tai Hun Kwon

Department of Mechanical Engineering, Pohang University of Science and Technology, San 31, Hyoja-dong, Namgu, Pohang, Gyeongbuk, Korea, thkwon@postech.ac.kr

ABSTRACT

Recently, nano structures are attracting more and more attention in various research fields of biomimetics, photonic crystals and so on. Such nano structures are commonly fabricated by using high-cost and low throughput procedures like e-beam lithography and nano imprinting lithography. Mass production methods, for example injection molding, are rarely utilized for nano structure replications due to a difficulty of making strong mold inserts. In this regard, the present study investigates the feasibility of a replication of nano dimple structures by injection molding. The present study proposes an efficient and easy mass production method for nano dimple structures by injection molding. Anodic aluminum oxide (AAO) technique is utilized to make a master with nano dimple structures, and then nickel mold insert is realized by using nickel electroforming (NE) technique with the AAO master. For a successful replication via injection molding, temperature is to be controlled with the help of namely rapid thermal control (RTC) system.

Keywords: nano structure, anodic aluminum oxide, nickel electroforming, injection molding, rapid thermal control system

1 INTRODUCTION

Due to the excellent replicable property and mass productivity, injection molding technique has been used in various research fields and industries. With development of micro chip [1], digital appliances and so forth, micro and nano scale replication techniques are attracting more attention in various research fields and industries. Their current and projected applications include mass storage equipments (Compact Disc, Digital Versatile Disc, Blue-Ray Disc, etc.), optical devices (LCD back-light unit, photonic crystals, etc.), biomimetic substrates (Gecko's foot, Self-cleaning surface [2], etc.), and biochemistry (cell culturing [3], tissue engineering, etc.).

Existing popular storage equipment, DVD, has about 1μm surface structure size. On the other hand, the next generation of high density storage equipment, Blue-Ray Disc, has below 400nm surface structure size. Interest on small and mass storage equipment is increasing more and more. For display appliances like LCD monitor, surface structure size and shape of back-light unit determine the video quality. Research for display appliances also focus on

large scale display by making large area of surface structure. In biomimetics, some organs of animals and plants have been found to have special functions with nano scale structures. On the foot skin of Gecko, crawling about on wall and ceiling, high aspect ratio nano pillars are observed. And on the surface of Lotus leaf which has self-cleaning effect, there are combined structure of micro particles and nano wax crystal. Several research fields focus on reproducing these special features. More uses of such nano structured templates could be found in cell culturing and tissue engineering for the purpose of separation, in-grouping, growing and so on. In many cases, nano structures are fabricated by means of e-beam lithography and nano imprinting lithography. Such processes, however, have high-cost and low throughput and limitation for large area application. On the other hand, Anodic aluminum oxide (AAO) process is quite economic since it needs just cheep materials, chemicals and devices. From AAO technique, somewhat arranged nano dimple or pore structures with 50~500nm pore-to-pore distance can be realized [4]. Also, shape and size of structures can be controlled by electrolyte type and electric current. Such AAO technique was used in recent researches like photonic crystal [5], nannochannel-array [6], bone implant [7], and so forth.

Mass production methods, for example injection molding, are rarely utilized for nano structure replications due to a difficulty in making strong mold insert. Nano structures fabricated by most lithography techniques and AAO process are composed of weak materials, like polymer, ceramic or aluminum, for using mold insert. In order to overcome this limitation, our research group has employed nickel electroforming technique [1]. Injection molding is the most typical mass production method. As a replication cycle time is shorter than any other replication processes like lithography and embossing, it have been used in various industries. Recently, injection molding for micro scale structures are achieved in plenty. Nano scale injection molding, however, still remains within a laboratory level [8].

The present study aims at a replication of nano structures by injection molding. Nickel mold insert is realized by using anodic aluminum oxide (AAO) and nickel electroforming techniques. In particular, mold insert temperature is controlled intensively during the injection molding process with the help of namely rapid thermal control (RTC) system [9].

2 EXPERIMENTAL

The overall replication procedure consists of three steps: i) basic nano structure template fabrication by means of AAO technique, ii) mold insert fabrication via nickel electroforming and iii) injection molding with the help of rapid thermal control system.

2.1 Fabrication of basic nano structure

To fabricate nano dimple structure, anodic aluminum oxide (AAO) technique was used. On the surface of 15mm x 15mm (width x length) pure aluminum template (99.999%, Good fellow), fabrication was realized step by step as described below. Figure 1 shows AAO fabrication process.

First, surface of template was electro-polished in 7℃ solution of perchloric acid and ethyl alcohol with 1:4 volume ratio. In this reaction, thin oxide layer was formed and melt in the solution. Therefore, surface of template became very smooth. Second, anodic oxidation was progressed in 0℃, 0.1M phosphoric acid as a electrolyte with 195 voltage for 16 hours. As shown in figure 1, disorganized alumina layer was formed from this reaction. Third, to remove the alumina layer, etching process was progressed in 65℃ solution of 9g chromium oxide, 20.2ml phosphoric acid and 500ml de-ionized water for 5 hours.

As shown in Figure 1(b), AAO template with nano dimple structure was fabricated. Nano patterns are well ordered and their pore to pore distance is about 500nm.

2.2 Fabrication of mold insert

To fabricate strong mold insert which withstands high pressure and rapid thermal change, nickel electroforming process was carried out on the nano structure template as a master. The whole nickel electroforming is shown in Figure 2(a) including a mold base and Figure 2(b) shows nickel electroformed mold insert and its surface SEM image.

2.3 Injection molding

In order to achieve a good trascriptability of nano structure in the injection molding process, one needs to set up and control temperature history during one cycle. In general molding case with micro scale or larger scale, melt and mold temperatures are fixed in adequate value which is usually given by polymer manufacturer. However, such a common processing condition (summarized in Table 1) turns out to be inadequate for replicating nano scale structures, mainly due to a fast solidification of molten polymer near the mold surface region. To resolve this issue, RTC (rapid thermal control) system is introduced particularly in the present injection molding process. RTC system increases locally and temporally the temperature of mold insert up to glass transition temperature of polymer, and consequently provides a more effective condition for the replication of nano structures with a better transcription quality. Figure 3 shows schematic view of the mold with RTC heater.

Melt temperature	285°C
Mold temperature	90°C
Filling time	0.20~0.40 sec
Packing time	2.0 sec
Packing pressure	140 ~ 160 MPa
Cooling time	15~25 sec

Table 1: Conventional injection molding processing condition with LUPOY

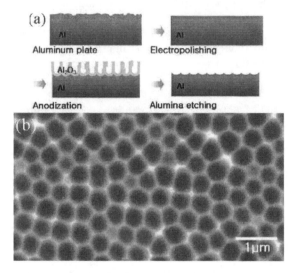

Figure 1: (a) Schematic diagram of AAO process. (b) SEM image of surface nano dimple structure of AAO template.

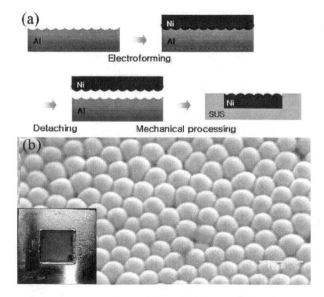

Figure 2: (a) Schematic diagram of nickel electroforming. (b) Image of mold insert and SEM image of surface nano lens-shape structure of mold insert.

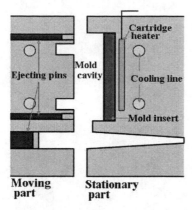

Figure 3: Schematic view of the mold with RTC heater [9]. Before injection of polymer melt, the heater increases the surface of mold insert. The temperature of the mold insert is measured by the thermocouple and monitored by the outside control box (not showed in the figure).

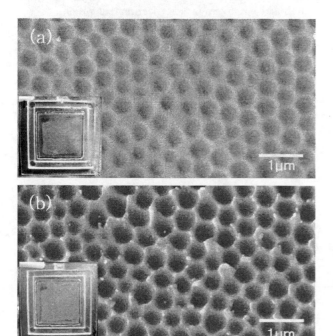

Figure 4: Image of replica which was made by (a) conventional case and (b) RTC system applied case, and their SEM image.

In the present study, injection molding experiment is carried out with RTC system and without RTC system (i.e. a conventional case) to find the effect of RTC on the transcription quality. As for the injection molding polymer, LUPOY (GP1000M, LG chem.) was used.

Figure 4(a) shows the image of replica which was made by the conventional case and its SEM image, and Figure 4(b) shows the images of RTC system applied case. As a means of measuring the transcriptability, we measured the dimension of structures in the molded article from both the conventional case and RTC system applied case: wall thickness of dimple of the latter is thinner than the former and depth of dimple of the latter is deeper than the former. From this result, RTC system provides a more effective condition for the replication of nano scale structures with better transcription quality.

For the detailed comparison, surface structures of basic AAO dimple template, nickel mold insert and replicas of injection molding were observed by AFM (Dimension3100 with Nanoscope Ⅴ in NCNT, POSTECH). Figure 5 shows the analysis result of basic AAO dimple template. The average height of nano dimple is about 120~140nm. Figure 6 shows the analysis result of nickel mold insert. Lens-shape structures are well arranged. Height of structure is about 120nm, close to the dimple depth. Figure 7 shows the analysis result of replica via injection molding with conventional condition. Dimple wall is a little bit blunt and Depth of dimple is about 100nm. Figure 8 shows the corresponding result from the replica of injection molding with RTC system. Dimple wall is sharp and some abnormal peaks are observed. These peaks might be formed by stretching during the demolding process. Depth of dimple is almost 120nm, close to the height of lens in the mold insert. As expected, from the result of AFM analysis, RTC system provides a more effective condition for the replication of nano scale structures with better transcription quality.

3 CONCLUSION

From the results of SEM images and AFM analysis, it is confirmed that the introduction of RTC system improves drastically the transcription quality of nano scale injection molding. As nano dimple structures are relatively simpler than any other structures, it was easy to be replicated by injection molding. For the cases of more complex structures, for example high aspect ratio structures, there may be problems in demolding process.

As interest in nano structures increases in various research fields of biomimetics, photonic crystals, cell culturing and so forth, this successful replication of nano structures via injection molding would open up windows widely to various applications of this technology towards interdisciplinary research fields of IT and BT.

4 ACKNOWLEDGMENT

The authors would like to thank the Korea Science and Engineering Foundation(KOSEF) grant funded by the Korea government(MOST) (R01-2005-000-10917-0), Defense Acquisition Program Administration and Agency for Defense Development (UD060049AD) and 'Development of nano patterning technique using the rapid thermal control system' project which is supported by LG electronics.

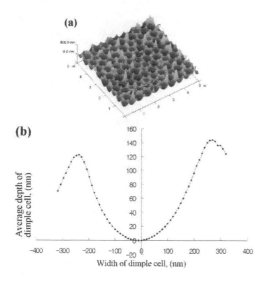

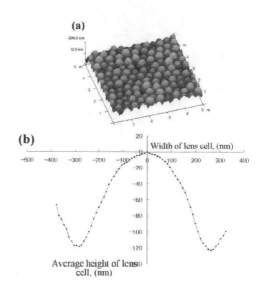

Figure 5: AFM analysis results of AAO dimple template; (a) 3D image, (b) average cross section data graph.

Figure 6: AFM analysis results of nickel mold insert; (a) 3D image, (b) average cross section data graph.

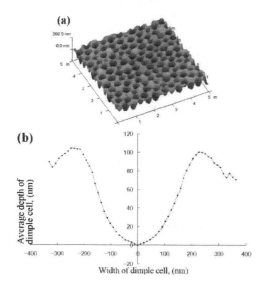

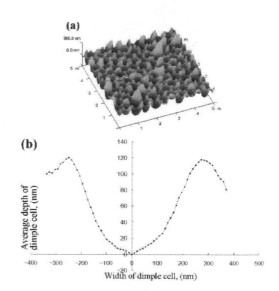

Figure 7: AFM analysis results of replica of injection molding without RTC; (a) 3D image, (b) average cross section data graph.

Figure 8: AFM analysis results of replica of injection molding with RTC system; (a) 3D image, (b) average cross section data graph.

REFERENCES

[1] D. S. Kim, S. H. Lee, T. H. Kwon and C. H. Ahn, Lab on a Chip, 5, 739~747, 2005

[2] S.Lee and T.H.Kwon, J.Micromech. Microeng., 17, 687-692, 2007

[3] A. Curtis and C. Wilkinson, Trens Biotechnol., 19, 97, 2001

[4] Masuda H, Ohya M, Asoh H, Nkao M, Nohtomi M and Tamamura T, Jpn. J. Appl. Phys., 38, L1403, 1999

[5] X. H. Wang, T. Akahane, H. Orikasa, and T. Kyotani, Appl. Phys. Lett., 91, 011908, 2007

[6] H. Masuda, H. Yamada, M. Satoh and H. Asoh, , Appl. Phys. Lett. 71, 2770 1997

[7] A. R. Walpole, E. P. Briggs, M. Karlsson, E. Pålsgård, P. R. Wilshaw, Mat.-wiss. u. Werkstofftech. 34, 1064, 2003

[8] H. Pranov , H. K. Rasmussen, N. B. Larsen and N. Gadegaard, polymer eng. & sci., 46, 160~171, 2005

[9] (Patent) LG electronics, The Korea Intellectual Property Office, (application) 10-2006-0043743, Dec. 26th, 2007

Rheological Characterization of Melt Compounded Polypropylene/clay Nanocomposites.

Mahmoud Abdel-Goad[1*], Bernd Kretzschmar[2], Rashid S. Al-Maamari[1], Petra Pötschke[2] and Gert Heinrich[2]

1: Petroleum and Chemical Engineering Department, College of Engineering, Sultan Qaboos University, P.O.Box 33, Al Khod 123 Muscat, Sultanate of Oman

2: Leibniz Institute of Polymer Research Dresden, Hohe Strasse 6, D-01069 Dresden, Germany

* Corresponding author: M. Abdel-Goad (E-mail: mahmoud@squ.edu.om)

ABSTRACT

Polypropylene/clay nanocomposites were prepared by using a two step melt compounding process on twin screw extruders. The PP-clay nanocomposites were rheologically investigated. The melt rheological measurements were performed using an ARES-rheometer in the dynamic mode at 220°C over frequency varying from 100 to 0.017 rad/s and in the parallel plate geometry with 25 mm diameter.

The results showed that the storage modulus and viscosity are found to be increased with the incorporation of clay into PP. The values of storage modulus and viscosity are significantly increased by the incorporation of modified clay and their rheological behavior changed completely.

The results of samples with modified clay showed an evidence of the exfoliation which could be detected rheologically.

Key words: PP/Clay Nanocomposites, Intercalation, Exfoliation, Rheology

1 INTRODUCTION

Polymer nanocomposites are nowadays the subject of intense research efforts owing to their various unique properties [1-2]. In recent years, polymer/clay nanocomposites have attracted great attention both in industry and academia in achieving various excellent properties of nanocomposites compared to conventional ones. These nanocomposites based on nanoscale-layered silicates exhibit remarkable improvement compared to the conventional micro-composites in various properties as reported by many authors [3-15]. Nanoscale-layered silicates does not use as only reinforcement material for polymers but also can change morphology of immiscible polymer blends. Nanocomposites based on clay and polypropylene have given currently considerable attention because of their potential in improvements the mechanical properties, heat resistance, gas permeability, and flammability barrier properties and the thermostability [16-27]. Clay-based nanocomposites have been studied for different polymers such as polypropylene, polystyrene, polyethylene, polyethylene oxide , poly (ε-caprolactone), and polyamide, etc. [28-36]. Polymer/clay nanocomposites have been widely studied using different preparation methods such as in-situ polymerization, solution blending and melt mixing to get the intercalated or exfoliated structures in homopolymers [37-43]

Melt rheology has been reported as a tool to characterize the polymer–clay nanocomposites [44]. It is known that melt rheology is very sensitive in characterization of the formation of percolated structures of anisotrop fillers in polymeric matrices. This was shown for different kind of fillers, like micrometer sized fibres, montmorillonite nanocomposites as for carbon nanotubes composites [45]. Therefore, it was the task of this study to characterize rheologically the polypropylene/clay nanocomposites with different weight ratios, detect the exfoliation with hope to determine the degree of the exfoliation using the rheology tool.

2 EXPERIMENTAL

Materials: In this work, Polypropylene/clay nanocomposites with an inorganic content of 5 wt% were prepared by melt compounding using a two step melt compounding process on a co-rotating twin screw extruder ZE 25 (Berstorff, Hannover). In the first step a masterbatch of an organically modified montmorillonite (Nanofil® 15 from Süd-Chemie, Moosburg) and a maleic anhydride-modified polypropylene (PP-g-MAH grade TPPP 2112 FA from Kometra, Merseburg) was compounded. In the second step this concentrate was diluted in the PP matrix to the nanocomposite with the focused clay content. This procedure results in optimum polypropylene nanocomposites performance concluded from comprehensive investigations of the compounding process [46]. Additionally, the influence of the process regime in the batch stage by combination of a reactive modification of PP-g-MA in situ with poly(alkyleneoxide)diamines with the compounding of the organoclay was investigated. Improved adhesion of this compatibilizer and the clay platelets in comparison to PP-g-MA, as no debonding could be observed in TEM images, and supported exfoliation of the clay using rheological methods was obtained [47].

Rheology: The PP/clay nanocomposites were rheologically investigated. The melt rheological measurements were performed using an ARES-rheometer (Rheometrics Scientific, USA) in the dynamic mode at 220°C over frequency varying from 100 to 0.017 rad/s and in the parallel plate geometry with 25 mm diameter. The strain amplitude was kept in the linear viscoelastic range.diameter and gap about 1mm.

3 RESULTS AND DISCUSSIONS

The rheological characterization of PP/clay nanocomposites in terms of the storage modulus, G', loss tangent and complex viscosity as function of frequency will be presented in this section.

Figure 1 shows an increase in the storage modulus, by the incorporation of clay into neat PP matrix. This enhancement in the dynamic modulus is significant, in particular, at low frequencies regime. As shown in Figure 1, the loading of 5wt% clay results in rising G' by factor around 4 at frequency 0.056 rad/s. This increase origins from the formation of physical network-like structure by the polymer chains and nano-filler. Additionally, the incorporation of high modulus material (clay) into PP leads to increase the dynamic modulus. However, the rheological behaviour does not change by the addition of 5 wt% unmodified clay. Since at low frequencies zone the PP/clay(5% non-modified) composite has liquid-like behavior as well as neat PP. The same weight of organomodified clay (5w%) changes the rheological behavior from liquid like to solid-like material . As shown in Figure 1 at low frequencies G' becomes nearly independent on the frequency at which second plateau appears in the low frequencies regime starting from ω 0.1 rad/s. This is an evidence of the exfoliation or rather partial exfoliation. This mechanism can be explained as; the diffusion of polymer molecules is favoured by making the galleries chemically compatible with the polymer. The comptability occurs by exchanging interlayers inorganic clay cations with organic cations (such as the alkylammonium cations) . Therefore, layer distance, gallery distance, of modified clay increases due to the modifier. Particles are distributed into single platelets . Clay platelets are distributed and dispersed into polymer matrix. The Compatibilizer (PP-g-MA) decreases the surface tension between silicate and polymer matrix. That enhances the dispersion, thus, the exfoliation.Meanwhile, the contribution of intercalation can not be ignored in such system. The values of storage modulus for PP/clay composite with modified clay increase by many order of magnitude. At ω 0.056 rad/s G' of PP/clay nanocomposite with 5wt% modified clay is greater by factor about 150 times than neat PP, meanwhile, its value higher by factor around 40 times than this composite with the same loading of non-modified nano-clay (5wt%). That reflects the development of the physical interaction between polymer chains and the layers of nano-clay.

The effect of reactive addition on the exfoliation was insignificant as shown in Figure 1, thus, G' shows nearly no improvement.

The same scenario was found in the melt viscosity. As shown in Figure 2, the viscosity increases by the incorporation of clay into neat PP, particularly, at low frequencies regime. At which the Newtonian behavior is clearly seen for both of neat PP and PP/clay composite with 5wt% non-modified clay. Modified 5wt% clay changes the behavior and values of viscosities drastically.

As shown in Figure 2, at low frequencies, the viscosity changes from Newtonian to shear thinning behavior. The increase of the viscosities values prove the physical network-like structure. Pronounced shear thinning has been reflected the nanodispersion. This is another evidence of the exfoliation which could be detected rheologically. 5wt% of organophilic montmorillonite in presence of PP-g-MAH which promotes clay exfoliation and sufficient interfacial adhesion between filler and PP. The well-distributed, very thin silicate packages within he PP matrix results in high degree of exfoliation. Shear thinning behavior with higher slope for better clay dispersion was previously observed for other nanocomposites in the literature [48].

Again the effect of reactive is invisible. This is, may be, related in somehow to the full exfoliation of nano-clay ,therefore, the influence of reactive addition is insignificant.

non-modified claysince h results in increasing G' In this article we are going to focus on The results showed that the dynamic mechanical moduli and viscosity increase with the incorporation of clay into PP and their values are found to increase with increasing the clay loadings particularly is significant at low frequencies. This increase origins from the formation of network-like structure, in addition to, the incorporation of high modulus material (clay) into PP.

Figure 3 shows the influence of clay feeding position on the viscosity of PP/clay composites with non-reactive modifier. As shown in this Figure there is , nearly, no difference between the viscosities of sample with side feeding clay and hoper feeding clay.

Here we can conclude that the prepared samples with reactive modifier, clay side feeder at 220°C show enhancing "slightly" the degree of exfoliation or rather it does not influence the values of the absolute viscosity.

The loss factor, tan δ, is plotted versus ω in Figure 4. tan δ relates the loss modulus to the storage modulus. It , somehow, indicates the viscoelasticity of the material. The addition of clay changes the viscoelasticity of PP. As shown in Figure 4, tan δ decreases by incorporation of clay into PP. That is noticeable in particular at low frequencies lower than 3 rad/s. This indicates the increase of the elasticity of the material. 5wt% loading of modified clay results in drop of tan δ at low frequencies which it becomes , nearly, independent on frequency along the measured frequencies. This is because of the development of the elastic component due to the high density of the physical interactions (Formation of elastic network-like structure). It is an indicator of exfoliation.

The effect of thermal treatment is clearly seen in Figure5. In this figure the viscosity increases by increasing the annealing time. As shown in this Figure the shear thinning exponent increases as well as the absolute values of the viscosity with increasing the annealing time up 6 hours at 220 C . followed by stability. This is, may be, because of enhancing of exfoliation. The degree of exfoliation was increased with increasing the annealing time at 220°C up to 6 hours followed by independent on the time. Here, the increasing in the degree of exfoliation

NSTI-Nanotech 2008, www.nsti.org, ISBN 978-1-4200-8503-7 Vol. 1

is attributed to the enhanced dispersion of clays. The dispersion of clay layers in the molten polymer depends on thermal diffusion of polymer molecules in the interlayer space galleries. The swelling effect of the clay layers up to a certain limit can be the reason.

4 CONCLUSION

In this work, polypropylene/clay nanocomposites were prepared by using a two step melt compounding process on twin screw extruders.

The materials were rheologically characterized. The rheological measurements are performed using an ARES-rheometer in the dynamic mode at 220°C over frequency varying from 100 to 0.017 rad/s .

The results showed that the the storage modulus and the viscosities are increased with the incorporation of clay into PP.

The exfoliation was detected rheologically and the results was given an evidence of the exfoliation for the samples with modified clay. However, the unmodified clay showed no sign of the exfoliation.

The change in the reactive modifier and clay feeding position did not influence the exfoliation.

The shear thinning exponent was increased with increasing the annealing time at 220°C up to 6 hours followed by independent on the time.

5 ACKNOWLEDGEMENTS

For financial support by Federal Ministry of Research and Education, Süd-Chemie AG, Basell Polypropylene GmbH, Germany

6 REFERENCES

1. J.W.Gilman, R.Harris and D.Hunter, 44th International Symposium, 1999, Long Beach, CA, PP 1408-1423.
2. A.B.Morgan, J.W.Gilman, R.Harris, C.L.Jackson, C.L.Wilkie and J.Zhu, Polym.Mater.Sci.Eng.2000, 83, PP 53.
3. S.S. Ray and M. Bousmina, *Macromol Rapid Commun* 26 (2005), pp. 1639–1646.
4. Brune DA, Bicerano J (2002) Polymer 43:369–387.
5. Y. Too, Ch Park, S.-G. Lee, K.-Y. Choi, D.S. Kim and J.H. Lee, *Macromol Chem Phys* 206 (2005), pp. 878–884.
6. P.B. Messersmith and E.P. Giannelis, *Chem Mater* 6 (1994), pp. 1719–1725.
7. Lloyd SM, Lave LB (2003) Environ Sci Technol 37:3458–3466.
8. B. Noval, *Adv Mater* 5 (1993), pp. 422–433.
9. A. Dasari, Z.-Z. Yu and Y.-W. Mai, *Polymer* 45 (2005), pp. 5986–5991.
10. S.S. Ray, M. Bousmina and A. Maazouz, *Polym Eng Sci* 46 (2006), pp. 1121–1129
11. Y. Kojima, A. Usuki, M. Kawasumi, A. Okada, A. Fujushima and T. Kurauchi et al., *J Mater Res* 8 (1993), pp. 1185–1189.
12. F.M. Mirabella Jr, *Dekker encyclopedia of nanoscience and nanotechnology*, Marcel Dekker, Inc, New York (2004) p. 3015–3030.
13. B.B. Khatua, D.J. Lee, H.Y. Kim and J.K. Kim, *Macromolecules* 37 (2004), pp. 2454–2459.
14. Reichert P, Nitz H, Klinke S, Brandsch R, Thomann R, Mu¨lhauptR (2000) Macromol Mater Eng 275:8–17
15. S.D. Burnside and E.P. Giannelis, *Chem Mater* 7 (1995), pp. 1597–1600.
16. E.Gianellis, Adv.Mater, 1996, 8, PP 29-35.
17. S.S. Ray and M. Bousmina, *Macromol Rapid Commun* 26 (2005), pp. 450–455.
18. Y. Wang, Q. Zhang and Q. Fu, *Macromol Rapid Commun* 24 (2003), pp. 231–235.
19. M.Y. Gelfer, H.H. Song, L. Liu, B.S. Hsiao, B. Chu and M. Rafailovich et al., *J Polym Sci B* 41 (2003), pp. 44–54.
20. D. Voulgaris and D. Petridis, *Polymer* 43 (2002), pp. 2213–2218.
21. W.S. Chow, Z.A. Mohd Ishak and J. Karger-Kocsis, *Macromol Mater Eng* 290 (2005), pp. 122–127.
22. D. Voulgaris and D. Petridis, *Polymer* 43 (2002), pp. 2213–2218.
23. A. Dasari, Z.-Z. Yu and Y.-W. Mai, *Polymer* 45 (2005), pp. 5986–5991.
24. H. Essawy and D. El-Nashar, *Polym Test* 23 (2004), pp. 804–807.
25. S. Mehta, F.M. Mirabella, K. Rufener and A. Bafna, *J Appl Polym Sci* 92 (2004), pp. 928–936
26. M. Maiti, A. Bandyopadhyay and A.K. Bhowmick, *J Appl Polym Sci* 99 (2006), pp. 1645–1656.
27. H. Lee, P.D. Fasulo, W.D. Rodgers and D.R. Paul, *Polymer* 46 (2005), pp. 11673–11689.
28. M.S. Wang and T.J. Pinnavaia, *Chem Mater* 6 (1994), pp. 468–474.
29. S.S. Lee and J.K. Kim, *J Polym Sci Part B Polym Phys* 42 (2004), pp. 246–252.
30. F.C. Chiu, S.M. Lai, J.W. Chen and P.H. Chu, *J Polym Sci Part B Polym Phys* 42 (2004), pp. 4139–4150.
31. R.A. Vaia, H. Ishii and E.P. Giannelis, *Chem Mater* 5 (1993), pp. 1694–1696.
32. R.A. Vaia and E.P. Giannelis, *Macromolecules* 30 (1997), pp. 8000–8009.
33. W.M. Choi, T.W. Kim, O.O. Park, Y.K. Chang and J.W. Lee, *J Appl Polym Sci* 90 (2003), pp. 525–529.
34. K.M. Lee and C.D. Han, *Polymer* 44 (2003), pp. 4573–4588.
35. R. Krishnamoorti and E.P. Giannelis, *Macromolecules* 30 (1997), pp. 4097–4102.
36. W. Loyens, P. Jannasch and F.H.J. Maurer, *Polymer* 46 (2005), pp. 903–914.
37. S. Choi, K.M. Lee and C.D. Han, *Macromolecules* 37 (2004), pp. 7649–7662.
38. S. Mehta, F.M. Mirabella, K. Rufener and A. Bafna, *J Appl Polym Sci* 92 (2004), pp. 928–936.

39. J. Park and S.C. Jana, *Macromolecules* 36 (2003), pp. 8391–8397.

40. J. Park and S.C. Jana, *Polymer* 45 (2004), pp. 7673–7679.

41. K.M. Lee and C.D. Han, *Macromolecules* 36 (2003), pp. 7165–7178.

42. X. Huang, S. Lewis, W.J. Brittain and R.A. Vaia, *Macromolecules* 33 (2000), pp. 2000–2004.

43. R.A. Vaia and E.P. Giannelis, *Macromolecules* 30 (1997), pp. 8000–8009.

44. R.Wagener and T.J.G.Reisinger, Polymer, 2003,44, PP 7513-7518.

45. M.Abdel-Goad, and P.Poetschke, IUPAC Paris 4-9.07.04 and PPS20, 20-24.06.04, Akron, USA.

46. *B. Kretzschmar, D. Pospiech, A. Leuteritz, D. Jehnichen, A. Janke, B. Tändler. Third World Congress Nanocomposites 2003. - San Francisco, 10.11. - 12.11.2003.*

47. *B. Kretzschmar, D. Pospiech, A. Leuteritz, M. Willeke, M. Abdel-Goad, D. Jehnichen, A. Janke, Nanocomposites 2004, 17 u. 18. März 2004, Brüssel, Belgien.*

48. K.M. Lee and C.D. Han, *Macromolecules* 36 (2003), pp. 804–815.

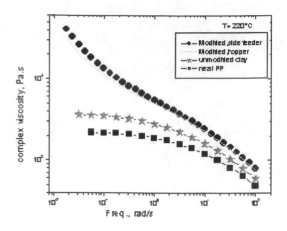

Figure 3. Frequency dependence of complex viscosity measured at 220 °C for the PP/clay nanocomposites.

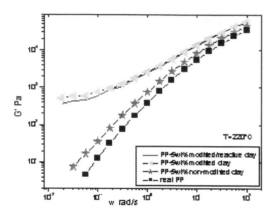

Figure 1. Frequency dependence of G' measured at 220 °C for the PP/clay nanocomposites.

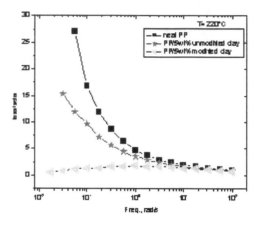

Figure 4. Frequency dependence of loss factor measured at 220 °C for the PP/clay nanocomposites.

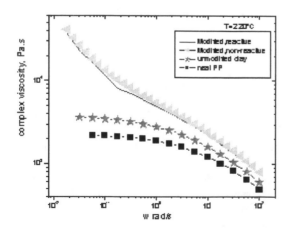

Figure 2. Frequency dependence of complex viscosity

$|\eta^*|$ measured at 220 °C for the PP/clay nanocomposites.

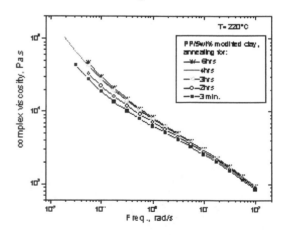

Figure 5: Effect of annealing time on the exfoliation process

NSTI-Nanotech 2008, www.nsti.org, ISBN 978-1-4200-8503-7 Vol. 1

Comparative thermodynamic study of functionalized homogeneous and multilayer latex particles

M. Corea[*], J. M. del Río[**]

[*] ESIQIE, Instituto Politécnico Nacional.
UPALM. Edificio Z-6. Col. San Pedro Zacatenco, C.P. 07738, México D.F. E-mail: mcorea@ipn.mx
[**] Programa de Aseguramiento para la Producción de Hidrocarburos. Instituto Mexicano del Petróleo
jmdelrio@imp.mx

ABSTRACT

In this work, two series of polymeric particles of poly(butyl acrylate-co-methyl methacrylate), P(BuA-MMA) functionalized with acrylic acid (AA) were synthesized by means of emulsion polymerization techniques. In both cases, the total carboxylic concentration was varied inside the particle between 0, 5, 10, 15 and 20 wt %. and number of particles was maintained constant in two systems. In the fist series, homogeneous latex particles were synthesized, whereas in second series, the latex particles were synthesized in three layers, where a gradient of carboxylic groups was generated inside the particle.

The behaviour of polymeric chains was determined by means of specific volume and adiabatic compressibility at infinite dilution, which were calculated from density and sound speed measurements at 30ºC.

Keywords: Emulsion polymerization, specific partial properties, functionalized polymers

1 INTRODUCTION

Functional groups are usually incorporated into polymeric latex by copolymerization with carboxylic acid comonomers[1]. These can impart colloidal stability, freeze-thaw stability, and improve film forming properties[2,3]. Another function of such monomers is as bonding agents in latex-based paper coatings. The carboxylic acid comonomer forms a major component of water soluble chains on the surface of the latex particle, proving both steric and electrostatic stabilization of the colloid (hence "electrosteric stabilization"). This surface coating of hydrophilic chains is often referred to as a "hairy layer"[4].

The applications of these kind of materials are in two main areas: (i) they can provide useful models for fundamental studies in colloidal science, physics and rheology[5], (ii) they can used in a broad range of applications, for example as binders in paints, adhesives, paper coatings, textiles, etc [6,7].

In recent years a new generation of carboxylic latex particles has been developed as supports in the biochemical and biomedical fields[8,9]. The carboxylic groups are able to form amine bonds with the amino groups of bioligands and they are frequently used to achieve protein binding[10,11]. The properties of these materials and their swelling process is affected by the synthesis process, quantity and location of functional groups inside the particle[12,13].

In this work two series of carboxylated latex particles synthesized by emulsion polymerization techniques were studied. Homogenous particles were synthesized in the first series while the second series, multilayer latex particles synthesized in four steps were prepared.

The analysis of behavior of polar and non polar groups of polymeric particle was made by means of specific partial properties, which indicated, the location of groups inside particle. In addition, a hydration process inside the particle was found.

1.1. Thermodynamics

The analysis of each latex was carried out, considering to latex as a three component system: water (component 1), non-polar groups (component 2) and polar groups (component 3). Now, two of these components (2 and 3) were group in an entity called fraction, which has internal composition. Therefore, the volume for a system composed by a component and a fraction is expressed as:

$$V = V(m_1, m_F, t_{f3}) \qquad (1)$$

where m_1, is the mass of the water and m_F the mass of polymeric particle:

$$m_F = m_2 + m_3 \qquad (2)$$

where m_2 and m_3 are the masses of non-polar and polar groups respectively. The variable t_{f3} is defined as:

$$t_{f3} = \frac{m_3}{m_F} \qquad (3)$$

and it is a measure of the composition of the fraction. The limits at infinite dilution of component 2 (non polar groups) and component 3 (polar groups) are taken as follows:

$$\lim_{\substack{t_F \to 0 \\ t_{f3} = t_{f3}^c}} v_{2;1,3}(t_F, t_{f3}) = v_{2;1,3}(0, t_{f3}^c) \equiv v_{2;1,3}^{\Delta}(t_{f3}^c) \quad (4)$$

and

$$\lim_{\substack{t_F \to 0 \\ t_{f3} = t_{f3}^c}} v_{3;1,2}(t_f, t_{f3}) = v_{3;1,2}(0, t_{f3}^c) \equiv v_{3;1,2}^{\Delta}(t_{f3}^c) \quad (5)$$

In equations (4) and (5) the concentration of the fraction F tends to zero while its composition is kept constant.

Specific partial properties by non-polar and polar groups can be interpreted in terms of:

$$v^o_{2;13} = v^o_M + \Delta v^o_h \qquad (6)$$

where $v_M{}^o$ is the intrinsic volume of the solute molecule, in which solvent molecules cannot penetrate, and is $\Delta v^o{}_h$ and it is named as hydration term.

In order to calculate the specific partial adiabatic compressibility at infinite dilution is

$$k^o_{3;1,2} = \left(\frac{\partial v^o_{3;1,2}}{\partial P} \right)_S = k^o_M + \Delta k^o_h \quad (7)$$

It is important to point out that $v^o{}_M$ and $k^o{}_M$ are positive and $\Delta v^o{}_h$ and $\Delta k^o{}_h$ are negative. The swelling process involves the repulsion of charges and a process of hydration inside the particle. In this way the repulsion between charges will increase the terms $v^o{}_M$ and $k^o{}_M$ while the process of hydration will decrease the terms $\Delta v^o{}_h$ and $\Delta k^o{}_h$.

2 EXPERIMENTAL

2.1 Materials

The monomers butyl acrylate (BuA), methyl methacrylate (MMA), and acrylic acid (AA) (National Starch & Chemical) were commercial grade and were used as received. Sodium dodecylbenzene sulfonate (SDBS) and potassium persulfate (from Aldrich) were reactive grade and were employed as surfactant and initiator, respectively; both were used without purification. The dispersion medium was distilled water.

2.2 Latex preparation

The carboxylated poly(BuA-MMA) samples were prepared via emulsion polymerization. All reactions were carried out in a semicontinuous reactor consisting of a jacketed rector and a feeding tank. A continuous flow of pre-emulsion material was ensured by a dosing pump. The reactor consisted of a 1-L stirred glass reactor under a dynamic flow of N_2 and at a temperature of 70 °C, controlled by a thermal bath. The stirring rate was adjusted to 250 rpm. In all cases, a seed of poly(BuA) was synthesized as a first stage and the number of particles was maintained constant. The AA total content in the latex was varied from 0 to 20 wt.%. The pH during reactions was kept at a value lower than 4 to ensure the incorporation of acrylic acid. The homogenous particles were carried out in two steps, including the seed, while the multilayer latex particles were prepared by four consecutive polymerization sequences. The formulation used to prepare both latex is presented in table 1 and 2.

Component	Content (g)	
	Stage 1	Stage 2
BuA	6.0	-
MMA		51.0
AA		3.0
Potassium sulfate	0.14	0.7
Tensoactive	0.02	2.52
Water	193.0	293.0

Table 1. Polymerization recipe for homogenous latex particle

Component	Content (g)			
	Stage 1	Stage 2	Stage 3	Stage 4
BuA	6.0	-	-	-
MMA	-	15.0	18.0	18.0
AA	-	2.0	1.0	-
Potassium sulafate	0.14	0.23	0.23	0.24
Tensoactive	0.02	0.76	0.76	1.0
Water	193.0	97.0	97.0	99.0

Table 2. Polymerization recipe for multilayer latex particles

2.3 Densimetry and utrasound speed

A DSA 5000 Anton Paar density and speed of sound analyzer was used. The samples were prepared as follows: A solution stock of 1.74 wt.% of polymer was made, and it was diluted to concentrations of 0.05, 0.1, 0.15, 0.2, 0.25, 0.3 and 0.35 wt.% of polymer. This procedure was made for each latex synthesized. The samples were degassed before to use. The measurements were made at 30 °C.

3 RESULTS AND DISCUSSION

The specific partial volume and adiabatic compressibility of non-polar groups ($v^\Delta_{2;1.3}$) and polar groups ($v^\Delta_{3;1.2}$) were calculated with following equations:

$$v^\Delta_{2:1,3}(t_{f3}) = v^o_{F;1}(t_{f3}) - \left(\frac{dv^o_{F;1}}{dt_{f3}} \right) \times t_{f3} \qquad (6)$$

$$v^\Delta_{3:1,2}(t_{f3}) = v^o_{F;1}(t_{f3}) + \left(\frac{dv^o_{F;1}(t_{f3})}{dt_{f3}} \right) \times (1 - t_{f3}) \qquad (7)$$

$$k^\Delta_{2:1,3}(t_{f3}) = k^o_{F;1}(t_{f3}) - \left(\frac{dk^o_{F;1}}{dt_{f3}} \right) \times t_{f3} \qquad (8)$$

$$k_{3;1,2}^{\Delta}(t_{f3}) = k_{F;1}^{o}(t_{f3}) + \left(\frac{dk_{F;1}^{o}(t_{f3})}{dt_{f3}} \right) \times (1 - t_{f3}) \quad (9)$$

where $v_{F;1}^{o}$ and $k_{F;1}^{o}$ were specific volume and the specific adiabatic compressibility of fraction (polymeric particle) of system, which were calculated as:

$$v = v_{F;1}^{o} + (v_1 - v_{F;1}^{o})t_1 \quad (10)$$

where v is the specific partial volume, v_1 is the specific partial volume of water in its pure state and t_1 is the fraction mass of component 1. In the same way, $k_{F;1}^{o}$ was calculated from specific adiabatic compressibility, k using the equation:

$$k = k_{F;1}^{o} + (k_1 - k_{F;1}^{o})t_1 \quad (11)$$

The specific partial volume (v) of latex samples was calculated from the density data, measurement at 30 °C by the following equation:

$$v = \frac{1}{\rho} \quad (12)$$

and the specific adiabatic compressibility (k) was calculated from the experimental density and sound speed data, measurement at 30 °C using the following equation:

$$k = \left(\frac{1}{\rho u} \right)^2 \quad (13)$$

The specific partial volume and specific adiabatic compressibility of non-polar and polar groups are shown in Figure 1.

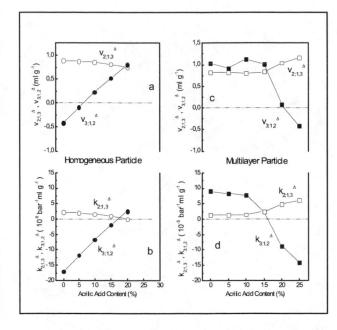

Figure 1. Specific partial thermodynamics properties of non-polar and polar groups of homogenous (panel a and b) and multilayer latex particles (panel c and d).

In the case of homogenous latex particles, both, the volume and adiabatic compressibility of polar groups have negative values when the acrylic acid concentration is low. This means that polar groups have a great hydration and this behavior is only possible if the carboxylic groups are on the surface of particle. When the acrylic acid concentration is increased, the value of thermodynamic properties is more positive. This is explained, when the carboxylic groups are located in the interior of particle, that is, that polar groups in the surface are more hydrated that polar groups in the interior of particle. While the non-polar groups are hydrated when the polar groups concentration is increased. For this reason, the thermodynamic properties decreased when the acrylic groups are raised.

The values of the specific partial thermodynamic properties of polar groups in the multilayer latex particles, suggest that a low concentration of polar groups, the hydration of these groups is not important. However, when the acrylic acid concentration is increased, the hydration of polar groups is higher. The results of thermodynamic properties of non-polar groups indicate that they suffer a dehydration when 15 wt.% of the acrylic acid concentration is reached. This behavior is interpreted in terms of the hydration of multilayer latex particle is located in the bound of each layer.

4 CONCLUSIONS

A simple methodology was developed by means of measurements of density and sound speed for the calculation of specific partial thermodynamic properties. The methodology employed to indicate the behavior of polar and non-polar groups in the polymeric chain.

The polar groups in the homogenous latex particle are located from the surface to the interior of the particle, while in the multilayer latex particle are located in the bound of each layer. In the same way, the hydration in the multilayer particle is not homogenous, because is located in the bound of each layer.

REFERENCES

[1] Koh, A. Y. C.; Mange, S.; Bothe, M.; Leyrer, R. J.; Gilbert, R. G.; Polymer, 2006, 47, 1159-1165.

[2] Lee, D.-Y.; Kim, J.- H.; Journal of Applied Polymer Science,1998, 69, 543-550.

[3] Mahdavian, A. R.; Abdollahi, M. ; Reactive & Functional Polymers, 2006, 66, 247-254.

[4] Coger, L.; Gilbert, R. G.; Macrmolecules, 2000, 33, 6693-6703.

[5] Innerlohinger, J.; Wyss, H. M.; Glatter, O. *J. Phys. Chem. B* **2004**, *108*, 18149-18157.

[6] Pedraza, E. P.; Soucek, M. D.; Polymer, 2005, 46, 11174-11185.

[7] Yuan, X.-Y.; Dimonie, V. L.; Sudol, E. D.; El-Aasser, M. S.; Macromolecules, 2002, 35, 8346-8355.

[8] Okubo, M.; Yonehara, H.; Yamashita, T. *Colloid Polym. Sci.* **2000**, *278*, 1007-1013.

[9] Santos, A. M.; Elaissari, A.; Martinho, J. M. G.; Pichot, C. ; Colloid Polym. Sci. **2004**, 282, 661-669.

[10] Menshikova, A. Y.; Evseeva, T. G.; Skurkis, Y. O.; Tennikova, T. B.; Ivanchev, S.S.; Polymer, **2005**, 46, 1417-1425.

[11] Sivakumar, M.; Panduranga Rao, K.; Reactive & Functional Polymers, **2000**, 46, 29-37.

[12] Ramos, J., Forcada, J., Polym, **2006**, 47, 1405-1413.

[13] Gualtieri, G., M., Gobran, R., H., Nien, Y.,-H., Kalidindi, S., R., J. of Appl. Polym. Scien., **2001**, 79, 1653-1664.

Synthesis and electrochemical characterization of sponge-like nickel nanoparticles

E. Ramírez-Meneses[*], M.A. Domínguez-Crespo[*], H. Dorantes- Rosales[**], A.M. Torres-Huerta[*], V. Montiel-Palma[***], G. Hernández-Tapia[****]

[*]Centro de Investigación en Ciencia Aplicada y Tecnología Avanzada, CICATA-IPN Unidad Altamira, Km. 14.5 Carretera Tampico-Puerto Industrial, C.P. 89600, Altamira, Tamaulipas. México. esramirez@ipn.mx, mdominguezc@ipn.mx, atorresh@ipn.mx

[**]Departamento de Metalurgia, Escuela Superior de Ingeniería Química e Industrias Extractivas-IPN, C.P. 07300 México, D.F. hdorantes@ipn.mx

[***]Centro de Investigaciones Químicas, Universidad Autónoma del Estado de Morelos, Av. Universidad 1001, Colonia Chamilpa, C.P.62201 Cuernavaca, Morelos, México.

[****]Gerencia de Catalizadores y Proceso, Instituto Mexicano del Petróleo, Eje Central Lázaro Cárdenas norte 152, C.P 07730 México, D.F. ghernand@imp.mx

ABSTRACT

Nickel and its alloys are the most studied electrode materials. Nickel exhibits a high initial electrocatalytic activity in the hydrogen evolution reaction (HER). Enhancement of cathodic activity of nickel for HER has been carried out by the formation of nanostructured nickel materials. In this context, the use of an organometallic precursor, able to decompose in mild conditions in the presence of a reducing gas has arisen as a suitable method of synthesis of these species. In this work, we investigated the synthesis of nickel nanoparticles starting from $Ni(COD)_2$, (COD=cycloocta-1,5-diene) and decomposed under H_2 atmosphere in THF in presence of 1,3-diaminopropane as stabilizer. TEM studies revealed the formation of characteristic particles resulting from the agglomeration of initially obtained nanocrystallites. The activity of the dispersed catalysts composed of Ni/C with respect to the HER was investigated using steady state polarization measurements.

Keywords: hydrogen evolution reaction, nickel, nanoparticles, organometallic.

1 INTRODUCTION

The hydrogen evolution reaction (HER) is the main reaction in the water electrolysis and has been one of the most studied electrochemical reactions. Although, platinum presents the highest activity for the HER, new electrode materials have been investigated, aiming at the reduction of the cost associated with the electrocatalysts development. Among these materials, nickel exhibits good corrosion resistance in aggressive environments and catalytic activity for the HER [1]. However, in order to improve its activity, resulting in reduced overpotentials for the HER, the real surface and the intrinsic activity of the electrode material must be increased.

Nickel nanoparticles less than 100 nm in primary particle diameter have attracted considerable interest in the last years due to their various applications such as catalysis superconductors, electronic, optical mechanic devices, magnetic recording media, and so on, [2,3]. Metal nanoparticles have generally been produced by both gas and liquid phase processes. Among the gas phase methods, hydrogen reduction of the metal chloride [4], laser-assisted gas phase photonucleation [5], and hydrogen plasma metal reaction [6] have been used. Alternatively, γ-ray irradiation [7], borohydride reduction of metal salts [8], sonochemical and thermal decomposition metal complexes [9] are the most useful techniques to synthesize metal nanostructures in liquid phase. Additionally, an organometallic approach using olefinic complexes as a source of metal atoms has shown an interesting alternative to control shape, size and dispersion of nanostructures [10,11]. The method leads to the access to monodispersed particles of very small size (1-2 nm) and allows for the performance of coordination chemistry at their surface. Therefore, the use of organometallic precursors is now well established as a method to obtain size and surface state controlled nanoparticles in mild conditions (room temperature and 1-3 bar of a reactive gas). Additionally, depending on their skeleton and functional groups, the stabilizers can interact more o less with the surface of the particles and then favour or not the growth of the particles in a privileged direction.

The present study was undertaken in order to investigate the electrochemical properties of Ni-base nanoparticles synthesized by an organometallic approach containing 1,3-diaminopropane as stabilizer in different Ni compositions. The particle size and shape of the resultant nanoparticles were characterized by Transmission Electron Microscopy (TEM), HR-TEM and electron diffraction pattern. The electrocatalytic activities of the powders for HER were investigated by cyclic voltammetry (CV), polarization curves and ac impedance.

2 EXPERIMENTAL PROCEDURE

2.1 Synthesis of metallic nanoparticles

The reaction was carried out, as a standard procedure in a Fischer–Porter bottle. In a typical procedure, a solution of THF (50 ml) including 50 mg of the precursor Ni(COD)$_2$ (Aldrich) and 1,3-diaminopropane (99%, Aldrich) 2 (Ni$_{2DAP}$) or 5 (Ni$_{5DAP}$) equivalents per Ni atom was reacted at 70 °C for 20 h, under dihydrogen pressure (3 bar), in Fischer-Porter bottle leads to the formation to dark gray colloids. The reaction is slow at lower temperatures [12]. The obtained solutions were purified by hexane washings. Finally, the resulting solution was evaporated in vacuum until the residue was completely dry.

Specimens for TEM analysis were prepared by the slow evaporation of a drop of each colloidal solution after the purification process deposited onto a carbon covered copper grid. TEM experiments were performed on a JEOL-1200 EX electron microscope, operating at 120 KV. The HRTEM study was carried out on a JEOL 2010 FasTem field emission transmission electron microscope with a resolution of 2.1 Å.

2.2 Catalysts preparation

The support of the electrode was a glassy carbon rod (5.0 mm in diameter). The base of the rod was polished with a cloth and alumina powder (ca. 0.3 μm). The working electrodes were prepared by attaching an ultrasonically re-dispersed catalyst (stabilized nickel nanoparticles) suspension containing a 4:1 ratio vulcan carbon black/ total metal (20 wt.%) powders in deionized water onto the glassy carbon. After drying under a high purity argon flow at room temperature, the deposited catalyst layer was then covered with ≈ 4 μl of a diluted aqueous Nafion solution and finally, the electrode was immersed in a nitrogen purged electrolyte to record the electrochemical measurements. Pt black and bulk were also analyzed as a reference. The electrodes were carefully prepared in order to obtain reproducible electrode surfaces and comparable electrocatalytic results.

3 RESULTS AND DISCUSSION

3.1 TEM and HR-TEM studies

TEM-bright field images of Ni nanoparticles synthesized in presence of 2 equiv. of 1,3-diaminopropane are shown in Figure 1. The micrograph (Figure 1a) reveals the formation of sponge-like particles with an average size of 138 nm. These particles are formed by smaller nanoparticles. The corresponding selected area diffraction pattern (SADP) was inserted on the top right in Figure 1a. The indexing of the ring diffraction pattern confirms the presence of the nickel with the following interplanar distances: 2.00, 1.73, 1.11, 1.02 and 0.77 Å. These distances correspond to the (111), (200), (311), (222) and (420) planes, respectively. Previous works for this type of particles have been observed in the chemistry of ruthenium [13], platinum [14], palladium [15] and recently for Ni obtained by microwave-assisted synthesis in presence of PVP [16]. The effect of higher stabilizer concentration (5 equiv.) on the dispersion is showed in Figure 1b. In this case, nanocrystalline sizes ranging between 5 to 14 nm were observed. It seems that higher concentrations of stabilizer induced an homogeneous distribution of the smaller nanoparticles.

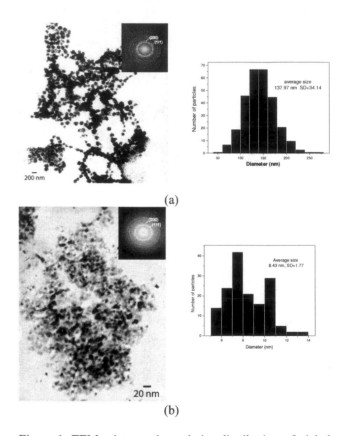

Figure 1. TEM micrographs and size distribution of nickel nanoparticles synthesized in THF from Ni(COD)$_2$ in presence of 1,3-diaminopropane (a) 2 eq. and (b) 5 eq.

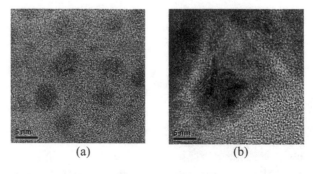

Figure 2. High-resolution TEM images of Ni nanoparticles synthesized in the presence of 5 eq. 1,3-diaminopropane.

The corresponding SADP in Figure 1b was also inserted. Five rings were indexed and they showed the following interplanar distances: 2.00, 1.79, 1.25, 1.04 and 1.28 Å. They are related to the (111), (200), (220), (311) planes of fcc Nickel and (113) plane of rhombohedra nickel oxide respectively. The presence of nickel oxide is attributed to the sample preparation process. The nanocrystallites showed in Figure 1b. were observed by HRTEM and they are showed in Figure 2a-b. These figures confirmed the well dispersed nickel nanoparticles with isolate cluster of around 5 nm in different zones.

Hence, the obtained results show the effect of the molar ratio of Ni/ $H_2NCH_2CH_2CH_2NH_2$ on the particle size. It causes a porous structure of the spherical Ni nanoflower or compacted Ni semispherical particles. The mean size of Ni nanoflowers (sponge-like) decrease gradually with the increase the amount of the stabilizer in the reaction medium. This may be ascribed to the influence of the nucleation rate and subsequently growing of the Ni structures. With the increase of the 1,3-diaminopropane, the interaction rate became faster, more dispersion can be reached inhibiting the growth of nanostructures and led to obtain smaller nanoclusters.

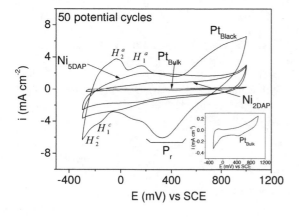

Figure 3. Cyclic voltammograms on Ni electrodes in 0.5 M H_2SO_4 (v=20 mV s^{-1}) after 50 potential cycles, different concentrations of stabilizer.

Figure 3 shows the cyclic voltammograms after 50 potential cycles of Ni catalysts and are compared with commercial Pt_{Black} and Pt_{Bulk} in 0.5 M H_2SO_4 at v=20 mV s^{-1} using a potential sweep of -300 to 1000 mV *vs* SCE. Pt_{Black} shows typical hydrogen adsorption desorption peaks in the potential range between -250 and 200 mV, and HER begins at about -300 mV vs SCE. Figure inset show the corresponding Pt_{Bulk} electrochemical behavior. The CVs for electrodes with different equiv. per nickel atom, i.e. 2 and 5 equiv. exhibited similar shape within evaluated potential range. It can be seen from the voltammogramms

that the charge densities associated with the redox peaks decrease depend on the amount of stabilizer.

In spite of many researches in the literature [17] have shown dissolution of Ni in acid media due noble metals are not completely inert at anodic potentials in these media, the obtained voltamogramms are nearly constant after only a few potential cycles and at 50 cycles the changes becoming small and dissolution have not been appreciated. This observation may be attributed to the presence of the stabilizer at surface of Ni particles, which interact with Ni nanostructures retarding its dissolution. FTIR spectra (not show here) indicate the presence of $H_2NCH_2CH_2CH_2NH_2$ interacting in the system Ni/stabilizer. Then these results can be related to the absorption mechanism of the stabilizer, the amine molecules can be adsorbed on Ni surface to form a film through coordinate bond which seems to serve as a block barrier inhibiting the Ni dissolution.

The cathodic polarization behavior (Tafel plots) after 50 potential cycles obtained for hydrogen evolution of the Ni particles having different stabilizer contents is shown in Figure 4. Both Ni electrodes are far of the electroactivity of commercial Pt_{Black}, but they are better than Pt_{bulk}.

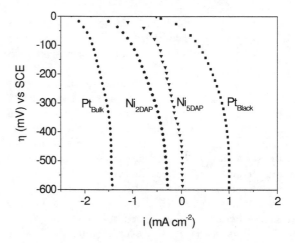

Figure 4. Cathodic polarization curves in 0.5 M H_2SO_4 (v=0.5 mV s^{-1}) of nickel nanoparticles synthesized from $Ni(COD)_2$ and compared with commercial Pt_{Black} and Pt_{Bulk}.

The major Tafel parameters are provided in Table 1. The obtained slopes (b_c) are very variety ranging between 86 and 285 mV dec^{-1}. The presence of high concentrations of the stabilizer in the Ni nanostructures increases the electrode activities which can be correlated with the particle size, as described above in TEM measurements. Then, Ni electrode having 5 equiv. was found to have higher electrocatalytic activity during HER than the other Ni catalyst (2 equiv.). The least active electrode was Pt_{Bulk} while the most active material was commercial Pt_{Black}. Table 1 also shows current densities at similar overpotentials. The higher current densities obtained for Ni_{5DAP} than this for Ni_{2DAP} at these overpotentials is only attributed to the influence of the stabilizer on the growth of the Ni nanostructures which evidently affect their catalytic activity.

Table 1: Kinetic parameters obtained from the mass-corrected Tafel Plots for hydrogen evolution reaction in 0.5 M H_2SO_4.

Parameter	Electrode			
	Pt_{bulk}	Pt_{black}	Ni_{5DAP}	Ni_{2DAP}
b_c (mV decada^{-1})	128	86	285	180
Log i_o (mA cm^{-2})	1.0E-3	1.6E-1	6.3 E-2	1.5E-2
i_1 (η=100 mV) (mA cm^{-2})	1.4E-2	2.2	3.3E-1	9.3E-2
i_2 (η=300 mV) (mA cm^{-2})	2.9E-2	7.2	6.8E-1	2.9 E-1
i_3 (η=500 mV) (mA cm^{-2})	3.2E-2	10.0	1.0	0.5

I_0 = exchange current density. i_1, i_2 and i_3 are current densities at different overpotentials.

CONCLUSIONS

The use of an organometallic approach lets to obtain size and surface state controlled nanoparticles in mild conditions. 1,3-diaminopropane-stabilized Ni nanoparticles were synthesized from $Ni(COD)_2$ in organic medium. An appropriated amount of stabilizer lets to obtain an uniform distribution of Ni nanoparticles with sponge-like structure or compacted Ni semispherical particles. Both samples displayed nearly monodispersed Ni nanoparticles through the specific coordination of the amine; this interaction seems to serve as a block barrier inhibiting the Ni dissolution in acid media. An important increased in the electrocatalytic activity during HER was observed as the molar $Ni/H_2NCH_2CH_2CH_2NH_2$ ratio was increased. The activities of these electrodes are far from the values displayed for commercial Pt, however by adjusting the kind of stabilizer containing long alkyl chains smaller clusters can be reached and increased the electrochemical performance.

ACKNOWLEDGMENT

The authors would like to thank SNI, CONACyT-México and SIP-IPN for financial support.

REFERENCES

[1] A. Kellenberger, N. Vaszilcsin, N. B. Waltraut, N. Duteanu, International Journal of Hydrogen Energy, 32, 3258, 2007.

[2] G. Schmid, L.F. Chi, Adv. Mater. 10, 515, 1998.

[3] K.Z. Chen, Z.K. Zhang, Z.L. Cui, D.H. Zuo, D.Z. Yang, Nanostruct. Mater. 8, 205, 1997.

[4] K. Otsuka, H. Yamamoto, A. Yoshizawa, Jpn. J. Chem., 6, 869, 1984.

[5] H. He, R.H. Heist, B.L. McIntyre, T.N. Blanton, Nanostruct. Mater. 8, 879, 1997.

[6] X.G. Li, T. Murai, A. Chiba, S. Takahashi, J. Appl. Phys. 86, 1867, 1999.

[7] F. Wang, Z. Zhang, Z. Chang, Mater. Lett. 55, 27, 2002.

[8] V. Srinivasa, S.K. Barika, B. Bodoa, D. Karmakarb, T.V. Chandrasekhar Raob, J. Magnetism and Magnetic Mater. 320, 788, 2008.

[9] Z.H. Zhou, J. Wang, X. Liu, H.S.O. Chan, J. Mater. Chem., 11, 1704, 2001.

[10] E. Ramirez, L. Eradès, K. Philippot, P. Lecante and B. Chaudret., J. Adv. Funct. Mater. 13, 17, 2219, 2007.

[11] O. Margeat, F. Dumestre, C. Amiens, B. Chaudret, P. Lecante and M. Respaud., Progress in solid State Chemistry., 33, 2-4, 71, 2005.

[12] N. Cordente, M. Respaud, F. Senocq, M. J. Casonove, C. Amiens and B. Chaudret. Nano Lett., 1, 565, 2001.

[13] K. Pelzer, O. Vidoni, K. Philippot, B. Chaudret, V. Colliere, Adv. Funct. Mater. 13, 118, 2003.

[14] Y. Song, Y. Yang, C.J. Medforth, J. Pereira, A.K. Singh, H. Xu, Y. Jiang, C.J. Brinker, F. van Swol, J.A. Shelnutt, J. Am. Chem. Soc., 126, 635, 2004.

[15] E. Ramirez, S. Jansat, K. Philippot, P. Lecante, M. Gómez, A. Masdeu-Bultó and B. Chaudret, J. Organomet. Chem., 689, 4601, 2004.

[16] W. Xu, K. Yong Liew, H. Liu, T. Huang, C. Sun, Y. Zhao, Mater. Lett. (2008), doi:10.1016/j.matlet.2007.12.057.

[17] D.A.J. Rand and R. Woods. Electroanalytical Chem. Interfacial Electrochem., 35, 209, 1972.

Magnetic properties of iron-filled multiwalled carbon nanotubes

N. Aguiló-Aguayo, J. García-Céspedes, E. Pascual and E. Bertran

FEMAN Group, IN₂UB, Departament de Física Aplicada i Òptica, Universitat de Barcelona.
C/ Martí i Franquès, 1, E-08028, Barcelona, Catalonia, Spain

ABSTRACT

Magnetic properties of iron-filled multiwalled carbon nanotubes have been investigated. Hysteretic and temperature dependent magnetic responses reveal a superparamagnetic behaviour at temperatures above 124 °C. The saturation magnetization (M_S) of the nanowires at 5 K is found to be 152 emu/g similar to the expected bulk value of Fe_3C. Analysis of field-cooled (FC) and the zero-field-cooled (ZFC) curves show a high magnetic anisotropy of the nanowires. Langevin function has been fitted using data at 300 K. The average domain size estimated has been $V \approx 4.18 \cdot 10^{-20}$ cm³ corresponding to spherical domains with radius $r \approx 2.2$ nm, below the value from TEM observations $r \approx 7$ nm, suggesting the presence of pseudo-single domains, with multi-domain-like and single-domain-like of radius $r \approx 2.2$ nm. The temperature dependence of coercivity give us an estimated value of blocking temperature $T_B \sim 397$ K. Possible uses of these material point to magnetic resonance imaging and other biomedical applications.

Keywords: Carbon nanotubs, magnetic properties, superparamagnetism, ZFC-FC

1 INTRODUCTION

In the last decades, superparamagnetic nanoparticles and carbon nanotubes have been a focus of interest in drug delivery, cancer treatment and diagnosis, and Magnetic Ressonance Imaging (MRI), among others novel biomedical applications.[1]

In the present article, magnetic behaviour of iron filled multiwalled carbon nanotubes has been studied from superparamagnetic approach. In addition, the structure and morphologic characteristics of iron-filled multiwalled nanotubes were determined by transmission electron microscopy (TEM) and selected area electron diffraction (SAED). The cause-effect relations with magnetic behaviour have been established.

2 EXPERIMENTAL

CNTs were produced by injection chemical vapor deposition (ICVD) using a quartz tube placed inside a tubular furnace (set up at 760 °C) with a laminar flow of Ar dragging a precursor vapor. The substrates were copper foils of 2 x 3 cm approximately and 100 μm thick. Another small quartz substrate of 10x15x2 mm was placed together for comparison after the process. The growth precursors involved were prepared in a liquid solution of ferrocene in toluene (mass percentage of 8.76 %) and released into the tube at a constant rate of 5.6 cm³/h. The solution was vaporised thanks to a furnace located just after the feedthrough, which is set up at 180 °C (the vaporisation temperature of ferrocene and toluene are 175 and 110 °C, respectively), and was carried through the main furnace by a gas mixture of Ar and H_2 ($\Phi Ar : \Phi H_2 = 250:2.5$; –flux values in standard cm³/min–). The pressure inside the tube was slightly over atmospheric, thus a pumping system was unnecessary.

Prior to CNTs deposition on copper, a 20 nm thick TiN barrier layer was deposited by magnetron sputtering, in order to avoid catalyst diffusion.

For a full description of the experimental set-up for CNTs production and a systematic study of the different deposition parameters influence see [2].

For SEM observations, a Hitachi H-4100FE operated at 25 kV were used. A Philips CM30 operated at 300 kV was used for TEM, HRTEM and SAED structural characterisation. A superonducting-quantum-interference-device (SQUID) magnetometer was used to study the magnetic behaviour of iron-filled multiwalled carbon nanotubes in the temperature range 5-300 K and in fields up to 55 kOe.

3 RESULTS AND DISCUSSION

3.1 Structure properties

CNTs grown on TiN-coated Cu show a partial alignment perpendicular to the substrate surface, due to a crowding effect [3]. The estimated length is between 20 and 30 μm, according to SEM observations. TEM observation allowed a more accurate study of the nanostructures. In figure 1 CNTs grown on TiN-coated Cu are shown. The image reveals a concatenation of multiwalled carbon capsules.

Electron diffraction patterns obtained from small areas of the samples revealed a crystalline structure corresponding to Fe_3C (figure 1b), that is a material commonly found in CNTs synthesized by CVD [4, 5], and is associated with the deactivation of Fe as a catalyst [6, 7]. However, Fe_3C might not be the only material encapsulated inside the nanostructure bodies, as confirmed by X-ray diffraction (figure 2).. Because of the great amount of catalyst material encapsulated in the sample, it

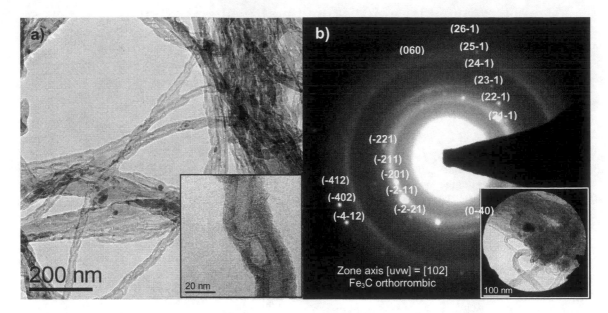

Figure 1. TEM analysis of CNTs obtained over TiN-coated Cu (a). It is observed a bamboo-like structure, with an important quantity of catalyst (iron based) material encapsulated on it. The inset in figure a) show HRTEM micrographs of the graphitic multiwalls. SAED patterns in b) are associated with the Fe₃C phases present into nanoparticles.

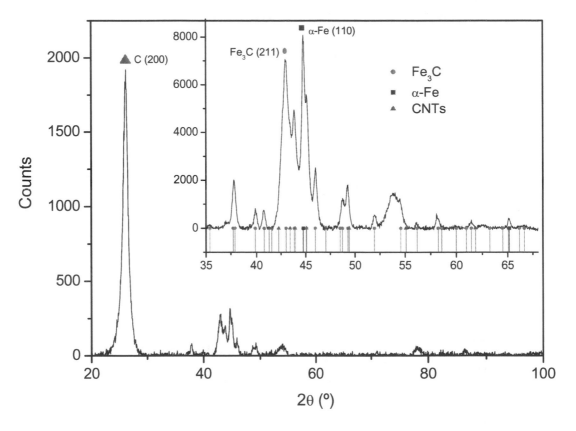

Figure 2. X-ray diffraction pattern of CNTs grown on copper. α-Fe, Fe₃C phases were found in addition to carbon phases.. Only main peaks have been labeled.

NSTI-Nanotech 2008, www.nsti.org, ISBN 978-1-4200-8503-7 Vol. 1

was possible to identify at least two different contributions among the many diffraction peaks, apart from C. The most feasible phases present into nanoparticles are Fe_3C and α-Fe (Ferrite).

3.2 Magnetic properties

Superconducting quantum interference device (SQUID) magnetometry studies have revealed the magnetic behaviour of the iron-filled multiwalled carbon nanotubes. The TEM studies have shown that the average size of the Fe grains was about 15 nm. The total mass of our sample was about ~ 85 μg. To measure it, a known mass of CNTs was oxidized in air at 600 °C for 30 minutes, converting all the carbon material into CO_2. Then, the remaining iron oxide was weighted. From this resulting mass, the iron content was extracted by considering all the material as Fe_2O_3. The corresponding apparent saturation magnetization was 152 emu/g at 5 K, this value is similar to the estimated saturation magnetization for a bulk sample of Fe_3C (~ 169 emu/g at 0 K), since it is the most predominating component, as X-ray diffraction and SAED results had predicted.

Figures 3 shows the M(H) curves versus H at different temperatures 5, 75, 150, 225 and 300 K. As the temperature is increased, the magnetization curve exhibits less hysteresis due to the superaparamagnetic behaviour is being achieved [9]. The inset of figure 3 represents the temperature dependence of saturation magnetization (M_S) and remanent magnetization (M_r), both values are decreasing with the increase of temperature in accordance with the above mentioned.

The ZFC-FC magnetization curves are shown in figure 4. When a field of 100 Oe is applied, we observed a maximum of magnetization located above 330 K, then the blocking temperature will be situated around this value, but this method is not accurate enough to determine the specific value of blocking temperature, T_B. However, those curves give information about the anisotropy of the sample [10]. In this case, the anisotropy of the sample is low since FC magnetization remains almost constant with the temperature.

For biomedical applications, it is very interesting to know the behaviour of the magnetic nanoparticles inside the human body, so that, we will study the magnetization curve at 300 K. This curve can be fit to a Langevin function L using the following relation [11].

$$\frac{M}{M_S} = L\left(\frac{\mu H}{k_B T}\right) = \coth\left(\frac{\mu H}{k_B T}\right) - \frac{k_B T}{\mu H} \qquad (1)$$

Where M_S is the saturation magnetization, μ is the effective moment given by the product $M_S \cdot \langle V \rangle$ and $\langle V \rangle$ is the average particle volume.

The effective moment value $\mu \approx 5.44 \cdot 10^{-17}$ emu is obtained using data for 300 K and fitting magnetization by

least-squares (figure 5). Using the effective moment and the known value of saturation magnetization of Fe_3C (the predominating material of our sample) at 0 K which is $\sigma_0 = 1302$ emu/cm^3 [12], we could estimate the diameter corresponding to spherical particles which is about $D \approx 4.3$ nm. This value is lower than the average size found in TEM observations, $D \approx 15$ nm, therefore, the present system suggests that we have a situation of pseudo-single domains, with regions of multi-domain-like and other regions single-domain-like of 4.3 nm with superparamagnetic behaviour.

To estimate the value of the blocking temperature, T_B, the dependence of the coercivity with temperature is plotted (figure 6). The coercivity depends on the anisotropy of the sample and its value is related with the difference between field-cooled (FC) and zero-field-cooled (ZFC) susceptibilities. We represent the coercivity H_C as the average along the positive and negative H axis, $H_C(T) = 1/2(|H_{C+}| + |H_{C-}|)$, for each temperature. The temperature dependence of the coercivity is fitted by least-square to the following equation [13]:

$$H_C = H_{Ci}\left[1 - \left[\frac{T}{T_B}\right]^{1/2}\right] \qquad (2)$$

We found a zero temperature coercivity of 2792 Oe and a blocking temperature of 397 K , greater than 330 K in accordance with the ZFC-FC mesurements.

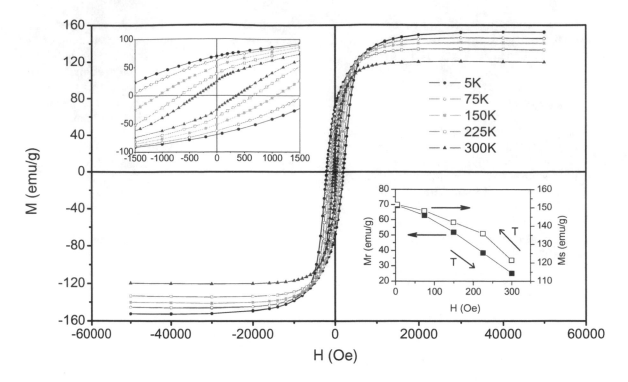

Figure 3. Magnetization versus applied field at temperatures of 5, 75, 150, 225 and 300 K for the Fe-filled MWCNTs. The inset shows the temperature dependence of remanent magnetization (close squares) and saturation magnetization (open squares) calculated from M(H) plots at different T.

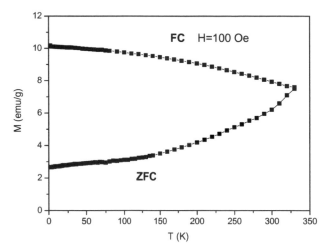

Figure 4. Magnetic susceptibility curves for ZFC and FC measured at low magnetic field of 100 Oe.

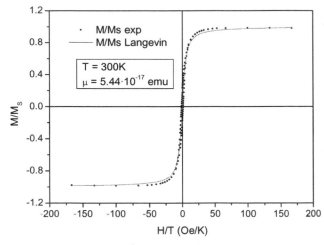

Figure 5. Magnetization versus H/T at 300 K fitted to a Langevin function, eq. (1). The effective moment found was $\mu \approx 5.44 \cdot 10^{-17}$ emu corresponding to a domain size of spherical particles with a diameter about 4.3 nm.

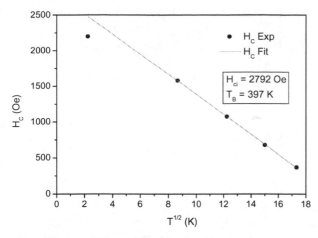

Figure 6. Temperature dependence of coercivity H_C. The H_C versus $T^{1/2}$ law was fitted to find the blocking temperature T_B.

4 CONCLUSIONS

A densely packed mat of partially aligned, bamboo-like CNTs have been grown on both faces of a TiN-coated Cu sheet, with an average size of about 30 μm, by means of ferrocene injection CVD.

Catalyst particles within CNTs have been identified as Fe_3C by SAED, although X-ray diffraction has shown other possible materials involved, as α- Fe and γ-Fe.

Magnetic properties of iron-filled multiwalled carbon nanotubes have been investigated using superconducting-quantum-interference-device magnetometry (SQUID). Hysteretic and temperature dependent magnetic responses reveal a superparamagnetic behaviour at temperatures above 397 K. The saturation magnetization is slightly different from the expected value of α-Fe due to the presence fraction of other components such as Fe_3C and γ-Fe in accordance with X-ray diffraction results. Magnetization data at 300K follows the Langevin function, which indicates that the diameter of the spherical particle domain is about 7nm, below the value from TEM observations suggesting the presence of pseudo-single domains.

5 ACKNOWLEDGMENTS

This study was supported by projects CSD2006-12 and DPI2006-03070 of MEDU of Spain. The authors thank Serveis Científico-tècnics of the Universitat de Barcelona (SCT-UB) for measurement facilities. J.G-C. thanks to Electrical Engineering and Materials Science & Metallurgy Departments of University of Cambridge for the use of their facilities for CNTs sample fabrication, specially to Prof. W.I.Milne, Dr. K.B.K.Teo and Dr. I.Kinloch.

REFERENCES

[1] C. Singh, M. S. P. Shaffer, A. H. Windle, Carbon 41, 359, 2003.

[2] C. S. S. R. Kumar, Wiley-VCH, 448, 2007.

[3] S. Fan, M. G. Chapline, N. R. Franklin, T. W. Tombler, A. M. Cassell, H. Dai, Science 283, 512, 1999.

[4] A. Oberlin, M. Endo, T. Koyama, Journal of Crystal Growth 32, 335, 1976.

[5] L. Sun, F. Banhart, Krasheninnikov, J. A. Rodríguez-Manzo, M. Terrones, P. M. Ajayan, Science 312, 1199, 2006.

[6] K. Hernadi, A. Fonseca, J. B. Nagy, D. Bernaerts, A. A. Lucas, Carbon 34 (10), 1249, 1996.

[7] S. Herreyre, P. Gadelle, P. Moral, J. M. M. Mollet, Journal of physical chemical solids 58 (10), 1539, 1997.

[8] H. Kim, M. J. Kaufman, W. M. Sigmund, Journal of Materials Research 18 (5), 1104 , 2003.

[9] D.L. Leslie-Pelecky, R. D. Rieke, Chemical Materials 8, 1770-1783, 1998.

[10] P. A. Joy, P. S. Anil Kumar, S. K. Date, Journal Physics: Condensed matter 10, 11049-11054, 1998.

[11] M. E. McHerny, S. A. Majetich, J. O. Artman, M.DeGaef, S. W. Staley, Physical Review B 49, 11358-11363, 1994.

[12] T. Mühl, D. Elefant, A. Graff, R. Kozhuharova, A. Leonhardt, I. Mönch, M. Ritsch, P. Simon, S. Groudeva-Zotova, C. M. Schneider, Journal of Applied Physics 93 (10), 7894-7896, 2003.

[13] S. Gangopadhyay, G. C. Hadjipanayis, Physical Review B 45, 9778-9787, 1992.

Synthesis and Characterization of Nanocrystalline Eu^{3+}-doped Gd_2O_3

S. Jáuregui-Rosas[1,2], O. Perales-Pérez[3*], M. Tomar[1], W. Jia[1], O. Vásquez[1] and E. Fachini[4]

[1]Department of Physics, University of Puerto Rico-Mayagüez Campus, Mayagüez, PR 00681
[2]Departament of Physics, Universidad Nacional de Trujillo, Trujillo - Perú
[3]Department of Engineering Science & Materials, University of Puerto Rico-Mayagüez Campus, Mayagüez, PR 00681. *ojuan@uprm.edu
[4]Chemistry Department, University of Puerto Rico-Rio Piedras Campus, San Juan, PR 00931

ABSTRACT

In the present work, nanocrystalline $Gd_{2-x}Eu_xO_3$ phosphors have been synthesized by sol-gel method. The effect of Eu^{3+} concentration and annealing temperature on structural and luminescence properties of nanocrystalline powders were investigated. X-ray diffraction analyses showed that cubic Eu-doped Gd_2O_3 was formed and exhibited an average crystallite size ranging from 29nm to 41nm when the annealing temperature varied from 750 to 950°C, respectively. Photoluminescence (PL) measurements verified the presence of all transitions of Eu^{3+} dopant, being the $^5D_0$ to $^7F_2$ transition the most intense. It was also found that, on a common weight basis, the PL intensity was strongly dependent on both the annealing temperature and dopant concentration. The highest PL intensity was observed for 'x'=0.15. The observed quenching in luminescence would result from exchange interactions. Preliminary results on the PL characterization of nanocrystalline $Gd_{2-x}Eu_xO_3$ thin films are also presented.

Keywords: Sol-Gel, Eu^{3+}-doped Gd_2O_3, nanocrystalline materials, luminescence, quenching.

1 INTRODUCTION

Nanosize materials are the focus of attention due to their novel and interesting size-dependent functional properties [1]. In particular, nanosize rare earth (RE)-doped phosphors have become one of the most promising materials because of their fundamental and potential applications including solid sate lasers, lighting and immunoassays systems [2,3]. The luminescence in phosphor materials is strongly dependent on both, the type of host and the nature of the dopant species. Then, any effort on the synthesis of RE sesquioxides (RE_2O_3), which are considered excellent host lattices for trivalent ions, is justified. One of these oxides, Gd_2O_3, is an inorganic insulator with a band gap of 5.4eV [4] and exhibits cubic structure at room temperature. The fact that Gd^{3+} is a well known contrast agent for magnetic resonance imaging, (MRI), increases the attractiveness of using RE-doped Gd_2O_3 nanophosphor as fluorescence and MRI labels [5]. In turn, Europium (Eu^{3+}) ions produce a red luminescence characterized by sharp PL peaks associated to the intra-4f shell transitions $^5D_0 \rightarrow ^7F_J$. This feature enables nanosize Eu^{3+}-doped Gd_2O_3 phosphors to be considered a promising biomarker material. Despite of this fact, there is a lack of systematic studies focused on the synthesis and characterization of this type of nanocrystalline material. Among the various synthesis approaches for nanocrystalline RE-oxides, the sol-gel technique exhibits the advantage of allowing a precise compositional control and homogeneity under moderately low-temperature conditions. Typical sol-gel methods involve the use of chelating agents that may greatly influence the final properties of the materials [6].

Under the above premises, the present work investigated the effects of the annealing temperature and atomic fraction of Eu^{3+} ions ('x') on the structural and luminescence properties of nanocrystalline Eu^{3+}-doped Gd_2O_3. This phosphor was synthesized by using a simple sol-gel method that did not require the use of any chelating agent.

2 EXPERIMENTAL

2.1 Materials

Required weights of the hydrate acetate salts of Gadolinium and Europium (99%, Alfa Aesar) were dissolved in 2-ethylhexanoic acid (98%, Alfa-Aesar). All reagents were used without further purification. Unlike other related reports, no chelating agent was used in our case.

2.2 Synthesis of Nanocrystalline Powders

A modified sol-gel method and subsequent annealing were used to prepare crystalline $Gd_{2-x}Eu_xO_3$ nanophosphor, with 'x' ranging from 0.01 to 0.30. Suitable amounts of Gd(III) and Eu(III) acetates were dissolved in 2-ethylhexanoic acid. A solid intermediate was obtained after evaporating the solvent at 200°C for 72h. In order to develop the desired structure, the solid intermediate was annealed at different temperatures (750, 850 y 950°C) for 2h and cooled down in air. The annealing temperatures were selected based on previous TG-DTA analyses of the precursor salts [7]. Annealed solids were grounded and submitted to characterization.

2.3 Characterization Techniques

Structural characterization of the solids were carried out in a Siemens D500 x-ray diffractometer (XRD) using the Cu-Kα radiation. Fourier Transform Infrared (FTIR) measurements were undertaken in a MIRacle TM ATR FTS 1000 spectrometer in the transmittance mode. Photoluminescence spectra were collected using a Spectrofluorometer FluoroMax-2 with a 150mW continuous ozone-free Xe lamp as the excitation source.

3 RESULTS AND DISCUSSION

3.1 XRD Analyses

Figure 1 shows the XRD patterns of the intermediate before and after annealing at different temperatures. The solids were prepared for a Eu atomic fraction, 'x', of 0.05. Only very sharp peaks corresponding to cubic Gd_2O_3 phase (JCPDS 12-0797) were identified in the annealed samples. Similar results were observed at different 'x' values [7].

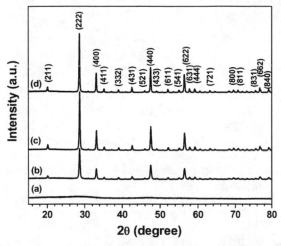

Figure 1: XRD patterns of $Gd_{2-x}Eu_xO_3$ (x=0.05) solids before (a) and after annealing at 750°C (b), 850°C (c) and 950°C (d). All peaks correspond to cubic Gd_2O_3.

The enhancement in the peak sharpness by increasing the annealing temperature would suggest the enlargement of the crystal size in the oxide.

As figure 2 shows, the average crystallite size - estimated by using the Scherre's equation- were 29nm, 36nm and 41nm when the solids were annealed at 750°C, 850°C and 950°C, respectively. This enlargement in crystallite size with annealing temperature can be explained by the improvement of the inter-grain diffusional processes, involved with grain growth, at higher temperatures. Figure 2 also suggest the negligible influence of Eu^{3+} ion content on the crystallite size. This observation agrees with the results reported by Pires, *et al.* [8] in Eu-Gd_2O_3 synthesized by precipitation.

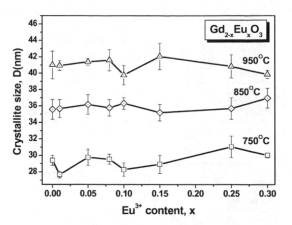

Figure 2: Effect of annealing temperature and Eu^{3+} content, 'x', on the average crystallite size of $Gd_{2-x}Eu_xO_3$ nanocrystalline powders.

x	*a* (Å)		
	750°C	850°C	950°C
0.00	10.816±0.002	10.817±0.001	10.813±0.001
0.05	10.809±0.004	10.799±0.006	10.816±0.001
0.10	10.818±0.002	10.816±0.001	10.806±0.005
0.15	10.827±0.002	10.822±0.001	10.821±0.001
0.30	10.826±0.002	10.830±0.001	10.825±0.001

Table 1. Variation of lattice parameter of Eu^{3+}-doped Gd_2O_3 powders annealed at different temperatures.

Table 1 shows the negligible variation of the lattice parameter '*a*' with the annealing temperature of the oxide produced at different 'x' values. Nevertheless, a small but noticeable increment on the lattice parameter was observed when the 'x' value varied from 0.05 to 0.30. This variation of the lattice parameter can be attributed to the actual substitution of Gd ions (0.94Å, ionic radii) by Eu ions (0.95Å) in the Gd_2O_3 lattice. The lattice parameter for pure Gd oxide (x = 0.00) is in good agreement with the bulk value (10.813Å).

3.2 FT-IR Measurements

The F-IR spectra of the solids produced at different annealing temperatures and x = 0.08 are presented in figure 5. The intense bands centered at 1535cm^{-1} and 1419cm^{-1} can be assigned to symmetric and asymmetric C=O stretching vibration modes in acetate species, respectively, contained in the intermediate. The spectra of the annealed powders only showed a sharp band at 540cm^{-1} that is assigned to the vibration of the Gd-O bond in the Gd_2O_3 structure. Furthermore, the absence of the acetate bands in the annealed samples suggests the complete thermal decomposition of the intermediate into the oxide structure. Similar spectra were obtained in other doped samples.

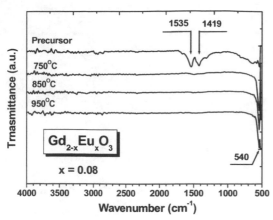

Figure 3: FT-IR spectra of Eu^{3+}-doped Gd_2O_3 (x=0.08) before (precursor) and after annealing at different temperatures.

3.3 Photoluminescence Measurements

Figure 4 shows the room temperature (RT) excitation spectra for Eu-doped Gd_2O_3 (x=0.15) annealed at different temperatures. The large band around 250nm corresponds to the $O^{2-} \rightarrow Eu^{3+}$ charge transfer band (CTB). Other weak bands at 274nm and 312nm are related to Gd^{3+} transitions promoted by the $Gd^{3+} \rightarrow Eu^{3+}$ energy transfer process. The weak narrow bands above 330nm are assigned to transitions of Eu^{3+} [9]. The rise of the CTB intensity by increasing the annealing temperature is attributed to the enhancement on crystallinity of the solids, as suggested by XRD. Figure 5 shows the RT emission spectra of the solids synthesized at different 'x' values obtained using the CTB (250nm) as the excitation wavelength. Figure 6 summarizes the variation of the relative luminescence intensity at 611nm with 'x' on a common sample-weight basis. As seen, the maximum intensity was obtained at x=0.15. This 'x' value is smaller than that reported in powders with similar crystallite size (41nm) but synthesized by liquid-phase reaction (x=0.20) [10]. Quenching became evident for 'x'>0.15.

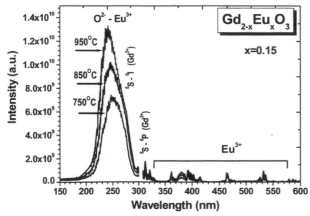

Figure 4: Effect of annealing temperature on excitation spectra of $Gd_{2-x}Eu_xO_3$ powders. The 611nm line was used for monitoring.

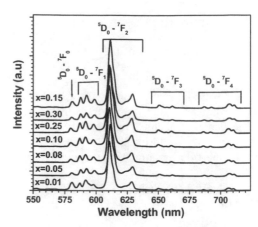

Figure 5: RT-Emission spectra $Gd_{2-x}Eu_xO_3$ powders synthesized at different 'x' values. The samples were annealed at 750°C for 2 hours.

In order to analyze the type of interactions involved with the observed quenching of luminescence, the mutual interactions constant, 's', was determined from the following relationship between the relative emission intensity (I) and the atomic fraction of the dopant species, 'x' [11]:

$$I \propto x^{\left(1 - \frac{s}{3}\right)} \tag{1}$$

Figure 7 shows the linearized plot of equation (1). The corresponding 's' value was estimated at of 3. This value is in excellent agreement with the one reported by Li and Hong [11]. The same authors proposed that the electric multipole resonant transfer between nearby Eu^{3+} ions and exchange interactions should be involved with the observed quenching phenomena.

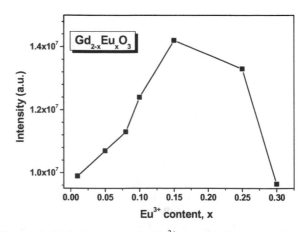

Figure 6: Relation between Eu^{3+} atomic fraction, 'x', and luminescence intensity at 611nm under excitation with 250nm radiation. The intensity values were measured from annealed samples on a common weight basis. The quenching phenomenon is evidenced by the drop in the luminescence intensity for 'x'>0.15.

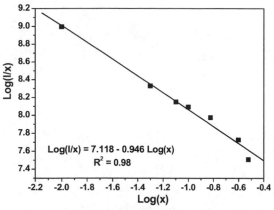

Figure 7: Log(I/x)-Log(x) plot for $Gd_{2-x}Eu_xO_3$ nanocrystalline powders synthesized at 750°C.

3.4 Characterization of Thin Films.

Based on the above preliminary results, the deposition of thin films of $Gd_{2-x}Eu_xO_3$ (x=0.08) onto fused quartz was attempted. The details of the experimental procedure will be published elsewhere [12]. The film exhibited the preferential growth of the (400) crystallographic plane (figure 8). The corresponding average crystallite size and lattice parameter were 22nm and 10.775Å, respectively. Figure 9 shows the RT PL-emission spectra obtained under different wavelength excitation. All emission bands correspond to typical intra shell $^5D_0$-^7F_J transitions of Eu^{3+} ions. The spectra also suggest that the Gd_2O_3 absorption band excitation (229nm) produces the strongest and brightest red emission compared with the conventional direct excitation wavelength (393nm). The inset shows the excitation spectrum of the thin film produced after thermal treatment at 750°C for 1h. Again, the CTB from O^{2-} to Eu^{3+}, the weak bands at 274nm and 312nm (related to internal Gd^{3+} transitions) and those weak bands representing the direct absorption of Eu^{3+} were observed. However, unlike nanocrystalline powders, a strong band at 229nm was detected. This band can be consequence of the absorption by Gd_2O_3 host [12].

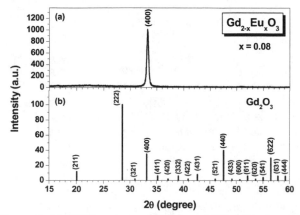

Figure 8: (a) XRD pattern of Eu-doped Gd oxide (x= 0.08) thin film; (b) standard pattern for Gd_2O_3 cubic phase.

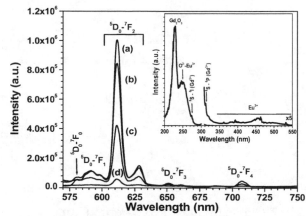

Figure 9: RT Emission spectra of $Gd_{2-x}Eu_xO_3$ (x=0.08) thin film, annealed at 750°C for 1h. The spectra were collected under different excitation wavelengths: (a) 229nm, (b) 248nm, (c) 274nm and (d) 393nm. The inset shows the excitation spectra obtained when the 611nm line was used.

4 CONCLUDING REMARKS

Powders and films of Eu^{3+}-doped Gd_2O_3 have been successfully synthesized by sol-gel method without the use of any chelating agent. The average particle size varied from 29nm to 41nm when the intermediate was annealed in the temperature range between 750°C and 950°C. PL characterization evidenced the formation of luminescent centers that facilitated the energy transfer from host Gd to Eu ions. The subsequent quenching of the luminescence for 'x'>0.15 was attributed to exchange interactions. PL analyses of the films suggested that excitation by the Gd_2O_3 absorption band would be more effective than direct excitation to obtain red luminescence.

5 ACKNOWLEDGMENTS

This material is based upon work supported by the National Science Foundation under Grant No. 0351449.

REFERENCES

[1] L. Brus, J. Phys. Chem., 90, 2555, 1986.
[2] G. Blasse, Chem. Mat., 1, 294, 1989; Chem. Mat., 6, 1465, 1994.
[3] M. Nichkova, et al., Anal. Biochem. 369, 34, 2007.
[4] H. Guo, et al., Appl. Surf. Scien., 230, 215, 2004.
[5] D. Dosev, et al., Appl. Phys. Lett. 88, 011906, 2006.
[6] C. Louis, et al., J. Solid State Chem., 173, 335, 2003.
[7] S. Jáuregui-Rosas, O. Perales-Perez, W. Jia, M. S. Tomar et al., Presented at MRS Spring Meeting, 2008.
[8] A.M. Pires, et al., J. Alloys Comp., 344, 276, 2002.
[9] M. Buijs, et al., J. Lumin. 37, 9, 1987.
[10] C.-S. Park, et al., J. Lumin. 118, 199, 2006.
[11] Y. Li and G. Hong, J. Lumin. 124, 297, 2007.
[12] S. Jáuregui-Rosas, O. Perales-Perez, W. Jia, M. S. Tomar et al., In preparation.

Biologically synthesised Silver nanoparticles using a Bacterial Culture Isolated from the riverine bank of Ganges in India

Naheed Ahmad*, Seema Sharma** and Ashok K Ghosh***

* University Department of Botany, Patna University, Patna 800003
naheedshamsi@yahoo.co.in
**Department of Physics, A N College, Patna 800013
seema_sharma26@yahoo.com
*** Department of Environment and Water Management, A N. College, Patna 800013
ashokghosh51@hotmail.com

ABSTRACT

Nanotechnology is a highly energized discipline of science and technology. One of the major challenges of nanotechnology is the synthesis of nanomaterials with a wide range of chemical compositions and sizes. Nanobiotechnology is a promising novel field specially for biodiverse countries like India.whose diversity is an asset which can be harnessed .Recently material scientists have been viewing with interest this diversity , particularly microorganisms as possible ecofriendly nanofactories..

In this paper, biosynthesis of nanomaterials is carried out from the microbial diversity obtained from the state of India, Bihar .Bacillus species isolated locally from the region has been used to synthesize silver nanoparticles at room temperature.

KEYWORDS

nanobiotechnology,nanofactories,microbial diversity,biosynthesis

INTRODUCTION

Nanotechnology is a highly energized discipline of science and technology , which is gaining importance in the new millennium . Nanomaterials are of considerable interest due to their unusual optical [1], chemical [2], petrochemical [3]and electronic properties[4] . These nanoscale materials have gained importance due to their potential applications in optics,biomedical sciences ,mechanics ,magnetics and energy science. Hence one of the various challenges of nanotechnology is the synthesis of nanomaterials with a wide range of chemical composition and sizes. In the current

world there is a growing need and awareness to develop green technologies .

Biological systems , are in general masters of ambient condition of chemistry and hence have ability to synthesize and sequester inorganic materials from nano to macroscale. Recently material scientists have been viewing with interest biological systems particularly microorganisms as possible

ecofriendly nanofactories [5-7]..Nanobiotechnology is a promising field specially for biodiversity rich countries like India [8] . Biodiversity, though the planets greatest asset remains paradoxically the least developed [9]. Microorganisms occupy diverse habitats, they develop unique and sometimes bizarre ways to survive hostile toxic environments. Metals are ubiquitous and many are necessary for the survival of organisms but even essential elements like Fe, Zn, Mg, etc. are toxic to organisms beyond certain limits. Bacteria react to most metals by reduction of ions or by the formation of water insoluble complexes. Microorganisms that tolerate heavy exposure to metals can be used for synthesis of nanoscale materials.However the exact mechanism leading to the formation of nanoparticles is not fully understood .Formation can be either intracellular or extra cellular . The ability of microorganisms to grow in presence of high metal concentrations might be a result of specific mechanisms of resistance. such mechanisms include efflux systems , alteration of solubility and change of toxicity in redox state of the metal ion, extracellular complexation and lack of specific metal transport systems [10 -11].

Early studies have revealed the use of various microorganisms for nanosynthesis. Bacteria *Bacillus subtilis* 168 was reported to reduce Au+ to nanoscale dimensions [6,7]. Several bacterial strains like *Psedomomas stutzeri* Ag259 were found to be silver resistant and were able to produce nanosize silver[12].Other workers have used the microbial diversity to produce Magnetite , Silica , Titania from fungi *Fusarium oxysporum and Vericillium* [13-14].*Klebsiella aerogens* was manipulated to produce Cds nanoparticles extracellularly[15]. In addition to gold and silver nanoparticles synthesis of semiconductors like CdS , ZnS and PbS have been obtained from bioorganisms *Clostridium thermoaceticum* precipitates was observed to precipitate CdS at the cell surface as well as the medium from $CdCl_2$ in presence of cystenine hydrochloride in the growth media [16].The monodispersity of silver / gold nanoparticles produced either intra or extracellularly by these bioorganisms is not very high or inferior to those obtained by the conventional chemical methods[17]

In this paper, biosynthesis of nanomaterials is carried out from the microbial diversity obtained from the state of India, Bihar (rich fertile alluvial Gangetic plain).Bacillus species which was identified as *Baccilus cresus* has been used to synthesize silver nanoparticles which was isolated from the Gangetic belt (riverine belt of Ganges river) was seen to synthesize silver nanoparticles extracellularly.This is the first report when the microbial diversity of the Gangetic plain has been tapped for nanosynthesis.

METHODOLOGY

Soil samples were collected from the riverine belt of Ganges in Bihar . Serial dilutions were done and plated on N.A plates. The plates were incubated for suitable time and distinct looking colonies were picked up and grown till pure cultures were obtained . Preliminary Silver resistance of various strains was done . The cultures were screened by spot inoculation of Silver nitrate of varying concentrations. The concentrations used were 25ppm .50 ppm ,75ppm , 100ppm . 150 ppm 200ppm 250ppm, 300 ppm …600ppm 800ppm till 1000ppm . Strains which could tolerate 800ppm of silver was selected for further study .

The selected silver tolerant strain was inoculated in YES media and incubated at room temperature until the absorbance of the culture was between 0.8-0.9 . silver at a concentration of 800 ppm was added and cultures were incubated further . Cell free supernatant was obtained by centrifugation of the cultures at 8000rpm for 10 min at room temperature.The absorbance scan of the supernatant was recorded to check for the characteristic nano silver peak. Cultures which showed best peak were selected for further studies. Optimization of the cultures was done to get the maximum yield.

Optical absorbance of the silver nanoparticles was taken on UV visible spectrophotometer (Shimadzu UV 2450), in range of 200-800 nm .XRD measurements of silver nanoparticles was done on Rigaku Miniflex and TEM of the silver nanoparticles was performed on carbon coated copper grids .

RESULTS AND DISCUSSION

The results revealed that the soil sample contains a diversity of unique bacteria which are silver tolerant. The bacterial culture which was selected for further study was identified as *Bacillus cereus* (fig 1).It was found that *Bacillus cereus* could tolerate silver upto 800ppm which was added to the media in form of silver nitrate.

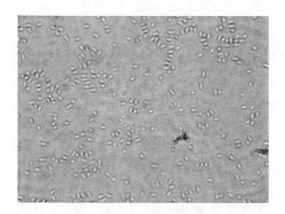

Fig 1 showing isolates of *Bacillus cereus*

Reduction was obtained in all three sets viz, when silver nitratre was added to the stationary phase , to the culture suspended in water and to the cell free suspension .recovery of silver was found to be maximum and easier in the first set hence this protocol was followed for further synthesis.Exposure to sunlight in all the three media was followed by colour change .Increase in room temperature was associated with reduction in time for the synthesis of nanoparticles. The absorbance scan of silver nanoparticles exhibited a sharp plasmon peak at 450 nm, which is characteristic of monodispersed Ag .

These particles were found to be stable in the solution on storage up to one month with no change in the absorbance maxima .The magnitude , peak wave length and spectral band width of the plasmon resonance associated with nanoparticle are dependent on particle size , shape and material composition as well as the local environment.

X-ray powder diffractogram of the biosynthesized nanosilver exhibits Braggs reflection due to (111), (200),(220)(311)and (222) corresponding FCC type bulk silver .(fig2).

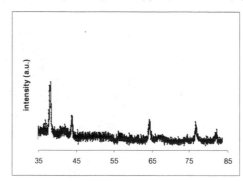

Fig. 2. XRD pattern of Silver nanoparticles

The diffraction peaks are broad around the their bases indicating that the silver particles are in nanosizes.

The TEM pictures (fig3) obtained reveal that the silver nanoparticles are in the 10 nm range.

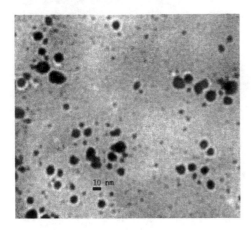

Fig. 3 Transmission Electron Microscopy
Micrograph of silver nanoparticles
synthesized by **Bacillus cresus**

Particles smaller than these could not be resolved .Aggregates of 0.25-05µ were observed with SEM image, probably indicating that the particles are aggregating to form larger particles. It was observed from EDAX measurements that the sample contained about 75% Ag .

Though the exact mechanism is still obscure it is probable that that certain enzymes are released which have reducing action . It is known that silver cations are highly are highly reactive and tend to bind with elctron donor groups containg sulphur, nitrogen or oxygen of the proteins . Thus it is seen that the microbial biodiversity may provide cheap viable and green ways for the production of nanoparticles each unique as environment it comes from.

CONCLUSION

In summary, a brief overview of the use of micro-organisms in the biosynthesis of Ag nanoparticles has been described. In the context of the present drive to develop green technologies in materials' synthesis, this method is a viable alternative to the physical and chemical methods currently in vogue. Extra-cellular secretion of enzymes offers the advantage of obtaining large quantities in a relatively pure state, free from other contaminating proteins associated with the organism. The process can be extended to the synthesis of nanoparticles of different chemical compositions, shapes and sizes by suitable identification of the enzymes secreted by the micro-organisms.

ACKNOWLEDGEMENT

The authors wish to thank University Grants Commission, Govt. of India for providing financial support under major research project scheme

REFERENCES

[1] A.Krolikowska, A Kudlski., A Michota, J. Bukowska (2003) SERS studies on Structure of thioglycolic acid monolays on silver and gold. Surf Sci.532:227-232.

[2] A Kumar. S Mandal, P.R Selvakanan, R Parischa, A B Mandale, M Sastry (2003) investigations into into interaction between surface bound Alkylamines and gold nanoparticles .Langmuir 19:6277-6282.

[3] N Chardrasekharan , PV Kamat (2000)- Improving photochemical performance by nanostructured TiO2 films by absorption of gold particles .J.Phys. Chem104:10851-10857.

[4] G Peto, G L.Paszti Z Molnar, Geszti , A Beck, L Guczi,(2000). Electronic structure of gold nanoparticles deposited in SiO2 /Si100.Mater SCi.Eng 19:95-99.

[5] G Southham and T.J Beveridge *Geochim .Cosmochim* Acta (1996)60, 4369-4379.

[6] T.J Beveridge and R.G.E Murray,. (1980) J.Bacteriol 1980 141, 867-887

[7]. D Fortin and T.J Beveridge (2000) in Biomineralisation –From Biol to Biotech &Medical applications (ed Baeuerien E.) Wiley VCH Weinghiem p700.

[8] E.O Wilson eds (1988) Biodiversity . Washington D.C natl. Acad .10.Bull A.T,

[9] H Goodfellow & J.H Slater (1992) Annu Review of Microb.46:219-52.

[10] S. Silver (1996) Bacterial resistance to toxic metal ions – Review Gene 179:9-19

[11] Beveridge T,J. , Hughes M.N ,Lee H,leung K.T., Poole R.K. , Savvaidis I, Silver S, Trevor JT (1977) Metal microbe interaction :contemperory approaches .Adv,Microb. Physiol. 38:178-243.

[12.] T Klaus, R Joerger. E Ollson, C-G Granqvist (1999)Silver based crystalline nanoparticles , microbially fabricated .Proc. natl. Acad Sci USA 96:13611-13614

[13]. A Ahmad, P Mukherjee, D Mandal, S.Senapati, MI Khan ,R Kumar, M Sastry (2002)Enzyme mediated extracellular synthesis of CDS nanoparticles by the fungus Fusarium oxysporium .J. Am Chem Soc. 124:12108-12109

[14] A Ahmad, MI Khan , S Senapati, R Kumar, M Sastry, MI Khan, (2003a) of monodisperse gold by anovel extremophilic actinomycete Thermonospora sp. Langmuir19:3550-3553

[15]. A Ahmad, P Mukherjee , S Senapati, D. Mandal, M.I Khan ,Kumar R, M Sastry (2003c)Extracellular synthesis of silver nanoparticles using the fungus *Fusarium oxysporum* Coll. Surf B:313-318

[16] D P Cunningham , LL Lundie (1993) Precipitation of cadmium y *Clostridium thermoaceticum.*Appl. Environ. Microbiol 59:7-14 .

[17] D Mandal., M.E Bolander, D Mukhopadya, G. Sarkar and P Mukherjee.(2006)The use of microorganisms for the formation of metal nanoparticles and their applications .Appl. Micriobiol. Biotechnol. 69:485-492

Comparative Study of Effect of Various Reducing Agents on Size and Shape of Gold Nanoparticles

Abhishek Kumar[1] and Nikhil Dhawan[2]

[1]Materials Science and Engineering Department, Stanford University, CA 94305, USA, akpec@stanford.edu
[2]Department of Metallurgical Engineering, Punjab Engineering College, Chandigarh 160012, India, dhawan.nikhil@gmail.com

ABSTRACT

Gold nanoparticles are synthesized by reduction of gold ions by a reducing agent in the presence of a capping agent. Many properties (like electromagnetic, optical and catalytic) vary with size and shape of the nanoparticles. In this paper, a comparative study of the effect of various reducing agents on size and shapes of gold nanoparticles has been made. Gold nanoparticles formed by hydrazine sulphate as a reducing agent have branched or the so-called sea urchin shape resembling to the shape that of the familiar sea animal. This shape was synthesized for the first time. TEM images depicting the morphological and structural features of nanoparticles formed by different techniques have been shown along with some of their applications.

Keywords: Electromagnetic properties, Shape, Reducing agent, synthesize.

1 INTRODUCTION

Nanoparticles have attracted considerable interest because of their unique optical, electromagnetic and catalytic properties that differ from the bulk ones. The origin of these properties is their large surface to volume ratio and from the coherent oscillation of the conduction electrons that can be induced by interactive electromagnetic fields. Recently, gold nanoparticles have found their large application in medical technology. Gold nanoparticles are being used to deliver protein-based drugs, and are of particular utility because the particles can carry multiple active groups. For example, they can be designed with targeting groups that direct treatment to a particular cell type, and also carry pharmacologically active groups that exert a therapeutic effect. One of the keys to understanding the size-dependent properties and applications of nanoparticles is generating libraries of particles with diverse sizes for physical study. Since many properties vary with the size and shape of the novel metal particles, the focus of many studies has been on the methods for controlling the size and shape of gold nanoparticles. The aim of this work has been to compare the effect of various reducing agents on size and shape of gold nanoparticles formed.

2 EXPERIMENTAL

Five identical samples of gold ions were prepared as follows. In each of the five conical flasks, 500 μL of 0.01 M $HAuCl_4$ and 500 μL of 0.01 M sodium citrate were added to 19 mL, 18 Ω deionized water. Sodium citrate acts as a capping agent, which prevents agglomeration of gold nanoparticles when they are formed. Each of the gold ion solution was reduced in a different way using different reducing agents. 500 μL of 0.1 M freshly prepared hydrazine sulphate (Sample A), 500 μL of 0.1 M freshly prepared $NaBH_4$ (Sample B), 500 μL of 30% hydrogen peroxide (H_2O_2) (Sample C) and 500 μL of 0.1 M potassium ferrocynide (Sample D) were added to four different samples and shaked gently for 7-10 seconds till the color changed indicating nanoparticles formation. Sample E was prepared by addition of 500 μL of 0.1 M ascorbic acid and heating at 60°C for 20 minutes. In yet another experiment involving synthesis of gold nanoparticles, 1000 ppm $KAu(CN)_2$ was reduced by 0.1 M freshly prepared $NaBH_4$ (Sample F). Each of the samples was left undisturbed after which they were analyzed by transmission electron microscopy (TEM). 2-3 drops of each solution were placed on carbon-coated copper grids for a transmission electron micrographic examination.

3 RESULTS AND DISCUSSION

The morphology of gold nanoparticles along with their size and shape has been outlined in the form of a grid with a short description for each morphology type. The nanoparticles formed have large surface area and hence large surface energy associated with them. As a result, they tend to agglomerate in order to lower their surface area and attain stable state. Use of a capping agent like sodium citrate is thus essential to prevent immediate agglomeration. The gold nanoparticles synthesized with the help of various reducing agents were characterized by TEM within 1 hr. of their synthesis so as to achieve non-agglomerated morphology of gold nanoshapes.

4 GOLD NANOPARTICLES GRID

A.

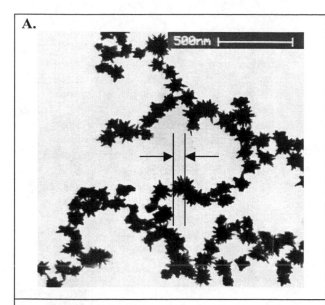

B.

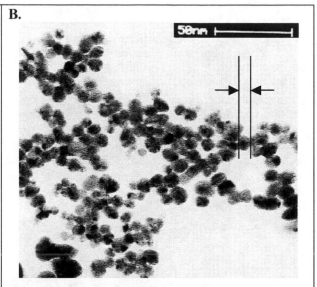

Reducing Agent: Hydrazine sulphate
Shape: Sea urchin
Size: 100-125 nm

Gold nanoparticles formed by this technique have approximately 18-20 needles erupting from a central core of 100-125 nm size nanoparticle. The length of longest needle observed is 70 nm and that the shortest needle is 10 nm long. The general aspect ratio of needles is 4:1. Such needles are assumed to be formed due to the ongoing competition between the hydrazine and the sulphate group towards the gold ion to be reduced, gold particles being attracted by sulphate group to some extent. These nanoparticles may act as active seeds for branched nanorods and nanowires synthesis.

Reducing Agent: Sodium borohydride
Shape: Nearly spherical
Size: 5-7 nm

Sodium borohydride being one of the strongest reducing agent reduces gold ions at the very instant it is put into solution containing gold ions and capping agent. Hence 5-7 nm size nanoparticles are synthesized. Citrate ions act as capping agents and prevents agglomeration upto certain extent. Within 5 hours of preparation the pink color of solution disappears indicating agglomeration of gold nanoparticles. These spherical gold nanoparticles are used as seeds to synthesis variable aspect ratio nanorods in the presence of a directing agent. The nanoparticles together with nanorods are used to detect various cancers and ulcers in human body.

C.

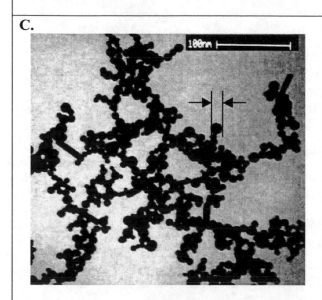

D.

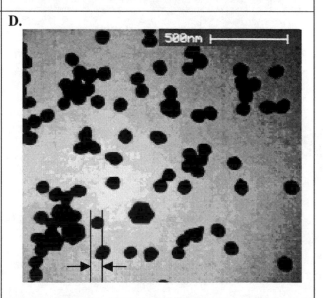

Reducing Agent: 30% Hydrogen peroxide
Shape: Spherical
Size: 5 nm

Synthesis of gold nanoparticles by using hydrogen peroxide showed smallest nanospheres (size: 5 nm) among all reducing agents used. There were also some large nanospheres of size range 15-20 nm. Apart from spherical structures nanorods upto length 50 nm and diameter 5 nm (aspect ratio = 8:1) were also observed. Hydrogen peroxide is a strong bleaching agent and can reduce gold ions very effectively. Nanospheres of this size range are best suited for biomedical applications. The exact reason behind the formation of nanorods is not known since no directing agent or micellar template was used.

Reducing Agent: Potassium Ferrocynide
Shape: Spherical nanospheres and plates
Size: 65-75 nm

Comparatively large nanoparticles of gold are produced using potassium ferrocynide as reducing agent. Potassium ferrocynide being weak reducing agent reduces gold ions slowly as compared to that reduced by sodium borohydride and hydrogen peroxide. Hence gold particles have more time to increase their size. Apart from nanospheres again some nanoplates are seen the exact reason behind formation of which is still to be known. Plates having 4 and 6 sides are easily seen. The length of each side of nanoplate is approximately 60 nm. The solution color changes to bluish-pink indicating nanoparticle formation. Blue color is due to use of potassium ferrocynide as reducing agent.

E.

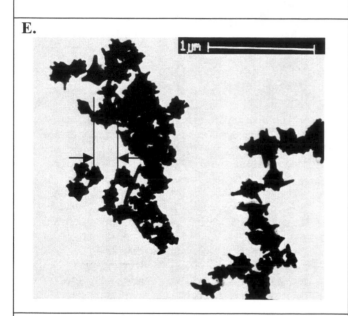

F.

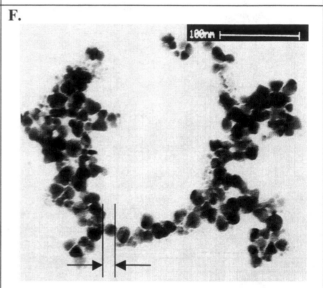

Reducing Agent: Ascorbic acid
Shape: Distorted shaped along with needles
Size: 200-225 nm

Ascorbic acid is a weak reducing agent and cannot reduce gold ion at room temperature or without any surfactant. Hence reduction was carried out at 60°C for about 20 min. The shape of gold nanoparticles is distorted with general size range 200-225 nm. These nanoparticles formed have high tendency to agglomerate and hence reduce surface energy to attain stable state. The exact mechanism and nature of formation of these nanoparticles by ascorbic acid reduction is yet to be understood. Due to their large size, these may also be termed as "pseudo gold nanoparticles". Some long needles of aspect ratio upto 10:1 were also observed.

Reducing Agent: Sodium borohydride (Reduction of gold cyanide ions)
Shape: Spherical
Size: 10-15 nm

For comparing relative affinities of chloride and cyanide ligands with gold core, an attempt to reduce gold cyanide ions with sodium borohydride was made. 1000-ppm solution of gold cyanide ions when reduced by 0.1 M borohydride solution (freshly prepared) gave gold nanoparticles, which were of size range 10-15 nm (bigger than those synthesized from gold chloride ions under same set of conditions). It may be assumed that cyanide has larger affinity with gold atom as compared to chloride. Moreover, large ranges of nanoparticles were found to be synthesized.

5 APPLICATIONS

Gold has attracted much attention in catalyst research and industrial applications. It has been demonstrated that gold nanoparticles (5-8 nm) dispersed on metal oxides can exhibit high catalytic activities for various types of reactions, like the epoxidation of propene and the low temperature oxidation of CO. The use of carbon nanotubes as potential catalyst supports is now attracting the interest of the catalytic community with evidence of unique metal/support interactions resulting in quite distinct catalytic behavior [4].

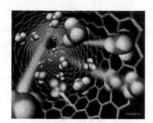

Figure 1: Au nanoparticles homogeneously dispersed on the surfaces of the CNTs.

These are also very good at scattering and absorbing light and this property is harnessed in a living cell to make cancer detection easier (Figure 2). Gold nanoparticles have 600 percent greater affinity for cancer cells than for non-cancerous cells [5]. Gold nanoparticles have also found applications in the field of electronics like digital data storage that requires them to form ordered arrays (Figure 3).

Figure 2: Binding gold nanoparticles (35-50 nm) to a specific antibody for cancer cells makes cancer detection easier.

6 CONCLUSION

Tiny billionth-of-a-meter sized clusters of gold atoms - gold "nanoparticles" — are being widely studied by scientists. They have many useful potential applications, from carriers for cancer-treatment drugs to digital data storage. We have provided a comparative study, in the form of a table for better understanding, of the effect of various reducing agents on the size and shapes of gold nanoparticles. Gold nanoparticles formed by hydrazine sulphate as a reducing agent have branched or the so-called sea urchin shape resembling to the shape that of familiar sea animal. This shape was synthesized for the first time and may find applications in various biomedical or related fields. The exact mechanism and nature of these branched gold nanoparticles is still under scrutiny at our research laboratory and full effort is being made to control the aspect ratio of needles erupting from the central core along with control in core size.

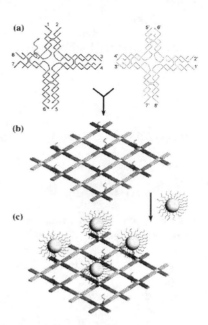

Figure 3: (a) Two-tile system of nanogrids; (b) DNA nanogrid (c) Ordered gold nanoparticles (~100 nm) on DNA grid [6].

ACKNOWLEDGEMENTS

We are grateful to Professor Philip Jennings and Dr. Eddy Poinern of Murdoch University, Perth 6150, Western Australia without whom this research would not have been possible. The TEM images shown in this work were provided by Dr. Peter Fallon of Biological Sciences, Murdoch University. His assistance is fully appreciated.

REFERENCES

[1] Govindaraj, A., Satishkumar, B. C., Nath, M. and Rao, C.N.R., "Metal Nanowires and Intercalated Metal Layers in Single-Walled Carbon Nanotube Bundles," Chem. Matter., **12**, 202 (2000).

[2] Jana, N. R., Gearheart, L. and Murphy, C. J., "Wet Chemical Synthesis of High Aspect Ratio Cylindrical Gold Nanorods," J. Phys. Chem. B, **105**, 4065 (2001).

[3] C.J. Murphy and N.R.Jana, "Controlling the Aspect Ratio of Inorganic Nanorods and Nanowires", Adv. Mater., Vol. 141), (2002), pp. 80–81.

[4] Wei Lu, Xicheng Ma1, Ning Lun, Shulin Wen, "Decoration of Carbon Nanotubes with Gold Nanoparticles for Catalytic Applications", Mat. Res. Soc. Symp. Proc. Vol. 820.

[5] Alex W. H. Lin, Nastassja A. Lewinski, Jennifer L. West, Naomi J. Halas and Rebekah A. Drezek, "Optically tunable nanoparticle contrast agents for early cancer detection: model-based analysis of gold nanoshells", J. Biomed. Opt. 10, 064035 (2005).

[6] James J. Storhoff, Anne A. Lazarides, Robert C. Mucic, Chad A. Mirkin, Robert L. Letsinger, and George C. Schatz, "What Controls the Optical Properties of DNA-Linked Gold Nanoparticle Assemblies?", J. Am. Chem. Soc. **2000**, 122, 4640-4650.

Shape-Controlled Synthesis and Catalytic Behavior of Supported Platinum Nanoparticles

Changkun Liu, Bing Zhou

Headwaters Technology Innovation, LLC. 1501 New York Ave., Lawrenceville, NJ 08648
cliu@htinj.com

ABSTRACT

A 5 wt% Pt/C catalyst with around 20 nm cubic platinum particles was prepared through a conventional preparation method (i.e. precursor impregnation, reduction, and calcination) by choosing of hydrophobic solvent in the impregnation procedure. These 20 nm cubic Pt particles showed high selectivity (ca. 99.4%) for the hydrogenation of o-chloronitrobenzene.

Keywords: nano-particle, shape control, supported catalyst, hydrogenation

1 INTRODUCTION

Supported metal catalysts are among the most important catalysts, being used on a large scale for refining of petroleum, conversion of automobile exhaust, hydrogenation of carbon monoxide, hydrogenation of fats, and many other processes. Since the researches show that shape control technologies may contribute the improvements to the final catalyst performance, the chemical industries are eager to apply this technology into commercial production [1-4]. However, in the last 5-10 years the efforts to use prevailing colloid shape control methods in industry were frustrated by many problems.

The most efficient method to precisely metal shape control was the colloid method by using functional polymer as capping agent or protective agent during the metal was reduced [5-9]. The stable metal colloidal solution was normally prepared in very low concentration. Impregnation of this well-defined metal nanopaticles colloid with supports is the convenient way to prepare supported catalyst with well metal shape-controlled. However in large-scale production or in order to prepare high loading supported catalyst, there is large amount of water needs to be very carefully evaporated off after the impregnation step. This process is not only energy and time consumed but also in the risk of nano-particles aggregation and lost of the morphological control. This inherent problem form colloid method itself restricts this method to be applied in large-scale catalyst production.

Traditionally, supported metal catalysts are typically made by impregnation of a porous supports with an aqueous solution of a metal salt, followed by reduction and calcination. This preparation method now is predominant in industry large-scale catalyst production but is always thought to lose ability for metal shape control.

In this study we disclose that precise shape control of metal particle also can be achieved by traditional metal preparation method. By choosing of Pt acetylacetonate as precursor and methanol/ toluene as solvent in the supports pretreatment and precursor impregnation, evenly dispersed 20nm cubic Pt nanoparticles with 5 wt% loading on carbon was available. This method may be reproduced in most of supported metal catalyst preparation, if metal acetylacetonate analogues are available, without changing extant industrial device and procedure.

2 EXPERIMENTS

Preparation of 20 nm cubic 5% Pt/C catalyst (Cubic-Pt)

6.0 g activated carbon was soaked with 30 ml methanol for 12 hours. After removal of excess methanol, the carbon solid was dispersed in 60 ml toluene and 0.605 g platinum (II) acetylacetonate in 20 ml toluene was added. The total suspension was kept rotated on rotary evaporator. Toluene was slowly evaporated out in 20mmHg vacuum and at 85 °C. The activated carbon with platinum complex was dried in vacuum oven at 80 °C for 6 hours and then reduced in 5% H_2 in N_2 flow at 100°C for 4 hours. After that gas flow was switched to nitrogen and let the system cool to room temperature.

General procedure of hydrogenation of 2-chloronitrobenzene

A typical hydrogenation procedure was conducted as follows: 4.04 g 2-chloro-nitrobenzene catalyst and proper amount of catalyst was dispersed in 60 ml ethanol. The suspension was added into a 300 ml stainless steel autoclave with glass liner equipped with a mechanically stir blade, a pressure gauge, a gas inlet tube attached to a hydrogen source and a cooling circler connected to the temperature controller. The autoclave was purged by nitrogen for three times. When the temperature stabilized at 25°C, the vessel was pressurized to 150 psi with hydrogen. During the reaction mixture was vigorously stirred and was temporarily stopped for sample collection when no pressure decrease was observed. The product was analyzed on gas chromatography Agilent 6890 equipped with a FID detector and Rtx-5 Amine column.

3 DISSCUSION

The HRTEM characterization of this carbon supported catalyst is shown in Figure 1. Figure 1 shows that very uniform Pt microcrystals are imbedded in the support. The size of the supported Pt crystallites was estimated to be 14.5nm from the broadening of the Pt (111) peak in X-ray diffraction (at $2\theta \approx 40^{\circ}$). The d_{TEM} of 20nm is in fairly good agreement with d_{XRD} of 14.5nm.

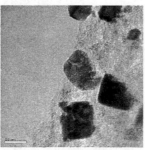

Figure 1. Left: Low-magnification TEM image of 5% wt. Pt /C, showing a uniform size distribution of the supported particles. Right: High-magnification image highlighting the cubic shape of the Pt nano–crystallites.

These cubic Pt nanocrystals supported on carbon, marked as **Cubic-Pt**, exhibit extra performance in catalytic hydrogenation of halo-nitrobenzene when compared to commercial available catalyst **DG-Pt**. **DG-Pt** is a commercial product from Degussa AG which composition is 5% Pt on carbon with bismuth doped as promoter.

Hydrogenation of o-chloronitrobenzene to o-chloroaniline is a tedious process and often yields many de-chloride byproducts during the reaction [10]. The results as shown in Table 1 indicated that catalyst **Cubic-Pt** took 10 h to almost complete the reaction, meanwhile catalysts **DG-Pt** only needed about 6 h. The product analysis showed that **Cubic-Pt** yielded the lowest concentration of de-chloride byproduct aniline (0.28% in 10 h) as compared to catalyst **DG-Pt** (0.48% in 6 h).

cubic 5% Pt/C; **DG-Pt**: 5% Pt/C with Bi doping bought from Degussa AG. [b] o-CAN = o-chloroaniline, AN = aniline, o-HOCAN = o-hydroxylamine-chlorobenzene. Other byproducts or intermediates with concentration lower than 0.10% are exclusive from calculation.

Efforts to prepare smaller size cubic Pt crystals on carbon as well as to optimize reaction parameters are ongoing. We believe that not only smaller size cubic Pt will increase hydrogenation rates but also more lateral boundary Pt atoms will result in a higher selectivity.

REFERENCES

1. B. C. Gates, Chem. Rev., 1995, 95, 511.
2. I. Balint, A. Miyazaki, K-i. Aika, Chem. Commun., 2002, 1044.
3. J. M. Rickard, L. Genovese, A. Moata and S. Nitsche, J. Catal., 1990, 121, 141.
4. A. Miyazaki, I. Balint, K-i. Aika, J. Catal., 2001, 204, 364.
5. V. F. Puntes, K. M. Krishnan, A. Paul Alivisatos, Science, 2001, 291, 2115.
5. T. S. Ahmadi, Z. L. Wang, T. C. Green, A. Henglein, M. A. EI-Sayed, Science, 1996, 272, 1924.
7. Y. G. Sun, Y. N. Xia, Science, 2002, 298, 2176.
8. T. Teranishi, R. Kurita, M. Miyake, J. Inorgan. Organometal. Polymers, 2000, 10, 145.
9. A. Miyazaki, Y. Nakano, Langmuir, 2000, 16, 7109.
10. X. X. Han, R. X. Zhou, X. M. Zheng, H. Jiang, J. Mol. Catal. A: Chem., 2003, 193, 103.

Catalysts [a]	Reaction Time (h)	Selectivity [b]			Conversion (%)
		o-CAN	AN	o-HOCAN	
Cubic-Pt	4.5	66.53	0.13	33.33	100
	6	86.16	0.19	13.65	100
	8.5	97.10	0.23	2.67	100
	10	99.41	0.28	0.31	100
DG-Pt	4.5	90.73	0.39	8.88	100
	6	98.94	0.48	0.59	100

Table 1: Hydrogenation of o-CNB with Pt/C catalysts
Reaction conditions: T = 25 °C; P_{H2} =10 bar; 4.04 g o-chloronitrobenzene in 60ml ethanol; 0.02g 5%Pt/C (substrate/catalyst = 5,050:1); [a] Catalyst, **Cubic-Pt**: 20 nm

Deposition and Characterization of Platinum Nanoparticles on Highly Orientated Pyrolytic Graphite

Nora Elizondo Villarreal[*,**], Carlos Novar[**], Ji Ping Zhou[**], Lorena Álvarez Contreras[***] and Ran Tel Vered[****]

[*]Department of Physics, Universidad Autónoma de Nuevo León, San Nicolás de los Garza, N. L., C.P. 66451, México, nelizond@mail.uanl.mx
[**]Department of Chemical Engineering and Texas Materials Institute, University of Texas at Austin, TX 78712, USA, jpzhou@ccwf.cc.utexas.edu, norae@che.utexas.edu
[***]Department of Chemical Materials, Research Center on Advanced Materials, S. C., Miguel de Cervantes 120, Complejo Ind. Chih., 31109 Chihuahua, Chih., México, lorena.alvarez@cimav.edu.mx
[****]Institute of Chemistry, The Hebrew University of Jerusalem, Jerusalem, 91904, Israel, telvered@gmail.com

ABSTRACT

Pt nanoparticles were synthesized by polyol method and then deposited on a highly orientated pyrolytic graphite (HOPG) substrate. They also were obtained by direct reduction of chloroplatinic acid on HOPG in a H_2 flow and by immersing the graphite electrode into a platinum plating solution, 0. 1 mM H_2Cl_6Pt and 0.1 M H_2SO_4 solution. The density of particles as well as their dimension depend in this case on the time of immersion in the hexachloroplatinic solution. Pt nanoparticles formed for these methods are mainly quasi hemispherical small Pt nanoparticles in size. The characterization of these systems was studied by high angle annular dark field (HAADF) technique and transmission electron microscope (HRTEM). The combination of these experimental techniques allowed, the determination of the structure and the distribution of Pt nanoparticles on HOPG. A rotation between the layers of HOPG was observed in the direction perpendicular to the basal plane by effect of heating in H_2.

Keywords: platinum nanoparticles, characterization, deposition, transmission electron microscope

INTRODUCTION

Platinum is considered one of the best electrocatalyst for low temperature reactions in a H_2/O_2 fuel cell. Metallic nanoparticles are of great interest because of the modification of properties observed due to size effects, modifying the catalytic, electronic, and optical properties of the monometallic nanoparticles. Interest in platinum nanoparticles derives mostly from the importance of highly dispersed platinum in catalysis. An "ideal" model system for investigating a particle size effect in electrochemical reactions-such as the methanol oxidation and oxygen reduction reactions- would possess all of the following characteristics: (1) Platinum nanocrystals should be size and shape monodisperse. (2) Nanocrystals should be dispersed on, and electrically connected to, a technologically relevant support surface that facilities spectroscopic characterization of the particles and of adsorbed intermediates. For many electrocataysis reactions the preferred support material is graphite. (3) Individual platinum particles on this support should be well-separated from one another. (4) The structure of the platinum nanocrystals on the support surface should be accessible both before and following the involvement of these particles in the catalytic process of interest. (5) Supported and platinum nanoparticles should be stable for days [1-5].

In this paper, a colloidal method of synthesis has been proposed to obtain metallic nanoparticles; the polyol method has been reported to produce small nanoparticles as the final product, easily changing composition and surface modifiers [6-9].

The gas-phase method of platinum salt particles disposed on a graphite surface by H_2 was used in this work. This method has the potential to yield nanocrystals that are disposed in direct contact with a substrate surface [10].

We describe also, an electrochemical method for preparing dispersions of platinum nanocrystals on a graphite basal plane surface involving the pulsed potentiostatic deposition of platinum from dilute $PtCl_6^{2-}$ using large overpotentials ($E_{overvoltage} \approx 500$ mV) [11-13].

A novel approach to characterize this kind of particles is based on the use of HAADF technique, in a high resolution transmission electron microscope (HRTEM), which allowed us the observation of the elements due to atomic number, densities, or the presence of strain fields due to differences in lattice parameters, structure, the presence of surfactants or any other surface modifier besides the size of the particle [14].

The HOPG is described as consisting of a lamellar structure The freshly cleaved surface consists of atomic steps and steps of several or dozens of atomic layers. The crystallographic planes do have a definite structure and the height of a single step is 0.34 nm [15]. The Moiré patterns were observed and rotations between the first and second layers of HOPG in the direction perpendicular to the basal plane by effect of heating Pt nanoparticles on HOPG in H_2 flow [14, 16].

EXPERIMENTAL SECTION

The polyol method was followed to obtain platinum metallic nanoparticles passivated with poly(vinylpyrrolidone) (PVP). Hexachloroplatinic (IV) acid (H_2PtCl_6) hydrate (99.99%), and poly (N-vinyl-2-pyrrolidone) (PVP-K30, MW = 40000) were purchased from Sigma Aldrich, and 1,2-ethylenediol (99.95%) was purchased from Fischer Chemicals; all the materials were used without any further treatment.

A 0.4 g sample of Poly (N-vinyl-2-pyrrolidone) (PVP) was dissolved in 50 mL of 1,2-ethylenediol (EG) under vigorous stirring, heating in reflux, until the desired temperature was reached (working temperatures ranged from 100 to 190 °C in increments of 10 °C). For the Pt metallic nanoparticles, a 0.1 mM EG-solution of the metal precursor was added to the EG-PVP solution, with continuous agitation for 3 h in reflux. When preparing the Pt metallic nanoparticles, the following criterion was used: after complete dissolution of PVP in EG, 2 mL of an EG solution of H_2PtCl_6 (0.05 M) was added to the EG-PVP solution in a period of 1 h. The reaction was carried out for 3 h at constant temperature. For this work the Pt nanoparticles presented the smaller average size for a synthesis performed at 130 °C. These Pt nanoparticles in a solution of ethanol were impregnated on HOPG and dried in a oven at 80°C.

The Pt nanoparticles preparation on HOPG by direct reduction of Platinum salts in H_2 flow on HPOG (the gas phase method) consisted in the partial oxidation of the support in a muffle furnace at 600°C for 24 hours. Then the impregnation of preoxidized HOPG with chloroplatinic acid solution in a four to one mixture of benzene to ethanol (absolute). The metal concentration was adjusted to produce the desired total metal loading (10 Wt % Pt). The amount of solvent was fixed using 50 ml/g of HOPG. A mixture of salt solution and HOPG was shaken while nitrogen was bubbled through the suspension at flow rate of 200-500 cc/min until to solution evaporated to dryness, i.e., after 40-60 hr for a 10 g sample in 500 ml of solution.

The samples prepared by these methods (Pt nanoparticles on HOPG) were heated in a H_2 flow at a temperature range from 450 °C to 1000 °C during time intervals from 2hr to 5 hr. Samples were then exfoliated with a scotch tape for TEM observation.

Pt nanoparticles were electrochemically deposited on a highly ordered pyrolytic graphite (HOPG, grade-1) which was obtained from SPI Supplies (West Chester, PA). Platinum deposition was carried out by immersing the potential in which electroless platinum deposition was not observed), followed by stepping the potential of the

graphite surface to a deposition potential of -0.6 V, for 100 ms. Following the application of the deposition pulse, the

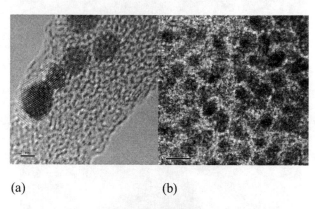

(a) (b)

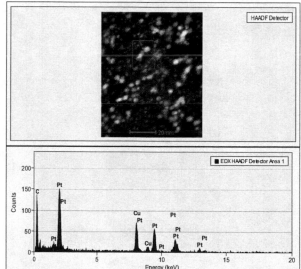

(c)

Figure 1. A microscopy images by HRTEM, shows a platinum nanoparticle synthesized by polyol method at: (a) and (b) synthesized at 145 °C. (c) HAADF shows of Pt nanoparticles and EDX HAADF analyze of Pt nanoparticles.

electrode potential was stepped back to 0.2 V, and the working electrode was removed from the plating solution. All electrochemical experiments were performed using a CH Instruments potentiostat model CHI 900B (CH Instruments, Austin, TX). A platinum coil (d=0.5 mm) and a Hg/Hg_2SO_4 were used as the counter and reference electrodes, respectively.

The Pt nanoparticles on HOPG for the electron microscopy analysis were prepared over lacey carbon TEM grids. HAADF images were taken with a JEOL 2010F microscope in the STEM mode, with the use of a HAADF detector with collection angles from 50 mrad to 110 mrad

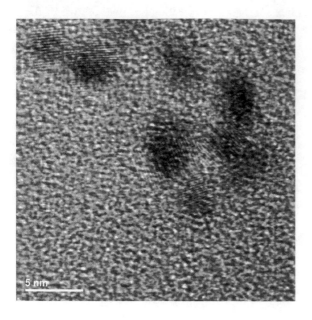

Figure 2. HRTEM image of Pt nanoparticles on HOPG synthetized by polyol heated in H$_2$ flow at 450°C.

Figure 4. HRTEM image of Pt nanoparticle synthesized on HOPG by the electrochemical method.

RESULTS AND DISCUSSION

By the polyol method with ethylene glycol as solvent-reductor, was possible to obtain monometallic platinum nanoparticles with narrow size distributions in systems with small particles (2-4 nm) and different structures depending on the temperature of reaction. The structure of a platinum nanoparticle is cubic, face centered has can be seen from Figure 1 (a). The monometallic synthesis of Pt nanoparticles by itself showed a distinctive morphology of quasi hemispherical small Pt nanoparticles, which does not depend on the temperature of reaction has can be seen from Figure 1 (a) and (b). In the figure (c) a image of HAADF of Pt nanoparticles with its respective analyze is shown.

The Pt nanoparticles synthesized by polyol method, were deposited successfully on the HOPG has can be seen from figure 2.

In the case when the Pt nanoparticles were synthesized by direct reduction of platinum salt in H$_2$ flow on HPOG we obtained nanoparticles with a considerable size distributions in systems of Pt nanoparticles and different structures depending on the temperature of reaction [17-18]. We obtained good results also in the preparation of Pt nanoparticles on HOPG by this method as can be seen in figure 3.

By immersing the graphite electrode into a platinum plating solution were obtained Pt nanoparticles in a narrow particle size distribution also for mean crystallite diameters smaller than 4 nm.

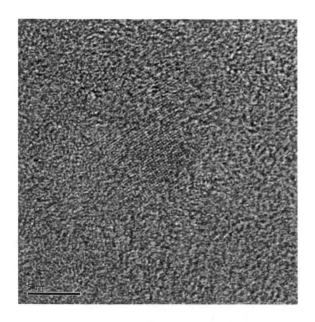

Figure 3. HRTEM image of Pt nanoparticle synthesized on HOPG graphite by direct reduction of chloroplatinc acid impregnated on HOPG and heated in H$_2$ flow at 950 °C.

It is important to observe that several patterns of diffraction of the samples Pt nanoparticles on HOPG presented rotations of some degrees between the layers of HOPG in the direction perpendicular to the basal plane by effect of heating these samples in H_2. Also Moire patterns were observed in some of these samples. Honeycomb structures were observed on the HOPG surface.

For preparing supported platinum nanoparticles on HOPG graphite for investigations of electrocatalysis, the advantage of the polyol method is the small size of the particles and the narrow distribution sizes of them. The second method have the potential to yield nanocrystals that are disposed in direct contact with HOPG; however, in neither case has it been possible to achieve good particle size monodispersity for platinum across a wide range of particle sizes.

Electrochemical deposition resulted an effective method to obtain directly nanoscale platinum particles on HOPG with a narrow distribution sizes.

ACKNOWLEDGMENT

Authors would like to acknowledge Professor Allen J. Bard for their support.

REFERENCES

[1] J. V. Zoval, J. Lee, S. Gorer, R. M. Penner, J. Phys. Chem. B, 102, 1166-1175, 1998. [2] D. Snow, M. Major and L. Green, Microelectronic Engr. 30, 969, 1996.

[2] J. M. Thomas, R. Raja, B. F. G. Johnson, S. Hermans., M. D. Jones, T. Khimyak, *Ind. Eng. Chem. Res.* 2003, *42*, 1563.D. Snow, M. Major and L. Green, Microelectronic Engr. 30, 969, 1996.

[3] L. M. Bronstein, D. M. Chernyshov, I. O. Volkov, M. G. Ezernitskaya, , P. M. Valetsky, V. G. Matveeva, E. M. Sulman, *J. Catal. 196*, 302, 2000.

[4] S. W. Han, Y. Kim, K. Kim, *J. Colloid Interface Sci.*, *208*, 272,1998.

[5] T. Freelink, W. Visscher, J. A. R. van Veen, J. Electroanal. Chem. 382, 65, 1995.

[6] M. M. Alvarez, J. T. Khoury, G. Schaaff, M. N. Shafigullin, I. Vezmar, R. L. Whetten, *J. Phys. Chem. B 101*, 3706, 1997.

[7] A. Henglein, *J. Phys. Chem. B*, *104*, 2201, 2000.

[8] K. C. Grabar, R. G. Freeman, M.B. Hommer, M. J. Natan, Anal. Chem. 34, 735, 1995.

[9] B. E. Baker, N. J. Kline, P. J. Treado, M. J. Natan, J. American Chemical Society, 118, 8721, 1996.

[10] C. H. Bartolomew, M. Boudart, J. Catalysis, 25, 173, 1972.

[11] M. T. Reetz, W. Helbig, J. Am. Chem. Society, 116, 7401, 1992.

[12] P. Allongue, E. Souteyrand, J. Electroanal. Chem., 362,79, 1993.

[13] H. Cachet, M. Froment, E. Souteyrand, Denning C. J. Elecrochem. Soc. 139, 2920, 1992.

[14] D. B. Williams, C. B. Carter, *Transmission Electron Microscopy*, Plenum Press, New York, 1996.

[15] L. Pauling, The Nature of the Chemical Bond, p. 235, 3rd. Edition 1960.

[16] H. Beyer, M. Müller, T. Schimmel, Applied Physics A, Science and Prosessing, 68,2, 0947-8396, 1999.

[17] A. Tomita and Y. Tamai, J. Phys. Chem., 78, 2254, 1974.

[18] R. T. Baker, R.S. Sherwood and E. O. Derouane, J. Cat. 75, 382,1982.

Effects of the Shape of Nanoparticles on the Crystallization Behavior of Poly(ethylene terephthalate) Nanocomposites in the Flow Fields

B. C. Kim[*], K. H. Jang[*], W. G. Hahm[**], T. Kikutani[***], and J. Lee[****]

[*]Hanyang University, Seoul, Korea, bckim@hanyang.ac.kr
[**]Korea Institute of Industrial Technology, Gyeonggi-do, Korea, wghahm@kitech.re.kr
[***]Tokyo Institute of Technology, Meguro-ku, Tokyo, Japan, tkikutan@o.cc.titech.ac.jp
[****] Tire Reinforcement Team, Production R&D Center, R&DB Labs., Hyosung Corporation, Anyang, Gyeonggi-do, Korea, jlee@hyosung.co.kr

ABSTRACT

The effects of the shape of nanoparticles on the crystallization behavior of high molecular weight poly(ethylene terephthalate) (PET ; inherent viscosity of 0.98 g/dl) in the shear and elongational flow fields were investigated. Two nanoparticles of different shape, the spherical polyhedral oligomeric silsesquioxanes (POSS) and the platelike surface-treated nanoclay, were selected. In general, both nanoparticles played the role of nucleating agent for PET in the shear flow field. Of the two nanoparticles, the clay proved to be the more effective nucleating agent in the shear low field. In the extensional flow field, however, the two nanoparticles affected the crystallization behavior of PET in different ways. The POSS promoted the crystallization of PET a little whereas the clay suppressed it. This discrepancy seemed to result from the different shape of the two nanoparticles.

keywords: PET, nanoclay, POSS, high-speed spinning, heterogeneous nucleation.

1   INTRODUCTION

Recently, polymer/inorganic nanocomposite materials have attracted considerable attention due to their improved properties, mainly concerning the thermal and mechanical properties [1-6]. Among them, polymer/silicate nano-scaled composites such as nano-sized clay and POSS contained nanocomposites are regarded as most desirable ones [7,8]. Owing to their high aspect ratio nanoclay is considered as an effective reinforcement when they are fully exfoliated. For improved dispersion in the polymer matrix and good interfacial adhesion organo-modified-mica is most widely used [9,10]. Cage-like structured poly(hedral oligomeric silsesquioxane) can be incorporated into polymer systems aiming at nanostructured polymeric materials whose properties bridge the property space between organic polymer and inorganic nanoparticle [11,12]. As the polymer matrix, poly(ethylene terephthalate) (PET) is commonly used [13,14].

Although many researches have been carried out to verify synergetic effects of nanoparticles on the physical properties of general-purpose PET (inherent viscosity of 0.5 to 0.7), there has been little research which compares the effects of clay and POSS on the high speed spinning process and physical properties of resultant fibers. Further, there is little literature on the high speed spinning process of high molecular weight PET (HMW-PET) and its nanocomposites irrespective of the practical significance in current tire-cord industry. In this paper, we investigated the effects of organoclay(clay) and trisilanol isobutyl-POSS(POSS) on the high-speed spinning process and the physical properties of spun fibers in relation with particle type and level.

2   EXPERIMENTAL

2.1   Materials

The organically modified Mica (SOMASIF, CO-OP Chemical Co, Ltd.) whose aspect ratio was 250 by L/D, C_{12}PPh-Mica, was synthesized by using ion exchange reaction between dodecyl triphenyl phophonium chloride(C_{12}PPh-Cl$^-$) and Na$^+$-Mica. POSS nanoparticles of average particle size 7 nm in the form of white powder was obtained from Hybrid Plastics Co and used as received. The organically modified clay was incorporated into PET by *in-situ* polymerization process to produce composite contained 1 wt% clay (PET/clay1). By using ultrasonicator, the clay was dispersed in ethylene glycol in the form of slurry. For this polymerization, 1.3 mole ratio of terephthalic acid to ethylene glycol was combined in a 2L 3156-SS stirred autoclave, fitted with reflux condenser, inert gas purge, and distillate removal, and reacted at 265 °C in a two-step process that has been reported previously [15]. The PET/POSS nanocomposites were prepared by melt mixing using internal mixer for 2 min at 265 °C with PET pellets and POSS nanoparticles. Before melt mixing, PET was dried under vacuum state. The content of POSS was 1 (PET/POSS1) and 2 wt% (PET/POSS2). PET and its nanocomposites were controlled to have an intrinsic viscosity of 1.05.

2.2   Preparation of nanocomposite fibers by high-speed melt spinning process

PET and its nanocomposite chips were completely dried in a vacuum oven overnight. High-speed melt spinning was carried out by using a single screw extruder equipped with a spinneret of a single hole of 1 mm diameter and a gear pump. Throughput rate and spinning temperature were controlled to 6.0 g/min and 300 °C, respectively. The quenching-air unit was not applied and the filaments were spun at room temperature over the range of the take-up speeds 1 to 4.5 km/min. The distance from the spinneret to the winder was around 300 cm.

2.3 Measurement of physical properties

On-line measurement of filament diameter in the melt spinning processes was performed using a back-illumination-type optical diameter monitor (Zimmer OHG model 460A/10). The signals of diameter were acquired every 1 ms (1 kHz) by using an A/D converter (Keyence NR-2000) and a computer. A series of on-line measurements were carried out by changing the measuring position along the filament with an interval of 100 mm in the melt spinning process. Diameter profiles of the spinning line were obtained by averaging 3000 data signals acquired during 2 seconds. After the evaluation of average diameter, the coefficient of variation (CV) was calculated.

Morphology was examined by scanning electron microscope (SEM, JEOL, JSM-6700F). Each fracture of the fibers was prepared by putting PET and PET nanocomposite fibers into epoxy resin and hardened epoxy resin was cut with a blade. The surfaces of cross-section of filament were coated with platinum under vacuum using a coating sputter.

Thermal properties were studied by using differential scanning calorimetry (DSC, TA instruments, DSC2010) under a nitrogen flow over the temperature range 30 to 300 °C. The weights of samples varied from 7 to 8 mg and the heating rate was 15 °C/min and cooling rate was 10 °C/min.

Mechanical properties were measured by using Instron 4465 testing tester according to ASTM D638. Yarn/fiber testing type was performed and the 2.5 N load cell was used. The gauge length and the crosshead speed were set to 30 mm and 30 mm/min, respectively.

3 RESLUTS AND DISCUSSION

3.1 Morphology by SEM

Fig. 1 and 2 show the cross section morphologies of the high-speed melt spun fibers. POSS nanoparticles show fine dispersion in the PET matrix while the clay nanoparticles do not. The poor dispersion of clay seems to come from re-agglomeration during spinning. In the case of clay, the clay is especially prone to align during elongational flow because the clay used in this study has plate shape and very large aspect ratio about 250 by L/D. POSS also gives rise to some re-agglomeration during spinning although it is reported that POSS can be dispersed into polymer matrix at

molecular level, which is, however, much less notable than in the case of clay [16, 17]. In general, the extent of re-agglomeration is decreased with increasing take-up speed.

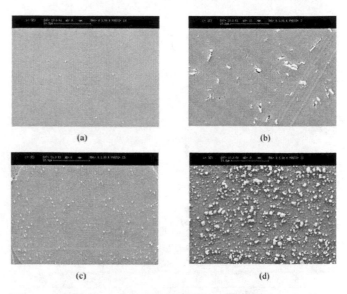

Figure 1: SEM images of (a) PET fiber, (b) PET/clay1, (c) PET/POSS1 and PET/POSS2 nanocomposite fibers at 2km/min take up speed.

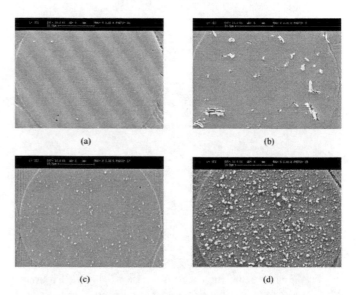

Figure 2: SEM images of (a) PET fiber, (b) PET/clay1, (c) PET/POSS1 and PET/POSS2 nanocomposite fibers at 3km/min take up speed.

3.2 Crystallization behavior

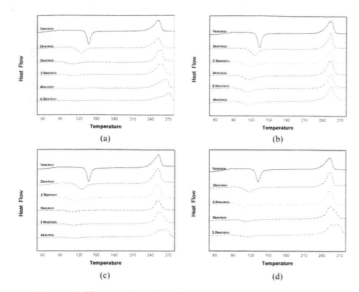

(a)

(b)

(c)

(d)

Figure 3: First heating thermograms of (a) PET fibers, (b) PET/clay1, (c) PET/ POSS1 and (d) PET/POSS2 nanocomposite fibers at various take up speeds.

Sample	Take up speed (km/min)	T_c (°C)	T_m (°C)	ΔH_c (J/g)
PET	1	136.1	252.7	28.3
	2	123.0	252.7	20.8
	3	107.9	256.1	9.4
	3.5	109.8	256.0	3.4
	4	110.4	256.6	2.5
	4.5	111.3	270.2	2.6
PET/Clay1	1	133.8	253.3	25.7
	2	126.0	253.6	18.8
	2.5	119.8	253.6	17.7
	3	114.8	253.7	16.6
	3.5	112.4	253.5	14.5
	4	113.2	254.1	13.2
PET/POSS1	1	136.3	252.9	24.3
	2	123.7	252.4	19.7
	2.5	113.9	252.9	18.6
	3	108.1	251.7	5.3
	3.5	109.1	254.5	4.9
	4	110.9	255.5, 266.8	3.8
PET/POSS2	1	132.1	252.2	23.9
	2	116.5	253.5	17.1
	2.5	110.6	252.7	11.4
	3	114.5	253.7	7.3
	3.5	101.0	252.4 , 266.1	5.2

Table 1: Thermal properties of PET and PET nanocomposite fibers with various take up speed

Fig. 3 shows the first heating scans of HMW-PET and its nanocomposite fibers at various take up speeds. The results are given in Table 1. During high-speed melt spinning, crystallization as well as orientation takes place when the spinning speed is higher than a critical value, e.g. 4.5 km/min for commodity PET. The stress-induced crystallization produces oriented crystals. In the case of nanocomposites, this would be more noticeable because of the nucleating effects of the nanoparticles incorporated. However, the clay nanoparticles suppress the crystallization of PET although they promote it at the static state as shown in Fig. 4 and Table 2.

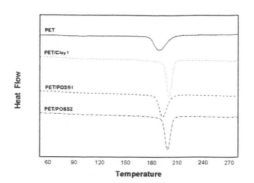

Figure 4: Cooling thermograms of PET, PET/clay1, PET/POSS1 and PET/POSS2.

Sample	T_c (°C)	ΔH_c (J/g)
PET	188.7	30.9
PET/clay1	200.6	39.9
PET/POSS1	192.8	34.2
PET/POSS2	198.6	36.4

Table 2: Crystallization behavior of PET and PET nanocomposites on the cooling condition

It is not possible to clearly account for why clay suppresses the crystallization and orientation of PET irrespective of its nucleation effects on PET. However it may be elucidated by Peterlin concept [18]. That is, the oriented clay nanoparticles may interrupt the crystallization of PET as schematically depicted in Fig. 5. In the case of POSS, however, spherically shaped POSS may not interrupt to form oriented crystals. Further, POSS is a low density material with caged-structure. Hence it would affect the spinning behavior of the nanocomposites much less. Consequently, it slightly improves the crystallization of PET. Glass transition temperature (T_g) band becomes broader with the increase of take up speed. The existence of the cold-crystallization peaks suggests that non-crystalline region of polymer still undergoes crystallization during heating process [19].

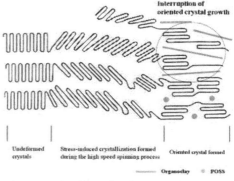

Figure 5: Suggested schematic diagram of crystal formation during high-speed spinning process based on Peterlin model.

As given in Table 1, the cold-crystallization temperature (T_c) of all fibers decreases with increasing take up speed. This indicates that the onset temperature of stress-induced crystallization decreases with increasing spinning speed. The T_c of PET/POSS fibers is similar to or slightly lower than that of PET fibers while that of PET/clay fibers is higher by 3 to 4 °C at all take up speeds. This also supports that clay interrupts stress-induced crystallization. Furthermore, the enthalpy of T_c peaks decreases with increasing take up speed at all heating scans. Comparing with fibers of low take up speed, highly drawn fibers have low degree of unoriented amorphous region in polymer matrix which can be altered into crystalline region. By the way, in the case PET/clay, the heating scans show little changes of heat of fusion (ΔH) with increasing take up speed, comparing with PET fibers. Furthermore, PET/POSS fiber shows very unique melting behavior. Although heating scan of PET fiber at 4km/min shows inconspicuous double T_m peaks, PET/POSS1 and PET/POSS2 show obviously double peaks at 4 and 3.5km/min, respectively. The double peaks are particularly notable at lower spinning speed and at higher POSS content. The double peaks of T_m suggest absence of one dominant crystals [19]. It can be said that the first T_m peak represents the melting temperature of PET crystals and the second one represents that of larger and oriented crystals.

4 CONCLUSION

Introducing nanoparticles had a very significant influence on the high-speed melt spinning process of HMW-PET. In the case of clay, it suppressed the crystallization of PET. On the other hand, POSS slightly enhanced it. Consequently, POSS improved the mechanical properties of PET fiber but clay had a negative effect on the mechanical properties of PET fiber.

REFERENCES

[1] E. P Giannelis, Appl. Organometallic Chem. 12, 675, 1998.
[2] E. P. Giannelis, Adv. Mater. 8, 29, 1996.
[3] P. B. Messersmith and E. P. Giannelis, Chem. Master. 5, 1064, 1993.
[4] J. W. Cho and D. R. Paul, Polym. 42(3), 1083, 2001.
[5] T. J. Pinnavaia, Sci. 220, 365, 1983.
[6] J. W. Gilman, Appl. Clay Sci. 15, 31, 1999.
[7] G. Z. Zhang, T. Shichi and K. Takagi, Mater. Lett. 57, 1858, 2003.
[8] G. H. Guan, C. C. Li and D. Zhang, J. Appl. Polm. Sci. 95, 1443, 2005.
[9] J. H. Chang, M. K. Mun and I. C. Lee, J. Appl. Polm. Sci. 98, 2009, 2005.
[10] J. H. Chang and M. K. Mun, J. Appl. Polm. Sci, 100, 1247, 2006.
[11] B. X. Fu, L. Yang, R. H. Somani, S. X. Zong, B. S. Hsiao and S. Philip, J. Polym. Sci. Polym. Phys. 39, 2727, 2001.
[12] Y. Zhao and D. A. Schiraldi, Polym. 46, 11640, 2005.
[13] S. H. Hwang, S. W. Paeng, J. Y. Kim and W. Huh, Polym. Bull. 49, 329, 2003.
[14] D. Wang, J. Zuh, Q. Yao and C. A. Wiklie, Chem. Mater. 14, 3837, 2002.
[15] R. M. Kroegel, D. M. Collard, C. L. Liotta and D. A. Schiraldi, Macromolecules 31, 765, 1998.
[16] J. J. Schwab, Sr. Reinerth, J. D. Lichtenhan, Y. Z. An, S. H. Philips and A. Lee, Polym. preprints 42, 48, 2001.
[17] J. Zeng, S. Kumar, S. Iyer, D. A. Schiraldi and R. I. Gonzarez, High Performance Polymers 17, 409, 2005.
[18] F. W. Billmeyer, "Textbook of Polymer Science", 3rd edition, John Wiley & Sons, New York, 1984.
[19] Y. P. In, P. KP, L. Tianxi and H. Chaobin, Polym. Int. 53, 1283, 2004.
[20] E. Giza, H. Ito, T. Kikutani and N. Okui, J. Macromol. Sci. –Phys. B39 (4), 546, 2000.

Effect of thermal oxidation on titanium oxides characteristics

M.V. Diamanti*, S. Codeluppi*, A. Cordioli*, MP. Pedeferri*

*CMIC Dept, Politecnico di Milano
Via Mancinelli 7, 20131 Milan, Italy, mariavittoria.diamanti@polimi.it

ABSTRACT

Nanocrystalline TiO_2, in the form of anatase or rutile, is one of the most important and used photocatalysts because of the excellent efficiencies of conversion of chemical species, and its stability in a wide range of environments and conditions. TiO_2 layers can be grown on titanium surfaces by means of anodizing processes: the so obtained oxides have thickness ranging from a few nanometers to tenths of micrometers, and present either amorphous or semicrystalline structure, depending on process parameters. When a homogeneous oxide with maximum thickness of the order of hundreds of nanometers is formed, a wide range of interference colors can appear on the oxide surface, showing different hues and saturations. Similar effects are obtained when the oxidation process consists of a thermal treatment, which can be performed at different temperatures in various atmospheres. This work is aimed at the exploration of the parameters involved in thermal oxidation and of their influence on the growing oxide characteristics, with particular reference to the surface thickness that can be achieved and the contingent generation of crystal phases, in particular anatase and rutile, which are responsible for photoinduced properties.

Keywords: anatase, anodizing, reflectance, thermal treatment, X-ray diffraction.

1 INTRODUCTION

While the exploiting of solar radiation has been extensively studied together with its applications in supplying energy and promoting chemical reactions (photoinduced processes), the role of semiconductor materials, such as titanium dioxide, only gained importance both for research studies and for industrial applications in the last twenty years [1, 2]. Titanium dioxide, in the crystalline form of anatase or rutile, is the focus of the most part of studies concerning photocatalysis, specially because of the high redox power of its electron-hole pair, which grants good efficiencies of conversion of chemical species, but also because of its commercial availability and its stability in many solvents under irradiation. The main drawback of titanium dioxide is the band gap, equal to 3,02 eV for rutile and 3,20 eV for anatase: therefore, photocatalytic performances are achieved only if the semiconductor is irradiated with UV light [3].

TiO_2 films with controlled morphology and structure at micro and nanometric scale can be obtained by means of anodic polarization: oxides can present smooth surface, or peculiar morphological features such as nanotubes. By increasing the thickness of the native oxide up to a few hundreds of nanometers, the surface acquires particular colors due to interference phenomena taking place at the metal-oxide and oxide-air interfaces: colors are determined by the oxide thickness [4].

While the oxides produced by anodizing have been extensively studied [5, 6], less information is available concerning thermal oxidation. Titanium oxides obtained by techniques alternative to anodizing, e.g. by thermal treatment or by pulsed laser, may present either an amorphous structure or the presence of crystal phases: the oxide stoichiometry principally consist of TiO_2, but non-stoichiometric phases ($Ti_{1+x}O_{2-y}$) and sub-stoichiometric phases (Ti_2O_3, TiO, Ti_2O, Ti_6O) may also form. In case of conspicuous alteration of the oxide structure from amorphous to crystalline, the interference color is lost and consequently the aesthetical qualities of the surface [7, 8]. Nevertheless, the formation of a small fraction of nanocrystals with non-stoichiometric composition in oxide containing prevalently anatase or rutile may enhance to some extent the photoactivity of the oxide, since these phases can modify the electronic structure of the resulting complex oxide by creating accessible energy levels in the oxide band-gap, therefore reducing the energy necessary to activate the semiconductor. This principle is also exploited when TiO_2 is doped with chemical elements (e.g. nitrogen or carbon) which enter the crystal structure in interstitial or substitutional positions, thus altering the electronic structure and reducing the band-gap: this method is used to shift light absorption from UV towards visible light [9].

In this research, the effect of thermal treatment both on bare titanium and on anodized titanium will be investigated: air and nitrogen atmospheres will be used. Investigations will focus on the oxide color and thickness and on the different structures achievable in the tested atmospheres.

2 MATERIALS AND METHODS

2.1 Materials preparation

Rectangular specimens (20 x 30 mm) were cut out of a sheet of commercial purity (grade 2) titanium, 0.5 mm thick. Surface preparation involved only a degreasing step with acetone. Samples were thermally treated in air and in nitrogen atmospheres, with temperatures ranging from 400°C to 700°C and treatment time ranging from 0.5 hours to 6 hours. A first series of tests was performed on non-anodized titanium; subsequently, the same treatments were repeated on specimens anodized either in sulfuric acid 0.5 M or with a two-step anodizing process described elsewhere [10], with cell potentials of 10 V, 20 V, 90 V and 100 V (anode-to-cathode potential differences, being the cathode a titanium net). These values were chosen in order to investigate the behavior of oxides with different initial thicknesses (Table 1). Anodizing was performed by applying a constant current density (20 mA/cm^2) by means of a galvanostat.

Cell potential (V)		10	20	90	100
Thickness (nm)	A	37	51	161	169
	2A	40	58	196	218

Table 1: Oxide thickness after anodizing, measured by reflectance. A: anodized in H$_2$SO$_4$; 2A: two-step anodizing.

2.2 Oxides characterization

The thickness of the oxide layer was derived from reflectance curves: in fact, thickness can be calculated on the basis of the position of maxima and minima in the reflectance spectrum (Fig. 1), which in turn are determined by Bragg's law for constructive and destructive interference. Reflectance data were elaborated with the FTM (Film Thickness Module) software; the software provided the refractive index dispersion curve for TiO$_2$ necessary to interpolate the reflectance curves. The evolution of colors with oxidation was also considered.

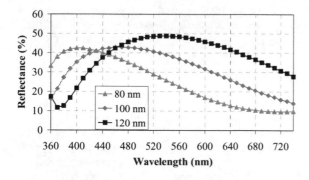

Figure 1: Reflectance spectra for 80, 100 and 120 nm thick oxides: as thickness increases reflectance peaks increase in number and shift towards higher wavelengths.

Concerning oxides photoactivity, X-ray diffraction patterns were acquired in order to evidence any crystal phase formed during treatments. Titanium anodized with a two-step procedure proved to be amorphous for any anodizing potential and current density applied; on the contrary, high anodizing potentials (90 V and 100 V) reached in H$_2$SO$_4$ electrolytes led to a partial crystallization of the oxide, with the formation of anatase nanocrystals in the amorphous matrix (Fig. 2) [5]. Therefore thermal oxidation was performed on oxides which had not only different thickness, but also different structure.

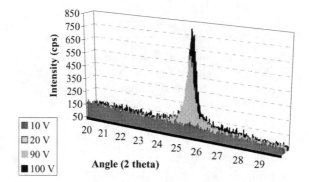

Figure 2: X-ray diffraction peak of anatase (25°) for the oxides obtained in H$_2$SO$_4$ 0.5 M.

3 RESULTS AND DISCUSSION

3.1 Oxide thickness

The thermal treatment in air of bare titanium led to the formation of nanometric oxides of different thicknesses, depending on temperature and duration of the treatment itself (Fig. 3). In particular, the applied temperature is more relevant than the treatment time: in fact, treatments performed at 450, 500 and 550°C all cause the growth of an oxide with thickness ranging from 30 to 60 nm, while at 600°C the oxide thickness increases more evidently, and also the effect of time becomes more relevant.

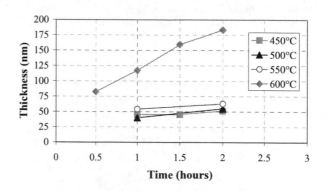

Figure 3: Relationship among treatment temperature, time and thickness achieved on bare titanium.

On anodized titanium, the achieved oxide thicknesses for the different temperatures and durations, for specimens anodized with cell potential of 100 V both in sulfuric acid and with a two-step anodizing, are reported as example; for the other applied cell potentials, analogous trends were observed (Table 2, 3).

	0.5 h	1 h	1.5 h	2 h
400°C	170	170	173	170
450°C	173	176	184	184
500°C	185	187	185	189
550°C	191	197	200	206
600°C	203	233	233	237

Table 2: Oxide thickness after thermal treatment in air of titanium anodized in H_2SO_4 with cell potential of 100 V, as a function of time and temperature.

	0.5 h	1 h	1.5 h	2 h
400°C	217	222	218	218
450°C	222	219	225	226
500°C	232	239	245	243
550°C	261	269	252	263
600°C	283	279	286	279

Table 3: Oxide thickness after thermal treatment in air of titanium treated with two-step anodizing, with cell potential of 100 V, as a function of time and temperature.

It is easily noticed that the influence of the treatment duration on oxide growth is almost negligible, compared to the effect of temperature, which allows an increase in oxide thickness of more than 20% with respect to the anodic oxide. For thin anodic oxides (10 V and 20 V), the percent increase was even more pronounced, exceeding 100%.

Thermal treatments were also performed at 650°C and 700°C: in these cases the conspicuous conversion of the amorphous structure to crystal phases led to the loss of the interference color. Reflectance spectra attested the degradation of the interference characteristics, with the disappearance of reflectance peaks; for this reason the oxide thickness could not be calculated for temperatures higher than 600°C.

The relationship between oxide thickness and temperature of thermal treatment is reported for both anodizing processes (Fig. 4). A first observation concerns the initial thicknesses of anodic oxides: for any applied potential, two-step anodizing generated thicker oxides. This is due to the higher homogeneity of the oxide on the whole surface: in fact, oxides grown in H_2SO_4 are not perfectly uniform in color and thickness at the microscale [5].

The effect of temperature is the same in the two anodizing conditions, that is, a non-linear increase of thickness, more pronounced as temperature increases; nevertheless, the oxides grown with two-step anodizing exhibit a lower tendency to a further oxidation as temperature increases. This is attributed to the higher barrier effect performed by these oxides towards oxidation, which in turn is imputable to the superior homogeneity.

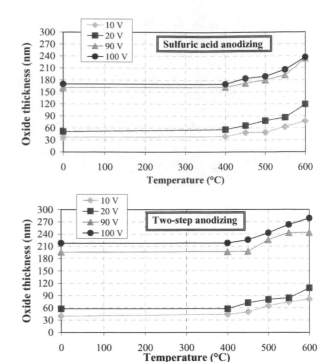

Figure 4: Oxide thickness variation with temperature for titanium anodized in H_2SO_4 and with two-step anodizing, for 2 hour thermal treatments.

Thermal treatments performed in N_2 atmosphere led to a less pronounced increase in oxide thickness; yet, a variation was noticed though the chosen atmosphere was considered to be inert, and this is due to O_2 impurities present in the N_2 gas used in these tests. A higher purity atmosphere wasn't considered since a further oxidation wasn't considered to affect the tests negatively. Thickness of thin anodic oxides increased by 60% on average for treatments performed at 600°C; thicker oxides increased by less than 20% in same conditions. The oxidation kinetics are similar to those observed in air.

3.2 Crystal phases

Thermal treatments performed in air on bare titanium produced the formation of rutile at high temperature (600°C or more). Surfaces treated at lower temperature presented completely amorphous oxides, probably on account of the particularly low oxide thickness. On previously anodized samples the presence of crystal phases was noticed at lower temperatures: anatase starts to crystallize at 500°C in very thin amorphous oxides (10 V) while the temperature necessary to its formation drops to 400°C with the increase of oxide thickness. Rutile appears at 600°C as for bare titanium. At 700°C the anatase component decreases or even disappears and a very intense peak of rutile is displayed (Fig. 5). Results obtained for titanium anodized in H_2SO_4 at 100 V are presented as example.

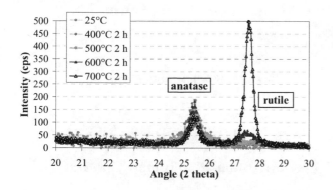

Figure 5: X-ray diffraction peaks of anatase (25°) and rutile (27.5°) for the oxide grown in H_2SO_4, cell potential: 100 V, thermally treated in air for 2 hours at various temperatures.

Macrocrystalline rutile is the thermodynamically stable structure of TiO_2, while anatase is a metastable phase which owes its stability at room temperature to the almost null transformation kinetics. However, thermodynamic stability is particle-size dependent, and at particle diameters below ca. 14 nm, anatase is more stable than rutile [3]. Therefore the observed behavior, i.e. the first appearance of anatase peak for low temperatures followed by the co-presence of the two structures and the final disappearance of anatase in favor of rutile phase, can be explained by considering the initial nucleation of nanocrystals, with average dimensions lower than 15 nm and thus with anatase structure, which increase in dimensions with increasing temperature, therefore gradually converting to rutile.

Finally, treatments performed in nitrogen atmosphere caused the formation of several crystal phases, besides anatase and rutile which are present as well. A non-stoichiometric oxide (Ti_6O) was observed on bare titanium and for 10 V anodizing, that is, for very thin oxides, at 600°C or more. This is ascribed to the low oxidative power of the N_2 atmosphere, which is only due to O_2 impurities, and therefore the shortage of oxygen can lead to the formation of titanium oxides with lower oxygen content.

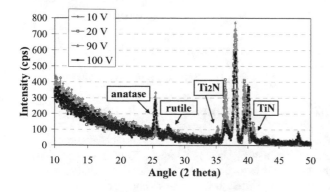

Figure 6: X-ray diffraction patterns of the oxides grown in H_2SO_4 thermally treated in N_2 at 700°C for 2 hours: crystal phases observed are labeled on the pattern.

The detection of titanium nitrides was considered particularly interesting: in fact, while the formation of a crystal phase containing nitrogen has no influence on band gap modification, it is surely indicative of the occurred adsorption of nitrogen in TiO_2 (Fig. 6). As mentioned in the introductive part, nitrogen adsorption is the physical phenomenon responsible for generation of admitted levels in the prohibited band gap, and therefore for band gap decrease, which cause the shift of TiO_2 light absorption from UV wavelengths towards visible light.

4 CONCLUSIONS

The presented research showed the possibility of creating oxides with different thicknesses and structures by means of thermal oxidation. While the behavior of titanium when subjected to thermal treatment was predictable, a particular focus was placed on the analysis of the behavior of anodized titanium to such treatments, since the presence of an oxide layer was supposed to slow down a further oxidation. Any oxide layer, either thin or thick, amorphous or semicrystalline, present on titanium surface actually restrains oxidation. While the duration of thermal treatment was proved to have a negligible influence on both oxide thickening and crystallization, the effect of temperature is definitely fundamental in determining thickness and structure, not only in terms of the quantity of a certain phase present in the oxide, but also (and mainly) for the type of crystal phases and the dimensions of the nanocrystals. Moreover, the presence of nitrogen in the atmosphere allowed the formation of titanium nitrides, which is considered as indicative of the adsorption of nitrogen atoms inside the oxide structure.

REFERENCES

[1] A. Fujishima, K. Honda, Nature, vol. 238 (1972), 37-38.
[2] S.N. Frank, A.J. Bard, J. Am. Chem. Soc., vol. 99 (1977), 303-304.
[3] O. Carp, C.L. Huisman, A. Reller, Prog. Solid State Chem., vol. 32 (2004), 33-177.
[4] U.R. Evans, Proceedings of the Royal Society of London. Series A, vol. 107 (1925), 71-74.
[5] M.V. Diamanti, MP. Pedeferri, Corros. Sci., vol. 49 (2007), 939-948.
[6] J.L. Delplancke, M. Degrez, A. Fontana, R. Winand, Surf. Technol., vol. 16 (1982), 16.
[7] P. Kofstad, High temperature corrosion. London and New York, Elsevier Applied Science (1988).
[8] L. Lavisse, D. Grevey, C. Langlade, B. Vannes, Appl. Surf. Sci., vol. 186, Issue 1-4 (2002), 150-155.
[9] L. Wan, J.F. Li, J.Y. Feng, W. Sun, Z.Q. Mao, Mat. Sci. Eng. B, vol. 139 Issue: 2-3 (2007), 216-222.

Maintaining Clean Surfaces in Cryogenic Measurement Environments

Jeffrey Lindemuth

Lake Shore Cryotronics, Westerville, OH 43021, jlindemuth@lakeshore.com

ABSTRACT

Surfaces of nanoscale materials of all types must be kept clean to properly characterize their behavior. Materials that are characterized at cryogenic temperatures and in vacuum present additional challenges to maintaining clean surfaces. Any residual contaminants will cryo-pump onto the cold surface of the system, including the sample. We describe a method of independent cooling of the cryostat and the sample. In this method the sample is kept at an elevated temperature, driving off condensates on the surface, while the cryostat is cooled to base temperature. All contaminants in the vacuum are cryo-pumped onto the cold surfaces. These surfaces are maintained at the lowest temperature possible. At this point the sample is cooled to base temperature. Analysis, including residual gas analysis, of the vacuum and surface shows that the condensations on the surface of the sample are reduced to a minimal level.

Keywords: probe station, cryogenic measurements, surface preparation, split flow cryostat

1 INTRODUCTION

In a cryogenic probe station the sample is often in vacuum and the system, including the sample, is cooled to cryogenic temperatures using a flow cryostat. Of course the vacuum is never ideal. As the system cools, the contaminants in the vacuum will cryopump onto the cold surfaces in the system [1]. The most common contaminants are water vapor, nitrogen and oxygen.

As the temperature of a surface cools below the freezing point of these gases, they can form a layer of ice on the surface of the sample. This layer of contamination will, at best, greatly affect any measurements of the sample, and at worst, destroy the sample. The measurement of the transport properties of many materials and devices of current research interest must be done at variable temperatures and with the sample in vacuum [2].

This paper is a description of using a split flow cryostat to control the rate of cryo-pumping on the sample. The sample is a simple piece of Kapton® film. The hydrophilic nature of the Kapton® film makes it an ideal model system. The Lake Shore Model HVTTP6 cryogenic probe station was used.

2 METHOD AND DISCUSSION

2.1 Description of Split Flow Cryostat

A split flow cryostat is a continuous flow cryostat with the ability to independently cool the radiation shields and the sample stage. The radiation shields are arranged as two stage radiation shields. This allows a cold inner shield and a warmer outer shield, and it also provides a radiation barrier between the room temperature vacuum chamber and the inner cold shield. With the split flow design, the sample stage can be warmer or colder than the coldest radiation shield. Keeping the sample warmer than the cold radiation shield makes the radiation shield the cryo-pump in the system. This comes with the small penalty of some radiation heating of the cold radiation shield by the sample. However, the solid angle of the sample seen by the radiation shield is small.

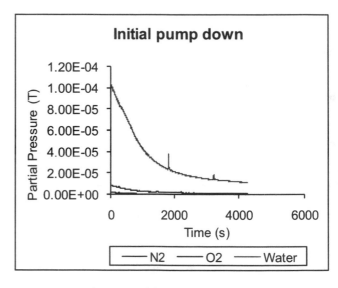

Figure 1: Initial pump down curve

2.2 Initial Sample Preparation

A Kapton® sample was used as the sample in this procedure. It was placed in the HVTTP6 probe station and the system was evacuated. A residual gas analyzer (RGA) was also connected to the cryostat chamber.

The initial pump down curve, with the system at room temperature, is shown in figure 1. Within about 1 hour of starting the pump down, the partial pressure for water is 20 µTorr

2.3 Raise the Sample Temperature for Bake-Out

Initially, the sample was heated to 400 K. Figure 2 is the RGA output vs. time. Notice the increase in water vapor as the sample heats up. This is the water vapor on the Kapton® sample driven off the surface of the Kapton®.

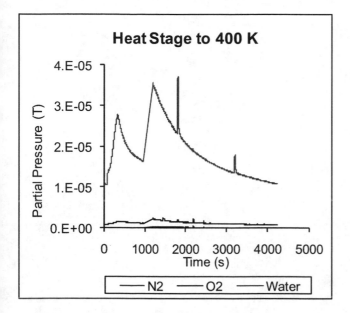

Figure 2: RGA partial pressure as the sample is warmed to 400 K

2.4 Using Split Flow Cryostat to Cool the System

Now the sample and sample stage are at 400 K. They are the warmest surfaces in the vacuum space. The HVTTP6 probe station uses a split flow cryostat. This design allows the radiation shields in the system to be cooled while keeping the sample stage at an elevated temperature. Nearly all of the contaminants will cryo-pump onto the radiation shields. Figure 3 shows the partial pressure recorded by the RGA during the cool down of the system.

Notice in figure 3 that the partial pressure of water has decreased from 10 µTorr at the end of the data in figure 2 to 3 µTorr in figure 3. This decrease in the partial pressure is due to the cryo-pumping effects of the cold radiation shields.

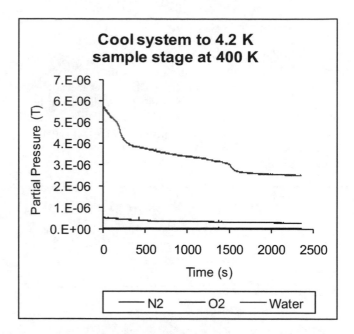

Figure 3: Partial pressure as the system is cooled to 4.2 K. Sample is at 400 K.

2.5 Cooling the Sample to 4.2 K

Now that everything in the system, except the sample stage and sample, are at 4.2 K, the split flow cryostat can be used to cool the sample. During this cooling of the sample the cold radiation shields are kept at 4.2 K. Figure 4 show the partial pressure during this cooling of the sample. Notice water partial pressure is nearly constant during the cool down. This is an indication that water was not cryo-pumped onto the sample during the cool down.

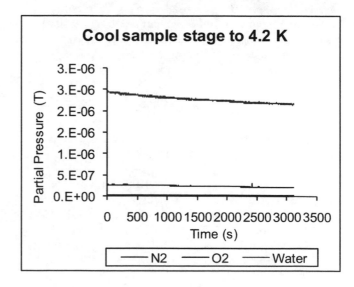

Figure 4: Partial pressure as the sample cools to 4.2 K

2.6 Warming the Sample

At this point the sample can be raised to any desired temperature for measurement. Using the split flow cryostat, it is possible to keep the radiation shields as the coldest surface in the system. Figure 5 is the RGA data from the sample. The sample is back to 400 K, the radiation shields are at 4.2 K. Note that the partial pressure of water remains at 2 µTorr. Compare this plot of the sample at 400 K to the data in figure 2.

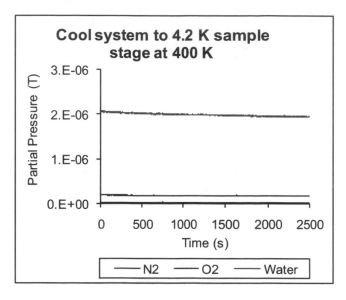

Figure 5: Partial pressure as the sample warms back to 400 K

3 CONCLUSIONS

By using a split flow cryostat in a cryogenic probing station, contaminates from the vacuum reaching the surface of a sample can be greatly reduced. The split flow cryostat allows the sample to be the warmest surface in the vacuum, while eliminating any cryo-pumping onto the sample.

REFERENCES

[1] Cryogenic scanning Hall-probe microscope with centimeter scan range and submicron resolution, Rafael B. Dinner, M. R. Beasley, and Kathryn A. Moler, Rev. Sci. Instrum. 76, 103702 (2005).

[2] Influence of source-drain electric field on mobility and charge transport in organic field-effect transistors, B. H. Hamadani, C. A. Richter, D. J. Gundlach, R. J. Kline, I. McCulloch, and M. Heeney, J. Appl. Phys. 102, 044503 (2007).

A Study about Surface Chemical-Physics Mechanisms
Occurring for Interacting Metal Nanostructures with gas NO$_2$

S. di Stasio[*][1], V. Dal Santo[**]

[*] Istituto Motori CNR, Aerosol & Nanostructures Laboratory, Via Marconi 8 – 80125 Napoli, Italy
[**] Istituto di Scienze e Tecnologie Molecolari CNR, Via Golgi 19 – 20133 Milano, Italy

ABSTRACT

Zinc nanostructures synthesized with different morphologies from the same evaporation/condensation technique are studied with concern to surface reactivity to NO$_2$ by Diffuse Reflectance Infrared Fourier Transformed Spectroscopy (DRIFTS) at variable temperatures. Synthesis of nanopowders is obtained, according to previous work, by gas flow thermal evaporation at 540 °C of bulk Zn grains. Two types of Zn powders are fabricated and characterized by SEM, TEM, XRD in experiments. The first one is constituted by grains ($\sim$10 μm) originated by the stratification of smaller aggregates ($\sim$200 nm) and isolated primary particles ($\sim$50 nm) born in the gas flow. The second one is constituted mainly by hollow Zn nanofibers with external and internal diameter about 100 and 60 nm. Comparison is made between the nanostructured powders with respect to commercial Zn standard dust. The Zn hollow nanofibers when exposed to NO$_2$ are found to exhibit dramatic reactivity, which is not observed either in the case of clustered aggregate zinc or of commercial Zn dust powder.

Keywords: chemical sensors, nanofibers, surface science, DRIFTS, CVD

1 INTRODUCTION

Transition element and relative oxides are used as nanostructured materials for gas-sensing manufacturing, owing to the properties of selectivity and sensitivity to different pollutants, hydrogen, humidity and ammonia. Common applications can be listed for Palladium, Platinum and TiO2, CuO, to cite a few. In particular, Zinc oxide is attractive material in that it exhibits a number of interesting features for instance it is classified as a wide-band-gap semiconductor (3.3 eV) with property of transparency in the visible region, piezoelectricity, photoluminescence. Possible applications include phosphor in flat panel displays and design of light emitting diodes when used as a substrate of gallium nitride (GaN) owing to the fact that ZnO does match very well the GaN lattice, expecially in its thin film form [1].

In previous paper by our group [2] we reported the receipt to fabricate hollow Zinc nanofiber by a evaporation-condensation route followed by deposition of metal vapour on quartz substrate.

The generation of nanocrystalline zinc nanostructures by thermal evaporation was also reported in the past. Yumoto et al [3] fabricated Zn crystals with metal bars as raw material. Recknagle et al [4] performed evaporation of zinc in partial vacuum environment followed by condensation of metal vapours by supersonic jet expansion.

NO$_2$ is one of the ever-present pollutants because it is generated both in the case of combustion of both hydrocarbons and hydrogen. Potentially it is responsible of formation of acid rains. It is harmful for human health at very low concentrations. For instance, Italian legislation fixed NO$_2$ concentration in urban environment at 0.1 ppm (alarm level) and 0.2 ppm (attention level).

For the reasons above it is attractive for scientists to speculate about the applicability of fabrication of low-cost sensors of nitrogen di-oxide for application to pollutant detection- In previous work [5] we demonstrated the feasility of sub-ppm NO$_2$ sensors made of Zn/ZnO material. In particular, we compared, at parity of chemical composition, the performance of powders with different morphological and structural properties at variable temperature, from room temperature (RT) to about 100 °C.

The aim of this communication is to represent an investigation about the physics-chemistry mechanisms involved in such dramatic effects. Apparently the Zn powders do differ in their surface-to-volume ratios. On this basis, we performed a series of experiment involving the surface of such materials with Diffuse Reflectance Fourier Transform Spectroscopy (DRIFTS).

2 EXPERIMENTS

Nanostructured powders are synthesized according to a gas-phase route described in detail in previous works [2,6]. Aerosol particles are collected for XRD analysis on high-efficiency quartz fiber filters. For characterization at SEM and TEM, Zn nanoparticles are deposited, by a thermal precipitator and a vacuum impactor, on optical glasses 18×18 mm (Menzel-Glaser) and 3 mm/400 mesh carbon-

---

[1] Corresponding author: s.distasio@im.cnr.it

coated copper grids (Agar Scientific), respectively. Zinc fibers are collected as a skein of sponge-like material at the exit of the expansion nozzle after run times typically of one hour. Scanning Electron Microscope is LEICA Cambridge 360 Field Emission SEM. X-Ray Diffraction (XRD) measurements, discussed elsewhere [6] are carried out on a PW1710 Philips rotating anode X-ray diffractometer.

Commercial Zinc dust powder used for comparison in XRD analysis is obtained by Fluka Chemical and used as received. DRIFT (Diffuse Reflectance Infrared Fourier Transform) spectra are recorded on a Digilab FTS-60 equipped with a KBr beamsplitter and a MCT detector operating between 400 and 4000 cm^{-1}, using an Harrick reaction chamber with KBr windows, and an Harrick DRA-2C1 diffuse reflectance accessory. Spectra are measured in Kubelka-Munk (K-M) units. Backgrounds are recorded by using dry KBr powder. All the samples are treated according to the following procedure: a) Calcination under O_2 (30%)/He flow (10 mL/min) at 200°C for 1 h (heating rate 5°C/min). b) Cooling down to RT in He flow (10 mL/min). c) Replacement of He flow with a NO_2 (1000 ppm vol ca.)/He flow (10 mL/min). d) Heating-up to 300°C at 10°C/min rate, holding the temperature for 6 min at Room Temperature (RT), 50, 100, 150, 200, 250, 300°C. DRIFT spectra are recorded 2 min after the reaching of the plateau temperature. e) cooling-down in NO_2/He flow. After cooling at RT, a comparative spectra are recorded for each sample to check with the spectra at the beginning of the heating treatment.

3   RESULTS AND DISCUSSION

Gas sensing properties are studied for standard Zn dust powder (not shown) and for Zn stratified nanoparticle aggregates (Fig. 1 top) and Zn hollow nanofibers (Fig. 1, middle and bottom).

Fig. 2 shows the DRIFTS spectra of stratified nanoparticle aggregates, synthesized by evaporation-condensation scheme [2] at T_{evap}=540 °C when exposed to the NO_2/He flux at different test temperatures (RT to 300°C). A doublet located at 1628 and 1600 cm^{-1} is visible in all the temperature range investigated, thus decreasing at the increase of temperature, which could be ascribed to R and P branches of NO_2 free (gas). A broad band located at 1300 cm^{-1} is evident in low (< 250°C) temperature spectra and decreases raising the temperature disappearing almost completely at 300°C. Around 1500 cm^{-1} another band very broad and feeble seems to be present. The analysis of bands evolution of Fig. 4 is complicated by the fact that, at increasing the temperature, a negative feature is observable between 1550 and 1300 cm^{-1}. Thus, the two peaks at 1300 and 1500 cm^{-1} could be ascribed to two vibrational modes of mono- or bidentate NO_3^- species [7] adsorbed on the ZnO thin film surface formed on the Zn surface, even if for a definitive attribution the presence of eventual absorption bands in the 970-1040 cm^{-1} region (not observable in our experiments) should be checked. At the increase of

temperature the band at 1300 cm^{-1} first grows till 150°C. Thereafter, it becomes less pronounced and it almost disappears at 300°C. The band is restored by cooling down the sample at RT in NO_2/He flow.

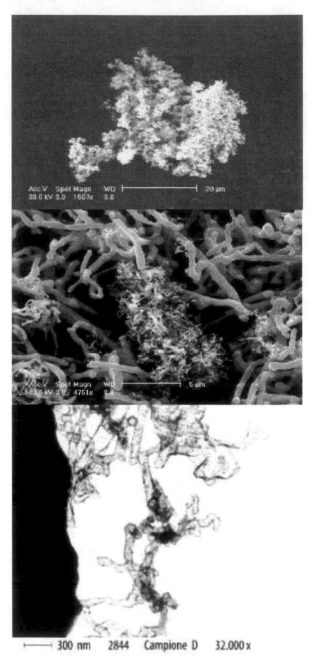

Figure 1: Top) Stratified Zn aggregates synthesized by evaporation at 560°C of bulk Zn grains and deposition of Zn vapors ; Middle) SEM image of a sponge of hollow Zn fibers lying between nested thick nanorods, as obtained by metal vapors deposition in the same experiment of Top); Bottom) TEM micrograph of hollow nano fibers.

Fig. 3 are the DRIFT spectra of Zn hollow nanofibers grown by the deposition within the quartz reactor orifice [2] of metallic zinc vapors obtained at T_{evap}=540°C. The

NSTI-Nanotech 2008, www.nsti.org, ISBN 978-1-4200-8503-7 Vol. 1

differences of amplitude (about one order magnitude) of the IR absorption bands between Zn aggregates and Zn nanofibers is probably due to the different amount of diffused IR radiation emerging from the sample resulting from different packing of sample cup and/or different nanostructure (micro-crystalline aggregates vs. mono-crystalline rods [2]).

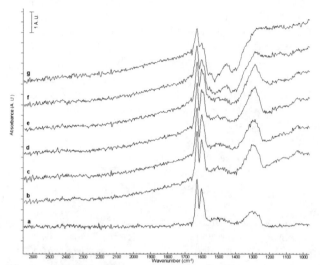

Figure 2: DRIFT spectra (background subtracted) of stratified aggregates exposed t NO$_2$/He with cauliflower-like morphology, relatively to wavenumbers 860-2640 cm^{-1} at increasing temperatures. a) RT; b) 50°C; c) 100°C; d) 150°C; e) 200°C; f) 250°C; g) 300°C.

Comparative analysis of spectra in the case of aggregates and hollow nanofibers shows dramatic differences.

In the case nanofibers (Fig. 2), we note that at RT no bands attributable to adsorbed species are observed. Only the doublet located at 1600 – 1628 cm^{-1}, due to NO$_2$ (gas) free is present. The presence of NO$_2$ feebly adsorbed on Zn^{2+} sites can not be excluded since the IR adsorption of such species fall in the 1642-1605 cm^{-1} range that is masked by gas phase NO$_2$ adsorption. At increasing the temperature from RT to 300 °C, the NO$_2$ doublet disappears and a broader shoulder between about 1900 and 1500 cm^{-1} grows up. At the mean time also a relatively narrow peak located at 2500 cm^{-1} appears. Contemporarily a new, complex spectral feature does appear between 1400 and 1200 cm^{-1}.

Fig. 4 shows the DRIFT spectra obtained for commercial bulk Zinc dust. At RT the spectra are markedly different with respect to those of Zn aggregates and hollow fibers. The 1600 – 1628 cm^{-1} doublet of NO$_2^-$ gas absorption is observed also in this case, but three intense bands located at 1510, 1315 and 1023 cm^{-1} are visible, which are increasing with the time of contact with NO$_2$. These bands can be attributed to adsorbed bidentate nitrates NO$_3^-$ [6]. At increasing the temperature from RT to 300°C, these 1510, 1315, 1023 cm^{-1} absorption bands exhibit an initial growth till 100°C, followed by a progressive depression, which can be interpreted as the process of thermal decomposition of adsorbed nitrates. In the meanwhile a negative feature

located at 1650 1760 cm^{-1} develops. After cooling down to RT the sample in NO$_2$/He flow all the bands grow up again.

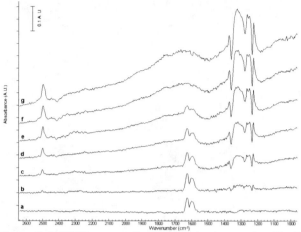

Figure 3: DRIFT spectra (background subtracted) of hollow nanofibers exposed t NO$_2$/He at 860-2640 cm^{-1}. a) RT; b) 50°C; c) 100°C; d) 150°C; e) 200°C; f) 250°C; g) 300°C.

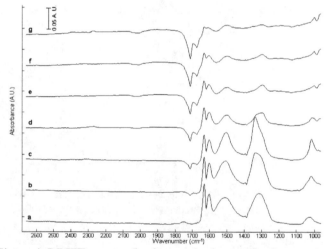

Figure 4: DRIFT spectra (background subtracted) of standard bulk Zn dust exposed t NO$_2$/He at 860-2640 cm^{-1}. a) RT; b) 50°C; c) 100°C; d) 150°C; e) 200°C; f) 250°C; g) 300°C.

An interpretative scheme for surface reactions yielding nitrites and nitrates has been reported previously [6,8]. Our previous experimental work on current response measurements [5] by Zn nanopowders constituted of nanofibers and nanoparticle aggregates exposed to NO$_2$ revealed relative abundance of conductive electrons in the case of nanofibers at room temperature. Similarly to present work, also in the previous study the samples were pre-treated, by exposure to humid synthetic air for 24 h at 200 °C. In particular, the current response of nanofibers at RT, 50%RH and 0.4 ppm NO$_2$ resulted almost 1 and 4 order of magnitudes higher with respect to nanoparticle aggregate and standard Zn powders, respectively. The essential point is that, even at low temperature, a thin film of ZnO is probably formed soon on the sample powders. The current response of the nanofibers samples decreases at the increasing temperature from RT to 150 °C.

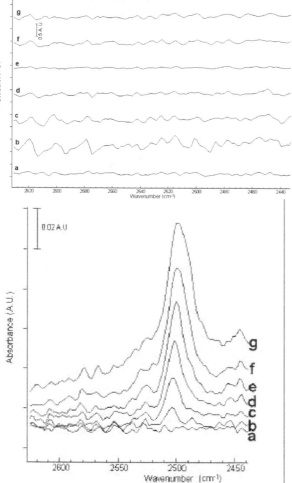

Figure 5: Zoom of DRIFTS at 2440-2650 cm^{-1} relatively to (Top) stratified Zn aggregates, (Bottom) hollow nanofibers.

A supplementary mechanism can be advocated to justify the previous experimental results [5]. Kase et al [9] proposed a particular process of NO chemisorption by the formation of the so-called *pseudo-oxygen defects*. Again the mechanism is cited according to the presumable formation of a thin film of ZnO on the surface of nanofiber and nanoparticle aggregates samples. It may describe the observed changes of electrical conductivity for an n-type semiconductor, which undergo contemporary NO chemisorption. In our case we propose at RT a similar mechanism occurring for NO$_2$ at ZnO film interface of nanopowders leading to adsorbed NO$_3^-$ and releasing of conductive electrons from the surface oxygen defects. This process, according to the cited paper [9] is most probable to occur at RT and thwarted at higher temperatures at which nitrates are not stable.

Figs. 5 shows the DRIFT difference spectra of cluster versus nanofiber samples at 2440 to 2630 cm^{-1}. When compared with spectra about 1500 to 1800 cm^{-1}, we observe that, at increasing temperatures, the nanofibers spectra

exhibit the rising of a narrow peak at 2500 cm^{-1} and, contemporarily, the progressive smoothing and disappearing of the doublet at 1600-1628 cm^{-1}, thus indicating probable consumption due to further oxidation of Zn to ZnO and/or desorption of NO$_2$ at increasing temperatures from RT to 200 °C. The 2500 cm^{-1} band in Fig. 5-bottom (nanofibers) may be attributed according to the literature [10] to the presence of adsorbed polymeric species of type (N$_x$O$_y$)$_z$. A structural peculiarity of the hollow fibers, which should be accounted for in the discussion of surface reactivity effects, is the huge surface area with respect to the other samples. According to a discussion reported alsewhere [6] the ratio between the specific surface (cm^2/g) of hollow nanofiber with respect to Zn standard powder can result higher than 300 and the ratio between the specific surface (cm^2/g) of hollow nanofibers with respect to stratified aggregates powder will be about 40 and 160 for clusters of 10 and 5 μm, respectively.

4   CONCLUSIONS

Zinc powders constituted by nanoparticles aggregates and hollow nanofibers have been studied with respect to NO$_2$ reactivity by DRIFT spectroscopy and compared with bulk Zn dust available in commerce. DRIFT spectra of zinc aggregates and standard dust show the formation of nitrates (NO$_3^-$) and nitrites (NO$_2^-$) adsorbed species, that are stable at RT, decompose as the temperature increases, and can be again formed at RT after their decomposition. Zinc nanofibers, external and internal diameters about 100, 60 nm, do exhibit a peculiar reactivity with respect to test gas NO$_2$. DRIFT spinctra , are characterized by two distinctive absorption bands at 2500 cm^{-1} and 1600-1628 cm^{-1}, which are observed to grow up and fall off at correlated rates, respectively, when temperature is increased from RT to 300°C This feature is not at all observed in the case of the other samples. This peculiarity can be explained as nanoscale effect related to the much higher specific surface (cm^2/g) and/or to a confinement effect of hollow nanofibers with respect to Zn aggregates and commercial Zn standard.

REFERENCES

[1] RF Xiao, XW Sun, HB Liao, N Cue, HS Kwok, J. Vac. Sci. Technol. A **15**, 2207 (1997).
[2] S di Stasio, *Chem. Phys. Lett.* 393, 498 (2004).
[3] H Yumoto, RR Hasiguti, J. Cryst. Growth 75, 289 (1986).
[4] S Recknagle, Q Xia, JN Chung, CT Crowe, H Hamilton, GS Collins, NanoStructured Materials 4, 103 (1994).
[5] C. Baratto, G. Sberveglieri, A. Onischuk, B. Caruso, S. di Stasio, Sensors and Actuators B 100, 261 (2004).
[6] S. di Stasio, V. Dal Santo V, Appl. Surface Sci., **253** 2899 (2006).
[7] K. Hadjiivanov, Catal. Rev Sci. Eng 42, 71 (2000).
[8] A. Chiorino, G. Ghiotti, F. Prinetto, M.C. Carotta, D. Gnani, G. Martinelli, Sensors and Actuators B 58, 338 (1999).
[9] K. Kase, M. Yamaguchi, T. Suzuki, K. Kaneko, J. Phys. Chem. 99, 13307(1995).
[10] K. Hadjiivanov, Catal. Lett. 68, 157 (2000).

Nanostructure LB Films of Novel OPV Compounds

B. Jiménez-Nava, V. Álvarez-Venicio, M. Gutiérrez-Nava and M.P. Carreón-Castro*

Departamento de Química de Radiaciones y Radioquímica, Instituto de Ciencias Nucleares, Universidad Nacional Autónoma de México, pilar@nucleares.unam.mx

ABSTRACT

We studied the formation of Langmuir monolayers of two series of amphiphilic monodendrons π-conjugated from first to third generation number with different length of terminal chains C_3H_7 and $C_{12}H_{25}$, the characterization of the monolayer was made through isotherms and images obtained by Brewster Angle Microscopy (BAM) to confirm the phase transition, the six compounds formed Langmuir monolayers stable and reproducible, the stability and homogeneity of the monolayers increases with the increment in the generation number. In addition, we observed that when the length of terminal chains is increases from C_3H_7 to $C_{12}H_{25}$ the stability has a notable increment for the second and third generation number. We had showed that the monodendron the second generation with terminal chains $C_{12}H_{25}$ has the best stability and reproducibility.

Keywords: Langmuir, monolayer, monodendrons, OPV, Brewster microscopy.

1. INTRODUCTION

Langmuir films, composed of amphiphilic molecules adsorbed at the air-water interface, continue to attract widespread attention due to their importance as models for two dimensional structures and phase transitions, as model biological membranes, and as precursors for technologically important Langmuir-Blodgett films [1-4].

There are various modern applications requiring highly ordered coatings of nanoscale thickness with smart behavior and electronic properties, microelectronic systems, microfluidic devices, micro- and nanopatterning, and optical data storage devices [5]. The search for new materials with a suitable combination of properties, microstructure and intermolecular interactions attracts attention to dendrimers. Dendrimers are materials with cascade, tree-like architectures, generation-dependent sizes in the nanometer range and capability of self-assembling into superstructures [6]. Dendritic macromolecules with well-defined geometrical sizes and surface functionality can be suggested as promising candidates for functional coatings. Numerous examples of dendrimer synthesis with varying chemical nature have been published in the last decade [7].

However, there is limited information on dendrimeric microstructures in the bulk state, and few studies are devoted to their behavior at solid surfaces [8, 9].

One particular class of dendrimers, monodendrons, with a focal functional group and semi-spherical shape are interesting candidates for organized monolayer formation and surface functionalization because of the specific shape and dendritic nature of the shell. We believe that Π-conjugated dendrons are a promising class of organic materials for the fabrications of organic light emitting diode (OLED's). They can be thought of as consisting of a conjugated light-emitting core, conjugated branches, and surface groups. Such materials can be designed so that the core defines the color of light emission, the surface groups control the processing properties, and the branches allow transport of charge to the core. Chemi-or-physisorption may offer an important route to the fabrication of dendrimer assemblies at interfaces [10].

The general approach of this investigation is to form and study Langmuir monolayers formed on the air–water interface, organized through self-ensambled imposed by terminal chains and examination of the stability of these monolayers deposited on solid substrates. In this publication, we focus on Langmuir monolayers of amphiphilic monodendrons of the first three generations G1-C_3H_7, G2-C_3H_7, G3-C_3H_7, G1-$C_{12}H_{25}$, G2-$C_{12}H_{25}$ and G3-$C_{12}H_{25}$. (Fig.1). Compounds with an alcohol polar head group were used to estimate the effect of the number of generation and the length of terminal alkyl chains in the formation. Isotherm superficial pressure (Π) vs. Molecular Area (A) and Brewster Angle Microscopy were used to study the stability and the morphology of the Langmuir monolayers. We made a simulation of the monodendron configuration.

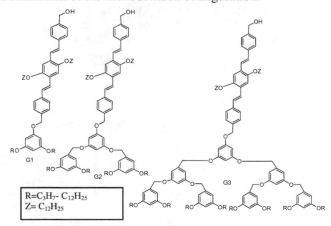

Figure 1: Dendrimeric compounds used for monolayer formation. GN corresponds to dendron of N generation; C_NH_N is the length of the alkyl terminal-chains.

2. EXPERIMENTAL

The six compounds were synthesized as described elsewhere [11]. Solutions at $\approx$ 1 mg/mL concentration were prepared, using HPLC chloroform (Aldrich, 99.9% pure). The dendrimer solution was spread over ultrapure water ρ=18.2 MΩ•cm used for the subphase was obtained from a Milli-DIPAK/ Milli-Q185 ultrapurification system from Millipore and allowed to stay for 15 min for residual solvent evaporation. The monolayer was then compressed at 10mm min^{-1} until the desired surface pressure was reached and deposited onto a hydrophilic surface of glass slide at a lift speed of 5 mm•min^{-1} at room temperature. Successive layers were deposited onto glass substrates (type Z). We typically spread 80–180 µL of solution on the water surface for isotherm, BAM images or Langmuir-Blodgett films.

The monitoring the monolayer formation was made by Π-A isotherms and Brewster angle microscopy (BAM) that are used to characterize the behavior of a monolayer film at the air-water interface. The trough employed for the formation and deposition of the Langmuir monolayers is a KSV 5000 system 3. The trough was set in a flexiglas enclosure soas to be protected from drafts and dust, the temperature was controlled to ± 0.1°C. All the isotherms presented here were taken at 20°C. The BAM images were gotten with a miniBAM plus of Nanofilm Technology Gmbh and the Langmuir films were formatted on a NIMA trough. Π-A isotherms were collected and *in situ* during acquisition of BAM images, surface pressure was measured by means of a platinum Wilhelmy plate.

Molecular semiempirical models of all compounds were built considering one molecule in the vacuum and taking into account just the hydrophilic and hydrophobic character of dendrimeric molecules and finding favorable energy were searched using Spartan 4.0 program. The simulation of a possible orientation of the monodendrons in the water-air interface in base to the Π-A isotherms and the calculated area thought the software Spartan.

3. RESULTS AND DISCUSSION

All compounds studied here were spread on an air–water interface to form a Langmuir monolayer. The lateral compression of the monolayers resulted in a gradual increase of the surface pressure that resembled classic amphiphilic behavior [12]. We got Π–A isotherms, BAM images of monodendrons of first to third generation with C_3H_7 chains (fig. 2).

An increase in the generation number caused significant shift in the Π–A isotherm to a higher cross-sectional area per molecule (A) and higher superficial pressure of phase transitions [13]. The BAM images show that with an increase in the generation number the homogeneity is higher and that the monolayers collapse forming multilayers. In addition, the BAM images show that the monolayers are formed for the self-assembled of molecules because we can watch the same images when we spread and when we decompress the monolayer, when the generation number is increase the self-assembled monolayers are more stable and homogeneous. At higher surface pressure (14 mN•m^{-1}), the initial stages of monolayer collapse become visible, manifesting itself as multiple islands of second layer formed on top of the primary monolayer [14].

The Π–A isotherms together with the BAM images provide the area per molecule of the liquid-condensed and the superficial pressure ideal for the deposit on a solid substrate of the Langmuir monolayers.

The Hysteresis curves (fig. 3b) show that the second generation monodendron with C_3H_7 terminal chains form the monolayers more reproducible and stable of the three monodendrons with C_3H_7 terminal-chains, it is understood like the effect of the C_3H_7 terminal chains is stronger than the sterics effects in the case of the second generation monodendron and weaker than the sterics effects of the rigid part of the third generation monodendron, It can be explain because even when the third generation monodendron has a bigger number of C_3H_7 chains, it has a bigger rigid structure too and the C_3H_7 chains are too short to have enough effect over the rigid part like in the second generation monodendron. The Π–A isotherms and the modeling show the area per molecule occupied for every monodendron in the liquid-condensed phase.

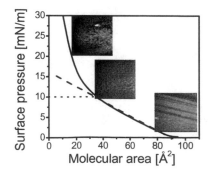 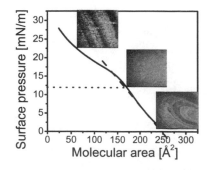 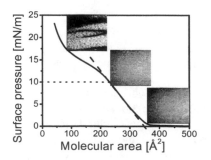

Figure 2: Pressure-area (Π-A) isotherms and BAM images of dendrimers at 25°C. (a) G1-C3H7 1.75mg/mL, 80µL. (b) G2-C3H7 1.18mg/mL, 100µL (c) G3-C3H7 0.93mg/mL, 100µL.

NSTI-Nanotech 2008, www.nsti.org, ISBN 978-1-4200-8503-7 Vol. 1

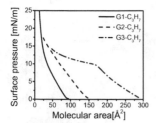 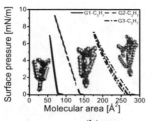

(a)   (b)

Figure 3: Comparison between the isotherms and hysteresis curves of the monodendrons with C_3H_7 terminal chains from first to third generation number.

For the monodendrons G1, G2 and G3 with C_3H_7 terminal-chains, we can observe that when increases the generation number (fig. 3a) the phase transitions are more marked and there for the monolayer is more stable. We can se the increment of molecular area with the generation number because the rigid part is increasing with the generation number and the increment in the slope of the phase transition can be understand like an increase in the self-assembled but to the third generation monodendron the hysteresis curve is wider than the second generation.

For the monodendrons with $C_{12}H_{25}$ terminal-chains we can observe an increase in the generation number caused significant shift in the Π-A isotherm (fig. 6) to a higher cross-sectional molecular area and higher superficial pressure of phase transitions [15]. The BAM images show phase images (Fig. 6) witch ones are extremely homogeneous with minor phase contrast over several micrometers across illustrating uniform composition of these monolayers that with an increase in the generation numbers the homogeneity is higher and that the monolayers collapse forming multilayers. In addition, the BAM images show that the monolayers are formed for the self-assembled of molecules because we can see the same images when we spread and when we decompress the monolayer, when the generation number is increased the self-assembled monolayers are more stable and homogeneous [16]. At higher surface pressure (20 mN•m^{-1}), the initial stages of monolayer collapse become visible, manifesting itself as multiple crest of second layer formed on top of the primary monolayer. The Π-A isotherms together with the BAM images provide the area per molecule of the liquid-condensed and the superficial pressure ideal for the deposit on a solid substrate of the Langmuir monolayers. The Π-A isotherms and the modeling show the area per molecule occupied for every monodendron in the phase of liquid-condensed.

Through the hysteresis curves (fig. 4b) we can get important information about the stability and reproducibility, there for, like is show in the hysteresis curves the reproducibility is good to the three monodendrons with terminal chains $C_{12}H_{25}$ and we can say that the monodendron of second generation number shows the best stability and reproducibility. The molecular semiempirical models show the most probable configuration of the moieties witch have a good congruence between the molecular area calculated for the

molecular semiempirical model and the area obtained from the isotherm.

For the monodendrons G1, G2 and G3 with $C_{12}H_{25}$ terminal-chains (fig. 4a), we can observe that when increases the generation number the change of phase transitions are more visible and there for the monolayer is more stable. We can se the increment of molecular area with the generation number because the rigid part is increasing with the generation number and the increment in the slope of the phase transition can be understand like an increase in the self-assembled.

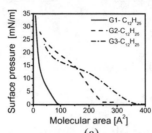 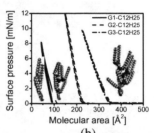

(a)   (b)

Figure 4: Comparison between the isotherms and hysteresis curves of monodendrons with $C_{12}H_{25}$ terminal chains.

The molecular area presents a different value with the increment in the chains length from C_3H_7 to $C_{12}H_{25}$ for the same generation (fig. 5), this can be understood like exits a number of changes witch begins to have an effect over the stability and the homogeneity of the monolayer, the chart of the first generation monodendrons show a little change, it can be explain like the two $C_{12}H_{25}$ chains have a little effect in the orientation of the monodendrons that the C_3H_7 chains do not have [17]. The figure 8b shows a big difference between the molecular areas of the second generation monodendrons ~100 Å^2 respect to the monodendron with C_3H_7 chains [18]. The difference of the surface pressure of the phase transition between the monodendrons of second generation number with C_3H_7 and $C_{12}H_{25}$ terminal chains, show that in this case the effect of the long chains is bigger than in the firs case, for this reason we get a monolayer more stable and homogenious, so we can expect that if we duplicate the number of chains again, we are going to get increment the stability and homogeneity, but if we see the figura 5c, we can observe that it does not happen, the reason why it did not happen, is because we increase the number of chains, but we increase the number of generation and the rigid part of the monodendrons is bigger, and the effect of the chains is not enough to get one monolayer as stable and homogenious as the monolayer of the second generation monodendron with $C_{12}H_{25}$ chains [19].

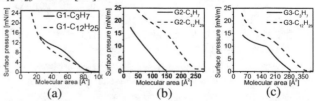

(a)   (b)   (c)

Figure 5: Comparison between the same generation number monodendrons with different length of terminal chains.

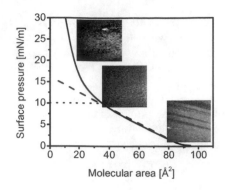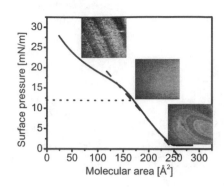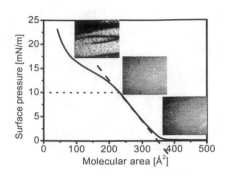

Figure 6: Pressure-area (Π-A) isotherms and BAM images of dendrimers at 25°C. (a) G1-C12H25 1.15mg/mL, 150μL. (b) G2-C12H25 0.5mg/mL, 171μL and (c) G3-C12H25 0.55mg/mL, 150μL.

4. CONCLUSIONS AND PERSPECTIVES.

We have form Langmuir monolayers of six new monodendrons with π-conjugated units. The isotherms and the BAM images confirmed the formation of the monolayer, the phase transitions and show that the length and number of terminal chains has a important effect in the stability. We had showed that the monodendron the second generation with terminal chains $C_{12}H_{25}$ has the best stability and reproducibility. There for, we can say that the generation number of the monodendrons is important factor to take into a count in the formation of Langmuir monolayers, but is too important to consider the length and number of terminal chains to get monolayers with the best stability and reproducibility too. The molecular semiempirical models showed a very good congruence with the molecular area obtained from the isotherms less than 10 $Å^2$.

We are working in the transference of the monolayers onto a glass substrate to form Langmuir-Blodgett films and will be able to study the electronics proprieties and fabrication of devices.

ACKNOWLEDGEMENTS

This work is supported by the PAPIIT-Projects from DEGAPA-UNAM IN118808. The authors also acknowledge to Martin Cruz Villafañe, Magda Sierra and Yazmin A. Valdez Hernández for their technical support.

REFERENCES

[1] G.L. Gaines, Insoluble Monolayers at the Liquid-Gas Interfaces (Interscience, Ney York, 1966).
[2] V.M. Kaganer, H. Möhwald, and P. Dutta, Rev. Mod. Phys. 71, 779 (1999).
[3] R.M. Kenn, C. Bohm, A.M. Bibo, I.R. Peterson, H. Möhwald, J. Als-Nielsen, and K. Kjaer, J. Phys. Chem. 95, 2092 (1991).
[4] D.B. Zhu, C. Yang, Y.Q. Liu, and Y. Xu, Thin Solid Films 210, 205 (1992).
[5] D. Quin, Y.N. Xia, J.A. Rogers, R.J. Jackman, X.M. Zhao, G.M. Whitesides, Top. Curr. Chem. 194 (1997) 1.
[6] J.M. Frechet, Science 263 (1994) 1711.
[7] D.A. Tomalia, Adv. Mater. 6 (1994) 529.
[8] A.W. Bosman, H.M. Janssen, E.W. Meijer, Chem. Rev. 99 (1999) 1665.
[9] V.V. Tsukruk, Adv. Mater. 10 (1998) 253.
[10] K. Ichimura, S.-K. Oh, M. Nakagawa, Science 288 (2000) 1624.
[11] V. Alvarez-Venicio, B. Jiménez-Nava, M.P. Carreón-Castro, I. Audelo Méndez, A. Acosta Huerta, E. Rivera, M. Gutiérrez-Nava, Polymer, Submitted (2007).
[12] A. Ulman, An Introduction to Ultrathin Organic Films, Academic Press, San Diego CA, 1991.
[13] S. Peleshanko, A. Sidorenko, K. Larson, O. Villavicencio, M. Ornatska, D.V. McGrath, V.V. Tsukruk, Thin Solid Films, 406 (2002) 233-240.
[14] Wen-Jung Pao, Fan Zhang, and Paul A. Heiney, Phys. Rev. E 67, 021601 (2003).
[15] A. Sidorenko, C. Houphouet-Boigny, O. Villavicencio, M. Hashemzadeh, D.V. McGrath, V.V. Tsukruk, Langmuir 16 (2000) 10569.
[16] F. Cardinali, J-L. Gallani, S. Schergna, M. Maggini, and J-F. Nierengarten, Tetrahedron Letters 46 (2005) 2969-2972.
[17] V.V. Tsukruk, V.V. Shilov, Structure of Polymeric Liquid Crystals, Naukova Dumka, Kiev 1990.
[18] David C. Tully; Jean M. J. Fréchet, Chem. Commun., 2001, 1229-1239.
[19] I. Bury, Bertrand Donnio, J-L. Gallani, and D. Guillon, Langmuir (2007), 23, 619-625.

Nanostructured films of novel azocompounds

Y. A. Valdez-Hernández*, E. Rivera**, M. P. Carreón-Castro*

* Instituto de Ciencias Nucleares, UNAM, México, yvaldez@nucleares.unam.mx
**Instituto de Investigaciones en Materiales, UNAM, México, riverage@iim.unam.mx
*Instituto de Ciencias Nucleares, UNAM, México, pilar@nucleares.unam.mx

ABSTRACT

The formation of Langmuir and Langmuir-Blodgett films from azocompounds like the azopolymers used in this work, that have different length of chains of ethylene glycol shows that these azocompounds are suitable to form thin films by the Langmuir-Blodgett technique, their spectroscopic and morphologic characterization confirm that the monolayer formation is stable and homogeneous in the interface air/water according the images obtained by Brewster's angle microscopy (BAM) and their transference onto solid substrates shows a linear increase multilayer according the UV-Visible spectra, that indicates homogeneity.

Keywords: Langmuir films, Langmuir-Blodgett films, azocompounds, azopolymers, Brewster microscopy.

1. INTRODUCTION

Actually many applications have been developed for azobenzenes exploring their important properties of photoisomerization and photonic, now their incorporation in polymers to form azopolymers is also very interesting, according to increase the potential optoelectronic and photonic applications [1, 2, 3, 4]. For study the applications the azopolymers are used in the solid state usually forming thin films built by casting or spin-coating methods. Techniques that allow control at the molecular level, such as the Langmuir-Blodgett (LB) and the Layer by Layer (LbL) methods, may lead to obtain nanostructured films with controllable thickness and architecture [5]. In the LB method, an insoluble layer of the material of interest is spread on an aqueous subphase and is compressed by constant speed, after the solid order (2-D) is obtained the layer is transferred onto a solid substrate by dipping, this substrate across the film with a constant speed. Repeated dipping on the same substrate leads to fabricate multilayers [6].

The study of the monolayer in the interface air/water is interesting too, and let us monitories the formation of the monolayer with a Brewster's Angle Microscope (BAM) that shows the behavior of the layer and all the phase transitions during the compression are observed.

2. EXPERIMENTAL

The synthesis and characterization of series *pnPEGMAN* (n= 2, 3, 4, and 6) were described in reference [6]. The general structure is shown in Fig. 1

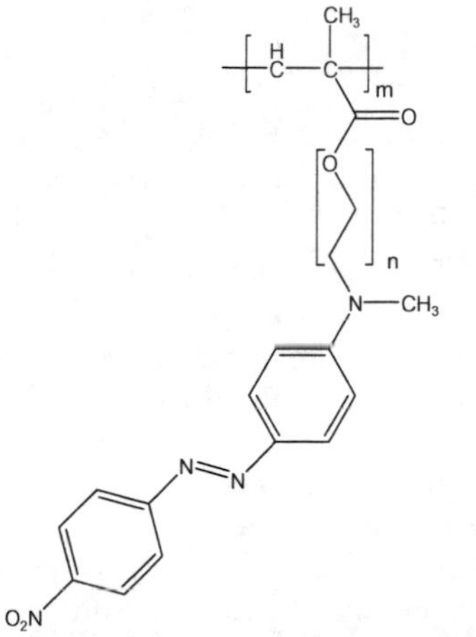

Figure 1: General structure of the azopolymers

Monolayer experiments and LB deposition were carried out at a subphase temperature of 22°C with KSV-5000 system 3 by Ltd Instruments. Ultra pure water supplied by a Milli-D coupled to a Milli-Q 185 purification system was used to prepare subphase.

Monolayers were obtained by spreading a solution of the azopolymer with a standard concentration of $1 mg \cdot mL^{-1}$ in chloroform (spectra grade, Aldrich) on a ultra pure water surface. During the isotherm experiments, monolayers were compressed at a barrier speed of $10 mm \cdot min^{-1}$. For the deposition onto substrates glass slides (Corning) were used pre-cleaned using sulfochromic mixture. The LB films were **Z** type in all cases.

UV-Visible measurements of LB films were obtained with a Varian CaryWin 100 spectrophotometer.

BAM images were obtained during compression monolayer in NIMA 622D2 through coupled to a miniBAM plus of Nanofilm Technology GmbH.

3. RESULTS

3.1 Langmuir monolayers formation and characterization

The surface pressure-area (Π-A) isotherms of the azocompounds are shown in Fig. 2.

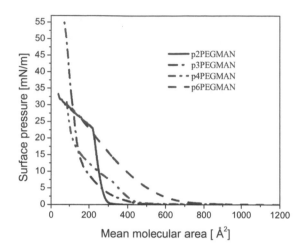

Figure 2: Surface pressure-mean molecular area isotherms of p*n*PEGMAN.

All the surface pressure-area isotherms are characterized by an initial long plateau region with a slow rise in different mean molecular areas, according to the size of the azopolymer, we observe a major area for azopolymer with ethylene glycol chain of 6, and the minor area is for the azopolymer with the ethylene glycol chain of 2.

For p6PEGMAN the slow rise begins at a mean molecular area about 680 Å² followed by a step rise in surface pressure with a limiting mean molecular area of about 520 Å². The extrapolation through the **x** axis of the portion of the curves corresponding at the solid state, around 15 mN·m⁻¹ yields and average surface area per molecule of 520 Å² for p6PEGMAN 380 Å² for p4PEGMAN, 180 Å² for p3PEGMAN and 340 Å² for p2PEGMAN. The low value for p3PEGMAN may be attributed to the formation of aggregate for the asymmetric length of the chain of ethylene glycol. The Langmuir films exhibited an excellent stability during the compression, we prove the reversibility of the monolayers under compression and decompression until surface pressure 15 mN·m⁻¹ and all the Langmuir films are stable and reversible.

The topography of the Langmuir films were obtained by BAM imaging at different surface pressures, where the film topography during the compression of a *pnPEGMAN* show a stable monolayer in all the cases.

The p2PEGMAN monolayer is illustrated in Fig. 3; We can see a high order of pre-compression in the film according the figure 3-a where the surface pressure is 0.2 mN·m⁻¹ the domains in Fig. 3-b were observed at 4 mN·m⁻¹ and 290 Å² when the compression has been started, with compression the domains were integrated to the film forming a stable and homogeneous monolayer until surface pressure at 33 mN·m⁻¹; after the mechanical collapse we observe fractures in the monolayer in the Fig. 3-f and the domains start again.

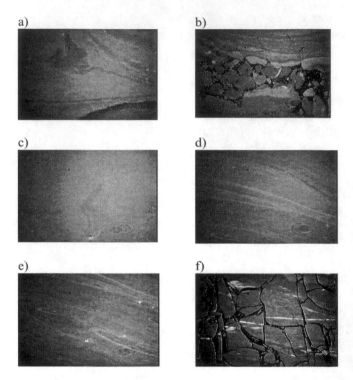

Figure 3: BAM images for p2PEGMAN at a)0 mN/m, b) 4 mN/m, c) 22 mN/m, d) 28mN/m, f) during decompression, collapse inducted.

3.2 Langmuir-Blodgett films fabrication and characterization.

Langmuir films were transferred by using LB technique onto glass slides in all the cases; the transfer ratio is 1± 0.1.

The LB films were examined by the UV-Visible spectroscopy; absorption spectra of p6PEGMAN from one to 20 layers are shown in Fig. 4, and the maximum absorption wavelength (λ) is 487 nm. Apparently no J-aggregation is present.

The absorbance at 479 nm was plotted against number of transferred layers and the relationship between absorbance and layer number was found to be linear with a linear correlation at 0.999.

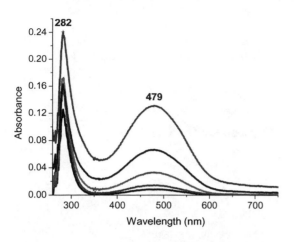

Figure 4: UV-Visible spectra of the LB Films of **p6PEGMAN** (from bottom to top: 1, 2, 5, 10 and 20 layers)

In the table 1 a resume of the maximums of absorption in UV-Visible are reported for all the azocompounds; and we can observe that the p4PEGMAN shows a red-shifted compared with the others azopolymers.

Azopolymer	Absorption LB films (nm)
p2PEGMAN	468
P3PEGMAN	469
P4PEGMAN	423
P6PEGMAN	479

Table 1: Resume of the maxims absorption of 20 layers from LB films and solution in chloroform

4. CONCLUSIONS

We have demonstrate that **pnPEGMAN** series forms molecular nanostructures with highly ordered and stability in monolayers; and the stability and homogeneity when are deposited onto solid substrates with the LB technique are excellent. We can conclude that the chain of ethylene glycol determine the mean molecular area and the area was decreased when the chain decreased too.

REFERENCES

[1]. Rau H. In: Rabek JK, editor. Photochemistry and photophysics, col. 2. Boca Raton, FL: CRC Press; 119, 1990.

[2]. Natansohn A, Rochon P. Chem Rev; 102:4139-4175, 2002.

[3]. Rivera E, Carreón-Castro M. P, Buendía I, Cedillo G, Dyes and Pigments; 68:217-226, 2006.

[4]. Rivera E, Carreón-Castro M. P, Rodríguez L, Cedillo G, Fomine S, Morales-Saavedra O, Dyes and Pigments; 74:396-406, 2007.

[5]. Oliveira Jr ON, He J-A, Zucolotto V, Balasubramanian S, Li L, H.S.N., et al. IN: Kumar J, Nalwa HS, editors. Handbook of polyelectrolytes and their applications. Los Angeles: American Scientific publishers; 1-37, 2002.

[6]. Oliveira Jr ON, Dos Santos Jr DS, Balogh DT, Zucolotto V, Mendoça CR, Adv Colloid Interface Sci, 116:179-192, 2005.

[7]. García-Tenorio T, Rivera E, Carreon-Castro MP. Morales-Saavedra O, Polymer 2008, submitted.

Acknowledgements:

The authors are grateful to DGAPA-UNAM (PAPIIT 118808 and 112203). They are pleased to acknowledge Fis. Luis Flores (FC-UNAM) and Martín Cruz (ICN-UNAM) for technical assistance.

Preparation and Characterization of Bi-2212 thin film using Pulsed Laser Deposition.

N. T. Mua [1], A.Sudaresan[2], C. N.R.Rao[2], C.R.Serrao[2], Shipra[2], T. D. Hien [1], and N. K. Man [1]

[1] International Training Institute for Materials Science (ITIMS),
Hanoi University of Technology,
No. 1, Dai Co Viet road, Hanoi, Vietnam

[2] Jawaharlal Nehru Centre for Advanced Scientific Research (JNCASR), Bangalore, India

Abstract

High-quality c-axis oriented Bi-2212 thin films have been grown ex situ on single crystal (100) surface of MgO substrates was produced by PLD. The properties of Bi-2212 thin films have been investigated. The films were characterized by SEM, FESEM, X-ray diffraction, resistivity, MVSH, and transport properties. The results show that these films are good quality with *c*-axis orientation and epitaxial growth, good electrical and magnetic properties .The critical current density J_c was found to be strongly dependent on the temperature and magnetic field. In optimized growth conditions the transport critical current density $Jc=6.3x10^6$ A/cm^2 in 0.5T magnetic field at 5 K, superconducting transition temperature $T_{c-onset}$ = 95K and T_{c-zero} =75K are obtained.

Keywords: *Bi-2212 thin film, pulsed laser deposition, superconductors.*

1. Introduction

High-temperature superconducting (HTS) thin films have a large critical current density Jc and small surface temperature resistance Rs in the microwave and millimeter wave region compared with that of normal conducting films. It can be used to design high-performance passive microwave devices, for example, filters, antennas, resonators, and delay lines [1-5]. The $Bi_2Sr_2Ca_{n-1}O_x$ system is thought to be one of the most promising high -Tc superconductor compounds use in applications ranging from power transmission cables to Josephson-junction -based electronic devices [10].

Pulsed laser deposition (PLD) or laser ablation is one of the most often used techniques for production of high-quality BSCCO thin films. The ablation process results in conservation of the target stoichiometry in the initially ablated cloud [6]. Ablated species from a target come into oxygen environment and transfer to the substrate. A relatively high oxygen background pressure can be used with this technique. This allows for the fabrication of oxide films with nearly perfect stoichiometry. Therefore, Bi-2212 thin films are very important for submillimeter wave devices. Several fabrication techniques of Bi-2212 films have been performed, such as pulse laser deposition (PLD), a molecular beam epitaxy (MBE) [7], a liquid phase epitaxy (LPE) [8], a metal chemical vapor deposition (MOCVD) [9]. The PLD method suitable for fabricating films with complex stoichiometry.

In this paper, we report the result of $Bi_2Sr_2Ca_1Cu_2O_{8+y}$ (BSCCO) thin films on MgO substrate prepared by the PLD method. Electrical properties such as the superconducting transition temperature (Tc) and the transport critical current density (Jc), surface morphology, crystal structure were investigated.

2. Experimental details

Precursor powder with Bi_2O_3, $SrCO_3$, $CaCO_3$ and CuO were prepared by solid satte reaction. The powders were calcined at 830^{0}C for 24 h in air. The calcinations were repeated three time at 830^{0}C for 24 h in order improve the homogeneity by intermittently grinding the powder. The calcined powder was compressed into a disk of 15 mm in diameter and 4mm. We have used the polyvinyl alcohol as a binder. The compacted powder disk was heat-treated at 845^0 C for 160 h.

The $Bi_2Sr_2Ca_1Cu_2O_{8+y}$ (BSCCO) thin films in this study were prepared by pulsed laser deposition (PLD).. Before deposition, the (100) MgO substrate was washed solution in an ultrasonic cleaner and subsequently by deionized water, acetone and ethanol and annealing at 800^0C. The excimer laser ($\lambda = 248$ nm) was operated at 300 mJ/pulse (the laser frequency is 3 Hz) for 30 min for each deposition.

The substrate is mounted on the heater in the cross- area of the plume caused by the beam and its temperature (Ts) was in the 730^0C. The ambient oxygen pressure was kept at 0.2 mbar. The films were in situ annealed at 680 ^{0}C for about 2 h and cooled down to room temperature for about 3 h in the ambient oxygen pressure of 800mbar. The thin films T84 , T92 and T93 were ex situ annealing at 845^0C for 3h, 2h and 1h. The films thickness was about 272 nm. The crystallographic texture of the film was studied by X-ray diffraction. SEM and FESEM.

Superconducting transition properties were measured by resistance, MVSH, and inductive measurements. The critical current density Jc of the film was calculated by model Bean:

$$J_c = \frac{60a|\Delta M|}{b(3a-b)} \; ; \text{ where a and b are the length and}$$

width (a>b, in cm) of the sample plane perpendicular the applied magnetic field. ΔM is the difference of the magnetization (emu/cm^3) between the field -up and field-down branches.

3. Results and discussion

Flatness of the thin films is importance in fabrication of layered structures and for research of basic properties of a few unit cell thick superconducting layers. SEM was used to investigate the influence of the energy of the laser pulses, the oxygen pressure and the temperature of the substrate on the surface morphology of the films. The SEM image of a film deposition on MgO at 0.2 mbar O_2, E = 300 mJ/pulse and substrate temperature Ts is about 730 ^{0}C is shown in figure 1.

Figure 1 shows a SEM image of the Bi-2212 film on MgO, with a smooth and dense microstructure.

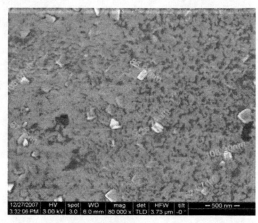

Figure 1. *SEM micrograph of the surface appearance of the BSCCO film on MgO substrate*

The presence of the (*00l*) peaks in the XRD patterns of the thin film corresponds to a well-crystallized single orthorhombic phase and *c*-axis-oriented film, as shown in figure 2. It shows that the BSCCO thin films of high structural quality have been epitaxially grown on single crystalline (100) MgO substrate. This reflects the perfection of the orientation of the different *c*-axis-oriented blocks of the film relative to the normal to the substrate.

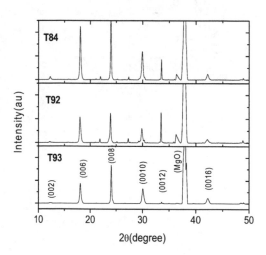

Figure 2. *XRD patterns of the Bi 2212 thin films on MgO substrate; T93 annealing at 845^0C/1h; T92 annealing at 845^0C/2h; T84 annealing at 845 ^{0}C/3h*

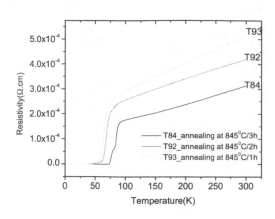

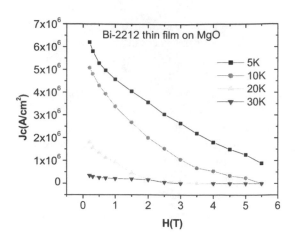

Figure 3. *The temperature dependent resistivity of Bi-2212 thin film on a MgO substrate*

Figure 4. *Current density of Bi-2212 thin film on MgO substrate depend on magnetic field at difference temperature*

The temperature dependent resistance, R(T), is shown in figure 3. The film exhibited good superconducting properties with $T_{c-onset}$ = 95K and T_{c-zero}=75K. When increasing time annealing, we find that value of $T_{c-onset}$ increasing.

The transport critical current density Jc is given by equation by model Bean: Jc = 60a ΔM / b(3a-b) = 6.3×10^6 A/cm^2 (where d=272 nm is the film thickness; and b=2.5 mm is the width of film, a= 2.8 mm is length of films).

This figure shows that the J_c values were found to increase initially with the decreasing of magnetic field. The results also indicate the enhanced H_{c1} values in these BSCCO samples. At magnetic field is 0.5T , J_c achieves maximum value is 6.5×10^6A/cm^2 at 5K in 0.5T

In the levitation effect of superconductors, two of the more important properties are critical current density J_c and the lower critical field H_{c1}. The repulsive force of sample depends on the value of J_c, and H_{c1} is the field of magnet. Because of J_c is achieved maximum at 0.5T, when the superconductor exhibits pinning force [9].

4. Conclusions

High quality epitaxial Bi-2212 thin films have been grown ex situ annealing by pulsed laser deposition (PLD) on single crystalline (100) MgO substrate. Our results show that the PLD method can be very useful in making Bi-2212 thin films with a smooth surface and good electrical, such as the transition temperature $T_{c-onset}$ = 95K and T_{c-zero}=75K, and critical current density Jc = 6.3×10^6 A/cm^2 in 0.5T magnetic field at 5 K, which are essential for the production of various superconducting devices with high performance.

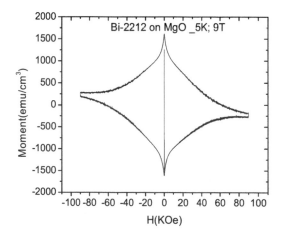

Figture 4. *Moment depend about magnetic field of thin film Bi-2212 on MgO substrate at 5K*

Acknowledgements

This work was supported by Jawaharlal Nehru centre for Advanced Scientific Research (JNCASR), Bangalore, India for providing supportive research environment for funding this work. One of the authors (N. T. Mua) would like to thank Dr. A.Sundaresan, Ph.D.Shipra, Ph.D.Claudy, Ph.D. Pranad, Ph.D Madhu and Ph.D Vengadesh in laboratory superconductivity and magnetism, Chemistry and Physics of Material Unit Bangalore, India for their help in prepare the thin films by PLD.

References

[1]. Mohan V. Jacob, Janina Mazierska and Michael Lorenz, Supercond. Sci. Technol. 16 (2003) 412–415.

[2]. N. D. Kataria, Mukul Misra, R. Pinto, Pramana-journal of physics 58 (2002) 1171–1177.

[3]. O. G. Vendik, I. B. Vendik and D. V. Kholodniak, Mater. Phys. Mech. 2 (2000) 15-24.

[4]. E. Moraitakisy, M. Anagnostouy, M. Pissasy, V. Psyharisy, D. Niarchosy and G. Stratakosz, Supercond. Sci. Technol. 11 (1998) 686–691.

[5]. E. Farber, J. P. Contour, G. Deutscher, Physica C 317–318 (1999) 550–553.

[6]. P. B. Mozhaev, F. Rönnung, P. V. Komissinskii, Z. G. Ivanov and G. A. Ovsyannikov, Physica C 336 (2000) 93-101.

[7]. H.Fujino, Y.Kasai, H.Ota,S.Migita, H. Yamamori, K. Matsumoto,S.Sakai, Physica C 362 (2001) 256

[8].T.Yasuda, T. Kawae, T.Yamashita, T.Uchiyama, I. Iguchi, M.Tonouchi, S.Takano, Physica C 378 (2002) 1265

[9]. K.Endo, H.Sato, K.Yamamoto, T.Mizukoshi, T.Yoshizawa, K.Abe, P.Badica, J.Itoh, K.Kajimura, H.Akoh, Physica C 372-376 (2002) 1075.

[10]. M.Nagao, M.Sato,H.Maeda, Appl.Phys.Lett,79 (2001) 2612

Welcome to Nanotech Germany

M. Rieland*, V. Müller**, F. Sicking*** and G. Bachmann***

*Federal Ministry of Education and Research, BMBF, Dept. 511, Nanomaterials, New Materials
53170 Bonn, Germany, martin.rieland@bmbf.bund.de
**International Bureau of the BMBF, DLR, Bonn, Germany, verena.mueller@dlr.de
***VDI Technologiezentrum GmbH, Duesseldorf, Germany, sicking@vdi.de, bachmann@vdi.de

ABSTRACT

Key technologies are "tickets" to the future. Nanotechnology is one of the most important key technologies. Nearly all branches of industry profit from the progress being made in nanotechnology research. And still, nanotechnology has a great wealth of innovative potential to exploit. Germany has set itself the goal of becoming a pacemaker for nanotechnological innovations. German researchers are right up front in third place in the international league table for patent applications. In 2008 and 2009, Germany is showcasing its nanotechnology performance internationally by the initiative "Welcome to Nanotech Germany - Research in Germany-Land of ideas". This initiative aims at stepping up cooperation between German and international research establishments and enterprises.

Keywords: nanotechnology, government funding, nano initiative, networks and infrastructure,

1 GENERAL DESCRIPTION OF R&D IN GERMANY

Germany is meeting the challenges of globalisation and international competition by investing in education and research. The Federal Government is determined to provide as much support as possible in this area – and has already been visibly successful. It's a fact: the High-Tech Strategy initiated by the Federal Ministry of Education and Research (BMBF) is having an impact. Targeted funding is to ensure that new impetus is given to turning research results into products, services and processes much more quickly, and thus creating new jobs. Up to 2009, altogether 15 billion euros of the budget have been earmarked for this investment programme in R&D. By 2010, state and private investment in research and development is projected to rise to three per cent of gross domestic product. Companies are counting on a growth rate of more than seven per cent for research and development in comparison with previous years. And the number of employees in companies involved in research and development has increased by 3.5 per cent. Cutting-edge research is not a prerogative of firms or the numerous universities in Germany; it also takes place at roughly 257 non-university research establishments, employing more than 70,000 staff. These establishments are incorporated in large organisations such as the Max Planck Society, the Helmholtz Association, the Leibniz Association and the Fraunhofer Society. This is where researchers find optimum conditions matched only at very few other institutions around the world. Which all goes to show: Germany puts strong efforts in R&D in order to meet future social and economic challenges.

2 GERMAN NANOTECH

One of the engines for future innovations is undoubtedly nanotechnology. Being a classic cross-cutting technology, nano developments build the starting point for progress in many economic sectors. The automotive industry, energy and environment, the health sector – in all these fields lessons learned from nanotechnology research set the trend for technological improvements. And the nano sector creates wealth: according to prognoses, a global market potential of up to a trillion euros in 2015 can be expected.

Self-cleaning windows, higher-performance tumour markers or miniature hard discs which can store just as much data as traditional hard discs – nanotechnology opens up manifold new opportunities. Indeed, our everyday lives have become unimaginable without nanotechnology.

Whether in sun creams, drinking water treatment plants or automotive paintwork, the tiny particles unfold their enormous potential with just the same degree of efficiency. Nearly all branches of industry profit from developments in nano research. Possible applications can be found in the optics sector, as well as in mechanical engineering, medical technology, and the chemical, electronic, textile and construction industries.

In the context of the "Nano Initiative – Action Plan 2010", the Federal Government has introduced a national funding concept spanning all policy areas. Part of the High-Tech Strategy for Germany, the Nano Initiative contributes to opening up new leading markets, networking business and science, and creating leeway for researchers and entrepreneurs.

Our main target is to expedite the transformation of ideas into new, marketable products, processes and services.

Germany must remain amongst the world's most research- and innovation-friendly nations. This holds true especiallay for the field of nanotechnology. By means of specifically created funding programmes the Federal Government is helping to establish Germany as leading nano location. 350 million euros public funding were invested in 2007 in R&D for this upcoming technology field. Only the USA and Japan invest more state funding in nanotechnology. Today, more than 650 companies in Germany with more than 50,000 employees are already developing, using or selling their nano products.

Today, Germany is right at the forefront of international nanoscience, not at least due to the fact, that the Federal Government realised the potential of this relatively young field of research very early on and systematically funded its development. For example, the amount of funds earmarked for nanotechnology by BMBF has increased more than five-fold since 1998.. In parallel, centres of competence have been established as supportive infrastructure. There- fore, the German research landscape offers students, aca- demics and companies from home and abroad the best possible prerequisites for success.

2.1 Nano-Life in Germany

Anyone, who is interested in doing nanotechnology re- search in Germany is highly welcome. Students and experts from abroad will profit not only from the most generous financial support for nanotechnology in Europe, but also enjoys a high standard of living. In the "country of poets and thinkers" the great philosophical and literary tradition continues to this day. The world's largest book fair takes place in Frankfurt am Main, with about 1,800 publishing houses presenting their works every year. And Germany also has a vast music and theatre scene. 135 publicly funded orchestras perform on 80 stages in the land of Bach and Beethoven. Cultural visitors come to Kassel for the "documenta", one of the most important exhibitions of international modern art worldwide, to Bayreuth for the "Richard-Wagner-Festspiele", or to Berlin for the Interna- tional Film Festival, the "Berlinale".

German countryside is surprisingly varied. There are 15 national parks and 90 natural preservation areas to investi- gate. Holidays in Germany might mean mountaineering in the Alps, walking in the legendary Black Forest or along the Wine Trail through the Palatinate or Baden, or simply relaxing on the white beaches of the Baltic or North Sea. And sport is a major feature, too. No other country has as many sports clubs as Germany – roughly 90,000, all told. In the big cities, major sporting events like inline skating or marathons take place regularly.

The society researchers experience in Germany is mul- ticultural and cosmopolitan: more than seven million for- eigners originating from all over the world live here. Dues- seldorf at the Rhine River is the European centre Japanese companies have chosen – there are Japanese restaurants, book shops and schools. In Hamburg, the Portuguese feel

at home, and Berlin is the favourite adopted home town for Turkish immigrants in Germany. An estimated 25,000 Chinese study at German universities across the country, while amongst Polish students, Viadrina European Univer- sity in Frankfurt/Oder is particularly popular. And more than 22,000 companies with foreign participation have successfully established themselves in Germany.

By comparison with the rest of the world, Germany, with its stable, democratic, constitutional state, is one of the safest countries of all – in which to live, to invest and, of course, to do research.

2.2 The Federal Government´s Nano-Initiative

Being a cross-sectional technology, the nano sector en- joys a special status in this context: on the basis of the "Nano-Initiative –Action Plan 2010", the Federal Govern- ment wants, above all, to improve the interface between basic research and rapid implementation. Comprehensive measures are on the drawing board which should help to improve the exploitation of nanotechnology in Germany – and meet the new challenges. Eight ministries, each with its own nano representative, have put together the package of actions. The sooner a good idea makes the transition from laboratory scale to a ground-breaking product the better.

There's lots to be done: apart from expediting the proc- ess from idea to product, one major task is to introduce nanotechnology to more new sectors and companies. Branch-level industrial dialogues between representatives from politics, the economy and associations help to find areas of application for economic areas not covered so far. In order to introduce nanotechnology to new fields of ap- plications special funding measures have been imple- mented. The action plan promotes so-called leading inno- vations which the Federal Government hopes will generate a high degree of growth and employment. For example, in the field of lighting technology, these include the "Nano- Lux" project and the OLED Initiative: whereas NanoLux is essentially preparing the way for energy-efficient lightemit- ting devices in the automotive industry, the OLED Initia- tive wants to create the technological basis for organic light-emitting devices as a cheap, large-scale lighting op- tion.

Small and medium-sized enterprises in particular often prove to be especially innovation friendly. They receive support from the action plan by means of funding and structural measures, such as the "NanoChance" programme for start-ups. And the Federal Government is also pooling its activities with regard to potential risks. A steering group under the direction of the Federal Ministry for the Envi- ronment is assessing the chances and risks involved in dealing with nanomaterials. In the context of this national nanodialogue, experts from industry and society evaluate dangers as well as perspectives and also discuss tomor- row's research needs.

Alongside excellent basic research, innovation-friendly industry, and well trained young personnel, Germany can rely on its internationally oriented technology networks and centres of competence – for yet more new, cutting-edge achievements "made in Germany".

2.3 Talented People and Networking

Expertise is the key to economic success. And innovation often begins with shrewd promotion of young talents. Targeted support already kicks in at school where nano topics and experimental learning arouse the curiosity of even the very young. In addition to this, the Federal Government's measures include science competitions in the nano field as well as information aimed at young people, telling them about promising new fields of employment and new educational opportunities. These are coordinated with the needs of industry. Fairs, workshops and surveys also help to identify new qualification trends and integrate them into tailored educational opportunities in higher and vocational education.

Success is usually a question of good teamwork and optimum communications. Thus, as a further measure in addition to promoting research, industry and young people, the Federal Government is pooling specialised knowledge in nanotechnology and also accelerating innovative processes by strategic information exchange. A network of so-called centres of competence, currently comprising nearly 500 participants, links universities, research institutes, companies, financial service providers, consultants and associations. These networks work in different subject areas both at the national level and in regional clusters. Apart from trends and current developments, these centres also focus on training and further education.

Currently, there are several centres of competence in the field of nanotechnology in Germany working along the value-added chain. In addition, a number of clusters have established themselves which bundle their activities nationally, regionally or even locally. A number of largely independent university networks complement the excellence teams at the local level. Examples of the largest networks are:

CC-NanoChem / NanoBioNet – Networks for Chemical Nanotechnology / Nanobiotechnology

The CC-NanoChem centre of excellence started up in 1998 at the Institute for New Materials in Saarbruecken, specialising in chemical nanotechnology. This is one of the decisive key areas needed for new materials and products as well as for new manufacturing and processing technologies. New processes make it possible to shape materials precisely at the level of atoms and molecules: the physical properties of materials can be tailored for change, for example. The NanoBioNet network followed in 2002. It focuses on nanobiotechnology – the interface between nanostructure science and biology. Both networks offer their members a forum for exchange between universities, research centres, small, medium-sized and large enterprises, as well as consultants and venture capitalists. And there is

expert assistance for start-ups, too. Even school students can come here to experience the things nanotechnology can do. Experimental kits and nano nights familiarise young people with this future technology.

Nanotechnology Centre of Competence "Ultrathin Functional Films"

Ultrathin films are often a key element in nanotechnology. Their field of use ranges from microelectronics and optics via medicine to wear protection. In Saxony, 51 companies, ten university institutes, 22 research establishments and five associations have bundled their know-how and got together to form a network. They have one common goal: to exploit industrial application potential. The Fraunhofer Institute for Material and Beam Technology (IWS) Dresden coordinates this network, which was chosen as the National Centre of Competence for the field of ultrathin functional films by the Federal Ministry of Research.

ENNaB – Excellence Network NanoBioTechnology

Gifted junior researchers and industrial enterprises working in the nano- and biotechnology fields are the groups targeted by ENNaB – Excellence Network NanoBioTechnology - at the university in Munich. The network understands its role as the link between university and industry – between basic research and industrial application. Its aims include training competent research and management personnel for universities and industry, and thus creating a solid basis for the commercial implementation of new, innovative applications in the field of nanobiotechnology.

HanseNanoTec Competence Centre

The HanseNanoTec competence centre is the point of contact for all researchers, entrepreneurs, financial service providers and funding organisations in the Hamburg region working in the field of nanotechnology. It bundles the Hanseatic city's expertise in this sector and initiates and promotes cooperation with supraregional and international partners. The objective is to optimise efficient acquisition of basic insights and swift transformation of nanotechnology knowledge into products, manufacturing processes and services.

UPOB – Ultraprecise Surface Figuring

The Brunswick nanotechnology competence centre is a cluster amalgamating production technologies, machines and machine components, metrology, sensor technology and materials. As the respective methods used differ significantly, the network is divided up into four core areas: mechanical/chemical processing, ion beam and plasma processing, optical processing and related topics, and characterisation of surfaces. This is a perfect way of bundling capabilities and presenting them to the public.

NSTI-Nanotech 2008, www.nsti.org, ISBN 978-1-4200-8503-7 Vol. 1

CCN – Competence Centre for Nanoanalytics

The main tasks of this network at the Centre for Nanotechnology in Muenster include the continued development of nanoanalytic measuring methods, their adaptation to technological demands and their standardisation. Special probe and raster scan techniques are amongst the network's methodological focus areas.

NanoMat

Nanotechnology materials are the focus of NanoMat, coordinated by the research centre in Karlsruhe. They coordinate research projects on "the synthesis and study of nanostructured metals and ceramics and the functions resulting from their nanoscale nature". Three research centres of the Helmholtz Association, ten universities with natural science and engineering departments, a Max Planck institute, a Leibniz institute, a Polish Academy of Sciences institute, three Fraunhofer institutes, the Society for Chemical Engineering and Biotechnology (DECHEMA), and four large corporate groups participate in the network.

NanOp – Competence Centre for the Application of Nanostructures in Optoelectronics

The NanOp centre of competence in Berlin, a network comprising universities, research establishments for applied and basic research, industrial enterprises, and banks as well as venture capital firms, is engaged in research and development in nanotechnologies for application in new and revolutionary products based on nanooptoelectronics. Close cooperation between all the partners involved in the network and fast transfer of know-how produce synergy effects.

3 ACTING RESPONSIBLE

Progress opens up enormous opportunities but might also incorporate inherent dangers. It is important not to play down these risks without actually exaggerating them. Germany puts strong efforts to investigate, to assess and to monitor possible toxic potential in nanomaterials. Results are discussed with the public in an open discourse.

Within the "Nano Initiative – Action Plan 2010" BMBF has started a funding initiative called NanoCare, which addresses this issue in a very efficient and exemplary manner. Here, the Federal Government, academia and industry are developing new, standardised procedures to improve assessment of the danger potential of nanoparticles. The public is actively involved.. The NanoCare cluster is combining the projects NanoCare, INOS and TRACER:

- The NanoCare project focuses on examining primary particles and agglomerates and the way they behave in biological media and systems. State-of-the-art analytical methods from the most diverse institutes are used to characterise the nanoparticles.

- The INOS project is geared to developing methods based on in vitro testing to evaluate the danger potential of engineered nanoparticles. The hazard analysis is based on a comprehensive investigation of the behaviour of nanoparticles in various cell culture media and the changes they undergo.

- Carbon nanotubes (CNT) and carbon nanofibres (CNF) are already considered to be some of the 21st century's key materials. In central fields of technology, such as chemistry, the automotive industry, aeronautics and aerospace, they are achieving technological breakthroughs. Emerging industrial manufacture of carbon nanotubes is likely to open up other fields of technology and pave the way for a multitude of mass applications. The objective of the TRACER project is to draw up recommendations for handling potential end products during manufacture, processing and use.

Dealing so openly with the opportunities and risks inherent in these miniscule particles is creating a positive climate in Germany which favours nanotechnological developments. On the basis of such social approval, the huge potential of this cross-discipline technology can be exploited to the full – which puts wind in the sails of researchers in their everyday work. And this is necessary, because nanotechnology is needed to face the new challenges of the 21st century. Climate change is a prominent example. There is a strong and further increasing demand for new energy technologies in order to meet the CO_2-emission reduction goals. Innovations in fields like energy production, energy efficiency or energy storing heavily rely on nanotechnology. Furthermore, new, workable solutions also have to be found for the continuing pollution of the earth by industrial society's waste products.

Thus, nanotechnology is one of the most promising pacemakers for the future. From 2025, at the latest, the vast majority of all significant new developments will be related to this field of technology. There are not a few people who think the tiny particles with the big impact will point the way out of today's technological cul-de-sac. But, at the same time, there is the need to make the risks associated with the particles calculable. This is the only way of ensuring that the future dividends inherent in the nanocosmos will pay off one day.

WEB LINKS

www.research-in-germany.de/nano
www.bmbf.de/en/nanotechnologie

Highlights of the R&D activities in the nanotechnology area performed by the Italian Interuniversity Consortium on Materials Science and Technology (INSTM)

T. Valente and D. Gatteschi

INSTM Consortium, Florence, Italy
Sapienza University of Rome & University of Florence, tvalente@instm.it

ABSTRACT

The Italian Interuniversity Consortium on Material Science and Technology (INSTM) is actually participated by 44 Italian Universities thus realizing an integrated scientific network among the community of chemists and engineers. INSTM is involved in many strategic areas for innovative devices development, such as: molecular materials for electronics and photonics; polymeric, composite, metallic and ceramic materials for structural and/or functional applications; nano-materials; nano-bio-materials and protective coatings. Research and development activities are performed through projects funded by public and private Italian institutions and International institutions, such as the European Commission. In the recent years much effort has been put to develop fundamental and applied knowledge in the area of nanotechnology and nanostructured materials. In this frame INSTM is actually coordinating three European Networks of Excellence and is participating in different Integrated Projects. A summary related to some of the on-going R&D activities is reported.

Keywords: magnetic materials, polymer nanocomposites, catalytic materials, ceramics

1 INTRODUCTION

Among the most relevant INSTM actions in the framework of nanotechnology and nanostructured materials they can be evidenced:

- activities related to molecular approach to nanomagnets and multifunctional materials, performed by partners grouped in the network of excellence MAGMANet, funded by the EU in the 6th Framework programmme;

- activities related to nanostructured and functionally polymer-based materials and nanocomposite, performed by partners grouped in the network of excellence Nanofunpoly, funded by the EU in the 6th Framework programme;

- activities related to integrated design of catalytic nanomaterials for a sustainable production, performed by partners grouped in the network of excellence Idecat, funded by the EU in the 6th Framework programme;

- activities related to structural ceramic nanocomposites for top-end functional applications, performed by partners

inside the integrated project Nanoker, funded by the EU in the 6th Framework programme.

An overview of the main objectives for the above mentioned projects will be reported, together with some relevant obtained results. The aim is to give a contribution to the state of the art knowledge on nanotechnology, to define future scenarios and directions of R&D strategies.

2 NANOMAGNETS AND MULTIFUNCTIONAL MATERIALS

MAGMANet is a Network of Excellence which connects more than 20 nodes from 10 different European countries (Italy, France, Germany, Spain, Portugal, Switzerland, The Netherlands, UK, Poland and Romania). Leading institution of the network is INSTM. The goal is that of integrating the participating Laboratories in order to make them more competitive in the field of "Molecular approach to Nanomagnets and Multifunctional Materials", it is a recent research area which is well inserted in the frame of molecular materials. Information can be obtained at the following address: http://www.magmanet-eu.net. The main research topics are collected in Fig. 1.

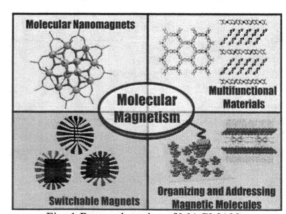

Fig. 1 Research topics of MAGMANet

2.1 Some highlights of R&D activities [1-2]

One of the key features of MAGMANet is the development of the so called Single Molecule Magnets, SMM, and Single Chain Magnets, SCM. The former are individual molecules which at low temperature behave as tiny magnets, and the latter are polymers whose magnetization relaxes slowly at low temperature. One of the possible uses, beyond advancement of basic science, is as memory elements. In order to do this it is necessary to organize the SMM on suitable surfaces. Encouraging results have been obtained using Langmuir Blodgett films, self assembled monolayers and dispersion in polymers. The relative efficiencies of the three techniques have been evaluated using magneto-optics techniques, which have shown how the environment dramatically influences the magnetic properties of the organized molecules. Magneto-optical techniques have also been used on metal alloys nanoparticles, showing that the magnetic properties can be tuner by interaction with visible light opening exciting perspectives. Other interests are addressed to the development of magnetic nanoparticles using molecular techniques. For instance it has been reported the use of undecanoic acid for the formation of monolayers of Cobalt Ferrite on silicon.

3 NANOCOMPOSITES AND NANOSTRUCTURED POLYMERS

The main objective of NANOFUN-POLY (http://www.nanofun-poly.com) is to generate a Network of Excellence designed to become the European organization on Multifunctional Nanostructured Polymers and Nanocomposite Materials. This objective will be reached through a trans-disciplinary partnership of 120 scientists combining excellence in different scientific areas, where the synergy of international excellence and multidisciplinary approaches will lead to develop and spread knowledge in innovative functional and structural polymer-based nanomaterials and their sustainable technologies. Applications that will benefit from NANOFUN-POLY concern strategic industrial sectors which can be competitive only by using advanced technologies: optoelectronics and telecommunications, packaging, agriculture, building construction, automotive and aerospace, etc. The NANOFUN-POLY Consortium consists of 29 partners (12 core partners, 17 Satellite Partners). Leading institution of the network is INSTM.

3.1 Some highlights of R&D activities: carbon nanotubes and nanocomposites [3-4]

Electrophoretically deposited single-walled carbon nanotube (SWCNT) films on a transparent conducting surface are used as electrodes for the electrodeposition of a π-conjugated polymer formed by the oxidative coupling of fluorine units. This method provides a uniform coverage of the conducting surface with respect to SWCNTs chemically assembled on a gold substrate.
The conductivity of the PF/C-SWCNTs/FTO sample increases and becomes ohmic with the thickness of the SWCNTs (Fig. 2). This result is consistent with a better percolation of the deposited carbon nanotubes.

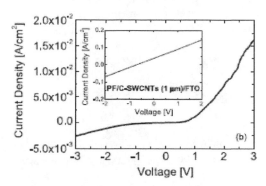

Figure 2. I–V characteristic of PF/C-SWCNTs/FTO (inset for PF/C-SWCNTs (≈1μm)/FTO sample)

Such enhanced conductivity of polymeric chains can be attributed to the entrapped nanotubes and nanotube bridging (Fig.3). The one-dimensional structure of CNTs may also induce and promote oriented polymerization, hence yielding an enhanced supramolecular order and higher conductivity. By combining the attractive properties of SWCNTs and polyfluorene, these nanocomposites open additional opportunities to achieve electrical contacts in nano-to micro-devices.

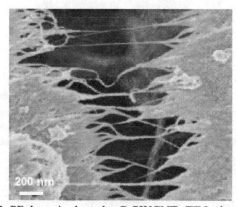

Figure 3. PF deposited on the C-SWCNTs/FTO electrode.

In the area of nanocomposite fabrication, a methodology showing how plasma fluorinated single-walled carbon nanotubes (SWNTs) reacted with a primary aliphatic amine (i.e.butylamine, BAM) hardener (BAM-SWNTs) was investigated, in order to prepare an integrated nanotube composite material. The grafting of butylamine onto CF4 plasma treated SWNTs was used to obtain a cross-linked epoxy nanocomposite. This methodology allowed to obtain a BAM-SWNTs/epoxy nanocomposite (Fig. 4) and experimental results showed that BAM-SWNTs acted as a catalyst, with interactions of cross-linking between the epoxy and amino-functionalized SWNTs during the cure reaction. Amino functionalized nanotubes had a better dispersion in the polymer matrix and the obtained nanocomposites presented an improvement in mechanical strength with respect to those prepared with un-functionalized SWNTs.

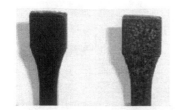

BAM-SWNTs/epoxy   BAM-u-SWNTs/epoxy   (a)

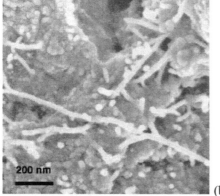

200 nm

(b)

Fig.4. (a) BAM-SWNTs/epoxy and BAM-u-SWNTs/epoxy nanocomposites; (b) image of BAM-SWNTs/epoxy nanocomposites

4   CATALYTIC NANOMATERIALS

The development of high-performance and conceptually innovative catalytic nanomaterials is of high impact for industry and for Europe sustainable future. In Europe, several excellent research teams exist in catalysis. IDECAT Network of Excellence (http://idecat.unime.it) aims are related to the creation of a coherent framework of research, know-how and training between the various catalysis communities (heterogeneous, homogeneous, bio-catalysis) with the objective of achieving an integration between the main European Institutions in the area. IDECAT integrates into a more general strategy of restructuring/reshaping the catalysis research in Europe and focuses its research actions on synthesis and mastering of nano-objects, the materials of the future for catalysis, integrating the concepts common also to other nanotechnologies, bridging the gap between theory and modeling, surface science, and kinetic\applied catalysis as well as between heterogeneous, homogeneous and biocatalytic approaches, integrated design of catalytic nanomaterials. The structure of IDECAT is based on a group of 37 laboratories from 17 Institutions, gathering over 500 researchers with a broad multidisciplinary expertise covering most of the aspects of catalysis. Leading institution of the network is INSTM.

4.1   Some highlights of R&D activities: catalytic membranes [5-6]

The use of layered materials (layered perovskite, anionic clays, pillared clays) in catalytic reactions, with their structure consisting of stacked sheets, represent an interesting opportunity for developing new materials with a tailored nanodesign, controlled accessibility to the site sand properties, tuneable pore size and volume, and high surface area. The evaluation of the scientific literature over the period 2000–2006 (English written), evidences that nearly 20,000 papers have been published on clays, layered perovskites (LP) pillared clays (PILC) and hydrotalcite (HT) materials, of which about 85% were dealing on catalysis. From an additional analysis it can be concluded that the LP and PILC materials are still mainly at the lab scale development stage, while HT and especially clays find a broad range of applications. They can offer possibility to develop new processes for environmental protection, selective oxidation and refinery/biorefinery and thus are subject of R&D investigations.

5   NANOCERAMICS AND n-COATINGS

The main objective of the Integrated Project NanoKer [7] (http://www.nanoker-society.org) is to find ceramic material solutions which allow the industrial application of knowledge-based nanoceramics and nanocomposites for top-end functional and structural applications. The industrial exploitation of nanostructured ceramics rely to the successful consolidation of these materials which preserve their nanostructure. Traditional processing techniques still show strong limitations in retaining conventional nanoparticles as the starting materials. Therefore, in addition to new material solutions, the full added-value chain of ceramic manufacturing has to be revisited. The technological objectives and expected breakthroughs of IP NANOKER will consist of new multifunctional materials with outstanding hardness, fracture resistance and fracture toughness operating in chemically and physically aggressive environments, and of new multifunctional materials processed into knowledge-based, industrially applicable nanoceramics and nanocomposites with added multifunctionality, e.g. biocompatible functions and very long lifetime, optical properties, tribochemical functions and excellent electrical conductivity, nanostructured coatings with tribological and barrier functions, etc.

IP Nanoker is carrying out research activities in many strategic fields of industrial application, such as Hip, knee and dental implants with life spans superior to the actual ones; new-concept bone substitutes; radiation windows for satellite guidance; satellite mirrors with high stability and reduce surface roughness; polycrystalline lasers of high efficiency; components and nanostructured coatings for engine in aeronautics; conductive nanoceramics to be machined by EDM technologies; metal-ceramic materials of extreme hardness for cutting tools and finally high-creep resistant ceramic nanocomposites.

22 European Partners are involved in this project and 5 of them are Italian. INSTM is a partner of the project.

5.1 Ceramic oxide-oxide nanocomposites development [8]

The research aim is to develop micro-nano and nano-nano alumina-based composites, in particular yttrium-aluminium garnet (YAG)–alumina nanocomposites which are promising materials for optical, electronic and structural applications. INSTM research units are involved in the wet chemical synthesis of nanocomposites powders whose full densification as a nanoscaled material is pursued by using conventional sintering routes coupled to particular mechanical and/or thermal powder pre-treatments. One of the main results up to now achieved is the production of an alumina-YAG (50 vol%) nano-nanocomposite material by coupling an optimised powder pre-treatment to an extensive mechanical activation, performed by a conventional wet milling. The set-up of this procedure allowed the Consortium INSTM to deposit an Italian Patent and now an European Patent is pending. After that powder compacts fully sintered (>98%) at very low sintering temperature, ranging between 1370 and 1420°C. As a consequence, a very fine microstructure was obtained in which α-alumina and YAG grains were lower than 300 nm in size (Fig.5).

Fig.5 - TEM micrographs of a 50 vol% alumina-YAG composite, after natural sintering.

5.2 New scaffolds for bone substitution based on hydroxyapatite (HAp) nanopowders [9]

This activity is aimed at finding possible solutions for the increasing need for advanced ceramic scaffolds for bone substitution, showing improved mechanical performances and biological properties in terms of cell adhesion and bone ingrowth. One of the major objectives is to achieve a suitable control of the porosity features in terms of total porosity, pore size distribution, pore distribution from the surface to the bulk of the component from the nano up to the micro-scale. To reach this goal, first of all, it is necessary to produce and characterize HAp nanopowders and then to control their sintering behavior in order to

control the final nano-microstructural features. Hydroxyapatite (HAp) nanopowders were synthesised following different precipitation routes and the pivotal role of the type of the preparation process on the thermal stability of HAp powders as well as on their sintering behavior and final fired microstructure (Fig.6) was clearly pointed out during the preliminary study through nanostructured scaffolds.

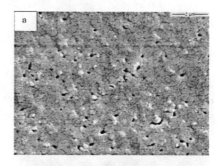

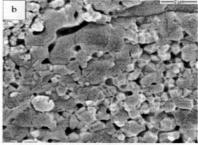

Fig. 6 – HAp materials prepared by precipitation (a) from calcium hydroxide and phosphoric acid solution and (b) from calcium nitrate and diammonium hydrogen phosphate solutions, and then sintered at 1050°C for 3 hrs.

5.3 Synthesis of pure YAG nanopowders [10]

Although a lot of investigations on pure alumina preparation have underlined the strong influence of some process parameters (temperature, pH, rate of addition and nature of the precipitation agent) on the phase evolution and on the properties of the synthesized alumina, when wet chemical syntheses and particularly precipitation are used, similar studies on pure yttrium-aluminium garnet (YAG) were lacking in the literature. YAG powders were therefore synthesized using a reverse-strike precipitation, by adding an aqueous solution of yttrium and aluminium chlorides to dilute ammonia while monitoring the pH to a constant value of 9. After precipitation, the gelly product was washed several times; precipitation and washing procedures were performed at three different temperatures, namely at 5, 25 and 60°C. After drying, the powders were calcined at different temperatures and times. Phase evolution was investigated by X-ray analysis; the evolution of crystallites formation and growth as a function of the temperature was followed by TEM observations (Fig. 7).

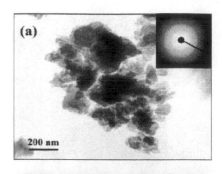

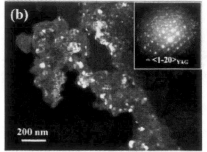

Fig. 7 - TEM of YAG powder synthesized at 25°C: (a) dried at 60°C, (b) pre-treated at 950°C

From this investigation it was possible to demonstrate a relevant influence of the co-precipitation temperature on the phases appearance, crystallization path and final homogeneity of these powders.

5.4 Coatings by nanostructured powders [11]

Three main deposition methods are actually under investigation: high velocity oxy-fuel (HVOF), suspension plasma spraying (SPS) and air plasma spraying (APS), for fabrication of wear and/or high temperature resistant coating. In the case of APS the materials under investigation are Al2O3/TiO2/ZrO2, Y2O3/ZrO2, CaO/ZrO2/SiO2, Cr2O3. With respect to coatings fabricated with the same methodology but with conventional microstructured powders improvements in terms of microhardness, fracture toughness, wear resistance and thermal cycling resistance have been observed. An example of coating microstructure is reported in Fig. 8.

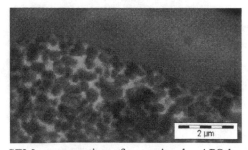

Fig. 8 – SEM cross section of a coating by APS based on Al2O3/TiO2/ZrO2 powders.

Even in the case of WC-Co HVOF coatings, improvements in terms of fracture toughness, microharndess and wear resistance have been observed. In both cases, a strict control of spraying temperature distribution, injection properties as well as jet properties is required in order to retain nanostructured areas inside the coatings. For these reasons an off-line modeling procedure of the jet and of the particle-jet interactions has been developed to address processing parameters optimization.

6 CONCLUSIONS

INSTM is actually involved in different research topics dealing with nanotechnology and nanostructured materials. National, European and International collaborations has been established not only on the scientific and/or academic side, but with the participations of industries and potential end-users. Results obtained up to now are relevant in terms of new generated knowledge. The transfer phase is just at the beginning and will be the area to which efforts will be finalized in the near future.

References
[1] C. Altavilla, E. Ciliberto, A. Aiello, C. Sangregorio, D. Gatteschi, J. Chem. Mater, 19, 24, 5890-5895 (2007).

[2] L. Bogani, L. Cavigli, M. Gurioli, R.L. Novak, M. Mannini, A. Caneschi, F. Pineider, R. Sessoli, M. Clemente-León, E. Coronado, A. Cornia, D. Gatteschi, Adv. Mat, 19, 22, 3906-3911 (2007)

[3] L. Valentini, D. Puglia, F. Carniato, E. Boccaleri, L. Marchese, J.M. Kenny, Comp. Sci. Tech., 68, 1008-1014 (2008)

[4] L. Valentini, F. Mengoni, L. Mattiello, J.M. Kenny, Nanotechnology, 18, 115502 (5pp) (2007)

[5] S. Abate, S. Perathoner, C. Genovese, G. Centi, Desalination 200, 760-761 (2006)

[6] G. Centi, S. Perathoner, Micr. Mes. Mater. 107, 3-15 (2008)

[7] NANOTEC IT Newsletter, No.8, June 2007, pp. 17-20

[8] P. Palmero, L. Montanaro, Advances in Science and Technology, vol. 45, pp. 1696-1703, Trans Tech Publications, Switzerland (2006)

[9] A. Bianco, I. Cicciotti, M. Lombardi, L. Montanaro, G. Gusmano, J. Therm. Anal. Calor. 88, 237-243 (2007).

[10] P. Palmero, A. Simone, C. Esnouf, G. Fantozzi, L. Montanaro, J. Eur. Cer. Soc. 26, 941–947 (2006)

[11] C. Bartuli, T. Valente, F. Casadei, M. Tului, J of Materials: Design and Application, Part L, Vol. 221, 175-185 (2007).

The importance of R&D alliances to sustain a nano-biotechnology ecosystem

J. Wauters

Flanders Investment and Trade, New York, NY, USA, jan.wauters@fitagency.com
Flanders Nano Bio Alliance, Larchmont, NY, USA, jan.wauters@flandersnanobio.org

ABSTRACT

The breakthroughs in nanoelectronics, nanotechnology and biotechnology are resulting in a growing convergence of nano with bio, as they both 'work' at the nanoscale. Nanobiotechnology will push biotech and nanotech R&D centers and companies from all across the spectrum to work more closely together. Consequently, the importance of regional, state, national and global initiatives cannot be overestimated. Regional cluster organizations are well suited to establish this as they are the basic drivers behind local and global ecosystems. This paper highlights international trends and initiatives in developing a nanobio ecosystem, from scientific research down to marketable products. The key issue at stake will be the discrepancy between different business models used in nanotech/nanoelectronics industries versus biotech and pharma industries.

Keywords: nanotechnology, biotechnology, nanobiotechnology, business model, convergence

1 INTRODUCTION

Biotechnology is without doubt poised to become a crucial technology driver in the 21st century. The fundamental knowledge of human – and other - life forms downto its fundamental building blocks, and the functions these building blocks perform with respect to health and comfort, is starting to revolutionize our human evolution more profoundly than ever. Today, bioscience breakthrough after breakthrough are made everywhere in the world. Started as one of the first technologies ever discovered and developed, think e.g. of the brewing process of beer or the fermentation process of cheese, "modern" biotechnology took off with the birth of genetic engineering. The discovery of the double helix structured DNA by Watson and Crick in 1953 marked the beginning of a new era in science and technology. This discovery created a vibrant biotechnology industry, beginning with the birth of the leading biotech company Genentech after the development of the recombinant DNA technique. Many companies have followed since then, often spun out from research institutes and large corporations. Today over 5,000 biotech companies exist worldwide. Their applications range from new biology-based medicines over agriculture to industrial biotech such as biofuels.

At the same time, the increased understanding of matter at the nanometer scale has led to a virtually infinite range of new nanotechnologies that can be applied in a wide range of applications. By manipulating matter at the atomic or molecular level, physical properties of materials can be altered to desired levels, and nano-scaled devices can be manufactured. Although many new nanotechnologies have been developed so far, we are only witnessing the very start of a nanotechnology era.

2 CONVERGENCE

Interestingly, nanotechnologies operate on the same scale as many biotechnologies, since biotech essentially deals with molecular structures which have nano-size building blocks. Consequently, scientists have started to investigate the use of nanotechnologies in biotechnology. As an example, DNA is a long polymer molecule, consisting of millions of repeating units, called nucleotides. One nucleotide is less than 1 nm long and the diameter of DNA is only 2 nm wide. Proteins, the engines in living organisms, measure between 1 and 20 nm in size, arranged in 3 dimensions.

Taking into account that the building blocks of living organisms are in the nano dimensions, it was inevitable that nanotechnology would be applied in biotechnology, hence the birth of "nanobiotechnology". More and more nanotechnology domains, from nanomaterials over nanoparticles and nanodevices to nano-imaging techniques, are being applied to biotechnology to understand biological processes fundamentally. In this way, scientists hope to improve medical diagnosis and disease treatments. Nanobiotechnologies hold the promise to significantly improve the discovery of drugs, to speed up drug development and increase their effectiveness.

3 ECOSYSTEM

The success of a high-tech industry in a country or region depends on many factors. When comparing different technology clusters worldwide, it becomes clear that if more than 2 of these factors are missing, growth will not be above average and will not result in a sustainable industry.

The far most important success factor of technology industries is related to research and development. It has been proven that the presence of R&D centers and excellent university research is a driving factor of new high-tech industrial activities: continuous flow of new know-how into the market through tech transfer and the creation of spin-off companies, and the presence and flow of well-educated and specialized human capital.

Other factors include the existence of a set of government support instruments such as financial incentives, tax support to high-tech industries, the presence of infrastructure, the presence of venture capital and last but not least, the presence of all aspects of the tech value chain, networking as much as possible.

4 R&D COOPERATION

The development of new applications based on emerging technologies such as nanotechnology, are increasingly research intensive and necessitate heavy and time-consuming investments. Many industries based on nanotechnologies and nanoelectronics such as computers and cell phones are faced with ever shorter product life cycles and shrinking time-to-market. In order to be able to bring their new products timely to market with enough return on investment, companies increasingly need to set up partnerships with other companies and research organizations to develop the needed technologies, in time and cost efficiently.

Take e.g. the semiconductor industry. The semiconductor industry started some 50 years ago with the discovery of the transistor. Since then, this industry has succeeded to develop new generations of so-called integrated circuits (IC) with a cycle of 2-3 years, following the so-called Moore's Law, named after co-founder Gordon Moore who predicted this evolution more than 40 years ago.

Today, semiconductor technology has dived well into the nanometer scale, bringing integrated circuits to market in which the transistors exhibit feature sizes of a few nanometer long. They constitute wat is called top-down nanotechnology, meaning that nano-sized dimensions are made starting from macroscopic dimensions in which deposition and lithographic techniques are so advanced they can make these tiny structures.

For these semiconductor technologies to be developed timely and cost efficient, industry has been forced to cooperate, not only in manufacturing, but more and more so in development and even research. This industry is used to work with research cooperation schemes, within the sequence of a value chain as well as between competitors. Within the value chain, more and more alliances have been created between supplier and customer, e.g. equipment supplier and IC manufacturer. More remarkable has been the cooperation between competitors. The key to success in this kind of cooperation scheme is the degree to which technology is distant from the market, i.e. the technology to be investigated should be pre-competitive and as generic as possible (but not fundamental research, since that has no commercial end point, and hence is of no commercial interest to companies). To put it differently, the technology is a basic technology that every player needs for their subsequent product development, but it does not constitute a characteristic that differentiates one player from its competitor.

This pre-competitive R&D cooperation scheme has proven its value, and is fairly common amongst semiconductor organizations. The most well-known example of such R&D alliance is IMEC. IMEC, an independent R&D center based in Leuven, Belgium, has managed to group the world's largest industry research partnership in semiconductor technology research, with members such as Intel, Texas Instruments, Samsung, Panasonic, and many more. Their business model [1], once unique, is now copied more and more throughout the world. It is based on co-ownership without accounting to one another. This means that all partners will co-own the intellectual property of the results of the research program, and can use it for their own commercialization purposes. The program also allows for some results to be exclusively owned by a certain partner, typically more partner-specific outcome.

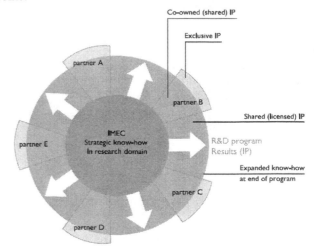

Figure 1 : R&D cooperation model of IMEC's industry partnership (source: IMEC).

The advantages of such joint R&D programs are manyfold: sharing of cost, sharing of infrastructure such as labs, sharing of expertise and sharing of risk. Since the results are pre-competitive, they do not constitute a risk of diluting the intellectual property rights of the product development that follows the research program, as this product development typically is carried out in-house and results in exclusive intellectual property rights.

5 BIO AND PHARMA

In life sciences, up till today, the situation has been much different. Biotechnology is a relatively young player in this field when compared to pharmaceutical companies. For many years, pharma's success has been based on the commercialization of so-called 'blockbuster' drugs, drugs that are used by a large population to treat a chronic condition, and have sales exceeding 1 billion USD. Many people do not realize this, but blockbuster drugs are really not that effective: they might be efficacious in only 40% of the general population. This also means that the development of such drugs has a long road to go as a sufficiently large number of clinical trials have to be carried out to show clear efficacy. As a result, development costs are extremely high, amounting up to 802 million USD in 2000, according to the Tufts Center for the Study of Drug Development. Nearly half of the total costs in drug development are due to the time value of money: it takes between 8 to 12 years for an experimental drug to get to the market. This, combined with the fact that a mere 10% of all drug candidates ever reaches the market, leads to the high risk, high cost picture of the pharma business [2]. No wonder that this market is dominated by a few large and well-established players, forming an effective barrier to new entrants.

With the emergence of biotechnology, the search for the right drug has changed. Thanks to molecular sciences and nanobiotechnology, industry is starting to be able to define diseases much more accurately, how they function at the molecular level and as a part of the biological system. The biological approach is based on a fundamental understanding of disease and is derived from human proteins or antibodies, in contrast with the typically used mice and other animals.

But there is a drawback. The insight in biological processes at the genomic and proteomic level requires understanding our genes and proteins, and this is quite a challenge. It is estimated that the human body contains over 500,000 different proteins, more than 10 times the number of genes. And it gets worse. Many diseases are related to more than one gene, and knowing how they work together is essential. It is like a geographical map. We have mapped the entire human genome, so we have the map, but we still do not know well how to read it. And a map is static, we do not see the functioning of all parts in the map.

What does this mean for R&D? Research will have to make radical changes to be able to cope with these new domains to turn these advances into targeted drugs. Thanks to genomics, researchers are starting to find out to predict which drugs would work best on which patients. This segmentation has significant consequences for the pharma business. Instead of deploying a blockbuster drug for a large population, these new targeted drugs will fragment the market based on probable responses for segmented patient groups. On the other hand, genomics and proteomics will enable pharma to define diseases much more accurately and develop much more efficacious drugs for smaller patient populations. It will increase market opportunities also in other ways. Patients will be served with comprehensive drug therapies that really work for them, starting from diagnostics, molecular markers for defining the disease state from which they suffer, monitoring mechanisms and targeted treatment using these new drugs.

6 CONVERGING BUSINESS MODEL?

As these markets become more fragmented, pharma companies will need to decide which disease family they want to concentrate their research on. This means that they will first build disease knowledge, which is generic in nature and should be allowed to be used by others too. Alternatively, it would mean that it would allow research partnerships based on pre-competitive molecular science to be set up. This industry should start to cooperate in research and development, much as the semiconductor industry has done in the last 2 decades. It will need to cope with the increasing complexity associated with many disease areas and product types to work in.

The resulting disease knowledge should lead to target knowledge, ending in a therapeutic molecule, new biomarkers and molecular diagnostics tools and methods. Most of the disease research will be carried out at or in collaboration with academia and research centers. As soon as it becomes clear how the disease works and what business can be made out of it in terms of diagnostics, targeted treatment and monitoring, pharma companies will take the lead. These companies will need to communicate with regulators much earlier as they will need to deploy proof of concept studies of treatments earlier in the development process.

With the emergence of nanobiotechnology, the challenge of cooperation is even higher. While the nanotechnology industry, and the semiconductor industry in particular, is used to work in true R&D cooperation schemes, biotech and life sciences still aren't. Typically, the development of new drugs, be it chemical or biological, is still a matter of exclusive intellectual ownerships. In essence, any cooperation between these 2 worlds, nanotech and biotech, would result in a clash of business models, and hence, inhibit the growing convergence on the fundamental scientific level.We believe it will not be a clash as the evolution of new drugs into the biotech scene, based on a more fundamental understanding of diseases, will indeed allow for more R&D partnerships to be formed within the life sciences domain as well as in nanobio, since new developments will need a combination of generic pre-competitive technology to be developed together with proprietary research for drug and therapeutic developments. Increased interaction to learn from each other's world is crucial in order to ripe the true benefits of nanobiotechnology. Those regions who will be able to join these different worlds together, will be the next technology pioneers in the world.

REFERENCES

[1] J. Van Helleputte, IMEC, private communication, 2007

[2] NanoBusinessLife magazine, Volume 2, p.22, 2007

Leveraging the Nation's Nanotechnology Research Centers:
True Interdisciplinary Buildings Foster a Seamless Transition from Lab Bench to Consumer

AUTHORS: Ahmad Soueid1, George B. Adams III2, and Michael Schaeffer3

1 Principal/Senior Vice President, HDR Architecture, Inc., ahmad.soueid@hdrinc.com
2 NCN Associate Director for Programs, Purdue University, gba@purdue.edu
3 Deputy Project Director, CFN, Brookhaven National Laboratory, mike@bnl.gov

ABSTRACT:

Spearheaded by the National Nanotechnology Initiative, the United States government has funded the construction and operation of several new research centers located at national laboratories. Additionally, state and local initiatives and private donors have funded higher education institutions to build a new breed of laboratories and cleanrooms supporting the nation's infrastructure for nanoscale research. Now operational, the Center for Functional Nanomaterials at Brookhaven National Laboratory and the Birck Nanotechnology Center at Purdue University are two examples of such facilities that offer private industry incubator space and/or operate as open access centers for use of their laboratory equipment. Interdisciplinary buildings are designed to provide a built environment that delivers "high end " laboratory space, true interdisciplinary research environments, and integration with the commercialization process.

keywords: facility design, economic development, infrastructure, nanotechnology, metrology

1. INTRODUCTION:

The ultimate goal of the National Nanotechnology Initiative (NNI) is new technology and industry that benefits society. NNI Program Component Areas include "Establishment of user facilities, acquisition of major instrumentation, and other activities that develop, support, or enhance the National scientific infrastructure for the conduct of nanoscale science, engineering, and technology R&D" [1]. The Birck Nanotechnology Center (BNC) in Discovery Park at Purdue University and the Center for Functional Nanomaterials (CFN) at the Brookhaven National Laboratory (BNL) are key examples of the many new nanotechnology research facilities. BNC and CFN successes owe much to their architectural designs and the support those designs provides for managing the challenges of nanotechnology research and its commercialization process.

This paper presents the BNC, CFN, and its sister facility, the Center for Integrated Nanotechnologies (CINT) at Sandia National Laboratories, as case studies to illustrate how architectural design contributes to programmatic success and to provide a more detail understanding of the opportunities available for leveraging the investment represented by these facilities.

2. INTELLECTUAL FUSION:

Nanotechnology is part of today's electronic devices, automobile catalytic converters, infection-fighting bandages, and leading-edge sports gear. It will be part of tomorrow's biomedical devices, pharmaceuticals, renewable energy and many, many other technologies. These and other end-user products are the combination of technologies from many disciplines. Their development requires the integration of basic science discipline knowledge from chemistry, physics, biology, and other fields; with engineering design that combines principles from electrical, mechanical, chemical, materials, and other fields; all taken forward by diverse entrepreneurial teams with access to leading-edge prototyping and characterization facilities.

Because nanotechnology is so diverse and pervasive a multidisciplinary team in proximity is a necessity for effective research, development, and commercialization. An interdisciplinary approach from "lab bench to business" is critical for successful economic development.

Scientific collaboration can be encouraged and enabled by architecture. Purdue University demonstrated the execution of this concept by merging 145 faculty members from 36 schools and departments including science, engineering, agriculture, pharmacy, and the liberal arts to create a "Birck Nanotechnology Community". Forty-five of these faculty, about 200 graduate students, and 30 technical and support staff members reside in the BNC building.

The building layout fosters dynamic and interactive collaboration spaces within the building while keeping within the 20-25% ratio of circulation space and without adding a cost premium to the overall project. There are many interaction spaces: seating in public areas, 10 small conference rooms, an office design that encourages leaving doors open, open-format student offices immediately

HDR

adjacent to laboratories, glass vision walls for many conference rooms and labs.

These spaces are intended to foster the informal meetings that are leading to collaborations, and they do. However, with this architectural support in place, BNC management has enacted planned personnel policies including placing faculty from different departments in adjacent offices and assigning graduate students working on different, but potentially synergistic research projects to the same multi-person office.

These graduate students report high satisfaction with close proximity to those working on different projects. Further, they note that high degree of interaction among students from different research projects leads to the rapid diffusion of best practices for experimentation and scientific instrument operation and, thus, more rapid progress.

Faculty research groups with aligned interests but based in different departments would, if housed in different buildings, formerly have rarely or never interacted. After spending time in the collaboration-supporting spaces of the BNC building, several such groups have entered into highly productive mergers. Figure 1 illustrates a portion of the web of such collaborations as of Dec. 2007.

★ Collaborated Publications

Figure 1: Each research group is represented by a disk with diameter proportional to the number of collaborations. Shorter, thicker lines between hubs indicate closer collaboration as measured by jointly authored publications [2].

The goal of providing the best support possible for the personnel involved in interdisciplinary nanoscale research and empowering a cultural change from individual laboratory facilities to shared spaces containing shared equipment thus directly informed BNC and CFN architectural design in anticipation of the desired operational policies.

3. QUALITY OF SPACE:

Recruitment of talent and grant funding are two major indicators of a successful program. Within the design and construction of appropriately specified "high end" laboratory space, both BNL and Purdue have recruited leading researchers as anchors to their programs. The facilities' aesthetically pleasing designs also provide natural mitigation of environmental "contaminants" affording dramatically better technical spaces and greatly improving researcher efficiency. The page of photographs shows the aesthetic aspects and collaboration spaces of CFN, BNC, and CINT. Researchers quickly feel at home in these buildings.

The CFN provides state-of-the-art capabilities with an emphasis on atomic-level tailoring to achieve desired properties and functions. CFN is a science-based user facility with the overarching theme of addressing challenges in energy security, consistent with the Department of Energy mission. Like its four sister DOE nanoscale science research centers, CFN works collaboratively with university, industry, and government laboratory researchers seeking breakthroughs in energy research as enabled by nanocatalysis, biological and soft materials, and electronic materials.

The $81M CFN facility (construction and major equipment) opened on May 21, 2007. The building features cleanrooms; general, dry, and wet laboratory space; office space for users and staff; and a wide range of materials-focused scientific fabrication tools and characterization instrumentation.

There is great promise for nanotechnology-based products comprising, for example, a multidisciplinary combination of NEMS/MEMS and biological molecules and/or organisms to create sensing/diagnostic devices with environmental and health care applicaitons. Nano-MEMS devices require a semiconductor-style cleanroom for fabrication. Working with biological molecules and/ or viable organisms and prototype materials, structures, or devices for health care use requires a pharmaceutical-grade cleanroom with biosafety containment. These two cleanrooms are not compatible in design or operational protocols [2]. These conflicting needs provide one example of the challenges of nanotechnology facility design. For the BNC, the design of the cleanrooms begins with a structure for a semiconductor fabrication cleanroom and modifies a portion of this area for the bio-pharma cleanroom. The semiconductor clearnroom requires the more complex infrastructure, so an incorporating bio-pharma section within the fabrication envelope is simpler design. Providing adjacency of the two spaces allows for materials transfer and enables research.

The BNC building was completed in October 2005 at a cost of $58M, of which $42M came from three generous private gifts. The 187,000 sq. ft. building includes the

Birck Nanotechnology Center
Purdue University

Birck Nanotechnology Center
Purdue University

Center for Integrated Nanotechnologies (CINT) Core Facility
Sandia National Laboratories

Center for Functional Nanomaterials
Brookhaven National Laboratory

Center for Integrated Nanotechnologies (CINT) Core Facility
Sandia National Laboratories

Nick Merrick © Hedrich Blessing

Scifres Nanofabrication Laboratory (SNL), a 25,000 sq. ft. cleanroom with research space rated at 1, 10 and 100 microparticles per cubic foot, an extraordinary level of cleanliness.

Within the SNL is a bio/pharma cleanroom that facilitates research at the juncture of nano- and bio-technologies. This unique capability puts the BNC at the forefront of facilities designed for research at this important frontier. Another 22,000 sq. ft. of dedicated laboratories outside the SNL provide specialized spaces with features such as temperature control to ±0.01 °C, floating inertial-mass floors for vibration isolation (NIST A-1 standard), and shielding from electromagnetic interference. BNC process water is the purest in the world, exceeding that of the best used for the manufacture of silicon integrated circuits.

The BNC project came in on time, on budget, and exceeds its extremely challenging technical specifications. Careful monitoring to ensure construction per design by the project management team lead to laboratories that exceed design goals by (1) a factor of 10 in cleanroom airborne particle count, (2) a factor of 30 in 60 Hz magnetic field generated by the building electrical power system, and (3) a factor of two in cleanroom floor vibration (yielding the lowest vibration for an elevated cleanroom floor ever measured by the project vibration consultant, a major vibration consulting firm to the worldwide semiconductor industry). This commitment to quality means that the scientific instruments in the facility routinely reach the limits of their performance capabilities.

The BNC design is supporting research in nanoelectronics, nanophotonics, nanobiotechnology and nanomedicine, energy conversion, heat transfer, surface science, nanomaterials, nanomanufacturing, MEMS/NEMS and micro/nanofluidics, and nanometrology. More information about BNC research activity is available [3] at www. nanoHUB.org.

CFN, BNC, and CINT planned for flexibility in their facility designs. Planning for the future and change reduces cost and down time.

4. GATEWAY TO INDUSTRY

Purdue University is a land grant university. Economic development for the State of Indiana is an important element of Purdue's mission. BNC, working with the Burton D. Morgan Center for Entrepreneurship, Purdue's Office of Technology Commercialization, and the Purdue Research Park has contributed to 702 new jobs with an overall increase in payroll of more than $36 million from 2003-2007. Companies that emerge from the BNC may find a place to grow in the nearby Purdue Research Park, currently home to 7 start-up companies in the Purdue Research Park based on nanotechnologies from BNC researchers.

The BNC laboratories are open to outside users. Companies making use of BNC facilities include ExxonMobil, Honda, Med Institute, and Cummins. BNC also features Nanotech Accelerator lab space for lease by companies investigating the promise of nascent nanotechnologies.

With an estimated 300 users from academia, industry and national laboratories, Brookhaven CFN serves as a hub to nanoscience studies in our national energy challenges.

REFERENCES

[1] "The National Nanotechnology Initiative: Strategic Plan", National Science and Technology Council, Committee on Technology, Subcommittee on Nanoscale Science, Engineering, and Technology, www.nano.gov.

[2] "Progress Report to Lilly Endowment, Inc.", France A. Córdova, Discovery Park, Purdue University, Dec. 31, 2007.

[3] John R. Weaver, "A Design for Combining Biological and Semiconductor Cleanrooms for Nanotechnology Research", Journal of the IEST, V. 48, No.1, pp. 75-82, 2005.

[4] Sands, Timothy D. (2007), "BNC Annual Research Symposium: Welcome and Overview", http://www. nanohub.org/resources/2622/.

Bridge a Gap between Nanotechnology R&D, Business and Public

Mizuki Sekiya, Soonhwa An, Saori Ishizu

Working Group on Strategic Area of Nanotechnology,
Technology Information Department,
National Institute of Advanced Industrial Science and Technology (AIST)
1-3-1, Kasumigaseki, Chiyoda-ku, Tokyo 100-8921, Japan,
mizuki-sekiya@aist.go.jp

ABSTRACT

I outline a new research project "Developing knowledge-based platform to support nanomaterials R&D for public acceptance" which has been started September 2007 with focus on National Institute of Advanced Industrial Science and Technology (AIST). We conduct research "Use of information and communication for facilitating public acceptance", which is one of the research components of a new research project. This research project aims to provide useful method for healthy development of nanotechnology. In the research, we deliberate communication tools and will recommend effective and suitable ways of providing information for public and society about nanotechnology R&D.

Keywords: public engagement, social implications, policy recommendations

1 INTRODUCTION

Initiatives to address the social implications of nanotechnologies have steadily progressed since 2004 in Japan. Meanwhile, there has not been a significant progress in practical applications for core nanotechnologies, which would normally be the most important and major pursuit. Exaggerated expectations for nanotechnologies may have moderated somewhat, and Japanese nanotechnology has now entered a stage of sound efforts toward risk management for nanoparticles, the development of nanotechnology standards, and a steady approach toward practical applications of fundamentally useful core technologies for nanotechnologies.

2 PUBLIC AND NANOTECHNOLOGY

We understand importance of public engagement into R&D process from the very beginning. Because if nanotechnology R&D goes well as researchers, investors and policy makers anticipate, economic and social structure will be transformed and our everyday life will change dramatically and may permanently. Nanotechnology products, although the definition still uncertain, already prolific in market, as cosmetics, sporting materials, or food ingredients [1]. Public attitude toward nanotechnology is not bad in Japan currently. Many consider nanotechnology will be useful to their life and to society [2]. But we are concerned that favorable attitude toward nanotechnology is based not on knowledge of actual nanotechnology R&D, but on surreal images about what nanotechnology may realize. Advancing nanotechnology R&D on such unstable and unreliable foundation is not healthy. This is clear when you look at other so-called emerging technologies closely. For example, genetically-modified organisms are not fully accepted in Japan or Europe. An advantage for addressing social implications of nanotechnology R&D is that it is allowed to learn from many precedential example technologies which were also expected to change society.

3 RESEARCH ON SOCIAL IMPLICATIONS OF NANOTECHNOLOGIES IN JAPAN

Every actor who takes part in nanotechnology R&D is responsible for providing information without concealing risks or exaggerating benefits of nanotechnology. Since nanotechnology R&D is involved in various research fields and multi-disciplinary in nature, single ministry or research institute can not properly manage issues originated from nanotechnology R&D or nanomaterials. This is true for activities for addressing social implications of nanotechnology. Under this concept, we started open forum "Nanotechnology and society" in 2004, where various stakeholders participated. However, our way of addressing social implications of nanotechnology R&D was not fully accepted in Japan in the beginning.

Some factors that may have delayed efforts to address societal implications are that, even though risk assessment and risk management are expected topics in the context of public acceptance, the word "risk" takes on a strongly negative connotation when translated into Japanese, people tend to fixate on the issue of "safety" rather than "risk," and the net effect is a resistance to conduct research for risk assessment and risk management.

Our patient effort has gradually borne fruit. We conducted novel endeavor with three other public research institutes under the different regulatory ministries, which was unprecedented cooperation. We studied ways to help

NSTI-Nanotech 2008, www.nsti.org, ISBN 978-1-4200-8503-7 Vol. 1

nanotechnology truly accepted in society and benefits people. After one year project, we submitted policy recommendations to the government, public institutes and private industry [3]. The policy recommendations were reflected in the Third Science Basic Plan which was prepared by the Council for Science and Technology Policy and has been in force since April 2006. The Third Science Basic Plan clearly stipulates that the government is responsible for addressing social implications of nanotechnologies. As a result, under the Third Science Basic Plan, the Developing Nanotechnologies and Engaging the Public is newly added to the Coordination Program of Science and Technology Projects [4]. The project that I introduce in this paper is a supplementary program for this new framework.

4 NEW RESEARCH PROJECT

The new research project has been launched in September 2007 and will be completed by the end of fiscal 2009. It aims to develop useful index for knowledge-based platform which supports nanomaterials R&D in industry. This research project is headed by University of Tokyo. AIST and National Institute for Materials Science participate in the project by making full use of the characteristics of each institute [Figure 1].

AIST contributes by studying communication with public and identify social trend to recommend an effective and suitable way of providing information and meaningful engagement of public into nanotechnology R&D process.

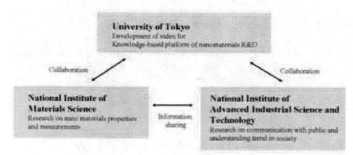

Figure 1: Implementation structure of the project

5 CONCLUSION

Nanotechnologies have the potential to offer solutions to a variety of problems facing the world we live in today. Advancing nanotechnology R&D to reach this ultimate goal is not easy task. Our new research project contributes to the issues by utilizing our knowledge and experience that acquired through past and on going research and related activities.

REFERENCES

[1] Nanotechnology-claimed consumer products inventory in Japan, Research Center for Chemical Risk Management of AIST http://staff.aist.go.jp/kishimoto-atsuo/nano/index.htm

[2] Fujita Y. & S. Abe (Nanotechnology Research Institute, AIST), 2005. Report of survey of questionnaire paper for nanotechnology and society.

[3] AIST, NIMS, NIES, NIHS, 2006. Research Project on Facilitation of Public Acceptance of Nanotechnology, Summary and Policy Recommendations.

[4] Council for Science and Technology Policy, Science and Technology Basic Plan http://www8.cao.go.jp/cstp/english/basic/index.html

Europe's First Institute of Piezoelectric Materials & Devices

Markys G Cain*

*National Physical Laboratory, Hampton Road, Teddington, UK, markys.cain@npl.co.uk

ABSTRACT

Europe's first organization dedicated to piezoelectric materials and devices has been launched by a multidisciplinary consortium of researchers and industrialists. The UK's National Physical Laboratory (NPL) is a founding member, and the institute is headed up by Danish materials manufacturer, Ferroperm.

Taking piezo into the mainstream:
- New institute signals Europe's commitment to piezo technologies
- Europe's first organization dedicated to piezoelectric materials and devices has been launched by a multidisciplinary consortium of researchers and industrialists.

Keywords: Piezoelectric, Institute, European, Organization, Network

1 INTRODUCTION

The European Institute of Piezoelectric Materials and Devices (Piezo Institute) was created by the EC-funded MIND Network of Excellence and incorporates expertise from countries including the UK, France, Italy, Denmark, Slovenia, Switzerland, Spain and Latvia. Members are currently being recruited from Europe and beyond.

The Piezo Institute will develop piezo-based sensors and other applications. It will be the European hub of expertise and resources in piezo technologies, offering research, resources, education and training. Its expertise includes ferroelectricity, electrostriction and pyroelectricity in materials including ceramics, single crystals, polymers and composites.

Founding members have skills in chemistry and process engineering, solid state physics, materials characterisation and measurement science, micro and nano technology, MEMS, numerical modeling and the design and manufacture of devices.

This expertise and associated technical resources are being committed to the new institute, which will integrate European expertise in piezoelectric materials and devices. The Piezo Institute aims to be financially independent and fully functional by 2009.

The institute will enable Europe to be more competitive. It will open doors to industry and make academic knowledge widely available for commercial applications in sectors including healthcare, automotive and defence.

2 BACKGROUND

"The science of piezoelectricity has been known for more than a century," notes Dr Markys Cain, a principal research scientist and knowledge leader at the UK's National Physical Laboratory, a founding member of the Piezo Institute. "The institute is Europe's recognition that there is now far greater potential for piezo applications in healthcare, transport, energy harvesting and environmental protection. It will help us to keep up with the rapid pace of piezo development in other parts of the world."

The Piezo Institute offers research and consultancy in chemistry and process engineering, solid state physics, materials characterisation, metrology, standards and the manufacture and testing of piezo devices.

2.1 Applying piezo phenomena

The aims of the new institute include miniaturisation and integration of piezo structures into multifunctional devices, such as sensors for monitoring radiation, temperature and biological threats. New transducers for medical imaging, underwater acoustics and telecommunication systems are improving sensing performance and environmental efficiency. Innovative piezo-based devices enable implants for health monitoring and targeted drug delivery, and disposable probes to inspect and treat cardiovascular disease.

Innovative devices being considered by piezo researchers in the emerging institute include implanted micro-tools for health monitoring and drug delivery, and wireless energy harvesting systems.

2.2 Transport

Piezoelectric materials are the basic component of many automotive sensors, including accelerometers and infra-red detectors. Improvement of sensor performance will reduce fuel consumption and improve safety. Piezoelectricity is becoming a leading technology in diesel fuel injection. The high pressures generated by piezo injectors improve efficiency of engines and reduce pollution. The Piezo Institute aims to develop self-testing structures with embedded sensors to monitor the effects of age such as micro-defects, cracks, delaminations and fatigue-related flaws.

2.3 Education and training in piezo

Education and training are among the core functions of the new Piezo Institute, which will help to enhance UK and European competitiveness by preparing students for scientific and industrial leadership in the field.

Europe's first Masters degrees in piezoelectric materials and devices will be offered by the institute and its partner universities. More than 15 Masters programmes in nine countries have already been identified by the Piezo Institute joint education programme. Students with a Bachelors or equivalent degree in a field related to materials science, chemistry, physics or engineering are eligible, and there are opportunities to work as interns at leading European companies.

2.4 Supporting PhDs and beyond

Piezo doctorates are being created at leading universities, with supervision from Piezo Institute scientists as well as part time employment and funding opportunities.

The institute is committed to continuous development of piezo-related skills, and provides industrial training once students enter the workforce. The training covers ferro and piezoelectric fabrication, micro and nanotechnology and piezo applications – all essential skills for scientists and engineers working in materials science, chemistry, physics, electronics, mechanics and systems engineering.

3 WHAT IS PIEZOELECTRICITY?

Piezoelectricity is the ability of certain materials to generate an electric charge in response to mechanical stress. They also have the opposite effect – the application of electric voltage produces mechanical strain in the materials. This makes piezoelectric materials effective in sensors and transducers used in the automotive and healthcare industries, and for environmental monitoring.

4 FOUNDING MEMBERS OF THE INSTITUTE

Ferroperm Piezoceramics A/S (Denmark)
University of Tours – LUSSI (Université François Rabelais – CNRS) (France)
Laboratoire de Céramique (Ecole Polytechnique Fédérale de Lausanne)
Jozef Stefan Institute (Slovenia)
Cranfield University (UK)
Instituto de Ciencia de Materiales de Madrid (Consejo Superior de Investigaciones Científicas) (Spain)
SIEMENS Corporate Technology – Erlangen (SIEMENS AG) (Germany)
Institute of Solid State Physics of the University of Latvia
National Physical Laboratory (UK)
Centro Ricerche Fiat (Italy)

Table 1: Founding members of the Network

5 CONCLUSIONS

Europe's first organization dedicated to piezoelectric materials and devices has been launched by a multidisciplinary consortium of researchers and industrialists.

In this presentation, the broad remit of the new Institute will be described and the ways in which individuals, universities or companies may benefit from membership is outlined. http://www.piezoinstitute.com

Figure 1: The Piezo Institute Logo

6 ACKNOWLEDGEMENTS

The author and the MIND European Network of Excellence wishes to recognize the sponsorship and support of the European Commission: MULTIFUNCTIONAL & INTEGRATED PIEZOELECTRIC DEVICES: Proposal / Contract NoE 515757-2: NMP-2003-3.4.2.1.1. Understanding Materials Phenomena (under NMP-2. Knowledge-based Multifunctional Materials, NMP-2.1 Development of fundamental knowledge).

NanoEnergy- Technological and economical impact of nanotechnology on the energy sector

W. Luther

VDI Technologiezentrum GmbH
Zukünftige Technologien Consulting, Düsseldorf, Germany, luther@vdi.de

ABSTRACT

According to a prognosis of the International Energy Agency the worldwide energy demand will reach more than 18.000 MTOE (Million Tons of Oil Equivalent) in the year 2030 which corresponds to a 50 % increase from today´s energy consumption. In respect of narrowing fossil energy resources and potential side effects of climate warming carbon dioxide emissions it becomes obvious that for a sustained longterm economical prosperity a fundamental renewal of the energy sector will be necessary. As a key and cross-sectional technology, nanotechnology has the potential to facilitate vital technological breakthroughs in the energy sector and, in this way, to make substantial contributions to a sustainable energy supply. The spectrum of possible applications for nanotechnologies ranges from continuous short and medium-term improvements in the use of conventional and regenerative energy sources to entirely new approaches to energy supply and utilisation in the long term. Nanotechnological innovations have an impact on the whole added value chain in the energy sector. Increases in efficiency and process innovations are possible in all subsectors, from primary energy production to energy transformation, transmission and storage and its utilisation by the end customer.

Keywords: nanomaterials, energy efficiency, CO_2-reduction, renewable energy sources, market potentials

1 PRIMARY ENERGY UTILISATION

Nanotechnologies offer substantial advantages in the utilisation both of conventional energy sources (fossil and nuclear fuels) and of renewable forms of energy such as geothermics, sun, wind, water, tides or biomass. Thanks to more wear-resistant nano-coated drilling tools, for example, the service life and efficiency of plants for the recovery of mineral oil and natural gas deposits or of geothermal heat can be optimised and, in this way, costs can be saved. Other examples are high-performance nanomaterials for lighter and more stable rotor blades in wind and tidal power plants, and wear and corrosion resistant coatings for mechanically stressed components (bearings, gears etc.). Nanotechnologies will play an essential role, in particular, in the increased use of solar energy through photovoltaics. Efficiency increases can be achieved in conventional crystalline silicon solar cells, for example, by using anti-reflective coatings for better light exploitation. First to profit from nanotechnology, however, is the further development of alternative cell types such as thin-film solar cells (of silicon or other material systems like copper/indium/selenium, for example), pigment-based solar cells or polymer solar cells. Because of their inexpensive materials and manufacturing process and their flexibility, the latter have a high potential, particularly for the supply of mobile electronic devices. Medium-term development targets here are an efficiency of approx. 10 % and a service life of several years, whereby the nanotechnologies can contribute, for example, to the optimisation of the coating design and the nanomorphology of the organic semiconductors in the component structures. In Germany a €360 million research initiative funded by the German Research Ministry as well as companies like BASF, Bosch, Merck and Schott has been launched to develop organic photovoltaic devices. The world market of thin film solar cells is expected to grow from $800 million in 2007 to $2 billion in 2010.

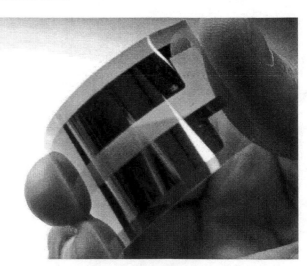

Fig. 1: Prototype of a flexible organic solar cell
(© Zentrum für Angewandte Energieforschung)

2 ENERGY TRANSFORMATION

Due to energy losses only a low percentage of primary energy sources is transformed into utilisable energy like electricity, heat and kinetic energy. Improvements in efficiency can be achieved, for example, by means of nano-scale protective heat and corrosion coatings for turbine blades of lightweight materials (e.g. titanium aluminides) for more efficient gas turbines in power plants or aircraft engines. The electrical power produced in the conversion of chemical energy by fuel cells can be increased by using nanostructured electrodes, catalysts and membranes, leading to economical applications in automobiles, in buildings or for the operation of mobile electronics in the future. The world market for fuel cells is expected to grow from around $1 billion in 2010 to over $20 billion in 2020. A promising future is also likely for thermo-electrical energy transformation. Using nanostructured semiconductors with optimised boundary surface design, gains in efficiency can be achieved which could open the way for their widespread application in the utilisation of waste heat, in automobiles for example, or even of human body heat for wearable electronics in textiles.

3   ENERGY STORAGE

The use of nanotechnologies for the improvement of electrical energy storage devices like batteries and supercapacitors is proving extremely successful. Because of a high cell voltage and the outstanding energy and output density, the lithium ion technology is regarded as one of the most promising variants for power storage. Nanotechnology can significantly increase the performance and safety of lithium ion accumulators. This applies in particular for the new ceramic, but still flexible, separators and high-performance electrode materials. In Germany an innovation alliance for the development of high performance Li-ion batteries for applications in hybrid and electrical vehicles and for stationary energy storage has been launched funded with €360 million from the German Research Ministry and industrial players. In the long term, hydrogen also appears to be a promising energy store for an environmentally friendly power supply. Apart from the necessary infrastructural adjustments, the efficient storage of hydrogen is seen as one of the critical factors for success in the progress towards a possible hydrogen economy. Present materials do not meet the requirements of the automobile industry, for example, with its demands for an H_2 storage capacity of up to ten per cent by weight. Here, various nanomaterials, some based on nanoporous organometallic compounds, offer development potentials which appear to be economically realisable, at least for the operation of fuel cells in mobile electronic devices.

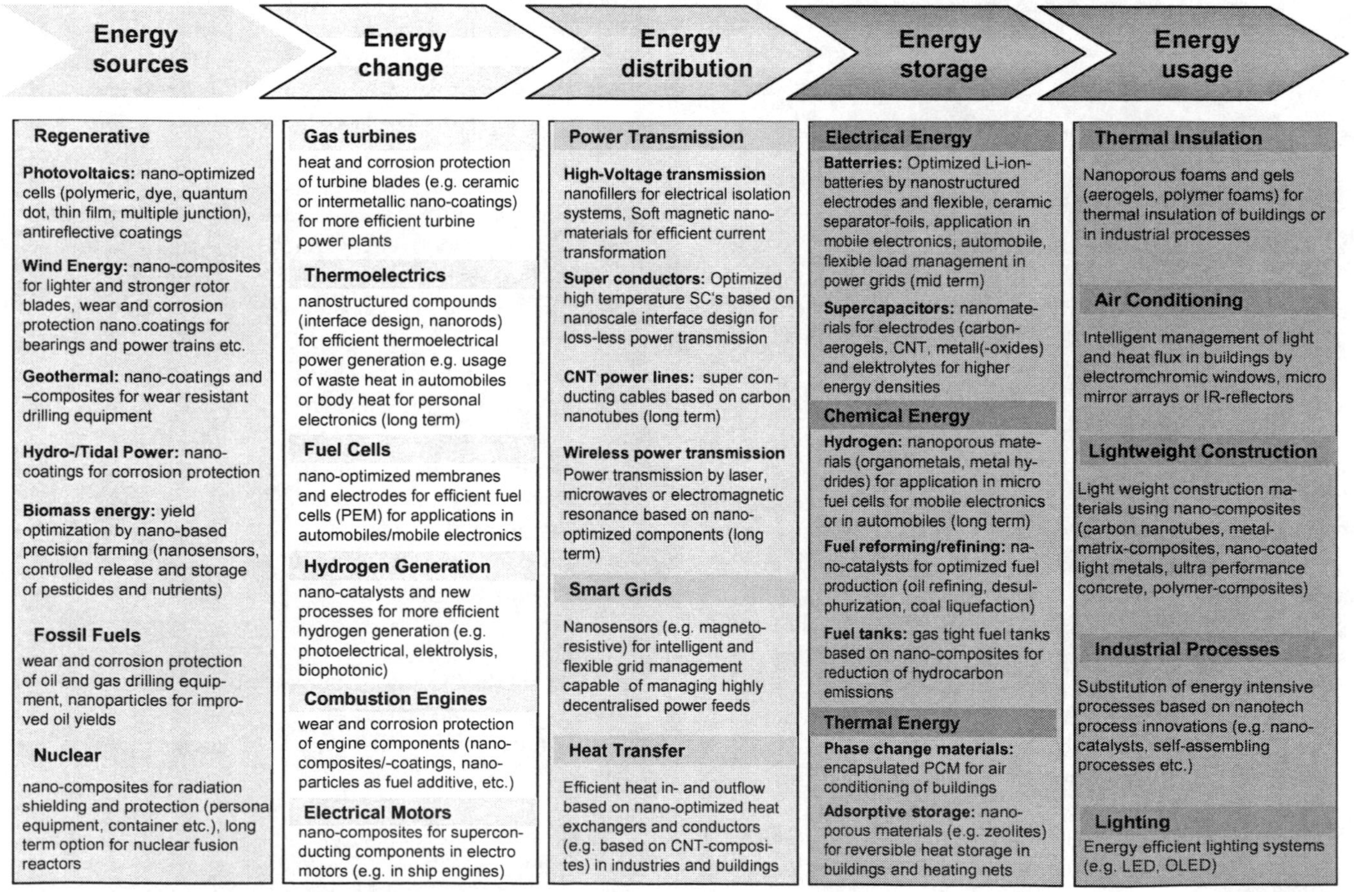

Fig 2: Overview on potential nanotechnology applications in the energy sector along the whole supply chain. (© VDI TZ)

With a view to reducing the energy losses in power transmission, it is hoped that the unusually high electrical conductivity of nanomaterials like carbon nanotubes can be used for power cables. There are also nanotechnological approaches to the optimisation of superconducting materials for loss-free transmission. In the long term, options also exist for power transport without cables, using lasers, microwaves or electromagnetic resonance, for example. For power distribution in the future, power networks will be needed which offer dynamic load and fault management, a demand led power supply with flexible price mechanisms and the possibility of power input from a large number of decentralised, renewable power sources. Nanotechnologies could make significant contributions to the realisation of this vision, for example through nano-sensoric and power electronics components capable of handling the extremely complex control and monitoring of such power networks.

Fig. 3: Nanoporous polymer foams for thermal insulation of buildings have a big potential for energy savings. (© BASF)

5   ENERGY SAVING

With the aim of a sustainable energy supply, in parallel with the opening of new energy sources, the efficiency of energy utilisation must naturally also be improved and unnecessary energy consumption avoided. This affects all branches of industry and also private households. Nanotechnologies offer a large number of approaches to energy saving. Examples are the reduction of fuel consumption in automobiles through the use of light construction methods based on nanocomposites, optimisation of fuel combustion by means of lighter, more wear-resistant engine components and nanoparticular fuel additives or even nanoparticles for optimised tyres with lower rolling resistance. There are also great energy saving potentials in building technology, for example, through the use of nanoporous thermal insulation, which could be applied with advantage particularly in the energy saving remediation of old buildings. In general, the control of the flow of light and heat through nanotechnological components such as switchable glass, for example, is a promising field for reducing the energy requirement in buildings.

6   CONCLUSIONS

Nanotechnology has the potential to be a key technology on the path to a sustainable CO_2-neutral power supply in the future. Their potential applications are extremely varied. For numerous firms in Germany, a location with pronounced strengths in the nanotechnologies and regenerative energy technologies, attractive market opportunities will open up. However, the success of nano-based innovations will strongly depend on public funding programmes, subsidies for regenerative energies and other framework conditions like regulation, the development of energy and raw material costs as well as evolving social demands resulting from consequences of the climate change. Germany has initiated several research initiatives to press ahead with nanotechnological developments in the energy sector to address the ambitious climate protection goals adopted by the European Commission. In addition to increased support for research and innovation, their realisation also demands coordinated action by all those involved in the added value chain. As an important step to utilise the potential of nanotechnologies in the energy sector it is necessary to establish a dialogue between all the actors with a role to play, transcending sectors and specialist scenes. To further this intersectoral dialogue the Hessian Ministry of Economics, Transport, Urban and Regional Development has launched several initiatives like the NanoEnergy conference which will take place the second time in September 2008 as well as an information brochure pointing out the innovation potentials in detail.

REFERENCES

[1] Luther, W.: Nanotechnology applications in the energy sector", Hessen Agentur, 2008 in press
[2] Cientifica: „Nanotechnologies and Energy", white-paper, Cientifica, London, 2/2007
[3] Nanoforum: "Nanotechnology helps solve the world's energy problems" April 2004, www.nanoforum.org

Internetlinks:
www.bmbf.de/de/nanotechnologie.php
www.hessen-nanotech.de
www.zukuenftigetechnologien.de/ENG/index.php

MEMS and NEMS Research for Education

Husak,M. - Jakovenko,J.

Department of Microelectronics
Faculty of Electrical Engineering, Czech Technical University in Prague
Technicka 2, CZ – 166 27 Prague 6, CZECH REPUBLIC
husak@feld.cvut.cz

ABSTRACT

Education in the branch of Microsystems is described in the paper. There are methods of education in the connection to research in the core of the paper. Methods are focused on active involvement of students in research work. Research student output are using in educational process. Individual tasks are parts of solutions of large research projects. Individual work contributes to their professional forming. The research in the area of nanoelectronics (MEMS and NEMS) is discussed. Several workplaces are used where students work on defined tasks. The use of special instrument, microscope and other nanotechnology workplaces in the different institutes is necessary. The education involves nanomaterials and smart materials for sensor and smart micro and nanosystems. Knowledges are become during research and individual student research work.

Keywords: Microsystems, education, MEMS, NEMS, smart materials, micro, nano

1 INTRODUCTION

MST represent an interdisciplinary area (typically interconnection of mechanical and electrical domains). Therefore analogy between quantities from various energy domains is implacable in models. Using analogy, non-electrical quantities can be converted to electrical ones. Solutions of electrical models are well elaborated and very good tools for modeling and simulation of their properties are developed. The integrated micro-sensors, micro-actuators and micro-systems are ever more being applied in various areas of common life – from the control of heating and air-conditioning in buildings to the constantly rising applications of micro-systems in medicine. Various analyzers are applied in environmental protection, based on special bio-chemical sensors. In parallel to the development of medical integrated systems there comes the need to follow their influence on the bio-system of the human body, etc. The micro-technologies are being augmented by the new nano-technologies (bringing new improved possibilities in developing device structures). They play an especially important role in the development of new materials and improvements of their properties.

The goal of the education process is research and education in the area of new types of intelligent integrated micro and nano sensors structures and actuators including electronic circuits for data signal processing and transfer. For the sensor realization are used micro technology resources together with nano technologies namely in the area of materials and chemical sensors and biosensors structures. Educational process includes modeling, properties simulation of RF MEMS and MEMS structure, development of active integrated strain gauges and wireless, Bluetooth and ZigBee data signal transfer, development of sensors using polymeric electronic, research of new opto-chemical sensors, development of micro and nano sensors for chemical and biochemical applications, build-in intelligence of integrated sensors systems, electro-magnetic compatibility in integrated circuits structures and bio-systems. The design, modeling, simulation, measurement and diagnostics in the laboratory are used for the pick-up student knowledges. The practical outputs of the student project are sensors and microsystems samples, instruments intended for biochemistry, medicine, living environment, food processing industry, and measurement and communication technology, models of sensor and structure. Some courses, that are introducing problems of nano and microelectronics, include in their practical part student research work. Students participate in solving partial task of larger research projects. Students present their research work results in the frame of individual courses, for examples: Sensors for electronics, Microelectronics, Micro and nanosystems, Physical electronics, etc. Interconnection of student education and research activities are depict on figure 1.

Talented students having interest in research work can work in specialised laboratories. These young researchers are in this way included in research teams and usually get small part of the project to solve [2]. For example, students can be invited into the laboratory of semiconductors in order to take a closer look with atomic force microscope (AFM). The microscope is used for three dimensional imaging of nano objects and surfaces in nano scale. The students can learn several AFM techniques that provide not only surface imaging (for example measurement of quantum dots InGa/GaAs can be demonstrated) in contact and semicontact regimes but also measurement of electrical properties of surface (spreading resistance, local distribution of potential), local anodic oxidation of material's surface, nanomanipulation etc. [1].

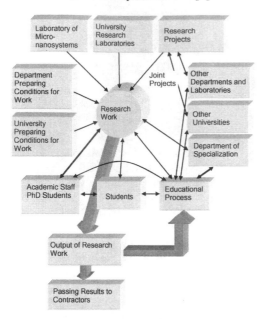

Figure 1: Interconnection of student education and research activities.

2   STUDENT'S RESEARCH ACTIVITIES IN EDUCATION

Students' involvement in research activities is performed in many ways that differ in a number of basic factors. Students are free in their decisions; they are not forced to participate in this work. However, it is recommended to do so. The research activities, projects and students' research activities are discussed on many levels, namely in education, during discussions with students, etc. Students get extended information about potential involvement in research activities. In curricula, there is a number of specialized courses utilising principle of project-oriented education. In these courses, the students get frequently task specifications that represent partial projects being parts of larger research projects. Topics of Master theses are usually based on actually solved projects in the Department. However, there is a disadvantage of long realization time of such a task. Nominal interval from defining topic of the Master thesis to its finishing is two years, but in recent years relatively many students ask for postponing the delivery. On the other side, if the student is good and teacher's supervision works well, the results acquired in the thesis may be usable in research projects. Efficiency of results applicability is very heterogeneous, in average it can be estimated to be up to 30 per cent. Here we are facing the problem that not all students have the sense of responsibility to deliver good work. There are students who just try to do their work "somehow" – to satisfy the basic requirements with as little effort as possible.

3   STUDENT'S RESEARCH ACTIVITIES IN EDUCATION

They acquire this knowledge in several ways, namely studying specialised courses, studying literature and active knowledge acquisition outside university. Let us mention an example of knowledge acquisition in the area of nanoelectronics, quantum electronics, microelectronics, or micro-nanosystems, fundamentals of micro and nanosystems (principles of operation of nanotubes, nanowires, nanoFET, semiconductor elements, circuits, electronic blocks and measurement methodology) are included. Courses constitute basic knowledge in the area of nanoelectronics, nanotechnology, electronics and microelectronics. They prepare students for further study of Micro and Nanosystems. Basic study levels are illustrated in figure 2. In the courses, laboratory equipment for microsystem design and diagnostics is used. Structure of micro-nanosystem courses is in figure 3.

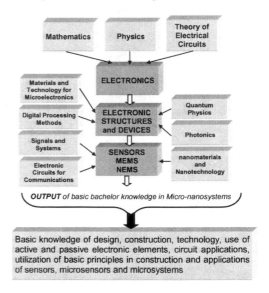

Figure 2: Basic courses supporting further education in micro-nanosystems.

Different themes are using as support of education in that subjects. Many themes relate to design of new types of smart integrated microstructures of microsensors, microactuators, including electronic circuits for processing and transmission of sensor signals meant for applications in environmental protection, biomedical applications, measurement techniques and other areas. For example some themes: Fast RF MEMS microwave signal switch (modeling of mechanical and thermal properties of the structure, use of equivalent electrical parameters to be created models, use of the CoventorWare program), RF sensor micro- and nano-structure for measurement of microwave radiation absorption. Sensor system with active integrated strain gauge, Integrated sensor structures with wireless data transmission, Polymer electronic structures and their applications in sensors, etc.

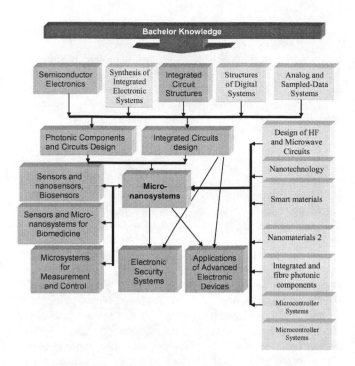

Figure 3: Concept of basic master microsystems courses.

The aim of the student project is the design, development, simulation and optimization of the properties of basic electronic structures, shape, signal processing, etc. The creation of models of the mechanical and thermal behavior of the structure, their application in the CoventorWare simulator. Comparison of models fit to prepared samples. Theme of Effects of integrated circuit architectures on biological environment is very suitable for the solving of the minimization of risks (of unwanted interaction) between the architecture of microelectronic devices and systems, and the biological environment.

4 WORKPLACES DETERMINED FOR STUDENT RESEARCH WORK

Workplace disposes of ECAD design center equipped with many workstations with implemented world software standards, like CADENCE and CoventorWare, etc. Software serve for design, modelling and simulation of integrated circuits and micromechanical microstructures are used. Technology service allows us usage of different IC technologies. The laboratory of diagnostics and microsystems is determined for measurement and testing of realized microstructures, both on-chip and encased samples of microsystems and microsensors. The laboratory is equipped with a number of instruments. Optical microscope workplace is used for inspection of micromechanical structures. The microscope is equipped with documentation subsystem for creation of graphical records that is composed of serial interface, camera for movement sensing under the objective and sublimation printer for printing quality images of microsystem structures. For processing of

graphical information from the microscope, a PC for image saving and software for graphics processing is used. The workplace is completed with reproduction equipment. Monitors are used for displaying manipulation with samples, monitoring moving events in microsystems, and demonstrations for students. A recorder is used for recording these events. Workplace for temperature measurements is intensively used for research activity, mostly microsensor calibration. The core of the workplace for development of temperature microsystems with high resolution and accuracy determined especially for biomedical purposes is represented by exact temperature calibrator 140SE-RS. The core of the workplace for high-speed collection of sensor data is represented by portable high-speed multichannel digitizer OMB-WAWEBOOK-512 with OMB-WBK20 interface that is determined for fast multichannel collection of sensor data and their evaluation on a connected PC. The workplace for biochemical measurement on microsystem structures is determined for testing of properties of biochemical microstructures, realized on semiconductor base (e.g. pH measurement at ISFET structures). The workplace of surface assembly allows manipulation and soldering of microelements and realization of professional electronic modules. It is a supporting workplace that is used for realization of additional works during development of measuring systems and devices for diagnostics and measurement.

5 FLOW OF THEME SOLVING

Many levels of the model are using as support of education approaches to complex problem solving may differ and depend on nature of solved MST [3]. System model (model realizing system function) may have several levels that are placed in different MST levels. Model on the level of energy domain can be used to consider MST operation in different energy domains. Equivalent models between energy domains can be used for many applications. Models are based on equivalence of discrete elements and their behaviour described by mathematic expressions in various energy domains [4]. The best-known equivalence is utilized between electrical, mechanical and thermal domains. In these domains individual equivalences are elaborated well – table 1.

Mechanical	Thermal	Electrical	
Mass M $F' = M\dfrac{d(v)}{dt}$	Capacitance C $Q' = C\dfrac{dT}{dt}$	Inductor L $V' = L\dfrac{d(i)}{dt}$	Capacitor C $i' = C\dfrac{d(V)}{dt}$
Spring k $F' = k\int v \cdot dt$	Capacitance C $T' = \dfrac{1}{C}\int Q \cdot dt$	Capacitor C $V' = \dfrac{1}{C}\int i \cdot dt$	Inductor L $i' = \dfrac{1}{L}\int V \cdot dt$
Damper b $F' = bv$	Resistance R $Q' = \dfrac{1}{R}(T_2 - T_1)$	Resistor R $V' = Ri$	Resistor R $i' = \dfrac{1}{R}V$

Table 1: Mechanical, thermal, and electrical analogies.

Material and structure models describe properties of intelligent materials and structures are utilized. Then modeling their properties as input parameters for higher-level models is necessary. Physical models are necessary to utilize new materials and new structure properties. Ideative model is used by the student at first, model creates an ideative model of a MST with input information, output functions, and inner logical functions. Soft model of MST using PC and libraries of electronic components and blocks. Realization of HARD model is a successive step and is used for verification of basic functions of the designed SOFT model. It is possible to use available elements for realization of the HARD model. This model illustrates characteristics and behaviour of the designed MST model. It is instructive for education, it is possible to demonstrate its behaviour and basic characteristics. There are close connections between SOFT and HARD models. The students can develop a real functional model. Design of MST is the most difficult part and follows after previous steps. Micro-models are developed from these macro-models according to the rules for design of integrated MST (technology, materials, software, etc.). Macro-model properties are compared with simulated and modeled properties of real MST. For these purposes, suitable tools are utilized (CoventoWare, CADENCE, etc.). Students usually design very simple MST; more complex models are realized in the frame of Master or PhD theses. Individual models and their interconnections are illustrated in figure 4. [4].

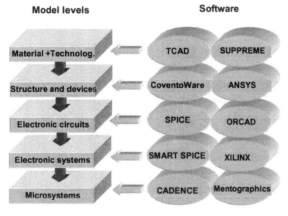

Figure 4: Flow of model levels.

When developing a system for pressure measurement, wireless information transmission and information processing, interconnection scheme (figure 5) can be used. Structure of integrated one-chip LC circuit realized on Si substrate is shown in Figure 6. In the middle part of the structure there is a capacitor with flexible membrane.

The capacitor measures external pressure. Integrated inductor is a part of integrated LC circuit. Inductor is produced by evaporation of metal spiral around capacitor. Besides L and C parameters, the structure displays a number of parasitic parameters caused by construction. Model on material and structural level can be used for modeling of basic properties of the structure as well.

6   CONCLUSIONS

Contributions of students' involvement in research activities: Motivation increase, Extension of specialised knowledge, Linking theory and practice directly during study, Students are often source of new ideas, ways of problem solving, unconventional approaches to solving given problems, Students see direct link between educational process and research work, Acquisition of deeper knowledge, Students become research team members, Mutual cooperation of students. New approach - a new method for education in the area of nanotechnology, narrow connection research and education, use of the nanotechnology equipment (AFM, nanolitography, etc.), The effort is targeted at the connection of the research an education of students in the university, connection different workplaces from the different institutes with branch of nanotechnology and nanoelectronics including nanomaterials.

7   ACKNOWLEDGEMENT

This research has been supported by the research program No. MSM6840770015"Research of Methods and Systems for Measurement of Physical Quantities and Measured Data Processing" of the CTU in Prague and partially by the Czech Science Foundation project No. 102/06/1624 "Micro and Nano Sensor Structures and Systems with Embedded Intelligence"

REFERENCES

[1] Dragoman,M., Dragoman,D., Nanoelectronics, Principles and Devices, Artech House 2006

[2] Husak,M., Student Education in Microsystems, *Proc. of iMEMS'2001*, Singapore 2001, pp.549-558.

[3] Romanowicz,B.F.: Methology for the modeling and simulation of microsystems. Kluwer Academic Publisher, Dordrecht 1998.

[4] Husak,M.: Microsystem Modeling – from Macromodels to Microsystem Design. The Nanotechnology Conference (NanoTech2003), February 23-27, 2003, San Francisco, California, USA, Proceedings vol.1 pp. 272 - 275.

Designing the new engineer and scientist

J. De Wachter*, J. Wauters**

* IMEC, Leuven, Belgium, jo.dewachter@imec.be
** Flanders Nano Bio Alliance, Larchmont, NY, USA, jan.wauters@flandersnanobio.org

ABSTRACT

Nano-bio science and technology is reaching a high level of technical expertise and knowledge. It opens up completely new ways of design, production and manufacturing. The deep understanding of materials and living matter evolves into the design and creation of totally new products and industries. Nano-bio know-how requires expertise in biology, chemistry, physics, engineering, medicine, agriculture … in other words an interdisciplinary expertise. In the USA and Europe the first interdisciplinary labs pop-up. But not only do we need more interaction between people with various backgrounds in the labs, we also need more interaction between labs and industries, the creative art scenes, the humanities, the society… There is lots of expertise showing great potential, but we have got to share our knowledge over the borders of our expertise fields. And then try to integrate all these different ideas into what we are doing.

Keywords: nanotechnology, biotechnology, nanobiotechnology, scientist, engineer, outreach

1 INTRODUCTION

A research & development lab involved in nano-bio research does not only need to set up interdisciplinary lab teams of chemists, biologists, engineers, physicists,… It must also actively involve people from the social sciences and other interested communities from outside the 'academic world'. Because the science and research is not only on materials but also on living matter. In the current nano-bio lab, we are not only talking about a new memory cell for an i-pod, but also about human cells connected with electronics. And there is more. At the nanoscale there is no difference between the building blocks of living or dead matter. It's the same thing we are dealing with: atoms and molecules.

So we have the opportunity with all this nano-bio knowledge to build great new things, but we can also mess it up terribly, because it's about the building blocks of everything around us, like DNA and even smaller molecules, … So from the start of this kind of research we must involve people with various intellectual views. Not only tech experts, also interested people with totally different backgrounds.

2 INTERDISCIPLINARY RESEARCH

That is new to IMEC, the nanoelectronics and nanotechnology center of Leuven, Belgium. Now, most of the people at IMEC are experts working in pilot lines and electronics labs. And that is very much OK, because IMEC has been able to transfer lots of knowledge to the market and will continue to do so for many years to come. But for the nano-bio research we need a broader approach. That's why IMEC launched the Ad!dictLab-IMEC nano research projects [1]. How do Ad!dictLab members, i.e. artists, designers, architects, … think about nano and how do they imagine future opportunities? What is important to them, even if they don't know a lot of the scientific expertise behind nano-bio? Or maybe they do? Then, we want to confront these inspirations, these ideas, with IMEC researchers and with the rest of the world. A real confrontation. Discussions, dialogs, confrontations between engineers, scientists, artists, designers, interested public,… about nano.

We have got to understand that with all this nano-bio expertise piling up, we really have the chance to do it right, right a way. But that is not easy, it is not like following the ongoing consumer markets. The drivers of the nano-bio research & development shouldn't be consumerism and militarism. And this interdisciplinary approach is also absolutely not about lowering standards and norms, on the contrary. We can and we want to build great new things. But we will have to do it in an interdisciplinary way. Why is that important? Because we have the chance now to invent things that can completely change the way we manufacture products. As was stated recently in a New York Times article about Olin College [2]: "Engineering has traditionally been focused on doing it right, but not on what's the right thing to do. That means designing products that are environmentally friendly and that respond to the needs of the people using them and not just what the purchasing department wants." That's the spirit, that's the vision we need.

3 TRUE EXPLORERS

So on the one hand, we want to do research on things that are going to be of a lot of value to the society, i.e. applied science. But also here we have got to do it in a long term vision. This means we cannot rush into the production of nano-bio consumer goods that are not designed properly.

On the other hand: there always must remain focus on deep level science, basic science.

So it is not all about useful knowledge or useful science or science=technology. We also want to make a point here that investing in science-for-the-sake-of science is also a must do. We know some great scientists and they are active in what they call basic science. We don't see products coming out of their research the coming decades. But the overall knowledge gained is so fantastic (about nature, the universe), it's priceless, that it is worth to invest in that. These great scientists are also the ones that do care about other views and visions in other fields than their own. And they go into discussions, not to impose their world view but to learn from others. They are using those dialogs to confront themselves and their own ideas. They question themselves constantly. And then magic happens. Great new knowledge originates. What these scientists are doing is an association exercise with other worlds and the result is great. They are not just deducting in a linear way. They are what we call the true, genuine explorers. Ask a true explorer to explain what she/he is doing to a kid, and he/she will manage. Because true explorers have that greater world view, that universal picture in their head, these association patterns with other elements and other ideas. And it is true that when we look at great scientists and their work, the difference between basic science, applied science and technology disappears. Because we can all appreciate the enormous value they create. We would like to make a reference here to what L.M. Branscomb and G. Holton describe as Jeffersonian science [3,4]. That's indeed the new science mode that has to be supported.

4 DEEP-LEVEL COAL MINING

So, interdisciplinary research is not at all about lowering science standards and norms. On the contrary, it will bring back more intellectuals to the labs. It's like deep level-coal mining. The deepest shafts are interconnected with other shafts. Because of the fact that they are interconnected they can dig deeper, because fresh air comes in through other shafts and new knowledge can find its way up more easily. Also, they don't have to surface all the time in order to share their views. The shafts that are not interconnected stop at a low depth.

So why is all this important? Because when we start to produce things, to make and sell products, we must realize that mankind is here to stay on this earth for a very, very long time. So we better do it right, right a way.
Also we have to remember that in the course of history of life on earth, extinction is the rule and survival is the exception [5].
And now, finally, with all this nano-bio knowledge and expertise piling up, we really have the chance to do it right. Let's do it. Let's sit down together, build deep knowledge and make great things.

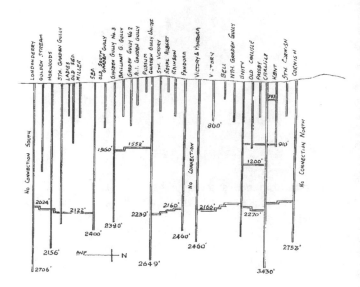

Fig. 1 : example of mine: deep shafts and horizontal connections illustrate interdisciplinary research (source: The Carlisle Gold Mine, History and family tree of John Jewell, www.jewel.asn.au).

5 CONCLUSION

Successful Nano-bio Outreach builds on 3 foundations:

5.1 Bring science and technology inspiration and passion in education through kids, teachers and parents

Regarding IMEC, IMECEXPO's classroom [6] and demonstrations currently serve as a location base for this. The IMECEXPO project is a joint initiative of IMEC and RVO-Society [7]. RVO-Society wants to stimulate youngsters for technology and science by means of original educational kits and social projects.

IMECEXPO has to be rebuilt to carry a more universal viewpoint : saving man and earth through knowledge, i.e. nano-bioscience. To establish this, we have to bring in new fresh ideas from outside imec. A first step is the AddictLab project.
A new project, "Youngsters' lab 16-18y" integrated in a new science gallery building, must be set up to complete this inspiring bring-science-to-(kids, teachers, parents)-life experience.
The underlying vision of this educational outreach effort is the belief in science and knowledge gaining as the key to human survival on earth and in the universe. Especially with the current growing beliefs among people in creationism and religious fundamentalism, it must be set clear that the only way forward is to continue to gain more knowledge and understanding. The quest for 'veritas' must continue. We have to advance science and technology. But we have to realize that if we produce something it has to

benefit the world and the universe in a big way, not only economically. Three pitfalls have to be avoided: Violence (aggressive power over people, non-democratic hierarchy, religious fundamentalism), Money (take the money and run attitude) and Ego (being vain, striving to become famous) Success is important, but not by means of power, money and ego. Success as a measure for bringing real value to humans and the world [8].

5.2 Interdisciplinary nano-bio research needs wide-angle interested researchers

This means bringing back the genuine, true explorer/scientist spirit and inspiration to current research institutes and universities (Jeffersonian science).

Ideas and visions by other community groups like art and design can be used as a trigger to start up discussions with nanotech researchers about their work. It also serves the need to look at nano science and technology from another angle or from a more universal viewpoint. The first goal of true research must not be to make money, or to become famous, but to gain knowledge, to go for truth, to build value and to strive for preservation of the human species on a healthy earth. And nanoscience and technology holds this potential. Nanoscience can give us the knowledge to create 100% eco-efficient industries.

If this switch in mindset succeeds among researchers, then existing research labs can restart into new groundbreaking research projects with revived human spirits.

5.3 Build strong business community relations, diverse networking

Building strong and intimate relations with industries that are reaching out to nano-bio science but have no clear current connection with it. These industries are e.g. the chemical, the biomedical, and the pharmaceutical industries. All of them are looking for new ideas and new markets, expecting a lot from nano-bio R&D. A new eco-efficient industry is the ultimate goal of nano-bio production and manufacturing. It will enlighten the current industrial production from the burdens of waste, energy spills, EH&S (environment, health and safety risks), overconsumption,...

REFERENCES

[1] www.addictlab.com
[2] J. Schwartz, "Re-engineering Engineering", , The New York Times Magazine, September 30, pp. 94-97, 2007
[3] www.branscomb.org
[4] G. Holton, "What kinds of science are worth supporting? A new look, and a new mode" in The Great Ideas Today (Chicago, Ill.: Encyclopedia Brittanica), 106-136, 1998.
[5] C. Sagan, A. Druyan, "The varieties of Scientific Experience", Penguin Press, 2006
[6] www.imecexpo.be
[7] www.rvo-society.be
[8] R. St John, "8 to be great: the 8 traits that lead to great success", Train of Thought Arts Inc, 2005

Nanotechnology in Pharmaceutical Education in USA

Yashwant V. Pathak* and Ajoy Koomer**
Department of Pharmaceutical Sciences
Sullivan University College of Pharmacy
2100 Gardiner Lane, Louisville, KY 40205, USA
*ypathak@sullivan.edu , **akoomer@sullivan.edu

ABSTRACT

Nanotechnology is on its way to make a big impact in Biotech, Pharmaceutical and Medical diagnostics sciences. Pharmaceutical education in USA is also taking significant steps in incorporating courses as well as offering specialization in nanotechnology and its applications in Pharmaceutical scenario. This paper discusses status of nanotechnology and its application in educational curriculum in US pharmacy schools. We surveyed the curriculum offered at various levels in Pharmacy schools which covers the nanotechnology and its application in pharmacy, using two methods. It was observed that none of the pharmacy colleges have a separate course at the Pharm.D. level on nanotechnology, though good numbers of them have incorporated chapters in some of the courses with different titles. On graduate level many colleges have reported to have courses addressing nanotechnology in their curriculum.

Keywords: nanotechnology, pharmacy curriculum, education

1 INTRODUCTION

The nanotechnology and related development in sciences is leading to technological revolution in many fields of sciences. Even though the nano particles have been in nature since time immemorial, the importance of utilizing this technology is growing significantly in last two decades and making strides in different fields. It has enormous potential to influence our lives and has been employed in many areas including electronics, defense, pharmaceuticals, consumer products, environment, medicine and many more. In United States, Japan, India, Europe and many countries several initiatives have been undertaken and members of the private as well as public sectors are intensifying their researches in this field. It is also a novel way to create intellectual properties for the private sectors [1]. Hundreds and millions of dollars are committed to the research and development in this field. This is going to change our traditional practices of design, synthesis, and analysis and manufacture different products in various fields. This impact is very well felt by

the pharmaceutical industry as well as field of medicine. Several new techniques are being introduced in the field of drug delivery systems, targeting the drug, prosthetics, using these techniques to overcome the blood brain barriers in drug treatments, and many other fields including the diagnostics and analytical techniques. The area of Nanomedicine is developing very fast. Nanomedicine is an emerging field of medicine with novel applications. Nanomedicine is a subset of nanotechnology, which uses tiny particles that are less than one-millionth of an inch in size. In nano medicine, these particles are much smaller than the living cell. Because of this, nano medicine presents many revolutionary opportunities in the fight against all types of cancer, neurodegenerative disorders and other diseases. This change is creating a challenge to the academic community in the field of sciences to educate the students with necessary knowledge, understanding and skills to interact and provide leadership in the emerging nanotechnology [2].

Pharmaceutical education throughout the Americas is responsible for preparing students to begin the practice of pharmacy as vital members of the health care team or to assume other roles where pharmacists' knowledge and skills are required. Pharmaceutical education prepares students to become informed citizens in a changing health care environment. It is responsible for generating and disseminating new knowledge about Pharmaceuticals and pharmaceutical services, and about the role of the pharmacist in the unique health care system. Pharmaceutical education provides students with the values necessary to serve society as caring, ethical, learning professionals and enlightened citizens. It provides students with scientific fundamentals and fosters attitudes necessary to adapt their careers to changes in health care needs over a lifetime. It also encourages students prior to and after graduation to take active leadership roles in shaping policies, practices and future directions of the Profession and national health priorities. Pharmaceutical education both at the undergraduate and graduate level is responsible to the profession and society for generating knowledge about pharmaceuticals, pharmaceutical products, therapeutic actions and rational drug use through basic and applied research. It promotes the pharmaceutical sciences by fostering graduate education and research. Pharmaceutical education continually evaluates its mission, goals, objectives and outcomes throughout the Americas and relative to specific, unique conditions in individual countries. Based on the advances in pharmaceutical education and research, the necessary changes will be implemented.

The 2004 CAPE Educational Outcomes are intended to be the target toward which the evolving pharmacy curriculum should be aimed [3]. Many schools have used the 1998 CAPE Educational Outcomes as the template for curricular mapping endeavors. To facilitate these endeavors and to stimulate communication between schools, AACP held annual assessment institutes (from 1998-2003) in exchange for feedback regarding how faculty members were using the document. Hence, many schools adopted the 1998 Educational Outcomes as the definitive set of abilities that each student would conceivably possess upon graduation. Before publication of the 2004 CAPE Educational Outcomes, schools were using the 1998 document to reevaluate their curricula and implement outcome assessment strategies as a means of continuous quality improvement. Even with the adoption of the 2004 CAPE Educational Outcomes, some schools may elect to refer to the more comprehensive 1998 version for more detailed direction in determining student abilities.

Because of the increasing knowledge regarding disease and drug therapy, it is impossible for didactic and experiential pharmacy education to adequately cover all aspects of disease management and pharmaceutical care [4]. It is imperative that the pharmacy curricula be dynamic and able to quickly respond to the expected advancement of patient- and population-centered pharmacy practice. The 2004 CAPE Educational Outcomes addresses these expected changes and allows individual schools to maximize their local resources while ensuring that basic principles of pharmaceutical care are learned. However, these outcomes must also be attainable. For example, the outcomes listed in the Systems Management and Public Health sections may be beyond what can be attained in a PharmD program.

In addition to the 1998 and 2004 CAPE documents, several additional pharmacy educational and competency statements either have been published recently or are in the midst of completion. The National Association of Boards of Pharmacy (NABP) recently revised the Competency Statements for the North American Pharmacist Licensure Examination (NAPLEX) [5]. The 3 competency areas described in the NAPLEX Blueprint correlate with the 2004 CAPE Educational Outcomes:

- Area 1: Assure safe and effective pharmacotherapy and optimize therapeutic outcomes
- Area 2: Assure safe and accurate preparation and dispensing of medications
- Area 3: Provide health care information and promote public health

According to the CAPE outcomes the focus of Pharmaceutical education has changed from product oriented to patient oriented teaching [6] and more stress has been given in clinical pharmacy practices and its training so that the pharmacist becomes an integral part of the patient care team in health sciences. Hence, it is interesting to see how the present day pharmaceutical education incorporates the recent advances in the field of nanotechnology in their curriculum and courses.

2 METHODS
We used two techniques to get the information. We have developed a small questionnaire and were sent to all the Deans of the Colleges of Pharmacy (106) using survey monkey. We received 20 responses while more than 25 e mails were bounced back as the dean's e-mail accounts were not receiving the unsolicited e mails. The second method was to visit the website of each college and look at their course and curriculum and look for the nanotechnology courses being offered or incorporated in any of the courses being taught at Pharm. D. level.

3 OBSERVATION AND RESULTS
3.1 Current status of Nanotechnology in Pharmaceutical Education in USA
Some of the observation we had seen from the results are:

1. The academic community in Pharmaceutical education in USA is reacting slowly to prepare the work force for the emerging opportunities for the students.
2. None of the 106 colleges have separate course on nanotechnology or related topics in their Pharm.D. Curriculum.
3. Majority of the colleges have incorporated one chapter on few courses, interestingly the courses are taught with different titles such as Compounding and manufacturing II, Novel drug delivery systems, Pharmaceutics I or II, Dosage form technology, but none of the courses were exclusively dedicated to nanotechnology.
4. At a graduate level the scenario was different many graduate programs have been offering courses in nanotechnology. The titles of the courses which were taught at masters or Ph.D. levels were application of

nanotechnology in drug delivery systems, protein and peptide drug delivery systems with reference to nanotechnology and its application in this field, controlled release , introduction to nanotechnology, nano crystal technology and polymeric nano particle synthesis, biotechnology based drug dosage forms, advanced drug delivery systems.

5. Graduate level courses also offered elective projects in nanotechnology, elements of nano science and nano technology, seminars in nano science and seminars in antibody mediated drug delivery systems based on nanotechnology.

6. A good number of universities were offering research opportunities at master's and Ph.D level in nanotechnology and its applications in Pharmaceutical sciences. But there were no structured courses in the curriculum per say.

3.2 Current status of Nanotechnology Education world wide

1. Currently there are very few universities world wide imparting specialized education in nanotechnology. The web search provided following countries with courses and degrees offered in nanotechnology. The number in bracket shows the number of universities offering the nanotechnology education programs leading to full degrees. Brazil(2), Mexico(2), Czech Republic(2), Denmark(40), France(10), Germany(4), Israel(1), Italy(1), Netherlands(2), Norway(1), Spain(1), Sweden(20), Switzerland(1), United Kingdom(8), Turkey(1), United States(30), Australia/New Zealand(13), Canada(4), India (17), Singapore(1) and Thailand (2).

2. The primary mission of these universities is to conduct research and develop in the area of nanotechnology and nano science, even today there is less focus on nanotechnology education.

3. Some research centers are supporting associate and certificate programs as part of their activities in conjunction with other degrees they are offering.

4. Most interesting part all over the world is the faculty members in this area are supporting their laboratories and researches though their grant funding and encouraging the graduate student s to get involved in this type of research.

3.3 Some of the suggested courses for the Colleges of Pharmacy in USA as well as world wide for Nanotechnology can be:

1. Generic methodologies for the nanotechnologies
2. Nanoscale magnetic materials and their applications in medicines and devices

3. Processing and properties and characterization of inorganic nano materials
4. Macromolecules delivery using nanotechnology
5. Bio nanotechnology
6. Nanotoxicology
7. Bionanomaterials
8. Nano Polymeric Products for pharmaceutical applications
9. Techniques to manipulate interfaces and surfaces at nano level.
10. Self assembling nanostructures and its applications in Pharmacy
11. Nanostructures for pharmaceutical applications like aerogels, carbon nano tubes, dendrimers, magnetic molecules, metallic nano particles, nano clays, nano crystals, quantum corrals, nano wires
12. Fabrication of nanostructures
13. Characterization techniques for nano structures
14. Nanoapplications in Biosciences
15. Nanoparticulate drug delivery systems

These can be offered as:
1. Certificate course for the Pharm.D.B.S students with at least 15 credit hours which can be taken while studying for their professional degree there can be four courses to be taken to complete the certificate course as follows
2. Masters level as specialization in nanotechnology
3. PhD level graduate programs in Nanotechnology in Pharmaceutical sciences.

3.4 Major constraints for implementing nanotechnology in pharmaceutical education:

1. Lack of resources and materials especially in teaching nanotechnology at undergraduate and professional level
2. Applied aspect needs to be well documented and resources need to be developed to incorporate the recent advances in the field
3. As the focus of the total pharmaceutical education is pharmacy practice based, we need to develop courses with application in practice and understanding the importance of these techniques in practice settings
4. Majority of the universities do not have facilities for the research as the equipments involved in manufacturing and characterizing are very expensive, hence need to develop courses where these need not be used.
5. There is a need for educating the educators at various levels about nanotechnology and also creates interests in students about nanotechnology.

4 CONCLUSIONS

Nanotechnology has great future in pharmaceutical education, as the science behind will be more and more applied; the courses will appear in the curriculum. The pharmaceutical education will have to gear up for the growing needs of the society in this area by incorporating topics appropriately.

REFERENCES

[1] National Science Foundation's national nanotechnology initiative http://www.nsf.gov/home/crssprgm/nano 2001.

[2] M Uddin and AR Chowdhary, Nanotechnology education, http:/www.action bioscience.org/education, Paper presented at the International conference on Engineering education, in Oslo, Norway, 2001.

[3] Susan P Bruce, Amy Bower, Emily Hak, and Amy H Schwartz, American Journal of Pharm. Education, 15, 70-74, 2006.

[4] T. Schwinghammer, American Journal of Pharm. Education, 13, 68-71, 2004.

[5] J Cerulli and M Malone, American Journal of Pharm. Education, 12, 67-69, 2003

[6] LL Maine, American Journal of Pharm. Education, 13, 68-71, 2004.

Partnering in Nanotechnology Ventures: Critical Decision Factors

J.L. Woolley*

* Santa Clara University, Leavey School of Business
500 El Camino Real, Santa Clara, CA, USA, jwoolleylin@scu.edu

ABSTRACT

Business executives are often faced with the decision of whether to optimize the firm's performance by partnering with another company or exploring growth alone. Although nanotechnology firms are no exception, the high level of tacit knowledge requisite in these ventures presents complicated decision factors that must be considered. This study discusses the traditional decision factors inherent in partnering decisions as well as factors specifically important to nanotechnology firm executives such as knowledge sharing, intellectual property use and protection, accelerated learning and innovation, and leveraged growth opportunities.

Keywords: interorganizational relationships, partnering, alliances, business

1 INTRODUCTION

Partnering between organizations has long been of interest to top management teams in efforts to optimize performance. Traditionally, organizations have partnered to share the costs associated with a project, tap into another organization's specific strengths including distribution, R&D, facilities, etc., and reduce overall risk. Research on interorganizational relationships has shown that they can help build legitimacy [1], support resources [2], provide access to resources [1, 3], and improve the likelihood of survival of a firm [4]. Several organizational researchers have shown that interorganizational relationships such as partnerships can improve the performance of an organization [1, 4, 5, 6, 7, 8].

Conversely, there are several disadvantages of engaging in partnerships. Working with another organization is inherently difficult. Each organization brings a differing culture, level of commitment, and set of expectations into the partnership than can either complement or conflict with the other organizations. If the partnership is based on research and development, this adds complexity to the relationship, especially since knowledge transfer is difficult to manage [9]. Additionally, the assignment of costs and benefits is often to the advantage of the more powerful entity in the relationship. In some cases, partnerships can even create a future local or even global competitor.

Operating in nanotechnology brings additional complexity to the decision to partner. Nanotechnology firms tend to be small and new or large incumbents.

Partnerships can be especially beneficial to new or small firms that lack resources and legitimacy in the industry [10]. At the same time, new or small firms have less power than established incumbents and this imbalance can lead to uneven negotiations and lack of equity. The large incumbents in nanotechnology have more power and, thus, more choices when deciding if, when, and with whom to partner. Additionally, all nanotechnology firms have distinguishing characteristics that create unique decision factors when considering partnerships. These include the varying state of the technology itself, the high level of tacit knowledge required to use nanotechnology, the small size of the technology, the inter-disciplinary nature, and the need for specialized facilities. These lead to critical issues when considering partnering such as knowledge sharing, intellectual property use and protection, accelerated learning and innovation, and opportunities for leveraged growth.

2 CRITICAL ISSUES IN NANOTECHNOLOGY PARTNERING

2.1 Knowledge sharing

As mentioned, partnerships are seen as an opportunity to gain knowledge from another organization. But knowledge sharing can go beyond simply discussing domain information and innovation techniques. For example, Niosi found that the majority of 60 new Canadian biotechnology firms studied gained complementary knowledge from the strategic alliances with which they participated [11]. Complementary knowledge builds on the organizations existing domain and extends both its knowledge stocks and capabilities. Thus, knowledge sharing can lead to improved performance for both parties.

In nanotechnology, the knowledge domain tends to be much broader and well integrated than in other high-technology areas. For instance, nanotechnology is seen as a convergence of biotechnology, information technology, physics, and chemistry. For knowledge sharing to be most effective, participants must have a broad understanding of the different domains involved. With regards to nanotechnology, this understanding may be difficult to find in one person. A team of multi-disciplined individuals may best serve the organization in its knowledge sharing efforts. Therefore, the knowledge sharing dynamic of partnerships requires organizations to consider the type of knowledge to share and the people who will take part in the exchange.

2.2 Intellectual property protection

Intellectual property protection is often a concern in partnerships. What is different about nanotechnology is that the technology may not be patentable or ready to patent. Additionally, much of the technology is contained in tacit knowledge by the researchers in the firm. Therefore, it is not only intellectual property that is at stake, but, more importantly, trade secrets that are not patentable. Thus, the unique nature of nanotechnology creates cause for concern over intellectual property and trade secret protection.

Adding to the complexity of the situation is the fact that nanotechnology is dealing with, by definition, things not visible to the naked eye. When problems arise concerning intellectual property, it is difficult to ascertain since physical evidence is almost impossible to obtain or analyze.

Nanotechnology includes innovations in both products and processes. Product innovations are those that directly lead to marketable goods and are relatively easy to identify. On the other hand, process innovations are not visible outside of the firm and are often more important than products because processes underlie the development and production of key competencies, and goods. When entering a partnership, it is often expected that some knowledge will be transferred. What causes difficulty is distinguishing between what is included in the purview of the partnership and how to separate and exclude knowledge outside this relationship. For example, if you are sharing product technology regarding thin layer deposition products you may inadvertently disclose your process technology.

Conversely, partnerships offer an opportunity to share intellectual property and trade secrets outside of market transactions. Traditionally, when intellectual property is exchanged between organizations, very specific contracts are written to "protect" each party from misuse. Partnerships allow a relationship to form around a specific issue or opportunity, and each organization can apply the knowledge it deems appropriate to the issue. This knowledge is not predetermined and therefore allows more flexibility and creativity.

When creating a partnership, first you must weight the possibility of losing some of your intellectual property or trade secret advantages. Second, you should examine the potential for the creative use of your IP or trade secrets that benefit both parties.

2.3 Accelerated learning and innovation

In addition to sharing knowledge, nanotechnology partnerships provide an opportunity to accelerate the learning and innovation. Interorganizational teams from each firm represent a variety of knowledge and resources. When talented individuals with a greater diversity of experience and perspectives are brought together and anchored on a specific issue, learning can take place at a much faster pace [12]. Powell, Koput, and Smith-Doerr argue that knowledge sharing in the biotechnology industry occurs in interorganizational collaborations which lead to learning outside of individual firms [13]. A partnership can accelerate learning when interorganizational teams take the opportunity to advance beyond the simple exchange of knowledge stocks. For example by comparing problems, simultaneous parallel problem solving can occur to which both parties can benefit.

Partnership opportunities to accelerate learning outside of the organization are attractive for several reasons. First, there are few places in the world that offer clear and consistent education for nanotechnology. Nanotechnology organizations wishing to obtain a particular talent or knowledge domain have few choices from which to hire talented individuals. Second, organizations that endeavor to teach employees on the job are constrained by the time, talent, and facilities that this requires which often come at the expense of product and process innovation. Third, partnerships enable learning based on the organizations existing products, processes, innovations, and problems. Therefore, partnerships offer a valuable opportunity for accelerated learning and innovation.

2.4 Leveraged growth opportunities

Partnerships provide access not only to resources outside of an organization including knowledge, physical assets, and talent, but also to growth opportunities by leveraging the resources of the partner's resources and network. By using the assets, knowledge, and other resources of the other organization to support your own organization's strategy, you can achieve what John Hagel calls "leveraged growth" [14]. Leveraged growth is the mobilization of resources outside the organization to expand. For example, this allows you to move into new markets without purchasing or owning assets specific to the expansion. Leveraged growth allows both organizations to focus on their respective capabilities, which leads to further growth, improved efficiency, and potentially, higher profits.

Nanotechnology is inherently expensive. Opportunities to leverage the resources outside the organization must be well considered. Often this takes the form of partnering with educational facilities or nanotechnology incubation organizations with access to the expensive equipment necessary to perform experiments and product development. In the consideration of partnerships, an organization must weigh the opportunity for leveraged growth through the resources of the partnering organization.

3 IMPLICATIONS AND DISCUSSION

This article discusses non-traditional benefits of partnerships and some of the decision factors that are associated. A partnership allows organizations to exchange ideas, and knowledge in a more creative forum. While intellectual property and trade secrets are often heavily protected, sharing some of this information in a partnership may lead to faster and better use of that IP. The opportunity to accelerate learning and innovation offered by certain partnerships may prove more valuable than the drawbacks. And lastly, partnerships provide access to a wider range of resources that may lead to leveraged growth opportunities.

When considering a partnership, each of these benefits must be weighed carefully. While it is useful to determine which of these benefits is most useful and likely, one must also examine the potential partner's resources and capability of enabling these benefits to be realized. All potential partners are not created equal. First, one must evaluate the potential partner's resources or access to resources that would prove useful for your organization. Second, while large firms may have the resources necessary, they may not be willing to share these resources. Third, the partnering organization's level of commitment will vary. Fourth, the culture of the potential partner will dictate your interactions. Make sure that it is compatible with your own team's culture and study. Fifth, if you decide to enter the partnership, make the sure that the goals and expectations of the union are well conveyed and documented.

REFERENCES

[1] J. A. C. Baum and C. Oliver, "Institutional Linkages and Organizational Mortality," Administrative Science Quarterly, 36, 187-218, 1991.

[2] A. H. Vandeven and G. Walker, "The Dynamics of Interorganizational Coordination," Administrative Science Quarterly, 29: 598-621, 1984.

[3] T. E. Stuart, H. Hoang, and R.C. Hybels, "Interorganizational endorsements and the performance of entrepreneurial ventures," Administrative Science Quarterly, 44, 315-349, 1999.

[4] W. Mitchell and K. Singh, 1996, "Survival of businesses using collaborative relationships to commercialize complex goods," Strategic Management Journal, 17, 169-195, 1996.

[5] J. A. C. Baum, T. Calabrese, and B. Silverman, "Don't go it alone: Alliance network composition and startups' performance in Canadian biotechnology," Strategic Management Journal, 21, 267-294, 2000.

[6] J. Hagedoorn, and J. Schakenraad, "The Effect of Strategic Technology Alliances on Company Performance," Strategic Management Journal, 15, 291-309, 1994.

[7] K. Singh and W. Mitchell, "Precarious collaboration: Business survival after partners shut down or form new partnerships," Strategic Management Journal, 17, 99-115, 1996.

[8] K. Singh and W. Mitchell, "Growth dynamics: The bidirectional relationship between interfirm collaboration and business sales in entrant and incumbent alliances," Strategic Management Journal, 26, 497-521, 2005

[9] Hagedoorn, Cloodt, van Kranenburg, 2006

[10] A. L. Stinchcombe, "Social structure and organizations," In J. March (Ed.), Handbook of Organizations, Rand McNally, 142-193, 1965.

[11] J. Niosi, "Alliances are not enough explaining rapid growth in biotechnology firms," Research Policy, 32, 737-750, 2003.

[12] J. Hagel and J.S. Brown, "Creation Nets" Harnessing the Potential of Open Innovation," http://www.edgeperspectives.com, 2006.

[13] W.W. Powell, K.W. Koput, and L. SmithDoerr, "Interorganizational collaboration and the locus of innovation: Networks of learning in biotechnology," Administrative Science Quarterly, 41, 116-145, 1996.

[14] J. Hagel, "Leveraged Growth: Expanding Sales Without Sacrificing Profits," Harvard Business Review, 80, 5-12, 2002.

Why Does One Nano Lab Cost More Than Another?

W. Wilson* and A. Gregg**

* Wilson Architects Inc., Boston, MA
** AGI Abbie Gregg, Inc., Tempe, AZ

Abstract

Construction costs of nano labs reflect the differential between the required performance criteria and the in situ conditions. Nano retrofits in existing buildings range from $300 to $1000 NSF, depending on the isolation required. New construction can be prepared for nano occupancies with strategic investments to the shell and core construction. Topics covered include decisions for: siting / utilities; foundation; super structure; high bay; low vibration; clean room; construction delivery; and testing and commissioning. Examples from seven current projects will be used – three renovations and four new construction.

Keywords

low vibration labs, clean rooms, high bay labs, shell and core decisions, fit-out decisions, construction cost

The label "nanotechnology" covers a host of research applications, some of which require vastly more costly infrastructure than others. Basic engineering issues, research themes, and design choices determine the budget of the nanotech facility.

The relative cost of nano labs is based primarily on five factors: 1) the level of performance that the research requires; 2) the means by which it is achieved; 3) the proportion of high performance space to the overall program; 4) the "baseline" conditions of the site; and 5) the builder's experience with this type of construction.

For the purposes of this manuscript, the most frequently asked questions in the design of a nano lab are examined. They are organized into themes, and include recommendations about which direction is most advantageous. When the cost difference between one solution and another is known, it is so stated. When the cost difference is not immediately available, general guidance is given. The building's end user, or the user representative, should participate in answering these questions because they are in the best position to evaluate whether or not the end justifies the means.

Instrument-Driven Science

Nano science is different from conventional bench science because of the reliance on tools and instrumentation needed to work at atomic scales. The operating requirements of the instruments used to conduct nano experiments are restrictive and severe. They simply do not work with precision and high resolution if they are jiggled, bumped or disturbed in any way at the scale of the atom.

Many manufacturers of sophisticated instruments, particularly in high resolution tunneling electron microscopy, won't guarantee results unless the spaces conform to their specifications. Research in microelectronics and materials involving etching, lithography, masking and deposition all require clean and stable environments in which precise patterns can be imparted onto a variety of substrates for testing and analysis. The instrument-driven sciences, therefore, rely on the stability of the space in which they are housed. Performance is key.

High Performance Space

Nano labs are "high performance spaces". That description starts to suggest the unique qualities of these labs and begins to differentiate them from typical laboratories. When the term "performance" is used, it refers to the stability and uniformity of the temperature, humidity, electro-magnetic fields, and vibration within the space.

Also included are issues of special construction such as high-bay spaces, and labs with pits. These are often needed for condensed matter physics in which probes are inserted into super-cold magnetic fields contained in dewars.

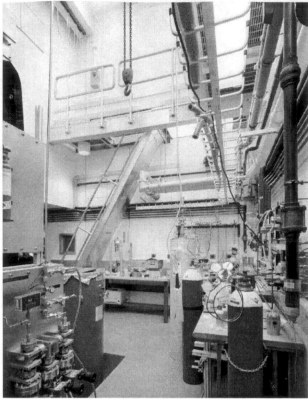

High Bay Lab
Harvard University, Cambridge, MA

And, perhaps the most costly space to create are clean room environments, where many of the high performance needs converge and tend to act against each other. Therefore, "high performance spaces" are generally characterized in four ways:

- **Low vibration labs.** The performance is defined by VC (vibration class) curves which plot velocity of displacement against frequency. To achieve VCE and lower (less than 125 micro inches/sec) inertia blocks are utilized and isolated from surrounding floors and structure by specialized springs.

- **Clean rooms.** The performance is defined by Class 1 through Class 100,000. The number refers to the number of particles of a specified size allowed in a volume of space. A Class 100 space has less than 100 particles greater in size than 0.5 microns per cubic foot. The lower the class number, the cleaner the space. These spaces are achieved with continuous air circulation using HEPA and ULPA filtration, specialized wall panel and flooring systems and gowning protocols to avoid the introduction or circulation of particulates. Support equipment, such as pumps, are often in a separate chase area or subfab area to limit particle generation in the clean room.

- **High bay labs.** These typically range from 20 to 25 feet in clear height. They are often outfitted with a 2 ton crane at the top to allow for the raising and lowering of dewars and chambers into specific experimental positions.

- **Cryogenic labs.** These labs are similar to high bay labs, except that the required height is achieved through the use of pits rather than high bay. The height is required to allow long experimental probes to be raised and lowered into low temperature liquids housed in dewars. The use of powerful magnets in the dewars precludes the use of magnetic rebar in the surrounding area.

In programming "high performance space", standard "Room Data Sheets" with "Room Condition Sheets" are often included to get at the true performance needs. These "room conditions" provide the first opportunity to start to gauge the relative cost of one space to another.

Clean Room – LISE Building
Harvard University, Cambridge, MA

Means and Methods

However, room conditions only go so far when it comes to cost. The next major cost factor is tied up in the means and methods by which the designer achieves the required stability. To achieve temperature and humidity stability, individual air handling units may be required for each experiment. For EMI stability, conduit may need to be shielded, wiring may need to be twisted and non-magnetic construction materials may be required. For vibration control, isolation blocks on springs may be installed. For clean environments, multiple air changes through HEPA filtration is needed. The list of problems and solutions is long and extensive and often expensive.

Program Distribution

Because achieving high performance space can be costly and difficult, the relative proportion of high performance space to the overall space program is the next major cost driver. When evaluating the cost of a nano lab at one institution versus another, it is important to understand the proportion, because it can dramatically skew the price comparison.

In a comparison of six recent projects, the proportion of high performance space was found to range between 20 and 70 percent of the overall program (called out as "special" and "clean room" in the figure below). The difference in cost of facilities is reflective of this range.

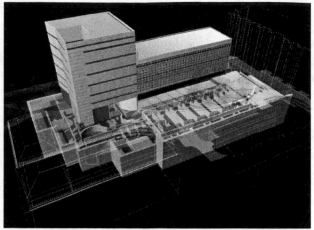

3-D Rendering-underground nano portion of LISE
Harvard University, Cambridge, MA
Drawing by Wilson Architects
Design Credit: Rafael Moneo with Wilson Architects

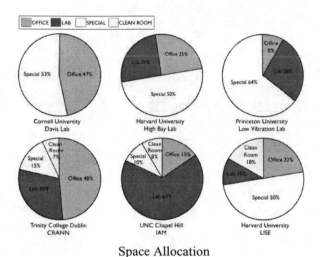

Space Allocation

Section of the LISE Building
Harvard University, Cambridge, MA
Drawing by Wilson Architects
Design Credit: Rafael Moneo with Wilson Architects

Site Conditions

Another great cost factor for nano labs is the baseline condition of the site that is chosen for the facility. Clearly, if a nano facility is placed next to a heavily trafficked road, a railway or subway, the baseline vibration and electromagnetic fields that will need to be mitigated is much higher and costlier than a quieter site. The need for low vibration and cryogenic pit space often dictates a slab on grade location which can generate the need for a large, deep basement footprint (as referred in the following two figures). The existing site utilities (relocation costs), the soil conditions (slurry walls) and the adjacency of other buildings (underpinning) greatly affect the budget.

Builder's Experience

Ultimately, the final judge of how much a space costs is the builder who successfully bids on the work. The same design will be priced differently by different bidders based on their experience with this type of project. An experienced builder will most likely price the project close to its actual worth, because they understand the complexity and the specialty trades necessary to build and commission the work. An inexperienced builder will tend to either grossly overestimate the work because of a "fear factor" about the restrictive requirements, or will underestimate it because of lack of understanding of the complexity. It is important to evaluate the bids with this in mind, and properly "scope" each bidder in a post-bid meeting.

Risk Management: Is the cure worse than the disease?

Once the room conditions, the means for meeting them, and the costs are identified, it is time to engage the users and project managers in a discussion of what to do next. In this case, it becomes a question of risk management, as the efficacy of many of the solutions is either apocryphal ("they did it this way at X institution, and they never had a problem"), or academic ("our calculations show that you will experience X perturbations...etc.), or just intuitive ("I don't care what your calculations show, there is no way I'm going to put the instrument there.."). As a researcher, or an owner, you are forced to make cost decisions based upon your intuitive sense of whether or not the effort of the solution matches the criticality of the criteria.

There have been clients who have chosen to give up on specific stability criteria once they saw the cost and draconian means that were taken to meet it. These same clients have chosen to move forward with costly solutions for some of the criteria because they believe in the necessity and legitimacy of the stated need.

Vibrationally isolated piping and conduit to experiment.
J.C. Davis Group Nanotechnology Lab
Cornell University, Ithaca, NY

Recommendations

Having been through a number of these "risk management" discussions about scope and cost, a list of recommendations and general guiding principles has been assembled. By organizing cost choices thematically, we have identified 14 menus of choices for achieving high performance space. A "check" is next to solutions that clients have chosen to use, and a "dollar sign" symbol is next to additional investments. They are additional not because they are necessarily without merit, but because they trigger costs that may be prohibitive.

Conclusion

The costs of nano labs are relative and dependent upon a number of factors, from the needs of the science, the experience of an architect and engineer, to the comfort level of a builder. Therefore, estimating the cost of a nano lab will not likely be as precise as the science that will go on inside it.

Index of Authors

Index of Keywords